Biochemistry

THE JONES AND BARTLETT SERIES IN BIOLOGY

AQUATIC ENTOMOLOGY
Patrick McCafferty,
Arwin Provonsha

BASIC GENETICS, *Second Edition*
Daniel L. Hartl

BIOCHEMISTRY
Robert H. Abeles,
Perry A. Frey,
William P. Jencks

BIOLOGICAL BASES OF HUMAN AGING AND DISEASE
Cary S. Kart, Eileen K. Metress,
Seamus P. Metress

THE BIOLOGY OF AIDS, *Second Edition*
Hung Fan, Ross F. Conner,
Luis P. Villarreal

CELL BIOLOGY: ORGANELLE STRUCTURE AND FUNCTION
David E. Sadava

CELLS: PRINCIPLES OF MOLECULAR STRUCTURE AND FUNCTION
David M. Prescott

CONCEPTS AND PROBLEM SOLVING IN BASIC GENETICS: A STUDY GUIDE
Rowland H. Davis,
Stephen G. Weller

CROSS CULTURAL PERSPECTIVES IN MEDICAL ETHICS: READINGS
Robert M. Veatch

EARLY LIFE
Lynn Margulis

ELECTRON MICROSCOPY
John J. Bozzola, Lonnie D. Russell

ELEMENTS OF HUMAN CANCER
Geoffrey M. Cooper

ESSENTIALS OF MOLECULAR BIOLOGY, *Second Edition*
David Freifelder,
George M. Malacinski

EVOLUTION
Monroe W. Strickberger

EXPERIMENTAL TECHNIQUES IN BACTERIAL GENETICS
Stanley R. Maloy

FUNCTIONAL DIVERSITY OF PLANTS IN THE SEA AND ON LAND
A.R.O. Chapman

GENERAL GENETICS
Leon A. Snyder, David Freifelder,
Daniel L. Hartl

GENETICS OF POPULATIONS
Philip W. Hedrick

THE GLOBAL ENVIRONMENT
Penelope ReVelle, Charles ReVelle

HANDBOOK OF PROTOCTISTA
Lynn Margulis, John O. Corliss,
Michael Melkonian, and
David J. Chapman, Editors

HUMAN GENETICS: A MODERN SYNTHESIS
Gordon Edlin

HUMAN GENETICS: THE MOLECULAR REVOLUTION
Edwin H. McConkey

HUMAN ANATOMY AND PHYSIOLOGY COLORING WORKBOOK AND STUDY GUIDE
Paul D. Anderson

HUMAN BIOLOGY
Donald J. Farish

THE ILLUSTRATED GLOSSARY OF PROTOCTISTA
Lynn Margulis,
Heather I. McKhann, and
Lorraine Olendzenski, Editors

INTRODUCTION TO HUMAN DISEASE, *Third Edition*
Leonard V. Crowley

MAJOR EVENTS IN THE HISTORY OF LIFE
J. William Schopf, Editor

MEDICAL BIOCHEMISTRY
N.V. Bhagavan

MEDICAL ETHICS
Robert M. Veatch

METHODS FOR CLONING AND ANALYSIS OF EUKARYOTIC GENES
Al Bothwell,
George D. Yancopoulos,
Fredrick W. Alt

MICROBIAL GENETICS
David Freifelder

MOLECULAR BIOLOGY, *Second Edition*
David Freifelder

MOLECULAR EVOLUTION
E.A. Terzaghi, A.S. Wilkins,
D. Penny

ONCOGENES
Geoffrey M. Cooper

ORIGINS OF LIFE: THE CENTRAL CONCEPTS
David W. Deamer,
Gail Raney Fleischaker

PLANT NUTRITION: AN INTRODUCTION TO CURRENT CONCEPTS
A.D.M. Glass

POPULATION BIOLOGY
Philip W. Hedrick

VERTEBRATES: A LABORATORY TEXT
Norman K. Wessels

WRITING A SUCCESSFUL GRANT APPLICATION, *Second Edition*
Liane Reif-Lehrer

Biochemistry

ROBERT H. ABELES
Brandeis University

PERRY A. FREY
University of Wisconsin

WILLIAM P. JENCKS
Brandeis University

Jones and Bartlett Publishers
Boston *London*

Editorial, Sales, and Customer Service Offices

Jones and Bartlett Publishers
One Exeter Plaza
Boston, MA 02116

Jones and Bartlett Publishers International
PO Box 1498
London W6 7RS
England

Library of Congress Cataloging-in-Publication Data

Abeles, Robert H.
Biochemistry / Robert H. Abeles, Perry A. Frey, William P. Jencks.
p. cm.
Includes bibliographical references and index.
ISBN 0-86720-212-2
1. Biochemistry. I. Frey, Perry A. II. Jencks, William P., 1927– . III. Title.
QP514.2.A24 1992
574.19′2—dc20 92-6344
CIP

Developmental and copy editor: Patricia Zimmerman
Production: Patricia Zimmerman
Design: Deborah Schneck
Illustrator: Intergraphics
Typesetter: Doyle Graphics
Cover design: Hannus Design Associates
Printing and binding: Courier Westford
Cover printer: Henry N. Sawyer

Cover illustration: Zif268 bound to a segment of double helical DNA.
[Used with permission. N.P. Pavletich and C.O. Pabo, "Zinc Finger-DNA Recognition: Crystal Structure of a Zif268-DNA Complex at 2.1Å." *Science, Washington D.C.* 252(1991): 809. Copyright 1991 by the American Association for the Advancement of Science.]

Printed in the United States of America
96 95 94 93 92 10 9 8 7 6 5 4 3 2

Brief Contents

Contents

Dedicated to Fritz Lipmann and Frank H. Westheimer,
who brought chemistry into modern biochemistry

Preface

The traditional approach to teaching biochemistry encourages an emphasis on learning biochemical facts before addressing biochemical principles and phenomena. There comes a time when the accumulation of knowledge requires that the practice of teaching ascend to a different plane, where descriptive knowledge is too voluminous for anyone to master and where enough is known to allow the student to focus on generalities and principles. At such a time, it becomes possible to explain why metabolism works as it does, why the biological macromolecules have the structures and reactivities that they exhibit, and how living cells can carry out their essential functions. We believe that biochemistry can now be taught with emphasis on a few chemical principles that rationalize large bodies of biochemical facts. This textbook represents our effort to present biochemistry in this way.

The method that we adopt is to present the chemical principles and facts that pertain to a biochemical system and then to explain the system itself. This textbook is organized so as to encourage a natural transition from organic chemistry to biochemistry, because most students study biochemistry after having taken organic chemistry. Chapter 1 explains this method briefly and contains a short review of cell biology to remind the student of the setting in which biochemistry takes place. Chapter 1 also describes some biochemical phenomena, without details, to exemplify and define biochemistry and its relation to chemistry and cell biology. Chapter 2 deals with the biochemical formation and cleavage of carbon–carbon bonds, the chemistry of which will be familiar to students who have just taken organic chemistry. Chapters 3 through 6 explain enzymes and how they work, in a way that is unique to this book. We adopt the case-study approach for the presentation of principles by considering chymotrypsin and trypsin in detail. These are the most thoroughly studied of all enzymes, and most of the underlying principles governing catalysis by enzymes can be explained in the context of what is known about them. Special emphasis is given in Chapter 5 to the importance of entropy and multiple weak binding interactions for catalysis by enzymes. The importance of entropy and binding energy in most biochemical phenomena is introduced here and reiterated in later chapters dealing with the structure and function of proteins and nucleic acids. Chapter 6 describes other proteolytic enzymes and the important cellular and organismic functions of proteases in processes such as blood clotting, digestion, and the activation of enzymes.

Chapter 7 describes the structures of proteins; it follows the early chapters on enzymes, in which many concepts of protein structure have been introduced without details and which include such basic information as the structures of amino acids and the peptide bond. The introduction of this information simplifies the more-detailed presentation in Chapter 7, which is followed in Chapter 8 by an examination of the principles that govern the structures of proteins and nucleic acids.

In Chapter 9, the Gibbs free energy changes of biochemical reactions are defined and explained in simple terms. Several basic concepts have been introduced in earlier chapters, the most fundamental of which is the coupling of two processes. An example is a conformational change in a macromolecule that is brought about by binding a small molecule. Coupling was described earlier in terms of equilibrium constants; Gibbs free energy is introduced and related to entropy and enthalpy, which have also been introduced in earlier chapters as concepts but without rigorous definitions.

Inasmuch as catalysis of acyl group transfers by proteases has been discussed in early chapters and protein structures are described in Chapter 7, Chapter 10 proceeds logically with protein biosynthesis and the importance of acyl activation in this process. Protein biosynthesis is the first biochemical pathway to be described in this textbook; it is also the medium by which we introduce nucleic acids, which have essential functions in this pathway. Enough information about the structures of nucleic acids is given to allow the principles of their actions in protein biosynthesis to be explained.

In Chapter 11, we introduce the basic chemistry of phosphotransfer reactions and the mechanistic enzymology of phosphotransferases and nucleotidyltransferases. The importance of these reactions in the assembly of coenzymes and nucleic acids, as well as membrane lipids, is presented here. In Chapter 12, we describe the structures of nucleic acids, along with their biosynthesis and processing, the chemistry of which entails nucleotidyl transfer reactions of the type introduced in Chapter 11. Nucleic acid biosynthesis and processing are the second and third biochemical pathways that are described in this textbook.

In Chapter 13, the organization and regulation of genes are described in sufficient detail to explain the basic principles of gene regulation in prokaryotic and eukaryotic cells. Several of the best-understood systems of prokaryotic and eukaryotic gene regulation are introduced to exemplify the principles and show diversities in the regulation of different systems. Some of the structural motifs in proteins that regulate genes also are introduced in Chapter 13. Several molecular cloning strategies are delineated in Chapter 14, as are some applications of cloned genes and recombinant DNA in biotechnology and basic biochemical research. Site-directed mutagenesis and its importance to biochemistry are explained here.

The basic chemistry of glycosyl group transfer reactions and the biosynthesis and breakdown of polysaccharides are described in Chapter 15. The assembly of *N*-linked oligosaccharides in glycoproteins is also presented.

Chapters 16, 17, and 18 delineate the basic chemistry and biochemistry of the oxidation-reduction of NAD^+ and FAD, decarboxylation, carboxylation, addition, and elimination. These chapters round out the chemical information that will enable the student to understand fatty acid biosynthesis and degradation, which are the subjects of Chapter 19. The isomerization reactions in Chapter 20 complete the reaction types that are required to understand carbohydrate metabolism in the glycolytic pathway and the tricarboxylic acid cycle, which are presented in Chapter 21. Knowledge of the oxidative metabolism of fatty acids and carbohydrates will help the student to understand how the reduced end products, NADH and $FADH_2$, are utilized in the terminal electron-transport and oxidative phosphorylation pathways to produce ATP, the principal mediator for energy exchange in biochemistry, as explained in Chapter 22.

Chapter 24 deals with biochemical oxygenation and oxidation reactions that utilize molecular oxygen. Several of the most-important biochemical oxygenation reactions in the metabolism of amino acids and neurotransmitters and in the metabolism of lipids serve to exemplify these important processes. Chapter 25 describes one-carbon metabolism and the insertion of one-carbon units in the synthesis of biological molecules. These two chapters round out the information that is required to explain the metabolism of amino acids and nucleotides and the biosynthesis of heme, which are the subjects of Chapters 26 and 27. The nutritional sources of nitrogen and the disposition of excess nitrogen in higher animals also are described in these chapters.

In Chapter 28, we outline the metabolism of complex lipids, including cholesterol, steroid hormones, sphingolipids, and lipoproteins. We emphasize the chemistry that is unique to cholesterol biosynthesis and its regulation. We also describe cholesterol homeostasis in human beings in sufficient detail to enable the student to understand its importance in heart disease. We describe membrane structures and some important principles of their functions in Chapter 29, and this opens the way for a detailed discussion of vectorial transport phenomena in Chapter 30, which concludes this textbook.

Our present understanding of biochemistry is unveiled in this book in three phases. The first fourteen chapters deal with the function, structure, and metabolism of proteins and nucleic acids, the most-important biological macromolecules. Chapters 15 through 27 take up intermediary metabolism, with a special emphasis on the chemical basis of metabolism. The book concludes with an examination of transport phenomena and membranes in Chapters 27 through 30. We introduce the basic chemistry that is required to understand each biochemical system, and we then describe the system and its function before proceeding to more-complex chemistry and other systems.

In general, the biochemistry in this textbook pertains to phenomena that are common to all cells. Specialized subjects such as hormone action, enzyme kinetics, coenzymes, and nutrition are not dealt with in detail in chapters exclusively about them; however, essential aspects are explained in connection with metabolism or the functions of proteins and other molecules. We attempt to focus on the main body of biochemical knowledge and to present it in a way that emphasizes the chemical principles that rationalize biochemistry.

As in most human endeavors, a few exceptions to our own rule for treating biochemistry common to all cells can be found in this book. For example, photosynthesis takes place only in certain plant cells and microorganisms; however, photosynthesis is so important to the existence of all forms of life that it should be included in every biochemistry course. Another example is muscle contraction, which is important in many complex organisms but is not common to all cells. However, it is a familiar and intensively studied example of a fundamental phenomenon in biochemistry: the use of biochemical free energy to do mechanical work. It is one of the biochemical systems described in the last chapter on coupled vectorial processes. Other examples are blood clotting in Chapter 6; antibody structure and function in Chapters 6, 7, and 13; and oxygen binding by hemoglobin in Chapter 10. Thus, while emphasizing the biochemistry that is common to all cells as the core knowledge of this textbook, we include specialized and particularly important phenomena within the appropriate chapters. With a few exceptions, the material set apart from the main text in boxes can be considered optional and need not be included in a one-semester course.

In learning any body of knowledge, every student faces the question of how to know when he or she has mastered the material. A related question is how to study the subject matter and gain a mastery of it. One can read a textbook any number of times without being confident of mastery. An effective way to study and evaluate understanding is to work problems that deal with the subject matter. Students who have a basic understanding of a subject can answer questions about it and can even pose new problems for further study.

The problems at the end of the chapters in this textbook are intended to assist the student in comprehending the most-important principles that are presented in each chapter. Some of the problems are in fact exercises that should assist the student in learning important biochemical components and processes. Other problems give the student an opportunity to apply basic chemical principles to the solution of biochemical problems. The principles required are to be found within the chapter under study or in preceding chapters. Because we present biochemical phenomena in the textbook within the framework of accumulating chemical principles, the problems for a given chapter are designed on the assumption that the information in preceding chapters has been mastered by the student. Therefore, should the order of chapter coverage be altered in a given course, the instructor may find it necessary to provide additional information to students before assigning some of the problems in the later chapters. Answers to the end-of-chapter problems can be found at the end of the book.

We would like to thank the reviewers, whose names are given on the next two pages, for reviewing the chapters and for making helpful suggestions. We would also like to thank Christopher Miller for contributing the chapter on biological membranes. Finally, we wish to acknowledge the late David Freifelder for his help and encouragement in getting the book started.

Robert H. Abeles
Perry A. Frey
William P. Jencks

Reviewers

WILLIAM S. ALLISON
Department of Chemistry, University of California, San Diego

THEODORE ALSTON
Department of Anesthesiology, Massachusetts General Hospital

FRANK T. BAYLISS
Department of Biology, San Francisco State University

STEPHEN BENKOVIC
Department of Chemistry, The Pennsylvania State University, University Park

KONRAD BLOCH
Department of Chemistry, Harvard University

PAUL O. BOYER
Department of Chemistry and Biochemistry, University of California, Los Angeles

KENNETH L. BROWN
Department of Chemistry, The University of Texas at Arlington

THOMAS C. BRUICE
Department of Chemistry, University of California, Santa Barbara

CHARLES H. CLAPP
Department of Chemistry, Bucknell University

W. WALLACE CLELAND
Institute for Enzyme Research, University of Wisconsin

GEORGE COHEN
Graduate Department of Biochemistry, Brandeis University

MICHAEL A. GOLDMAN
Department of Biology, San Francisco State University

HARRY GRAY
Department of Chemistry, Beckman Institute, California Institute of Technology

GORDON A. HAMILTON
Department of Chemistry, The Pennsylvania State University, University Park

RALPH A. JACOBSON
Department of Chemistry, California Polytechnic State University, San Luis Obispo

GEORGE L. KENYON
Department of Chemistry, Pharmaceutical Chemistry, and Pharmacology, University of California, San Francisco

JACK F. KIRSCH
Department of Biochemistry and Molecular Biology, University of California, Berkeley

JEREMY KNOWLES
Department of Chemistry, Dean of the Faculty of Arts and Sciences, Harvard University

FRED KULL
Department of Biological Science, State University of New York, Binghamton

I. ROBERT LEHMAN
Department of Biochemistry, Stanford University

ROWENA G. MATTHEWS
Department of Biological Chemistry, The University of Michigan Medical School

PAUL R. ORTIZ DE MONTELLANO
Department of Chemistry, Pharmaceutical Chemistry, and Pharmacology, University of California, San Francisco

C. DALE POULTER
Department of Chemistry, University of Utah

ROBERT O. POYTON
Department of Molecular, Cellular, and Developmental Biology, University of Colorado, Boulder

DAVID M. PRESCOTT
Department of Molecular, Cellular, and Developmental Biology, University of Colorado, Boulder

DANIEL V. SANTI
Department of Biochemistry, School of Medicine, University of California, San Francisco

KENNETH SAUER
Department of Chemistry, University of California, Berkeley

PAUL SCHIMMEL
Department of Biology, Massachusetts Institute of Technology

WILLIAM M. SCOVELL
Department of Chemistry, Bowling Green State University

JULES A. SHAFER
Department of Biological Chemistry, The University of Michigan Medical School

DAVID N. SILVERMAN
University of Florida College of Medicine

RICHARD B. SILVERMAN
Department of Chemistry, Biochemistry, Molecular Biology, and Cell Biology, Northwestern University

PAUL A. SRERE
Pre-Clinical Sciences, Veterans Administration Medical Center, Dallas

JOSEPH J. VILLAFRANCA
Center for Biomolecular Structure and Function, The Pennsylvania State University, University Park

Biochemistry and Life

1

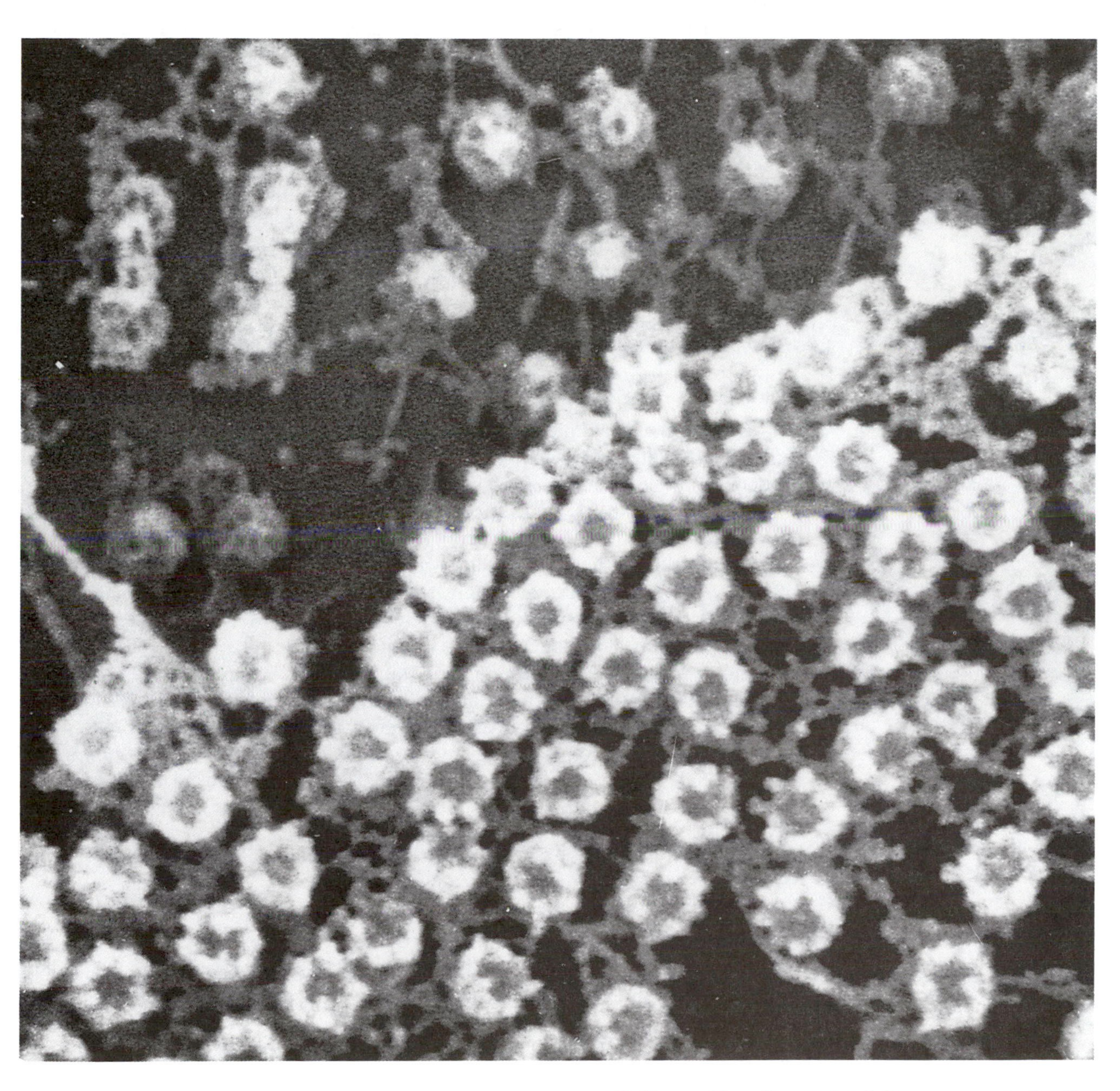

Nuclear surface of an isolated Xenopus oocyte showing nuclear pores. (Courtesy of Hans Ris.)

Our understanding of how living systems work has grown enormously in the past 150 years — much more than anyone could have imagined at the beginning of this period. This growth has been responsible for a change in our view of the natural world that is at least as revolutionary as the changes that have been brought about by the extraordinary advances in physics and astronomy in the same period. The revolution in biology is largely a consequence of the development of a new way of approaching and describing the operation of living systems—a new language—that makes possible the analysis of biological processes in chemical terms, rather than in vague and vitalistic terms. It is hard for an educated person today to comprehend the extent of the aura of mystery that surrounded anything connected with the word "life" 150 years ago. The mystery has been disappearing from one area after another as our understanding has increased at an exponential rate.

The new understanding of life is described largely in the language of biochemistry. Owing to the enormous advances in the past 60 years, biochemistry now serves the basic role in describing biological phenomena that anatomy did a century ago. Our understanding of biology has grown from a descriptive level based on gross and microscopic anatomy to the point at which the basic chemical nature of the majority of biological materials is known, many of the biochemical pathways for the synthesis and degradation of these materials have been worked out, and the mechanisms for the regulation of some of these pathways are becoming understood. We are now extending our understanding from the biochemistry of homogeneous systems to the mechanism of action of heterogeneous systems that convert chemical energy into other kinds of energy, such as mechanical energy in muscle contraction and osmotic work in the transport of ions and molecules across membranes. This record of progress leads more and more biologists and biochemists to believe that most biological phenomena will eventually be understood in chemical terms.

Biochemistry may be defined as the science that is concerned with the structures, interactions, and transformations of biological molecules. By another definition, biochemistry is the chemistry of life. The two definitions are similar, because chemistry is concerned with the structures, interactions, and transformations of molecules in general, and biological molecules are the molecules of life. In this textbook, we shall describe most of the molecules of life in general terms. We shall also describe many specific molecules. We shall emphasize their interactions and transformations and how these processes support life. Owing to the complexities of living systems, it is not possible to cover all of the known biological molecules in a single textbook. However, it is possible to explain the chemical principles governing the interactions and transformations of biological molecules; and those principles will be emphasized in this textbook.

In this chapter, we begin by introducing and reviewing the biology of living cells, the setting in which most biochemical phenomena take place. We also define the science of biochemistry and explain how this textbook is organized to present the major biochemical subjects.

Cells

The biological cell is the basic unit of life and the setting for most biochemical phenomena. In unicellular life forms, a single cell is capable of self-maintenance and reproduction, the most-essential life processes. In *organisms,* animals and plants, the individual cells have these same capabilities, but they also support the functions of other cells and are in turn supported by other cells to the advantage of the whole organism. Multicellular organisms are capable of more, in some cases much more, than self-maintenance and reproduction. The highest animals can alter their environments, to either their benefit or their detriment, whereas most unicellular forms thrive or die in the environments in which they find themselves.

Biochemical Energy

All cellular functions require energy. Therefore, one of the most-fundamental biochemical functions of any cell is the production of energy in forms that are useful for maintaining the cell and ensuring its survival and reproduction. A more-accurate statement is that a cell converts nutrients, its sources of energy, into biochemically useful forms, and then uses the energy for its survival. A great part of biochemistry, and of this textbook, is concerned with the chemistry by which available energy sources are used by cells to produce biochemically useful energy. Another large part of this textbook deals with how this energy is used by cells in their essential functions of self-maintenance and reproduction.

The most-important chemical form of energy in most cells is ATP, *adenosine 5′-triphosphate.* ATP is a *high-energy molecule* found in all cells. The energy of ATP is used in cells as the *chemical free energy* that is made available by cleavage of the phosphoanhydride bonds shown in color in the adjoining structural formula for ATP.

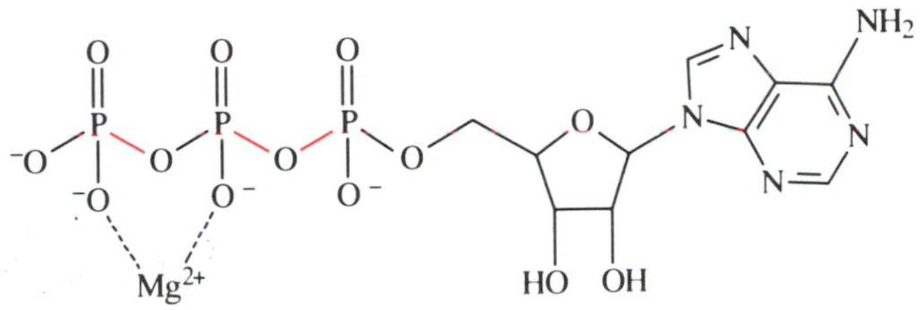

Adenosine triphosphate (ATP)

(See Chapter 9 for a discussion of chemical free energy.) ATP is shown as a coordination complex with Mg^{2+} because most of the ATP in cells exists in the form of MgATP. Moreover, in most ATP-dependent cellular processes, it is MgATP that reacts rather than ATP itself. This is well known to biochemists, who, in referring to cellular ATP, regard it as understood that ATP in cells is MgATP. In this textbook, we shall follow this convention and refer to ATP with the understanding that it is complexed with Mg^{2+} in its biochemical reactions. ATP is not the only important form of chemical free energy in cells, and it is not the only high-energy molecule. However, it is by far the single most-important form of readily available free energy in cells.

Cell Maintenance and Reproduction

Cells maintain themselves by regenerating any components that become damaged or are turned over by normal cellular processes. The macromolecular cellular components consist of *deoxyribonucleic acid* (DNA), *ribonucleic acids* (RNA), *proteins, lipids, complex carbohydrates,* and conjugates of these molecules. The chemistry of these molecules and their origins and functions will be described in considerable detail in this textbook. The nucleic acids contain and process the genetic information that specifies the structures of all cellular macromolecules. The proteins carry out many of the most-important functions required for maintaining and replicating cells. The cellular functions carried out by proteins range from the catalysis of biochemical reactions by enzymes, to muscle contraction by myosin and actin, to the transport of ions and molecules across cell membranes, to the maintenance of structure by fibrous proteins such as collagen and keratin. The lipids are major components of cell membranes, which ensure the structural and biochemical integrity of cells. The complex carbohydrates are important in maintaining cell structure and function, often as conjugates with lipids and proteins, but also in their own right as the principal constituents of such diverse substances as *chitin* in the exoskeletons of insects and heparin, a blood anticoagulant. Complex carbohydrates also play an important role in the recognition of cells by other cells and molecules.

The reproduction of cells differs from ordinary maintenance in that it requires the coordinated reproduction of all the working cellular components to produce an exact replica of a cell. Cellular reproduction in biology may be either *meiosis*, the division of germ cells, or *mitosis,* somatic cell division; in either case, it is a highly organized process that is initiated by the replication of DNA, the genetic material, and completed by the synthesis of other cellular components.

Eukaryotic and Prokaryotic Cells

Cellular organisms are classified in biology as *eukaryotes* or *prokaryotes.* The two classes differ in several respects but most fundamentally in that a eukaryotic cell has a nucleus and a prokaryotic cell has no nucleus. Eukaryotic cells are generally larger than prokaryotic cells and more complex in their structures and functions. All complex multicellular organisms, including all animals and plants, as well as many unicellular species, are eukaryotes; whereas all bacteria are prokaryotes.

Figure 1-1 is a graphic representation of the principal features of a typical eukaryotic cell. Figures 1-2 through 1-4 are electron micrographs of sections through several eukaryotic cells. The cell is encased by the plasma membrane; the cytoplasm and nucleus are inside. The *plasma membrane* is composed of a double layer, a *bilayer,* of lipids and also a number of proteins (see Chapter 29 for membrane structure and Chapter 7 for protein structure). This membrane is impermeable to most small and large molecules, but many of the proteins in the membrane are receptors and transport proteins for particular molecules that must be recognized by the cell. Such molecules may transmit signals to the cell by binding to receptors or they may be transported into or out of the cell. The

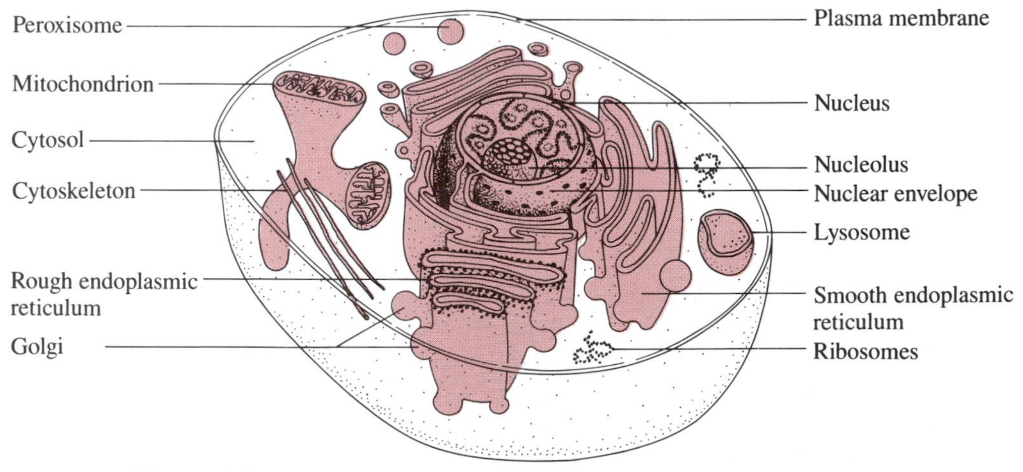

Figure 1-1

A Graphic Representation of a Eukaryotic Cell. The principal cellular compartments are the cytoplasm and the nucleus, enclosed by the plasma membrane. The organelles of the cytoplasm are the mitochondria, endoplasmic reticulum, lysosomes, peroxisomes, and Golgi apparatus. The liquid part of the cytoplasm is the cytosol.

plasma membrane is very difficult to visualize in electron micrographs of eukaryotic cells because it is so thin relative to the overall cellular dimensions. The *cytosol* is the liquid part of the cytoplasm. It consists of the water and water-soluble components, including proteins, salts, minerals, and cellular nutrients. The cytosol also contains the many small molecules that take part in the biochemistry of the cell. Cellular biochemistry includes many chemical pathways, the *metabolic pathways,* for the *anabolism* (biosynthesis) and *catabolism* (biodegradation) of molecules. The metabolic pathways are collectively the *metabolism* of the cell, and the biochemical intermediates are *metabolites.* The soluble metabolites are important components of the cytosol. Most of the principal features of eukaryotic cells are visible in the electron micrograph of a liver cell shown in Figure 1-2.

The *nucleus* contains the chromosomes, which consist mainly of DNA and *histones,* the proteins bound to DNA in chromosomes. The nucleus is enclosed

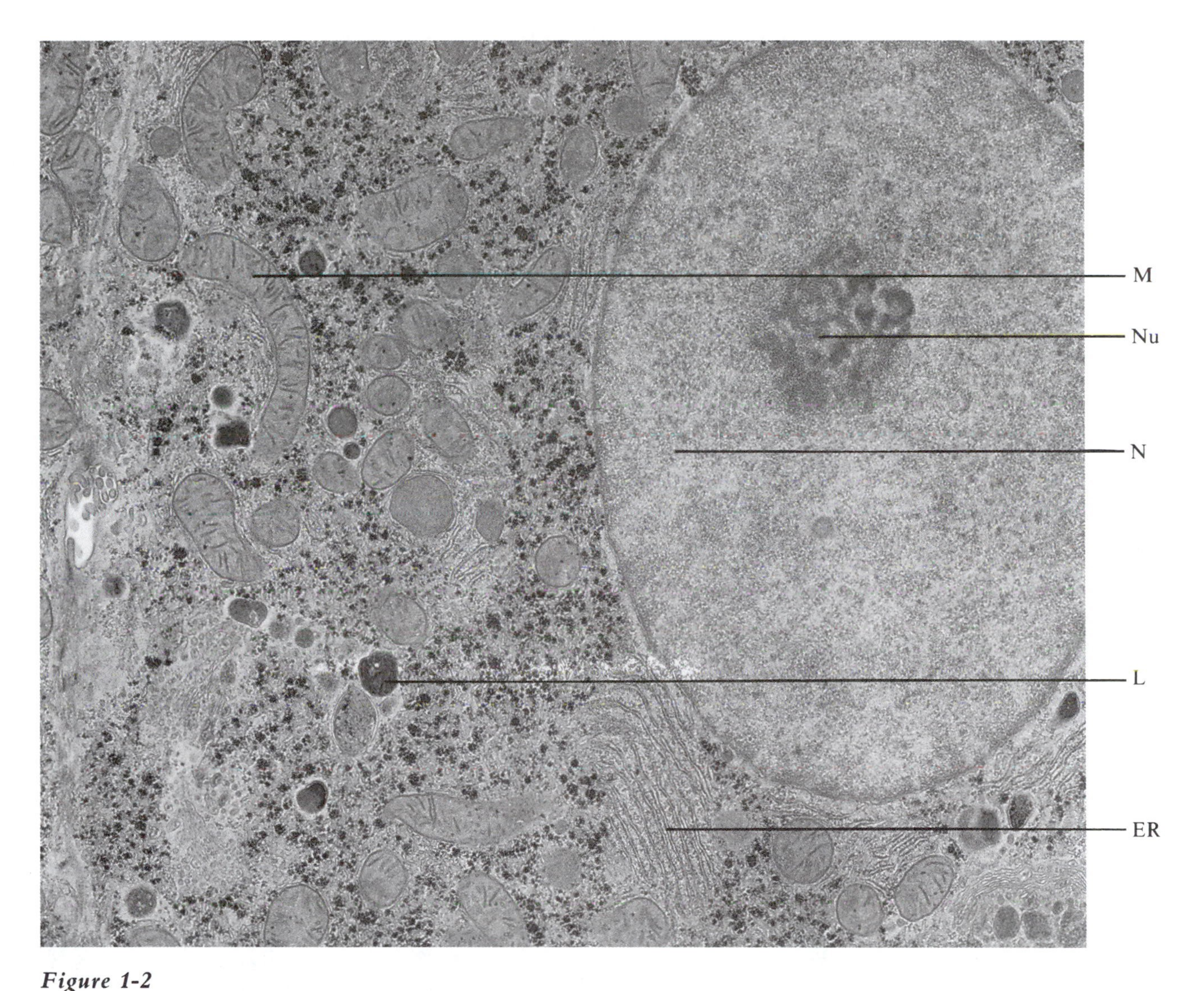

Figure 1-2

Electron Micrograph of a Section Through a Rat Liver Cell. Many of the principal parts of the liver cell are visible in this micrograph. Several are identified as follows: N, nucleus; Nu, nucleolus; ER, endoplasmic reticulum; L, lysosome; M, mitochondrion. (Courtesy of Keith Porter.)

by the *nuclear envelope,* a porous membrane that defines the nuclear boundary. Also inside the nucleus is the *nucleoplasm,* which contains the enzymes and other proteins that process the genetic information. Encoded within the structure of DNA is the genetic information specifying the structures of all the molecules in the organism. The *nucleolus* in the nucleus produces the *ribosomes,* the ribonucleoprotein particles with overall molecular weights of several million. Ribosomes are large as measured on the scale of molecular sizes, but they can traverse the nuclear envelope and enter the cytoplasm through the nuclear pores. The nuclear pores are gaps in the nuclear envelope that are shown at high magnification in Figure 1-3. In the cytoplasm, the ribosomes play a central role in the biosynthesis of proteins, which is described in Chapter 10. The genetic information in DNA is first transcribed into molecules of RNA by a process that is also described in Chapter 10. That part of the genetic information that specifies protein structures is transcribed into a special species of RNA, *messenger RNA* (mRNA). The mRNA migrates to the cytoplasm where it directs the synthesis of proteins by the ribosomes.

In addition to the cytosol, the cytoplasm contains the mitochondria, the endoplasmic reticulum, the lysosomes and peroxisomes, the ribosomes, the Golgi apparatus, and the cytoskeleton. Because their principal function is the production of ATP, *mitochondria* are the energy factories of the cell. ATP

Figure 1-3

Nuclear Pores in the Nuclear Envelope. The nuclear pore complexes in this electron micrograph are on the nuclear surface of an isolated *Xenopus* oocyte nucleus. The pores allow RNA and ribosomes to cross the nuclear envelope into the cytoplasm. (Courtesy of Professor Hans Ris.)

production in the mitochondria is catalyzed by enzyme complexes that are embedded in the inner mitochondrial membrane. This process is described in Chapter 22. Mitochondria are visible in most electron micrographs of eukaryotic cell sections, and numerous mitochondrial images can be seen in Figures 1-2 and 1-4. Mitochondrial images in a single micrograph of a cell section can give a misleading concept of the nature of a mitochondrion; indeed, for many years an incorrect picture of mitochondrial morphology was generally accepted, based in part on their appearance in cell sections. Three-dimensional reconstructions of mitochondria, derived from images in serial sections from single cells, show that they are large cellular organelles and that there are only a few per cell. In fact, certain yeast cells contain only a single mitochondrion. Many mitochondrial images appear in cell sections because the mitochondria are serpentine, and often have branches. Therefore, because the arms of a single mitochondrion wind throughout the cell, the sectioning process cuts a given mitochondrion in many places, creating many images of a single organelle.

The *endoplasmic reticulum* is a system of membranes in which protein biosynthesis and other important metabolic processes take place. This organelle contains a large number of ribosomes, which appear approximately spherical in low-resolution electron micrographs, in cells in which protein biosynthesis is the major function of the endoplasmic reticulum. The ribosomes give the organelle a rough, studded appearance, so that it is known as *rough endoplasmic reticulum.* Endoplasmic reticulum that is engaged primarily in steroid hormone biosynthesis contains few ribosomes, and this organelle is not rough; it is known as *smooth endoplasmic reticulum.* Many cells contain both rough and smooth endoplasmic reticulum, and both types are shown in Figure 1-4.

The *lysosomes* and *peroxisomes* are small vesicular bodies containing biodegradative enzymes, which decompose foreign substances that invade the cell. Peroxisomes and lysosomes are visible in Figure 1-2. The lysosomes contain hydrolytic enzymes, whereas the peroxisomes contain oxidative enzymes. These enzymes also digest the components of the cell itself as a part of the normal turnover of cellular constituents and in the event of cell death.

The *Golgi complexes* are the sites at which newly biosynthesized proteins are chemically processed and given signals that target them for transport outside the cell or to specific sites within the cell. Many proteins that are biosynthesized in the endoplasmic reticulum are converted into *glycoproteins* by attachment of oligosaccharide units, which serve as signals for directing them to the Golgi complexes. This process is described in Chapter 15.

The *cytoskeleton* of a eukaryotic cell maintains the cellular structure and plays a central role in bringing about the morphological changes that accompany cell division. The cytoskeleton consists of proteins such as *tubulin* and *actin,* which aggregate to form the *microtubules* and *microfilaments,* respectively. In muscle cells, actin filaments interact with filaments of myosin to bring about muscle contraction (Chapter 30).

Eukaryotic cells are highly diverse among species; they are found in many unicellular animals and microorganisms, as well as in all multicellular animals and plants. Eukaryotic cell diversity extends to intraspecies diversity in complex animals and plants, an obvious example being the diversity of cells in a human being. The human body consists of more than two hundred different cell types, each specialized for a particular function to support the body as a whole. Cellular diversity in higher animals and human beings can be appreciated by considering the muscle cells, which contain large amounts of the contractile proteins myosin and actin; the skin cells, which contain large amounts of the tough structural protein keratin; the fat cells of adipose tissue, which store potential energy in the form of fat; the bone cells, which contain large amounts of calcium phosphate and hydroxylapatite; and the lung cells, which are rich in

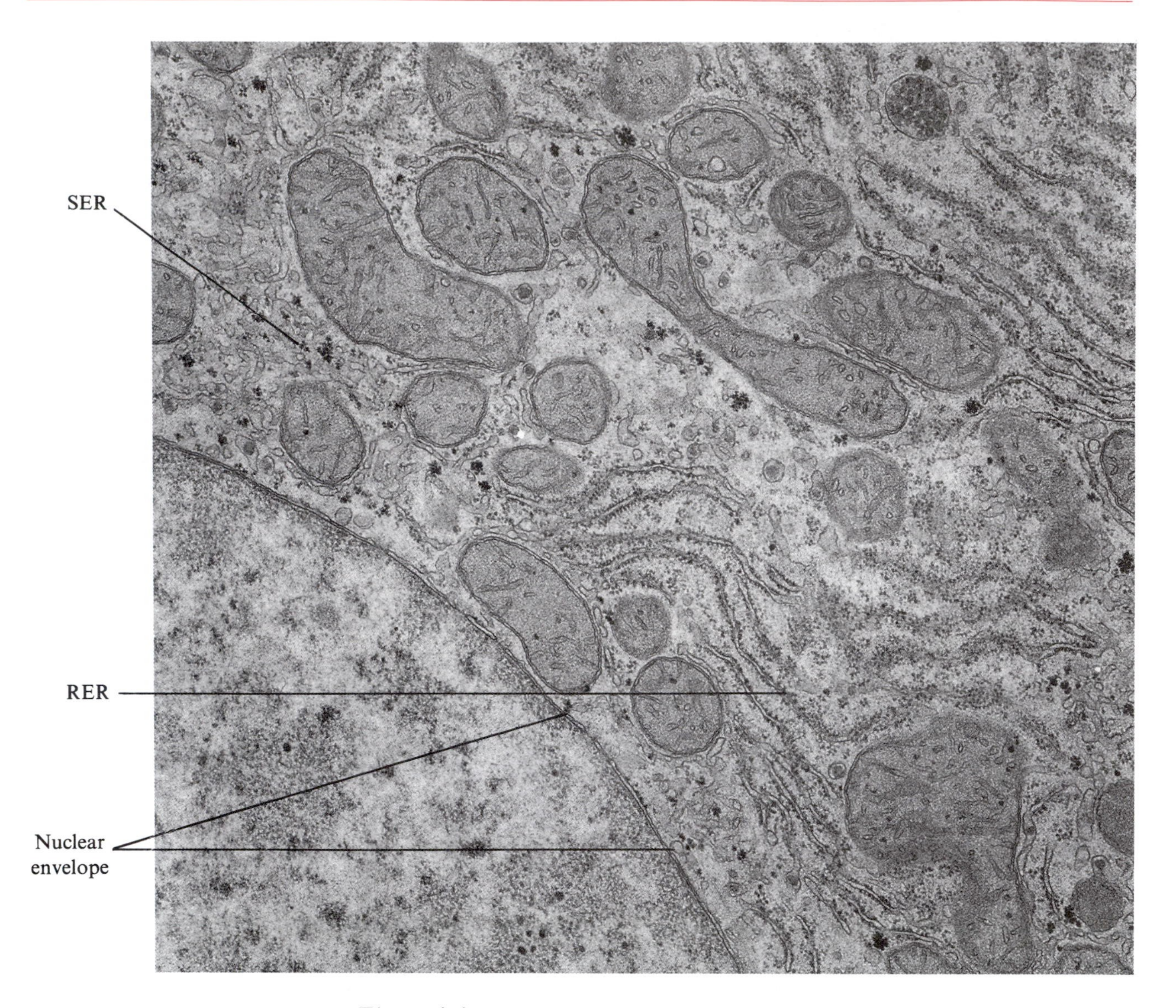

Figure 1-4

Smooth and Rough Endoplasmic Reticulum. Both smooth and rough endoplasmic reticulum are visible in this highly magnified electron micrograph of a section through a rat liver cell. The smooth endoplasmic reticulum (SER) is near the top of the micrograph and the rough endoplasmic reticulum (RER) is in the bottom half. The dark spots associated with the RER are ribosomes. (Courtesy of Keith Porter.)

the rubberlike protein elastin. Eukaryotic cell diversity extends to the protective coverings surrounding the cells. All eukaryotic cells have a plasma membrane. Certain eukaryotic cells, including yeast and plant cells, also have tough outer walls, such as that of the dividing yeast cell shown in the electron micrograph of Figure 1-5. Cell walls are composed of cross-linked polymeric materials. Plant cells also have cell walls or are implanted in a matrix of tough cellulose, as shown in Figure 1-6.

Despite the diversities among eukaryotic cells, certain structures and functions are similar for all of them because all must generate their own energy in the form of ATP. They must also undergo self-replication and maintenance. The underlying principles and mechanisms supporting these functions are

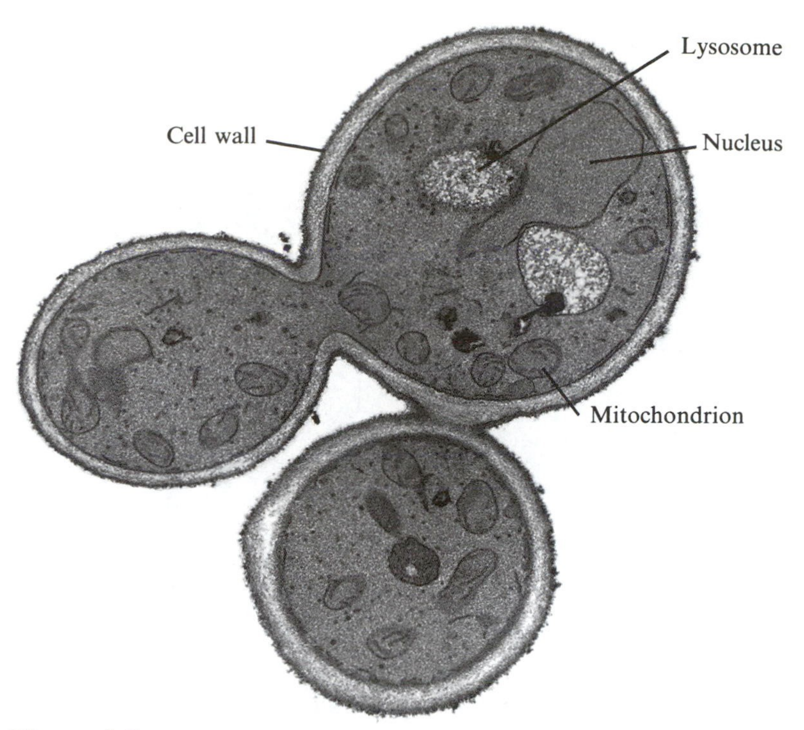

Figure 1-5

A Section Through a Budding Yeast Cell. The thick cell wall stands out in this electron micrograph, and the mitochondria, lysosomes, and nucleus also are visible. (This micrograph was obtained by the late Dr. Barbara Stevens and was provided through the courtesy of Professor Nicole Gas.)

similar in all cells, and these principles and mechanisms are the main focus of the succeeding chapters in this textbook. Some of the most-important functions of subcellular organelles are illustrated in Figure 1-7 for the nucleus, the mitochondria, the endoplasmic reticulum, and the lysosomes and peroxisomes. Nutrients entering the cell are partly metabolized in the cytosol and further metabolized in the mitochondria, which produce most of the ATP for the cell. The ATP is used as the main source of energy for the biosynthesis of cellular constituents. The nucleolus in the nucleus produces ribosomes, which are found throughout the cytoplasm and are particularly associated with the endoplasmic reticulum. Messenger RNA for protein biosynthesis is produced in the nucleus by transcription from a segment of DNA and migrates through the cytosol to the endoplasmic reticulum, where the genetic code is translated into a molecule of protein. The enzymes in the lysosomes and peroxisomes degrade cellular debris by hydrolysis in the lysosomes and oxidation in the peroxisomes. The molecules produced by intracellular digestion are often recycled to resynthesize cellular constituents.

Prokaryotic cells, the bacteria, differ from eukaryotic cells in many ways, most fundamentally in that they contain no nuclear envelope. There are other major differences in structure and organization. For example, bacteria contain no mitochondria or endoplasmic reticulum or other subcellular organelles. There are many types of bacteria, the most well known and thoroughly studied of which is *Escherichia coli (E. coli),* an inhabitant of the intestinal tract of all animals, including human beings. A section through another bacterium, a dividing *Sporosarcina ureae* cell, is shown in Figure 1-8. The dark, outer coating is the cell wall, and the constrictions in the cell envelope are the points at which the cell is dividing, first into two and then into four cells.

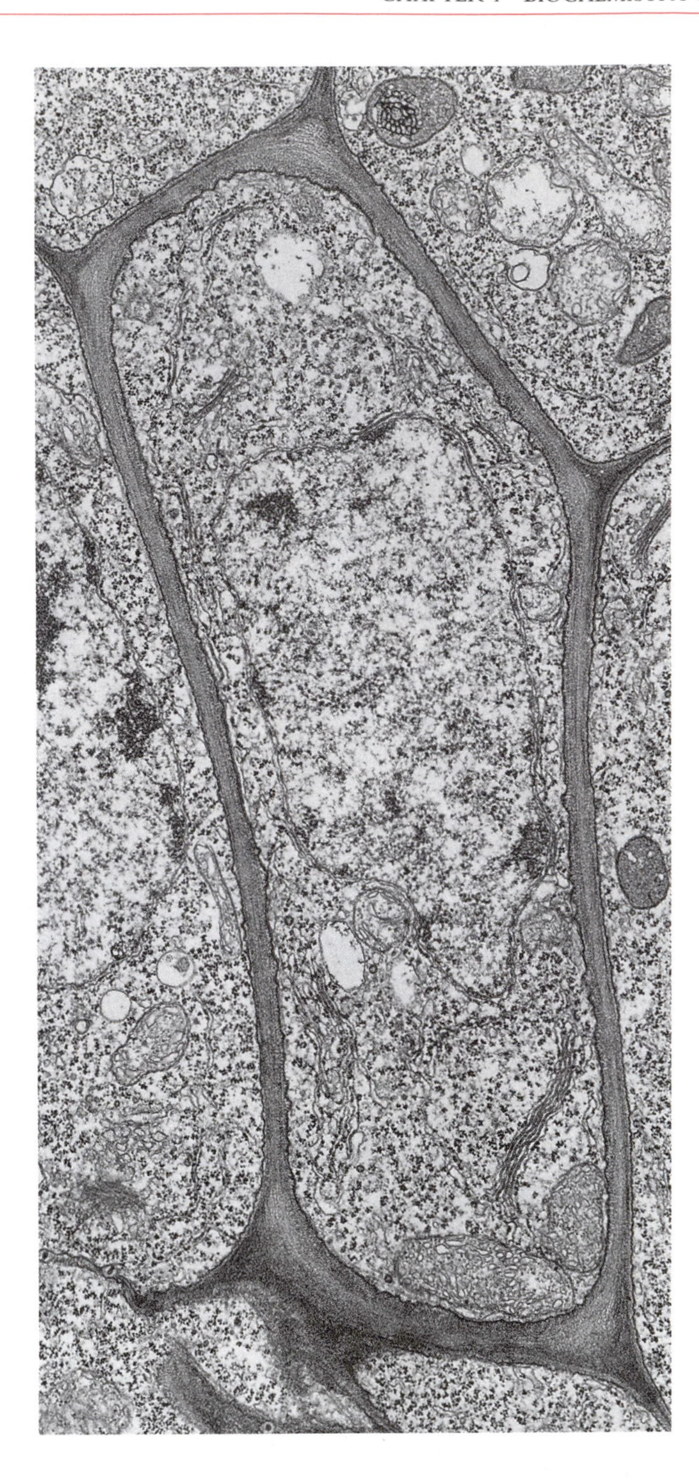

Figure 1-6

Plant Cells: An Electron Micrograph of a Section Through the Tip of a Root from the Bean *Phaseolus vulgaris*. The nucleus, plasma membrane, endoplasmic reticulum, and mitochondria are visible in this micrograph. Note that these cells do not have distinct cell walls. Rather, the cells are implanted into a matrix of tough cellulose that serves as a collective wall for all the cells. (Courtesy of Eldon Newcomb.)

E. coli and other bacteria have an outer membrane (a tough wall), a plasma membrane, and a space (the periplasmic space) separating the wall and plasma membrane. The outer membrane is composed of lipids, lipoproteins, and a few proteins, including *porins*, which allow molecules of molecular weight not higher than 600 to diffuse across the membrane. The cell wall is a thick, tough, and highly porous structure that is composed of a complex of cross-linked polysaccharides, proteins, and lipids. Polysaccharides are polymers of sugars (Chapter 15). The wall maintains the shape of the cell. The plasma membrane is similar in principle and function to the plasma membrane of a eukaryotic cell. In addition to serving as an impermeable barrier containing specific transport

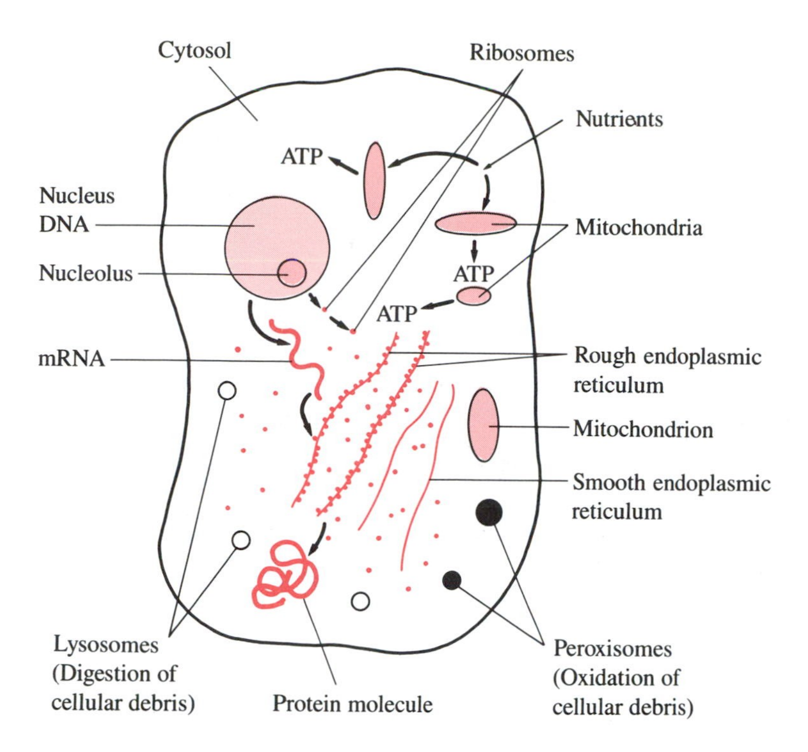

Figure 1-7

Functional Drawing of a Eukaryotic Cell. Nutrients enter the cell and are used as sources both of energy and of building materials for the biosynthesis of macromolecules. Nutrients that enter the cell are metabolized in lysosomes, peroxisomes, and the cytosol. The energy-yielding metabolism is then completed, with ATP formation, in the mitochondria. Ribosomes that are produced in the nucleus by the nucleolus pass through the nuclear envelope and migrate to the nearby endoplasmic reticulum. Messenger RNA, which is produced by transcription from DNA in the nucleus, migrates through the nuclear pores and reaches the endoplasmic reticulum, where it is translated into a protein by the ribosomes.

systems, the bacterial membrane also houses the ATP-producing apparatus that is analogous to the mitochondrial membrane systems of eukaryotic cells. The production of ATP by membrane-bound systems is discussed in Chapter 22. The cytoplasm consists of the cytosol and ribosomes, the ribonucleoproteins that play a central role in protein biosynthesis. Bacterial metabolism, including protein biosynthesis, takes place mainly in the cytosol. The genetic information in the DNA is transcribed into molecules of RNA, as described in Chapter 10, and the resulting messenger RNA (mRNA), ribosomal RNA (rRNA), and precursors of transfer RNA (tRNA) interact in the cytoplasm and bring about the synthesis of proteins (Chapter 10).

Because eukaryotic cells are complex and contain nuclei, mitochondria, endoplasmic reticulum, lysosomes, peroxisomes, Golgi complexes, and a cytoskeleton, they are much larger than prokaryotic cells and contain more DNA. Prokaryotic cells range from 0.5 to a few micrometers in diameter or length, depending on their shapes, and reproduce rapidly. *E. coli* typically undergo cell division every 20 to 40 minutes, depending on growth conditions. Eukaryotic cells are typically more than ten times as large in linear dimensions and thousands of times as large in volume as prokaryotic cells. This may be responsible for their longer doubling times—many hours to several days—in culture.

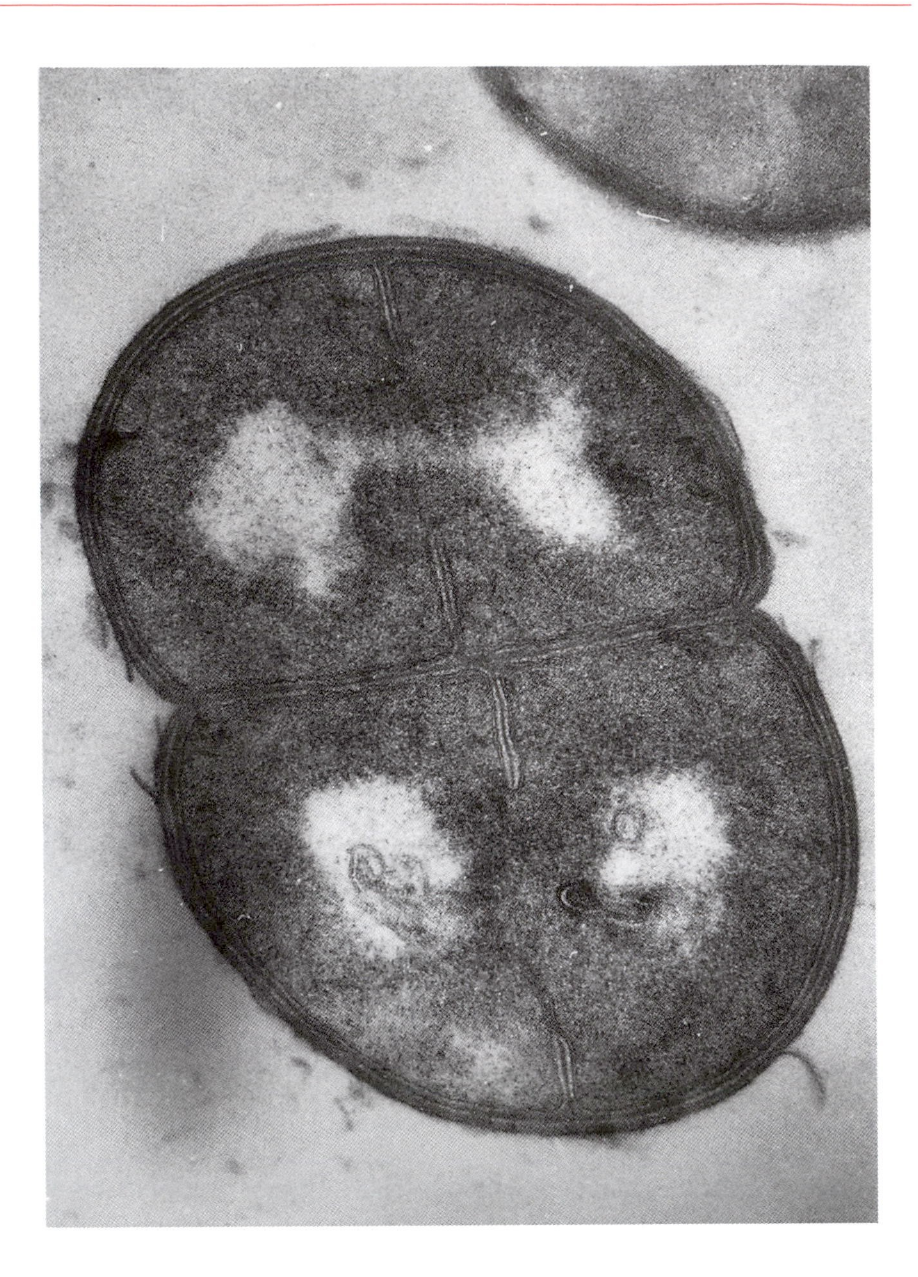

Figure 1-8

Electron Micrograph of a Section Through Dividing Bacteria. This is a thin section of *Sporosarcina ureae,* which is a gram positive coccus. One complete septum has formed between daughter cells, and growing septa are forming within each daughter cell at 90° to the original division plane. The cells are approximately 1 μm in diameter. (Micrograph courtesy of T. J. Beveridge, Department of Microbiology, University of Guelph, Guelph, Ontario, Canada.)

Multicellular Organisms

Higher animals, plants, and human beings consist of organized multicellular systems, in which each system is itself supported and maintained by intercellular interactions and in which each system interacts with other systems to support the organism as a whole. In human beings and animals, the various multicellular systems are the organs. The major organs of human beings and higher animals include the central nervous system, the liver, the heart and circulatory system, the lungs, the skin, the skeleton, the muscles, the digestive system, the reproductive organs, and the secretory glands. Each organ is composed of cells that must stick together through intercellular interactions to form and maintain the structure of the organ. Intercellular interactions within an organ entail molecular recognitions and attractions that are poorly understood and are a subject of considerable interest to cell biologists. Intercellular interactions have a biochemical basis that will become defined as our knowledge of cell biology and biochemistry increases.

Hormones and Communications

The various human or animal organs must function in harmony to support the life of the creature as a whole. For example, it would not do for the digestive system to operate in the absence of food, because the digestive juices are themselves corrosive to the digestive tract. The corrosiveness of digestive fluids is well understood by victims of peptic ulcers, which may result from the overproduction of acid and the digestive enzyme pepsin. The digestive process is discussed in more detail later in this chapter. The harmonious functions of the organs are controlled by the brain and the peripheral nervous system.

The commands from the brain to the various organs follow several routes. The central nervous system connects the brain to the muscles through the nerves and controls muscular contraction. The connections between the brain and other organs are mediated by the secretory glands and the circulatory system. The secretory glands produce and release signal molecules, *hormones,* that are carried to specific organs and control their functions. Hormone secretion is of two types: secretion through ducts directly into the target organs and secretion into the bloodstream, which carries the hormones to the target organs. Hormones that are carried by the bloodstream circulate throughout the body to the cells, most of which do not recognize them. Only the specific target organs recognize the hormones intended for them, because only they have the recognition factors, the specific *receptors,* on their cell surfaces. The receptors are protein molecules embedded in the plasma membranes of cells that recognize and bind hormones and other specific signal molecules.

Figure 1-9 illustrates the sites of the major endocrine glands of the human body. *Endocrine glands* secrete hormones into the bloodstream; the major glands are the *pituitary glands,* the *thyroid,* the *adrenal glands,* the *pancreas,* and the *gonads* (testis and ovary). Many of the endocrine glands are activated by

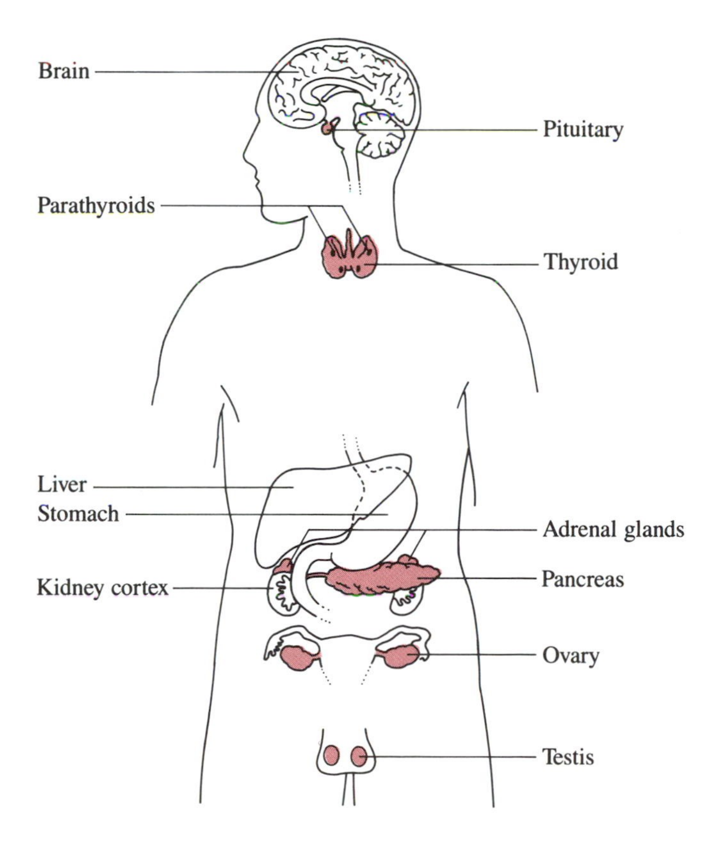

Figure 1-9

Endocrine Glands of the Human Body. The hypothalamus of the brain releases factors into the pituitary, stimulating it to secrete hormones into the bloodstream. The pituitary hormones are absorbed by specific receptors on the surfaces of target cells. The target cells for many pituitary hormones are in the other endocrine glands, which secrete other hormones into the blood stream. Some of the pituitary hormones are absorbed by organs and stimulate specific functions. The hormones that are secreted by the thyroid, adrenal glands, pancreas, testis, and ovary also stimulate the function of specific organs after they are taken up by receptors on the cell surfaces of those organs. Table 1-1 lists some of the hormones, the glands that secrete them, and some of the functions that they stimulate.

releasing factors secreted by the *hypothalamus* on command from the central nervous system. The hypothalamus is located next to the pituitary glands at the base of the brain, and it secretes the releasing factors directly into the pituitary glands. The pituitary glands are the master endocrine glands because they secrete hormones that control several other endocrine glands. The names and general functions of some of the hormones secreted by glands are listed in Table 1-1. The hormones in Table 1-1 include examples from the classes of known hormones, but this table is far from comprehensive in scope. Many more hormones are known, and those produced by the endocrine system are by no means the only signal molecules. Many other intercellular signal molecules are known, and the known signal molecules certainly constitute only a fraction of the total number in a complex animal.

As molecules, the hormones fall into the following three chemical types: the steroid hormones, hormones derived from amino acids, and the polypeptide hormones. Representative examples of steroid hormones and those derived from amino acids are shown in Figure 1-10, together with the structure of the amino acid *tyrosine,* which is related as a biosynthetic precursor to several amino acid hormones. The steroid hormones shown are the sex hormones testosterone (male) and estradiol (female), which are secreted by the gonads and stimulate the development of secondary sex characteristics, as well as sexual function. The biosynthesis of sterols, which are precursors of steroid hormones, is described in Chapter 28. Amino acids are discussed in succeeding chapters, and the biosyntheses of representative amino acids are presented in Chapter 26. The structural formulas for the hormones *thyroxine* and *epinephrine* (adrenaline), both of which are derived from the amino acid tyrosine, are shown in

Table 1-1

The major endocrine glands and the hormones that they secrete

Endocrine Gland	Hormones	Structural Class	Stimulation Function
Anterior pituitary	Oxytocin	Nonapeptide	Uterine contraction, milk flow
	Vasopressin	Nonapeptide	Blood pressure, water balance
Posterior pituitary	Corticotropic (ACTH)	Polypeptide	Adrenal steroid synthesis
	Growth hormone	Polypeptide	General biosynthesis Growth-factor release
	Luteinizing hormone	Polypeptide	Ovary functions
	Follicle-stimulating hormone (FSH)	Polypeptide	Ovulation
	Prolactin	Polypeptide	Milk synthesis
	Thyrotropin	Polypeptide	Thyroid
Adrenal cortex	Glucocorticoids	Steroids	Diverse, protein synthesis, control of metabolism, etc.
Adrenal medulla	Epinephrine	Tyrosine derivative (Figure 1-10)	Energy production Heart function
Thyroid	Thyroxine	Tyrosine derivative	Many cellular effects
Pancreas	Insulin	2 polypeptides	Glucose absorption
	Glucagon	Polypeptide	Release of glucose from cellular glycogen
Ovary	Estrogens	Steroids (Figure 1-10)	Development and function of secondary sex organs
Testis	Androgens	Steroids (Figure 1-10)	Development and function of secondary sex organs

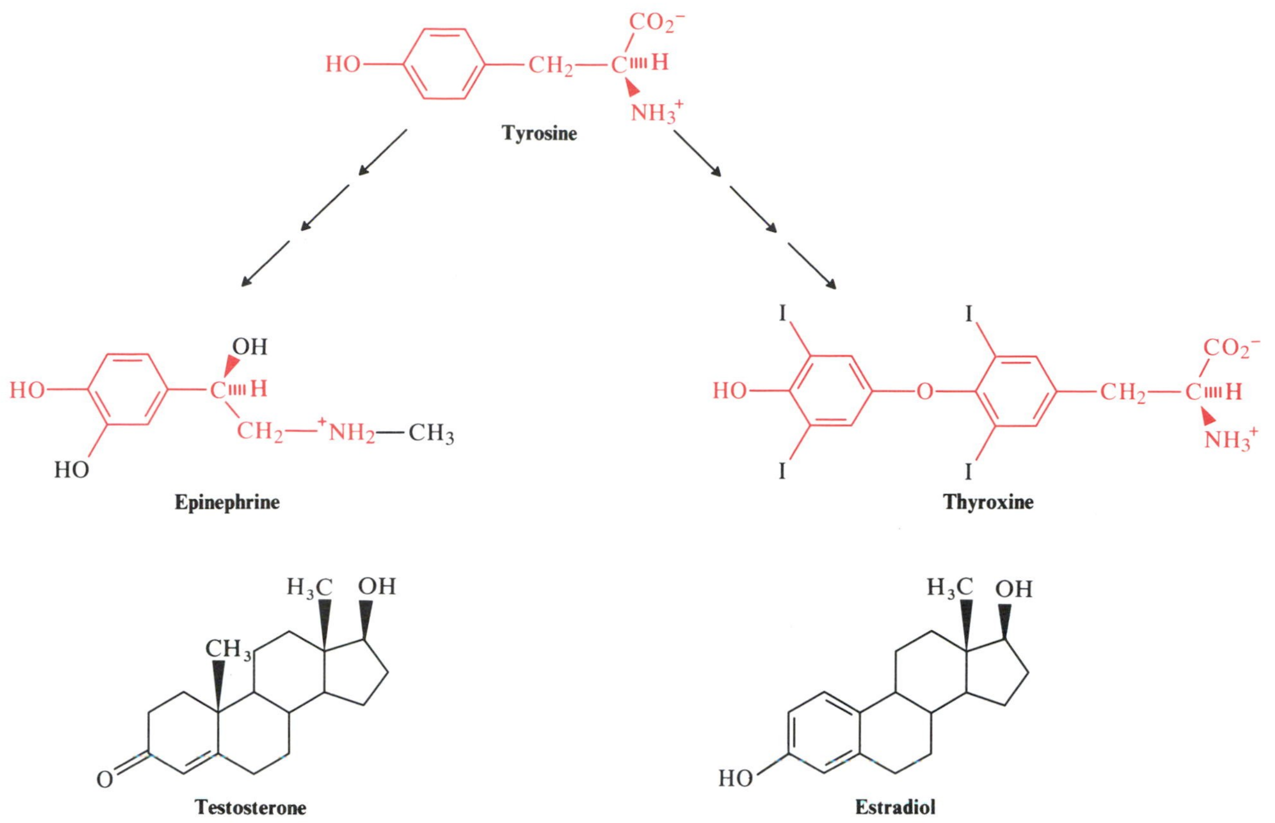

Tyrosine

Epinephrine

Thyroxine

Testosterone

Estradiol

Figure 1-10

The Structures of Some Hormones. Thyroxine and epinephrine are derivatives of the amino acid tyrosine. Thyroxine, which is secreted by the thyroid gland, stimulates many cellular functions, including energy utilization. The two steroid hormones shown are the sex hormones testosterone and estradiol.

Figure 1-10 and compared with the structure of tyrosine. Thyroxine is secreted by the thyroid gland and stimulates many cellular processes. Epinephrine is secreted by the adrenal medulla and stimulates the catabolism of carbohydrates in muscle cells. Polypeptides are polymers of amino acids and are discussed in Chapter 7. The master endocrine glands, the pituitaries, secrete only polypeptide hormones, and the releasing factors secreted by the hypothalamus to the pituitaries also are polypeptides.

Upon binding its cognate hormone, a cell-surface receptor transmits a signal into the cell and switches on a specific cellular process. The hormone often may not enter the cell; so there must be a mechanism, a *signal transduction,* by which the presence of the hormone is translated into a signal that is recognized inside the cell, where the relevant biochemical changes take place. As a result of recent research, the mechanisms of signal transduction from cell surfaces into cells are becoming understood in chemical terms. In signal transduction, the binding energy that is made available by the interaction of a receptor with a signal molecule, such as a hormone, is used to activate some biochemical process. The binding energy is usually used to drive a change in the conformation of a receptor, which may activate an enzyme or an ion channel. The use of binding energy to drive conformational changes is discussed in Chapter 5, and the means by which the signal from the hormone epinephrine is transmitted to the energy-producing systems of cells is described in Chapter 21.

Digestion

A biochemical process that all people experience is the digestion of foods. Certainly, a failure in digestion is a memorable experience for everyone. Digestion breaks down the complex molecules in foods to the simple molecules that can be used in metabolism for the generation of energy and the biosynthesis of cellular components. Digestion of food is moderately well understood in biochemical terms for human beings and animals. Digestion in biochemistry extends to the breakdown *within cells* of complex compounds to simple molecules that can be either excreted or used in cellular metabolism. Both of these digestive processes are essential to the well being of complex animals.

Digestion of Food

The major nutrients in food that require digestion are proteins, starches, and fats, all of which are large molecules that are digested by hydrolysis into smaller molecules. Proteins are long-chain polymers of α-amino acids, and starches are polymers of sugars, whereas fats are *triglycerides* or *phospholipids*, which are glycerol esters of fatty acids. (The chemical structures of triglycerides and phospholipids are given in Chapter 11 and those of sugars and starches are in Chapter 15.) The digestive reactions are catalyzed by enzymes in the digestive tract. Enzymes are themselves proteins and are often defined as biological catalysts.

The first enzyme encountered by food during ingestion into the mouth is *salivary amylase*. This enzyme initiates digestion by catalyzing the partial hydrolysis of starches into shorter oligomers. Its main importance may be to protect teeth from decay by digesting adhering food particles into soluble molecules that flow easily into the stomach, thereby removing potential nutrients for bacteria. Food quickly reaches the stomach, which normally contains 0.01 M to 0.1 M hydrochloric acid (pH 2 to 1*) and the enzyme *pepsin*, a member of the class of enzymes known as proteases, which catalyze the hydrolytic breakdown of proteins to smaller polypeptides. Unlike most enzymes, pepsin is stable and catalytically active at the low pH values in the stomach, where the hydrochloric acid disrupts the natural three-dimensionally folded structures of most proteins by converting them into unfolded, *denatured* forms. Denaturation exposes proteins to cleavage by pepsin and other proteases encountered further along the digestive tract. Hydrochloric acid and pepsin evidently are not truly essential for digestion, however, because people who have had their stomachs removed can still digest foods reasonably well.

Most of the digestive enzymes are in the small intestine, where they are secreted after being synthesized in the pancreas or the lining of the intestine. Many of these enzymes are produced and secreted in the form of proenzymes, or *zymogens*, which are inactive and will not digest the tissues in which they are synthesized and stored. Other enzymes activate them when they reach the small

*The pH is (approximately) the negative logarithm of the hydrogen ion concentration; that is, $pH = -\log[H^+]$, so that pH 2 is 10^{-2} M hydrogen ion. At neutrality, the pH is 7, corresponding to 10^{-7} M hydrogen ion. See the Appendix for definitions of pH, pK_a, and their relations.

intestine. The activated proteases include *trypsin, chymotrypsin, elastase, carboxypeptidases,* and *aminopeptidases,* all of which attack proteins and cleave them into smaller fragments. Trypsin, chymotrypsin, and elastase are known as *serine proteases,* in recognition of the importance of the amino acid serine in their mechanism of action. The actions of these enzymes and the role played by serine are discussed in detail in Chapters 3 through 5.

Nutrient fats are cleaved by *lipases* and *phospholipases,* which have the remarkable property of bringing about the hydrolytic breakdown of water-insoluble molecules after attaching themselves to the fat particles. This process is facilitated by bile, a natural solution of detergents that is synthesized in the liver, stored in the gall bladder, and then secreted into the small intestine.

After proteins are hydrolyzed, the free amino acids enter the cells lining the intestine through specific transport systems. There also are transport systems for the individual sugars derived from the hydrolysis of carbohydrates. Some of these carrier systems are capable of *active* transport, in which the transported molecules are moved against a concentration gradient by the utilization of energy from ATP or some other source. Other systems bring about *passive* transport by a process of facilitated diffusion, in which molecules move down a concentration gradient and no energy is utilized in the transport process. In facilitated diffusion, specific carrier molecules within cell outer membranes bind the molecules and relay them across the membranes into the cells. Some of these transported molecules, especially neutral molecules such as sugars, undergo phosphorylation inside the cells. An example of such a sugar phosphate is glucose-6-phosphate, the chemical structure of which is shown in the margin. The negative charges on glucose-6-phosphate prevent it from leaking out of the cells, which makes it possible to maintain a high concentration of this energy source within a cell. Phosphates in general are negatively charged in neutral solutions, and they are retained within cells because the cell membranes are impermeable to most ions.

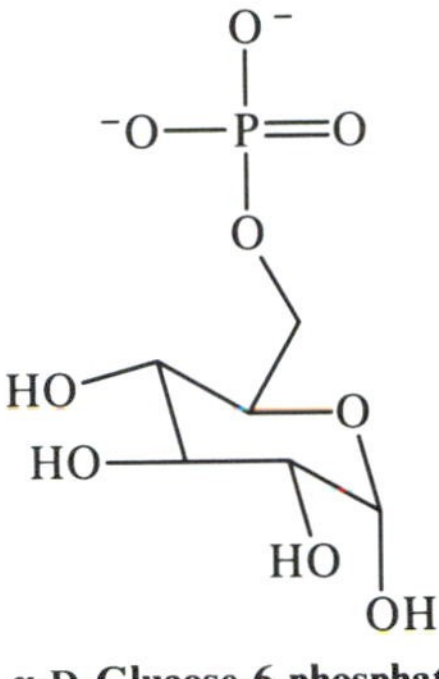

α-D-Glucose-6-phosphate

The absorption of fatty acids and partially digested fat globules probably occurs by simple diffusion across the cell membrane. The resynthesis of fats from the fatty acid molecules transported into the cell diminishes their intracellular concentration and maintains a concentration gradient across the membrane that facilitates the continued diffusion of fatty acids into the cell. The fats may undergo secretion into the lymphatic circulation as *chylomicron particles* or be incorporated into *lipoproteins* and transported into the lymph or bloodstream. Chylomicrons and lipoproteins are complexes of fats, other lipids, and proteins and are described in Chapter 28.

Why Digest Food?

Why is digestion necessary? Hydrolysis of nutrient polymers to monomers or dimers, transport of these molecules into cells, and resynthesis of similar or identical polymers inside the cells require a large number of enzymes. Moreover, in the hydrolysis of polymers, a significant amount of energy is lost; and the resynthesis of new polymers requires the input of more energy that might be saved by the absorption and utilization of intact polymers.

The immediate reason that intact polymers from nutrients are rarely if ever used directly is that the cells are impermeable to polymers. The impermeability of cells to polymers is essential for the maintenance of cell structure and the control of metabolism. A basic problem that must be solved by every cell is the maintenance of an internal environment that is very different from the extracellular environment. It must concentrate and store foodstuffs and other essential chemicals; it must also minimize the entrance of undesirable chemicals.

The construction of large polymers and aggregates from small molecules is a fundamental part of the development and maintenance of cells and tissues. The impermeability of membranes to polymers is an important design feature for keeping enzymes and other proteins inside cells, where they carry out their metabolic functions. Sugars are stored as large polymers of starch or *glycogen* (animal starch), and fats are stored as insoluble droplets and particles that are also retained inside cells for later use as needed. When energy is required, these stored particles and polymers can be converted into small molecules that can be used for metabolism and eventual production of ATP. It is important for these storage forms to be large polymers or particles. If these materials were stored inside a cell as many small, soluble molecules and somehow prevented from crossing membranes, a large osmotic pressure would develop within the cell that no cell wall could resist, and the cell would explode.

Another reason for the digestion of polymers is that living organisms go to great lengths to preserve their own particular identities. The individuality of a species and of an organism is expressed in the specific structures of the large polymers within their cells, and it is important to destroy such structures in nutrients derived from other species before these nutrients are incorporated into the cells of an ingesting organism.

Intracellular Digestion

Particles and droplets of potential nutrients or toxic substances in the interstitial spaces between cells are absorbed into cells, and there are mechanisms for digesting these substances within the cells. The absorptive processes, *endocytosis,* are known as *phagocytosis* for the absorption of particles and *pinocytosis* for the absorption of extracellular fluid. The mechanism of endocytosis entails the initial invagination of the cell membrane containing the particle or droplet. The invagination closes upon itself to form a small vesicle in the interior of the cell, as is illustrated in Figure 1-11. The contents of the resulting vesicles are not directly utilized by the cell. Instead, the vesicles fuse with the lysosomes, which contain a collection of digestive enzymes that attack polymers and degrade them to their monomers. Digestion at the cellular level protects the cell against foreign materials while salvaging useful nutrients.

Many of the digestive enzymes in lysosomes are proteases, which are collectively known as *cathepsins.* Lysosomes contain some forty different hydrolytic enzymes, which attack proteins, polysaccharides, nucleic acids, and lipids. The interior of lysosomes is more acidic than the cytoplasm, and this acidity facilitates the actions of lysosomal processes. Lysosomes are variously called the "hyena" and the "garbage can" of the cell; they are intracellular scavengers that play the important role of digesting large molecules that have been brought into the cell by pinocytosis or phagocytosis.

Other intracellular scavengers are the peroxisomes, which contain oxidizing enzymes that attack a wide variety of ingested molecules by oxidative rather than hydrolytic chemistry. Peroxisomes are morphologically similar to lysosomes but carry out oxidations rather than hydrolyses, as mentioned earlier. These particles are especially abundant in *leukocytes,* the white cells of the blood, which ingest and then digest bacteria and other foreign agents in the bloodstream and tissues.

The lysosomes and peroxisomes, as lines of defense, provide partial protection against foreign materials and microorganisms, although they occasionally fail. One unhappy consequence of failure can be testified to by those who have suffered from food allergies and certain types of food poisoning. More-

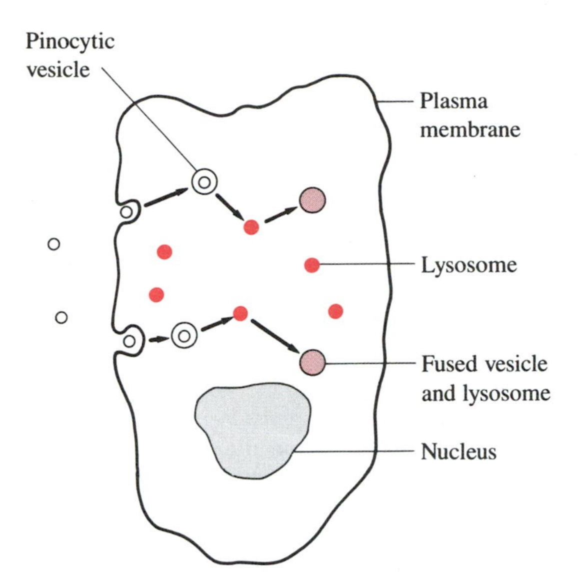

Figure 1-11

Absorption of Fat Globules into a Eukaryotic Cell. The small fat bodies are attracted to the plasma membrane, which forms invaginations that eventually seal off from the membrane to form vesicles in the process of *pinocytosis*. The pinocytic veiscles consist of an outer membrane, which is part of the cell membrane that separated when the invagination was sealed off, and the fat globule contained within the vesicle. The pinocytic vesicles undergo fusion with lysosomes, so that the vesicle and lysosomal membranes are fused together and the fat globules are inside the lysosomes. The lysosomal enzymes digest the fat globule. The fragments of plasma membrane fused with the lysosomal membrane are recycled to the plasma membrane.

serious consequences arise from invasion by microorganisms that cause infectious disease.

The digestive functions of lysosomes and peroxisomes are not limited to the breakdown of foreign substances. In fact, defense against foreign bodies may be a secondary function of these organelles. The principal function may be the digestion of cellular constituents during the normal turnover of cellular components. In eukaryotes, the lifetimes of individual proteins are measured in hours or days. Cellular proteins, especially enzymes, are constantly biosynthesized and degraded. The difference between the rates of biosynthesis and degradation determines the intracellular concentration of an enzyme. Controlled variation in these rates provides a mechanism for regulating the concentration of a particular enzyme within a cell. The degradative process is carried out by enzymes within the lysosomes, whereas the biosynthesis of proteins is carried out in the endoplasmic reticulum, as described earlier in this chapter. The concentrations of other cellular constituents are similarly regulated. In addition, when a cell dies, the lysosomal membranes break and the released enzymes solubilize the cell debris. This process is aided by the leukocytes. Proper lysosome function is very important to the health of cells and multicellular organisms, and their functions can be compromised when they absorb indigestible substances (Box 1-1).

BOX 1-1

THE HEALTH CONSEQUENCES OF LYSOSOMAL LOADING

The importance of lysosomes is highlighted by the consequences of their malfunctions in human cells. The effects of several genetic diseases arise from the absence of a lysosomal enzyme. Some of these diseases are not immediately lethal, but they have serious consequences that lead to misery, anguish, and death. Tay-Sachs disease, for example, is caused by the hereditary deficiency of a lysosomal enzyme that normally degrades complex intracellular lipopolysaccharides known as *gangliosides* or *glycosphingolipids*. Breakdown of these molecules by hydrolysis is an essential biochemical function in the turnover of cellular constituents. If the lipid is not degraded, it accumulates in the lysosomes of the brain cells of infants with Tay-Sachs disease and causes mental deterioration and death. Although the disease is uncommon, the recessive gene that causes it has been estimated to be present in 3% of the Jewish population in New York City.

The expression *lysosome loading* refers to the collective phenomena leading to the accumulation of undigestible matter within lysosomes. Undigested matter may be cellular substances that would normally be digested but remain undigested for want of an enzyme that should be present but is not, as in Tay-Sachs disease. Alternatively, undigested matter in lysosomes may be foreign substances that are simply not subject to biochemical degradation within cells. Over time, such matter can become voluminous in certain types of cells. Most of the molecules in a brain cell (an exception being DNA) turn over rapidly, from hundreds to more than a hundred thousand times in the lifetime of an individual. However, nerve cells themselves do not divide significantly because brain is not a regenerative tissue, and so they are never replaced. Therefore, although the lysosomal membranes turn over constantly, the indigestible contents of nerve-cell lysosomes continue to accumulate throughout life. This means that over time the lysosomes become loaded with indigestible foreign matter. The phenomenon of lysosome loading is a significant factor in aging, because excessive loading in lysosomes has deleterious effects on cell function and can lead to cell death. If a means could be found to induce nerve cells to divide slowly, without harming brain function, the problem of lysosome loading might be overcome, because the dying cells could be replaced by new cells with new lysosomes.

A well-known consequence of lysosome loading is the damage to lung cells that results from the inhalation of toxic materials, such as smoke. Excess smoke can cause the serine protease *elastase* to be released from lysosomes when they become overloaded and rupture. This has a seriously debilitating consequence to the lungs. Normal respiration depends on the elastic properties of the protein elastin, which brings about passive contraction of the lung in exhalation after an active inhalation. Cleavage of elastin by elastase causes a loss of the elasticity of the lung and leads to a defect in respiration that appears clinically as emphysema.

Viruses

Animal, plant, and bacterial cells are susceptible to invasion and infection by virus particles. Viral infections undermine the normal functions of cells, which often undergo lysis and cell death as the ultimate consequence of viral infection. Virus particles can drastically alter the intracellular biochemistry to the detriment of the cell and the advantage of the virus. Diseases of animals that are caused by viruses include the common cold, measles, hepatitis, poliomyelitis, and acquired immune deficiency syndrome (AIDS). Plant viruses cause various diseases in plants. Some virus particles are put to beneficial use, however, in recombinant DNA research and in the biotechnology industry. Some of the practical biochemical uses of bacterial viruses are described in Chapter 14.

A virus is a complex particle consisting of a molecule of a nucleic acid, either DNA or RNA, that is encased within a coating of proteins. The nucleic acid contains the genetic information that specifies the structures of the molecules contained within the virus. However, viruses do not have a metabolism, do not absorb nutrients, cannot produce ATP, contain only a few enzymes, and are incapable of independent self-replication. Virus particles vary in size and morphology from a millionth of a centimeter in diameter for the smallest spherical viruses to a hundred thousandth centimeter in length for filamentous viruses. Spherical viruses are not actually spherical in shape, but rather

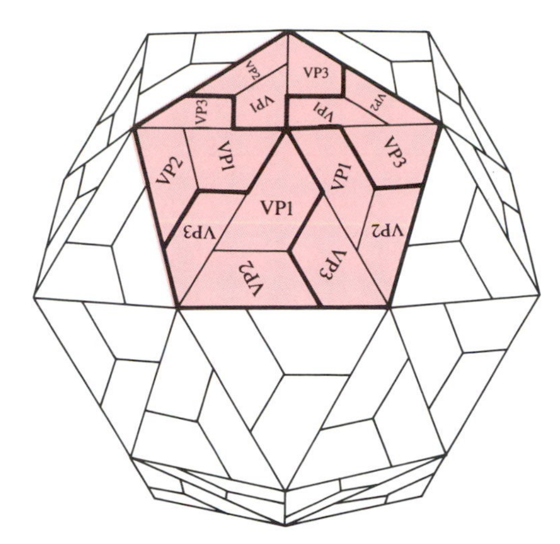

Figure 1-12

The Icosahedral Structure of a Small Virus Particle. The icosahedral structure is typical of small RNA viruses such as human rhinovirus, the virus of the common cold, tomato bushy stunt virus, and southern bean mosaic virus. The protein coat assemblage shown here is for human rhinovirus 14. Each virion shell contains sixty protomers consisting of four different proteins: VP1, VP2, VP3, and VP4. Each facet of the icosahedron is composed of one subunit each of VP1, VP2, and VP3. VP4 is an internal structural protein. The RNA is encased by the protein shell. The overall morphology of this virus is known from electron microscopy, and the molecular structure has been determined by X-ray crystallography. (From M. G. Rossman et al. Adapted with permission from *Nature*, vol. 317, p. 145. Copyright © 1985 Macmillan Magazines Limited.)

polyhedral, as shown by the icosahedral structure in Figure 1-12. The icosahedral structure is typical of small viruses. The molecular weight of the common cold virus depicted in Figure 1-12 is about 8.5×10^6, and the diameter is about 300 Å (1 Å, an angstrom, is 10^{-10} meter, or 0.1 nanometer). These dimensions are comparable to those of the largest known multienzyme complexes, such as the pyruvate dehydrogenase complex described in Chapter 17. Other virus particles are much larger and can be much more complex in structure.

Although they are not free living forms, virus particles have the capacity to enter and infect living cells. Upon infecting a cell, a virus particle subverts many normal cellular functions and turns the cellular machinery to the process of replicating the virus. The first stage of this process is the replication of the viral genome, which is generally catalyzed by specific viral enzymes acting in concert with host enzymes. After the viral genome has been replicated, it is transcribed and translated into the viral molecules by the enzymes and biosynthetic machinery of the host cell. The ATP and molecular building blocks for this biosynthesis are supplied by the metabolism of the host. The newly synthesized viral molecules are then assembled into new virus particles.

The life cycles of different viruses vary. The sole function of *lytic viruses* is self-replication within the cells that they infect, and typical phases in the life cycles of many lytic viruses are illustrated in Figure 1-13. An electron micrograph of a typical virus particle appears in Figure 1-14. Viral replication produces many copies of the virus, which become physically overwhelming. Under this stress, the cell undergoes lysis, dies, and spills out hundreds of virus particles. The free virus particles then invade many other cells and repeat their life cycle. Many of the symptoms of viral diseases are caused by energy depletion and cell death.

Some viruses remain dormant within cells for various periods of time, even years, before undergoing replication and completing the infectious cycle. *Lysogenous viruses* do not undergo replication immediately after they infect a cell. Infection by a lysogenous virus normally leads to the initial incorporation of the viral genetic information into the host chromosome, so that the viral genes become integrated into the host genome. The viral genes may then direct the production of viral particles, leading ultimately to cell lysis. Alternatively, the viral genes may reside in a dormant state within the host genome for a long period of time before becoming activated. In either case, integration of the viral genes into the host genome is not normally reversible. The AIDS virus is an example of a lysogenous virus that may reside in a dormant state for many years before becoming active. A few viruses carry *oncogenes*, the genes that cause infected cells to be transformed into malignant tumor cells (Box 1-2).

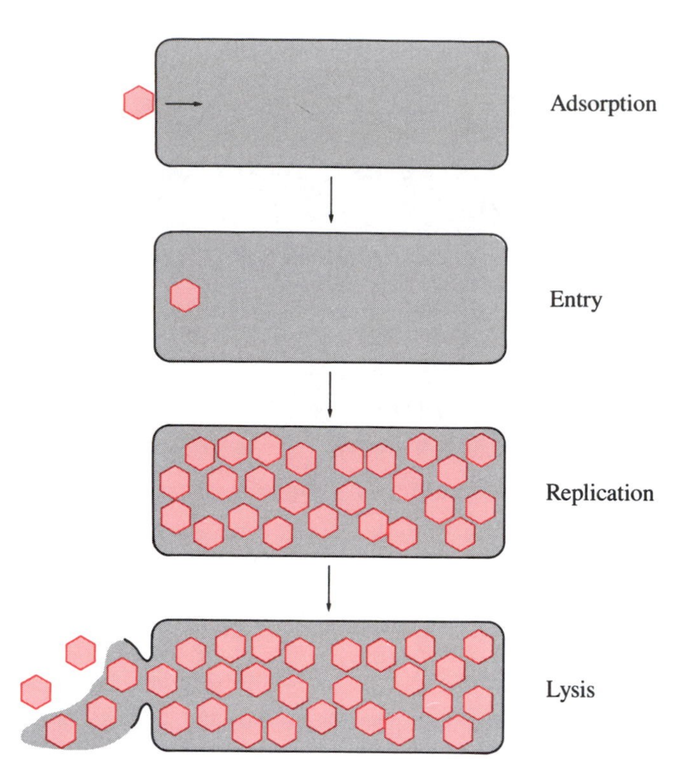

Figure 1-13

The Life Cycle of a Lytic Virus. Lytic viruses become attached to the cell membrane and penetrate into the cell. Once inside the cell the virus takes over much of the biosynthetic machinery for nucleic acids and proteins and uses it, together with cellular energy sources and molecular building blocks, to proliferate the virus particles. When the cell becomes filled with virus particles, it undergoes lysis and spills out hundreds or thousands of viruses.

BOX 1-2

ONCOGENIC VIRUSES AS CARRIERS OF ONCOGENES

Oncogenic genes are mutant forms of mammalian genes rather than essential parts of the viral genome. Oncogenic viruses contain these extraviral genes as a result of the life history of their ancestors. A cancerous cell that became infected by a virus might have transmitted an oncogene to the virus particles that are produced during viral proliferation. Alternatively, a normal host gene that is susceptible to mutation to an oncogene, a *proto-oncogene,* might have been transferred to an infecting virus, undergoing mutation during the transfer process. Virus particles that contain oncogenes act as *vectors* for the oncogenes by carrying them into cells and, perhaps, into other individual organisms of the host species. The oncogene can then be incorporated into the host genome and, upon activation, can transform the host cells into cancerous cells. Transformation does not require the entire viral genome to be incorporated into the host chromosome; incorporation of the oncogene itself is sufficient to cause transformation. It is very likely that oncogenes are also generated from proto-oncogenes by mutations induced by *carcinogens,* cancer-causing environmental factors, after being transferred into the host cell.

In many cases where there is reason to suppose that a virus is a causative agent in a cancer, the evidence is sketchy and inconclusive. One problem with obtaining information about the possible involvement of a virus in cancer is that the incidence of transformation is extremely low in cultured cells infected with viruses that have been shown to carry oncogenes. It is even more difficult to connect a human tumor with a viral vector, because a cancerous cell will normally not be infected with any virus, even though one of its ancestors may originally have been infected by a viral vector that introduced an oncogene into its genome.

Very few human cancers are caused by viruses. Human lymphotrophic T-cell virus I (HTLV-I) causes T-cell leukemia in people. The T-cells are a part of the immune system. A related virus, HTLV-III, infects T-cells, as well as other cells, and causes AIDS.

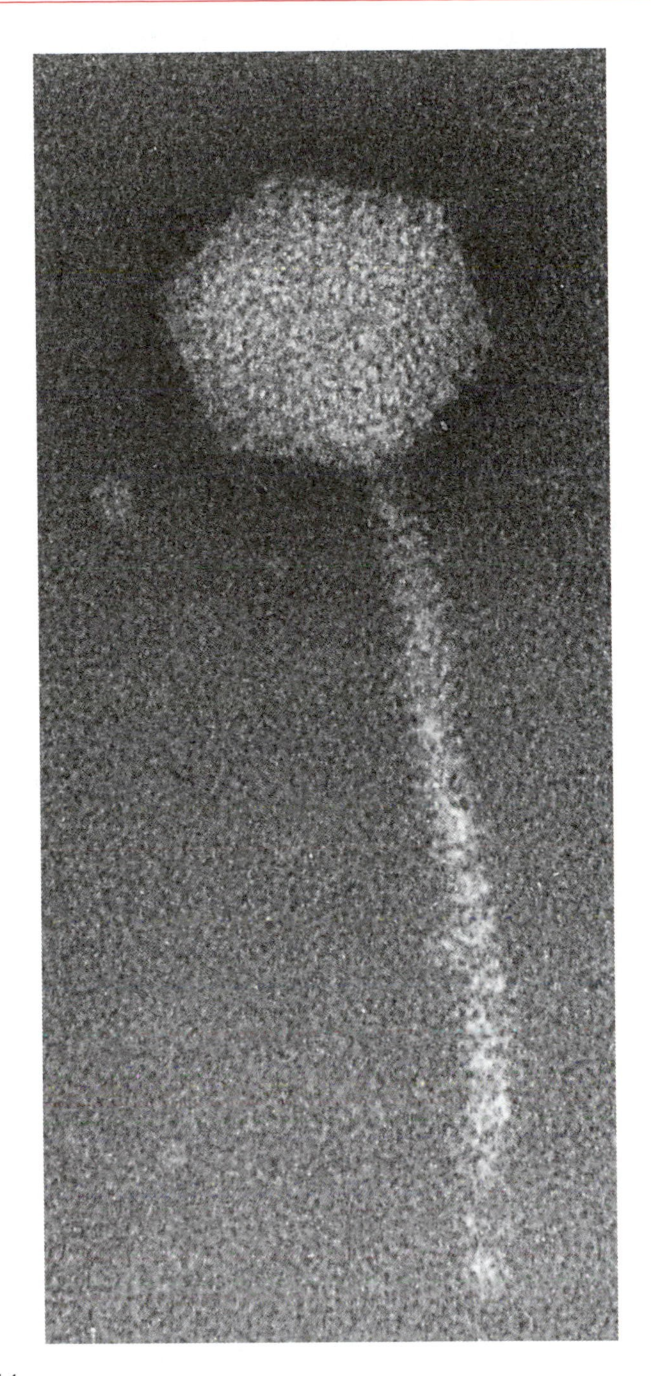

Figure 1-14

Electron Micrograph of Bacteriophage λ at 650,000-fold Magnification.
(Micrograph courtesy of Harold Fisher and originally published in *An Electron Micrographic Atlas of Viruses* by Robley Williams and Harold C. Fisher. Charles C. Thomas Publishers.)

Bacteriophages are bacterial viruses. Bacteriophages are very important in the biotechnology industry and in the recombinant DNA methods used in molecular biology and biochemistry. The *E. coli* bacteriophage M13 is widely used for determining the coded information in DNA by a process that is described in Chapter 12. Still other bacteriophages are used as vehicles for the cloning of genes and for the overproduction of enzymes and other proteins in *E. coli* cells.

The Chemical Basis of Biochemistry

The principal objective of research in biochemistry is to obtain an understanding of the workings of living cells and organisms in chemical terms. The structures of biological molecules, their transformations, and their interactions leading to biological phenomena are the principal subjects of biochemistry. With few exceptions, biological molecules are much more complex than typical molecules in organic chemistry. The molecular complexities of biochemistry introduce a challenge to biochemical experimentation, but at the same time make it exceptionally interesting. The reward of success in these experiments is an enhanced understanding of life.

Chemical Reactions in Biochemistry

The life-giving processes in metabolism, ATP production, cellular replication and repair, intercellular and interorgan communications, and the regulation of metabolism are brought about by chemical transformations of complex molecules and interactions between small and large molecules. The complexities of biochemical compounds and their reactions may appear overwhelming at first, even for someone who has mastered the organic chemistry of norbornanes and their rearrangements. Consideration of these complexities leads one to an appreciation of the extraordinary accomplishments of the chemists and biochemists who elucidated the structures of biochemical compounds and the chemical pathways through which they are formed and react.

However, most biochemical *reactions* are not as complex as they may at first appear. For example, the reactions illustrated in Figure 1-15 show that the apparent complexities arise entirely from the structures of the reactants. The chemical reactions themselves are simply the elementary reactions of organic chemistry. In each reaction, only the functional groups that are highlighted in color undergo any chemical change. The reactions themselves are unremarkable in the chemical sense. However, when they are strung together in metabolic pathways, such as those discussed in protein biosynthesis (Chapter 10), nucleic acid biosynthesis and processing (Chapter 12), fatty acid and carbohydrate

Figure 1-15

Typical Biochemical Reactions Involving Complex Reactants and Products. The functional groups shown in color are the only groups that undergo chemical changes.

1. S_N2 DISPLACEMENT

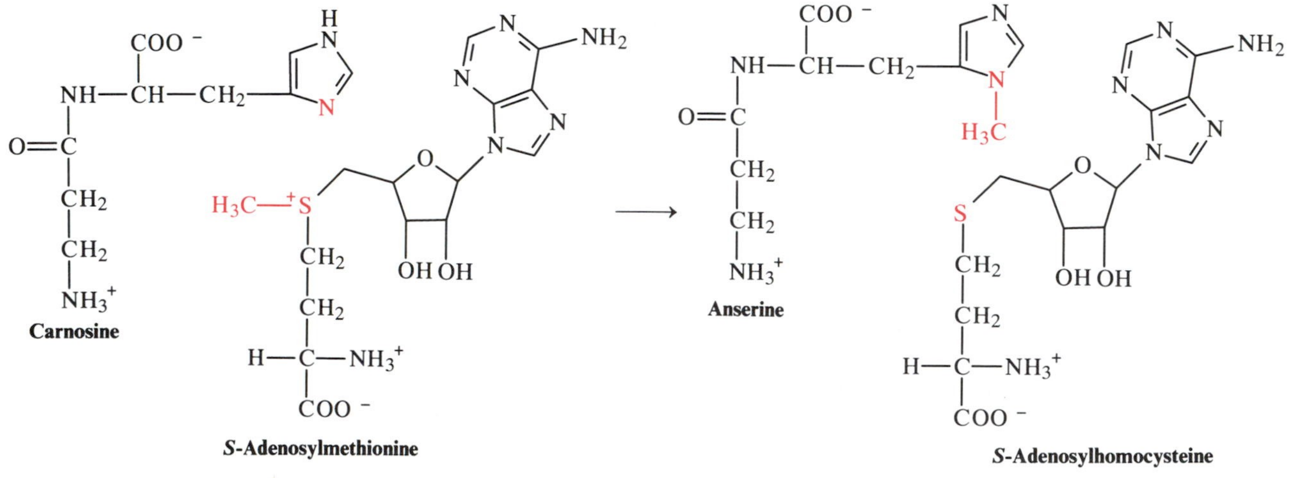

Carnosine

S-Adenosylmethionine

Anserine

S-Adenosylhomocysteine

Figure 1-15 (continued)

2. ALDOL CONDENSATION

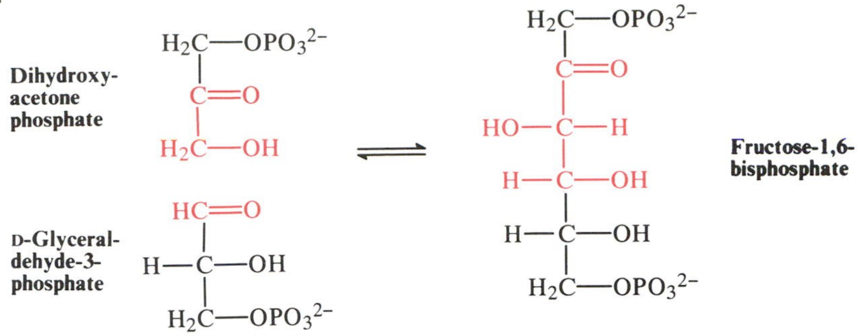

Dihydroxyacetone phosphate

$H_2C{-}OPO_3^{2-}$, $C{=}O$, $H_2C{-}OH$

D-Glyceraldehyde-3-phosphate

$HC{=}O$, $H{-}C{-}OH$, $H_2C{-}OPO_3^{2-}$

$\rightleftharpoons$

Fructose-1,6-bisphosphate

$H_2C{-}OPO_3^{2-}$, $C{=}O$, $HO{-}C{-}H$, $H{-}C{-}OH$, $H{-}C{-}OH$, $H_2C{-}OPO_3^{2-}$

3. PHOSPHORYLATION

Adenosine triphosphate (ATP) + **α-D-Glucose** $\rightleftharpoons$ **Adenosine diphosphate (ADP)** + **α-D-Glucose-6-phosphate**

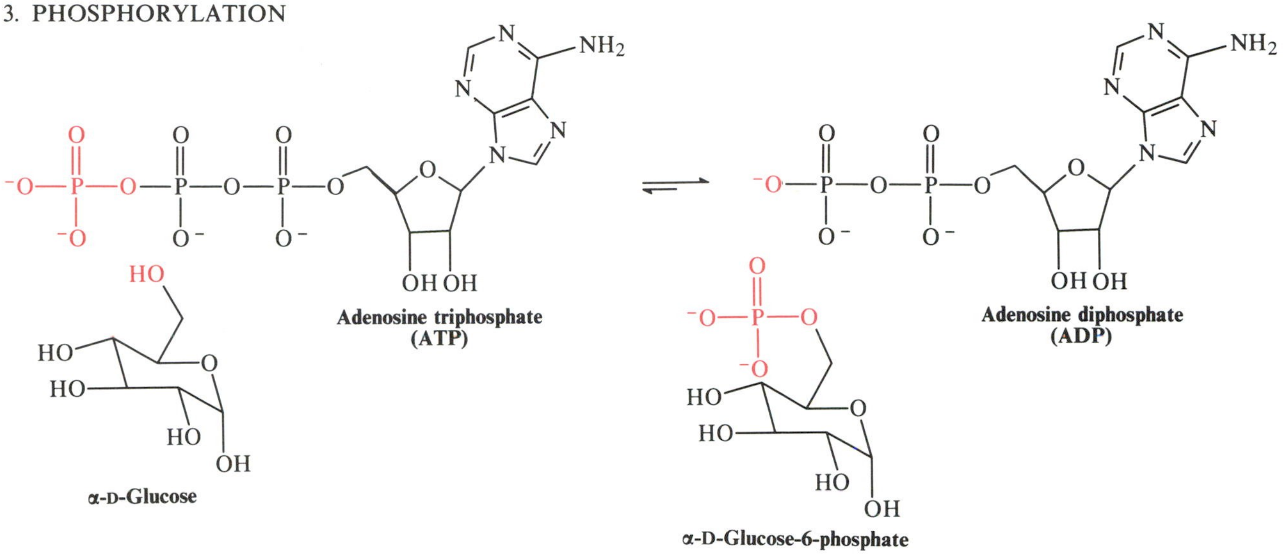

4. AMIDE SYNTHESIS

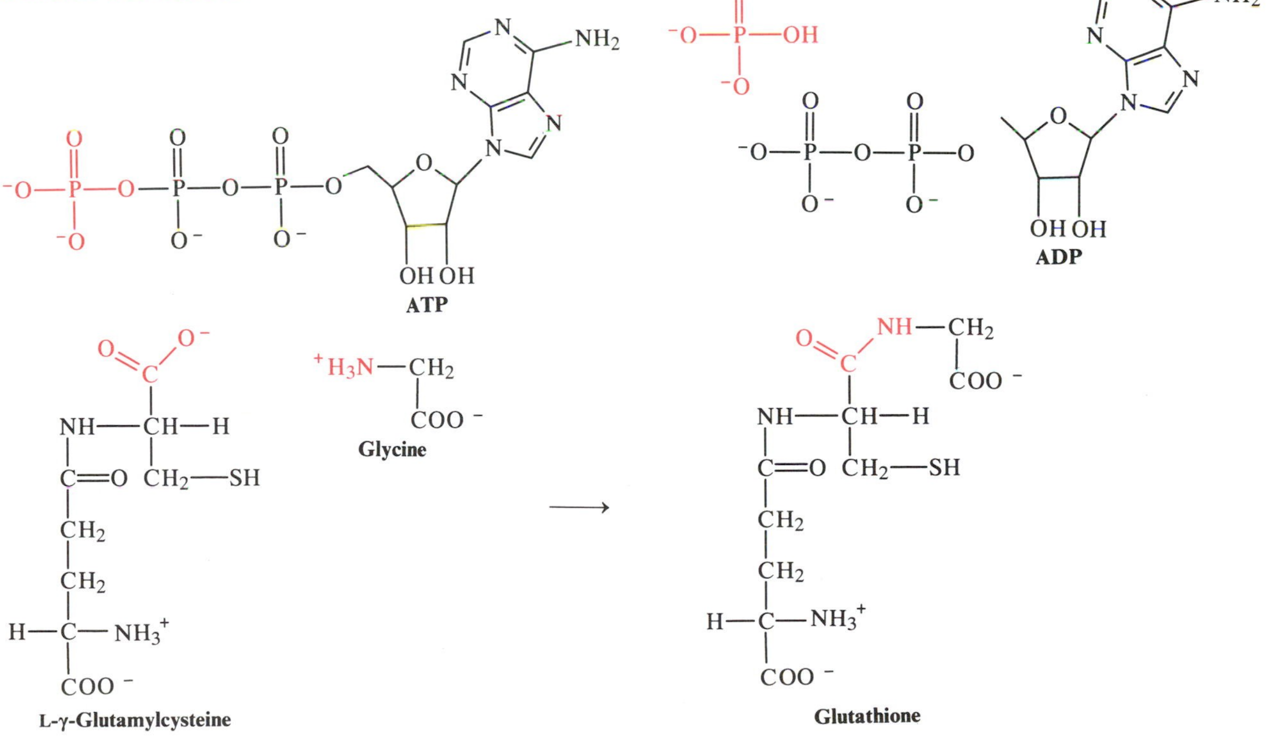

ATP + **L-γ-Glutamylcysteine** + **Glycine** $\longrightarrow$ $^{-}O{-}PO_2{-}OH$ + **ADP** + **Glutathione**

metabolism (Chapters 19 and 21), amino acid metabolism (Chapter 26), and nucleotide metabolism (Chapter 27), they lead to remarkable phenomena.

Figure 1-15 includes specific biochemical examples of the following organic reactions:

Reaction 1 is the alkylation of a nitrogen atom by S_N2 displacement on a carbon atom of an alkylsulfonium ion.

$$-\mathrm{N(H)}: \quad \mathrm{CH_3{-}S^+{-}} \longrightarrow -\ddot{\mathrm{N}}{-}\mathrm{CH_3} \quad \ddot{\mathrm{S}}{-} + \mathrm{H^+}$$

Reaction 2 is an aldol condensation.

$$\mathrm{O{=}C(H)} \quad \mathrm{H{-}C{-}C({=}O){-}} \longrightarrow \mathrm{HO{-}C{-}C{-}C({=}O){-}}$$

Reaction 3 is a phosphorylation.

$$-\mathrm{C{-}OH} \quad {}^-\mathrm{O{-}P({=}O)(O^-){-}O{-}P({=}O)(O^-){-}O{-}} \longrightarrow -\mathrm{C{-}O{-}P({=}O)(O^-){-}O^-} \quad \mathrm{HO{-}P({=}O)(O^-){-}O{-}}$$

Reaction 4 is the formation of an amide from an amine and a carboxylic acid that is activated by an acid anhydride, in this case an anhydride of phosphoric acid.

$$-\mathrm{C({=}O){-}O^-} \quad {}^-\mathrm{O{-}P({=}O)(O^-){-}O{-}P{-}} \xrightarrow{\;-{}^-\mathrm{O{-}P{-}}\;} -\mathrm{C({=}O){-}O{-}P({=}O)(O^-){-}O^-} \xrightarrow{\;-\mathrm{NH_2}\;} -\mathrm{C({=}O){-}NH{-}} \quad {}^-\mathrm{O{-}P({=}O)(O^-){-}OH}$$

Each reaction in Figure 1-15 is straightforward when attention is focused on the groups that actually undergo chemical changes. The reactions appear complex only when one's attention is diffused throughout the complex structures to functional groups that do not undergo any chemical changes.

The Importance of Coenzymes

The reactions in Figure 1-15 are catalyzed by enzymes acting alone, the only other factor being the Mg^{2+} that is coordinated to ATP in nearly all of its biochemical reactions. Other reactions are more complex in the organic chemical sense and are catalyzed by enzymes with the assistance of *coenzymes,* which are complex organic or organometallic molecules that possess special chemical or physical properties not found in proteins and which facilitate difficult chemical transformations. Many coenzymes are derivatives of the water-soluble *vitamins,* which are essential trace nutrients for animals and human beings. The water-soluble vitamins are not themselves coenzymes, but they become coenzymes after they are biochemically activated by specific chemical modifications. Several water-soluble vitamins are shown in Figure 1-16. These molecules are essential in the diets of higher animals and human beings, because they are not synthesized by higher organisms. They are also essential for the growth of lower eukaryotes, such as *Tetrahymena,* which require all of the vitamins in Figure 1-16. Diets that are deficient in these substances cause visible lesions such as dermatitis and, eventually, more-serious

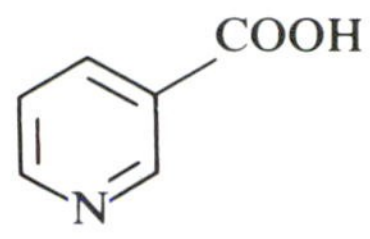

Niacin
(Nicotinic acid)

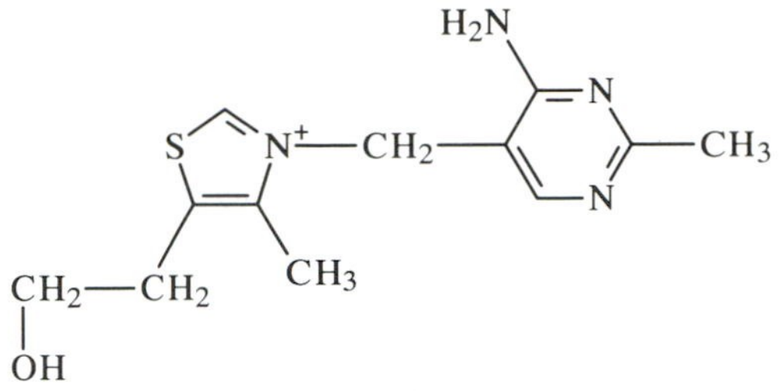

Thiamine

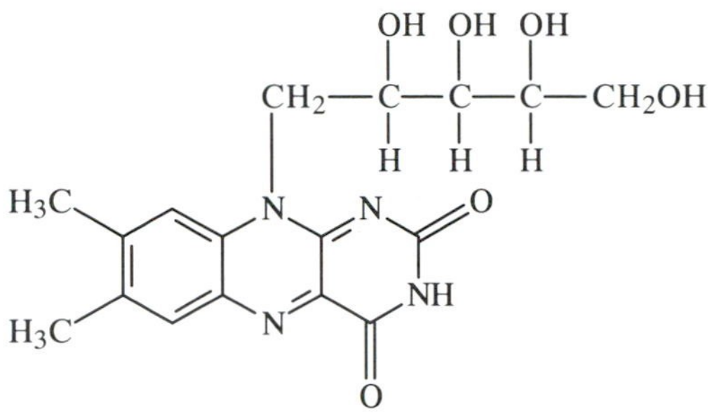

Riboflavin

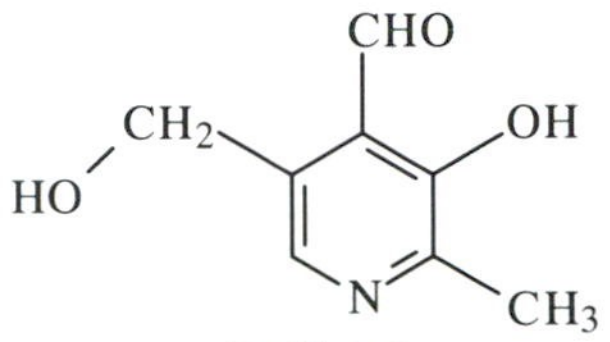

Pyridoxal

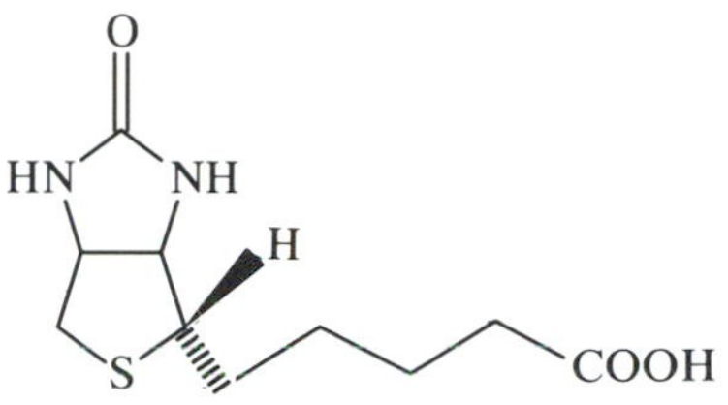

Biotin

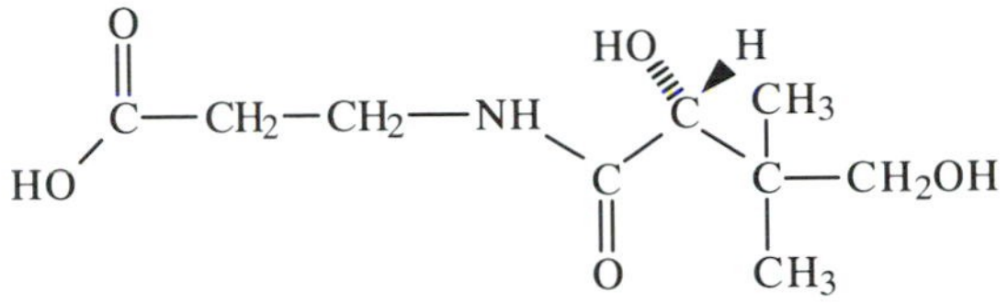

Pantothenic acid

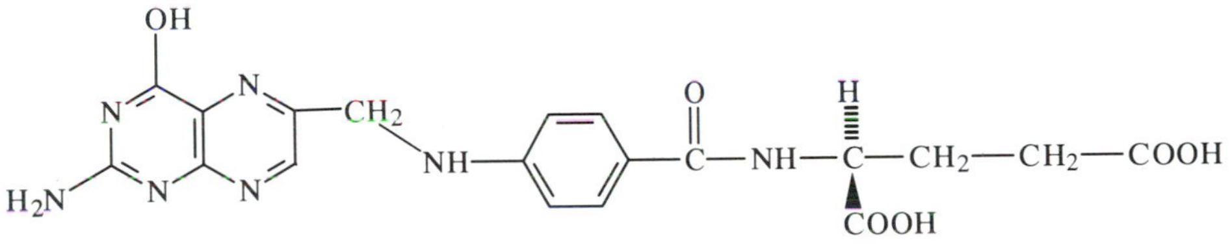

Folic acid

Figure 1-16

Structures of Water-soluble Vitamins. The niacin family of vitamins also includes *niacinamide,* or nicotinamide, and the reduced form of pyridoxal, *pyridoxine,* is another form of the vitamin that is nutritionally equivalent to pyridoxal. The organometallic vitamin B_{12}, *cyanocobalamin,* is not shown here but appears in Chapter 20. Vitamin C, *ascorbic acid,* also is not shown but appears in Box 1-3.

consequences, including mental retardation and other developmental defects. The vitamins are biosynthesized by bacteria and plants and are valuable constituents of vegetables, fruits, dairy products, and meats.

The biochemically active form of thiamine is *thiamine pyrophosphate* (TPP, Chapter 17), that of niacin is *nicotinamide adenine dinucleotide* (NAD^+,

Chapter 16), and that of pyridoxal is *pyridoxal-5′-phosphate* (PLP, Chapters 2 and 18). Riboflavin is the central component of *flavin adenine dinucleotide* and *flavin mononucleotide* (FAD and FMN, Chapter 16). Pantothenic acid is a central constituent of *coenzyme A* (Chapter 2). Folic acid is the nutrient form of *tetrahydrofolate,* which is discussed in Chapter 25. And *biotin* is covalently bonded to certain enzymes in its biochemically active form (Chapter 17).

Other coenzymes are complex organic and organometallic compounds that are not derivatives of vitamins. These include *heme,* which acts as a coenzyme in oxygenation reactions (Chapter 24) and is the essential O_2-binding factor in hemoglobin (Chapter 10). *Lipoic acid* is an essential component of α-ketoacid dehydrogenase complexes (Chapter 17). And iron-sulfur cluster compounds are important components of proteins that catalyze oxidation-reduction processes (Chapters 22 and 23).

Despite the chemical complexities of coenzyme-dependent biochemical reactions, the rule that only small parts of the reacting molecules undergo chemical change holds also for these reactions. For example, the reaction shown in Figure 1-17 involves complex molecules in which only one atom undergoes a chemical change, yet the reaction is a difficult one that requires the niacin coenzyme NAD^+.

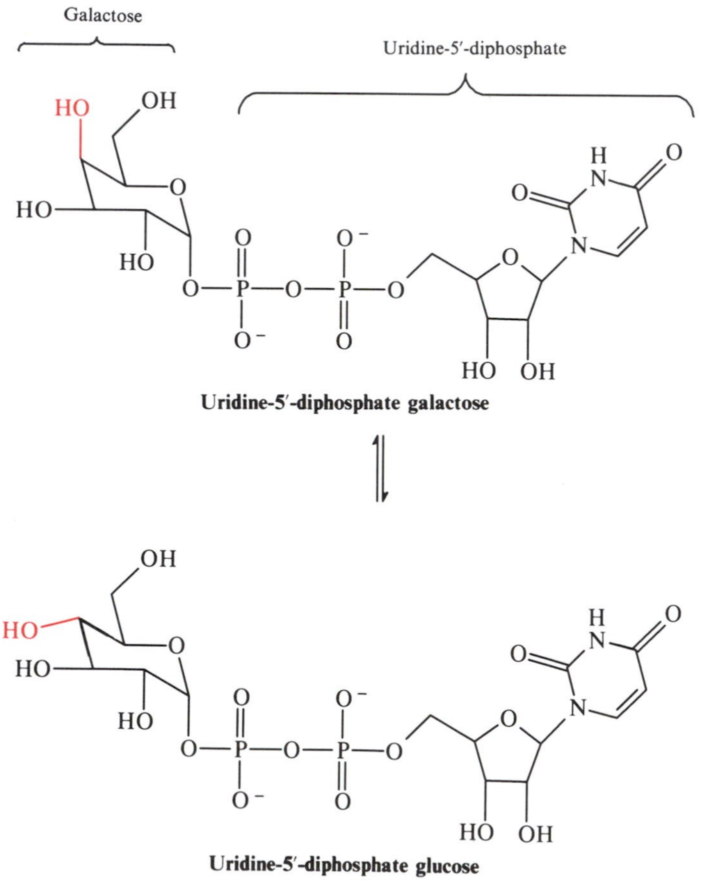

Figure 1-17

The Interconversion of the Sugars Galactose and Glucose. This apparently simple reaction is catalyzed by uridine-5′-diphosphate galactose 4-epimerase, an enzyme that requires the action of NAD^+, a niacin coenzyme.

BOX 1-3

VITAMINS

Ascorbic acid *(vitamin C)* and vitamin B_{12} are two water-soluble vitamins that are not shown in Figure 1-16. Ascorbic acid is an excellent reducing agent and is required for certain enzymatic hydroxylation reactions. Vitamin B_{12}, *cyanocobalamin,* is activated to a coenzyme form; its structure and function are described in Chapters 20, 25, and 26.

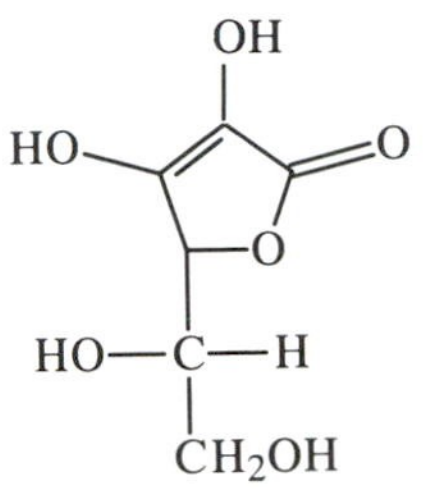

L-Ascorbic acid (Vitamin C)

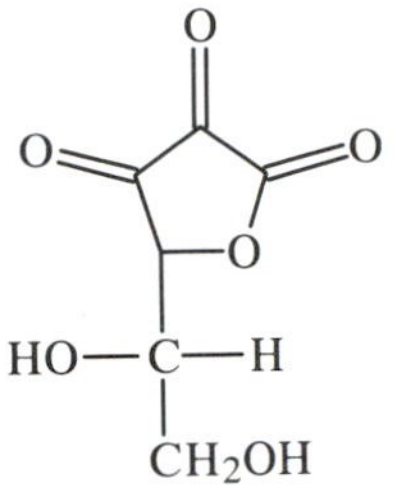

L-Dehydroascorbic acid

Vitamins are compounds that are required in the metabolism of human beings and higher animals but are not biosynthesized in their bodies. Therefore, nearly all of them are required in the diets of all vertebrates in order to maintain good health. Rats and a few other mammals can biosynthesize vitamin C and do not require it in their diets. The vitamins were originally discovered by nutritional biochemists and physicians, who found that certain disorders could be cured by either the dietary administration or the injection of naturally occurring compounds, which came to be known as vitamins. The disorders caused by dietary deficiencies of the vitamins are listed in the table.

Disorders that are cured by vitamins

Vitamin	Disorder
Ascorbic acid (vitamin C)	Scurvy
Niacin (niacinamide)	Pellagra
Thiamin	Beriberi
Riboflavin	Dermatitis
Pyridoxal (pyridoxine)	Numerous
Vitamin B_{12}	Pernicious anemia
Biotin	Seborrhea (skin disorder)
Vitamin D	Rickets
Vitamin K	Hemorrhage
Vitamin A (retinol)	Night blindness

The last three vitamins entered in the table, vitamins D, K, and A, are fat-soluble vitamins rather than water-soluble vitamins. The structures of fat-soluble vitamins are shown below. Vitamin A is part of the principal visual pigment and is derived from carotene in the diet. Vitamin D is required for calcium deposition in the bones. Vitamin K is required for the blood-clotting process (Chapter 6).

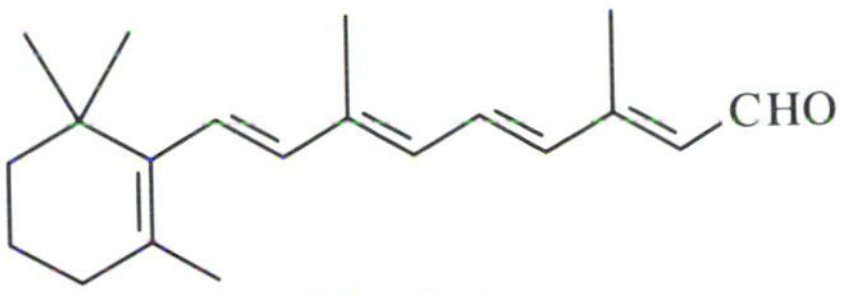

Vitamin A

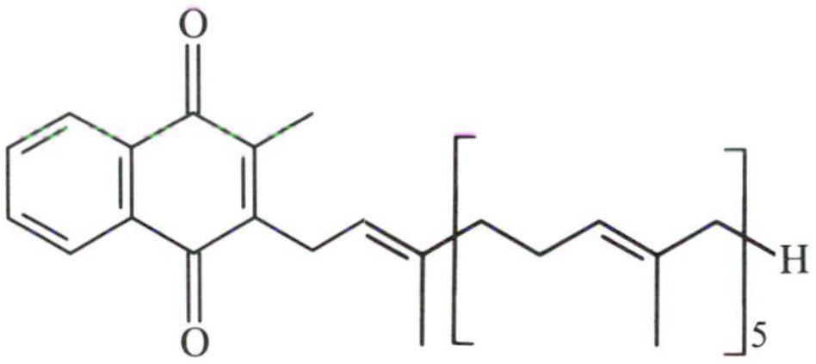

Vitamin K_2

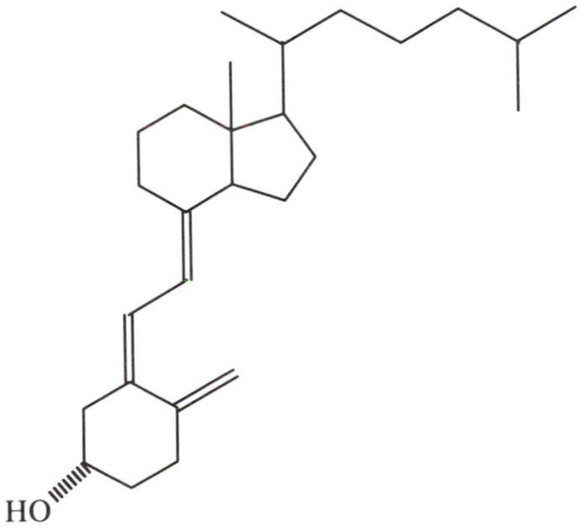

Vitamin D_3

Complex Structures of Biological Molecules

Most biological molecules such as those in metabolic pathways have complex structures such as those shown in Figures 1-15 and 1-17. Inasmuch as a biochemical reaction involves only a small part of a reacting molecule, one might inquire why these molecules are so complex. One answer may be that cells are themselves structurally and functionally complex and are composed of complicated molecules. One might suppose that most of the functional groups in molecules such as those in Figure 1-17 play no role in the chemical reaction of any one group. This is consistent with the fact that complex molecules are biosynthesized or degraded one step at a time in metabolism. Thus, the chemical change in a given step of metabolism entails a single chemical reaction at a single functional group of a reactant, similar to a reaction in one step of a long chemical synthetic scheme.

The foregoing explanation for the existence of complex biological molecules is satisfactory as a beginning, but it is only the most obvious and superficial reason. Higher-order rationalizations become necessary as one probes more deeply into the *raison d'être* for many biochemical processes. As an example, the reaction in Figure 1-17 is the only means by which the common sugars galactose and glucose can be interconverted in cells. The unchanged parts of the molecules seem to be excess baggage. However, careful investigations have shown that the nonreacting parts of the molecules play crucial roles in their interconversion. It turns out that any enzyme catalyzing the transformation of galactose into glucose would need unreactive pendant groups as binding anchors to prevent the reaction intermediate from escaping during the catalytic process. In the metabolism of galactose, the uridine-5′-diphosphate is enzymatically attached to the sugar before its conversion into the glucose derivative, *and it is then removed*. The mechanism of this reaction is discussed in more detail in Chapter 20. In addition, the apparently unreactive parts of the molecules in Figure 1-17 participate even more directly in the catalytic process through their interactions with the enzyme. The details of the latter interactions are still unknown; however, the underlying principle of the use of binding energy derived from unreactive parts of molecules in catalyzing their reactions is explained in Chapter 5.

Biochemical Reactions and Interactions

Biochemistry as a science includes all of the biochemical reactions of metabolism; all of the macromolecular interactions; all of the guided movements of ions, metabolites, and signal molecules; and all of the molecular structures of living cells. This is an enormous amount of information that can no longer be included in a single textbook. However, enough is now known that it is possible to present biochemistry as more than a collection of facts. Underlying principles are coming into focus that organize large bodies of facts about related phenomena in biochemistry. Some of these principles are mature—for example, the underlying chemistry of many biochemical reactions—and others are still in the process of development—such as the use of binding energy between a macromolecule and a ligand in enzyme catalysis and signal transduction. In this textbook, we shall emphasize the principles of biochemistry and illustrate them with examples and a limited amount of descriptive biochemistry.

Enzymes are the most extensively studied, the most complex, and the most generally important macromolecules in biochemistry. Some of the principles of enzyme catalysis—specifically, the utilization of binding energy—also explain other biochemical phenomena such as signal transduction. Therefore, we begin the discussion of biochemistry in this textbook with enzymes. Examples of the important classes of reactions that are catalyzed by enzymes are introduced in Chapter 2, without structural detail, to explain how enzymes catalyze the reactions of carbon–carbon bond formation and cleavage. These reactions will be familiar to students who have completed a course in organic chemistry. The way that enzymes work is then examined in Chapters 3, 4, and 5, using the case-study approach with the serine proteases chymotrypsin and trypsin, the most thoroughly studied of all enzymes, as the case in point. The structures of proteins in general are then described in Chapters 7 and 8. Chapter 9 presents biochemical energetics in a formal way, and the synthesis and biosynthesis of proteins is in Chapter 10. After the introduction of nucleic acid structure in Chapter 10, the biochemistry of phosphoryl transfer and phosphoryl ester transfer is given in Chapter 11, followed by nucleic acid biochemistry in Chapter 12. Chapters 13 and 14 describe the principles and mechanisms underlying the regulation of genes and the use of these mechanisms in recombinant DNA technology. The biochemistry of glycosyl transfer, carboxylation, decarboxylation, addition and elimination reactions, and oxidation-reduction coenzymes is presented in Chapters 15 through 18. It is remarkable how few reaction types are needed to understand a complex metabolic pathway such as fatty acid biosynthesis and degradation or carbohydrate metabolism, which are discussed in Chapters 19 and 21, respectively. The biochemistry of ATP formation by animals, microorganisms, and plants is described in Chapters 22 and 23. Biological oxygenations are described in Chapter 24 and one-carbon metabolism in Chapter 25. Nitrogen metabolism is in Chapters 26 and 27, followed by complex lipids in Chapter 28 and membrane structure and function in Chapter 29. Chapter 30 concludes the text with a description of energy transduction in vectorial transport, muscle contraction, and related phenomena.

This textbook emphasizes principles while including the biochemistry that is common to most cells. Much detailed biochemistry is not specifically covered, not only to avoid the appearance of greater complexity than is needed to achieve a good understanding of principles, but also to minimize the size of this volume. Students who have mastered the principles should be able to comprehend new or unfamiliar biochemical information with ease and satisfaction.

Summary

Biochemistry is the chemistry of life, and the biological cell is the basic unit of life. Eukaryotic cells are the complex cells of higher unicellular forms, animals, and plants; whereas prokaryotic cells are bacteria. The most-essential functions of most cells are self-maintenance and reproduction, both of which require energy. The most-important form of biochemical free energy is ATP, which exists in cells as the complex MgATP, and a large part of the biochemistry of cells is directed toward the production of ATP. Cell maintenance and energy production are accomplished through the processes of anabolism (biosynthesis) and catabolism (biodegradation), which are collectively known as metabolism. Cell replication begins with the biosynthesis of the macromolecules: first, the genetic molecules of DNA; then, RNA; and, finally, proteins, lipids, and complex carbohydrates.

Eukaryotic cells are encased by a plasma membrane that contains the cytosol, nucleus, endoplasmic reticulum, ribosomes, mitochondria, lysosomes, peroxisomes, and cytoskeleton. The cytosol contains most of the soluble components, including enzymes, metabolites and salts. The nucleus contains DNA and the enzymes required to transcribe the information within DNA into RNA and to keep DNA in repair. The RNA migrates into the cytosol and is both a carrier of genetic information and a major participant in the biosynthesis of proteins, which is carried out by ribosomes in the endoplasmic reticulum. The mitochondria generate most of the ATP for the cell. The lysosomes and peroxisomes contain hydrolytic and oxidative enzymes, respectively, that carry out the breakdown of cellular constituents and foreign matter. Turnover of cellular constituents is a normal process of cellular regulation, repair, and renewal. The cytoskeleton maintains the cellular shape and participates in organizing the cellular organelles.

Multicellular organisms consist of a diversity of cells. In higher organisms, such as animals, the cells having related functions are associated as organs, which include the central nervous system, heart and circulatory system, lungs, liver, stomach and digestive system, skin, muscles, skeleton, and secretory glands. The organs are controlled by the brain through nerves and the secretory glands, which on command from the brain secrete hormones, or signal molecules, into the bloodstream. The hormones bind to receptors on the surfaces of cells in specific organs and activate the cells by signal transduction to carry out specific functions.

Digestion is important in the biochemistry of whole animals and of individual cells. In animals and human beings, digestion of the macromolecules in foodstuffs to small molecules that can be used for biosynthesis and energy production is carried out by the digestive organs. In cells, the lysosomes and peroxisomes digest cellular debris and foreign matter to molecules that can be excreted or used for the resynthesis of cellular constituents.

Virus particles infect cells and subvert their biochemistry to the proliferation of the virus particles. Viruses are the infectious agents in many diseases of human beings, animals, and plants. Infection of a cell by a lytic virus leads to cell lysis and the release of hundreds or thousands of virus particles. Infection of a cell by a lysogenous virus entails the incorporation of the viral genome into the host chromosomes. The integrated viral genome can then direct the proliferative reproduction of virus particles. A few viruses carry cancer-causing oncogenes.

The long sequences of biochemical reactions in the metabolic pathways occur step by step in discrete chemical reactions catalyzed by enzymes. Most biochemical molecules are structurally complex, but many individual reactions are straightforward reactions of organic chemistry. More-complex or more-difficult reactions require the actions of coenzymes as well as enzymes. Many coenzymes are derived from water-soluble vitamins.

ADDITIONAL READING

Prescott, D. M. 1988. *Cells.* Boston: Jones and Bartlett Publishers.

de Duve, C. 1984. *A Guided Tour of the Living Cell,* 2 vols. New York: Scientific American Books.

Rossman, M. G., Erickson, J. W., Frankenberger, E. A., Griffith, J. P., Hecht, H.-J., Johnson, J. E., Kamer, G., Luo, M., Mosser, A. G., Rueckert, R. R., Sherry, B., Vriend, G. 1985. *Nature* 317: 145.

PROBLEMS

1. Explain each of the following in one sentence.

(a) organism
(b) ATP
(c) enzyme
(d) prokaryote
(e) mitochondrion
(f) peroxisome
(g) cell wall
(h) cytoskeleton
(i) organ
(j) cell
(k) DNA
(l) protein
(m) plasma membrane
(n) ribosome
(o) endoplasmic reticulum
(p) metabolite
(q) virus
(r) endocrine system
(s) eukaryote
(t) RNA
(u) hormone
(v) lysosome
(w) lipid
(x) nucleus
(y) mRNA
(z) bacteriophage

2. What are the principal phases of the infection of a cell by a lytic virus? How do lysogenous viruses differ from lytic viruses?

3. What are the functions of coenzymes?

4. How do prokaryotes differ from eukaryotes?

5. What are signal molecules and why are they necessary?

6. What is lysosome loading and why is it important?

7. Should viruses be regarded as a form of life. Why? Why not?

8. What are the principal cellular functions that are common to all cells? What kinds of functions are carried out only by specialized cells?

9. What is the Golgi complex and what is known about its function?

10. What is a definition of biochemistry? How is it related to chemistry and biology?

11. How is cellular shape maintained in various kinds of cells?

12. What are vitamins and why are they important?

13. What is the importance of DNA and RNA to cells? To viruses?

14. What are the functions of lysosomes and peroxisomes? How do they differ?

15. What is signal transduction and why is it important?

Carbon-Carbon Bond Formation and Cleavage

2

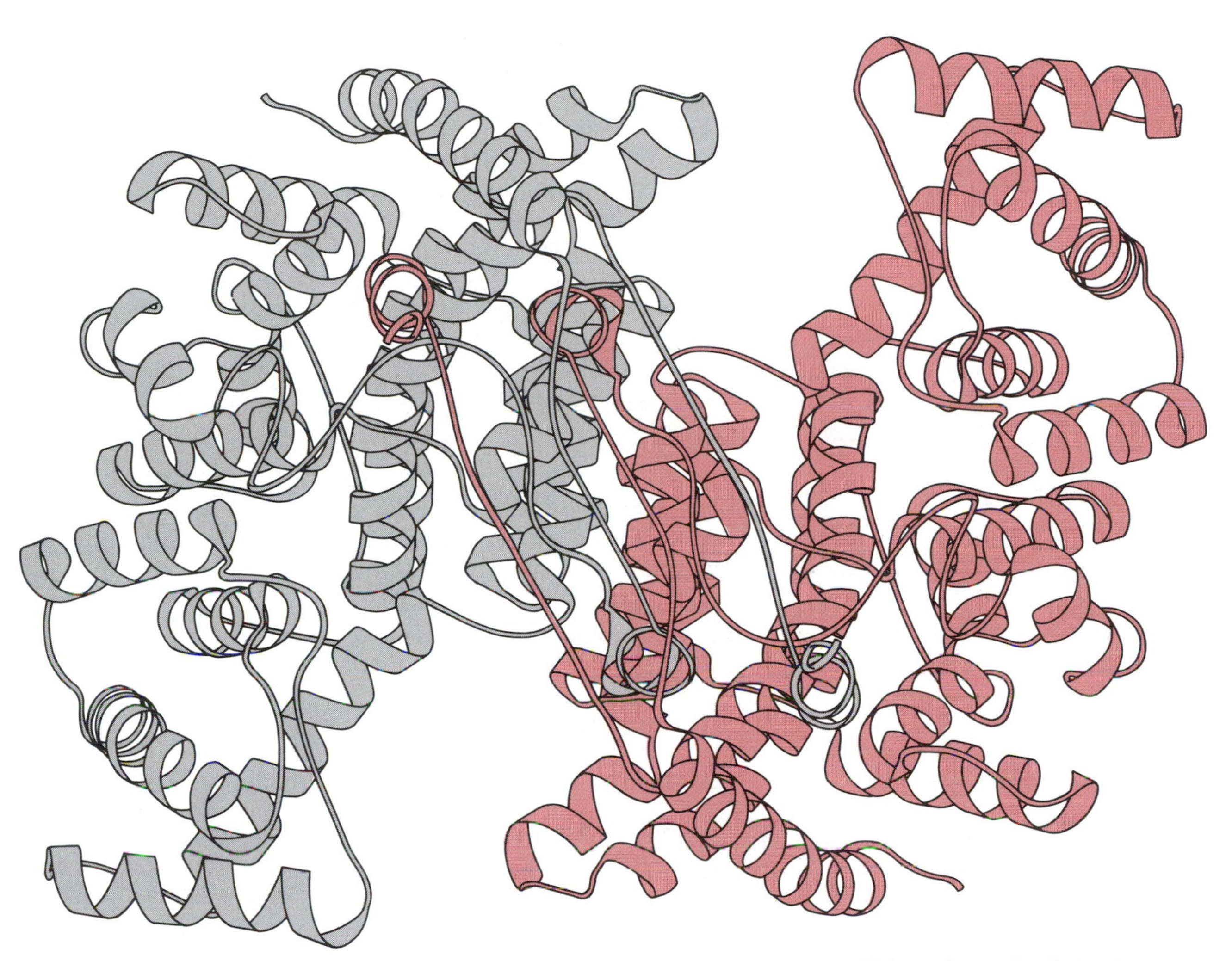

A ribbon diagram of the polypeptide chains in citrate synthase. The enzyme consists of two identical subunits, each of which has a catalytic site, that are bound together into a dimeric structure. (After a drawing courtesy of James Remington.)

Living organisms continually degrade organic compounds and synthesize new ones. Consider how much reconstruction of this type is required when, for example, a boiled egg is converted into a component of your body. In the course of this chemical transformation, a large number of carbon–carbon bonds are degraded and reformed. The mechanisms through which these bond-forming and bond-breaking reactions generally proceed are well known in organic chemistry. An understanding of these mechanisms makes it relatively easy to understand many of the metabolic pathways in which compounds are synthesized or degraded.

In this chapter, we shall first consider the chemical mechanisms through which C–C bonds are formed, and then see how biological systems utilize these mechanisms. In the course of this discussion, we will encounter several of the molecules that are important in biochemistry and learn something about how they work. These molecules include: amino acids; proteins, which are constructed as chains of amino acids; enzymes, which are proteins that make reactions proceed with extraordinary speed and specificity; coenzymes, which participate in the action of enzymes by mechanisms that are now reasonably well understood; vitamins, some of which are converted into coenzymes; and ATP, which is the "common currency" of biochemical energy that can be used to drive chemical reactions and to do work, as in muscle contraction. All of these compounds will be examined in greater detail in subsequent chapters.

Chemistry of the Synthesis and Cleavage of the Carbon–Carbon Bond

Let us begin with an outline of the basic organic chemistry required to make and break a carbon–carbon bond. The two possible ways of forming and cleaving carbon–carbon bonds are:

1. The combination of two carbon radicals or, in the reverse direction, the fragmentation of the bond into two radicals, as shown in equation 1.

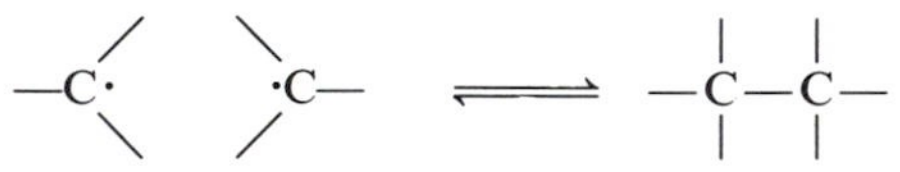

$$-\overset{\diagup}{\underset{\diagdown}{C}}\cdot \quad \cdot\overset{\diagdown}{\underset{\diagup}{C}}- \rightleftharpoons -\overset{|}{\underset{|}{C}}-\overset{|}{\underset{|}{C}}- \tag{1}$$

2. The combination of a positively charged carbon atom, a *carbocation,* with a negatively charged carbon atom, a *carbanion,* shown in equation 2. In the reverse direction, the carbon–carbon bond undergoes fragmentation into a carbocation and a carbanion. Biological systems make use of the ionic mechanism in most C–C bond formations and cleavages.

$$-\overset{+\diagup}{\underset{\diagdown}{C}} \quad \overset{\diagdown}{\underset{\diagup}{\bar{C}}}- \rightleftharpoons -\overset{|}{\underset{|}{C}}-\overset{|}{\underset{|}{C}}- \tag{2}$$

A carbocation is extremely unstable unless it is incorporated into a molecule in which it is stabilized. A *carbonyl group* commonly supplies the carbon atom that can react as a carbocation. The chemical basis for this is explained by the structure of a carbonyl group in an aldehyde, ketone, ester, or lactone, which can be described as a resonance hybrid of the following structures.

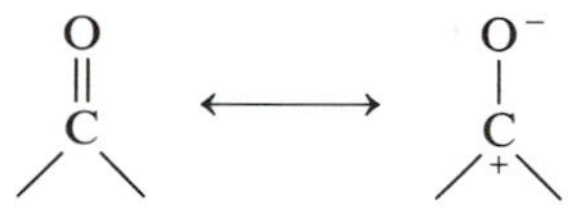

$$\overset{O}{\overset{\|}{C}} \longleftrightarrow \overset{O^-}{\overset{|}{C^+}}$$

The carbonyl group has a large dipole moment, which shows that there is considerable positive charge on the carbon atom, approximately one-half of a positive charge in a ketone, and an equal amount of negative charge on the oxygen atom.

The other component of the reaction in equation 2 is a carbanion, which is also extremely unstable unless the negative charge is stabilized. In biochemical reactions, stabilization is usually brought about by an adjacent electron-deficient atom. In most cases, the carbanion is attached to a functional group that allows the negative charge to be delocalized by resonance. This distributes the negative charge over several atoms in addition to the carbon, thus obviating the need to have a high charge density on carbon while allowing it to react as a carbanion. The best-known activating group in both chemistry and biochemistry is, again, the carbonyl group. This group allows delocalization of the negative charge onto the oxygen atom of the enolate anion, where it is relatively stable:

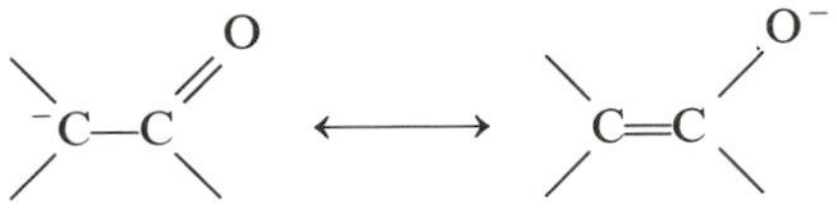

$$^{-}C{-}C{=}O \longleftrightarrow C{=}C{-}O^{-}$$

Thus, carbonyl groups can provide both the nucleophilic and the electrophilic components for carbon–carbon bond formation and cleavage as in the aldol reaction, the mechanism of which is illustrated in Figure 2-1. Because of these properties, carbonyl groups are commonly present in biochemical

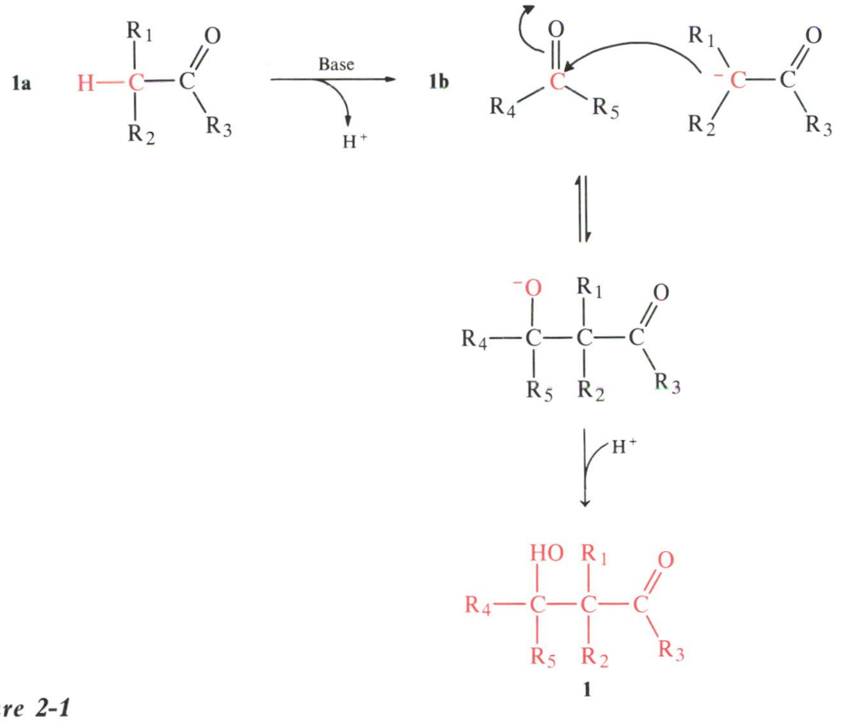

1a $H{-}C(R_1)(R_2){-}C({=}O)R_3$ —Base→ (H^+)

1b $R_4R_5C{=}O$ + $^{-}C(R_1)(R_2){-}C({=}O)R_3$

⇌

$R_4{-}C(^{-}O)(R_5){-}C(R_1)(R_2){-}C({=}O)R_3$

↓ H^+

$R_4{-}C(HO)(R_5){-}C(R_1)(R_2){-}C({=}O)R_3$

1

Figure 2-1

The Mechanism of the Aldol Condensation Reaction. In the aldol condensation, a proton bonded to a carbon atom α to a carbonyl group in **1a** is acidic enough to be removed, leading to the formation of a carbanion. The carbon atom of a carbonyl group in compound **1b** is electropositive; attack on this carbon by the carbanion leads to carbon–carbon bonding in the addition compound. Protonation of the initial adduct produces the product, **1**.

$CH_3—CH_2—CH_2—CO_2^-$
Butyrate

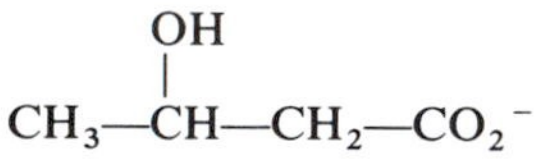

$CH_3—CH(OH)—CH_2—CO_2^-$
β-Hydroxybutyrate

compounds, and biological systems frequently form carbon–carbon bonds by use of the aldol condensation.

Aldol condensation leads to the general structure **1** in Figure 2-1; and, whenever we encounter a biological molecule that is represented by this structure, we can conclude that it most likely was biosynthesized through an aldol condensation of the precursors **1a** and **1b**. Conversely, if a compound is to be degraded in a biological system through a retro-aldol condensation, the β-hydroxyketone grouping in structure **1** must either be present or be introduced into the molecule. For example, consider the catabolic degradation of butyrate. We know that butyrate cannot itself undergo carbon–carbon bond cleavage; however, β-hydroxybutyrate can undergo a retro-aldol reaction. After reading Chapters 16 and 18, you will know what reactions are available to convert butyrate into β-hydroxybutyrate and will then be able to deduce the pathway by which biological systems degrade butyric acid.

Carbon–Carbon Bond Formation by Aldolase: D-Fructose-1,6-bisphosphate

An example of a biological aldol condensation is the formation and degradation of D-fructose-1,6-bisphosphate (fructose-1,6-P_2), as shown in equation 3.

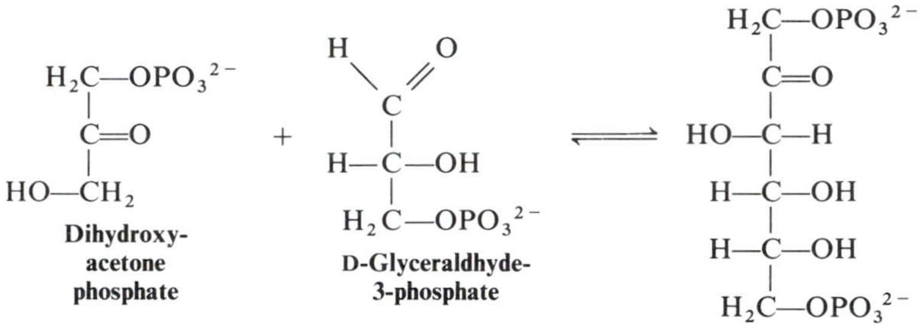

$$\underset{\textbf{Dihydroxyacetone phosphate}}{H_2C(OPO_3^{2-})—C(=O)—CH_2—OH} + \underset{\textbf{D-Glyceraldhyde-3-phosphate}}{HC(=O)—C(H)(OH)—H_2C—OPO_3^{2-}} \rightleftharpoons \underset{\textbf{D-Fructose-1,6-bisphosphate}}{H_2C(OPO_3^{2-})—C(=O)—C(HO)(H)—C(H)(OH)—C(H)(OH)—H_2C—OPO_3^{2-}} \quad (3)$$

Fructose is a sugar that plays an important part in carbohydrate metabolism, and its role will be discussed in more detail in Chapter 21. Its structure is shown in equation 3 in the open-chain form to facilitate discussion of the chemistry of its reactions. In solution, however, this compound exists mainly (98%) in a cyclic form, as do most other sugars.

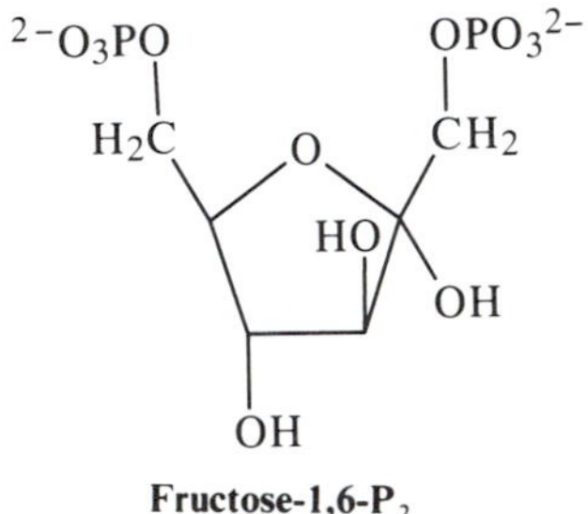

Fructose-1,6-P_2 (cyclic form)

Fructose-1,6-P_2 is formed reversibly from D-glyceraldehyde-3-phosphate (glyceraldehyde-3-P) and dihydroxyacetone phosphate (dihydroxyacetone-P) in the cytoplasm of the cell. The three molecules participating in the reaction are phosphorylated. This is true of many biologically important compounds. As mentioned in Chapter 1, phosphorylated compounds, and most other charged molecules, cannot enter or leave a cell easily because they cannot cross the cell membrane without some special transport mechanism. Therefore, after a compound is phosphorylated inside the cell, it cannot escape. Phosphorylation is also important in other ways that will be described later.

If one were to prepare a solution of glyceraldehyde-3-P and dihydroxy-acetone-P, little if any fructose-1,6-P_2 would be formed, even after a long time. Some way must be found to accelerate the reaction. Therefore, before continuing the discussion of how fructose-1,6-P_2 is formed, we shall consider how chemical reactions can be accelerated in biological and nonbiological systems.

Transition States, Catalysis, and Enzymes

In most cases, a chemical reaction does not occur every time two molecules collide. Usually, many collisions and the proper orientation of the two reacting molecules are required before a new bond will form. This means that a barrier prevents the occurrence of a reaction on each collision, and the molecules must have enough energy to overcome this barrier in order to react. A large population of molecules has a distribution of molecules with different amounts of energy; and, when two molecules with enough energy collide in the correct orientation, the new bond is formed and the reaction takes place. When bond breaking occurs in the reverse direction, the energy barrier must also be surmounted. An energy diagram, shown in Figure 2-2, is useful for describing the formation and breaking of bonds. This diagram shows the energetic barrier to reaction that must be overcome for bond formation ($\Delta G_f^‡$) or cleavage ($\Delta G_r^‡$) to occur. The relative energies of the starting materials and products (ΔG_{react}) determine the relative concentrations of the reactant and product at equilibrium. The progress of the reaction is often defined by a *reaction coordinate,* which is a measure of the formation and breaking of the chemical bonds.

It is useful to define the *transition state* indicated by the double dagger (‡) in Figure 2-2. It is the structure ($AB^‡$) of the reacting complex at the highest

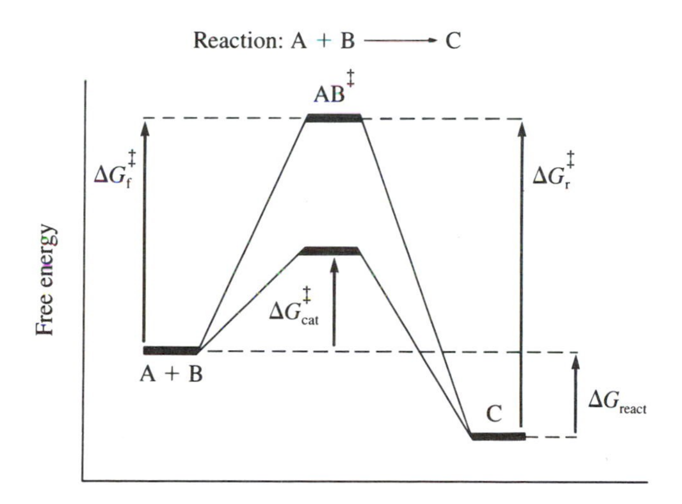

Figure 2-2

Energy Diagram for a Chemical Reaction. The free-energy levels are indicated for the reactants (A + B) and product (C), as well as for the transition state ($AB^‡$), for the uncatalyzed and catalyzed reaction. The defined free-energy changes are: $\Delta G_f^‡$ and $\Delta G_r^‡$, energy of activation for the uncatalyzed forward and reverse directions, respectively; ΔG_{cat}, energy of activation for the catalyzed reaction in the forward direction; ΔG_{react}, free-energy change for the reaction of A with B to form C.

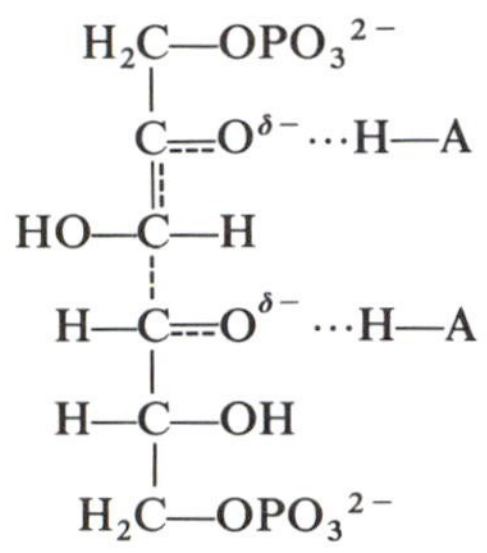

$H_2C—OPO_3^{2-}$
$C{=}O^{\delta-}\cdots H—A$
$HO—C—H$
$H—C{=}O^{\delta-}\cdots H—A$
$H—C—OH$
$H_2C—OPO_3^{2-}$

Figure 2-3

Possible Structure of the Transition State for the Aldol Condensation of Dihydroxyacetone-P and Glyceraldehyde-3-P. The structure illustrates a possible transition state for the addition of the enolate carbanion of dihydroxyacetone-P to the aldehyde group of glyceraldehyde-3-P. In this transition state, the partial bonds are indicated by dashed lines and the hydrogen bonds by dotted lines. The two acidic groups AH stabilize the transition state by hydrogen bonding to the partially negatively charged oxygens.

point in the energy diagram. The transition state is not a real chemical compound because it collapses to give reactants or products with no barrier, so that it does not exist for an appreciable time. Nevertheless, the concept of a transition state is useful because it offers a simple way to describe what is happening in chemical terms at the highest energy point along the reaction path. The energy difference between reactants and the transition state determines the rate of the forward reaction, and the difference in energy between the product and the transition state determines the rate of the reverse reaction.

Figure 2-3 represents a reasonable structure of the transition state for the condensation reaction of glyceraldehyde-3-P with dihydroxyacetone-P. In this structure, the carbon–carbon bond between the two reactants is neither fully formed nor broken, and the two carbonyl groups have partially developed charges and partial double-bond character. The partial negative charges are stabilized through hydrogen bonding by the acids, HA. This structure can be transformed with no barrier to give either the product, fructose-1,6-P_2, or the reactants, glyceraldehyde-3-P and dihydroxyacetone-P.

Two things are necessary in order to reach this transition state and allow the reaction to occur. First, the reacting molecules must have enough energy. At any one time, only a small fraction of the colliding molecules in solution have sufficient energy to reach the transition state. The rate of reaction can be increased by increasing the temperature, which will increase the average kinetic energy of the molecules, so that a larger fraction of colliding molecules will have sufficient energy to reach the transition state. Second, the colliding molecules must be in precisely the correct relative orientation to form the transition state. That the second requirement will be met is extremely improbable, because the requirements to form a new bond are very strict. The two molecules cannot be in the wrong position by even a fraction of an angstrom if the reaction is to proceed at an optimal rate. It is even more unlikely that catalyzing acid and base molecules will be in the correct positions to donate and accept protons at the right moment. This low probability is defined by *entropy,* a subject that will be discussed in greater detail in Chapters 5 and 9.

Most biochemical reactions do not occur at significant rates without the assistance of a *catalyst,* which increases the reaction rate without being consumed by the reaction. Catalysts accelerate chemical reactions by stabilizing the transition state; that is, by decreasing the energy that is required to reach the transition state. In the transition state of Figure 2-3, acids HA are catalysts. They lower the energy of the transition state by stabilizing the negative charges on oxygen.

Cells contain extraordinarily efficient catalysts known as *enzymes,* which can accelerate chemical reactions by factors ranging from 10^{12} to 10^{15}, or even more. Enzymes differ from nonbiological catalysts in that they are highly specific. For example, the condensation of glyceraldehyde-3-P and dihydroxyacetone-P discussed earlier is catalyzed by the enzyme *aldolase,* which will catalyze only the condensation of the D-stereoisomer of glyceraldehyde-3-P with dihydroxyacetone-P. It will not catalyze the condensation of glyceraldehyde-3-P and dihydroxyacetone at a significant rate. Nor will it catalyze the condensation of the L-stereoisomer (mirror image) of glyceraldehyde-3-P with dihydroxyacetone-P. (Figure 2-4 shows the two stereoisomers, or enantiomers, of alanine, an α-amino acid.)

The degree of specificity of different enzymes is variable. Some enzymes will tolerate moderate changes in the structure of substrates that react in the enzyme-catalyzed reaction. The basis of the catalytic efficiency and selectivity or specificity of enzymes is a problem of major interest to chemists and biochemists. It will be discussed more extensively in Chapters 4 and 5.

Amino Acids, Proteins, and Enzymes

Before considering how enzymes work, we need to know something about their structures. All enzymes are proteins, which are polymers of L-α-amino acids. The general structure of an L-α-amino acid is illustrated by any one of the three stereochemical formulations shown in the margin. The three stereochemical representations are shown to aid in visualizing the arrangement of substituents on the α-carbon for the L-stereoisomer. The carbon bearing the nitrogen is asymmetric so that two stereoisomers, or *enantiomers,* exist, the L- and D-isomers. The interchange of any two substituents gives the D-isomer, as is illustrated for the L- and D-isomers of alanine in Figure 2-4. They are mirror images of each other. Frequently, the enantiomers are designated as *(R)* and *(S)**. Using this convention, we designate L-alanine as *(S)*-alanine.

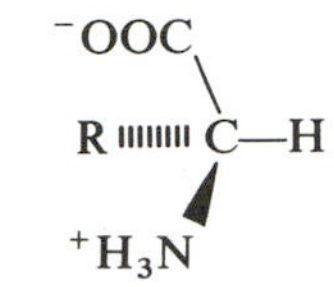

L-α-Amino acid (three-dimensional)

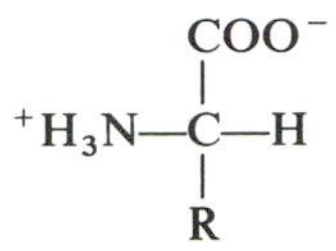

Fischer projection

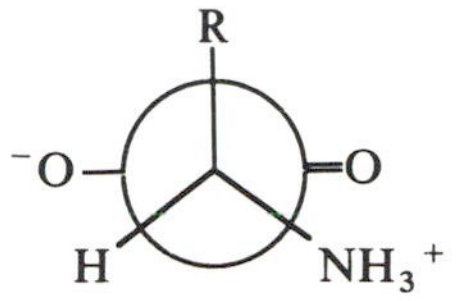

Newman projection

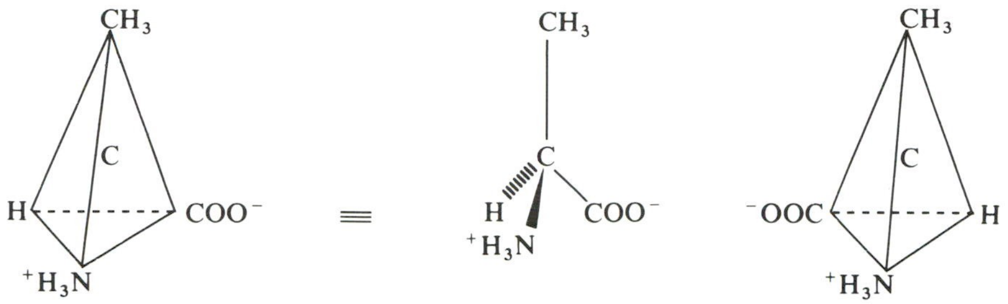

L-Alanine

D-Alanine

Figure 2-4

Structures of L- and D-Alanine.

Only the L-isomers of amino acids are components of proteins. The simplest amino acid is glycine, which lacks an alkyl substituent on the α-carbon and has instead a second hydrogen. The alkyl substituents of a few hydrophilic and hydrophobic amino acids are shown below in Table 2-1. (The structures of the twenty amino acids found in proteins are shown on the inside front cover of this book.) Amino acids with hydrophilic substituents tend to be water soluble.

Table 2-1

Structures of alkyl α-carbon substituents for a few amino acids

HYDROPHILIC SUBSTITUENTS		HYDROPHOBIC SUBSTITUENTS	
Substituent	**Amino Acid**	**Substituent**	**Amino Acid**
$—CH_2OH$	Serine	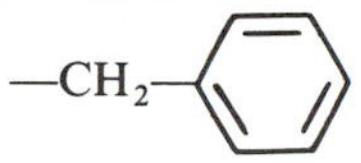 $—CH_2—C_6H_5$	Phenylalanine
$CH(CH_3)OH$	Threonine	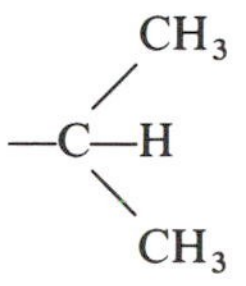	
$—CH_2\text{-}CO_2^-$	Aspartate	$—CH(CH_3)_2$	Valine
$—(CH_2)_4\text{-}NH_3^+$	Lysine		

*For the definition of *R* and *S* nomenclature, see, for instance, E. Eliel, *Stereochemistry of Carbon Compounds*.

These substituents can form hydrogen bonds, and several have positive or negative charges in neutral solutions. The hydrophobic substituents are nonpolar, uncharged, and cannot form hydrogen bonds; and these amino acids tend to have a low solubility in water.

All amino acids have at least two functional groups, the *α-carboxyl group* and the *α-amino group.* Because carboxyl groups can react with amino groups to form amides, amino acids can combine to form long-chain polymers; that is, *proteins.* A short polymer of amino acids is a *peptide* or *polypeptide.* A protein consists of amino acids connected to one another by amide bonds, which are known as *peptide bonds* when they connect amino acids.

$$\begin{array}{ccccccccccccccccccccccccc}
 & & {}^{+}H_3N & & O & & & & H & & O & & & & H & & O & & & & H & & O & & \\
 & & | & & \| & & & & | & & \| & & & & | & & \| & & & & | & & \| & & \\
R_1 & - & C & - & C & - & N & - & C & - & C & - & N & - & C & - & C & - & N & -\cdots\cdots\cdots- & C & - & C & - & O^- \\
 & & | & & & & | & & | & & & & | & & | & & & & | & & | & & & & \\
 & & H & & & & H & & R_2 & & & & H & & R_3 & & & & H & & R_n & & & &
\end{array}$$

A protein

An ionized carboxyl group (COO^-) is at the C-terminal end and a protonated amino group (NH_3^+) is usually at the N-terminal end of a protein. A peptide chain typically contains one hundred or more amino acids, so that the molecular weight of proteins can range from 10,000 (approximately sixty amino acid residues) to several hundred thousand. Many important components of biological systems, in addition to enzymes, are proteins.

The long peptide chains of proteins might be expected to fold up randomly into a mixture of an enormous number of three-dimensional structures. In fact, native proteins are folded up into precise and well-defined structures. As mentioned earlier, the alkyl substituent groups are either hydrophobic or hydrophilic; and there is a marked tendency for the hydrophilic groups to appear on the water-exposed surfaces of proteins, with the hydrophobic groups in the interior or on membrane-exposed surfaces. Figure 2-5 shows how the peptide chain of citrate synthase is folded. The mechanism by which citrate synthase catalyzes carbon–carbon bond formation is a principal subject of this chapter. The three-dimensional structure of this protein was determined by x-ray crystallography. We see that the protein chain is fixed in a very definite configuration; and, for citrate synthase, there are two chains in each molecule; that is, the protein consists of two *subunits.*

When a protein is heated, dissolved in certain solvents, or exposed to strong acid or base, it loses this highly organized structure and is converted into a mixture of poorly defined structures that are sometimes called *random coils.* This loss of the native structure is known as *denaturation* and is almost always accompanied by loss of biological activity. The boiling of an egg is an example of protein denaturation. When the egg is heated, the proteins unfold and eventually form a solidified mass in the hard-boiled egg. The biological competence is lost even with brief heating; that is, the egg can no longer give rise to a chick. Similarly, enzymes lose their ability to catalyze reactions when they are denatured. Denaturation occurs when the numerous weak bonds that hold the protein in its native structure are broken by heat or a denaturing solvent. The nature of these weak bonds will be considered in Chapter 5.

We can now ask how an enzyme in its native structure can catalyze a chemical reaction. The enzyme is folded so that there is a pocket, or cleft, on the surface, called the *active site,* into which the reacting molecules (substrates) can bind and undergo the chemical transformation that is catalyzed by the enzyme. Substrates bind at the active site because the structures of specific substrates interact with complementary structures in the active site. Thus, negative charges

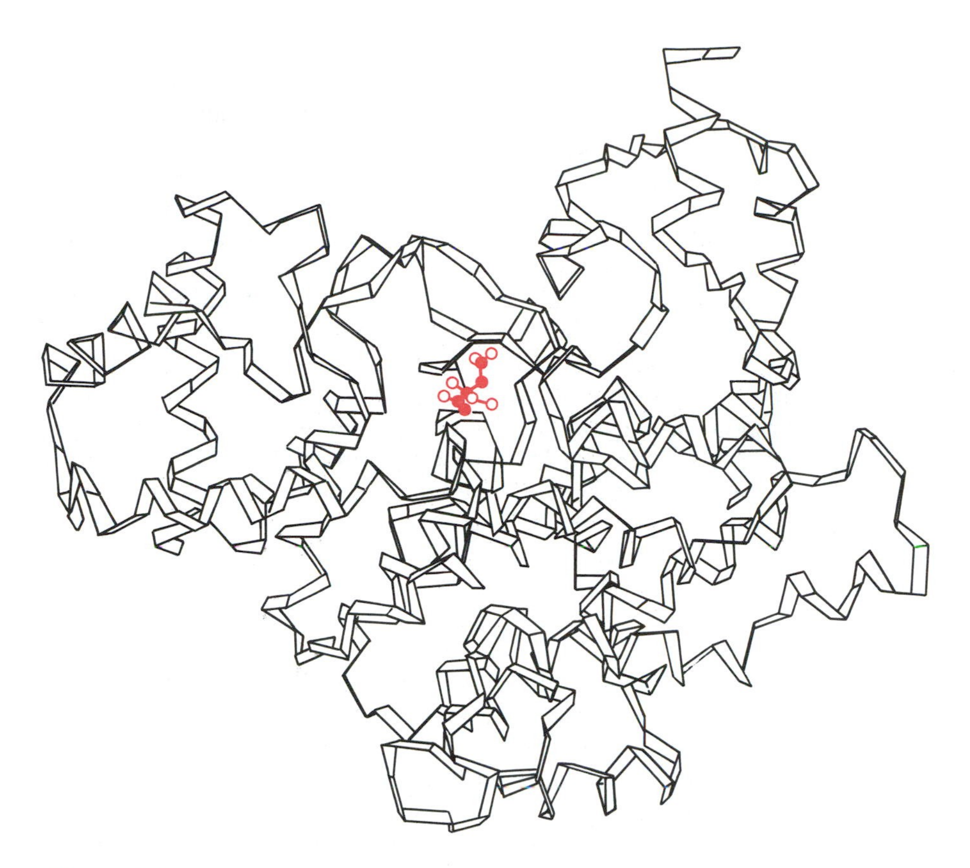

Figure 2-5

The Structure of Citrate Synthase. The polypeptide chain of a protein such as citrate synthase is folded into a definite arrangement of peptide bonds. Each peptide linkage is a planar unit defined by the α-carbon, the carbonyl carbon and oxygen atoms, and the amide nitrogen and hydrogen atoms. One means of illustrating the folding motif in a polypeptide is to represent each peptide linkage by a segment of tape connecting the α-carbon atoms of two amino acids. This allows the polypeptide chain to be traced without being obscured by the side-chain R groups. Such a representation of one of the two identical peptide chains in the enzyme citrate synthase is illustrated in this figure. In citrate synthase, each chain has a catalytic site and catalyzes the formation of a carbon–carbon bond between oxaloacetate and acetyl CoA (equation 16). The drawing shows a molecule of substrate (in color) bound at the active site. (Courtesy of James Remington.)

on the substrate associate with positive charges on the enzyme, nonpolar hydrophobic groups with complementary hydrophobic groups, and hydrogen-bond donors with hydrogen-bond acceptors. In addition, functional groups at the active site that participate in catalysis may also contribute to binding the substrate.

What might an active site that catalyzes a simple aldol condensation such as equation 4 look like?

$$R_1{-}C({=}O){-}H + H_3C{-}C({=}O){-}R_2 \longrightarrow R_1{-}\underset{H}{\overset{OH}{C}}{-}CH_2{-}\overset{O}{\overset{\|}{C}}{-}R_2 \qquad (4)$$

In the hypothetical active site of Figure 2-6, the R_1 and R_2 groups of the substrates fit into complementary regions of the active site, which will be hydrophobic if R_1 and R_2 are hydrophobic and will contain polar or charged groups if R_1 and R_2 contain such groups. In addition, in the active site there is a —COOH group, of an aspartate or glutamate residue, and an NH_3^+ group provided by a lysine residue (Table 2-1), which form hydrogen bonds to carbonyl groups on the substrates. These groups also facilitate the formation of the C–C bond: the COOH group polarizes the carbonyl group of the aldehyde, increasing its carbocationic character, which facilitates bond formation with the carbanion; the NH_3^+ group forms a hydrogen bond to the carbonyl group of the other substrate and facilitates formation of the enolate anion. A carboxylate group is in the proper position to abstract a proton and produce the carbanion.

Several aspects of enzyme catalysis are now apparent. The enzyme binds the two substrates in exactly the correct position for bond formation; it uses the binding energy in order to facilitate the close approximation of the reactants. Furthermore, once the reactants are bound at the active site, they are also in close contact with the acids and bases required for catalysis. Thus, the necessity for a highly improbable collision of two substrates with three catalysts is eliminated.

We can also discern some of the properties of enzymes that are responsible for specificity toward particular substrates. Clearly, if a substrate is too large, it will not fit into the active site. However, it is more difficult to understand why the enzyme will not work well with a small substrate. Small substrates can diffuse into the active site, but they will fit loosely and, therefore, will not be aligned properly for reaction. For example, if the smaller substituent R_1 of a substrate fits more deeply into the R_1-binding site, then its carbonyl group will not be aligned properly for hydrogen bonding. Conversely, if the small substrate

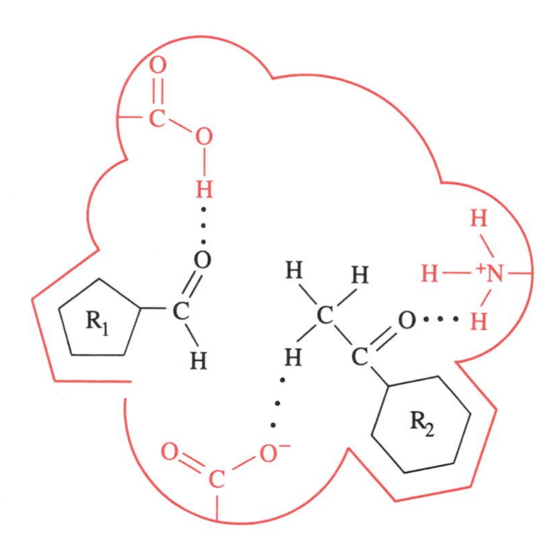

Figure 2-6

A Hypothetical Active Site for an Aldol Condensation. A hypothetical active site for catalyzing the aldol condensation of equation 4 is illustrated here, showing specific binding subsites for substituents R_1 and R_2 and three catalytic amino acid residues. The complementarity between the binding subsites and substituents R_1 and R_2 is illustrated pentagonally for R_1 and hexagonally for R_2. Catalytic groups COOH, COO^-, and NH_3^+ assist in binding the substrates through hydrogen bonds; groups COOH and COO^- also catalyze the reaction by mediating the proton-transfer steps of the aldol condensation.

forms a hydrogen bond, then its R_1 group will not interact optimally with its binding region. In turn, the binding energy required for the proper alignment of the substrate molecules will be lost. The problem of enzyme specificity will be discussed in more detail in Chapter 5.

From the foregoing discussion, we might expect that the condensation between glyceraldehyde-3-P and dihydroxyacetone-P would occur by reaction of a carbocation provided by glyceraldehyde-3-P and the carbanion derived from dihydroxyacetone-P. Equations 5 and 6 give a more-detailed description of this reaction. In the first step, a proton is removed from dihydroxyacetone-P by a base, A_2^-, to form the carbanion, the enolate anion, as shown in equation 5.

$$\begin{array}{c} H_2C{-}OPO_3^{2-} \\ | \\ C{=}O\cdots HA_1 \\ | \\ HO{-}C{-}H \\ | \\ H \\ \quad {}^{-}A_2 \end{array} \quad \rightleftharpoons \quad \begin{array}{c} H_2C{-}OPO_3^{2-} \\ | \\ C{-}O^{-}\cdots\cdots HA_1 \\ \| \\ HO{-}C{-}H \\ \\ HA_2 \end{array} \qquad (5)$$

This reaction might be helped by an acid, HA_1, that can stabilize the developing negative charge on the oxygen atom of the enolate by forming a hydrogen bond to it. The acid can also donate a proton to the enolate, but the reactive form is the enol-anion that is shown.

The formation of the carbon–carbon bond by addition of the carbanion to the carbonyl group of glyceraldehyde-3-P is shown in equation 6.

$$\begin{array}{c} H_2C{-}OPO_3^{2-} \\ | \\ C{-}O^{-}\cdots\cdots HA_1 \\ \| \\ HO{-}C{-}H \\ \qquad HA_2 \\ H \quad\; O \\ \backslash \;\, /\!/ \\ C \\ | \\ H{-}C{-}OH \\ | \\ H_2C{-}OPO_3^{2-} \end{array} \quad \rightleftharpoons \quad \begin{array}{c} H_2C{-}OPO_3^{2-} \\ | \\ C{=}O\cdots\cdots HA_1 \\ | \\ HO{-}C{-}H \\ | \\ H{-}C{-}OH\cdots\cdots {}^{-}A_2 \\ | \\ H{-}C{-}OH \\ | \\ H_2C{-}OPO_3^{2-} \end{array} \qquad (6)$$

This reaction might be assisted by an acid, HA_2, that stabilizes the negative charge on the carbonyl group to which the carbanion adds and eventually donates a proton to it to give the stable alcohol product. If no acid were available, the condensation reaction would lead to the formation of an unstable alkoxide ion, which can decompose to give glyceraldehyde-3-P and dihydroxyacetone-P. Although this mechanism looks plausible from the organic chemical view, the enzyme *aldolase,* which catalyzes the reactions in mammalian cells, has adopted a more-powerful way of forming a carbanion from dihydroxyacetone-P—namely, by forming a protonated imine from the carbonyl group.

The Mechanism of Action of Aldolase

We now consider catalysis by aldolase of the condensation of dihydroxyacetone-P and glyceraldehyde-3-P to form fructose-1,6-P_2. The most-difficult part of the reaction is the removal of the proton from dihydroxyacetone-P to form the carbanion that undergoes aldol condensation with the carbonyl group of glyceraldehyde-3-P. Abstraction of this proton to form a carbanion near neutrality is extremely unfavorable energetically. How then does the enzyme facilitate proton abstraction?

The enzyme has, in the course of evolution, produced a mechanism that forms a carbanion-equivalent at the active site that is much more stable and easier to form than the enolate carbanion. The enzyme utilizes the ε-amino group of a lysine residue at the active site ($E—NH_2$) to replace the oxygen of the carbonyl group of the substrate to form an *imine*, which is partially protonated at neutral pH, as is shown in equation 7.

$$E{-}NH_2 + \underset{HO-CH_2}{\overset{H_2C-OPO_3^{2-}}{C{=}O}} + H^+ \rightleftharpoons E{-}{}^+NH{=}\underset{HO-CH_2}{\overset{H_2C-OPO_3^{2-}}{C}} + H_2O \qquad (7)$$

This protonated imine greatly facilitates abstraction of the proton because the positive charge on the nitrogen atom makes abstraction of the positively charged proton more favorable energetically. This is equivalent to saying that, because of the positive charge on the nitrogen, very little negative charge accumulates on the carbon atom when a proton is abstracted.

We can now write a complete mechanism (Figure 2-7) for the reaction catalyzed by aldolase, in which E represents the active site of the enzyme and B is a base at the active site. According to this mechanism, dihydroxyacetone-P first condenses with the $—NH_2$ group ($E—NH_2$ of lysine) on the enzyme to form an imine. The formation of the imine also requires the participation of acids and bases at the active site, which have been omitted from Figure 2-7 to reduce its complexity. Subsequently, the base at the active site abstracts a proton to form an enamine (equation 8), which is a carbanion-equivalent with the electron pair localized mainly on the uncharged nitrogen atom.

$$B{:}\quad \underset{H-C-}{\overset{|}{C}}{=}\overset{H}{N^+}{-}R \;\underset{}{\overset{\pm BH^+}{\rightleftharpoons}}\; \left[\overset{\overset{H}{|}}{N}{-}R\text{ on }C{=}C \longleftrightarrow \overset{\overset{H}{|}}{N^+}{-}R\text{ on }C{=}N^+,\ {}^-C \right] \qquad (8)$$

Enamine

In one resonance form of the enamine, however, the electron pair is on carbon, so that enamines react as carbanions. This carbanion-equivalent can readily condense with the carbonyl group of glyceraldehyde-3-P. The condensation is facilitated by an acid, BH^+, which protonates the carbonyl group. BH^+ may or may not be the acid that is formed when B: abstracts a proton from dihydroxyacetone-P to give the enamine. The product of the condensation is the enzyme-bound imine of fructose-1,6-P_2. The imine then undergoes hydrolysis to regenerate the enzyme and release the product. Hydrolysis, again, includes participation of functional groups at the active site (not shown).

Experimental Support for the Proposed Mechanism of Action of Aldolase: Hydrogen Exchange

Let us consider some of the experimental evidence from which the mechanism of action of aldolase was deduced. The enzyme and dihydroxyacetone-P were allowed to react in deuterium oxide (D_2O, or ^2H_2O) instead of H_2O. Deuterium

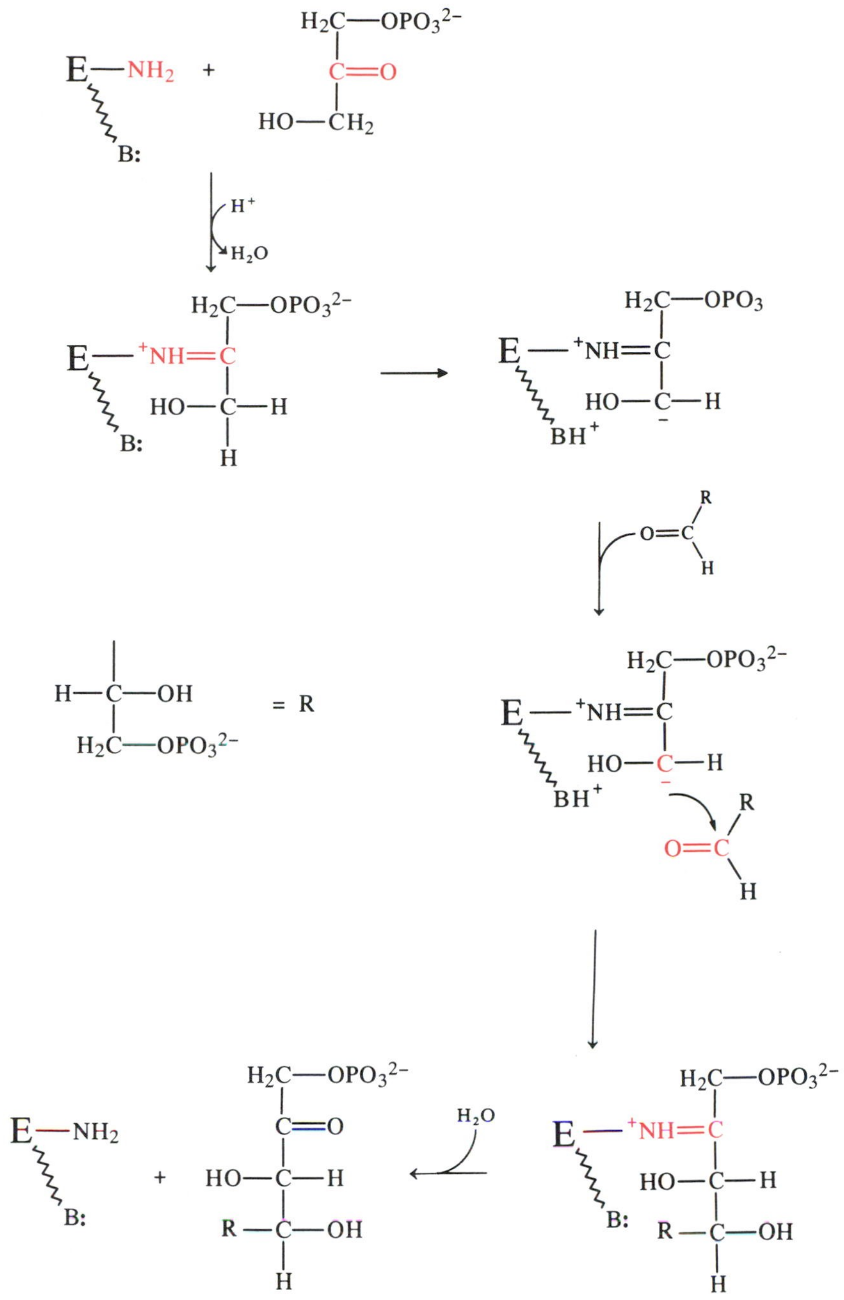

$$E{-}NH_2 + H_2C(OPO_3^{2-}){-}C{=}O{-}CH_2OH$$

$\xrightarrow{H^+,\ -H_2O}$

$$E{-}^{+}NH{=}C(H_2C{-}OPO_3^{2-}){-}C(HO)(H){-}H \quad \longrightarrow \quad E{-}^{+}NH{=}C(H_2C{-}OPO_3){-}\bar{C}(HO){-}H \ (BH^+)$$

$$H{-}C(OH){-}H_2C{-}OPO_3^{2-} = R$$

$\xrightarrow{O{=}C(R)H}$

$$E{-}^{+}NH{=}C(H_2C{-}OPO_3^{2-}){-}\bar{C}(HO){-}H \ (BH^+) + O{=}C(R)H$$

$\longrightarrow$

$$E{-}^{+}NH{=}C(H_2C{-}OPO_3^{2-}){-}C(HO)(H){-}C(R)(OH){-}H$$

$\xrightarrow{H_2O}$

$$E{-}NH_2 + H_2C(OPO_3^{2-}){-}C{=}O{-}C(HO)(H){-}C(R)(OH){-}H$$

Figure 2-7

Mechanism of the Aldolase Reaction. Each step including proton transfer is catalyzed by acidic and basic groups, although many of these processes are not explicitly shown. In the first step, the active-site amino group reacts with the carbonyl group of dihydroxyacetone-P with elimination of water to form a protonated imine. The proton from carbon-3 of the iminium ion is abstracted by an enzymatic base to form the enamine, the carbanion-equivalent for the condensation step. The enamine adds as a carbanion to the carbonyl group of the aldehyde substrate, and the imine then undergoes hydrolysis to release the product.

is a heavy isotope of hydrogen, with an atomic weight of 2, and it serves as a useful label for hydrogen because it can be analyzed by NMR (nuclear magnetic resonance) or MS (mass spectrometry). The dihydroxyacetone-P was then reisolated and equilibrated with H_2O to remove rapidly exchangeable ^2H (^2H on the OH-group—see Box 2-1); and its deuterium content was determined. It was found that one deuterium atom had been incorporated into the hydroxymethylene group of dihydroxyacetone-P.

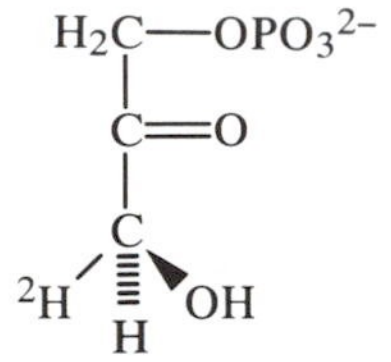

$H_2C—OPO_3^{2-}$
C=O
C
^2H H OH

(S)-[3-^{2}H]Dihydroxy-acetone-P

The reaction sequence of Figure 2-8 will account for the incorporation of this deuterium. Dihydroxyacetone-P initially reacts with the amino group (E—NH_2) of the active-site lysine to form an imine. A base (B) at the active site of the enzyme abstracts a proton from the imine to produce a carbanion (recall that the structure shown in Figure 2-8 is one of several resonance forms). The proton on the base (BH^+) exchanges rapidly with solvent $^2H^+$ to give B^2H^+. The proton released from BH^+ is essentially lost in a sea of $^2H^+$ and the probability that it will ever return to dihydroxyacetone-P is very low. B^2H^+ now reprotonates the enamine of dihydroxyacetone-P to form the deuterated imine of dihydroxyacetone-P. The exchange of solvent $^2H^+$ into dihydroxyacetone-P, catalyzed by aldolase, supports the hypothesis that the enzyme can catalyze formation of the equivalent of a carbanion, the enamine, as required by the mechanism described in Figure 2-7.

BOX 2-1

RAPIDLY AND SLOWLY EXCHANGING HYDRONS

A hydron is any isotope of hydrogen, which are: protium, ^1H; deuterium, ^2H; and tritium, ^3H, with atomic weights of 1, 2, and 3, respectively. Deuterium is a stable, "heavy" isotope, whereas tritium is a radioactive, heavy isotope that decays with the emission of β-particles and a half-life of 12.7 years. Both deuterium and tritium are often used as tracers to follow the reaction pathways of hydrons in biochemical and chemical reactions.

To understand the use of these tracers, one must know how long they will stay attached to a molecule. Organic molecules contain three types of hydrons:

1. Hydrons that for all practical purposes do not exchange with solvent hydrons.

2. Hydrons that exchange within seconds or minutes.

3. Hydrons that exchange slowly, with half-times of hours at room temperature in the absence of acids or bases.

When an alcohol such as methanol (CH_3OH) is mixed with heavy water (^2H_2O or ^3H_2O) and then reisolated—say, by fractional distillation—the hydron of the methanolic OH group will be isotopically labeled. However, the carbon-bound hydrons of the methyl group will be unlabeled; that is, they will not have exchanged with the hydrons of heavy water. Carbon-bound hydrons usually do not exchange rapidly with solvent hydrons. However, hydrons that are bound to heteroatoms such as O, N, or S generally exchange very rapidly with solvent hydrons.

Hydrons bound to carbon adjacent to a carbonyl group exchange slowly with solvent hydrons. This exchange is accelerated by acids or bases. For example, the methyl hydrons of acetone will exchange very slowly with hydrons of water; however, if acetone is dissolved in ^2H_2O in the presence of 0.1 M NaOH, all of the methyl hydrons of acetone will be rapidly replaced by ^2H.

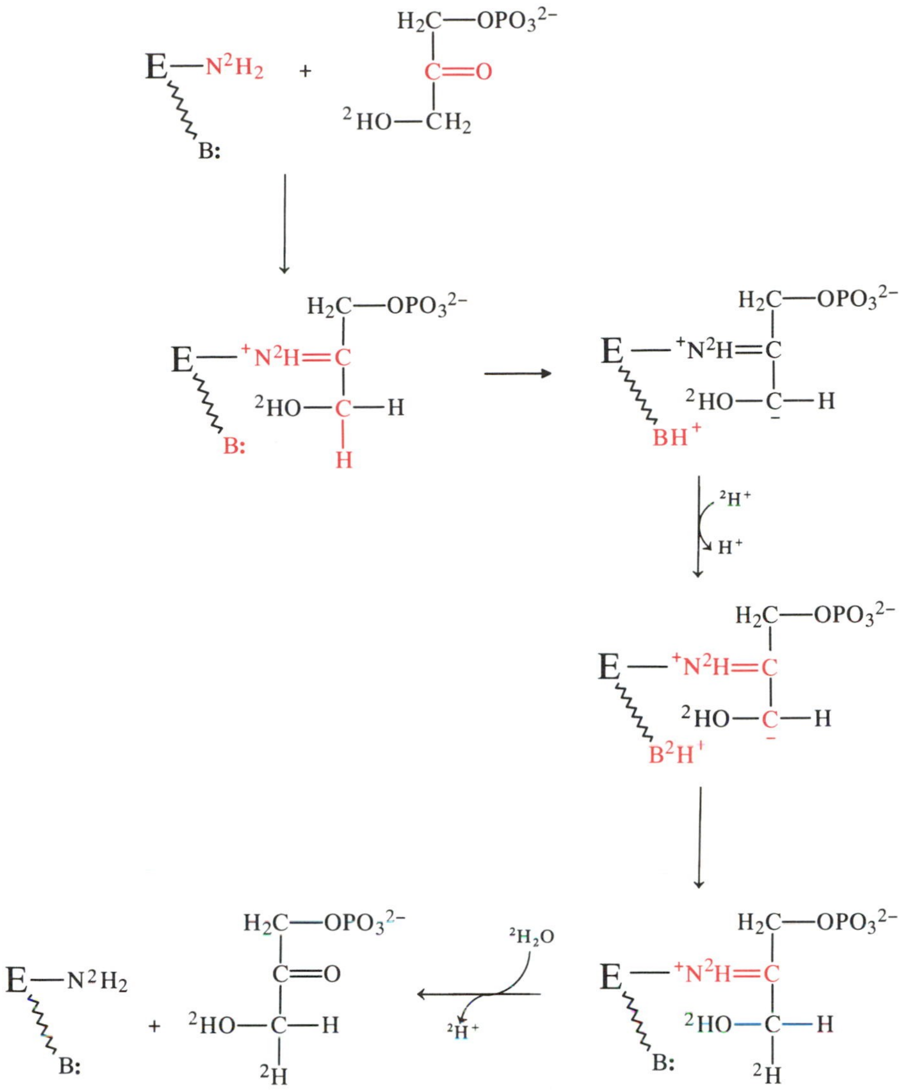

Figure 2-8

Aldolase-catalyzed Exchange of Deuterium into Carbon-3 of Dihydroxyacetone-P. The mechanism of deuterium exchange is closely related to the condensation mechanism in Figure 2-7. The early steps of enamine formation are the same as in the condensation, but the aldehyde cosubstrate is absent. The protonated enzymatic base quickly exchanges its proton with the deuterium in the solvent (^2H_2O), and it then transfers deuterium to carbon-3 of the enamine, a process that is the reverse of proton abstraction. Upon hydrolysis of the imine, deutero-dihydroxyacetone-P is released from the enzyme. All deuterium atoms bonded to N or O quickly exchange with protons in water (H_2O) during the purification of [3-^{2}H]dihydroxyacetone-P, but the 3-deuterium is bonded to carbon and is stable to exchange.

Other aspects of this experiment require further comment. Note that only one of the four carbon-bound hydrogens of dihydroxyacetone-P was exchanged. However, all four hydrogens are adjacent to the carbonyl group and, therefore, should be subject to exchange. If we were to put dihydroxyacetone-P into ^2H_2O under basic conditions, all four hydrogens would undergo exchange. Why, then, does the enzyme catalyze the exchange of only one hydrogen?

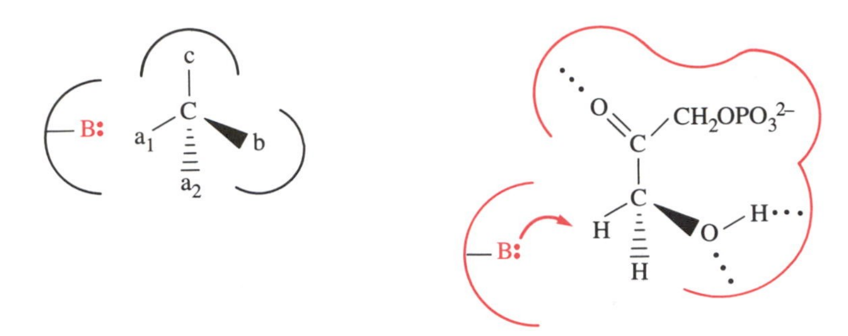

Figure 2-9

Stereospecific Proton Abstraction from Prochiral Centers in Active Sites. In the structure at the left, immobilization of groups *b* and *c* by enzymatic binding orients group a_1 into position to react with the base, whereas group a_2 is out of position and cannot react. In the structure at the right, two functional groups of dihydroxyacetone-P are immobilized by binding to aldolase. This allows one hydrogen but not the other on carbon-3 to be abstracted by the enzymatic base.

This case exemplifies an important difference between catalysis by enzymes and catalysis in solution: the positions of acids and bases in enzyme catalysis are precisely defined, whereas in solution acids and bases move freely and can interact at any point with the substrate. The location of the base at the active site determines which of the hydrogens will be exchanged. The base cannot make contact with the hydrogens on the carbon bearing the phosphate; therefore, they will not exchange. In contrast, when the exchange is catalyzed by a base in solution, the base moves around freely and can abstract any of the four hydrogens. These precisely defined positions account, in part, for the high specificity of enzyme-catalyzed reactions.

It is now apparent why only the hydrogens on the hydroxymethylene group are subject to exchange. Why is only one of these hydrogens exchanged? Two of the four substituents of the hydroxymethylene group, the two hydrogens, are identical. This structure can be represented as follows: Caabc (a = H, b = OH, c = $COCH_2OPO_3^{2-}$). One might think of the two hydrogens (the two a's) as equivalent; that is, indistinguishable. This, however, is not correct. The left-hand structure in Figure 2-9 diagrams the interaction of a Caabc-type molecule with a hypothetical active site. Assume that substituents c and b fit into specific regions of the active site and are immobilized by binding interactions. In this case, substituent a_1 is required to interact specifically with base B. If the interactions of c and b with their speciric sites are maintained, then a_2 can never interact with B; that is, the enzyme will catalyze only the exchange of a_1, but never of a_2. The right-hand structure in Figure 2-9 illustrates how similar binding interactions of dihydroxyacetone-P in an active site lead to stereospecific interaction of one hydrogen with a base.

Many molecules can be represented as Caabc, and a carbon atom with this arrangement of groups is called *prochiral.* It is important to remember that, in all cases, the “a” substituents are not equivalent and that enzymes will always specifically recognize one of them. In summary, then, we now understand why aldolase catalyzes the exchange of only one of the four hydrogens of dihydroxyacetone-P.

Evidence for Imine Formation: Reduction by $NaBH_4$

According to the mechanism of action of aldolase presented in Figure 2-7, an imine is formed between the substrate (dihydroxyacetone-P or fructose-1,6-P_2)

and an amino group at the active site. What is the evidence for this? If one adds sodium borohydride ($NaBH_4$), a reducing agent that transfers the equivalent of a hydride ion (H^-), to a buffered solution containing enzyme and dihydroxyacetone-P, the enzyme becomes catalytically inactive; that is, it loses the ability to catalyze both the aldol condensation and the exchange of the hydrogen on dihydroxyacetone-P with solvent deuterium. If $NaBH_4$ is added to enzyme alone, the catalytic activity is unaffected. The presence of *both* $NaBH_4$ *and* dihydroxyacetone-P is therefore required to inactivate the enzyme. Inactivation is due to the reduction by $NaBH_4$ of the imine formed between the enzyme and dihydroxyacetone-P (Figure 2-7). This reduction leads to the conversion of the imine into a secondary amine, which is stable to hydrolysis (equation 9). Consequently, the active site is permanently modified and catalysis of the aldol condensation can no longer occur.

$$E{-}\overset{+}{N}H{=}C(CH_2OPO_3^{2-}){-}C(OH)H_2 \xrightarrow{NaBH_4} E{-}NH{-}CH(CH_2OPO_3^{2-}){-}C(OH)H_2 \tag{9}$$

The reaction with $NaBH_4$ can be further characterized by repeating the reduction using dihydroxyacetone-P in which one of the carbons has been labeled with ^{14}C (see Box 2-2). After this $NaBH_4$ reduction, ^{14}C is covalently bound to the enzyme protein. When this protein is hydrolyzed with HCl, a new amino acid is found, N^6-dihydroxypropyl-L-lysine, which corresponds to the reduced imine formed from lysine and dihydroxyacetone-P, except for the phosphate group that is lost during hydrolysis.

$$^-O(O{=})C{-}CH(NH_3^+){-}CH_2{-}CH_2{-}CH_2{-}CH_2{-}\overset{+}{N}H_2{-}CH(CH_2OH)_2$$

N^6-Dihydroxypropyl-L-lysine

This shows that lysine is the amino acid at the active site of aldolase that provides the amino group for the formation of an imine with dihydroxyacetone-P.

BOX 2-2

TRACERS FOR CARBON

Carbon-14 (^{14}C, atomic weight 14) is a radioactive isotope of carbon that emits low-energy β-radiation, which can be readily detected. Carbon-14 may be introduced into organic molecules in order to detect the molecule, or that part of the molecule that contains the isotope, after a reaction has occurred. For example, in our experiment in which aldolase is reduced with $NaBH_4$ in the presence of [^{14}C]dihydroxyacetone-P, we can separate the protein from unreacted $NaBH_4$ and [^{14}C]dihydroxyacetone-P by precipitation and wash it extensively. Exposure to a radiation-detecting device will show that ^{14}C is associated with the protein. We can, therefore, conclude that dihydroxyacetone-P, or at least the ^{14}C-labeled carbon of dihydroxyacetone-P, is attached to the protein. Proteolytic enzymes catalyze the partial hydrolysis of proteins by cleaving specific peptide bonds. The products resulting from digestion by many proteolytic enzymes are polypeptides that can be purified and characterized (Chapter 7). One of these polypeptides from the reduction of aldolase contains the ^{14}C from [^{14}C]dihydroxyacetone-P attached to the amino acid lysine. Because the amino acid sequences for aldolase and the [^{14}C]polypeptide are known, we know exactly where the active site is located.

There is also a heavy, stable isotope of carbon, ^{13}C (atomic weight 13) that can be detected by NMR or MS, which is also useful for tracing reaction pathways of carbon.

Further Evidence for Imine Formation: Oxygen Exchange

Additional evidence in support of the proposed mechanism is provided by the following experiment in which $H_2{}^{18}O$ is used. Ordinary H_2O, as well as dihydroxyacetone-P, contains mainly ^{16}O. Oxygen-18 is a stable (i.e., nonradioactive) isotope of oxygen, and $H_2{}^{18}O$ is prepared by separating the 0.2% of ^{18}O that is found in nature from the more-abundant ^{16}O. The enzyme-catalyzed condensation between dihydroxyacetone-P and glyceraldehyde-3-P is carried out in $H_2{}^{18}O$. The product, fructose-1,6-P_2, is isolated and found to contain ^{18}O in the carbonyl group. This shows that the oxygen (^{16}O) that was originally in the carbonyl group of dihydroxyacetone-P is no longer present in the carbonyl group of fructose. This is exactly what is required by the imine mechanism of equation 7. Formation of an imine from dihydroxyacetone-P and the lysine amino group leads to elimination of the carbonyl oxygen as $H_2{}^{16}O$. When the imine formed between fructose-1,6-P_2 and lysine is hydrolyzed, a new oxygen atom (^{18}O) is introduced from the solvent.

Metal-dependent Aldolase Reactions

Not all enzymes that catalyze aldol condensations between aldehydes and ketones, and related reactions, utilize the imine mechanism. Enzymes that do not utilize this mechanism require a metal, generally Mg^{2+}, for activity. The Mg^{2+} brings two positive charges close to the carbonyl group and stabilizes formation of the enolate (equation 10).

$$\text{B:}\ \ \underset{\text{H}}{\underset{|}{\text{CH}_2}}\text{—}\overset{\overset{\text{Mg}^{2+}}{\text{O}}}{\overset{\|}{\text{C}}}\text{—CH}_2\text{—} \rightleftharpoons \text{BH}^+\ \ \text{CH}_2\text{=}\overset{\overset{\text{Mg}^{2+}}{\vdots}}{\overset{^-\text{O}}{\overset{|}{\text{C}}}}\text{—CH}_2\text{—} \tag{10}$$

Carbon-Carbon Bond Formation Through Other Mechanisms

The Claisen Condensation: Thiolase and Coenzyme A

Another reaction that leads to the formation of C–C bonds is the Claisen condensation. In this reaction, a carbanion reacts with the carbonyl group of an ester to form a ketone according to equation 11; several enzymes catalyze related reactions.

$$\text{R}_1\text{—}\overset{\text{O}}{\overset{\|}{\text{C}}}\text{—}\underset{\text{R}_4}{\underset{|}{\overset{\text{R}_5}{\overset{|}{\text{C}^-}}}} + \text{R}_3\text{O—}\overset{\text{O}}{\overset{\|}{\text{C}}}\text{—R}_2 \rightleftharpoons \text{R}_1\text{—}\overset{\text{O}}{\overset{\|}{\text{C}}}\text{—}\underset{\text{R}_4}{\underset{|}{\overset{\text{R}_5}{\overset{|}{\text{C}}}}}\text{—}\underset{\text{OR}_3}{\underset{|}{\overset{\text{O}^-}{\overset{|}{\text{C}}}}}\text{—R}_2 \rightleftharpoons \text{R}_1\text{—}\overset{\text{O}}{\overset{\|}{\text{C}}}\text{—}\underset{\text{R}_4}{\underset{|}{\overset{\text{R}_5}{\overset{|}{\text{C}}}}}\text{—}\overset{\text{O}}{\overset{\|}{\text{C}}}\text{—R}_2 + \text{R}_3\text{O}^- \tag{11}$$

However, the enzymatic reactions do not use oxygen esters, but use a thioester derived from coenzyme A, the structure of which is shown in Figure 2-10. (Coenzyme A is frequently abbreviated as CoASH and esters of coenzyme A as RCOSCoA.)

It is advantageous to use thioesters in condensation reactions because the carbonyl carbon atom has more positive character than the carbonyl in the

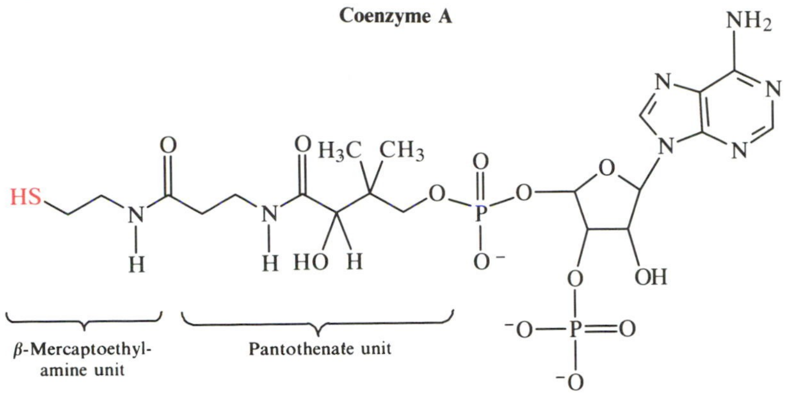

Coenzyme A

HS — β-Mercaptoethylamine unit — Pantothenate unit

Figure 2-10

The Structure of Coenzyme A.

corresponding oxygen esters. Consider the resonance forms for an oxygen ester in Figure 2-11. The contribution from form II tends to decrease the positive charge on the carbon, whereas forms I and III tend to increase the positive charge. Corresponding resonance forms can be written for a thioester. However, for the thioester, the contribution from II is believed to be less important, whereas I and III may be more important than for the oxygen ester. Hence, the carbonyl carbon of the thioester is more positive than that of the oxygen ester. Positive charge on carbon will make it easier for a nucleophilic reagent, such as a carbanion, to attack the carbonyl group. It will also make it easier to remove a proton from the adjacent carbon atom to form a carbanion.

Thiolase is an enzyme that catalyzes a Claisen condensation (equation 12).

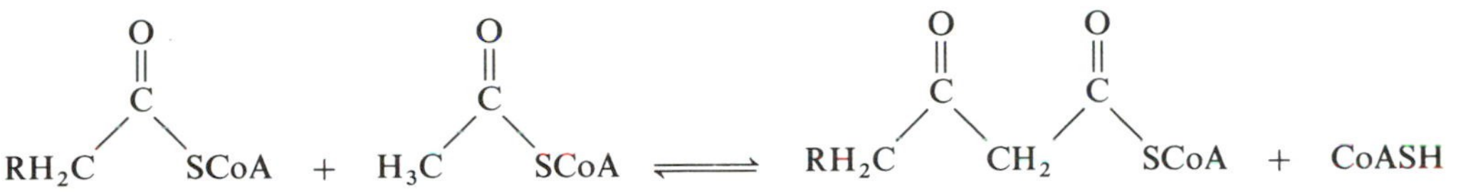

$$RH_2C{-}\overset{O}{\overset{\|}{C}}{-}SCoA + H_3C{-}\overset{O}{\overset{\|}{C}}{-}SCoA \rightleftharpoons RH_2C{-}\overset{O}{\overset{\|}{C}}{-}CH_2{-}\overset{O}{\overset{\|}{C}}{-}SCoA + CoASH \qquad (12)$$

$$R = CH_3{-}(CH_2)_n \; (n = 0, 1, 2, \ldots, 20)$$

The probable sequence of reactions catalyzed by thiolase is shown in Figure 2-12, which shows only the essential intermediates. The active-site functional groups, with the exception of an —SH group, are omitted.

In the reaction catalyzed by thiolase (Figure 2-12), the CoA-thioester ($RCH_2COSCoA$) initially undergoes an ester interchange reaction with a sulfhydryl (—SH) group of a cysteine residue at the active site to form an enzymatic thioester and release CoASH. A carbanion derived from acetyl CoA then adds to the carbonyl group of the enzyme-derived thioester to give a tetrahedral adduct, which collapses to form final products. This reaction, like the reaction catalyzed by aldolase, includes a covalent intermediate between the enzyme and the substrate. The first evidence for the presence of the essential —SH group came from experiments with iodoacetic acid (ICH_2COOH) and iodoacetamide (ICH_2CONH_2). These compounds react with —SH groups and

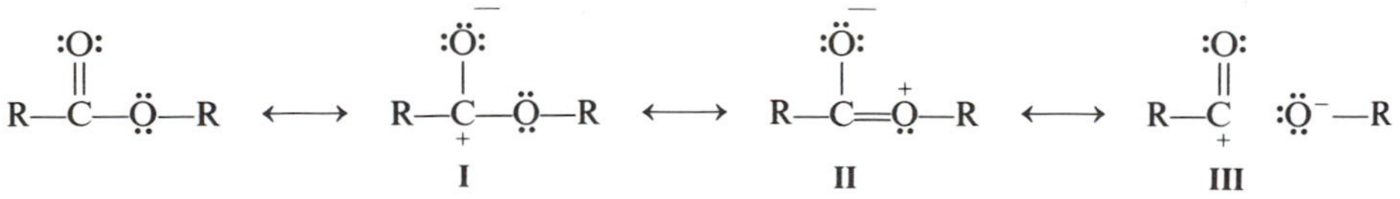

$$R{-}\overset{:O:}{\overset{\|}{C}}{-}\ddot{O}{-}R \longleftrightarrow \underset{\text{I}}{R{-}\overset{:\ddot{O}:^-}{\overset{|}{C^+}}{-}\ddot{O}{-}R} \longleftrightarrow \underset{\text{II}}{R{-}\overset{:\ddot{O}:^-}{\overset{|}{C}}{=}\overset{+}{O}{-}R} \longleftrightarrow \underset{\text{III}}{R{-}\overset{:O:}{\overset{\|}{C^+}}\;:\ddot{O}^-{-}R}$$

Figure 2-11

Resonance Forms for an Oxygen Ester.

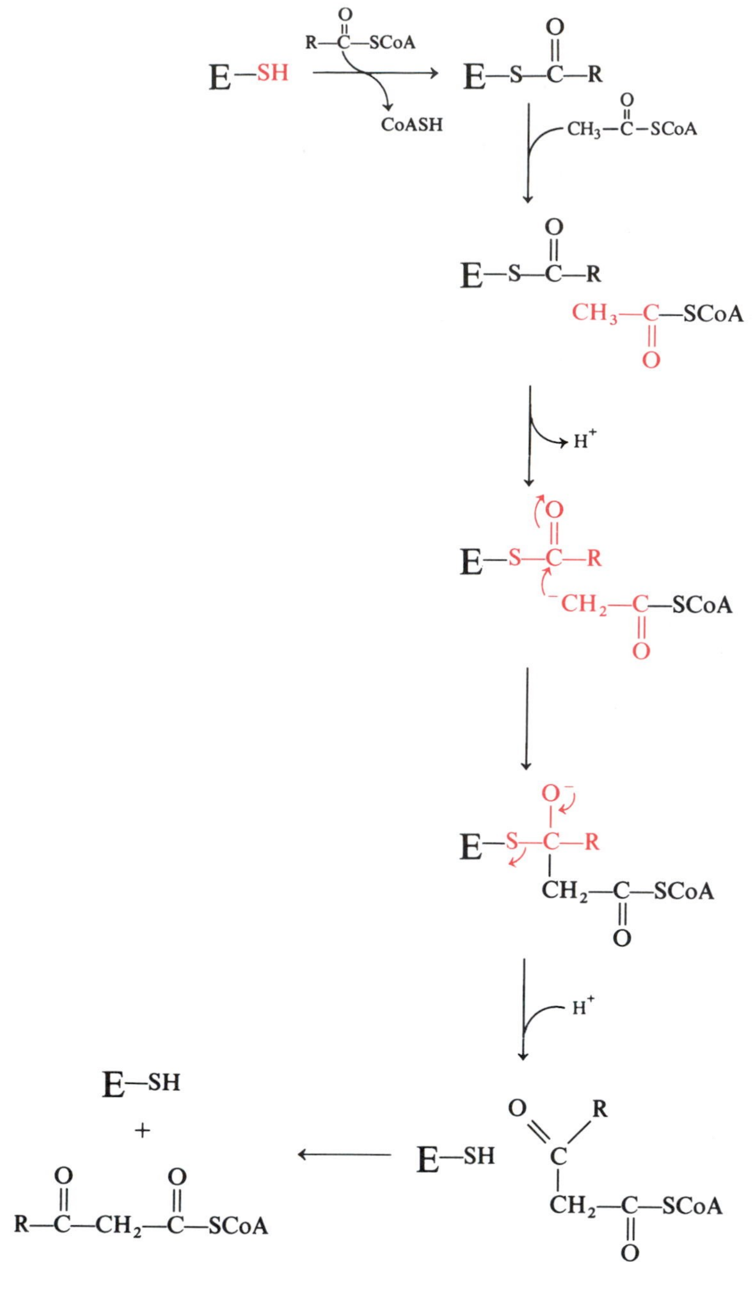

Figure 2-12

The Mechanism of the Thiolase Reaction. The mechanism of the thiolase reaction begins with the acylation of the sulfhydryl group of the active-site cysteine by an acyl CoA, accompanied by the release of CoASH. A molecule of acetyl CoA is then bound to the active site, and it undergoes enolization to the carbanion, which subsequently attacks the carbonyl group of the enzymatic acyl thioester. The resulting tetrahedral adduct eliminates the enzymatic thiol group to form the final condensation product. The functions of other catalytic groups that are required to facilitate proton transfers are omitted to simplify the illustration; however, each proton-transfer step is catalyzed by an acid or a base on the enzyme. Although represented as a unidirectional process for simplicity of discussion, the reaction is readily reversible and slightly favors cleavage of β-ketoacids; hence, the name thiolase or β-ketothiolase.

are frequently used to identify —SH groups in proteins by their reactions according to equations 13 and 14.

$$E\text{—}S^- + CH_2(I)\text{—}\overset{O}{\overset{\|}{C}}\text{—}O^- \longrightarrow E\text{—}S\text{—}CH_2\text{—}\overset{O}{\overset{\|}{C}}\text{—}O^- + I^- \quad (13)$$

$$E\text{—}S^- + CH_2(I)\text{—}\overset{O}{\overset{\|}{C}}\text{—}NH_2 \longrightarrow E\text{—}S\text{—}CH_2\text{—}\overset{O}{\overset{\|}{C}}\text{—}NH_2 + I^- \quad (14)$$

Addition of either of these reagents to thiolase inactivates the enzyme. The rate of inactivation is reduced in the presence of substrate. This suggests that inactivation occurs at the active site and that an —SH group is present at the active site. When thiolase is inactivated with ^{14}C-iodoacetate, one mole of iodoacetate becomes covalently attached per mole of enzyme. Acid hydrolysis of the labeled enzyme yields a ^{14}C-labeled derivative of cysteine, *S*-[^{14}C]carboxymethylcysteine.

$$^-O_2C\text{—}\underset{NH_3^+}{\overset{H}{C}}\text{—}CH_2\text{—}SH$$

L-Cysteine (Cys, C)

$$^-O_2C\text{—}\underset{NH_3^+}{\overset{H}{C}}\text{—}CH_2\text{—}S\text{—}{}^{14}CH_2\text{—}CO_2^-$$

***S*-[^{14}C]Carboxymethylcysteine**

The most-convincing evidence for the intermediate formation of a thioester of acetate on the enzyme comes from the following experiment. Treatment of thiolase with ^{14}C-labeled acetyl CoA followed by dialysis, which removes all small molecules (Box 2-3), produces a ^{14}C-labeled protein. From the amount of

BOX 2-3

SEPARATIONS OF SMALL MOLECULES FROM LARGE MOLECULES

Dialysis is a method by which small molecules or ions are separated from large molecules such as proteins, RNA, and DNA. In this procedure, a solution of the large and small molecules is enclosed in a bag made of a semipermeable membrane, which is permeable to small molecules but not to large molecules. The bag is suspended in a large volume of a buffer. The small molecules inside the bag diffuse through the membrane into the outer buffer solution until their concentrations inside and outside the bag are equal. Initially, owing to the large volume of the outer buffer, the probability that the small molecules will reenter the bag is small, because the concentration outside is small; but, at equilibrium, the concentrations inside and outside the bag are equal, and the rate of diffusion across the membrane is the same in both directions. Large molecules such as proteins also have a strong tendency or potential for diffusing through the bag into the outer buffer, but diffusion of large molecules is blocked by the small size of the pores in the semipermeable membrane. Therefore, proteins will remain inside the bag. This procedure can be used not only to remove undesired small molecules from a protein, but also to change the buffer in which the protein is dissolved.

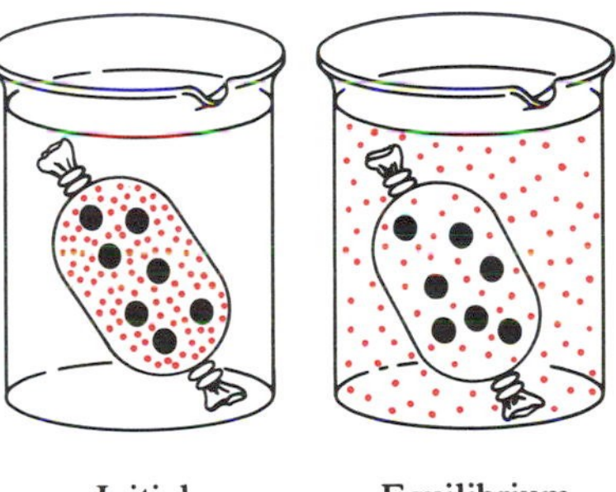

Gel permeation chromatography also separates small molecules from large molecules by the effects of pore permeability. In chromatography, the pores are spaces within beads packed into a column rather than within a membrane. Gel permeation chromatography is described in Chapter 7. For some purposes, gel chromatography is more efficient and convenient than dialysis, and for other purposes dialysis can be more convenient.

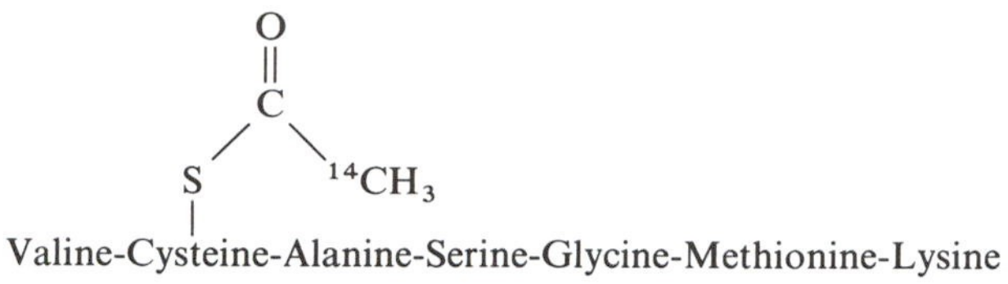

O
‖
C
S $^{14}CH_3$
|
Valine-Cysteine-Alanine-Serine-Glycine-Methionine-Lysine
Peptide A

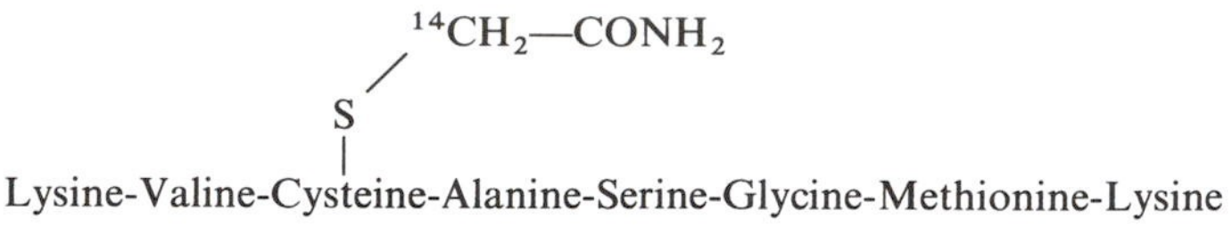

$^{14}CH_2$—$CONH_2$
S
|
Lysine-Valine-Cysteine-Alanine-Serine-Glycine-Methionine-Lysine
Peptide B

Figure 2-13

The ^{14}C-labeled Active-Site Peptides of Thiolase. The peptides derived by systematic degradation of the covalent [^{14}C]acetyl thiolase and the inactivated enzyme prepared by alkylation with [^{14}C]iodoacetamide are illustrated. The amino acids in the peptides are indicated by their names and arranged by convention from left to right in the N-terminal to C-terminal directions. The structure of cysteine is given in the text and those of valine and lysine appear in Table 2-1. The structures of the other amino acids may be found among those of the twenty common amino acids inside the front cover of this book. The methods by which amino acid sequences of peptides are determined are described in Chapter 7.

radioactivity associated with the protein, it can be calculated that one mole of ^{14}C-labeled compound is associated with one mole of protein. When the ^{14}C-labeled protein is degraded with a proteolytic enzyme, peptide A in Figure 2-13, which contains a ^{14}C-labeled acetyl group, can be isolated. (Proteolytic enzymes degrade proteins by catalyzing the hydrolysis of peptide bonds. See Chapters 3 through 6.)

This experiment establishes directly that reaction of acetyl CoA with the enzyme leads to acetylation of one SH group, as proposed in the mechanism shown in Figure 2-12. Inactivation of the enzyme with ^{14}C-iodoacetamide and proteolytic degradation of the ^{14}C-protein leads to peptide B in Figure 2-13. Note that peptides A and B contain the same amino acid sequence, with an additional lysine in peptide B. Iodoacetamide, therefore, reacts with the same cysteine residue that is acetylated by acetyl CoA. These and other results, not described here, strongly support the proposed mechanism.

Both the substrate (acetyl CoA) and the covalent thiolase-substrate complex are thioesters. Therefore, this covalent adduct does not have any special chemical properties, such as a larger intrinsic susceptibility of the carbonyl group to nucleophilic attack, compared with the substrate. The advantage in forming a covalent adduct with the enzyme, if there is one, must lie in some other aspect of catalysis. The reader might speculate what the advantage of forming a covalent bond with the enzyme might be (hint: see Chapters 5 and 11).

Structural Requirements for C–C Bond Formation and Cleavage

We now know that two basic enzymatic reactions—that is, those related to aldol condensations and those related to Claisen condensations—lead to the formation of carbon–carbon bonds. Biological molecules formed through these

Aldol condensation

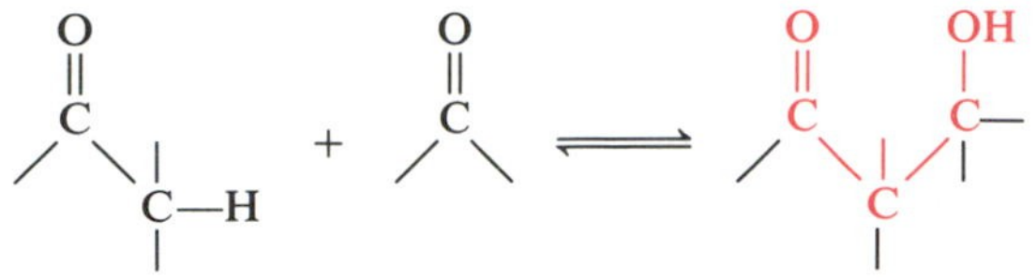

Claisen condensation

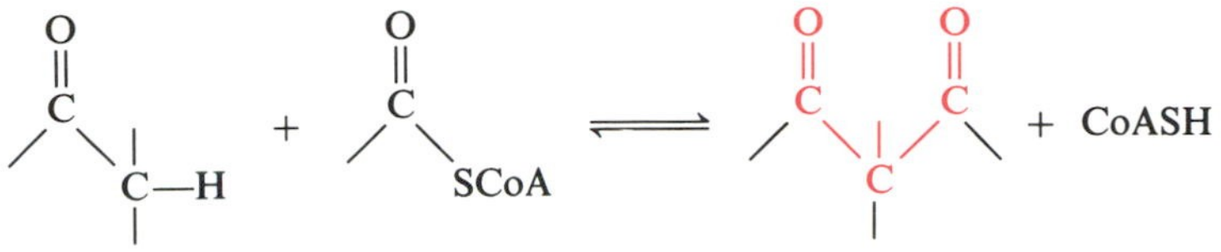

Figure 2-14

Structural Requirements for C–C Bond Formation and Cleavage.

reactions have characteristic structural features I and II shown in Figure 2-14. If we encounter a biological molecule that contains structure I, we can conclude that it was very likely formed by an aldol condensation. We can then also make reasonable guesses about the structures of the precursors. Consider the following example. Certain microorganisms synthesize 2-deoxy-D-ribose-5-phosphate.

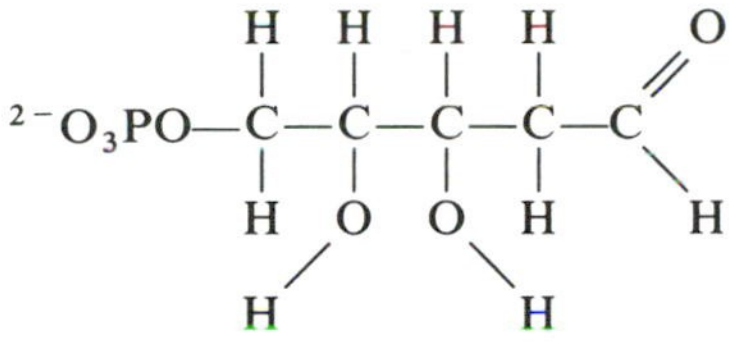

2-Deoxy-D-ribose-5-phosphate

This molecule contains configuration I in Figure 2-14. It is derived through an aldol condensation, and the precursor molecules are glyceraldehyde-3-P and acetaldehyde. Analogous conclusions can be drawn when we encounter molecules that contain configuration II.

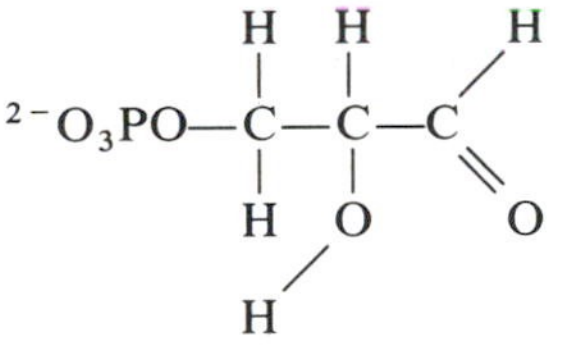

Glyceraldehyde-3-P 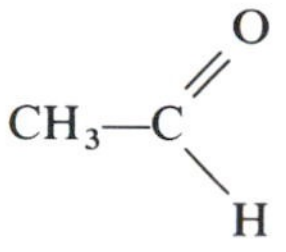 **Acetaldehyde**

The realization that molecules undergoing formation and degradation of C–C bonds have characteristic structures helps us to predict strategies for the degradation and synthesis of compounds. For example, butyryl CoA is metabolized to two molecules of acetyl CoA (equation 15). How can this be accomplished?

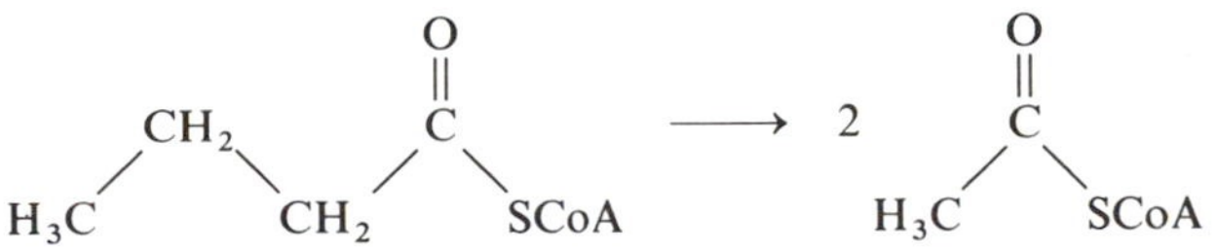

$$H_3C{-}CH_2{-}CH_2{-}C(=O){-}SCoA \longrightarrow 2\ H_3C{-}C(=O){-}SCoA \qquad (15)$$

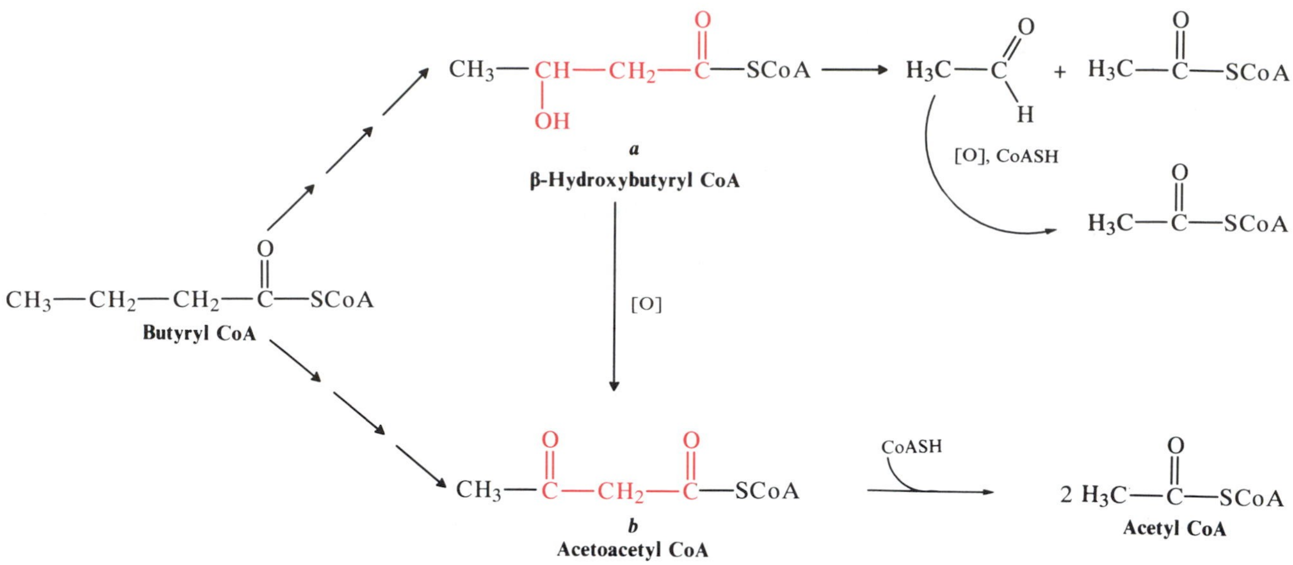

$$CH_3{-}CH_2{-}CH_2{-}\overset{O}{\overset{\|}{C}}{-}SCoA$$

Butyryl CoA

$$CH_3{-}\underset{OH}{\underset{|}{CH}}{-}CH_2{-}\overset{O}{\overset{\|}{C}}{-}SCoA \longrightarrow H_3C{-}CHO + H_3C{-}\overset{O}{\overset{\|}{C}}{-}SCoA$$

a

β-Hydroxybutyryl CoA

$$H_3C{-}CHO \xrightarrow{[O],\ CoASH} H_3C{-}\overset{O}{\overset{\|}{C}}{-}SCoA$$

$\downarrow$ [O]

$$CH_3{-}\overset{O}{\overset{\|}{C}}{-}CH_2{-}\overset{O}{\overset{\|}{C}}{-}SCoA \xrightarrow{CoASH} 2\ H_3C{-}\overset{O}{\overset{\|}{C}}{-}SCoA$$

b

Acetoacetyl CoA

Acetyl CoA

Figure 2-15

Possible Routes for the Degradation of Butyryl CoA. Butyryl CoA could, in several steps, be oxygenated either to β-hydroxybutyryl CoA (*a*) or to acetoacetyl CoA (*b*), either of which could be cleaved into two molecules of acetyl CoA. In cells, butyryl CoA is in fact converted into acetoacetyl CoA, and β-hydroxybutyryl CoA is an intermediate on the route to acetoacetyl CoA (Chapter 19).

From the products, it is apparent that C–C cleavage occurs between C-2 and C-3. But butyryl CoA does not contain a configuration characteristic of C–C cleavage. Therefore, we must conclude that the metabolism of butyryl CoA consists of a series of steps leading to a structure in which cleavage can occur. For example, conversion of butyryl CoA into either β-hydroxybutyryl CoA or acetoacetyl CoA, as illustrated in Figure 2-15, would introduce the appropriate configurations. Compound *a* contains configuration I (Figure 2-14), and *b* configuration II. Either compound can therefore undergo C–C cleavage and the appropriate products will be formed. As we continue our discussion of basic biochemical reactions, the reader will understand how butyryl CoA can be converted into either *a* or *b*. In fact, butyryl CoA is degraded in biochemical systems by the path in which acetoacetyl CoA is an intermediate; β-hydroxybutyryl CoA is also an intermediate in the formation of acetoacetyl CoA (Chapter 19).

Citrate Synthase

Another important reaction in which a C–C bond is formed is the condensation between acetyl CoA and oxaloacetate catalyzed by the enzyme citrate synthase (equation 16).

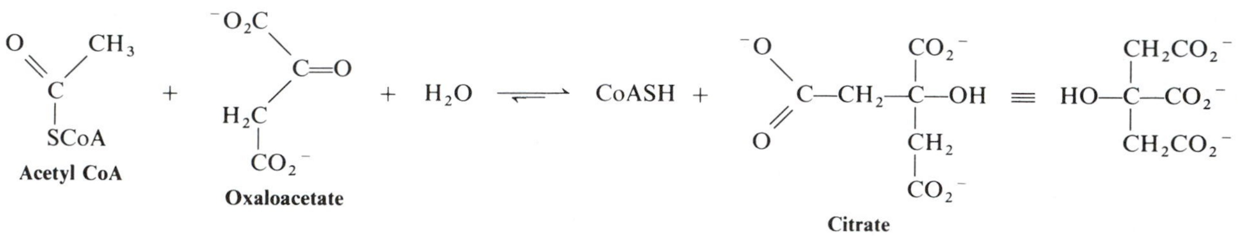

$$\underset{\text{Acetyl CoA}}{CH_3{-}\overset{O}{\overset{\|}{C}}{-}SCoA} + \underset{\text{Oxaloacetate}}{^-O_2C{-}\overset{O}{\overset{\|}{C}}{-}CH_2{-}CO_2^-} + H_2O \rightleftharpoons CoASH + \underset{\text{Citrate}}{^-O_2C{-}CH_2{-}\underset{CH_2CO_2^-}{\underset{|}{\overset{CO_2^-}{\overset{|}{C}}}}{-}OH} \equiv HO{-}\underset{CH_2CO_2^-}{\underset{|}{\overset{CH_2CO_2^-}{\overset{|}{C}}}}{-}CO_2^- \quad (16)$$

Citrate synthase is present in all living organisms and catalyzes the initial step of the citric acid cycle, a major metabolic pathway that will be discussed in Chapter 21. In biological systems, this enzyme functions primarily in the synthetic direction.

Note that, in the reaction catalyzed by citrate synthase, one of the reactants is a CoA-ester. In the products, however, there is no ester linkage and free CoASH is formed. This reaction, therefore, includes the hydrolysis of a thioester, as well as C–C bond formation. The reaction products and chemical considerations suggest the minimal reaction sequence that is shown in Figure 2-16.

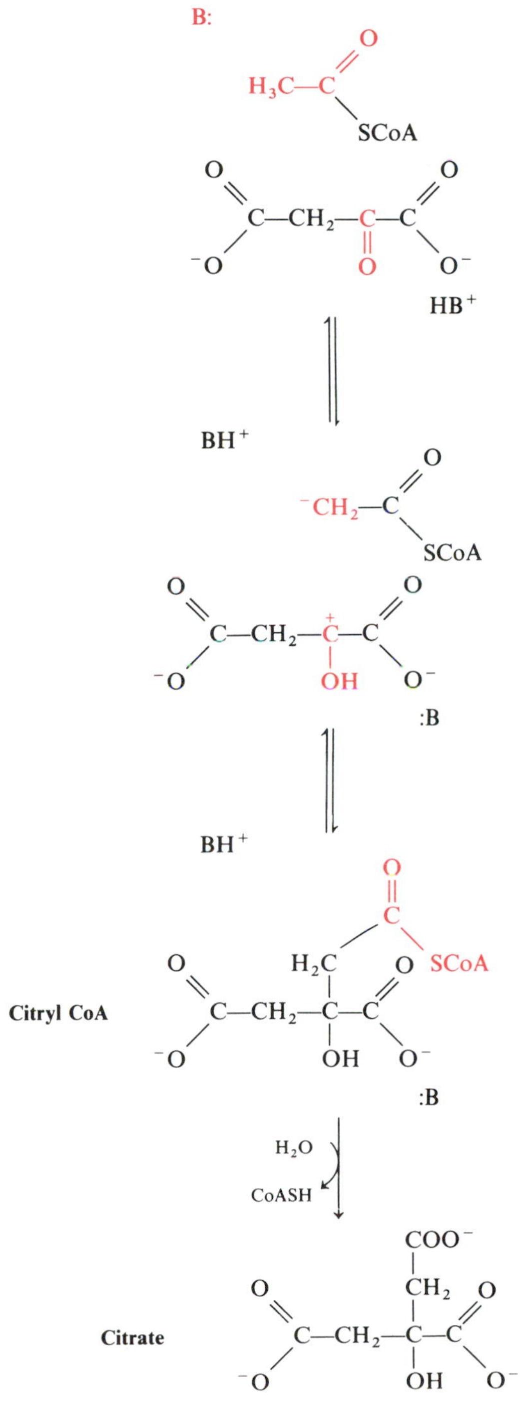

B: $H_3C-C(=O)SCoA$ + $^-O-C(=O)-CH_2-C(=O)-C(=O)-O^-$ HB$^+$

⇌

BH$^+$ $^-CH_2-C(=O)SCoA$ + $^-O-C(=O)-CH_2-\overset{+}{C}(OH)-C(=O)-O^-$:B

⇌

BH$^+$

Citryl CoA: $^-O-C(=O)-CH_2-C(OH)(CH_2-C(=O)SCoA)-C(=O)-O^-$:B

↓ H_2O, CoASH

Citrate: $^-O-C(=O)-CH_2-C(OH)(CH_2COO^-)-C(=O)-O^-$

Figure 2-16

The Mechanism of the Citrate Synthase Reaction. The condensation of acetyl CoA with oxaloacetate begins with the abstraction of a proton from acetyl CoA by an enzymatic base to form a carbanion, which attacks the carbonyl carbon of oxaloacetate. The resulting citryl CoA undergoes hydrolysis to citrate and CoASH.

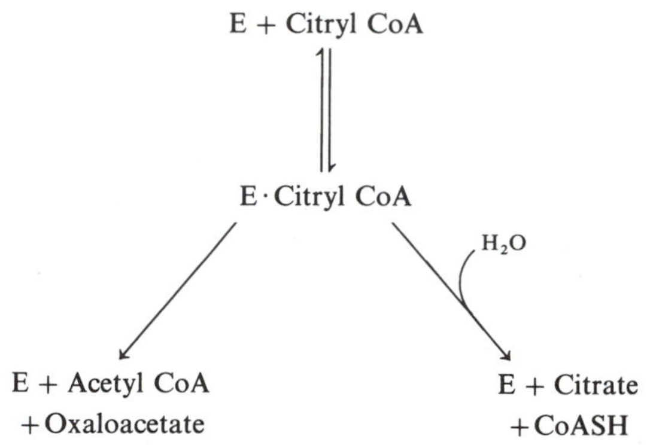

E + Citryl CoA

⇅

E·Citryl CoA

H_2O

E + Acetyl CoA + Oxaloacetate   E + Citrate + CoASH

Figure 2-17

Evidence for Citryl CoA in the Citrate Synthase Reaction Mechanism. When synthetic citryl CoA was combined with citrate synthase, the enzyme catalyzed its decomposition to a mixture of acetyl CoA, oxaloacetate, citrate, and CoASH. This was consistent with citryl CoA being an intermediate in the reaction, because the formation of acetyl CoA and oxaloacetate would represent reversal of citryl CoA formation in Figure 2-16, and citrate formation would represent hydrolysis of citryl CoA in the forward direction of Figure 2-16.

Although citryl CoA is an intermediate in the proposed reaction sequence, no free citryl CoA can be detected in the course of the catalytic reaction. If it is a participant, it must be very tightly bound to the enzyme so that its conversion into either starting material or product must be much faster than its release from the enzyme.

Citryl CoA has been synthesized and it was found that it is converted by the enzyme into acetyl CoA and oxaloacetate, as well as into citrate (Figure 2-17). This finding is good evidence for the intermediate involvement of citryl CoA, in agreement with the mechanism depicted in Figure 2-16. Other aspects of the mechanism in Figure 2-16 are much more difficult to demonstrate (see Box 2-4).

Citrate Lyase and ATP: Activation of Carboxyl Oxygen by ATP

Citrate synthase functions in the synthetic direction in most biological systems. *Citrate lyase,* on the other hand, is used to degrade citrate to acetyl CoA and oxaloacetate according to equation 17.

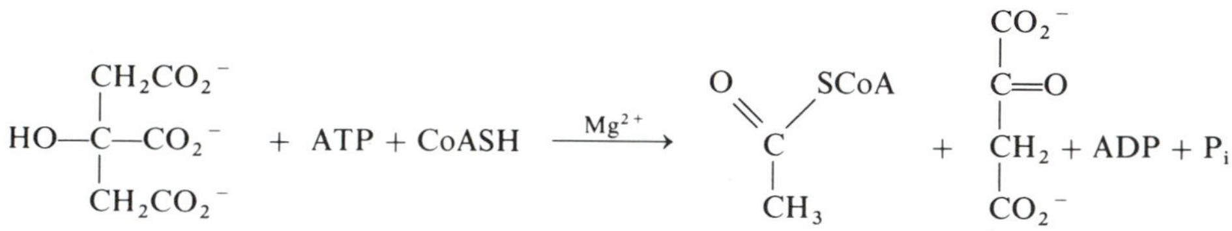

$$HO\text{—}C(CH_2CO_2^-)_2\text{—}CO_2^- + ATP + CoASH \xrightarrow{Mg^{2+}} CH_3\text{—}C(=O)\text{—}SCoA + {}^-O_2C\text{—}C(=O)\text{—}CH_2\text{—}CO_2^- + ADP + P_i \quad (17)$$

This reaction is more complicated than the other reactions presented so far. Not only is a carbon–carbon bond broken, but ATP is hydrolyzed to adenosine diphosphate (ADP), and a thioester is formed. Thus, ATP, the most-important high-energy molecule in biological systems (Chapter 1), is a participant in this reaction.

BOX 2-4

SUBSTRATE SYNERGISM IN PROTON ABSTRACTION FROM ACETYL COENZYME A

The mechanism proposed for the action of citrate synthase requires that a base at the active site abstract a proton from acetyl CoA. In principle, it should be easy to demonstrate this. One would allow the enzyme to react with acetyl CoA containing a heavy isotope of hydrogen (^2H or ^3H) in the acetyl group, and the appearance of the heavy isotope in the solvent, or its disappearance from acetyl CoA, would be analyzed. Alternatively, the reaction would be carried out in heavy water (^2H_2O or ^3H_2O), and one would determine whether the enzyme can catalyze the replacement of the α-hydrogens on acetyl CoA by deuterium or tritium.

Exchange experiments of this kind have already been described in connection with aldolase. When these experiments were carried out with citrate synthase, no hydrogen exchange between solvent hydrogens and the α-hydrogens of acetyl CoA could be detected. This was a surprising and confusing result.

Important insight into the mechanism of action of this enzyme was gained through an observation of H. Eggerer and his colleagues. These investigators showed that tritium is released from [^3H]acetyl CoA into the solvent when the labeled substrate is combined with the enzyme *together with malate,* a structural analog of oxaloacetate that is unreactive as a substrate. They obtained the results given in the adjoining table.

^3H released from [^3H]acetyl CoA catalyzed by citrate synthase

Additions to Citrate Synthase	^3H Released to H_2O (dpm*)
[^3H]Acetyl CoA	200
[^3H]Acetyl CoA + L-Malate	6440
[^3H]Acetyl CoA + D-Malate	210
[^3H]Acetyl CoA + D, L-Malate	2530

*dpm = disintegrations per minute, a measure of the amount of ^3H.

$$\begin{array}{c} CO_2^- \\ | \\ HO{-}C{-}H \\ | \\ CH_2 \\ | \\ CO_2^- \end{array} \qquad \begin{array}{c} CO_2^- \\ | \\ O{=}C \\ | \\ CH_2 \\ | \\ CO_2^- \end{array}$$

L-Malate **Oxaloacetate**

These experiments show that citrate synthase does indeed catalyze the exchange of the α-protons of acetyl CoA but only in the presence of L-malate. L-Malate is a *substrate analog;* that is, it is a molecule that is similar to the true substrate oxaloacetate but that cannot participate in the condensation reaction with acetyl CoA. A reasonable interpretation of this experiment is that malate brings about a conformational change in citrate synthase that enables the active site to become catalytically active, so that abstraction of a proton from the α-position of acetyl CoA can occur. In the normal catalytic process, the interaction with oxaloacetate, rather than L-malate, presumably brings about the change to a catalytically active conformation. The data show that only one of the two enantiomers of malate, L-malate, is active. L-Malate presumably occupies a portion of the substrate and product binding site. The OH group of malate probably occupies the site at which the OH group of citrate is bound. Only in the L-isomer is the OH group oriented correctly relative to the two carboxyl groups so that the steric requirements of the binding site are met.

A situation analogous to that seen with citrate synthase is found with many two-substrate enzymes (see Chapter 11, pyruvate kinase). The second substrate is required to bring the enzyme into a catalytically active state. Citrate synthase is probably the first enzyme for which this effect was clearly demonstrated. This kind of behavior is called *substrate synergism* because binding of one substrate (or part of a substrate) activates the enzyme for reaction with another substrate.

Biological systems generally obtain energy for their activities by oxidizing organic molecules (Figure 2-18). However, the energy that is used is not obtained directly from oxidation. Instead, energy derived from oxidation is first converted into ATP, which is an anhydride that contains three phosphates (see Chapter 1). ATP is then used as a "fuel" for the majority of energy-requiring (endergonic) processes in biology. When ATP is used as an energy source, it is usually hydrolyzed to ADP + P_i (inorganic phosphate); however, in some reactions it is cleaved to give AMP and PP_i (pyrophosphate).

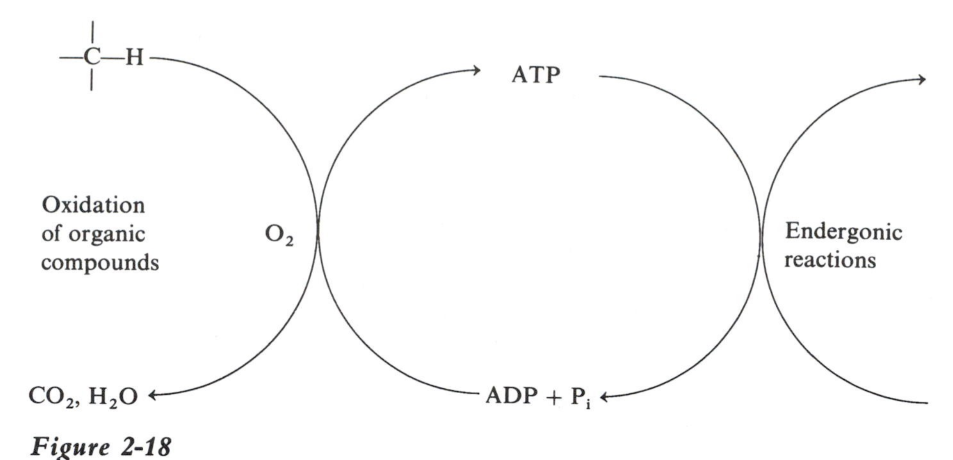

Figure 2-18

Generation of ATP and Its Utilization for Energy-requiring Reactions.

Endergonic reactions do not proceed spontaneously; they require energy input to make them occur. Examples of energy-requiring processes that use ATP are biosynthetic reactions, muscle contraction, and the transport of metabolites against concentration gradients. The generation and utilization of ATP will be discussed extensively in many chapters of this textbook.

We now return to citrate lyase and consider the function of ATP. However, the mechanism of the reaction catalyzed by citrate lyase is likely to be very similar to the mechanism of the reaction catalyzed by citrate synthase. Citrate will be broken down to oxaloacetate and a carbanion derived from acetate (equation 18).

$$\mathrm{HO{-}C(CH_2CO_2^-)_2{-}CO_2^-} \longrightarrow \mathrm{^-CH_2CO_2^-} + \mathrm{O{=}C(CH_2CO_2^-){-}CO_2^-} + \mathrm{H^+} \qquad (18)$$

The reaction, as written in equation 18, includes the formation of an acetate carbanion, which is very unfavorable energetically owing to the presence of two negative charges on acetate. If the negative charge on the carboxylate group were removed—for example, by esterification—carbanion formation would be greatly facilitated. However, esterification of citrate is an energy-requiring process. To esterify citrate, the carboxylate oxygen of citrate must be removed and replaced by RS^- or RO^-. Removal of the carboxylate oxygen is energetically very unfavorable, because O^{2-} is an extremely poor leaving group. This problem is overcome by the use of ATP to phosphorylate the carboxylate group; that is, to form an anhydride between citrate and phosphate. The phosphoryl group is an excellent leaving group and can readily be displaced by CoASH. We now have citryl CoA; and carbon–carbon bond cleavage is no longer a problem, according to the mechanisms just described. *The primary function of ATP in biochemical reactions is to activate oxygen in organic compounds so as to convert it into a better leaving group* (Box 2-5).

There is experimental evidence that citryl phosphate is an intermediate in the reaction catalyzed by citrate lyase. Figure 2-19 shows a reasonable reaction sequence for the reaction catalyzed by citrate lyase. All aspects of the mechanism have not been definitely elucidated and, when further experimental results become available, it may have to be modified. According to this mechanism, two enzyme-bound intermediates are formed: citryl phosphate and citryl CoA. These

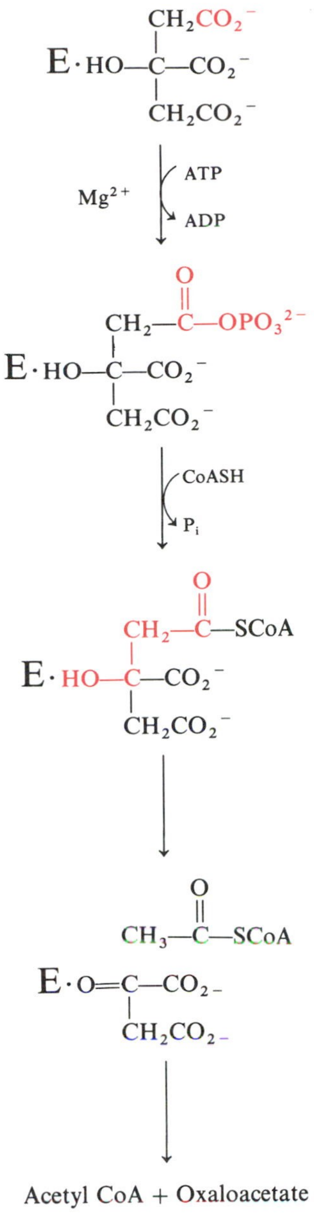

$$E\cdot HO{-}\overset{CH_2CO_2^-}{\underset{CH_2CO_2^-}{C}}{-}CO_2^-$$

Mg^{2+} ATP → ADP

$$E\cdot HO{-}\overset{CH_2{-}\overset{O}{\overset{\|}{C}}{-}OPO_3^{2-}}{\underset{CH_2CO_2^-}{C}}{-}CO_2^-$$

CoASH → P_i

$$E\cdot HO{-}\overset{CH_2{-}\overset{O}{\overset{\|}{C}}{-}SCoA}{\underset{CH_2CO_2^-}{C}}{-}CO_2^-$$

$$\overset{O}{\overset{\|}{CH_3{-}C{-}SCoA}}$$
$$E\cdot O{=}\underset{CH_2CO_2^-}{C}{-}CO_2^-$$

Acetyl CoA + Oxaloacetate

Figure 2-19

A Hypothetical Mechanism of Action for Citrate Lyase. All chemical steps take place at the active site through direct interactions of the enzyme-bound substrates. The reaction begins with the phosphorylation of citrate by ATP to form citryl phosphate, with ADP being released from the enzyme. Citryl phosphate then reacts with CoASH to form citryl CoA and release inorganic phosphate. Finally, citryl CoA undergoes a carbon–carbon cleavage reaction to produce acetyl CoA and oxaloacetate, which are released from the enzyme. The final cleavage is strictly analogous to the first step of the citrate synthase reaction. Citrate cleavage enzyme differs from citrate synthase in that it catalyzes the ATP-dependent formation of citryl CoA from citrate and CoA, whereas citrate synthase catalyzes the hydrolysis of citryl CoA to form citrate.

BOX 2-5

MECHANISM OF CARBOXYL ACTIVATION BY ATP IN THE CITRATE LYASE REACTION

In another experiment pertaining to citrate lyase, citrate labeled in the carboxyl groups with ^{18}O was the substrate, and the phosphate produced in the reaction was found to contain one ^{18}O that must have been derived from citrate. This result provided direct evidence for the proposed mechanism of carboxyl activation by ATP.

The transfer of ^{18}O from a carboxylate to a phosphate that is derived from unlabeled ATP is typically found in reactions in which ATP activates a carboxyl group; we shall encounter many examples of this type of oxygen transfer in ATP-dependent reactions.

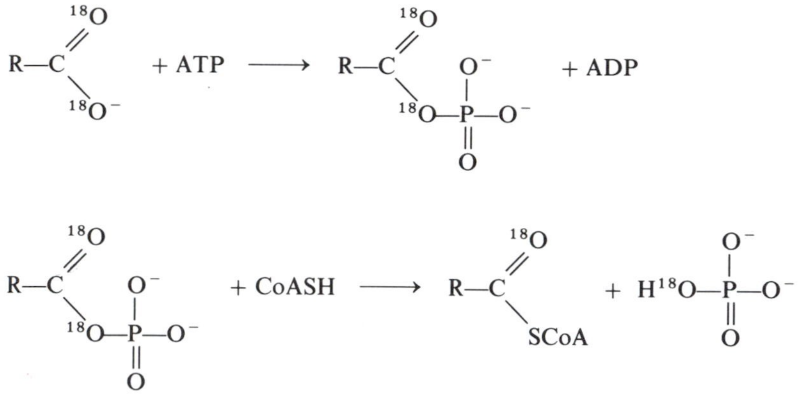

$$R\text{—}C(={}^{18}O)\text{—}{}^{18}O^- + ATP \longrightarrow R\text{—}C(={}^{18}O)\text{—}{}^{18}O\text{—}P(=O)(O^-)\text{—}O^- + ADP$$

$$R\text{—}C(={}^{18}O)\text{—}{}^{18}O\text{—}P(=O)(O^-)\text{—}O^- + CoASH \longrightarrow R\text{—}C(={}^{18}O)\text{—}SCoA + H^{18}O\text{—}P(=O)(O^-)\text{—}O^-$$

compounds are tightly bound to the enzyme and are not released into the solution in the course of the reaction. It is known, however, that the enzyme can convert citryl phosphate into acetyl CoA and oxaloacetate in the presence of CoA, but without ATP. Citryl CoA is converted into products in the absence of both CoA and ATP.

The mechanism shown in Figure 2-19 has common features with the mechanism of citrate synthase (Figure 2-17). The carbon–carbon bond is cleaved to form acetyl CoA and oxaloacetate, presumably with the intermediate formation of the carbanion of acetyl CoA (not shown in Figure 2-19). The major difference between citrate cleavage enzyme and citrate synthase is that the former utilizes ATP to activate the carboxyl oxygen, whereas the latter uses acetyl CoA to form citryl CoA.

Serine Transhydroxymethylase-Pyridoxal Phosphate-Tetrahydrofolic Acid

Several amino acids also can undergo aldol-type condensations. For example, serine transhydroxymethylase catalyzes the reaction in Figure 2-20. The enzyme is nonspecific for the R group and acts on (also catalyzes the synthesis of) both the threo and erythro isomers of L-α-amino-β-hydroxyacids. As the name implies, the enzyme can act upon serine to produce formaldehyde and glycine (R = H in Figure 2-20).

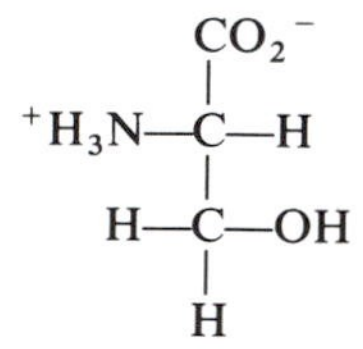

$$CO_2^-\text{—}C(H)(^+H_3N)\text{—}C(H)(H)\text{—}OH$$

L-Serine (Ser, S)

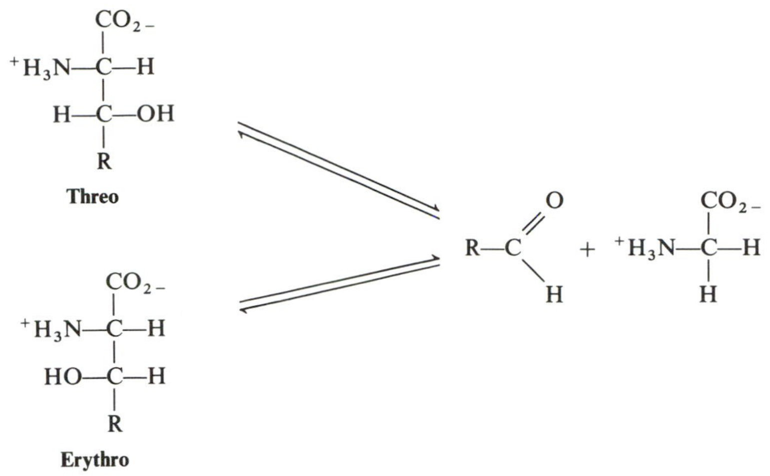

Threo: $^+H_3N{-}C(CO_2^-)(H){-}C(H)(OH){-}R$

Erythro: $^+H_3N{-}C(CO_2^-)(H){-}C(HO)(H){-}R$

$$\text{Threo} \rightleftharpoons R{-}CHO + {}^+H_3N{-}CH_2{-}CO_2^- \rightleftharpoons \text{Erythro}$$

Figure 2-20

Carbon–Carbon Bond Cleavage Reactions Catalyzed by Serine Transhydroxymethylase.

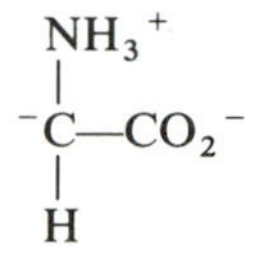

The chemical basis for the reaction catalyzed by serine transhydroxymethylase is similar to that for reactions already discussed. The carbon–carbon bond is cleaved to give a carbanion and the carbonyl group of an aldehyde. However, a carbanion that is stabilized only by COO^- and NH_3^+ is very unstable. Biological systems have found a very effective way of facilitating the formation of an α-carbanion from amino acids such as serine. Many enzymes that form an α-carbanion from an amino acid use a coenzyme, pyridoxal-5′-phosphate (PLP), to activate the amino acid. When a coenzyme is used, the catalytically active species is a complex formed between the enzyme and the coenzyme. This complex is known as a *holoenzyme,* whereas the same enzyme without a coenzyme is called the *apoenzyme.*

$$^-C(NH_3^+)(H){-}CO_2^-$$

Unstable carbanion

As mentioned in Chapter 1, many coenzymes are derived from water-soluble vitamins (generally the B-vitamins: B_6, B_{12}, etc.), and PLP is derived from vitamin B_6 (Figure 2-21). Deficiency of a vitamin leads to nonfunctional enzymes and the interruption of metabolic pathways. These interruptions have serious physiological consequences and may be fatal.

Before returning to the mechanism of action of serine transhydroxymethylase, we shall consider how PLP facilitates the formation of a carbanion of

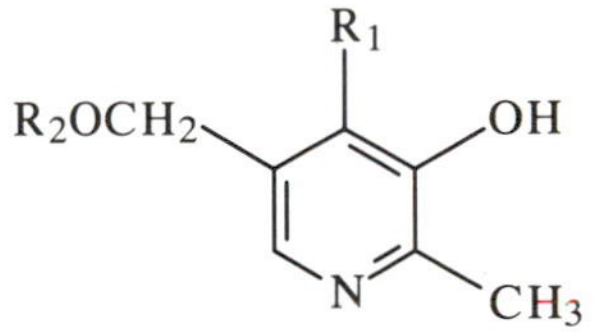

Pyridine ring bearing R_1, R_2OCH_2, OH, and CH_3 substituents:

$R_1 = CH_2OH$ Pyridoxine (vitamin B_6)
$R_1 = CHO$ Pyridoxal
$R_1 = CH_2NH_2$ Pyridoxamine

$R_1 = CHO$, $R_2 = PO_3^{2-}$: Pyridoxal-5′-phosphate
$R_1 = CH_2NH_2$, $R_2 = PO_3^{2-}$: Pyridoxamine-5′-phosphate
} Coenzymes

Figure 2-21

Compounds Related to Vitamin B_6.

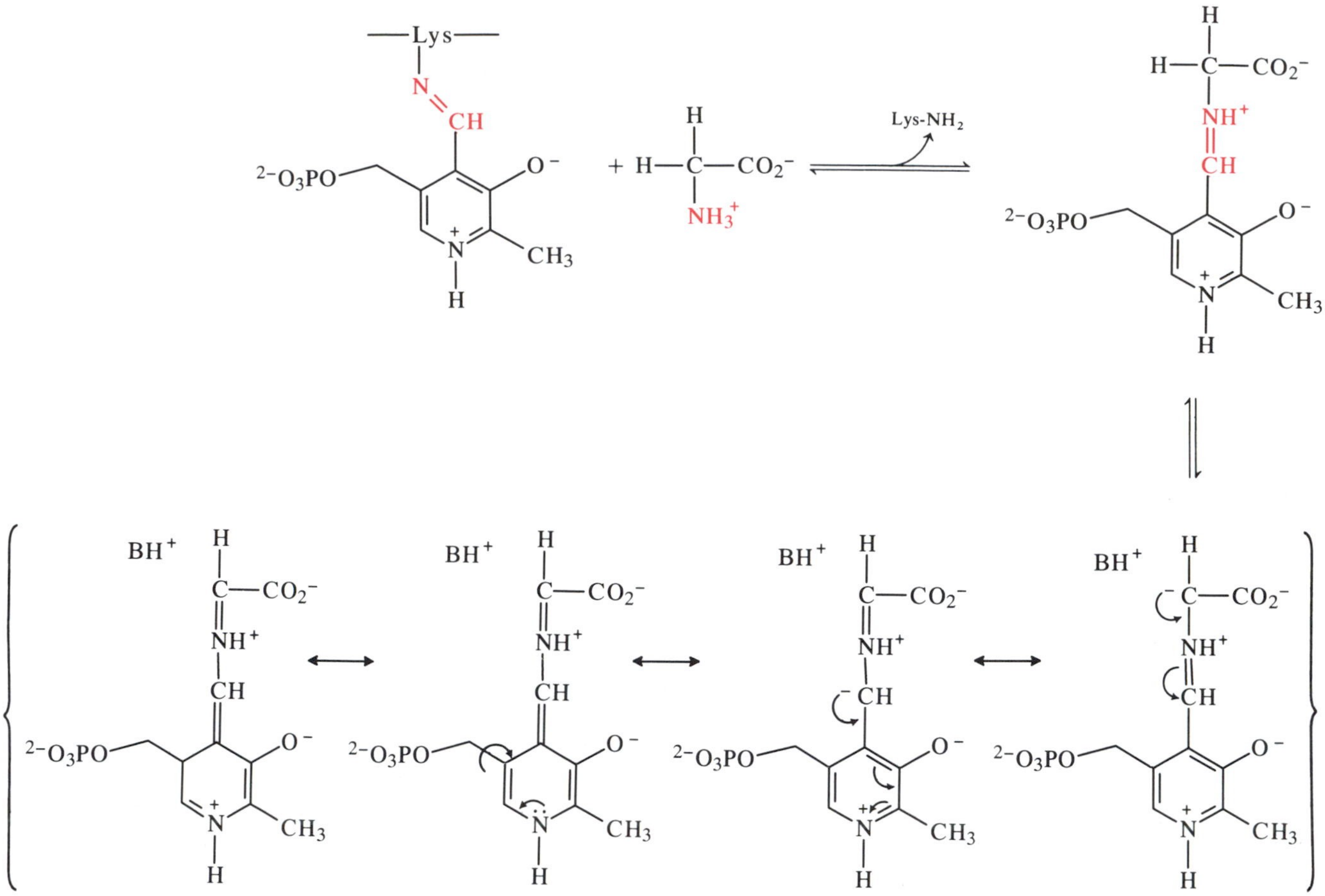

Figure 2-22

Carbanion Stabilization by Pyridoxal Phosphate. The carbanion generated by abstraction of a proton from the α-carbon of the imine formed between PLP and an α-amino acid is stabilized through delocalization of the electron pair throughout the pyridinium ring of PLP. This delocalization is represented by the resonance forms shown at the bottom of the figure.

an amino acid (Figure 2-22). The initial step, as in all reactions involving PLP, is the formation of an imine (Schiff base) between the amino acid and PLP. Proton abstraction from the α-carbon of the amino acid leads to the formation of a relatively stable carbanion-equivalent. The carbanion is stable because the negative charge is highly dispersed throughout the conjugated system; that is, many stable resonance forms exist (Figure 2-22).

In all enzymes that utilize PLP as a cofactor, PLP forms an imine with the amino group on the side chain of a lysine residue at the active site of the enzyme Figure 2-23). When the amino acid substrate enters the active site, it reacts with this imine to displace the lysine amino group and form a new imine, as shown in Figure 2-23. Note that the free aldehyde of PLP is not an intermediate in this reaction. This process is a *transimination,* but it has sometimes been called a "trans-Schiffization" after Schiff base, the common name for the imine linkage. Concomitant with the formation of the Schiff base with the substrate, the free lysine amino group is released. There is evidence that this amino group participates as the base catalyst in subsequent events.

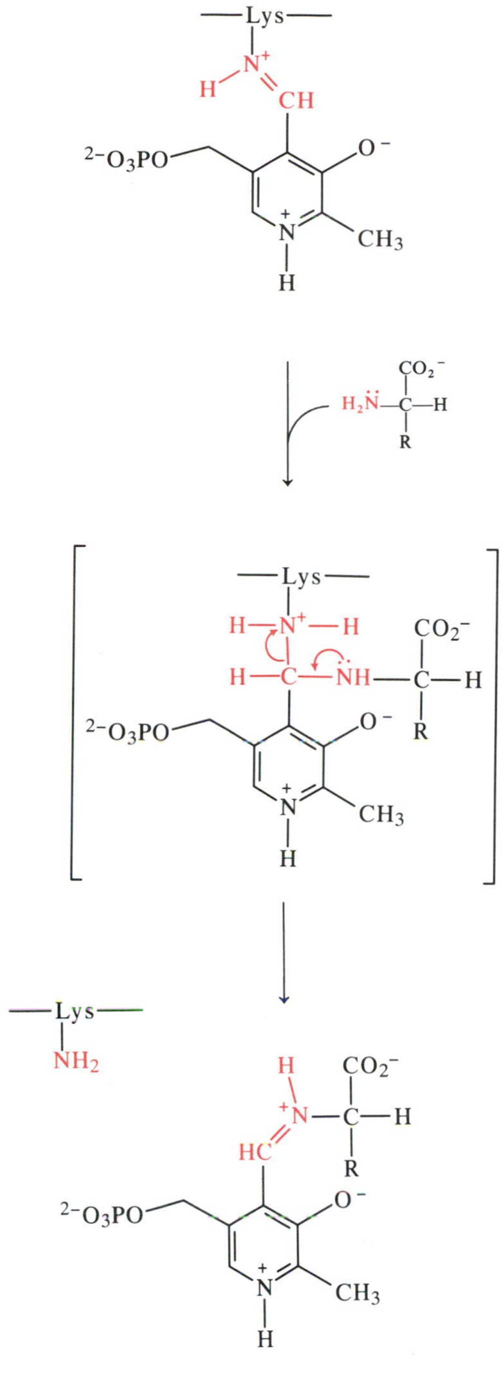

Figure 2-23

A Probable Intermediate in Imine (Schiff Base) Formation. The transimination between enzyme-bound PLP and an α-amino acid is probably a two-step process that includes the bracketed tetrahedral addition intermediate.

A mechanism for the reaction catalyzed by serine transhydroxymethylase is shown in Figure 2-24. This mechanism should be familiar, because it is a reverse aldol condensation. Cleavage of the C–C bond gives an aldehyde and the carbanion of glycine, which is still attached to PLP as the imine. This carbanion is stabilized by resonance with the pyridoxal cofactor, and it is subsequently protonated to give the imine of glycine and PLP. Glycine is then

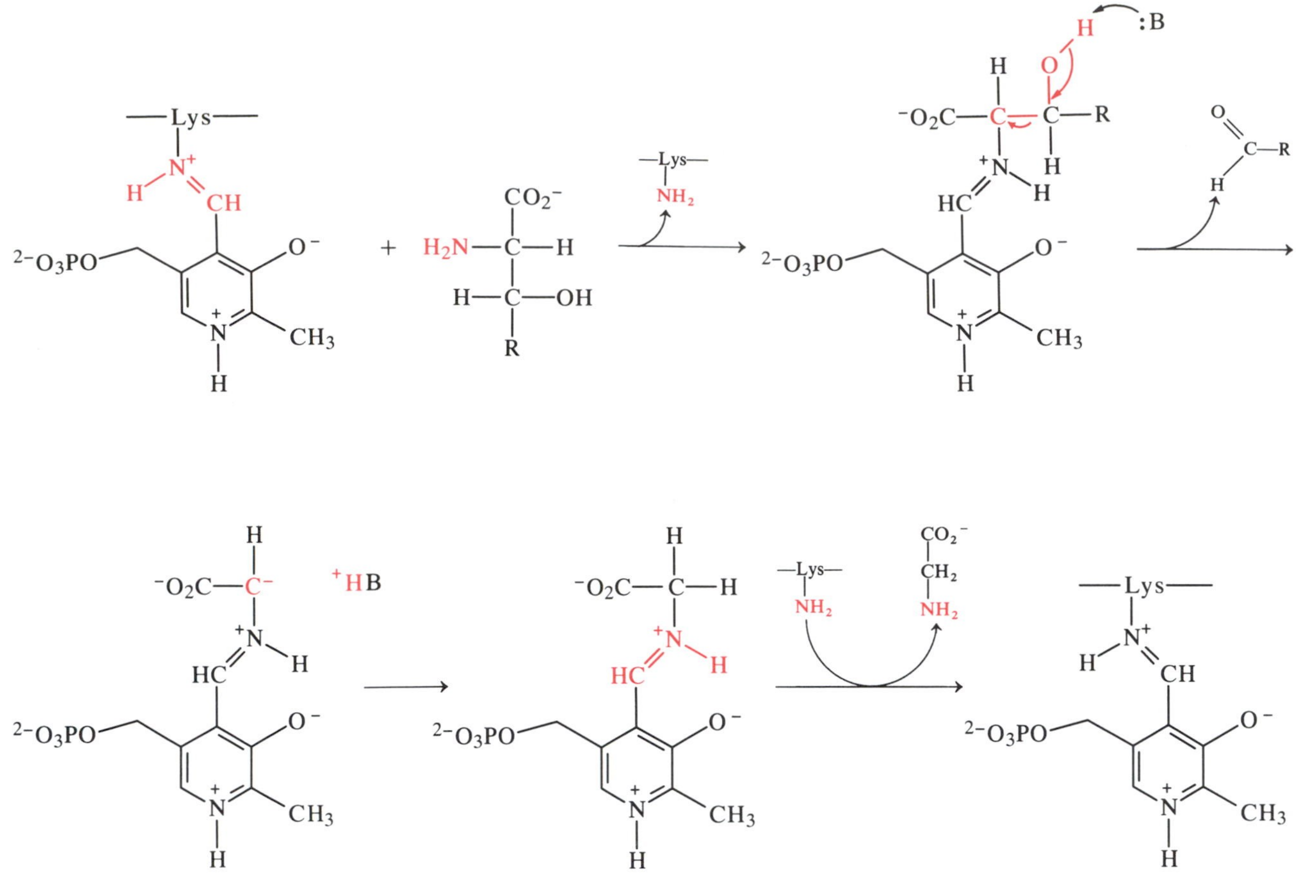

Figure 2-24

The Mechanism of Action of PLP in the Serine Transhydroxymethylase Reaction. The imine formed between serine and PLP is subject to C–C cleavage between carbon-2 and carbon-3 owing to the stabilization of the resulting carbanion by PLP. This is essentially an aldol reaction, in which a stabilized carbanion undergoes reversible addition to the carbonyl group of a ketone or an aldehyde.

displaced by transimination with the side-chain amino group of the active-site lysine to release glycine and regenerate the lysine-PLP-imine.

The action of serine transhydroxymethylase on serine leads to the formation of formaldehyde and glycine. However, formaldehyde is a toxic substance and is never present in its free form in biological systems. When the enzyme acts on serine, another coenzyme molecule, tetrahydrofolic acid, shown in Figure 2-25, participates in the reaction. This cofactor reacts with the formaldehyde very rapidly, before it can leave the active site, to form the adduct N^5, N^{10}-methylene tetrahydrofolate (Figure 2-25), which is then released from the enzyme. The overall reaction then is equation 19.

$$\text{Serine} + \text{Tetrahydrofolate} \rightleftharpoons \text{Methylene tetrahydrofolate} + \text{Glycine} \quad (19)$$

The reaction is reversible and can produce serine, particularly in certain bacteria that utilize one-carbon compounds as their sole source of carbon. However, in animals the reaction normally produces methylene tetrahydrofolate, an important metabolite that is used in many biochemical reactions of one-carbon compounds, in particular the biosynthesis of nucleotides. We will consider these reactions in more detail in Chapters 25 and 26, and it is not necessary to learn the complete structure of tetrahydrofolate at this time.

In the reverse direction (condensation), methylene tetrahydrofolate rather than free CH_2O is utilized. This compound probably undergoes hydrolysis at the active site to deliver CH_2O. It is remarkable that tetrahydrofolate is not required for the reaction of any other substrate with this enzyme. Thus, serine transhydroxymethylase will catalyze the condensation of acetaldehyde or benzaldehyde with glycine in the absence of tetrahydrofolate.

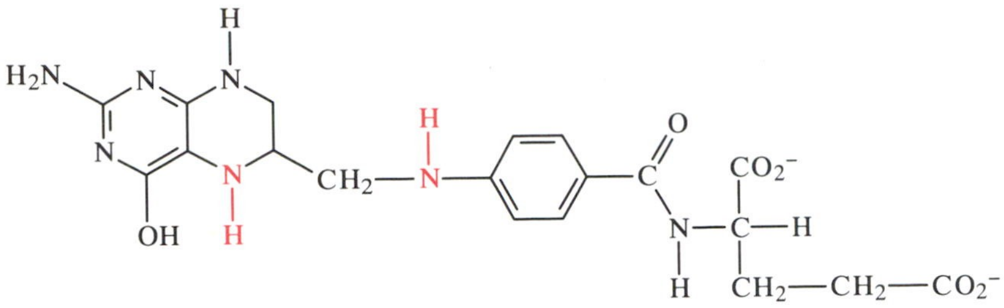

Tetrahydrofolate

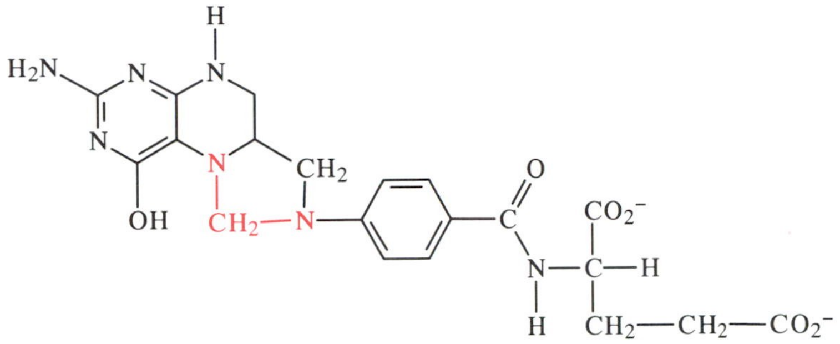

N^5,N^{10}-Methylene tetrahydrofolate

Figure 2-25

The Structures of Tetrahydrofolate and N^5, N^{10}-Methylene Tetrahydrofolate (Methylene Tetrahydrofolate). The highlighted methylene group in methylene tetrahydrofolate is derived from formaldehyde produced in the cleavage of serine by serine transhydroxymethylase. This carbon is potentially available for reaction as a *formaldehyde-equivalent* and participates in several important reactions of one-carbon metabolism (Chapter 25).

Summary

This chapter presents an overview of several of the concepts, compounds, and reactions that are important in biochemistry. Enzymes utilize many of the same mechanisms that are well known in organic chemistry for the synthesis and degradation of organic compounds. For example, the formation and cleavage of carbon–carbon bonds usually occur through aldol and Claisen condensations or the reverse, cleavages. In these reactions, bond formation is by interaction of a carbon atom that has carbocationic character, generally a carbonyl group, with a carbon atom that has carbanionic character. The carbanion is frequently stabilized as an enolate by an adjacent carbonyl group.

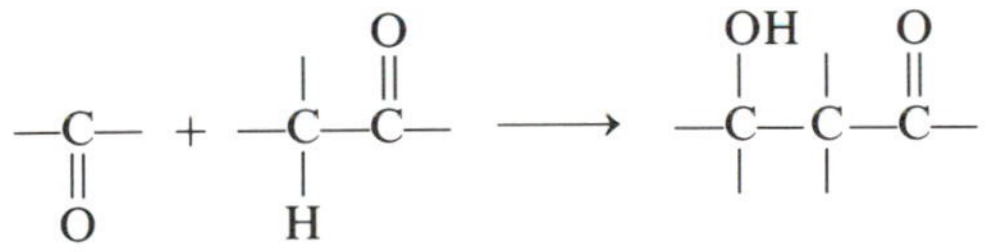

Aldol condensation

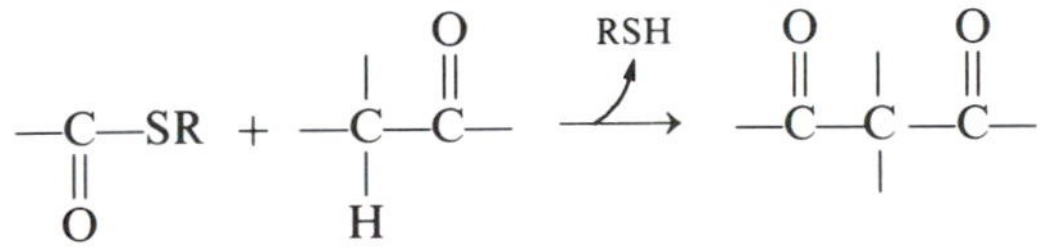

Claisen condensation

In the Claisen condensation, a thioester derived from coenzyme A (CoASH) is frequently used.

When amino acids participate in reactions in which carbon–carbon bonds are made or broken, pyridoxal phosphate participates as a coenzyme. This cofactor forms an imine with the amino acid that facilitates formation of a carbanion-equivalent.

Some additional classes of compounds and processes that are important for understanding biochemistry were briefly introduced, including amino acids, proteins, catalysis, activation energy, enzymes, and isotopes.

ADDITIONAL READING

ELIEL, E. 1962. *Stereochemistry of Carbon Compounds.* New York: McGraw-Hill.

POPJAK, G. 1970. Stereospecificity of enzymatic reactions. In *The Enzymes*, vol. 2, 3d ed., pp. 115–215. New York: Academic Press.

HORECKER, B. L., ORESTES, T., and LAI, C. Y. 1973. Aldolases. In *The Enzymes*, vol. 7, 3d ed. New York: Academic Press.

SNELL, E. E., and DI MARI, S. 1973. Schiff base intermediates in enzyme catalysis. In *The Enzymes*, vol. 7, 3d ed. New York: Academic Press.

KARPLUS, M., BRANCHAUD, B., and REMINGTON, J. 1990. Proposed mechanism for the condensation reaction of citrate synthase: 1.9-Å structure of the ternary complex with oxaloacetate and carboxymethyl coenzyme A. *Biochemistry* 29: 2213–2219.

PROBLEMS

1. Complete the following:

(a) $CH_3—CH_2—COSCoA + CH_3—CH_2—COSCoA \longrightarrow$

(b) $CH_3—CHOH—CHO + CH_3—COSCoA \longrightarrow$

(c) 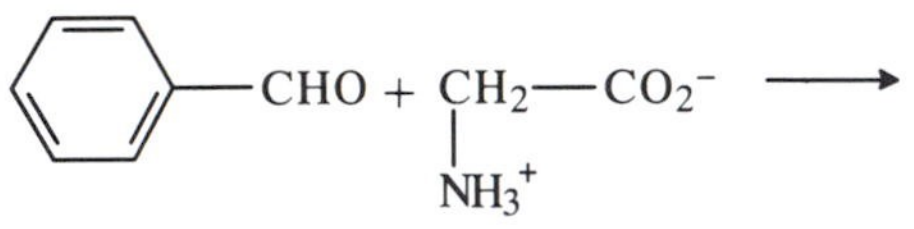 $C_6H_5—CHO + CH_2(NH_3^+)—CO_2^- \longrightarrow$

(d) 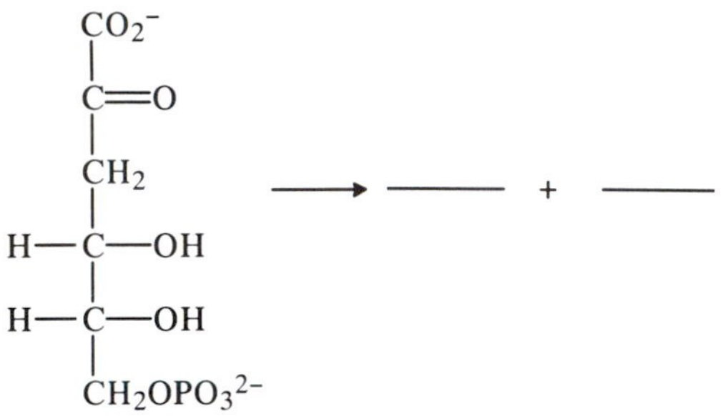 $CO_2^-—C(=O)—CH_2—CH(OH)—CH(OH)—CH_2OPO_3^{2-} \longrightarrow$ ______ + ______

(e) 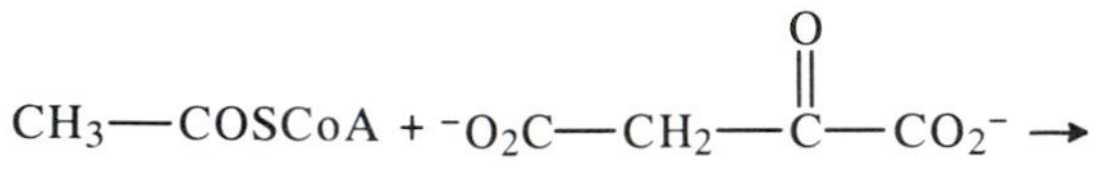 $CH_3—COSCoA + {}^-O_2C—CH_2—C(=O)—CO_2^- \longrightarrow$

2. An enzyme catalyzes:

$o\text{-}H_2N—C_6H_4—C(=O)—CH_2—CH(NH_3^+)—CO_2^- + H_2O \longrightarrow o\text{-}H_2N—C_6H_4—C(=O)—OH + H_3C—CH(NH_3^+)—CO_2^-$

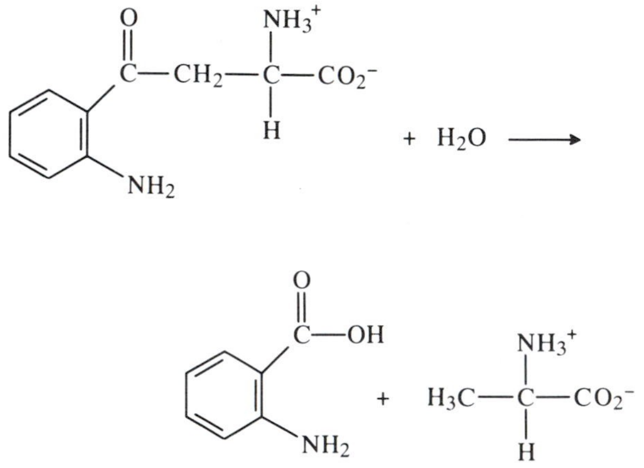

What cofactor is most likely involved? Write a mechanism for the reaction.

3. The enzyme that catalyzes reaction *d* in problem 1 is inactivated by bromopyruvate ($BrCH_2—CO—CO_2^-$). Write a mechanism for this inactivation.

4. In the reaction catalyzed by aldolase, an imine is an intermediate. The active-site lysine that takes part in imine formation has an unusually low pK_a.

(a) How does this low pK_a facilitate the reaction?

(b) What structural feature of the enzyme could bring about this low pK_a?

5. A microorganism uses

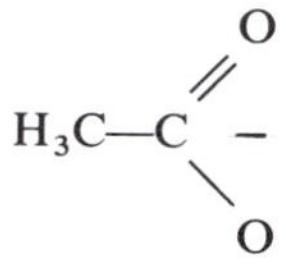

$H_3C—C(=O)—O^-$

to synthesise aromatic compounds. When acetate labeled at C-1($CH_3—COO^-$) with ^{14}C is administered, the following compound can be isolated and is labeled as indicated:

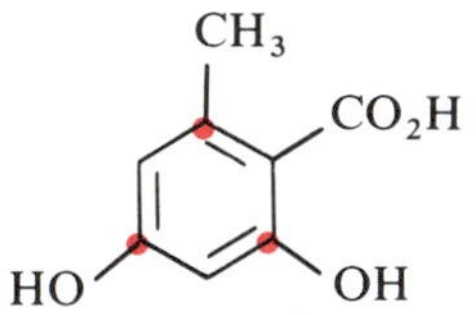

(Structure: benzene ring bearing CH_3, CO_2H, OH, and HO substituents, with labeled ring carbons marked)

Orsellinic acid

Write a probable reaction sequence.

6. An enzyme uses

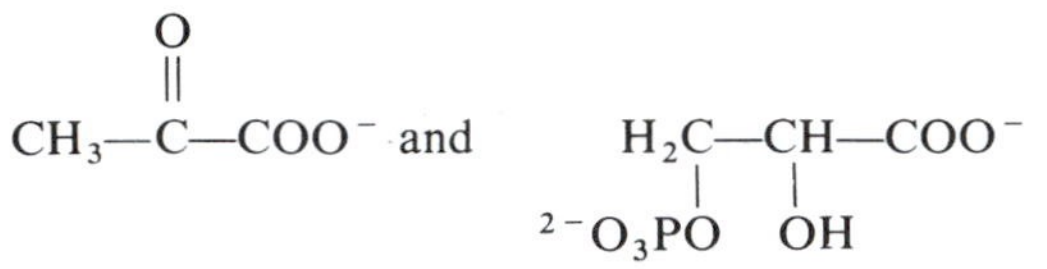

$CH_3—C(=O)—COO^-$ and $H_2C(OPO_3^{2-})—CH(OH)—COO^-$

as substrates and catalyzes the synthesis of a six-carbon compound. When the enzyme is exposed to $NaBH_4$ in the presence of pyruvate ($CH_3—CO—COO^-$), it is inactivated and the carbon atoms of pyruvate become covalently linked to the enzyme. What is the probable product of the reaction? What is the mechanism of the reaction?

Introduction to Chymotrypsin: A Case Study for Catalysis by Enzymes

3

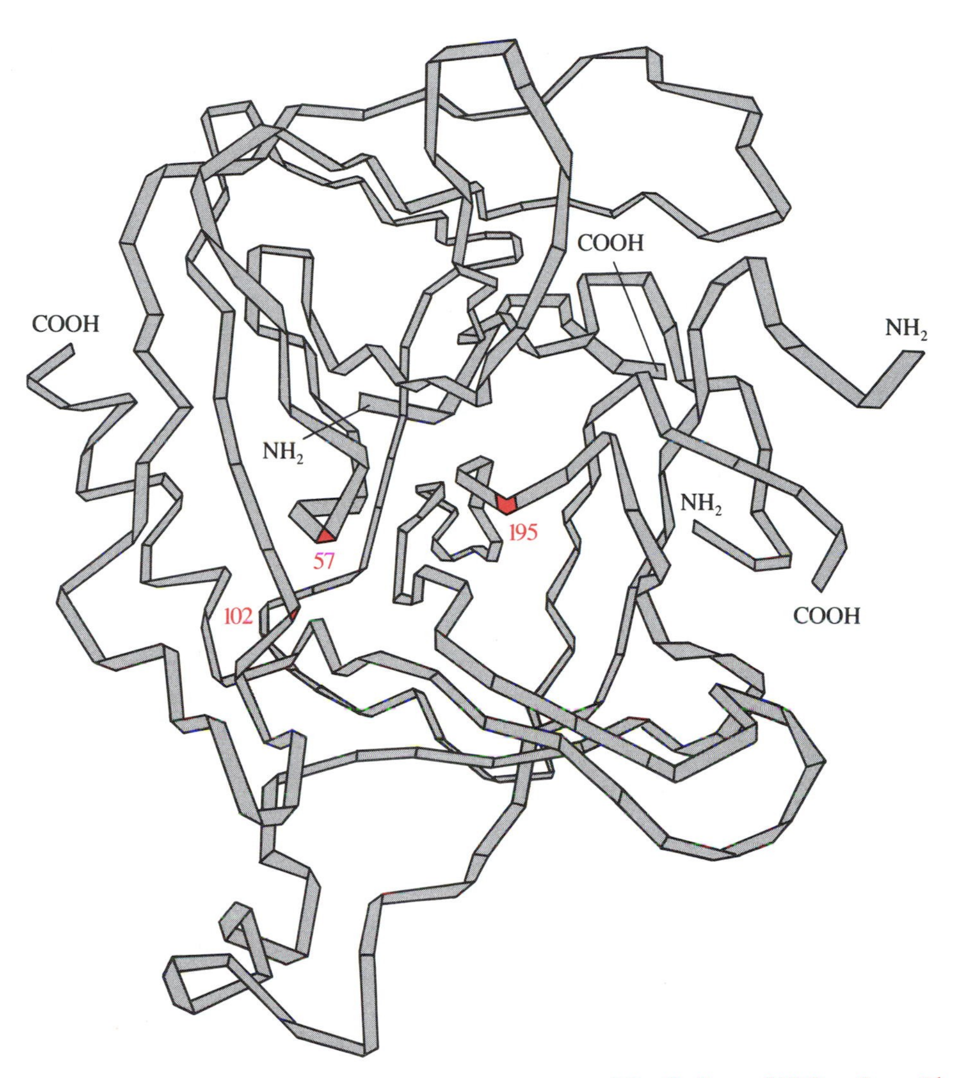

A model of three folded polypeptide chains of α-chymotrypsin, showing the positions of the important amino acids at the active site.

From an overview of the chemistry and biochemistry of the most-important reactions in which carbon–carbon bonds are formed, we now turn to the question of how these and other reactions are brought about in biological systems. This demands an understanding of enzymes, the catalysts that make these reactions occur with extraordinary speed and accuracy.

Proteolytic enzymes, or *proteases,* are the enzymes that are best known to most people. They include digestive enzymes that cleave the proteins of ingested food into amino acids that can be absorbed from the intestine. Closely related enzymes used in meat tenderizers and detergents serve a similar function of breaking up proteins and other large protein-containing organic polymers and particles. Some may think that proteases are not very interesting, but they are important for many functions in living cells and body fluids: they play crucial roles in such diverse processes as blood clotting, immunological reactions, and the production of hormones (Chapter 6). Recently, they have even been shown to play an important role in the transmission of information that controls the growth and functioning of cells in complex organisms, including mammals.

These enzymes are given detailed consideration in this book because of their biochemical importance and because they catalyze reactions that are relatively simple in the chemical sense. Many proteases are available in large quantities and were among the first enzymes to have been isolated and characterized. Chymotrypsin, the most thoroughly studied of all enzymes, is described in this and subsequent chapters. Chymotrypsin can serve as a case study that illustrates how the properties of an enzyme can be characterized and how an enzyme can accelerate reaction rates by enormous factors, much larger than are obtained with most chemical catalysts.

The Specificities of Chymotrypsin and Other Proteases

Inasmuch as the digestion of proteins requires the complete hydrolysis of peptide chains into their constituent amino acids, one might expect that in nature a single enzyme would have evolved that could cleave the peptide bond of any amino acid in a protein and carry out the whole process singlehandedly. In fact, evolutionary selection has led to a considerable number of different proteases that carry out protein digestion, each with a different specificity for cleavage of a peptide chain at a position that is occupied by a particular kind of amino acid. This suggests that there must be some advantage for the catalytic process that arises from the interaction of an enzyme with the specific R-group of an amino acid at the cleavage point. The specificities of the proteases exemplify a general phenomenon—*the catalytic activities of the great majority of enzymes depend on a recognition between specific groups of the substrates and the enzymes that catalyze their reactions.*

Chymotrypsin is a digestive enzyme that is available in large quantities, in pure form, and at low cost. Although much is known about its mechanism of action, we still cannot account quantitatively for the extraordinary increases in the rates of hydrolysis of acyl compounds that it brings about.

The Hydrolysis of Peptides

Proteases catalyze the hydrolysis of peptide bonds (equation 1), in which a water molecule reacts with a peptide carbonyl group and the bond to the amino group of the next amino acid in the chain is cleaved. This is a special case of amide hydrolysis (equation 2), which appears to be a simple reaction.

$$^{+}H_3N\text{—}\underset{R_1}{\overset{H}{C}}\text{—}\overset{O}{\overset{\|}{C}}\text{—}\underset{H}{N}\text{—}\underset{R_2}{\overset{H}{C}}\text{—}CO_2^- + H_2O \longrightarrow {}^{+}H_3N\text{—}\underset{R_1}{\overset{H}{C}}\text{—}CO_2^- + {}^{+}H_3N\text{—}\underset{R_2}{\overset{H}{C}}\text{—}CO_2^- \tag{1}$$

$$R\text{—}\overset{O}{\overset{\|}{C}}\text{—}NHR' + H_2O \longrightarrow R\text{—}\overset{O}{\overset{\|}{C}}\text{—}O^- + {}^{+}H_3N\text{—}R' \tag{2}$$

However, it is important to realize that the hydrolysis of an amide comprises not only the formation of a C–O bond and the cleavage of a C–N bond, but also the removal of two protons from water and the addition of two protons to the leaving nitrogen atom at physiological pH values.

Chymotrypsin, trypsin, and elastase are three proteases that are synthesized in the pancreas in the form of inactive proenzymes, or zymogens (e.g., chymotrypsinogen and trypsinogen), so that they do not digest the tissue in which they are formed or stored, as noted in Chapter 1. In certain disease states, they are released and become activated, and the resulting enzymes cause serious damage to the tissue; pancreatitis is a painful and often fatal disease. Each zymogen is secreted into the intestine after the ingestion of food and is then converted into an active state through the action of other protease molecules, which cleave the zymogen at specific peptide bonds. This leads to changes in the three-dimensional conformation of the protein that convert it into a catalytically active structure.

The serine proteases are *endopeptidases* that attack the peptide bond of certain amino acids within a peptide chain, which cleaves the chain into shorter segments by hydrolysis. They differ from *exopeptidases*, which cleave the terminal amino acids from the ends of the peptide chain (Figure 3-1). The nature of the reactions that are catalyzed by these enzymes was first determined by analysis of the amino acids at the ends of the new peptide chains that are formed upon cleavage of a protein. Chymotrypsin is an endopeptidase that produces peptide chains that may have almost any amino acid at the amino-terminal end; however, all of the peptides contain amino acids with bulky, nonpolar, hydrophobic R-groups at the newly formed carboxyl-terminal end. An examination of the specificity and mechanism of action of chymotrypsin and related enzymes illustrates the structures of most of the important amino acids and how the structures of some of these amino acids can be related to their function. These structures are summarized on the inside front cover of this book.

Substrate Specificities of Serine Proteases

Chymotrypsin is active toward peptides that contain *phenylalanine* and *tyrosine*, which are derived from alanine by the substitution of a phenyl or phenolic group for a hydrogen atom. It is also very active toward peptides of *tryptophan*, in

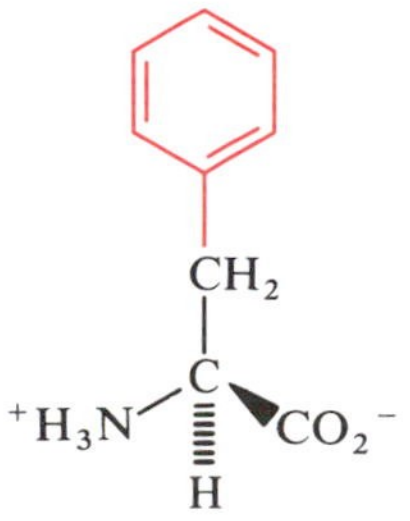

CH_2 / $^{+}H_3N$—C—CO_2^- / H

L-Phenylalanine (Phe, F)

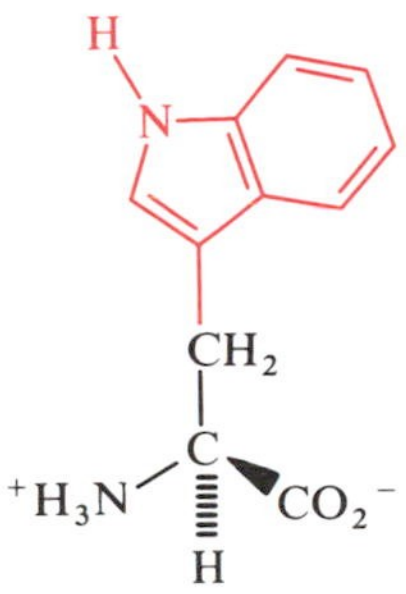

H, N; CH_2 / $^{+}H_3N$—C—CO_2^- / H

L-Tryptophan (Trp, W)

OH; CH_2 / $^{+}H_3N$—C—CO_2^- / H

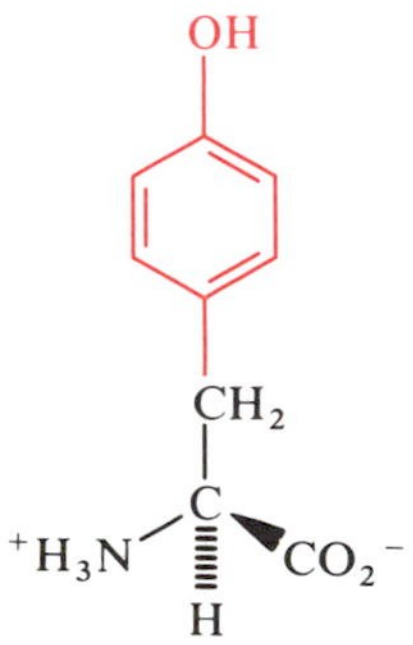

L-Tyrosine (Tyr, Y)

Figure 3-1

The Specificities of Some Proteases for Hydrolysis of Peptide Bonds.

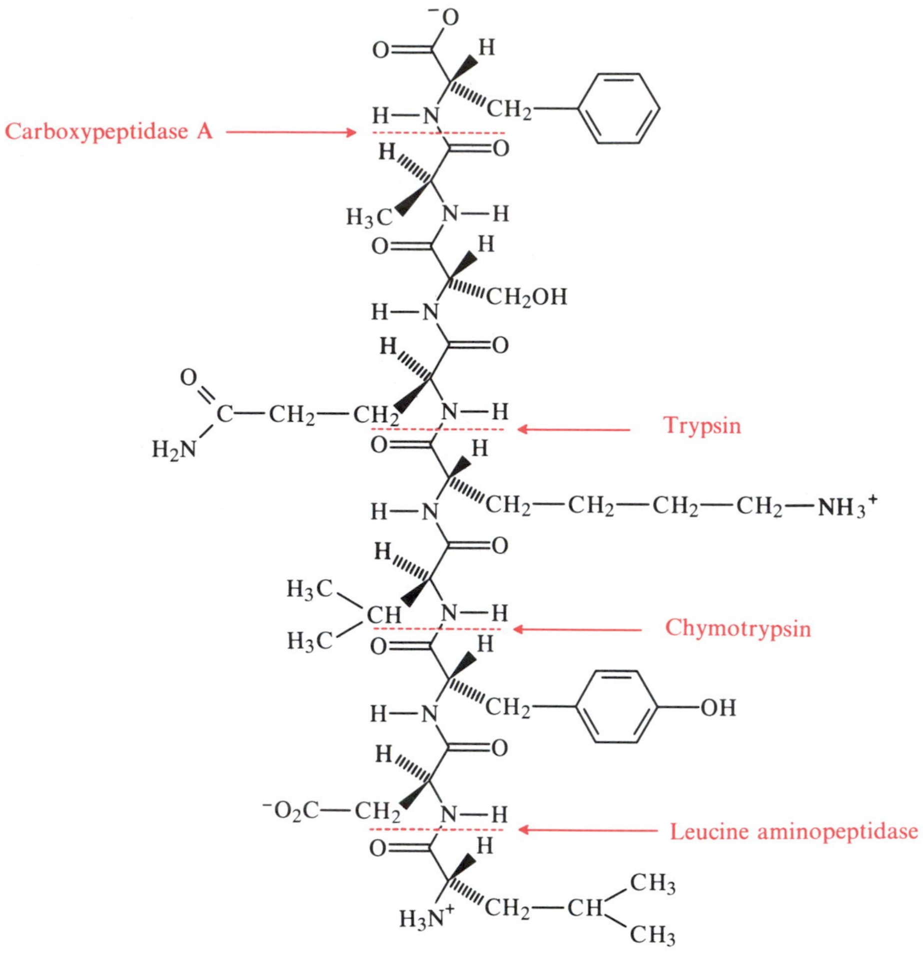

Carboxypeptidase A

Trypsin

Chymotrypsin

Leucine aminopeptidase

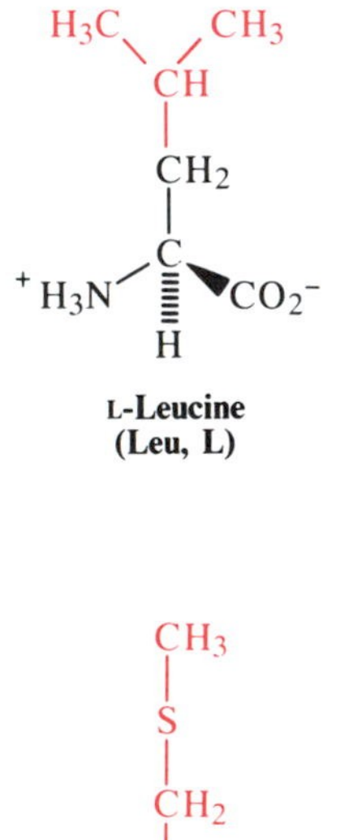

L-Leucine (Leu, L)

which indole substitutes for a methyl hydrogen of alanine. The hydrocarbon side chain of *leucine* and the nonpolar side chain of *methionine* result in cleavage at these amino acids, but *isoleucine* and *valine* residues are much less susceptible to hydrolysis. Leucine is a 2-propyl and methionine is a CH_3SCH_2-derivative of alanine, whereas isoleucine and valine are methylethyl and dimethyl derivatives of alanine. Chymotrypsin slowly attacks peptides of *isoleucine* and *valine*, which have branched hydrocarbon side chains. Thus, if a peptide chain is exposed to chymotrypsin, it will be cleaved first into peptides with tryptophan, tyrosine, or phenylalanine on the C-terminal ends. Formation of peptides with C-terminal leucine or methionine proceeds more slowly, peptides with C-terminal isoleucine and valine are formed still more slowly, and peptides with other C-terminal amino acids appear very slowly indeed.

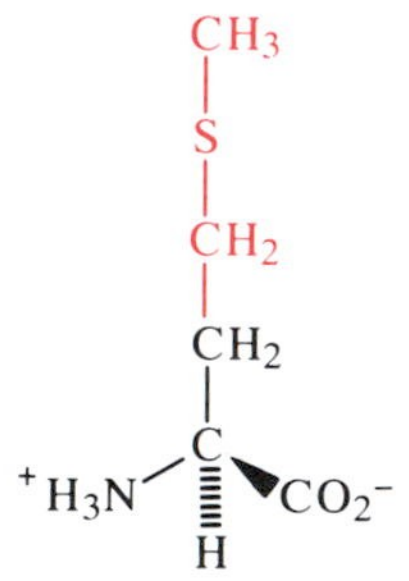

L-Methionine (Met, M)

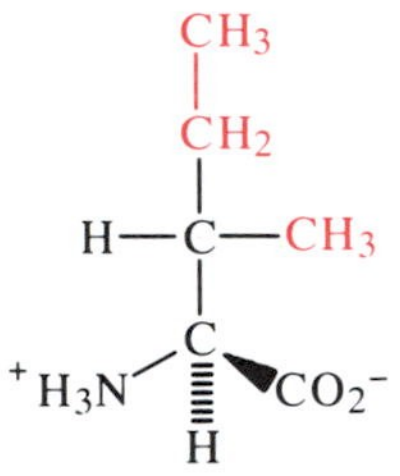

L-Isoleucine (Ile, I)

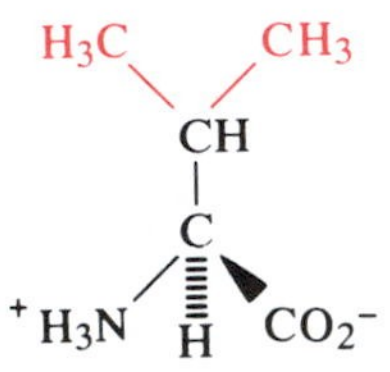

L-Valine (Val, V)

Ever since the discovery of chymotrypsin's high selectivity toward hydrophobic amino acid residues, it has been assumed that the side chains of these amino acids must bind specifically to the enzyme in such a way that the peptide bonds are in exactly the correct position for the enzyme to exert its catalytic action upon the acyl groups that are cleaved. The structural basis for this "lock and key" specificity has become clear through one of the major advances of this century: the determination of the complete three-dimensional structure of crystalline chymotrypsin by X-ray diffraction analysis.

The structure is illustrated in Figure 3-2. The structural model shows that this enzyme is a globular protein that is made up of three intricately folded polypeptide chains A, B, and C; it is not highly elongated like a fibrous protein. Space-filling molecular models of the structure show that the catalytically functional site, the active site, contains a "hydrophobic pocket" that is lined with relatively nonpolar, hydrophobic amino acids. The phenyl or indole rings of phenylalanine, tyrosine, or tryptophan fit very nicely indeed into this pocket.

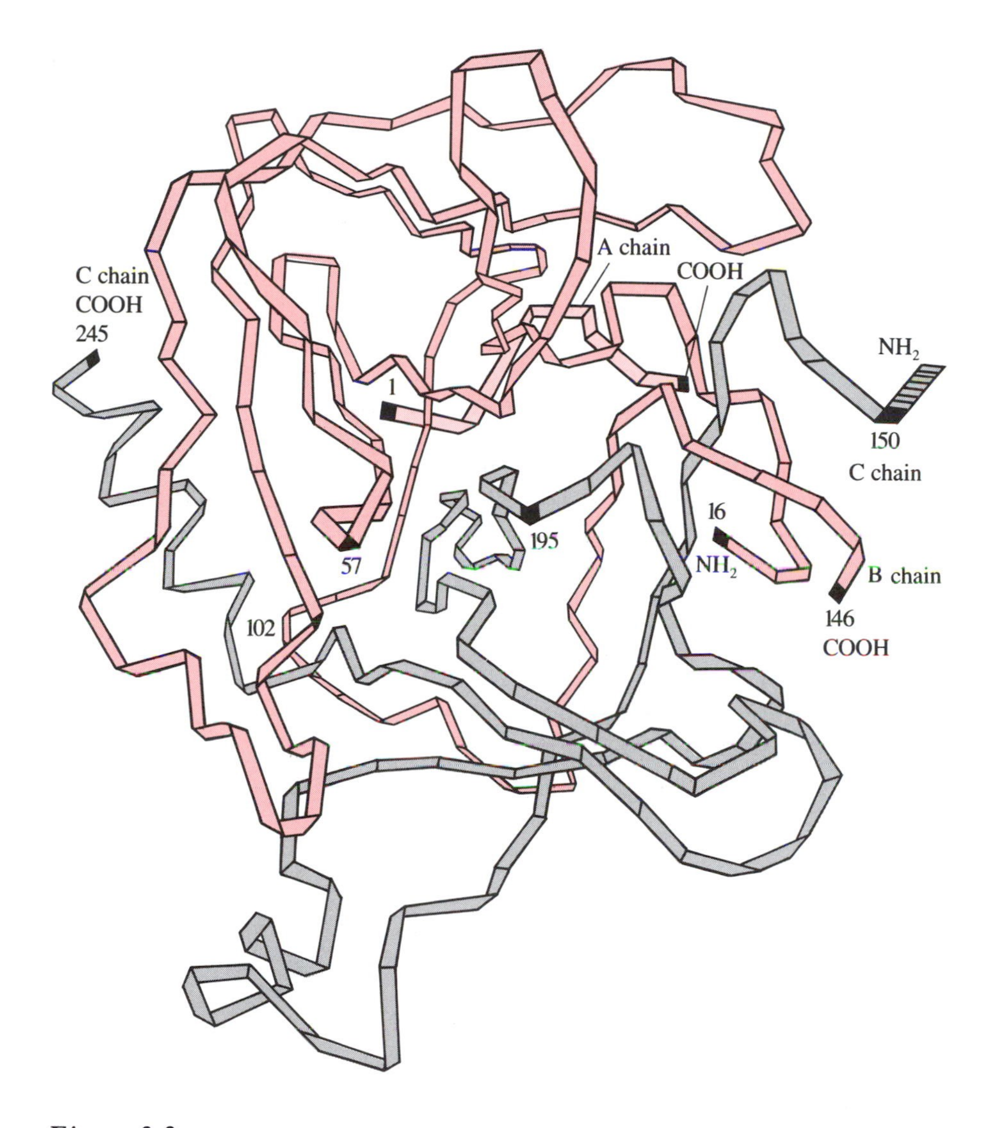

Figure 3-2

The Three-dimensional Structure of Chymotrypsin, Determined by X-ray Crystallography. Shown are the peptide chains of α-carbons in chymotrypsin, as well as the positions of the active-site residues Ser-195, His-57, and Asp-102.

The fit is illustrated diagrammatically in Figure 3-3A. This specificity pocket provides a "handle" that keeps the substrate bound and properly oriented with respect to the catalytic groups at the active site. We shall describe these groups and their functions in subsequent chapters.

Small molecules will also bind in this pocket, as is shown for indole in Figure 3-3B. Unreactive substrate analogs of this kind compete with the substrate for the pocket, so that they inhibit the binding and hydrolysis of substrates such as tryptophan peptides. We shall describe this kind of inhibition quantitatively in Chapter 4.

Chymotrypsin is far more tolerant of variations in the structures of the amino acids in the leaving group, X, than in the acyl group.

$$R_1\text{—C(=O)—NH—C(H)(}RH_2C\text{)—C(=O)—X}$$

R = phenyl, p-hydroxyphenyl, or indole

$$X = \text{—N(H)—C(H)(}R'H_2C\text{)—C(=O)—}X', \quad \text{—NHR}'', \quad \text{—SR}'', \quad \text{—O—C(=O)—R}'', \quad \text{—Cl}, \quad \text{—O}^-$$ (see Box 3-1)

In fact, if chymotrypsin is provided with the acyl group of one of its specific amino acids, such as phenylalanine, tyrosine, or tryptophan, it will catalyze the hydrolysis of almost any acyl derivative that is subject to chemical hydrolysis. Esters and thioesters are considerably more reactive than amides and peptides, as they are in nonenzymatic, base-catalyzed hydrolysis. Anhydrides and acyl chlorides are also cleaved rapidly.

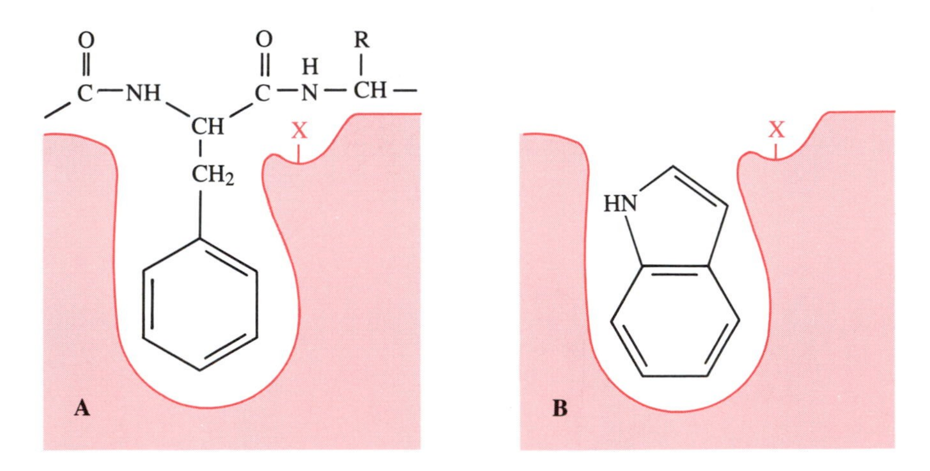

Figure 3-3

The Specificity Pocket of Chymotrypsin. Part A shows a phenylalanyl side chain fitted into the pocket and held by hydrophobic bonding, with the hydroxyl group of Ser-195 (—X) near the acyl-carbonyl group with which it reacts. Part B shows indole, a competitive inhibitor, in the binding pocket.

BOX 3-1

CHYMOTRYPSIN-CATALYZED EXCHANGE OF CARBOXYL-OXYGEN ATOMS

Chymotrypsin catalyzes the exchange of the carboxyl-oxygen atoms of unesterified substrates with the oxygen atoms of water. This is measured by adding the enzyme and substrate to water that contains the heavy oxygen isotope ^{18}O and measuring the incorporation of the labeled oxygen atom into the carboxylate group of the substrate. This is called a *virtual reaction* because no net chemical reaction takes place, except for the replacement of ^{16}O by ^{18}O.

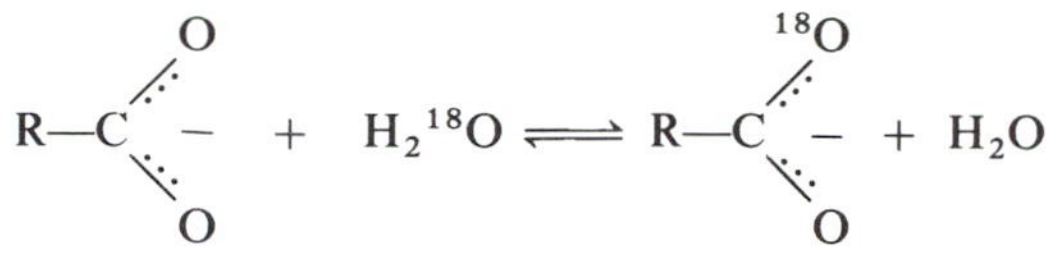

$$R{-}CO_2^- + H_2{}^{18}O \rightleftharpoons R{-}C({}^{18}O)O^- + H_2O$$

Although the exchange reaction is usually studied near neutral pH, where the carboxyl group exists as the carboxylate ion, the dependence of the rate of exchange on pH indicates that the protonated form, R-COOH, is the reactive species. This is not surprising, because carboxylic acids are far more reactive than carboxylate ions toward nucleophilic attack at the carbonyl group; and the enzyme-catalyzed exchange reaction of the free acid is essentially the same as the hydrolysis of an ordinary ester substrate.

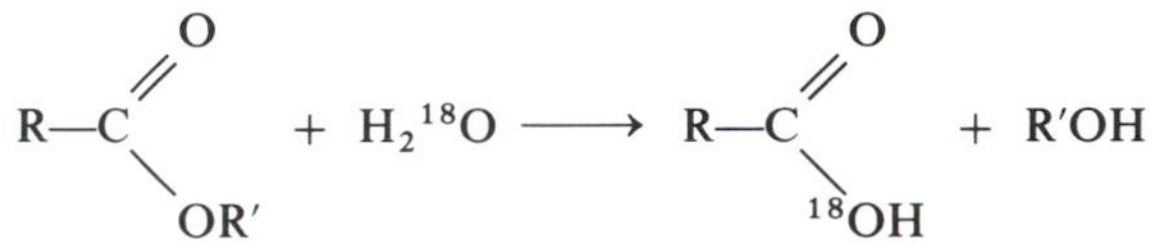

$$R{-}C(=O){-}OR' + H_2{}^{18}O \longrightarrow R{-}C(=O){-}{}^{18}OH + R'OH$$

$R' = H$ or alkyl

Thus, in the exchange reaction, the enzyme accepts the undissociated acid, in which $R' = H$ in the above reaction, as the substrate. A low concentration of the undissociated carboxylic acid is always present at equilibrium in solution, even at neutral pH.

Trypsin is a serine protease that is very similar to chymotrypsin in structure and reaction specificity, but its substrate specificity is for cleavage at the *arginine* and *lysine* residues in the peptide chain. The basic groups in the side chains of these amino acids are positively charged at physiological pH values. The pK_a of the guanidinium group of arginine is 12.5 and that of the ammonium group in the side chain of lysine is 10.5, so that both of these side chains are protonated at the pH of 8.0 in the small intestine.

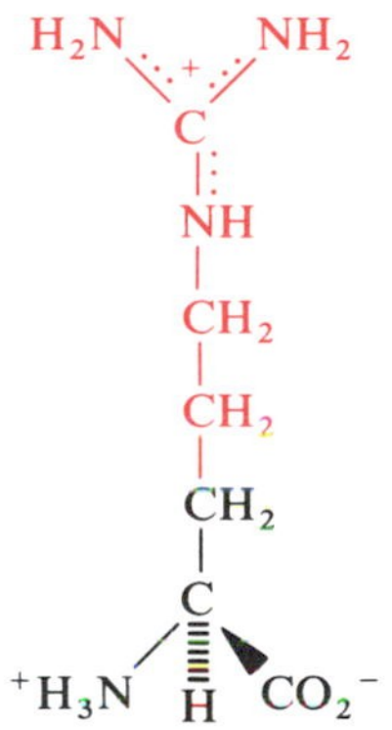

It was long supposed that the structural basis for this specificity would be found to be the binding of the positively charged side chains of substrates to a negative charge in the active site of trypsin. The negative charge could be supplied by the side-chain carboxylate group of one of the acidic amino acids, *aspartate* or *glutamate*. This expectation has been confirmed by the complete three-dimensional structure of trypsin, which has been determined by X-ray crystallography.

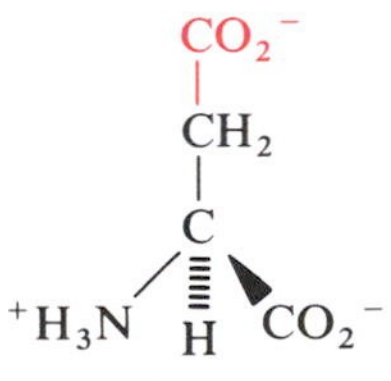

$$(H_2N)_2C^+{-}NH{-}CH_2{-}CH_2{-}CH_2{-}CH(^+H_3N)CO_2^-$$

L-Arginine (Arg, R)

$$^-O_2C{-}CH_2{-}CH(^+H_3N)CO_2^-$$

L-Aspartate (Asp, D)

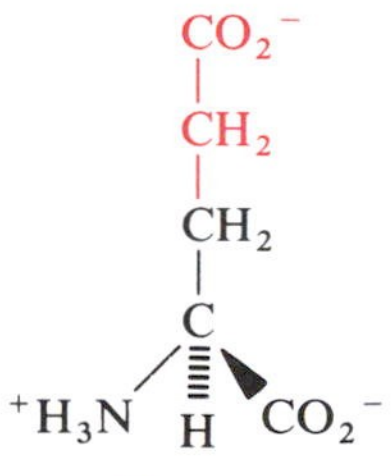

$$^-O_2C{-}CH_2{-}CH_2{-}CH(^+H_3N)CO_2^-$$

L-Glutamate (Glu, E)

$$^+H_3N{-}CH_2{-}CH_2{-}CH_2{-}CH_2{-}CH(^+H_3N)CO_2^-$$

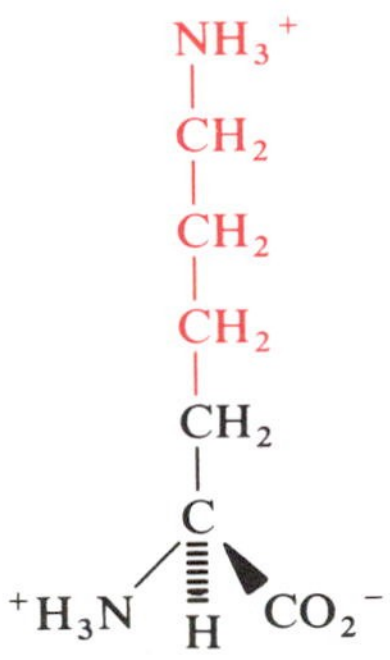

L-Lysine (Lys, K)

The substrate-binding pocket at the active site of trypsin is similar to that of chymotrypsin except for the presence of a negative charge at the bottom of the pocket (Figure 3-4A). The charge is supplied by the β-carboxylate group of an aspartate residue in the peptide chain of trypsin, which interacts with the positive charges of specific amino acid substrates. Curiously, there is a water molecule between the positive charge of the substrate side chain and the

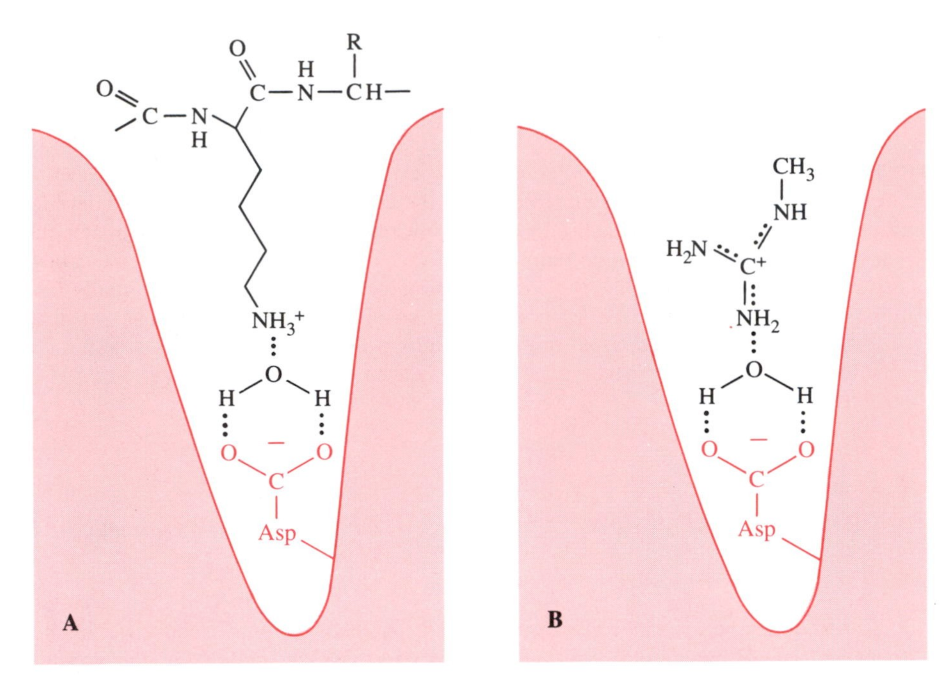

Figure 3-4

The Specificity Pocket of Trypsin. In part A, a lysyl side chain is fitted into the pocket and held by an ionic bond to an aspartate anion at the base of the pocket. In part B, a methylguanidinium ion, a competitive inhibitor, is bound in the pocket.

carboxylate group in the pocket. The importance of this electrostatic interaction is revealed not only in the substrate specificity, but also in the nature of trypsin inhibitors. The activity of trypsin toward specific substrates is inhibited by cationic molecules, such as the methylguanidinium ion, which bind to the anionic site in the pocket in competition with the side chains of specific substrates (Figure 3-4B).

Elastase is a less-specific serine protease that cleaves the acyl group of amino acids with small side chains, such as alanine, and exhibits low activity toward substrates for chymotrypsin or trypsin. The X-ray structure analysis has shown that elastase is similar to chymotrypsin and trypsin in many ways; it differs in that the binding pocket is sterically blocked, so that large amino acid side chains cannot fit into the active site. The elastase in leukocytes is released into lung tissues that are damaged by smoking or asbestos and cleaves elastin, an elastic structural protein in the lung; this is an important cause of emphysema (see also Box 1-1).

We shall see that the binding of substrates is not a passive aspect of catalysis, but rather an active function in bringing about rate enhancement. This raises the question of how elastase can function effectively with less binding than trypsin or chymotrypsin to substrate side chains. Elastase compensates for its lack of binding in a pocket by binding the peptide chain to a series of subsites that fit the amino acid residues neighboring the cleavage site (Figure 3-5).

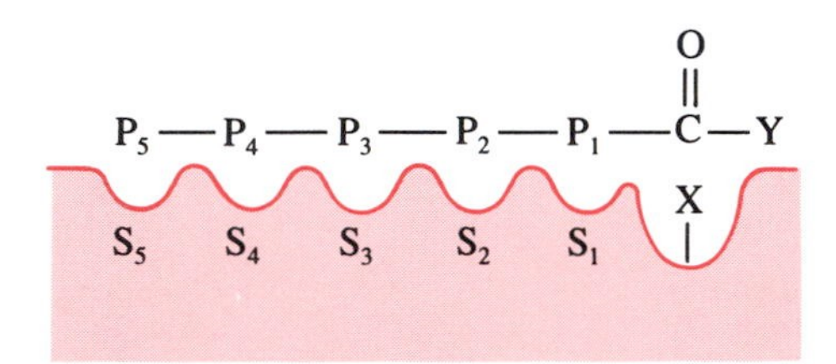

Figure 3-5

Multiple Binding Interactions Between a Polypeptide Chain and Elastase. Aminoacyl residues are designated P_1 through P_5, and complementary binding subsites of elastase are designated S_1 through S_5. The active-site serine is symbolized by —X and the leaving group by —Y.

Elastase has five subsites, S_1 through S_5, which bind five amino acid residues, P_1 through P_5, and help to bring the peptide bond that is to be cleaved into the correct position relative to the catalytic groups at the active site. The interactions of amino acid residues with these subsites leads to tighter binding of the peptide, as should be expected. But binding to these sites also has the remarkable effect of bringing about a large increase in the catalytic activity of the active site. Analogous subsites exist in chymotrypsin and trypsin, but they are less important for the catalytic activity of these enzymes.

Exopeptidases

Carboxypeptidases and aminopeptidases are exopeptidases that remove amino acids one at a time from the carboxyl-terminal and amino-terminal ends of a peptide chain, respectively (Figure 3-1). The substrate specificities of these enzymes are, accordingly, toward the carboxylate or the protonated amino groups of the terminal amino acids that are cleaved. A carboxypeptidase is a *metalloenzyme* that contains a tightly bound metal ion, such as Zn^{2+}, which is essential for its catalytic activity. A Zn^{2+} ion facilitates the hydrolysis of a peptide in two ways: it polarizes the carbonyl group (**1a**), which aids the nucleophilic attack of water, and it binds the attacking water molecule so that it can easily lose a proton to form hydroxide ion, which is a strong nucleophile (**1b**).

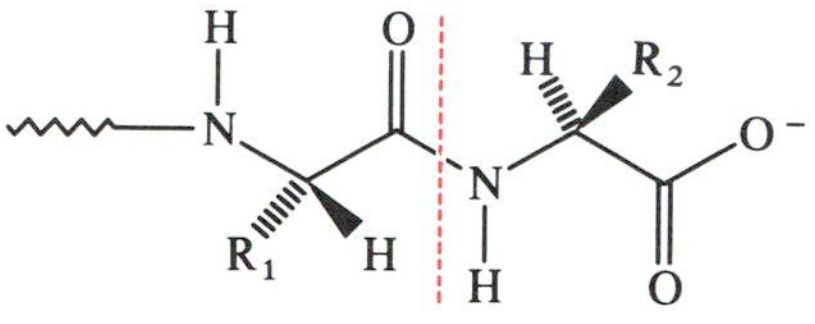

Carboxyl-terminal end

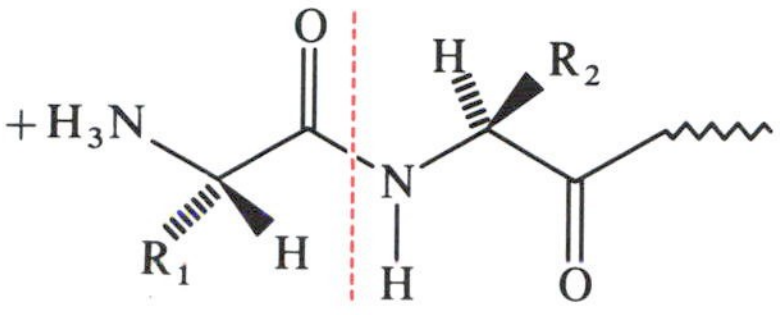

Amino-terminal end

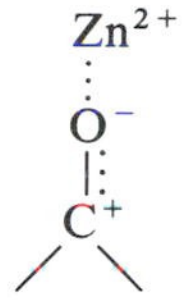

Zn^{2+}···O^-–C^+

1a

Zn^{2+}–HOH ⇌ Zn^{2+}–HO^- + H^+

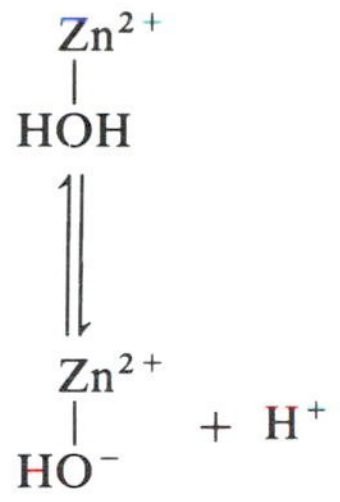

1b

Carboxypeptidase A is selective for carboxyl-terminal amino acids with large hydrophobic side-chain R-groups, such as phenylalanine, tyrosine, and tryptophan. Carboxypeptidase B, on the other hand, exhibits selectivity for the basic amino acids arginine and lysine at the carboxyl terminus.

Pancreatic carboxypeptidase A is the most thoroughly studied carboxypeptidase. Its activity is inhibited by reagents that react with the guanidinium group of arginine. Its three-dimensional structure, determined by X-ray crystallography, suggests that a guanidinium group of an arginine residue supplies a site at which the carboxylate group of a peptide substrate can be bound through a cyclic, bifunctional, hydrogen-bonded ion pair (**2**). This specific interaction is ideally suited for holding the substrate in the correct position for reaction at the active site. Increasing evidence implicates arginine as a specific amino acid for binding carboxylate and phosphate groups in many enzymes and other proteins.

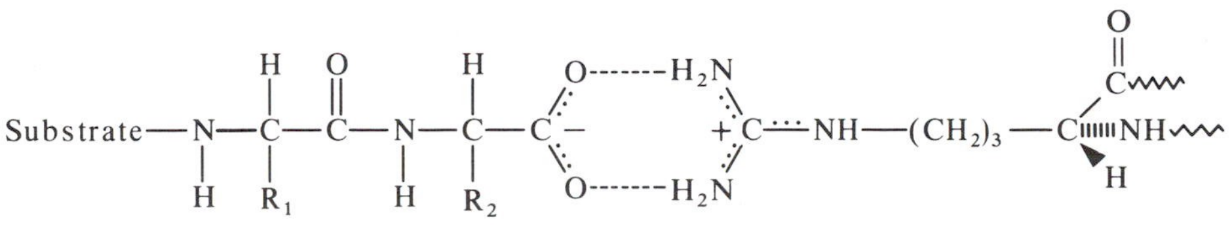

Substrate—N(H)—CH(R_1)—C(=O)—N(H)—CH(R_2)—$C(O^-)O$ ···H_2N/H_2N =C^+—NH—$(CH_2)_3$—CH(C=O)NH

2

BOX 3-2

ANGIOTENSIN-CONVERTING ENZYME

Angiotensin, a potent hormone that causes an increase in blood pressure, is an octapeptide. It is formed from proangiotensin by removal of a carboxyl-terminal dipeptide. Proangiotensin and angiotensin contain the amino acids histidine (His) and proline (Pro) in addition to other amino acids that you have already seen.

H_2N-Asp-Arg-Val-Tyr-Ile-His-Pro-Phe-His-Leu-CO_2H

Proangiotensin

H_2N-Asp-Arg-Val-Tyr-Ile-His-Pro-Phe-CO_2H

Angiotensin

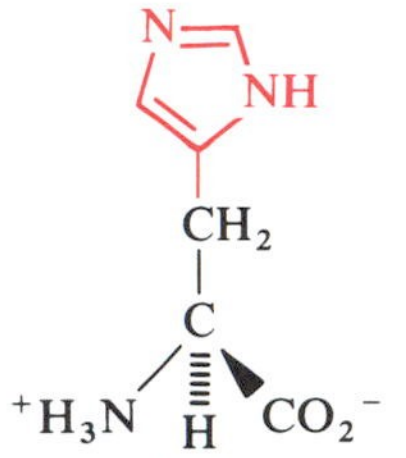

L-Histidine (His, H)

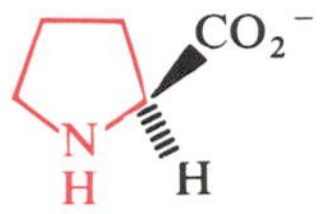

L-Proline (Pro, P)

The proangiotensin is cleaved from α_2-globulin, which is secreted by the liver. This cleavage is brought about by renin, a kidney enzyme.

Angiotensin-converting enzyme is a carboxydipeptidase. It is similar to carboxypeptidase except that the zinc atom is further from the C-terminal end of the binding site, so that a dipeptide rather than a single amino acid is cleaved from the end of the peptide. There has been extensive research into the development of inhibitors for angiotensin-converting enzyme, because such inhibitors have the potential of being effective drugs for lowering the blood pressure of patients with hypertension.

Byers and Wolfenden showed that benzylsuccinate is a strong inhibitor of carboxypeptidase. This compound can be regarded as a *bisubstrate analog* because a single molecule contains groups that are present on two molecules of substrate, or the two molecules of product in this case. Benzylsuccinate has a carboxylate group corresponding to the terminal carboxylate of a peptide and another one corresponding to the carboxylate group at the site of peptide cleavage.

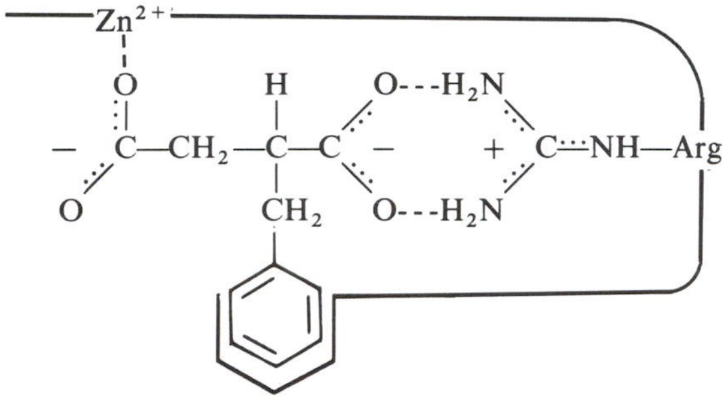

This is a strong inhibitor because a single molecule that contains two binding groups often binds very much more strongly than the two separate molecules.

Comparatively little is known about aminopeptidases. The best-known example is leucine aminopeptidase from the intestine, which exhibits high activity toward leucine as the N-terminal amino acid but is also quite active toward most other N-terminal amino acids.

The Rates of Enzyme-catalyzed Reactions

Chymotrypsin and all other enzymes are catalysts, as defined in Chapter 2: they are agents that increase reaction rates without themselves undergoing permanent chemical change. Our description of enzyme catalysis up to this point has been essentially qualitative. We have described how the substrate specificities of chymotrypsin and trypsin are determined by the structures of the substrates and the enzymes. To proceed further toward an understanding of enzyme specificity and catalysis, we must define the activity of enzymes quantitatively. The rate of an enzyme-catalyzed reaction usually levels off at high substrate concentration, when the enzyme is saturated with substrate, so that we must have a way of describing these rates at different substrate concentrations; we must be able to

BOX 3-2 (continued)

A group of workers at Squibb Pharmaceuticals prepared an inhibitor that was designed to resemble the hydrolysis products of proangiotensin. Angiotensin-converting enzyme could bind a peptide in a position such that the peptide carbonyl group at the site of cleavage can interact with the catalytic zinc ion, as illustrated in the following diagram.

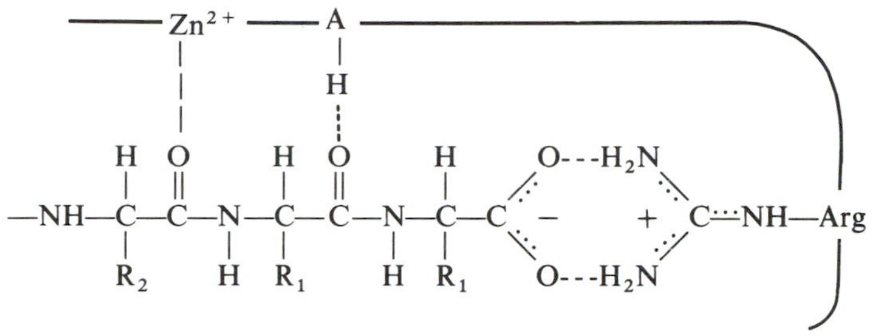

The inhibitor can bind similarly, with a strong interaction between the thiol anion and the zinc. This compound is an extremely potent inhibitor of the enzyme, with a dissociation constant of approximately 2×10^{-8} M. It has been widely used for the control of blood pressure under the trade name of Captopril.

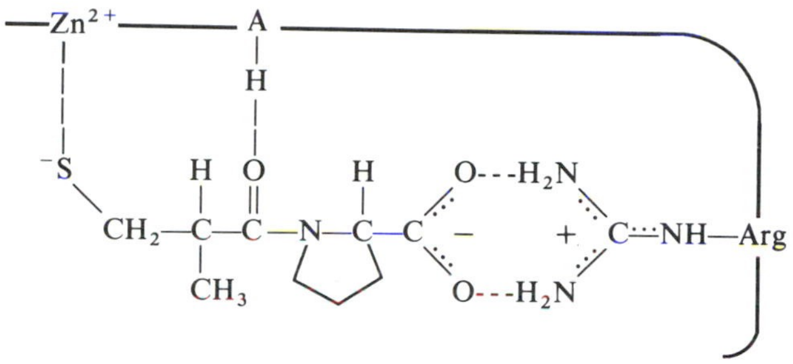

Notice that the C-terminal proline of the inhibitor does not correspond to the terminal amino acid of angiotensin. It is interesting that the inhibitor is able to bind so well with the ring at the carboxyl terminal position.

describe the kinetics of enzyme-catalyzed reactions. Knowledge of the rate constants for catalysis of reactions by enzymes under different conditions and of the structures of enzymes, as determined by X-ray crystallography, chemical modification, and other techniques, allows us to obtain some understanding of how enzymes bring about catalysis.

Proteolytic enzymes induce extraordinary increases in the rate of peptide-bond hydrolysis—canned beef is stable for years, but proteins are cleaved in minutes or seconds in the presence of bacteria, the enzymes in meat tenderizers, or purified enzymes. To make progress toward understanding the actions of proteases such as chymotrypsin, researchers have subjected these enzymes to structure-function examination by using small, chemically simple substrates that could be systematically varied in structure and studied quantitatively. To evaluate the relative activity of chymotrypsin toward different substrates requires that we describe this activity as a function of substrate concentration. This is done by measuring the rate of the reaction at different substrate concentrations in the presence of a constant amount of enzyme. Enzyme activity is usually measured as an *initial rate* for four reasons: to avoid the effects of the reverse reaction, to avoid curvature in the rate as the substrate concentration decreases, to avoid inhibition of the reaction by accumulated products, and to avoid the effects of time-dependent changes in enzyme activity that might be caused by the assay conditions.

Saturation Kinetics

Chymotrypsin, like most other enzymes and catalysts in general, shows a leveling off, or saturation, of its catalytic activity as the substrate concentration is increased. This is illustrated in Figure 3-6, in which v is the initial rate of hydrolysis of the substrate and [S] is the concentration of substrate, which is very much larger than the concentration of enzyme. The saturation effect is observed as a maximal rate of reaction at high substrate concentration. It is explained by postulating that the enzyme and substrate combine as an enzyme-substrate complex (E·S). At saturation, all of the enzyme is in the form of E·S, so that further increases in substrate concentration do not increase the concentration of E·S or the rate.

The hyperbolic curve in Figure 3-6 can be characterized by two numbers, the maximal velocity at a saturating concentration of substrate, V_m, and the initial slope of the curve at low substrate concentrations, V_m/K_m. The value of K_m is also equal to the concentration of substrate that gives half-maximal activity. It is called the *Michaelis constant* and is equal to the ratio of the maximal velocity and the initial slope, $V_m/(V_m/K_m)$. V_m and K_m are *experimental* quantities that define the hyperbolic curve in Figure 3-6; they have no unique mechanistic meaning.

Equation 3 describes the simplest of several different kinetic models that are consistent with this saturation behavior.

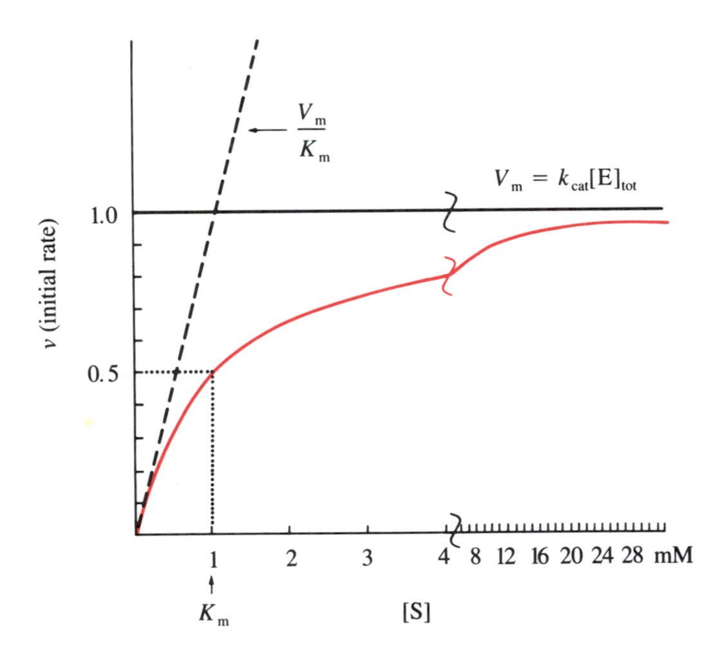

Figure 3-6

Dependence of Velocity on Substrate Concentration for a Simple One-Substrate Enzyme-catalyzed Reaction. The curve shows the increase in velocity (v) with increasing substrate concentration ([S]). The scale on the abscissa is broken to compress the display of data. Note that the velocity approaches a saturation value, which is illustrated in the graph by the horizontal line and is the maximal velocity V_m (or k_{cat}). The K_m is the concentration of substrate at half-saturation. The slope of the dashed line is V_m/K_m.

$$\mathrm{E + S} \underset{k_{-1}}{\overset{k_1}{\rightleftharpoons}} \mathrm{E \cdot S} \xrightarrow{k_2} \text{Products} + \mathrm{E} \tag{3}$$

The interpretation of V_m according to this *model* is straightforward—it is simply the rate at which the enzyme-substrate complex breaks down to products and free enzyme. According to equation 3, the observed velocity, or rate, depends on the concentration of E·S (equation 4). At high substrate concentration, the enzyme is saturated with substrate, all of the enzyme is converted into E·S, and the observed rate of product formation is independent of the substrate concentration as described by equation 5.

$$v = k_2[\mathrm{E \cdot S}] \tag{4}$$

$$V_m = k_2[\mathrm{E}]_{tot} \tag{5}$$

Thus, the observed rate at substrate saturation is determined by the rate constant k_2. If the concentration of enzyme is doubled in the presence of a large excess of substrate, the concentration of E·S and the rate also are doubled; that is, the observed rate is directly proportional to the enzyme concentration, and the reaction is *first order* with respect to the enzyme. The rate of a first-order reaction is proportional to a first-order rate constant, such as k_2, which has units of reciprocal time: $\sec^{-1}$ or $\min^{-1}$. Because the rate is independent of the substrate concentration at saturation, k_2 is a characteristic property of most enzymes as long as the concentration of substrate is much larger than that of the enzyme.

The rate of product formation is proportional to the concentration of E·S and is always given by equation 4. At the substrate concentration that gives half-maximal activity ($[S] = K_m$), exactly half of the enzyme is in the form of E·S. Half-maximal activity, therefore, corresponds to half-maximal concentration of E·S during the period of rate measurement.

When the substrate concentration is far below the K_m, so that the fraction of enzyme present as the E·S complex is very small, the rate of product formation increases linearly with increasing substrate concentration. The initial slope of the curve in Figure 3-6 is equal to V_m/K_m. Because the rate increases linearly with substrate concentration, we can say that the reaction is first order with respect to substrate under these conditions. The rate also increases linearly with enzyme concentration, so that the reaction is also first order with respect to enzyme. Thus, the rate under these conditions is described by equation 6, in which $k' = k_2/K_m$.

$$v = k'[\mathrm{E}][\mathrm{S}] = \frac{k_2}{K_m}[\mathrm{E}][\mathrm{S}] \tag{6}$$

Equation 6 is a rate equation for a *second-order* reaction because the reaction is first order with respect to each of two reagents, [E] and [S]. The units for a second-order rate constant such as k' or k_2/K_m are $M^{-1} \cdot \sec^{-1}$ or $M^{-1} \cdot \min^{-1}$.

The observed first-order rate constant under conditions of substrate saturation is usually referred to as k_{cat} rather than k_2, because in more-complex kinetic models it may represent a complex of several rate constants. The second-order rate constant at very low substrate concentrations is then k_{cat}/K_m. *Thus, the leveling off of the rate with increasing substrate concentration (Figure 3-6) represents a transition from a second-order dependence on both the reactants, [E] and [S], at very low substrate concentrations, to a first-order dependence on [E] alone at high substrate concentrations. The curve is characterized by any two of three variables: (1) the first-order rate constant,* k_{cat}*; (2) the second-order rate constant,* k_{cat}/K_m*; and (3) the substrate concentration at half-maximal velocity,* K_m.

These constants are useful for characterizing the relative activities of an enzyme toward different substrates. The first-order rate constant, k_{cat}, is a measure of how active the enzyme is in bringing about the reaction of a substrate once it is bound to the active site in the E·S complex. The second-order rate constant, k_{cat}/K_m, is a measure of the catalytic effectiveness of the free enzyme in dilute solution that includes the specific binding of the substrate to the enzyme. The K_m is a measure of the dependence of the rate on substrate concentration. It is frequently assumed that K_m is the equilibrium constant for dissociation of the E·S complex to E + S, which measures the affinity of the enzyme for binding the substrate, but we shall see that this often is not correct.

Investigators disagree on whether k_{cat} or k_{cat}/K_m is the more-useful rate constant to use for comparing the activities of different enzymes or substrates. The second-order rate constant, k_{cat}/K_m, is sometimes said to be the more representative of physiological conditions because the K_m values of most enzymes are larger than the physiological concentrations of their substrates. It is often not clear, however, what the intracellular concentrations of substrates really are, owing to intracellular compartmentalization. And some of the important enzymes in the principal metabolic pathways appear to be present at concentrations that are comparable to or larger than the concentrations of their substrates; they may even be higher than the K_m values. In such cases, k_{cat} may be the more-significant parameter for the evaluation of relative activities under physiological conditions.

When an enzyme is catalyzing the reactions of two different substrates in the same solution, the relative activity toward the two substrates is determined by the substrate concentrations and the ratio of the values of k_{cat}/K_m for the two substrates. This is true even if the concentrations of the substrates are larger than K_m, so that the enzyme is saturated.

Evaluation of V_m and K_m

The determination of accurate experimental values for V_m and K_m is more difficult than might be supposed. It is usually easier to obtain an accurate value for V_m/K_m, which is measured at low concentrations of substrate. It is often difficult to estimate the substrate concentration that gives half-maximal velocity from the nonlinear plot of initial rates against substrate concentrations (Figure 3-6) because the maximal velocity cannot be determined from such a plot, unless the rates are measured at concentrations of substrate that extend to more than ten times the K_m. Limited solubility or limited availability of substrates often makes this impossible. The investigator should never omit the preparation of such a plot, however, because it provides the most-direct picture of the experimental data, especially the scatter of the points.

We shall show in the next section that the dependence of the rate on substrate concentration according to the kinetic model of equation 3 and Figure 3-6 is described by equation 7, the *Michaelis-Menten equation.*

$$v = \frac{V_m[S]}{K_m + [S]} = \frac{V_m}{\frac{K_m}{[S]} + 1} \tag{7}$$

To make the best use of the available experimental data, the observed rates are usually plotted according to a linearized form of this equation. Equation 7 can be converted into a linear form in several ways, of which the simplest is to take the reciprocal, equation 8.

$$\frac{1}{v} = \frac{1}{V_m} + \frac{K_m}{V_m} \cdot \frac{1}{[S]} \tag{8}$$

This equation is linear in $1/v$ and $1/[S]$, and a plot of $1/v$ against $1/[S]$ is linear for data that are consistent with equations 3 and 7. The ordinate intercept of this plot is $1/V_m$ and the slope is K_m/V_m (Figure 3-7). The value of K_m also may be obtained from the negative intercept on the abscissa, which is $-1/K_m$ (Figure 3-7). This double-reciprocal correlation is known to biochemists as a *Lineweaver-Burk plot*. Although this plot is widely used, it suffers from the weakness of all reciprocal plots in that the data are weighted very differently at high and low substrate concentrations; consequently, it is not highly accurate and may be misleading.

The *Eadie-Hofstee plot* of v against $v/[S]$ is more satisfactory for the estimation of V_m (Figure 3-8). This plot is based on equation 9, which is obtained from equation 8 by multiplying both sides by v and rearranging.

$$v = V_m \cdot \frac{v}{[S]} \cdot K_m \tag{9}$$

The value of V_m is obtained directly from the ordinate intercept and the value of $-K_m$ from the slope of the plot. The weakness of this plot is that the substrate concentration does not appear directly on either the abscissa or the ordinate; this makes it difficult to relate the points on the plot directly to the experimental results.

For accurate determination of V_m and K_m, the observed rate measurements should be treated statistically, with appropriate weighting for the

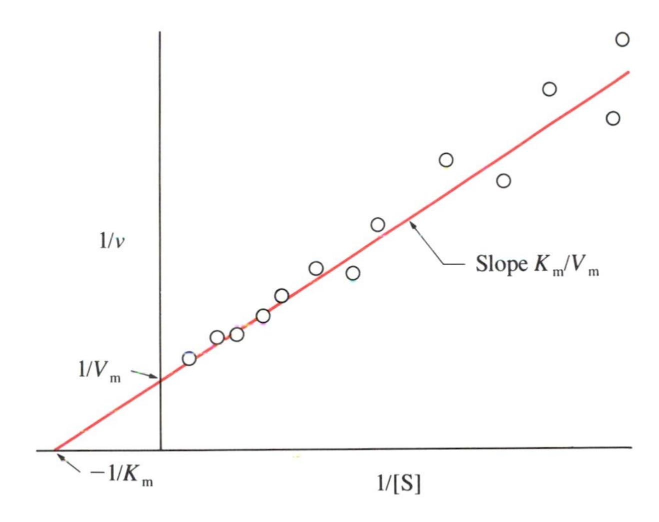

Figure 3-7

A Lineweaver-Burk Plot of $1/v$ Against $1/[S]$.

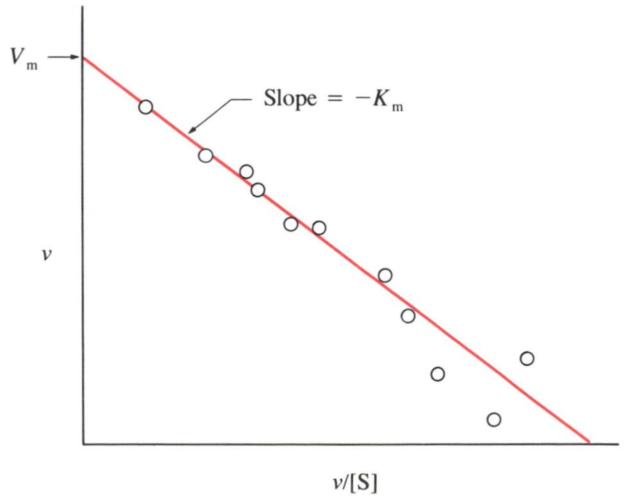

Figure 3-8

An Eadie-Hofstee Plot of v Against $v/[S]$.

uncertainty of the data at different substrate concentrations. Computer programs are available for such calculations.

The most-useful comparisons of enzyme activity are based on k_{cat} and k_{cat}/K_m. These rate constants refer to the molar concentration of enzyme molecules or, if there are known to be several sites in each enzyme molecule, to the concentration of active sites in the solution. The k_{cat} is also known as the *turnover number* of the enzyme, because it gives the number of catalytic cycles that are completed by the enzyme per unit of time under the assay conditions. Values of k_{cat} and k_{cat}/K_m can be obtained only for homogeneous enzymes of known molecular weight and concentration, for which the number of active sites per enzyme molecule is also known.

If an enzyme is impure or the number of active sites per molecule is unknown, the specific activity may be described by *international units,* the amount of enzyme that produces 1 micromole of product per minute under standard conditions, or the *katal* (kat), the amount of enzyme that reacts with 1 mole of substrate per second to give product under standard conditions. Standard conditions must be defined for each enzyme. Although the katal is recommended by the International Union of Biochemistry, it is rarely used.

The Specificity of Chymotrypsin

We are now in a position to make a quantitative evaluation of the activity of chymotrypsin toward different substrates. For example, the kinetic constants for catalysis by chymotrypsin of the hydrolysis of *N*-acetyl-L-tyrosyl glycinamide (AT-GA), and *N*-acetyl-L-tyrosyl ethyl ester (AT-OEt), are shown in Table 3-1. The values of k_{cat} show that the enzyme catalyzes hydrolysis of the ester 390 times as fast as the peptide when it is saturated with substrate.

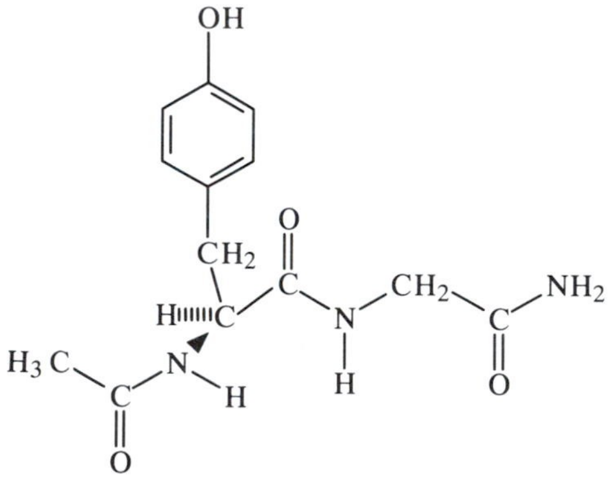

AT-GA 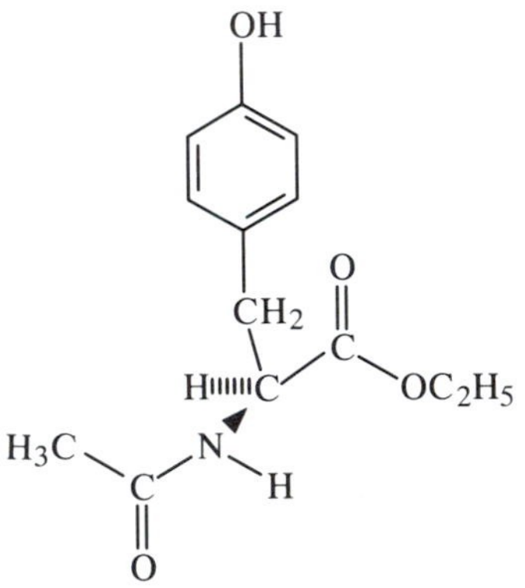 **AT-OEt**

Table 3-1

Kinetic constants for the activity of α-chromotrypsin toward *N*-acetyl-L-tyrosine-glycinamide (AT-GA) and *N*-acetyl-L-tyrosine ethyl ester (AT-OEt) at 25° and pH 7.9

	SUBSTRATE		
	AT-GA	**AT-OEt**	**Ratio (Ester/Peptide)**
$k_{cat}(sec^{-1})$	0.50	193	390
$K_m(M)$	0.023	0.0007	0.03
$k_{cat}/K_m(M^{-1}\,sec^{-1})$	22	280,000	12,700

However, the values of the second-order rate constant, k_{cat}/K_m, show that at very low substrate concentration the ester is cleaved 12,700 times as fast as the peptide. Both comparisons show that the activity of the enzyme parallels the higher chemical reactivity of esters than that of amides toward base-catalyzed hydrolysis. The difference between the comparisons of k_{cat} and k_{cat}/K_m is due to the 33-fold smaller K_m value for the ester, which means that the enzyme becomes saturated by a much lower concentration of the ester than of the peptide. One might at first think that this lower K_m means that the enzyme binds the ester more tightly than the peptide. However, we will return to these data later and find that this is not the case. This is one of many examples in which K_m is *not* a measure of how tightly the substrate binds to the enzyme.

Derivation and Interpretation of the Michaelis-Menten Equation

A Steady-State Kinetic Derivation

The kinetic model of equation 3, and other kinetic models, can be treated quantitatively by steady-state kinetics when the concentration of substrate is much larger than that of the enzyme.

$$E + S \underset{k_{-1}}{\overset{k_1}{\rightleftharpoons}} E \cdot S \xrightarrow{k_2} \text{Products} + E \tag{3}$$

Steady-state kinetics can be applied to a system when the change in the concentration of an intermediate species in the course of a reaction is very small compared with the change in the concentration of reactants and products. This situation describes the *steady-state approximation.* It is a satisfactory approximation for an initial rate measurement of a reaction in the presence of a large amount of substrate and a very small, catalytic amount of enzyme, because the enzyme must turn over many times in order to bring about a significant change in the concentration of substrate. The rate of formation of the E·S complex must be equal to the rate of its breakdown for turnover to occur, so that the concentration of the E·S complex is constant after the first few turnovers. Over a long observation period, much longer than an initial rate measurement, the steady-state level of E·S will gradually decline because the concentration of substrate decreases as the reaction proceeds.

The steady-state rate equation for the kinetic model of equation 3 is obtained from the rate equations for the individual steps of the reaction. According to this model, the rate of formation of the E·S complex, v_f, depends on the concentrations of enzyme and of substrate, so that it is given by equation 10.

$$v_f = k_1[E][S] \tag{10}$$

The E·S complex can break down to give products with the rate constant k_2, but it can also break down to give back E and S with the rate constant k_{-1}. Therefore, the total rate at which E·S breaks down, v_b, is given by equation 11, which describes the sum of these two rates.

$$v_b = (k_{-1} + k_2)[E \cdot S] \tag{11}$$

The steady-state requirement that the rates for the formation and breakdown of the E·S complex must be equal is given by equation 12.

$$k_1[E][S] = (k_{-1} + k_2)[E \cdot S] \tag{12}$$

This requirement is often stated in the equivalent form that the rate of change in the concentration of the steady-state intermediate is approximately zero.

This equation can be related to experimentally measurable quantities with a *conservation equation*, equation 13, and a *definition* of K_m, the Michaelis constant,

$$[E]_{tot} = [E] + [E\cdot S] \tag{13}$$

$$K_m = \frac{k_{-1} + k_2}{k_1} \tag{14}$$

as the ratio of the rate constants for breakdown and formation of the E·S complex, equation 14. According to the kinetic model of equation 3, the rate of product formation, $d[P]/dt$, depends only on the concentration of E·S and the value of the rate constant k_2, as shown in equation 15.

$$v = \frac{d[P]}{dt} = k_2[ES] \tag{15}$$

$$v = \frac{k_2[E]_{tot}}{\frac{K_m}{[S]} + 1} = \frac{V_m}{\frac{K_m}{[S]} + 1} \tag{16}$$

Simple algebra then gives the Michaelis-Menten equation (equation 16), which specifies the rate at any concentration of substrate.

Equation 16 has the properties that are required to account for saturation curves such as that in Figure 3-6. When the substrate concentration is much larger than K_m, the substrate term drops out and the rate becomes $v = V_m$, which is independent of substrate concentration. When the substrate concentration is much less than K_m, the rate becomes $v = (V_m/K_m)[S]$; that is, it becomes directly proportional to the substrate concentration and is second-order overall. When $[S] = K_m$, the observed rate is $v = V_m/2$, half the maximal rate.

Interpretation of the Michaelis-Menten Equation

Rate equations are most useful and interesting when they can be interpreted in terms of the mechanism of a reaction. If an enzymatic reaction follows the model of equation 3, its mechanism can be interpreted in two ways, which differ with respect to the behavior, or the partitioning, of the intermediate E·S. The partitioning is controlled by the relative values of k_2 and k_{-1}.

In the simplest case, the E·S complex is at equilibrium with the free enzyme and substrate. This is the interpretation that most people find familiar and intuitively reasonable. It describes the case in which the E·S complex dissociates to E and S many times for each time that it goes forward to products; that is, $k_{-1} \gg k_2$. When E·S is at equilibrium with free E and S, its concentration is described by an equilibrium constant K_s (equation 17).

$$K_s = \frac{[E][S]}{[E\cdot S]} = \frac{k_{-1}}{k_1} \tag{17}$$

It can easily be shown that, when equilibrium obtains, $K_s = k_{-1}/k_1$ because at equilibrium the rate of E·S formation ($k_1[E][S]$) is equal to the rate of its breakdown in the *reverse direction* ($k_{-1}[E\cdot S]$). In this case, in which $k_{-1} \gg k_2$, the definition of the Michaelis constant K_m (equation 14) reduces to that for the definition of K_s (equation 17); that is, $K_m = K_s$. The dependence of the observed rate on the concentration of substrate is then determined by the dissociation constant of the E·S complex.

However, an alternative interpretation of the kinetic model of equation 3 gives rise to saturation kinetics without invoking the dissociation constant of the E·S complex at all. This situation is sometimes regarded as a textbook curiosity, learned and soon forgotten. However, it deserves understanding because it describes the behavior of many, perhaps most, enzymes. It also describes the operation of metabolic pathways that include irreversible steps, as well as catalysis of a number of chemical reactions, such as hydrogenation or rearrangements, that involve irreversible binding of a molecule to a metallic catalyst.

Suppose that the rate of dissociation of E·S back to E and S is negligibly small; that is, $k_{-1} \ll k_2$. When this is so, essentially every molecule of E·S that is formed will go on to give products, the back reaction can be neglected, and the kinetic model of equation 3 reduces to equation 18.

$$\mathrm{E} + \mathrm{S} \xrightarrow{k_1} \mathrm{E \cdot S} \xrightarrow{k_2} \text{Products} + \mathrm{E} \tag{18}$$

We are left then with a simple system of two consecutive irreversible reactions, an initial second-order reaction that depends on the concentrations of both substrate and free enzyme, equation 19, and a subsequent first-order reaction that depends on the concentration of only E·S, equation 20.

$$v = k_1[\mathrm{E}][\mathrm{S}] \tag{19}$$

$$v = k_2[\mathrm{E \cdot S}] \tag{20}$$

In the steady state, the rates for the two steps of the overall reaction that are defined by equations 19 and 20 are equal to each other and to the observed initial rate, at a given concentration of substrate. By the same algebraic manipulations employed for equation 16, it can be shown that the initial rate is given by equation 21.

$$v = \frac{k_2[\mathrm{E}]_{\mathrm{tot}}}{\dfrac{k_2/k_1}{[\mathrm{S}]} + 1} \tag{21}$$

This is the Michaelis-Menten equation for the case in which $k_{-1} \ll k_2$ in equation 14; in this case, $k_{\mathrm{cat}} = k_2$ and $k_{\mathrm{cat}}/K_{\mathrm{m}} = k_1$.

Now, if we start with a very low concentration of substrate, every molecule of E·S that is formed will go on to products rapidly, and the observed rate will increase linearly as the substrate concentration is increased. This corresponds to the initial slope in Figure 3-6. The observed rate under these conditions is determined by the rate of formation of the E·S complex through the second-order reaction that is described by equation 19; this reaction is slow and rate determining when the substrate concentration is very low. However, as the substrate concentration is increased, the rate of formation of E·S will increase until it becomes faster than the breakdown of the E·S complex to products. When this happens, the E·S complex will begin to pile up, and at a sufficiently high concentration of substrate all of the enzyme will exist as the E·S complex. The rate of product formation is then determined entirely by the rate of breakdown of the E·S complex to products, according to the first-order reaction of equation 20. Because a further increase in substrate concentration cannot cause any further increase in the concentration of E·S or in the rate, the reaction has reached a limiting maximum velocity. This is shown by the horizontal line in Figure 3-6.

What we see in this system is a transition from a second-order reaction that determines the rate at low substrate concentration to a first-order reaction that determines the rate at high substrate concentration. As the substrate concentration is increased, there is a change in rate-determining step from the second-order to the first-order reaction. The change in rate-determining step occurs because the second-order reaction depends on the concentration of substrate, but the first-order reaction does not. This situation results in exactly the same shape of the substrate "saturation" curve in Figure 3-6, as is observed for the equilibrium case. The difference is that the saturation behavior is caused by a *change in rate-limiting step,* whereas in the case described earlier (equation 17) it is caused by the *equilibrium binding of the substrate*. Equilibrium conditions exist when $k_{-1} \gg k_2$, and it is only when this is the case that $K_m = K_s$.

For the situation in which saturation behavior is caused by a change in rate-determining step and $k_2 \gg k_{-1}$, the k_{-1} term in the definition of K_m can be neglected and K_m reduces to the ratio of the two rate constants k_2/k_1 (see equation 21). The second-order rate constant for the reaction at low substrate concentration, k_2/K_m or k_{cat}/K_m, is then equal to k_1, in agreement with equation 19. In this situation, the K_m corresponds to the substrate concentration at which the change in rate-determining step from k_1 to k_2 is halfway complete and both steps are equally rate determining.

This is sometimes called the Haldane interpretation of Michaelis-Menten kinetics after J. B. S. Haldane, who originally described steady-state kinetics for enzymes. Recent work has shown that the binding of substrates is completely or partially irreversible much more often than had previously been supposed, so that this interpretation applies to many enzymatic reactions.

These two limiting interpretations of saturation kinetics can be described by the energy diagrams in Box 3-3. Some reactions of substrates with enzymes occur at an encounter-controlled rate, which represents a special case of the Haldane interpretation, as described in Box 3-4.

BOX 3-3

RATE-DETERMINING STEPS

The two interpretations of Michaelis-Menten kinetics can be described by the adjoining reaction diagrams. When $k_{-1} \gg k_2$ (Case 1), the E·S complex breaks down to give back E + S much more often than it overcomes the larger barrier to break down to products. This means that E, S, and E·S are interconverted rapidly and are at equilibrium, whereas the breakdown of E·S with the rate constant k_2 is the rate-determining step. When $k_{-1} \ll k_2$ (Case 2), the E·S complex breaks down to give products much more often than it can overcome the larger barrier to dissociate back to E + S. This means that formation of the E·S complex is essentially irreversible and that the rate-determining step at low substrate concentrations is the binding of substrate with the rate constant k_1. If $k_{-1} = k_2$, the barriers for the breakdown of E·S in the two directions are equal and both steps are partially rate determining.

These diagrams are drawn to describe the reaction under the conditions in which the substrate concentration is far below K_m. Diagrams of this kind are often useful for visualizing the kinetic behavior of reactions, but they are frequently misleading and must be used with great caution, especially for enzymatic reactions. They are most useful for illustrating first-order processes, such as the dissociation of E·S to reactants and to products with the first-order rate constants k_{-1} and k_2, respectively. They can also be used for *pseudo first-order* reactions, in which the observed rate constant is first order because all of the reacting molecules except one are present in great excess, so that there is no significant change in their concentration in the course of the reaction.

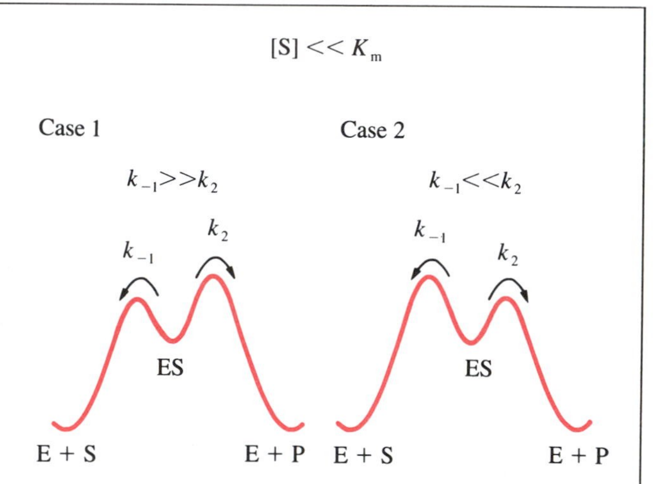

Reaction diagrams to show binding of a substrate to an enzyme at equilibrium, followed by rate-limiting formation of products (Case 1) and irreversible binding of substrate, followed by rapid formation of products (Case 2).

BOX 3-4

DIFFUSION-CONTROLLED REACTIONS

A few enzymes catalyze reactions of dilute substrates at diffusion-controlled rates. These enzymes have evolved to maximum possible efficiency, at least for catalyzing reactions at substrate concentrations below K_m. This phenomenon represents a special case of the second interpretation of Michaelis-Menten kinetics (Case 2 in Box 3-3), because it means that every encounter of the enzyme with a substrate results in product formation, so that the rate of dissociation of the E·S complex to release substrate must be neglibible; that is, $k_{-1} \ll k_2$.

The maximum rate constant for the diffusion-controlled encounter of a substrate with the active site of an enzyme is on the order of $10^9\ M^{-1} \cdot s^{-1}$. Several enzymes are known to bind their substrates more slowly, with rate constants ranging from 10^7 to $10^8\ M^{-1} \cdot s^{-1}$. These lower measured rate constants can be accounted for by postulating that only a small fraction of the enzyme or substrate is in the correct conformation, orientation, or ionization state for binding, so that only from 1% to 10% of the diffusional encounters result in productive binding. An alternative explanation is that formation of the correct E·S complex, $E \cdot S_2$ below, requires a conformational change with the rate constant $k_{1'}$ immediately following the initial diffusional encounter.

$$E + S \underset{k_{-1}}{\overset{k_1}{\rightleftharpoons}} E \cdot S_1 \underset{k_{-1'}}{\overset{k_{1'}}{\rightleftharpoons}} E \cdot S_2 \xrightarrow{k_2} \text{Products}$$

Cleland has given the ratio k_2/k_{-1} the appropriate name of the *stickiness ratio*. When this ratio is small, the substrate nearly always dissociates from the Michaelis complex, so that E·S is at equilibrium; when it is large, the substrate sticks to the enzyme and nearly always goes on to products.

Why Enzymes Are Large Molecules

Why are enzymes such large molecules? Chymotrypsin and trypsin have molecular weights of about 25,000 and are among the smaller enzymes. Other enzymes range in molecular weight from about 14,000 to 8,000,000. The very largest enzymes consist of many individual polypeptide chains that are associated in large complexes. However, the molecular weight of individual polypeptide chains can be as much as 100,000, and even more in some cases.

There is no definitive answer to the question at this time, but there are several plausible possibilities.

One possible answer is that many enzymes and other proteins are constructed from a series of *domains*, which have well-defined structures of their own and interact with each other in the intact enzyme. These domains can have different functions, such as the binding of different substrate molecules that react with each other or regulation of enzyme activity to provide metabolic control. In some cases, these domains have developed separately through evolution and are combined in an enzyme molecule to provide a particular combination of catalytic or regulatory properties. This kind of assembly line and interconvertibility of parts was rediscovered by automobile manufacturers very recently, compared with the time scale of evolution. As with automobiles, some of these parts may be larger than is really necessary and thereby contribute to the weight of the final product.

It is critically important for catalysis that there be a very strong and exact fit between the enzyme and its specific substrates in the transition state of the reaction that is catalyzed. This exact fit and strong binding stabilize the transition state and are largely responsible for catalysis, as will be discussed in more detail in Chapter 5. Exact fit requires that the enzyme have an exact, well-defined structure; a floppy enzyme would have too many possible conformations to provide an exact fit. At the same time, the enzyme must be flexible enough to allow the substrates to bind and the products to dissociate. This combination of qualities is possible if there are domains of well-defined structure that can move relative to each other in a large protein. The structures of many enzymes suggest that they can exist in an open conformation that permits binding and dissociation and in a closed conformation that brings about catalysis.

Other conformational changes control the activities of enzymes. Enzyme activities may be regulated by covalent modifications, such as phosphorylation and dephosphorylation of serine, threonine, or tyrosine, and by interactions with metabolites at second, or *allosteric, sites,* which are separate from active sites. These interactions induce conformational changes that activate or inhibit the enzyme. Such control of enzyme conformation may require a large structure.

Finally, there are several proteins that contain more than a single enzyme activity, often for catalysis of reactions that are metabolically related. These multienzyme proteins have molecular weights of as much as several hundred thousand; they may arise from fusion of the genes that code for the synthesis of the individual enzymes.

Summary

Catalysis of the hydrolysis of peptides and proteins by chymotrypsin and other proteases shows specificity for the site of cleavage that is determined by the binding of amino acids and their side chains at complementary sites on the enzyme. Carboxypeptidases and aminopeptidases are metalloenzymes that cleave amino acids from the carboxyl- and amino-terminal ends of the peptide chain, respectively. The activity of enzymes follows saturation kinetics and is characterized by k_{cat} and k_{cat}/K_m, which are first- and second-order rate constants for turnover at saturation and in very dilute solution, respectively. The values of k_{cat} and k_{cat}/K_m may be evaluated from Lineweaver-Burk or Eadie-Hofstee plots or by weighted statistical analysis. The saturation kinetics can arise either from binding of substrates at equilibrium ($k_{-1} > k_2$) or from irreversible formation of the E·S complex at high substrate concentrations ($k_1[S] > k_2$); this occurs when $k_{-1} < k_2$.

ADDITIONAL READING

Boyer, P. D., ed. 1971. *The Enzymes,* vol. 3, 3d ed. New York: Academic Press.

Cleland, W. W. 1970. Steady state kinetics. In *The Enzymes,* vol. 2, 3d ed., ed. P. D. Boyer, pp. 1–66. New York: Academic Press.

Cornish-Bowden, A. and Wharton, C. W. 1988. *Enzyme Kinetics.* Oxford: IRL Press.

Matthews, B. W. 1988. Structural basis of the action of thermolysin and related zinc peptidases. *Accounts of Chemical Research* 21: 333–340.

Tsukada, H., and Blow, D. M. 1985. Structure of α-chymotrypsin refined at 1.68 Å resolution. *Journal of Molecular Biology* 184: 703–711.

PROBLEMS

1. (a) Derive the Michaelis-Menten equation in terms of $[E]_{tot} + [S] + K_m$.

(b) Reduce the equation to the limiting cases at high and low substrate concentration when $k_{-1} > k_2$ and $k_2 > k_{-1}$. What is the meaning of K_m in these situations?

2. *N*-Acetyl-L-valine methyl ester is hydrolyzed by 10^{-5} M chymotrypsin with $v/\mu M\ s^{-1} = 0.40$, 0.50, 0.58, 0.70, 1.0, and 1.3 in the presence of 0.075, 0.087, 0.10, 0.22, and 0.63 M substrate, respectively. What are the values of K_m, V_{max}, and k_{cat} for this enzyme?

3. Can you design an inhibitor of elastase that might help to prevent emphysema? What would be a likely problem in the clinical use of this inhibitor?

4. What is the behavior of a system in which E·S is unreactive and the product is formed only from a second-order reaction of E + S? Why does substrate saturation not prove that the reaction proceeds through E·S?

5. Derive the rate equation for the following scheme when $k_1 \gg k_2$.

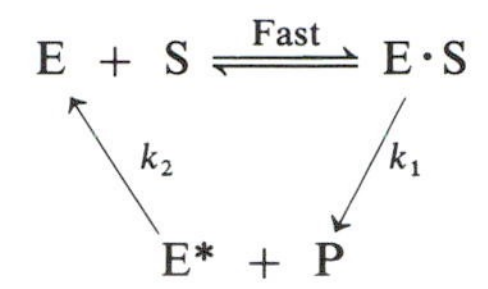

(Apply the steady-state approximation to E*.) Note that for a cyclic system $k_2[E^*] = k_1[E \cdot S]$. What is the rate-determining step and the predominant form of the enzyme at substrate saturation?

6. Draw the structures of the following pairs of amino acids and explain the differences between them: Asp and Lys; T and E; F and L; Arg and Y; Met and Ser.

7. What interactions between enzymes and substrates account for the difference in substrate specificities of trypsin and α-chymotrypsin?

Catalysis by Chymotrypsin: The Catalytic Pathway

4

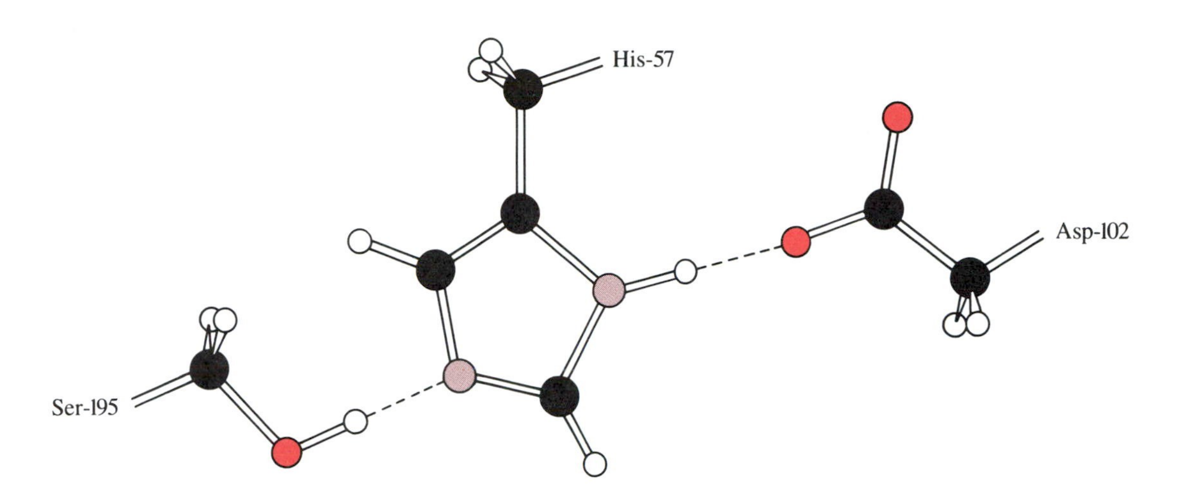

The hydrogen-bonding network linking serine-195, histidine-57, and aspartate-102 in chymotrypsin.

Investigators followed two main lines of research in studying the mechanism of action of chymotrypsin. First, they determined the discrete chemical reactions in the overall catalytic process and identified the chemical intermediates on the catalytic pathway. Then, they studied the chemical mechanisms by which the enzyme is able to increase the rates of the reaction steps in the catalytic pathway. These two aspects are related but different, and in general it is essential to determine the catalytic pathway before addressing the mechanisms of rate enhancement for individual steps. Therefore, we shall start with an examination of the catalytic pathway for chymotrypsin in this chapter.

The elucidation of the reaction pathway for catalysis by chymotrypsin has shown that the basic mechanism can be described by straightforward chemistry, without invoking mysterious forces. Hydrolysis proceeds by the formation of a covalently bonded acyl-enzyme, and the formation and hydrolysis of this acyl-enzyme are assisted by catalysis of the proton transfers that are required to give stable products. The dependence on pH of k_{cat}, k_{cat}/K_m, and K_m and the reactions with an active-site-directed inactivator identify the group at the active site that is responsible for catalysis of proton transfer. The reaction is inhibited by indole and other competitive inhibitors that resemble specific substrates and compete with them for binding to the active site.

Covalent Catalysis

Discovery of the Acyl-Enzyme

In 1953, Hartley and Kilby examined catalysis by chymotrypsin of the hydrolysis of *p*-nitrophenyl acetate (PNPA, equation 1).

$$H_3C{-}C({=}O){-}O{-}C_6H_4{-}NO_2 + H_2O \longrightarrow CH_3COO^- + {}^-O{-}C_6H_4{-}NO_2 + 2H^+ \quad (1)$$

PNPA **PNP⁻**

This substrate has an advantage over more-specific substrates in that the *p*-nitrophenolate ion product is bright yellow, with a strong absorption at 400 nm, so that the course of the reaction can be easily followed spectrophotometrically. PNPA is a very poor substrate for chymotrypsin because the acyl group does not have the hydrophobic and acylamino substituents that are required for efficient catalysis by this enzyme. This might have been thought to be a disadvantage, but it turned out to be essential for the success of their experiment.

Hartley and Kilby found that, in the presence of a large excess of PNPA, the release of *p*-nitrophenolate ion was linear with time, as expected. But they also noted that, when the increase in absorbance at 400 nm was extrapolated back to zero time, it did not extrapolate to zero absorbance (Figure 4-1). Many investigators would have dismissed this as an artifact, which could have been caused by a small amount of hydrolysis or a yellow impurity in their PNPA preparation. However, Hartley and Kilby followed up their observation and showed that the reaction proceeds with an initial "burst" of *p*-nitrophenol release, which requires the presence of enzyme, followed by the usual linear, zero-order release of product from turnover of the enzyme. The burst corresponded to one mole of *p*-nitrophenol for each mole of enzyme. This result suggests that the burst represents a chemical reaction of PNPA with the enzyme

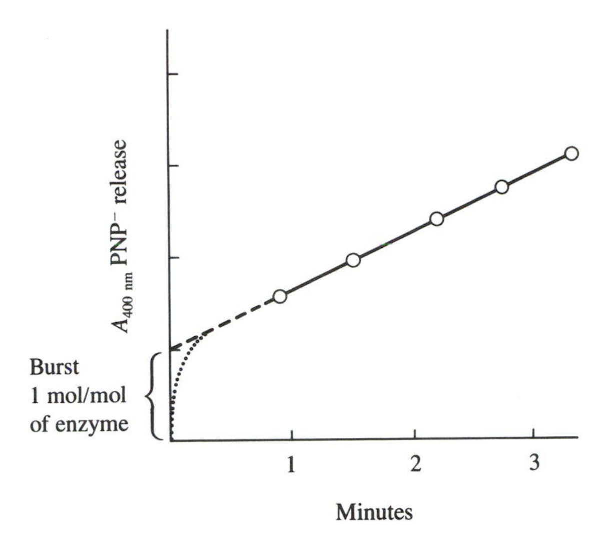

Figure 4-1

The Chymotrypsin-catalyzed Hydrolysis of *p*-Nitrophenyl Acetate Shows an Initial Burst of *p*-Nitrophenolate Release. The initial burst of PNP^- is the first step of the mechanism (equation 2a), in which PNPA reacts with the enzyme, acyl-enzyme is formed, and PNP^- is released. The second phase is the turnover of the acyl-enzyme, which undergoes a relatively slow hydrolysis according to equation 2b and is immediately reacylated by PNPA.

and that the overall reaction proceeds in two phases: an initial rapid reaction of the substrate with enzyme, which gives a stoichiometric burst of *p*-nitrophenol release, followed by a slower steady-state reaction that is responsible for the subsequent release of *p*-nitrophenol. Hartley and Kilby suggested the most-probable explanation of this phenomenon—that the hydrolysis of PNPA proceeds in two steps. The first step is an initial rapid acylation of some group on the enzyme by PNPA, with the release of one equivalent of *p*-nitrophenol in the burst (equation 2a). The second step is the slow hydrolysis of this acyl-enzyme, followed by fast reacylation of the free enzyme by PNPA, which accounts for the slow turnover phase of *p*-nitrophenolate production (equation 2b).

$$\text{HO—E} + \text{PNPA} \xrightarrow{\text{Fast (burst)}} H_3C\text{—}C(=O)\text{—O—E} + PNP^- \qquad \text{(2a)}$$

$$H_3C\text{—}C(=O)\text{—O—E} + H_2O \xrightarrow{\text{Slow (turnover)}} \text{HO—E} + CH_3CO_2^- \qquad \text{(2b)}$$

This is a most-important conclusion. It means that the mechanism of the chymotrypsin-catalyzed hydrolysis of this substrate is an example of *covalent catalysis,* a catalytic pathway in which a part of the substrate forms a covalent bond with a group on the enzyme to give an intermediate chemical species. In a second step, the intermediate undergoes a further reaction to form product and regenerate the original form of the enzyme.

The burst observed by Hartley and Kilby in their experiments could have been detected only with a poor substrate like PNPA. A poor substrate requires the use of a large amount of enzyme for each rate measurement in order to obtain a reasonable overall rate. The large amount of enzyme that was used gave an observable burst and made it possible to correlate the size of the burst with the amount of enzyme used. With good substrates, the amounts of enzyme that are required to give easily measurable rates would have been too small to give an observable burst. More recently, the burst has been observed with *p*-nitrophenyl esters of good, specific substrates in the presence of large amounts of enzyme by using rapid reaction techniques and by carrying out the reaction at low pH values or very low temperatures, conditions under which the enzyme is much less active.

Covalent catalysis is one of a small number of chemical mechanisms that are known to be utilized by enzymes to accelerate reactions; we will examine other mechanisms in subsequent chapters. The discovery of covalent catalysis is an important experimental milestone in the conceptual development of biochemistry because it imparts chemical reality to the enzyme-substrate complex; it reinforces the optimism of those who believe that catalysis by enzymes can ultimately be explained in chemical terms. For many years, from early in the development of biochemistry to the middle of this century, doubts were expressed even about the existence of enzyme-substrate complexes, because they could not be isolated. It was seriously suggested that, notwithstanding the observation of saturation kinetics, enzymes may exert their actions at a distance from their substrates. Even today, the vitalist tradition reappears periodically in the form of suggestions that an understanding of enzymatic catalysis will require the discovery of new and mysterious forces that are different from those currently known to chemists. Although one cannot deny the possibility of some future discovery of mysterious forces, it is encouraging to find that increasingly sophisticated investigations of enzymatic catalysis turn up increasingly complete chemical pictures of catalysis that invariably can be accommodated by well-established chemical principles. The detection and characterization of the covalent acyl-enzyme intermediate of chymotrypsin is a particularly important example of such a biochemical advance.

Another important early step toward conceptualizing enzymes as definite chemical entities was their crystallization. *Urease,* which catalyzes the hydrolysis of urea, was crystallized in the 1930s; and, a few years after the discovery of burst kinetics, an acyl-enzyme intermediate of chymotrypsin was crystallized. This provided an even stronger chemical basis for covalent catalysis and the concept of the enzyme-substrate complex. The crystallization was accomplished with the relatively stable trimethylacetyl-enzyme at low pH, where the hydrolysis of all acyl-enzyme forms of chymotrypsin is slow. The knowledge that an enzyme-substrate complex could be crystallized and put in a bottle encouraged biochemists to proceed with the examination of the chemical properties of such intermediates.

Criteria for Covalent Catalysis

Three criteria must be satisfied in order to prove that the mechanism of action of an enzyme includes covalent catalysis.

1. *Isolation and characterization.* The covalent enzyme-substrate complex must be isolated, chemically characterized, and shown to be present during turnover.
2. *Chemical reactivity.* It must be shown that the intermediate is formed and reacts further at rates that account for the rate of the overall catalytic process.

3. *Generalization.* When covalent enzyme-substrate intermediates are first observed by the use of poor substrates, it is important to demonstrate that physiological substrates react by the same mechanism.

The importance of the third criterion is exemplified by the use of PNPA as a substrate for chymotrypsin. Although it is a poor substrate, it has a high intrinsic chemical reactivity and might react with the enzyme to form an intermediate that is not on the catalytic pathway for normal substrates. Therefore, it is important to repeat the experiment with specific substrates. Proof by these criteria requires that the enzyme be available in large quantities; that the intermediate be stable enough to be characterized, chemically and kinetically; and that fast reaction instrumentation be available for measuring the rates of formation and further reaction of an intermediate.

Indirect Criteria for Covalent Catalysis

Because of the difficulty in satisfying all of the direct criteria in many cases, the conclusion that an enzymatic reaction proceeds through covalent catalysis is often based in whole or in part on indirect evidence. Studies of chymotrypsin exemplify several of these methods.

The kinetic constants for the chymotrypsin-catalyzed hydrolysis of a series of derivatives of *N*-acetyl-L-tryptophan are shown in Table 4-1. The three esters undergo hydrolysis at essentially the same maximal rate, whereas the amide is hydrolyzed much more slowly and with a much larger K_m value. This difference is similar to that between acetyl-L-tyrosine esters and amides, discussed in Chapter 3.

It is significant that the maximum rates of hydrolysis for the three esters are identical, within experimental error. In nonenzymatic hydrolysis, a *p*-nitrophenyl ester reacts several orders of magnitude faster than an ethyl or methyl ester; so the enzyme must operate by some mechanism that brings about a leveling of these rates. This mechanism is one in which the enzyme promotes the formation of a *common intermediate* and the subsequent reaction of this intermediate is rate determining for all three esters.

There are several possible explanations for the formation of a common intermediate in enzyme-catalyzed reactions. In this reaction, the intermediate is a common acyl-enzyme, as illustrated by the scheme in Figure 4-2. All of the esters react quickly to form the same acyl-enzyme intermediate, AcTrp-E, which

Table 4-1

Kinetic constants for the chymotrypsin-catalyzed hydrolysis of *N*-acetyl-L-tryptophan derivatives

***N*-Acetyl-L-tryptophan-X**	k_{cat} **(s^{-1})**	$K_m \times 10^3$ **(M)**
X = OCH_2CH_3	27	0.097
OCH_3	28	0.095
O–C_6H_4–NO_2 (*p*-nitrophenyl)	31	0.002
NH_2	0.026	7.3

Source: G. W. Schwert and M. A. Eisenberg, *J. Biol. Chem.* 179 (1949): 665.

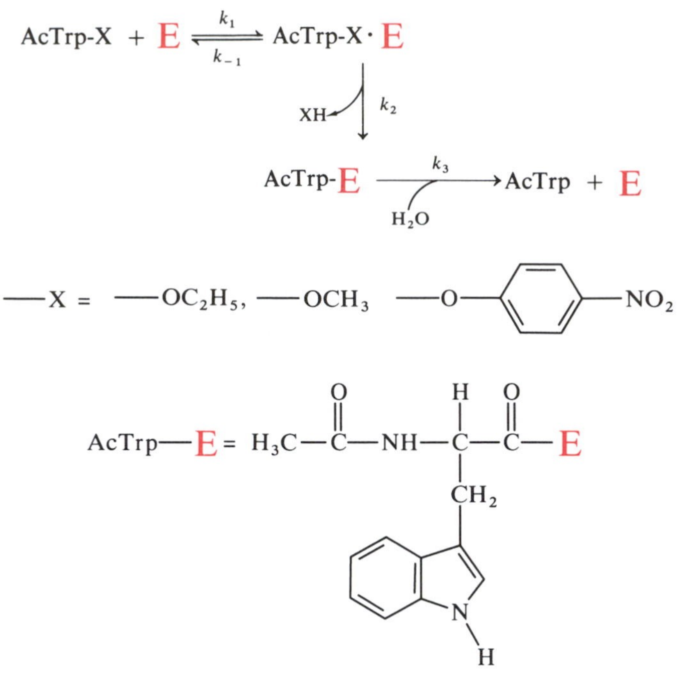

$$\text{AcTrp-X} + \text{E} \underset{k_{-1}}{\overset{k_1}{\rightleftharpoons}} \text{AcTrp-X}\cdot\text{E} \xrightarrow[-\text{XH}]{k_2} \text{AcTrp-E} \xrightarrow[\text{H}_2\text{O}]{k_3} \text{AcTrp} + \text{E}$$

$—X = —OC_2H_5$, $—OCH_3$, $—O—C_6H_4—NO_2$

AcTrp—E = $H_3C—C(=O)—NH—CH(CH_2\text{-indolyl})—C(=O)—E$

Figure 4-2

The Mechanism of Chymotrypsin-catalyzed Hydrolysis of *N*-Acetyl-L-tryptophan Derivatives. All of the tryptophan derivatives in Table 4-1 bind to the enzyme to form Michaelis complexes that differ with respect to the nature of group X. The Michaelis complexes react to form a covalent *N*-acetyl-L-tryptophanyl-enzyme (AcTrp-E) with the release of XH. The common intermediate undergoes hydrolysis with the rate constant k_3 to form the free enzyme and *N*-acetyl-L-tryptophan.

undergoes hydrolysis in the rate-determining step. Therefore, the value of k_{cat} for all the esters is determined by the rate of this second step. A corollary of this rate-determining step is that, for the ester substrates, the E · S complex is almost exclusively in the form of the acyl-enzyme, which accumulates owing to its slow rate of hydrolysis; very little of the noncovalent Michaelis complex is present.

The reaction mechanism in Figure 4-2 provides a more-complete description of the interaction of chymotrypsin with the *N*-acetyl-L-tryptophan substrates in Table 4-1; group X in Figure 4-2 is the leaving group. For the ester substrates, XH is ethanol, methanol, or *p*-nitrophenol; for the amide, it is NH_3. The substrate binds to the enzyme in the k_1 step, which is rapid and reversible (k_{-1} is $> k_2$). The enzyme is acylated, and the leaving group is expelled in the k_2 step; the acyl-enzyme reacts with water to give hydrolysis products in the k_3 step.

The values of K_m and k_{cat} for the kinetic model in Figure 4-2 are given by equations 3 and 4, respectively, which we will not derive here.

$$K_m = \left[\frac{k_{-1} + k_2}{k_1}\right]\left[\frac{k_3}{k_2 + k_3}\right] \tag{3}$$

$$k_{cat} = k_2\left[\frac{k_3}{k_2 + k_3}\right] \tag{4}$$

The esters are very reactive acylating agents so that the rate constant for acylation, k_2, is large—much larger than the rate constant for deacylation, k_3. Equation 4 then simplifies to $k_{cat} = k_3$, so that the deacylation step, k_3, is rate

BOX 4-1

PARTITIONING OF A COMMON INTERMEDIATE

Most of the indirect methods for detecting the existence of covalent catalysis are based on observations of the behavior of a common intermediate that is formed from several different substrates. Catalysis of the hydrolysis of *N*-acetyl-L-tryptophan esters by chymotrypsin illustrates the special case in which the reaction of the common intermediate is the rate-determining step of the overall reaction, so that the same maximum rate is observed when this intermediate is formed from several different esters. More generally useful techniques are based on the *partitioning* of a common intermediate to give two different products, rather than its rate of reaction, because the partitioning of a common intermediate for reaction through two different pathways must be the same under a given set of experimental conditions regardless of whether formation or breakdown of the intermediate is rate determining.

Although there is strong evidence for an acyl-enzyme intermediate in the hydrolysis of esters by chymotrypsin, there is no guarantee that peptides and amides are cleaved by the same mechanism. For many years, the possibility was seriously considered that the hydrolysis of these much less reactive substrates proceeds by the direct attack of water, without the formation of an acyl-enzyme.

The acyl-enzyme mechanism was finally demonstrated by Fastrez and Fersht, who showed that the same partitioning behavior is observed for the common intermediate that is formed in the hydrolysis of *N*-acetyl-L-phenylalanine esters and amides by chymotrypsin.

$$\begin{array}{l} \text{AcPhe-OCH}_3 + \text{E} \\ \text{AcPhe-NHR}^1 + \text{E} \\ \text{AcPhe-NHR}^2 + \text{E} \end{array} \longrightarrow \text{AcPhe-E} \begin{array}{l} \xrightarrow{k_{HOH}} \text{AcPhe-OH} + \text{E} \\ \xrightarrow{k_N[R^3NH_2]} \text{AcPhe-NHR}^3 + \text{E} \end{array}$$

Partitioning of a common acyl-enzyme intermediate between water and 0.05 M alanine amide in the chymotrypsin-catalyzed cleavage of AcPhe-X

Leaving Group	AcPheAlaNH$_2$ Produced (%)
X = —OMe	69
—NH—C_6H_4—N^+Me_3	67
—NH—C_6H_4—NMe_2	63
—NHAlaNH$_2$	67

J. Fastrez and A. R. Fersht, *Biochemistry* 12 (1973): 2025.

The partitioning was measured by determining the ratio of transfer of the acyl group to 0.05 M alanine amide (R^3NH_2 in the adjoining scheme) to give a new peptide, and to water, to give hydrolysis (see the table). The alanine amide, $H_2NCH(CH_3)CONH_2$ ($AlaNH_2$), is an efficient acceptor for the acyl group because it binds to the enzyme in the site that is occupied by the leaving group during the hydrolysis of a peptide. The results in the table show that the same ratio of transfer to water and to the amide acceptor is found regardless of whether the substrate is the methyl ester, an anilide, or a peptide of *N*-acetyl-L-phenylalanine. It is unlikely that these very different substrates would show the same relative reactivities toward water and alanine amide if these reagents were to attack the substrate directly. These results are expected, however, if the same acyl-enzyme is an intermediate in the reactions of all the substrates.

limiting for ester hydrolysis. The value of k_3 is the same for all of the esters, because it represents the hydrolysis of the common intermediate, AcTrp-E.

The fact that the amide substrate, *N*-acetyl-L-tryptophan amide, reacts with a maximal rate that is only one-thousandth that of the esters and has a larger K_m can be explained by the same mechanism with an acyl-enzyme intermediate (Figure 4-2). With the amide substrate, the same acyl-enzyme is formed and undergoes hydrolysis with the same rate constant as the esters.

However, amides are less reactive than esters, and the small value of $k_{cat} = 0.026\ s^{-1}$ for this substrate (Table 4-1) means that the rate constant for *formation* of the acyl-enzyme, k_2, is much smaller than the value of $k_2 = 29\ s^{-1}$ for reactions of esters. Thus, the formation of the acyl-enzyme becomes rate determining for amide substrates of chymotrypsin because the rate constant k_2 for acyl-enzyme formation from the amide is much smaller than k_3, the rate constant for hydrolysis of the acyl-enzyme. Therefore, equation 4 reduces to $k_{cat} = k_2$ for *N*-acetyl-L-tryptophan amide, and for other amides as well. Because k_2 is the rate-determining bottleneck in the reaction of amides, the enzyme piles up behind this step and exists almost entirely as the noncovalent Michaelis complex of AcTrp-$NH_2 \cdot$E when it is saturated with substrate.

Evidence for the formation of a common acyl-enzyme intermediate from several different substrates can also be demonstrated by showing that the intermediate undergoes partitioning to different products and that the same ratio of products is formed when the intermediate is formed from different starting materials. Several examples are described in Box 4-1.

pH-Dependence and Histidine

Having learned that the catalysis by chymotrypsin is an example of covalent catalysis, we are in a position to inquire about the chemistry of the acylation and deacylation steps. The first question is: What is the nature of the covalent bond that is formed between the acyl group and the enzyme? We shall approach this question somewhat indirectly by first considering evidence that implicates the imidazole ring of histidine as an essential catalytic group at the active site. The imidazole group, like several other functional groups in proteins, exhibits both nucleophilic and basic properties. We will see that, in the development of mechanistic models for the action of chymotrypsin, the decision of which functional role is played by histidine is not as simple as it might seem.

The Dependence of Rate on pH

A typical pH-rate profile for the hydrolysis of an uncharged peptide by chymotrypsin is shown in Figure 4-3A. This kind of bell-shaped dependence on pH has been found for many enzyme-catalyzed reactions. Such pH-rate profiles can provide important information about the mechanisms of enzymatic reactions; however, it is usually difficult to interpret this information unambiguously as evidence for a specific catalytic mechanism. The final assessment of the meaning of pH-rate data generally requires additional independent evidence.

The reader should by now be aware that a meaningful statement about the pH-dependence or any other aspect of enzyme activity must specify whether activity at substrate saturation, k_{cat}, or activity at low substrate concentration, k_{cat}/K_m, is meant. It is likely that pH-rate profiles will be different at different substrate concentrations because k_{cat}, K_m, and k_{cat}/K_m usually have different dependencies on pH. The curve in Figure 4-3A is incomplete and could be misleading because it does not specify the concentration of substrate. It could describe the pH-dependence of either k_{cat} or k_{cat}/K_m or some combination of them. It is not unlikely that a particular concentration of substrate at which the rate was measured would saturate the enzyme at one pH value but not at another.

This is, in fact, the case for chymotrypsin. The curve in Figure 4-3A is actually a composite of two curves, as is illustrated in Figures 4-3B and 4-3C. The pH-dependence of the curve for maximum activity, k_{cat} (Figure 4-3B), is

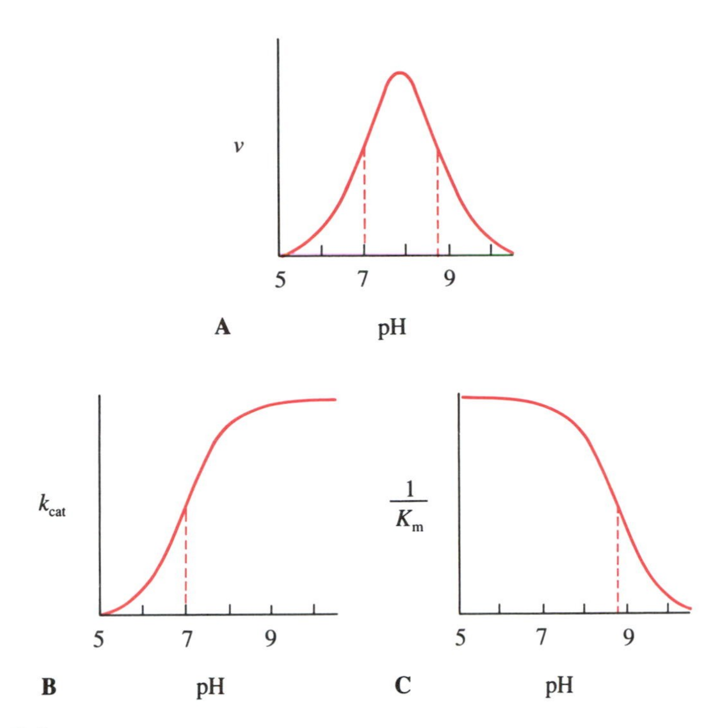

Figure 4-3

Dependence on pH of v (Part A), the Velocity at Low Substrate Concentration, k_{cat} (Part B), and $1/K_m$ (Part C) for Chymotrypsin-catalyzed Reactions.

sigmoid and resembles the titration curve of an acid with a pK_a of approximately 7; at high pH, the *maximum* rate becomes independent of the pH. The sigmoid pH-dependence of $1/K_m$ is the converse of that for k_{cat} and arises from the ionization of an acid with a pK_a of approximately 8.7 (Figure 4-3C). For the hydrolysis of peptides by chymotrypsin, K_m is a true dissociation constant, equal to K_s, so that the pK_a of 8.7 for $1/K_m$ describes an ionization that controls the binding of substrate by the enzyme to form the E·S complex; we will return to this pK_a in Chapter 5.

The dependence of k_{cat} on pH ordinarily arises from the ionization of a group in the E·S complex that is required for activity. This group is usually part of the active site of the enzyme, but it could also be a group on the bound substrate. For chymotrypsin, the pH-dependence of k_{cat} shows that the rate at which the E·S complex reacts is maximal when a group with a pK_a of 7 is in the unprotonated form and that the enzyme is inactive when this group is protonated. This pH-dependence is described by the mechanistic scheme in Figure 4-4, in which $K_a = [ES][H^+]/[ESH^+] = 10^{-7}$ M.

$$\begin{array}{ccc} E\cdot S & \xrightarrow{k_{cat}} & \text{Products} \\ \pm H^+ \Big\updownarrow K_a & & \\ {}^+HE\cdot S & & \end{array}$$

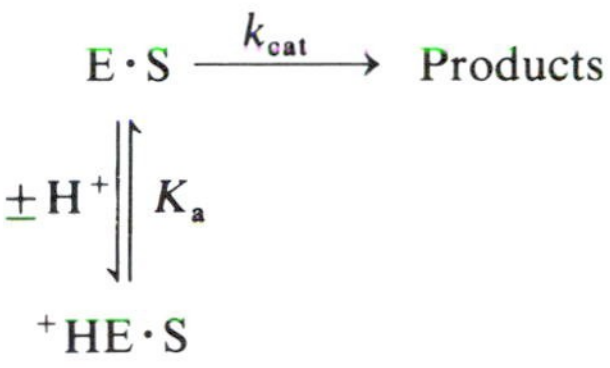

Figure 4-4

A Kinetic Scheme That Explains the pH-Dependence of k_{cat} for Chymotrypsin. The kinetic model shown rationalizes the k_{cat}-pH profile in Figure 4-3 by postulating that the enzyme-substrate complex contains a functional group that ionizes and must be in its unprotonated form for the enzyme to exhibit activity. The enzyme form $^+$HE·S is inactive, whereas the form E·S undergoes reaction to form products. K_a is the acid dissociation constant for $^+$HE·S.

The dependence of k_{cat}/K_m on pH ordinarily arises from the ionization of a group in the free enzyme or the substrate that is required for activity. It is easy to distinguish these possibilities because the pK_a values for ionization of the substrate can be measured by titration. The pH-rate profile for peptide hydrolysis by chymotrypsin that is shown in Figure 4-3A in fact describes the dependence of k_{cat}/K_m on pH. At low pH values, it shows the increase in k_{cat} with increasing pH; at high pH, it shows the increase in K_m (or decrease in $1/K_m$) with increasing pH. This pH-dependence can be assigned to ionizations of the free enzyme because the peptide substrates do not ionize in this pH region.

Because k_{cat}/K_m is a second-order rate constant that pertains only to the free enzyme and substrate, its dependence on pH can be described by the kinetic

$$\begin{array}{c} \text{E (inactive)} \\ \pm H^+ \Big\Updownarrow K_1 \\ EH^+ + S \xrightarrow{k_{cat}/K_m} \text{Products} \\ \pm H^+ \Big\Updownarrow K_2 \\ EH_2^{2+} \text{ (inactive)} \end{array}$$

Figure 4-5

A Kinetic Scheme That Explains the pH-Dependence of k_{cat}/K_m for Chymotrypsin. This kinetic model refers to the data in Figure 4-3C, the pH dependence of $1/K_m$. In this scheme, the enzyme has two ionizing functional groups that control its activity. Form EH^+ binds the substrate and is active, whereas form E does not bind the substrate and so cannot be active. It is this ionization, with the acid dissociation constant K_1, that controls binding and accounts for the $1/K_m$-pH profile. Enzyme form $EH_2^{\ +}$ also is inactive; and, because its acid dissociation constant is the same as that for the k_{cat}-pH profile in Figure 4-3B, the two ionizations are thought to represent the acid-base behavior of the same group ($^+HE \cdot S$ in Figure 4-4).

scheme in Figure 4-5, in which $K_1 = 10^{-8.7}$ M and $K_2 = 10^{-7}$ M. The value of K_2 is the same as K_a for the E·S complex and presumably represents the ionization of the same group with a pK_a of 7.0. The generalizations that the pH-dependence of k_{cat} arises from ionizations of E·S and that the pH-dependence of k_{cat}/K_m arises from ionizations of free E and S hold for the great majority of enzymatic reaction schemes, with a few exceptions for complex kinetic mechanisms.

The dependence on pH of $1/K_m = 1/K_s$ for catalysis of peptide hydrolysis by chymotrypsin (Figure 4-3C) arises from the ionization of a group on the free enzyme that is required for binding the substrate. Binding is maximal when this group is protonated and falls off with deprotonation at pH values above the pK_a of 8.7. However, maximal enzyme activity (V_m) can be obtained by increasing the substrate concentration to a sufficiently high level to overcome the unfavorable K_m, as shown by the pH-independent curve for k_{cat} at high pH.

In the general case, in which K_m is a complex constant composed of rate constants as well as equilibrium constants (as in the chymotrypsin-catalyzed hydrolysis of esters), the pH-dependence of K_m is likely to include contributions from the pH-dependence of these kinetic constants. Because of this complication, it is usually desirable to consider the pH-dependence of the simpler constants, k_{cat} and k_{cat}/K_m before attempting to analyze the effects of pH on K_m.

The Dependence of k_{cat} on pH

We shall now consider how to interpret the dependence of k_{cat} on pH; we will return in Chapter 5 to the ionization of the free enzyme (pK_a = 8.7) that controls substrate binding. Because the maximum activity of the enzyme follows the titration curve of an acid with a pK_a of 7 and increases with increasing pH, we can say that the rate of the rate-determining step is proportional to the fraction of some ionizable group with a pK_a of 7 that is in the basic form in the enzyme-substrate complex (Figure 4-4). We can assign this ionization to a group in the enzyme that is required for catalysis and is probably at the active site, because the uncharged substrate has no ionizations in this pH range.

The most likely candidate for this pK_a is the protonated imidazole group of a histidine residue (equation 5); this group, in either histidine or peptides of histidine, generally exhibits a pK_a between 6 and 7.

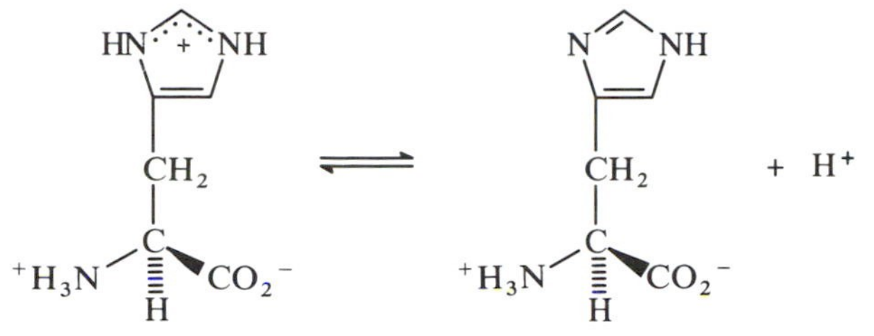

$$\text{HN}\overset{+}{}\text{NH (imidazolium)}-CH_2-C(H)(^{+}H_3N)(CO_2^-) \rightleftharpoons \text{N}\text{NH (imidazole)}-CH_2-C(H)(^{+}H_3N)(CO_2^-) + H^+ \qquad (5)$$

L-Histidine
(His, H)

A less likely candidate is the protonated amino group of an N-terminal amino acid, which usually has a pK_a near 8 (equation 6), or a carboxylate group, which normally has a pK_a near 5.

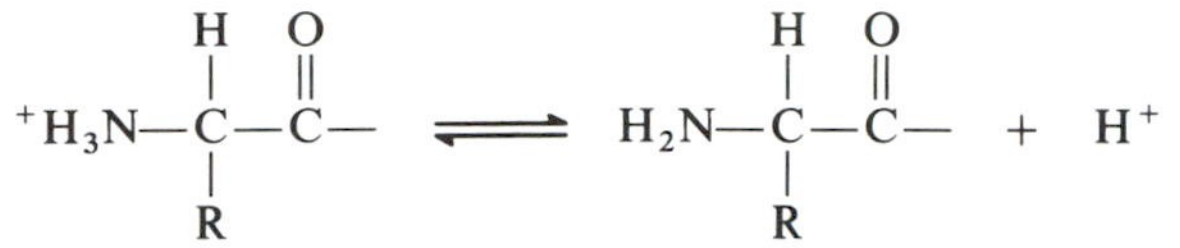

$$^{+}H_3N-\underset{R}{\overset{H}{C}}-\overset{O}{\overset{\|}{C}}- \rightleftharpoons H_2N-\underset{R}{\overset{H}{C}}-\overset{O}{\overset{\|}{C}}- + H^+ \qquad (6)$$

We can conclude that the basic form of an imidazole group in the active site of chymotrypsin probably plays a role in catalysis; that is, in the formation of an acyl-enzyme from a peptide substrate. Other evidence confirms this conclusion for chymotrypsin, but in general it is not possible to definitively identify a group at the active site from a pH-rate profile alone. There are a number of ways in which a pK_a value can be perturbed by the microenvironment, and other factors can influence the interpretation of pH-rate profiles; some of these are described in Box 4-2.

BOX 4-2

INTERPRETATIONS OF pH-RATE PROFILES

There are several reasons that the interpretation of pH-rate profiles may not be easy even if the pH-dependencies of K_m and k_{cat} are separated.

First, the imidazole may not participate directly in the catalytic process, but protonation of the imidazole ring may cause a conformational change that makes the enzyme catalytically inactive.

Second, the dissociation with a pK_a of 7 might represent the perturbed dissociation of a group that normally has a higher or lower pK_a. The protonated 6-amino group of a lysine residue, for example, ordinarily dissociates with a pK_a of 10.4, but the pK_a of this group in the active site of the enzyme acetoacetate decarboxylase is reduced to 6.0. This decrease in pK_a is caused by electrostatic destabilization of the perturbed lysine-6-ammonium group by adjacent positive charges, which favors dissociation of the proton:

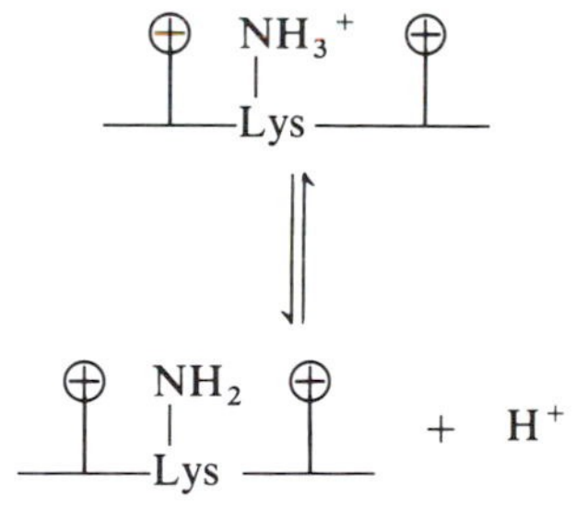

$$\oplus \quad NH_3^+\text{-Lys} \quad \oplus \rightleftharpoons \oplus \quad NH_2\text{-Lys} \quad \oplus + H^+$$

Alternatively, a pK_a of 7 could represent the ionization of a carboxylic acid group that has its pK_a perturbed upward by a poor ion-solvating environment or by adjacent negative charges. This has been shown to occur in several enzymes. (*Continued on next page.*)

BOX 4-2 (continued)

INTERPRETATIONS OF pH-RATE PROFILES

Third, every pH-rate profile that suggests a requirement for a particular group has a kinetically ambiguous interpretation that involves the conjugate acid (or base) of that group and some other ionization. For example, the activity of chymotrypsin might require not imidazole base, but instead the imidazolium ion along with the basic form of a very weak acid, such as water or an alcohol (ROH). The rate law and hence the pH-rate profile for this alternative interpretation are identical with those for imidazole and free ROH (see the equation below). The identity follows simply from inserting the ionization constants $K_{\mathrm{ImH^+}}$ and K_{ROH} for imidazolium ion and ROH into the rate law. There is *no way* in which these two interpretations can be distinguished by simple kinetic methods; however, there are other reasons for rejecting this interpretation for chymotrypsin.

$$v = k[\mathrm{ImH^+}][\mathrm{RO^-}] = k\frac{K_{\mathrm{ROH}}}{K_{\mathrm{ImH^+}}}[\mathrm{Im}][\mathrm{ROH}]\frac{[\mathrm{H^+}]}{[\mathrm{H^+}]}$$

$$= k'[\mathrm{Im}][\mathrm{ROH}]$$

$$K_{\mathrm{ImH^+}} = \frac{[\mathrm{Im}][\mathrm{H^+}]}{[\mathrm{ImH^+}]} \qquad K_{\mathrm{ROH}} = \frac{[\mathrm{RO^-}][\mathrm{H^+}]}{[\mathrm{ROH}]}$$

Fourth, the apparent pK_a of the E·S complex may not represent the ionization of a group at all, but may instead represent a change in the rate-determining step. An example of this is found in the nonenzymatic reaction of hydroxylamine with acetone to form acetoxime, which follows an apparent pK_a at pH 3.2 (red line in Figure A). The rate constant, k, refers to the concentration of NH_2OH (not NH_3OH^+). However, none of the reactants undergo ionization near pH 3.2.

The explanation comes from the different dependence on pH of the two steps in the reaction.

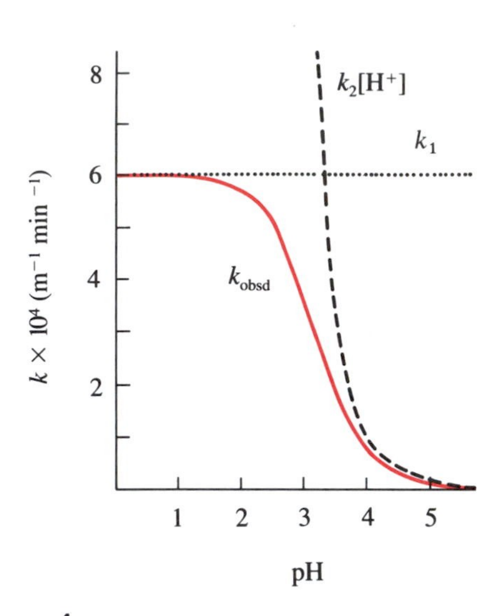

Figure A

The effect of pH on the formation of acetoxime from acetone and hydroxylamine. The red line is the observed pH-rate profile, with an apparent pK_a at pH 3.2. The dashed black line is the rate constant of the second step. The observed rate levels off at low pH values because attack of free hydroxylamine, with the rate constant k_1, becomes rate limiting. The observed rate constant, k, refers to hydroxylamine as the free base.

At low pH, the attack of free hydroxylamine on acetone is rate determining (dotted line in the pH-rate profile). At high pH, the rate-limiting step is the acid-catalyzed dehydration of the addition compound. The rate increases as the pH decreases until almost every molecule of addition compound that is formed undergoes rapid dehydration ($k_2[\mathrm{H^+}] > k_{-1}$) and the k_1 step becomes rate limiting.

$$\mathrm{HO{-}NH_2} + \mathrm{C{=}O} \underset{k_{-1}}{\overset{k_1}{\rightleftharpoons}} \mathrm{HO{-}NH{-}C{-}OH} \xrightarrow{k_2[\mathrm{H^+}]} \mathrm{HO{-}N{=}C} + \mathrm{H_2O}$$

Identification of Histidine in the Active Site

The best chemical evidence for the presence of a histidine residue in the active site came from the irreversible inactivation of chymotrypsin by *p*-toluenesulfonyl phenylalanine chloromethyl ketone (TPCK). This compound is a structural analog of an *N*-acylphenylalanine substrate, so that it is recognized by the enzyme and binds specifically at the active site like the normal substrate.

This specificity is shown by the fact that inactivation by TPCK is prevented by the presence of specific substrates or competitive inhibitors. 2-Chloroketones are active alkylating agents, so that TPCK can act as a "Trojan horse." It binds at the active site, in disguise as a substrate, and then alkylates a nearby nucleophile to block the site and destroy enzyme activity.

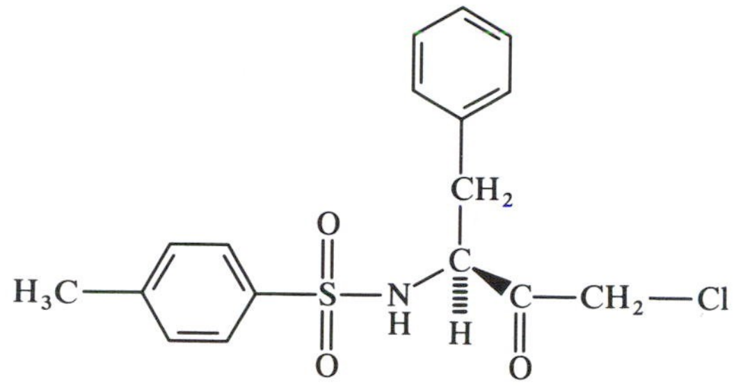

p-Toluenesulfonyl phenylalanine chloromethyl ketone (TPCK)

Inactivation of chymotrypsin by radioactive [^{14}C]TPCK gave a radioactive protein containing one mole of covalently bound [^{14}C]TPK per mole of enzyme. Chemical degradation of this radioactive enzyme by a combination of oxidative and hydrolytic procedures gave a radioactive [^{14}C]carboxymethylhistidine derivative that was formed from the histidine residue at position 57 of the peptide chain.

This type of *active-site-directed inactivator* is particularly useful because it takes advantage of the specific binding affinity of an enzyme toward its substrates. This affinity causes the inactivator to react with, and therefore to label, a specific group at the active site. Other chemically similar groups in other parts of the enzyme react much more slowly. Reagents of this kind are also important tools for examining the chemistry of the binding sites of many kinds of receptor proteins. They are of particular interest in physiology and medicine because they provide a method for selectively inactivating a particular enzyme or receptor without affecting other proteins.

All of this chemical evidence that implicates an imidazole group at the active site has been confirmed by determination of the X-ray crystal structure of chymotrypsin. Structure determinations of the enzyme and of complexes of the enzyme with substrate analogs show directly that the imidazole group of histidine-57 is at the substrate-binding site.

The Nature of the Acyl-Enzyme Intermediate

We now have evidence that establishes beyond a reasonable doubt that there is an imidazole group (His-57) in the active site of chymotrypsin and that protonation of this group is responsible for the pK_a of 7 for k_{cat} in the pH-rate profile of chymotrypsin. Could this imidazole be the nucleophilic group in chymotrypsin that reacts with the substrate to form the acyl-enzyme? Chemical evidence suggests that imidazole might play such a catalytic role. It is often useful to examine chemical reactions in aqueous solution that may serve as models for enzyme-catalyzed reactions in order to characterize the chemistry of reactions that may be taking place at the active site of the enzyme. However, model reactions must be evaluated with skepticism because they may, in fact, have little or no relation to the enzyme-catalyzed reaction. We shall review the chemical properties of imidazole as a nucleophilic catalyst and then show that, although imidazole is a good catalyst for reactions of reactive substrates, it is *not*, in fact, the nucleophilic catalyst in chymotrypsin.

Imidazole as a Nucleophilic Catalyst

The imidazole molecule is an effective catalyst of the hydrolysis of PNPA and other reactive acyl compounds with good leaving groups. Imidazole reacts rapidly with PNPA to give *N*-acetylimidazole as an intermediate, which subsequently undergoes hydrolysis (Figure 4-6). The reaction shows a burst of PNP^- release and follows the same dependence on pH as k_{cat} for the enzymatic reaction. Thus, imidazole catalyzes the same reaction of PNPA as does chymotrypsin, also by covalent catalysis; the reaction with imidazole appears to be a "model reaction."

Two chemical properties of imidazole are responsible for its special competence as a nucleophilic catalyst. First, it is a highly reactive nucleophile toward activated esters such as PNPA, many orders of magnitude more reactive than water, the ultimate acyl acceptor in hydrolytic reactions. Second, the intermediate acylimidazoles react with water several orders of magnitude faster than does PNPA itself. Very few nucleophiles exhibit both of these properties, so that imidazole is an unusually effective nucleophilic catalyst.

Furthermore, both imidazole and chymotrypsin catalyze the transfer of acyl groups to acceptors other than water by the mechanism shown in Figure 4-7. In these reactions, the same acyl intermediates that are formed in hydrolysis—the acyl-enzyme in the case of chymotrypsin and acylimidazole with imidazole—react with other nucleophilic molecules to form stable products. Thus, both imidazole and chymotrypsin catalyze the reactions of activated esters with amines to form amides (Figure 4-7), with alcohols to give other esters, and with thiols to give thioesters.

These properties of imidazole, as well as the presence of histidine-57 in the active site of chymotrypsin, led several investigators to suggest that histidine-57 is the nucleophilic catalyst in the enzyme. In fact, it is not. This exemplifies the importance of the criteria presented earlier in this chapter for establishing

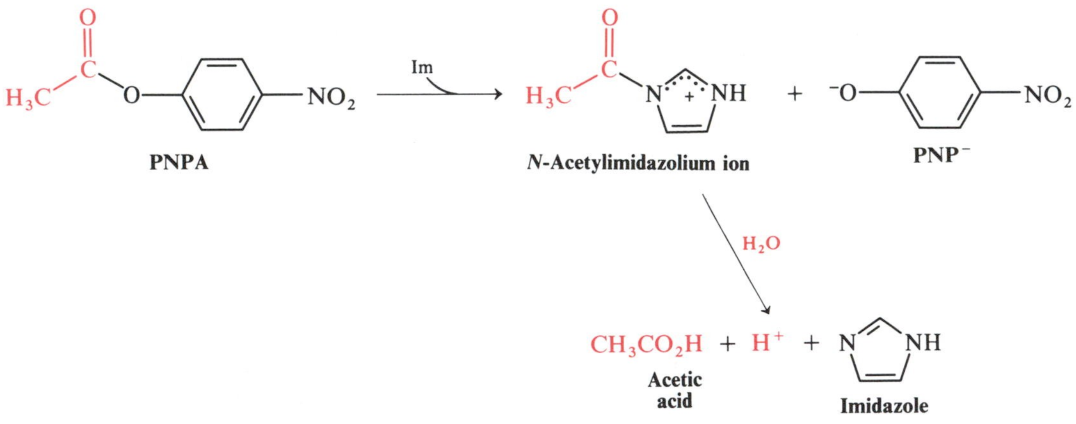

PNPA
Im
N-Acetylimidazolium ion
PNP^-
H_2O
$CH_3CO_2H + H^+ +$
Acetic acid
Imidazole

Figure 4-6

The Mechanism of Imidazole-catalyzed Hydrolysis of PNPA. Imidazole-catalyzed hydrolysis of PNPA is kinetically and mechanistically similar to hydrolysis of PNPA by chymotrypsin. Imidazole acts as a nucleophilic catalyst that reacts with PNPA in the first step to form PNP^- and a covalent intermediate, which undergoes hydrolysis in the second step.

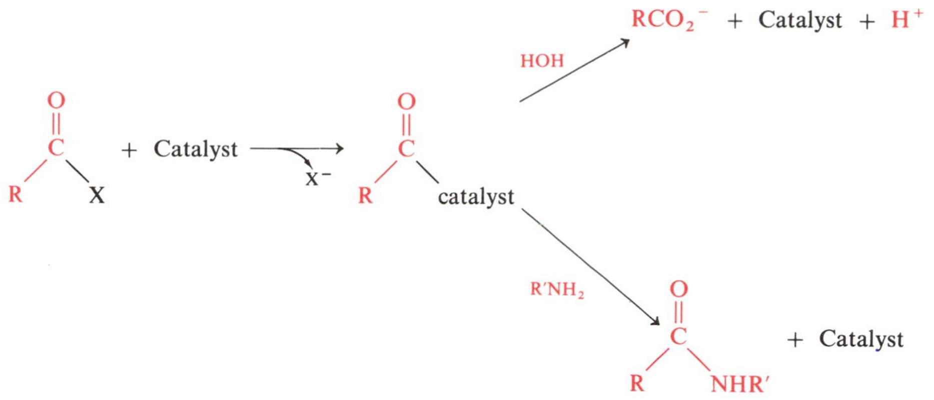

$$R\text{—}\overset{O}{\overset{\|}{C}}\text{—}X + \text{Catalyst} \xrightarrow{-X^-} R\text{—}\overset{O}{\overset{\|}{C}}\text{—catalyst}$$

$$R\text{—}\overset{O}{\overset{\|}{C}}\text{—catalyst} \xrightarrow{HOH} RCO_2^- + \text{Catalyst} + H^+$$

$$R\text{—}\overset{O}{\overset{\|}{C}}\text{—catalyst} \xrightarrow{R'NH_2} R\text{—}\overset{O}{\overset{\|}{C}}\text{—}NHR' + \text{Catalyst}$$

Figure 4-7

Partitioning of Acyl Intermediates in Hydrolysis Catalyzed by Chymotrypsin and Imidazole. Both chymotrypsin and imidazole catalyze the hydrolysis of activated acyl compounds by displacing the leaving group, —X, to form an acyl-catalyst intermediate. Hydrolysis of the intermediate to the acid, with regeneration of the catalyst, takes place in the second step. The intermediate in both cases can be partitioned into two products by including an amine in the reaction mixture. The amine competes with water to accept the acetyl group from the intermediate.

covalent catalysis—the intermediate must be isolated and chemically characterized, and it must be shown that it is formed with normal substrates. The next section will show that the isolated intermediate *does not* contain an acylimidazole. In fact, imidazole has not been found to be a nucleophilic catalyst in any acyl-transferring enzyme. It is a nucleophilic catalyst in several enzymes that catalyze phosphoryl group transfer, which will be described in Chapter 11.

Serine-195 as the Nucleophilic Catalyst

The first clue to the discovery of the chemical structure of the acyl-enzyme intermediate came, surprisingly, from studies of war gases and insecticides. One class of these compounds is a series of reactive phosphorylating agents that exhibit widely varying degrees of toxicity toward mammals and insects, depending on their mode of absorption into the organism and other factors. The best known of these compounds is diisopropylphosphorofluoridate (DFP), but many other uncharged, fully esterified phosphate or phosphonate derivatives have similar toxicities. These compounds exert their toxic effects by inactivating acetylcholinesterase.

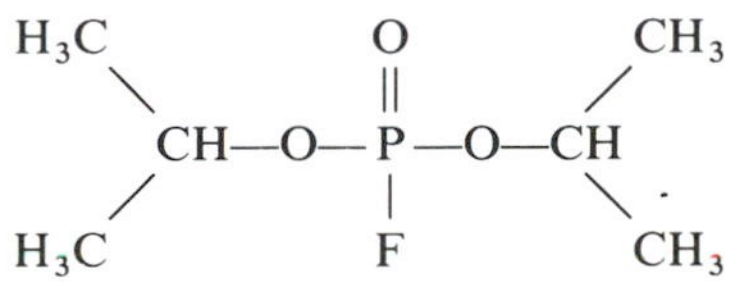

$$(H_3C)_2CH\text{—}O\text{—}\underset{F}{\underset{|}{\overset{O}{\overset{\|}{P}}}}\text{—}O\text{—}CH(CH_3)_2$$

Diisopropylphosphorofluoridate (DFP)

$$H_3C\text{—}\overset{O}{\overset{\|}{C}}\text{—}O\text{—}CH_2\text{—}CH_2\text{—}\underset{CH_3}{\underset{|}{\overset{CH_3}{\overset{|}{N^+}}}}\text{—}CH_3$$

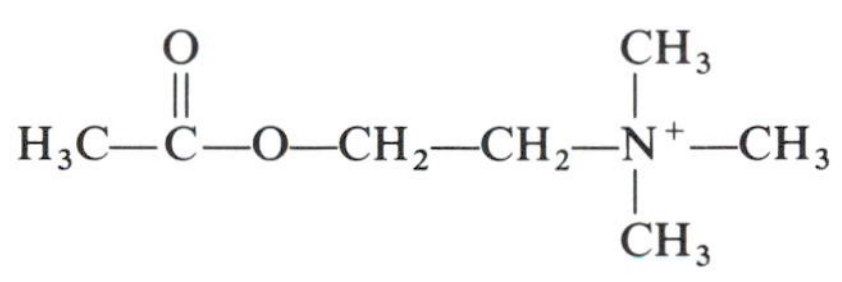

Acetylcholine

Acetylcholine is one of the several neurotransmitter substances that are stored in vesicles at nerve endings and released into the *synapse,* or neuromuscular junction, when an action potential reaches the nerve ending. The molecules of acetylcholine then combine either with a specific acetylcholine receptor to stimulate some further process or with acetylcholinesterase to undergo hydrolysis. The extremely fast cleavage of acetylcholine by this enzyme, which is essentially diffusion controlled, ensures that the duration of the stimulus will be short. If acetylcholinesterase is inhibited, cleavage of acetylcholine is prevented, and the receptor continues to respond. This quickly leads to a severe physiological imbalance resulting in death unless an antidote that antagonizes the action of acetylcholine, such as atropine, is administered.

The mechanisms of action of acetylcholinesterase and chymotrypsin are closely related, and both enzymes are inactivated by DFP, although inactivation of chymotrypsin does not have such drastic physiological consequences. This inactivation, like that by specific chloroketones, is stoichiometric, covalent, and retarded by the presence of substrates or reversible inhibitors. It is also difficult, though not impossible, to reverse. The reaction with DFP that is labeled with ^{32}P leads to the incorporation of one equivalent of radioactive inhibitor into the covalent structure of the protein (equation 7);

$$\text{iPrO—P(=O)(OiPr)—F} + \text{HO-Ser-E} \longrightarrow \text{iPrO—P(=O)(OiPr)—O-Ser-E} + F^- + H^+ \qquad (7)$$

CH_2—OH, C, ^+H_3N, H, CO_2^-

L-Serine (Ser, S)

CH_2—$O^{32}PO_3^{2-}$, C, ^+H_3N, H, CO_2^-

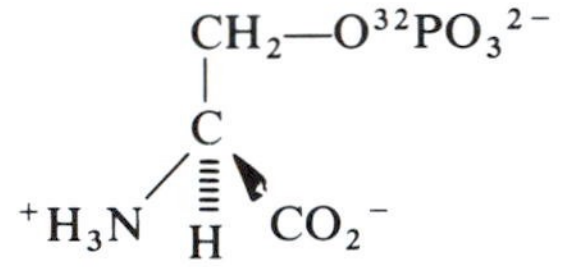

[^{32}P]Phosphoserine

this radioactive protein may be isolated and chemically degraded. The site of phosphorylation in both enzymes has been identified as the hydroxyl group of a serine residue through isolation and identification of radioactive [^{32}P]phosphoserine after hydrolysis of the inactivated enzyme. Thus, inactivation of chymotrypsin by DFP proceeds by formation of a hydrolytically stable diisopropylphosphoryl-enzyme from the reaction of DFP with a serine hydroxyl group (equation 7). By analogy, normal substrates probably react with the same serine group to form the corresponding acyl-enzyme species, which are labile to hydrolysis and undergo cleavage to products.

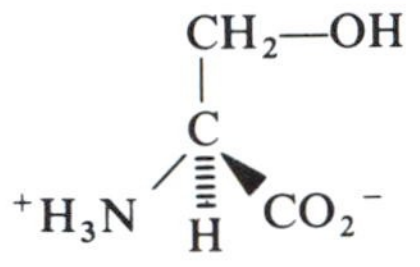

This analogy is valid. Characterization of the acetyl-enzyme that is formed during the reaction of an acetyl ester is complicated by the fact that it cannot be subjected to hydrolytic degradation in acid without loss of the acetyl group. However, a peptide containing *O*-acetylserine can be isolated by digesting the acetyl-chymotrypsin intermediate with a proteolytic enzyme under mild conditions. Furthermore, the X-ray crystal structure of chymotrypsin shows that serine-195 in the peptide chain of chymotrypsin is located in the active site, with its hydroxyl group near the imidazole group of histidine-57 and the hydrophobic specificity pocket.

The proximity of serine-195 and histidine-57 within the active site, in contrast with the large distance between them in the amino acid sequence, perfectly exemplifies the importance of the three-dimensional structure of the enzyme, which brings together amino acid functional groups of several peptide segments through a specific folding motif for the peptide chain.

The importance of the three-dimensional structure at the active site is demonstrated by the experiment illustrated in Figure 4-8. The native acetyl-enzyme undergoes rapid hydrolysis at neutral pH as a consequence of the properties of the active site that make chymotrypsin an effective catalyst. However, chymotrypsin, and proteins in general, can be unfolded into a denatured state by the addition of concentrated urea, which destabilizes the folded peptide and disrupts the active-site structure. The unfolded acyl-enzyme loses its special reactivity and undergoes hydrolysis at the same slow rate as an ordinary acetyl ester. When the urea is dialyzed away, the enzyme spontaneously refolds to its native, active state and the acyl-enzyme regains its full

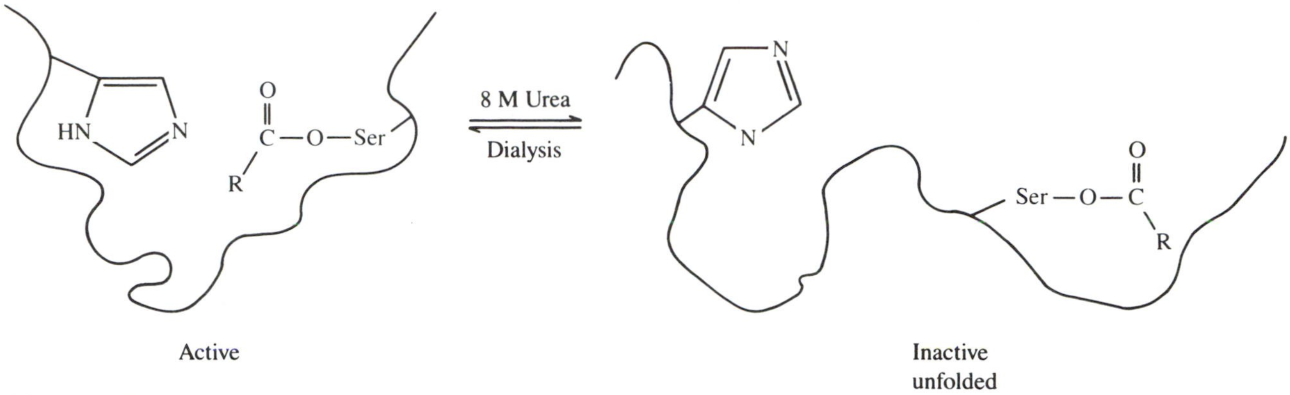

Active

8 M Urea

Dialysis

Inactive unfolded

Figure 4-8

Denaturation of an Acyl-Enzyme of Chymotrypsin in Urea Causes Loss of Its Catalytic Ability. Urea causes unfolding of the peptide chain. Removal of urea by dialysis allows refolding of the enzyme and restores activity.

reactivity. This experiment proves that the special properties of the active site not only are a consequence of the chemistry of the peptide chain itself, but also require folding of the chain into the correct conformation with imidazole next to the serine. The almost identical rate of hydrolysis of the acetyl-enzyme, in the presence of urea, with that of a simple acetate ester confirms that the acetyl-enzyme is indeed a serine ester; it is not an artifact that is brought about by migration of the acetyl group from imidazole to serine under the conditions employed for degrading the acetyl-enzyme to an acetylserine peptide. Acetylimidazole undergoes hydrolysis one thousand times as fast as *O*-acetylserine and denatured acetyl-chymotrypsin.

General Acid-Base Catalysis and Histidine-57

If the hydroxyl group of serine-195 is the active-site nucleophile for covalent catalysis by chymotrypsin, what then is the role of the imidazole group in histidine-57? The function of imidazole is to act as a catalyst for the proton transfers that must take place in the hydrolysis of a peptide; that is, as a general acid-base catalyst. If an acyl-enzyme were to be formed directly from a peptide and the serine hydroxyl group without proton transfer, the immediate product would be an amine anion and an O-protonated ester (equation 8).

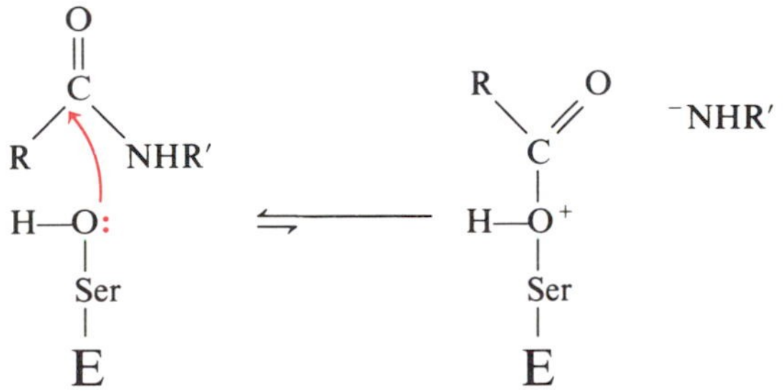

$$R{-}C(=O){-}NHR' + H{-}\ddot{O}{-}Ser{-}E \rightleftharpoons R{-}C(=O){-}O^{+}(H){-}Ser{-}E \quad {}^{-}NHR' \qquad (8)$$

Such species are so unstable that they could not possibly be formed as intermediates. If they should form, they would instantly react to regenerate the reactants. Their instabilities are apparent from the pK_a value for the ionization of an amine to an anion, which is approximately 35, and that for the dissociation of a proton from a protonated alcohol, which is less than zero. The imidazole group of histidine-57 can avoid the formation of these unstable species: initially, by acting as a proton acceptor to avoid formation of the O-protonated ester and, then, by donating the proton to the leaving amine to avoid the formation of the amine anion.

The phenomenon of general acid-base catalysis is often puzzling to those who have not yet encountered it because it is not consistent with the descriptions of acid and base catalysis that are usually presented in elementary chemistry courses. Most of us learned that the rate of an acid- or base-catalyzed reaction is determined by the concentration of hydrogen ion or hydroxide ion but is independent of the total buffer concentration, because the pH of a buffer solution does not change appreciably with increasing buffer concentration when the ratio of the acid and base components of the buffer is held constant. This is shown in Figure 4-9A for a base-catalyzed reaction that takes place in imidazole buffers at pH 7.0 and 7.3. The rate is twice as fast at pH 7.3 as at pH 7.0 because the concentration of hydroxide ion is twice as large at pH 7.3; and the rates at both pH values are constant at increasing buffer concentration. This reaction follows the rate law of equation 9 and is said to be *specific base catalyzed*; that is, the rate depends on the concentration of the specific base HO^- and not on the concentration of any other base in the solution. A reaction that depends only on the concentration of hydrogen ions (equation 10) is said to be *specific acid catalyzed.*

$$v = k[\text{substrate}][\text{HO}^-] \tag{9}$$

$$v = k[\text{substrate}][\text{H}^+] \tag{10}$$

The surprising thing is that the rates of some base-catalyzed reactions do not remain constant at a given pH value but increase as the buffer concentration is increased. This is shown in Figure 4-9B for a different base-catalyzed reaction in the presence of imidazole buffers. The reaction is the hydrolysis of *N,O*-diacetylserinamide, a serine ester with blocked amino and carboxyl groups that serves as a model for the acetylserine intermediate in the chymotrypsin-catalyzed hydrolysis of *p*-nitrophenyl acetate. The rate of hydrolysis of this ester increases more rapidly at the higher pH value, at which more of the imidazole buffer is present as imidazole free base, and the slopes of the lines are directly proportional to the fraction of the basic species of imidazole in the buffer. This result shows that the rate of the reaction depends not only on the concentration

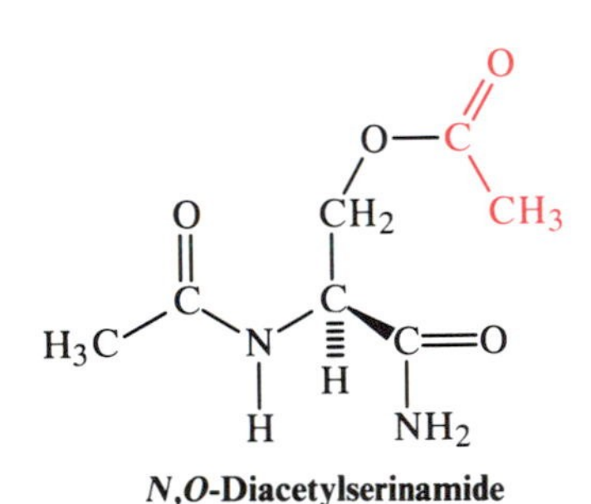

***N,O*-Diacetylserinamide**

Specific base catalysis | General base catalysis

Rate | pH 7.3 | pH 7.0

A [Imidazole buffer]

Rate | pH 7.3 | pH 7.0

B [Imidazole buffer]

Figure 4-9

The Dependence of Reaction Rates on the Concentration of Buffer for Specific Base Catalysis (A) and for General Base Catalysis (B).

of the specific base, hydroxide ion, but also on the concentration of the base imidazole. The observed rate of the reaction can be described by the rate law of equation 11.

$$v = k_0[\text{ester}] + k_1[\text{ester}][\text{HO}^-] + k_{\text{Im}}[\text{ester}][\text{Im}] \tag{11}$$

The reaction is also catalyzed by the basic species of other buffers, so that we must conclude that it is catalyzed by buffers in general. That is, it is *general base catalyzed.* The contributions of water (the k_0 term) and hydroxide ion (the k_1 term) to the observed rate are determined from the intercepts at zero buffer concentration for different pH values, as shown in Figure 4-9B.

Similarly, a reaction that proceeds at a rate that is proportional to the concentration of the acidic species of the buffer is said to be *general acid catalyzed.* The rate law for a reaction with general acid catalysis by imidazole buffers would include the term:

$$k_{\text{ImH}^+}[\text{substrate}][\text{ImH}^+]$$

General acid-base catalysis is the second general mechanism by which enzymes can bring about catalysis. General acid or base catalysis has little or no significance for the synthetic chemist, who can add strong acids or hydroxide ion to reaction mixtures to bring about the necessary proton transfers; but both are critical to enzymes because there is no way to bring about a large change in the concentration (or activity) of hydrogen or hydroxide ions at an active site under ordinary physiological conditions. The existence of general acid-base catalysis is indispensable for the action of enzymes at neutral pH, because it provides a mechanism for bringing about the necessary proton transfers without utilizing hydrogen or hydroxide ions, which exist at concentrations of only about 10^{-7} M under physiological conditions. Such low concentrations of these ions could not account for the observed rates of many enzyme-catalyzed reactions that require proton transfer, even if the proton transfer were diffusion controlled.

The functional groups of enzymes that can act as acid or base catalysts include the —COO^- groups of aspartate and glutamate (pK_a ~ 4.5), the imidazole group of histidine (pK_a = 7), the —SH group of cysteine (pK_a = 8–9), the protonated amino groups of lysine (pK_a = 10.7), and the N-terminal amino group of the peptide chain (pK_a = 8). Of these groups, imidazole is the best fitted for a catalytic role because of its pK_a of 7—a weaker base would be less reactive, whereas a stronger base would be fully protonated at pH 7 and therefore less readily available at physiological pH. Similarly, the imidazolium ion with its pK_a of 7 is the strongest acid and most-effective general acid catalyst at neutral pH. These statements are generally true; nonetheless, our earlier cautionary statements about microenvironmental effects on the pK_a's of active-site residues apply. In particular, it is common for carboxyl groups in proteins to exhibit values of pK_a near 6 or 7 because of the low polarity of their microenvironment in the active site.

CH_2—SH, C, ^+H_3N, CO_2^-, H

L-Cysteine
(Cys, C)

The Mechanism of Chymotrypsin Acylation

A reasonable mechanism for the action of chymotrypsin that utilizes imidazole as both a base and an acid catalyst is shown in Figure 4-10. As the hydroxyl group of serine attacks the peptide bond in step 1, the imidazole group of histidine accepts the proton from serine, which avoids the formation of an unstable O-protonated intermediate. The attack by serine probably gives a

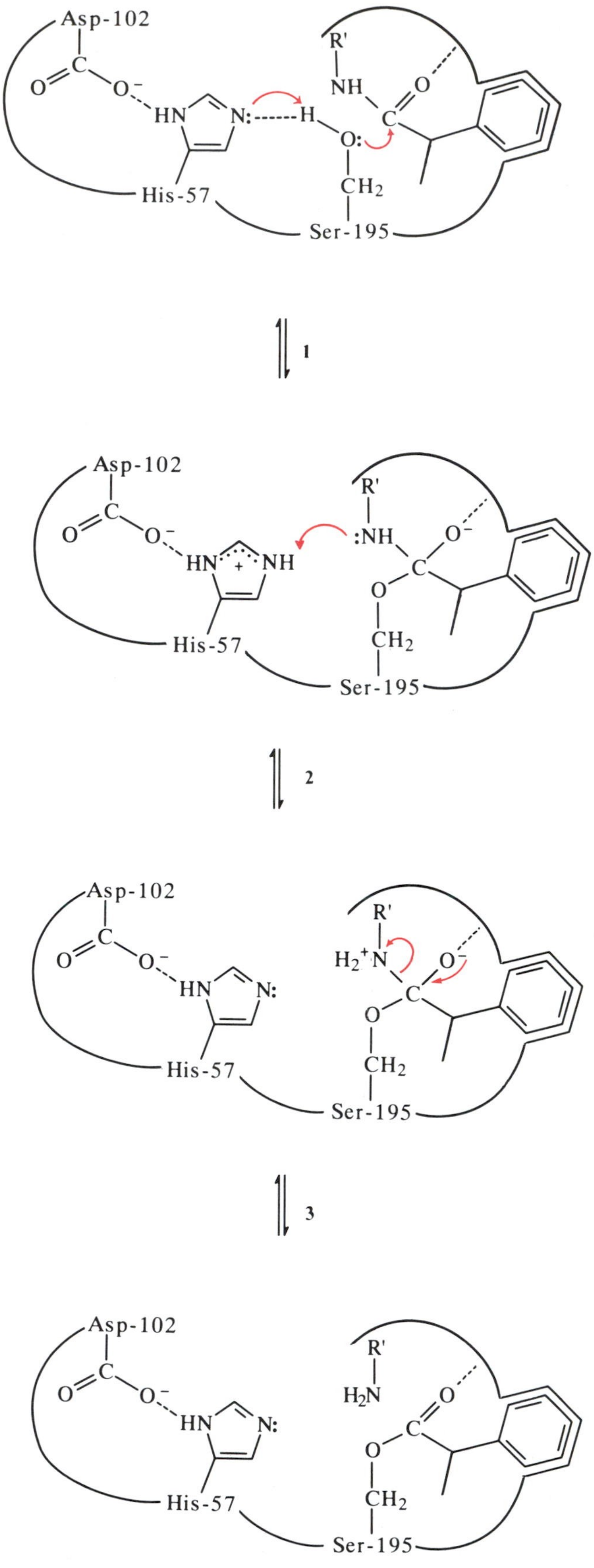

Figure 4-10

A Mechanism for Catalysis of the Acylation Step by Chymotrypsin. His-57 acts as a general base in abstracting a proton from the hydroxyl group of Ser-195 as it attacks the acyl group (1) and the protonated His-57 acts as a general acid to donate a proton to the leaving amino group of the tetrahedral addition intermediate (2) so that the C–N bond can break (3).

tetrahedral addition intermediate, such as is known to be formed in many similar nonenzymatic reactions of acyl compounds. In the breakdown of this intermediate to form the acylserine ester (step 2), the protonated imidazole that was generated in the first step acts as a general acid catalyst, donating a proton to the leaving group and facilitating the reaction by ensuring that the leaving group departs as a neutral amine (step 3) and not as an amine anion.

Different mechanisms are shown in Figure 4-10 for general base catalysis of the formation of the tetrahedral addition intermediate and general acid catalysis of its breakdown. Proton removal by imidazole from the serine hydroxyl group as it attacks the peptide carbonyl group is shown as a single step, in a concerted mechanism, whereas proton donation to the leaving amine occurs before C–N cleavage, in a two-step mechanism. In general, catalysis of this kind usually occurs in a concerted mechanism when the intermediate that could be formed without proton transfer is very unstable, whereas it tends to follow a step-by-step mechanism when moderately stable intermediates, such as a protonated amine, can be formed.

Hydrolysis of the acylserine intermediate may take place by an analogous series of reactions, with the imidazole acting to remove a proton from the attacking water molecule and to donate a proton to the serine oxygen atom as it is expelled from the tetrahedral addition intermediate. The driving force for general acid-base catalysis in all of these steps arises from the avoidance of extremely unstable ionic forms of intermediates.

There is evidence in addition to the proximity of histidine and serine in the active site of chymotrypsin that supports the role of imidazole as a general acid-base catalyst. This additional evidence is that the hydrolysis of several substrates and acyl-enzyme intermediates proceeds more slowly in D_2O than in H_2O by a factor of about three. General base-catalysis of the nonenzymatic hydrolysis of esters, such as *N,O*-diacetylserinamide, and of other acyl compounds also proceeds about three times more slowly in D_2O than in H_2O. Solvent deuterium isotope effects such as these are typical of reactions in which a proton (or deuteron) is transferred in the reaction. In D_2O, the protons on the serine hydroxyl group and on the nitrogen atoms of imidazole will be rapidly exchanged for deuterons from the solvent, as is the case for almost all protons bound to nitrogen, oxygen, or sulfur (see Box 2-1, Chapter 2). The transfer of a deuteron in a chemical reaction is usually from two to ten times as slow as proton transfer. The solvent deuterium isotope effects in both enzymatic and nonenzymatic ester hydrolysis are consistent with general base catalysis by imidazole, in which a proton undergoes transfer to and from imidazole in the two reaction steps. Nucleophilic catalysis by imidazole, as in equation 12, gives little or no isotope effect in D_2O.

HN–imidazole–N: + C(=O)(OR) ⟶ HN–imidazolium(+)–N–C(=O) + $^-$OR (12)

This interpretation of solvent isotope effects with chymotrypsin and other enzymes is probably correct, but an uncertainty remains because of the possibility that the structure of the enzyme may be slightly different in D_2O and H_2O. It is conceivable that a small difference in enzyme structure, rather than a direct isotope effect on the reaction, is responsible for the observed decrease in rate in D_2O.

The Quantitative Examination of Enzyme Inactivation and Inhibition

Before proceeding further, we should consider how to deal quantitatively with the simplest types of enzyme inactivation and inhibition. We have already seen examples of both inactivators and inhibitors in the irreversible inactivation of chymotrypsin by *p*-toluenesulfonyl phenylalanine chloromethyl ketone (TPCK) earlier in this chapter and in the reversible, competitive inhibition of chymotrypsin by indole and of trypsin by methylguanidinium ion (Chapter 3). The conclusion that these molecules act exclusively at the active site could not have been reached from qualitative observations; it is important to characterize inactivation and inhibition quantitatively. In particular, there are several different classes of inhibition that can be distinguished by quantitative studies, and each type of inhibition has a specific meaning in terms of inhibitor-enzyme interactions.

Irreversible Inactivation and Pseudo First-Order Kinetics

The reaction of an enzyme with an inactivator such as TPCK is irreversible and proceeds to completion in the presence of a large molar excess of the inactivator. The disappearance of enzyme activity follows first-order kinetics under these conditions, according to the rate law of equation 13,

$$v = k_{obsd}[E] = k_{obsd}(\text{activity}) \qquad (13)$$

and a plot of log(activity) against time—that is, activity against time on semilogarithmic graph paper—gives a straight line (Figure 4-11). The first-order rate constant may be obtained by measuring the half-time ($t_{1/2}$) of the inactivation and calculating k_{obsd} from the relationship $k_{obsd} = \ln 2/t_{1/2} = 0.693/t_{1/2}$.

If the first-order plot is not linear and curves upward, it is likely either that the enzyme preparation contains different classes of active sites that are

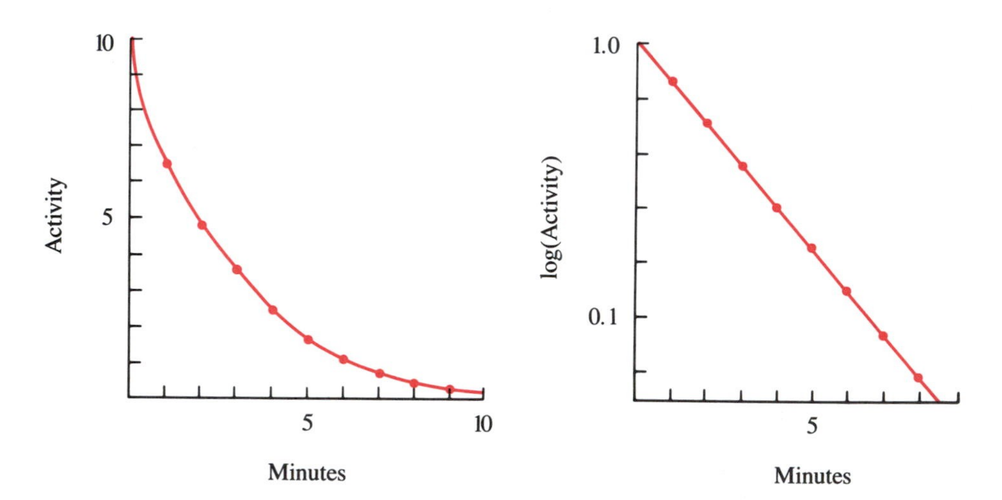

Figure 4-11

Pseudo First-Order Kinetics for the Inactivation of Chymotrypsin by TPCK. The graph at the left shows the decrease in activity with time. A first-order decay gives a straight line when plotted on semilogarithmic graph paper, as shown in the graph of the same data at the right.

inactivated with different rate constants or that reaction with the inactivator does not result in complete loss of enzyme activity. In the latter case, a plot of $(A - A_{\infty})$ against time on semilogarithmic graph paper will be linear (A_{∞} is the activity that remains when the reaction with the inactivator is complete). This result is often obtained if the inactivator causes a conformational change that decreases, but does not abolish, enzyme activity. It shows that the group that reacts with the inactivator is not essential for catalytic activity.

The inactivation reaction is *pseudo first order*, rather than true first order, because the rate depends on the concentration of the inactivator, as well as the concentration of enzyme. The disappearance of enzyme activity follows first-order kinetics because the inactivator is present in a large excess, compared with the enzyme, so that its concentration remains essentially constant during the reaction. The treatment of the dependence of the inactivation rate on the concentration of the inactivator is analogous to the treatment of the dependence of enzyme activity on substrate concentration, except that pseudo first-order rate constants, k_{obsd}, are used instead of initial rate measurements.

It is essential to distinguish between the *inactivation* of an enzyme (sometimes called irreversible inhibition) and reversible inhibition. Inactivation is usually caused by the formation of a covalent bond with some group at the active site of the enzyme. Reversible inhibition is relieved by dilution or dialysis, which removes inhibitor from the enzyme. Confusion can arise because the *rate* of inactivation is often decreased by the presence of substrate, which protects the enzyme by binding to the active site. However, reaction with an inactivator will usually proceed to completion, even in the presence of substrate, because it is irreversible. This differs from competitive inhibition, which is ordinarily *reversed* rapidly by the addition of a high concentration of substrate.

Reversible Inhibition by Competitive Inhibitors

The amount of inhibition by a reversible inhibitor is dependent upon binding of the inhibitor to the enzyme at equilibrium. This equilibrium is usually reached rapidly, so that the amount of inhibition is constant as long as the experimental conditions are maintained constant. The terms *competitive, noncompetitive,* and *uncompetitive* inhibition are reserved for reversible inhibitors, which give time-independent inhibition that can be described by equilibrium constants. Some confusion may arise from the fact that many irreversible inactivators that resemble the substrate bind to the active site and act as reversible, competitive inhibitors before they react irreversibly to inactivate the enzyme (see Box 4-3). Thus, the first task in the experimental evaluation of an inhibitor is to determine whether the inhibition that is being observed is time dependent and whether it can be rapidly reversed by removing the inhibitor or diluting the solution.

In this chapter, we will examine competitive inhibition as an example of inhibition in which binding of the inhibitor is rapid and reversible. Noncompetitive and uncompetitive inhibition are usually observed with two-substrate enzymes and will be described when the kinetics of these enzymes are examined in Chapter 11.

The dependence on substrate concentration of the initial rate for the chymotrypsin-catalyzed hydrolysis of *N*-acetylphenylalanine-alanine amide in the presence of two increasing concentrations of inhibitor, $[I]_1$ and $[I]_2$, is shown in Figure 4-12A. At low concentrations of substrate, the rate is decreased in the presence of the inhibitor because some of the free enzyme binds to the inhibitor and is no longer available for the binding and hydrolysis of the

BOX 4-3

THE KINETICS OF ENZYME INACTIVATION

It is of interest to know the second-order rate constant, k_i, for the reaction of an enzyme with an inactivator in very dilute solution in order to compare the effectiveness of different inactivators. We also wish to know the value of the dissociation constant, K_i, for the reversible dissociation of the enzyme-inactivator complex (E·I) that forms before the irreversible inactivation step.

If the inactivator resembles the substrate and binds to the active site, the rate of the inactivation reaction will show saturation behavior, just as the reaction of a substrate shows saturation behavior during catalysis.

$$E + I \underset{K_i}{\rightleftharpoons} E \cdot I \xrightarrow{k_2} E_{inactive}$$

The second-order rate constant for inactivation in a solution of inactivator that is very dilute relative to the K_i value (but still in large excess over enzyme) is given by $k_i = k_2/K_i$, which is analogous to the second-order rate constant k_{cat}/K_m for the reaction of a very dilute solution of the substrate. The first-order rate constant, k_2, for reaction of the bound inactivator with the enzyme is analogous to k_{cat}, and the dissociation constant, K_i, of the noncovalent E·I complex is analogous to K_s. A comparison of K_i for the inactivator with K_s for a substrate shows how closely the noncovalent binding of the inactivator resembles that of the substrate.

Some inactivators do not bind strongly enough in a noncovalent complex to give saturation behavior, so that they can be evaluated only in terms of the second-order rate constant, k_i. The value of k_i is then given directly by the slope of a plot of the observed pseudo first-order rate constants for inactivation against the concentration of inactivator, because $k_i = k_{obsd}[I]$. If there is saturation behavior, the values of K_i, k_2, and k_i can be evaluated from a Lineweaver-Burk or Eadie-Hofstee plot, using k_{obsd} and [I] instead of v and [S].

substrate. This situation is described in Figure 4-13, in which K_i is defined as the dissociation constant for the enzyme-inhibitor (E·I) complex. However, if the substrate concentration is increased enough, most of the inhibition can be overcome because the equilibrium is forced toward the binding of substrate by competition between the substrate and the inhibitor for the free enzyme. At a higher concentration (lower curve), there is more inhibition because more of the enzyme is tied up in the E·I complex, but the inhibition can still be overcome at sufficiently high concentrations of substrate. Thus, the inhibitor serves to inhibit the enzyme by reducing the binding of substrate. This decreases the velocity and k_{cat}/K_m at low substrate concentrations and increases the observed K_m value.

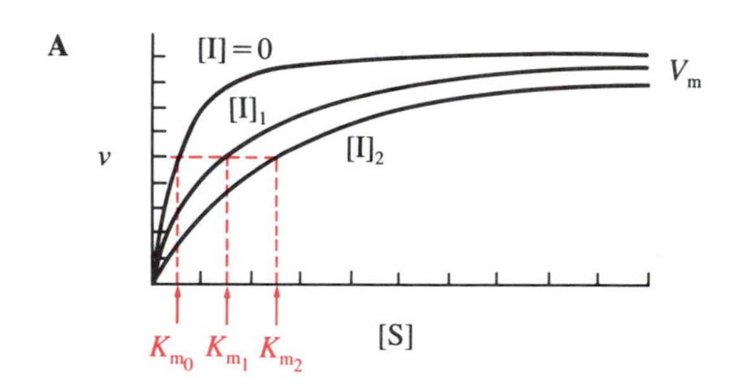

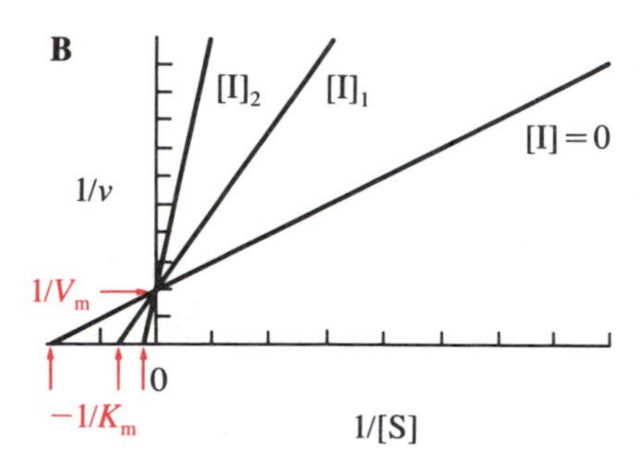

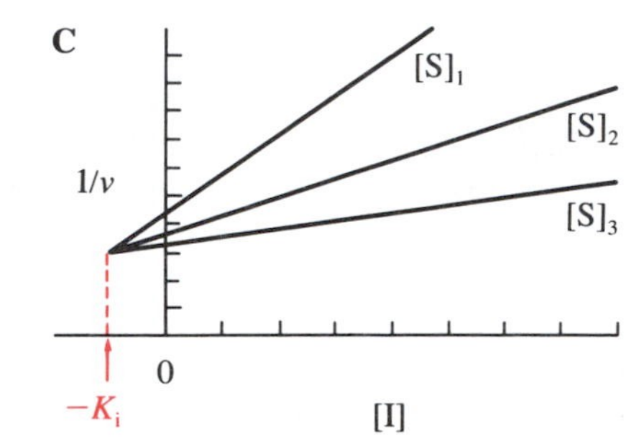

Figure 4-12

Graphs of Experimental Data for Competitive Inhibition of an Enzyme. Part A shows the effect of an inhibitor on the saturation curve of v versus [S]; part B, the Lineweaver-Burk plots of the same data; part C, the plot of $1/v$ versus [I].

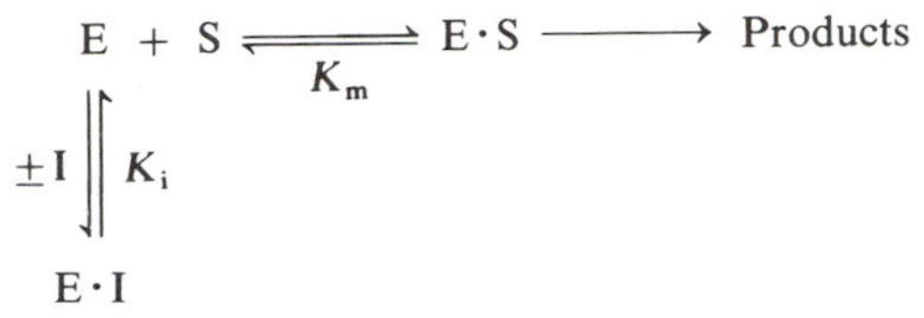

$$\begin{array}{ccc} E + S & \underset{K_m}{\rightleftharpoons} & E \cdot S \longrightarrow \text{Products} \\ \pm I \Big\Updownarrow K_i & & \\ E \cdot I & & \end{array}$$

$$K_i = \frac{[E][I]}{[E \cdot I]}$$

Figure 4-13

Kinetic Scheme for Inhibition of an Enzyme by a Reversible Competitive Inhibitor.

The V_m and k_{cat} values are unchanged because at sufficiently high substrate concentration it is still possible to convert all of the enzyme into the active E·S complex.

This behavior is shown in a double reciprocal Lineweaver-Burk plot of the effect of substrate concentration on the rate at the two inhibitor concentrations in Figure 4-12B. The ordinate intercept at infinitely high substrate concentration, $1/V_m$, is not affected by the inhibitor, but the observed K_m increases with increasing inhibitor concentration, so that the observed rate is decreased at the lower substrate concentrations. In a third type of plot, of $1/v$ against [I] at different substrate concentrations (Figure 4-12C), the lines intersect in the upper-left quadrant. This behavior exemplifies classical *competitive inhibition.*

The rate expression for competitive inhibition according to the scheme of Figure 4-13 is given in equation 14, which is the same as the simple Michaelis-Menten equation except that the K_m term is multiplied by the term $(1 + [I]/K_i)$ (equation 15).

$$v = \frac{V_m[S]}{K_m(1 + [I]/K_i) + [S]} = \frac{V_m[S]}{K_{m(obsd)} + [S]} \tag{14}$$

$$K_{m(obsd)} = K_m(1 + [I]/K_i) \tag{15}$$

This term describes simple saturation behavior when a molecule binds to an enzyme or a receptor. Here, it shows how the observed K_m value is increased because the binding of substrate has to compete with the binding of inhibitor (Figure 4-13). It is apparent from equation 14 that V_m is independent of the concentration of inhibitor, and this is because the inhibitor has no effect on the rate of breakdown of the E·S complex. The value of the dissociation constant of the enzyme-inhibitor complex, K_i, may be obtained from the observed K_m values in the presence of inhibitor (Figure 4-12B, equation 15), from the point of intersection of the lines in Figure 4-12C, or directly from equation 14.

Competitive inhibition is understood most clearly by considering the two important kinetic constants for enzyme catalysis: k_{cat}/K_m, for the second-order reaction of free E with S, and k_{cat}, for the first-order reaction of the E·S complex. A competitive inhibitor binds to E and not to E·S. Therefore, it inhibits the reaction of E with the second-order rate constant k_{cat}/K_m by decreasing the concentration of free E. It has no effect on the reaction of E·S, with the first-order rate constant k_{cat}. The inhibition of E must arise from an increase in $K_{m(obsd)}$ because k_{cat} is constant and k_{cat}/K_m decreases.

Summary

Chymotrypsin catalyzes the hydrolysis of peptides and esters by covalent catalysis. The hydroxyl group of serine-195 at the active site of the enzyme forms a covalent bond with the acyl group of the substrate to give an acyl-enzyme intermediate. Formation and hydrolysis of the acyl-enzyme is assisted by the imidazole group of histidine-57, which provides general acid-base catalysis of the proton transfers that are required in these reactions. The inactivator *p*-toluenesulfonyl phenylalanine chloromethyl ketone, TPCK, resembles a substrate molecule and alkylates the catalytic imidazole group. Indole is a competitive inhibitor because it competes with the substrate for binding to the active site and increases K_m without changing k_{cat}.

ADDITIONAL READING

ONG, E. B., SHAW, E., and SCHOELLMANN, G. 1964. An active center histidine peptide of α-chymotrypsin. *Journal of the American Chemical Society* 86: 1271–1272.

JENCKS, W. P. 1987. General acid-base catalysis. In *Catalysis in Chemistry and Enzymology*, ch. 3. New York: Dover. Tests for covalent catalysis.

JENCKS, W. P. 1976. Enforced general acid-base catalysis of complex reactions and its limitations. *Accounts of Chemical Research* 9: 425–432.

PROBLEMS

1. Catalysis by chymotrypsin of the hydrolysis of *N*-acetyl-Gly-Phe-Ala-NH_2 is inhibited by Ala-NH_2, but this inhibition is not competitive with the concentration of *N*-acetyl-Gly-Phe-Ala-NH_2. What is the mechanism of this inhibition?

2. Explain the mechanistic interpretation of k_{cat}, k_{cat}/K_m, and K_m for the hydrolysis of *p*-nitrophenyl acetate catalyzed by chymotrypsin.

3. You have isolated an unknown proteolytic enzyme from a strain of bacteria. What would you do to characterize this enzyme, in a fairly short time and without use of complicated instrumentation?

4. An enzyme, ω-amidase (E), cleaves the amide group of glutamine. It also catalyzes the hydrolysis of monoesters and other derivatives of dicarboxylic acids. If an additional nucleophile, such as hydroxylamine, is added, this nucleophile reacts to form the expected product, the hydroxamic acid for hydroxylamine, so that the overall reaction follows the course:

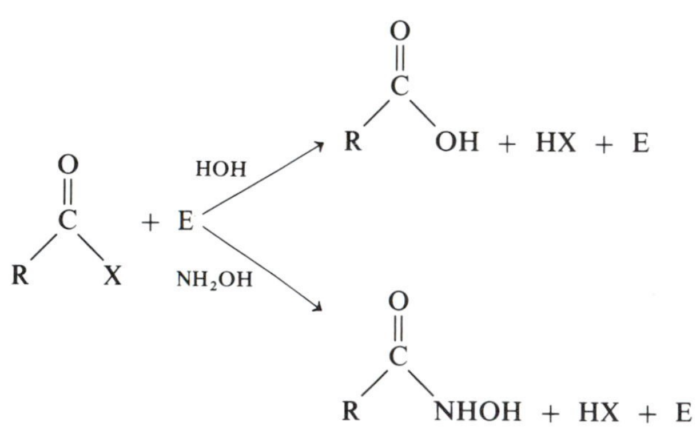

The following observations were made:

(a) In the presence of added hydroxylamine, the rates of the different observed reactions for *p*-methylphenyl glutarate behave as shown in graph A.

(b) However, for methyl glutarate, the same experiment gives the results shown in graph B.

(c) The same maximum rate is reached when a sufficiently high concentration of hydroxylamine or methylamine is the acyl-group acceptor.

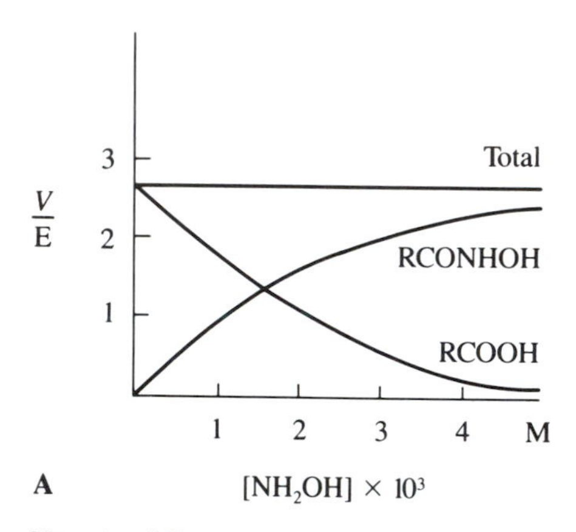

Note the different scales.

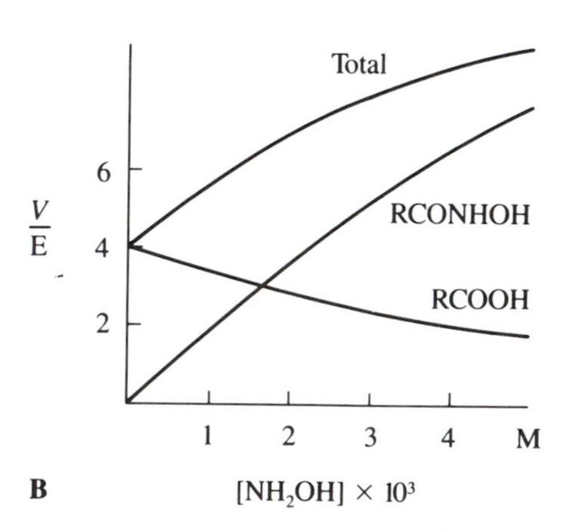

How many conclusions can you reach from this evidence? Can you deduce anything from the relative amounts of hydrolysis product and hydroxamic acid that are formed at the different hydroxylamine concentrations from the different substrates? What do you conclude from the additional observation that three substituted phenyl esters of glutarate exhibit the same maximum rate of hydrolysis?

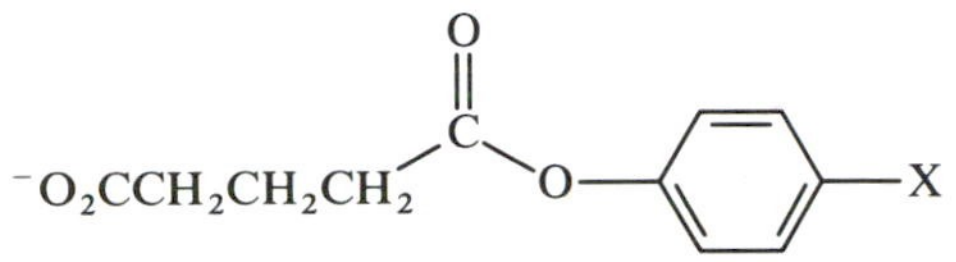

$$^{-}O_2CCH_2CH_2CH_2-\overset{\overset{O}{\|}}{C}-O-C_6H_4-X$$

Substituted phenyl esters of glutarate

5. It was found that the hydrolysis of a series of ethyl, phenyl, and nitrophenyl (NP) esters by papain (E) occurs with the kinetic parameters given in the table below. The chemical reactivity of these esters with a thiol anion in the absence of enzyme is also shown. The values of log $1/K_m$ increase with the chemical reactivity of the esters as shown in the graph. Interpret the behavior of the values of k_{cat} and K_m.

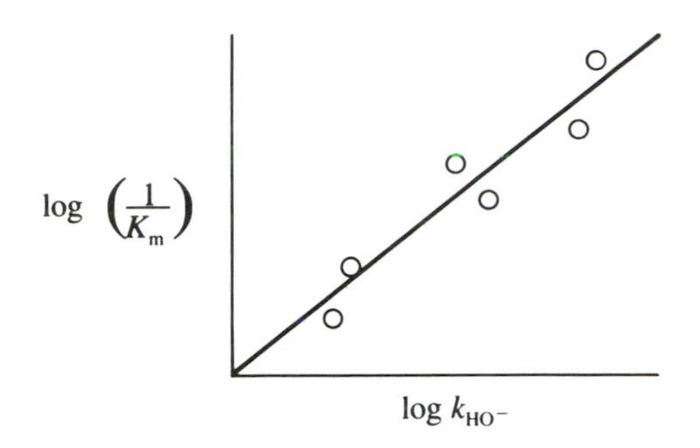

(k_{HO^-} for alkaline hydrolysis of ester)

Substrate: $PhCH_2OCONHCH_2\overset{\overset{O}{\|}}{C}-X$	ENZYMATIC		NONENZYMATIC
	$10^5\, K_m/M$	k_{cat}/s^{-1}	$HOEtS^- +$ Ester $k_2/M^{-1}s^{-1}$
X 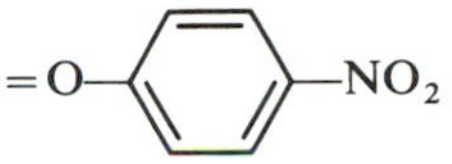 = O–C_6H_4–NO_2 (para)	0.93	2.73	19,500
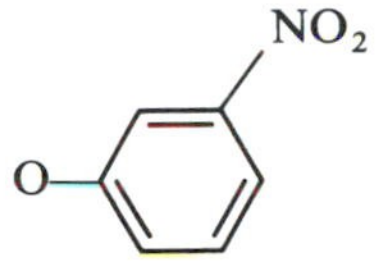 O–C_6H_4–NO_2 (meta)	1.89	2.18	11,000
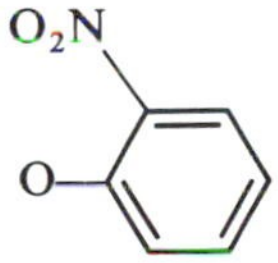 O–C_6H_4–NO_2 (ortho, O_2N)	15.2	2.14	12,800
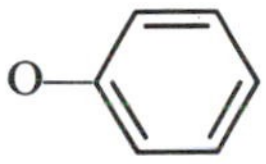 O–C_6H_5	10.7	2.45	531
$O-CH_2CH_3$	514	1.96	—

Catalysis by Chymotrypsin: The Importance of Binding Energy

5

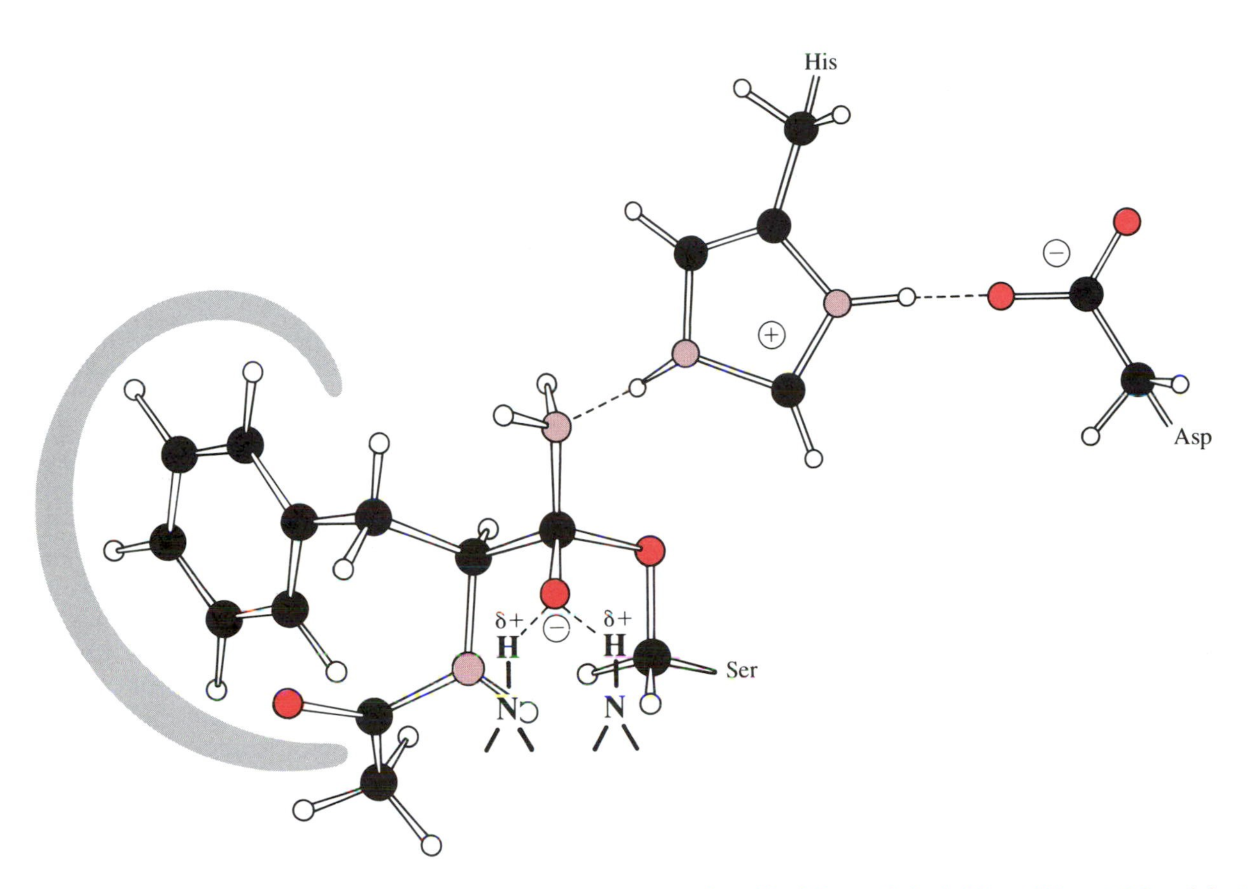

A model of the tetrahedral adduct of N-acetylphenylalanine amide with the β-hydroxyl group of serine-195 at the active site of chymotrypsin, showing two peptide hydrogen bonds to the oxyanion and the hydrogen bond from protonated histidine-57 to the leaving group.

Almost anything interesting that happens in living systems involves specific *binding interactions* between molecules. The noncovalent binding of a small molecule to a specific site on a large molecule often brings about an interaction with the binding site that makes something happen. Well-known examples include the binding of hormones and neurotransmitters to receptors that control the functioning of the neuromuscular system, the heart, the digestive tract, and other organs. In this chapter, we examine the role of these binding interactions in catalysis by enzymes. Our understanding of the utilization of binding energy for catalysis is limited; however, it is generally better than our understanding of the utilization of binding energy in other systems because enzymes have been studied quantitatively in great detail and the three-dimensional structures of many enzymes are now known.

Many properties of enzymes cannot be fully explained by the simple chemical events that are known to take place during catalysis, such as those described in Chapter 4. These properties include substrate specificity, conformational changes, and, especially, the extraordinary increases in the rates of chemical reactions that are brought about by enzymes. Substrate specificities may be explained, in part, by the fit of specific substrates into their binding pockets, as described in Chapter 3. But many enzymes that catalyze reactions of specific substrates are inactive toward simpler substrates that bind well enough but do not react.

Furthermore, studies of nonenzymatic "model reactions" for nucleophilic catalysis and general acid-base catalysis have shown that chemical mechanisms generally increase reaction rates by factors of as much as about 10^3. This is important, but it is not nearly enough to account for the rate increases by factors ranging from 10^8 to 10^{14} or more that are brought about by many enzymes. It suggests that the binding interactions between enzymes and their specific substrates at active sites are responsible for very large increases in reaction rate.

In this chapter, we will examine some of the mechanisms by which binding interactions between an enzyme and a specific substrate can contribute to specificity and catalysis. These mechanisms include bringing about an increase in the probability of reaction by holding the reacting groups in exactly the correct position for reaction, and destabilizing the bound substrate so that its reactivity is increased. Specific substrates can also bring about changes in enzyme conformation, and they can bind in more-productive conformations than nonspecific substrates. These mechanisms for catalysis can be summarized by the statement that an enzyme utilizes binding interactions with specific substrates to stabilize the transition state much more than it stabilizes the ground state of the enzyme-substrate complex.

Conformational Changes

Many enzymes undergo a change in conformation upon binding substrates, and these conformational changes may play an important role in catalysis. However, there are only a few cases in which the nature of these conformational changes has been determined, and even fewer in which the changes can be evaluated quantitatively. In this section, we will examine a conformational change that is essential for bringing chymotrypsin into an active state. This has been analyzed in some detail and serves as a model for conformational changes in other systems that are less well understood.

The Effect of pH on K_m for Chymotrypsin

We return now to the question of why the activity of chymotrypsin falls off at high pH values, as described by the bell-shaped pH-rate profile for k_{cat}/K_m. We know from Chapter 4 (Figure 4-3) that the decrease in activity at low pH values represents a decrease in k_{cat}, which is caused by protonation of the imidazole group that is required for general base catalysis, and that the decrease in activity at high pH arises from an increase in K_m. We also know that the acylation step is rate determining for the hydrolysis of peptide substrates, so that $K_m = K_s$ for peptides. Therefore, the decrease in rate at high pH is caused by a decrease in the binding of substrates. Why should this be?

Competitive inhibitors of chymotrypsin, such as benzyl alcohol, show the same decrease in binding at high pH as substrates. This is expected, because these inhibitors are structurally similar to substrates and bind to the same site. When the binding of an inhibitor is measured at high pH values and the pH is held constant by automatic titration, it is found that the binding of each molecule of inhibitor, I, is accompanied by the uptake of one proton according to equation 1.

$$E + I + H^+ \rightleftharpoons {}^+HE \cdot I \qquad (1)$$

When the same experiment is carried out at lower pH values, less than one proton is taken up. The decrease in the amount of proton uptake with decreasing pH follows a sigmoid titration curve with a pK close to 9, shown in Figure 5-1; this is the same as the pK for the increase in substrate binding.

This behavior shows that substrates and inhibitors bind to a particular ionic form of the enzyme and that the basic form, which exists at high pH, does not bind substrates or inhibitors. If the binding is forced to occur by the addition of high concentrations of inhibitor, a proton must be taken up in order to convert the enzyme into the correct protonation state for binding.

This situation is described by the scheme in Figure 5-2 for the binding of a substrate, S. An ionizing group of the enzyme with a pK_a close to 9 controls binding. When the enzyme is in the protonated form, EH^+, S binds tightly but, when the enzyme is unprotonated, S binds very weakly. Therefore, the observed binding of S decreases as the fraction of the protonated form decreases. However, if enough substrate is added, binding can still be forced to go to

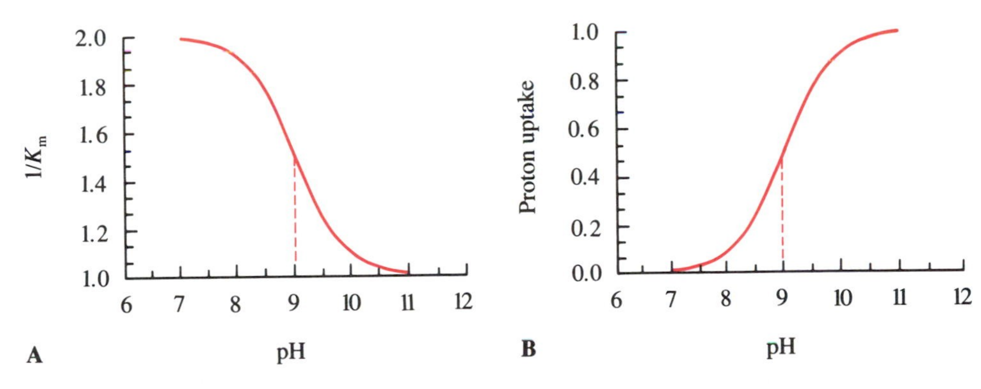

Figure 5-1

The dependence on pH of (A) $1/K_m$ for catalysis of the hydrolysis of a peptide by chymotrypsin; and (B) the proton uptake that accompanies the binding of benzyl alcohol, an inhibitor, to chymotrypsin at high pH values.

$$\begin{array}{ccc} {}^{+}HE + S & \rightleftharpoons & {}^{+}HE \cdot S \\ \Big\updownarrow K_a & & \\ E + H^{+} & & \end{array}$$

Figure 5-2

Scheme Accounting for the Effect of pH on Substrate Binding to Chymotrypsin. The scheme is an expansion of the center line in Figure 4-5, showing that a protonated form of the enzyme, ^{+}HE, binds the substrate S to form the active Michaelis complex $^{+}HE \cdot S$, whereas the unprotonated form E cannot bind S. This scheme explains the alkaline leg of the pH rate profile in Figure 5-1 (and Figure 4-3). Note that the enzyme form E can be forced to bind a proton by a high concentration of S. This is because S binds only to ^{+}HE, which is present at a low concentration even above the pK_a; and a high concentration of S can draw all of the enzyme into the form $^{+}HE \cdot S$.

completion by drawing the equilibrium in the direction of $^{+}HE \cdot S$ formation (or $^{+}HE \cdot I$ for an inhibitor). Figure 5-2 shows that this binding can occur only when one proton is taken up with each molecule of substrate to give $^{+}HE \cdot S$.

Figure 5-2 is a simple example of a *coupled equilibrium system,* in which the binding of one ligand (the substrate) is coupled to the binding of another (the proton), and saturation by substrate drives the binding of a proton. In this particular case, proton uptake is the key experimental manifestation of the phenomenon of coupled binding because the proton is one of the binding ligands. In many other systems, a species other than the proton is the second ligand, but the same principle of coupled binding applies.

Why does the protonated form of chymotrypsin bind an uncharged substrate so much better than the unprotonated form? The pK_a near 9 probably represents the dissociation of a protonated amino group, which is expected to dissociate in this pH range. However, protonation of an amino group in the binding site gives a positively charged group, $—NH_3^{+}$, which would be expected to decrease rather than increase the binding of an uncharged substrate or inhibitor. This suggests that the amino group is not in the binding site.

The reason for increased binding at low pH is that there is a difference in the structure of the enzyme at high and low pH; that is, protonation of the amino group causes a *conformational change* of the protein. There is strong, independent evidence linking conformational changes and changes in binding of small molecules to chymotrypsin with changes of pH. Indeed, the conformational change that occurs upon protonation is essentially the same as that which occurs when the inactive zymogen, chymotrypsinogen, is converted into active chymotrypsin.

Activation of Chymotrypsinogen

Chymotrypsinogen is synthesized in the pancreas and secreted into the small intestine as the zymogen, which is a single peptide chain that has almost no catalytic activity. This protects surrounding tissues against digestion by the enzyme. When the zymogen reaches the small intestine, it is attacked by active trypsin and by chymotrypsin itself, which converts it first into π-chymotrypsin, then into δ-chymotrypsin, and finally into α-chymotrypsin by successive cleavages of four peptide bonds. This process brings about the excision of two dipeptides, as shown schematically in Figure 5-3, and the conversion of the single peptide chain of chymotrypsinogen into three shorter chains with two new amino-terminal end groups, isoleucine and alanine.

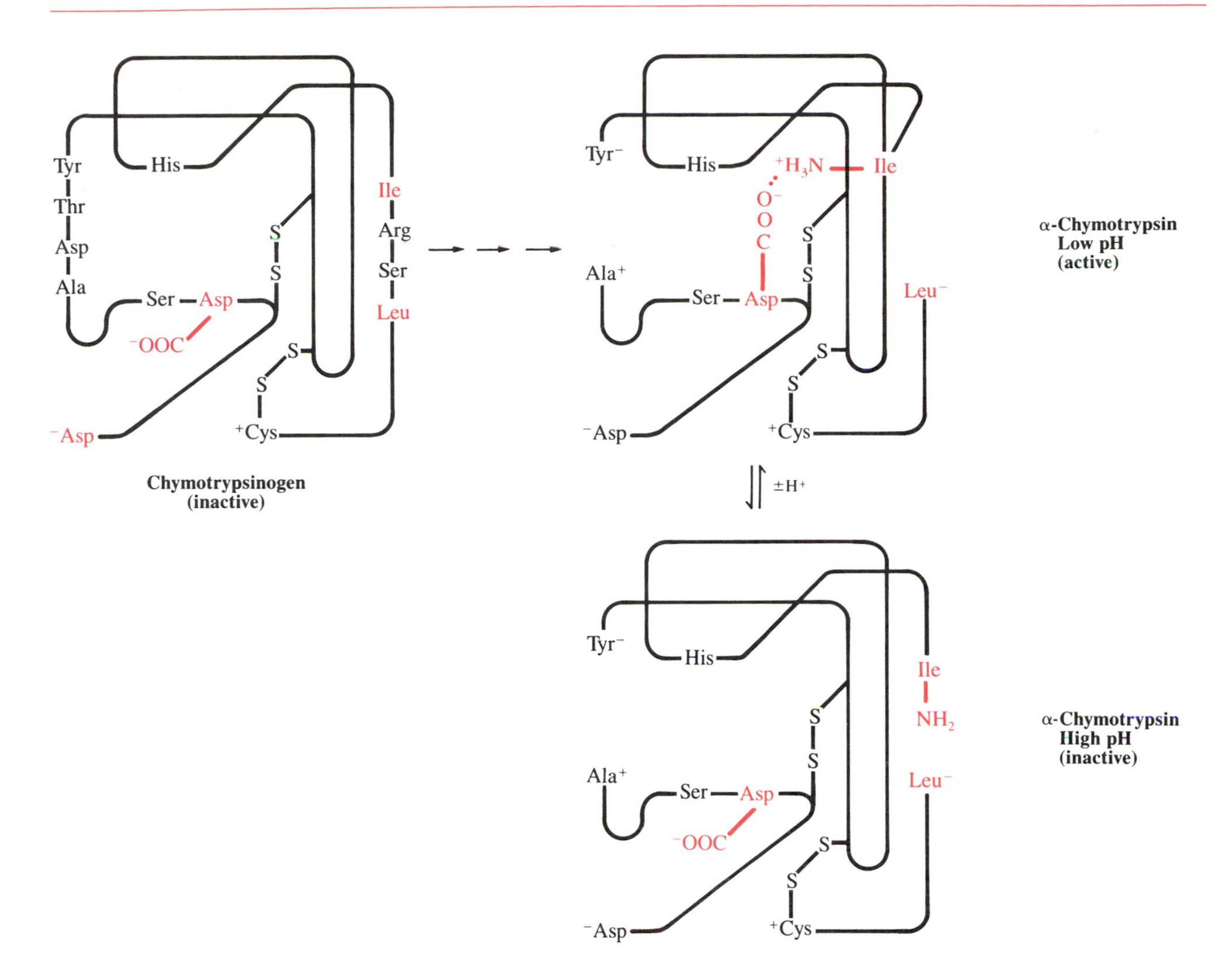

Figure 5-3

Schematic Drawings to Describe the Conversion of Chymotrypsinogen into Active Chymotrypsin. An N-terminal isoleucine amino group is released and forms an ion pair with an aspartate carboxylate group at the active site. Deprotonation of the Ile-16-NH_3^+ at high pH breaks the ion pair and converts the enzyme into an inactive form that does not bind substrate.

Figure 5-3 also shows several disulfide cross-links that help to prevent the peptide chains of chymotrypsin from falling apart when the enzyme is activated. These stabilizing cross-links are formed by the oxidation of cysteine residues in the peptide chain to give disulfide bridges. The disulfide of two cysteines is called *cystine,* which is a common cross-linking agent in extracellular proteins.

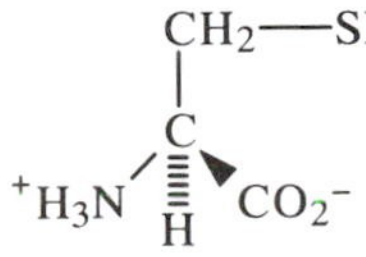

L-Cysteine (Cys, C)

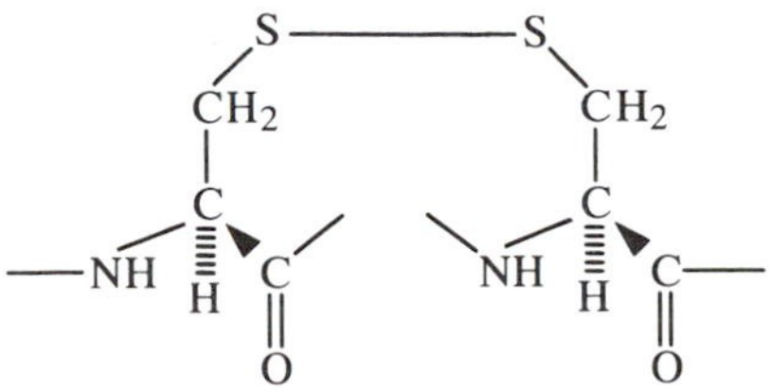

L-Cystine

An Ion Pair Causes the Change in Conformation

When the bond between isoleucine-16 and arginine-17 is cleaved in the intestine, the new N-terminal isoleucyl amino group immediately picks up a proton to form the charged ammonium group, —NH_3^+. This forms an ion pair with the β-carboxylate group, —COO^-, of an aspartate residue that is located near the other end of the peptide chain. The formation of this ion pair is one of the primary events that brings about the conformational changes required to convert the enzyme from an inactive form into an active form, as shown in Figure 5-3.

Asp-194 Ile-16

The ion pair linking isoleucine-16 and aspartate-194 is an *electrostatic interaction* that is critical for maintaining the three-dimensional structure of the active enzyme. In chymotrypsinogen, the amino group of isoleucine-16 is part of the peptide chain, so that it cannot form an ionic bond with aspartate-194, and the zymogen cannot bind substrates. The active enzyme binds substrates at neutral pH. But, when the pH is increased and the isoleucyl —NH_3^+ group loses a proton, the ionic bond with aspartate-194 is broken and the enzyme loses its ability to bind substrates and inhibitors because the —COO^- group of aspartate blocks the binding pocket (Figure 5-4). The same result can be brought about by acetylation of the amino group of the N-terminal isoleucine to give structure **1**. The acetylated amino group in this structure cannot be protonated at neutral pH and cannot form an ionic bond with aspartate-194.

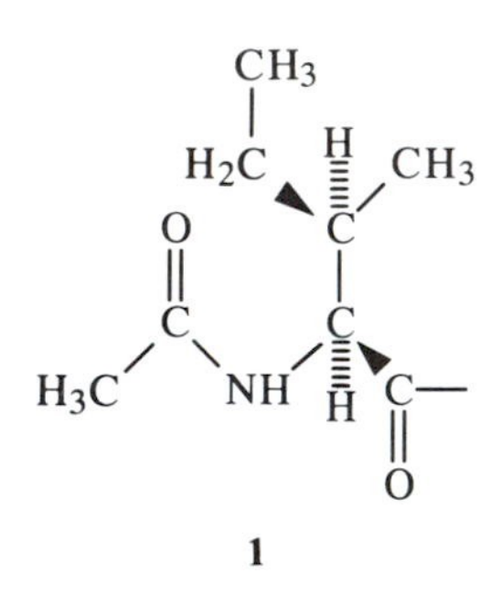

1

These conclusions are based in part on the three-dimensional crystal structures of chymotrypsin and chymotrypsinogen that were determined by X-ray diffraction. The structures do not indicate that activation requires a large *overall* conformational change; instead, there are several small changes. The most-important changes are a twisting of the aspartate side chain and a small movement of the peptide chain. These changes open the hydrophobic binding pocket, by shifting the obstructing group, and allow specific substrates to bind properly to the active site.

Substrate Binding and the Conformational Change

The binding of a substrate, or an inhibitor, to chymotrypsin at high pH provides a relatively simple example of one of the most-important processes in biochemistry: the utilization of binding energy between a small molecule and a macromolecule, or between two macromolecules, to make something happen that otherwise would not happen. The binding energy that arises from the noncovalent interactions between the substrate and its binding site in protonated chymotrypsin is *utilized* to force the enzyme into its active form when a sufficient concentration of substrate is added to the unprotonated, inactive form of the enzyme at high pH. This phenomenon is known in enzymology as *induced fit*—the process by which the binding of a specific substrate brings about a change in the conformation of an enzyme and converts it into a catalytically active form.

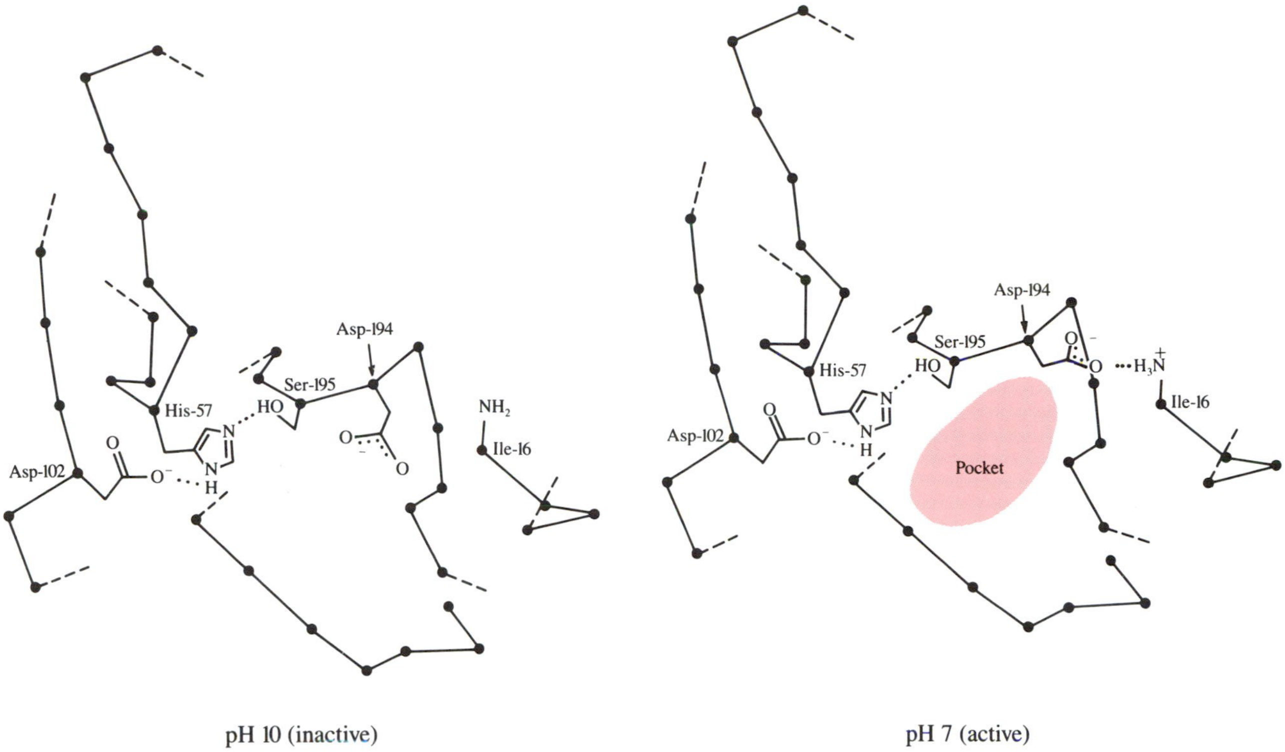

pH 10 (inactive) pH 7 (active)

Figure 5-4

At high pH, the β-carboxylate group of aspartate-194 blocks the specificity pocket of chymotrypsin. At neutral pH, the protonated isoleucine amino group forms an ion pair with the aspartate carboxylate group, which opens the pocket and allows the hydrophobic side chains of specific substrates to bind. The diagrams also show the carboxylate group of aspartate-102, which holds the imidazole group of histidine-57 in place. The imidazole can then accept a proton from the hydroxyl group of serine-195 when this hydroxyl group attacks the substrate.

Chymotrypsin is one of the very few systems in which the nature of the conformational change is known at the molecular level and in which it is possible to make a quantitative analysis of the utilization of binding energy to bring about this change.

A General Model: The Thermodynamic Box

In the general case, these induced conformational changes are described by a box of equilibria, such as that defined by the binding scheme in Figure 5-5. An enzyme or a receptor protein may exist in a resting, inactive state, P, in the absence of a specific ligand, L; the concentration of the activated state, P*, is small or insignificant. This means that the equilibrium constant K^* is much smaller than one ($K^* \ll 1.0$). When the ligand binds to the protein, an increased fraction of the protein is in the activated state, $P^* \cdot L$; that is, $K_L > K^*$. When K_L is >1.0, most of the protein will exist in the active form in the presence of sufficiently high concentrations of ligand.

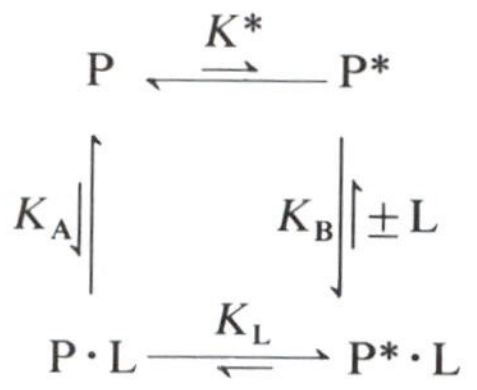

$$\begin{array}{ccc} P & \overset{K^*}{\rightleftharpoons} & P^* \\ K_A \updownarrow & & \updownarrow K_B \quad \pm L \\ P \cdot L & \underset{}{\overset{K_L}{\rightleftharpoons}} & P^* \cdot L \end{array}$$

Figure 5-5

The Thermodynamic Box of Equilibria for Coupling of the Binding of a Ligand to a Protein Conformational Change. In this system, a protein exists in two conformations, P and P*, which exhibit different binding properties. A ligand, L, binds weakly to P and strongly to P*. The equilibrium between the two conformational states favors P in the unliganded state, but the high affinity of P* for L causes P*·L to be the favored form of the liganded protein. These facts are explained by coupling between the equilibria for ligand binding and for the change in protein conformation, as defined by equations 2 and 3.

A straightforward analysis of the equilibrium constants in Figure 5-5 shows that they are related by equations 2 and 3.

$$K^* K_B = K_A K_L \tag{2}$$

$$\frac{K_L}{K^*} = \frac{K_B}{K_A} \tag{3}$$

Because $K_L > K^*$, it follows that $K_B > K_A$. This means that the ligand binds more strongly to P* than to P. It is this strong binding to P* that draws the equilibrium toward the active form P*·L when the ligand is present.

For chymotrypsin, the ligand can be regarded as either the proton or the substrate. Binding of the proton by the inactive state (E) converts it into the active state ($^+$HE), which binds the substrate. Conversely, binding of the substrate converts the inactive enzyme into a state that binds a proton to form $^+$HE·S. When several ligands bind or when several interacting subunits of one macromolecule bind the ligand, the situation is more complicated mathematically but is the same in principle. The reactions shown in Figure 5-5 represent a thermodynamic *box;* we will see other examples of such boxes later. A quantitative analysis of the conformational change is presented in Box 5-1.

Allosteric Enzymes

Induced conformational changes have widespread significance in biochemistry. The most widely known examples may be the *allosteric proteins* that exhibit cooperative or anticooperative binding of ligands because binding of a ligand causes a change in the structure of the protein. Many of these proteins have several subunits that interact with each other. Hemoglobin, for example, is a four-subunit protein that binds four molecules of O_2 with *cooperativity.* In hemoglobin, and other proteins that exhibit cooperative binding effects, the binding of a ligand induces a conformational change. In hemoglobin, the conformational changes are transmitted from one subunit to another through subunit-subunit interactions. These conformational changes convert the subunits from a state with a low affinity for oxygen into a high-affinity state, so that the affinity becomes much larger as more oxygen molecules bind (Chapter 10).

In allosteric enzymes, the ligand-induced conformational changes are manifested in changes of K_m or k_{cat} upon binding of substrates, activators, or inhibitors. Ligands that cause such conformational effects are called *effectors.* When the binding of one molecule, such as O_2 or a substrate, causes a

BOX 5-1

A QUANTITATIVE ANALYSIS OF THE CONFORMATIONAL CHANGE IN CHYMOTRYPSIN

The conformational change upon protonation of the isoleucine-16 amino group of chymotrypsin has been demonstrated by taking advantage of the fact that the dye proflavin undergoes a change in its absorption spectrum when it binds to the active site of chymotrypsin, so that the amount of binding may easily be measured spectrophotometrically. The change in absorbance at 465 nm that results from this change with a final pH of 6.84 is shown as the dashed line in the adjoining figure.*

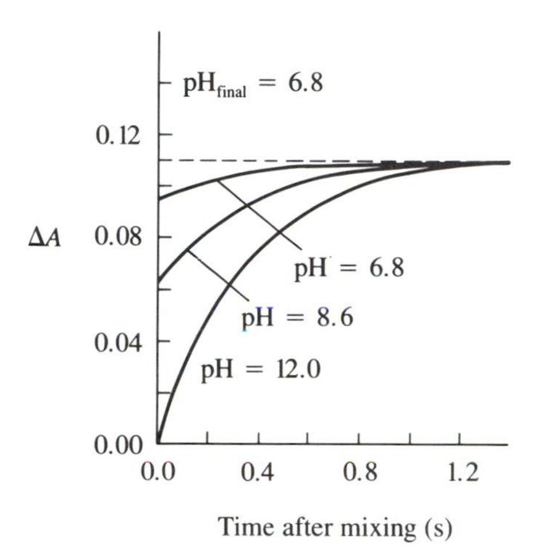

*After A. R. Fersht and Y. Requena, *J. Mol. Biol.* 60 (1971): 279.

Now, if a sample of enzyme is allowed to stand at pH 12 and is then rapidly mixed with proflavin and a buffer that brings it to pH 6.84, in a stopped flow apparatus, the change in absorbance that is caused by proflavin binding does not take place immediately. Instead, it follows a first-order course with a half-time of a fraction of a second, as shown in the lower curve of the figure. However, if the enzyme is initially in the active form at pH 6.84, most of the change in absorbance takes place immediately upon mixing with proflavin. This is followed by a small further change, with the same half-time as in the experiments at high pH.

These results show that a time-dependent change must take place before proflavin can bind to the high-pH form of the enzyme. Because binding to the active form of the enzyme at low pH takes place immediately, this change must be the change of the inactive, high-pH conformation to the active, low-pH conformation. The reaction can be described by equation a,

$$E + H^+ \underset{}{\overset{\text{Fast}}{\rightleftharpoons}} EH^+ \xrightarrow{k} E^*H^+ \qquad \text{(a)}$$

in which E^*H^+ represents the active, low-pH form of the enzyme that binds proflavin rapidly, and k is the first-order rate constant for the change in conformation of the protonated inactive enzyme, EH^+.

The size of the time-dependent changes in absorbance in the figure provides a measure of the amount of enzyme that was initially in the inactive form at each pH value and underwent a change to the active form at pH 6.84. The amounts of inactive enzyme decrease with decreasing pH with a midpoint at pH 8.8, as expected from the pH-dependence for the binding of substrates and inhibitors. At neutral pH, a small amount, approximately 15%, of the enzyme is initially in the inactive form and is converted into the active form in the presence of excess proflavin, as shown by the upper curve of the figure. This is well below the pK of 8.8 for conversion into the inactive form of the enzyme, so that we can conclude that 15% of the protonated enzyme is in the inactive form in the absence of substrate or inhibitor. Thus, the equilibrium constant, K_1 (equation b),

$$E^*H^+ \overset{K_1}{\rightleftharpoons} EH^+ \qquad \text{(b)}$$

for the conversion of protonated enzyme from the active into the inactive form, is $0.15/0.85 = 0.18$.

The effect of pH on binding in this system provides a relatively simple example of the kind of coupled equilibria that are important for many systems in which small molecules bind to proteins. The coupling between the equilibrium constants for conformational changes and ionizations can be described by the thermodynamic box in equation c.

$$\begin{array}{ccc} E^*H^+ & \overset{K_a^*}{\rightleftharpoons} & E^* + H^+ \\ K_1 \updownarrow & \searrow K_0 & K_2 \updownarrow \\ EH^+ & \overset{K_a}{\rightleftharpoons} & E + H^+ \end{array} \qquad \text{(c)}$$

This box contains five equilibrium constants, but the system is described completely by only three constants; the others are readily calculated once these three are known.

The diagonal dissociation constant K_0 for the ionization of E^*H^+ to $E + H^+$ is the same as the product of the equilibrium constants for its component steps: $K_a^* K_2$ for the upper path through E^*, and $K_1 K_a$ for the lower path through EH^+. Therefore, the ratios K_1/K_2 and K_a^*/K_a are identical, so that any difference in the equilibrium constants K_1 and K_2 is balanced by a difference between K_a^* and K_a. (Box 5-1 continued on next page.)

BOX 5-1 (continued)

A QUANTITATIVE ANALYSIS OF THE CONFORMATIONAL CHANGE IN CHYMOTRYPSIN

The value of pK_a can be calculated from the observed values of $pK_0 = 8.8$ and $K_1 = 0.18$. The resulting value of $pK_a = 7.9$ represents the true dissociation constant of the isoleucine-NH_3^+ group in the inactive form of the enzyme, EH^+, in which there is no ion pair. This is a normal value for an α-amino group in a peptide chain. The *observed* pK is higher than this because the equilibrium is pulled toward the protonated form by formation of the ion pair in E^*H^+. No binding is seen at high pH, so that the value of K_2 must be >20 and pK_a^*, for the isoleucyl ammonium ion in the ion pair, is >10. This shows that the ion-paired species cannot lose a proton and that the concentration of the active form of the enzyme is not significant unless protonation and ion pair formation have taken place. The ratio of the active to the inactive form of the enzyme changes by more than 20/0.18, or by more than 100-fold, when isoleucine is protonated and the ion pair is formed. Conversely, because the equilibria are coupled, we can say that the binding of substrate and the resulting conformational change bring about a decrease of more than 100-fold in the acid dissociation constant of the enzyme.

This is an interesting system for analysis of the utilization of binding energy because the structural basis for the energy balance is known. The important point is that the binding energy of the substrate is *utilized* to force the conformation of the inactive enzyme to change to that of the active enzyme at high pH. The observed binding represents the energy for binding to the active form of the enzyme minus the energy that is required to convert the inactive form of the enzyme into the active form. In this example it was shown directly that this change is a consequence of the more-favorable binding energy and tighter binding of the substrate to the active, low-pH form of the enzyme. It is more difficult to analyze conformational changes in most other systems, because they are more complex.

The maximum interaction energy that can be obtained for the binding of a ligand to one form of a macromolecule is called the *intrinsic binding energy.* If the binding of a ligand causes a change in the conformation of a macromolecule, the *observed* binding energy is always less favorable than the intrinsic binding energy.

conformational change that increases the binding of another molecule of O_2 or a substrate, the system is said to show *positive cooperativity;* if it decreases the binding of another molecule, the system shows *negative cooperativity*.

Information Transfer

A particularly important role of ligand-induced conformational changes, especially in higher organisms, is in the transmission of information. The information transfer systems are just beginning to be identified and understood. One example is the receptor that responds to acetylcholine. Acetylcholine that is released at a nerve ending combines with a specific receptor, and this forces the receptor to change its conformation. This touches off a chain of further reactions that result in the generation of an action potential in another nerve fiber, contraction of a muscle, or one of many other physiological processes. How these processes occur is presently under intensive investigation. Other examples of the same kind of process include the actions of most hormones and of many drugs, which bring about conformational changes upon combination with specific receptors and initiate different sequences of events.

Regulation of the synthesis of certain enzymes and other proteins in bacteria is brought about by the binding of a metabolite, an *inducer,* to a *repressor* molecule, a regulatory protein that binds to DNA and prevents the biosynthesis of specific proteins (Chapter 13). Binding of inducer causes a conformational change in the repressor that greatly reduces its affinity for DNA and leads to its dissociation from its cognate regulatory element of DNA. This allows the synthesis of a specific messenger RNA (mRNA), coded by a neighboring segment of DNA. Because mRNA codes for the biosynthesis of a specific protein, the repressor inhibits the synthesis of that protein. Comparable systems in higher organisms are similar in principle but are generally more complex and are less well understood at this time.

Mechanisms for Enzyme Catalysis by Utilization of Binding Energy

We now pose some more direct and quantitative questions about how chymotrypsin and other enzymes bring about such extraordinary rate accelerations with specific substrates. To do this, we must first define the magnitude of the rate accelerations that are characteristic of enzyme-catalyzed reactions.

Rate Acceleration Factors

The rate accelerations that are brought about by enzymes are large, so large that it is often difficult to estimate them accurately. One of the problems that must be overcome is the selection of an appropriate nonenzymatic reaction for comparison with the enzymatic reaction. It is not very useful, for example, to compare k_{cat} for an enzymatic reaction with the second-order rate constant for the H^+-catalyzed reaction of the same substrate to give the same product. This is because H^+ is a much stronger acid than any group on the enzyme and k_{cat} is a first-order rate constant that cannot be compared directly with a second-order rate constant. The best comparison for a pH-independent value of k_{cat} is with the pH-independent first-order rate constant for the corresponding nonenzymatic reaction under the same conditions.

A better approach is to compare k_{cat}/K_m, the second-order rate constant for the enzymatic reaction, with the second-order rate constant for the corresponding nonenzymatic reaction between the substrate and a group known to be at the active site of the enzyme. This comparison includes the effect of substrate binding on the rate acceleration.

Any valid comparison requires knowledge of the catalytic pathway for the enzyme, so that a suitable nonenzymatic model reaction can be chosen. A serious problem in comparing rates is that the enzymatic reaction often proceeds by a different mechanism than that followed by the nonenzymatic reaction. This usually means that the comparison provides only a lower limit for the rate acceleration factor. A related problem is that the rates of many nonenzymatic reactions cannot be measured at all, because they are so slow that side reactions interfere with rate measurements. In these cases, lower limits for the rate acceleration or estimates that are based on related reactions are the best that can be achieved.

A reasonable comparison can be made for chymotrypsin. The first-order rate constant for the pH-independent hydrolysis of ethyl acetate in water is approximately $10^{-10}\ s^{-1}$, whereas k_{cat} for the pH-independent hydrolysis of *N*-acetyl-L-tyrosine ethyl ester bound to the active site of chymotrypsin at pH 8 is approximately $10^{2}\ s^{-1}$ at 25°C. These rate constants correspond to half-times of about 85 years and 0.01 second, respectively. These two esters do not differ much in their chemical reactivity, so that a quantitative understanding of the mechanism of catalysis by chymotrypsin requires an explanation for a rate acceleration factor on the order of 10^{12}. This is a very large number; it used to be even larger than the national debt of the United States. Different enzymes bring about widely different amounts of catalysis, but the factor of 10^{12} is typical of the rate increases that must be explained in order to understand the mechanism of catalysis by enzymes.

Catalysis from Noncovalent Binding Interactions

We now consider the question of how much the rate of a reaction can be increased as the result of noncovalent binding interactions between a specific substrate and the active site of an enzyme. The chymotrypsin-catalyzed hydrolysis of *N*-acetyl-L-tyrosine ethyl ester, a specific substrate, proceeds with rate constants of $5300\ s^{-1}$ and $195\ s^{-1}$ for the formation and hydrolysis of the acyl-enzyme intermediate, respectively. However, the corresponding rate constants for the chymotrypsin-catalyzed hydrolysis of ethyl acetate are so small that they cannot be readily measured. The value of k_3 for the hydrolysis of the acetyl-enzyme that is formed during the hydrolysis of *p*-nitrophenyl acetate is $7 \times 10^{-3}\ s^{-1}$, and the rate constant for the formation of the acetyl-enzyme from ethyl acetate must be much smaller than this.

Now, ethyl acetate is a smaller molecule than a specific substrate, such as *N*-acetyl-L-tyrosine ethyl ester, and can bind to the active site, but the enzyme shows little or no catalytic activity toward ethyl acetate. *Therefore, the rate constants quoted above show that the increase in acylation rate that is brought about from the noncovalent interactions of the enzyme with N-acetyl-L-tyrosine ethyl ester, compared with ethyl acetate, amounts to a factor of more than* 10^7.

How can the noncovalent interactions between the specific substrate *N*-acetyl-L-tyrosine ethyl ester and chymotrypsin cause a rate increase of more than 10 millionfold? We should make it clear before proceeding further that the answer to this question is not known in detail at the present time, for this or any other enzyme. However, several mechanisms that certainly contribute to such rate increases are known; and in a few cases it is possible to estimate the magnitude of the rate increase that can be brought about by a particular mechanism.

We have seen that covalent catalysis and general acid-base catalysis are two of the factors that contribute to the rate acceleration that is brought about by chymotrypsin. However, examination of nonenzymatic reactions indicates that the rate increases from general acid-base catalysis and nucleophilic reactions of the serine hydroxyl group, compared with water, can account for only about 10^4 of the total rate acceleration of $\sim 10^{12}$ by chymotrypsin that must be explained. We must look for other mechanisms to explain the large rate increases that are observed.

The same questions can be asked about most enzyme-catalyzed reactions. A particularly simple and challenging example is catalysis of bimolecular S_N2 displacement, such as the transfer of a methyl group from a sulfonium ion to a sulfide to give a sulfonium ion, as shown in equation 4.

$$(H_3C)(R)S^+{-}CH_3 + :S(CH_3)(R) \rightleftharpoons (H_3C)(R)S: + H_3C{-}{}^+S(CH_3)(R) \quad (4)$$

This simple reaction, which will be described in Chapter 25, is catalyzed by an enzyme that brings about an increase in reaction rate that is comparable to the rate increase brought about by chymotrypsin, but there is no possibility of general acid-base catalysis of this reaction; and nucleophilic catalysis is not useful because the sulfur atom of the sulfide is already a strong nucleophilic reagent. How can an enzyme accelerate a simple reaction of this kind?

Many enzymologists expected that determination of the three-dimensional structures of enzymes by X-ray diffraction of crystals would explain their catalytic activity, either through known mechanisms or through new and unexpected mechanisms. These structures have yielded valuable information; however, our most-optimistic expectations have not been realized. For example, the X-ray crystal structure of chymotrypsin confirmed the presence in the active site of the imidazole group of histidine-57 and the hydroxyl group of serine-195, which had already been identified by chemical methods, but it showed no indication of any remarkable new mechanism. The structure did show that the imidazole ring of histidine-57 is oriented properly in the active site by hydrogen bonding to the carboxylate group of aspartate-102 and that the hydroxyl group of serine-195 is oriented in the active site in position to form a hydrogen bond with histidine-57.

What is apparent from X-ray studies of chymotrypsin and many other enzymes is that specific substrates fit snugly into an active site that is located in a groove, cleft, or cavity and provides a well-defined structure with many contacts between the enzyme and its substrate. Many enzymes even have flaps, or domains, that fold down so as to engulf the substrate almost completely in the active site. This observation and the large rate increases for *specific* substrates support the idea that the binding interactions themselves supply the driving force that is responsible for much, or probably most, of the catalytic effectiveness of enzymes.

Intramolecular Reactions

Before examining more-complicated proposals, we should consider what can be gained from the most obvious and elementary mechanism that an enzyme can utilize to increase reaction rates—bringing the reacting and catalyzing groups together in the correct position to react. This simple mechanism can bring about much larger rate increases than was expected only a few years ago.

Consider two molecules, A and B, that can react with each other to form a product, such as an ester that reacts with a serine hydroxyl group to form an acyl-serine. In order for these molecules to react, they must first come together in *exactly* the correct relative positions to form a new covalent bond, as illustrated schematically in Figure 5-6. In the nonenzymatic reaction, this complex is

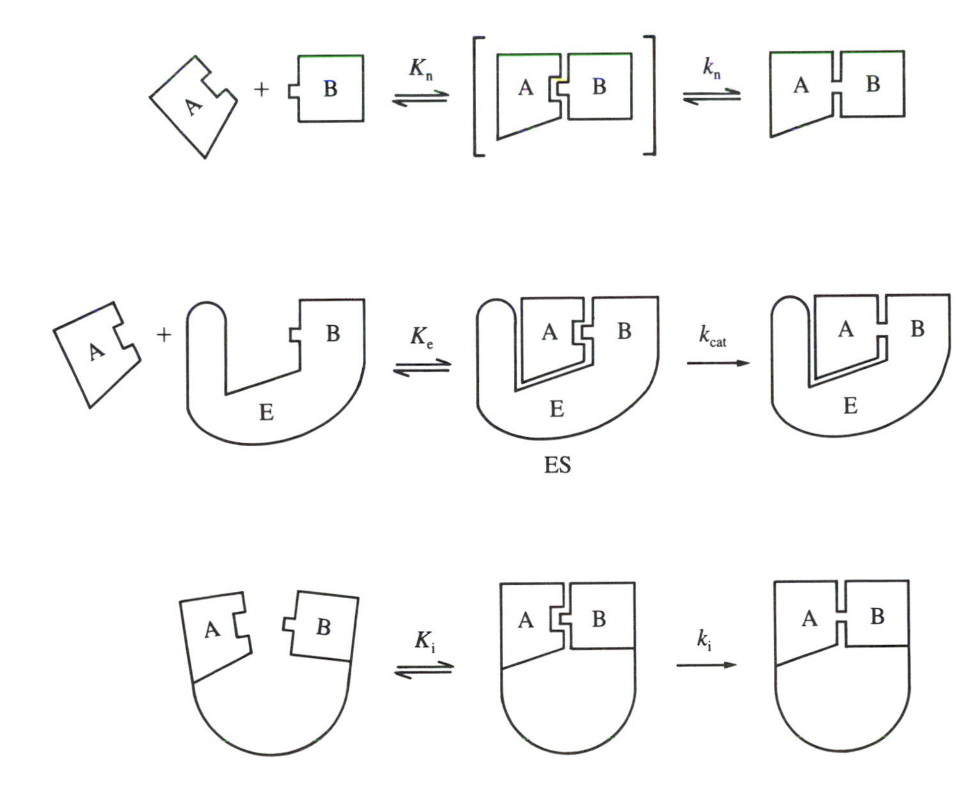

Figure 5-6

Top: A bimolecular reaction requires that the molecules A and B come together in precisely the correct position for reaction, with the equilibrium constant K_n, before they react, with the rate constant k_n. *Center:* An enzyme-catalyzed reaction requires binding of the substrate, with the association constant K_e, and reaction of the E·S complex, with the rate constant k_{cat}. *Bottom:* An intramolecular reaction also requires that A and B come together in precisely the correct position for reaction, with the equilibrium constant K_i, before they react with the rate constant k_i. If $k_i = k_n$, the advantage of the intramolecular compared with the bimolecular reaction is equal to K_i/K_n.

formed by random motions and collisions of A and B, with the equilibrium constant K_n, and it reacts to give products with the rate constant k_n. An enzyme that catalyzes the reaction of A with B utilizes the binding forces for its specific substrates to bring A and B together at the active site, in the ES complex, with a binding constant $K_e = 1/K_s$; they then react to give products, with a rate constant k_{cat}. Part of the catalysis that is brought about by the enzyme represents the utilization of the specific binding interactions at the active site to bring the reactants into a favorable position for reaction with each other in the enzyme-substrate complex, as shown in the center of Figure 5-6.

We now consider the question of *how much* of a rate increase can be brought about simply by bringing two reactants together at the active site. The most-direct estimation of this rate increase comes from a comparison of the rates of intermolecular and intramolecular reactions in solution, such as the upper and lower reactions in Figure 5-6. In the intramolecular reaction, the chemist brings the reacting groups together by linking them covalently, thereby constructing a simple model for the action of an enzyme in bringing them together at the active site (Figure 5-6, center).

The intramolecular reaction of a substituted phenyl succinate (Figure 5-7, top) is 100,000 times faster than the bimolecular reaction of acetate ion with the corresponding phenyl acetate, with all compounds at a concentration of 1 M (Figure 5-7, bottom). In both cases, the first product is the anhydride, which undergoes hydrolysis rapidly. The ratio of the rate constants for the mono- and bimolecular reactions of $k_1/k_2 = 10^5$ M has units of molarity because the first-order and second-order rate constants have units of s^{-1} and $M^{-1}s^{-1}$, respectively. This ratio is called the "effective molarity," because it is the concentration of acetate that would be required to make the bimolecular reaction of phenyl acetate (at 1 M) proceed as fast as the intramolecular reaction. This effective molarity of 10^5 M is much larger than any real

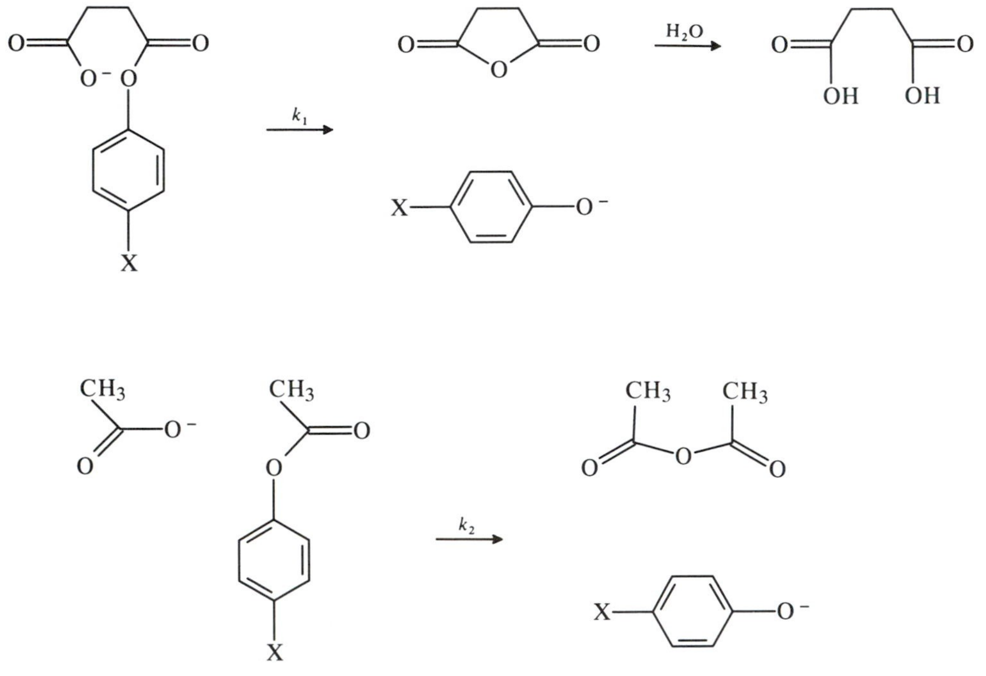

Figure 5-7

The intramolecular reaction of phenyl succinate (top) is approximately 10^5 times faster than the bimolecular reaction of 1 M acetate ion with phenyl acetate (bottom). The ratio k_1/k_2 is approximately $10^5\ s^{-1}/M^{-1}s^{-1} = 10^5$ M.

concentration that can be imagined (solid potassium acetate is 16 M). Therefore, we are forced to the remarkable conclusion that incorporating the carboxylate group into the ester molecule increases the reaction rate to a value that is about 10,000 times larger than it could be in a solution in which the phenyl ester is completely surrounded by carboxylate ions.

The Importance of Entropy

We now address the question of how is it *possible* that an intramolecular reaction can proceed at a rate that corresponds to the physically impossible concentration of 10^5 M for one of the reactants? The effect cannot be attributed to structural constraints in phenyl succinate that force the carboxylate ion into contact with the ester because the succinate ester has rotational freedom about three bonds, as shown in structure **2**.

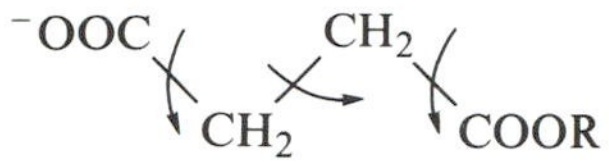

^{-}OOC — CH_2 — CH_2 — COOR

2

The rapid reaction rate of phenyl succinate can be understood on the basis of *entropy,* or the *probability* that the reaction will occur. The rate of a reaction is determined by two factors: the probability for the reaction and the activation energy (or heat) that is needed to make the reaction proceed. The likelihood that the acetate ion in the lower reaction in Figure 5-7 will find itself in *exactly* the right position and orientation to react with *p*-nitrophenyl acetate and form a new covalent bond is very small, so that the probability of the intermolecular reaction is very low. Even at a concentration of 1 M, or 10 M, the reacting molecules can exist in an enormous number of different positions that are not the exactly correct position required for the reaction to occur.

The probability that the reactants are in exactly the correct position for reaction corresponds to the equilibrium constant K_n of Figure 5-6, and is measured by the entropy of activation, $\Delta S^\ddagger$. Entropy can be regarded as a measure of probability, and a reaction with a large negative entropy of activation has a low probability of taking place. (Entropy also has a low probability of being easily conceptualized, and we shall return to it later.) The energy or heat needed for the reaction to proceed is measured by the enthalpy of activation, $\Delta H^\ddagger$, and will appear in the rate constant k_n of Figure 5-6.

Another way to think about this is that the carboxylate group in the intramolecular reaction of phenyl succinate (Figure 5-7, top) is held near the ester by covalent bonds, so that it cannot diffuse away and has many opportunities to react in a given time. On the other hand, the acetate ion in the intermolecular reaction of Figure 5-7 must find its way from some random location in solution to exactly the correct position for reaction with the ester; it always has the freedom to diffuse away instead of reacting. Thus, the intermolecular reaction has a much smaller probability for reaction and a more negative $\Delta S^\ddagger$ compared with the intramolecular reaction. Once the molecules are in the correct relative positions for reaction, the chemistry is essentially the same for the intermolecular and intramolecular systems, so that $k_n = k_i$ for the upper and lower reactions of Figure 5-6. The rate acceleration of 10^5 M means that the equilibrium constant for bringing the reactants into exactly the correct position in the intramolecular reaction, K_i in Figure 5-6, is 10^5 times larger than that for the bimolecular reaction, K_n in Figure 5-6.

The observed rate increase of 10^5 M for the intramolecular reaction in Figure 5-7 shows that an enzyme can accelerate a reaction by this amount if it can hold two molecules or reacting groups together as closely as they are held in phenyl succinate. If an enzyme can be even more efficient and hold the reacting groups even more exactly in the correct position, it can cause a correspondingly larger rate increase by further increasing the probability of reaction; that is, by reducing the amount of entropy that must be lost in order for the reaction to occur. However, there is a limit to the rate increase that can be obtained by this means. This limit is imposed by the maximum amount of entropy that can be lost, which corresponds to a rate factor of approximately 10^8 M for two reacting molecules or groups.

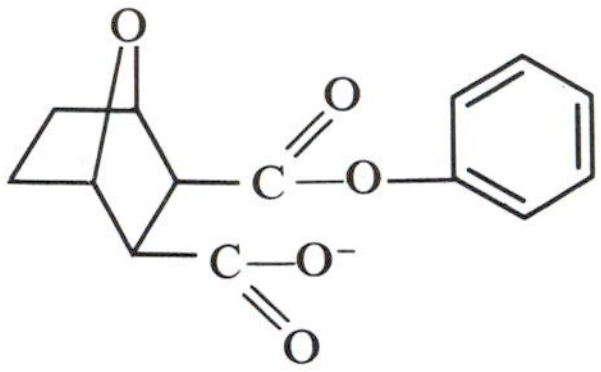

This maximum rate constant ratio has been observed for the bicyclic compound **3**, in which rotation of the C–C bonds of phenyl succinate is prevented and the carboxylate group is always in the correct position to react with the ester. In compound **3** the reaction appears to have a maximum probability and requires a minimal loss of entropy to reach the transition state.

Some intramolecular reactions show rate increases of more than 10^8 M compared with the corresponding bimolecular reactions. In the transition states of these reactions, strained reactants benefit from relief of unfavorable interactions. Strain has frequently been postulated in enzyme-catalyzed reactions, but few clear examples are known. The main advantage from holding reacting groups tightly together at the active site is probably to bring about loss of entropy; molecules must be held very firmly indeed in exactly the correct position to react if a large rate increase is to be obtained from loss of entropy.

Enzymes generally cannot hold reacting groups in position as accurately as they are held by the covalent bonds in intramolecular model systems, which are more rigid than enzymes. This accounts for the fact that the maximum advantage that has been observed from bringing reacting groups together at the active site of an enzyme is about 10^5 M, rather than the limiting value of 10^8 M. However, an additional rate increase can be realized if there is more than one group at the active site that participates in catalysis. The imidazole ring of histidine-57 is such an additional group in chymotrypsin; it catalyzes the reaction of serine-195 with the bound substrate and the reaction of an attacking water molecule with the acyl-enzyme intermediate. Each additional catalytic group might give an additional rate increase ranging from 10^2 to 10^4 M. This factor is much smaller than 10^8 M, because these groups usually catalyze proton transfer reactions. Such reactions have intrinsically high probabilities and entropies of activation, so that the rate increase that can be obtained by making them intramolecular is smaller.

We can conclude that one effect of substrate binding at the active site of an enzyme is to increase the probability for the reaction and decrease the requirement for loss of entropy, and this can account for large rate increases in the reactions of specific substrates. Nonspecific substrates for chymotrypsin, such as ethyl acetate, have low reactivities compared with specific substrates like *N*-acetyl-L-tyrosine ethyl ester, because they are not held tightly in the correct position at the active site. Intramolecularity induced by substrate binding also provides one explanation for catalysis by enzymes of simple S_N2 displacement reactions, such as the transfer of a methyl group from a sulfonium ion to a sulfide (equation 4). However, it is not yet possible to make an exact quantitative estimation of the catalytic advantage that can be attributed to this effect for the specific substrates of a particular enzyme.

Intramolecularity and the Chelate Effect

The advantage of favorable probability and entropy is also the basis for the *chelate effect*, which is an equilibrium rather than a rate effect. It is important in all fields of chemistry.

We shall draw an illustrative example from inorganic chemistry. Acetate ion has a low affinity for calcium ion, with $K_{eq} = 3\ M^{-1}$ (equation 5).

$$CH_3CO_2^- + Ca^{2+}(H_2O)_6 \xrightleftharpoons{K_{eq}} CH_3CO_2^- \cdot Ca^{2+}(H_2O)_5 + H_2O \quad (5)$$

However, ethylenediaminetetraacetic acid (EDTA), a molecule that can be regarded as four acetate ions incorporated into the same molecule (Figure 5-8), binds Ca^{2+} with an equilibrium constant of $10^{10.6}\ M^{-1}$. The probability that an acetate ion and a calcium ion will come together from dilute solution is small, and the probability that two acetate ions and one calcium ion will come together is much smaller. Each acetate ion that binds to calcium releases one coordinated water molecule. The coordination of EDTA contrasts sharply with that of acetate because much of the binding is intramolecular, although the chemical properties of the coordinating groups are very similar.

The probability for the binding of one carboxylate group of EDTA to calcium is similar to that for acetate, as illustrated in the first binding step of Figure 5-8. However, the probability for the second, third, and fourth carboxylate groups to bind is much larger, because of intramolecularity. Figure 5-8

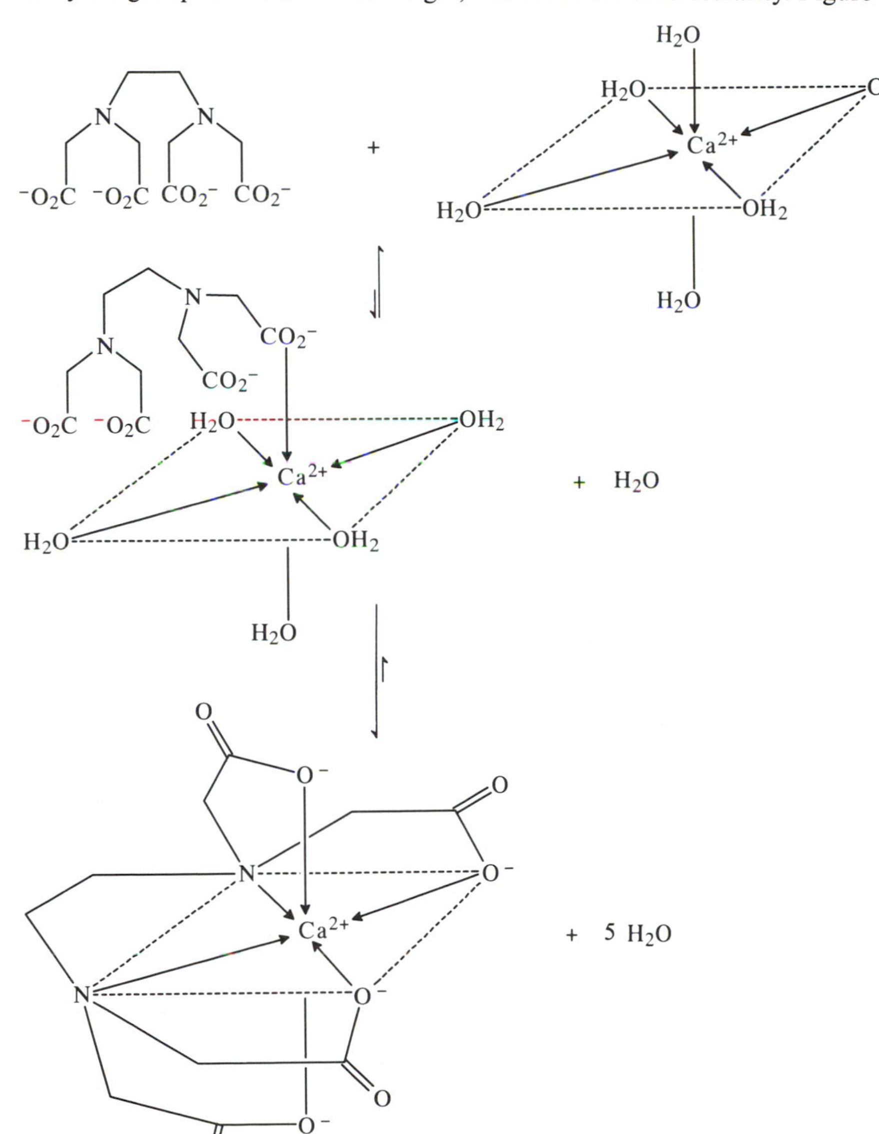

Figure 5-8

Chelation of Calcium Ion by EDTA. Formation of the first bond between EDTA and calcium ion is a bimolecular reaction with an unfavorable equilibrium constant. However, once this bond has formed, the additional bonds form in intramolecular reactions, with relatively little loss of entropy. Therefore, these steps are very favorable and pull the reaction toward completion.

shows that this process has a high probability because it does not immobilize any additional EDTA molecules in the solution, and it results in the release of five water molecules that are free to move through the solution. This increases the entropy of the system. Once one acetate group of EDTA is bound to the ion, the probability that the other acetate groups will bind to form a chelated structure is several orders of magnitude larger than in the intermolecular reaction.

Stated another way, the binding of a calcium ion to four acetate ions requires the loss of entropy of four acetate ions that had been free to move independently through the solution. If acetate ion were the ligand, the first binding step would be similar to that for EDTA in Figure 5-8, but the binding of three more acetate ions in the second phase would immobilize these ions with a large loss of entropy. In contrast, the binding of EDTA requires loss of entropy for only a single molecule of EDTA, so that it is much more probable. A more-complete description of how entropy is responsible for chelation and the rapid rates of intramolecular reactions is given in Box 5-2.

BOX 5-2

MORE ABOUT ENTROPY

In order for two molecules, A and B, to come together to form a product, A-B, or a transition state, $[A\text{-}B]^{\ddagger}$, they must lose their freedom to move about and rotate independently. These losses are improbable happenings, and the amount of improbability is measured by the entropy of activation, $\Delta S^{\ddagger}$. The freedom of a molecule to move about and be located anywhere in a given volume is measured by its *translational entropy*. For 1 M standard states in the gas phase, the translational entropies, S_{trans}, of A, B, and A-B are typically close to 30 entropy units (e.u., cal K^{-1} mol^{-1}) for each molecule, so that the formation of a transition state or product requires the loss of ~30 e.u., as shown in the adjoining table.

Translational, rotational, and internal entropies

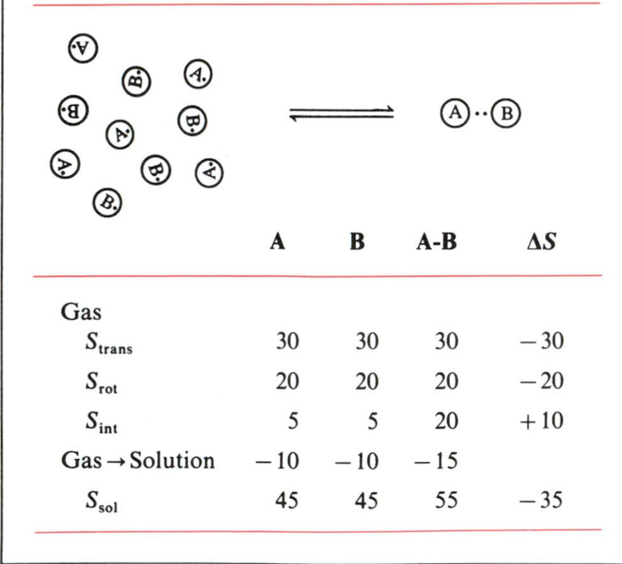

	A	B	A-B	ΔS
Gas				
S_{trans}	30	30	30	−30
S_{rot}	20	20	20	−20
S_{int}	5	5	20	+10
Gas → Solution	−10	−10	−15	
S_{sol}	45	45	55	−35

The freedom of a molecule to rotate into any position about its three axes is measured by its *rotational entropy*, S_{rot}. The rotational entropy is more variable than the translational entropy, but if A, B, and A-B are typical nonlinear organic molecules it is about 20 e.u. No more entropy can be lost than is present in the reactants, so that this sets an upper limit of about −50 e.u. on the total entropy change.

The observed loss of entropy will be less than the maximum loss if the transition state or product has a large *internal entropy*. This arises from internal motions, such as low frequency vibrations or rotations about bonds that are absent in the reactants. For example, if we assume that the internal entropy is 5 e.u. for A and for B, and 20 e.u. for A-B, there is a gain of $S_{int} = +10$ e.u. in the reaction. The freedom of movement and entropy of molecules are smaller in solution than in the gas phase by 10 to 15 e.u. (for a 1 M standard state in both phases). The total loss of entropy that is required for the formation of A-B from A and B in solution in the example shown in the table is 35 e.u., which corresponds to a factor of 10^8 M. In general, the maximum entropy change from losses of translational and rotational entropy for reactions in solution is from −35 to −40 e.u., which corresponds to factors ranging from 10^8 to 10^9 M. However, much smaller changes are possible if the transition state or product is loose or floppy and, therefore, has a large internal entropy.

These entropy changes account for all known examples of chelate effects and rate accelerations in intramolecular reactions that do not involve strain. Intramolecular reactions do not require the loss of translational or overall rotational entropy. They usually require only the loss of the entropy from internal rotations, which are typically about 4 e.u. each.

BOX 5-3

CATALYSIS BY UNREACTIVE GROUPS ON ENZYMES: THE "ANION HOLE" IN CHYMOTRYPSIN

The correct positioning of catalytic groups at the active site of an enzyme can give rise to significant catalysis even for groups that have no detectable catalytic activity in solution. For example, there is an "anion hole" in the active site of chymotrypsin and related enzymes that is formed from two peptide bonds. The N–H groups of these bonds have a partial positive charge that stabilizes the developing negative charge on a substrate molecule when it is attacked by the serine hydroxyl group, as shown below. Simple peptides are not effective catalysts for reactions of this kind in aqueous solution. However, catalysis by such groups is possible when they are positioned correctly at the active site of the enzyme, because they are already in the correct position to stabilize the developing negative charge by hydrogen bonding. The optimal stabilization is obtained when the tetrahedral addition species is formed, with an increase in length of the C–O bond and a full negative charge on the oxygen atom.

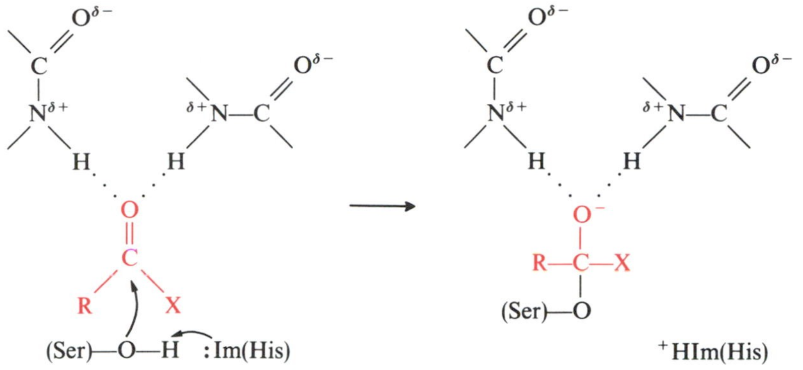

Chelation is responsible for the tight binding of metal ions to enzymes and other proteins. The liganding groups are usually the carboxylate groups of aspartate or glutamate, the nitrogen atoms of imidazole or lysine, the oxygen atoms of peptide bonds, and the thiol anion of cysteine.

Intramolecularity and the chelate effect are important for many biochemical systems, including the binding in enzyme-substrate complexes. Many biochemical substrates, such as nucleotides, contain polar, ionic, hydrogen-bonding, and hydrophobic groups in their structures that interact with complementary groups in the E·S complexes. Any one such interaction may be relatively weak; but, once one has been formed, all of the others become intramolecular and are much more stable. This effect also accounts for the extraordinarily high binding affinities of proteins with other molecules in many biological systems, such as the binding of antibodies to antigens. It can also bring about catalysis by properly located groups at the active site of an enzyme that would not cause catalysis in solution (Box 5-3).

The Importance of Binding for Specificity and Catalysis

Specificity is perhaps the most important and characteristic property of enzymes. It is the property that most clearly separates enzymes from synthetic catalysts and enzyme models. Specificity is usually thought of in terms of the good fit of a specific substrate into the active site, according to some sort of "lock and key" model, and catalysis is considered separately, in terms of the chemistry that occurs after the substrate has bound. However, on closer examination, it becomes clear that the separation of catalytic activity into specific binding and

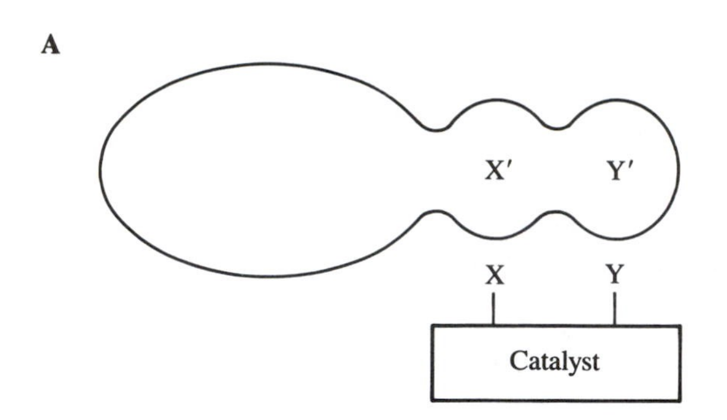

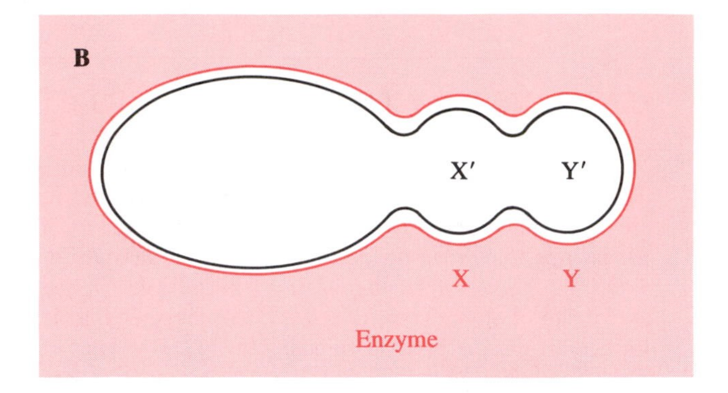

Figure 5-9

A chemical catalyst may catalyze a reaction by using groups X and Y to interact with groups X′ and Y′ of the substrate. An enzyme may have these same catalytic groups, but it can also utilize the noncovalent binding interactions with a specific substrate to increase the reaction rate of the substrate.

chemical catalysis is artificial and misleading. We have seen that chymotrypsin reacts much faster with a specific substrate than with a smaller substrate and that the catalytic activity of most enzymes toward their specific substrates is much greater than it is toward nonspecific substrates in the E·S complex, as measured by k_{cat} for the reaction of bound substrate. This is one of many pieces of evidence that show that the unique ability of enzymes to interact with the substituent groups of their specific substrates contributes directly to their catalytic activity. The binding forces between enzymes and specific substrates are *utilized* to bring about rate accelerations.

This difference between enzymes and most chemical catalysts is summarized schematically in Figure 5-9. A chemical catalyst may have groups, such as X and Y, that react with groups in the substrate, X′ and Y′, to bring about a rate increase (Figure 5-9A). The active site of an enzyme may contain these same catalytic groups, such as the hydroxyl group of serine and the imidazole group of a histidine residue in chymotrypsin, but it also has a three-dimensional structure that interacts with many other parts of a specific substrate. It utilizes the noncovalent interactions with these groups to bring about further rate increases (Figure 5-9B). These rate increases can be very large indeed, by factors of as much as $\sim 10^{10}$, and may account for the greater part of the total catalysis. These interactions bring about a *utilization* of the binding energy between the enzyme and these groups on specific substrates. The same kind of utilization of binding energy is required in the ligand-receptor interaction of hormones, neurotransmitters, and other molecules that control the activities of an organism. An example of the use of hydrophobic binding interactions in catalysis by chymotrypsin is described in Box 5-4.

Mechanisms for Catalysis Through Binding

There are four ways that an enzyme can utilize binding interactions with a specific substrate to increase the observed reaction rate of the bound substrate.

1. Binding interactions with a specific substrate may be utilized to fix the substrate tightly in precisely the correct position at the active site, so

BOX 5-4

HYDROPHOBIC BINDING AT THE ACTIVE SITE OF CHYMOTRYPSIN

We have seen that nonpolar, hydrophobic substituents of specific substrates for chymotrypsin give high activities because these substituents fit into the specificity pocket of this enzyme. A quantitative estimation of the importance of binding to the specificity site of hydrophobic groups can be obtained by comparing the rates of hydrolysis of a series of *N*-acetylaminoacyl esters with the partition coefficients for distribution of molecules containing the hydrophobic substituents, R, between water and octanol. These partition coefficients provide a quantitative measure of the hydrophobic character of a series of hydrocarbon substituents of increasing size. A plot of $\log(1/K_s)$ for the different substrates against the logarithm of the partition coefficient has a slope of 1.0. This shows that the increase of substrate binding can be accounted for by hydrophobic binding to the active site. What is not expected is that a similar plot of $\log k_2$ against the logarithm of the partition coefficient also has a large slope of 1.2. The slope of the plot for the second-order rate constant, k_{cat}/K_m, is therefore 2.2. This shows that the binding interactions of the hydrophobic substituent with the active site cause a large increase in the rate of hydrolysis of the substrate even after it is bound to the enzyme. It also shows that the total increase in rate that can be obtained from these binding interactions is much larger than is predicted by this measure of hydrophobic interactions.

$$H_3C{-}\overset{\overset{\displaystyle O}{\|}}{C}{-}NH{-}\overset{\overset{\displaystyle COOR'}{|}}{\underset{\underset{\displaystyle R}{|}}{C}}{-}H$$

***N*-Acetylaminoacyl ester**

that it can react with a high probability and a small loss of entropy. This was described earlier.

2. The binding energy of a specific substrate may be utilized to *destabilize* the substrate in the E·S complex, so that it reacts faster. This destabilization can be through geometric strain, electrostatic interactions, or desolvation of reacting groups at the active site. Some mechanisms for destabilization are described in the next section.
3. A specific substrate can bring about a change in the conformation of an enzyme that converts it into a catalytically active state. This is sometimes called *induced fit*.
4. The binding energy of a specific substrate may be utilized to induce it to bind productively in the active site, whereas a nonspecific substrate may bind nonproductively in an incorrect position. This can decrease k_{cat} for poor substrates, but it has no effect on k_{cat}/K_m and it does not contribute to catalysis.

These mechanisms may overlap—for example, the geometric strain of a substrate from binding to an enzyme is unlikely to occur without some change in the conformation of the enzyme. And the tight fixation of a substrate to cause maximal loss of entropy is likely to be accompanied by compression and strain. Nevertheless, it is useful to consider the mechanisms separately for purposes of description and analysis.

Destabilization Mechanisms

Destabilization or strain in the enzyme-substrate complex is the first mechanism that was suggested for catalysis by enzymes; it has been popular ever since. We will see that in a general sense destabilization is an essential feature of catalysis by enzymes, but in its more restricted sense of catalysis through physical strain or compression it is probably less important than has often been believed.

There are three mechanisms for destabilization that can increase the reaction rate of substrates in the E·S complex:

1. The best-known mechanism is *geometric destabilization,* in which binding of a substrate to an enzyme pushes, pulls, bends, or twists reacting groups in such a way as to make them react faster. However, there are only a few well-documented examples of catalysis by this mechanism in enzymatic reactions; this kind of destabilization is better represented in nonenzymatic reactions. It has frequently been suggested that nucleophilic displacement reactions occur by compression of the nucleophile against the substrate, in order to assist bond formation. But it is hard to distinguish this mechanism from exact fixation and a decrease of entropy, which also requires that the reactants be held firmly against each other. The most important role of physical compression is probably to hold the reacting groups very firmly in exactly the right position for reaction, with a large loss of entropy, as described earlier.

Geometric destabilization is more likely to be important for reactions that require bending, with changes in bond angles. Catalysis by this mechanism occurs when binding groups at the active site interact weakly or unfavorably with the bound substrate but interact favorably in the transition state of the catalyzed reaction. The bound substrate then has a low stability and the transition state is stabilized when the bond angles of the bound substrate have changed enough to allow favorable interactions with the active site.

2. *Electrostatic destabilization* of charged groups on the bound substrate and the active site can bring about a rate increase if the reaction results in loss of charge on the substrate. We have seen an example of electrostatic destabilization in the decrease of 4.7 units, from 10.7 to 6.0, in the pK_a of the protonated 6-amino group of a lysine residue in the active site of acetoacetate decarboxylase. This decrease in pK_a is caused by destabilization of the charged RNH_3^+ group by one or more adjacent positive charges in the active site (see Box 4-2).

3. *Destabilization by desolvation* of charged groups on substrates bound in the active site is a particularly appropriate mechanism for an enzyme to utilize in order to bring about a rate increase, because it does not require a rigid enzyme. Charged groups are very strongly stabilized by interaction with the solvent in aqueous solution. The transfer of some ions from the gas phase to water is favorable by more than 100 kcal mol^{-1}, which corresponds to stabilization in water by a factor of more than 10^{70}. Therefore, charged groups become much less stable, and hence more reactive, in an environment that does not provide good stabilization by solvation.

Decarboxylation of a Pyruvate Adduct of Thiamine Pyrophosphate

The decarboxylation of compound **4** provides an excellent example of catalysis by desolvation (equation 6).

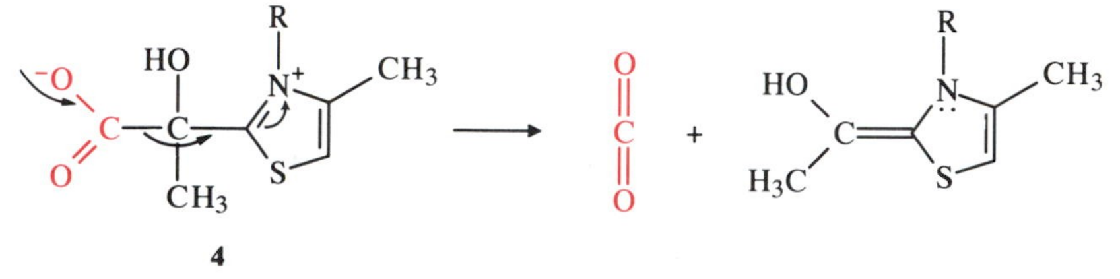

$$\mathbf{4} \longrightarrow O{=}C{=}O \; + \; \ldots \qquad (6)$$

This compound is a model for the adduct of pyruvate with the coenzyme thiamine pyrophosphate, TPP, which will be described when decarboxylation is discussed in Chapter 17.

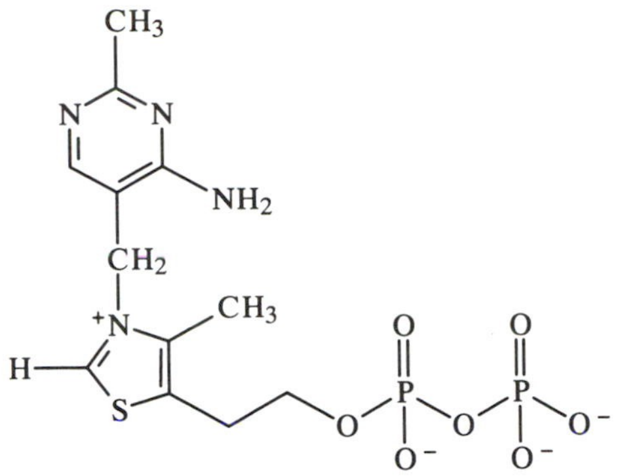

Thiamine pyrophosphate (TPP)

Compound **4** undergoes decarboxylation from 10^4 to 10^5 times faster in ethanol than in water and reacts even more rapidly in aprotic solvents, which cannot form hydrogen bonds with the —COO^- group. The reaction involves the conversion of a doubly charged reactant into uncharged products; and the large rate increase in ethanol is caused, at least in part, by the fact that the carboxylate group is strongly solvated by water but is much less stable and is correspondingly more reactive in ethanol.

The enzyme pyruvate decarboxylase catalyzes decarboxylation of the pyruvate adduct of TPP and has been shown to have a hydrophobic active site. Evidence for this is provided by nonpolar dyes that bind to the active site and serve as indicators for hydrophobic regions. It is likely that the poor solvating environment of this active site increases the rate of decarboxylation by this kind of destabilization mechanism. Nonpolar solvents and the active site provide a poor solvation and a hostile environment for the charges on compound **4** but are more hospitable to the transition state; the active site can grasp the uncharged product with a strong attractive force. This causes an increase in rate because there is a more-favorable interaction of the active site with the transition state than with the bound carboxylate adduct of TPP in the ground state and, therefore, a smaller barrier for the reaction.

This relative stabilization is even larger for compound **5**, which is a derivative of thiamine pyrophosphate with an uncharged ring.

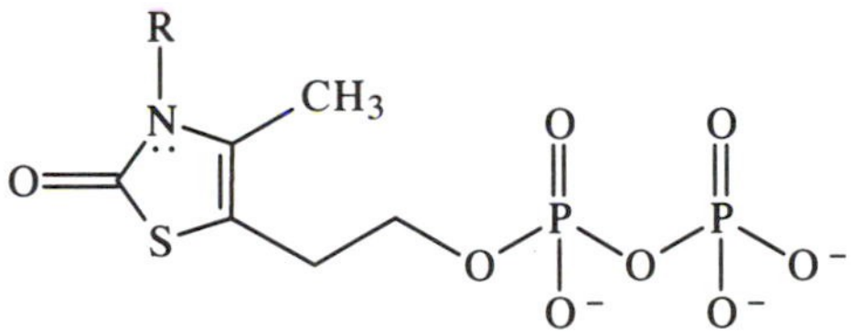

5

Thiamine thiazolone pyrophosphate

This compound binds strongly to pyruvate decarboxylase, with a dissociation constant of $<5 \times 10^{-10}$ M. This is much stronger binding than is observed for TPP itself, which has $K_{diss} = 10^{-5}$ M. Compound **5** lacks the carboxylate group of compound **4** and is uncharged, so that it is not destabilized by a nonpolar environment and binds very strongly to the active site. *The very strong observed binding means that there is a large amount of binding energy available for a planar, uncharged group at the active site.* The same binding energy is potentially available for compound **4** and for TPP itself, but the observed binding is weaker because of the destabilization of the charged groups.

Transition States, Transition State Analogs, and Catalysis by Enzymes

We have seen that it is useful to think of chemical and enzymatic reactions in terms of transition states, the hypothetical species at the highest point of the energy barrier for the reaction. Catalysis by enzymes decreases this barrier for the reaction of the enzyme-substrate complex, as shown in Figure 5-10. This decrease represents a stabilization of the transition state when it is bound to the enzyme, compared with the transition state in solution. In other words the enzyme has a very high affinity for binding the transition state. A transition state has no real lifetime or existence as a chemical species because it breaks down to reactants or products faster than a bond vibration, in 10^{-13} to 10^{-14} second, but it has a definable structure.

The decarboxylation of compound **4**, which has two charges, gives a product with no charge, as shown in equation 6, so that the transition state for this reaction, structure **6**, will have less charge than compound **4** and more charge than the product.

6

Compound **5**, thiamine thiazolone pyrophosphate, is called a *transition state analog* because its strong binding represents the same kind of very strong binding that is observed for the transition state, owing to its decreased charge

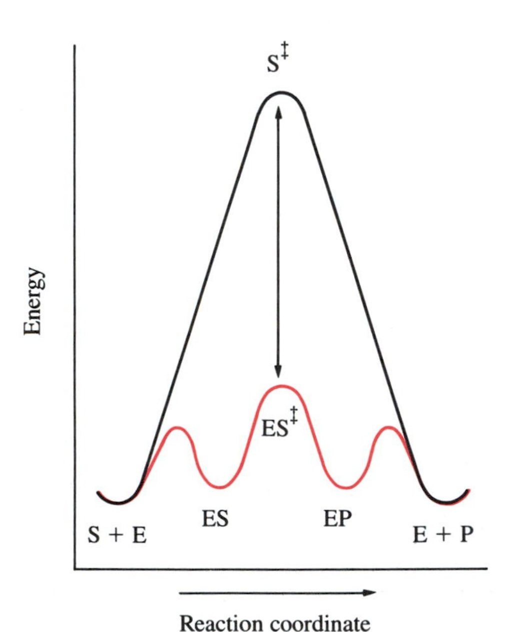

Figure 5-10

An enzyme reduces the barrier for a reaction by stabilizing the transition state for reaction of the E·S complex. This causes a large increase in the reaction rate of E·S compared with S.

and decreased requirement for charge solvation. The strong binding of the transition state analog gives an approximate indication of the stabilization that would be available for binding a substrate molecule if there were no destabilization by poor solvation.

Transition state analogs are an attractive tool for probing the nature of the active site and the mechanism of catalysis by enzymes. Strong binding of a molecule that resembles a postulated transition state supports the hypothesis that the reaction proceeds through that transition state. However, it does not prove the hypothesis; and conclusions from such comparisons should be treated cautiously, because strong binding may occur for many other reasons.

Transition state analogs are powerful inhibitors of enzymes because of their strong binding to the active site. Therefore, they may be useful if it is desired to inhibit an enzyme for the investigation or treatment of a disease, when the activity of the enzyme is important in the disease.

The stabilization, or strong binding, of the transition state means that the active site is complementary to the transition state. The structures of the substrate and the transition state are different, so that there will be less-favorable binding of the substrate. The imperfect, weaker binding of the substrate corresponds to the destabilization or strain relative to the transition state that we have been considering. This *difference* in the stabilization, or binding, of substrates and the transition state is the basis for enzyme catalysis. This was first pointed out clearly by Linus Pauling.

Requirements for Catalysis

Stabilization of *both* the bound substrate *and* the transition state will not give catalysis; there must be stabilization of the transition state and less stabilization, or a relative destabilization, of the bound substrate. The reason for this is shown in Figure 5-11. First, consider reaction A, which has a certain energy barrier in

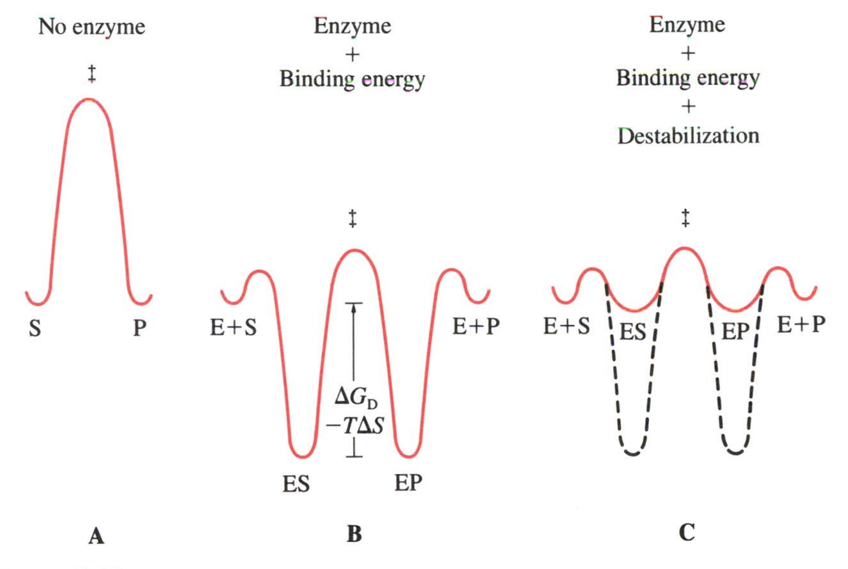

Figure 5-11

If an enzyme stabilizes the E·S complex and the transition state for the reaction of S by the same amount, there is no catalysis, because the barrier for the reaction of E·S (B) is the same as the barrier for the reaction of S (A). The enzyme becomes an effective catalyst when it stabilizes the transition state much more than the E·S and E·P complexes (C). This is brought about by destabilization mechanisms (ΔG_D) and loss of entropy ($-T\Delta S$) in the E·S and E·P complexes.

the absence of the enzyme. Then, consider system B, in which the transition state and the complexes of enzyme with substrate (E • S) and with product (E • P) are all stabilized by the same amount. This system has a large stabilization of the transition state. However, the enzyme is useless because the barrier for the reaction of E • S to reach the transition state is exactly the same as the barrier for the reaction of S in the absence of enzyme.

In order to have catalysis, it is necessary to *destabilize* the E • S and E • P complexes, or to prevent their stabilization, as shown in Figure 5-11C. This can be done by the induction of strain, desolvation, or other mechanisms, as indicated by ΔG_D in Figure 5-11B.

We have seen that an important part of the effective barrier for the reaction involves bringing the reacting groups together in exactly the correct position to react, with a loss of entropy. This part of the barrier can be overcome by bringing the reactants into nearly the correct position for reaction in the E • S complex. This is improbable and corresponds to a decrease in entropy, as shown by the $-T\Delta S$ term in Figure 5-11B.

The destabilization of the E • S complex, G_D, and the loss of entropy, $-T\Delta S$, correspond to the increase in the Gibbs energy of the E • S complex in Figure 5-11C that makes it easier for the E • S complex to reach the transition state and form products. A more formal description of these relationships will be presented in Chapter 8.

The situation is summarized by the energy diagram in Figure 5-12. The left and right sides of Figure 5-12 show that the barrier for the reaction of the E • S complex is much smaller than the barrier for the reaction of the free substrate. It follows that the stabilization of the transition state, S‡, by the enzyme in E • S‡, which is shown by the upper arrow, is much larger than the stabilization of the substrate upon binding to the enzyme to give E • S, shown by the lower arrow. If the stabilization of the bound substrate were as large as the stabilization of the transition state, as illustrated by the energy of ES′ in Figure 5-12, the enzyme would not catalyze the reaction because the barrier for the reaction would be just the same for ES′ as for S.

In order for the enzyme to work, there must be *both* stabilization of the transition state *and* much less stabilization of the E • S complex. In other words,

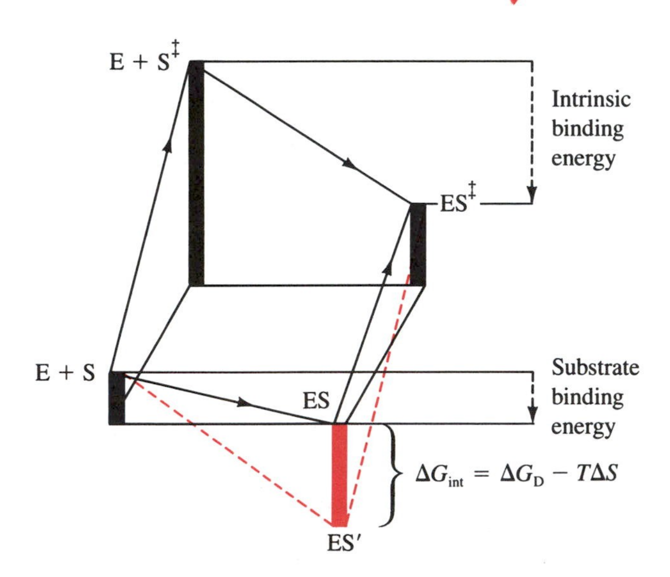

Figure 5-12

An Energy Bar Diagram That Describes Catalysis by an Enzyme. The barrier for ES to reach the transition state and react is much smaller than that for S.

there must be a relative destabilization of the E·S complex, compared with the transition state, ES‡. The difference between the amount of stabilization of the transition state and of the bound substrate is the *interaction energy,* ΔG_{int}. The interaction energy arises from simple destabilization, ΔG_D, and from loss of entropy, $-T\Delta S$, because of the low probability of binding the substrate in exactly the correct position for reaction in the transition state. Thus, the primary role of desolvation, strain, conformational changes and entropy loss is to destabilize the E·S complex, so that it can be converted into the transition state easily and the situation shown in Figure 5-11B is avoided.

Nonproductive Binding

The fourth general mechanism by which the value of k_{cat} can be increased for specific substrates compared with nonspecific substrates is *nonproductive binding.* If a specific substrate binds in the correct position, catalysis will be efficient and k_{cat} will be large; but, if small, incomplete substrates bind incorrectly in various ways, only a small fraction of the several species of E·S complexes will contain the substrate bound in the correct position to react. Therefore, the poor substrates will have small values of k_{cat}. Thus, nonproductive binding results in relatively large values of k_{cat} for specific substrates compared with nonspecific substrates. However, it does not contribute to catalysis for reactions of good substrates and it has no effect on k_{cat}/K_m, the second-order rate constant for the reaction of E and S.

An interesting example of nonproductive binding is observed with D- and L-amino acid esters and is described in Box 5-5. The nonphysiological D-enantiomers show "negative specificity" for k_{cat}, because the addition of specific side chains increases the fraction of L-substrate molecules that bind productively, but increases the fraction of D-substrate molecules that bind nonproductively.

Conformational Changes and Catalysis

Many enzymes undergo a change in conformation when their specific substrates bind at the active site. This change may bring the catalytic groups at the active site into the correct position to interact with reacting groups of the substrate. A conformational change of this kind is described by the term induced fit, as noted earlier in this chapter. We have seen an example of such a conformational change when a substrate or inhibitor binds to chymotrypsin at alkaline pH and forces the enzyme to change from the inactive to the active conformation by taking up a proton and forming an ion pair (Asp-194—COO^- · ^+H_3N—Ile-16).

In many enzymes, the active site is located in a pocket or crevice that may be lined by amino acids that are located in several different domains or subunits of the enzyme. In some enzymes a flap folds down over the substrate when it binds, so that the substrate is almost completely surrounded by the enzyme. Binding sites of this kind are well suited to provide the largest possible interaction of groups in the enzyme with a specific substrate. These interactions can provide a large amount of binding energy for stabilizing the transition state. They can also hold the substrate in place very firmly in the correct position to react, with a large loss of entropy, and can cause destabilization of the ES complex so that the transition state can be reached more easily. Conformational changes may be required in order to bring about this strong interaction with the substrate; they are certainly required when a flap folds down to cover the bound substrate.

BOX 5-5

NEGATIVE SPECIFICITY

Hydrolysis of the L- and D-enantiomers of the *p*-nitrophenyl esters of *N*-acylamino acids catalyzed by chymotrypsin was studied by Ingles and Knowles. These reactions occur with rate-limiting deacylation of the acylenzyme intermediate at substrate saturation and are faster for the L-substrates than for the D-substrates. In fact, derivatives of D-amino acids, which are nonphysiological, are generally inhibitors of proteolytic enzymes.

Figure A shows the values of k_{cat} for hydrolysis of the glycine derivative and several more specific D- and L-substrates. The rate constants have been corrected for small differences in the chemical reactivity of the substrates. The hydrolysis of the L-substrates increases in the order from glycine, to leucine, to phenylalanine, to tryptophan, but rates of the D-substrates *decrease* in the same order. This has been called *negative specificity*—substituents that bind favorably at the active site and increase the reaction rate of L-substrates cause a decrease in the rate with D-substrates.

This behavior can be explained by progressively better *productive binding* of specific L-substrates and progressively more *nonproductive binding* of the enantiomeric (mirror image) substrates when substituents that bind well to the active site are added. Figure B (on facing page) shows schematically how a specific *N*-benzoyl-L-phenylalanine substrate will bind productively in the active site. A smaller derivative of glycine can bind productively, with the binding constant K_{B1}, but is also likely to bind in one or more nonproductive modes (K_{B2}, K_{B3}), in which the reactive groups are not properly aligned in the active site. Figure B shows binding of the complete substrate, but similar behavior can occur with the acyl-enzyme.

The substituents on the specific L-amino acid substrates in Figure B decrease the amount of binding in nonproductive modes, so that the fraction of nonproductive binding is decreased and the observed catalytic constant increases, as described by equations a and b (other factors can also contribute to this rate increase).

$$K_{B(obsd)} = K_{B1} + K_{B2} + K_{B3} \cdots \tag{a}$$

$$k_{cat(obsd)} = k_c \frac{K_{B1}}{K_{B1} + K_{B2} + K_{B3}} \tag{b}$$

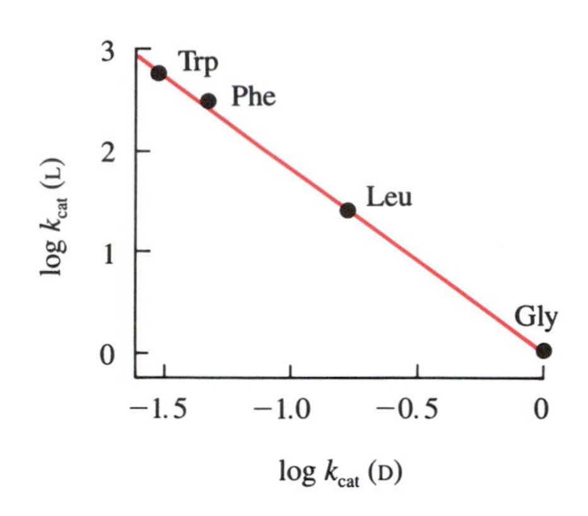

Figure A

Plot of log k_{cat} for D- versus L-acyl-α-chymotrypsins. (After D. W. Ingles and J. R. Knowles, *Biochem. J.* 104 (1967): 369.

The same substituents on the D-amino acid derivatives *increase* the amount of nonproductive binding and cause a corresponding decrease in the catalytic constant.

Although nonproductive binding decreases k_{cat}, it has no effect on k_{cat}/K_m (equation c).

$$\frac{k_{cat(obsd)}}{K_m} = k_{cat(obsd)} K_{B(obsd)} = k_c K_{B1} = k_c/K_{m(productive)} \tag{c}$$

On the other hand, too much movement of the enzyme decreases the effectiveness of catalysis because it implies that the enzyme can exist in a large number of different conformations, only one of which is optimal for catalysis. The ideal enzyme binds and stabilizes the transition state in a single conformation and undergoes the minimal number of conformational changes that are needed in order to bind the substrate in this conformation. There is seldom

BOX 5-5 (continued)

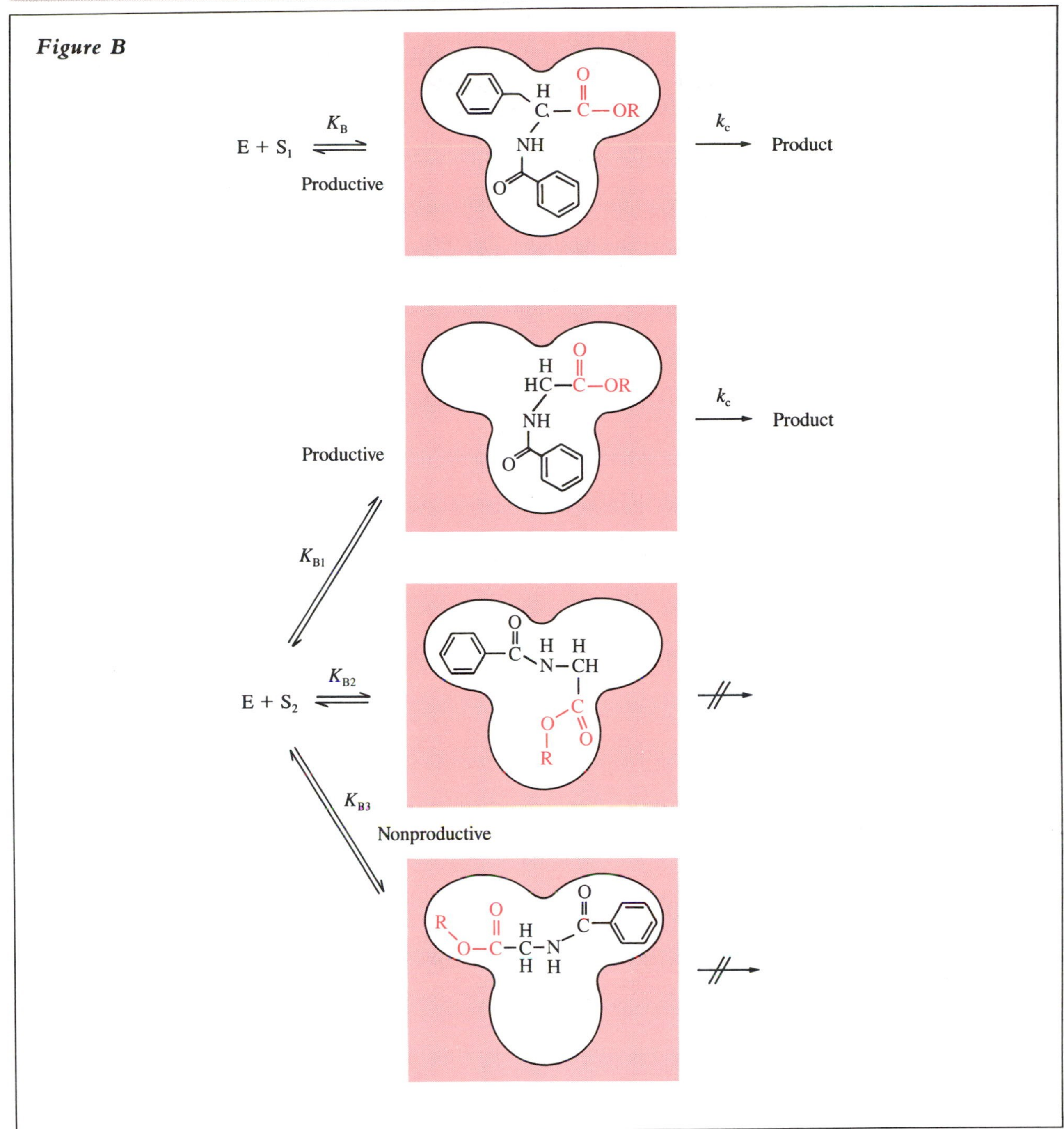

an advantage for an enzyme to be initially in an inactive conformation and then undergo a change to an active conformation when the substrate binds. The enzyme is most effective when it is in the active conformation initially, so that binding energy does not have to be used up in order to bring it into the active conformation.

Summary

The specific binding interactions between substrates and binding sites on enzymes are responsible for a large fraction of the rate increases that are brought about by enzymes. Similar specific binding interactions control many physiological functions through the binding of hormones, neurotransmitters, and other regulatory molecules to receptors. This specific binding increases reaction rates within the active site of enzymes by holding the reacting groups in precisely the correct position to react, with a large loss of entropy, and by destabilizing the bound substrates so that they can reach the transition state more easily. The binding interactions are fully expressed to stabilize the transition state. Binding energy can also be used to cause a change in the conformation of an enzyme or a receptor. These changes can be described by interaction energies and a thermodynamic box.

ADDITIONAL READING

FERSHT, A. R., and REQUENA, Y. 1971. Equilibrium and rate constants for the interconversion of two conformations of α-chymotrypsin. *Journal of Molecular Biology*, 60: 279.

JENCKS, W. P. 1975. Binding energy, specificity, and enzymic catalysis: the Circe effect. *Advances in Enzymology*, 43: 219.

JENCKS, W. P. 1987. Economics of enzyme catalysis. *Cold Spring Harbor Symposium*, 52: 65.

PROBLEMS

1. Draw a hypothetical active site for the reaction of a sulfide with a sulfonium ion (equation 4) and describe the mechanisms that the enzyme might utilize to catalyze this reaction.

2. Acetate ion binds Ca^{2+} to give $AcO^- \cdot Ca^{2+}$ with an association constant of $K_a = 10\,M^{-1}$, whereas malonate dianion, $^-OOCCH_2COO^-$, binds Ca^{2+} more strongly with $K_a = 500\,M^{-1}$. Explain, and calculate the effective molarity of the second carboxylate group of malonate.

3. Write the equations for the binding of a hormone to the inactive form of a hormone receptor and to the active form that sets off a physiological response. Connect the two equations to make a thermodynamic box and draw an energy diagram to show how the receptor is converted into the activated state by binding of the hormone.

4. The reaction of chymotrypsin with one equivalent of acetic anhydride at pH 9 causes irreversible loss of enzyme activity. Why does this occur?

5. Acetylcholinesterase catalyzes the hydrolysis of acetylcholine with a second-order rate constant that is close to diffusion controlled, but catalysis of the hydrolysis of ethyl acetate is extremely slow. Describe probable reasons for this difference.

More Hydrolytic Enzymes, Blood Clotting, and the Antigen-Antibody Response

6

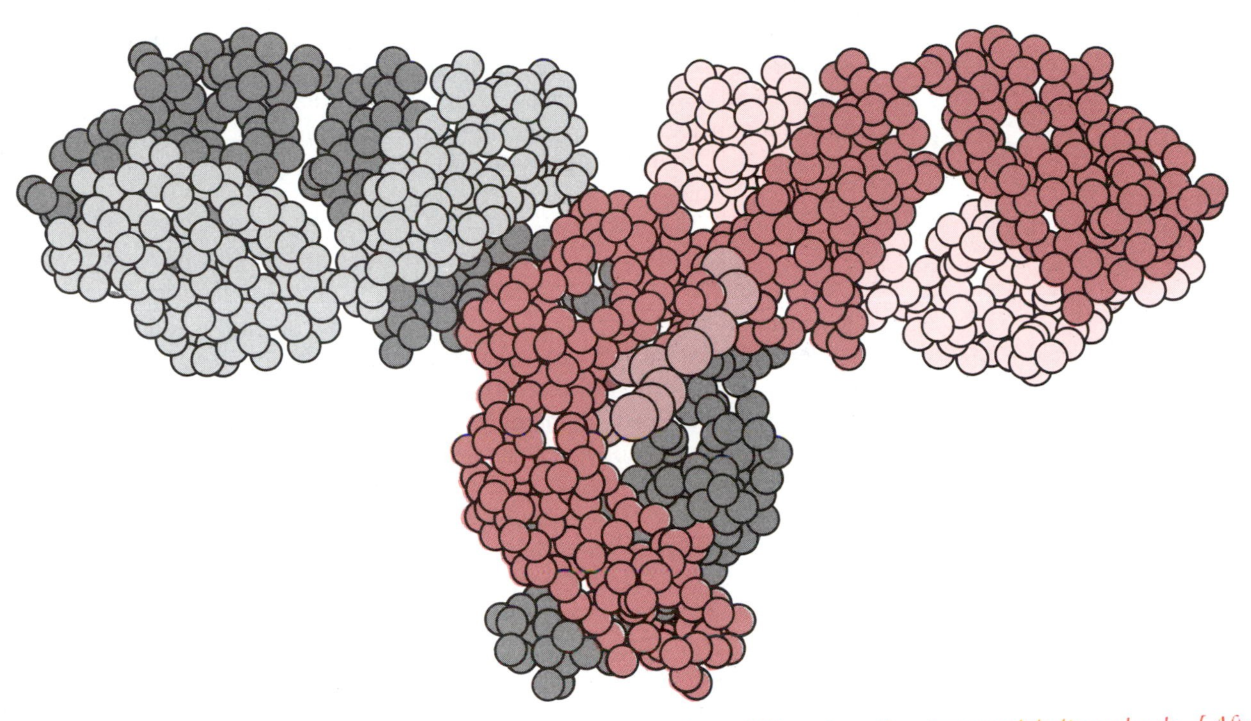

Space-filling view of an immunoglobulin molecule. [After E. W. Silverton, M. A. Navia, and D. R. Davies. Proc. Natl. Acad. Sci. USA 74 (1977): 5142.]

In this chapter, we describe several more enzymes that catalyze the hydrolysis of acyl compounds and briefly examine two systems that are used by a complex organism to respond to dangerous events in the environment: the blood-clotting cascade, which responds to injury to blood vessels, and the immune system, which responds to invasion by bacteria, viruses, and other foreign matter. Hydrolytic enzymes play important roles in both of these responses.

We have examined the properties of chymotrypsin in some detail in order to illustrate principles for the catalytic action of enzymes and some of the methods by which enzymes can be characterized. It might be assumed that chymotrypsin and other proteolytic enzymes are not of outstanding interest because the digestion of food is not the most exciting topic in biochemistry. However, it has become increasingly apparent in recent years that proteolytic enzymes and related enzymes that catalyze the hydrolysis of other acyl compounds play critical roles in the regulation and operation of many important and interesting physiological processes.

Enzymes That Work at an Interface

The great majority of the enzymes that have been studied thoroughly are water-soluble enzymes, which are relatively easy to characterize quantitatively. However, a large fraction of biochemical reactions do not occur in aqueous solution; they take place on membranes or in other heterogeneous media. Many of these enzymes are not well characterized because it is technically much more difficult to obtain quantitative information about enzymes that are not water soluble or that react with insoluble substrates.

Pancreatic lipase serves as a relatively simple example of an enzyme that acts at an interface. This enzyme plays an important role in digestion by cleaving insoluble fats. Lipase does not catalyze the hydrolysis of soluble substrates that are cleaved by ordinary enzymes; instead, it reacts with emulsions (Figure 6-1) and monomolecular films of fatty acid esters to introduce a water molecule from the aqueous phase into the ester group in the lipid phase. For example, it does not attack methyl butyrate that is dissolved in aqueous solution, but it cleaves an emulsion of methyl butyrate that is formed at higher concentrations than the

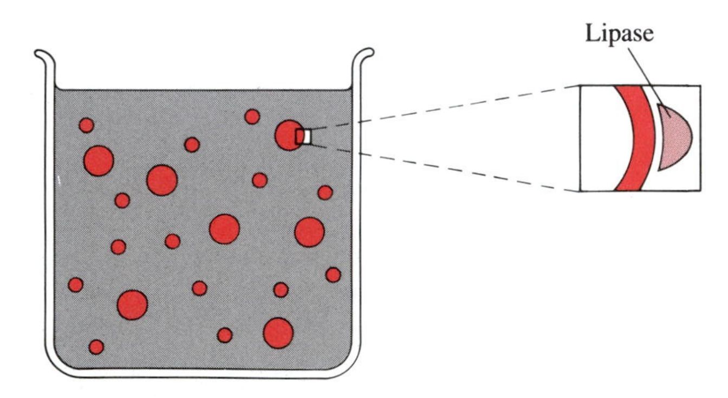

Figure 6-1

Pancreatic Lipase Acts at the Water-Lipid Interfaces of Particles in Emulsions. An aqueous emulsion consists of small particles of water-insoluble matter dispersed in water; this is illustrated in the drawing as a collection of tiny droplets dispersed throughout an aqueous phase. Pancreatic lipase acts at the surfaces of particles in aqueous emulsions of lipids—that is, at water-lipid interfaces. The observed activity of a fixed amount of lipase is, therefore, dependent on the surface area to which it has access. A few large particles will expose less surface than many small particles consisting of the same amount of lipid.

solubility limit of this ester in water. The identification of pancreatic lipase as a serine enzyme was delayed for many years because its activity is not inhibited by solutions of DFP (diisopropylphosphorofluoridate, Chapter 4) in water. However, when the concentration of DFP is increased above its solubility limit, so that it forms an emulsion in water, it reacts with the enzyme and causes inactivation.

The activity of lipases is strongly influenced by the presence of bile salts and other detergents, including protein molecules, that change the properties of the aqueous-organic interface and make it possible for the enzyme and water to interact with the acyl groups of water-insoluble esters. Bile is synthesized in the liver, stored in the gall bladder, and secreted through the bile duct into the intestine where it assists the digestion of fats. Bile contains effective detergents or surfactants that emulsify fat into a dispersed form with a large surface area, which is readily attacked by lipases. These detergents are bile salts such as *glycocholate* and *taurocholate,* shown in Figure 6-2. These compounds consist of cholic acid, which provides a hydrophobic, lipophilic moiety derived from cholesterol, and glycine or taurine (an aminosulfonate), which provides a charged, hydrophilic moiety.

It is still not clear how to examine the properties of enzymes such as pancreatic lipase quantitatively and in the detail that has been achieved with soluble, homogeneous enzyme systems. The biochemist who studies reactions in heterogeneous systems is likely to be faced with such challenges as a reaction rate that depends on the surface area of the insoluble substrate, rather than its concentration, and on the rate at which the incubation mixture is shaken. The activity of lipase is often examined by measuring the decrease in surface area of a monomolecular film of substrate that is spread over water, but this is a difficult technique and it is not applicable to many insoluble substrates.

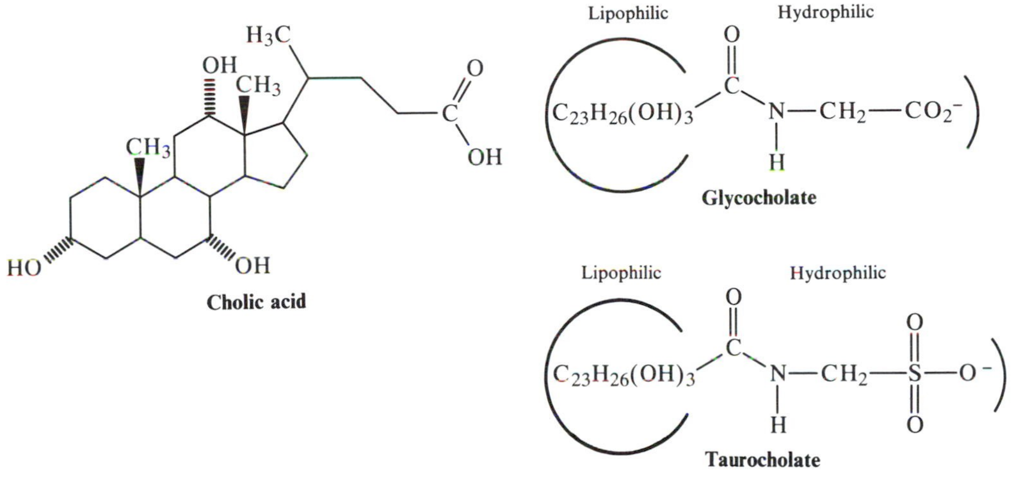

Cholic acid

Lipophilic Hydrophilic

$C_{23}H_{26}(OH)_3-C(=O)-NH-CH_2-CO_2^-$

Glycocholate

Lipophilic Hydrophilic

$C_{23}H_{26}(OH)_3-C(=O)-NH-CH_2-SO_2-O^-$

Taurocholate

Figure 6-2

Structures of Cholic Acid, Glycocholate, and Taurocholate. Cholate esters are hydrophobic molecules, largely hydrocarbon in nature, which have little solubility in water. Glycocholate is a glycinamide of cholic acid and taurocholate is an analogous sulfonate. Both glycocholate and taurocholate have hydrophobic and hydrophilic ends that interact with like molecules when exposed to an aqueous medium. The hydrophobic ends tend to associate with each other and with other hydrophobic molecules and to avoid exposure to water, whereas the hydrophilic ends tend to interact with water. Glycocholate and taurocholate are *emulsifying agents,* because they tend to facilitate the dispersion of water-insoluble lipids into small particles in aqueous media. The interiors of these particles are hydrophobic havens for lipid molecules, and the surfaces are defined by the interface between water and the solvated (hydrated) hydrophilic ends of taurocholate and glycocholate.

Other Hydrolytic Enzymes

There are many other serine enzymes that catalyze hydrolysis and several other classes of enzymes that utilize different chemistry at the active site to catalyze hydrolysis. The most-important of these classes are the sulfhydryl hydrolases, which have a cysteine instead of serine residue at the active site.

Subtilisin BPN′ is a bacterial serine protease having an active site almost identical with that of chymotrypsin but a very different structure in the remainder of the enzyme. The probable reason for this is described in Box 6-1.

Shorter-chain esters of fatty acids and other compounds, including acetylcholine, are cleaved by a wide variety of *esterases* that are found in most tissues and in serum. The functions of these enzymes are incompletely understood. Some, but by no means all, of them are serine enzymes.

More-exotic, but not necessarily less important, serine enzymes include acrosin and cocoonase. Acrosin is released from the acrosomes of sperm when the sperm approaches the ovum. It makes penetration into the ovum possible by cleaving proteins in the zona pellucida, a protective layer of protein and carbohydrate that surrounds the ovum (Figure 6-3). As the reader might guess, cocoonase cleaves protein fibers (which may be silk) in order to permit the escape of a mature butterfly or moth from its cocoon.

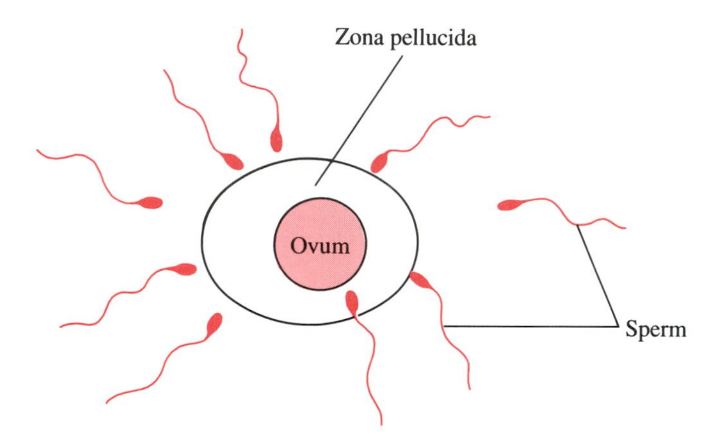

Figure 6-3

Sperm release acrosin when they encounter the protective zona pellucida that surrounds the ovum.

BOX 6-1

CONVERGENT EVOLUTION OF A SERINE ENZYME

Subtilisin BPN′ is a bacterial enzyme that is almost identical in its basic catalytic properties with the chymotrypsin family of enzymes. Subtilisin is of particular interest because X-ray crystallography has shown that the three-dimensional structure of its active site is almost exactly the same as that of chymotrypsin. The relative positions of the serine hydroxyl, histidine imidazole, and aspartate carboxylate groups, as well as a series of subsites for binding adjacent amino acid residues of a peptide substrate, are almost identical with those in chymotrypsin. On the other hand, the structure of the rest of the enzyme is entirely different from that of chymotrypsin—the amino acid sequence and the three-dimensional fold of the peptide chain are so different from those of the chymotrypsin family that it does not appear possible that these enzymes could have evolved from a common ancestor.

This enzyme seems, therefore, to provide an example of *convergent evolution* in biochemistry. Apparently the particular three-dimensional arrangement of serine, histidine, and aspartate residues in the active site of chymotrypsin is so effective for catalysis that the identical arrangement of catalytic groups has been chosen by evolutionary selection on at least two separate occasions from the large number of possible structures of an active site. Selection of the same structures for the peptide-binding subsites, in addition to the amino acids that participate more directly in catalysis, is particularly interesting because it suggests that the role of these subsites in holding the peptide chain in place makes an important contribution to the catalytic process.

Sulfhydryl Hydrolases

Nature has made a relatively small structural change, the substitution of a cysteine for the serine residue at the active site of chymotrypsin and related enzymes, to create another important class of acyl-hydrolyzing enzymes, the sulfhydryl hydrolases. The sulfhydryl hydrolases in mammals include several intracellular proteolytic enzymes (cathepsins), but the best-known members of this class are the plant proteases such as papain, bromelin, and ficin (from the papaya, pineapple, and fig, respectively). These enzymes are highly susceptible to inactivation by oxidation of the sulfhydryl group to inactive disulfides, so that it is often necessary to reduce enzyme preparations by reaction with sulfhydryl compounds to obtain maximum activity. Cleavage of a mixed disulfide containing cysteine by a thiolate (mercaptide) ion is shown in equation 1.

$$\text{Cys—S—S—R}' + {}^{-}\text{S—R} \rightleftharpoons \text{Cys—S}^{-} + \text{R}'\text{—S—S—R} \quad (1)$$

The process leading to the activation of a cysteine proteinase is illustrated in Figure 6-4.

The mechanism of action of these enzymes is closely similar to that of the serine hydrolases. The mechanism includes covalent catalysis through an acyl-enzyme intermediate, which is a thiol ester of cysteine rather than an oxygen ester of serine. The existence of an acyl-enzyme intermediate has been demonstrated by showing that a series of esters with different leaving groups but the same acyl group all undergo hydrolysis at the same maximum rate. This suggests that there is a common acyl-enzyme intermediate, which is cleaved in the rate-determining step, k_{cat}, as illustrated in Figure 6-5. The acyl enzyme has been directly observed after the addition of a thiono ester of hippuric acid to the

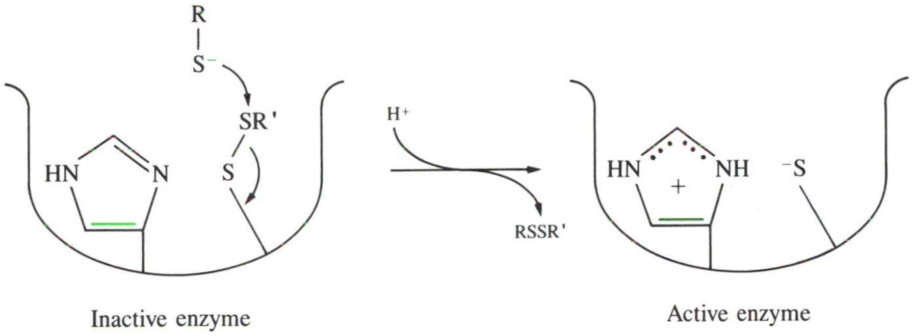

Inactive enzyme — Active enzyme

Figure 6-4

Activation of Oxidized Papain by Thiolate Anions.

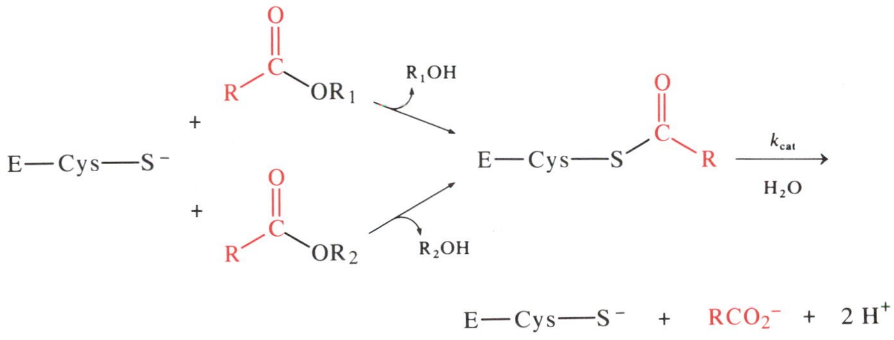

$$\text{E—Cys—S}^{-} + \text{RCOOR}_1 \text{ (or RCOOR}_2) \xrightarrow{-\text{R}_1\text{OH (or R}_2\text{OH)}} \text{E—Cys—S—C(=O)—R} \xrightarrow[\text{H}_2\text{O}]{k_{cat}} \text{E—Cys—S}^{-} + \text{RCO}_2^{-} + 2\,\text{H}^{+}$$

Figure 6-5

Formation and Decomposition of a Common Acyl-Enzyme Intermediate in Papain-catalyzed Hydrolysis of Substrates with Identical Acyl Groups.

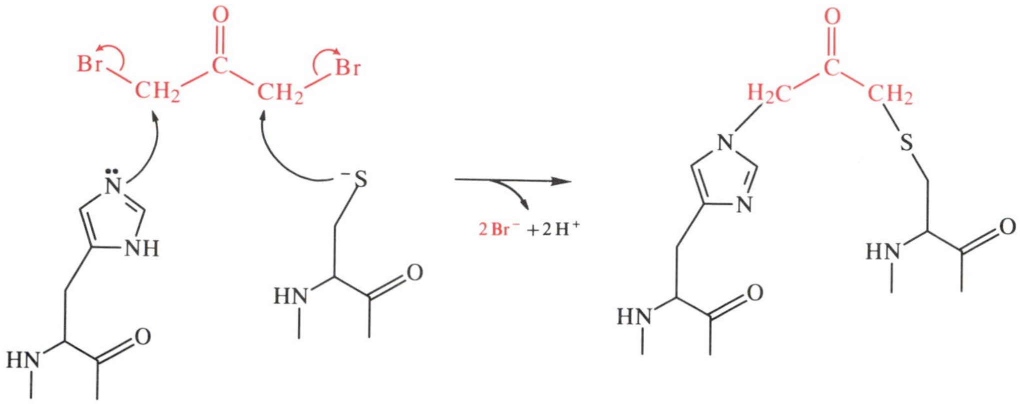

$2\,Br^- + 2\,H^+$

Figure 6-6

Dibromoacetone forms a cross-link by reacting with imidazole and thiol groups in the peptide chain of papain.

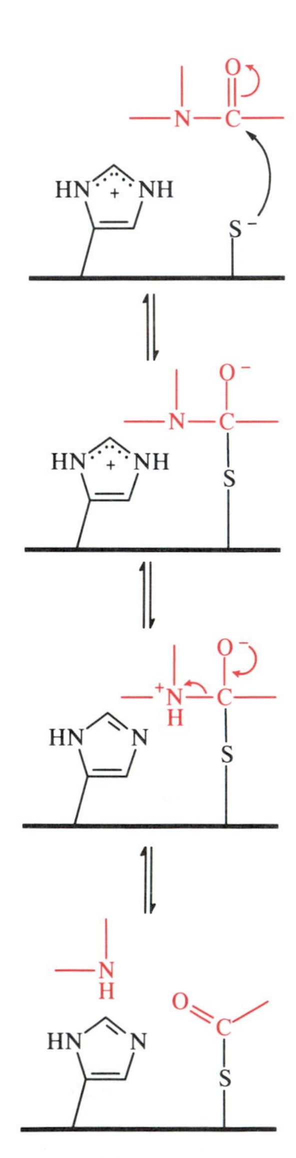

Figure 6-7

The Mechanism of Acylation of Papain.

enzyme, which reacts according to equation 2 to form an acyl-enzyme with the characteristic absorption spectrum of a dithioester.

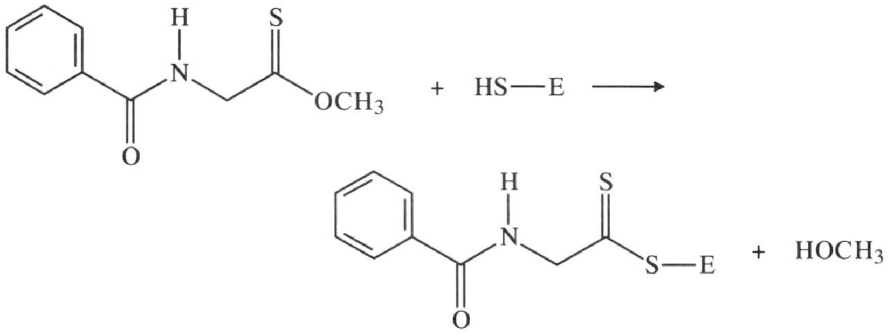

$$+ \; HS\text{—}E \longrightarrow \qquad + \; HOCH_3 \qquad (2)$$

λ_{max} 313 nm

There is also an imidazole group of a histidine residue near the nucleophilic sulfhydryl group at the active site. The spatial relationship of these two groups was first demonstrated by treating papain with a bifunctional alkylating agent, 1,3-dibromoacetone. This compound reacts with both the thiol group of the cysteine and the imidazole ring of histidine to form a cross-linked product that can be isolated after hydrolysis of the protein (Figure 6-6). The juxtaposition of the imidazole and sulfhydryl groups was later confirmed by X-ray structure determination. The imidazole group presumably acts to abstract a proton from the attacking sulfhydryl group and to donate a proton to the leaving amine group of the substrate, performing a function similar to that of the imidazole group of chymotrypsin (Figure 6-7).

There is evidence that the proton transfer from the cysteine —SH to imidazole takes place before the thiol anion attacks the peptide acyl group. The pH-rate profile of k_{cat} for papain is a broad plateau with inflection points that represent ionizations with pK_a values of 4.2 and 8.2 (Figure 6-8). This means that there is a low-pH form of the enzyme-substrate complex, $EH_2{}^+$, that is catalytically inactive; an active form, EH, that is formed by the loss of a proton from a group with a pK_a of 4.2; and another catalytically inactive form, E^-, that is formed by loss of a second proton from a group with a pK_a of 8.2. These ionizations represent dissociations of the thiol and imidazole groups, as shown in Figure 6-9. The acidic, diprotonated species, EH_2, has no base to accept the proton when the thiol attacks the substrate; and the basic, unprotonated species, E^-, is inactive, perhaps because there is no proton available to protonate the leaving group.

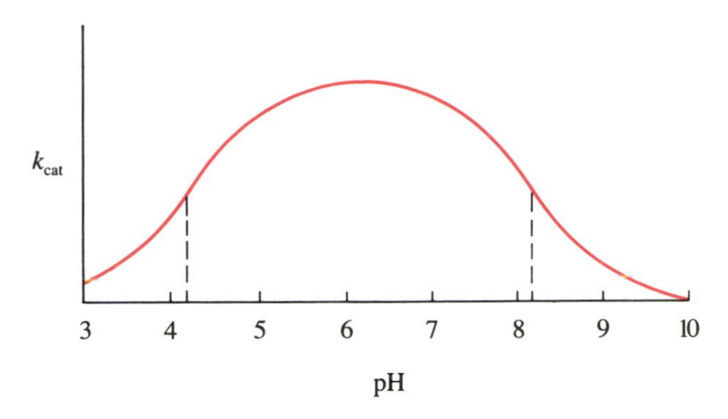

Figure 6-8

The Bell-shaped Dependence on pH of k_{cat} for Catalysis of Peptide Hydrolysis by Papain.

Thus, the active form of the enzyme at the maximum of the pH-rate profile contains either uncharged imidazole and thiol groups or the thiol anion and protonated imidazole. The pH-rate profile cannot distinguish between these two structures, but the nuclear magnetic resonance spectrum has shown that the enzyme exists predominantly as the ion pair at pH 7, with the proton on imidazole; that is, K_z in Figure 6-9 is > 1.0. This is surprising, because imidazole is less basic than a thiol anion in aqueous solution. Evidently, there are electrostatic interactions in the active site of the enzyme that stabilize the ion pair.

Aspartyl Proteases: Pepsin and Chymosin

The other major class of proteolytic enzymes, in addition to the serine enzymes, the sulfhydryl enzymes, and metalloenzymes such as carboxypeptidase (Chapter 3), consists of the aspartyl proteases. The best-known member of this group is pepsin, which is active in the strongly acidic environment of the stomach near

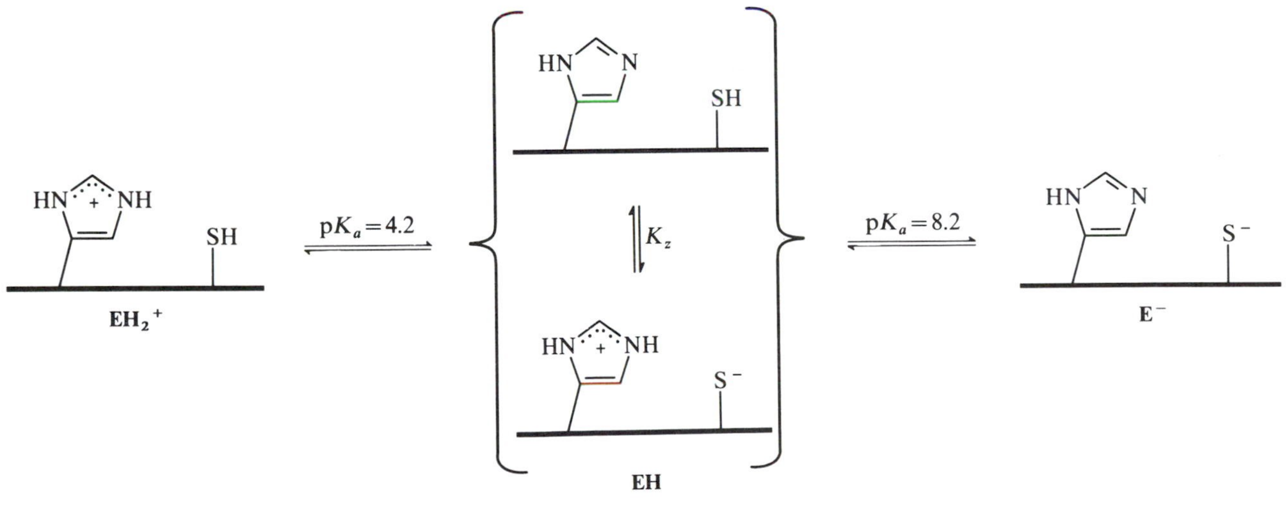

Figure 6-9

The dissociation of protonated imidazole and thiol groups accounts for the maximum in the pH-rate profile for hydrolysis by papain.

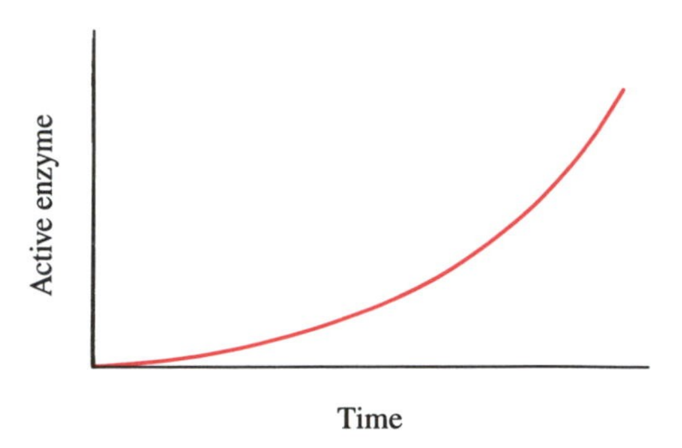

Figure 6-10

The Autocatalytic Activation of Pepsinogen by Pepsin.

pH 2. The closely related enzyme chymosin (rennin) is well known as the active principle of rennet, the enzyme preparation that catalyzes the coagulation of milk in the preparation of cheese and junket. This coagulation is a consequence of the specific cleavage of a phenylalanine–methionine bond in κ-casein, one of the proteins of milk. Chymosin is found in the fourth stomach of the calf and has properties very similar to those of pepsin.

These enzymes are synthesized as the inactive proenzymes pepsinogen and prochymosin. The activation of pepsinogen to pepsin is catalyzed by active pepsin, which results in the removal of several peptide fragments and a decrease in molecular weight from 41,000 to 34,000 for active pepsin (equation 3).

$$\underset{M_r = 41{,}000}{\text{Pepsinogen}} \xrightarrow{\text{Pepsin}} \underset{M_r = 34{,}000}{\text{Pepsin}} + \text{Peptides} \tag{3}$$

Thus, it provides an example of an *autocatalytic reaction,* because the product of the reaction, pepsin, is also a catalyst for the reaction. Therefore, the rate of formation of active enzyme increases progressively with time as the concentration of active pepsin increases (Figure 6-10). However, it has recently been shown that, at low concentrations or when the separated molecules are bound to a solid support, pepsinogen itself will slowly undergo spontaneous activation in the absence of pepsin. Evidently, a proteolytic activity of the pepsinogen molecule toward itself is developed at low pH that is sufficient to initiate the activation, without assistance from other molecules of pepsinogen or pepsin.

The activity of pepsin follows a bell-shaped pH-rate profile, with inflection points near pH 1.5 and 4.2 for k_{cat} (the pK_a values of the E · S complex) and near pH 1 and 4.7 for k_{cat}/K_m (the pK_a values of the free enzyme). These pK values represent the ionization of carboxylic acid groups on aspartate residues that are required for activity. The requirement for free carboxyl groups in the active site has been shown by inactivating the enzyme with reagents that react rapidly with carboxyl groups, such as the alkylating agent *p*-bromophenacyl bromide (equation 4), followed by chemical identification of the product.

$$\text{E—C(=O)—O}^- + \text{Br—CH}_2\text{—C(=O)—C}_6\text{H}_5 \longrightarrow \text{E—C(=O)—O—CH}_2\text{—C(=O)—C}_6\text{H}_5 + \text{Br}^- \tag{4}$$

X-ray crystallography of pepsin and closely related enzymes has shown the presence of two aspartyl β-carboxyl groups at the active site that account for the observed pH-rate profile. The low pK_a value, in the range of 1.0 to 1.5, can be accounted for by stabilization of the complex between the carboxylic acid and the carboxylate ion at the active site by hydrogen bonding, as is shown in Figure 6-11. The species with two nonionized COOH groups is much less stable

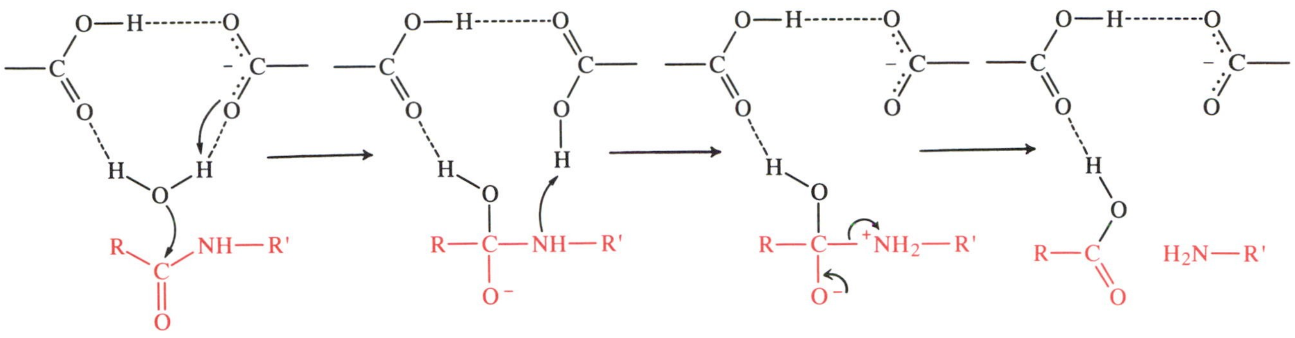

Figure 6-11

A Mechanism for Catalysis by Two Aspartyl β-Carboxyl Groups at the Active Site of Pepsin. The two aspartyl β-carboxyl groups at the active site are associated by hydrogen bonding, with one in the acid form (—CO_2H) and the other in the ionized form (—CO_2^-). These groups are thought to catalyze the hydrolysis of peptides bound at the active site by first coordinating, through hydrogen bonding, with the attacking water molecule and then stabilizing the tetrahedral addition intermediate and facilitating its cleavage to products.

and readily undergoes dissociation of a proton to form the hydrogen-bonded complex, with a water molecule that bridges the acid and the anion.

It is likely that the carboxylate anion and the carboxylic acid act as a proton acceptor and donor, respectively, to catalyze hydrolysis of the peptide bond. In a possible mechanism shown in Figure 6-11, the carboxylate group acts as a base to remove a proton from the water molecule that attacks the peptide bond; the other proton of water is transferred to a carboxyl oxygen atom and then to the nitrogen atom of the peptide bond to facilitate cleavage of the C–N bond.

Regulation: The Blood-clotting Cascade

There are two general classes of regulation of functions in living organisms. First, a plant, animal, microorganism, human being, or any other living system must maintain an internal environment that is nearly constant and, second, the organism must be able to respond to changes in such a way as to avoid irreversible injury or death. Enzymes, membranes, metabolic processes, and other systems can function effectively and maintain a constant internal environment if there is a constant source of available energy that can be utilized for the necessary activities of the organism. This condition is called *homeostasis*. It is brought about by a large number of complex, interacting regulatory systems that respond to changes in conditions in such a way as to maintain the internal environment remarkably constant.

One example is the regulation of the concentration of glucose in the blood by the polypeptide hormone *glucagon*. When glucose is used up in metabolism and the concentration of blood glucose decreases, the α-cells of the pancreas release glucagon into the blood stream. The glucagon binds to receptors in the liver and initiates a cascade of reactions that result in the release of glucose from the liver and a return of the blood glucose to its normal concentration (see Chapter 21).

Initiation of Blood Clotting

The second class of regulation is the response to a stimulus that brings about a *change* in the concentration or the activity of molecules in response to a change in conditions, such as an injury or an infection. These responses to external stimuli require changes that are appropriate to the condition: they must be turned on when they are needed and, equally important, they must be turned off when conditions return to normal. A well-known example is the clotting of blood, which must take place rapidly in order to stop bleeding from a wound but must be strictly limited to the site of injury in order to maintain the supply of blood to other tissues. Blood clots are an important component of the blockage of arteries that causes myocardial infarction and stroke, whereas insufficient clotting of the blood can cause cerebral hemorrhage.

In this section, we will examine the blood-clotting system as an example of this second class of regulation. Most of the enzymes that control blood clotting are serine proteinases that activate clotting factors.

It is perhaps more remarkable that blood does not clot in the circulatory system than that it *does* clot under a large variety of different circumstances, because clotting is rapidly initiated by any of a large number of conditions; it is difficult to prevent the clotting of blood once it is removed from the circulatory system unless inhibitors of the clotting pathway are added. The smooth wall of blood vessels that is formed by endothelial cells does not initiate clotting, but glass and many other surfaces will initiate clot formation within a few minutes. Clotting that is initiated by surfaces is classically known as the "intrinsic pathway." When blood leaks out of a blood vessel into the tissue, it encounters a large number of compounds that activate clotting by the "extrinsic pathway" within seconds. These compounds include phospholipids in membranes and collagen, a structural protein of connective tissue. The intrinsic and extrinsic pathways are outlined in Figure 6-12. This remarkable cascade of reactions deserves appreciation but not memorization.

When a blood vessel is ruptured, there is an immediate contraction of the vessel that reduces the outflow of blood. This is rapidly followed by the aggregation of *platelets* that adhere to the damaged site and may form a plug that is strong enough to stop the flow of blood out of a small vessel. There is also interaction with collagen, phospholipids, and *tissue factor,* a lipid-containing protein, or lipoprotein, that stimulates clot formation. The fragile platelets break up and release *thromboplastin,* a poorly defined mixture of phospholipids and proteins that activate the sequence of reactions in Figure 6-12 at several different points. They also provide a nucleus upon which fibrin is precipitated to form the clot.

The important property of the cascade of reactions in Figure 6-12 is that the catalytic activation of clotting factors by other, activated factors *amplifies* the clotting process. This amplification changes a system that does not clot at all under normal conditions to a state in which clotting occurs in seconds. An important contribution to this amplification is provided by feedback loops, in which activated factors themselves activate different factors that occur earlier in the cascade. For example, the inactive enzyme prekallikrein is converted by activated factor XII into active kallikrein, which speeds up the activation of factor XII (Figure 6-12). The primary goal of the cascade is to produce active *thrombin,* and thrombin accelerates the formation of more thrombin from prothrombin by activating factor VIII and factor V. Factor VIII is the antihemophilic factor. It is a glycoprotein—that is, a protein that contains carbohydrate—and the importance of its role in clotting is shown by the serious bleeding tendency of hemophiliacs. The hereditary absence of this protein in some people has been significant both for the development of our understanding of human genetics and for history.

The activation of prothrombin to thrombin takes place on a phospholipid membrane in a complex that also contains factors V and X and another activator protein, APC. This complexation increases the effective concentration

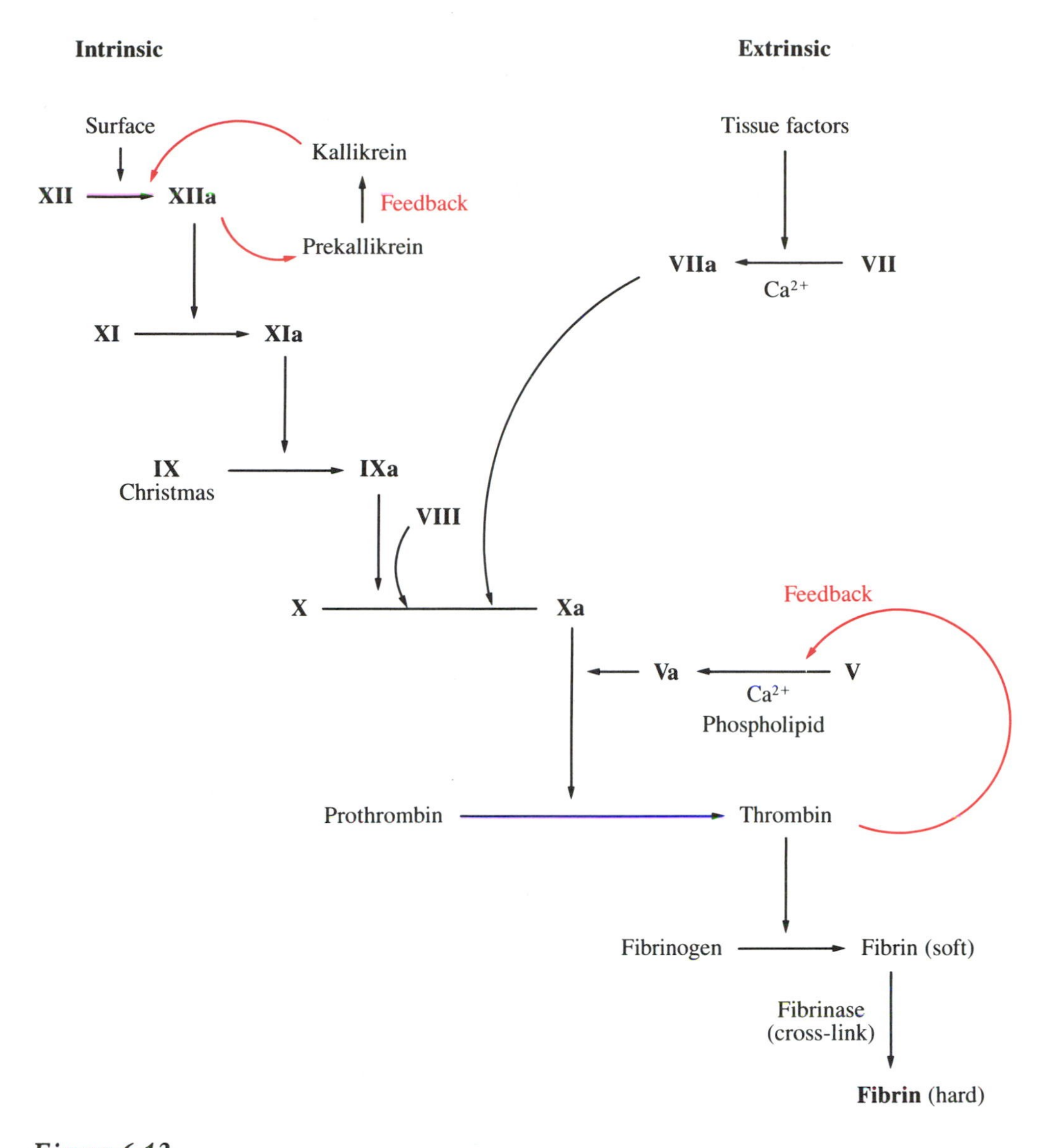

Figure 6-12

Blood clotting occurs through a cascade of reactions catalyzed by serine proteases. Serine proteases acting in a cascade bring about a rapid amplification of an intrinsic or extrinsic stimulus for blood clotting.

of the proteins relative to each other and has been estimated to accelerate the activation process by a factor of 10^5. *Calcium ion* is required for this and several other reactions in the cascade because it serves as a bridge between a binding site on the clotting factors and the negative charge of the phospholipid membrane.

The blood clot is finally formed when the active thrombin attacks fibrinogen to give fibrin. Fibrinogen is a large, long protein of molecular weight 340,000 that is soluble in water and remains in solution in blood plasma. It is a dimer of two subunits, each of which is constructed from three chains, α, β, and γ; the dimer is held together by disulfide bonds that connect the two subunits and parts of the individual chains (Figure 6-13). Its solubility is maintained by two regions of the protein, A and B, that have a high density of negative charge. Clotting occurs when thrombin, a serine protease that closely resembles trypsin,

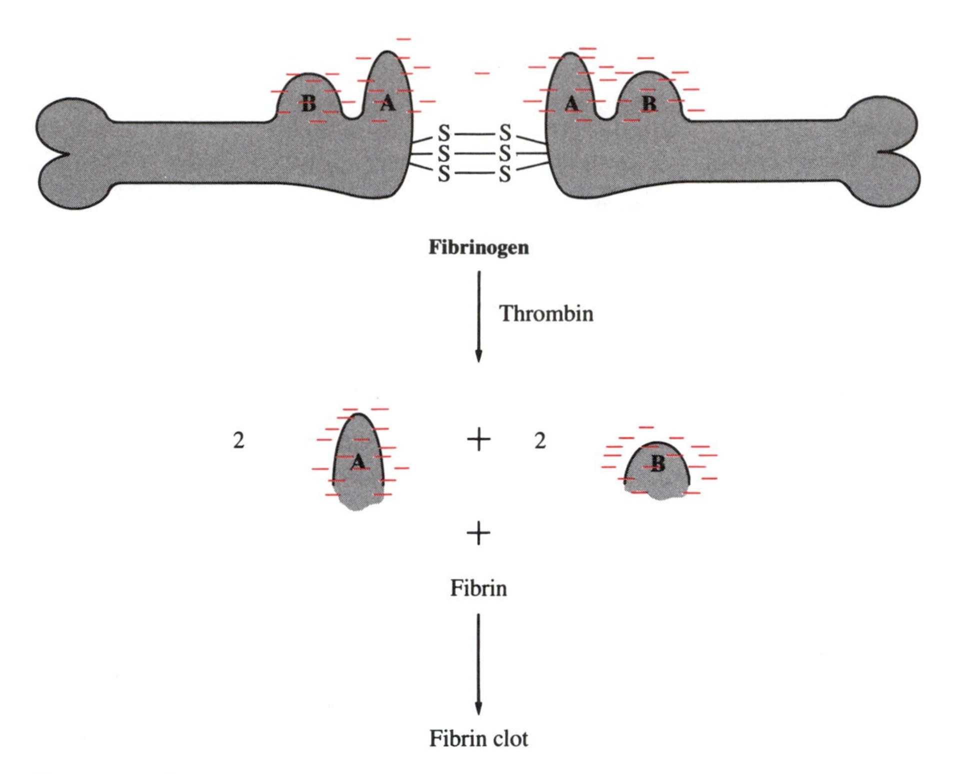

Figure 6-13

The soluble fibrinogen molecule consists of peptide chains and subunits that are held together by disulfide bonds. When the negatively charged peptides A and B are cleaved off by thrombin, the solubility decreases and fibrin precipitates to form a clot.

cleaves off the fibrinopeptides A and B by catalyzing the hydrolysis of a peptide bond to arginine. This causes a drastic decrease in solubility, owing to the loss of the hydrophilic domains, so that the fibrin molecules aggregate in a staggered arrangement to form the insoluble clot.

The final step in the formation of a blood clot is the conversion of the soft clot of precipitated fibrin into a hard clot, in which the fibrin molecules are interconnected by a network of covalent cross-links. This cross-linking is brought about by the enzyme transglutaminase, or *fibrinase,* which catalyzes the attack by a lysine ε-amino group of one fibrin molecule on the terminal δ-amide group of a glutamine residue in another fibrin molecule to displace ammonia (Figure 6-14). This connects the two fibrin molecules through a new amide bond between the side chains of glutamate and lysine.

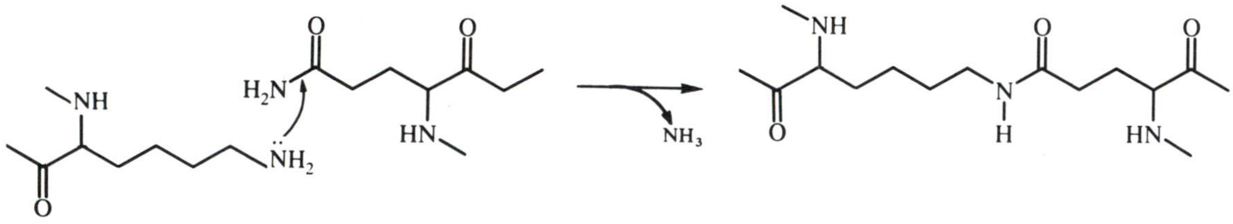

Figure 6-14

Fibrinase-catalyzed Formation of an ε-Lysyl-δ-Glutamyl Cross-Link in Fibrin.

Prevention of Blood Clotting

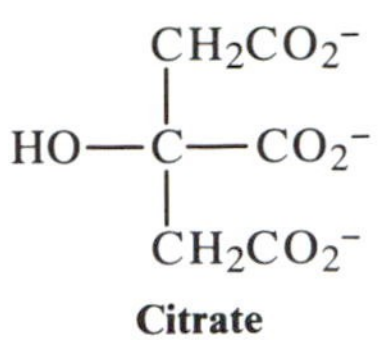

The removal of free calcium ions by combination with the chelating agent EDTA (Chapter 5) or the chelator *citrate*, shown in the margin, provides an effective and widely utilized method of preventing blood clotting.

The role of calcium can also be blocked in a quite different way that is frequently used to decrease the probability of clot formation in patients with coronary thrombosis and similar circulatory diseases. The binding of calcium to prothrombin and several other factors in the clotting cascade is mediated by a recently discovered amino acid in these proteins, *γ-carboxyglutamate,* which is shown in Figure 6-15. The bis-carboxyl groups on the side chains of carboxyglutamate residues are similar in structure to EDTA and citrate and serve as effective calcium chelating agents. Carboxyglutamate is produced by the addition of carbon dioxide to glutamate residues in the prothrombin molecule, through a reaction that requires *vitamin K* (Figure 6-15) and oxygen. Vitamin K (the "Koagulation vitamin") is a naphthoquinone with a side chain of four or five isoprenoid units; it is one of the fat-soluble vitamins. The mechanism of this very interesting carboxylation reaction is unknown and is currently under active investigation.

The role of calcium in the clotting process is blocked if the synthesis of carboxyglutamate is prevented. This can occur in vitamin K deficiency, which is rare, but it can also be induced by the administration of dicumarol or warfarin (Figure 6-16), which are structural analogs of vitamin K that antagonize its normal function in promoting carboxylation of glutamate residues in prothrombin. Dicumarol and warfarin are medically significant anticoagulants, and warfarin is widely used as a rat poison. Other mechanisms for the prevention and control of blood clots are described in Box 6-2.

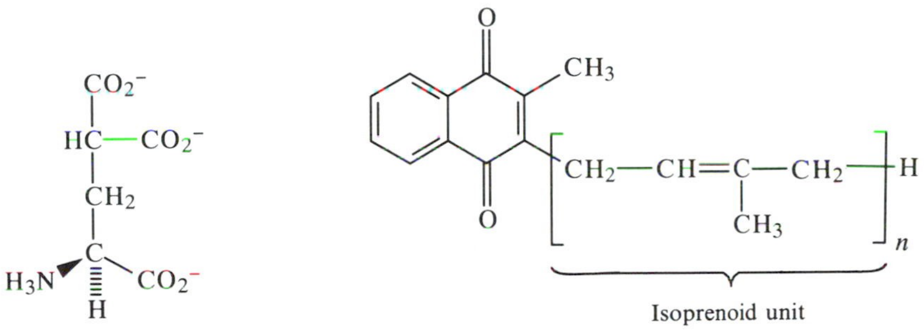

γ-Carboxyglutamate

Isoprenoid unit

Vitamin K

Figure 6-15

The Structures of γ-Carboxyglutamate and Vitamin K.

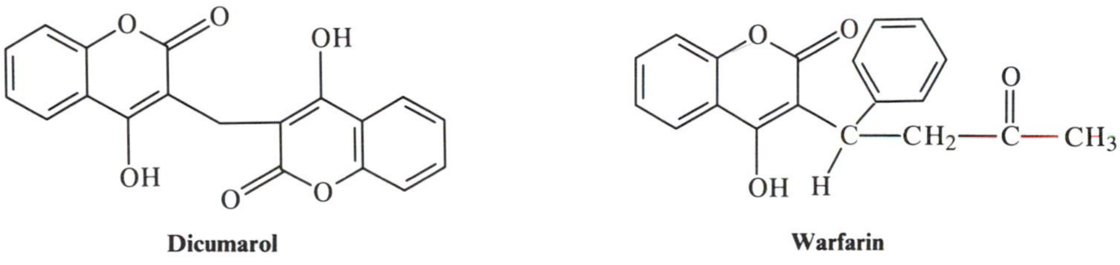

Dicumarol

Warfarin

Figure 6-16

The Structures of the Anticoagulants and Vitamin K Antagonists Dicumarol and Warfarin.

PREVENTION OF BLOOD CLOTTING

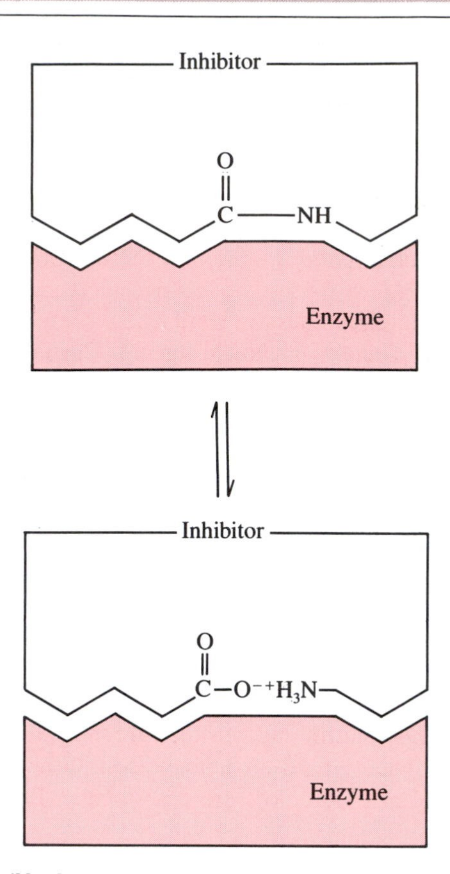

Reversible cleavage of a peptide bond in a protease inhibitor.

It is at least as important to prevent blood clotting in the wrong place at the wrong time as to initiate it when it is needed. This prevention depends on the smoothness of the vascular walls, the sensitivity of the platelets, and the existence in serum of potent *inhibitors* of serine proteases that block the action of thrombin and other components of the blood-clotting system. These inhibitors play a complementary role to the autocatalytic, feedback-stimulated cascade of reactions that initiates clotting because they prevent a clot from spreading beyond the site of injury. The clotting process is delicately balanced so as to provide an all-or-nothing, localized sequence of reactions—any minor initiation or spread of clotting is blocked by inhibitors and a clot is formed only when a stimulus initiates the cascade of reactions strongly enough to overcome this inhibition and produce a burst of fibrin production in a localized region.

Inhibitors of serine proteases are widely distributed in animals and plants and play an important role in preventing the digestion of tissues by endogenous enzymes. They are found in the pancreas and pancreatic duct, for example, where they protect against any proteolytic enzymes that might become activated before they reach the small intestine. A deficiency of inhibitor in the lung leads to digestion of the lung tissue by elastase with a resulting loss of its natural elasticity and the development of emphysema.

A protease inhibitor is itself a protein, with a structure that is closely complementary to the active site and subsites of a particular enzyme, so that it binds tightly to the enzyme and prevents its attack on its usual substrates. Many of these inhibitors are substrates for the proteases, in a sense, because a peptide bond of the inhibitor is cleaved at the active site by the enzyme. However, the cleaved peptide does not dissociate from the protease, and the three-dimensional structure of the inhibitor remains intact. The newly formed carboxylate and amino groups are held next to each other, so that with some inhibitors the enzyme also catalyzes the reverse reaction, the resynthesis of the peptide bond with the expulsion of water. After a short time, an equilibrium is set up and the bound inhibitor consists of an equilibrium mixture of the cleaved and virgin forms.

Clotting is also prevented by *heparin,* a sulfated polysaccharide. Heparin causes a large increase in the ability of one of these protease inhibitors, antithrombin, to bind to thrombin and prevent clotting. Heparin is synthesized by mast cells, which are found around most blood vessels. Other sulfated polysaccharides with a high density of negative charge have a similar anticoagulant effect.

A second, more slowly acting defense mechanism against excessive blood clotting is provided by the enzyme *plasmin* (fibrinolysin), which breaks up clots by cleaving fibrin in the clot. Plasmin is still another serine protease, which exists in the blood in the form of an inactive precursor, plasminogen, that is converted into plasmin upon the cleavage of an arginyl–valine bond by a proteolytic enzyme. This activation may be stimulated by stress, epinephrine, or damage to the wall of the blood vessel.

It has recently been shown that plasmin and a DFP-sensitive protease that activates plasminogen to plasmin are associated in some way with the transformation of cells into a malignant state. Inactivation by DFP (diisopropylphosphorofluoridate) indicates that the enzyme is a serine protease (Chapter 4). Normal cells treated with plasmin undergo morphological changes similar to those seen in transformed cells, and plasmin appears to be required for the morphological changes that take place in transformed cells. The significance of this correlation is under active investigation.

R = H or SO_3^-

Heparin

Antibodies and Antigens

The immune system provides a second example of response by a complex organism to a threat from the outside world. This remarkable system responds to attack by bacteria, viruses, and other foreign invaders by forming antibodies that bind to antigens on the foreign materials and initiate reactions that kill foreign organisms. The operation of the immune system is analogous to the blood-clotting system in that it involves amplification through cascades of successive reactions. It is also similar in that it is important for it *not* to operate unless there is a need to respond to the presence of foreign materials. *Autoimmune diseases* are caused by a failure of this protective regulatory mechanism and can result in serious or fatal consequences when the immune system of an animal or person attacks its own molecules as a result of mistakenly recognizing them as foreign substances.

Antibodies are synthesized by B lymphocytes (B cells), a particular class of leukocyte, or white blood cell, that are found in lymph nodes, the thymus gland, and the blood. When there is an invasion by a foreign organism or material, B lymphocytes accumulate in large numbers at the site of invasion and initiate an inflammatory response that usually destroys or contains the invader. Lymphocytes recognize a particular structure, or *epitope,* on a foreign molecule, an *antigen,* and respond by synthesizing large numbers of daughter cells that produce antibodies that bind with very high affinity to molecules having this specific structure.

Antigen–antibody binding may be compared to the interaction of the active site of an enzyme with the transition state for the reaction of its specific substrates; both are very tight and specific. In fact, antibodies have recently been prepared with binding specificity toward transition state analogs, molecules that structurally resemble the transition states for particular reactions. These antibodies bind the analogs tightly; they also catalyze the chemical reactions for which they are transition state analogs. Such antibodies are *catalytic antibodies,* or "abzymes," which increase the rate of the reaction that proceeds through the transition state that is stabilized by binding tightly to the antibody (see Box 6-3).

Antibodies are formed against any of an extraordinarily large number of molecular structures. This diversity protects an organism against attack by almost any sort of foreign invader. In fact, there are few chemical structures that are not capable of stimulating the formation of specific antibodies under favorable conditions.

BOX 6-3

CATALYTIC ANTIBODIES

An antibody binds tightly to a particular region of an antigen, the epitope, which constitutes a small part of a macromolecule, typically a protein. The diversity of possible antibodies allows virtually any part of a protein, or even another molecule chemically conjugated to a protein, to be recognized as an epitope for a particular antibody and to elicit the proliferation of the cells that produce that antibody. Small molecules that, when bound to proteins, stimulate the formation of antibodies that bind them tightly are known as haptens. An antibody that binds to a particular hapten can be prepared by attaching the hapten to a large molecule and immunizing a rabbit with this hapten-antigen.

An antibody that is made to a transition state analog as the hapten will bind this transition state analog tightly. An enzyme catalyzes a reaction by stabilizing a transition state that is formed from a substrate that is bound in the active site, so that an antibody that binds and stabilizes a transition state should also be an enzyme that catalyzes the reaction that proceeds through that transition state.

This has recently been found to be the case for several different classes of reaction. One example is catalysis of the isomerization of chorismate to prephenate, which is a step in the synthesis of phenylalanine and tyrosine that occurs by a Claisen rearrangement:

BOX 6-3 *(continued)*

CATALYTIC ANTIBODIES

^-O_2C, CO_2^-, O, HO, H, H

Chorismate

$\longrightarrow$ [^-O_2C, O, CO_2^-, HO]‡ $\longrightarrow$

^-O_2C, CH_2—C, CO_2^-, O, HO, H

Prephenate

An antibody was prepared with binding specificity for a molecule that closely resembles the transition state of this reaction:

H, ^-O_2C, O, CO_2^-, HO

This antibody is, in fact, an active catalyst of the synthesis of prephenate from chorismate. The antibody stabilizes the transition state and increases the rate of the reaction by a factor of 10^4 compared with the uncatalyzed reaction.

This wide range of specificities could be brought about in either of two ways:

1. The lymphocyte could bind the antigen, build an antibody molecule that binds tightly to the antigen, and then synthesize many more molecules of this specific antibody.
2. Each of a very large number of lymphocytes could synthesize an antibody with a different structure. An antigen will bind to the particular cell that makes its specific complementary antibody and stimulate this cell to make more copies of that antibody and to divide into more cells that make the same antibody.

The second of these mechanisms may seem unlikely because it requires the presence of an extraordinarily large number of different lymphocytes in order for there to be at least one cell among them able to recognize a particular antigen. Nevertheless, this has been shown to be the correct mechanism (Chapter 13). There are so many different specificities of different lymphocytes that there is almost always a cell that recognizes a foreign antigen, and contact with this antigen has been shown to stimulate the cell to synthesize the complementary antibody and to divide. This is the *clonal mechanism* of antibody selectivity: the stimulated cell divides to form a clone of identical cells that continue dividing. This provides a large number of cells that synthesize and release the specific antibody.

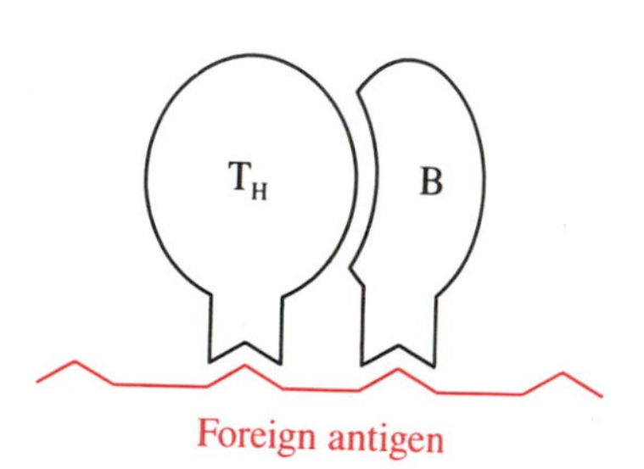

Figure 6-17

Activation of Antibody Production. Binding of a B cell to a specific antigen activates the cell to produce more antibody to the antigen and to divide. This process is stimulated by a helper T cell, T_H, which also binds to the specific antigen.

The activation of specific B lymphocytes is brought about with assistance from antibody presenting cells (APC) and T lymphocytes (T cells). The APC cells are stored in lymph nodes and the inner layers of the skin, where they can respond to invading foreign materials. The APC cells bind the antigen and present it to both T cells and B cells; the T_H cells are "helper cells" that are also specific for the antigen and stimulate the activity of the B cells (Figure 6-17).

An effective antigen must be attached to a large molecule to activate lymphocytes and stimulate antibody formation. Therefore, immunization is often carried out by the injection of antigen that is attached to a large molecule, together with insoluble materials, called *adjuvants,* that stimulate the movement of lymphocytes to the site of immunization and the antibody response.

The Structure of Antibodies

The structure of a typical antibody molecule is shown schematically in Figure 6-18. This structure was determined by X-ray diffraction of a crystalline antibody. The molecule has the overall shape of the letter Y, with two flexible arms that contain the specific binding sites for an antigen. These arms contain regions that have different amino acid sequences in different antibodies, so that each forms a structure that is specific for a particular antigen. These are the *variable regions* of the antibody and are designated F_{ab}, in which the subscript indicates that they contain the functional region of the antibody that binds a specific antigen. The remaining regions of the antibody molecule have essentially the same structure for antibodies of differing specificity and are designated F_c, in which the subscript indicates a constant structure. The antibody is constructed from two heavy chains, with a molecular weight ranging from 50,000 to 77,000, and two light chains, with a molecular weight of approximately 25,000.

When an animal or human being is infected with a microorganism, or is immunized with inactivated bacteria or viruses, the first specific antibody that is generated in response to the infection is IgM. This molecule is a polymer that contains five basic units which are connected by disulfide bonds and arranged in a cyclic structure, as shown in Figure 6-19. From 1 to 2 weeks later, the concentration of IgM decreases and the concentration of IgG increases. Figure 6-18 shows a typical structure of IgG.

It is significant that each molecule of antibody has two or more binding sites for antigens, whereas bacteria, viruses, carbohydrates, and many proteins have several identical antigenic sites. When only one antigenic site is on a molecule, it is called a hapten. A hapten will combine with an antibody, but does not give precipitation. However, an antigen with several sites that combine with the binding sites of a specific antibody forms a polymer or interlocking network that grows to a large size and precipitates from solution (Figure 6-20). These precipitates are then attacked by several classes of white blood cells that can ingest and digest them.

This binding to multiple sites is another example of the advantage of intramolecularity in chemical and biological reactions. The multiple binding sites produce a chelate effect and greatly increase the stability of the antigen-antibody complex because of the favorable entropy of binding.

A second defense mechanism against foreign organisms is provided by the *complement system,* which is a cascade of proteolytic reactions that is analogous to the blood-clotting cascade. The cascade is triggered by the binding of a six-headed molecule of complement, C1, to a polymeric antigen-antibody aggregate.

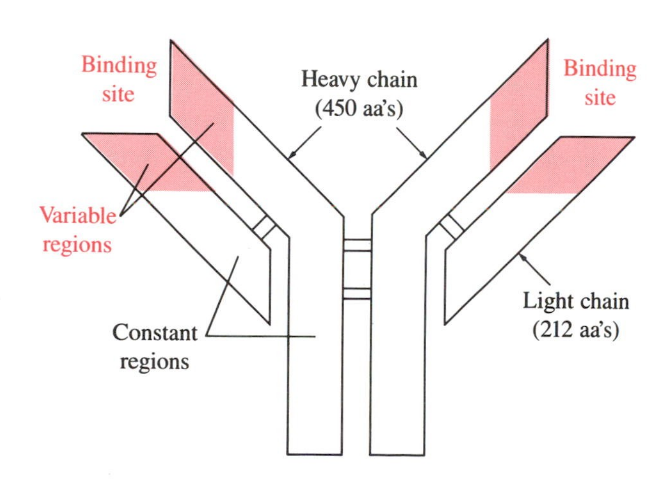

Figure 6-18

Schematic Drawing of an Immunoglobulin G (IgG) Antibody Molecule. The drawing shows the arrangement of the light and heavy chains and of the constant and variable regions of the amino acid sequence.

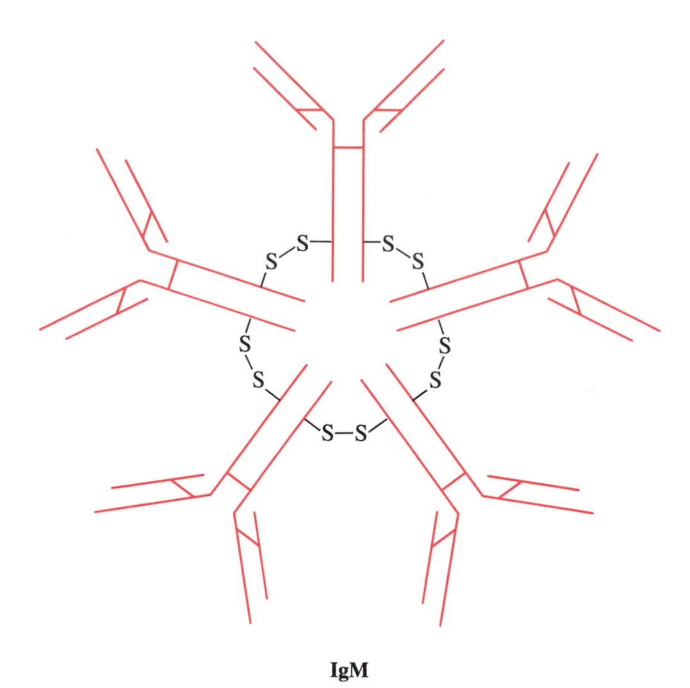

Figure 6-19

Schematic Drawing of an IgM Antibody Molecule. The structure of the individual subunits, which are connected by disulfide bonds, is similar to that of IgG antibodies.

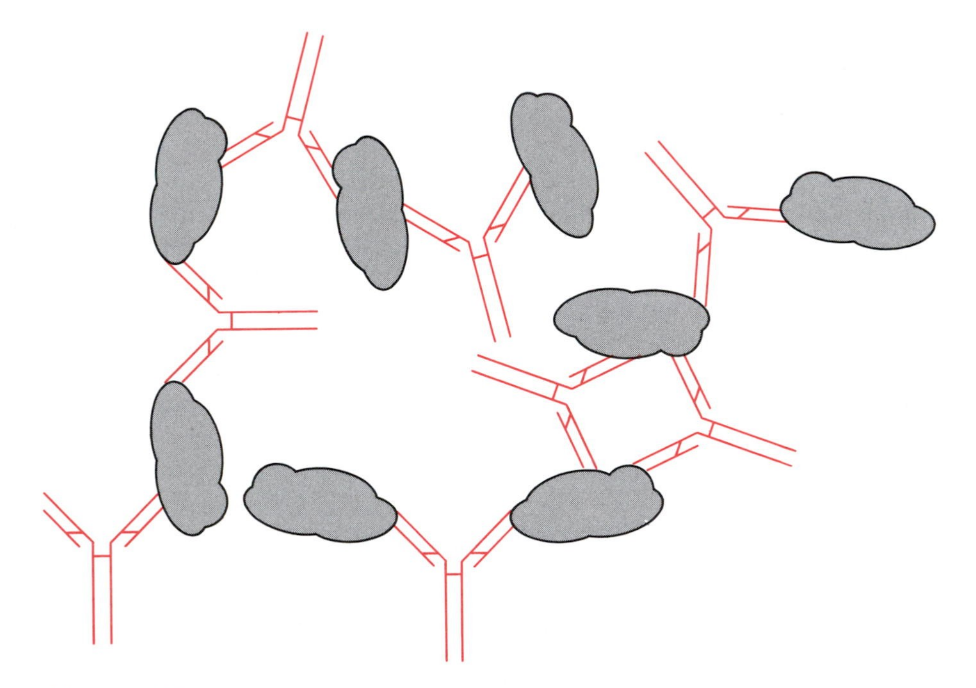

Figure 6-20

The Formation of a Precipitate in an Antigen-Antibody Reaction. The reaction of antibody with a polyvalent antigen, such as a bacterium or virus, forms an aggregate that may precipitate out of solution and activate the complement cascade.

The C1 molecule does not bind effectively to individual antibody molecules, but the entropic advantage from the chelate effect of the antibody molecules in a precipitate or aggregate provides the driving force for the binding of the six heads of C1 to six F_c units of antibody molecules in an aggregate. The binding causes a conformational change in C1 that activates a protease activity in a subunit of C1, which in turn activates a different protease activity in a different subunit of C1. These enzymes are typical serine proteases that are inactivated by DFP. The cascade continues when the activated protease activates another complement enzyme, C4, which is followed by sequential activation of C2, C3, and C5. The ability of each enzyme to activate molecules of the next enzyme in the series results in a rapid activation of the response. Finally, C5b, an activated fragment of C5, combines with several additional components of complement, C6, C7, and C8, to form a complex that binds to the microorganism or foreign cell and digests an area of the cell membrane. This forms a hole in the membrane through which the cytoplasm leaks out of the cell, killing the microorganism or cell. In addition, the activated C3 and C5 molecules, which have a carboxyl-terminal arginine residue at the activation site, stimulate additional responses to the foreign organism. These responses include the attraction of white blood cells to the site of invasion and activation of these cells to attack and digest the foreign material.

Finally, it is essential that the antibody response should be used only against foreign proteins and other antigens, not against the proteins and tissues of the individual making the antibody. Occasionally, the mechanism for this self-tolerance breaks down and an individual develops autoimmune disease. This large and complex subject is described briefly in Box 6-4.

BOX 6-4

SELF-TOLERANCE, AUTOIMMUNE DISEASES, AND THYROID HORMONE

Under normal conditions, a person's own proteins and other antigens do not generate antibodies that react with them; the immune response is generated only against foreign antigens. The mechanisms by which this self-tolerance is brought about are complex and still not completely understood, but the overall pattern is known. Tolerance toward one's own antigens is developed in the embryo by the destruction of specific helper T cells that are required for the development of the immune response for those particular antigens. There are also suppressor T cells that are specific for particular antigens and inhibit the response of B cells and helper T cells to those antigens.

Failure of this self-tolerance is believed to be responsible for several autoimmune diseases. One of the earliest to be recognized is Hashimoto's thyroiditis, in which antibodies are generated against *thyroglobulin.* This protein is synthesized in the thyroid gland, where it is cleaved to produce *thyroxine,* an iodine-containing hormone that is released from the thyroid gland and regulates the overall rate of metabolism. In Hashimoto's disease, autoantibodies are formed that react with thyroglobulin in the patient's thyroid gland. This leads to the destruction of thyroglobulin, which causes hypothyroidism and a reduced metabolic rate of the patient.

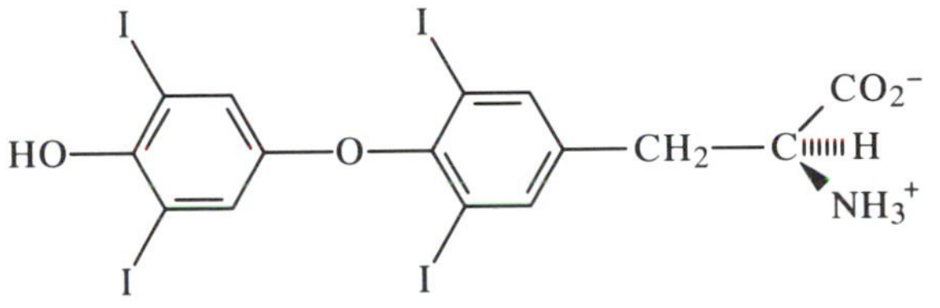

Thyroxine
(L-3,3′,5,5′-Tetraiodothyronine)

There is evidence that autoimmunity is responsible for several other diseases that include lupus erythematosus, in which autoantibodies react with many tissues throughout the body; rheumatoid arthritis, in which they react with joint tissues; and pernicious anemia, in which they are directed against proteins in the stomach that are required for the uptake of vitamin B_{12}. This vitamin is required for the synthesis of red blood cells (see Chapters 20 and 25).

Summary

Hydrolytic enzymes are important for many physiological functions in addition to the digestion of foodstuffs. Enzymes that catalyze hydrolysis utilize several different mechanisms for catalysis that include general acid-base catalysis of the attack by water or the hydroxyl group of serine and attack by a thiol anion of cysteine in an ion pair. They can bring about hydrolysis of insoluble molecules at the interface between aqueous and organic phases, which is important in many physiological systems. Blood clotting and the complement system of the immune response are two examples of physiological responses by an organism to different challenges. Both of these systems are cascades of enzymatic reactions that rapidly amplify the response. It is equally important that these responses be turned off when they are not needed, to avoid thrombosis of the blood vessels and the development of autoimmune diseases.

ADDITIONAL READING

WALSH, C. 1979. *Enzymatic Reaction Mechanisms.* San Francisco: Freeman, ch. 3.

ROITT, I. M., BROSTOFF, J., and MALE, D. 1985. *Immunology.* St. Louis: Mosby.

MANN, K. G. 1987. The assembly of blood clotting complexes on membranes. *Trends in Biochemical Sciences* 12: 229–233.

ZWAAL, R. F. A., and HENKER, H. C., eds. 1986. *Blood Coagulation.* New York: Elsevier.

JACKSON, D. Y., JACOBS, J. W., SUGASAWARA, S. H., REICH, S. H., BARTLETT, P. A., and SCHULTZ, P. G. 1988. An antibody-catalyzed Claisen rearrangement. *Journal of the American Chemical Society*, 110: 4841.

HILVERT, D., and NARED, K. D. 1988. Stereospecific Claisen rearrangement catalyzed by an antibody. *Journal of the American Chemical Society*, 110: 5593.

PROBLEMS

1. Explain why the activity of pepsin at pH 3 decreases when the pH is increased or decreased.

2. Pancreatic lipase does not hydrolyze a dilute solution of phenyl propionate, CH_3CH_2COOPh, in water. What would you do to make lipase hydrolyze this ester?

3. The activity of bromelin, a proteolytic enzyme from pineapple, is found to decrease with time after it is dissolved in water. What are two likely reasons for the decrease in activity, and how would you prevent this decrease or regain the activity?

4. How would you test whether a proenzyme activates itself?

5. Suppose that you are going to set up the world's first blood bank. What would you do to prevent the stored blood from clotting? Consider several possibilities.

6. Explain why it is useful for an antibody molecule to have several identical binding sites for its specific antigen.

7. Design a molecule that could be attached to a protein and used as an antigen to stimulate the formation of an antibody that has catalytic activity for a reaction.

The Structure of Proteins

7

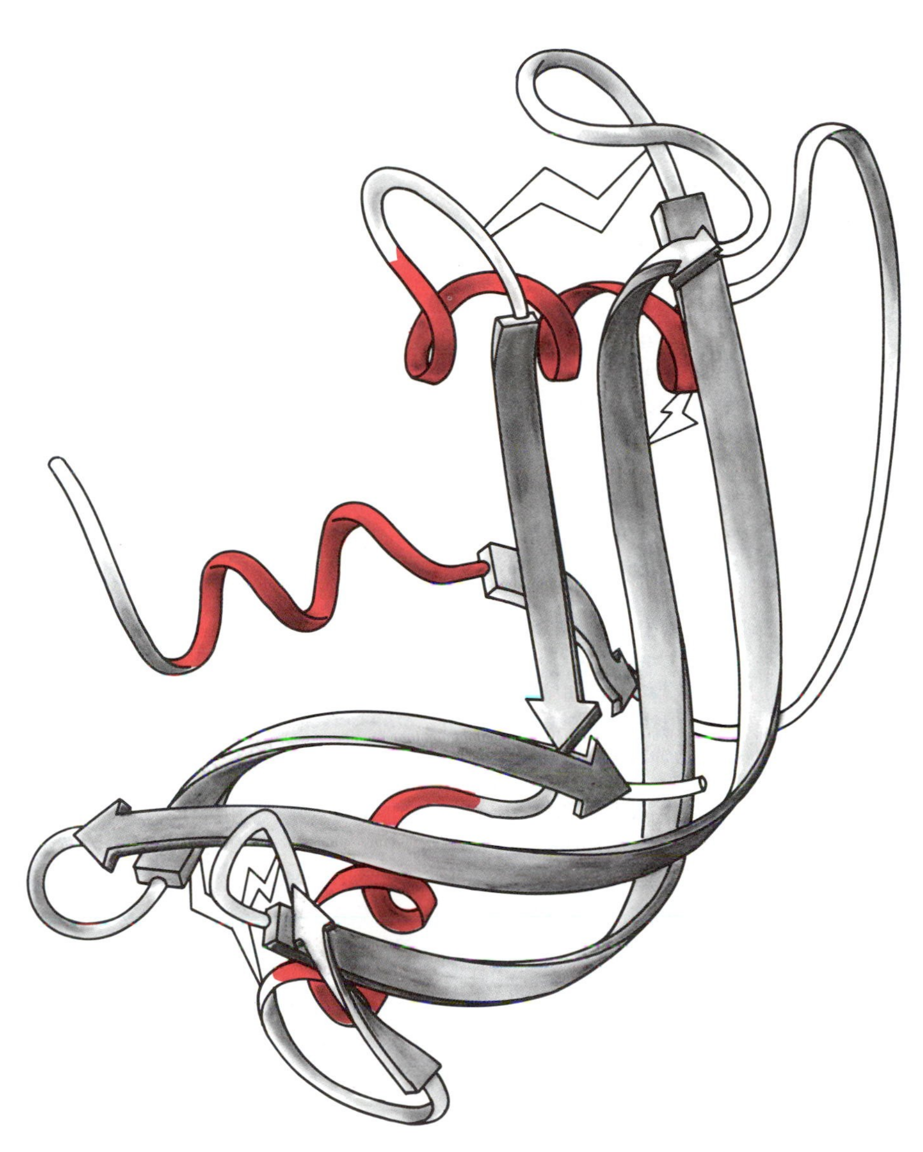

Ribbon diagram of the polypeptide chain in ribonuclease A.

We have seen that proteins are large molecules that do extraordinary things. It is hard to imagine how life could exist without proteins—although there has been some speculation that life on earth may have been initiated by the self-replication of nucleic acids, which eventually developed the ability to direct the synthesis of proteins. Many proteins are enzymes that catalyze reactions with a speed and specificity that has not been approached by synthetic catalysts; protein receptors respond to hormones and neurotransmitters by initiating complex physiological responses; other proteins recognize foreign molecules and respond in complex ways that protect an organism from damage; proteins cause contractile tissues such as muscles to move—the list goes on to include most of the processes that we associate with a living organism.

We have already encountered proteins both as enzymes and as substrates that are cleaved by enzymes. In this chapter, we ask the question: How is it possible to characterize and understand the properties of such an enormous and complex organic molecule as a protein? Because of their large size and unusual properties, especially their low stability under the conditions that are used for characterizing most organic molecules, it was necessary to develop entirely new experimental methods for the characterization of proteins. These methods constitute a new field of chemistry: protein chemistry.

One of the major advances of this century is the determination of the structure of proteins—first, the sequence of amino acids in the peptide chain and, then, the three-dimensional structure by X-ray diffraction of crystalline proteins. These structures are extraordinarily complex, but we will see that some simplification is introduced by the existence of α-helices and β-sheets, which are ordered substructures in different parts of the protein molecule. The remainder of the protein is usually described as a "random coil," although it is neither random nor a coil; it does, however, have an almost identical structure in different molecules of a particular protein. It might better be called an "ordered jumble." The structure of collagen, a structural protein of connective tissue, corresponds to a triple helix and includes covalent cross-links that give it great tensile strength.

The properties of proteins have been determined by a large variety of chemical and physical methods. The molecular weight can be determined most accurately from the amino acid sequence or by sedimentation at very high speeds in an ultracentrifuge, but approximate molecular weights are often determined either by the rate of movement of the unfolded protein in an electric field in the presence of a charged detergent or by passage through a column that separates large molecules according to their size. The charge of a protein determines its rate of movement in an electric field by electrophoresis and depends mainly on the state of ionization of the amino acid side chains. The charge also makes possible the separation and purification of proteins by chromatography on ion-exchange resins and by electrophoresis or isoelectric focusing. Proteins can also be purified by precipitation with concentrated ammonium sulfate or organic solvents and by affinity chromatography on columns that interact with specific sites on the protein.

In the next chapter, we examine the question of how proteins are held in their active, native structure under physiological conditions. What are the forces that hold proteins in their native structure, and how are these forces disrupted when a protein is placed in an unfavorable environment that causes it to unfold into an inactive, denatured state?

The Covalent Structure of Proteins

We have already seen that proteins are constructed from long chains of α-amino acids that are connected by peptide bonds (Figure 7-1). When a chain of amino

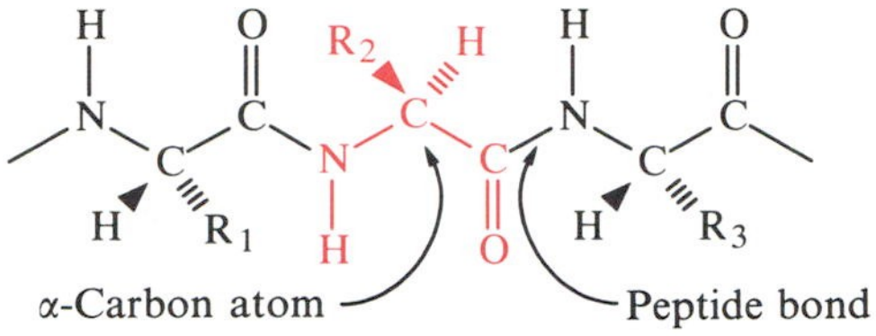

α-Carbon atom — Peptide bond

Figure 7-1

A Segment of a Peptide Chain.

acids is synthesized in the cell, it folds up spontaneously into the complex structure of the native, functional protein. The first step in characterizing the structure of a protein is to determine the sequence of the amino acids in the peptide chain of the protein.

The determination of the sequence of amino acids in the peptide chains of insulin by Sanger in 1953 is a major milestone in biochemistry because it is the first instance in which the structure of such a complex biological molecule was defined in chemical terms. This problem of describing the chemistry of a macromolecule, like many such problems, had seemed almost impossible to solve until it was solved by Sanger; now its solution has become an almost routine procedure (though by no means a trivial one).

Insulin is a protein hormone that is synthesized in the pancreas and released into the bloodstream. It regulates the concentration of glucose in the blood by controlling the uptake of glucose into cells. The glucose can be used immediately as a source of energy or stored as a polysaccharide, glycogen. The problem for Sanger was to determine the sequence of the twenty-one amino acids in the A chain and the thirty amino acids in the B chain of insulin (Figure 7-2). The two chains are connected by two disulfide cross-links between cysteine residues, and there is an additional disulfide cross-link in the A chain. Most other proteins are much larger than insulin; for chymotrypsin and trypsin, for example, it is necessary to determine the sequence of more than two hundred amino acids.

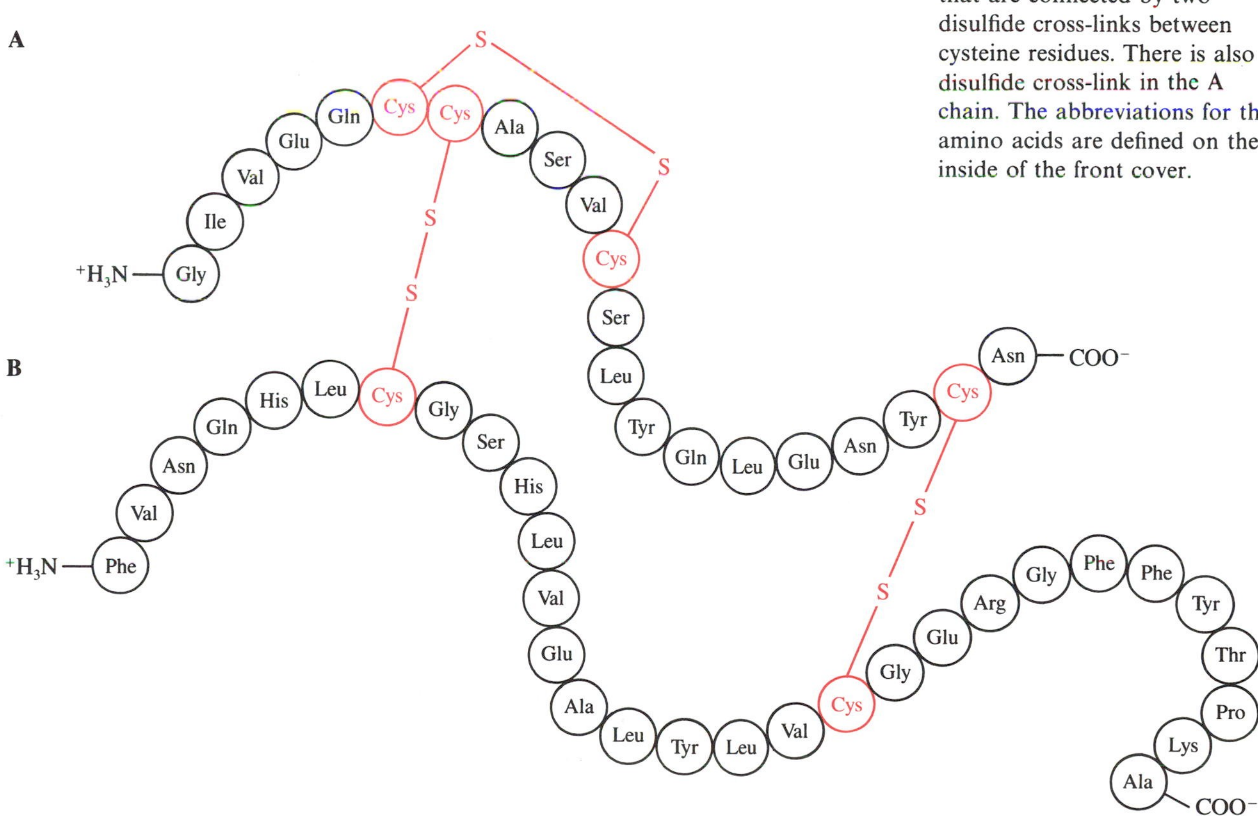

Figure 7-2

The Primary Structure of Insulin. Insulin consists of two chains of amino acids, A and B, that are connected by two disulfide cross-links between cysteine residues. There is also a disulfide cross-link in the A chain. The abbreviations for the amino acids are defined on the inside of the front cover.

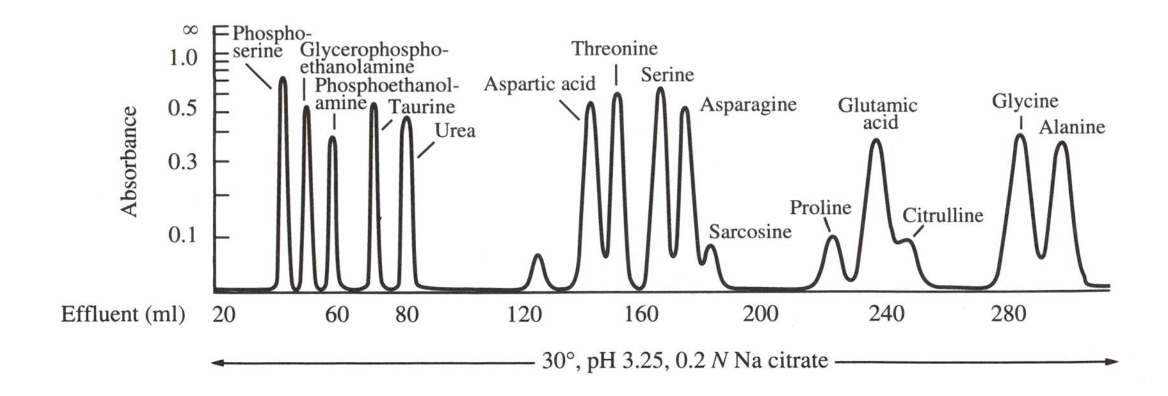

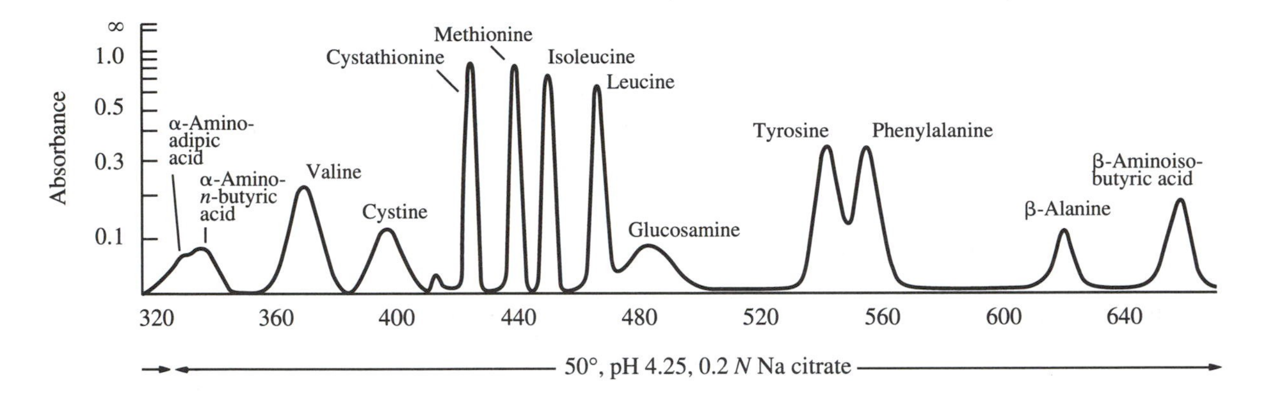

Figure 7-3

Analysis of a Mixture of Amino Acids by Ion-Exchange Chromatography. A different column is used for basic amino acids, such as lysine and histidine. [After D. H. Spackman, W. H. Stein, and S. Moore. *Anal. Chem.* 30 (1958): 1190.]

Quantitative Amino Acid Analysis

The total amino acid composition of a purified protein or peptide can be determined by *ion-exchange chromatography* after hydrolysis of the peptide chain in acid. The different amino acids bind to an ion-exchange resin in a chromatography column with different affinities, by interacting with the charged groups and the resin backbone of the column, so that they are eluted from the column at different times when a large volume of an eluting fluid is passed through the column.

Figure 7-3 illustrates the results that are obtained with an automatic amino acid analyzer that separates the amino acids of a protein hydrolysate on ion-exchange columns and analyzes the concentrations of the different peaks. This procedure is described in Box 7-1.

Cleavage to Peptides and Analysis of the Amino Acid Sequence

The first step in the sequence determination of a protein is to cleave the peptide chain into smaller peptides and then separate homogeneous samples of these peptides. Trypsin is especially useful for this initial cleavage, because of its specificity for lysine and arginine residues. A polypeptide chain containing five

BOX 7-1

ION-EXCHANGE CHROMATOGRAPHY OF AMINO ACIDS

A protein or peptide is cleaved into its amino acids by hydrolysis in acid at elevated temperature.

Any disulfide bonds in the protein are cleaved by reduction with borohydride and then blocked or oxidized before hydrolysis, as shown in equations a, b, and c, so that cystine residues in the protein will appear as derivatives of cysteine in the analysis. A mixture of amino acids is added to a chromatography column that is packed with an insoluble polymer to which are attached charged groups, such as substituted ammonium ions (Figure 7-3). A counter-ion, such as chloride, is present in the aqueous phase surrounding the polymer.

(a) Cystine ($-CH_2-S-S-CH_2-$) $\underset{O_2}{\overset{BH_4^-}{\rightleftharpoons}}$ Cysteine ($-CH_2-SH$) + Cysteine ($HS-CH_2-$)

(b) $-CH_2-S^-$ + $I-CH_2-COO^-$ (Iodoacetate) $\longrightarrow$ $-CH_2-S-CH_2-COO^-$ + I^-

(c) $-CH_2-SH$ + HCO_3H (Performic acid) $\longrightarrow$ $-CH_2-SO_3^-$

The eluting fluid contains dilute salt and buffer, and the ionic strength and pH of this fluid may be changed during the development of the column by the gradual addition of a more-concentrated salt solution containing acid, base, or a buffer with a different pH. A different column that is packed with a negatively charged ion-exchange resin is used to separate the "basic," positively charged amino acids: arginine, lysine, and histidine. A separation of amino acids by ion-exchange chromatography is shown in Figure 7-3.

Analysis of the amino acid fractions that are eluted from the column gives the relative concentrations of the amino acids in the protein. The number of residues of each amino acid in the protein is easily calculated if the molecular weight of the protein is known. We will see later how the molecular weight is determined.

A separate spectrophotometric analysis must be carried out for tryptophan, which is destroyed by acid hydrolysis of the peptide chain. It is also necessary to correct for losses during hydrolysis of some amino acids, such as serine, by carrying out the analysis after different times of hydrolysis and extrapolating to zero time.

---

such residues, for example, will be cleaved by trypsin into six shorter peptides (Figure 7-4). The shorter peptides are then separated and analyzed by ion-exchange chromatography or other techniques. The peptide chain can also be cleaved at specific amino acids by chemical methods, such as the reaction with cyanogen bromide that is described in Box 7-2.

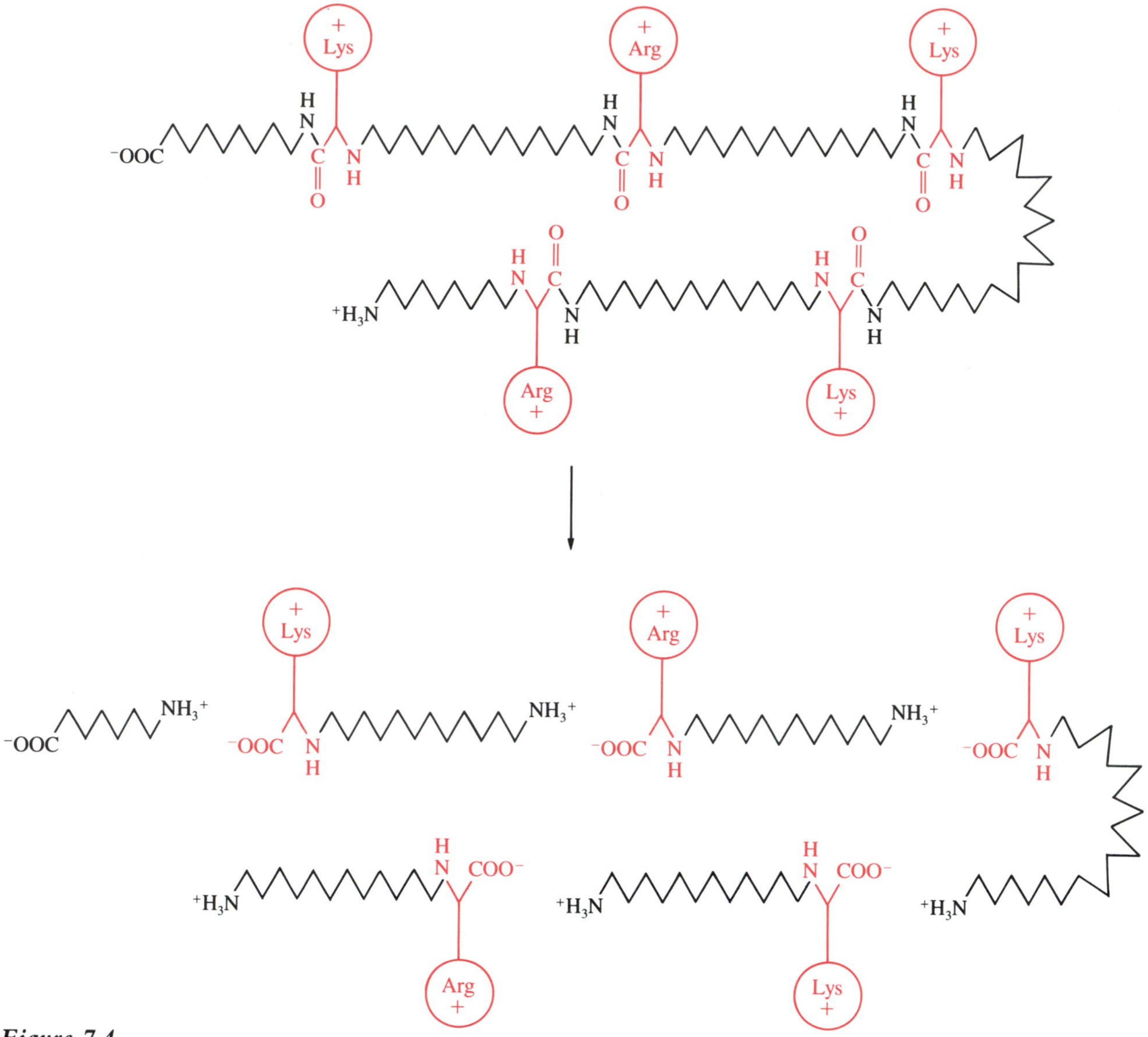

Figure 7-4

Trypsin cleaves a peptide chain specifically at positively charged residues to give smaller peptides with carboxyl-terminal lysine or arginine residues.

The amino acid sequence of the isolated peptides is then determined by the sequential cleavage of amino acids from the carboxyl-terminal and amino-terminal ends of each peptide. This can be accomplished by the use of exopeptidases, which are specific for the amino- or carboxyl-terminal ends of peptide chains, or by chemical methods. Carboxypeptidase successively cleaves amino acids from the carboxyl-terminal end of the peptide (equation 1),

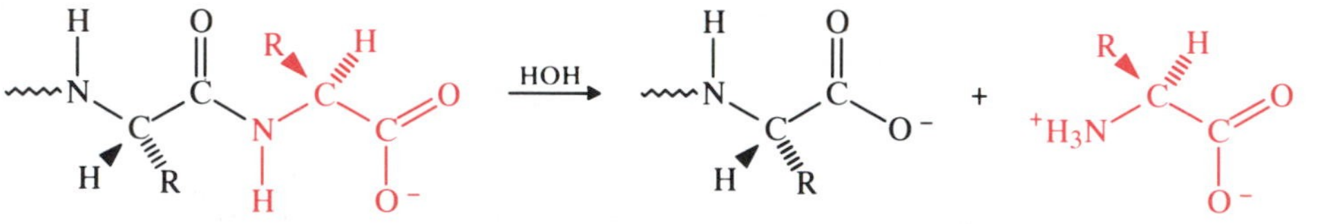

(1)

and it is possible to determine the sequence of amino acids for several residues into the peptide by following the time course for the release of the amino acids. Some chemical methods for determining the sequence of amino acids in peptides are described in Box 7-3.

BOX 7-2

A CHEMICAL METHOD FOR THE SPECIFIC CLEAVAGE OF A PEPTIDE CHAIN

The *cyanogen bromide* procedure is one of several chemical methods that provides an alternative to trypsin for cleaving a peptide chain into shorter peptides. Cyanogen bromide, N≡CBr, cleaves the chain specifically at methionine residues and gives a manageable number of peptides, because most proteins contain only a small number of methionine residues. The probable mechanism of action of cyanogen bromide is to activate the methionine side chain toward intramolecular nucleophilic attack by the adjacent peptide carbonyl group. This gives an unstable imidate that is rapidly hydrolyzed to give homoserine lactone.

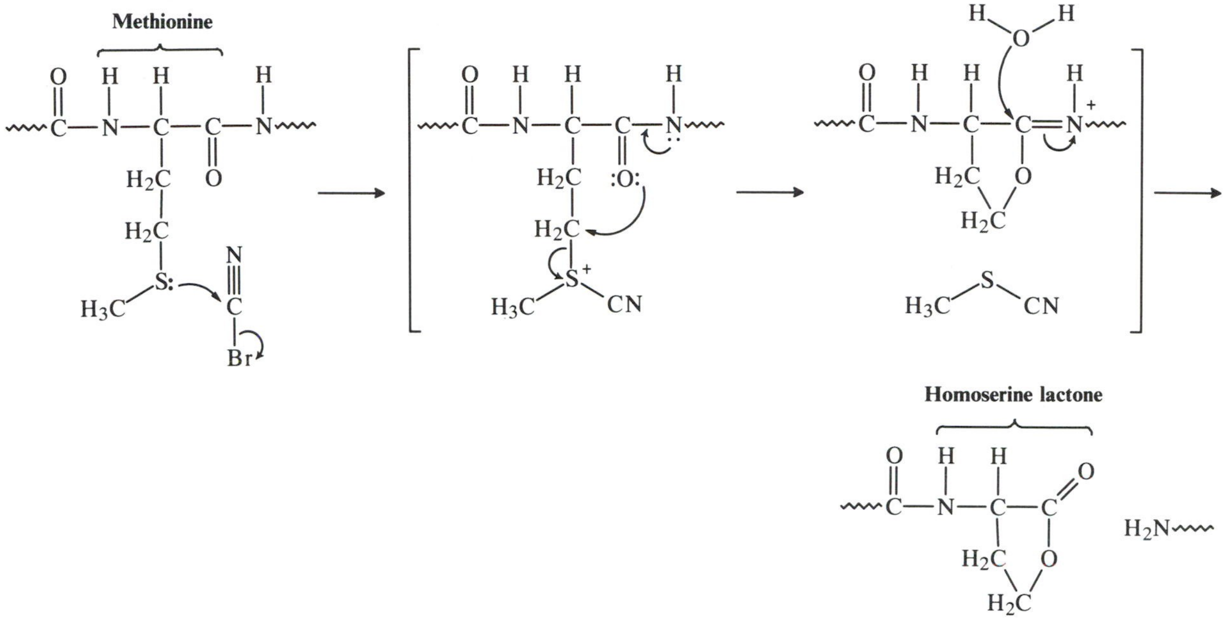

Some puzzlement was caused for a time by the finding that certain proteins appear to have no N-terminal amino acids, as measured by the fluorodinitrobenzene reaction (Box 7-3). It was found that the reason for this is that the terminal amino acid in these proteins is itself acylated, although not by an α-amino acid, so that the reaction of the terminal amino group with fluorodinitrobenzene is blocked. The acyl group may be acetic acid, formic acid (**1**, R = CH_3 or H), or the α-carboxyl group of glutamate, which forms the cyclic amide 5-oxoproline (this compound is also known as 5-oxopyrrolidine-2-carboxylic acid or pyroglutamic acid).

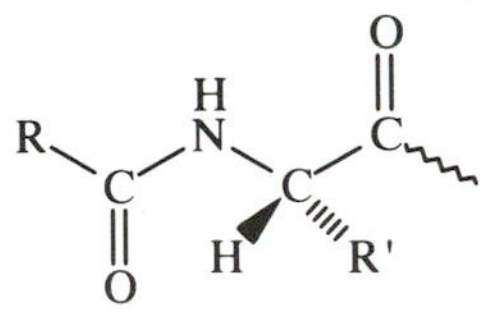

1

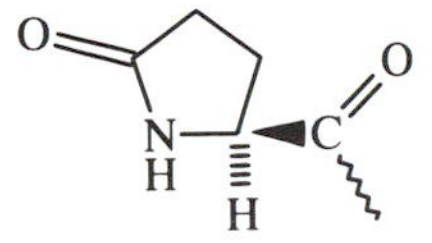

5-Oxoproline

BOX 7-3

CHEMICAL ANALYSIS OF AMINO-TERMINAL AND CARBOXYL-TERMINAL AMINO ACIDS

The most-useful chemical method for the analysis of peptide sequences is the reaction of N-terminal amino acids with phenylisothiocyanate, the *Edman degradation* (Figure A). This reaction removes amino acids sequentially from the N-terminal end of the chain as their phenylthiohydantoin (PTH) derivatives. In the first step of the reaction, isothiocyanate undergoes nucleophilic attack by the terminal amino group of the peptide to give a substituted thiourea. This step is carried out in dilute base, so that the electron pair of the free amino group is not blocked by protonation. Upon treatment with weak acid, the terminal amino group of the thiourea attacks the peptide bond of the terminal amino acid in a facile intramolecular reaction to give the phenylthiohydantoin derivative of the original N-terminal amino acid. This may be identified by chromatography, by comparison with standard phenylthiohydantoin derivatives of known amino acids. Cleavage of the peptide bond gives a new N-terminal amino acid that may be identified by repetition of the whole procedure. The Edman degradation has been developed to the point that it can be used successfully to determine the sequence of as many as sixty amino acids, starting from the N-terminal end of a protein chain.

Hydrazinolysis provides a chemical method, analogous to the action of carboxypeptidase, for identifying the amino acid on the carboxyl-terminal end of a peptide. Anhydrous hydrazine reacts with the peptide bonds of

Figure A

One cycle of Edman degradation of a peptide.

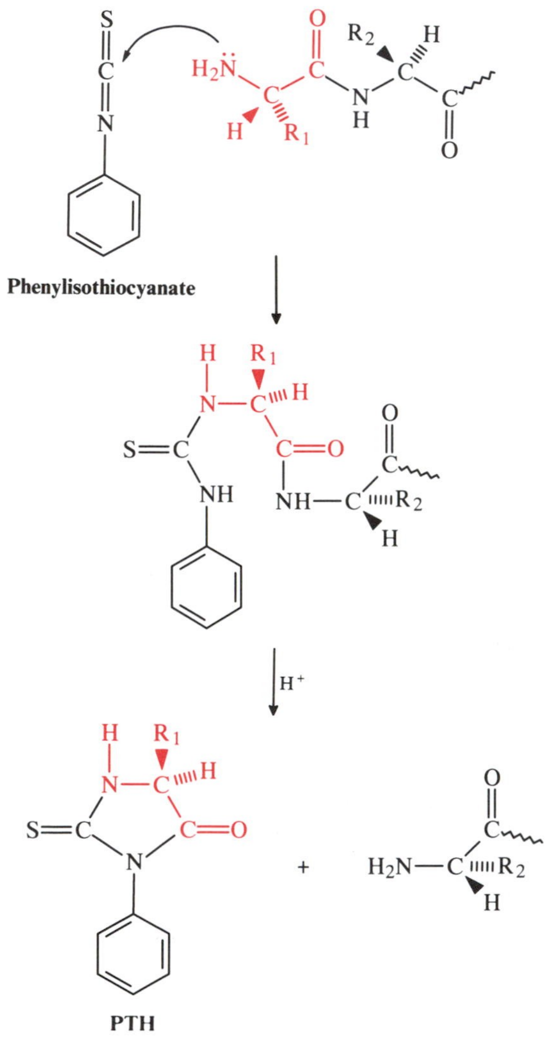

Phenylisothiocyanate

H^+

PTH

BOX 7-3 (continued)

Figure B
Hydrazinolysis of a peptide.

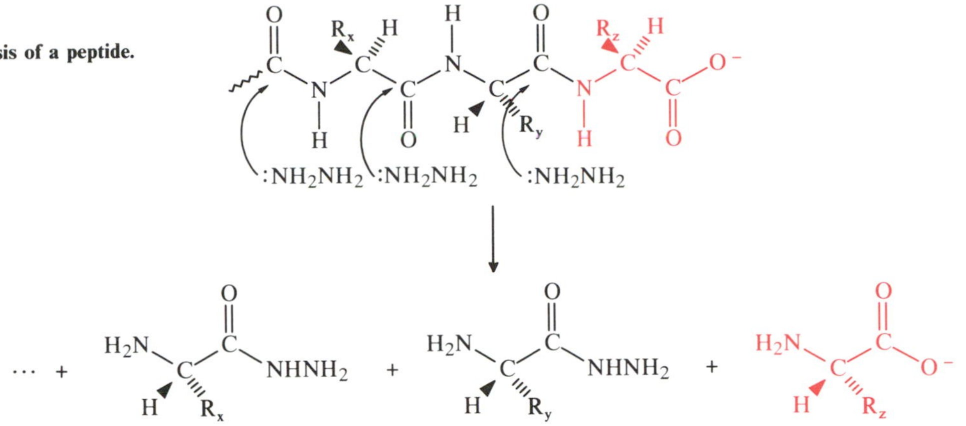

Figure C
Reaction of a peptide with Sanger's reagent, FDNB.

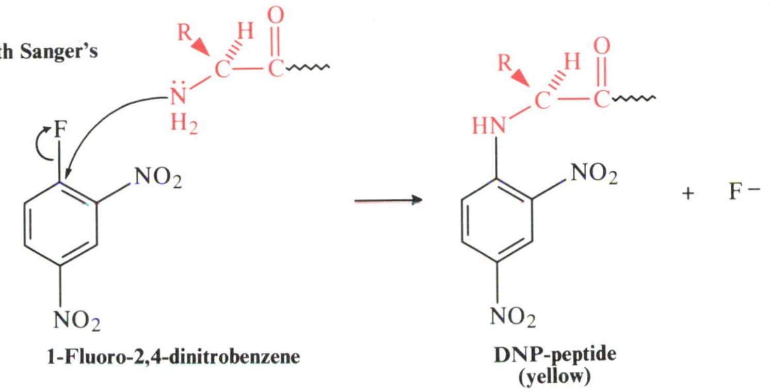

1-Fluoro-2,4-dinitrobenzene

DNP-peptide (yellow)

the amino acids in a peptide at 100°C to form the corresponding hydrazides. The unactivated carboxylate group at the carboxyl-terminal end of the peptide chain does not react, as illustrated in Figure B, so that the free amino acid is released and may be identified by chromatography.

The amino-terminal end of a peptide chain may also be identified by reaction with 1-fluoro-2,4-dinitrobenzene (Sanger's reagent, FDNB) or other electrophilic reagents. The free amino group reacts through *nucleophilic aromatic displacement* of fluoride ion to give the yellow dinitrophenyl (DNP) derivative (Figure C). The yellow color facilitates the location and purification of the peptide by chromatography; and, after acid hydrolysis of the peptide, the N-terminal DNP-amino acid may be easily identified by chromatography.

Other reagents that give stable derivatives of N-terminal amino acids are *dansyl chloride,*

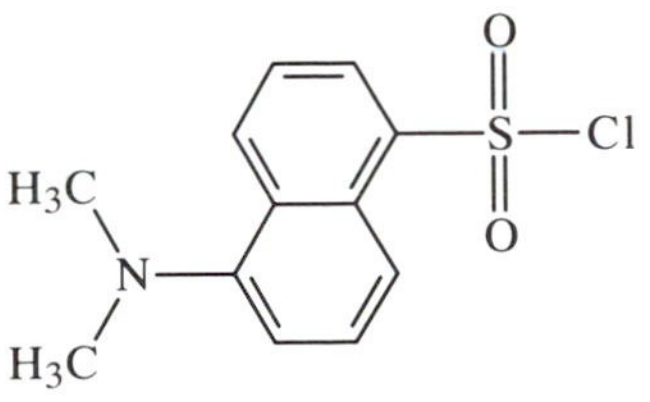

Dansyl chloride

which gives a strongly fluorescent product, and *dabsyl chloride,*

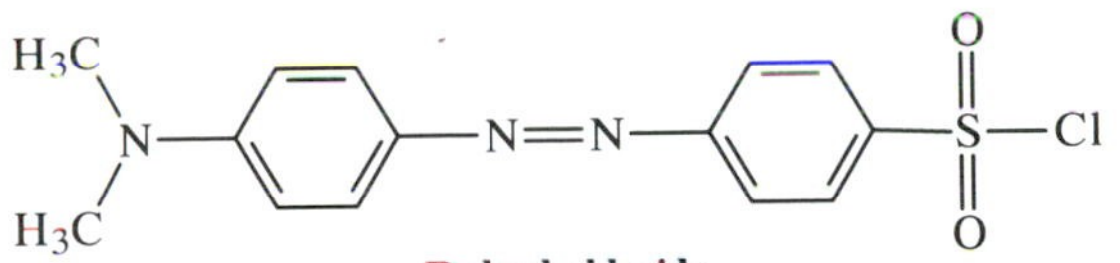

Dabsyl chloride

which gives a highly colored product that is easily visualized after chromatography.

Analysis of the N-terminal amino acids of a sample of protein or peptide is a useful method for determining the number of peptide chains and for detecting the presence of impurities in the sample. Insulin, for example, has two N-terminal residues, a glycine on the A chain and phenylalanine on the B chain (Figure 7-2). Impurities in the preparation usually give a variable amount of additional N-terminal amino acids.

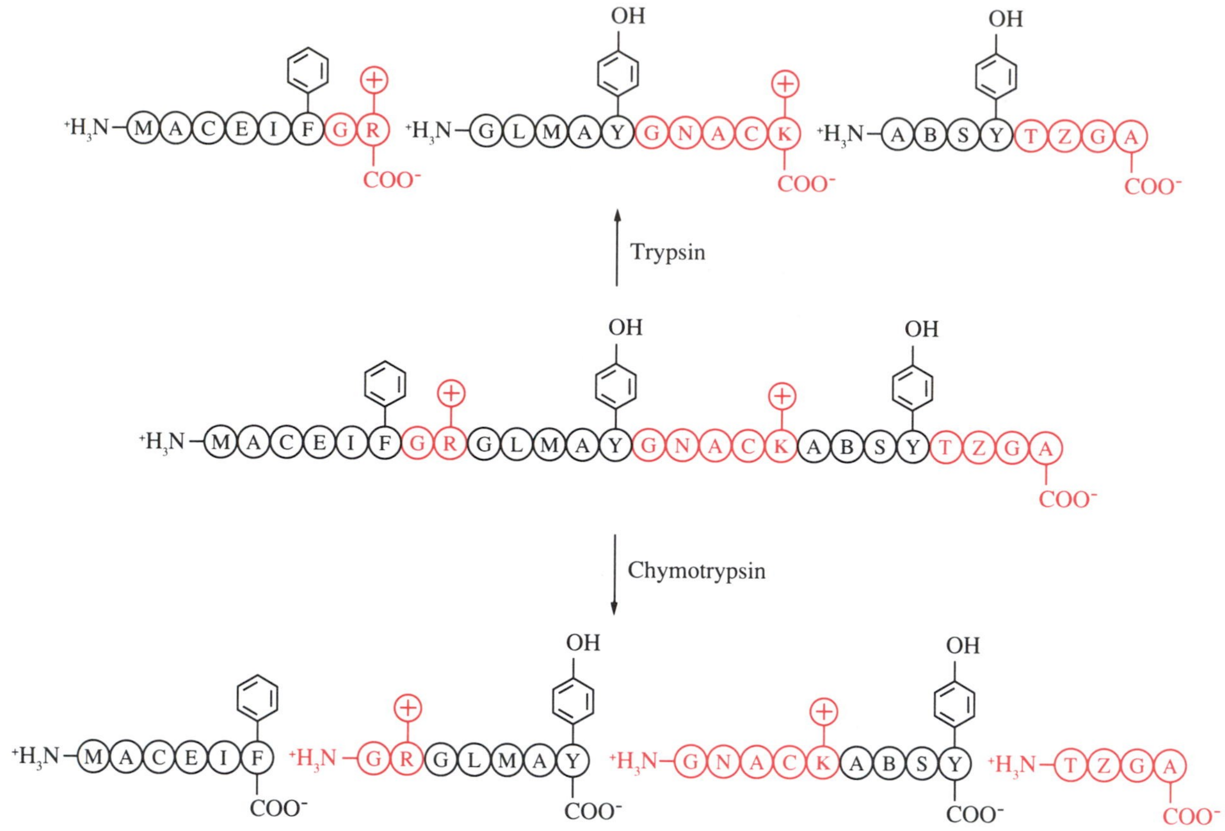

Figure 7-5

Cleavage of a peptide chain at different positions by the specific proteases trypsin and chymotrypsin gives different peptides. The sequence of the original peptide can be deduced from the sequences of the short peptides by matching the regions of overlapping sequence in the peptides.

Alignment of Peptides

Once the amino acid sequence has been determined for the individual peptides of a protein, the problem remains of determining the relative positions of the peptides in the protein. The positions of the amino- and carboxyl-terminal peptides are easily decided by comparing the amino- and carboxyl-terminal sequences of the intact protein with those of the individual peptides. The positions of the other peptides are determined by cleaving the protein with different reagents to produce peptides that overlap the connecting points of the first set of peptides. For example, if the sequence has been determined first on the tryptic peptides, cleavage of the protein with the cyanogen bromide method will give a new series of peptides with C-terminal methionine residues. Still another series of peptides can be obtained by cleavage of the peptide chain with chymotrypsin, which gives peptides that generally result from attack at tyrosine, phenylalanine, or tryptophan residues. Comparison of the sequences of the peptides that are prepared by two different methods of specific cleavage will generally give the sequence of the entire peptide chain, because the sequence of a peptide prepared by one method will usually overlap the sequences of two adjacent peptides that were obtained by the other method. This is illustrated in Figure 7-5 for peptides obtained by cleavage with trypsin and with chymotryp-

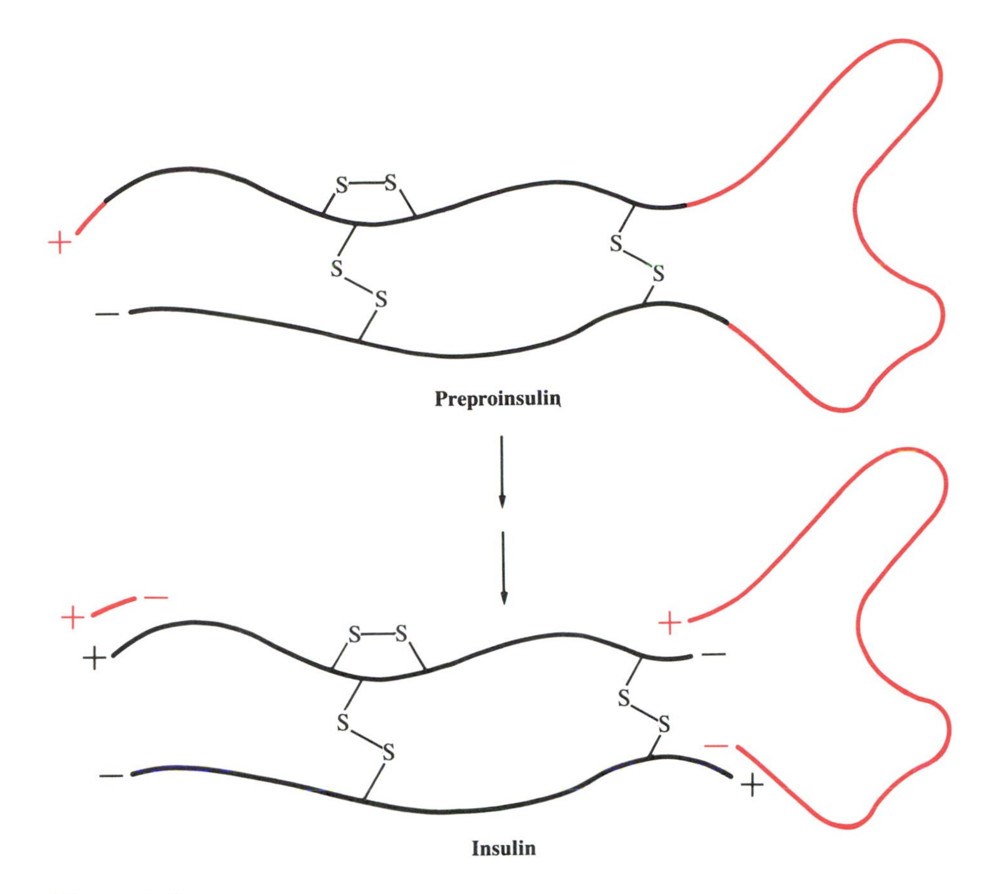

Figure 7-6

Insulin is formed from preproinsulin by processing; that is, by excision of peptides from the middle and the N-terminal end of the peptide chain.

sin. It is often possible to determine the relative positions of the original peptides from the amino acid composition of these overlapping peptides, without carrying out a complete sequence determination of each peptide.

In the past few years, the direct sequencing of proteins by chemical methods has been increasingly replaced by nucleotide sequencing of DNA, which carries the information for the amino acid sequence of the protein in its sequence of nucleic acid bases: a sequence of three bases provides the code for each amino acid of the protein. Relatively simple methods have been developed for the sequencing of nucleic acids, so that reliable amino acid sequences for proteins can be obtained by preparing and isolating the nucleic acid that carries the code for a particular protein and determining its sequence. These techniques will be described in Chapter 12. This method of DNA sequencing gives the sequence of the complete protein chain as it is synthesized in the cell. This is often different from the sequence of a protein that is isolated after it has been excreted from the cell, because proteins generally undergo *processing* before they are released into the extracellular space. The newly synthesized protein generally has a *leader sequence* at the N-terminal end that interacts with a protein complex in the cell membrane to facilitate extrusion of the peptide chain into the extracellular space. The leader sequence is then cleaved to give the mature protein. For example, insulin (Figure 7-6) is synthesized in the β-cells of the pancreas as a single chain of *preproinsulin,* which is shortened first to proinsulin and then to the two chains of active insulin by processing with specific proteolytic enzymes. Such processing is unnecessary and does not occur with many proteins that are retained within a cell, as are most bacterial proteins. The synthesis of proteins will be described more completely in Chapter 10.

The Primary Structure Determines the Three-dimensional Structure of Proteins

The structure of a protein molecule is extraordinarily complex. A peptide chain of two hundred amino acids can exist in an enormous number of different structures because there are several different possible positions for rotation of the chain at each amino acid in the chain. However, a native protein usually exists in a single structure that is so highly ordered that it can be crystallized and examined by X-ray diffraction. This means that there must be a large number of specific interactions between the amino acids in the chain that hold the protein in its native, active structure. When the protein is synthesized in the cell, it folds up spontaneously into its native three-dimensional structure. We will see later in this chapter how this structure can be divided into substructures, which make up the *secondary* and *tertiary* structures of the protein.

The Size and Shape of Proteins

The large size and irregular shape of protein molecules make it difficult to determine their molecular weight and structure in solution by the techniques that are usually used for small molecules. Consequently, a number of new and ingenious techniques have been developed for measuring the properties of proteins and other large molecules. Methods that utilize osmotic pressure and other colligative properties of solutions are widely used to determine the molecular weight of small molecules, but they are difficult to apply to proteins both because of their large size (and therefore the low molar concentrations that are attainable) and because of the presence of accompanying small counterions in most protein solutions that balance the charge of the protein. The methods that have been developed for proteins depend to various extents on the size and the density of the molecule, and there are difficulties of one kind or another with all of them when the protein has an unusual shape or composition, especially if it contains a significant amount of carbohydrate or lipid.

Sedimentation

Large, heavy particles tend to settle to the bottom of a solution. This process is too slow to affect proteins in the earth's gravitational field, but the centrifugal field developed by modern ultracentrifuges, as much as 500,000 times the force of gravity, will sediment proteins and other macromolecules in a few minutes or hours. The sedimentation is faster for larger molecules and can be used to estimate the molecular weight. The molecular weight cannot be obtained from the sedimentation rate alone because this rate depends on the density and shape, as well as the weight, of the protein. However, it can be measured from the sedimentation and diffusion rates if the density is known or can be estimated. A sedimentation pattern for the ultracentrifugation of bovine serum mercaptalbumin is shown in Figure 7-7 and the use of ultracentrifugation to determine the molecular weight is described briefly in Box 7-4.

An important practical application of ultracentrifugation is that it enables us to follow the dissociation of proteins into subunits—we usually want to know the number and nature of the subunits of which a protein is composed and the conditions under which it dissociates into subunits. Dissociation may be brought about by diluting the protein, by increasing the salt concentration so as to decrease the electrostatic attraction between subunits, or by increasing the net charge of the subunits so that they repel one another, by increasing or decreasing the pH. The number of charged groups changes with pH as —COO^- groups are protonated at low pH or the —NH_3^+ groups on lysine side chains are deprotonated at high pH.

An approximate measure of the molecular weight may be obtained without an analytical ultracentrifuge by sedimenting a protein in a *sucrose gradient* with a preparative ultracentrifuge. A sucrose gradient can be prepared by carefully adding a concentrated solution of sucrose to the bottom of a centrifuge tube and then adding progressively more dilute solutions, which have a lower density. This is usually done by gradually mixing water into the concentrated solution of sucrose that is being added. The protein is mixed with standard proteins of known molecular weight and the mixture is layered carefully over a sucrose solution. During ultracentrifugation, the different proteins will sediment as discrete bands or zones, at different rates that depend on their size and weight. The sucrose gradient stabilizes these bands so that they can be removed sequentially from the centrifuge tube and their relative sedimentation coefficients determined. The method works well if the standard proteins are similar in shape and composition to the unknown. It has the

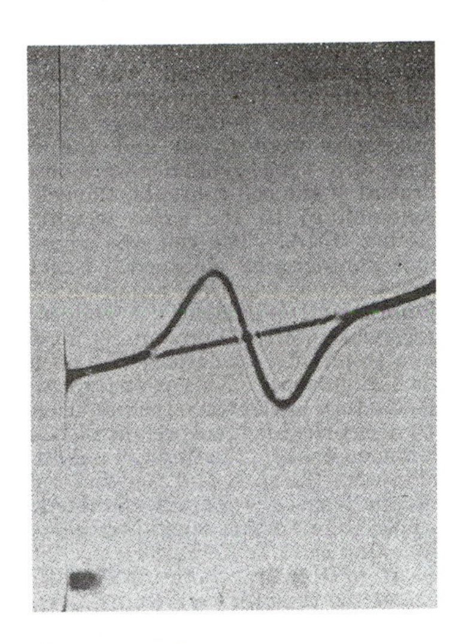

Figure 7-7

Ultracentrifugation of Bovine Serum Mercaptalbumin in 2.59 Molal Cesium Chloride. The protein is at sedimentation equilibrium and the concentration was measured by Schlieren optics in this experiment. [Reprinted with permission from J. B. Ifft and J. Vinograd. *J. Phys. Chem.* 66 (1962): 1990. Copyright 1962 American Chemical Society.]

BOX 7-4

ULTRACENTRIFUGATION

The observed rate of sedimentation per unit field of force for a molecule in the ultracentrifuge is the *sedimentation coefficient, s,* which represents a balance between the driving force for sedimentation and the resistance to movement from friction as described by equation a.

$$s = \frac{M(1 - \bar{v}\rho)}{f} \qquad \text{(a)}$$

The driving force for sedimentation depends on the difference between the densities of the solvent, ρ, and the protein, ρ_{prot}; on the molecular weight of the protein, M; and on the centrifugal field of force. The density of the protein is ordinarily expressed as its reciprocal, the *partial specific volume,* $\bar{v} = 1/\rho_{prot}$, which has a value close to 0.74 for ordinary proteins. The centrifugal field of force is given by $\omega^2 x$, in which ω is the angular velocity in radians per second and x is the distance from the center of rotation.

It is possible to determine the molecular weight by ultracentrifugation, from the sedimentation coefficient, s, and equation a, if the *frictional coefficient, f,* is known. This can be determined by measuring the rate of diffusion of the protein molecule, because the *diffusion coefficient, D,* depends inversely on the frictional coefficient according to equation b.

$$D = \frac{RT}{f} \qquad \text{(b)}$$

(R is the gas constant and T is the absolute temperature.) Diffusion coefficients are usually determined by measuring the spreading of an initially sharp boundary between a protein solution and solvent as a function of time.

BOX 7-4 (continued)

ULTRACENTRIFUGATION

Inserting equation b into equation a gives the well-known *Svedberg equation,* c, which permits the determination of molecular weight from measurements of sedimentation and diffusion rates.

$$M = \frac{RTs}{D(1 - \bar{v}\rho)} \tag{c}$$

The frictional coefficient, f, is a function of the average asymmetry and hydration of the protein molecule; it increases upon unfolding of the peptide chain. This accounts for the high viscosity of proteins that have been denatured in urea or guanidine hydrochloride in the presence of a thiol, such as mercaptoethanol, that cleaves the disulfide bonds. A fully denatured protein has a decreased sedimentation coefficient, because of its increased frictional coefficient, and exhibits the properties that are expected for a long-chain, flexible, linear polymer.

The most widely used ultracentrifugal method for the accurate measurement of molecular weights at the present time avoids the necessity for separate determinations of sedimentation and diffusion coefficients by measuring the distribution of the protein in the centrifuge cell at *sedimentation equilibrium* after prolonged centrifugation. This equilibrium distribution depends both on the rate of sedimentation, which pulls the protein toward the bottom of the ultracentrifuge cell, and on the rate of diffusion, which results in an upward movement from the more-concentrated protein at the bottom of the cell. By centrifugation at low speeds in a cell with a very short path length (0.1 mm), this sedimentation equilibrium may usually be reached in less than a day. The molecular weight determined by the method of sedimentation equilibrium is given by equation d.

$$M = \frac{2RT \ln c}{\omega^2 r^2(1 - \bar{v}\rho)} \tag{d}$$

in which c is the concentration of protein at a given distance r from the axis of rotation; in practice, the molecular weight is obtained from the slope of a plot of $\ln c$ against r^2.

Sedimentation and diffusion coefficients are often measured by taking advantage of the difference in refractive index of a protein and the solvent, so that the change in refractive index at the boundary between the sedimenting protein and the solvent can be observed as a *Schlieren pattern,* a measure of the change in refractive index with distance. Sedimentation equilibrium measurements are usually carried out with the much more sensitive *interference optics,* which permits the use of lower protein concentrations.

The sedimentation and diffusion rates of proteins often change significantly with changing protein concentration because of aggregation or activity coefficient effects, so that it may be necessary to carry out a series of measurements and extrapolate the results to zero protein concentration. This correction is much less important for measurements that are made at the low protein concentrations that are made possible by the use of interference optics.

---

advantage that sensitive techniques, such as ultraviolet spectrophotometry or measurement of enzymatic activity, may be used to measure the positions of the bands.

SDS Electrophoresis

A much simpler method for estimating the molecular weight of proteins and protein subunits, although it does not have a very solid theoretical basis, is by electrophoresis in a stabilized medium of polyacrylamide gel in the presence of sodium dodecylsulfate (SDS), an anionic detergent, after cleavage of disulfide cross-links. This method of protein separation by polyacrylamide gel electrophoresis is often abbreviated SDS-PAGE. *Electrophoresis* is the movement of a molecule in an electric field. The detergent, SDS, contains a long hydrocarbon chain of twelve carbon atoms, which interacts with the relatively nonpolar interior of most proteins to unfold the native structure, and a charged sulfate group that remains exposed to the solvent and gives a large net negative charge to the protein-detergent complex.

Polyacrylamide is prepared by the polymerization of acrylamide, through a radical chain reaction that is initiated by the addition of a trace of a radical-forming reagent (equation 2).

$$n\,H_2C{=}C(H)(C{=}O)NH_2 \xrightarrow{\text{Radical initiator}} \sim CH(C{=}O\,NH_2)-CH_2-CH(C{=}O\,NH_2)-CH_2-CH(C{=}O\,NH_2)-CH_2\sim \quad (2)$$

Acrylamide **Polyacrylamide**

Polyacrylamide is a useful polymer because it is strongly hydrophilic, so that it can be used with aqueous solutions and does not denature proteins. The tightness, or the "pore size," of the gel can be controlled by adding varying amounts of a bifunctional acrylamide, *methylene bis-acrylamide.*

$$CH_2{=}CH-C(=O)-NH-CH_2-NH-C(=O)-CH{=}CH_2$$

Methylene bis-acrylamide

This reagent can be incorporated into two chains, so that it connects the chains by cross-linking them and produces gels with different degrees of tightness, or pore size. These gels have different resistances to the movement of large compared with small molecules.

Now, if a band of this protein-detergent solution is placed in a polyacrylamide gel and a voltage is applied to the system, the negatively charged complex will move toward the anode at a rate that has been found to be inversely proportional to its molecular weight. The molecular weight of a protein is estimated by comparing its observed mobility after electrophoresis and staining of the protein band with the mobility of proteins of known molecular weight. If the protein consists of several peptide chains, this molecular weight is usually that of the subunit, because the detergent is a denaturing agent. Figure 7-8 shows the separation of a series of proteins of different molecular weight by SDS gel electrophoresis and Figure 7-9 shows the correlation of the molecular weight of a series of proteins with their mobility in SDS gel electrophoresis.

The rate of movement of the protein is a function of its charge and size, which determine the number of anionic SDS molecules that bind to it, the strength of the applied electric field, and the resistance to movement in the gel. The method works because the amount of SDS bound to each gram of protein (about 0.4 g/g at 5×10^{-4} M SDS) is almost constant for different proteins, and the SDS-protein particles take on a more or less constant rodlike shape. The cross-linked polyacrylamide gel serves a dual purpose. First, it provides a stable medium for electrophoresis that is easy to handle and minimizes problems from diffusion and convection currents. Second, it provides resistance to movement of the charged complexes, which is larger for the larger complexes and helps to achieve a separation according to molecular weight by a kind of filtration process. A schematic illustration of the movement of a protein-detergent complex in gel electrophoresis is shown in Figure 7-10.

If antibodies can be obtained that are specific for a particular protein of interest, that protein can be identified in an SDS-polyacrylamine gel at extremely low concentration, even in crude mixtures, by *Western blotting.* The proteins are transferred from the gel to an overlaid sheet of filter paper, by "blotting," and allowed to react with the specific antibody. The position of the protein-antibody complex is then determined by adding a second antibody that is specific for reaction with the first antibody (this antibody and the first antibody are obtained from different animal species). The second antibody is labeled with radioactivity or with an attached fluorescent molecule so that its

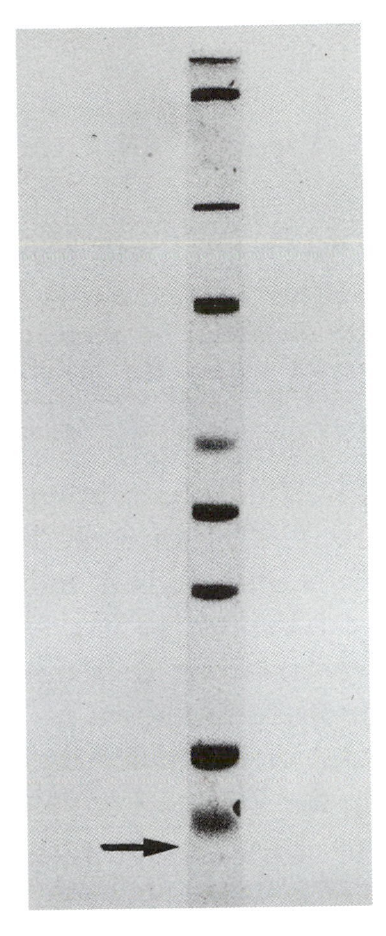

Figure 7-8

SDS-Polyacrylamide Electrophoresis of a Mixture of Proteins with Different Molecular Weights. The proteins were initially at the top of the gel and moved downward upon electrophoresis. The proteins and their molecular weights, in descending order, are myosin (200,000), phosphorylase *a* (100,000), bovine serum albumin (68,000), ovalbumin (43,000), glyceraldehyde-3-phosphate dehydrogenase (36,000), carbonic anhydrase (29,000), myoglobin (17,200), and cytochrome *c* (11,700). The arrow shows the position of a dye that was used as a marker. The proteins were stained after electrophoresis. (From K. Weber and M. Osborn. In *The Proteins,* vol. 1, 3d ed., ed. H. Neurath, R. L. Hill, and C.-L. Boeder. New York, Academic Press, 1975, p. 181.)

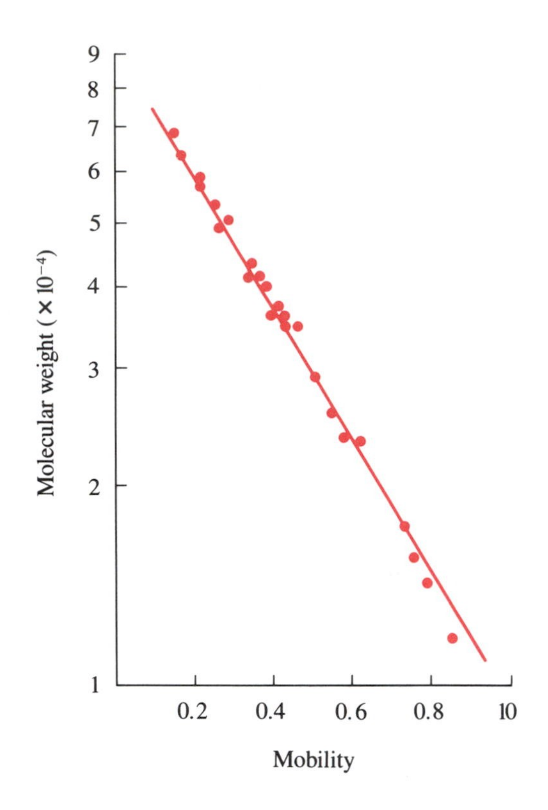

Figure 7-9

Correlation of the Molecular Weight of a Series of Proteins with Their Mobility in SDS Gel Electrophoresis. The proteins have molecular weights ranging from 14,000 to 70,000. (From K. Weber and M. Osborn. In *The Proteins,* vol. 1, 3d ed., ed. H. Neurath, R. L. Hill, and C.-L. Boeder. New York: Academic Press, 1975, p. 182.)

position on the gel can be located. The location of the fluorescent antibody is determined in a dark room by irradiation with light in the near-ultraviolet region of the spectrum. The location of radioactivity is determined with an *autoradiogram,* in which the dried paper is placed on a film that is exposed by the radioactive antibody and gives a dark band at the position that corresponds to the protein of interest.

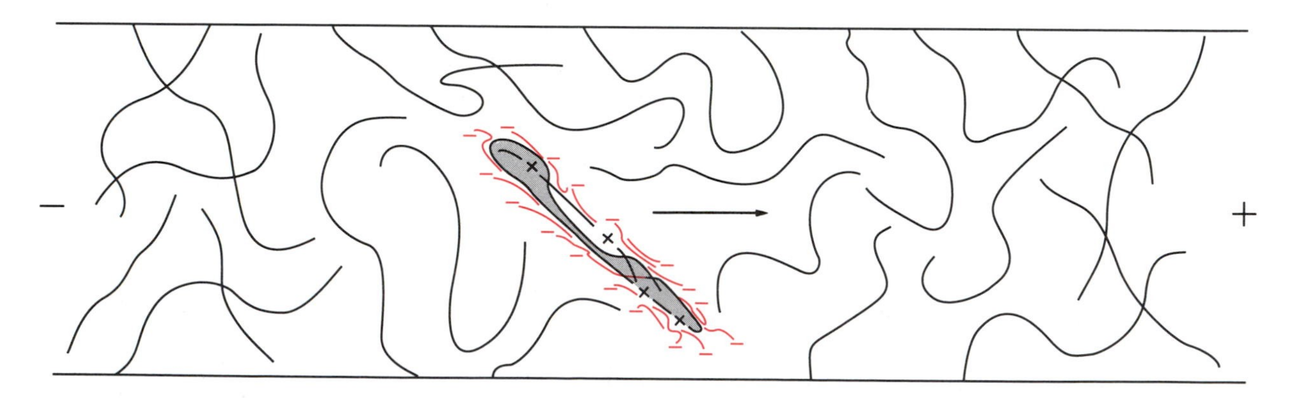

Figure 7-10

The negatively charged complex of a denatured protein and the anionic detergent SDS moves toward the anode in an electric field. Resistance to this movement by the gel is larger for larger proteins.

Molecular Exclusion Chromatography

Still another physical method for the determination of the size of proteins and their subunits is provided by molecular exclusion chromatography, which is widely used for the separation of proteins of different size. When proteins are passed through a column that is packed with a cross-linked hydrophilic polymer of polyacrylamide, dextran (a polysaccharide), or a similar material, they emerge in the effluent approximately in the order of decreasing molecular size. Contrary to what one might guess initially, the *larger* molecules appear first. By varying the degree of cross-linking, the range of molecular weight in which separation takes place may be varied from 10^2 to 10^6 or more. If the column has been calibrated with molecules of known molecular weight, the elution position of a protein may be used to obtain an estimate of its size and approximate molecular weight.

The method is based upon a simple, but intriguing, principle. Proteins and other molecules in solution occupy a finite volume that usually may be neglected relative to the volume adjacent to the surface or walls of the container. However, if a stringy polymer such as a dextran is added to the solution, the space that is available to the protein becomes restricted, simply because the protein and the long polymer cannot occupy the same volume at the same time. This volume exclusion effect is more significant for large than for small molecules and can even be used as a method for forcing proteins to precipitate out of solution. Thus, if a protein is equilibrated with a dextran or polyacrylamide gel phase, the protein will be excluded from the gel phase to some extent and will have a higher concentration in the aqueous phase (Figure 7-11). A small molecule that occupies less volume will be excluded much less and will have a similar concentration in the gel and aqueous phases.

Now, if the gel is transferred to a column and a mixture of protein and small molecules is passed through the column, the protein will have a relatively short path length through the column, because it is excluded from much of the volume, and will pass through quickly. However, the small molecule has to follow a longer path, through a larger volume of solvent, so that it appears later in the effluent. Similarly, large and small protein molecules will have different excluded volumes and retention times on the column, so that they will appear as separate peaks in the effluent.

The difference between the movement of a large and a small molecule may be compared to the difference between the movement of a large boat down a

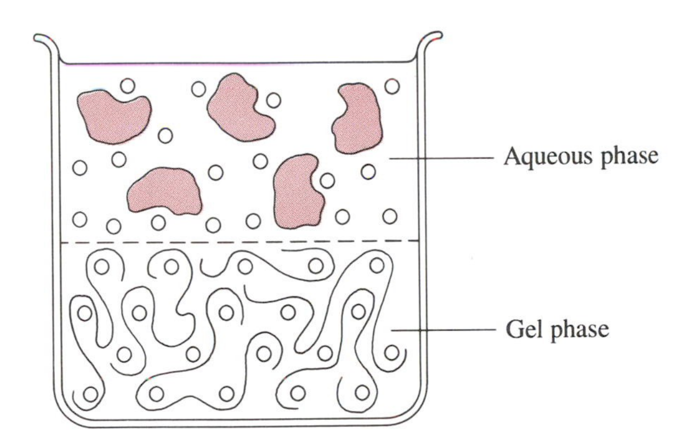

Figure 7-11

Large molecules of protein (red) are excluded from a gel phase because the gel and the protein cannot occupy the same space. However, small molecules (circles) are found in both the gel phase and the aqueous phase because they can fit between the strands of the gel.

relatively straight river and the movement of a small boat that explores a network of small streams in a swamp. The large molecules go directly through the gel, whereas the small molecules follow a longer, meandering path through the interstices of the gel.

The gel is often prepared in the form of small beads to improve its flow characteristics, and it is easy to see that a large molecule that is excluded from the beads will flow through the column faster than a small molecule that can equilibrate with the gel volume in the interior of the beads (Figure 7-12). The volume required for the appearance of a given protein, therefore, depends on its size and the extent to which it is excluded from the gel phase; hence the name "molecular exclusion chromatography." The commonly used names, "gel filtration" and "molecular sieving" are misleading because the process is not a filtration—the largest molecules are not held back, but pass through rapidly.

The elution positions of different proteins are usually described by a constant, K, that relates the measured elution position, V_e, to the void volume of the column, V_0, and the total volume, V_t (equation 3).

$$K = \frac{V_e - V_0}{V_t - V_0} \tag{3}$$

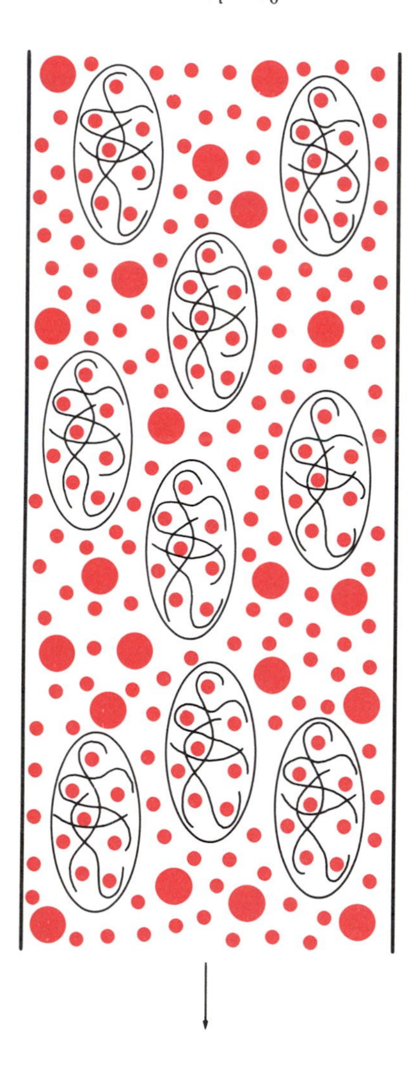

Figure 7-12

Large molecules of protein pass rapidly through a column containing beads of a stringy gel because they slide past the gel particles, whereas small molecules have to wander through the particles and take a longer route through the column.

The total volume, V_t, is the volume that is available to small molecules, including the solvent. The void volume, V_0, is the volume available to large molecules that are completely excluded from the gel phase. The void volume is often determined from the elution position of dextran blue, a large dye-containing polymer. Figure 7-13 shows an example of the separation of a series of proteins by molecular exclusion chromatography.

Analytical methods utilizing these properties of gels do not provide an exact measure of the molecular weight of proteins that have different shapes or compositions, because they depend on the *volumes* of the solution occupied by the macromolecule in question and by the gel fibers. Gel chromatography provides a measure of the *Stokes radius* and, hence, of the diffusion coefficient of proteins. In other words, it measures the "effective" size of the protein for diffusion or viscosity, the radius that the molecule would have if it were a sphere. For this reason, the elution position of native proteins can be related empirically only to the molecular weights of protein standards of similar overall structure. The fact that many globular proteins have a more or less similar shape and composition makes possible a preliminary estimation of the molecular weight of an unknown protein by this method by comparing its elution position with that of appropriate standards, and the technique has been widely utilized for this purpose. *Globular proteins* have a small or moderate ratio of length to width; *fibrous proteins* are elongated proteins that usually have a structural function. Figure 7-13 shows the relation between elution position and molecular weight

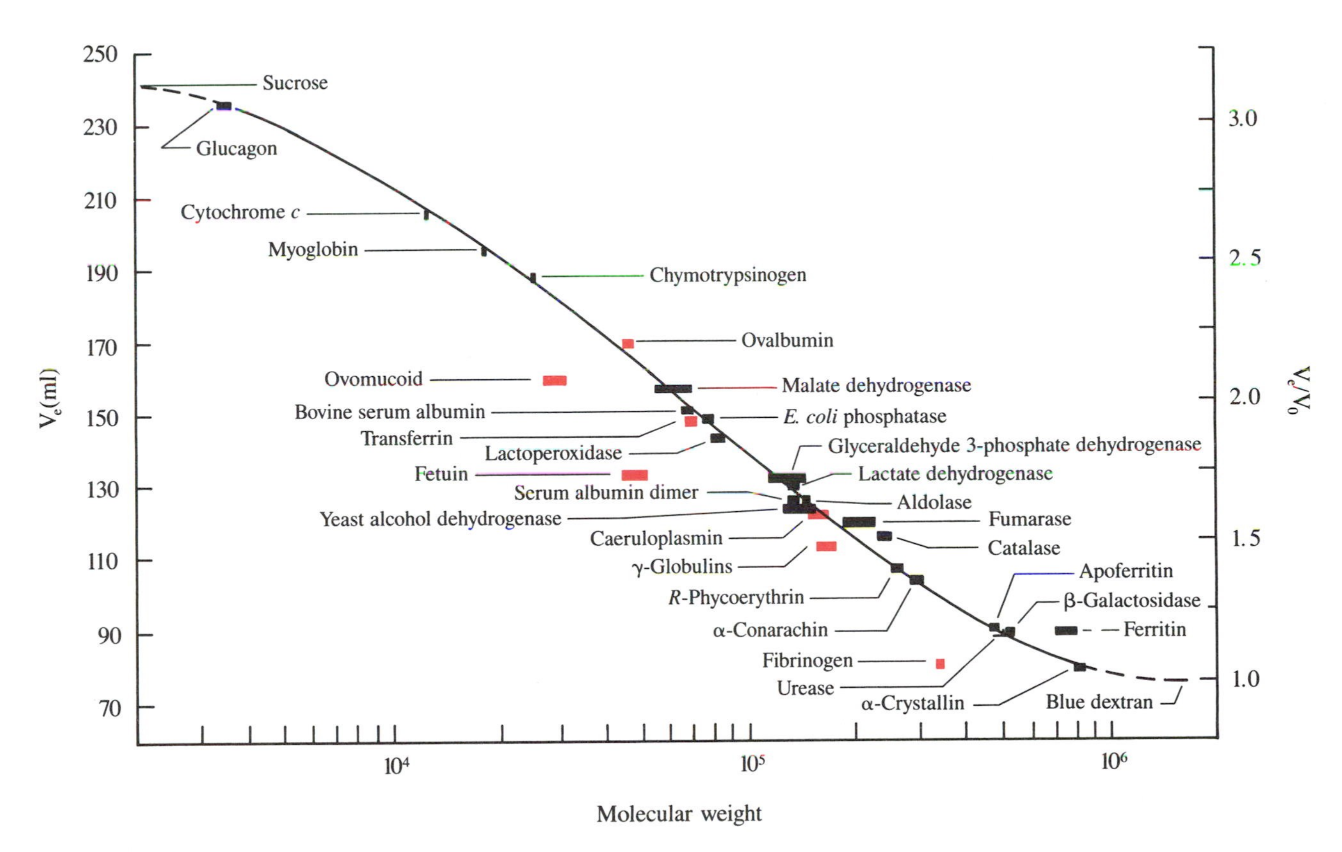

Figure 7-13

Selectivity Curve for Gel Chromatography with Sephadex G-200. The volume of solvent required to elute a protein is a function of the molecular weight of the protein. The red bars represent glycoproteins. [From P. Andrews. *Biochem. J.* 96 (1965): 597.]

for several globular proteins. The method is particularly useful because the elution position can be determined by measurement of the absorption spectrum, enzymatic activity, or other specific properties of the protein of interest. Therefore, it is often possible to obtain an estimate of the molecular weight of a particular protein even if it is contaminated by the presence of other proteins.

A more-reliable estimate of the molecular weight may be obtained by molecular exclusion chromatography of fully denatured proteins on a column of agarose gel, in the presence of concentrated guanidine hydrochloride and a thiol. Denaturation unfolds the protein and destroys its three-dimensional structure, so that differences in behavior caused by these specific structural differences disappear. Therefore, one cannot use enzymatic activity or other properties that depend on the structure of the native protein to assay proteins with this technique.

Glycoproteins and lipoproteins do not give reliable results with this technique or with most other techniques for the determination of molecular weight. Such proteins can be characterized by analysis of their chemical composition and by ultracentrifugation.

Gel columns also provide a simple and useful method for measuring the binding of small molecules to proteins, as described in Box 7-5.

BOX 7-5

MEASUREMENT OF THE BINDING OF SMALL MOLECULES TO PROTEINS

Agarose gels and other cross-linked gels provide a simple and useful method for measuring the binding of small molecules to enzymes and other proteins. For example, they can be used for measuring the binding of an inhibitor or a substrate to an enzyme. When the gel is added to a mixture of enzyme and inhibitor, the protein is totally excluded from the gel phase, but small molecules will be distributed between the gel and aqueous phases. Measurement of the concentration of the inhibitor in the two phases in the presence of different concentrations of protein, therefore, provides a measure of the amount of the molecule that is bound to the protein, because the protein-bound inhibitor is only in the aqueous phase. The dissociation constant can be calculated from these concentrations of inhibitor and the known concentration of protein.

This system can be transferred to a column that contains the gel and has previously been equilibrated with a known concentration of an enzyme inhibitor by passing a solution of inhibitor through the column. When a protein is added to the column, it will pass through rapidly, because it is excluded from much of the volume of the column, and will bind the inhibitor during its passage. If the total concentration of inhibitor in the effluent from the column is monitored, a plateau level representing the concentration of free inhibitor will be observed initially in tubes 1 through 18 of the adjoining illustration, followed by a peak in tubes 19 through 30 that corresponds to the additional inhibitor that is bound to the protein. Later, a second, negative peak in the inhibitor concentration appears, in tubes 35 through 50. This corresponds to a deficit of inhibitor; it represents inhibitor that was initially in the column but that bound to the protein and moved faster through the column to give the first peak. The amount of bound inhibitor may then be calculated from these concentrations and the concentration of protein. The dissociation constant of the enzyme-inhibitor complex can be calculated from the concentration of protein and amount of inhibitor that is bound at different concentrations of inhibitor.

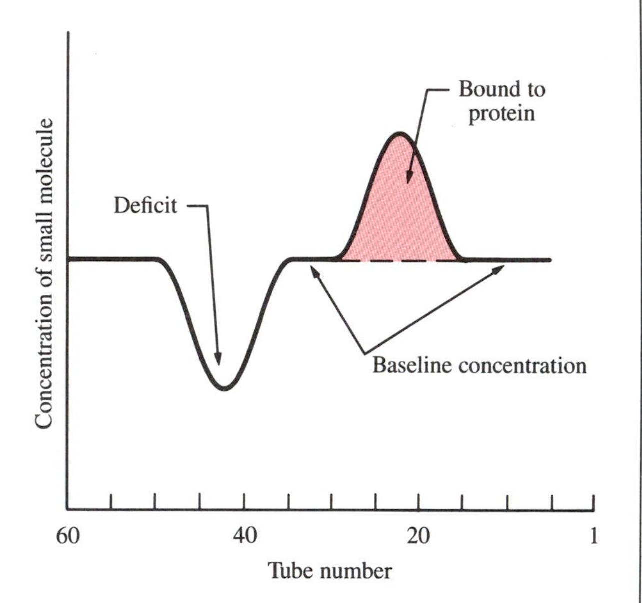

The Charge of Proteins

The charges on the outside of a protein help to keep soluble proteins in solution and are responsible for many important properties of proteins. They also make possible the separation and characterization of proteins by several different methods. The charges are provided by amino acids with charged side chains and by phosphate, sulfate, and other charged groups that may have been attached to the protein. Most of the charges are from carboxylate groups of the side chains of aspartate and glutamate residues and from the protonated ε-amino groups of lysine and guanidine groups of arginine side chains. The carboxylic acid groups of the side chains generally have pK values ranging from 4 to 5, so that they are dissociated to carboxylate ions at neutral pH, whereas the protonated ε-amino group of lysine and the guanidinium group of arginine have pK values near 10 and 12, respectively, so that these groups exist in the charged, protonated form at neutral pH.

pK_a Values

The usual ranges of pK_a values for dissociable side chains of amino acids are summarized in Table 7-1. The terminal α-carboxyl and α-amino groups have relatively low pK_a values near 3 and 8, respectively, because of the electron-withdrawing effect of the nearby peptide bonds. The protonated imidazole group of histidine has a pK_a ranging from 6 to 7, the sulfhydryl group of cysteine has a pK_a value near 9, and the phenolic hydroxyl group of tyrosine has a pK_a near 10. Protonated amide and peptide groups dissociate below p$K_a = 0$ (the proton is on the oxygen atom of the amide) and the hydroxyl groups of serine and threonine dissociate with a pK_a ranging from 13 to 14, outside the range at which most proteins are stable. All of these pK_a values can be shifted by

Table 7-1

Typical pK_a values for dissociating groups in proteins

Group	Formula	pK_a
Protonated amide and peptide	$-C(OH){\cdots}N^{+}(H)-$	−2–0
Terminal α-carboxyl	—COOH	3
Aspartate β-carboxyl	—COOH	4
Glutamate γ-carboxyl	—COOH	5
Histidine imidazole	imidazolium ring (HN ⁺ NH)	6–7
Terminal α-amino	$-NH_3^+$	8
Lysine ε-amino	$-NH_3^+$	10
Arginine guanidinium	$-NH{\cdots}C^{+}(NH_2)_2$	12
Cysteine sulfhydryl	—SH	9
Tyrosine hydroxyl	—OH	10
Serine, threonine hydroxyl	—OH	13–14

perturbations that are caused by neighboring charged groups, which stabilize ionic species of opposite charge and destabilize those of the same charge, and by partial burying or shielding from the aqueous solvent, which destabilizes ionized species of either charge, as described in Chapter 4.

Because of the large number of dissociable groups of varying pK_a and these perturbations, the titration curves of proteins are usually irregular, smooth curves that provide an indication of the total number of dissociable groups in the major classes but do not distinguish individual amino acids. The difference between the number of ionizable groups in the carboxylic acid region and the number of aspartate and glutamate residues that are determined by amino acid analysis of the protein provides an estimate of the number of glutamine and asparagine residues in the protein, because these groups do not dissociate in this range and are hydrolyzed under the conditions of the amino acid analysis (equation 4).

$$R-C(=O)-NH_2 \xrightarrow[H_2O]{H^+} RCOOH + NH_4^+ \quad (4)$$

Proteins that have a large number of basic amino and guanidine groups of lysine and arginine residues are classically known as *basic proteins.* This term can be confusing because these groups are protonated and positively charged at neutral pH. Therefore, they are really acidic groups under physiological conditions (equation 5).

$$\text{Lys-NH}_3^+ \rightleftharpoons \text{Lys-NH}_2 + H^+ \quad (5)$$

These proteins carry a large net positive charge and many of them, such as the *histones* and *protamines,* bind electrostatically to negatively charged DNA and play an important role in maintaining the structure of chromosomes.

Isoelectric and Isoionic Points

At most pH values, a protein molecule will carry a net charge because the number of positively charged ionizing groups does not exactly balance the number of negatively charged groups. However, if the pH is lowered or raised, the fraction of positively charged groups is increased or decreased, respectively, and at some pH value there will be no net charge. This is the *isoelectric point,* which occurs at a characteristic pH value for every protein. The isoelectric point is close to the *isoionic point,* the pH value that is obtained from prolonged dialysis of a protein against water so that all counterions are removed and the positive and negative charges of the protein balance each other. A protein usually has its minimum solubility at the isoelectric point, because there is no net charge that helps to keep it in solution. Furthermore, precipitation into a solid phase at the isoelectric point does not require the accompanying precipitation of strongly hydrated counterions. It is likely to have a compact structure at the isoelectric point, because there is a minimum electrostatic repulsion of protonated imidazole groups or other groups with the same charge on the surface.

Electrophoresis

The net charge on a protein at a given pH value determines the direction and the rate of its movement by electrophoresis in an electric field, toward the anode if the charge is negative and toward the cathode if it is positive. Most electrophoresis today is *zone electrophoresis,* in which a band of protein migrates in a stabilizing medium that prevents disruption by convection currents. These stabilizing media include filter paper moistened by buffer; polyacrylamide or starch gels, which can be stained to identify the positions of protein bands after drying; and a sucrose gradient in a vertical column, which permits the separation of different proteins on a preparative scale. Zone electrophoresis is now the most commonly used method for determining the homogeneity of a protein preparation.

The classical method of electrophoresis, which is still preferred for careful quantitative work, is *moving boundary electrophoresis,* in which the movement of proteins into the solvent from a sharp boundary between the protein solution and solvent is measured, usually by the Schlieren refractive index method. As in ultracentrifugal sedimentation, the change in protein concentration at a boundary, dc/dr, gives rise to a Schlieren peak that defines the position of the boundary.

Electrophoresis patterns of human serum proteins are shown in Figure 7-14. The large, rapidly moving peak (in part A), which comprises about 60% of the total protein, is serum albumin; most of the smaller peaks constitute the various globulin fractions, which include antibody molecules. The results of electrophoresis of serum on paper and on starch gel are shown for comparison. The isoelectric point of a protein may be determined by measuring its mobility

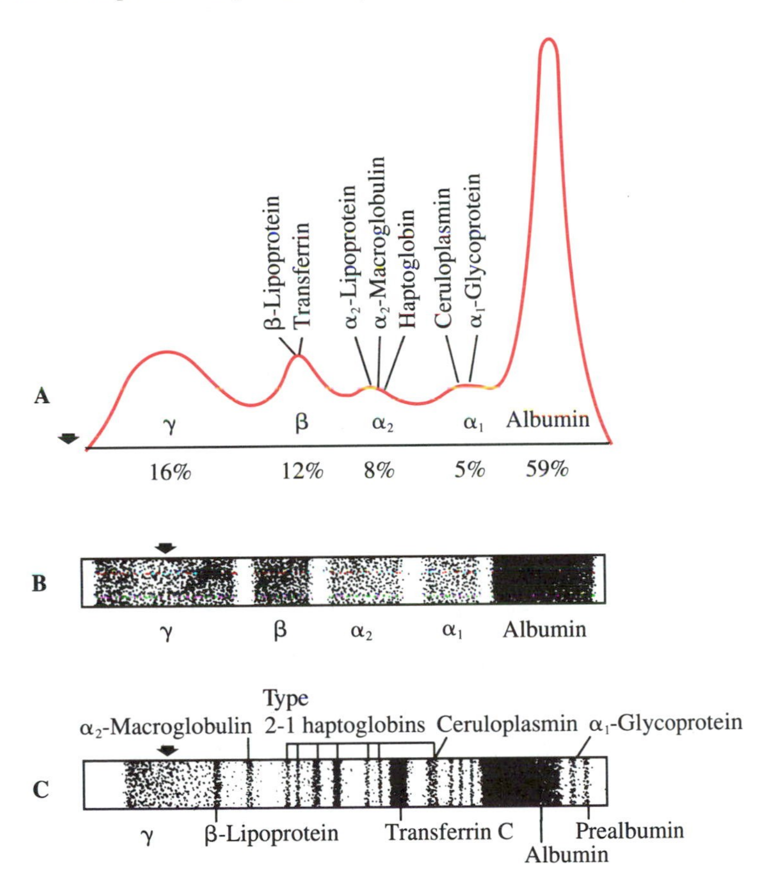

Figure 7-14

Schematic representation of the electrophoretic pattern of normal human serum in pH 8.6 buffer as obtained by: (A) moving boundary electrophoresis, (B) paper electrophoresis, (C) starch gel electrophoresis. The broad vertical arrow indicates the starting point in each case. (From F. W. Putnam. In *The Plasma Proteins,* vol. 1, 2d ed., ed. F. W. Putnam. New York: Academic Press, 1975, p. 18.)

upon electrophoresis in a series of buffers of different pH. There will be no net movement of a protein in a buffer of the same pH as the isoelectric point of the protein, so that a plot of protein mobility against the pH of different buffers will cross the origin at the isoelectric point.

The charge of proteins has also been used for the purification of proteins by the techniques of ion-exchange chromatography and isoelectric focusing. These methods are described in the next section.

The Purification of Proteins

The separation and isolation of pure proteins has occupied a large fraction of the working time of biochemists since biochemistry began; it is a far more difficult problem than the isolation of most small molecules. Proteins are delicate and are often subject to irreversible denaturation upon exposure to nonphysiological conditions—sometimes even upon dilution and exposure to air-water or water-glass interfaces. Furthermore, proteins are usually difficult to crystallize and may not be completely homogeneous when they have been crystallized. The purification of enzymes and other proteins is, therefore, an art in itself.

Precipitation

The classical method for protein purification is by fractional precipitation, and the most widely used precipitant is concentrated ammonium sulfate. By varying the concentration of ammonium sulfate, the pH, the temperature, and the method of addition of ammonium sulfate, it is possible to obtain a considerable degree of purification of a desired protein with ammonium sulfate precipitation. However, the separation of different proteins into different fractions is seldom complete, so that it is necessary to choose between a high yield and a high degree of purification. The interaction of proteins with ammonium sulfate and other salts is described in Box 7-6. Proteins can also be selectively precipitated by the addition of organic solvents, such as alcohol or acetone, but it is necessary to do this carefully at low temperature because these solvents also cause proteins to undergo denaturation. These techniques were developed extensively during World War II for the fractionation of blood plasma into stable protein fractions that could be stored and utilized under difficult conditions.

We have already seen that proteins are least soluble near their isoelectric point, which may be useful in purification, and that they can be forced out of solution by the addition of long hydrophilic polymers, such as polyethylene glycol, that tend to take up the space that the protein needs in solution. Fractionation can sometimes be brought about by precipitation with oppositely charged polymers, such as protamines for acidic, anionic proteins; the positively charged protamines combine with the anionic protein to give an insoluble complex.

Chromatography

The purification of proteins has been revolutionized in the past three decades by the development of chromatographic techniques that are based on the differing physical and chemical properties of different proteins. These methods are capable of providing far better resolution and purification than can be obtained by classical fractionation procedures. This improved resolution is inherent in chromatographic techniques because of the repeated fractionation events that occur as the protein passes down a chromatographic column.

BOX 7-6

SALTING OUT AND THE HOFMEISTER SERIES

One of the writers of this text asked his laboratory instructor in biochemistry at a well-known medical school, "How does ammonium sulfate salt out proteins?" The instructor smiled knowledgeably and said, "I would be glad to explain it to you, but I don't have the time just now." The question remains unanswered, in spite of the extensive research, calculation, and speculation that has been carried out by many workers in the past century. The instructor, unfortunately, has disappeared.

The precipitation of proteins by organic solvents can be explained by destabilization of charged groups in an aqueous-organic solvent, compared with water, but low concentrations of salts usually *increase* the solubility of proteins by stabilizing these charged groups. The precipitation of proteins is one example of the *salting out* of organic molecules that is generally observed in concentrated salt solutions.

Hofmeister noted in the last century that there are large differences in the effects of different salts on the solubility of proteins and that the order of effectiveness for salting out proteins is Na_3 citrate $> Na_2SO_4 \sim K_2HPO_4 > (NH_4)_2SO_4 >$ Na acetate $> NaCl > NaNO_3$. This is in marked contrast with the effect of increasing ionic strength on electrostatic interactions at low salt concentration, which is independent of the nature of the salt. Other salts, such as LiBr and LiI, extend the series in the opposite direction and actually *increase* the solubility of proteins. The order of this *Hofmeister series* is followed for the effects of moderately concentrated salts on a remarkably large number of different phenomena in aqueous solutions.

The same salts that increase protein solubility also tend to cause proteins to dissociate into subunits and to unfold into a denatured state at high salt concentrations, whereas those that salt out proteins also tend to protect them against denaturation and dissociation. All of these phenomena can be understood in terms of exposure of the protein to the aqueous environment—salt solutions that interact favorably with the protein, such as LiBr, will tend to increase the amount of the protein surface that is exposed to the solvent and therefore increase solubility, dissociation into subunits, and denaturation, whereas salts that interact unfavorably with the surface will tend to decrease solubility, dissociation, and unfolding, so that the minimum surface area is exposed to the solvent. In general, these two types of salts exhibit different behavior in concentrated solutions, depending on whether they interact more strongly with the peptide chain of the protein or with the solvent.

The order of salting-out effectiveness in the Hofmeister series generally follows the order of increasing strength of interaction of the ions, especially the anions, with water and is correlated with their hydration energy, their basicity, and (inversely) their ionic radius. It is intuitively reasonable that a solution in which there is a very strong interaction between the water and the salt will resist the insertion into the solution of a weakly interacting solute, such as an organic molecule or protein, because the salt will prefer to interact with the water rather than with the solute. Thus, salting out can be regarded as a kind of "squeezing out" of solutes because of the strong mutual interaction, internal cohesion, or internal pressure of the salt solution.

Salts that interact very weakly with water molecules, more weakly than do other water molecules, have the opposite effect. They tend to decrease the overall cohesion of the solvent and thereby favor the insertion of organic molecules into the solvent. Ions that interact weakly with water, such as iodide and perchlorate, may then interact favorably with dipolar organic groups, such as peptide bonds, and facilitate their exposure to the solvent. There has been some success in putting these notions on a quantitative basis with treatments based on internal pressure or the scaled particle theory, but the properties of concentrated salt solutions are still far from being completely understood.

The separation of proteins on the basis of their charge is carried out by *ion-exchange chromatography*. The charged groups of the ion exchanger are usually attached to a flexible, hydrophilic polymer such as cellulose, because the charged groups on the more-rigid cross-linked polymers that are used for the separation of small molecules, such as amino acids, do not interact well with proteins. Commonly used ion-exchange materials for proteins include carboxymethyl cellulose (CMC) and diethylaminoethyl (DEAE) cellulose, which provide carboxylate groups and protonated amino groups to interact with positively and negatively charged proteins, respectively. Proteins are selectively eluted from columns of these materials with a gradient of increasing salt concentration, which decreases the electrostatic interaction between the protein and the resin, or with a pH-gradient that decreases the net charge on the protein. The separation of UDP-galactose 4-epimerase on a column of DEAE Sephadex is shown in Figure 7-15.

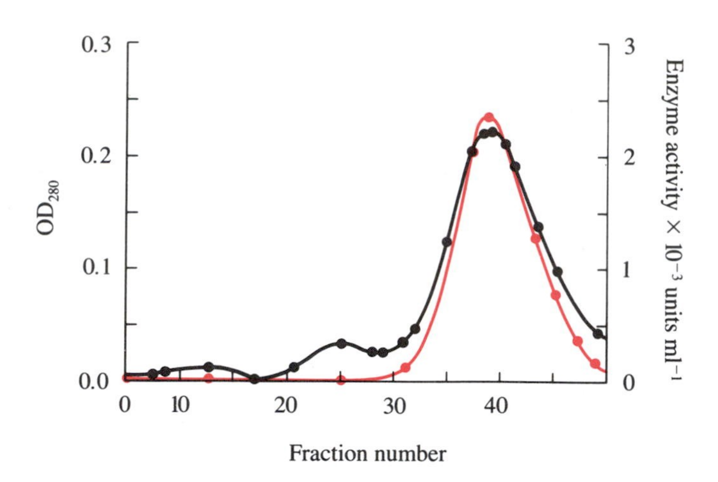

Figure 7-15

Purification of UDP-galactose 4-epimerase by DEAE Sephadex Chromatography. Enzymatic activity is indicated in red; the protein concentration, from the optical density (OD) at 280 nm, is indicated by the black curve. [After D. B. Wilson and D. S. Hogness. *J. Biol. Chem.* 239 (1964): 2469.]

Isoelectric Focusing

The characteristic isoelectric point of proteins has recently been utilized for the development of the highly sensitive method of *isoelectric focusing* for protein analysis and separation. The method consists of the application of an electric field to a solution that contains the proteins and a mixture of small, multiply charged molecules with differing isoelectric points in a medium that is stabilized with sucrose or a polyacrylamide gel. One end of the system contains a strongly basic solution and the other a strongly acidic one. The small molecules migrate in one or the other direction as long as they have a net charge but ultimately stop moving when their net charge becomes zero. This sets up a gradient of pH from one end of the tube to the other. A protein molecule behaves similarly and migrates in one direction or the other until all of it collects in a sharp band at the pH in the gradient that corresponds to the isoelectric point of the protein. The separation of hemoglobins from different species by isoelectric focusing is shown in Figure 7-16. A two-dimensional separation of a series of proteins from wheat is shown in Figure 7-17. The proteins were first separated in one direction by

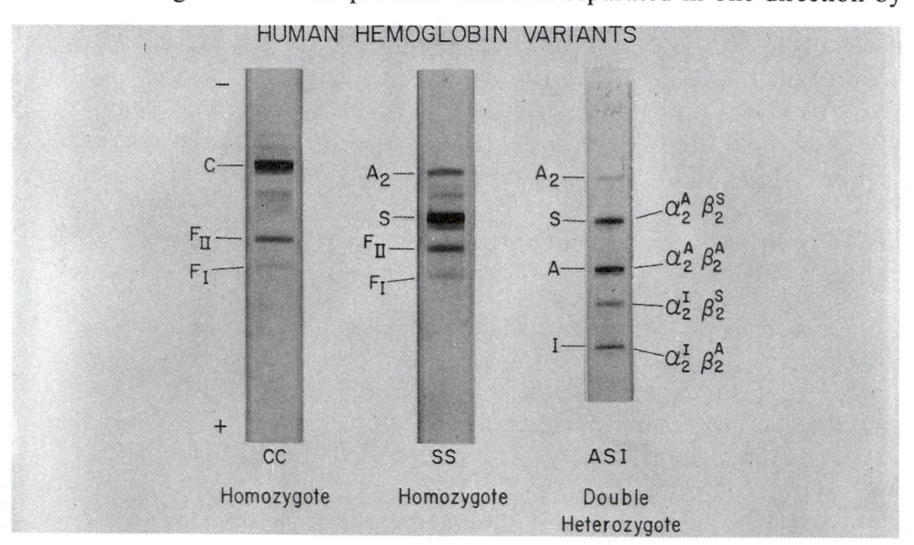

Figure 7-16

Isoelectric Focusing of Hemoglobins from a Number of Different Species. The results show that the isoelectric points of the hemoglobins are different in different species, and some species have several different hemoglobins. (Courtesy of James W. Drysdale.)

electrofocusing in a starch gel and were then separated in the perpendicular direction by electrophoresis.

Because few proteins have exactly the same isoelectric point, this is an extremely sensitive method for separating and identifying proteins. In fact, the method is so sensitive that a number of proteins, including crystalline proteins, that had been considered to be homogeneous by other criteria have been found to separate into several bands when they were subjected to isoelectric focusing. Because the alteration or substitution of a single charged amino acid can cause a detectable change in the isoelectric point of a protein, this is not as surprising as it might be, but it does raise difficulties of analysis and interpretation for a number of proteins.

If a particular enzyme exists in different forms with small chemical or structural differences in the native protein, the different isomers are called *isozymes*. It is often difficult to distinguish true isozymes from different forms of the protein that arise in the course of the isolation procedure—by partial proteolytic digestion, for example, or from partial dissociation of the protein into subunits.

The ideal method for protein separation and purification takes advantage of *specific* chemical properties of the protein in question, rather than general physical properties that are likely to overlap with those of other proteins. For example, certain enzymes undergo dissociation into subunits in the presence of substrates or inhibitors. Such a protein can easily be purified by separation according to molecular weight with molecular exclusion chromatography, first in the absence and then in the presence of the substrate or inhibitor.

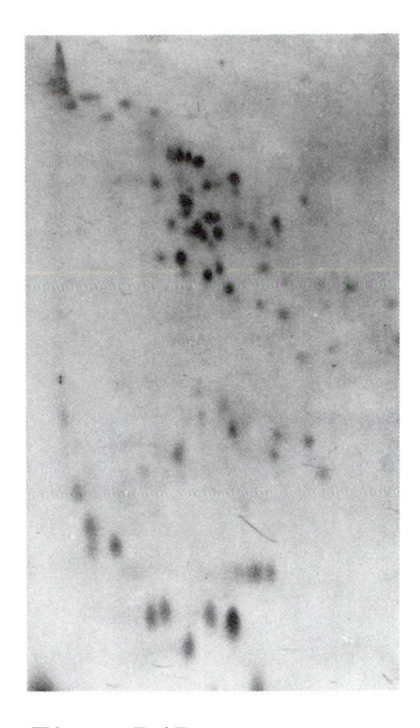

Figure 7-17

Two-dimensional Fractionation of Gliadin Proteins from Wheat. The proteins were first separated by starch gel electrofocusing in the range pH 5 (left) to 9 (right) and then by electrophoresis in aluminum lactate buffer, pH 3.2. (Courtesy of Colin W. Wrigley.)

Affinity Chromatography

Another technique for protein purification, *affinity chromatography*, has recently been developed that is based directly upon the utilization of specific interactions between proteins and ligands, such as the interaction between enzymes and their substrates or inhibitors. A molecule that binds specifically to a particular protein is attached covalently to a hydrophilic polymer, which can then be used for purification of the protein by either a batch-by-batch or a column procedure.

For example, chymotrypsin can be separated from trypsin by chromatography on a column containing agarose to which trypsin inhibitor has been attached, as shown in Figure 7-18. The chymotrypsin comes off the column readily, but the trypsin binds to the trypsin inhibitor. It is eluted only at a low pH because acid weakens the binding of trypsin to the inhibitor.

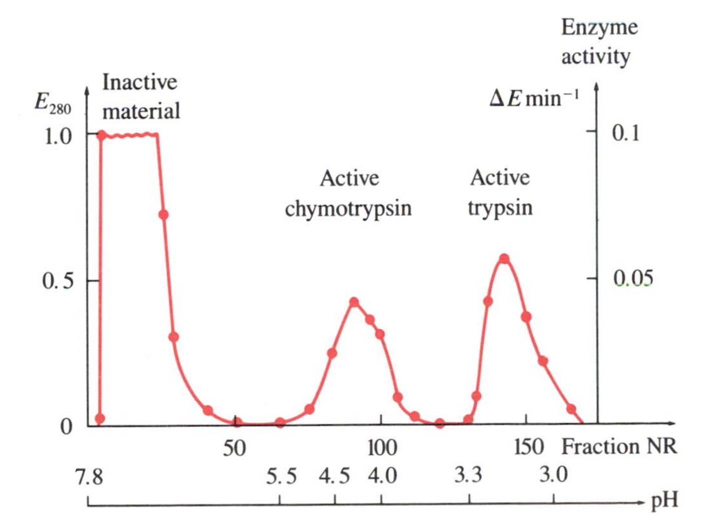

Figure 7-18

Separation of Chymotrypsin and Trypsin on a Column Prepared with Soybean Trypsin Inhibitor Agarose. The trypsin comes off the column later than chymotrypsin because it binds for a time to the trypsin inhibitor and is retarded. (From J. Porath and T. Kristiansen. In *The Proteins,* vol. 1, 3d ed., ed. R. C. Hill, C. L. Boeder, and H. Neurath. New York: Academic Press, 1975, p. 135.)

The Secondary Structure of Proteins

Proteins are synthesized by the sequential addition of activated amino acids to the growing peptide chain (Chapter 10) and generally fold up into the structure of the native protein as they are formed. Furthermore, many proteins can be completely unfolded in the presence of a denaturing agent such as urea or guanidine hydrochloride, as will be discussed in the next chapter, and will refold into their active, native structure when the denaturing agent is removed. These observations establish the important conclusion that the complex three-dimensional structure of a native protein is determined by its *primary structure,* the sequence of amino acids in the peptide chain. The structure may be modified by additional steps after synthesis, as in the activation of chymotrypsinogen to chymotrypsin, but most modifications of this kind do not cause large changes in the arrangement of amino acids in most of the protein.

How can we describe a molecule that is as complicated as a protein? It is difficult to reach conclusions and generalizations by examining the amino acid sequence or even the extraordinarily complex three-dimensional structure of a protein that has been determined by X-ray diffraction. Fortunately, description is simplified by the existence of regularities in the structures at a higher level of organization than the primary structure. A large fraction of the peptide chain of most proteins is folded into regions of more or less regular *secondary structure* in such a way as to satisfy three basic requirements:

1. The highly polar C=O and N—H groups of peptide bonds are usually hydrogen bonded to proton donors and acceptors in the protein, if they are not hydrogen bonded to water, to prevent a net loss of hydrogen bonding and solvation of these highly polar groups when the protein folds up. Most peptide groups satisfy this requirement by hydrogen bonding to other peptide groups. The peptide bond is stabilized by resonance with electron donation from the nitrogen atom to the carbonyl group. There is a strong dipole, with approximately 0.5 positive charge on nitrogen and 0.5 negative charge on the oxygen atom. This resonance and dipole favor the formation of a hydrogen bond and inhibit rotation around the C–N bond, because of its partial double-bond character. Many of the peptide hydrogen bonds inside most proteins are arranged in periodic structures, or networks, that make up these regular secondary structures.

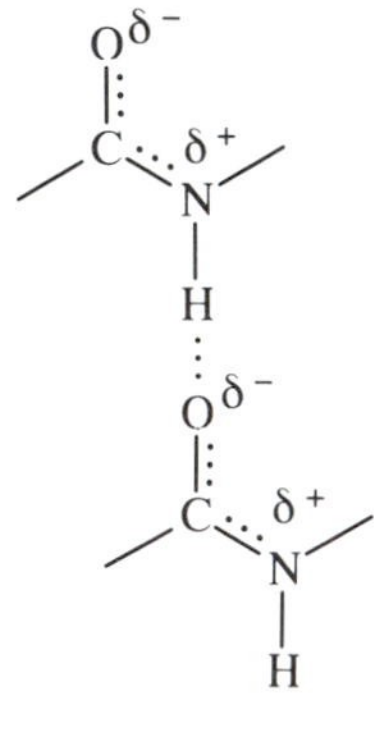

2. The peptide chain will fold so that almost all charged groups and most of the polar hydroxyl and amide groups on amino acid side chains are in contact with water. The solvation energies of charged groups are enormous. Therefore, desolvation and burying of these groups in the interior of the protein is energetically forbidden, unless they form an ion pair with oppositely charged groups or they are solvated internally by several polar groups. On the other hand, the side chains of leucine, isoleucine, valine, methionine, and phenylalanine are nonpolar and tend to be buried in the interior of the protein, where they are not in contact with water.

3. The peptide chain must fold in such a manner that the side chains of different amino acids do not occupy the same space, and the bond and rotational angles along the peptide chain are not strained. These requirements seem simple and obvious, but they play a major role in determining protein structure in ways that are still only partly understood.

Three regular secondary structures that satisfy these requirements are the α-helix, the β-pleated sheet, and the β-turn. Regions of a protein that do not fit clearly into one of these structures are usually labeled "random coil." In fact, native proteins have a highly ordered structure, as they must have if they crystallize and give high-resolution diffraction patterns, and random coils are seldom coiled. However, no one has yet produced a better name that has succeeded in replacing the nonrandom, noncoiled "random coil."

The Alpha-Helix

The best-known regular secondary structure is the *α-helix* (Figure 7-19), which was identified by Linus Pauling in 1951. The backbone peptide chain is arranged in a helix with 3.6 amino acid residues in each turn, in such a way that each peptide N—H group is hydrogen bonded to the peptide carbonyl group that is

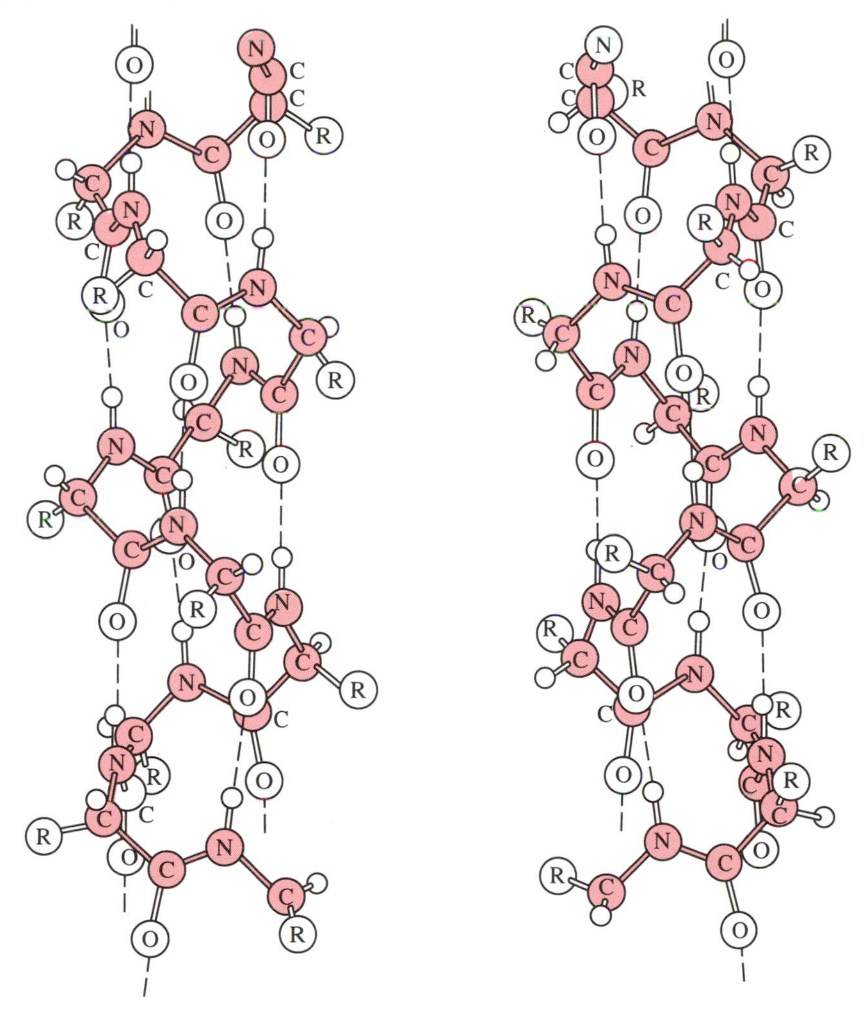

Figure 7-19

Drawings of Left-handed (A) and Right-handed (B) α-Helical Forms of a Polypeptide Chain Containing L-Amino Acids. (After B. W. Low and J. T. Edsall. In *Currents in Biochemical Research,* ed. D. E. Green. New York: Wiley-Interscience, 1956, p. 398.)

four residues further along the chain. The distance along the helix between each turn, the *pitch* of the helix, is 5.5 Å, or 0.55 nm. Segments of α-helical structure of variable length are frequently found in the folded peptide chains of globular proteins. The prototype of a globular protein is an approximately spherical, soluble protein, but the term is frequently applied to other proteins, including some that are highly elongated. Most enzymes are globular proteins. The α-helix is also found in some structural fibrous proteins, such as the keratin of hair.

Beta-Sheets

The β-pleated sheet structure (Figure 7-20) is more difficult to visualize. *Beta-pleated sheet* structures can be formed either in a parallel direction, in which adjacent peptide chains run in the same direction, or in an antiparallel direction, which occurs when a peptide chain doubles back on itself. The antiparallel β-sheet is more common. The β-structure is formed by hydrogen bonding of the N—H and C═O groups of one amino acid to complementary groups of an adjacent chain. In a β-sheet structure, this is followed by hydrogen bonding of the N—H and C═O groups of the next amino acid to complementary groups of

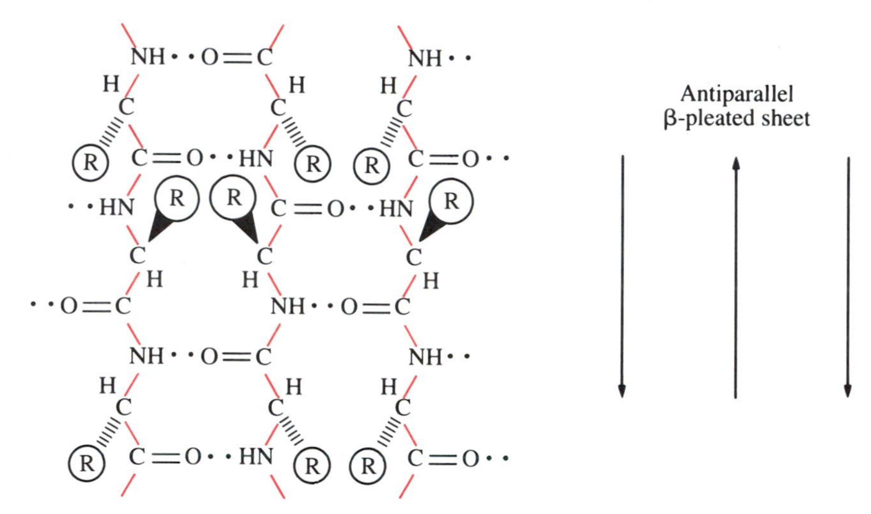

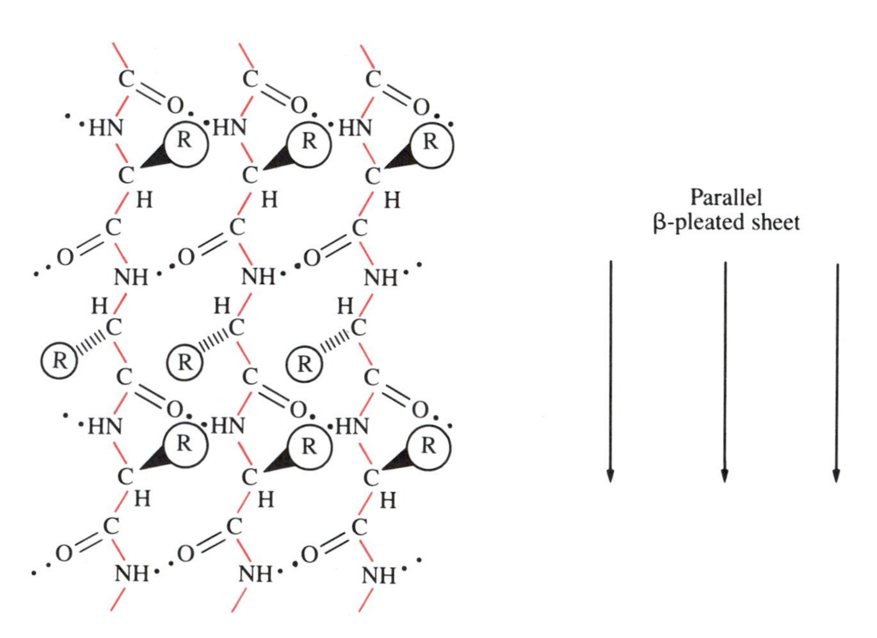

Figure 7-20

The β-Pleated-Sheet Structure of Proteins. Note the difference in the arrangement of the side chains, R, between the structure in which the direction of the peptide strands is antiparallel and that in which it is parallel.

an adjacent chain on the other side. The side chains (R) of the amino acids protrude alternately from the two sides of the sheet, and the β-sheet has a tendency to curl slightly to accommodate the structures of the L-amino acids in the interacting peptide chains. The hydrogen-bonded structure of the pleated sheet is relatively strong and rigid, so that it is an important part of many structural, fibrous proteins, such as silk fibroin. The structure is frequently strengthened by interaction of nonpolar, hydrophobic R groups of amino acids on adjacent chains, which pack tightly against each other.

Beta-Turns

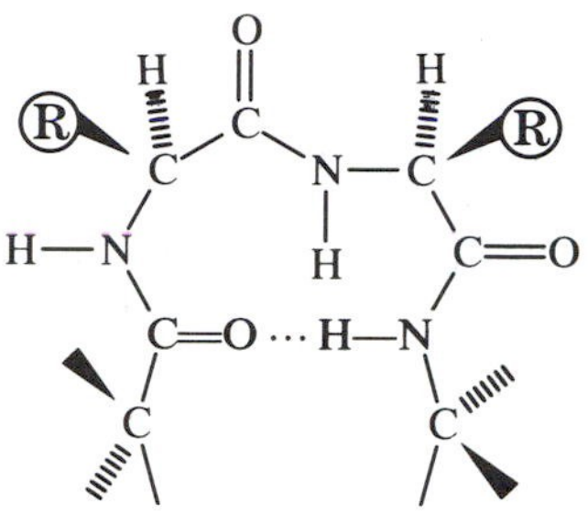

The *β-turn* is of particular interest because it was discovered only recently—which demonstrates that there are always interesting things to find for those who have the curiosity and imagination to look for them. The typical globular protein is formed of segments of peptide chain, including α-helices and β-sheets, that must fold back into the interior when they reach the surface of the protein. This folding region has been found to have a characteristic structure of four amino acids that make up the β-turn. The amino acids in the β-turn characteristically include small glycine residues, proline residues that permit an acute angle in the turn, and one or more hydrophilic amino acids, such as asparagine, serine, or amino acids with charged side chains. The hydrophilic amino acids in the β-turn have highly polar side chains that interact favorably with the aqueous environment on the outside of the protein. This makes an important contribution to determining the three-dimensional structure of the protein and helps to keep it in solution. These groups also serve as sites of attachment for phosphate and sugar groups that are added to certain proteins in the final stage of their synthesis. In spite of its small size, the β-turn is an important structural element that may involve as many as 30% of the amino acids in a globular protein.

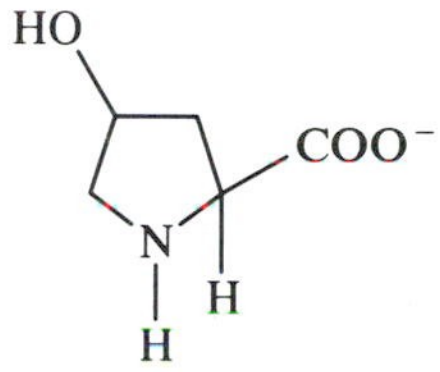

β-Turn

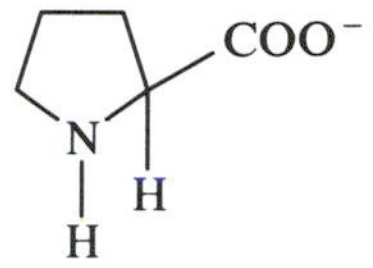

L-Proline (Pro, P)

4-Hydroxyproline

There is a tendency for certain amino acids to be found preferentially in α-helical or in β-sheet structures, as was noted for β-turns. However, the preference is by no means absolute—it is a statistical preference, which can be used to predict the secondary structure of a region of the peptide chain from the amino acid sequence of that region with approximately 60% accuracy. We still do not know how to predict the way that a peptide chain of known sequence will fold to give the specific three-dimensional structure of a native protein.

Collagen

A specialized kind of secondary structure is found in *collagen,* an important structural protein and the most-abundant protein in most animals. It is the principal insoluble protein of *connective tissue,* which holds together the bodies of mammals and other animals. Tendons, for example, are composed almost entirely of collagen.

Collagen has an unusual triple-helical structure that is formed from three intertwined peptide strands and is largely responsible for the great strength of this protein. The strands are composed of 33% glycine residues and include large amounts of alanine and the cyclic amino acids proline and hydroxyproline. Every third amino acid is a glycine residue, and the small steric requirements of glycine, along with the unusual bond angles of proline and hydroxyproline, make the three-stranded helical structure possible. This structure is closely packed and an amino acid that is larger than glycine would distort or break up the helix. The hydroxyl group of hydroxyproline presumably contributes to the stability of the triple helix by hydrogen bonding to other residues, possibly through an intermediate water molecule.

BOX 7-7

INTERACTIONS WITH POLARIZED LIGHT

Light that is polarized in a plane may be regarded as being composed of equal amounts of right and left circularly polarized light, the vectors of which add up to give the plane-polarized light. Now, if this light interacts with an asymmetric, optically active molecule that has some degree of right or left "handedness," the two forms of circularly polarized light will interact with the molecule differently (Figure A), just as a right-handed corkscrew can move more easily through a channel made by a right-handed corkscrew than by a left-handed one. The greater *refraction* of one form of circularly polarized light by this molecule, because of a relatively strong interaction of one electrical component of the light with the asymmetrically oriented electron density of the molecule, slows the transmission of this form and thereby causes a net rotation of the plane of the plane-polarized light. This is *optical rotation.* The greater *absorption* of one form compared with the other is measurable, as the "ellipticity," and constitutes *circular dichroism.* Both the optical rotation and the circular dichroism depend on wavelength and become larger as the maximum of the absorption band of the optically active transition is approached. However, at the absorption maximum, the circular dichroism, Θ, is at a maximum (Figure B, top), whereas the optical rotation decreases sharply, crosses the baseline, and changes sign, to give a "Cotton effect" (Figure B, center).

Now, if these optically active chromophores (or even optically inactive chromophores) are themselves arranged in a regular, *asymmetric* pattern, such as a right- or left-handed helix, the optical transitions can interact with each other and give rise to a greatly enhanced optical rotation or circular dichroism. This is the case for the α-helix of a protein and for the helix of nucleic acids. These asymmetries give rise to characteristic optical rotatory dispersion (ORD) or circular dichroism (CD) spectra, as shown in curve 1 of Figure C for the CD spectrum of poly-L-lysine. The spectrum of the random polypeptide chain is shown in curve 3. Because several absorption bands are optically active (there is no simple relation between the extinction coefficient of an absorption band and its influence on optical rotation or circular dichroism), the observed spectra represent a composite from several transitions that differ both in sign and in wavelength. A different pattern is found for the β-structure, as shown in curve 2 of Figure C. The amounts of these different structures in the protein can be estimated by measuring the intensity or the wavelength dependence of an observed pattern. The widely used parameter b_0 is a measure of this wavelength dependence (dispersion) of the optical rotation and is an approximate

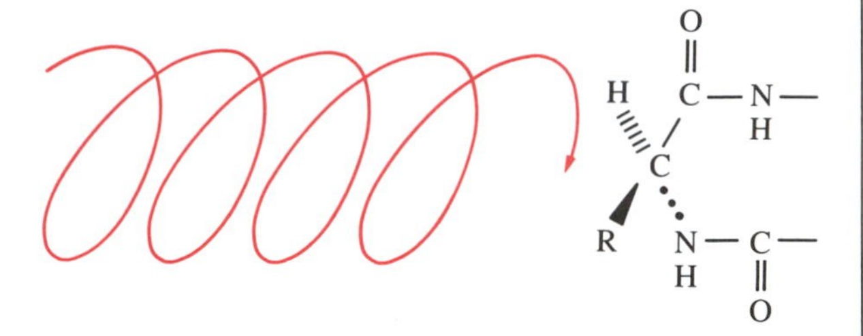

Figure A

An asymmetric, optically active molecule, such as an α-amino acid, will interact differently with right and left circularly polarized light. It will rotate the plane of polarized light.

Vitamin C and Collagen. Studies with hydroxyproline containing a radioactive label have shown that it is not incorporated into collagen. Instead, there is an enzyme, a hydroxylase, that oxidizes some of the proline residues in collagen to hydroxyproline with molecular oxygen and a reducing agent, after the proline has been incorporated into the peptide chain. The hydroxylase contains iron that helps to catalyze the hydroxylation and this iron can be oxidized to an inactive form.

Ascorbic acid (vitamin C) is required to keep this iron in its active, reduced state. Consequently, a deficiency of vitamin C leads to inadequate hydroxylation of proline and to *scurvy,* in which defective collagen leads to weakness or lesions in joints, blood vessels, and skin. Scurvy was a cause of serious disability and death among sailors for many years until it was found that limes and other citrus fruits that contain a high concentration of ascorbic acid will prevent scurvy. Limes were first widely adopted by the British navy and gave rise to the name "limeys," for British sailors, but there was a delay of several years before citrus fruits were generally made available to sailors.

Much of the strength of mature collagen and another structural protein, *elastin,* is provided by covalent cross-links between the triple helices. These cross-links are formed by condensation of the amino groups of lysine or hydroxylysine side chains with the oxidation products of lysine and hydroxy-

BOX 7-7 (continued)

index of the amount of α-helix in a protein; several fully helical polypeptides and proteins have b_0 values near −600. Estimates of helix content from CD or ORD spectra show a moderately good (although far from perfect) correlation with the results of X-ray structural analysis.

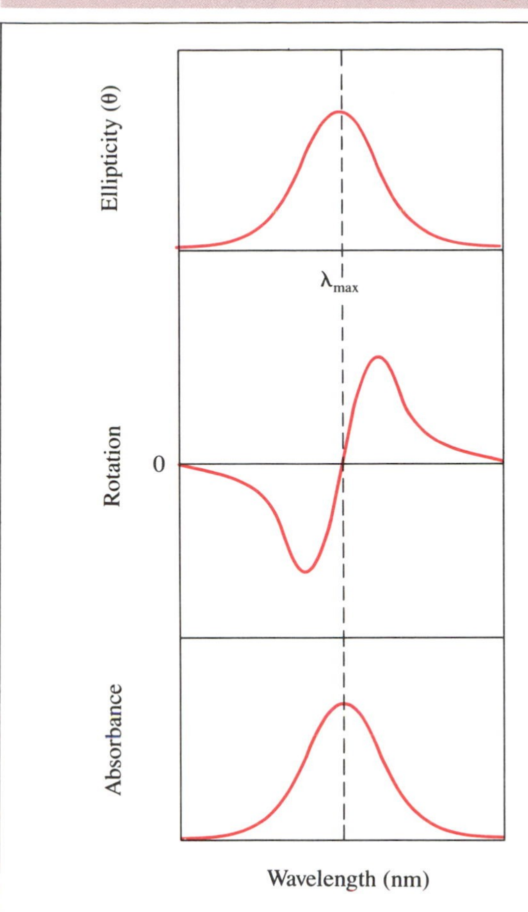

Figure B
The ellipticity is maximal at the same wavelength as the absorption maximum, but the rotation of polarized light reaches a maximum and minimum above and below the wavelength of maximum absorption.

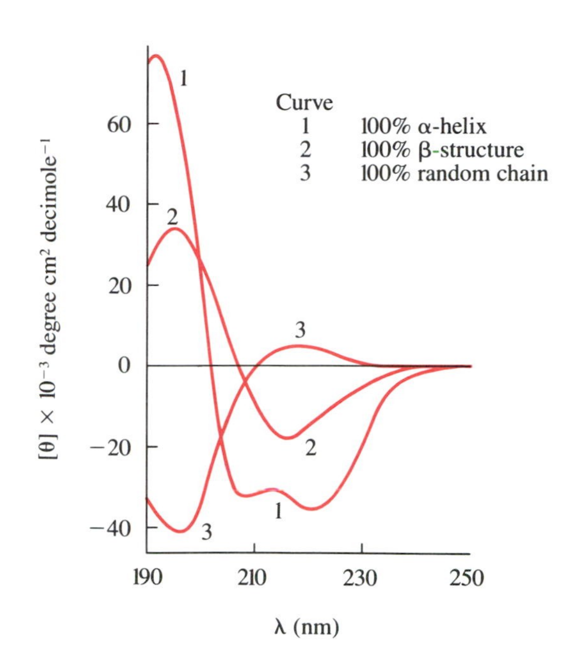

Figure C
The circular dichroism spectra of poly-L-lysine in different conformations; Θ is the ellipticity. The α-helix is poly-L-lysine at pH 10.5, the β-structure is formed after heating at pH 10.5, and the random structure is poly-L-lysine near neutral pH. (Courtesy of G. Fasman.)

lysine residues on adjacent chains through a complex series of reactions to form several different types of stable covalent bonds. Lysinonorleucine is one example of a cross-linking amino acid. Hydroxylysine is formed by hydroxylation of lysine residues by lysyl hydroxylase, a copper-containing enzyme, after their incorporation into the peptide chain of collagen and is also deficient in scurvy.

$H_3N^+—CH_2—CH(OH)—CH_2—CH_2—CH(NH_3^+)—COO^-$

5-Hydroxylysine

$—NH—CH(C=O)—(CH_2)_4—NH—(CH_2)_4—CH(C=O)—NH—$

Lysinonorleucine

Optical Rotation of Proteins

Optically active L-amino acids that are located in the α-helices and β-sheets of proteins interact with polarized light in different ways, so that the *circular dichroism* or *optical rotatory dispersion spectra* of proteins may be used to estimate the amounts of these secondary structures in a given protein with a fair degree of accuracy. These techniques are described in Box 7-7.

The Tertiary Structure of Proteins

The next higher level of organization of a protein, its *tertiary structure,* is the folding of the different segments of helices, sheets, turns, and the remainder of the peptide chain into the three-dimensional structure of the native protein or protein subunit. It is difficult to make generalizations about tertiary structures because of the large variety of folding patterns in different proteins, but a few general patterns have emerged recently.

The driving force for the formation of tertiary structures is essentially the same as that for secondary structures. The strong solvation of charged groups requires that the protein fold in such a way that almost all of these groups are on the outside of the protein, where they are exposed to water. Exceptions are found in ion pairs between substructures inside the protein and in ion channels, which consist of a group of α-helices with charged groups and other hydrophilic groups on one side of the helix; the helices are arranged so that the polar groups project into the inside of the channel, where they can be solvated by water in the channel and interact with transported ions (see Chapter 29). Hydrophobic, nonpolar side chains of amino acids tend to avoid the aqueous exterior and are usually buried in the inside of the protein.

Figure 7-21 is a diagrammatic representation of the structure of a small enzyme, *ribonuclease.* The elements of secondary structure are indicated by helices and by flattened arrows for chains that are segments of β-sheet structures. This method of illustrating secondary and tertiary structures was developed by Jane Richardson and has been very helpful in showing the structural pattern of an entire protein molecule. It is difficult to interpret the

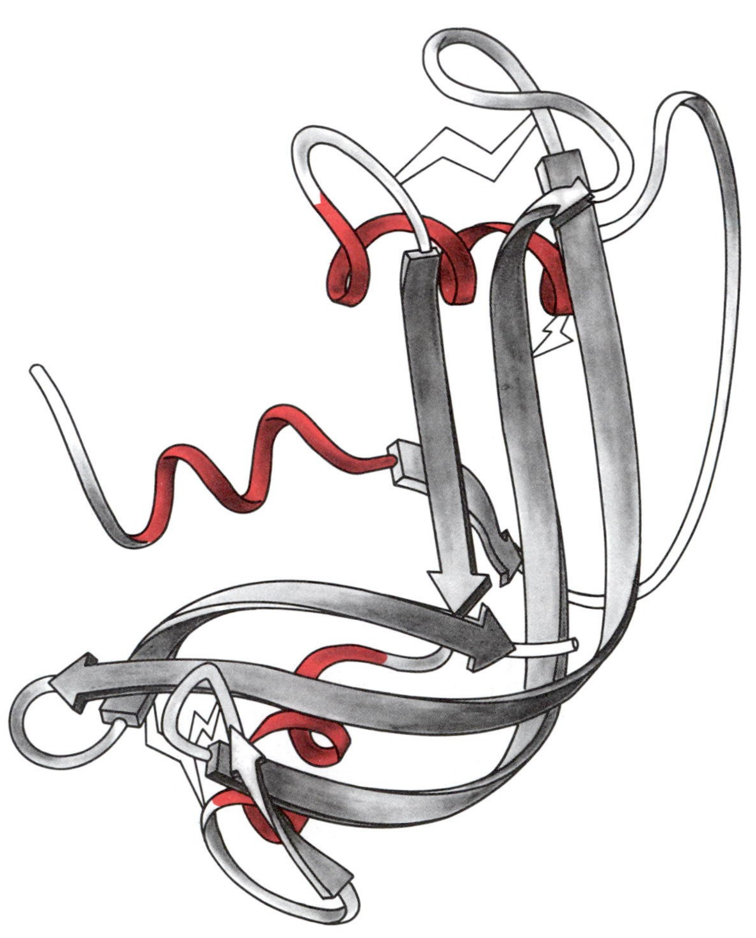

Figure 7-21

The Tertiary Structure of Ribonuclease A. The red coils represent α-helices and the flattened arrows represent protein chains that are in antiparallel β-sheets.

structure of a protein in a simple way by examining directly the complex structure that is obtained by X-ray diffraction of a crystalline protein.

A number of proteins are constructed of individual structured units, or domains, that are connected by more or less flexible regions. The connecting regions may serve as "hinges" to allow movement of the domains relative to one another, as in myosin, the major contractile protein of muscle, and are often the primary sites of proteolytic attack on the protein. The active site of an enzyme is frequently located in a groove or cleft that may be formed from adjacent domains. This makes possible a maximal interaction of the substrate with the binding site of the enzyme and, in some enzymes, there is a still larger interaction when the domains on each side of the cleft come together and surround the substrate almost completely.

There is also evidence that structurally similar regions, or domains, form the binding sites in a number of different enzymes for particular classes of compounds, such as ATP and other nucleotides. This has led to the suggestion that nature has mass-produced particular structures for use in different enzymes by transferring whole regions of genetic information from one DNA molecule to another, so that it was not necessary that the binding site for each substrate in these classes of compounds should evolve independently in each enzyme.

Quaternary Structure and Antibodies

The final level of structural organization of a protein is the interaction of different peptide chains with each other. Many of the larger soluble proteins, and some insoluble proteins, are composed of two or more peptide chains or subunits, which may be either identical or different. These subunits can sometimes be separated in concentrated salt solutions or under mild denaturing conditions. This level of structural organization is known as *quaternary* structure.

Why is it advantageous for protein subunits to combine to form large molecules? The reason is obvious for structural proteins that combine in an organized fashion to form a large structure, such as tubulin in microtubules and actin in microfilaments (Chapters 1 and 30). For hemoglobin, enzymes, and other biochemically active proteins, the combination of subunits into whole molecules makes the complete structure stronger and more protected from proteolytic attack and, in some cases, makes possible a functional interaction between the subunits that may greatly increase the usefulness of the molecule. These subunit interactions are responsible, for example, for the cooperative behavior of the subunits of hemoglobin that gives an optimal oxygen saturation curve and for allosteric effects in enzymes that are important for the regulation of enzyme activity and metabolism.

The interactions between different subunits occur through contacts of nonpolar, hydrophobic regions, through hydrogen bonding of proton donors to proton acceptors, and through electrostatic interactions between charged side chains of amino acids. Charged groups seldom provide stabilization of the interior structure of the protein, but frequently form specific interactions with other subunits. These electrostatic interactions also play a critical role in transmitting information from one subunit to another in order to cause conformational changes during the oxygenation of hemoglobin, and in the cooperative behavior of other multisubunit proteins.

The structure of antibodies (Chapter 6) provides an example of how several of these different kinds of structures can be fitted together to give a complex functional protein. Figure 7-22 shows that the structure of the most-common circulating antibody in the blood, immunoglobulin G (IgG), consists of twelve identifiable regions that are connected by peptide chains, disulfide bonds,

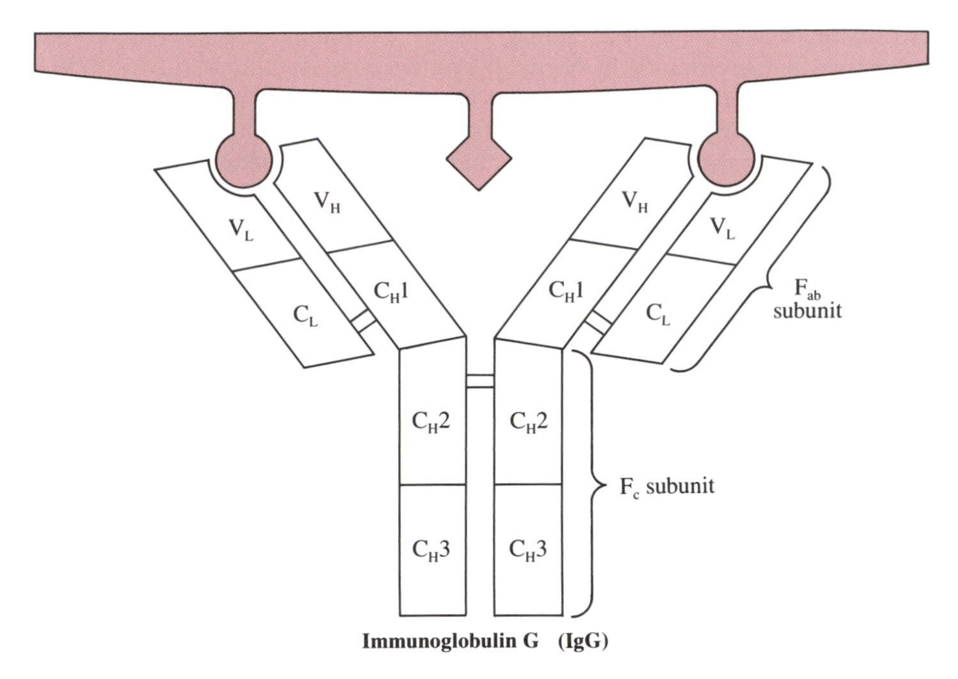

Figure 7-22

Schematic Representation of the Subunit Structure of an Antibody, Immunoglobulin G (IgG). The double lines represent disulfide bonds that connect the subunits, the subscripts H and L refer to the heavy and light chains, and C and V refer to the regions of constant and variable structure, respectively, in different antibody molecules. This antibody molecule is bound to two specific antigen sites on the outside of a cell. [After J. D. Capra and A. B. Edmundson. *Scientific American,* 236 (1977): 52. Copyright © 1977 by Scientific American, Inc.]

and noncovalent interactions. Cleavage of the disulfide bonds gives two large, heavy chains, each containing four connected regions, and two light chains with two regions each. The two specific binding sites that react with the complementary sites on an antigen are formed cooperatively from portions of the ends of both the heavy and light chains, V_H and V_L, so that different structures of individual subunits provide many combinations of subunits with different specificities. The regions of variable and constant amino acid sequence are labeled V and C, respectively, in Figure 7-23. The structures and folding patterns of different regions in the antibody molecule are sufficiently similar to one another to suggest that they evolved from a common precursor. For example, both the constant and the variable regions of the light chain consist of structures based on two sheets of β-structure, one composed of four and the other of three antiparallel strands, as shown by the broad arrows in Figure 7-23. Each of these sheets is slightly twisted. If the structure of the region of constant amino acid sequence is rotated, it is apparent that its overall structure is very similar to that of the variable region. The variable region has additional loops of amino acids of highly variable sequence that make up the specific combining site.

The way that these subunits fit together to form the overall structure of the entire IgG molecule is shown in Figure 7-24. The light and heavy chains tend to

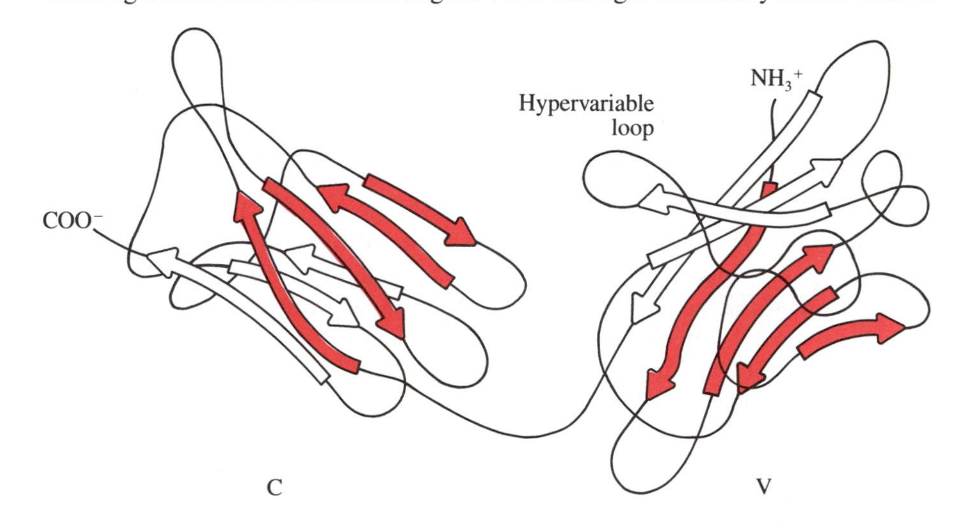

Figure 7-23

The light chain of an IgG antibody folds into a constant (C) and a variable (V) domain. The β-sheets are indicated by the broad arrows. Four β-sheets in each domain are shown in color to indicate their relative positions in the two domains. [After J. D. Capra and A. B. Edmundson. *Scientific American,* 236 (1977): 54. Copyright © 1977 by Scientific American, Inc.]

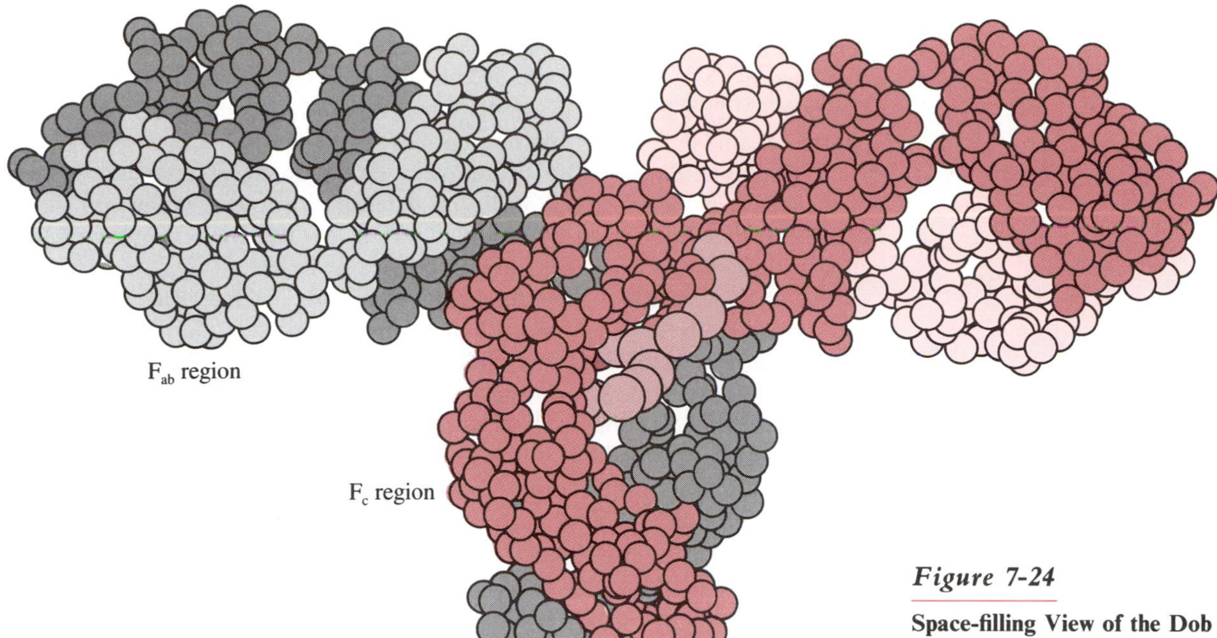

Figure 7-24

Space-filling View of the Dob Ig Molecule. One complete heavy chain is in dark pink and the other is dark gray; the two light chains are lightly shaded. The large spheres represent the individual hexose units of the complex carbohydrate. In this view, the twofold axis of symmetry is vertical. A crevice is seen between the C_H2 of the heavy, dark pink chain and the C_l domain of the F_{ab} on the left. [After E. W. Silverton, M. A. Navia, and D. R. Davies. *Proc. Natl. Acad. Sci. USA* 74 (1977): 5142.]

coil around each other to form a Y-shaped molecule that has a well-defined structure in the regions in which the different subunits are fitted together but is flexible in the "hinge" region at the bends in the "Y." The two arms of the Y that contain the combining sites constitute the F_{ab} regions and can be separated from the rest of the molecule by brief treatment with proteolytic enzymes, which attack the relatively exposed hinge region. The remainder of the molecule is the F_c fragment at the center of the molecule. This is the region in which carbohydrate is attached.

Most other antibody molecules preserve this same kind of basic structure but differ in the number of subunits and combining sites. Some of them are composed essentially of several IgG units that combine to form a large molecule with a large number of combining sites. For example, the IgM molecule contains five units, each with a structure similar to that of IgG. The entropic advantage from intramolecularity in a multivalent antibody of this kind means that it can bind very tightly indeed to a multivalent antigen, such as a virus, to form a polymeric network. This large aggregate may precipitate out of solution, and initiate the complement pathway (Chapter 6), or be attacked by macrophages.

The kind of fitting together of different structural regions that is found in immunoglobulins is characteristic of other soluble proteins, including enzymes. However, most globulins differ from immunoglobulins in that they contain one or more regions of α-helical structure.

Summary

The amino acid composition of proteins can be determined by ion-exchange chromatography after hydrolysis of the protein in acid. The amino acid sequence of the peptide chain can be determined by cleaving the chain into shorter peptides and determining the sequence in each peptide by step-by-step degradation from the N-terminal end with the Edman procedure and from the C-terminal end with carboxypeptidase. The overall sequence can be determined by matching the sequences of peptides that were obtained from specific cleavage by two different procedures, such as cleavage at basic amino acids with trypsin and at methionine residues with cyanogen bromide. The sequence of the amino acid chain that was initially synthesized can be determined from the nucleotide sequence of the DNA that carries the code for that protein, as described in Chapter 12. This sequence is usually longer than that of the protein because the peptide chain is processed by the removal of peptides after its synthesis.

The molecular weight of proteins can be estimated by sedimentation in an ultracentrifuge, by electrophoresis of denatured protein bound to sodium dodecyl sulfate in a polyacrylamide gel (SDS-PAGE), by molecular exclusion chromatography, or from the amino acid sequence.

The movement of the protein in an electric field by electrophoresis is determined by the state of ionization of its amino acid side chains, the N- and C-terminal amino acids, and any charged groups that have been added to the protein after its synthesis. Isoelectric focusing in a pH gradient provides a sensitive method for the analysis and purification of proteins. Purification is more commonly carried out by fractional precipitation with ammonium sulfate, by chromatography on positively or negatively charged ion-exchange resins such as DEAE cellulose or carboxymethyl cellulose (CMC), by high performance liquid chromatography (HPLC), and by affinity chromatography.

Proteins contain regions of regular secondary structure consisting of α-helices, β-sheets, and β-turns. The structural protein collagen exists as a triple helix that is made possible by the presence of a large number of glycine and proline residues. Covalent cross-linking of these helices makes collagen an extraordinarily strong structural protein. The secondary structure of proteins can be estimated by optical rotatory dispersion (ORD) or circular dichroism (CD). The tertiary structure is defined by the amount of α-helix, β-sheet, and "random coil," as well as by the arrangement of these structures. The quaternary structure is defined by the number and relation of the different subunits in a multisubunit protein.

ADDITIONAL READING

Creighton, T. E. 1983. *Proteins: Structures and Molecular Principles,* New York: Freeman.

Creighton, T. E. 1987. Stability of alpha-helices. *Nature* 326: 547–548.

Louie, G., Thao, T., Englander, J. J., and Englander, S. W. 1988. Allosteric energy at the hemoglobin beta chain C terminus: studies by hydrogen exchange. *Journal of Molecular Biology* 201: 755–764.

McCammon, J. A., and Karplus, M. 1983. The dynamic picture of protein structure. *Accounts of Chemical Research* 16: 187–193.

Neurath, H., Hill, R. L., and Boeder, C.-L., eds. 1975. *The Proteins,* vols. 1–5, 3d ed. New York: Academic Press.

Richardson, J. S. 1981. Protein anatomy. *Advances in Protein Chemistry* 34: 168–339.

Wuthrich, K., Billeter, M., and Braun, W. 1984. Polypeptide secondary structure determination by nuclear magnetic resonance observation of short proton–proton distances. *Journal of Molecular Biology* 180: 715–740.

PROBLEMS

1. Explain the procedure of isoelectric focusing.

2. What will happen to the rate of movement of a protein through a column of DEAE cellulose as the pH is changed?

3. Explain the difference between the methods of SDS electrophoresis and molecular exclusion chromatography for separating proteins on the basis of their molecular weight.

4. Explain how the isoelectric point and the isoionic point are related to the p*K* values of ionizing groups on proteins.

5. How much can you say about the mechanism of salting out of proteins?

6. Which amino acids are commonly found in a β-turn, and why is this so?

7. Discuss the stability and strength of collagen.

Forces and Interactions in Aqueous Solution

8

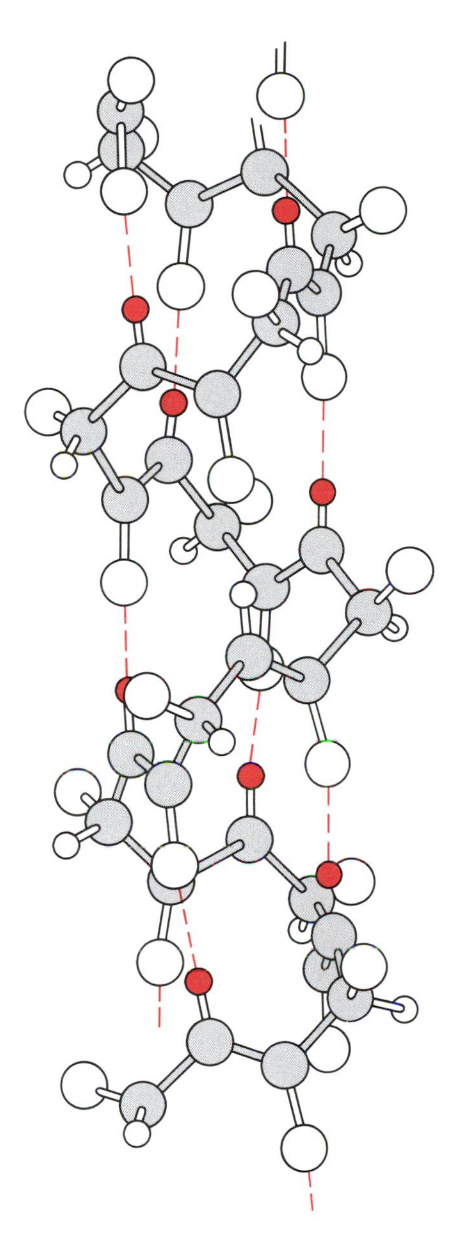

A ball-and-stick model of an α-helix in a protein. The polypeptide chain is shown in gray and the hydrogen bonds are in red.

Most of the important and interesting processes that take place in living systems are brought about by binding interactions between molecules. These include the reactions of enzymes with substrates, the interaction of hormones, drugs, and neurotransmitters with their receptors, and the binding of chains of DNA to each other in a double helix that mediates the transfer of genetic information from one generation to the next. The forces that are responsible for these interactions also bring about the folding of proteins into their native structure. When a protein is heated or exposed to nonaqueous solvents or solvent mixtures, its native structure is generally lost and it becomes denatured.

In this chapter, we examine some of these forces and binding interactions, with particular emphasis on their role in maintaining the three-dimensional structure of native proteins. The single most-important fact to remember about these interactions is that the cells of living systems function in an aqueous environment, so that the binding forces between molecules always represent a *competition* between the interaction of a molecule with water and with the other molecule. This fact has made it difficult to obtain a quantitative understanding of binding interactions in biological systems. However, we have some understanding of the general nature of these interactions and of their role in maintaining the structure and activity of proteins and nucleic acids. A second important generalization about these binding interactions is that the individual interactions between two groups are weak, but the binding energy that can be obtained from the cooperative binding of several groups can be very large indeed; it is large enough to mediate most of the specific interactions that take place in living systems.

The Denaturation of Proteins

The structure of proteins in their native state was described in Chapter 7. One source of insight into the nature of the forces that hold proteins in their native structure is to examine the conditions that cause them to unfold into a random, denatured structure.

Proteolytic enzymes cleave the peptide chains of proteins that they encounter in the digestive tract, intracellular vesicles, or elsewhere much faster than they digest themselves. This is a curious fact because proteolytic enzymes are themselves proteins. Although they do inactivate themselves by autolysis (self-digestion) on long standing in their activated form, this is generally a slow process compared with the rate at which they cleave other proteins in the digestive tract.

There are several reasons for this difference in rate, but the simplest and probably the most important is simply that proteolytic enzymes have a well-defined three-dimensional structure in their native state. Most of the peptide bonds that are susceptible to proteolytic cleavage are buried in the interior of this structure and are not accessible to attack by proteolytic enzymes. Proteins in food, on the other hand, are usually converted into a denatured, partially unfolded state by cooking and by exposure to acid in the stomach, so that the peptide bonds are much more easily attacked. If a protein gets all the way to the slightly alkaline small intestine (pH about 8) without denaturation, it can usually be attacked sequentially at exposed residues by several different exopeptidases and endopeptidases, so that it is gradually decomposed and can no longer maintain its native structure.

The importance of this native structure is nicely illustrated by the behavior of proteolytic enzymes themselves. Trypsin and chymotrypsin, like other proteins, are converted into an unfolded, denatured state by high concentrations of the denaturing agents urea and guanidine hydrochloride. If a low concentration of urea or guanidine hydrochloride (the chloride salt of guanidinium ion) is added to trypsin, the enzyme's rate of autolysis is greatly increased and it is rapidly inactivated. However, if trypsin is added to a *concentrated* solution of guanidine hydrochloride, it is not attacked, because all of the enzyme molecules are converted into an unfolded, inactive state and have no proteolytic activity (Figure 8-1). If the denaturing agent is then removed (under acidic conditions, in which proteolysis does not take place), the enzyme refolds to its native state and

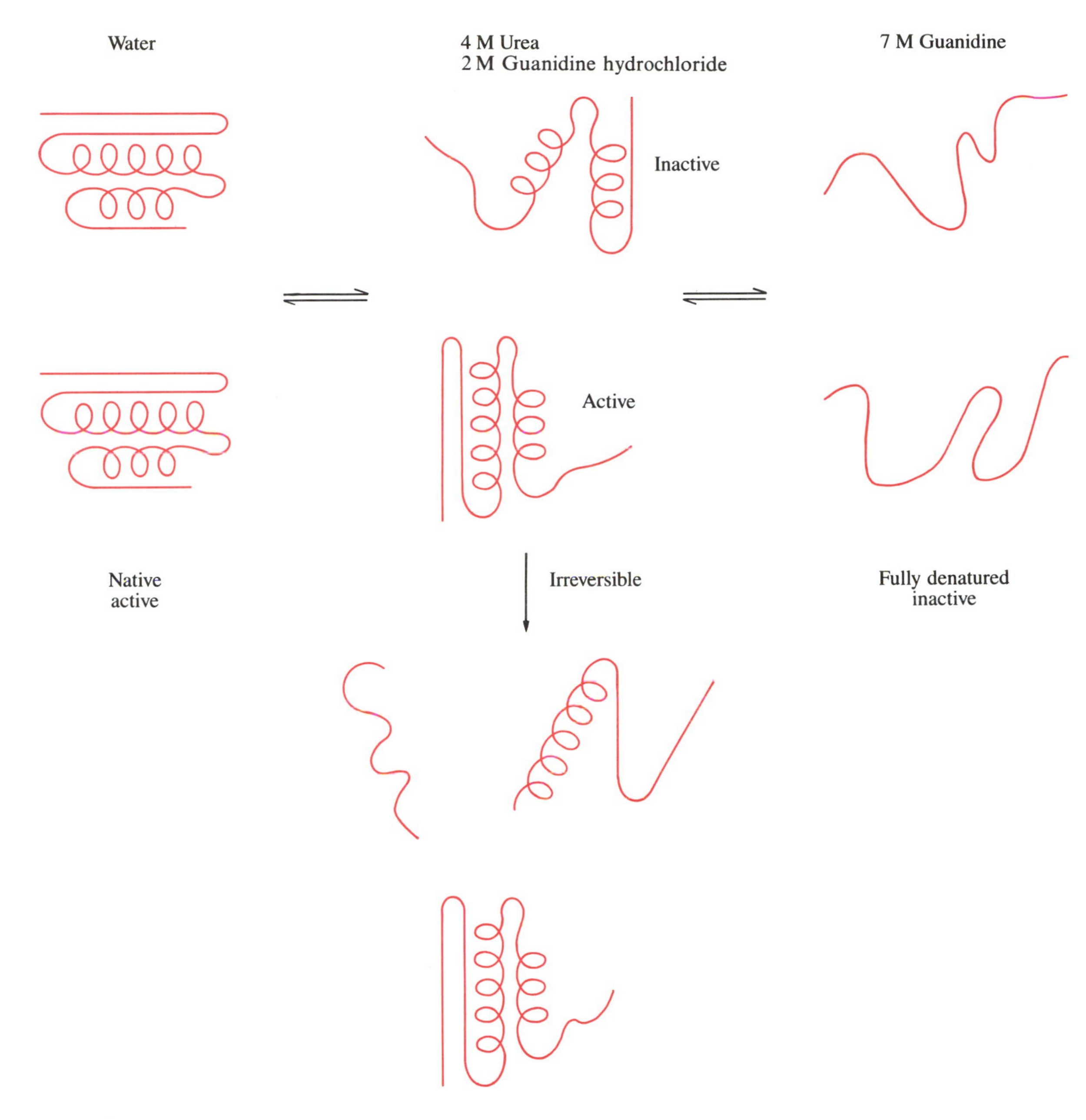

Figure 8-1

Denaturation of Trypsin. Trypsin is stable for some time in water, and it is still active after it has been denatured in 7 M guanidine hydrochloride, allowed to stand, and then diluted into water. However, in 2 M guanidine hydrochloride or 4 M urea, it is rapidly inactivated. In these solvents, some of the molecules are partially unfolded by the denaturing agent, and these molecules are cleaved rapidly by other molecules of the enzyme that are still active.

is fully active. Inactivation takes place at the low concentration of denaturing agent because some of the molecules unfold enough to be attacked by the remaining molecules that are still active (Figure 8-1, center). Under these conditions, digestion takes place until all of the enzyme activity is lost.

H_2N–C(=O)–NH_2

Urea

H_2N···C(+)(–NH_2)···NH_2

Guanidinium ion

Urea and guanidine hydrochloride denature these enzymes and other proteins because they interact favorably with the peptide chain in the interior of the native protein. They "dissolve out" the peptide chain, so that it prefers to be exposed to the solvent instead of remaining in the folded structure of the native protein. They do this because they interact more favorably with the peptide groups and the relatively nonpolar side chains of the amino acids in the interior of the protein than does water. In the absence of the denaturing agent, the peptide groups and the amino acid side chains interact more favorably with each other than with water, so that the protein folds up into its native three-dimensional structure.

Probability, Entropy, and Denaturation

It is extraordinarily improbable that a protein should exist in its unique native structure. The protein is made up of a chain of amino acids that are connected through peptide bonds and this chain can rotate into a very large number of different positions, in addition to the particular position that is characteristic of the native protein. Rotation about the bonds of the amino acids in the peptide chain is not free; but, even if each amino acid can rotate into just three different positions, a chain of 100 amino acids could take up $3^{100} = 10^{47}$ different positions. Thus, from the standpoint of probability it is much more likely that a protein will unfold into its denatured state, in which it can take up a very large number of different positions, than that it will stay in the native state.

This high probability of unfolding into a disordered state (Figure 8-2) is a manifestation of entropy, which was discussed in Chapter 5. The entropy, S, can be defined as a logarithmic measure of the number of accessible states, w, in which the system can exist. This is described by equation 1 of statistical

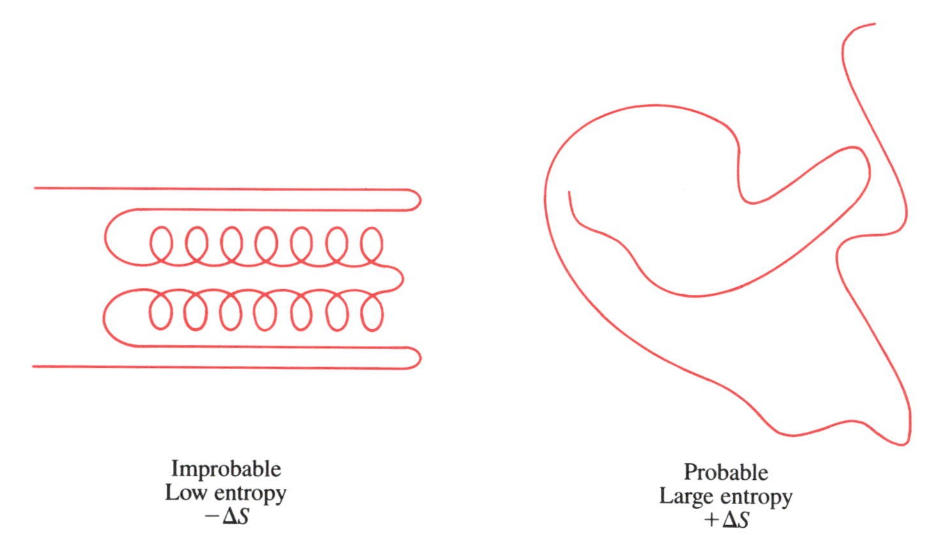

Figure 8-2

The Probabilities for the Native and Denatured States of a Protein. The peptide chain of a native protein (left) has a particular arrangement and structure. A denatured protein (right) can take up any of a very large number of different structures, one of which is shown here. This corresponds to a large entropy.

mechanics, in which k is Boltzmann's constant = 1.98 cal K^{-1} mol^{-1}, and ln is the natural logarithm.

$$S = k \ln w \tag{1}$$
$$= 1.98 \times 2.303 \log w$$

The entropy for the reversible denaturation of the small enzyme ribonuclease is 285 cal K^{-1} mol^{-1}, or 285 e.u., for example, which is very much larger than the entropy changes of less than ± 20 e.u. that are usually observed for reactions of small organic molecules. Equation 1 shows that this increase in entropy corresponds to a probability factor of $w = 10^{62}$ that favors the denatured over the native state. In other words, the reaction behaves as if the denatured protein could exist in 10^{62} more different conformations, compared with the native protein. However, this number cannot be assigned entirely to the unfolding of the protein because there will also be changes in the entropy of the solvent upon denaturation.

The only reason that all proteins do not exist in the denatured state, therefore, is that there are strong noncovalent forces from hydrogen bonding and other interactions that hold the protein together in its native structure. These forces are sufficient to overcome the strong disruptive influence of entropy and disorder.

The complex, ordered structure of a native protein is one manifestation of the state of extraordinarily high order and low entropy that is characteristic of living systems. This point has been nicely described by Schrödinger in a short book: *What Is Life?* Another system with a high information content and a correspondingly low entropy is the sequence of bases in nucleic acids. This sequence carries the information that is needed to synthesize the components of a cell, including its proteins.

Temperature Dependence, Enthalpy, and Thermodynamics

Under physiological conditions, the very large probability and entropy of protein denaturation is counterbalanced by an even larger stabilizing influence of the *enthalpy* for conversion of the protein from the denatured into the native state. For the reaction in the reverse direction, this change in enthalpy is sometimes called the heat or energy of denaturation. The enormous favorable entropy of denaturation of $\Delta S = 285$ cal K^{-1} mol^{-1}, or 285 e.u., for the enzyme ribonuclease is opposed by an enthalpy change of $\Delta H = 95$ kcal mol^{-1}, which is sufficient to keep the protein in its native state at physiological temperatures.

The balance between the contributions of entropy and enthalpy to the observed stability of the protein is given by the standard Gibbs energy change of the reaction, $\Delta G°$, according to equation 2.

$$\Delta G° = \Delta H - T\Delta S \tag{2}$$

The standard Gibbs energy change is a logarithmic function of the equilibrium constant of the reaction according to equation 3 (R is the gas constant). Equation 3 shows that a positive value of $\Delta G°$ corresponds to an unfavorable equilibrium constant for denaturation, K_D.

$$\Delta G° = -RT \ln K_D \tag{3}$$

Thus, when $\Delta G°$ is negative a reaction will tend to proceed in the forward direction ($K_D > 1.0$) and if it is positive it will tend to proceed in the reverse direction ($K_D < 1.0$). The Gibbs energy is sometimes called the free energy, ΔF, especially in the older literature.

The entropy term in equation 2, ΔS, is multiplied by the absolute temperature, K, so that the contribution of the entropy term that favors denaturation at 27°C, or $(27+273) = 300\,\text{K}$, is $-T\Delta S = -300\,\text{K} \times 285\,\text{cal}\,\text{K}^{-1}\,\text{mol}^{-1} = -85{,}000\,\text{cal}\,\text{mol}^{-1}$, or $-85\,\text{kcal}\,\text{mol}^{-1}$. This is overcome by the unfavorable increase of enthalpy upon denaturation of $\Delta H = 95\,\text{kcal}\,\text{mol}^{-1}$, so that the standard Gibbs energy change for denaturation, $\Delta G°$, is positive; it is equal to $95 - 85 = +10\,\text{kcal}\,\text{mol}^{-1}$.

Thus, denaturation at 27°C is unfavorable by $+10\,\text{kcal}\,\text{mol}^{-1}$; the conversion of denatured into native enzyme, in the reverse direction, is favorable by $-10\,\text{kcal}\,\text{mol}^{-1}$, as shown in Figure 8-3.

At 300 K, the value of $\Delta G° = 10\,\text{kcal}\,\text{mol}^{-1}$ gives a value of $K_D = 0.00000005 = 5 \times 10^{-8}$, from $\Delta G° = -RT \ln K = -1.98 \times 300 \times 2.303 \log K$, so that the native state is favored over the denatured state of the enzyme by a factor of 2×10^7. *Thus, $\Delta G°$ is a measure of the actual, net driving force for a reaction.* Not infrequently, as in this case, it represents a small difference between two larger numbers.

The enthalpy, or heat, of a reaction can be measured either directly, in a calorimeter, or indirectly, from the dependence on temperature of the equilibrium constant for the reaction. These methods will be described in Chapter 9.

The large ΔH for protein denaturation accounts for the stability of this protein over a large range of temperature. Most proteins are stable over a range of temperature as the temperature increases ($K_D < 1.0$) and then suddenly undergo complete conversion into the denatured form over a small range of temperature; they are entirely in the denatured form at temperatures above this range ($K_D > 1.0$). An example is shown in Figure 8-4. The midpoint of the denaturation curve, at which the protein is 50% denatured, is the *melting temperature*, T_m. The amount of denaturation can be determined by measuring the change in ultraviolet absorbance or optical rotation or by following the large

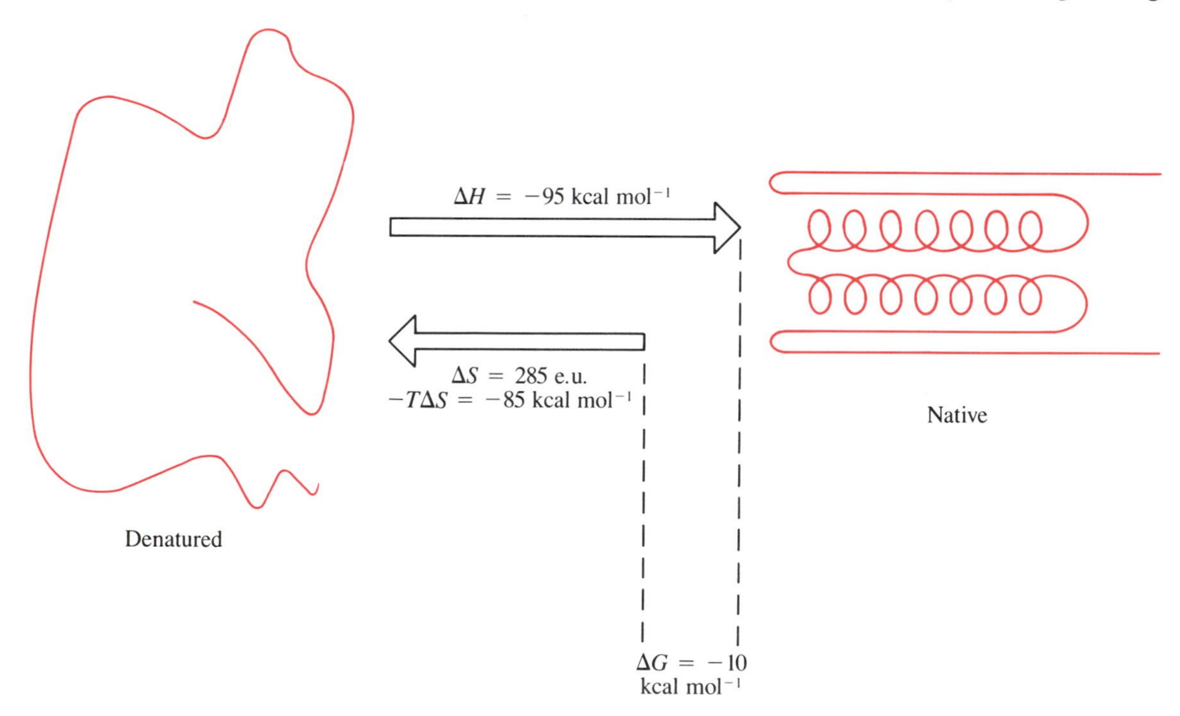

Figure 8-3

The Thermodynamic Basis for the Structure of a Protein. The native structure of a protein results from a counterbalance of the unfavorable entropy and the favorable enthalpy. The stabilization by enthalpy is slightly greater than T times the entropy increase upon denaturation, $T\Delta S$.

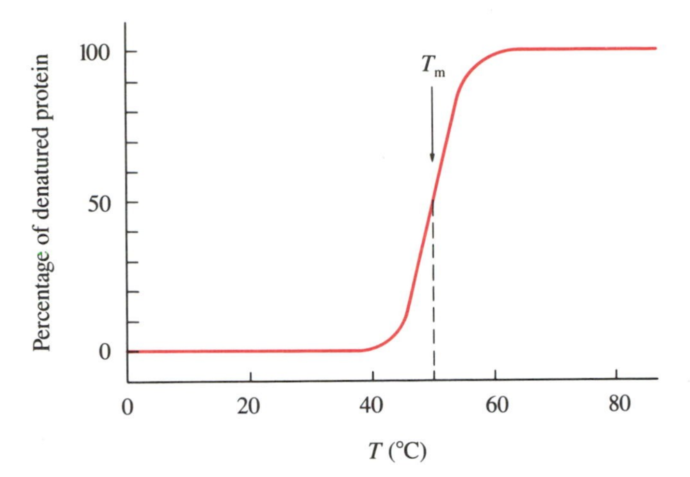

Figure 8-4

Thermal Melting of Proteins. Most proteins are unfolded or rearranged into an inactive, denatured state at high temperatures. The transition usually occurs over a small range of temperature and is 50% complete at the "melting temperature," T_m.

increase in viscosity as the compact native protein unfolds into a random, denatured structure.

It is important to distinguish the Gibbs energy change of a reaction, ΔG, from the enthalpy or energy change of the reaction, ΔH. We are accustomed to reactions that proceed spontaneously when they give off a large amount of heat, as in an explosion; and in the early days of chemistry it was thought that the evolution of heat is a measure of the tendency of a reaction to proceed spontaneously. Thermodynamics is concerned largely with the fact that this is not so. Protein denaturation is the ultimate example of a reaction that can be driven by an increase in probability and entropy in spite of a strongly *unfavorable* enthalpy change. The denaturation of a protein usually takes up a large amount of heat, so that it is the reverse of an explosion.

The enthalpy, or heat, of a reaction has generally been measured in calories by biochemists and most physical chemists. We are now in the midst of a conversion from calories into joules, following the recommendations of a series of international commissions, so that the literature read by the next generation will contain both calories and joules. We will use both units in this book, but will generally use calories in the text. It is helpful to remember that 4.2 joules = 1 calorie.

Hydrogen Bonds

For many years, almost any biochemist would say with some assurance that the most-important force that maintains proteins in their native structure is *hydrogen bonding* and that the denaturation of proteins by heat is brought about by breaking these hydrogen bonds with a large, unfavorable, enthalpy change. The pendulum then swung far toward the view that only *hydrophobic interactions* between nonpolar groups provide significant stabilization to proteins, and that hydrogen bonding provides little, if any, stabilization. At the present time, it appears that both hydrogen bonding and hydrophobic interactions provide important contributions to protein stability. We will consider interactions between nonpolar groups in the next section.

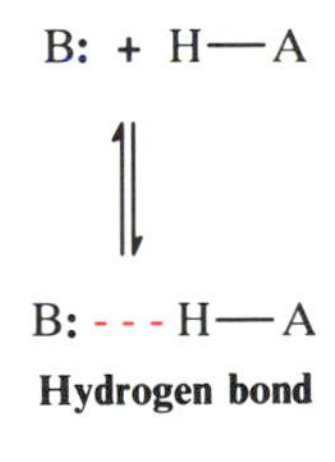

Hydrogen bond

It has been difficult to evaluate the relative contributions of hydrogen bonding and other forces for two reasons: first, as already noted, binding forces under physiological conditions represent a *competition* between binding to water and binding to some other group, not the net formation of a new bond in a vacuum; and, second, the results of studies on small model compounds in aqueous solution cannot be applied directly to proteins because the different parts of a protein are held together by a covalent peptide backbone, whereas small molecules are not.

Two basic experimental facts are clear. First, there is no doubt that hydrogen bonds *exist* in native proteins. We have seen that almost every peptide bond is hydrogen bonded to another peptide group or some other proton donor or acceptor, and many of them are organized in secondary structures such as the α-helix, the β-sheet, and β-turns. Second, the *net* breaking of these hydrogen bonds in the absence of water is very unfavorable, with an unfavorable energy change ranging from about 3 to 5 kcal mol^{-1}. This accounts for the great stability of proteins in the absence of water. In fact, proteins that are protected from water in seeds and spores can withstand high temperatures, often above the boiling point of water, without becoming denatured. This stability is important for the successful reproduction of many species of plants and microorganisms. Under most conditions, a protein will not break its internal hydrogen bonds and unfold unless it can form new hydrogen bonds to water or to some other solvent. In other words, water is itself a strong denaturing agent, although it is not quite strong enough to unfold proteins under physiological conditions.

Hydrogen bonds are now believed to be almost entirely electrostatic in nature; theoretical chemists have concluded that there is little or no contribution of covalent bonding to hydrogen bonds. The electrostatic interaction arises from dipoles or charges. For example, an ion pair of a carboxylate group of glutamate and a protonated amino group of lysine can form a hydrogen-bonded ion pair:

$$R{-}C({=}O){-}O^{-}\cdots H{-}N^{+}H_2{-}R'$$

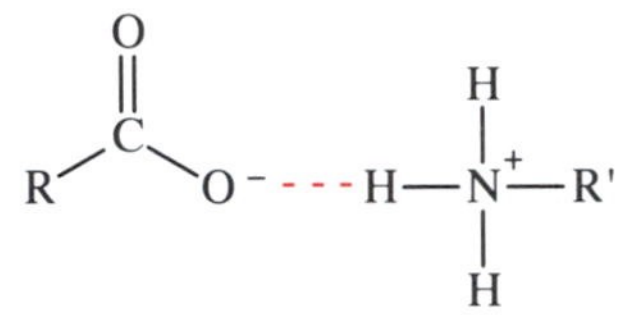

We have already described the function of a hydrogen-bonded ion pair in chymotrypsin between a carboxylate group of aspartate and the protonated α-amino group of isoleucine that converts the enzyme into an active form when chymotrypsinogen is activated to chymotrypsin (Chapter 4).

Most hydrogen bonds in the interior of proteins are formed from interactions between dipoles, rather than charged groups. Hydrogen is less electronegative than oxygen, so that the hydroxyl group of serine has a dipole that interacts favorably with the lone-pair electrons on the imidazole group of histidine at the active site of chymotrypsin:

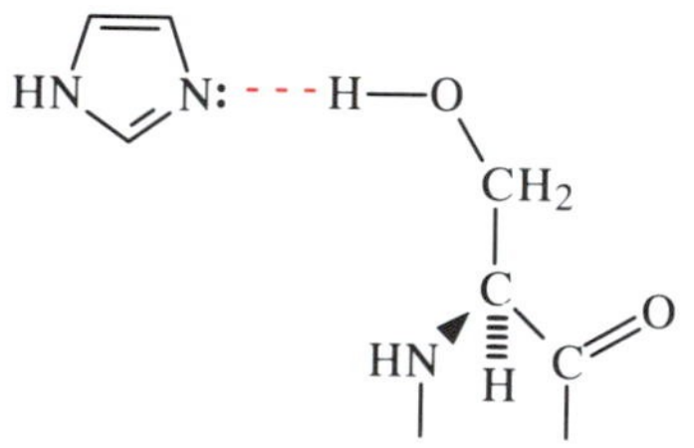

$$\text{HN(imidazole)N:}\cdots H{-}O{-}CH_2{-}C_{\alpha}H(NH{-})(C{=}O)$$

Peptide and other amide groups have a very large dipole because of the strong electron-withdrawing power of the carbonyl group and the strong electron-donating power of nitrogen. This causes movement of electrons into the carbonyl group and produces about 0.5 negative charge on the carbonyl oxygen atom and about 0.5 positive charge on the NH of the peptide. The resulting dipole is responsible for the hydrogen bonding of peptide groups in proteins and for a large fraction of the favorable enthalpy change that provides a driving force for the folding of proteins into their native structure.

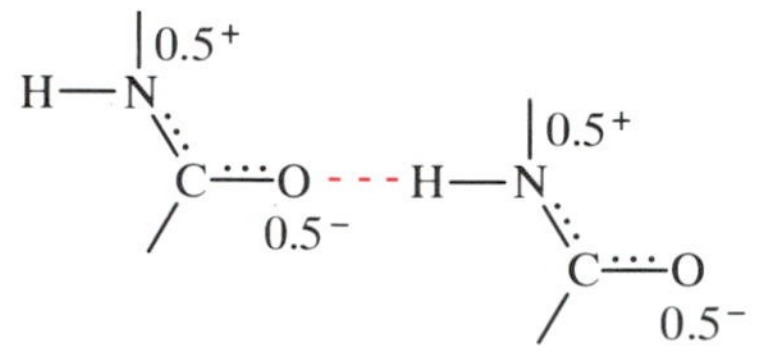

$$H{-}N^{0.5+}{\cdots}C{\cdots}O^{0.5-}\cdots H{-}N^{0.5+}{\cdots}C{\cdots}O^{0.5-}$$

The stability of a given hydrogen bond in a folded protein arises partly from the *favorable* interaction between the two hydrogen-bonded peptide groups and partly from the very *unfavorable* situation that would arise if the polar peptide groups were to separate and find themselves in the hydrophobic environment of the interior of a protein, with no hydrogen bonding. These hydrogen bonds are found in the α-helices and β-sheets that make up much of the secondary structure of proteins, and nearly all of the peptide groups in proteins form hydrogen bonds with other parts of the protein or the solvent.

The strength of a hydrogen bond in a nonpolar environment of ~3 to 5 kcal mol^{-1} (equation 4) is not relevant to most biochemical systems, because it is unusual for a hydrogen bond to be lost under physiological conditions.

$$\text{B:} + \text{H—A} \rightleftharpoons \text{B:}\cdots\text{H—A} \qquad (4)$$

The single point that is most important to remember about hydrogen bonds in biochemistry is that the "breaking" of a hydrogen bond hardly ever causes a net loss of hydrogen bonding; it nearly always represents the *exchange* of one hydrogen bond for another, usually for a hydrogen bond to water (Figure 8-5). The denaturation of a protein, for example, includes the exchange of peptide–peptide and water–water hydrogen bonds for two peptide–water hydrogen bonds (equation 5).

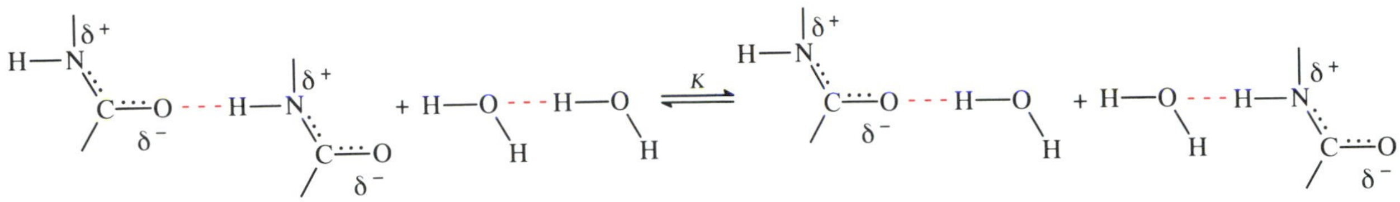

$$\text{H—}\overset{\delta^+}{\text{N}}\text{—C}\overset{\delta^-}{=}\text{O}\cdots\text{H—}\overset{\delta^+}{\text{N}}\text{—C}\overset{\delta^-}{=}\text{O} + \text{H—O(H)}\cdots\text{H—O(H)} \overset{K}{\rightleftharpoons} \text{H—}\overset{\delta^+}{\text{N}}\text{—C}\overset{\delta^-}{=}\text{O}\cdots\text{H—O(H)} + \text{H—O(H)}\cdots\text{H—}\overset{\delta^+}{\text{N}}\text{—C}\overset{\delta^-}{=}\text{O} \qquad (5)$$

A serious challenge to the primary role of hydrogen bonding in maintaining protein structure arose from the observation by Klotz and Franzen that hydrogen bonding between molecules of *N*-methylacetamide, which is a model for the peptide groups of proteins, can be observed in nonaqueous solvents but is very weak in aqueous solution. In fact, the infrared spectra of mixtures containing different concentrations of *N*-methylacetamide and water show no evidence for hydrogen bonding between the amide molecules as long as there are enough water molecules present to form hydrogen bonds to the carbonyl and N—H groups. This shows that there is competition between water and other amide molecules for hydrogen bonding to these groups, according to equation 5,

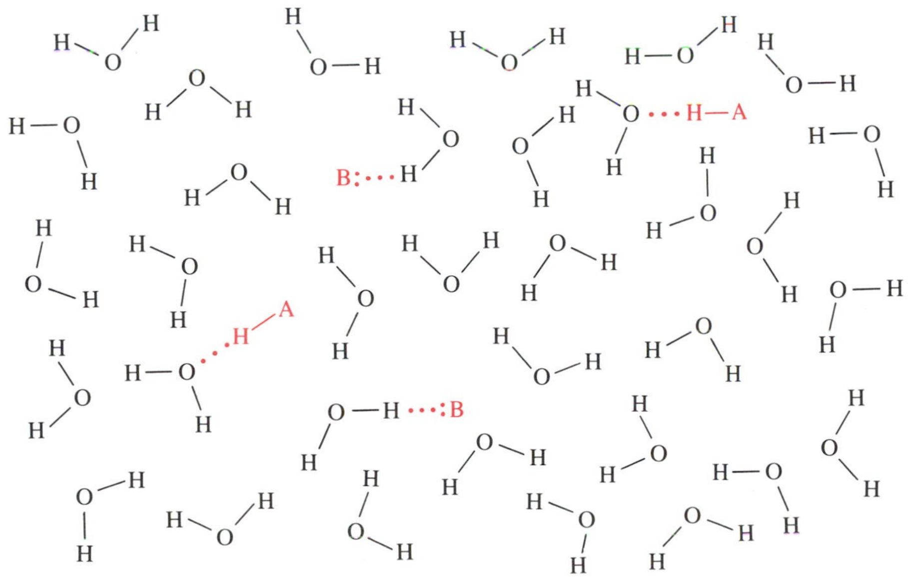

Figure 8-5

Hydrogen Bonding Between Water and Other Polar Molecules That Undergo Hydrogen Bonding.

and that hydrogen bonding between an amide and water is more favorable than between two amides. The favorable equilibrium constant of equation 5 means that, unless other factors are important, hydrogen bonding should not give a favorable equilibrium constant for the folding of proteins into their native structure in the presence of water.

There *are* other factors that are important. First, there are other forces, such as those arising from interactions between nonpolar groups, that make a significant contribution to protein stability and can help to stabilize a structure that includes hydrogen bonds. Second, and probably more important, is the contribution of *cooperativity* in hydrogen bonding that results from the covalent connections between peptide groups in the protein backbone.

The equilibrium constant for breaking a hydrogen bond between two peptide bonds in a partly folded peptide chain (Figure 8-6A) is much smaller than that for breaking a hydrogen bond between two molecules in dilute solution (Figure 8-6B) because the two molecules will diffuse apart and have a low probability of finding each other in the solution. The reaction in the partly folded peptide chain is favorable because the hydrogen-bonding groups are held near to each other, so that they have a higher probability of forming an additional hydrogen bond. The formation of hydrogen bonds during the folding of a protein is more probable than formation of hydrogen bonds between separated molecules because it requires less loss of entropy; it does not involve

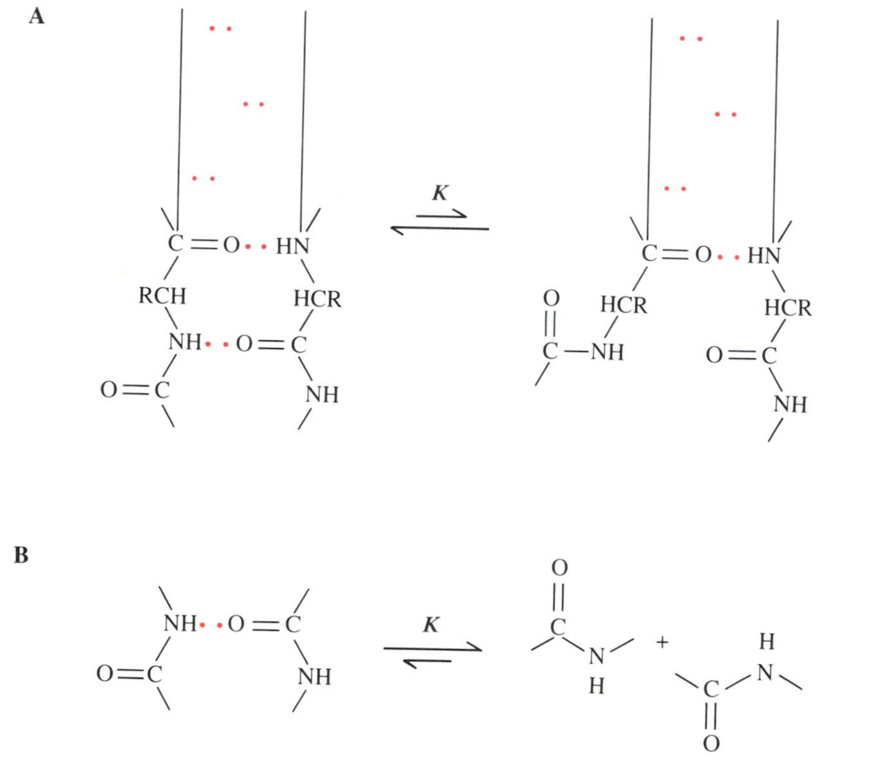

Figure 8-6

Equilibrium Constants for Intramolecular and Intermolecular Peptide Hydrogen Bonds. The equilibrium constant for hydrogen-bond formation between peptide groups is likely to be favorable if the groups are in an ordered structure (A), such as an α-helix or a β-sheet, because the formation of the hydrogen bond is an intramolecular reaction. The equilibrium constant for hydrogen-bond formation between isolated peptide groups (B) is unfavorable because the separated groups will diffuse away from each other. They have a low probability of reforming a hydrogen bond unless their concentration is very high.

the loss of entropy that is required for hydrogen bonding between two separated molecules. The formation of a hydrogen bond between two separate molecules requires that the individual molecules lose their freedom to be in any of a very large number of different positions in the solution, the translational entropy, and their freedom to be in a large number of different rotational states, the rotational entropy. The folding of a protein into its native structure is made possible by the same kind of induced intramolecularity that plays an important role in catalysis by enzymes, as explained in Chapter 5.

The stability of a native protein arises from the sum of the contributions of the covalent bonds of the peptide chain, hydrogen bonding, nonpolar interactions, and some less-important interactions including ionic bonds. These contributions are often reinforced in extracellular proteins by the cross-linking of peptide chains by disulfide bonds. The interactions must take place in such a way as to satisfy the strict steric demands from the space-filling properties and permissible bond angles of the constituent amino acids. Just how these contributions can be added up to account for the spontaneous folding of an unfolded peptide chain in water into its native structure is still too difficult a problem to solve with any degree of certainty. The difficulty of the problem is apparent from the fact that the favorable value of $\Delta G^{\circ} = -10\,\text{kcal mol}^{-1}$ for the folding of ribonuclease into its native structure represents a small difference between the much larger favorable enthalpy term of $\Delta H = -95\,\text{kcal mol}^{-1}$ and the unfavorable entropy term of $T\Delta S = -85\,\text{kcal mol}^{-1}$ for folding of the protein. We will see later that even these numbers are not direct measures of bond strengths and entropies of folding, because the observed thermodynamic parameters are perturbed strongly by interactions with the aqueous solvent.

Nonpolar and Electrostatic Interactions

The discovery of regular hydrogen-bonded structures in proteins and nucleic acids led to the conclusion that hydrogen bonding is the principal force that is responsible for maintaining the structure of these macromolecules. However, it later became increasingly apparent that the tendency of all but the most-polar molecules to clump together and avoid water (Figure 8-7) also provides an important driving force for maintaining the structure of macromolecules in aqueous solution.

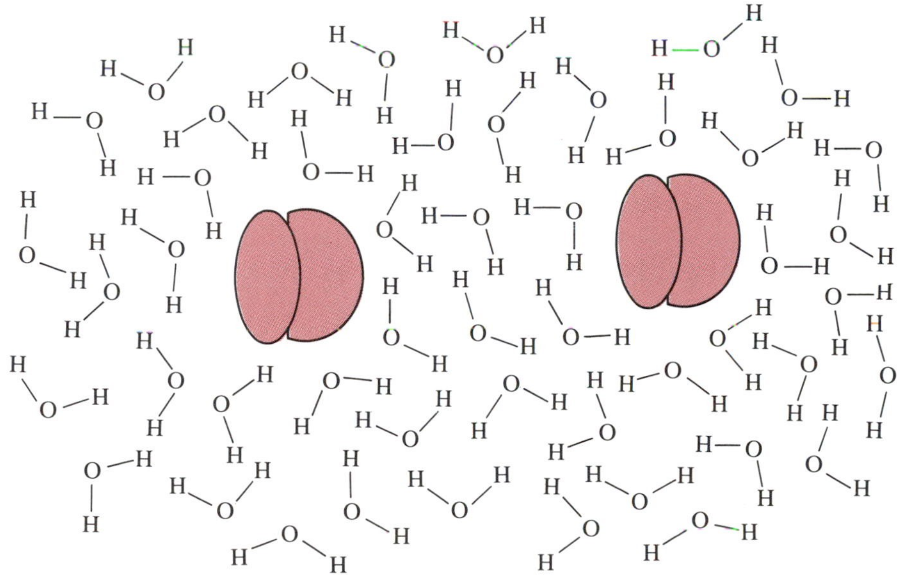

Figure 8-7

Self-association of Water and of Nonpolar Molecules. Nonpolar molecules suspended in water tend to associate with each other, rather than with water molecules, and water molecules associate together through hydrogen bonds.

The structure, and often the activity, of biological macromolecules depends in large part on the most simple and important principle of solution chemistry—that like attracts like. Nonpolar molecules tend to associate with other nonpolar molecules and polar, hydrogen-bonding molecules, especially water, tend to associate with other hydrogen-bonding molecules. The most important of the many unusual properties of water is probably that it is a small molecule, so that it contains a very large number of hydrogen bonds per unit of volume. Although the individual hydrogen bonds are not unusually strong, their high density and the strong dipolar interactions of water give this solvent a very large *cohesive energy density*, which is larger than that of any other liquid except mercury. Therefore, it is hard to insert any molecule into water unless the molecule can interact strongly with water through hydrogen bonds or electrostatic interactions. This high density of hydrogen bonds is the basis for many of the extraordinary properties of water that make life, as we know it, possible. These properties have been cited many times, but their importance cannot be emphasized too strongly. It is still a worthwhile and relevant experience to read L. J. Henderson's classic monograph on this subject, *The Fitness of the Environment*, which was first published in 1913.

Everyone is familiar with the tendency of oil in water to coalesce into droplets, on a macroscopic scale, and on a molecular scale it is common for dyes and other large organic molecules in aqueous solution to combine as dimers or larger aggregates. We have already described how the nonpolar side chains of amino acids on adjacent chains stabilize the β-pleated-sheet structure of proteins (Chapter 7); they also help to hold adjacent layers of pleated sheet together in structural proteins and interact to stabilize α-helices and the overall three-dimensional structure of native proteins. The double helix of DNA is fitted together through hydrogen bonds in the well-known Watson-Crick base-paired structure, as will be described in Chapter 12, but the preference of the nucleic acid bases to stack upon each other rather than to remain exposed to water appears to be an equally important driving force that maintains the structure of native nucleic acids.

There is no satisfactory name for the tendency of solutes in water to stick together unless they interact with water more strongly than they interact with each other. The term "hydrophobic interaction" is most descriptive but leads to some confusion because it is widely used to describe a particular kind of interaction in water, which will be discussed shortly. The more-general term *nonpolar interaction* is useful because these solutes generally have less-important polar interactions per unit of volume than does water. However, some of these molecules are themselves moderately polar and may even have larger dipole moments than that of water.

Micelles

Micelles provide a useful, although crude, model for soluble, globular proteins (Figure 8-8). Micelles are formed from soaps and other detergents that have a hydrophilic group attached to a long hydrophobic chain.

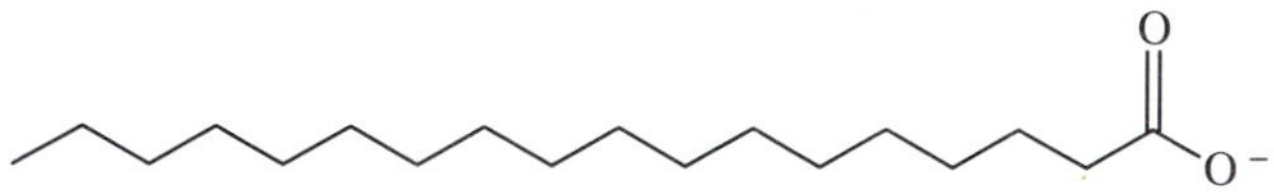

O
O⁻

Stearate anion (18 carbon atoms)

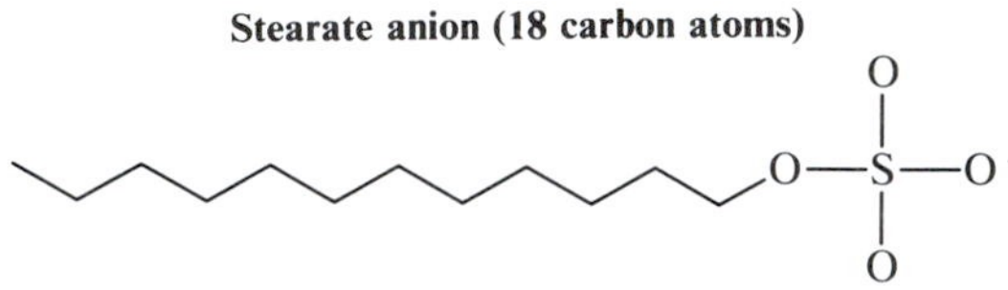

O
O—S—O⁻
O

Dodecyl sulfate (12 carbon atoms)

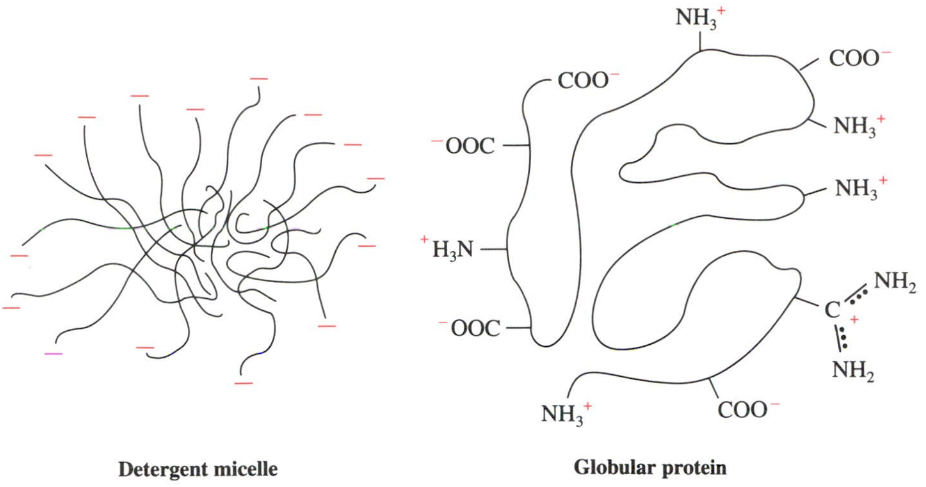

Detergent micelle **Globular protein**

Figure 8-8

Hydrophobic Interactions in Micelles and Proteins. Molecules of an anionic detergent in water tend to clump together in micelles, in which the nonpolar chains interact with each other and the charges are solvated by water. The micelle stays in the solution because of these charges. A native protein is similar to a micelle in that nearly all of the charged groups are on the outside, exposed to the solvent, and most of the uncharged groups are on the inside, with less or no exposure to the solvent.

Soaps are anions of long-chain fatty acids, such as stearic acid, and detergents have a charged or polar head group attached to a hydrocarbon tail, as in dodecyl sulfate. The hydrocarbon chains of these molecules clump together into an aggregate that has relatively little contact with water, whereas the polar head groups keep the aggregate suspended in water. The ability of these aggregates to take up oils and other water-insoluble substances accounts for the usefulness of soaps and other detergents as cleansing agents.

Soluble proteins are similar to micelles in that they are kept in solution by charged groups and other polar groups on the outside, which surround a less-soluble interior. The model is a crude one because proteins have a unique three-dimensional structure that gives an X-ray diffraction pattern, like a crystalline solid, whereas the interior of a micelle is composed of hydrocarbon chains from many detergent molecules that can take up many different positions and structures, as in a liquid.

Micelles and most proteins lose their native structure when alcohols or other organic solvents are added to water, because aqueous-organic solvents are better solvents than water for the relatively nonpolar groups in the interior of the structures. These cosolvents decrease the tendency of nonpolar groups to remain in the interior of the protein and maintain its native structure. They cause denaturation by "dissolving out" the interior structure, because exposure of the nonpolar groups to the solvent becomes more favorable.

Many detergents denature proteins by the same mechanism—the nonpolar tails of the detergent molecules bind to the nonpolar interior of the protein and cause it to unfold, whereas the charged head groups keep the denatured complex in solution. We have seen an example of this in the detergent-protein complexes that are formed with sodium dodecyl sulfate in SDS-electrophoresis. The process is essentially the formation of a mixed micelle containing protein and detergent. Some detergents owe their excellent antiseptic properties to their ability to disrupt proteins and membranes and thereby kill microorganisms.

Estimates of the relative strength of nonpolar interactions between molecules in water may be obtained as described in Box 8-1.

BOX 8-1

ESTIMATION OF THE STRENGTH OF NONPOLAR INTERACTIONS

As for hydrogen bonding, it is difficult to obtain a reliable quantitative estimation of the contribution of nonpolar interactions to the stability of proteins. An indication of the *relative* strength of these interactions for different amino acid side chains can be obtained from measurements of the partitioning ratios of molecules containing these side chains between water and an immiscible solvent, such as octanol or ethyl acetate, which may be regarded as a crude model for the interior of a protein. Nonpolar groups on a molecule will favor its transfer to an organic solvent, and the distribution of a solute when it is shaken or otherwise equilibrated between water and the organic solvent provides a measure of the amount by which the solute prefers the nonaqueous to the aqueous environment. The distribution coefficient can be used to calculate the Gibbs energy of transfer from water to the organic solvent, ΔG_T, from the ratio of the concentrations of the molecule in the organic solvent and in water after equilibration, C_{org}/C_{HOH}:

$$\Delta G_T = -RT \ln \frac{C_{org}}{C_{HOH}}$$

Surprisingly good estimates of the strength of nonpolar interactions have also been obtained simply by correlating the stability of a protein with the surface area or volume of the protein that is exposed to the aqueous solvent or is buried in the interior. These correlations are successful because the different properties of different amino acid side chains average out if a sufficiently large number of amino acids are present. The partitioning of a molecule between octanol and water provides a measure of its hydrophobic character. Hydrophobic molecules will tend to enter the organic phase when the two phases are shaken.

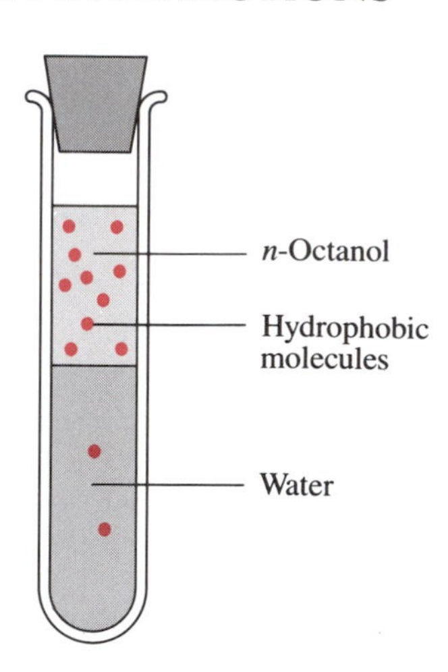

The Basis of Nonpolar Interactions

Two general kinds of explanations have been proposed to account for nonpolar interactions. The first focuses on the properties of water and describes the tendency of nonpolar groups to come together as a kind of squeezing out that results from the hydrogen bonding of water molecules to one another; it is likely to be accompanied by changes in the hydrogen-bonded structure of liquid water. The second focuses on the attraction of nonpolar groups for each other, which will tend to bring them together if it is larger than their attraction for water. We will conclude that both of these factors are important—a protein unfolds when the solvent-protein interactions become stronger than the sum of the solvent-solvent interactions and the peptide-peptide interactions in the folded protein.

Hydrophobic Interactions

The first explanation is called the *hydrophobic interaction* or “hydrophobic bond.” “Interaction” is more descriptive than “bond” because the interaction is not bonding in the usual sense. The low solubility of hydrocarbons and other nonpolar molecules in water shows that the interaction of water with nonpolar molecules is very unfavorable. The hydrophobic interaction is a consequence of the high polarity and strong hydrogen-bonding ability of liquid water, especially at low temperature. This causes a resistance to the insertion of nonpolar molecules or surfaces because water molecules interact more favorably with each other than with a nonpolar molecule or surface. Therefore, nonpolar groups tend to interact with each other, so that they have the least possible exposure to water and the water can interact most strongly with itself.

We might expect that the insertion of a nonpolar molecule into water would result in the breaking of hydrogen bonds between water molecules. However, this does not always occur. The water molecules at a nonpolar surface cannot form hydrogen bonds or dipolar interactions with the surface, but they can form hydrogen bonds with other water molecules at the surface. When this happens, some of the possible structures that they would take up in liquid water are energetically forbidden. Therefore, the water molecules are more ordered—they can exist in a smaller number of positions. This situation corresponds to a decrease in the entropy of the system, $-\Delta S$, and an increase in Gibbs energy, $+\Delta G$, because it is improbable; it would not happen spontaneously in liquid water, and the water molecules would like to return to the greater freedom that they had in pure water. The loss of entropy is partly compensated by a negative (favorable) change in enthalpy, $-\Delta H$, from the formation of hydrogen bonds, but the net change in Gibbs energy remains positive. This positive change in Gibbs energy describes the great difficulty of dissolving oil in water.

It was believed for a number of years that these negative changes in ΔS and ΔH were the result of a net increase in the number of hydrogen bonds of the water molecules around the nonpolar molecule. However, it has recently become possible to calculate the energy of systems of this kind with the aid of modern computers and computational techniques. The results of "Monte Carlo" calculations of the lowest energy states for a large number of different arrangements of water molecules surrounding a hydrocarbon have led to the conclusion that the number of hydrogen bonds is approximately the same for water molecules around the hydrocarbon as for those in the bulk solution. These calculations indicate that a favorable change in enthalpy arises from interaction of the water molecules with the dissolved hydrocarbon by dipole-induced dipole and dispersion forces (which will be described shortly). The entropy is decreased because the water molecules can exist in a smaller number of different positions in the region occupied by the nonpolar molecule.

When two nonpolar molecules in water come together in a single solvent cavity, the unfavorable interactions with water are reduced because the surface area of the hydrophobic molecules that is in contact with water is decreased (Figure 8-9). This results in a *decrease* in the number of water molecules surrounding the solutes with a low entropy and, therefore, an overall favorable change in Gibbs energy, $-\Delta G$. The increase in the number of different positions in which water molecules can find themselves corresponds to an increase in the entropy, or probability, of the system, and is believed to be the factor that provides the main driving force for the association of nonpolar molecules or surfaces with each other in aqueous solutions. It is accompanied by an increase in enthalpy, $+\Delta H$, because there are fewer molecules in contact with the nonpolar surface.

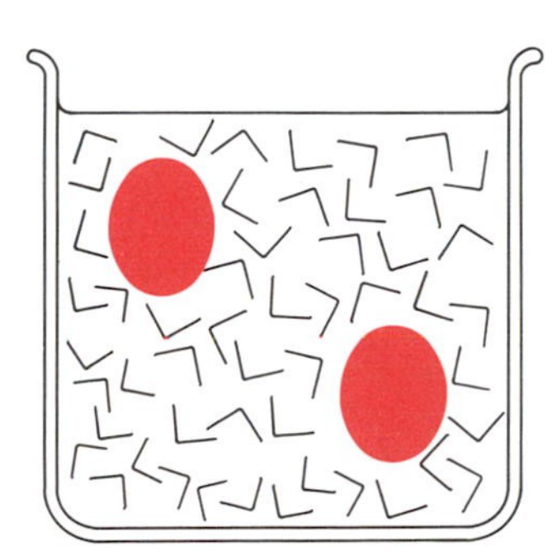

Hydrophobic solute

$-\Delta S$
$-\Delta H$
$+\Delta G$

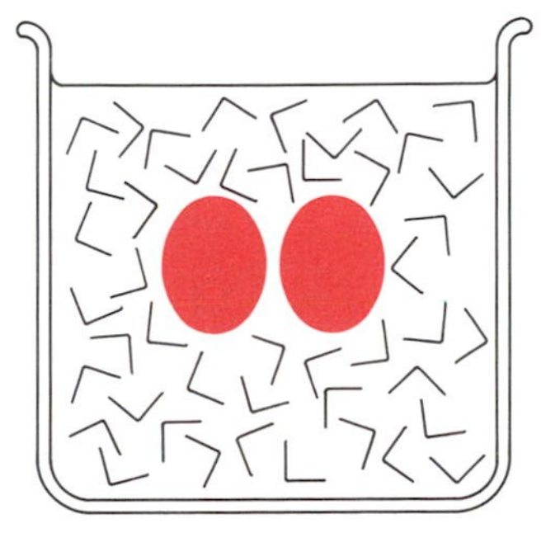

Hydrophobic interaction

$+\Delta S$
$+\Delta H$
$-\Delta G$

Figure 8-9

The Changes in Entropy and Enthalpy of Water Caused by Hydrophobic Molecules. The insertion of hydrophobic molecules into water is unfavorable ($+\Delta G$) because water molecules are less free to take up different positions in the neighborhood of the solute ($-\Delta S$); this is partially compensated by favorable interactions between the water molecules and the solute ($-\Delta H$). When hydrophobic molecules come together these changes are partially reversed and there is a favorable hydrophobic "interaction."

The folding up of a protein to form hydrophobic interactions in the interior of the folded protein is one example of this process, because it decreases the exposure of nonpolar surfaces to water. The formation of most chemical bonds gives off heat; there is a negative change of enthalpy and the reaction is, therefore, more favorable at low temperature. However, the *positive* enthalpy change for hydrophobic interactions means that the interaction becomes stronger with *increasing* temperature. This is probably the most-dramatic experimental manifestation of hydrophobic interactions. It is true up to about 60°C; above this temperature, the enthalpy change for interaction of nonpolar groups in water becomes zero and then negative; at high temperatures, the strength of interactions between nonpolar groups in water decreases with increasing temperature and, with few exceptions, the great majority of proteins undergo denaturation.

Consequently, many proteins have maximum stability at temperatures ranging from about 20° to 40°C. Above this range, they undergo denaturation because of the positive enthalpy of denaturation that was described for ribonuclease; at lower temperatures, they become less stable because of a weakening of hydrophobic interactions. Denaturation breaks hydrophobic interactions, which can lead to the ordering of water molecules that come into contact with exposed nonpolar surfaces (Figure 8-9) and to a decrease in entropy and enthalpy. The observed changes in enthalpy and entropy for protein denaturation at low temperatures, therefore, represent a *balance* between the negative changes in these parameters from breaking hydrophobic interactions and the positive changes from unfolding and breaking hydrogen bonds.

Cold Denaturation

In a few proteins, the weakening of hydrophobic interactions with decreasing temperature is large enough to cause the protein to undergo denaturation near 0°C. In these proteins, the decrease in enthalpy from the breaking of hydrophobic interactions is larger than the increase in enthalpy from the breaking of other interactions, such as hydrogen bonds, so that the net enthalpy of denaturation is negative and denaturation becomes more favorable with decreasing temperature. Cold denaturation is particularly disconcerting to the biochemist who carefully protects an enzyme from denaturation by keeping it cold at all times during purification, only to find that the enzyme has lost its activity, whereas a less-careful colleague may keep it at room temperature and obtain an active enzyme.

Cold denaturation may also arise from the dissociation of subunits, which can be caused by the breaking of electrostatic, as well as hydrophobic, interactions; electrostatic bonds also show a small or negative temperature dependence. This is because the strength of the electrostatic interaction itself, between the oppositely charged groups, has very little dependence on temperature, but the hydrogen bonding of water to the charged groups of the separated ions (equation 6) gives a negative change in enthalpy and is favored at low temperature.

$$\text{—C(=O)—O}^- \cdots \text{H—}\overset{+}{\text{N}}\text{H}_2\text{—} + 5\,H_2O \underset{}{\overset{-\Delta H}{\rightleftharpoons}} \text{—C}(\text{O}\cdots\text{HOH})_2^- + (\text{H}_2\text{O}\cdots)\text{H—}\overset{+}{\text{N}}(\text{H}\cdots\text{OH}_2)_2\text{—} \qquad (6)$$

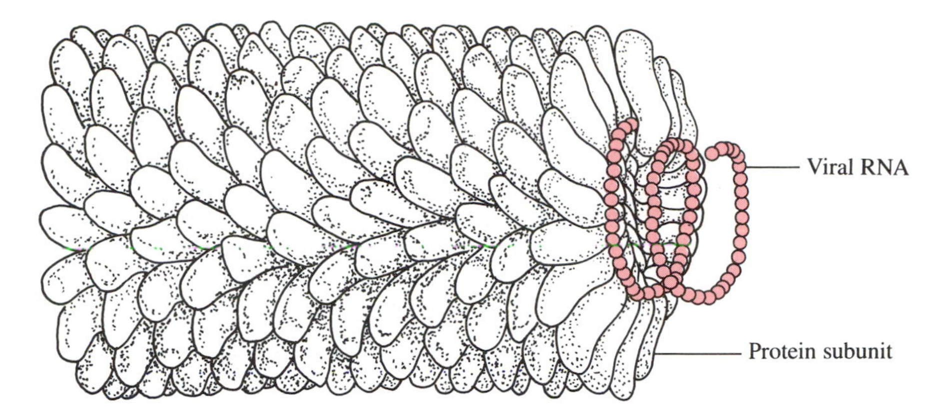

Figure 8-10

Helical Polymerization of Subunits in Tobacco Mosaic Virus. (After a drawing by Donald L. D. Caspar.)

Tobacco mosaic virus consists of a large number of identical protein molecules (molecular weight 17,000), which surround a coil of RNA and protect it against damage. This linear polymer of nucleotides, which carries the genetic information of the virus, will be described in Chapters 10 and 12. The protein molecules form a large, helical structure and the RNA is coiled in a groove inside that structure (Figure 8-10). Under suitable conditions of pH and ionic strength, the protein subunits aggregate spontaneously into this helical structure, even in the absence of RNA. The assembly process is reversible, and the structure reverts to the separated subunits with *decreasing* temperature. The structure of this protein has been determined by X-ray diffraction and provides the explanation for this unusual temperature dependence. The contact regions between subunits along the helical structure consist of alternating polar and hydrophobic regions, and the contact regions between adjacent layers of the helix consist largely of charged regions that provide favorable electrostatic interactions between oppositely charged side chains of amino acids. The inverse temperature dependence is accounted for by the breaking of hydrophobic interactions and electrostatic bonds between the subunits, both of which have this unusual dependence on temperature.

Water "structure" has often been invoked as a significant factor that determines the behavior of aqueous solutions. There is no doubt that there are differences in the amounts of hydrogen bonding and the arrangement of water molecules under different conditions, but water structure itself probably does not have a large influence on the behavior of molecules in aqueous solution (see Box 8-2).

Dispersion Interactions

The second general explanation for nonpolar interactions in water is based on the attraction of nonpolar groups for each other. Oil droplets and nonpolar liquids are generally held together by the weak attractive forces of van der Waals or London *dispersion interactions* between any two molecules that are in contact. Although these forces are weak for organic molecules, they are often still weaker for water.

The strong hydrogen bonds and dipole interactions of water lead to a degree of structuring and resist close packing of the water molecules, so that water is an unusually open liquid with a large amount of empty space between the water molecules. One of the most-important properties of liquid water is its fraction of occupied volume, which is only 0.36. In a typical protein, the fraction of occupied volume is 0.75, which is more than twice as large and is the same as

BOX 8-2

WATER "STRUCTURE"

It is often said that there is an increase in water "structure" around nonpolar molecules and surfaces and that this structure is a kind of iceberg. Although any such structure almost certainly bears little resemblance to an iceberg and has an exceedingly short lifetime, there is direct evidence for an ordering of water molecules in the neighborhood of nonpolar solutes. One kind of evidence is the time that is required for water molecules to reorient themselves in the presence of nonpolar solutes. These reorientation times can be measured by dielectric relaxation and by nuclear magnetic resonance; they are found to increase in the presence of organic solutes. A more easily observed manifestation of the same phenomenon is the increase in viscosity that occurs upon the addition of alcohols, dioxane, or dimethylformamide to water. When dimethylformamide is added to water, for example, the viscosity increases to a maximum value that is nearly three times as large as the viscosity of pure water. However, with further addition of dimethylformamide, the viscosity decreases and finally reaches the viscosity of pure dimethylformamide, which is less than that of water.

Changes in water "structure" are often proposed as explanations for various events that happen in aqueous or partly aqueous solutions. Such changes in water structure are, in a sense, an "ideal" explanation for almost any event, because there is almost nothing that can be done to water that does not result in a change in some property of the solution that can be attributed to a change in water structure.

The widespread use of this "explanation" is also its downfall. If almost anything changes water structure, there is virtually no predictive value to a hypothesis that is based on changes in water structure. Furthermore, a small change in the ratio of structured to unstructured water molecules will not provide much driving force for a reaction because the system is initially at equilibrium, with $\Delta G = 0$, and a small change in this ratio will give only a small change in ΔG. Karl Popper has said that the difference between myth and science is that a scientific hypothesis can be proved to be wrong; it is subject to the test of *falsification,* but a myth is never proved wrong. It follows that explanations based on changes in water structure that cannot be rejected by the test of falsification must be classed as myths.

that for closely packed spheres. Most organic liquids have intermediate values. The density of proteins is 1.35, one-third larger than that of water. The strength of van der Waals interactions depends largely on the amount of matter that is in close contact, so that these interactions are stronger between two organic molecules, or two parts of a protein, than between the water and the organic molecule or protein. Therefore, the organic molecules or parts of a protein will tend to interact favorably with each other.

This kind of interaction has a *negative* enthalpy change, like most chemical bonds. The negative enthalpy change that is found for the dimerization or aggregation of nucleic acid bases, dyes, and many other relatively large molecules in water suggests that this kind of direct attraction is important. Many of the molecules that show such interaction, such as nucleic acid bases, are moderately polar and many contain aromatic rings. The breaking of these dispersion interactions and of hydrogen bonds is responsible for the positive change in enthalpy that is observed for the denaturation of most proteins, especially at high temperatures. Dispersion and dipole-dipole interactions are described in Box 8-3.

Advocates of the "squeezing out," hydrophobic interaction approach have often shown a tendency to minimize or ignore the contribution of attractive interactions, and physical chemists who are concerned with dispersion forces have shown the same tendency toward the hydrophobic interaction approach. The two approaches are not mutually exclusive and it is likely, as is often the case in such matters, that both views are correct. On the one hand, water is a strongly dipolar, hydrogen-bonded liquid that resists the insertion of anything that cannot interact strongly with it; on the other hand, proteins are large, dense molecules with a high polarizability, so that the peptide chain is likely to fold up and interact with itself more strongly than it interacts with small water molecules.

The final conclusion must be that it is the *difference* between the dense, polarizable protein and the hydrophilic, hydrogen-bonded aqueous solvent that is responsible for the folding of proteins in water. Any description of the problem eventually returns to the simple conclusion that "like attracts like."

BOX 8-3

DIPOLE–DIPOLE AND DISPERSION INTERACTIONS

The short-range attractive interactions in proteins and most other biochemical systems are dominated by dipole–dipole, dipole–induced-dipole, and dispersion (induced-dipole–induced-dipole) interactions.

Peptide groups and other groups of atoms with differing electronegativities have a significant dipole moment. The dipole of a peptide group may be attributed to the large electronegativity of oxygen and electron donation by resonance from the nitrogen to the carbonyl group.

0.5^+ H—N ⋯ C⋯O 0.5^-

The arrow that describes the dipole is conventionally drawn pointing toward the negative end of the dipole. The dipole of a peptide or amide is large; there is approximately 0.5 negative charge on the oxygen and 0.5 positive charge on the nitrogen atom. We have already seen that these charges are responsible for the formation of hydrogen bonds between peptide groups in a protein, as the result of a favorable dipole–dipole interaction.

δ^+ H—N ⋯ C⋯O δ^- - - - H—N δ^+ ⋯ C⋯O δ^-

In an α-helix, a large number of these dipoles point in the same direction, so that the α-helix has a large dipole moment. However, the charges at each end of the helix of +0.5 and −0.5 are approximately the same as those of an individual peptide bond.

The electron cloud in a molecule is not fixed; it is *polarized*, which means that its electrons can move in response to a nearby charge. Therefore, a molecule that is near a dipole will respond to the dipole with a redistribution of charge, to form a complementary dipole. This gives a favorable dipole–induced-dipole interaction.

Now, if two molecules with no net dipole are next to each other, a small, transient dipole will develop in one of them at some instant, simply because of the random redistribution of electron density. This will then induce a transient dipole of opposite sign in the adjacent molecule, which gives rise to a net attractive force. This is an induced-dipole–induced-dipole interaction, or *London dispersion interaction.*

All three of these dipole interactions are short range and depend critically on the position and separation of the groups or molecules. The strength of each interaction decreases with the sixth power of the distance, so that the relative position of the molecules and the fit between them is critical. Induced dipoles, or dispersion interactions, also depend on the polarizability of the interacting molecules.

An approximation for the energy, E, of London dispersion interactions is given by:

$$E = \frac{3\alpha_1\alpha_2}{2d^6} \times \frac{I_1 I_2}{I_1 + I_2}$$

in which α_1 and α_2 are the polarizabilities of the two molecules, I_1 and I_2 are their ionization potentials, and d is the distance between them. In practice, the sixth-power dependence on this distance and the importance of an exact fit between the molecules make the strengths of these interactions very difficult to estimate even in the most-favorable cases; they are almost impossible to calculate accurately for aqueous solutions. The problem is that it is necessary to evaluate the difference between solute-solute, water-water, and solute-water interaction energies for systems in which the exact geometries of the interactions are not known. We are left with the qualitative conclusion that it is highly probable that these interactions are stronger in the interior of a tightly packed native protein molecule than they are for the interaction of an unfolded protein with the less-dense aqueous solvent.

BOX 8-3 (continued)

Hexanitrodiphenylamine

The acidity of hexanitrodiphenylamine is larger in aqueous-organic solvents than in water. The dissociation of uncharged acids, such as acetic acid, is usually inhibited by the addition of organic solvents to water because the dissociation results in the development of two charges and these charges are less stable in the solvent mixtures than in water.

$$CH_3COOH \rightleftharpoons CH_3COO^- + H^+$$

However, the acidity of hexanitrodiphenylamine *increases* by a factor of 30 in 24% acetone, compared with water. This increase may be attributed to the favorable dispersion interaction of the resonance-stabilized, colored, polarizable hexanitrodiphenylamine anion with acetone, which is considerably larger than with water.

Electrostatic Interactions

Repulsive forces between charged groups can cause denaturation of a protein, especially at low ionic strength. If the salt concentration is low, the charges on proteins are not shielded by a cloud of ions of opposite charge, so that interactions between charges become much more important than at high ionic strength. Some proteins undergo denaturation at low ionic strength, or even at physiological salt concentrations, because groups of amino acid side chains with the same charge on the outside of the protein repel each other and cause unfolding, whereas groups of opposite charge attract each other to form new electrostatic interactions in the same protein and with other protein molecules. This results in denaturation and precipitation. The ionic strength, μ, is defined by equation 7,

$$\mu = \frac{\Sigma c_i Z_i^2}{2} \tag{7}$$

in which c_i is the concentration and Z_i is the charge of each ion (equation 7). For a salt of two singly charged ions, such as NaCl, the ionic strength is equal to the concentration of the salt.

Most proteins are denatured by acid or base largely because of electrostatic repulsion of side chains that have the same charge. Figure 8-11 shows how

Figure 8-11

Protein Denaturation in Acids and Bases. The addition of acid to a protein solution decreases the pH and protonates negatively charged carboxylate groups. This increases the net positive charge on the protein and may cause the protein to unfold when the electrostatic repulsion between the positively charged groups becomes significant. The addition of base can cause unfolding by bringing about electrostatic repulsion between negative charges.

a decrease in the pH can result in the protonation of several carboxylate side chains that breaks ion-pair bonds and causes dissociation of a dimeric protein. A further decrease in pH will lead to the protonation of most or all of the carboxylate groups, so that the repulsion of the positive charges causes unfolding of the peptide chain. Analogous results can be brought about by removing the protons from $—NH_3^+$ groups and increasing the net negative charge of the protein at high pH.

BOX 8-4

DENATURATION IN MIXED SOLVENTS

Urea and guanidine hydrochloride are the standard, and among the best, denaturing agents for proteins. These compounds were originally thought to act by forming hydrogen bonds to the peptide bonds of the protein that are stronger than the hydrogen bonds to other peptide bonds or water. This causes the peptide bonds in the interior of the protein to become exposed to the solvent. However, the mechanism must be more complicated because few, if any, compounds are better at hydrogen bonding than water and there is no evidence that urea and guanidinium ion form stronger hydrogen bonds than do peptides.

In order to examine this problem critically, it is useful to consider the denaturation process in more detail. A protein that has been fully denatured by urea or guanidine hydrochloride is essentially a randomly oriented polypeptide chain in solution. In denaturation, the compact structure of the native protein unfolds into a random form that results in a greatly increased exposure of the interior of the protein to the solvent. This increased exposure to the solvent provides the driving force for denaturation that is brought about by changing the solvent. We can look at the denaturation as a "dissolving out" of the interior of the protein by the denaturing solvent.

The mechanism of action of different denaturing agents can be examined by determining their effectiveness in dissolving small molecules that resemble the different groups that are in the interior of a protein. Organic solvents such as alcohols and dioxane, for example, increase the solubility of hydrocarbons and other nonpolar molecules in water, but they do not increase the solubility of the polar molecule acetyltetraglycine ethyl ester (ATGEE), which is a model for the polar peptide groups in the interior of proteins. Therefore, denaturation by such organic solvents can be attributed to a breaking up or dissolving out of the nonpolar groups in the interior of proteins but not to a favorable interaction with polar groups. In fact, organic solvents do denature

Acetyltetraglycine ethyl ester (ATGEE)

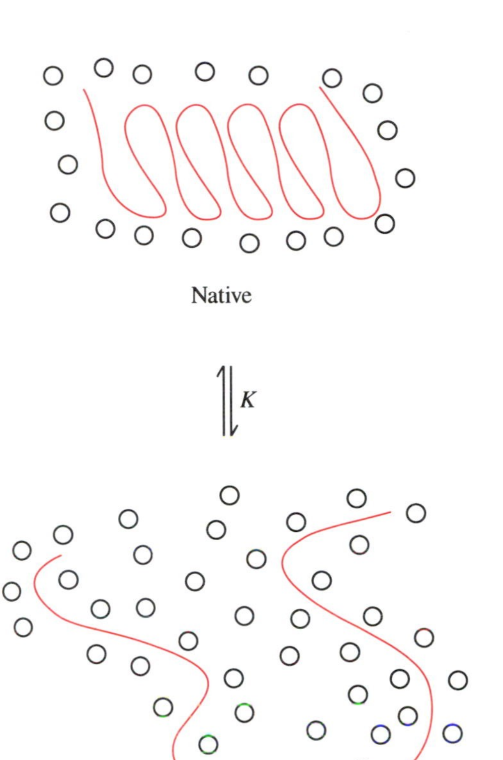

Denatured

proteins, largely by disrupting their tertiary structure, but do not unfold them completely and usually *increase* the stability of hydrogen-bonded secondary structures such as the α-helix.

Urea and guanidine hydrochloride have the remarkable property of interacting favorably with *both polar and nonpolar groups,* as shown by their ability to increase the solubility of both nonpolar solutes and ATGEE. This property enables them to unfold a protein completely by dissolving out all of the groups in the interior of the protein. Their effectiveness is a consequence of the fact that it is easier to insert almost any organic molecule into aqueous urea or guanidine hydrochloride than into water, so that there is a favorable interaction with these denaturants even if there is no increase in hydrogen bonding. Alcohols and other organic solvents are less-effective denaturants because they

BOX 8-4 (continued)

DENATURATION IN MIXED SOLVENTS

have an *unfavorable* influence on highly polar, hydrogen-bonding groups.

These matters are dealt with quantitatively by using the change in solubility to calculate the Gibbs energy of transfer of a molecule from water to another solvent such as aqueous urea. A negative free energy of transfer, $-\Delta G_{tr}$, corresponds to an increase in solubility and means that the molecule prefers to be in the urea solution. The change in solubility can be described as a change in the activity coefficient of the molecule, taking the activity coefficient in dilute aqueous solution as 1.0. The activity of a molecule, a, is related to its concentration, c, by its activity coefficient, f, according to:

$$a = fc$$

The activity and activity coefficient are measures of the tendency of a molecule to go somewhere or do something—if the activity or activity coefficient is high, the molecule has a strong tendency to precipitate out of solution, undergo a chemical reaction, or transfer to a more-favorable environment. A low activity and activity coefficient mean that the molecule is less active—it is content to be where it is and has less tendency to do these things.

At equilibrium the activity of a molecule in two different phases must be the same, so that it has no tendency to move from one to the other. Therefore, the activity of a molecule in a saturated solution that is at equilibrium with a solid is the same as the activity of the solid. This is described by:

$$a_{solid} = a_h = a_u = f_h c_h = f_u c_u$$

for the activity of a molecule in water, a_h, and in aqueous urea, a_u. Therefore, an increased solubility in the urea solution, $c_u > c_h$, corresponds to a decrease in the activity coefficient of the molecule in the urea solution compared with the activity coefficient in water, which is defined as 1.0. This is described by:

$$\frac{c_u}{c_h} = \frac{f_h}{f_u} = \frac{1.0}{f_u}$$

The increased solubility represents a proportionally decreased activity coefficient. These relations are most useful when the saturated solution is very dilute, so that the solute does not alter its own solubility and activity coefficient.

The increased solubility of a molecule in aqueous urea means that its transfer from water to aqueous urea is favorable and the Gibbs energy of transfer to aqueous urea, ΔG_{tr}, is negative, as described by:

$$\Delta G_{tr} = -RT\ln\left(\frac{S_u}{S_h}\right) = -RT\ln\left(\frac{f_h}{f_u}\right)$$

in which S_h and S_u are the solubilities in water and urea, respectively. The same relations should hold for a group on a protein, so that, if a group is more soluble in urea, its transfer from water to urea has a negative, favorable change in free energy and, if it is in the interior of the native protein, it will tend to become exposed to urea and denature the protein.

This conclusion is described quantitatively by:

$$K_D = \frac{a_D}{a_N} = \frac{f_D c_D}{f_N c_N}$$

in which the subscripts D and N refer to denatured and native protein, respectively. The thermodynamic equilibrium constant of a reaction is based on the activities, not the concentrations, of the reactants; and the activities are defined in such a way that the equilibrium constant remains constant when the ionic strength or other properties of the solvent change. Therefore, the increase in the concentration of denatured protein in urea, c_D, must be accounted for by a decrease in the activity coefficient of the denatured protein, f_D. This decrease occurs because the groups in the interior of the native protein that become exposed to solvent upon denaturation have a lower activity coefficient in urea than in water.

Thus, the different free energies of transfer of different molecules from water to aqueous urea and to other solvents can be used to estimate the free energies of transfer of similar groups on the inside of a protein to these solvents and to compare the abilities of these solvents to cause protein unfolding by interacting favorably with these groups. For example, the substitution of methyl groups for the protons of urea decreases the hydrogen-bonding ability of the molecule and its favorable interaction with ATGEE but increases its interaction with nonpolar compounds. Substitution of methyl groups on urea decreases its denaturing activity toward some proteins but increases it toward others. This difference can be correlated with differences in the ratios of polar to nonpolar groups in the interior of the native protein that become exposed to the solvent upon denaturation.

Salts of large, poorly solvated anions, such as lithium bromide and sodium perchlorate, tend to dissolve, dissociate, and denature proteins. These salts also increase the solubility of ATGEE. Therefore, their effects on proteins can be attributed to a favorable interaction with strongly polar groups, such as the peptide bond, and the absence of the strongly unfavorable interaction with nonpolar groups that is observed with salts of sulfate and other strongly solvated anions.

Nucleic Acids

Nucleic acids are similar to proteins in that they are held in their native structure by hydrogen bonding, which is between the acidic and basic groups on complementary base pairs in the double helix, and by noncovalent interactions between the bases that are stacked above each other in the double helix. The structure and properties of nucleic acids are described in Chapter 12.

Summary

The strength of binding interactions in aqueous solution always represents a *difference* between the strengths of interaction of a molecule with a binding site and with water. Hydrogen bonds between two small molecules are weak, because they are easily disrupted by water. However, cooperative hydrogen bonds provide stability to the native structures of proteins and nucleic acids because the formation of most of the bonds is intramolecular and requires little loss of entropy. Heating causes breaking of the hydrogen bonds and denaturation of the protein as the native structure is disrupted.

Nonpolar or "hydrophobic" interactions arise from favorable dispersion interactions between nonpolar side chains of amino acids and from a decrease in the unfavorable exposure of nonpolar groups to water when they interact with each other in the native structure of a protein. This results in the release of water molecules that were not free to move about and rotate freely, with a resulting increase of entropy. Nonpolar interactions generally become stronger with increasing temperature, up to 60°C, and some proteins become denatured at low temperatures, at which these interactions are weakened. Proteins may undergo denaturation at low ionic strength, or at high or low pH values, because of electrostatic repulsion between substituents with the same charge. Urea and guanidine hydrochloride denature proteins because they interact favorably with polar peptide and amide groups and with nonpolar side chains of amino acids. This causes the protein to unfold when the interactions of the peptide chain with the solvent become favorable enough to overcome the hydrogen bonding and nonpolar interactions that hold a protein in its native conformation. There is a large increase in entropy upon unfolding of a native protein into a disordered structure with an enormous number of different possible conformations. This provides a large driving force for denaturation of the protein.

ADDITIONAL READING

HENDERSON, L. J. 1958. *Fitness of the Environment: An Inquiry into the Biological Significance of the Properties of Matter*. Boston: Beacon Press.

SCHRÖDINGER, E. 1967. *What Is Life? The Physical Aspect of the Living Cell and Mind and Matter*. Cambridge: University Press.

PRIVALOV, P. L., and GILL, S. J. 1988. Stability of protein structure and hydrophobic interaction. In *Advances in Protein Chemistry* vol. 39, ed. C. B. Anfinsen, Jr., J. T. Edsall, F. M. Richards, and D. S. Eisenberg. New York: Academic Press, pp. 191–234.

KAUZMANN, W. 1959. Some factors in the interpretation of protein denaturation. In *Advances in Protein Chemistry*, vol. 14, ed. C. B. Anfinsen, Jr., M. L. Anson, K. Bailey, and J. T. Edsall. New York: Academic Press.

SCHELLMAN, J. A. 1987. The thermodynamic stability of proteins. *Annual Review of Biophysics and Biophysical Chemistry* 16: 115–137.

JORGENSEN, W. L., GAO, J., and RAVIMOHAN, C. 1985. Monte Carlo simulations of alkanes in water: hydration numbers and the hydrophobic effect. *Journal of Physical Chemistry* 89: 3470–3473.

PROBLEMS

1. A protein is found to be 30% denatured at pH 7 and 25°C in the presence of 4 M urea. What is ΔG for denaturation under these conditions? Report your answer in calories and joules.

2. Solutions of chymotrypsin at pH 7 gradually lose activity on standing. Explain why this occurs and why the rate of activity loss changes with increasing temperature (assume that the activity is always assayed at the same temperature as the initial incubation).

3. The entropy of unfolding of a peptide is $50\,\text{cal}\,K^{-1}\,\text{mol}^{-1}$. If this represents an increase in the number of states in which the peptide and surrounding solvent molecules exist upon denaturation, how large is this increase?

4. A protein undergoes reversible denaturation at pH 8.0 with $\Delta H = 37{,}000\,\text{cal}\,\text{mol}^{-1}$ and $\Delta S = 111$ e.u. What is its "melting temperature"; that is, at what temperature will it be 50% denatured?

5. The temperature of a solution of protein is suddenly increased to 70°C. The protein then undergoes denaturation over the next few minutes. What happens to the temperature of the solution?

6. An esterase has been isolated from liver and was found to undergo denaturation at pH 7 with increasing temperature. The value of ΔH is $54{,}000\,\text{cal}\,\text{mol}^{-1}$ and ΔS is $200\,\text{cal}\,K^{-1}\,\text{mol}^{-1}$ for denaturation. Will this protein be in the native or the denatured state at 40°C? What is the equilibrium constant for its denaturation at this temperature?

7. It has been suggested that proteins are similar to micelles in some respects. How might this suggestion be tested experimentally?

Energy

9

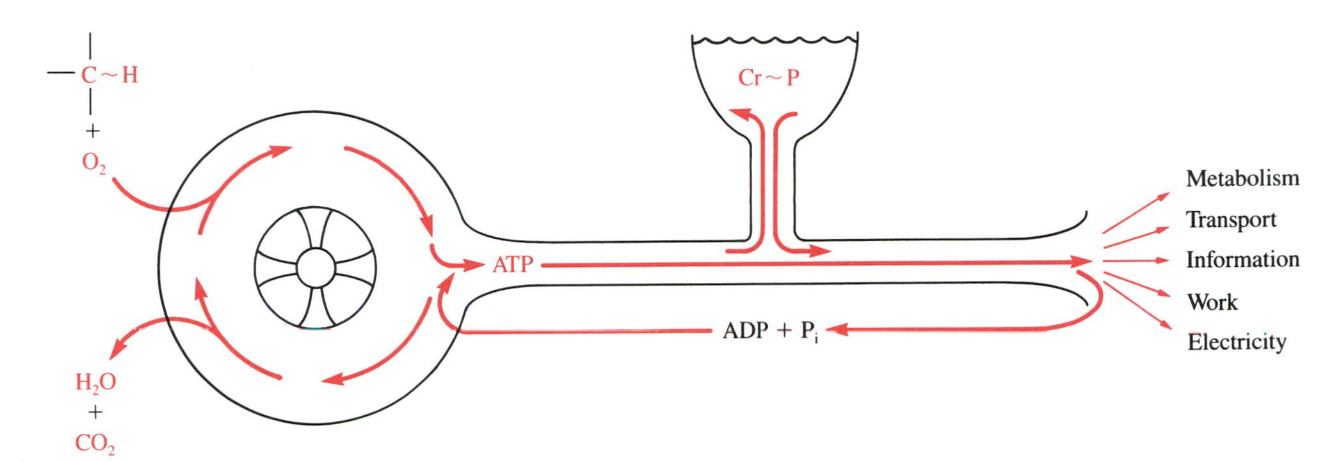

Lipmann's metabolic dynamo, 1941. The energy that is available from the oxidation of the carbon–hydrogen bonds of foodstuffs to carbon dioxide and water is used to form an energy-rich bond in ATP from ADP and inorganic phosphate, P_i. This energy is utilized to drive the different reactions and processes that take place in a living organism. Creatine phosphate, Cr ~ P, provides a reservoir of energy-rich phosphate bonds that can rapidly regenerate ATP from ADP.

Enzymes can catalyze a reaction so as to make it proceed rapidly; but, if the reaction is not energetically favorable, it will not take place unless a source of energy is available that supplies an additional driving force for the reaction. We need to know not only how fast a biochemical reaction can proceed in the presence of the appropriate enzyme, but whether, and how far, it will proceed. Understanding the energetics of the reaction is largely a matter of understanding its equilibrium constant, which may seem to be a simple matter. However, equilibrium constants are not as simple as they may seem, and experienced biochemists, as well as students, have had difficulty with this subject.

Historically, a meaningful understanding of the energetics of biochemical reactions lagged well behind the discovery of the reactions themselves. An understanding in chemical terms of how living systems work became possible when the chemical structures of organic compounds became known in the nineteenth century. The chemical pathways for the metabolism of compounds in biological systems were worked out, beginning at that time, after it had been shown that fermentation and glycolysis could be studied in a relatively simple, cell-free system. However, knowledge of these pathways did not identify the driving force that makes them proceed and did not explain how they provide useful energy to the cell.

The Energetics of Living Systems: The "Metabolic Dynamo"

One of the most-important functions of metabolism is to generate the energy that is needed in order for a living system to function. The *overall* energy balance of living systems was examined in detail near the turn of the century by measuring the total amount of energy that was taken up by an organism, in the form of foodstuffs and oxygen, and the amount of heat that was given off over a period of time.

However, such studies tell us very little about what is actually happening. The measurement of total energy input and output, or the *basal metabolic rate,* may be compared to studying the input and output of a slot machine. It shows what goes in and what comes out, and one can watch the wheels go around. But it tells us nothing about what goes on inside, about how the foodstuffs and oxygen that are taken up are converted into useful chemical processes and work.

Our present understanding of the energetics of biological systems arose as much from a new way of thinking about these systems as it did from new discoveries. This occurred in the 1940s and can be traced in large part to Fritz Lipmann and Hermann Kalckar. In his classic review "Metabolic Generation and Utilization of Phosphate Energy" in 1941, Lipmann described biochemistry as a metabolic dynamo in which the energy that is obtained from the oxidation of organic compounds is converted into useful energy, in the form of the "high energy" phosphoanhydride bond of ATP (see the illustration on page 235). The "energy rich" bonds are indicated by the squiggles, $\sim$, in this illustration; squiggles could also be included in ATP and ADP to indicate that these compounds contain high-energy phosphoanhydride bonds. The squiggle means that useful energy can be released when these bonds are cleaved by hydrolysis or oxidation. In this chapter, we will see how the energy changes that take place in the reactions of living systems can be described and understood.

Aerobic organisms do not use oxygen to *burn* carbohydrates and fats and give off heat; instead, they capture and store the energy that is available from the oxidation of these compounds through a series of controlled chemical reactions. These reactions release the energy in steps that make it available to drive other chemical reactions. Most of these steps are reversible and the overall process can proceed with very high efficiency. It has no resemblance whatever to a steam engine.

These steps are carried out in the metabolic pathways that are represented by the wheel of the metabolic dynamo. These pathways constitute a large part of our present knowledge of biochemistry. Some useful energy is made available from the synthesis of ATP when sugars are broken down through *glycolysis* or *fermentation,* which do not require oxygen. However, in aerobic organisms most of the energy is made available through a quite different mechanism called *oxidative phosphorylation.* We will examine these processes in detail in Chapters 21 and 22. In muscle and some other tissues, the small pool of energy in the form of ATP is backed up by a reservoir of energy that is stored as *phosphocreatine,* a high-energy compound that is at equilibrium with ATP, as is shown in the drawing on page 235 and in equation 1.

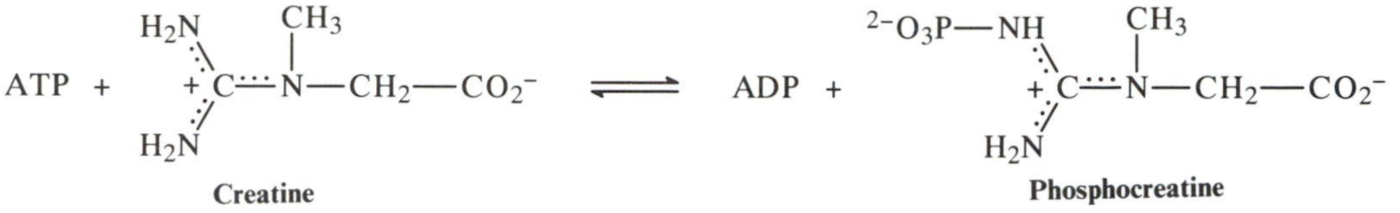

$$\text{ATP} + \underset{\textbf{Creatine}}{(H_2N)_2C^{+}{\cdots}N(CH_3){-}CH_2{-}CO_2^-} \rightleftharpoons \text{ADP} + \underset{\textbf{Phosphocreatine}}{(^{2-}O_3P{-}NH)(H_2N)C^{+}{\cdots}N(CH_3){-}CH_2{-}CO_2^-} \quad (1)$$

(Reminder: In this book, ATP represents MgATP unless otherwise stated.)

The chemical energy that is stored in ATP and other high-energy compounds is used to provide the driving force for the great majority of processes associated with life. These processes include not only the many metabolic pathways that are essential for a living system, but also many other processes that perform work, such as muscle contraction and the active transport of small molecules and ions. We are just beginning to understand how some of these processes work. This energy is also used to make DNA and RNA, which carry the genetic code for every living organism, so that it is also used to provide the *information* that is passed on from one generation to the next.

Group Transfer Reactions

The chymotrypsin-catalyzed hydrolysis of an ester or peptide, shown in Figure 9-1, is a *group transfer reaction,* in which an acyl group is transferred to water. Such reactions are usually irreversible, under physiological conditions in aqueous solution, because the high concentration of water and the low concentration of protons drive the reaction toward hydrolysis. However, chymotrypsin will also catalyze group transfer to other acceptors, such as alcohols and amines, forming esters or amides (Figure 9-1). These and other group transfer reactions are likely to be reversible, so that it is necessary to consider the *extent* to which they will proceed at equilibrium, as well as the *rate* at which they can occur.

Protein synthesis is a group transfer reaction that increases the length of a peptide chain by the addition of an amino acid ester of tRNA to a peptide ester of tRNA (equation 2).

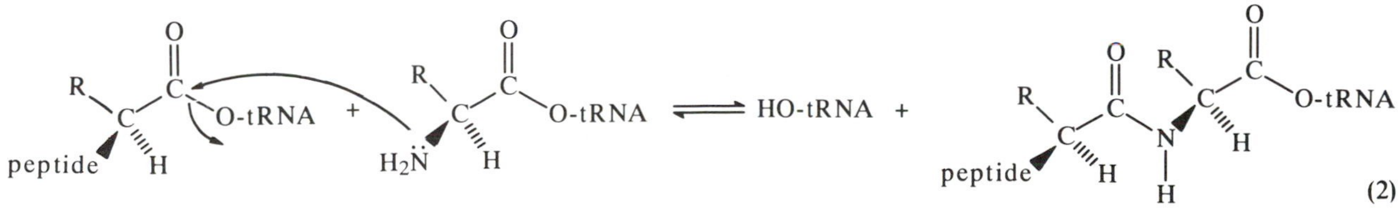

$$\text{peptide}{-}CHR{-}C(=O){-}O\text{-tRNA} + H_2N{-}CHR{-}C(=O){-}O\text{-tRNA} \rightleftharpoons \text{HO-tRNA} + \text{peptide}{-}CHR{-}C(=O){-}NH{-}CHR{-}C(=O){-}O\text{-tRNA} \quad (2)$$

The aminoacyl groups are activated by bonding to a hydroxyl group of transfer RNA (tRNA), the structure of which is described in Chapter 10. The amino group of the amino acid ester reacts with the acyl group of the peptide ester,

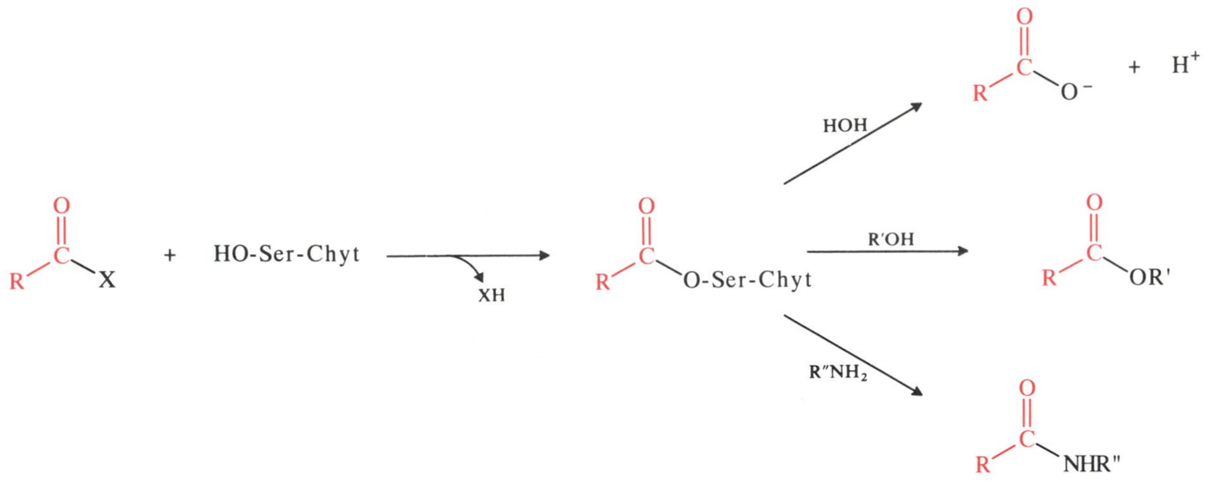

$$RC(=O)X + \text{HO-Ser-Chyt} \xrightarrow{-XH} RC(=O)\text{O-Ser-Chyt}$$

$$RC(=O)\text{O-Ser-Chyt} \xrightarrow{HOH} RC(=O)O^- + H^+$$

$$RC(=O)\text{O-Ser-Chyt} \xrightarrow{R'OH} RC(=O)OR'$$

$$RC(=O)\text{O-Ser-Chyt} \xrightarrow{R''NH_2} RC(=O)NHR''$$

Figure 9-1

Group Transfer Catalyzed by α-Chymotrypsin. Chymotrypsin will catalyze group transfer to alcohols and amines, in addition to water, by catalyzing the transfer of the acyl group of the acyl-enzyme to other acceptors. The extent to which this occurs depends on the energetics of the system in which chymotrypsin acts. Important factors are the concentrations of acceptors other than water, the thermodynamic stabilities of the products, the pK_a values of the alternative acceptors, and the compatibility of an alternative acceptor with the active site.

which displaces tRNA and extends the peptide chain by one amino acid. We need to know how the activated aminoacyl esters can be formed and whether or not their reaction with an amino group to form a peptide will proceed spontaneously.

"Energy Rich" Compounds and Gibbs Energies

The tendency of a reaction to proceed is measured by its equilibrium constant. An equilibrium constant larger than 1 means that the reaction will tend to proceed in the forward direction when all of the reactants and products are present at a concentration of 1.0 molar (1 M). This concentration is often taken as the *standard state* for the comparison of different reactions, so that they can be compared under the same conditions. However, for many purposes, particularly for reactions with very large or very small equilibrium constants, it is not convenient to describe the tendency of a reaction to proceed by its equilibrium constant. It is often more useful to use a logarithmic expression.

In biochemistry, the tendency for a reaction to proceed is usually described by the *standard Gibbs energy change* of the reaction at pH 7, $\Delta G^{\circ\prime}$, as shown in equation 3.

$$\Delta G^{\circ\prime}\,(\text{pH } 7) = -RT \ln K \tag{3}$$

The $\Delta G^{\circ\prime}$ (pH 7) is a logarithmic measure of the equilibrium constant and refers to the *total* concentrations of the reactants and products at equilibrium, including all ionic species of the indicated reactants. However, it describes the reaction only at pH 7.0; it does not describe any protons that are given off or taken up in the reaction. The proportionality constant, RT, is the product of the gas constant R, and the absolute temperature, T. The value of the gas constant is $R = 1.98$ cal mol^{-1} deg^{-1}, and the absolute temperature is given in the Kelvin scale (K). Absolute zero is 0 K and −273°C (Celsius); 0°C is 273 K.

The value of $\Delta G^{\circ\prime}$ (pH 7) is a measure of the useful energy that can be obtained from a biochemical reaction when the reactants (except for water) are in their standard state of 1.0 M total concentration at pH 7. The standard state for liquid water is taken as 1.0. The Gibbs energy change is often called the *free energy change*, $\Delta F^{\circ\prime}$; it is a measure of the amount of energy that can be obtained from a reaction that is available to do work or to drive another reaction. Equation 3 shows that a negative value of $\Delta G^{\circ\prime}$ corresponds to an equilibrium constant that is larger than 1; a large negative value of $\Delta G^{\circ\prime}$ means that a large amount of energy is given off in the reaction and can be made available to drive synthetic reactions or to do work.

Compounds with a large negative standard free energy change for their reaction are frequently called "energy rich" or "high energy" compounds, or compounds with a high "group potential." The ΔG values are measured in *calories or joules*, J. Several international bodies have accepted the joule as the standard but calories are still in wide use, so that it is important to remember the interconversion factor of 4.2 joules = 1 calorie.

A number of energy-rich compounds and some other metabolites that are important in biochemical reactions are shown on a scale of Gibbs energy in Figure 9-2. The scale shows the standard Gibbs energy change for *hydrolysis* of the compounds at pH 7, $\Delta G^{\circ\prime}$ (pH 7), in kilocalories mol^{-1} and in kilojoules

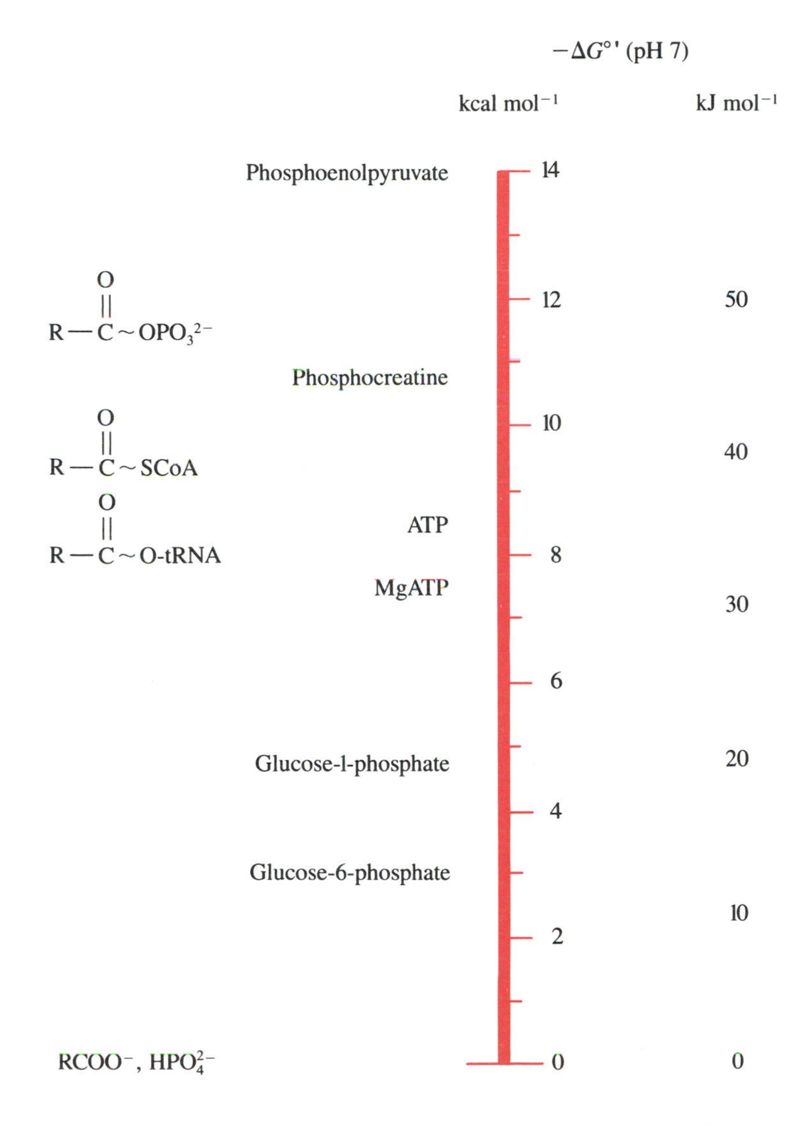

Figure 9-2

Gibbs Energy Scale for the Hydrolysis of Some Compounds of Biochemical Interest.

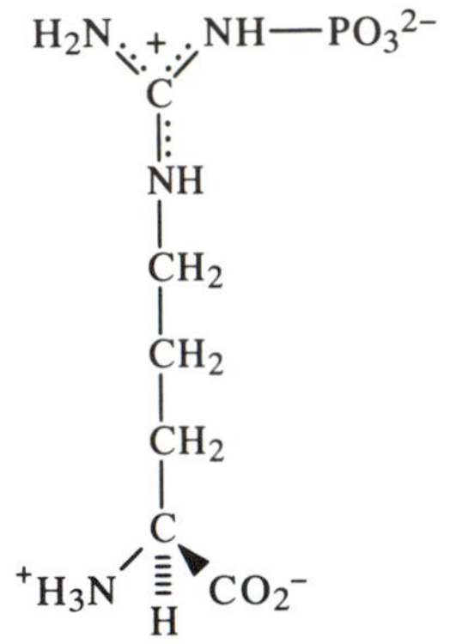

Phosphoarginine

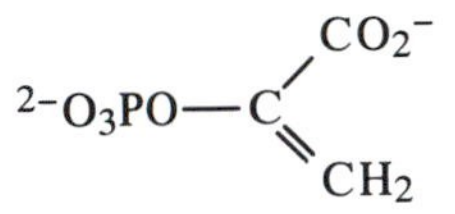

Phosphoenolpyruvate (PEP)

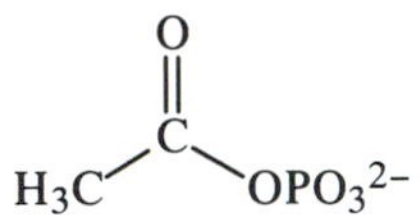

Acetyl phosphate

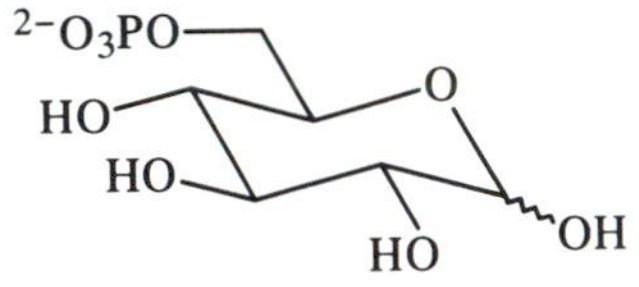

Glucose-6-phosphate

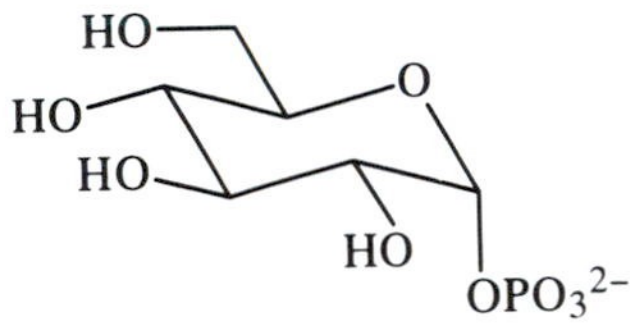

α-D-Glucose-1-phosphate

mol^{-1}. The most widely used and best known of these compounds is ATP, which has a $\Delta G^{\circ\prime}$ value of $-8500\,cal\,mol^{-1}$ at pH 7. This corresponds to an equilibrium constant for hydrolysis to ADP and P_i of $K = 10^{6.25} = 1{,}778{,}000$ M. ATP has been called the "common currency" for the transfer of chemical energy in the cell. Other nucleoside triphosphates have a very similar Gibbs energy for hydrolysis, but are used for a smaller number of reactions.

Phosphocreatine (equation 1) has a more-negative value of $\Delta G^{\circ\prime} = -10{,}800\,cal\,mol^{-1}$ for hydrolysis at pH 7 that enables it to serve as a reservoir of chemical energy, as noted earlier. Phosphocreatine contains a phosphorylated guanidinium group, which is thermodynamically unstable. The same function is served in invertebrates by phosphoarginine, which has a very similar phosphorylated guanidinium group. Earthworms and a few other organisms use other phosphorylated guanidinium compounds.

Phosphoenolpyruvate (PEP) and several acyl phosphate anhydrides are phosphate compounds that are even higher in energy (Figure 9-2). They are formed in glycolysis and fermentation and can serve as good sources for the synthesis of ATP because their reaction with ADP to make ATP is favorable thermodynamically. Their *formation* from ATP is *unfavorable* and can give only an extremely low yield of product because they are much higher in energy than ATP. Simple monoesters of phosphate, such as glucose-6-phosphate, are "low energy" compounds with a small group transfer potential. Their free energy of hydrolysis is only about $-3000\,cal\,mol^{-1}$, so that they are formed readily and almost irreversibly from ATP. Glucose-1-phosphate is slightly higher in energy than glucose-6-phosphate. Finally, the hydrolysis products of all of these phosphate compounds have a $\Delta G^{\circ\prime}$ of zero, by definition.

A more-general definition of a high-energy, or energy-rich, compound is given by equation 4:

$$\text{High-energy compound} + \underset{\substack{\uparrow \\ \text{Readily available}}}{X} \xrightarrow{\text{Physiological conditions}} \text{Products} \qquad \underset{\substack{\uparrow \\ \text{Large negative}}}{-\Delta G^{\circ\prime}} \tag{4}$$

A high-energy compound is a compound that will react under physiological conditions with a substance that is readily available in the environment to give products, with a large negative change in Gibbs energy. The most-common usage of this definition is to describe the hydrolysis reactions of phosphate and acyl compounds, such as ATP and aminoacyl tRNA. However, for aerobic organisms, an equally important high-energy compound is any carbohydrate or fat containing a C–H bond that can be oxidized through a series of reactions to give carbon dioxide and water. This is indicated by the wheel of the metabolic dynamo on page 235. Oxygen is readily available in the environment, so that this oxidation is the source of energy for all aerobic organisms and, ultimately, for most living organisms on earth. It provides a very large amount of energy, enough to make several molecules of ATP from each C–H bond. Oxidation-reduction reactions are measured on a scale of oxidation-reduction potentials that will be described shortly.

Concentrations of Reactants

The standard Gibbs energy change at pH 7 ($\Delta G^{\circ\prime}$, pH 7; equation 3) describes the amount of Gibbs energy that is available from a reaction that occurs at pH 7.0 when the total concentrations of the reactants and products (except for H^+ and HO^-) are at a standard state of 1 M. However, it is not an accurate measure of the amount of available energy that can be obtained from a reaction under physiological conditions, because biochemical reactions do not occur at a standard state of 1 M. The amount of available energy from the reaction of equation 5 is described by the more-general equation 6.

$$\mathrm{A} + \mathrm{B} \xrightleftharpoons{K_{eq}} \mathrm{C} + \mathrm{D} \tag{5}$$

$$\Delta G = \Delta G^\circ + RT \ln \frac{[\mathrm{C}][\mathrm{D}]}{[\mathrm{A}][\mathrm{B}]} \tag{6}$$

This equation includes a term for the concentrations of the reactants and products; physical chemists will recognize this as an entropy term that corresponds to the entropy of concentration (or dilution). When the concentration of a reactant, such as ATP, is high compared with that of the hydrolysis product, ADP, the concentration term in equation 6 will become smaller. The reaction will then have a more-negative Gibbs energy change, ΔG, so that it will have a greater tendency to proceed forward.

It is useful to consider two special cases of the general equation 6. First, it reduces to equation 7 when all of the reactants and products are at a standard state of 1 M.

$$\Delta G = \Delta G^\circ + RT \ln \frac{1 \times 1}{1 \times 1} = \Delta G^\circ + 0 = \Delta G^\circ \tag{7}$$

The available Gibbs energy of the reaction, ΔG, is then equal to the standard Gibbs energy change of the reaction, ΔG°.

Second, when the reaction is at equilibrium, there is no driving force for it to proceed in either the forward or the reverse direction, and the Gibbs energy change is zero:

$$\Delta G = 0 = \Delta G^\circ + RT \ln \frac{[\mathrm{C}]_{eq}[\mathrm{D}]_{eq}}{[\mathrm{A}]_{eq}[\mathrm{B}]_{eq}} = \Delta G^\circ + RT \ln K_{eq} \tag{8}$$

The ratio of the concentrations of reactants and products at equilibrium is the equilibrium constant, so that this equation reduces to the definition of the standard Gibbs energy change:

$$\Delta G^\circ = -RT \ln K_{eq} \tag{9}$$

Figure 9-3 shows the dependence of the Gibbs energy change of a reaction on the ratio of the concentrations of reactants and products when $K_{eq} = 1.0$. When the concentrations of reactants and products in the reaction $\mathrm{A} \rightleftharpoons \mathrm{B}$ are in the standard state of 1 M, so that [A] = [B], then the concentration term in the

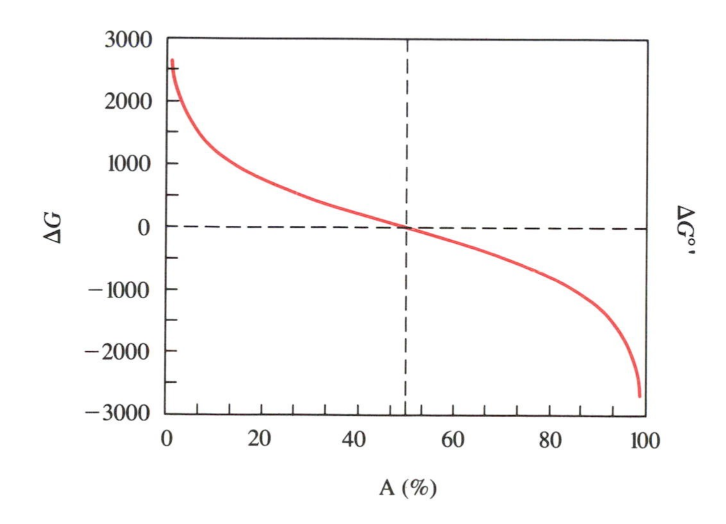

Figure 9-3

The Dependence of ΔG on the Fraction of Product. For a reaction $\mathrm{A} \leftrightarrows \mathrm{B}$, the relation between ΔG and the percentage of A is plotted. This figure refers to a reaction that takes place in aqueous solution at a specified pH with $K_{eq} = 1$; that is, ΔG is zero at 50% B, and $\Delta G^{\circ\prime}$ at that pH also is zero (see equations 9 and 10). For a different reaction with a different value of K_{eq}, the $\Delta G^{\circ\prime}$ will have a value other than zero (see equation 9).

equation is zero ($\ln 1/1 = \ln 1 = 0$) and the Gibbs energy change is equal to the standard Gibbs energy change (equation 10).

$$\Delta G = \Delta G^\circ + RT \ln \frac{[B]}{[A]} = \Delta G^\circ \tag{10}$$

However, when the concentration of reactants is larger than that of products, the value of ΔG decreases or becomes more negative and the driving force for the reaction becomes more favorable.

This effect of concentrations can be large. It is very important under physiological conditions, especially for reactions in which one molecule gives two molecules of products. Consider the hydrolysis of ATP, which has a value of $\Delta G^{\circ\prime}(\text{pH } 7) = -8500 \text{ cal mol}^{-1}$ for a standard state of 1 M. For typical physiological conditions of 10^{-3} M ATP, 10^{-4} M ADP, and 10^{-2} M P_i (inorganic phosphate), this gives a much more negative value of ΔG; namely, $-13{,}200 \text{ cal mol}^{-1}$ (equation 11):

$$\begin{aligned}\Delta G &= \Delta G^{\circ\prime}(\text{pH } 7) + RT \ln \frac{[\text{ADP}][\text{P}_i]}{[\text{ATP}]} \\ &= -8500 + 1360 \log \frac{10^{-4} \times 10^{-2}}{10^{-3}} = -13{,}200\end{aligned} \tag{11}$$

Most of this difference arises from the fact that two molecules of dilute products are formed from one molecule of reactant. The value of 1360 in equation 11 is RT times the conversion factor from natural to base 10 logarithms (2.303) for a temperature of 25°C. It is a convenient number to remember. At 37°C, this value is 1420.

Figure 9-4 shows the relation between concentrations and a logarithmic term, the pH, for the ionization of an acid:

$$\text{HA} \overset{K_a}{\rightleftharpoons} \text{H}^+ + \text{A}^- \tag{12}$$

This relation is described by the well-known Henderson-Hasselbach equation:

$$\text{pH} = \text{p}K_a + \log \frac{[\text{A}^-]}{[\text{HA}]} \tag{13}$$

which relates the pH to the $\text{p}K_a$ of an acid and gives rise to the titration curve in Figure 9-4. Again, when the concentrations of the acid and the base are equal, the concentration term cancels out and the variable term (the pH) becomes

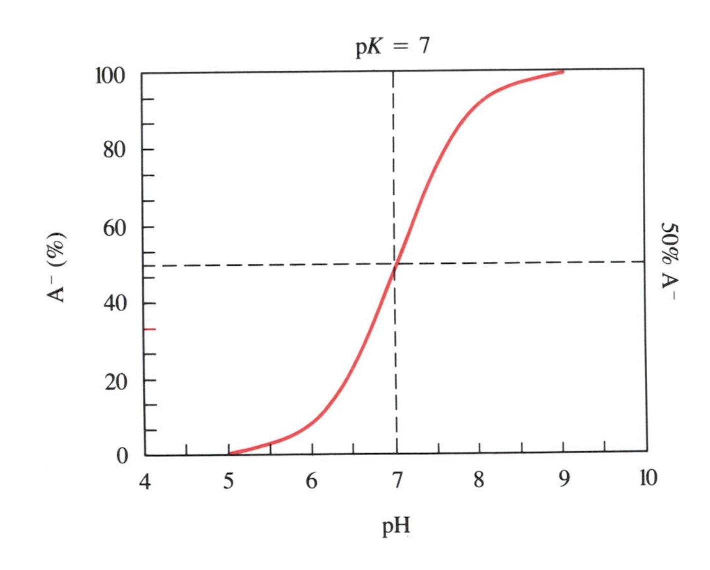

Figure 9-4

The Dependence of pH on the Fraction of Base, A^-. The relation between pH and the fraction of A^- for the dissociation of an acid HA with a $\text{p}K_a$ value of 7 is plotted in the figure (see equation 13). Note that the concentrations of A^- and HA are equal when the pH is the same as the $\text{p}K_a$ (7 in the figure). For an acid with a $\text{p}K_a$ value other than 7, the curve will be shifted on the horizontal axis.

equal to the constant term (the pK of the acid). This graph has the same form as the one in Figure 9-3, except that it is rotated by 90°.

Oxidation-reduction reactions are described by equation 14,

$$E = E_0'(\text{pH } 7) + \frac{RT}{n\text{F}} \ln \frac{[\text{oxidant}]}{[\text{reductant}]} \tag{14}$$

in which E is the potential, in volts. Equation 14 is plotted in Figure 9-5. Again, the reaction is described by a sigmoid curve that relates the ratio of the concentrations of oxidant and reductant to the oxidation-reduction potential, E, as described by equation 14. The standard term, E_0' (pH 7), is called the *standard oxidation-reduction potential*. The concentration term here includes the number of electrons that are transferred in the oxidation-reduction reaction, n, and the Faraday, F, which is equal to 23,062 cal V^{-1}, or 96,486 J V^{-1}. At 25°C, the concentration term in this equation reduces to approximately 0.06 log {[oxidant]/[reductant]} for a one-electron transfer. Note that the curve is flatter when two electrons are transferred ($n = 2$). The larger number of electrons provides a sort of *cooperativity*. Two electrons must be transferred in order for the reaction to occur, and the change in potential is more abrupt for a given change in concentration ratio.

The Faraday is important because it is the conversion factor between the oxidation-reduction scale and the Gibbs energy scale. For example, a two-electron oxidation reaction with a favorable change in E_0' of 0.3 V has a Gibbs energy change of $2 \times -0.3 \times 23{,}062 = -13{,}840$ cal $\text{mol}^{-1} = -13.84$ kcal mol^{-1}.

The oxidation-reduction potentials of two compounds provide a measure of the driving force for electron transfer between the compounds and determine the extent to which electrons can be transferred from one to the other compound. Thus, they are important for the electron-transfer reactions in the oxidation of metabolites and in photosynthesis, which are described in Chapters 22 and 23, respectively.

These three equations, for ΔG, pH, and E, have essentially the same form. Each of them provides a measure of the dependence of a *variable* driving force or potential (ΔG, pH, or E) on a constant *standard* term ($\Delta G^{\circ\prime}$, pK, or E_0') and a *logarithmic* term for the concentrations of reactants and products. These relations are summarized in Table 9-1.

The Gibbs energy that is available from the hydrolysis of ATP and other compounds must also be corrected for changes in the concentrations of reactants and products that are brought about by complexation with metals.

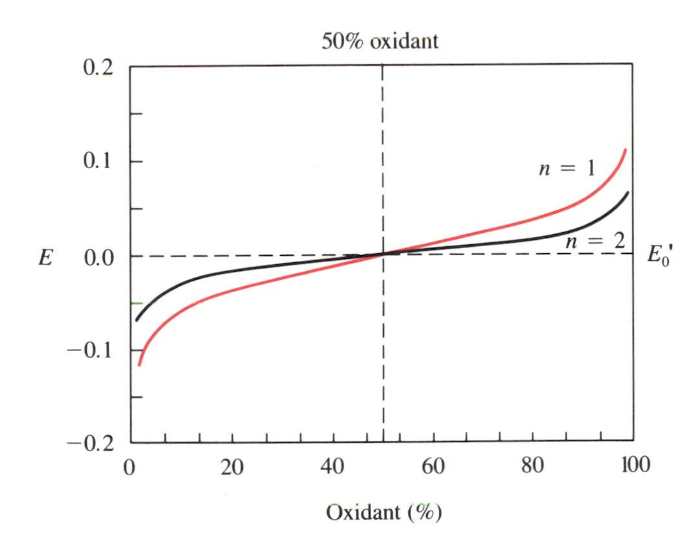

Figure 9-5

The Dependence of E on the Fraction of Oxidant. The lines are calculated from equation 14 for an oxidation-reduction reaction with a value of E_0' of zero. The two curves refer to redox reactions in which $n = 1$ (one-electron) or $n = 2$ (two-electrons). The curves will be displaced on the vertical axis if the E_0' value for a reaction is other than zero.

Table 9-1

Comparative concentration dependencies of Gibbs energy changes, pH changes, and oxidation-reduction potentials

Variable		Standard Term		Logarithmic Concentration Term
ΔG	=	$\Delta G^\circ, \Delta G^{\circ\prime}$	+	$RT \ln \frac{[\text{Products}]}{[\text{Reactants}]}$
Chemical potential		$-RT \ln K$ $-RT \ln K_{app}$		
pH	=	pK_a	+	$\log \frac{[A^-]}{[HA]}$
Hydrogen ion potential		$-\log K_a$		
E	=	$E_0{}'$	+	$\frac{RT}{nF} \ln \frac{[\text{Oxidant}]}{[\text{Reductant}]}$
Oxidation-reduction potential				

The most important of these is magnesium, which complexes significantly with ATP and ADP at physiological concentrations. The value of $\Delta G^{\circ\prime}$ (pH 7) for the hydrolysis of ATP in the presence of 1 mM magnesium ions is -7.6 kcal mol^{-1}. This is significantly less negative than the value of -8.5 kcal mol^{-1} in the absence of magnesium ion.

BOX 9-1

CONCENTRATIONS AND ACTIVITIES

The great majority of equilibrium constants used in biochemistry, and the corresponding Gibbs energies, are based on the *concentrations* of the reactants and products. They often refer to standard conditions of 0.1 M potassium chloride and 25°C. However, these equilibrium constants often change when the ionic strength is changed, especially if the reactants are charged.

Physical chemists use more-general equilibrium constants and Gibbs energy changes that are defined by the *activities* rather than the concentrations of the reactants and products. The activity is equal to the concentration of a compound at zero ionic strength, as the concentration of the compound approaches zero; this corresponds to an ideal solution. The relation between the activity, a, and the concentration, c, of a compound, b, under a particular set of experimental conditions is given by

$$a_b = f_b c_b$$

in which f is an *activity coefficient* (some disrespectful biochemists call it a fudge factor). If the equilibrium constant is defined in terms of activities, rather than concentrations, the activity coefficients are the terms that are necessary in order to make the observed concentrations of the reactants at equilibrium fit that equilibrium constant under a particular set of reaction conditions.

We have seen in Chapter 8 that activity coefficients are useful for describing both the denaturation of proteins in the presence of urea and the solubility of compounds in mixed solvents. Activities and activity coefficients can be determined by a number of physical techniques, in addition to measurements of equilibrium constants and solubilities. They contribute to the chemical potential, or tendency to react, or escaping tendency of a compound under a particular set of conditions.

For charged molecules at low or moderate ionic strengths, the activity coefficients can be calculated from one of the several forms of the Debye-Hückel equation. These equations provide a measure of the stabilizing effect of ions on charged molecules in solution from electrostatic interactions. An approximate form of the equation for dilute solutions in water is

$$\log f_{\pm} = \frac{(Z^+Z^-)0.50\sqrt{\mu}}{1 + \sqrt{\mu}}$$

in which $f_{\pm}$ is the mean activity coefficient of the ions, Z is the charge of each ion, and μ is the ionic strength.

Experimental and General Gibbs Energies: $\Delta G^{\circ\prime}$ (pH 7) and ΔG°

The quantity $\Delta G^{\circ\prime}$(pH 7) is an *apparent,* or *experimental,* Gibbs energy of hydrolysis at pH 7.0 that is extremely useful for describing the driving force for different reactions under physiological conditions. It is particularly useful for comparing different compounds and reactions under physiological conditions, as shown in Figure 9-2; and it is the quantity that is usually used by biochemists to compare reactions of different compounds. However, it is not a fundamental physical quantity and it tells us nothing about the driving force for reactions under different experimental conditions or about the reasons why different reactions have different values of $\Delta G^{\circ\prime}$(pH 7). In order to deal with these questions, it is essential to describe reactions by a more-general quantity, ΔG°, and to understand the difference between $\Delta G^{\circ\prime}$(pH 7) and ΔG°. These two standard Gibbs energy changes are defined as follows:

$\Delta G^{\circ\prime}$(pH x) is an experimental, or apparent, standard Gibbs energy change that is obtained from an experimental, or apparent, equilibrium constant, K_{exp} or K_{app}, at a particular pH value:

$$\Delta G^{\circ\prime}(\text{pH } x) = -RT \ln K_{app}$$

Both $\Delta G^{\circ\prime}$(pH x) and K_{app} are defined by:

1. The *total, stoichiometric* concentrations of each reagent, including all ionic species of that reagent (e.g., $[RCOO^-] + [RCOOH]$).
2. A chemical equation that does not include any hydrogen ions and hydroxide ions that may participate in the reaction; that is, it is not balanced with respect to H^+ or HO^- ions.

This convention is *specific*. It is true only at the specified pH value, if H^+ or HO^- ions are included in the reaction. It is based upon the *total* concentration of each reagent, as might be estimated experimentally by an analytical procedure for measuring total phosphate concentration, for example. It is useful for comparing standard Gibbs energies of hydrolysis of different compounds under physiological conditions, as shown in Figure 9-2.

ΔG° (ΔF° in the older literature) is the standard Gibbs energy change that is obtained from the equilibrium constant K:

$$\Delta G^{\circ} = -RT \ln K$$

Both ΔG° and K are based on:

1. The concentrations of *particular ionic species* of all reagents.
2. A chemical equation that is balanced with respect to mass and charge, including any H^+ or HO^- ions that are included in the equilibrium equation.

This convention is *general* and is true at any pH value. However, it cannot be used directly for comparing the free energies of hydrolysis for different compounds under physiological conditions.

Both $\Delta G^{\circ\prime}$(pH x) and ΔG° refer to dilute aqueous solutions and take the activity of liquid water as 1.0. Hybrids of $\Delta G^{\circ\prime}$(pH x) and ΔG° have sometimes been used. However, they are confusing and are not recommended.

The usual procedure for dealing with these quantities experimentally is, first, to measure the equilibrium constant and the value of $\Delta G^{\circ\prime}$ at some pH value, by using the total, stoichiometric concentrations of reactants and products.

This value can then be converted into a more-general ΔG° that is defined by particular ionic species of each reactant and includes any H^+ or HO^- ions that are involved in the equilibrium.

BOX 9-2

INTERCONVERSION OF $\Delta G^{\circ\prime}$ (pH x) AND ΔG°

The Gibbs energy change for a reaction

$$A + B \rightleftharpoons C + D$$

when the reactants and products are not at the standard state concentration, ΔG, can be obtained from ΔG° according to the following (equation 6 in the text):

$$\Delta G = \Delta G^{\circ} + RT \ln \frac{[C][D]}{[A][B]} \quad (6)$$

in which the reactants and products include hydrogen and hydroxide ions and particular ionic species are specified. This equation can be used to calculate the standard Gibbs energy according to one convention if the value for another convention is known. Analogous equations can be written for other reactions.

For example, the hydrolysis of glycine ethyl ester can be described by:

$$\Delta G = \Delta G^{\circ} + RT \ln \frac{[H_3N^+CH_2COO^-][H^+][HOEt]}{[H_3N^+CH_2COOEt]} \quad (6a)$$

This equation describes the reaction to give the ionized carboxylate group and a hydrogen ion as products and can be used to calculate ΔG at any pH.

The Gibbs energy change of a reaction at a particular pH value, x, can also be obtained from $\Delta G^{\circ\prime}$ (pH x)

$$\Delta G = \Delta G^{\circ\prime}(\text{pH } x) + RT \ln \frac{[C]_{tot}[D]_{tot}}{[A]_{tot}[B]_{tot}}$$

in which hydrogen and hydroxide ions are omitted and the concentrations refer to total, stoichiometric concentrations of the reactants and products.

$\Delta G^{\circ\prime}$ (pH x) and ΔG° are related by ΔG, which is the same for both conventions under a particular set of experimental conditions.

For example, the value of $\Delta G^{\circ\prime}$ for the hydrolysis of glycine ethyl ester at pH 8 and 37°C may be calculated from ΔG° (equation 6a) by solving for ΔG under conditions in which $\Delta G = \Delta G^{\circ\prime}$ (pH 8); that is, under standard state conditions for $\Delta G^{\circ\prime}$ (pH 8). This is done by inserting the activity of hydrogen ions at pH 8 and the fraction of 1 M total reactants and products that are in the ionic form specified by equation 6a at pH 8:

$$\begin{aligned}\Delta G &= \Delta G^{\circ\prime}(\text{pH } 8) \\ &= \Delta G^{\circ} + RT \ln \frac{(1.0)(10^{-8})(1.0)}{(0.50)} \\ &= 1370 + 1420(-8 + 0.3) = -9560\end{aligned}$$

Note that 10^{-8} M is inserted for the hydrogen ion activity and 0.5 for the fraction of glycine ethyl ester that is present as the cation, because the pK_a of the ammonium group of glycine ethyl ester is 8.0. No correction is necessary for glycine because the pK_a of glycine is 9.7 and almost all of the 1 M glycine is present as the zwitterion at pH 8. This expression gives a value of $-9{,}560$ cal mol^{-1} for $\Delta G^{\circ\prime}$ (pH 8) at 37°C.

The value of $\Delta G^{\circ\prime}$ at a different pH value can then be obtained by calculating $\Delta G^{\circ\prime}$ for this pH value from ΔG° and the fraction of each reactant that is in the specified ionic form.

An example of this procedure is given in the following section.

Ester Hydrolysis: A Case Study

The history of the energetics of ester hydrolysis provides a striking example of the importance of understanding the correct use of Gibbs energies. On the one hand, it was shown in the nineteenth century that the equilibrium constant for the hydrolysis of an ester is close to 1.0. Therefore, one might reasonably conclude that there is no large driving force for the hydrolysis of an ester. On the other hand, it was found in the middle of the twentieth century that esters of amino acids and tRNA are high-energy compounds, with a large driving force for hydrolysis. This apparent contradiction caused considerable confusion and, in fact, delayed the development of our understanding of protein synthesis. It is important to understand the reasons for this situation, both for understanding the use of Gibbs energies in practice and for understanding the reasons that some compounds are "high energy."

The problem can be described by using the Gibbs energy change for the hydrolysis of glycine ethyl ester as an example. The standard Gibbs energy change for the hydrolysis of glycine ethyl ester varies from $+560$ cal mol^{-1}, when it is described according to one convention, to -6860 cal mol^{-1}, when it is described according to a different convention. According to the first convention, this hydrolysis is a slightly unfavorable reaction; whereas, according to the second convention, glycine ethyl ester is an energy-rich compound with a very favorable Gibbs energy change for hydrolysis. Let us see why these Gibbs energies are so different.

We will examine four different conventions that describe the standard Gibbs energy change for the hydrolysis of glycine ethyl ester. All four are correct ways to describe the reaction and give the same value of ΔG for the reaction under the same reaction conditions. However, different conventions can be useful for describing the reaction under different conditions.

Convention 1. K and ΔG° Are Based on the Molarity of All Reactants. According to this convention, water is treated just like any other reactant. For a reaction in liquid water, the concentration of liquid water, 55.5 M, is used. The reaction can be described by equation 15.

$$H_3N^+\!-\!CH_2\!-\!\overset{O}{\overset{\|}{C}}\!-\!OEt + H_2O \underset{}{\overset{K_1}{\rightleftharpoons}} H_3N^+\!-\!CH_2\!-\!\overset{O}{\overset{\|}{C}}\!-\!OH + EtOH \qquad (15)$$

At 39°C, the equilibrium constant of the reaction is 0.4, which corresponds to a value of $\Delta G^\circ = +560\,\text{cal mol}^{-1}$, or $+2350\,\text{J mol}^{-1}$, at 39°C (equation 16).

$$K_1 = \frac{[\text{RCOOH}][\text{HOEt}]}{[\text{RCOOEt}][\text{HOH}]} = 0.4 \qquad (16)$$

It is reasonable that the replacement of ethanol by water should have an equilibrium constant close to 1, because HOH and HOEt are structurally very similar. This convention is the one that is most often used by chemists and is the convention that was used for measurement of the equilibrium constant in a mixed solvent of ethanol and water.

Convention 2. In Dilute Aqueous Solution, the Activity of Water Is Taken to Be 1.0. For reactions in dilute aqueous solution, it is convenient to describe water in terms of an *apparent* concentration or activity of 1.0 instead of 55.5 M. The equilibrium constant then describes the concentrations of the reactants that are likely to be measured experimentally. The concentration of water is simply removed from the equilibrium expression, equation 17,

$$K_2 = \frac{[\text{RCOOH}][\text{HOEt}]}{[\text{RCOOEt}]} = K_1[\text{HOH}] = 55.5\,\text{M} \times K_1 = 22\,\text{M} \qquad (17)$$

which shows that this equilibrium constant, K_2, is equal to 55.5 K_1 or 22 M. This corresponds to a considerably more favorable value of $\Delta G^\circ{}_2 = -1900$ cal mol^{-1}, or $-8000\,\text{J mol}^{-1}$. The difference between this value and that of Convention 1 corresponds to the molarity of water, 55.5 M, and is equal to $1420 \times \log_{10} 55.5 = -2480\,\text{cal mol}^{-1}$. The high concentration of water drives the hydrolysis toward completion in dilute aqueous solutions. This convention is followed for the great majority of biochemical reactions, because they are carried out in dilute aqueous solutions. However, the convention that is used is not always specified, and this can cause confusion.

Convention 3. Hydrolysis to Give Ionized Products. It is equally valid, and sometimes more useful, to express the hydrolysis reaction in terms of the ionized products, RCOO^- and H^+, according to the reaction

$$H_3N^+\!-\!CH_2\!-\!\overset{O}{\overset{\|}{C}}\!-\!OEt + H_2O \overset{K_3}{\rightleftharpoons} H_3N^+\!-\!CH_2\!-\!\overset{O}{\overset{\|}{C}}\!-\!O^- + H^+ + HOEt \qquad (18)$$

The equilibrium constant, K_3, for this reaction is

$$K_3 = \frac{[RCOO^-][H^+][HOEt]}{[RCOOEt]} = K_2K_a = 0.11\ M^2 \tag{19}$$

which equals $K_2K_a = 0.11\ M^2$, in which K_a is the ionization constant for the acid:

$$K_a = \frac{[RCOO^-][H^+]}{[RCOOH]} \tag{20}$$

The standard Gibbs energy change for the reaction according to Convention 3 is $\Delta G^\circ{}_3 = -RT \ln K_3 = +1360\ \text{cal mol}^{-1}$, or $+5720\ \text{J mol}^{-1}$. The small equilibrium constant and the positive value of ΔG° show that the reaction is unfavorable. This is because the products are $RCOO^-$ and H^+, and the ionization of RCOOH to $RCOO^-$ and H^+ is unfavorable at a standard state of 1 M for all reactants and products.

Convention 4. $\Delta G^{\circ\prime}$ at a Given pH Only, Based on Ratios of Total Reagent Concentrations. Conventions 1 through 3 give values of ΔG° that are independent of the pH, because they include specified ionic species and any H^+ ions that are given off in the reaction. However, we have seen that it is useful to compare the Gibbs energies for different reactions under the same conditions. This can be done by making use of a pH-dependent Gibbs energy, $\Delta G^{\circ\prime}$(pH x), which is a standard Gibbs energy change that is correct only at a particular pH value. It is based upon the *total* concentration of every reactant, including all ionic species of that reactant.

At pH 6, glycine exists almost entirely as the zwitterion, whereas the ester exists as the cation. The apparent equilibrium constant is

$$K_{app} = \frac{[H_3N^+CH_2COO^-][HOEt]}{[H_3N^+CH_2COOEt]} = \frac{K_3}{[H^+]} = \frac{0.11}{10^{-6}} = 1.1 \times 10^{+5} \tag{21}$$

This equilibrium constant is the same as that for Convention 3, except that the hydrogen ion is omitted. The value of K_{app} is then $K_3/10^{-6} = 1.1 \times 10^{+5}$ M at pH 6. This gives a value of $\Delta G^{\circ\prime}$ (pH 6) of $-6860\ \text{cal mol}^{-1}$, or $-28{,}800\ \text{J mol}^{-1}$. At pH 5, the equilibrium constant would be $1.1 \times 10^{+4}$ M.

These different conventions are summarized in Table 9-2. They show how four different conventions give four widely different values for the standard free energy of hydrolysis. Each of them is valid and is useful for a different purpose.

This and the example at pH 8 (see Box 9-2) show why glycine ethyl ester and other amino acid esters are high-energy compounds under physiological conditions. We will see in Chapter 10 that the tRNA ester of a peptide donates its acyl group to the amino group of the tRNA ester of an amino acid to form a new peptide bond that is much more stable (equation 2); its Gibbs energy of hydrolysis is much less favorable. Thus, the synthetic reaction is strongly downhill, with a negative change in Gibbs energy, so that the synthesis of the peptide is driven forward and is essentially irreversible. One of the most-important functions of ATP in biochemistry is to make synthetic reactions and other processes irreversible, so that they can take place readily and can be subject to control.

In summary, we can identify three factors that make the standard Gibbs energy for the hydrolysis of esters much more negative under physiological conditions than might be expected from equilibrium constants that were measured under other conditions, such as the conditions that might be used by a chemist to determine the equilibrium constant.

1. The high concentration of liquid water, 55 M, tends to drive the hydrolysis reaction forward and gives a negative contribution to ΔG°. This is described by Convention 2, as compared with Convention 1.

Table 9-2

Equilibrium constants and standard Gibbs free energy changes for the hydrolysis of glycine ethyl ester according to different conventions

Convention 1: $[H_2O] = 55\,M$

$$\frac{[RCOOH][HOEt]}{[RCOOEt][H_2O]} = K_1 = 0.4 \qquad \Delta G^\circ = +560\ \text{cal mol}^{-1} = +2350\ \text{J mol}^{-1}$$

Convention 2: $a_{H_2O} = 1.0$

$$\frac{[RCOOH][HOEt]}{[RCOOEt]} = K_2 = 22\,M \qquad \Delta G^\circ{}_2 = -1900\ \text{cal mol}^{-1} = -8000\ \text{J mol}^{-1}$$

Convention 3: Ionized Products

$$\frac{[RCOO^-][H^+][HOEt]}{[RCOOEt]} = K_3 = 0.11\,M^2 \qquad \Delta G^\circ{}_3 = +1360\ \text{cal mol}^{-1} = +5720\ \text{J mol}^{-1}$$

Convention 4: At pH 6

$$\frac{[RCOO^-][HOEt]}{[RCOOEt]} = K_{app}(\text{pH } 6.0) = 1.1 \times 10^{+5} \qquad \Delta G^{\circ\prime} = -6860\ \text{cal mol}^{-1} = -28{,}800\ \text{J mol}^{-1}$$

2. Ionization of an acid product *pulls* the reaction toward hydrolysis at physiological pH. This can be thought of in two ways:

First, the concentration of RCOOH is very low at physiological pH. Therefore, ΔG calculated from equation 17 and $\Delta G^\circ{}_2$ of Convention 2 is very negative.

Second, the concentration of H^+ is very low at physiological pH. Therefore, ΔG calculated from equation 18 and $\Delta G^\circ{}_3$ of Convention 3 is very negative.

3. The value of ΔG for the hydrolysis of a compound that gives two products will be much more favorable in dilute solution under physiological conditions than at a standard state of 1 M, as described above for ATP (equation 11).

What Makes a Reaction Go? Enthalpy and Entropy

For most of recorded history, it was believed that the factor that determines whether a reaction will proceed spontaneously or not is the evolution of heat. Explosions and forest fires are not reversible.

Nevertheless, this belief is wrong. One of the important developments of the nineteenth century was the demonstration that some reactions proceed spontaneously with the *uptake* of heat and the conclusion that this can be accounted for by the entropy, or probability, of the reaction. One example of an entropy-driven reaction is the denaturation of a protein (Chapter 7); others are the dissolving of a salt in water and the melting of ice. The driving force for a reaction is not the evolution of heat, or the change in enthalpy, but the change in Gibbs energy. The change in Gibbs energy is the basis of everything that has been described in this chapter up to this point. The Gibbs energy is often called the *free energy* because it is the amount of energy that is available to do work.

The Gibbs energy change is made up of the heat, or *enthalpy*, change of a reaction and the *entropy* term, $T\Delta S$:

$$\Delta G = \Delta H - T\Delta S \tag{22}$$

Physical chemists are likely to write equation 22 in the form of

$$\Delta H = \Delta G + T\Delta S \tag{23}$$

and interpret the enthalpy change in terms of changes in Gibbs energy and entropy. The choice depends on what is most important. Biochemists are usually interested in the driving force of a reaction—namely, ΔG. The enthalpy and entropy changes are interesting and important, but in biochemical systems it is often difficult or impossible to interpret experimental values of these parameters in physical terms. When a solute is added to an aqueous solution or a protein molecule is perturbed, there are often large changes in ΔH and $T\Delta S$ that offset each other; therefore, there is little change in ΔG. Changes in the arrangement of water molecules in the solvent or around the protein may cause large, compensating changes in ΔH and $T\Delta S$ but have little influence on the driving force of the reaction, ΔG. Because almost anything that is added to an aqueous solution is likely to cause such changes in the "structure" of the solvent, it is often difficult to ascribe observed changes in ΔH and $T\Delta S$ to changes in the physical properties of a protein.

The enthalpy change of a reaction may be determined in two ways:

1. The amount of heat that is evolved in the course of a reaction can be measured with a calorimeter. Lavoisier used a calorimeter to measure the heat given off by a guinea pig in a closed system containing ice. The heat given off by the guinea pig caused melting of the ice to water, which was measured. With modern calorimeters, more-exact measurements can determine the release of extremely small amounts of heat.

2. The enthalpy change can be determined by measuring the temperature dependence of the equilibrium constant of a reaction according to the van't Hoff equation:

$$\frac{d \ln K}{dT} = -\frac{\Delta H}{RT^2} \tag{24}$$

This equation states that the change of the natural logarithm of the equilibrium constant with temperature provides a measure of ΔH; R is the gas constant. The value of ΔH is usually obtained from a logarithmic plot of the equilibrium constant against the reciprocal of the temperature according to

$$\frac{d \ln K}{d(1/T)} = -\frac{\Delta H}{R} \tag{25}$$

The slope of the resulting line gives $-\Delta H/R$.

If the data are obtained carefully it is often found that the line is not straight. This means that the enthalpy change is itself temperature dependent, so that there is a significant heat capacity term, ΔC_p.

$$\Delta C_p = \frac{d\Delta H}{dT} \tag{26}$$

The value of ΔC_p can be obtained from careful measurements of the temperature dependence of ΔH.

It is sometimes convenient to determine ΔH (in cal mol^{-1}) directly from equation 27

$$\log\left(\frac{K_2}{K_1}\right) = \frac{-\Delta H}{4.575}\left(\frac{1}{T_2} - \frac{1}{T_1}\right) \tag{27}$$

by using values of the equilibrium constant determined at temperatures T_1 and T_2. Energies of activation of a reaction, E_a, may be obtained similarly from the slopes of logarithmic plots of the rate constant against $1/T$ or from an integrated equation corresponding to equation 27.

We encountered entropy earlier as an important driving force for catalysis by enzymes and for the denaturation of proteins. Entropy is one of the most-difficult, and one of the most-important, subjects to understand in all of physical chemistry. Entropy has variously been described as disorder, chaos, a measure of the number of states in which a system can exist, a state with a high probability, and the inability to specify the exact position and properties of a particle. In statistical mechanics, it is defined according to the expression

$$S = k \ln w \tag{28}$$

in which k is Boltzmann's constant and w is the number of equivalent states that are available to the system.

The entropy increases when the products of a particular reaction are more probable, or can exist in a larger number of equivalent states, compared with the reactants. It also increases when there is no change of enthalpy, but work has to be done to reverse the reaction. For example, the entropy increases when a salt solution is diluted with water or when solutions of potassium chloride and sodium chloride are mixed. Both of these processes are difficult to reverse; they can be reversed only by expending Gibbs energy.

The denaturation of proteins provides a dramatic example of an increase of entropy in biochemistry. As described in Chapter 7, the denaturation of ribonuclease has the enormous values of $\Delta H = 95\,\text{kcal mol}^{-1}$ and $\Delta S = 285\,\text{cal mol}^{-1}\,\text{K}^{-1}$ (or 285 e.u.). These values are very much larger than those observed for most reactions in aqueous solution. They show that the increased probability of an unfolded protein provides an enormous driving force toward denaturation. This is counteracted in the native protein by an extremely unfavorable enthalpy change for the unfolding reaction. The positive entropy change arises from the extremely large number of different structures that the unfolded protein can take up, compared with the native protein.

It may be more informative to consider states of low entropy, or reactions that proceed with a negative entropy change. A loss of entropy is required for the precise location of two molecules next to each other, such as two substrates that are bound in exactly the correct position to react at the active site of an enzyme. It is extremely improbable that two molecules would find themselves in this particular position by chance alone, and the negative entropy change is a measure of this improbability. As discussed in Chapter 5, this is the basis for the very large rate accelerations of intramolecular reactions and for a large fraction of the rate acceleration that is brought about by enzymes.

Information is a form of negative entropy. An important example of this is the particular arrangement of nucleic acid bases that carries genetic information from one generation to the next. Schrödinger has written a fascinating book, *What Is Life?*, that is concerned largely with entropy. The conclusion is that life is an extraordinarily improbable state with a high degree of order and a high information content. It requires a large amount of energy to construct and maintain this state of low entropy. Death, on the other hand, represents a large increase in entropy and has a notably high probability; the expenditure of large amounts of energy by the human species has not produced a way to avoid it.

Energy-rich Compounds

Table 9-3 shows some of the compounds that can participate in biochemical group-transfer reactions. The compounds are arranged in groups according to

Table 9-3

"Energy rich" compounds that may undergo group transfer

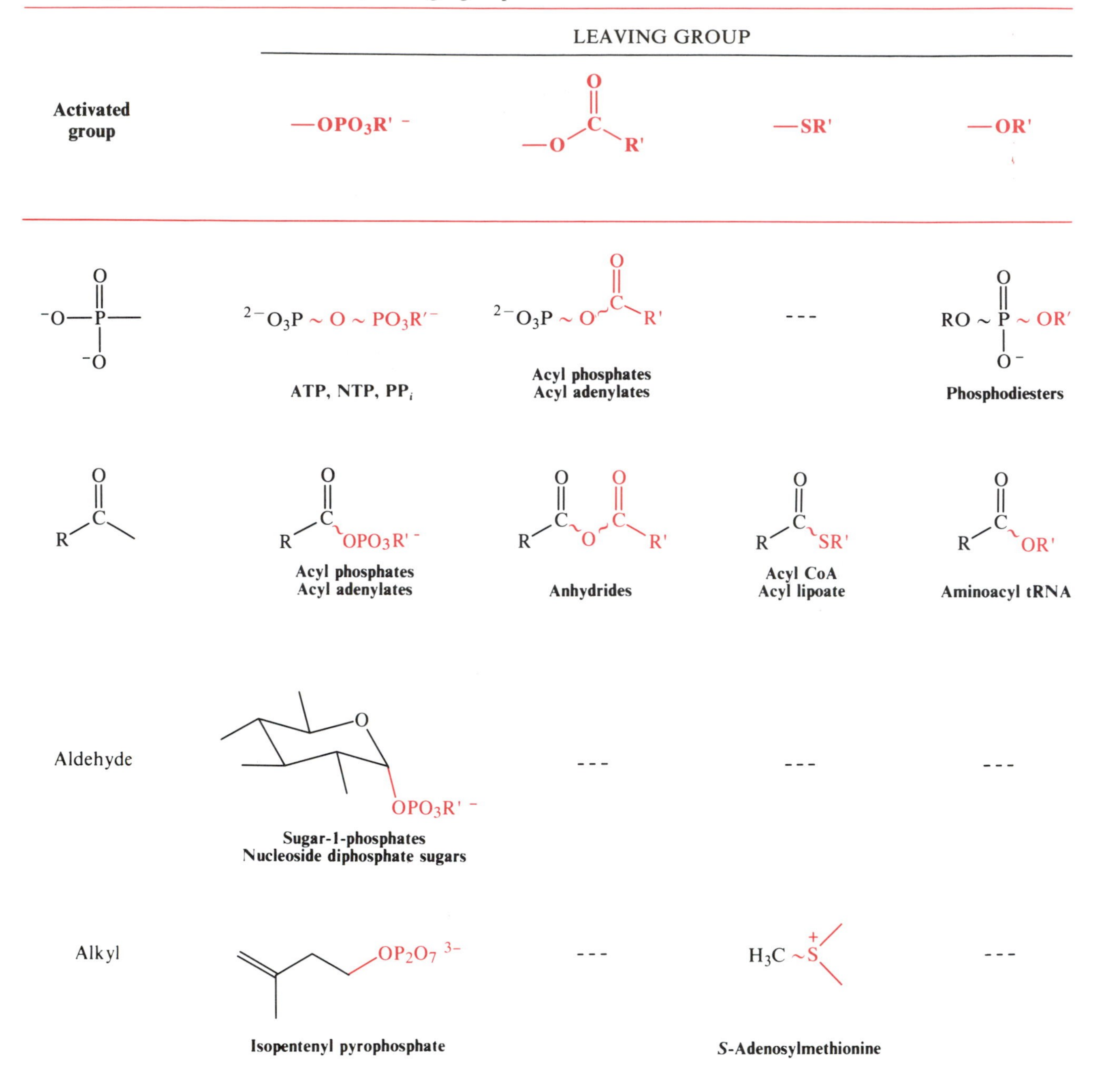

Activated group	LEAVING GROUP: $—OPO_3R'^-$	$—O—C(=O)—R'$	$—SR'$	$—OR'$
$^{-}O—P(=O)(O^-)—$	$^{2-}O_3P \sim O \sim PO_3R'^-$ **ATP, NTP, PP$_i$**	$^{2-}O_3P \sim O—C(=O)—R'$ **Acyl phosphates Acyl adenylates**	---	$RO \sim P(=O)(O^-) \sim OR'$ **Phosphodiesters**
$R—C(=O)—$	$R—C(=O) \sim OPO_3R'^-$ **Acyl phosphates Acyl adenylates**	$R—C(=O) \sim O—C(=O)—R'$ **Anhydrides**	$R—C(=O) \sim SR'$ **Acyl CoA Acyl lipoate**	$R—C(=O) \sim OR'$ **Aminoacyl tRNA**
Aldehyde	$OPO_3R'^-$ **Sugar-1-phosphates Nucleoside diphosphate sugars**	---	---	---
Alkyl	$OP_2O_7^{3-}$ **Isopentenyl pyrophosphate**	---	$H_3C \sim S^+$ ***S*-Adenosylmethionine**	---

their oxidation state. Derivatives of acids are generally the most energy rich and reactive. Both the energy-rich nature and the reactivity tend to decrease as the state of oxidation decreases.

Most of these compounds are of great importance in metabolism and the functioning of a cell because they can donate groups for the synthesis of other molecules. They have a high group *potential,* so that group transfer is likely to be thermodynamically favorable. It is important to know why these compounds are energy rich, but our understanding of the reasons why particular compounds are effective group donors is incomplete at the present time. We have seen that, in some cases, large favorable free energies for hydrolysis under physiological

Table 9-3

LEAVING GROUP		
$-NH_2R'$	$H_2N{\cdots}C^{+}(NH_2){\cdots}N<$	**Other**
$^{2-}O_3P \sim NH_2R'$ **Phosphoramidate**	$^{2-}O_3P \sim HN{\cdots}C^{+}(NH_2){\cdots}N<$ **Phosphocreatine Phosphoarginine**	$^{2-}O_3P \sim O-C(COO^-)=CH_2$ **Phosphoenolpyruvate**
$R-C(=O)-NHR'$ **Peptides Gln, Asn**	$R-C(=O)-N$(imidazole) **Acylimidazole**	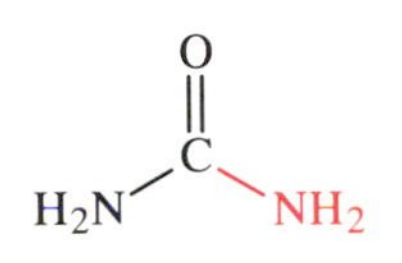 **Urea**
R'–ribose–N+ (nicotinamide, $CONH_2$), HO OH **NAD$^+$**	---	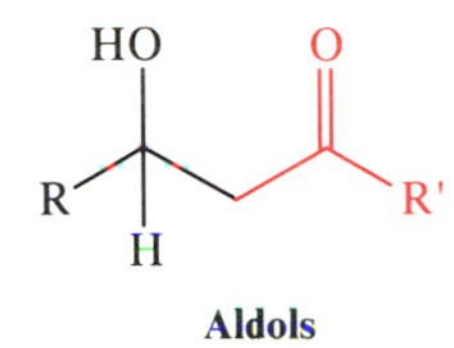 **Aldols**
H–N, N–CH_3 **N^5-Methyltetrahydrofolate**	---	---

conditions arise from the formation of fairly strong acids during hydrolysis, followed by dissocation of these acids with the release of a proton. The low acidity at neutral pH then pulls the reaction toward hydrolysis. The hydrolysis products are often stabilized by resonance more than the starting materials, as in the formation of carboxylate ions. This is one of the reasons for the high acidity of these compounds that is responsible for the large negative free energies of hydrolysis. However, it is difficult to make generalizations, because the list of factors that can influence the values of $\Delta G^{\circ\prime}$ (pH 7) includes the relative bond energies of the reactants and products, solvation effects, electrostatic repulsion and attraction, relative resonance stabilization of reactants and products, and the ionization or isomerization of reactants and products. For example, the hydrolysis of phosphoenolpyruvate is strongly favorable because the final product, pyruvate, is much more stable than the initial enol product:

$$^{2-}O_3PO{-}C(COO^-){=}CH_2 \xrightarrow{H_2O} {}^{2-}O_3POH + HO{-}C(COO^-){=}CH_2 \rightleftharpoons O{=}C(COO^-){-}CH_3$$

Summary

The energy that is required by living systems is provided by a "metabolic dynamo" that converts the energy that is available from the metabolism of foodstuffs into "energy rich" compounds, such as ATP and phosphocreatine. These compounds provide the energy that drives chemical reactions and other processes in the organism. What follows is a summary of how the energetics of these reactions can be described.

The useful energy that can be obtained from a reaction, ΔG, depends on the concentrations of reactants and products and the standard Gibbs energy change of the reaction, ΔG°, according to

$$\Delta G = \Delta G^\circ + RT \ln \frac{[C][D]}{[A][B]}$$

for the reaction

$$A + B \rightleftharpoons C + D$$

The value of ΔG° depends on the logarithm of the equilibrium constant:

$$\Delta G^\circ = -RT \ln K$$

There are two types of equilibrium constants and Gibbs (free) energies:

1. Experimental: K_{app}, K', $\Delta G^{\circ\prime}$
These hold at a given pH *only* and refer to the *total* stoichiometric concentrations of reactants and products (excluding H^+ and HO^-).
2. General: $K, \Delta G^\circ$
These are true at any pH value and refer to the concentrations (or activities) of *stated ionic species.* They are defined by a balanced equation and the equilibrium constant includes any protons that are taken up or released in the reaction.

Several equations (and several ΔG° values) are often possible; for example,

$$ATP^{4-} (+H_2O) \rightleftharpoons ATP^{3-} + HPO_4{}^{2-} + H^+ \quad \text{(a)}$$

$$ATP^{3-} (+H_2O) \rightleftharpoons ADP^{2-} + H_2PO_4{}^- \quad \text{(b)}$$

Equation a corresponds fairly closely to the composition of a solution at pH 8, equation b at pH 5. But either could be used to calculate ΔG at either pH by insertion of the correct concentrations (activities) of H^+ and the stated ionic species at that pH.

The value of $\Delta G^{\circ\prime}$ at a particular pH value can be obtained from the value at a different pH as follows:

Convert the observed $\Delta G^{\circ\prime}$ at one pH into ΔG°, which refers to concentrations of specified ionic species and includes H^+ or HO^-. ΔG° is good at any pH value.

Convert ΔG° back into $\Delta G^{\circ\prime}$ at the desired pH value.

The driving force for a reaction

$$A + B \rightleftharpoons C + D$$

at a given pH value is given by

$$\Delta G_{obs} = \Delta G^{\circ\prime} + RT \ln [C][D]/[A][B]$$

in which the *total* concentration of each reactant, including all ionic species, is used.

The quantity ΔG° describes a balanced equation, including any protons that are involved in the reaction, and refers to the activities or concentrations of particular ionic species.

There are two special cases:

1. At equilibrium: $\Delta G = 0$ and

$$\begin{aligned}\Delta G^\circ &= -RT \ln \frac{[C][D]}{[A][B]} \text{(at equilibrium)} \\ &= -RT \ln K\end{aligned}$$

2. In the standard state:

$$RT \ln \frac{1 \times 1}{1 \times 1} = 0;$$

$$\Delta G = \Delta G^\circ$$

The Gibbs energy, ΔG, enthalpy (or energy or heat), ΔH, and entropy, ΔS, of a reaction are related by

$$\Delta G = \Delta H - T\Delta S$$

The enthalpy can be measured by determining the equilibrium constant at several temperatures, according to

$$\frac{d \ln K}{d(1/T)} = -\frac{\Delta H}{R}$$

or

$$\log \frac{K_2}{K_1} = -\frac{\Delta H}{4.575}\left(\frac{1}{T_2} - \frac{1}{T_1}\right)$$

The pH and the oxidation-reduction potential of a solution, E (in volts), depend on a standard term and the logarithm of a concentration ratio:

$$\text{pH} = \text{p}K_a + \log \frac{[\text{Base}]}{[\text{Acid}]}$$

$$E = E_0'(\text{pH } 7) + \frac{RT}{nF} \ln \frac{[\text{Oxidant}]}{[\text{Reductant}]}$$

in which n is the number of electrons transferred and $F = 96{,}500$ coulombs is the faraday, the electrical charge of one mole of electrons.

ADDITIONAL READING

LIPMANN, F. 1941. Metabolic generation and utilization of phosphate bond energy. *Advances in Enzymology* 1:99–162.

FASMAN, G. 1975–1977. *Handbook of Biochemistry and Molecular Biology,* 3d ed. Physical Chemistry I, Recommendations. Cleveland: CRC Press, p. 93.

CORNELL, N. W., LEADBETTER, M., and VEECH, R. L. 1979. Effects of free magnesium concentration and ionic strength on equilibrium constants of the glyceraldehyde phosphate dehydrogenase and phosphoglycerate kinase reactions. *Journal of Biological Chemistry* 254: 6522–6527.

LAWSON, J. W. R., and VEECH, R. L. 1979. Effects of pH and free Mg^{2+} on the K_{eq} of the creatine kinase reactions and other phosphate hydrolyses and phosphate transfer reactions. *Journal of Biological Chemistry* 254: 6528–6537.

VEECH, R. L., LAWSON, J. W., CORNELL, N. W., and KREBS, H. A. 1979. Cytosolic phosphorylation potential. *Journal of Biological Chemistry* 254: 6538–6547.

MOROWITZ, H. J. 1978. *Foundations of Bioenergetics*. New York: Academic Press.

HINZ, H.-J. 1986. *Thermodynamic Data for Biochemistry*. New York: Springer-Verlag.

PROBLEMS

1. The equilibrium constant for the reaction of ADPβS and 1,3-diphosphoglycerate to give ATPβS and 3-phosphoglycerate is 400. What is $\Delta G^{\circ\prime}$ for this reaction?

2. What is ΔG for the hydrolysis of ATP in a muscle that has been contracting and cleaving ATP, so that the concentration of ATP is 10^{-4} M, ADP is 10^{-3} M, and P_i is 0.04 M?

3. Calculate the standard Gibbs free energy change for the hydrolysis of ATP at pH 6.5 based on the following ionization equilibria:

(a) $H_2PO_4^- \rightleftharpoons HPO_4^{2-} + H^+$ $\quad pK' = 6.5$

(b) $ATP^{3-} \rightleftharpoons ATP^{4-} + H^+$ $\quad pK' = 6.0$

(c) $ADP^{2-} \rightleftharpoons ADP^{3-} + H^+$ $\quad pK' = 6.0$

4. Calculate the approximate value of the equilibrium constant for the hydrolysis of ATP at pH 6.5.

5. Calculate $\Delta G^{\circ\prime}$ (pH 7) from K_2 *and* from K_3 for the hydrolysis of glycine ethyl ester.

6. Draw an approximate curve of $\Delta G^{\circ\prime}$ against pH between pH 0 and pH 9 for the hydrolysis of glycine ethyl ester.

7. Estimate $\Delta G^{\circ\prime}$ and K for the transfer of a phosphoryl group from ATP to glucose to give glucose-1-phosphate and ADP at pH 7.0, from Figure 9-2.

8. Estimate the equilibrium constant for the transfer of a phosphoryl group from phosphoenolpyruvate to ADP to give ATP at pH 7.0.

Protein Biosynthesis and Acyl Activation

10

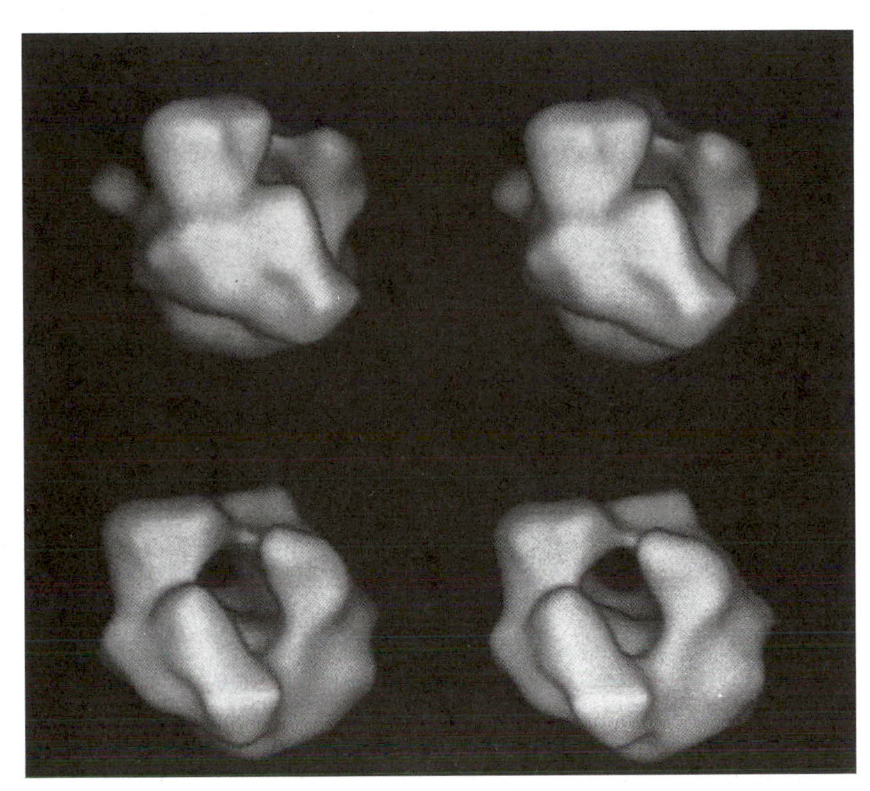

Reconstruction images of a ribosome from E. coli. This merged reconstruction consists of stereo pairs viewed from two directions (see Figure 10-12 for details). (Courtesy of P. Penczek, M. Radermacher, and J. Frank.)

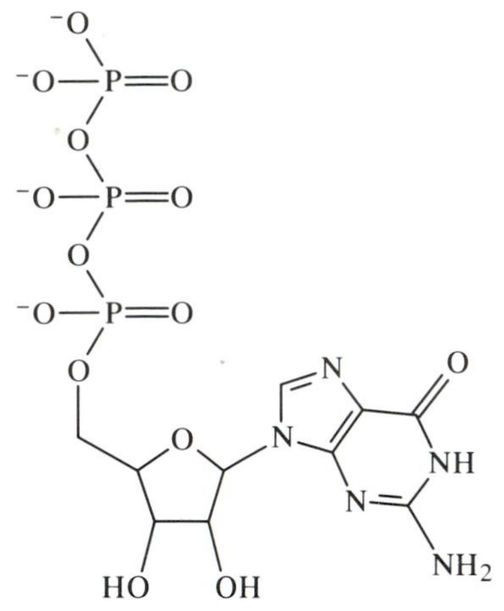

GTP

We now consider how cells assemble and link amino acids into proteins. The free energy of ATP and its analog guanosine triphosphate (GTP) is used in several ways to accomplish the biosynthesis of complex molecules such as proteins and DNA, and we shall point out at each stage of protein biosynthesis the importance of this free energy to the overall process. Some aspects of protein biosynthesis are better understood than others. For example, the chemical mechanism by which the protein biosynthetic machinery uses the energy of ATP to form the peptide bond is much better understood than that by which it uses the energy of GTP for other purposes. We shall begin by explaining how amino acids are activated and why activation is necessary for the formation of peptide bonds. We shall then explain how the genetic information in DNA specifies the amino acid sequence of a protein; how this information is transmitted to the protein synthesizing machinery of a cell; and, finally, how it is used to biosynthesize a protein.

Acyl Activation

In order to aminoacylate the amino terminus of a peptide or a protein, the carboxylate group of the amino acid must first be activated so that the synthesis will occur readily and irreversibly. One means of chemically activating a carboxylate group is by reaction with an anhydride, an acid halide or, as in equation 1a, a carboethoxyhalide. (The carboxylate oxygen atoms are shown in color to aid in tracing them).

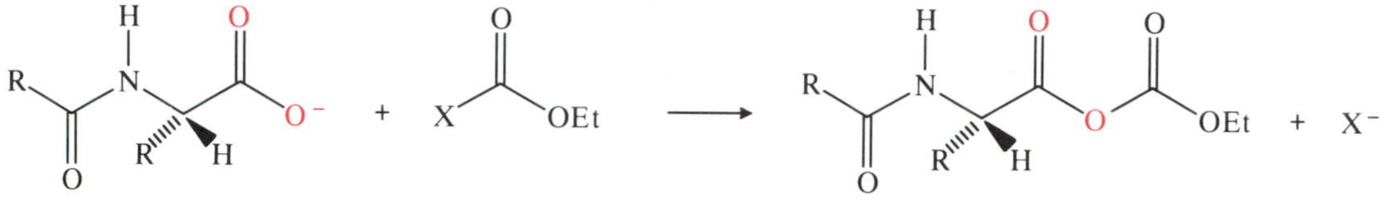

(1a)

The activated amino acid then can react with the amino group at the N-terminus of a peptide chain to form a new peptide bond. This is shown in equation 1b.

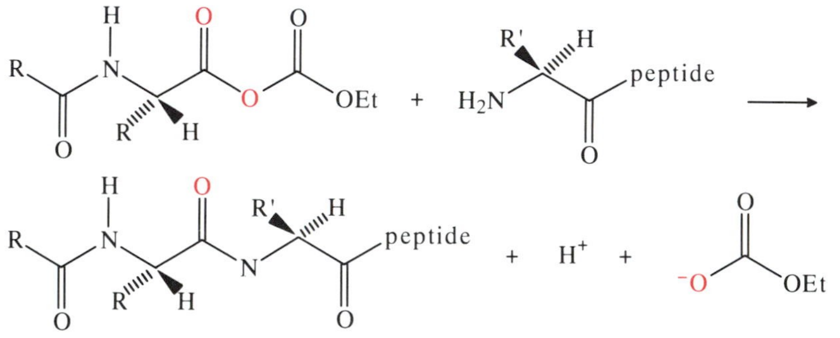

(1b)

The amino group of the activated amino acid must be protected to prevent self-condensation, and this protection is provided by the R—C═O group in equations 1a and 1b.

Why is activation required? It is required because an unactivated carboxylate group is unreactive toward nucleophilic attack. To make a peptide from the carboxylate group of an amino acid (RCO_2^-) by direct reaction with the amino group of another amino acid, it would be necessary to expel O^{2-} as the leaving group; but O^{2-} is much too poor (unstable) as a leaving group to allow a reaction to occur. Therefore, an activating group must be attached to an oxygen of the carboxylate group to make it a good leaving group. This is what happens in protein biosynthesis.

Another way to view activation is to think of peptide-bond formation as a dehydration. (The importance of ATP as a dehydrating agent was explained in

Chapter 2.) If there were no activating agent, one product of peptide synthesis would be a water molecule containing an oxygen atom from the carboxylate group of the amino acid. An oxygen atom is not easily removed from a carboxylic acid; and, if it were somehow removed, it would tend to add to the product to give back starting materials. The oxygen can be extracted irreversibly from the carboxylate group by forming a covalent bond to another group, a dehydrating agent such as an anhydride, to make the oxygen into a good leaving group. The oxygen atom will then appear as a part of the dehydrating agent, rather than as water (equations 1a and 1b).

Other carboxylic acids in cells are activated by exactly the same chemical principle exemplified by equations 1a and 1b, with ATP acting as the activator and dehydrating agent. As a phosphoanhydride, the principal role of ATP in metabolism is to act as a dehydrating agent to activate acyl groups, phosphoryl groups, and other molecules, to make their further reactions possible.

Activation by ATP is manifested both kinetically and thermodynamically. It plays a *kinetic* role by making reactions occur fast enough to be useful; it also plays a *thermodynamic* role by making reactions possible that would tend to proceed in the reverse direction were it not for the large driving force provided by this high-energy compound. ATP is a remarkably effective anhydride in cells because its spontaneous hydrolysis is very slow, occurring over a period of weeks, whereas other potential activating compounds, such as carboxylic acid anhydrides, are hydrolyzed within minutes.

Activation of Acetate

The conversion of acetate into acetyl CoA is a simple example of biological acyl group activation by ATP. Some bacteria contain an acetate kinase that catalyzes the phosphorylation of acetate by ATP to give acetyl phosphate, as shown in equation 2.

$$\text{AMP}-O-P(=O)(O^-)-O-P(=O)(O^-)-O^- \cdot Mg^{2+} + {}^-O-C(=O)-CH_3 \xrightleftharpoons{\text{Acetate kinase}} {}^{2-}O_3P-O-C(=O)-CH_3 + \text{MgADP} \qquad (2)$$

Acetyl phosphate is the anhydride of acetic acid and phosphoric acid. It is activated to transfer either an acetyl or a phosphoryl group to a nucleophilic reagent. Therefore, it can be considered either a high-energy acyl compound or a high-energy phosphoryl compound.

The acetate kinase reaction is reversible. In fact, the value of $\Delta G^{\circ\prime}$ for equation 2 is $+3\ \text{kcal mol}^{-1}$, and the reaction proceeds in the reverse direction outside a bacterial cell. Within a cell, however, the reaction does not exist in isolation. A second enzyme, phosphotransacetylase, is found in the same cells and catalyzes the transfer of the activated acetyl group to the sulfur atom of coenzyme A to give acetyl CoA, as shown in equation 3.

$$H_3C-C(=O)-O-PO_3^{2-} + \text{HS}-\text{CoA} \xrightleftharpoons{\text{Phosphotransacetylase}} H_3C-C(=O)-S-\text{CoA} + HOPO_3^{2-} \qquad (3)$$

The value of $\Delta G^{\circ\prime}$ for this reaction is $-3\ \text{kcal mol}^{-1}$, so that it proceeds in the forward direction. The sum of the $\Delta G^{\circ\prime}$ values for these two reactions is near zero; therefore, the two occurring together allow acetyl CoA to be formed. Further favorable reactions of acetyl CoA tend to decrease the intracellular concentration of acetyl CoA, and this makes the two reactions more favorable.

BOX 10-1

THE DISCOVERY AND IMPORTANCE OF COENZYME A

Coenzyme A is the active carrier of acetyl and other acyl groups in biological systems. The function of CoA in the activation of acyl groups for carbon–carbon bond formation is delineated in Chapter 2, and its central role in the metabolism of fatty acids and carbohydrates is described in Chapters 19 and 21. In fact, CoA is the most generally important activating group for acyl group transfer reactions in living cells.

The discovery of CoA came about as a result of the curiosity of F. Lipmann about how acyl groups could be activated for biochemical reactions. He chose to study a reaction that might not seem to have great importance—the acetylation of sulfanilamide to give acetylsulfanilamide, which is then excreted.

In the course of studying this obscure reaction, Lipmann discovered and purified CoA as a factor that

$$H_3C{-}C({=}O){-}SCoA + H_2N{-}C_6H_4{-}S(O_2){-}NH_2 \longrightarrow H_3C{-}C({=}O){-}NH{-}C_6H_4{-}S(O_2){-}NH_2 + HSCoA$$

was present in extracts and was required to acetylate sulfanilamide. Lipmann's examination of this reaction led to the discovery of acetyl coenzyme A by F. Lynen. Acetyl coenzyme A is the acetylating agent for sulfanilamide and many other more important biological compounds.

The important chemistry here is, again, activation and dehydration. The oxygen atom of acetate is converted into a good leaving group when it becomes the bridge atom in the anhydride with phosphate (equation 2). Attack by the nucleophilic sulfur displaces this oxygen atom, and it appears in the inorganic phosphate. The path of the oxygen atom can be followed by using acetate labeled with ^{18}O, as shown by colored oxygen atoms in equations 2 and 3.

In mammalian systems, and eukaryotes in general, the activation of acetate to acetyl CoA occurs according to the same general principle but through a different series of reactions that does not include the formation of free acetyl phosphate, which is an unstable high-energy anhydride. Acetyl phosphate is so high in energy that it is hard to form from ATP (Chapter 9); moreover, it is a reactive acylating and phosphorylating reagent that may react nonenzymatically with other compounds in a cell.

In eukaryotes, including mammals, the activation of acetate is catalyzed by a single enzyme that gives pyrophosphate and AMP, rather than P_i and ADP, as the final products, according to equation 4.

$$H_3C{-}C({=}O){-}O^- + ATP + CoASH \xrightleftharpoons{\text{Acetyl CoA synthetase}} H_3C{-}C({=}O){-}SCoA + AMP{-}O + PP_i \quad (4)$$

This reaction, which is catalyzed by acetyl CoA synthetase, proceeds through a mixed anhydride of acetic acid and adenylic acid (AMP) through the reaction pathway shown in Figure 10-1. Acetate attacks the α-phosphorus of ATP, instead of the γ-phosphorus, to form acetyl adenylate (acetyl AMP), with expulsion of inorganic pyrophosphate. Acetyl adenylate remains tightly bound to the enzyme and reacts quickly with the sulfur atom of coenzyme A, which is also bound to the active site, to give acetyl CoA as the product. The enzyme stabilizes acetyl adenylate both thermodynamically and kinetically, so that its formation at the active site occurs with an equilibrium constant close to one. The formation of this intermediate entails the attachment of an oxygen atom in acetate to the phosphorus atom of AMP, and this bridging oxygen appears in the product AMP.

Acetyl adenylate is thermodynamically high in energy and kinetically slow to form in the absence of an enzyme to stabilize it and facilitate its formation.

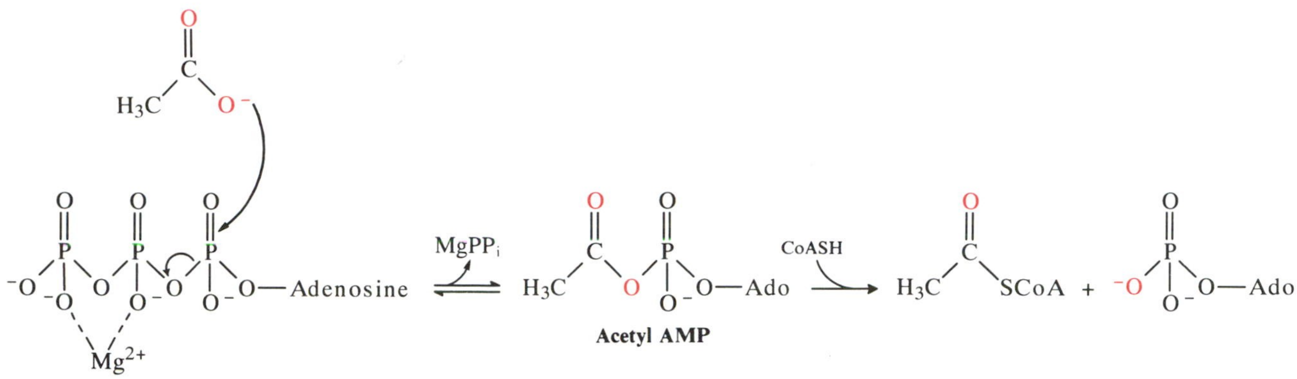

Figure 10-1

The Pathway of the Acetyl CoA Synthetase Reaction. Shown are the sequence of reaction steps at the enzymatic active site and the structure of acetyl adenylate. Acetyl adenylate remains enzyme bound throughout the reaction. Note that labeled oxygen from acetate appears in AMP.

Catalysis of its formation is one of the functions of acetyl CoA synthetase. The binding interactions between the enzyme and acetyl adenylate stabilize the intermediate thermodynamically, making it possible to form at the active site. These binding interactions also hold acetyl adenylate in place and facilitate its reaction with coenzyme A. Tight binding of acetyl adenylate by the enzyme also protects the cell against its potentially toxic effects; acetyl adenylate is highly reactive and would rapidly and indiscriminately acetylate amines and thiols if it were to diffuse freely within a cell.

The acetyl CoA could, in principle, be produced at the active site of acetyl CoA synthetase either by a two-step mechanism through a mixed anhydride, Figure 10-1, or by a fully concerted mechanism in which the thiolate anion of coenzyme A ($CoAS^-$) attacks the carboxylate group of acetate ($-CO_2^-$) at the same time that ATP abstracts an oxygen dianion (O^{2-}) from the same carboxylate group. There is no precedent for a concerted reaction mechanism of this type in organic chemistry. Nevertheless, this mechanism was favored by a number of investigators for some years. Three experiments are relevant to determining the mechanism:

1. The mechanism in Figure 10-1 predicts that isotopically labeled pyrophosphate will exchange into ATP in the presence of acetate but in the absence of coenzyme A. When ATP reacts with acetate to form the acetyl adenylate, pyrophosphate is released. If there is radioactive pyrophosphate in the solution, it will react with the mixed anhydride to regenerate acetate and ATP, incorporating the radioactive pyrophosphate into ATP. This exchange reaction is observed for some, but not all, enzymatic reactions of this class.

The failure of some enzymes to catalyze this exchange reaction was at first accepted as evidence that those reactions occur through a concerted mechanism. However, an alternative explanation is that some enzymes exhibit *substrate synergism;* that is, they do not catalyze any reactions unless all of their substrates are present. Presumably, the binding of an additional substrate molecule causes the enzyme to change into a conformation that is active for catalysis, even if that substrate molecule does not directly participate in the particular step being catalyzed. Thus, acetyl CoA synthetase may exist in a catalytically inactive conformation until CoA binds and converts it into the active conformation. Examples of substrate synergism exhibited by enzymes catalyzing reactions of different types are citrate synthase (Chapter 2) and pyruvate kinase (Chapter 11).

2. The mixed anhydride acetyl adenylate was synthesized, and acetyl CoA synthetase was shown to catalyze its reaction with pyrophosphate to make ATP and to catalyze its reaction with coenzyme A to make acetyl CoA.

3. By the use of large amounts of enzyme, it was shown that the acetyl adenylate is formed from ATP and acetate and that it accumulates as an enzyme-bound intermediate.

These experimental results are consistent with the step-by-step mechanism of Figure 10-1; but they do not rule out the possibility that, when all three substrates are bound, the enzyme catalyzes the reaction by a one-step, concerted mechanism.

Activation of Amino Acids

Aminoacyl tRNA synthetases catalyze the activation of amino acids for protein biosynthesis by the reaction pathway shown in Figure 10-2. This is essentially the same mechanism as that used by acetyl CoA synthetase to activate acetate, except that tRNA is the acceptor of the activated aminoacyl group instead of CoA. The activated amino acids differ from most other activated acids in biochemistry in many respects, one of which is that they are not thioesters with coenzyme A. An activated aminoacyl group is an ester with a hydroxyl group of a transfer RNA (tRNA). As shown in Figure 10-2B, the aminoacyl group rapidly migrates back and forth between the 2′- and 3′-positions of an adenosyl moiety of tRNA in an intramolecular reaction, even in the absence of the activating enzyme.

Transfer RNA is a family of small ribonucleic acids, each of which is specific for a particular amino acid and acts as the activating group and carrier for this amino acid between the activating enzyme and the protein biosynthesizing machinery. The chemical structures of nucleic acids will be described in the next section. Each species of tRNA has a specific structure and accepts a specific amino acid; the individual species of tRNA are known as $tRNA^{Ala}$, $tRNA^{Lys}$, and so forth, to designate which amino acid a particular tRNA molecule accepts for activation. The activation of each amino acid with its cognate tRNA is catalyzed by a specific aminoacyl tRNA synthetase; that is, alanine tRNA synthetase catalyzes the activation only of alanine and utilizes only $tRNA^{Ala}$ as the activating tRNA in equation 5.

$$\text{Alanine} + \text{ATP} + \text{tRNA}^{\text{Ala}} \longrightarrow \text{Alanyl tRNA}^{\text{Ala}} + \text{AMP} + \text{PP}_i \quad (5)$$

As with acetyl CoA synthetase, the ATP acts as an activating agent to form a mixed anhydride, an aminoacyl adenylate, and to extract the oxygen from the carboxyl group of the amino acid so that it appears in AMP. This can be shown by using ^{18}O-labeled amino acids and tracing the ^{18}O into the products. The fate of ^{18}O is shown in Figure 10-2A. The activated amino acid is an ester that can react with the amino group of another amino acid to form a peptide bond and to lengthen a peptide chain.

Aminoacyl tRNA synthetases catalyze the exchange of [^{32}P]pyrophosphate into ATP to form $AT^{32}P$, but the exchange occurs only in the presence of the amino acid for which the synthetase is specific. This exchange provides evidence for the activation mechanism in Figure 10-2A. For example, alanine tRNA synthetase can catalyze this exchange only in the presence of alanine, and no other amino acid supports the exchange. This is because the exchange mechanism requires the formation of the aminoacyl adenylate, as shown in Figure 10-2A; and alanine tRNA synthetase catalyzes the formation of alanyl adenylate but not any other aminoacyl adenylate.

Direct evidence for this mechanism is the fact that aminoacyl adenylate intermediates can be isolated from several aminoacyl tRNA synthetases.

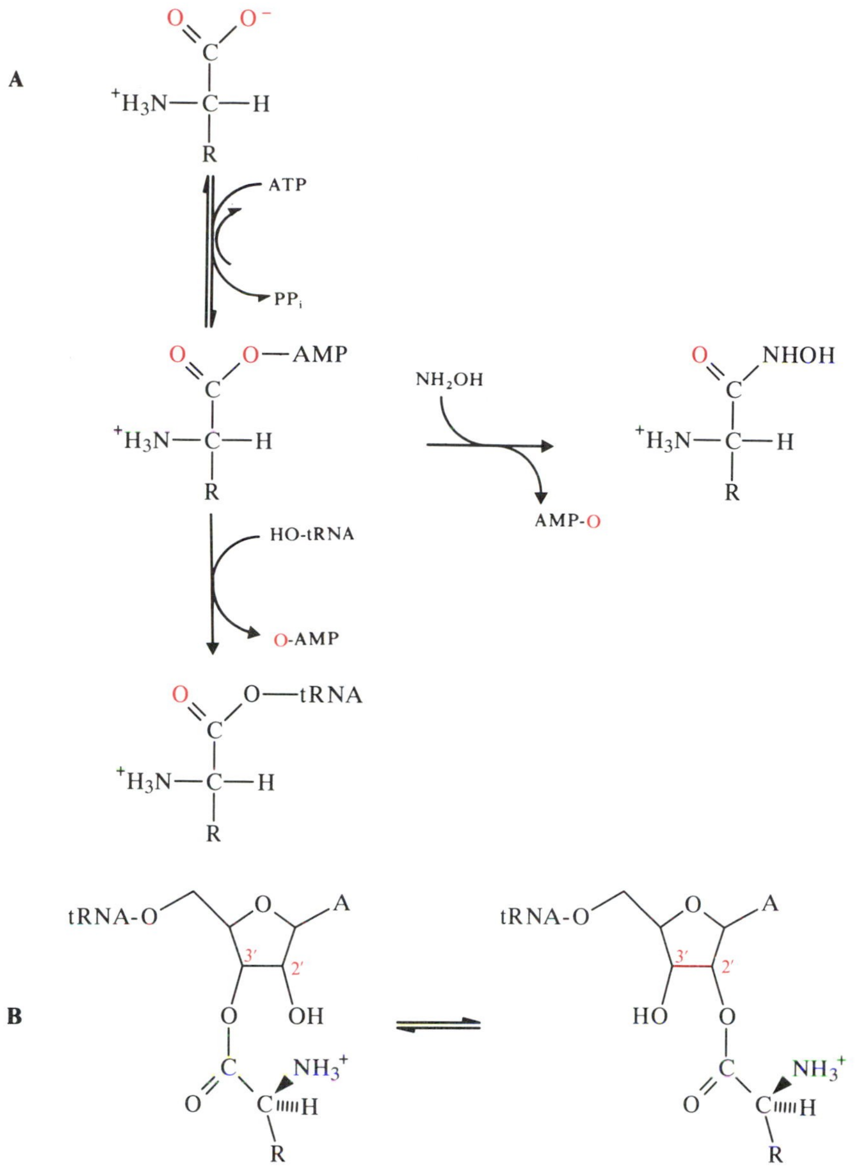

Figure 10-2

The Mechanistic Pathway of the Aminoacyl tRNA Synthetase Reactions. At least one aminoacyl tRNA synthetase is present in every cell for each amino acid. Each aminoacyl tRNA synthetase is specific for one amino acid and for a single species of tRNA, the *cognate* tRNA, for that amino acid. In the first step of part A, the amino acid reacts with the α-phosphorus of ATP to form pyrophosphate and an aminoacyl adenylate, which does not dissociate from the enzyme but reacts in the second step with tRNA to produce the aminoacyl tRNA. The aminoacyl tRNA then dissociates from the active site. The aminoacyl adenylate intermediate can also react with hydroxylamine (NH_2OH) in a nonphysiological reaction that is convenient for detecting the presence of the aminoacyl adenylate. The aminoacyl group migrates back and forth rapidly (part B) between the 2′- and 3′-hydroxyl groups of the terminal adenosine residue of tRNA.

However, it is difficult to demonstrate the formation of the aminoacyl adenylate with some of these enzymes, either by [^{32}P]PP_i-ATP exchange or by isolation of the intermediate, because they exhibit activity only in the presence of the specific tRNA required to activate the amino acid. This is another example of substrate synergism. In the presence of tRNA, however, the intermediate does not accumulate and cannot be observed, because it is rapidly converted into the aminoacyl tRNA.

Additional indirect evidence for the formation of aminoacyl adenylates as intermediates is that they can be trapped by reaction with added hydroxylamine,

BOX 10-2

PROOFREADING IN PROTEIN BIOSYNTHESIS

It is important to prevent the synthesis of inactive proteins, which may be formed by the misincorporation of amino acids in the course of protein biosynthesis. Errors are minimized in protein biosynthesis by several mechanisms in the process of translation, two of which are carried out largely by the aminoacyl tRNA synthetases. These enzymes bind and activate amino acids with a high degree of specificity; they also correct any misactivation events through proofreading mechanisms.

Specificity

Aminoacyl tRNA synthetases select the correct amino acids by taking advantage of enzyme specificity. Both the binding and the catalytic activity can be much more favorable for specific compared with nonspecific amino acids. It is particularly easy to discriminate against amino acids that are larger than the correct amino acid because larger substrates can be effectively excluded from the active site. For example, valine tRNA synthetase excludes isoleucine in comparison with valine by a factor of at least 60,000.

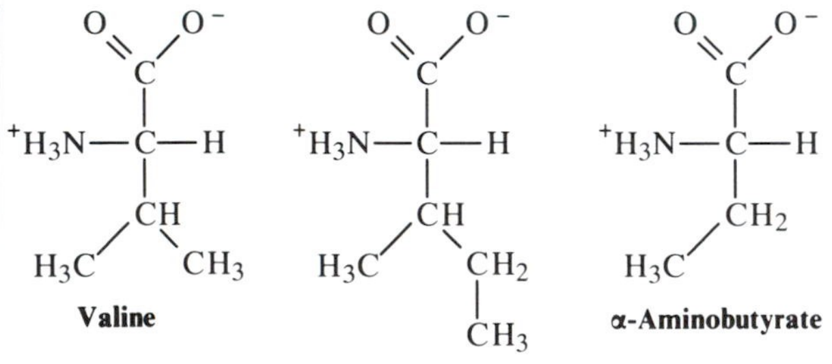

$^{+}H_3N-CH(COO^-)-CH(CH_3)-CH_3$

Valine

$^{+}H_3N-CH(COO^-)-CH(CH_3)-CH_2-CH_3$

Isoleucine

$^{+}H_3N-CH(COO^-)-CH_2-CH_3$

α-Aminobutyrate

It is much more difficult, however, to obtain specificity for molecules that are smaller than the correct substrate.

Proofreading

Although binding and catalysis in the activation of amino acids provide one chance for screening, there is a limit to the effectiveness of this screening, which depends upon the specific binding energies between the enzyme and the amino acid. Additional opportunities for specificity are afforded by proofreading processes, which remove incorrect products. Proofreading by aminoacyl tRNA synthetases is an important means by which these enzymes guard against misincorporation of amino acids into a protein. In their proofreading function, aminoacyl tRNA synthetases catalyze the hydrolysis of misactivated amino acids that adventitiously appear, albeit with low frequencies. Hydrolysis can occur at either or both steps of the activation process; that is, the formation of an incorrect aminoacyl adenylate and the formation of an incorrect aminoacyl tRNA.

An example of proofreading is the action of valine tRNA synthetase on α-aminobutyrate. This enzyme will activate α-aminobutyrate to α-aminobutyryl adenylate, which lacks one methyl group in comparison with valyl adenylate. However, the enzyme catalyzes the hydrolysis of α-aminobutyryl adenylate ten times as fast as the overall rate of valine activation, thereby effectively preventing the accumulation of a significant amount of the incorrectly activated amino acid. This has been called a "sieve" mechanism because it screens out incorrect molecules that are smaller than the correct substrate.

A misactivated amino acid may mistakenly be transferred to an incorrect tRNA. A second proofreading function of aminoacyl tRNA synthetases corrects these mistakes. An example is the mischarging of valine by isoleucine tRNA synthetase. The alkyl side chain of valine is one methylene group shorter than that of isoleucine, and isoleucine tRNA synthetase will on rare occasion accept valine, activate it to valyl adenylate, and mischarge $tRNA^{Ile}$ to valine $tRNA^{Ile}$. When this occurs, isoleucine tRNA synthetase catalyzes the hydrolysis of valine $tRNA^{Ile}$.

Proofreading exemplifies the general principle that a certain amount of specificity can be obtained in one step, but a second reaction that occurs after an irreversible step can provide additional specificity, so that the overall specificity is the product of the two specificity ratios.

All proofreading mechanisms cost energy because ATP is used up to make the incorrect product. This loss is well worthwhile in order to prevent an error. If one incorrect amino acid is incorporated into an enzyme molecule or some other protein, that protein may be unable to perform its function or to operate correctly. In this event, all of the energy that was used to synthesize the defective protein is wasted, and the protein may even be harmful to the cell. Consequently, there are mechanisms to prevent or correct mistakes in every step of protein biosynthesis; and these mechanisms work very effectively to synthesize proteins with a very low error rate, even lower than that of Japanese automobile manufacturers.

NH_2OH, which reacts readily with activated acyl groups to give a hydroxamic acid (Figure 10-2A). Hydroxamic acids give a red color in the presence of $FeCl_3$ and are useful for assaying activated acyl groups.

The most-important result for establishing a step-by-step mechanism is the demonstration that both the formation of the mixed anhydride and its reaction with tRNA occur fast enough to account for the observed turnover of the enzyme. This has been accomplished by the use of rapid reaction techniques and large quantities of enzyme. In other words, the aminoacyl tRNA intermediate has been shown to be *kinetically competent.*

The importance of aminoacyl tRNA synthetases in the biosynthesis of proteins is *much greater* than simply the chemical activation of amino acids. This is a remarkable statement in view of the fact that activation is essential. Nevertheless, the statement is correct because aminoacyl tRNA synthetases also mediate the translation of the genetic code, a function that is at least as important as activation. The specificities of aminoacyl tRNA synthetases for activating amino acids are, collectively, the first stage in the translation of the genetic code. We shall return to this subject in a later section of this chapter.

Information Transfer

The organized structure and function of a cell represents a tremendous amount of information, all of which must be processed each time a cell is replicated, and portions of which must be processed in cell repair. All of this information is stored in the structure of DNA. To be of use, the information in the DNA must be accessible to the machinery of a cell. One of the most-important cellular functions is the biosynthesis of proteins, the information for which resides in DNA. The process by which the information in DNA is used to produce proteins requires acyl activation and group transfer; it also requires phosphoryl group transfer in the processing of information in DNA. In this section, we introduce the structures of nucleic acids and outline the flow of genetic information from DNA into other molecules.

RNA and DNA

The amino acid sequences in proteins, and therefore their structures, are specified by the structure of DNA. You may be familiar with the transfer of genetic information from DNA (deoxyribonucleic acid) to RNA (ribonucleic acid) to proteins:

$$\text{DNA} \longrightarrow \text{RNA} \longrightarrow \text{Proteins}$$

With very few exceptions, genetic information is *stored* and *transmitted* from one generation of any species to the next in the form of DNA. This information is *expressed* in the functions of biomolecules by first being *transcribed* into RNA. One means by which RNA acts to express genetic information is by carrying the code for protein structure to the protein biosynthetic machinery, where it is *translated* into the amino acid sequences of proteins.

An important exception to the preceding rule is found in *retroviruses,* which store their genetic information in the form of RNA instead of DNA. Viruses are simple particles that cannot reproduce themselves as free life forms but must rely on the enzymes in the biosynthetic machinery of host cells to be reproduced. Because the protein biosynthetic machinery in the host does not recognize RNA as a storage form of genetic information, the genetic information in the RNA of retroviruses is expressed by first being transcribed into DNA by *reverse transcription.* It is then again transcribed into the species of RNA that carries genetic information into the host protein biosynthetic machinery. The functions of *reverse transcriptase,* the enzyme that catalyzes reverse transcription, are further described in Chapter 12.

Nucleotides and 3′,5′-Phosphodiesters

Nucleic acids are composed of nucleotides that are chemically linked in specific, linear sequences. Thus, nucleic acids are often called polynucleotides. *Nucleotides* are composed of heterocyclic bases, shown in Figure 10-3, bonded to phosphorylated forms of the sugars ribose and 2′-deoxyribose, as shown in Figure 10-4. RNA is composed of ribonucleotides, which contain ribose; and DNA is composed of deoxyribonucleotides, which contain 2′-deoxyribose. We have already discussed a few ribonucleotides, ATP, GTP, ADP, and AMP. The other nucleotides are structurally related to these nucleotides but differ in important ways.

The two chemical differences between RNA and DNA are the structures of the heterocylic bases (Figure 10-3) and the presence of ribose in ribonucleotides and deoxyribose in deoxyribonucleotides. RNA contains the purines adenine (A) and guanine (G) and the pyrimidines uracil (U) and cytosine (C). DNA contains adenine, guanine and cytosine, as does RNA, but DNA contains thymine (T) instead of uracil, which is found only in RNA. The structures and names of the nucleic acid bases must be committed to memory. There is no way to avoid this. It may be helpful to remember the structures of the pyrimidine and purine rings. Thymine, which is found only in DNA, is simply uracil with a methyl group attached to carbon-5. A few additional, chemically modified forms of these bases appear in one species of RNA, transfer RNA, which is described in a later section of this chapter. The heterocyclic bases are biosynthesized in complex multistep biosynthetic pathways that are described in Chapter 27.

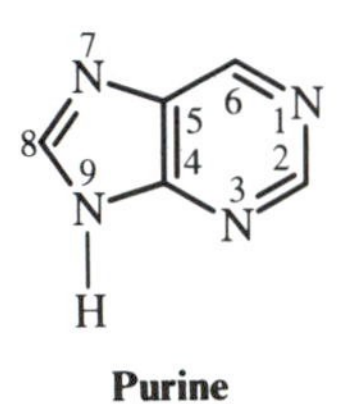

Purine

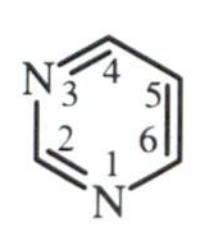

Pyrimidine

The structures and names of a few nucleosides and nucleotides are given in Figure 10-4. The nucleosides consist of heterocyclic bases that are bonded to carbon-1′ of ribose or 2′-deoxyribose, always in the stereochemical configuration shown, and the nucleotides are phosphorylated nucleosides. The most-common nucleotides are phosphorylated at either the 5′- or the 3′-position of the ribose or 2′-deoxyribose ring, because in nucleic acids the phosphodiester bridges span these two positions of the sugar rings, as described shortly. However, phosphorylation of ribose rings at the 2′-OH group is also possible and does occur in a few instances in biochemistry. An example is $NADP^+$, nicotinamide adenine dinucleotide phosphate, which is related to NAD^+ and is described in Chapter 16.

In nucleic acids, one nucleotide is linked to another by a phosphodiester bridge between the 3′-OH of one and the 5′-OH of the next; that is, by 3′,5′-phosphodiester linkages. Representative segments of RNA and DNA are shown in Figure 10-5, on page 268. In RNA and DNA, the nucleotides are always linked 3′- to 5′-, so that the polymers are invariably linear.

Figure 10-3

The Structures of the Common Heterocyclic Bases in RNA and DNA. The names and accepted abbreviations are given beneath the structures. A, T, G, and C are found in DNA and A, U, G, and C are in RNA. The numbering in the rings shown for adenine is the same for guanine and that for uracil is the same for thymine and cytosine.

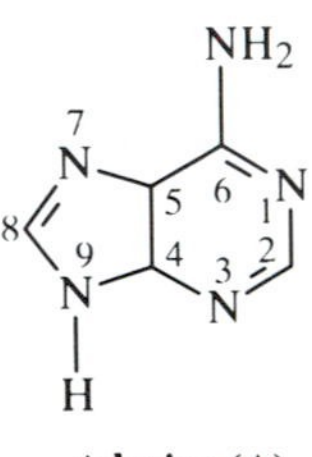

Adenine (A)

Guanine (G)

Uracil (U)

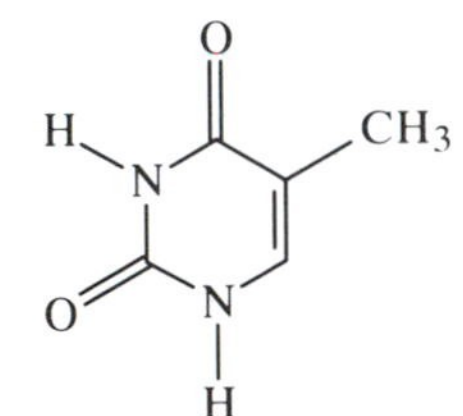

Thymine (T)

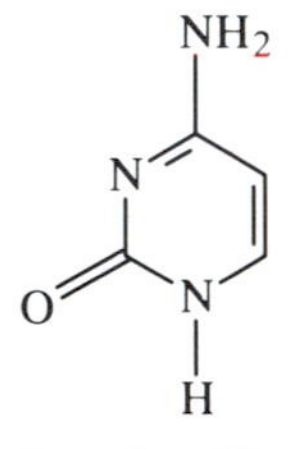

Cystosine (C)

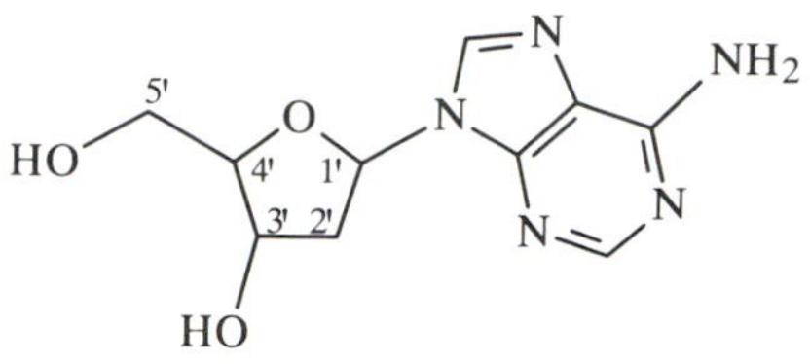

Deoxyadenosine (dAdo)

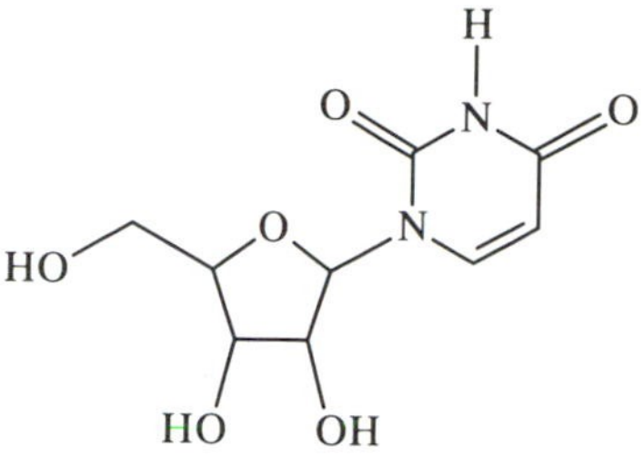

Uridine (Urd)

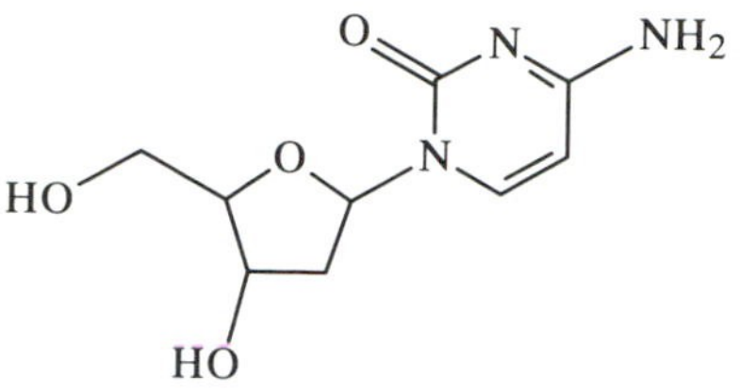

Deoxycytidine (dCyd)

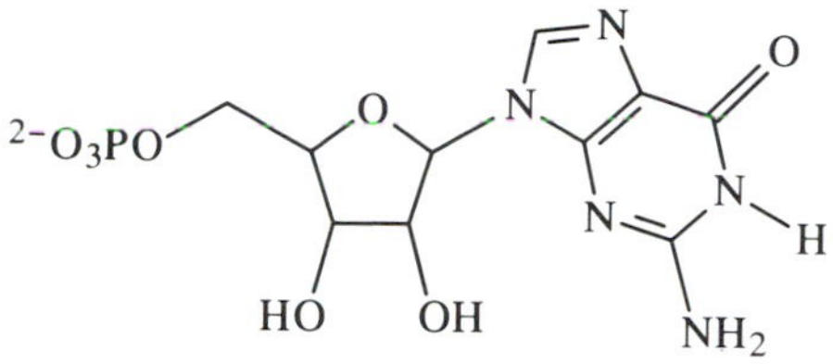

Guanosine 5′-phosphate (GMP)

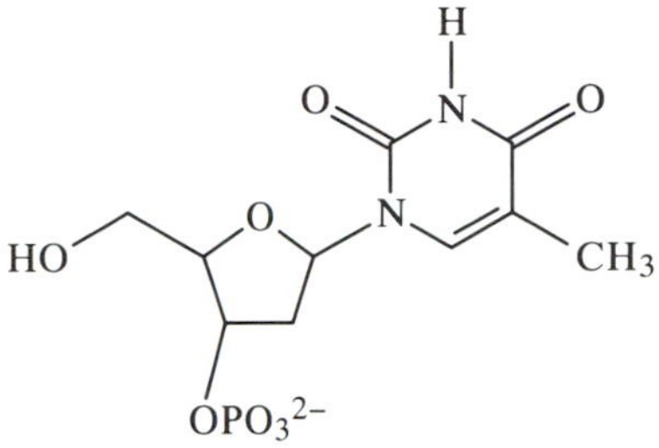

Thymidine 3′-phosphate (3′-TMP)

Figure 10-4

The Structures of Some Nucleosides and Nucleotides. The accepted names and abbreviations are given beneath the structures. The names and three-letter abbreviations of the ribonucleosides are: adenosine, Ado; guanosine, Guo; uridine, Urd; thymidine, Thd; and cytidine, Cyd. The names of the deoxynucleosides are the same but with the prefix deoxy. The abbreviations of the deoxynucleosides are: dAdo, dGuo, dUrd, dThd, dCyd. The nucleotides are named as the nucleosides plus designation of the phosphorylation level and position of phosphorylation, as in the following: guanosine 5′-monophosphate (GMP), thymidine 3′-monophosphate (3′-TMP), and adenosine 5′-triphosphate (ATP). The atom numbers in the ribose and deoxyribose rings are primed to distinguish them from the atom numbers in the heterocyclic bases.

The Structure of DNA

DNA is contained within chromosomes (Chapter 12) and, in microorganisms, in extrachromosomal elements such as plasmids. Chromosomal DNA molecules are large, having molecular weights in the millions. RNA molecules are much smaller, with molecular weights ranging from 6000 to more than 1 million. DNA, which stores and transmits genetic information, is not metabolically active except in cell division. The structure of DNA is an important factor in stabilizing it against adventitious chemical reactions that could threaten its structure and the integrity of the genetic information. Native DNA is double stranded, with the two strands wound about each other in a double helix. The segments of RNA and DNA single strands shown in Figure 10-5 illustrate the polarity that exists in a polynucleotide. The strands have an intrinsic directionality, with a 5′ end and a 3′ end, shown in Figure 10-5 as projecting upward and downward, respectively. In the double helix, the polarities of the two strands are antiparallel, and the heterocyclic bases are paired between them by hydrophobic bonding and hydrogen bonds. Figure 10-6 shows the hydrogen bonding between adenine and thymine and between cytosine and guanine.

The rules for base pairing (AT and GC) are closely followed in double-stranded DNA. Therefore, the base sequences in the two strands are complementary. *Complementarity* in this case is a form of redundancy that is an

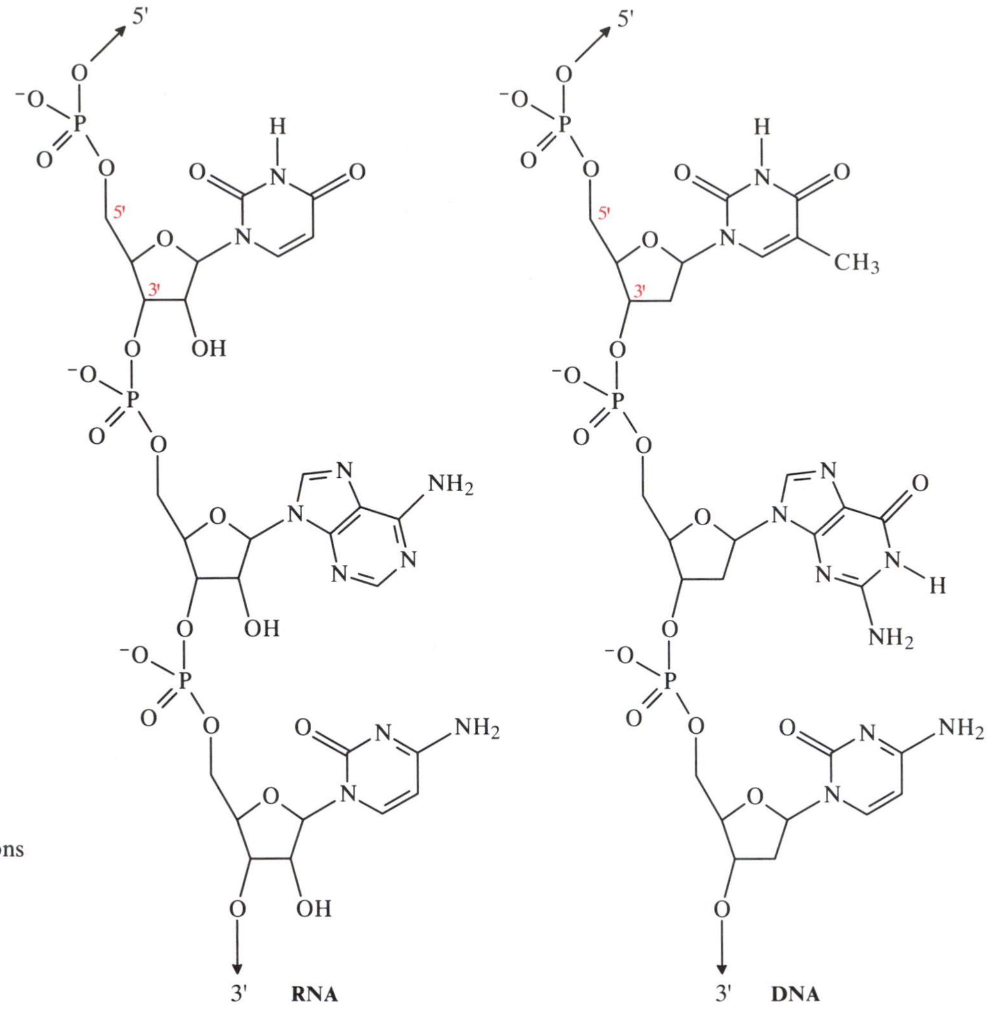

Figure 10-5

The Structures of 3′,5′-Phosphodiester Linkages in RNA and DNA. Abbreviations for these sequences are 5′-...pUpApC...-3′ and d(5′-...pTpGpC...-3′) or 5′-...UAC...-3′ and 5′...TGC...3′.

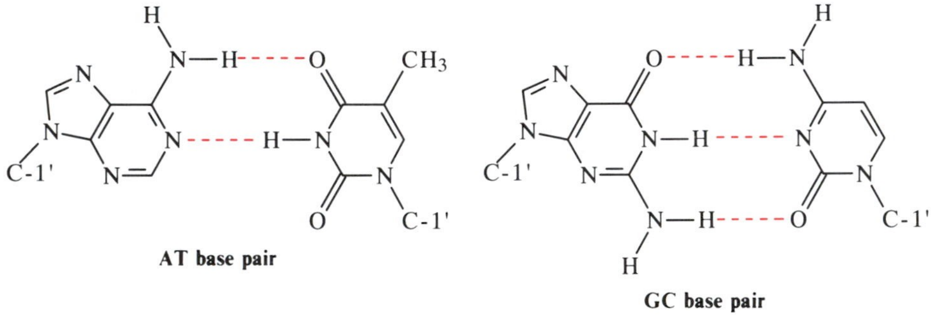

Figure 10-6

The Hydrogen Bonds in the Base Pairs AT and GC.

important factor in preserving the informational content of DNA. The information in DNA is the nucleotide sequence, and the presence of a complementary sequence ensures that if part of the sequence information is somehow lost in one strand it can be recovered from the complementary sequence in the other strand by repair processes. The structure, replication, and repair of DNA are described in Chapter 12.

Messenger RNA (mRNA)

RNA, in contrast with DNA, is metabolically active in protein biosynthesis. RNA exists in several forms, each of which has a specific function in the synthesis of proteins. The amino acid sequences of proteins are specified by the sequences of bases in segments of DNA. The code relating a base sequence to an amino acid sequence is based on triplets of nucleotide bases. Within a DNA segment that specifies a protein, an individual triplet base sequence codes for the insertion of a particular amino acid into a specific position of the protein. The DNA supplies the information for the synthesis of mRNA, which in turn determines the sequence in which amino acids that have been attached to specific tRNA molecules are incorporated into a growing peptide chain in protein synthesis.

In the transfer of sequence information from DNA to mRNA, the base-pairing rules are followed, and only one strand of the DNA, designated the *sense strand,* is transcribed into mRNA. When one strand of the DNA in equation 6 acts as a template to synthesize mRNA, the mRNA is composed of ribonucleotides in the sequence complementary to that strand.

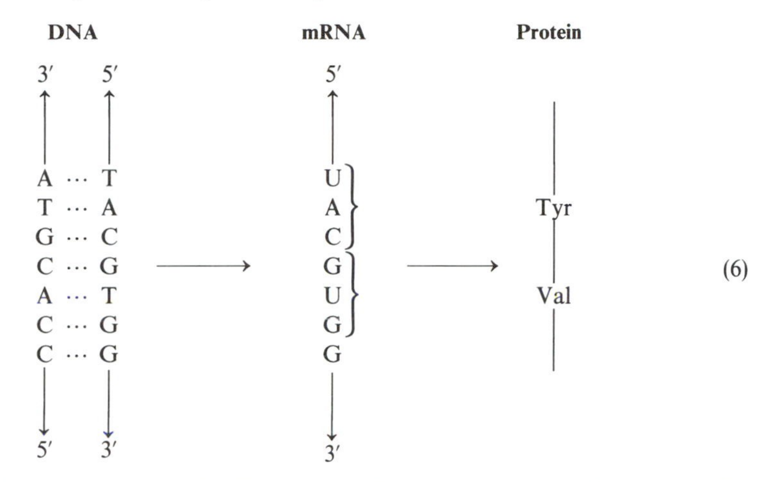

(6)

Three-base segments of mRNA code for the incorporation of specific amino acids in the protein biosynthetic machinery, which is described in the next sections. In equation 6, the segment of mRNA codes for the incorporation of tyrosine and valine into a protein. The genetic code comprises triplets of nucleotides, most of which specify the incorporation of particular amino acids. The triplets in mRNA are the *codons,* reading 5′ to 3′, and the *code words* are the amino acids specified by the triplets. The genetic code is summarized in Table 10-1 and inside the back cover of this book.

Table 10-1
The genetic code

Amino Acid	Codons	Amino Acid	Codons
Phe	UUU, UUC	Ser	UCU, UCC, UCA, UCG, AGU, AGC
Tyr	UAU, UAC	Leu	UUA, UUG, CUU, CUC, CUA, CUG
Cys	UGU, UGC	Pro	CCU, CCC, CCA, CCG
Trp	UGG	Arg	CGU, CGC, CGA, CGG, AGA, AGG
His	CAU, CAC	Ile	AUU, AUC, AUA
Gln	CAA, CAG	Thr	ACU, ACC, ACA, ACG
Asn	AAU, AAC	Val	GUU, GUC, GUA, GUG
Lys	AAA, AAG	Ala	GCU, GCC, GCA, GCG
Glu	GAA, GAG	Gly	GGU, GGC, GGA, GGG
Asp	GAU, GAC	(Termination)	UAA, UGA, UAG
Met (initiation)	AUG		

In bacteria, mRNA is produced in the nucleoid and is essentially ready to be translated into an amino acid sequence as it diffuses through the cytoplasm. In a eukaryotic cell, mRNA is produced within the nucleus and must be transported into the cytoplasm and the membranes of the endoplasmic reticulum. To protect it in transit, and to provide recognition signals, eukaryotic mRNA is modified by capping at the 5′ end and polyadenylylation at the 3′ end. Eukaryotic mRNA also contains noncoding sequences that are removed before decoding. These processes are described in Chapter 12.

Transfer RNA (tRNA)

Transfer RNA is the species of RNA that is utilized to activate amino acids for protein biosynthesis and to begin the process of translating the genetic code. All species of tRNA are single strands of ribonucleic acid containing segments that have a high degree of base complementarity with downstream segments; this allows intrastrand base pairing and the formation of loops when the strand folds back upon itself, as illustrated in Figure 10-7. Consequently, tRNA contains a high degree of secondary structure consisting of base-paired double-helical segments. The secondary structures are similar in all species of tRNA, as deduced from the base-pairing rules and the nucleotide sequences of tRNA molecules and verified by X-ray diffraction.

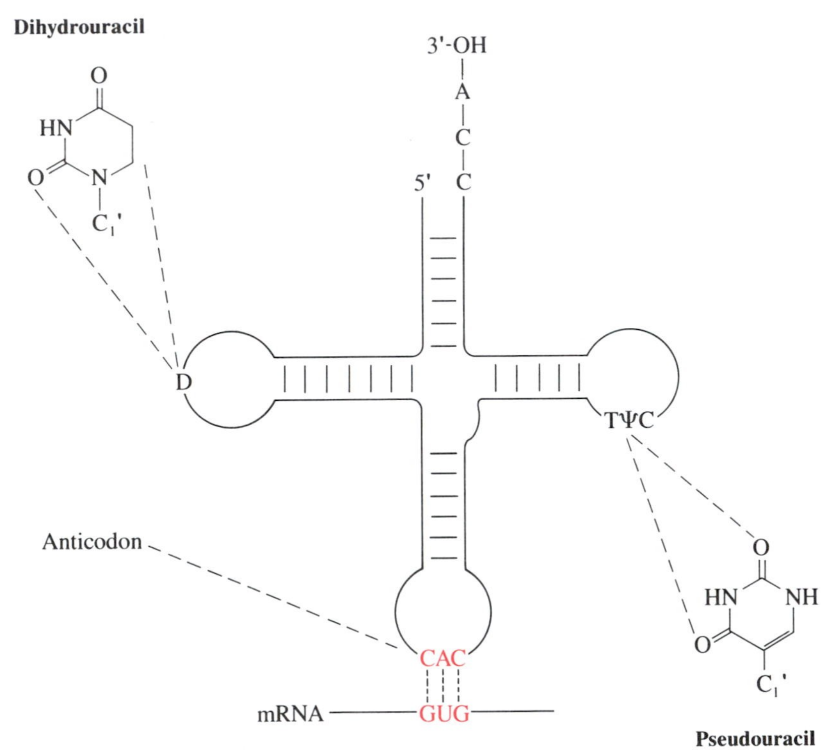

Figure 10-7

The Secondary Structure of tRNA. The arms are base-paired segments in which the chain turns back on itself and forms double-helical segments because of self-complementary nucleotide sequences. The loops are intervening regions with little or no possibility of base pairing. The loop at the bottom is the anticodon loop. The anticodon binds directly to the codon on mRNA. The other two loops always contain the same unusual bases with which they are identified. The unusual base pseudouracil is always flanked by cytosine and ribothymidine, which has ribose as the sugar instead of deoxyribose as in thymidine. The CCA-end is the 3′ end at the top and is the site at which an amino acid is attached.

Figure 10-7 shows that the secondary structures of all species of tRNA resemble a clover leaf, in which the four stems represent double-stranded helical segments composed of complementary base sequences and hydrogen-bonded bases. The three loops contain no internal base complementarity, and the 3′ end of the molecule always has the sequence CCA. One loop, the *anticodon* loop, contains a three-base sequence that directly participates in translating the genetic code. The activated amino acid is attached on the 2′- or 3′-hydroxyl group of the 3′-terminal adenosine.

Transfer RNA molecules contain several modified bases found in particular segments of the structure. The functions of these bases are not yet fully understood. Two of the modified bases are dihydrouracil (D) and pseudouracil (Ψ), which are also shown in Figure 10-7. These two bases appear in specific locations and are used to designate the D-loop and stem and the TΨC-loop and stem. Each species of tRNA, such as $tRNA^{Phe}$, $tRNA^{Ala}$, and so forth, has a slightly different nucleotide sequence and length, but all are about 75 nucleotides long and have the major secondary structural features shown in Figure 10-7. Transfer RNA molecules are not planar as in the diagram in Figure 10-7; rather, they have a tertiary structure, shown in Figure 10-8, in which the TΨC-loop is folded over onto the D-loop. The two loops are held together by interloop hydrogen-bonded base pairs, and these are known as tertiary base pairs.

The unusual heterocyclic bases in tRNA are introduced by enzymatic modification of certain conventional bases in newly synthesized tRNA, rather than by the incorporation of unconventional nucleotides into the polymer. The modified bases include N^6-isopentenyladenosine, 5-methylcytosine, N^6-dimethyladenosine, shown in Figure 10-9, as well as dihydrouridine and pseudouridine. The methyl-donor molecule for the methylated bases is *S*-adenosylmethionine (Chapter 25). The isopentenyl groups arise from isopentenyl pyrophosphate (Chapter 28). Pseudouridine is derived from uridine by an interesting and mechanistically uncharacterized isomerization, in which the *N*-ribosyl bond is cleaved and reformed to carbon-5 of the ring, as illustrated in Figure 10-10.

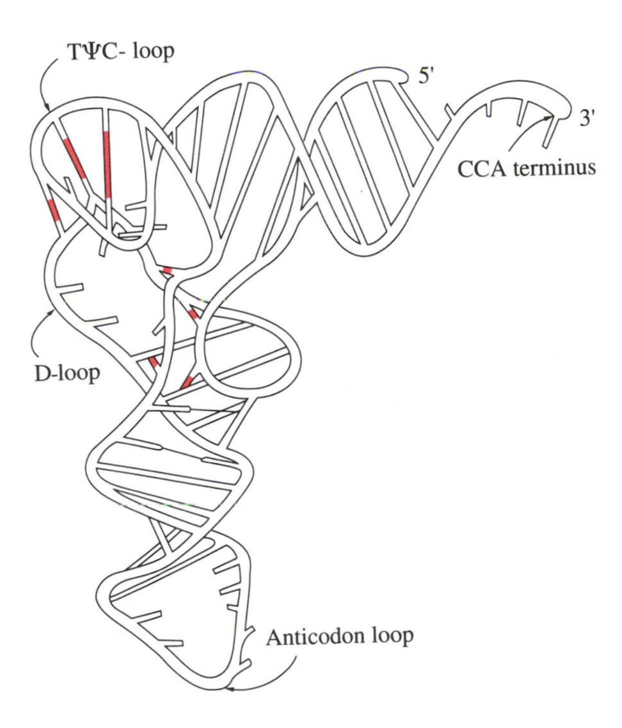

Figure 10-8

The Tertiary Structure of tRNA. This representation of the three-dimensional structure of yeast phenylalanyl tRNA emphasizes the relations among secondary structural elements. Note that the TΨC-loop is folded over onto the D-loop. The two loops are held together by interloop pairing of bases.

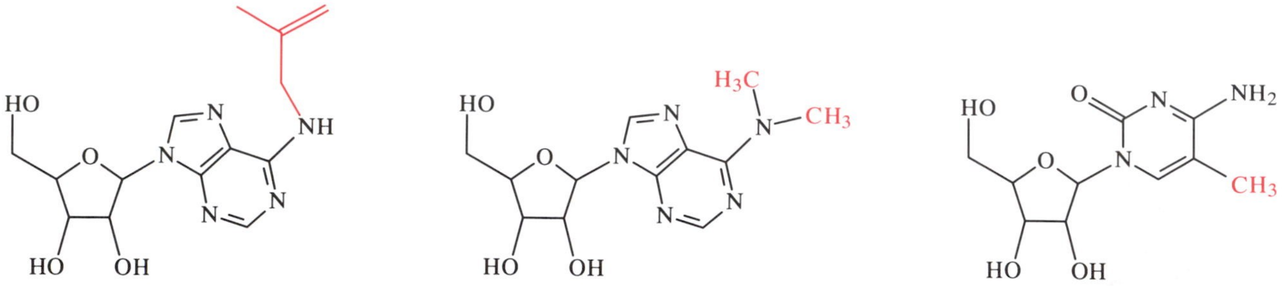

N^6-Isopentenyladenosine

N^6-Dimethyladenosine

5-Methylcytidine

Figure 10-9

Structures of Some Modified Bases Found in tRNA.

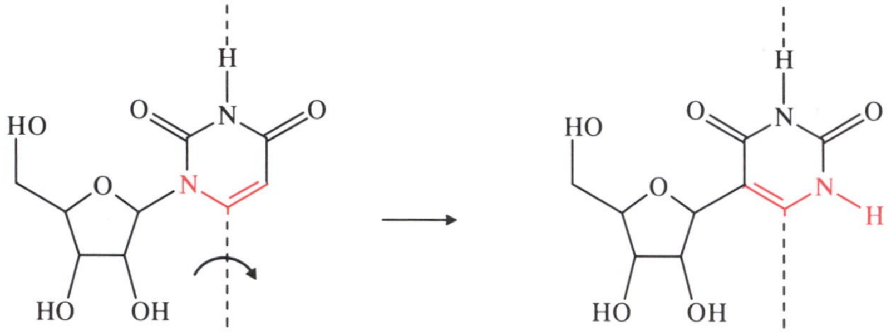

Uridine

Pseudouridine (Ψ)

Figure 10-10

The Structural Relation Between Uridine and Pseudouridine.

The Genetic Code

The specific interaction of an aminoacyl tRNA with mRNA that transmits the amino acid code from DNA to tRNA arises from the anticodon, a three-base sequence in the anticodon loop of the tRNA, shown in Figure 10-7. The anticodon is paired by specific hydrogen bonding to a codon of complementary bases in mRNA. The base sequence of a codon specifies a particular amino acid for incorporation into a protein. The codon GUG ($5' \rightarrow 3'$) in equation 6 and Figure 10-7 is the code for valine and is complementary to the anticodon CAC ($3' \rightarrow 5'$) in $tRNA^{Val}$. A CGU-sequence on mRNA codes for arginine and is complementary to the anticodon GCA on the $tRNA^{Arg}$ shown below.

```
mRNA (codon)             5'----C----G----U----3'
                               :    :    :
tRNA^Arg (anticodon)     3'----G----C----A----5'
```

The four bases in DNA can be combined in sixty-four different three-base sequences ($4 \times 4 \times 4$), because there are four ways to choose each base. Therefore, sixty-four codons are potentially available for twenty amino acids. A few codons serve as start and stop signals, but most are amino acid codons. This makes the genetic code *degenerate,* because there are more codes than there are amino acids; that is, more than one triplet can code for a particular amino acid. The specificity of each codon resides mainly in the first two bases of the triplet. The third base is less specific and may be one of two or even four bases without a change in specificity. The third base is known as the *wobble base.* For example, valine, which is shown in equation 6 as being coded by GUG in mRNA, may also be coded by GUA. The cognate anticodons in the two species of tRNA that are used to activate valine are CAC and CAA, respectively. Phenylalanine is coded by UUU, and the synthesis of polyphenylalanine directed by synthetic poly U was an important step in deciphering the coding language for amino acids (Box 10-3). The codons are matched with the amino acids that they designate in Table 10-1.

BOX 10-3

DISCOVERY OF THE GENETIC CODE

The first information about the genetic code came from simple experiments using crude and fractionated extracts of *Escherichia coli*. An extract of bacterial cells consisted mainly of the cytosol and ribosomes in the buffer used to extract broken cells. The ribosomes could be separated from soluble enzymes by high-speed centrifugation. An extract, designated S-30, was the supernatant fluid obtained from centrifugation of a suspension of broken cells in buffer at 30,000 × g in a high-speed centrifuge. Further centrifugation of the extract in a preparative ultracentrifuge at 100,000 × g gave an S-100 pellet, which consisted mainly of ribosomes, and an S-100 supernatant fluid consisting of the cytosol diluted slightly with buffer. The S-100 fluid contained all of the soluble proteins in the cell, including the aminoacyl tRNA synthetases and all of the other protein factors required in protein biosynthesis. Many early investigations of the genetic code were carried out with these extracts and ribosomal pellets supplemented with synthetic polyribonucleotides or oligoribonucleotides and radioactive amino acids.

The earliest experiments utilized S-30 extracts and polyribonucleotides, together with radioactive amino acids. In the first successful experiment, M. Nirenberg added poly(U) to the extract, together with ATP, GTP, and a ^{14}C-labeled amino acid; after an incubation period, he added acid to precipitate any [^{14}C]polypeptide that had been produced. Inasmuch as the ^{14}C-amino acids are soluble in acid, radioactivity counts in the precipitate signaled the formation of a polypeptide, which is insoluble in acid. With poly(U) as the polyribonucleotide, radioactivity appeared in the precipitate only when the amino acid was ^{14}C-phenylalanine. No other radioactive amino acid gave this result. And no other polyribonucleotide gave this result with ^{14}C-phenylalanine. Therefore, the code for phenylalanine incorporation into a polypeptide had to be UUU in mRNA. This was the first and simplest codon assignment in the genetic code. In similar experiments, poly(A) stimulated the production of acid-insoluble radioactivity from ^{14}C-lysine, and poly(C) stimulated the formation of poly[^{14}C]proline. Thus, AAA and CCC were shown to be the codons for lysine and proline, respectively.

Other codon assignments were more difficult to make, because they consisted of mixed bases. Progress was made through the use of copolymers of ribonucleotides containing two nucleotides to stimulate the formation of acid-insoluble products from ^{14}C-labeled amino acids. These copolymers stimulated the formation of acid-insoluble radioactivity from more than one amino acid. For example, a random copolymer of A and C contained the codons AAA, ACA, AAC, CAA, CAC, ACC, CCA, and CCC; and such a copolymer stimulated the incorporation of radioactivity into acid precipitates when the amino acid was ^{14}C-labeled lysine, proline, threonine, aspartate, histidine, or glutamine. Therefore, because the codons AAA and CCC had already been assigned to lysine and proline, those for threonine, aspartate, histidine, and glutamine must have been among the other six permutations and combinations of A and C. These experiments narrowed the choices for other codon assignments and supported the final assignments.

Further progress was made by observing the binding of specific species of [^{14}C]aminoacyl $tRNA^{aa}$ to ribosomes. In these experiments, use was made of the S-100 supernatant fluid, which contained the soluble enzymes and transfer RNA, and the S-100 pellets, which contained the ribosomes. Both fractions were required for protein biosynthesis. The soluble fraction was used to prepare activated amino acids such as [^{14}C]phenylalanyl $tRNA^{Phe}$, which could be shown to bind to ribosomes in the presence of the synthetic triplets UUU and UUC, but not with other triplets (and not with the dimer UU). Thus, both UUU and UUC were assigned as codons for phenylalanine.

The final assignments depended upon the work of H. G. Khorana, who pioneered the chemical synthesis of polynucleotides with defined sequences. Khorana developed chemical methods for synthesizing specific oligonucleotides, which could be polymerized to produce block copolymers with definite nucleotide sequences. A block copolymer is one in which a sequence is repeated in blocks, such as ACUACUACUACU...; this example contains the three triplet codons ACU, CUA, and UAC in the three possible reading frames. Block copolymers were slightly more complex as templates for polypeptide formation than homo polymers but not as complex as random copolymers. Thus, the ordinary copolymer of A and C discussed earlier contained eight codons, but the three-letter block copolymer above contains only three.

An example of the progress made with block copolymers is the result obtained using the polymer with the perfectly alternating sequence UCUCUCUCUC.... This copolymer has only two codons: UCU and CUC. When it was used to stimulate polypeptide synthesis in the extracts of *E. coli*, a ^{14}C-polypeptide was produced containing only serine and histidine. This polypeptide could be produced only when *both* serine *and* histidine were present, and no polypeptide was formed when only one of the two was present. Owing to the sensitivity of peptide bonds linking serine to other amino acids, the serine-histidine linkage could be selectively cleaved by partial acid hydrolysis. This led to the formation of a single dipeptide, serine-histidine. Other dipeptides such as histidine-serine, histidine-histidine, or serine-serine were not produced. Therefore, the polypeptide must itself have been a perfectly alternating block copolymer of serine and histidine. This was an important result in that not only did it help to define the codons for serine and

BOX 10-3 (continued)

DISCOVERY OF THE GENETIC CODE

histidine, but it also showed that mRNA is read continuously, with no spaces between codons. Other experiments using block copolymers with two and three bases, coupled with analysis of the polypeptides produced, allowed the triplet codes to be assigned.

The foregoing experiments were consistent with triplet and nonoverlapping codes. Moreover, elegant genetic experiments gave results that were also most consistent with a triplet nonoverlapping code. The code is the same for prokaryotes and eukaryotes but may not be quite universal. Mitochondria in eukaryotic cells contain a small amount of DNA, as well as ribosomes and other machinery for protein biosynthesis; and they biosynthesize a few mitochondrial proteins. The genetic code within mitochondria is almost the same as the "universal code," but recent experiments indicate that it may differ very slightly from the universal code.

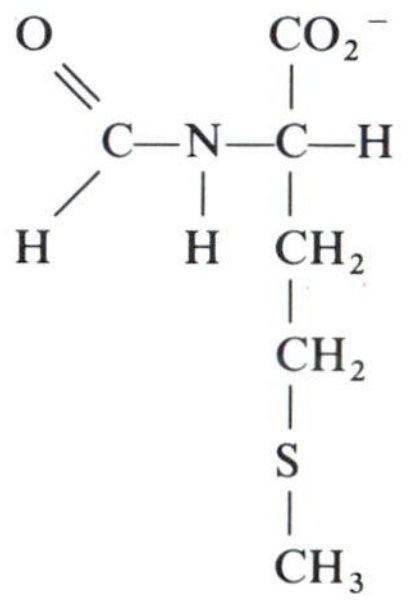

```
O         CO2−
 \\        |
  C—N—C—H
 /   |   |
H    H  CH2
         |
        CH2
         |
         S
         |
        CH3
```

***N*-Formylmethionine**

A particularly important codon is AUG, which, in bacteria, codes for the incorporation of *N-formylmethionine* into a protein. *N*-Formylmethionine is the first amino acid to be incorporated into the N-terminal position of a protein and is the amino acid that initiates polypeptide synthesis. The triplet AUG is, therefore, the *initiation codon* for protein biosynthesis. The importance of this codon cannot be overemphasized. Because the translation of the base sequences in mRNA depends on reading triplet codes, the question of where this reading should begin, or of how to establish the *reading frame* for translation, is crucially important in determining how the base sequence is to be read. Shown here is a nucleotide sequence and the three possible reading frames.

5′. . . GUC AUG CUC UAG GAA ACG CAC CUA UAA GCU GAU CGU . . . 3′

5′. . . G UCA UGC UCU AGG AAA CGC ACC UAU AAG CUG AUC GU . . . 3′

5′. . . GU CAU GCU CUA GGA AAC GCA CCU AUA AGC UGA UCG U . . . 3′

The three reading frames define three different sequences of codons and amino acids. Recognition of the initiation sequence AUG in the first reading frame by *N*-formyl-Met-$tRNA^{fMet}$ establishes the reading frame for subsequent amino acid codons and, in the same stroke, starts the protein biosynthetic reaction.

N-Formylmethionyl $tRNA^{fMet}$ arises in prokaryotes by *N*-formylation of a particular species of activated methionine, methionyl $tRNA^{fMet}$. The formyl group is added to the amino acid after it has been activated to its specific rRNA ester. However, the formyl group, and often the methionine itself, may be removed after the protein has been synthesized. In eukaryotes, protein synthesis is generally initiated with methionine, and the initiator codon AUG codes for methionine rather than *N*-formylmethionine. The chemistry underlying *N*-formylation requires the vitamin folic acid and is analogous to all folate-dependent transformylations, which are described in Chapter 25.

The Biosynthesis of Proteins

Protein biosynthesis begins with the activation of amino acids as the specific species of aminoacyl tRNA. Peptide-bond formation in the biosynthesis of proteins is carried out by an extraordinarily intricate machine, the ribosome, which assembles a protein molecule from all the species of aminoacyl tRNA. The

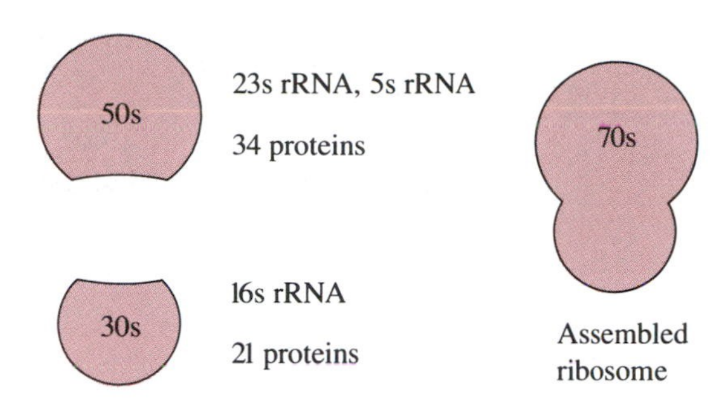

Figure 10-11

The Composition of the Ribosome from *E. coli*. The ribosome from *E. coli* consists of three species of rRNA and 55 proteins in two subunits. The larger 50s subunit contains the 23s and 5s rRNA and 34 proteins; the smaller 30s subunit contains the 16s rRNA and 21 proteins. The two subunits become associated into a ribosome in the process of initiation, which is described in Figure 10-13.

amino acids are assembled in the correct order by matching the nucleotide sequence information in a molecule of mRNA, which was obtained from the nucleotide sequence of DNA, with the anticodon triplet of each aminoacyl tRNA. The ribosome builds a protein molecule from 150 amino acids in from 10 to 200 seconds in eukaryotes and even faster in bacteria.

The bacterial ribosome is a particle that is built around the framework of a third variety of RNA, ribosomal RNA (rRNA). The assembled and functioning ribosome is a large particle with a sedimentation coefficient of 70s in the ultracentrifuge. It is composed of two smaller particles of 30s and 50s consisting of rRNA and proteins, shown in Figure 10-11. There are approximately 34 protein molecules, 23s rRNA, and 5s rRNA in the 50s subunit; and there are 21 proteins and 16s rRNA in the smaller 30s particle. A reconstruction image of an *E. coli* ribosome is shown in Figure 10-12, on page 276.

We will describe bacterial protein biosynthesis here. Eukaryotic protein synthesis is similar, but the various factors have different designations and slightly different properties.

The synthesis of a polypeptide requires chain initiation, chain elongation with a translocation step to prepare for the next amino acid, and finally chain termination. The process is controlled by a series of protein factors that regulate individual steps through reaction sequences that are driven forward by the hydrolysis of GTP.

Chain Initiation

Bacterial protein synthesis is initiated by the assembly of 30s and 50s ribosomal subunits with mRNA according to the sequence of events outlined in Figure 10-13. Formation of the initiation complex includes the actions of several proteins, the initiation factors known as IF-1, IF-2, and IF-3, which are initially bound to the 30s ribosomal subunit. Factors IF-1 and IF-2 facilitate the binding of mRNA and *N*-formylmethionyl $tRNA^{fMet}$ to the 30s subunit. Factor IF-3 then dissociates from the 30s subunit, and the 50s subunit binds in a process that leads to the hydrolysis of GTP and the release of IF-1 and IF-2. This completes the formation of the initiation complex, in which the anticodon of *N*-formylmethionyl $tRNA^{fMet}$ is bound to the codon AUG on mRNA, with the *N*-formylmethionyl group in the P-site, the peptidyl transfer site of the completed

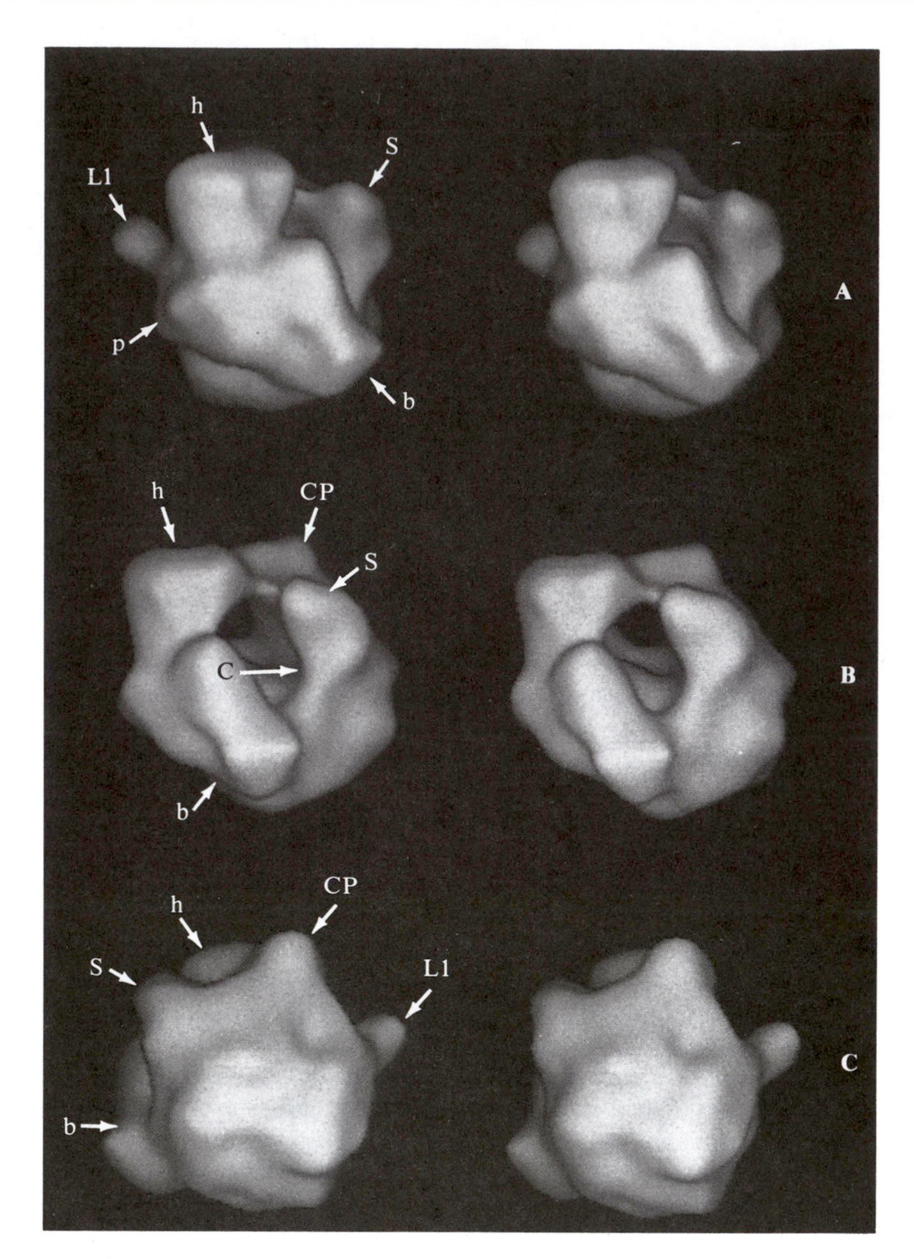

Figure 10-12

A Reconstruction Image of a Ribosome from *E. coli.* This is a merged reconstruction, presented as stereo pairs from three viewing directions. In part A, the ribosome is oriented in such a way that the 30s subunit is in front. Part B is a view of the ribosome after a 40° rotation of part A around an axis lying vertical in the plane of the image. Part C is a view after a 150° rotation of part A around the same axis. For the 30s subunit, p = platform, h = head, b = main body. For the 50s subunit, L1 = L1 ridge, CP = central protuberance, S = stalk base (the stalk is not visible in the merged reconstruction because of its high mobility). (Reconstruction image courtesy of P. Penczek, M. Radermacher, and J. Frank.)

ribosome. Initiation-complex formation is a partially understood process, in which the initiation factors bring about hydrolysis of GTP. This hydrolysis and similar reactions in later steps provide the driving force and control for the steps that keep the process moving in the forward direction. The initiation factors also

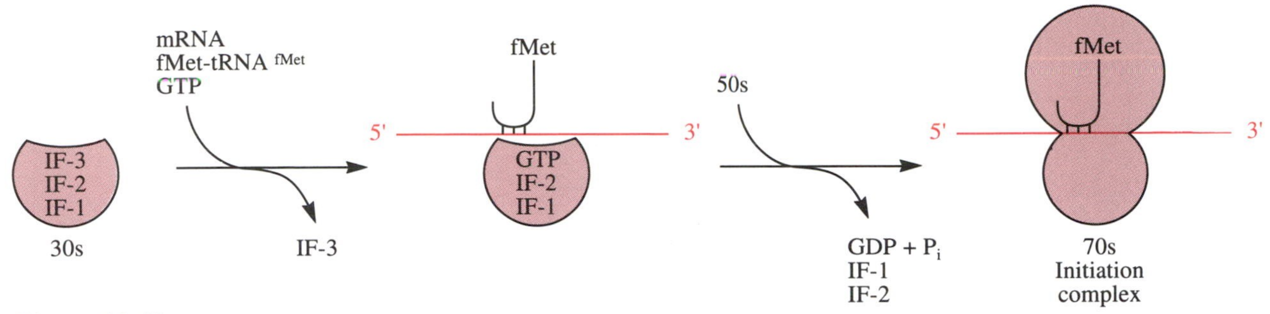

Figure 10-13

Assembly of the Initiation Complex. The 30s ribosomal subunit is initially bound to three proteins, initiation factors IF-1, IF-2, and IF-3, which are not part of the ribosome itself but which facilitate assembly of the initiation complex. In the first step of initiation, the subunit binds mRNA and *N*-formylmethionyl tRNA^{fMet}, and the formation of the complex is accompanied by the dissociation of IF-3. In the second step, the complex binds the 50s subunit in a process that is accompanied by the dissociation of IF-1 and IF-2 and the hydrolysis of GTP. The assembled initiation complex is primed to bind other species of aminoacyl tRNA and form peptide bonds in chain elongation, which is described in Figure 10-14.

ensure that only *N*-formylmethionyl tRNA will bind productively to the ribosome and the initiation codon.

Chain Elongation

Figure 10-14 outlines the series of elongation steps that add amino acids to the chain and move the apparatus along the mRNA molecule. These processes incorporate amino acids into the polypeptide in the order specified by the base sequence in mRNA according to the reading frame established in the initiation complex. The elongation process begins with a molecule of the aminoacyl tRNA that has an anticodon corresponding to the next codon downstream from the initiation codon AUG, where *N*-formylmethionyl tRNA^{fMet} is bound in the *P-site,* the peptidyl transfer site. The second aminoacyl tRNA binds through its anticodon to the next codon in the mRNA sequence at a second ribosomal site, the *A-site* (amino acid site), in a process that is controlled by elongation factors (EF) and the hydrolysis of another molecule of GTP. After the second aminoacyl tRNA is bound, the new peptide bond is formed by reaction of the amino group of the second aminoacyl tRNA in the A-site with the activated *N*-formylmethionyl group in the P-site. This releases the tRNA^{fMet}, which dissociates and vacates the P-site. The ribosome must then slide along the mRNA by the distance corresponding to three bases so that the second aminoacyl tRNA can enter the P-site and vacate the A-site. The process is repeated when a third aminoacyl tRNA binds to the next codon in the A-site, and it continues as long as codons are recognized by species of aminoacyl tRNA. The hydrolysis of GTP molecules drives the process in the forward direction.

The elongation process summarized in Figure 10-14 begins with the binding of aminoacyl tRNA at the A-site by an interesting sequence of steps that are outlined in Figure 10-15. The binding of GTP to elongation factor EF-T causes it to dissociate into two subunits, EF-Ts and the complex EF-Tu · GTP. The aminoacyl tRNA binds to EF-Tu · GTP and gives TC, a ternary complex of aminoacyl tRNA, Tu, and GTP. The complex TC then combines with the ribosome, RS, to give RS · TC. This binding affords an opportunity for matching

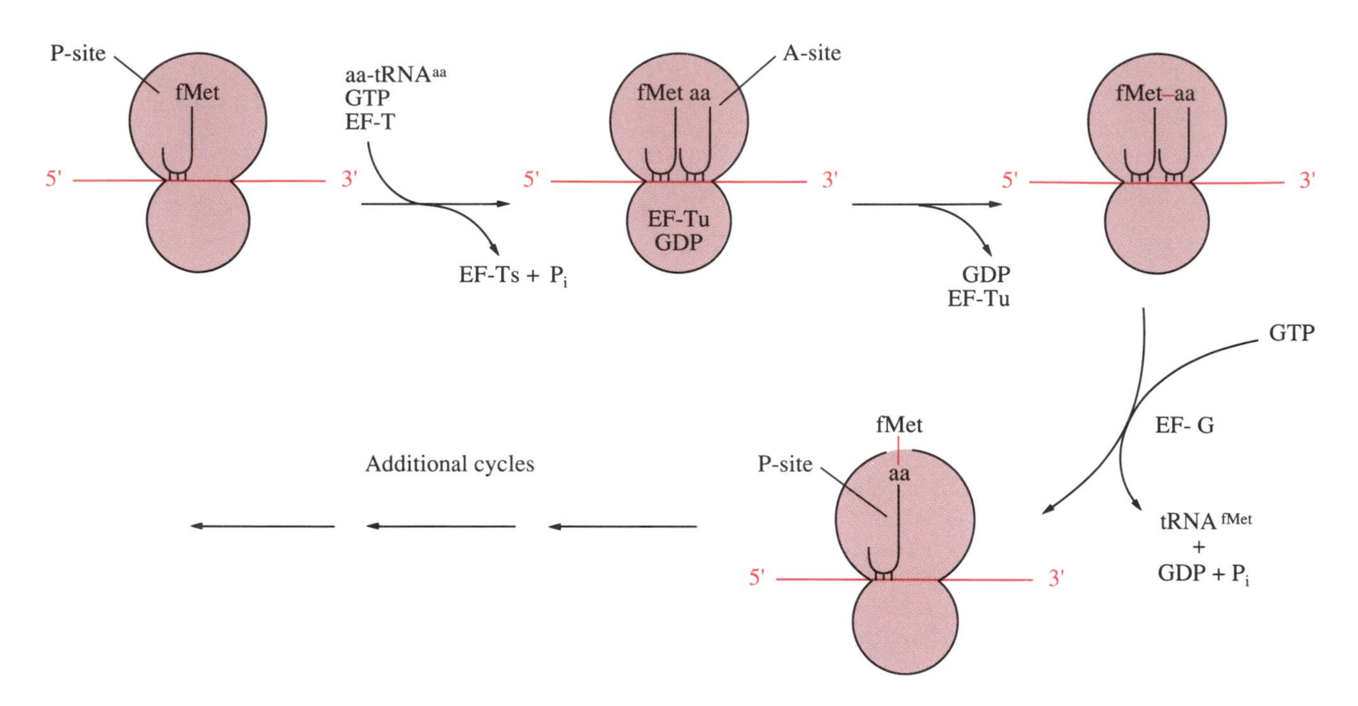

Figure 10-14

Elongation of a Peptide Chain by the Assembled Ribsome. The assembled ribosome has two binding sites for species of aminoacyl tRNA: the P-site for peptidyl transfer and the A-site for binding the incoming aminoacyl tRNA. In the initiation complex, *N*-formylmethionyl $tRNA^{fMet}$ is bound at the P-site with its activated carboxyl group in position to react with the amino group of an incoming aminoacyl tRNA.

In the first elongation step, a specific aminoacyl tRNA, designated aa-$tRNA^{aa}$, binds to the A-site adjacent to *N*-formylmethionyl $tRNA^{fMet}$. This species of aa-$tRNA^{aa}$ is the one carrying the anticodon complementing the codon on the 3′ side of the initiation codon AUG, to which *N*-formylmethionyl $tRNA^{fMet}$ is paired. This step is facilitated by elongation factor EF-T and GTP, with the release of EF-Ts from EF-T and P_i from GTP.

In the next step, the peptide bond is formed by reaction of the amino group of aa-$tRNA^{aa}$ in the A-site with the activated carboxyl group of *N*-formylmethionyl $tRNA^{fMet}$ in the P-site, accompanied by the dissociation of elongation factor EF-Tu, derived from EF-T, along with GDP.

In the last step, $tRNA^{fMet}$ is dissociated and the peptidyl $tRNA^{aa}$ (fMet-aa-$tRNA^{aa}$) migrates to the P-site in a process that requires the action of elongation factor EF-G and the hydrolysis of GTP. The elongation cycle is then repeated with another aminoacyl tRNA binding to the A-site and the peptidyl tRNA in the P-site.

the anticodon of the aminoacyl tRNA with its corresponding codon on mRNA. The wrong anticodon usually will not bind in this step. Productive binding of the correct aminoacyl tRNA then activates the elongation factor Tu to hydrolyze GTP, which is an irreversible step. The complex is then partitioned between two processes: (1) the reaction of the aminoacyl tRNA to form a new peptide bond, which is governed by the rate constant k_p; and (2) the dissociation of the complex without reaction, with the rate constant k_{off}. The elongation factors acting with GTP guide the correct species of aminoacyl tRNA into position and assure fidelity of translation.

Whenever a bound aminoacyl tRNA is incorrectly matched with the mRNA, it dissociates rapidly; that is, k_{off} is large, so that no mistake in peptide-

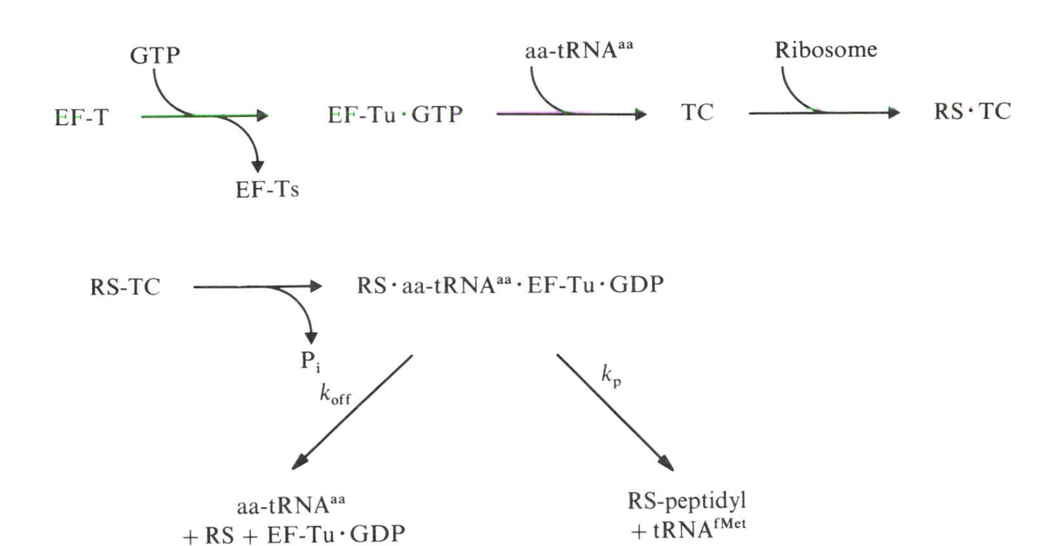

Figure 10-15

Control of Elongation by Elongation Factor EF-T. Elongation factor EF-T consists of two subunits, EF-Ts and EF-Tu, each of which has a specific role in facilitating elongation. Upon binding GTP, EF-Ts dissociates from EF-T, and the complex EF-Tu·GTP binds an aminoacyl tRNA to form the complex TC. A ribosome engaged in elongation—for example, the initiation complex—binds complex TC concomitant with the hydrolysis of GTP and release of P_i. The resulting complex of ribosome, aminoacyl tRNA, EF-Tu, and GDP has two fates: (1) peptidyl transfer and release of tRNA from the peptidyl tRNA (tRNA^{fMet} in the figure) controlled by the rate constant k_p or (2) dissociation of the aminoacyl tRNA controlled by the rate constant k_{off}. When the correct aminoacyl tRNA is bound, $k_p > k_{off}$ and peptidyl transfer occurs. When the incorrect aminoacyl tRNA is bound $k_p < k_{off}$, and the complex dissociates. The k_p/k_{off} processes constitute a second screen ensuring fidelity of translation.

bond formation is made even if the wrong aminoacyl tRNA happens to be bound at first. This second step is an example of proofreading; it is a check on the fidelity of translation following the irreversible step of GTP hydrolysis. This sequence of reactions provides *two* opportunities to select the correct aminoacyl tRNA and avoid incorporation of an incorrect amino acid. The hydrolysis of GTP separates the two steps, so that each step reduces the possibility of error by a large factor.

Another means by which an incorrect amino acid can be incorporated is through a mistaken aminoacyl activation event, in which the wrong amino acid is adventitiously ligated to a species of tRNA. If, for example, isoleucine tRNA synthetase mistakenly produced valyl tRNA^{Ile}, the ribosome would be blind to this mistake and incorporate valine in place of isoleucine. This almost never happens, because aminoacyl tRNA synthetases have their own editing functions that essentially eliminate such mistakes. The two main editing functions of aminoacyl tRNA synthetases are described in Box 10-2.

Peptidyl transfer generates a peptide attached to the tRNA in the A-site of the ribosome. To carry out another cycle of elongation, the peptidyl tRNA must be translocated into the other tRNA binding site on the ribosome, the P-site, so that the amino acid site is free to bind another aminoacyl tRNA and repeat the cycle. This interesting process is brought about with the help of another elongation factor, EF-G, and a molecule of GTP that is hydrolyzed to GDP and P_i (Figure 10-14). This reaction may be somewhat analogous to muscle contraction, because it involves movement and is driven by the hydrolysis of a high-energy phosphoanhydride bond. Protein synthesis then proceeds through repetition of the elongation step, attaching amino acids sequentially to the peptide chain according to the sequence of codons in mRNA and translocating the growing peptide chain back and forth between the amino acid and peptide sites of the ribosome.

Chain Termination

Elongation continues until the ribosome encounters a *nonsense codon* on the mRNA, which is a codon that is not complementary to the anticodon of any aminoacyl tRNA. This is the signal for termination of peptide synthesis. The ester bond that attaches the completed peptide chain to tRNA is hydrolyzed in a reaction that is assisted by three termination factors, R1, R2, and R3, and is probably catalyzed by peptidyl transferase, as shown in Figure 10-16. The protein, ribosome, tRNA, and mRNA then dissociate from one another, and the system is ready to synthesize another molecule of protein, beginning with initiation, the assembly of the initiation complex. The termination process in eukaryotes is similar, except that it involves the hydrolysis of another molecule of GTP.

The Direction of Peptide-Chain Synthesis

At first glance, it may seem curious that protein synthesis occurs by the addition of the acyl group of an entire peptide chain in peptidyl tRNA to the amino group of an aminoacyl tRNA molecule, rather than by the addition of the activated acyl group of aminoacyl tRNA to the amino group on the growing peptide chain. From a purely chemical standpoint, the latter approach would seem to be a simpler means of producing the peptide bond. Indeed, in the chemical synthesis of polypeptides, this strategy is employed (Box 10-4).

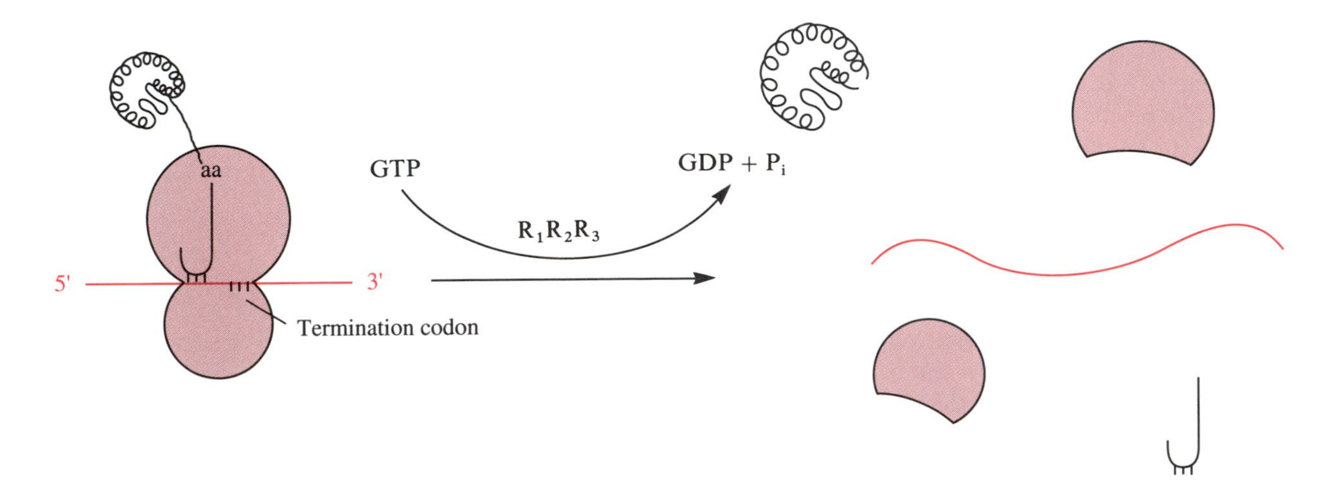

Figure 10-16

Termination of the Elongation Phase in the Biosynthesis of a Protein. Elongation ceases when the ribosome encounters a nonsense codon that is not recognized by a species of aminoacyl tRNA. Termination factors then gain access and facilitate the dissociation of the ribosome and the cleavage of the protein from the peptidyl tRNA. The process requires GTP and results in its hydrolysis to GDP and P_i.

BOX 10-4

THE CHEMICAL SYNTHESIS OF POLYPEPTIDES

Synthesis of the peptide bond requires activation of the carboxyl group of one amino acid followed by transfer of the aminoacyl group to the α-amino group of a second amino acid, as in the biosynthesis of a peptide bond. The differences between the chemical synthesis and the biosynthesis of a peptide bond are the use of activating molecules other than ATP and the use of protective groups in the chemical synthesis. Both of these differences arise from the fact that the reaction microenvironment and the specificity conferred by enzymatic interactions in biosynthesis cannot be replicated in nonenzymatic reactions. The activated amino acids in biosynthesis are aminoacyl tRNA molecules, in which the amino acids are activated as esters. Although esters are activated, they are not so reactive as to undergo nonenzymatic side reactions with free amino groups in cells at rates that would cause problems for the cell. At the peptidyl transfer site, however, enzymatic catalysis is brought to bear, and aminoacyl transfer is catalyzed between the proper molecules, with specificity being achieved through specific binding interactions that focus catalysis on the correct functional groups. If the aminoacyl group were more chemically reactive than an ester, the biosynthetic process would be undermined by side reactions. In the chemical synthesis of the peptide bond, the activated carboxyl group must be much more reactive than an ester because enzymatic catalysis is not available to enhance the rate; and protective groups are required to prevent the otherwise inevitable side reactions.

Several methods for activating the carboxyl group have been used for the chemical synthesis of peptide bonds, but the most widely used method employs dicyclohexylcarbodiimide (DCC). The reaction scheme in Figure A exemplifies the use of this reagent and of the *t*-butyloxycarbonyl (*t*-Boc) group as an amine protective group in the synthesis of a peptide bond. The amino

Figure A

t-Boc-Ala

DCC

BOX 10-4 (continued)

THE CHEMICAL SYNTHESIS OF POLYPEPTIDES

group of the activated amino acid must be protected to prevent its reaction with a second molecule of the activated amino acid. The carboxyl group of the reacting amino acid in the second step must also be protected against side reactions, generally as the benzyl (Bz) ester. The protective groups must be removed to complete the synthesis. All reactions are carried out in organic solvents.

The reaction of DCC with carboxylic acids produces an adduct with the carboxyl oxygen atom bonded to the diimide carbon atom of DCC. This activates the carboxyl group toward reactions with nucleophiles, such as amino groups. For this reason, the amino group of the amino acid must be protected in some way, such as with the *t*-Boc group. The free amino group of a second amino acid, such as glycine benzyl ester, will then react to displace the activating group, forming a peptide bond and dicyclohexylurea. Note that dicyclohexylurea differs from DCC by the addition of water to the carbodiimide group. Thus, DCC provides a means of removing the elements of

Figure B

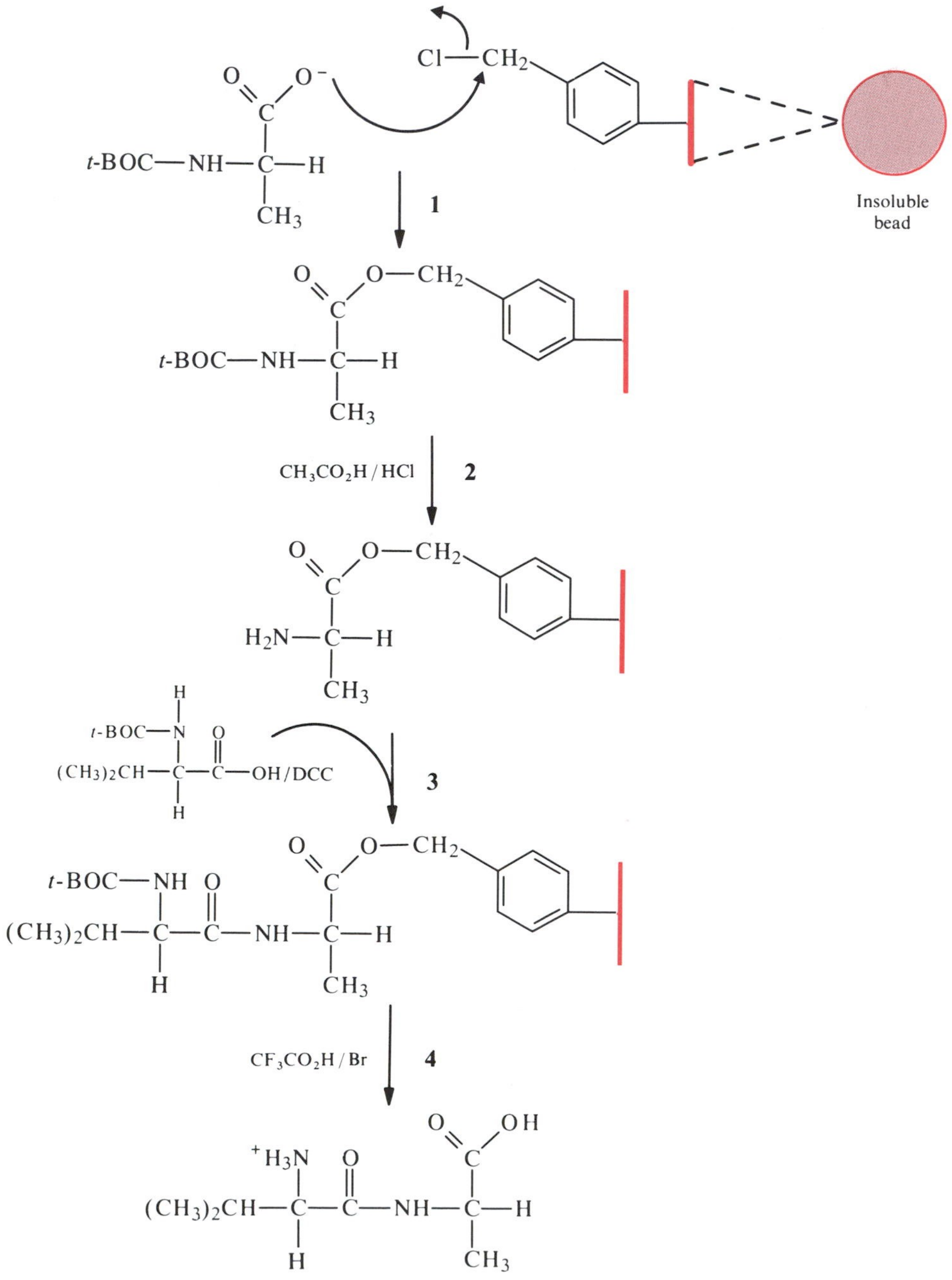

BOX 10-4 (continued)

water from the carboxyl and amino groups and driving the reaction forward. Deprotection of the condensed product to remove the *t*-Boc and Bz groups produces the dipeptide alanyl glycine.

Multiple cycles of such synthesis can be used to synthesize oligopeptides. The method is laborious, however, because the intermediates must be purified after each cycle. B. Merrifield greatly simplified the problem of purification in repetitive cycles by introducing the solid-phase method. The chemical principles and procedures are similar to those described here, but the growing polypeptide chains are covalently bonded to insoluble beads, so that the intermediates can be purified between cycles simply by filtering and washing the beads. Porous polystyrene beads are excellent supports for this type of synthesis, and each bead can hold many growing polypeptide chains. A typical sequence of solid-phase polypeptide-synthesis reactions is illustrated in Figure B. Phenyl rings of polystyrene can be chloromethylated to benzyl chloride groups that can alkylate a *t*-Boc-amino acid to form a *t*-Boc-aminoacyl benzyl moiety on the bead in step **1** (initiation). In step **2**, the amino group is deprotected by treatment with acetic acid and HCl. The beads are filtered, washed, and then subjected to the peptide synthesis in step **3**, in which another *t*-Boc-amino acid is activated with DCC and forms a peptide bond with the free amino group on the aminoacyl benzyl bead. This generates a *t*-Boc-dipeptidyl benzyl bead that can be deprotected by a repetition of step **2** and elongated by reaction with another *t*-Boc-amino acid and DCC in a repetition of step **3**. Iteration of steps **2** and **3** (elongation), with filtration and washing between steps, allows the synthesis of a long polypeptide. In the scheme presented here, the synthesis is curtailed (termination) at the dipeptide stage by treatment with acid, which brings about deprotection and release from the bead to generate the dipeptide valyl alanine.

On closer inspection, it is clear that the biosynthetic mechanism is the most reasonable one. It is important to match the amino acid at the *growing end* of the peptide chain with the new amino acid, according to the code of the mRNA. There would be no such direct matching if the activated amino acid on the aminoacyl tRNA were to add to the amino-terminal end of the peptide chain because the N-terminal amino acid is attached to the peptide, not to tRNA. The enzyme would have to hold on to the N-terminal end of the peptide by noncovalent binding interactions. The occasional escape of the peptide would then give rise to an incomplete protein with no means for recognition and specific reattachment to the synthetic machinery of the ribosome.

Inhibitors of Protein Synthesis

Several antibiotics and toxins act by blocking protein synthesis in bacteria or the host. The antibiotic chloramphenicol, for example, blocks protein synthesis in bacteria by inhibiting the catalysis of covalent bond formation by peptidyl transferase. Diphtheria toxin blocks protein synthesis in susceptible animals by inactivating EF-G through an extraordinary mechanism. Diphtheria toxin consists of several subunits, one of which brings about its passage into the cell through the cell membrane. Another subunit is an enzyme that catalyzes a reaction of the coenzyme NAD^+ (Chapter 16) with a subunit of the translocation factor, EF-2. The nicotinamide ring of NAD^+ is displaced and the remainder of the molecule, ADP-ribose, is attached through carbon-1 of ribose to the factor EF-2, which inactivates EF-2 and blocks protein synthesis (equation 7).

$$NAD^+ + \text{EF-2} \longrightarrow \text{ADP-ribose-EF-2} + \text{Nicotinamide} \qquad (7)$$

The antibiotic puromycin inhibits protein synthesis by inducing early termination of peptide synthesis. Puromycin resembles a molecule of phenylalanine attached to the 3′-terminal adenylyl moiety of tRNA closely enough to bind to the amino acid site (A-site) of the ribosome and react with peptidyl tRNA to make a new peptide bond. However, protein synthesis is immediately

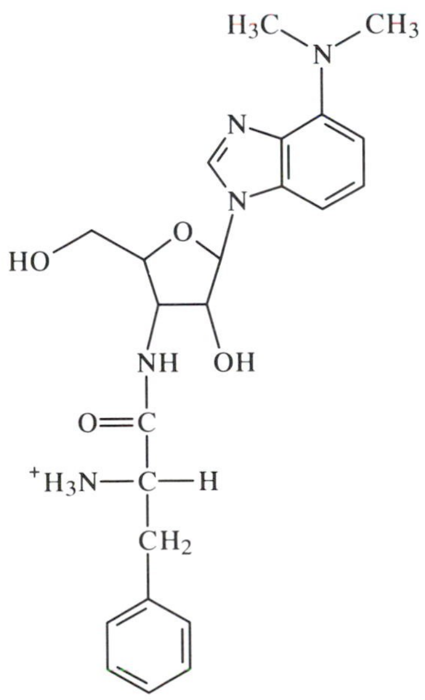

Puromycin

stopped when this happens because, in puromycin, the acyl group of phenylalanine is attached to 3′-aminoribose as the unreactive amide, rather than as a reactive ester, and the amino acid is not activated for further chain lengthening. Synthesis of the peptide chain then ceases.

Folding and Processing

Folding of the peptide chain into the correct three-dimensional structure of a protein and a series of processing steps that modify the covalent structure of many proteins are just as important for the proper functioning of proteins as is their synthesis with the correct amino acid sequence. It is clear that the primary sequence of amino acids carries information that determines the folding of the peptide chain, and some proteins can be unfolded in urea or guanidine hydrochloride and will refold into the correct three-dimensional structure if the denaturing agent is removed (Chapter 4). But this is not true for all proteins. It may be that some proteins fold into the correct three-dimensional structure a little bit at a time, as they come off the ribosome, to give a final structure that is different from the structure that would have been formed if the synthesis had been in the reverse direction, from the C-terminal end. The active structure may also differ from that formed when the fully denatured, unfolded protein is allowed to refold. In other words, a native active protein may be in a *metastable state* that is different from the most thermodynamically stable state. It is not yet clear whether this happens.

Almost all proteins are modified by one or several forms of chemical processing after the synthesis of the peptide chain. In bacteria, the formyl group of formylmethionine, often the methionine itself, and sometimes additional amino acids are cleaved from the N-terminal end of the peptide. There are analogous deletions in eukaryotes. Acetyl groups or other groups may be added to the N-terminal end of the peptide and, when this occurs, no N-terminal amino acid is found upon N-terminal amino acid analysis.

One or a series of sugar residues may be added to the hydroxyl group of serine or threonine or to the amide group of asparagine or glutamine (Chapter 15). The sugars serve as important carriers of information about the function of the protein. The nature of this information transfer is just beginning to be understood.

Serine, threonine, and tyrosine may also be phosphorylated on their hydroxyl groups. Some phosphoproteins contain large numbers of phosphate esters of this kind. It is now becoming clear that a number of enzymes and other proteins are regulated through phosphorylation by protein kinases (Chapter 21).

Hydroxylation by molecular oxygen can introduce new amino acids into a completed peptide chain. It is by this mechanism, for example, that the hydroxyproline residues of collagen are biosynthesized.

The processing of proteins and the secretion of extracellular proteins is brought about in the endoplasmic reticulum and the Golgi apparatus. Newly synthesized proteins that will ultimately be translocated to membranes of the cell or cell organelles or be secreted to the outside of the cell contain hydrophobic *leader sequences* that allow the proteins to pass through membranes as they are being translated, as noted for insulin in Chapter 6. The leader sequences are subsequently removed by proteolysis after the proteins have passed through the membranes. The amino acid sequences contain information that determines which membranes they pass through and what is their final destination. The detailed mechanisms of these important processes are just beginning to be understood.

Molecular Recognition in the Translation of the Genetic Code

We have seen that the translation of the genetic code in DNA into the amino acid sequence of a protein begins with mRNA, more than twenty species of tRNA, and the amino acids. The translating machinery consists of the twenty aminoacyl tRNA synthetases and the ribosomes. The first stage of translation is the activation of amino acids through the actions of aminoacyl tRNA synthetases, which vary in size and quaternary structure (Table 10-2). Each aminoacyl tRNA synthetase matches one amino acid with one species of tRNA *in addition* to catalyzing the activation itself. This mode of activation, which uses specific activating species for specific amino acids, stands in sharp contrast with the chemical synthesis of proteins, in which a single activating group is generally used for all amino acids (Box 10-4).

Aminoacyl tRNA Synthetases

The process of matching an amino acid with a particular species of tRNA requires the recognition of specific structural features in each molecule. The recognition of an amino acid (molecular weight ~125) by an aminoacyl tRNA

Table 10-2

Sizes and quaternary structures of aminoacyl tRNA synthetases

Enzyme	Organism	Quaternary Structure	Amino Acids* per Subunit
Alanine	*E. coli*	α_4	875
Aspartate	Yeast (cytoplasm)	α_2	557
Glycine	*E. coli*	$\alpha_2\beta_2$	303 (α) 689 (β)
Glutamine	*E. coli*	α	551
	Yeast	Unknown	809
Glutamate	*E. coli*	α	471
Histidine	*E. coli*	α_2	424
	Yeast (cytoplasm)	Unknown	526
	Yeast (mitochondria)	Unknown	546
Isoleucine	*E. coli*	α	939
Methionine	*E. coli*	α_2	677
	Yeast (cytoplasm)	α_2	751
Threonine	*E. coli*	α_2	642
	Yeast (cytoplasm)	Unknown	734
	Yeast (mitochondria)	Unknown	462
Phenylalanine	*E. coli*	$\alpha_2\beta_2$	307 (α) 795 (β)
Tryptophan	*E. coli*	α_2	334
	Yeast	α_2	374
Tyrosine	*E. coli*	α_2	424
	Yeast (mitochondria)	Unknown	492

*Primary translation product released from the ribosome.

synthetase through the characteristic structure of its side-chain R-group is more easily comprehended than the recognition of a large molecule of tRNA (molecular weight ~25,000). Indeed, the recognition sites between species of tRNA and their cognate aminoacyl tRNA synthetases are still not thoroughly known, although this subject has been actively investigated for more than 20 years.

The most-obvious recognition site for a given species of tRNA by its cognate aminoacyl tRNA synthetase is the anticodon, which is at the opposite end of the molecule from the CCA-end to which the amino acid will be attached. It is clear that methionyl tRNA synthetase and a few others utilize the anticodon as a part of the recognition interaction. Recognition between species of tRNA and aminoacyl tRNA synthetases is idiosyncratic, however, because in other cases recognition by the synthetase is not at all dependent upon the anticodon. In these species of tRNA, a change in the anticodon has no effect on the amino-acid-accepting properties of the tRNA. Therefore, these species of tRNA must have other recognition sites that interact specifically with their cognate aminoacyl tRNA synthetases. It has recently been shown, for example, that the recognition between $tRNA^{Ala}$ and the alanine tRNA synthetase from *E. coli* is dominated by the presence of a wobble base pair (GU) in the amino acid acceptor stem. All known species of $tRNA^{Ala}$ contain the G3 · U70 wobble base pair, and the substitution of this pair into other species of tRNA confers upon them the capacity to accept alanine either *in vivo* or *in vitro*. In contrast with the location of the anticodon, the G3 · U70 base pair is in the amino acid acceptor stem and near the end, so that the recognition site and accepting site are close together. The role of the base pair G3 · U70 is shown most clearly by the fact that alanine tRNA synthetase catalyzes the activation of alanine to the CCA-end of a microhelix that contains only the acceptor stem of $tRNA^{Ala}$, shown here. The numbers of the bases in the microhelix correspond to those in $tRNA^{Ala}$.

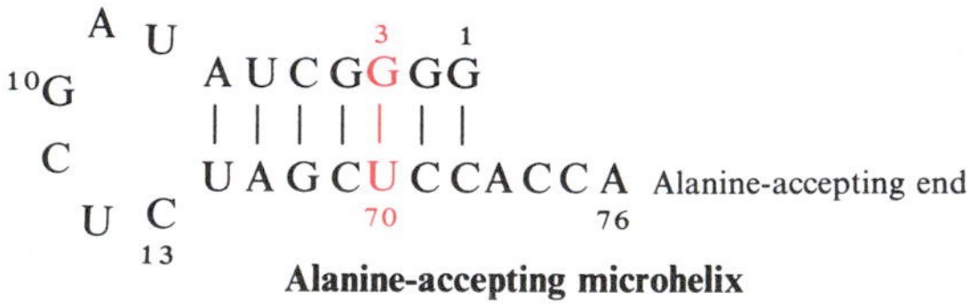

```
        A U          3   1
  10G        A U C G G G G
             | | | | | | |
    C        U A G C U C C A C C A  Alanine-accepting end
     U C               70        76
       13
```

Alanine-accepting microhelix

Ribosomes

The second stage of translation is the assembly of activated amino acids into proteins by the ribosomes. The complexities of ribosome function are outlined in Figure 10-14 and consist of chain initiation, chain elongation, and chain termination. These functions require the coordinated actions of initiation factors, mRNA, and *N*-formylmethionyl $tRNA^{fMet}$ for initiation; the elongation factors and more than twenty species of aminoacyl tRNA for chain elongation; and termination factors for chain termination. The GTP hydrolyzed at each stage of ribosome action presumably provides the energy for the actions of the initiation, elongation, and termination factors. All three processes include some sort of physical organization, reorganization, or disassembly of the protein biosynthetic complex.

The ribosomal complex matches the anticodons that identify the more than twenty species of aminoacyl tRNA with the codons of mRNA; it then catalyzes peptide-bond formation. The complex does not recognize aminoacyl structure and will insert an incorrect amino acid if confronted with a mischarged aminoacyl tRNA such as alanine $tRNA^{Phe}$. Ribosomal recognition consists of identifying chain-initiation sites and chain-termination sites and facilitating the matching of codons with anticodons. Their catalytic functions consist of peptide-bond formation and, in cooperation with other factors, the energy-requiring physical transformations required for initiation-complex formation, chain elongation, and disassembly at chain-termination sites.

Errors in Protein Biosynthesis: Hemoglobin and Cooperativity

Before 1950, our understanding of genetics centered on the Mendelian inheritance of spots, colors, and other traits. Since then, our understanding of the nature of double-stranded DNA and its basic role in genetics has fueled a revolution in genetics and biology. This revolution made it possible to understand genetics in chemical terms. Today, most high-school science students know that one gene codes for one enzyme, one triplet of bases in DNA codes for one amino acid, and a change in one base of DNA can give a mutation that results in the incorporation of a different amino acid and the synthesis of a protein with altered properties. Such mutations can be brought about by errors in replication, by ionizing radiation, and by carcinogenic chemicals.

Sickle-Cell Anemia

There is no better illustration of the consequences of genetic errors than the hemoglobin molecule, the oxygen carrier of the red blood cell. The hemoglobin in red cells carries O_2 by way of the blood stream from the lungs to the tissues, releases the O_2, and is returned to the lungs to absorb more O_2. Hemoglobin is a four-subunit protein, with two types of similar but not identical subunits ($\alpha_2\beta_2$) and an overall molecular weight of 64,500.

More than one hundred genetically variant forms of human hemoglobin are known. Sickle-cell anemia is caused by a point mutation in DNA that leads to the substitution of a valine for glutamate in position 6 of the β-subunit of hemoglobin, a change in which a nonpolar amino acid side chain is substituted for a negatively charged side chain. Point mutations are usually caused by a single base change in the gene that specifies the structure of that protein, such that the code for one amino acid is changed to a code for a different amino acid. For example, the mutation Glu $\rightarrow$ Val can be brought about by a codon change of GAA or GAG in mRNA to GUA or GUG. These changes correspond to mutation of a T to an A at one position of the gene for the β-subunit of hemoglobin.

The mutation Glu $\rightarrow$ Val in sickle-cell hemoglobin reduces the solubility of deoxyhemoglobin enough to cause it to precipitate in the red blood cell. When red cells pass through capillaries in the tissues, oxygen is removed from oxyhemoglobin and taken up by the tissues. Sickle-cell deoxyhemoglobin that is formed in the red cell tends to precipitate. This stiffens the red cell so that it cannot be sufficiently deformed to allow it to squeeze through the capillary. Capillary blockage causes a further loss of oxygen, so that a chain reaction is initiated that can lead to severe circulatory disturbances and pain. The outcome can be fatal.

The Binding of Oxygen by Myoglobin and Hemoglobin

In the oxygenation of hemoglobin, oxygen binds reversibly to iron (Fe^{2+}) in heme, *iron protoporphyrin IX*, which is bound to each of the four subunits.

Iron protoporphyrin IX

Heme is a porphyrin ring, a macrocycle composed of four pyrrole rings that are connected by one-carbon bridges, with ferrous iron bound to the four nitrogen atoms. The biosynthesis of this molecule is described in Chapter 26.

Myoglobin is a much simpler oxygen-binding protein than hemoglobin, because it consists of a single subunit (molecular weight 16,900) with one molecule of heme. Myoglobin serves the important function in muscle of binding O_2 and storing it in an available but immobilized form, whereas hemoglobin binds O_2 in the lungs and carries it to cells, including muscle cells, through the blood stream. The binding of O_2 to myoglobin follows a simple hyperbolic saturation curve with increasing oxygen pressure, as is shown in Figure 10-17A. The shape of the binding curve for myoglobin is the same as that of the enzyme-substrate saturation curve of initial rate plotted against substrate concentration.

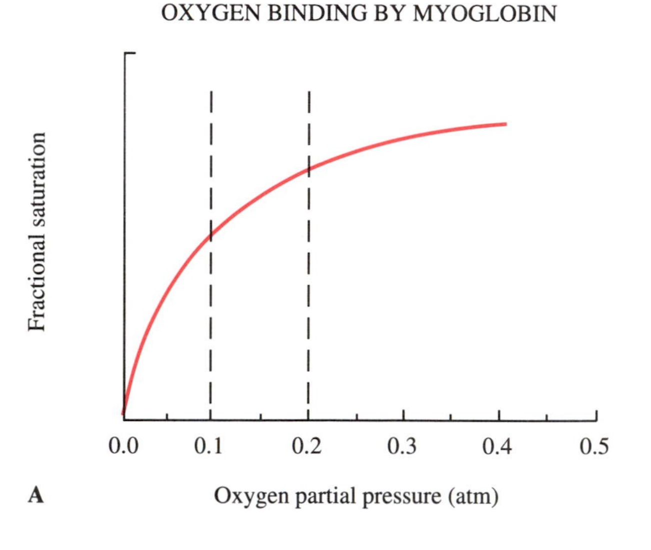

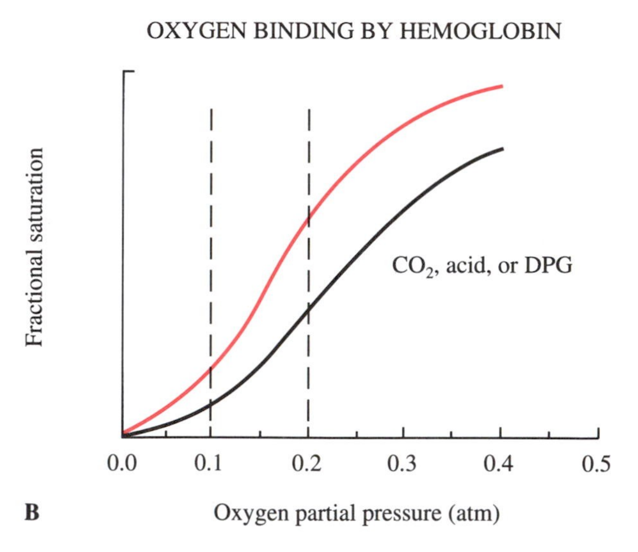

Figure 10-17

Oxygen-binding Curves for Myoglobin and Hemoglobin. The oxygen-binding curve for myoglobin (A) is a simple saturation curve. Myoglobin is a monomeric protein that binds only one molecule of oxygen. The oxygen binds to the Fe in the heme of myoglobin and, because all myoglobin molecules are alike and display a single affinity for oxygen, the binding follows a simple hyperbolic saturation curve. Hemoglobin is tetrameric ($\alpha_2\beta_2$) and contains four hemes per molecule, one in each subunit. Hemoglobin binds oxygen cooperatively and displays a sigmoid binding curve (B). Note that, in the presence of CO_2 or diphosphoglycerate (DPG) or at low pH, the binding curve remains sigmoid but is displaced to the right. This is because CO_2 and DPG interact with hemoglobin in such a way as to decrease its affinity for oxygen.

The O_2-saturation curve for hemoglobin is sigmoid, as shown in Figure 10-17B; it is very different from that for myoglobin. This sigmoid curve is physiologically useful because it allows hemoglobin to pick up and discharge large amounts of oxygen at the different partial pressures of O_2 that exist in lungs and tissues. The sigmoid shape of the curve gives hemoglobin the capacity to bind very different amounts of oxygen within a relatively narrow range of O_2 partial pressures, as indicated by the dashed lines in Figure 10-17B. A simple hyperbolic curve gives a much smaller difference and a much smaller release of oxygen, as illustrated for myoglobin in Figure 10-17A. Myoglobin can release more oxygen when the partial pressure of oxygen in the tissues becomes very low, which happens under conditions of rapid oxygen utilization. Thus, myoglobin serves as a reservoir for the binding of oxygen in muscles, whereas hemoglobin is adapted to be an efficient *transporter* of oxygen.

Cooperativity in Hemoglobin

The sigmoid O_2-binding curve reveals the cooperative binding of oxygen by hemoglobin. Cooperativity simply means that, when one molecule of oxygen is bound by a molecule of hemoglobin, additional molecules of oxygen are more tightly bound to that molecule of hemoglobin. A molecule of hemoglobin has the capacity to bind as many as four molecules of oxygen, one for each heme that is bound to the four subunits.

Cooperativity arises from interactions among the α- and β-subunits of hemoglobin that depend on their state of oxygenation. The mechanism of cooperative oxygen binding by hemoglobin has been studied intensively for more than fifty years and is now well understood. The details are complex, but the basic principle is simple. The stable structures of deoxyhemoglobin and oxyhemoglobin are different, as is indicated schematically by the circles and boxes in Figure 10-18. The circles represent the stable conformation of deoxyhemoglobin, which has a low affinity for oxygen, and the squares represent the structure of the high-affinity conformation of hemoglobin that exists whenever oxygen is bound to the heme in at least one of the subunits.

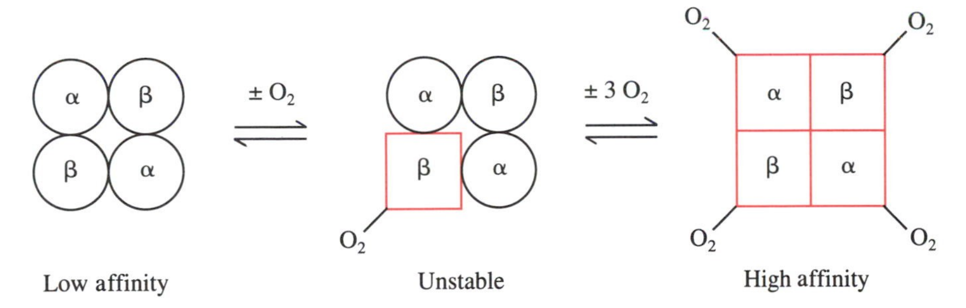

Figure 10-18

A Schematic of the Effect of Oxygen on the Structure of Hemoglobin. The circles and squares symbolize two conformational states of the α- and β-subunits of hemoglobin. The circles symbolize a low-affinity state and the squares a high-affinity state. The two states differ significantly in structure and in their intersubunit interactions, as well as in their oxygen affinities. The binding of one oxygen to one subunit in the tetrameric protein alters the conformation of that subunit and alters intersubunit interactions. This strains the other subunit structures and induces them to undergo structural changes and assume the high-affinity state. This transition is responsible for the cooperative binding behavior of hemoglobin. Thus, the form in which one or more oxygen molecules is bound to hemoglobin exhibits enhanced binding affinity for oxygen, leading to the sigmoid binding curve and cooperative oxygen binding.

Hemoglobin in the absence of oxygen has a low observed affinity for oxygen because all of the subunits are in the low-affinity conformation. When an oxygen molecule binds to a subunit of a molecule of hemoglobin, an unstable situation is created that is very likely to be reversed (Figure 10-18). This accounts for the weak binding of oxygen at low O_2 partial pressures in Figure 10-17B. However, as the oxygen pressure increases, more oxygen binds until most of the molecules have at least one oxygen bound to a subunit. The high-affinity state of the four subunits now becomes more stable and the molecule flips into the high-affinity conformation, indicated by the squares in Figure 10-18. The intermediate states are unstable because the low-affinity conformations (circles) and the high-affinity conformations (squares) would have to coexist in the same molecule, and they do not fit well together. The best available experimental evidence indicates that the conformational change is essentially completed by the time one molecule of oxygen is bound to a molecule of hemoglobin. This is a good thing for hemoglobin, because with only one molecule of oxygen bound it is in a favorable state to bind three more with high affinity in the steep part of the binding curve of Figure 10-17B.

An Insertion Mutant of Hemoglobin

Not all mutant proteins contain changes of a single amino acid. Other kinds of mutations are also possible, a rare kind being the insertion of an additional amino acid into an internal position of a protein. An example of this is hemoglobin Catonsville, which exhibits altered O_2-binding properties. Hemoglobin Catonsville differs structurally from normal hemoglobin in that it contains an additional glutamate inserted between proline-37 and threonine-38 in the α-subunit. This insertion appears at the interface of α- and β-subunits and causes the mutant to bind O_2 with increased affinity and decreased cooperativity, so that the O_2-binding curve is displaced to the left of that in Figure 10-17B. The increased O_2 affinity inhibits O_2 release to the tissues and causes pallor and jaundice. The molecular defect in hemoglobin Catonsville has only recently been identified.

Oxygen-Binding Energies and Conformational Changes

The energetics of the conformational change brought about by the binding of oxygen to a subunit of hemoglobin can be described by the thermodynamic box shown in Figure 10-19. The overall change in Gibbs free energy between the low-affinity form of hemoglobin (circle) and the high-affinity form with bound oxygen (square-O_2) is the same by either the clockwise or the counterclockwise binding pathway. The difference in energy of any two sides of the box is the interaction energy ΔG_{int} (equations 8a and 8b).

$$\Delta G_1 + \Delta G_2 = \Delta G_3 + \Delta G_4 \qquad (8a)$$

$$\Delta G_1 - \Delta G_3 = \Delta G_2 - \Delta G_4 \qquad (8b)$$

Figure 10-20 shows the energy-level diagram for this system. There is a relative destabilization of the low-affinity form of hemoglobin with oxygen bound and of the high energy form in the absence of oxygen, which is a consequence of the interaction energy. If there were no interaction energy, the solid lines on opposite sides of the box would be parallel.

The change from one state to another is brought about by the binding of oxygen, which pulls the iron atom into the plane of the porphyrin ring (Figure 10-21). In the absence of oxygen, the iron is held outside the plane of the ring by coordination of iron to the imidazole ring of a histidine residue. When the iron is

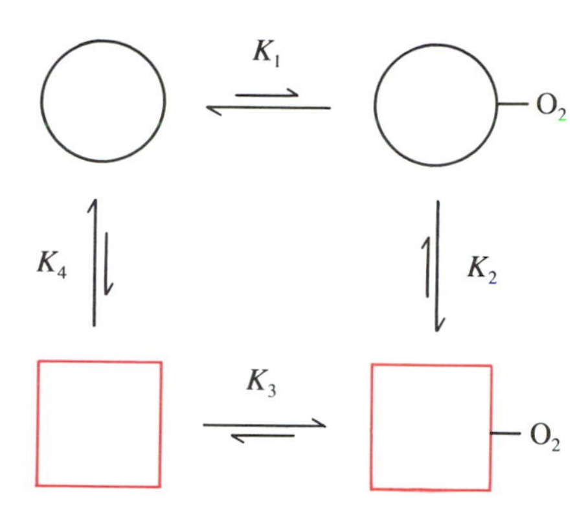

Figure 10-19

A Thermodynamic Box That Describes the Energetics for the Conformational Transitions Between the High- and Low-Affinity Structures and the Binding of Oxygen to the Subunits of Hemoglobin. The relative energy levels for the four forms are illustrated in Figure 10-20.

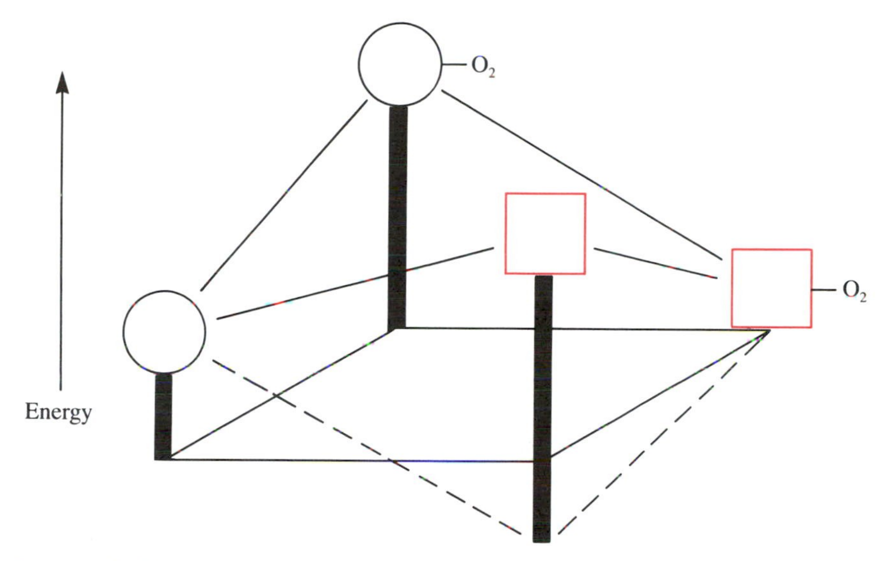

Figure 10-20

An Energy-Level Diagram Relating the Energies of the Low- and High-Affinity Conformations of Hemoglobin with and without Bound Oxygen. The relative energy levels of hemoglobin subunits are shown for the two conformations, the low-affinity (circles) and high-affinity (squares) states, in both unliganded and oxygen-liganded forms. Note that the relative energies of the two conformations are reversed in the unliganded and oxygen-liganded states. Unliganded hemoglobin normally exists in the low-affinity state, but oxygen binding raises its energy above that of the high-affinity state. This causes the structure to relax to the high-affinity state which, with oxygen bound, is lower in energy than the unliganded high-affinity form.

drawn into the porphyrin ring plane by its coordination to oxygen, it brings the histidine residue with it. This movement of histidine sets off a chain of structural changes throughout the protein that is analogous to the motions of a series of connected gears and levers that rearranges the entire structure of the subunit. These changes result from the perturbation of a large network of hydrogen bonds, electrostatic bonds, and van der Waals contacts extending throughout the subunit. These changes alter the shape of the subunit enough so that its interactions with its neighboring subunits are disturbed, inducing an unstable state that remains destabilized until all of the subunits have changed to the new conformation of oxyhemoglobin.

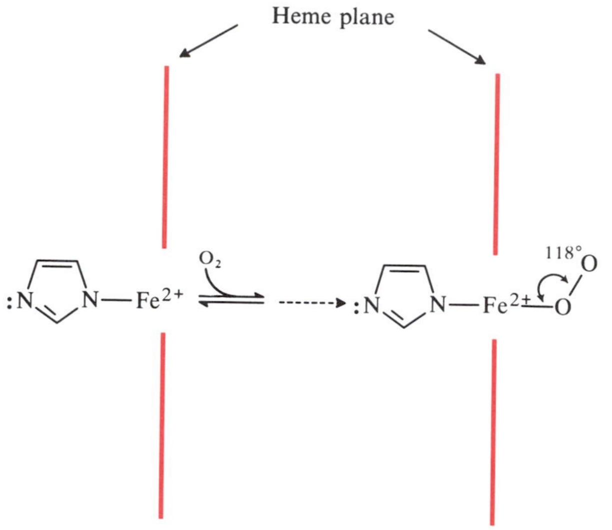

Heme plane

:N N—Fe^{2+} $\overset{O_2}{\rightleftharpoons}$ ------▸ :N N—Fe^{2+}—O—O 118°

Figure 10-21

The Effect of Oxygen Binding on the Structure of Heme and Its Histidine Ligand. In the unliganded form of hemoglobin, the iron is pulled out of the heme plane by coordination with a histidine of the protein, because of the steric interactions between the heme ring and the imidazole ring of histidine. Upon binding oxygen, the iron is drawn into the ring plane, and the imidazole ring of histidine is pulled along with the iron. This small movement of histidine is transmitted to other groups of the protein through its H-bonding network and causes the structure to relax to a high-affinity conformation. This change in structure destabilizes the interactions of this subunit with the other subunits and induces them also to change into the high-affinity conformation.

The interactions between different subunits are brought about, at least in part, through ion pairs between charged side chains of amino acids in the different subunits. Many of the mutations in hemoglobin involve the loss of these charged groups or changes in them, which disturb the interactions between subunits and alter cooperativity. A change of one amino acid and one charge can significantly decrease the cooperativity and the oxygen-carrying ability of hemoglobin, resulting in clinical symptoms. One hundred fifty different single amino acid mutations have so far been identified. Most of these mutations can be accounted for by a single base change in DNA.

Hemoglobin is an allosteric protein in which the binding of a ligand to one subunit can result in changes in structure and behavior of a different subunit. Analogous behavior is found in regulatory enzymes. The activities of many regulatory enzymes are controlled by the binding of substrates, inhibitors, or activators to sites on multiple subunits. The changes in activity are reflected in either K_m or V_m or both. An important example of such an enzyme is aspartate transcarbamoylase, which is discussed in Chapter 27.

Summary

The carboxylate group of a carboxylic acid must be activated by bonding to a good leaving group in order to form an acyl derivative of another molecule. The activated forms of most biological carboxylic acids in the cell are esters of coenzyme A (CoA), which are produced by the actions of enzymes such as acetyl

CoA synthetase. These enzymes utilize ATP as the energy source to activate the carboxylate oxygen atom of substrates such as acetate for reaction and displacement in a later step by the thiol group of CoA to form the CoA ester. ATP acts as a dehydrating agent in these reactions, and it plays the same role in the activation of amino acids for protein biosynthesis.

Each amino acid required for protein biosynthesis is activated by an aminoacyl tRNA synthetase, which catalyzes its ligation to the 2′- or 3′-hydroxyl group at the CCA-end of a specific species of transfer RNA (tRNA). In the resulting aminoacyl tRNA, the carboxyl group of the amino acid is chemically activated for peptide-bond formation, and the amino acid is matched with the species of tRNA that can direct its incorporation into the correct sequence positions of all the proteins in the cell.

The information for the biosynthesis of proteins flows from DNA to RNA to the protein biosynthetic machinery of the cell. The sequence of amino acids in a protein is specified by the sequence of nucleotides in a gene in DNA. This information is transcribed into the complementary nucleotide sequence of messenger RNA (mRNA) according to the base-pairing rules that are similar to those followed in the pairing of heterocyclic bases in double-stranded DNA. In DNA, adenine (A) in one strand is paired with thymine (T) in the other, and guanine (G) is paired with cytosine (C). The base-pairing rules for transcription are, from DNA to RNA, A with uracil (U), T with A, G with C, and C with G.

The genetic code consists of codons—triplets of nucleotide sequences in mRNA that allow the overall nucleotide sequence to be translated into an amino acid sequence. Each codon specifies a particular function in translation. One codon specifies the start point for initiation in reading the code, most of the codons specify the insertion of amino acids in the correct order during elongation of the peptide and two codons specify the stopping point for termination. In translation, each amino acid is activated by esterification to a specific tRNA, which contains an anticodon complementary in sequence to the codon specifying the incorporation of that amino acid.

Protein biosynthesis is carried out by ribosomes, which consist of ribosomal RNA (rRNA) and more than fifty small proteins. The assembly and functions of ribosomes are directed by the actions of proteins known as initiation factors, elongation factors, and termination factors. The initiation complex consists of the two major ribosomal subunits, mRNA, and *N*-formylmethionyl $tRNA^{fMet}$ bound through its anticodon to the initiation codon on mRNA (5′-AUG-3′). Other species of aminoacyl tRNA are then bound by pairing their anticodons with the succeeding codons along the mRNA during peptide elongation; and as each is bound the peptide bond is formed between the amino group of the newly bound aminoacyl tRNA and the activated carboxyl group of the growing peptidyl tRNA, with release of a species of tRNA. In termination, chain growth is stopped at a nonsense codon that fails to bind the anticodon of any species of tRNA. The energy requirements for initiation-complex formation, elongation, and disassembly of the elongation complex during termination are derived from GTP, which undergoes hydrolysis to GDP and P_i coupled with the actions of proteins that facilitate these processes.

Changes in the nucleotide sequences in genes are mutations that appear as mistakes in the amino acid sequences of proteins. Human hemoglobin is an example of a protein for which more than one hundred fifty mutant variants are known. Hemoglobin consists of four subunits—two α-chains and two β-chains, each of which contains a molecule of heme. Heme is a macrocyclic ligand consisting of four substituted pyrrole rings linked by methine bridges, with the pyrrole nitrogens coordinated to Fe^{2+}. Hemoglobin reversibly binds O_2 as a ligand to Fe^{2+} in heme. Each molecule of hemoglobin can bind as many as four molecules of O_2, one to each heme; and binding is cooperative. In cooperative

O_2-binding by hemoglobin, the binding of O_2 to heme in a subunit increases the affinities of heme in other subunits for O_2. Cooperativity is mediated through subunit conformational changes that are induced by the coordination of O_2 to Fe^{2+} in heme.

Sickle-cell anemia results from a point mutation in DNA that alters the genetic code and leads to the replacement of glutamate by valine at position 6 of the β-subunit in hemoglobin. This reduces the solubility of the deoxy form of hemoglobin; precipitation of deoxyhemoglobin in the red blood cells alters the physical properties of blood cells and reduces their permeability in capillaries. Many other mutant forms of hemoglobin are known to exist.

ADDITIONAL READING

Ochoa, S., and Mazumder, R. 1974. Polypeptide chain initiation. In *The Enzymes*, vol. 10, 3d ed., ed. P. D. Boyer. New York: Academic Press, pp. 1–51.

Lucas-Lenard, J., and Beres, L. 1974. Protein synthesis: peptide chain elongation. In *The Enzymes*, vol. 10, 3d ed., ed. P. D. Boyer. New York: Academic Press, pp. 53–86.

Tate, W. P., and Caskey, C. T. 1974. Polypeptide chain termination. In *The Enzymes*, vol. 10, 3d ed., ed. P. D. Boyer. New York: Academic Press, pp. 87–118.

Freifelder, D. 1987. *Molecular Biology*, 2d ed. Boston: Jones and Bartlett.

Schimmel, P. 1987. Aminoacyl tRNA synthetases: general scheme of structure-function relationships in the polypeptides and recognition of transfer RNA's. *Annual Review of Biochemistry* 56: 125–158.

PROBLEMS

1. Acetyl CoA synthetase also catalyzes an exchange reaction that is different from that discussed in the text but is consistent with the mixed anhydride intermediate. What is this reaction and what experimental conditions should be used to make its observation significant?

2. Draw the chemical structural formulas for dAdo, UDP, 3′-AMP, Cyd, dTDP, dGDP, Guo, dThd, dCDP.

3. What purely chemical advantage does an aminoacyl tRNA have over an aminoacyl adenylate or an aminoacyl phosphate as an activated amino acid for protein biosynthesis?

4. Define the following terms:

(a) Nucleoside
(b) CDP
(c) TTP
(d) Genetic code
(e) Codon
(f) Initiation
(g) Transcription
(h) Translation
(i) Anticodon
(j) Elongation
(k) Ribosome
(l) Nucleotide
(m) 5′ end
(n) D loop
(o) EF-T
(p) IF-3
(q) P-site
(r) 16s rRNA
(s) Activation
(t) $tRNA^{Leu}$
(u) mRNA
(v) Termination
(w) Puromycin
(x) Mutation

5. The list in question 4 includes several protein biosynthetic processes. Aminoacyl tRNA synthetases are important in two of these processes. Which two processes are they and what is the role played by the aminoacyl tRNA synthetases in each one?

6. What nucleotide sequences in DNA and mRNA would code for the following amino acid sequence?

Met-Arg-Val-Val-Ala-Cys-Ser-Phe-Asp-Pro-His-Asn

7. What is meant by cooperative O_2 binding by hemoglobin? What mechanism explains cooperative O_2 binding by hemoglobin? Why is O_2 binding by myoglobin not cooperative?

8. What is the structure and biological function of heme?

9. How is binding energy used by hemoglobin to drive the change in conformation from that displaying low affinity for O_2 to that displaying high affinity for O_2?

10. What is the biological value of cooperative O_2 binding by hemoglobin?

Phosphoryl Group Transfer

11

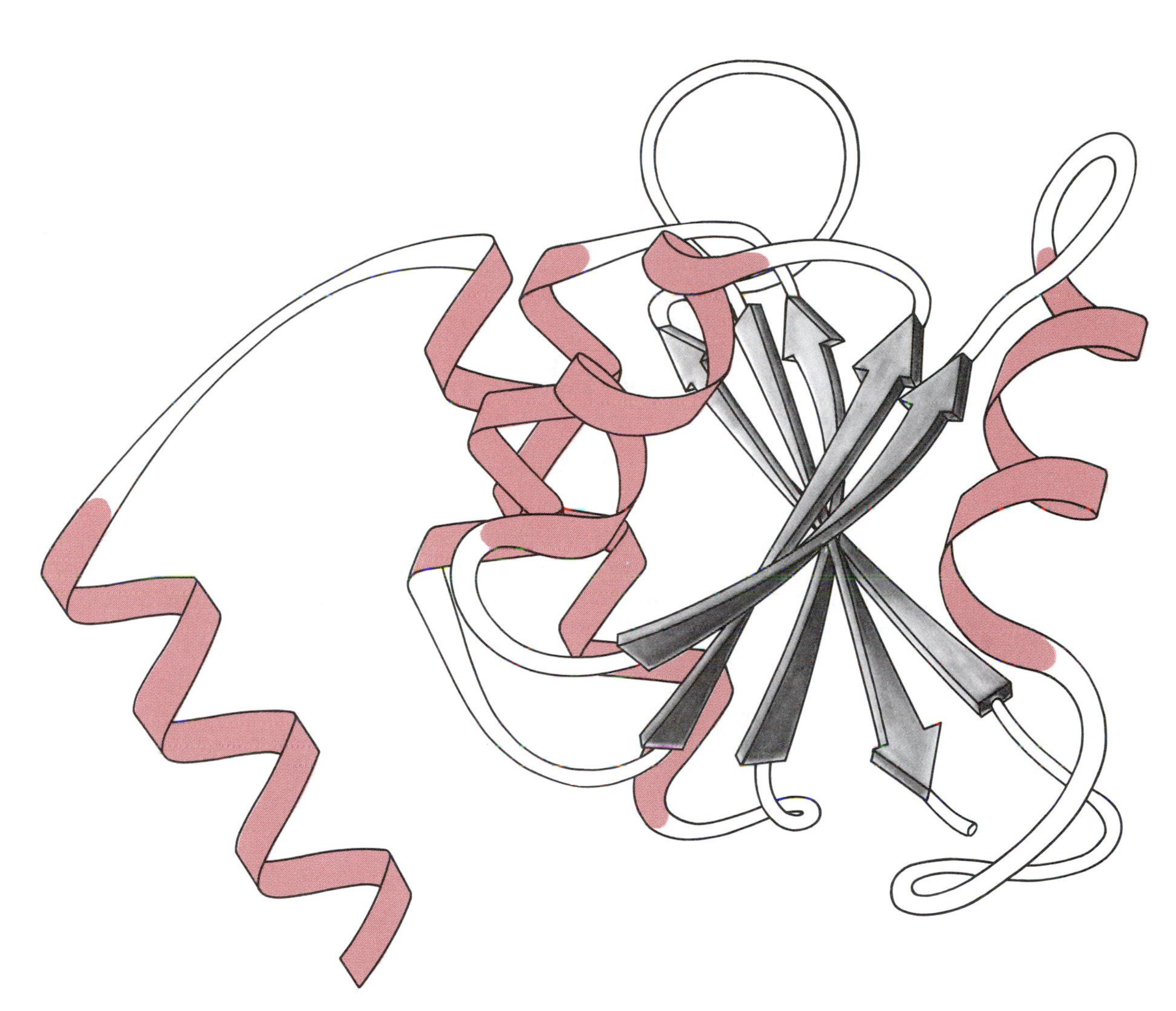

A ribbon diagram of the substrate-binding domain in pyruvate kinase. (Adapted from a drawing by Jane Richardson.)

All biosynthetic pathways include reactions in which two molecules are joined together into a third molecule. In many of these reactions, phosphoryl, acyl, or glycosyl groups are ligated to other molecules, and the reactivities of these groups are normally increased by attachment to good leaving groups that can be displaced in the ligation step. A general formulation of activation and ligation is given by equations 1a and 1b, in which X—is a phosphoryl, acyl, or glycosyl group; $RO \sim PO_3^{2-}$ is a high-energy phosphorylating agent—usually ATP; and :YH is a nucleophilic substrate.

$$X\text{—}OH + RO{\sim}PO_3^{2-} \rightleftharpoons X\text{—}OPO_3^{2-} + R\text{—}OH \quad (1a)$$

$$X\text{—}OPO_3^{2-} + :YH \rightleftharpoons X\text{—}Y + HOPO_3^{2-} \quad (1b)$$

These reactions are reversible, but $\Delta G^{\circ\prime}$ for the sum of equations 1a and 1b is usually negative and large, making the overall process virtually irreversible. We have already seen several examples of acyl group activation in Chapter 10. The ultimate purpose of such phosphoryl group transfer reactions is to provide control and a strong chemical force to drive biosynthetic reactions. Not all group transfer reactions are irreversible, however; many are nearly isoenergetic and have the important function of regulating the pools of biosynthetic intermediates that will be used for the production of end products in virtually irreversible steps.

In this chapter, we consider the chemistry and enzymology of phosphoryl and phosphoryl ester transfer reactions.

Chemistry of Phosphoryl Transfer

In all transfers of phosphoryl, phosphoryl ester, and pyrophosphoryl groups, the phosphorus reacts as an electrophilic species being transferred from one nucleophile to another, as in the hydrolysis of glucose-1-phosphate catalyzed by acid phosphatase (equation 2).

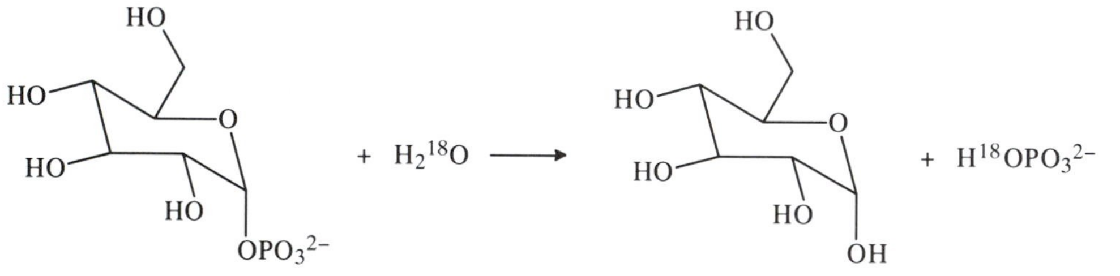

HO, HO, HO, HO, O, OPO_3^{2-} (glucose-1-phosphate) $+ H_2{}^{18}O \longrightarrow$ HO, HO, HO, HO, O, OH (glucose) $+ H^{18}OPO_3^{2-}$ (2)

This reaction was carried out in $H_2{}^{18}O$ by M. Cohn, who showed that the ^{18}O appears in the phosphate rather than in the glucose produced in the reaction. All enzymatic phosphoryl transfers proceed in this way, with cleavage of the phosphorus–oxygen bond and the addition of a nucleophile to phosphorus. Many other reactions of phosphate compounds proceed with carbon–oxygen bond cleavage and the addition of a nucleophile to carbon. These are reactions of compounds that have been *activated* by phosphorylation, such as acyl group transfer and glycosyl group transfer reactions.

Hydrolysis of Phosphate Esters

The simplest phosphoryl group transfer reactions are those in which the phosphoryl groups are transferred to water. The chemistry of the hydrolysis of phosphoric esters is not as well understood as that of the hydrolysis of analogous carbon compounds. The mechanisms of the hydrolysis of a few types of phosphate esters are moderately well known. The phosphomonoester monoanions undergo hydrolysis through the *dissociative transition state* shown

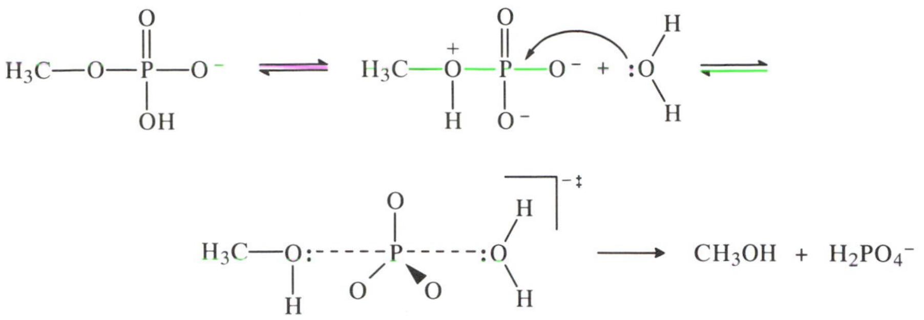

Figure 11-1

Hydrolysis of Methyl Phosphate Monoanion Through a Dissociative Transition State. The reactive species of methyl phosphate is the *minor* form, in which the proton is bonded to the leaving oxygen atom. In the transition state, the water molecule and methanol are weakly bonded to the phosphorus. No discrete addition intermediate is formed.

in Figure 11-1. This reaction is characterized by the importance of bond cleavage in the transition state. There is little bonding either to the attacking water molecule or to the departing methanol in the transition state. Extensive studies show that the minimum requirements for this mechanism are that the phosphoryl group must carry two negative charges and the departing group must be a good leaving group. In the hydrolysis of methyl phosphate, these requirements can be met by one form of the monoanion, in which the proton is bonded to the oxygen that bridges the methyl group and phosphorus (Figure 11-1); and it is this species that is thought to be reactive, although it is no doubt the minor component of the equilibrium. Some phosphomonesters may be unable to form the reactive species shown in Figure 11-1 before P–O bond cleavage, and the reaction may proceed with concerted transfer of the proton to the leaving alcohol. The dianionic forms of phosphomonoesters and related compounds also react by a mechanism with a dissociative transition state if they have a sufficiently good leaving group that does not require protonation to leave. The dianion of methyl phosphate does not undergo hydrolysis at an appreciable rate because methoxide (CH_3O^-) is a poor leaving group.

The hydrolysis of *phosphodiester monoanions* cannot proceed by the dissociative mechanism because the phosphoryl group of a diester cannot have a double negative charge to provide the driving force needed to expel the leaving group. This is an important fact in biochemistry. If the phosphodiester linkage could be cleaved by water at a significant rate, DNA and the genetic information that it carries would soon be lost, with disastrous consequences for the cell and for life itself. Phosphodiester anions do react very sluggishly at very high temperatures with *hydroxide* ions through an *associative transition state,* illustrated in Figure 11-2. This transition state is characterized by a high degree of bonding between the phosphorus and both the entering and the leaving groups. The rate of this reaction is very slow under any conditions, particularly in neutral solutions where the concentration of hydroxide is so low that the reaction is not observable at physiological temperatures. (The estimated half-time is 80 million years at pH 7 and 37°C.)

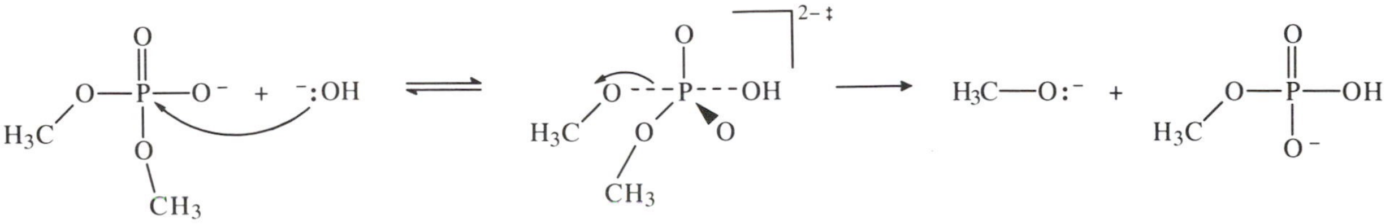

Figure 11-2

Base-Catalyzed Hydrolysis of Dimethyl Phosphate Through an Associative Transition State. This is an extremely slow reaction that cannot be observed at body temperatures in neutral solutions. The displacement of methoxide by hydroxide proceeds through a transition state in which both hydroxide and methoxide are fairly strongly bonded to phosphorus.

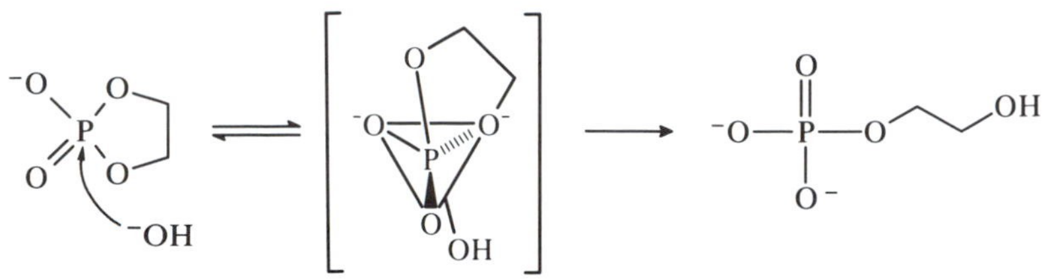

Figure 11-3

The Mechanism of Base-catalyzed Hydrolysis of Five-membered Cyclic Phosphodiesters. In the hydrolysis of five-member-ring cyclic phosphates, a discrete pentavalent intermediate shaped like a trigonal bipyramid is formed and decomposes to products.

Hydrolysis of Cyclic Phosphodiesters

The five-member-ring cyclic phosphodiesters are a special class of diesters that undergo hydrolysis 10 million times as fast as the noncyclic or six-member-ring cyclic analogs. These molecules react by the extreme associative mechanism, in which a discrete pentavalent intermediate is formed with single bonds linking phosphorus with both the entering and the leaving groups. This is shown in Figure 11-3. The intermediate is a trigonal bipyramid, with the attacking nucleophile entering an apical position. This intermediate decomposes by the departure of the leaving group from an apical position.

The driving force for the formation of the pentavalent intermediate in this case is that its formation relieves the bond-angle strain that exists in five-member-ring cyclic phosphodiesters. This is an example of rate enhancement through relief of strain in the transition state leading to the intermediate. Cyclic phosphodiesters with six ring atoms react at rates comparable to those for noncyclic phosphodiesters.

Nucleophilic Catalysis

The hydrolysis of acyl phosphates and phosphomonoesters with good leaving groups is catalyzed by nucleophiles such as pyridine. Pyridine reacts initially as the attacking nucleophile, as in Figure 11-4, to displace the leaving group and form an *N*-phosphorylpyridinium zwitterion (carrying both negative and positive charges). The intermediate subsequently reacts with water to form phosphate and regenerate pyridine.

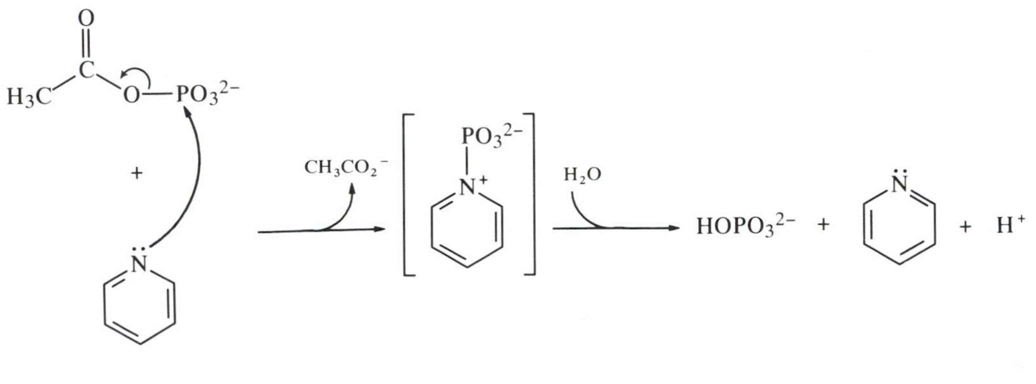

Figure 11-4

Nucleophilic Catalysis of the Nonenzymatic Hydrolysis of Acetyl Phosphate by Pyridine.

Enzymatic Phosphoryl Transfer

Enzymatic phosphoryl group transfer mechanisms are poorly understood, although very recent evidence indicates that the basic chemical mechanisms are similar to the nonenzymatic counterparts. As a rule, little is known about transition states and whether the reactions are associative or dissociative. For reactions of phosphodiesters, the dissociative transition state appears to be forbidden because of the impossibility of generating a double negative charge on the phosphate. A pentavalent intermediate may be formed in one case (ribonuclease, Chapter 12), because a five-member-ring cyclic phosphate is an intermediate.

An important fact in many enzymatic phosphotransfer reactions is the high negative charge associated with the triphosphate in ATP under physiological conditions. The negative charges tend to shield each phosphorus against reaction with incoming nucleophiles, and this property is one reason that ATP is kinetically stable in the cell. ATP is neutralized at the active sites of enzymes by (1) coordination with metal ions and (2) ion pairing with arginine guanidinium and lysine ammonium ions placed within the active sites in precisely the optimal locations for close ionic and hydrogen-bonding interactions with the phosphate oxygens. Typical neutralizing interactions are illustrated in Figure 11-5, which is a general structural representation of ATP interacting with Mg^{2+}, a lysine ε-ammonium ion, and an arginine guanidinium ion at the active site of an enzyme. Such interactions are the rule in active sites of phosphotransferases and other enzymes that catalyze nucleophilic substitution on phosphorus. In fact, ATP in cells exists largely in a complex with Mg^{2+} or other metal ions, which neutralize two of the negative charges; and these complexes are the true substrates for almost all enzymatic reactions of nucleoside di- and triphosphates. The hydrogen-bonded ion pair between the guanidinium group of arginine and an oxygen of the γ-phosphoryl group of MgATP neutralizes the second negative charge on the γ-phosphate and constitutes a binding interaction. The hydrogen-bonded ion pair between the lysine ε-ammonium ion and an oxygen of the α-phosphoryl group neutralizes that group and contributes to binding.

Although the overall charge of ATP is neutralized in the structure shown in Figure 11-5, the oxygen atoms of the triphosphate still carry their negative charges; and the γ-phosphoryl group can still react through a dissociative transition state. The structure in Figure 11-5 is illustrative, but it is not a definitive or universal representation of enzyme-ATP binding interactions.

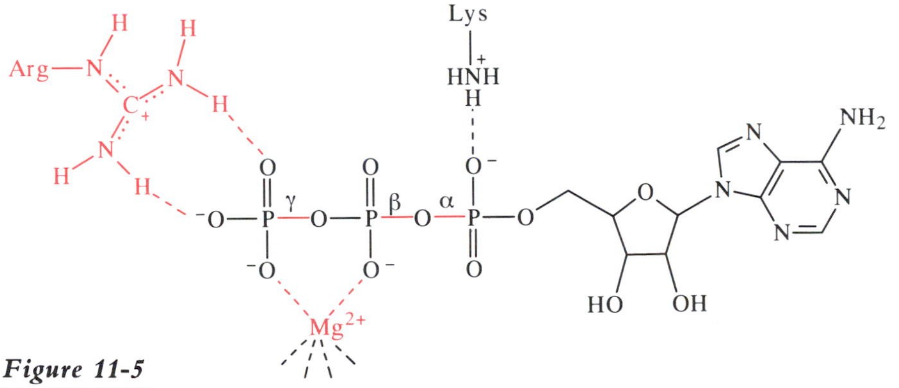

Figure 11-5

An Illustration of Typical Binding Interactions Between MgATP and a Binding Site. Hydrogen-bonded ionic binding interactions between anionic phosphoryl oxygens and the guanidinium and ammonium groups of arginine and lysine, respectively, are shown. The γ-bond is cleaved in phosphoryl group transfer reactions, the α-bond is cleaved in phosphoryl ester transfer, and the β-bond is cleaved in pyrophosphoryl transfer reactions.

Many variant binding modes are possible, but all probably include two or more of the interactions shown in Figure 11-5.

Figure 11-5 also shows the bonds cleaved in reactions catalyzed by various enzymes. The γ-bond is cleaved in phosphotransferase (kinase) reactions and the α-bond is cleaved in nucleotidyltransferase reactions. A few enzymes, such as phosphoribosyl pyrophosphate synthase (Chapter 15), catalyze cleavage of the β-bond, with transfer of the pyrophosphoryl group. Chemical precedents indicate that cleavage of the α- and β-bonds should proceed through associative transition states, whereas cleavage of the γ-bond of ATP should be through a dissociative transition state. The γ-phosphoryl group, with two negative charges and a good leaving group, fulfills the structural requirements to form a dissociative transition state, whereas the α- and β-phosphoryl groups do not satisfy these requirements.

Phosphoryl Transfer Reactions: Phosphotransferases

Adenylate Kinase

The transfer of phosphoryl groups between different nucleotides and other molecules such as creatine phosphate (or arginine phosphate) is important for utilizing and replenishing the cellular pool of energy-rich phosphate compounds. The best-known enzyme of this class is adenylate kinase, which is found in especially high concentration in muscle and was formerly known as myokinase. A kinase is a phosphotransferase that catalyzes the transfer of a phosphoryl group, most often the terminal phosphoryl (the γ-phosphate) of ATP, to an acceptor molecule. Adenylate kinase catalyzes the phosphorylation of adenylic acid, AMP in equation 3, by MgATP to give ADP and MgADP.

$$\text{AMP} + \text{MgATP} \rightleftharpoons \text{ADP} + \text{MgADP} \tag{3}$$

When ATP is depleted by cellular energy requirements—for example, in muscle contraction or in producing phosphorylated metabolites—the resulting ADP can be used to produce more ATP by dismutation catalyzed by adenylate kinase; that is, by the reverse of equation 3. Inasmuch as a phosphoanhydride is both cleaved and formed, this reaction is approximately isoenergetic and reversibility is assured.

The kinetic behavior of adenylate kinase exemplifies a particular variant of a widely observed pattern for multisubstrate enzymes, in which all of the substrates must bind to the enzyme before any product can dissociate. This is often called *sequential kinetics,* which, for a two-substrate enzyme such as adenylate kinase, refers to the fact that the two substrates are bound in sequence by the enzyme and both must be bound in a ternary complex with the enzyme before any product is released.

In a study of the steady-state kinetics of a multisubstrate enzyme, the initial rates are measured as a function of the concentration of a single substrate while the concentrations of all other substrates are held constant. This generates a saturation curve for one substrate at constant levels of cosubstrates, and the data describe a straight line in a double-reciprocal (Lineweaver-Burk) plot. The measurements are repeated at several fixed concentrations of a second substrate and plotted together, as illustrated in Figure 11-6 for adenylate kinase. The lines intersect at the left of the ordinate in the graph, which is typical of sequential kinetics and is commonly observed with two-substrate enzymes. This behavior is described by equation 4, the rate equation for adenylate kinase, in which [ATP] refers to [MgATP].

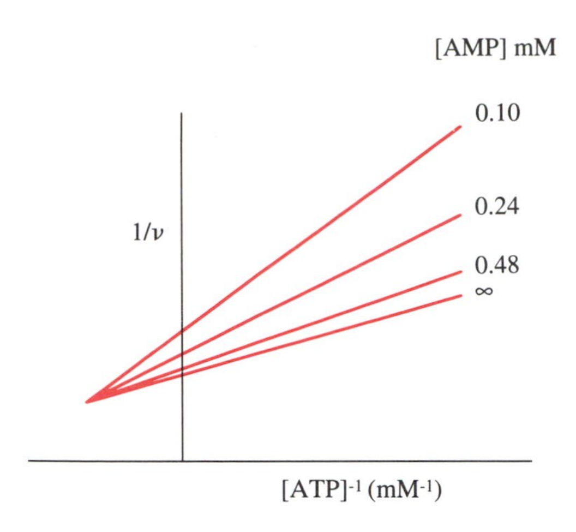

Figure 11-6

Lineweaver-Burk Plots of Initial Rates in the Adenylate Kinase Reaction. This pattern of lines exemplifies the double-reciprocal plots of kinetic data for a sequential kinetic pathway—in this case, the equilibrium random pathway in the adenylate kinase reaction. The concentration of AMP is held constant, whereas that of MgATP is varied for each line. Each line corresponds to a different fixed concentration of AMP.

$$v = \frac{V_m[\text{ATP}][\text{AMP}]}{[\text{ATP}][\text{AMP}] + K_{s\text{ATP}}[\text{AMP}] + K_{s\text{AMP}}[\text{ATP}] + K'_{s\text{ATP}}K_{s\text{AMP}}} \quad (4)$$

For adenylate kinase, the two substrates are bound randomly and are always in equilibrium with the ternary complex. Therefore, the K_m values for MgATP and AMP are the dissociation constants $K_{s\text{ATP}}$ and $K_{s\text{AMP}}$. The kinetic pathway is represented in shorthand form by the scheme in Figure 11-7, which follows a convention that simplifies the writing of complex kinetic pathways by connecting the free enzyme and various binary and ternary complexes with lines instead of reversible arrows. The dissociation constants for ATP and AMP binding to the free enzyme (K_s) or to the binary complexes E·AMP or E·MgATP (K_s') may or may not be equal. Under conditions in which initial

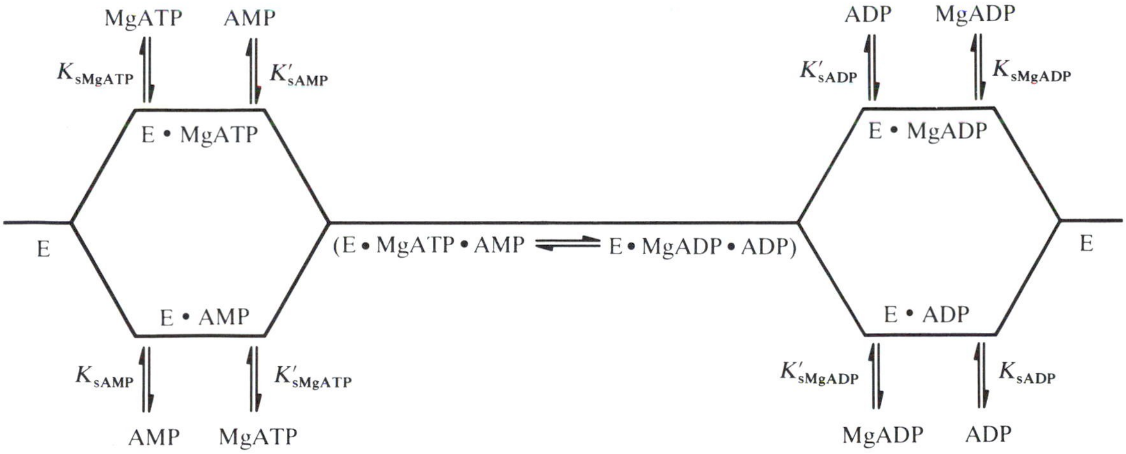

MgATP AMP ADP MgADP
$K_{s\text{MgATP}}$ $K'_{s\text{AMP}}$ $K'_{s\text{ADP}}$ $K_{s\text{MgADP}}$
E • MgATP E • MgADP
E (E • MgATP • AMP ⇌ E • MgADP • ADP) E
E • AMP E • ADP
$K_{s\text{AMP}}$ $K'_{s\text{MgATP}}$ $K'_{s\text{MgADP}}$ $K_{s\text{ADP}}$
AMP MgATP MgADP ADP

Figure 11-7

The Equilibrium Random Binding Kinetic Mechanism for Adenylate Kinase. This is a shorthand way of illustrating the kinetic mechanism. The solid lines connect the free enzyme, E, with the various binary and ternary enzyme-substrate and enzyme-product complexes. All binding steps are shown as reversible for purposes of generality, and the constants, K_s, are dissociation constants. Each step is reversible and at equilibrium when all components are present at significant concentrations. For initial rate measurements in the absence of any product, all of the product-releasing steps are virtually irreversible, and the concentrations of the binary enzyme-product complexes are nearly zero.

forward rates are measured in the absence of the products, the reaction is virtually irreversible; and the dissociation constants for ADP and MgADP in Figure 11-7 do not appear in the rate equation (equation 4).

Because the substrate-binding steps for adenylate kinase are at equilibrium, the rate-limiting step is the conversion of one ternary complex into the other. This does not necessarily mean that the phosphoryl transfer step limits the rate, however, because conformational changes may also occur and have an effect; and the interconversion of ternary complexes could include more than a single step. Such additional steps would not appear in the overall kinetics; that is, their existence would not alter the rate equation.

One kind of additional step that could intervene in the interconversion of ternary complexes is the transfer of the phosphoryl group to a nucleophilic group of the enzyme, forming a covalent phosphoryl-enzyme as an intermediate. Powerful evidence against this hypothesis is the reaction of a substrate having a chiral [^{18}O]thiophosphoryl group, (S_p)-[γ-^{18}O]ATPγS, in place of the γ-phosphoryl group of ATP (equation 5). Transfer of the [^{18}O]thiophosphoryl group to AMP proceeds with overall *inversion* of configuration at phosphorus.

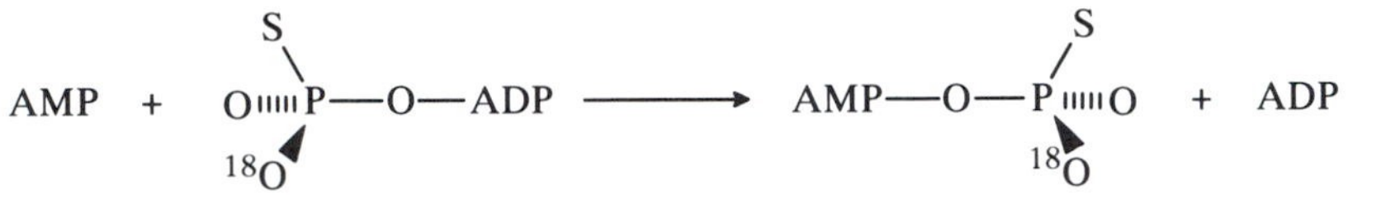

$$\text{AMP} + \text{O}\cdots\text{P}(\text{S})(^{18}\text{O})\text{—O—ADP} \longrightarrow \text{AMP—O—P}(\text{S})(^{18}\text{O})\cdots\text{O} + \text{ADP} \quad (5)$$

This is consistent with a single transfer of the terminal [^{18}O]thiophosphoryl group, because each enzymatic phosphoryl transfer proceeds with inversion of configuration at phosphorus. If the thiophosphoryl group had been transferred first to an enzymatic group, it would have undergone inversion in that step. A subsequent transfer to AMP would have inverted it again and resulted in overall retention of configuration, contrary to fact. Therefore, the transfer probably proceeds directly from MgATP to AMP within the ternary complex, without nucleophilic catalysis by the enzyme.

The chemistry of the adenylate kinase reaction presumably consists of simple nucleophilic attack on the terminal phosphate of MgATP by the phosphate group of AMP, transferring the phosphoryl group and displacing MgADP, as illustrated in Figure 11-8. In the reverse direction, which is physiologically important for replenishing ATP, two molecules of ADP bind to the enzyme, one as MgADP, and undergo the reverse transfer.

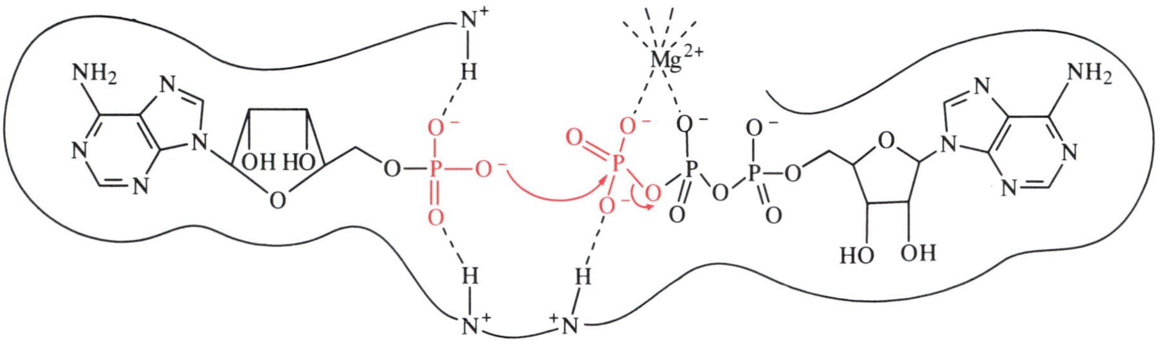

Figure 11-8

A Conceptualization of the Adjacent Substrate Binding Sites of Adenylate Kinase. The mechanism of the adenylate kinase reaction requires the substrates AMP and MgATP to be bound at adjacent sites. The phosphoryl group is transferred directly from MgATP to AMP.

Structural analyses of adenylate kinases from several species support the foregoing interpretation of mechanistic information. The structures are a source of much additional insight into the actions of this enzyme, however, by showing that binding energy between each substrate and its cognate binding site is used to bring the sites into adjacent positions through large conformational changes in the enzyme. The free enzyme is in an open conformation with the substrate-binding sites separated in different domains. With both AMP and MgATP bound, the enzyme is in its closed and active conformation, which excludes water and catalyzes phosphoryl transfer between nucleotides.

These structural differences between free adenylate kinase and the closed active form are shown in Figure 11-9. Part A shows the α-carbon backbone of the porcine enzyme, and part B shows the backbone structure with diadenosine pentaphosphate (Ap_5A) bound to the two substrate-binding subsites. Ap_5A is a two-substrate analog that spans the binding sites, and it is a very potent inhibitor. The two conformations are very different, in part because the subsites are drawn together when Ap_5A binds. The conformational change does not result simply from the fact that Ap_5A holds the two subsites together, however, because a major part of the change is brought about by the binding of AMP alone, as shown in Figure 11-10. The structure in part B shows that the AMP-binding lobe is closed with AMP in its binding site; the ATP-binding lobe also closes when Ap_5A binds.

Diadenosine pentaphosphate (Ap_5A)

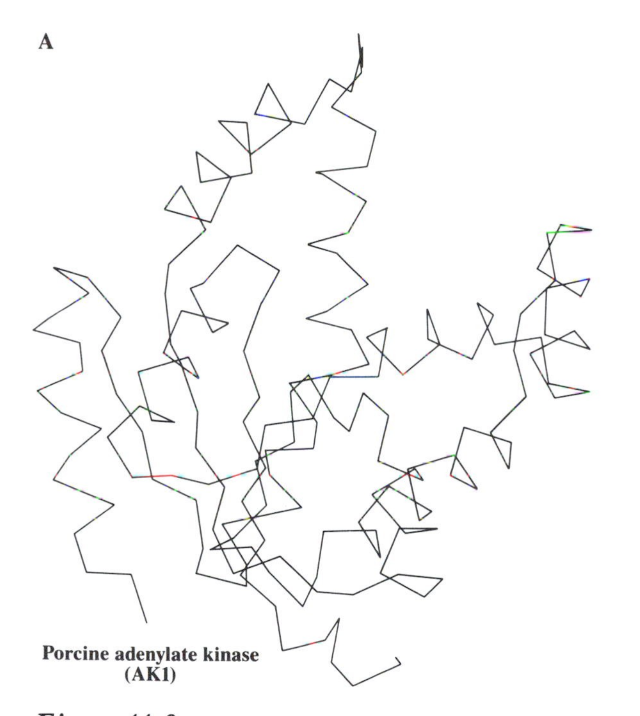

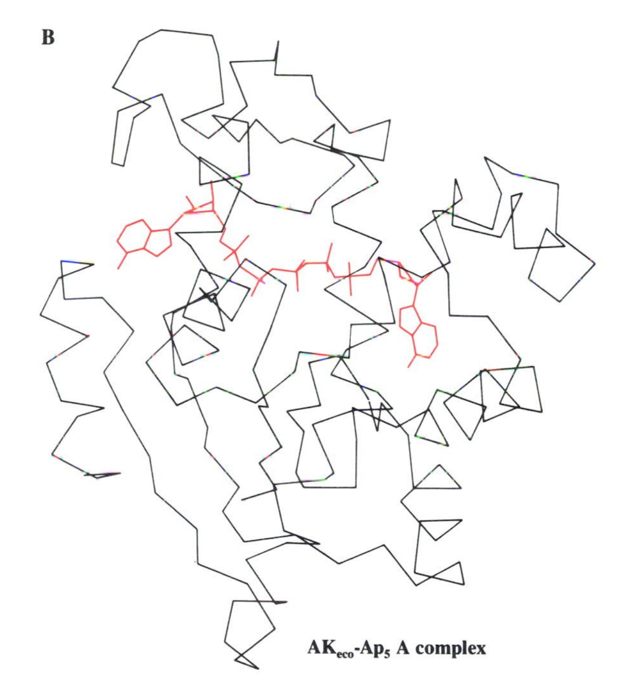

Figure 11-9

The Structural Change in Adenylate Kinase Brought about by Binding Ap_5A to the Enzyme. Shown are α-carbon backbone models for (A) porcine adenylate kinase (AK1) and (B) the complex between diadenosine pentaphosphate (Ap_5A), shown in red, and *E. coli* adenylate kinase (AK_{eco}). The polypeptide chains of these two species of adenylate kinase are folded similarly. Note the two major lobes and the large cleft in AK1 (part A). The binding sites for AMP and MgATP are in the two domains roughly defined by the lobes and cleft. The binding of Ap_5A (red structure in part B) to the two substrate-binding sites of AK_{eco} draws them together and partly closes the cleft. [Adapted from D. Dreusicke and G. E. Schulz. *J. Mol. Biol.* 199 (1988): 359–371; and C. W. Müller and G. E. Schulz. *J. Mol. Biol.* 202 (1988): 909–912.]

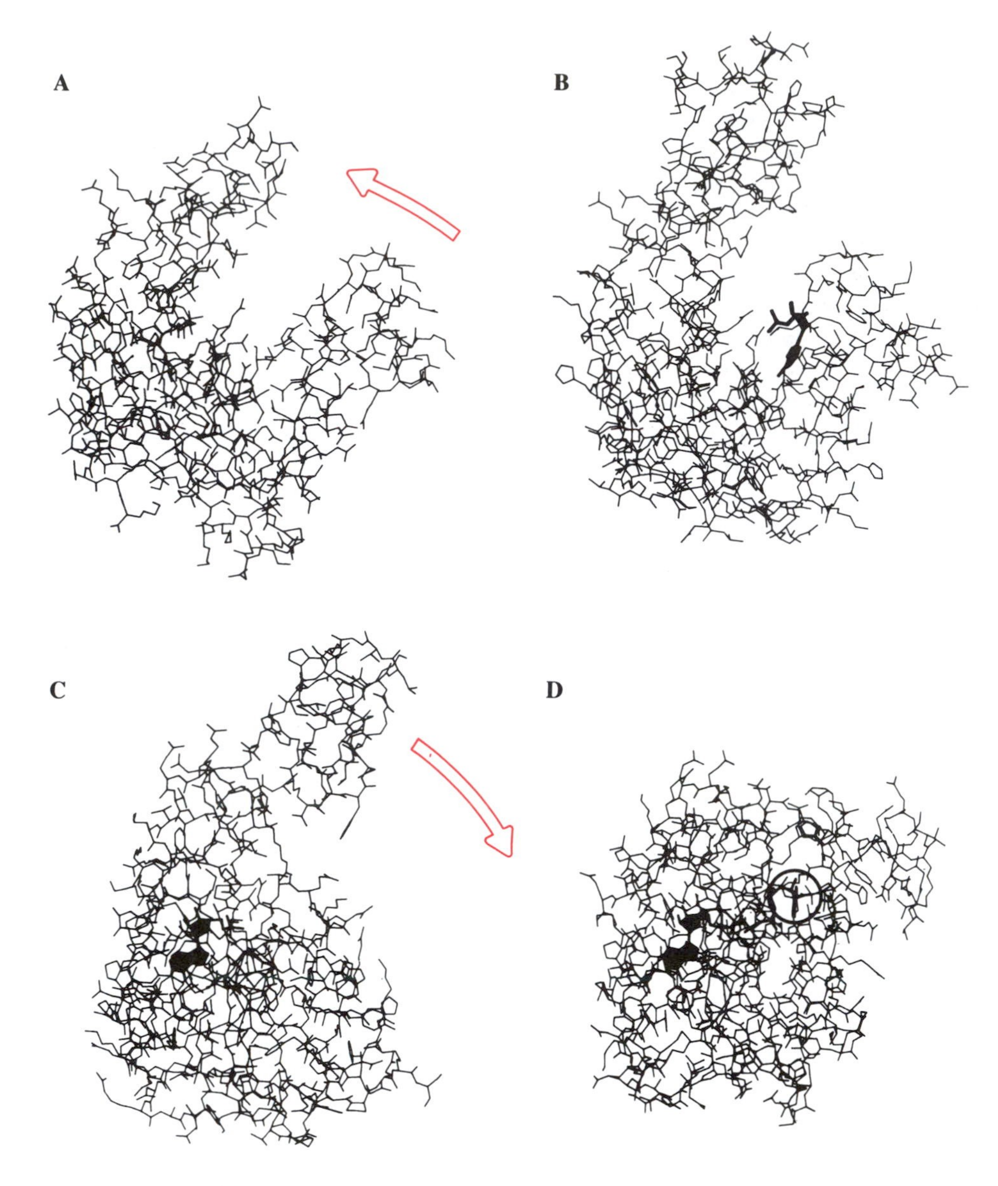

Figure 11-10

A Structural Model for Step-by-Step Conformational Changes Brought about by Substrates Binding to Adenylate Kinase. Parts A through D are interpretive models of the conformational changes brought about in adenylate kinase by the binding of AMP and Ap_5A. Structures are available for the porcine muscle adenylate kinase (AK1), shown in part A; the complex of AMP with bovine mitochondrial adenylate kinase (AK3), shown in parts B and C; and the complex of Ap_5A with *E. coli* adenylate kinase (AK_{eco}), shown in part D. The models depict all atoms except for hydrogens. The domain movements are indicated by the arrows. Part A is a model of AK1 with no bound ligand, showing a deep cleft poised to accept the substrates. Part B is a model of AK3 with bound AMP viewed as superimposed in AK1. AMP, at the right-hand side of the cleft, is drawn in heavy lines. The movement of the AMP-binding domain is indicated by the arrow in part A. Part C shows the model in part B rotated by 90° about a vertical axis. Part D is a model of AK_{eco} with bound Ap_5A. The two-substrate analog Ap_5A is drawn in heavy lines, the adenine and ribose rings of one adenosine (AMP-site) being shown in solid black, and the other adenosine (ATP-site) encircled. The movement of the ATP-binding domain is indicated by the arrow in part C. [Adapted from G. E. Schulz, C. W. Müller, and K. Diederichs. *J. Mol. Biol.* 213 (1990): 627–630.]

The structure of the ternary complex **E·AMP·MgATP** is probably similar to the structure with $MgAp_5A$ bound at the active site, and this closed structure is the active form that catalyzes phosphoryl group transfer between nucleotides, while excluding water. Thus, the enzyme can bind MgATP alone, but the complex **E·MgATP** is not in the active conformation and does not catalyze phosphoryl transfer to water.

Nucleoside Diphosphate Kinase

Nucleoside diphosphate kinase catalyzes the transfer of the terminal phosphoryl group of ATP to a nucleoside diphosphate, such as GDP in equation 6.

$$\text{MgATP} + \text{MgGDP} \rightleftharpoons \text{MgADP} + \text{MgGTP} \tag{6}$$

In cells, nucleoside diphosphate kinase equilibrates high-energy phosphate groups among nucleotides so that if a nucleoside triphosphate becomes depleted it can be replenished by phosphotransfer from ATP. It is especially important for maintaining GTP at concentrations required for protein biosynthesis (Chapter 10). An important mechanistic fact about this enzyme is that it also catalyzes the exchange of radiochemically labeled ADP with ATP to give labeled ATP, which is nearly the same reaction. The significance of this exchange reaction will shortly become clear.

In spite of the apparent chemical similarity of the nucleoside diphosphate kinase reaction to the adenylate kinase reaction, the two enzymes work by entirely different mechanisms. This is revealed by the steady-state kinetic pattern, which for nucleoside diphosphate kinase is a series of parallel lines in the Lineweaver-Burk plots of Figure 11-11, rather than the intersecting lines found with adenylate kinase. The parallel line plots are described by equation 7,

$$v = \frac{V_m[\text{ATP}][\text{GTP}]}{[\text{ATP}][\text{GTP}] + K_{m\text{ATP}}[\text{GTP}] + K_{m\text{GTP}}[\text{ATP}]} \tag{7}$$

which differs from equation 4 for the adenylate kinase reaction in that it lacks the constant term in the denominator. The slope of a Lineweaver-Burk plot is K_m/V_m, and for a single line in a multisubstrate pattern, in which the fixed cosubstrate may not be at a saturating concentration, it is $K_{m\,app}/V_{m\,app}$ for apparent values. The parallel lines mean that the slope is a constant for each varied substrate and independent of the fixed concentration of the cosubstrate; that is, $K_{m\,app}/V_{m\,app}$ equals K_m/V_m.

The parallel line pattern in primary Lineweaver-Burk plots is characteristic of the steady-state *Ping-Pong* mechanism. In this mechanism, a group (**X—**) is transferred from a donor molecule to the enzyme in a first half-reaction (on page 306), which proceeds independently of the presence of the acceptor substrate. The group is then transferred from the enzyme to the acceptor in a second half-reaction.

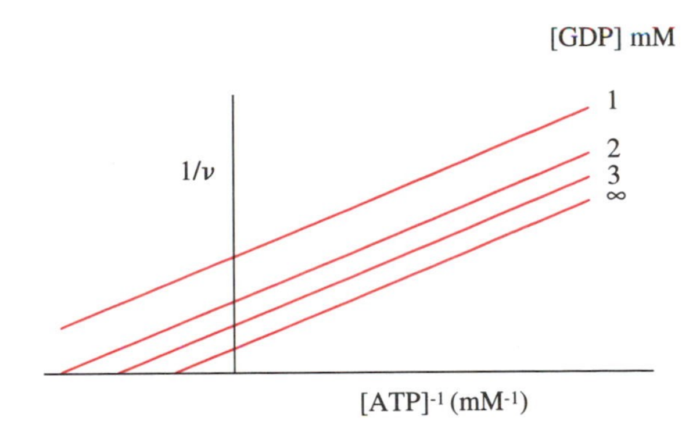

Figure 11-11

Lineweaver-Burk Plots of Initial Rates in the Nucleoside Diphosphate Kinase Reaction, an Example of a Ping-Pong Kinetic Mechanism.

$$E + A{-}X \rightleftharpoons E{-}X + A$$
$$E{-}X + B \rightleftharpoons E + B{-}X$$

The overall kinetic pathway for nucleoside diphosphate kinase is illustrated in short form by the scheme in Figure 11-12. Note that the reaction of ATP with the enzyme to form ADP and an intermediate phosphoryl-enzyme proceeds independently of the presence of GDP. Indeed, this kinetic pathway excludes GDP from the enzyme until ADP has departed. Therefore, in contrast with the sequential pathway, ternary complexes are specifically excluded as intermediates. The K_m values for substrates in equation 7 are not dissociation constants, but rather are determined by the rate constants in Figure 11-12 because this is a steady-state kinetic pathway.

Some enzyme kineticists initially disapproved of the term Ping-Pong, but it survives because it aptly describes the reaction. This is shown by the half-reactions of equations 8a and 8b. (The nucleotides react as their magnesium complexes.)

$$\text{ATP} + \text{E} \rightleftharpoons \text{E}\cdot\text{ATP} \rightleftharpoons \text{E—P} + \text{ADP (ping)} \quad (8a)$$
$$\text{E—P} + \text{GDP} \rightleftharpoons \text{E—P}\cdot\text{GDP} \rightleftharpoons \text{E} + \text{GTP (pong)} \quad (8b)$$

The "ping" is the transfer of the phosphoryl group to the enzyme and the release of the first product, ADP, in the first half-reaction. The "pong" is the transfer of the phosphoryl group from the enzyme to GDP to give GTP. The second reaction is the pong of the first reaction because it is chemically the same reaction in reverse, except for the identity of the nucleotide base. The reverse relationship is more obvious and perfect for the exchange reaction of ^{32}P from [β-^{32}P]ADP into ATP (equation 9), which you should write out as half-reactions to illustrate the point.

$$\text{ATP} + \text{ADP*} \rightleftharpoons \text{ATP*} + \text{ADP} \quad (9)$$

This exchange reaction is an important clue that the pathway is likely to be Ping-Pong.

Ping-Pong kinetics is an example of a steady-state enzyme kinetic pattern that suggests something specific and important about the *chemical* reaction mechanism. It implicates an intermediate composed of the enzyme and the transferred group, the phosphoryl group if the enzyme is nucleoside diphosphate kinase.

The *interpretation* of a kinetic pattern as Ping-Pong kinetics is not always straightforward. It is often difficult to decide whether Lineweaver-Burk plots are really parallel lines, and several factors can perturb kinetic behavior. Apparently

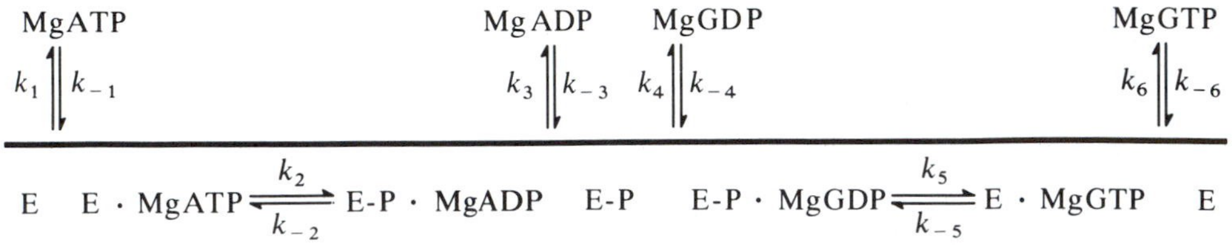

MgATP (k_1 / k_{-1}) MgADP (k_3 / k_{-3}) MgGDP (k_4 / k_{-4}) MgGTP (k_6 / k_{-6})

E E · MgATP $\underset{k_{-2}}{\overset{k_2}{\rightleftharpoons}}$ E-P · MgADP E-P E-P · MgGDP $\underset{k_{-5}}{\overset{k_5}{\rightleftharpoons}}$ E · MgGTP E

Figure 11-12

The Ping-Pong Kinetic Mechanism for Nucleoside Diphosphate Kinase. In the first half-reaction, MgATP binds to the enzyme and phosphorylates a histidine residue in the active site; MgADP then undergoes dissociation from the enzyme. In the second half-reaction, MgGDP is bound to the phosphoryl-enzyme (E-P) and phosphorylated to MgGTP. The substrate- and product-binding steps may be either equilibrium or steady-state processes and, for purposes of generality, are shown here as steady-state steps with associated rate constants.

parallel lines are sometimes found not to be parallel when they are examined carefully. However, the Ping-Pong pathway can always be verified by other, more-definitive tests and experiments, all of which are suggested by the Ping-Pong pathway itself.

Half-reactions 8a and 8b and the scheme in Figure 11-12 are statements of the fact that the formation of a phosphoryl enzyme can be demonstrated directly by reaction of the enzyme with ATP labeled with ^{32}P in the terminal phosphoryl group, [γ-^{32}P]ATP, in the absence of a nucleoside diphosphate. This intermediate can be isolated by molecular exclusion chromatography through a Sephadex column that separates proteins from smaller molecules such as nucleotides. The radioactive [^{32}P]phosphoenzyme will be separated from ADP and unreacted [γ-^{32}P]ATP, directly showing the formation of a phosphoenzyme (Figure 11-13). Reaction of this [^{32}P]phosphoenzyme with GDP

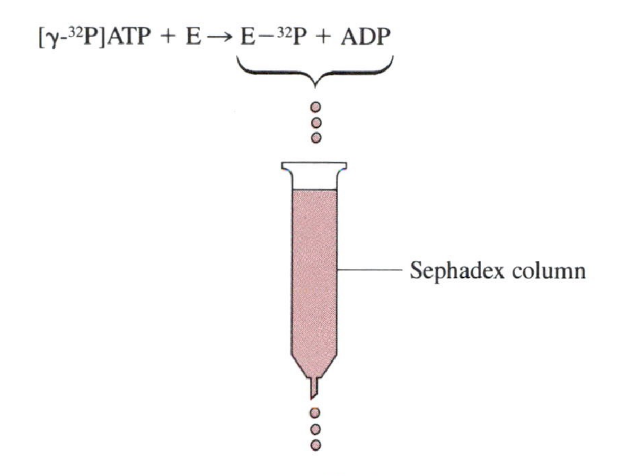

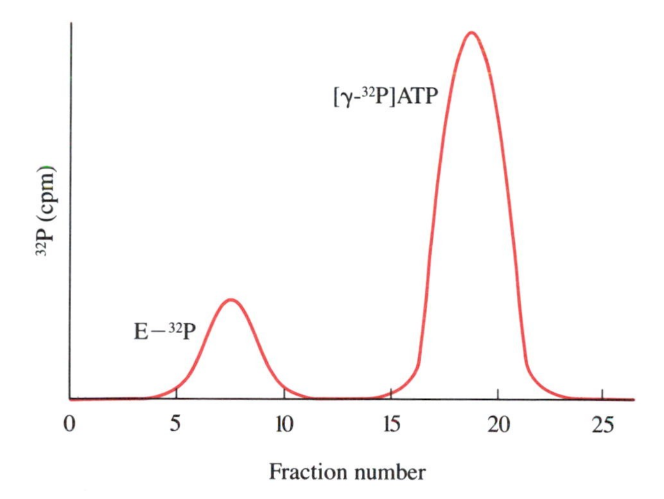

Figure 11-13

Isolation of a [^{32}P]Phosphoryl-Enzyme. The Sephadex G-25 column separates molecules on the basis of differences in molecular weight. Because the [^{32}P]phosphoryl-enzyme has a molecular weight greater than 25,000, whereas that of ATP is less than 700, the two are efficiently separated. The first peak eluted from the column is the protein, which contains ^{32}P because it has been phosphorylated by [γ-^{32}P]ATP. The second peak contains excess [γ-^{32}P]ATP and the ADP formed in the phosphorylation of the enzyme, but no protein.

and rechromatography through the same column gives unlabeled protein and [γ-^{32}P]GTP in place of E-^{32}P and [γ-^{32}P]ATP.

To be quite sure of the meaning of the experiment in Figure 11-13, it is important to demonstrate that the ^{32}P associated with the enzyme is covalently bonded to the enzyme. This can be done by denaturing E-^{32}P and observing that the radioactivity remains protein bound. It is not uncommon for a purified enzyme to contain acceptor substrate in a noncovalently bound form that is phosphorylated by ATP and not removed by molecular exclusion chromatography. Denaturation of E-^{32}P will dissociate such noncovalently bonded species.

Alkaline hydrolysis of the phosphoenzyme form of nucleoside diphosphate kinase gives a phosphohistidine with phosphate on the imidazole ring (equation 10).

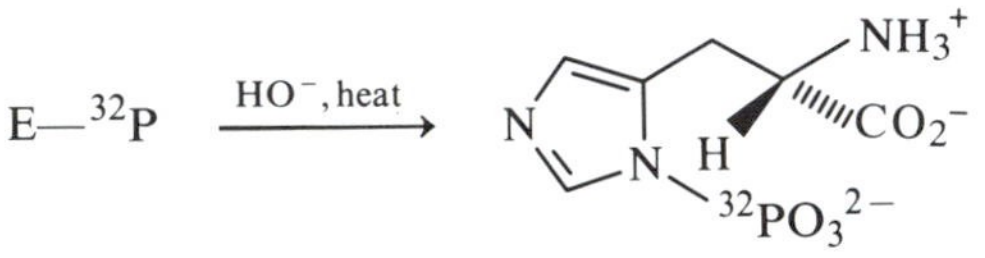

$$\text{E—}^{32}\text{P} \xrightarrow{\text{HO}^-,\text{heat}} \text{N-}^{32}\text{PO}_3^{2-}\text{-histidine}\ (\text{NH}_3^+,\ \text{CO}_2^-,\ \text{H}) \tag{10}$$

In this and several other enzymatic reactions, the imidazole group acts as a nucleophilic catalyst of phosphoryl transfer. (It is not a nucleophilic catalyst of acyl transfer in chymotrypsin and other enzymatic acyl transfer reactions.) Powerful support for this role is provided by the stereochemical course of the phosphoryl transfer catalyzed by nucleoside diphosphate kinase. The chiral [^{18}O]thiophosphoryl analog of ATP (equation 5) reacts as a substrate and transfers the [^{18}O]thiophosphoryl group to a nucleoside diphosphate with overall *retention* of configuration about phosphorus. This stereochemistry is required by the double-displacement mechanism, in which the formation of the intermediate E-P proceeds with inversion and the transfer to acceptor in the second-half reaction proceeds with a second inversion. This results in overall retention of configuration, as is illustrated in Figure 11-14.

Adenylate kinase and nucleoside diphosphate kinase catalyze very similar reactions. Why should they act by such different mechanisms? The answer to this question lies in the structures of the phosphoryl group acceptors and the *principle of economy in the evolution of binding sites.* Nucleoside diphosphate kinase catalyzes a symmetrical reaction in which the acceptor substrates are structurally very similar; and it does so through the efficient utilization of a single binding site that can fit and productively bind any nucleoside triphosphate (Figure 11-15). Phosphoryl transfer to the enzyme produces the phosphoryl-enzyme and a nucleoside diphosphate, which dissociates and can be replaced by any other nucleoside diphosphate. Reverse phosphoryl transfer forms the product. With this mechanism, only one binding site that contains a suitable nucleophilic functional group is required. The essential function of the nucleophilic catalyst is to accept the transferred group and conserve the bond energy during the interchange of acceptors. The evolution of a single binding site

$$\text{E-His:}\ \ \text{O}\cdots\text{P(S)(}^{18}\text{O)—O—ADP} \xrightarrow{\text{ADP}} \text{E-His—P(S)(}^{18}\text{O)}\cdots\text{O} \xrightarrow{\text{GDP}} \text{E-His:}\ \ \text{O}\cdots\text{P(S)(}^{18}\text{O)—O—GDP}$$

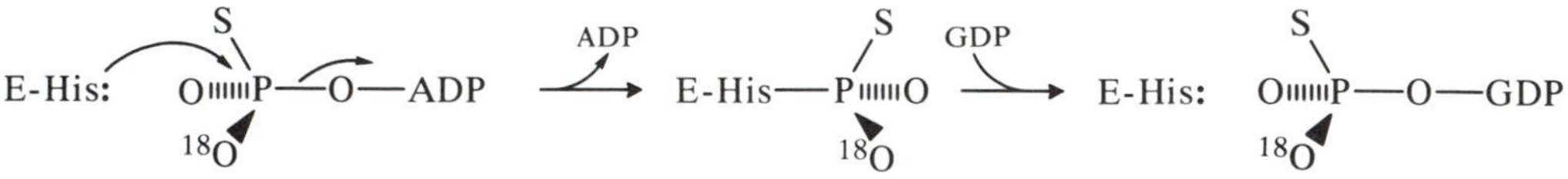

Figure 11-14

The Stereochemical Course of Thiophosphoryl Transfer by Nucleoside Diphosphate Kinase. The configuration of the chiral [^{18}O]thiophosphate group is inverted twice, once as it is transferred to the active-site histidine and again when it is transferred to GDP. Therefore, there is retention of configuration in the *overall* reaction.

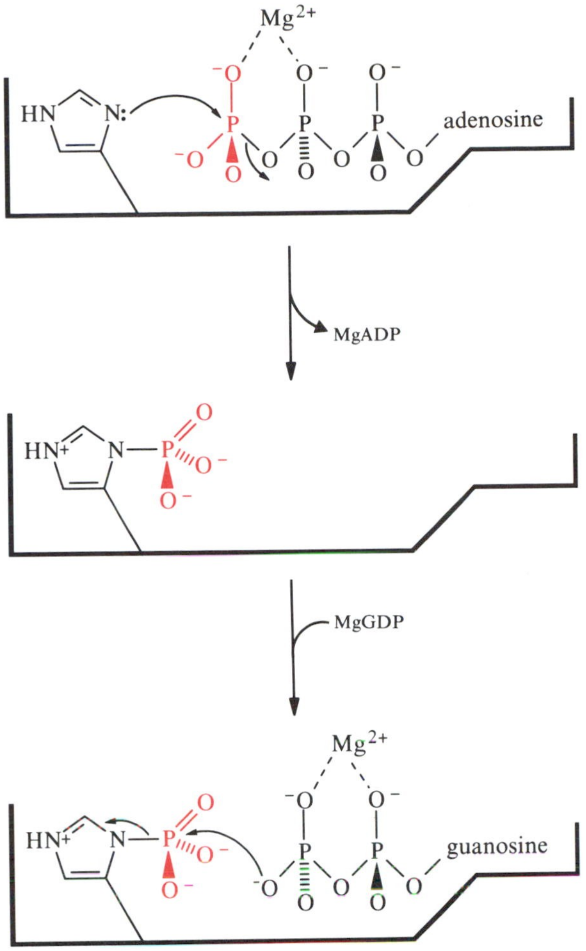

Figure 11-15

Economy in the Evolution of Binding Sites. Nucleoside diphosphate kinase is thought to have a single binding site that may be occupied by either MgATP or MgGTP. In the phosphoryl-enzyme, the site may be occupied by MgADP or MgGDP. The evolution of this binding mode, and the associated reaction mechanism, is thought to be governed by the principle of economy in the evolution of binding sites.

should be a favored process compared with the evolution of two binding sites, and this may have been the important factor in the evolutionary selection of the Ping-Pong mechanism for nucleoside diphosphate kinase.

The nucleoside diphosphate kinase mechanism in its simplest conception would not be applicable to the adenylate kinase reaction and was not selected in the course of evolution. The adenylate kinase reaction, although chemically similar, is asymmetric with respect to the structures of the phosphoryl acceptors AMP^{2-} and $MgADP^{-}$, which differ substantially both in structure and in charge. A nucleoside diphosphate kinaselike binding site for adenylate kinase could not recognize and productively bind AMP^{2-} or $MgADP^{-}$ in a single acceptor subsite. Therefore, adenylate kinase evolved with two adjacent binding sites, one for MgATP and the other for AMP (or MgADP and ADP in the reverse reaction). The sites overlap in the space traversed by the phosphoryl group in flight (Figure 11-8). With both substrates bound in adjacent sites, phosphoryl transfer can proceed directly from donor to acceptor, obviating the need for a nucleophile.

Other examples of chemically similar group transfers that proceed by these two mechanisms are known and can be rationalized in the same way. It is rare for the two acceptors in a group transfer reaction to be similar enough to bind effectively to the same site; therefore, the Ping-Pong mechanism with a shared binding site for acceptors is the exception rather than the rule in two-substrate group transfer reactions. Covalent enzyme-substrate intermediates are more commonly found in more-complex reactions; for example, those having three or more substrates.

ATP Production from Higher-Energy Phosphates

Phosphoryl transfer from ATP to carboxylate ions and certain other groups is thermodynamically uphill because acyl phosphates are considerably higher in energy than ATP (Chapters 9 and 10). Therefore, most such reactions proceed in the reverse direction to synthesize ATP from an acyl phosphate and ADP under physiological conditions. Several biological phosphates are higher in energy than ATP and serve as sources of high-energy phosphate for the production of ATP in response to cellular needs. The most important among them are bis-1,3-phosphoglycerate, phosphoenolpyruvate (PEP), and creatine phosphate; and for each of these molecules there is a phosphotransferase that catalyzes its reaction with ADP to produce ATP. All of these enzymes catalyze direct, single-step phosphoryl transfers within ternary complexes by sequential kinetic mechanisms analogous to that of adenylate kinase. All of them catalyze the phosphoryl transfer with inversion of configuration at phosphorus in the P-chiral substrates [γ-^{18}O]ATPγS or [γ-^{17}O, ^{18}O]ATP.

We have seen in Chapter 10 that the phosphorylation of acetate is catalyzed by acetate kinase (equation 11).

$$H_3C{-}C({=}O){-}O^- + {}^{17}O{\cdots}P({}^{18}O)(O){-}O{-}ADP \rightleftharpoons H_3C{-}C({=}O){-}O{-}P({}^{18}O)({}^{17}O)(O) + ADP \qquad (11)$$

The reaction is energetically unfavored in the direction of equation 11, but it can be drawn in the forward direction in *E. coli* by coupling with phosphotransacetylase to form acetyl CoA when the bacteria are grown on acetate. Phosphoryl transfer catalyzed by acetate kinase is accompanied by inversion of configuration at chiral phosphorus, which is consistent with kinetic and other evidence that indicates a single displacement mechanism.

Phosphoglycerate kinase catalyzes the phosphorylation of 3-phosphoglycerate by ATP to give bis-1,3-phosphoglycerate (equation 12).

$$O{=}C(O^-){-}C(H)(OH){-}CH_2OPO_3^{2-} + ATP \rightleftharpoons O{=}C(O{-}PO_3^{2-}){-}C(H)(OH){-}CH_2OPO_3^{2-} + ADP \qquad (12)$$

The reaction normally proceeds in the reverse, energetically favorable direction in the cell to synthesize ATP. Phosphoglycerate kinase is an important enzyme in the metabolism of virtually all cells because it catalyzes an important step in the glycolytic pathway for glucose utilization by producing one of the ATP molecules synthesized in glycolysis and fermentation (Chapter 21).

Creatine kinase catalyzes the reversible transfer of phosphate between ATP and creatine to give creatine phosphate and ADP (equation 13).

$$H_3C{-}N(CH_2CO_2^-){-}C(NH{-}PO_3^{2-}){=}NH_2^+ + ADP \rightleftharpoons H_3C{-}N(CH_2CO_2^-){-}C(NH_2){=}NH_2^+ + ATP \qquad (13)$$

Creatine phosphate, which is present at high concentrations in muscle and some

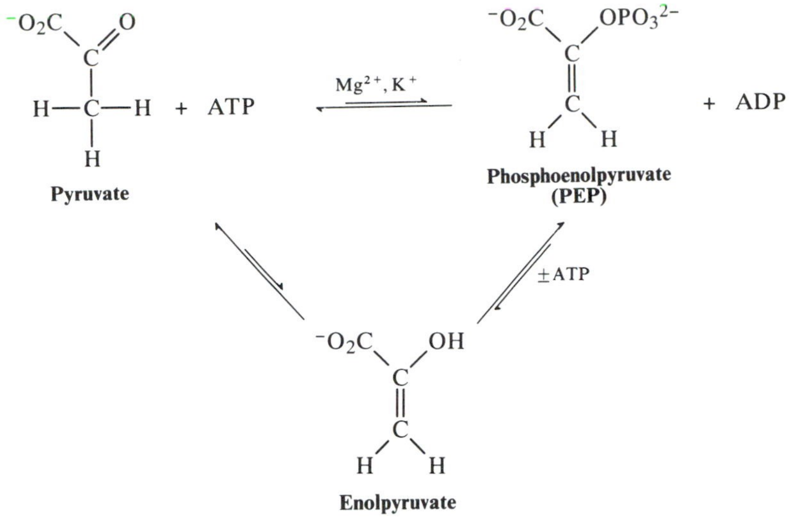

$$\underset{\textbf{Pyruvate}}{{}^-O_2C{-}CO{-}CH_3} + ATP \xrightarrow{Mg^{2+},\,K^+} \underset{\textbf{Phosphoenolpyruvate (PEP)}}{{}^-O_2C{-}C(OPO_3^{2-}){=}CH_2} + ADP$$

$$\textbf{Pyruvate} \rightleftharpoons \underset{\textbf{Enolpyruvate}}{{}^-O_2C{-}C(OH){=}CH_2} \xrightleftharpoons{\pm ATP} \textbf{PEP}$$

Figure 11-16

Chemical Steps in the Pyruvate Kinase Reaction.

other tissues, is a *phosphagen,* meaning that it can generate high-energy phosphate. It is itself a high-energy phosphate, with a standard free energy of hydrolysis of $-9\,\text{kcal mol}^{-1}$ at pH 7 (Chapter 9), and is a ready source of phosphoryl groups for generating ATP through the action of creatine kinase. As such, it serves an important cellular function as a reservoir of high-energy phosphoryl groups for ATP production in rapid response to cellular needs. A similar enzyme, arginine kinase, catalyzes the similar reaction of ATP and the guanidino group of arginine in invertebrates to give phosphoarginine. These enzymes also act through sequential catalytic pathways and ternary complexes to transfer the phosphoryl group directly between substrates bound at adjacent sites.

The phosphorylation of pyruvate by ATP to give PEP consists of two chemically distinct processes, which are shown in Figure 11-16. They are the enolization of pyruvate to enolpyruvate and the phosphorylation of enolpyruvate to PEP. The overall reaction is highly unfavorable for PEP production because PEP, with a standard free energy of hydrolysis of $-14\,\text{kcal mol}^{-1}$, is a compound of much higher energy than ATP (Chapter 9). PEP, like bis-1,3-phosphoglycerate, is a product of glycolysis and fermentation that serves as a phosphoryl donor for the synthesis of ATP.

The reason that PEP is energy rich is that enolpyruvate is much less stable than pyruvate itself, as is generally true of enols relative to the corresponding ketones. Consider the *reversal* of the pyruvate kinase reaction in Figure 11-16 to form pyruvate and ATP. This is the overwhelmingly favored direction for pyruvate kinase. The transfer of a phosphoryl group from PEP to an acceptor is a low-energy process. However, the enolpyruvate produced by phosphoryl transfer is highly unstable relative to pyruvate, and the ketonization of enolpyruvate provides a strong driving force for phosphoryl transfer and the formation of pyruvate.

Synthesis of Low-Energy Phosphate Compounds from ATP

Biosynthetic reactions of phosphate compounds are often downhill, thermodynamically favorable reactions that lead to the almost irreversible formation of the desired products. The most-common phosphorylation reaction in biosynthetic pathways is the phosphorylation of hydroxyl groups by ATP to give phosphomonoesters. A few examples are listed in Table 11-1. All these reactions proceed by sequential kinetic mechanisms through ternary complexes and all

Table 11-1

Phosphorylation of alcohols by ATP

Enzyme		Product
Hexokinase	⟶	Glucose-6-phosphate
Glycerokinase	⟶	Glycerol-1-phosphate
Choline kinase	⟶	Phosphocholine
Phosphofructokinase	⟶	Fructose-1,6-bisphosphate
Protein kinases	⟶	Phosphoproteins (Ser-P, Thr-P, Tyr-P)
Polynucleotide kinase	⟶	Polynucleotide 5′-phosphates

proceed by direct phosphoryl transfer from ATP to the alcohol substrates with inversion of configuration at chiral phosphorus, as in the adenylate kinase reaction.

The phosphorylation reactions in Table 11-1 are important cellular events that are crucial for the health and survival of cells for the following reasons:

1. The phosphorylation of glucose and other alcohols has the effect of trapping glucose-6-phosphate and other phosphorylated compounds inside the cell because anions cannot diffuse through membranes. If these molecules were not phosphorylated, some other means would have to be invented to prevent them from escaping by diffusion through the cellular membranes.
2. Phosphorylation is also a means of providing a binding "handle" that facilitates further interactions of these compounds with enzymes and other macromolecules involved in their metabolism. Ionic binding is often an important component of the interactions of small molecules with enzymes and receptors.
3. Some of these phosphate esters are eventually converted by metabolism into high-energy compounds, such as PEP, that can regenerate ATP.
4. The phosphate groups serve as good leaving groups in elimination reactions and in nucleophilic displacements on carbon. Phosphates, pyrophosphates, and triphosphates are the leaving groups at several points in the biosynthesis of sterols, steroids and terpenes, *S*-adenosylmethionine, and adenosylcobalamin (Vitamin B_{12} coenzyme).
5. Many phosphorylated compounds are biosynthetic intermediates. An example is phosphocholine (Table 11-1), an intermediate in the biosynthesis of phosphatidyl choline, a component of cellular membranes.
6. Other roles of phosphorylation include the regulation of such cellular functions as the actions of protein kinases that catalyze phosphorylations of proteins. Many enzymes are regulated by phosphorylation-dephosphorylation mechanisms, in which the phosphorylation of serine or threonine activates or inactivates the enzyme and dephosphorylation reverses the effect. For example, glycogen phosphorylase is activated by the phosphorylation of a serine hydroxyl group, whereas the pyruvate dehydrogenase complex is inhibited by serine phosphorylation (Chapter 21). Phosphorylations of serine, threonine, and tyrosine play extremely important roles in regulating many cellular functions, including the expression of hormonal effects on the control of cell growth.

Phosphomonoester Transfer

Many phosphomonoester transfers are catalyzed by nucleotidyl transferases and by ATP-dependent synthetases that activate molecules by transfer of the adenylyl (AMP) group of ATP. Acetyl CoA synthetase in yeast and mammalian cells is an example of the latter type and was discussed in Chapter 10 in connection with acyl activation. These enzymes are important for activating sugars in polysaccharide and complex-carbohydrate biosynthesis, for coenzyme biosynthesis, and for many fundamental reactions of protein and nucleic acid biosynthesis. Other phosphomonoester transfers are important in the biosynthesis of complex lipids. Examples of these reaction types will be described in this section.

UDP-glucose Pyrophosphorylase

Uridine diphosphate glucose pyrophosphorylase catalyzes the transfer of the UMP group from MgUTP to α-D-glucose-1-phosphate (Glc-1-P) to form UDP-glucose and $MgPP_i$, as illustrated in Figure 11-17. The kinetic pathway for this reaction is a special case of sequential kinetics, in which substrates are bound and products are released in the compulsory order specified in Figure 11-18. In this sequential pathway, UTP binds to the enzyme before glucose-1-phosphate, and $MgPP_i$ is released before UDP-glucose. It is a special case because no degree of randomness in binding is detectable by kinetic measurements. The compulsory ordered binding pathway lies at one pole of the possible sequential pathways, whereas the equilibrium random binding pathway followed in the adenylate kinase reaction lies at the opposite pole. Many intermediate variants are possible, among which one of the pathways is favored under given reaction conditions, but others may be favored with changing conditions such as pH or the type of buffer used.

Equation 14 is the rate equation for this kinetic pathway, and it has the same form as that for the equilibrium random binding pathway (equation 4).

$$v = \frac{V_m[\text{UTP}][\text{Glc-1-P}]}{[\text{UTP}][\text{Glc-1-P}] + K_{m\text{Glc-1-P}}[\text{UTP}] + K_{m\text{UTP}}[\text{Glc-1-P}] + K_{s\text{UTP}}K_{m\text{Glc-1-P}}} \tag{14}$$

However, $K_{m\text{UTP}}$ and $K_{m\text{Glc-1-P}}$ for the ordered steady-state kinetic pathway are not equal to the dissociation constants for UTP and Glc-1-P, as they are in the

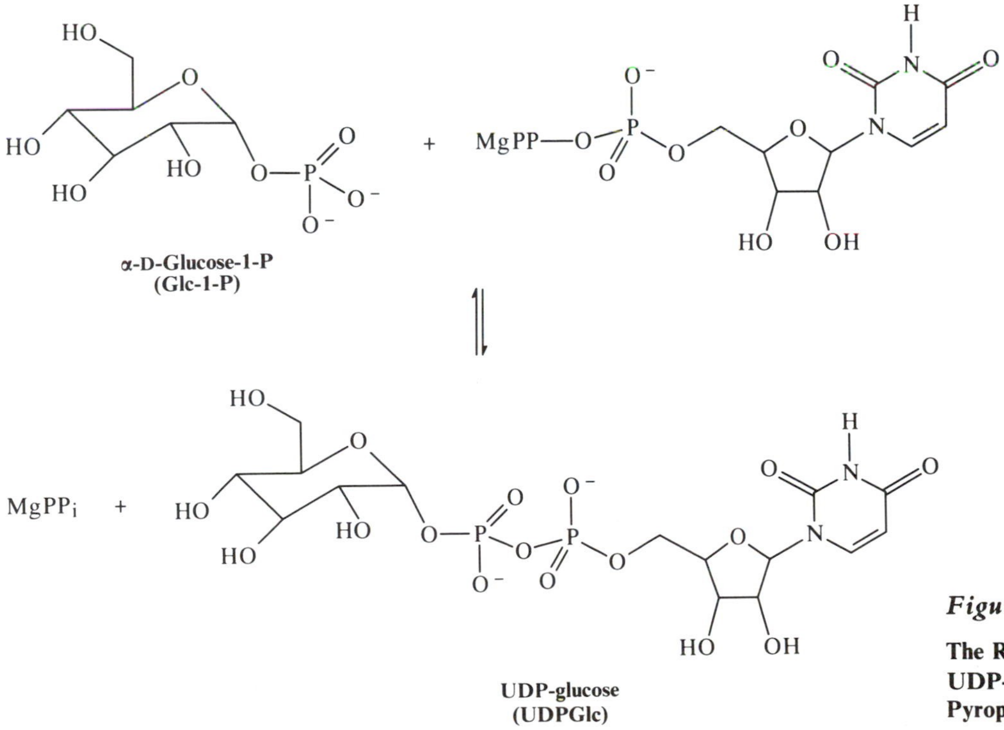

α-D-Glucose-1-P
(Glc-1-P)

MgPP—O—P(O)(O⁻)—O— ... + MgPPi

UDP-glucose
(UDPGlc)

Figure 11-17

The Reaction Catalyzed by UDP-glucose Pyrophosphorylase.

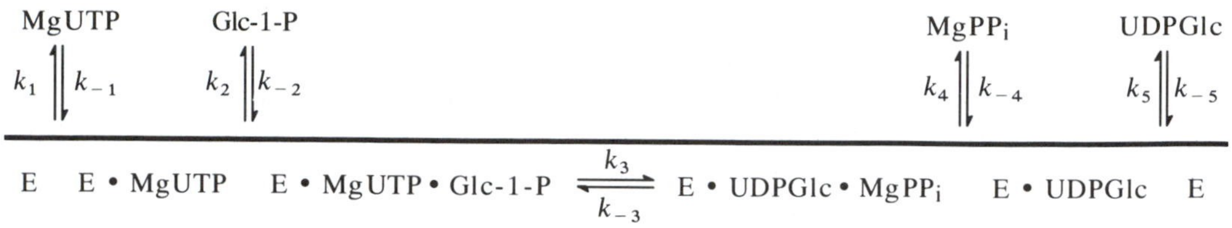

$$
\begin{array}{ccccccccc}
 & \text{MgUTP} & & \text{Glc-1-P} & & \text{MgPP}_i & & \text{UDPGlc} & \\
 & k_1 \downarrow\uparrow k_{-1} & & k_2 \downarrow\uparrow k_{-2} & & k_4 \uparrow\downarrow k_{-4} & & k_5 \uparrow\downarrow k_{-5} & \\
\text{E} & \text{E}\cdot\text{MgUTP} & & \text{E}\cdot\text{MgUTP}\cdot\text{Glc-1-P} \underset{k_{-3}}{\overset{k_3}{\rightleftharpoons}} & \text{E}\cdot\text{UDPGlc}\cdot\text{MgPP}_i & & \text{E}\cdot\text{UDPGlc} & & \text{E}
\end{array}
$$

Figure 11-18

The Steady-State Ordered Kinetic Mechanism for UDP-Glucose Pyrophosphorylase. UDP-glucose and glucose-1-phosphate are abbreviated as UDPGlc and Glc-1-P, respectively.

equilibrium random binding pathway. Instead, they are determined by the kinetic rate constants in Figure 11-18 and can be expressed in terms of those constants. The only dissociation constant that appears in equation 14 is K_{sUTP} in the last term of the denominator.

The fact that the form of equation 4 for the equilibrium random pathway and that of equation 14 for the steady-state ordered pathway are identical exemplifies the complexities of multisubstrate enzyme kinetics. The equilibrium random and steady-state ordered binding pathways cannot be distinguished by the form of their rate equations or the appearance of Lineweaver-Burk plots of initial rates with no products present. One means by which they may be distinguished is by analysis of the inhibitory effects of varied concentrations of products on the kinetic plots and rate equations. This subject is presented in a later section in this chapter.

The chemical mechanism of the UDP-glucose pyrophosphorylase reaction is analogous to that of adenylate kinase, a direct group transfer, except that the transferred group is the uridine-5′-phosphoryl group rather than the phosphoryl group. The transfer proceeds with inversion of configuration at phosphorus with a P-chiral substrate containing sulfur at P^1 and with UMPS instead of UMP as the transferred group. This is consistent with a single-displacement mechanism and direct transfer. No covalent E-UMP can be isolated. However, when the enzyme is incubated with UTP or UDP-glucose, it binds one molecule per molecule of enzyme very tightly, so that a simple radiochemical experiment with [^{14}C]UTP gives a false positive result that might be interpreted as the formation of E-UMP. Moreover, the partially purified enzyme contains tightly bound UTP and UDP-glucose, and for this reason it catalyzes an exchange reaction characteristic of Ping-Pong kinetics. Removal of the nucleotides by treatment with charcoal abolishes the exchange.

The UDP-glucose pyrophosphorylase reaction is the most-important means by which glucose is activated as a glucosyl donor for the biosynthesis of polysaccharides and complex carbohydrates. UDP-glucose is the glucosyl donor for the biosynthesis of starch and glycogen; it is also an intermediate in the biosyntheses of a variety of other nucleotide sugars, including UDP-galactose, UDP-glucuronic acid, and UDP-xylose. UDP-glucose pyrophosphorylase is one of a class of nucleotide sugar pyrophosphorylases that produce activated sugars for biosynthesis of complex carbohydrates, including ADP-glucose in plants, GDP-mannose, UDP-*N*-acetylglucosamine, CDP-glucose, and dTDP-glucose, among others.

Galactose-1-P Uridylyltransferase

Galactose-1-phosphate uridylyltransferase catalyzes the transfer of a UMP group from UDP-glucose to galactose-1-P according to Figure 11-19. The

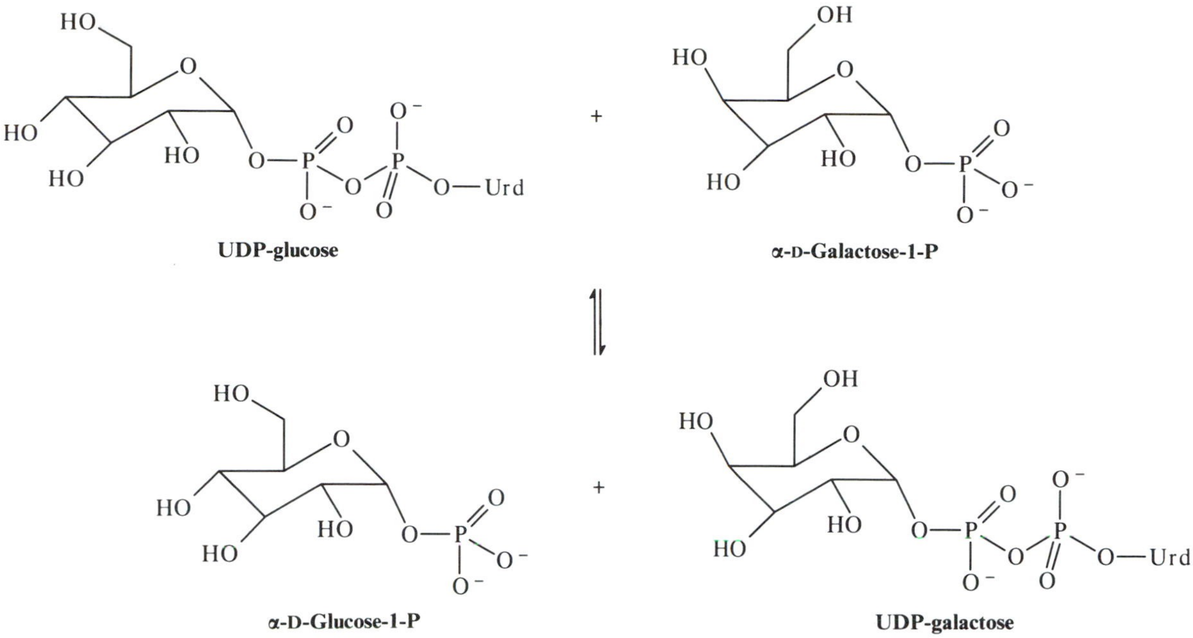

UDP-glucose + α-D-Galactose-1-P ⇌ α-D-Glucose-1-P + UDP-galactose

Figure 11-19

The Reaction Catalyzed by Galactose-1-P Uridylyltransferase. Galactose-1-phosphate and UDP-galactose are abbreviated as Gal-1-P and UDPGal, respectively.

reaction is formally similar to the UDP-glucose pyrophosphorylase reaction; however, the kinetics and chemical reaction mechanism are entirely different. The galactose-1-P uridylyltransferase reaction follows the Ping-Pong kinetic pathway in Figure 11-20 in which a covalent E-UMP is the intermediate. The Lineweaver-Burk plots are parallel lines described by a rate equation having the same form as that for the nucleoside diphosphate kinase reaction. In the intermediate, UMP is bonded to the imidazole ring of a histidine residue. And, in the overall reaction, configuration at chiral phosphorus is retained, as a consequence of the two inversions in the two transfer steps.

The mechanistic difference between the UDP-glucose pyrophosphorylase and galactose-1-P uridylyltransferase reactions is rationalized on the basis of the principle of economy in the evolution of binding sites, as invoked earlier for the adenylate kinase and nucleoside diphosphate kinase reactions. The galactose-1-P uridylyltransferase reaction (Figure 11-19) is nearly symmetrical with respect to acceptor substrates, galactose-1-P and glucose-1-P, which are sterically similar and have the same charge. This allows the enzyme to work with a single binding site containing a nucleophilic catalyst, histidine-166 in the *E. coli*

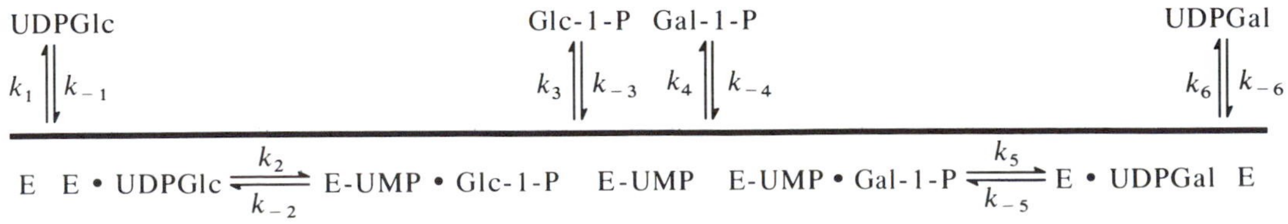

UDPGlc (k_1, k_{-1}) Glc-1-P (k_3, k_{-3}) Gal-1-P (k_4, k_{-4}) UDPGal (k_6, k_{-6})

$$\text{E} \quad \text{E} \cdot \text{UDPGlc} \underset{k_{-2}}{\overset{k_2}{\rightleftharpoons}} \text{E-UMP} \cdot \text{Glc-1-P} \quad \text{E-UMP} \quad \text{E-UMP} \cdot \text{Gal-1-P} \underset{k_{-5}}{\overset{k_5}{\rightleftharpoons}} \text{E} \cdot \text{UDPGal} \quad \text{E}$$

Figure 11-20

The Ping-Pong Kinetic Mechanism for Galactose-1-P Uridylyltransferase.

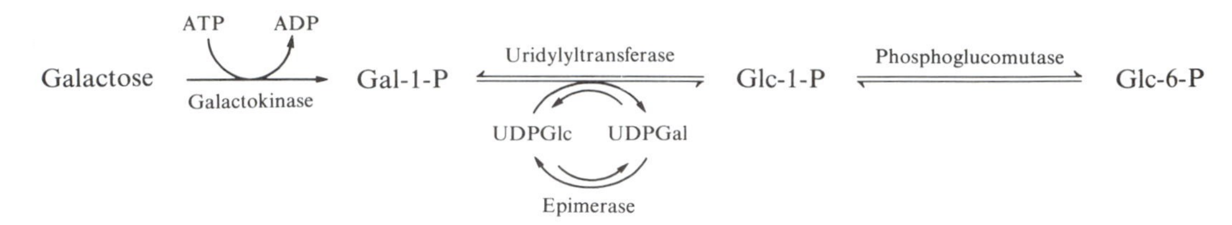

Figure 11-21

The Leloir Pathway of Galactose Metabolism.

enzyme. In contrast, the UMP-acceptors in the UDP-glucose pyrophosphorylase reaction (Figure 11-17) are PP_i and glucose 1-P, which are structurally and electrostatically too different to use the same binding site. Therefore, through evolution, two acceptor binding sites appeared, both adjacent to the uridylyl binding site and obviating the need for a nucleophilic group because direct transfer is possible in ternary complexes.

Galactose metabolism is of special interest because galactose is activated by UDP-glucose rather than by UTP. Moreover, a defect in galactose 1-P uridylyltransferase is responsible for galactosemia in human beings, an autosomal recessive trait that blocks utilization of galactose and leads to its accumulation to toxic levels. The condition causes irreversible neurological damage and early cataract formation among other problems. Withdrawal of lactose and galactose from the diet prevents the development of symptoms.

Galactose is metabolized in plants, microorganisms, and animals by the Leloir pathway in Figure 11-21. Phosphorylation by galactokinase produces galactose-1-P, which is then converted into UDP-galactose and glucose-1-P. Glucose-1-P is metabolized through the glycolytic pathway (Chapter 21). UDP-glucose acts catalytically because the enzyme UDP-galactose 4-epimerase catalyzes its regeneration from UDP-galactose. This replenishes UDP-glucose for another iteration of the pathway. Thus, the Leloir pathway converts galactose into glucose 1-P, using one ATP in the process. ATP utilization in galactose metabolism is in no sense inefficient relative to glucose metabolism because glucose also must be phosphorylated by ATP (Chapter 21).

The Leloir pathway also enables the production of UDP-galactose from glucose through the action of UDP-galactose 4-epimerase on UDP-glucose, which is produced in the absence of galactose from glucose 1-P and UTP in the UDP-glucose pyrophosphorylase reaction. The UDP-galactose 4-epimerase reaction presents mechanistic difficulties because it requires nonstereospecific hydride transfer at a single active site. The reaction mechanism is described in Chapter 20.

Determination of Kinetic Pathways in Multisubstrate Enzymatic Reactions

The similarity of rate equations 4 and 14 for the equilibrium random binding pathway of adenylate kinase and the steady-state ordered binding pathway of UDP-glucose pyrophosphorylase exemplifies one of the difficulties in the kinetic analysis of multisubstrate enzymatic reactions. The only information that can be obtained from the primary kinetic analysis is the form of the rate equation, and that form usually does not characterize the binding pathway, much less the chemical mechanism. An important exception is the Ping-Pong pathway, where the form of the rate equation indicates that there is a covalent enzyme-substrate intermediate in the chemical mechanism.

In a multisubstrate reaction, the steady-state kinetics is analyzed by initial rate measurements, varying one substrate concentration at a time while holding the others constant. In this way, for a single set of measurements, all the complex rate laws reduce to the same form, that of the simple one-substrate Michaelis-Menten rate law. This is easily seen by factoring the reciprocal form of equation 4, for adenylate kinases, in such a way as to conform with the fact that, in a particular experiment, AMP is varied while ATP is held constant. Equation 15 is this form of the rate equation for the adenylate kinase reaction, shown earlier as equation 4.

$$\frac{1}{v} = \frac{1}{V_m}\left(1 + \frac{K_{sATP}}{[ATP]}\right) + \frac{K_{sAMP}}{V_m}\left(1 + \frac{K'_{sATP}}{[ATP]}\right)\frac{1}{[AMP]} \tag{15}$$

With [ATP] held constant, the factors containing [ATP] are multipliers of $1/V_m$ and K_s/V_m, and the equation has the same form as the reciprocal of the Michaelis-Menten equation for a one-substrate reaction (equation 8 in Chapter 3), in which AMP is the substrate. By varying [AMP] at several different constant values of [ATP], several *apparent* values of the maximum velocity and Michaelis constant can be obtained. The apparent values, $V_{m\,app}$ and $K_{m\,app}$, are defined by equations 16 and 17.

$$\frac{1}{V_{m\,app}} = \frac{1}{V_m}\left(1 + \frac{K_{sATP}}{[ATP]}\right) \tag{16}$$

$$\frac{K_{m\,app}}{V_{m\,app}} = \frac{K_{sAMP}}{V_m}\left(1 + \frac{K'_{sATP}}{[ATP]}\right) \tag{17}$$

Notice that values of V_m and K_{sATP} can be obtained from a plot of $1/V_{m\,app}$ versus 1/[ATP] where, according to equation 16, the intercept will be $1/V_m$ and the slope will be K_{sATP}/V_m. The value of K'_{sATP} can be obtained by plotting equation 17 in a similar way.

It is often possible to distinguish equilibrium random binding from steady-state ordered binding in bisubstrate reactions by studying the effects of inhibitors on the form of the rate equation. In particular, products acting as inhibitors create different and characteristic inhibition patterns for these two pathways; in fact, product inhibition patterns can characterize and distinguish other pathways for three-substrate and higher-order reactions as well.

Cleland's Rules for Enzyme-Inhibitor Interactions

Cleland's rules enable one to deduce the types of enzyme-inhibitor binding interactions that lead to inhibition simply by inspecting the plots of $1/v$ versus 1/[varied substrate] obtained in the presence and absence of the inhibitor. An inhibitor may be an unreactive molecule related in structure to a substrate, or it may be a reaction product that binds to the enzyme in place of a substrate and reverses the flow of chemical events in catalysis. When the inhibitor is a reaction product, its interaction with the enzyme gives an inhibition pattern that is characteristic of the kinetic pathway for enzyme-substrate and enzyme-product binding; and these inhibition patterns for several inhibitors can be useful for deducing the kinetic binding pathway. Cleland's rules are the following:

1. An inhibitor that binds to the same enzyme form as does the *varied* substrate, or to a form that is connected only by reversible steps to that form, increases the slope of the Lineweaver-Burk plot.
2. An inhibitor that binds to a different enzyme form from that to which the varied substrate binds increases the intercept of the double-reciprocal plot.

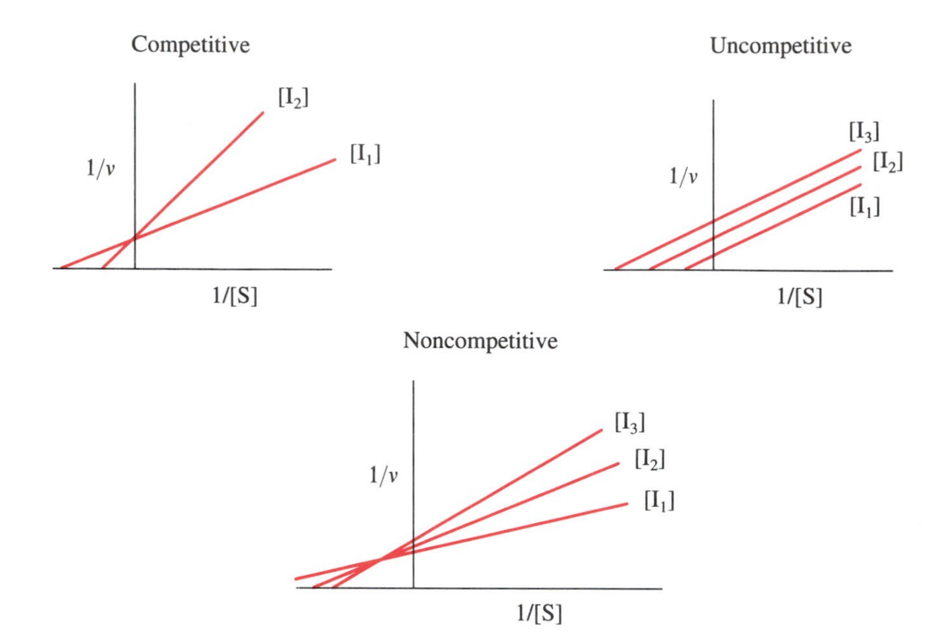

Figure 11-22

Competitive, Uncompetitive, and Noncompetitive Inhibition Patterns in Double-Reciprocal Plots.

Thus, an inhibitor that binds only according to rule 1 will affect only the slopes in the double-reciprocal plot, and this corresponds to *competitive inhibition* (Chapter 4). An inhibitor that binds only according to rule 2 affects only the intercepts; that is, the lines will be parallel, and a strictly parallel inhibition pattern is known as *uncompetitive inhibition.* An inhibitor that binds in accord with *both* rule 1 *and* rule 2 exhibits both slope and intercept effects; that is, the lines will have different slopes and different intercepts and will converge at the left of the ordinate. This type of inhibition is known as *noncompetitive inhibition.* The three types of inhibition patterns are illustrated in Figure 11-22.

BOX 11-1

PRODUCT INHIBITION IN BISUBSTRATE GROUP TRANSFER REACTIONS

In bisubstrate group transfer reactions, there are two substrates and two products; and the presence of either product inhibits the reaction by binding to the enzyme. Any such interaction with a product tends to reverse the reaction and reduce the rate. This phenomenon is expressed in the rate equation in a specific way. Although rate equations for inhibition will not be presented here, the interactions and inhibition patterns described herein are expressed by appropriate rate equations that can be straightforwardly derived using standard methods.

In Chapter 4, we showed that competitive inhibition is expressed in the rate equation in the form of the factor $(1 + [I]/K_i)$ as a multiplier of K_m; with increasing values of [I], this term increases the slope of a double-reciprocal plot (Figure 11-22). In uncompetitive inhibition, the inhibition factor appears as a multiplier of V_m, which increases the intercept of a double-reciprocal plot (Figure 11-22). In noncompetitive inhibition, both V_m and K_m are multiplied by such terms, and the inhibitor increases both the slope and the intercept.

In inhibition experiments, the enzyme-inhibitor interactions lead to modified rate equations that incorporate multiplier terms in [I] such as those in the following equations, in which K_{is} designates an inhibitory effect on the slope and K_{ii} an effect on the intercept of a double-reciprocal plot.

Competitive inhibition	$K_{m\,app} = K_m(1 + [I]/K_{is})$
Uncompetitive inhibition	$V_{m\,app} = V_m(1 + [I]/K_{ii})$
Noncompetitive inhibition	Both of the above equations

A difference between single-substrate and multisubstrate reactions is that, with more than one substrate, a particular inhibitor can be competitive when one substrate is varied, but it can be uncompetitive or noncompetitive when a different substrate is varied. This fact is the basis for the utility of these rules in analyzing kinetic pathways.

Application of Cleland's rules to the idealized kinetic pathways introduced here for bisubstrate group transfer reactions exemplifies the use of product inhibition patterns for characterizing kinetic pathways and rate

BOX 11-1 (continued)

equations for complex enzymatic reactions. In an equilibrium random binding pathway such as that for adenylate kinase in Figure 11-7, each product ADP and MgADP binds to E, as does each substrate according to rule 1. Each product can *in principle* also bind to a different enzyme form, a binary enzyme-product complex; but, in a rapid equilibrium binding pathway, these *product* binary complexes do not exist at significant levels in the absence of a pool of products because the rate-limiting step is the interconversion of ternary complexes and product-dissociation steps are at equilibrium. Therefore, each product binds only according to rule 1, and competitive inhibition is exhibited by each product versus each corresponding substrate in the idealized equilibrium random pathway.

The other sequential pathway that we have considered is the steady-state ordered binding mechanism for UDP-glucose pyrophosphorylase (Figure 11-18). Consider UDP-glucose (UDPGlc), which binds only to E, acting as an inhibitor. When MgUTP is the varied substrate, UDPGlc binds according to rule 1 and exhibits competitive inhibition. When glucose-1-P is the varied substrate, UDPGlc binds according to both rule 1 and rule 2 and is a noncompetitive inhibitor. $MgPP_i$ binds only to E·UDPGlc, which is not the same as E or E·MgUTP to which the substrates bind; so rule 2 applies. However, E·UDPGlc is connected by reversible steps to all other enzyme forms, so that rule 1 also applies. Therefore, inhibition by $MgPP_i$ is noncompetitive with respect to both substrates.

An important special case for inhibition by $MgPP_i$ is that in which the fixed concentration of glucose-1-P is so high that virtually no E·MgUTP exists in the steady-state, owing to the fact that it has all been captured by glucose-1-P to form the ternary complex. In this special case, the reversible connection between E·UDPGlc and E is broken. Because there is no pool of free UDPGlc, its dissociation from E·UDPGlc is irreversible in an initial rate measurement; therefore, there is also no reversible connection to E in the forward direction. In this case, inhibition by $MgPP_i$ versus MgUTP follows only rule 2 and is uncompetitive. The observation of uncompetitive inhibition in this special case is a particularly strong indicator of the ordered binding pathway.

Unreactive substrate or product analogs also are often used as inhibitors to characterize kinetic pathways. These molecules, which are sometimes referred to as dead-end inhibitors, bind to the enzyme in place of a substrate or product and effectively extract it from the productive catalytic pathway. The competitive inhibitors for trypsin and chymotrypsin described in Chapter 4 are examples of this type of inhibitor. Inhibition by these molecules is also subject to analysis by the Cleland rules.

It is important to recognize that enzyme-substrate and enzyme-inhibitor interactions can be complex; and the idealized pathways defined in Figures 11-7, 11-12, 11-18, and 11-20 do not always prevail. Moreover, complex kinetic mechanisms are not always unambiguously determined by product inhibition studies alone. Kinetic analyses should never be lightly undertaken. Three of the enzymes cited as examples in this chapter—nucleoside diphosphate kinase, UDPGlc pyrophosphorylase, and galactose-1-P uridylyltransferase—exhibit the idealized kinetic behavior defined in Figures 11-12, 11-18, and 11-20 at substrate and product concentrations below and a fewfold higher than K_m values. At higher inhibitor concentrations, other interactions almost always come into play; and these interactions also may be useful for analyzing kinetic behavior. However, adenylate kinase does not behave exactly as illustrated by Figure 11-7, although all of the interactions shown do in fact occur. The problem that arises is that the substrates and products are similar enough that *additional* binding interactions occur, in which AMP (and ADP) bind to the site at which MgATP (and MgADP) normally bind. The resulting dead-end complexes are inhibited forms. You can apply the inhibition rules to these interactions to determine how they affect the kinetic behavior.

Biosynthesis of Nucleotide Derivatives

Phosphoadenosine Phosphosulfate (PAPS)

PAPS, the biochemical sulfating agent, is 3′-phosphoadenosine 5′-phosphosulfate. Its biosynthesis proceeds in two steps catalyzed by an adenylyltransferase and a kinase. In equation 18, the adenylyltransferase catalyzes the displacement of PP_i from ATP by SO_4^{2-} to form adenosine 5′-phosphosulfate (APS).

$$\text{ATP} + {}^{-}\text{O}-\overset{\overset{\text{O}}{\|}}{\underset{\underset{\text{O}}{\|}}{\text{S}}}-\text{O}^{-} \rightleftharpoons \text{Ado}-\text{O}-\overset{\overset{\text{O}}{\|}}{\underset{\underset{\text{O}^{-}}{|}}{\text{P}}}-\text{O}-\overset{\overset{\text{O}}{\|}}{\underset{\underset{\text{O}}{\|}}{\text{S}}}-\text{O}^{-} + \text{PP}_i \quad (18)$$

APS

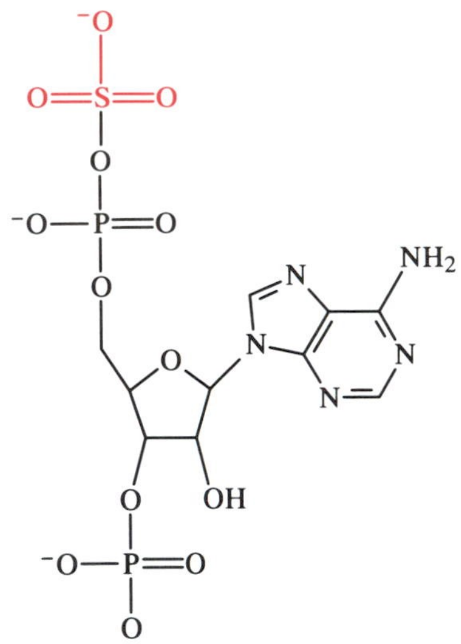

3′-Phosphoadenosine-5′-phosphosulfate (PAPS)

APS is then phosphorylated on the ribosyl-3′-OH by the kinase in equation 19.

$$\text{APS} + \text{ATP} \rightleftharpoons \text{PAPS} + \text{ADP} \quad (19)$$

The overall process requires two molecules of ATP to activate one sulfate. In fact, three phosphoanhydride bonds from two ATP's must be used to produce PAPS. The first step is very unfavorable because bisulfate (HSO_4^-) is a much stronger acid ($pK_a = 2$) than phosphate ($pK_a = 7$ and 12). The reaction is partially drawn to the right by the pyrophosphatase-catalyzed hydrolysis of PP_i. However, a further strong boost is made available by the phosphorylation of the 3′-OH in APS to PAPS in the second step, which is a steeply downhill reaction.

PAPS is the physiological donor of sulfate groups and the substrate for sulfotransferases in biosynthesis and detoxification. Phenols, for example, are detoxified by sulfation to sulfate esters and excreted. PAPS is the sulfate dónor in the biosynthesis of sulfated, complex polysaccharides.

Cyclic AMP

Adenosine 3′,5′-cyclic phosphate (cyclic AMP, cAMP, or 3′,5′-AMP) is synthesized by intramolecular transfer of the adenylyl (AMP) group within ATP to itself; that is, by attack of the 3′-OH group on the α-phosphorus of ATP to expel pyrophosphate, as illustrated in Figure 11-23. Adenylyl cyclase is a membrane-bound enzyme that catalyzes this reaction and is controlled by interactions with hormones. It is activated by binding interactions with a variety of receptors, which are in turn activated by binding interactions with hormones or other effector molecules. The cellular concentration of cyclic AMP is also modulated by cyclic AMP phosphodiesterases, which catalyze its hydrolysis to AMP. Cyclic AMP phosphodiesterases are themselves subject to regulation.

Cyclic AMP is a "second messenger" that mediates the actions of hormones, which are secreted into the blood stream by glands on command from the hypothalamus (Figure 11-24; see also Chapter 1). Hormonal messages are often relayed by the receptors in cell membranes to adenylyl cyclase, also in

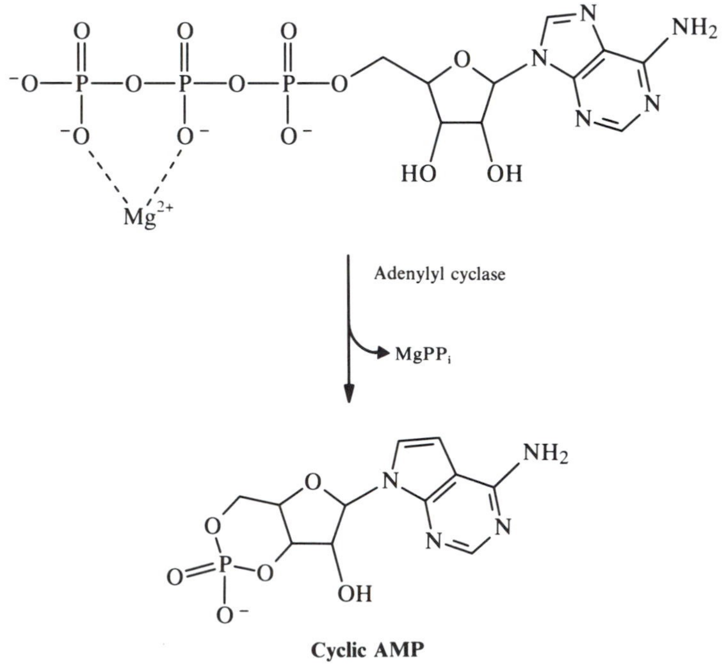

Adenylyl cyclase

$MgPP_i$

Cyclic AMP

Figure 11-23

The Conversion of ATP into Cyclic AMP Catalyzed by Adenylyl Cyclase.

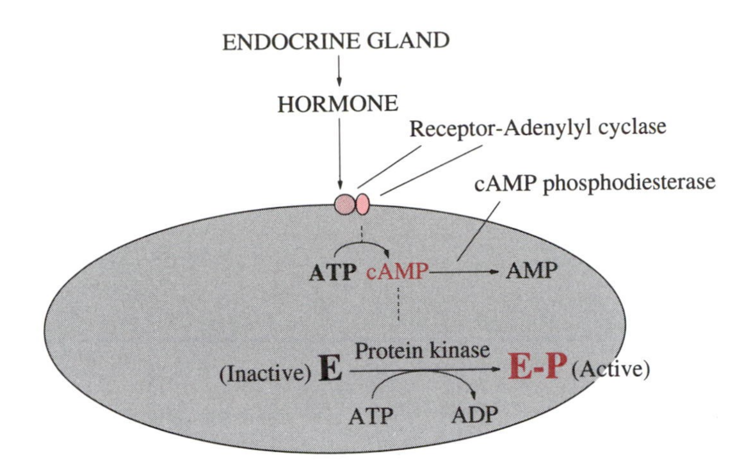

Figure 11-24

Cyclic AMP as a Second Messenger of Hormone Action.

the cell membrane, the activtion of which produces a burst of cyclic AMP. The cyclic AMP "turns on" a variety of cellular activities such as the degradation of glycogen (Chapter 21). Cyclic AMP often acts by activating a protein kinase, which in turn catalyzes phosphorylation of a protein, thereby activating or inactivating that protein. A protein phosphatase, which may itself be subject to regulatory interactions with effectors, subsequently cleaves the phosphate group from the protein and returns it to its original state.

Biosynthesis of Fats and Phospholipids

Phospholipids contain two long-chain fatty acids and a phosphodiester attached to the three-carbon backbone of glycerol (Figure 11-25). One of the fatty acids is often either palmitic (hexadecanoic) or stearic (octadecanoic) acid (16 or 18 carbons), and the central fatty acid is usually an unsaturated fatty acid such as oleic acid. The phosphate is esterified to glycerol and either serine, choline, ethanolamine, or *myo*-inositol, which are designated —X in Figure 11-25. The phospholipid is named by the prefix *phosphatidyl* preceding the name of the alcohol, —X. Common examples are phosphatidyl serine and phosphatidyl choline. The compound lacking an alcohol —X is a *phosphatidic acid.*

Phospholipids are the principal constituents of membranes (Chapter 29). With large hydrophobic fatty acyl groups at one end and an electrostatically charged and hydrophilic group at the other, they are detergents. In aqueous media, they aggregate into bilayers with the hydrophobic groups inside and the charged head groups exposed to water. The bilayers undergo closure to essentially spherical vesicles (Chapter 29). Variations in the alcohol, —X, and the fatty acids affect the properties of phospholipids and cell membranes. A choline deficiency can give rise to fatty liver, probably because it has a role in the transport of fats.

Phospholipids are synthesized by the transfer of a phosphoryl ester group from a phosphoanhydride to an alcohol. This is the type of reaction that occurs in the formation of cyclic AMP from ATP, and it is the foundation for the biosynthesis of 3′,5′-phosphodiester linkages in nucleic acids (Chapter 12).

Phospholipid biosynthesis occurs in two ways: (1) the activated phosphate ester is transferred to the free hydroxyl group of a bis-fatty acylglycerol or diacylglycerol, a *diglyceride* (Figure 11-26); (2) alternatively, the activated phosphatidyl group in a *CDP-diglyceride* is transferred to the alcohol, —X (Figure 11-27).

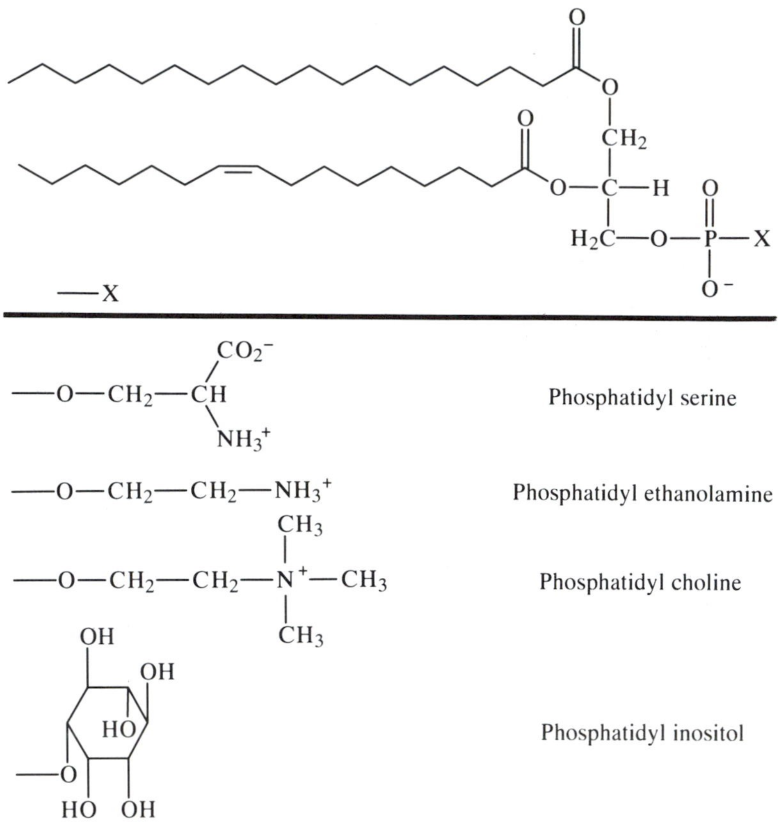

—X	
$—O—CH_2—CH(CO_2^-)(NH_3^+)$	Phosphatidyl serine
$—O—CH_2—CH_2—NH_3^+$	Phosphatidyl ethanolamine
$—O—CH_2—CH_2—N^+(CH_3)_3$	Phosphatidyl choline
—O—inositol	Phosphatidyl inositol

Figure 11-25

Structures of Phospholipids.

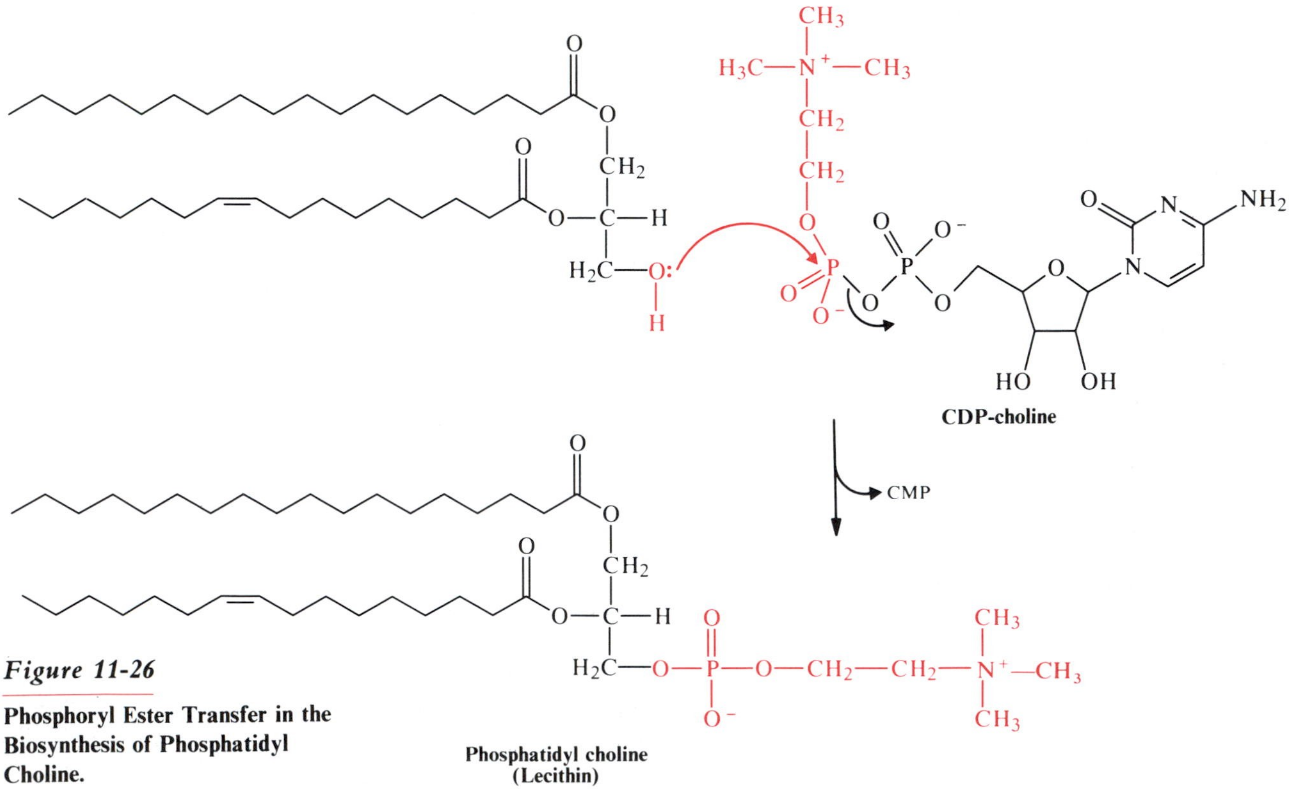

CDP-choline

CMP

Phosphatidyl choline (Lecithin)

Figure 11-26

Phosphoryl Ester Transfer in the Biosynthesis of Phosphatidyl Choline.

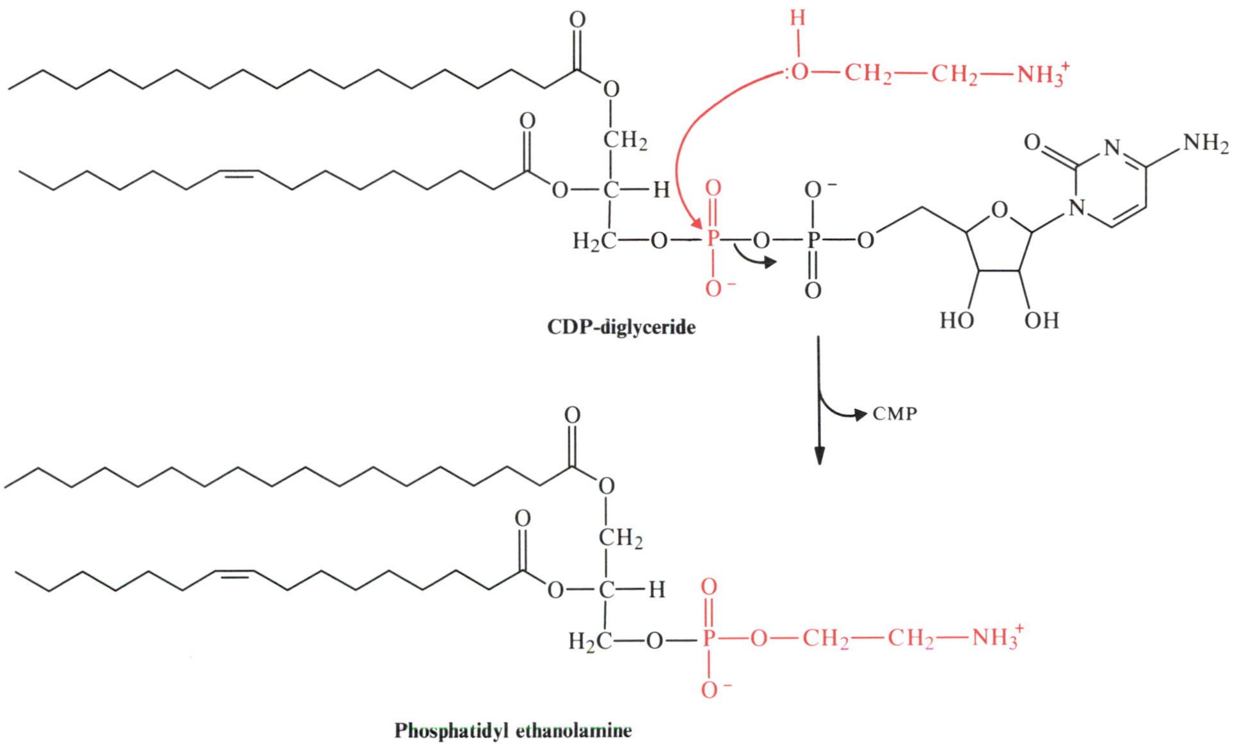

CDP-diglyceride

CMP

Phosphatidyl ethanolamine

Figure 11-27

Phosphoryl Ester Transfer in the Biosynthesis of Phosphatidyl Ethanolamine.

The route to a phospholipid through the CDP-activated alcohol choline, shown in Figure 11-26, is the predominant pathway in most animals. Phosphocholine is activated as a phosphoanhydride in CDP-choline. The 3-hydroxyl group of a diglyceride attacks the phosphate of phosphocholine to displace CMP, giving phosphatidyl choline, which is commonly called lecithin.

In some bacteria and mammals, phosphatidic acid is activated by phosphoanhydride bonding in CDP-diglyceride, and the phosphatidyl group is transferred to the hydroxyl group of an alcohol to form the corresponding phospholipid, as illustrated in Figure 11-27 for the reaction with ethanolamine. Phosphatidyl ethanolamine appears in membrane lipids and is a precursor of phosphatidyl choline, which can be formed by methylation of the amino group.

The activated esters CDP-choline and CDP-diglyceride are produced by nucleotidyl (NMP) transfers that are analogous to those in the UDP-glucose pyrophosphorylase reaction. Phosphocholine from choline kinase (Table 11-1) reacts with CTP, displacing PP_i and forming CDP-choline in equation 20.

$$\text{Phosphocholine} + \text{CTP} \rightleftharpoons \text{CDP-choline} + PP_i \qquad (20)$$

CDP-diglycerides are produced in similar reactions of phosphatidic acids with CTP (equation 21).

$$\text{Phosphatidic acid} + \text{CTP} \rightleftharpoons \text{CDP-diglyceride} + PP_i \qquad (21)$$

The biosynthetic pathways for phospholipids and neutral fats are summarized in Figure 11-28. Glycerol is phosphorylated by ATP, which traps it within the cell as glycerophosphate. It is then acylated on the two free hydroxyl groups by *acyl transfer* from two molecules of fatty acyl CoA to give a phosphatidic acid. The fatty acids are activated by fatty acid activating enzymes

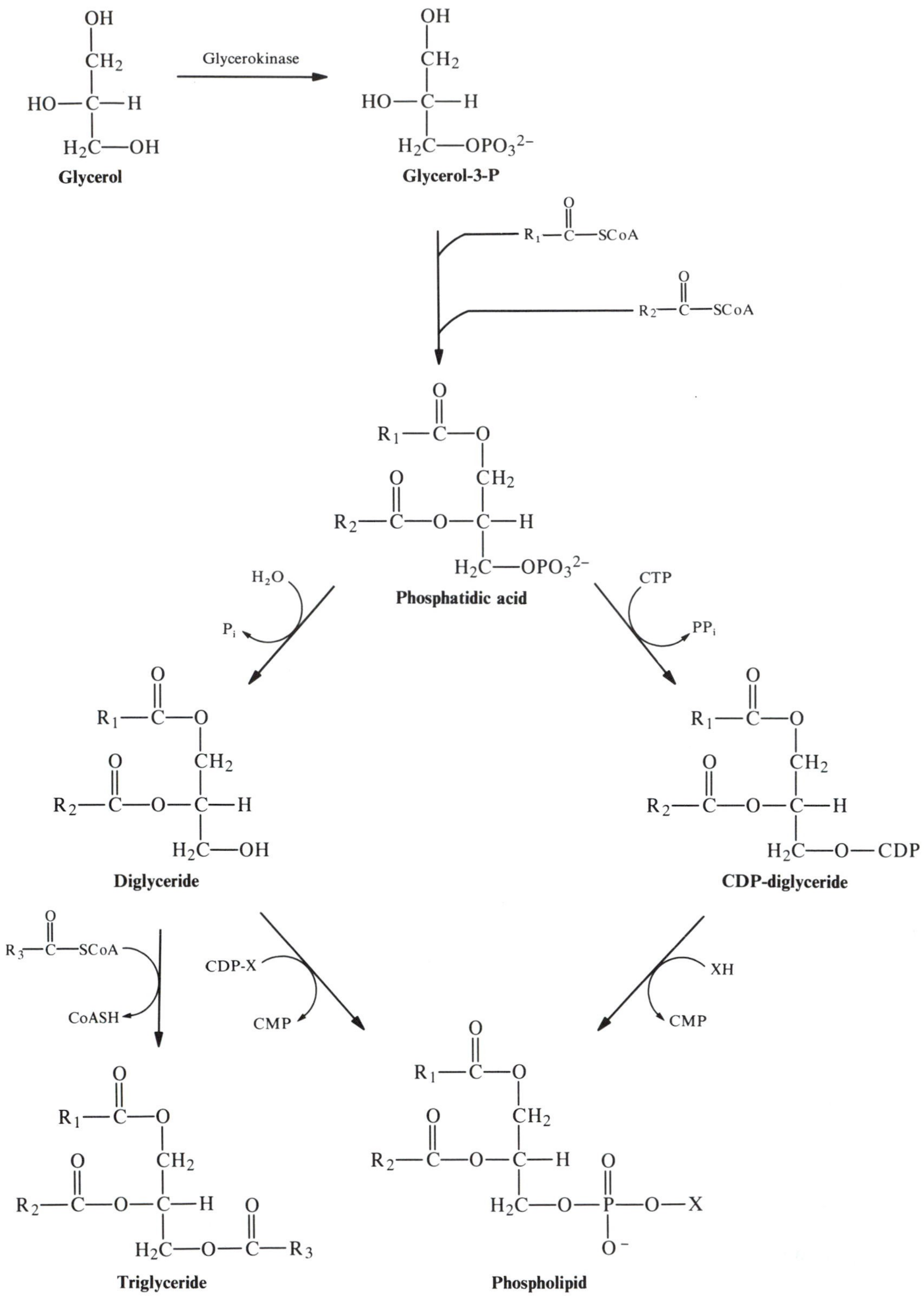

Figure 11-28

Summary of Phospholipid Biosynthesis.

through the same mechanism as that of acetyl CoA synthetase and aminoacyl tRNA synthetases (Chapter 10). Phospholipids are then synthesized by the two routes described earlier, either by activation of the phosphatidic acid to a CDP-diglyceride, followed by transfer of the phosphatidyl group to the appropriate alcohol, or by hydrolysis of phosphatidic acid to a diglyceride and reaction with an activated alcohol, CDP—X, to give the phospholipid.

It is apparent that the diglyceride is a key player in this scheme. It is also the precursor of neutral fats—the tris-fatty acylglycerols, or *triglycerides*—which are the principal lipids in body fat. Diglycerides are thereby also key players in multibillion dollar programs for weight control. All that is needed to synthesize a neutral triglyceride is acylation of a diglyceride by a third fatty acyl CoA.

Phosphohydrolases

Phosphohydrolases catalyze phosphotransfer and phosphomonoester transfer to water. Their biochemical importance is less well understood than that of kinases and nucleotidyl transferases. They certainly are involved in both the degradation and the biosynthesis of nucleotides and nucleic acids. Some of them act to maintain the pool of inorganic phosphate in cells. They also play important roles in the regulation of metabolism by catalyzing the hydrolysis of such phosphorylated molecules as fructose-1,6-bisphosphate, glucose-1,6-bisphosphate, cyclic AMP, and phosphorylated proteins (Chapter 21).

Alkaline Phosphatase

Alkaline phosphatases are phosphomonoesterases that catalyze the hydrolysis of essentially all phosphomonoesters, according to equation 22.

$$R{-}O{-}PO_3^{2-} + H_2O \longrightarrow R{-}OH + HOPO_3^{2-} \quad (22)$$

The enzyme is present in animals and microorganisms and is named for the fact that its optimum pH is above pH 7. Acid phosphatases have pH optima below pH 7. The substrate specificity of an alkaline phosphatase is limited to the requirement that substrates have a phosphoryl group bonded to oxygen, sulfur, or nitrogen—that is, the substituent R in equation 24 can be almost anything, including the adenosyl 5′-diphosphoryl group in ATP, the phosphoryl group in pyrophosphate, any alkyl or aryl group, a sugar, and even a polynucleotide.

One explanation for the absence of specificity becomes clear from the reaction mechanism and the kinetic properties of this enzyme. The kinetic pathway for the phosphohydrolase activity of alkaline phosphatase from *E. coli* is delineated in the upper pathway of Figure 11-29. In the first half of the

Figure 11-29

The Kinetic Mechanism for Phosphomonoester Hydrolysis and Phosphotransfer Catalyzed by Alkaline Phosphatase.

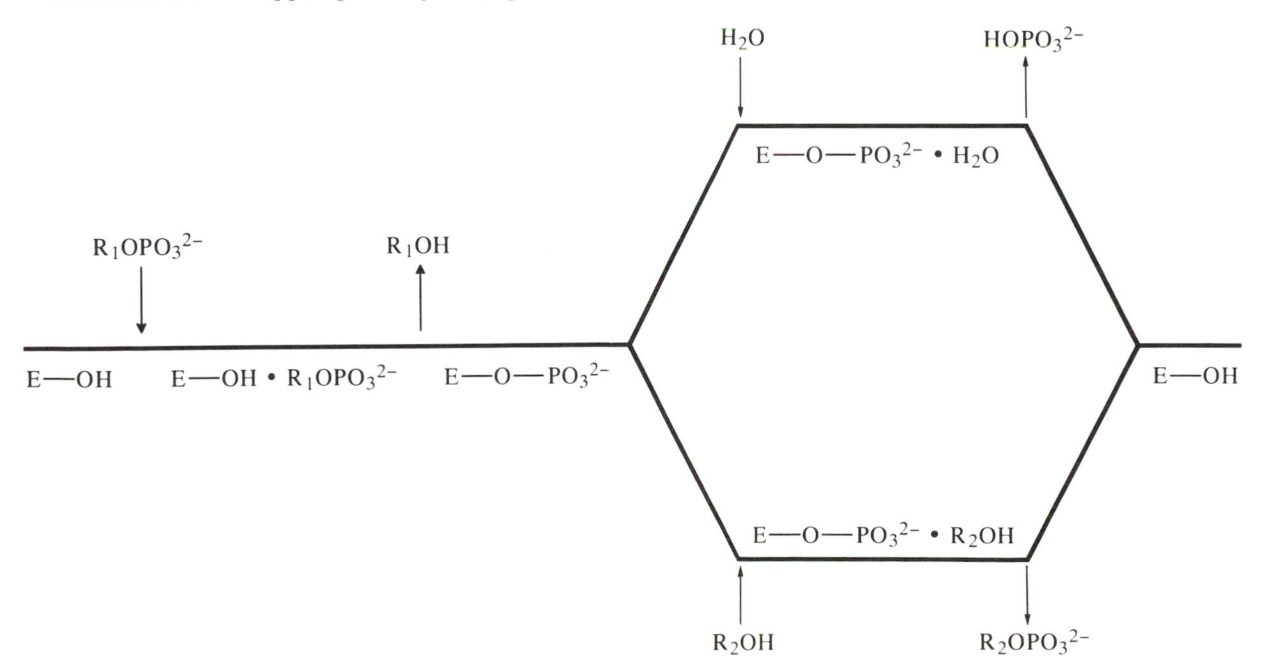

reaction, a phosphomonoester binds to the active site and phosphorylates a serine hydroxyl group of the enzyme; the product alcohol then dissociates from the phosphoryl-enzyme. In the second half of the reaction, water binds in place of the product alcohol, accepts the phosphoryl group, and regenerates the free enzyme. The first half of the reaction is almost never rate limiting; the second half is the same for all substrates and limits the rate. Therefore, nearly all phosphomonoester substrates react with the same value for k_{cat}.

The lower pathway in Figure 11-29 is the phosphotransferase pathway for alkaline phosphatase. The enzyme catalyzes the transfer of the phosphoryl group from one alcohol to another in the presence of a high concentration of a second alcohol. The phosphotransferase activity results from the fact that phosphorylation of the enzyme by substrates is reversible and the fact that the enzyme is nonspecific for substrates.

The activity of alkaline phosphatase depends on the presence of a metal ion such as Zn^{2+} or Mg^{2+}. Considerable evidence, obtained in experiments utilizing nuclear magnetic resonance, implicates the metal in binding the substrate by showing that the metal is coordinated to the phosphoryl group of substrates and the phosphoryl-enzyme.

Pyrophosphatases

Pyrophosphatases act in all cells to catalyze the hydrolysis of Mg-pyrophosphate into two molecules of inorganic phosphate. The reaction mechanism differs from that of alkaline phosphatase in that it does *not* occur through a two-step phosphoryl-transfer mechanism involving a covalent phosphoryl-enzyme. Cleavage occurs by attack of H_2O on the enzyme-bound $MgPP_i$.

Inorganic pyrophosphatases are important in metabolism, because many biosynthetic reactions that are thermodynamically either unfavorable or only slightly favorable produce pyrophosphate. Examples include acyl activation reactions (Chapter 10); nucleotide sugar biosynthesis, such as the UDP-glucose pyrophosphorylase reaction; phospholipid biosynthesis; and nucleic acid biosynthesis (Chapter 12). The action of inorganic pyrophosphatase in the hydrolysis of pyrophosphate makes all of these biosynthetic reactions energetically more favored than they would otherwise be, because the free energy of hydrolysis of pyrophosphate is about $-7\,\text{kcal mol}^{-1}$. This is always a significant and often the decisive factor in driving biosynthetic reactions.

Summary

Phosphate compounds are important in cells for the following reasons: First, phosphate esters and phosphoanhydrides are ionic under physiological conditions, which prevents them from diffusing across membranes and traps them inside cells and within cellular organelles. Second, the phosphate and pyrophosphate groups are good leaving groups in group transfer reactions, which allows them to serve as activating groups for acyl, glycosyl, and alkyl groups. The phosphorylated groups are transferred by ligases to acceptors, with displacement of the phosphate group, in numerous biosynthetic ligation reactions. Third, certain high-energy phosphates produced in metabolism, such as creatine phosphate, phosphoenolpyruvate, and 1,3-diphosphoglycerate, serve as sources of high-energy phosphoryl groups for ATP biosynthesis. Fourth, phosphate esters and phosphoanhydrides are kinetically stable under physiological conditions; but, when bound to active sites of enzymes, they become reactive through the catalytic effects of enzyme-substrate interactions. These properties of phosphates allow the cell to exercise control over the reactions of activated phosphate compounds.

Kinases catalyze phosphoryl transfer from nucleoside triphosphates, usually ATP, to various acceptor molecules. High-energy phosphates such as phosphoenolpyruvate, 1,3-diphosphoglycerate, creatine phosphate, and arginine phosphate can generate ATP from ADP in the presence of the specific kinases. Other kinases activate molecules for conversion into important metabolic intermediates. Kinases such as adenylate kinase and nucleoside diphosphate kinases maintain balance in the nucleotide pools and ensure availability of ATP. Several kinases catalyze phosphorylations of proteins that are regulated by their phosphorylation states.

Nucleotidyltransferases catalyze nucleoside phosphoryl transfers to a variety of acceptor molecules to form metabolic intermediates, coenzymes, and nucleic acids. These transfer reactions are required for the activation of sugars as nucleoside diphosphate sugars; the biosynthesis of DNA and RNA (Chapter 12); the biosynthesis of complex lipids; and the biosynthesis of cyclic AMP.

Kinases and nucleotidyltransferases catalyze nucleophilic attack by acceptors on the γ-phosphate or α-phosphate of a nucleotide. An important aspect of nucleotide binding to enzymes is the neutralization of negative charges on the phosphate groups undergoing reaction. Charge neutralization is effected through coordination with divalent metal ions and with the ammonium and guanidinium groups of the enzymes. The enzymes catalyze phosphoryl and nucleotidyl transfers either (a) by direct transfer from donor to acceptor within ternary enzyme-substrate complexes in sequential kinetic pathways or (b) by double-displacement, Ping-Pong kinetic pathways that include the formation of covalent phosphoryl-enzyme or nucleotidyl-enzyme intermediates. The principle of economy in the evolution of binding sites is a dominant factor guiding the evolution of group-transfer enzymes. This principle accounts for the fact that a few enzymes catalyze phosphoryl group transfer by two-step double-displacement mechanisms through phosphoryl-enzyme intermediates and Ping-Pong kinetic pathways, whereas most of them proceed by single-step direct transfers between substrates bound in ternary complexes generated by sequential kinetic pathways.

ADDITIONAL READING

KNOWLES, J. R. 1980. *Annual Review of Biochemistry* 49: 877.

WESTHEIMER, F. H. 1987. *Science* 235: 1173.

BENKOVIC, S. J., and SCHRAY, K. 1973. *The Enzymes*, vol. 8, 3d ed., ed. P. D. Boyer. New York: Academic Press, p. 201.

FREY, P. A. 1989. *Advances in Enzymology and Related Topics in Molecular Biology* 62: 119.

FREY, P. A. *The Enzymes*, vol. 20, 3d ed., ed. P. D. Boyer and D. S. Sigman. New York: Academic Press, forthcoming.

CLELAND, W. W. 1970. *The Enzymes*, vol. 2, 3d ed., ed. P. D. Boyer. New York: Academic Press, p. 1.

HERSCHLAG, D., and JENCKS, W. P. 1990. *Biochemistry* 29: 5172.

PROBLEMS

1. Figure 11-20 shows the kinetic pathway for the reaction catalyzed by galactose-1-P uridylyltransferase. Using Cleland's rules, predict the product inhibition patterns for UDP-galactose and glucose-1-P acting as product inhibitors of the forward reaction.

2. Dephosphocoenzyme A is coenzyme A (Chapters 2 and 10) lacking the 3-phosphoryl group in the ribosyl ring. Dephosphocoenzyme A is produced by the action of an enzyme that catalyzes the reaction of phosphopantetheine with ATP (reaction a). The coenzymes NAD^+ and FAD

(Chapter 16) are produced in similar reactions of nicotinamide mononucleotide and flavin mononucleotide with ATP (reactions b and c). Nucleoside phosphotransferase catalyzes reaction d, and adenosine kinase catalyzes reaction e.

(a) Phosphopantetheine + ATP $\rightleftharpoons$ Dephospho-CoA + PP_i

(b) Nicotinamide mononucleotide + ATP $\rightleftharpoons$ NAD^+ + PP_i

(c) Flavin mononucleotide + ATP $\rightleftharpoons$ FAD + PP_i

(d) dAMP + dGuo $\rightleftharpoons$ dAdo + dGMP

(e) Ado + ATP $\rightleftharpoons$ AMP + ADP

By what type of kinetic reaction pathway are these reactions most likely to occur? How would one determine whether these reactions proceed by single-displacement or double-displacement mechanisms?

3. What is the importance of the following enzymes in living cells?

(a) Pyruvate kinase and 3-phosphoglycerate kinase.

(b) Creatine kinase and arginine kinase.

(c) Hexokinase and glycerokinase.

(d) UDP-glucose pyrophosphorylase and choline kinase.

(e) Adenylate kinase and nucleoside diphosphate kinase.

4. What is meant by dissociative and associative transition states in phosphoryl transfer? What types of phosphate compounds undergo phosphoryl transfer by these mechanisms?

5. Explain the types of experimental evidence that support the assignment of a double-displacement phosphoryl transfer mechanism in enzymatic reactions. Contrast these lines of evidence with those that support the assignment of single-displacement mechanisms.

6. According to the mechanism of Figure 11-16 for the pyruvate kinase reaction, the enzyme should catalyze the enolization of pyruvate to enolpyruvate or the enolate of pyruvate independently of phosphoryl transfer—that is, in the absence of ATP, according to the following equation:

$$E + H_3C{-}C({=}O){-}CO_2^- \rightleftharpoons E \cdots C(H_3C)(=O)CO_2^- \rightleftharpoons E \cdots C(O^-)(=CH_2)CO_2^- + H^+$$

The reverse of this process carried out in D_2O would lead to the incorporation of D^+ to form 3-deuteropyruvate. Nevertheless, the enzyme does not catalyze the exchange of deuterium from D_2O into pyruvate. In the presence of inorganic phosphate, however, the enzyme catalyzes this exchange at a fast rate. How can this property of pyruvate kinase be most reasonably explained?

Nucleic Acids

12

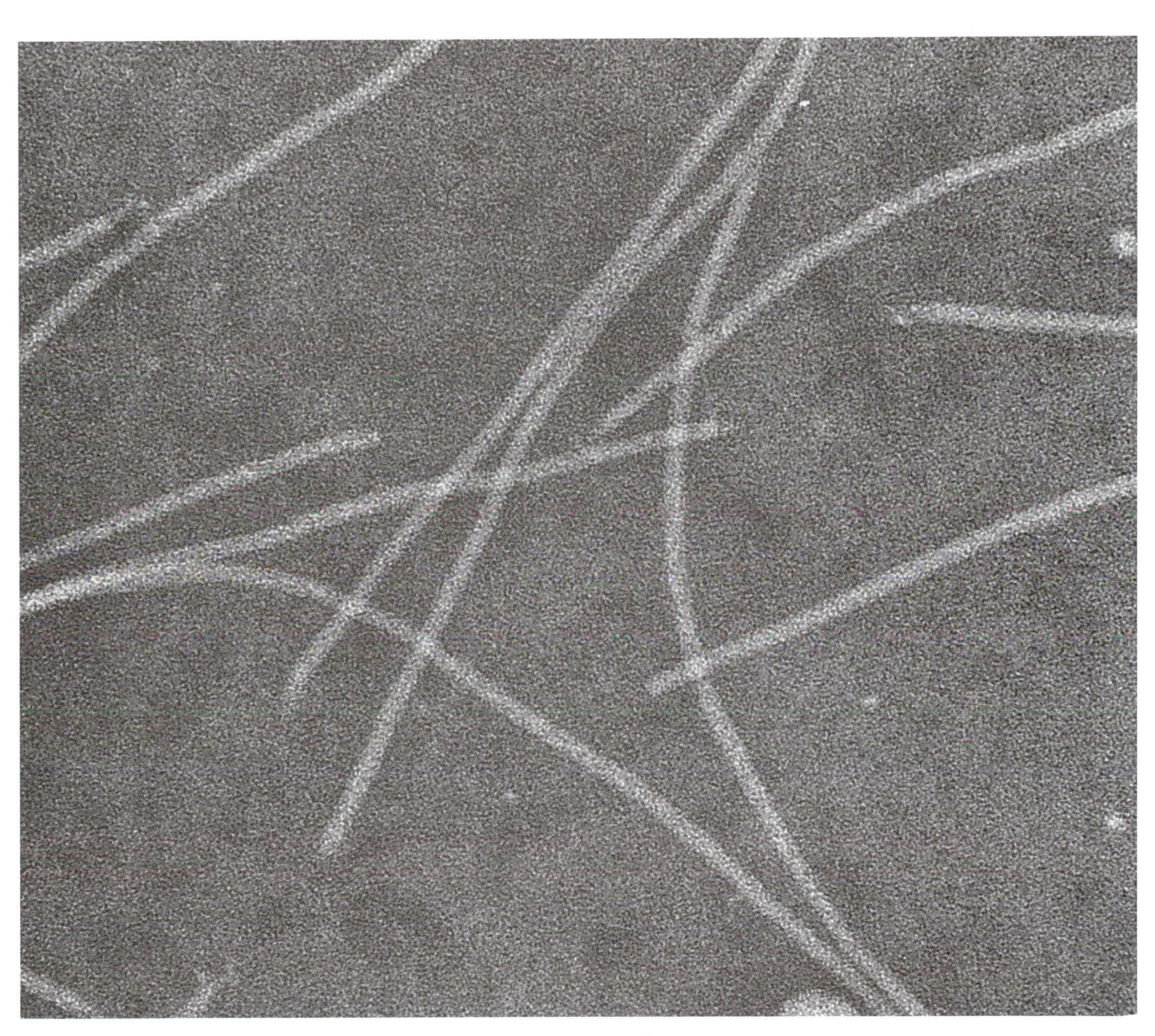

Electron micrographic images of the filamentous E. coli bacteriophage fd. Its morphology and life cycle are similar to those of M13, a commonly used DNA sequencing vector. (Micrograph courtesy of Harold Fisher and originally published in An Electron Micrographic Atlas of Viruses by Robley Williams and Harold C. Fisher. Charles C. Thomas Publishers.)

The essential function of DNA in cells and organisms is to *store and preserve genetic information.* The essential function of all species of RNA is to *mediate the expression* of the genetic information in DNA. The most-basic aspects of the chemical structures of ribonucleic acids (RNA) and deoxyribonucleic acids (DNA) and their functions in protein biosynthesis are described in Chapter 10. The structures of the heterocyclic bases and 3′,5′-phosphodiester linkages in the nucleic acids will not be repeated here, but they can be found in the inside back cover of this volume. In this chapter, we shall describe the structures of the nucleic acids, as well as their biosynthesis and processing, in greater detail.

DNA Structure

The Double Helix

The chemical structure of DNA as a linear polymer of 2′-deoxynucleotides linked through 3′,5′-phosphodiester bonds has been set forth in Chapter 10, as were the structures of the hydrogen-bonded base pairs AT and GC. Watson and Crick elucidated the physical structure of this polymer in 1953 as a two-stranded double helix. They showed in model-building experiments that a double-helical structure, illustrated in Figures 12-1 and 12-2, accounts for all of the properties of DNA. Their model gained immediate acceptance because it explained the chemical and physical properties of DNA and it provided a beautiful rationale

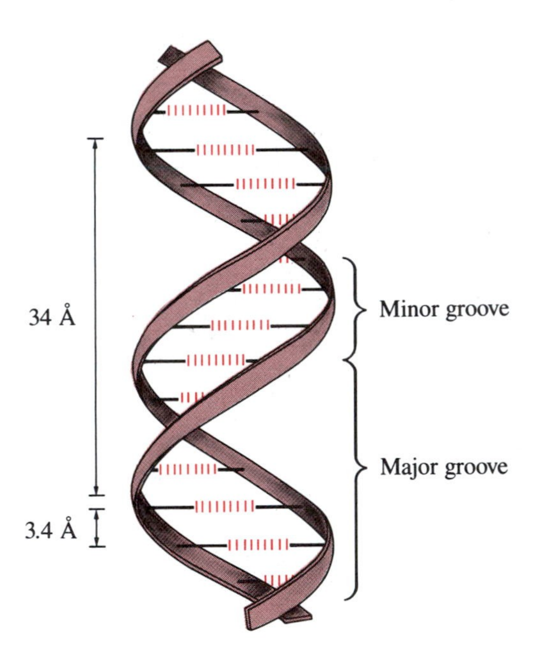

Figure 12-1

The Helical Twist of Double-stranded B-DNA. The ribbon strands describe the helical paths followed by the 2′-deoxyribose-3′,5′-phosphodiester backbone of the DNA, and the hydrogen-bonded bases project from the deoxyribosyl-C-1′ into the center of the helix. The strand association is such that it leads to two helical grooves in DNA, the major and minor grooves, that are best visualized in the space-filling models in Figure 12-2. In the B-helix, the rise is 34 Å for a complete helical turn, the repeating unit; and each base pair is separated from its neighbors by 3.4 Å along the helical axis, so that there are ten base pairs in the repeating unit. A-DNA differs from B-DNA in the rise per helical turn and the spacing between base pairs, as shown by the space-filling models in Figure 12-2.

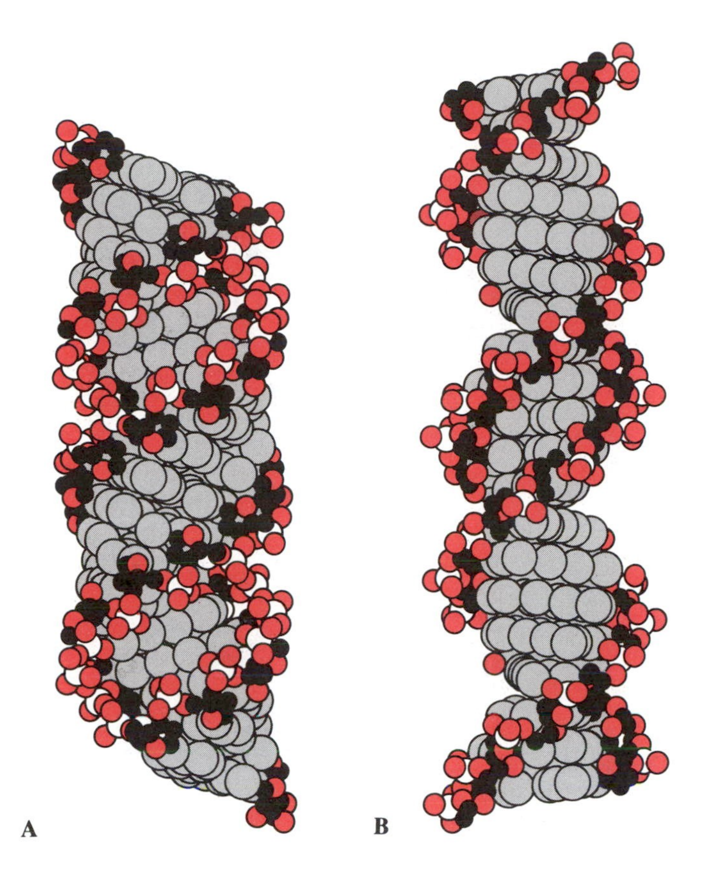

Figure 12-2

Space-filling Models of A- and B-DNA. Part A is a space-filling model of a segment of A-DNA, and part B a model of B-DNA. Atoms of the bases are shown as large circles; atoms of the deoxyribose phosphate backbones are shown as smaller circles, with carbon atoms being smallest, oxygen atoms intermediate and shaded, and phosphorus atoms larger and unshaded. [After R. D. Wells, T. C. Goodman, W. Hillen, G. T. Horn, R. D. Klein, J. E. Larson, U. R. Müller, S. K. Neuendorf, N. Panayotatos, and S. M. Sturdivant. Adapted with permission from *Progress in Nucleic Acid Research and Molecular Biology* 24 (1980): 167–267. Copyright CRC Press, Inc., Boca Raton, Florida.]

for the most-fundamental biological properties of DNA. Furthermore, the double helix made it possible to advance and test hypotheses regarding how DNA is replicated in the course of cell division and how the genetic information in DNA can be used to direct the biosynthesis of the various species of RNA and proteins. It also became clear how the structure can protect the genetic information through its intrinsic stability and by facilitating the repair of damaged DNA.

The models shown in Figure 12-2 are A and B double helices of DNA. DNA exists in several helical forms, and A and B helices are the most-common right-handed forms (α-helices). The B-helix, the predominant form in dilute aqueous solutions, is 20 Å across and each helical turn is 34 Å along the longitudinal axis. There are ten base pairs in each helical turn, the longitudinal spacing between base pairs is 3.38 Å, and each base pair consists of a base from each strand. The regularity of the structure is assured by the fact that, although the base pairs differ in chemical structure, each pair is formed from a purine and a pyrimidine, with the result that the AT and GC pairs have nearly the same dimensions. The A-helix is the predominant form in an apolar solvent, such as ethanol-water, which decreases the activity of water and dehydrates the DNA. The A-helix is also about 20 Å across, actually about 3% wider than the B-helix, but it differs in that the longitudinal spacing between base pairs is only 2.56 Å. The length of the repeating unit is 28 Å, shorter than that of the B-helix, with eleven base pairs per helical turn.

The models in Figure 12-2 clearly show the overall conformational differences between A- and B-DNA. The most-fundamental structural basis for this is the difference in the conformations of the deoxyribose rings in the two forms. These conformations are shown in Figure 12-3. Bond rotational freedom in the deoxyribose ring is severely limited by the small size of the ring, so that only two conformations are predominant, the (C)3′-*endo* and the (C)2′-*endo*

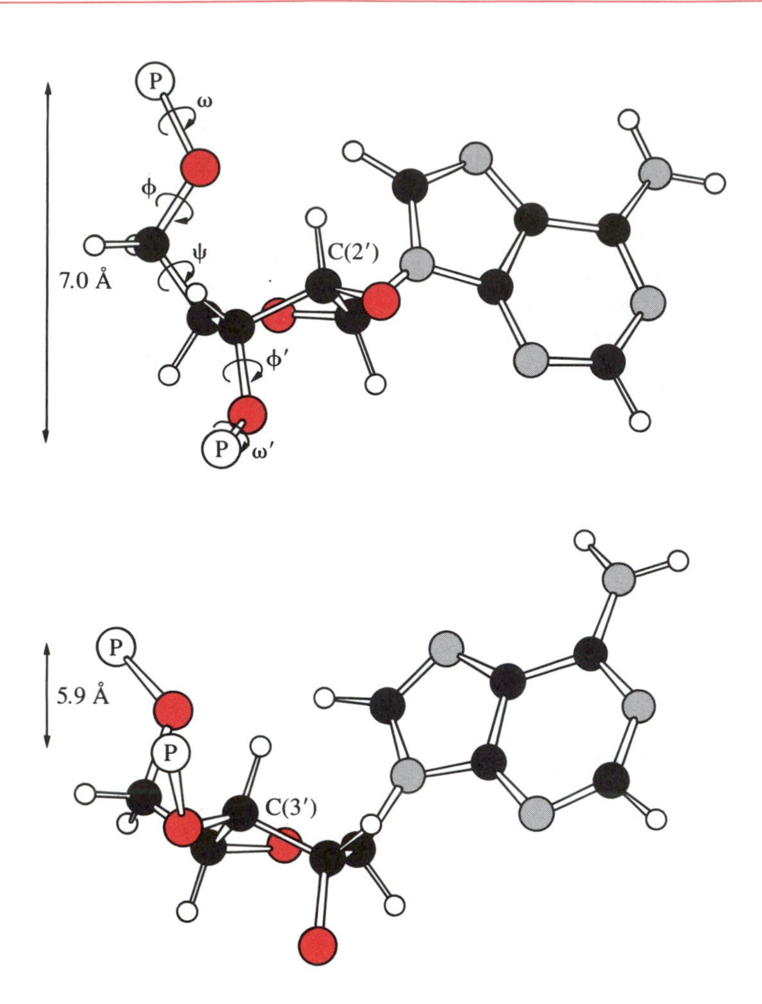

Figure 12-3

The 3′-*endo*- and 2′-*endo*-Deoxyribosyl Conformers in A- and B-DNA. In the (C)2′-*endo* conformation of deoxyribose, the ring is puckered so that the 2′ carbon projects above the ring plane on the same side as the heterocyclic base. In the (C)3′-*endo* conformation, the ring is puckered so that the 3′ carbon projects above the ring. Because the 3′ carbon is bonded to the 3′ oxygen in the phosphodiester bond, the difference in the positions of the 3′ carbon of these two conformations has a substantial effect on the spacing of the phosphorus atoms. The spatial difference is transmitted to the phosphorus atoms through the free rotations about the bonds C-3′—O-3′ (Φ′), 3′O—P (ω′), C-4′—C-5′ (Ψ), C—5′—O-5′ (Φ), and O-5′—P (ω). [From M. Sundaralingam, in *Structure and Conformation of Nucleic Acids and Protein-Nucleic Acid Interactions*, ed. M. Sundaralingam and S. T. Rao. Baltimore: University Park Press, 1974, pp. 487–524.]

conformations in Figure 12-3. The prefix *endo*- designates which of the ring puckered conformers projects carbons 2′ or 3′ on the same side of the ring plane as the heterocyclic base. Note that the ring conformation greatly affects the spacing between the two phosphodiester groups to which each deoxyribose is attached. In the (C)3′-*endo* conformation, the spacing between phosphorus atoms is 5.9 Å whereas, in the (C)2′-*endo* conformation, the spacing is 7.0 Å. These spacings control the distances between paired bases in DNA, so that A-DNA has the (C)3′-*endo* and B-DNA the (C)2′-*endo* conformation in the deoxyribose rings.

Figure 12-4 illustrates the hydrogen bonding and dimensional similarity in the AT and GC base pairs. Variant forms of DNA differ in the number of bases per turn and in the length of the repeating unit; all forms are interconvertible with variations in cation, ionic strength, temperature, and base composition. The most-prevalent forms are α-helices; however, a left-handed double helix, the Z-helix, is also known and is favored by high GC content and high ionic strength. The biological significance of Z-DNA is not known.

Antiparallel Strands in DNA

Directionality in nucleic acid chains arises from the chemical structure of the deoxyribose and the phosphodiester internucleotide linkage, which links the 5′- and 3′-oxygens of neighboring nucleotides. By convention, a given strand is thought of as being oriented in the 5′-to-3′ direction, and base sequences are read in the 5′-to-3′ direction.

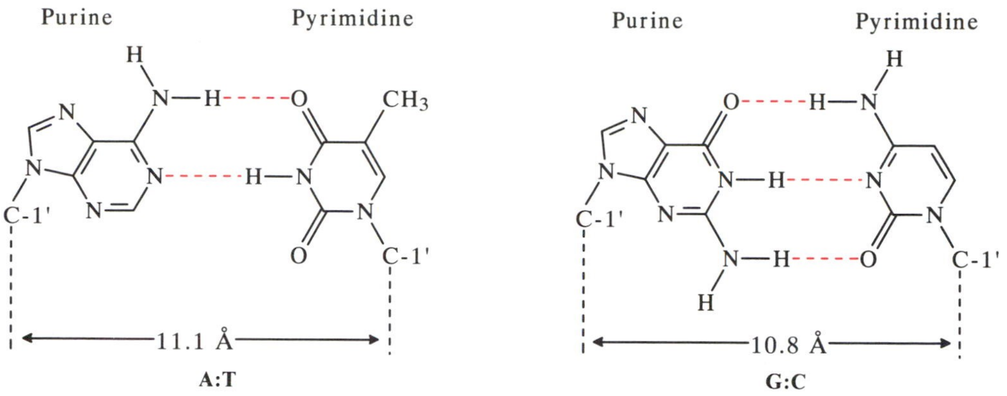

Purine Pyrimidine

11.1 Å

A:T

Purine Pyrimidine

10.8 Å

G:C

Figure 12-4

Structures of the Base Pairs AT and GC. Shown are the base pairs AT and GC with their hydrogen bonds and lateral dimensions. The bases at opposing positions of the two strands are associated by hydrogen bonds. Each base pair consists of a purine and a pyrimidine, either AT or GC, and the base pairs are stacked along the helical axis. The base pairs are shown as they would appear if the DNA helix were viewed down its longitudinal axis through the center. Note that the two pairs have similar dimensions and can fit similarly into the helical space, and that the GC pair is held by three hydrogen bonds, whereas the AT pair has only two.

The two strands in double-helical DNA are antiparallel; that is, they are oriented in opposite directions relative to each other. This is shown in the shorthand drawing of Figure 12-5, in which the left-hand strand is oriented downward (5′ to 3′) and the right-hand strand is oriented upward (5′ to 3′). Note that the bases are paired between the strands according to the base-pairing rules (AT and GC), and their antiparallel directions lead to complementarity in the base sequences. Thus, the sequence (5′ to 3′) in the strand at the left is ACTGA, whereas that at the right is TCAGT.

AT and GC Base Pairs

The bifunctional hydrogen bonds linking complementary bases on the two strands of DNA are made possible by the delocalization of electrons through resonance, which produces dipoles on the amide and amidine segments of the

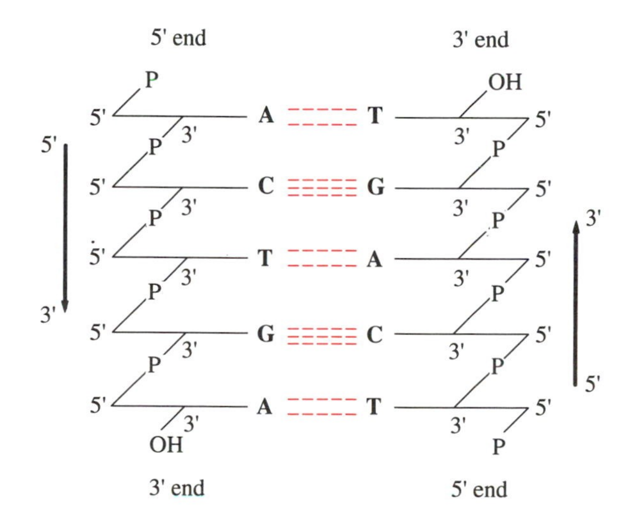

Figure 12-5

Antiparallel Strands in DNA. The antiparallel orientation of the two DNA strands in the double helix is illustrated in this drawing, which shows the strand at the left oriented downward in the 5′-to-3′ direction, whereas that at the right is oriented upward in the 5′-to-3′ direction.

bases. Delocalization of electrons is illustrated by the resonance forms of the amidine (**1**) and amide (**2**) segments of adenine and thymine, respectively.

1 **2**

In the base pair AT (or AU), the partial positive and negative charges on nitrogen and oxygen make possible the formation of the cyclic, hydrogen-bonded structures of the base pairs shown in Figure 12-4. In the GC base pair, analogous resonance effects make it possible to form three hydrogen bonds between amide and amidine segments. The three hydrogen bonds in the GC pair make this structure more stable than the AT pair, which has only two hydrogen bonds. Consequently, nucleic acids that have a large content of GC base pairs are correspondingly more stable and require a higher temperature to fall apart into separated, denatured single strands.

Base Stacking and Melting in DNA

The bases in the two strands of DNA that are twined around each other in the double helix interact through mutually parallel base-to-base hydrophobic bonds, which decreases their absorption of ultraviolet (UV) light. (All nucleic acid bases absorb light at 260 nm; adenine, thymine, and uracil have absorption maxima near 260 nm. Cytosine has an absorption maximum near 275 nm; and guanine absorbs maximally near 250 nm with a shoulder at 280 nm.) Parallel association is known as *base stacking,* and the decrease in UV absorbance is the *hypochromic effect.* When the double helix unfolds upon heating and the strands come apart, the UV absorbance increases; and this change can be used to measure the amount of unfolding. When the temperature of DNA in solution is increased gradually in small increments, the structural changes are reversible and highly cooperative. Because of this cooperativity, a plot of A_{260} against temperature is a smooth curve with a sharp transition over a narrow temperature range. The sharp transition signals structural melting of the DNA. The midpoint temperature at which the change is half maximal is the melting temperature, T_m. The sharp transition between the helical and melted states is illustrated in Figure 12-6, which shows that the hypochromicity changes dramatically over a narrow temperature range, the midpoint of which is T_m.

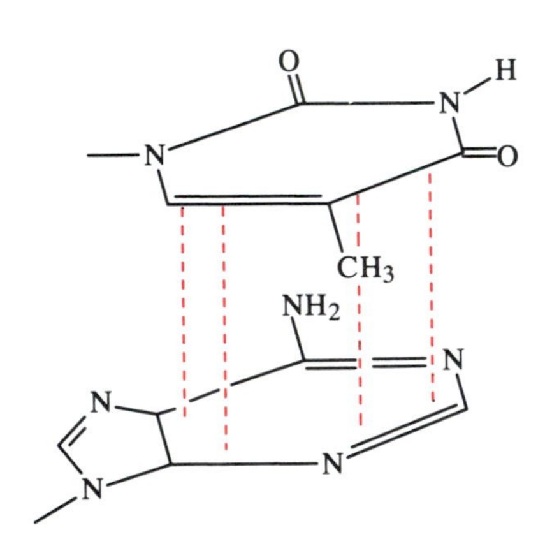

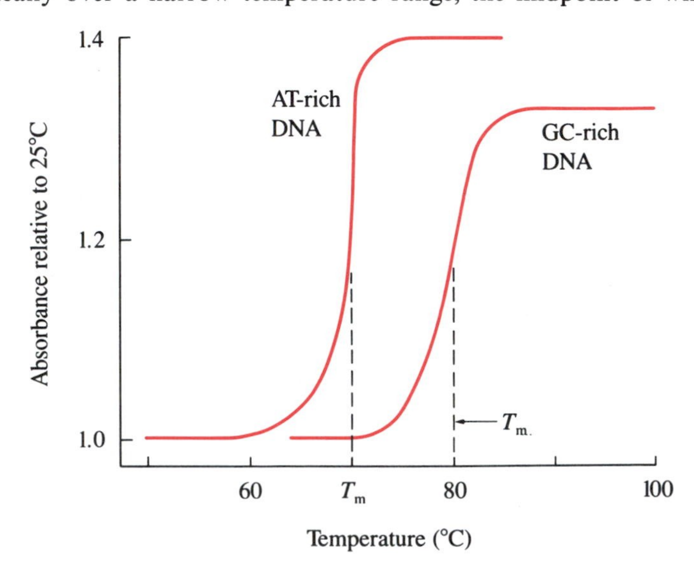

Figure 12-6

The Melting Curves for AT-rich and GC-rich Double-helical DNA. The transition between double-stranded native and strand-separated denatured DNA can be detected by changes in absorbance at 260 nm (ΔA_{260}), changes in circular dichroism, or changes in viscosity. The midpoint of the transition is T_m and is at a higher temperature for the GC-rich DNA than for the AT-rich DNA because of the greater stability of the GC base pair.

Melting is accompanied by changes in other physical properties, such as the rotation of polarized light in the double helix, the circular dichroism, and the viscosity. Any of these changes can be used to measure the unfolding of DNA in the way that A_{260} is used in Figure 12-6.

Both the hydrogen bonds and the stacking interactions of the bases in nucleic acids supply energy for maintaining the structure of the molecule, but it is now believed that the stacking interactions are quantitatively more important. However, the hydrogen bonds play the critical role in providing the specificity for information transfer. This is made possible largely because an incorrect hydrogen bond is extremely *unstable*. If the bases are mismatched, hydrogen bonds cannot be formed either to the complementary base or to water, and the system is destabilized.

The factors contributing to the structures of proteins, described in Chapter 8, are also important in determining the structures of nucleic acids, including the helical forms of DNA. Hydrogen bonding determines the specificity of base pairing, the stacking of base pairs along the longitudinal axis of the double helix is essentially through hydrophobic interactions that stabilize the helix, and the negatively charged phosphodiesters are exposed on the surface of the helix where they interact with cations or the positively charged groups in DNA-binding proteins.

Renaturation of DNA

The melting curves in Figure 12-6 represent denaturation of DNA in experiments in which the temperature is gradually increased. However, the denaturation of DNA is *reversible;* when a heated solution is allowed to cool very gradually, the double-helical structure can be regenerated. The cooling process must be gradual to be reversible because, in a cooling solution, numerous imperfectly base-paired segments of DNA are transiently formed, and quick cooling can trap many of them and lead to a mixture of incorrect structures. By cooling slowly, many possible paired structures can be sampled, and sufficient time allows the most-stable one to be formed.

The generation of double-helical structures by gradually cooling heated samples of DNA, which is known as *annealing,* is usually done for purposes other than simply renaturing denatured DNA. There are many important applications of the annealing process in genetic engineering, some of which are described in Chapter 14. Moreover, this is a powerful means for comparing samples of DNA from different sources for their relatedness. The more closely the nucleotide sequences in two samples of DNA are related, the greater will be the observed hypochromic effect upon annealing. In an experiment, specimens of DNA from related species are obtained and processed into single-stranded samples. The heated samples are mixed and allowed to anneal. When the two species are closely related, their nucleotide sequences are similar and a high degree of hypochromicity is observed. When the species are dissimilar, the nucleotide sequences are dissimilar, leading to poor pairing and a lower hypochromic effect. This technique is used for comparing and testing genetic relationships among species.

Correlation of Physicochemical Properties with the Double Helix

The following chemical and physical properties of DNA guided Watson and Crick in formulating the double helix:

1. DNA consists of nucleotides linked through phosphodiester bridges.
2. Samples of DNA from different species contain the same heterocyclic bases but in different amounts that characterize the species.

3. In all species of DNA, the stoichiometric amounts of heterocyclic bases are related as A + G = C + T and A = T and G = C.
4. DNA isolated from all species is viscous but melts sharply to a less viscous form at a T_m that is characteristic of the species.
5. The higher the GC content, the higher the T_m.
6. Melting leads to an increase in the UV absorption of DNA, the hypochromic effect.

All of these properties are explained by the double-helical structure of DNA. The base composition is elegantly explained by base pairing between helices, wherein the AT and GC base pairs are nearly the same size and span the distance between the helical strands. Viscosity is explained by the linearity and length of the helical structure. The phenomenon of melting is explained by the separation of helical strands at temperatures high enough to disrupt the interchain attractive forces. The hypochromic effect is caused by the base-stacking interactions perturbing the UV absorption, which is relieved upon melting. The GC base pair is held together by three hydrogen bonds more strongly than is the AT base pair by two hydrogen bonds, which accounts for the fact that species of DNA with higher GC contents exhibit greater heat stability and higher values of T_m.

In general, eukaryotic DNA is much longer than prokaryotic DNA, which in turn is much longer than viral DNA. These relations originate with the differences in the numbers of genes carried by DNA obtained from these sources.

Superhelical DNA

B-DNA and the other common α-helical forms of DNA are highly regular structures in their most-stable forms. DNA is not structurally rigid, however, and it can undergo conformational and other structural changes. One aspect of regularity in DNA structure is the number of base pairs per turn of the helix, which is ten for B-DNA. Some species of DNA vary in the number of base pairs per turn owing to physical constraints imposed upon the helix, either by binding proteins or by the closure of a DNA molecule into a circle. DNA that is subjected to these perturbations may be underwound—that is, contain more than ten base pairs per helical turn—or be overwound—with fewer than ten base pairs per turn. Superhelical DNA is either underwound or overwound, and the DNA topoisomerases and DNA gyrases catalyze the interconversions of relaxed and superhelical DNA.

A simple way to conceptualize superhelicity is to consider circular DNA, which is the predominant form of DNA in bacteria and viruses. A relaxed molecule of circular DNA can lie flat with exactly ten base pairs per turn all around the circle, as is illustrated in Figure 12-7A. If one strand is broken,

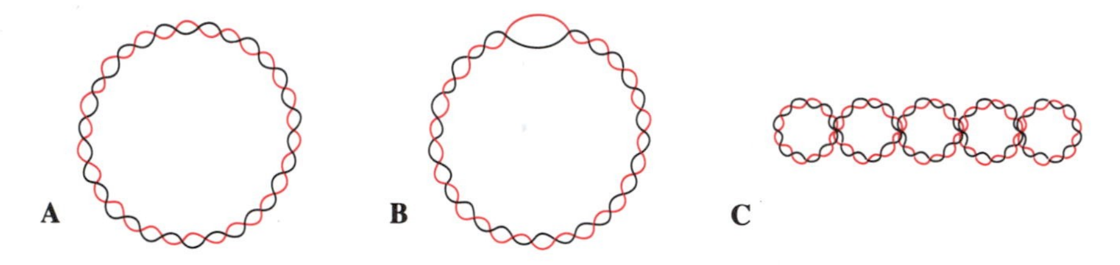

Figure 12-7

Relaxed and Superhelical Circular DNA. The DNA in part A is relaxed; that in part B is underwound with the strain localized in the loop; and the superhelical structure in part C is the underwound molecule from part B, in which the strain is spread throughout the structure.

unwound by one or more turns, and then rejoined, a nonhelical segment such as that in Figure 12-7B will be formed. Unwinding introduces strain into the molecule because the unordered segment is less stable than it would be if it were helical. Because DNA is not conformationally static, the instability introduced by unwinding can be accommodated by structures other than one in which the instability is localized in one unordered segment. When the strain is distributed throughout the structure, there will be more than ten base pairs per turn of the helix, and the instability will be expressed as a twisting of the helix along its longitudinal axis, as in Figure 12-7C. A similar condition will result from overwinding, except that the twist in the superhelix will be in the opposite direction.

In Figure 12-7, the circular helix is unwound by four complete turns. Therefore, when the strain is uniformly distributed around the circle, there are four twists in the superhelix. All possible forms are in equilibrium, so that there will be species present with a one-turn gap in the helix and three twists in the superhelix and so forth. The superhelical forms are in equilibrium with those having gaps in the helix; however, the twists are favored at higher temperatures. The gaps tend to appear preferentially in regions of high AT content, where the helix is least stable.

The biological importance of superhelices is not fully known, although there is confidence that they are important in linear as well as circular DNA. For one thing, the DNA in chromatin is underwound to accommodate the histones, which are the basic proteins around which the DNA is wound in the *nucleosomes* of eukaryotic chromosomes (Figure 12-8). For another thing, the process of DNA replication introduces positive supercoiling in DNA, and relaxation of supercoiling is a part of the replication process. Finally, the gaps that appear transiently in the helices of underwound DNA, especially in AT-rich regions, may be important signals for enzymes that process DNA, some of which act specifically on supercoiled DNA.

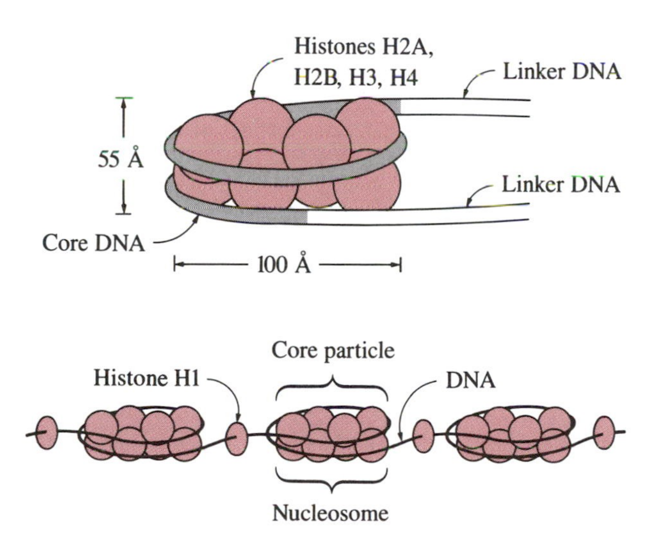

Figure 12-8

Structure and Composition of Nucleosomes. A nucleosome in eukaryotic chromatin consists of DNA wound around histones, which are basic proteins rich in lysine and arginine, in a beadlike structure. The histone composition is one H1 (MW 21,000), two H2a (MW 14,500), two H2b (MW 13,700), two H3 (MW 15,300) and four H4 (MW 11,300). Histone H1 is bound to the linker region of DNA that links nucleosomes to one another, as shown at the bottom.

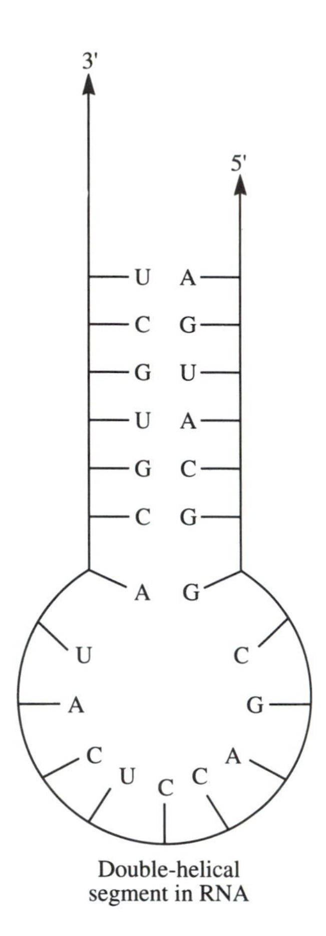

Double-helical segment in RNA

RNA Structure

The basic chemical linkages in RNA and DNA are similar, but the structure of RNA differs from that of DNA in several ways. The three main differences are:

1. Most RNA is single stranded rather than double stranded.
2. Both RNA and DNA contain adenine, guanine, cytosine, and a fourth base—uracil in RNA and thymine in DNA.
3. RNA contains ribose, whereas DNA contains 2′-deoxyribose.

The chemical linkages between the ribose rings and the bases and the phosphodiester linkages between the nucleotides are the same in the two polymers.

Although most RNA is single stranded, the double helix is also an important structural motif in RNA. The double helices in RNA are not as long as those in DNA, and for this reason many of them are less stable than those in DNA. The clearest and most thoroughly studied example of double-helical structure in RNA is tRNA, the structure of which is described in Chapter 10 (see Figures 10-7 and 10-8). Double-helical structure arises in a molecule of RNA when the strand turns back upon itself and base pairing occurs between antiparallel complementary or quasicomplementary segments of base sequences within the same strand. Inasmuch as short quasicomplementary sequences in a given strand usually contain a few mismatched base pairs, such as the G-U pair shown in the margin, and the paired segments are in any case usually less than twelve base pairs in length, the helical segments of RNA are less stable than the double helix in DNA. An exception to this is tRNA, in which the D-loop and TΨC-loop are held together by tertiary base pairs that stabilize the tertiary structure (Figure 10-8).

Chemical Properties of Nucleic Acids

DNA and RNA are structurally similar, and in some respects they have similar chemical properties. However, the 2′-hydroxyl group of RNA gives it a chemical reactivity that is not present in DNA. The following sections delineate and compare the principal chemical properties of DNA and RNA that are important in the characterization of these molecules.

Hydrolysis of RNA and DNA

Both RNA and DNA can be cleaved to their constituent molecules by hydrolysis in strong acid (pH < 1, heat). The products of acid hydrolysis are the heterocyclic bases, ribose or 2′-deoxyribose, and phosphate in equimolar amounts; that is, 1 base : 1 ribose : 1 phosphate. The pyrimidines are much more resistant to hydrolytic cleavage from ribose than are the purines, which can be removed selectively by hydrolysis at pH 4. The purines are more sensitive to hydrolysis because both adenine and guanine can be protonated at pH values near 4, with protonation mainly on N-7 of guanine and N-3 of adenine, and the protonated bases are very good leaving groups from the sugar, as shown in Figure 12-9. Removal of adenine and guanine by hydrolysis in dilute acid without cleavage of the phosphodiester linkages produces *apurinic* DNA. The phosphodiester linkages are cleaved by heating in strong acid. RNA and DNA have similar reactivities in acidic solutions.

In alkaline solutions, RNA is much more reactive than DNA. DNA is stable to hydrolysis in neutral and alkaline solutions; however, RNA is easily hydrolyzed to a mixture of ribonucleoside 2′- and 3′-phosphates in dilute

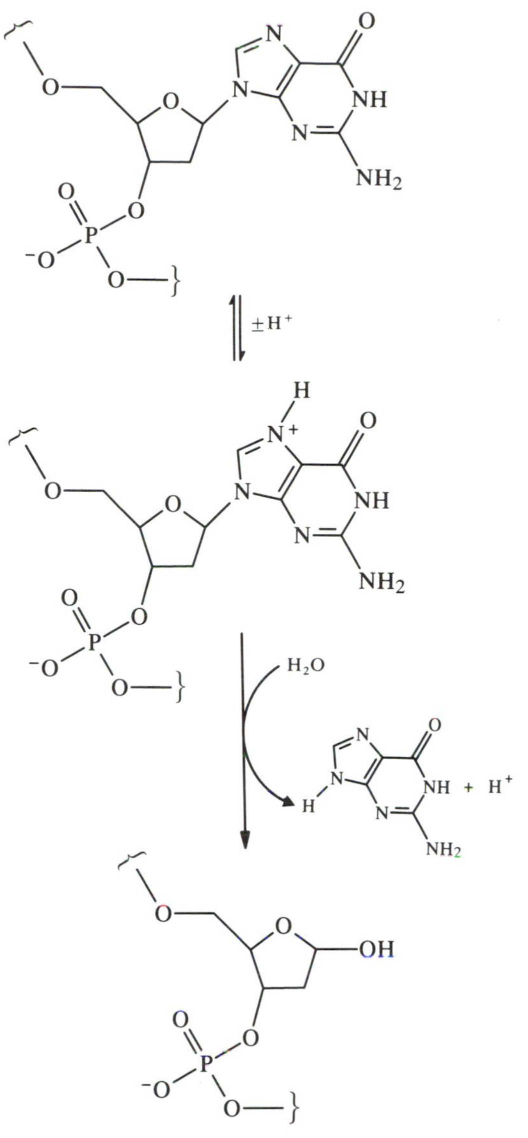

Apurinic DNA

Figure 12-9

Hydrolytic Cleavage of Guanine from a Polynucleotide in Dilute Acid. Purines can be selectively removed from nucleic acids by acid hydrolysis under mild conditions. Both adenine and guanine can be protonated by dilute acid, with protonation at N-7 of guanine and N-1 of adenine. The protonated forms are good leaving groups from ribose carbon-1 and can be displaced by water, with release of adenine or guanine and the formation of apurinic DNA. Shown here is the acid-catalyzed removal of a guanine ring from a segment of DNA.

solutions of NaOH (0.1 M NaOH at 37°C). This is because dilute bases react with the 2′-hydroxyl groups of RNA to deprotonate them to 2′-oxyanions, which quickly attack the phosphodiester linkages and cleave them to 2′,3′-cyclic phosphates, as shown in Figure 12-10. The cyclic phosphates undergo rapid hydrolysis to the 2′- and 3′-phosphates under the reaction conditions for the reasons given in Chapter 11.

Why should the 2′-oxyanion react with the phosphodiester so much faster than the hydroxide ion even though the hydroxide is present at the higher concentration? The answer to this question explains the lability of RNA and the stability of DNA, which has similar phosphodiester linkages but no 2′-OH groups. The reason that the 2′-oxyanion reacts so rapidly is that its reaction is intramolecular and is a highly probable event; that is, very little entropy must be lost for the reaction to occur. The reaction of hydroxide ion with the same phosphodiester bonds is a highly improbable event that requires the loss of a great deal of entropy to extract the hydroxide ion from solution and to fix it in the correct orientation to react with the phosphodiester. Therefore, DNA is stable in dilute base, whereas RNA is labile. This is an important fact that favors DNA as the form in which genetic information is preserved.

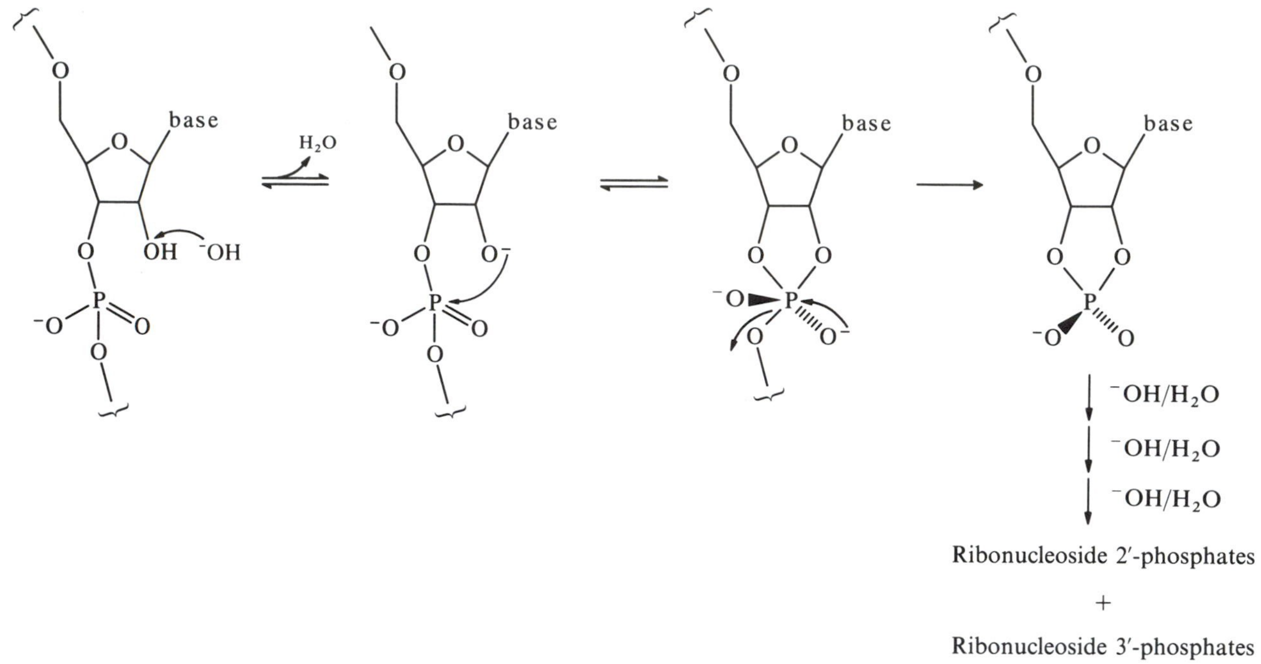

$^{-}OH/H_2O$

Ribonucleoside 2′-phosphates

+

Ribonucleoside 3′-phosphates

Figure 12-10

The Mechanism of Base-catalyzed RNA Hydrolysis. In RNA, the 2′-hydroxyl groups are adjacent to the phosphodiester linkages and, in the presence of dilute base, a fraction of them are deprotonated. The nucleophilic 2′-oxyanions can attack the neighboring phosphorus to form an intramolecular pentavalent adduct, which can collapse with the elimination of an oxyanion, either in the reverse direction or in the forward direction, with cleavage of the phosphodiester linkage by elimination of the 5′-oxyanion of the neighboring nucleotide unit. This process can occur at every phosphodiester linkage in RNA and accounts for the base lability of RNA; it cannot take place in DNA.

Other Chemical Cleavages of DNA

Apurinic DNA is cleaved by dilute base, in contrast with intact DNA, which is not cleaved by bases. Cleavage of apurinic DNA by hydroxide ions is an elimination reaction that proceeds by C–O cleavage between carbon-3′ and the 3′-oxygen of ribose at the apurinic sites according to the mechanism in Figure 12-11. Once an adenine or guanine is removed from a nucleotide site in DNA, carbon-1′ becomes a hemiacetal and is in equilibrium with the open-chain aldehyde form shown in Figure 12-11. The 2′-protons in the aldehyde are activated by the carbonyl group and so are acidic enough to be abstracted by hydroxide, with elimination of the phosphate from carbon-3′ and cleavage of the chain.

Another means of removing adenine and guanine from DNA or RNA is by alkylation of the bases, followed by heating. Alkylation labilizes the bonds linking these bases to the ribose rings for the same basic reason that protonation labilizes them in Figure 12-9; that is, alkylation creates a quaternary ammonium group, and the positive charge makes the base a good leaving group. Alkylation of an adenine nucleotide site in DNA by dimethyl sulfate and cleavage of N^3-methyladenine is illustrated in Figure 12-12. Dimethyl sulfate methylates N-7 of guanine and, in the cleavage, N^7-methylguanine is cleaved from the ribose ring. In both cases, the purine is removed, creating an apurinic site that is subject to hydroxide-catalyzed elimination of the 3′-phosphate, with cleavage of the chain.

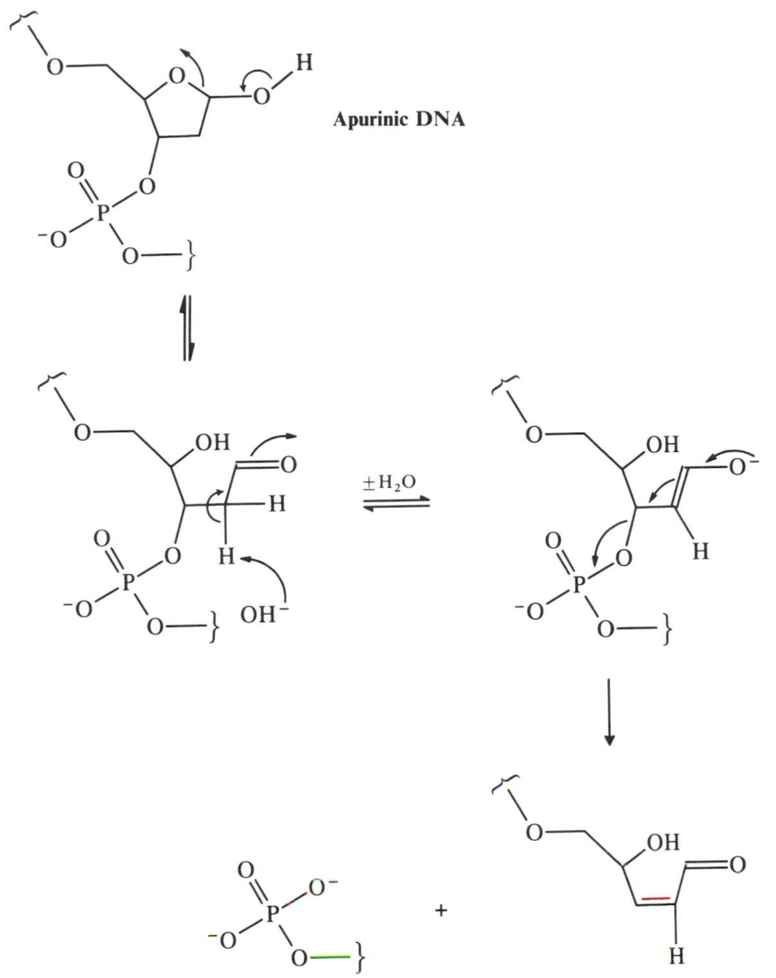

Apurinic DNA

$\pm H_2O$

Figure 12-11

Base-catalyzed Cleavage of Apurinic DNA. Ribose carbon-1′ at apurinic sites in DNA is a hemiacetal that is in equilibrium with the ring-opened aldehyde form. The interconversion of the hemiacetal and aldehyde forms is catalyzed by acids and bases. In the aldehyde form, the 2′-protons are acidic owing to the inductive electron-withdrawing effect of the aldehyde carbon and the resonance stabilization of the 2′-carbanion as an enolate ion by the aldehyde group. The enolate anion generated by dilute base can eliminate the 3′-phosphate to cleave the phosphodiester linkage.

By the application of alternative chemistry, the thymine and cytosine rings also can be removed from DNA. The pyrimidines are subject to addition of nucleophiles, and the adducts generally undergo secondary reactions that can lead to cleavage of the base from 2′-deoxyribose in DNA. A particularly useful example of this type of reaction is the treatment of DNA with hydrazine, which initially adds to C-5 of thymine and cytosine and ultimately leads to fragmentation and cleavage of the rings according to Figure 12-13.

The 2′-deoxyribosyl hydrazone produced in hydrazinolysis of DNA (Figure 12-13) also can undergo elimination of the 3′-phosphate, but this requires a much higher concentration of base than does apurinic DNA. Piperidine is used for this purpose in the reaction shown in Figure 12-14. The reactions in Figures 12-12 through 12-14 are used in the Maxam-Gilbert

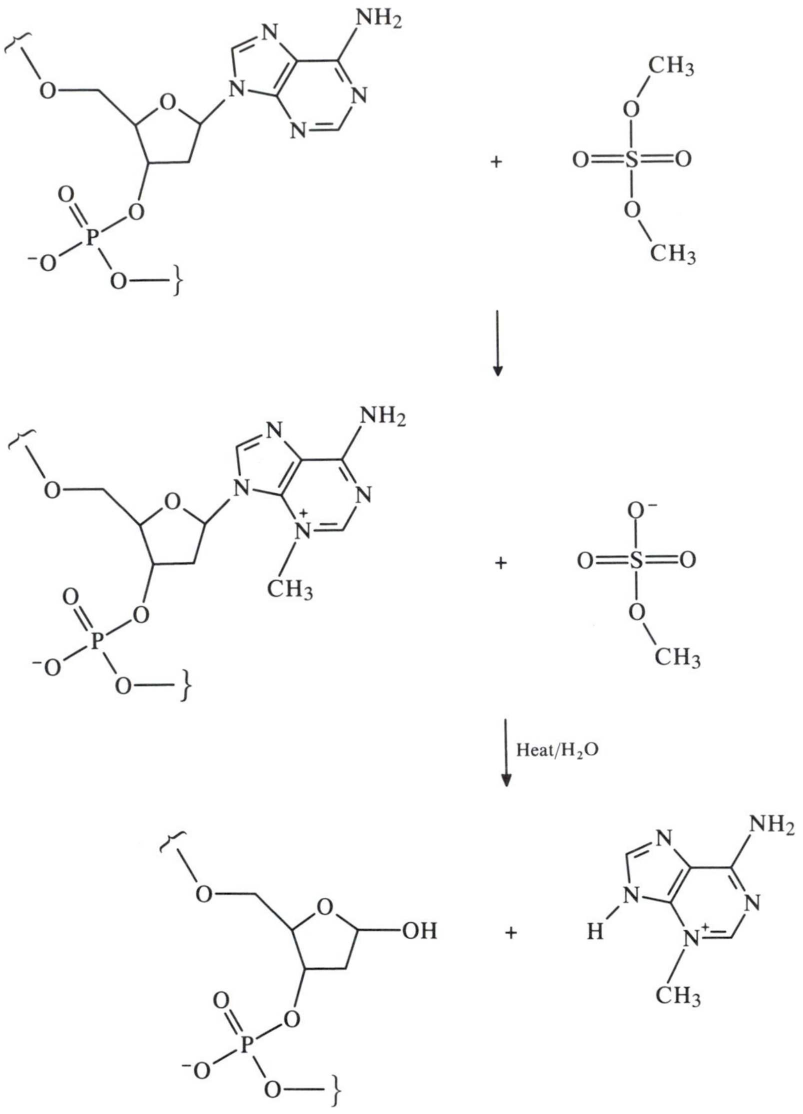

Figure 12-12

Methylation and Cleavage of DNA. Methylation of a nitrogen of a heterocyclic base has the same electronic effect as protonation of the base; that is, it introduces an additional covalent bond. In the methylation of adenine or guanine, the additional bond converts the nitrogen into a quaternary ammonium ion. The adenine ring undergoes methylation at N-3 to form a methylated quaternary ammonium ion, a structure that is analogous to that of the protonated adenine ring. Methylation facilitates the displacement of N^3-methyladenine from ribose-C-1′ by water, just as protonation of N-3 facilitates the cleavage of adenine, to form an apurinic site. Guanine undergoes similar methylation at N-7, and N^7-methylguanine also is displaced by water to form an apurinic site. This method of introducing apurinic sites into DNA is used in the Maxam-Gilbert sequencing method described later in this chapter.

method for determining the nucleotide sequences in segments of DNA, which is described in a later section of this chapter.

At first glance, it might appear that the elimination reactions leading to chain cleavage in Figures 12-12 and 12-14 could also be carried out on the

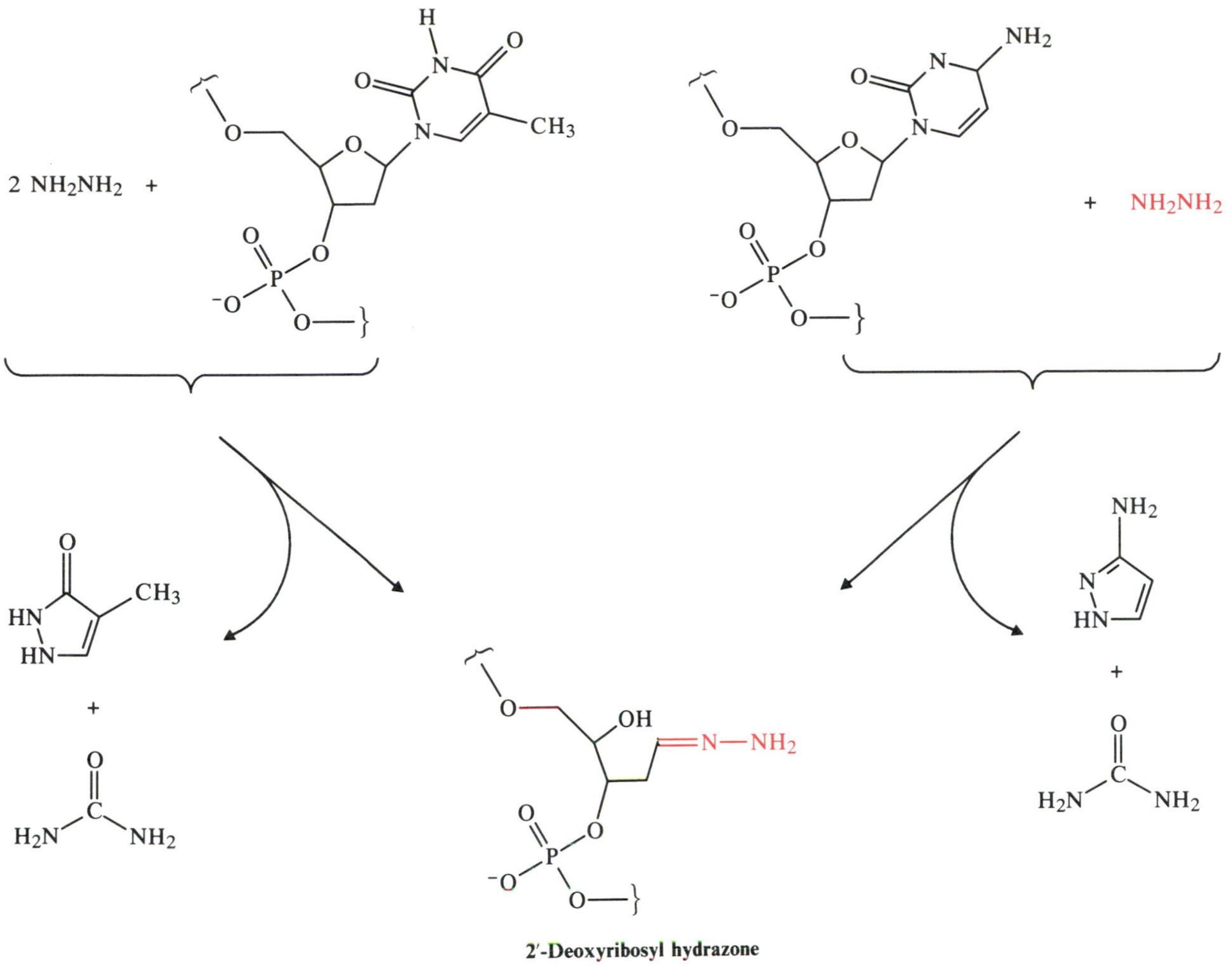

2 NH_2NH_2 + ... + NH_2NH_2

2′-Deoxyribosyl hydrazone

Figure 12-13

Removal of Pyrimidines from DNA by Reaction with Hydrazine. Pyrimidine rings undergo addition of nucleophiles at carbon-6 to saturate the 5, 6 double bond. This addition can destabilize the ring under certain conditions. (It can also facilitate heterocyclic ring modification catalyzed by enzymes. One example is the biosynthesis of dTMP from dUMP, which is described in Chapter 27.) An example of pyrimidine ring destabilization by nucleophilic addition that has been put to practical use is the reaction of hydrazine (NH_2NH_2) with cytosine and thymine rings in DNA. Hydrazine initially undergoes nucleophilic addition to carbon-6, and the adducts can decompose through a series of reactions into the products shown, with the formation of apyrimidinic sites in DNA. Under the reaction conditions, the aldehyde groups of the ribose ring in the apyrimidinic sites react with excess hydrazine to form the corresponding hydrazones at these sites. This method of depyrimidination is used in the Maxam-Gilbert nucleotide sequencing method.

analogous species of RNA; however, the 2′-OH group interferes and prevents a high yield of cleavage. This is because α-hydroxy aldehydes undergo a base-catalyzed rearrangement to the corresponding α-hydroxy ketones, which, for apurinic RNA, do not undergo the elimination at carbon-3′. (This type of rearrangement is explained for enzymatic reactions in Chapter 20). Therefore, the chemistry in Figures 12-12 through 12-14 is not practical for determining the nucleotide sequence of RNA.

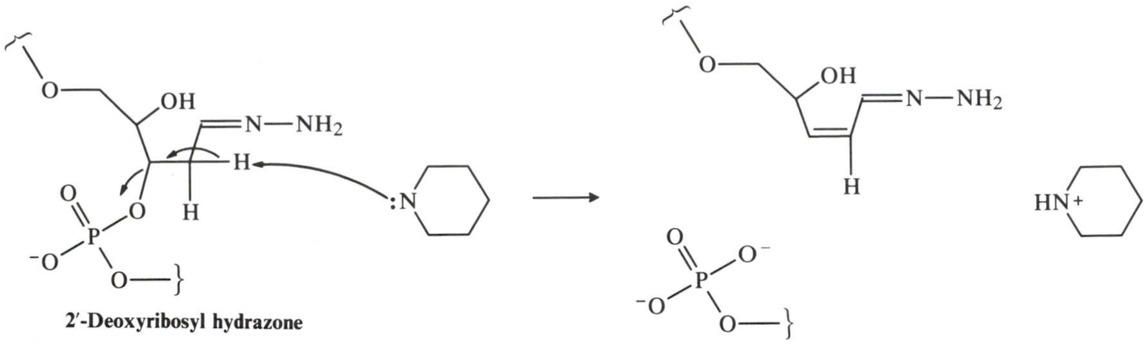

2′-Deoxyribosyl hydrazone

Figure 12-14

Cleavage of 2′-Deoxyribosyl Hydrazone DNA by Piperidine. The hydrazones in apyrimidinic sites are less reactive in base-catalyzed elimination of 3′-phosphates than are the apurinic sites, although the elimination is catalyzed by a very high concentration of hydroxide; that is, in strongly alkaline aqueous solutions. Piperidine is a general base that can be used for this purpose in high concentrations without generating high alkalinity. Piperidine is used as shown to cleave phosphodiester bridges at apyrimidinic sites of DNA that have been introduced in the removal of pyrimidines by hydrazine in the Maxam-Gilbert sequencing method.

Nucleases

Nucleases are enzymes that catalyze the hydrolysis of RNA or DNA, with cleavage of the phosphodiester linkage. *Endonucleases* are nucleases that cleave at internal positions of nucleic acids to form smaller oligonucleotides; *exonucleases* cleave the terminal nucleotide from one end of a nucleic acid, shortening it by one nucleotide. There are 5′-exonucleases, which attack the 5′ end of a polynucleotide, and 3′-exonucleases, which attack the 3′ end. In this section, we consider the mechanisms of action of a few endonucleases. Nucleases function in the cell to turn over nucleic acids, to provide a means of salvaging nucleotides from turnover, to repair lesions and mistakes in DNA, and to degrade foreign nucleic acids.

Pancreatic Ribonuclease A

The most thoroughly studied nuclease is pancreatic ribonuclease A, or RNase A, a small enzyme consisting of a single polypeptide chain 124 aminoacyl residues in length and having a molecular weight of 14,000. It catalyzes the hydrolysis of phosphodiester linkages in RNA with cleavage of the RNA into oligonucleotides. RNase A exhibits considerable selectivity for cleaving at pyrimidine sites in RNA to generate fragments terminating with uridine or cytidine 3′-phosphates according to equation 1, in which the pyrimidine site is C for cytidine and X, Y, and Z can be any ribonucleotide residues.

$$5'\text{-}\ldots\text{XpYpCpZp}\ldots\text{-}3' + H_2O \longrightarrow 5'\text{-}\ldots\text{XpYpC—}3'\text{-}PO_3^{2-} + 5'\text{-HO—Zp}\ldots 3' \quad (1)$$

The mechanism of action of RNase A takes advantage of the most-important chemical property distinguishing RNA from DNA: the sensitivity of RNA to alkaline hydrolysis. The mechanism, as it is currently understood, is given in Figure 12-15, which shows the mechanism by which RNA is cleaved into fragments with 2′,3′-cyclic nucleotides at the 3′ ends. The active site contains two histidines (His-12 and His-119) that are believed to act as general acid-base catalysts in the mechanism. Lysine-41 also is a critical residue that binds the phosphodiester group of the substrate. In the first step, histidine-12 abstracts the

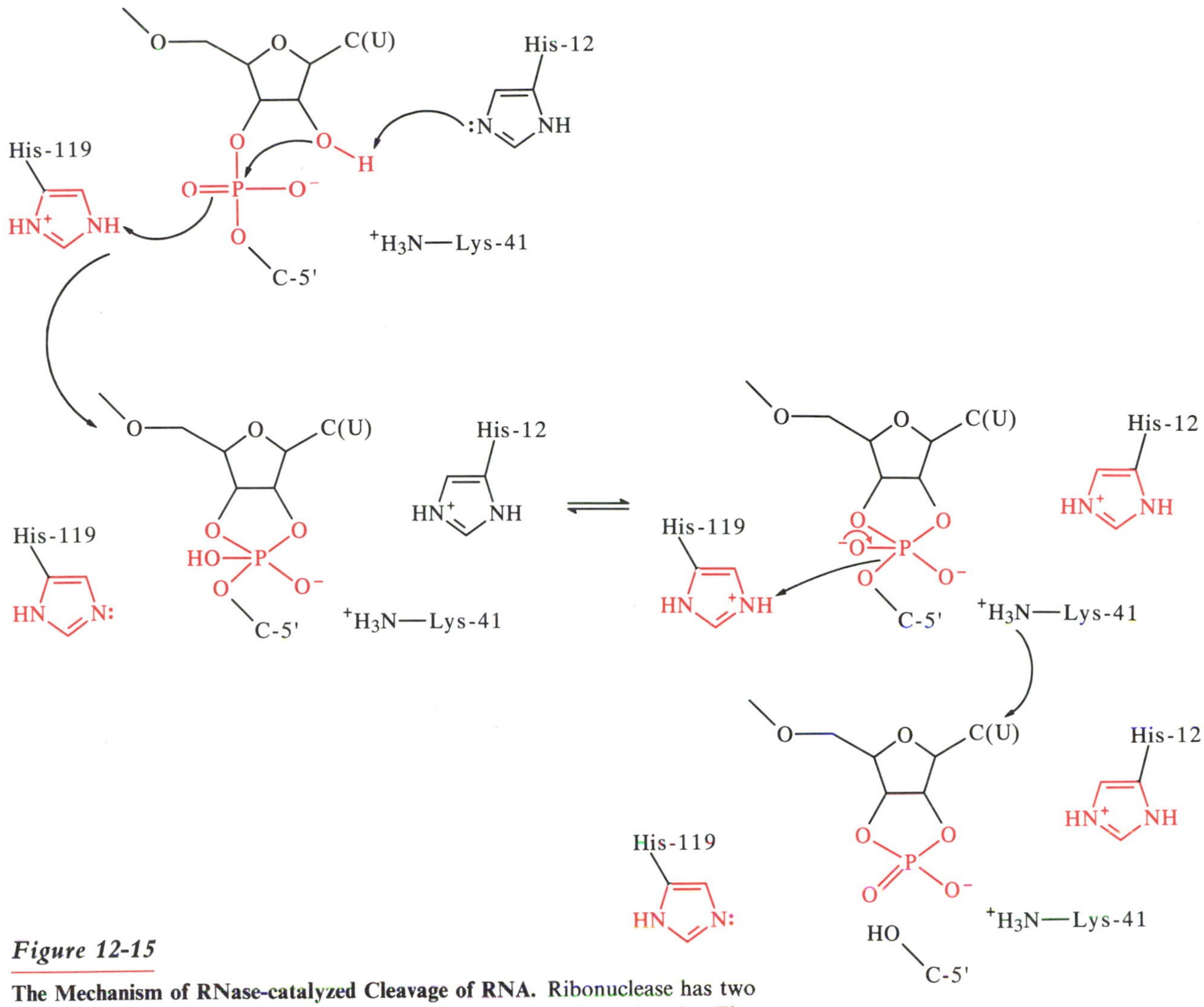

Figure 12-15

The Mechanism of RNase-catalyzed Cleavage of RNA. Ribonuclease has two histidines (His-12 and His-119) and one lysine (Lys-41) at the active site. The mechanism by which RNase cleaves RNA includes the intermediate formation of 2′,3′-cyclic phosphodiesters at the 3′ ends of cleaved fragments. The enzyme also catalyzes the hydrolysis of the cyclic phosphodiester ends to 3′-phosphates. The histidine side chains function as general acid-base catalysts in the formation and hydrolysis of the 2′,3′-cyclic phosphodiester ends. The exact mechanism of action of RNase is not known, but from the structure of the enzyme it appears that histidine-12 catalyzes the reaction of the 2′-hydroxyl group with the phosphodiester to form the pentavalent adduct; histidine-119 catalyzes the decomposition of this adduct to the 2′,3′-cyclic phosphodiester, with the elimination of the 5′-hydroxyl group of the neighboring nucleotide. In this mechanism, histidine-12 functions as a general base to form the initial pentavalent adduct, and histidine-119 functions as a general acid in catalyzing the departure of the 5′-oxygen as a hydroxyl group from the pentavalent adduct. Lysine-41 appears to participate in binding the phosphodiester group through ionic bonding.

2′-hydroxyl proton as the 2′-oxygen attacks the phosphorus in the phosphodiester linkage to form a trigonal bipyramidal addition intermediate. The increased negative charge associated with the equatorial oxygens of the addition intermediate is stabilized by the positively charged ε-ammonium group of lysine-41. In the second step, the protonated imidazolium ion of histidine-119 catalyzes the departure of the 5′-oxygen to cleave the chain and form the 2′,3′-

cyclic nucleotide. The hydrolysis of the cyclic nucleotide presumably proceeds by the microscopic reverse of this mechanism, in which the attacking water molecule takes the place of the 5′-OH group. Histidine-119 in its basic form abstracts a proton from the attacking water. The resulting trigonal bipyramidal intermediate decomposes by departure of the 2′-oxygen catalyzed by proton transfer from the imidazolium group of histidine-12. This mechanism is consistent with the stereochemical course of the reaction of the chiral phosphorothioate *endo*-uridine 2′,3′-cyclic phosphorothioate, which proceeds with inversion of configuration at phosphorus.

The first experimental evidence for the importance of histidine-12 and histidine-119 in the action of RNase was obtained in chemical-modification experiments utilizing iodoacetate and iodoacetamide. These molecules are alkylating agents that react with many nucleophiles. Nucleophiles displace iodide from these molecules and form stable bonds to the methylene carbon, as is shown in equation 2, in which the nucleophile is symbolized by X: and iodoacetate is the alkylating agent.

$$\text{R—X:} + \text{H}_2\text{C(CO}_2^-\text{)—I} \longrightarrow \text{R—X—CH}_2\text{—CO}_2^- + \text{I}^- \quad (2)$$

In proteins, the most-reactive nucleophile with iodoacetate and iodoacetamide is the thiol group of cysteine. However, in RNase A, all four cysteines are oxidized and cross-linked as disulfide linkages, which are unreactive with these reagents. The histidine imidazole ring is second in reactivity toward alkylating agents among amino acid side chains. In RNase A, both histidine-12 and histidine-119 react with iodoacetate.

RNase A reacts with [^{14}C]iodoacetate and is inactivated in the process. The inactive enzyme contains ^{14}C corresponding to one molecule of covalently bonded inactivator per molecule of enzyme. Complete acid hydrolysis of the inactivated ^{14}C-labeled protein gives [^{14}C]carboxymethylhistidine. From analysis of the position of the [^{14}C]carboxymethylhistidine in the amino acid sequence, part of the alkylation is known to occur by alkylation of histidine-12 and part by alkylation of histidine-119. Therefore, alkylation of either residue inactivates the enzyme, and a given molecule could react with iodoacetate at histidine-12 or histidine-119 but not at both sites. These facts suggest that both histidine-12 and histidine-119 are in the active site of the enzyme and both probably participate in catalyzing the reaction. Thus, alkylation of either histidine physically blocks alkylation of the second, and alkylation of either one blocks interactions with substrates. This interpretation is verified by the X-ray crystal structure of the enzyme; Figure 12-16 illustrates the structure of ribonuclease S, showing the active-site residues. Ribonuclease S differs from ribonuclease A by the cleavage of the peptide bond between residues 20 and 21, and it is catalytically active when the S-peptide (residues 1 through 20) is present.

RNase A is very reactive with iodoacetate but much less reactive with iodoacetamide. This is interesting because iodoacetamide is intrinsically more reactive as an electrophile than iodoacetate. The reason for the faster reaction of iodoacetate is that it is anionic and can bind at the active site in place of the phosphodiester anion, whereas iodoacetamide is electrostatically neutral and is not bound by the active site.

Partial digestion of RNase A with the proteolytic enzyme subtilisin cleaves the enzyme between residues 20 and 21, producing RNase S and the S-peptide—residues 1 through 20. RNase S is inactive—it lacks histidine-12—but is activated by the addition of the S-peptide, which binds to the cleaved protein and reconstitutes the enzymatic activity.

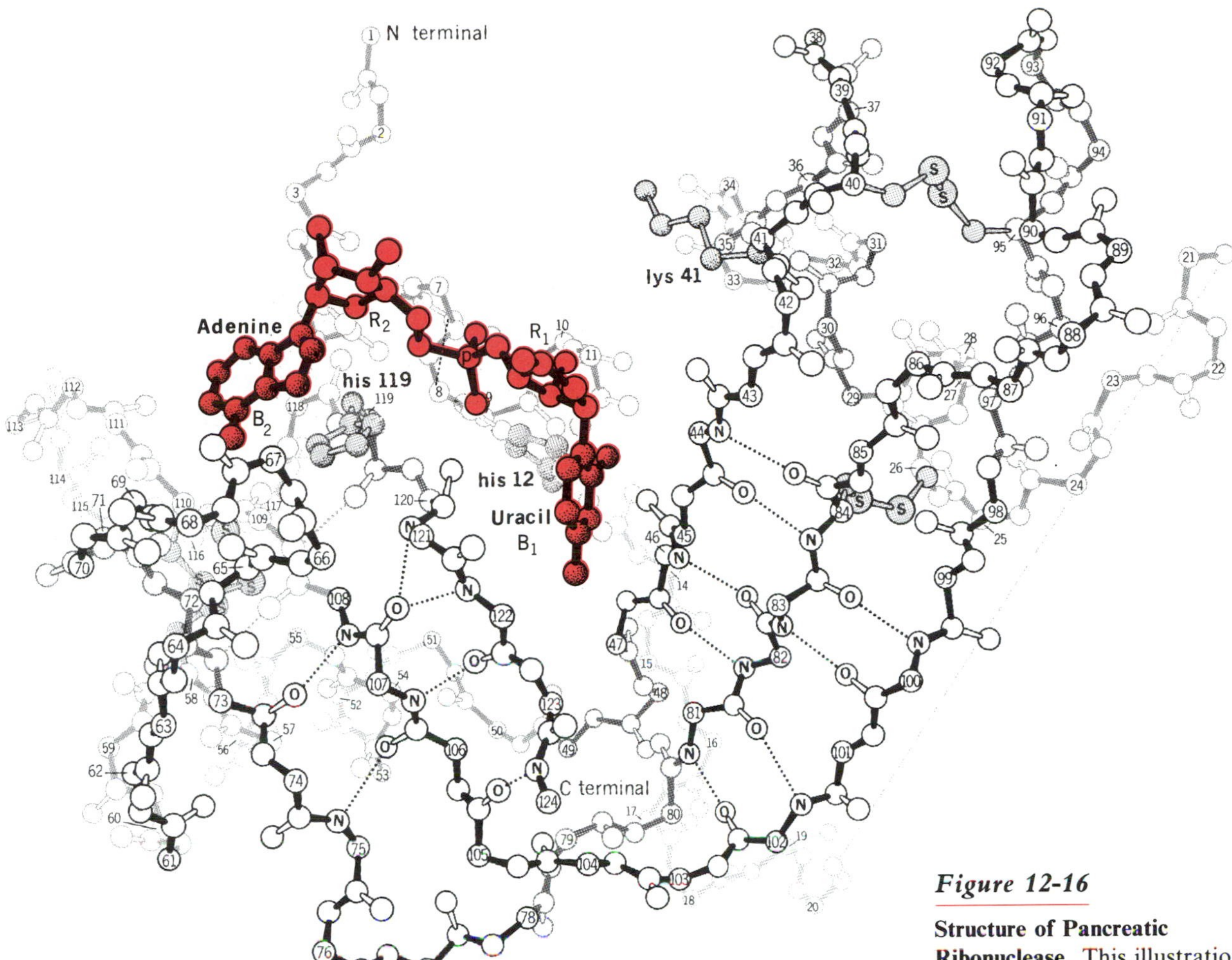

Figure 12-16

Structure of Pancreatic Ribonuclease. This illustration of the three-dimensional structure of RNase S shows the proximity of the catalytic residues (His-12 and His-119) in the active site. Lysine-41 also is highlighted in the figure. A dinucleotide, ApU, is shown bound to the active site. The dashed line connects residues 20 and 21, which have been cleaved in the transformation of ribonuclease A into ribonuclease S. (Copyright © Irving Geis.)

Ribonuclease A was the first enzyme to be studied in both native and denatured forms. C. B. Anfinsen showed in the 1950s that the enzyme could be completely denatured and then renatured to its original activity. This proved that the structure of the protein is determined by the amino acid sequence and not by other factors, such as the pairing of cysteines in disulfide linkages. In these experiments, RNase was reduced and denatured by the combined use of 2-mercaptoethanol and the denaturing agents urea and guanidine hydrochloride. It was then allowed to refold by gradually removing the denaturants by dialysis, while preventing reoxidation of the cysteine sulfhydryl groups. The enzyme regained its structure and activity. Upon reoxidation of the sulfhydryl groups, the same disulfide linkages as those originally present were regenerated.

In related experiments, the cysteines were reoxidized before allowing the native three-dimensional structure to form. This led to unnatural disulfide linkages, and removal of the denaturant did not allow the molecules to return to their native structures. However, the addition of a trace of 2-mercaptoethanol facilitated the unscrambling of the disulfides and regeneration of the correct disulfides. The corrected enzyme was fully active.

These experiments proved that no factors extrinsic to the amino acid sequence are required to generate the native structure of RNase A. It is often said that these experiments prove that the native structure of RNase A is the most-stable structure for the protein. This statement is not quite true. The experiments prove that the native structure is the most stable of those that are kinetically accessible in the renaturation experiments and in the cell. It can be the most-stable structure if all possible structures are kinetically accessible on the time scale of the folding process. It has so far not been possible to determine whether the folding process allows all possible structures to be sampled.

Snake-Venom Phosphodiesterase

Snake-venom phosphodiesterase is noteworthy because it catalyzes the hydrolysis of phosphodiester bonds in nucleic acids by an entirely different mechanism than that of RNase A and gives different products. Its mechanism of action allows it to cleave both DNA and RNA. For these reasons, it is widely used in biochemical laboratories for preparative and analytical purposes. Its cleavage specificity is for the purine (especially adenine) sites in RNA or DNA, and it gives oligonucleotides with 5′-phosphate and 3′-OH ends (equation 3).

$$5'\text{-}\ldots\text{WpXpApYp}\ldots 3' + H_2O \longrightarrow 5'\text{-}\ldots\text{WpX—}3'\text{-OH} + 5'\text{-}{}^{2-}O_3P\text{—ApYpZp}\ldots 3' \quad (3)$$

Apart from its specificity, very little is known about snake-venom phosphodiesterase. The enzyme catalyzes the cleavage of nucleic acids with *retention* of configuration at phosphorus. This means that the reaction probably proceeds by a double-displacement reaction mechanism through a covalent nucleotidyl-enzyme intermediate (equations 4a and 4b).

$$5'\ldots\text{WpXpApXpYp}\ldots 3' + \mathbf{E} \rightleftharpoons 5'\ldots\text{WpX—}3'\text{-OH} + \mathbf{E}\text{—pApXpYp}\ldots 3' \quad (4a)$$

$$\mathbf{E}\text{—pApXpYp}\ldots 3' + H_2O \longrightarrow \mathbf{E} + 5'\text{-}{}^{2-}O_3P\text{—ApXpYp}\ldots 3' \quad (4b)$$

Restriction Endonucleases

Bacteria contain endonucleases that catalyze the hydrolysis of double-stranded DNA at specific sites but act only on DNA that is foreign to the bacteria in which they exist. Their generic name, *restriction endonucleases,* derives from the fact that they have the effect of protecting the bacteria from invasive foreign DNA, such as the DNA of bacteriophages. Thus, the growth of a particular bacteriophage may be *restricted* in a particular bacterial host because of the presence of a restriction endonuclease that degrades the phage DNA. The bacterial DNA is protected from cleavage by the presence of modified bases at the cleavage sites for which the restriction enzyme is specific.

The best-known example of a restriction enzyme is Eco RI (*E. coli* restriction enzyme I). Like all restriction enzymes, Eco RI cleaves DNA at specific sequences known as *palindromes,* which are sequences containing a twofold axis of symmetry. DNA contains many such sequences four to six base pairs in length. The cleavage specificity of Eco RI is shown in equation 5.

$$\begin{array}{c} 5'\text{-}\ldots\text{G-A-A-T-T-C}\ldots\text{-}3' \\ |\;\;|\;\;|\;\;|\;\;|\;\;| \\ 3'\text{-}\ldots\text{C-T-T-A-A-G}\ldots\text{-}5' \end{array} + H_2O \longrightarrow \begin{array}{c} 5'\text{-}\ldots\text{G} \\ | \\ 3'\text{-}\ldots\text{C-T-T-A-A} \end{array} + \begin{array}{c} \text{A-A-T-T-C}\ldots\text{-}3' \\ | \\ \text{G}\ldots\text{-}5' \end{array} \quad (5)$$

The dimeric structure of the enzyme is in keeping with the symmetry of the palindromic base sequence at the cleavage site. The X-ray crystal structure of the dimeric enzyme in complex with its specific sequence is in accord with the expectation of a symmetrical interaction of each subunit with one of the strands. The specificities of several other restriction enzymes are given in Chapter 14.

The DNA of *E. coli* is methylated at adenine-N^6 in all the sequences such as that in equation 5, at which Eco RI would otherwise cleave. The presence of the N^6-methyl groups at these positions protects the DNA of *E. coli* from being cleaved by Eco RI.

Almost nothing is known about the chemical mechanism by which the restriction endonucleases catalyze hydrolysis of DNA. It is known to be different from those of RNase A and snake-venom phosphodiesterase. Unlike the RNase A reaction, it cannot involve a 2′-OH group and a cyclic intermediate because DNA contains no 2′-OH groups. And, unlike snake-venom phosphodiesterase, Eco RI catalyzes cleavage with inversion of configuration at phosphorus.

Different bacteria contain different sets of restriction enzymes that vary in cleavage specificities. Eukaryotic cells do not have restriction enzymes, and eukaryotic DNA is not protected from them. Therefore, the restriction enzymes are indispensable tools in recombinant DNA technology. More than two hundred specific enzymes are available for cleaving a particular species of DNA into defined sets of double-stranded oligomers, including whole genes. These enzymes have made gene cloning possible (see Chapter 14).

The enormous variety of specificities available has also facilitated the analysis of nucleotide sequences in genes. In nucleotide sequencing, which is described later in this chapter, a few hundred base pairs can be sequenced in a single experiment. Because most genes are longer than this, it is necessary to sequence several sets of fragments from a given gene and look for overlapping sequences. This can always be done by using several restriction endonucleases to cleave a gene into several sets of fragments, just as proteolytic enzymes and cyanogen bromide are used with proteins to produce two or more sets of peptides in amino acid sequencing.

DNA Replication

The reactions leading to the biosynthesis of DNA and RNA are undoubtedly among the most-studied, the most-complicated, and the most-important examples of nucleotidyl transfer. One might argue that they are the most-important reactions in living systems; thus, they are primary subjects of courses in molecular biology. The fundamental aspects of the chemistry of DNA replication will be described in this section, followed by RNA biosynthesis in the next section.

DNA replication is the most-fundamental and most-critical part of chromosome replication, which takes place only in the course of cell division. DNA replication is moderately well understood today as the result of concentrated research in the past 25 years.

Semiconservative Replication

A basic question regarding the replication of DNA was whether the process of producing two exact replicas of double-stranded DNA required the retention or conservation of one or both strands of the preexisting DNA in a dividing cell. Three distinct possibilities could be considered, as illustrated in Figure 12-17. First, two entirely new molecules of DNA, each identical with the parent, might be synthesized. This would be unconservative replication. Second, replication might proceed with conservation of both parent DNA strands in one molecule and two new strands in the new DNA molecule. This would be conservative replication. Third, replication might proceed by a process that conserves the two parent strands, with one parent strand and one newly synthesized strand in each of the two molecules of DNA. This would be semiconservative replication.

Meselson and Stahl succeeded in proving that DNA replication is semiconservative. In their experiment, they used ^{15}N, a stable (unradioactive) heavy isotope of nitrogen, and density gradient ultracentrifugation to show that (1) all molecules of newly synthesized DNA are identical and (2) each molecule contains one parent strand and one newly synthesized strand of DNA. To do this, they grew bacteria on a defined growth medium, using NH_4^+ or $^{15}NH_4^+$

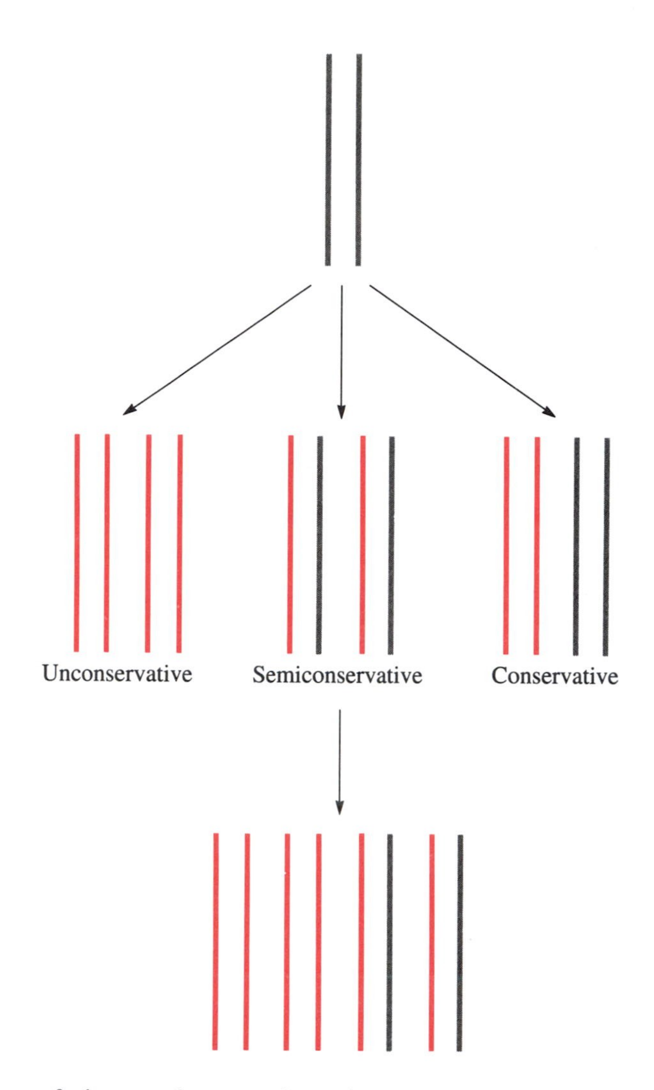

Figure 12-17

The Possible Modes of Chain Synthesis in DNA Replication: Conservative, Semiconservative, and Unconservative. The parent DNA and the conserved strands in the daughter molecules are shown in color. In unconservative replication, both daughter strands would be newly synthesized from nucleotides. In semiconservative replication, each daughter molecule would contain one strand from the parent and o newly synthesized strand. In conservative replication, one daughter molecule would consist of both parent strands and the other would consist of newly synthesized strands. Meselson and Stahl showed that replication is semiconservative by the method illustrated in Figure 12-18. In the second generation, semiconservative replication leads to two granddaughters with newly synthesized strands and two with one conserved and one newly synthesized strand, so that the ratio of conserved to newly synthesized strands in the second generation is 1 to 3.

as the source of nitrogen for growth, and analyzed the buoyant density of the bacterial DNA that they isolated from the bacteria at different stages of growth.

In *density gradi nt ultracentrifugation,* a gradient of CsCl is set up inside a centrifuge tube that continuously increases in density from the top to the bottom of the tube. A sample of unknown density is layered on the top of the gradient and then subjected to high-speed centrifugation. The sample migrates in the centrifugal field until it reaches a point in the density gradient at which its buoyant density matches the density of the gradient. At this point, the centrifugal force experienced by the sample is counterbalanced by the buoyancy of the CsCl gradient, and the sample ceases to migrate. The cell position of this equilibrium point is a sensitive measure of the buoyant density of the sample.

The buoyant density of [^{15}N]DNA is larger than that of DNA containing only naturally occurring nitrogen. (Nitrogen in nature consists of 99.64% ^{14}N and 0.36% ^{15}N.) Therefore, the equilibrium position of a sample of DNA in a density gradient is dependent on its ^{15}N content. Meselson and Stahl chose ^{15}N as the heavy isotope for reasons of convenience and economy, because $^{15}NH_4^+$ was available, inexpensive, and a convenient nitrogen source for bacterial growth. In their experiment, they grew bacteria with $^{15}NH_4^+$ in the growth medium for many generations, so that the DNA would contain essentially only ^{15}N. They isolated the [^{15}N]DNA from a sample of the bacteria, and they shifted the bacteria to a growth medium that was identical except that the new medium contained NH_4^+ instead of $^{15}NH_4^+$. After one generation time and two generation times in the new medium, they again isolated samples of DNA

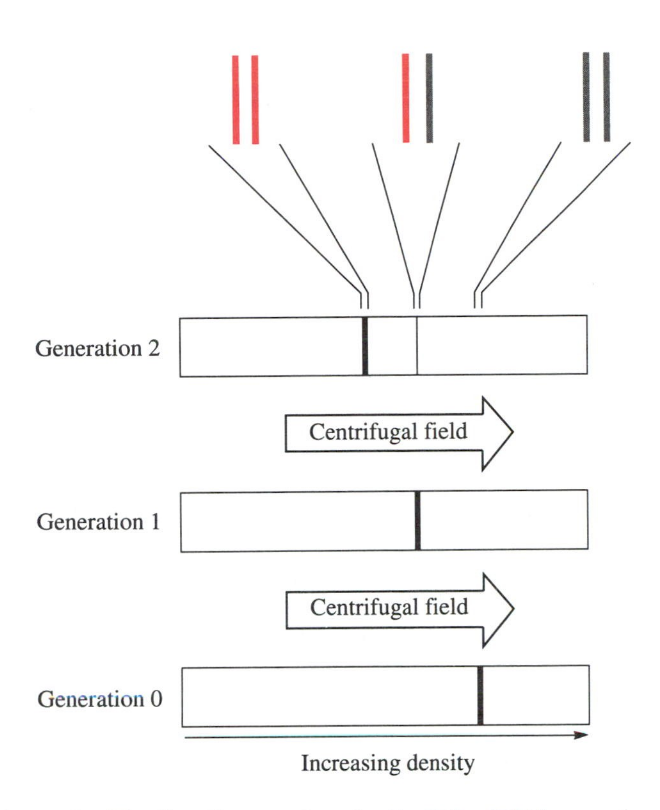

Figure 12-18

Proof of Semiconservative DNA Replication. Meselson and Stahl obtained the following results in their demonstration of semiconservative DNA replication in bacteria. Illustrated are density gradient centrifugation cells with the centrifugal field and density gradient oriented from left to right. [^{15}N]DNA from the parent cells migrates to the furthest point in the density gradient, corresponding to ^{15}N in both strands. DNA from the first generation of growth on $^{14}NH_3$ migrates to a single position, with a lower buoyant density, corresponding to one conserved parent strand (^{15}N) and one newly synthesized strand (naturally occurring N, >99% ^{14}N). DNA from the second generation migrates to two positions, one corresponding to the same species as the second generation and the second to a position of lowest buoyant density corresponding to naturally occurring N, in a ratio of 1:3.

from the bacteria. Finally, they analyzed all of the DNA samples by density gradient centrifugation.

The three samples of DNA migrated differently in the density gradients, with the results illustrated in Figure 12-18, and the differences could be interpreted only in terms of semiconservative replication. The [^{15}N]DNA migrated to the position of highest density. After one generation of growth in unenriched NH_4^+, none of the DNA migrated to the position of either [^{15}N]DNA or unlabeled DNA, but all of it migrated to a position intermediate between the two. This position corresponded to a *hybrid* DNA containing 50% ^{15}N-enrichment. The absence of any fully ^{15}N-enriched DNA and any unenriched DNA excluded both the conservative and the unconservative models of replication. After two generations of growth in unenriched NH_4^+, two bands of DNA appeared in a ratio of 1:3, the first of which had the same density as the hybrid DNA in the first generation and the second of which corresponded to unenriched DNA. This also supported semiconservative replication.

Although the observation of hybrid DNA was *consistent* with one strand being fully enriched in ^{15}N and the other being unenriched, proof of this required another experiment, because the hybrid could not be distinguished by its buoyant density from a species in which both strands were 50% enriched. Meselson and Stahl further examined this question by denaturing the hybrid DNA and allowing it to renature. The separated strands in denatured DNA could reform double-stranded DNA with different partners. They then repeated the gradient ultracentrifugation with the renatured sample and observed all three species of DNA—fully ^{15}N-enriched, unenriched, and hybrid DNA—proving that the original sample of hybrid DNA consisted of one fully ^{15}N-enriched strand and one unenriched strand.

DNA Polymerases

The discovery of DNA polymerase by A. Kornberg is a classic example of the biochemical methods by which many enzymes have been discovered and purified. Kornberg determined that replication somehow involved the sequential linkage of nucleotides into the polymer. The chemical problem of effecting the linkage should most obviously be overcome by the use of deoxynucleoside triphosphates (dNTP) as substrates of the polymerization system, because they could be linked together by sequential attack of a deoxyribosyl 3′-hydroxyl group on the α-phosphorus of a dNTP with displacement of PP_i in a nucleotidyl transfer. This would be energetically favorable and further driven by inorganic pyrophosphatase-catalyzed hydrolysis of PP_i. Kornberg did not know whether the substrates would be deoxynucleoside 5′-triphosphates or deoxynucleoside 3′-triphosphates, but the 3′-triphosphates had not been discovered in cells; therefore, the 5′-triphosphates seemed to be the most-probable substrates. However, many additional possibilities existed, and all had to be taken into account. Most alternative possibilities would have required more-complex polymerization enzymes than would the polymerization of 5′-dNTPs, however, and such enzymes were never discovered.

The first requirement for discovering an enzyme by biochemical methods is to have or invent a reliable method for measuring the amount of enzyme; that is, an enzyme assay. By use of an assay, the amount of an enzyme in an extract or purification fraction can be measured. This measurement allows the effectiveness of an attempted purification step to be assessed. Kornberg began by developing an assay for an enzyme that would catalyze the polymerization of dNTPs into a polynucleotide. His assay had the advantages of simplicity, accuracy, and reliability. In developing it, Kornberg considered the known chemical properties of DNA and used the best-known and most-reliable one as the basis for the assay, which is that nucleic acids in general and DNA in particular are insoluble in acidic solutions. Because dNTPs are soluble in acid, he reasoned that an enzyme that would catalyze the polymerization of $dNT^{32}Ps$ would create a $[^{32}P]DNA$ that could be precipitated by the addition of acid, leaving the unreacted $dNT^{32}Ps$ in solution. The radioactivity of $[^{32}P]DNA$ in the resulting precipitate could be accurately measured.

In his early attempts to detect such an enzyme in crude extracts of *E. coli*, Kornberg observed a few hundred counts of radioactivity in precipitates from reaction mixtures containing millions of counts in the $dNT^{32}Ps$. The precipitated radioactivity appeared only under certain conditions and remained in the precipitates after repeated washing with acid. Equation 6 defines the reaction that Kornberg observed; that is, a reaction in which the enzyme produces a polydeoxyribonucleotide from the *four dNTPs* in the presence of Mg^{2+}, a *primer*, and a *template.*

$$n\,(\text{dATP, dCTP, dGTP, dTTP}) \xrightarrow[\text{Mg}^{2+}]{\text{Template/Primer}} \text{poly d(ACGT)} + n\ \text{PP}_i \qquad (6)$$

The reaction is thermodynamically favored and further driven by the action of inorganic pyrophosphatase, which catalyzes the energetically downhill hydrolysis of pyrophosphate produced in the DNA polymerase reactions. Equation 6 is catalyzed by the DNA polymerase that Kornberg first observed and purified, DNA polymerase I, and by all other DNA polymerases that have since been discovered. Two polymerases, DNA polymerase I and DNA polymerase III, play important roles in replication. *DNA polymerase III* is primarily a replication enzyme, whereas *DNA polymerase I* is important in both replication and repair of damaged DNA.

The template and primer are crucial to the action of all DNA polymerases. These enzymes cannot initiate the biosynthesis of a polymer but can only *extend* existing polymers; the role of the primer is to serve as the starting point for polymer growth. In DNA replication, specialized enzymes generate specific primers for initiating the action of DNA polymerases. This makes excellent sense for a replication system that must replicate accurately, but only under controlled conditions and at the appropriate times. It would not do for DNA polymerases to have the capacity to start new chains without control. In laboratory experiments, the primer must be supplied by the experimenter. The template has the independent but equally central role of specifying the sequence with which deoxynucleotides are incorporated into the new polymer. The new sequence is determined by base pairing between incoming dNTP molecules and the bases in the template.

The basic reaction catalyzed by DNA polymerases is illustrated in Figure 12-19. The template is the conserved chain of DNA, which specifies the

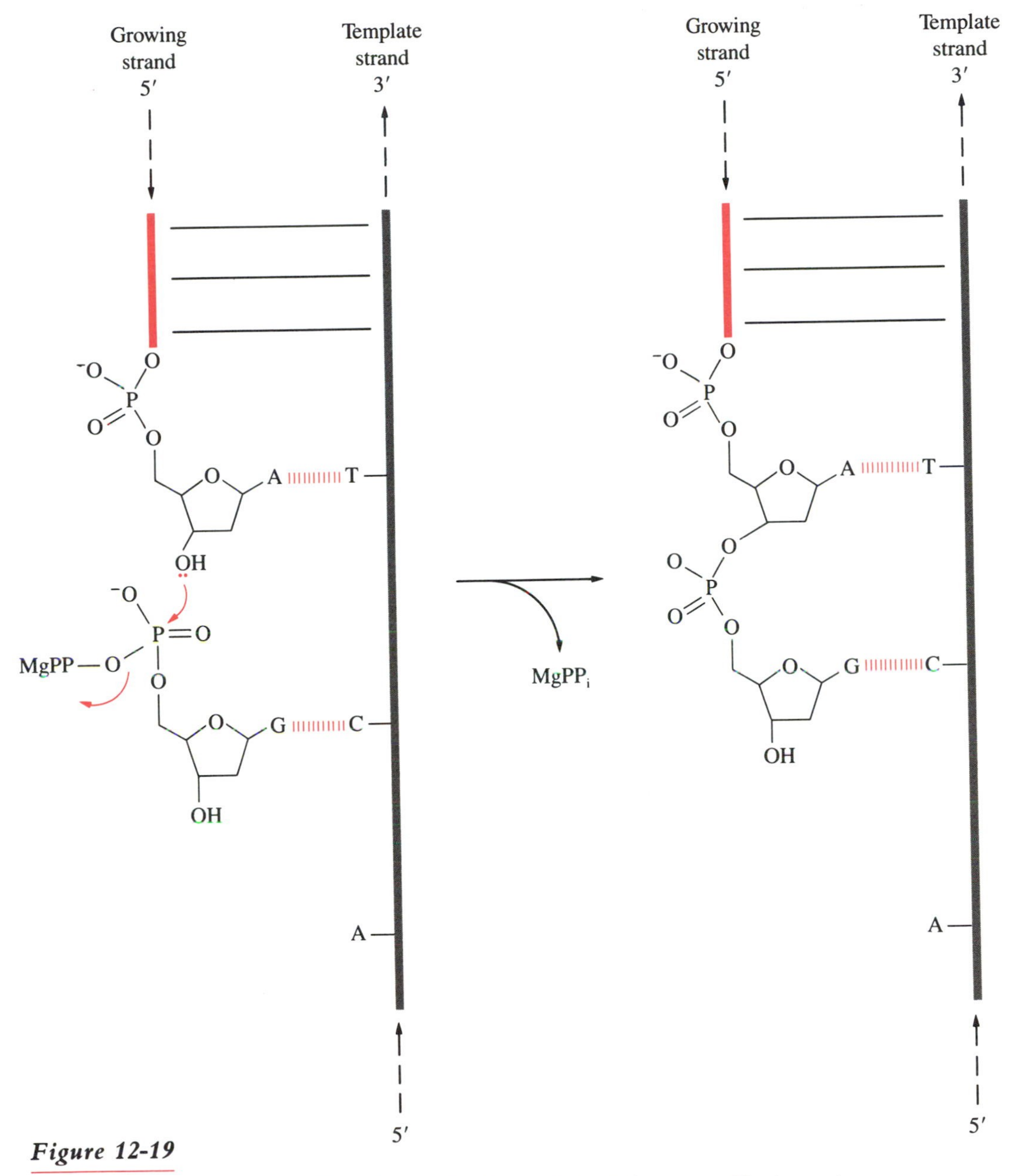

Figure 12-19

The Chemical Mechanism of DNA Polymerization by DNA Polymerase. The formation of a 3′,5′-phosphodiester linkage between deoxyadenosine and deoxyguanosine illustrates the basic reaction catalyzed by all DNA polymerases. In the reaction shown, the guanine ring of dGTP in the growing strand is hydrogen bonded to the cytosine ring of dCMP in the template strand, and the 3′-hydroxyl group of deoxyadenosine at the terminus of the growing strand displaces pyrophosphate from dGTP to form the phosphodiester bridge between the nucleotides. In the next step, dTTP will bind by base pairing with the dAMP residue in the next position of the template strand, and a new phosphodiester bond will be generated by the same mechanism.

nucleotide sequence that is to be synthesized in the new or growing chain. The chain grows in steps such as that shown, in which DNA polymerase catalyzes the transfer of a dGMP group from dGTP to the 3′-hydroxyl group of the deoxynucleoside residue at the 3′ end of the growing DNA chain. The 3′-hydroxyl group reacts as a nucleophile at P_α of dGTP, displacing pyrophosphate and forming the phosphodiester bond. Only the complementary deoxynucleoside triphosphate, dGTP in Figure 12-19, can bind productively to give chain elongation. In the next step of elongation, dTTP will bind, complementing dA, and the 3′-hydroxyl of the dGMP that is added in the step shown will attack P_α of dTTP, displacing PP_i and forming the next phosphodiester bond.

The polymerization catalyzed by DNA polymerases is perfectly adequate for producing a single strand of DNA, but it cannot replicate double-stranded DNA. The problem is that DNA chains are polar and the two strands in DNA are antiparallel. The mechanism in Figure 12-19 is unidirectional, allowing for chain growth in only the 5′ to 3′ direction. Therefore, in replication the two strands must be replicated either simultaneously from the opposite ends or by some processes other than that shown in Figure 12-19.

DNA Replication Forks

Electron micrographs show that the two strands are replicated simultaneously from a fork, and the fork moves in a single direction, 5′ to 3′ for one strand and 3′-to-5′ for the other (Figure 12-20). Chain growth in the 3′-to-5′ direction would be possible in principle by utilizing deoxynucleoside 3′-triphosphates and allowing attack by the 5′-hydroxyl end, but no such nucleotides or enzymes with the required specificity have been discovered. Thus, the antiparallel nature of DNA, the mechanism of Figure 12-19, and the observation of replication forks require that one of the strands in a fork be synthesized in the direction *away* from the fork.

The first clue to the mechanism by which polymerization of one strand proceeds in the direction opposite to that of fork movement was obtained by Okazaki, who observed the release of short DNA fragments, about a thousand nucleotides in length, when replication was interrupted. These were named Okazaki fragments and shown to be short segments synthesized as fragments of the *lagging strand*, as shown in Figure 12-21. The lagging strand is synthesized in the 3′-to-5′ direction from the Okazaki fragments, which are themselves produced in bursts of 5′-to-3′ chain growth catalyzed by DNA polymerase III. The *leading strand* is synthesized as a continuous polymer in the 5′-to-3′ direction, also by DNA polymerase III.

DNA Primase

The mechanism of discontinuous polymerization in the lagging strand is a complex affair that requires three enzymes in addition to DNA polymerase III, all of which catalyze NMP transfers. The first is a special RNA polymerase, *DNA primase*, that produces very short RNA oligomers that are complementary to short sequences in the lagging DNA strand. We shall see that RNA polymerases in general, unlike DNA polymerases, do not require primers to start polymerization. DNA primase catalyzes the formation of complementary RNA fragments by nucleotidyl transfer through the chemical mechanism shown in Figure 12-19, but it utilizes the ribonucleotides ATP, CTP, GTP, and UTP and does not need a primer. The complementary oligoribonucleotides then prime DNA polymerase III, which extends them with deoxyribonucleotides and generates the Okazaki fragments. A primase also generates a short RNA oligomer at the origin of replication and primes DNA polymerase III to synthesize the leading strand.

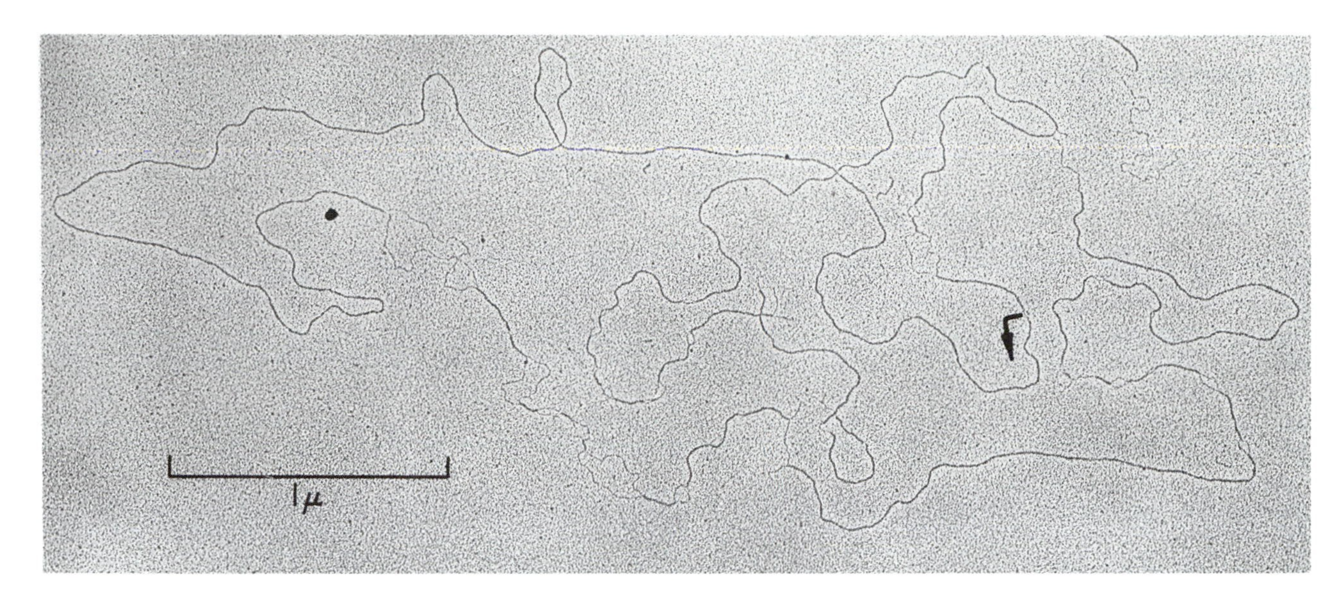

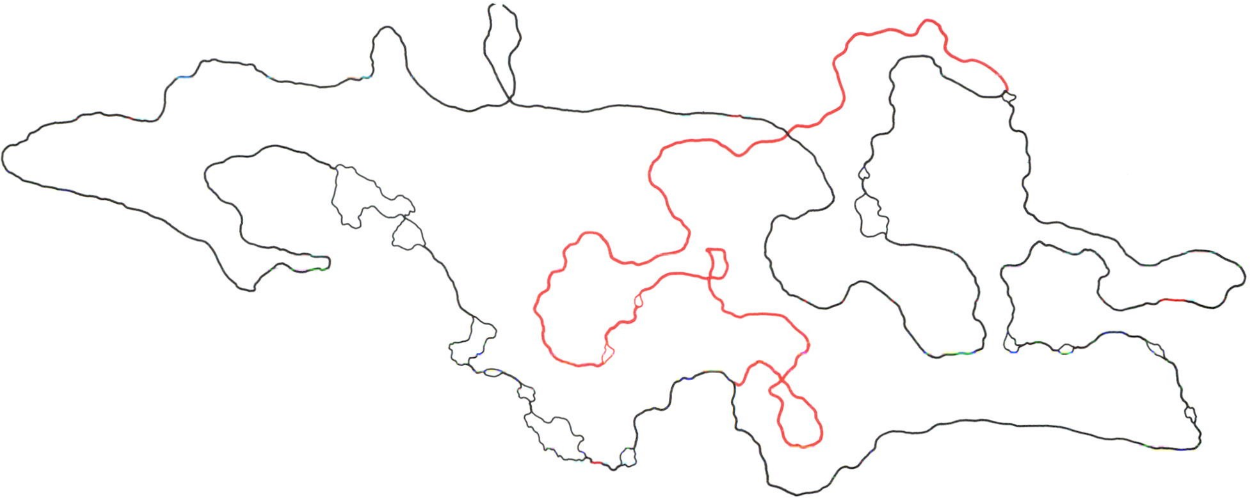

Figure 12-20

Electron Micrograph of a Replication Fork. This electron micrograph shows the replicative intermediate of bacteriophage λ (48,514 base pairs). Replication is bidirectional from a unique site and generates a looped structure known as a θ structure. The replicative intermediate was isolated from intracellular DNA and partially denatured so that the branch points could be positioned with respect to the physical and genetic map of this phage. The branch points and cross bar of the θ structure are highlighted in the interpretive drawing below the micrograph. The arrow in the electron micrograph indicates the deduced position of the ends of the linear molecule as it exists within the phage before infection. (Micrograph courtesy of Ross Inman.)

DNA polymerase I, the second enzyme that is required for discontinuous biosynthesis of the lagging strand, is the utility infielder of DNA replication. This enzyme is a DNA polymerase but is also an exonuclease. It exhibits both 3′,5′- and 5′,3′-exonucleolytic activities. Its 5′,3′-exonucleolytic activity is called into play to remove the short RNA sequences that primed the Okazaki fragments. The gaps remaining between Okazaki fragments are then filled in by the polymerase activity of DNA polymerase I, using the 3′ ends as primers and the parent strand as the template.

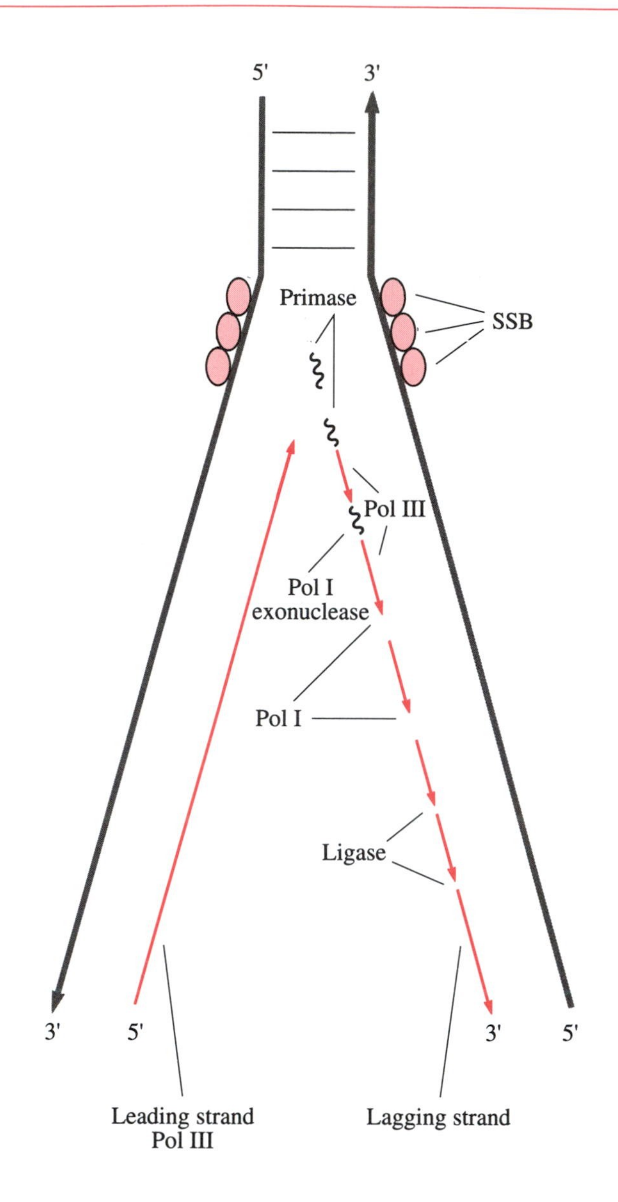

Figure 12-21

The Mechanism of DNA Replication. DNA replication requires the coordinated actions of many enzymes, including but probably not limited to DNA polymerase I (Pol I), DNA polymerase III (Pol III), DNA ligase, DNA primase, topoisomerases, helicases, and single-strand-binding proteins (SSB). The basic process is illustrated here. Pol III catalyzes the continuous 5′-to-3′ synthesis of the leading strand, using one parent strand as the template. The discontinuous 3′-to-5′ synthesis of the lagging strand is catalyzed by DNA primase, Pol I, and DNA ligase. Primase begins the process by producing short oligoribonucleotides that serve to prime the Pol III catalyzed 5′-to-3′ synthesis of Okazaki fragments, using the other parent strand as the template. Pol I removes the RNA primers from the 5′ ends of the Okazaki fragments through its 5′-exonuclease activity, and it fills the gaps between the Okazaki fragments through its polymerase activity. The filled fragments are finally sealed together by DNA ligase.

DNA Ligase

When the gaps are filled, the lagging strand remains discontinuous because the polymerases cannot catalyze phosphodiester-bond formation between the 5′-phosphate and 3′-hydroxyl ends of the mature Okazaki fragments. This process is catalyzed by *DNA ligase,* an ATP-dependent (or NAD^+-dependent) enzyme with specificity for closing "nicks" in DNA. The enzyme is nonspecific with regard to nucleotide sequences, but it is specific for nicks in double-stranded DNA with 5′-phosphate and 3′-OH ends.

DNA ligases contain a lysyl residue at the active site that plays a crucial role in the reaction mechanism, which is outlined in Figure 12-22. The lysine reacts with ATP in the first step to displace PP_i and generate an AMP-enzyme. (It is interesting that a DNA ligase from a bacteriophage utilizes NAD^+ as the AMP donor in place of ATP.) The AMP-enzyme then binds nicked DNA and transfers the AMP group to the 5′-phosphoryl group in the nick. The resulting ADP-ribosyl entity in the nick activates the 5′-phosphoryl group and facilitates nucleophilic attack by the 3′-hydroxyl group to displace AMP and form the

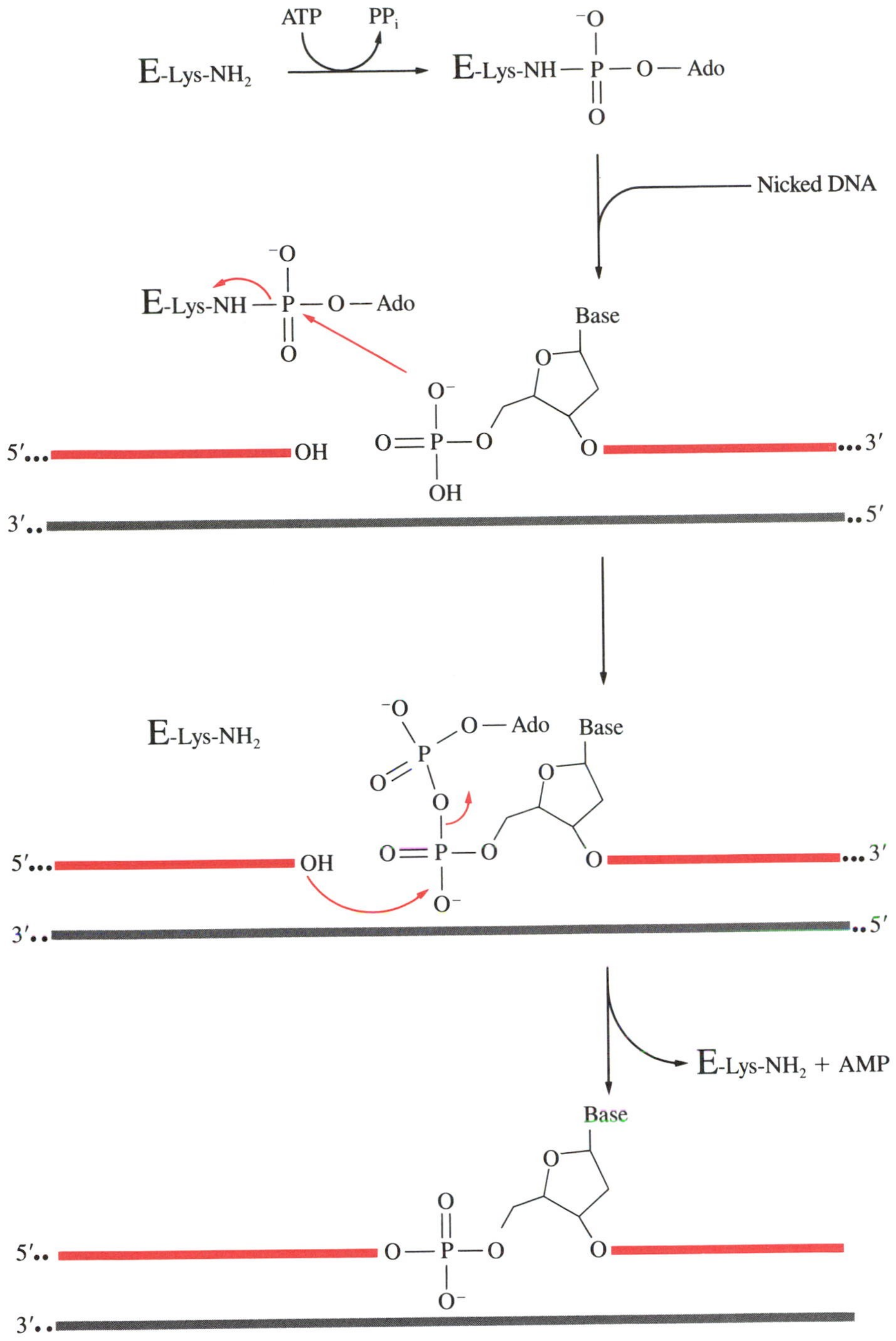

Figure 12-22

The Chemical Mechanism by Which DNA Ligase Seals Nicked DNA. DNA ligase first reacts with ATP to form a covalent enzyme-AMP complex (adenylyl-enzyme), with AMP activated by bonding to a lysine ε-amino group. This form of the enzyme binds nicked DNA and repairs the nick by transferring the AMP group to the 5′-phosphate group to form a phosphanhydride. This activates the 5′-phosphate by supplying a good leaving group, AMP, which is displaced by the 3′-hydroxyl group in the nick, thus forming the 3′,5′-phosphodiester bond.

phosphodiester linkage. This linkage is created by the extraction of an oxygen atom from the 5′-phosphate of the DNA nick and its replacement by the 3′-hydroxyl as a ligand to phosphorus. The dehydrating agent is a phosphoanhydride bond of ATP (or NAD^+), and the leaving group is AMP.

Helicases and DNA-binding Proteins

Figure 12-21 shows how the actions of DNA polymerases, DNA primase, and DNA ligase catalyze replication; however, several additional proteins are required for this process, and it is not clear that all of them are yet known. One part of the replication process that cannot be catalyzed by the enzymes so far described is the separation of DNA strands. The enzymes clearly cannot replicate DNA that is tightly wound up into a double helix, and the strands must be separated at the replication fork to allow the actions of DNA polymerases and DNA primase to proceed. Strand separation is accomplished by the actions of helicases and single-strand-binding proteins. The *helicases* separate DNA strands at replication forks by mechanisms that are unknown in detail but that include the binding of the enzyme to DNA and its migration by physical translocation into the fork in a process that separates the strands. Strand separation requires energy input, and helicase action is accompanied by ATP hydrolysis. Thus, helicases are DNA-dependent ATPases and utilize the energy of ATP to separate the strands of DNA. One energy-requiring function of helicases is the migration of the enzyme along the DNA with the replication fork.

Strands separated by helicase action will not remain separated for long unless they are held apart by some means. The *single-strand-binding proteins* play an essential role in replication by preventing the separated strands from coming together again. The structures of these proteins are such that they bind specifically to single-stranded DNA and not to double-stranded DNA. The binding energy is used in this way to prevent separated strands from coming together. Thus, the helicases catalyze strand separation using the energy provided by ATP hydrolysis, and the single-strand-binding proteins trap the separated strands and hold them apart by the use of binding energy.

Topoisomerases

Topoisomerases catalyze the introduction and relaxation of superhelicity in DNA. Several types of enzymes with varying specificities are known to be important in the replication of DNA, as well as in the repair, genetic recombination, and transcription of DNA. The simplest topoisomerases, designated topoisomerase I, relax superhelical DNA, a process that is energetically spontaneous. The *gyrases,* which are known as topoisomerase II, catalyze the energy-requiring and ATP-dependent introduction of negative superhelical twists into DNA. In DNA replication, topoisomerases I and II have the important function of relaxing the positive superhelicity that is introduced ahead of the replicating forks by the action of helicases. In addition, gyrases introduce negative twists into segments of DNA that allow single-strand regions to appear. These regions are important means of facilitating the actions of other enzymes that can bind only to single-stranded DNA.

The chemical mechanisms by which topoisomerases act include the transient cleavage and reformation of phosphodiester bonds in DNA, as illustrated in Figures 12-23 and 12-24. The enzyme binds DNA, and a nucleophilic amino acid, generally tyrosine in topoisomerases, reacts with a phosphodiester linkage in one strand to displace either the 5′-OH or the 3′-OH of one of the two nucleotides at that position of the strand. This process links the enzyme to the cleaved strand through a new phosphodiester. The resulting nick allows the double-stranded DNA to undergo topological changes, to relaxed DNA if the enzyme is topoisomerase I; the nick is then resealed by the displacement of the tyrosyl hydroxyl group of the enzyme (Figure 12-24) by the 5′-OH or the 3′-OH end of the nicked strand, and the enzyme is released. If the enzyme is DNA gyrase (topoisomerase II), it binds to DNA and cleaves one strand by covalent-bond formation; it then introduces negative superhelical twists into the nicked DNA by an unknown mechanism, in a process that is energetically driven by the hydrolysis of ATP. The nick is then resealed by reversal of the nicking reaction and release of the enzyme.

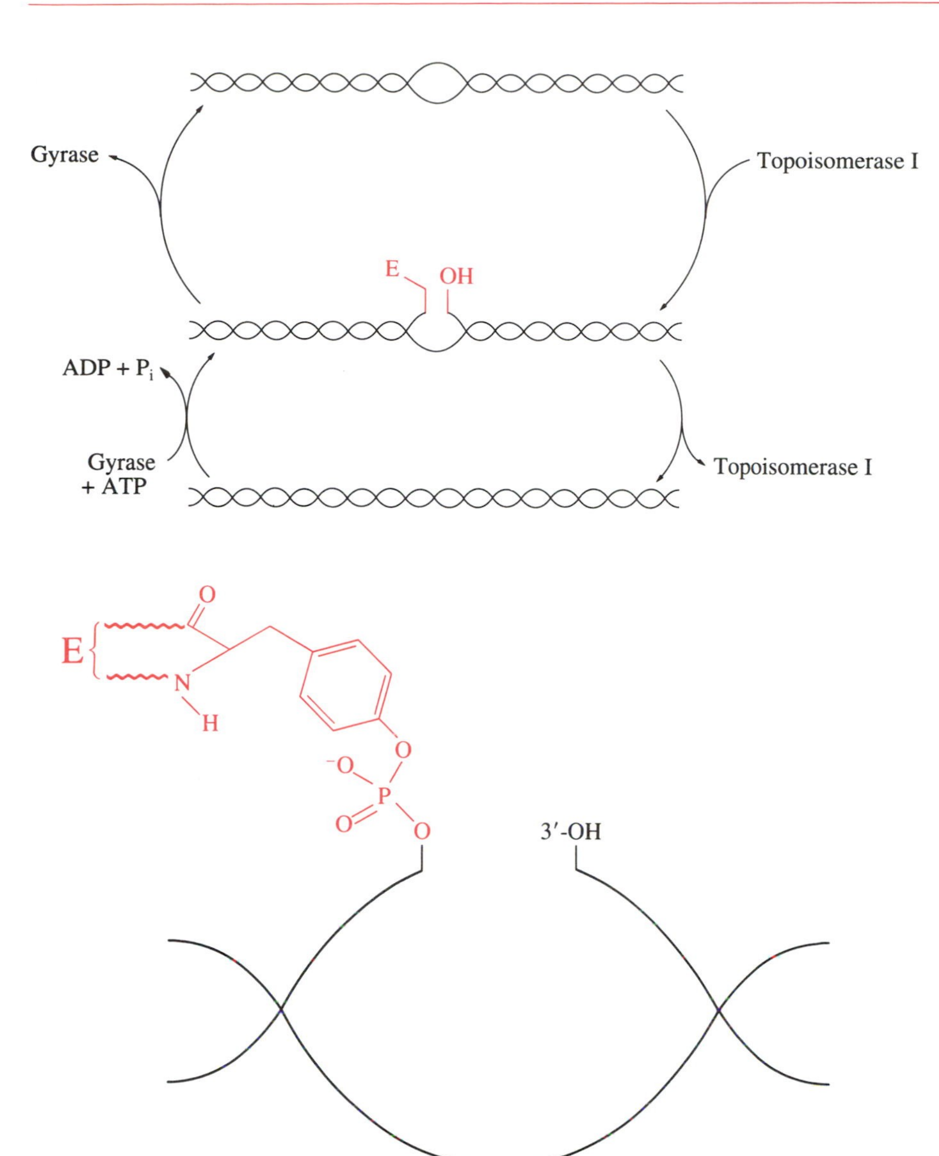

Figure 12-23

The Mechanism of Action of DNA Topoisomerases. Topoisomerase I acts by utilizing a tyrosyl side chain to cleave a phosphodiester bond in superhelical DNA, forming a covalent bond between the enzyme and one strand of the DNA. The DNA is then free to relax, and the phosphodiester bond can reform with cleavage of the DNA from the protein. DNA gyrase (topoisomerase II) cleaves DNA and reforms the phosphodiester bonds in a similar way, but the covalent enzyme-DNA complex can introduce positive superhelical twists into the DNA by an unknown mechanism, the energy for which is supplied by the hydrolysis of ATP.

Figure 12-24

The Chemical Structure of DNA Topoisomerase Intermediates.

BOX 12-1

DNA REPAIR

The protection of the structural integrity of DNA is almost as important to an organism as life itself. Many kinds of damage to DNA are possible, beginning with mistakes in replication and extending to degradation by invasive environmental agents. The molecular consequences of these events are varied, from simple mistakes in nucleotide sequence introduced in the course of replication to strand cleavage or chemical destruction of heterocyclic bases by ultraviolet light or chemical agents. A few kinds of damage simply alter the nucleotide sequence and result in mutation, a fundamental event in evolution. A mutation may or may not strengthen the survivability of the affected organism but is, in either case, required for evolution. Simple mutations are not subject to repair by an organism.

There are a number of mechanisms for repairing DNA that protect the genetic information and probably limit the rate of aging. Only a few of these mechanisms

BOX 12-1 (continued)

DNA REPAIR

are known. One is the action of DNA polymerase I, which has a proofreading function in replication. Using its exonuclease activities, it removes any nucleotides that happen to be incorporated incorrectly as mismatches owing to slight imperfections in the recognition of "correct" base pairs by DNA polymerase III.

A second system in *E. coli* works with DNA polymerase I to remove damaged segments and replace them with corrected sequences. This system consists of three proteins, uvr A, uvr B, and uvr C, that recognize the conformational irregularities caused by slight imperfections in base pairing. These proteins serve a search-and-destroy function in which they bind randomly to DNA and then migrate along the double strand in search of irregularities. The search phase requires energy because it requires physical movement; the source of energy is a phosphoanhydride bond of ATP, which undergoes hydrolysis by an ATPase action catalyzed by one of the proteins. Upon recognition of an imperfection, a nuclease activity is called into play that cleaves a phosphodiester bond by hydrolysis at a certain distance from the site of the lesion in that strand. DNA polymerase I removes the offending segment through its exonuclease activity; it then replaces the nucleotides through its polymerase activity, using the 3′-OH end of the repairing strand as the primer and the base sequence in the other strand as the template. The repair is finally completed by DNA ligase-catalyzed closure.

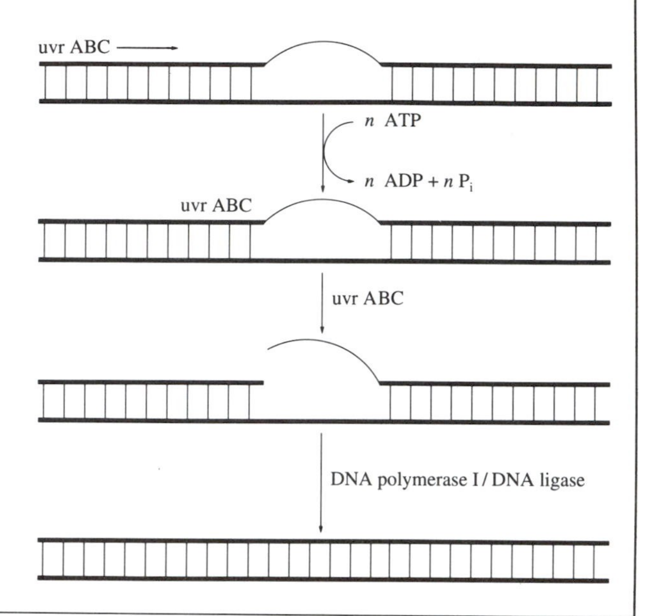

BOX 12-2

CHEMICAL SYNTHESIS OF OLIGODEOXYNUCLEOTIDES

Much research has been directed toward the discovery and development of chemistry for synthesizing oligodeoxynucleotides of defined nucleotide sequence. The importance of this research cannot be overemphasized, because the availability of such oligodeoxynucleotides is important for the cloning of genes and essential for the preparation of specific, single-site mutants of proteins (site-directed mutagenesis), which is described in Chapter 14. Two methods have been widely used: the phosphotriester method developed by H. G. Khorana and the phosphite triester method introduced by R. L. Letsinger. The first synthetic gene, the gene specifying the nucleotide sequence in phenylalanyl tRNA, was assembled by Khorana and his associates from oligodeoxynucleotides that they had synthesized by the phosphotriester method.

The most widely used method, that employed in most of the automated DNA synthesizers, is the solid-phase phosphite triester method described here. The solid-phase technology offers the same advantages to oligodeoxynucleotide synthesis as those described in Chapter 10 for polypeptide synthesis—namely, the operational simplicity with which chemical intermediates can be purified by filtration and washing of the solid particles. In oligodeoxynucleotide synthesis, the solid particles are silica based, either controlled-pore glass or silica particles.

Protective groups are important in oligodeoxynucleotide synthesis, as they are in polypeptide synthesis and most other chemical synthetic procedures involving polyfunctional molecules. Deprotection is, therefore, an essential part of the synthetic procedure, and much research in many laboratories has been required to develop suitable protection and deprotection strategies. In oligodeoxynucleotide synthesis, the points of chemical protection are the 5′-hydroxyl groups of the deoxynucleoside-3′-phosphoramidites added to the growing chain; one oxygen in the phosphoramidite group; and the exocyclic amino groups of the bases A, G, and C.

BOX 12-2 (continued)

CHEMICAL SYNTHESIS OF OLIGODEOXYNUCLEOTIDES

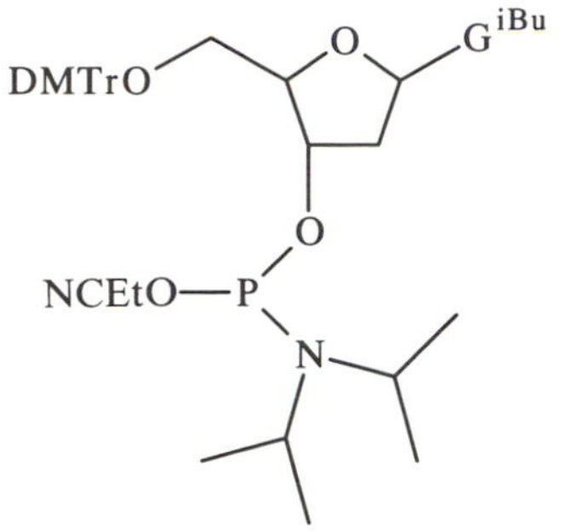

DMTrO, G^{iBu}, O, O, NCEtO—P, N

A protected deoxynucleoside 3′-phosphoramidite

The most widely used protective groups are shown in Figure A (below). They are 4,4′-dimethoxytriphenylmethyl (DMTr) for the 5′-hydroxyl groups, benzoyl for bases A and C, isobutyryl for base G, and β-cyanoethyl (NCEt) for the phosphoramidite group.

The silica particles are chemically treated to allow a 5′-protected deoxynucleoside to be attached through esterification of its 3′-hydroxyl group with an activated carboxyl group of a spacer molecule attached to the particle. The oligodeoxynucleotide is then built up from the immobilized deoxynucleoside using the chemistry shown in Figure B (on page 362). In step **1**, the 5′-hydroxyl group of the immobilized nucleoside is exposed by deprotection using dichloroacetic acid (DCA) in methylene chloride. In step **2**, the 5′-hydroxyl group reacts with a protected deoxynucleoside-3′-phosphoramidite in a tetrazole-catalyzed condensation to form a phosphite triester linkage. The phosphite triester is then oxidized in step **3** to a phosphate triester. The sequence is then iterated, beginning with deprotection of the 5′-hydroxyl group of the newly added deoxynucleotide.

The protective groups are selected in such a way that they can be removed at two stages in the synthesis. The DMTr groups are removed from the 5′ ends of the growing chains in each cycle of nucleotide addition by the use of dichloroacetic acid in methylene chloride, a procedure that does not interfere with the other protective groups. All other protective groups are removed and the completed oligonucleotide is removed from the solid support by reaction with NH_3. In the sequence shown in Figure B, the starting nucleoside is 5′-O-DMTr-protected benzoyl-A attached to the solid support through its 3′-oxygen and a succinamide spacer. The first step is detritylation, the second is coupling with the 5′-protected 3′-β-cyanoethyl diisopropylphosphoramidite of isobutyryl-G to form the internucleotide linkage to the phosphite triester, and the third step is oxidation to the phosphotriester. The sequence can be repeated with the incorporation of other internucleotide linkages as desired. The process can be terminated at any desired point by detritylation of the protected 5′-hydroxyl followed by treatment with ammonia to remove all other protective groups and cleave the oligodeoxynucleotide from the support, as illustrated in Figure B in the release of the dinucleotide GpA.

Figure A

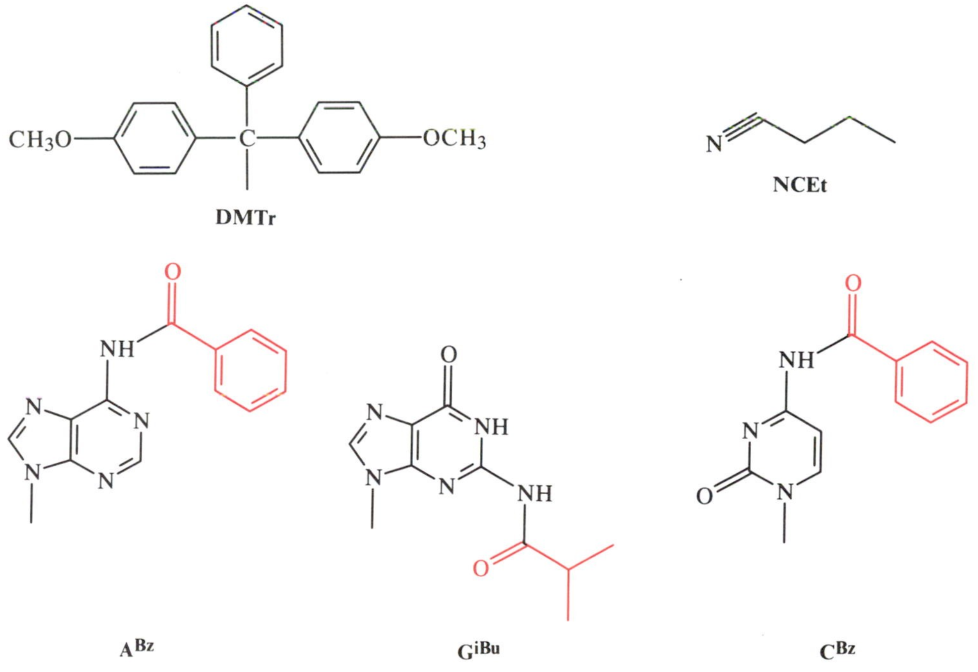

CH_3O, C, OCH_3

DMTr

N

NCEt

O, NH, N, N, N, N

A^{Bz}

O, N, NH, N, N, NH, O

G^{iBu}

O, NH, N, O, N

C^{Bz}

BOX 12-2 (continued)

CHEMICAL SYNTHESIS OF OLIGODEOXYNUCLEOTIDES

Figure B

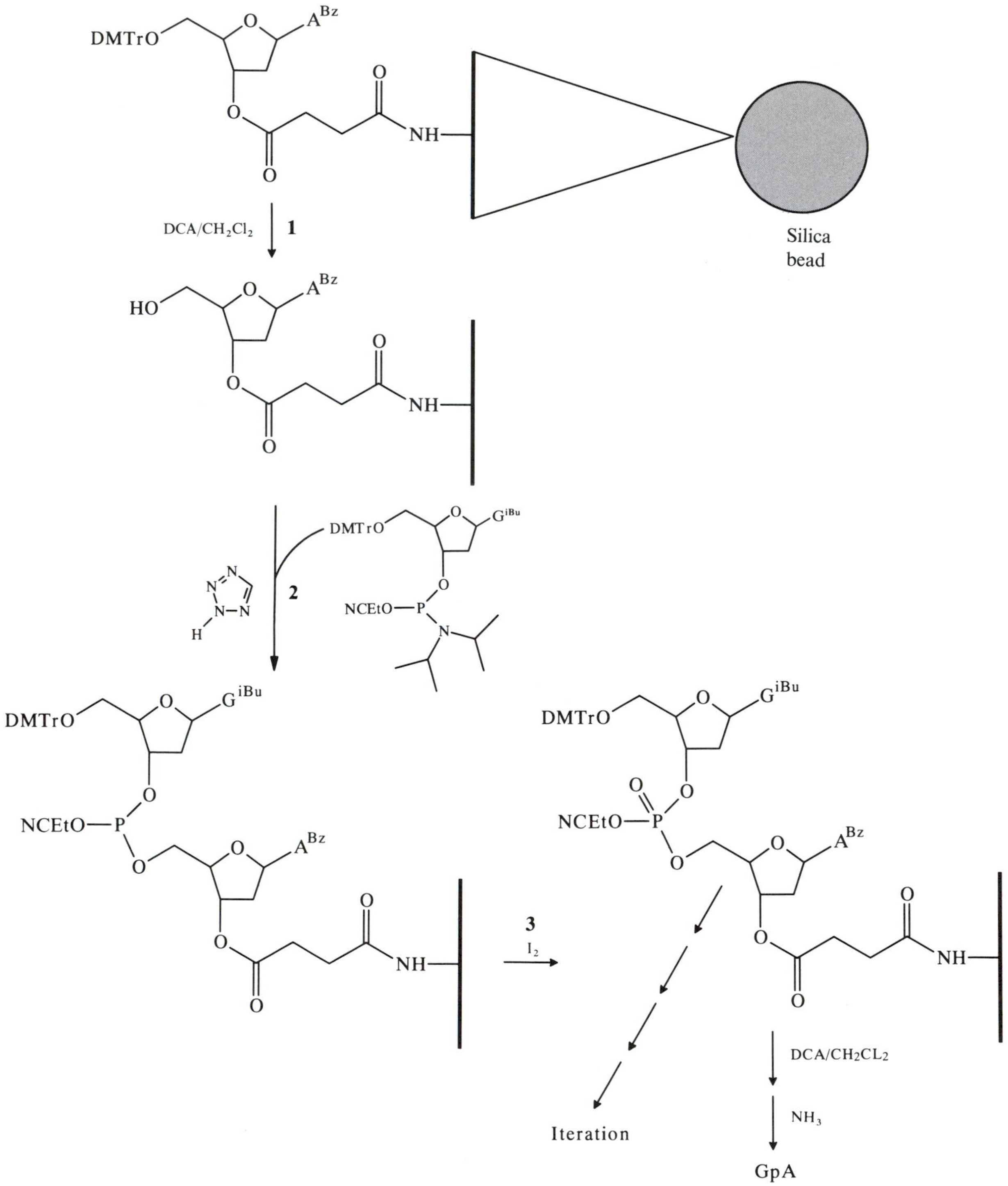

The phosphotriester method is similar in principle to the phosphite triester approach, except that the coupling chemistry involves only tetravalent phosphates and oxidation to the phosphate level is not required. Most of the protective strategies and many of the protective groups were originally worked out in the phosphotriester method.

Nucleotide Sequence Analysis

The nucleotide sequences of DNA in genes are determined by application of the same basic sequencing principles used for amino acid sequence analysis of proteins. These principles are:

1. Cleave the DNA into two or more sets of overlapping fragments.
2. Determine the nucleotide sequences of the fragments by controlled degradation and end-group analysis.
3. Align the fragments by the use of overlapping sequences.

In nucleotide sequencing, the defined fragments are produced by digesting the gene with restriction endonucleases. The gene fragments are then isolated by agarose gel electrophoresis, which separates fragments on the basis of size. The nucleotide sequences in the fragments can be determined by either of two procedures, the Maxam-Gilbert chemical procedure or the dideoxy method of Sanger.

Maxam-Gilbert Sequencing

In the Maxam-Gilbert procedure, a double-stranded DNA fragment from 100 to 200 base pairs in length is first treated with alkaline phosphatase to remove the 5′-phosphates from the ends of both strands; it is then again phosphorylated by the use of polynucleotide kinase and [γ-^{32}P]ATP. The ^{32}P-labeled strands are separated by denaturation and gel electrophoresis, and the ^{32}P-label is used for analytical purposes to detect DNA fragments in agarose electrophoretic gels throughout the sequencing procedure.

Each strand is treated chemically by the methods outlined in Figures 12-9 and 12-11 through 12-14 to cleave the chain selectively at certain sites. For example, cleavages at a *small percentage* of a random selection of A and G sites can be effected by limited treatment with dimethyl sulfate to methylate N-7 of G and N-3 of A at a few purine sites. Hydrolysis of the methylated bases creates apurinic sites at these positions, which undergo base-catalyzed elimination of the 3′-phosphate and cleavage of the diester linkage. Methylation conditions are selected that lead, on average, to cleavage at only one point in a given strand.

The A and G sites are distinguished by treating the methylated DNA in such a way as to cleave the strands selectively. The methylated DNA is divided into two aliquots, one of which is heated in neutral solution to hydrolyze both N^7-methylguanine and N^3-methyladenine, and the apurinic sites are cleaved by alkaline elimination to give the (A + G) fragments. The second sample of methylated DNA is hydrolyzed by treatment with dilute acid at ambient temperature, which removes only the N^3-methyladenine, and these apurinic sites are cleaved by alkaline elimination to give the A fragments.

The fragments from both strands are separated by agarose gel electrophoresis under conditions that allow oligonucleotides differing in length by one nucleotide to be separated. From the sizes of the oligonucleotides, the spacings between A and G sites in the original strand can be deduced. Comparison of the (A + G) fragments with the A fragments allows the positions of A and G in the sequence to be distinguished.

In separate experiments conducted at the same time, the DNA strand is treated with hydrazine to remove C and T at a few positions; it is then heated with piperidine to eliminate the 3′-phosphate and cleave the chain, giving (C + T) fragments. Depyrimidination can be made selective for C by hydrazinolysis in 2 M NaCl to produce C fragments. The two sets of fragments are separated by agarose gel electrophoresis and, from the spacings of the (C + T) fragments compared with the C fragments, the positions of C and T in the strand can be deduced.

The chemical cleavage conditions are selected to give an average of one cleavage point on a given strand, with the points of cleavage being determined by the method used: G and A in the (A + G) fragments, only A in the A fragments, and so forth. Therefore, each base in the strand can be located by the pattern of cleavages as analyzed by agarose gel electrophoresis. The migration distance for each fragment is determined by the length of the fragment, as measured by the number of nucleotide residues between the point of cleavage and the 5′-^{32}P end. The positions of the ^{32}P-labeled fragments on the electrophoretic gels are located by radioautography of the gels, which gives images of the radioactive bands on a photographic film. Figure 12-25 illustrates how the partial sequence ... GAA ... can be deduced from the relative migration positions of selected (A + G) and A fragments of an oligodeoxynucleotide. The same sequencing principle applies to the interpretation of the patterns created by (C + T) and C fragments. Figure 12-25 illustrates only four of the cleavage points and the corresponding fragments; however, in a real experiment, all possible fragments are produced, and all of the sequence can be deduced from their migration positions on the sequencing gel.

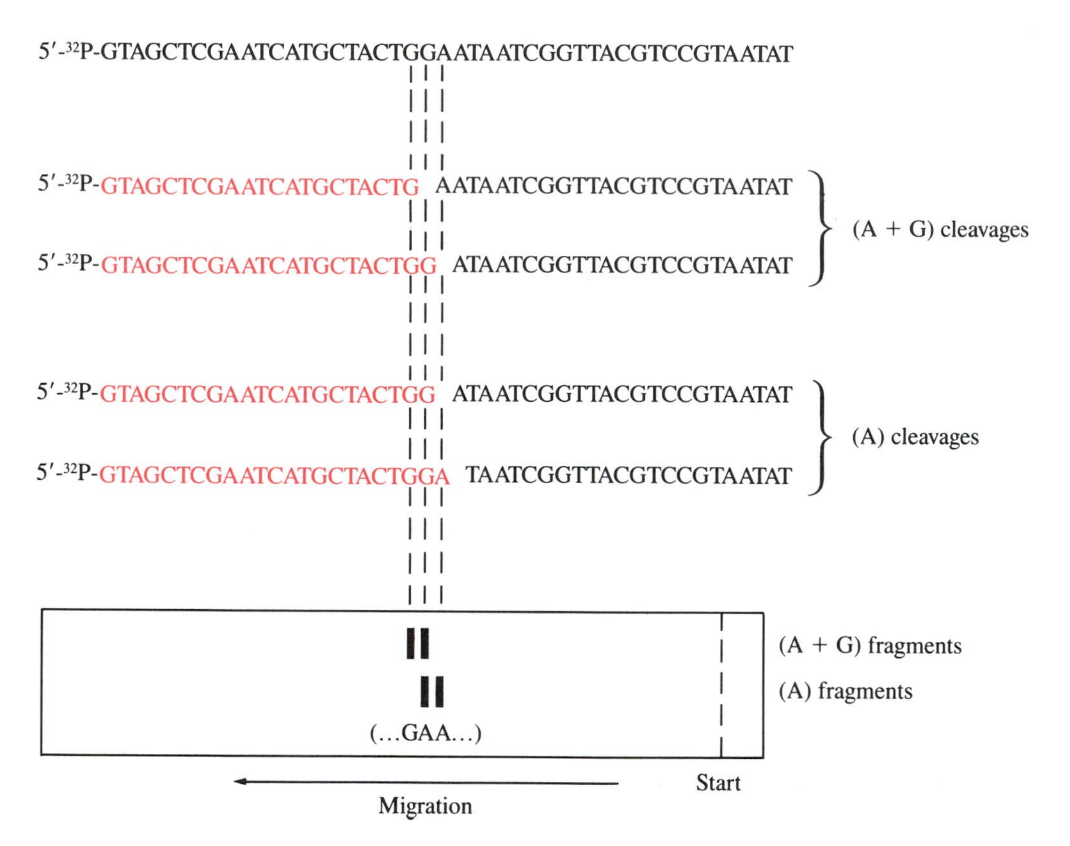

Figure 12-25

A Diagram of a Maxam-Gilbert Sequencing Gel. The chemistry by which the single-stranded DNA is cleaved into fragments is described in the text. The fragments are separated by agarose gel electrophoresis, which can separate fragments differing in length by one nucleotide. The 5′ ends of the fragments on the gel are labeled with ^{32}P, which can be detected by radioautography, and appear as bands on the gel. Nonradioactive bands from the 3′ ends are present but not detected. The small band pattern shown here for demonstration purposes corresponds to the sequence GAA. In a real experiment, many bands are present, and sequences of several hundred nucleotides can be read from the band pattern.

Sanger Dideoxy Sequencing

Several methods based on the partial replication of DNA by the use of DNA polymerase are available, using the principle introduced by Sanger. In these methods, oligodeoxynucleotides are produced by controlled DNA polymerase catalyzed synthesis on a template rather than by degradation. Various chain-termination methods are used to arrest DNA synthesis by DNA polymerase at known positions in the sequence—for example, at all A positions or all T positions, and so forth. Agarose gel electrophoresis is used to separate the resulting single-stranded oligodeoxynucleotides, which differ in length by one nucleotide or a few nucleotides.

The dideoxy method is the simplest of the various means by which the end groups are specified in partial replicative sequencing. A single-stranded DNA fragment is cloned into a specific site of a single-stranded DNA bacteriophage, usually an M13 bacteriophage, in which the phage nucleotide sequence is known downstream from the 3′ end of the cloned fragment. A synthetic oligodeoxynucleotide with a sequence complementary to that near the 3′ end of the cloned fragment is hybridized to the bacteriophage DNA. This serves as a primer for DNA polymerase I, as illustrated in Figure 12-26. A DNA polymerase reaction is then carried out by using the four dNTPs, one of which contains ^{32}P at the α-position, and the cloned fragment is the template for more than two hundred

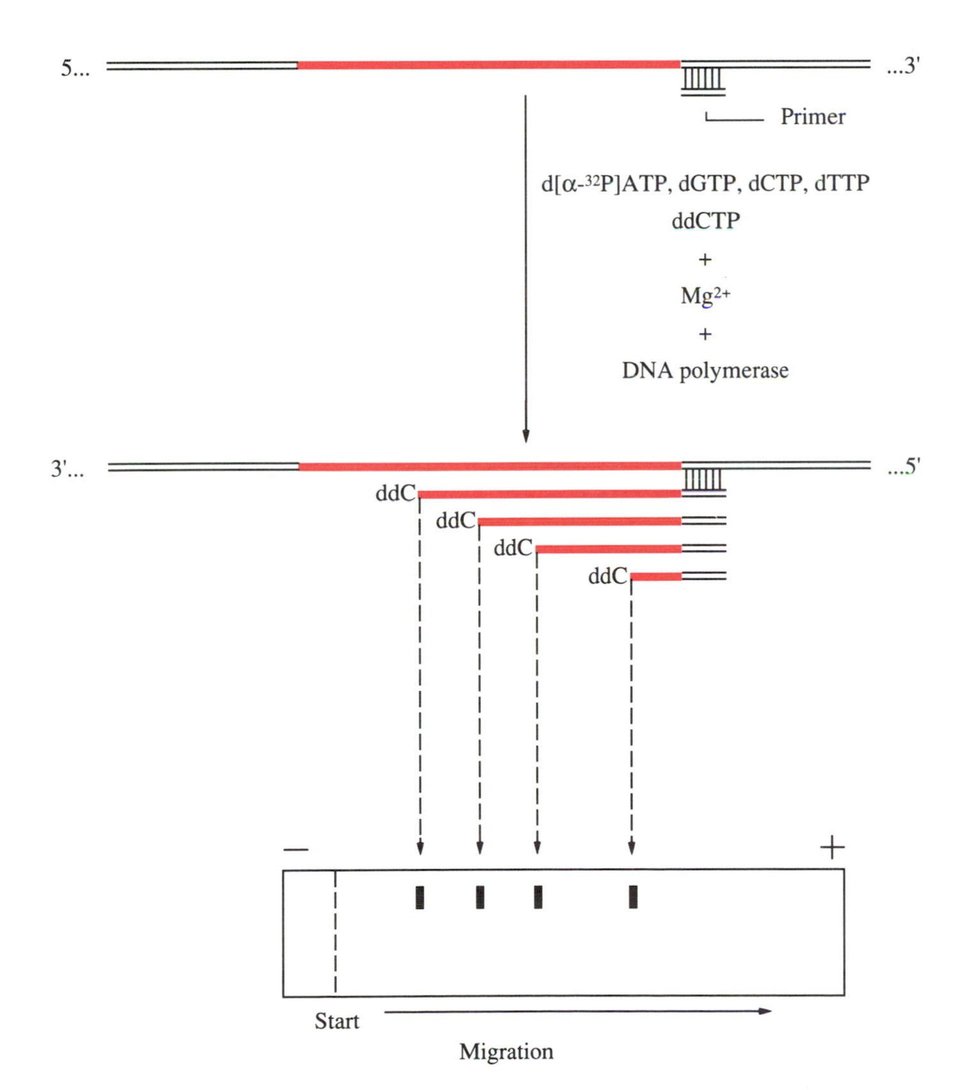

Figure 12-26

Sanger Dideoxy Sequencing of DNA. The method by which 2′,3′-ddCTP is used to produce DNA fragments in the limited replication sequencing method of Sanger is as follows. A single strand of DNA is cloned into a single-stranded vector such as bacteriophage M13, the nucleotide sequence of which is known. A short oligodeoxynucleotide having a sequence complementary with that of the vector just upstream from the 3′ end of the cloned fragment is hybridized to the vector and serves as the primer for a DNA polymerase reaction. Included in the polymerase reaction is a small amount of ddCTP along with all four dNTP molecules, one of which is labeled with ^{32}P in the α-position. Chain growth stops each time that ddC is incorporated, and this phenomenon leads to the production of a series of fragments of various lengths, *all of which terminate in ddC*. These terminations correspond to the positions of dG in the template. The fragments are separated by agarose gel electrophoresis, and the nucleotide sequence is read from the band pattern. An example of such a pattern is illustrated in Figure 12-27.

cycles. The polymerization is interrupted at selected sites by inclusion in the reaction mixture of a single 2′,3′-dideoxynucleoside 5′-triphosphate (ddNTP). The ddNTPs are substrates for DNA polymerase, but they arrest the polymerization reaction by generating 3′-deoxy ends, which cannot be elongated. Therefore, all oligonucleotides produced in the presence of ddCTP will terminate in ddC on the 3′ ends, as illustrated in Figure 12-26. When the oligonucleotides are separated, the differences in length will correspond to the spacing of the *complementary* base guanine in the template sequence. A separate experiment is conducted with each ddNTP, and the gel separations are run side by side. The oligonucleotide bands are detected by radioautography of the ^{32}P label in the newly synthesized oligonucleotides. The sequence of the template can be read directly from the gels when the products of four polymerization reactions are separated side by side on an electrophoretic gel. A typical gel pattern and the corresponding template nucleotide sequence are shown in Figure 12-27.

2′,3′-Dideoxynucleoside triphosphate

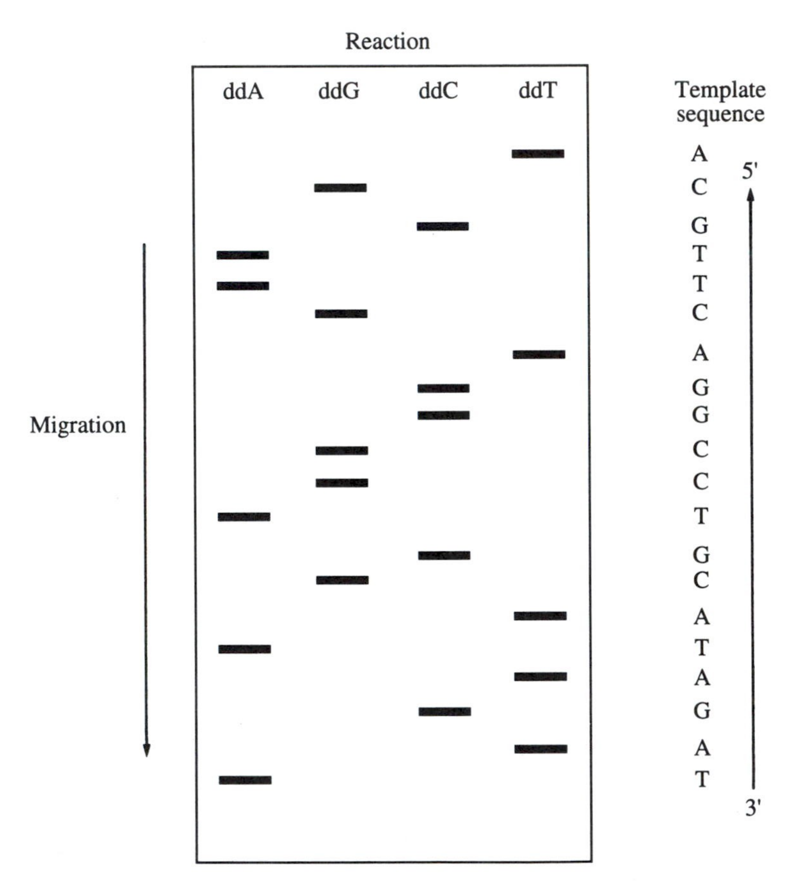

Figure 12-27

Correlation of a Fragment Pattern with Template Sequence in a Sanger Dideoxy Sequencing Gel.

Nucleotide Sequences in RNA

RNA cannot be sequenced directly by either the Maxam-Gilbert or the Sanger method. However, a fragment of RNA can be transcribed in reverse by the use of reverse transcriptase (Box 12-3) to produce a complementary fragment of DNA, which can then be subjected to deoxynucleotide sequencing by either the Maxam-Gilbert or the Sanger method. The RNA sequence is then deduced from the nucleotide sequence of the (+) strand of the DNA. The overall procedure is illustrated in Figure 12-28.

BOX 12-3

REVERSE TRANSCRIPTASE AND REPLICATION OF RETROVIRUSES

The genetic information in retroviruses is RNA rather than DNA (Chapter 1), and RNA cannot be replicated by the DNA replication complexes of cells. Moreover, RNA polymerase requires DNA as the template and cannot transcribe or replicate RNA. Therefore, the retroviruses have special replication systems that allow their genomes to be expressed and preserved. The genomic RNA of *bacterial* retroviruses, such as Qβ and MS2, acts as mRNA and directs the host protein biosynthetic machinery to produce the viral proteins. One of the Qβ-viral proteins (molecular weight = 55,000), the biosynthesis of which is directed by the viral genome, forms a complex with two of the bacterial protein biosynthetic components, EF-Tu and EF-Ts (Chapter 10); the complex is the enzyme Qβ replicase, which catalyzes the specific replication of the viral genome. With genomic viral RNA serving as the template, Qβ replicase catalyzes the 3′-to-5′ polymerization of nucleoside 5′-triphosphates to form a complementary (minus strand) polymer, which in turn serves as the template for Qβ replicase-catalyzed synthesis of the coding (plus strand) genomic RNA. Qβ replicase is, therefore, an RNA-dependent RNA polymerase.

Lysogenous retroviruses that infect eukaryotes, including the oncogenic viruses and the AIDS virus, have their genomes incorporated into the host DNA as a step in their life cycles. These viruses produce reverse transcriptase, which exhibits unique enzymatic activities allowing it to produce double-stranded DNA by using the viral genomic RNA as the template; the double-stranded DNA is subsequently integrated into the host chromosome. Reverse transcriptase exhibits three major enzymatic activities, which it utilizes to produce a double-stranded DNA in which the (+) strand contains the DNA version of the nucleotide sequence in the viral genomic RNA. These enzymatic activities correspond to an RNA-dependent DNA polymerase, which catalyzes the polymerization of the four species of dNTP using RNA as the template and primer; RNase H, a nuclease that acts specifically to degrade RNA in RNA-DNA hybrids; and single-strand DNA-dependent DNA polymerase.

A long terminal repeat sequence at each end of the viral genome serves a signal role in the process. A specific tRNA binds near this sequence at the 5′ end and serves as the primer for RNA-dependent DNA polymerase activity in step 1 of the illustration on page 368. In step 2, the RNase H activity removes the 5′ end of the genomic RNA (+) strand, which frees the 3′ end of the newly synthesized (−) strand of DNA to hybridize with the 5′ end of a second molecule of viral genomic RNA in step 3. Reverse transcriptase then elongates the DNA strands in both directions in steps 4 and 5, while the RNase continues to degrade the RNA portions of the DNA-RNA duplexes so produced. This leaves a short segment of the DNA version of the terminal repeat hybridized to the 3′ end of the newly synthesized (−) strand of DNA, and this segment rehybridizes to the same repeat sequence near the 5′ end in step 6, where it serves as the primer for a final round of DNA-dependent DNA polymerization in step 7 to produce a double-stranded DNA with the terminal repeats at both ends in both the (+) and (−) strands. The finished DNA is then integrated into the host chromosome and ultimately expressed in viral replication.

The terminal repeat sequences clearly play a crucial role in the production of the double-stranded DNA version of the viral genome. They are also thought to be equally important in the integration of the finished viral DNA into the host chromosome, which is as yet not fully understood.

BOX 12-3 (continued)

REVERSE TRANSCRIPTASE AND REPLICATION OF RETROVIRUSES

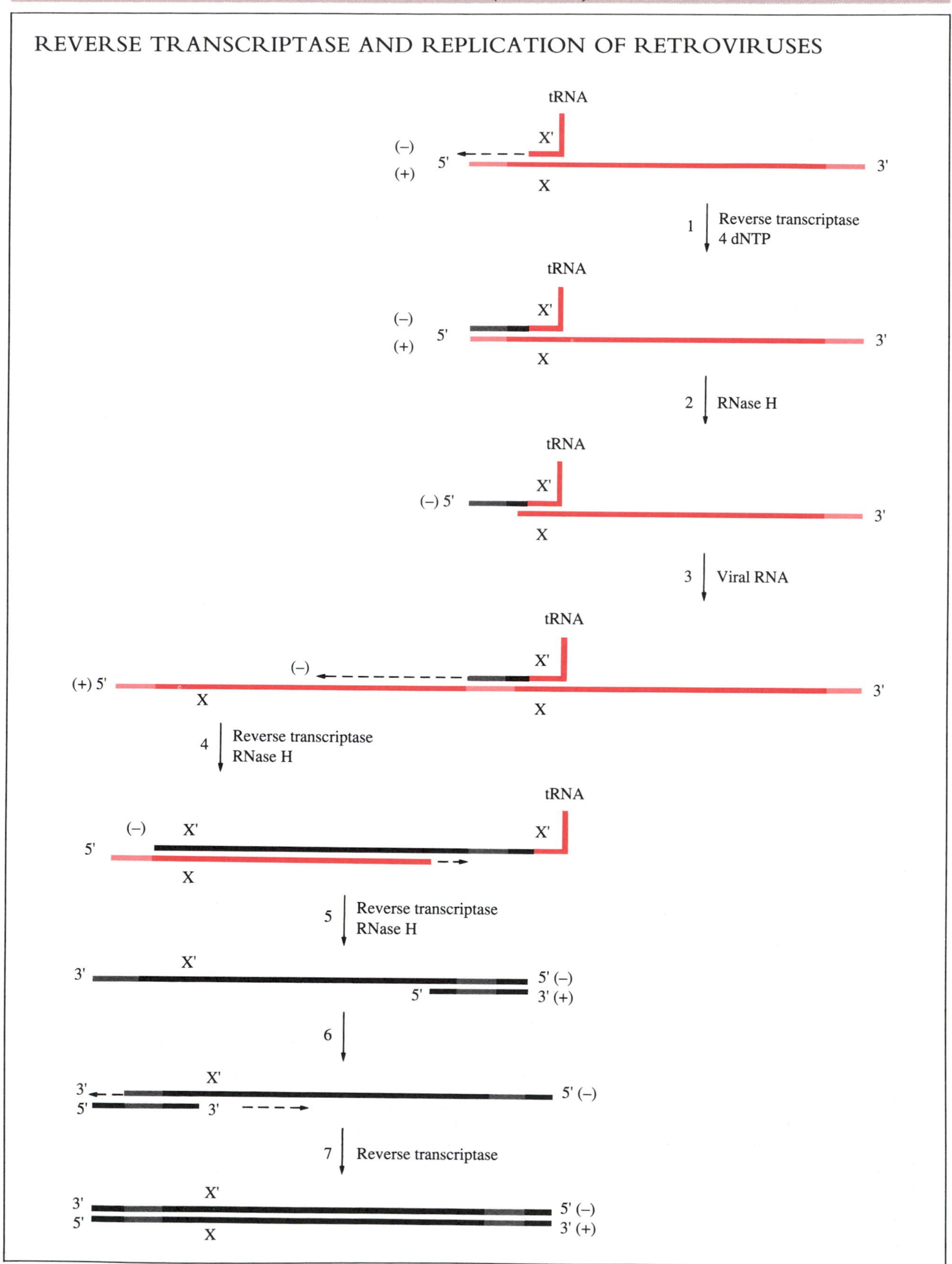

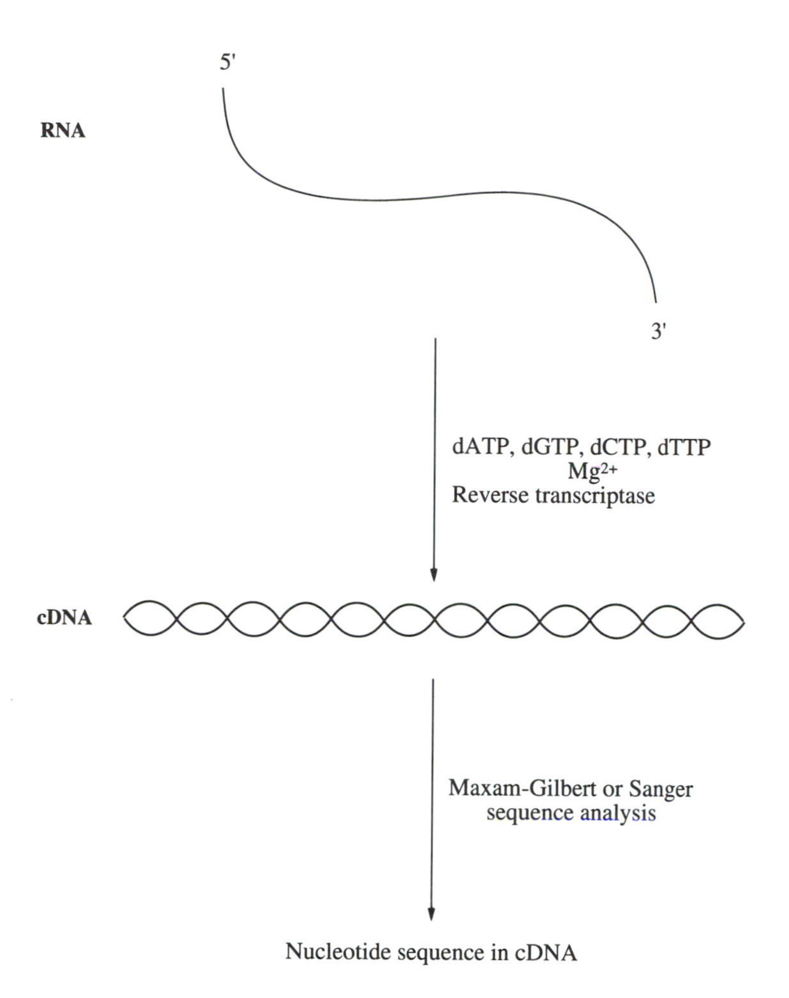

Figure 12-28

Nucleotide Sequences in RNA.

RNA Biosynthesis

RNA Polymerases

RNA polymerases differ fundamentally from DNA polymerases in their substrate specificities and the absence of a requirement for a primer, but the basic chemical reaction in RNA polymerization (equation 7) is essentially the same as that in DNA polymerization and is illustrated in Figure 12-19.

$$n\,(\text{ATP, CTP, GTP, UTP}) \xrightarrow[\text{Mg}^{2+}]{\text{DNA template}} \text{poly (ACGU)} + n\,\text{PP}_i \qquad (7)$$

RNA polymerase requires the Mg^{2+} complexes of the four ribonucleoside 5′-triphosphates as NMP donor substrates, as well as template DNA to specify the sequence of nucleotide incorporation. The action of this enzyme is termed transcription because the information in the template sequence is transcribed into the nucleotide sequence of RNA. Bacteria contain a single RNA polymerase that produces RNA transcripts for mRNA, rRNA and tRNA. Eukaryotic cells contain three RNA polymerases in the nucleus: one in the nucleolus that produces pre-RNA (precursor-RNA) for rRNA; another in the nucleoplasm that produces transcripts for mRNA; and a third, also in the nucleoplasm, that produces pre-tRNA. Mitochondria of eukaryotic cells contain a fourth RNA polymerase to transcribe the mitochondrial DNA.

RNA polymerases are multiprotein complexes whose subunit composition varies with the state of the transcription process. The RNA polymerase of *E. coli* consists of a four-subunit core enzyme that interacts with another type of subunit. The core enzyme contains three types of subunits, two subunits designated α of uncertain function, a subunit designated β that is the catalytic subunit, and a subunit designated β' that binds template DNA. A separate subunit designated σ combines reversibly with the core enzyme and recognizes sequences similar to 3′-TATAAA-5′ and guides the core enzyme to the *promoters,* the sites at which RNA polymerase binds to DNA. (TATAAA is a *consensus* or average of the sequences actually found and will normally differ by no more than one base from an actual sequence.) The molecular weight of the RNA polymerase from *E. coli* ($\alpha_2\beta\beta'\sigma$) is 450,000.

Once bound to a promoter, RNA polymerase is self-priming and begins polymerization at a template start site on the 3′ side of the sequence TATAAA by binding either ATP or GTP. The enzyme utilizes the 3′-OH group of ATP or GTP as the primer for the first polymerization cycle. Therefore, the 5′ end of the new RNA chain is a triphosphate originating with ATP or GTP as shown in equation 8, in which the identities of X and subsequent nucleotides are specified by the template DNA.

$$\text{pppA} + \text{XTP} \xrightarrow{\text{Mg}^{2+}} \text{pppApX} + \text{PP}_i \qquad \text{X} = \text{A, C, G, U} \tag{8}$$

In prokaryotes, the mRNA produced in this process is used in translation to specify amino acid sequences in proteins (Chapter 10).

RNA Processing

Most RNA transcripts produced by RNA polymerases are not in their final forms as mRNA, rRNA, or tRNA but undergo a variety of structural alterations on the way to becoming mature species capable of carrying out their cellular functions. Only mRNA of prokaryotes is in its mature form as an initial transcript. Many of the changes that initial RNA transcripts undergo are nucleotidyl transfers, but some are alterations in the structures of the heterocyclic bases. RNA processing is currently an active and rapidly advancing field of research.

An example of RNA processing is the production of rRNA. The initial transcripts for ribosomal RNA in both prokaryotes and eukaryotes are very long and carry the sequences for both the large and the small ribosomal subunits. These pre-RNA molecules are cleaved by specific nucleases, acting in several stages, to the final mature species of rRNA. Much remains to be learned about RNA processing. A few patterns are known and some specific systems are described in the following sections.

Transfer RNA

The structure of tRNA is described in Chapter 10. All species of tRNA have the sequence . . . CCA-3′-OH at the 3′ end, which accepts the aminoacyl groups in protein biosynthesis and maintains their activated states until they are transferred to the growing peptide chains. All have similar but nonidentical secondary structures, though varying significantly in overall nucleotide sequences. All have unusual heterocyclic bases that have been generated by modification of the bases originally incorporated into the initial RNA transcript, and the initial transcripts are always longer than the mature structures. The three-dimensional structures are also similar, but not identical, and the anticodon sequences are always found in the anticodon loops of mature tRNA.

The genes specifying the various species of tRNA are much longer than the finished tRNA molecules. In *E. coli,* the initial transcripts are cleaved from both ends by specific endonucleases to approximately the length of the final tRNA (about 70–90 nucleotides). A few additional nucleotides are removed from the 3′ end by exonucleases. In some cases, this process stops at the CCA end but, in others, part of this sequence is also removed and then restored by the action of a terminal nucleotidyltransferase utilizing ATP and CTP. This enzyme has the important function of ensuring that all species of tRNA have the correct 3′-terminal sequence.

The unusual heterocyclic bases in tRNA are introduced by enzymatic modification of bases in the nuclease-trimmed tRNA. The modified bases include N^6-isopentenyladenosine, 5-methylcytosine, N^6-dimethyladenosine, 5,6-dihydrouridine, and pseudouridine (Figures 10-7 and 10-9), as well as others.

Capping and Polyadenylylation of Eukaryotic mRNA

Eukaryotic mRNA is further processed in a variety of ways, many of which entail NMP transfer. Two of these are 5′ capping and 3′ polyadenylylation. In 5′ capping, the 5′-triphosphate end is modified shortly after the initiation of transcription. The process begins with the removal of the terminal phosphate by phosphatase action (equation 9).

$$\text{pppApXpY}\ldots\text{pZ} + \text{H}_2\text{O} \xrightarrow{\text{Mg}^{2+}} \text{ppApXpY}\ldots\text{pZ} + \text{P}_i \tag{9}$$

The capping enzyme then catalyzes GMP transfer from GTP to the diphosphate end (equations 10a and 10b).

$$\text{E-Lys} + \text{GTP} \longrightarrow \text{E-Lys-GMP} + \text{PP}_i \tag{10a}$$

$$\text{E-Lys-GMP} + \text{ppApXpY}\ldots\text{pZ} \longrightarrow \text{GpppApXpY}\ldots\text{pZ} + \text{E-Lys} \tag{10b}$$

The capping reaction proceeds in two steps with a nucleotidyl-enzyme as a covalent intermediate, in which the GMP group is first transferred to the ε-amino group of a lysine residue. In the second step, the 5′-pyrophosphoryl group of mRNA accepts the GMP group from the GMP-enzyme to form capped mRNA. The guanine ring is subsequently methylated by *S*-adenosylmethionine on N-7 and later on the 2′-OH of the guanosyl ribose.

In polyadenylylation, the 3′-OH end of mRNA undergoes a series of sequential additions of AMP that add from 50 to 100 adenylyl units to the 3′ end according to equation 11.

$$\text{m}^7\text{GpppApXpY}\ldots\text{pZ} + n\ \text{ATP} \longrightarrow \text{m}^7\text{GpppApXpY}\ldots\text{pZ(pA)}_n + n\ \text{PP}_i \tag{11}$$

Capping and polyadenylylation protect eukaryotic mRNA against metabolic degradation; they also introduce structural signals that facilitate the transport of mature mRNA from the nucleus to the endoplasmic reticulum, where it directs protein biosynthesis.

RNA Splicing

The sizes of many species of RNA transcripts are reduced by *splicing* out *intervening sequences* that are absent from the mature forms. The functions and significance of the intervening sequences are generally not well understood. All the splicing mechanisms include nucleotidyl transfer reactions, some of which proceed without catalysis by an enzyme.

Eukaryotic pre-mRNA contains many noncoding *introns* (intervening sequences) separating the *exons* (Figure 12-29). The introns are spliced out in the

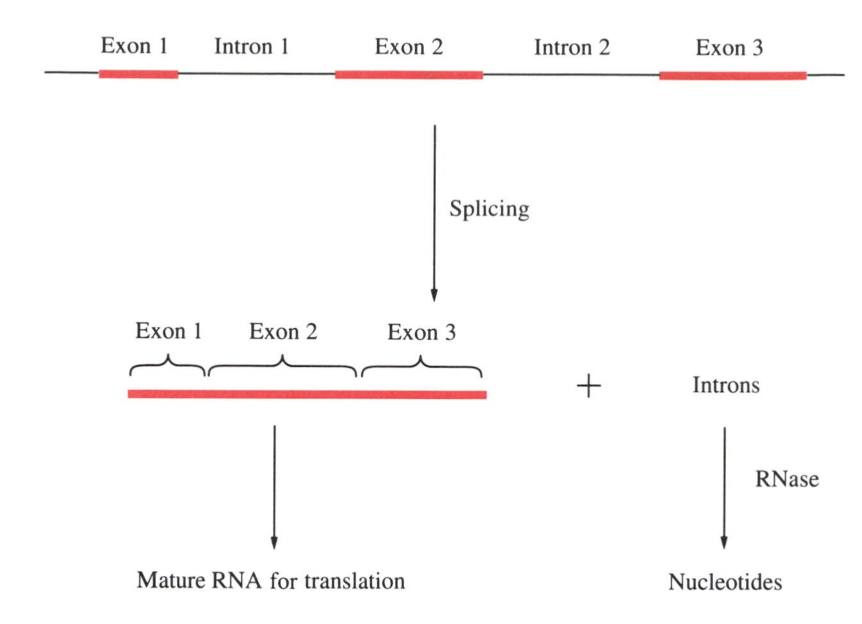

Figure 12-29

Exons and Introns. Exons are the coding regions and introns the noncoding regions of mRNA. The introns are spliced out with joining of the exons before translation of the sequence into a protein.

most-important steps of mRNA processing. Splicing is catalyzed by complexes of snRNA (small nuclear RNA) and proteins (40S–60S) known as *spliceosomes.* The splicing mechanism entails the two nucleotidyl transfer steps shown in Figure 12-30, in which the 2′-OH group of an AMP-residue near the 3′ end of an intron initially attacks the phosphodiester linkage at the 5′ end of the intron, cleaving the bond to form a *lariat* of the intron and a 3′-OH at the end of the upstream exon. This 3′-OH group then attacks the phosphodiester linking the downstream exon to the intron, cleaving out the intron and joining the two exons. The beauty of this scheme is that both the nucleotidyl transfers are isoenergetic, so that no energy input is required. The process is iterated until all of the exons have been joined and the introns excised.

Some splicing reactions require energy to remove intervening sequences. This is the case in the processing of some species of eukaryotic tRNA, in which the intervening sequences are removed hydrolytically by the actions of endonucleases. The resulting 5′-OH and 3′-phosphate ends of the tRNA are then rejoined by the action of *RNA ligase,* which uses ATP to drive the rejoining reaction. RNA ligase acts by the same chemical mechanism as DNA ligase (Figure 12-22), but it differs in that the "nick" is really a cleavage point in single-stranded RNA, the two segments being held together for sealing by secondary structure in the RNA, as illustrated in Figure 12-30. Thus, the secondary structure of RNA ensures that the correct ends are brought together in the ligation. If the chains were allowed to become separated, fragments from several species of processed RNA could become randomized during splicing.

T. Cech first discovered that some species of pre-RNA can undergo *self-splicing* to remove intervening sequences. The process is illustrated in Figure 12-31. In pre-RNA of *Tetrahymena,* the 3′-OH of GMP attacks the phosphodiester linkage at the 5′ end of the intervening sequence and cleaves it to

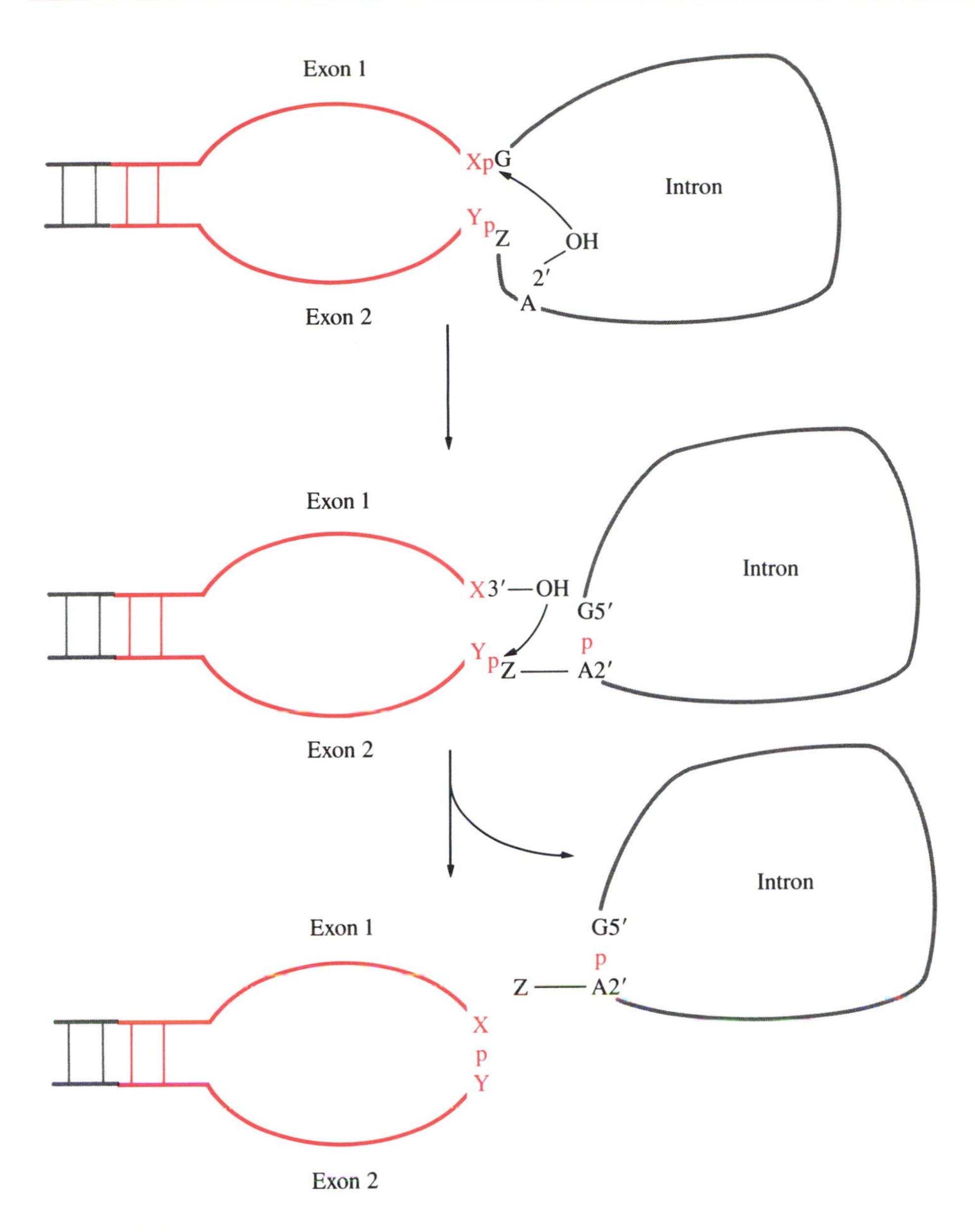

Figure 12-30

Spliceosome-catalyzed Excision of Introns. The 2′-hydroxyl group of an A residue near the 3′ end of an intron attacks the phosphodiester bond at the splice site and cleaves it, forming a 2′,5′-phosphodiester linkage with the G residue at the 5′ end of the intron. The resulting 3′-hydroxyl group of exon 1 attacks the phosphodiester linkage between residues at the junction of exon 2 and the intron, cleaving it to form a 3′,5′-phosphodiester linkage between exons 1 and 2. The intron is released as the lariat structure shown, in which the 3′ end is residue Z and the 5′ end is in a 2′,5′-phosphodiester linkage with an A residue near the end of the intron.

generate a 3′-OH end on the splicing fragment and a 5′-phosphate end on the intervening sequence. The newly created 3′-OH then attacks the phosphodiester linking the 3′ end of the intervening sequence to the 5′ end of the other splicing segment, cleaving out the intervening sequence and sealing together the two splicing segments. Again, this scheme consists of two essentially isoenergetic

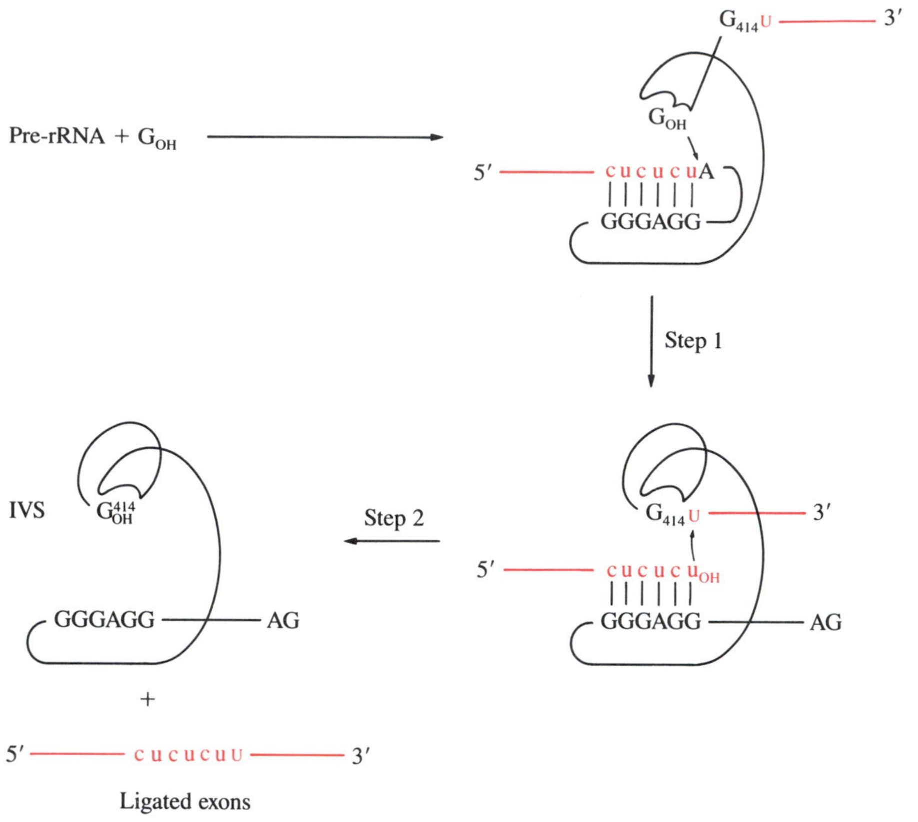

Figure 12-31

Self-splicing RNA. Self-splicing RNA adopts a reactive tertiary structure determined by its nucleotide sequence. The structure is illustrated here with bold lines and lowercase letters designating the exon. The uppercase letters designate residues in the intervening sequence (IVS) that is spliced out. This structure binds guanosine or GTP (G_{OH}) and catalyzes the reaction of its 3′-hydroxyl group with a phosphodiester linkage, UpA, cleaving it to form a new 3′,5′-linkage between G and the A residue. The resulting 3′-hydroxyl group (U-3′-OH) attacks a second phosphodiester linkage between G_{414}pU, forming a new phosphodiester to the U residue and cleaving away the IVS. The excised IVS has G_{414} at the 5′ end and G_{OH} at the 3′ end. [Adapted with permission from D. Herschlag and T. Cech. *Biochemistry* 29 (1990): 10159–10171, Figure 1A. Copyright 1990 American Chemical Society.]

steps and requires no external energy source. The pre-RNA undoubtedly exists in a specific tertiary structure that binds GMP and facilitates the nucleotidyl transfer reactions.

Summary

DNA is the repository of genetic information and its carrier from one generation to the next, and RNA mediates the expression of the genetic information in DNA. DNA is double stranded and helical and can be reversibly denatured by heating slowly to temperatures higher than T_m, the temperature at which sharp melting occurs. RNA is single stranded, but it also contains segments of helical structure, which result from the chain turning back on itself and undergoing regional hybridization where quasicomplementarity exists between two segments of the strand.

The nucleotide sequences of genes in DNA can be determined by the use of restriction endonucleases to produce overlapping fragments, followed by sequence analysis of the fragments, using either the Maxam-Gilbert chemical sequencing method or one of the Sanger partial replicative methods such as the dideoxy method.

DNA replication takes place in the course of cell division; it is semiconservative; and it is catalyzed by the coordinated actions of DNA polymerases, DNA primase, DNA ligase, helicases, topoisomerases, and single-strand-binding proteins. The transcription of nucleotide sequences of genes within DNA into RNA is catalyzed by RNA polymerases. The mRNA transcripts in prokaryotes are directly translated into protein sequences. The pre-mRNA in the nuclei of eukaryotes is further processed before being used in translations. mRNA processing consists of 5′ capping, 3′ polyadenylylation, and splicing out the introns and fusing the exons to make mRNA that can be translated in protein biosynthesis. The initial RNA transcripts for tRNA and rRNA undergo further processing in both prokaryotes and eukaryotes. This processing consists of trimming ends and splicing out intervening sequences, as well as base modification in tRNA.

ADDITIONAL READING

KORNBERG, A., and BAKER, D. 1991. *DNA Replication,* 2d ed. New York: Freeman.

WELLS, R. D., GOODMAN, T. C., HILLEN, W., HORN, G. T., KLEIN R. D., LARSON, J. E., MÜLLER, U. R., NEUENDORF, S. K., PANAYOTATOS, N., and STURDIVANT, S. M. 1980. *Progress in Nucleic Acid Research and Molecular Biology* 24: 167–267.

SUNDARALINGAM, M. 1974. In *Structure and Conformation of Nucleic Acids and Protein-Nucleic Acid Interactions,* ed. M. Sundaralingam and S. T. Rao. Baltimore: University Park Press, pp. 487–524.

DICKERSON, R. E., DREW, H. R., CONNER, B. N., WING, R. M., FRATINI, A. V., and KOPKA, M. L. 1982. *Science (Washington, D.C.)* 216: 475.

SAENGER, W. 1984. *Principles of Nucleic Acid Structure.* New York: Springer-Verlag.

FREIFELDER, D. 1987. *Molecular Biology,* 2d ed. Boston: Jones and Bartlett Publishers.

PROBLEMS

1. Compare the structure of DNA with that of proteins. What structural principles are similar for the two classes of molecules? Which structural motifs in the two classes of molecules are analogous and which are not? Compare the structural dynamics for proteins and DNA.

2. Compare the structures of proteins and RNA as in question 1; that is, in terms of structural principles, motifs, and dynamics.

3. The nucleotide unit in a nucleic acid chain contains segments that differ in internal degrees of freedom. Identify these segments and explain how they differ with respect to internal degrees of freedom.

4. The chemical transformations in the Maxam-Gilbert and Sanger dideoxy nucleotide sequencing methods differ in many ways. The sequencing principles are similar in the two methods. What are the underlying sequencing principles, and by what means are these principles applied in the two methods?

5. What proteins are known to be required for the replication of DNA? What is the essential function of each protein in the replication process?

6. Why is DNA a better molecule for storing genetic information than RNA?

7. The transcription of DNA into RNA is catalyzed by RNA polymerase. What are the essential steps in the process by which RNA polymerase transcribes a particular segment of DNA into a complementary segment of RNA?

8. How might 5′ capping and 3′ polyadenylylation protect eukaryotic mRNA from enzymatic degradation during transit from the nucleus to the cytoplasm?

9. A-DNA, with eleven base pairs per helical turn, is more compact than B-DNA, which has ten base pairs per turn. Why is A-DNA more compact along the helix axis than B-DNA, and why are the two forms almost the same diameter?

10. What are the functions of nucleases in cells?

13

Gene Regulation

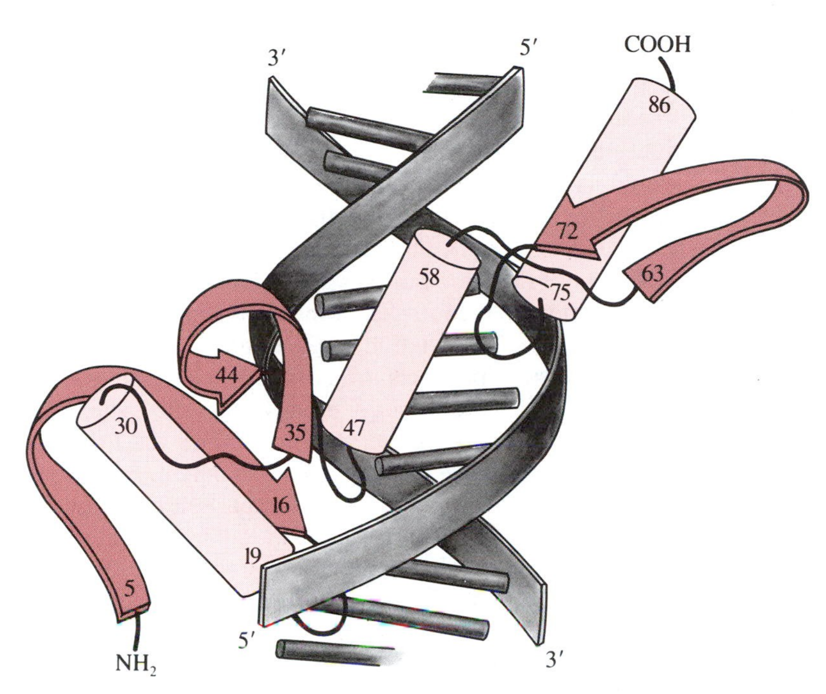

A diagram of the zinc finger protein Zif 268 bound to a segment of DNA. Each zinc finger consists of a segment of α-helix (cylinder) and a segment of β-sheet (arrow) held together by coordination to a Zn^{2+} ion. [Adapted from Figure 2 in N. P. Pavletich and C. O. Pabo, Science (Washington, D.C.) 252, 809 (1991). Copyright 1991 by the American Association for the Advancement of Science.]

Gene expression is the display of genetic traits in cells or organisms. Genes are segments of DNA that express traits by specifying the structures of proteins and nucleic acids and regulating the timing of their biosynthesis. All genes specify the structures of macromolecules, some of which regulate the expression of other, related genes. The genes in DNA are interspersed with other, relatively short nucleotide sequences known as *regulatory elements* that play important roles in controlling gene expression. Genes that specify the amino acid sequences of proteins are examples of structural genes whose expression in protein biosynthesis was delineated in Chapter 10. Other structural genes specify the nucleotide sequences of pre-rRNA and pre-tRNA, the production and posttranscriptional processing of which were outlined in Chapter 12.

Although protein and nucleic acid biosyntheses are understood in general, much remains to be learned, and even less is known about the mechanisms by which the expression of genes is regulated. Regulation is necessary because uncontrolled expression would lead to the production of unneeded macromolecules in even the simplest cell. Moreover, uncontrolled gene expression would defeat the differentiation of cells in the development of a complex organism, because all of its cells contain the same genetic information in the DNA but express it differentially. For example, the muscle and liver cells of rabbits have the same genes in DNA, but they have very different morphologies and functions. Uncontrolled gene expression in a rabbit embryo would not allow cellular differentiation in the course of development.

Some gene products are needed all of the time—for example, those that are responsible for the energy needs of the cell. Others are required only under certain conditions or at certain times. The rule, therefore, is that *a gene is expressed in the production of its product only when that product is needed by the cell.* The mechanisms by which this important outcome is achieved are complex and varied and only partly understood. However, the most-basic principles underlying gene regulation are known for prokaryotes, and these systems are a major focus of this chapter. In principle, the expression of a gene could be regulated at any or all points along the route to the production of the macromolecule. In fact, the most-efficient regulatory point is at the level of gene transcription into RNA, and most of the regulatory mechanisms function to control transcription. However, the rate of mRNA translation is also significant in regulating protein biosynthesis. Systems for eukaryotic gene regulation are much more complex, although some of the same principles apply.

Regulation of Prokaryotic Genes

Certain enzymes are needed at all times and under all conditions in bacteria and are produced *constitutively;* that is, in constant amounts at all times. They are constituents of the cells. Other enzymes are needed only under certain metabolic conditions, and their synthesis is induced under these conditions. Transcription of the genes specifying the structure of these enzymes is the principal point at which regulation occurs.

Metabolically related bacterial genes often appear in the genome as clusters or *operons* of genes and regulatory elements. The structural genes in operons are expressed together. In this section, we describe the structures and regulation of three of the best-known bacterial operons.

Regulation of the Lactose Operon

The Structure of the Lactose Operon. The first gene regulation system to be successfully analyzed at the molecular level, by F. Jacob and J. Monod, is that controlling the uptake and breakdown of lactose in *E. coli.* A cluster, or *operon,* of genes and regulatory elements named the *lac* operon consists of structural genes and two regulatory elements. A separate, regulatory gene controls the expression of the structural genes in the operon. This is the most thoroughly studied and best understood of all the gene-regulation systems, and elements of this system are widely used in recombinant DNA technology. Bacteria contain many other operons consisting of related genes and regulatory elements. All have certain regulatory features in common, and the *lac* operon exemplifies the most-general features; however, most other operons exhibit some additional features that promote the unique cellular functions of the enzymes produced. The galactose and tryptophan operons, discussed in later sections, are examples of this diversity.

E. coli can utilize many compounds as sources of carbon and energy for growth, one of which is lactose, a disaccharide consisting of glucose and galactose linked by a glycosidic bond. Lactose is a β-galactoside, because the glycosidic linkage between carbon-1 of the galactosyl moiety and carbon-4 of glucose has the β-equatorial configuration (Figure 13-1). The opposite, α-axial configuration is found in other glycosides (Chapter 15).

To utilize lactose for growth, *E. coli* must first absorb it from the growth medium and then break it down into simpler molecules that can be metabolized to produce energy and the carbon-containing building blocks for the biosynthesis of all the molecules in the cell. The *lac* operon contains the structural genes for the enzymes required to absorb lactose and break it down into glucose and galactose, which are in turn further metabolized by other systems.

When lactose is absent from the growth medium, there is little need for the lactose-processing enzymes, and the structural genes specifying these proteins are transcribed only in small amounts. The transcription of these genes is controlled by a regulatory gene and by two short regulatory elements in the *lac* operon, the *operator* and the *promoter*. The regulatory gene and regulatory elements act as an on-off switch in controlling the expression of the structural genes for lactose utilization.

An illustration of the alignment of structural genes and regulatory elements in the *lac* operon of *E. coli* is given in Figure 13-2. The segments in color are the regulatory gene, *lac I*, and the regulatory elements *lac P*, the

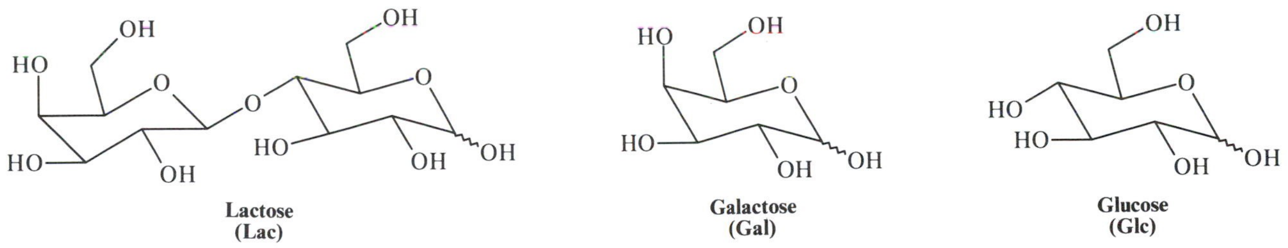

Lactose (Lac)

Galactose (Gal)

Glucose (Glc)

Figure 13-1

Structures of Lactose, Galactose, and Glucose. Lactose is the β-glycoside formed by linking C-1 of galactose to the hydroxyl group on C-4 of glucose. The general name *glycoside* refers to compounds in which an alkyl substituent is attached to the anomeric oxygen (see below) of the cyclic form of a sugar. In lactose, C-4 of glucose is attached to the anomeric oxygen at C-1 of galactose, and it is a *galactoside*. It is known as a β-galactoside because the configuration of the glycoside linkage is equatorial with the galactosyl group in the conformation shown here. Cyclic forms of D-galactose and D-glucose are also shown. The wavy bonds linking C-1 of the cyclic sugars to the hydroxyl group refer to the fact that there are two cyclic forms differing in configuration at C-1: the β-form with the hydroxyl group equatorial and the α-form with the hydroxyl group axial. These isomers are known as anomers, and the exocyclic oxygen is the anomeric oxygen.

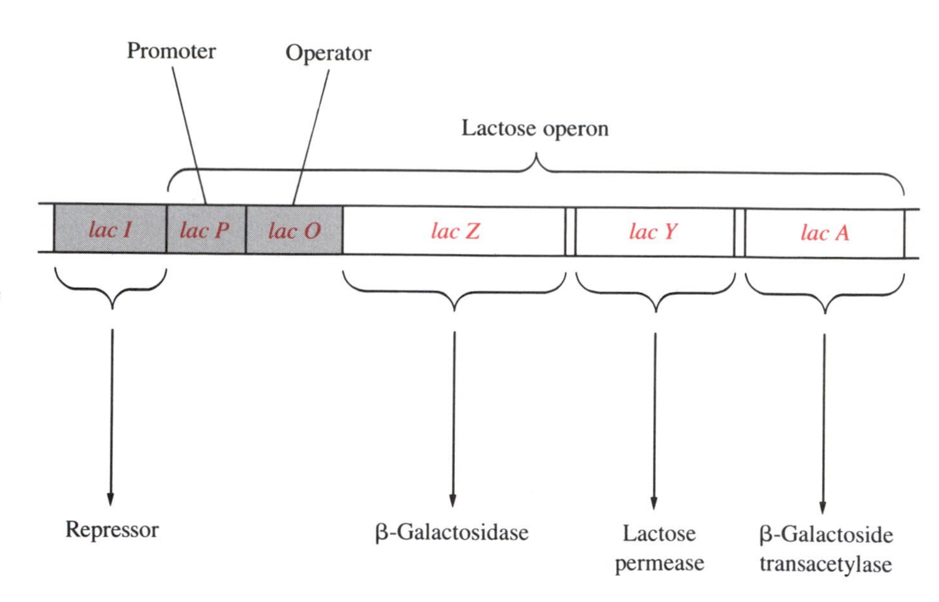

Figure 13-2

The Structure of the Lactose Operon in *E. coli.* The *lac* operon consists of three structural genes and two regulatory elements. The structural genes are *lac Z, Y,* and *A,* which code for the enzymes *β*-galactosidase, lactose permease, and *β*-galactoside transacetylase, respectively. The regulatory gene *lac I* codes for the repressor, a protein that inhibits the transcription of the structural genes by binding to the regulatory element *lac O,* the operator. The other regulatory element, *lac P,* is the site at which RNA polymerase binds in order to transcribe the structural genes for protein biosynthesis.

promoter (Chapter 12), and *lac O,* the operator, which control the expression of the structural genes *lac Z, lac Y,* and *lac A. Lac Z* specifies the amino acid sequence of *β*-galactosidase, which catalyzes the hydrolysis of lactose to galactose and glucose; *lac Y* specifies lactose permease, which facilitates the transport of lactose from the medium into the *E. coli;* and *lac A* specifies the amino acid sequence of *β*-galactoside transacetylase, which participates in the metabolism of *β*-galactosides other than lactose. *β*-Galactosidase and *β*-galactoside transacetylase are located in the cytosol; however, lactose permease is located in the membrane, where it binds and transports lactose from the growth medium into the cytosol.

Negative Regulation. *Lac I,* which is not in the *lac* operon, codes for the regulatory protein Lac repressor, which operates the on-off switch for the transcription of the structural genes. The Lac repressor is a tetrameric protein that binds very tightly to the operator segment of the operon and blocks the transcription of the structural genes by RNA polymerase, which must bind to the promoter. Because the operator is located between the promoter and the structural genes, the repressor bound to the operator is a barrier to transcription. Thus, whenever the repressor binds to the operator there is almost no transcription of the structural genes and almost no biosynthesis of the enzymes required for the absorption and metabolism of lactose. This constitutes negative control because, in its normal state, the *lac* operon is switched off by the presence of the repressor, and it is turned on only when *allo*-lactose or another inducer is present.

Very low levels of *β*-galactosidase, permease, and transacetylase are produced in the repressed state of the *lac* operon, only one or a few molecules of each per cell. These proteins cannot support the metabolism of lactose, but we shall see that they too are crucial for the regulation of the operon.

When lactose is present in the growth medium of *E. coli, and is needed for growth,* the few molecules of permease in the membrane bind and transport a small amount into the cell. Most of this lactose is hydrolyzed to glucose and galactose by the action of *β*-galactosidase. However, *β*-galactosidase exhibits a second catalytic activity in addition to the hydrolysis of lactose: the isomerization of lactose to *allo*-lactose, in which the galactosyl moiety is bonded to oxygen-6 rather than to oxygen-4 of glucose.

***allo*-Lactose**

A few molecules of lactose are converted into *allo*-lactose, which derepresses the *lac* operon by binding to the *lac* repressor and inducing it to dissociate from the operator. Derepression of the *lac* operon is illustrated in Figure 13-3. This is another example of the use of binding energy in biochemistry (Chapter 5). When *allo*-lactose is absent, the repressor binds tightly to the operator; however, in the presence of *allo*-lactose the repressor dissociates from the operator. *Allo*-lactose dramatically alters the properties of the repressor by changing its conformation to one that has little affinity for the operator. When the repressor dissociates from the operator, RNA polymerase can bind to the promoter and transcribe the structural genes, leading to the biosynthesis of large quantities of the lactose-metabolizing enzymes. The derepression of genes by molecules like *allo*-lactose is known as *induction,* and the derepressing molecules are *inducers* of the biosynthesis of specific proteins.

Note in Figures 13-2 and 13-3 that the operator does not block the transcription of *lac I,* which codes for the repressor. The Lac repressor is always produced, but whenever the inducer is present it binds the repressor and the operon is derepressed. As soon as the inducer is metabolized, as *allo*-lactose eventually will be, the repressor will again bind to the operator and block the transcription of the structural genes.

Positive Regulation. The *lac* operon is also subject to positive regulation; that is, transcription is turned on by a molecule binding to the promoter. Positive control is applied by molecules that are participants in the more-generalized control phenomenon known as *catabolite repression,* in which the expression of several operons is prevented by the presence of high levels of glucose. Catabolite repression draws its name from the fact that genetic studies show that the effect is not brought about by glucose itself but rather by a catabolite of glucose. *E. coli* grows well on glucose, and in its presence the bacterium preferentially utilizes it rather than other carbohydrates that may be present. The mechanism by which

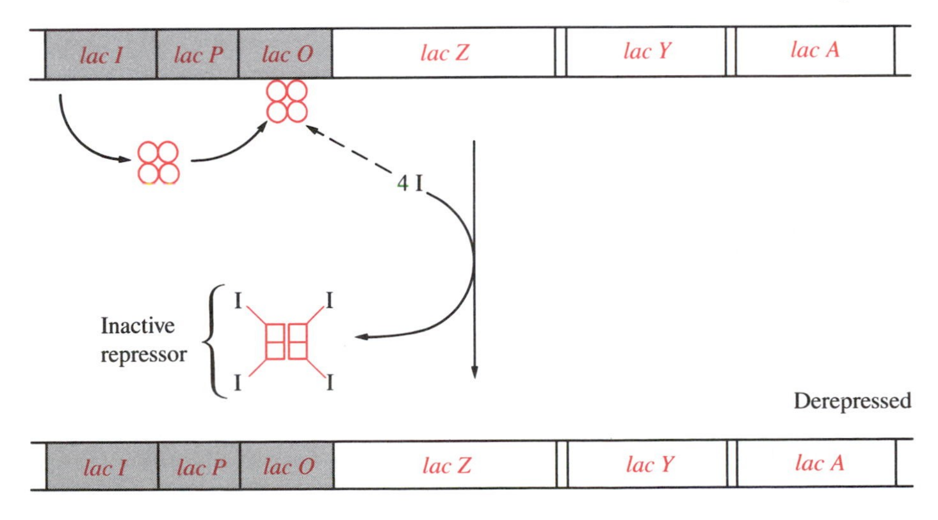

Figure 13-3

Negative Regulation of the Lactose Operon in *E. coli.* The *lac* operon is normally repressed in the absence of lactose, so that the enzymes of lactose metabolism are produced only at low levels. When lactose is present, a small amount is converted into the inducer *allo*-lactose, designated "I" in the figure, which binds to the repressor and prevents it from binding to the operator.

the preferential utilization of glucose is brought about is one that prevents the transcription of operons such as *lac*, even in the presence of inducers, but allows transcription whenever glucose is absent and inducers are present.

Figure 13-4 shows how positive control is mediated by the cyclic AMP binding protein known as the *catabolite activation protein* (CAP). This protein by itself exerts little or no effect on the *lac* operon; however, when it binds cyclic AMP, it acquires the property of binding and activating promoters. The *lac* promoter is activated to bind RNA polymerase whenever the complex of CAP and cyclic AMP is bound to it and repressor is absent. Thus, in the absence of

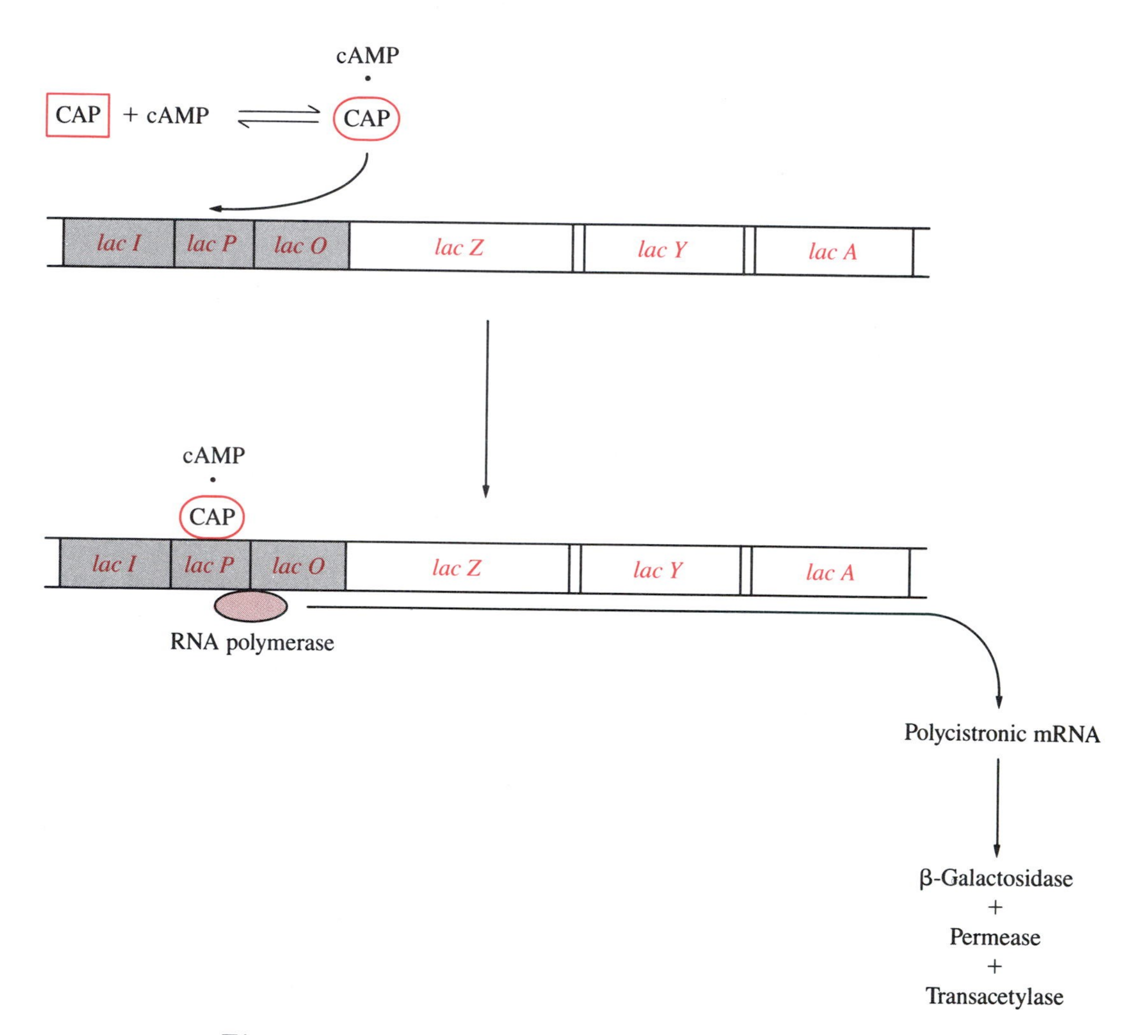

Figure 13-4

Positive Control of the Lactose Operon by cAMP and CAP Protein. Transcription of the *lac* operon is inefficient unless the complex formed between cAMP and the catabolite activation protein (CAP) is bound to the promoter. When the promoter is activated by the complex cAMP·CAP, RNA polymerase binds to the promoter-operator elements of the operon and transcribes the structural genes, *lac Z, Y,* and *A,* which are transcribed into a single polycistronic mRNA. The mRNA contains all of the nucleotide sequence information, including the start and stop codons, for the biosynthesis of β-galactosidase, permease, and transacetylase. As long as the inducer is present to keep the repressor from binding to the operator (Figure 13-3), and the complex of cAMP and CAP is bound to the promoter, these enzymes are produced in large amounts.

the complex CAP·cAMP, the *lac* operon is poorly transcribed, but with CAP·cAMP bound to the promoter transcription is facilitated; it occurs whenever the repressor is not bound to the operator.

When glucose is present in *E. coli,* the cyclic AMP levels are low, little cAMP·CAP is available, and the transcription of *lac* and several other operons is poor. At low levels of glucose, cyclic AMP levels are high and cAMP·CAP is available and bound to *lac P* and other promoters; however, *lac* is not transcribed as long as the repressor is bound to the operator. When glucose is absent and lactose is present, β-galactosidase produces *allo*-lactose, which binds to the repressor and induces the production of the enzymes required for lactose utilization.

The effects of high and low levels of glucose and lactose on the regulation of the *lac* operon and the production of the enzymes for lactose utilization are listed in Table 13-1. Note that induced, high-level synthesis of these enzymes occurs only under conditions in which *both* positive regulation and negative regulation favor transcription of the structural genes; that is, low levels of glucose and high levels of lactose. Under all other conditions, the enzymes for lactose utilization are produced at low levels.

Coordinate Translation of Polycistronic mRNA. In the derepressed state, in which polycistronic mRNA for the structural genes is produced and translated, β-galactosidase, permease, and transacetylase are not produced in equal amounts. They are said to be *coordinately* derepressed, because they are produced in successively decreasing amounts according to the order in which they are translated from polycistronic mRNA; that is, they are translated in the relative amounts β-galactosidase > permease > transacetylase. This regulation is at the level of translation (Chapter 10), and it results from two phenomena that are illustrated in Figure 13-5. First, there are short segments of RNA

Table 13-1

Effects of nutrients on transcription of the lac operon

[Glucose]	[Lactose]	[cAMP]	*lac O-P* Region	Levels of Lac Enzymes
High	Low	Low	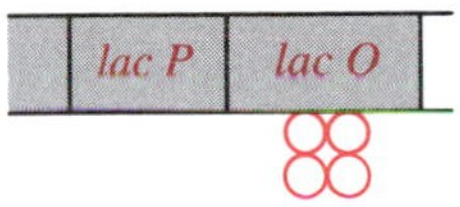 *lac P* *lac O*	Low
High	High	Low	 *lac P* *lac O*	Low
Low	Low	High	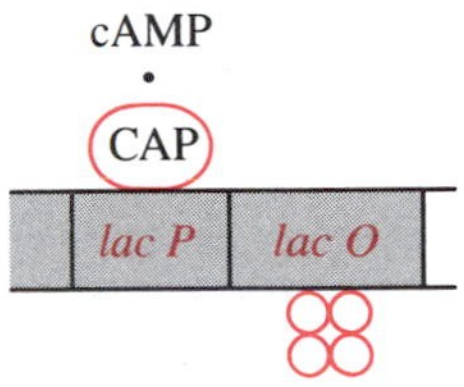 cAMP • CAP *lac P* *lac O*	Low
Low	High	High	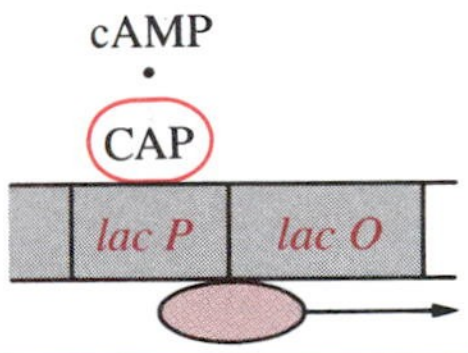 cAMP • CAP *lac P* *lac O*	Induced

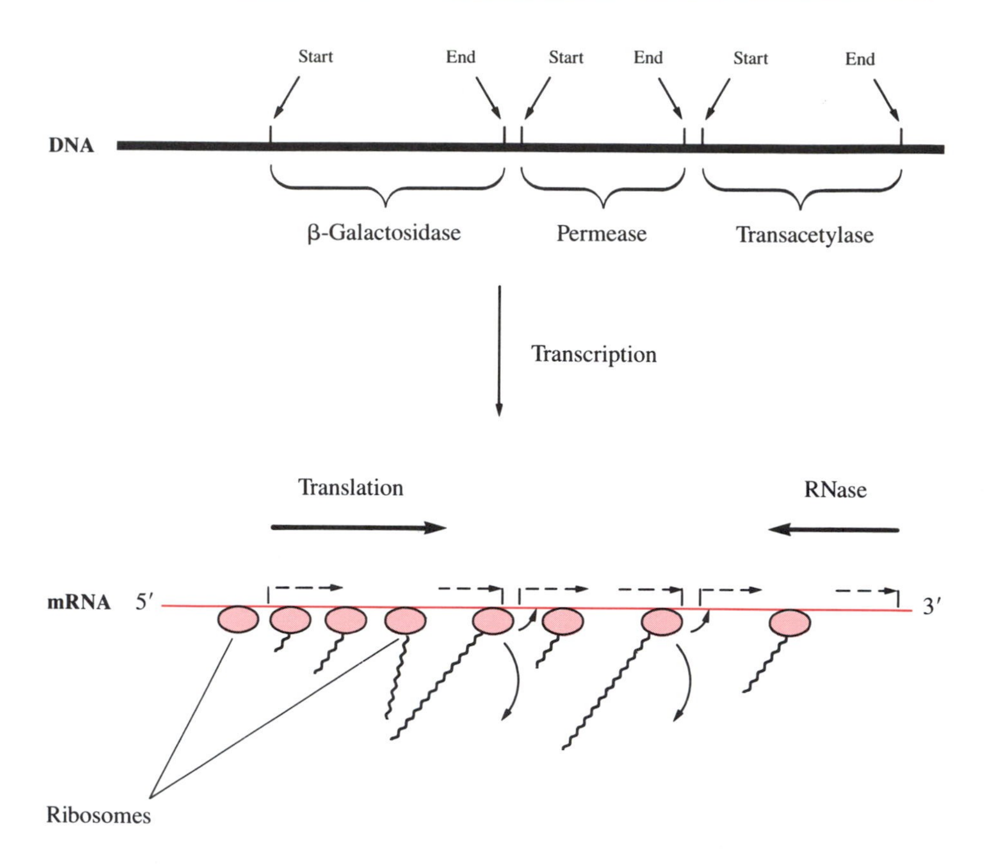

Figure 13-5

Coordinate Translation of Polycistronic mRNA. The best ribosome-binding site for translation precedes the initiation codon for the first cistron to be translated. After the stop codons for the first and second cistrons, there are short segments of mRNA that allow the ribosomes the opportunity either to dissociate or to reinitiate at the next start codon. Although binding of ribosomes at the start codons of the second and third cistrons can occur, it is less probable than at the start of the first cistron. In addition, RNases preferentially degrade mRNA from the 3′ end in *E. coli*. Therefore, less of the second cistron and even less of the third is translated relative to the first.

between the stop codon of each cistron and the start codon of the next. (A cistron is a nucleotide sequence in DNA that encodes the amino acid sequence of a protein. See Chapter 10.) These noncoding segments allow the translating ribosome to undergo the chain-termination process and dissociate from the mRNA (Chapter 10). A new initiation complex can be assembled at any of the start codons, but the best ribosome-binding site for initiation of translation precedes the cistron for β-galactosidase. Second, the polycistronic mRNA is subject to degradation by nucleases, which in *E. coli* preferentially attack from the 3′ end of mRNA and so preferentially degrade the cistrons in the order transacetylase > permease > β-galactosidase.

Discovery of the Lactose Repressor. Most of the original evidence for the existence of operons, in which the expression of several related genes is tightly coupled, was genetic in nature. The biochemical evidence for the enzymes associated with the *lac* operon was fairly straightforward to obtain. For example, some mutants lacked all three enzyme activities and could not grow on lactose. These were mutants that had deletions in *lac Z* or mutants that had particular defects in the regulatory elements—for example, a mutated promoter that would not bind RNA polymerase. However, certain mutants always produced large amounts of all three enzymes. These were mutants with a defective operator that could not bind the repressor or they were mutants that were deficient in the Lac repressor. They produced the enzymes of lactose utilization constitutively; that is, as constituents of the cell, as if they were required for cell growth under all conditions.

The biochemical characterization of mutants that always produced the Lac enzymes was not an easy matter because the repressor does not catalyze any reaction that could be used to detect its presence. To observe the presence of a functioning repressor, it proved to be necessary to detect binding of inducers by the repressor. Lactose itself was not an inducer, and *allo*-lactose was not known at the time. Fortunately, an artificial inducer, isopropylthiogalactoside (IPTG), was known and could be prepared in a radiochemically labeled form. [Methylthiogalactoside (MTG) was also found to be an inducer.] The radiochemical IPTG-binding assay made it possible to detect the presence of repressor in extracts and partially purified fractions from *E. coli.*

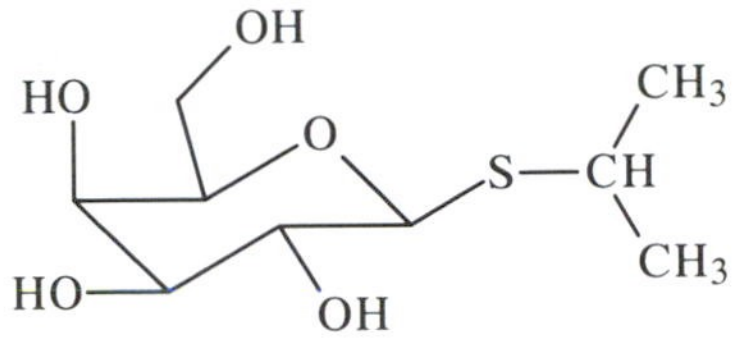

IPTG

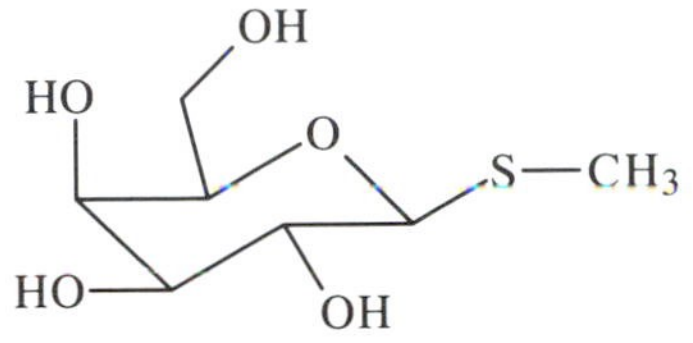

MTG

By use of the IPTG-binding assay, the existence of a molecule with the property of the hypothetical Lac repressor—that is, the ability to bind [^{14}C]IPTG—could be proved. This property was detected in extracts from cells that exhibited derepression of the enzymes for lactose utilization. At first, it was not known whether the repressor was a protein or a nucleic acid or perhaps a complex carbohydrate. However, the [^{14}C]IPTG-binding molecule was soon shown to be destroyed by heat or proteolytic enzymes, even in partly purified preparations. These properties were most consistent with the repressor being a protein, as proved to be the case once it was purified to homogeneity.

IPTG was an important molecule for the identification and purification of the Lac repressor. However, it and MTG have a continuing importance in biotechnology. The molecules are particularly valuable for use in switching on the *lac* operon and keeping it on even when there is no lactose or need for the lactose metabolizing system. They are much more metabolically stable than *allo*-lactose and persist in *E. coli,* while keeping the *lac* operon turned on even in the absence of any lactose. This property of IPTG is widely used in biotechnology, along with the regulatory elements of the *lac* operon, to control genes other than those of the *lac* operon that may be cloned in *E. coli* (Chapter 14).

Structure of the Lactose Operator. The nucleotide sequence of the *lac* operator exhibits symmetry that complements the presumed symmetry of the tetrameric repressor. The operator sequence is shown in Figure 13-6, in which the bases that define a quasi twofold axis of symmetry are highlighted. Note that a protein (repressor) with paired identical subunits can in principle bind to such a sequence in the manner illustrated in Figure 13-6B. This mode of binding once

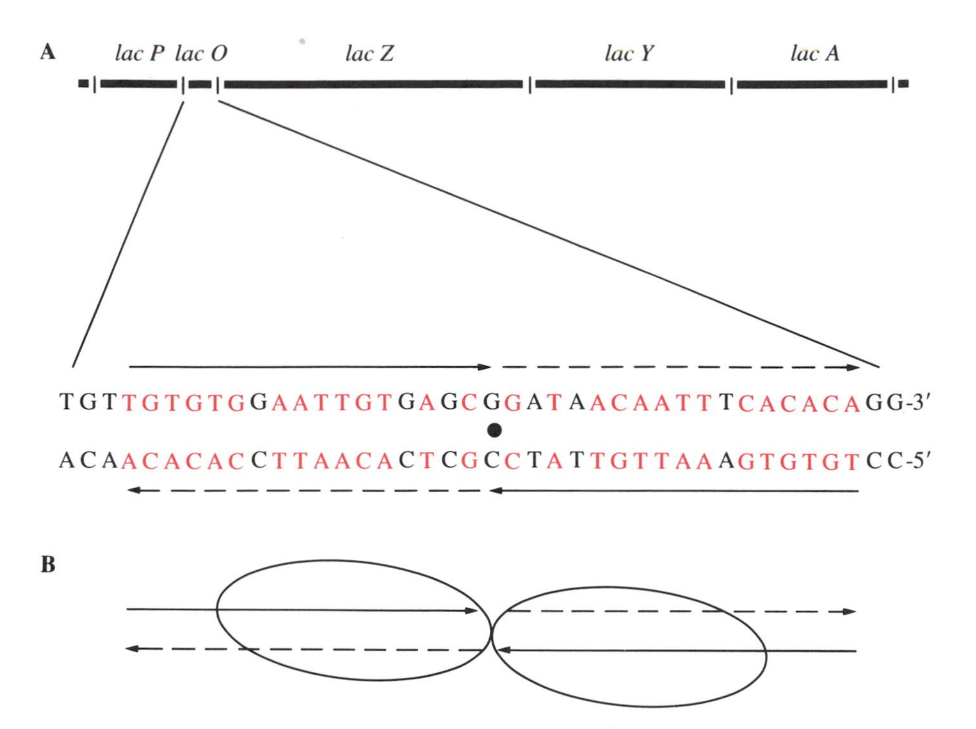

Figure 13-6

Symmetry in the Lactose Operator. In this nucleotide sequence of the *lac* operator in *E. coli* (Part A), the bases that define a quasi twofold axis of symmetry are in color; the axis is perpendicular to the plane and indicated at the center marked with the black spot. Rotation about this axis through 180° reproduces the sequence shown in color. This property of the operator sequence complements the even-numbered subunit composition of the repressor to which it binds. Associated pairs of subunits can in principle bind to this sequence in the manner illustrated in part B.

again exploits the chelate effect and allows very strong binding of the repressor through multiple, almost identical contacts.

Regulation of the Galactose Operon

The galactose operon (*gal*) in *E. coli* is a cluster of genes and regulatory elements that contain the information for production of enzymes required to utilize or produce galactose. The regulation of the *gal* operon is similar in some respects to that of the *lac* operon, but it differs in other ways. The reason for the difference is that the survival of *E. coli* depends on the ability of the bacterium to *produce* galactose for the production of its cell wall when galactose is absent from the medium. Regulation of the *lac* operon provides only for the utilization of lactose.

Galactose Utilization in *E. coli* and Eukaryotes. Galactose catabolism in bacteria and eukaryotes essentially entails the conversion of galactose into glucose-1-P through the actions of three enzymes in the Leloir Pathway, galactokinase, galactose-1-P uridylyltransferase, and UDP-galactose 4-epimerase (Figure 13-7). Glucose-1-P is then broken down by the same pathways through which glucose is utilized (Chapter 21). The Leloir Pathway was described in Chapter 11, along with the functions of the kinase and uridylyltransferase; the mechanism of action of the epimerase will be described in Chapter 20. The genes encoding the three required enzymes in *E. coli* are *galE, galT,* and *galK,* for epimerase, transferase, and kinase, respectively; the clustering of these genes in the *gal* operon is similar to that of *lac Z, lac Y,* and *lac A* in the *lac* operon.

The regulatory mechanisms described herein for the *lac* operon also apply to regulation of the *gal* operon. However, owing to a requirement for balance between preferential glucose metabolism through catabolite repression on the one hand and the special requirements of cells for a supply of galactose on the other hand, the regulation of the *gal* operon is slightly more complex. Galactose utilization by *E. coli* is outlined in Figure 13-7, where the importance of the Leloir enzymes in both the breakdown and biosynthesis of galactose is illustrated. Galactose in the growth medium is largely subject, in the absence of

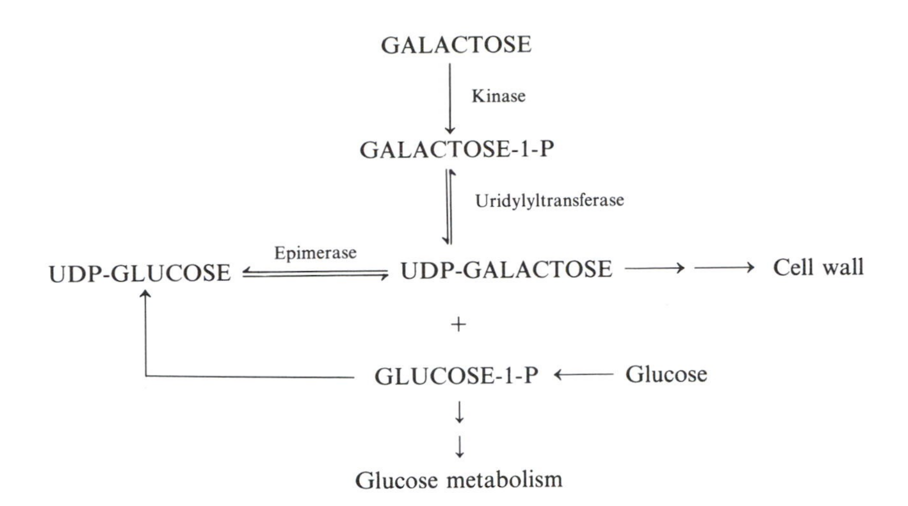

Figure 13-7

Galactose Metabolism in *E. coli.* The scheme shows how the enzymes of the Leloir Pathway are required both for breaking galactose down as an energy and carbon source and for producing UDP-galactose from glucose-1-P for cell wall biosynthesis. All three enzymes are required to convert galactose into glucose-1-P; however, only UDP-galactose 4-epimerase is required to convert UDP-glucose into UDP-galactose. UDP-glucose is produced from glucose-1-P by the action of UDP-glucose pyrophosphorylase (Chapter 11).

glucose, to degradation as a source of energy and carbon; however, part of the UDP-galactose is used for biosynthesis of the bacterial cell wall. In the absence of galactose in the growth medium, UDP-galactose is still required for cell walls; it can be obtained only from UDP-glucose by the action of UDP-galactose 4-epimerase, one of the three enzymes of the *gal* operon. Thus, the regulation of the *gal* operon must allow for the breakdown of galactose in the growth medium but, owing to catabolite repression, only in the absence of glucose; it must also allow for the conversion of glucose into UDP-galactose for cell-wall biosynthesis when no other source of galactose is available. The added complexity exemplifies the diversities in gene regulation that exist, even in simple cells such as *E. coli*.

Galactose and its metabolism are also crucial in all eukaryotic cells. Galactose can serve as an energy source in eukaryotes in the same way as in bacteria, by conversion into glucose-1-P through the Leloir pathway; it is also required in eukaryotes as a major component of the oligosaccharide moieties of glycoproteins and glycolipids.

Structure and Regulation of the Galactose Operon. The structure of the *gal* operon is shown schematically in Figure 13-8. It consists of the three structural genes, *galE*, *galT*, and *galK*, as well as an operator and a *bifunctional* promoter region. A major difference from the *lac* operon is that the regulatory gene, *galR*, is far away from the operon. Another major difference is that the promoter region functions as if it were *two* promoters rather than one. Regulation of the *gal* operon is less well understood than that of the *lac* operon.

In negative regulation, the *gal* repressor, the product of *galR*, binds to the operator-promoter region and inhibits transcription of the *gal* operon. Enzyme synthesis is induced by galactose in either the absence or the presence of glucose, but in the absence of glucose induction is much higher. Genetic studies of promoter mutants indicate the existence of two functional regions: one that responds to activation by cAMP·CAP (positive regulation) and another that allows transcription to proceed in the absence of cAMP. Two species of mRNA

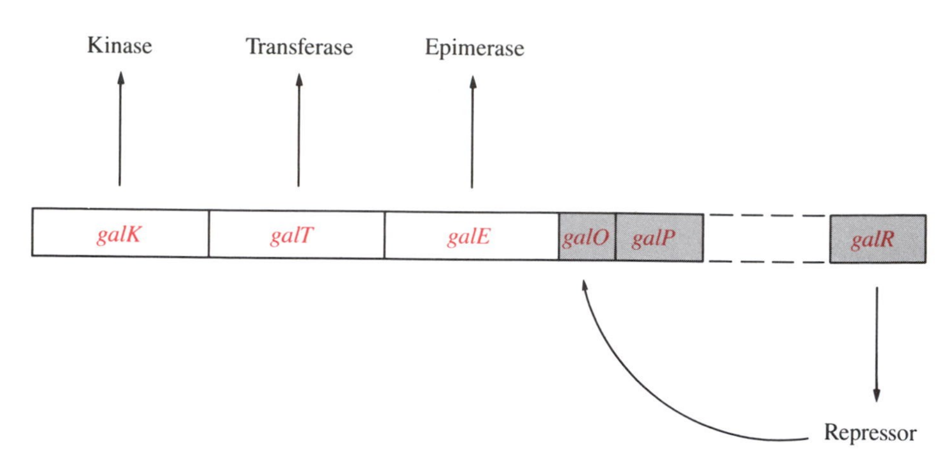

Figure 13-8

The Structure of the Galactose Operon in *E. coli.* This operon differs from the *lac* operon in that there are two promoter functions; that is, the promoter region functions as two promoters, one of which is activated by the CAP-cAMP complex and the other is not.

are produced in transcription controlled by the two promoter regions, both of which are polycistronic and encode the three enzymes. However, one species is five nucleotides longer than the other, showing that transcription begins from slightly different sites under the control of the two promoter regions. Thus, positive regulation of the *gal* operon differs from that of the *lac* operon in that transcription is only partly regulated by the cAMP·CAP system.

Owing to the bifunctional nature of the promoter, the complex cAMP·CAP can bind and promote the initiation of transcription; however, RNA polymerase can also bind at a nearby site and initiate transcription with less efficiency in the absence of cAMP. Therefore, catabolite repression is only partly effective in repressing the expression of the *gal* genes. This state of affairs is in the interest of effective cell growth under all conditions, because cell-wall biosynthesis requires UDP-galactose, which can be produced from glucose only when UDP-galactose 4-epimerase is present. Inasmuch as the cells must be able to synthesize this enzyme for growth under all conditions, the existence of a secondary promoter function even in the absence of cAMP·CAP—that is, under conditions of catabolite repression—is important to the survival of the cell.

The Gal enzymes are, therefore, constitutively produced under all conditions in significant amounts, greater than those of the Lac enzymes, to support cell-wall biosynthesis. They are induced by galactose and artificial inducers such as 6-deoxygalactose under all conditions, but they are induced in greater amounts if the glucose level is low than they are under conditions of catabolite repression at high levels of glucose.

Regulation of the Tryptophan Operon

Bacteria produce tryptophan and other amino acids when they are not available in the growth medium. The tryptophan (*trp*) operon encodes the enzymes required to produce tryptophan for cellular needs. The regulatory apparatus for this operon senses the needs of the cell for tryptophan and responds by controlling the amounts of the biosynthetic enzymes produced by the cell.

Tryptophan Biosynthesis. Chorismate is a common intermediate in the biosynthesis of aromatic amino acids and is converted in five steps into tryptophan (Chapter 26). The structural genes in the *trp* operon code for the biosynthesis of these enzymes. The system that regulates the *trp* operon senses the presence of tryptophan and responds by shutting down the production of the enzymes required for tryptophan biosynthesis. Because this is a biosynthetic system, the regulation of the mechanism by which the *trp* operon is repressed differs from that of the *lac* operon, which is a biodegrative system.

Structure and Regulation of the Tryptophan Operon. Figure 13-9 illustrates the structural genes and the regulatory elements of the *trp* operon. The structural genes *trpA*, *trpB*, *trpC*, *trpD*, and *trpE* encode the amino acid sequences of the enzymes required to produce tryptophan from chorismate (Chapter 26). The regulatory elements are the promoter, the operator, and an *attenuator*. The repressor protein is specified by the gene *trpR* at a separate location.

The *trp* operon is subject to two types of regulation, repression and attenuation, both of which depend on the tryptophan concentration in the cell (Figure 13-9). Transcription is repressed when the complex of repressor protein and tryptophan bind to the operator. This type of regulation works as an on-off switch, and transcription is blocked when the repressor protein is saturated with tryptophan at high concentrations of the amino acid. At lower concentrations of tryptophan, transcription is possible but the rate is regulated by the attenuator in such a way that it decreases with increasing tryptophan concentrations.

The exact mechanism by which the attenuator modulates the rate of transcription is not known, but the process is thought to entail control of transcription termination by a leader sequence within the attenuator. Transcription and translation of the operon are kinetically coupled; that is, a ribosome binds to the mRNA and begins translating shortly after RNA

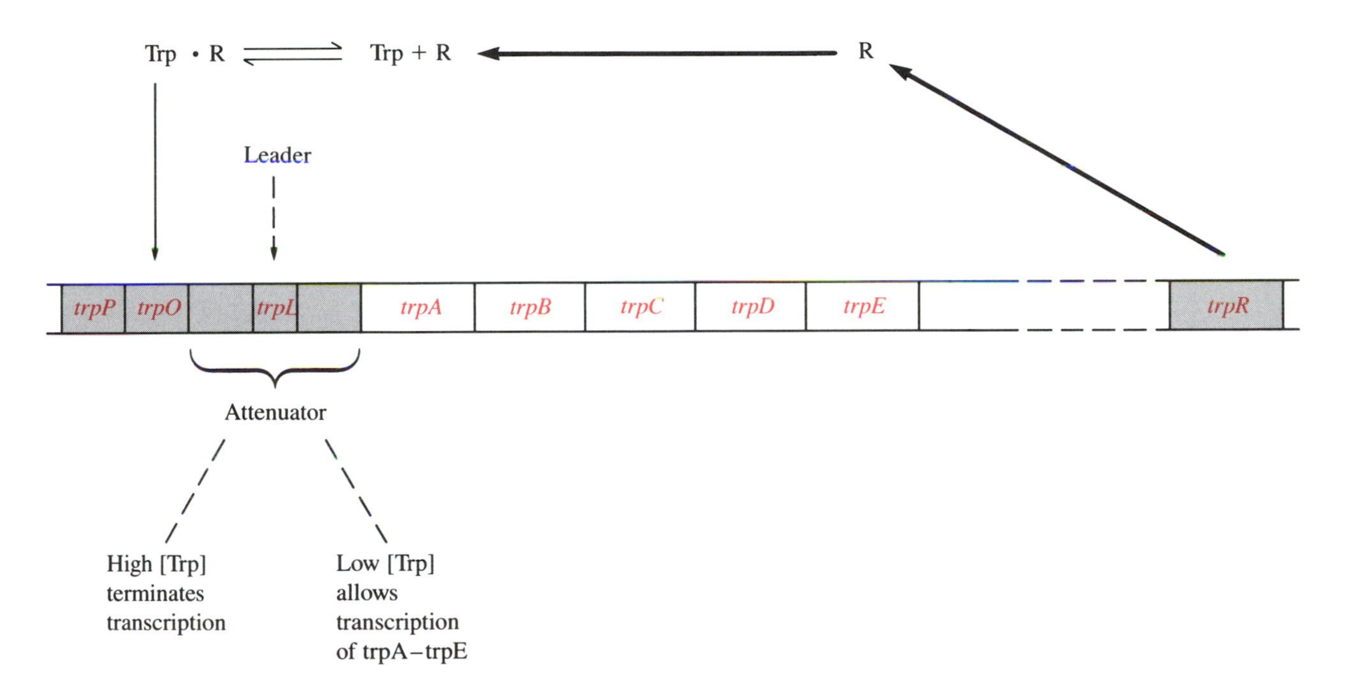

Figure 13-9

Structure and Regulation of the Tryptophan Operon. The *trp* operon consists of structural genes *trpA* through *trpE*, which encode the enzymes required for tryptophan biosynthesis from chorismate; a promoter; an operator; and an attenuator. The attenuator contains a leader, which encodes a fourteen amino acid peptide containing two tryptophan residues. The repressor is encoded by a separate gene *trpR* well removed from the operon. The repressor protein R does not itself block transcription. Instead, tryptophan reversibly binds to the repressor, and the complex of repressor and tryptophan binds to the operator and represses transcription. Thus, at high concentrations of tryptophan, the *trp* operon is not transcribed. At somewhat lower concentrations of tryptophan, the attenuator modulates the rate of transcription.

polymerase begins transcription and before it has passed through the attenuator. The leader specifies a fourteen amino acid peptide that contains a Trp-Trp sequence. The small amount of Trp-$tRNA^{Trp}$ that is available at low concentrations of tryptophan limits the rate at which the ribosome can translate the leader, and this allows a smooth downstream transcription right through the five structural genes, which are then translated by the ribosome. However, high tryptophan concentrations increase the amount of available Trp-$tRNA^{Trp}$, so that the rate of leader translation is very fast, and this somehow leads to a high frequency of transcription termination in the attenuator. There is a septa-U sequence between the leader and the start codon for *trpA* that is the site of termination. Termination of transcription at this site blocks the expression of the structural genes.

Regulation of Eukaryotic Gene Expression

The expression of eukaryotic genes is much less well understood than the expression of prokaryotic genes. Some of the same principles of repression and induction also apply to eukaryotic gene regulation; however, eukaryotic systems differ in several respects from prokaryotic systems. For example, eukaryotic genes are not organized in operons. This fact has several consequences for the regulation of gene expression in eukaryotes compared with prokaryotes. One is that related genes are not regulated together in the same way as they are in operons, and another is that each species of mRNA transcribed from a eukaryotic gene codes for only a single protein—that is, there are no polycistronic species of mRNA. Because of these differences, the regulation of related genes is generally more complex than in prokaryotes, where they can be regulated together. Moreover, coordinate derepression of related genes cannot occur by the mechanisms described for the enzymes of an operon, because this type of regulation requires translation of polycistronic mRNA. Other differences are the following:

1. Although eukaryotic genes are not organized in operons that are transcribed into polycistronic RNA, related genes are often organized in families that are related through coordinated regulation of their transcription. This in itself makes eukaryotic gene regulation more complex than prokaryotic regulation.
2. Eukaryotic genes consist of exons, the coding sequences, interspersed with introns that are spliced out of pre-mRNA, pre-rRNA and pre-tRNA (Chapter 12). The role of introns is not thoroughly understood, and the splicing process may be related to the regulation of gene expression.
3. Histones are bound to eukaryotic genes and may play a role in gene regulation, probably at the stage of cellular differentiation.
4. Eukaryotic DNA contains many apparently meaningless, highly repetitive sequences interspersed among the genes. These may be truly meaningless or they may have some currently obscure function.
5. Finally, eukaryotic genes can proliferate within a genome, resulting in a large number of copies of the same gene. This generally leads to high levels of expression of that gene.

In this section, we shall describe a few of the currently accepted principles of eukaryotic gene expression. Eukaryotic gene regulation is a much larger subject, however, in that it is the basis for developmental biology and cellular diversity in differentiated organisms. Because all germ cells of a differentiated organism contain the same DNA, cellular diversity entails the permanent exposure of selected genes to high expression in a given cell type. Cells of other types must have unique patterns of gene exposure. The details of the means by which genes are activated for expression in the course of development are beyond the scope of this textbook. However, biochemical evidence indicates that the chromatin structure around the genes specifying the most highly expressed proteins is much less compact and much more exposed to external interactions than chromatin around unexpressed genes. Exposure presumably facilitates the transcription process. The most highly expressed proteins in a specialized cell are generally those that are most important to the function of a given cell type. In muscle cells, for example, these proteins would include myosin and actin (Chapter 30).

Eukaryotic Regulatory Elements

Some of the principles underlying the regulation of gene expression in prokaryotes are called into play in eukaryotes. For example, gene expression is regulated primarily at the transcriptional level through the action of promoters located upstream of a gene. However, operators and repressors, which block transcription in prokaryotes, do not regulate eukaryotic promoters. In eukaryotes, gene regulation is primarily positive; that is, *transcription factors* activate transcription. Transcription factors are proteins that bind to eukaryotic promoters and facilitate transcription by RNA polymerase.

Figure 13-10 shows a generic regulatory region upstream of the gene for a typical eukaryotic protein. The regulatory elements include a promoter and often an *enhancer*. Eukaryotic promoters contain within them some or all of the nucleotide sequences in Figure 13-10, a TATA box, a CAAT box, and a GC box. A given promoter may lack one of these sequences. The transcription factors recognize these sequences and bind to them. When all the transcription factors are present, the promoter is recognized by RNA polymerase and transcription proceeds. The enhancers are nucleotide sequences from fifty to two hundred base pairs in length located either upstream or downstream from the gene.

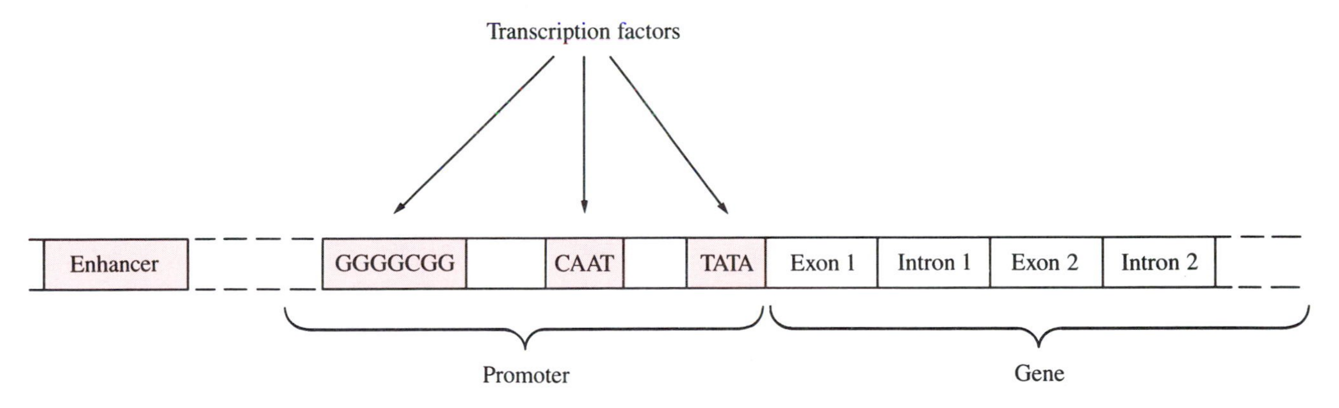

Figure 13-10

Regulatory Elements of a Eukaryotic Gene. Transcription of a eukaryotic gene is typically controlled by a promoter that contains at least two of the short sequences shown, a TATA box, a CAAT box, and a GC box. Transcription factors activate the promoter by mechanisms that include recognition of these sequences. RNA polymerase II in the nucleus binds to the activated promoter and transcribes the gene. The mRNA produced by transcription contains the exons and introns of the gene, and the introns are spliced out before 5′ capping and 3′ polyadenylylation of mRNA (Chapter 12). An enhancer sequence may be either upstream or downstream of the promoter and gene. Enhancers increase the frequency of transcriptional initiation.

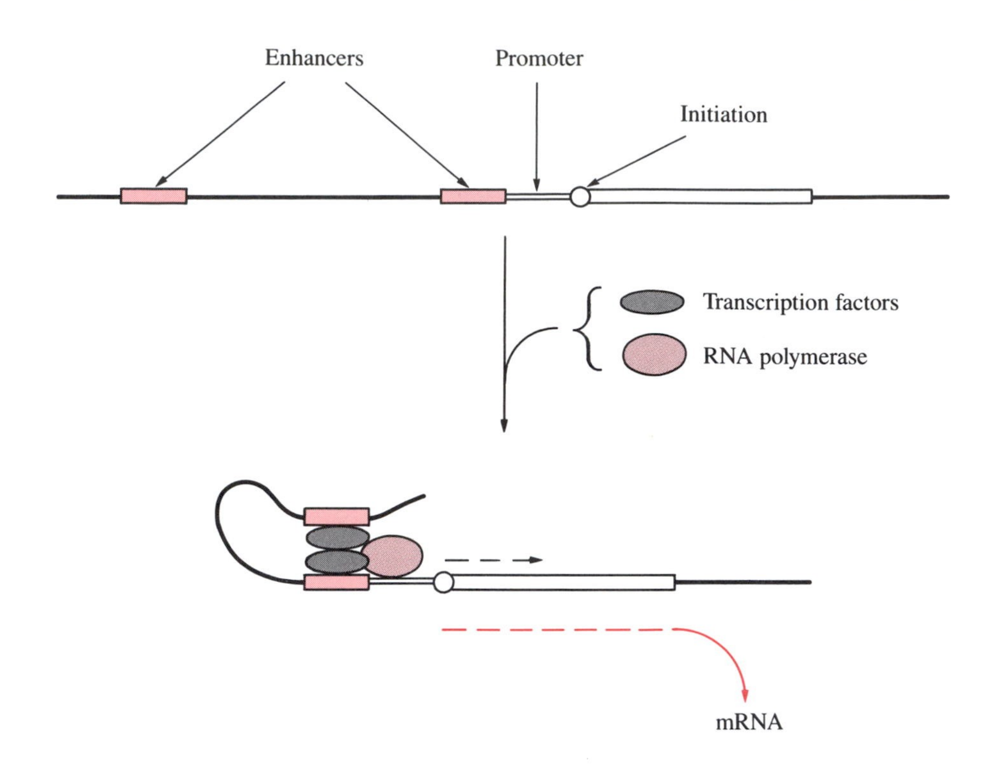

Figure 13-11

Interactions of Enhancers with Transcriptional Factors. Enhancer sequences hundreds or thousands of base pairs upstream or downstream of a promoter may enhance the frequency of transcription by binding transcriptional factors, which signal the points at which RNA polymerase binds and initiates transcription. In this figure, enhancer sequences are near and distal from a promoter and transcription initiation site for a gene. Transcription is initiated when transcription factors bind to the enhancers and promoter, signalling RNA polymerase to bind and initiate transcription. Transcription factors bound to the enhancers also bind to each other, thus creating a loop in the DNA at the point of transcription initiation.

Enhancers greatly increase the frequency with which transcription is initiated and may be separated by a thousand base pairs or more from the gene and promoter.

Activation of transcription by enhancers hundreds of base pairs upstream or downstream from a gene is possible because DNA is not a rigid or perfectly linear molecule. Sequences that are well separated in the linear structure can be brought together by the interactions of DNA binding proteins such as transcription factors, which may bind to each other as well as to separated nucleotide sequences of DNA. These interactions produce looplike structures in the DNA, in which both an enhancer sequence and a promoter are bound to transcription factors (Figure 13-11).

The primary RNA transcript contains all of the exons and introns in the gene and undergoes the processing described in Chapter 12 before leaving the nucleus—namely, 5′ capping, 3′ polyadenylylation, and splicing out of introns. The mature mRNA is then transported to the endoplasmic reticulum, where it is translated.

Regulation of Galactose Metabolism in Yeast

Galactose metabolism is as important in eukaryotes as in prokaryotes. Galactose is a source of energy equivalent to glucose through the Leloir pathway (Figure 13-7), and it is required for the biosynthesis of glycoproteins and glycolipids in eukaryotes.

The regulation of the genes specifying the enzymes of the Leloir pathway in yeast differs from that in *E. coli.* Figure 13-12 shows how the presence of galactose as a nutrient leads to the expression of the genes. In yeast, the genes specifying the Leloir enzymes are on one chromosome and designated *GAL1*, *GAL7*, and *GAL10*, for the kinase, transferase, and epimerase, respectively. *GAL2* is on a different chromosome and codes for a protein required to transport galactose into the cell. *GAL4* and *GAL80* are regulatory genes on other chromosomes. The protein specified by *GAL4* activates the transcription

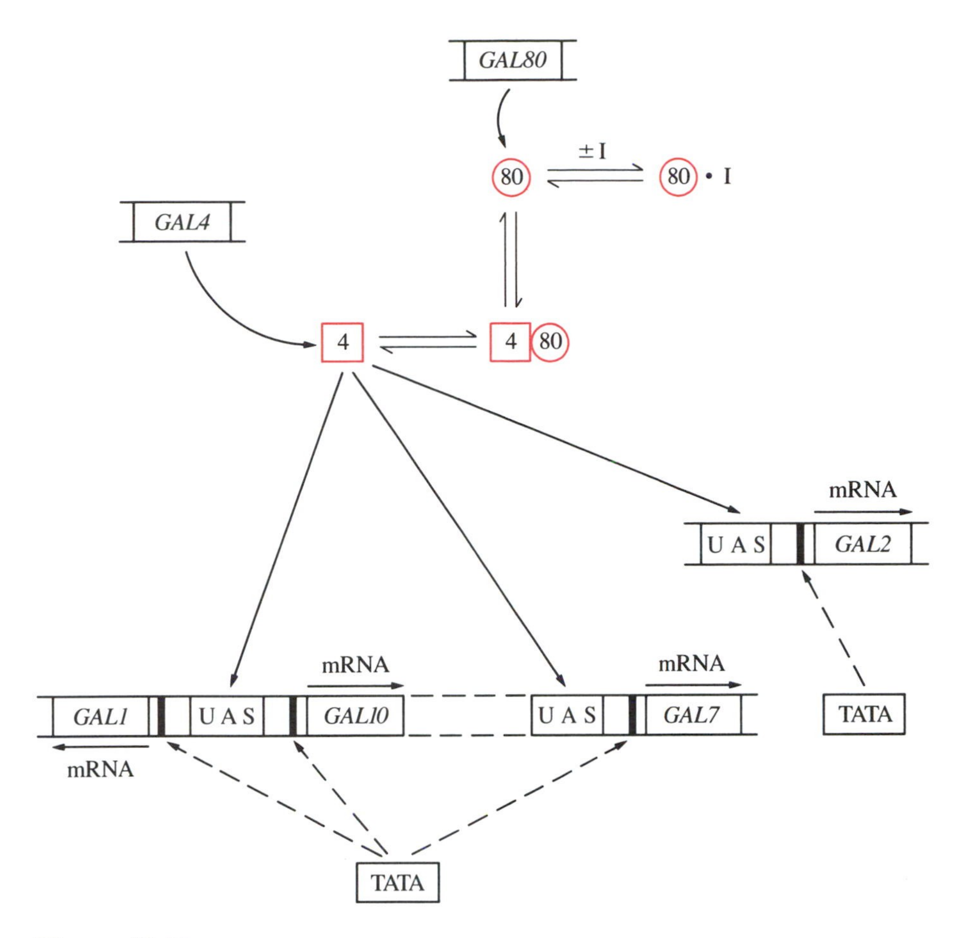

Figure 13-12

Regulation of the Transcription of mRNA Specifying Enzymes of Galactose Metabolism in Yeast. The structural genes for galactokinase, galactose-1-P uridylyltransferase, and UDP-galactose 4-epimerase are *GAL1*, *GAL7*, and *GAL10*, respectively, and are located on one chromosome. *GAL2*, on a different chromosome, is the structural gene for a protein that promotes the uptake of galactose. *GAL4* and *GAL80* on still other chromosomes specify proteins that regulate the expression of *GAL1*, *GAL7*, and *GAL10*. The GAL4 protein is a transcription activator that binds to upstream activating sequences (UAS). The GAL80 protein normally binds to the GAL4 protein and prevents it from activating transcription. Induction of the Leloir enzymes by galactose is caused by the binding of an inducer to the GAL80 protein, which causes it to dissociate from the GAL4 protein. The GAL4 protein then activates the transcription of *GAL1*, *GAL7*, *GAL10*, and *GAL2*.

of *GAL1*, *GAL7*, and *GAL10* by binding to the upstream activation sequences (UAS), which are located within about one hundred pairs upstream from the TATA boxes of the promoters. The GAL80 protein normally forms a complex with the GAL4 protein, and the complex does not activate transcription. The GAL4 and GAL80 proteins are constitutively produced, so that *GAL1*, *GAL7*, and *GAL10* are normally not transcribed at a high level of expression. (In this state, sufficient UDP-galactose 4-epimerase must be produced to allow for the biosynthesis of the small amount of galactose required for glycoprotein biosynthesis.)

Galactose induces the production of the Leloir enzymes by more than a thousandfold. Either galactose or an inducer derived from galactose binds to the

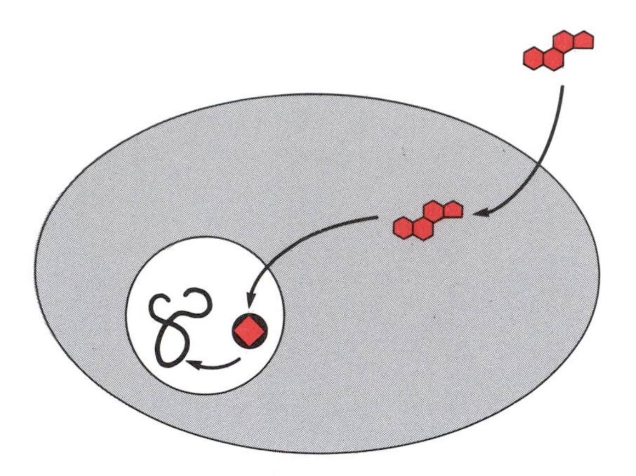

Figure 13-13

Hormonal Activation of Gene Transcription. The steroid hormones diffuse into the nuclei of target eukaryotic cells and bind to receptors that activate transcription of specific genes in the chromosomes.

GAL80 protein and prevents it from inactivating the GAL4 protein. This allows the GAL4 protein to bind to the UAS elements and activate the transcription of *GAL1, GAL7,* and *GAL10.*

Hormonal Regulation of Gene Expression

Hormones are signal molecules that are recognized by target cells through receptors that bind them and facilitate their actions. Some hormones such as epinephrine bind to receptors in the plasma membrane and activate preexisting enzymes through the mediating actions of second messengers (Chapters 1 and 21). Other hormones activate transcription of specific genes, generally by a process similar to that illustrated in Figure 13-13. Many hormones are hydrophobic molecules that penetrate cell membranes and enter cells freely. Once inside its target cell, a hydrophobic hormone can be transported into the nucleus, where it activates the transcription of a gene or gene family to produce certain proteins required for a specific cell function. Examples are the steroid hormones, including androgens, estrogens, progesterone, and glucocorticoids. These hormones diffuse through cell membranes and cytoplasm and bind to specific receptors in nuclei. Transcription of specific genes is increased in rate and, at least in some cases, the mRNA produced is stabilized. Such hormone receptors may be regarded as hormonally activated transcription factors.

Structures of Proteins That Regulate Gene Expression

It is reasonable to expect proteins that bind and regulate the transcription of genes to have structural features that complement the structure of DNA. Recent evidence supports this notion. One of the chain-folding motifs that has been recognized in several DNA-binding proteins is known as the *helix-turn-helix* motif, which is exemplified by the Cro protein of λ bacteriophage. Cro protein is the λ repressor; that is, it represses the transcription of bacteriophage λ genes that have been integrated into the bacterial chromosome (Chapter 14). The Cro protein contains 66 amino acid residues, it binds to DNA as a dimer, and it can be dimeric in solution under certain conditions. The α-carbon backbone of one subunit is shown in Figure 13-14. Note that amino acids 1 through 6 form part of a β-sheet, and residues 7 through 14 form a helix projecting upward at a slight angle to the right. This is followed by a turn, another helix of residues 15 through 23 perpendicular to the plane of Figure 13-14, and another turn followed by a helix of residues 27 through 36 approximately in the plane of the figure. Residues 40 through 54 fold back to the N terminus and complete the β-sheet with

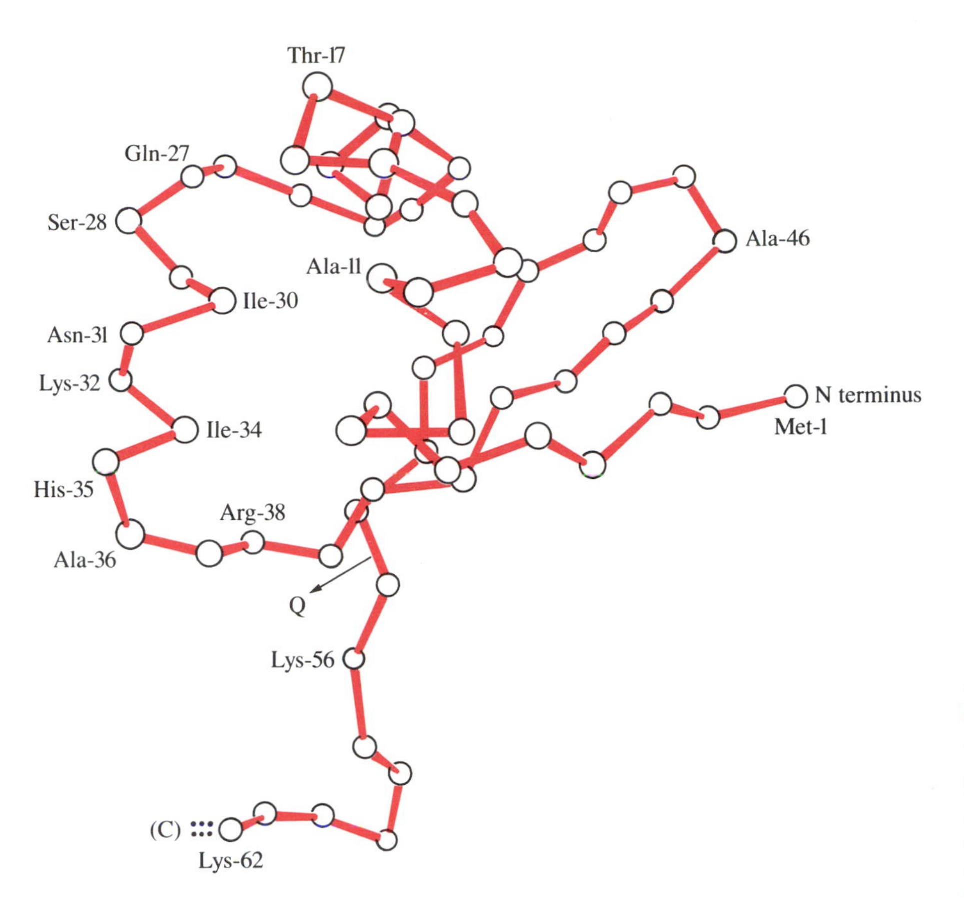

Figure 13-14

The α-Carbon Backbone of the Bacteriophage λ Repressor (Cro Protein). (After W. F. Anderson, D. H. Ohlendorf, Y. Takeda, and B. W. Matthews. Adapted with permission from *Nature*, vol. 290, p. 754. Copyright © 1980 Macmillan Magazines Limited.)

residues 1 through 6. From approximately residue 56 to the C terminus, the chain is thought to interact with the C terminus of a second subunit in the dimeric structure.

The helix-turn-helix motif is thought to allow binding to DNA by projection of one of the helices into the major groove of DNA. This is illustrated for the Cro protein in the stereo drawing of Figure 13-15. Note that when one helix in a subunit of a dimeric Cro protein is aligned in the major groove at the optimal angle for binding, the same helix in the other subunit is also positioned at the correct angle to bind to the major groove downstream from the first subunit. The binding mode illustrated in Figure 13-15 is consistent with chemical data showing which nucleotides in DNA are in contact with Cro protein. The side chains of the amino acids in the complex, which are not shown in Figure 13-15, are in the correct positions to form hydrogen bonds with DNA.

Another protein structural motif frequently found in eukaryotic DNA binding proteins is that of the *zinc finger proteins.* The first such protein to be discovered and characterized with respect to amino acid sequence is transcription factor IIIA (TFIIIA) of *Xenopus.* The zinc finger proteins contain in their amino acid sequences the periodic arrangement of histidine and cysteine residues X_3–Cys–X_{2-4}–Cys–X_{12}–His–X_{3-4}–His–X_4, in which the X's are other amino acids. The imidazole and thiol groups of histidine and cysteine side chains are good ligands for Zn^{2+}. They are spaced in such a way as to allow them to condense around Zn^{2+}. This coordination creates loops of amino acid residues projecting from each zinc ion, as illustrated in Figure 13-16—hence, the name zinc finger. These loops are thought to bind to the major groove of DNA. The GAL4 protein in yeast, the transcription factor that regulates the transcription of the genes specifying the enzymes of galactose metabolism, is one of hundreds of zinc finger proteins.

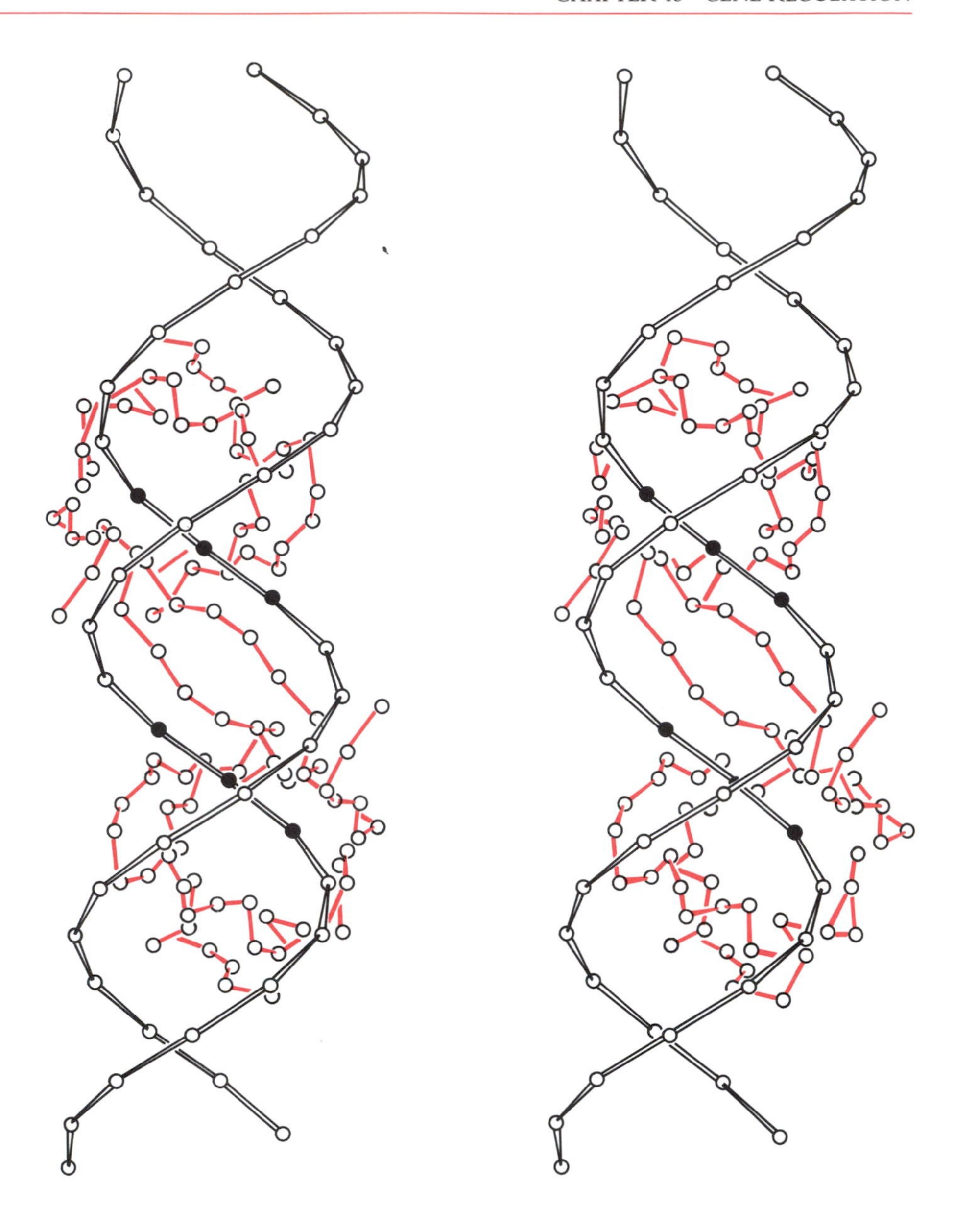

Figure 13-15

The Presumed Structure of Dimeric Cro Protein Bound to DNA. (After W. F. Anderson, D. H. Ohlendorf, Y. Takeda, and B. W. Matthews. Adapted with permission from *Nature*, vol. 290, p. 754. Copyright © 1980 Macmillan Magazines Limited.)

The crystal structure of the DNA-binding domain of the zinc finger protein Zif268 bound to a segment of DNA has recently been solved. The structure verified the hypothesis that the zinc fingers bind to the major groove of DNA, as illustrated in Figure 13-17.

Diversity in Proteins Through Variations in Exon Splicing

Families of similar but different proteins in an organism may arise from a single gene. Examples are the isoforms of myosin and troponin T, in muscle cells, and fibronectins, which are cell-to-cell adhesion glycoproteins. Alternative splicing modes for processing pre-mRNA into species of mRNA encoding different but related proteins are set forth in Figure 13-18. A gene may contain more than one promoter, any one of which may be utilized in transcription. The spliced species of mRNA will differ with respect to which exons are spliced together. A gene

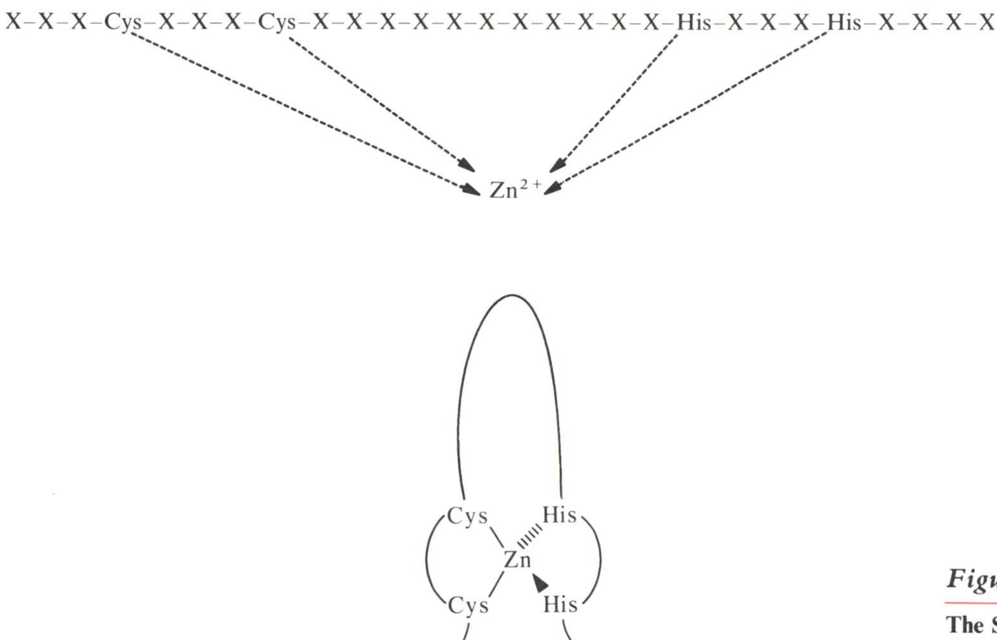

Figure 13-16

The Structural Basis for Zinc Fingers.

may contain more than one point at which a primary transcript is cleaved and polyadenylylated, and this too can lead to alternative splicing of exons. It is possible for an exon to be spliced out with an intron, so that the mRNA is missing an exon. It is also possible for an intron to be retained in the processed mRNA, in which case the protein will have an insert in its sequence. Clearly, a large variety of structures can be formed through alternative splicing of pre-mRNA, and the possibilities are not exhausted by the examples in Figure 13-18.

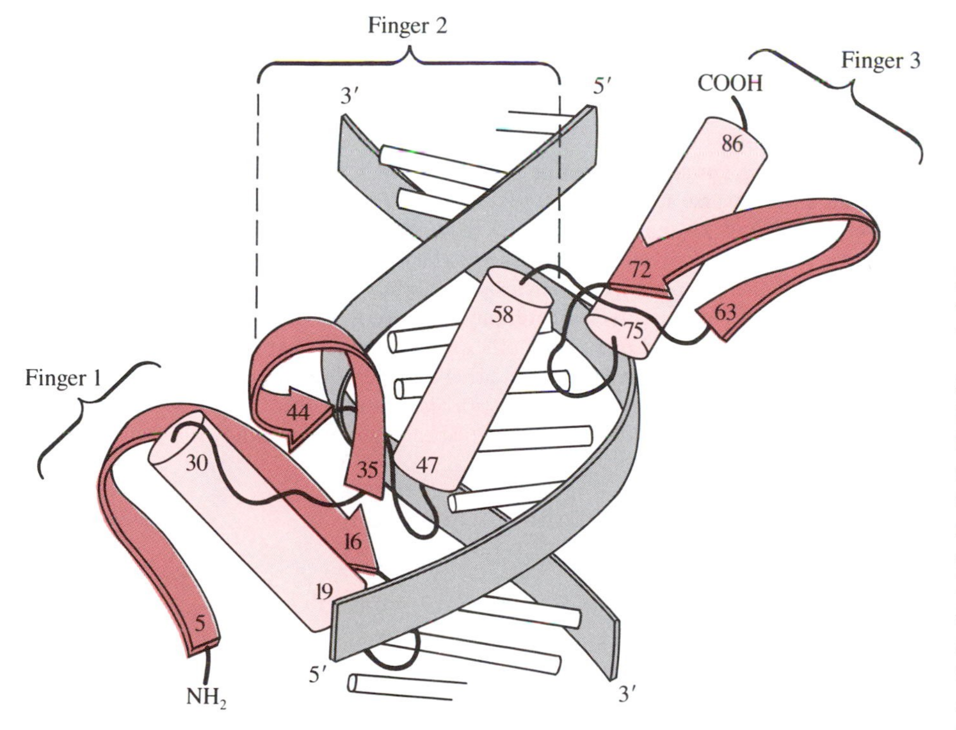

Figure 13-17

A Graphic Illustration of the X-ray Crystal Structure of the Zinc Finger Protein Zif268 Bound to DNA. Each zinc finger consists of an α-helix, shown as a cylinder, and a β-strand, shown as an arrow ribbon, held together through coordination to zinc. [After N. P. Pavletich and C. Pabo, *Science (Washington, D.C.)* 252 (1991): 809–817, Fig. 2. Copyright 1991 by the American Association for the Advancement of Science.]

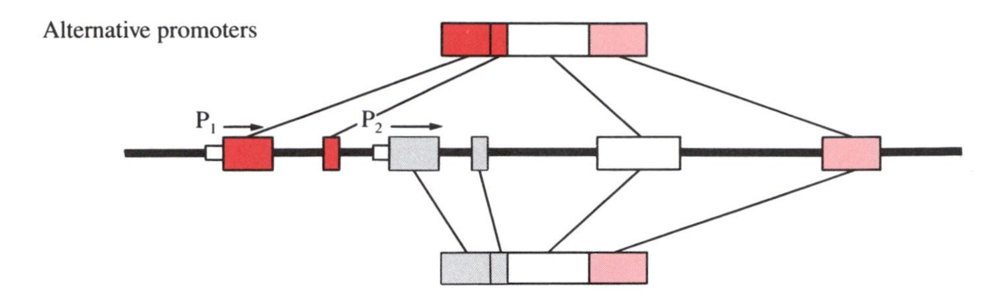

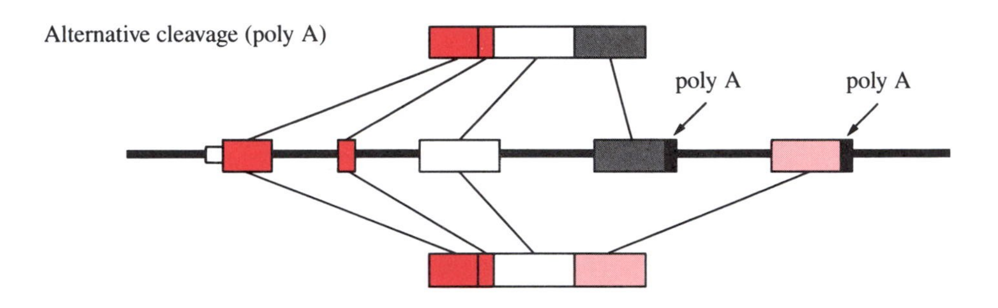

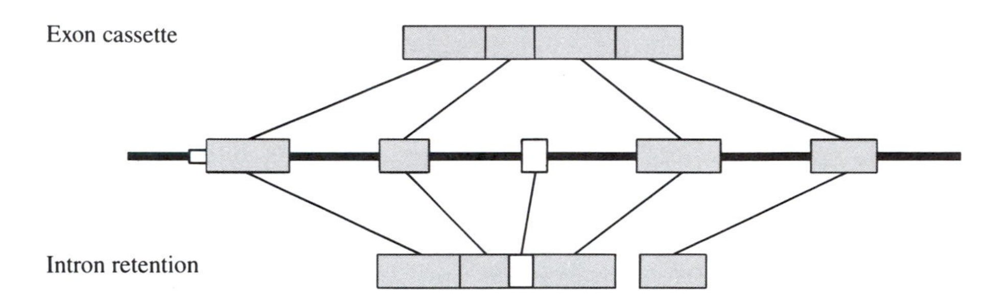

Figure 13-18

Protein Diversity Through Alternative Splicing of mRNA. Shown are four possibilities for diverse species of mRNA to arise from a single primary gene transcript. The mRNA may contain more than one promoter and undergo splicing in such a way as to eliminate one or the other promoter from the processed mRNA. This will have the effect of eliminating different exons from the mature mRNA. Alternatively, there may be more than one cleavage and polyadenylylation site, and one or the other may be eliminated by alternative splicing. In some cases, an exon may be either retained or spliced out and eliminated from the processed mRNA. It can also happen that a short intron may be either retained or spliced out of two species of mature mRNA.

Antibody Diversity Through Variable Recombination of Genes

An individual human being can produce more than 10 million different antibodies, each of which recognizes and binds a specific antigen. This diversity arises from the diverse population of proliferated B lymphocytes in the lymphatic system. The molecular basis for diverse antibody production by B lymphocytes is of considerable interest to biomedical scientists. The basic question is how the immunoglobulin genes in a single cell type can be so diverse. It cannot be that there are more than 10 million immunoglobulin genes in B lymphocytes, with each cell expressing only one of the genes, because there are not 10 million genes in a whole animal. The answer to this question is that the DNA in B lymphocytes is not quite identical from cell to cell; it differs in the immunoglobulin genes.

The most-important information comes from the nucleotide sequences of immunoglobulin genes in diverse cells. Comparisons of the nucleotide sequences strongly suggest that an immunoglobulin gene in a given cell is derived from a parent that has undergone genetic recombination in such a way as to lose segments of DNA. In *genetic recombination,* DNA is cleaved and recombined in a way that generates a different nucleotide sequence; it manifests itself in many ways, including the inheritance of genetic traits. In the production of B lymphocytes in the course of embryological development, recombination of the immunoglobulin genes leads to losses of different segments by different cells, so that each cell has a different nucleotide sequence in each immunoglobulin gene. A population of cells is, thereby, programmed to produce a population of different antibody molecules in the adult.

Only those parts of the immunoglobulin genes that encode the variable regions of the heavy, H, and light, L, chains of antibodies are subject to frequent recombination. The structural organization of an immunoglobulin (IgG), described in Chapter 6, is also illustrated in Figure 13-19. The variable regions of the H and L chains form the antibody-antigen combining site and consist of residues 1 through 109 in both chains. The constant regions are residues 110 through 446 of the H chain and 110 through 214 of the L chain.

Figure 13-20 illustrates how recombination of the embryonic *L* gene leads to many variant genes specifying L chains differing in sequence in the variable

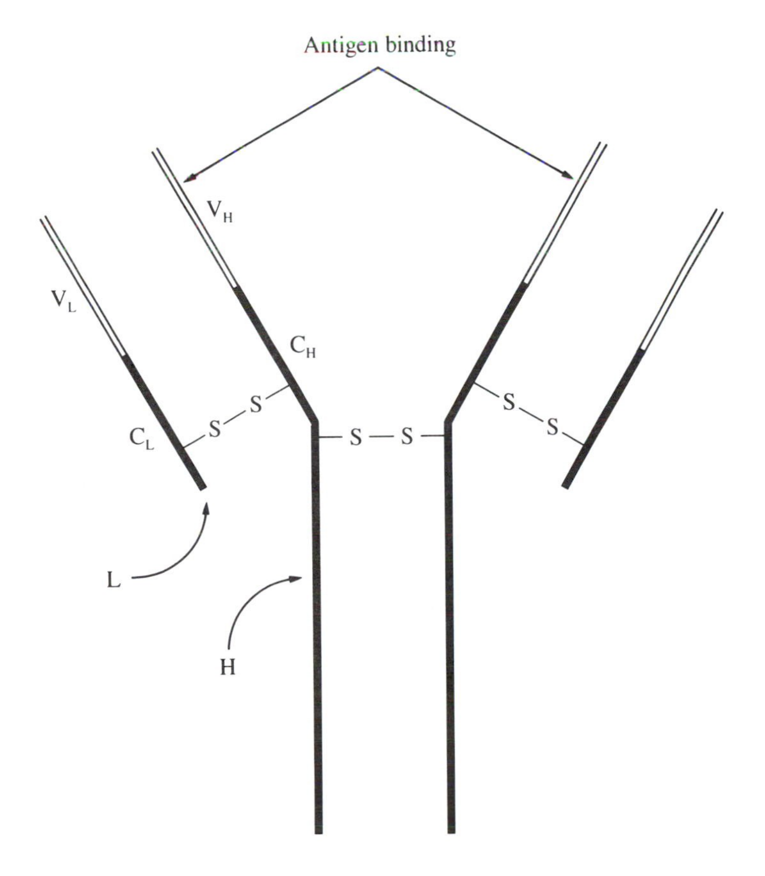

Figure 13-19

The Arrangement of Heavy and Light Chains in Immunoglobulins (IgG). In each IgG, there are two heavy (H) chains and two light (L) chains, and each chain contains variable (V_H and V_L) and constant (C_H and C_L) regions. The constant regions of all IgG molecules have the same amino acid sequence, but the variable regions differ from one antibody to another. The antigen-binding site is formed by the interaction of the variable regions of the H and L chains, with two binding sites per antibody molecule.

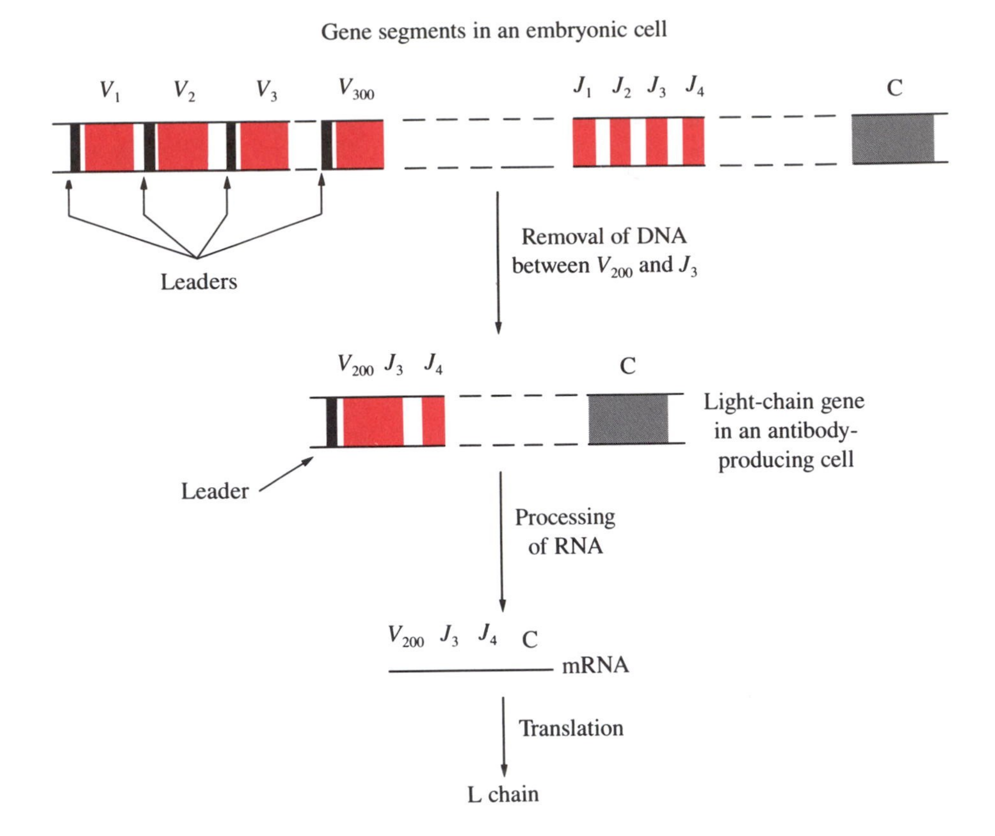

Figure 13-20

Recombination of Embryonic DNA Segments into a Gene for an L Chain of IgG. Embryonic DNA contains about three hundred *V* segments, four *J* segments, and a C segment. These segments are combined in various ways to form an *L* gene. In the recombination, the DNA between a *V* and a *J* segment is deleted, and that upstream from the recombined *V* segment is also deleted. In this illustration, the segment V_{200} is fused with J_3, but any other fusion is also allowed in other cells. In the mRNA, the exons for the variable region are V_{200}, J_3, and J_4, and that for the constant region is C.

region. There are two types of L chains, κ and λ, both of which have constant and variable regions. Recombination in the variable region of the two types is similar and Figure 13-20 describes the κ type. In the embryo, there are many DNA segments from which the gene for a light chain can be assembled, about three hundred designated *V*, four designated *J*, and a single one designated *C*. Each gene for a light chain consists of a single *V* segment, one to four *J* segments, and the *C* segment. The combined *V* and *J* segments encode the variable region and the *C* segment the constant region. The *V* and *J* segments are assembled by recombination of one *V* with one *J* segment, fusing them and deleting the intervening DNA. The DNA at the left of the fused *V* segment is also deleted. In Figure 13-20, the segment V_{200} is recombined with the segment J_3, producing the variable coding region $V_{200}J_3J_4$. In the mRNA, the coding sequences J_3 and J_4 are separated by an intron, and another intron separates J_4 from *C*. The introns are spliced out in mRNA processing.

There are about 3000 possible sequences for an L chain. This number arises from the 1200 ways to combine one *V* segment with a *J* segment (300×4) multiplied by the number of possible amino acids at the site of recombination, which is 2.5 on average. An analogous recombination scheme for H chains

BOX 13-1

POLYCLONAL AND MONOCLONAL ANTIBODIES

An antigen elicits the production of antibodies that recognize, bind, and mark it as a foreign molecule for attack by macrophages and neutrophils. An antigen causes the proliferation of B lymphocytes that produce antibodies with binding specificity for the antigen. A given antibody recognizes a small segment of a macromolecule such as a protein, and a given protein elicits the production of many different antibodies. The antibodies produced in response to a given antigen are known as *polyclonal antibodies.* Each B lymphocyte produces a single antibody, so that polyclonal antibodies with binding specificities for different parts of a single protein are produced by different B lymphocytes. The polyclonal antibodies bind to the foreign protein and cover it, marking it for degradation. Many proteins on the surfaces of bacteria or viruses can act as antigens and elicit antibody production, so that a whole bacterium or virion may be marked with thousands of antibody molecules. The macrophages and neutrophils recognize the clusters of antibodies on marked bodies and destroy the bodies in a complex series of oxidative and digestive processes.

Each B lymphocyte produces a particular immunoglobulin, and more than 10 million types of B lymphocytes are possible. In a given individual, only a few hundred or thousand types of B lymphocytes are present in substantial numbers at a given time, the others being present in very low and ineffective numbers. When a new antigen is sensed by the immune system, a new group of B lymphocytes is stimulated to proliferate and produce antibodies that recognize and bind the invading molecules. This increases the diversity of antibodies in the bloodstream. The mechanism by which the immune system recognizes a foreign antigen and stimulates the proliferation of new B lymphocytes is partly known and is currently under very intensive investigation.

Monoclonal antibodies are those produced by a single B lymphocyte. It is not possible to culture B lymphocytes for many generations owing to their mortality. Myeloma cells are lymphocytes that produce myeloma immunoglobulins and are immortal; therefore, they can be cultured indefinitely. They exhibit an important property; they can be induced to undergo cell fusion with normal B lymphocytes to produce cells that are immortal and that produce antibodies specified by the normal B lymphocytes. These cells can be cloned and cultured to produce very large amounts of pure, chemically homogeneous antibodies of any type produced by normal B lymphocytes. These are monoclonal antibodies.

Monoclonal antibodies are used for many purposes in basic research and biotechnology. Their specific binding properties make them valuable for use in purifying proteins by affinity chromatography. A tiny amount of a protein can yield partial amino acid sequence information that allows one to synthesize an oligopeptide fragment. Such a fragment can even be synthesized from the translated amino acid sequence derived from a cloned gene (Chapter 14). This peptide can be attached to a larger protein and used to generate cells that produce antibodies specific for the fragment. Fusion of these cells with myeloma cells and selection for those that produce the desired antibody allows one to clone fusion cells producing the monoclonal antibody. These antibodies can be attached to solid supports and used as the chromatographic adsorbent for purifying the protein from crude extracts. Monoclonal antibodies are also used to detect specific molecules in simple binding assays.

allows about 5000 different amino acid sequences to be encoded. Therefore, the potential number of different variable regions in an IgG molecule are 3000×5000, or 1.5×10^7. In this way, more than 10 million possible amino acid sequences can be created from a few hundred segments of embryonic DNA.

Summary

Gene expression is regulated by mechanisms that ensure the production of proteins and nucleic acids only under conditions in which they are needed by the cell. In bacteria, related genes are often found together in operons—clusters of related genes together with regulatory elements, the operators and promoters. In negative control, transcription of the genes in an operon is turned on when a repressor is displaced from the operator by an inducer. Positive control exerted in catabolite repression is brought about by binding of the cAMP·CAP complex to the promoter, which stimulates transcription by RNA polymerase.

Eukaryotes do not contain operons. In eukaryotic gene expression, each gene is separately transcribed under the control of promoters, enhancers, and transcription factors. The promoters typically contain sequences such as TATA boxes, CAAT boxes, and GC boxes. The transcription factors bound to promoters function as signals for RNA polymerase to bind and initiate transcription. Each gene may be controlled by the cooperative actions of more than one transcription factor, and other signal molecules such as hormones often regulate the functions of transcription factors by binding to them and altering their properties.

Diversity in related eukaryotic proteins is achieved in several ways. Genes within the chromosome may be exposed or shielded at various stages of the development of a cell or organism. Second, alternative splicing in the processing of mRNA can lead to variations in the nucleotide sequences of mature mRNA. In the development of embryonic B lymphocytes, gene recombination produces B lymphocytes that produce as many as 15 million different antibodies.

DNA-binding proteins include those with domains such as the helix-turn-helix or zinc finger. These proteins bind DNA by projecting structural domains into the major groove of helical DNA.

ADDITIONAL READING

FREIFELDER D, 1987 *Molecular Biology,* 2d ed. Boston: Jones and Bartlett.

HARTL D. L., 1991. *Basic Genetics,* 2d ed. Boston: Jones and Bartlett.

PRESCOTT, D. M. 1988. *Cells.* Boston: Jones and Bartlett.

PAVLETICH, N. P., and PABO C. O. 1991. *Science (Washington, D.C.)* 252: 809.

ANDERSON, W. F., OHLENDORF, D. H., TAKEDA Y., and MATTHEWS, B. W. 1980. *Nature* 290: 754.

PROBLEMS

1. How does constitutive synthesis of the enzymes in the *lac* operon of *E. coli* contribute to regulating the expression of this operon?

2. Suppose that an enzyme has the capacity to excise a U from a particular nucleotide sequence in mRNA. Suggest three ways in which this enzyme might participate in the regulation of gene expression.

3. A class of nucleotide sequences in mRNA known as Shine-Delgarno sequences has the property of binding ribosomes more tightly than others. How might this property be exploited to enhance the level of the expression of a gene?

4. Explain why some *E. coli* mutants with deletions in *lac Z* fail to produce permease and transacetylase.

Recombinant DNA and Protein Engineering

14

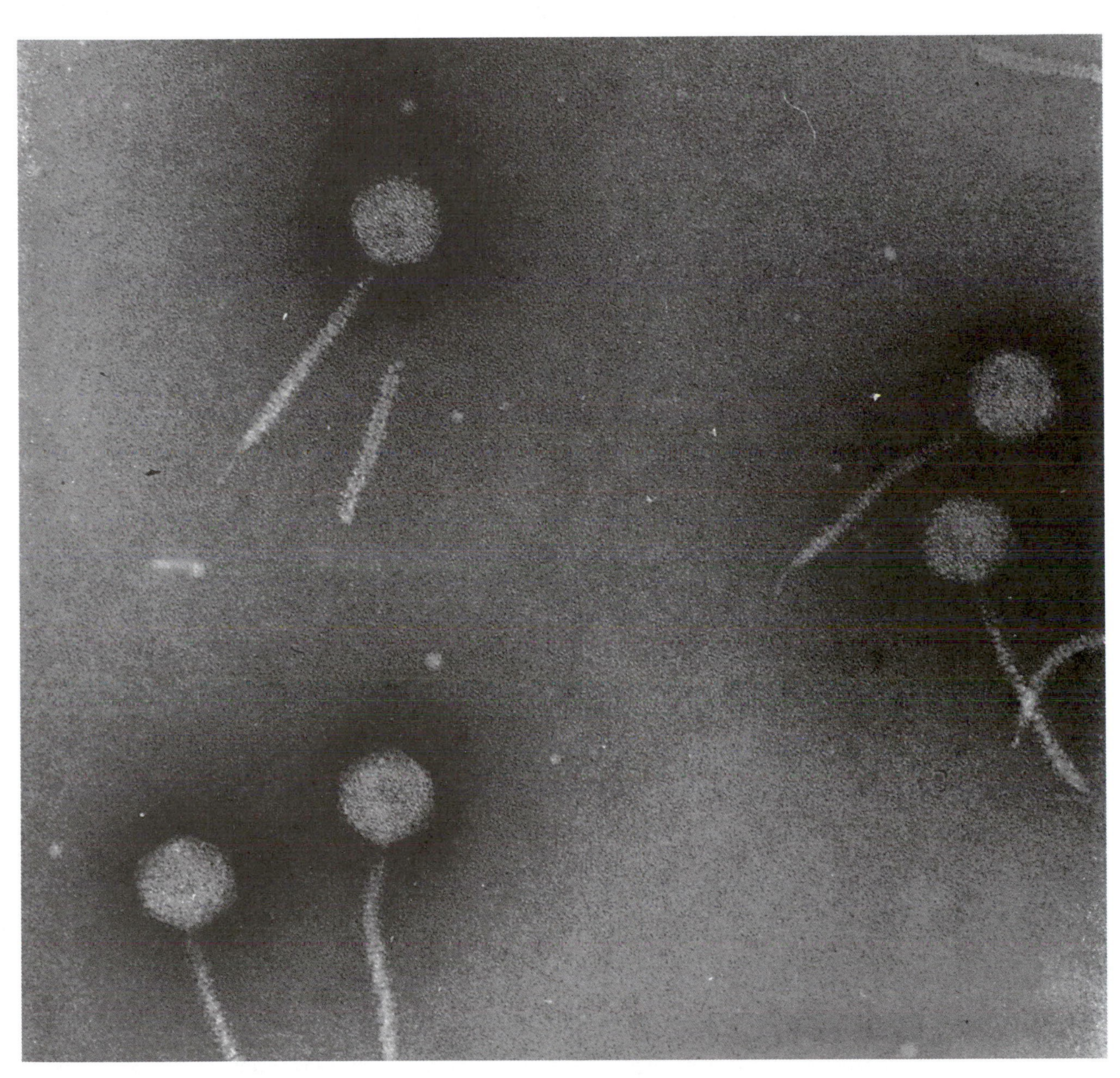

Electron micrographic image of bacteriophage λ at 250,000-fold magnification (Courtesy of Harold Fisher and originally published in An Electron Micrographic Atlas of Viruses by Robley Williams and Harold C. Fisher. Charles C. Thomas Publishers.)

Gene regulation is well enough understood to allow a few of the systems to be exploited for practical purposes in the large-scale production of proteins and other natural products for industrial and research applications. This is an important and growing activity in the biotechnology industry. Manipulation of the gene-expression systems requires recombinant DNA technology, a subject of this chapter. This technology also allows the production of specific mutant forms of proteins by site-directed mutagenesis, in which alterations in the amino acid sequences of proteins can be programmed into structural genes. The functional properties of such mutant proteins can give valuable information about the mechanism of action of the normal protein; the structural properties of the mutants give equally valuable information about the factors governing the folding of a protein into its three-dimensional structure.

Recombinant DNA

Recombinant DNA is any species of DNA that has been produced by subdividing natural DNA and then joining the segments to form a new species of DNA. Recombination is most often the natural process of *genetic recombination* that occurs constantly and leads to the transmission of genetic traits, to biological diversity, and to the evolution of species. Recent advances in DNA biochemistry have made it possible to use the chemistry of recombination to carry out directed fragmentation and recombination of DNA in the laboratory, and this has allowed genes to be cloned, studied, and even systematically modified. These methods and their current and potential utility are the subjects of this section. Genetic engineering can be carried out in a few bacterial systems and increasingly in yeast. In this chapter, we shall limit the discussion to the use of *E. coli* in genetic engineering, and we shall begin by describing how genes are carried into *E. coli* by cloning vehicles (vectors).

Genes and Vectors

Most genetic engineering in *E. coli* is geared to manipulating the expression of genes that are not normally a part of the genome of *E. coli*. The DNA of *E. coli* is always expressed whenever the cells are used in genetic engineering; however, in most cases, the *engineered genes* are inserted within cloning vehicles, or *vectors*, by methods to be described shortly. Vectors are small elements of DNA that can be introduced into *E. coli* cells and replicated. They contain their own genes that control their own replication; in genetic engineering, additional genes and the necessary transcription-regulation elements are artificially introduced into the vector DNA. These accessory genes carried by the vectors can then be expressed as proteins within *E. coli*. *Gene cloning* consists of ligating a gene into a vector, introducing the vector into a host cell, and isolating the host cells containing and expressing that gene.

Plasmid pBR322. The vectors used in *E. coli* are of two general types, plasmids and bacteriophages. *Plasmids* are molecules of extrachromosomal, circular DNA that appear in most bacteria and some eukaryotic species. Plasmids vary in many basic properties, including whether they are capable of controlled self-transmission from one cell to another, whether they can undergo replication to only a few or many copies per cell, and whether they naturally carry genes in addition to those required for their self-replication.

The *E. coli* plasmid pBR322, for which a genetic map is illustrated in Figure 14-1, has been engineered and used in many ways as a vehicle for genetic cloning and expression. It has a molecular weight of 2.9×10^6; it can be

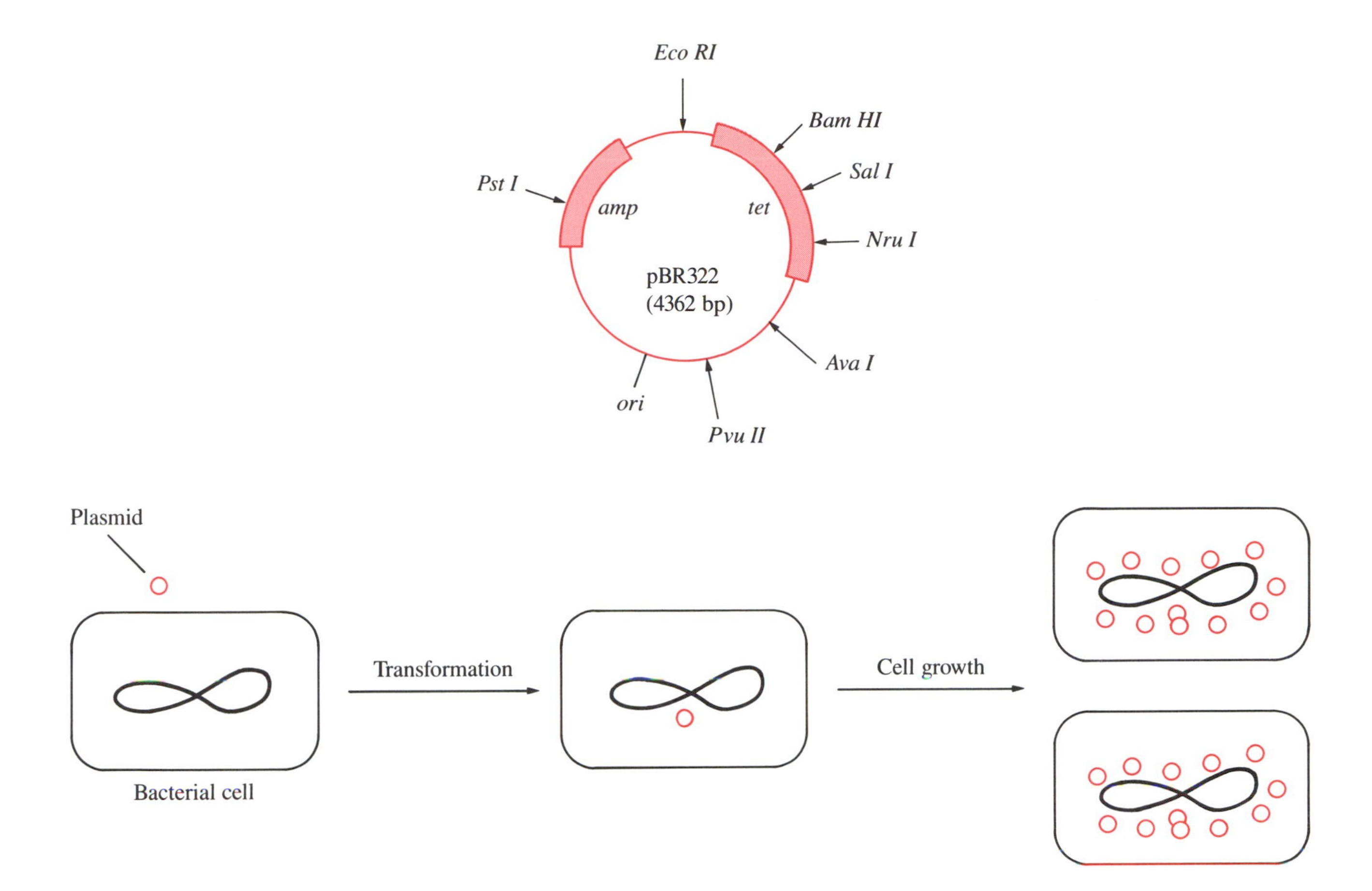

Figure 14-1

Plasmid pBR322. This *E. coli* plasmid carries genes (*tet* and *amp*) that encode tetracycline- and ampicillin-inactivating enzymes. The replication of this plasmid in *E. coli* requires host enzymes but is controlled by the plasmid, with the origin of replication at the position marked *ori*. The points at which base sequences recognized by restriction endonucleases are located are indicated by arrows. The sequence and cleavage specificities of several restriction enzymes are given in Table 14-2. Cells of *E. coli* that do not contain the plasmid can acquire it at a very low frequency through exposure and environmental stress. *E. coli* can acquire the plasmid at a high frequency by *transformation*, a process in which the cells are treated with $CaCl_2$ to enhance the permeability of the cell walls and then exposed to a large number of plasmids. A few plasmids enter cells and, as the cells mature and divide, the plasmids are replicated within the cells. pBR322 is a high-copy-number plasmid.

introduced into a host cell by the process of transformation; it has a high copy number; and it carries drug-resistance genes (resistance is generally to tetracycline and ampicillin, tet^+ and amp^+ in genetic notation). It also has several palindromic sequences, marked by the arrows in Figure 14-1, that are cleavage sites for specific restriction endonucleases. These properties make pBR322 an ideal vector for use in producing and using recombinant DNA. Because of the plasmid's large copy number, a gene introduced into pBR322 by recombination, together with a suitable promoter-operator to facilitate its transcription in *E. coli*, can be induced to produce a large amount of the protein specified by the gene. In effect, *E. coli* that has been transformed by this plasmid, which normally replicates to about fifty copies per cell, becomes a tiny reactor that can be induced to produce a large amount of a protein specified by the accessory gene carried within the plasmid DNA, irrespective of whether the cell has any need for that protein. *E. coli* cells transformed by such vectors as pBR322 are *overexpression* systems for the cloned gene.

The drug-resistance genes in the plasmid provide a valuable cloning tool, because they encode the enzymes that inactivate the drugs; by this means, they confer drug resistance to any *E. coli* that has been transformed by the plasmid. The value of drug resistance in cloning will become apparent in subsequent sections.

pBR322 can be isolated from harvested bacteria by suitable treatment of an extract. Plasmid DNA is generally separated from chromosomal DNA, initially by precipitation of chromosomal DNA with detergent and then by centrifugation in a CsCl gradient. pBR322 is not self-transmissible with high frequency under laboratory conditions; however, it can be induced to enter *E. coli* with a high frequency by the process of *transformation.* Exposure of *E. coli* to $CaCl_2$ under specific conditions reversibly weakens the cell walls and allows the plasmid to enter the cell. Upon transfer to a suitable growth medium, the transformed cells grow with replication of the plasmid and expression of any genes that they carry, such as those for the production of tetracycline- and ampicillin-inactivating enzymes.

Bacteriophage M13. The other general type of vectors for *E. coli* are the bacteriophages, including M13, which was briefly described in Chapter 12 as a sequencing vector, and phage λ, a lysogenic phage, the life cycle of which includes the integration of its DNA into the bacterial chromosome. The infection of bacterial cells with phages is operationally a simple process, consisting essentially of mixing phages with appropriate host cells; however, the detailed mechanisms of infection are complex and vary with the phage and host cell. *Transfection* is the introduction of phage DNA into a host, and this method is often employed when using phage DNA as a vehicle for transmitting genes that have been cloned into phage DNA into a bacterial host. Both M13 and λ are widely used in recombinant DNA research; however, M13 is almost universally used for nucleotide sequence analysis by the partial replication methods (dideoxy sequencing, Chapter 12) and for mismatched-oligonucleotide-induced mutagenesis. M13 is less suitable than pBR322 for certain other applications in genetic engineering. In practice, derivatives of M13 and pBR322 are used together in many research applications. Other vectors are listed in Table 14-1.

The life cycle of M13 is illustrated in Figure 14-2. M13 phages are among the very few that infect cells, undergo replication, and are extruded from the cells without causing cell lysis or death. M13 is a filamentous phage, containing a

Table 14-1

E. coli cloning vehicles currently in use

Plasmids	Phages
pBR322 (*tet*$^+$, *amp*$^+$)	M13mp8 (contains *lac* o/p, *lac Z*, and a polylinker)
pBR325 (*tet*$^+$, *amp*$^+$, *cam*$^+$)	λgt4 · λB (thermally induced)
pSC101 (*tet*$^+$)	$\lambda\Delta$z1 (cloning site in *lac Z* gene)
ColE1 (*imm*E1)	

Note: *tet*$^+$, tetracycline resistance; *amp*$^+$, ampicillin resistance; *cam*$^+$, chloramphenicol resistance; *imm*E1, plasmid kills bacteria that are not infected with a ColE plasmid by production of colicin B, a protein that kills sensitive bacteria.

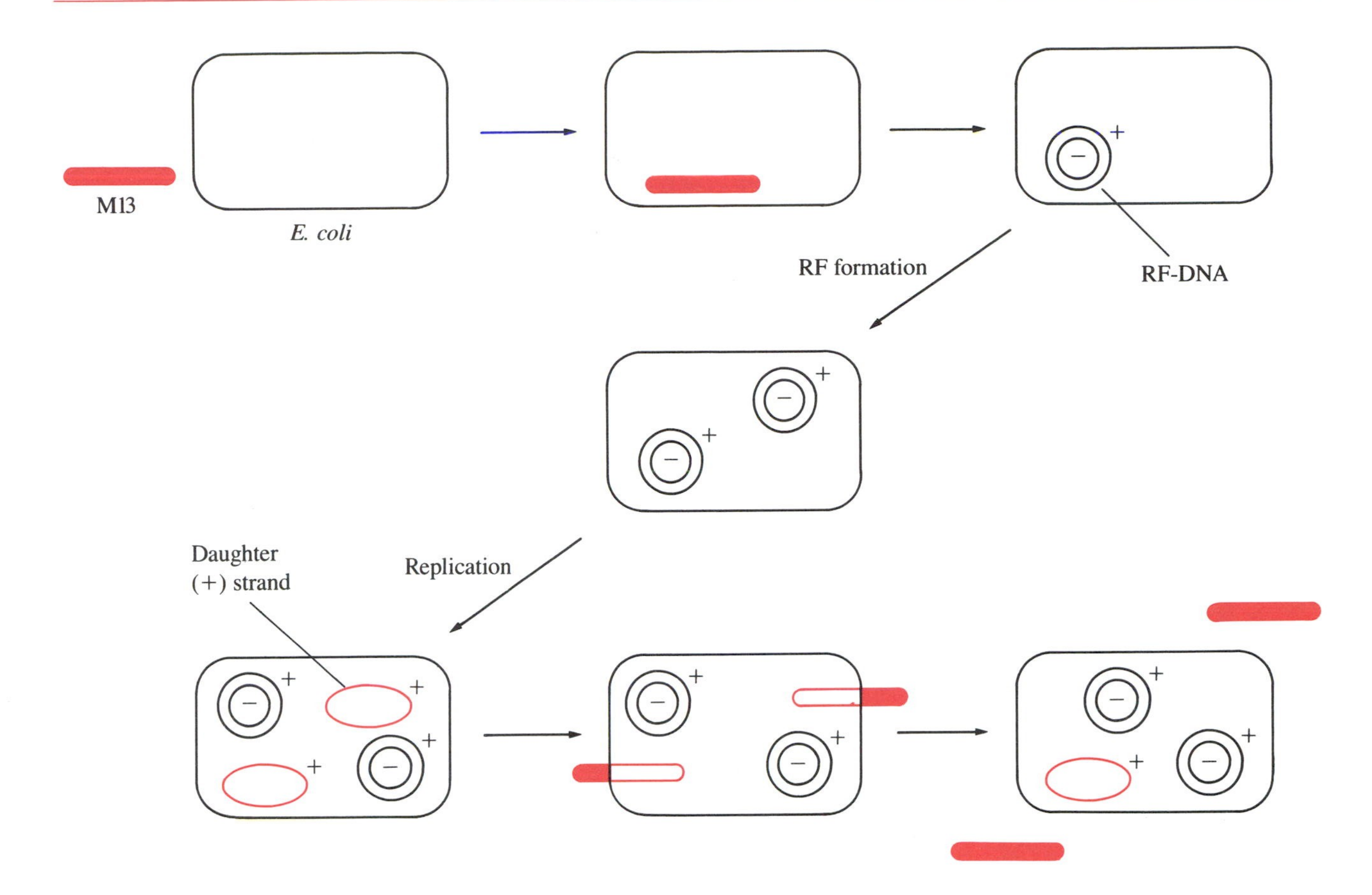

Figure 14-2

Events in the Life Cycle of Phage M13. M13 is a filamentous phage that infects *E. coli,* is replicated within the cell, and is extruded from the cell without causing cell lysis. It is one of the few bacteriophages that do not cause lysis. In the initial stage, the intact phage is absorbed through the F-pilus of a cell and lodged on the inside membrane surface, where some of the coat proteins are removed. The single-stranded (+) circular DNA is simultaneously converted into the double-stranded replicating form (RF) by synthesis of the complementary (−) strand. The RF is replicated to several copies per cell. The (−) strand in RF codes for the synthesis of the phage proteins by the bacterial protein biosynthetic system. The RF undergoes a special replication (rolling circle, Figure 14-3) culminating in the production of single (+) strands that are coated with a phage protein. These coated (+) strands migrate to the bacterial membrane, where they exchange the binding protein acquired in the course of replication for their final coating of three proteins and are extruded through the bacterial wall.

single strand of circular DNA, and it enters a cell spontaneously at the F-pilus of a cell. (Cells that do not express F-pili are not infected.) At the inner-membrane surface of a cell, it loses its coat protein and is simultaneously converted into the *replicative form* (RF-1), which is double stranded. The RF-1 can be replicated and is the form that encodes the phage proteins. Transcription of the newly synthesized (−) strand in the RF form and protein biosynthesis are carried out by the host enzymes.

Replication of any bacteriophage requires both the biosynthesis of the coat proteins and the replication of the phage genome, which, in M13 and certain other phages, is circular, single-stranded DNA. Generation of the (+) strand exemplifies the diversity of replication modes for phage DNA and viruses in general. The mechanisms of viral genome replication vary considerably, although the rolling-circle mechanism shown in Figure 14-3 is a common mode for single-stranded DNA phage.

Rolling-circle replication begins with the introduction of a nick in the (+) strand at a specific location, which is brought about by the nucleophilic action of an enzyme that becomes covalently attached to the 5′-phosphoryl group. This process creates a primer, the 3′-OH end, from which DNA polymerase III can catalyze elongation. A single-stranded DNA binding protein encoded by the (−) strand associates with the original (+) strand and aids in its displacement from the (−) strand. The new replication complex is known as an RF-2 complex, and replication begins with the production of (+) strands from the (−) template

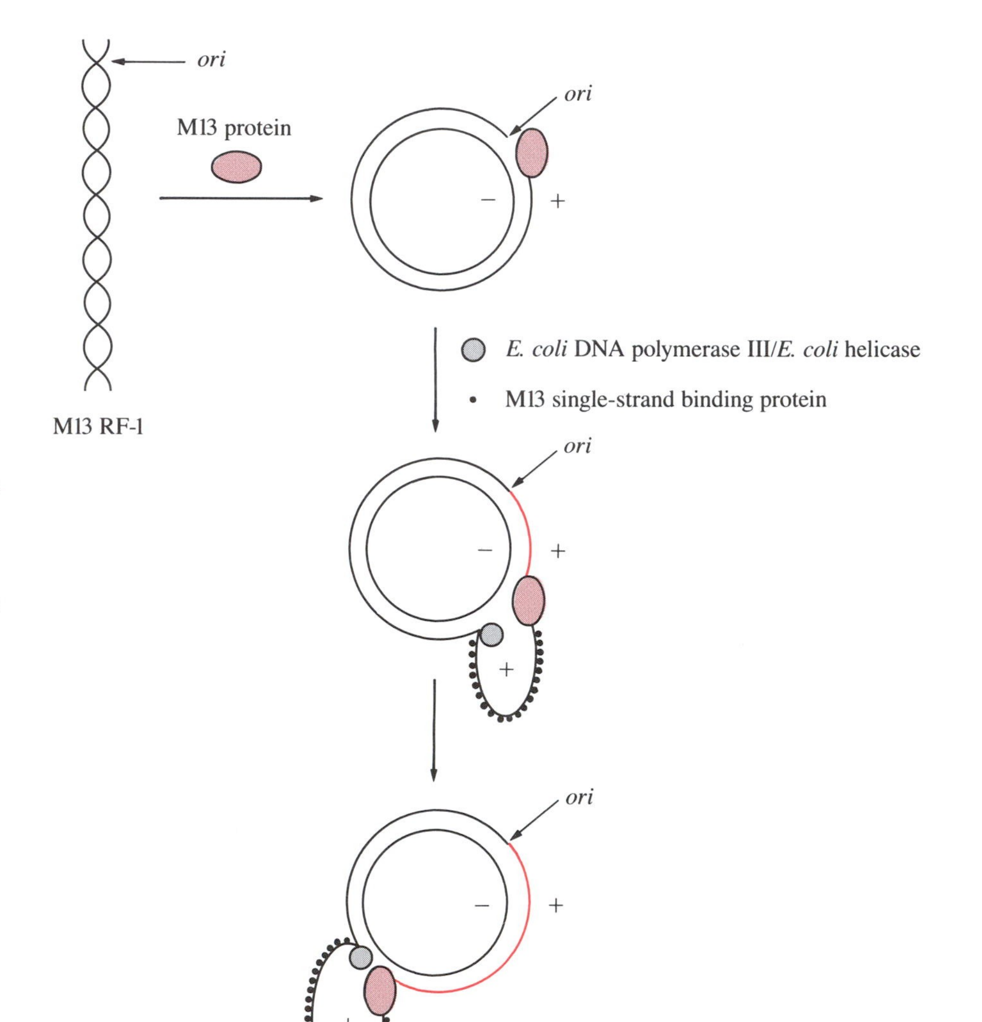

Figure 14-3

The Rolling-Circle Replication of RF Forms of Phage DNA. The molecular details by which the (+) strand of the RF form is replicated by the rolling-circle mechanism is less well understood for M13 than for ΦX174, another single-stranded DNA *E. coli* bacteriophage. The main events, as illustrated here, begin with cleavage of the (+) strand at a specific origin of replication by the action of a phage protein that cleaves the phosphodiester by forming a covalent bond with the 5′-phosphoryl group. This relaxes any supercoiling and releases the 3′-OH group to serve as the primer for DNA polymerase III. Polymerization proceeds with the (−) strand as the template to generate a new (+) strand and displace the original (+) strand. The original (+) strand is looped out and stabilized by binding to a phage protein. Polymerization continues around the (−) strand and is followed by circularization and closing of the coated (+) strands. The molecular details by which the rolling circles are closed to coated (+) strands are not fully known.

catalyzed by DNA polymerase. By processes that are not yet clearly defined, the new (+) strands are circularized and make their way to the cell membrane, where the single-stranded DNA binding protein is displaced by the coat proteins. Extrusion of the complete phage completes the life cycle.

Other single-stranded bacteriophages are replicated by analogous or related processes; however, nearly all of them lead to cell lysis. The only ones that do not are a few filamentous phages of *E. coli.* This and other properties of M13 make it useful for recombinant DNA research. The life cycle of M13 allows large amounts of specialized DNA to be produced and isolated as single-stranded DNA from the phage particles. Recall from Chapter 12 that nucleotide sequencing by the partial-replication methods utilizes single-stranded DNA carrying the sequence of interest. Moreover, the RF form also can be isolated, so that a recombinant segment of DNA, such as a gene, that was cloned into the phage DNA earlier also is available in double-stranded form. By methods shortly to be explained, the recombinant DNA can be excised from the RF-1 and inserted into other vectors, such as derivatives of pBR322, which are more convenient for other purposes.

The isolation of single (+) stranded DNA (ssDNA) and duplex RF-1 DNA of M13 is relatively straightforward. To obtain ssDNA, the phages are simply isolated from a growing culture of infected *E. coli* by centrifugation of the cells. The supernatant fluid contains the phages, and the cells contain the RF-1 form of phage DNA. To obtain (+) strands from phages, the coat proteins are removed from the phages by treatment with a detergent. The ssDNA is separated from the detergent by selective precipitation, by the use of salts and organic solvents, and the ssDNA is dissolved in aqueous solution and extracted with phenol to remove any residual proteins. The ssDNA may be brought to a high stage of purity by ultracentrifugation in a gradient of CsCl. The duplex RF-1 form is isolated from the cells by disruption of the cell walls and separation from chromosomal DNA by the same methods used for plasmid DNA. The RF-1 form can also be used in the same way as plasmid DNA to transform *E. coli.*

Cloning DNA into Vectors

The principles underlying molecular cloning are elegantly simple. Any method by which a fragment of double-stranded DNA can be inserted into a vector such as a plasmid or virus without diminishing the infectivity or transformation capacity of the vector can be the basis for molecular cloning. One widely used method depends entirely on the specificities of restriction endonucleases and DNA ligase (Chapter 12). Consider the plasmid pBR322 in Figure 14-1, which contains several restriction sites. The sites cleaved by restriction endonucleases offer a means for introducing a new segment of DNA. The simplest method for achieving this is outlined in Figure 14-4. Cleavage of the vector by Eco RI opens (linearizes) the vector specifically at one site and produces *staggered ends.* Any segment of DNA with complementary staggered ends will hybridize with the linearized vector. For example, DNA from a species other than *E. coli* can be isolated and cut into fragments by digestion with Eco RI, creating fragments with the complementary staggered ends; and some of the fragments may contain intact genes for proteins from that species. When the fragments are mixed with the linear plasmid DNA and allowed to hybridize, circular double-stranded and nicked hybrids will be obtained. Because each nick contains a 5′-phosphate and a 3′-OH end, the nicks can be sealed by the action of DNA ligase. (Polynucleotide kinase and ATPase are included to ensure that all the 5′ ends are phosphorylated.) This forms a class of plasmid containing DNA inserts between the *Eco RI* sites. After integration into the vector by DNA ligase, the DNA fragment is said to be cloned into the vector.

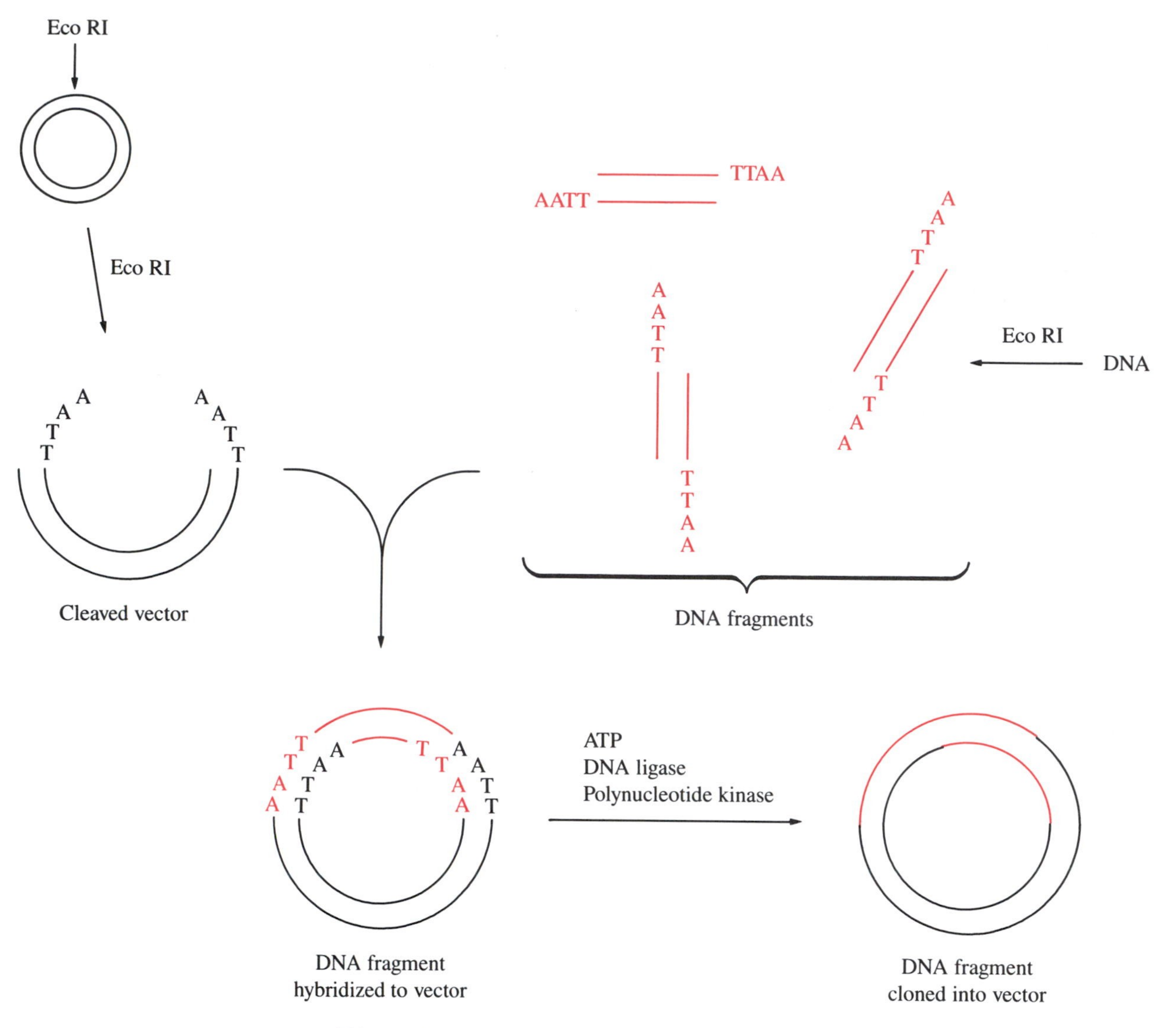

Figure 14-4

Molecular Cloning of DNA Fragments into a Vector. Digestion of plasmid pBR322 with Eco RI cleaves it at a single site and creates staggered ends, with complementary nucleotide sequences in the overhanging ends. These sequences are illustrated in the linear plasmid, or cleaved vector. Similar digestion of chromosomal DNA by Eco RI produces fragments of DNA, also with staggered ends that are complementary to the ends of the linear plasmid DNA. Mixture of the DNA digest with the linear plasmid produces nicked circular hybrids containing the plasmid DNA and a fragment. Treatment of the hybrids with DNA ligase and ATP seals the nicks by ligating the 3′-OH ends to the 5′-phosphate ends. (Polynucleotide kinase is included in the ligation reaction to ensure that the 5′ ends are phosphorylated.) The closed circular DNA contains all of the plasmid plus the DNA fragment, which is said to be cloned. *E. coli* transformed by this new plasmid will produce many copies of it, including the DNA fragment, which may contain a complete gene.

The nucleotide sequences at the hybridization sites before ligation are shown in Figure 14-5. Note that a 180° rotation of the DNA fragment within the plane of the figure will allow the opposite orientation of the DNA fragment to be hybridized to the vector, and this also can be integrated into the vector by DNA

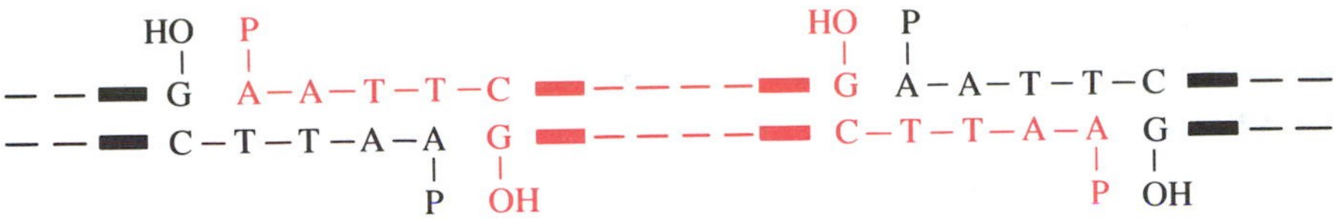

```
        HO  P                               HO  P
        |   |                               |   |
-- ■ G   A-A-T-T-C ■-----■ G   A-A-T-T-C ■ --
-- ■ C-T-T-A-A   G ■-----■ C-T-T-A-A   G ■ --
             |   |                     |   |
             P   OH                    P   OH
```

Figure 14-5

Nucleotide Sequences at the Nicks in the Circular DNA of Figure 14-4. The specificity of Eco RI leads to staggered ends with the nucleotide sequence AATT-3′-OH in the overhang. These ends will undergo base pairing with one another, as illustrated here for the example shown in Figure 14-4. The central segment shown in color represents a DNA fragment obtained by digestion of chromosomal DNA by Eco RI. Note that the two ends are identical with each other and with those of the linear plasmid. The hybrid can be sealed at the nicks by DNA ligase. Notice also that the DNA fragment can hybridize with the plasmid in two ways. Rotation of the fragment by 180° within the plane of the page while holding the plasmid ends in place will allow the rotated fragment to form similar base pairs with the opposite ends of the linear plasmid. These, too, can be sealed with DNA ligase.

ligase. Thus, any fragment that can be cloned into the vector will be cloned in either of the two possible orientations. In the initial cloning of a segment of DNA, the fact that a desired fragment may appear in two orientations is not a problem and may be advantageous. A gene within such a fragment may be expressed in either orientation if its regulatory elements are also contained within the fragment and if these elements allow the expression of the gene in *E. coli.*

In many applications, it is useful to use two restriction enzymes to define the boundaries of DNA fragments that are to be cloned into a vector. In Figure 14-1, the sites of pBR322 cleaved by the restriction endonucleases Eco RI and Bam HI (or any other pair of sites) offer a means for introducing a new segment of DNA in a specific orientation within the vector. The sequences and cleavage specificities given in Table 14-2 are as illustrated in Figure 14-6, which shows how a molecular cloning experiment can lead to the incorporation of a segment of DNA between the *Eco RI* and *Bam HI* sites. If pBR322 is treated with Eco RI and Bam HI, a segment of DNA will be excised, and the remaining linear

Table 14-2

Specificities of some restriction enzymes

Enzyme	Bacterium	Cleavage Specificity
Eco RI	*Escherichia coli*	G |AATT C C TTAA| G
Bam HI	*Bacillus amyloliquefaciens* H	G |GATC C C CTAG| G
Pst I	*Providentia stuartii*	C TGCA| G G |ACGT C
Sal I	*Streptococcus albus* G	G |TCGA C C AGCT| G
Sma I	*Serratia marcescens*	CCC|GGG GGG|CCC

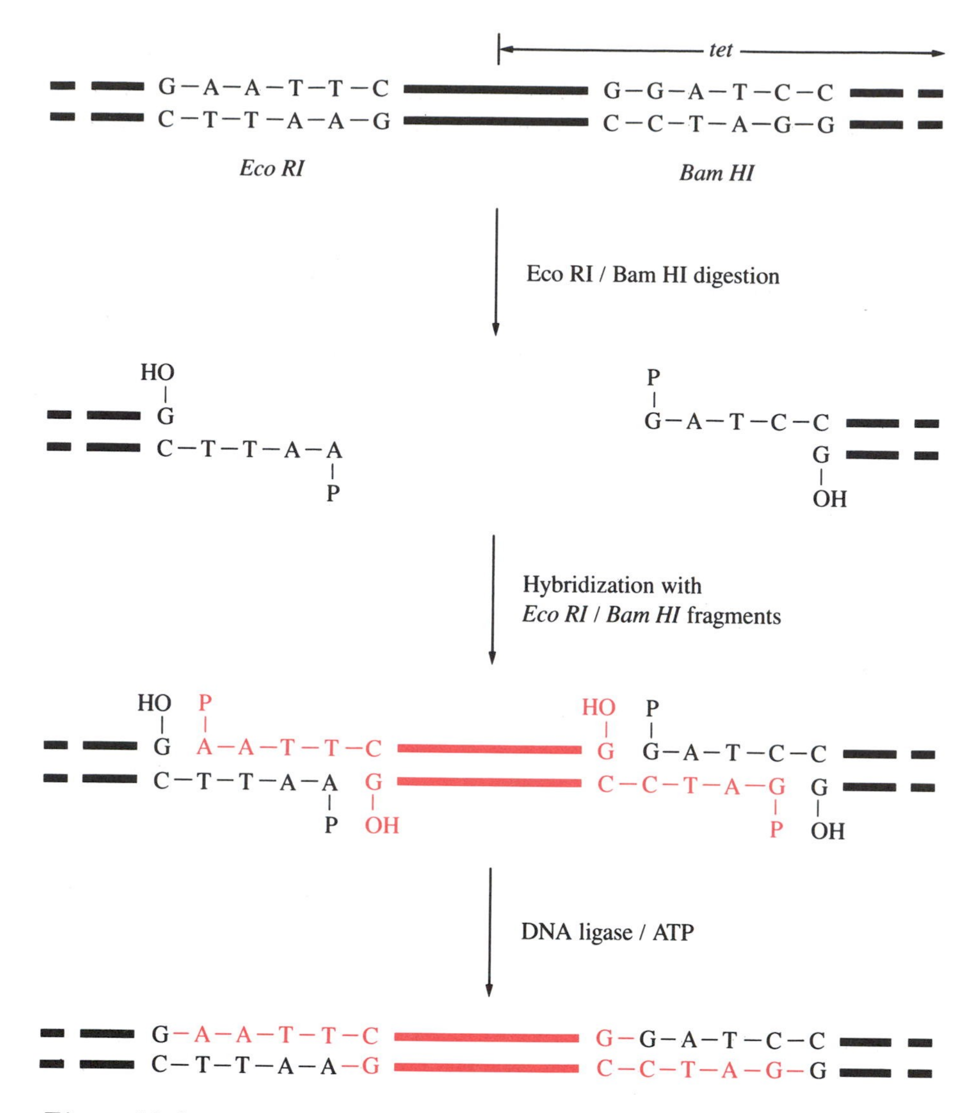

Figure 14-6

Molecular Cloning of a Restriction Fragment Between Two Restriction Sites of pBR322. Digestion of pBR322 by Eco RI and Bam HI excises a fragment of the plasmid and creates staggered ends that are complementary to those of DNA fragments obtained by digestion of another species of DNA by Eco RI and Bam HI. The staggered ends of the linearized plasmid and those of the fragments can be joined by the use of DNA ligase to form hybrids. In practice, the 5′-phosphate ends of the plasmid and DNA fragments are removed by the use of alkaline phosphatase before being isolated. The 5′-phosphates are then restored by the use of polynucleotide kinase in the last step, when the complementary staggered ends are joined by DNA ligase.

plasmid DNA will contain two *different* staggered ends, each of which contains a 5′-phosphate and a 3′-OH end. The linear plasmid DNA can be isolated from the small excised fragment and from uncut circular DNA by gel electrophoresis. The linear DNA will undergo hybridization with any segment of double-stranded DNA that contains complementary staggered ends. Such segments will be created from a sample of DNA that has been digested with both Eco RI and Bam HI. Figure 14-7 gives a physical overview of the linearization by Eco RI and Bam HI, hybridization with complementary fragments, and sealing by DNA ligase.

The recombinant plasmids contain intact origins of replication and so can be transformed into *E. coli* and replicated. In this experiment, the recombinant plasmid will no longer confer tetracycline resistance to the host cell, owing to partial deletion of the *tet* gene by digestion with Eco RI and Bam HI; however, the *amp* gene remains intact and will confer ampicillin resistance. Thus, all *E. coli* that are transformed by the recombinant plasmid will grow in media containing ampicillin, but not in media with tetracycline; that is, they are amp^+ tet^- in genetic notation. This property of the recombinant plasmid offers a means of screening a transformed bacterial culture for cells that carry the circular plasmid.

The method outlined in Figures 14-6 and 14-7 is illustrative of one of many methods that have been used to insert fragments of DNA into a cloning vector. A complete description of the available methods is beyond the scope of this textbook. (You will be given an opportunity to devise methods in the problems at the end of the chapter.)

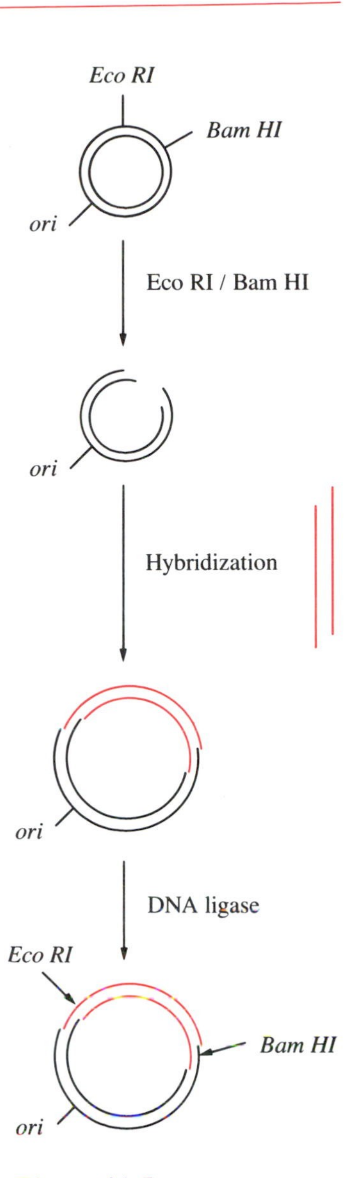

Figure 14-7

Physical Overview of Molecular Cloning in Figure 14-6.

Engineered Cloning Vectors

The usefulness of pBR322 and the RF-1 form of M13 can be enhanced by introducing special restriction sites into their structures. This can be done by excising small fragments by digestion with restriction enzymes and inserting synthetic fragments having any desired sequence. Many commercial vectors of this type have been prepared. For example, the commercial forms of M13 have unique restriction sites between which genes with complementary staggered ends can be inserted. These can then be transfected into *E. coli* and amplified by growth of the host. Both the new RF-1 and the mature phage containing the (+) strand can be isolated. Similarly, variants of pBR322 can be prepared with many specific restriction sites in a small segment, a *polylinker region,* for use in cloning DNA fragments with any of a large variety of different complementary ends. An example of such a construction is shown in Figure 14-8. This vector contains six

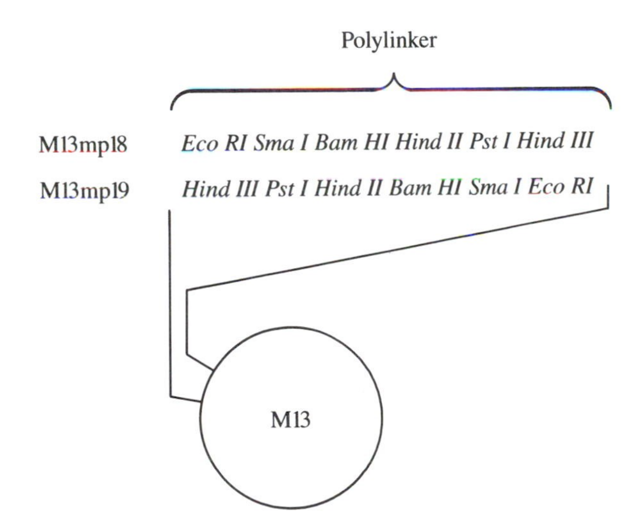

Figure 14-8

Polylinkers Engineered into Phage M13. Fragments of duplex deoxyoligonucleotides with base sequences corresponding to clusters of unique sites for cleavage by restriction endonucleases can be synthesized and cloned into phages or plasmids. Such nucleotide sequences are known as polylinker regions. Shown here are two versions of M13 containing the same polylinker in opposite orientations. The polylinker contains cleavage sites for six restriction endonucleases, and any one or a pair of restriction sites can, in principle, be used for cloning.

restriction sites from the *Eco RI* site to the *Hind III* site that can be used to clone a fragment with ends complementary to those created by cleavage of the vector with any two of the restriction enzymes. The vector is shown in two versions, with opposite orientations of the polylinker region. This vector enables the investigator to use any two of a variety of restriction enzymes to prepare fragments of chromosomal DNA. A particular gene in chromosomal DNA may, for example, happen to be cleaved at a point within the gene by Eco RI, so that this enzyme will not be useful for cloning. If the gene is not cleaved by either Hind III or Sal I, these enzymes can be used to digest the chromosomal DNA to obtain fragments that can be cloned between the *Hind III* and *Sal I* sites.

A problem that often arises with a cloned gene is that the regulatory genes and elements that control the transcription of the gene in its normal chromosomal setting are often not present or functional when the gene is cloned into a vector. In this case, the cloned gene may be intact in the vector but not expressed in the host cell. This problem is generally overcome by placing the gene's transcription under the control of a host regulation system. A convenient system, among others, in *E. coli* is the *lac* promoter-operator sequence, which is not very large and can be conveniently synthesized and cloned into one end of a polylinker region of a plasmid (Figure 14-9). When present in a multicopy plasmid, this promoter will bind RNA polymerase and promote the transcription of downstream genes as if it were the start site for β-galactosidase, the main limitation being that the Lac repressor in *E. coli* will normally bind to the operator and prevent transcription. However, there is often insufficient re-

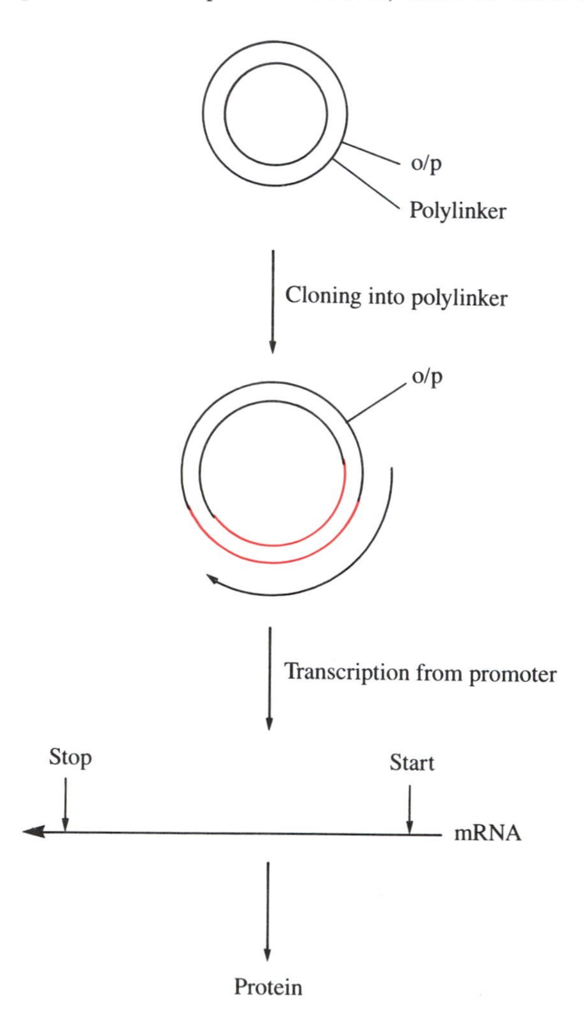

Figure 14-9

Construction of an Expression Vector for Cloned Genes. Operator-promoter (o/p) sequences are often cloned into expression vectors in front of polylinker regions, into which genes can be cloned. Transcription of the cloned genes can then be controlled by the o/p sequence, which is normally one of the regulatory elements of the host cell. For example, if the *lac* o/p is cloned ahead of the polylinker region, transcription of the cloned gene will be regulated by the *lac* repressor and can be derepressed by IPTG.

pressor in *E. coli* to occupy all of the *lac* operators in a multicopy plasmid (10–100, depending on growth conditions); moreover, the *lac* operator can be derepressed by IPTG or MTG in the growth medium. In this way, the transcription of a cloned gene can be artificially induced in *E. coli*. Many other strategies are available for regulating the expression of foreign genes in *E. coli*, and analogous strategies are available for controlling the expression of genes cloned into yeast vectors.

Cloning Eukaryotic Genes

Eukaryotic genes present several special cloning problems. A major obstacle is that eukaryotic genes normally contain introns (noncoding sequences) interspersed with the exons (coding sequences, Chapter 12). In eukaryotic cells, the introns are postranscriptionally spliced out of pre-mRNA to form mRNA for translation into a protein; however, the splicing systems are not normally present in host cells for cloning vectors. Therefore, such a gene may be successfully cloned and transcribed, but expression as a protein may be impossible in the host owing to the presence of introns.

The problem of introns may be overcome through the use of reverse transcriptase, the complex viral enzyme that transcribes the RNA genome of a retrovirus into DNA in a virus-infected cell (Box 12-3). If the processed mRNA for a protein can be isolated from a eukaryotic cell, it can be transcribed into a *complementary DNA* (cDNA) by reverse transcriptase (Chapter 12). The use of reverse transcriptase for this purpose is illustrated in Figure 14-10. Eukaryotic mRNA is more stable than prokaryotic mRNA and can often be isolated by methods that will not be described here. The cDNA will not be the normal eukaryotic gene, because it lacks the coding sequences for introns, but it contains the sequences for the exons in the correct order and will code for the desired protein if it can be cloned into a vector and transcribed in a microorganism.

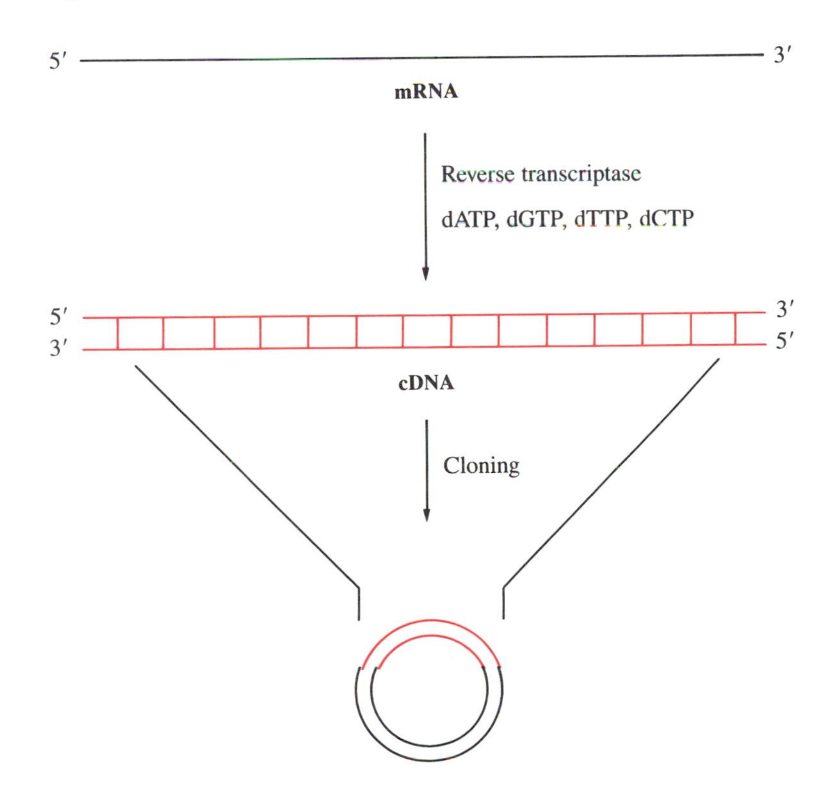

Figure 14-10

The Production of cDNA from mRNA by Use of Reverse Transcriptase. Eukaryotic mRNA lacking the introns of eukaryotic genes can be a substrate for reverse transcriptase to produce cDNA, which can be cloned into a suitable vector for expression.

cDNA contains blunt ends and must be cloned into vectors by methods other than those in Figures 14-4 to 14-7. One method is to use an adapter, as in Figure 14-11. An *adapter* is a synthetic molecule of double-stranded DNA that is blunt at one end but has an overhang at the other end corresponding in sequence to a restriction site. DNA ligase, which normally catalyzes the sealing of nicks in DNA, can also be used for *blunt-end ligation,* in which two blunt ends are brought together and ligated. The conditions under which blunt-end ligation may be brought about differ from those for ordinary nick-sealing, in that a high level of DNA ligase and a high concentration of one blunt-ended fragment are required to attain a useful rate. With cDNA and two adapters, a statistical distribution of products is formed, a fraction of which will contain the adapters in the correct orientation for cloning the cDNA between two restriction sites in a vector. If the cloning vehicle contains regulatory elements that will promote transcription of the cDNA, the cells carrying the vector can be screened for those that express the correct protein.

The second major problem with cloned eukaryotic genes is that functional proteins in eukaryotes often have undergone posttranslational modifications that are not brought about in host cells for vectors. Thus, a protein may be produced but be inactive owing to the fact that it is not processed beyond translation. One type of posttranslational processing is proteolytic activation, as in the activation of zymogens such as chymotrypsinogen and procarboxypeptidase (Chapters 5 and 6). Another is the attachment of oligosaccharides to glycoproteins (Chapter 15). Problems such as these must be surmounted on a case-by-case basis. Proteolytic activation can sometimes be carried out *in vitro,* and yeast cells can sometimes serve as hosts for the production of glycoproteins. Moreover, the carbohydrate moieties of glycoproteins often do not participate in their functions and are not required for some applications. For example, enzymes that are glycoproteins do not require the carbohydrate for activity. However, the activities of other types of proteins may require the pendant oligosaccharides, which often function as recognition elements in complex organisms.

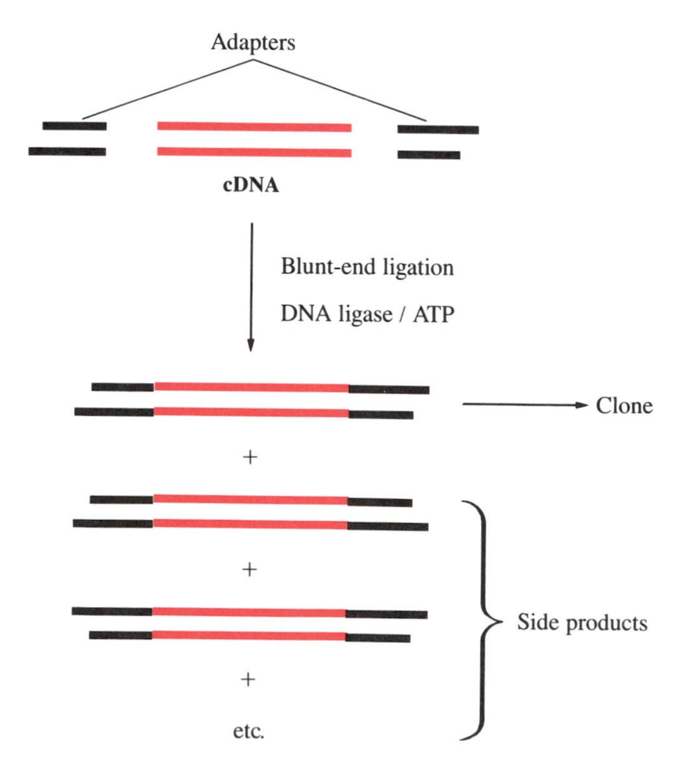

Figure 14-11

Adapting cDNA with Staggered Ends for Cloning. cDNA has blunt ends, which are not suitable for cloning into a vector. One method for creating staggered ends is to introduce adapters. An adapter is a synthesized fragment of DNA that contains the desired staggered ends, as well as blunt ends. The blunt ends of the adapters can be joined to those of cDNA by blunt-end ligation, in which an excess of the fragment and a high level of DNA ligase are used.

Screening and Selection of Cells Expressing Cloned Genes

A general problem that must be worked out in gene cloning is the identification and isolation of cells that contain and express the gene of interest. This is done by screening cells that produce the desired protein. Screening and selection methods are many and varied and, to a considerable degree, tailored to the protein of interest. A few general aspects are mentioned here.

The initial criterion for selection of cells carrying a gene cloned into a vector such as pBR322 is that the transformed cells will exhibit drug resistance; that is, they will grow in a medium containing ampicillin or tetracycline or both if the gene has been cloned into the *Eco RI* site or between the *Eco RI* and *Bam HI* sites. When a DNA fragment is cloned into the *Eco RI* site according to Figure 14-4, transformed *E. coli* will grow in the presence of either ampicillin or tetracycline.

When a gene is cloned between the *Eco RI* and *Bam HI* sites according to Figure 14-7, transformed *E. coli* will be resistant to ampicillin, but they will not grow in media containing tetracycline owing to the cleavage of the *tet* gene at the *Bam HI* site. A very dilute suspension of cells is *plated,* or spread onto the surface of agar containing a growth medium supplemented with ampicillin, and cells that are viable will grow in colonies, each of which originated with a single cell. Thus, each colony will be a clone containing *E. coli* that carry the plasmid, because other cells cannot grow with ampicillin. Only cells that have been transformed with a circular plasmid containing the intact gene for ampicillin resistance will grow in the presence of ampicillin. The plasmid in all such cells should carry a cloned segment of DNA between the restriction sites.

The next screening stage is to identify the colonies that express the protein of interest. In a few kinds of cloning experiments, the target gene will have been purified essentially to homogeneity or even chemically synthesized, and most colonies will test positive for the desired cloned DNA. However, in most cases very few colonies will express the gene of interest, because many different fragments from a DNA digest will have been cloned into the plasmid. One must then determine which of the many colonies of drug-resistant bacteria express the protein of interest. This can be done by the application of various tests. For example, if the target gene is an enzyme, its activity can be detected by an appropriate microanalytical technique. Alternatively, if antibodies to the protein are available, its presence in a colony can be detected by microimmunological techniques. Once a colony of cells is shown to produce the protein, a few of the cells can be picked up and allowed to grow in a suitable medium. Large-scale cell growth from such a culture can produce the protein in large amounts.

Site-directed Mutagenesis

The nucleotide sequence of a gene cloned into a bacterial vector can be altered in specific ways and expressed in a host cell to produce a mutant nucleic acid or protein. The functional properties of the altered macromolecule, compared with those of the wild-type molecule, give information about the relation of structure to function. The preparation of such mutant macromolecules is *site-directed mutagenesis,* an important recent development in biochemistry. Before the development of site-directed mutagenesis, structure-function studies of biological macromolecules were limited to measurements of the effects of alterations in the structures of small substrates and regulator molecules on macromolecular function, as well as to the effects of chemical modifications on the properties of macromolecules. Recent developments in chemical and molecular biological methods have made it possible to introduce virtually any desired change in the nucleotide sequence of a cloned gene. In this section, we shall briefly describe how this methodology allows one to produce a *point mutation,* a specific change in an amino acid at any point in the primary structure of a protein. Other mutation types, such as deletions and insertions, can also be introduced specifically. The functional properties and structures of these mutants constitute important information for use in learning the mechanisms of action of proteins.

Mutagenesis Directed by Mismatched Oligonucleotides

The widely used method for site-directed mutagenesis outlined in Figure 14-12 depends on the use of a chemically synthesized oligodeoxynucleotide, from fifteen to twenty nucleotides in length, that is complementary to a segment of the (+) strand of a gene cloned into a commercial version of M13 such as M13mp18. A suitable mismatched oligodeoxynucleotide will contain one or two mismatched nucleotides at its center but be perfectly complementary to the target sequence at all other positions. The mismatched oligodeoxynucleotide will hybridize to the (+) strand and serve as a primer for elongation of the mutant (−) strand by DNA polymerase. When polymerization is complete, the nicked duplex DNA can be sealed by the action of DNA ligase. Selective degradation of the (+) strand by any method, followed by production of a new (+) strand from the template mutant (−) strand, will give a mutant duplex that can be used as the RF-1 form of M13 to transform *E. coli.* By growing the transformed *E. coli,* one can produce and isolate the mutant RF-1 form of the vector and excise the mutant gene by digestion with the appropriate

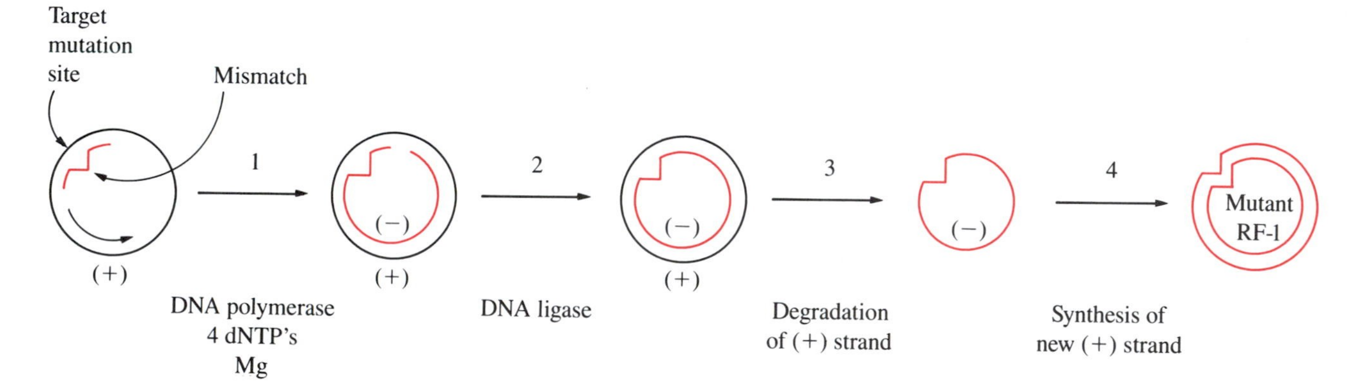

Figure 14-12

Site-directed Mutagenesis by the Use of a Mismatched Oligonucleotide. The (+) strand of a single-stranded vector such as M13 carrying the gene of interest is annealed with a synthetic oligonucleotide that has the sequence of the (−) strand on each side of the target site for mutagenesis. At the point of the desired mutation, the oligonucleotide carries the desired sequence, which constitutes a mismatch with the (+) strand. The hybrid will form if the oligonucleotide is about twenty nucleotides in length. The oligonucleotide is then extended in step **1** by the action of a DNA polymerase, and the ends are joined by DNA ligase in step **2**. In step **3**, the (+) strand is largely degraded by any of several methods, two of which are described in Figures 14-13 and 14-14. Synthesis of a new (+) strand in step **4**, with the mutant (−) strand serving as the template, produces a mutant duplex vector carrying the mutation in both strands. Expression of the mutated vector in a suitable host will generate the mutant protein or nucleic acid.

restriction endonucleases. The isolated gene can then be recloned into a suitable derivative of pBR322 or another vector for large-scale overexpression of the protein in *E. coli.*

Various methods are available for degrading the (+) strand of the mismatched RF-1 in Figure 14-12, one of which is outlined in Figure 14-13. In this method, which was originally introduced by T. Kunkel, the M13 is grown in a particular strain of *E. coli* that lacks two enzymes, the consequence of which is that this strain allows dUTP to be incorporated into DNA in place of dTTP. One of these enzymes is a dUTPase, encoded by *dut,* that normally catalyzes the cleavage of dUTP; in its absence, dUTP levels are high enough for dUTP to compete with dTTP for incorporation into DNA. The other deficient enzyme is an *N*-glycosylase, encoded by *ung,* that normally recognizes misincorporated U in DNA and catalyzes the hydrolytic removal of the uracil ring. This creates a lesion in the DNA that is recognized by the DNA repair systems, which excise the lesions and resynthesize the affected strands. This strain of *E. coli* can grow and produce M13 in which the DNA contains U in place of T with a low frequency. The U-containing (+)-strand is purified from the phage and converted into mismatched RF-1 *in vitro* by steps **1** and **2** in Figure 14-12. When

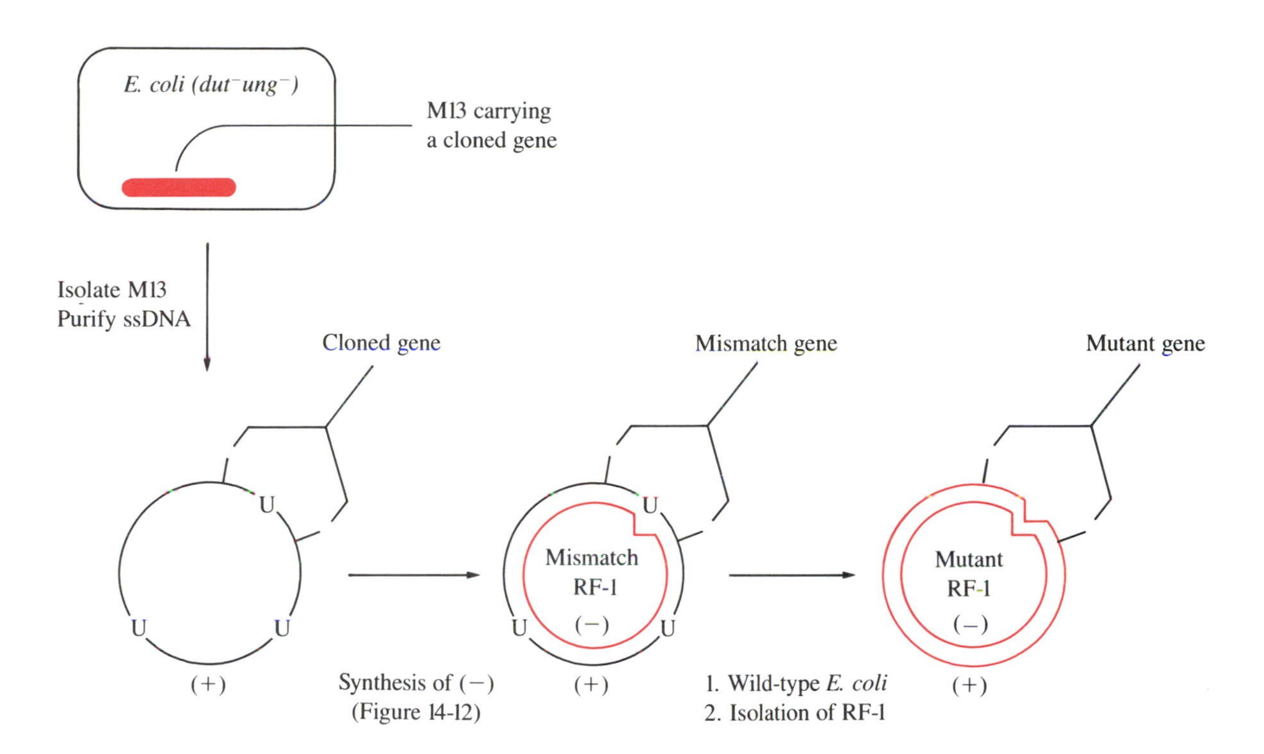

Figure 14-13

The Use of U-containing (+) Strands from M13 in Site-directed Mutagenesis. A strain of *E. coli* is defective in preventing and repairing the misincorporation of dUMP into DNA because of the absence of two enzymes, a dUTPase and a glycosylase. M13 grown in this strain contains U at a significant percentage of T-sites. When this (+) strand is used as the template for mismatch-oligonucleotide-mutagenesis according to steps **1** and **2** in Figure 14-12, an infective duplex carrying U in the (+) strand is generated. This duplex can be used to transform *E. coli* that is wild type with respect to the genes coding for the dUTPase and glycosylase. In wild-type *E. coli* transformed with the U-containing RF-1 form of M13, the (+) strand is largely destroyed and a new (+) strand is synthesized by the repair process. The newly synthesized (+) strand contains the mutation at the site of the mismatch.

the U-containing RF-1 is used to transform a strain of *E. coli* containing the enzymes that recognize and excise misincorporated U, the (+) strands will be destroyed and replaced with newly synthesized strands *in which the mismatch has also been repaired by introduction of the mutation in the* (+) *strand.* The resulting mutant RF-1 can be isolated, and the mutant gene can be excised by appropriate restriction digestion.

Another method by which the (+) strand can be selectively destroyed is by the use of a sulfur-containing analog of a deoxynucleoside triphosphate for the elongation of the mismatched oligonucleotide catalyzed by DNA polymerase in step **1** of Figure 14-12. The procedure was introduced by F. Eckstein and is outlined in Figure 14-14. The mismatched duplex RF-1 will contain thiophosphodiester bridges at selected sites instead of phosphodiesters. When such a bridge is present at a site recognized by a certain restriction endonuclease, the thiophosphodiester in the (−) strand will resist cleavage, whereas the (+) strand will be cleaved at that site, creating a nick. Subsequent digestion by an exonuclease will destroy most of the (+) strand. When it has been substantially degraded, the (+) strand is again synthesized, using DNA polymerase, by elongation from the ends of residual segments. Elongation through the mismatch site will introduce the new base into the resynthesized (+) strand and give the mutant RF-1, with the thiophosphodiester bridges still present in the (−) strand. This RF-1 will transform *E. coli* and express the mutant phenotype.

Cassette Mutagenesis

Cassette mutagenesis is a direct and simple method for altering the nucleotide sequence of a gene. It is most effective for specially engineered genes that contain regularly spaced, unique restriction sites throughout the nucleotide sequence. *Engineered genes* are chemically synthesized versions of natural genes, in which the nucleotide sequence has been altered in such a way as to introduce unique restriction sites at regular intervals without altering the translated amino acid sequence of the protein. Such alterations in the nucleotide sequence depend on the degenerative triplet codes for amino acids. For example, as shown in Table 10-1, the mRNA codons for alanine are GCA, GCG, GCU, and GCC. Thus, a naturally occurring gene coding for alanine at a specific location in a protein will contain a triplet sequence 3′-CGX, and this can be changed by altering X in such a way as to create a unique restriction site at that point in the gene, while retaining a code for alanine at that position in the protein chain.

Natural genes and artificial genes can be chemically synthesized by the use of automatic oligodeoxynucleotide synthesizers, which can prepare oligodeoxynucleotides up to about 75 nucleotides in length (Chapter 12). Full-length genes for smaller species of RNA, such as tRNA, can be synthesized one strand at a time. However, for the longer genes that encode proteins, the synthesis must be carried out in steps. Oligonucleotides corresponding in sequence to segments of

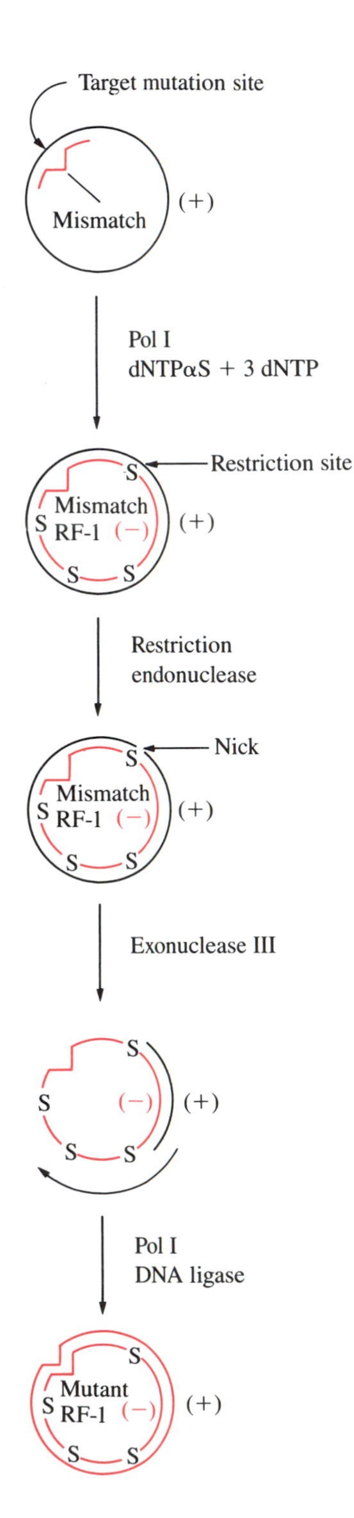

Figure 14-14

The Use of Thiophosphodiester Bridges for Selective Destruction of (+) Strands in Site-directed Mutagenesis. The thiophosphodiester bridges resist the action of nucleases such as restriction endonucleases and exonucleases and thereby protect the (−) strand during the nucleolytic degradation of most of the (+) strand of the mismatched RF-1 form of M13. When most of the (+) strand is digested, the remaining segment serves as the primer for the resynthesis of a (+) strand, in which the resynthesized strand is fully complementary to the (−) strand and carries the mutation.

both strands in a gene are synthesized. The lengths of these strands are planned in such a way as to allow them to form hybrid duplexes with staggered ends. This is illustrated in Figure 14-15. Pairs of these fragments with complementary staggered ends can be annealed and ligated by DNA ligase. It is often possible to

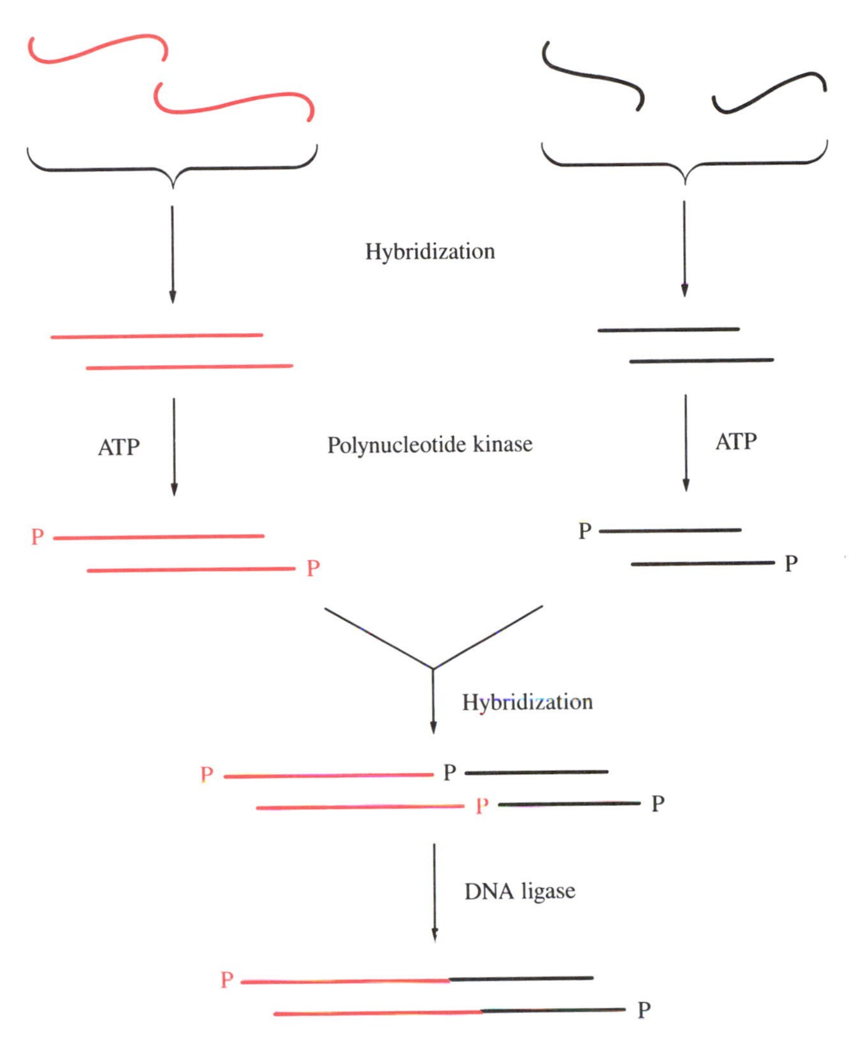

Figure 14-15

The Basic Synthetic Scheme for the Assembly of Chemically Synthesized Gene Fragments. The synthesis of a large gene is a step-by-step process that begins with the chemical synthesis of many relatively short segments of single-stranded oligodeoxynucleotides. Each oligodeoxynucleotide carries the nucleotide sequence of a segment of one strand in the gene. Pairs of oligonucleotides with complementary segments of the two strands in the gene are hybridized and isolated as short duplexes. The original synthetic design of complementary single-stranded segments is such that, when two are hybridized, they have staggered ends. The 5′ ends are then phosphorylated by polynucleotide kinase. The staggered ends allow two double-stranded fragments to be hybridized together through complementary base pairing, and they can then be joined through the action of DNA ligase. This figure illustrates how just one such ligation can be brought about in the synthesis of a gene. In practice, many such joining reactions are required, and the process can be organized in various ways. For example, several double-stranded fragments with unique complementarities in the ends can often be ligated in a single reaction, as shown in Figure 14-16.

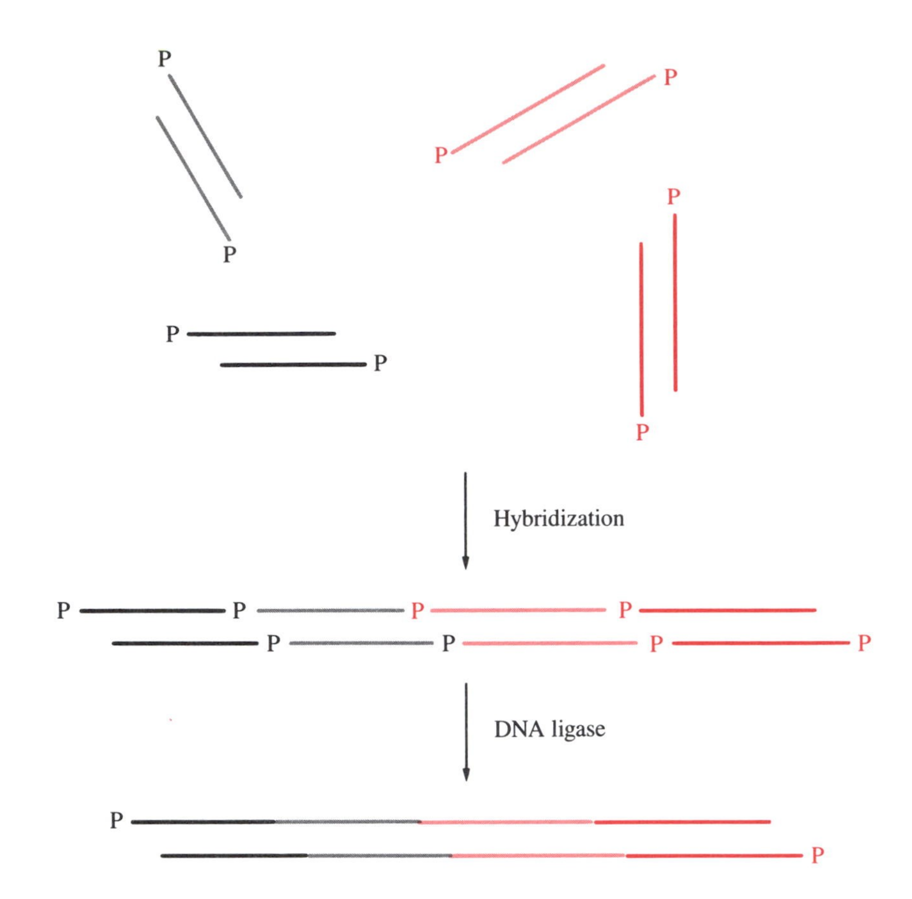

Figure 14-16

Ligation of Several Double-stranded Fragments in the Assembly of a Chemically Synthesized Gene. In the assembly of a gene from fragments synthesized chemically, several double-stranded fragments can often be assembled in one reaction. Four fragments are illustrated at the top of the figure in four colors to symbolize their different nucleotide sequences. The ends of a given fragment will hybridize to ends of other fragments in specific ways—that is, pink to red and grey but not to black, and so forth. The fragments will then pair up in a single alignment, and can be joined by DNA ligase to form a larger segment of the gene. The assembled segment itself has staggered ends and can be joined by the same procedure to other segments to form the complete gene.

assemble several duplex fragments into a larger segment of the gene in a single reaction, as illustrated in Figure 14-16. Systematic assembly by this procedure allows whole genes to be synthesized.

In a planned synthesis, an artificial gene may be designed to contain unique restriction sites at regular intervals, and it will be bounded on each end with convenient restriction sites for cloning into a suitable vector.

Once an engineered gene is cloned into a high-expression vector containing suitable regulatory elements, the vector can be transformed into a suitable strain of *E. coli* to produce large amounts of the protein. The gene can be mutated by purely chemical means by following the strategy in Figure 14-17. A segment of the gene is excised by the use of two restriction enzymes that cleave uniquely at the sites most closely flanking the point of mutation, and the linear vector is isolated. A replacement fragment is chemically synthesized with the desired nucleotide sequence and cloned into the linear vector by the procedures described in Figures 14-4 to 14-7. The circular vector will contain the mutant gene and the same expression system used for the natural protein. This method has the advantage that any desired mutation may be introduced into the gene; however, for the procedure to be most useful, an engineered gene with appropriately spaced and unique restriction sites must first be chemically synthesized.

Importance of Site-directed Mutagenesis

The value of site-directed mutagenesis for studying the structure-function relations in proteins and nucleic acids can hardly be overemphasized. The technique allows any single change in the amino acid sequence of a cloned

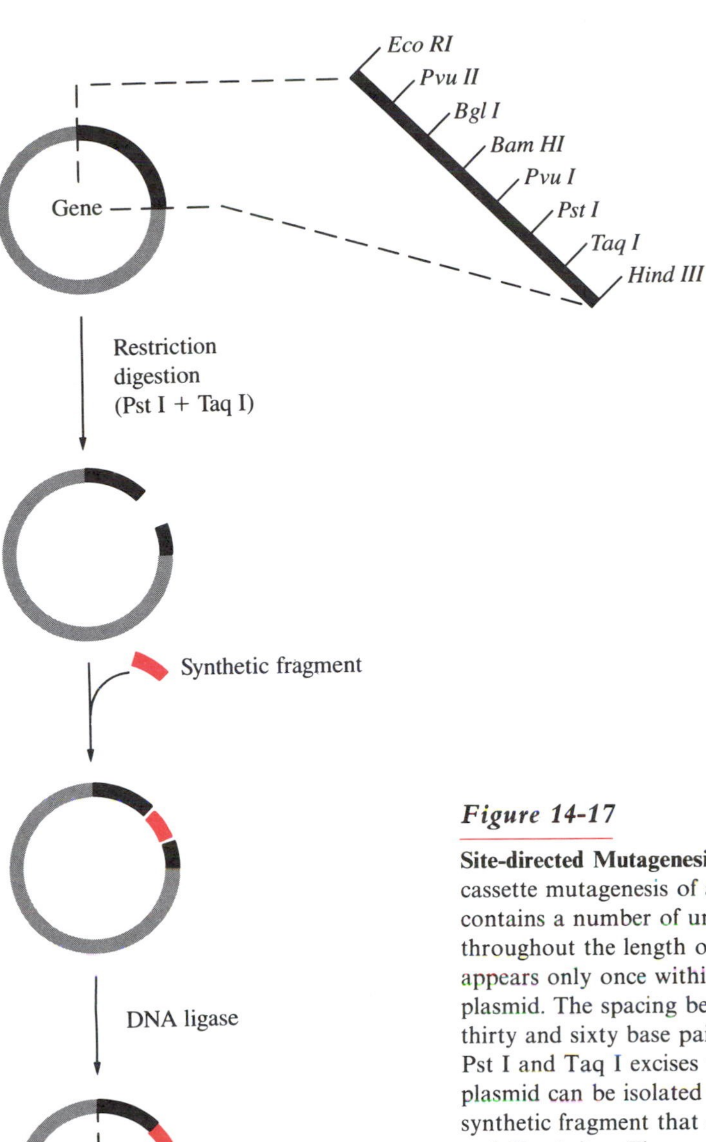

Figure 14-17

Site-directed Mutagenesis by the Cassette Technique. This figure illustrates cassette mutagenesis of an engineered gene cloned into a plasmid. The gene contains a number of unique restriction sites spaced approximately uniformly throughout the length of the gene. The uniqueness of these sites is that each appears only once within the gene and does not appear elsewhere in the plasmid. The spacing between unique restriction sites typically ranges between thirty and sixty base pairs. Digestion of the gene with the restriction enzymes Pst I and Taq I excises the fragment between these two sites. The linear plasmid can be isolated from the excised fragment and annealed with a new synthetic fragment that contains the staggered ends characteristic of the *Pst I* and *Taq I* sites. The synthetic fragment may have any length and nucleotide sequence; however, for site-directed mutagenesis, it will have the same length as the excised fragment, and the nucleotide sequence will differ in only one or two positions to change the coding for a single amino acid. The annealed fragment is finally ligated into the plasmid with DNA ligase.

protein to be brought about, and the functional consequences of this change can be determined in the mutant protein. By this technique, it has been found that tyrosine-198 of carboxypeptidase A (Chapter 6) is not critical for the function of the enzyme, because the mutant with phenylalanine at this site is almost as active as the wild-type enzyme. Tyrosine-198 had been proposed, on the basis of other chemical studies, to play a significant role in catalysis. The mutation of glutamate-165 at the site of triose phosphate isomerase (Chapter 21) to aspartate by this technique caused a dramatic decrease in activity. Because there was no significant change in the structure of the protein, this result supported the proposition that the γ-carboxylate group of the active-site glutamate acts as a general base. In the aspartate mutant, the β-carboxylate group was relatively ineffective because of being displaced a short distance from its normal position. The serine proteases (Chapters 3 through 5) containing alanine in place of the active-site serine or histidine exhibit only traces of activity. And the active-site nucleophilic histidine in galactose-1-P uridylyltransferase (Chapter 11) has been identified as histidine-166 by site-directed mutagenesis. Site-directed mutagenesis can be applied to the analysis of protein structure and function in many other ways, and more will undoubtedly be discovered.

Site-directed mutagenesis is also important for studying nucleic acid function. The sequences of promoters with respect to function have been extensively studied by this technique. And the importance of nucleotide sequences in the splicing functions of nucleic acids are being investigated by systematically varying the sequences near splice sites.

BOX 14-1

THE POLYMERASE CHAIN REACTION

The polymerase chain reaction (PCR) is an important recent development in nucleic acid chemistry that allows tiny amounts of double-stranded DNA to be amplified a billionfold within a short time. The amount of a specific segment of DNA available for study is often small, and the cloning techniques described in this chapter allow small segments such as genes to be amplified by incorporating them into cloning vectors and then growing the vectors in host cells. The polymerase chain reaction makes it possible to amplify DNA *in vitro* in a very short time and with a minimum of work.

The minimum information required to allow the PCR to be applied to the amplification of a targeted segment of DNA are nucleotide sequences flanking both sides of the target. The steps in the PCR are illustrated in Figure A. Two primers, oligodeoxynucleotides about twenty nucleotides in length and complementary to sequences flanking the two strands of the target, are synthesized by the method described in Chapter 12. The sequences of these primers are designed so that the 3′ ends will be oriented toward the target when they are annealed to the separated DNA strands adjacent to the target sequences. The sample DNA is heated in step **1** to separate the strands, and both primers are added and allowed to anneal to the strands at the complementary locations flanking the target segments. With the primers in place and the sample cooled, the 3′ ends of the primers are extended beyond the target sequences in step **2** by DNA polymerase. Note that the sequences beyond the target in the extended primers are complementary to both primers. This allows another cycle of denaturation, annealing of primers, and polymerization; the system can be cycled in this way any number of times. After a few cycles, the principal product after annealing is the double-stranded target segment flanked by the primer sequences.

With recent methodological improvements, as many as twenty-five cycles of PCR can be carried out in slightly more than one hour. Each cycle can double the amount of the target, so that the final amplification will be 2^n in n cycles. Therefore, a 67 millionfold amplification is possible in about an hour. One of the recent methodological advances is the use of a heat-stable DNA polymerase from *Thermus aquaticus,* Taq polymerase. With this polymerase, it is not necessary to cool the reaction mixture to 37°C for the chain-extension step, and it is not necessary to add fresh polymerase with each cycle. The reaction is carried out in a single vessel with no additions of reactants or enzyme between cycles. A cycle consists of changes in temperature. A reaction mixture consisting of a target, the two primers, the four deoxynucleotide triphosphates, Mg^{2+}, buffer, salts, and Taq polymerase is heated to 95°C for 15 seconds to denature the target, cooled to 54°C for 15 seconds to anneal the primers, and held at 72°C for 30 seconds for primer extension by the polymerase. The process can be automated, and commercial PCR machines are available.

The PCR facilitates many kinds of DNA analysis and cloning that would otherwise be time consuming or impossible. The PCR facilitates gene cloning, prenatal analysis of fetal DNA for inherited defects, diagnosis of infectious disease, taxonomic analysis of genes in DNA from closely related species, forensic analysis of DNA samples, and site-directed mutagenesis. Many other uses in DNA chemistry will no doubt be found as well.

BOX 14-1 (continued)

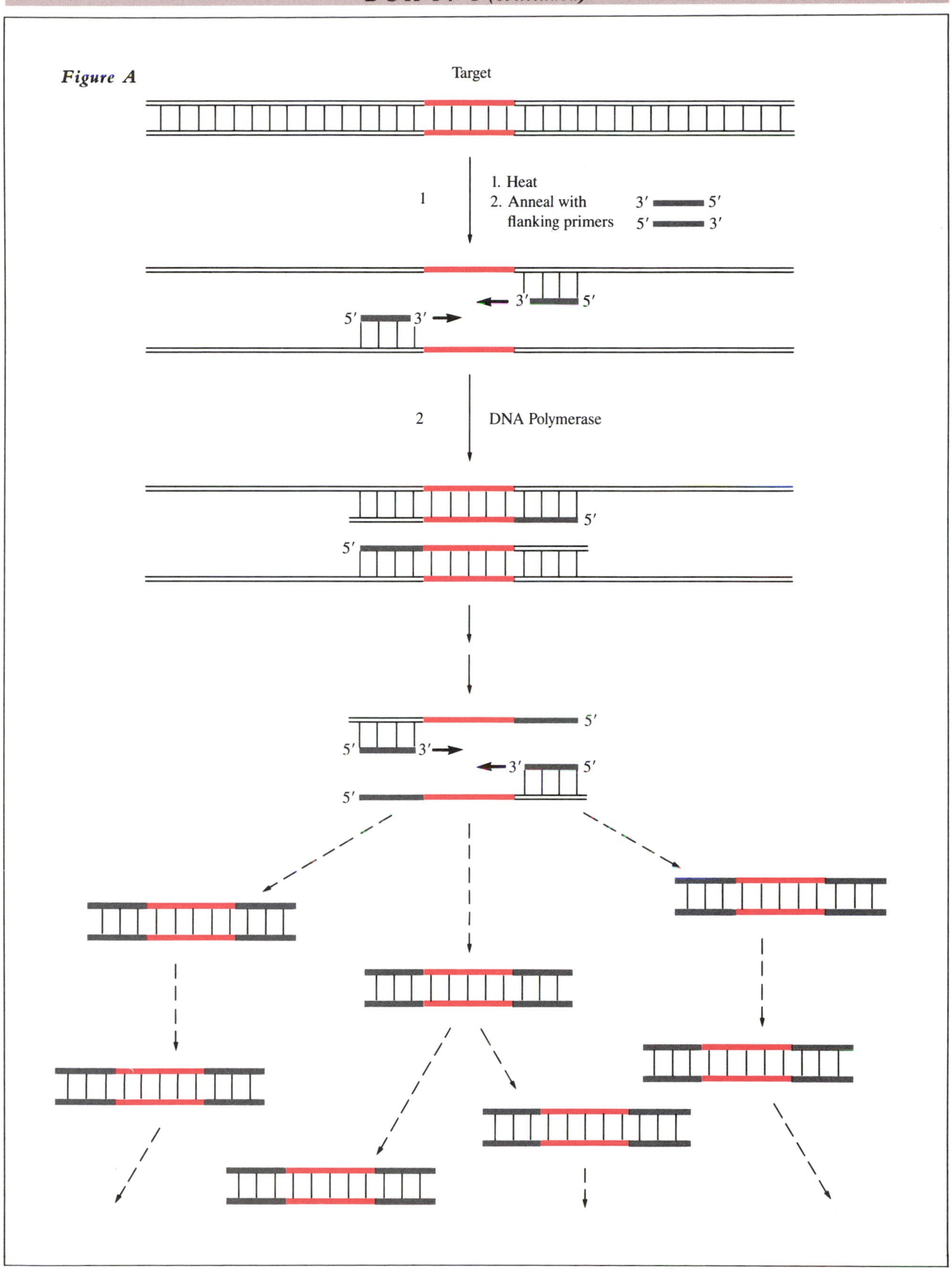

Summary

Recombinant DNA is any species of DNA that has been cleaved from its chromosomal DNA and inserted into another segment of DNA. In the molecular cloning of genes, segments of chromosomal DNA are initially produced by digestion with restriction endonucleases. These segments are ligated between compatible sites of a cloning vehicle, or vector, such as plasmid pBR322 or phage M13. The vectors carrying the cloned segments are then either transformed or transfected into *E. coli*, and the cells are screened for the desired gene product. Any cell that produces the gene product is assumed to have acquired the cloned gene through the vector. When the vector is a multicopy plasmid and carries host regulatory elements adjacent to the start site for the cloned gene, the gene product can be overproduced by the transformed cells.

In site-directed mutagenesis, defined mutations can be introduced specifically into a gene that has been cloned into a suitable expression vector. Site-directed mutagenesis of proteins is a valuable means of systematically studying structure-function relations.

ADDITIONAL READING

Wu, R., Grossman, L., and Moldave, K., eds. 1989. *Recombinant DNA Methodology.* New York: Academic Press.

Kunkel, T. A. 1985. *Proceedings of the National Academy of Sciences U.S.A.* 82: 488.

Nakamaye, K. L., and Eckstein, F. 1986. *Nucleic Acids Research* 14: 9679.

Sambrook, J., Fritsch, E. F., and Maniatis, T. 1989. *Molecular Cloning: A Laboratory Manual.* Cold Spring Harbor Laboratory Press.

White, T., Arnheim, N., and Erlich, H. 1989. The polymerase chain reaction. *Trends in Genetics* 5: 179.

Mullis, K. B. 1990. The unusual origin of the polymerase chain reaction. *Scientific American,* April, p. 56.

PROBLEMS

1. By the method of molecular cloning in Figure 14-4, a single staggered cut is introduced into a cloning vector by a single restriction enzyme. The restriction fragments generated from chromosomal DNA by this same enzyme are then allowed to hybridize with the cleaved vector and are sealed with DNA ligase. There are some practical problems with this method. What are these problems? (Optional) How might it be possible to overcome these problems?

2. One method for generating fragments of DNA from chromosomal DNA is to subject it to shearing force, as by forcing the solution rapidly through a narrow-bore syringe needle. Under controlled conditions, the fragment sizes can be fairly uniform, with strand cleavages randomly distributed. The ends are generally blunt rather than staggered, however, so that ligation into a cloning vector is difficult. Suggest a means by which terminal deoxynucleotidyltransferase can be used to facilitate the cloning of such fragments into vectors. (Terminal transferase catalyzes the addition of a polydeoxynucleotide to the 3′-OH end of a DNA primer.) Which other well-known enzyme (in addition to DNA ligase) should be used with terminal transferase to further facilitate the process?

3. When engineered vectors such as the one in Figure 14-9 are used as expression vectors, they are controlled by the o/p cloned ahead of the polylinker cloning site. Suppose that the *lac* o/p is used for a multicopy *E. coli* vector and a gene for RNA ligase is cloned into the polylinker. How should the growth medium for the transformed *E. coli* be formulated in order to achieve a high expression of RNA ligase? (Bacterial growth media normally contain salts of metals, phosphate, a source of nitrogen, and a source of carbon for energy, as well as any other special compounds that may be needed.)

4. A class of nucleotide sequences in mRNA known as Shine-Delgarno sequences have the property of binding ribosomes more tightly than others. How might this property be exploited in an expression vector to enhance the level of the expression of a cloned gene?

5. Galactose-1-P uridylyltransferase, an enzyme in the *gal* operon, catalyzes the reaction of UDP-glucose with galactose-1-P to form UDP-galactose and glucose-1-P by a Ping-Pong mechanism through a covalent intermediate in which the UMP-moiety of UDP-glucose is bonded to a histidine residue. The *E. coli* enzyme contains fifteen histidine residues and, of the possible site-directed mutants in which asparagine is substituted for histidine, thirteen are active and two (N164 and N166) are inactive. What can be concluded about the mechanism of the reaction? Why is asparagine a good choice as a replacement for histidine in this experiment?

6. Consider again the enzyme in question 5. The site-directed mutant protein G166 catalyzes the following reaction, but the mutant G164 does not. What can be concluded about the mechanism of action of this enzyme?

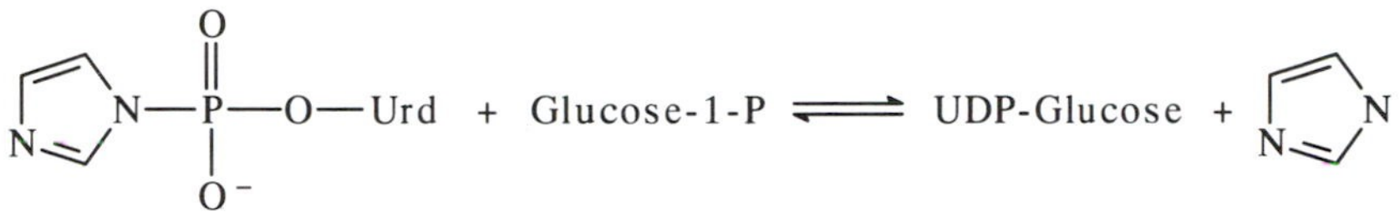

$$\text{Im-N}-\overset{\displaystyle O}{\overset{\|}{\underset{\underset{\displaystyle O^-}{|}}{P}}}-O-\text{Urd} + \text{Glucose-1-P} \rightleftharpoons \text{UDP-Glucose} + \text{Imidazole}$$

Glycosyl Group Transfer

15

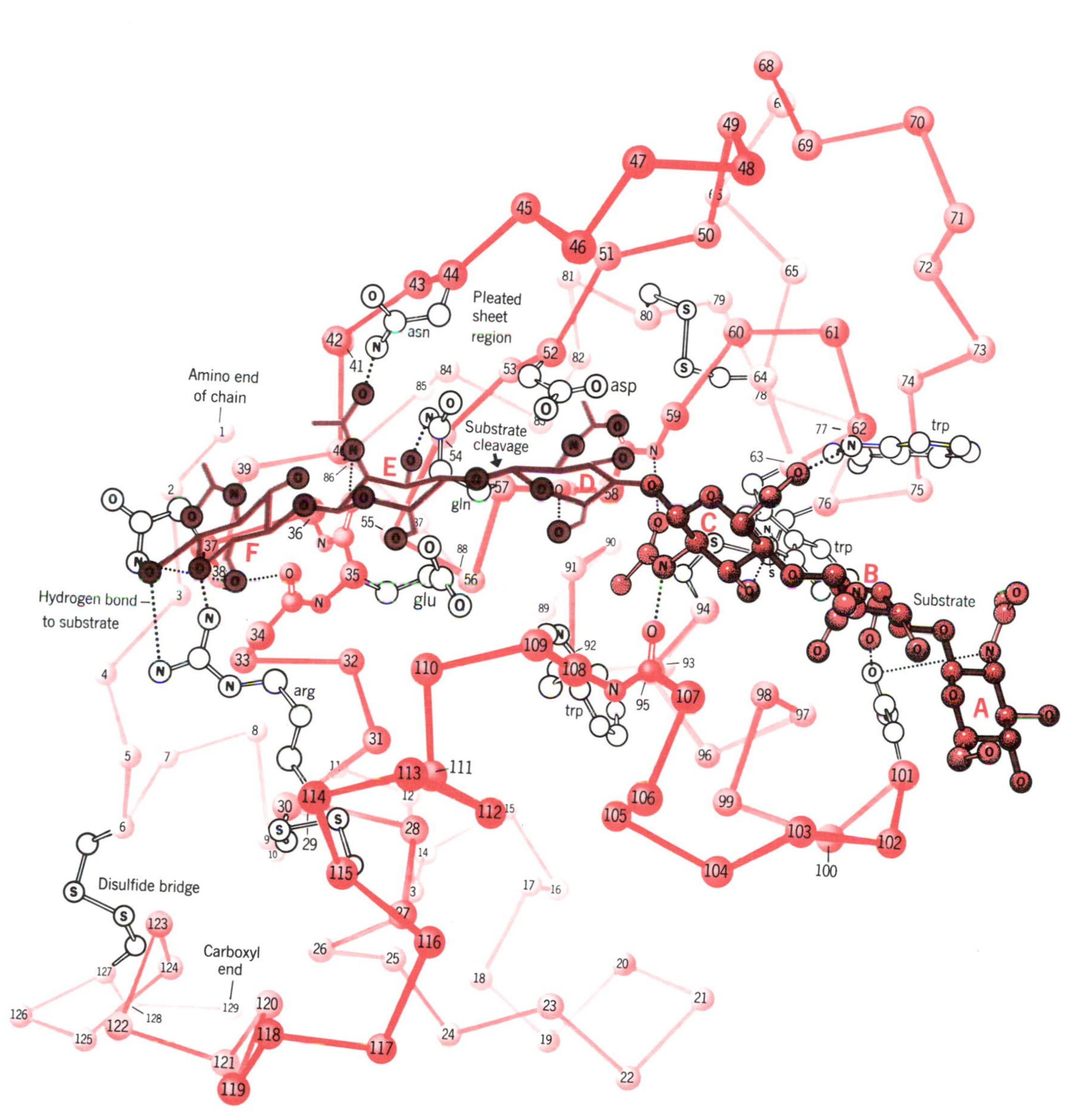

Ball-and-stick model showing the main chain in chicken lysozyme. The hexasaccharide substrate is shown in color. The binding sites for rings A through F are inferred from model building and from the binding of smaller oligosaccharides. (Copyright © Irving Geis.)

Sugars and sugar derivatives are ubiquitous in biological systems. Sugars and polysaccharides are major energy sources for cells and organisms. The sugars ribose and 2-deoxyribose are important components of nucleic acids and nucleotides. Complex oligosaccharides are components of glycoproteins and glycolipids, and they are important cellular recognition factors and signals. The biosynthesis of complex carbohydrates requires the activation of sugars into chemical states that allow the sugar glycosyl units to be transferred to appropriate molecules, which include nucleoside bases, proteins, other sugars, and polysaccharides.

In the activation of sugars for the biosynthesis of glycosyl derivatives such as glycoproteins and complex carbohydrates, a sugar is converted either into a nucleotide derivative, such as UDP-glucose or UDP-galactose, or into a phosphate ester, such as 5-phosphoribosyl-1-pyrophosphate (PRPP). The production of nucleotide sugars is described in Chapters 11 and 13; other nucleotide sugars such as UDP-*N*-acetylglucosamine, UDP-*N*-acetylgalactosamine, and GDP-mannose are produced in analogous reactions.

In glycosyl transfer reactions, a sugar derivative undergoes the transfer of the sugar unit as a *glycosyl* group to some nucleophilic acceptor molecule, which may be a nucleoside base, an amino acid side chain in a protein, a hydroxyl group of another sugar or carbohydrate, a phosphate, or often water. In this chapter, we shall describe the basic chemical process of glycosyl transfer. We shall also describe some specific examples of biological glycosyl transfer processes that are important in the production and degradation of complex carbohydrates.

The Chemistry of Glycosyl Transfer

Glycosyl compounds are derivatives of sugars in their cyclized forms. The *glucosyl* group of UDP-glucose is highlighted in the structure shown below. The sugars in glycosyl compounds are either *pyranoses*, such as the glucose in UDP-glucose, or five-member-ring *furanoses*, such as the fructosyl ring in sucrose. Naturally occurring glycosyl compounds have substituents of many types, including other glycosyl groups in polysaccharides, nucleotides in nucleotide sugars, lipids in glycolipids, and proteins in glycoproteins. The principal subject of this chapter is the assembly and degradation of these molecules through glycosyl group transfer reactions.

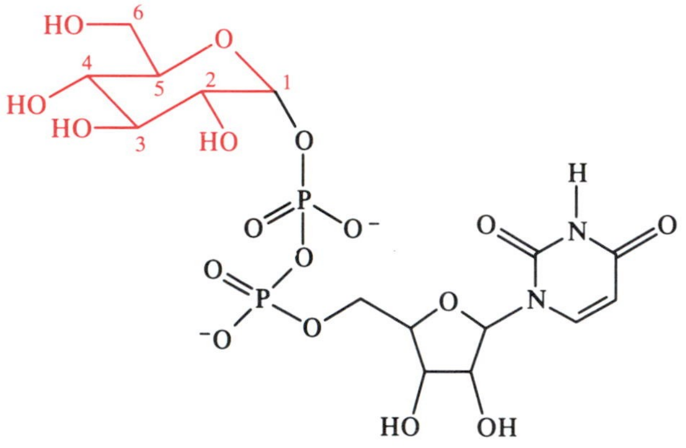

UDP-glucose

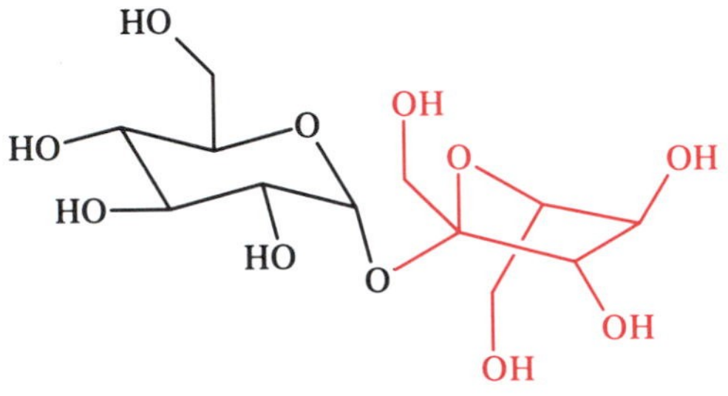

Sucrose

Reactions in which glycosyl groups are transferred to water are believed to occur in step-by-step mechanisms through extremely unstable glycosyl carbocationic intermediates, as illustrated in Figure 15-1 for the acid-catalyzed hydrolysis of methyl-α-glucoside. Stabilization of the carbocation by electron donation from the adjacent oxygen atom is illustrated in the margin.

Glycosides are extremely stable toward hydrolysis, but they can be cleaved when there is a very good leaving group at C-1. Protonation of the leaving oxygen atom in Figure 15-1 gives methanol as the leaving group, which is rapidly expelled. In the absence of acid, the leaving group would be methoxide ion (CH_3O^-), which is very unstable and a poor leaving group. If there is a good leaving group and a strong nucleophile, glycosyl groups can be transferred

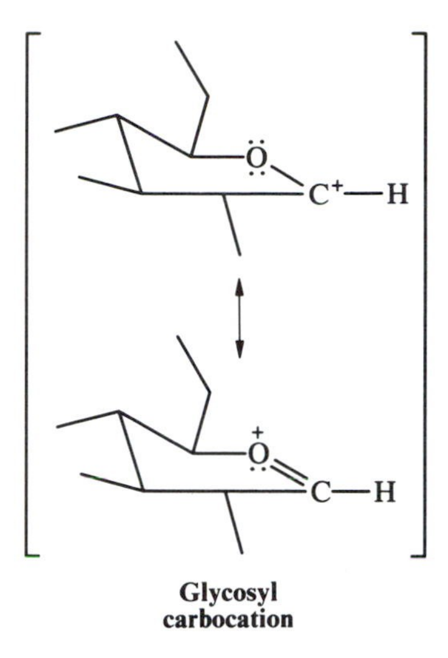

Glycosyl carbocation

Methyl-α-glucoside

α-D-Glucose

β-D-Glucose

Figure 15-1

Acid-catalyzed Hydrolysis of Methyl-α-glucoside. Glycosides are subject to acid-catalyzed hydrolysis but are stable in basic solutions. Acid-catalyzed hydrolysis requires protonation of the glycoside by an acid. The protonated —OCH_3 is the leaving group. Loss of methanol produces the glycosyl cation, which is immediately captured by water.

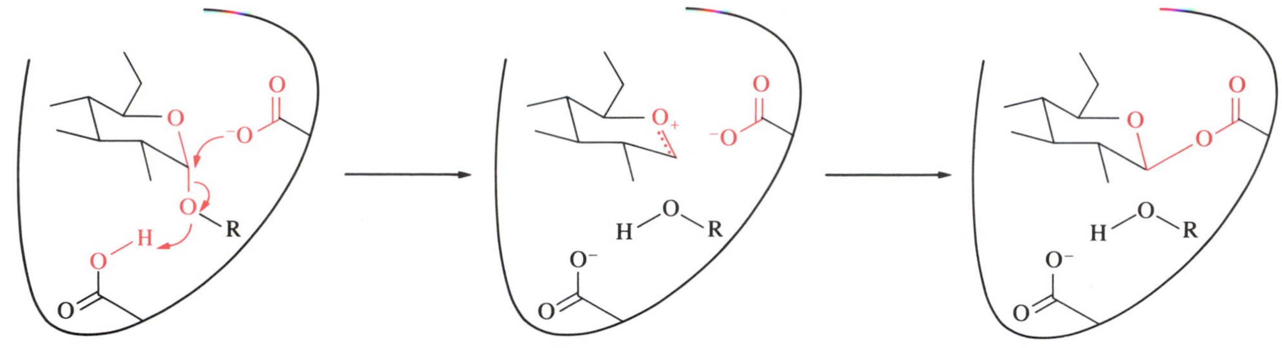

Figure 15-2

Possible Mechanistic Role of Acidic Residues at the Active Sites of Glycosyl Transfer Enzymes. Two carboxyl groups at the active sites of glycosyl transfer enzymes can catalyze the reactions by acid catalysis in the cleavage of the glycoside linkage. This accounts for the action of one carboxylic acid group. A second carboxyl group is presumably a carboxylate ion, which either facilitates the formation of a glycosyl cation by stabilizing it electrostatically or acts as a nucleophilic catalyst by forming a bond to the glycosyl group.

through a bimolecular displacement mechanism; that is, in a single concerted step in which the nucleophile displaces the leaving group in one step with no intermediate and with inversion of configuration at C-1. In general, it is not known whether enzymatic catalysis of glycosyl transfer occurs through a two-step mechanism, with an intermediate glycosyl carbocation, or by a concerted nucleophilic displacement of the leaving group.

In an enzymatic glycosyl transfer reaction, the enzyme must catalyze the departure of the leaving group and the attack by a nucleophile on C-1 of the glycosyl group. The nucleophile may be a group on the enzyme, which leads to the formation of an intermediate glycosyl-enzyme, or it may be a nucleophilic group of the acceptor substrate. Several enzymes that catalyze glycosyl transfer, including glycosidases, have the side-chain carboxyl groups of glutamate or aspartate at the active site. A protonated carboxyl group may protonate the leaving oxygen atom, as illustrated in Figure 15-2; and a carboxylate anion may stabilize a transient carbocation electrostatically. Alternatively, a carboxylate anion may serve as an acceptor for the glycosyl group and form an intermediate glycosyl-enzyme.

Glycosidases

Glycosidases catalyze the hydrolysis of glycoside linkages in oligosaccharides and polysaccharides. These enzymes include the amylases, which are essential for the digestion of starch by higher animals. They also include the α-glycosidases and β-glycosidases, which catalyze crucial steps in the processing of the oligosaccharide moieties of glycoproteins.

Lysozyme from the white of hens' eggs is the most thoroughly studied glycosidase. It catalyzes the hydrolysis of the glycoside linkage indicated in Figure 15-3. Bacterial cell walls are constructed of polysaccharides consisting of alternating *N*-acetylmuramyl and *N*-acetylglucosaminyl units linked by α-1,4-glycosyl bonds and cross-linked by oligopeptides. Muramic acid is 3-*O*-lactylglucose, and glucosamine is 2-deoxy-2-aminoglucose; the *N*-acetyl derivatives of these sugars are shown in Figure 15-3. Lysozyme catalyzes the cleavage of the glycosidic bonds at C-1 of *N*-acetylmuramyl units.

A model of the three-dimensional structure of lysozyme, as determined by X-ray crystallography, is shown in Figure 15-4. The active site contains two acidic amino acids, glutamate-35 and aspartate-52. The structure of the enzyme with bound oligosaccharides suggests that these two acidic residues are situated on opposite sides of the glycosidic linkage in the bound substrate, similar to the arrangement shown in Figure 15-5. The mechanism of lysozyme action is not known, but the structure suggests that substrate binding entails the use of

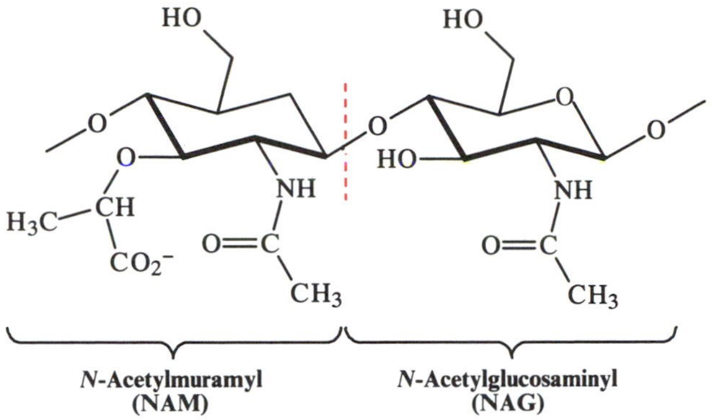

HO HO

O O O

O HO

NH NH

H_3C—CH

O=C O=C

CO_2^- CH_3 CH_3

N-Acetylmuramyl (NAM) **N-Acetylglucosaminyl (NAG)**

Figure 15-3

Specificity of Lysozyme. Lysozyme catalyzes the cleavage of the glycoside linkage between *N*-acetylmuramic acid and *N*-acetylglucosamine in alternating polymers of these sugars. The bond that is cleaved is indicated by the dashed line.

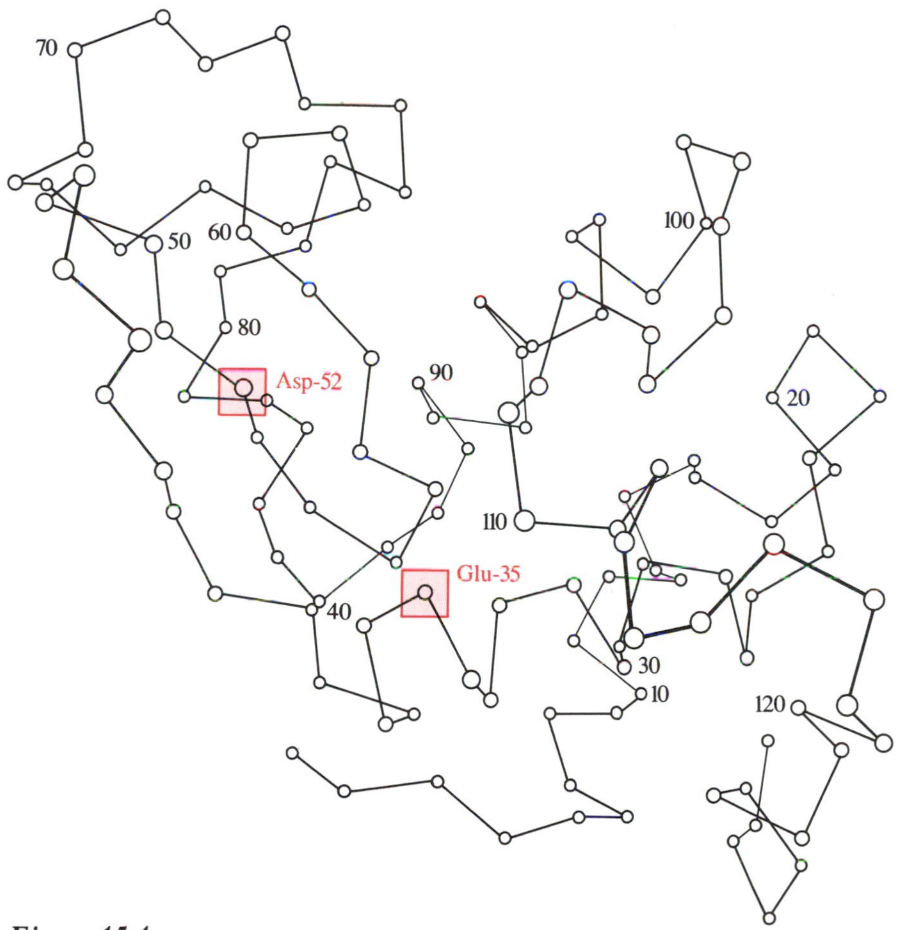

Figure 15-4

Three-dimensional Structure of Lysozyme. This perspective drawing of the α-carbon chain of lysozyme from hen egg whites gives a view into the cleft in which substrates and inhibitors bind. The α-carbon atoms of residues Glu-35 and Asp-52 are highlighted and shown to occupy opposed positions in the cleft. The side chains of these residues are thought to interact with opposite sides of the substrate, somewhat as is illustrated in Figure 15-5. (After T. Imoto, L. N. Johnson, A. C. T. North, D. C. Phillips, and J. A. Rupley. In *The Enzymes*, vol. 7, 3d ed., ed. P. D. Boyer. New York: Academic Press, 1972, p. 665.)

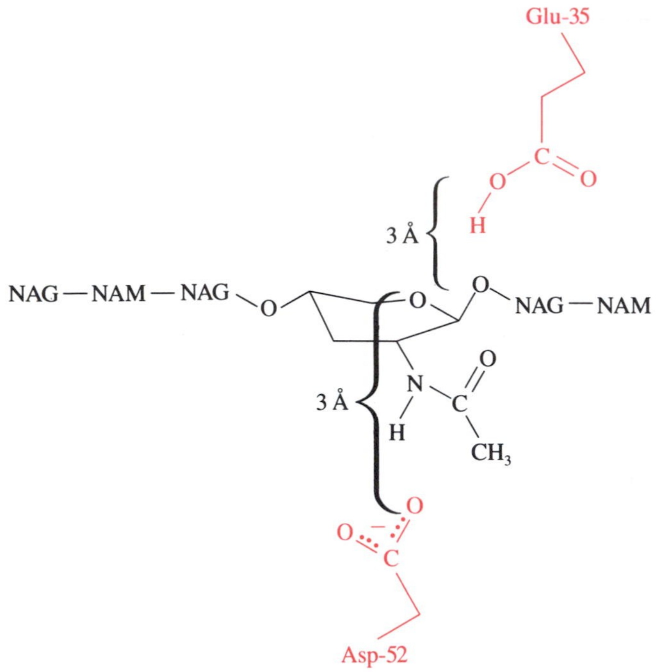

Figure 15-5

Disposition of Glutamate-35 and Aspartate-52 Relative to the Substrate Glycosyl Group at the Active Site of Lysozyme. [After W. N. Lipscomb, *Proc. Robert A. Welch Found. Conf. Chem. Res.* 15 (1972): 151.]

binding energy at a number of sites along the polymer to distort the sugar ring at the scissile glycoside linkage toward planarity. This would stabilize either a carbocationic intermediate or a transition state that resembles a carbocation. The trigonal bipyramidal transition state for displacement of the leaving group by a carboxylate ion is planar about the carbon-1 atom and would be stabilized by the same binding interactions that would stabilize a carbocation. Such a displacement would lead to a covalent glycosyl-enzyme, which would then undergo hydrolysis in a second step. The overall stereochemistry in either mechanism results in retention of configuration at C-1, which is the observed steric course. If the mechanism includes a carbocationic intermediate, the enzyme shields one side of the carbocation from reaction with water. In this case, the enzyme structure would have to be such as to prevent the β-carboxylate group of aspartate-52 from forming a full covalent bond with the glycosyl intermediate at the cationic C-1, while allowing this same group to stabilize the carbocationic intermediate through an electrostatic interaction.

Synthesis of Polysaccharides and Oligosaccharides

UDP-glucose and other nucleotide sugars serve as glycosyl donors in many biological glycosyl transfer reactions. The leaving group can be either UDP^{3-} or UDP^{2-}, which are good leaving groups. They are good leaving groups because they are stable ions in water. The nucleotide sugars, though not especially reactive in the absence of catalysts, are responsive to enzymatic catalysis. Thus, the molecules possess potential *kinetic* reactivity. They are also *thermodynamically* activated for glycosyl transfer reactions, again because of the stability of UDP as a leaving group. The reaction of an ordinary alkyl glycoside with a sugar or other alcohols is nearly isoenergetic, but the reaction of UDP-glucose with another sugar is a downhill reaction. These processes are compared

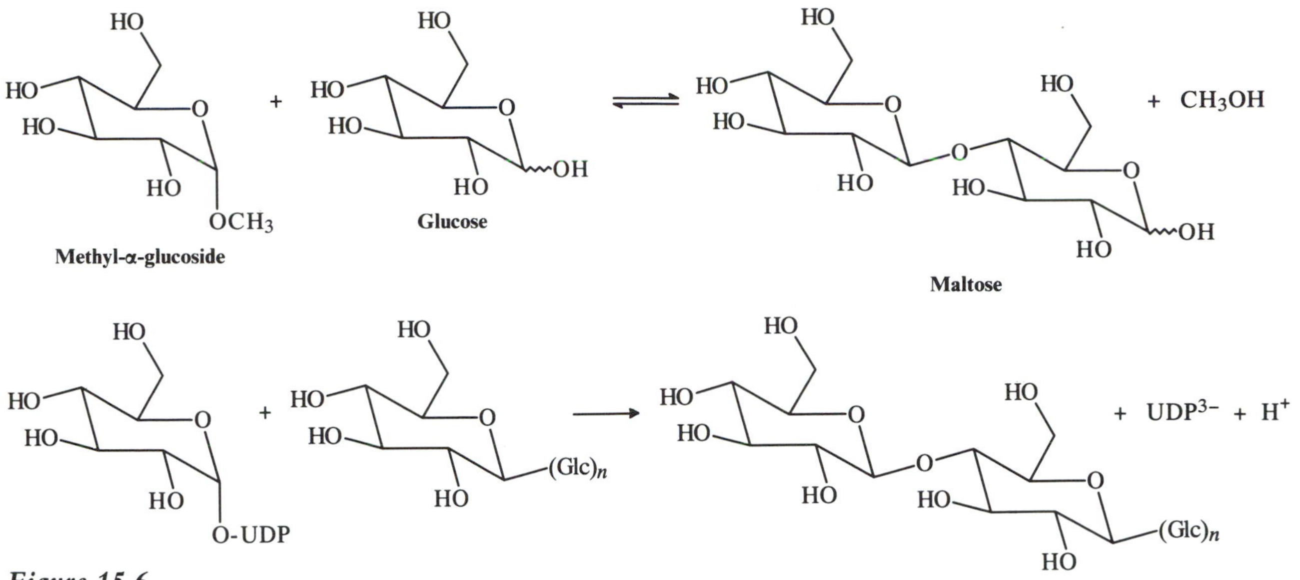

Figure 15-6

Synthesis of a Glycoside. The nonenzymatic synthesis of the disaccharide maltose in a hypothetical nonenzymatic reaction of methyl-α-glucoside would have an equilibrium constant near 1. The known enzymatic synthesis of the glucoside linkages in glycogen using UDP-glucose as the glucosyl donor is energetically downhill.

in Figure 15-6 for the hypothetical production of maltose from methyl-α-glucoside and glucose and the biochemical reaction of UDP-glucose with glycogen. The reaction of methyl-α-glucoside is essentially isoenergetic because methanol is the coproduct with maltose. With UDP-glucose, however, the coproducts are a proton and the trianion of UDP. The low concentration of protons at pH 7 pulls the reaction forward.

Glycogen Synthetase

Glycogen is a branched polymer of glucose in which glucosyl groups are joined by α-1,4-glucosyl and α-1,6-glycosyl linkages, as shown in Figure 15-7. It is found in substantial amounts in muscle, where it serves as the primary storage form of glucose residues in animals. Polymerization of glucose as glycogen traps it within the cell and does not tie up a large amount of water. An equivalent concentration of glucose itself would interact with water molecules and cause a large increase in the osmotic pressure inside a cell.

The basic α-1,4-glucosyl linkage is generated by the enzyme glycogen synthetase, which utilizes UDP-glucose as the glucosyl donor and glycogen itself as the acceptor (equation 1).

$$\text{UDP-glucose} + (\text{Glucose})_n \longrightarrow \text{UDP} + (\text{Glucose})_{n+1} \qquad (1)$$

The acceptor group in glycogen is the 4-OH of the terminal glucose residue. Thus, the reaction leads to the extension of a glycogen chain by addition of one glucose residue to the nonreducing end. All of the glucosyl residues in glycogen originate with UDP-glucose and are introduced as α-1,4-glucosyl units by glycogen synthetase.

The α-1,6-glucosyl linkages in glycogen are introduced by a branching enzyme that catalyzes the isomerization of an α-1,4 to an α-1,6 linkage as illustrated in Figure 15-8. This is an intramolecular glycosyl transfer of an oligoglucosyl segment about seven units in length to an internal 6-OH group.

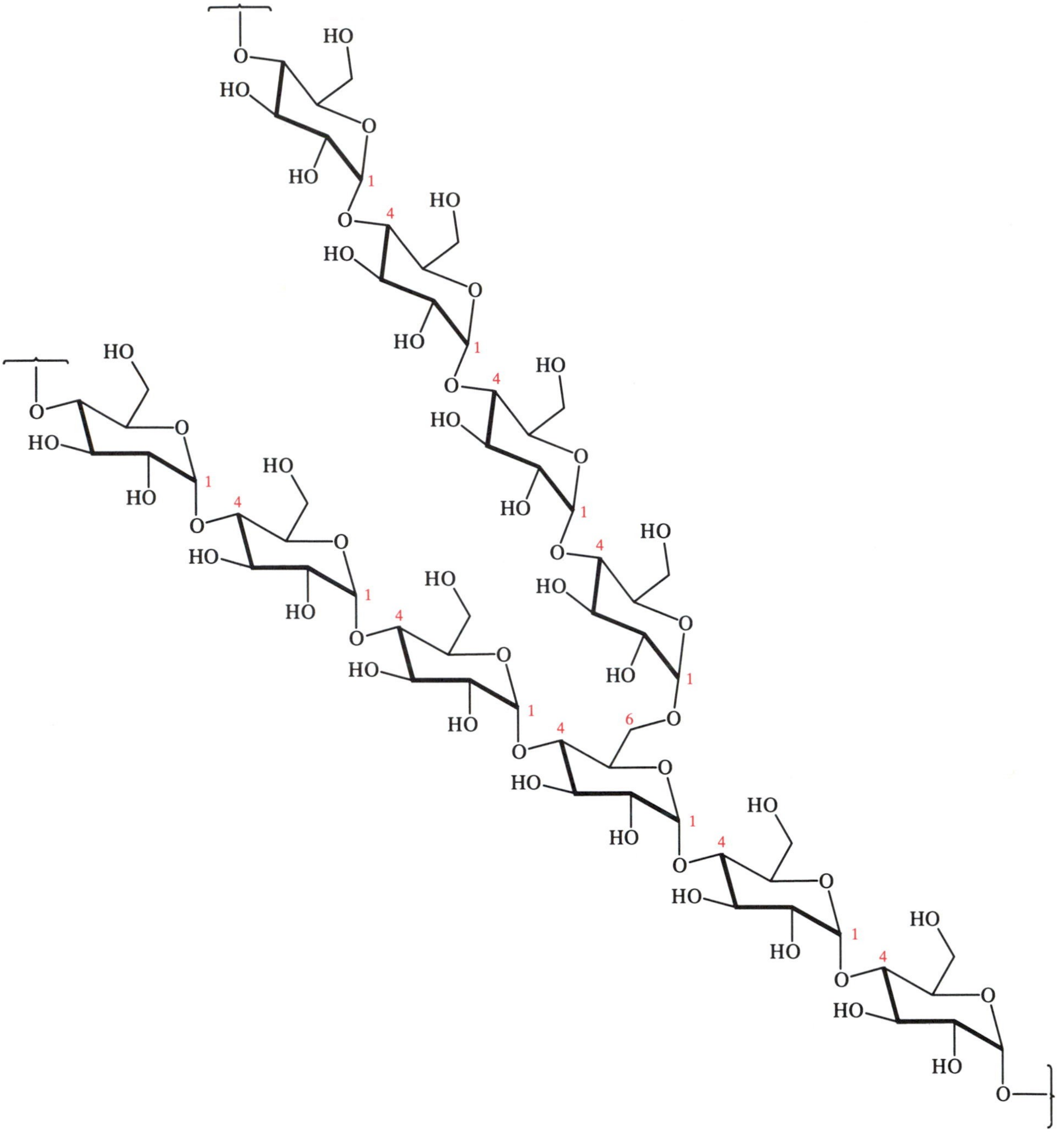

Figure 15-7

Structure of Glycogen Showing α-1,4- and α-1,6-Glucosidic Linkages.

The branching reaction increases the number of end glucosyl residues and makes the structure more compact. The physiological importance of adding terminal glucosyl residues by this process is that it increases the number of points at which glycogen can be degraded by phosphorylase, which degrades glycogen by acting on terminal glucosyl residues, as is explained in the next section of this chapter.

Plant starch is similar to glycogen but is produced from ADP-glucose, the adenosine analog of UDP-glucose. It is less branched than glycogen and so tends to form bundles of parallel hydrogen-bonded chains. Consequently, it is

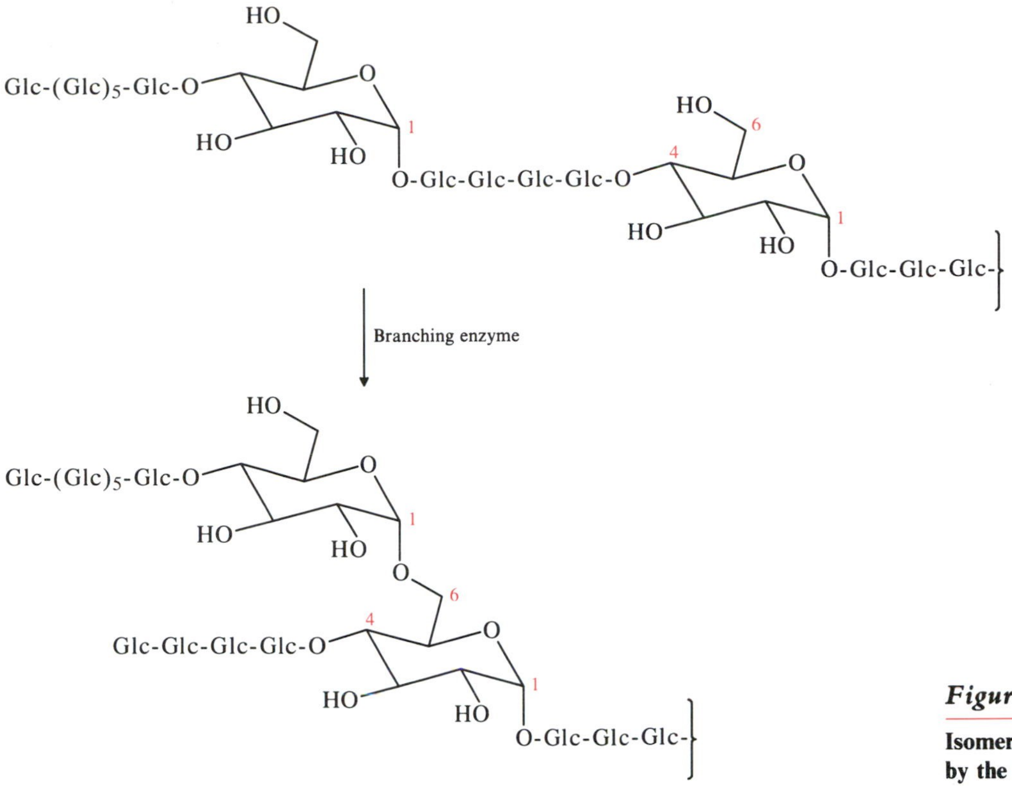

Figure 15-8

Isomerization of Linear Glycogen by the Branching Enzyme.

less soluble and its glycosidic bonds are less accessible for reaction than those of glycogen. The chains of glycogen do not pack together well and tend to form intrachain hydrogen bonds in helical segments.

Glycogen Phosphorylase

The breakdown of glycogen to α-D-glucose-1-phosphate for utilization by the cell is catalyzed by glycogen phosphorylase according to equation 2.

$$(\text{Glucose})_n + \text{P}_i \rightleftharpoons \alpha\text{-D-Glucose-1-P} + (\text{Glucose})_{n-1} \quad (2)$$

As is illustrated in Figure 15-9, a glucosyl group from a nonreducing end of

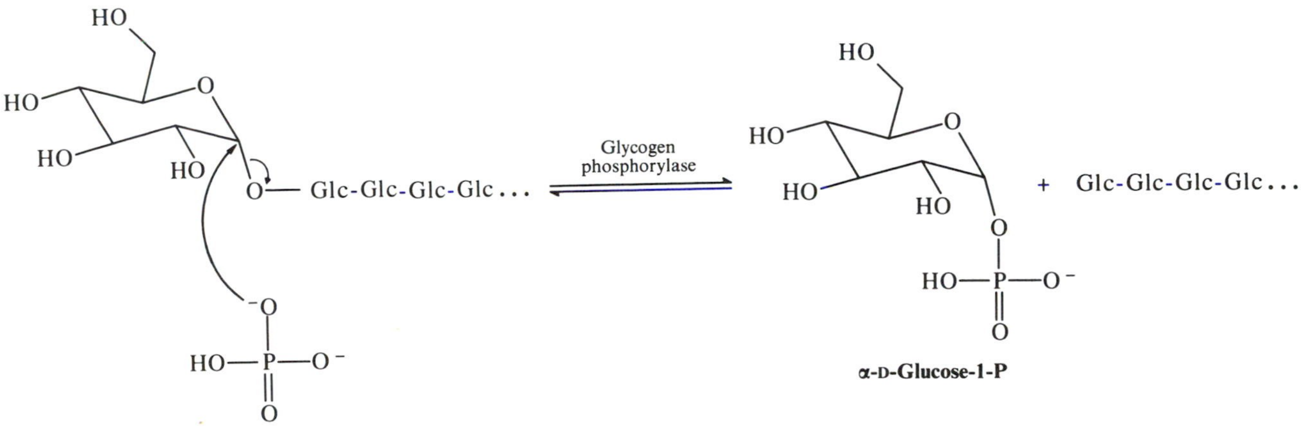

Figure 15-9

Phosphorolytic Cleavage of Glycogen Catalyzed by Glycogen Phosphorylase.

glycogen is transferred to inorganic phosphate with retention of configuration to produce α-D-glucose-1-P and shorten the glycogen chain by one glucosyl residue. Iteration efficiently generates a large amount of glucose-1-P from glycogen. The reaction is reversible and slightly favors glycogen synthesis; however, it is drawn in the forward direction by the presence of phosphoglucomutase. This enzyme catalyzes the isomerization of glucose-1-P to glucose-6-P, which is energetically downhill and supplies glucose in a form that can be utilized as a source of chemical free energy for the production of ATP in glycolysis and the citric acid cycle (Chapters 21 and 22).

Cellulose

Cellulose also is a polymer of glucose in plants, but it differs from plant starch in the stereochemistry of the glycosidic linkages. Cellulose is a linear, unbranched polymer in which the linkages are all β-1,4, as illustrated in Figure 15-10. Cellulose is synthesized mainly from ADP-glucose, which has an α-glucosyl group like UDP-glucose; however, the glucoside linkages in cellulose are β-1,4, so that the reaction proceeds with inversion of configuration at C-1 of glucose. This apparent minor difference in structure between starch and cellulose has enormous chemical, biochemical, industrial, and ecological consequences. The pages of this book and other cellulose-containing materials have a defined structure because cellulose is a water-insoluble polymer. The β-1,4 linkages allow for the intrachain hydrogen bonding shown in Figure 15-10 and for numerous interchain H bonds conferring insolubility and high tensile strength.

Animals generally cannot digest cellulose, although cattle and other ruminants have microorganisms in their digestive tracts that cleave cellulose to degradation products that the animals can utilize to produce energy. This symbiotic relation allows the animals to live on grasses. It remains a problem to understand how Nebuchadnezzar was able to subsist on grass for a considerable

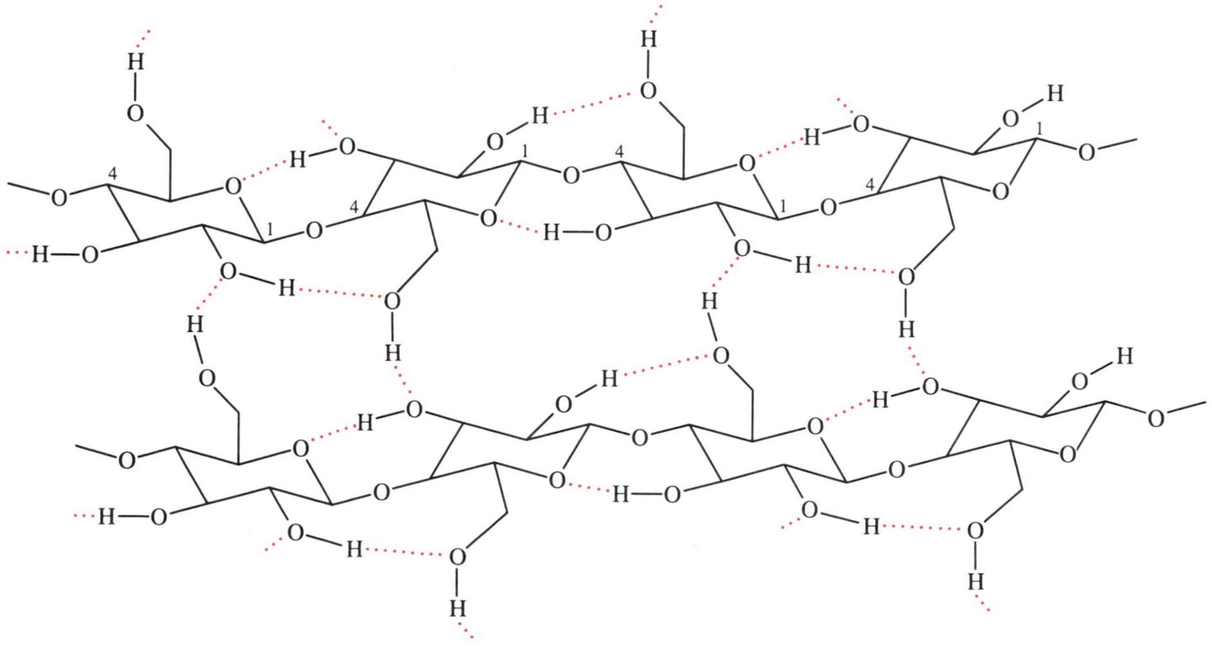

Figure 15-10

Structure of Cellulose Showing β-1,4-Glucosidic Linkages.

period. If a way could be found to duplicate this feat today, so that cellulose could be used as a foodstuff for human beings, starvation and famines could be prevented, because there is more cellulose in the world than any other organic compound.

Biosynthesis of Disaccharides

Several important disaccharides are biosynthesized by glucosyl transfer from UDP-glucose to a hydroxyl group of an acceptor sugar. Glucosyl transfer from UDP-glucose to the 2-OH group of the cyclic hemiketal of fructose-6-P gives sucrose phosphate, which is hydrolyzed by a phosphatase to sucrose, as shown in Figure 15-11. The reaction proceeds with overall retention of configuration at C-1 of glucose.

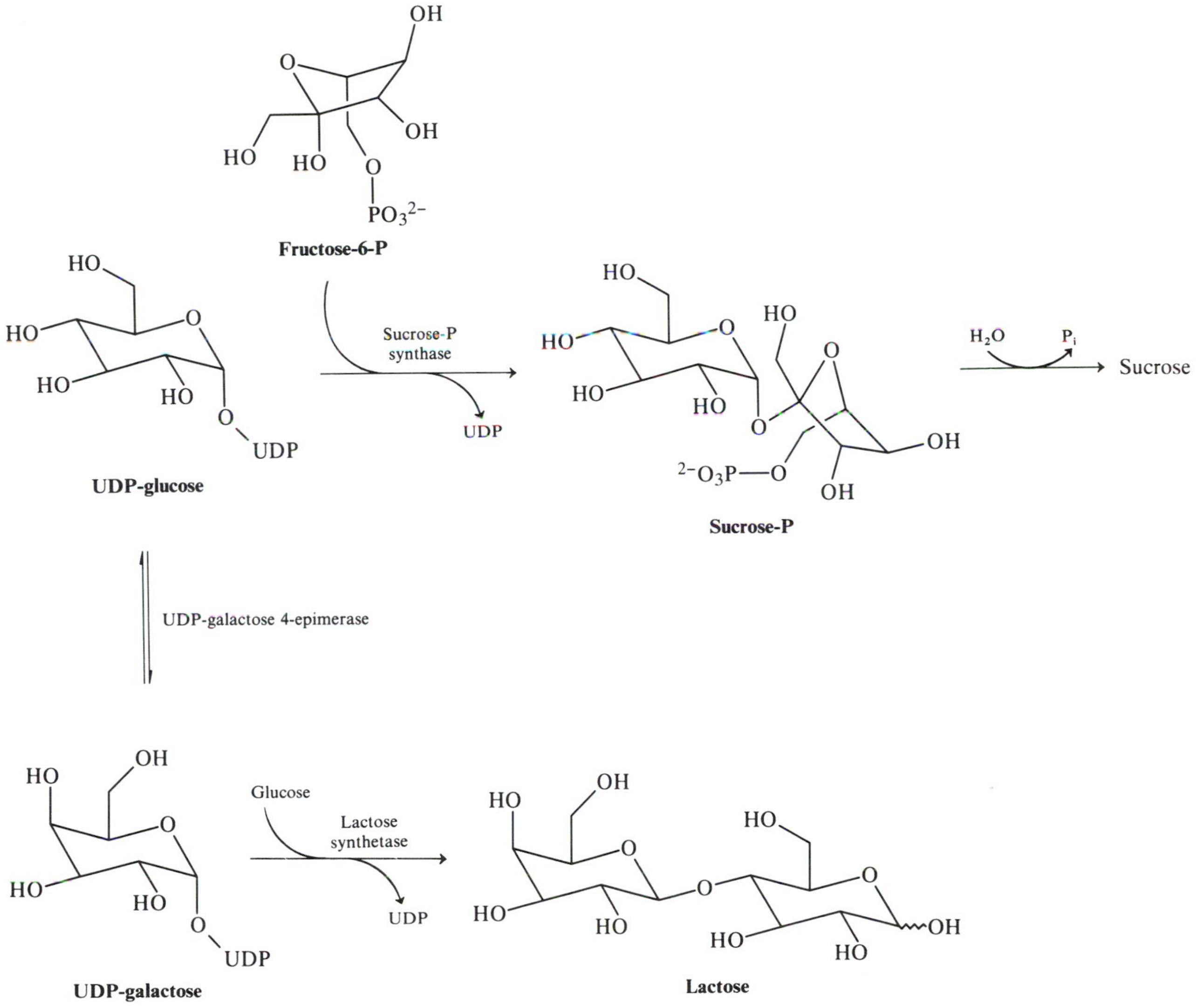

Figure 15-11

Enzymatic Synthesis of Sucrose and Lactose. The biosynthesis of sucrose takes place only in plants. Lactose is biosynthesized in mammary glands.

Figure 15-12

The Mechanistic Reaction Pathway in Catalysis by Sucrose Phosphorylase.

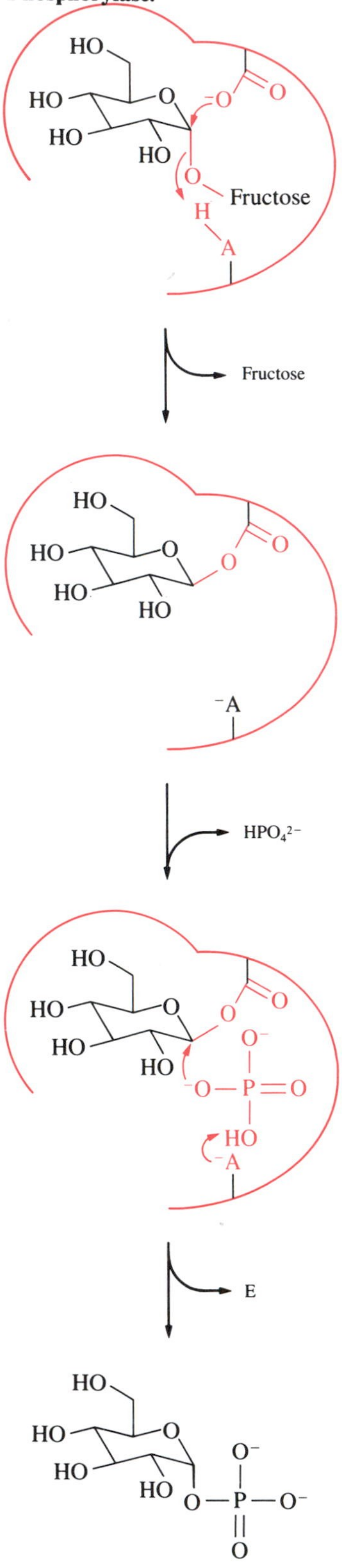

α-D-Glucose-1-P

The analogous reaction of UDP-galactose with glucose, also shown in Figure 15-11, is catalyzed by lactose synthetase and gives lactose with overall inversion of configuration at C-1 of galactose. Lactose synthetase from milk is a complex of two proteins, galactosyl transferase and α-lactalbumin, which exist in an association-dissociation equilibrium, according to equation 3.

$$\text{Lactose synthetase} \rightleftharpoons \text{Galactosyl transferase} + \alpha\text{-Lactalbumin} \quad (3)$$

The galactosyl transferase efficiently catalyzes galactosyl transfer from UDP-galactose to *N*-acetylglucosamine units on glycoproteins or to *N*-acetylglucosamine itself. Its activity toward glucose is very low, primarily owing to a very high value of K_m. However, in the complex with α-lactalbumin as lactose synthetase, it catalyzes lactose biosynthesis efficiently with a low K_m for glucose. Thus, the regulatory effect of α-lactalbumin is to reduce the K_m for glucose.

Sucrose Phosphorylase

In certain bacteria such as *Pseudomonas saccharophila,* sucrose is used as a nutrient and is degraded initially to fructose and glucose 1-P (equation 4).

$$\text{Sucrose} + P_i \rightleftharpoons \alpha\text{-D-Glucose-1-P} + \text{Fructose} \quad (4)$$

This reaction is catalyzed by the enzyme sucrose phosphorylase. In animals, nutrient sucrose is hydrolyzed to glucose and fructose. The sucrose phosphorylase reaction is of interest as a glycosyl transfer reaction that has been subjected to a thorough mechanistic analysis. It proceeds with overall retention of configuration at C-1 of the glucosyl group. The reaction follows the Ping-Pong kinetic pathway, in which the intermediate is a covalent glucosyl-enzyme.

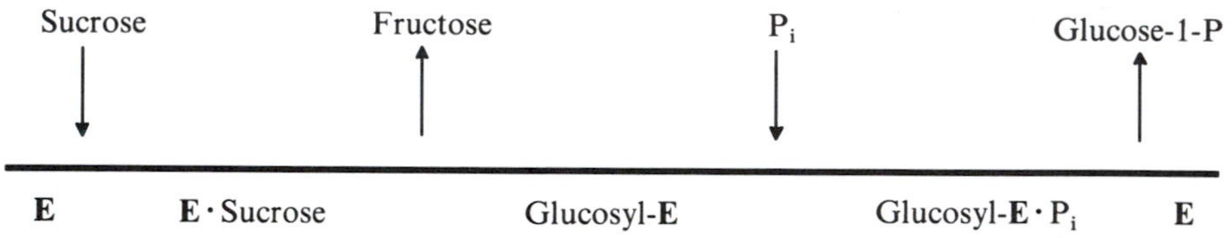

The glucosyl-enzyme can be isolated at low pH, where the enzyme activity is low, and in a denatured form. The glucosyl group in the intermediate is bonded to the carboxyl group of the protein, and the isolated intermediate has the *β* configuration at C-1 of the glucosyl group. The glucosyl groups in sucrose and glucose-1-P have the α configuration at C-1, so that the configuration of the glucosyl group in the intermediate is inverted relative to the substrate and product. The overall reaction follows the course outlined in Figure 15-12, which shows how the double displacement at C-1 leads to overall retention of configuration.

Phosphoribosyl Pyrophosphate (PRPP)

The source of *β*-ribosyl groups for nucleotides is α-D-5-phosphoribosyl-1-pyrophosphate (PRPP). This compound is produced from ribose 5-P in an interesting reaction (equation 5) catalyzed by a pyrophosphokinase that transfers the pyrophosphoryl group from ATP to the anomeric hydroxyl group of α-D-ribose 5-P.

$$\text{Ribose-5-P} + \text{ATP} \xrightarrow[\text{Mg}^{2+}]{\text{PRPP synthetase}} \text{PRPP} + \text{AMP} + \text{H}^+ \quad (5)$$

The pyrophosphate group at C-1 of PRPP is an excellent leaving group for phosphoribosyl transfer reactions, and these are important steps in the biosynthesis of nucleotides. In nucleotides, the heterocyclic bases are bonded to C-1 of ribose-5-P or 2′-deoxyribose-5-P in the β configuration; that is, inverted relative to the α configuration in PRPP. All such linkages are formed from PRPP (Chapter 27).

One of the reactions in which PRPP is the ribosyl group donor is the first step of the purine biosynthetic pathway, the reaction of PRPP with glutamine catalyzed by 5-phosphoribosylamine synthase according to equation 6.

$$\text{PRPP} + \text{Gln} + H_2O \xrightarrow{\text{Phosphoribosylamine synthase}} \text{5-phosphoribosylamine} + \text{Glu} + \text{PP}_i \quad (6)$$

The amino group in 5-phosphoribosylamine originates with the amide group of glutamine. The amido nitrogen displaces pyrophosphate from PRPP to form the β-riboside; that is, with overall inversion of configuration. In the process, the amido-NH_2 is hydrolytically removed from glutamine, which is converted into glutamate. The enzyme will also accept NH_3 as the acceptor substrate in place of glutamine. The reaction mechanism apparently involves a single displacement of pyrophosphate at C-1 of the ribosyl group, in contrast with the sucrose phosphorylase reaction.

Dolichol Phosphate and Complex Oligosaccharides

An important part of protein biosynthesis in eukaryotic cells is the attachment of complex oligosaccharides to proteins in the production of glycoproteins. These oligosaccharides are recognition groups whose functions are only partly understood; they are added to the polypeptide chains, initially in the endoplasmic reticulum, and serve as signals to direct the proteins to the Golgi apparatus. The oligosaccharides are processed in the Golgi complex by a variety of reactions that remove some of the sugars and add others.

A typical oligosaccharide in *N*-glycosidic linkage to a protein is shown in Figure 15-13. In glycoproteins, the oligosaccharides are linked either through asparagine, the *N*-glycosyl linkage, or through the 3-OH of serine or threonine, the *O*-glycosyl linkage.

The biosynthesis of complex oligosaccharides and their attachment to asparagine residues of proteins proceeds by an interesting series of phosphomonoester and glycosyl transfers and a phosphatase step, in which dolichol phosphate plays a central role.

$$H_3C{-}\overset{CH_3}{\overset{|}{C}}{=}CH{-}CH_2{-}\left[CH_2{-}CH{=}\overset{CH_3}{\overset{|}{C}}{-}CH_2\right]_n{-}CH_2{-}\overset{CH_3}{\overset{|}{C}H}{-}CH_2{-}CH_2{-}OPO_3^{2-}$$

Dolichol phosphate
(n = 15–19)

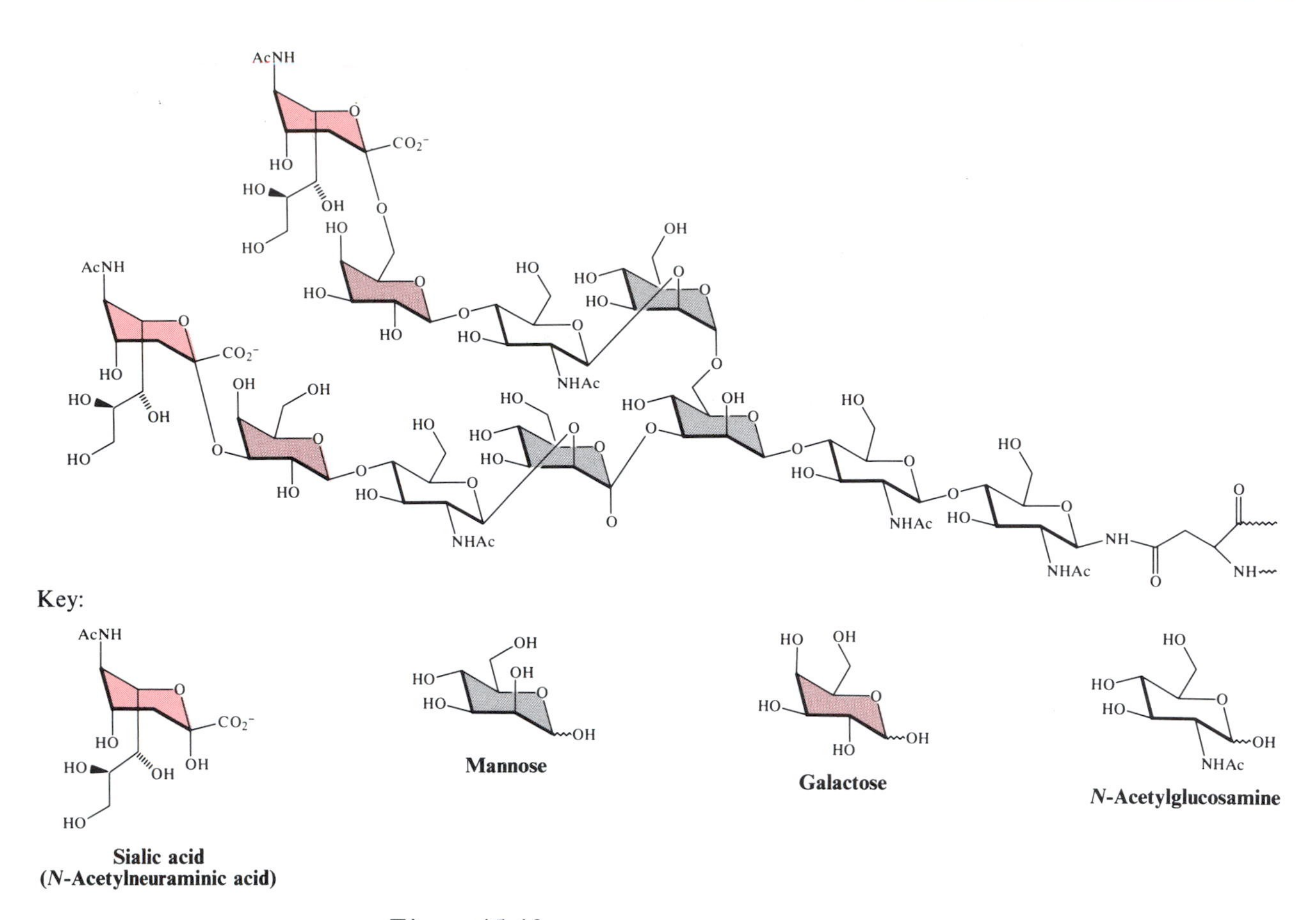

Figure 15-13

Structure of a Typical Oligosaccharide Attached to a Protein.

The *N*-linked oligosaccharides are assembled while attached to dolichol phosphate,which is anchored in the membrane of the endoplasmic reticulum. The oligosaccharide in Figure 15-13 contains four *N*-acetylglucosaminyl (NAG) units, three mannosyl (Man) units, two galactosyl (Gal) units, and two sialyl (*N*-acetylneuraminyl, or NeuNAc) units. The sugars originate with the nucleoside diphosphate derivatives UDP-*N*-acetylglucosamine (UDP-NAG), UDP-galactose (UDP-Gal), UDP-glucose (UDP-Glc), and GDP-mannose (GDP-Man). The activated form of mannose is always the GDP derivative. The sugars are connected by 1,2-, 1,3-, 1,4-, and 1,6-glycosyl linkages in either α- or β-anomeric configurations. Other oligosaccharides also contain glucose, glucuronic acid, and sialic acid among other sugars. With the number of possible glycosyl linkages and sugars that are available for the assembly of oligosaccharides, many structures are possible.

The biosynthesis of oligosaccharide groups for attachment to asparagine residues takes place in the endoplasmic reticulum, the site of protein biosynthesis. The process begins with the assembly of a precursor oligosaccharide on dolichol-P and the transfer of the oligosaccharyl group to a nascent polypeptide chain as it grows from a ribosome. The composition of the precursor oligosaccharide, which is shown in Figure 15-14, is $NAG_2Man_9Glc_3$. The glycosyl groups shown in color in Figure 15-14 are donated to the growing oligosaccharide by the nucleotide derivatives UDP-NAG and GDP-Man. The other glycosyl groups are donated by dolichol-P-Glc and dolichol-P-Man, which in turn are produced from dolichol-P and the nucleotide sugar according to equations 7 and 8.

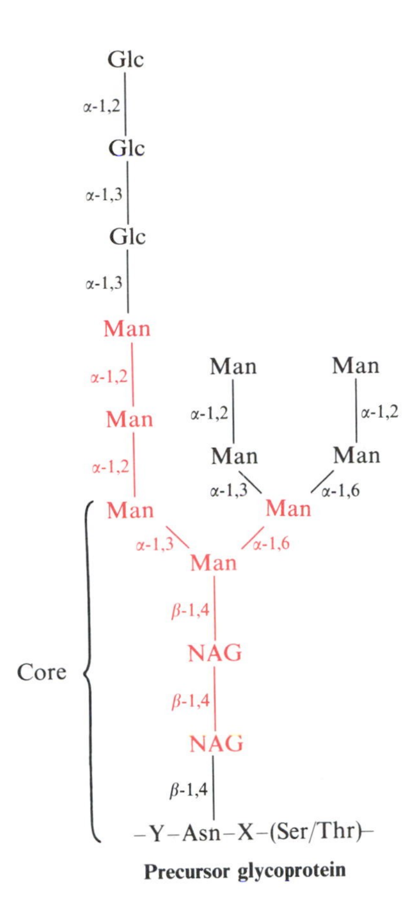

Figure 15-14

Structure of the Precursor Oligosaccharide in the Endoplasmic Reticulum. The oligosaccharide $NAG_2Man_9Glc_3$ is assembled on dolichol phosphate in the endoplasmic reticulum and transferred to an asparagine residue in a sequence [Asn–X–Ser (or Thr)] in a nascent polypeptide chain being synthesized on a ribosome. The glycosyl groups shown in color are derived from UDP-NAG or GDP-Man, and the others are derived from dolichol–P-Glc or dolichol-P-Man. The assembly of the precursor oligosaccharide in the endoplasmic reticulum is illustrated in Figure 15-15. The core (NAG_2 Man_3) survives later processing in the Golgi complex, where some or all of the peripheral glycosyl groups are removed by specific glycosidases. New glycosyl groups are also added in the Golgi to give a diversity of structures, one of which is shown in Figure 15-13.

$$\text{Dolichol—P} + \text{UDP-Glc} \rightleftharpoons \text{Dolichol—P—Glc} + \text{UDP} \tag{7}$$

$$\text{Dolichol—P} + \text{GDP—Man} \rightleftharpoons \text{Dolichol—P—Man} + \text{GDP} \tag{8}$$

Figure 15-15 illustrates how the precursor oligosaccharide in Figure 15-14 is assembled while attached to dolichol phosphate. The polyprenyl portion of dolichol phosphate is hydrophobic and firmly anchored in the membrane of the endoplasmic reticulum. The phosphate group is ionized and exposed at the membrane surface, where it interacts with enzymes. In the first step, the phosphate group accepts an *N*-acetylglucosamine phosphate (NAG-P) in a phosphomonoester transfer reaction with UDP-NAG. This is not a glycosyl transfer and not an NMP transfer but a glycosyl phosphoryl transfer, in which UMP is released and a pyrophosphoryl linkage is introduced between dolichol and NAG. There follows a series of glycosyl transfers in which NAG is added from UDP-NAG and five mannoses are added from GDP-Man. The remaining glucosyl and mannosyl groups are added from dolichol-P-Glc and dolichol-P-Man to complete the assembly of the precursor complex saccharide.

The oligosaccharide is then transferred to the carboxamido nitrogen of an asparagine residue in a growing polypeptide chain. This transfer is catalyzed by an oligosaccharyl transferase that recognizes the asparaginyl sequences Asn-X-Ser (or Thr) in nascent polypeptide chains. In these sequences, the residue X can be any amino acid except proline and aspartate. The dolichol diphosphate released is then hydrolyzed by a phosphatase to dolichol phosphate, which can start another cycle of oligosaccharide biosynthesis.

The glycopeptide is transferred to the Golgi complex, where it is further processed as it traverses the compartments. In this processing, some of the

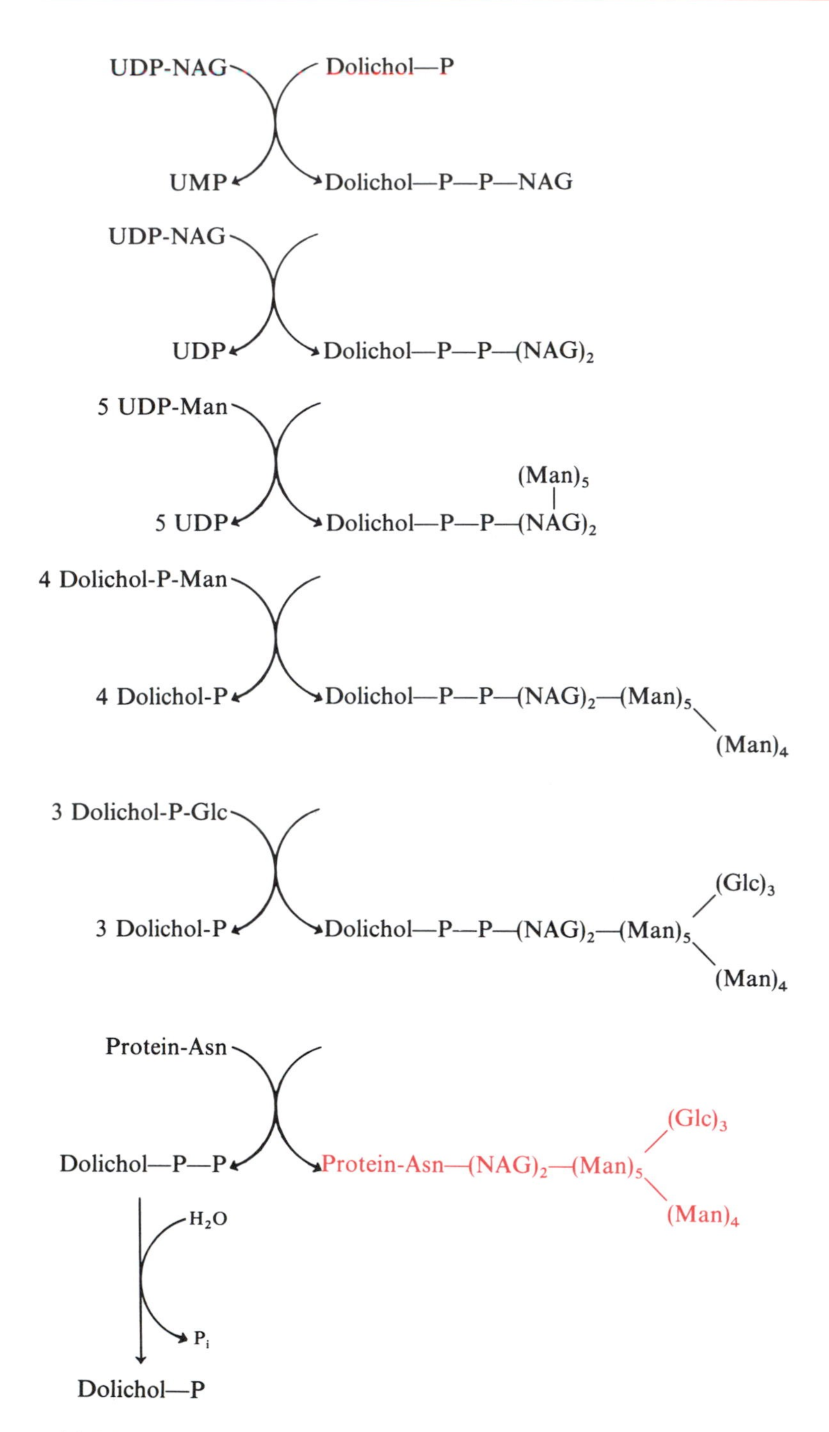

Figure 15-15

Assembly of an *N*-linked Oligosaccharide on a Glycoprotein. The oligosaccharide in Figure 15-14 is assembled on dolichol phosphate in the endoplasmic reticulum by the glycosyl transfer reactions shown here. The process is begun by transfer of an *N*-acetylglucosaminyl-1-phosphoryl group from UDP-*N*-acetylglucosamine to dolichol phosphate to form dolichol diphosphate *N*-acetylglucosamine. Another *N*-acetylglucosaminyl group and five mannosyl groups are added from nucleotide derivatives to the growing oligosaccharide. The remaining mannosyl units and three glucosyl groups are added from the corresponding dolichol phosphosugars. An oligosaccharyl transferase finally transfers the completed oligosaccharide to the peptide. The oligosaccharide on the glycoprotein is processed further in the Golgi complex.

glycosyl groups are removed by the actions of specific glycosidases, and other glycosides, such as those shown in Figure 15-13, are added. The glycosides in the core structure NAG_2Man_3, shown in Figure 15.14, are not removed in this process; in many cases, only a few of the other mannosyl groups are removed. This variability, together with the variability of the glycosides added in the Golgi complex, allows a high degree of diversity in the structures of the oligosaccharides in glycoproteins.

Summary

Glycosides are cyclic derivatives of sugars, in which a glycosyl group is attached to another group. The attached groups in glycosides may be other sugars in complex carbohydrates, a nucleotide in nucleotide sugars, a protein in glycoproteins, or a lipid in glycolipids. The derivatives are produced in enzymatic glycosyl group transfer reactions. The mechanisms of these reactions may include either extremely unstable glycosyl carbocationic intermediates or covalent glycosyl-enzyme intermediates. Examples of glycosyl transfer reactions include the biosyntheses and degradations of disaccharides and carbohydrates, as well as the biosynthesis of 5′-phosphoribosyl pyrophosphate, the donor of ribosyl groups in nucleotide biosynthesis.

The *N*-linked oligosaccharides attached to glycoproteins produced in eukaryotic cells are biosynthesized in the endoplasmic reticulum. The oligosaccharides are assembled on membrane-bound dolichol diphosphate by a series of glycosyl transfer reactions. The precursor oligosaccharide is transferred from the dolichol diphosphate oligosaccharide to an asparagine residue in a nascent polypeptide chain on a ribosome in the course of protein biosynthesis. This glycosyl transfer is catalyzed by oligosaccharyl transferase. The oligosaccharide is further processed in the Golgi complex.

ADDITIONAL READING

Imoto, T., Johnson, L. N., North, A. C. T., Phillips, D. C., and Rupley, J. A. 1972. Vertebrate lysozymes. In *The Enzymes,* vol. 7, 3d ed., ed. P. D. Boyer. New York: Academic Press, p. 665.

Wallenfels, K., and Weil, R. 1972. β-Galactosidase. In *The Enzymes,* vol. 7, 3d ed., ed. P. D. Boyer. New York: Academic Press, p. 617.

Mieyal, J. J., and Abeles, R. H. 1972. Disaccharide phosphorylases. In *The Enzymes,* vol. 7, 3d ed., ed. P. D. Boyer. New York: Academic Press, p. 515.

Kornfeld, R., and Kornfeld, S. 1985. *Annual Review of Biochemistry* 54: 631.

Hubbard, S. C., and Ivatt, R. J. 1981. *Annual Review of Biochemistry* 50: 555.

Ginsberg, V., and Robbins, P, eds. 1984. *Biology of Carbohydrates.* New York: Wiley.

Lennarz, W. J., ed. 1980. *The Biochemistry of Glycoproteins and Proteoglycans.* New York: Plenum.

PROBLEMS

1. Draw the structures of five disaccharides other than lactose consisting of one molecule of glucose and one molecule of galactose. Suggest names for these molecules. How many structures of this type are possible?

2. Why are the glycosidic bonds between sugars subject to acid-catalyzed hydrolysis and not to base-catalyzed hydrolysis?

3. Glucose in dilute aqueous solutions consists of about 37% α-D-glucose and 62% β-D-glucose, with less than 1% of the open-chain and hydrated forms. Explain why there is so little hydrate despite the fact that the concentration of water is 55 M.

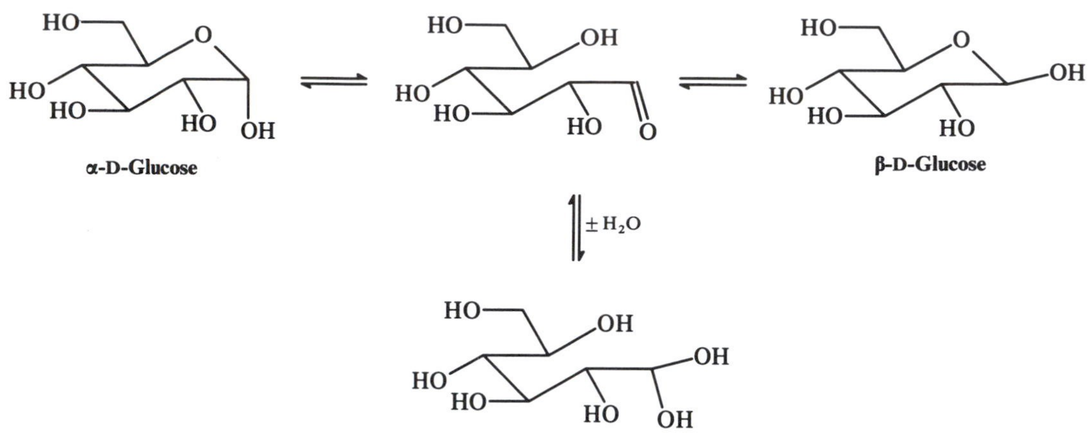

α-D-Glucose ⇌ ⇌ β-D-Glucose

$\pm H_2O$

4. Consider the hydrolysis of a β-galactoside by β-galactosidase as an example of a glycosyl transfer. There is evidence suggesting that this reaction might proceed by way of a covalent galactosyl-enzyme as an intermediate; however, the evidence is inconclusive, and the reaction might include a galactosyl carbocationic intermediate. Show with simple drawings how binding interactions with the enzyme could stabilize the transition state for the formation of a carbocationic intermediate. Assume that there is a carboxylic acid and a carboxylate group in the active site similar to glutamate-35 and aspartate-52 in lysozyme. Show with similar drawings how enzyme-binding interactions could stabilize a trigonal bipyramidal transition state for the formation of a covalent galactosyl-enzyme, in which the galactosyl group is attached to one of the acidic residues in the active site. Compare the enzyme-binding interactions for these two mechanisms. What information could allow you to decide which mechanism should be excluded?

5. Glycogen and other polysaccharides contain branching points. One means of characterizing such polymers is to determine the fraction of branched glycoside residues relative to the total number of glycoside units. Explain how an exhaustively methylated sample of glycogen (all OH groups converted into OCH_3) could be chemically analyzed in such a way as to give accurate information about the number of branches in the polymer.

6. Draw the structure of the oligosaccharide part of the precursor oligosaccharide shown in Figure 15-14.

Biological Oxidation: NAD$^+$ and FAD

16

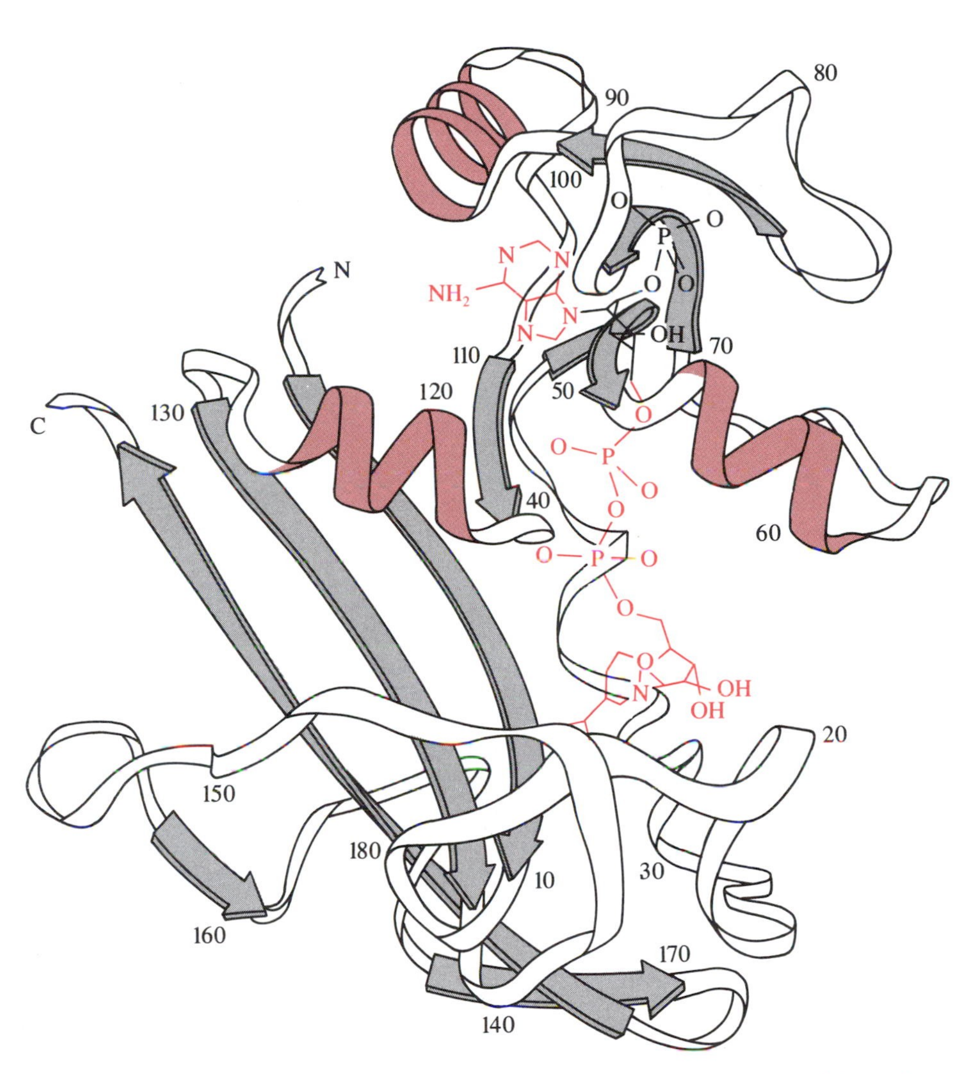

A ribbon diagram of dihydrofolate reductase with NADPH and trimethoprim bound at the active site. (Adapted from "Dihydrofolate Reductase" by J. Kraut and D. A. Matthews. In Biological Macromolecules and Assemblies, vol. 3, ed. F.A. Jurnak and A. McPherson. New York: Wiley-Interscience, © 1987, Fig. 4b. Adapted by permission of John Wiley & Sons, Inc.)

Many biological reactions are *redox* (oxidation-reduction) reactions. These are reactions in which electrons are transferred between molecules. Redox reactions are not only important in metabolic and catabolic processes, but also central to the reactions that lead to the generation of ATP. In this chapter, we shall consider two types of molecules, flavins and pyridine nucleotides, that play an important role in redox reactions. We shall describe the mechanism by which electron transfer takes place.

Oxidation Means Loss of Electrons

The reader is undoubtedly familiar with the concept of oxidation and reduction. However, before we begin to consider biological oxidations, we will review the basic chemistry of oxidation and reduction. Consider equations 1 and 2.

$$\underset{\downarrow\, 2\times 2e^-}{2Cu} + \underset{\uparrow\, 2\times 2e^-}{O_2} \longrightarrow 2CuO \qquad (1)$$

$$\underset{\downarrow\, 3\times 4e^-}{3CH_3CH_2OH} + \underset{\uparrow\, 4\times 3e^-}{4MnO_4^-} + 4K^+ + H^+ \longrightarrow 3CH_3CO_2^- + 4MnO_2 + 4K^+ + 5H_2O \qquad (2)$$

In both reactions, oxidation and reduction take place. In equation 1, copper is oxidized by O_2 and, in equation 2, ethanol is oxidized by $KMnO_4$. In both equations, the substance oxidized loses electrons and the oxidizing agent gains electrons, with the numerical loss of electrons indicated by the downward dashed arrows and the gain of electrons indicated by the upward dashed arrows. Reduction is the reverse of oxidation, so that the substance reduced gains electrons and the reducing agent loses electrons. In equation 1, Cu is the reducing agent and is oxidized, whereas O_2 is the oxidizing agent and is reduced.

With inorganic species such as Cu and MnO_4^-, it is relatively easy to decide whether electrons are lost or gained and how many electrons are involved. This is because inorganic species are assigned oxidation states, and the loss or gain of electrons is indicated by the change in oxidation state. Thus, in equation 1, the oxidation state of Cu changes from 0 to +2 and Cu is oxidized, whereas, in equation 2, the oxidation state of Mn changes from +7 to +4, and Mn is reduced.

The concept of oxidation states is much less frequently used with organic molecules (see Box 16-1 for a discussion of the use of oxidation states in organic molecules); therefore, it is difficult to decide whether an organic reaction involves oxidation or reduction. The best way to overcome this uncertainty is to remember the relative oxidation states of the various functional groups in organic molecules. Fortunately, this is not difficult to do because there are relatively few functional groups. Table 16-1 lists the important functional groups and indicates which interconversions result in oxidation or reduction.

BOX 16-1

OXIDATION STATES OF ORGANIC MOLECULES

The use of numerical oxidation states for carbon in organic molecules is complicated by the fact that most organic molecules contain several carbon atoms, each of which may be in a different oxidation state. The oxidation state of carbon in such a molecule could be expressed for each carbon atom, but this would be a cumbersome way to describe the oxidation state of a molecule with more than two carbon atoms. The oxidation state of carbon can be computed as an "average" number for all of the carbon atoms. This method can lead to fractional oxidation states. Another way to express the oxidation state for organic compounds is to consider the sum of oxidation states for the carbon atoms in a molecule.

The rules for assigning oxidation states to atoms other than carbon in organic molecules are as follows:

1. That of hydrogen is assigned $+1$.
2. Those of oxygen and sulfur are assigned -2.
3. Those of halogens are assigned -1.
4. That of nitrogen is -3.

The following procedure can be used for computing the sum of oxidation states for carbon atoms in organic molecules that are not phosphate, sulfate, or nitrate esters. For these molecules, the sum of oxidation states for carbon should be calculated for the parent alcohol. (The following method must be modified if it is to be used for phosphonate, sulfonate, nitro, and nitroso compounds.)

The sum of the oxidation states for all carbon atoms in an organic molecule or ion can be computed by considering the molecular formula and overall charge. The oxidation state for the atoms in a molecule or ion must add algebraically to the overall charge on the molecule or ion. To calculate the sum of oxidation states for carbon atoms, algebraically add the oxidation states for the hydrogen, halogen, nitrogen, oxygen, and sulfur atoms. Subtract this number from the overall charge on the molecule or ion to obtain the sum of the oxidation states for all of the carbon atoms in the molecule.

The application of these rules to some one-carbon molecules gives the following oxidation states for carbon: methane (CH_4), -4; methanol (CH_4O), -2; formaldehyde (CH_2O), 0; formic acid (CH_2O_2), $+2$; and carbon dioxide (CO_2), $+4$. These numbers show that the oxidation states increase in two-electron increments in the progression from methane to carbon dioxide.

Consider a more-complicated species such as the pyruvate ion ($H_3C—CO—COO^-$). The sum of oxidation states for hydrogen and oxygen is $3(+1)+3(-2)= -3$. Subtracting this sum from the overall charge gives $(-1)-(-3) = +2$, which is the sum of the oxidation states for the three carbon atoms. You can verify that the same result is obtained for pyruvic acid, which has the formula $C_3H_4O_2$ and a charge of zero. Consider lactate ion ($H_3C—CHOH—COO^-$). The sum of noncarbon oxidation states is $5(+1) + 3(-2) = -1$. The sum of oxidation states for the three carbon atoms is $-1-(-1) = 0$. Comparing this with the sum of the carbon oxidation states in pyruvate ($+2$) shows that pyruvate is oxidized by two electrons relative to lactate. Consider the carbon in *n*-propanol, C_3H_8O. The sum of oxidation states for hydrogen and oxygen is $8(+1) + (-2) = +6$, and for the carbon atoms it is $0-6 = -6$. This means that propanol is reduced by 8 electrons relative to pyruvate and by 6 electrons relative to lactate. Consider propylamine, C_3H_9N. The sum of oxidation states for hydrogen and nitrogen is $9(+1) + (-3) = +6$. The sum of oxidation states for carbon is $0 - 6 = -6$, the same as for propanol.

This method can lead to rather large numbers for the carbon oxidation states of molecules having many carbon atoms. However, it always allows a comparison of the oxidation states of molecules having the same number of carbon atoms. To determine the relative oxidation states of molecules having different numbers of carbon atoms, compute the average oxidation state per carbon from the sum of carbon oxidation states for each molecule. This will often give fractional oxidation states, but these fractions still allow a comparison of the oxidation states of the molecules. For example, for pyruvate, the average oxidation state of carbon is $+2/3$, which means that pyruvate is intermediate in oxidation state between formate ($+2$) and formaldehyde (0). This is as it should be because pyruvate contains a carboxyl group, a ketonic group, and a methyl group. For lactate, the average oxidation state of the carbon atoms is 0, placing it at the same overall oxidation level as formaldehyde.

Balancing Redox Equations

The oxidation states of molecules are useful in balancing chemical equations for redox reactions. One may know the nature of the organic substrates and products of a reaction, and any reaction can be described by a balanced chemical equation. However, the balanced chemical equation may not be obvious for a complex reaction. For example, in a process that entails a series of chemical steps, the overall stoichiometry may not be obvious, but the balanced chemical equation will always give the correct stoichiometry. A simple example is given here to show how oxidation states can be used to balance biochemical redox equations.

Consider a reaction that is catalyzed by a cell extract, such as the transformation of malate

BOX 16-1 (continued)

OXIDATION STATES OF ORGANIC MOLECULES

($C_4H_4O_5^{2-}$) into pyruvate ($C_3H_3O_3^-$) and carbon dioxide (CO_2). The first question that must be answered is whether this is a redox reaction. To answer this question, the sum of carbon oxidation states in malate must be compared with the sum of carbon oxidation states in pyruvate *plus* carbon dioxide. For malate, the sum of carbon oxidation states is $(-2) - [4(+1) + 5(-2)] = +4$. For pyruvate and carbon dioxide, the sums of carbon oxidation states in the two molecules are, as already calculated, +2 and +4 for a total of +6. The products are more highly oxidized than the reactants, so that it is a redox process. The second question is: "What is the balanced chemical equation?" From the available information, the balanced oxidation half-reaction can be written as:

$$C_4H_4O_5^{2-} \rightleftharpoons C_3H_3O_3^- + CO_2 + 2e^- + H^+ \quad \text{(a)}$$

Note that half-reactions must be balanced both electrostatically and chemically by the inclusion of electrons and hydrogen ions. In this case, two electrons had to be added to the right side to balance the oxidation states of carbon, which are +4 on the left and +6 on the right. This led to an electrostatic imbalance, which was corrected by the inclusion of a hydrogen ion on the right side. This hydrogen ion also chemically balanced the hydrogen on the two sides of the equation. Equation a describes the oxidation part of the overall reaction, but there must also be a reduction half-reaction. For a process that takes place in a crude extract, the nature of the oxidizing agent might not be known; so we can represent it in a general form as X, which might represent FAD or NAD^+. The reduction half-reaction can be written as:

$$X + 2H^+ + 2e^- \rightleftharpoons XH_2 \quad \text{(b)}$$

The sum of the oxidation half-reaction (equation a) and the reduction half-reaction (equation b) is:

$$C_4H_4O_5^{2-} + X + 2H^+ \rightleftharpoons C_3H_3O_3^- + CO_2 + XH_2 + H^+ \quad \text{(c)}$$

Note that the equation is balanced chemically and electrostatically. It tells us that there must be an oxidizing agent that accepts two electrons in this reaction. In fact, the reaction is catalyzed by malic enzyme, which utilizes $NADP^+$ as the oxidizing agent and produces $NADPH + H^+$, as well as pyruvate and carbon dioxide.

As a second example, consider the transformation of fumarate ($C_4H_2O_4^{2-}$) into malate ($C_4H_4O_5^{2-}$) in a crude extract. As always, the first question is whether this is a redox reaction. The sum of oxidation states for carbon in fumarate is $-2 - [2(+1) + 4(-2)] = +4$. This is the same as the sum of the carbon oxidation states for malate calculated earlier. Therefore, the transformation of fumarate into malate is not a redox process, and no oxidizing or reducing agent is required. The balanced chemical equation is:

$$C_4H_2O_4^{2-} + H_2O \rightleftharpoons C_4H_4O_5^{2-} \quad \text{(d)}$$

The reaction is catalyzed by the enzyme fumarase.

Table 16-1

Relative oxidation states of functional groups

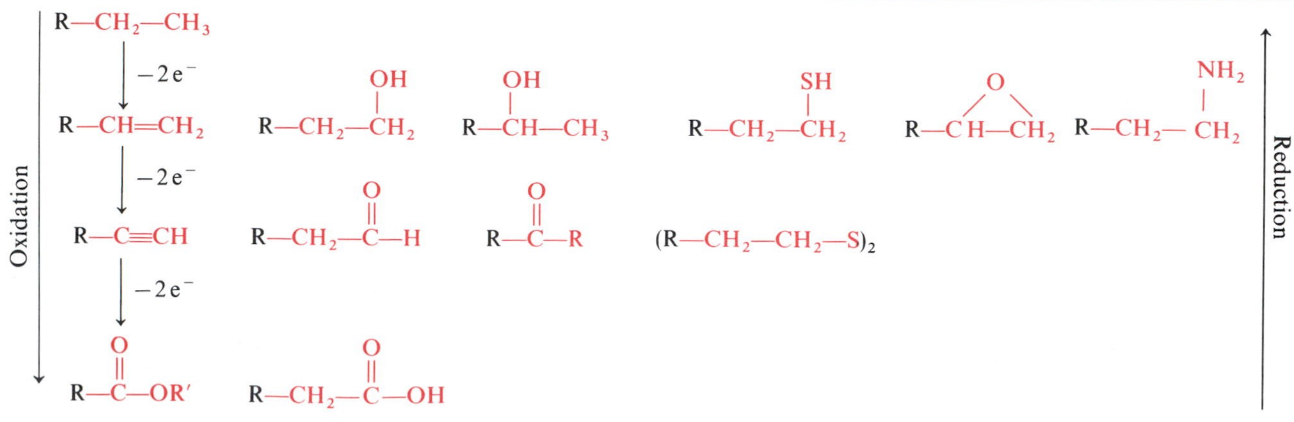

(Oxidation ↓ downward; Reduction ↑ upward)

R—CH$_2$—CH$_3$					
↓ $-2e^-$					
R—CH=CH$_2$	R—CH$_2$—CH$_2$(—OH)	R—CH(—OH)—CH$_3$	R—CH$_2$—CH$_2$(—SH)	R—CH—CH$_2$ (epoxide, —O—)	R—CH$_2$—CH$_2$(—NH$_2$)
↓ $-2e^-$					
R—C≡CH	R—CH$_2$—C(=O)—H	R—C(=O)—R	(R—CH$_2$—CH$_2$—S)$_2$		
↓ $-2e^-$					
R—C(=O)—OR′	R—CH$_2$—C(=O)—OH				

Glyoxalase-catalyzed Intramolecular Oxidation-Reduction

The reaction catalyzed by glyoxalase is perhaps one of the simplest redox reactions because it is intramolecular. Glyoxalase catalyzes the conversion of the thiohemiacetal of pyruvaldehyde (CH_3—CO—CHO) to *S*-lactylglutathione according to the reaction sequence in Figure 16-1. This is an intramolecular redox process in which C-1 is oxidized and C-2 is reduced. C-1 is converted from the oxidation state of an aldehyde into that of an ester, and C-2 is converted from a ketone into an alcohol.

The two mechanisms in Figure 16-2 can be considered for the glyoxalase reaction; in both mechanisms, B_1 and B_2 represent bases at the active site of the enzyme. In mechanism I, hydride ion is transferred directly from C-1 to C-2. According to this mechanism, the hydride transfer is facilitated by the presence of two bases at the active site, one of which removes a proton from the hydroxyl group on C-1 and leads to increased charge density on oxygen, thus providing a driving force for hydride expulsion. The other base protonates the carbonyl group and therefore makes the carbon more positive and a better hydride acceptor.

In mechanism II, the transfer of protons and electrons takes place in separate steps. A base (B_2) abstracts a proton from C-1. This proton is reasonably acidic because of the presence of the carbonyl group. Proton abstraction leads to the formation of an enol (electron transfer). Enol formation is facilitated by the acid B_1H^+, which provides a proton to the carbonyl group, thereby preventing a buildup of negative charge on the oxygen as the enol is formed. Finally, the proton is transferred from B_2H^+ to C-2 in a process that is also catalyzed by B_1, which accepts the proton that is released as the carbonyl group is formed. An important difference between the two mechanisms is that, in mechanism I, hydrogen is directly transferred as a hydride (H^-) from C-1 to C-2; whereas, in mechanism II, hydrogen is first transferred as a proton (H^+) to an active-site base and from there to C-2, again as a proton.

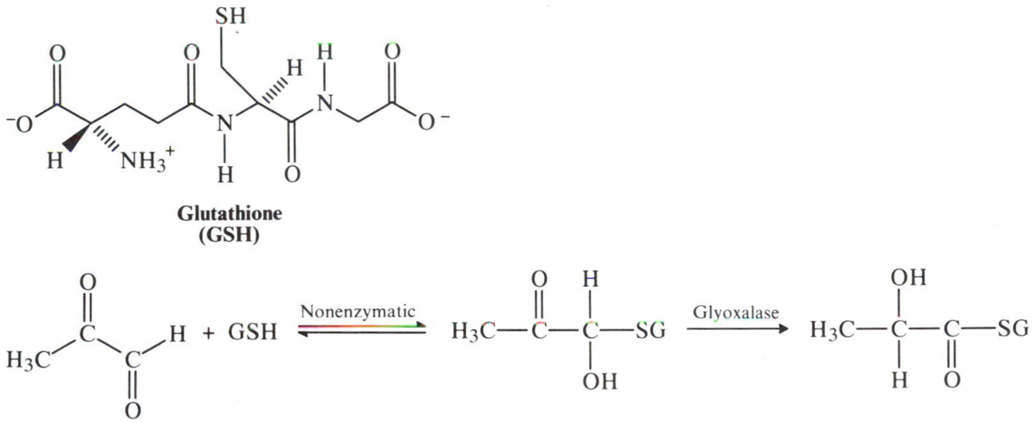

Glutathione (GSH)

$$H_3C\text{—}CO\text{—}CHO + GSH \xrightleftharpoons{\text{Nonenzymatic}} H_3C\text{—}CO\text{—}CH(OH)\text{—}SG \xrightarrow{\text{Glyoxalase}} H_3C\text{—}CH(OH)\text{—}CO\text{—}SG$$

Figure 16-1

An Intramolecular Redox Reaction Catalyzed by Glyoxalase. Glutathione is a tripeptide of glycine, cysteine, and glutamate, in which glutamate is linked through its γ-carboxyl group. Glutathione and pyruvaldehyde are in equilibrium with their adduct, a thiohemiacetal that is the substrate for glyoxalase I. The enzyme catalyzes the oxidation of the thiohemiacetal and the concomitant reduction of the ketonic group to form *S*-lactylglutathione. This is an intramolecular redox reaction.

Mechanism I

Mechanism II

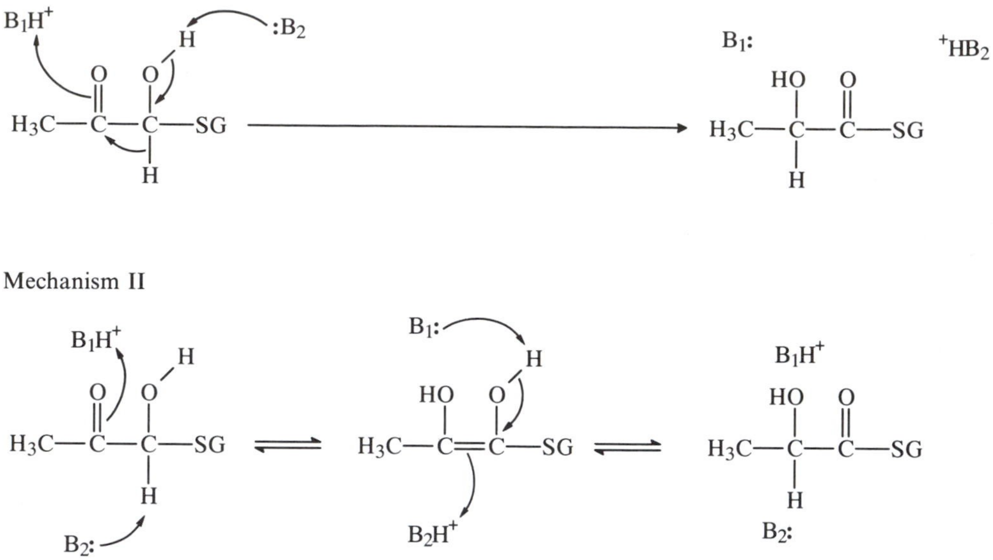

Figure 16-2

Hypothetical Mechanisms for the Glyoxalase Reaction. In mechanism I, a base (B_2) at the active site abstracts the hydroxyl proton, and the conjugate acid of a second base (B_1H^+) donates a proton to the carbonyl group to provide the driving force for a 1,2-hydride shift. In mechanism II, the active-site base (B_2) abstracts the α-proton at the same time that B_1H^+, the conjugate acid of the second base, donates a proton to the carbonyl group. This produces the enediol intermediate, which can ketonize in either direction to regenerate the substrate or form the product.

What experiment can be carried out to distinguish between the two mechanisms? Suppose that the reaction were carried out in ^2H_2O. What would happen in ^2H_2O if the reaction proceeded by mechanism I? In mechanism I, a hydride is transferred from C-1 to C-2, and at no stage of the reaction does this hydrogen have an opportunity to exchange with solvent protons. Therefore, if a hydride mechanism is operative, we would expect no deuterium incorporation from the solvent. In mechanism II, however, the hydrogen migrates twice as a proton and spends some time on a base at the active site (B_2H^+). We do not know the nature of B_2, but it must be one of the following: —COO^-, —S^-, —imidazole, or —NH_2. After the proton has transferred to any of these bases, it will be readily exchangeable with deuterium from the solvent. In D_2O, therefore, the protonated base B_2H^+ can lose its proton before it is transferred to C-2 by an exchange mechanism such as that in Figure 16-3. If reaction 3 is much faster than reaction 2, we can isolate a product containing deuterium at C-2 when the reaction is carried out in D_2O. In other words, the hydrogen originally bonded to C-1 of the starting material is not present in the reaction product. (Recall the discussion of exchangeable and nonexchangeable hydrogen in Chapter 2.)

To recapitulate, when the reaction is carried out in ^2H_2O, no deuterium will be in the product if a hydride transfer occurs by mechanism I, whereas deuterium will be found in the product if an intermediate enol is involved, as in mechanism II. Therefore, this experiment appears to provide a clear-cut distinction between the two mechanisms.

Remember, however, that we assumed that the rate of exchange is fast relative to product formation (reaction 2). What happens if this condition is not

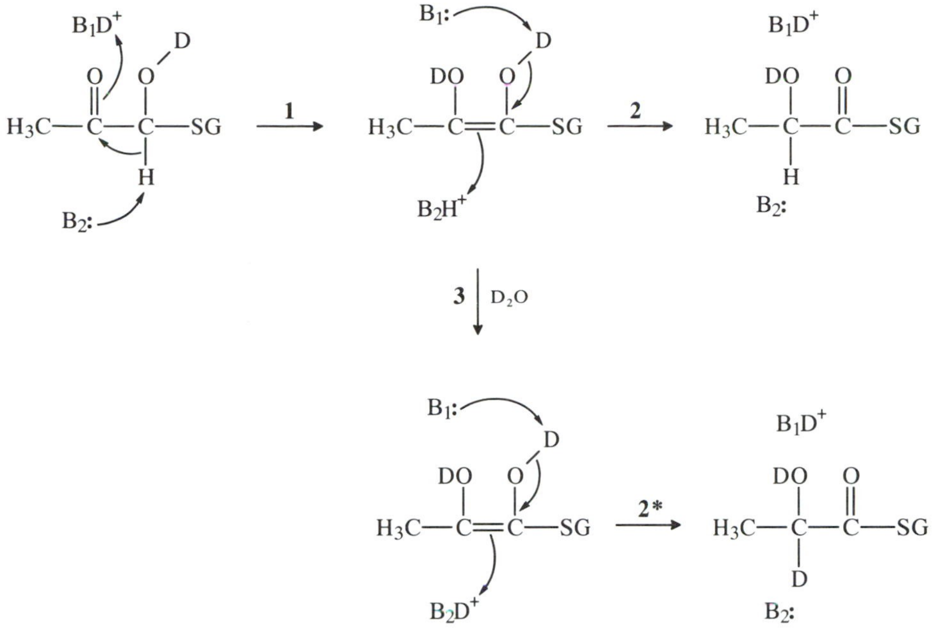

Figure 16-3

Glyoxalase-catalyzed Proton Exchange Through the Enediol Mechanism. When carried out in D_2O, the glyoxalase reaction results in the incorporation of a significant amount of deuterium in the product. This cannot occur by hydride transfer (mechanism I in Figure 16-2), but it can be explained by enolization (mechanism II). The reaction scheme shows how the enzymatic base B_2H^+ can exchange its proton with deuterium in D_2O when reaction 3 competes with reaction 2. To the extent that reaction 3 occurs, deuterium is incorporated into the product.

met? Consider the situation when exchange with solvent is slow relative to product formation (reaction 3 is slower than reaction 2). In this case, the hydrogen on B_2H^+ would not have time to exchange with solvent deuterons before product formation; it would be transferred to C-2 of the product, so that deuterium incorporation from the solvent would not be observed even if mechanism II were operative. Under these conditions, the experiment could not distinguish between mechanisms I and II. For the intermediate situation, where reactions 2 and 3 proceed at comparable rates, some deuterium would be incorporated, but not a full gram-atom per mole of product. Reactions in which the exchange with solvent is slow relative to the transfer process are frequently encountered in enzymatic reactions.

What conclusion can be drawn from these isotope exchange experiments? If the experiment is done in ^2H_2O and the heavy isotope of hydrogen is incorporated into the product, one can conclude that hydride transfer is *not* involved. If no heavy hydrogen is incorporated, a definite conclusion cannot be reached. The observation would be consistent with hydride transfer, but also with a proton transfer in which the proton undergoing transfer is not subject to exchange.

The glyoxalase reaction has actually been carried out in D_2O, and some deuterium (less than one atom) was incorporated into the product. It was, therefore, concluded that a hydride mechanism is not involved. Glyoxalase represents a case in which exchange with solvent proceeds at a rate that is comparable with the rate of proton transfer.

Pyridine and Flavin Nucleotides in Biological Redox Processes

Most biological oxidation-reduction reactions are not intramolecular processes, but are bimolecular, two-substrate reactions involving an oxidizing and a reducing substrate. In nonbiological reactions, many molecules can serve as oxidizing or reducing agents and, for the oxidation or reduction of a given molecule, one can generally use a number of different oxidizing or reducing agents. Biological systems differ in that they use relatively few molecules for the oxidation or reduction of the many substrates in metabolism. The most-important molecules used as biological oxidizing or reducing agents are pyridine nucleotides and flavin coenzymes.

The pyridine nucleotide coenzymes are nicotinamide adenine dinucleotide (NAD^+) and nicotinamide adenine dinucleotide phosphate ($NADP^+$), the structures of which are shown in Figure 16-4. In both coenzymes, all of the chemistry takes place in the nicotinamide part. When NAD^+ (or $NADP^+$) serves as an oxidizing agent, the nicotinamide ring is reduced by the addition of

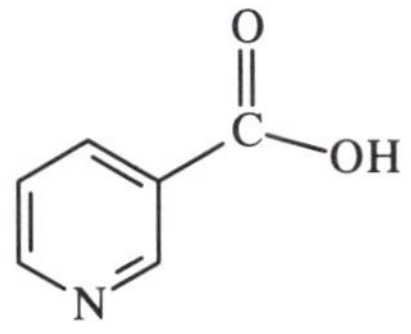

Niacin (Nicotinic acid)

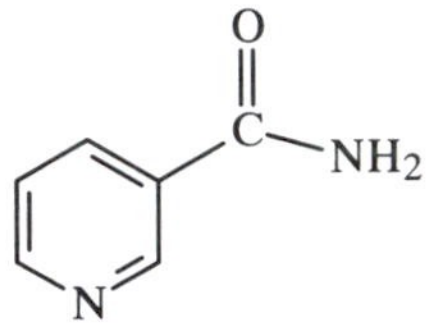

Niacinamide (Nicotinamide)

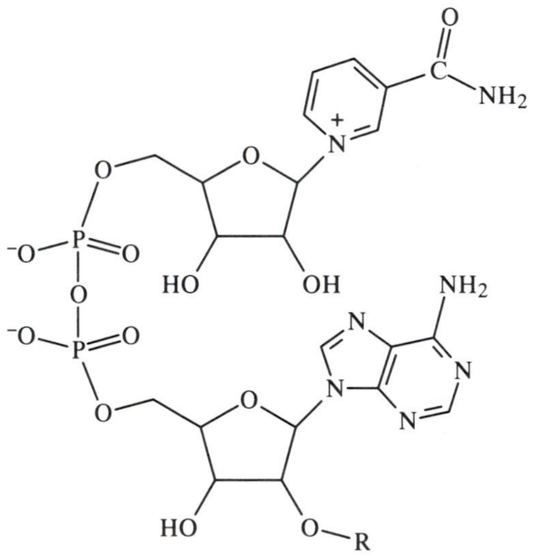

NAD^+ (R = H)
$NADP^+$ (R = PO_3^{2-})

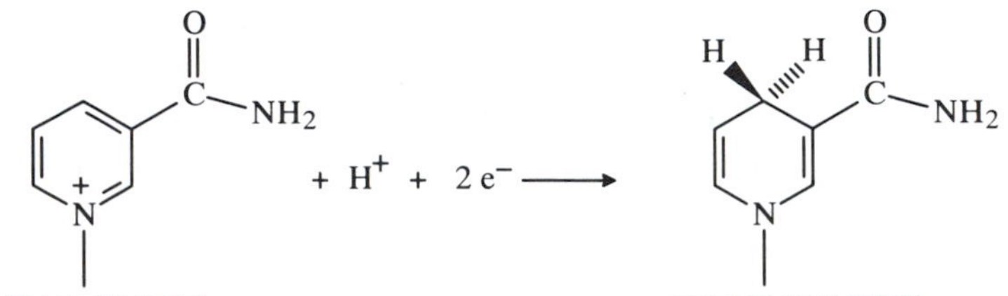

NAD (NADP) + H^+ + $2e^-$ ⟶ NADH (NADPH)

Figure 16-4

The Structures of Niacin and Pyridine Nucleotide Coenzymes. NAD^+ is nicotinamide adenine dinucleotide, and $NADP^+$ is the 2′-phosphoryl form of NAD^+. NADH and NADPH are the reduced forms, which contain the 1,4-dihydronicotinamide ring.

a hydride-equivalent (H^-) exclusively to the 4-position. The reduced form of the coenzyme is known as NADH (or NADPH).

NAD^+ and $NADP^+$ are derived from the vitamin niacin, the deficiency of which leads to a disease known as pellagra in human beings and black tongue in dogs. In the early 1900s, pellagra was found among children in the United States, and, in 1920, it was recognized that pellagra is caused by a dietary deficiency. The deficiency cannot be attributed solely to a low level of niacin in the diet, however, because most organisms, including higher animals, can biosynthesize nicotinic acid from the amino acid tryptophan. Dietary insufficiency of the vitamin, therefore, results only when *both* niacin *and* protein intake are low.

The flavin coenzymes are flavin adenine dinucleotide (FAD) and flavin mononucleotide (FMN), the oxidized and reduced forms of which are shown in Figure 16-5. The reduced forms are known as $FADH_2$ and $FMNH_2$, re-

Riboflavin (Vitamin B_2)

FMN (Flavin mononucleotide)

FAD (Flavin adenine dinucleotide)

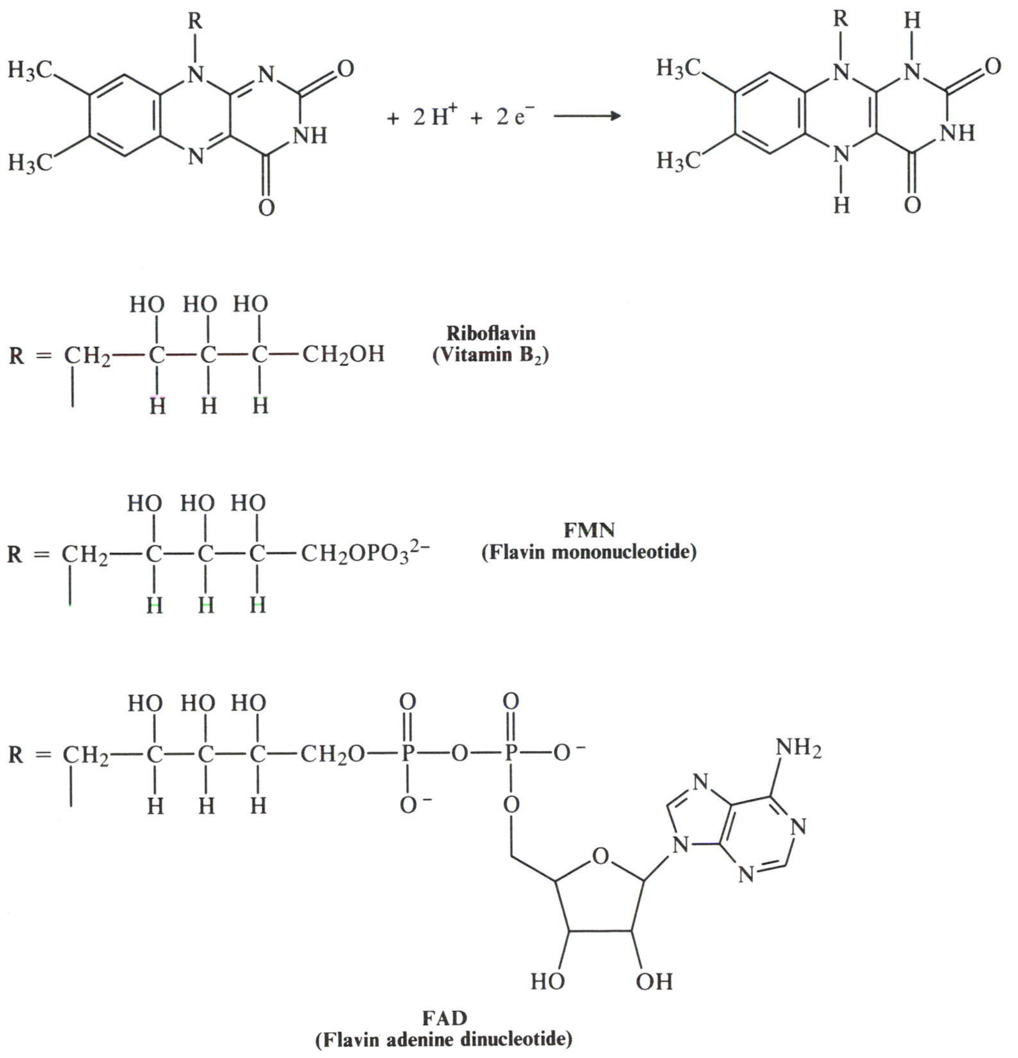

Figure 16-5

The Structures of Riboflavin and Flavin Coenzymes. The isoalloxazine ring of riboflavin and riboflavin coenzymes is shown at the top in the oxidized and two-electron-reduced forms. FAD is flavin adenine dinucleotide, and FMN is flavin mononucleotide. The reduced forms are known as $FADH_2$ and $FMNH_2$, respectively.

spectively. The flavin coenzymes are derived from the vitamin riboflavin (vitamin B_2), and all of the redox chemistry takes place in the flavin part of the coenzyme molecules.

Flavins and pyridine nucleotides participate in redox processes involving many different functional groups. We shall cite two representative examples among the many reactions in which these coenzymes participate. NAD^+ is the coenzyme for the oxidation of alcohols by the enzyme alcohol dehydrogenase (equation 3).

$$CH_3{-}CH_2{-}OH + \text{(pyridinium ring, } \overset{+}{N}{-}\text{, with } {-}\overset{O}{\overset{\|}{C}}{-}NH_2\text{)} \rightleftharpoons CH_3{-}CHO + \text{(1,4-dihydropyridine ring, } N{-}\text{, with H H at C-4 and } {-}\overset{O}{\overset{\|}{C}}{-}NH_2\text{)} + H^+ \quad (3)$$

Examples of FAD-dependent enzymes are provided by D-amino acid oxidase and L-amino acid oxidase, which catalyze the oxidation of D- and L-α-amino acids, respectively, according to equation 4.

$$R_1{-}\underset{NH_2}{\underset{|}{CH}}{-}CO_2^- + \text{(oxidized flavin: } H_3C, H_3C, N{-}R_2, N, N, NH, O, O\text{)} \longrightarrow R_1{-}\underset{NH}{\underset{\|}{C}}{-}CO_2^- + \text{(reduced flavin: } H_3C, H_3C, N{-}R_2, N{-}H, N{-}H, NH, O, O\text{)} \quad (4)$$

These enzymes also catalyze the oxidation of $FADH_2$ by O_2, which regenerates FAD and produces hydrogen peroxide (equation 5).

$$FADH_2 + O_2 \longrightarrow FAD + H_2O_2 \quad (5)$$

The overall reaction catalyzed by D- or L-amino acid oxidase is then equation 6.

$$R{-}\underset{NH_2}{\underset{|}{CH}}{-}CO_2^- + O_2 \longrightarrow R{-}\underset{NH}{\underset{\|}{C}}{-}CO_2^- + H_2O_2 \quad (6)$$

When the reaction is carried out under anaerobic conditions, reduced FAD accumulates.

Many substrates are acted upon by enzymes that require flavin coenzymes

Table 16-2

Functional groups found in substrates for enzymes that use pyridine nucleotide or flavin coenzymes

$$\text{>}\overset{H}{\overset{|}{C}}{-}OH \rightleftharpoons \text{>}C{=}O \qquad {-}S{-}S{-} \rightleftharpoons 2{-}SH \qquad \text{>}\underset{H}{C}{-}O^- \rightleftharpoons \text{>}\underset{^-O}{C}{=}O$$

$$\text{>}\overset{H}{\overset{|}{C}}{-}NH_2 \rightleftharpoons \text{>}C{=}NH \qquad \text{>}\underset{H}{C}{=}O \rightleftharpoons \text{>}\underset{RO\ (RS)}{C}{=}O \qquad {-}\overset{H}{\underset{|}{\overset{|}{C}}}{-}\overset{H}{\underset{|}{\overset{|}{C}}}{-}\overset{O}{\overset{\|}{C}}{-} \rightleftharpoons {-}\underset{|}{C}{=}\underset{|}{C}{-}\overset{O}{\overset{\|}{C}}{-}$$

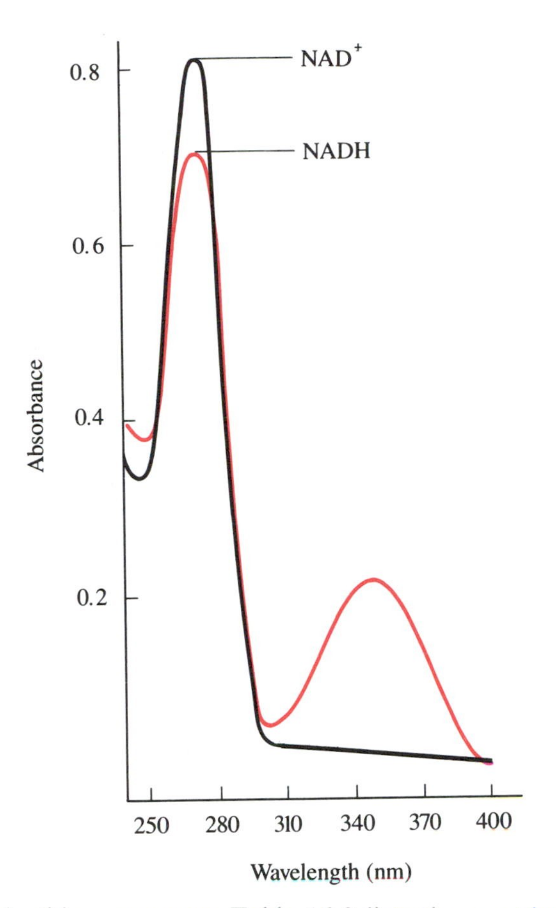

Figure 16-6

Ultraviolet-Visible Absorption Spectra of NAD$^+$ and NADH.

or pyridine nucleotide coenzymes. Table 16-2 lists the most-important functional groups that are acted upon by these coenzymes.

The progress of a dehydrogenase reaction in which NADH or NADPH is produced can be followed conveniently by measuring the light absorbed at 340 nm, which is the wavelength at which NADH (or NADPH) has an intense and characteristic absorption band. Because NADH (or NADPH) is kinetically unreactive with O_2, many of the dehydrogenation reactions can be carried out *in vitro* under aerobic conditions. The ultraviolet-visible absorption spectra of oxidized and reduced NAD$^+$ are presented in Figure 16-6.

The Structures of Dehydrogenases

M. Rossman first succeeded in determining the three-dimensional structure of a dehydrogenase, the NAD$^+$-dependent lactate dehydrogenase. The NAD$^+$-binding domain of this enzyme has turned out to exemplify a structural motif that has been found in many enzymes that bind NAD$^+$, FAD, or other nucleotides, and in some other enzymes as well.

The folding of the polypeptide chain in lactate dehydrogenase is shown by the drawing in Figure 16-7, which also shows the location of the binding site for NAD$^+$. The NAD$^+$-binding domains in lactate dehydrogenase and alcohol dehydrogenase are compared in Figure 16-8. The structure consists of a twisted β-sheet surrounded by α-helices. The chain passes alternately through the sheet, turns outward into a helix, and again enters the sheet. The slight twist in the β-sheet is propagated with each passage of the chain. This structural motif is also found in FAD-dependent dehydrogenases.

Figure 16-7

Folding of the Chain in the Nucleotide-binding Domain of Lactate Dehydrogenase. The location of the binding site for NAD^+ is indicated by the structure of NAD^+ (in red). (After a drawing courtesy of Hazel Holden.)

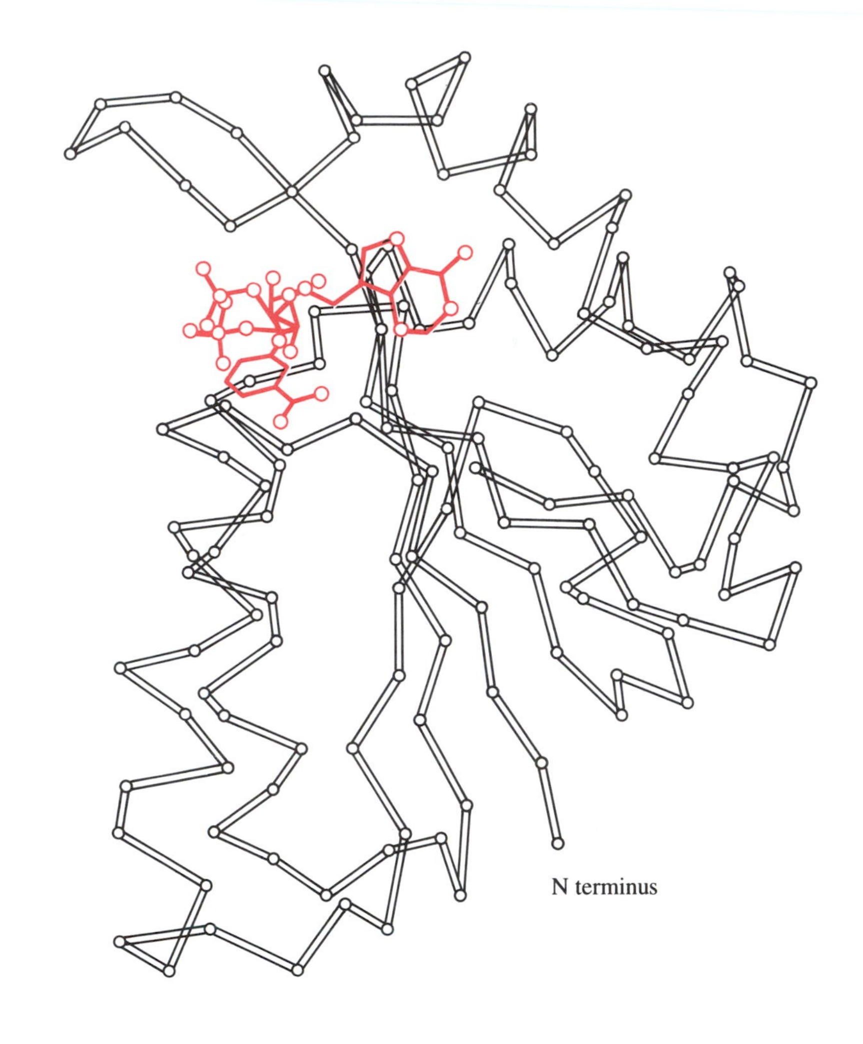

Figure 16-8

Comparison of the Nucleotide-binding Domains of Lactate and Alcohol Dehydrogenases. The chain segments forming the β-sheet are shown as arrows, and those in α-helices are shown as coiled ribbons. (After drawings by Jane Richardson.)

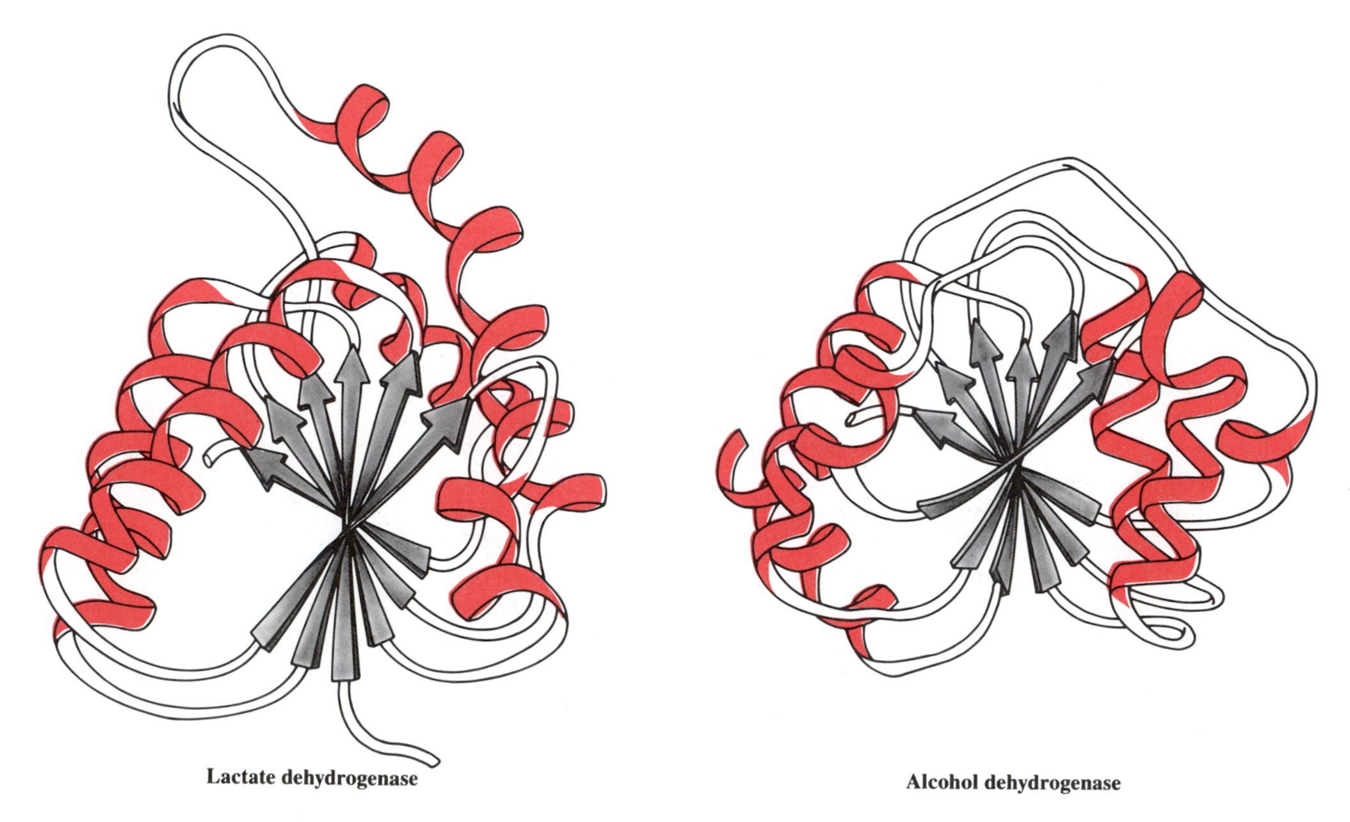

Reduction of Pyridine Nucleotide and Flavin Coenzymes: The Transfer of Two Electrons and One or Two Hydrogens

How does electron transfer occur? One can envision the following two limiting processes: (1) both electrons are transferred simultaneously; or (2) the electrons are transferred in two discrete steps. The classification of redox processes into these two categories is undoubtedly an oversimplification. The electrons may in some cases be transferred sequentially, but in very rapid succession, so that the intermediate has an extremely short lifetime. In such cases, it may well be difficult to distinguish between one-step and two-step electron transfer. Nevertheless, it is useful to discuss redox reactions in terms of these two classifications because it facilitates our thinking about these processes, as well as our experimental approach.

We shall consider the two-electron transfer first. This transfer could be concomitant with the transfer of one hydrogen, in which case a hydride ion (H^-) would be transferred. A hydride ion is an extremely reactive species and cannot exist in aqueous solution. Many biochemists believe that reactions in which pyridine nucleotides are electron donors or acceptors proceed by hydride transfer; however, this has not been conclusively proved. A nonenzymatic reaction in which a hydride is probably transferred is the Cannizzaro reaction (equation 8).

$$2\ C_6H_5{-}CHO \xrightarrow{OH^-} C_6H_5{-}CH_2OH + C_6H_5{-}COO^- + H^+ \qquad (8)$$

Benzaldehyde — Benzyl alcohol — Benzoate

In this reaction, two molecules of benzaldehyde react under strongly basic conditions. One molecule of benzaldehyde serves as electron donor and is converted into benzoic acid, and the other molecule serves as electron acceptor and is converted into benzyl alcohol. The reaction probably includes the initial addition of OH^- to benzaldehyde to give the hydrate, which is ionized under the basic conditions used (equation 9).

$$C_6H_5{-}CHO + OH^- \rightleftharpoons C_6H_5{-}CH(O^-)(OH) + H_2O \qquad (9)$$

The hydrate then transfers a hydride to another molecule of benzaldehyde (Figure 16-9). The Cannizzaro reaction is particularly favorable for a hydride transfer. In the hydrate of benzaldehyde, two electronegative atoms, one of them negatively charged, are attached to the carbon from which the hydride is transferred. Therefore, considerable "electron pressure" exists, which facilitates the expulsion of the negatively charged H^-. The hydride is transferred to a carbonyl carbon atom,

$$C_6H_5{-}CH{=}O \longleftrightarrow C_6H_5{-}C^{+}H{-}O^-$$

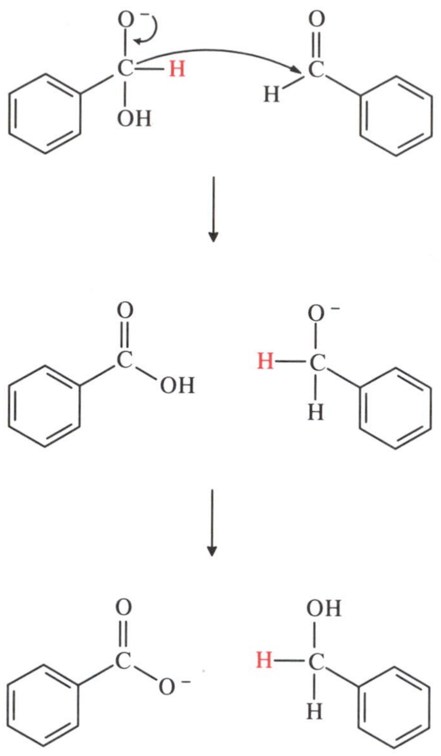

Figure 16-9

The Mechanism of Hydride Transfer in the Cannizzaro Reaction. Hydroxide ion is added to the carbonyl carbon of benzaldehyde to form the hydrate anion, and the anion reacts with a second molecule of benzaldehyde by transferring a hydride to the carbonyl carbon. This is followed by an immediate proton transfer from benzoic acid to benzyl alcohol anion, which is a much stronger base than benzoate.

which carries considerable positive charge and, therefore, is a good acceptor for the negatively charged H^-. Thus, optimal conditions for hydride transfer are twofold: (1) a high electron density on the carbon that donates the hydride and (2) a highly electron deficient carbon, which accepts the hydride ion.

When pyridine nucleotides act as electron donors or acceptors, these conditions are met, at least to some extent. Consider the oxidation of ethanol by NAD^+ as catalyzed by alcohol dehydrogenase (equation 3). There is considerable positive charge density at carbon-4 of the nicotinamide ring, so that NAD^+ can function as a hydride acceptor.

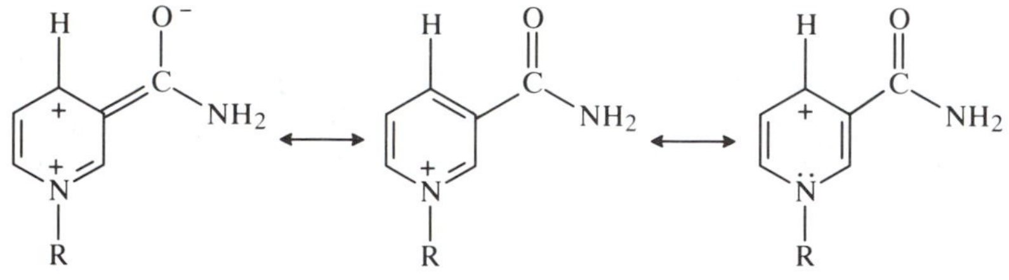

Many alcohol dehydrogenases have a general base in the active site that abstracts the hydroxylic proton of the substrate at the same time the hydride is transferred to carbon-4 of the nicotinamide ring (equation 10). This mechanism avoids the formation of a discrete hydride ion (H^-), which is very reactive and cannot exist in water. It also avoids the formation of a substrate oxyanion.

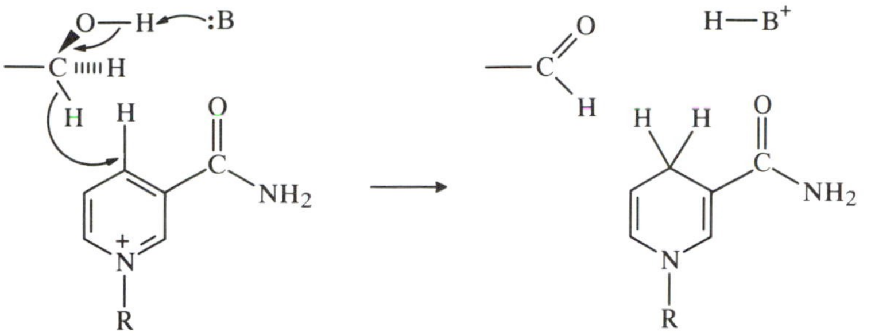

(10)

Lactate dehydrogenase has a histidine residue at the active site that is thought to act as the general base. The interactions of the substrates and histidine at the active site of lactate dehydrogenase are illustrated in Figure 16-10.

Liver alcohol dehydrogenase contains Zn^{2+} at the active site. The hydroxide (OH^-) bound to Zn^{2+} could act as the base that facilitates the removal of the proton from the alcoholic hydroxyl group, as in structure I. Alternatively, the hydroxyl group of ethanol might be directly coordinated to Zn^{2+}, which would greatly reduce its pK_a and allow it to exist as the alkoxide ion, as in structure II. These two structures differ most fundamentally in the intervention of a water molecule between the substrate and Zn^{2+} in structure I. In either case, the metal acts to facilitate hydride transfer.

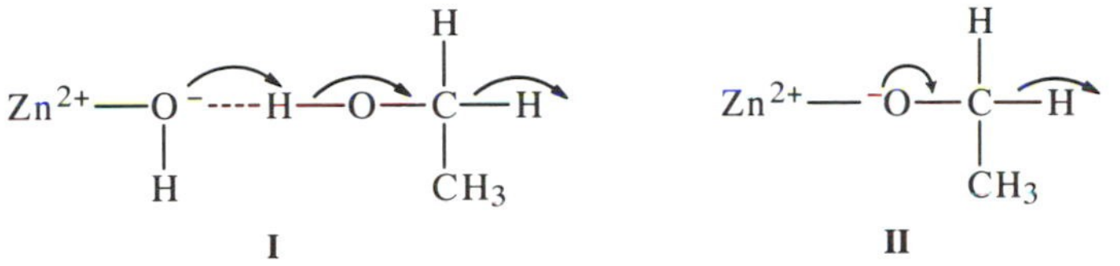

Two-electron transfers can also occur independently of the transfer of a hydrogen atom. An example of this type of mechanism is provided by mechanism II of the glyoxalase reaction (see Figure 16-2).

A two-electron transfer mechanism has also been proposed for oxidations in which flavins participate. For example, consider the oxidation of an amino

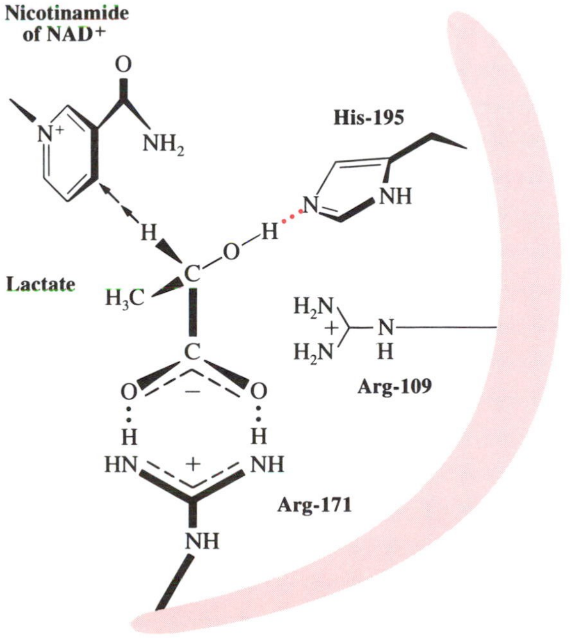

Figure 16-10

Histidine as a General Base at the Active Site of Lactate Dehydrogenase. Histidine-195 is positioned so that it can act as a general base to abstract a proton from the hydroxyl group of lactate in the transfer of hydrogen from carbon-2 to NAD^+. (Adapted, with permission, from J. J. Holbrook, A. Liljas, S. J. Steindel, and M. G. Rossman. In *The Enzymes,* vol. 11, 3d ed., ed. P. D. Boyer, New York: Academic Press, 1975, p. 243.)

acid by D-amino acid oxidase. It is proposed, with the support of considerable experimental evidence, that the first step is the abstraction of a proton from the substrate by a base in the active site to form a carbanion in step 1 of Figure 16-11. It is likely that a positive charge on the enzyme interacts with the

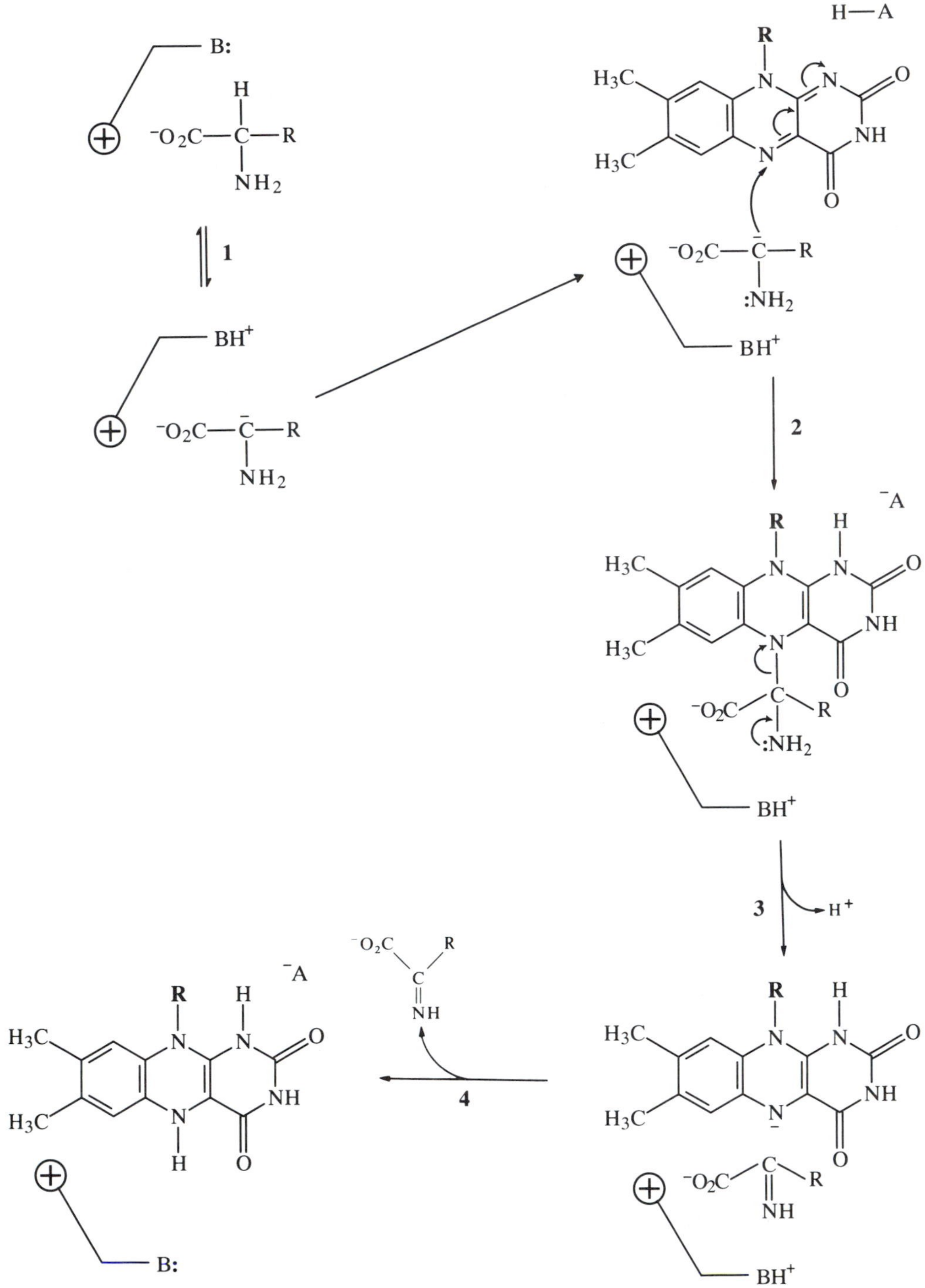

Figure 16-11

A Two-Electron Transfer Mechanism for D-Amino Acid Oxidase. In this mechanism, the enzyme initially catalyzes substrate-α-carbanion formation in step 1, and this carbanion adds to N-5 of the flavin coenzyme in step 2. Decomposition of the resulting adduct in step 3 generates the imino acid and a flavin anion, which is protonated by the enzyme in step 4.

carboxylate group of the amino acid to further facilitate carbanion formation. The carbanion then reacts with the flavin to form a covalent bond in step 2, and electron transfer takes place in step 3, subsequent to bond formation. Finally, the product is released and the flavin is protonated in step 4. According to this mechanism, proton transfer and electron transfer are separate steps.

The mechanism of flavin action in Figure 16-11 is by no means proven, although powerful evidence supporting carbanion formation has been obtained; but it is one of several possible mechanisms that have been considered. The electrons are transferred in pairs in Figure 16-11, but this need not be the case; they can be transferred in two discrete steps. If so, intermediates containing unpaired electrons (radical intermediates) are formed. This allows an alternative mechanism for D-amino acid oxidase, one that includes radical intermediates. For example, the flavin-catalyzed oxidation of amino acids could involve proton abstraction and substrate-carbanion formation, as in the mechanism shown in Figure 16-11. Electrons from the carbanion could then be transferred to flavin in two steps, as in Figure 16-12. An intermediate stage in the reaction consists of the formation of a radical pair, a substrate-derived radical and a flavin radical derived from the transfer of one electron and one proton to the flavin in step 1 of Figure 16-12. The second electron is transferred in step 2 to form the dihydroflavin. In an alternative radical mechanism, a hydrogen atom is transferred directly from the substrate to the flavin without prior carbanion formation. A number of investigators have proposed radical mechanisms for both flavin- and pyridine-nucleotide-catalyzed reactions.

It should be kept in mind that radical intermediates proposed for biological reactions are not "free" radicals. These radicals are formed within the confines of the active site and are not free to interact with other molecules of the environment.

Figure 16-12

A Mechanism for D-Amino Acid Oxidase in Which There Are Two One-Electron Transfer Steps. The carbanion initially generated as shown in Figure 16-11 transfers one electron to the flavin, and the enzyme protonates the ring in step 1 to form a substrate radical and the cation radical form of the flavin. In the second step, the substrate radical transfers another electron to the flavin cation radical to form the imino acid and dihydroflavin.

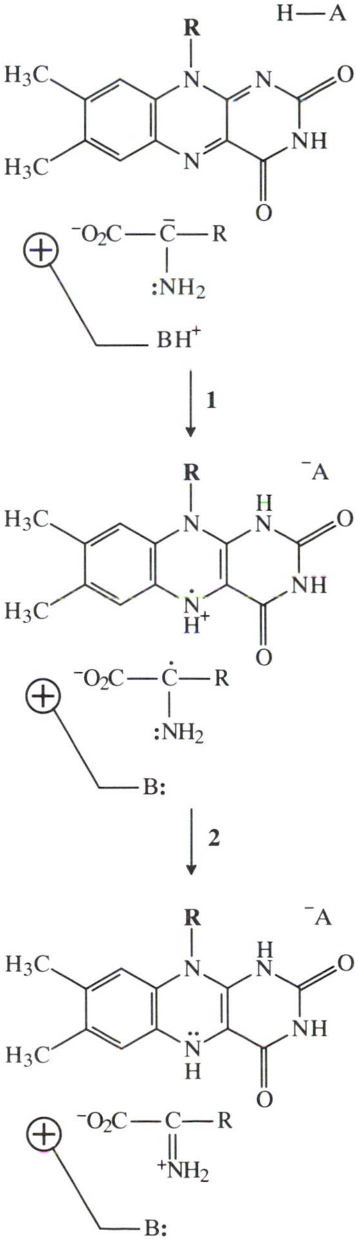

Spectroscopic Identification of Radical Intermediates

It is apparent from the foregoing discussion that an important distinction between one-step and two-step redox processes is the intermediate formation of radical species; that is, radical species are formed only when electron transfer is in two discrete steps. How can radical intermediates be detected? In many cases, radical intermediates have characteristic optical (electronic absorption) spectra and should, therefore, be detectable by the observation of spectral changes in the course of the reaction. Detection by optical spectroscopy has been successful in some cases, but complications are sometimes encountered owing to the fact that the spectrum of a species often depends upon the microenvironment. Radicals at the active site of an enzyme are in a unique environment, which may have unpredictable effects on the spectrum. Moreover, spectra are not absolutely unique, so that different species may give rise to similar spectra. For example, the optical spectrum observed when D-amino acid oxidase acts on its substrate was at one time considered to be indicative of a flavin radical intermediate. Subsequent experiments, however, failed to yield supporting data for radical formation. It is now agreed that the species that gives the "radicallike spectrum" in this reaction is actually not a radical. Finally, only radicals in which the unpaired electron is delocalized through a conjugated system of π-bonds will give rise to an optical spectrum.

Another very useful technique for detecting radical intermediates is ESR (electron spin resonance) spectroscopy. ESR spectroscopy is analogous to NMR spectroscopy in that electromagnetic radiation is absorbed when the sample is placed in a magnetic field. The two differ most fundamentally in that the energy absorption by an *unpaired electron* is measured in ESR, whereas the energy absorption by a *magnetic nucleus* is measured in NMR. Energy absorption is in the microwave frequency range for ESR and in the radiowave frequency range for NMR.

ESR measurements are carried out by placing the molecule in a magnetic field, applying microwave electromagnetic radiation, and measuring absorption of microwave energy as a function of the magnetic field. In practice, the first derivative of the change in absorption is recorded.

ESR is useful for detecting and studying most species that contain one or more unpaired electrons, including organic radicals and transition metals with unpaired electrons. Figure 16-13 shows an ESR spectrum of a copper-containing protein (conalbumin) in which the four left-hand signals arise from the copper atom and the signal at the right arises from the interactions of copper with nitrogen atoms. The magnetic field strength at which the energy absorption takes place is determined by the nature of the atom containing the unpaired electron and by its environment. This technique, therefore, allows not only the determination of the number of unpaired electrons, but also the identification of the atom on which the electron is located. In addition, it supplies information about the environment of the radical.

Detection of a radical does not necessarily implicate it as an intermediate in the reaction under consideration. The radical may be formed through a side reaction and may not be an essential intermediate. To establish that a radical intermediate is formed in a reaction sequence, its kinetic competence must be established; that is, it must be shown to be formed rapidly enough and to react to form the product rapidly enough to function as an intermediate on the main reaction pathway.

Unfortunately, ESR spectroscopy also has limitations. When intermediates consist of radical pairs, it may not be possible to detect an ESR signal. Thus, failure to detect an ESR signal cannot be used to eliminate the possibility of intermediate radicals. Although no ESR signals have been detected in reactions catalyzed by D-amino acid oxidase or alcohol dehydrogenase, the

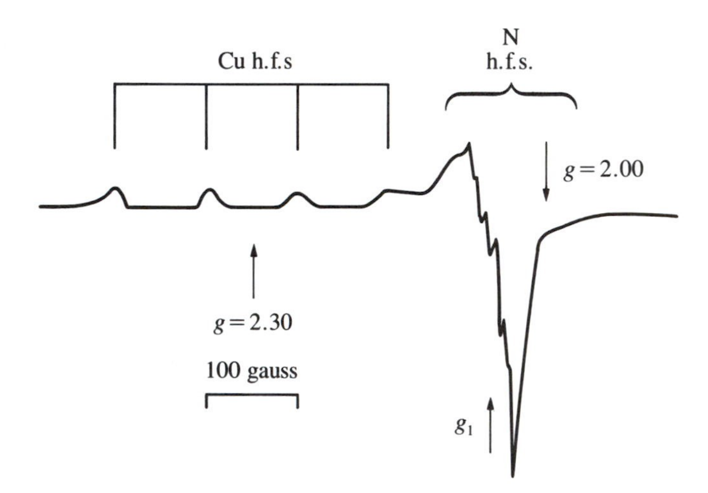

Figure 16-13

The ESR Spectrum of the Copper Protein Conalbumin. The terms h.f.s. (hyperfine splitting) and g provide information concerning the environment of the unpaired electron. For example, for organic radicals, generally $g = 2.00$. For a more-detailed discussion of ESR spectroscopy see, for instance, *Biological and Biochemical Applications of E.S.R.* by D.S.E. Ingram, Plenum.

possibility of radical processes cannot be ruled out. The uncertainty regarding the mechanism of action of flavins and pyridine nucleotides applies to the majority of biological redox reactions.

Direct Hydrogen Transfer in Reactions of NAD$^+$ and NADP

In the early 1950s, F. Westheimer and B. Vennesland began what has now become a classical study of the mechanism of action of alcohol dehydrogenase. They considered the following question: Does the oxidation of ethanol catalyzed by alcohol dehydrogenase proceed with direct hydrogen transfer from ethanol to NAD$^+$? The oxidation of [1,1-$^2H_2$]ethanol was carried out in H_2O (equation 11) and the deuterium content of the resulting reduced NAD$^+$ was determined.

$$H_3C-C(^2H)_2-OH + \text{NAD}^+ \text{(pyridinium, 4-H, 3-CONH}_2\text{, N-R)} \xrightarrow{\text{Alcohol dehydrogenase}} H^+ + H_3C-C(^2H)=O + \text{NADH (4-}^2\text{H, 4-H, 3-CONH}_2\text{, N-R)} \quad (11)$$

It was found to contain one atom of deuterium per molecule. [Note: Reduced NAD$^+$ (NADH) has a deuterium atom, derived from [1,1-$^2H_2$]ethanol, and a hydrogen atom (originally present in NAD$^+$) in the C-4 position.] When this reaction was carried out in ^2H_2O with undeuterated ethanol, no deuterium was found in NADH, consistent with the results obtained with deuterated ethanol in H_2O. These experiments established that the oxidation of ethanol by NAD$^+$ proceeds with "direct" hydrogen transfer; that is, at no stage in the course of the reaction is the hydrogen subject to exchange with solvent protons. The reverse reaction, the reduction of acetaldehyde by NADH, also occurs with "direct" hydrogen transfer.

These results are consistent with a reaction in which the hydrogen from C-1 of ethanol is transferred to NAD$^+$ as a hydride ion (H^-) or hydrogen atom ($H\cdot$). For the reasons discussed earlier, this interpretation is equivocal; that is, a proton transfer, with slow exchange of the intermediate with solvent, cannot be excluded. However, hydrogen transfer experiments have been done with many NAD$^+$-dependent enzymes, and in all cases hydrogen is transferred directly between substrate and NADH with no incorporation of solvent hydrogen. It seems exceedingly unlikely that, in *all* reactions involving NAD$^+$ or NADP$^+$, exchange of the intermediate with solvent protons would be slow relative to the product-forming step; therefore, the conclusion that H^- or $H\cdot$ is transferred seems justified.

Reactions Involving NAD$^+$ Are Stereospecific

As shown above in equation 11, alcohol dehydrogenase catalyzed reduction of NAD$^+$ by [1,1-$^2H_2$]ethanol leads to the formation of reduced NAD$^+$ containing ^2H at C-4. In the enzymatic reoxidation of the reduced NAD$^+$ (NAD^{2}H) by acetaldehyde in Figure 16-14, *all* of the deuterium is transferred to ethanol. This result seems surprising at first, because there are two hydrogens, one ^1H and one ^2H, in the 4-position of NADH; and one might expect either the ^1H or the ^2H to

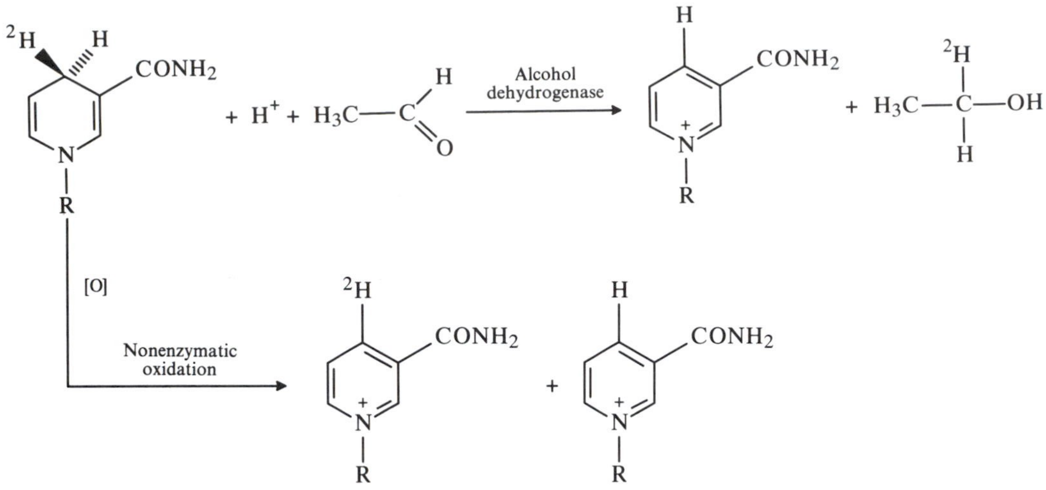

Figure 16-14

Stereospecific Enzymatic Oxidation of NAD^{2}H. In the upper reaction, the sample of NAD^{2}H, produced in the alcohol dehydrogenase catalyzed reduction of NAD$^+$ by dideuteroethanol in equation 11, is reoxidized by the action of alcohol dehydrogenase with an excess of acetaldehyde. All of the deuterium is transferred to acetaldehyde to form monodeuteroethanol and NAD$^+$. In the lower reaction, the same sample of NAD^{2}H is reoxidized nonstereospecifically in a nonenzymatic oxidation to form both NAD$^+$ and deutero-NAD$^+$.

be transferred. This is indeed the case in the nonenzymatic reoxidation in Figure 16-14; only in the enzymatic reaction is the deuterium specifically removed from NAD^{2}H.

The explanation of this phenomenon lies in the asymmetry of the dihydronicotinamide ring of NADH. Because of the presence of two substituents on the nicotinamide ring, the carboxamide group and the ADP-ribose moiety, the two faces of the ring are not equivalent. The active site of the enzyme contains specific binding points for the carboxamide group and the ADP-ribose, and these binding interactions orient the nicotinamide ring in the active site. Figure 16-15 illustrates this principle. The three-dimensional structure of the enzyme is such that the substrate also is bound at a specific site adjacent to the nicotinamide ring and can approach only from one side of the ring. Consequently, the same hydrogen that is added in the reductive process is removed in the reverse, oxidative reaction. This is in contrast with most nonenzymatic reactions, in which oxidizing agents are freely diffusible in solution, have access to both sides of the dihydronicotinamide ring, and can remove hydrogen from either side.

The hydrogen of NADH that is removed by alcohol dehydrogenase catalyzed reduction of acetaldehyde is referred to as H_a, and the other hydrogen on C-4 is H_b. Not all NAD$^+$- or NADP$^+$-dependent enzymes transfer hydrogen to the same side of the nicotinamide ring (Figure 16-15). Thus, some enzymes specifically transfer H_a and others H_b, and they are referred to as having A-side or B-side specificity. When deuterium is in the H_a position of NADH (Figure 16-15), C-4 of the nicotinamide ring is chiral and has the configuration *R*. Therefore, the H_a-hydrogen is in the pro-*R* position of C-4, and H_b is in the pro-*S* position.

Surprising results were obtained when the stereospecificity of the reaction catalyzed by alcohol dehydrogenase was investigated with respect to ethanol. The reactions of equations 12 through 14, all catalyzed by alcohol dehydrogenase, were carried out.

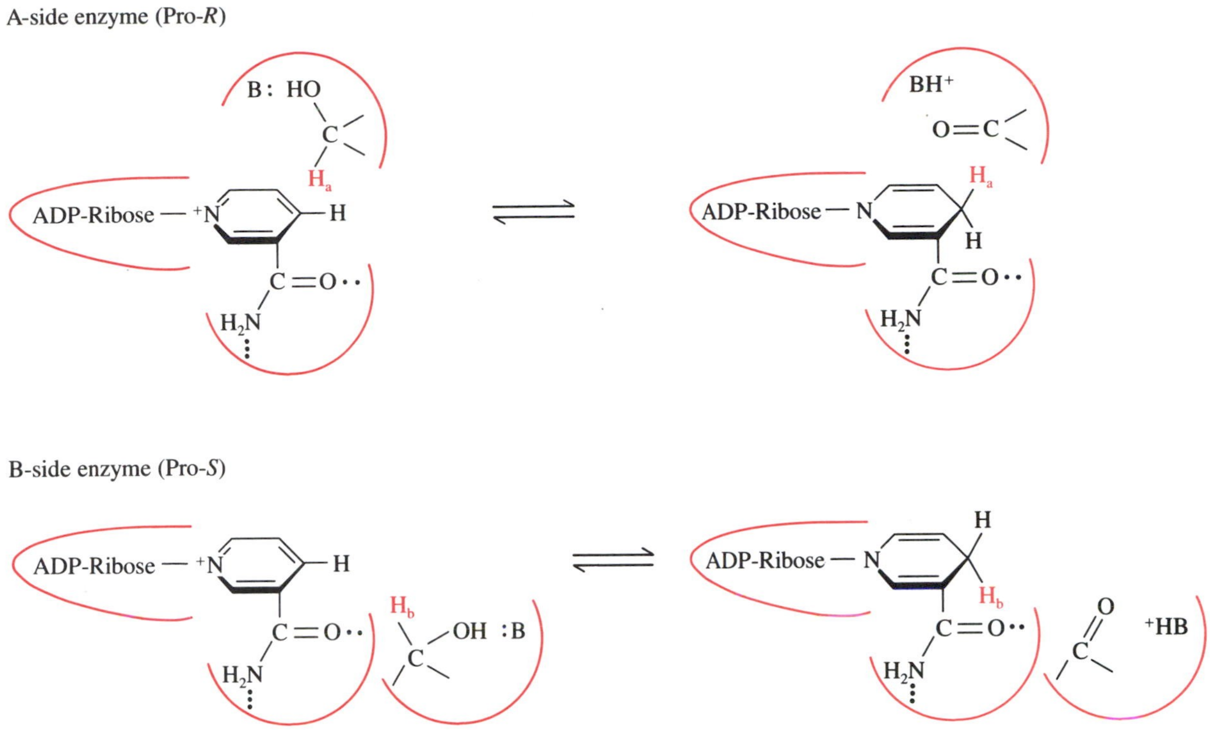

A-side enzyme (Pro-*R*)

B-side enzyme (Pro-*S*)

Figure 16-15

A Model for Stereospecific Enzymatic Hydrogen Transfer to NAD$^+$. Specific binding sites for the ADP-ribose moiety and carboxamide group of NAD$^+$ orient the nicotinamide ring relative to the substrate in its adjacent binding site in such a way that the hydrogen that is transferred has access to only one side of the nicotinamide ring. For alcohol dehydrogenase, this is designated the A-side; and in NADH the hydrogen at C-4 that projects above the A-side is designated H_a. Many dehydrogenases share A-side specificity with alcohol dehydrogenase, but a number of others bind their substrates on the opposite side of the nicotinamide ring, the B-side, and can transfer hydrogen only to the B-side. This hydrogen in NADH is designated H_b. H_a is often designated pro-*R* because the absolute configuration at carbon-4 is *R* when there is deuterium (^2H) on the A-side. H_b is then designated pro-*S*.

$$CH_3\text{—}C^2H_2\text{—}OH + NAD^+ \xrightarrow[\text{dehydrogenase}]{\text{Alcohol}} CH_3\text{—}C^2HO + NAD^2H + H^+ \qquad (12)$$

$$CH_3\text{—}CHO + NAD^2H + H^+ \xrightarrow[\text{dehydrogenase}]{\text{Alcohol}} H_3C\text{—}\underset{^2H}{\overset{H}{C}}\text{—}OH + NAD^+ \qquad (13)$$

$$H_3C\text{—}\underset{^2H}{\overset{H}{C}}\text{—}OH + NAD^+ \xrightarrow[\text{dehydrogenase}]{\text{Alcohol}} CH_3\text{—}CHO + NAD^2H + H^+ \qquad (14)$$

Alcohol dehydrogenase was used to prepare reduced NAD$^+$, containing ^2H at C-4, by reduction of NAD$^+$ with dideuteroethanol (equation 12). The NAD^{2}H so obtained was then used to reduce CH_3CHO to [1-^{2}H]ethanol (equation 13), which contained one deuterium per molecule. This monodeuteroethanol was again reoxidized with NAD$^+$ (equation 14), with the result that *all* of the deuterium from the C-1 position of ethanol was transferred back to NAD$^+$.

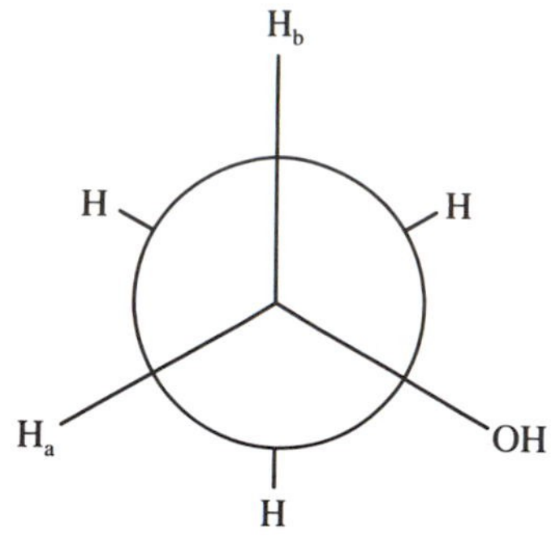

Figure 16-16

A Newman Diagram of Ethanol, Showing the Prochiral Center. The view of ethanol in this diagram is along the carbon–carbon bond, with C-1 at the front and C-2 behind. The three substituents H_a, H_b, and OH are shown projecting outward from C-1. Monodeuteroethanol produced by alcohol dehydrogenase catalyzed reduction of acetaldehyde by NAD^2H in equation 13 has deuterium in place of H_a, which creates a chiral center at C-1. This molecule is (*R*)-[1-^{2}H]ethanol.

The same deuterium that had been introduced into NAD^2H in the reduction of equation 13 was removed in the reoxidation of equation 14. This means that the two C-1 hydrogens of ethanol are not equivalent, and one of them is specifically removed during the oxidation. This may at first be surprising because we tend to think of ethanol as a symmetric molecule. But an explanation is readily apparent when we consider that the structure of ethanol can be represented as Caabc, in which a = H, b = CH_3, and c = OH. We pointed out in Chapter 2 that the two "a" substituents in such molecules are not sterically equivalent, and they react stereospecifically at enzymatic active sites. When [1-H^2]ethanol is prepared enzymatically by reduction of unlabeled acetaldehyde with NAD^2H, an ethanol molecule is produced in which H_a in the Newman diagram of Figure 16-16 is replaced by deuterium. This molecule is designated as (*R*)-[1-^{2}H]ethanol. Monodeuteroethanol has four different substituents at C-1 (CH_3, ^1H, ^2H, OH). It is, therefore, an asymmetric molecule, and the stereoisomers are optically active.

The situation at C-1 of ethanol is formally similar to that at C-4 of the dihydronicotinamide ring in Figure 16-14. Both carbons are of the steric type Caabc, in which the two chemically identical substituents are H; in both cases, the two hydrogens are transferred stereospecifically at the active site of alcohol dehydrogenase. Carbons of the steric type Caabc are known as prochiral centers, because they become chiral when one of the two chemically identical substituents is changed, as by replacement with another isotope or a different chemical group.

The structural models in Figure 16-17 show how stereospecific hydrogen transfer from the prochiral C-1 of ethanol can be explained by the structure of the enzymatic active site. Structure A, which undergoes reaction in Figure 16-17, leads to the transfer of the pro-*R* hydrogen from ethanol to NAD^+. The alternative structure, B, which would allow the pro-*S* hydrogen to be transferred, appears to be impossible owing to steric interference between the methyl group of ethanol and the phenyl ring of phenylalanine-93 in the enzyme.

In another experiment that further clarified the stereospecificity of alcohol dehydrogenase, (*R*)-[1-^{2}H]ethanol, prepared enzymatically, was derivatized with *p*-toluenesulfonyl chloride (TsCl), according to equation 15.

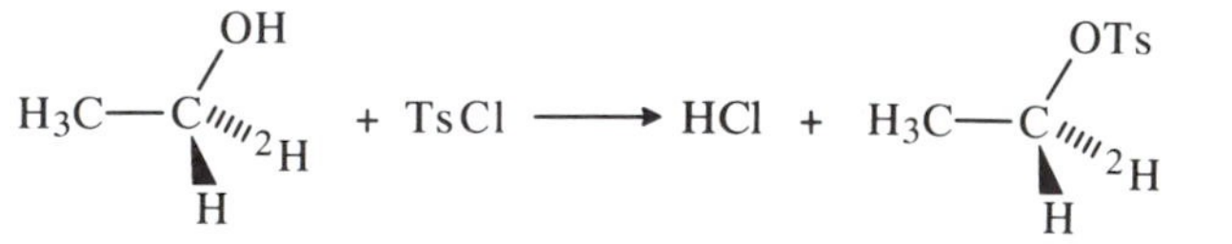

$$H_3C-C(OH)(H)(^2H) + TsCl \longrightarrow HCl + H_3C-C(OTs)(H)(^2H) \qquad (15)$$

The *p*-toluenesulfonate was then treated with base and converted into the alcohol (equation 16).

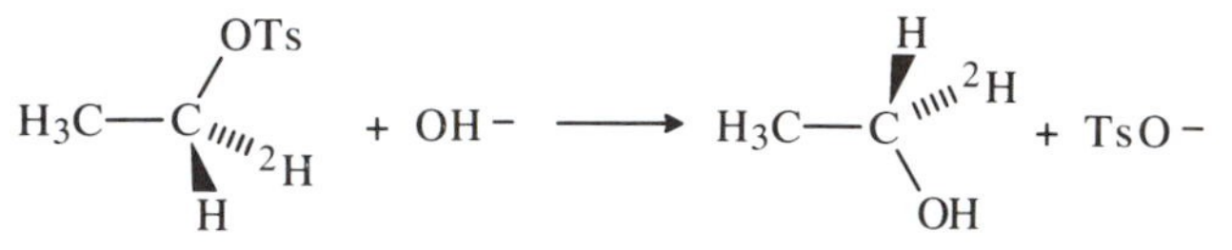

$$H_3C-C(OTs)(H)(^2H) + OH^- \longrightarrow H_3C-C(H)(^2H)(OH) + TsO^- \qquad (16)$$

Because the reaction with OH^- is a displacement reaction, which causes inversion of configuration at C-1, (*S*)-[1-^{2}H]ethanol was obtained. When this sample of [1-^{2}H]ethanol was subjected to oxidation by NAD^+, catalyzed by alcohol dehydrogenase, the resulting NADH contained *no* deuterium (equation 17).

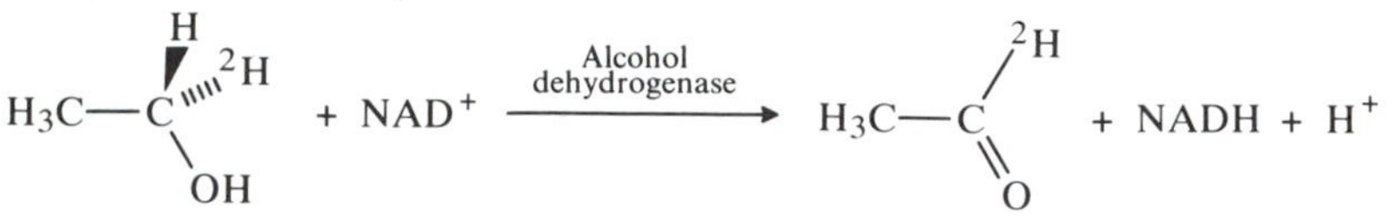

$$H_3C-C(H)(^2H)(OH) + NAD^+ \xrightarrow{\text{Alcohol dehydrogenase}} H_3C-C(^2H){=}O + NADH + H^+ \qquad (17)$$

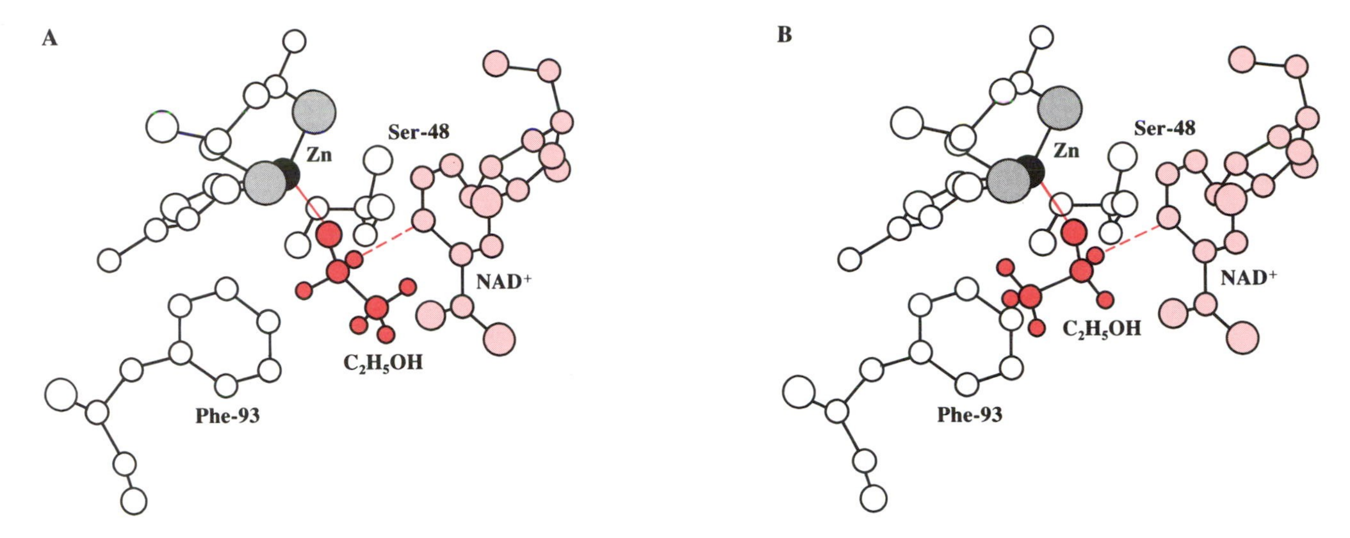

Figure 16-17

A Structural Model for Stereospecific Hydrogen Transfer at the Active Site of Alcohol Dehydrogenase. In structure A, ethanol is modeled into the active site of alcohol dehydrogenase in such a way as to allow the hydroxyl group to be coordinated to Zn^{2+} and the 1-pro-*R* hydrogen to be transferred to NAD^+. This is probably similar to the actual structure. Structure B shows how ethanol would have to be oriented in the active site to allow coordination to Zn^{2+} and transfer of the 1-pro-*S* hydrogen to NAD^+. Note that in structure B there is severe steric interference between the methyl group of ethanol and Phe-93. (After H. Eklund and C. Bränden, Alcohol dehydrogenase. In *Biological Macromolecules and Assemblies,* vol. 3, ed. F. A. Jurnak and A. McPherson. New York: Wiley-Interscience, © 1987, p. 119. Adapted by permission of John Wiley & Sons, Inc.)

Recall that enzymatic oxidation of (*R*)-[1-^{2}H]ethanol resulted in the transfer of deuterium to NADH (equation 14).

This latter experiment further illustrates the point mentioned earlier that the selectivity towards a specific hydrogen displayed by alcohol dehydrogenase is not due to the presence of an isotope. This experiment shows that, when (*S*)-[1-^{2}H]ethanol is used, only hydrogen is transferred and not deuterium; therefore, the enzyme is specific for the spatial position of the hydrogen and not for the heavy isotope.

Thiohemiacetals as Intermediates in the Dehydrogenation of Aldehydes

Aldehyde Dehydrogenase

Many enzymes that catalyze the oxidation of aldehydes and utilize pyridine nucleotide coenzymes have —SH groups that are essential for catalysis. An example is liver aldehyde dehydrogenase, a rather nonspecific enzyme that catalyzes the oxidation of many aldehydes to the corresponding carboxylic acids. The initial step in the catalytic mechanism, which is illustrated in Figure 16-18, is believed to result in thiohemiacetal formation between the enzyme and the substrate. The thiohemiacetal is then oxidized by NAD^+ to produce an

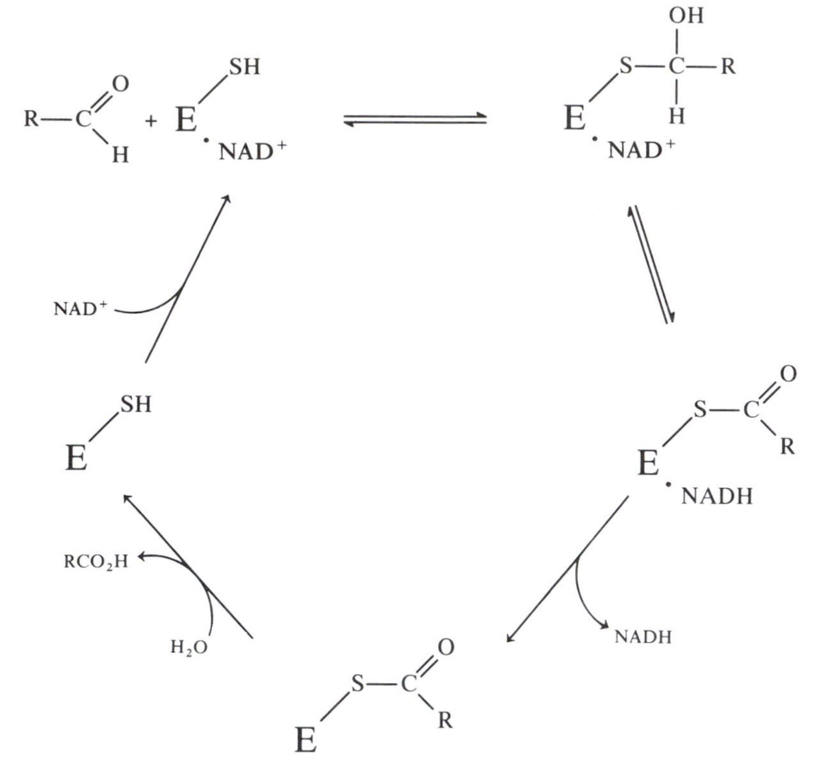

$$R{-}CHO + E(SH)\cdot NAD^+ \rightleftharpoons E(S{-}CH(OH){-}R)\cdot NAD^+ \rightleftharpoons E(S{-}C(=O){-}R)\cdot NADH \xrightarrow{-NADH} E{-}S{-}C(=O){-}R \xrightarrow{H_2O,\ -RCO_2H} E{-}SH \xrightarrow{NAD^+} E(SH)\cdot NAD^+$$

Figure 16-18

The Mechanism of the Aldehyde Dehydrogenase Reaction. The substrate aldehyde binds to the enzyme as a thiohemiacetal with a cysteinyl —SH group at the active site, and this adduct transfers a hydride to NAD^+, which is bound in an adjacent site. The resulting enzyme-acylthioester undergoes hydrolysis to the carboxylic acid, NADH is released, and the catalytic cycle is repeated.

enzyme-bound thioester and NADH. Transfer of hydride from the thiohemiacetal to NAD^+ is reminiscent of the Cannizzaro reaction in that hydrogen is transferred from a carbon bonded to two heteroatoms. The enzyme-bound thioester is then hydrolyzed to produce the acid. Note that this reaction is wasteful, because the intermediate thioester (a high-energy compound) is hydrolyzed. As we shall soon see, not all enzymes are so uneconomical, and in some cases the high-energy bond is preserved.

Glyceraldehyde-3-phosphate Dehydrogenase

The oxidation of glyceraldehyde-3-P catalyzed by glyceraldehyde-3-phosphate dehydrogenase according to equation 18 is a reaction in which the intermediate high-energy bond (thioester) is not wasted.

$$\begin{array}{c} O{=}C{-}H \\ | \\ H{-}C{-}OH \\ | \\ H_2C{-}OPO_3^{2-} \end{array} + HOPO_3^{2-} + NAD^+ \longrightarrow \begin{array}{c} O{=}C{-}OPO_3^{2-} \\ | \\ H{-}C{-}OH \\ | \\ H_2C{-}OPO_3^{2-} \end{array} + NADH + H^+ \qquad (18)$$

1,3-Diphospho-glyceric acid

This enzyme plays a key role in the metabolism of all organisms (see Chapter 21). The product of the reaction, 1,3-diphosphoglyceric acid, is a high-energy compound, which reacts with ADP to form ATP and 3-phosphoglycerate in a reaction catalyzed by phosphoglycerate kinase (Chapter 11). This is one of the few instances in which the coupling of an oxidation process to the formation of a high-energy bond is mechanistically well understood.

We shall briefly review some of the experimental facts that have been obtained with glyceraldehyde-3-phosphate dehydrogenase. One of the first questions one might ask is: What is the role of P_i? This was investigated by P. Boyer and S. Velick and their associates in the 1950s. If one uses catalytic amounts of enzyme and omits P_i from the reaction mixture, no product is formed. An experiment was done with a relatively large amount of enzyme, so that enzyme-bound intermediates could be detected, with the results shown in Figure 16-19. At time zero ($t = 0$), glyceraldehyde-3-P, NAD^+, buffer, and glyceraldehyde-3-phosphate dehydrogenase were mixed, and a rapid increase in absorbance at 340 nm was observed; that is, NADH was formed. However, the reaction stopped when all the enzyme-bound NAD^+ had been reduced. At this point, only a small fraction of the available *free* NAD^+ and substrate had been consumed. At the time indicated in Figure 16-19, P_i was added, and more NADH was rapidly formed. The amount of NADH formed was equivalent to the amount of P_i added and considerably larger than the amount of enzyme present.

What can be concluded from this experiment? Clearly, P_i is needed in order to have catalysis; that is, to produce more moles of product than the number of moles of enzyme present. In addition, oxidation can occur as evidenced by NADH formation in the absence of P_i. These experimental results are consistent with the mechanism in Figure 16-20. Glyceraldehyde-3-P initially reacts with the —SH group of a cysteine in the enzyme to form a thiohemiacetal. The thiohemiacetal then reduces the enzyme-bound NAD^+, and an enzyme-thioester is formed. Up to this point, the reaction resembles that described earlier for aldehyde dehydrogenases. However, in the absence of P_i, the glyceraldehyde-3-phosphate dehydrogenase cannot react further; that is, the enzyme-thioester does not undergo hydrolysis. (It actually undergoes hydrolysis at a very slow rate, relative to the rate at which phosphorolysis proceeds, and

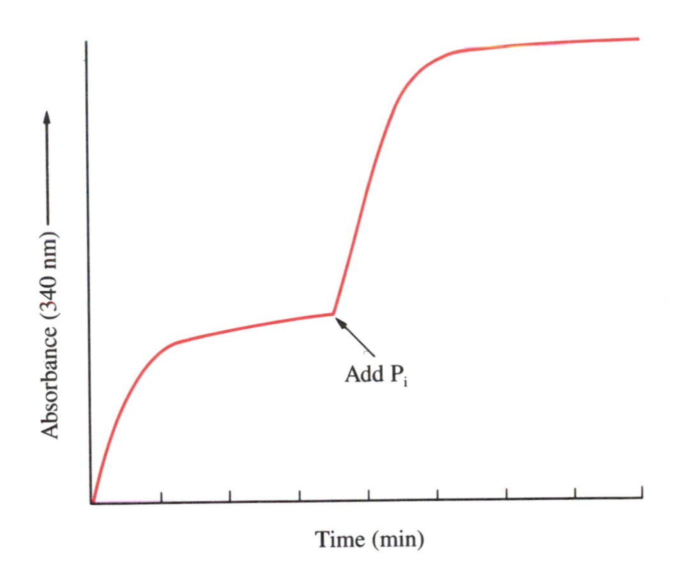

Figure 16-19

Spectrophotometric Analysis of the Mechanism of NADH Formation by Glyceraldehyde-3-phosphate Dehydrogenase. At time zero, glyceraldehyde-3-P is added to a solution containing enzyme, NAD^+, and a phosphate-free buffer. This initiates a burst of NADH formation that is equivalent to the amount of enzyme in the solution and appears as an increase in A_{340}. After a short time, NADH formation slows down and nearly reaches a plateau. Addition of P_i elicits a second burst of NADH. This course of NADH production is consistent with the mechanism in Figure 16-20.

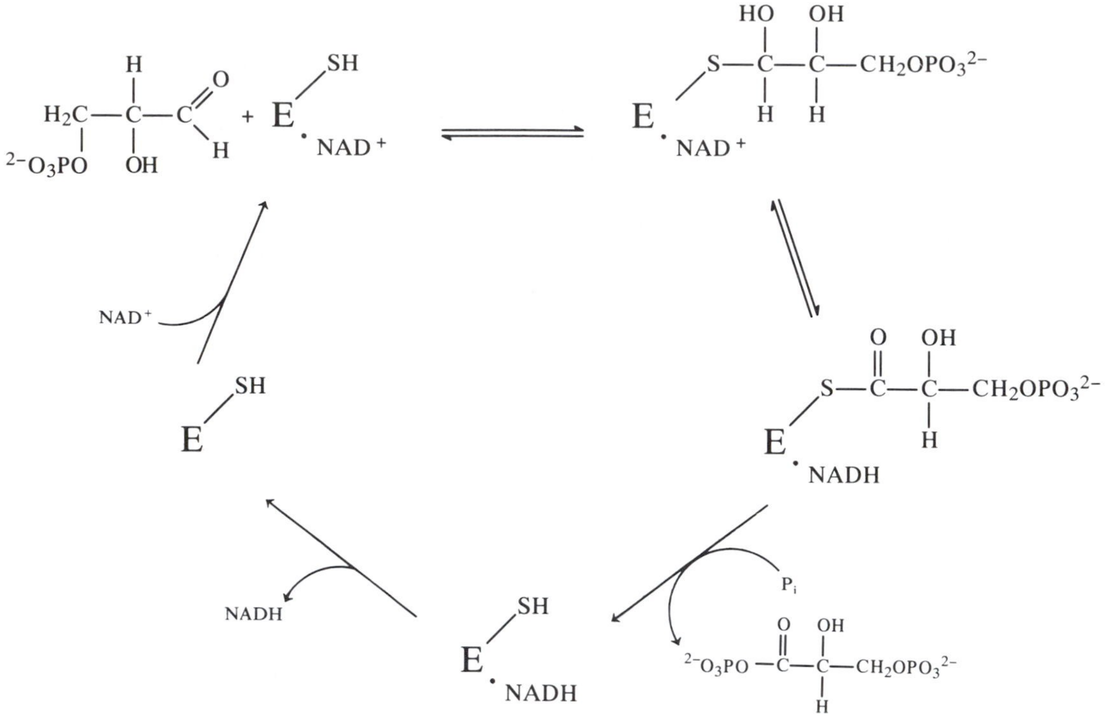

Figure 16-20

The Mechanism of the Glyceraldehyde-3-phosphate Dehydrogenase Reaction. The mechanism is chemically analogous to that of the aldehyde dehydrogenase reaction, except that the enzyme-acylthioester does not undergo rapid hydrolysis. This form of the enzyme reacts with P_i instead of water to form an acyl phosphate, 1,3-diphosphoglycerate (or acetyl phosphate in equation 21), NADH is released, and the catalytic cycle is repeated.

NADH is formed very slowly, following the initial fast NADH production.) The reaction up to this point accounts for the formation of a limited amount of NADH, which is formed in the absence of P_i. When P_i is added, the enzyme-bound thioester reacts to form an acyl phosphate, 1,3-diphosphoglycerate, which is released from the enzyme. NADH can then be released, and the enzyme can catalyze the oxidation of another substrate molecule.

What is the evidence for the participation of an —SH group and for thioester formation? Several chemical modification experiments implicated a cysteinyl —SH group at the active site. First, the participation of an —SH group in the catalytic process was indicated by the observation that glyceraldehyde-3-phosphate dehydrogenase could be inactivated by compounds such as *p*-chloromercuribenzoate or iodoacetamide, which are known to react with —SH groups (equations 19 and 20).

$$R{-}S^- + I{-}CH_2{-}\overset{O}{\overset{\|}{C}}{-}NH_2 \longrightarrow R{-}S{-}CH_2{-}\overset{O}{\overset{\|}{C}}{-}NH_2 + I^- \quad (19)$$

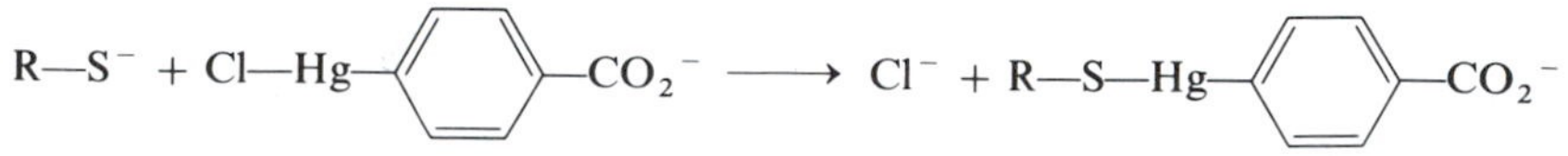

$$R{-}S^- + Cl{-}Hg{-}C_6H_4{-}CO_2^- \longrightarrow Cl^- + R{-}S{-}Hg{-}C_6H_4{-}CO_2^- \quad (20)$$

Furthermore, the rate of inactivation could be decreased by glyceraldehyde-3-P but not by P_i or NAD$^+$. This suggested that the substrate and iodoacetamide react with the same —SH group.

Second, glyceraldehyde-3-phosphate dehydrogenase, from rabbit muscle, was inactivated with ^{14}C-iodoacetamide. The inactivated enzyme contained ^{14}C that was not removed by dialysis, confirming that iodoacetamide was strongly attached to the enzyme. The inactivated enzyme was degraded with proteolytic enzymes, and a peptide containing ^{14}C was isolated and purified. The amino acid sequence of this peptide was then determined and found to contain two cysteinyl residues, only one of which had reacted with iodoacetamide. The partial amino acid sequence of the peptide containing the [^{14}C]carboxamidomethyl group was found to be:

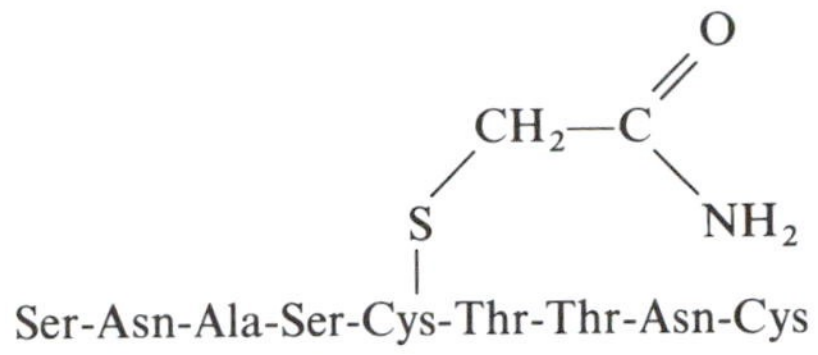

$$\text{Ser-Asn-Ala-Ser-Cys(S-CH}_2\text{-C(=O)NH}_2\text{)-Thr-Thr-Asn-Cys}$$

Before explaining how it is known that the cysteine residue identified by chemical modification is the same one present in the intermediate enzyme-thioester, we must briefly describe the reaction of glyceraldehyde phosphate dehydrogenase with acetaldehyde and acetyl phosphate. The enzyme catalyzes the oxidation of acetaldehyde according to equation 21 at a slow rate.

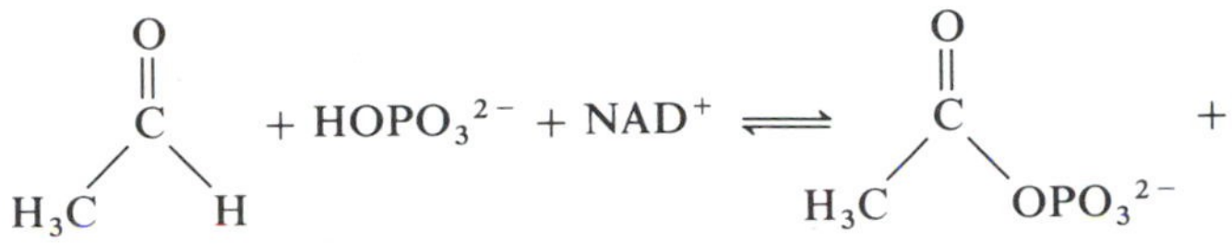

$$H_3C\text{-}C(=O)\text{-}H + HOPO_3^{2-} + NAD^+ \rightleftharpoons H_3C\text{-}C(=O)\text{-}OPO_3^{2-} + NADH + H^+ \qquad (21)$$

This reaction is believed to proceed by the same mechanism as the oxidation of glyceraldehyde-3-P; that is, the mechanism in Figure 16-20 involving the formation of an enzyme-thioester intermediate, in which acetaldehyde replaces glyceraldehyde-3-P. In addition, the enzyme catalyzes the hydrolysis of acetyl phosphate, in the absence of NADH, to acetic acid and P_i. This hydrolytic reaction is closely related to the oxidation of acetaldehyde. It represents a partial reversal of the reaction, in which acetyl phosphate reacts with the enzyme to form a thioacetyl-enzyme, the same acetyl-enzyme that is formed in the oxidation of acetaldehyde. In the presence of NADH, the thioester would be reduced and acetaldehyde would eventually be formed; however, in the absence of NADH, the thioacetyl-enzyme undergoes hydrolysis and acetic acid is released.

By making use of the acetyl phosphate reaction, investigators conducted a third experiment, which showed that the same cysteine residue that is alkylated by iodoacetamide also participates in forming the thioester intermediate when glyceraldehyde-3-phosphate dehydrogenase reacts with its substrate. When glyceraldehyde-3-phosphate dehydrogenase was allowed to react with [^{14}C]acetyl phosphate, ^{14}C-labeled enzyme could be isolated; the enzyme so obtained was degraded to peptides by the use of proteolytic enzymes. Isolated from these peptides was a ^{14}C-labeled peptide, which contained an *S*-acetyl group in the following amino acid sequence.

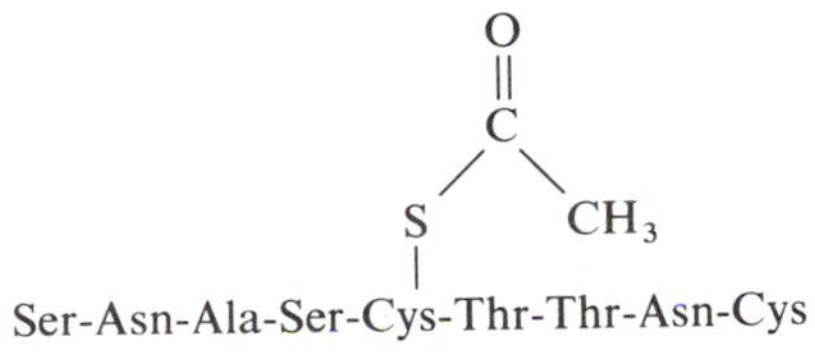

$$\text{Ser-Asn-Ala-Ser-Cys(S-C(=O)-CH}_3\text{)-Thr-Thr-Asn-Cys}$$

Notice that *S*-acetylcysteine in this peptide is part of the same amino acid sequence as the cysteine residue that was labeled by iodoacetamide. These experiments confirm that iodoacetamide and the substrate react with the same cysteine-SH group, and they provide strong support for the proposed mechanism of action of glyceraldehyde-3-phosphate dehydrogenase.

Glutathione Reductase

Several enzymes utilize both nicotinamide coenzymes and flavin coenzymes in catalyzing oxidation-reduction reactions. An example is glutathione reductase, the structure of which is known; the active site is shown in Figure 16-21. This enzyme is a flavoprotein that catalyzes the reduction of glutathione by NADPH according to equation 22.

$$\text{G—S—S—G} + \text{NADPH} + \text{H}^+ \rightleftharpoons 2\text{G—SH} + \text{NADP}^+ \quad (22)$$

Glutathione (G-SH)
(γ-Glutamylcysteinylglycine)

As shown in Figure 16-21, NADPH is bound at the active site adjacent to and nearly parallel with FAD, which intervenes between NADPH and a disulfide group. NADPH reduces FAD, which in turn reduces the disulfide; and the disulfide reduces oxidized glutathione (G–S–S–G) to glutathione.

Glutathione is an important reductant in a number of cellular reactions, among which are the maintenance of the sulfhydryl groups of proteins in their reduced states, the reductive detoxification of hydrogen peroxide, and the detoxification of xenobiotics through the formation of glutathione thioesters. Oxidized glutathione produced in some of these processes is reduced by NADPH through the action of glutathione reductase.

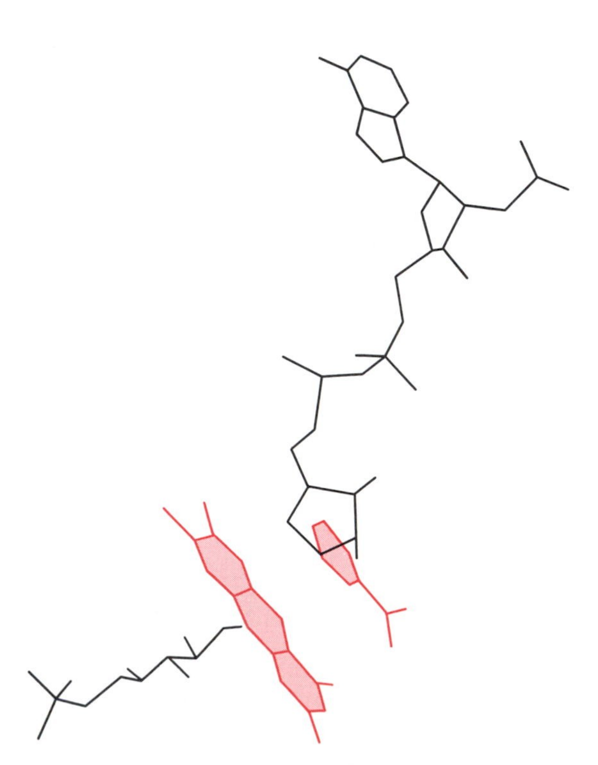

Figure 16-21

NADP$^+$ and FAD at the Active Site of Glutathione Reductase. [Adapted with permission from E. F. Pai, P. A. Karplus, and G. E. Schulz. *Biochemistry* 27 (1988): 4465–4474, Fig. 4. Copyright 1988 American Chemical Society.]

NAD^+ as an Analytical Tool

Because of the high extinction coefficient of NADH at 340 nm ($\varepsilon_{340} = 6220\,cm^{-1}\,M^{-1}$), NAD^+ is widely used for analytical purposes. For example, the concentration of ethanol in a solution can be determined by adding alcohol dehydrogenase and NAD^+ and spectrophotometrically measuring the amount of NADH produced, as indicated by the increase in A_{340}. The amount of NADH is stoichiometrically equivalent to the amount of alcohol initially present and, from Beer's Law, the change in A_{340} is a direct measure of the amount of NADH produced. Beer's Law is given by equation 23,

$$A_\lambda = \varepsilon_\lambda \cdot [\text{chromophore}] \cdot 1 \tag{23}$$

$$\Delta A_{340} = 6620[\text{NADH}] \tag{24}$$

in which "l" is the optical path length in centimeters, usually 1 cm for conventional cuvets.

From this, it can be seen that the change in A_{340} is related to the concentration of NADH by equation 24, when an optical cell with a 1-cm path length is used. The use of alcohol dehydrogenase to determine the concentration of ethanol is an example of one of many ways in which NAD^+-dependent enzymes can be used to determine the concentration of compounds.

The utility of NAD^+ as an analytical tool was greatly extended by the development of the coupled assay. We shall illustrate this approach by describing the determination of glucose. No known NAD^+-dependent enzyme acts on glucose. However, glucose-6-phosphate dehydrogenase is an $NADP^+$-dependent enzyme that catalyzes the oxidization of glucose-6-P to 6-phosphogluconolactone, and the enzyme hexokinase utilizes ATP to convert glucose into glucose-6-P. The two reactions may be coupled in series, so that with a combination of the two enzymes, glucose can be determined spectrophotometrically as NADPH. The coupling of these two reactions for this purpose is illustrated in Figure 16-22, where it is shown that, for each glucose molecule oxidized, one molecule of NADPH is produced. Because the value of ε_{340} for NADPH is the same as that for NADH, equation 24 can also be used for calculating the amount of glucose from the measured ΔA_{340}.

Many coupled systems have been developed for the determination of biologically important compounds. Some of these systems use two or three coupling enzymes in addition to the NAD^+-dependent enzyme.

Glucose

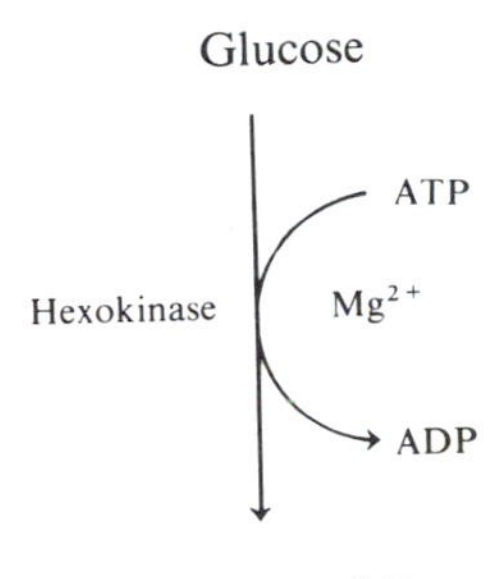

Glucose-6-P

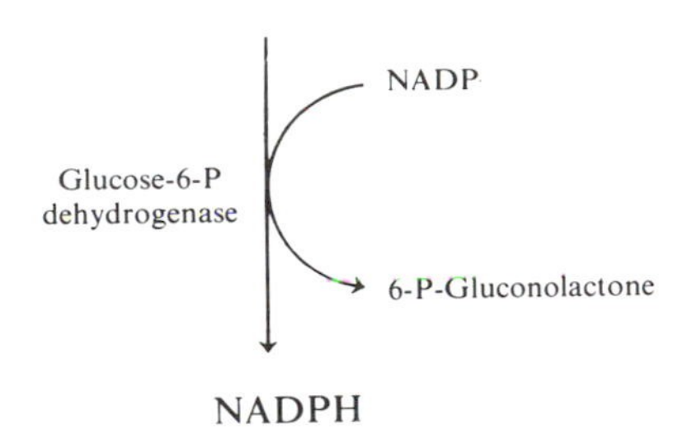

NADPH

Figure 16-22

Coupled Assay of Glucose Using Hexokinase and Glucose-6-phosphate Dehydrogenase. When the hexokinase and glucose-6-phosphate dehydrogenase reactions are coupled, the amount of NADPH produced is stoichiometrically equivalent to the amount of glucose oxidized. The NADPH can be quantitatively determined by the change in absorbance at 340 nm (ΔA_{340}), using equation 23. This is the most convenient and widely used method for accurately determining glucose.

Summary

Niacin and riboflavin coenzymes are widely important in mediating electron and hydrogen transfer in biological oxidation-reduction reactions. The nicotinamide coenzymes NAD^+ and $NADP^+$ are reduced by direct, enzyme-catalyzed transfer of a hydrogen atom and two electrons from substrates to form NADH and NADPH, respectively. The flavin coenzymes FAD and FMN also are often reduced by two electrons and hydrogen to $FADH_2$ and $FMNH_2$; however, one-electron transfer processes are known for the flavin coenzymes, and the enzymatic transfer of two electrons may occur in two one-electron steps in many flavin reactions.

Direct hydrogen transfer in enzymatic reactions of NAD^+ and $NADP^+$ occur stereospecifically, both with respect to nicotinamide carbon-4 in the coenzyme and with respect to the hydrogen-donor atoms of substrates.

Dehydrogenation of an alcohol or an amine by an NAD^+- or $NADP^+$-dependent enzyme includes hydride transfer to form the reduced coenzyme and a carbonyl compound or imine. However, the oxidation of an aldehyde is initiated by the addition of an enzymatic —SH group to the aldehyde to form a thiohemiacetal, followed by hydride transfer to NAD to form NADH and an enzyme-thioester. The acyl group of the enzyme-thioester is subsequently transferred to water or another nucleophile such as phosphate or coenzyme A.

ADDITIONAL READING

BOYER, P. D., ed. 1972. *Oxidation-Reduction: Part A. The Enzymes,* vol. 6, 3d ed. New York: Academic Press.

HARRIS, J. L., and WATERS, M. 1976. Glyceraldehyde-3-phosphate dehydrogenase. In *The Enzymes,* vol. 13, 3d ed., ed. P. D. Boyer. New York: Academic Press, p. 1.

WILLIAMS, C. H. 1976. Flavin-containing dehydrogenases. In *The Enzymes,* vol. 13, 3d ed., ed. P. D. Boyer. New York: Academic Press, p. 90.

EKLUND, H., BRÄNDEN, C.-I. 1987. Alcohol dehydrogenase. In *Active Sites of Enzymes,* ed. F. A. Jurnak and A. McPherson. Biological Macromolecules and Assemblies, vol. 3. New York: Wiley-Interscience, p. 73.

WALSH, C. T. 1976. *Enzymatic Reaction Mechanisms.* San Francisco: Freeman.

PROBLEMS

1. The glyoxalase reaction is carried out in H_2O with the substrate shown below containing ^2H at C-1.

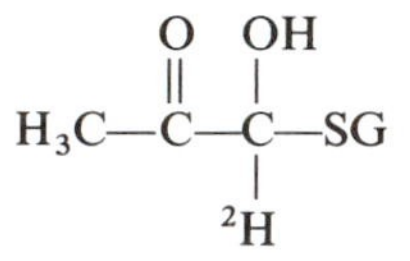

$$H_3C-\overset{O}{\overset{\|}{C}}-\overset{OH}{\overset{|}{\underset{^2H}{\underset{|}{C}}}}-SG$$

Predict the ^2H content of the product for a hydride mechanism and for an enol mechanism, in which the conjugate acid of the enzymatic base exchanges either rapidly or slowly with the solvent.

2. The compound shown below is an extremely effective inhibitor of glyceraldehyde phosphate dehydrogenase.

$$\underset{^{2-}O_3P-O}{H_2C}-\overset{H}{\underset{HO}{C}}-\overset{H}{\underset{O-PO_3^{2-}}{C}}-CHO$$

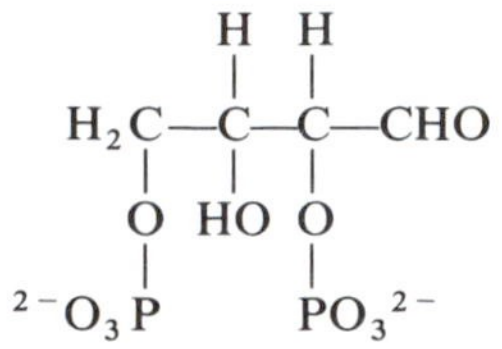

It inhibits the enzyme essentially irreversibly. Suggest a rationale for why this inhibitor is so potent.

3. Suggest a mechanism for the dehydrogenation of histidinol according to the following equation:

$$\text{Histidinol} + 2\,NAD^+ + H_2O \longrightarrow \text{Histidine} + 2\,NADH + 2\,H^+$$

Histidinol differs from histidine in that the carboxyl group of histidine is a hydroxymethyl group.

4. How would you synthesize a sample of [nicotinamide-4-^{2}H]NAD^+?

Decarboxylation and Carboxylation

17

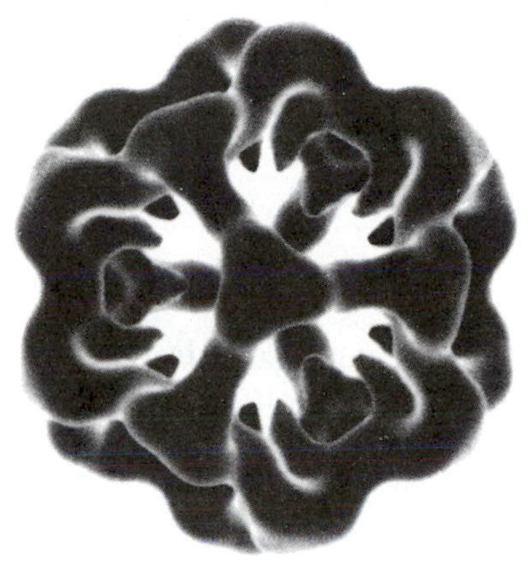
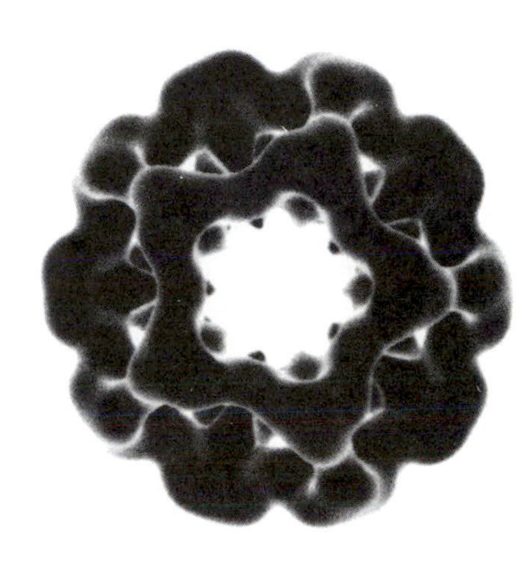

Three-dimensional structure of the truncated core of the yeast pyruvate dehydrogenase complex determined by cryo-electron microscopy The truncated form (residues 181–454) of the dehydrolipamide acetyltransferase component (E_2) of the pyruvate dehydrogenase complex was expressed in Escherichia coli and exhibited catalytic activity similar to that of the wild-type enzyme. The 22 Å reconstructions show the characteristic two-, three-, and fivefold views of the pentagonal dodecahedron structure. The vertices of this sixty subunit complex consist of trimers, and the structure serves as the scaffolding to which the other components of the complex (E_1, E_3, and protein X) are attached. E_2 from E. coli consists of twenty-four subunits and has cubic symmetry. (Courtesy of Timothy S. Baker, Xiao-D. Niu, Lester J. Reed, John P. Schroeter, and James K. Stoops, manuscript in preparation.)

Degradative aerobic metabolism converts organic compounds into carbon dioxide and water. Carbon dioxide, a major product of aerobic metabolism, generally arises from the decarboxylation of α-ketoacids and β-ketoacids produced in the course of catabolism. Carboxylic acids in cells are also produced in biosynthetic (anabolic) reactions, frequently in carboxylation reactions. In these reactions, carbon dioxide is added to an organic molecule. Although carboxylation reactions are the reverse of decarboxylation, the metabolic processes by which carboxylation occurs are not the reverse of the corresponding decarboxylation reactions. Synthetic biological reactions generally do not proceed through the reversal of the corresponding degradative reaction sequences.

In this chapter, we shall describe the chemical mechanisms by which carboxylation and decarboxylation reactions occur, and then we shall consider the enzymes and coenzymes that catalyze these reactions.

Decarboxylation

β-Ketoacids Readily Undergo Decarboxylation

Decarboxylation of a carboxylic acid leads to the formation of CO_2 and a carbanion according to equation 1. Carbon dioxide is a stable molecule, whereas the carbanion is a high-energy species that cannot exist for long under physiological conditions. Therefore, the main barrier to decarboxylation is the formation of the carbanion; and decarboxylation will be facilitated when a mechanism exists to stabilize the carbanion produced by decarboxylation.

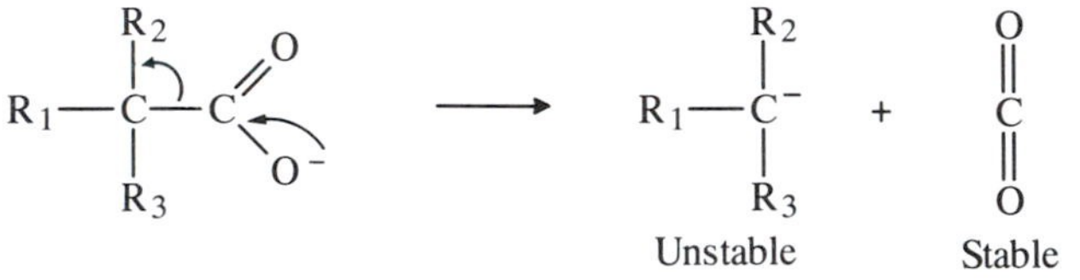

$$R_1-C(R_2)(R_3)-C(=O)O^- \longrightarrow \underset{\text{Unstable}}{R_1-C^-(R_2)(R_3)} + \underset{\text{Stable}}{O=C=O} \quad (1)$$

How can this be accomplished? If the carbanion is adjacent to an electron-deficient group such as the carbonyl group in a ketone, ester, aldehyde, or carboxylic acid, it will be stabilized by delocalization of the electron pair. Hence, β-ketoacids readily undergo decarboxylation according to equation 2.

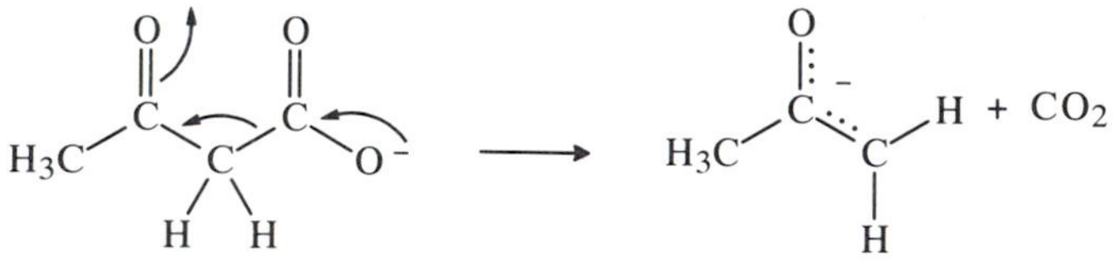

$$H_3C-C(=O)-CH_2-C(=O)O^- \longrightarrow H_3C-C(\overset{\delta-}{O})\cdots CH-H + CO_2 \quad (2)$$

Carboxylic acids that have no carbonyl group in the β-position are stable to decarboxylation under physiological conditions (equation 3).

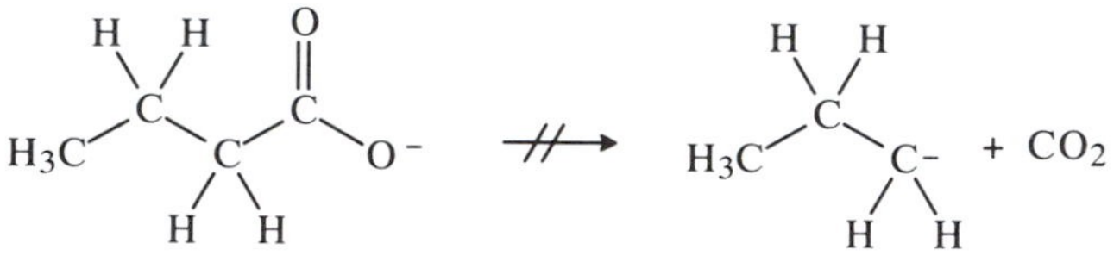

$$H_3C-CH_2-CH_2-C(=O)O^- \not\longrightarrow H_3C-CH_2-CH_2^- + CO_2 \quad (3)$$

Molecules such as acetic acid or butyric acid in equation 3 undergo decarboxylation only under extreme conditions, such as fusion with solid NaOH.

β-Ketoacids Frequently Participate in Biological Decarboxylation

How can a decarboxylation reaction be catalyzed? Decarboxylation of a β-ketoacid entails the formation of an enolate ion such as that shown in equation

2. This carbanion is greatly stabilized by delocalization of the electrons; nevertheless, an enolate is not a particularly stable intermediate under physiological conditions. The degree to which a β-carbonyl group stabilizes a carbanion is represented for aqueous solutions by the pK_a values for alkanes and ketones, which for an ordinary alkane is estimated to be about 50 ($\Delta G^\circ = +68\ \text{kcal mol}^{-1}$ at 25°C) and for acetone is about 20 ($\Delta G^\circ = +27\ \text{kcal mol}^{-1}$ at 25°C).

$$\text{R—H} \underset{}{\overset{pK_a \approx 50}{\rightleftharpoons}} \text{R}^- + \text{H}^+ \quad (4)$$

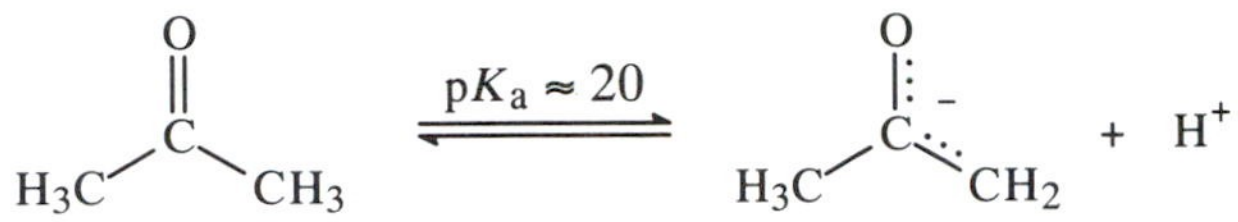

$$\text{H}_3\text{C—C(=O)—CH}_3 \overset{pK_a \approx 20}{\rightleftharpoons} \text{H}_3\text{C—C(}\overset{..}{=}\text{O)}\cdots\text{CH}_2^- + \text{H}^+ \quad (5)$$

Thus the β-carbonyl group stabilizes a carbanion by about $40\ \text{kcal mol}^{-1}$, but an enolate is still quite unstable in neutral aqueous solutions. Therefore, any interaction with an enzyme that stabilizes the negative charge will be helpful in catalyzing decarboxylation.

An enzyme-bound enolate can be stabilized by a positively charged entity, as illustrated in Figure 17-1, such as the proton of an acidic group or the positive charge of a metal ion placed near the carbonyl oxygen. In Figure 17-1A, the enolate is protonated to the enol form concomitant with decarboxylation, so that the enolate anion is never produced. The enol is much more stable than the

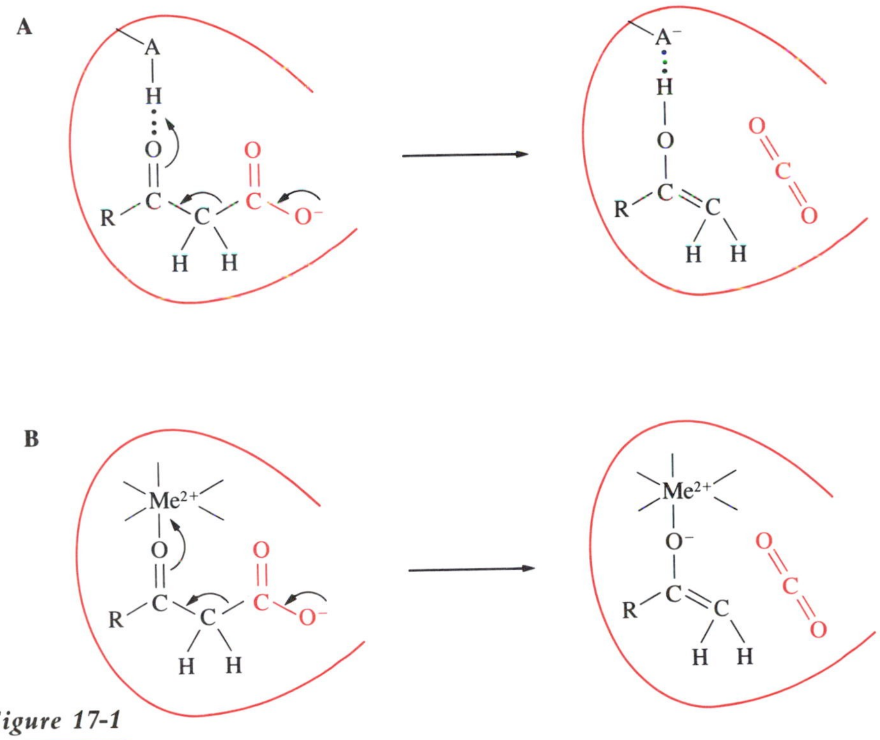

A

B

Figure 17-1

Stabilization of Enolate Anions at Active Sites. Part A illustrates how an enzymatic acid can donate a hydrogen bond to the β-carbonyl group of a β-ketoacid and donate a proton to the enolate resulting from decarboxylation. The hydrogen bond in the enzyme-substrate complex polarizes the carbonyl group by increasing the partial positive charge on the carbonyl oxygen, and proton transfer stabilizes the enolate resulting from decarboxylation. Part B shows how a metal ion can polarize the β-carbonyl group through coordination bonding; it also stabilizes the enolate.

enolate, and it is thought that the enol is the intermediate in enzymatic reactions rather than the enolate. In Figure 17-1B, a metal ion stabilizes the enolate through an electrostatic interaction. Stabilization of the enolate lowers the activation energy for the reaction and increases the rate. Conversion of the β-carbonyl group into a protonated imine also facilitates decarboxylation by the mechanism shown in equation 6.

$$R'(H)N^{+}{=}C(R)-CH_2-C(=O)O^- \longrightarrow R'(H)N-C(R){=}CH_2 + CO_2 \qquad (6)$$

The pK of an imine is near 7, so that under physiological conditions the imine-nitrogen can be positively charged and acts as a very effective electron sink. Decarboxylation leads to the formation of an enamine, a lower-energy species than an enolate. With a carbonyl group in the β-position, a partial positive charge can reside on the oxygen through the intrinsic polarity of the carbonyl group and the effects of hydrogen bonding with an enzymatic acid (Figure 17-1). In contrast, the nitrogen of a protonated imine carries a full positive charge, and the β-iminium ion facilitates decarboxylation even more effectively than does a carbonyl group. Therefore, if the keto group of a β-ketoacid is converted into a protonated imine, the rate of decarboxylation will be greatly enhanced.

Enzymatic Decarboxylation of a β-Ketoacid

An example of enzymatic decarboxylation is provided by acetoacetate decarboxylase, an enzyme isolated from a microorganism, which catalyzes the reaction of equation 7. How does the enzyme catalyze this reaction?

$$H_3C-C(=O)-CH_2-C(=O)O^- + H^+ \longrightarrow \underset{\textbf{Acetone}}{H_3C-C(=O)-CH_3} + CO_2 \qquad (7)$$

Acetoacetate decarboxylase brings about decarboxylation through imine formation, as shown in Figure 17-2. Acetoacetate first reacts with the ε-NH_2 group of a lysine residue at the active site to form an imine. The imine then undergoes decarboxylation to form an acetone imine, which undergoes hydrolysis to acetone and the regenerated enzyme.

The initial phase of acetoacetate decarboxylation is very similar to that of the aldolases discussed in Chapter 2; that is, the reaction of the carbonyl group of acetoacetate with the ε-amino group at the active site. Imine formation between aldehydes and ketones occurs by the two-step mechanism shown in Figure 17-3. The amino group adds as a nucleophile to the carbonyl group to form a tetrahedral adduct, which then undergoes acid catalyzed dehydration to the imine. This is a fast and reversible reaction; it is also efficiently catalyzed by enzymes when it is a compulsory step in the enzymatic mechanism.

Several experiments have provided evidence for imine formation between acetoacetate decarboxylase and acetoacetate. In one experiment, the decarboxylation of acetoacetate was carried out in $H_2{}^{18}O$, and the acetone produced was isolated and found to contain ^{18}O. The fact that acetone produced through the decarboxylation of acetoacetate in $H_2{}^{18}O$ acquired ^{18}O meant that the

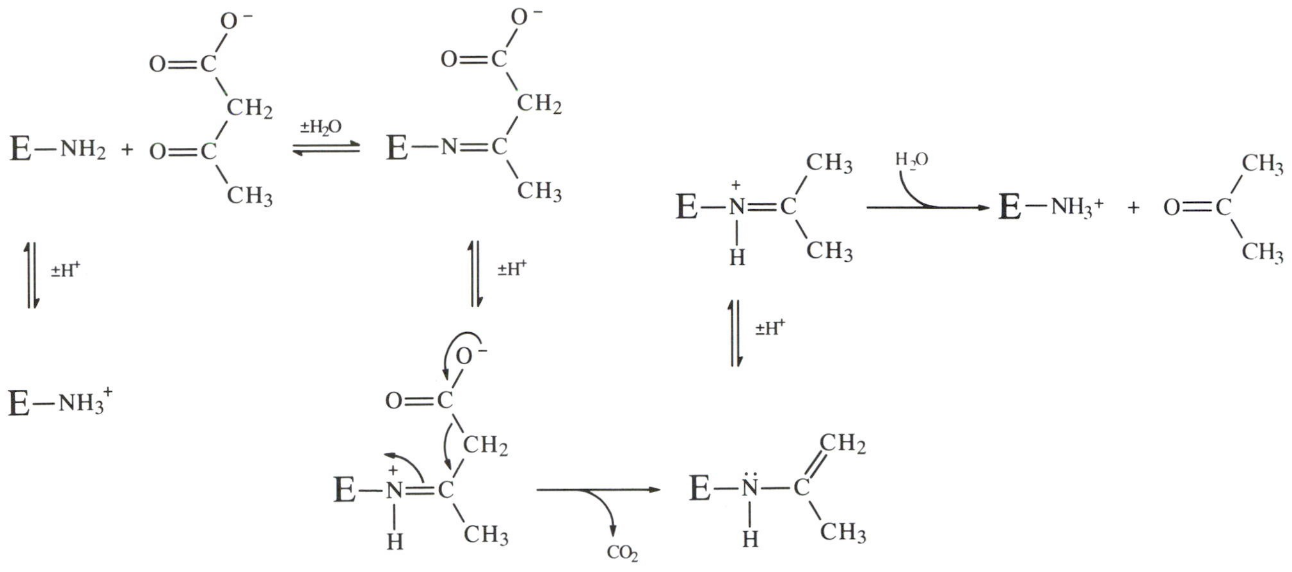

Figure 17-2

The Mechanism of Decarboxylation by Acetoacetate Decarboxylase.

reaction mechanism must have led to the loss of ^{16}O originally in the β-carbonyl group of acetoacetate.

The loss of ^{16}O from acetoacetate and the incorporation of ^{18}O from the solvent into acetone is exactly what is predicted by the imine mechanism. When acetoacetate reacts with the enzyme to form the imine, it loses its β-carbonyl oxygen in the form of water (Figures 17-2 and 17-3). The acetone produced is derived from the hydrolysis of an imine, in the course of which oxygen from the solvent is incorporated into acetone. If acetoacetate had been decarboxylated directly, without imine formation, the oxygen of the carbonyl group of acetoacetate would be present in the acetone produced. Thus, incorporation of solvent oxygen into acetone provides evidence for the imine mechanism.

Direct evidence for the imine mechanism was obtained by chemical trapping of the intermediate imine. Sodium borohydride ($NaBH_4$) was added to acetoacetate decarboxylase and acetoacetate. Within a short time, the enzymatic activity was lost. In control experiments, $NaBH_4$ added to acetoacetate decarboxylase in the absence of acetoacetate did not inactivate the enzyme.

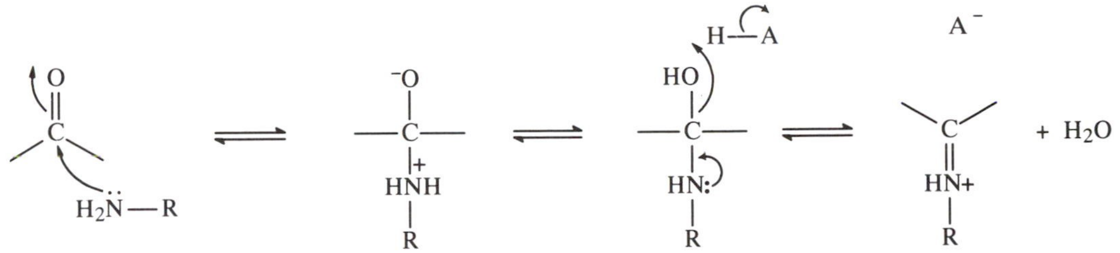

Figure 17-3

The Mechanism of Imine Formation Between a Ketone and a Primary Amine. The first step is addition of the unprotonated amino group to the carbonyl group, followed by proton transfer to convert the initial zwitterionic addition intermediate into the more-stable neutral adduct. The dehydration of this adduct to the imine is acid catalyzed. Enzymes catalyze this process when imines are compulsory intermediates.

Acetoacetate decarboxylase is inactivated by $NaBH_4$ only in the presence of the substrate. Why? It is known that $NaBH_4$ reduces protonated imines; therefore, it is possible that $NaBH_4$ reduces one or both of the intermediate protonated imines, as illustrated in Figure 17-4. Reduction of the protonated imine by $NaBH_4$ produces a secondary amine, in which the C—N bond is stable. Consequently, the lysine residue at the active site becomes chemically modified to a stable secondary amine, and it can no longer react with substrate molecules to form an imine, which makes the enzyme catalytically inactive. Therefore, the inactivation observed with $NaBH_4$ is also consistent with an imine mechanism.

The proposal that an imine intermediate is involved would be on a much firmer basis if we knew more about the chemistry of the reaction with $NaBH_4$. It would be very useful to know the following: Has a secondary amine in fact been formed after reduction with $NaBH_4$? If a secondary amine has been formed, which amino acid residue of the protein is modified? Is the secondary amine produced from the substrate or product-derived imine in Figure 17-4?

In principle, one could get the answers to these questions by hydrolyzing the $NaBH_4$ inactivated enzyme and separating a "new" amino acid from the hydrolysate—that is, one that was not present before treatment with $NaBH_4$. The structure of the "new" amino acid could then be determined. The purification of the new amino acid would be difficult, because only one out of many amino acids would be chemically modified; the total amount of material available for such experiments is often small because enzymes are often not available in large quantities.

To reduce the experimental difficulties, a different approach was used. The $NaBH_4$-inactivation was carried out in the presence of acetoacetate labeled with ^{14}C in the β-carbon (CH_3—CO—$^{14}CH_2$—COO^-). When the inactivated enzyme was isolated and freed from excess substrate, it had ^{14}C radioactivity associated with it, which indicated that it was covalently bound to all or part of the substrate molecule.

The inactivated enzyme was hydrolyzed, and the resulting amino acids were separated by chromatography. The isolation and characterization of the

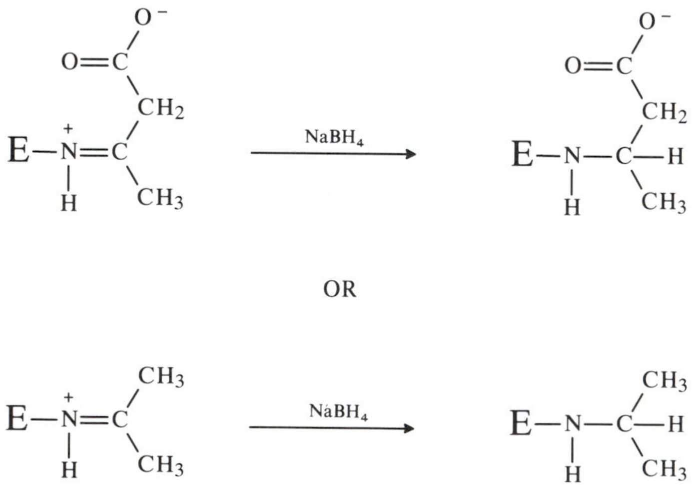

$$\text{E—}\overset{+}{\text{N}}\text{(H)=C(CH}_3\text{)—CH}_2\text{—C(=O)O}^- \xrightarrow{NaBH_4} \text{E—N(H)—C(H)(CH}_3\text{)—CH}_2\text{—C(=O)O}^-$$

OR

$$\text{E—}\overset{+}{\text{N}}\text{(H)=C(CH}_3\text{)(CH}_3\text{)} \xrightarrow{NaBH_4} \text{E—N(H)—C(H)(CH}_3\text{)(CH}_3\text{)}$$

Figure 17-4

Protonated Imines of Acetoacetate Decarboxylase That Are Subject to Reduction by $NaBH_4$. Both the enzyme-acetoacetate imine at the top and the enzyme-acetone imine at the bottom are potentially reducible by $NaBH_4$. In practice, the acetone-imine is the one that is reduced to a secondary amine, leading to inactivation of the enzyme.

hydrolytic degradation product was simplified by the fact that it was radioactive, which facilitated its detection and analysis. A radioactive compound was separated from the other amino acids, and it was shown to have the following structure:

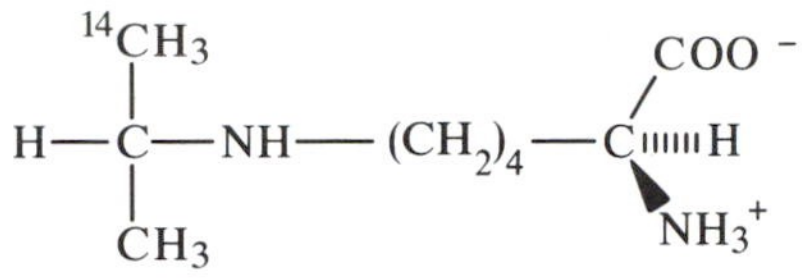

$$\mathrm{H{-}\overset{^{14}CH_3}{\underset{CH_3}{C}}{-}NH{-}(CH_2)_4{-}\overset{COO^-}{\underset{NH_3^+}{C}}{\cdots}H}$$

N^{ε}-[^{14}C]Isopropyl-L-lysine

This molecule was produced through the reduction by $NaBH_4$ of the imine formed between the ε-NH_2 group of lysine and acetone. This experiment established that a lysine residue participates in imine formation and that only the imine resulting from decarboxylation is trapped by $NaBH_4$. The location of this lysine residue in the amino acid sequence of an oligopeptide encompassing the active site was determined by the methods described in Chapter 7.

Decarboxylation of β-Hydroxyacids

A number of enzymes catalyze decarboxylations of carboxylic acids that do not contain β-keto groups (Figure 17-5). Note that NAD^+ or $NADP^+$ participates

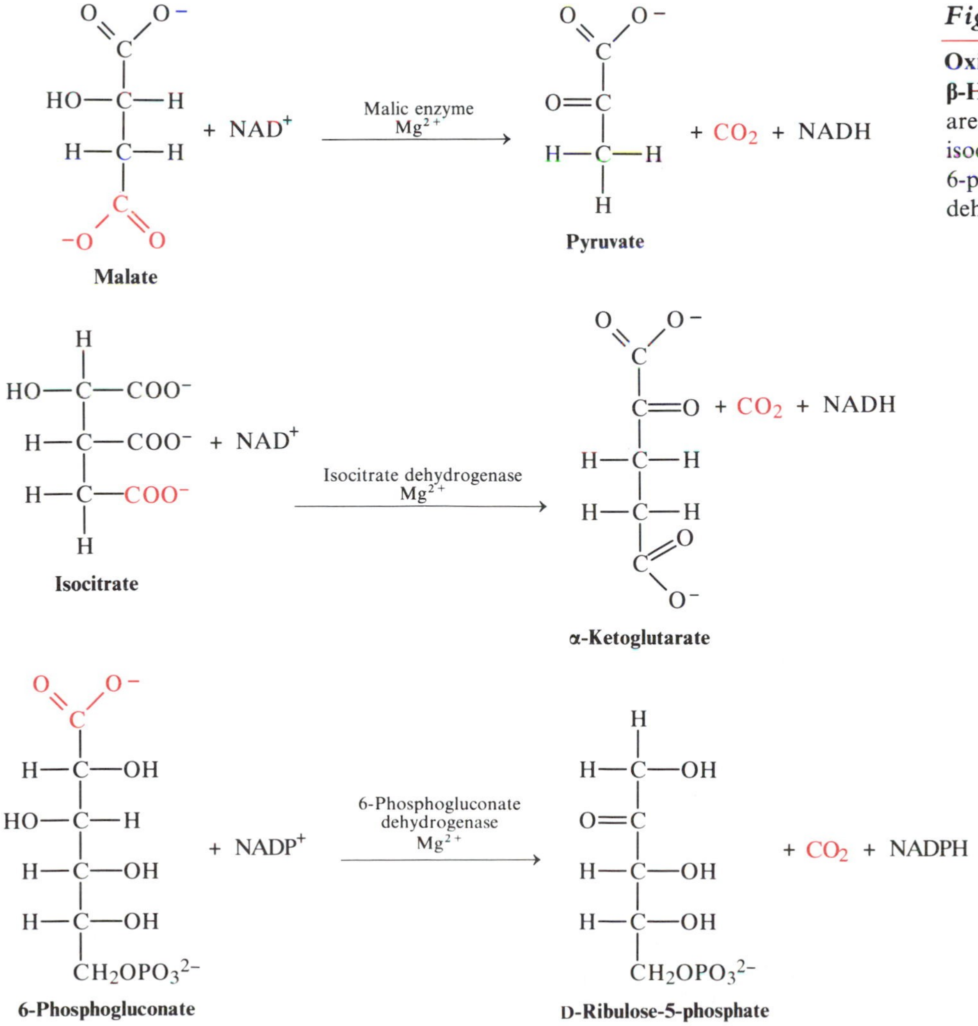

$$\mathrm{^-OOC{-}CH(OH){-}CH_2{-}COO^-} + NAD^+ \xrightarrow[\text{Mg}^{2+}]{\text{Malic enzyme}} \mathrm{^-OOC{-}CO{-}CH_3} + CO_2 + NADH$$

Malate → **Pyruvate**

$$\mathrm{HC(OH)(COO^-){-}HC(COO^-){-}CH_2{-}COO^-} + NAD^+ \xrightarrow[\text{Mg}^{2+}]{\text{Isocitrate dehydrogenase}} \mathrm{^-OOC{-}CO{-}CH_2{-}CH_2{-}COO^-} + CO_2 + NADH$$

Isocitrate → **α-Ketoglutarate**

$$\mathrm{^-OOC{-}CH(OH){-}CH(OH){-}CH(OH){-}CH(OH){-}CH_2OPO_3^{2-}} + NADP^+ \xrightarrow[\text{Mg}^{2+}]{\text{6-Phosphogluconate dehydrogenase}} \mathrm{CH_2OH{-}CO{-}CH(OH){-}CH(OH){-}CH_2OPO_3^{2-}} + CO_2 + NADPH$$

6-Phosphogluconate → **D-Ribulose-5-phosphate**

Figure 17-5

Oxidative Decarboxylation of β-Hydroxyacids. The reactions are catalyzed by malic enzyme, isocitrate dehydrogenase, and 6-phosphogluconate dehydrogenase.

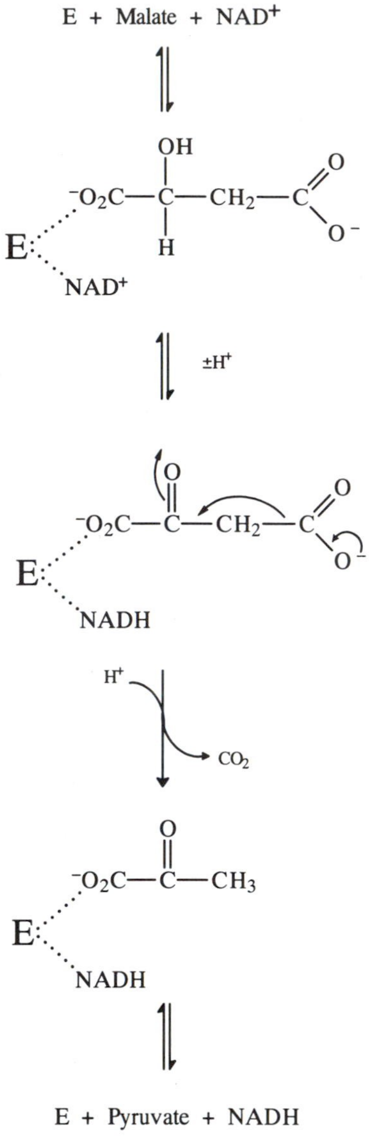

in all of these reactions, and an oxidation occurs. This suggests that oxidation precedes decarboxylation, so that the actual species that undergoes decarboxylation is a β-ketoacid; this is known to be the case for the reaction catalyzed by malic enzyme. A probable mechanism for the decarboxylation of malate is shown in Figure 17-6. The decarboxylation of isocitrate and 6-phosphogluconate is believed to be accomplished through similar mechanisms, in which the enzyme-bound substrate is first oxidized to the β-ketoacid, with NAD^+ or $NADP^+$ serving as the hydrogen acceptor. The β-ketoacid undergoes decarboxylation without being released from the enzyme.

The following experimental evidence supports the mechanism in Figure 17-6:

1. The enzymes catalyze the decarboxylation of the presumed β-ketoacid intermediates in the absence of NAD^+ (or $NADP^+$).
2. When the presumed β-ketoacid intermediates are added to the enzymes in the presence of NADH (NADPH), they are reduced to the corresponding β-hydroxyacids.

Thus, the malic enzyme catalyzes the dehydrogenation and decarboxylation in discrete steps; that is, dehydrogenation and decarboxylation do not occur in concert but are distinct reactions.

The enzymes that catalyze the dehydrogenations and decarboxylations of β-hydroxyacids do not form imines before decarboxylation. Instead, they require a divalent cation; and it is likely that this cation facilitates the decarboxylation through coordination with the β-carbonyl group, as illustrated in Figure 17-1. The metal provides positive charge to help stabilize the carbanion resulting from decarboxylation.

Figure 17-6

The Mechanism of the Oxidation and Decarboxylation of Malate Catalyzed by Malic Enzyme. The malic enzyme catalyzes the dehydrogenation of malate by NAD^+ to oxaloacetate and NADH. Enzyme-bound oxaloacetate undergoes β-decarboxylation to form pyruvate and CO_2, which are released from the enzyme along with NADH.

α-Ketoacids Also Are Subject to Decarboxylation

The decarboxylation of α-ketoacids occurs frequently in biological systems. It is not obvious that α-ketoacids should decarboxylate readily, because decarboxylation of these acids would not produce a stabilized carbanion. As we shall see, these compounds undergo a chemical modification before decarboxylation, which converts them into structures resembling β-ketoacids. This chemical modification is facilitated by a coenzyme, thiamine pyrophosphate (TPP), which is derived from the vitamin thiamine (vitamin B_1, Chapter 1).

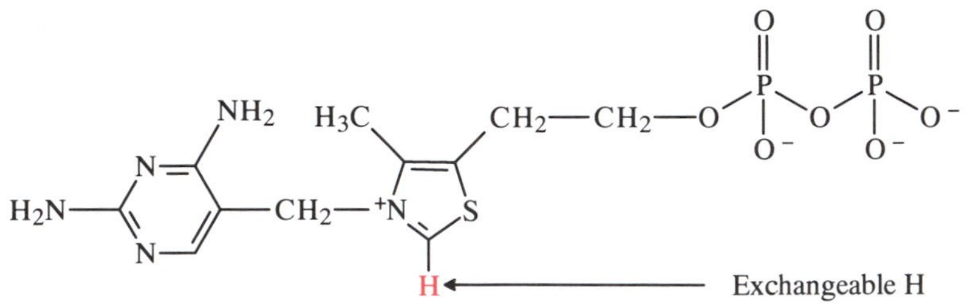

Thiamine pyrophosphate (TPP)

TPP is the cofactor in the decarboxylation of all α-ketoacids. As we shall see subsequently, these decarboxylations are of considerable importance in metabolism. Deficiency of Vitamin B_1 leads to a disease known as beri-beri, which causes damage to the heart and the peripheral nervous system. The disease is particularly prevalent in areas where rice is a major food. The situation is aggravated when rice is polished, because thiamine is primarily located in the outer layer of the rice kernel.

How does TPP function in decarboxylation of α-ketoacids? This question vexed biochemists for a long time. TPP is a rather complex molecule and can undergo a variety of chemical reactions. These properties of TPP led workers to propose a large number of incorrect mechanisms, utilizing the various chemical properties of TPP. The problem was finally solved by R. Breslow, who observed that the hydrogen bonded to carbon-2 of the thiazolium ring is readily exchangeable with solvent protons. This observation led him to postulate the mechanism that is now generally accepted. L. O. Krampitz applied Breslow's mechanism to the enzymatic reactions.

We shall first describe the mechanism for yeast pyruvate decarboxylase, which catalyzes the reaction of equation 8.

$$H_3C{-}C(=O){-}COO^- + H^+ \longrightarrow H_3C{-}C(=O){-}H + CO_2 \qquad (8)$$

The first step in the enzymatic mechanism is the reaction of pyruvate with enzyme-bound TPP to give an adduct according to equation 9.

$$H_3C{-}C(=O){-}COO^- + \text{R}{-}{}^+\text{N(thiazolium, C2–H)S} \rightleftharpoons \text{R}{-}{}^+\text{N(thiazolium)S}{-}C(OH)(CH_3){-}C(=O)O^- \qquad (9)$$

This adduct is well suited to decarboxylation. Note its structural similarity to a β-ketoacid. The structure —C=N$^+$—R is a much more effective electron sink than the carbonyl group, owing to the positive charge.

Loss of CO_2 from the thiamine-pyruvate adduct leads to the formation of a carbanion, shown in Figure 17-7, which becomes protonated to give hydroxyethyl-TPP (HETPP). HETPP undergoes an elimination reaction, which yields acetaldehyde and the TPP anion. In the final step, the anion is protonated to regenerate TPP. The following evidence supports the mechanism in Figure 17-7:

1. HETPP has been isolated from enzymatic reactions and has been chemically synthesized.
2. HETPP can function as an intermediate in the reaction catalyzed by pyruvate decarboxylase and in other reactions in which TPP and pyruvate take part.

Pyruvate-TPP

β-Ketoacid

An enzyme in yeasts, plants, and some microorganisms catalyzes the somewhat more complex reaction of equation 10, which is on the biosynthetic pathway to valine.

$$2\,H_3C{-}C(=O){-}COO^- + H^+ \longrightarrow H_3C{-}C(=O){-}C(OH)(CH_3){-}COO^- + CO_2 \qquad (10)$$

Acetolactate

The initial steps in the reaction are those shown for pyruvate decarboxylation, but the reaction diverges in that the HETPP anion is not protonated; instead it reacts with a second molecule of pyruvate, as illustrated in Figure 17-8, by

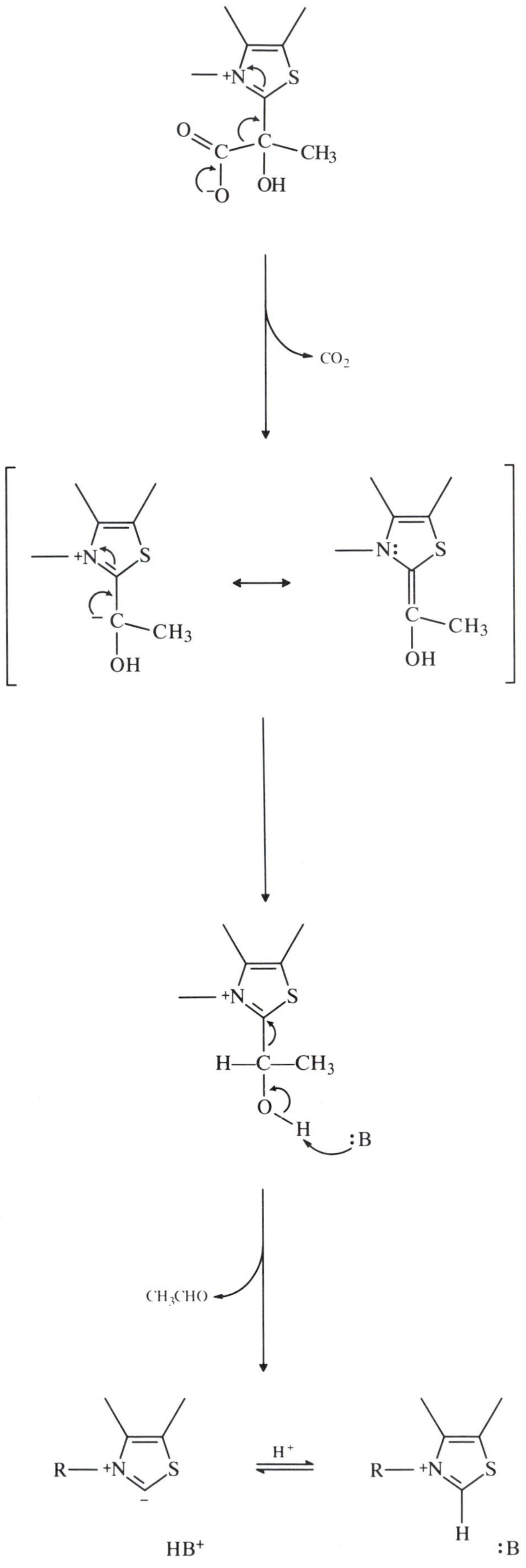

CO_2

CH_3

OH

CH_3CHO

R— +N S

HB+

H+

:B

Figure 17-7

The Mechanism by Which Pyruvate Decarboxylase and TPP Catalyze the Decarboxylation of Pyruvate. The adduct formed between TPP and pyruvate reacts as a β-ketoacid to undergo decarboxylation to hydroxyethylidene-TPP, a resonance stabilized carbanionic species, which is then protonated at carbon-1 to form HETPP. Elimination of TPP from HETPP produces acetaldehyde and TPP.

adding as a nucleophile to the carbonyl group of pyruvate. The resulting adduct then undergoes elimination of the TPP zwitterion in a process that is analogous to the production of acetaldehyde in Figure 17-7.

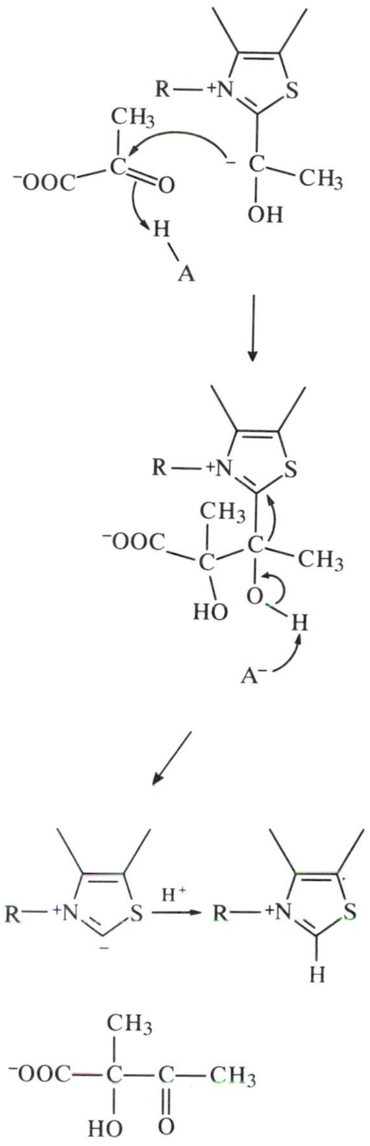

Oxidative Decarboxylation of α-Ketoacids

In aerobic organisms, pyruvate is not converted into acetaldehyde and CO_2; instead it is oxidatively decarboxylated to give acetyl CoA and CO_2 according to equation 11, in which the oxidizing agent is NAD^+.

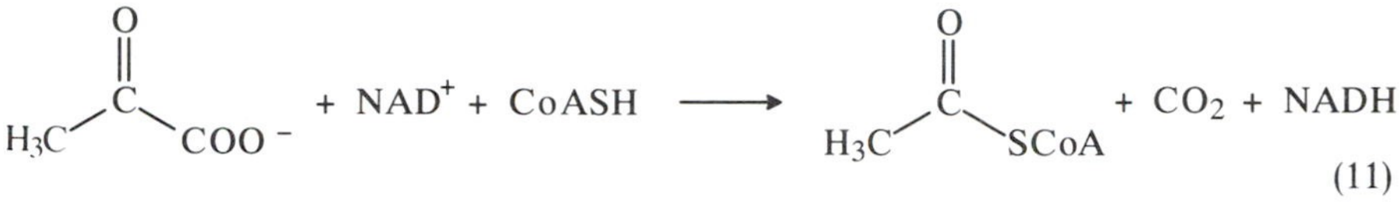

$$H_3C-C(=O)-COO^- + NAD^+ + CoASH \longrightarrow H_3C-C(=O)-SCoA + CO_2 + NADH \quad (11)$$

This reaction gives rise to the high-energy compound acetyl CoA, which is utilized in a number of metabolic reactions. The importance of this reaction in the metabolism of carbohydrates and fatty acids is explained in Chapters 19 and 21.

The oxidative decarboxylation of pyruvate is a multistep process that is carried out by an enzyme complex, the pyruvate dehydrogenase complex, consisting of three different enzymes, each of which participates in catalyzing the reaction. These enzymes are pyruvate dehydrogenase (E_1), dihydrolipoyl transacetylase (E_2), and dihydrolipoyl dehydrogenase (E_3). The complex also contains the coenzymes TPP, FAD, and lipoamide, a coenzyme that we have not yet encountered. Lipoic acid is covalently bound to dihydrolipoyl transacetylase through an amide linkage with the ε-NH_2 group of one of the lysine residues.

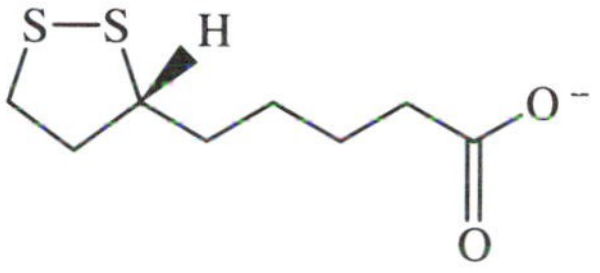

Lipoate 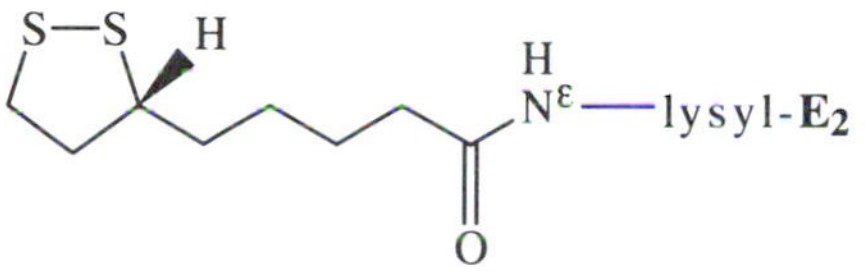 **Lipoyl-E_2**

The first enzyme in the pyruvate dehydrogenase complex to react with the substrate is pyruvate dehydrogenase (E_1). This enzyme catalyzes the decarboxylation of pyruvate, already described for pyruvate decarboxylase in Figure 17-7. Decarboxylation of the pyruvate-TPP adduct leads to the *anion* of HETPP, hydroxyethylidene-TPP, a resonance stabilized species shown in Figure 17-7, which reacts with a lipoyl group bound to the next enzyme in the complex, dihydrolipoyl transacetylase (E_2), according to Figure 17-9. In this reaction, TPP is regenerated and S^8-acetyldihydrolipoamide is formed.

The reaction between hydroxyethylidene-TPP and lipoamide includes an oxidation-reduction, as well as the transfer of the two-carbon moiety to lipoamide. The hydroxyethylidene group becomes oxidized and lipoamide is reduced. Acetyl-TPP is transiently generated in this process; however, the exact mechanism by which this reaction occurs is not known.

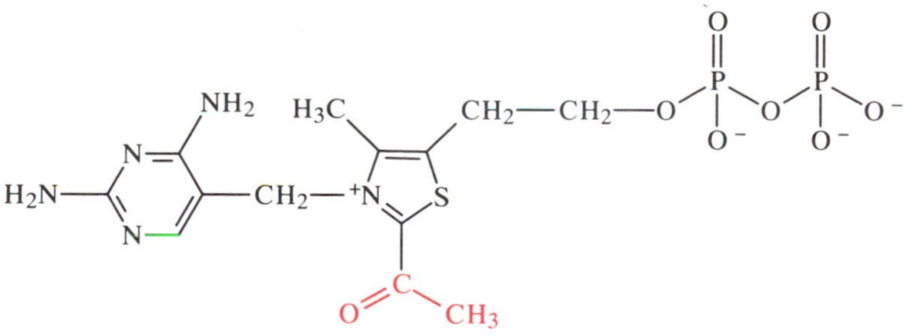

Acetyl-TPP

Figure 17-8

The Mechanism of Acetolactate Formation. Enzyme-bound hydroxyethylidene-TPP resulting from the decarboxylation of pyruvate reacts as a carbanion and adds to the carbonyl group of the second molecule of pyruvate to form an acetolactate-TPP adduct. Elimination of TPP from this latter species produces acetolactate and TPP.

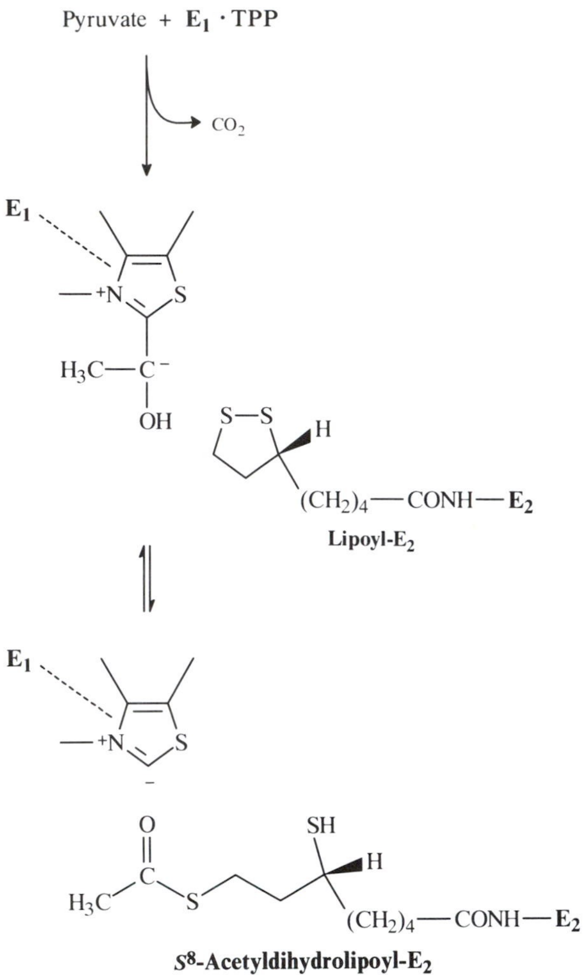

Pyruvate + $E_1 \cdot TPP$

CO_2

E_1

H_3C—C⁻ OH

S—S H

$(CH_2)_4$—CONH—E_2

Lipoyl-E_2

E_1

O

SH H

H_3C C S

$(CH_2)_4$—CONH—E_2

S^8-Acetyldihydrolipoyl-E_2

Figure 17-9

Reductive Transacetylation Catalyzed by Pyruvate Dehydrogenase (E_1). The intermediate hydroxyethylidene-TPP from the decarboxylation of pyruvate reacts with a lipoyl moiety bound to dihydrolipoyl transacetylase (E_2) to form S^8-acetyldihydrolipoyl-E_2 and TPP. Acetyl-TPP is transiently formed in this process, the exact mechanism of which is unknown.

Dihydrolipoyl transacetylase then catalyzes acetyl group transfer to CoA according to equation 12.

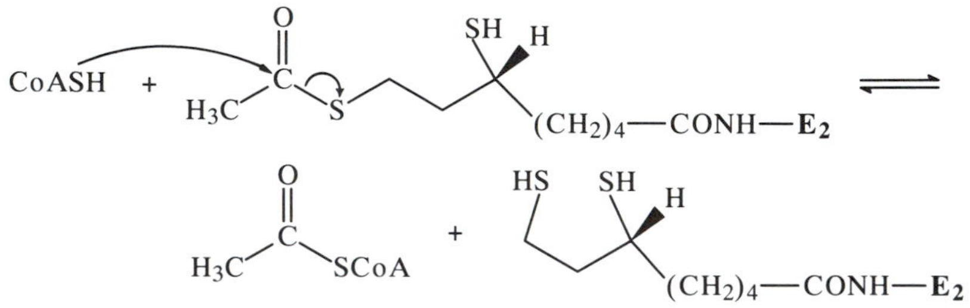

$$\text{CoASH} + \text{H}_3\text{C}\text{—C(=O)—S—}\ldots\text{(SH)(H)}(\text{CH}_2)_4\text{—CONH—E}_2 \rightleftharpoons \text{H}_3\text{C—C(=O)—SCoA} + \text{HS}\ldots\text{(SH)(H)}(\text{CH}_2)_4\text{—CONH—E}_2 \quad (12)$$

This leads to the formation of reduced lipoyl groups and acetyl CoA, which is released from the enzyme complex.

Reduced lipoamide must be oxidized to bring it back to its original oxidation state in order to catalyze another reaction cycle. The reoxidation of reduced lipoamide is catalyzed by dihydrolipoyl dehydrogenase, a flavoprotein, according to equation 13. The reduced flavin resulting from the oxidation of lipoamide is reoxidized by NAD^+ (equation 14).

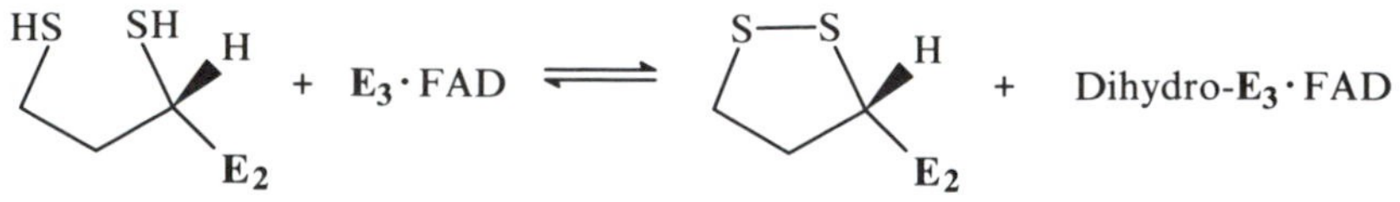

$$\text{HS, SH (H) E}_2 + \text{E}_3\cdot\text{FAD} \rightleftharpoons \text{S—S (H) E}_2 + \text{Dihydro-E}_3\cdot\text{FAD} \quad (13)$$

$$\text{Dihydro-}\mathbf{E_3}\cdot\text{FAD} + \text{NAD} \rightleftharpoons \mathbf{E_3}\cdot\text{FAD} + \text{NADH} + \text{H}^+ \qquad (14)$$

The pyruvate dehydrogenase complexes from mammalian and bacterial sources have complex structures. The architecture of these complexes has been elucidated largely through the work of L. J. Reed and his colleagues. The *E. coli* complex has a molecular weight of about 5 million. The enzymes are assembled in an ordered mosaic in a cubic morphology, as illustrated in Figure 17-10. The electron microscopic images in Figure 17-11 reveal the cubic morphology of the complex. Dihydrolipoyl transacetylase forms the core of this structure and the other enzymes are arranged around this core.

It is possible to dissociate this complex into the component enzymes. At pH 9.2, E_1 spontaneously dissociates and can be separated from the subcomplex E_2E_3. Upon brief treatment with 4 M urea, the subcomplex E_2E_3 dissociates into E_3 and the 24-subunit core of E_2, which can be separated by chromatography. The complex spontaneously undergoes reassembly when the components are mixed in the correct proportions.

In the course of catalysis, no reaction intermediates are released from the pyruvate dehydrogenase complex. It is, therefore, essential that the enzymes interact with one another within the complex. The two-carbon fragment of

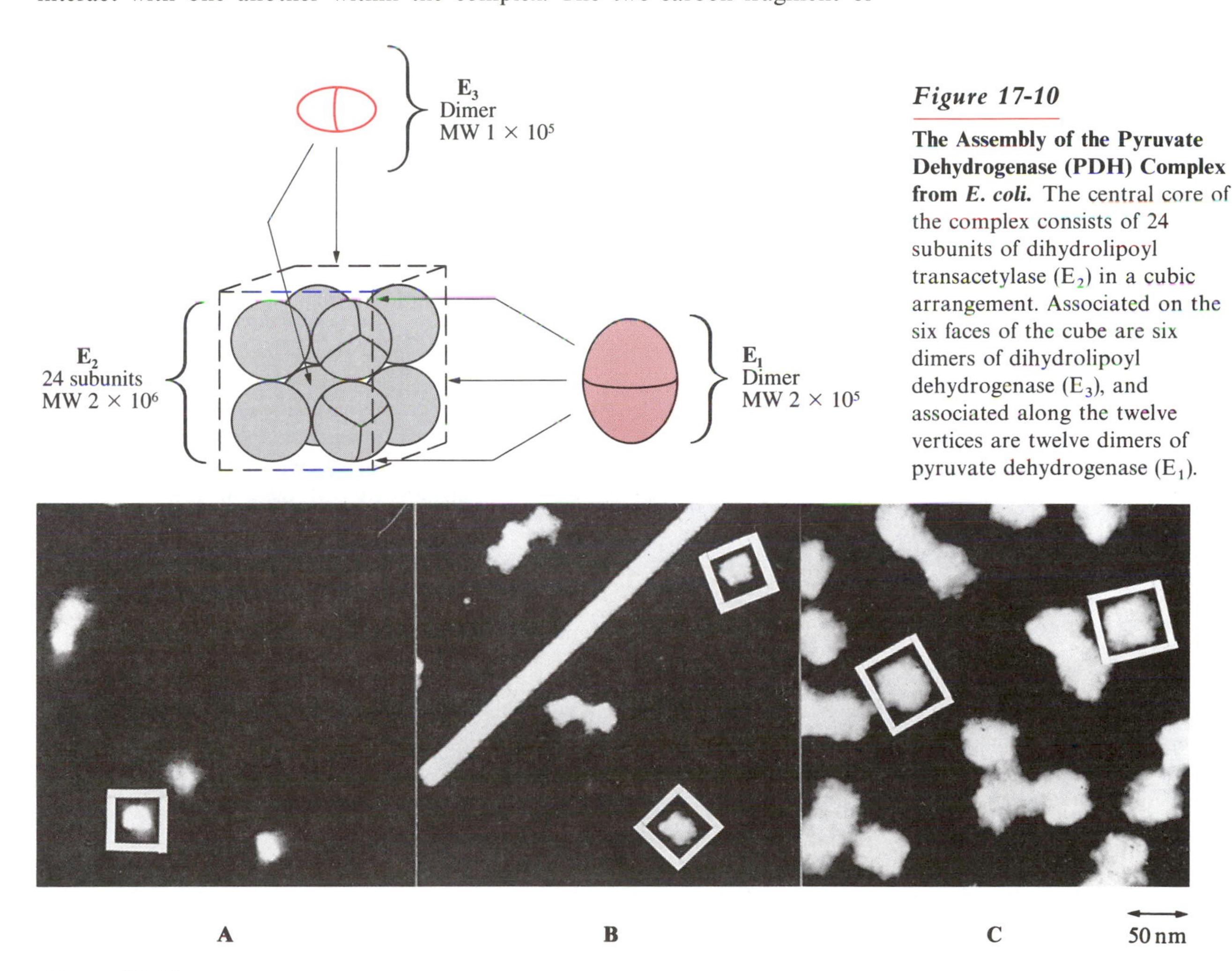

Figure 17-10

The Assembly of the Pyruvate Dehydrogenase (PDH) Complex from *E. coli*. The central core of the complex consists of 24 subunits of dihydrolipoyl transacetylase (E_2) in a cubic arrangement. Associated on the six faces of the cube are six dimers of dihydrolipoyl dehydrogenase (E_3), and associated along the twelve vertices are twelve dimers of pyruvate dehydrogenase (E_1).

Figure 17-11

Scanning Transmission Electron Microscopic Images of the PDH Complex from *E. coli*. (A) E_2; (B) E_2E_3 subcomplex; (C) PDH complex. In part B, the linear structure is tobacco mosaic virus, an internal size standard. (Courtesy of H. Yang. J. Hainfeld. and J. Wall.)

hydroxyethylidene-TPP (Figure 17-7) must be passed from thiamine on E_1 to lipoamide bound to E_2 (Figure 17-9), the S^8-acetyldihydrolipoamide must transfer its acetyl group to CoA at an active site on E_2 (equation 12), and the resulting dihydrolipoyl moiety must be able to interact with FAD associated with E_3 (equation 13). The three active sites are far apart in molecular space, $\geqslant 50$ Å between the TPP-site on E_1 and FAD on E_3. The structural flexibility of the hydrocarbon side chains of lipoamide and lysine in the lipoyl-N^{ε}-lysyl moiety bound to E_2 is undoubtedly important in these interactions. Owing to free rotation about most of the bonds in the lipoyl-N^{ε}-lysyl side chains, the system acts as a flexible arm in facilitating the movement of the reactive sulfur-containing part of the coenzyme among the active sites.

The pyruvate dehydrogenase complex from mammalian sources is similar to the bacterial complex, but it contains a larger number of constituent enzymes. The E_2-core consists of 60 subunits in a polyhedral arrangement. Moreover, the mammalian complex contains additional regulatory enzymes, a protein kinase and protein phosphatase, that regulate the activity of the complex by a process of phosphorylation and dephosphorylation of serine residues in E_1. Regulation by covalent modification is described in Chapter 21.

In addition to the pyruvate dehydrogenase complex, other complexes catalyze the decarboxylation and dehydrogenation of α-ketoacids. These complexes are structurally very similar to the pyruvate dehydrogenase complex. For example, a complex exists in bacteria, as well as in higher organisms, that catalyzes the decarboxylation and dehydrogenation of α-ketoglutarate according to equation 15.

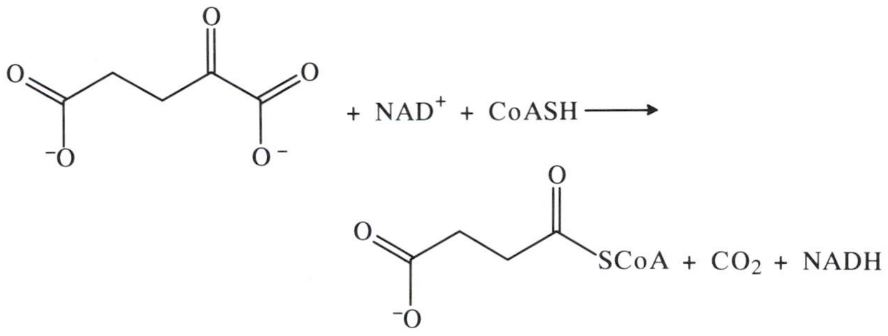

$$\text{α-ketoglutarate} + NAD^+ + CoASH \longrightarrow \text{succinyl-SCoA} + CO_2 + NADH \quad (15)$$

Figure 17-12

Metabolism of Branched-Chain Amino Acids. The amino acids leucine, isoleucine, and valine are degraded by conversion into the corresponding α-ketoacids, which are substrates for the branched-chain α-ketoacid dehydrogenase complex.

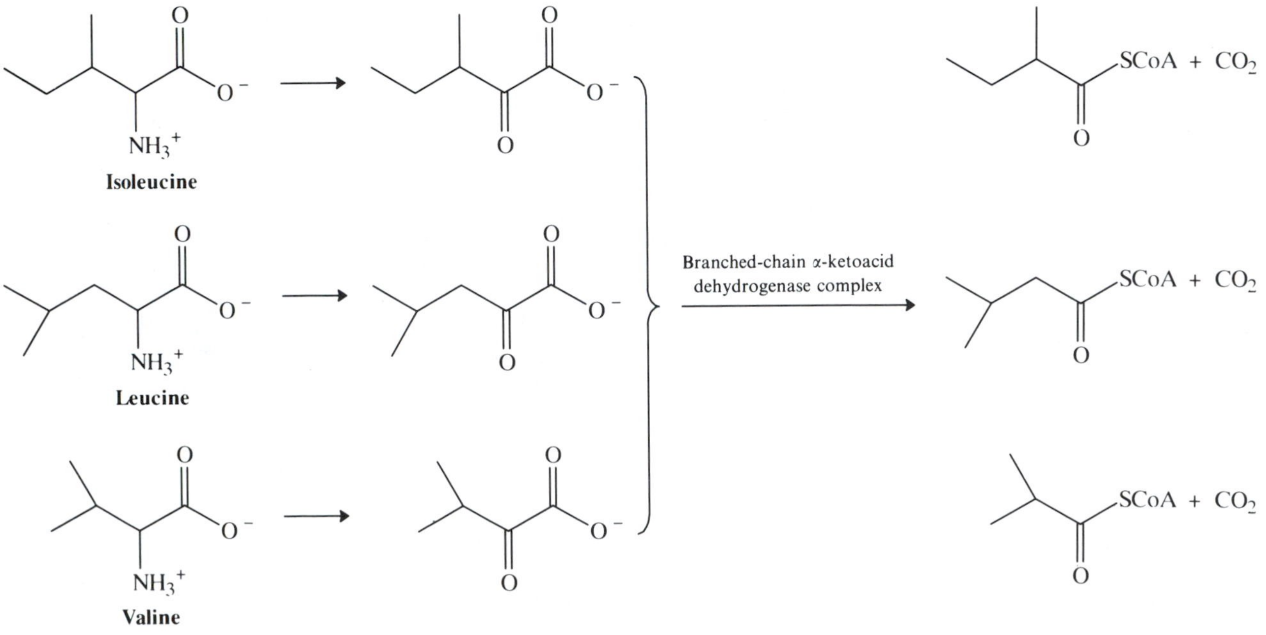

BOX 17-1

MAPLE SYRUP URINE DISEASE AND LACTIC ACIDEMIA

In the genetic disease known as maple syrup urine disease (the disease is so called because the urine and other body secretions emit an odor similar to that of maple syrup), the branched-chain α-ketoacid dehydrogenase complex is inactive, and the α-ketoacids accumulate and are excreted in the urine. The disease usually manifests itself near the first week of an infant's life and results in feeding difficulties and neurological symptoms. Unless therapy is started, infants suffering from this disease survive at most for a few months. The disease is treated by placing the infant on a diet in which the intake of branched-chain amino acids is reduced.

Genetic deficiencies are also the cause of *lactic acidemia,* a human condition caused by the inactivity of the pyruvate dehydrogenase complex. Failure of the body to metabolize pyruvate to acetyl CoA leads to its reduction by NADH through the action of lactate dehydrogenase to large amounts of lactate, which is excreted. The metabolic pathways to pyruvate and lactate are described and explained in Chapter 21.

The branched-chain α-ketoacid dehydrogenase complex catalyzes the decarboxylation of α-ketoacids, which are metabolically derived from the amino acids leucine, isoleucine, and valine. This process is illustrated in Figure 17-12. Some people are unable to metabolize α-ketoacids derived from branched amino acids. This condition is described in Box 17-1.

The Decarboxylation of Amino Acids

An important reaction in amino acid metabolism is the decarboxylation of amino acids according to equation 16:

$$R{-}CH(NH_3^+){-}COO^- + H^+ \longrightarrow R{-}CH_2{-}NH_3^+ + CO_2 \qquad (16)$$

Decarboxylases for many amino acids are found in microorganisms. In higher organisms, amino acid decarboxylation gives rise to biologically important molecules. For example, the decarboxylation of dihydroxyphenylalanine (DOPA) gives rise to dopamine according to equation 17.

$$\text{DOPA} + H^+ \longrightarrow \text{Dopamine} + CO_2 \qquad (17)$$

DOPA: $(HO)_2C_6H_3{-}CH_2{-}CH(NH_3^+){-}COO^-$; Dopamine: $(HO)_2C_6H_3{-}CH_2{-}CH_2{-}NH_3^+$

Dopamine is a neurotransmitter and a precursor of norepinephrine, another neurotransmitter in the central and peripheral nervous system. Dopamine and norepinephrine are also intermediates in the biosynthesis of the hormone epinephrine in chromaffin cells of the adrenal medula.

In Parkinson's disease, there appears to be a deficiency of dopamine in the brain. This deficiency cannot be corrected simply through the administration of dopamine, which does not pass through the "blood-brain barrier." Relief can be obtained in many cases by the administration of L-DOPA, which penetrates the blood-brain barrier and is converted into dopamine in the brain through the action of DOPA decarboxylase.

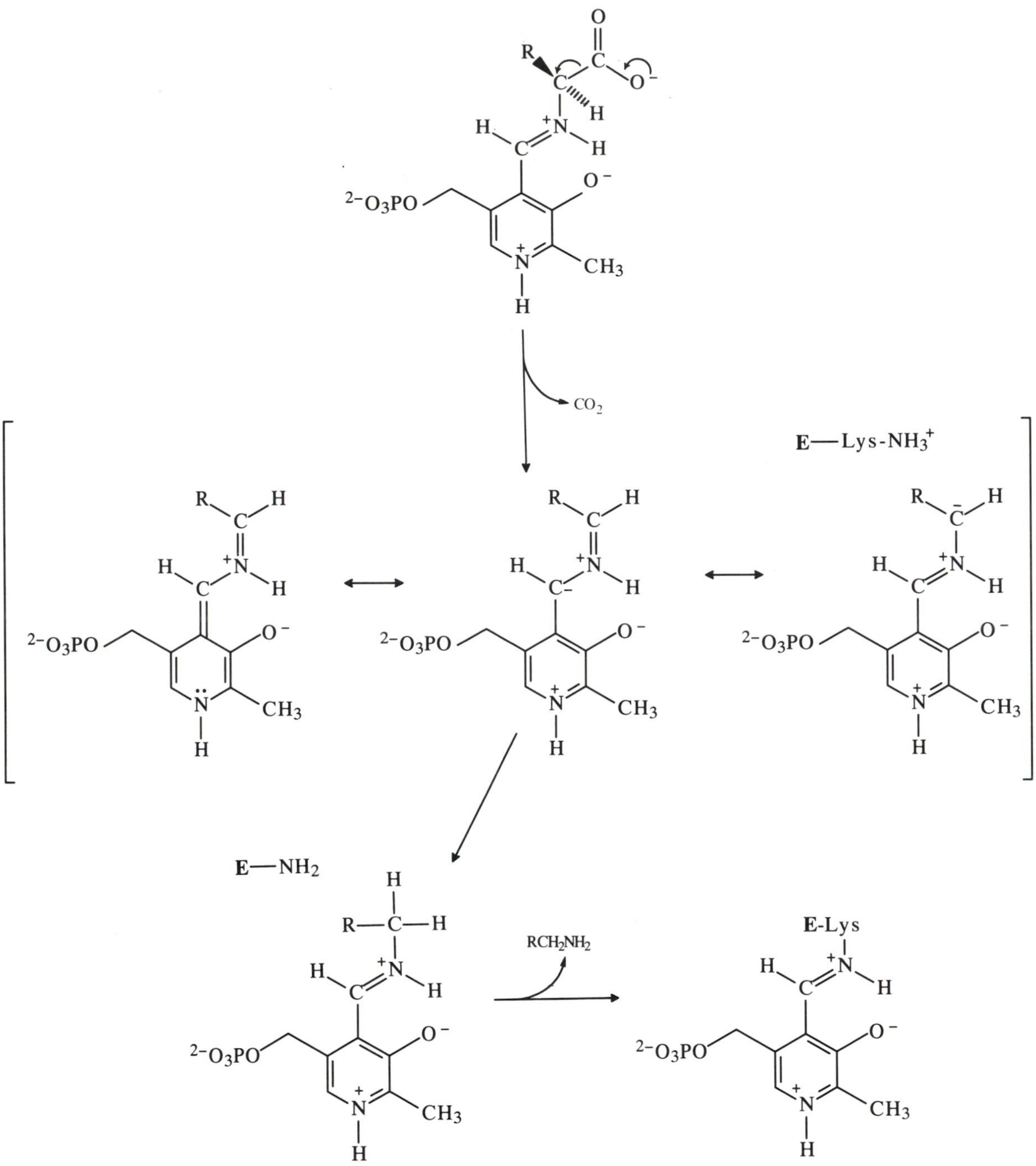

Figure 17-13

The Mechanism of PLP-catalyzed Decarboxylation of Amino Acids.

Glutamate is decarboxylated by glutamate decarboxylase to γ-aminobutyrate (GABA) according to equation 18.

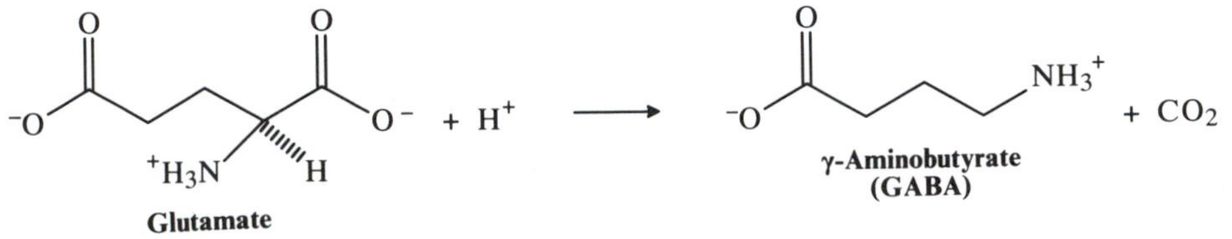

$$\text{Glutamate} + H^+ \longrightarrow \gamma\text{-Aminobutyrate (GABA)} + CO_2 \quad (18)$$

GABA is found almost exclusively in the central nervous system and probably functions as an inhibitory neurotransmitter.

Ornithine decarboxylase is a widely distributed enzyme that markedly increases in activity during cellular reproduction. It is thought that spermine, the product of the decarboxylation of ornithine (equation 19), as well as polyamines derived from spermine, are required for DNA replication, although the exact role of these compounds is unknown.

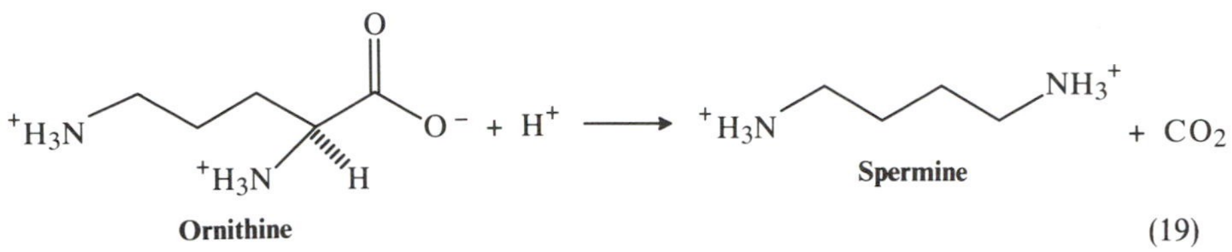

Ornithine + H^+ ⟶ **Spermine** + CO_2 (19)

It is apparent that the decarboxylation of amino acids presents a chemical problem because no electron sink is available to stabilize the carbanion resulting from the decarboxylation. From Chapter 2, we know that pyridoxal phosphate (PLP) stabilizes a carbanion on the α-carbon atom of amino acids. That is exactly what is needed here, a means to stabilize the carbanion formed when the amino acid undergoes decarboxylation.

With these considerations in mind, the mechanism of decarboxylation shown in Figure 17-13 is reasonable; most amino acid decarboxylases depend for their activity on PLP. The initial step in the mechanism is the reaction of the amino group of the substrate with the aldehyde group of PLP to form an aldimine. As explained in Chapter 2 (Figure 2-23), PLP is itself bound to enzymes through aldimine formation with the ε-NH_2 group of a lysine residue at the active site. This binding mode facilitates aldimine formation between PLP and an amino acid. Aldimine formation with PLP is the "transimination" shown in Figure 17-14, and it differs mechanistically from the mechanism of imine formation in Figure 17-3. The first step of transimination is nucleophilic addition of the amino group to a protonated imine, which carries a full positive charge on nitrogen and is highly reactive. The analogous step in the mechanism of imine formation in Figure 17-3 is addition of an amine to a carbonyl group, which is also reactive as an electrophile but does not carry a full positive charge. A carbonyl group is less reactive than a protonated imine. Thus, PLP bound as a protonated aldimine is highly reactive in aldimine formation with amino acids.

Decarboxylation of the PLP-amino acid aldimine leads to a resonance stabilized carbanion, shown in Figure 17-13. This carbanion is more stable than that derived from β-ketoacids; hence, PLP is a very effective catalyst for the decarboxylation of amino acids. In the last stage of the reaction, the carbanion resulting from the decarboxylation is protonated, probably by a proton furnished by a general acid at the active site, perhaps by the protonated lysine. The product-PLP aldimine then reacts with the ε-NH_2 group of lysine in a reverse transimination to regenerate the enzyme-PLP aldimine and release the product.

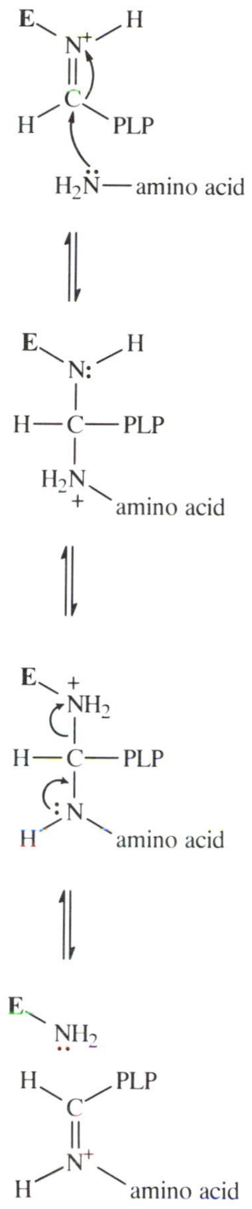

E—N^+(H)=C(H)—PLP + $H_2\ddot{N}$—amino acid ⇌ E—N(H)—C(H)(PLP)—$\overset{+}{N}H_2$—amino acid ⇌ E—$\overset{+}{N}H_2$—C(H)(PLP)—N(H)—amino acid ⇌ E—$\ddot{N}H_2$ + H(PLP)C=N^+(H)—amino acid

Figure 17-14

The Mechanism of Aldimine Formation Between Enzyme-bound PLP and Amino Acids.

Carboxylation

There are several important biological reactions in which carboxylic acids are formed through carboxylation reactions. In all carboxylations, a carbanion reacts with CO_2 or, in some cases, an enzyme-bound form of CO_2. This process is described by equation 20.

$$\mathrm{R{-}\overset{|}{\underset{|}{C}}{}^{-} + O{=}C{=}O \longrightarrow R{-}\overset{|}{\underset{|}{C}}{-}C(=O)O^{-}} \quad (20)$$

(R = a group that stabilizes a carbanion)

Therefore, only molecules that can form relatively stable carbanions are substrates for carboxylation reactions.

As we already know, the groups most commonly found in biological molecules that stabilize carbanions are the carbonyl groups of esters and ketones; and all biological carboxylations occur α to these groups. As we shall see, there are two types of carboxylation reactions, those that require the coenzyme biotin and those that do not. We shall first discuss the carboxylations that do not require a coenzyme.

Phosphoenolpyruvate (PEP) Is a Substrate for Carboxylation

The enzyme phosphoenolpyruvate carboxykinase, an important enzyme in carbohydrate metabolism, catalyzes the carboxylation of PEP to oxaloacetate according to equation 21.

$$\mathrm{H_2C{=}C(OPO_3^{2-}){-}C(=O)O^{-} + GDP + CO_2 \rightleftharpoons {}^{-}O{-}C(=O){-}CH_2{-}C(=O){-}C(=O)O^{-} + GTP} \quad (21)$$

Phosphoenolpyruvate **Oxaloacetate**

The enzyme is widely distributed in mammals and microorganisms. In one microorganism (*Propionibacterium shermannii*), an interesting variant of this reaction occurs in which phosphate reacts in place of GDP and pyrophosphate is formed in place of GTP. In this bacterium, PP_i can serve as a high-energy compound in place of ATP or GTP.

$$\mathrm{H_2C{=}C(OPO_3^{2-}){-}C(=O)O^{-} + P_i + CO_2 \rightleftharpoons {}^{-}O{-}C(=O){-}CH_2{-}C(=O){-}C(=O)O^{-} + PP_i} \quad (22)$$

The exact mechanism of the carboxylation in these reactions is unknown, but the available data suggest that the enzyme initially catalyzes a reaction between phosphoenolpyruvate and GDP (or P_i) to give the enzyme-bound enolate of pyruvate and GTP (Figure 17-15). GTP is then released and the enzyme-bound enolate reacts with CO_2 to form oxaloacetate. If this mechanism is correct, the enzyme must be able to stabilize the intermediate enolate in some way. A likely possibility is that the enzyme can supply positive centers that interact with the negative centers of the enolate. Exactly how the enolate is stabilized is not known.

Ribulose-1,5-bisphosphate Carboxylase

Another enzyme that catalyzes a carboxylation reaction is ribulose-1,5-bisphosphate carboxylase. This enzyme catalyzes CO_2 fixation in photosynthesis according to equation 23.

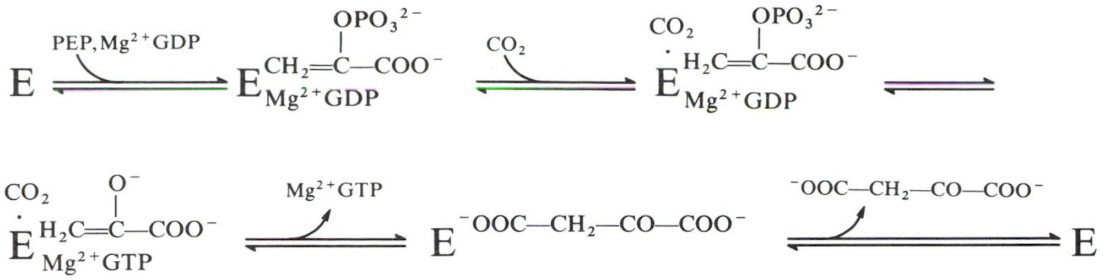

$$E \xrightleftharpoons{PEP, Mg^{2+}GDP} E\begin{matrix} CH_2{=}C(OPO_3^{2-}){-}COO^- \\ Mg^{2+}GDP \end{matrix} \xrightleftharpoons{CO_2} E\begin{matrix} CO_2 \\ \cdot\, H_2C{=}C(OPO_3^{2-}){-}COO^- \\ Mg^{2+}GDP \end{matrix} \rightleftharpoons$$

$$E\begin{matrix} CO_2 \\ \cdot\, H_2C{=}C(O^-){-}COO^- \\ Mg^{2+}GTP \end{matrix} \xrightleftharpoons{Mg^{2+}GTP} E\ ^-OOC{-}CH_2{-}CO{-}COO^- \xrightleftharpoons{^-OOC{-}CH_2{-}CO{-}COO^-} E$$

Figure 17-15

The Mechanism of Carboxylation by PEP Carboxykinase. Phosphoenolpyruvate and the Mg^{2+}-complex of GDP bind to form a ternary complex. Phosphoryl transfer from phosphoenolpyruvate to MgGDP produces MgGTP and the enolate of pyruvate, which is released, and the enzyme binds CO_2. The enolate reacts as a nucleophile in an addition reaction with the electrophilic carbon of CO_2 in the carboxylation step to form oxaloacetate, which is released in the last step to regenerate the free enzyme.

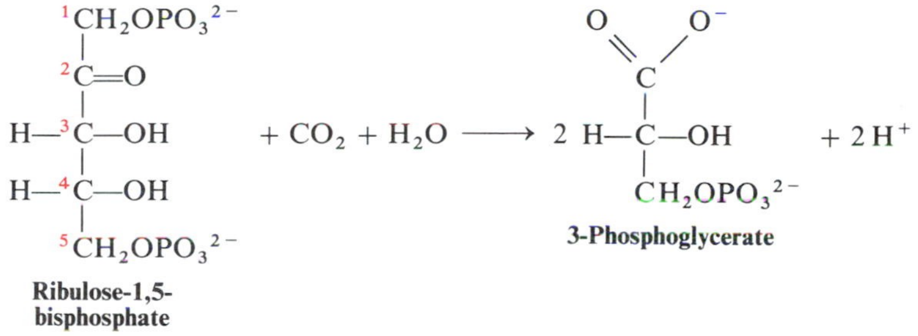

$$\underset{\textbf{Ribulose-1,5-bisphosphate}}{\begin{matrix} {}^1CH_2OPO_3^{2-} \\ | \\ {}^2C{=}O \\ | \\ H{-}{}^3C{-}OH \\ | \\ H{-}{}^4C{-}OH \\ | \\ {}^5CH_2OPO_3^{2-} \end{matrix}} + CO_2 + H_2O \longrightarrow 2\ \underset{\textbf{3-Phosphoglycerate}}{\begin{matrix} O{=}C{-}O^- \\ | \\ H{-}C{-}OH \\ | \\ CH_2OPO_3^{2-} \end{matrix}} + 2\,H^+ \qquad (23)$$

The enzyme is in the green leaves of plants and is the most-abundant protein in the world. The reaction includes not only a carboxylation, but also carbon–carbon bond cleavage. How does this reaction occur? Because the final products are two molecules of 3-phosphoglycerate, it is very likely that its immediate precursor is a six-carbon compound that can be cleaved into these two molecules. Structure A in equation 24 represents such a molecule.

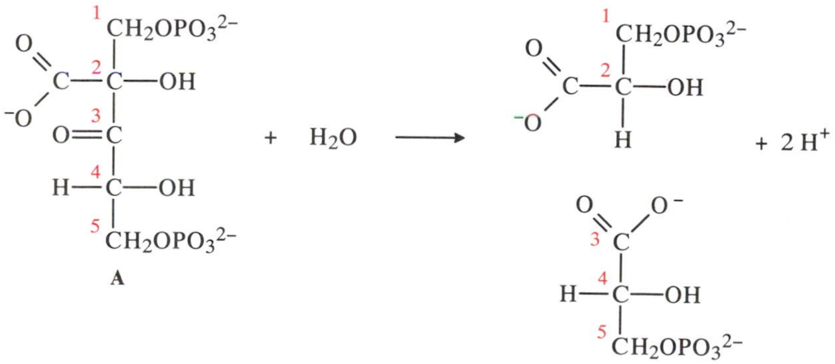

$$\underset{\textbf{A}}{\begin{matrix} {}^1CH_2OPO_3^{2-} \\ | \\ ^-O(O{=})C{-}{}^2C{-}OH \\ | \\ O{=}{}^3C \\ | \\ H{-}{}^4C{-}OH \\ | \\ {}^5CH_2OPO_3^{2-} \end{matrix}} + H_2O \longrightarrow \begin{matrix} {}^1CH_2OPO_3^{2-} \\ | \\ ^-O(O{=})C{-}{}^2C{-}OH \\ | \\ H \end{matrix} + \begin{matrix} O{=}{}^3C{-}O^- \\ | \\ H{-}{}^4C{-}OH \\ | \\ {}^5CH_2OPO_3^{2-} \end{matrix} + 2\,H^+ \qquad (24)$$

It can be cleaved between C-2 and C-3 to give two molecules of 3-phosphoglycerate.

How do we obtain molecule A? It can be obtained by carboxylation of carbanion B in equation 25.

Figure 17-16

The Mechanism by Which Ribulose-1,5-bisphosphate Is Carboxylated by the Action of Ribulose Bisphosphate Carboxylase. The first chemical step is enolate formation at carbon-3, through base-catalyzed proton abstraction, and this is followed by proton transfer in the second step to the carbonyl group. The 3-hydroxyl group of the resulting enediol is deprotonated to form a carbanion at carbon-2, which undergoes nucleophilic addition to CO_2 to form the intermediate 2-carboxyrib-3-ulose-1,5-bisphosphate. The hydrate of this intermediate is generated and, by deprotonation of the hydrate and cleavage between carbons 2 and 3, forms two molecules of 3-phosphoglycerate, with the intermediate formation of the enolate at carbon-2.

(25)

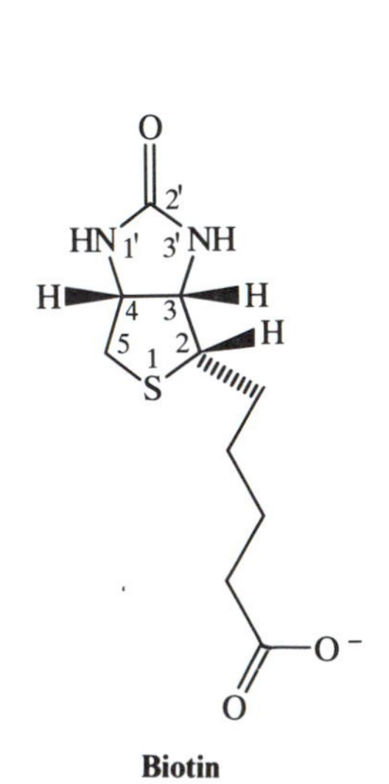

Biotin

Elementary chemical reactions can convert ribulose-1,5-bisphosphate into carbanion B. We are now in a position to write a complete reaction sequence for the reaction catalyzed by ribulose bisphosphate carboxylase (Figure 17-16). The structure of ribulose bisphosphate carboxylase is shown in Figure 17-17.

Biotin-dependent Carboxylation Reactions

Biotin is considered to be a vitamin but, unlike the other vitamins, there is no nutritional requirement for biotin in animals because it is synthesized by intestinal bacteria. Biotin deficiencies, therefore, are rare. However, a deficiency

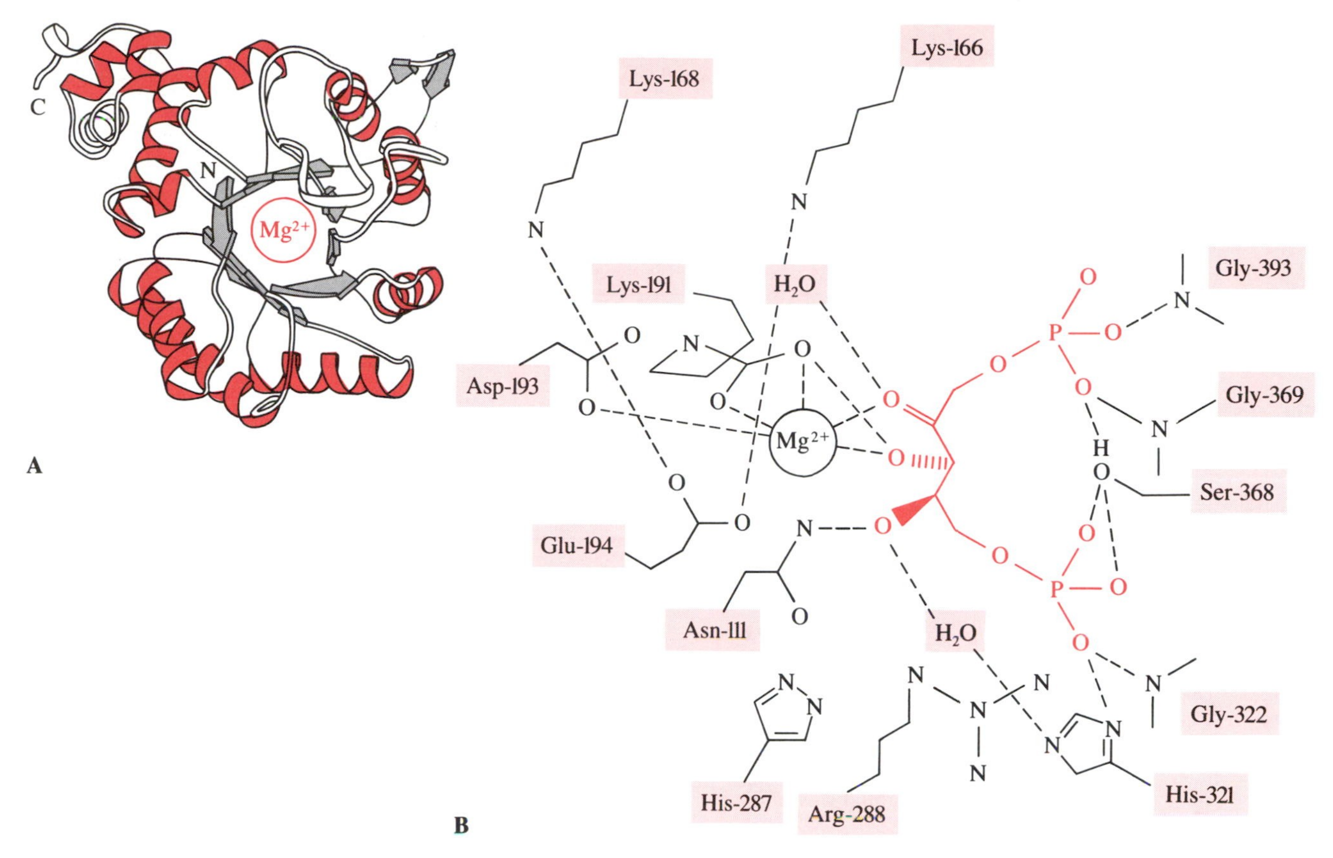

Figure 17-17

The Structure of Ribulose Bisphosphate Carboxylase. The ribbon drawing of the peptide chain of ribulose bisphosphate carboxylase from the photosynthetic bacterium *Rhodospirillum rubrum* (part A) shows the location of Mg^{2+} at the active site. The schematic view of ribulose bisphosphate bound at the active site (part B) shows its interactions with enzymatic residues. [After T. Lundqvist and G. Schneider. Crystal structure of activated ribulose-1,5-bisphosphate carboxylase complexed with its substrate, ribulose-1,5-bisphosphate. *J. Biol. Chem.* 266 (1991): 12604–12611.]

can be *induced* in experimental animals by feeding the protein avidin, which is obtained from eggs. This protein binds biotin extremely tightly and prevents it from carrying out its coenzymatic function.

Biotin is covalently bound to carboxylases through its carboxyl group, which forms an amide linkage with the ε-NH_2 group of a lysine residue.

Examples of biotin-dependent carboxylation reactions are shown in Figure 17-18. Note that carboxylation always takes place adjacent to a carbonyl group, which stabilizes the required carbanion. These reactions differ from the carboxylation reactions described earlier in two important respects: (1) they are coupled to the hydrolysis of ATP; and (2) the carboxylating agent is HCO_3^- rather than CO_2.

The participation of ATP in these reactions suggests that the hydrolysis of ATP is somehow coupled to the carboxylation. How is this accomplished? Before answering this question, we shall consider the carboxylation reaction in more detail. These reactions proceed in two steps, the partial reactions of equations 26 and 27.

$$HCO_3^- + \text{Enzyme} + \text{ATP} \rightleftharpoons \text{Enzyme—}CO_2 + \text{ADP} + P_i \quad (26)$$

$$\text{Enzyme—}CO_2 + S \rightleftharpoons S\text{—}CO_2 + \text{Enzyme} \quad (27)$$

Biotin
Lysine
Carboxylase

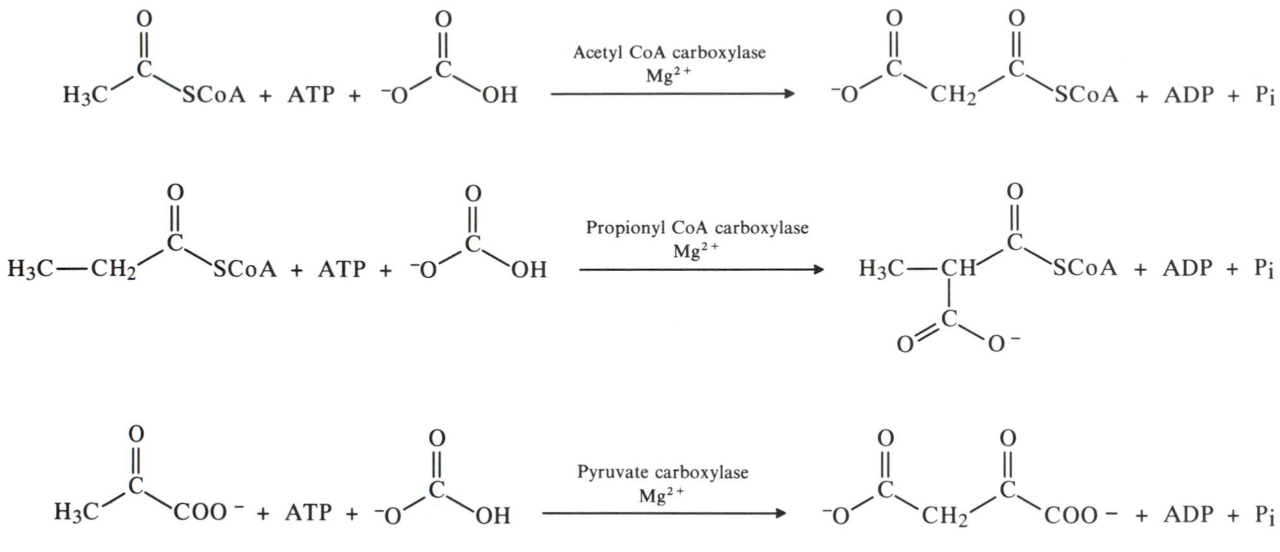

$$H_3C\text{—}C(=O)\text{—}SCoA + ATP + {}^{-}O\text{—}C(=O)\text{—}OH \xrightarrow[\text{Mg}^{2+}]{\text{Acetyl CoA carboxylase}} {}^{-}O\text{—}C(=O)\text{—}CH_2\text{—}C(=O)\text{—}SCoA + ADP + P_i$$

$$H_3C\text{—}CH_2\text{—}C(=O)\text{—}SCoA + ATP + {}^{-}O\text{—}C(=O)\text{—}OH \xrightarrow[\text{Mg}^{2+}]{\text{Propionyl CoA carboxylase}} H_3C\text{—}CH(COO^{-})\text{—}C(=O)\text{—}SCoA + ADP + P_i$$

$$H_3C\text{—}C(=O)\text{—}COO^{-} + ATP + {}^{-}O\text{—}C(=O)\text{—}OH \xrightarrow[\text{Mg}^{2+}]{\text{Pyruvate carboxylase}} {}^{-}O\text{—}C(=O)\text{—}CH_2\text{—}C(=O)\text{—}COO^{-} + ADP + P_i$$

Figure 17-18

Three Biotin-dependent Carboxylation Reactions. In all biotin-dependent carboxylation reactions, the carboxylating substrate is bicarbonate, and ATP supplies the free energy required to dehydrate bicarbonate. Biotin maintains the carboxyl group in a chemically reactive form as N^1-carboxybiotin (see equations 26 and 27).

The existence of these two partial reactions was established by the following experiment. Acetyl CoA carboxylase was incubated with $H^{14}CO_3^-$ and ATP in the absence of acetyl CoA. The reaction mixture was then passed through a Sephadex column and effluent fractions were collected. (Recall gel permeation chromatography from Chapter 7. Sephadex is a gel permeation medium that separates molecules according to size. The largest molecules are excluded from the gel and pass through the column most rapidly, whereas the small molecules penetrate the gel and are retarded.) In the experiment under discussion, Sephadex chromatography separated the enzyme from $H^{14}CO_3^-$ and ADP, with the results shown in Figure 17-19. Radioactivity was associated with the protein, which suggested that ^{14}C in $H^{14}CO_3^-$ had in some way become covalently attached, or is at least very tightly bound, to the enzyme; that is, enzyme-$^{14}CO_2$ was formed and could be isolated.

When ATP was omitted from the reaction mixture, no radioactivity was associated with the protein. ATP was, therefore, required to form enzyme-$^{14}CO_2$. Moreover, ADP and P_i were formed in amounts equivalent to enzyme-$^{14}CO_2$. This experiment established the occurrence of the first partial reaction (equation 26).

When enzyme-$^{14}CO_2$, which had been isolated using the Sephadex column, was combined with acetyl CoA, [^{14}C]malonyl CoA was formed; and all of the radioactivity was removed from the protein. This experiment provided evidence for the second of the two partial reactions, equation 27.

Because enzyme-$^{14}CO_2$ could be isolated, its structure could also be determined. The nature of enzyme-CO_2 was established largely through the work of F. Lynen and D. Lane and their collaborators, who showed that CO_2 is not directly bonded to the enzyme but is bonded to biotin. CO_2 is covalently bonded to the N-1 atom of biotin; that is, opposite the alkyl side chain.

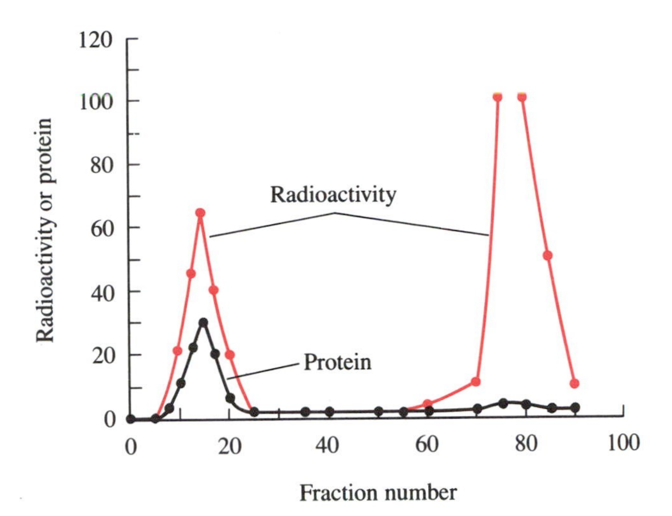

Figure 17-19

Formation of Carboxylated Acetyl CoA Carboxylase. The reaction mixture, containing acetyl CoA carboxylase, $H^{14}CO_3^-$, and ATP but no acetyl CoA, was placed on a Sephadex column. The effluent from the column was collected in small fractions. Each fraction was assayed for radioactivity (red point) and protein content (black point). Protein eluted in fractions 10 through 20. Note that radioactive material emerged with the eluted protein. $H^{14}CO_3^-$, or a compound derived from $H^{14}CO_3^-$, must be strongly associated or covalently bound to the protein. A large peak of radioactive material eluted in fractions 70 through 90. This was where small molecules eluted from the column, and it contained excess $H^{14}CO_3^-$. The original reaction mixture contained more HCO_3^- than enzyme.

We now return to the original question concerning the function of ATP. In synthetic reactions, ATP generally activates oxygen so that it can be displaced or eliminated. We, therefore, expect that ATP functions in the same manner in this reaction and will activate the oxygen of bicarbonate to facilitate its displacement. This is the case, as shown by the fact that, when the reaction is carried out with $HC^{18}O_3^-$, the resulting P_i contains ^{18}O. Clearly, an oxygen atom from HCO_3^- is transferred to P_i.

It is not definitely known how ATP activates HCO_3^- but the reactions in Figure 17-20 account for all the experimental facts and is chemically reasonable. The proposed intermediate, carboxyphosphate, has never been isolated or synthesized. Its existence is, therefore, not conclusively established.

The role of Mg^{2+} in the transfer of a phosphoryl group from ATP to HCO_3^- is somewhat speculative. It is proposed that Mg^{2+} is complexed with two oxygens of ATP, and in this way neutralizes two negative charges and contributes to the maintenance of electrical neutrality. Evidence has been

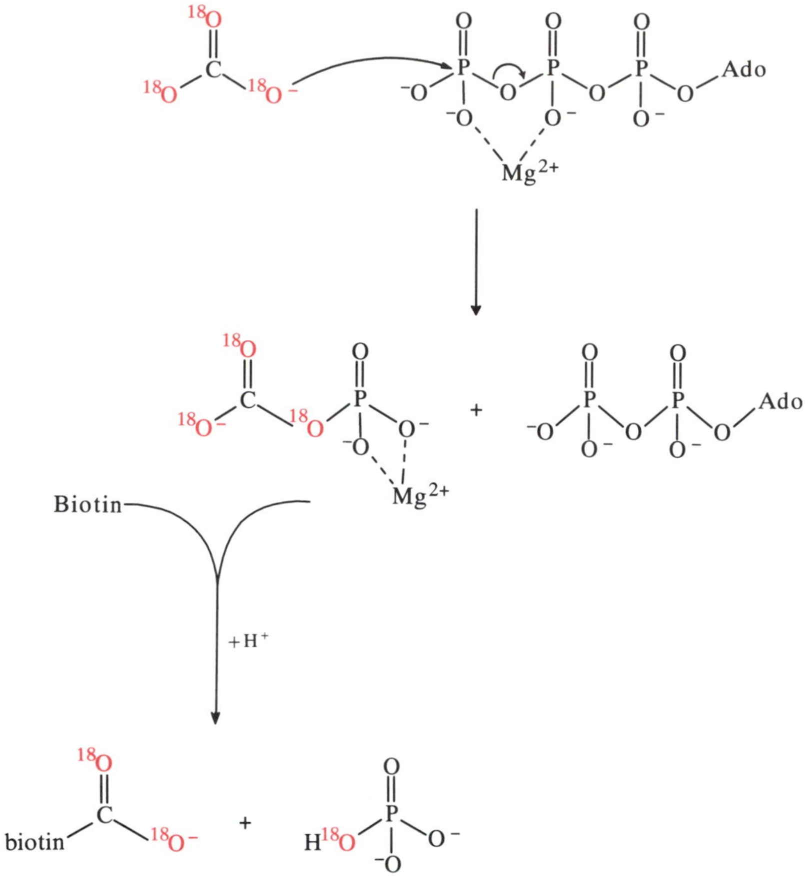

Figure 17-20

The Formation of a Hypothetical Intermediate, Carboxyphosphate, in the ATP-dependent Carboxylation of Biotin.

obtained recently that suggests that this is generally the role of Mg^{2+} in reactions of the terminal phosphoryl group of ATP. A more-detailed mechanism for the carboxylation of acetyl CoA is shown in Figure 17-21.

Acetyl CoA Carboxylase from *E. coli* Consists of Three Different Subunits

The composition of acetyl CoA carboxylase from *E. coli* is extremely interesting. The enzyme can be resolved into three separate subunits, and each subunit catalyzes one or more of the reactions in Figure 17-21. The three subunits are biotin carboxylase (molecular weight 95,000); a carboxyl carrier protein (CCP, molecular weight 22,500), which contains covalently bound biotin; and a carboxytransferase (molecular weight 130,000). Biotin carboxylase catalyzes reactions 1 through 3 of Figure 17-21, the HCO_3^-- and ATP-dependent carboxylation of biotin bonded to the carboxyl carrier protein. It can also catalyze the carboxylation of free biotin. Carboxytransferase catalyzes the transfer of CO_2 from CCP-CO_2 to acetyl CoA—reactions 4 and 5 in Figure 17-21; it also catalyzes the reverse reaction, the carboxylation of CCP by malonyl CoA.

This type of "modular" design is not present in the vertebrate enzyme. The avian enzyme, for example, is a high molecular weight polymer consisting of protomers of molecular weight 410,000. All of the catalytic functions appear to be located on one polypeptide chain.

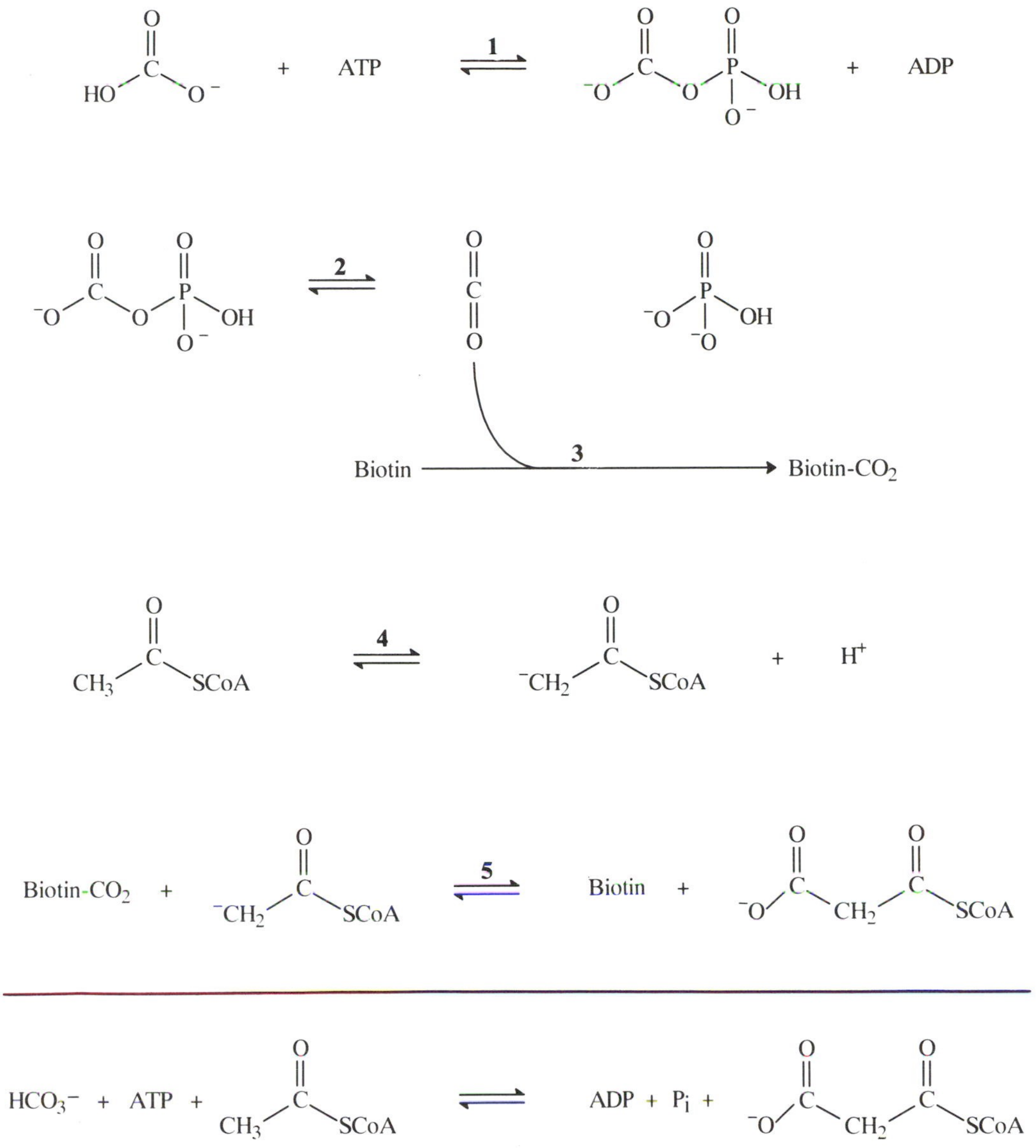

$$HO-\overset{\displaystyle O}{\overset{\|}{C}}-O^- + ATP \overset{1}{\rightleftharpoons} {}^-O-\overset{\displaystyle O}{\overset{\|}{C}}-O-\overset{\displaystyle O}{\overset{\|}{\underset{\displaystyle O^-}{\underset{|}{P}}}}-OH + ADP$$

$${}^-O-\overset{\displaystyle O}{\overset{\|}{C}}-O-\overset{\displaystyle O}{\overset{\|}{\underset{\displaystyle O^-}{\underset{|}{P}}}}-OH \overset{2}{\rightleftharpoons} O=C=O \quad {}^-O-\overset{\displaystyle O}{\overset{\|}{\underset{\displaystyle {}^-O}{\underset{|}{P}}}}-OH$$

$$\text{Biotin} \xrightarrow[\quad CO_2 \quad]{3} \text{Biotin-}CO_2$$

$$CH_3-\overset{\displaystyle O}{\overset{\|}{C}}-SCoA \overset{4}{\rightleftharpoons} {}^-CH_2-\overset{\displaystyle O}{\overset{\|}{C}}-SCoA + H^+$$

$$\text{Biotin-}CO_2 + {}^-CH_2-\overset{\displaystyle O}{\overset{\|}{C}}-SCoA \overset{5}{\rightleftharpoons} \text{Biotin} + {}^-O-\overset{\displaystyle O}{\overset{\|}{C}}-CH_2-\overset{\displaystyle O}{\overset{\|}{C}}-SCoA$$

$$HCO_3^- + ATP + CH_3-\overset{\displaystyle O}{\overset{\|}{C}}-SCoA \rightleftharpoons ADP + P_i + {}^-O-\overset{\displaystyle O}{\overset{\|}{C}}-CH_2-\overset{\displaystyle O}{\overset{\|}{C}}-SCoA$$

Figure 17-21

Steps in the Acetyl CoA Carboxylase Reaction. Steps 1 through 3 are catalyzed by biotin carboxylase of *E. coli*, with the biotin bonded to the carboxyl carrier protein. The carboxyltransferase catalyzes reactions 4 and 5. All five reactions are catalyzed by the avian single-chain enzyme (molecular weight 410,000).

Summary

Biological molecules that undergo enzymatic decarboxylation have the following general structure:

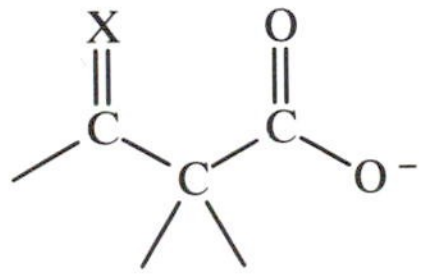

$$-\overset{\displaystyle X}{\overset{\|}{C}}-\underset{/\ \backslash}{C}-\overset{\displaystyle O}{\overset{\|}{C}}-O^-$$

in which X can be O or NH. Decarboxylation leads to the formation of a carbanion, hence the requirement for —C=X, which stabilizes the carbanion.

$$\mathrm{-C(=X)-C(\!<\!)-C(=O)-O^-} \longrightarrow \mathrm{-C(=X)-C^-(\!<\!)} + CO_2$$

An example of enzymatic decarboxylation is the conversion of acetoacetate into acetate and CO_2, catalyzed by the bacterial enzyme acetoacetate decarboxylase (Figure 17-2). A lysine residue at the active site of the decarboxylase reacts with the substrate to convert the β-carbonyl group into a protonated imine, which undergoes decarboxylation. The imine is positively charged under physiological conditions; it therefore stabilizes the intermediate carbanion more than a carbonyl group, which can at best acquire a fraction of a positive charge.

Amino acid decarboxylases utilize the cofactor pyridoxal phosphate (PLP), which facilitates the decarboxylation by forming an imine with the amino acid (Figure 17-14).

In enzymatic decarboxylation of α-keto acids, the cofactor thiamine pyrophosphate (TPP) is utilized. Formation of an adduct between the cofactor and the substrate creates an electron sink that allows the decarboxylation to proceed. An example of the decarboxylation of an α-ketoacid is the conversion of pyruvate into acetaldehyde and CO_2, which takes place in microorganisms and plants. A more-complex reaction is the oxidative conversion of pyruvate, NAD^+, and CoA into acetyl CoA, NADH, and CO_2. This reaction is catalyzed in animals, microorganisms, and plants by an enzyme complex consisting of three enzymes: pyruvate dehydrogenase (E_1), dihydrolipoyl transacetylase (E_2), and dihydrolipoyl dehydrogenase (E_3). TPP is associated with E_1 and, in addition, the coenzymes lipoamide and FAD are required. The latter coenzymes are bound to E_2 and E_3, respectively.

Enzymes also catalyze carboxylation reactions in which CO_2 is added to a carbanion. The carboxylation of ribulose bisphosphate and PEP does not require the participation of a cofactor. The carboxylation of other molecules requires the cofactor biotin. An example of an enzyme that catalyzes a biotin-dependent carboxylation is acetyl CoA carboxylase. This carboxylation consists of two partial reactions:

1. HCO_3^-, ATP, and biotin react to form biotin—CO_2, ADP, and P_i.
2. The CO_2 group of biotin—CO_2 is transferred to the carbanion of the substrate (—$CH_2COSCoA$ in the case of acetyl CoA carboxylase).

ADDITIONAL READING

BOYER, P. D., ed. 1972. Carboxylation and decarboxylation. In *The Enzymes*, vol. 6, 3d ed. New York: Academic Press.

KNOWLES, J. R. 1989. The mechanism of biotin-dependent enzymes. *Annual Review of Biochemistry* 58: 195.

MIZIORKO, H. M., and LORIMER, G. H. 1983. Ribulose-1,5-bisphosphate carboxylase-oxygenase. *Annual Review of Biochemistry* 52: 507.

WALSH, C. T. 1976. *Enzymatic Reaction Mechanisms*, San Francisco: Freeman.

LUNDQVIST, T., and SCHNEIDER, G. 1991. Crystal structure of activated ribulose-1,5-bisphosphate carboxylase complexed with its substrate, ribulose-1,5-bisphosphate. *Journal of Biological Chemistry* 266: 12604.

Rose, I. A., and Mullhofer, G. 1965. The position of carbon–carbon cleavage in the ribulose diphosphate carboxydismutase reaction. *Journal of Biological Chemistry* 240: 1341.

Hamilton, G. A., and Westheimer, F. H. 1959. On the mechanism of the enzymatic decarboxylation of acetoacetate. *Journal of the American Chemical Society* 81: 6332.

Reed, L. J. 1974. Multienzyme complexes. *Accounts of Chemical Research* 7: 43.

PROBLEMS

1. 3-Phenyl-3-bromopropionate readily undergoes decarboxylation. What products will be formed, and what is the mechanism of decarboxylation?

C_6H_5—CH(Br)—CH_2—COO^-

3-Phenyl-3-bromopropionate

2. Anion **1** undergoes decarboxylation more slowly than its acid form **2**. Explain why this is so.

RO—C(=O)—C(=O)—C(H_3C)(CH_3)—C(=O)—C(=O)—O^-

1

RO—C(=O)—C(=O)—C(H_3C)(CH_3)—C(=O)—C(=O)—O—H

2

3. Metal ions accelerate the nonenzymatic decarboxylation of **3** much more than that of **1** in problem 2. Why?

$^-$O—C(=O)—C(=O)—C(H_3C)(CH_3)—C(=O)—C(=O)—O^-

3

4. Exposure of acetoacetate decarboxylase to $NaBH_4$ in the presence of acetoacetate leads to reductive trapping only of the acetone-derived imine and not the imine derived from acetoacetate. This fact can be rationalized in several ways. Suggest two mechanistic rationales for this observation.

5. In addition to catalyzing carboxylation, ribulose-1,5-bisphosphate carboxylase catalyzes an oxygenation:

Ribulose-1,5-P_2 + O_2 ⟶

$H_2C(OPO_3^{2-})$—COO^- + 3-Phosphoglycerate

Write a mechanism for this reaction.

This reaction, by forming phosphoglycolic acid, reduces the efficiency of CO_2 fixation. There is considerable interest in preventing this reaction.

6. What kind of exchange experiment would you carry out to gain evidence for the first partial reaction catalyzed by acetyl CoA carboxylase?

7. An enzyme catalyzes the decarboxylation of phenylalanine to phenylethylamine.

C_6H_5—CH_2—C(H)(NH_3^+)—COO^- ⟶

C_6H_5—CH_2—CH_2—NH_3^+ + CO_2

Phenylethylamine

(a) Write a mechanism for this reaction.
(b) When phenylethylamine is added to the enzyme in ^2H_2O, deuterium is incorporated at C—1. Maximally, one deuterium atom is incorporated. Explain this observation.

8. Why does the following compound not undergo decarboxylation?

(7-oxabicyclic structure bearing COO^-)

9. What is the role of ATP in carboxylation reactions?

10. Propose a compound that may be a good inhibitor of ribulose bisphosphate carboxylase.

11. (a) Write a mechanism of action for propionyl CoA carboxylase.
(b) When propionyl CoA is added to propionyl CoA carboxylase in ^2H_2O in the absence of ATP, no deuterium is incorporated into the α-carbon of propionyl CoA. Is this observation consistent with your mechanism? Explain.

12. Prephenic acid is an intermediate in the bacterial biosynthesis of aromatic amino acids. It can undergo two types of metabolic transformations:

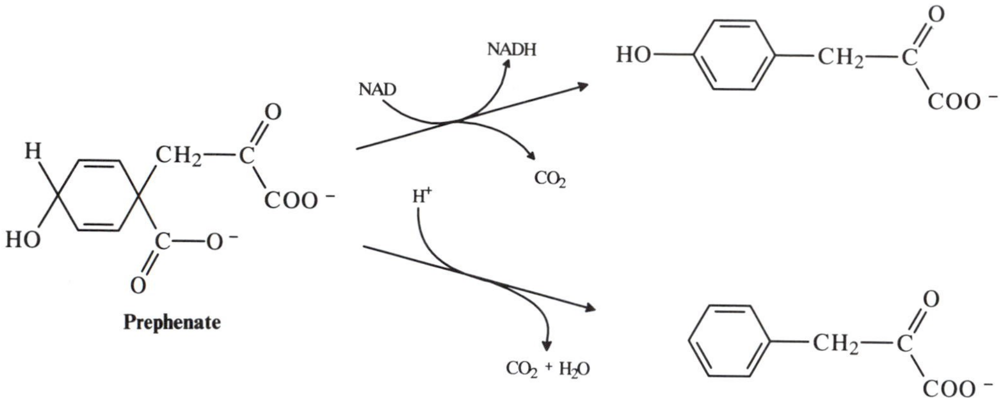

NAD → NADH, CO_2

H^+ → CO_2 + H_2O

Prephenate

Write mechanisms for both decarboxylations.

Addition, Elimination, and Transamination

18

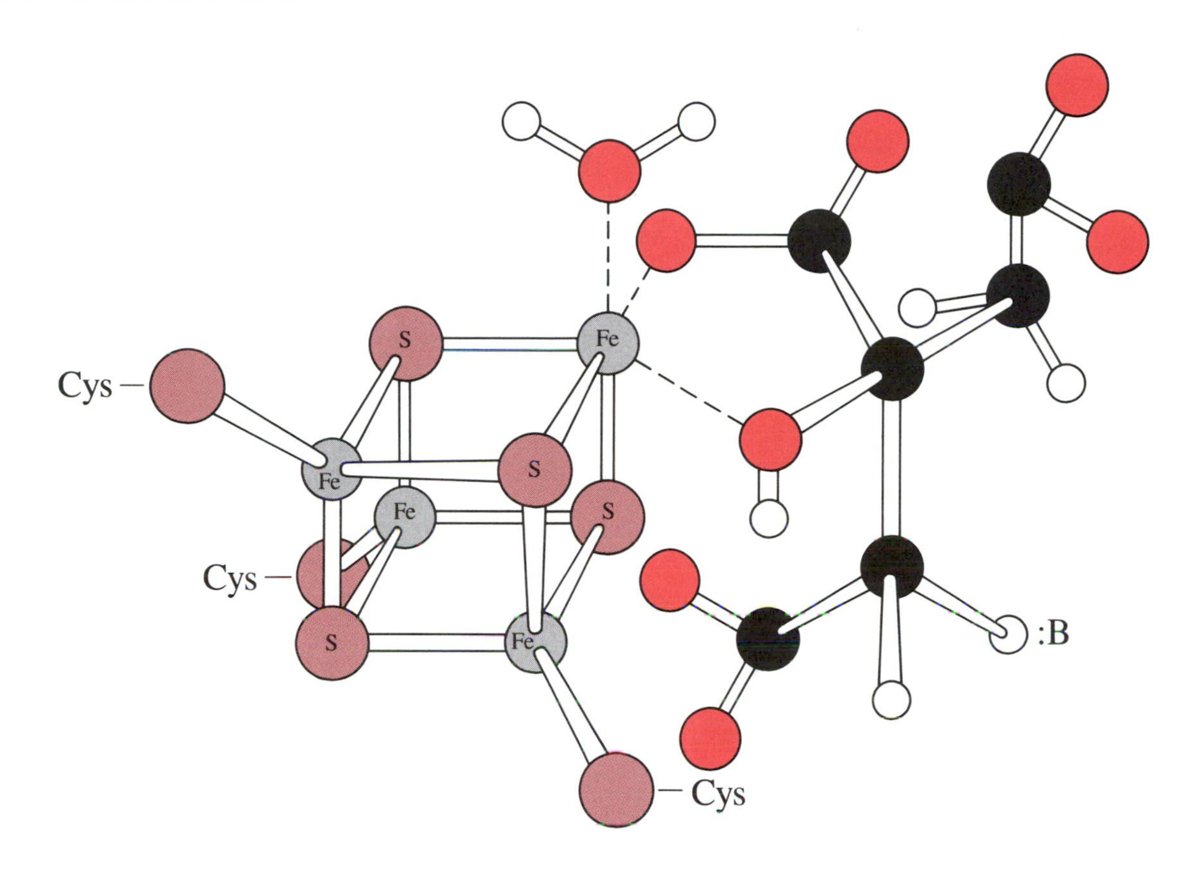

A structural model of the iron-sulfur cluster in aconitase, with the substrate bound to one iron ion.

The biosynthesis and biodegradation of molecules in metabolism frequently require the *addition* or *elimination* of the functional groups at various stages of the overall process. Recall the importance of the carbonyl group in facilitating carbon–carbon bond formation and cleavage (Chapter 2) and in decarboxylation and carboxylation reactions (Chapter 17). These are essential processes in metabolism. The carbonyl groups that make these reactions possible are often introduced through the dehydrogenation of alcohols, amines, or aldehydes by NAD or FAD (Chapter 16) or through their oxidation by O_2 (Chapter 24). Hydroxyl groups destined for oxidation to carbonyl groups are often introduced into biological molecules by the addition of water to an alkene. Conversely, in catabolic pathways the amino groups of amino acids or hydroxyl groups of carbohydrates must often be eliminated. Thus, addition and elimination reactions are essential steps in both anabolic and catabolic pathways.

It is often important in metabolism to convert an amine into a ketone because of the useful activating effect of the ketonic carbonyl group. This is frequently accomplished by a kind of molecular barter, an exchange transaction known as transamination, in which the carbonyl group is acquired by exchange with an amine according to equation 1.

$$R_1-\underset{\underset{O}{\|}}{C}-COO^- + R_2-\underset{\underset{NH_3^+}{|}}{CH}-COO^- \rightleftharpoons R_1-\underset{\underset{NH_3^+}{|}}{CH}-COO^- + R_2-\underset{\underset{O}{\|}}{C}-COO^- \quad (1)$$

In this chapter, we present the chemical basis for addition, elimination, and transamination reactions in metabolism.

Addition and Elimination Reactions Are Generally α,β to a Carbonyl Group

Addition and elimination reactions in the synthesis or degradation of biological molecules occur in substrates that contain a carbonyl group β to a leaving group, with at least one hydrogen on the α-carbon atom:

$$-\underset{\underset{X}{|}}{\overset{|}{C}}-\underset{|}{\overset{\overset{H}{|}}{C}}-\overset{\overset{O}{\|}}{C}- \rightleftharpoons -\overset{|}{C}=\underset{|}{C}-\overset{\overset{O}{\|}}{C}- + XH \quad (2)$$

$$X = OH, NH_3^+, SH, OR, SR$$

These substrates are generally thioesters, carboxylic acids, ketones, or aldehydes; and the proton α to the carbonyl group is eliminated with the leaving group, X, in the β-position. Specific examples are the reactions of equations 3 and 4, which are catalyzed by β-methylaspartase and β-hydroxyacyl CoA dehydratase, respectively.

$$\text{L-}\beta\text{-Methylaspartate} \rightleftharpoons \text{Mesaconate } ({}^-O_2C-C(CH_3)=CH-CO_2^-) + NH_4^+ \quad (3)$$

L-β-Methylaspartate **Mesaconate**

$$CH_3(CH_2)_n-CH(OH)-CH_2-\overset{\overset{O}{\|}}{C}-SCoA \rightleftharpoons CH_3(CH_2)_n-CH=CH-\overset{\overset{O}{\|}}{C}-SCoA + H_2O \quad (4)$$

Eliminations Proceed Through either Two-Step or Concerted Mechanisms

Before turning to specific enzymatic eliminations, let us briefly consider the general mechanisms by which nonenzymatic elimination reactions occur (Figure 18-1). These reactions can proceed either through two-step processes (E1cB or E1), in which either H^+ or X^- is initially eliminated, or by a concerted process (E2), in which the bonds to H and X are simultaneously broken.

The E1cB mechanism is a two-step process in which the first step is the abstraction of a proton, as illustrated in Figure 18-1A. In the second step, the free electron pair of the carbanion provides the driving force for the expulsion of the leaving group, X^-. This mechanism is possible only when the carbanion can be stabilized through interaction with an electron-withdrawing center. Molecules that undergo E1cB elimination, therefore, generally have the structure shown in the margin. This structure, as pointed out in the preceding section, is found in many biological molecules that undergo elimination reactions. It is, therefore, tempting to assume that biological elimination reactions proceed through E1cB mechanisms, and a number of enzyme-catalyzed elimination reactions are known to do so; however, insufficient evidence is available to conclude that this is true for all enzymatic elimination reactions.

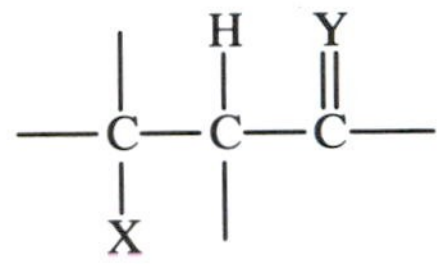

$Y = O, NH_2^+$
$X = OH, NH_3^+, SH, OR, SR$

The E1 mechanism (Figure 18-1B) also is a two-step process. The leaving group is eliminated as X^- in the first step to form a carbocation as the intermediate. The carbocation facilitates the loss of a proton and formation of the double bond.

The E2 mechanism in Figure 18-1C is a concerted process, in which the proton is abstracted by a base at the same time that the leaving group departs.

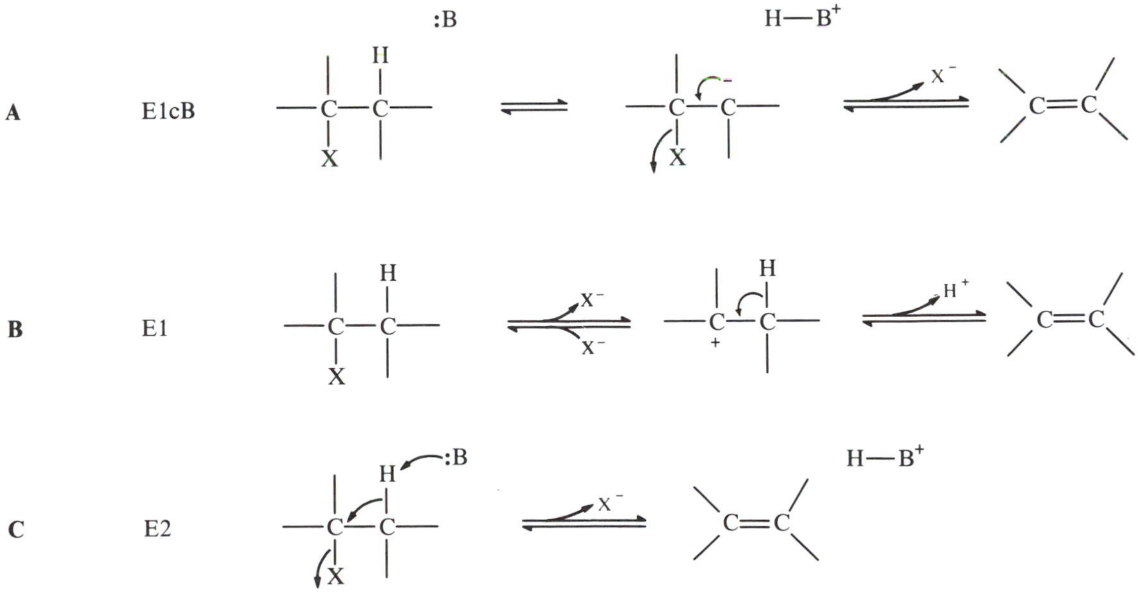

A E1cB

B E1

C E2

Figure 18-1

The Three Elimination Mechanisms E1cB, E1, and E2. Mechanism E1cB in part A includes the formation of a carbanion intermediate by loss of a proton in the first step. The carbanion eliminates the leaving group, X^-, in the second step. In the mechanism E1 in part B, X^- is eliminated in the first step, to form a carbocationic intermediate, which loses a proton in the second step. Mechanism E2 (part C) is a concerted process, in which a base abstracts the proton at the same time that the group X^- leaves.

There is no intermediate because the dissociation of the carbon–hydrogen bond and the expulsion of the leaving group occur simultaneously. The carbon skeleton in the transition state may have a predominantly carbanionic or carbocationic character, depending upon the extent to which bonds have been broken when the transition state is reached. Because of this, the distinction between two-step mechanisms and concerted mechanisms can become "fuzzy." It is, therefore, sometimes difficult, even in nonenzymatic reactions, to determine which mechanism is followed. Some of the experimental criteria used to determine which mechanism is employed will be discussed in connection with specific enzymatic reactions.

Evidence For the Carbanion Mechanism in Biological Elimination Reactions

We will begin the discussion of elimination reactions with the enzyme β-methylaspartase, which catalyzes the reaction of equation 3. In considering the mechanism, it is useful to decide whether breaking the C–H bond is rate determining. As we have already seen, this can be done by using a substrate molecule in which the hydrogen that is eliminated is replaced by deuterium. If a deuterium kinetic isotope effect is seen—that is, V_{max} for the deuterated compound is smaller than that for the nondeuterated compound—it can be concluded that C–H cleavage is rate determining.

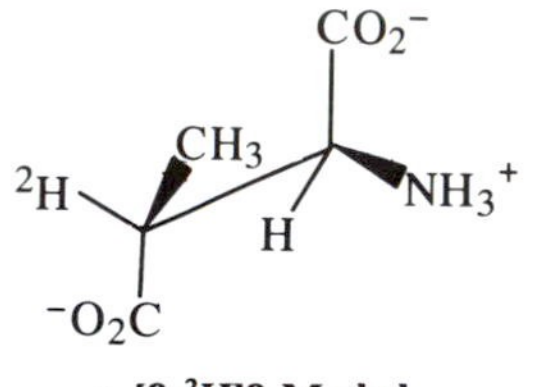

L-[β-^{2}H]β-Methylaspartate

When [β-^{2}H]β-methylaspartate is the substrate, no deuterium kinetic isotope effect is observed over a wide pH range; therefore, C–H cleavage cannot be rate determining.

In another experiment, [β-^{2}H]β-methylaspartate was added to β-methylaspartase, and the reaction was allowed to proceed until from 10% to 15% of the substrate had been converted into product. Under these conditions, essentially no reverse reaction would occur; that is, no β-methylaspartate would be formed by addition of NH_4^+ to mesaconate. The remaining substrate was reisolated from the reaction mixture and analyzed for deuterium. It was found that a significant fraction of the deuterium originally present in the substrate had been replaced by hydrogen; that is, the enzyme catalyzed the exchange of the β-^{2}H with the protons of water.

What is the significance of this observation? It is that there must be an intermediate stage in the reaction at which the β-hydrogen of the substrate has been abstracted, but the final reaction products, mesaconate and NH_4^+, have not been formed. One can further conclude that the reaction intermediate is converted back into the substrate, β-methylaspartate, at a rate that cannot be much slower than its conversion into product. (If that were not the case, we would not see the exchange reaction.)

The E1cB mechanism shown in Figure 18-2 is consistent with the experimental results, which show that breaking of the C–N bond (step 3) is rate determining. The absence of a deuterium kinetic isotope effect on V_{max} in the reaction of [β-^{2}H]β-methylaspartate is consistent with this mechanism because the rate-determining step does not include the breaking of a C–H bond. The observed exchange of the substrate β-^{2}H with water protons also is consistent with this mechanism. In the catalytic process, the deuterium-labeled substrate is converted through steps 1 and 2 into a carbanion through abstraction of the β-^{2}H by the base on the enzyme. This deuterium-labeled base is now subject to exchange with solvent protons. Step 3 in the forward direction is slow relative to the rates of steps 2 and 1 in the reverse direction. Therefore, the complex of

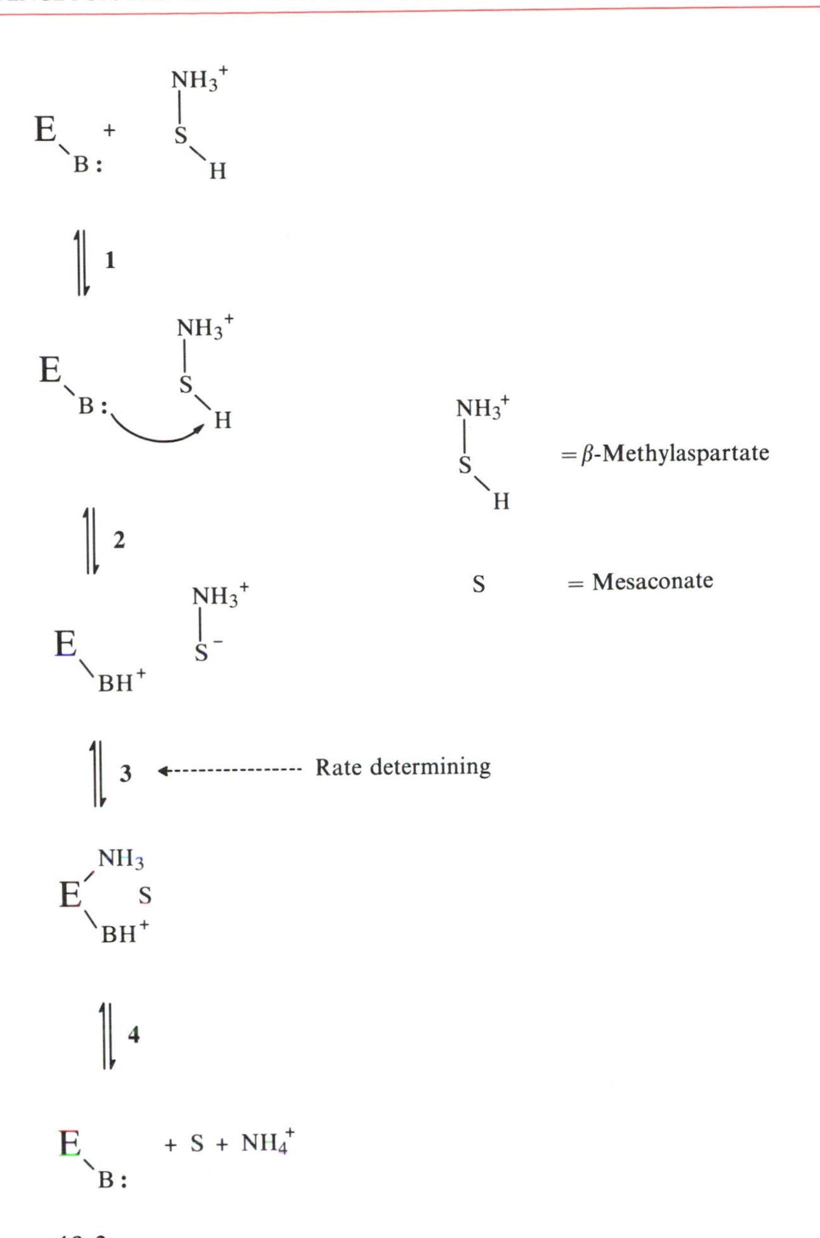

Figure 18-2

The Mechanism of Action of β-Methylaspartase. The mechanism shown is consistent with the experiments described in the text. The enzyme binds β-methylaspartate in step 1, and the β-proton is abstracted by a base at the active site in step 2. The elimination of NH_3 occurs in step 3, the rate-limiting step, and the products are released in step 4. Steps 1 and 2 are faster in the reverse direction than the forward direction of step 3, so that the enzyme catalyzes the exchange of the β-H of β-methylaspartate with solvent hydrogen faster than the overall reaction proceeds to product formation.

enzyme and carbanion is converted back into β-methylaspartate, now lacking deuterium, more frequently than it is converted into the product. The interpretation of the experimental results is actually more complex than this, as explained in Box 18-1.

BOX 18-1

ALTERNATIVE MECHANISTIC POSSIBILITIES FOR β-METHYLASPARTASE

The mechanism presented in the text is clearly consistent with the experimental results. But other possibilities should be considered. Recall that the conclusion from the isotope experiments was that breaking the C–H bond precedes the rate-determining step. The rate-determining step need not *necessarily* be the breaking of the C–N bond; it could, for example, be the release of either of the reaction products from the enzyme. Let us consider a concerted bond-cleavage mechanism, in which the release of either NH_4^+ or mesaconate from the enzyme is rate determining in the following scheme (S is mesaconate and H—S—NH_3^+ is β-methylaspartate):

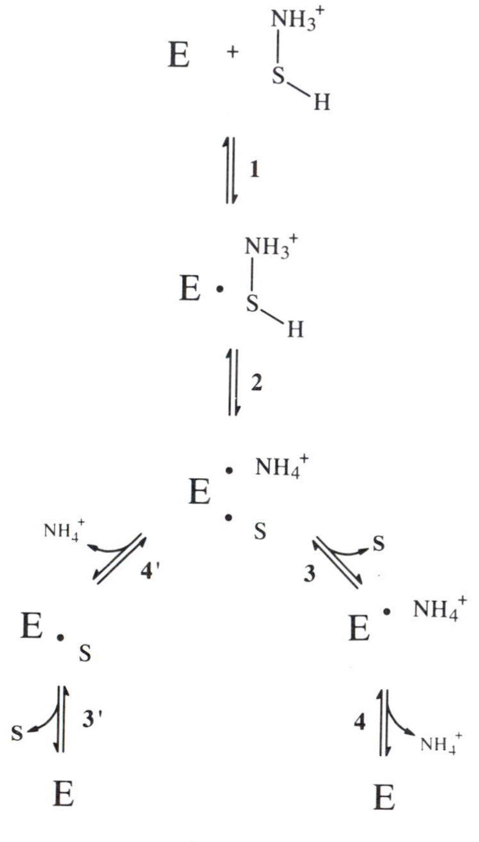

$$E + H\text{—}S\text{—}NH_3^+ \underset{}{\overset{1}{\rightleftharpoons}} E\cdot H\text{—}S\text{—}NH_3^+ \overset{2}{\rightleftharpoons} E\cdot S\cdot NH_4^+$$

$$E\cdot S\cdot NH_4^+ \overset{4'}{\rightleftharpoons} E\cdot S + NH_4^+ \qquad E\cdot S \overset{3'}{\rightleftharpoons} E + S$$

$$E\cdot S\cdot NH_4^+ \overset{3}{\rightleftharpoons} E\cdot NH_4^+ + S \qquad E\cdot NH_4^+ \overset{4}{\rightleftharpoons} E + NH_4^+$$

Two alternatives are shown, in which either S (mesaconate) or NH_4^+ is released from the enzyme first. Consider the situation in which the release of NH_4^+ is rate determining in step 4, and S is released in step 3 before NH_4^+. Either the $E\cdot NH_4^+$ can undergo dissociation in step 4 to E and NH_4^+ in a unimolecular process or it can react with S in a bimolecular process, the reverse of step 3, to give the complex $E\cdot S\cdot NH_4^+$ and eventually free β-methylaspartate by reversal of steps 2 and 1. If the concentration of S is increased sufficiently, step 3 will become as fast as or faster than step 4. (It is a second-order process, the rate of which is proportional to the concentration of S.) Therefore, when the conversion of β-methylaspartate into mesaconate is carried out in the presence of a sufficiently high concentration of [^{14}C]mesaconate, [^{14}C]β-methylaspartate should be found. When this experiment was carried out, [^{14}C]β-methylaspartate did not appear. Therefore, this mechanism can be eliminated.

A similar experimental approach was used to eliminate the mechanism in which release of S is rate determining and NH_4^+ is released before S. What experiment would you carry out that would provide evidence for or against the mechanism in which the release of S is rate determining?

The experiments described here, as well as other evidence that we have not discussed, strongly support the E1cB mechanism for β-methylaspartase.

Other examples of enzyme-catalyzed elimination of H_2O are the reactions catalyzed by enolase and fumarase (equations 5 and 6).

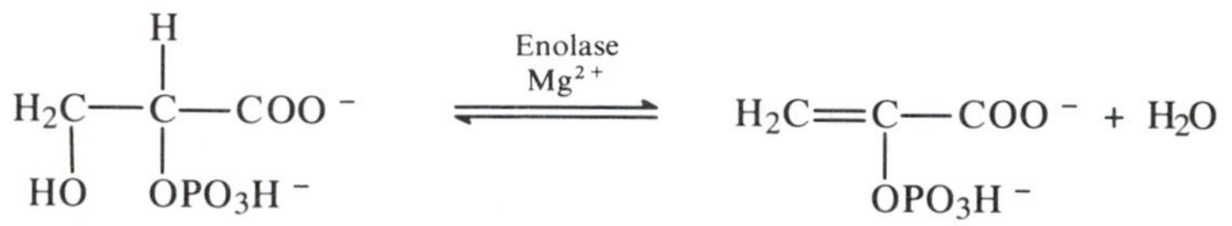

$$\underset{\textbf{2-Phosphoglycerate}}{H_2C(HO)\text{—}C(H)(OPO_3H^-)\text{—}COO^-} \xrightleftharpoons[]{\text{Enolase},\ Mg^{2+}} \underset{\textbf{Phosphoenolpyruvate}}{H_2C{=}C(OPO_3H^-)\text{—}COO^-} + H_2O \qquad (5)$$

$$^{-}OOC{-}\underset{H}{\overset{H}{C}}{-}\underset{H}{\overset{OH}{C}}{-}COO^{-} \underset{}{\overset{\text{Fumarase}}{\rightleftharpoons}} \text{(H)(}^{-}OOC\text{)C=C(COO}^{-}\text{)(H)} + H_2O \quad (6)$$

L-Malate — Fumarate

Both of these enzymes participate in the important metabolic pathways of glycolysis and the tricarboxylic acid (TCA) cycle, which we shall describe in Chapter 21. The mechanism of action of both enzymes has been extensively studied, and considerable evidence has been obtained supporting the E1cB mechanism for enolase. The mechanism of action of fumarase, on the other hand, is less clear.

Aconitase Brings about Isomerization Through the Elimination and Addition of H_2O

Another reaction in which H_2O is eliminated is the interconversion of citrate, isocitrate, and *cis*-aconitate catalyzed by aconitase (Figure 18-3). At equilibrium, the percentage of each acid is 88.4%, 7.5%, and 4.1%, respectively. This reaction presents some interesting problems.

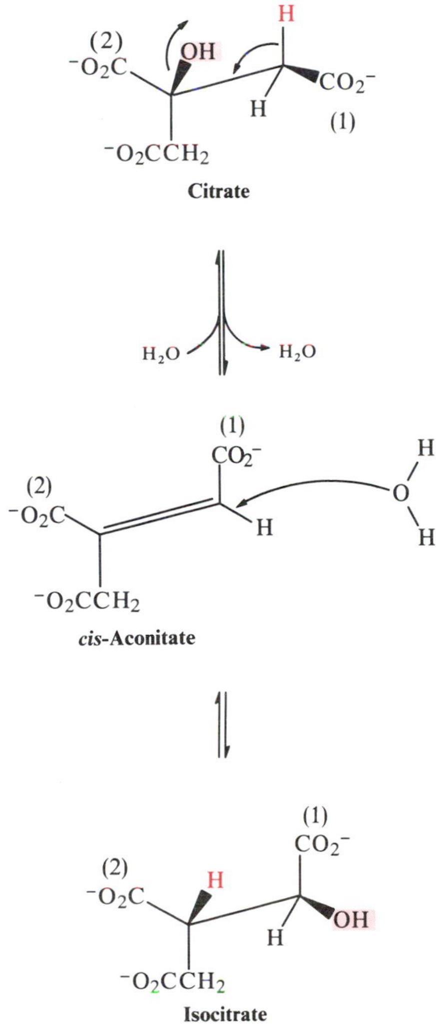

Figure 18-3

Aconitase-catalyzed Dehydration of Citrate and Hydration of *cis*-Aconitate. The elimination of OH is β and elimination of H is α to the carbonyl moiety of the carboxylate group (1) in the dehydration of citrate; and, in the formation of isocitrate, the addition of OH is β and of H is α to the carbonyl of the carboxyl group (2).

The conversion of citrate into isocitrate is an important reaction in the metabolic degradation of carbohydrates and fats through the TCA cycle (Chapter 21). Recall from Chapter 17 that the CO_2 produced in the catabolism of carbon sources is generated by oxidative decarboxylation of β-hydroxyacids and α-ketoacids. Citrate is the first intermediate in the TCA cycle, but it is a tertiary alcohol and so cannot be oxidized as either a β-hydroxyacid or an α-ketoacid. Aconitase catalyzes the rearrangement of citrate to isocitrate by an elimination-addition mechanism, and isocitrate is then oxidatively decarboxylated to an α-ketoacid (α-ketoglutarate), which also is oxidatively decarboxylated (Chapters 17 and 21). Thus, aconitase plays an essential role in the degradation of carbohydrates and fats.

The reactions catalyzed by aconitase further exemplify the generalization that elimination and addition reactions occur α, β to a carbonyl group. The reaction catalyzed by aconitase includes an addition and an elimination of H_2O. First, H_2O is eliminated from citrate to form *cis*-aconitate. The —OH group eliminated is initially β to a carboxyl group (labeled 1 in Figure 18-3) and the proton is α to the same carboxyl group. To form isocitrate, H_2O is added to *cis*-aconitate. Here, the —OH group is again added β to a carboxyl group (labeled 2 in the diagram) and the proton is added α to the same carboxyl group.

Chemical logic leads one to conclude that the interconversion of citrate and isocitrate requires the intermediate *cis*-aconitate. Is *cis*-aconitate necessarily a *free* intermediate in the enzymatic interconversion of citrate and isocitrate? An answer to this question was found through an experiment devised by Dickman and Speyer in 1956. They carried out the enzymatic conversion of isocitrate into citrate in 10% D_2O, as well as the conversion of *cis*-aconitate into citrate in 10% D_2O; and they measured the deuterium content of citrate produced in each experiment. They found that citrate produced from *cis*-aconitate contained more than three times as much deuterium as that formed from isocitrate. The fact that citrate derived from isocitrate contained less deuterium than that derived from *cis*-aconitate meant that *free* *cis*-aconitate cannot be an intermediate.

It is very likely that *enzyme-bound* *cis*-aconitate is the intermediate in the conversion of citrate into isocitrate. The *cis*-aconitate can be released from the enzyme occasionally to produce free *cis*-aconitate; however, the interconversion of citrate and isocitrate proceeds predominantly through *cis*-aconitate that remains bound to the enzyme. This is illustrated in Figure 18-4, where the reversible dissociation of *cis*-aconitate is an infrequent event relative to the conversion of enzyme-bound *cis*-aconitate into isocitrate.

The experiment of Dickman and Speyer gives us additional information about the reaction. Recall that the citrate produced from isocitrate in 10% D_2O contained less than one-third the deuterium found in citrate produced from *cis*-aconitate; that is, very little deuterium from the solvent is found in citrate generated directly from isocitrate. This tells us that the proton that is abstracted from isocitrate in the formation of enzyme-bound *cis*-aconitate is the same proton that is returned to *cis*-aconitate when it is hydrated to citrate. Hence, the hydrogen derived from isocitrate must be conserved; that is, it does not undergo equilibration with solvent protons (deuterons) during the conversion of isocitrate into citrate (Figure 18-5). Additional experimental support for direct hydrogen transfer appears in Box 18-2.

The elimination of the hydroxyl group in dehydration reactions such as those catalyzed by aconitase, fumarase, and enolase requires acid catalysis because hydroxide ion is a poor leaving group. Proton donation by a general acid at the active site to the leaving hydroxide would convert it into water in the transition state, and this may be one means by which enzymes catalyze the elimination of hydroxide (Figure 18-6). A general acid in a protein would be a

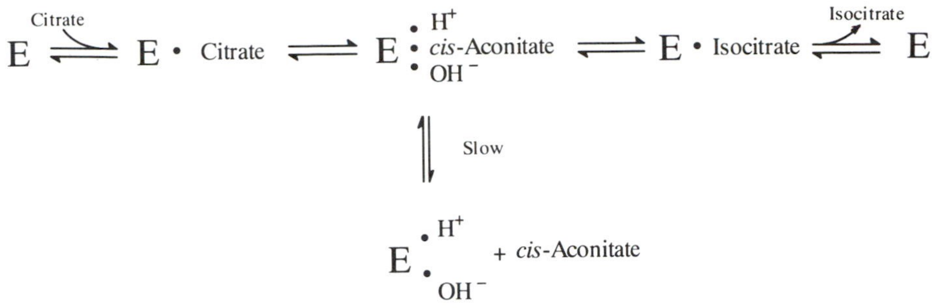

Citrate
E ⇌ E • Citrate ⇌ E : H^+ *cis*-Aconitate OH^- ⇌ E • Isocitrate ⇌ E
Isocitrate

Slow

E : H^+ OH^- + *cis*-Aconitate

Figure 18-4

The Role of *cis*-Aconitate in the Aconitase-catalyzed Conversion of Citrate into Isocitrate. *cis*-Aconitate is formed in the dehydration of citrate at the active site of aconitase. *cis*-Aconitate is an enzyme-bound intermediate that is hydrated to isocitrate most of the time; however, an infrequent but observable event is the reversible dissociation of *cis*-aconitate from the active site. Thus, *cis*-aconitate is generated by the enzyme and is equilibrated over time with citrate and isocitrate, but *free* *cis*-aconitate is not a compulsory intermediate in the conversion of citrate into isocitrate.

Isocitrate

Citrate

Slow

E + OH^- + H^+ +

cis-Aconitate

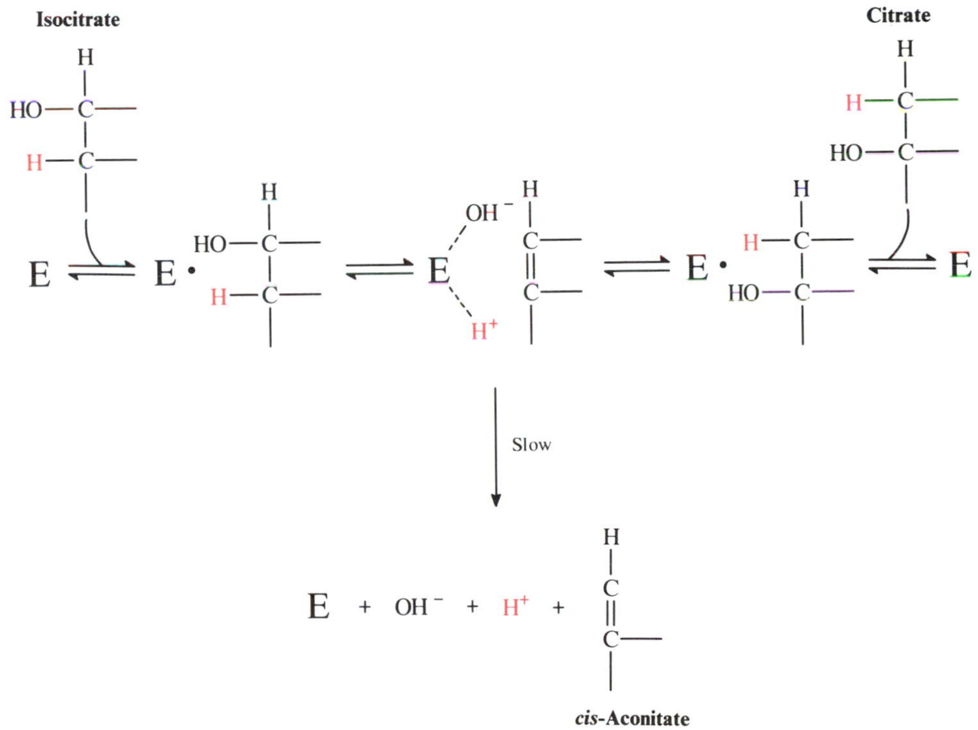

Figure 18-5

Aconitase-catalyzed Direct Proton Transfer. The proton abstracted from citrate by a general base at the active site of aconitase is transferred back to aconitase to form isocitrate in the hydration step without undergoing exchange with water protons to a significant extent. Whenever *cis*-aconitate is dissociated from the active site, the abstracted proton is free to exchange with solvent protons.

BOX 18-2

ACONITASE-CATALYZED DIRECT PROTON TRANSFER WITHOUT SOLVENT EXCHANGE

Direct hydrogen transfer in the reaction catalyzed by aconitase was demonstrated unequivocally by the use of [3-^{3}H]isocitrate as a substrate. When this compound was converted into citrate, most of the tritium originally present in isocitrate was found in citrate.

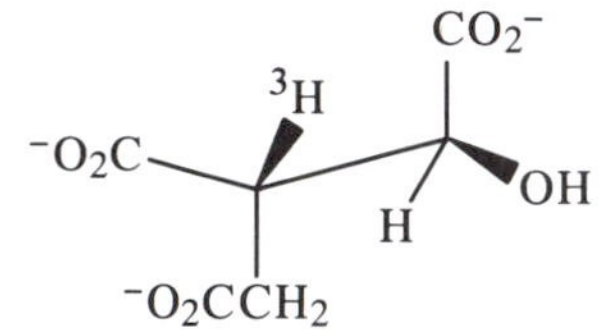

[3-^{3}H]Isocitrate

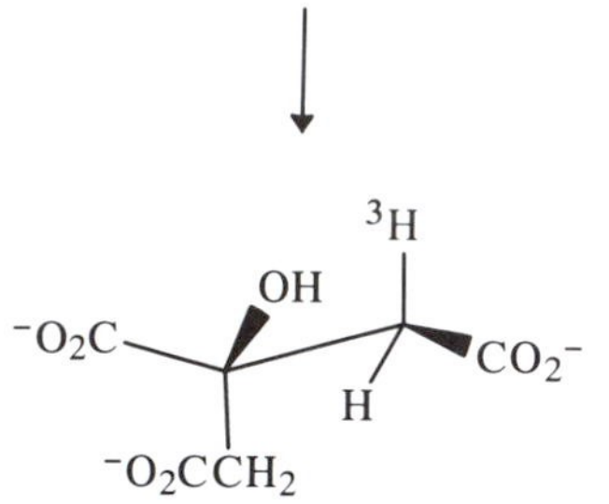

[4-^{3}H]Citrate

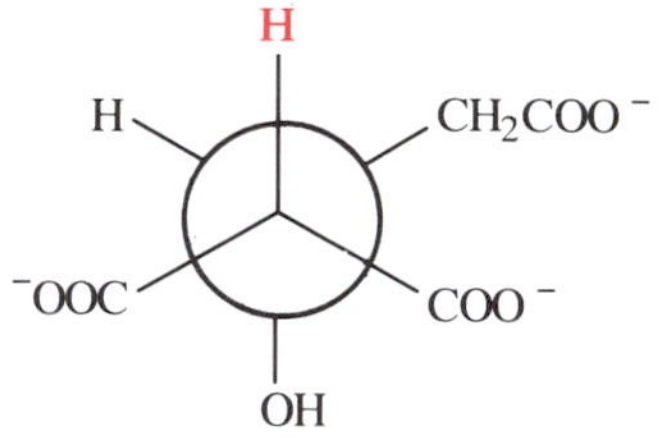

Inasmuch as *cis*-aconitate is an intermediate (enzyme-bound), the hydrogen must be transferred as a proton and not in a 1,2-hydride shift. We have already discussed reactions in which an enzyme-bound proton does not equilibrate with solvent protons, and we have considered the possible mechanisms by which this can occur.

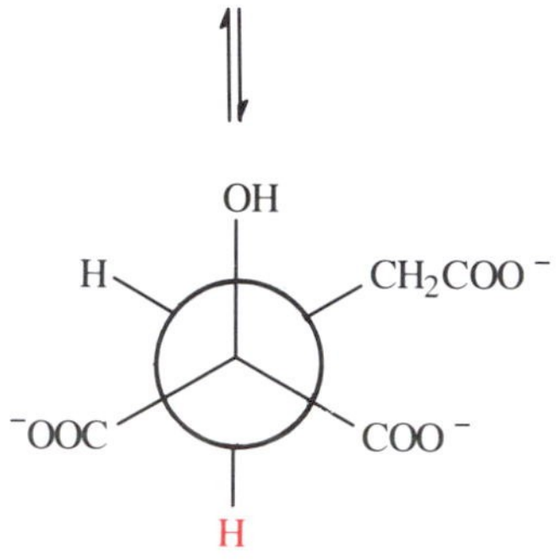

Concomitant with proton transfer, OH^- also is transferred. When isocitrate labeled with ^{18}O in the hydroxyl group is converted into citrate, no ^{18}O is found in citrate. Therefore, in contrast with the proton, OH^- is not transferred directly but instead equilibrates with water.

The conclusion that the hydrogen originally present in the substrate is found in the reaction product leads to an interesting stereochemical problem. The following Newman diagrams show the stereochemical course of the reaction.

Citrate

Isocitrate

Note that hydrogen is transferred from one face of the molecule to the opposite face. How is this accomplished? One possibility is that the base at the active center that abstracts the proton is structurally flexible enough to transport the proton from one face of the molecule to the other. Examples are known in nonenzymatic reactions in which intramolecular proton transfers occur by migration across a molecule over a considerable distance. These reactions have been said to proceed by "guided tour mechanisms." The other possibility is that the substrate molecule rotates in the course of the reaction. Several ingenious mechanisms have been proposed for such a rotation, and some of them have been given colorful names, such as the "Baylor twist." In general, these mechanisms use the carboxyl groups of the substrate as anchor points. If one carboxyl group is disengaged from the enzyme when *cis*-aconitate is formed, the molecule can pivot on the other two without dissociating from the enzyme. At this time, however, the mechanism by which the hydrogen is transferred is not understood.

Bronsted acid—that is, a proton donor such as —COOH (Glu, Asp, or the C-terminus) or the conjugate acid of a basic group (NH_3^+ of Lys or the N-terminus or the imidazolium of His).

Another means of stabilizing hydroxide is by coordination to a metal, which can act as a Lewis acid by binding unshared electron pairs of hydroxyl groups. Dehydration by aconitase is thought to exemplify the function of a metal as a Lewis acid to stabilize hydroxide ion as a leaving group. The enzyme contains an iron-sulfur cofactor that forms a complex with substrates through

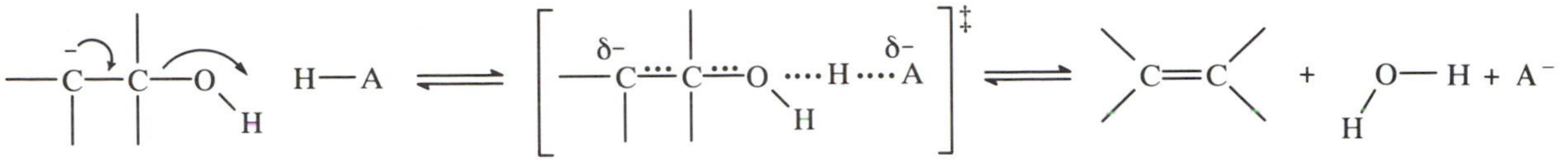

$$-\overset{|}{\underset{|}{C}}-\overset{|}{\underset{|}{C}}-O\diagdown_{H}\quad H-A \rightleftharpoons \left[-\overset{\delta-}{\overset{|}{\underset{|}{C}}}\cdots\overset{|}{\underset{|}{C}}\cdots O\cdots H\cdots \overset{\delta-}{A}\right]^{\ddagger} \rightleftharpoons \;\rangle C{=}C\langle\; + \; H-O-H + A^-$$

Figure 18-6

General Acid Catalysis in the Elimination of a Hydroxyl Group. The hydroxide group (HO^-) is a poor leaving group. A general acid in the active site of an enzyme facilitates the departure of the HO^- group by proton transfer in the transition state; this converts it into water, a stable species and an excellent leaving group.

coordination with the leaving hydroxyl group. The iron cofactor in aconitase, illustrated in Figure 18-7, is an iron-sulfur cluster, an organometallic complex consisting of iron and inorganic sulfide (Fe_3S_4) bound to the protein by coordination to three cysteine residues. In electron-transfer proteins (Chapter 22), iron-sulfur clusters ordinarily consist of equal numbers of inorganic sulfide atoms and cysteine ($Fe_2S_2Cys_2$ or $Fe_4S_4Cys_4$). The iron-sulfur cluster in purified aconitase differs in that it lacks one iron and one cysteinyl group, and this makes it reactive in binding Fe^{2+}. Aconitase is activated by Fe^{2+}, which enters the fourth iron-position of the cofactor and binds the substrate through coordination to a carboxylate group and probably also the hydroxyl group (Figure 18-7). Coordination with the iron in the iron-sulfur complex is thought to facilitate dehydration by stabilizing hydroxide ion as a leaving group.

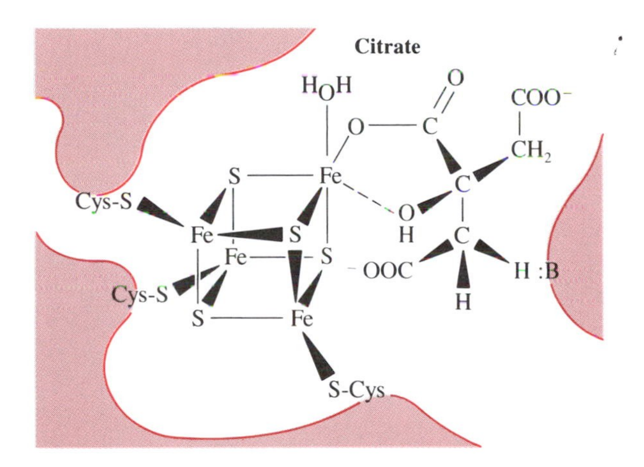

Figure 18-7

The Structure and Function of the Iron-Sulfur Cofactor at the Active Site of Aconitase. Purified aconitase contains an iron-sulfide cluster at the active site (Fe_3S_4) that is related to the iron-sulfide complexes in electron-transfer proteins. It differs in that one of the iron positions in the approximately cubic cluster is vacant. The enzyme is activated by Fe^{2+}, which binds to the cluster in the vacant position. This Fe^{2+} binds to the substrate by coordination to a carboxylate group and appears to facilitate the elimination of the OH group in the formation of *cis*-aconitate. The iron also stabilizes the hydroxide group and facilitates its readdition to *cis*-aconitate to form isocitrate or citrate. The iron-hydroxide complex rapidly exchanges hydroxide with water, so that ^{18}O in [^{18}O]citrate is lost to solvent in the conversion into isocitrate. [Adapted with permission from M. M. Werst, M. C. Kennedy, A. L. P. Houseman, H. Beinert, and B. M. Hoffman. *Biochemistry* 29 (1990): 10533, Fig. 10. Copyright 1990 American Chemical Society.]

In the complex of aconitase and *cis*-aconitate, the hydroxide is believed to be coordinated to iron, which stabilizes it as the nucleophile undergoing addition to the carbon–carbon double bond in the formation of isocitrate.

β-Hydroxydecanoyl Thioester Dehydrase Catalyzes an Elimination and Isomerization Reaction

Another example of an enzymatic dehydration is provided by the β-hydroxydecanoyl thioester dehydrase from *E. coli*. This enzyme catalyzes the elimination of H_2O from β-hydroxydecanoyl CoA to form the corresponding α, β-unsaturated decanoyl CoA, and it catalyzes the isomerization of the double bond from the α, β- to the β, γ-position, as shown in Figure 18-8. The β, γ-unsaturated fatty acid, as we shall see subsequently, plays an important role in the biosynthesis of other unsaturated fatty acids in this organism.

The elimination reaction presumably follows an E1cB mechanism, as discussed earlier. A base (B_1) at the active site abstracts the α-proton (Figure 18-9). An acidic group (B_2H) hydrogen bonded to the —OH group facilitates its departure through proton donation in the transition state to form H_2O as the leaving group (Figures 18-6 and 18-9).

Two bases in the active site probably also participate in the isomerization. One base abstracts a proton from the γ-position, and the other adds a proton to the α-position. Abstraction of a proton from the γ-position of the α, β-unsaturated intermediate can be either a discrete step or concerted with proton

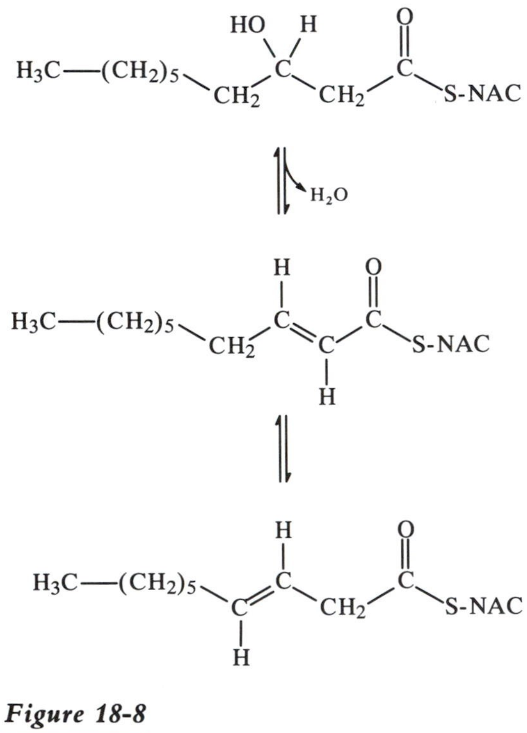

S-NAC = *N*-Acetylcysteamine

Figure 18-8

Dehydration and Isomerization Catalyzed by β-Hydroxydecanoyl Dehydrase. The natural substrate for this enzyme is β-hydroxydecanoyl CoA; however, the enzyme accepts other related thioesters such as the *N*-acetylcysteamine ester shown. The enzyme catalyzes the dehydration of the substrate to the α,β-unsaturated ester; it also catalyzes the isomerization of the double bond to the β,γ-position.

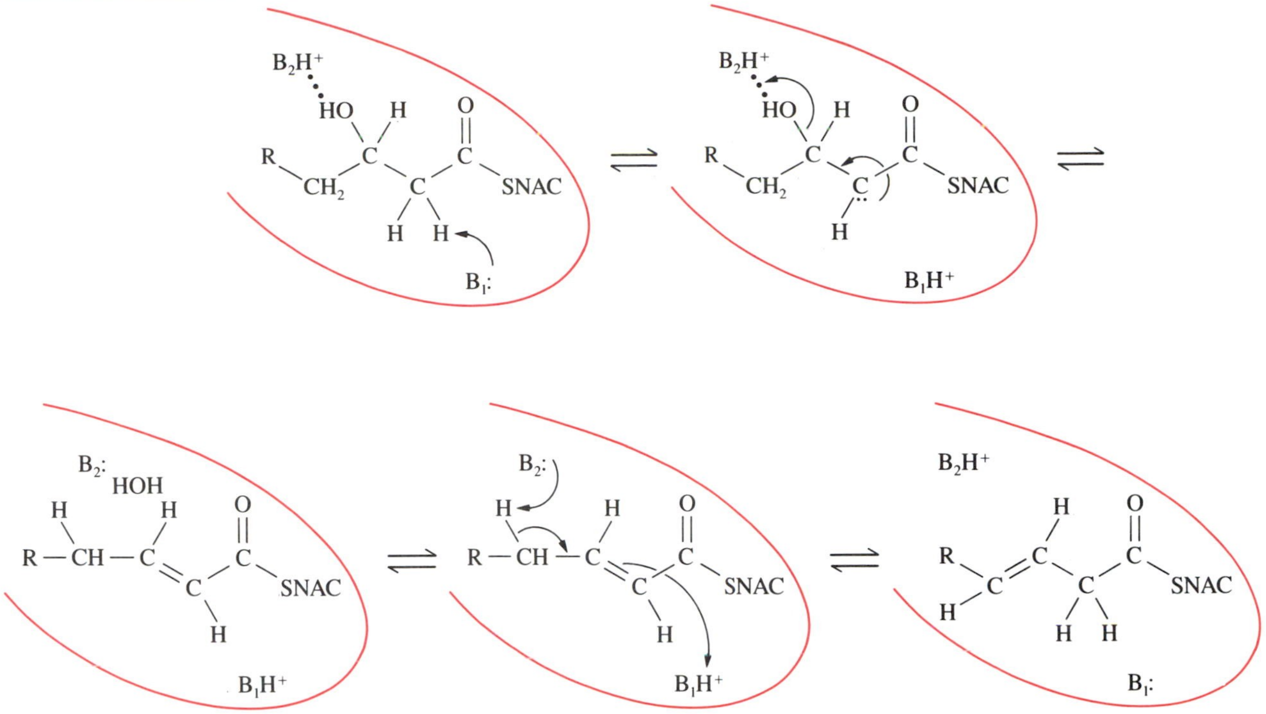

Figure 18-9

The Mechanism of Dehydration and Isomerization Catalyzed by β-Hydroxydecanoyl Thioester Dehydrase. The dehydration and isomerization can be catalyzed at the active site by the actions of two general acid-base groups. In the mechanism shown, base B_1 abstracts a proton α to the carbonyl group to form a carbanion intermediate. In the second step, the carbanion intermediate eliminates the hydroxyl group from the β-carbon, a process that is catalyzed by the general acid B_2H^+, which donates a proton to the leaving hydroxyl to form water and the α,β-unsaturated acyl thioester. In the isomerization, the base B_2 abstracts a proton from the γ-carbon and the acid B_1H^+ donates a proton to the α-carbon. The enzyme normally acts on CoA esters *in vivo* but is also active with other thioesters such as *N*-acetylcysteamine.

donation to the α-position. The intermediate carbanion that would result from proton abstraction from the γ-position is a resonance-stabilized species,

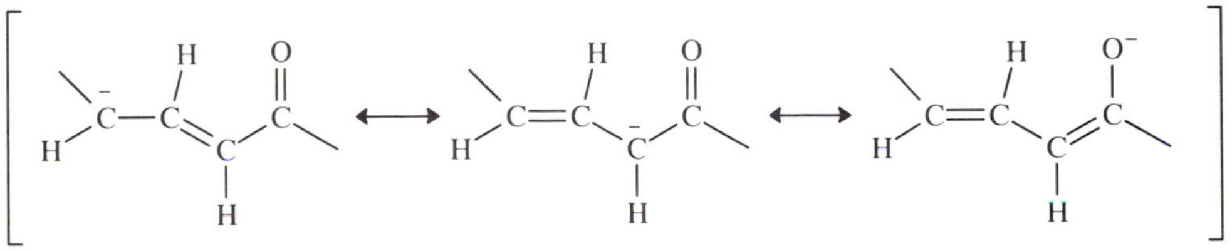

in which the electron pair is delocalized through the double bond and the carbonyl group. It is possible that the same functional groups participate in the hydration-dehydration and the isomerization reactions.

α,β-Dihydroxyacids Can Be Converted into β-Ketoacids Through Elimination Reactions

The elimination reactions discussed so far include formation of a double bond as a result of the elimination of water or NH_3. Closely related elimination reactions in which substrates contain two adjacent —OH groups lead to the formation of ketones. An example is the conversion of 6-phosphogluconate into 2-keto-3-deoxy-6-phosphogluconate (equation 7) catalyzed by the enzyme 6-phosphogluconate dehydratase.

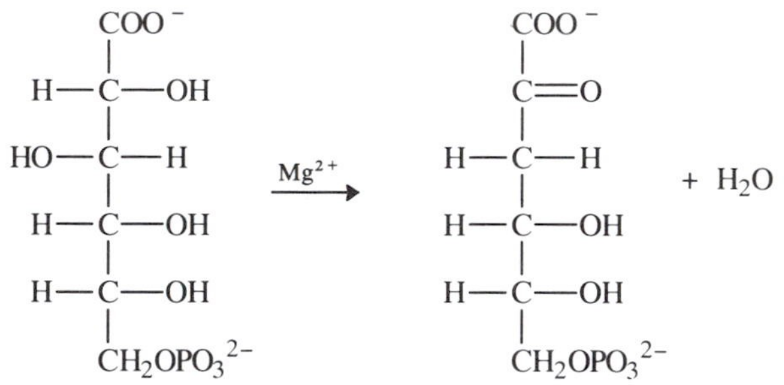

$$\begin{array}{c} COO^- \\ | \\ H-C-OH \\ | \\ HO-C-H \\ | \\ H-C-OH \\ | \\ H-C-OH \\ | \\ CH_2OPO_3^{2-} \end{array} \xrightarrow{Mg^{2+}} \begin{array}{c} COO^- \\ | \\ C=O \\ | \\ H-C-H \\ | \\ H-C-OH \\ | \\ H-C-OH \\ | \\ CH_2OPO_3^{2-} \end{array} + H_2O \qquad (7)$$

6-Phosphogluconate **2-Keto-3-deoxy-6-phosphogluconate**

In this reaction, the hydroxyl group at C-2 is oxidized to a carbonyl group, and the hydroxyl group at C-3 is reduced to CH_2. There is no net oxidation or reduction, and no additional electron acceptor or donor is utilized. This type of reaction is very useful because it allows a carbonyl group to be introduced into a molecule without the use of an oxidizing agent. This is particularly important for anaerobic organisms.

From earlier discussions, you will undoubtedly see why the introduction of carbonyl groups is important both in biosynthetic and in degradative reactions. The reaction catalyzed by 6-phosphogluconate dehydratase is involved in the degradative metabolism of 6-phosphogluconate by *Pseudomonas saccharophila.* At this point, you should be able to predict which compounds are produced in the degradation of 2-keto-3-deoxy-6-phosphogluconate. Once this problem has been worked out, it will be apparent why the introduction of a carbonyl group at C-2 is essential.

Some aspects of the probable mechanism of action of 6-phosphogluconate dehydratase are shown in Figure 18-10. The elimination of water is proposed to proceed by an E1cB mechanism (see Box 18-3). A base at the active site removes a proton from C-2 to form a carbanion, which is stabilized by the adjacent carbonyl group of the Mg^{2+}-complex. Next the OH group is eliminated. This is undoubtedly facilitated by an acid at the active site, because OH^- is a very poor

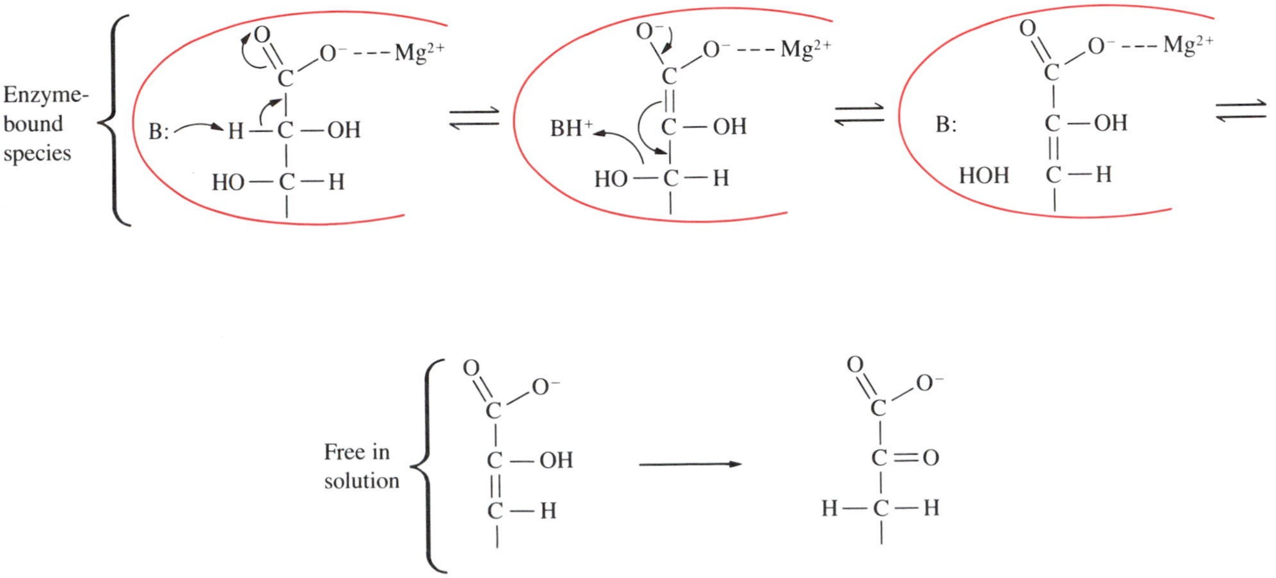

Figure 18-10

The Catalytic Mechanism of 6-Phosphogluconate Dehydrase in the Elimination of Water. The dehydration mechanism is thought to be similar to that in Figure 18-10 to form the α,β-unsaturated product, as is shown in the upper part of this figure as an enzymatic process. A difference is that the substrate is not a thioester but instead has the carboxyl group complexed with Mg^{2+} to neutralize the charge. The enzymatic product is the enol of 2-keto-3-deoxy-6-phosphogluconate, which undergoes ketonization after being released from the enzyme.

BOX 18-3

EXPERIMENTAL SUPPORT FOR THE E1cB MECHANISM OF ACTION FOR 6-PHOSPHOGLUCONATE DEHYDRASE

Experimental support for the mechanism of Figure 18-10 was provided by the following experiment. The conversion of 6-phosphogluconate into 2-keto-3-deoxy-6-phosphogluconate was carried out in ^3H_2O and not allowed to go to completion. The unreacted 6-phosphogluconate was reisolated, and tritium was found in the C-2 position. The tritium could not have been introduced through overall reversal of this irreversible reaction. Therefore, tritium must have been introduced at some intermediate stage in the reaction.

A reasonable mechanism for the introducton of tritium is through the carbanion, which can react in either of two ways: (1) it can become reprotonated to the starting material, 6-phosphogluconate; or (2) the carbanion can be converted into the reaction product, 2-keto-3-deoxy-6-phosphogluconate. If the former happens, and if the conjugate acid of the base that abstracts the C-2 proton allows this proton to equilibrate with solvent protons, then tritium will be incorporated from the solvent into the unreacted substrate. The observation that tritium exchange into the substrate occurs is, therefore, consistent with a carbanion mechanism, in which the rate of conversion of the carbanion back into the substrate occurs nearly as fast as (or faster than) its conversion into the product.

leaving group. Elimination leads to the formation of an enol at the active site, which undergoes ketonization spontaneously after being released from the enzyme.

Pyridoxal Phosphate as the Cofactor for Elimination Reactions with Amino Acids as Substrates

There are many enzymes that catalyze elimination and addition reactions in which amino acids are substrates. These enzymes frequently require pyridoxal-5′-phosphate (PLP) as the coenzyme. There are two groups of reactions in this category: β-elimination and replacement reactions and γ-elimination and replacement reactions.

We will first discuss β-elimination and replacement reactions (Figure 18-11), which are important in the metabolism of amino acids. β-Elimination reactions can generate α-ketoacids, which are intermediates in the metabolism

β-Elimination reactions:

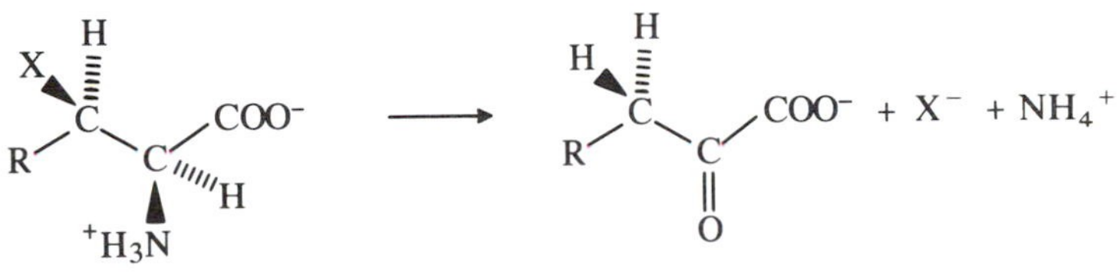

β-Replacement reactions:

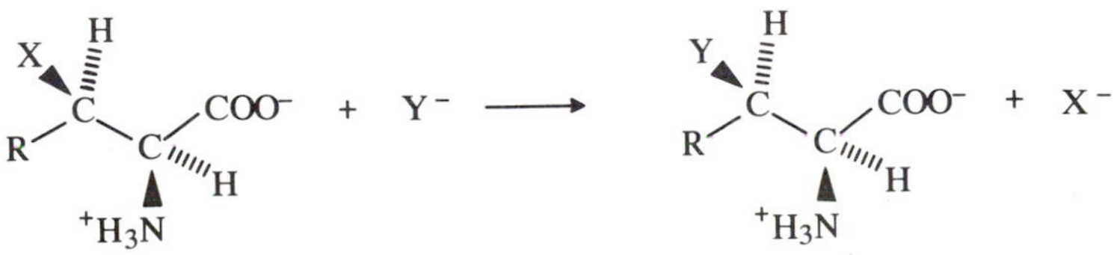

Figure 18-11

α-Elimination and α-Replacement Reactions of Amino Acids.

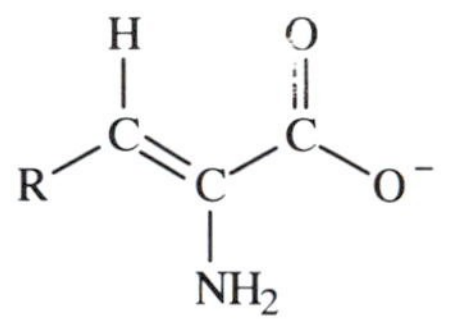

of almost all amino acids. In view of our earlier discussion of the mechanism of action of PLP, the mechanism of these elimination reactions should be evident. However, a mechanism for the β-elimination reaction is presented in Figure 18-12.

The β-elimination reaction is similar to other reactions in which PLP plays a role, in that initially an aldimine is formed. This facilitates abstraction of the α-proton from the substrate. The electron pair made available by proton abstraction expels a leaving group, X^-, to form an enamine (II). The steps subsequent to the formation of structure II in Figure 18-12 are somewhat speculative, and other reaction sequences have been written. For instance, intermediate III could undergo hydrolysis directly to produce PLP and the enamine shown in the margin. This enamine can undergo isomerization (tautomerization) to the imine (IV) in Figure 18-12, which will finally be hydrolyzed to the keto acid. Experimental results require that the conversion of the enamine into imine take place within the active site of the enzyme. The mechanism shown in Figure 18-12 includes enzyme-catalyzed protonation of the β-carbon before the decomposition of the product-PLP adduct; it, therefore, appears somewhat more attractive than a mechanism in which the enamine is

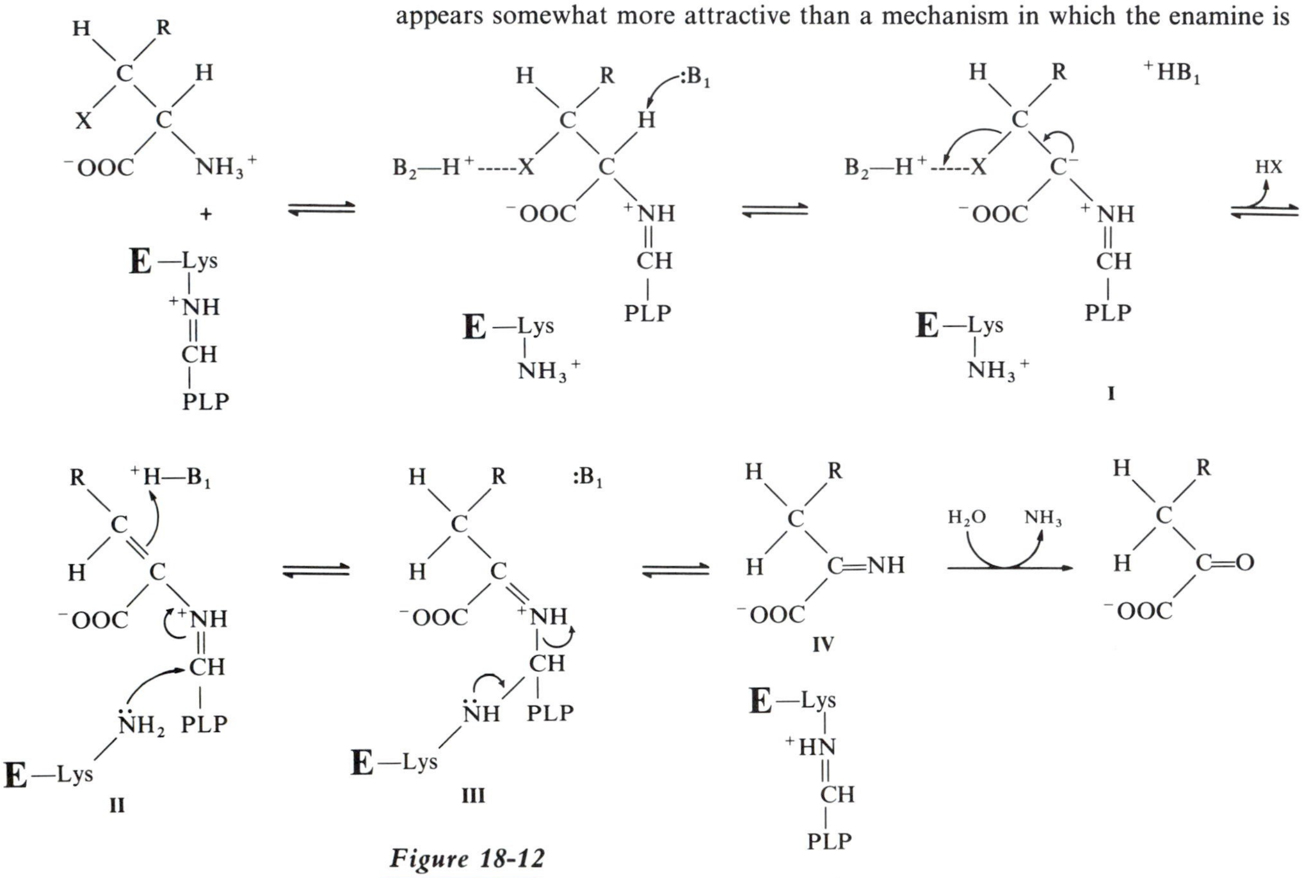

Figure 18-12

The Mechanism of PLP-dependent α,β-Elimination Reactions. In the first step, the enzyme-PLP complex forms an aldimine with the amino group of the amino acid. This activates the α-proton of the amino acid, which is abstracted by a base to form the α-carbanion (I). The carbanion expels the leaving group, X, from the β-carbon to form the α,β-unsaturated intermediate (II). Addition of the active-site lysyl-ε-amino group to the pyridoxaldimine carbon of structure II, and protonation of the substrate γ-carbon by a general acid produces the imine complex (III). The substrate imine (IV) is eliminated from complex III by decomposition of the tetrahedral adduct at the C-4′ of PLP. The substrate imine (IV) undergoes hydrolysis to the α-ketoacid, either in an enzymatic process at the active site or in solution after being released from the enzyme.

released as the product and allowed to undergo tautomerization and hydrolysis in solution. It should also be pointed out that it is unknown at this time whether the actual product released from the enzyme is the ketone or the imine.

A specific example of a β-elimination reaction is provided by L-serine dehydratase (equation 8).

$$HOCH_2\text{—}C(H)(^{+}H_3N)\text{—}COO^- \longrightarrow H_3C\text{—}C(=O)\text{—}COO^- + NH_4^+ \quad (8)$$

The mechanism for the β-replacement reaction is very similar to that for the β-elimination reaction. β-Replacement reactions proceed to intermediate II in Figure 18-12. At this point, intermediate II does not undergo hydrolysis but instead reacts with another nucleophile, Y. This reaction is essentially a reversal of the elimination except that X is replaced by Y.

Elimination and replacement can also occur at the γ-position of an amino acid (equations 9 and 10).

γ-Elimination:

$$XH_2C\text{—}CH_2\text{—}C(H)(^{+}H_3N)\text{—}COO^- + H_2O \longrightarrow H_3C\text{—}CH_2\text{—}C(=O)\text{—}COO^- + NH_4^+ + X^- \quad (9)$$

γ-Replacement:

$$XH_2C\text{—}CH_2\text{—}C(H)(^{+}H_3N)\text{—}COO^- + Y^- \longrightarrow YH_2C\text{—}CH_2\text{—}C(H)(^{+}H_3N)\text{—}COO^- + X^- \quad (10)$$

The mechanism for a γ-elimination shown in Figure 18-13 (on the next page) is, in many respects, similar to that in Figure 18-12 for β-elimination. As in the β-replacement reaction, γ-replacement can occur essentially by a reversal of the elimination reaction.

In some respects, the mechanism for the γ-elimination differs from other PLP mechanisms discussed so far. Here, PLP is used to introduce a protonated imine at the α-carbon of the amino acid (II in Figure 18-13). This is an electron-deficient center, which enhances the acidity of the β-hydrogen and facilitates its abstraction. The electron pair of the resulting β-carbanion expels the leaving group from the γ-carbon.

Transaminases, PLP-dependent Enzymes That Catalyze the Interconversion of Ketoacids and Amino Acids

Another important reaction of amino acids in which PLP participates is transamination, in which an amino group and a carbonyl group are interchanged between two molecules (equation 11).

$$^{+}H_3N\text{—}C(H)(R)\text{—}COO^- + O{=}C(R_1)\text{—}COO^- \rightleftharpoons O{=}C(R)\text{—}COO^- + {}^{+}H_3N\text{—}C(H)(R_1)\text{—}COO^- \quad (11)$$

Many amino acids and keto acids can participate in these reactions. Oxaloacetate and α-ketoglutarate are the ketoacids most frequently involved in transaminations. An abbreviated mechanism of this reaction is represented by

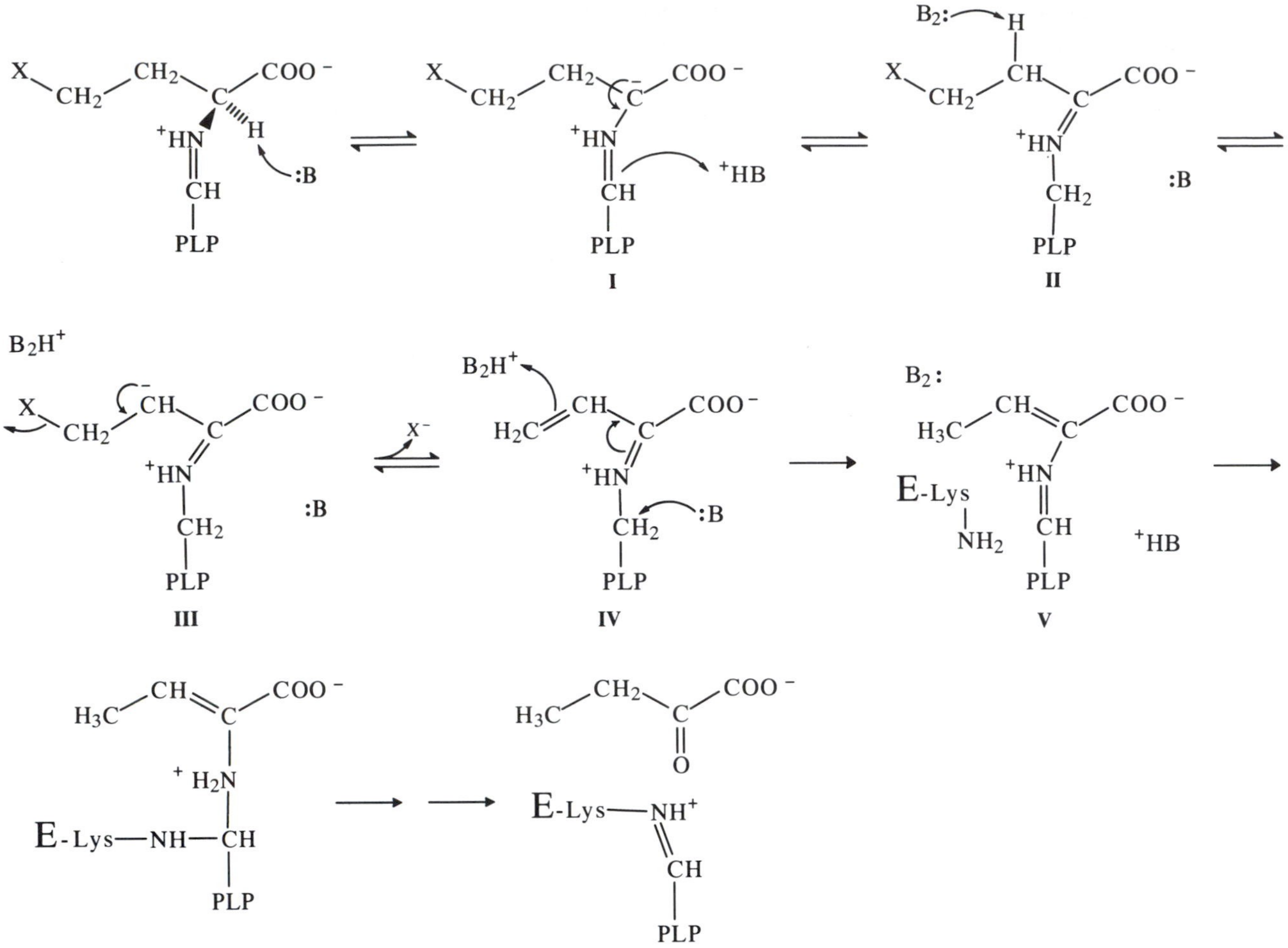

Figure 18-13

The Mechanism of PLP-dependent β,γ-Elimination Reactions. The initial steps to form an α-carbanion complex (I) between PLP and the substrate are similar to those for the α,β-elimination in Figure 18-13. Complex I is protonated on the pyridoxaldimine carbon (C-4′) by a general acid to form complex II, in which the imine carbon is the α-carbon of the amino acid. The protonated α-imino group activates a β-proton to abstraction by a base, to form the β-carbanion (III), which expels leaving-group X^- to form the β,γ-unsaturated complex (IV). Protonation of the γ-carbon of complex IV and abstraction of a proton from pyridoximine C-4′ by general acid and base groups produces complex V, which is decomposed in several steps to the enzyme-PLP complex and the α-ketoacid.

equations 12 and 13, in which PLP-CHO is pyridoxal-5′-phosphate and PLP-$CH_2NH_3^+$ is pyridoxamine-5′-phosphate.

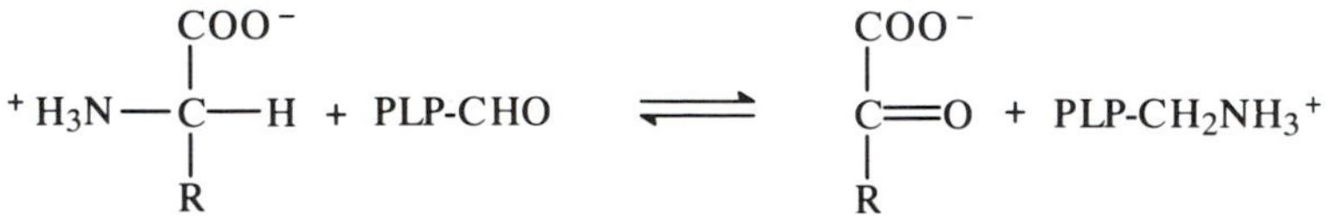

$$^+H_3N-\underset{R}{\overset{COO^-}{C}}-H + \text{PLP-CHO} \rightleftharpoons \underset{R}{\overset{COO^-}{C}}{=}O + \text{PLP-}CH_2NH_3^+ \quad (12)$$

$$\underset{R_1}{\overset{COO^-}{C}}{=}O + \text{PLP-}CH_2NH_3^+ \rightleftharpoons {}^+H_3N-\underset{R_1}{\overset{COO^-}{C}}-H + \text{PLP-CHO} \quad (13)$$

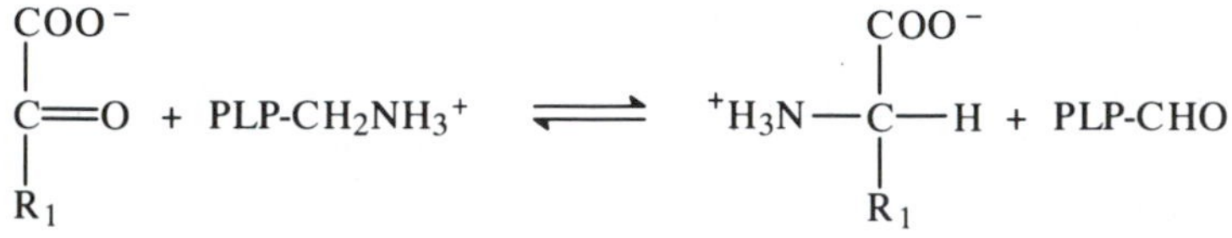

The reaction consists of two parts. First the amino acid [R-$CH(NH_3^+)COO^-$] reacts with enzyme-bound PLP to form pyridoxamine-5′-phosphate and the ketoacid product (R-$COCOO^-$). This ketoacid is released from the enzyme. The ketoacid substrate (R_1COCOO^-) then reacts with the enzyme-bound pyridoxamine-5′-phosphate to regenerate PLP and form the amino acid product [$R_1CH(NH_3^+)COO^-$].

Considerable experimental evidence supports this mechanism, including, first, that transaminases generally show a kinetic pattern consisting of parallel lines when v^{-1} is plotted against $[S]^{-1}$ at several fixed concentrations of the second substrate ("Ping Pong" kinetics). Recall that Ping Pong kinetics is observed in a two-substrate reaction when one substrate reacts with the enzyme to give a "modified" enzyme and a product that is released from the enzyme (Chapter 11). The modified enzyme then reacts with the second substrate. This regenerates the original enzyme and produces the second product. With transaminases, an amino acid reacts with enzyme-bound PLP to produce a ketoacid (product) and enzyme-bound pyridoxamine-5′-phosphate ("modified" enzyme). After release of the ketoacid product, the ketoacid substrate reacts with enzyme-bound pyridoxamine-5′-phosphate to produce the amino acid product and regenerate enzyme-bound PLP. Second, the partial reactions (equations 12 and 13) can be demonstrated by the isolation and characterization of the complex of enzyme with pyridoxamine-5′-phosphate when sufficient enzyme is available. An example is alanine transaminase, which catalyzes the reaction of equation 14:

$$\underset{\textbf{Alanine}}{{}^+H_3N{-}CH(CH_3){-}COO^-} + \underset{\alpha\textbf{-Ketoglutarate}}{{}^-OOC{-}CH_2{-}CH_2{-}C({=}O){-}COO^-} \rightleftharpoons \underset{\textbf{Pyruvate}}{CH_3{-}C({=}O){-}COO^-} + \underset{\textbf{Glutamate}}{{}^-OOC{-}CH_2{-}CH_2{-}CH(NH_3^+){-}COO^-} \quad (14)$$

When the enzyme and alanine are allowed to react, in the absence of α-ketoglutarate, one mole of pyruvate per mole of enzyme is produced; and the enzyme-bound PLP is converted into pyridoxamine-5′-phosphate. The steps in the reaction between PLP and the amino acid to produce pyridoxamine-5′-phosphate and the ketoacid are very similar to other PLP reactions, which have already been described.

The foregoing PLP-dependent reactions are part of many biosynthetic and catabolic processes, particularly the interconversion of amino acids. Through transamination reactions, amino acids are converted into ketoacids. This is generally the first step in the degradation of an amino acid.

Summary

The majority of biological elimination and addition reactions take place α,β to a carbonyl group. This suggests that these reactions include a carbanionic intermediate in an E1cB mechanism because the carbonyl group can stabilize an α-carbanion by about 40 kcal mol^{-1}. Formation of the carbanion (abstraction of the α-proton) may or may not be rate determining. β-Methylaspartase catalyzes a rapid exchange of the α-hydrogen of the substrate with solvent protons (Figure 18-2). Abstraction of the α-proton is, therefore, not rate limiting.

The enzyme aconitase catalyzes the conversion of citrate into isocitrate. This reaction formally involves the transfer of a hydroxyl group to an adjacent carbon (Figure 18-3). In fact, the transfer is accomplished by elimination of H_2O to form *cis*-aconitate and readdition of H_2O in the opposite sense to generate isocitrate.

Intramolecular oxidation/reduction can also be achieved through an elimination reaction:

$$-\underset{HO}{\overset{|}{C}}-\underset{OH}{\overset{H}{C}}-\overset{O}{\overset{\|}{C}}- \xrightarrow{-H_2O} >C=\underset{OH}{C}-\overset{O}{\overset{\|}{C}}- \longrightarrow -\underset{H}{\overset{|}{C}}-\underset{\|}{\overset{|}{C}}-\overset{O}{\overset{\|}{C}}-$$

Elimination of H_2O from a diol leads to an enol, which undergoes tautomerization to a ketone. An example is the conversion of 6-phosphogluconate into 2-keto-3-deoxy-6-phosphogluconate.

Amino acids also undergo elimination reactions and replacement reactions:

$$H_2\underset{X}{C}-\underset{H}{\overset{NH_3^+}{C}}-COO^- \xrightarrow{-HX} H_2C=\overset{NH_3^+}{C}-COO^- \longrightarrow H_3C-\overset{O}{\overset{\|}{C}}-COO^-$$

$$H_2C=\overset{NH_3^+}{C}-COO^- \xrightarrow{+HY} H_2\underset{Y}{C}-\underset{H}{\overset{NH_3^+}{C}}-COO^-$$

In these reactions, an α-carbanion is initially formed, followed by elimination of the β-leaving group (X). The resulting enamine undergoes hydrolysis to give the ketone as the final product. Replacement reactions are similar, except that HY adds to the intermediate enamine. The cofactor PLP participates by forming an aldimine (Figure 18-12) that enhances the acidity of the α-hydrogen.

There are also reactions in which a group in the γ-position is eliminated or replaced. In these cases the aldimine initially formed between the amino group of the substrate and PLP undergoes isomerization to an imine in the α-position of the amino acid:

$$B_2{:}\curvearrowright H;\quad X-CH_2-\overset{H}{CH}-\underset{\|\;{}^{+}HN-CH_2-PLP}{C}-COO^-$$

This enhances the acidity of the β-H, thereby facilitating the elimination of X.

ADDITIONAL READING

BLOCH, K. 1971. β-Hydroxydecanoyl thioester dehydrase. In *The Enzymes,* vol. 5, 3d ed., ed. P. D. Boyer. New York: Academic Press, p. 441.

WALSH, C. T. 1976. *Enzymatic Reaction Mechanisms.* San Francisco: Freeman.

BRAUNSTEIN, A. E., and GORYACHENKOVA, E. V. 1984. The β-replacement: specific pyridoxal-P-dependent lyases. In *Advances in Enzymology and Related Areas of Molecular Biology,* vol. 56. New York: Wiley-Interscience, p. 1.

DAVIS, L., and METZLER, D. E. 1972. Pyridoxal-linked elimination and replacement reactions. In *The Enzymes*, vol. 7, 3d ed., ed. P. D. Boyer. New York: Academic Press, p. 33.

JANSONIUS, J. N., and VINCENT, M. G. 1984. Structural basis for catalysis by aspartate aminotransferase. In *Active Sites of Enzymes*, ed. F. A. Jurnak and A. McPherson. *Biological Macromolecules and Assemblies*, vol. 3. New York: Wiley-Interscience, p. 187.

PROBLEMS

1. In the E1cB mechanism, either proton removal or the departure of X^- can be rate determining. Suggest an experiment to distinguish between the two possibilities (for a nonenzymatic reaction).

2. In one mechanism of PLP-dependent β-elimination, intermediate III in Figure 18-13 undergoes hydrolysis to a product-related enamine, which then undergoes hydrolysis to an imine (IV in Figure 18-13). Experimental evidence indicates that if this occurs it must happen at the active site and not in solution; that is, the enamine is not released as the enzymatic product. What evidence could indicate that this is the case?

3. The bacterial enzyme cystathionine synthase catalyzes the following reaction:

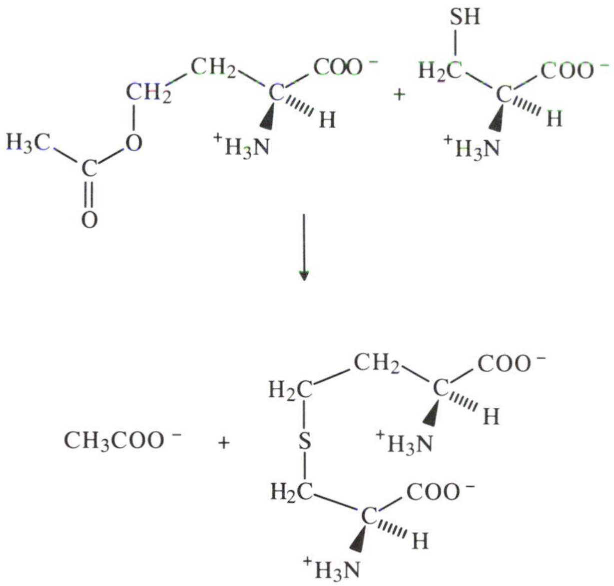

$H_3C{-}C(=O){-}O{-}CH_2{-}CH_2{-}CH(^+H_3N)COO^-$ + $HS{-}H_2C{-}CH(^+H_3N)COO^-$

$\downarrow$

CH_3COO^- + $^+H_3N(COO^-)CH{-}CH_2{-}H_2C{-}S{-}H_2C{-}CH(^+H_3N)COO^-$

Write a mechanism for the reaction. At which atom of the product would you expect to find ^2H if the reaction were carried out in [^2H] H_2O? Would you expect any ^2H in the starting material? Explain your answer.

4. An interesting phenomenon was observed in studies of exchange reactions catalyzed by the enzyme in problem 3. When the enzyme acts on α-aminobutyrate in D_2O, both β-hydrogens exchange with solvent deuterium at nearly identical rates. When the enzyme acts on α-aminopentenoate, one of the two β-hydrogens exchanges ten times as fast as the other. How do you explain these results?

5. When alanine and [^{14}C]pyruvate are incubated with alanine transaminase, [^{14}C]alanine is produced. Also, when alanine is added to transaminase in ^2H_2O, [α-^2H]alanine is produced. Explain how these two observations confirm the two-step transamination mechanism.

6. Predict which compounds are produced in the catabolic degradation of 2-keto-3-deoxy-6-phosphogluconate to three-carbon metabolites. How can *one* of the three-carbon products be further degraded to one- and two-carbon fragments? (Further degradation of the other three-carbon product is explained in Chapter 21.)

Fatty Acids

19

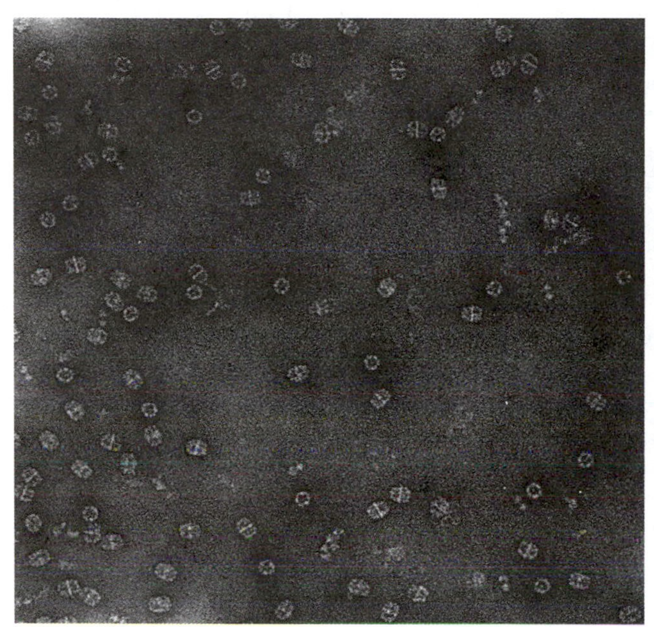

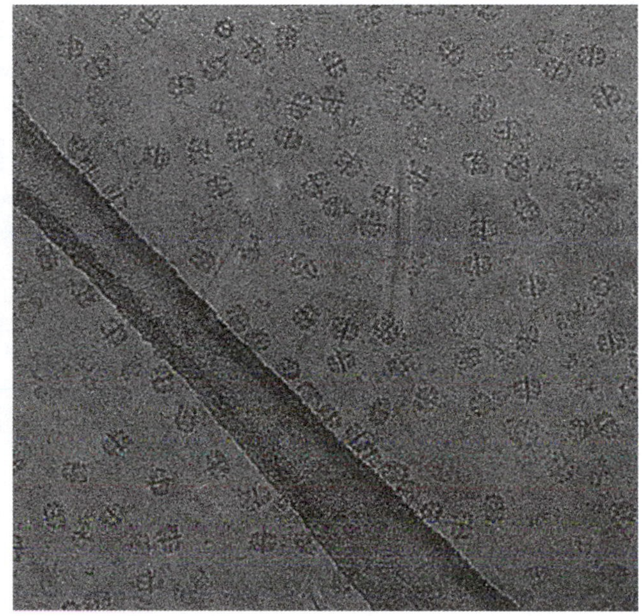

Negative stain (left) and cryo-electron (right) microscope fields of the yeast fatty acid synthase ×150,000. The enzyme (molecular weight = 2.5×10^6) is a complex of two multifunctional proteins, α and β (molecular weights = 208,000 and 220,000, respectively), that are organized in an α6β6 complex. The structure has a shape similar to a prolate ellipsoid with major and minor axes of 250 and 220 Å, respectively. The six α subunits constitute the high-density protein band that bisects the major axis of the structure, and the six archlike β subunits are attached three on either side of this central core. As a result of surface interactions, the frozen-hydrated images consist primarily of side views of the structure, whereas the stain images supported by Butvar film consist of a nearly equal population of side and end views (ring-shaped image) of the structure. The good concordance of the images indicates that the structure has been well preserved by both methods used to visualize it. (Courtesy of James K. Stoops.)

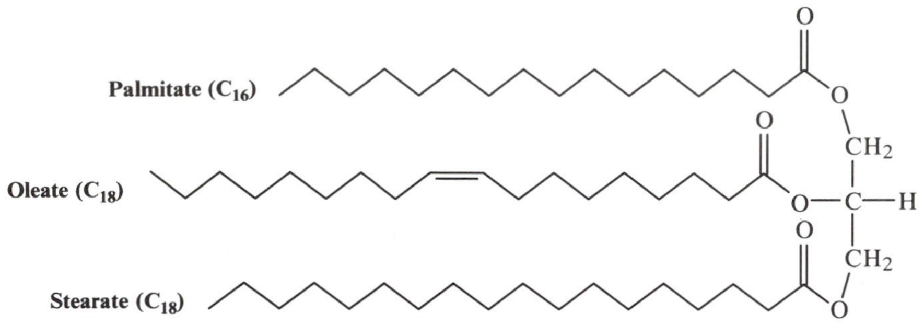

Palmitate (C_{16})
Oleate (C_{18})
Stearate (C_{18})

Figure 19-1

The Structure of a Typical Triglyceride. The triglycerides are biosynthesized by acyl group transfer from various species of fatty acyl CoA to glycerol (Chapter 11). The triglyceride shown here was biosynthesized by acyl group transfer from palmitoyl CoA, oleyl CoA, and stearyl CoA to the hydroxyl groups of glycerol. Triglycerides are degraded by the action of lipases, which catalyze their hydrolysis to fatty acids and glycerol.

Fatty acids are molecules of the general structure C_nH_m—COOH, in which C_nH_m is a hydrocarbon attached to a carboxylate group. Living cells synthesize and metabolize an enormous variety of fatty acids; however, most are linear molecules with an even number of carbon atoms ranging from 2 (acetic acid) to 24 (lignoceric acid). They may be as simple as saturated alkyl carboxylates or they may contain double bonds, as in the unsaturated fatty acids. Complex and rare fatty acids contain branches or even cyclopropyl rings in the hydrocarbon segment.

The sodium salts of long-chain fatty acids are detergents (the main component of soap) and are highly damaging to the structural integrity of cell membranes. Therefore, these molecules are not present in free form in the cell but rather are combined with other molecules as complex derivatives such as fatty acyl CoA, phospholipids, and glycerides. The biosyntheses of these compounds are described in Chapters 10 and 11.

Fatty acids serve three major functions in cells.

1. Fatty acids are the basic building material for complex molecules such as steroids and prostaglandins.
2. Fatty acids are constituents of phospholipids and glycolipids, which are the main structural components of cell membranes.
3. Fatty acids are the basic constituents of triglycerides (Chapter 11), which are a reservoir of stored energy. The oxidation of hydrocarbon chains to CO_2 yields more energy per unit of mass than does the oxidation of polysaccharides (just as the burning of gasoline yields more energy than the burning of an equivalent weight of wood).

As a source of energy, fatty acids are stored in triglycerides (Figure 19-1) and are released through the action of lipases (esterases) when required for energy production. In this chapter, we shall discuss the degradation and biosynthesis of fatty acids and related compounds.

Fatty Acid Degradation: Sequential Removal of Two-Carbon Units as Acetyl CoA

Classic experiments carried out by Knoop in 1904 yielded important information concerning fatty acid catabolism. Knoop fed dogs straight-chain fatty acids in which the ω-carbon atom was bonded to a phenyl group. He then isolated the urinary excretion products and measured the amounts of benzoic acid and phenylacetic acid. He found that when the number of methylene groups in ω-phenyl fatty acids was uneven—that is, n in the structure was an odd number—the main excretion product was phenylacetate. When n was an even number, however, benzoate was excreted.

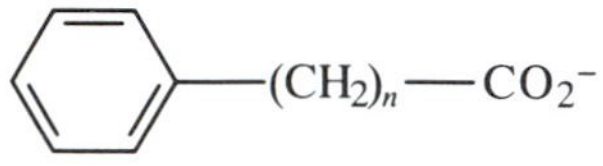

C_6H_5—$(CH_2)_n$—CO_2^-

ω-Phenyl fatty acid ($n = 3, 4, \ldots$)

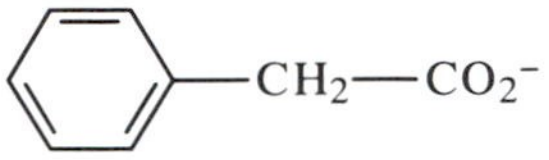

C_6H_5—CH_2—CO_2^-

Phenylacetate

C_6H_5—CO_2^-

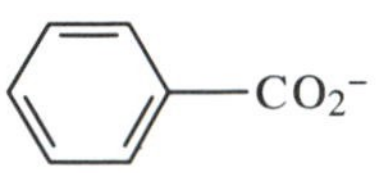

Benzoate

On the basis of these experiments, Knoop postulated that fatty acids are degraded by the sequential removal of two-carbon units, starting from the carboxyl end of the fatty acid molecule. Subsequent work confirmed Knoop's postulate. Knoop's experiment was ingenious in its use of a "labeled" molecule to study a metabolic process. In this case, the fatty acid was labeled with a phenyl group; today, isotopes such as ^{14}C are used as labels.

Through what reaction sequence could a two-carbon fragment be removed from a fatty acid? From the discussion of carbon–carbon bond-breaking reactions in Chapter 2, we know that the bond between the C_α and C_β of a fatty acid molecule cannot be broken. The fatty acid must, therefore, be modified before bond cleavage can occur. Conversion of the fatty acid into the corresponding β-keto fatty acyl CoA would provide a molecule that can undergo carbon–carbon bond cleavage by action of the enzyme thiolase (equation 1, Chapter 2).

$$R{-}CH_2 \wr CH_2{-}C(=O){-}O^- \qquad (\beta \;\wr\; \alpha)$$

$$R{-}C(=O){-}CH_2{-}C(=O){-}SCoA + CoASH \longrightarrow R{-}C(=O){-}SCoA + H_3C{-}C(=O){-}SCoA \qquad (1)$$

How is a fatty acid converted into a β-keto thioester? The fatty acid is converted into the CoA ester first and a carbonyl group must eventually be introduced in the β-position. Direct oxidation of a $-CH_2-$ group generally does not occur. However, a double bond can be introduced by dehydrogenation between C_α and C_β (Chapter 17). A hydroxyl group can then be introduced by the addition of H_2O to the double bond (Chapter 18). The β-hydroxyacyl CoA is oxidized to the β-ketoacyl CoA ester. The β-keto ester is a substrate for thiolase, which catalyzes the carbon–carbon bond cleavage according to equation 1. The overall process of fatty acid activation and oxidation to a β-ketoacyl CoA ester is shown by equations 2 through 5, and the β-ketoacyl CoA is cleaved to acetyl CoA and a fatty acyl CoA by the action of thiolase (equation 1).

1. Activation of $-COO^-$ (acyl CoA synthase)

$$R{-}CH_2{-}CH_2{-}C(=O){-}O^- + CoASH + ATP \longrightarrow R{-}CH_2{-}CH_2{-}C(=O){-}SCoA + AMP + PP_i \qquad (2)$$

2. Dehydrogenation to an α,β-unsaturated fatty acyl CoA (acyl CoA dehydrogenase)

$$R{-}CH_2{-}CH_2{-}C(=O){-}SCoA + FAD \longrightarrow (R)(H)C{=}C(H){-}C(=O){-}SCoA + FADH_2 \qquad (3)$$

3. Hydration of the double bond (enoyl CoA hydratase)

$$(R)(H)C{=}C(H){-}C(=O){-}SCoA + H_2O \longrightarrow R{-}C(HO)(H){-}CH_2{-}C(=O){-}SCoA \qquad (4)$$

4. Oxidation of the alcohol group (L-3-hydroxyacyl CoA dehydrogenase)

$$R{-}C(HO)(H){-}CH_2{-}C(=O){-}SCoA + NAD^+ \longrightarrow R{-}C(=O){-}CH_2{-}C(=O){-}SCoA + NADH + H^+ \qquad (5)$$

5. Cleavage of a carbon–carbon bond (β-keto thiolase)

$$R{-}C({=}O){-}CH_2{-}C({=}O){-}SCoA + CoASH \longrightarrow R{-}C({=}O){-}SCoA + H_3C{-}C({=}O){-}SCoA \quad (1)$$

This process is repeated until the fatty acid is completely degraded. Note that all of the reactions have been described in preceding chapters. Fatty acids containing an even number of carbon atoms are converted into acetyl CoA molecules, whereas those containing an odd number of carbon atoms are converted into acetyl CoA and a single propionyl CoA molecule. Acetyl CoA and propionyl CoA are eventually oxidized to CO_2 and H_2O through the tricarboxylic acid cycle (Chapter 21).

A problem arises in the degradation of certain unsaturated fatty acids such as the C_{16} acid, palmitoleate, which contains a double bond between C-3 and C-4. The degradation pathway of equations 2 through 5 requires a double bond between C-2 and C-3, the formation of which is prevented by the presence of a double bond between C-3 and C-4. Another enzyme, an isomerase, shifts the double bond between C-2 and C-3; that is, to the α,β-position (equation 6).

$$H_3C{-}(CH_2)_5{-}CH{=}CH{-}CH_2{-}C({=}O){-}SCoA \rightleftharpoons H_3C{-}(CH_2)_6{-}CH{=}CH{-}C({=}O){-}SCoA \quad (6)$$

Note that the isomerization of equation 6 changes the position of the double bond and also converts the *cis* double bond into the *trans* configuration. The α,β-unsaturated fatty acyl CoA then undergoes the conventional degradation, beginning with hydration to the β-hydroxyacyl CoA according to equation 4.

Reactions represented by equations 1 through 5 are known as the β-oxidation of fatty acids because the oxidation steps take place at the β-carbon of the fatty acyl CoA. The process is often referred to simply as β-oxidation or the β-oxidation pathway.

Energetics of Fatty Acid Breakdown

Biological organisms use two different kinds of molecules for long-term energy storage: polysaccharides and fatty acids. Of these, fatty acids are by far the most-efficient form of stored energy. To understand this, let us consider the fate of a single molecule of stearoyl CoA, a saturated fatty acid with an 18-carbon alkyl chain. This molecule is broken down through eight "rounds" of the cycle described in the preceding section, resulting in nine acetyl CoA molecules. The overall process is described by equation 7.

$$\text{Stearoyl CoA} + 8\,\text{CoA} + 8\,\text{NAD}^+ + 8\,\text{FAD} + 8\,\text{H}_2\text{O} \longrightarrow 8\,\text{H}^+ + 9\,\text{Acetyl CoA} + 8\,\text{NADH} + 8\,\text{FADH}_2 \quad (7)$$

The reaction products afford two opportunities for generating energy in the form of ATP: (1) the oxidation of NADH and $FADH_2$ through the electron-transport chain and oxidative phosphorylation to produce ATP, and (2) the metabolism of acetyl CoA to CO_2 through the tricarboxylic acid (TCA) cycle, which also produces NADH and $FADH_2$ for oxidative phosphorylation and ATP production. The TCA cycle and oxidative phosphorylation will be described in Chapters 21 and 22.

The overall result of stearoyl CoA oxidation is described by equation 8, which shows that the complete oxidation of 1 mole of stearoyl CoA produces 148 moles of ATP.

$$\text{Stearoyl CoA} + 26\,O_2 + 148\,ADP^{3-} + 148\,H_2PO_4^- \longrightarrow 165\,H_2O + 18\,CO_2 + \text{CoASH} + 148\,ATP^{4-} \quad (8)$$

This reaction should be compared to the oxidative metabolism of eighteen carbons in the form of carbohydrate—for example 3 moles of a hexose such as glucose, according to equation 9.

$$3\,C_6H_{12}O_6 + 18\,O_2 + 114\,ADP^{3-} + 114\,H_2PO_4^- \longrightarrow 132\,H_2O + 18\,CO_2 + 114\,ATP^{4-} \quad (9)$$

Thus, for an equimolar production of CO_2, fatty acid breakdown yields about 30% more ATP than does sugar metabolism. More importantly, the fatty acid is about half the weight of an equimolar amount of sugar. Therefore, the fatty acid is a far more efficient storage form of energy for an animal, which must carry it around.

An additional benefit of fatty acid oxidation is the amount of water produced. Note from equations 8 and 9 that more water is produced than ATP. The water produced in excess of ATP results from the oxidation of fatty acid or sugar. For each gram of fatty acid oxidized, a little more than a milliliter of water is produced and, for each gram of sugar, about 0.6 ml of water is produced. Thus, the camel can set out onto the desert in the summer, and the bear into its cave for the winter, comforted by the knowledge that the vast stores of fat that they carry will provide not only energy but water as well.

All this a consequence of the fact that the free energy of combustion of hydrocarbons is about twice as large as that for carbohydrates. Similarly, gasoline is a better fuel than wood, which is largely cellulose.

Fatty Acid Biosynthesis

Fatty acids could, in principle, be synthesized through reversal of the process by which they are degraded (equations 1 through 5). This is not the case, however. The enzyme system for fatty acid synthesis differs in many respects from the degradative system, although it is based on the same chemical principles. The two systems exist in different compartments of eukaryotic cells; fatty acid synthesis occurs in the cytoplasm, whereas degradation takes place in the mitochondrion.

Initiation of Fatty Acid Synthesis

S. Wakil established that malonyl CoA plays an important role in the biosynthesis of fatty acids. He observed that fatty acid biosynthesis in cell extracts requires the presence of *bicarbonate*, and that malonyl CoA is an intermediate. Malonyl CoA is derived from the carboxylation of acetyl CoA, which is catalyzed by acetyl CoA carboxylase (equation 10).

$$H_3C{-}C({=}O){-}SCoA + ATP + {}^-O{-}C({=}O){-}OH \longrightarrow {}^-O{-}C({=}O){-}CH_2{-}C({=}O){-}SCoA + ADP + P_i \quad (10)$$

Malonyl CoA

Two carbons of each fatty acid chain are furnished by acetyl CoA; however, all

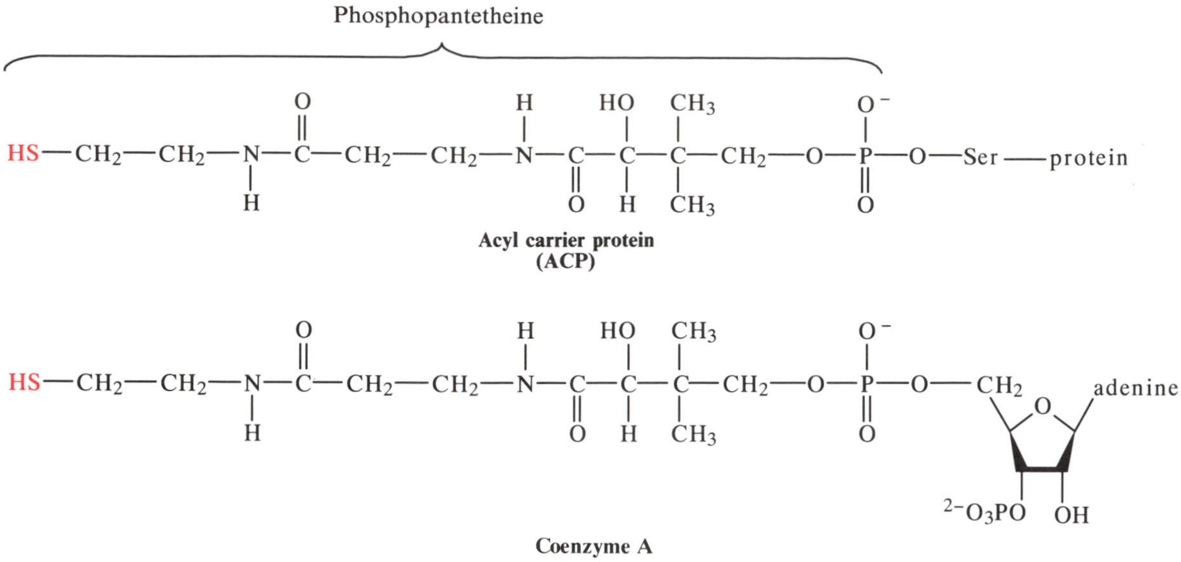

Phosphopantetheine

$$HS{-}CH_2{-}CH_2{-}N(H){-}C(=O){-}CH_2{-}CH_2{-}N(H){-}C(=O){-}C(H)(OH){-}C(CH_3)_2{-}CH_2{-}O{-}P(=O)(O^-){-}O{-}Ser{-}protein$$

Acyl carrier protein (ACP)

$$HS{-}CH_2{-}CH_2{-}N(H){-}C(=O){-}CH_2{-}CH_2{-}N(H){-}C(=O){-}C(H)(OH){-}C(CH_3)_2{-}CH_2{-}O{-}P(=O)(O^-){-}O{-}CH_2{-}\text{ribose(adenine)}(3'\text{-}OPO_3^{2-})(2'\text{-}OH)$$

Coenzyme A

Figure 19-2

The Structures of Acyl Carrier Protein and Coenzyme A. Both ACP and coenzyme A contain the phosphopantetheine moiety. In ACP, the phosphopantetheine is bonded to the β-hydroxyl group of a serine residue; whereas, in coenzyme A, it is bonded to the 5′-hydroxyl group of adenosine-3′-phosphate.

of the other carbon atoms are furnished by malonyl CoA, which is the building block for fatty acid biosynthesis.

In fatty acid biosynthesis, the malonyl and acetyl groups of malonyl CoA and acetyl CoA are first transferred to the thiol group of phosphopantetheine in a protein known as acyl carrier protein (ACP). In this protein, a phosphopantetheine moiety is covalently attached to a serine residue and takes the place of coenzyme A, as shown in Figure 19-2. Recall that phosphopantetheine is also a component of coenzyme A (Figure 19-2). The transfer of the acyl moieties of acetyl CoA and malonyl CoA to ACP are described by equations 11 and 12.

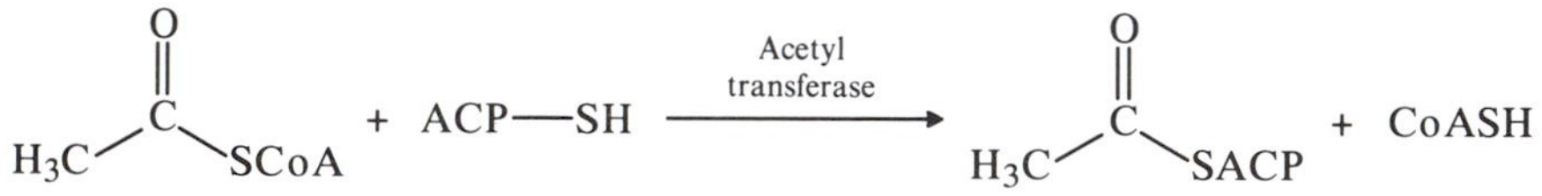

$$H_3C{-}C(=O){-}SCoA + ACP{-}SH \xrightarrow{\text{Acetyl transferase}} H_3C{-}C(=O){-}SACP + CoASH \quad (11)$$

$$^-O{-}C(=O){-}CH_2{-}C(=O){-}SCoA + ACP{-}SH \xrightarrow{\text{Malonyl transferase}} {}^-O{-}C(=O){-}CH_2{-}C(=O){-}SACP + CoASH \quad (12)$$

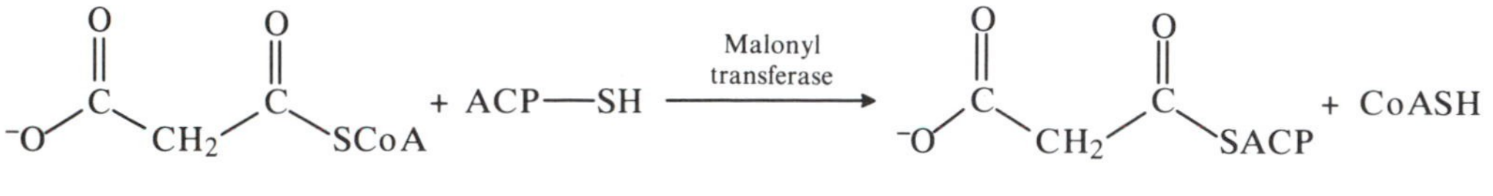

Extension of the Alkyl Chain

Next, a condensation reaction, catalyzed by acyl malonyl ACP condensing enzyme, takes place to produce acetoacetyl ACP and ACP—SH according to equation 13.

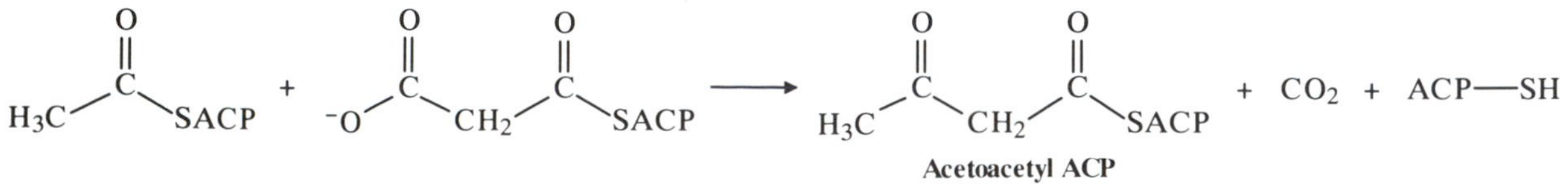

$$H_3C{-}C(=O){-}SACP + {}^-O{-}C(=O){-}CH_2{-}C(=O){-}SACP \longrightarrow \underset{\text{Acetoacetyl ACP}}{H_3C{-}C(=O){-}CH_2{-}C(=O){-}SACP} + CO_2 + ACP{-}SH \quad (13)$$

The acyl malonyl ACP condensing enzyme catalyzes the reaction of equation 13 through an intermediate acyl enzyme, in which the acyl group is bonded to the β-thiol group of a cysteine residue of the condensing enzyme, as illustrated in Figure 19-3. The first step is acyl group transfer from acetyl ACP to the cysteine–SH group and release of ACP. In the next step, the acetyl-enzyme reacts with malonyl ACP to form acetoacetyl ACP.

This condensation, like other condensations of this type explained in Chapter 2, includes the addition of a carbanion to a carbonyl group. The reaction sequence shown in Figure 19-4 differs from other condensation reactions that we have seen, in that the carbanion is produced through decarboxylation of malonyl ACP and not through dissociation of a proton.

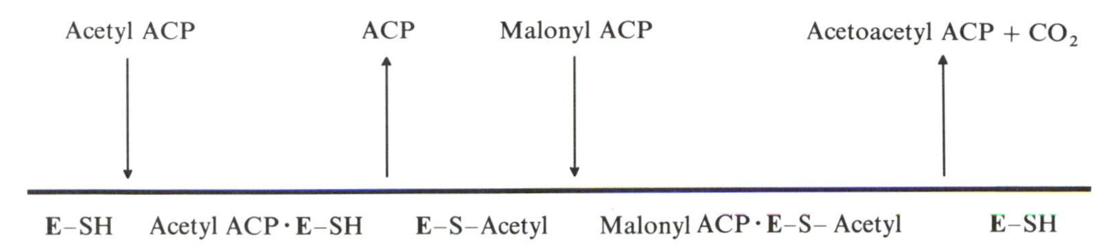

Figure 19-3

The Mechanism of Group Transfer by Acyl Malonyl ACP Condensing Enzyme. In the first condensation step of fatty acid biosynthesis, acetyl ACP undergoes acyl group transfer with the enzyme to form an acetyl enzyme and ACP. In the second step, the acetyl enzyme undergoes condensation with malonyl ACP to form acetoacetyl ACP. In subsequent steps catalyzed by other enzymes, acetoacetyl ACP is reduced to butyryl ACP, which in turn acylates the condensing enzyme through the same process as acetyl ACP to form a butyryl enzyme. The butyryl enzyme is then extended by two carbons in the second step of this scheme; that is, by reaction with malonyl ACP. The acyl malonyl ACP condensing enzyme will accept any acyl ACP not longer than sixteen carbons as an acylation substrate and extend it by two carbons.

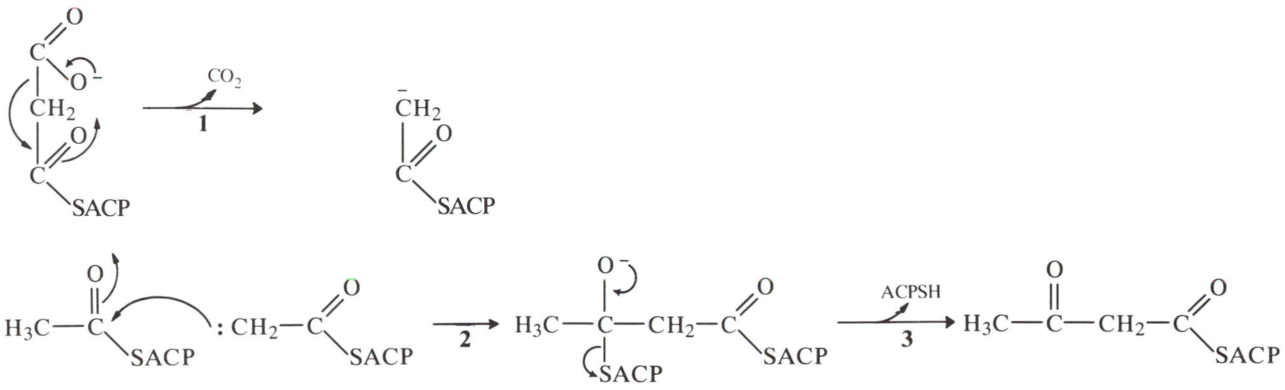

$^{-}O_2C{-}CH_2{-}C(=O){-}SACP \xrightarrow[1]{CO_2} {}^{-}CH_2{-}C(=O){-}SACP$

$H_3C{-}C(=O){-}SACP + :CH_2{-}C(=O){-}SACP \xrightarrow{2} H_3C{-}C(O^-)(SACP){-}CH_2{-}C(=O){-}SACP \xrightarrow[3]{ACPSH} H_3C{-}C(=O){-}CH_2{-}C(=O){-}SACP$

Figure 19-4

The Mechanism of Carbon–Carbon Condensation Catalyzed by Acyl Malonyl ACP Condensing Enzyme. In the formation of acetoacetyl ACP, the acetyl enzyme binds malonyl ACP and catalyzes its decarboxylation to the carbanion of acetyl ACP in step 1. In step 2, the carbanion adds to the carbonyl group of the acetyl enzyme to form an addition intermediate, which eliminates the thiol group of the enzyme in the third step to form acetoacetyl ACP. In subsequent elongation steps, the acetyl enzyme is replaced by butyryl enzyme, hexanoyl enzyme, and so forth; the carbanion undergoes addition to the carbonyl groups of these acyl enzymes.

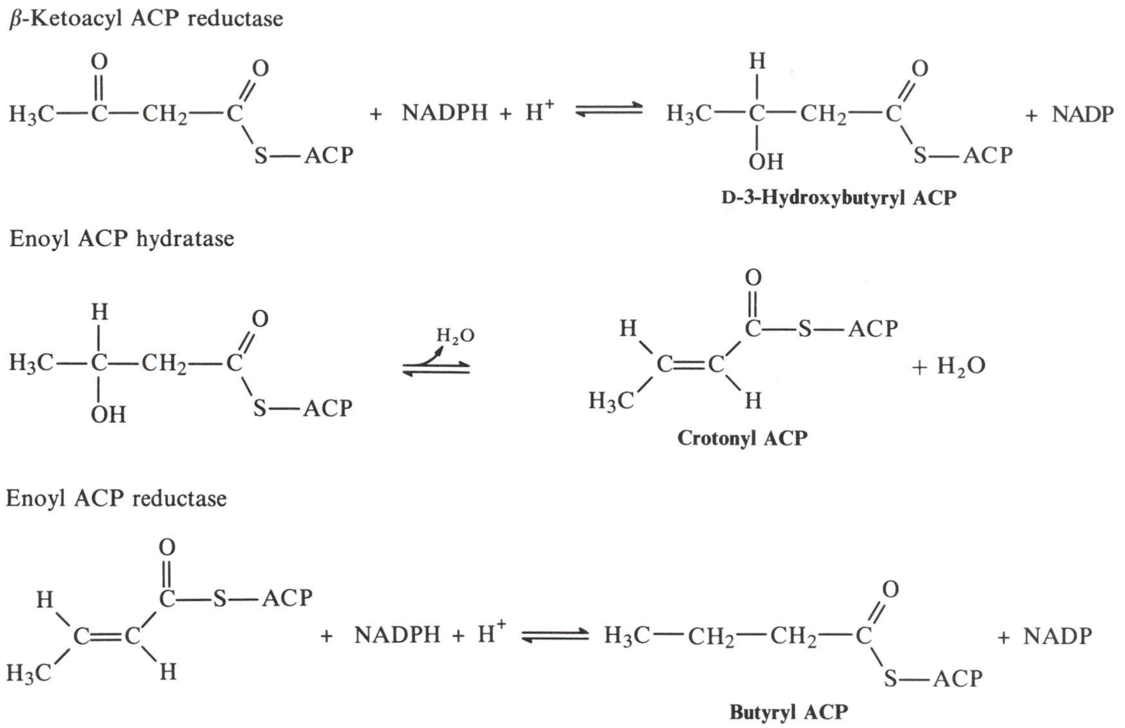

β-Ketoacyl ACP reductase

$$H_3C-\overset{O}{\overset{\|}{C}}-CH_2-C(=O)-S-ACP + NADPH + H^+ \rightleftharpoons H_3C-\overset{H}{\underset{OH}{C}}-CH_2-C(=O)-S-ACP + NADP$$

D-3-Hydroxybutyryl ACP

Enoyl ACP hydratase

$$H_3C-\overset{H}{\underset{OH}{C}}-CH_2-C(=O)-S-ACP \overset{H_2O}{\rightleftharpoons} \underset{H_3C}{\overset{H}{}}C{=}C\underset{H}{\overset{C(=O)-S-ACP}{}} + H_2O$$

Crotonyl ACP

Enoyl ACP reductase

$$\underset{H_3C}{\overset{H}{}}C{=}C\underset{H}{\overset{C(=O)-S-ACP}{}} + NADPH + H^+ \rightleftharpoons H_3C-CH_2-CH_2-C(=O)-S-ACP + NADP$$

Butyryl ACP

Figure 19-5

Reduction of Acetoacetyl ACP to Butyryl ACP in Fatty Acid Biosynthesis. β-Ketoacyl ACP reductase catalyzes the reduction of acetoacetyl ACP by NADPH to D-3-hydroxybutyryl ACP, which is then dehydrated by the action of enoyl ACP hydratase to crotonyl ACP. Reduction of crotonyl ACP to butyryl ACP by NADPH is catalyzed by a flavoprotein, enoyl ACP reductase.

Acetoacetyl ACP is reduced to butyryl ACP through the reactions shown in Figure 19-5. This sequence is very similar to that by which β-oxidation in fatty acid degradation occurs in the reverse direction. The principal difference is that NADPH is used in the reductive steps rather than NADH. As a general rule, NADPH is utilized in reductive biosynthetic (anabolic) pathways, whereas NADH is produced in catabolic pathways.

In the next alkyl chain extension cycle, butyryl ACP reacts with malonyl ACP to form β-ketohexanoyl ACP. This compound undergoes reduction, dehydration, and reduction through the reactions in Figure 19-5 and is converted into hexanoyl ACP. Extension of the carbon chain continues until the C_{16} fatty acid thioester, palmityl ACP, is formed.

Termination of Alkyl Chain Growth

Palmityl ACP does not undergo further condensation but is hydrolyzed to palmitate, the final product of fatty acid synthesis. The stoichiometry of the overall process is shown in equation 14.

$$\text{Acetyl CoA} + 7\,\text{Malonyl CoA} + 14\,\text{NADPH} + 20\,H^+ \longrightarrow \text{Palmitate} + 6\,H_2O + 7\,CO_2 + 14\,\text{NADP}^+ + 8\,\text{CoASH} \qquad (14)$$

Interactions of Enzymes in Fatty Acid Synthesis

In *E. coli* and other bacteria, the enzymes of fatty acid biosynthesis can be purified as individual enzymes and are thought to act sequentially in the biosynthesis of fatty acids. In yeast and higher animals, fatty acid synthesis takes place within relatively large multienzyme complexes that contain not only all of the enzyme activities required for the synthesis of fatty acids in *E. coli*, but also ACP.

It has so far not been possible to separate the individual enzymes from any of the eukaryotic fatty acid synthetase complexes. For instance, the molecular weight of the yeast complex is 2,300,000. It consists of two different types of protein chains, α and β, each of which contains several enzyme activities. There are six copies of each chain in a particle of the complex, which may be designated $\alpha_6\beta_6$; and the α chain contains an ACP domain.

The fatty acid synthase complex from pigeon liver has a molecular weight of about 500,000 and consists of two identical polypeptide chains, each of which contains all seven enzymatic activities, as well as an ACP domain. The seven activities are those responsible for catalyzing the reactions of equations 11 through 13, the reactions in Figure 19-5, and the hydrolysis of palmitoyl ACP to palmitate. Other animal fatty acid synthase complexes appear to be similar to the pigeon liver complex.

We shall describe the function of the yeast complex in further detail to explain how these complexes work. The complex contains a number of sulfhydryl (—SH) groups, which are necessary for catalytic activity. The sulfhydryl groups fall into two categories on the basis of their reactivities toward iodoacetate. One class of sulfhydryl groups readily reacts with iodoacetate and belongs to the condensing enzyme. (The condensing enzyme from *E. coli* also contains a sulfhydryl group that is essential for catalytic activity.) The second class of sulfhydryl groups is relatively insensitive to iodoacetate and belongs to the ACP of the fatty acid synthase complex. In addition to the sulfhydryl groups, there are two serine hydroxyl groups to which the substrate and intermediates become covalently attached.

Figure 19-6 is a schematic illustration of how these hydroxyl and sulfhydryl groups participate in the overall reaction. Initially, the acyl and malonyl groups of acetyl CoA and malonyl CoA are transferred to serine-OH groups of the complex, as indicated in the diagram. The acetyl group is transferred, first to the ACP-sulfhydryl group (not shown) and then to the sulfhydryl group of the condensing enzyme. The malonyl moiety is then transferred to the ACP-sulfhydryl group. The condensation reaction then takes place and ACP-bound acetoacetate is formed. The acetoacetate is then reduced to butyryl ACP through exactly the same three-step reaction sequence as that in *E. coli* (Figure 19-5). Three enzymes take part in the conversion of AcAc-ACP into butyryl ACP. The acyl group attached to the ACP domain can be transported from one enzyme domain to the next. ACP thus serves as a flexible arm, 20 Å in length, that transports the acyl group from one enzyme domain to another. The resulting butyryl group bonded to ACP is transferred to the sulfhydryl group of the condensing enzyme, which frees ACP for acylation by

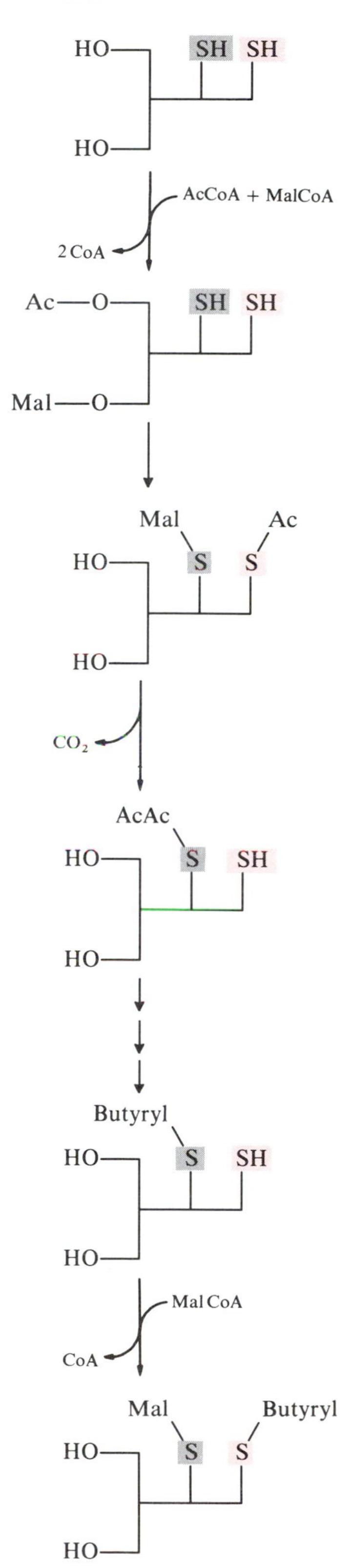

Figure 19-6

Acylation and C–C Bond Formation by the Condensing Enzyme and ACP Domains in Animal Fatty Acid Synthase Complexes. Two serine hydroxyl groups of the condensing enzyme are acylated by acetyl CoA and malonyl CoA. The acetyl group is then transferred to a sulfhydryl group of the condensing enzyme, and the malonyl group is transferred to the sulfhydryl group of the ACP domain. The condensing enzyme catalyzes the decarboxylation of the malonyl group and the addition of the resulting carbanion to the acetyl group by the mechanism of Figure 19-4. The resulting acetoacetyl ACP is reduced to butyryl ACP by the sequence of reactions in Figure 19-5. The butyryl group is then shifted to the sulfhydryl group of the condensing enzyme, and another malonyl CoA reacts with the ACP domain to acylate the sulfhydryl group. In subsequent steps, another carbon–carbon bond is formed by decarboxylation of malonyl ACP and condensation of the carbanion with the butyryl group.

malonyl CoA. The system is thereby positioned for another cycle of chain growth, beginning with the decarboxylation of malonyl ACP and condensation of the carbanion with the butyryl group to form β-ketohexanoyl ACP. The cycle is repeated until palmitoyl ACP is formed and undergoes hydrolysis to palmitate, which is finally released from the complex and frees it for another cycle of palmitate formation.

High concentrations of palmitate in the cell would be destructive because of its properties as a surfactant. Palmitate is immediately activated to palmitoyl CoA (equation 2), as are all fatty acids, including those released from glycerides by the action of lipases.

An important property of this system is that all of the reaction intermediates remain covalently attached to the enzyme complex. Only the final product is released. This arrangement is an obvious advantage in the synthesis of a large molecule.

Control of Fatty Acid Biosynthesis

Because the major building block of fatty acids is malonyl CoA, modulation of its synthesis is a logical means of controlling fatty acid synthesis. The biotin-dependent enzyme, acetyl CoA carboxylase (Chapter 17), catalyzes the carboxylation of acetyl CoA to malonyl CoA. Acetyl CoA carboxylase is subject to allosteric regulation by several molecules. Citrate strongly activates acetyl CoA carboxylase; it also causes the enzyme to undergo polymerization into long fibrous structures. For example, the molecular weight of rat liver acetyl CoA carboxylase increases from 5.2×10^5 in the absence of citrate to 4.8×10^6 when citrate is present at high concentrations. The polymerized form is catalytically active. Palmitoyl CoA, on the other hand, the product of fatty acid synthesis, inhibits malonyl CoA production and thus reduces the rate of fatty acid production.

Regulation of acetyl CoA carboxylase through activation by citrate and inhibition by palmitoyl CoA is a logical means of balancing the activities of major metabolic pathways in response to cellular needs. The scheme in Figure 19-7 illustrates these relations. The TCA cycle, which is described in Chapter 21, oxidizes acetyl CoA to CO_2; and it utilizes the electrons derived from the oxidation to reduce NAD^+ and FAD to NADH and $FADH_2$, respectively. The reduced coenzymes are subsequently reoxidized by oxygen in the terminal electron-transport pathway, and the energy released is used to produce ATP by oxidative phosphorylation (Chapter 22). The first step of the TCA cycle is the condensation of acetyl CoA with oxaloacetate to form citrate, which is catalyzed by citrate synthase (Chapter 2). Thus, the presence in a cell of a high citrate concentration signals a high activity of the TCA cycle, presumably resulting from a high level of acetyl CoA. Excess acetyl CoA that is not needed for energy production through the TCA cycle and oxidative phosphorylation can be used for fatty acid synthesis, and the high citrate concentration facilitates this process by activating acetyl CoA carboxylase to produce malonyl CoA.

Palmitate produced by fatty acid biosynthesis is quickly converted into palmitoyl CoA for incorporation into glycerides or phospholipids (Chapter 11). The presence of a high concentration of palmitoyl CoA is a signal of excess fatty acids in the cell. Therefore, it is reasonable for palmitoyl CoA to act as an inhibitor of fatty acid biosynthesis by blocking the carboxylation of acetyl CoA to malonyl CoA.

Another aspect of control in fatty acid synthesis is the fact that short-chain fatty acids are produced *in vivo* in many animals despite the fact that the purified synthase produces only palmitate *in vitro*. The shorter-chain fatty acids appear

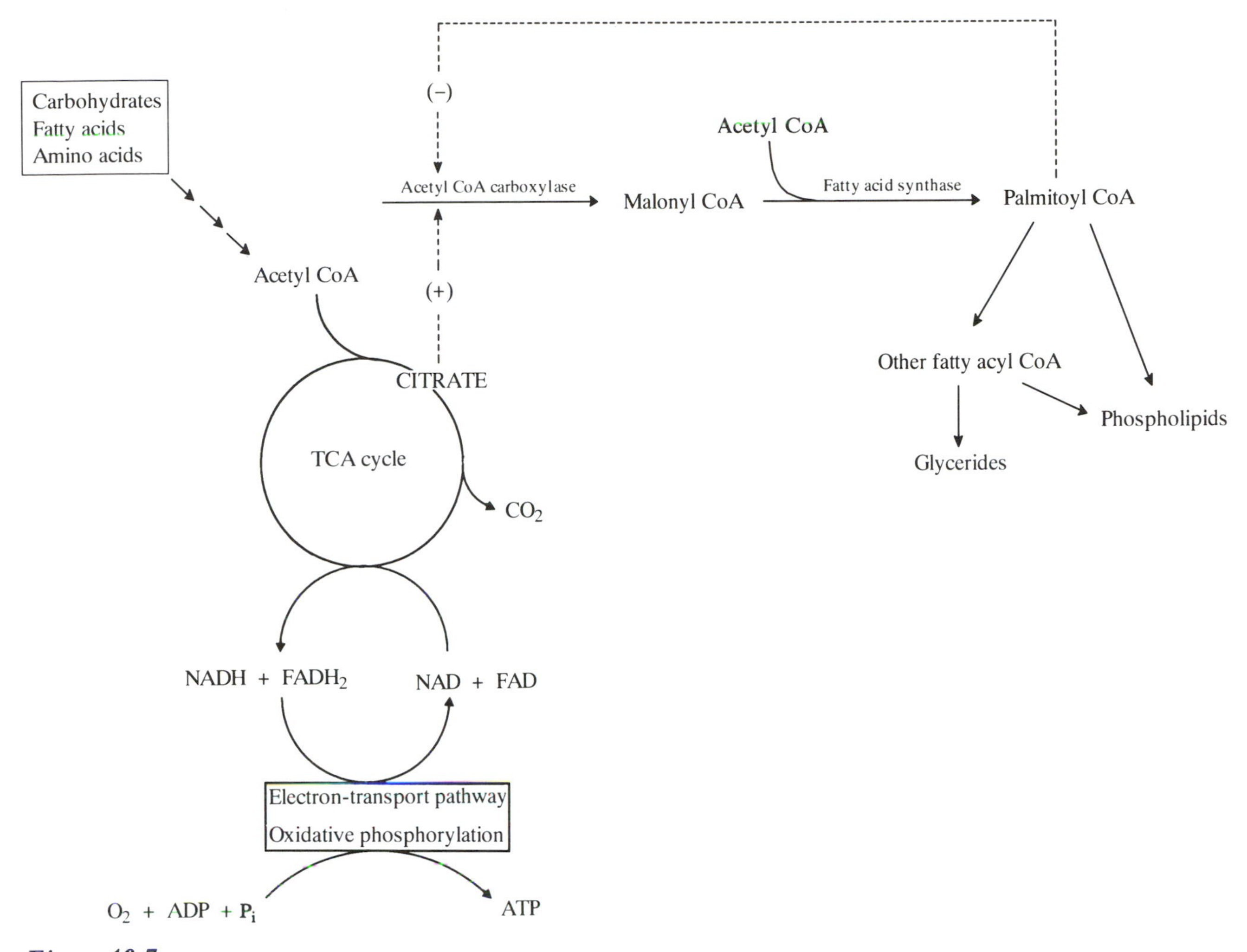

Figure 19-7

Regulation of Acetyl CoA Carboxylase by Citrate and Palmitate. This figure illustrates the relations between fatty acid biosynthesis and the TCA cycle in *E. coli*, both of which utilize acetyl CoA, and how the control of fatty acid biosynthesis is related to the operation of the two pathways. High citrate levels result from high activity in the TCA cycle in response to a high concentration of acetyl CoA. The high citrate concentration activates acetyl CoA carboxylase, which converts excess acetyl CoA into malonyl CoA for fatty acid biosynthesis. Palmitate, which is produced by fatty acid synthase, modulates fatty acid biosynthesis by inhibiting acetyl CoA carboxylase. Under normal cellular conditions, palmitate is converted into palmitoyl CoA and other fatty acyl CoA's for phospholipid and glyceride biosynthesis. Acetyl CoA is derived from the catabolism of carbohydrates, amino acids, and fatty acids. The oxidation of acetyl CoA through the TCA cycle produces NADH and $FADH_2$ (Chapter 21). The NADH and $FADH_2$ are reoxidized in the electron-transport pathway, and the electrons derived from this process are transferred to O_2 (Chapter 22). The energy derived from oxidation-reduction in the electron-transport pathway is used for ATP production.

to be produced by premature hydrolysis of the growing acyl ACP. This is brought about by regulation of the thioesterase activity of the fatty acid synthase complex, in which regulatory proteins bind to the complex and alter the specificity of the thioesterase.

Modification of Long-Chain Fatty Acids

The sequence of synthetic reactions discussed earlier ends at palmitate, a sixteen-carbon saturated fatty acid. But, if we analyze the fatty acid content of phospholipids in a "typical" organism, we will find a complex collection of chemical species consisting of fatty acids with one or more double bonds and chain lengths in excess of sixteen carbons, as well as palmitate. This is because most organisms contain enzymes that modify palmitoyl CoA by introducing double bonds or elongating the hydrocarbon chain or both.

Fatty Acid Chain Elongation

In addition to possessing the systems that synthesize palmitate from acetyl CoA and malonyl CoA, most organisms contain enzyme systems that can elongate fatty acids. The elongation system utilizes acetyl CoA as the two-carbon unit of extension, and the carbon–carbon bond-forming reaction is essentially a reversal of the β-ketothiolase reaction in the degradation of fatty acids (equation 1). The elongation of palmitoyl CoA by two carbon atoms is described by equation 15.

$$\text{Acetyl CoA} + \text{Palmitoyl CoA} + 2\,\text{NADPH} + 2\,\text{H}^+ \longrightarrow \text{Stearoyl CoA} + \text{H}_2\text{O} + 2\,\text{NADP}^+ + \text{CoASH} \qquad (15)$$

Unsaturated Fatty Acids

In aerobic organisms, the introduction of double bonds is carried out by enzymes located on internal membranes. An example of this kind of reaction is the conversion of palmitoyl CoA into palmitoleyl CoA (equation 16).

$$CH_3\text{—}(CH_2)_5\text{—}CH_2\text{—}CH_2\text{—}(CH_2)_7\text{—}C(=O)\text{—}SCoA + NADPH + O_2 + H^+ \longrightarrow$$

$$CH_3\text{—}(CH_2)_5\text{—}CH{=}CH\text{—}(CH_2)_7\text{—}C(=O)\text{—}SCoA + NADP^+ + 2\,H_2O \qquad (16)$$

There are several aspects of this reaction that warrant discussion. Most importantly, molecular oxygen is the primary oxidant. This is very different from our earlier encounter with chain unsaturation in fatty acid metabolism, the first step in fatty acid breakdown. In that case, the oxidation of the hydrocarbon was powered by the reduction of FAD in a conventional redox reaction (equation 3). Why is the present case different? Why should the cell go to the considerable trouble of coupling this double-bond formation with the reduction of molecular oxygen? The answer, as usual, lies in the chemistry of the process. In fatty acid breakdown, the double bond formed is α,β to a carbonyl group, and thus the carbon from which the hydrogen atom is abstracted is "activated"; that is, it is relatively acidic ($pK_a \sim 24$) compared with the unactivated C–H bond ($pK_a \sim 50$). However, in the unsaturation of palmitate, the carbons from which the protons must be removed are not activated, and so a much more powerful oxidizing agent than FAD is needed. Such an agent is molecular oxygen.

Because of this requirement, another problem arises. The reduction of O_2 to H_2O requires four electrons (and four hydrogens), but only two are available from palmitoyl CoA. Therefore, an additional source of electrons is needed, and this is the reason for the participation of NADPH in this reaction. This is quite an eccentric situation: an oxidation reaction that requires NADPH. Oxidases that utilize NADPH as a cosubstrate are known as mixed-function oxidases, and their mechanisms of action are discussed in Chapter 24.

Mammals do not have desaturation enzymes that introduce double bonds into fatty acids beyond the C-9 position. Therefore, they cannot synthesize such fatty acids as linoleic (18:2 *cis* Δ9, Δ12) or linolenic acid (18:3 *cis* Δ9, Δ12, Δ15).* These fatty acids must be furnished in the diet and are, therefore, known as essential fatty acids. They are important in mammalian metabolism because they may be converted into other important fatty acids by elongation and introduction of additional double bonds.

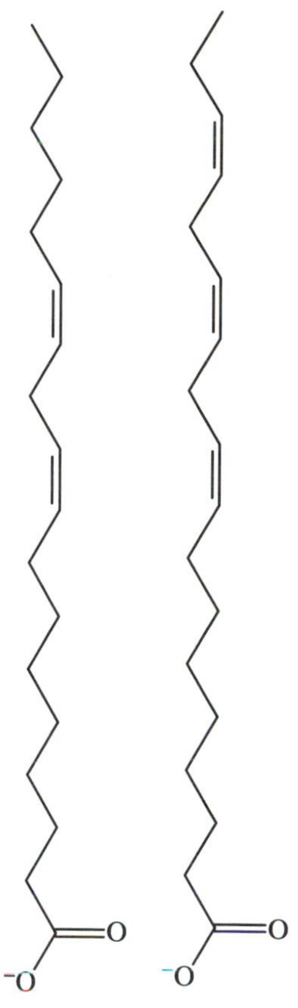

Cellular Compartmentation of Fatty Acid Metabolizing Systems

Fatty acids are synthesized in the cytoplasm, where they are "fixed" as triglyceride droplets or constituents of membranes. But the enzymes required to break down these fatty acids for the purpose of energy generation are located inside the mitochondrion. Because the inner membrane of this organelle is impermeable to fatty acyl CoA esters, a "carrier" system exists to move these compounds into the mitochondrion. This system depends on carnitine, which is acylated by fatty acyl CoA in a reaction catalyzed by a carnitine acyltransferase (equation 17).

$$R{-}\overset{O}{\overset{\|}{C}}{-}SCoA + \underbrace{(CH_3)_3\overset{+}{N}{-}CH_2{-}CH(OH){-}CH_2{-}COO^-}_{\text{Carnitine}} \longrightarrow \underbrace{(CH_3)_3\overset{+}{N}{-}CH_2{-}CH(O{-}\overset{O}{\overset{\|}{C}}{-}R){-}CH_2{-}COO^-}_{\text{Acylcarnitine}} + CoASH \qquad (17)$$

The resulting acylcarnitine is transported across the mitochondrial membrane by a specific transport protein. Once inside the mitochondrion, the acylcarnitine reacts with CoASH to again form fatty acyl CoA and carnitine. The carnitine freed inside the mitochondrion is transported back to the cytosol, making it available for reacylation by another fatty acyl CoA (Figure 19-8).

A similar problem of compartmentation exists for acetyl CoA. Malonyl CoA, the building block for fatty acid synthesis, is derived from acetyl CoA, which is synthesized inside the mitochondrion from pyruvate through the action of the pyruvate dehydrogenase complex (Chapter 17). But acetyl CoA destined for fatty acid synthesis must be transported into the cytosol, where acetyl CoA carboxylase and the fatty acid synthase are found.

*Nomenclature of fatty acids: 18:0 refers to an 18-carbon fatty acid with no double bonds; 18:2 refers to an 18-carbon acid with two double bonds; 18:2 is Δ9, Δ12—that is, with double bonds between carbons 9 and 10 and between carbons 12 and 13.

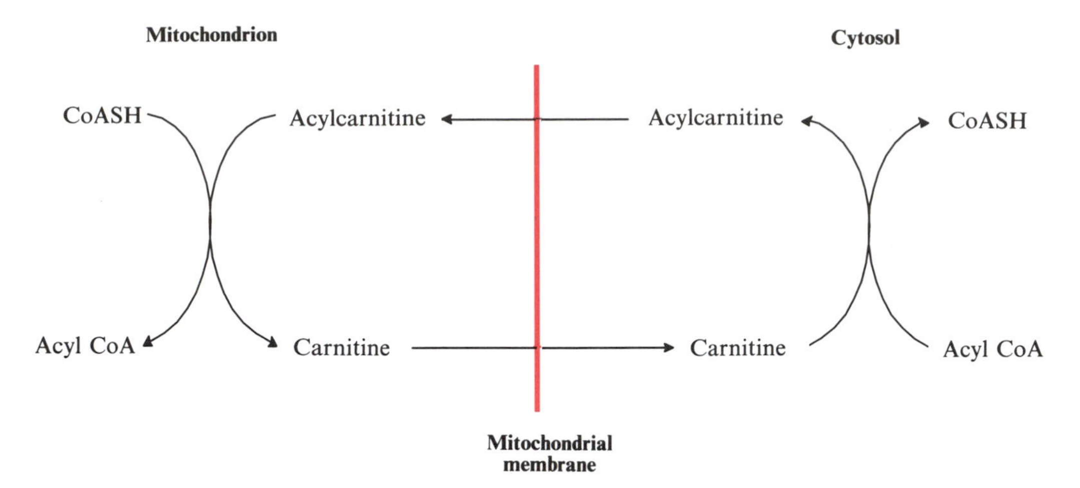

Figure 19-8

The Role of Carnitine in the Transport of Acyl Groups into the Mitochondrion. Cytosolic and mitochondrial carnitine acyltransferases catalyze acyl group transfer between carnitine and CoA on both sides of the mitochondrial membrane. A carnitine transport system embedded within the membrane facilitates the diffusion of acylcarnitine and carnitine between the cytosol and the mitochondrion. The transport system consists of a receptor protein that can bind carnitine or acylcarnitine on either side of the membrane, escort it to the opposite side, and release it there. The mechanisms by which this and other analogous transport systems act are unknown.

Like the long-chain fatty acyl CoAs, acetyl CoA cannot penetrate the mitochondrial membrane and must, therefore, be exported by another means. This is accomplished through the condensation of acetyl CoA with oxaloacetate according to equation 18 to form citrate inside the mitochondrion.

$$\mathrm{CH_3{-}C({=}O){-}SCoA} + \underset{\textbf{Oxaloacetate}}{\mathrm{^-O_2C{-}C({=}O){-}CH_2{-}CO_2^-}} + \mathrm{H_2O} \rightleftharpoons \mathrm{CoASH} + \underset{\textbf{Citrate}}{\mathrm{^-O{-}C({=}O){-}CH_2{-}C(OH)(CO_2^-){-}CH_2{-}CO_2^-}} \tag{18}$$

The mechanism of this reaction is described in Chapter 2. Citrate can diffuse into the cytosol, where it is converted back into acetyl CoA and oxaloacetate through the action of citrate lyase according to equation 19.

$$\mathrm{HO{-}C(CO_2^-)(CH_2CO_2^-)_2} + \mathrm{ATP} + \mathrm{CoASH} \xrightarrow{\mathrm{Mg^{2+}}} \mathrm{CH_3{-}C({=}O){-}SCoA} + \mathrm{^-O_2C{-}C({=}O){-}CH_2{-}CO_2^-} + \mathrm{ADP} + \mathrm{P_i} \tag{19}$$

The mechanism of the citrate lyase reaction is described in Chapter 2.

The scheme in Figure 19-9 illustrates the process by which acetyl CoA in the mitochondrion is converted into fatty acids in the cytosol. Because the cleavage of citrate requires ATP, it costs the cell one molecule of ATP to export one molecule of acetyl CoA from the mitochondrion for the purpose of fatty acid synthesis.

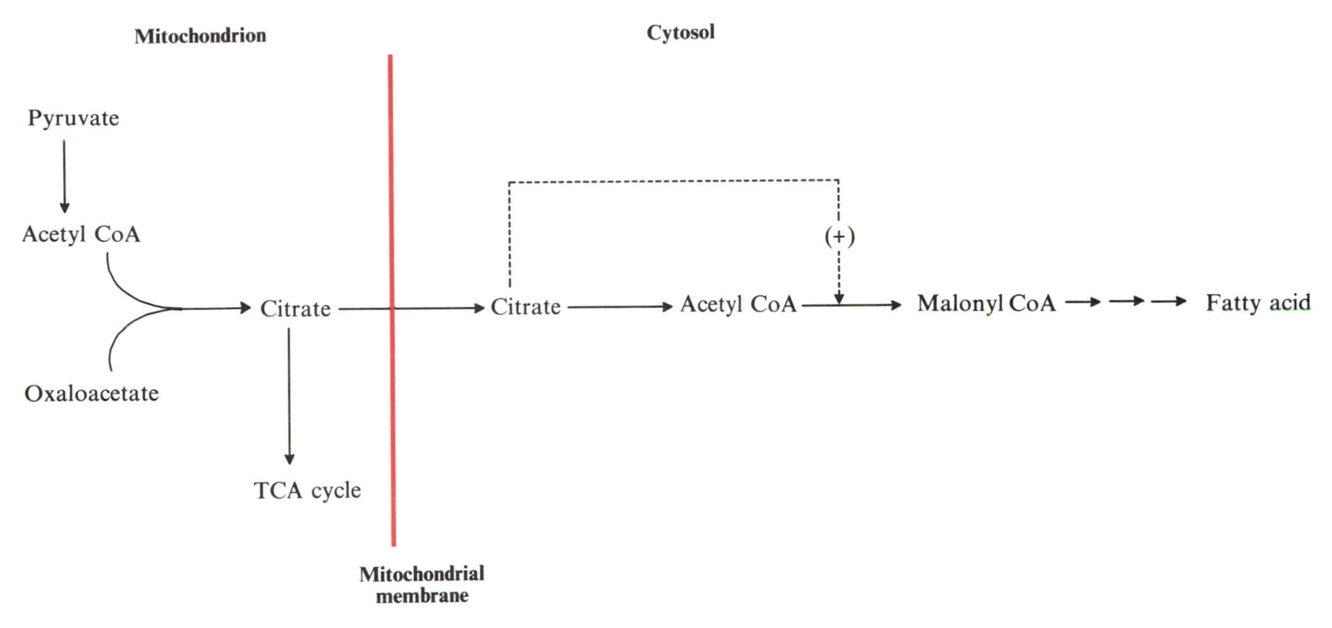

Figure 19-9

Partitioning of Acetyl CoA and Citrate for Energy Production and Fatty Acid Biosynthesis. Acetyl CoA produced in the mitochondria of eukaryotic cells is condensed with oxaloacetate to citrate by citrate synthase (Chapter 2). The citrate is partitioned between the TCA cycle where it is used for energy production and transport outside the mitochondrion. Under conditions in which excess citrate is produced, the overflow is transported to the cytosol, where it is used to produce fatty acids. Citrate is both the fuel for and the activator of fatty acid biosynthesis. It is cleaved to acetyl CoA for carboxylation to malonyl CoA, and it is a specific activator for acetyl CoA carboxylase, as indicated by (+).

Acetyl CoA Can Also Give Rise to Ketone Bodies

Under certain conditions, acetyl CoA can be produced by fatty acid degradation more rapidly than it can be utilized in other metabolic processes. When this occurs, acetyl CoA is converted by enzymes into acetoacetate, D-3-hydroxybutyrate, and acetone. These compounds are known as ketone bodies and are found primarily in the liver.

Ketone bodies are formed through the initial condensation of three molecules of acetyl CoA to produce 3-hydroxy-3-methylglutaryl CoA (HMG CoA). The first step in this reaction sequence is catalyzed by thiolase and produces acetoacetyl CoA according to equation 20.

$$H_3C{-}C({=}O){-}SCoA + H_3C{-}C({=}O){-}SCoA \rightleftharpoons H_3C{-}C({=}O){-}CH_2{-}C({=}O){-}SCoA + CoASH \qquad (20)$$

This reaction is the reversal of the final step in a cycle of fatty acid degradation (equation 1). The equilibrium of this reaction strongly favors degradation. However, this unfavorable equilibrium can be overcome by the next condensation step, which leads to the formation of HMG CoA according to equation 21.

$$CH_3COCH_2COSCoA + CH_3COSCoA + H_2O \longrightarrow {}^{-}O_2C{-}CH_2{-}C(OH)(CH_3){-}CH_2{-}CO{-}SCoA + CoASH \quad (21)$$

3-Hydroxy-3-methylglutaryl CoA (HMG CoA)

HMG CoA is then converted into acetyl CoA and acetoacetate by HMG CoA lyase according to equation 22.

$$ {}^{-}O_2C{-}CH_2{-}C(OH)(CH_3){-}CH_2{-}CO{-}SCoA \longrightarrow {}^{-}O_2C{-}CH_2{-}CO{-}CH_3 + H_3C{-}CO{-}SCoA \quad (22)$$

Acetoacetate

The overall reaction is the sum of equations 20 through 22; that is, the conversion of two moles of acetyl CoA into one mole of acetoacetate and two moles of CoASH (equation 23).

$$2\ \text{Acetyl CoA} + H_2O \longrightarrow \text{Acetoacetate} + 2\ \text{CoASH} + H^+ \quad (23)$$

Acetoacetate can be reduced to D-3-hydroxybutyrate or decarboxylated to acetone, as shown in Figure 19-10 along with the overall pathway by which ketone bodies arise.

Acetoacetate is primarily produced in the liver and serves as a fuel in many tissues. It is converted by the sequence of reactions in Figure 19-11 into acetyl CoA, which then enters the TCA cycle and is used for energy production (Chapter 21). Acetoacetate is normally used as an energy source, in preference to glucose, by heart muscle and the renal cortex. The preferred energy source for the brain is glucose, but under conditions of starvation the brain will also utilize acetoacetate.

BOX 19-1

HIGH-FAT WEIGHT-LOSS DIETS

In the 1970s, a popular weight-reduction diet called the "carbo-calorie diet" became popular in the United States. It was reported to allow the overweight dieter to eat all he or she wants, as long as the amount of total carbohydrate intake was kept under a certain specified limit. The dieter was encouraged to eat as much fat as desired, however. Thus, whole milk was preferred to skim milk as a drink, because more sugar per volume is contained in skim milk. In the same way, unlimited amounts of butter, meat, whipped cream, and fried foods were encouraged. Not surprisingly, this diet was welcomed by many, and it did in fact lead to substantial and rapid weight loss, even in the face of a high level of maintained food intake. One reason for the success of this diet was the fact that its enormous intake of fat combined with its low intake of sugar led to the production of ketone bodies, which have the additional effect of suppressing the appetite. Thus, such "ketogenic diets" are somewhat self-regulating.

The problem with this diet is that it leads to high levels of saturated fats in the bloodstream. As public consciousness began to focus on the relation between blood cholesterol and atherosclerosis, the popularity of ketogenic diets abruptly waned.

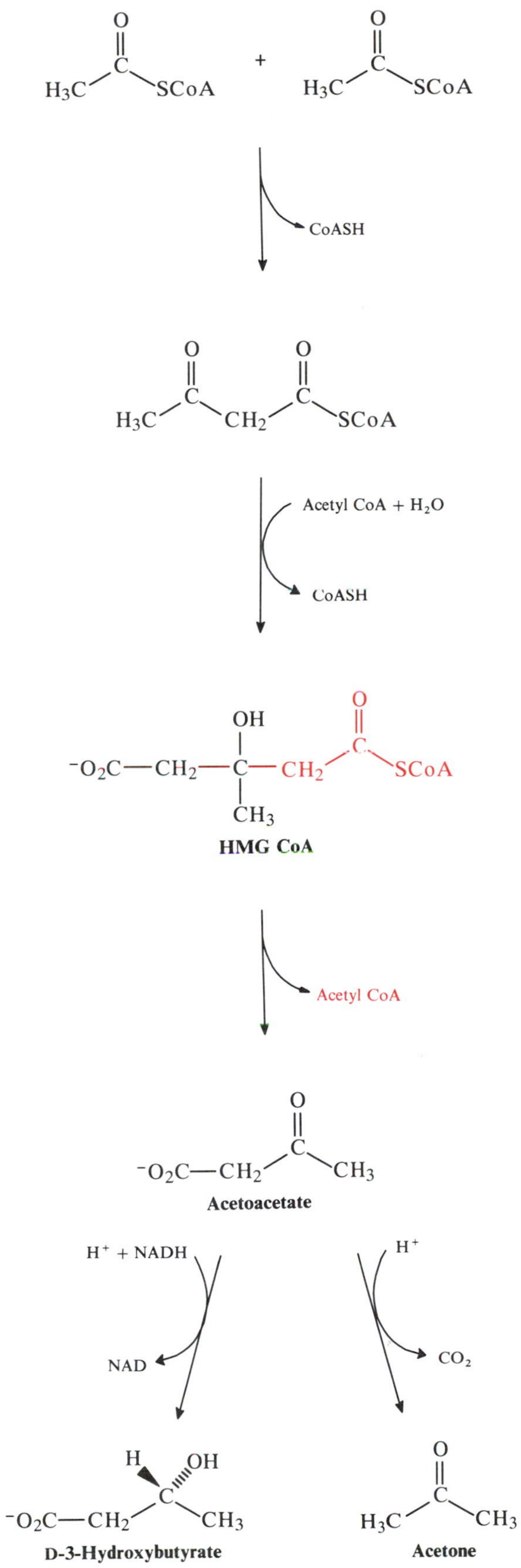

$H_3C{-}C({=}O){-}SCoA$ + $H_3C{-}C({=}O){-}SCoA$

↓ CoASH

$H_3C{-}C({=}O){-}CH_2{-}C({=}O){-}SCoA$

↓ Acetyl CoA + H_2O → CoASH

$^{-}O_2C{-}CH_2{-}C(OH)(CH_3){-}CH_2{-}C({=}O){-}SCoA$

HMG CoA

↓ Acetyl CoA

$^{-}O_2C{-}CH_2{-}C({=}O){-}CH_3$

Acetoacetate

H^+ + NADH → NAD

H^+ → CO_2

$^{-}O_2C{-}CH_2{-}CH(OH){-}CH_3$

D-3-Hydroxybutyrate

$H_3C{-}C({=}O){-}CH_3$

Acetone

Figure 19-10

The Pathway of Ketone-Body Production.

CoA transferase

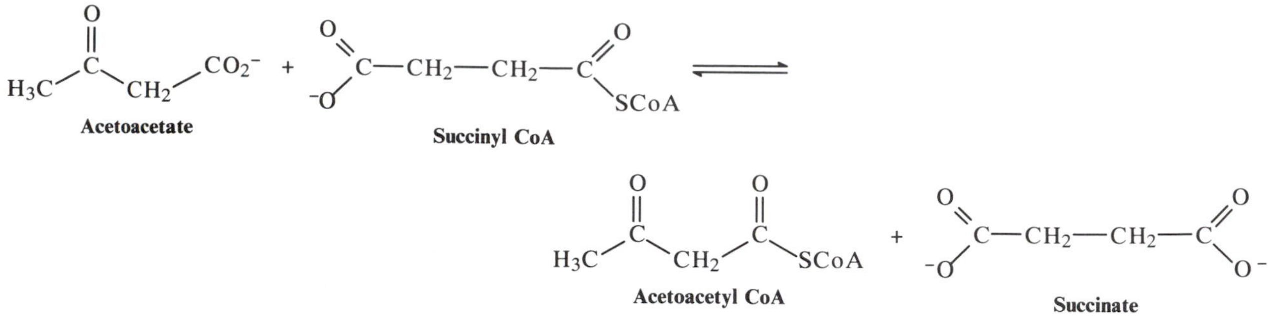

$$H_3C{-}\overset{O}{\overset{\|}{C}}{-}CH_2{-}CO_2^- + {}^-O{-}\overset{O}{\overset{\|}{C}}{-}CH_2{-}CH_2{-}\overset{O}{\overset{\|}{C}}{-}SCoA \rightleftharpoons H_3C{-}\overset{O}{\overset{\|}{C}}{-}CH_2{-}\overset{O}{\overset{\|}{C}}{-}SCoA + {}^-O{-}\overset{O}{\overset{\|}{C}}{-}CH_2{-}CH_2{-}\overset{O}{\overset{\|}{C}}{-}O^-$$

Acetoacetate + Succinyl CoA ⇌ Acetoacetyl CoA + Succinate

Thiolase

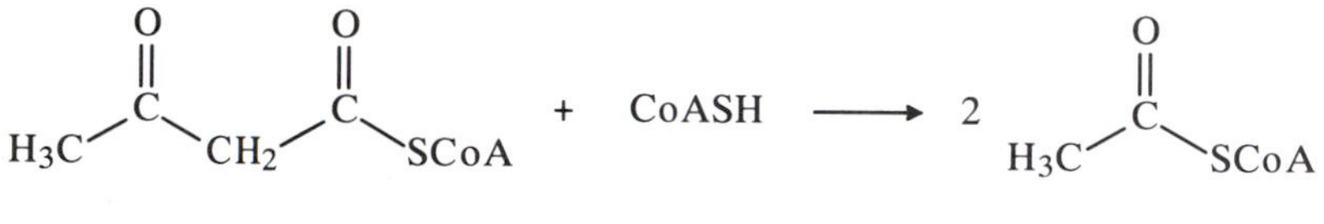

$$H_3C{-}\overset{O}{\overset{\|}{C}}{-}CH_2{-}\overset{O}{\overset{\|}{C}}{-}SCoA + CoASH \longrightarrow 2\ H_3C{-}\overset{O}{\overset{\|}{C}}{-}SCoA$$

Figure 19-11

Metabolism of Ketone Bodies to Succinyl CoA and Acetyl CoA. Succinyl CoA and acetyl CoA are further metabolized in the tricarboxylic acid cycle (Chapter 21).

Summary

Most fatty acids are straight-chain carboxylic acids with even numbers of carbon atoms. The hydrocarbon substituents either are saturated or contain one, two, or three double bonds. Both the degradation and biosynthesis of fatty acids proceed through chemically similar cycles of reactions, in which the alkyl chain is degraded or elongated by two-carbon units; however, the two processes are catalyzed by different systems of enzymes. The system that catalyzes fatty acid degradation consists of separate enzymes that are found in the mitochondria of eukaryotes, whereas the fatty acid synthase is a multienzyme complex that catalyzes biosynthesis in the cytosol.

In fatty acid degradation, a fatty acid is first activated to the corresponding fatty acyl CoA and then dehydrogenated by FAD to an α,β-unsaturated derivative and $FADH_2$. The α,β-unsaturated fatty acyl CoA is hydrated to a β-hydroxyacyl CoA and dehydrogenated by NAD^+ to NADH and a β-ketoacyl CoA, which is then cleaved by CoASH and thiolase to acetyl CoA and a new fatty acyl CoA two carbon atoms shorter than the original fatty acyl CoA. This cycle of reactions is repeated until the fatty acid is completely broken down to acetyl CoA.

Fatty acid synthesis begins with the carboxylation of acetyl CoA to malonyl CoA by acetyl CoA carboxylase. Malonyl CoA is the main building block for fatty acid biosynthesis, which is carried out by the fatty acid synthase multienzyme complex. The process is initiated by the transfer of the acetyl group of acetyl CoA to the acyl carrier protein (ACP) of the complex, followed by its further transfer to the acyl malonyl condensing enzyme. The ACP is then acylated by malonyl CoA to malonyl ACP. The acyl malonyl condensing enzyme catalyzes the decarboxylation of malonyl ACP to a carbanionic enolacetyl ACP and its condensation with the acetyl enzyme to form β-acetoacetyl enzyme. The β-acetoacetyl group is transferred to ACP and reduced in three enzymatic steps to butyryl ACP. In this process, the ACP acts as a carrier for transport of the acyl group to three enzymatic active sites within the complex. The reactions are reduction of β-acetoacetyl ACP by NADPH to β-hydroxylbutyryl ACP, dehydration to butenoyl ACP, and reduction by NADPH to butyryl ACP. The butyryl group is transferred to the acyl malonyl condensing enzyme, and the alkyl chain is extended in two-carbon units by repetition of this reaction sequence. The process ends with the hydrolysis of palmitoyl ACP to palmitate, the product of fatty acid synthase. Palmitate is activated to palmitoyl CoA, elongated to other saturated fatty acyl CoAs, and oxidized to unsaturated fatty acyl CoAs.

Cellular compartmentation of fats and fatty acid metabolizing enzymes in eukaryotic cells requires intracellular transport mechanisms to facilitate chemical communications and the regulation of fatty acid metabolism. Fatty acyl CoA from glycerides in the cytosol is transported into mitochondria by a carnitine-dependent system. Fatty acid biosynthesis in the cytosol requires acetyl CoA, which is produced primarily in the mitochondria. Mitochondrial acetyl CoA is condensed with oxaloacetate by citrate synthase; the citrate is transported into the cytosol, where it is cleaved by CoASH to acetyl CoA and oxaloacetate. Acetyl CoA is then carboxylated to malonyl CoA for fatty acid biosynthesis by acetyl CoA carboxylase.

Fatty acid biosynthesis is regulated in the cytosol by citrate and palmitate. The regulation site is acetyl CoA carboxylase, which is activated by citrate, the source of acetyl CoA for fatty acid biosynthesis, and inhibited by palmitate, the end product of fatty acid biosynthesis.

Enzymes in livers of higher animals lead to the formation of ketone bodies such as acetoacetate, β-hydroxybutyrate, and acetone from acetyl CoA under conditions of starvation. Acetoacetate is a fuel in many organs.

ADDITIONAL READING

VOLPE, J. J., and VAGELOS, P. R. 1973. Saturated fatty acid biosynthesis and its regulation. *Annual Review of Biochemistry* 52: 507.

WAKIL, S. J., and STOOPS, J. K. 1983. Structure and mechanism of fatty acid synthetase. In *The Enzymes,* vol. 16, 3d ed., ed. P. D. Boyer. New York: Academic Press, p. 3.

BLOCH, K., and VANCE, D. E. 1977. Control mechanisms in the synthesis of saturated fatty acids. *Annual Review of Biochemistry* 46: 263.

HOLLOWAY, P. W. 1983. Fatty acid desaturation. In *The Enzymes,* vol. 16, 3d ed., ed. P. D. Boyer. New York: Academic Press, p. 63.

JEFFCOAT, R. 1979. The biosynthesis of unsaturated fatty acids and its control in mammalian liver. *Essays in Biochemistry* 15: 1.

MCGARRY, J. D., and FOSTER, D. W. 1980. Regulation of hepatic fatty acid oxidation and ketone body production. *Annual Review of Biochemistry* 49: 395.

PROBLEMS

1. A mechanism for the condensation of acetyl ACP and malonyl ACP to acetoacetyl ACP is presented in Figure 19-4. Propose an alternative mechanism for this reaction. Suggest an experiment that will distinguish your alternative mechanism from that described in Figure 19-4.

2. The transport of fatty acyl CoA from the cytosol into the mitochondria of eukaryotic cells requires carnitine. Presumably the requirement for carnitine offers some advantage to cellular function that has led to the evolution of carnitine-dependent transport systems in mitochondrial membranes. Explain what this advantage might be. Why would the direct transport of fatty acyl CoA by a CoASH-dependent transport system in the mitochondrial membrane be disadvantageous?

3. The TCA cycle oxidizes acetyl CoA to CO_2 and to NADH and $FADH_2$. NADH and $FADH_2$ are used for the production of ATP (Chapter 21). The first step of the TCA cycle is the condensation of acetyl CoA with oxaloacetate to form citrate and CoASH, and citrate is subsequently oxidized back to oxaloacetate and CO_2 in the cycle. Citrate can also be transported outside the mitochondrion and cleaved to oxaloacetate and acetyl CoA for fatty acid biosynthesis. It would appear to be simpler for acetyl CoA destined for fatty acid biosynthesis to be transported *as such* across the mitochondrial membrane. This would also appear to be energy efficient because ATP would not then be required to cleave citrate to acetyl CoA in the cytoplasm. An acetyl CoA transport system could presumably have evolved to allow acetyl CoA to be transported, but no such system has appeared, presumably because its apparent advantages are offset by some disadvantages to the cell. What problems might be caused for a cell by such a transport system?

4. Valine and isoleucine can be used as energy sources and, in their catabolism, they are broken down to succinyl CoA, which is an intermediate in the TCA cycle. The breakdown of these amino acids entails a complex series of reactions in which a late intermediate is methylmalonyl CoA.

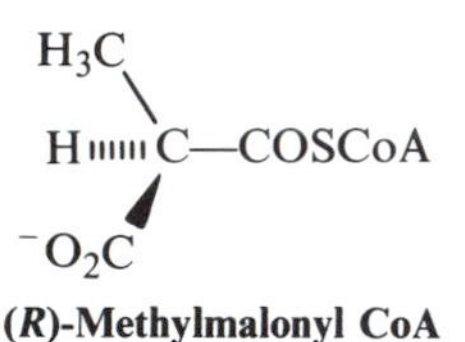

(*R*)-Methylmalonyl CoA

This compound is converted into succinyl CoA by a vitamin-B_{12}-dependent enzyme. When there is a defect in this enzyme, or in any other enzyme required to convert the vitamin into its biologically active form, methylmalonyl CoA accumulates to abnormally high concentrations in the cell. What problems might methylmalonyl CoA cause in fatty acid metabolism?

5. Propionyl CoA is a precursor of methylmalonyl CoA in the degradation of isoleucine and valine. It is carboxylated to methylmalonyl CoA by propionyl CoA carboxylase, a biotin-dependent enzyme. If this enzyme is defective, propionyl CoA will be the end product of leucine and valine breakdown. What problems in fatty acid metabolism might be caused by high levels of propionyl CoA?

6. The phosphopantetheine moiety of the acyl carrier protein (ACP) is thought to act as a “swinging arm” in transporting acyl groups from one active site in the fatty acid synthase to another. What properties of the phosphopantetheine moeity would allow it to function in this way? How many active sites in the fatty acid synthase would be serviced in this way by phosphopantetheine?

Isomerization

20

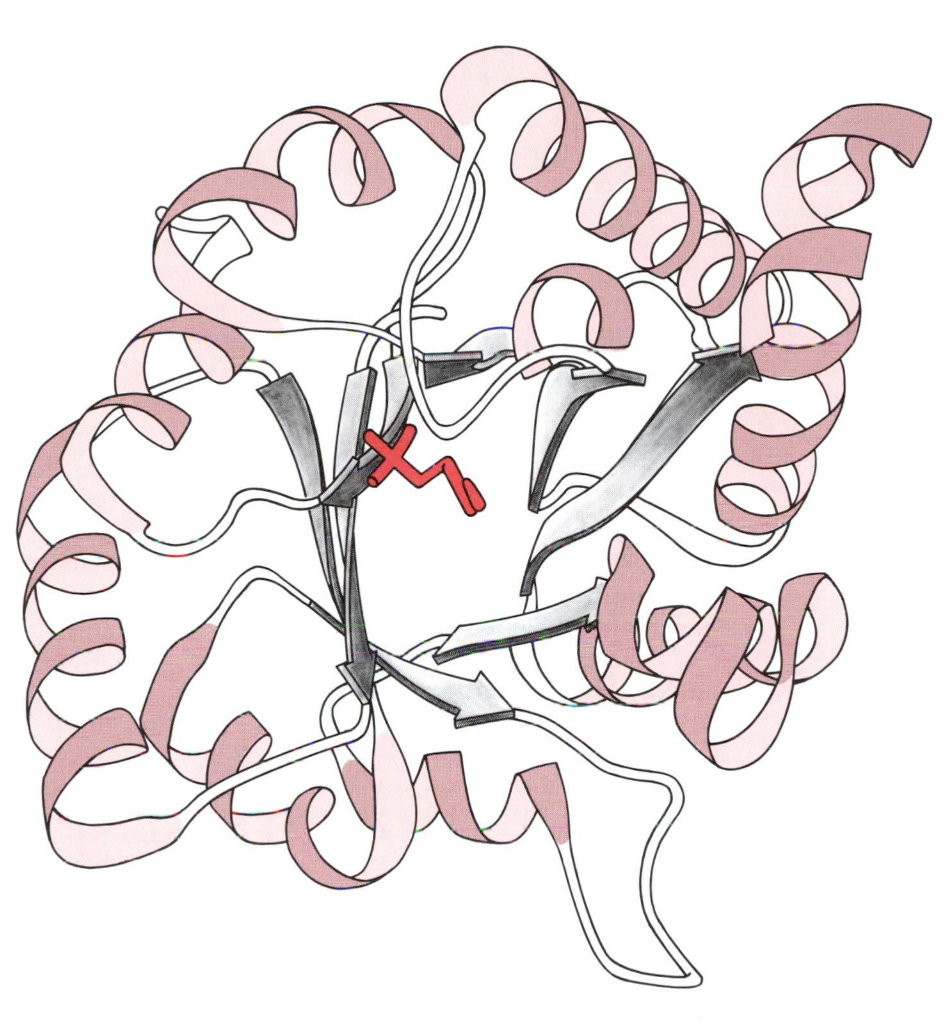

A ribbon diagram of triose phosphate isomerase showing a stick model of a substrate bound at the active site. (Adapted from a computer-generated drawing supplied by Daniel Peisach and James Griffith.)

In most catabolic and biosynthetic pathways, certain metabolic intermediates are not structurally suited for succeeding chemical steps. In some cases, the configuration around one carbon atom is incompatible with a later step that is required in the overall pathway. The metabolic pathway will then include an enzymatic racemization or epimerization that allows the correct configuration to be introduced. In other cases, an intermediate may arise that contains suitable functional groups to facilitate further reactions, but one of these groups is not in the best position within the molecule to potentiate a later step in the pathway. The metabolic pathway will then include a step that leads to the migration of this functional group to a position that allows it to facilitate a subsequent step. These isomerization reactions shuffle the functional groups and substituents in a molecule without breaking it down or oxidizing it. In this chapter, we consider the various types of enzymatic isomerizations that are essential steps in metabolic pathways. The isomerizations include racemizations, epimerizations, α-hydroxycarbonyl rearrangements, phosphomutase reactions, and rearrangements requiring vitamin B_{12}.

Racemization of Amino Acids

An amino acid racemase catalyzes the interconversion of the L- and D-enantiomers of a given amino acid. Racemases for most amino acids are found in microorganisms; an example is alanine racemase. This enzyme is important to certain bacteria that require D-alanine for the biosynthesis of their cell walls but find L-alanine in their environment. This enzyme is an important potential target for antibacterial drugs, because a compound that specifically inhibits alanine racemase would prevent bacterial cell division by blocking the biosynthesis of the cell wall. 3-Fluoroalanine inactivates the alanine racemase and is effective in killing bacteria. Unfortunately, it also has unacceptable side effects in animals and human beings and so is rarely used as a drug.

How can the racemization of alanine be achieved? Clearly, one of the substituents of the carbon atom to be racemized must be removed and then returned to the other side of the carbon, as shown in equation 1.

$$\text{a—C(b)(c)(d)} \longrightarrow \left[\text{a} \quad \text{C(b)(c)(d)}\right] \longrightarrow \text{c—C(b)(d)—a} \tag{1}$$

The substituent removed by alanine racemase and by amino acid racemases in general is the α-hydrogen. It is abstracted by an enzymatic base as a proton. This abstraction is facilitated by the coenzyme pyridoxal-5′-phosphate (PLP, Chapters 2 and 17). In fact, most enzymes that catalyze the racemization or epimerization of amino acids use PLP as the cofactor. A partial mechanism of racemization is shown in equation 2.

$$\text{}^{-}\text{O(O=)C—C(H)(R)—NH}_3^+ + \text{CHO—PLP} \underset{H_2O}{\overset{H_2O}{\rightleftharpoons}} \text{}^{-}\text{O(O=)C—C(H)(R)—NH}^+\text{=CH—PLP} \;\; (:B) \rightleftharpoons \text{}^{-}\text{O(O=)C—}\bar{\text{C}}\text{(R)—NH}^+\text{=CH—PLP} \;\; (H—B^+) \tag{2}$$

Initially, an aldimine (Schiff base) is formed between PLP and the amino acid. The formation of this imine facilitates abstraction of the α-hydrogen by a base at the active site, as explained in Chapter 2.

Subsequent steps in racemization present a problem. A proton must be returned to the side of the molecule *opposite* that from which it was removed, so that the configuration at the α-carbon is inverted. Given the close contacts between an active site and a bound substrate, it is not obvious how a proton could be transported from one side of a carbon to the opposite side. It might be accomplished in one of several ways:

1. The base that removed the proton may be structurally flexible enough to return the proton to the opposite face of the molecule.
2. The proton may be returned after the carbanion is rotated through 180°.
3. The enzymatic active site has two bases, one unprotonated and located on one side of the α-carbon atom and the other protonated (conjugate acid form) located on the other side of the α-carbon. The unprotonated base abstracts the α-proton and the protonated base adds a proton to the opposite side of the carbanion (equation 3).

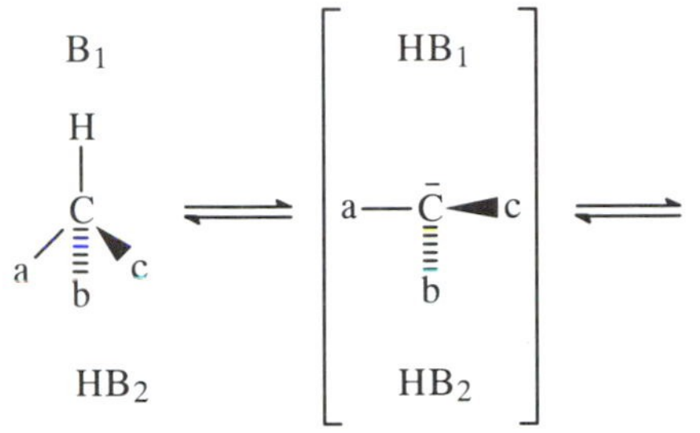

$$B_1 \cdots H\text{–}C(a)(b)(c) \cdots HB_2 \rightleftharpoons \left[HB_1 \cdots a\text{–}\bar{C}(b)(c) \cdots HB_2 \right] \rightleftharpoons HB_1 \cdots (a)(b)(c)C\text{–}H \cdots B_2 \qquad (3)$$

This mechanism (equation 3) probably applies in most cases. Regardless of the mechanism of racemization, the final step in the reaction sequence is hydrolysis of the imine and release of the product.

Epimerization of UDP-galactose

Organisms encounter a variety of carbohydrates that may serve as sources of energy through metabolic oxidation. However, relatively few carbohydrates can be utilized in the available metabolic pathways. Therefore, to utilize other sugars, an organism must convert the many carbohydrates that it encounters into the few that can be utilized for energy production. An important step in many of these conversions is the epimerization of carbon atoms; that is, inversion of configuration of one of several asymmetric carbon atoms.

An example of such a carbohydrate is galactose, which differs from glucose in the configuration of C-4. Galactose in the mammalian diet is derived from the hydrolysis of lactose, a component of milk. There is no energy-yielding reaction sequence for the breakdown of galactose, and it must be converted into glucose by epimerization at C-4 in order to be metabolized. Before epimerization, galactose is converted into UDP-galactose through the Leloir pathway, as described in Chapters 11 and 13. UDP-galactose is converted into UDP-glucose by UDP-galactose 4-epimerase according to equation 4.

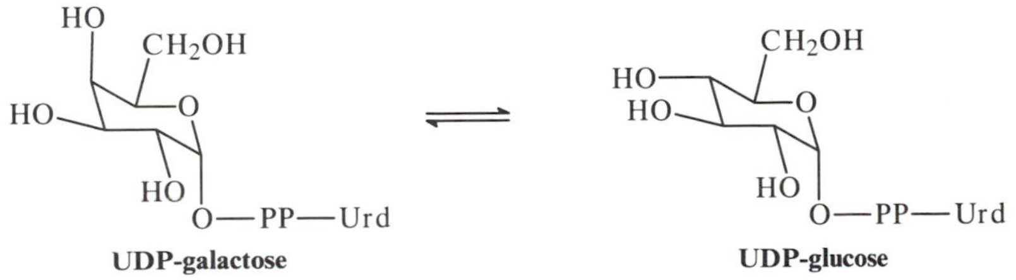

$$\text{UDP-galactose} \rightleftharpoons \text{UDP-glucose} \qquad (4)$$

UDP-galactose (HO, CH_2OH, HO, O, HO, O—PP—Urd) ⇌ UDP-glucose (HO, CH_2OH, HO, O, HO, O—PP—Urd)

To attain epimerization at C-4, one of the four bonds linking this carbon to the rest of the molecule must be broken. For example, removal of either the 4-hydroxyl group or the 4-hydrogen from one side of the ring and transfer to the other side would give epimerization. There is no obvious way in which the 4-hydrogen could be removed as a proton because the pK_a is very high. Removal of the 4-hydroxyl group also is difficult.

The problem has been solved as illustrated in Figure 20-1. The enzyme utilizes NAD^+ to remove the hydrogen from C-4 of UDP-galactose by oxidizing this carbon to the ketone level, resulting in the formation of UDP-4-ketoglucose and NADH. Both UDP-4-ketoglucose and NADH remain bound to the active site, and NADH again reduces the ketone group at C-4.

The hydrogen is returned to the face of 4-ketoglucose opposite that from which it was removed, and this brings about epimerization. The intermediate complex, containing the UDP-4-ketoglucose and NADH, is a very "tight" complex; that is, the release of UDP-4-ketoglucose occurs rarely, and the release of NADH has never been observed.

How can the hydrogen be transported from one side of the sugar ring to the other? Does the sugar remain stationary and the NADH move from one side of the sugar ring to the other? Or does the sugar move relative to NADH? There is considerable evidence suggesting that NADH remains stationary and the carbohydrate moiety of UDP-4-ketoglucose rotates, as illustrated in Figure 20-1. It is known that, in the formation of the enzyme • UDP-galactose complex, interactions between the enzyme and the UDP moiety of the substrate are of prime importance. The glucose part of the substrate is loosely bound, and glucose is free to rotate.

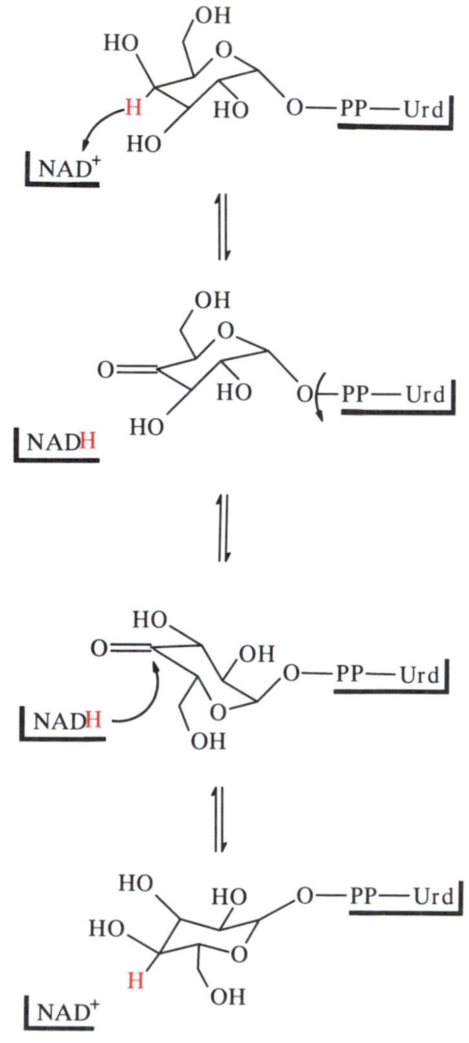

Figure 20-1

The Epimerization of UDP-galactose to UDP-glucose. UDP-galactose 4-epimerase contains NAD^+ tightly bound to the active site. In the first step, the NAD^+ is reduced and C-4 of the galactosyl group in UDP-galactose is oxidized. The 4-ketoglucosyl group then undergoes a rotation within the active site that allows the opposite face of the ring to approach NADH. The 4-keto intermediate is reduced to UDP-glucose concomitant with the oxidation of NADH to NAD^+. The product subsequently dissociates from the active site. In the mechanism of action of the microbial enzyme, NADH and UDP-4-ketoglucose normally do not dissociate from the active site during the catalytic cycle. UDP-galactose and UDP-glucose dissociate freely.

Part of the evidence for the loose binding of glucose is derived from studies of a model reaction. When the enzyme is allowed to react with glucose and UMP, glucose is oxidized at C-1 to gluconolactone and NAD^+ is reduced to NADH. The K_m for glucose is very large, in excess of 3 molar, and either α-D-glucose or β-D-glucose can reduce the NAD^+. When galactose is used in place of glucose to reduce the NAD^+, both C-1 and C-4 of galactose are oxidized at comparable rates. These facts indicate that sugars are weakly bound and can assume a number of orientations in the active site. This "freedom" of glucose is very likely utilized in the catalysis of epimerization.

Carbonyl-Alcohol Interconversion

An isomerization frequently encountered in metabolism is the interchange of a carbonyl and an adjacent hydroxyl group:

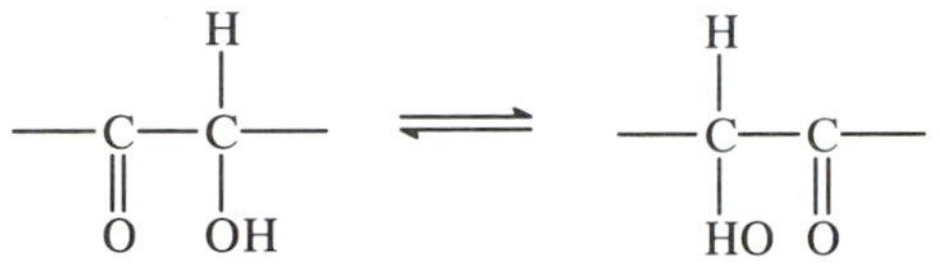

$$-C(=O)-CH(OH)- \rightleftharpoons -CH(OH)-C(=O)-$$

Note that the reaction does not change the oxidation state of the molecule. This is an important reaction because it changes the location of a carbonyl group. Recall that bonds between carbon atoms α and β to a carbonyl group are subject to cleavage (Chapter 2); therefore, the movement of a carbonyl group within a molecule alters the position at which it may undergo carbon–carbon bond cleavage.

A specific example of this type of isomerization is the interconversion of glucose-6-phosphate and fructose-6-phosphate by phosphoglucose isomerase (equation 5).

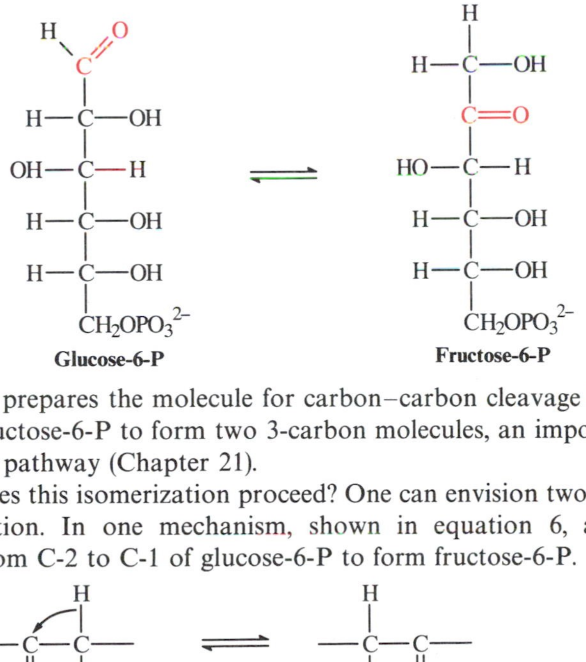

$$\underset{\textbf{Glucose-6-P}}{HC(=O)-CH(OH)-CH(OH)-CH(OH)-CH(OH)-CH_2OPO_3^{2-}} \rightleftharpoons \underset{\textbf{Fructose-6-P}}{CH_2(OH)-C(=O)-CH(OH)-CH(OH)-CH(OH)-CH_2OPO_3^{2-}} \quad (5)$$

This reaction prepares the molecule for carbon–carbon cleavage between C-3 and C-4 of fructose-6-P to form two 3-carbon molecules, an important step in the glycolysis pathway (Chapter 21).

How does this isomerization proceed? One can envision two mechanisms for this reaction. In one mechanism, shown in equation 6, a hydride is transferred from C-2 to C-1 of glucose-6-P to form fructose-6-P.

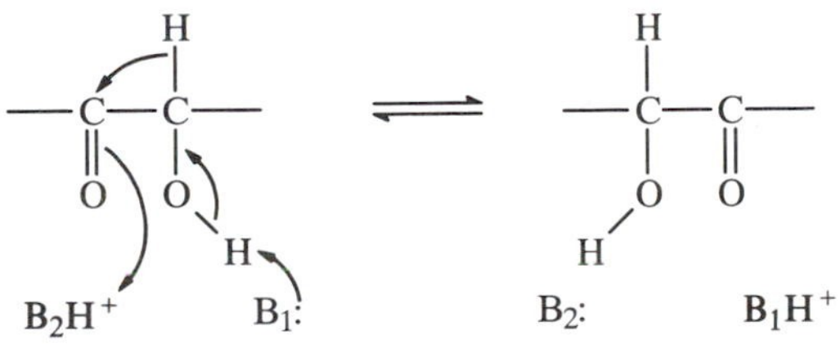

$$-C(=O)-CH(OH)- \;\; B_2H^+ \;\; B_1: \rightleftharpoons -CH(OH)-C(=O)- \;\; B_2: \;\; B_1H^+ \quad (6)$$

If this is the mechanism, and the reaction is carried out in isotopically labelled H_2O (either ^2H_2O or ^3H_2O), no deuterium (^2H) or tritium (^3H) will be bound to carbon in the product molecule.

The other mechanism is an enolization through an enediol intermediate according to equation 7.

$$
\begin{array}{c} B_1: \\ H \\ | \\ -C-C- \\ \| \quad | \\ O \quad OH \\ H \\ | \\ B_2 \end{array}
\rightleftharpoons
\left[\begin{array}{c} B_1: \\ H \\ -C{=}C- \\ | \quad | \\ O \quad O \\ H \qquad H \\ \bar{B}_2: \end{array} \right]
\rightleftharpoons
\begin{array}{c} B_1: \\ H \\ | \\ -C-C- \\ | \quad \| \\ HO \quad O \\ H \\ | \\ B_2 \end{array}
\qquad (7)
$$

Here the hydrogen also is transferred from C-2 to C-1, but it is not a direct transfer. Before arriving at its final destination, the proton makes an intermediate stop on a base at the active site of the enzyme—B_1 in equation 7. As pointed out in earlier chapters, bases B_1 and B_2 are functional groups (—SH, —OH, —NH_2, —COO^-, imidazole) of amino acid residues. Regardless of which of these groups is the base, a proton bonded to it may exchange rapidly with solvent protons. Therefore, we expect that, if the reaction is carried in ^2H_2O or ^3H_2O, isotopic hydrogen will be bonded to carbon in the product.

Many isomerizations of this type are known and have been tested for incorporation of hydrogen from the solvent. In the majority of cases, it is found that hydrogen from the solvent is incorporated into the product. Therefore, these reactions proceed by a mechanism involving intermediate enediols.

In the isomerization of glyceraldehyde-3-P and dihydroxyacetone-3-P by triose phosphate isomerase, B_1 in equation 7 is the side chain of glutamate-165 and B_2 is histidine-95.

With some enzymes of this type, little (less than 1 atom) or no hydrogen is incorporated into the product. In this case, either the enzyme is so constructed that H—B_1 is not readily accessible to the solvent—that is, there is a physical barrier for exchange—or the rate of proton transfer from H—B_1 to the product is much faster than the exchange with solvent protons. This problem was discussed in Chapters 2 and 16.

Carbon-Skeleton Rearrangements

There are also enzymes that catalyze isomerizations in which there is a change in the carbon skeleton. In the simplest of these reactions, there is migration of double bonds, as, for example, in the reaction catalyzed by isopentenyl pyrophosphate isomerase (equation 8).

$$
\underset{\textbf{Isopentenyl-PP}}{(H_3C)(H_2C{=})C-CH_2-CH_2-OPO_2O_6{}^{3-}} \rightleftharpoons \underset{\textbf{Dimethylallyl-PP}}{(H_3C)_2C{=}CH-CH_2-OPO_2O_6{}^{3-}} \qquad (8)
$$

This is a step in the biosynthesis of cholesterol (Chapter 28). Formally, the reaction entails the migration of both a double bond and a hydrogen. It is known that this hydrogen transfer is not a direct 1, 3 shift; that is, the conversion of isopentenyl-PP into dimethylallyl-PP in ^2H_2O or ^3H_2O leads to the incorporation of ^2H or ^3H from the solvent into the methyl group of dimethylallyl-PP.

All evidence available to date points to the mechanism shown in Figure 20-2. Initially, a base (B_1H^+) at the active site protonates the methylene carbon. This produces a carbocation, which facilitates transfer of a proton to B_2 and migration of the double bond.

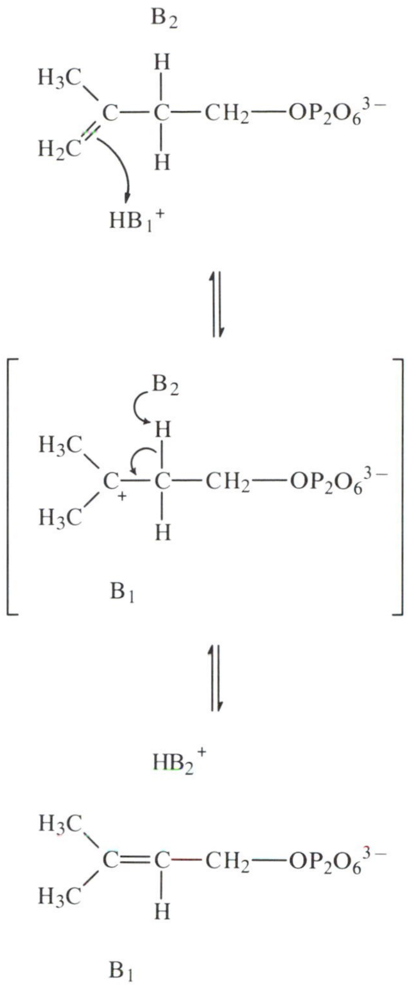

Figure 20-2

The Isomerization of Isopentenyl-PP to Dimethylallyl-PP. Isopentenyl-PP isomerase probably catalyzes its reaction by initially transferring a proton to the double bond of isopentenyl-PP to form a carbocationic intermediate that is stabilized by electrostatic interaction with a negatively charged group of the enzyme. The carbocationic intermediate is neutralized by abstraction of a proton to form the double bond in dimethylallyl-PP.

One could also write a carbanion mechanism that would be acceptable on chemical grounds. Several experimental facts argue against a carbanion or concerted mechanism, and further evidence for a carbocation mechanism is presented in Box 20-1.

BOX 20-1

EVIDENCE FOR A CARBOCATIONIC INTERMEDIATE IN THE ISOPENTENYL-PP ISOMERASE REACTION

When the CH_3 group in isopentenyl-PP is replaced by CF_3, the rearrangement is 20,000-fold slower. Because the CF_3 group is strongly electron withdrawing, one expects it to destabilize the intermediate carbocation; however, it would have little effect on or would stabilize a carbanion.

Moreover, $(CH_3)_2{}^+NH—CH_2—CH_2—OP_2O_7{}^{3-}$ is a very powerful inhibitor of the reaction ($K_i = 10^{-11}$). The protonated cationic amine is most likely an analog of the carbocationic intermediate in Figure 20-2, in which the positively charged nitrogen takes the place of the carbocation. Presumably, in the normal reaction, the high-energy carbocation is stabilized through an electrostatic interaction with a negative charge on the enzyme. Interaction of this same negative charge with the positively charged nitrogen of the inhibitor probably makes a major contribution to the strong affinity of the inhibitor for the enzyme. It should be noted that the interaction of the ammonium group with the enzyme is not solely responsible for the strong binding. For instance, conversion of the pyrophosphate moiety of the inhibitor into phosphate increases K_i by 10^5. Clearly, the simultaneous occurrence of multiple binding interactions is responsible for the tight binding of the inhibitor (see Chapter 5).

An example of a rearrangement in which carbon-atom migration occurs is the reaction catalyzed by acetohydroxy acid isomeroreductase (equation 9).

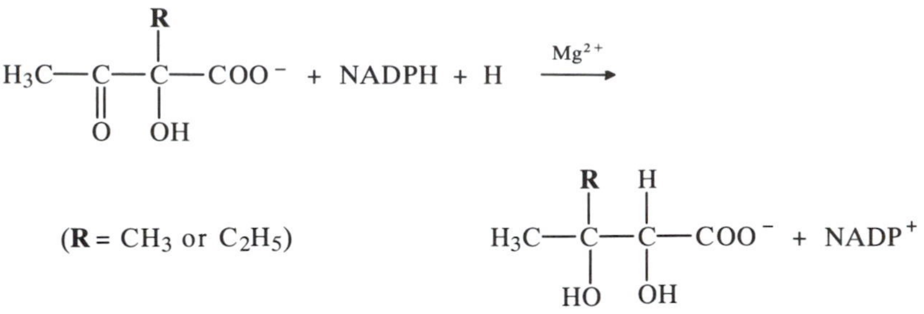

$$H_3C{-}\underset{\underset{O}{\|}}{C}{-}\overset{\overset{\textbf{R}}{|}}{\underset{\underset{OH}{|}}{C}}{-}COO^- + NADPH + H \xrightarrow{Mg^{2+}} H_3C{-}\overset{\overset{\textbf{R}}{|}}{\underset{\underset{HO}{|}}{C}}{-}\overset{\overset{H}{|}}{\underset{\underset{OH}{|}}{C}}{-}COO^- + NADP^+ \quad (9)$$

($\textbf{R} = CH_3$ or C_2H_5)

This is a reaction in the biosynthesis of leucine and valine in microorganisms and plants. The chemical mechanism probably includes the intermediate shown in Figure 20-3. The reaction is reminiscent of the pinacol rearrangement in that the R group migrates with its electrons to the adjacent carbon. In the pinacol rearrangement, a carbocation is generated and an alkyl group α to the cationic carbon undergoes a 1,2 rearrangement, as shown below. Migration of the R group is promoted through the electron pressure created by transfer of the proton from the α-OH group to a base and by the carbocationic character associated with the electropositive β-carbonyl carbon atom.

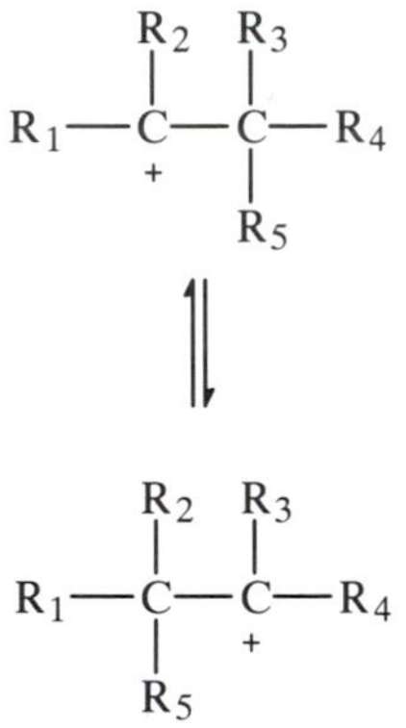

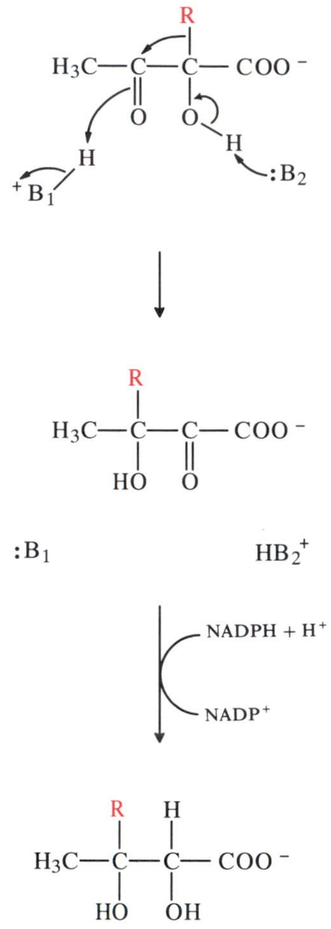

$$R_1{-}\overset{\overset{R_2}{|}}{\underset{+}{C}}{-}\overset{\overset{R_3}{|}}{\underset{\underset{R_5}{|}}{C}}{-}R_4 \rightleftharpoons R_1{-}\overset{\overset{R_2}{|}}{\underset{\underset{R_5}{|}}{C}}{-}\overset{\overset{R_3}{|}}{\underset{+}{C}}{-}R_4$$

$$H_3C{-}\underset{\underset{O}{\|}}{C}{-}\overset{\overset{R}{|}}{\underset{\underset{O-H}{|}}{C}}{-}COO^- \quad (^+B_1{-}H,\ :B_2) \longrightarrow H_3C{-}\overset{\overset{R}{|}}{\underset{\underset{HO}{|}}{C}}{-}\underset{\underset{O}{\|}}{C}{-}COO^- \quad (:B_1,\ HB_2^+) \xrightarrow[NADP^+]{NADPH + H^+} H_3C{-}\overset{\overset{R}{|}}{\underset{\underset{HO}{|}}{C}}{-}\overset{\overset{H}{|}}{\underset{\underset{OH}{|}}{C}}{-}COO^-$$

Figure 20-3

The Mechanism of the α-Hydroxy-β-ketoacid Isomeroreductase Reaction. The most-logical mechanism for the isomeroreductase reaction requires two steps: one in which the alkyl group migrates from the α-carbon to the β-carbon and the other in which the α-carbon is reduced by NADPH. The driving force for the alkyl migration is explained in the text.

Rearrangements Requiring Coenzyme B_{12}

The coenzyme form of vitamin B_{12} is required for several remarkable enzymatic rearrangements. The structure of the coenzyme B_{12} (*adenosylcobalamin*) is shown in Figure 20-4. The coenzyme contains a corrin ring, which is related to the porphyrin ring of heme. Recall that heme in hemoglobin binds molecular oxygen (Chapter 10). The biosynthesis of the porphyrin ring is described in Chapter 26, and the biosynthesis of corrin begins similarly but diverges. The history of the discovery of vitamin B_{12} is in Box 20-2.

The structure of corrin differs from that of porphyrin in heme in that the pyrroline rings are at a lower oxidation level than the pyrrole rings in porphyrin, so that corrin contains nine asymmetric carbon atoms, whereas heme contains none. Moreover, one of the methine bridges in porphyrin is missing in corrin. Adenosylcobalamin is a unique molecule in that the central cobalt atom coordinated with corrin is covalently bonded to the 5′-carbon of an adenosyl moiety. Vitamin B_{12} coenzyme is remarkable in that it is a naturally occurring

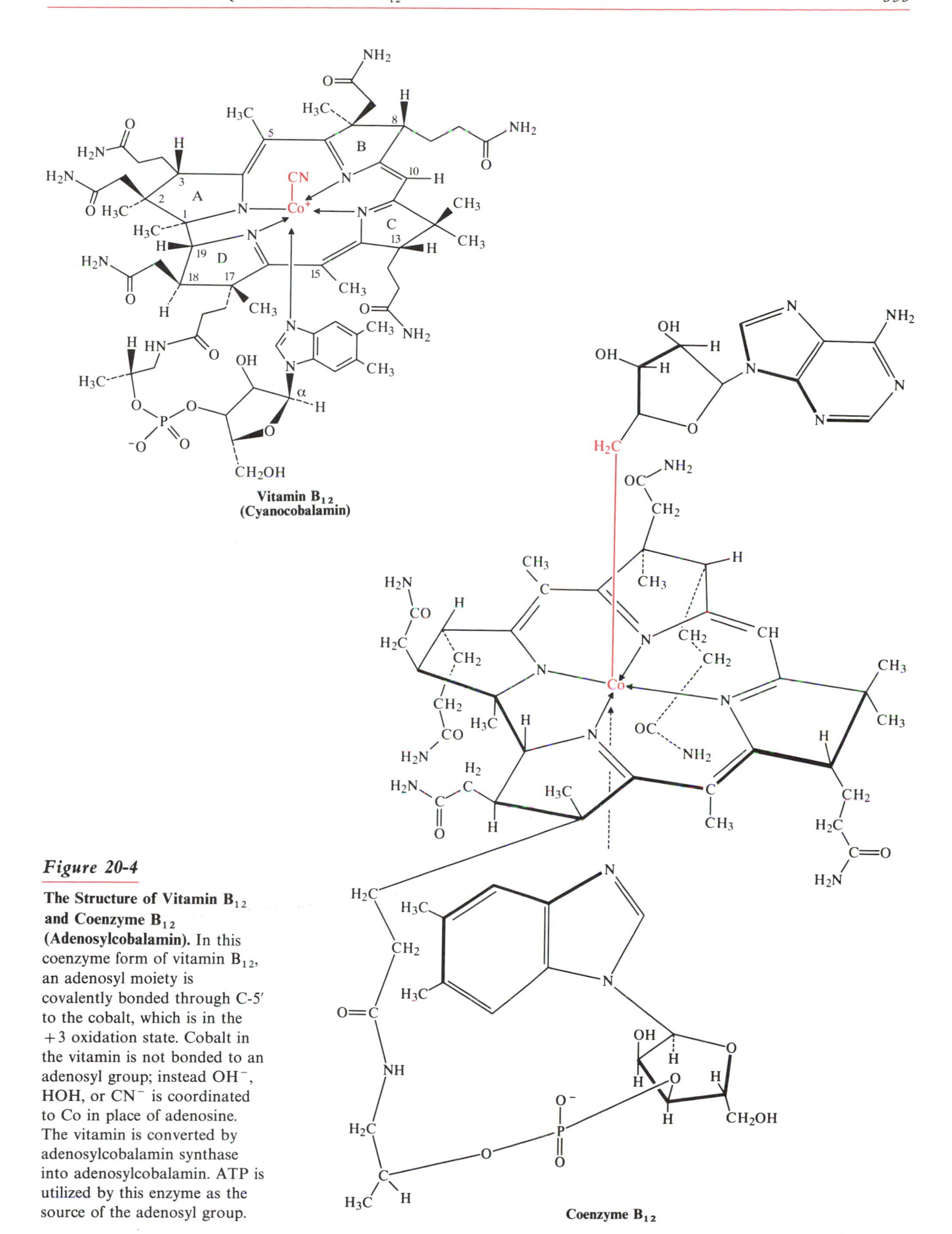

Figure 20-4

The Structure of Vitamin B_{12} and Coenzyme B_{12} (Adenosylcobalamin). In this coenzyme form of vitamin B_{12}, an adenosyl moiety is covalently bonded through C-5′ to the cobalt, which is in the +3 oxidation state. Cobalt in the vitamin is not bonded to an adenosyl group; instead OH^-, HOH, or CN^- is coordinated to Co in place of adenosine. The vitamin is converted by adenosylcobalamin synthase into adenosylcobalamin. ATP is utilized by this enzyme as the source of the adenosyl group.

BOX 20-2

VITAMIN B_{12}

It is instructive to briefly consider the history of the discovery of Vitamin B_{12}, which is closely tied to the study of a disease known as pernicious anemia. This disease has been recognized for more than 100 years. It is a macrocyclic anemia: red blood cells are abnormally large and few in number, from about 25% to 50% of the normal value. The bone-marrow cells become enlarged while still immature. At later stages in the disease, neurological symptoms develop and lead eventually to death.

Progress in the treatment of pernicious anemia first came from the discovery, in 1925, by Whipple and Robshiet-Robbins that feeding liver to anemic dogs accelerated red blood cell formation. Minot and Murphy followed up this discovery by administering raw or lightly cooked liver to human patients suffering from pernicious anemia. They found that from one-half to one pound of liver a day brought about remarkable recoveries, including a rapid improvement of the blood picture. Minot and Murphy described their success rather modestly, as follows:

> The dietetic treatment of pernicious anemia is of more importance than hitherto generally recognized. Forty-five patients with pernicious anemia are continuing to take a special diet. They have now been living from about six weeks to two years. Following the diet, all the patients showed a prompt, rapid, and distinct remission of their anemia, coincident with at least rather marked symptomatic improvement, except for pronounced disorders due to spinal cord degeneration. ... All the patients remain to date in a good state of health except three, who discontinued the diet; two rapidly improved on resuming it and the other had just commenced to gain. As the diet was advised for most of the patients less than eight months ago, enough time has not yet elapsed to determine whether or not the remission will last any longer than in other cases.

In 1934, Whipple, Minot, and Murphy were awarded the Nobel Prize in medicine and physiology for their discovery.

It was known that pernicious anemia is also associated with partial atrophy of gastric mucosa. This fact led Castle, in 1928, to propose that the stomach produces a substance that he called intrinsic factor. He further proposed that intrinsic factor acts upon a substance, termed extrinsic factor, in certain foods to produce a liver factor, which prevents pernicious anemia. This reasoning led to a new treatment of pernicious anemia. Juice from normal human stomachs was administered along with meat as a source of extrinsic factor. This treatment proved to be an effective treatment of pernicious anemia. Later, preparations derived from hog stomachs were used. We know today that Castle's original hypothesis was not quite correct. Intrinsic factor is a protein required for vitamin B_{12} to be absorbed from the intestine. Only a complex of B_{12} and intrinsic factor can be efficiently absorbed. The immediate cause of pernicious anemia is the lack of intrinsic factor and the consequent inability to absorb B_{12}. Today, pernicious anemia is treated by monthly injections of vitamin B_{12}.

The success of the treatment with nutrient liver indicated that liver probably contains a substance that cures pernicious anemia. Many laboratories undertook to isolate this substance. Initially, an extract was produced from liver that alleviated the disease. Instead of consuming large quantities of liver, patients could be healed with 1 mg of extract. The purification effect of the "liver factor," which was eventually called vitamin B_{12}, culminated in 1947, when two groups almost simultaneously obtained crystalline vitamin B_{12}. The crystalline vitamin was isolated by Karl Folkers and his associates at Merck and Company and by E. Lester Smith and his co-workers at Glaxo.

Another 10 years was required to completely elucidate the structure of vitamin B_{12}. Features of the structure were derived through classical chemical structure determination by the group at Merck and by Lord Todd and his colleagues. The final structure was obtained through X-ray crystallography by D. Hodgkin. Vitamin B_{12} was the first large molecular structure derived through X-ray crystallography. Hodgkin was awarded a Nobel Prize in chemistry in 1964.

A major contribution to our understanding of the role of vitamin B_{12} in metabolism came from the laboratory of H. A. Barker. He investigated the metabolism of an amino acid and discovered that a new coenzyme, which was a derivative of vitamin B_{12}, was required. D. Hodgkin and her co-workers established the structure of B_{12} coenzyme by X-ray crystallographic analysis.

In 1972, Woodward, Eschenmoser, and their co-workers achieved the complete chemical synthesis of vitamin B_{12}. Woodward had received the Nobel Prize in 1965 for his earlier work.

Although we have learned much about vitamin B_{12} since pernicious anemia was first described in 1870, many questions still remain unanswered. The biochemical basis of pernicious anemia is not completely understood: that is, What biochemical lesions are caused by B_{12} deficiency, and how are they related to clinical symptoms? Questions concerning the mechanism of action of vitamin B_{12} coenzyme in catalysis still remain.

The story of vitamin B_{12} is a typical example of how biochemical problems are solved. Initially, a phenomenon is observed in a living organism. Next, crude extracts are obtained in which, or with which, the phenomenon can be studied. Gradually, these extracts are refined and, eventually, a characterizable compound is isolated. Finally, the phenomenon originally observed in living organisms can be explained on a molecular basis.

compound that contains a carbon–cobalt bond. In addition to being covalently bonded to adenosine, the cobalt atom is octahedrally coordinated to the four pyrroline nitrogen atoms in the approximately planar corrin ring, and 5,6-dimethylbenzimidazole is in the lower coordination position of the cobalt. Vitamin B_{12}, the precursor of the coenzyme, does not contain the adenosyl moiety; instead water, hydroxide ion, or cyanide ion is coordinated to cobalt at the upper coordination position in place of adenosine (Figure 20-4).

Figure 20-5 shows some of the enzymatic reactions requiring adenosylcobalamin. The reactions shown by no means represent an exhaustive list of those that utilize vitamin B_{12} coenzyme. Others will be discussed in Chapters 25, 26, and 27. The reactions in Figure 20-5, as well as many other reactions requiring

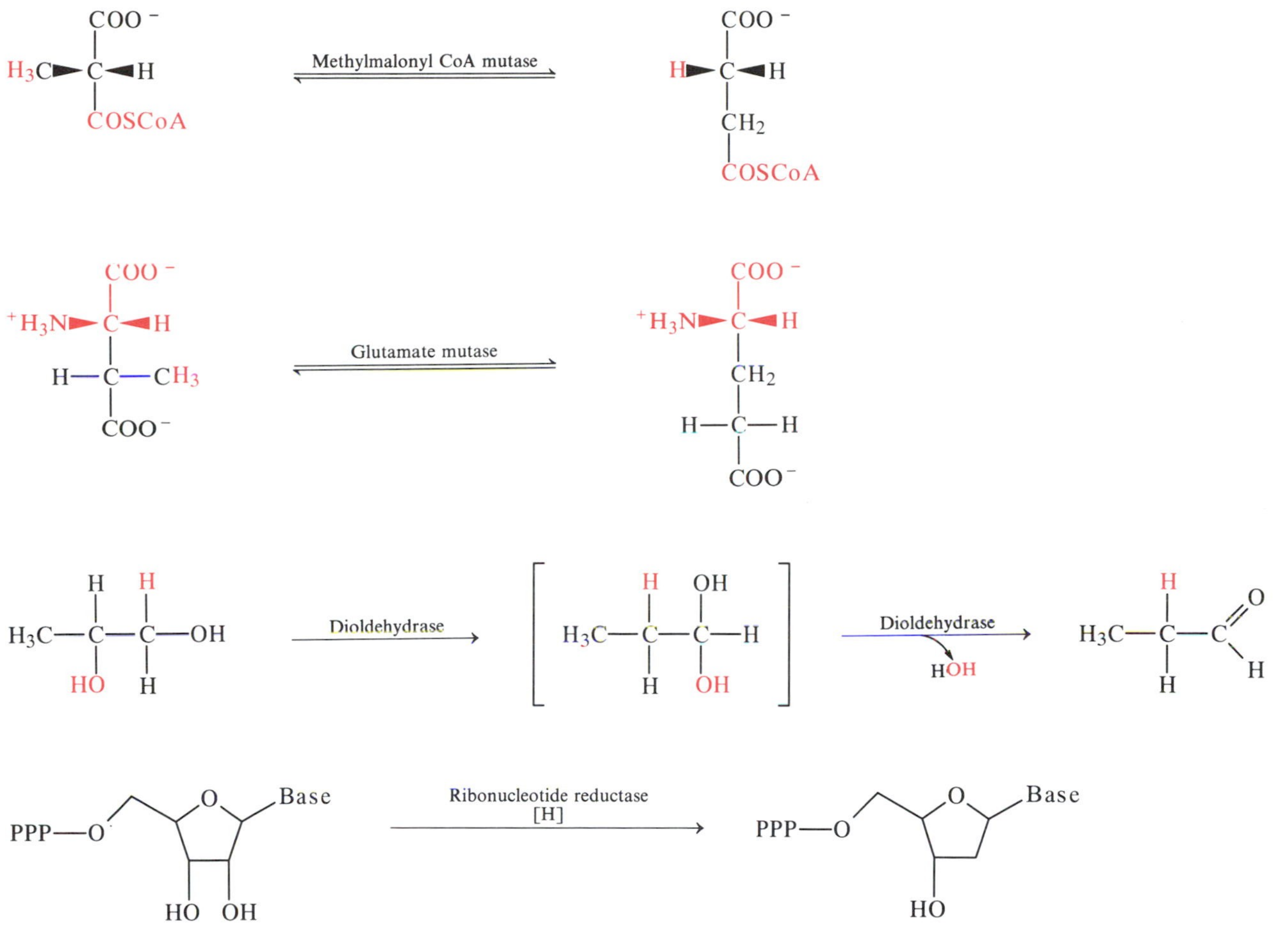

Methylmalonyl CoA mutase

Glutamate mutase

Dioldehydrase

Ribonucleotide reductase
[H]

Base = adenine, cytosine, or guanine

Figure 20-5

Examples of Reactions Requiring Adenosylcobalamin. All of the enzymes catalyzing these reactions either contain adenosylcobalamin or are activated by this coenzyme. Ribonucleotide reductase from *Lactobacillus leichmanii* requires adenosylcobalamin and utilizes ribonucleoside triphosphates as substrates; however, the ribonucleotide reductases in animals and *E. coli* do not utilize adenosylcobalamin. These enzymes contain another cofactor, a tyrosine radical, that is generated by the action of another cofactor, a μ-oxo-Fe_2 complex, in a reaction requiring O_2. The tyrosine radical is thought to initiate a radical chain reaction that leads to a substrate radical (Chapter 27).

B_{12} coenzyme, can be represented by the general formulation in equation 10,

$$\begin{array}{c}-\overset{|}{\underset{X}{\underset{|}{C_\beta}}}-\overset{|}{\underset{H}{\underset{|}{C_\alpha}}}-\end{array} \rightleftharpoons \begin{array}{c}-\overset{|}{\underset{H}{\underset{|}{C_\beta}}}-\overset{|}{\underset{X}{\underset{|}{C_\alpha}}}-\end{array} \qquad (10)$$

in which a group designated X and a hydrogen bonded to an adjacent carbon atom exchange places. For example, in the dioldehydrase reaction shown in Figure 20-5, X = OH, and in the methylmalonyl CoA mutase reaction X = COSCoA. Box 20-3 contains a more-detailed description of the reaction catalyzed by dioldehydrase.

BOX 20-3

THE ROLE OF ADENOSYLCOBALAMIN IN THE DIOLDEHYDRASE REACTION

The mechanism of action of dioldehydrase has been studied extensively, and the currently accepted mechanism is shown in Figure A. The coenzyme and all intermediates remain bound at the active site throughout the catalytic process. The initial combination of the substrate with the enzyme-coenzyme complex activates the coenzyme in such a way as to induce the cleavage of the carbon–cobalt bond in step 1. This process presumably entails a substrate-induced change in the structure of the enzyme, and the binding interactions between the altered enzyme and the coenzyme weaken the carbon–cobalt bond. The cleavage of this bond is homolytic; that is, one electron remains with Co, which becomes Co(II), and the other remains with the adenosyl radical. The adenosyl radical then abstracts a hydrogen atom from the substrate in step 2 and becomes 5′-deoxyadenosine (R—CH_3). This process converts the substrate into a radical, and in step 3 the hydroxyl group on C-2 migrates to C-1. The new radical is related to the product, and it abstracts a hydrogen atom from 5′-deoxyadenosine in step 4 to form propionaldehyde hydrate at the active site while regenerating the adenosyl radical. In step 5, the adenosyl radical recombines with Co(II) to regenerate the coenzyme, and propionaldehyde hydrate is dehydrated to propionaldehyde. After dissociation of propionaldehyde, the enzyme-coenzyme complex is ready for the next reaction cycle.

Figure A

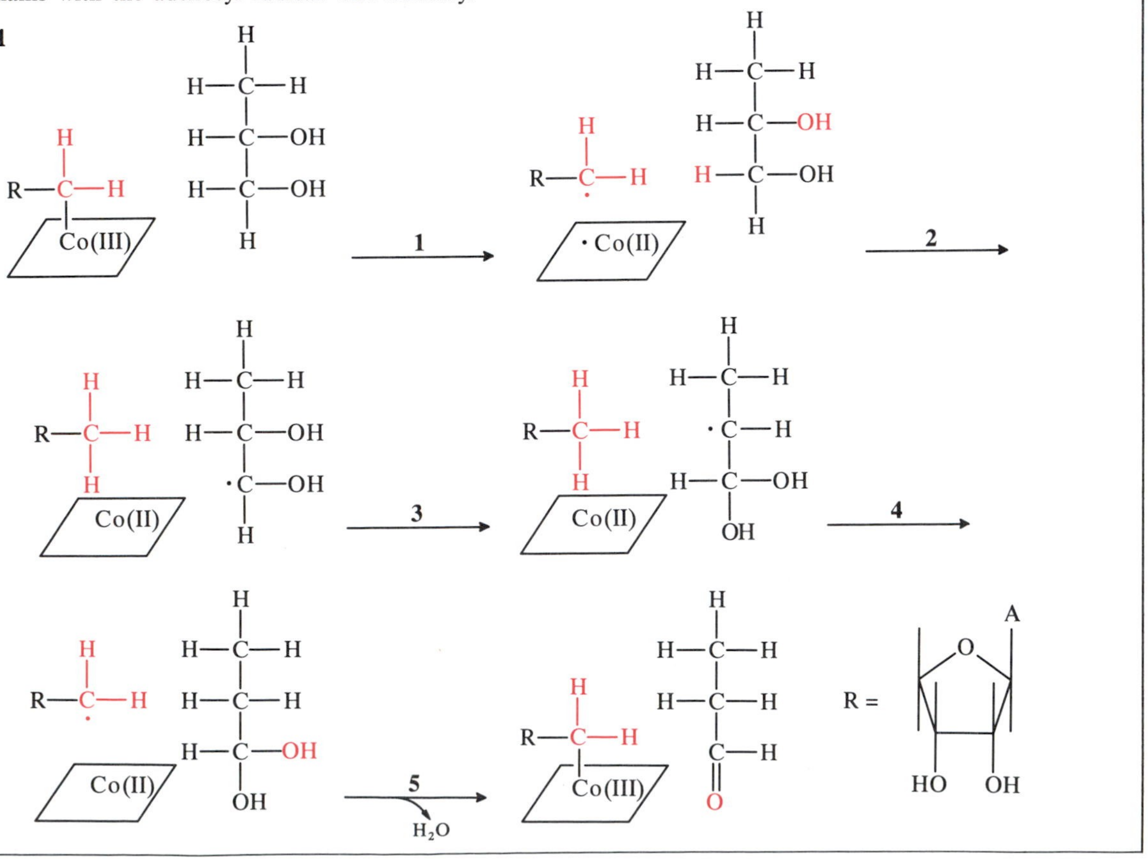

BOX 20-3 (continued)

THE ROLE OF ADENOSYLCOBALAMIN IN THE DIOLDEHYDRASE REACTION

The reactions catalyzed by methylmalonyl CoA mutase and dioldehydrase (and, for that matter, other reactions involving B_{12} coenzyme) probably proceed by analogous mechanisms, in all of which the essential function of the vitamin B_{12} coenzyme is to facilitate the process of abstracting a hydrogen atom (H·) from a substrate. The resulting radical related to the substrate is transformed into a radical related to the product, which abstracts a hydrogen atom from 5′-deoxyadenosine and becomes the product.

Phosphoryl Migration and Phosphomutases

The interconversion of glucose-1-P and glucose-6-P is catalyzed by phosphoglucomutase (equation 11),

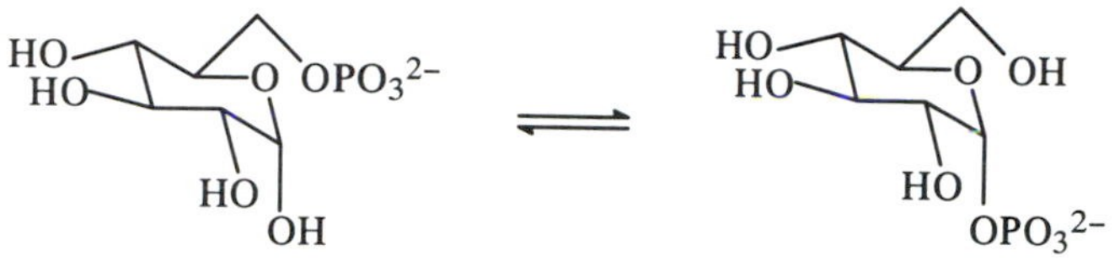

HO, HO, O, OPO_3^{2-}, HO, OH ⇌ HO, HO, O, OH, HO, OPO_3^{2-} (11)

Glucose-6-P **Glucose-1-P**

and the interconversion of 2-phosphoglycerate and 3-phosphoglycerate is catalyzed by phosphoglycerate mutase (equation 12).

$$\mathrm{COO^-{-}CH(OH){-}CH_2{-}OPO_3^{2-}} \rightleftharpoons \mathrm{COO^-{-}CH(OPO_3^{2-}){-}CH_2{-}OH} \quad (12)$$

3-Phosphoglycerate **2-Phosphoglycerate**

To maintain maximal activity, both enzymes require a cofactor. The cofactor for phosphoglucomutase is glucose-1,6-bisphosphate (glucose-1,6-P_2) and that for phosphoglycerate mutase is glycerate-2,3-bisphosphate (glycerate-2,3-P_2).

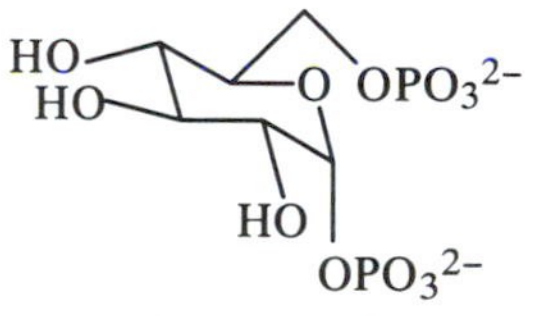

HO, HO, O, OPO_3^{2-}, HO, OPO_3^{2-} — **Glucose-1,6-P_2**

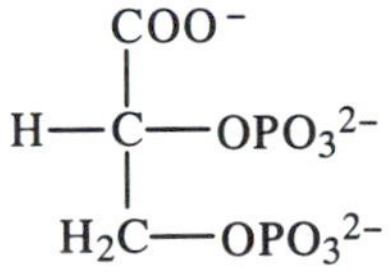

$\mathrm{COO^-{-}CH(OPO_3^{2-}){-}CH_2{-}OPO_3^{2-}}$ — **Glycerate-2,3-P_2**

The mechanism of action of phosphoglucomutase is illustrated in Figure 20-6. The active phosphoglucomutase is a phosphoenzyme, in which a serine residue is phosphorylated. Glucose-1-P combines with the phosphoenzyme and the phosphoryl group is transferred from the enzyme to the substrate. This forms a dephosphoenzyme and glucose-1,6-P_2. A phosphoryl group is then transferred from glucose-1,6-P_2 to the enzyme to regenerate the phosphoenzyme. When the phosphoryl group from C-1 of glucose-1,6-P_2 is transferred, glucose-6-P is formed and the isomerization is completed. If the phosphoryl group is transferred from C-6 of glucose-1,6-P_2 to the enzyme, the starting substrate, glucose-1-P, is regenerated. A metal ion is required for the phosphoglucomutase reaction, preferably Mg^{2+}, and considerable evidence indicates that the metal is coordinated both to the phosphoryl group and to amino acid side chains of the protein.

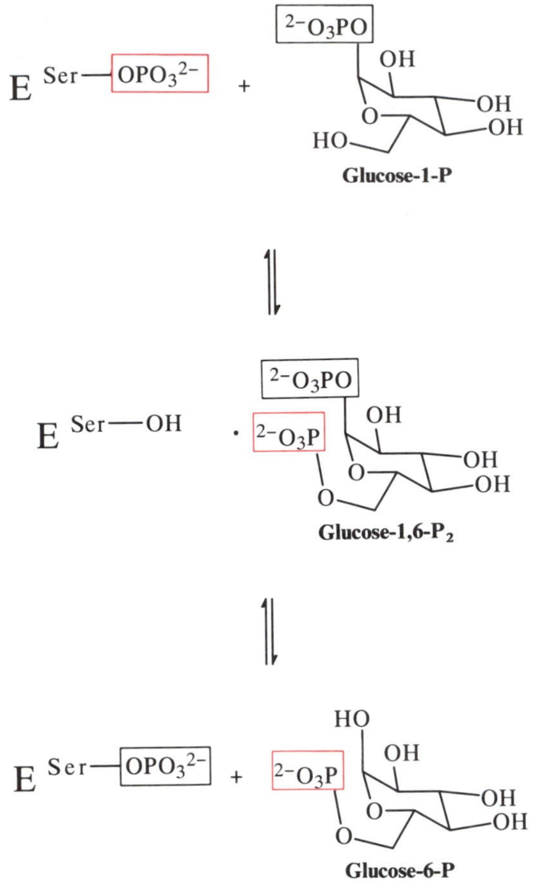

Figure 20-6

The Mechanism of Action of Phosphoglucomutase. Phosphoglucomutase contains a phosphorylated serine at the active site. When the enzyme binds glucose-1-P, it catalyzes the formation of glucose-1,6-P_2 and the dephosphoenzyme as an intermediate. The phosphoryl group is transferred from C-1 of glucose-1,6-P_2 to regenerate the phosphorylated serine of phosphoglucomutase. Glucose-6-P then dissociates from the active site. The reaction is reversible but favors glucose-6-P with a $K_{eq} = 17$.

Glucose-1-P

Glucose-1,6-P_2

Glucose-6-P

Part of the evidence for the mechanism described here is derived from experiments with glucose-6-P labeled with ^{32}P. When the enzyme reacts with this substrate, it is observed that after a few turnovers (actually, one is sufficient) the enzyme contains one labeled phosphate group per molecule of enzyme. When the ^{32}P-labeled enzyme so obtained is allowed to act on unlabeled glucose-1-P, the ^{32}P-label is lost from the enzyme and appears in the reaction product. This is exactly what is expected from the mechanism shown in Figure 20-6. Note that a product molecule never ends up with the phosphate that was present in the substrate molecule from which it was derived. The phosphate found in a product molecule is always derived from the enzyme, and the phosphate on the enzyme is derived from the substrate molecule that was previously processed.

What is the role of the cofactor glucose-1,6-P_2? The intermediate complex in Figure 20-6 can undergo an infrequent dissociation to release glucose-1,6-P_2 and produce the dephosphoenzyme (not phosphorylated) according to equation 13.

$$\mathrm{E \cdot Glucose\text{-}1,6\text{-}P_2} \rightleftharpoons \underset{\text{Dephosphoenzyme}}{\mathrm{E}} + \mathrm{Glucose\text{-}1,6\text{-}P_2} \qquad (13)$$

The dephosphoenzyme is catalytically inactive. The function of the glucose-1,6-P_2 is to prevent the accumulation of the dephosphoenzyme; that is, it shifts the equilibrium of equation 13 toward the active complex E · glucose-1,6-P_2, which can generate the phosphoenzyme and either glucose-1-P or glucose-6-P.

The structure of phosphoglucomutase is illustrated in Figure 20-7 as the α-carbon chain. Phosphoglucomutase is a large, single-chain protein with a molecular weight of 62,000 for the muscle enzyme. The structure is remarkable

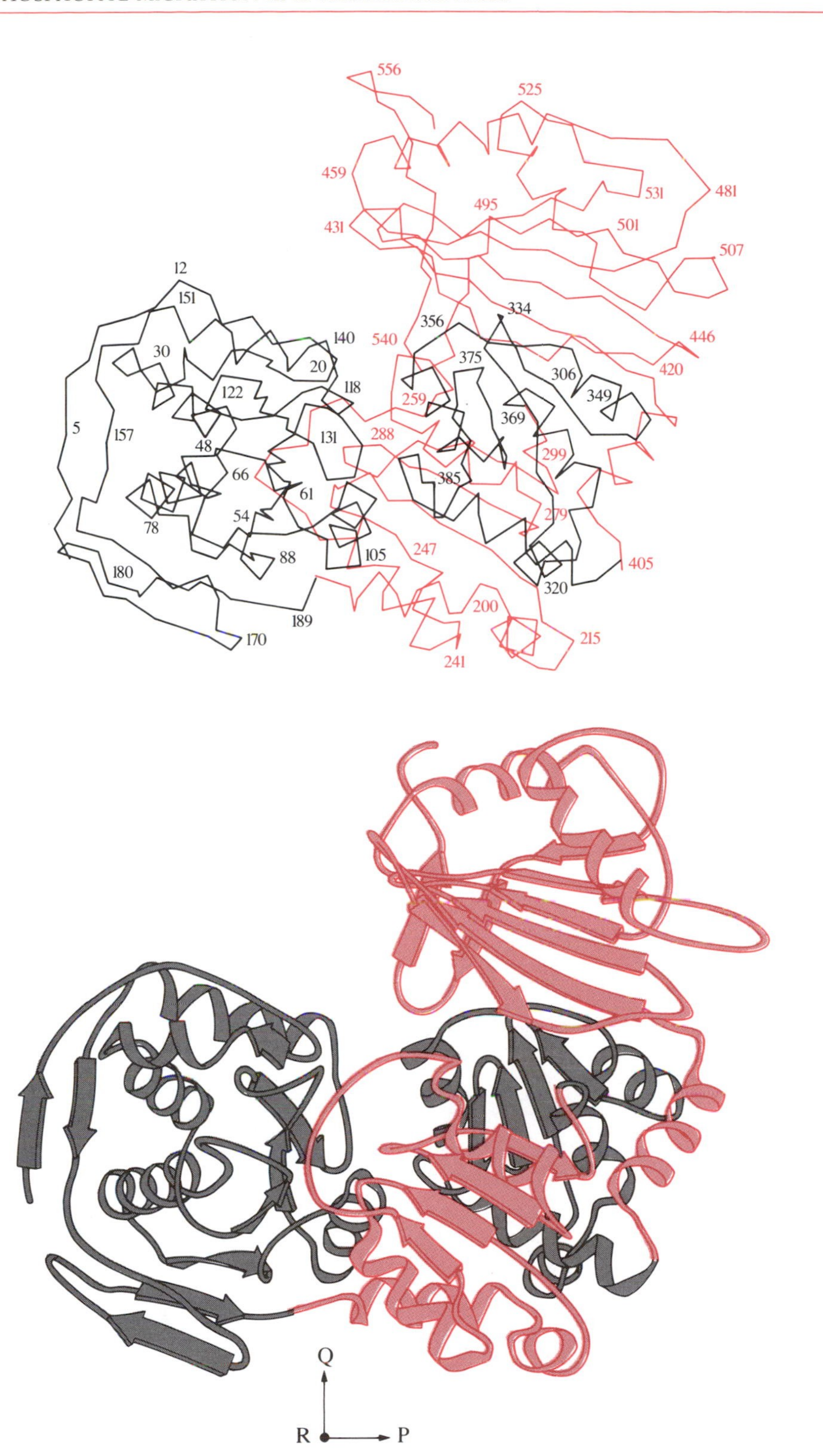

Figure 20-7

The α-Carbon Backbone of Muscle Phosphoglucomutase. Phosphoglucomutase is a large protein (molecular weight 62,000). The chain is folded into four domains that are indicated by different colors in these drawings. The upper drawing shows the α-carbon backbone of the protein. The lower drawing is a ribbon diagram of the phosphoglucomutase chain. (Adapted from J.-B. Dai, Y. Liu, W. J. Ray, Jr., and M. Konno. *Journal of Biological Chemistry*, in press.)

in that the chain is folded into four distinct domains, three of which are topologically similar. The active site is in a large cleft, as shown in Figure 20-8. The active site is remarkable for the fact that the substrate is much smaller than the cleft.

The mechanism of the phosphoglycerate mutase reaction is similar to that of phosphoglucomutase, except that the phosphoryl group in the phosphoenzyme is bonded to the imidazole ring of a histidine residue instead of to a serine hydroxyl group as in phosphoglucomutase.

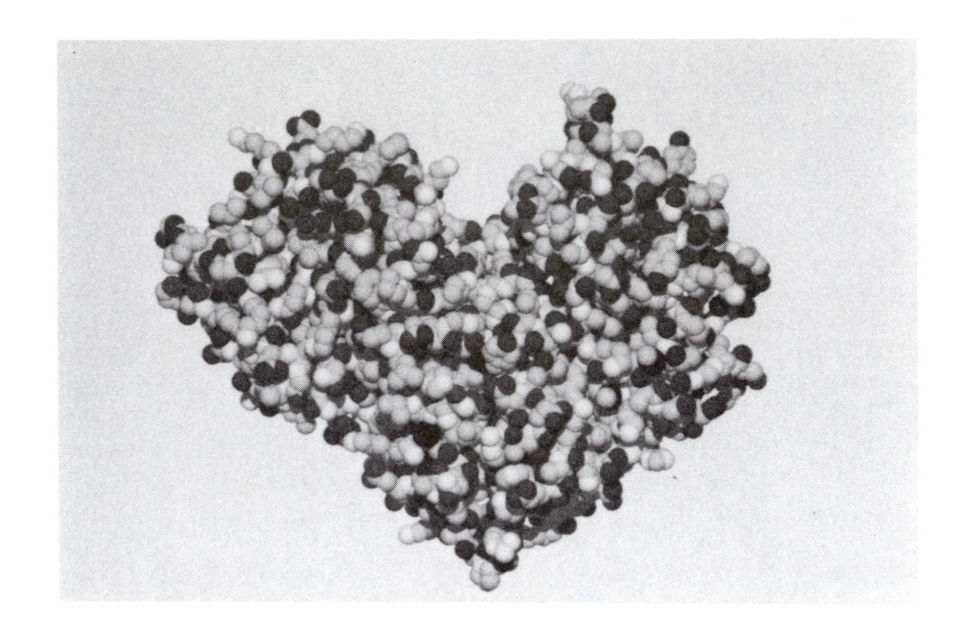

Figure 20-8

A Space-filling Model of Phosphoglucomutase Showing the Location of Phosphoserine in the Active Site at the Base of a Large Cleft. In this space-filling model of phosphoglucomutase, the β-O of serine-116 is at the base of the substrate-binding cleft. Serine-116 is phosphorylated in the active enzyme, and the phosphoryl group projects down into the cavity. (From J.-B. Dai, Y. Liu, W. J. Ray, Jr., and M. Konno. *Journal of Biological Chemistry*, in press.)

Not all phosphomutases function by the mechanism in Figure 20-6. Plant phosphoglycerate mutase operates through the intramolecular phosphoryl transfer mechanism of Figure 20-9, in which the enzyme is not phosphorylated initially. Upon combination with the 3-phosphoglycerate, the phosphoryl group is transferred from the substrate to the enzyme to form the phosphoenzyme and glycerate. Glycerate remains bound at the active site, and the phosphoryl group is then returned from the enzyme to the 2-OH of glycerate to form 2-phosphoglycerate.

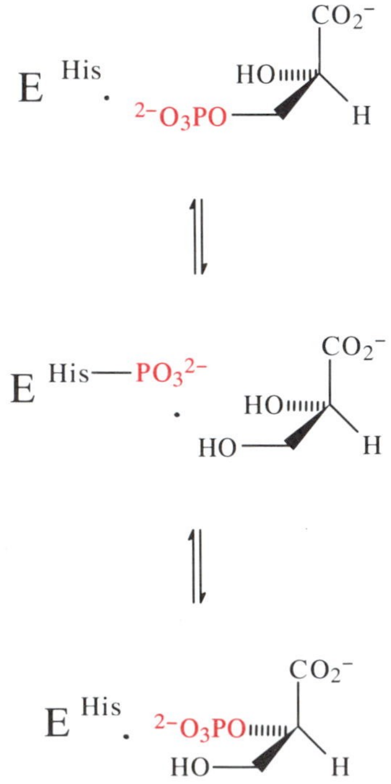

Figure 20-9

The Mechanism of Action of Plant Phosphoglycerate Mutase. Most phosphoglycerate mutases catalyze the interconversion of 3-phosphoglycerate and 2-phosphoglycerate by a mechanism analogous to that of phosphoglucomutase in Figure 20-6, except that the nucleophilic catalyst at the active site is a histidine instead of a serine. Shown here is the mechanism by which the plant phosphoglycerate kinase catalyzes the same reaction. The enzyme is normally *dephosphorylated,* and this form of the enzyme binds 3-phosphoglycerate to form the complex E · 3-P-glycerate shown at the top. The phosphoryl group is transferred to the enzyme, leading to the intermediate E—P · glycerate. The phosphoryl group from the enzyme is then transferred to the 2-OH group of glycerate, which yields the product complex E · 3-P-glycerate.

CO_2^-
E His · HO
$^{2-}O_3PO$ H

⇅

CO_2^-
E His—PO_3^{2-} · HO
HO H

⇅

CO_2^-
E His · $^{2-}O_3PO$
HO H

BOX 20-4

PHOSPHOGLUCOMUTASE: "SUBSTRATE-INDUCED RATE EFFECT"

Studies of phosphoglucomutase by W. Ray and coworkers have provided an interesting example of the use of the binding energy between an enzyme and a specific substrate for effective catalysis—that is, a "substrate-induced rate effect." There is little difference between the chemical reactivity of water, HOH, and that of the hydroxyl group at the 6-position of glucose-1-phosphate, ROH. But the rate constant for catalysis of phosphoryl transfer from the phosphoenzyme to glucose-1-phosphate bound to the active site is 3×10^{10} times as fast as transfer to water. The difference must be brought about by interaction between the enzyme and the bound substrate. Transfer to xylose is 70,000 times as fast as transfer to water, but transfer to xylose in the presence of phosphite is 2×10^9 as fast as transfer to water, almost as fast as the normal substrate. The addition of phosphite holds the five-carbon sugar in almost exactly the correct position for reaction in the active site. It is interesting that phosphite increases the reactivity with water by 580-fold and xylose-1-phosphate increases the reaction with water by a factor of 170,000. This shows that addition of parts of the specific substrate bring the enzyme into an active conformation that will rapidly transfer phosphate from the enzyme to water.

These differences in rate arise largely because of correct positioning of the hydroxyl group of the substrate, so that the probability of reaction is high; only a small amount of entropy must be lost in order to fix the reacting hydroxyl group in exactly the correct position for reaction (Chapter 5). However, the large increases in reaction rate with water in the presence of phosphite or xylose 1-phosphate show that the binding of portions of the specific substrate can force a change in the conformation of the enzyme that increases the reactivity of the phosphoryl group in the active site by a factor of as much as 100,000.

BOX 20-5

RACEMIZATION OF PROLINE

Enzymes that catalyze the racemization of amino acids generally use pyridoxal-5′-phosphate (PLP) as a cofactor. However, the enzymes proline racemase and hydroxyproline 2-epimerase do not require a cofactor. If the usual PLP mechanism were required for these reactions, as it is for alanine racemase, the PLP-substrate intermediate would be a secondary imine carrying a positive charge. Such an imine would be very effective in facilitating the abstraction of the α-proton, but it would also be highly unstable. No reactions are known in which a secondary, positively charged imine is formed with PLP.

In the simplest mechanism that one can envision, the α-proton of proline is removed to form a symmetric carbanion, which can be protonated at either face to give D- or L-proline.

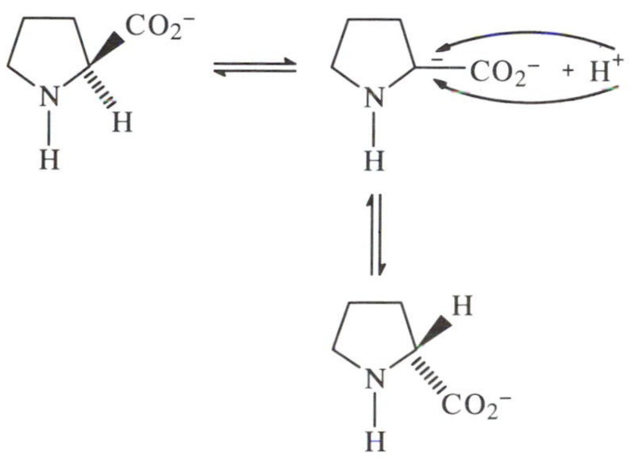

A minimal kinetic mechanism for the racemization of proline is:

$$E + P_H \underset{k_{-1}}{\overset{k_1}{\rightleftharpoons}} E \cdot P_H \underset{k_{-2}}{\overset{k_2}{\rightleftharpoons}} E\text{—}H^+ \cdot P^-$$

$$k_{-3} \upharpoonleft\downharpoonright k_3$$

$$E + P^H \underset{k_{-4}}{\overset{k_4}{\leftrightharpoons}} E \cdot P^H$$

Suppose that we tentatively make the following assumptions: the steps governed by k_1 and k_4 are fast, and the H^+ of $E\text{—}H^+ \cdot P^-$ rapidly exchanges with solvent protons. When L-proline is allowed to racemize completely [L-proline (P_H) is converted into DL-proline ($P^H + P_H$)] in ^2H_2O or ^3H_2O, the α-proton of proline is replaced by deuterium or tritium. This exchange is certainly consistent with the minimal kinetic mechanism shown above.

The results of the following experiment were more informative: L-proline (P_H) was added to the enzyme in ^2H_2O, and the reaction was allowed to proceed until 10% of L-proline had been converted into D-proline (P^H). The D- and L-proline were analyzed for deuterium, and the remaining L-proline was found to contain no deuterium. However, the newly formed D-proline contained one deuterium atom per molecule. This result could be consistent with the minimal kinetic mechanism shown

BOX 20-5 (continued)

RACEMIZATION OF PROLINE

only if $k_3 > k_{-2}$. In other words, the intermediate must have been unevenly partitioned in the forward and reverse directions and decomposed to P^H more often than to P_H.

This interpretation was easily tested. D-Proline (P^H) was added to the enzyme in ^2H_2O and the reaction was allowed to proceed until 10% of D-proline had been converted into L-proline. The deuterium distribution was again determined, with the result that essentially all the deuterium was found in L-proline. If $k_3 > k_{-2}$, as postulated earlier, deuterium would have been incorporated into D-proline in this experiment. This was not found; therefore, the minimal kinetic mechanism had to be modified.

The following mechanism of proline racemization is consistent with the experimental results presented here and with other experiments beyond the scope of this text.

$$E^H + P_H \underset{k_{-1}}{\overset{k_1}{\rightleftharpoons}} E^H \cdot P_H$$

$$k_{-2} \Big\uparrow\Big\downarrow k_2$$

$$E^H_H \cdot P^-$$

$$k_{-3} \Big\uparrow\Big\downarrow k_3$$

$$E_H + P^H \underset{k_{-4}}{\overset{k_4}{\rightleftharpoons}} E_H \cdot P_H$$

$$E_H \underset{k_{-5}}{\overset{k_5}{\rightleftharpoons}} E^H$$

The revised mechanism has a number of additional features. There are two forms of the enzyme, $\mathbf{E_H}$ and $\mathbf{E^H}$, which differ in the position of the proton and are specific for one of the proline enantiomers. $\mathbf{E^H}$ reacts only with P_H and $\mathbf{E_H}$ only with P^H. Clearly, $\mathbf{E^H}$ and $\mathbf{E_H}$ must be interconvertible. One additional condition is required; that is, the protons on the enzyme are not exchangeable with solvent protons when the substrate is bound. The protons in $\mathbf{E^H}$ and $\mathbf{E_H}$ readily exchange with solvent protons, but those in $\mathbf{E^H} \cdot P_H$, $\mathbf{E^H_H} \cdot P^-$ and $\mathbf{E_H} \cdot P^H$ do not exchange.

It is instructive to trace the fate of P^H as it reacts with the racemase in ^2H_2O. The protons of the enzyme will exchange with solvent deuterons.

$$E_H \xrightleftharpoons{2H_2O} E_{2H} \qquad E^H \xrightleftharpoons{2H_2O} E^{2H}$$

The deuterated enzyme then reacts with P^H to form a complex, which reacts to form the carbanion.

$$E_{2H} + P^H \rightleftharpoons E_{2H} \cdot P^H \rightleftharpoons E^H_{2H} \cdot P^-$$

$$\Big\uparrow\Big\downarrow$$

$$E^H \cdot P_{2H} \rightleftharpoons E^H + P_{2H}$$

This central complex can undergo one of two possible reactions. It can return to P^H, in which case it regains the same proton that it originally donated to the enzyme. Notice that the proline molecule so formed, P^H, will not contain deuterium, consistent with the experimental results. Alternatively, the carbanion can be converted into the opposite enantiomer. In undergoing this conversion, it will acquire one deuteron from the enzyme. This mechanism is consistent with experimental results: starting with L-proline (P^H) in ^2H_2O under initial velocity conditions, one finds no ^2H in L-proline, but every molecule of D-proline contains ^2H.

The simplest structural rationale for the kinetics of solvent proton exchange is that the enzyme contains two bases, which, in the enzyme-substrate complex, are positioned on opposite sides of the α-carbon of proline. The two forms of the enzyme, $\mathbf{E_H}$ and $\mathbf{E^H}$, differ with respect to which base is protonated. Chemical information indicates that the two bases may be cysteine residues.

Summary

Enzymatic racemization or epimerization at carbon can be thought of as a two-step process. In the first step, the center to be racemized is converted into a symmetric structure by removal of one of the substituents of the carbon atom (frequently, a proton). In the second step, the substituent is returned to the opposite side of the carbon from which it was removed. In amino acids, the α-hydrogen is abstracted as a proton, forming a carbanion (symmetric structure). The proton can be returned to the opposite face, which brings about the opposite configuration at the α-carbon of the amino acid. Pyridoxal-5′-phosphate is generally a cofactor that facilitates removal of the α-proton. Epimerization of sugars is exemplified by UDP-galactose 4-epimerase. The hydrogen bonded to C-4 of the galactosyl group is abstracted by NAD^+ and returned as a hydride ion by NADH. The symmetric intermediate is UDP-4-ketoglucose.

Another important isomerization is the interconversion of isomeric α-hydroxyketones or α-hydroxyaldehydes:

$$\begin{array}{c} \text{HO}\;\;\text{O} \\ |\;\;\;\;\| \\ \text{—C—C—} \\ | \\ \text{H} \end{array} \rightleftharpoons \begin{array}{c} \text{O}\;\;\text{OH} \\ \|\;\;\;\;| \\ \text{—C—C—} \\ \;\;\;\;\;| \\ \;\;\;\;\;\text{H} \end{array}$$

These reactions generally proceed through enediol intermediates.

$$\begin{array}{c} \text{HO}\;\;\text{OH} \\ |\;\;\;\;| \\ \text{—C=C—} \end{array}$$

Several isomerizations entail changes in the carbon skeleton of the substrates, such as the shift in the double bond exemplified by isopentenyl pyrophosphate isomerase. This enzyme participates in the biosynthesis of cholesterol and steroids. A carbon-skeleton rearrangement that is required in the biosynthesis of leucine and valine is catalyzed by acetohydroxy acid isomeroreductase. This enzyme catalyzes a 1,2 alkyl shift similar to the pinacol rearrangement. The vitamin B_{12} coenzyme adenosylcobalamin is required in a number of isomerization reactions, including carbon-skeleton rearrangements. These are probably radical reactions, in which the coenzyme functions as the radical initiator.

Phosphomutases catalyze the transfer of a phosphoryl group between two positions within a molecule. The mechanisms of these reactions frequently include a phosphoenzyme as an intermediate.

ADDITIONAL READING

Noltmann, E. A. 1972. Aldose-ketose isomerases. In *The Enzymes,* vol. 6, 3d ed., ed. P D. Boyer. New York: Academic Press, p. 272.

Ray, W. J., Jr., and Peck, E. J. 1972. Phosphomutases. In *The Enzymes,* vol. 6, 3d ed., ed. P. D. Boyer. New York: Academic Press, p. 408.

Barker, H. A. 1972. Coenzyme B_{12}-dependent mutases causing carbon chain rearrangements. In *The Enzymes,* vol. 6, 3d ed., ed. P. D. Boyer. New York: Academic Press, p. 509.

Frey, P. A. 1987. Complex pyridine nucleotide-dependent transformations. In *Pyridine Nucleotide Coenzymes: Chemical, Biochemical, and Medical Aspects,* vol. 2B, ed. D. Dolphin, R. Poulson, and O. Avramovic. New York: Wiley, p. 462.

Walsh, C. T. 1976. *Enzymatic Reaction Mechanisms.* San Francisco: Freeman.

Ray, W. J., Jr, Long, J. W., and Owens, J. D. 1976. An analysis of the substrate-induced rate effect in the phosphoglucomutase system. *Biochemistry* 15: 406.

Powers, U. M., Koo, C. W., Kenyon, G. L., Gerlf, J. A., and Kozarich, J. W. 1991. Mechanism of the reaction catalyzed by mandelate racemase: chemical and kinetic evidence for a two-base mechanism. *Biochemistry* 30: 9225.

PROBLEMS

1. Write a carbanion mechanism for the reaction catalyzed by isopentenyl pyrophosphate isomerase.

2. How does fluoroalanine inactivate alanine racemase?

3. Propose another possible mechanism of action of UDP-galactose 4-epimerase.

4. A bacterial enzyme catalyzes the following:

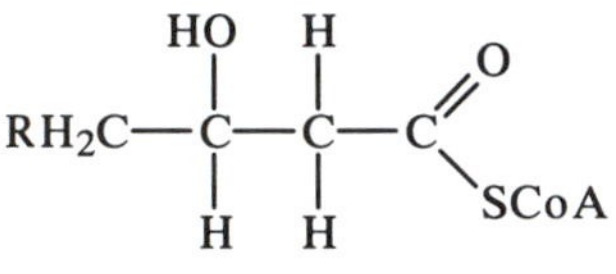

$RH_2C{-}C(OH)(H){-}C(H)(H){-}C(=O)SCoA$

A

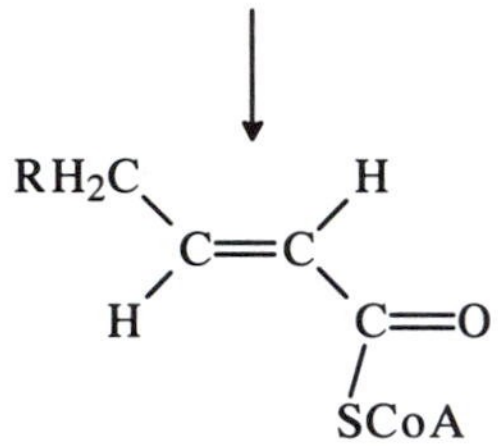

↓

$RH_2C{-}CH{=}CH{-}C(=O){-}SCoA$

B

+

$R{-}CH{=}CH{-}CH_2{-}C(=O)SCoA$

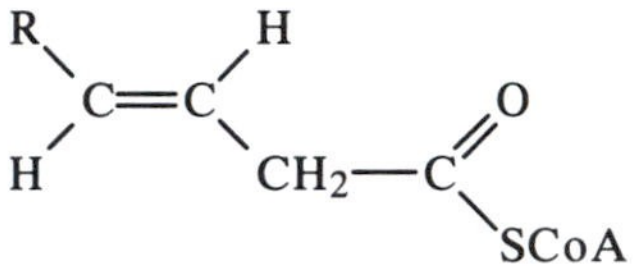

C

(a) Which of the following probably occurs?

$A \rightleftharpoons B \rightleftharpoons C$

$A \rightleftharpoons C \rightleftharpoons B$

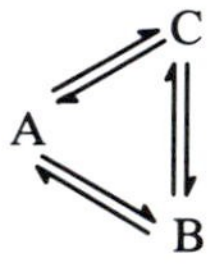

$A \rightleftharpoons C$, $C \rightleftharpoons B$, $A \rightleftharpoons B$ (cyclic)

(b) Write a mechanism for the reaction.

5. Propose a mechanism for the action of mandelate racemase.

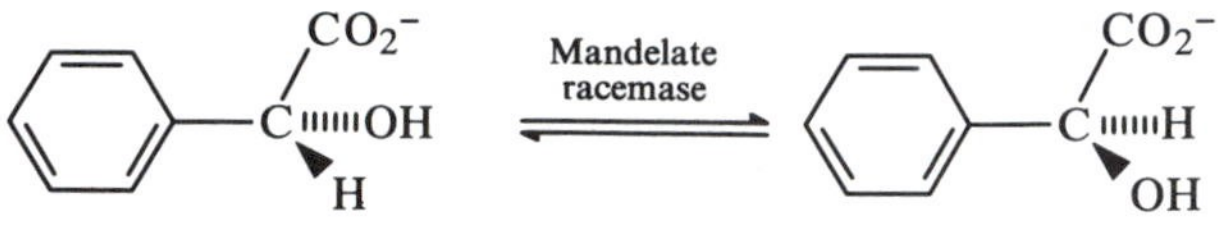

$C_6H_5{-}C(CO_2^-)(OH)(H) \underset{\text{racemase}}{\overset{\text{Mandelate}}{\rightleftharpoons}} C_6H_5{-}C(CO_2^-)(H)(OH)$

Glycolysis and the Tricarboxylic Acid Cycle

21

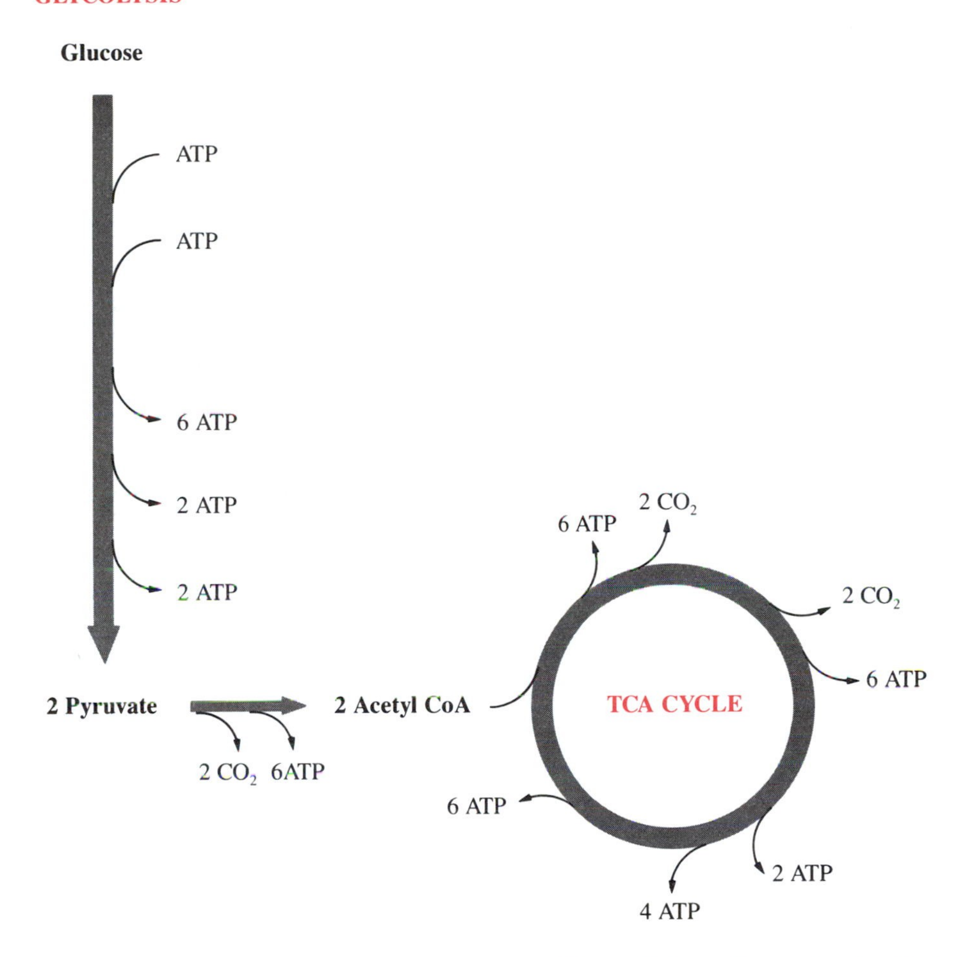

The oxidation of a molecule of glucose through the glycolytic pathway and the tricarboxylic acid (TCA) cycle leads to the net production of as many as thirty-eight molecules of ATP.

The central requirement for a cell to continue its day-to-day existence is a source of readily available energy—a fuel to run biosynthetic reactions, to power the storage of information in the genes, and to drive many specialized energy-requiring functions (e.g., motility; generation of heat, light, or electricity; excretion of unwanted chemicals or absorption of desirable ones). In this chapter, we shall describe how a cell derives energy by consumption of nutrient molecules in its surroundings. The common source of immediate energy in biological systems is ATP, and so the problem is: *How are nutrient molecules utilized to bring about the synthesis of enough ATP to meet the energy needs of the cell?*

The cell derives energy from nutrients by subjecting them to a series of chemical transformations that have, overall, a large negative free-energy change; that is, by metabolizing them. However, instead of allowing this free energy to be dissipated into the environment, the pathways of energy metabolism capture this free energy by forming ATP. In effect, the free energy inherent in the breakdown of nutrients is stored in a potential form—the high-energy phosphate bond of ATP—and may be drawn upon as needs arise in the cell. In this chapter and the next, we will explore how this process of energy capture is carried out.

As pointed out in Chapter 1, a remarkable thing about cellular metabolism is that, although the cells process a tremendous number of different molecules, only a small number of *types* of chemical reactions are required for metabolism. Thus, the chemical reactions in protein biosynthesis, starting from amino acids, consist of acyl group transfer and phosphoryl group transfer; the major nucleic acid biosynthetic processes are phosphoryl group transfer reactions (Chapters 10 and 12). Fatty acid biosynthesis requires carboxylation, acyl group transfer, decarboxylation, carbon–carbon bond formation, reduction, and dehydration (Chapter 19).

In considering glycolysis and the tricarboxylic acid cycle (TCA cycle), as well as other metabolic pathways, we will see that the sequence of reactions in a metabolic pathway is completely determined by the *chemical* problems that must be solved; like the synthetic organic chemist, nature chooses a sequence of reactions best designed to reach a desired product from a given starting material. As we describe the pathways of intermediary metabolism, it is important to keep in mind that these pathways, though seemingly complex, are not arbitrarily so—that is, each step in a reaction sequence has a good chemical reason for being there.

Energy Metabolism: The Grand Strategy

Most of this discussion will concern the pathways and reactions that lead to the generation of ATP; that is, the tactics used by a cell to supply itself with usable fuel. First, however, it is worthwhile to point out the overall strategy employed to achieve this goal, to prevent the enzymological trees from obscuring the thermodynamic forest. Except in photosynthesis (Chapter 23), all biological energy is derived from oxidative processes; in aerobic organisms, which we consider first, the final electron acceptor is O_2.

Let us begin by considering the general strategy for the metabolism of glucose. In aerobic organisms, glucose is oxidized to $CO_2 + H_2O$, and O_2 serves as the ultimate electron acceptor.

$$\text{Glucose} + O_2 \longrightarrow 6\,CO_2 + 6\,H_2O \qquad \Delta G^\circ = -690 \text{ kcal mol}^{-1}$$

In this process, a large amount of energy is liberated, which is used to generate ATP. The conversion of glucose into CO_2 does not occur in one step; instead, it

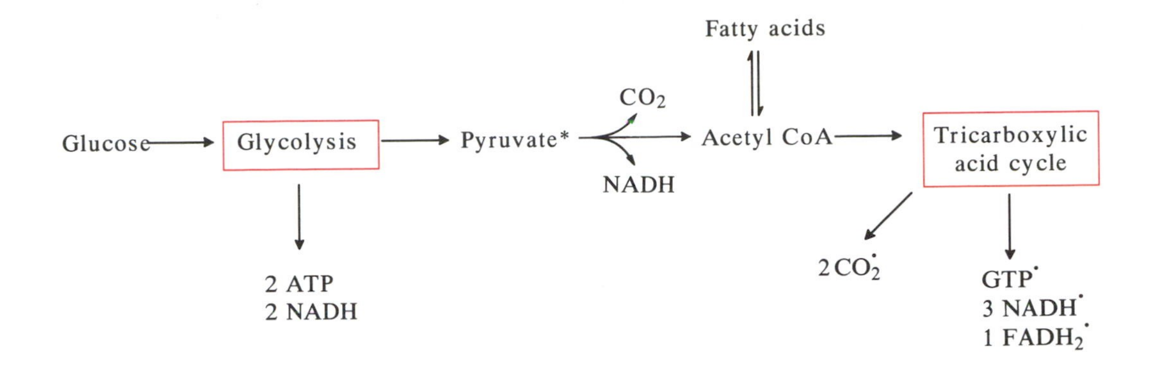

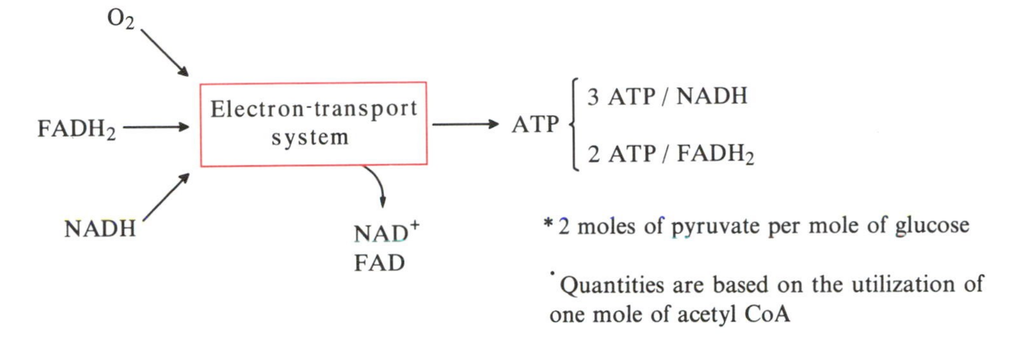

* 2 moles of pyruvate per mole of glucose

• Quantities are based on the utilization of one mole of acetyl CoA

Figure 21-1

Aerobic Glucose Metabolism.

is accomplished in stages through three major metabolic pathways, as shown schematically in Figure 21-1. A mole of glucose is first converted through a series of reactions known as *glycolysis* into 2 moles of pyruvate. Concomitantly, 2 moles of ATP and 2 moles of NADH are produced. The 2 moles of pyruvate are then converted into 2 moles of acetyl CoA and 2 additional moles of NADH by pyruvate dehydrogenase, a complex of several enzymes that was described in Chapter 17. Acetyl CoA is oxidized to CO_2 and H_2O through a series of reactions in the *tricarboxylic acid cycle,* also called the Krebs cycle or the citric acid cycle. The NADH and $FADH_2$ produced in these reactions is finally reoxidized to NAD^+ and FAD through the electron-transport chain. In the course of this oxidation, 3 moles of ATP per mole of NADH and 2 moles of ATP per mole of $FADH_2$ are produced. The three major metabolic pathways shown in Figure 21-1 are found, with some minor modifications, in the cells of all aerobic organisms.

Glycolysis, or a modification thereof, is also found in anaerobic organisms. When glycolysis occurs under anaerobic conditions, it is called fermentation. When glucose is fermented, pyruvate is not oxidatively decarboxylated to acetyl CoA but is converted into other compounds, such as ethanol. The fate of pyruvate in anaerobic organisms will be described later in this chapter.

A large fraction of the ATP that is produced in the aerobic metabolism of glucose comes from the oxidation of NADH and $FADH_2$ through the electron-transport chain. This ability to couple the oxidation of NADH and $FADH_2$ by O_2 to the formation of ATP makes aerobic metabolism exceedingly productive of useful energy and gives aerobic organisms an enormous advantage over anaerobic organisms. The three major metabolic pathways shown in Figure 21-1 are described in the following sections. We begin with glycolysis.

Glycolysis

Glycolysis is the process of converting glucose into pyruvate. The overall equation is:

$$2\,NAD^+ + C_6H_{12}O_6 + 2\,HPO_4^{2-} + 2\,ADP^{3-} \longrightarrow 2\,C_3H_4O_3 + 2\,ATP^{4-} + 2\,NADH + 2\,H_2O$$

in which 2 moles of ATP are produced per mole of glucose.

The reaction sequence by which glucose is converted into pyruvate is shown in Figure 21-2. It is one of many metabolic pathways that you will encounter. It is quite difficult to memorize these pathways, and there is no need to do so. If you understand the basic principles underlying a pathway, you will usually be able to readily "derive" it.

To illustrate this point, we shall analyze the fundamental principles underlying the glycolytic pathway and briefly describe the individual reactions in Figure 21-2. When thinking about glycolysis, it is helpful to divide it into two stages: in stage 1, glucose is converted into a three-carbon compound, glyceraldehyde-3-P, in reactions 1 through 5 in Figure 21-2; stage 2 is the conversion of glyceraldehyde-3-P into pyruvate. The breakdown of glucose to a *single* three-carbon species, glyceraldehyde-3-P in stage 1, has the advantage that one set of enzymes can be used in subsequent reactions; whereas, if two different fragments were produced, two sets of enzymes would be needed.

STAGE 1 OF GLYCOLYSIS

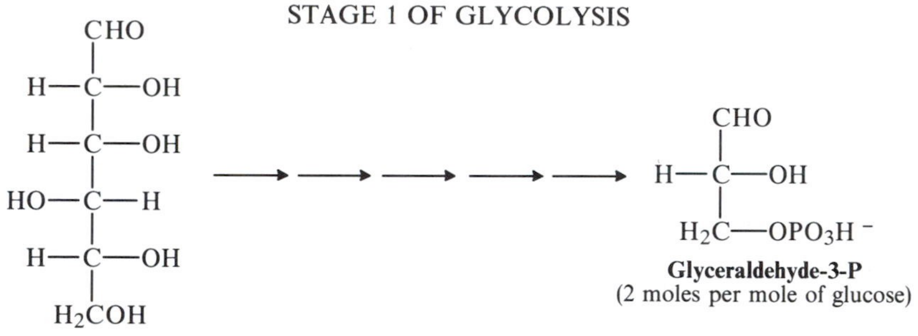

```
        CHO
        |
    H—C—OH
        |
    H—C—OH                                   CHO
        |                                     |
   HO—C—H    ——→ ——→ ——→ ——→ ——→        H—C—OH
        |                                     |
    H—C—OH                                H2C—OPO3H⁻
        |
      H2COH
```

Glucose

Glyceraldehyde-3-P
(2 moles per mole of glucose)

The first stage of glycolysis (stage 1) begins with the phosphorylation of glucose by ATP, which is catalyzed by hexokinase. Phosphorylation of the sugar tends to keep the sugar inside the cell, because cells are generally impermeable to phosphorylated compounds. Next, the phosphorylated sugar must be converted into two 3-carbon fragments, dihydroxyacetone-P and glyceraldehyde-3-P. To achieve this, the C–C bond between C-3 and C-4 of glucose must be broken. Carbon–carbon bond cleavage is generally α,β to a carbonyl group (Chapter 2). However, the carbonyl group of glucose is at the C-1 position, so that cleavage of the C–C bond would be between C-2 and C-3, and a 2-carbon and 4-carbon fragment would be formed. To obtain the desired 3-carbon fragments, the carbonyl group must be "moved" to C-2. This is exactly what is accomplished by phosphoglucose isomerase, which converts glucose-6-P into fructose-6-P (reaction 2, Figure 21-2). Phosphorylation of the fructose-6-P in reaction 3 gives rise to fructose-1,6-P_2. Cleavage of this diphosphorylated compound by fructose-1,6-P_2 aldolase in reaction 4 gives two phosphorylated 3-carbon fragments, glyceraldehyde-3-P and dihydroxyacetone-P. Glyceraldehyde-3-P and dihydroxyacetone-P are readily interconvertible (reaction 5), so that eventually one glucose molecule is converted to 2 moles of glyceraldehyde-3-P. This completes stage 1 of glycolysis.

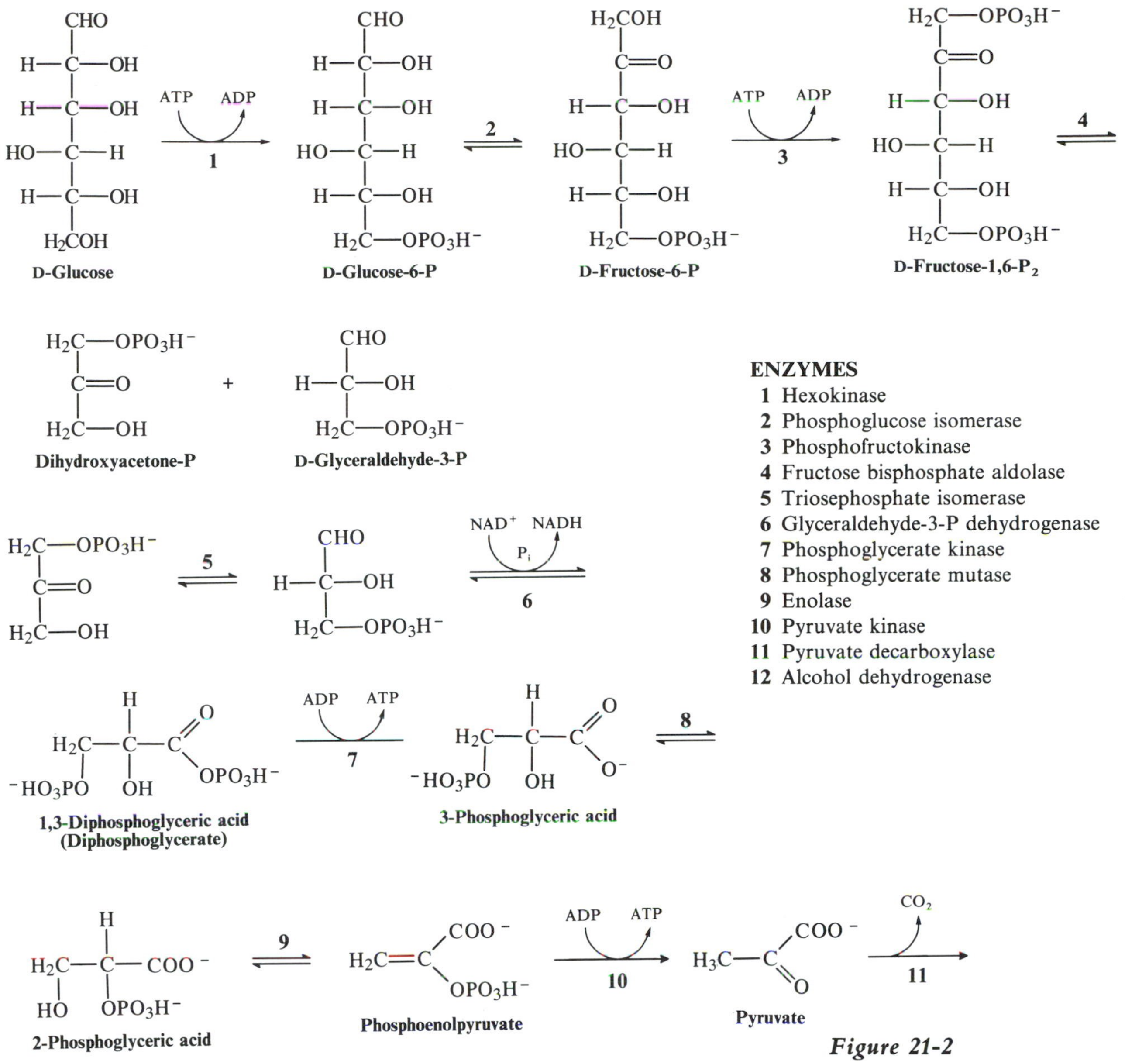

D-Glucose → (1; ATP → ADP) → D-Glucose-6-P ⇌ (2) D-Fructose-6-P → (3; ATP → ADP) → D-Fructose-1,6-P_2 ⇌ (4)

Dihydroxyacetone-P + D-Glyceraldehyde-3-P

Dihydroxyacetone-P ⇌ (5) Glyceraldehyde-3-P ⇌ (6; NAD^+ → NADH, P_i)

1,3-Diphosphoglyceric acid (Diphosphoglycerate) → (7; ADP → ATP) → 3-Phosphoglyceric acid ⇌ (8)

2-Phosphoglyceric acid ⇌ (9) Phosphoenolpyruvate → (10; ADP → ATP) → Pyruvate → (11; CO_2) →

Acetaldehyde (CH_3—CHO) ⇌ (12; NADH → NAD^+) Ethanol (CH_3—CH_2OH)

ENZYMES

1 Hexokinase
2 Phosphoglucose isomerase
3 Phosphofructokinase
4 Fructose bisphosphate aldolase
5 Triosephosphate isomerase
6 Glyceraldehyde-3-P dehydrogenase
7 Phosphoglycerate kinase
8 Phosphoglycerate mutase
9 Enolase
10 Pyruvate kinase
11 Pyruvate decarboxylase
12 Alcohol dehydrogenase

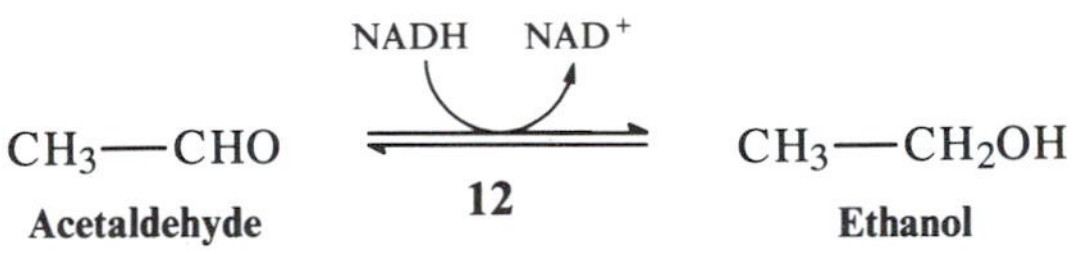

Figure 21-2

Glycolysis—Conversion of Glucose into Pyruvate. The breakdown of glucose to pyruvate is catalyzed by enzymes **1** through **10**. Note that 2 moles of pyruvate are formed from 1 mole of glucose. In anaerobic systems, pyruvate cannot be further oxidized. This illustration also shows how pyruvate can be converted into ethanol by enzymes **11** and **12** in anaerobic fermentation by yeast.

In the second stage, the 3-carbon compound glyceraldehyde-3-P is converted into pyruvate, with the formation of 2 moles of ATP per mole of glyceraldehyde-3-P.

STAGE 2 OF GLYCOLYSIS

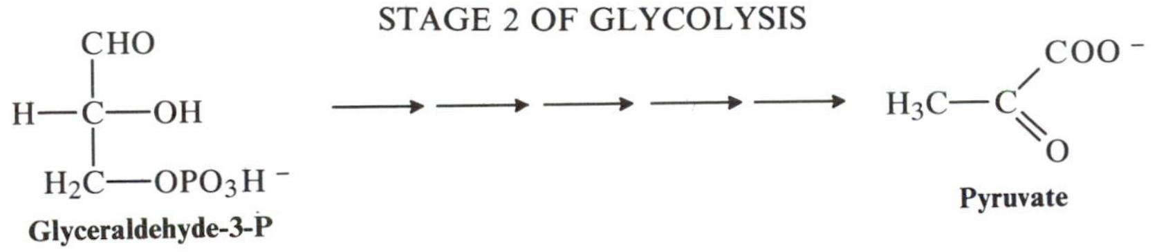

Glyceraldehyde-3-P → → → → → Pyruvate

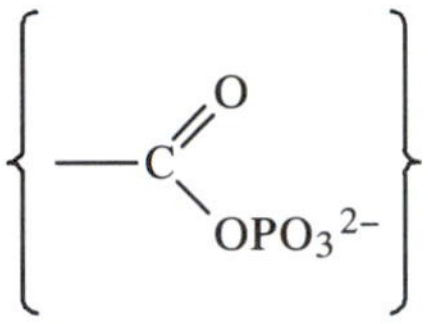

$$\left\{ -C\begin{matrix}{}^{\diagup\!\!\diagup O} \\ {}_{\diagdown OPO_3^{2-}}\end{matrix} \right\}$$

Stage 2 begins with the oxidation of glyceraldehyde-3-P with the uptake of phosphate to give 1,3-diphosphoglycerate. This leads to the formation of a compound containing a high-energy phosphoryl group. This is the first high-energy compound that is formed in the fermentation of glucose. The mechanism of this oxidation was described in Chapter 17. One phosphoryl group of diphosphoglycerate can be transferred to ADP to produce ATP (reaction 8) and 3-phosphoglycerate. At this point, two molecules of ATP have been produced per molecule of glucose metabolized. However, recall that two molecules of ATP have also been utilized (reactions 1 and 3); therefore, no net ATP formation has occurred up to this point.

The subsequent steps, reactions 7 through 10, are designed to convert the phosphoryl group of 3-phosphoglycerate into a high-energy phosphoryl group, so that another molecule of ATP can be obtained. This problem is solved by moving the phosphoryl group from C-3 to C-2; that is, the isomerization of 3-phosphoglycerate to 2-phosphoglycerate in reaction 8. Water can then be eliminated from 2-phosphoglycerate to form phosphoenolpyruvate, a high-energy compound, in reaction 9. Phosphoenolpyruvate can react with ADP to produce ATP and pyruvate. Note that the conversion of 2-phosphoglycerate into pyruvate is a type of elimination reaction that is frequently seen with dihydroxy compounds; it was described in Chapter 18. This reaction differs from others that we have seen in that the intermediate phosphorylated enol, phosphoenolpyruvate, is stable. In most other reactions of this type, the enol is not stable; the free enol spontaneously undergoes isomerization to the ketone.

One can better understand the rationale underlying the reaction sequence 3-phosphoglycerate → 2-phosphoglycerate → phosphoenolpyruvate → pyruvate by considering a possible hypothetical sequence:

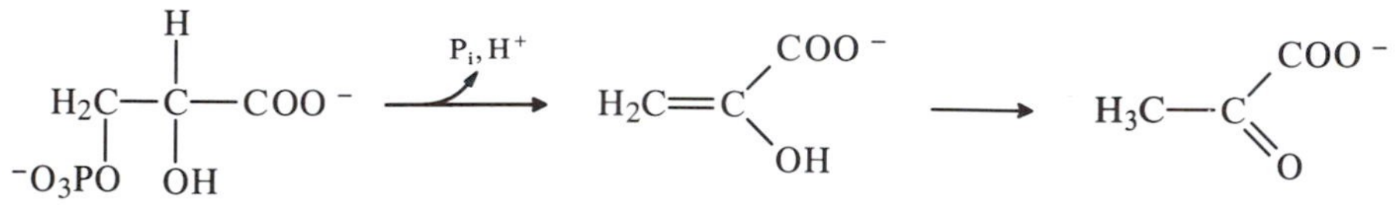

$$\underset{^{-}O_3PO\quad\;\; OH}{H_2C-\overset{H}{C}-COO^-} \xrightarrow{P_i,\,H^+} H_2C{=}C\begin{matrix}{}^{\diagup COO^-} \\ {}_{\diagdown OH}\end{matrix} \longrightarrow H_3C-C\begin{matrix}{}^{\diagup COO^-} \\ {}_{\diagdown\!\!\diagdown O}\end{matrix}$$

In this sequence, the isomerization step (3-phosphoglycerate → 2-phosphoglycerate) has been omitted. Phosphate is eliminated directly from 3-phosphoglycerate to produce the enol of pyruvate, which then undergoes ketonization to form pyruvate. Like reactions 8 through 10 in Figure 21-2, this reaction sequence would produce pyruvate from 3-phosphoglycerate, but the phosphoryl group would be lost, no high-energy intermediate would be formed, and no ATP would be produced.

The reaction of phosphoenolpyruvate with ADP to produce ATP in reaction 10 produces two molecules of ATP from each molecule of glucose metabolized (two molecules of phosphoenolpyruvate are formed from each molecule of glucose). *Two more molecules of ATP were formed from two molecules of 1,3-diphosphoglyceric acid, but two molecules of ATP were utilized to form fructose-1,6-bisphosphate. Therefore, a net total of two molecules of ATP has been synthesized from each molecule of glucose up to this point.*

The TCA Cycle

Before the tricarboxylic acid, or TCA, cycle is entered, pyruvate must first be converted into acetyl CoA. This reaction is catalyzed by the pyruvate dehydrogenase complex according to the following equation.

$$CH_3{-}CO{-}COO^- + NAD^+ + CoASH \longrightarrow CH_3{-}CO{-}SCoA + CO_2 + NADH$$

This reaction was described in Chapter 17. The stoichiometry of the metabolism of acetyl CoA through the TCA cycle is given by:

$$2\,H_2O + CH_3—CO—SCoA + GDP + P_i + 3\,NAD^+ + FAD \longrightarrow 2\,CO_2 + FADH_2 + GTP + 3\,NADH + 2\,H^+ + CoASH$$

The TCA cycle produces 2 moles of GTP, 6 moles of NADH, and 2 moles of $FADH_2$ from 2 moles of acetyl CoA (1 mole of glucose gives rise to 2 moles of acetyl CoA). NADH and $FADH_2$ are reoxidized by O_2 through the electron-transport chain. The oxidation of each mole of NADH produces 3 moles of ATP, and 2 moles of ATP are obtained from each $FADH_2$. Thus, the metabolism of each mole of glucose under aerobic conditions produces a total of 38 moles of ATP (or GTP, which is equivalent to ATP as a high-energy phosphate compound).

The reactions of the TCA cycle are shown in Figure 21-3. The first reaction is the condensation of oxaloacetate and acetylCoA; the final reaction regenerates oxaloacetate. Thus, only acetyl CoA is actually consumed in the process. In other words, oxaloacetate, and indeed the cycle as a whole, serves as a *catalyst* to convert acetyl CoA into CO_2. This type of cyclic process, a reaction sequence in which the original substrate is eventually regenerated, is found frequently in biochemical systems.

First Phase of the TCA Cycle: CO_2 Production

When we consider the rationale underlying the TCA cycle, it is important to keep in mind that the two carbon atoms of the acetyl moiety of acetyl CoA are converted into CO_2. As indicated in Chapter 17, the strategy used throughout biochemical systems to produce CO_2 is through decarboxylation of α-keto or β-keto acids. Hence, the first four reactions of the cycle concern the production of keto acids and their decarboxylation to produce 2 moles of CO_2 per mole of acetyl CoA.

The first reaction of the TCA cycle is the condensation of oxaloacetic acid and acetyl CoA, which leads to citric acid (reaction 1, Figure 21-3). This is one of the reactions described in Chapter 2 that results in the formation of a new C–C bond. Citric acid cannot be oxidized to a keto acid because the OH group is on a tertiary carbon atom. Isomerization of citric acid to isocitric acid (reaction 2) places the OH group on a secondary carbon atom, where it can be oxidized to a carbonyl group. This oxidation produces a β-keto acid, oxalosuccinate, which is then decarboxylated. You will recall that isocitrate dehydrogenase catalyzes this oxidation (reaction 3) as well as the decarboxylation of isocitrate (Chapter 17). Reaction 3 gives rise to an α-keto acid, α-ketoglutarate, which is decarboxylated through reaction 4. This reaction includes a condensation with coenzyme A to give succinyl CoA, a high-energy compound.

At this point, two molecules of CO_2 have been formed, along with two molecules of NADH. The carbon skeleton of glucose has now been completely broken down.

Second Phase of the TCA Cycle: Regeneration of Oxaloacetate

The second phase begins with the conversion of succinyl CoA, a compound with a high-energy thioester bond, into succinate (Figure 21-3, reaction 5). The energy inherent in this conversion is not thrown away but is trapped by the production

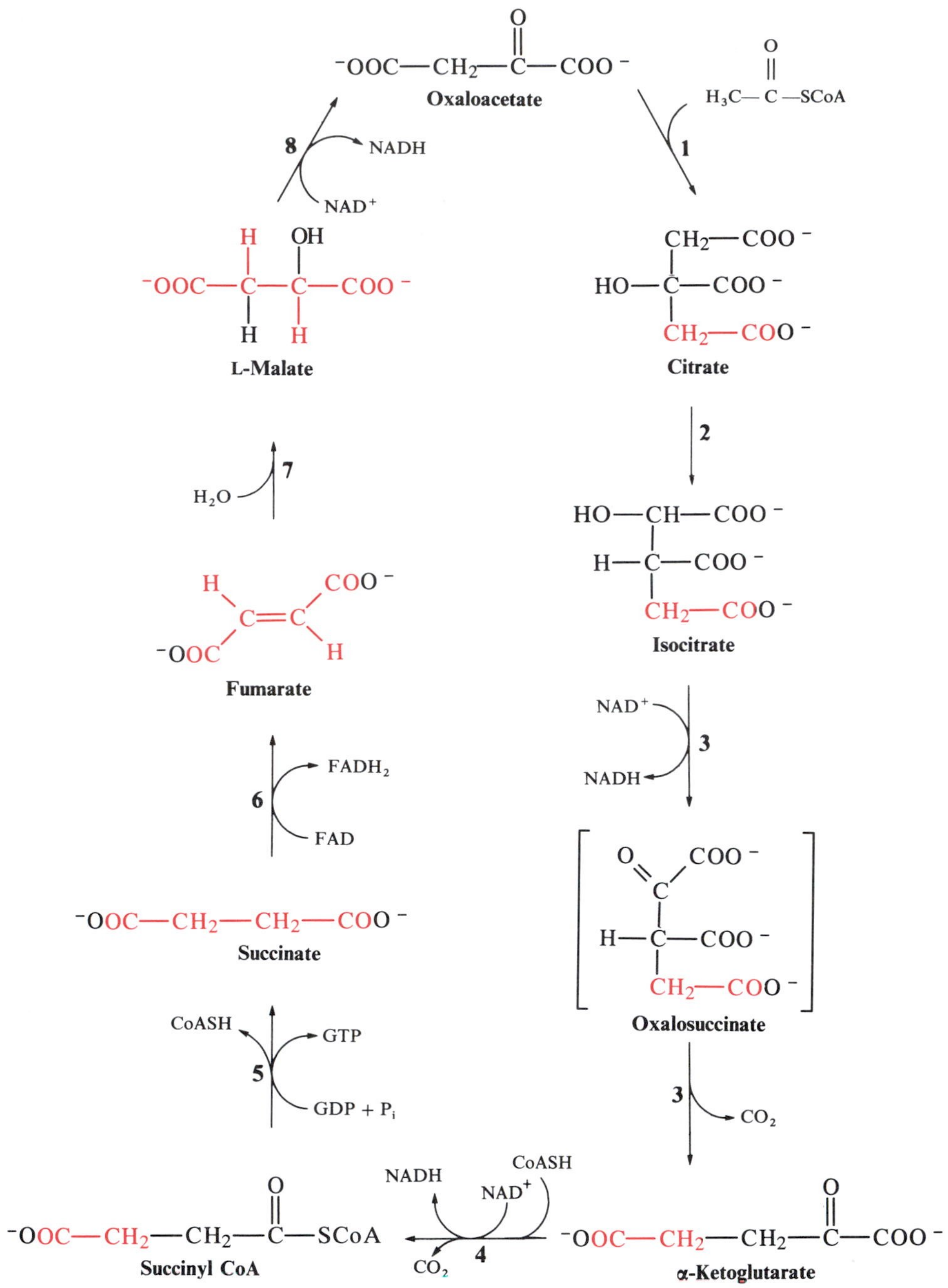

Figure 21-3

Tricarboxylic Acid Cycle (TCA cycle, or Krebs cycle). This scheme indicates the labeling pattern when isotopically labeled acetyl CoA undergoes one turn of the cycle; the red atoms are derived from acetyl CoA. The cycle is initiated by the condensation of oxaloacetate and acetyl CoA. When the cycle is completed, oxaloacetate is regenerated. The net result is: CH_3—CO—SCoA → $CO_2 + 2\,H_2O$ + CoASH + 2 NADH + $FMNH_2$ + GTP The enzymes are: **1**, citrate synthase; **2**, aconitase; **3**, isocitrate dehydrogenase; **4**, α-ketoglutarate dehydrogenase complex; **5**, succinyl CoA synthase; **6**, succinate dehydrogenase; **7**, fumarase; **8**, malate dehydrogenase.

of a molecule of GTP. This reaction is the reverse of an acyl-activation reaction in which a carboxylic acid is converted into a thiol ester by utilization of the energy of ATP or GTP. Succinate is now ready to be converted into oxaloacetate. This reaction sequence entails the oxidative formation of a double bond, hydration of this double bond, and finally oxidation of the resulting alcohol to a ketone, oxaloacetate. Note that two oxidation reactions are involved: the oxidation of succinate to fumarate and the oxidation of malate to oxaloacetate. Each of these reactions is used to produce reducing power—$FADH_2$ in the former case and NADH in the latter. It should be noted that this is essentially the same reaction sequence that is used to convert fatty acids into β-keto acids (Chapter 19).

With the regeneration of oxaloacetate, one turn of the cycle is now completed. One mole of acetyl CoA has been converted into 2 moles of CO_2; and, in the process, GTP, NADH, and $FADH_2$ have been produced. One mole of oxaloacetate has been used and then regenerated; this preserves the catalytic function of the cycle.

Experimental Evidence for the TCA Cycle

When ^{14}C-labeled compounds became available, it was possible to conduct additional experiments to test the validity of the reaction sequence in the TCA cycle. Consider, for example, the following experiment: acetyl CoA labeled in the carboxyl carbon atom is added to a system in which the TCA cycle is active. We then ask: What is the fate of the radioactive carbon atom in a *single turn* of the cycle? Will ^{14}C appear in CO_2 or will it be found in the various intermediates of the cycle? The first expectation, from examining the scheme of Figure 21-4, might be that 50% of the ^{14}C would be found in CO_2 after the first turn of the cycle. Acetyl CoA that is labeled with ^{14}C reacts with oxaloacetate to produce carboxyl-labeled citrate. Citrate is then converted through the actions of aconitase into isocitrate. It was assumed at first that, because citrate is a symmetric molecule, two species of isocitrate could be formed (paths *a* and *b*, Figure 21-4). In one species, the OH group is adjacent to the labeled carboxyl group (path *b*) and, in the other species, it is adjacent to an unlabeled carboxyl group, derived initially from oxaloacetate. Through subsequent reactions, the isocitrate formed through path *b* will give rise to $^{14}CO_2$ and that formed through path *a* will give rise to unlabeled CO_2. The results that were obtained from experiments of this kind were surprising. In the first turn of the cycle, no labeled CO_2 was produced from acetyl CoA labeled with ^{14}C in the carboxyl group.

The failure to observe $^{14}CO_2$ formation led to many proposals of schemes for the TCA cycle that did not include citrate. However, the problem was finally resolved by Ogston, who pointed out that the two terminal carboxyl groups are *not* equivalent for a citrate molecule that is bound to an enzyme. Figure 21-4 shows how the enzyme can distinguish between the two apparently equivalent carboxyl groups. When citrate is isomerized to isocitrate at the active site of the enzyme, the OH group is moved so that only the species in which the OH group is not adjacent to the labeled carboxyl group is formed (path *a*). Consequently, in the subsequent decarboxylation, only nonisotopic CO_2 is produced in the first turn of the cycle. This stereoselectivity was discussed in Chapter 2. You will recognize that citrate is a molecule of the form Caabd, in which the two "a" substituents are not equivalent; therefore, an enzyme that acts on the molecule can specifically select one of the "a" substituents. Further consideration of the TCA cycle leads to the conclusion that, in the *second* turn of the cycle, $^{14}CO_2$ will be produced from carboxyl-labeled acetyl CoA, and this has been shown to occur by experiment.

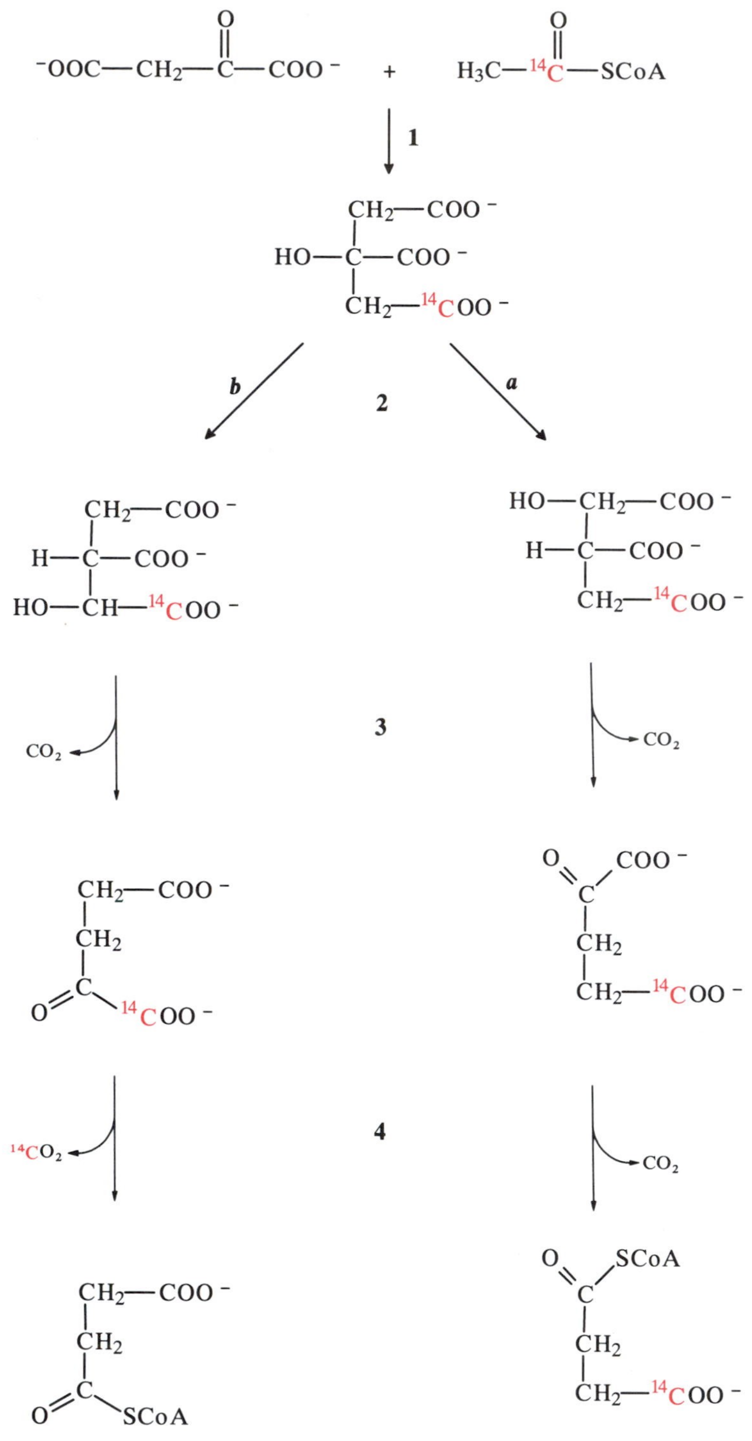

$^-OOC{-}CH_2{-}C(=O){-}COO^-$ + $H_3C{-}^{14}C(=O){-}SCoA$

1

$HO{-}C(COO^-)(CH_2{-}COO^-)(CH_2{-}^{14}COO^-)$

2 (b / a)

b: $CH_2{-}COO^-$ / $H{-}C{-}COO^-$ / $HO{-}CH{-}^{14}COO^-$

a: $HO{-}CH_2{-}COO^-$ / $H{-}C{-}COO^-$ / $CH_2{-}^{14}COO^-$

3

b: CO_2 → $CH_2{-}COO^-$ / CH_2 / $O{=}C{-}^{14}COO^-$

a: CO_2 → $O{=}C{-}COO^-$ / CH_2 / $CH_2{-}^{14}COO^-$

4

b: $^{14}CO_2$ → $CH_2{-}COO^-$ / CH_2 / $O{=}C{-}SCoA$

a: CO_2 → $O{=}C{-}SCoA$ / CH_2 / $CH_2{-}^{14}COO^-$

Figure 21-4

Condensation of [1-^{14}C]Acetyl CoA with Oxaloacetate. Path *a* assumes that "upper" —COO^- is lost as CO_2. Path *b* assumes that "lower" —COO^- is lost.

Upon first encountering the TCA cycle, students are often puzzled at what makes it continue to turn. If the beginning substrate, oxaloacetate, is regenerated at the end, why is this not some sort of perpetual motion machine? It is important to understand that the TCA cycle is powered to turn always in the same direction because of the constant removal of NADH and production of NAD^+ by the reduction of molecular oxygen. For example, the dehydrogenation of malate to form oxaloacetate (reaction 8, Figure 21-3) is not by itself a particularly exergonic reaction and should not by itself proceed forward. However, because O_2 is continually depleting NADH and producing NAD^+, this reaction is *driven* in the forward direction by the high "oxidation pressure" of NAD^+. If oxygen were removed, NADH would rapidly accumulate and the whole TCA cycle would simply stop turning, because all the NAD^+ would be converted into NADH.

The TCA Cycle Occurs in Mitochondria

In 1948, Kennedy and Lehninger discovered that isolated rat-liver mitochondria will catalyze the complete oxidation of pyruvate to carbon dioxide and water. Other cellular fractions, such as cytoplasm, nuclei, or microsomes, were inactive. Mitochondria contain all the enzymes of the TCA cycle. The glycolytic enzymes, on the other hand, are located in the cytosol. We will see later that a useful consequence of this *compartmentation* of pathways is that intermediates that are used in several different biochemical pathways can be selectively utilized by being transported into the appropriate compartment.

The TCA Cycle Interacts with Other Metabolic Pathways

So far, we have examined the role of the TCA cycle in the production of energy. Although this is one of its main functions, the cycle also serves as a "crossroad" for the interconversion of various metabolites. Several of the intermediates of the cycle can serve as precursors for other metabolic processes. For example, citrate can be converted through a reaction catalyzed by citrate lyase into acetyl CoA (Chapter 2) in the cytosol of eukaryotes.

$$\text{Citrate} + \text{ATP} + \text{CoA} \longrightarrow \text{Acetyl CoA} + \text{Oxaloacetate} + \text{ADP} + P_i$$

This acetyl CoA is then used for the synthesis of fatty acids (Chapter 19). Succinyl CoA is a precursor for heme biosynthesis. The keto acids, α-ketoglutarate and oxaloacetate, can be converted through transamination reactions into the amino acids glutamate and aspartate, respectively (Chapter 18).

There are other reactions through which additional metabolites can be fed into the TCA cycle, thus speeding up the turnover of the cycle. These reactions can serve as a mechanism for controlling the level of operation of the cycle, a topic to be discussed in more detail later in this chapter. For example, pyruvate carboxylase (Chapter 17), catalyzes the following reaction:

$$\text{Pyruvate} + HCO_3^- + \text{ATP} \longrightarrow \text{Oxaloacetate} + \text{ADP} + P_i$$

Thus, oxaloacetic acid can be supplied to the TCA cycle through the carboxylation of pyruvate. Because oxaloacetate serves as a catalyst in the cycle, the introduction of more oxaloacetate will speed up the rate at which acetyl CoA can be metabolized through the cycle.

The amino acids glutamate and aspartate can be utilized to provide energy by conversion into the keto acids α-ketoglutarate and oxaloacetate through the reverse of the transamination reactions. Degradation of other amino acids leads to the production of acetyl CoA, succinyl CoA, and fumarate, all of which are intermediates of the TCA cycle (Chapter 26). Thus the TCA cycle extracts energy from proteins, as well as carbohydrates. When there is a strong need for additional energy, as in the starvation of animals or human beings, the combined capabilities of the TCA cycle and the amino acid degradation pathways provide energy but cause emaciation.

Many Monosaccharides and Polysaccharides Can Enter Glycolysis

Although glucose plays a central role in glycolysis, other monosaccharides and the sugars in several disaccharides and polysaccharides that an organism may encounter also can enter the glycolytic cycle. Monosaccharides are converted into molecules that can directly enter the glycolytic pathway. They may be converted into one of the phosphorylated hexoses or trioses that are part of the glycolytic pathway. Polysaccharides are broken down by hydrolysis or phosphorolysis to monosaccharides that can either enter glycolysis directly or be converted into compounds that can do so.

We will consider several examples of the pathways by which sugars can enter glycolysis by way of illustration. An example of the metabolism of a polysaccharide is provided by glycogen (Chapter 15), a polymer of glucose that serves as a storage depot for glucose. This polymer is degraded through the action of glycogen phosphorylase to glucose-1-P.

$$(\text{Glucose})_n + \text{P}_i \longrightarrow \text{Glucose-1-P} + (\text{Glucose})_{n-1}$$

Glucose-1-P can then be converted by phosphoglucomutase into glucose-6-P, which can be utilized for glycolysis.

Glycogen phosphorylase is a tetramer that exists in two forms: an active phosphorylated form, phosphorylase *a*, and an inactive form, phosphorylase *b*. The active form differs from the inactive form in that it contains one phosphoserine residue per monomer. The active form (*a*) can be converted into the inactive form (*b*) through the action of a phosphatase. The *b* form can be converted into the *a* form through an ATP-dependent phosphorylation that is catalyzed by the enzyme phosphorylase kinase. These reactions provide a mechanism for regulation of the supply of glucose and, therefore, the supply of energy by the metabolism of glucose.

REGULATION OF GLYCOGEN PHOSPHORYLASE BY PHOSPHORYLATION AND DEPHOSPHORYLATION

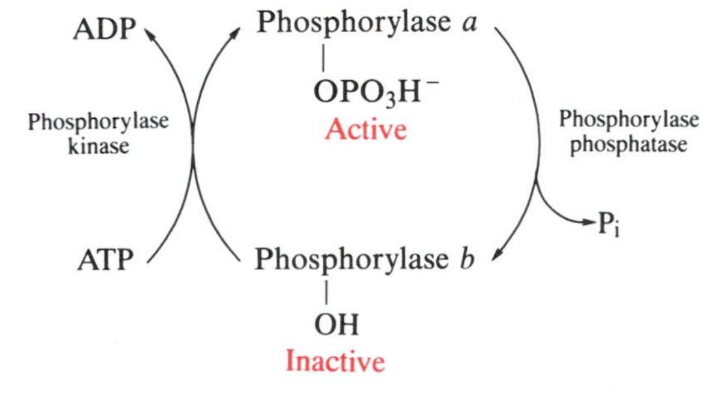

The metabolism of sucrose and lactose illustrates how disaccharides are processed in mammals. Sucrose is first hydrolyzed to glucose and fructose. Glucose can readily enter the glycolytic pathway. Fructose can be phosphorylated by hexokinase, but the affinity of hexokinase for fructose is low, relative to that for glucose. Therefore, this phosphorylation does not occur in the presence of high concentrations of glucose. In the liver, fructose can be converted into glyceraldehyde-3-P through the following reaction sequence. The reactions are catalyzed by: 1, fructokinase; 2, fructose-1-P aldolase; 3, triose kinase; and 4, triose phosphate isomerase.

CONVERSION OF FRUCTOSE INTO GLYCERALDEHYDE-3-P

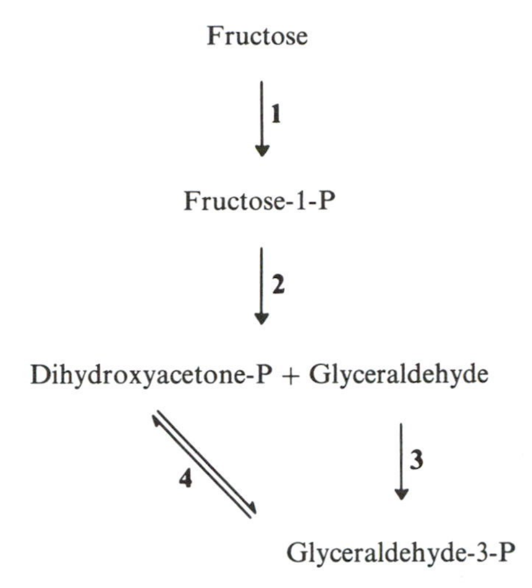

The disaccharide lactose is the major carbohydrate component of milk. It is hydrolyzed to glucose and galactose. Galactose is then converted into glucose-1-P through the action of three enzymes in the Leloir pathway, which we have seen in earlier chapters (Chapters 11, 13, and 15).

THE CONVERSION OF GALACTOSE INTO GLUCOSE

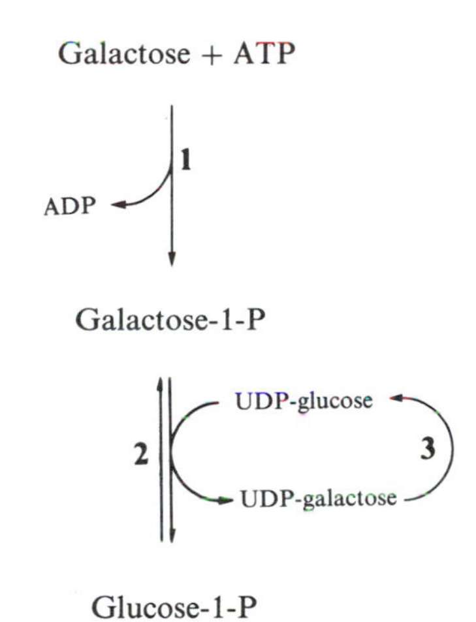

The overall reaction is ATP + galactose → ADP + glucose-1-P. The enzymes of the Leloir pathway are: 1, galactokinase; 2, hexose-1-P uridylyltransferase; and 3, UDP-galactose 4-epimerase.

The livers of infants suffering from galactosemia, a genetic disease, lack the enzyme hexose-1-phosphate uridyltransferase. Consequently, these infants cannot metabolize galactose, which is a component of lactose. Galactose accumulates in the blood of these infants and causes severe illness, including mental deficiency, by a mechanism that is not well understood. The condition can be alleviated by withdrawing milk, which contains lactose.

Another Pathway for Glucose Metabolism: NADPH and Pentoses

There are two types of pyridine nucleotide coenzymes, NAD^+ and $NADP^+$, which can be reduced to NADH and NADPH, respectively. In general, these coenzymes serve two different functions: (1) NADH is used primarily for energy production, and (2) NADPH is used primarily for reductive biosynthesis—for example, the synthesis of fatty acids and steroids. The metabolism of glucose through glycolysis and the TCA cycle produces NADH. The *pentose shunt* is another pathway for glucose metabolism and leads to the formation of five-carbon sugars and NADPH; it is also called the hexose monophosphate pathway. These five-carbon sugars are used for the biosynthesis of DNA, RNA, and coenzymes, including CoA, NAD^+, FMN, and ATP. This alternative pathway of glucose metabolism functions primarily in adipose tissue, the mammary gland, the adrenal gland, and the liver, all of which carry out reductive biosynthesis. The activity of the pentose shunt is very low in muscle, which uses glucose primarily for energy production and does not carry out reductive biosynthesis. The enzymes of the pentose shunt are found in the cytosol of the cell.

The pentose shunt consists of two phases. In the first phase, glucose-6-phosphate is converted into CO_2 and a pentose, ribulose-5-P, with the concomitant production of 2 moles of NADPH (Figure 21-5). As already indicated, the ribulose-5-P and NADPH are used for biosynthesis. In the second phase of the shunt, any ribulose-5-P that is not used for biosynthesis is converted into compounds that can be utilized in glycolysis.

How is the conversion of glucose-6-P into ribulose-5-P and CO_2 accomplished? From earlier discussions, we can predict the reactions required. Carbon

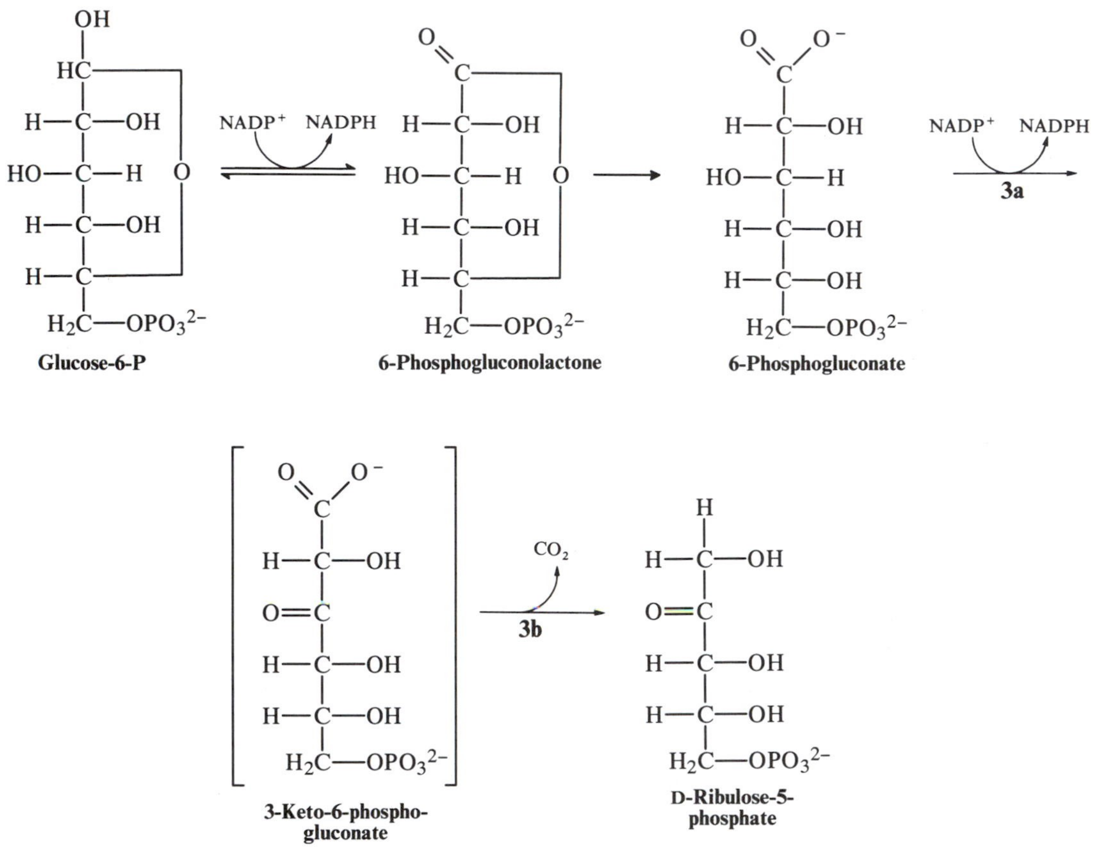

Glucose-6-P → ($NADP^+$ → NADPH) → 6-Phosphogluconolactone → 6-Phosphogluconate → ($NADP^+$ → NADPH, 3a) → [3-Keto-6-phospho-gluconate] → (CO_2, 3b) → D-Ribulose-5-phosphate

Figure 21-5

Conversion of Glucose-1-P into CO_2 and D-Ribulose-5-P. The enzymes are: **1**, glucose-6-P dehydrogenase; **2**, lactonase; and **3**, 6-phosphogluconate decarboxylase. The bracketed compound, 3-keto-6-phosphogluconate, is an enzyme-bound intermediate.

dioxide is generally produced by the decarboxylation of α- or β-ketoacids. Therefore, we expect that glucose-6-P will be converted into either an α- or a β-ketoacid, which is then decarboxylated. In fact, glucose-6-P is converted into a β-ketoacid through the reactions shown in Figure 21-5.

All of these reactions have been described before. Reaction 1 is the oxidation of an aldehyde to a lactone. In reaction 2, the lactone is hydrolyzed to the acid, a reaction equivalent to ester hydrolysis. Reactions 3a and 3b are catalyzed by a single enzyme, which first oxidizes position-3 of 6-phosphogluconolactone. This is a typical oxidation of the type, shown in the margin, that utilizes pyridine nucleotide coenzymes. This oxidation produces a β-ketoacid, which is a substrate for decarboxylation reactions. The β-ketoacid, 3-keto-6-phosphogluconolactone, is not released from the enzyme and is directly decarboxylated.

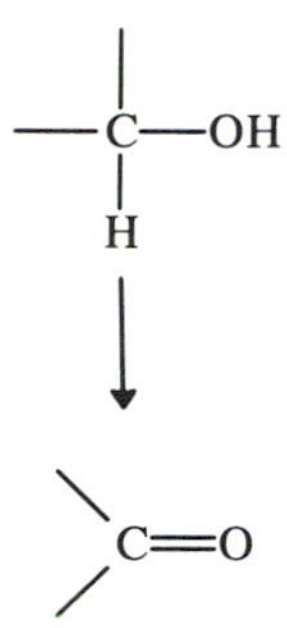

In the next stage of the pentose shunt, ribulose-5-P that is not used for biosynthetic purposes is converted into molecules that can enter the glycolytic pathway. One might think that this could be accomplished by converting the five-carbon sugar into glyceraldehyde-3-P and acetaldehyde. Such a conversion is possible in principle, but it would be wasteful because no pathways exist that allow acetaldehyde to be used for energy production. However, nature has devised an ingenious and effective metabolic sequence that begins with the two reactions shown in Figure 21-6. These two reactions are isomerizations of the five-carbon sugars similar to those described in Chapter 20. One is an aldose-ketose interconversion, and the other is epimerization at C-3.

D-Ribulose-5-phosphate ⇌ (1) D-Ribose-5-phosphate

D-Ribulose-5-phosphate ⇌ (2) D-Xylulose-5-phosphate

Figure 21-6

Enzymes of the Pentose Shunt. The enzymes are: **1**, phosphopentose isomerase; and, **2**, ribulose-5-P 3-epimerase.

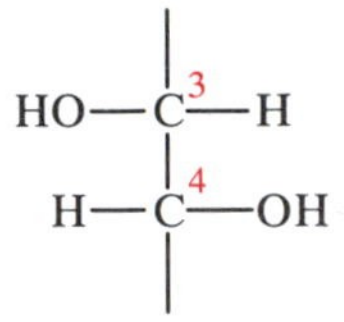

HO—C3—H
H—C4—OH

The next set of reactions is catalyzed by the enzymes transketolase and transaldolase (Figure 21-7). These enzymes require a ketose and an aldose as substrates. To provide the aldose, ribulose-5-P is converted into ribose-5-P (Figure 21-6). The epimerization of ribulose-5-P at C-3 to give xylulose-5-P is necessary to meet the specificity requirements of these enzymes; that is, both enzymes require a *trans* structure at C-3 and C-4. Ribose-5-P and xylulose-5-P produced in these isomerization reactions participate in the reactions of Figure 21-7. These reactions are catalyzed by the enzymes transketolase and transaldolase, which were described in Chapter 2. Recall that transketolase is an enzyme that requires thiamine pyrophosphate and transfers a two-carbon fragment from a donor ketose to an acceptor aldose. Transaldolase (a Schiff-base enzyme) requires no cofactor and transfers a three-carbon fragment from a donor ketose to an acceptor aldose. Through the action of transketolase and transaldolase, two pentose molecules are converted into a molecule of fructose-6-P, which can be utilized in glycolysis, and a molecule of erythrose-4-P, which

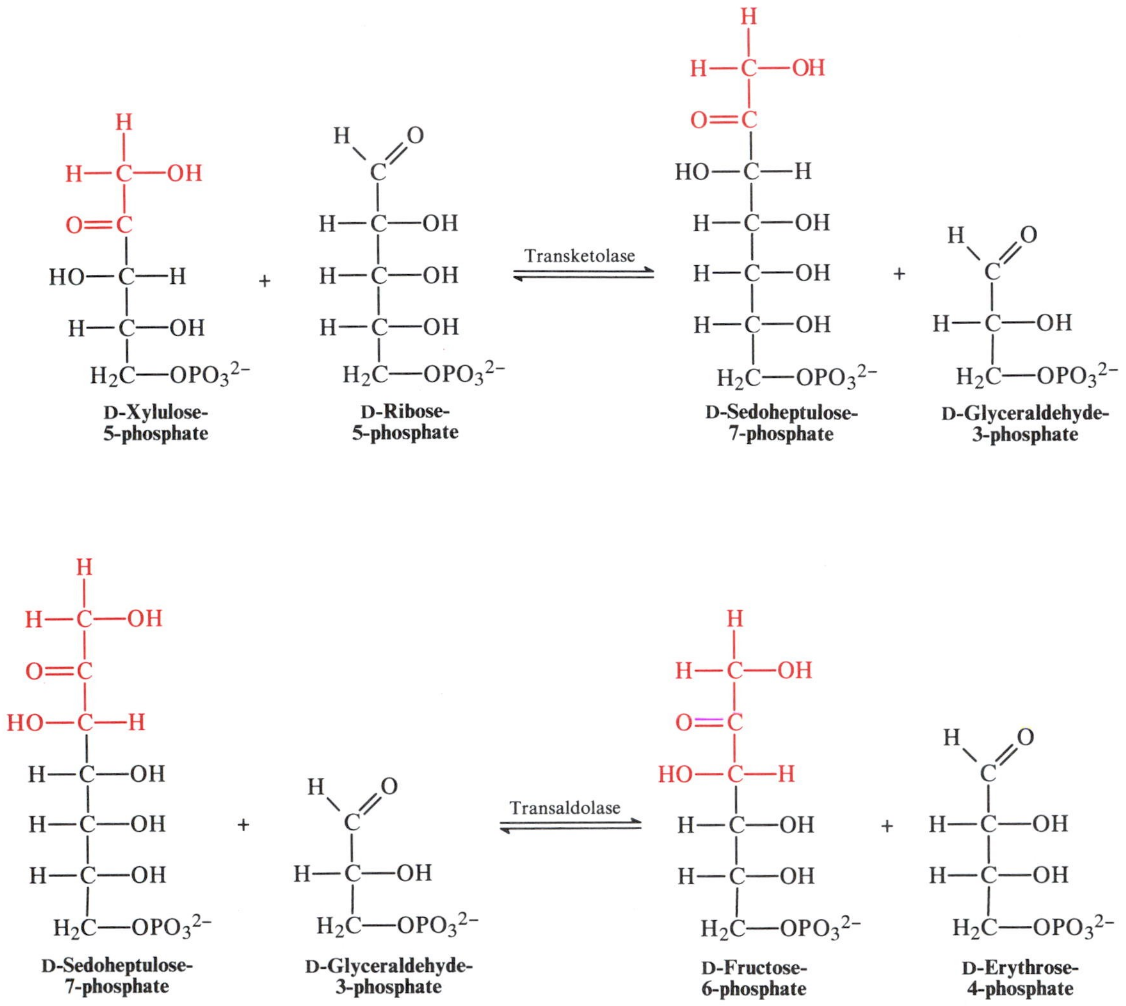

D-Xylulose-5-phosphate + D-Ribose-5-phosphate ⇌ (Transketolase) D-Sedoheptulose-7-phosphate + D-Glyceraldehyde-3-phosphate

D-Sedoheptulose-7-phosphate + D-Glyceraldehyde-3-phosphate ⇌ (Transaldolase) D-Fructose-6-phosphate + D-Erythrose-4-phosphate

Figure 21-7

Transketolase and Transaldolase. Transketolase transfers a two-carbon fragment, transaldolase a three-carbon fragment.

cannot enter glycolysis. Erythrose-4-P is converted into molecules that can enter glycolysis through the following transketolase reaction:

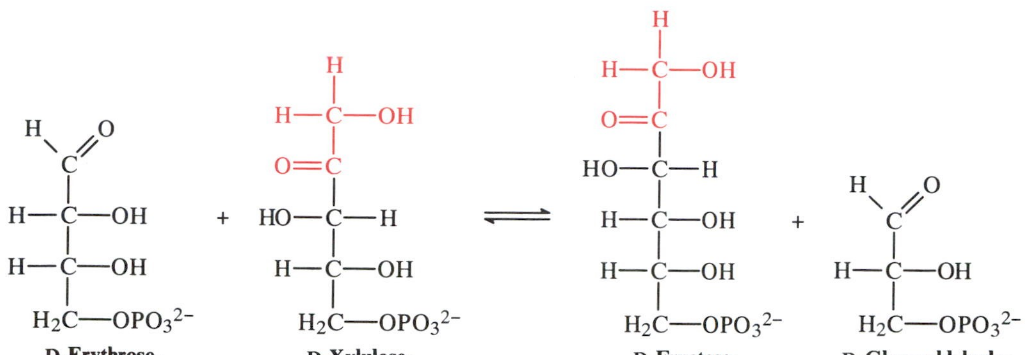

$$\text{D-Erythrose-4-phosphate} + \text{D-Xylulose-5-phosphate} \rightleftharpoons \text{D-Fructose-6-phosphate} + \text{D-Glyceraldehyde-3-phosphate}$$

Erythrose-4-P reacts with xylulose-5-P in a reaction catalyzed by transketolase to produce fructose-6-P and glyceraldehyde-3-P. Both of these molecules can enter glycolysis. The net reaction, which takes place in the second phase of the shunt, is:

$$3 \text{ Pentoses} \longrightarrow 2 \text{ Hexoses} + 1 \text{ Triose}$$

All of the pentose molecules that are not used in synthetic reactions can be completely converted into molecules that enter glycolysis, so that they can be utilized for energy production.

Control of the Rates of Major Metabolic Processes in Biological Systems

So far in this chapter, two major energy-producing pathways have been discussed: glycolysis and the tricarboxylic acid cycle. In an earlier chapter, we described the biosynthesis and degradation of fatty acids. In addition to these major pathways, there are a multitude of other metabolic cycles and pathways that serve to degrade and synthesize the many metabolites that are encountered by living organisms. A cell is a virtual network of metabolic activity. Almost any metabolite could enter a multitude of pathways and undergo any one of several metabolic fates. There must be some mechanism for control that decides, in response to environmental demands, which metabolic pathway predominates. The mechanism by which these metabolic controls operate is incompletely understood and is one of the most important problems in modern biochemistry. In this section, we cannot cover this area in detail, but we will present a few examples that illustrate some of the problems and mechanisms of control.

There are two major mechanisms by which metabolic pathways can be regulated:

1. By controlling the entry of a metabolite into the cell. For example, one of the major effects of insulin is to promote the entry of glucose into muscle and fat cells.
2. By controlling the activity of certain enzymes.

We will examine three different mechanisms for the control of enzyme activity.

Amount of Enzyme. The quantity of enzyme present can vary, in accord with the nutritional state of the organism or in response to the presence of certain compounds. Most enzymes, like almost all components of living organisms, are continuously synthesized and degraded. The quantity of enzyme that is present at a particular time is determined by its rate of synthesis and degradation. A change in the rate of either of these processes will change the amount of enzyme present in the cell. For example, the level of the enzyme carboxykinase, which catalyzes the following reaction,

$$H_2C{=}C(OPO_3^{2-})(COO^-) + CO_2 + GDP \longrightarrow {}^-OOC{-}CH_2{-}C({=}O)(COO^-) + GTP$$

is controlled by the nutritional state of the animal. This is one of the enzymes required for the conversion of pyruvate into glucose, a process that will be described later in this chapter. The concentration of this enzyme increases significantly when the animal is in a state of starvation. The concentrations of some enzymes increase in response to particular compounds. For example, the concentration of cytochrome P450, a hemoprotein, increases in the presence of a drug. This enzyme participates in the detoxification of foreign compounds, including barbiturates, certain drugs, and other foreign substances (Chapter 24). Administration of barbiturates causes a large increase in the concentration of this protein.

Presence of Small Molecules. Enzyme activity can be controlled by small molecules in the environment. The small molecule may be an enzyme's specific substrate or some other molecule. The simplest situation is that shown by the solid lines in Figure 21-8, an example of normal Michaelis-Menten kinetics. Here, the substrate does not actually control the turnover rate of the enzyme-substrate complex, although the rate of product formation does depend on the substrate concentration. A slightly more complex situation is found with enzymes that show substrate inhibition at high substrate concentrations, as shown by the dashed lines in Figure 21-8.

The next level of complexity entails allosteric enzymes, of which phosphofructokinase is an example. This enzyme plays an important role in controlling the rate of glycolysis. It catalyzes the following reaction:

$$\text{Fructose-6-P} + \text{ATP} \longrightarrow \text{Fructose-1,6-P}_2 + \text{ADP}$$

The kinetics observed with this enzyme are shown in Figure 21-9. Curves A and B at the right show that the enzyme activity follows a sigmoidal dependence on the concentration of fructose-1,6-bisphosphate, rather than the hyperbolic kinetics seen with enzymes that obey normal Michaelis-Menten kinetics. In

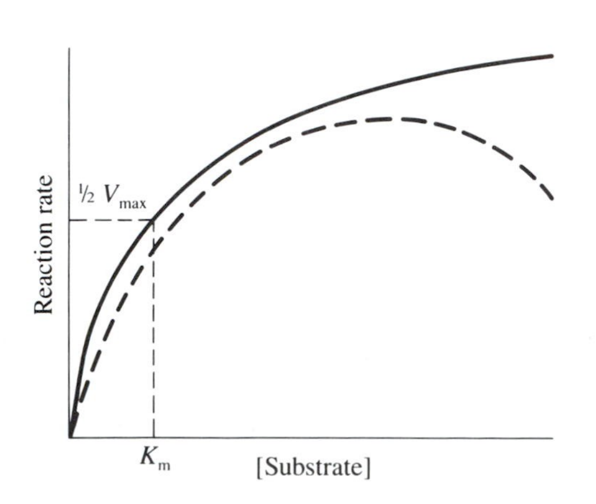

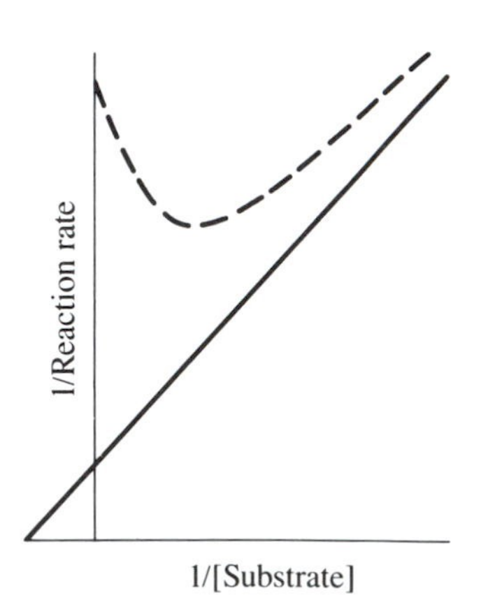

Figure 21-8

Reaction Kinetics. Normal Michaelis-Menton kinetics and substrate inhibition are shown by solid lines and dashed lines, respectively. Note that, at very high substrate concentrations, the rate of product formation decreases. The graph at the left is a plot of reaction rate against substrate concentration. The graph at the right uses the same substrate concentrations, but the inverse of the reaction rate is plotted against the inverse of substrate concentration.

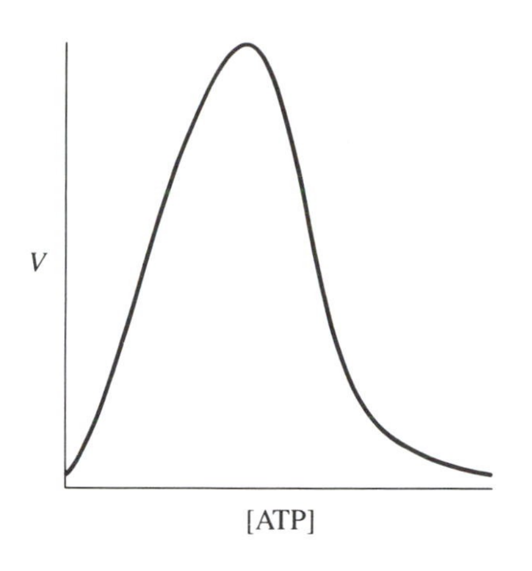

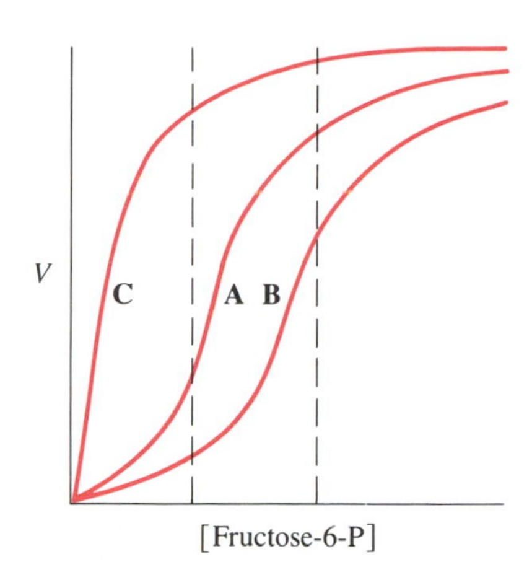

A→B: increasing ATP
C: A + AMP

Figure 21-9

Kinetics of Phosphofructokinase. At the left, reaction velocity is plotted against ATP concentration; fructose-6-P remains constant. This is an example of substrate inhibition. At the right, curve A is at a lower ATP concentration than curve B. These curves are examples of sigmoidal kinetics. Note the large change in rate with increasing substrate concentration. The change in rate with increasing fructose-6-P concentration in this range is much greater than that for a reaction that obeys simple Michaelis-Menten kinetics. Curve C results from the addition of AMP, other conditions being the same as for curve A. AMP is an activator.

certain ranges of substrate concentration, the sigmoidal curve is much more sensitive to substrate concentration than the hyperbolic curve. In the most-sensitive region of the hyperbolic curve (C), at low substrate concentration, the rate is directly proportional to substrate concentration. In the sigmoidal curves A and B, on the other hand, the rate varies exponentially with substrate concentration in the steep part of the curve. Therefore, a small change in substrate concentration will produce a very large change in the reaction rate. Enzymes that show this type of sigmoidal response have a tendency to maintain substrate concentrations below the level of maximal response and they are, therefore, more effective in maintaining the concentration of a substrate within a narrowly defined range than are enzymes that show normal Michaelis-Menten kinetics. The sigmoidal curves in Figure 21-9 are analogous to those described in Chapter 10 for oxygen binding to hemoglobin.

The left-hand curve in Figure 21-9 shows that the effect of the other substrate, ATP, is complex. If the concentration of ATP is increased at a constant concentration of fructose-1-P, one observes an increase in rate that is followed by a rate decrease, which represents substrate inhibition. ATP will eventually shut off the reaction. Inhibitory levels of ATP shift the sigmoidal curve for the dependance of the rate on the concentration of fructose-6-P to the right, and this shifts the region of the maximal sensitivity to substrate concentration. Qualitatively, the effect of high concentrations of ATP is to reduce the activity of the enzyme at all concentrations of fructose-6-P.

The rate of catalysis can be further controlled by effectors, which are molecules that are not substrates. Such a molecule can increase the rate of reaction (a positive effector) or decrease the rate of reaction (a negative effector). In curve C of the right-hand graph in Figure 21-9, we see the effect of the positive effector AMP on phosphofructokinase activity. AMP increases the rate of the reaction and abolishes the sigmoidal dependance of the rate on the substrate concentration. On the other hand, citrate and phosphoenolpyruvate decrease the activity of phosphofructokinase and are negative effectors of this enzyme.

ATP, AMP, and citrate frequently function as allosteric effectors. The rationale for this is easy to understand: when ATP levels are high, a cell's energy requirements are satisfied, glycolytic activity can be reduced, and metabolic activities can be used to fulfill requirements other than energy needs. Hence, high levels of ATP tend to shut off enzymes that control pathways in which ATP is formed. Citrate also provides a signal that energy production is not required.

When citrate levels are high, the TCA cycle operates at a high rate, which in turn leads to a high rate of ATP production. Thus, citrate and ATP frequently have the same effect on allosteric enzymes.

When ATP has been utilized, AMP is formed. High levels of AMP indicate low levels of ATP and hence the need for energy production. Thus, AMP frequently activates enzymes that participate in energy production, such as phosphofructokinase.

Another frequently encountered regulatory process is feedback, or end-product, inhibition. Here, the final product of a metabolic sequence turns off enzymes near the beginning of the sequence when enough of the final product has accumulated to satisfy the requirements of the cell.

Chemical Modification. Enzymes can be activated or deactivated through chemical modification. We have already discussed glycogen phosphorylase, which is activated by phosphorylation. Another example is the mammalian pyruvate dehydrogenase complex, which is inactivated by phosphorylation of a serine hydroxyl group of the component E_1 with ATP. This reaction is catalyzed by a kinase that is closely associated with the mammalian pyruvate dehydrogenase complex. The pyruvate dehydrogenase complex produces the acetyl CoA that enters the TCA cycle. This enzyme, therefore, provides a crucial entrance to energy production. At high ATP levels, which indicate low energy demands, the enzyme is phosphorylated and turned off. Here, again, we see ATP controlling an energy-producing pathway.

The pyruvate dehydrogenase complex of *E. coli* is not regulated by phosphorylation and dephosphorylation; instead, the component E_1 is inhibited by GTP, and the inhibition is relieved by ADP. Again, the enzyme from *E. coli* is inhibited when metabolic energy is abundant, and it is active when metabolic energy is low.

Several Metabolic Pathways Control Blood Glucose Levels

BIOSYNTHESIS OF GLYCOGEN (Glc_n)

Glucose-6-P

⇅

Glucose-1-P

↓ UTP → PP_i

UDP-glucose

↓ Glc_n

Glc_{n+1} + UDP

So far, we have discussed the various kinds of mechanisms that are available to a biological organism for the control of metabolic pathways. We shall next consider how these controls are used in the metabolism of a specific substance, glucose. Glucose is the major source of energy for the brain and is an important energy source for other organs such as the heart, kidney, and muscle. Therefore, maintenance of adequate blood glucose levels is extremely important. How is this control achieved?

Before considering the actual control mechanism, we shall consider how glucose is processed by the liver. Figure 21-10 summarizes the major avenues of glucose metabolism found in the liver. The liver can either remove glucose from the blood stream or release glucose into the blood. We shall first consider the fate of glucose when it is taken up by the liver. Glucose from the bloodstream, after entering the liver, is converted through the action of hexokinase or glucokinase into glucose-6-P. Glucose-6-P can then be processed through the glycolytic pathway, in which case glucose is used for energy production. If energy requirements are low, glucose-6-P can be converted into glycogen. Conversion of glucose-6-P into glycogen (Glc_n) occurs by the reaction sequence shown in the margin. The last enzyme in the sequence is glycogen synthetase. This enzyme exists in two forms, synthetase I and synthetase D, which are interconvertible. The D (dependent) form requires high levels of glucose-6-P for activity, whereas the I (independent) form is active without glucose-6-P. The I form is converted into the D form by phosphorylation, catalyzed by a protein

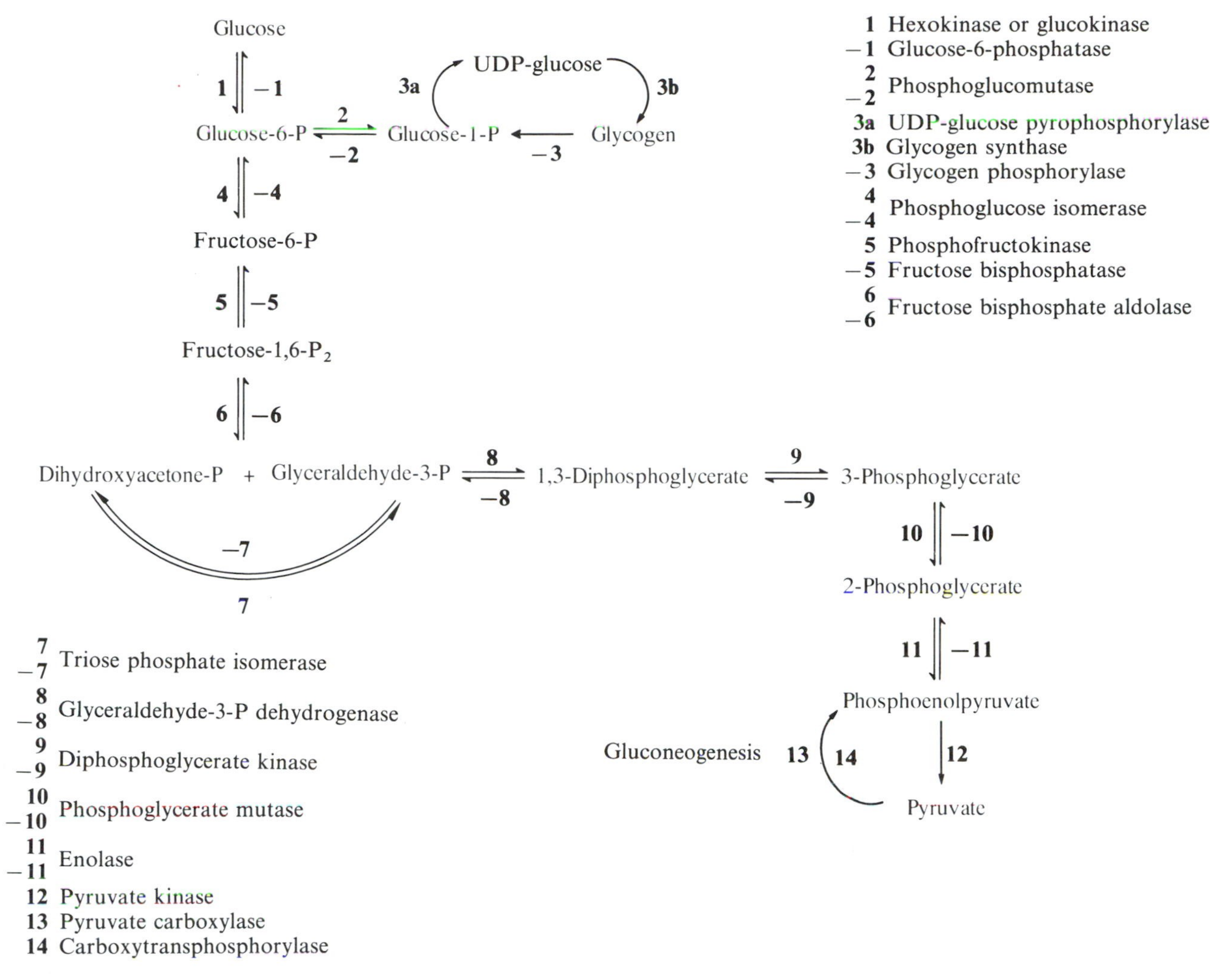

Figure 21-10

Metabolism of Glucose in the Liver. Blood glucose can be metabolized through glycolysis (reactions 1 through 12). Glucose-6-P can be diverted to glycogen (reactions 2, 3a, and 3b). This would lower blood sugar. Glycogen can give rise to glucose-6-P (reactions −3, and −2). Glucose-6-P can be hydrolyzed to glucose, which would raise blood sugar, or can be used for energy production through glycolysis. Gluconeogenesis (reactions 13 and 14) reverses the action of pyruvate kinase (PEP + ADP → Pyruvate + ATP), which is highly exergonic. Glucose is then synthesized, starting with pyruvate through reversal of glycolysis.

kinase. The D form can be converted into the I form through the action of a phosphatase.

REGULATION OF GLYCOGEN SYNTHETASE BY PHOSPHORYLATION AND DEPHOSPHORYLATION

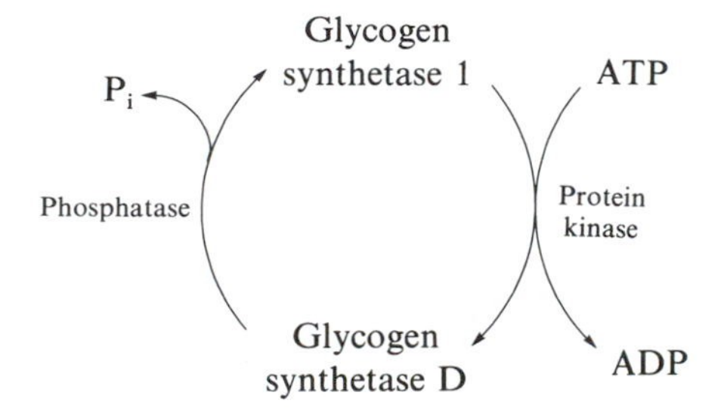

Glucose can be generated by the liver in two ways. First, glucose can be released from glycogen. The reaction sequence through which glycogen is broken down into glucose has been discussed. Second, glucose can be synthesized from lactate through gluconeogenesis, a process that is essentially the reversal of glycolysis. Although the same intermediates are formed in glycolysis and gluconeogenesis, some of the enzymes required differ for the two processes. Usually degradative pathways and biosynthetic pathways do not utilize the same enzymes. We have already seen an example of this in fatty acid degradation and biosynthesis.

In the conversion of lactate into glucose, lactate is first converted into pyruvate. Next, pyruvate must be converted into phosphoenolpyruvate. This cannot be done simply through the reversal of the reaction catalyzed by pyruvate kinase because the transfer of phosphate from phosphoenolpyruvate to ADP is highly exergonic and is essentially irreversible.

$$H_2C{=}C(OPO_3^{2-})(COO^-) + ADP \longrightarrow H_3C{-}C({=}O){-}COO^- + ATP$$

Therefore, a different set of reactions is used to convert pyruvate into phosphoenolpyruvate:

$$CH_3{-}CO{-}COO^- + CO_2 + ATP \rightleftharpoons {}^-OOC{-}CH_2{-}CO{-}COO^- + ADP + P_i$$

$$^-OOC{-}CH_2{-}CO{-}COO^- + GTP \rightleftharpoons CH_2{=}C(OPO_3^{2-})(COO^-) + GDP + P_i + CO_2$$

$$CH_3{-}CO{-}COO^- + GTP + ATP \rightleftharpoons CH_2{=}C(OPO_3^{2-})(COO^-) + GDP + ADP + P_i$$

Two enzymes are required: biotin-dependent pyruvate carboxylase, which utilizes the energy released by the hydrolysis of ATP to drive the formation of oxaloacetate from pyruvate and CO_2; and carboxytransphosphorylase, which catalyzes the conversion of oxaloacetate and GTP into phosphoenolpyruvate with the release of GDP and CO_2. Thus, the conversion of 1 mole of pyruvate into phosphoenolpyruvate is an expensive process because it requires 1 mole of ATP and 1 mole of GTP. For the conversion of phosphoenolpyruvate into fructose-1,6-P_2, gluconeogenesis employs the same enzymes as those used in glycolysis.

In the subsequent synthesis of glucose from phosphoenolpyruvate, the same intermediates are formed as in glycolysis, but in some cases gluconeogenesis employs different enzymes. These enzymes catalyze the glycolytic reactions in the reverse direction until fructose-1,6-P_2 is produced. Fructose-1,6-P_2 is then converted into fructose-6-P through the action of fructose-1,6-bisphosphatase. Recall that the reversal of this reaction, the conversion of fructose-6-P into fructose-1,6-P_2, is catalyzed by phosphofructokinase. Finally, the last step in glucose production, the conversion of glucose-6-P into glucose, is catalyzed by glucose-6-phosphatase. Here, again, the formation of glucose-6-phosphate from glucose is catalyzed by a kinase.

BOX 21-1

REGULATION OF GLUCONEOGENESIS AND GLYCOLYSIS BY FRUCTOSE-2,6-BISPHOSPHATE

We have pointed out that phosphofructokinase and fructose-1,6-bisphosphatase are important in controlling flux through the glycolytic system. The action of phosphofructokinase favors glycolysis and energy production, whereas that of fructose-1,6-bisphosphatase tends to favor glucose production and glycogen formation. Fructose-2,6-bisphosphate is important in regulating these enzymes.

The concentration of fructose-2,6-bisphosphate in the liver determines whether the phosphofructokinase–fructose-1,6-bisphosphatase system operates in the direction of glycolysis or gluconeogenesis.

Fructose-2,6-bisphosphate is an allosteric activator of phosphofructokinase, and it acts synergistically with AMP in opposing the inhibition of phosphofructokinase by ATP and citrate. Fructose-2,6-bisphosphate also competitively inhibits fructose-1,6-bisphosphatase, and it acts synergistically with AMP in the inhibition of fructose-1,6-bisphosphatase. Thus, a high concentration of fructose-2,6-bisphosphate activates glycolysis and inhibits gluconeogenesis. In contrast, gluconeogenesis is favored when the concentrations of ATP and citrate are high and that of fructose-2,6-bisphosphate is low.

Fructose-2,6-bisphosphate is produced by the bifunctional enzyme 6-phosphofructo-2-kinase/fructose-2,6-bisphosphatase, which catalyzes both of the following reactions.

$$\text{Fructose-6-phosphate} + \text{ATP} \rightarrow \text{Fructose-2,6-bisphosphate} + \text{ADP}$$

$$\text{Fructose-2,6-bisphosphate} + H_2O \rightarrow \text{Fructose-6-phosphate} + P_i$$

The activities of 6-phosphofructo-2-kinase/fructose-2,6-bisphosphatase are under hormonal control linked through the secondary messenger cAMP. The enzyme is subject to phosphorylation by cAMP-dependent protein kinase; phosphorylation of serine-32 inhibits the 6-phosphofructo-2-kinase activity and increases the fructose-2,6-bisphosphatase activity. Thus, at high concentrations of cAMP, the cAMP-dependent protein kinase keeps 6-phosphofructo-2-kinase/fructose-2,6-bisphosphatase in its phosphorylated state and allows gluconeogenesis to proceed. At low levels of cAMP, the dephosphorylated form of the enzyme produces a high concentration of fructose-2,6-bisphosphate, which inhibits gluconeogeneis and activates glycolysis. The intracellular concentration of cAMP is under hormonal control by glucagon and epinephrine.

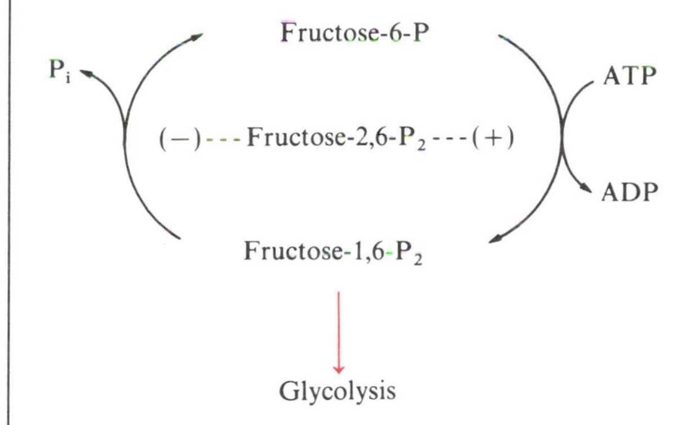

Gluconeogenesis
Fructose-6-P
P_i
ATP
ATP + Citrate---(−)
ADP
Fructose-1,6-P_2

Phosphofructokinase Is an Important Control Point in Glycolysis

We have outlined several pathways through which glucose metabolism occurs. We are now ready to consider some of the mechanisms by which the flux through these pathways is controlled. An important control point in glycolysis is the phosphorylation of fructose-6-P catalyzed by phosphofructokinase. If that reaction were shut off, glucose would be prevented from entering glycolysis. The liver could then release glucose into the bloodstream or convert glucose into glycogen. On the other hand, acceleration of the reaction catalyzed by phosphofructokinase would tend to shift glucose utilization towards glycolysis; that is, energy production.

We have already discussed phosphofructokinase and have pointed out that it is an allosteric enzyme: its catalytic activity can be controlled by various small molecules that act either as positive or as negative effectors. AMP is a positive effector and increases the catalytic efficiency of the enzyme. ATP, phosphoenolpyruvate, and citrate are negative effectors and decrease the rate at which fructose-1,6-P_2 is formed.

Fructose-1,6-diphosphatase hydrolyzes fructose-1,6-P_2 and, thus, in effect reverses the reaction catalyzed by phosphofructokinase. It would be futile to have both reactions proceed at maximal rates. It is therefore very reasonable that AMP is a positive effector of phosphofructokinase and a negative effector of the phosphatase.

The Synthesis and Breakdown of Glycogen Is Subject to Hormonal Control

The reactions in the synthesis and degradation of glycogen are summarized in Figure 21-11. Glycogen degradation occurs when an organism requires an increase in blood glucose levels. How does the organism signal to the liver that such a situation has occurred? This question is part of a more-general problem,

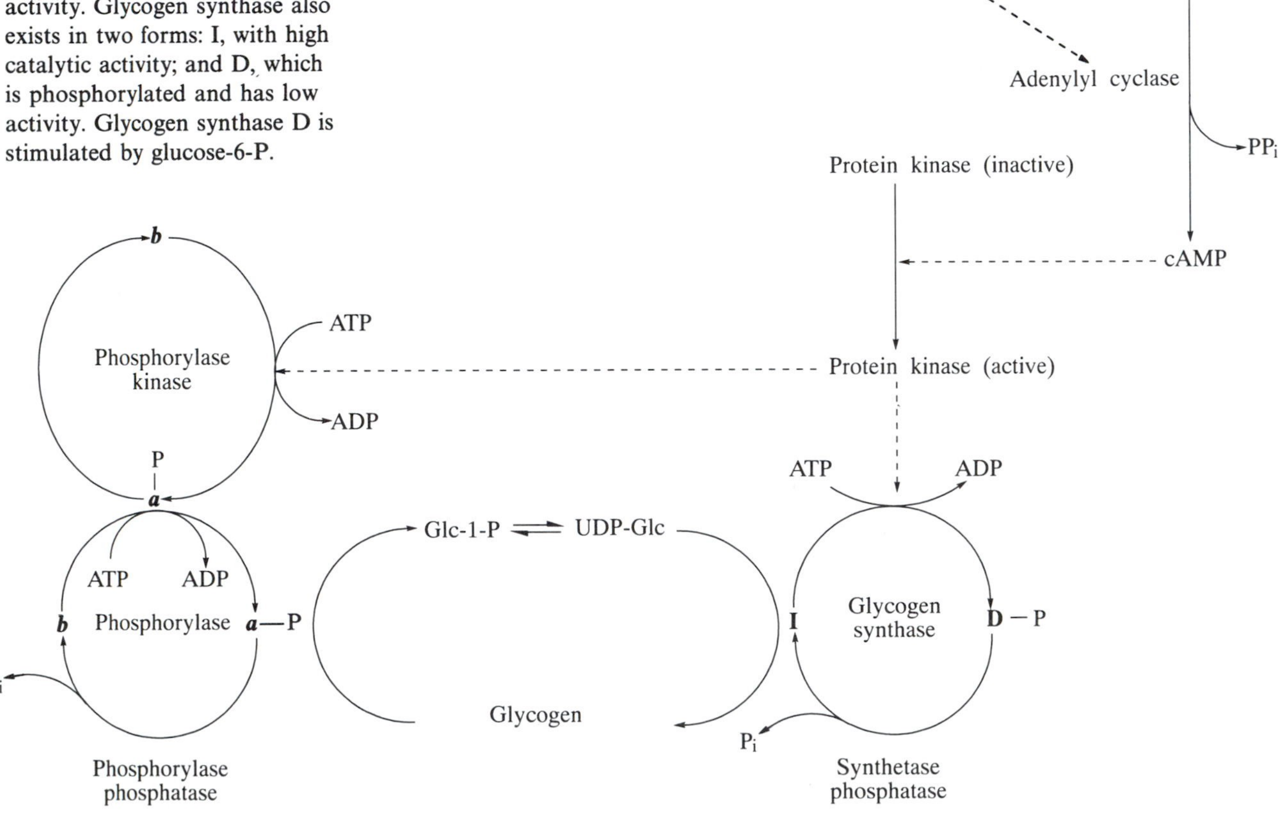

Figure 21-11

Glycogen Synthase and Glycogen Phosphorylase. Glycogen phosphorylase exists in two states. The phosphorylated enzyme, phosphorylase *a*, has high activity and is a tetramer. Phosphorylase *b* is not phosphorylated and is a dimer; this form has low catalytic activity. Glycogen synthase also exists in two forms: I, with high catalytic activity; and D, which is phosphorylated and has low activity. Glycogen synthase D is stimulated by glucose-6-P.

Epinephrine

Thyroxine
(L-3,5,3′,5′-Tetraiodothyronine)

Ser–Tyr–Ser–Met–Glu–His–Phe–Arg–Trp–Gly–Lys–Pro–Val–Gly–Lys–Lys–Arg–Pro–Val–Lys–
1 2 3 4 5 6 7 8 9 10 11 12 13 14 15 16 17 18 19 20

Val–Tyr–Pro–Asp–Ala–Gly–Glu–Asp–Gln–Ser–Ala–Glu–Ala–Phe–Phe–Pro–Leu–Glu–Phe
21 22 23 24 25 26 27 28 29 30 31 32 33 34 35 36 37 38 39

ACTH
(Adrenocorticotropin)

Figure 21-12

Structures of Various Hormones.

namely: How do organs communicate with each other? Communication occurs through hormones, which are produced by glands and sent through the bloodstream to target organs. These organs have receptors that can interact with the hormones. Only target organs with appropriate receptors will respond to the hormonal signal. The situation is analogous to transmitted radio signals: only receivers that pick up the correct wavelength can receive the signal. Once hormones interact with their target organ, they produce effects in one of several ways:

1. They alter the permeability of the cell. This type of effect has been discussed in connection with insulin.
2. They can affect the rate of enzyme-catalyzed reactions.
3. They can affect the rate of protein synthesis, including the synthesis of enzymes.

We described several types of hormones in Chapter 1, including thyroxine and epinephrine, which are derived from amino acids. Other hormones, such as ACTH (adrenocorticotropin) in Figure 21-12, insulin, and growth hormone, are polypeptides or proteins. Another group of hormones, described in Chapter 28, consists of steroids.

The hormones that give the signal for glycogen breakdown are ACTH (a polypeptide hormone), glucagon (a polypeptide hormone) and epinephrine (a hormone derived from phenylalanine). ACTH is produced by the anterior pituitary gland; glucagon is released by the α cells of the pancreas in response to low blood sugar; and epinephrine is produced by the adrenal gland in response to muscular activity. These hormones activate the enzyme adenylyl (adenylate) cyclase, which is located in the cell membrane. Adenylyl cyclase catalyzes the conversion of ATP into 3′,5′-cyclic AMP (cAMP, Figure 21-13). cAMP in turn triggers a series of events that eventually lead to glycogen breakdown and decreased glycogen synthesis from glucose, as shown in Figure 21-11.

We shall consider the control of glycogen degradation to release glucose. The immediate action of cAMP is to activate a protein kinase. This kinase is a tetramer (R_2C_2) consisting of two kinds of subunits: a regulatory subunit (R) and a catalytic subunit (C). The tetramer has low catalytic activity. The regulatory subunit can bind cAMP. When it does so, the enzyme undergoes dissociation and the active catalytic subunit is released.

Next, the activated protein kinase phosphorylates the enzyme phosphorylase kinase, using ATP as the phosphoryl donor. This phosphorylation converts phosphorylase kinase from an inactive form into an active form. In muscle, an additional mechanism exists that partly activates phosphorylase

ACTIVATION OF cAMP-DEPENDENT PROTEIN KINASE

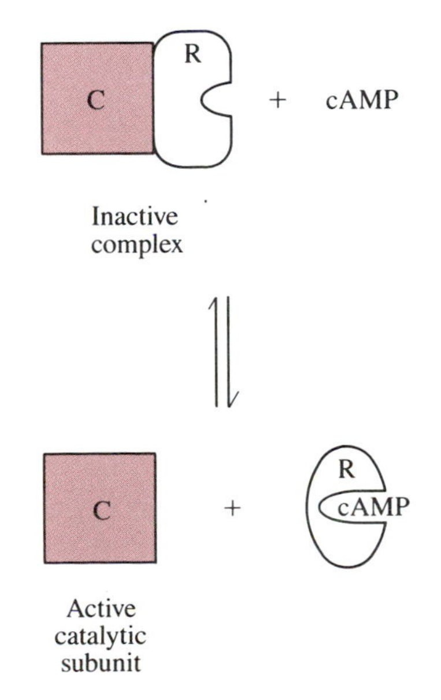

Figure 21-13

The Reaction Catalyzed by Adenylyl Cyclase.

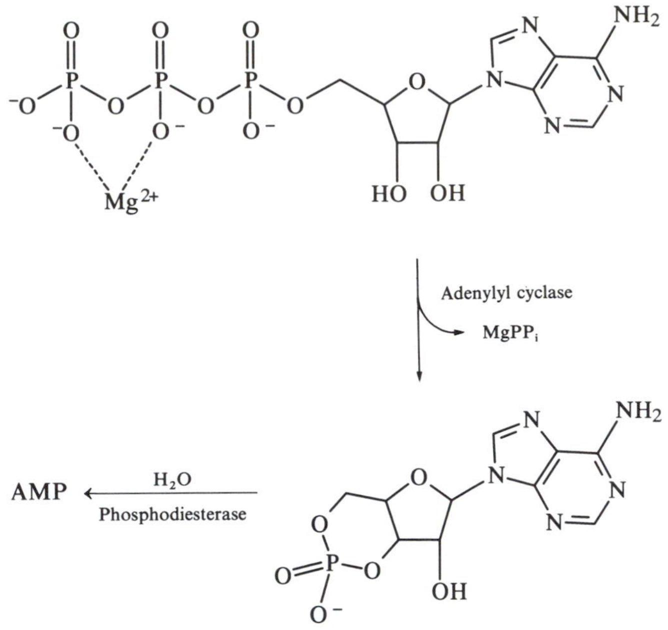

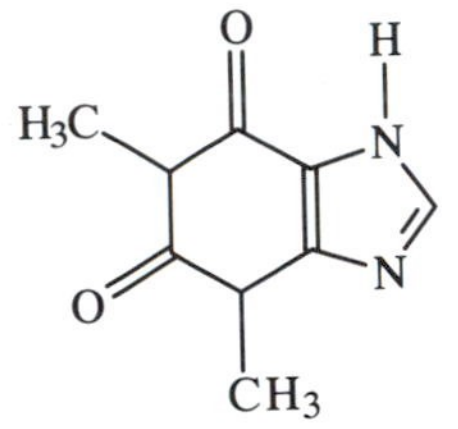

Theophylline

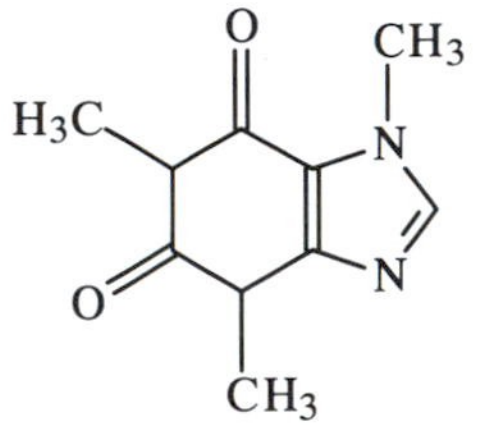

Caffeine

kinase. This activation is brought about through interaction with Ca^{2+}. This is of particular importance in muscle because muscle contraction is triggered by the release of Ca^{2+} (Chapter 30). The same signal, therefore, that triggers muscle contraction is also a signal for the release of glucose, which provides energy for muscular contraction. Phosphorylase kinase catalyzes the phosphorylation of a specific serine residue of glycogen phosphorylase, which converts this enzyme from the inactive (*b*) form into the active (*a*) form. It is reasonable to assume that there is also a system that deactivates protein kinase. Deactivation is brought about by the hydrolytic conversion of cAMP into AMP. This hydrolysis is catalyzed by a specific phosphodiesterase, an enzyme that is inhibited by the drugs theophylline and caffeine. These drugs, therefore, enhance the effects of hormones that act through cAMP. The active phosphorylase kinase is also deactivated by the action of a phosphatase. The active form of glycogen phosphorylase is deactivated through the action of a protein phosphatase. This returns the system to its resting state.

In summary, the activation of glycogen phosphorylase can be initiated through the action of several hormones. These hormones do not act directly on the target enzyme; instead, they act through a cascading series of intermediate enzymes. Through this system, the signal that is provided by a few molecules of hormone can trigger the release of a large amount of glucose.

The control of glycogen synthesis must be closely related to that of degradation. It would obviously be wasteful to allow two opposing processes to proceed independently. Indeed, the hormones that control the degradation of glycogen also control its biosynthesis. Again, a protein kinase is involved that is activated through the hormone/cAMP system. This kinase phosphorylates the active form of glycogen synthase (the I form) and thereby converts it into the relatively inactive form (the D form). The D form can be converted back into the I form through the action of a phosphatase. Thus, the initial action of these hormones activates glycogen degradation and deactivates glycogen synthesis.

The hormone/cAMP system just described controls several other important biological processes, in addition to glycogen synthesis and degradation. cAMP acts as a "second messenger" for a number of other hormones such as vasopressin, thyroxine, and the thyroid-stimulating and melanocyte-stimulating hormones. The hormone/cAMP system is effective because it serves as a biological amplification system. The concentrations of hormones in the blood are approximately 10^{-10} M, whereas the concentration of cAMP in the cell is 10,000 times as large, approximately 10^{-6} M.

Structure and Function of the Hormone Receptor

Several hormones can activate adenylyl cyclase and, hence, the production of cAMP. How is this accomplished? The system for the activation of adenylyl cyclase consists of three distinct components:

1. The hormone receptor
2. A G-protein
3. Adenylyl cyclase

Figure 21-14 shows a model for the interaction of these components. The hormone receptor is located at the extracellular surface of the plasma membrane. It is a large protein. For instance, the epinephrine receptor is a 75 kD protein. This receptor is also known as the β-adrenergic receptor to indicate that it binds a particular class of pharmacologically active compounds. Associated with the receptor is the G-protein. It consists of three different subunits, which are designated α, β, and γ; they have molecular weights of 45,000, 37,000, and 7000, respectively. The α subunit binds GDP. When the receptor is occupied by the hormone, GDP dissociates and GTP from the cytosol binds to the receptor. The GTP–G-protein complex then dissociates into an active GTP–αG subunit and the β and γ subunits. The GTP–αG subunit is the important messenger of the system because it activates adenylyl cyclase to catalyze the synthesis of cAMP from ATP. However, it is also a GTPase with a slow turnover of about 4 min^{-1}. It continues to activate adenylyl cyclase until the bound GTP is hydrolyzed to give GDP–αG-protein, which is inactive. The GDP–αG-protein subunit can then recycle through the receptor-hormone complex to regenerate the active GTP–αG subunit for as long as hormone is still bound to the receptor. The release of GDP and the binding of GTP are severalfold slower than the hydrolysis of GTP, so that these steps regulate the rate of recycling to the active GTP–αG subunit and the duration of action of the active adenylyl cyclase.

If the hydrolysis of GTP on the GTP–αG subunit is inhibited, it remains active and continues to activate adenylyl cyclase, so that the concentration of cAMP in the cell continues to increase. This occurs with nonhydrolyzable analogs of GTP, for example.

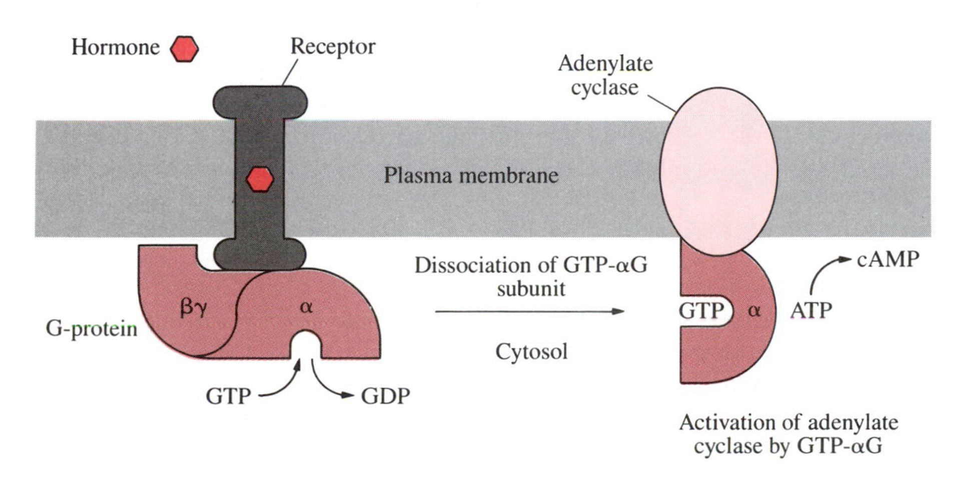

Figure 21-14

The cAMP-GTP Cycle.

BOX 21-2

ADDITIONAL G-PROTEIN SYSTEMS

Related G-proteins have recently been found in a number of other systems:

1. There is a G-protein that is inhibitory rather than stimulatory for adenylyl cyclase.

2. Another G-protein activates phospholipase C to catalyze the release of 1,4,5-inositol triphosphate (IP_3) from a phospholipid in the cell membrane.

Phosphatidylinositol-4,5-diphosphate

↓

IP_3 + **Diacylglycerol**

The IP_3 then binds to a receptor on the endoplasmic reticulum; this causes the release of calcium ions into the cytoplasm, which in turn regulates a number of cellular processes. The IP_3 is inactivated by hydrolysis of a phosphate group or through phosphorylation by ATP to give IP_4, an inactive form.

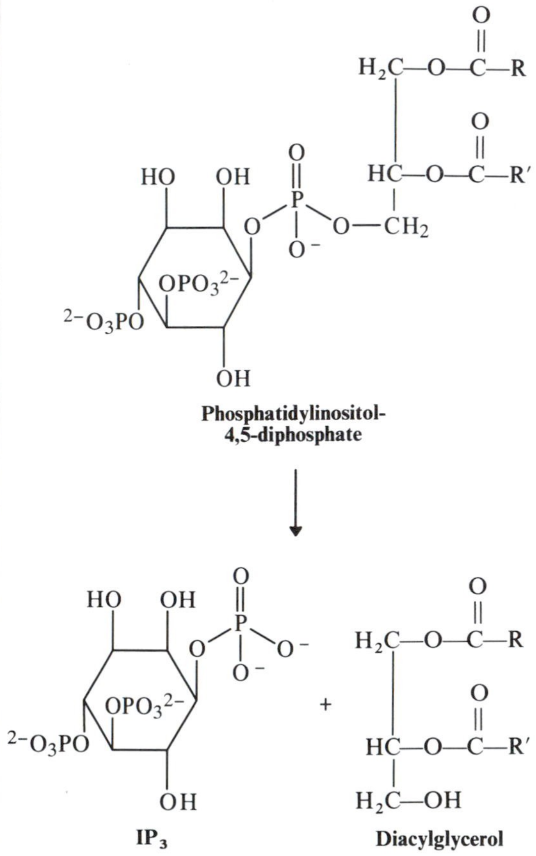

This hydrolysis also explains the action of cholera toxin, which blocks the hydrolysis of GTP in the GTP–αG subunit. Cholera toxin is a dimeric protein with a molecular weight of 87,000. One of its subunits is brought into the cell, where it catalyzes the transfer of the ADP-ribose portion of NAD^+ to a subunit of the G-protein and inhibits the GTPase activity of this subunit.

$$NAD^+ + GTP\text{—}\alpha G \text{ subunit} \xrightarrow{\text{Cholera toxin}} GTP\text{—}\alpha G \text{ subunit—ADP-ribose} + \text{Nicotinamide}$$

When GTP hydrolysis is blocked, the subunit remains in the active form and continues to synthesize cAMP. This continuing synthesis of cAMP in cells of the intestinal epithelium leads to a massive loss of sodium chloride and water from the intestine. This is the primary cause of death in cholera; however, it can be treated by the administration of large amounts of water and salt.

BOX 21-2 (continued)

ADDITIONAL G-PROTEIN SYSTEMS

3. The production of a nerve impulse from the retina when light is absorbed by rhodopsin in a rod is brought about by a G-protein and GTP. The mechanism of this reaction is described more fully in Chapter 29. Light absorption initiates a cascade of reactions that result in the closing of a channel and the development of an action potential:

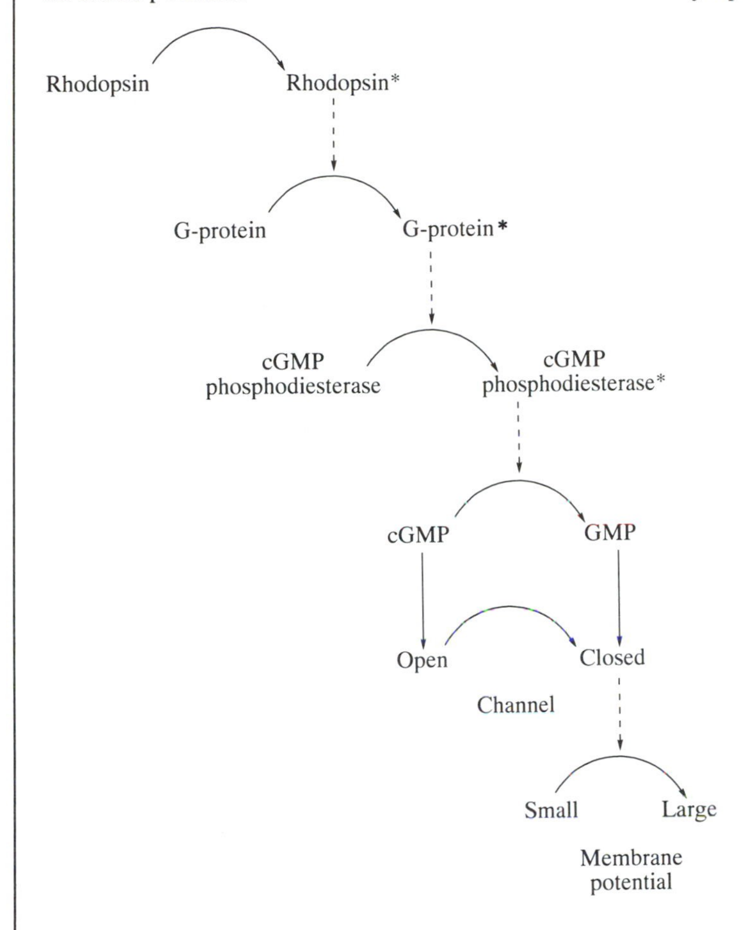

Light absorption activates the G-protein, which in turn activates a phosphodiesterase that cleaves cGMP to inactive GMP. In the dark, the cGMP acts on a channel in the cell membrane to keep it open. When the cGMP is cleaved, the channel closes, the membrane potential of the cell increases, and a signal is transmitted through a synapse to a nerve and eventually to the brain.

Receptors

As mentioned in Chapter 1, receptors are protein structures that are usually located on cell membranes. Each receptor can combine with a specific regulatory molecule and, as a result of this combination, the receptor is either activated or inactivated and some event takes place. Frequently, another protein or several proteins will be activated to perform a physiological function. With regard to the β-adrenergic receptors just discussed, a G-protein becomes activated.

A molecule that activates the receptor is called an agonist, and a molecule that combines with the receptor but does not activate it is an antagonist. Antagonists prevent the action of agonists. In a sense, receptors are very similar to enzymes in that they specifically combine with certain molecules. An enzyme then acts upon the molecule with which it has combined, whereas a receptor generally modifies another structure. Agonists are analogous to substrates for enzymes and antagonists to inhibitors.

How does the combination of a receptor with an agonist promote the initiation of a physiological event? The answer to this question is not well understood. One can speculate, however, by using enzymes as a model. It seems likely that the combination of agonist and receptor brings about a conformational change; that is, a change in the structure or charge distribution of the receptor. This conformational change may be transmitted to another molecule, which is thereby activated, or it may open or close a channel that allows ions to cross a membrane. This can produce an effect by changing the membrane potential or the ions may themselves act as an agonist that controls an activity of the cell. Thus, interaction of a hormone with a β-adrenergic receptor brings about a structural change in the receptor, which causes the G-protein to undergo a conformational change, which in turn allows the exchange of GDP and GTP.

Fermentation, Anaerobic Glycolysis

In the metabolic processes discussed so far, O_2 serves as an electron acceptor. There are, however, many microorganisms that exist in the total absence of O_2. How do these organisms obtain energy? Let us consider how yeast metabolizes glucose under anaerobic conditions. Yeast and other anaerobic organisms convert glucose into pyruvate. When a molecule of glucose is converted into two molecules of pyruvate, there is a net synthesis of two molecules of ATP. *This is the total yield of ATP that is obtained from a molecule of glucose under anaerobic conditions.* Thus, anaerobic metabolism is much less productive of useful energy than is aerobic metabolism. The enormously greater energy production of aerobic metabolism is made possible by the ability of aerobic organisms to couple the reoxidation of NADH by molecular oxygen to the synthesis of ATP.

In the process of converting glucose into pyruvate in glycolysis, 2 moles of NADH are produced from each mole of glucose. The concentration of NAD^+ in the cell is very low relative to the concentration of substrates that are metabolized. Soon, all of the NAD^+ would be converted into NADH and metabolism would stop. To avoid this, NAD^+ is continuously regenerated by the oxidation of NADH. Because no oxygen is used in anaerobic metabolism, a molecule found in the environment is used to reoxidize NADH. In the fermentation of glucose by yeast, acetaldehyde, which is derived from the decarboxylation of pyruvate, is used for this purpose. The reaction of acetaldehyde and NADH regenerates NAD^+ and gives rise to ethanol through the following reaction sequence.

$$\text{Glucose} \longrightarrow \xrightarrow[2\,NAD^+]{2\,NADH} \longrightarrow 2\,CH_3{-}CO{-}COO^- \xrightarrow[2\,CO_2]{} 2\,CH_3{-}CHO \xrightarrow[2\,NAD^+]{2\,NADH} 2\,CH_3{-}CH_2OH$$

Thus the end products of anaerobic glucose metabolism from fermentation in yeast are ethanol and CO_2. This is the basic process by which the alcohol of wine, beer, and other alcoholic beverages is produced.

Fermentation Can Give Products Other Than Ethanol

An interesting artificial modification of the ethanol fermentation was discovered by Neuberger. When HSO_3^- is added to the fermenting medium, ethanol production ceases. Instead, glycerol and the acetaldehyde-HSO_3^- adduct accumulate. The reason for this is that acetaldehyde is trapped by adduct formation with HSO_3^-.

$$H_3C{-}CHO + HSO_3^- \longrightarrow H_3C{-}CH(OH){-}SO_3^-$$

Therefore, acetaldehyde is not available for reoxidation of the NADH that is produced by glyceraldehyde-3-P dehydrogenase. Instead, the NADH is oxidized by dihydroxyacetone-P in a reaction catalyzed by glycerophosphate dehydrogenase.

$$CH_2OH{-}C(=O){-}CH_2OPO_3^{2-} + NADH + H^+ \longrightarrow CH_2OH{-}CH(OH){-}CH_2OPO_3^{2-} + NAD^+$$

Glycerol-3-P

This results in the formation of glycerol-3-P, which is subsequently hydrolyzed to glycerol by the action of phosphatases.

$$CH_2OH{-}CH(OH){-}CH_2OPO_3^{2-} + H_2O \longrightarrow CH_2OH{-}CH(OH){-}CH_2OH + HPO_4^{2-}$$

This sequence of reactions has been of considerable importance for the commercial production of glycerol. It supplied a significant portion of Germany's glycerol requirement for the manufacture of the explosive nitroglycerine in World War I.

It is beyond the scope of this book to discuss the many interesting modifications in fermentation pathways that exist among microorganisms. We will cite only one example. The end product of glucose fermentation in bacteria is generally not ethanol. In one type of fermentation, lactic acid is found. Lactic acid arises from the reduction of pyruvate.

$$H_3C{-}C(=O){-}COO^- + NADH + H^+ \longrightarrow H_3C{-}CH(OH){-}COO^- + NAD^+$$

In this reaction, pyruvate, instead of acetaldehyde, is used to reoxidize NADH.

The same process occurs in glycolysis in mammals and other aerobic organisms when it is necessary to form ATP more rapidly than it can be supplied by oxidative phosphorylation. This is the reason that lactic acid accumulates in the blood of athletes and others who exercise hard for a short period and exhaust their stores of ATP and creatine phosphate. The accumulation of this lactic acid can lower the pH of the blood and extracellular fluids and cause *acidosis*.

The Discovery of Glycolysis and the TCA Cycle

Because these processes are of fundamental importance and their investigation has greatly altered the way that we examine biological processes, we shall briefly review the history of their discovery.

The study of fermentation and glycolysis has a long history—in fact, it provided the foundation of modern biochemistry. In 1897, H. Buchner and E. Buchner discovered that cell-free extracts of yeast could convert sucrose into ethanol. This was the first demonstration that a biochemical process could occur outside a living cell; it was a major scientific breakthrough. This discovery was a subject of much controversy. It was believed at that time that biological transformations could occur only within living cells. The Buchners' discovery was thus contrary to accepted dogma. The eventual acceptance of the fact that biological reactions can be studied outside a cell has been the foundation of most of the subsequent development of biochemistry. It is now almost axiomatic that, in the investigation of any biochemical phenomenon, one must first find a cell-free system in which the process occurs.

The cell-free system has many advantages. For example, it avoids the problem of selective permeability that is encountered with intact cells. One can, therefore, add many substrates and inhibitors that might not enter the intact cells and study their effects. It also becomes possible to vary the concentration of these substances over a larger and more precisely defined range. In addition, interactions between various systems are reduced so that the properties of a system that is under investigation are not obscured by interactions with other systems.

People have often worried that the study of a biological system outside the cell will provide misleading information. The history of biochemistry has shown that this is rarely the case. On the contrary, the realization that cell-free systems can be utilized is responsible for much of the knowledge of modern biochemistry.

The Buchners' discovery laid the foundation for the next major advance in the understanding of fermentation. This was the discovery in 1905, by Harden and Young, that inorganic phosphate is required to maintain the production of ethanol from glucose in yeast extracts and that phosphate is consumed in the process. These investigators also isolated a phosphorylated hexose, fructose-1,6-bisphosphate, which is the key intermediate in fermentation. They succeeded in separating yeast juice into two components: a dialyzable, heat-stable component, which they called cozymase; and a nondialyzable heat-labile component, which they called zymase. Each component was inactive in fermentation, but the combination of the two components catalyzed the conversion of glucose into ethanol. Eventually, it was shown that zymase contains the enzymes that participate in fermentation and cozymase consists of the coenzymes ATP, ADP, and NAD. Thus, these early studies not only began to clarify the sequence of chemical transformations that glucose undergoes in the course of fermentation, but also led to the discovery of several of the most-important coenzymes and laid the foundation for the isolation of enzymes.

Subsequently, it was found that muscle extracts can convert glucose into lactate and that the reactions of this glycolysis are similar to those in fermentation. By 1940, the complete pathway of glycolysis, as carried out in muscle, was discovered largely through the work of G. Embden, O. Meyerhof, G. Cori, J. Parnus, and O. Warburg.

The Discovery of the TCA Cycle Is a Milestone in Modern Biochemistry

The TCA cycle was formulated in 1937 by Krebs. The work of many investigators provided the background for his discovery. Perhaps the foundation for the formulation of the TCA cycle was the discovery of the dehydrogenases by T. Thunberg, F. Batells, and L. S. Stern (1910–1920). These workers prepared anaerobic, minced tissue suspensions and demonstrated that these suspensions could catalyze the reduction of certain dyes by several of the organic acids that are known to be present in cells. These acids are succinic, malic, and citric acid. By 1930, more-sophisticated methods became available and it was possible to show that tissue suspensions can utilize O_2 to oxidize these acids to CO_2. In 1935, A. Szent-Gyorgyi was able to formulate the following important reaction sequence:

$$\text{Succinate} \longrightarrow \text{Fumarate} \longrightarrow \text{Malate} \longrightarrow \text{Oxaloacetate}$$

He made an additional important observation: the addition of malate or fumarate to tissue suspensions led to O_2 consumption that was far in excess of that required to metabolize the added dicarboxylic acid. He concluded that these acids must fulfill a *catalytic* role in the metabolism of some endogenous substance that was present in the tissue homogenate. The endogenous substrate was probably acetyl CoA derived from glycogen, which was present in the tissue preparations used. From our discussion of the TCA cycle, we know that malate and fumarate are precursors of oxaloacetate, which plays a catalytic role in the operation of the TCA cycle. We can, therefore, readily see why the addition of any of these compounds will accelerate the rate of acetyl CoA oxidation.

C. Martinus and F. Knoop, in 1936, succeeded in formulating an additional reaction sequence:

$$\text{Citrate} \longrightarrow \alpha\text{-Ketoglutarate} \longrightarrow \text{Succinate}$$

Important information, which was crucial in the final formulation of the citric acid cycle, was provided by experiments that were carried out with malonate, a competitive inhibitor of succinic dehydrogenase. Malonate inhibited the stimulation of O_2 consumption when di- or tricarboxylic acids were added to tissue homogenates, and succinate was found to accumulate under these conditions. Krebs concluded that succinate dehydrogenase plays a crucial role in the processes by which these acids stimulate oxygen consumption. An additional important observation was that succinate also accumulated when fumarate was added to these tissue preparations in the presence of malonate. These observations establish that a pathway must exist for the conversion of fumarate into succinate that does not include succinic dehydrogenase, because that enzyme is inhibited by malonate. These experiments suggested the occurrence of a cyclic process and were important in enabling Krebs to formulate what is now called the TCA cycle. We have seen that cyclic processes are critical for metabolism and synthesis in biochemistry. In this connection, it should be pointed out that, previously, Krebs had proposed the ornithine cycle, through which urea is formed (Chapter 26).

The TCA cycle as formulated in 1937 involved an initial step in the reaction between pyruvate and oxaloacetate to produce citrate. However, the nature of the two-carbon fragment that actually condensed with pyruvate was unknown. For the next ten years, biochemists worked on the problem of characterizing the two-carbon fragment. Finally, in 1948, it was identified as acetyl CoA.

Summary

All biological energy is derived from redox processes. In aerobic organisms, the final electron acceptor is O_2. For the oxidation of glucose to CO_2 and H_2O, $\Delta G^\circ = -690\ \text{kcal mol}^{-1}$. This energy is not released in a single reaction. Energy is released through a series of reactions and is converted into ATP in increments. Glycolysis is the first major metabolic process in the oxidation of glucose. In this process, 1 mole of glucose is converted into 2 moles of pyruvate, and 1 mole each of ATP and NADH are generated. Pyruvate is converted into acetate and then metabolized through the tricarboxylic acid cycle (TCA cycle) to CO_2 and H_2O. In this process, 3 moles of NADH, 1 mole of GTP, and 1 mole of $FADH_2$ are generated. NADH and $FADH_2$, generated through glycolysis and the TCA cycle, are oxidized through a series of enzymes (electron-transport system) to NAD^+ and FAD. In this process, 3 moles of ATP are generated per mole of NADH and 2 moles of ATP per mole of $FADH_2$.

Sugars other than glucose can enter glycolysis. This is achieved by their conversion into glucose or into one of the intermediates of glycolysis. For example, fructose, a product of sucrose hydrolysis, is converted through the action of three enzymes into glyceraldehyde-3-P, which then enters the glycolytic pathway.

Glucose can also be metabolized through the pentose shunt to CO_2 and five-carbon sugars. These can be used in the biosynthesis of nucleotides or, after further metabolic modifications, they can enter glycolysis. In addition, the pentose shunt produces NADPH, which is used in reductive biosynthesis.

In every living organism, there are many metabolic pathways. These pathways interact with one another, and the flux of metabolism responds to physiological stimulants. To assure effective interaction and response, the level of activity of the various pathways must be controlled. This is achieved through two major mechanisms: (1) control of entry of metabolites into cells; (2) control of activity of certain enzymes. Enzyme activity is controlled in three ways: (1) by the amount of enzyme present; (2) by reversible chemical modification; (3) by the presence of small molecules in the environment that can either inhibit or activate catalysis.

Molecules such as ATP, ADP, or citrate, which do not necessarily participate in a reaction, also can either accelerate or retard the reaction. It is easy to understand why these molecules are utilized to control enzyme reactions. High levels of ATP indicate that there is no need to generate ATP and, consequently, the rate of glycolysis can decrease. High levels of ATP inhibit phosphofructokinase and, consequently, the rate of glycolysis decreases. High levels of ADP signal a deficiency of ATP, which can be corrected by increased glycolysis. ADP accelerates the phosphorylation of fructose-6-P by phosphofructokinase and, hence, the rate of glycolysis. High levels of citrate indicate a high rate of TCA cycle activity, and the organism can afford to divert molecules that would normally enter the TCA cycle into other pathways. For example, acetyl CoA can be used for fatty acid synthesis rather than entering the TCA cycle.

The liver is largely responsible for the control of blood glucose levels. Glucose can be removed from the circulation either by glycolysis or by conversion into glycogen, a storage polymer of glucose. Glucose is incorporated into glycogen, first, by conversion into glucose-6-P and then into glucose-1-P. Glucose-1-P is transformed into UDP-glucose, which is the substrate for the incorporation of glucose into glycogen through the action of glycogen synthase. When blood glucose levels are low, the liver releases glucose, derived from glycogen, into the blood stream. This is accomplished through the action of glycogen phosphorylase, which catalyzes the reaction between glycogen and P_i to yield glucose-1-P. Glucose-1-P is converted into glucose-6-P which is hydrolyzed to free glucose. The action of these enzymes must be closely controlled. When synthesis is required, degradation must be turned off and vice versa. To accomplish this control, the liver must "sense" the body's need for glucose. In other words, a signal is required to inform the liver that glucose must be released to the blood stream or removed from the blood stream. Hormones are signal molecules that are used to communicate between organs. There are several types of hormones: those derived from amino acids (e.g., thyroxine and epinephrine, or adrenalin); those that are polypeptides or proteins (e.g., insulin); and those that are steroids.

Hormones that give the signal for glycogen breakdown are ACTH, glucagon (polypeptide hormones), and epinephrine, which is derived from phenylalanine. Upon release, these hormones activate AMP cyclase, which catalyzes the conversion of ATP into cyclic AMP (cAMP). Cyclic AMP triggers a series of events that lead to glycogen breakdown and decreased glycogen synthesis. cAMP initiates the process by activating protein kinase. The activated protein kinase phosphorylates phosphorylase kinase with ATP as the phosphate donor and thereby activates the enzyme. Phosphorylase kinase catalyzes the phosphorylation of a serine residue of glycogen phosphorylase and converts this enzyme from an inactive (*b*) form into an active (*a*) form. Phosphatases exist that dephosphorylate active glycogen phosphorylase and phosphorylase kinase. The action of these enzymes returns the system to its resting state.

Glycogen synthesis is subject to hormonal control by the same enzymes that control degradation. Again, a protein kinase is activated through the hormone/cAMP system. This kinase phosphorylates the active (I) form of glycogen synthase and converts it into the relatively inactive (D) form. The hormone/cAMP system leads to the deactivation of glycogen synthase and the activation of glycogen phosphorylase. The hormone/cAMP system also controls many other biological processes. cAMP is referred to as a "second messenger." It is a very powerful amplification system. One hormone molecule gives rise to approximately 10,000 cAMP molecules.

Hormonal stimulation of adenylyl cyclase utilizes a transmembrane receptor protein and the associated G-protein that is located on the cytoplasmic side of the receptor. The G-protein consists of α, β, and γ subunits. The α subunit binds GDP. When the hormone binds to the receptor, GDP dissociates and is replaced by GTP from the cytoplasmic side. The GTP–αG complex then dissociates and combines with adenylyl cyclase, whereupon the cyclase becomes active. GTP–αG continues to activate the cyclase until the bound GTP is hydrolyzed by the α subunit. The GDP–αG complex then associates with the receptor and the β and γ subunits; it can then undergo a new cycle.

For most of the metabolic reactions discussed in this chapter, O_2 serves as the final electron acceptor. Many microorganisms exist in the total absence of air and still must obtain energy for biological processes. Yeast, for instance, can metabolize glucose under anaerobic conditions. The same reactions are used as in glycolysis, and 2 moles of pyruvate are produced from each mole of glucose. Concomitantly, 2 moles of ATP are produced. This is the total yield of ATP from the anaerobic metabolism of glucose. Clearly, anaerobic metabolism is much less efficient than aerobic metabolism. The conversion of glucose into pyruvate also results in the formation of 2 moles of NADH per mole of glucose. NADH must be reoxidized to NAD^+, because NAD^+ is present only in catalytic amounts. An aerobic organism would use O_2 to oxidize NADH. Under anaerobic conditions, yeast uses acetaldehyde, derived from the decarboxylation of pyruvate, to regenerate NAD^+. This leads to the formation of ethanol. The metabolism of glucose under anaerobic conditions is called fermentation, which is only one example of anaerobic metabolism. Anaerobic metabolism is diverse, and anaerobic organisms can metabolize a multitude of substrates.

ADDITIONAL READING

Bioteux, A., and Hess, B. 1981. Design of glycolysis. *Phil. Trans. R. Soc. Lond.* B 293: 5–22. (A stimulating presentation of the regulation of glycolysis. Published in a symposium volume entitled *The Enzymes of Glycolysis: Structure, Activity, and Evolution*, which brings together X-ray crystallography, enzymology, and evolutionary biology.)

ROSE, I. A. 1981. Chemistry of proton abstraction by glycolytic enzymes (aldolase, isomerases, and pyruvate kinase). *Phil. Trans. R. Soc. Lond.* B 293: 131–144.

KNOWLES, J. R., and ALBERY, W. J. 1977. Perfection in enzyme catalysis: the energetics of triosephosphate isomerase. *Accounts of Chemical Research* 10: 105–111.

FRUTON, J. S. 1972. *Molecules and Life: Historical Essays on the Interplay of Chemistry and Biology.* Wiley-Interscience. (Includes a meticulously documented account of the elucidation of the nature of fermentation and how it led to enzyme chemistry.)

HINKLE, P. C., and MCCARTY, R. E. 1978. How cells make ATP. *Scientific American* 238(3): 104–123.

KNOWLES, J. 1990. Stabilization of a reaction intermediate as a catalytic device: definition of the functional role of the flexible loop in triosephosphate isomerase. *Biochemistry* 29: 3186.

PETSKO, G., 1991. Structure of the triosephosphate isomerase-phosphoglycolohydroxamate complex: an analogue of the intermediate on the reaction pathway. *Biochemistry* 30: 5821.

PILKUS, S. J., and EL-MAGHRABI, M. R. 1988. Hormonal regulation of hepatic gluconeogenesis and glycolysis. *Annual Review of Biochemistry* 57: 755.

PROBLEMS

1. Arsenate ($HAsO_4^{2-}$) is toxic because it can replace HPO_4^{2-} in many reactions. For example, in the presence of $HAsO_4^{2-}$, glyceraldehyde 3-P dehydrogenase converts glyceraldehyde-3-P into 1-arseno-3-phosphoglyceraldehyde, which undergoes hydrolysis rapidly. What is the effect of $HAsO_4^{2-}$ on the ATP yield from glycolysis?

2. How is the isotopic label distributed in ethanol and CO_2 when [3-^{14}C]glucose undergoes fermentation? Assume that the interconversion of dihydroxyacetone phosphate and glyceraldehyde-3-P is rapid.

3. What is the labeling pattern of oxaloacetate if $^{14}CH_3—COO^-$ is metabolized through the TCA cycle? Assume that the cycle turns once, three times, and a large number of times.

4. Explain why the addition of the dicarboxylic acid succinate, fumarate, or oxaloacetate enhances the rate of CO_2 production in liver homogenates.

5. We know that cells have a receptor for epinephrine. What experimental approach could be used to locate this receptor?

6. Could a cAMP analog that prevents the action of cAMP be pharmacologically useful? Justify your answer.

7. Propose a mechanism of action of phosphopentose isomerase and an experiment to test this mechanism.

The Electron-Transport Pathway and Oxidative Phosphorylation

22

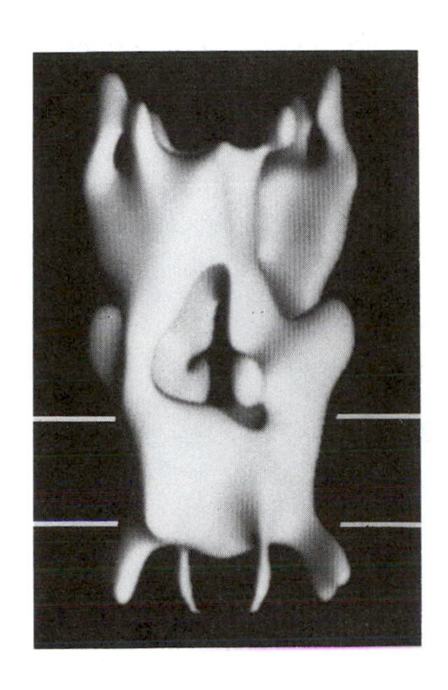
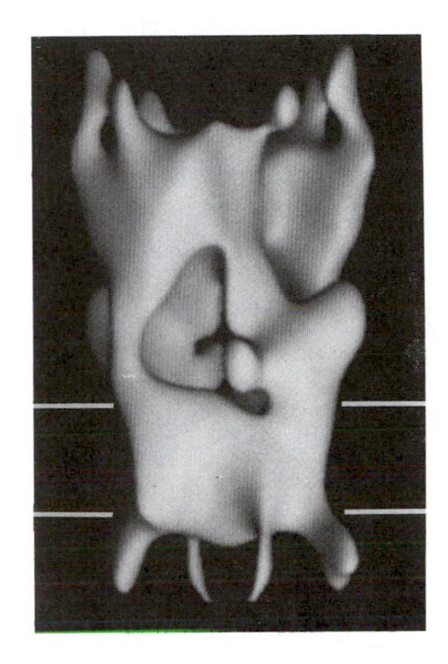

Reconstruction image of cytochrome oxidase. These are surface-shaded representations of one molecule of a beef heart mitochondrial cytochrome c oxidase dimer determined by electron cryo-microscopy of thin crystals frozen in vitreous (noncrystalline) ice. A stereo view is shown here. One functional unit (one-half of a dimer) has a molecular weight of 200,000 and consists of twelve or thirteen different polypeptides, two heme a moieties, and two copper ions. The two horizontal white lines delineate the position of the lipid bilayer in which the molecule is embedded. (Courtesy of T. G. Frey and R. Henderson.)

Only a small fraction of the total energy potentially available from the oxidation of glucose is made available as ATP and GTP in the *substrate-level phosphorylation* reactions of glycolysis and the TCA cycle, which produce six molecules of ATP per molecule of glucose (Chapter 21). Inasmuch as two molecules of ATP are used up in the hexokinase and phosphofructokinase steps of glycolysis, these two pathways yield a net of only four molecules of ATP per molecule of glucose. This corresponds to 32 kcal mol^{-1} of chemical free energy, which is very small compared with the 686 kcal mol^{-1} of heat released by the combustion of glucose.

Much of the energy not captured as ATP in glycolysis and the TCA cycle is conserved in the form of ten molecules of NADH and two molecules of $FADH_2$. The NADH is produced in the reactions catalyzed by pyridine nucleotide-dependent dehydrogenases acting on intermediates in glycolysis and the TCA cycle. The two molecules of $FADH_2$ are generated by the succinate dehydrogenase reaction and are bound to that enzyme. Succinate dehydrogenase is embedded in a very important membrane that is the principal subject of this chapter. The chemical events within this membrane generate about thirty-two molecules of ATP from the oxidation of ten molecules of NADH and two molecules of $FADH_2$ by O_2. This ATP is the main source of energy for most aerobic organisms. This chapter is concerned with the problem of how ATP is formed from the oxidation of NADH and $FADH_2$ through a series of reactions catalyzed by enzymes in the mitochondrial membrane.

The Site of Electron Transport and ATP Formation

It took a long time for biochemists to realize that the metabolic pathways do not provide most of the ATP directly. The glycolysis pathway was worked out in the 1930s, and attention became focused on energy production by this pathway and, later, by the TCA cycle. In 1941 it was realized, in particular by F. Lipmann and H. Kalckar, that the energy balance should be described in terms of the amount of high-energy phosphate that is formed as ATP and that most of the ATP is not produced by the reactions of glycolysis and the TCA cycle. At about this time, Belitzer in Russia and Kalckar showed that the oxidation of C–H bonds by O_2, brought about by the mitochondria, could give ATP from added ADP and P_i. It was learned later, largely through the work of Lehninger and Kennedy, that the most-important immediate source of C–H bonds in this process is the NADH generated by the TCA cycle and glycolysis.

The oxidation of NADH by O_2 and the concomitant production of ATP in eukaryotic cells take place in mitochondria and are catalyzed by membrane-bound enzymes. The overall process is known as *oxidative phosphorylation.* Figures 22-1 and 22-2 are electron micrographs showing the cristae of sectioned mitochondria from heart and liver cells. Figure 22-3 illustrates the inner and outer membranes, the matrix and intermembrane spaces, and the cristae. The inner membrane is the site of oxidative phosphorylation and is the membrane, referred to earlier, in which succinate dehydrogenase is embedded.

Oxidative phosphorylation can be divided into the reactions catalyzed by two systems: the electron-transport pathway and ATP synthase. The electron-transport pathway is a system of multienzyme complexes consisting of redox proteins and a quinone. This system catalyzes the oxidation of NADH by O_2 in a series of steps. The energy released by electron transfer through the electron-transport pathway is used by ATP synthase to produce ATP from ADP and P_i. In anaerobic bacteria, the oxidizing agent in the electron-transport pathway is

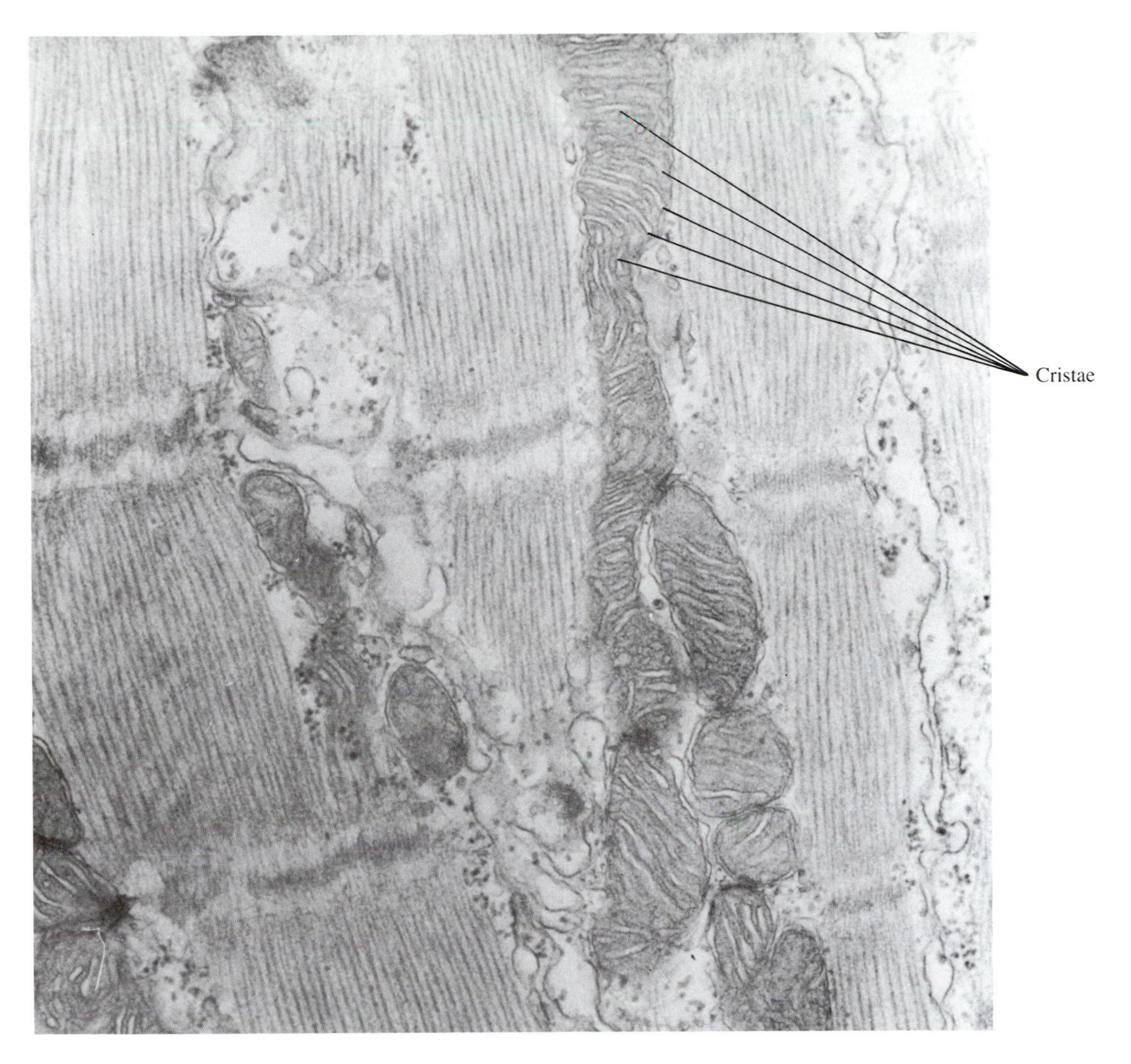

Figure 22-1

Electron Micrograph of Heart Mitochondria. This enlargement of an electron micrograph taken of a section through fish heart cells shows cross sections through mitochondria. The presence of many cristae in the inner membrane (see Figure 22-3) is characteristic of highly aerobic cells such as heart cells, which consume large amounts of energy in beating. (Micrograph courtesy of Enrique Valdivia.)

some molecule other than O_2, such as CO_2, NO_2^-, SO_4^{2-}, or another oxidized molecule. Systems utilizing molecules other than O_2 as the terminal electron acceptor will not be described here.

In eukaryotic cells, the electron-transport pathway and ATP synthase in the inner membrane of the mitochondria produce most of the ATP used by the cell. Mitochondria contain, in addition to the electron-transport pathway and

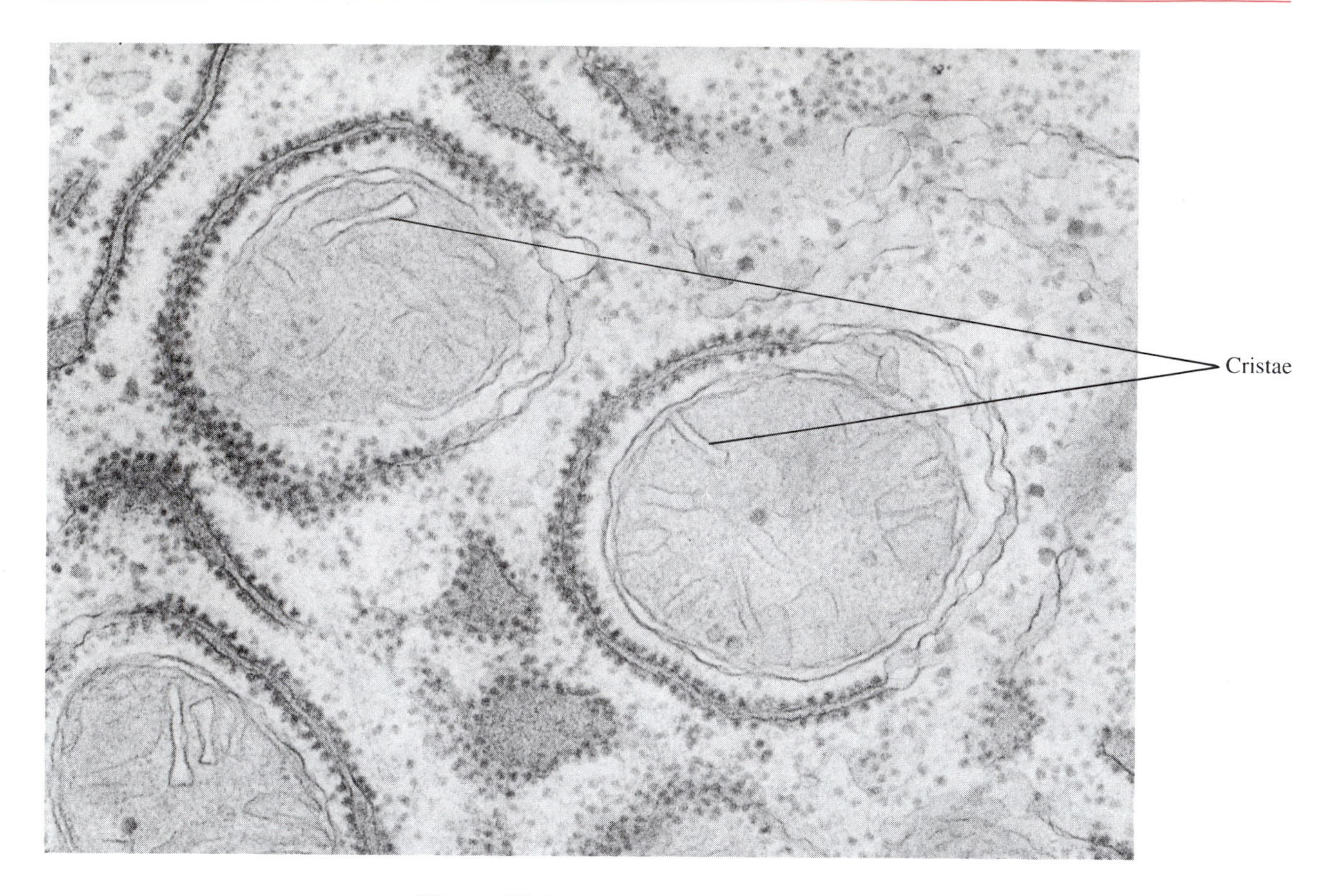

Figure 22-2

Electron Micrograph of Liver Mitochondria. This enlargement of an electron micrograph taken of a section through fish liver cells shows cross sections of mitochondria. The few cristae defined by the convolutions in the inner mitochondrial membrane are characteristic of cells that are not engaged in physical work and so are not highly aerobic. (Micrograph courtesy of Enrique Valdivia.)

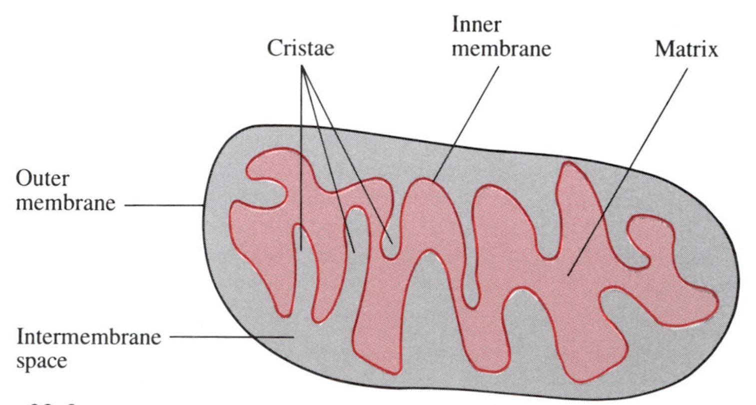

Figure 22-3

The Membrane Structure of a Mitochondrion. The outer membrane is permeable to many small cytosolic molecules. The inner membrane is impermeable to all molecules except those for which the membrane contains transport systems. Complexes I through IV of the electron-transport pathway are embedded in the inner membrane. ATP synthase also is in the inner membrane. The TCA enzymes, fatty acid degradation enzymes, mitochondrial DNA, and ribosomes are in the matrix space enclosed by the inner membrane.

ATP synthase, the enzymes of the TCA cycle, the pyruvate dehydrogenase complex, and the fatty acid degradation system. They also contain all of the systems required to transport pyruvate and the reducing electrons of NADH from the cytosol, where glycolysis takes place, into the mitochondrion. In addition, they contain a small amount of DNA, protein biosynthetic machinery, and several other transport systems, including the one that transports ADP into the mitochondrion and ATP out of it.

Bacteria and prokaryotes in general do not contain mitochondria; they produce ATP by oxidative phosphorylation within the bacterial membrane. The bacterial systems are remarkably similar to the eukaryotic systems. This is thought to signify an ancestral relationship between mitochondria and bacteria. According to this view, the mitochondria arose through a symbiotic relationship between early eukaryotic cells and intracellular bacteria. Through time, the eukaryotic cells became dominant and captured the bacteria as intracellular organelles. This is consistent with the fact that mitochondria contain their own specific DNA that encodes several important mitochondrial proteins. Most mitochondrial proteins are encoded by nuclear DNA, however, and must be transported into the mitochondria in the course of cell division.

Energetics of Electron Transport and ATP Formation

In mitochondria, the oxidation of NADH by O_2 and the formation of ATP from ADP and P_i (equations 1 and 2)

$$\text{NADH} + \tfrac{1}{2}\,\text{O}_2 + \text{H}^+ \xrightarrow{\text{ETP}} \text{H}_2\text{O} + \text{NAD}^+ \qquad (1)$$

$$\text{ADP}^{3-} + \text{HPO}_4{}^{2-} + \text{H}^+ \xrightarrow{\text{ATP synthase}} \text{ATP}^{4-} + \text{H}_2\text{O} \qquad (2)$$

are chemically distinct processes that are coupled; that is, they are mechanistically linked and interdependent under normal conditions. ATP formation depends on electron transport by virtue of the proton motive force generated by the action of the electron-transport pathway and used by ATP synthase as the source of energy for ATP production.

The energy potentially available from the oxidation of NADH by O_2 can be calculated from the reduction potentials (E_0') of the NAD^+/NADH and O_2/H_2O systems, which are -0.320 volt and $+0.815$ volt, respectively. Using these values, we can calculate $E_0' = 1.135$ volts for the oxidation of NADH by O_2 according to equation 1. From this and equation 3,

$$\Delta G^{\circ\prime} = -nFE_0' \qquad (3)$$

$$(n = 2 \text{ and } F = 23{,}000 \text{ cal V}^{-1}\text{ mol}^{-1})$$

which relates the difference in reduction potentials to the free-energy change (Chapter 9), we calculate $\Delta G^{\circ\prime} = -52\,\text{kcal mol}^{-1}$. This is sufficient free energy to produce 6 moles of ATP from each mole of NADH, because $\Delta G^{\circ\prime}$ for the hydrolysis of ATP is $-8.5\,\text{kcal mol}^{-1}$ and ATP hydrolysis is the reverse of ATP synthesis. Oxidative phosphorylation actually produces about half of the theoretical amount, or 3 moles of ATP per mole of NADH. This is excellent efficiency when compared with typical man-made machines. These calculations refer to standard conditions, which do not exist in cells. More energy is required to produce ATP at the low concentrations of ADP and P_i within cells; so the efficiency of the system is more than 50%.

The Electron-Transport Pathway

Mitochondrial Structure

The structure of the mitochondrion is largely defined by the two membranes illustrated in Figure 22-3, the outer mitochondrial membrane and the inner mitochondrial membrane. The outer membrane is permeable to most small molecules and ions of the cytosol. The inner membrane is largely impermeable to molecules and ions but contains specific transport systems for pyruvate, glycerophosphate, malate, acyl carnitine esters, and other molecules that participate in mitochondrial functions. The inner mitochondrial membrane also houses ATP synthase and the components of the electron-transport pathway. The inner membrane has a much larger surface area than does the outer membrane. Because it is contained within the outer membrane, it is highly convoluted into narrow folds known as *cristae,* shown in the micrographs of Figures 22-1 and 22-2 and illustrated in Figure 22-3. The space between the inner and outer membranes is the *intermembrane space.* Inside the inner membrane is the *matrix space.* The enzymes of the TCA cycle and the pyruvate dehydrogenase complex, as well as the enzymes of fatty acid degradation, are in the matrix space.

Electron transport in the inner membrane is carried out by a number of redox cofactors, also called electron carriers, such as iron-sulfur clusters, hemes, flavins, quinones, and copper ions. The electron carriers are bound to various proteins that facilitate electron transfer. D. E. Green and Y. Hatefi showed that the proteins are organized into complexes that catalyze different steps of the overall oxidation of NADH by O_2. Figure 22-4 shows the initial pathway of electron transfer from NADH or succinate, the transfer of electrons from one complex to the next, and finally the reduction of O_2 to water. The complexes can be isolated from membranes that have been disrupted by detergents. They are particulate preparations, with the proteins containing the redox cofactors embedded in membrane fragments. These complexes and the reactions that they catalyze are described in the following subsections.

The structures of some of the redox cofactors are shown in Figures 22-5, 22-6, and 22-7. Iron-sulfur (FeS) clusters (Figure 22-5) consist of Fe^{2+} and Fe^{3+} ions complexed with inorganic sulfide (S^{2-}) and cysteine residues of the proteins. Most FeS clusters contain an equal number of iron and sulfide ions,

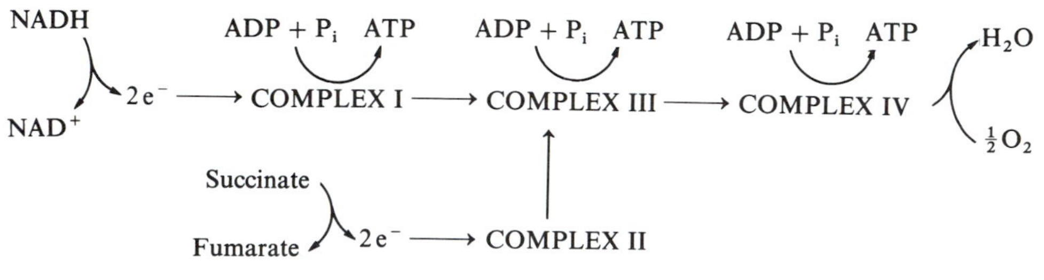

NADH → NAD$^+$: $2e^-$ ⟶ COMPLEX I (ADP + P_i → ATP) ⟶ COMPLEX III (ADP + P_i → ATP) ⟶ COMPLEX IV (ADP + P_i → ATP); $\frac{1}{2}O_2$ → H_2O

Succinate → Fumarate : $2e^-$ ⟶ COMPLEX II ⟶ COMPLEX III

Figure 22-4

The Pathway of Electron Transport in Mitochondria. The electron-transport complexes are assemblages of proteins and redox cofactors embedded in the inner mitochondrial membrane. Electrons generated by the dehydrogenation of NADH or succinate are transferred from one complex to another until they reach complex IV, cytochrome oxidase. Here, they are used to reduce O_2 to H_2O. ATP is produced as a result of the passage of electrons through complexes I, III, and IV.

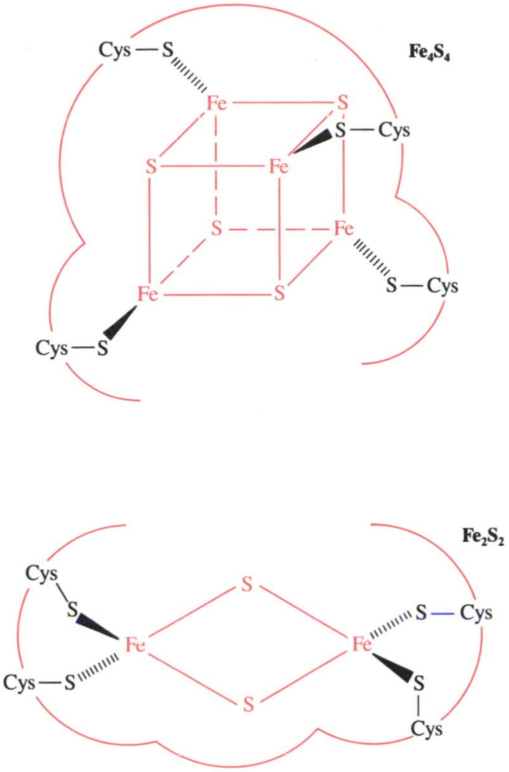

Fe$_4$S$_4$

Fe$_2$S$_2$

Figure 22-5

The Structures of Iron-Sulfur (FeS) Clusters. FeS clusters are one-electron carriers in the electron-transport pathway. The Fe atoms undergo reduction and oxidation between the +3 and +2 states. Inorganic sulfides (S^{2-}) are bound between Fe atoms. The cysteines (Cys) are part of the protein. FeS clusters are associated with proteins in the electron-transport complexes. They undergo transient one-electron reduction and oxidation while relaying electrons along the electron-transport pathway.

Ubiquinone (UQ)

$H^+ + 1e^-$

$H^+ + 1e^-$

Dihydroubiquinone (UQH_2)

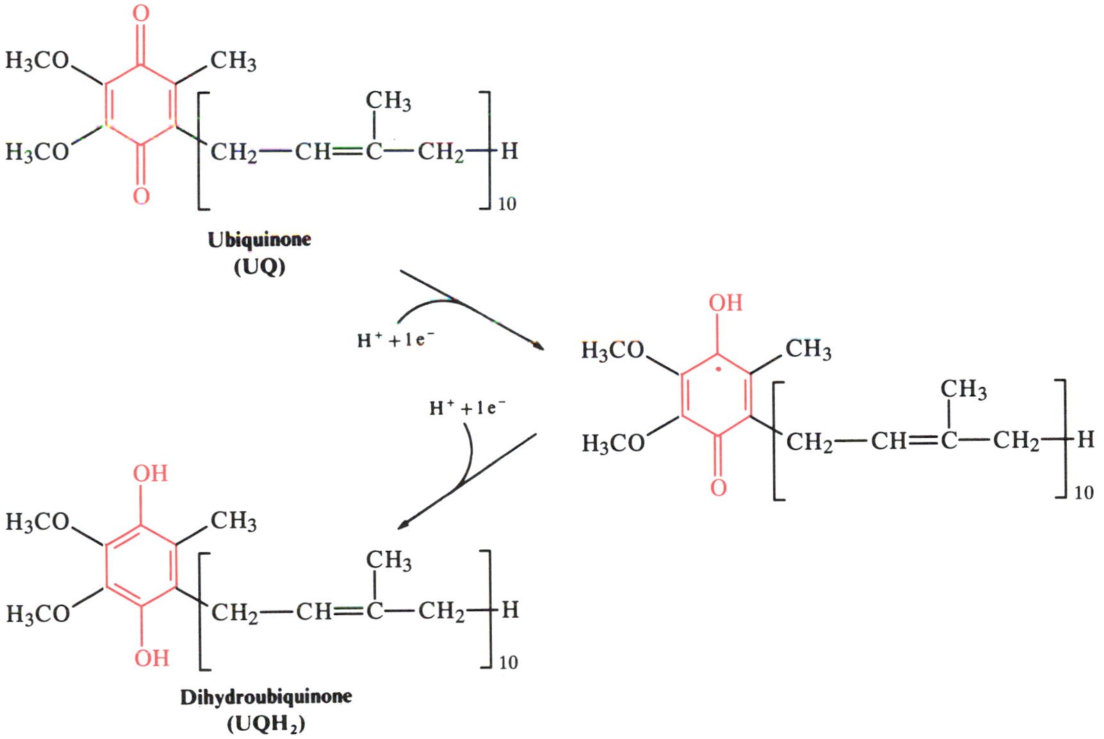

Figure 22-6

The Structures of Ubiquinone (UQ) and Dihydroubiquinone (UQH_2). Ubiquinone is a two-electron carrier that can undergo reduction and oxidation in two one-electron steps. The UQ/UQH_2 system relays electrons between complexes I and III and complexes II and III in the electron-transport pathway.

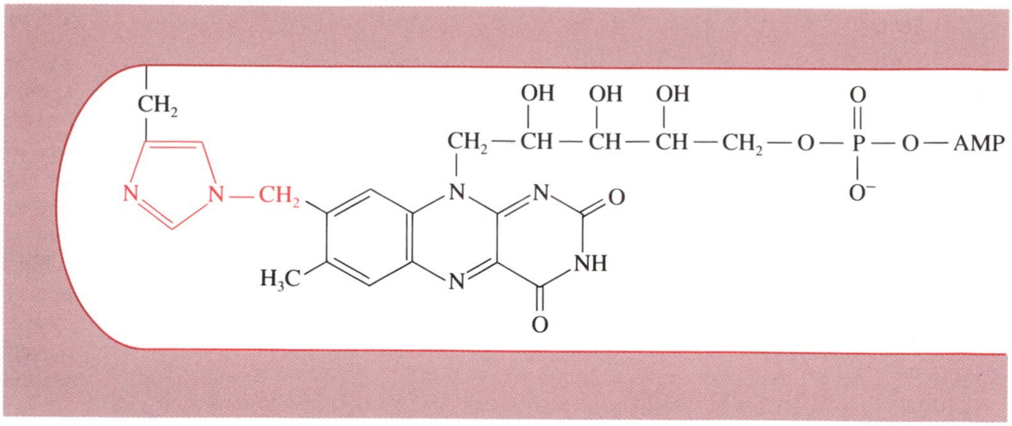

Figure 22-7

Covalent Attachment of FAD to Succinate Dehydrogenase (Complex II). FAD is covalently bonded to a histidyl imidazole ring in succinate dehydrogenase. The flavin is reduced to $FADH_2$ by succinate and is then reoxidized by ubiquinone in a process that includes the action of FeS cofactors in succinate dehydrogenase.

such as Fe_4S_4, which contains four of each. These cofactors undergo reduction and oxidation in one-electron steps; that is, they are one-electron carriers. Ubiquinone (Figure 22-6) is a two-electron carrier that undergoes reduction and oxidation in one-electron steps. FAD and FMN can undergo either two-electron or one-electron redox reactions (Chapter 16) and are sometimes covalently bound to proteins, as illustrated in Figure 22-7. The sizes, protein compositions, and cofactor contents of the four electron-transfer complexes in the electron-transport pathway are listed in Table 22-1.

Complex I: NADH Dehydrogenase

Complex I is known as NADH dehydrogenase (or NADH: ubiquinone oxidoreductase). It catalyzes the reduction of ubiquinone (UQ) by NADH according to equation 4.

$$NADH + UQ + H^+ \longrightarrow NAD^+ + UQH_2 \quad (4)$$

The structure of ubiquinone, also known as *coenzyme Q*, is shown in Figure 22-6. It consists of an electron-accepting quinone group attached to a hydrophobic polyisoprenoid tail that anchors the molecule to the membrane. Although UQ and UQH_2 are held in the membrane, they are mobile; they

Table 22-1

The electron-transfer complexes of the electron-transport pathway

Proteins	Molecular Weight	Cofactors	Number of Proteins
Complex I	1×10^6	Fe_4S_4, Fe_2S_2, FMN	26
Complex II	>100,000	Fe_4S_4, Fe_2S_2, FAD	4
Complex III	450,000	2 Cyt b, Cyt c_1	>6
Complex IV	200,000	Cyt a, Cyt a_3, Cu	>7

shuttle electrons from complex I to complex III of the overall pathway by lateral diffusion within the plane of the membrane.

The redox cofactors in complex I are FMN and FeS clusters. The FeS clusters are compulsory one-electron transfer centers, whereas NADH is a two-electron donor (Chapter 16). Therefore, direct electron transfer between an FeS center and NADH is not allowed. However, because flavins can undergo either one- or two-electron redox reactions, FMN can couple electron transfer between NADH and the FeS centers. FMN first undergoes a two-electron reduction by NADH to $FMNH_2$, possibly by a hydride transfer mechanism. The $FMNH_2$ can then reduce the FeS clusters in two 1-electron steps, with the formation of a flavin semiquinone as a transient intermediate. The reduced FeS clusters can then reduce UQ to UQH_2 in two 1-electron steps through a UQ-semiquinone radical (Figure 22-6). The capability of the FMN in NADH dehydrogenase to undergo both one- and two-electron transfers is fundamental to coupling the oxidation of NADH to the reduction of ubiquinone through the FeS clusters. There are several FeS clusters in complex I, and it is likely that electron transfer proceeds from one to another in additional intermediate steps. A reasonable reaction sequence for NADH dehydrogenase is summarized in equations 5a through 5c.

$$\text{NADH} + \text{E}\cdot\text{FMN} + \text{H}^+ \longrightarrow \text{NAD}^+ + \text{E}\cdot\text{FMNH}_2 \tag{5a}$$

$$\text{E}\cdot\text{FMNH}_2 + 2\,\text{FeS}_{\text{ox}} \longrightarrow \text{E}\cdot\text{FMN} + 2\,\text{FeS}_{\text{red}} + 2\,\text{H}^+ \tag{5b}$$

$$2\,\text{FeS}_{\text{red}} + \text{UQ} + 2\,\text{H}^+ \longrightarrow 2\,\text{FeS}_{\text{ox}} + \text{UQH}_2 \tag{5c}$$

Complex II: Succinate Dehydrogenase

Complex II is succinate dehydrogenase, also known as succinate: ubiquinone oxidoreductase, which was introduced in Chapter 21 as part of the TCA cycle. It is localized in the inner mitochondrial membrane. Succinate dehydrogenase is a flavoprotein that also contains FeS clusters and catalyzes the dehydrogenation of succinate to fumarate, with reduction of the flavin. The covalent attachment of FAD to the major subunit in this complex is illustrated in Figure 22-7. The electron acceptor for reduced succinate dehydrogenase is ubiquinone, and the reduction of ubiquinone by the reduced flavin may be chemically similar to the analogous process in complex I. A reasonable reaction sequence for complex II is summarized in equations 6a through 6c.

$$\text{Succinate} + \text{E}\cdot\text{FAD} \longrightarrow \text{Fumarate} + \text{E}\cdot\text{FADH}_2 \tag{6a}$$

$$\text{E}\cdot\text{FMNH}_2 + 2\,\text{FeS}_{\text{ox}} \longrightarrow \text{E}\cdot\text{FMN} + 2\,\text{FeS}_{\text{red}} + 2\,\text{H}^+ \tag{6b}$$

$$2\,\text{FeS}_{\text{red}} + \text{UQ} + 2\,\text{H}^+ \longrightarrow 2\,\text{FeS}_{\text{ox}} + \text{UQH}_2 \tag{6c}$$

Complex III: Cytochrome bc_1 Complex

Complex III, which is also known as the cytochrome bc_1 complex, catalyzes electron transfer from UQH_2 to cytochrome *c*, a soluble, redox-active heme-protein that associates reversibly with the inner membrane (equation 7).

$$\text{UQH}_2 + 2\ \text{Cyt}\ c\ (\text{Fe}^{3+}) \longrightarrow \text{UQ} + 2\ \text{Cyt}\ c\ (\text{Fe}^{2+}) + 2\,\text{H}^+ \tag{7}$$

The complex contains three other cytochromes and an iron-sulfur protein with a Fe_2S_2 center. The cytochromes in the electron-transport pathway are distinguished by their heme structures, the chemistry of the heme-protein interactions, their visible electronic spectra, and their reduction potentials. The spectra of oxidized and reduced cytochrome *b* are shown in Figure 22-8 and are typical of the spectra of cytochromes. Complex III contains two *b*-type cytochromes that

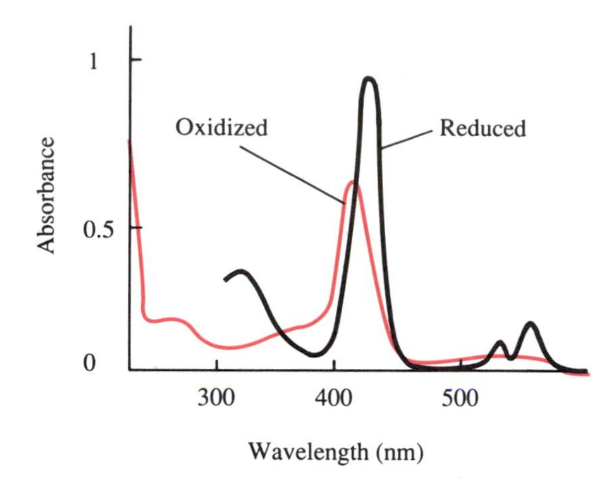

Figure 22-8

Visible Absorption Spectra of Reduced and Oxidized Cytochrome *b*.

consist of iron protoporphyrin IX (Figure 22-9) associated with one polypeptide with a molecular weight of 30,000. The *b*-type cytochromes exhibit reduction potentials E_0' of -0.03 volt and $+0.05$ volt, respectively. Complex III also contains cytochrome c_1 ($E_0' = 0.23$ volt), which, like all *c*-type cytochromes, contains heme *c* covalently linked to the protein by bonds between two cysteines of the protein and two vinyl groups of heme. Complex III also contains UQ-binding proteins and several proteins that do not interact with electron carriers.

It is not obvious how complex III or any other enzyme catalyzes the reaction of equation 7. This is because UQH_2 undergoes an overall two-electron oxidation in which the electrons are partitioned to two molecules of cytochrome *c*. Evidence indicates the existence of dual electron-transfer circuits within this complex, in which ubisemiquinone is an intermediate.

Complex IV: Cytochrome Oxidase

Complex IV is cytochrome oxidase, which catalyzes the reduction of O_2 to H_2O by reduced cytochrome *c* according to equation 8.

$$4\ \text{Cyt}\ c\ (\text{Fe}^{2+}) + \text{O}_2 + 4\,\text{H}^+ \longrightarrow 4\ \text{Cyt}\ c\ (\text{Fe}^{3+}) + 2\,\text{H}_2\text{O} \qquad (8)$$

Cytochrome oxidase consists of at least seven subunits ranging in molecular weight from 5000 to 50,000. Cytochrome oxidase contains two cytochromes with heme *a* as the prosthetic group (Figure 22-9). There are also two redox-active copper ions (Cu^+/Cu^{2+}) that are coupled to reduction of the cytochromes. Because of interactions between copper and the cytochromes, and their similar redox potentials, it is difficult to observe redox processes of the individual electron carriers. Each copper ion is paired with a cytochrome, and these pairs exhibit reduction potentials, E_0', of $+0.25$ volt and $+0.35$ volt, respectively.

The reaction catalyzed by cytochrome oxidase is complex because it requires the accumulation of four electrons from four molecules of reduced cytochrome *c* and the transfer of these four electrons, as well as four protons, to molecular oxygen to form two molecules of water (equation 8). The four

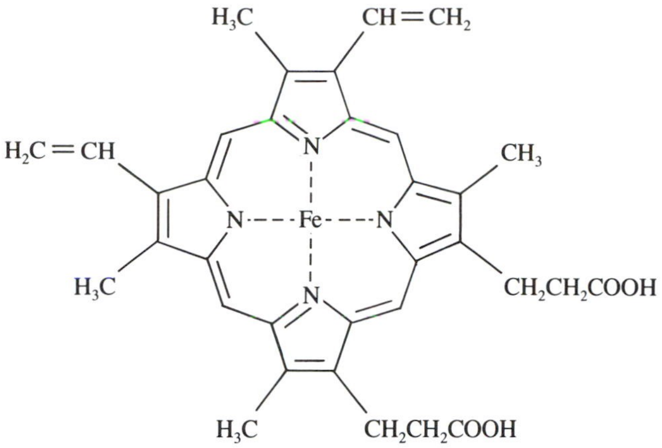

Iron protoporphyrin IX

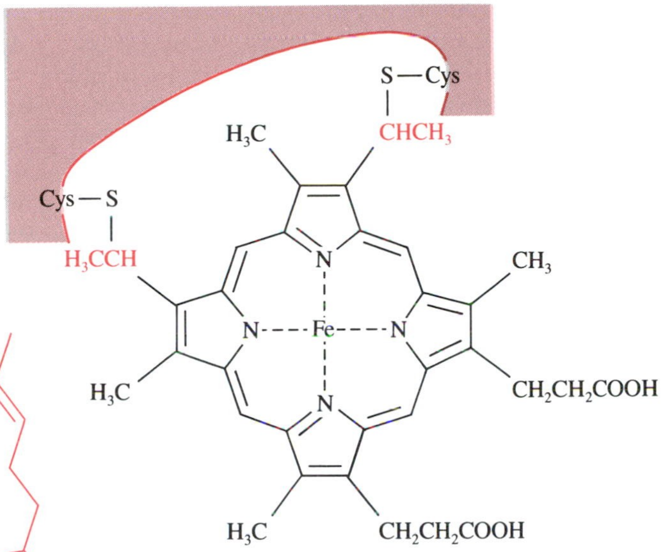

Heme *c*

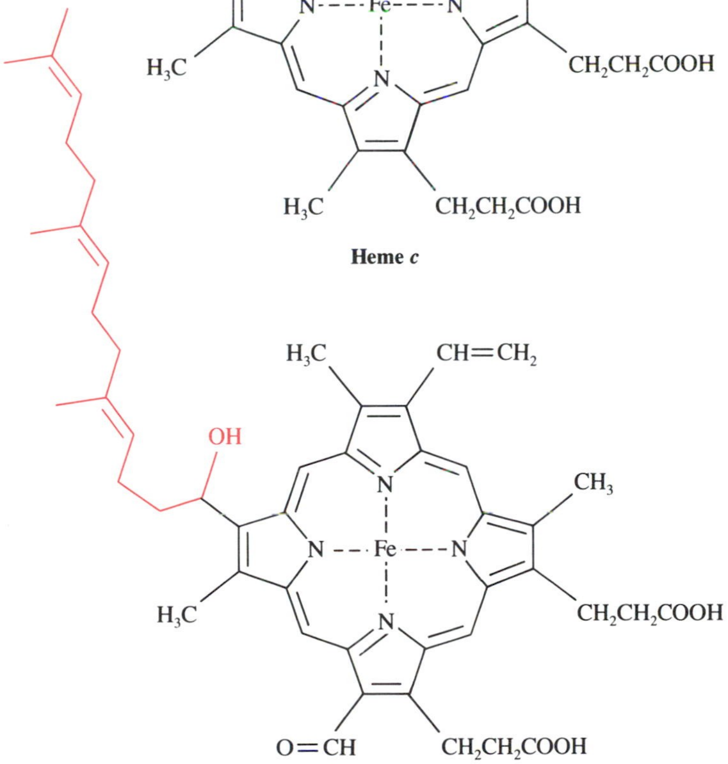

Heme *a*

Figure 22-9

Structures of Various Hemes in Cytochromes of the Electron-Transport Pathway. The structures shown are iron protoporphyrin IX, heme *c* found in cytochrome *c*, and heme *a* found in cytochrome oxidase (complex IV).

electrons reduce the two copper ions and the two hemes. The oxygen is believed to be bound between a copper ion and a heme iron, where it accepts the four electrons and adds four protons and is transformed into two water molecules. The protons are taken up from the mitochondrial matrix and contribute to the proton gradient that constitutes the driving force for oxidative phosphorylation.

Interactions among the Electron-Transfer Complexes

Two questions regarding the mechanism by which complexes I through IV carry out the oxidation of NADH or succinate by O_2 are: How do the complexes communicate with one another? In what order do they accept and relay electrons? The answer to the first question requires knowledge of the physical properties of the inner mitochondrial membrane. Biological membranes are films composed of phospholipids and other lipids that are not covalently cross-linked. The membranes have good tensile strength; their interiors are hydrophobic; and, because the molecules are not cross-linked, membranes are flexible. Under normal conditions, the interior of a membrane is liquid and allows lateral motion of membrane components, including membrane proteins. The fluidity of a membrane allows the UQ/UQH_2 system to migrate laterally within the membrane. This motion allows the UQ/UQH_2 system to transfer electrons from one complex to another. The UQ/UQH_2 system transfers electrons from complexes I and II to complex III. Cytochrome *c* is loosely associated with the membrane and relays electrons from complex III to complex IV. The properties of membranes are described in greater detail in Chapter 29.

When membranes are physically disrupted, the fragments close up into spherical structures known as *vesicles.* Membrane vesicles prepared from mitochondria contain the components of oxidative phosphorylation and can be used to study the processes of the electron-transport pathway and ATP synthesis.

The answer to the second question, regarding the sequence of electron-transfer steps, is summarized in Figure 22-4. Complex I transfers electrons from NADH to complex III, which relays them to complex IV; finally complex IV reduces O_2. Succinate is oxidized by complex II, which transfers the electrons to the main pathway by reducing complex III.

Complexes I through IV have been characterized over a long period through the efforts of many researchers, and for many years the relations and interactions among these complexes were not understood. It was known through the work of Warburg and Keilin that the electron-transport pathway contains cytochromes and flavins and that the system is complex and membrane bound. The cytochromes were distinguished by spectral analysis of crude membrane preparations by using inhibitors to block the electron-transport pathway at different points and then reducing the system with different substrates. Using this approach, Keilin and Chance were able to reduce the cytochromes selectively and obtain the first information about the order in which they are reduced in the passage of electrons through the pathway. These experiments worked on the following principle: Certain inhibitors of the electron-transport pathway were found to block electron transfer to particular electron acceptors. To determine the sequence in which the acceptors could be reduced by a particular reductant, all of the components were first reduced by a substrate such as NADH or succinate, in the absence of O_2. The complete visible absorption spectrum was recorded. An inhibitor was then added, and O_2 was admitted to oxidize all of the electron carriers beyond the site of inhibition, leaving those preceding the blockage site reduced. The spectral differences resulting from the use of different inhibitors allowed the chromophoric carriers to be assigned relative positions in the electron-transport pathway. Several of the inhibitors in Figure 22-10 were among those used.

These *crossover* experiments worked as illustrated in Figure 22-11. With the system fully reduced and anaerobic, the 100% reduced level was spectrally recorded for each carrier. The addition of an inhibitor such as antimycin A, followed by O_2, allowed cytochromes c, c_1, a, and a_3 to become oxidized, whereas NADH and $FMNH_2$ of NADH dehydrogenase remained reduced. With cyanide, all the electron acceptors remained reduced. These and related

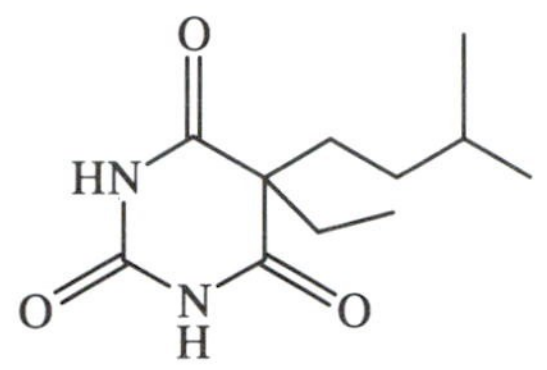

Amytal

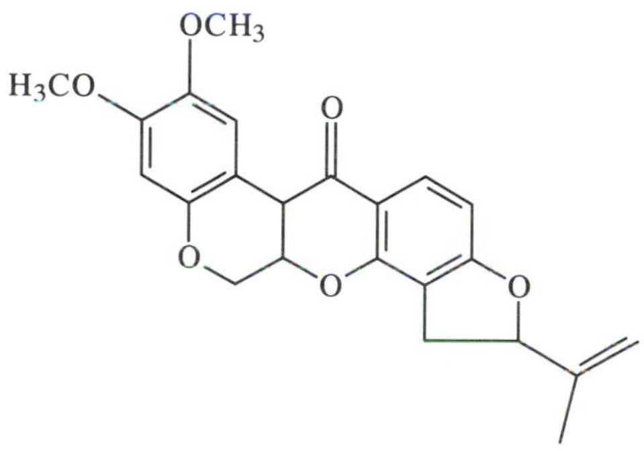

Rotenone

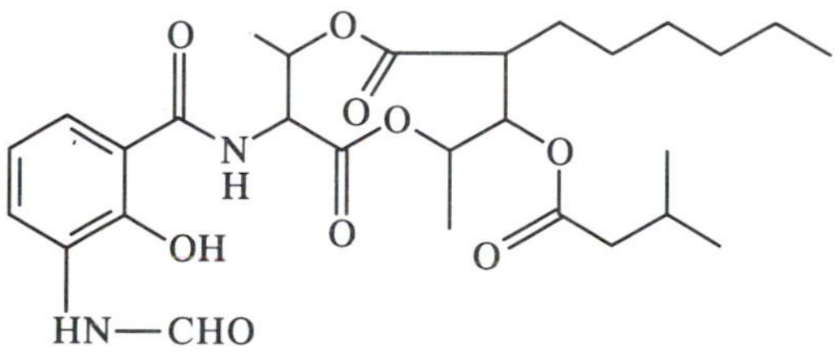

Antimycin

Figure 22-10

Inhibitors of the Electron-Transport Pathway. Shown are the structures of amytal, rotenone, and antimycin, all of which are inhibitors of the electron-transport pathway. Each inhibitor blocks the pathway at a different point. Amytal inhibits NADH dehydrogenase (complex I) and antimycin inhibits the cytochrome bc_1 complex (complex III).

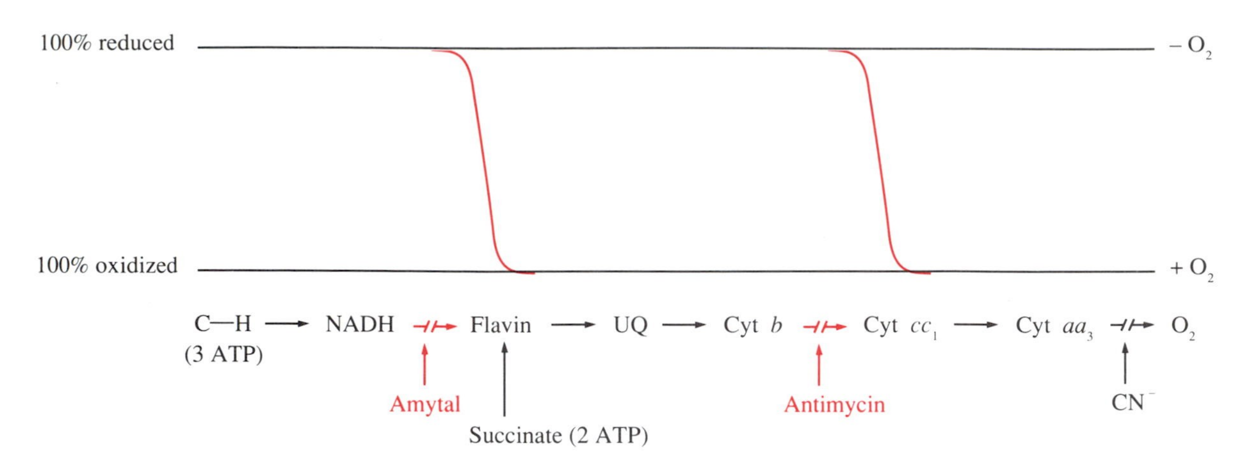

Figure 22-11

Typical Data Showing Inhibition of the Electron-Transport Pathway by Specific Inhibitors. In these experiments, all of the electron carriers in the electron-transport pathway were first reduced under anaerobic conditions by a substrate such as NADH. A specific inhibitor of electron transport was then added, and O_2 was subsequently admitted. The absorption spectrum of each electron carrier was recorded before and after admission of O_2. Shown are the transitions from fully reduced to fully oxidized for two electron carriers just following the site of inhibition by two inhibitors. When the system is inhibited by amytal, all of the initially reduced electron carriers, including the flavins, are reoxidized by O_2. This means that the flavins are at or near the beginning of the electron-transport pathway and that amytal inhibits a system at the beginning of the pathway. The flavins are now known to be redox components of NADH dehydrogenase and succinate dehydrogenase. When the inhibitor is antimycin, the *c*-type cytochromes are reoxidized by O_2, but the *b*-type cytochromes and flavins are not reoxidized. This means that the *c*-type cytochromes follow the *b*-type cytochromes and flavins in the pathway, and that antimycin inhibits the electron transfer from cytochrome *b* to cytochrome *c*.

experiments with other inhibitors and reducing substrates allowed the electron-transfer pathway to be outlined.

This electron-transfer pathway was directly proved by the isolation and characterization of complexes I through IV. Each complex was shown to be a segment of the pathway and to transfer electrons to the next complex. In addition, the reduction potentials of the isolated complexes could be measured. Inasmuch as electron transfer can proceed only from a reduced carrier to an oxidized carrier with a more-positive reduction potential, the order in which the complexes interact has to correspond to the reduction potentials. The measured potentials were found to be consistent with their assigned positions in the overall pathway, as illustrated in Figure 22-12. Two very important additional dividends of the isolation and characterization of the complexes were the discoveries of the essential carriers UQ and the iron-sulfur proteins.

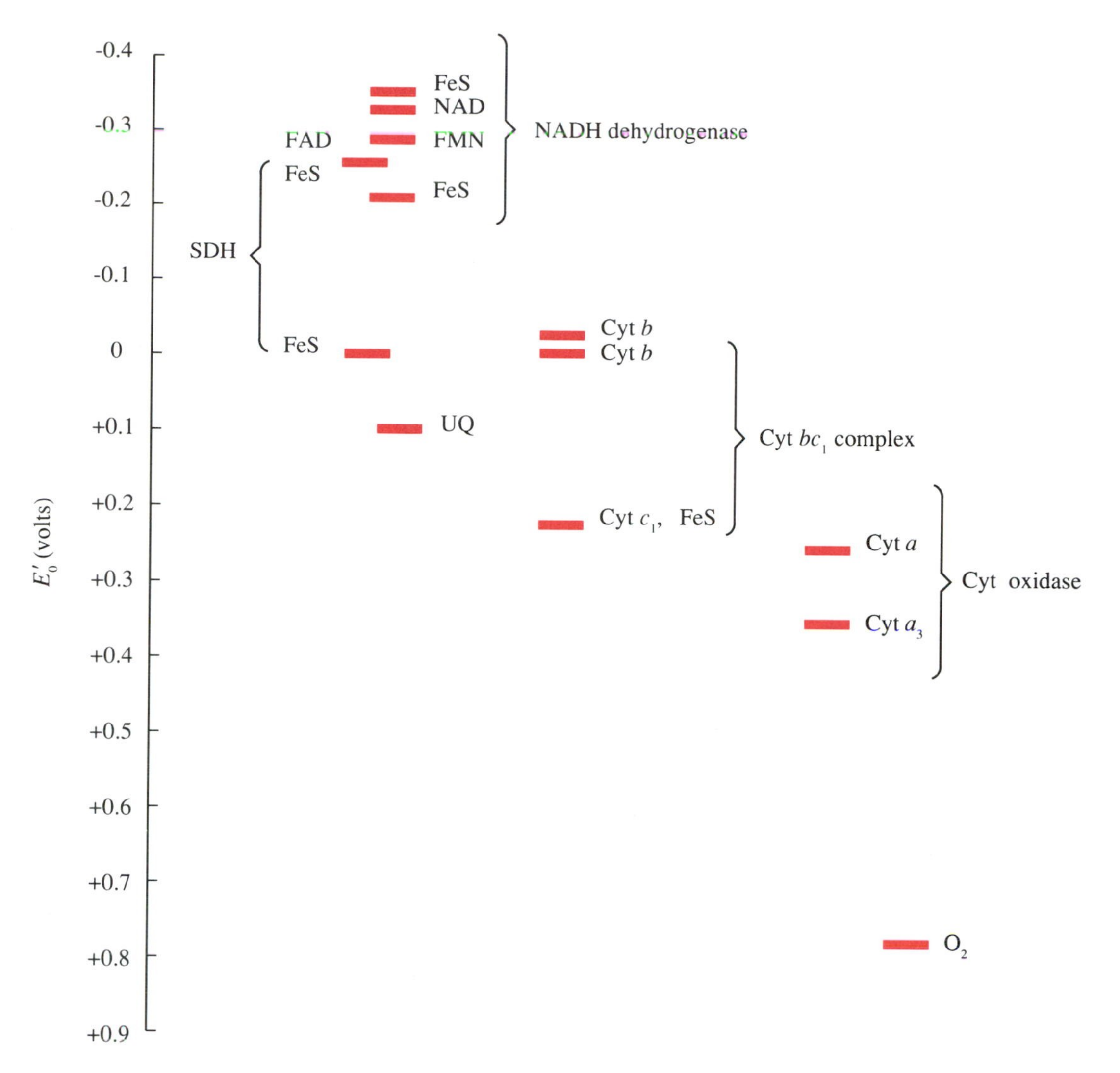

Figure 22-12

Reduction Potentials for the Redox Cofactors in the Electron-Transport Pathway. The redox cofactors in NADH dehydrogenase (complex I) have the least-positive reduction potentials and the cofactors in cytochrome oxidase (complex IV) have the most-positive potentials. This is consistent with their positions in the electron-transport pathway. The reduction potentials of the cofactors of succinate dehydrogenase (SDH, complex II) and the cytochrome bc_1 complex (complex III) are at intermediate values, consistent with their positions in the electron-transport pathway.

The Synthesis of ATP

Intact mitochondria produce ATP from ADP and P_i in a reaction that is strictly coupled to the electron-transport pathway; that is, ATP formation and electron transport are linked in such a way that the occurrence of one depends on the occurrence of the other. The mechanism by which these processes are coupled was a controversial subject for many years.

Coupling of Electron Transport and ATP Formation

The coupling between electron transport and the formation of ATP can be observed in simple experiments, such as those illustrated in Figure 22-13. Oxygen consumption is conveniently monitored by use of an O_2 electrode, and the rate of O_2 uptake corresponds to the rate of electron transport through the mitochondrial pathway. This rate depends on ATP synthesis, as shown by the experiment in Figure 22-13A, in which mitochondria consume O_2 at only a low rate until ADP is added and ATP formation commences, whereupon the rate of O_2 uptake increases. This experiment demonstrates that electron transfer to O_2 and the formation of ATP are coupled; as long as they are coupled, neither process can occur unless the other also occurs.

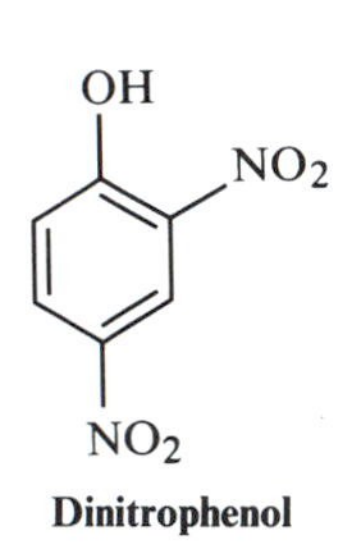

Dinitrophenol

The second experiment, shown in Figure 22-13B, also depends on measuring the respiration rate. Respiring and phosphorylating mitochondria display a particular rate of O_2 uptake. Certain molecules known as uncouplers have the property of preventing ATP formation without blocking respiration. Dinitrophenol is such a molecule. The addition of dinitrophenol to respiring and phosphorylating mitochondria *increases* the respiration rate but stops ATP synthesis. The fact that O_2 uptake is increased means that the nature of the coupling between ATP synthesis and electron transport is such that the rate of ATP formation limits the rate of electron transport. Although dinitrophenol prevents ATP formation, it does not inhibit the ATPase activity of ATP synthase. The coupling mechanism must explain this fact as well as the mechanism by which uncouplers act to dissociate electron transport from ATP formation. (The ability of dinitrophenol to prevent ATP formation, while allowing respiration, at one time made it a candidate as a weight-loss drug, because it would cause nutrient energy to be dissipated. In tests, dinitrophenol did cause weight loss. Unfortunately, it also caused blindness, which eliminated it from further consideration as a weight-loss drug.)

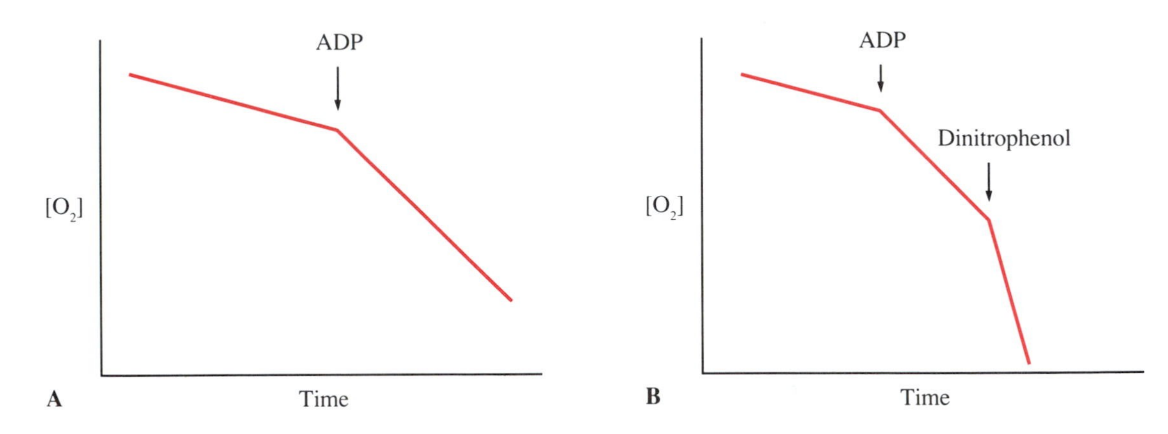

Figure 22-13

Experimental Demonstrations of Coupling Between Electron Transport and ATP Formation. In part A, the initial rate of O_2 uptake is recorded for respiring mitochondria in a phosphate buffer that contains no ADP. At the indicated time of ADP addition, the rate of O_2 uptake increases dramatically. This is because the rate of ATP formation is initially limited by the availability of ADP, and the rate of ATP formation in turn limits the rate of electron transport through the pathway, which in turn limits the respiration rate. In part B, the respiration rate at saturating ADP is increased by uncouplers of oxidative phosphorylation, such as dinitrophenol, which prevent ATP formation without blocking electron transfer or respiration. Parts A and B together show that ATP formation limits the respiration rate.

Certain molecules, including dicyclohexylcarbodiimide and oligomycin (Figure 22-14), inhibit both ATP formation and electron transport. Oligomycin inhibits ATP synthase, and dicyclohexylcarbodiimide irreversibly inactivates the enzyme. When ATP synthase is inactivated, electron transport also is blocked; but it can be restored by the addition of an uncoupling agent such as dinitrophenol. This is illustrated in Figure 22-15. The coupling mechanism must explain how uncoupling agents such as dinitrophenol work.

The coupling between ATP formation and electron transport to O_2 can be described by the ratios of ATP formed per O_2 consumed, the P/O or P/2 e^- ratios. These ratios have been measured by several techniques, one of which is illustrated in Figure 22-16. Mitochondria that are deficient in ADP are incubated with an electron donor such as pyruvate or succinate in the presence of an excess of P_i but no ADP. The rate of oxygen uptake is recorded, to obtain a basal rate, and a small amount of ADP is then added. This causes a brief burst of O_2 consumption that can be accurately measured. The ratio of the amount of ADP added to the amount of O_2 consumed in the burst is used to calculate the P/O value. The P/O ratio turns out to be about 3 for NADH, and for all electron donors that generate NADH within mitochondria, and about 2 for succinate and other donors that generate an $E \cdot FADH_2$ within the mitochrondrial membrane. The coupling mechanism must explain these ratios.

A hint of the nature of this mechanism is afforded by the properties of the Ca^{2+} transport system in mitochondria. Mitochondria have an active transport mechanism for pumping Ca^{2+} ions from the outside into the matrix. ATP is an energy source for driving the accumulation of calcium inside the mitochondrion. Calcium transport is also driven by oxidation of C–H bonds in the same molecules that support the generation of ATP. Calcium transport driven by ATP is blocked by oligomycin, which inactivates the ATP synthase. However, Ca^{2+} transport driven by C–H oxidation through the electron-transport pathway is not blocked by oligomycin. This means that Ca^{2+} transport that is driven by the oxidation of C–H bonds by the electron-transfer pathway *does not require the formation of ATP*. Therefore, both electron transport and ATP

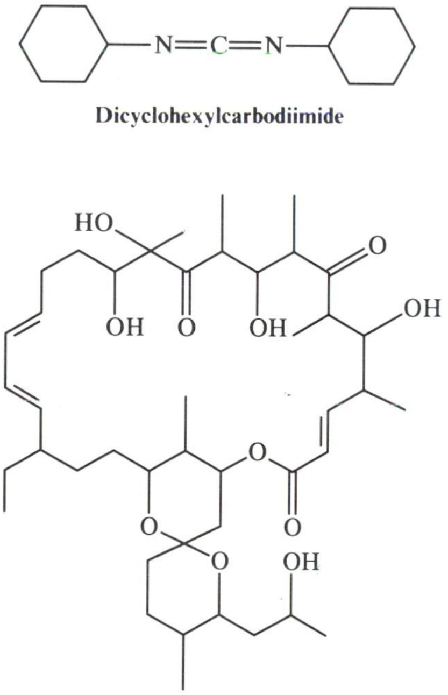

Figure 22-14

Structures of Inhibitors of ATP Synthesis. Dicyclohexylcarbodiimide and oligomycin inhibit ATP synthase. Dicyclohexylcarbodiimide inactivates the enzyme by reacting with a specific carboxyl group in F_1ATPase.

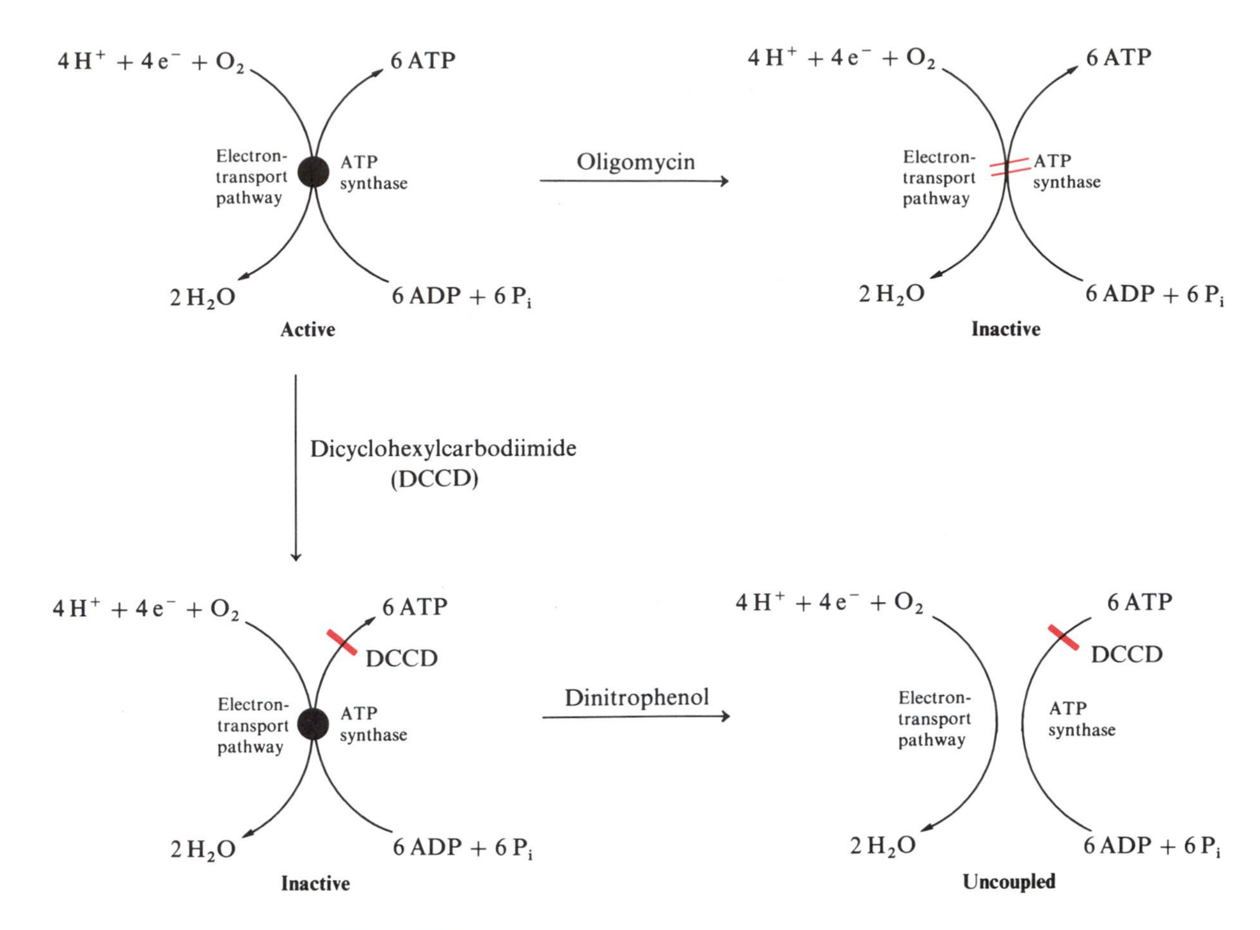

Figure 22-15

The Actions of Inhibitors and Uncouplers on Electron Transport and Oxidative Phosphorylation. In coupled electron transport and ATP synthesis, which is illustrated at the upper left, the actions of ATP synthase and the electron-transport pathway are coupled—that is, interdependent. When the action of ATP synthase is inhibited by oligomycin (upper right) or when the enzyme is irreversibly inactivated by dicyclohexylcarbodiimide (lower left), the electron-transport pathway is inhibited. Uncouplers such as dinitrophenol unlink the electron-transport pathway from the action of ATP synthase and allow it to proceed independently, as illustrated at the lower right.

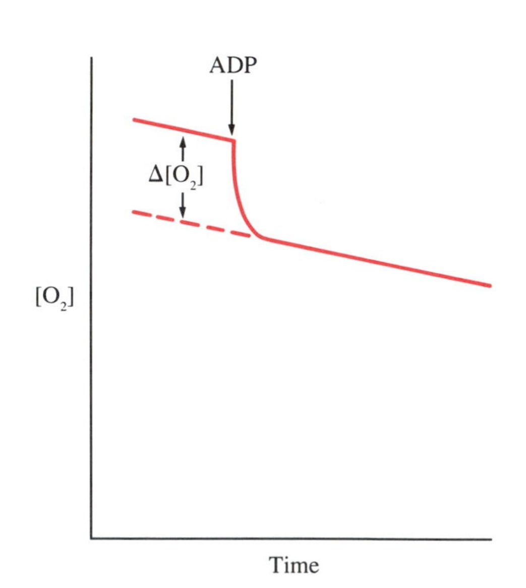

Figure 22-16

Measurement of the P/O Ratio for Mitochondria. The initial O_2-uptake rate for respiring mitochondria is recorded in the absence of added ADP, as described in Figure 22-13. A small, carefully measured amount of ADP is added. This causes a burst of O_2 uptake owing to the increased O_2 consumption accompanying a burst of ATP formation. The ratio of ADP added to O from O_2 consumed in the burst is the P/O.

hydrolysis must produce something, perhaps the same thing, that directly provides the energy for pumping Ca^{2+}. It is now known that both ATP hydrolysis and electron transport by mitochondria generate a pH differential and electrochemical potential difference between the inside and outside of the inner mitochondrial membrane.

Energy Coupling by a Proton Motive Force

The coupling mechanism is the means by which energy is harvested from electron transport and used to provide energy for ATP formation. P. Mitchell discovered in the early 1960s that mitochrondria contain a hydrogen ion pump that is driven by electron transport. He advanced the hypothesis that energy is released in the form of the work required to pump hydrogen ions from the matrix to the intermembrane space of mitochondria. The primary role of the oxidation reaction of oxidative phosphorylation is to build up an electrochemical gradient across the inner membrane of the mitochondrion. This is brought about by the uptake of protons from the matrix and their release into the intermembrane space by the oxidative reactions of the electron-transport complexes. The resulting pH differential and electrochemical potential is known as the *proton motive force,* and it provides the energy for the synthesis of ATP by ATP synthase. This is the chemiosmotic theory of energy coupling. The same proton motive force is used to support calcium transport and the transport of other metal ions, such as magnesium, and some metabolites.

Uncouplers of oxidative phosphorylation dissipate the energy of the ion gradients by making the mitochondria permeable to hydrogen ions, thereby discharging the electrochemical potential and pH differential. All uncouplers are, like dinitrophenol, lipophilic molecules that are soluble in the membrane and contain a weakly acidic functional group. Such molecules serve as transporters of hydrogen ions across the membrane, as illustrated in Figure 22-17 for dinitrophenol, thereby removing the pH gradient and the proton motive force.

The electrochemical potential also controls the rate of electron transport because most of the reactions are reversible; a large proton gradient can stop or even reverse the oxidative reactions. The uncouplers eliminate these restraints on electron transport by discharging the pH differential and electrochemical potential, allowing an increase in rate. The fact that all uncouplers act with

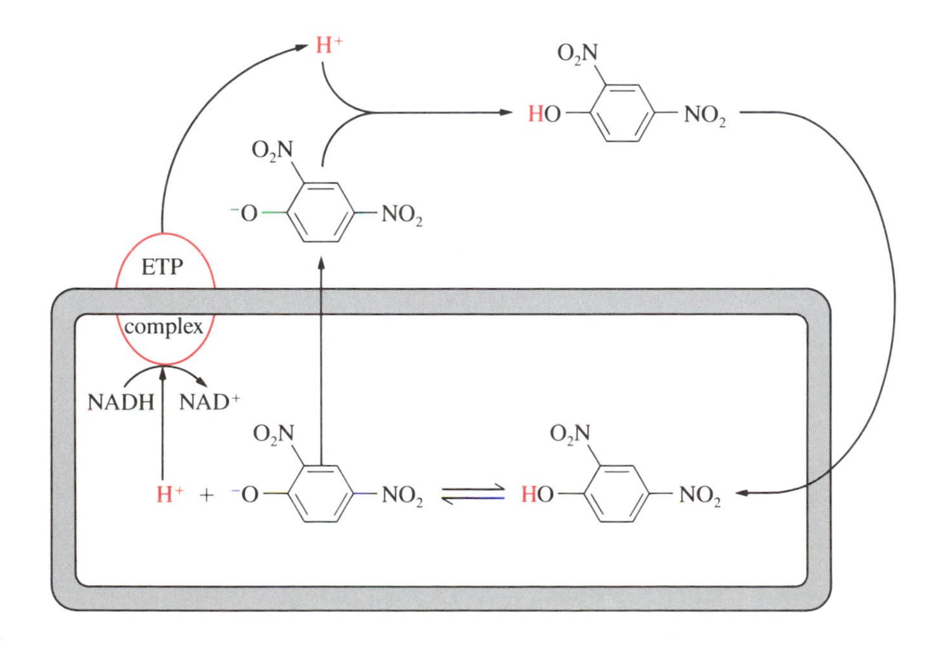

Figure 22-17

Dissipation by Dinitrophenol of the pH Gradient Created by an Electron-Transport Complex in the Inner Mitochondrial Membrane. Protons translocated across the inner mitochondrial membrane by the action of electron-transport complexes are returned to the mitochondrial matrix by dinitrophenol. The membrane is permeable to both dinitrophenol and the dinitrophenolate ion. This diffusibility allows dinitrophenol and dinitrophenolate to shuttle protons from outside the mitochondrion into the matrix, thus dissipating the pH gradient created by electron transport.

equal effectiveness on all sites of ATP formation is particularly nicely explained by the Mitchell hypothesis. No uncoupler acts preferentially on one site of coupling because all of the sites respond to the same proton motive force. The Mitchell hypothesis is supported by many experiments that directly demonstrate that complexes I, III, and IV pump hydrogen ions across the membranes.

The electrochemical gradient, or proton motive force, has two components: the gradient of pH across the membrane and the membrane potential (equation 9).

$$Y = \text{"proton motive force" (mv)} = \text{membrane potential} + 60\,\Delta\text{pH} \qquad (9)$$

The charge difference on the two sides of the membrane can be measured by the use of lipophilic cations that diffuse across the membrane. From the partitioning of such ions on the two sides at equilibrium, it is possible to show that the membrane potential for respiring, energy-coupled mitochondria is about -150 mV.

The Mitchell hypothesis is also consistent with the P/O ratios for NADH and succinate and the linkage in the electron-transport pathway outlined in Figure 22-18. Electrons from NADH (P/O = 3) pass through complexes I, III,

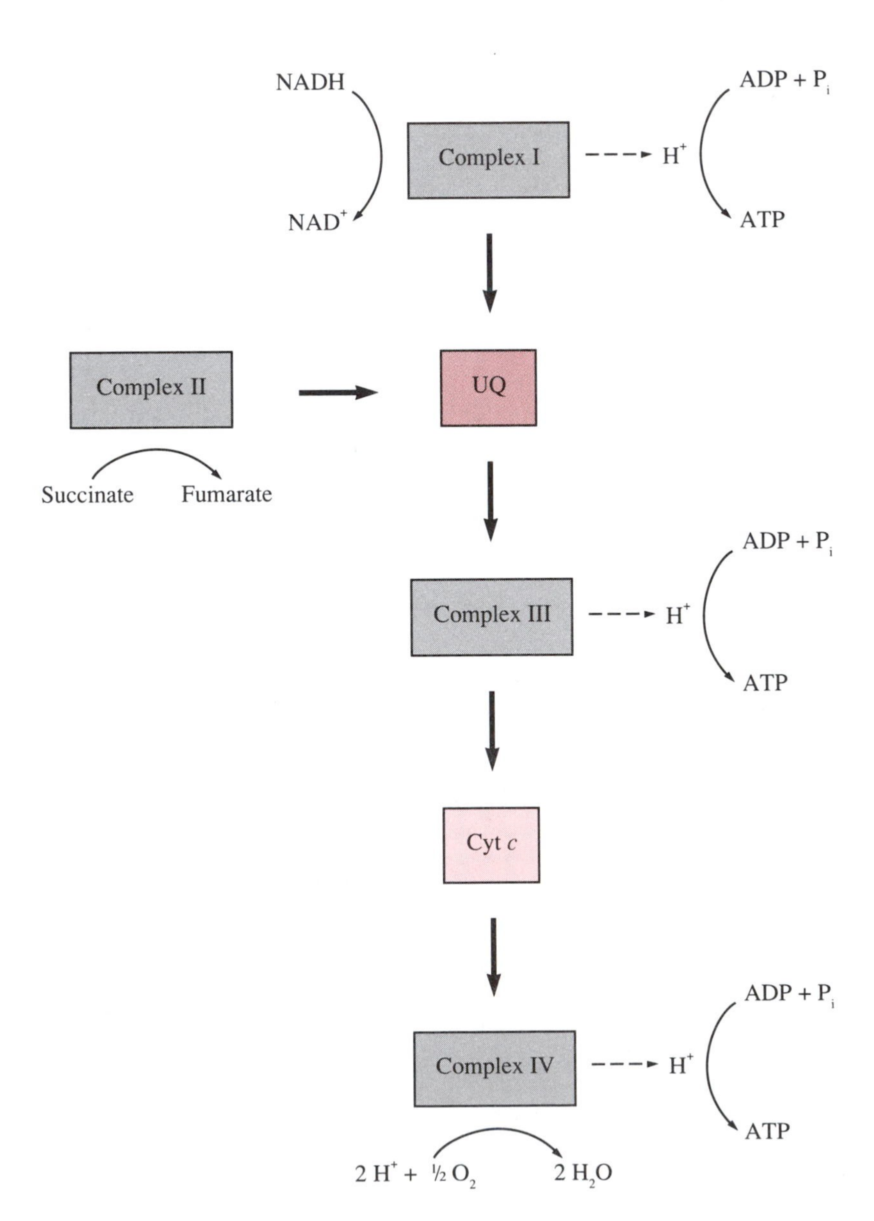

Figure 22-18

Sites at Which Electron Transfer Through the Electron-Transport Pathway Is Coupled with ATP Formation. Complexes I, III, and IV pump hydrogen ions across the inner mitochondrial membrane. These proton gradients provide the energy for ATP formation. Complex II does not pump hydrogen ions and so does not support ATP formation. Ubiquinone relays electrons from complexes I and II to complex III of the electron-transport pathway, and cytochrome *c* relays electrons from complex III to complex IV. These relations lead to the production of three molecules of ATP per NADH oxidized and two molecules of ATP per succinate oxidized.

and IV, each of which produces enough hydrogen ions to provide the energy for the formation of one molecule of ATP per electron pair. Complex II, succinate dehydrogenase, generates UQH_2, as does complex I, but it does not pump hydrogen ions. The UQH_2 generated by complex II transfers electrons to O_2 through complexes III and IV. Complex I is bypassed, so that the P/O ratio is 2 for succinate rather than 3.

The inhibitors of ATP synthase—dicyclohexylcarbodiimide and oligomycin—block ATP formation without interfering with the generation of electrochemical potentials and pH differentials. With ATP synthase inactivated, the proton gradient generated by electron transport cannot be used for ATP formation. A large gradient is built up and reaches a limit. The addition of an uncoupler to the inhibited system dissipates the proton gradient and allows electron transport to proceed.

F_1-F_0 ATPase Is the ATP Synthase

Racker purified a coupling factor from mitochondrial membranes that is required for ATP synthesis by oxidative phosphorylation and has ATP synthase activity. The purified ATPase, which is known as F_1, catalyzes the hydrolysis of ATP. In mitochondria, this ATPase is bound to a protein, F_0, that is embedded in the mitochondrial membrane. The F_0 fragment can act as a channel to allow protons to cross the membrane, but it cannot make ATP. When these two subunits are brought together, they form the F_1-F_0 ATPase of mitochondria and are able to synthesize ATP when protons cross the membrane. The F_1 ATPase is a complex of five proteins, α, β, γ, δ, and ε, with the composition $\alpha_3\beta_3\gamma\delta\varepsilon$ and an overall molecular weight of about 360,000. The F_1-F_0 ATPase resembles a mushroom, with the F_1 subunit protruding into the mitochondrial matrix and the F_0 stalk buried in the membrane. Figure 22-19 is an electron micrograph showing F_1 subunits of ATP synthase protruding from the mitochondrial inner membrane.

The F_1 ATPase has the remarkable property of catalyzing the reaction of equation 10.

$$ADP^{3-} + H_2PO_4^{\ -} + 3\,H^+_{\ out} \rightleftharpoons ATP^{4-} + H_2O + 3\,H^+_{\ in} \qquad (10)$$

In the forward direction, it synthesizes ATP from ADP and phosphate when three protons enter the mitochondrial matrix; and, in the reverse direction, it pumps three protons out of the mitochondrial matrix when one molecule of ATP is hydrolyzed. Thus, the enzyme catalyzes a *coupled vectorial process.* It catalyzes the movement of protons from one side of the membrane to the other side, a vectorial process, at the same time that it catalyzes a chemical process, the hydrolysis or synthesis of ATP. When a proton gradient is built up by oxidation reactions, the return of protons into the mitochondrial matrix results in ATP synthesis, as described in equation 10 and Figure 22-20. The proton gradient provides the energy that is needed for ATP synthesis, which is otherwise very unfavorable.

If there is no proton gradient, there is no driving force for ATP synthesis and the enzyme will operate in the reverse direction, to hydrolyze ATP and pump protons out of the mitochondrial matrix. If the proton gradient is removed by the presence of an uncoupler, such as dinitrophenol, ATP will be hydrolyzed, but no proton gradient will be built up.

The mechanism of ATP synthesis in oxidative phosphorylation by the F_1-F_0 ATPase is not known, but several steps of the reaction are understood.

Surprisingly, the hydrolysis of ATP at the active site of the enzyme has been shown to be reversible. Boyer showed that the F_1 ATPase will catalyze

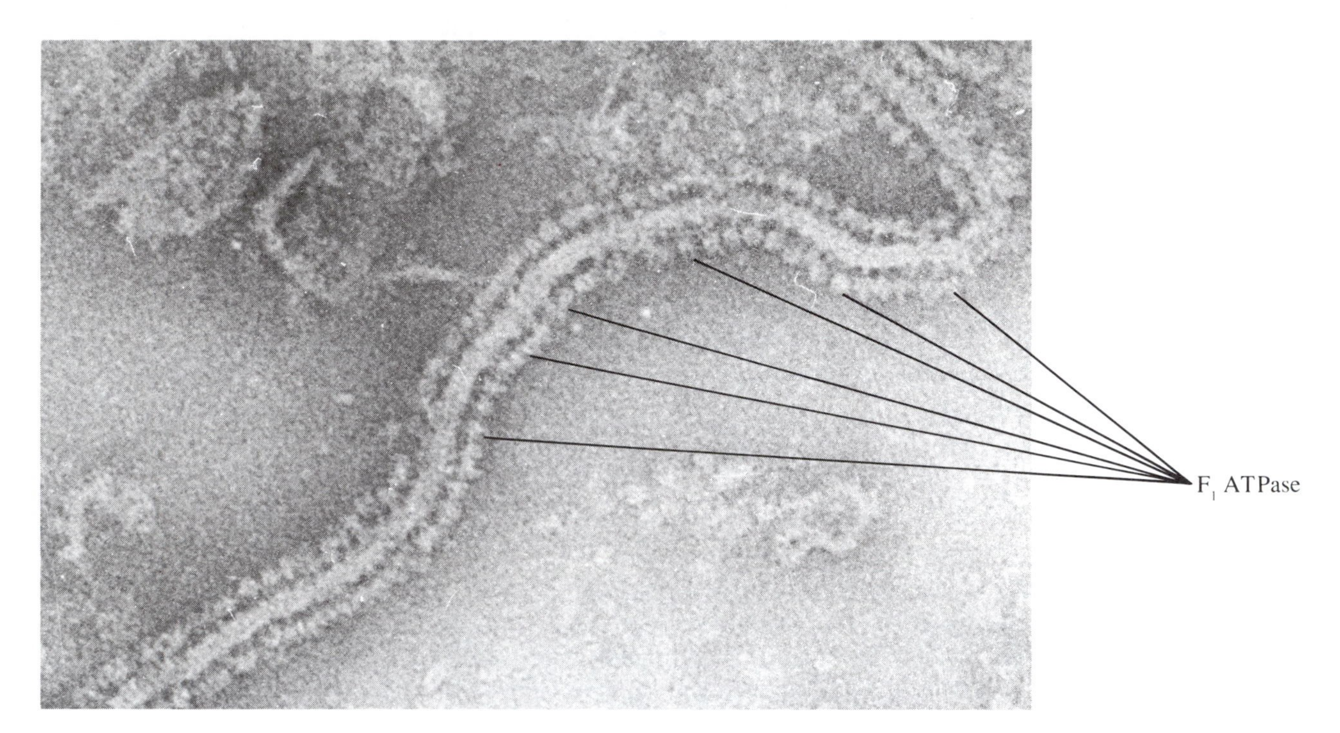

Figure 22-19

Electron Micrograph of a Mitochondrial Inner Membrane Showing the ATP Synthase. Shown in cross section is a part of the inner membrane of a mitochondrion that has been osmotically swelled so as to force two segments of the membrane together. The spherical structures are the F_1ATPase part of ATP synthase. F_1ATPase is anchored to the membrane by association with the F_0 part, which is embedded in the membrane. (Micrograph courtesy of Enrique Valdivia.)

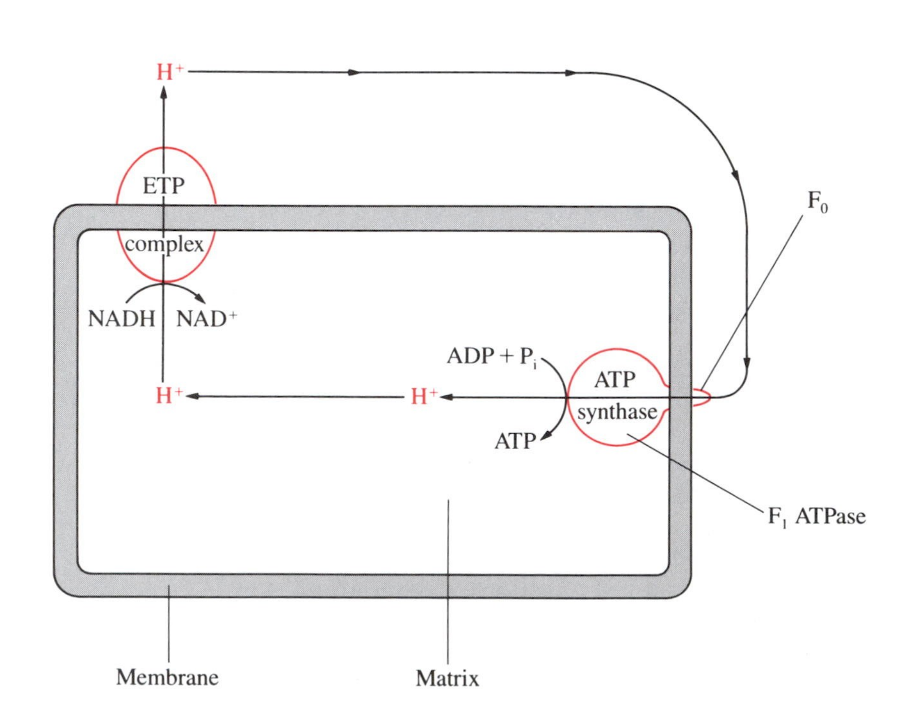

Figure 22-20

Proton Translocation in Oxidative Phosphorylation. Complexes I, III, and IV of the electron-transport pathway pump hydrogen ions from the matrix side of the inner mitochondrial membrane across the membrane in the course of electron transfer. The translocated hydrogen ions can re-enter the matrix by passing through the proton translocation channels of ATP synthase. The energy released is used by ATP synthase to produce ATP from ADP and P_i.

exchange of ^{32}P-inorganic phosphate into ATP at the same time that it hydrolyzes ATP (equation 11).

$$ATP + {}^{32}P_i \rightleftharpoons AT^{32}P + P_i \quad (11)$$

Reversal of the hydrolysis at the active site is also shown by the incorporation of ^{18}O from $H_2{}^{18}O$. One atom of ^{18}O must be incorporated upon hydrolysis; however, in the absence of the proton motive force, and at low ATP concentrations, the number of reaction reversals at the active site of ATP synthase for each P_i released increases markedly and as many as four water oxygens may appear in the P_i released (equation 12).

$$ADP{-}P(=O)(O^-){-}O^- + H_2{}^{18}O \xrightleftharpoons{F_1ATPase} ADP + H^{18}O{-}P(={}^{18}O)({}^{18}O^-){-}{}^{18}O^- \quad (12)$$

Penefsky has shown directly that the hydrolysis of ATP at the active site of the enzyme is reversible. When small amounts of ATP are added to a large amount of enzyme—so that all the ATP, ADP, and P_i is bound at the active site—it is found that the hydrolysis reaction at the active site proceeds only about 50% toward completion. Thus, the equilibrium constant for ATP hydrolysis at the active site is close to 1.0. This is very different from the equilibrium constant for the hydrolysis of MgATP to MgADP + P_i in solution, which is approximately 10^6 M.

The system can be described by the thermodynamic box in Figure 22-21. The large difference in the equilibrium constant for ATP hydrolysis in solution and at the active site of the enzyme means that K_1 in Figure 22-21 is 10^6 times as large as K_4. This requires that, on the other two sides of this thermodynamic box, K_3 must be 10^6 times as large as K_2; that is, the enzyme binds and stabilizes ATP very strongly compared with ADP and P_i. This shifts the equilibrium constant for ATP synthesis at the active site, compared with its value in solution, as illustrated in Figure 22-22. The ADP and phosphate at the active site are brought together in the correct position to react, with a large loss of entropy, so that ATP formation is far more probable than it is in solution. There may also be destabilization of the bound ADP and P_i, ΔG_D, that is relieved when ATP is synthesized.

Although the enzyme solves the problem of synthesizing ATP by strongly binding ATP, this raises another problem. As pointed out by Boyer, there must be a *binding change,* so that free ATP can be released from the enzyme. There must also be a way to prevent the reverse reaction, the hydrolysis of ATP, unless it is coupled to the pumping of three protons according to equation 10.

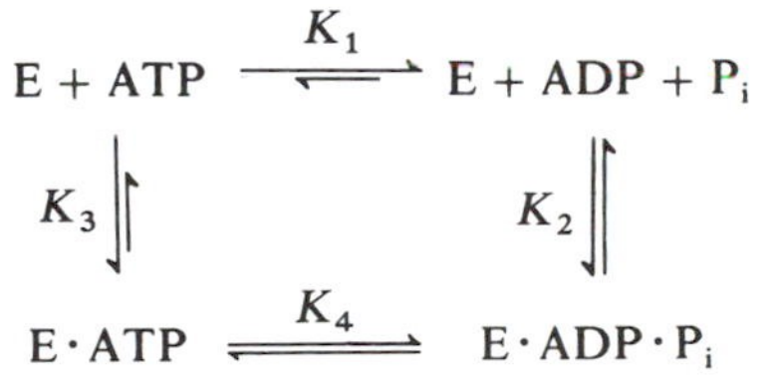

$$\begin{array}{ccc} E + ATP & \xrightleftharpoons{K_1} & E + ADP + P_i \\ K_3 \updownarrow & & \updownarrow K_2 \\ E \cdot ATP & \xrightleftharpoons{K_4} & E \cdot ADP \cdot P_i \end{array}$$

Figure 22-21

A Thermodynamic Box Describing the Binding and Hydrolysis of ATP by ATP Synthase. The equilibrium constant K_4 for the conversion of $E \cdot ATP$ to $E \cdot ADP \cdot P_i$ is about 1.0. Because the equilibrium constant for the hydrolysis of ATP to ADP + P_i (K_1) is large (about 10^6), the enzyme must bind ATP in such a way as to stabilize it relative to ADP and P_i. This stabilization leads to a large value for K_3; that is, to tight binding of ATP.

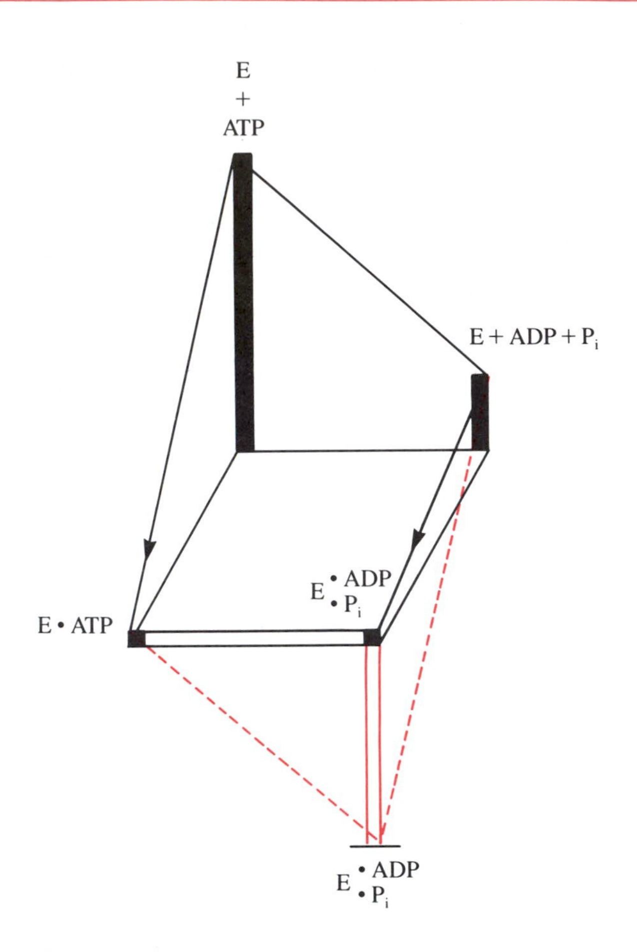

Figure 22-22

Free Energy Level Diagram for a Two-State Model of Conformational Coupling in ATP Formation. This energy diagram describes the thermodynamic box of Figure 22-21. It shows a plane of zero energy and bars that represent an increase in energy above zero or a decrease below zero. The arrows show that ATP binds much more strongly to the enzyme than do ADP and P_i. The back of the diagram shows the negative $\Delta G^{\circ\prime}$ for the hydrolysis of ATP to ADP and phosphate and the solid line in front shows the $\Delta G^{\circ\prime}$ close to 0 for the hydrolysis of ATP to ADP and P_i on the enzyme. The dashed lines show the $\Delta G^{\circ\prime}$ that would be observed if $\Delta G^{\circ\prime}$ for hydrolysis were the same at the active site as in solution. The interaction energy is the difference between the energy of $E \cdot ADP \cdot P_i$ (solid lines) and its energy if there were no interaction energy (dashed lines).

It is not known how these problems are solved, but the solution probably requires a mechanism that separates the chemical and binding-dissociation steps for ATP, ADP, and phosphate from the transport steps, as indicated in Figure 22-23. According to this model, the overall chemical reaction, which is represented by the red arrows, is separated into the binding-dissociation steps (top) and the chemical step (bottom). The transport process, which is shown by the black arrows, is separated into movement of the proton carrier between the two sides of the membrane with three bound protons for $E \cdot ADP \cdot P_i$ (left) and with no protons for $E \cdot ATP$ (right). When this cycle proceeds in a counter-

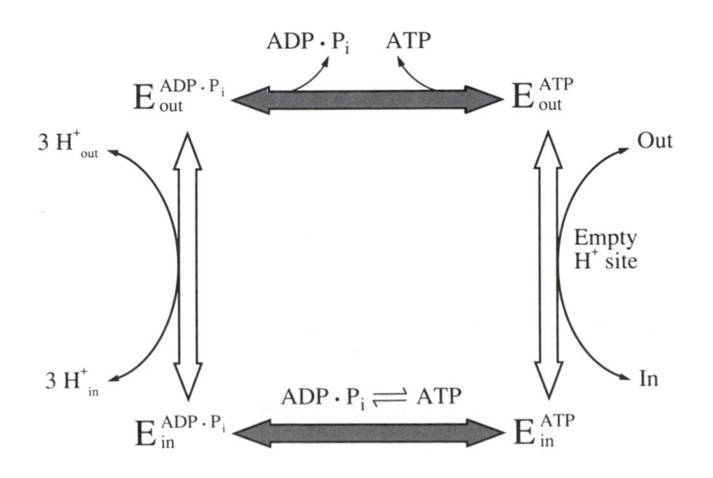

Figure 22-23

The Function of ATP Synthase. The function of ATP synthase may require the separation of the chemical synthesis of ATP from the binding and dissociation steps. The chemical reaction is represented by thick red arrows; it is separated into the binding and dissociation step at the top and the chemical transformation at the bottom. The proton-transport processes are represented by the thick black arrows. The system produces ATP when it operates in the counterclockwise direction. When the proton translocation site is empty, ATP is released from ATP synthase and ADP and P_i are bound (top). The movement of three protons through the translocation site is required to complete the cycle.

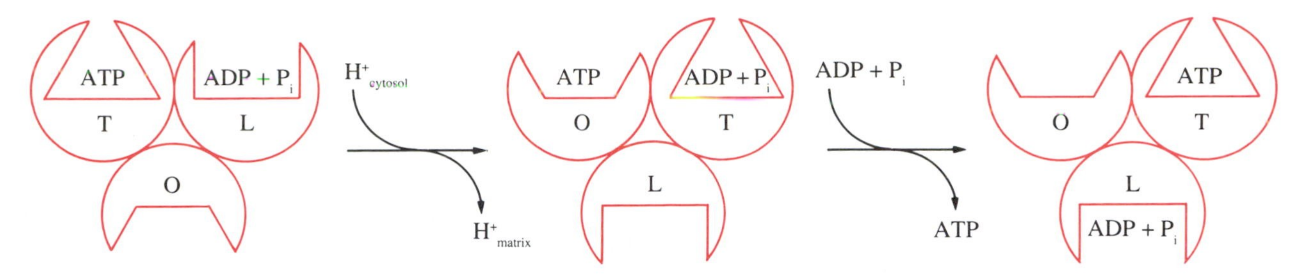

Figure 22-24

The Three-State Model for Conformational Coupling of ATP Formation to Translocation of Hydrogen Ions.

clockwise direction, it will synthesize one ATP for every three protons that enter the matrix. In the reverse direction, it will pump three protons out of the matrix for each ATP that undergoes hydrolysis.

The enzyme contains three β subunits with nucleotide-binding sites that interact strongly with one another. Binding of a nucleotide to one subunit accelerates the dissociation of nucleotides from an adjacent subunit. The role of these interactions is not yet clear, but a model proposed by Boyer is shown in Figure 22-24. According to this model, the three catalytic β subunits exist as T (tight), L (loose), and O (open) states. The T state binds nucleotides tightly and is catalytically active. The O and L states are inactive, but the O state allows binding and dissociation of nucleotides. Proton translocation causes a unidirectional switching of states among the three subunits, as shown in Figure 22-24.

Transport of Metabolites by Mitochondria

Several metabolites from metabolic pathways are substrates for the major oxidative pathways for energy production in mitochondria. The most-obvious examples are the pyruvate and NADH produced by glycolysis. Other examples are fumarate, succinate, and glutamate, all of which are generated in amino acid degradation or the urea cycle. To be used as substrates for the TCA cycle and the production of ATP, these molecules must be transported into the mitochondrion. The outer membrane is permeable to small molecules, but the otherwise impermeable inner membrane contains specific transport systems that exhibit high selectivity for these molecules. The carriers are membrane proteins that bind the molecules selectively and facilitate their diffusion across the membrane. The mechanisms by which these transport systems act are not understood.

Pyruvate and Intermediates of the TCA Cycle

The pyruvate transporter promotes the transfer of pyruvate from the intermembrane space into the matrix in exchange for hydroxide ion. This is an important system because pyruvate is the end product of aerobic glycolysis in the cytosol and the substrate for the pyruvate dehydrogenase complex in the mitochondrion. The tricarboxylic acid transporter transports citrate into the mitochondrion and a dicarboxylic acid carrier transports phosphate, fumarate, succinate, and malate. Glutamate and aspartate are exchanged between the two sides of the membrane by the glutamate carrier.

Electrons from Cytosolic NADH Are Transported by Shuttles

NADH produced by the oxidation of glyceraldehyde-3-phosphate in glycolysis cannot enter mitochondria. However, the reducing equivalents from cytoplasmic NADH are brought into mitochondria by two shuttle systems. The glycerol-3-phosphate shuttle operates through the actions of two different glycerol-3-phosphate dehydrogenases, one in the cytoplasm and the other embedded in the inner mitochondrial membrane, as shown in Figure 22-25. The cytosolic enzyme catalyzes the reduction of dihydroxyacetone by NADH according to equation 13.

$$\text{NADH} + \text{H}^+ + \text{Dihydroxyacetone phosphate} \rightleftharpoons \text{NAD}^+ + \text{Glycerol-3-phosphate} \tag{13}$$

Glycerol-3-phosphate diffuses through the outer mitochondrial membrane to the inner membrane, where it is dehydrogenated by a mitochondrial glycerol-3-phosphate dehydrogenase, a flavoprotein embedded in the inner membrane (equation 14).

$$\text{Glycerol-3-phosphate} + \text{E}\cdot\text{FAD} \rightleftharpoons \text{Dihydroxyacetone phosphate} + \text{E}\cdot\text{FADH}_2 \tag{14}$$

Dihydroxyacetone phosphate diffuses back into the cytoplasm to accept another electron pair from NADH. The reduced mitochondrial glycerol 3-phosphate dehydrogenase ($\text{E}\cdot\text{FADH}_2$) reduces ubiquinone in the inner membrane (equation 15).

$$\text{E}\cdot\text{FADH}_2 + \text{UQ} \rightleftharpoons \text{E}\cdot\text{FAD} + \text{UQH}_2 \tag{15}$$

UQH_2 in the inner membrane transfers electrons to complex III of the electron-transport pathway, so that complex I is bypassed, as in the case of succinate dehydrogenase. Therefore, only two moles of ATP are generated per electron pair from cytosolic NADH transported into the mitochondrion by the glycerol 3-P shuttle.

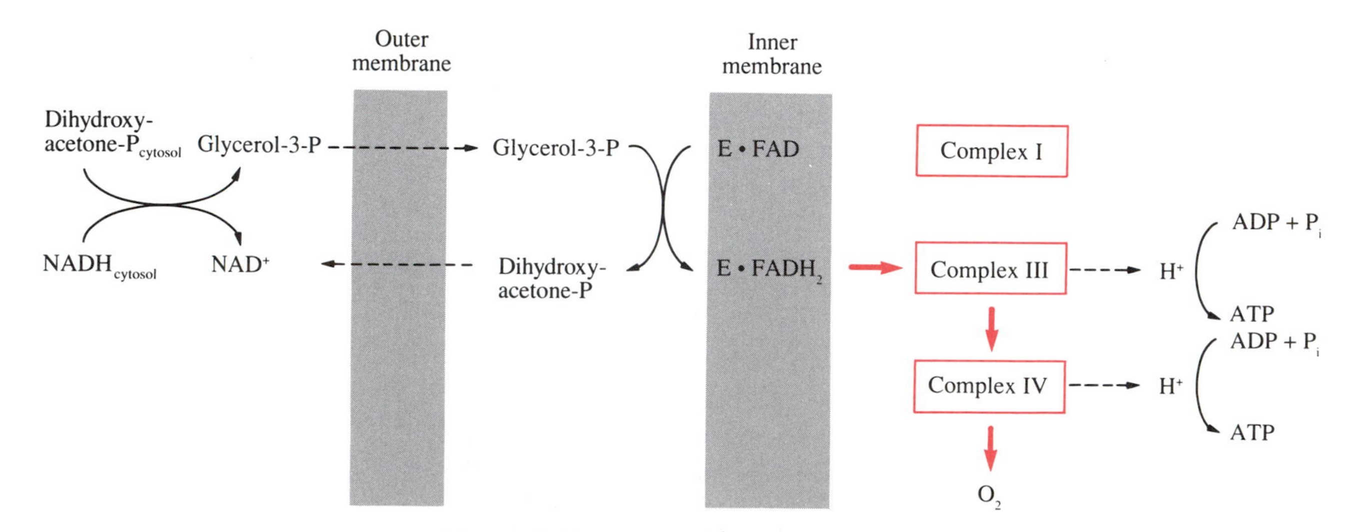

Figure 22-25

The Glycerol Phosphate Shuttle for the Transfer of Reducing Electrons from Cytosolic NADH to the Electron-Transport Pathway.

Heart and liver mitochondria use the malate shuttle, described by equations 16a through 16d,

$$\text{Oxaloacetate}_{(c)} + \text{NADH} + \text{H}^+ \rightleftharpoons \text{Malate}_{(c)} + \text{NAD}^+ \quad (16a)$$

$$\text{Malate}_{(c)} \rightleftharpoons \text{Malate}_{(m)} \quad (16b)$$

$$\text{Malate}_{(m)} + \text{NAD}^+ \rightleftharpoons \text{Oxaloacetate}_{(m)} + \text{NADH} + \text{H}^+ \quad (16c)$$

$$\text{Oxaloacetate}_{(m)} + \text{Glu}_{(m)} \rightleftharpoons \text{Asp}_{(m)} + \text{2-Ketoglutarate}_{(m)} \quad (16d)$$

to transport reducing equivalents from cytoplasmic NADH into the matrix. Cytoplasmic malate dehydrogenase catalyzes the reduction of oxaloacetate by NADH to malate. Malate diffuses to the inner membrane where it is transported into the matrix by the malate carrier. Mitochondrial malate dehydrogenase catalyzes its dehydrogenation to form NADH in the matrix, and the oxaloacetate, which cannot diffuse out, is cleared by a transamination reaction with glutamate to form aspartate and α-ketoglutarate. Aspartate is transported out of the mitochondrion and converted back into oxaloacetate in the cytosol. The malate shuttle has the overall effect of removing NADH from the cytoplasm and producing NADH in the matrix. Mitochondrial NADH is oxidized through complexes I, III, and IV, so that three molecules of ATP are formed for each electron pair transported by the malate shuttle.

When the malate shuttle operates, the complete aerobic degradation of glucose through glycolysis and the TCA cycle can, in principle, generate 38 moles of ATP per mole of glucose. When the glycerol 3-phosphate shuttle operates, only 36 ATP molecules can be formed for each glucose molecule oxidized.

The ADP-ATP Transporter

Mitochondria are essentially dynamos that generate ATP for all cellular processes; therefore, ATP formed in the matrix must be transported through the inner membrane into the cytoplasm. Moreover, the ADP generated in the cytosol must be transported into the matrix, where it can be used to form more ATP. ATP and ADP carry electrostatic charges of -4 and -3, respectively, and cannot diffuse across membranes, which are hydrophobic. A single transport system accomplishes both transport objectives through a one-to-one exchange of cytoplasmic ADP for mitochondrial ATP. The exchange is reversible but operates mainly to transport ADP into and ATP out of the mitochondrion, because ATP is continuously produced inside the mitochondrion, and utilized on the outside. The exchange of cytoplasmic $MgADP^-$ for matrix $MgATP^{2-}$ constitutes net migration of negative charge from the matrix to the cytoplasm. This is energetically downhill. The exchange reduces the membrane potential and, to the extent that it does so, it reduces the energy available for ATP formation. Recall that, of the total energy potentially available from the oxidation of glucose, about half actually goes into ATP. Part of the balance is used by the ATP-ADP transporter; and more is used to transport other species such as Ca^{2+} into the matrix.

The ATP-ADP transporter is a dimeric protein with a single nucleotide-binding site per dimer. It is embedded in the inner membrane and protrudes on both the matrix and the cytosolic sides. The transporter exhibits a higher affinity for $MgADP^-$ than for $MgATP^{2-}$ on the cytosolic side, whereas on the matrix side it has a higher affinity for $MgATP^{2-}$ than for $MgADP^-$. Figure 22-26 graphically represents the action of this transporter. The mechanism and conformational changes that bring about transport of the molecules between the two sides of the membrane are not known.

Figure 22-26

The Operation of the ADP-ATP Transporter in the Inner Mitochondrial Membrane.

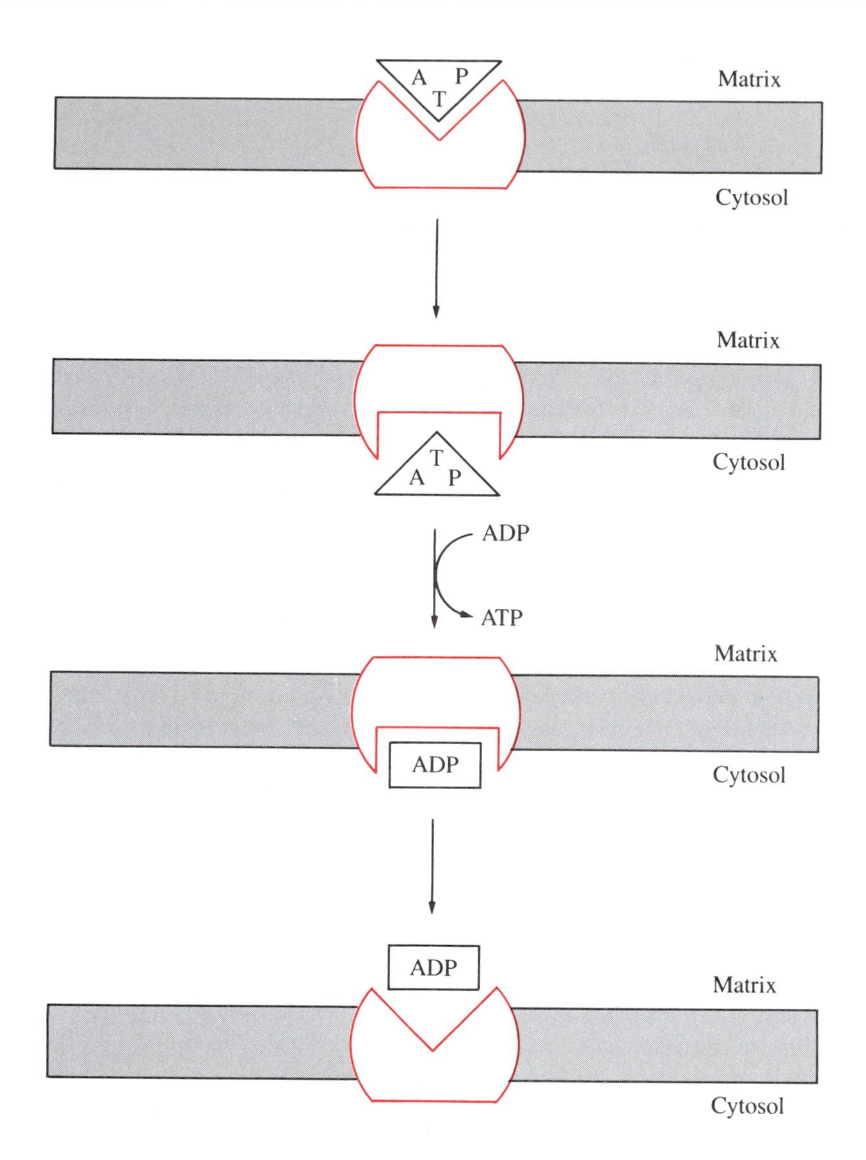

Summary

NADH produced in glycolysis and the TCA cycle is oxidized by oxygen through the electron-transport system located in the inner mitochondrial membrane of eukaryotic cells. Succinate in the TCA cycle also is oxidized by this system. These processes are carried out by a series of four electron-transport complexes (I, II, III, and IV) embedded within the membrane and ubiquinone (UQ). The complexes and UQ act in a specific sequence shown below.

$$\text{NADH} \longrightarrow \underset{\text{ADP} + \text{P}_i \,\to\, \text{ATP}}{\text{Complex I}} \longrightarrow \text{UQ} \longrightarrow \underset{\text{ADP} + \text{P}_i \,\to\, \text{ATP}}{\text{Complex III}} \longrightarrow \underset{\text{ADP} + \text{P}_i \,\to\, \text{ATP}}{\text{Complex IV}} \longrightarrow \text{O}_2$$

$$\text{Succinate} \longrightarrow \text{Complex II} \longrightarrow \text{UQ}$$

Complexes I, III, and IV couple electron transfer to the translocation of protons from the inside of the inner mitochondrial membrane—the matrix—to the intermembrane space. ATP synthase, which is also embedded in the inner membrane, couples the formation of ATP from ADP and P_i with the translocation of protons from the intermembrane space back into the matrix. Thus, the electron-transport chain harvests the energy from the oxidation of electron donors such as NADH and succinate and uses it to create an electrochemical potential and pH differential between the two sides of the inner mitochondrial membrane, and ATP synthase uses this energy to produce ATP from ADP and P_i. In prokaryotic cells, similar electron-transport systems and ATP synthases are located in the plasma membrane. A few energy-requiring cellular processes such as membrane transport are driven by direct use of electrochemical membrane potentials rather than by ATP.

Mitochondrial membranes contain transport systems for moving pyruvate, malate, and glycerol-P from the cytosol into the matrix. These systems have two main functions: they allow products of glycolysis and other catabolic pathways that occur in the cytosol to be transported into the mitochondrial matrix for further oxidation in the TCA cycle; and they allow reducing equivalents from NADH that are generated in the cytosol through glycolysis to be transported into the matrix for oxidation through the electron-transport system. The mitochondrial membrane also contains a transport system for shuttling ATP and ADP from one side of the membrane to the other.

ADDITIONAL READING

BOYER, P., CHANCE, B., ERNSTER, L., MITCHELL, P., RACKER, E., and SLATER, E. C. 1977. Oxidative phosphorylation and photophosphorylation. *Annual Review of Biochemistry* 46: 966.

CHANCE, B., and WILLIAMS, G. R. 1955. Respiratory enzymes in oxidative phosphorylation II: difference spectra. *The Journal of Biological Chemistry* 217: 395.

ERNSTER, L., ed. 1984. Bioenergetics. In *New Comprehensive Biochemistry*, vol. 9. Amsterdam: Elsevier.

MITCHELL, P., and MOYLE, J. 1965. Stoichiometry of proton translocation through the respiratory chain and adenosine triphosphatase systems of rat mitochondria. *Nature* 208: 147.

PENEFSKY, H. S., PULLMAN, M. E., DATTA, A., and RACKER, E. 1960. Partial resolution of the enzymes catalyzing oxidative phosphorylation II: participation of a soluble adenosine triphosphatase in oxidative phosphorylation. *The Journal of Biological Chemistry* 235: 3330.

SKULACHEV, V. P., and HINKLE, P. C., eds. 1981. *Chemiosmotic Proton Circuits in Biological Membranes*. Reading, Massachusetts: Addison-Wesley.

PROBLEMS

1. How does each of the following participate in the production of ATP?

(a) Electron transport complex I and II
(b) Ubiquinone
(c) ATP synthase
(d) Cytochrome oxidase
(e) Conformational coupling
(f) Cytochrome *c*

2. Why are three molecules of ATP produced in mitochondria per molecule of NADH oxidized by oxygen? Why are only two molecules of ATP produced per molecule of $FADH_2$ oxidized?

3. The energy of pH and ion gradients created across membranes is used for various purposes in cells. What are some of the uses of this energy?

4. How does dinitrophenol prevent the formation of ATP in mitochondria? How does dinitrophenol increase the rate of oxygen consumption by mitochondria?

5. How were dicyclohexylcarbodiimide and oligomycin important in studies of energy coupling in oxidative phosphorylation?

6. What are the main chemical communication systems between the mitochondrial matrix and the cytosol of a eukaryotic cell?

7. How is the energy generated by the action of the electron-transfer complexes of the inner mitochondrial membrane used to produce ATP?

8. How many molecules of ATP are produced by the oxidation of acetyl CoA in the mitochondrion? How many of these are produced by the action of ATP synthase?

9. What are the main components of ATP synthase and what are their functions?

10. The oxidation of glucose by eukaryotic cells produces ATP with a high energy efficiency. However, the efficiency is not 100%, because the free energy available from the ATP produced is about $-290\,\text{kcal mol}^{-1}$, compared with the $-686\,\text{kcal mol}^{-1}$ of heat that is available from the combustion of glucose. What happens to the energy from the oxidation of glucose in cells that is not used for the production of ATP?

Photosynthesis

23

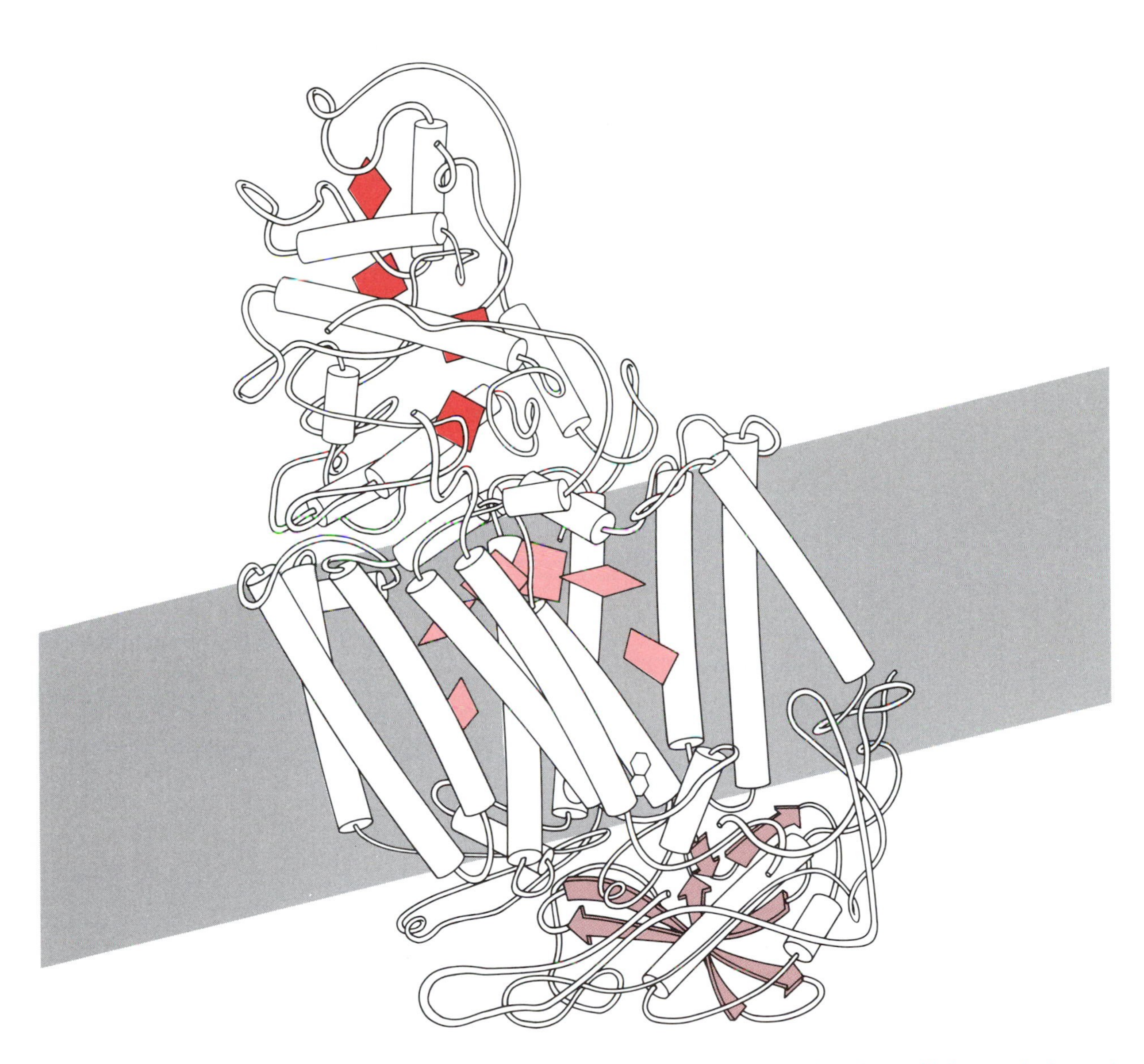

Schematic drawing of the bacterial photosynthetic reaction center from Rhodopseudomonas viridis. This remarkable system contains three subunits, each of which contains close to three hundred amino acids that are largely hydrophobic. Its eleven α-helices form a cylinder in the cell membrane, and additional structures project outward from both sides of the membrane. Cytochromes are shown in red; pheophytins in pink. *See page 641 for details.* *(After a drawing by Jane Richardson.)*

In the preceding chapter, we described the synthesis of high-energy phosphate compounds from the oxidation of metabolites in oxidative phosphorylation. The source of this energy is the carbon–hydrogen bond, which can be oxidized to carbon dioxide and water in the presence of molecular oxygen. These processes can be described by the metabolic dynamo (Figure 23-1), which was discussed in Chapter 9.

We now consider the problem of where this energy-rich C ~ H bond arises. The ultimate source of energy for life on earth is radiation from the sun. This energy is utilized to drive the formation of oxidizable substrates containing the C ~ H bond from carbon dioxide and water, which are the end products of oxidative metabolism. This is shown on the left side of Figure 23-1.

The overall reaction that occurs in photosynthesis by plants produces a 6-carbon sugar and 6 molecules of oxygen from 6 molecules of carbon dioxide and 12 molecules of water. Six molecules of water also are produced (equation 1).

$$6\,CO_2 + 12\,H_2O \longrightarrow C_6H_{12}O_6 + 6\,O_2 + 6\,H_2O \qquad (1)$$

We will see later that the water molecules that are formed are not the same ones that are used in this reaction. The reaction is thermodynamically uphill by 690,000 cal mol^{-1}. This energy is supplied by radiation from the sun. The overall reaction is the reduction of carbon dioxide, and the total reaction involves 24 electrons that reduce 6 molecules of carbon dioxide.

How do plants grow? There is a long and fascinating history of the development of our understanding of photosynthesis. Aristotle looked around him at the conditions under which plants grew and concluded that they grew by obtaining their nutrition from the humus. He did not do any experiments.

In 1648, von Helmont carried out an extraordinary experiment in which he weighed a bucket full of sand, then grew a willow tree in this sand, and weighed the resulting tree and sand (Figure 23-2). The result, which must have been astonishing at the time, was that the willow tree and sand (after drying) weighed far more than the original sand. This proved that the material constituting this tree must have come from the atmosphere.

In 1774, Priestley showed that a mouse died rapidly if put into a container in which a candle had burned until it went out. However, if the air in which the

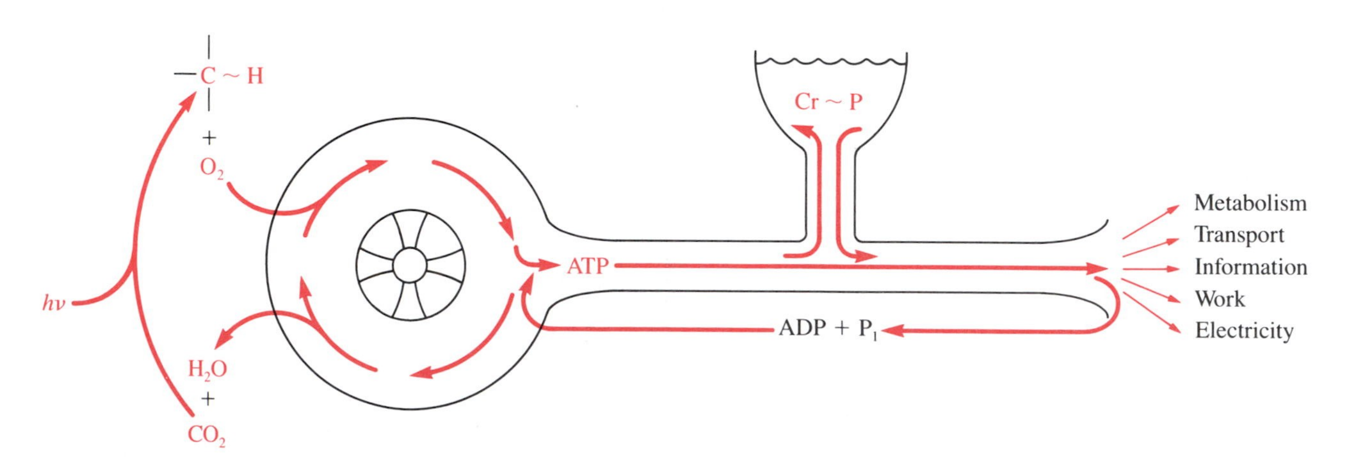

Figure 23-1

The Metabolic Dynamo. The energy of sunlight is utilized in photosynthesis by green plants to reduce carbon dioxide to a sugar and oxidize water to oxygen. The sugar contains energy-rich C ~ H bonds that are oxidized to carbon dioxide and water; this provides the energy that drives the metabolic dynamo. Creatine phosphate (Cr ~ P) provides a reservoir of high-energy phosphate bonds.

candle had burned was renewed in the presence of a green plant and light, the air regained the ability to support the life of a mouse (Figure 23-3). This was the first clear demonstration that oxygen is formed in photosynthesis.

More-recent work has progressively dissected the photosynthetic process into simpler systems that can be studied experimentally (Figure 23-4). Photosynthesis was first studied with plants, then with leaves, and then cells. The cells were broken into chloroplasts, which were broken into grana, which were disrupted further into thylakoid discs. All of these preparations are active and all contain intact membranes, so that there is always an inside and an outside.

Figure 23-2

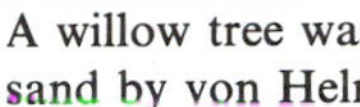

A willow tree was grown in sand by von Helmont. The mass of this tree must have come from water and the atmosphere.

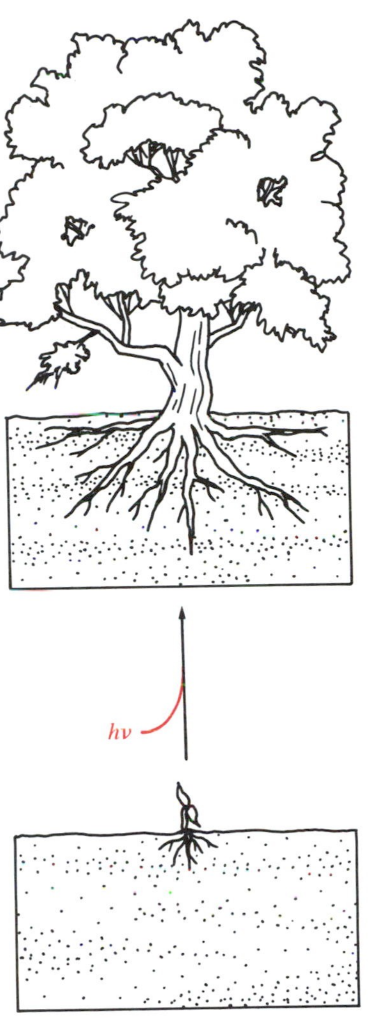

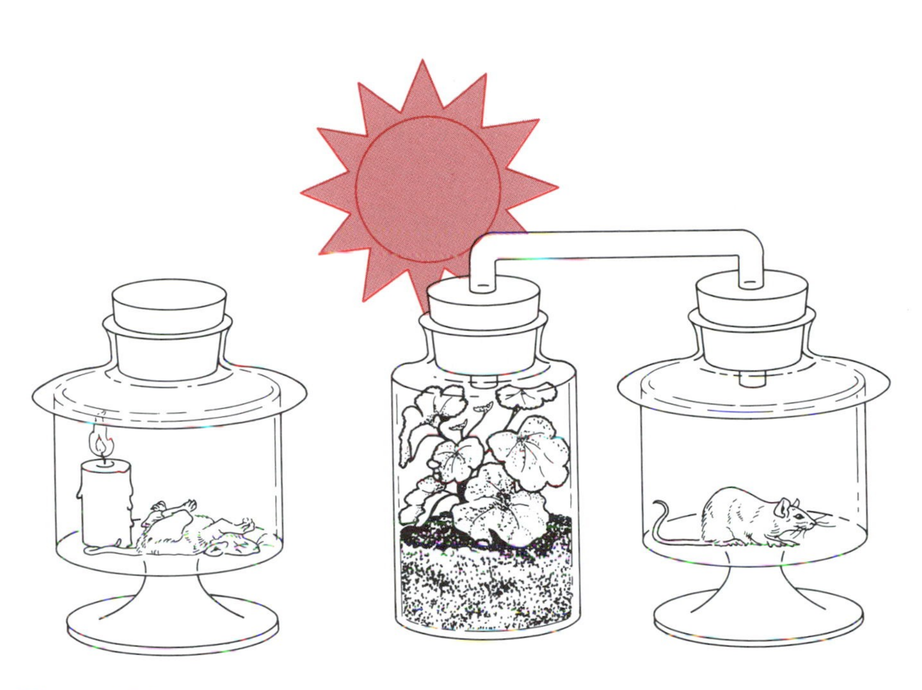

Figure 23-3

Priestley found that a mouse could not live in a container in which a candle had burned. However, a green plant that was exposed to sunlight was found to produce an atmosphere that supported life for a mouse.

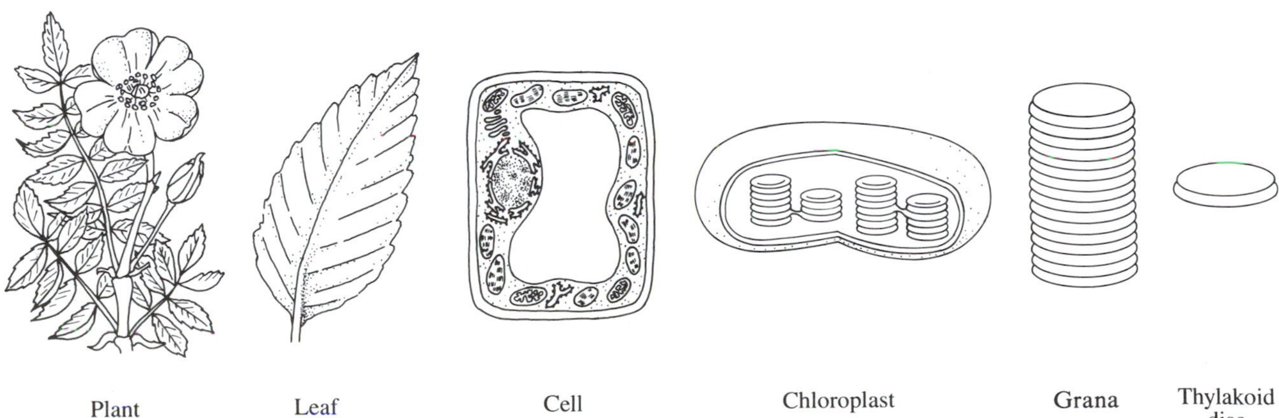

Figure 23-4

Systems That Can Carry Out Photosynthetic Reactions.

What Is Photosynthesis?

Many theories have been proposed to explain the biochemistry of photosynthesis. One possibility is that light splits carbon dioxide to form oxygen and reduces the carbon to carbohydrate (equation 2).

$$H_2O + CO_2 \xrightarrow{\text{Light}} CH_2O + O_2 \tag{2}$$

This is wrong. Another possibility is that light directly splits water into hydrogen and oxygen (equation 3).

$$2\,H_2O \xrightarrow{\text{Light}} 4\,H + O_2 \tag{3}$$

This is not right either (although water does supply the oxygen atoms of O_2).

The correct mechanism was proposed by van Neil; it is shown in equation 4, in which (CH_2O) represents a molecule of carbon dioxide that has been reduced to one carbon unit in a molecule of a sugar [$(CH_2O)_n$].

$$CO_2 + 2\,H_2O^* \xrightarrow{\text{Light}} (CH_2O) + H_2O + O_2^* \tag{4}$$

When the water on the left side of equation 4 is labeled with ^{18}O, it gives rise to labeled oxygen. This shows that the oxygen does come from the water, and not from splitting of the carbon dioxide. The oxygen from carbon dioxide appears in water. This is the reason that the water molecules that constitute reactants are different from those that constitute products in equation 1.

Many microorganisms do not use water but use other substrates that can be oxidized. An example is hydrogen sulfide, which can be oxidized to elemental sulfur (equation 5).

$$CO_2 + 2\,H_2S \xrightarrow{\text{Light}} (CH_2O) + H_2O + 2\,S \tag{5}$$

Thus, photosynthesis includes a series of reactions that incorporate carbon dioxide into carbohydrate, a process that requires reduction to form new C–H bonds and the oxidation of another molecule, such as water or hydrogen sulfide, that is readily available to the organism. The general equation for photosynthesis is shown in equation 6,

$$CO_2 + 2\,H_2A \xrightarrow{\text{Light}} (CH_2O) + H_2O + 2\,A \tag{6}$$

in which H_2A is a substance that can be oxidized to a product A.

However, the real progress in understanding the mechanism of photosynthesis was made only when the system was simplified at the biochemical level. The breakthrough came when the light-dependent reactions were separated from those that occur in the dark.

The light reaction was shown to bring about the reduction of ferricyanide to ferrocyanide (equation 7).

$$Fe^{3+}(CN)_6{}^{3-} \xrightarrow{\text{Light}} Fe^{2+}(CN)_6{}^{4-} \tag{7}$$

This is a simple one-electron reduction. It demonstrates that photosynthesis can produce reducing power directly. This is known as the Hill reaction, after its discoverer, R. Hill.

The physiologically important Hill reaction reduces the coenzyme $NADP^+$. The light reaction can reduce 2 molecules of $NADP^+$ to 2 molecules of NADPH (equation 8).

$$2\,NADP^+ + 2\,H_2O \xrightarrow{\text{Light}} 2\,NADPH + O_2 + 2\,H^+ \tag{8}$$

Oxygen is a by-product of the reaction.

The dark reaction results in the fixation of carbon. The dark reaction can be demonstrated by studying the time course of carbon fixation with flashes of light. Carbon dioxide is converted into carbohydrate by incorporation into organic compounds after the absorption of light. This involves the chemistry of carboxylation and reduction, using the reducing power of NADPH that is produced in the light reaction.

A third type of reaction, which occurs to some extent in both the light and the dark, is the synthesis of ATP from ADP and phosphate. This is photophosphorylation (equation 9).

$$\mathrm{ADP} + \mathrm{P_i} \xrightarrow{\text{Light}} \mathrm{ATP} \tag{9}$$

We will examine the light reaction first, then the dark reactions, and finally photophosphorylation.

The Light Reaction

The photochemistry of photosynthesis is initiated by the excitation of chlorophyll by light. *Chlorophyll* is a porphyrin that contains a magnesium ion, instead of the iron of hemes, and a long phytol side chain twenty carbon atoms in length (Figure 23-5). It differs from iron porphyrins in that one of the pyrrole rings is reduced. In the *bacteriochlorophyll* of photosynthetic bacteria, a second pyrrole ring is reduced, giving a chlorophyll that absorbs at higher wavelength. The chlorophyll molecule contains a long alcohol side chain and an additional ring that is attached to one of the pyrroles.

Some light is absorbed by chlorophyll directly. However, in green plants and photosynthetic bacteria, there are also several *accessory pigments* that absorb light and transfer the excitation energy to chlorophyll (Figure 23-6). The most important of these in plants are the *carotenoids*, which are long, unsaturated hydrocarbon chains (Figure 23-7). *Phycobilins*, which are porphyrins that have been opened up into a chain, are found in some photosynthetic bacteria. With all of these pigments, there is an extremely high probability that light will be absorbed over a wide range of wavelengths and the energy then transferred to chlorophyll.

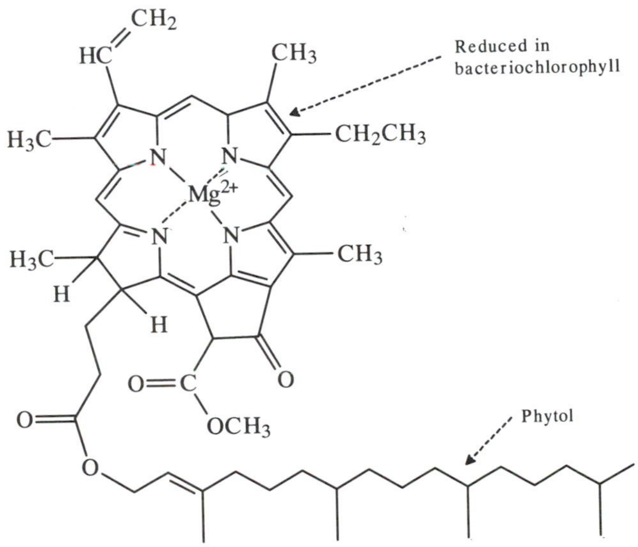

Reduced in bacteriochlorophyll

Phytol

Chlorophyll *a*

Figure 23-5
Chlorophyll *a*.

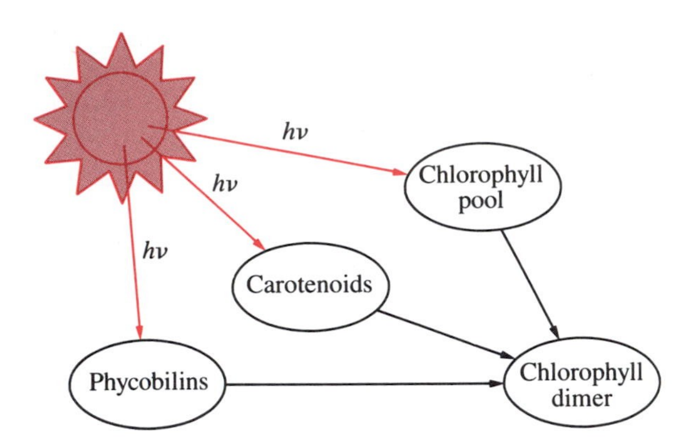

Figure 23-6

Much of the sunlight that is used for photosynthesis is absorbed by *accessory pigments*, such as carotenoids and phycobilins. The light energy is then transferred to a pool of chlorophyll molecules and, finally, to the active pair of chlorophyll molecules that provides the reducing power for photosynthesis.

CAROTENOIDS

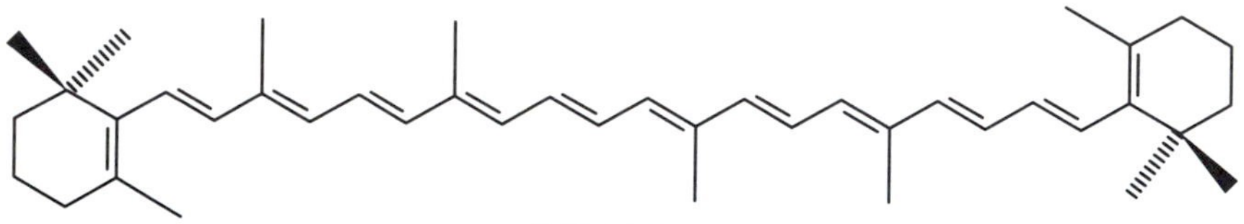

β-Carotene

XANTHOPHYLLS (oxygenated carotenoids)

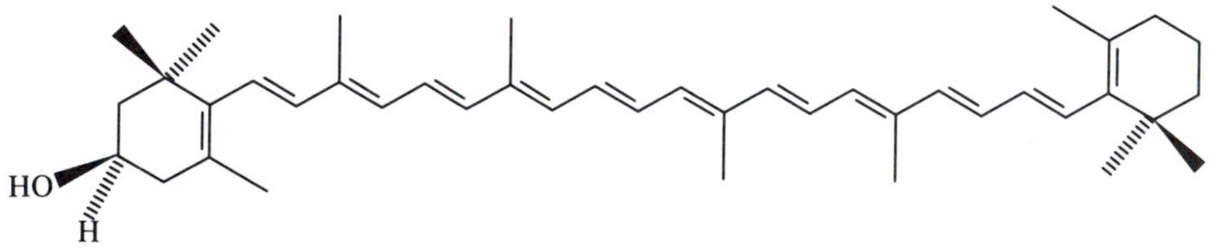

HO

H

Zeaxanthin

PHYCOBILINS (and related compounds)

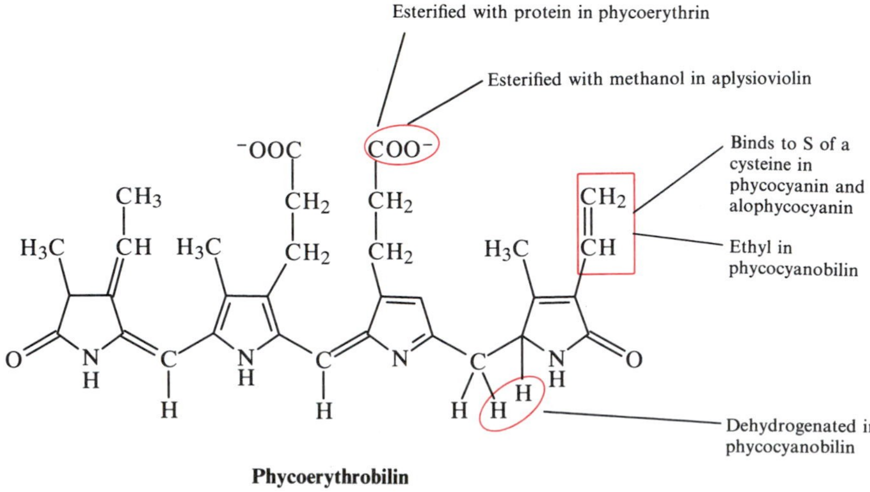

Esterified with protein in phycoerythrin

Esterified with methanol in aplysioviolin

Binds to S of a cysteine in phycocyanin and alophycocyanin

Ethyl in phycocyanobilin

Dehydrogenated in phycocyanobilin

Phycoerythrobilin

Figure 23-7

Accessory Pigments in Photosynthesis.

The chlorophyll molecules themselves form a pool. In most photosynthetic systems, the energy of a particular group or pair of chlorophyll molecules in the excited state is slightly lower than that of most of the chlorophyll molecules. These molecules will collect the excitation energy from the chlorophyll molecules in the pool (Figure 23-6). The lower energy for the excited state of the chlorophyll dimer is due to its slightly longer maximum wavelength of absorption, because the energy of a quantum of light decreases with increasing wavelength.

It has been shown by illuminating chloroplasts with flashes of increasing light intensity that the maximum energy that can be utilized in the flash is enough to give 1 oxygen molecule for each 3,000 chlorophyll molecules. Eight quanta of light are required to form 1 oxygen molecule, and so this corresponds to about 400 chlorophyll molecules for each quantum of light. It suggests that there is a pool of about 400 chlorophyll molecules that can be excited and transfer energy to each active dimer of chlorophyll molecules.

The absorption of light excites an electron from an occupied orbital of chlorophyll to an unoccupied orbital of higher energy (Figure 23-8). In the absence of an acceptor, this electron can return to the ground state and emit its energy as fluorescence. However, if there is an appropriate electron acceptor of slightly lower energy nearby, the excited electron can be transferred to this acceptor and reduce it. This prevents the loss of energy by fluorescence and, for this reason, the compounds that undergo this electron transfer are quenching reagents and are sometimes designated as Q.

The important point is that, when this excitation and electron transfer have taken place, there is an electron available that can be used for reduction

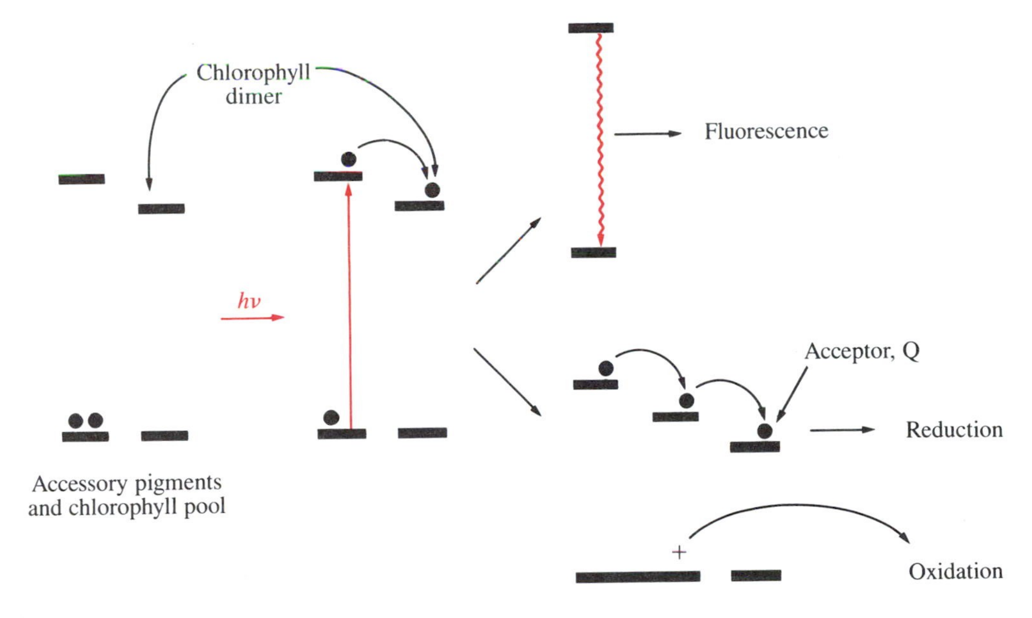

Figure 23-8

Light excites molecules in the chlorophyll pool or accessory pigments to an excited state, and this energy is rapidly transferred to the chlorophyll dimer, which has a slightly lower energy. If there is no electron acceptor present, the electron will return to the ground state and the energy will be emitted as fluorescence. If an electron acceptor is present, the electron is transferred to reduce the acceptor. Electrons from the reduced acceptor are then used to reduce other molecules and, ultimately, to reduce $NADP^+$ to NADPH. The oxidizing equivalent that is left behind after transfer of the electron is used for the oxidation of H_2O to O_2 in green plants.

when it, or a pair of electrons, is transferred to an appropriate reagent (Figure 23-8). There is also an electron deficiency, or "hole," that is left behind in the chlorophyll. This is an oxidizing agent, as indicated by the plus sign in Figure 23-8. It will accept electrons from other pigments or reducing agents to bring about oxidation. The formation of these oxidizing and reducing equivalents is the critical and important step in photosynthesis.

Pathways for Photosynthesis

The occurrence of photosynthesis in green plants is widespread and well known. However, it is a relatively recent occurrence on this planet—photosynthesis was brought about by bacteria and other microorganisms for millions of years before green plants appeared. Although there are several different pathways for photosynthesis in microorganisms, these pathways are generally simpler than photosynthesis in green plants, and we will examine an example of photosynthesis in bacteria first.

Bacterial Photosynthesis

A typical sequence of reactions for the light reaction in bacterial photosynthesis is shown in Figure 23-9. Light is used to excite a molecule of bacteriochlorophyll in a dimer that absorbs maximally at 870 nm. The maximum wavelength for absorption of bacteriochlorophyll is longer than that of plant chlorophyll because of the additional reduced pyrrole group in the chlorophyll ring (Figure 23-5). The excitation corresponds to a change of potential from approximately +0.45 volt to −0.45 volt. Thus, the excited electron is a good reducing agent.

The excited electron jumps very rapidly to *pheophytin*, which is a chlorophyll molecule that does not contain magnesium. This gives a radical pair. One electron is on the pheophytin, and an electron deficiency (+), or "hole," remains on the chlorophyll. The electron is then transferred to an iron-quinone complex. When two electrons are collected, they can be used to reduce $NADP^+$ to NADPH. This is the critical chemical reaction in the initial stage of photosynthesis because it produces a reduced carbon–hydrogen bond, C~H, in NADPH.

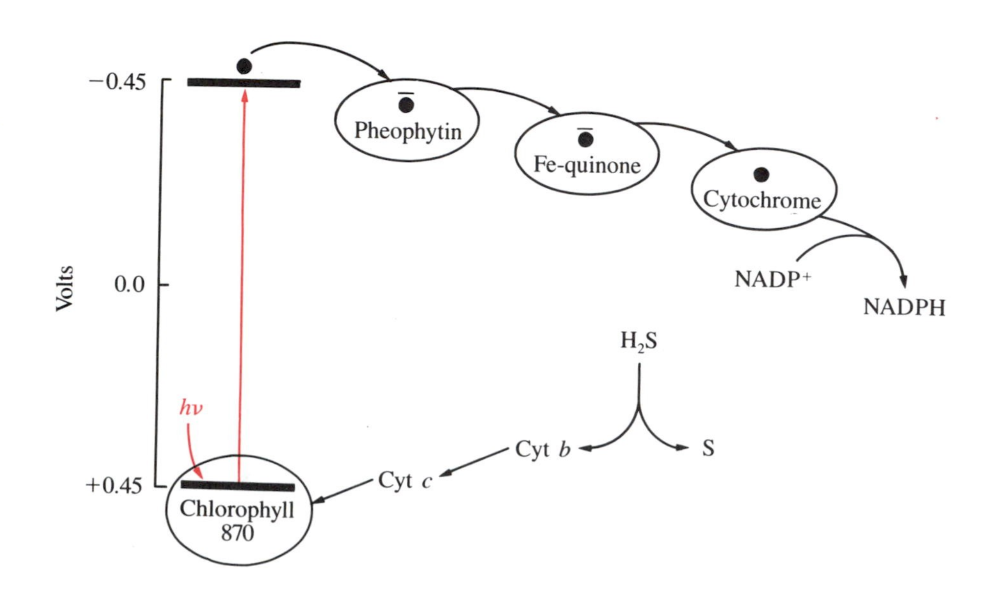

Figure 23-9

A Typical Series of Reactions in Bacterial Photosynthesis. When bacteriochlorophyll absorbs light, the excited electron (−) is transferred to bacteriopheophytin, then to an iron-ubiquinone complex, and then to a cytochrome. These electrons can be used to reduce water to hydrogen gas or to reduce $NADP^+$ to NADPH. The oxidizing equivalent (+) that is left behind can be reduced by H_2S, through cytochromes *b* and *c*, to give elemental sulfur.

The oxidizing equivalent that is left behind is used to oxidize an electron donor. For example, hydrogen sulfide can transfer electrons to a cytochrome chain consisting of cytochrome *b* and cytochrome *c*. Electrons can then be transferred to the oxidizing equivalent (the "hole") remaining in the chlorophyll to regenerate starting material. The oxidation product from hydrogen sulfide is elemental sulfur.

The structure of the photosynthetic reaction center of *Rhodopseudomonas viridus* is known from X-ray diffraction; it was determined by J. Diesenhofer, R. Huber, and H. Michel in 1984. The structure is shown schematically on page 633. It is immediately apparent that the location of the electron carriers in the protein is in the same order as the sequence of electron transfers after excitation by light. Light is absorbed by a special pair of chlorophyll molecules at the top of the structure. The excitation energy is rapidly transferred to pheophytins that are largely in the membranous region and then to menaquinone and ubiquinone. The transfer from excited menaquinone to ubiquinone may be facilitated by an Fe^{2+} ion between them. When two electrons are taken up by the quinone, it adds two protons to form the hydroquinone. The hydroquinone is a reducing agent for the cell. When it is oxidized, it delivers two protons that form a proton gradient, which can be used for photophosphorylation, as described later in this chapter. The cytochromes at the top of the molecule deliver an electron to the oxidized chlorophyll and allow a new reaction cycle to take place.

Photosynthesis in Plants

The site of photosynthesis in green plants is the *chloroplast* (see Figure 23-4), an organelle within the plant cell that has some resemblance to the mitochondrion in animal cells. The photosynthetic machinery is contained in the *grana*, which consist of stacks of interconnected discs. These discs are composed of interconnected membranes and together make up a distinct compartment of the cell, the *thylakoid*.

Photosynthesis is brought about by two photosystems in green plants. With two photosystems, there are two excitations, so that more energy is available to form NADPH and to oxidize water to molecular oxygen. This is important because the difference in potential between oxygen and NADPH is larger than that between elemental sulfur and NADPH. There are several kinds of evidence that two photosystems exist. One is that the maximum rate of photosynthesis that can be obtained by irradiation with light at 690 nm is considerably increased by light at 700–710 nm. This suggests that there are two systems, one of which can absorb light at 680–690 nm and the other at 700–710 nm, that complement each other in the overall reaction.

A scheme for plant photosynthesis is shown in Figure 23-10. This is sometimes called the Z system because the two excitations and the electron transfer between them form a (sideways) Z.

In this system, photosystem I, long-wavelength light excites active chlorophyll molecules with a maximum absorption at 700 nm, P700. The excited electron is rapidly transferred to a chlorophyll molecule with a slightly different structure, then to phylloquinone (Vitamin K), several membrane-bound ferredoxins, and a soluble ferredoxin, each of which contains an iron-sulfur cluster. Finally, two electrons are transferred to a flavoprotein enzyme and used to reduce $NADP^+$ to NADPH.

Some of the electrons from photosystem I, however, are used to reduce a quinone, plastoquinone, to the corresponding hydroquinone. This results in the uptake of two protons on one side of the thylakoid membrane and contributes to the buildup of a proton gradient across the membrane that is used for photophosphorylation, as described later in this chapter.

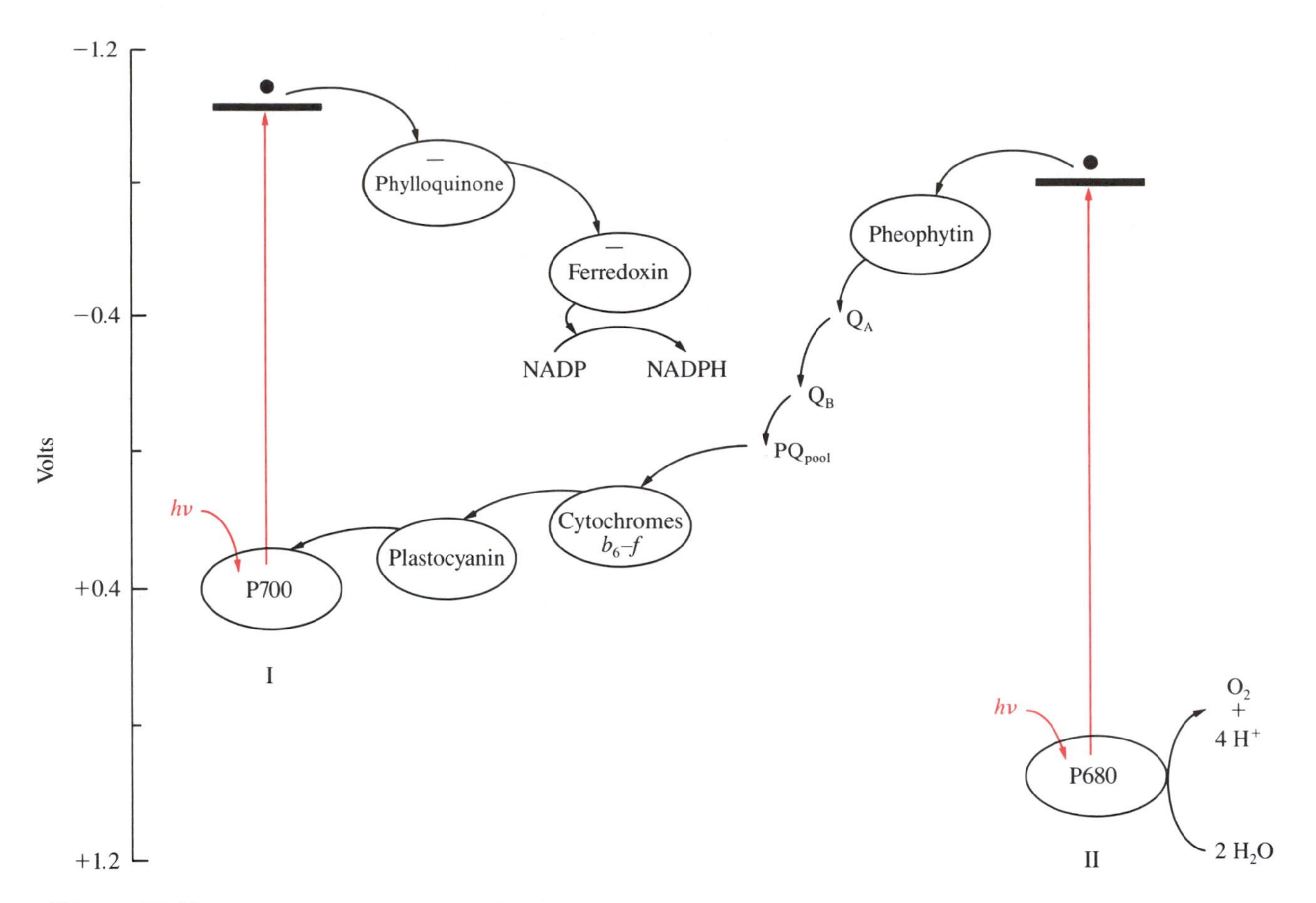

Figure 23-10

The Z Scheme for Photosynthesis in Green Plants Through Two Photosystems, P700 (I) and P680 (II). Excitation of P700 provides an electron that reduces several electron carriers, including phylloquinone and ferredoxin, and then contributes to the reduction of $NADP^+$ to NADPH. Excitation of P680 provides an electron that passes through pheophytin, several quinones, and the cytochrome b_6–f complex before reducing the P700, which was left in an oxidized state after excitation and electron transfer. Four oxidizing equivalents from the excitation of P680 are collected in a Mn-protein complex and are then used to oxidize two water molecules to O_2, with the release of four protons.

The remaining "hole," or oxidizing equivalent, in photosystem I is reduced by a reducing equivalent that is formed in photosystem II. Excitation of P680 chlorophyll in photosystem II by light of shorter wavelength gives a reducing equivalent that is rapidly transferred to pheophytin. It then travels down an electron-transport chain and is transferred to a pool of plastoquinone, PQ, which is a quinone with a long hydrophobic side chain and a structure similar to that of ubiquinone. The reduction of plastoquinone by two electrons results in the uptake of two protons from the outside of the chloroplast (equation 10).

$$\textbf{PQ} \xrightarrow[2\,H^+_{out}]{2\,e} \textbf{PQH}_2 \qquad (10)$$

Ⓡ = ($CH_2CH{=}C(CH_3)CH_2$)$_n$

Electrons then move through a cytochrome b_6–cytochrome f complex to plastocyanin, a copper-containing electron-transporting protein, and are then used to reduce the oxidizing equivalent that was formed in the first photosystem (Figure 23-10).

The oxidizing equivalent that is left behind in P680 is a strong oxidizing agent. When four of these oxidizing equivalents (⊕) are collected, they are used to oxidize water to molecular oxygen, as shown in equation 11.

$$2\,H_2O + 4\oplus \longrightarrow O_2 + 4\,H_{in}^{+} \qquad (11)$$

It is important to balance this equation and notice that four protons are formed in the oxidation. The enzyme that catalyzes this remarkable reaction contains four Mn atoms and collects the four oxidizing equivalents on Mn and, probably, on another site. The four Mn atoms and four alternating oxygen atoms are probably located at the corners of a distorted cube. It has been suggested that two water molecules bind to two of the Mn atoms and are oxidized to release oxygen and four protons (equation 11). This reaction is analogous to the reaction in the opposite direction that is catalyzed by cytochrome oxidase, which collects four reducing equivalents from the respiratory chain and uses them to reduce oxygen to water (Chapter 22). However, the mechanisms of the two reactions are quite different.

These reactions result in the accumulation of protons in the interior of the thylakoid space, which becomes more acidic than the surrounding stroma in the chloroplast by some 2 to 3 pH units. The probable sources of these protons are from the oxidation of water, as just noted, and from the plastoquinone system when the hydroquinone (equation 10) is oxidized by cytochrome *b* back to the quinone (equation 12).

$$PQH_2 + 2\,Cyt\,b_{ox} \longrightarrow PQ + 2\,Cyt\,b_{red} + 2\,H^{+} \qquad (12)$$

It appears that the enzymes catalyzing these reactions are arranged in such a way that the protons are released on only one side of the thylakoid membrane and thus build up a proton gradient.

One of the important inhibitors of photosynthesis is 3-(3,4-dichlorophenyl)-1,1-dimethylurea, or DCMU.

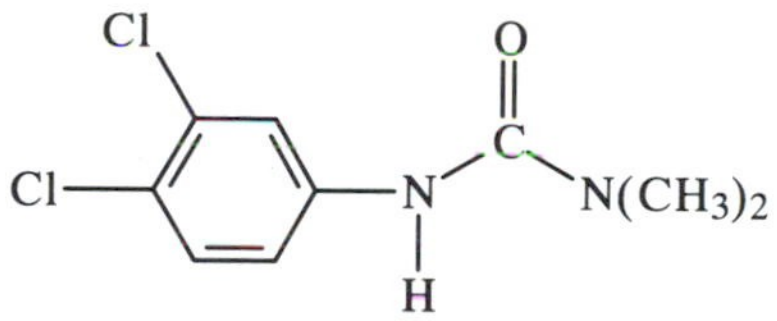

3-(3,4-Dichlorophenyl)-1,1-dimethylurea (DCMU)

This compound blocks photosynthesis by blocking electron transfer between photosystem II and cytochrome *f*, so that P700 in photosystem I cannot be activated by light to reduce $NADP^+$. Therefore, DCMU blocks the growth of plants and is an effective herbicide.

It is possible to reduce the oxidized P700 by ascorbic acid (vitamin C), as well as by the second photosystem. When this is done, only 2 quanta of light are required to provide the two electrons that give a molecule of NADPH. In the absence of ascorbic acid, both photosystems are required, so that a total of four excitations and 4 quanta of light are required to reduce one molecule of NADP to NADPH.

Dark Reactions: The Fixation of Carbon Dioxide

The synthesis of carbohydrate in photosynthesis requires that molecules of carbon dioxide from the atmosphere must be incorporated into a larger molecule and reduced by NADPH, so that they can become part of a glucose molecule. This does not require light directly; it is brought about in a series of "*dark reactions*" and is called carbon dioxide *fixation*. The end result is the synthesis of a reduced carbon atom, C~H, that can be oxidized to provide energy for living systems (see Figure 23-1). This series of reactions makes possible the existence of life on earth. It was discovered by Melvin Calvin, James Bassham, and Andrew Benson from 1945 to 1953.

The new carbon–carbon bond is formed by the addition of carbon dioxide to a five-carbon ketosugar, ribulose-1,5-bisphosphate, to produce two molecules of 3-phosphoglycerate, as shown in equation 13.

$$O{=}C{=}O + \underset{\text{Ribulose-1,5-bisphosphate}}{CH_2OPO_3^{2-}\text{–}C({=}O)\text{–}CH(OH)\text{–}CH(OH)\text{–}CH_2OPO_3^{2-}} + H_2O \longrightarrow \underset{\text{3-Phosphoglycerate}}{2\ {}^{-}O_2C\text{–}CH(OH)\text{–}CH_2OPO_3^{2-}} \quad (13)$$

This carboxylation reaction was described in Chapter 16; it occurs through enolization of the ketone to form an enediol, followed by addition of carbon dioxide to the enediol and hydrolysis of the resulting six-carbon sugar to give two molecules of phosphoglycerate.

The two molecules of 3-phosphoglycerate are then reduced and converted into fructose-6-phosphate, which can be metabolized to provide energy. Fructose-6-phosphate is synthesized by the reversal of several of the steps in glycolysis. Recall that in glycolysis fructose-6-phosphate is cleaved to give glyceraldehyde-3-phosphate and dihydroxyacetone phosphate. The phosphoglyceric acid that is formed in photosynthesis is activated by ATP to give 1,3-bisphosphoglycerate and is then reduced to glyceraldehyde-3-phosphate by the use of a molecule of NADPH that was formed in the light reaction (Figure 23-11). This is a critical step in photosynthesis because it incorporates into glyceraldehyde-3-phosphate the reducing equivalents that were formed in the light reaction by the reduction of $NADP^+$ to NADPH. One molecule of glyceraldehyde-3-phosphate is then isomerized to dihydroxyacetone phosphate, which condenses with another molecule of glyceraldehyde-3-phosphate to give fructose-1,6-bisphosphate. The overall result is that a molecule of carbon dioxide has been incorporated into a sugar and reduced to a form in which it can be utilized in metabolism to provide energy.

The next problem to consider is how the ribulose-1,5-bisphosphate that reacts with carbon dioxide is regenerated, so that carbon fixation can continue. The enzymes that catalyze this process are aldolase and transketolase, a thiamine pyrophosphate enzyme. These enzymes, together with triosephosphate isomerase and sedoheptulose-1,7-bisphosphatase, catalyze the overall reaction that is described by equation 14.

$$\text{Fructose-6-phosphate} + 3\ \text{Glyceraldehyde-3-phosphate} + 3\ \text{ATP} \longrightarrow 3\ \text{Ribulose-1,5-bisphosphate} + 3\ \text{ADP} + P_i \quad (14)$$

The reaction sequence that leads to the formation of ribulose 1,5-bisphosphate is described in Box 23-1 (see also Chapter 21).

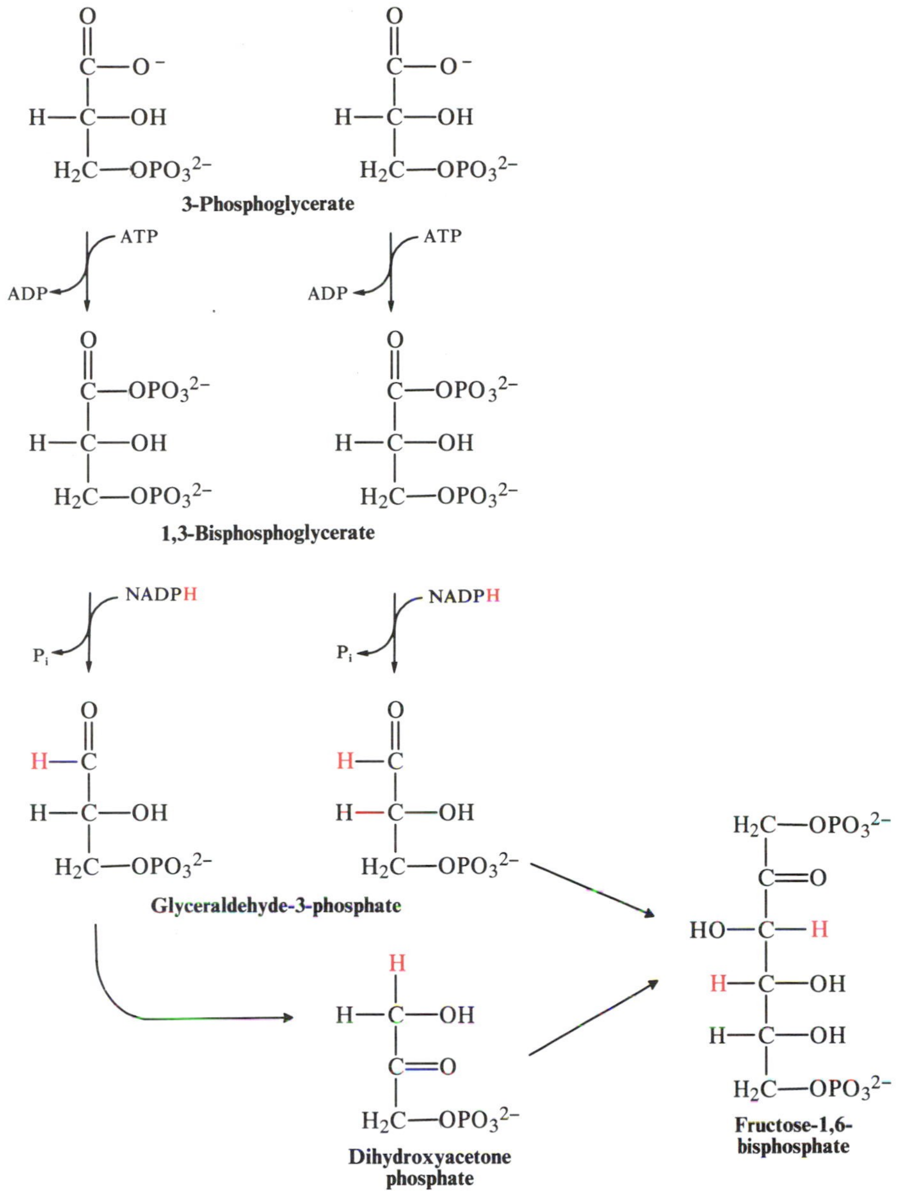

3-Phosphoglycerate

ATP

ADP

1,3-Bisphosphoglycerate

NADPH

P_i

Glyceraldehyde-3-phosphate

Dihydroxyacetone phosphate

Fructose-1,6-bisphosphate

Figure 23-11

The hydrogen of NADPH that is formed from the light reactions of photosynthesis is used to reduce 1,3-bisphosphoglycerate to glyceraldehyde-3-phosphate. Two molecules of glyceraldehyde-3-phosphate are condensed to give fructose-1-6-bisphosphate, after one of them has been isomerized to dihydroxyacetone phosphate.

BOX 23-1

THE PATHWAY FOR THE REGENERATION OF RIBULOSE-1,5-BISPHOSPHATE

Figure A (on page 646) shows how carbon-chain lengths are rearranged to form 5-carbon sugars from the 6-carbon and 3-carbon chains of fructose-6-phosphate and glyceraldehyde-3-phosphate. The first reaction forms 5-carbon and 4-carbon sugars from fructose-6-phosphate and glyceraldehyde-3-phosphate by transfer of a 2-carbon unit, which is catalyzed by transketolase. The second reaction is the addition to erythrose-4-phosphate of the three carbon atoms of dihydroxyacetone phosphate, which is formed by isomerization of glyceraldehyde-3-phosphate, to give sedoheptulose-1,7-bisphosphate. This reaction is catalyzed by transaldolase. Finally, a 2-carbon unit is transferred from sedoheptulose-7-phosphate to glyceraldehyde-3-phosphate to give two 5-carbon sugars, ribose-5-phosphate and xylulose-5-phosphate. (*Continued on pages 646 and 647.*)

BOX 23-1 (continued)

THE PATHWAY FOR THE REGENERATION OF RIBULOSE-1,5-BISPHOSPHATE

Figure A

The pathway for the regeneration of 5-carbon sugars from fructose-6-phosphate and glyceraldehyde-3-phosphate.

$$C_6 + C_3 \rightarrow C_5 + C_4$$

Fructose-6-phosphate + **Glyceraldehyde-3-phosphate** $\xrightarrow{1}$ **Xylulose-5-phosphate** + **Erythrose-4-phosphate**

$$C_4 + C_3 \rightarrow C_7$$

Erythrose-4-phosphate + **Dihydroxyacetone phosphate** $\xrightarrow{2}$ **Sedoheptulose-1,7-bisphosphate** $\rightarrow$ (P_i) **Sedoheptulose-7-phosphate**

$$C_7 + C_3 \rightarrow 2\,C_5$$

Sedoheptulose-7-phosphate + Glyceraldehyde-3-phosphate $\xrightarrow{1}$ **Ribose-5-phosphate** + **Xylulose-5-phosphate**

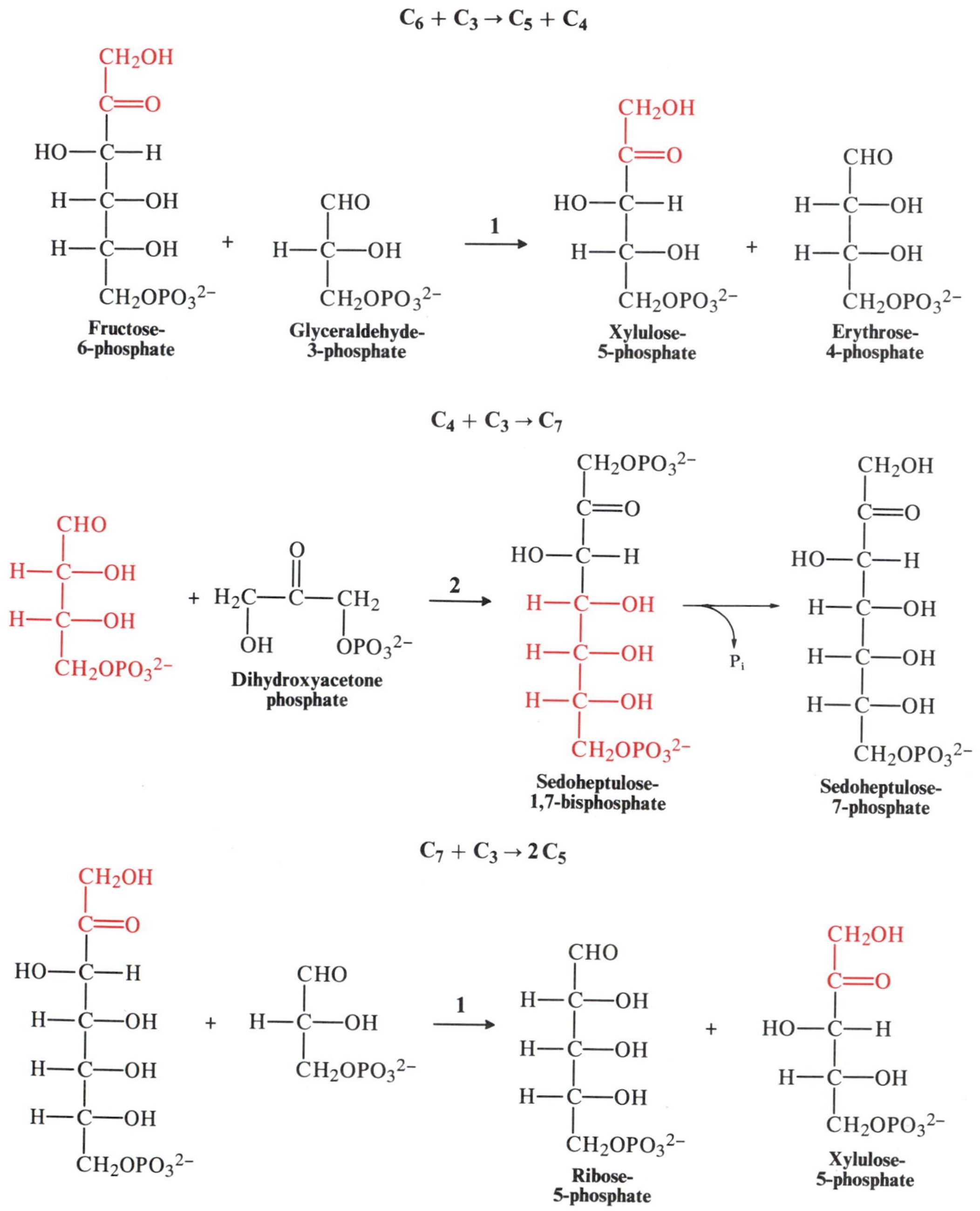

1. Transketolase
2. Transaldolase

BOX 23-1 (continued)

THE PATHWAY FOR THE REGENERATION OF RIBULOSE-1,5-BISPHOSPHATE

Figure B

The synthesis of ribulose-1,5-bisphosphate from ribose-5-phosphate and xylulose-5-phosphate.

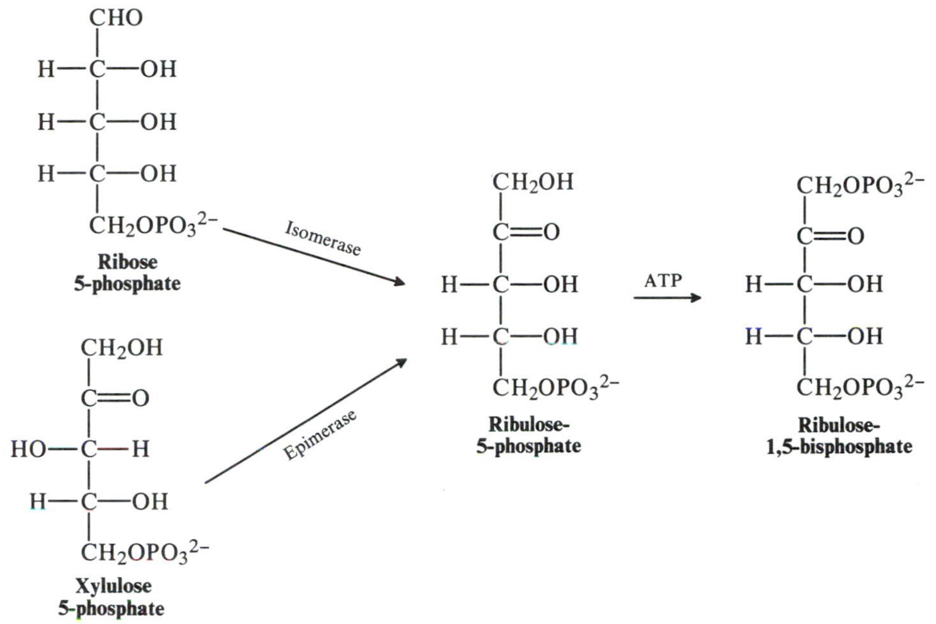

Ribose 5-phosphate

Isomerase

Xylulose 5-phosphate

Epimerase

Ribulose-5-phosphate

ATP

Ribulose-1,5-bisphosphate

Figure B shows how ribulose-1,5-bisphosphate is formed from these 5-carbon sugars by the action of an isomerase and an epimerase, followed by phosphorylation by ATP.

Photooxidation

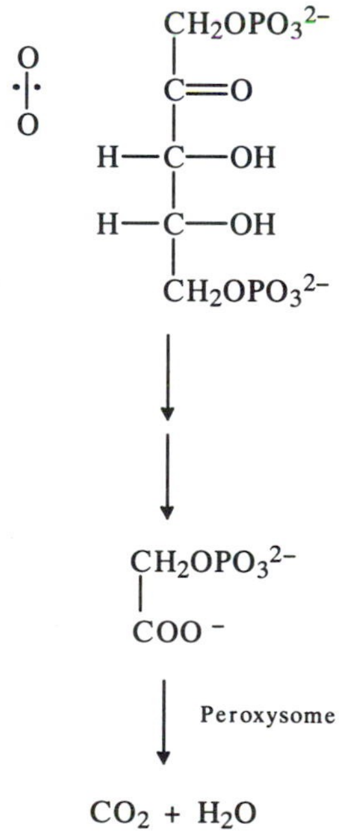

An important—and wasteful—side reaction of photosynthesis arises from the imperfect specificity of ribulose bisphosphate carboxylase for catalysis. This side reaction causes an enormous loss of the products of photosynthesis; if it could be avoided, there would be a large increase in the amount of food available for feeding the growing population of this planet. This side reaction is *photooxidation;* it is sometimes called photorespiration.

Photooxidation occurs because ribulose bisphosphate carboxylase cannot always distinguish between carbon dioxide and oxygen, which is present at a much higher concentration in the atmosphere. We have seen that ribulose bisphosphate carboxylase converts its substrate into an enediol, which provides a carbanion to attack carbon dioxide. Oxygen also reacts rapidly with carbanions, and it is not very different in size from carbon dioxide, so that a significant fraction of the enediol intermediates that are formed from ribulose-1,5-bisphosphate at the active site of the enzyme react with oxygen. This reaction cleaves the ribulose-1,5-bisphosphate and forms phosphoglycolate, as shown in the margin. The phosphoglycolate is oxidized further to carbon dioxide and water by a group of enzymes in *peroxysome* particles. The

Peroxysome

$CO_2 + H_2O$

unfortunate result of this reaction is not only the loss of the carbon dioxide molecule that would otherwise have been fixed, but also the destruction of a segment of the ribulose bisphosphate molecule. This constitutes a large loss of CH groups that could be oxidized to produce useful energy.

A mechanism for avoiding this waste has been evolved by a number of plants that can concentrate carbon dioxide in their cells through a "four-carbon cycle." These plants include sugar cane, corn, crab grass, and many weeds.

The four-carbon cycle is shown in Figure 23-12. All of the reactions in this cycle but one are closely analogous to those in the citric acid cycle. Phosphoenolpyruvate is carboxylated to oxaloacetate, which is then reduced by NADPH to give malate in bundle-sheath cells. The malate is transported to mesophyll cells, which surround the bundle-sheath cells, where it is oxidatively decarboxylated to regenerate NADPH and form pyruvate and carbon dioxide. The carbon dioxide then diffuses into the bundle-sheath cells, where it is used for photosynthesis without significant competition from oxygen.

The pyruvate is reconverted into phosphoenolpyruvate by an extraordinary enzyme, pyruvate phosphate dikinase. This enzyme catalyzes the phosphorylation of both pyruvate and inorganic phosphate by ATP to form phosphoenolpyruvate, pyrophosphate, and AMP, as shown in equation 15.

$$\text{Pyruvate} + \text{ATP} + \text{P}_i^* \longrightarrow \text{Phosphoenolpyruvate} + \text{PP}_i^* + \text{AMP} \quad (15)$$

Use of labeled inorganic phosphate has shown that the phosphate is incorporated into the pyrophosphate product. The pyrophosphate is then hydrolyzed rapidly to inorganic phosphate (equation 16).

$$\text{PP}_i^* + \text{H}_2\text{O} \longrightarrow \text{P}_i + \text{P}_i^* \quad (16)$$

The purpose of this curious reaction is to provide a large driving force for the synthesis of phosphoenolpyruvate, which is thermodynamically unfavorable when phosphoenolpyruvate is formed simply by the transfer of one phosphate group from ATP. Hydrolysis of the additional high-energy bond of pyrophosphate pulls the reaction effectively toward completion. This is a high price to pay, but it is worthwhile in order to avoid the losses caused by photooxidation.

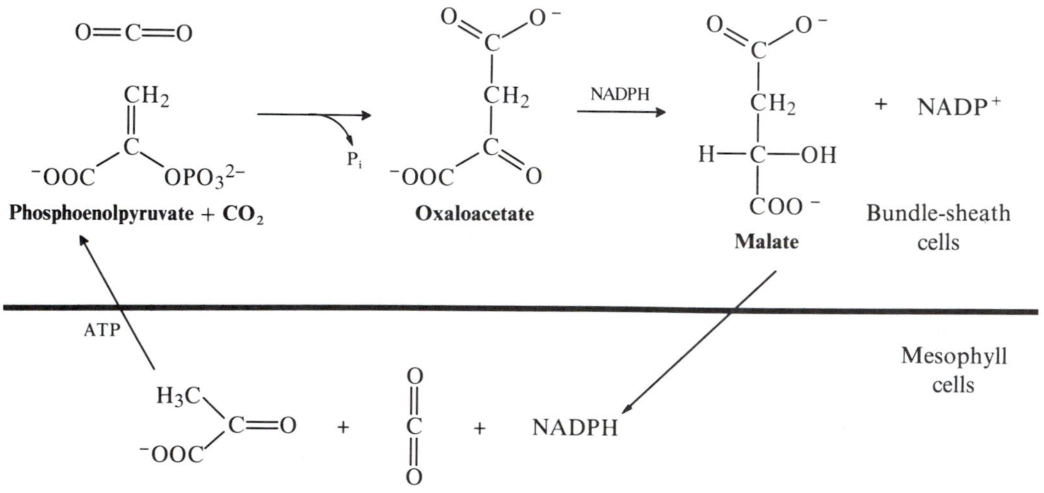

Phosphoenolpyruvate + CO_2
Oxaloacetate
Malate
Bundle-sheath cells
Mesophyll cells

Figure 23-12

The Four-Carbon Cycle. The four-carbon cycle utilizes the energy of ATP to form phosphoenolpyruvate and incorporate CO_2 into a four-carbon compound that is transported into mesophyll cells and decarboxylated in order to increase the concentration of CO_2 in these cells.

Photophosphorylation

The formation of energy-rich ATP from ADP and P_i by photophosphorylation was discovered by D. Arnon and is the second important reaction of photosynthesis (equation 17).

$$ADP + P_i \xrightarrow{h\nu} ATP \tag{17}$$

It is brought about with the energy that is provided by light, but it occurs in the absence of light, in a "dark reaction," after the light processes. The light reaction produces a proton gradient as a by-product of the reduction and oxidation of quinones and the oxidation of two H_2O molecules to O_2, as described earlier (equations 10 and 11), and the proton gradient provides the driving force for the synthesis of ATP (equation 18).

$$3\,H^+_{in} + ADP + P_i \rightleftharpoons 3\,H^+_{out} + ATP \tag{18}$$

The synthesis of ATP by photophosphorylation is important because ATP is required for photosynthesis. It is used to activate 3-phosphoglycerate to 1,3,-bisphosphoglycerate, so that the carboxylic acid group can be reduced to an aldehyde as shown in Figure 23-11, and it is required for the regeneration of ribulose-1,5-bisphosphate from ribulose-5-phosphate (Figure B, Box 23-1).

The synthesis of ATP coupled to proton transport was demonstrated in an important experiment carried out by Jagendorf in 1966. Chloroplasts were soaked in a succinic acid buffer at pH 4 until the succinic acid entered the particle and produced a low pH inside the chloroplast. The chloroplast was then placed in a buffer at pH 8. The flow of protons derived from the succinic acid inside to the more-alkaline solution outside was found to be coupled to the synthesis of ATP from ADP and P_i (Figure 23-13). The use of succinic acid buffers, rather than just an acidic solution with low buffering capacity, meant that a large number of protons were available to be transported down the pH gradient.

Quantitative studies have shown that the reaction proceeds with the stoichiometry shown in equation 18, with three protons transported across the membrane for each molecule of ATP that is formed. Strong evidence for this stoichiometry is provided by the dependence of the ratio of ATP to ADP and

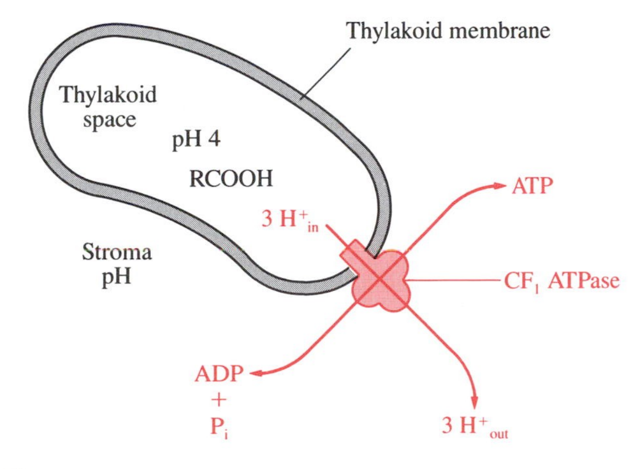

Figure 23-13

The transport of three H^+ ions from the inside to the outside of the thylakoid space through CF_1 ATPase is coupled to the synthesis of a molecule of ATP from ADP and P_i.

phosphate concentrations at equilibrium on the proton concentration outside the chloroplast. The equilibrium constant for the reaction of equation 18 is shown in equation 19.

$$K = \frac{[ATP][H^+]^3_{out}}{[ADP][P_i][H^+]^3_{in}} \tag{19}$$

If three protons are involved, the ratio of [ATP]/[ADP] at a constant concentration of P_i will change according to the third power of the hydrogen ion activity outside the chloroplast. This result was obtained experimentally.

Several additional types of evidence are consistent with this mechanism for photophosphorylation. The mechanism is essentially the same as that for oxidative phosphorylation (Chapter 22), except that the direction of the proton gradient between the inside and the outside of the membrane is opposite that of oxidative phosphorylation.

1. Jagendorf's results have been confirmed and shown to require a pH gradient of 2 or more pH units. Gradients of 4 pH units or more are commonly observed in illuminated chloroplasts. The pH gradient appears to be more important than the membrane potential in providing the proton motive force that drives ATP synthesis in plants.
2. Chloroplasts contain an ATPase, the CF_1F_0ATPase, with properties that are very similar to those of the F_1F_0ATPase of mitochondria (Chapter 22). The hydrolysis of ATP by this ATPase has been shown to pump protons into the chloroplast. Approximately 3.4 protons are pumped for each ATP that is cleaved. This reaction appears to be the reverse of the reaction in which ATP is synthesized.
3. Light has been shown to give rise to a proton gradient across the membrane. This can arise from the synthesis of oxygen from water that is catalyzed by the manganese enzyme and from the oxidation of reduced plastoquinone by cytochrome. The addition of ATP has been shown to produce a proton gradient that reverses this electron-transfer reaction; that is, it results in the transfer of electrons from reduced cytochrome *f* to plastoquinone, which forms reduced plastoquinone. This shows that the conversion of an oxidation-reduction reaction to a proton gradient and ATP is reversible, at least in part.
4. A flash of light has been shown to result in subsequent ATP synthesis in the dark. The amount of ATP that is formed is proportional to the amount of the proton gradient that is produced by the light.

However, all of the foregoing evidence does not *prove* that the proton gradient is directly responsible for the synthesis of ATP. It could also be explained by the synthesis of some other intermediate, "$\sim$X," which then gives rise to ATP. Direct evidence that light-driven proton transport and a proton gradient can synthesize ATP was provided by a simple but ingenious experiment conducted by Stoekenius and Racker. This experiment was in fact carried out with the F_1F_0 ATPase of mitochondria instead of the CF_1F_0 ATPase of chloroplasts, but the properties of the two enzymes are very similar.

Certain bacteria contain a remarkable protein pigment, *bacteriorhodopsin*, that pumps a proton across the bacterial membrane when it absorbs light. A similar pigment, *rhodopsin*, is responsible for the detection of light in vision, as described in Chapter 29. The molecule that is responsible for the color of rhodopsin and bacteriorhodopsin is a protonated imine of the aldehyde *retinal*.

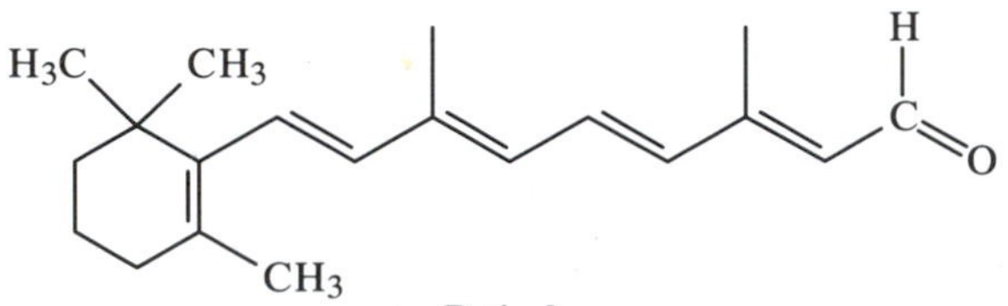

Retinal

The aldehyde adds to the ε-amino group of a lysine residue in bacteriorhodopsin to form an imine. The imine is buried in the interior of the protein, where it is surrounded by seven α-helices that pass from the inside to the outside of the membrane (Figure 23-14). However, the nitrogen atom of the imine can accept a proton from the carboxylic acid group of aspartate-96, which in turn can accept protons from the inside of the cell, as shown in Figures 23-14 and 23-15. When the pigment absorbs light, the C-13 double bond is isomerized from *trans* to *cis*. This rotates the imine and its protonated nitrogen atom so that the proton can be transferred to the carboxylate group of aspartate-85, from which it dissociates to the *outside* of the membrane. The imine then isomerizes back to the all *trans* structure, without transporting a proton, so that it can accept another proton from the inside of the membrane and repeat the cycle. The final result is that one proton is transferred from the inside to the outside of the membrane in each reaction cycle (Figures 23-14 and 23-15).

Stoekenius and Racker inserted both bacteriorhodopsin and the F_1F_0 ATPase into phospholipid vesicles. When these vesicles were illuminated in the presence of ADP and inorganic phosphate, ATP was synthesized (Figure 23-16). The light causes bacteriorhodopsin to pump protons into the vesicles, which

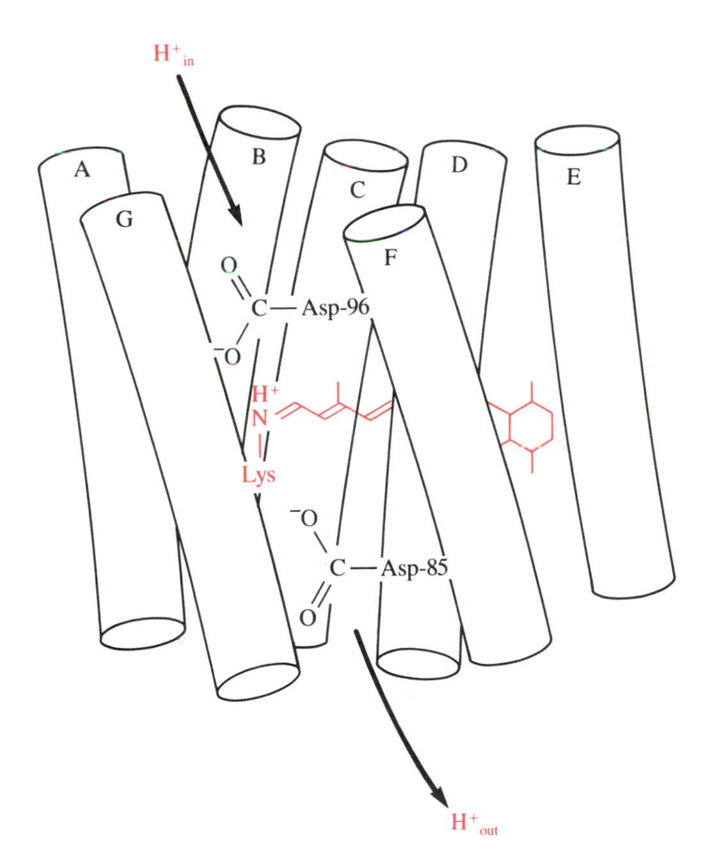

Figure 23-14

Bacteriorhodopsin is a protein in the cell wall of certain halobacteria, which exist in an environment with a high concentration of salt. The protein contains a molecule of retinal that is bonded to the nitrogen atom on the side chain of a lysine residue to form an imine (Figure 23-15) and is surrounded by seven transmembrane helices. A proton from the cytoplasm can be transferred to the carboxylate group of aspartate-96 and then to the nitrogen atom of the imine. Light causes isomerization of the imine, so that the proton is transferred to the carboxylate group of aspartate-85 and then to the outside of the cell (Figure 23-15).

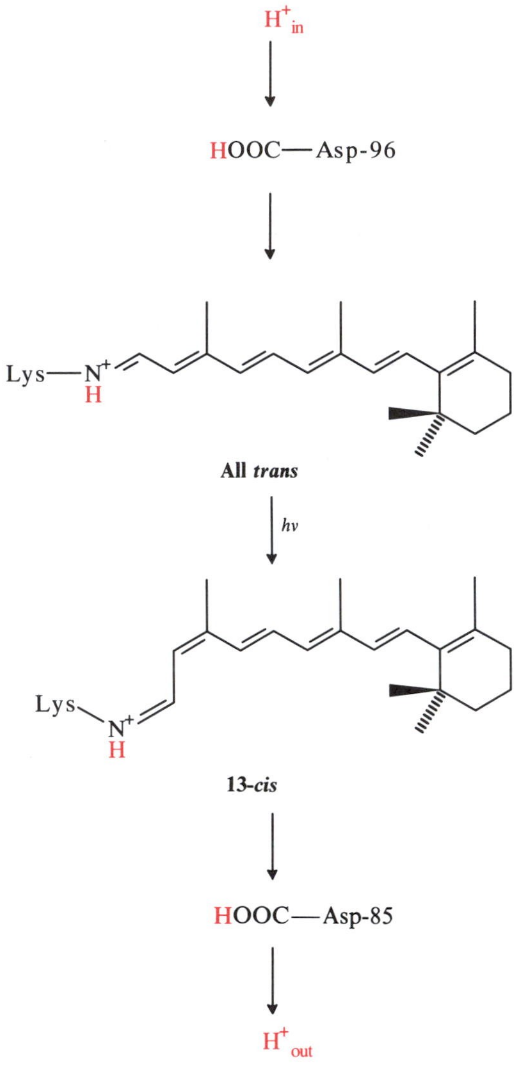

H^+_{in}

HOOC—Asp-96

Lys—N+ H

All *trans*

hv

Lys N+ H

13-*cis*

HOOC—Asp-85

H^+_{out}

Figure 23-15

Light absorption by bacteriorhodopsin drives the transport of a proton across a membrane. A proton from the inside of the membrane can protonate the carboxylate group of aspartate-96, which then donates a proton to the nitrogen atom of the imine formed from retinal and the amino group of a lysine residue. Light causes isomerization of the retinal imine from the all *trans* to the 13-*cis* conformation. This changes the position of the proton in such a way that it can be transferred to the carboxylate group of aspartate-85, from which it dissociates to the outside of the cell. The unprotonated retinal imine can then isomerize back to the all *trans* form, so that the cycle can be repeated.

creates a proton gradient. The ATPase then utilizes this proton gradient to synthesize ATP as the protons pass through the ATPase to return to the outside solution. This experiment shows that light can produce a proton gradient and the proton gradient can drive the synthesis of ATP from ADP in a simple, defined system that contains only the bacteriorhodopsin that creates the proton gradient and the ATPase, which utilizes the transport of protons down a concentration gradient to bring about the synthesis of ATP.

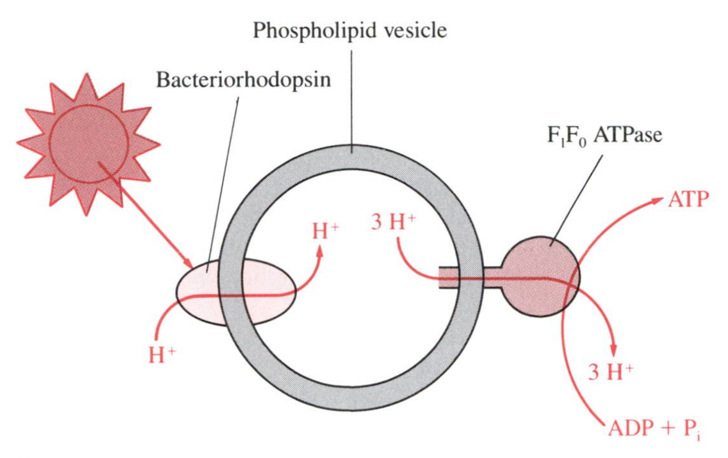

Figure 23-16

The Stoekenius-Racker experiment demonstrated that transport of protons across a membrane causes the synthesis of ATP from ADP and phosphate in a simple, purified system. Absorption of light by bacteriorhodopsin pumps protons into a phospholipid vesicle, and the F_1F_0 ATPase synthesizes ATP from ADP and inorganic phosphate when it transports three protons from the inside to the outside of the vesicle.

Summary

Photosynthesis provides the energy for life on earth by using the energy of sunlight to excite an electron of chlorophyll. Light is absorbed by chlorophyll and by a number of accessory pigments that absorb light at shorter wavelengths and transfer their excitation energy to chlorophyll. These excited electrons in chlorophyll reduce $NADP^+$ to NADPH, which provides the reducing equivalents that ultimately convert carbon dioxide into carbohydrate through a series of dark reactions. Excitation also produces an oxidizing equivalent. Green plants utilize four oxidizing equivalents and a manganese enzyme to synthesize molecular oxygen from water. Bacteria and other photosynthetic microorganisms may utilize the oxidizing equivalents in other reactions, such as the oxidation of hydrogen sulfide to sulfur. Most photosynthetic bacteria have only a single photosystem. However, green plants have two photosystems in the "Z system," which act sequentially to provide enough energy to form molecular oxygen and NADPH from water and $NADP^+$.

The formation of O_2 and H_2O and the oxidation of reduced plastoquinone by cytochrome *b* also release protons in the interior of the chloroplast. This builds up a pH gradient across the chloroplast membrane that is used to synthesize ATP from ADP and inorganic phosphate by photophosphorylation. The CF_1F_0 ATPase of chloroplasts is a proton pump that transports three protons into the chloroplast for each ATP that is cleaved. It couples the synthesis of ATP in photophosphorylation to the transport of three protons out of the chloroplast. Certain bacteria contain bacteriorhodopsin, a carotenoid-protein with retinal imine as its prosthetic group. Bacteriorhodopsin binds a proton from one side of the membrane to the nitrogen atom of retinal imine, undergoes isomerization of the retinal imine when it is excited by light, and then discharges the proton to the other side of the membrane; it is a light-activated proton pump.

Carbon dioxide is incorporated into carbohydrate and reduced in a series of dark reactions. Ribulose-1,5-bisphosphate carboxylase catalyzes the addition of carbon dioxide to the 2-position of ribulose-1,5-bisphosphate to give two molecules of 3-phosphoglycerate. In other dark reactions of photosynthesis, 3-phosphoglycerate is phosphorylated by ATP and phosphoglycerate kinase to 1,3-bisphosphoglycerate, which is reduced by NADPH and glyceraldehyde-3-phosphate dehydrogenase to glyceraldehyde-3-phosphate. Part of the glyceraldehyde-3-phosphate is used for carbohydrate biosynthesis. The ribulose-1,5-bisphosphate consumed in the carboxylation reaction is regenerated from glyceraldehyde-3-phosphate and fructose-6-phosphate by the actions of transketolase and transaldolase, together with triosephosphate isomerase and sedoheptulose-1,7-bisphosphatase.

ADDITIONAL READING

Wu, D., and Boyer, P. D. 1986. Bound adenosine-5′-triphosphate formation, bound adenosine-5′-diphosphate and inorganic phosphate retention, and inorganic phosphate oxygen exchange by chloroplast adenosine triphosphatase in the presence of Ca^{2+} or Mg^{2+}. *Biochemistry* 25: 3390.

Deisenhofer, J., Michel, H., and Huber, R. 1985. The structural basis of photosynthetic light reactions in bacteria. *Trends in Biochemical Science* 10: 243.

Brudvig, G. W., Beck, W. F., and de Paula, J. C. 1989. Mechanism of photosynthetic water oxidation. *Annual Review of Biophysics and Biophysical Chemistry* 18: 25.

Brudvig, G. W., and Crabtree, R. H. 1986. Mechanism for photosynthetic O_2 evolution. *Proceedings of the National Academy of Sciences* 83: 4586.

Foyer, C. H. 1984. *Photosynthesis.* New York: Wiley.

Govindjee, ed., 1982. *Photosynthesis*, vols. 1 and 2. New York: Academic Press.

Lundquist, T., and Schneider, G. 1989. Crystal structure of the complex ribulose-1,5-bisphosphate carboxylase and a transition state analogue, 2-carboxy-D-arabinitol-1,5-bisphosphate. *Journal of Biological Chemistry* 264: 7078.

PROBLEMS

1. Explain the utilization of water molecules in photosynthesis by green plants, as described in equation 1.

2. Why is the Hill reaction so important?

3. Explain the importance of accessory pigments in photosynthesis. What is the range of light absorption wavelengths that they should have to be effective? Chlorophyll *b* has a strong absorption maximum at 453 nm.

4. Explain how the size of the pool of chlorophyll molecules that can transfer energy to the active pair of chlorophyll molecules can be determined.

5. Why did the Z scheme of photosynthesis, consisting of two photosystems, evolve before human beings appeared on earth?

6. Why does ascorbic acid increase the yield of carbohydrate from photosynthesis at low light intensity?

Oxidations and Oxygenations

24

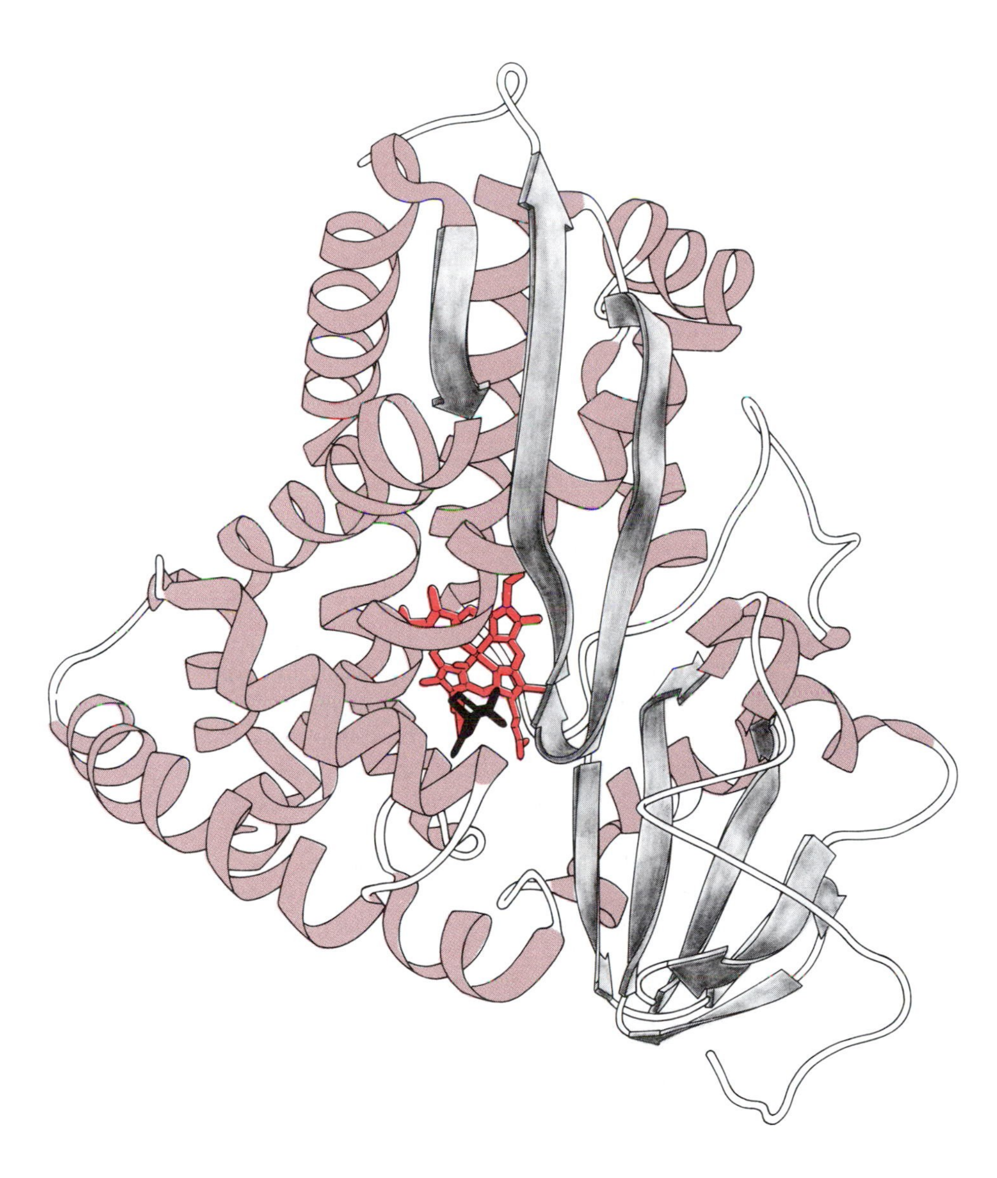

A ribbon drawing of the structure of cytochrome P450. The porphyrin ring (red) is visible in the center of the molecule and a molecule of camphor (black) is bound near the bottom of the porphyrin ring. (Adapted from a computer-generated drawing by Dan Peisach of a structure determined by Thomas Poulos and Joseph Kraut.)

In this chapter, we consider some of the reactions in which molecular oxygen participates directly in the oxidation of molecules. In oxidase reactions, oxygen is reduced but is not incorporated into another molecule; and, in oxygenation reactions, one or both oxygen atoms of oxygen are incorporated into a substrate molecule. We have already described hemoglobin, which binds oxygen reversibly to ferrous iron in a porphyrin (Chapter 10). This is a simple binding process and not an oxidation-reduction reaction. Cytochrome oxidase collects four electrons from the oxidation of substrates in the citric acid cycle and donates them to a molecule of oxygen to form two molecules of water, with the uptake of four protons (equation 1; see Chapter 22).

$$4e^- + 4H^+ + O_2 \longrightarrow 2H_2O \qquad (1)$$

It is an *oxidase* because it accepts electrons from another molecule and uses them to reduce oxygen. The manganese complex of photosynthesis catalyzes the reverse reaction (equation 2).

$$\text{Enzyme}^{4+} + 2H_2O \longrightarrow \text{Enzyme} + O_2 + 4H^+ \qquad (2)$$

It collects four oxidizing equivalents in the light reaction of green plants and uses them to form one molecule of oxygen from two water molecules, with the release of four protons (Chapter 23).

The reactions of equations 1 and 2 result in oxidation or reduction by the transfer of electrons from one molecule to another.

Oxygenases, on the other hand, bring about oxidation by the direct transfer of oxygen from molecular oxygen, O_2, to the substrate. A dioxygenase catalyzes the transfer of two oxygen atoms to the substrate, as in the oxidation of catechol to *cis,cis*-muconate catalyzed by pyrocatechase (equation 3).

$$\text{catechol (OH, OH)} + {}^{18}O_2 \longrightarrow \text{CO}^{18}\text{OH}, \text{CO}^{18}\text{OH} \qquad (3)$$

Pyrocatechase and several enzymes that catalyze similar reactions contain nonheme iron, Fe^{3+}, and are bright red; others contain Fe^{2+}. A monooxygenase catalyzes the transfer of only one oxygen atom. This uses only two of the four oxidizing equivalents of oxygen, so that a reducing agent is required to convert the remaining two oxidizing equivalents into water.

$\cdot\ddot{O}-\ddot{O}\cdot$

1

Molecular oxygen has two unpaired electrons.

Oxygen is one of the big guns of biochemistry. Oxygen itself reacts remarkably slowly for a molecule that exists in the triplet state with two unpaired electrons and is a diradical, **1**. Most of the biological reactions of oxygen are catalyzed by metalloenzymes, which usually contain iron or copper. This combination of oxygen and catalysts can bring about a number of unusual reactions of molecules that are ordinarily unreactive. If there is a reaction for which there is no obvious mechanism or precedent in the ordinary reactions of organic chemistry, it is likely that it entails either molecular oxygen or vitamin B_{12}.

The increase in the reactivity of oxygen that can be brought about by metal ions makes oxygen a powerful reagent for the detoxification of xenobiotic molecules. Metal catalysis also makes oxygen reactive with hydrocarbons, which are unreactive toward most chemical reagents. Several classes of microorganisms contain oxygenases that utilize oxygen to attack hydrocarbons and convert them into molecules that can be used by the microorganism or dispersed in water, so that they can be degraded by other organisms. This is the basis for the use of such microorganisms to disperse oil spills. Similar oxygenases in complex organisms such as people attack hydrocarbons, including some toxic molecules that are taken up from the environment. These reactions may afford a method for producing foodstuffs from hydrocarbons, which cannot be metabolized directly by most organisms.

Most oxidations by molecular oxygen are strongly favorable energetically; they have large negative values of $\Delta H°$ and $\Delta G°$. In some cases, this energy is large enough to be emitted in the form of visible light, as bioluminescence. There are very few biological reactions that release this much energy. Thus, the high reactivity of oxygen is accompanied by a strong thermodynamic driving force for its reactions.

Two-Electron Oxidations and Monooxygenases

Reduced flavins—$FADH_2$, for example—are produced in flavin-catalyzed oxidation reactions, such as the α,β-dehydrogenation of a fatty acyl CoA (equations 4 and 5).

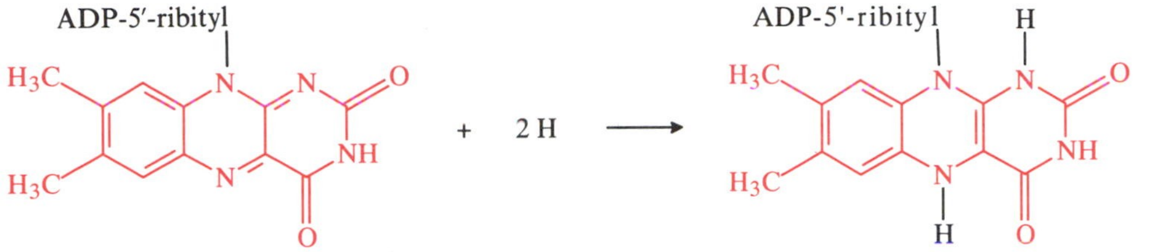

$$\text{FAD (ADP-5'-ribityl isoalloxazine)} + 2\,H \longrightarrow FADH_2 \quad (4)$$

FAD   FADH$_2$

$$RCH_2\text{—}CH_2\text{—}\overset{O}{\overset{\|}{C}}\sim SCoA + FAD \longrightarrow RCH\text{=}CH\text{—}\overset{O}{\overset{\|}{C}}\sim SCoA + FADH_2 \quad (5)$$

We have seen in Chapter 16 that the reduced flavin is reoxidized to FAD in the presence of oxygen, with the production of hydrogen peroxide as a by-product (equation 6) because oxygen has four oxidizing equivalents and the oxidation of $FADH_2$ is a two-electron oxidation.

$$FADH_2 + O_2 \longrightarrow FAD + HOOH \quad (6)$$

This enzyme is an oxidase, because the substrate is oxidized by the coenzyme, FAD, and molecular oxygen reacts only in the second step of the reaction, to regenerate FAD from $FADH_2$.

In mitochondria, however, the reduced flavin feeds electrons into the electron-transport chain. These electrons are then utilized for energy production through oxidative phosphorylation, as described in Chapter 22. The electrons are first transferred to an electron-transfer flavoprotein (ETF) and then to ubiquinone reductase (another flavoprotein) and to ubiquinone according to Figure 24-1 before they enter the mitochondrial electron-transport chain.

$$FADH_2 + ETF \longrightarrow FAD + ETFH_2 \xrightarrow{\text{Q (Ubiquinone)}} QH_2 \xrightarrow[\text{ATP}]{\frac{1}{2}O_2} Q + H_2O$$

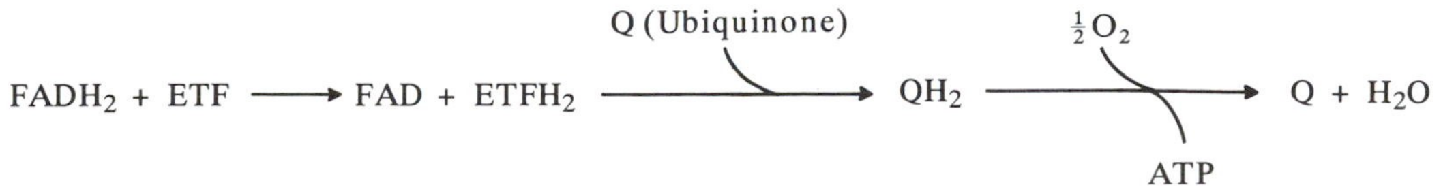

Figure 24-1

Electron Transfer from $FADH_2$ Derived from the Dehydrogenation of Acyl CoA to O_2 Through the Electron-Transport Pathway. $FADH_2$ reduces an electron-transport flavoprotein (ETF), which reduces ubiquinone. Reduced ubiquinone (QH_2) permeates the electron-transport membrane, and the electron-transport complexes catalyze its oxidation by O_2. ATP is produced by the electron-transport pathway in the oxidation of $FADH_2$ (Chapter 22).

Desaturases

The oxidation between the α- and β-carbon atoms of a thiol ester is relatively easy to carry out, because the carbonyl group activates the α-hydrogen atoms. However, the oxidation of stearic acid to oleic acid, which has a 9, 10 double bond near the middle of the hydrocarbon chain, is a much more difficult reaction because it requires attack on the center of the hydrocarbon chain of stearic acid, a saturated 18-carbon fatty acid, as shown in Figure 24-2. This reaction uses Fe^{2+} at the active site, but the oxidant is molecular oxygen. A molecule of $FADH_2$ reduces $(Fe^{3+})_2$ associated with the desaturase in a process that is mediated by cytochrome b_5, as shown in Figure 24-3. The ferrous ions reduce oxygen to a peroxide derivative, which then removes two hydrogen atoms from the 9- and 10-carbon atoms of the hydrocarbon to form the double bond and two molecules of water. This enzyme is known as a *desaturase* because it converts a saturated hydrocarbon into an unsaturated hydrocarbon. It is found on the inner surface of the endoplasmic reticulum membrane and is associated with an electron-transport system, which forms $FADH_2$ from FAD and provides the two additional electrons required to use up the four oxidizing equivalents of O_2.

The desaturase contains ferrous ions, Fe^{2+}, which provide the two reducing equivalents that are used for the overall reaction with oxygen, as shown in Figure 24-3. The Fe^{2+} probably reacts directly with oxygen in this process to produce a peroxide derivative, perhaps an iron-peroxide, that oxidizes the hydrocarbon. The reduced desaturase is regenerated from reduced cytochrome b_5, which is formed from reduced flavin by cytochrome b_5 reductase. The reduced reductase is regenerated from NADH by cytochrome b_5 reductase. This pathway provides the reducing equivalents that are required for a number of reactions of this type.

Several similar desaturases catalyze the formation of double bonds at different positions up to 9 carbons away from the acyl group. However, mammals cannot synthesize linoleic acid, an 18-carbon acid with double bonds in the 9 and 12 positions; its formal name is $\Delta^{9,12}$-octadecadienoic acid. We will see shortly that linoleic acid is required for the synthesis of arachidonic acid, which is a precursor of prostaglandins, thromboxanes, and leukotrienes. These compounds play important regulatory roles in animals. Therefore, linoleic acid is an *essential* unsaturated fatty acid that must be obtained in the diet.

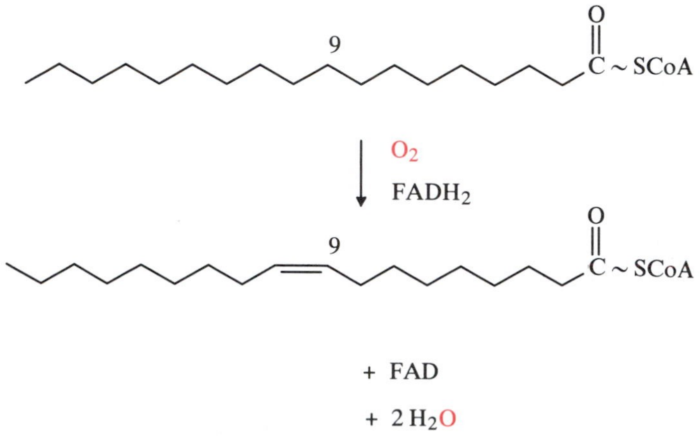

9 O C ~ SCoA

↓ O_2 $FADH_2$

9 O C ~ SCoA

+ FAD

+ 2 H_2O

Figure 24-2

Dehydrogenation of Stearyl CoA to Oleyl CoA by the Acyl CoA Desaturase.

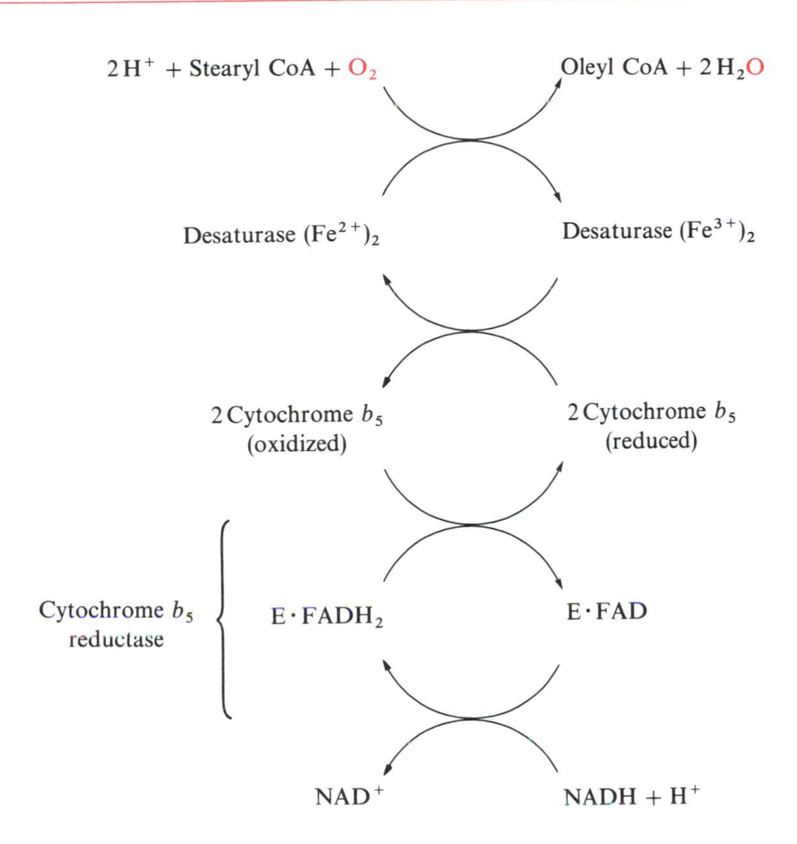

Figure 24-3

Electron Transfer to Fatty Acyl CoA Desaturase Through the Electron-Transport Chain. In the dehydrogenation of stearyl CoA to oleyl CoA by molecular oxygen at the active site of the desaturase enzyme, the Fe^{2+} is oxidized to Fe^{3+}. The iron must be reduced to allow another cycle of desaturation. Reduction is by cytochrome b_5 in the membrane, and oxidized cytochrome b_5 is reduced by cytochrome b_5 reductase and NADH.

Cytochrome P450

The best-known and probably the most-widespread monooxygenases are the cytochrome P450 class of enzymes. These oxygenases contain heme iron, which exhibits an absorption maximum near 450 nm when carbon monoxide is bound to the iron. This spectral band is the basis for the designation of these heme proteins as cytochrome P450s. Some of them may be regarded as the incinerators of the cell—they will attack almost anything, including drugs, pesticides, and carcinogens, and convert their substrates into a form that can be degraded further or excreted. However, they are also required for the biosynthesis of steroid hormones from cholesterol, a process in which methyl groups and methylene groups are hydroxylated. There are many P450 enzymes with different specificities that are important in biosynthesis for the formation of hormones, neurotransmitters, and other useful compounds. However, P450 enzymes can also convert harmless hydrocarbons into active carcinogens.

The cytochrome P450s catalyze the insertion of an oxygen atom into a carbon–hydrogen bond according to equation 7.

$$-\overset{|}{\underset{|}{C}}-H + O_2 + NADH + H^+ \longrightarrow -\overset{|}{\underset{|}{C}}-OH + H_2O + NAD^+ \quad (7)$$

This reaction is one example of a monooxygenase reaction, which is generally formulated according to equation 8.

$$S + O_2 + AH_2 \longrightarrow SO + H_2O + A \quad (8)$$

The high reactivity of oxygen that is brought about by its reaction with iron allows these enzymes to catalyze the reaction of oxygen with unreactive groups on drugs and other foreign molecules that find their way into an organism. For example, they are able to convert methyl groups into hydroxymethyl groups, which can react with acylating or sulfating agents to produce compounds that are rapidly excreted. The oxidation of methyl groups to hydroxymethyl groups is also required for the synthesis of steroids. The oxygenation of hydrocarbons by cytochrome P450s in bacteria can help to clean up oil spills and has a potential for the production of foodstuffs.

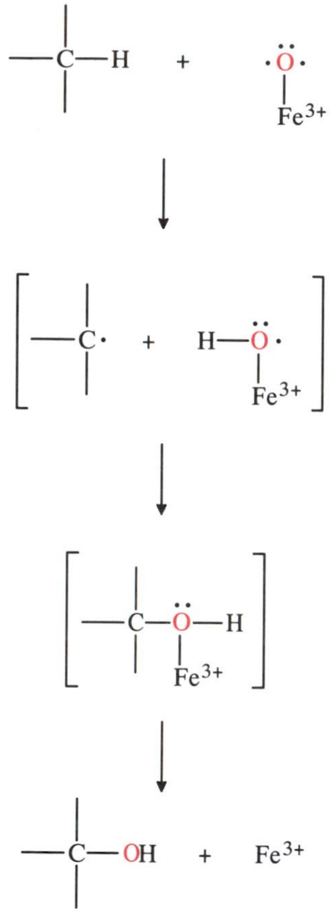

The P450 enzymes are able to attack hydrocarbons and other stable molecules by the use of a radical mechanism. They bring about oxygenation by the perferryl ion, FeO^{3+}, which may be regarded as an oxygen atom that is attached to a ferric ion in the active site of the enzyme, as shown in Figure 24-4. The exact electronic structure of this species is still a subject of intensive debate and research, and it is often represented in alternative ways such as Fe(V)═O or as $\cdot N^{+}$—Fe(IV)═O, in which $\cdot N^{+}$ represents a heme cation radical. In these two alternative structures, oxygen is shown with an octet of paired and bonding electrons. The important point for the present purpose is that oxygen in this species is highly reactive and can abstract a hydrogen atom from a substrate.

The reaction can be represented as in Figure 24-4, in which the perferryl ion is shown as an oxygen atom that is coordinated to Fe^{3+} through one of its electron pairs and behaves as if it were a diradical. This picture has the advantage that oxygen is known to be the chemically reactive species, and it reacts as a radical. In the hydroxylation of a hydrocarbon, it abstracts a hydrogen atom to form the corresponding carbon radical. The enzyme then rapidly donates the OH group to the radical to give the product and the Fe^{3+}-enzyme. There may be rearrangement of the carbon chain of the substrate, because certain radicals undergo rearrangement rapidly. Rearrangement is observed when it can occur within approximately 10^{-8} seconds, the time that is required for recombination of the radical with $FeOH^{3+}$.

If a heteroatom, such as S or N, is adjacent to carbon and is oxidized, it may react with FeO^{3+} to give the corresponding oxide. Alternatively, the FeO^{3+} may attack the C–H bond to form a carbon radical, which is followed by hydroxylation and then cleavage of the bond to the heteroatom to give a carbonyl compound, as shown in Figure 24-5. This is a common mechanism for the dealkylation of alkylamines.

Figure 24-4

Hydroxylation of a Hydrocarbon by a Cytochrome P450 Monooxygenase.

X = N or S

Figure 24-5

Oxygenation of Amines and Thioethers by a Cytochrome P450 Monooxygenase. An amine or thioether may be oxygenated either on the heteroatom itself or on the adjacent carbon atom. Hydroxylation of a carbon bonded to nitrogen produces a carbinolamine, which undergoes elimination of the amine to form the aldehyde. This is the mechanism by which cytochrome P450 dealkylates amines.

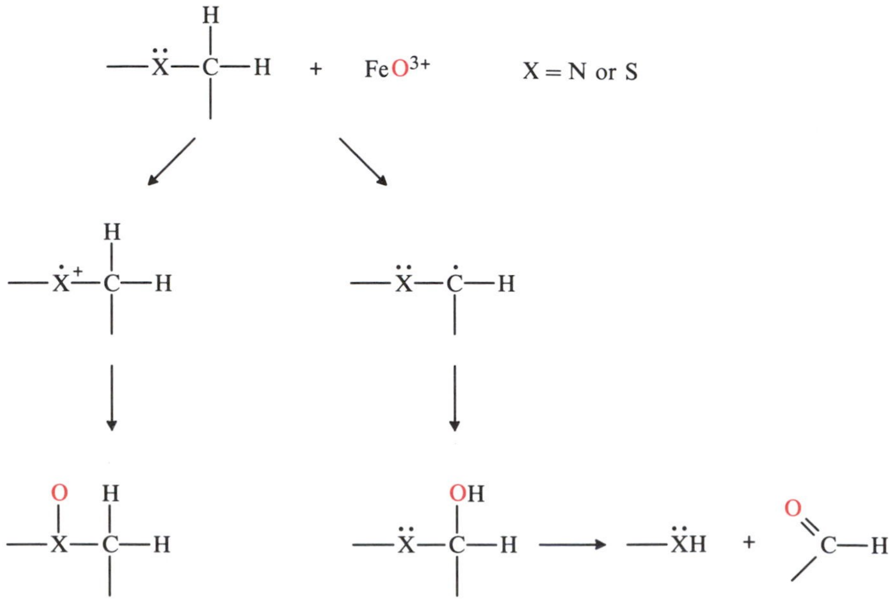

The active enzyme-FeO^{3+} species can be regenerated from an oxygen donor, such as iodosylbenzene (XO), as shown in the upper pathway of Figure 24-6. However, under physiological conditions, it is ordinarily regenerated from molecular oxygen, as shown in the lower pathway of Figure 24-6. The reaction requires two 1-electron reduction steps and the addition of two protons in order to form FeO^{3+} and H_2O from Fe^{3+} and O_2. The first reduction converts the enzyme-substrate complex from $E_S^{Fe^{3+}}$ into $E_S^{Fe^{2+}}$; the second reduction occurs after the addition of oxygen and gives $E_S^{FeO_2^+}$. The electrons can be supplied by an NADPH-cytochrome P450 reductase and cytochrome b_5. One of the oxygen atoms is removed as water, after the addition of two protons, which gives the active form of the enzyme-substrate complex, $E_S^{FeO^{3+}}$

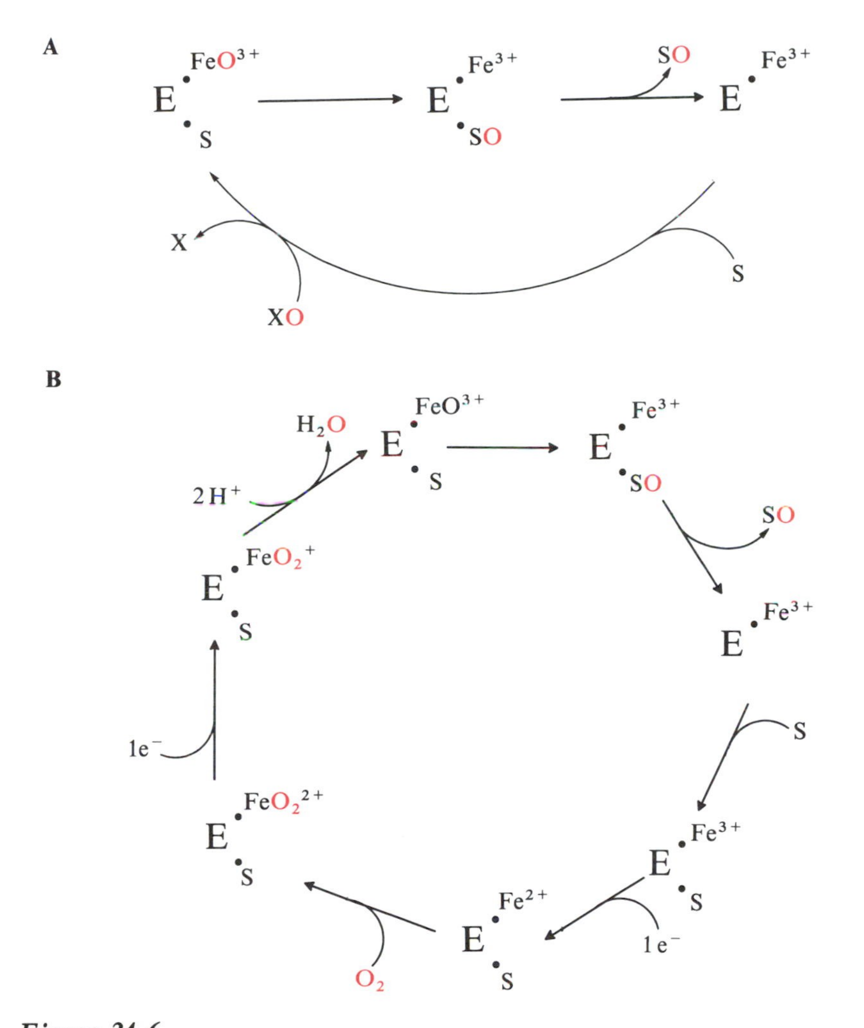

Figure 24-6

Oxygenation of Cytochrome P450. Oxygenation of a substrate by the perferryl form of cytochrome P450 leaves the heme with Fe^{3+} and no oxygen for another cycle of substrate oxygenation. Part A shows that iodosobenzene (XO) can oxygenate the heme-Fe^{3+} to the perferryl species, which can oxygenate another substrate molecule. Part B shows how the oxygenation normally proceeds in a series of steps beginning with the binding of a molecule of substrate. In the succeeding steps, a reducing agent donates one electron to reduce Fe^{3+} to Fe^{2+}, a molecule of O_2 is bound by Fe^{2+}, another electron is donated to the oxygenated heme, and two protons are finally required to dehydrate the reduced and oxygenated heme, regenerating the perferryl species with the substrate in position to be oxygenated in the next cycle.

The incinerator activity of monooxygenases can also give rise to toxic products, of which the most important are probably carcinogens. For example, the hydroxylation of benzpyrene gives an epoxide, which can then react with water to give a diol. A second epoxidation gives the diol epoxide shown in equation 9.

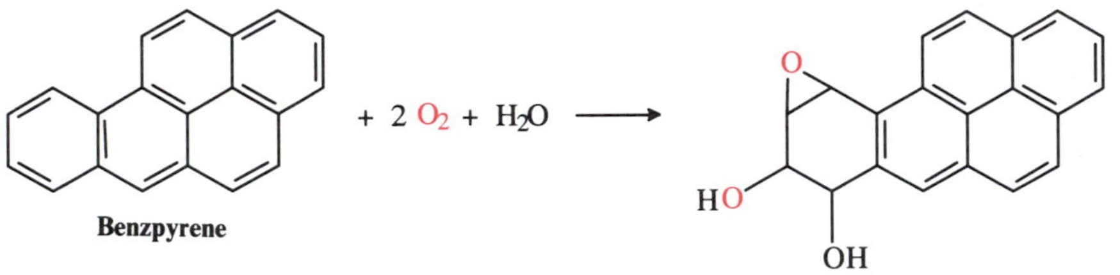

$$\text{Benzpyrene} + 2\,O_2 + H_2O \longrightarrow \text{diol epoxide} \quad (9)$$

This is one of the most-carcinogenic compounds known and has been called the "ultimate carcinogen." It is thought to act by intercalating into DNA and reacting covalently with the DNA to produce a mutation.

Phenylalanine Hydroxylase

The synthesis of tyrosine from phenylalanine and oxygen is catalyzed by phenylalanine hydroxylase according to equation 10.

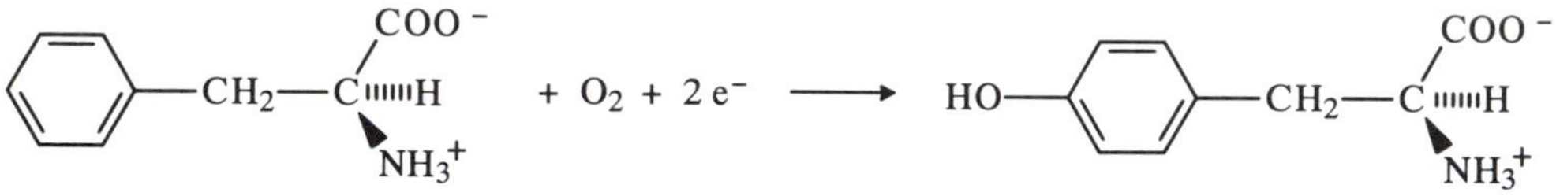

$$C_6H_5{-}CH_2{-}CH(COO^-)(NH_3^+) + O_2 + 2\,e^- \longrightarrow HO{-}C_6H_4{-}CH_2{-}CH(COO^-)(NH_3^+) \quad (10)$$

This enzyme is also a monooxygenase that uses only two of the four oxidizing equivalents of molecular oxygen in the formation of the tyrosine product. The other two oxidizing equivalents of O_2 are converted into water by a coenzyme, 5,6,7,8-tetrahydrobiopterin (Figure 24-7). The oxidized coenzyme, dihydrobiopterin, is converted back into tetrahydrobiopterin by reduction with NADH, catalyzed by dihydropteridine reductase.

Phenylalanine hydroxylase is of particular interest because of the consequences of its absence in children with *phenylketonuria.* This is the first metabolic disease that was shown to be caused by the absence of an enzyme in

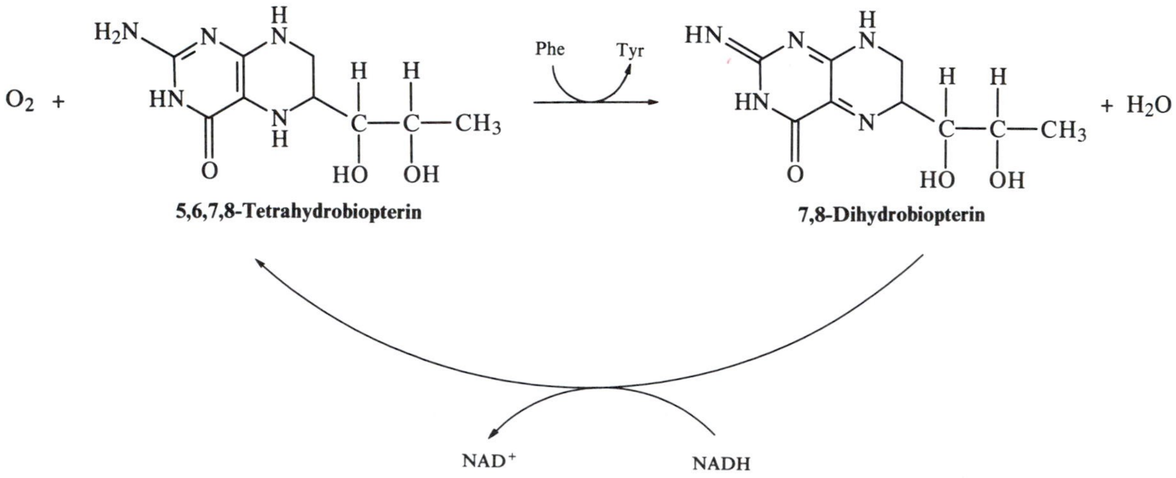

$$O_2 + \text{5,6,7,8-Tetrahydrobiopterin} \xrightarrow{\text{Phe} \rightarrow \text{Tyr}} \text{7,8-Dihydrobiopterin} + H_2O$$

(7,8-Dihydrobiopterin is reduced back to 5,6,7,8-Tetrahydrobiopterin: NADH → NAD^+)

Figure 24-7

The synthesis of tyrosine from phenylalanine catalyzed by phenylalanine hydroxylase utilizes the coenzyme 5,6,7,8-tetrahydrobiopterin.

affected people; it is the result of an inborn error of metabolism. When phenylalanine cannot be converted into tyrosine, its concentration increases in the tissues and blood and it is converted into phenylpyruvate by transamination. The phenylpyruvate is excreted in the urine; that is, there is phenylketonuria. The phenylpyruvate is toxic and gives rise to mental retardation if it is allowed to accumulate.

Infants are now screened for the presence of a high concentration of phenylalanine in their blood and are placed on a diet that is low in phenylalanine if they have phenylketonuria. This prevents the development of serious consequences from the disease and the children can generally tolerate a less-rigid diet when they become older.

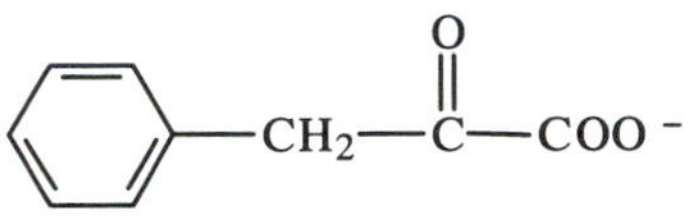

Phenylpyruvate

Serotonin

Tetrahydrobiopterin is also the coenzyme that provides reducing equivalents for the synthesis of serotonin by the hydroxylation of tryptophan (Figure 24-8). The initial product of hydroxylation, 5-hydroxytryptophan, is decarboxylated to give 5-hydroxytryptamine, which is better known as *serotonin.* A potent and extremely interesting compound, serotonin has many functions, a large fraction

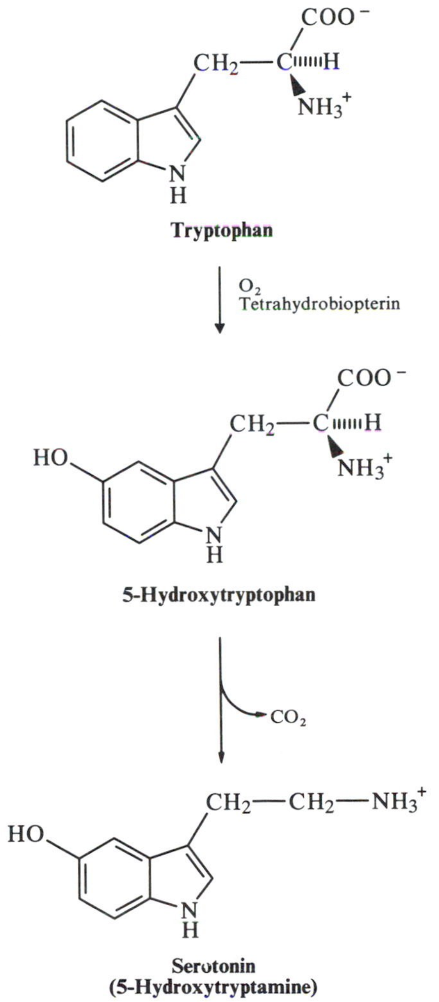

Tryptophan

O_2
Tetrahydrobiopterin

5-Hydroxytryptophan

CO_2

Serotonin (5-Hydroxytryptamine)

Figure 24-8

The Synthesis of Serotonin by 5-Hydroxylation and Decarboxylation of Tryptophan.

of which we would like to avoid. It stimulates the contraction of smooth muscles and causes vasoconstriction. This increases the blood pressure and causes headaches. Serotonin is a component of the venom of some toads and wasps. It also serves as a neurotransmitter that passes a signal from one nerve to another across a synapse. There is increasing evidence, but not yet decisive proof, that its metabolism is abnormal in manic-depressive psychosis. It is even possible that this metabolic disorder is the cause of such a psychosis.

The Synthesis of Catecholamines

Tyrosine hydroxylase catalyzes the hydroxylation of tyrosine to 3,4-dihydroxyphenylalanine, which is widely known as DOPA (Figure 24-9). This is the third reaction in which tetrahydrobiopterin serves as the cofactor. It gives rise to three

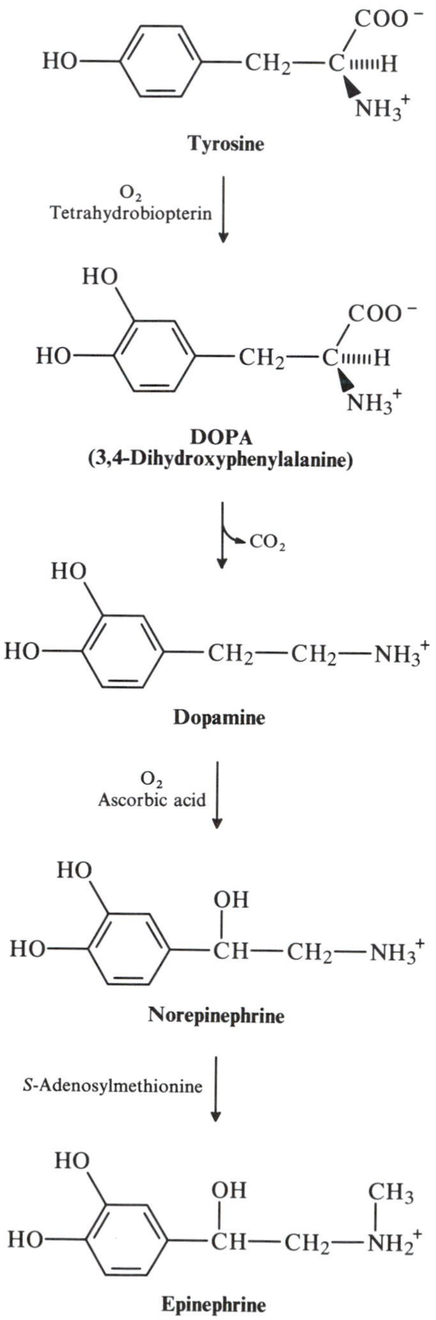

Figure 24-9

The Synthesis of Neurotransmitters and Hormones from Tyrosine.

Tyrosine

O_2
Tetrahydrobiopterin

DOPA
(3,4-Dihydroxyphenylalanine)

CO_2

Dopamine

O_2
Ascorbic acid

Norepinephrine

S-Adenosylmethionine

Epinephrine

important hormones or neurotransmitters: dopamine, norepinephrine, and epinephrine. The decarboxylation of DOPA gives dopamine, hydroxylation of dopamine gives norepinephrine, and methylation of norepinephrine on the amino group gives epinephrine (Figure 24-9). All these compounds contain *ortho*-hydroxyl groups on the benzene ring and are therefore known as catecholamines.

Dopamine is synthesized in the brain. In Parkinson's disease, the cells that release dopamine degenerate and the concentration of dopamine in the brain decreases. Parkinsonism is characterized by tremor, disturbance of posture, and muscular weakness. The administration of dopamine is not helpful, because dopamine does not cross the blood-brain barrier. However, L-DOPA does cross the blood-brain barrier and is decarboxylated to dopamine. It is given simultaneously with a DOPA decarboxylase inhibitor, which does not enter the central nervous system but prevents decarboxylation of L-DOPA in the liver. Administration of L-DOPA and the decarboxylase inhibitor ameliorates many of the symptoms of the disease.

The enzyme that catalyzes the hydroxylation of dopamine to norepinephrine, dopamine monooxygenase, contains two reduced copper ions that provide the two electrons needed to balance two of the oxidizing equivalents of O_2. These copper ions are reduced from the Cu^{2+} to the Cu^{+} oxidation state by ascorbic acid (vitamin C) and are reoxidized when one of the oxygen atoms of O_2 is reduced to water (equations 11a and 11b).

$$E_{Cu^+Cu^+} + \text{Dopamine} + O_2 + 2H^+ \longrightarrow E_{Cu^{2+}Cu^{2+}} + \text{Norepinephrine} + H_2O \quad (11a)$$

$$E_{Cu^{2+}Cu^{2+}} + \text{Ascorbic acid} \longrightarrow E_{Cu^+Cu^+} + \text{Dehydroascorbic acid} \quad (11b)$$

Histamine and γ-Aminobutyric Acid (GABA)

Two additional amines, histamine and γ-aminobutyric acid, are mentioned here because they play important roles in the regulation of physiological responses. Histamine is formed by decarboxylation of the amino acid histidine (equation 12), and γ-aminobutyric acid (GABA) is formed by decarboxylation of glutamate (equation 13).

$$\text{Histidine} + H^+ \longrightarrow \text{HN-imidazolyl}{-}CH_2{-}CH_2{-}NH_3^+ + CO_2 \quad (12)$$

Histidine (imidazolyl–CH_2–$C(COO^-)(H)(NH_3^+)$) → Histamine

$$^-OOC{-}CH_2{-}CH_2{-}C(COO^-)(H)(NH_3^+) + H^+ \longrightarrow {}^-OOC{-}CH_2{-}CH_2{-}CH_2{-}NH_3^+ + CO_2 \quad (13)$$

Glutamate → γ-Aminobutyric acid (GABA)

Both of these reactions are catalyzed by enzymes that utilize pyridoxal phosphate as the coenzyme.

Histamine is well known as a vasodilating agent that is released in allergic responses. It also stimulates the release of acid into the stomach by activating an ATP-dependent proton pump. This pump uses the energy that is released by the hydrolysis of ATP to pump protons into the stomach from cells in the stomach mucosa (Chapter 30). A histamine analog, cimetidine (Tagamet), binds to the histamine receptor and prevents its activation by histamine. It is an effective drug for the treatment of gastric and duodenal ulcers because it decreases acid secretion.

γ-Aminobutyric acid is secreted in the brain and acts as an inhibitory neurotransmitter. A severe deficiency of vitamin B_6, pyridoxal, is expected to inhibit the decarboxylation of glutamate to GABA by glutamate decarboxylase, which requires pyridoxal phosphate as a cofactor. This decrease in the concentration of GABA may account for the fact that convulsions are a characteristic symptom of pyridoxal deficiency.

Dioxygenases

We have already described pyrocatechase, an Fe^{3+}-enzyme that opens the benzene ring of catechol and incorporates both of the oxygen atoms of O_2 into the product, *cis, cis*-muconate (equation 3). Other dioxygenases insert oxygen into double bonds in the biosynthesis of vitamins and of signal molecules.

Retinal and Vitamin A

An important dioxygenase is the enzyme that catalyzes the formation of retinal and vitamin A from β-carotene (Figure 24-10). β-Carotene is a carotenoid, a hydrocarbon that contains forty carbon atoms and a long conjugated chain of double bonds. It is one of the several carotenoids that impart much of the color to tomatoes, carrots, and the leaves of many plants.

Figure 24-10

The Formation of Retinal and Vitamin A from β-Carotene.

β-Carotene

O_2

H C O

***trans*-Retinal**

O C H

***trans*-Retinal**

$NADP^+$ NADPH

H HO—C H

Retinol (Vitamin A)

O C OH

Vitamin A acid

H C O

11-*cis*-Retinal

Opsin

H H R—C=N^+—Lys

Rhodopsin

hν

trans-Retinal + Opsin

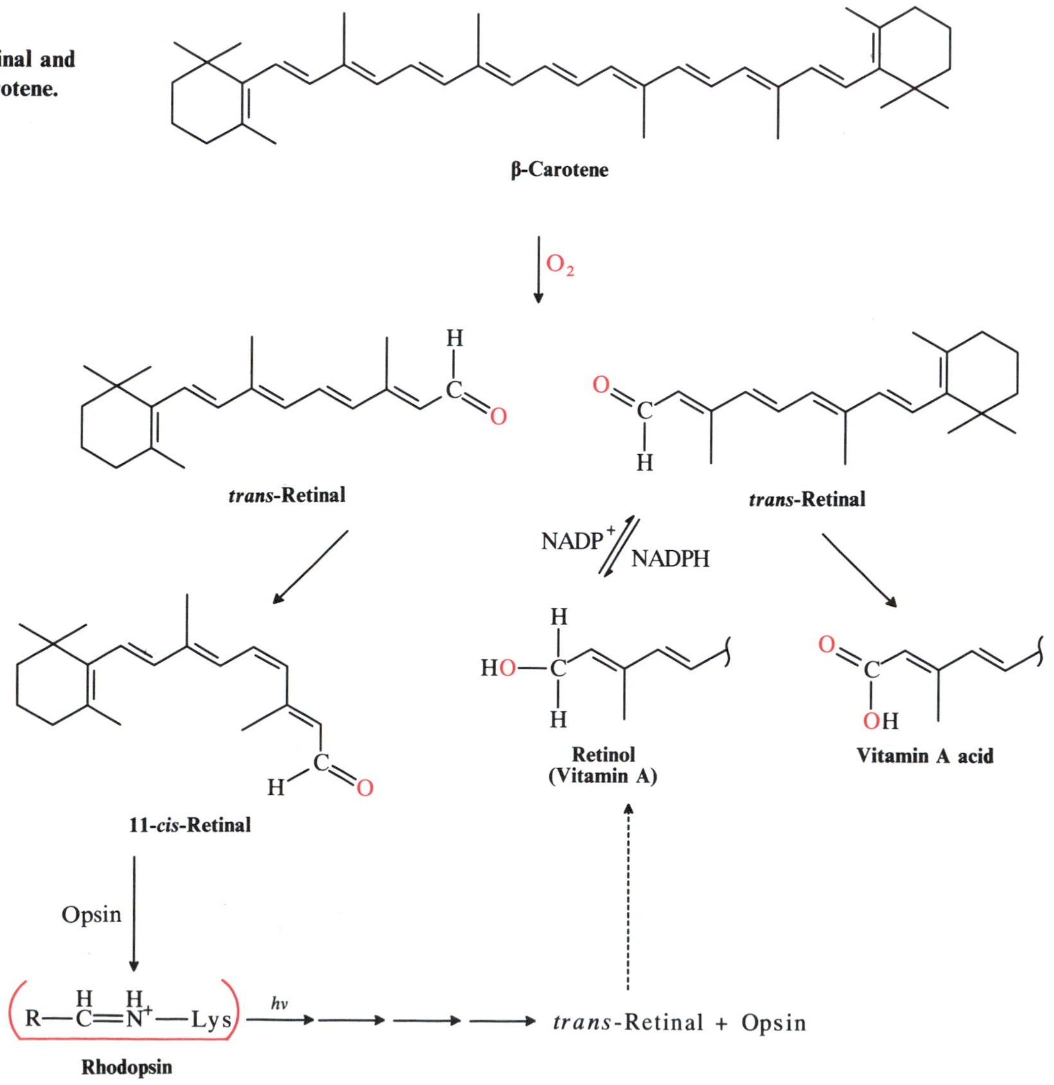

β-Carotene is cleaved by a dioxygenase that catalyzes the attack of both atoms of molecular oxygen at the center of the chain to form two molecules of *trans*-retinal, a conjugated aldehyde that contains twenty carbon atoms. Retinal is best known as the prosthetic group of rhodopsin, a carotenoid-protein that is the primary light-absorbing molecule responsible for vision. The vision process will be described in Chapter 29.

However, there is a much more important role for retinal, as well as the corresponding alcohol, retinol, and acid, retinoic acid. These compounds are the three oxidation states of vitamin A, which is required for the normal growth and development of human beings and other mammals. Vitamin A is usually dispensed by druggists as retinol, which is more stable than the aldehyde, but the different oxidation states are rapidly interconverted in the body.

Vitamin A is the least understood of the vitamins; the determination of its function is a major challenge to biochemists and physiologists. It is required for normal growth and differentiation and presumably plays a critical role in the regulation of these processes. The requirement for vitamin A to prevent night blindness is better understood because it is the source of the retinal in rhodopsin.

Prostaglandins

Another example of a dioxygenase is provided by cyclooxygenase in the synthesis of prostaglandins. Prostaglandins are members of a series of compounds with twenty carbon atoms, the eicosanoids (eikosi is Greek for twenty), which also include thromboxanes and leukotrienes. These compounds are extremely potent regulators of a large number of physiological responses including inflammation, blood clotting, fever, pain, the induction of labor, and blood pressure. All prostaglandins are synthesized from arachidonic acid, a twenty-carbon acid containing four *cis* double bonds, which is one of the acids found in phospholipid membranes. It is usually attached to the phospholipid at the 2-position of glycerol.

An appropriate stimulus causes a cell to release active phospholipase C or phospholipase A_2. These enzymes cleave phospholipids and cause the release of arachidonic acid (Figure 24-11). The enzyme cyclooxygenase (which is also known as prostaglandin endoperoxide synthase) then catalyzes the addition of two molecules of oxygen to arachidonic acid to form prostaglandin, PGG_2. One of the oxygen molecules adds across two double bonds to give a cyclic hydroperoxide and form a five-membered carbon ring. The second oxygen molecule adds to a double bond to form a hydroperoxide, which is reduced by glutathione to give a hydroxyl group. Opening of the endoperoxide ring can give PGE_2, the hydroxyketone that is shown in Figure 24-11, or PGD_2, a prostaglandin in which the positions of the ketone and hydroxyl groups of PGE_2 are exchanged. There are several other prostaglandins, which have slightly different structures.

Thromboxanes and Prostacyclins

PGG_2 can also be converted into thromboxanes and prostacyclins (Figure 24-12). The platelets of blood contain a thromboxane synthase that is normally inactive but becomes activated when platelets make contact with tissue fluids that do not ordinarily enter the blood vessel. The active thromboxane causes constriction of the blood vessels and aggregation of the platelets, which are initial steps in the blood-clotting cascade (Chapter 6). Prostacyclin I_2, on the other hand, is found in the lining of the blood vessels and causes their dilation and the inhibition of platelet aggregation. This compound helps to prevent blood clotting under normal conditions but is overcome by the activating

Figure 24-11

The Synthesis of a Prostaglandin (PGE_2) from Arachidonic Acid.

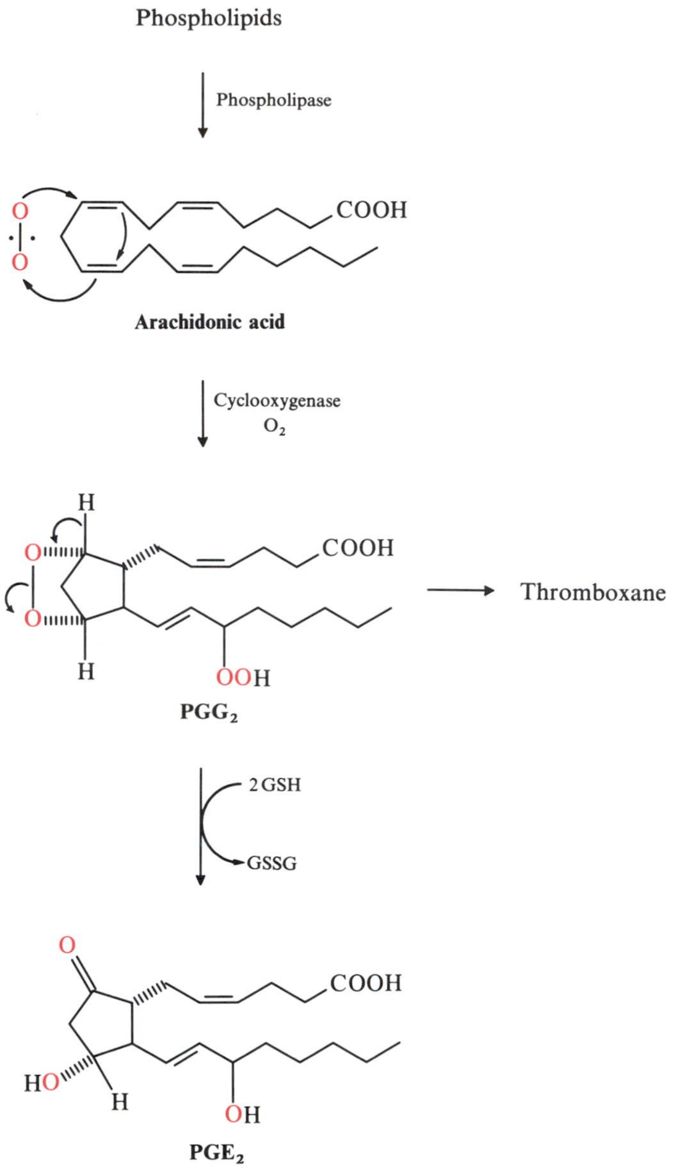

Phospholipids

Phospholipase

COOH

Arachidonic acid

Cyclooxygenase O_2

COOH

Thromboxane

OOH

$\mathbf{PGG_2}$

2GSH

GSSG

COOH

HO H OH

$\mathbf{PGE_2}$

factors of the blood-clotting cascade when the blood contacts tissue fluids. Many of these compounds have a very short half-life, so that their activity can be controlled precisely.

Aspirin (acetylsalicylate) exerts its remarkable effects largely by blocking cyclooxygenase and inhibiting the synthesis of prostaglandins and thromboxanes. Aspirin is a good acylating agent and inactivates cyclooxygenase by acetylating it on the hydroxyl group of a serine residue (equation 14).

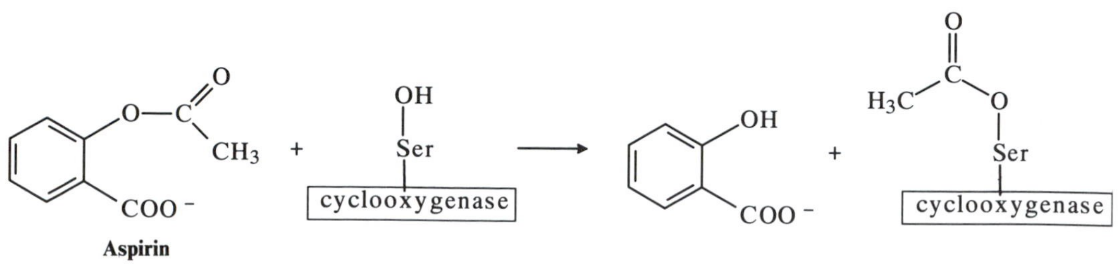

Aspirin + OH–Ser–cyclooxygenase → salicylate (COO^-, OH) + H_3C–C(=O)–O–Ser–cyclooxygenase (14)

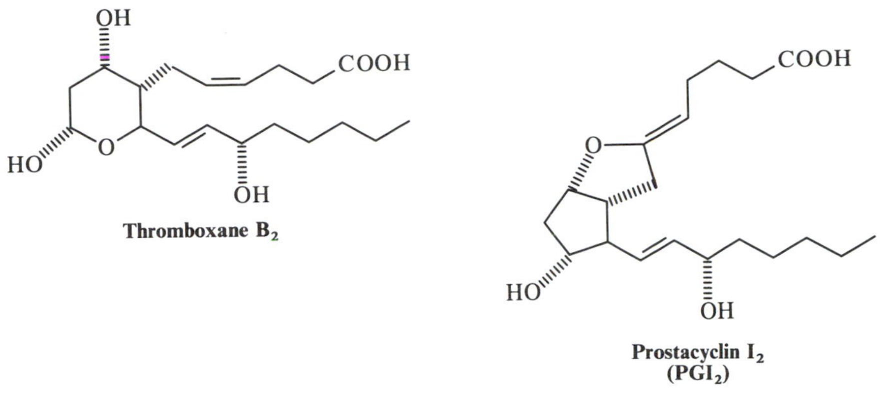

Figure 24-12

Structures of a Thromboxane and a Prostacyclin.

This inhibits the synthesis of prostaglandins and thromboxanes and decreases the inflammation and pain that are induced by these compounds. It also causes a modest decrease in the rate of blood clotting. This is why it is used in low doses to help prevent clotting in the blood vessels of the heart that results in myocardial infarction.

Leukotrienes

Arachidonic acid can also be converted into leukotrienes, which are members of a series of *acyclic*· compounds. Leukotrienes are extraordinarily potent activators of allergic responses and can cause fatal anaphylactic shock. They are direct mediators of the symptoms of asthma.

Leukotrienes are synthesized by the lipoxygenase-catalyzed addition of oxygen to position-5, −12, or −15 of arachidonic acid to form the corresponding hydroperoxide. The formation of 5-hydroperoxyeicosatetraenoic acid (5-HPETE) is shown in Figure 24-13. The hydroperoxide can undergo dehydration to the epoxide, which can be attacked by water to give a diol or by glutathione to give the glutathione adduct in a reaction catalyzed by glutathione *S*-transferase. Additional leukotrienes are formed by hydrolysis of the glutamate and glycine residues from the glutathione. The diol, leukotriene B_4, is a chemotactic agent that stimulates leukocytes (white blood cells) to move up the concentration gradient toward the source of the leukotriene.

The structures and activities of these remarkable compounds derived from arachidonic acid have been discovered only in the past few years and are still being investigated intensively in many laboratories. The reader will not be surprised to learn that a large number of investigators in medical schools, universities, research institutes, and pharmaceutical companies are attempting to determine the mechanisms of regulation and action of these compounds and to develop drugs that will control their effects.

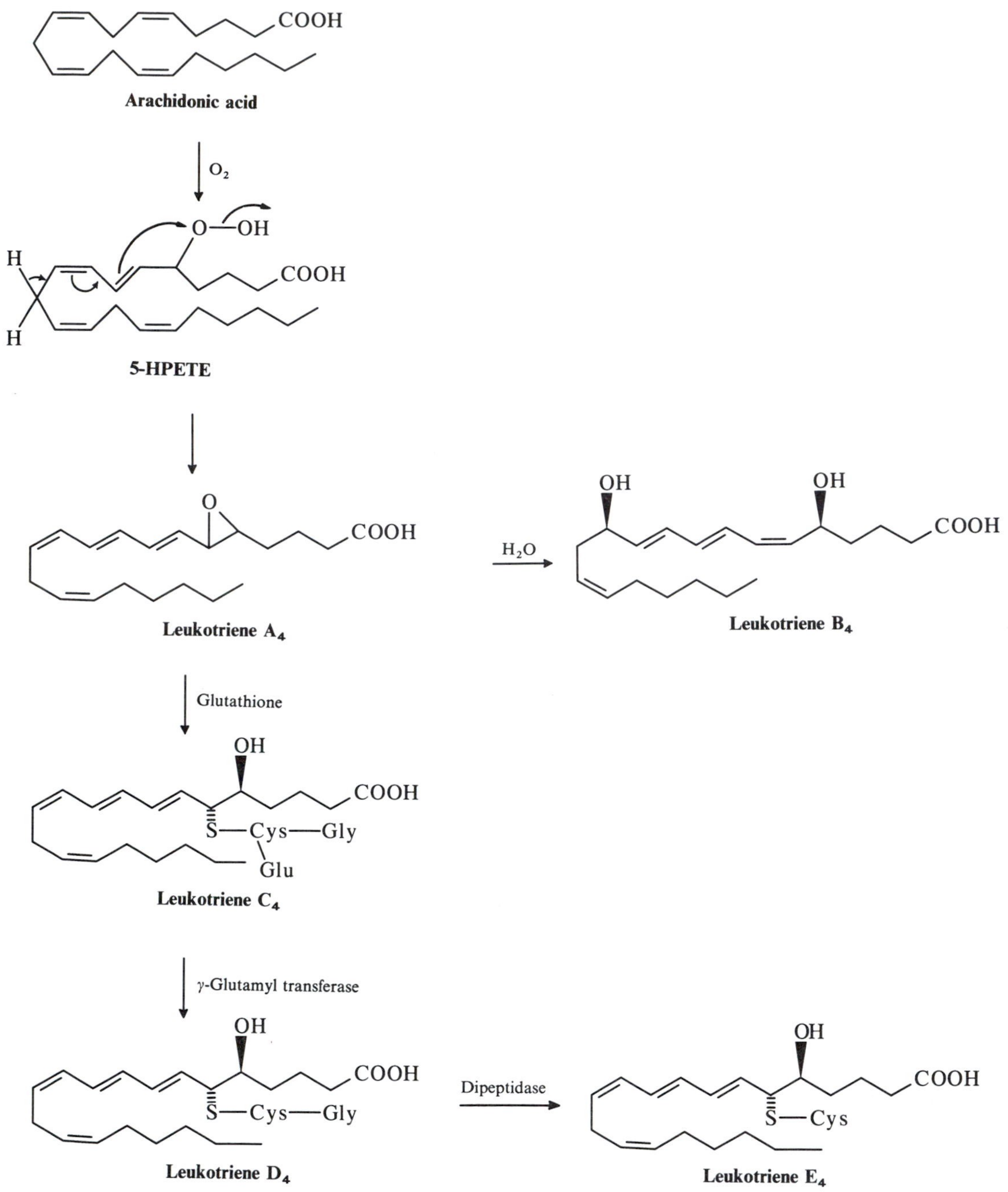

COOH

Arachidonic acid

O_2

O—OH

H

COOH

H

5-HPETE

O

COOH

H_2O

OH OH

COOH

Leukotriene A_4

Leukotriene B_4

Glutathione

OH

COOH

S—Cys—Gly

Glu

Leukotriene C_4

γ-Glutamyl transferase

OH

COOH

S—Cys—Gly

Dipeptidase

OH

COOH

S—Cys

Leukotriene D_4

Leukotriene E_4

Figure 24-13

The Synthesis of Leukotrienes from Arachidonic Acid.

BOX 24-1

DIOXYGENASES THAT UTILIZE AN α-KETOACID

Several oxygenases use one of the oxygen atoms of O_2 for oxygenation of their substrates and incorporate the other oxygen atom into α-ketoglutarate, which brings about an oxidative decarboxylation to succinate and CO_2 according to the following equation.

$$-\overset{|}{\underset{|}{C}}-H + {}^{18}O_2 + \alpha\text{-ketoglutarate } (COO^- - C{=}O - CH_2 - CH_2 - COO^-) \longrightarrow -\overset{|}{\underset{|}{C}}-{}^{18}OH + ({}^{18}O^- - C{=}O - CH_2 - CH_2 - COO^-) + CO_2$$

This is an alternative to the use of NADPH or other reducing agents for the removal of the oxidizing equivalents of O_2 that remain after a monooxygenation.

Prolyl and lysyl hydroxylases are required for the synthesis of collagen and elastin. These enzymes form hydroxyproline and hydroxylysine residues in the peptide chain according to the following equation.

$$\text{Prolyl residue (C(=O)NHR')} + {}^{18}O_2 + \alpha\text{-Ketoglutarate} \longrightarrow \text{4-}H^{18}O\text{-prolyl residue (C(=O)NHR')} + [{}^{18}O]\text{Succinate} + CO_2$$

The hydroxyproline is required to maintain the hydrogen-bonded structure of the native proteins. 5-Hydroxylysine is produced in a similar reaction and provides sites for glycosylation of the proteins.

$$\text{Lysyl residue } (RNH, H, C(=O)NHR', {}^{+}NH_3) + {}^{18}O_2 + \alpha\text{-Ketoglutarate} \longrightarrow \text{5-}{}^{18}OH\text{-lysyl residue} + [{}^{18}O]\text{Succinate} + CO_2$$

The enzymes contain iron that is kept in a reduced state by ascorbic acid and other reducing agents. A deficiency of ascorbic acid in scurvy decreases the activity of these enzymes and results in serious weakening of connective tissues because of a decrease in the stability of these connective-tissue proteins (Chapter 7).

Summary

Oxidases catalyze the oxidation of a substrate and reduction of oxygen to hydrogen peroxide or water. Oxygenases, on the other hand, catalyze the direct incorporation of molecular oxygen into the substrate. Monooxygenases catalyze the incorporation of one atom of oxygen into the substrate to give product. The other oxygen atom of O_2 is generally reduced to water by a reducing agent, such as reduced cytochrome b_5 in the endoplasmic reticulum. Dioxygenases catalyze the incorporation of both oxygen atoms of O_2 into the substrate. Metal ions, such as iron or copper, usually participate in catalysis by oxygenases. Cytochrome P450 oxygenases utilize heme iron and are thought to catalyze oxygenation by the perferryl ion, FeO^{3+}, through a radical mechanism. Tetrahydrobiopterin is the coenzyme that provides the reducing equivalents that are required for the reactions catalyzed by three important oxygenases: phenylalanine hydroxylase, which is absent in children with phenylketonuria; tryptophan hydroxylase, which is required for the synthesis of serotonin; and tyrosine hydroxylase, which catalyzes the formation of 3,4-dihydroxyphenylalanine (DOPA), the precursor of dopamine, norepinephrine, and epinephrine. These catecholamines are neurotransmitters and hormones, and a defect in the synthesis of dopamine is responsible for Parkinson's disease. A dioxygenase catalyzes the cleavage of β-carotene into two molecules of retinal, which is readily converted into two other forms of vitamin A: retinol and retinoic acid. Arachidonic acid is released from phospholipids by phospholipases and can be converted by oxygenases into prostaglandins, thromboxanes, prostacyclins, and leukotrienes. The synthesis of prostaglandins from arachidonic acid is initiated by cyclooxygenase, which is inhibited by aspirin; and leukotrienes are formed by lipoxygenase. These extremely potent compounds initiate a large number of responses to stress or injury, not all of which are desirable.

ADDITIONAL READING

Walsh, C. 1979. *Enzymatic Reaction Mechanisms.* New York: Freeman, chap. 11–16.

Guengerich, F. P., and MacDonald, T. L. 1990. Mechanisms of cytochrome P-450 catalysis. *FASEB Journal* 4: 2453–2459.

Lazarus, R. A., Dietrich, R. F., Wallick, D. E., and Benkovic, S. J. 1981. On the mechanism of action of phenylalanine hydroxylase. 1981. *Biochemistry* 20: 6834–6841.

Ortiz de Montellano, P. R., and Stearns, R. A. 1987. Timing of the radical recombination step in cytochrome P-450 catalysis with ring-strained probes. *Journal of the American Chemical Society* 109: 3415–3420.

Wilcox, D. E., Porras, A. G., Hwang, Y. T., Lerch, K., Winkler, M. E., and Solomon, E. I. 1985. Substrate analogue binding to the coupled binuclear copper active site in tyrosinase. *Journal of the American Chemical Society* 107: 4015–4027.

Kaufman, S. 1979. The activity of 2,4,5-triamino-6-hydroxypyrimidine in the phenylalanine hydroxylase system. *The Journal of Biological Chemistry* 254: 5150–5154.

PROBLEMS

1. Why do you think that evolution led to the use of aromatic amino acids as precursors for many regulatory compounds, such as epinephrine and serotonin?

2. Explain why exact regulation of the rate of phospholipid hydrolysis is important.

3. Why are large doses of aspirin used for the treatment of rheumatoid arthritis and rheumatic fever?

4. Suppose that you isolated an enzyme that catalyzes the formation of $RCOCH_2OH$ from $RCOCH_3$. How would you determine whether it is an oxidase or an oxygenase? What cofactor might it utilize if it were an oxidase? An oxygenase?

5. Explain why α-ketoglutaric acid is required for the action of certain oxygenases.

6. Can you explain how phenylpyruvate is formed from phenylalanine in phenylketonuria?

7. Write a plausible mechanism for the action of anthranilate hydroxylase, which contains Fe^{2+} and catalyzes the reaction:

$$\text{anthranilate (}NH_2\text{, }COO^-\text{)} + NADPH + O_2 + 3\,H^+ \longrightarrow \text{catechol (OH, OH)} + NADP^+ + NH_4^+ + CO_2$$

8. Give a rationale for the fact that α-ketoglutarate acts as a cosubstrate for prolyl hydroxylase? (Do not write a detailed mechanism.)

9. How do you think that 5-hydroxytryptophan is converted into serotonin?

10. An outbreak of convulsions in children was traced to inadequate vitamin B_6 in a commercial baby food. What would you treat these infants with immediately?

25

One-Carbon Metabolism

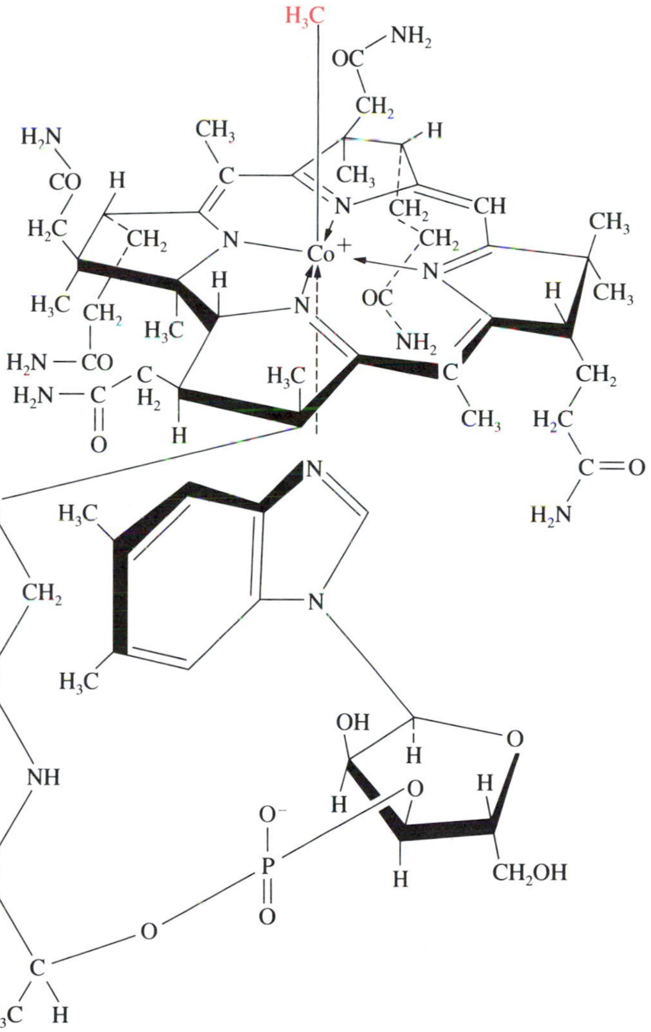

Structure of methylcobalamin.

By 1950, it was known through the work of DuVignaud and his colleagues that rats can metabolize methanol, formaldehyde, and formic acid. It was shown that these compounds can be metabolized to provide the methyl groups of choline, methionine, and other molecules. The study of the metabolism of methanol, formaldehyde, and formate opened up a new area of research—that of one-carbon metabolism.

The biochemistry of one-carbon compounds is a field to itself because of the unusual chemical properties of these compounds. Even the coenzymes that are utilized in one-carbon metabolism are generally different from those that react with other compounds. The differences arise largely because of the very different chemical reactivity of these compounds compared with most metabolites; their small size also may be significant. The carbonyl groups of formaldehyde and formic acid are more reactive than those of acetaldehyde and acetate compounds by factors of as much as 1000. This difference arises partly from steric hindrance to reaction, because of the bulk of the methyl group, and partly from hyperconjugation that stabilizes the acetyl derivatives, as shown by the resonance forms **1**, so that they are less reactive than formyl derivatives.

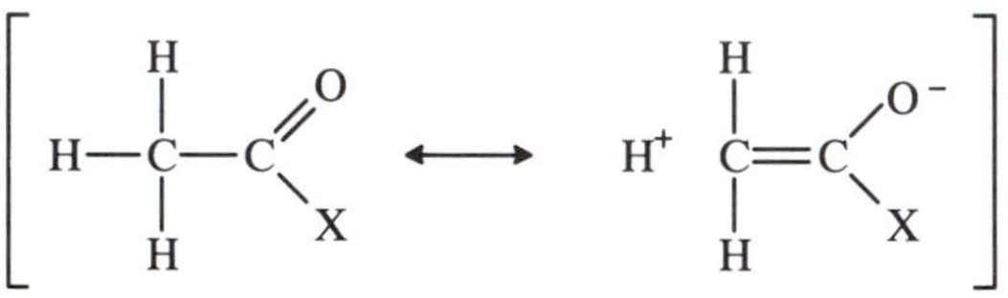

1

In this chapter, we will examine some of the biochemical reactions of formaldehyde and formic acid derivatives, as well as methyl transfer reactions in biochemistry. The reactions of bicarbonate, which frequently require the coenzyme biotin, are described in Chapter 17.

Tetrahydrofolic Acid

Tetrahydrofolic acid is the coenzyme that plays a central role in the metabolism of one-carbon compounds.

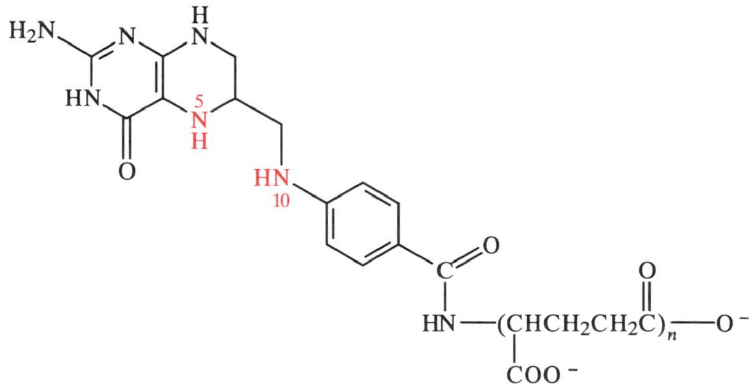

Tetrahydrofolic Acid

The vitamin folic acid is a precursor of tetrahydrofolic acid, in which the reduced piperazine ring of tetrahydrofolic acid is fully oxidized and aromatic.

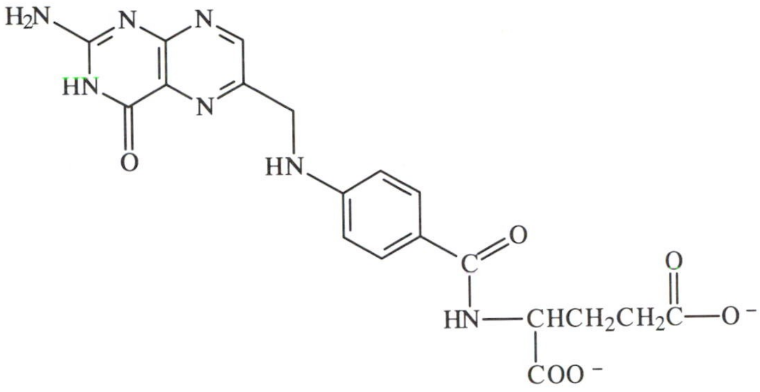

Folic Acid

In the most biologically reactive forms of tetrahydrofolate, the carboxyl group is linked to an oligoglutamate peptide, but the vitamin contains a single glutamate.

Human beings and other mammals do not biosynthesize folic acid, but it is required for the metabolism of one-carbon compounds. Therefore, folic acid is a vitamin that must be supplied by an external source. This requirement is met by a balanced diet; in addition, bacteria in the intestine synthesize folic acid and some of this is absorbed, so that a deficiency of folic acid is rare today.

The reactive part of tetrahydrofolic acid consists of the nitrogen atoms at the 5- and 10-positions, which can carry the one-carbon fragments. The pK_a of the protonated nitrogen-5 is 5, whereas that of the protonated nitrogen-10 is less than zero. The pK_a of an anilino nitrogen such as nitrogen-10 is normally about 3, but the positive charge on protonated nitrogen-5 below pH 5 destabilizes a second positive charge nearby on nitrogen-10 and depresses the pK_a to a value less than zero. Protonation of both nitrogen-5 and nitrogen-10 places two positive charges in close proximity; this is an unstable arrangement that can occur only when it is forced by very strongly acidic solutions.

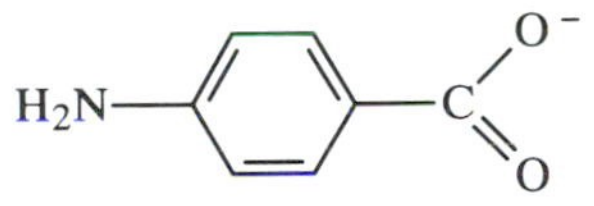

There is also a *p*-aminobenzoic acid group in tetrahydrofolic acid, which is attached to a short chain of glutamate residues. The glutamate residues help the binding of the molecule to enzymes, but the complete role of this side chain is not well understood.

***p*-Aminobenzoic acid anion**

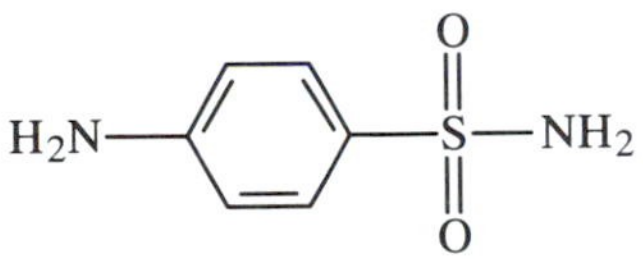

The pathway for the biosynthesis of tetrahydrofolic acid in bacteria is particularly interesting because it is responsible for the mechanism of action of sulfanilamide, the first widely used antibiotic. The sulfonamides and closely related compounds are also notable because they are one of the first classes of drugs for which a rational mechanism of action was deduced. Sulfanilamide closely resembles *p*-aminobenzoic acid, so that the bacterial enzymes that synthesize dihydrofolic acid incorporate sulfanilamide in place of *p*-aminobenzoic acid. This gives a product that is inactive because the enzymes that catalyze the addition of glutamate residues to the carboxylate group of *p*-aminobenzoic acid cannot add glutamate to the sulfonamide group of sulfanilamide. Furthermore, sulfanilamide can repress the biosynthesis of *p*-aminobenzoic acid. The consequence is that sulfanilamide and related antibiotics kill many bacteria that take it up and incorporate it into an inactive folic acid analog. However, it is almost harmless to people because human beings do not synthesize folic acid; they obtain it from their diet or from bacteria.

Sulfanilamide

The Synthesis of Folic Acid

Folic acid and its derivatives are synthesized through an unusual pathway that includes several curious and interesting reactions. We will not describe this pathway in detail (in fact, not all of the details of the pathway are understood), but the principal reactions are noteworthy.

The initial step of the reaction is the removal of a formic acid equivalent, HCOOH, from the five-membered ring of GTP. This is followed by cyclization of the ribose side chain to form a six-membered ring and the removal of two carbon atoms from this side chain to give intermediate **2** in Figure 25-1. This complicated process is catalyzed by the enzyme GTP cyclohydrolase. The

Figure 25-1

The Biosynthesis of Dihydrofolic Acid.

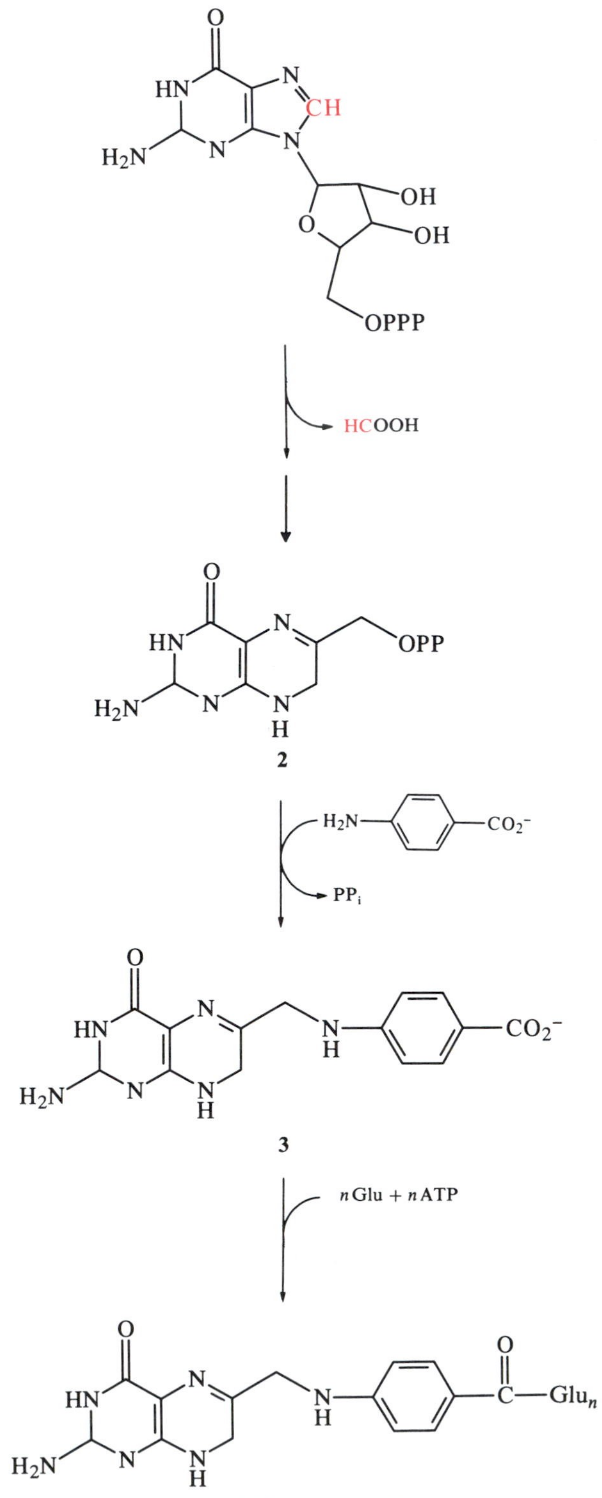

HCOOH

2

$H_2N-C_6H_4-CO_2^-$

PP_i

3

nGlu + nATP

Dihydrofolic Acid

reaction requires an Amadori rearrangement of the ribose side chain, which is an internal oxidation-reduction reaction that gives a ketone group at position 2 of the ribose. The ketone can then cyclize with the exocyclic nitrogen atom to give the *six-membered* ring of the product. A possible mechanism for this Amadori rearrangement is shown in the following equation.

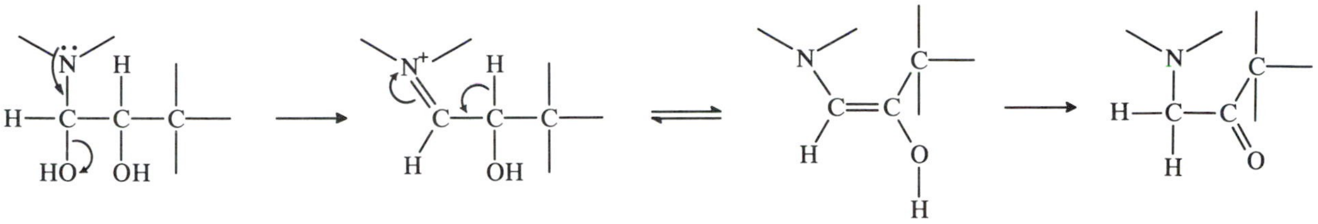

The two terminal carbon atoms of ribose are then removed and the resulting alcohol is pyrophosphorylated by ATP to give intermediate **2**. The pyrophosphate group is then replaced by *p*-aminobenzoic acid to give intermediate **3**.

Finally, several glutamate residues are attached successively to the carboxyl group of *p*-aminobenzoic acid to give the polyglutamate chain of the final product, dihydrofolic acid. Note that the dihydrofolic acid molecule in Figure 25-1 is oriented upside down compared with the tetrahydrofolic acid shown earlier. This is because the synthesis is described starting from GTP, which is written in its usual orientation with the carbonyl group at the top.

Metabolism of One-Carbon Groups

The reaction scheme in Figure 25-2 summarizes the main reaction pathways for interconverting the one-carbon fragments attached to tetrahydrofolic acid in *5-methyl tetrahydrofolate, methylene tetrahydrofolate, formyl tetrahydrofolate,* and *methenyl tetrahydrofolate.* These are the species of tetrahydrofolate that participate in one-carbon transfer reactions. The apparent complexities of these reactions are simplified by considering the three oxidation levels of the one-carbon groups separately. The reactions account for transfer of one-carbon units at the methyl, hydroxymethyl (formaldehyde), and formic acid levels of oxidation; these three levels are connected by enzymatic dehydrogenation reactions.

The most-interesting and, probably, the most-important of the reactions in Figure 25-2 are those in which the one-carbon unit is at the formaldehyde level of oxidation in the center of the scheme. The conversion of serine into glycine, catalyzed by the enzyme serine transhydroxymethylase, results in the release of a formaldehyde equivalent that is bound as a CH_2 group in methylene tetrahydrofolate, as described in Chapter 2. The formaldehyde equivalent is inserted between the nitrogen-5 atom of the ring and the nitrogen-10 atom of tetrahydrofolate. This key intermediate is *reduced* by NADPH to give 5-methyl tetrahydrofolate, which provides the methyl group of methionine. It can be *oxidized* by $NADP^+$ to give methenyl tetrahydrofolate, which is at the formic acid level of oxidation. This can undergo addition of water to give either 5-formyl tetrahydrofolate or 10-formyl tetrahydrofolate, both of which provide formyl groups in metabolism and biosynthetic reactions.

We will examine reactions at the formaldehyde level of oxidation first and then the reactions in which formic acid and methyl equivalents are transferred.

The Synthesis of Thymidylic Acid

The addition of a methyl group to deoxyuridylic acid (dUMP) to form thymidylic acid (TMP) is a particularly interesting and important reaction in

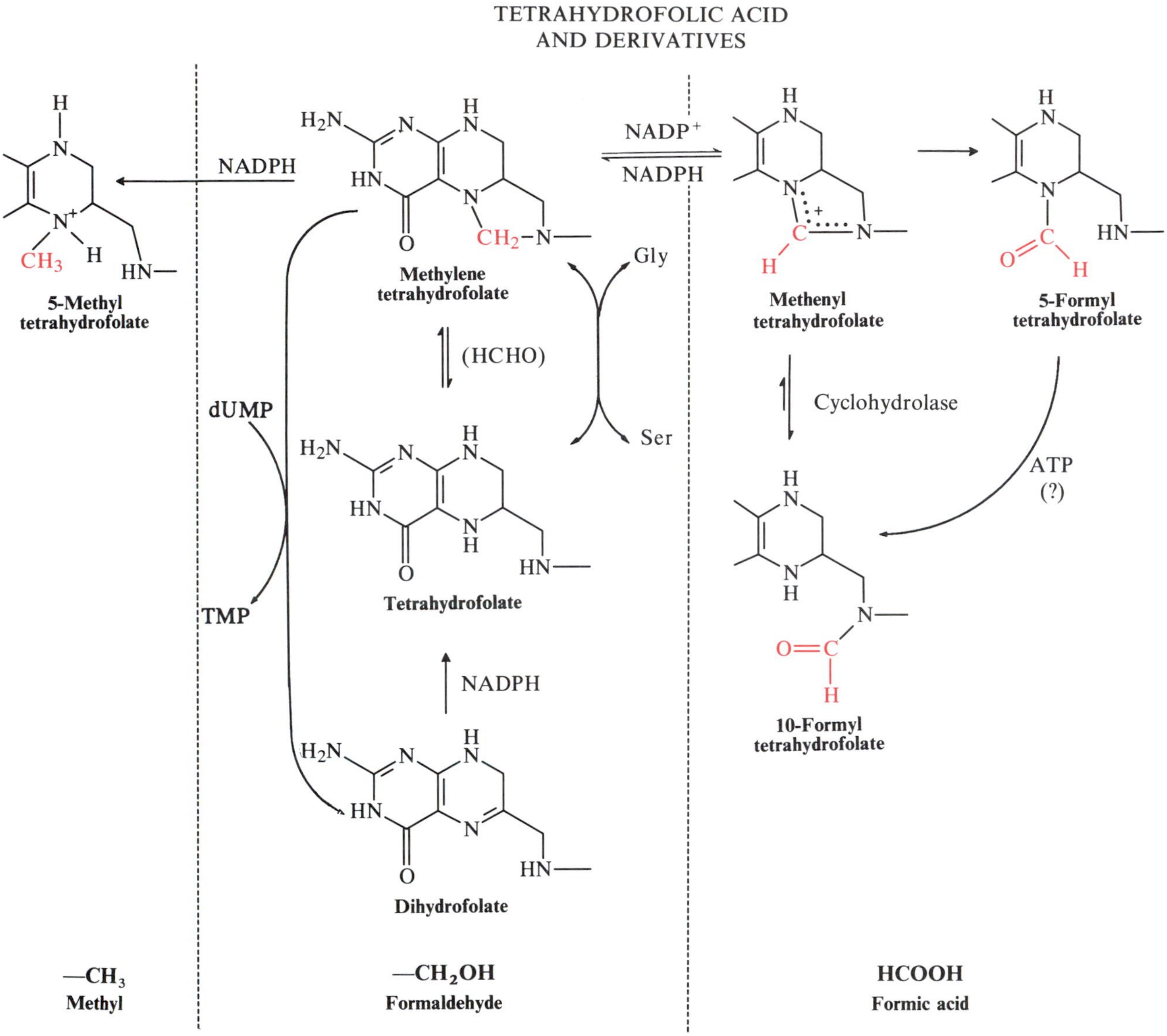

TETRAHYDROFOLIC ACID AND DERIVATIVES

5-Methyl tetrahydrofolate

Methylene tetrahydrofolate

Methenyl tetrahydrofolate

5-Formyl tetrahydrofolate

Tetrahydrofolate

10-Formyl tetrahydrofolate

Dihydrofolate

—CH_3
Methyl

—CH_2OH
Formaldehyde

HCOOH
Formic acid

Figure 25-2

The Three Oxidation States of Tetrahydrofolate Derivatives. The most highly reduced folate derivative is 5-methyl tetrahydrofolate shown at the left. The one-carbon unit of methylene tetrahydrofolate shown in the center is at the oxidation level of formaldehyde. Methylene tetrahydrofolate can be synthesized by reaction of tetrahydrofolate with formaldehyde. In the enzymatic synthesis, which is catalyzed by serine transhydroxymethylase in cells, the hydroxymethyl group of serine is transferred to tetrahydrofolate. The most highly oxidized one-carbon derivatives of tetrahydrofolate are 10-formyl tetrahydrofolate and methenyl tetrahydrofolate, which are shown at the right. These compounds are interconverted by the action of cyclohydrolase. The benzoyl polyglutamate part of tetrahydrofolate is omitted from the structures.

one-carbon metabolism. The methyl group is ultimately derived from the amino acid serine, and the product is important because TMP is required for the synthesis of DNA. Therefore, inhibition of TMP synthesis inhibits DNA synthesis. DNA synthesis is required for cell growth and multiplication, so that inhibition of DNA synthesis is one of the most-effective ways of controlling cancer by decreasing the rate of cell division. Several of the drugs that are widely used for cancer chemotherapy act by inhibiting enzymes that are required for TMP synthesis.

Figure 25-3 shows the overall reaction path for the synthesis of TMP from dUMP and methylene tetrahydrofolate. This curious and difficult reaction is more easily understood if it is divided into its two parts:

1. The methylene group is transferred from methylene tetrahydrofolate.
2. A hydrogen atom is transferred from tetrahydrofolate.

Together, these two processes give the methyl group of TMP, which is derived from a formaldehyde equivalent and a reducing equivalent. The reducing equivalent is supplied by NADPH, which reduces dihydrofolate to tetrahydrofolate. The formaldehyde equivalent comes from serine. Both of these reactions are unusually important and interesting.

A probable reaction sequence for the mechanism of action of thymidylate synthase is shown in Figure 25-4. In initial steps (numbered 1), a sulfhydryl group at the active site adds to the double bond of dUMP in a Michael-type addition reaction to form an enolate carbanion, and methylene tetrahydrofolate is converted into the iminium ion. This carbanion then attacks the methylene group of the highly reactive iminium ion in step 2. A proton is then removed from carbon-5 of dUMP in step 3 to regenerate the enolate anion, which expels nitrogen-5 of tetrahydrofolate in step 4 and forms a reactive methylidene group. Tetrahydrofolate then reduces the methylene group by transfer of a hydrogen in step 5, giving the enolate anion and dihydrofolic acid. Finally, the enolate anion expels the enzymatic thiolate group in step 6.

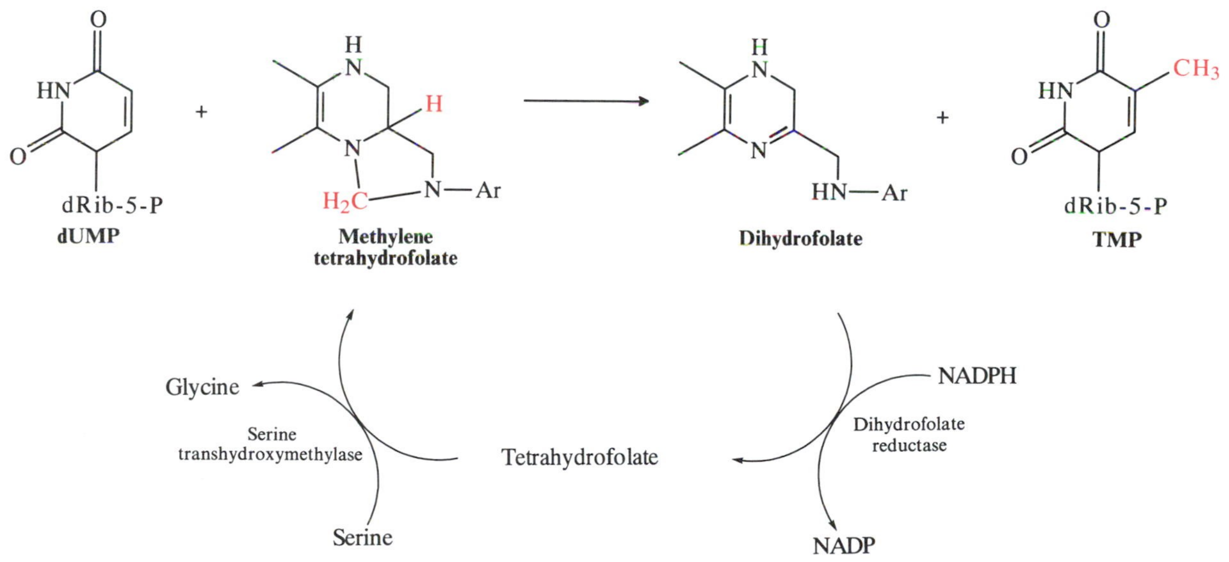

Figure 25-3

Interconversions of Folate Derivatives in the Biosynthesis of dUMP to TMP. Thymidylate synthase catalyzes the reaction of methylene tetrahydrofolate with dUMP to give TMP and dihydrofolate. Dihydrofolate is converted back into methylene tetrahydrofolate in two steps. In the first step, dihydrofolate is reduced to tetrahydrofolate by NADPH and dihydrofolate reductase. Serine transhydroxymethylase then catalyzes the reaction of tetrahydrofolate with serine to give methylene tetrahydrofolate and glycine. Thus, the source of the methyl group in thymine is the hydroxymethyl group of serine and a hydride from NADPH.

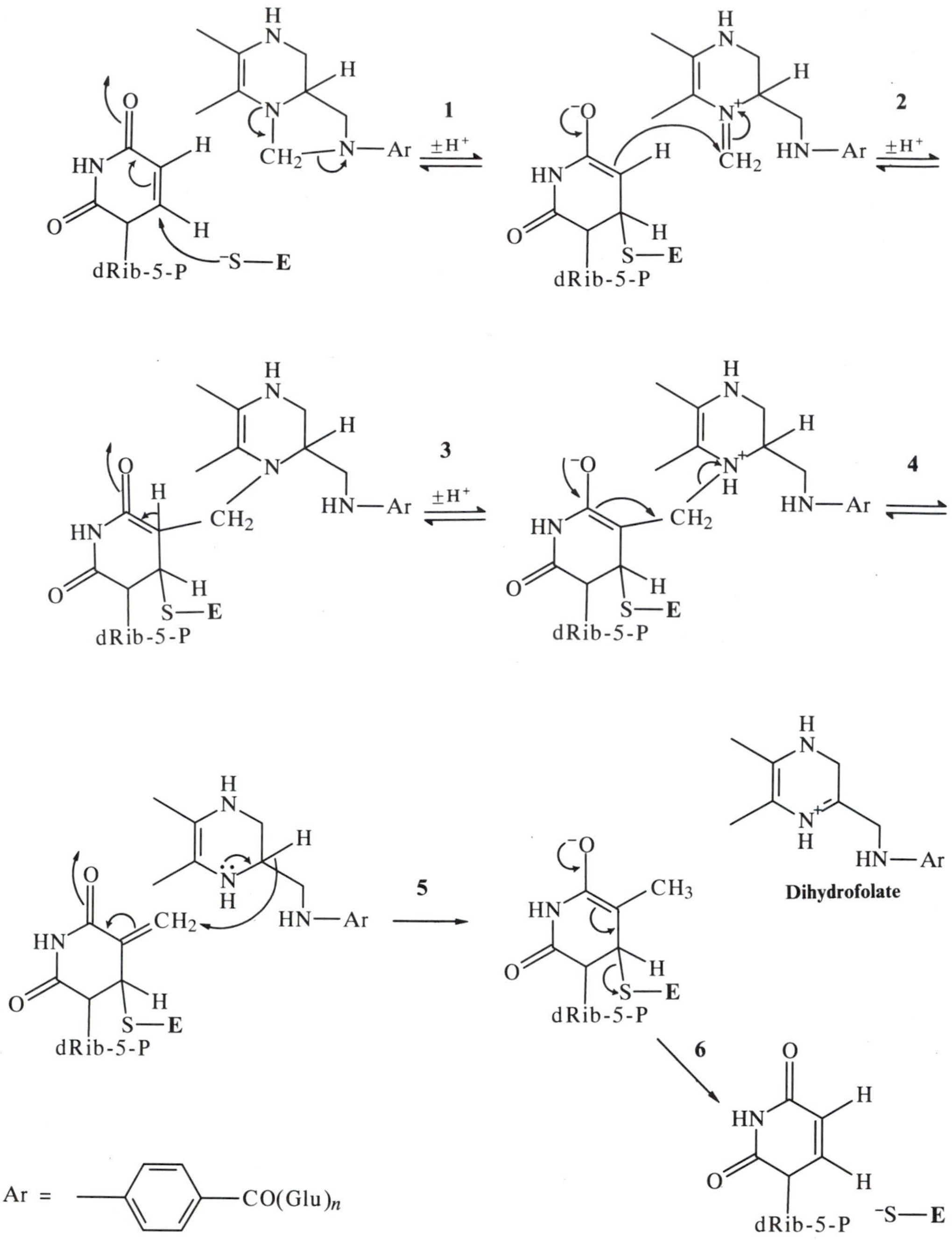

Figure 25-4

Steps in the Conversion of dUMP and Methylene Tetrahydrofolate into TMP and Dihydrofolate. The reaction of dUMP with methylene tetrahydrofolate to produce TMP and dihydrofolate is catalyzed by thymidylate synthase through the steps shown. The reaction begins with the Michael-type addition of a cysteinyl thiolate at the active site to carbon-6 of dUMP to form an enolate anion. In the next step, the enolate adds to the iminium ion form of methylene tetrahydrofolate to form an addition compound. The proton on carbon-5 is abstracted by an enzymatic base, and the resulting enolate ion eliminates tetrahydrofolate, which reduces the methylidene group to a methyl group of a third enolate intermediate and forms dihydrofolate. This last enolate eliminates the enzymatic thiolate group to form TMP.

Anticancer Drugs

The inhibition of DNA synthesis and the growth of rapidly dividing cells by several chemotherapeutic agents for cancer can be accounted for by specific inhibition of the reactions in Figures 25-3 and 25-4. 5-Fluorouracil (5-FU) is an analog of uracil that can be converted into 5-fluorodeoxyuridylate (FdUMP) in the body.

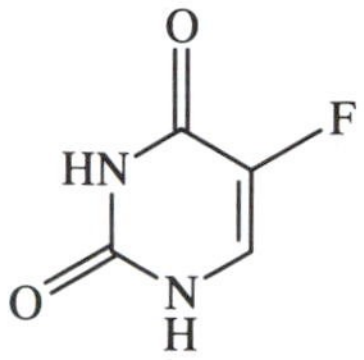

5-Fluorouracil (5-FU)

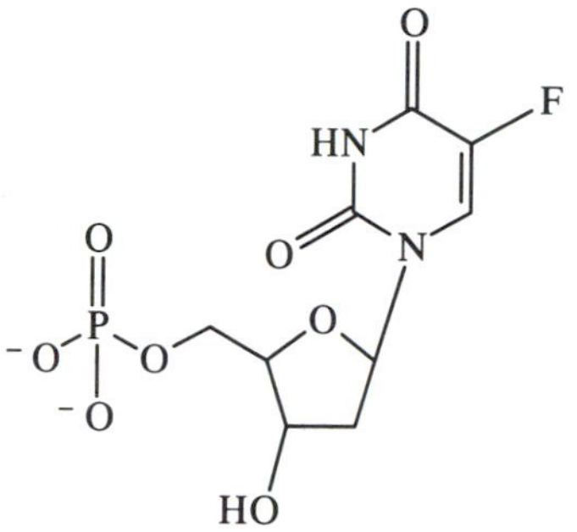

5-Fluorodeoxyuridylate (FdUMP)

FdUMP inactivates thymidylate synthase in an example of "suicide" inactivation. The enzymatic reaction of FdUMP proceeds just as it does with dUMP through steps 1 and 2 in Figure 25-4. However, step 3 cannot occur because at this point in the normal reaction a proton would be abstracted from carbon-5 and, in the analogous complex with 5-fluorouracil, the proton is replaced by a fluorine atom, which cannot be abstracted. Therefore no product can be released, the active site of the enzyme is blocked, TMP cannot be synthesized, and the synthesis of DNA comes to a halt.

The synthesis of DNA can also be inhibited by blocking the reduction of dihydrofolate to tetrahydrofolate catalyzed by dihydrofolate reductase, which prevents the formation of methylene tetrahydrofolate in the reaction cycle of Figure 25-3. Catalysis of the reduction of dihydrofolate by dihydrofolate reductase is strongly inhibited by aminopterin and by methotrexate (amethopterin), which is the N^{10}-methyl analog of aminopterin.

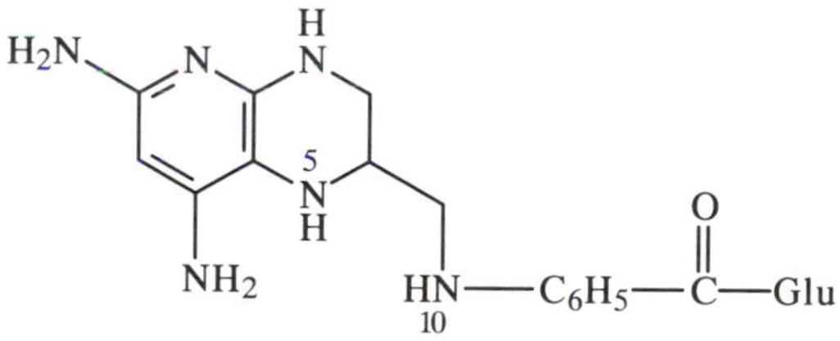

Aminopterin

These inhibitors have extraordinarily high affinities for dihydrofolate reductase, with dissociation constants ranging from 10^{-10} to 10^{-11} M. The protonated inhibitors bind to the dihydrofolate-binding site of the enzyme. The high specificity of these inhibitors for this enzyme makes them clinically useful for cancer chemotherapy.

Reaction of Formaldehyde with Tetrahydrofolate

Formaldehyde is a highly reactive, toxic compound that does not exist at significant concentrations in normal tissues. It can be formed by the oxidation of methanol and is a factor in the severe toxicity of methanol. An important role of tetrahydrofolate in the reaction cycle of TMP synthesis is, therefore, to carry a formaldehyde equivalent, without forming free formaldehyde.

Formaldehyde does, however, react rapidly with tetrahydrofolate in the laboratory, according to the reaction scheme in Figure 25-5. The nitrogen-5 atom of tetrahydrofolate attacks the carbonyl group to give a hydroxymethyl adduct, which undergoes dehydration to give the extremely reactive key intermediate X. This intermediate is a stabilized carbocation. It has never been observed directly, because it undergoes very rapid ring closure to methylene tetrahydrofolate. However, it can be regenerated from methylene tetrahydrofolate and is presumably the reactive species for transfer of formaldehyde equivalents from methylene tetrahydrofolate.

In the pathway of one-carbon metabolism, the reduction of this intermediate by NADPH is catalyzed by a flavin enzyme and gives 5-methyl tetrahydrofolate. The reactions of this compound will be discussed in the section on methionine biosynthesis. The reduction reaction is interesting because it includes the incorporation of a proton from the solvent into the methyl group instead of direct transfer of hydrogen from NADPH. The mechanism of this reduction is not yet understood.

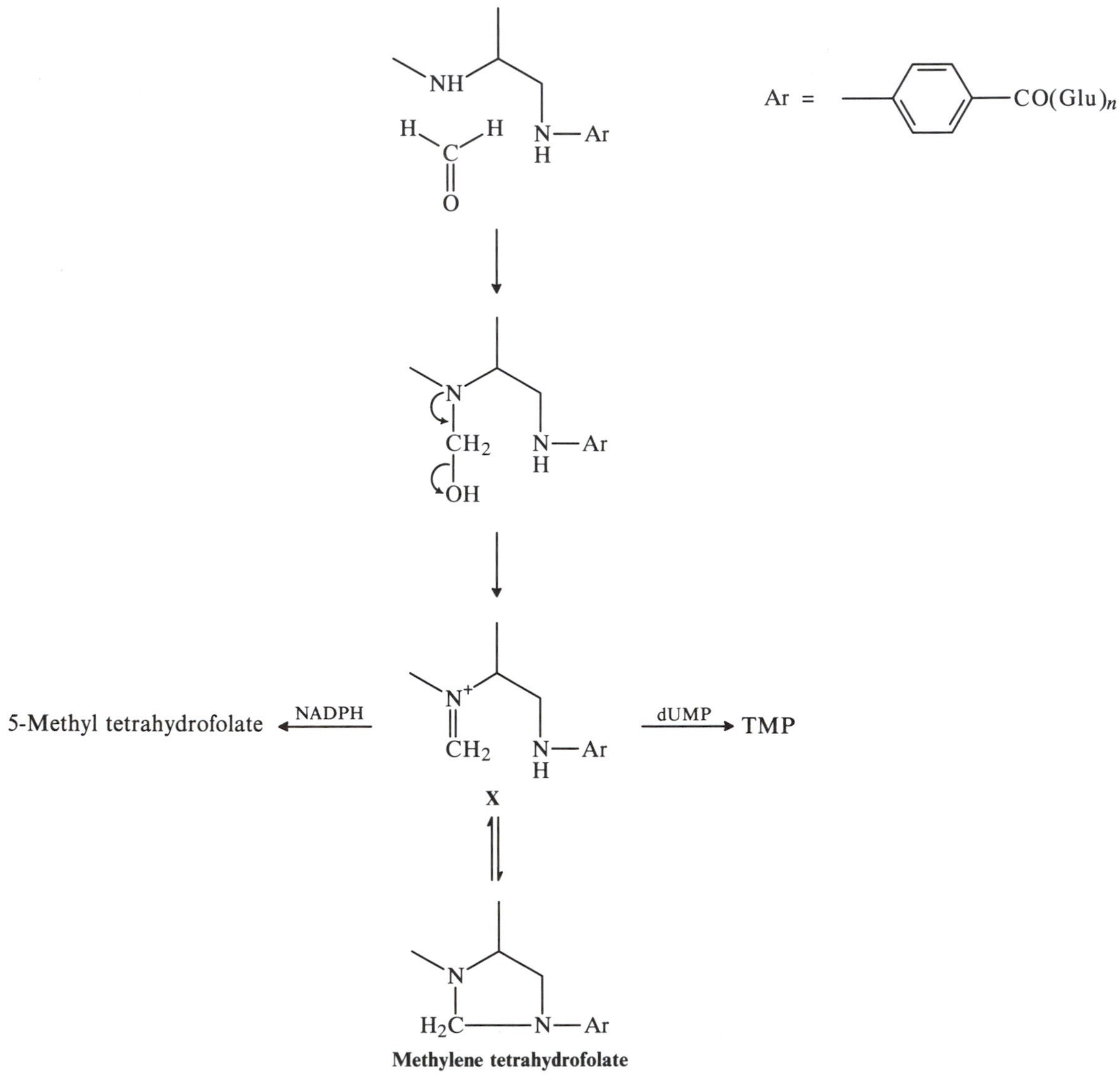

Ar = —C₆H₄—CO(Glu)$_n$

5-Methyl tetrahydrofolate ← NADPH — X — dUMP → TMP

Methylene tetrahydrofolate

Figure 25-5

Steps in the Reaction of Formaldehyde with Tetrahydrofolate to Form Methylene Tetrahydrofolate.

Formate Activation and Transfer

Activated formyl groups can be obtained by the NADP-dependent oxidation of methylene tetrahydrofolate to methenyl tetrahydrofolate, as shown in Figure 25-2, or by the activation of formic acid by ATP to form 10-formyl tetrahydrofolic acid. This is a typical ATP-dependent acyl activation reaction (Chapter 9), which gives ADP and phosphate as products and proceeds through an enzyme-bound formyl phosphate intermediate (Figure 25-6).

Methenyl tetrahydrofolate can also be converted into 5-formyl tetrahydrofolate (Figure 25-2). 10-Formyl tetrahydrofolate is a formyl group donor for several important reactions including the synthesis of *N*-formylglutamate. It reacts with a specific methionyl tRNA to form formylmethionyl tRNA, which initiates synthesis of the peptide chain in protein biosynthesis (Chapter 10). It is also the formyl group donor in two reactions of purine biosynthesis: the synthesis of formamidoimidazole carboxamide ribotide and of formylglycinamide ribotide. These reactions are described in Chapter 27.

It is interesting that several of these reactions in purine biosynthesis take place in complexes that engage in several enzyme activities. For example, the enzyme that synthesizes formylglycinamide ribotide also has cyclohydrolase activity, which generates 10-formyl tetrahydrofolate from methenyl tetrahydrofolate (Figure 25-2).

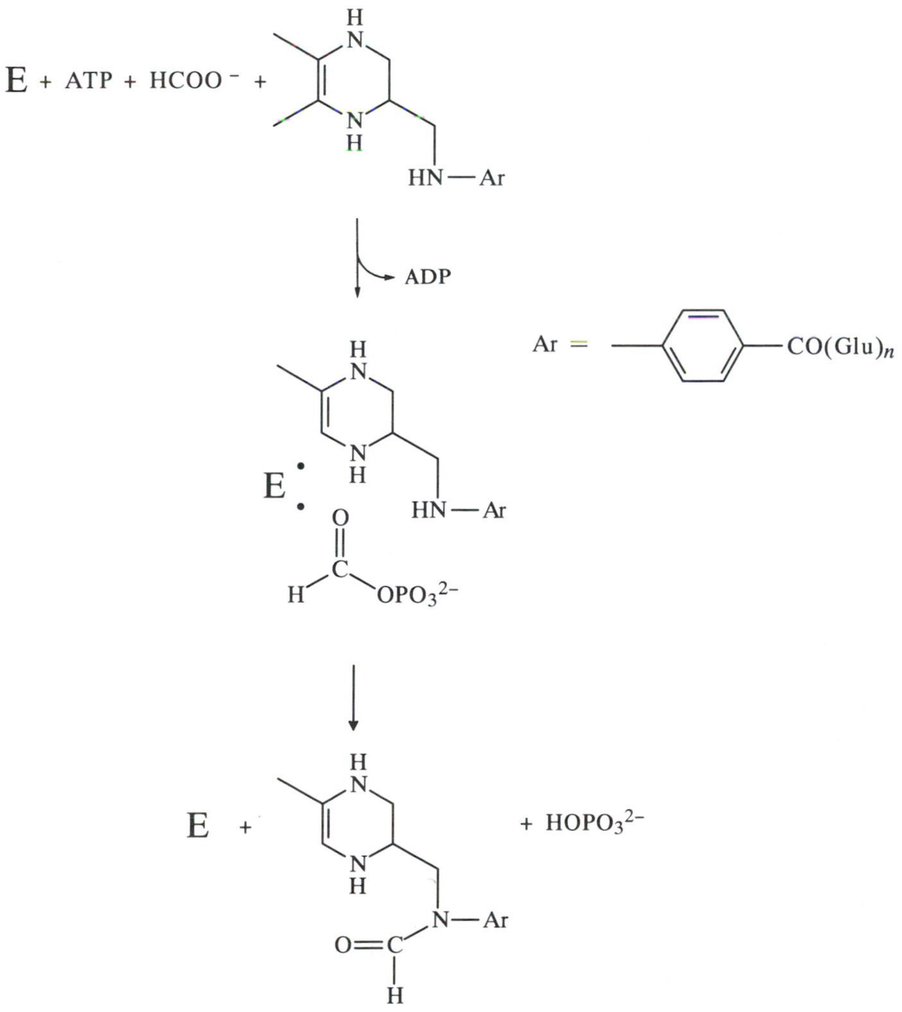

Figure 25-6

Formyl Phosphate as an Intermediate in the Formation of 10-Formyl Tetrahydrofolate. 10-Formyl tetrahydrofolate synthase catalyzes the activation of formate by ATP to formyl phosphate at the active site. Formyl phosphate does not dissociate from the active site but donates the formyl group to nitrogen-10 of tetrahydrofolate.

Methyl Transfer

The occurrence of methylation reactions in biological systems has been recognized for a long time. In 1950, Challenger wrote an extensive review on methylation reactions, in which he gave an interesting explanation of the toxicity that had been discovered in certain green wallpapers. The green dye that was then in use for the preparation of these wallpapers contained arsenic, which was methylated by bacteria that grow in the paste used to attach the wallpaper to the wall. This methylation produced $(CH_3)_3As$, a volatile compound, which was inhaled by people and was responsible for the toxicity. The methylation of arsenic is interesting but not of major biological importance. However, there are many important biological reactions in which molecules are methylated on N, O, or S atoms. These reactions occur both in biosynthetic (anabolic) and in catabolic processes, including the methylation of DNA and the methylation of the ethanolamine component of complex lipids.

S-Adenosylmethionine: The Source of Methyl Groups

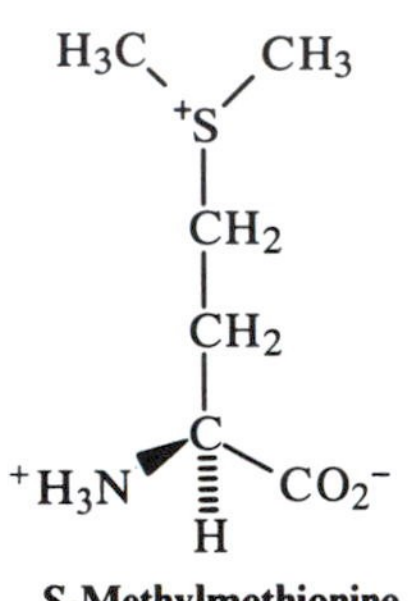

***S*-Methylmethionine**

***S*-Adenosylmethionine**

What is the source of the methyl groups in methylation reactions? It was known in 1950 that methionine could serve as a methyl donor. It was also realized that methionine was not sufficiently reactive to be a good methyl donor because the homocysteine mercaptide anion is a poor leaving group. It was suggested that *S*-methylmethionine might be the methyl donor. If this compound served as the methyl donor, methionine would be the leaving group, which is more attractive on chemical grounds than a mercaptide anion. A few experiments appeared to support the notion that *S*-methylmethionine is the biological methyl donor; however, it proved not to be the case. Cantoni discovered the true methyl donor, *S*-adenosylmethionine, in 1955. *S*-Adenosylmethionine is the methyl donor in the vast majority of biological methylation reactions. An example of such a methylation reaction is the formation of epinephrine shown in Figure 25-7.

All methylation reactions requiring *S*-adenosylmethionine appear to be simple S_N2 displacements. An important contribution of the enzyme in these reactions is probably the proper alignment of the substrate for the displacement. The contribution of the enzyme may, therefore, be largely entropic. The transfer of a methyl group from *S*-adenosylmethionine to an acceptor produces *S*-adenosylhomocysteine according to Figure 25-7. *S*-Adenosylhomocysteine is a potent inhibitor of all reactions in which a methyl group is transferred from *S*-adenosylmethionine to an acceptor; therefore, it is important to prevent the accumulation of *S*-adenosylhomocysteine in cells. This is accomplished through the action of the enzyme *S*-adenosylhomocysteinase, which catalyzes the conversion of *S*-adenosylhomocysteine into adenosine and homocysteine (Figure 25-8). The equilibrium of this reaction lies in the direction of synthesis, but *in vivo* it is shifted in the degradative direction by removal of the reaction products. Homocysteine is converted into methionine and adenosine into inosine. The formation of methionine will be described in the next section.

S-Adenosylmethionine is produced by *S*-adenosylmethionine synthase in a very interesting reaction; ATP reacts with methionine and inorganic triphosphate is formed according to Figure 25-9. The triphosphate is then hydrolyzed by the same enzyme to PP_i and P_i. The formation of the sulfonium compound proceeds by nucleophilic attack of the methionine sulfur atom on the 5′ carbon of ATP to displace triphosphate. This is one of two reactions in which a displacement of this kind is known to occur, the other being the formation of adenosylcobalamin. The hydrolysis of the triphosphate drives the reaction to the right; it makes the reaction highly exergonic in the synthetic direction.

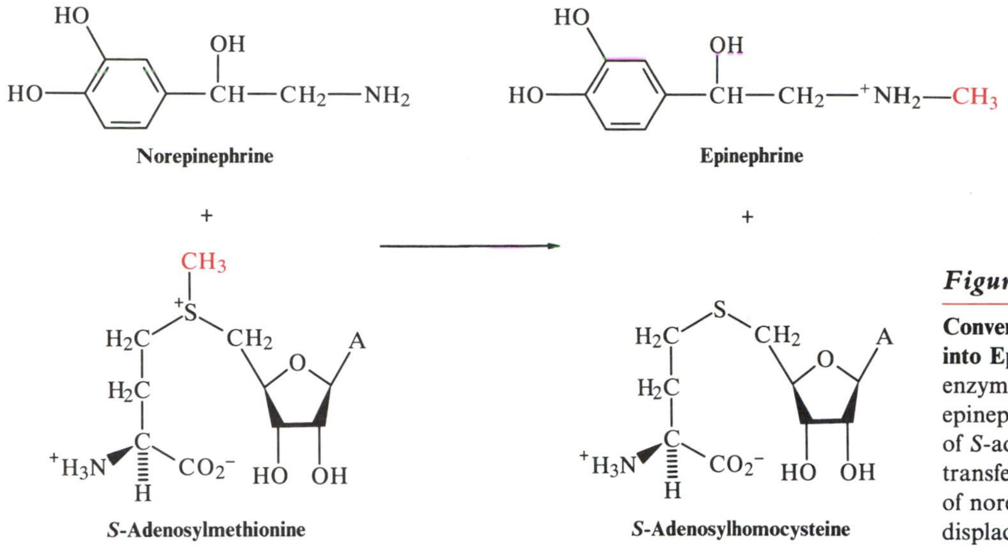

Norepinephrine + S-Adenosylmethionine → Epinephrine + S-Adenosylhomocysteine

Figure 25-7

Conversion of Norepinephrine into Epinephrine. In the enzymatic formation of epinephrine, the methyl group of *S*-adenosylmethionine is transferred to the amino group of norepinephrine in an S_N2 displacement.

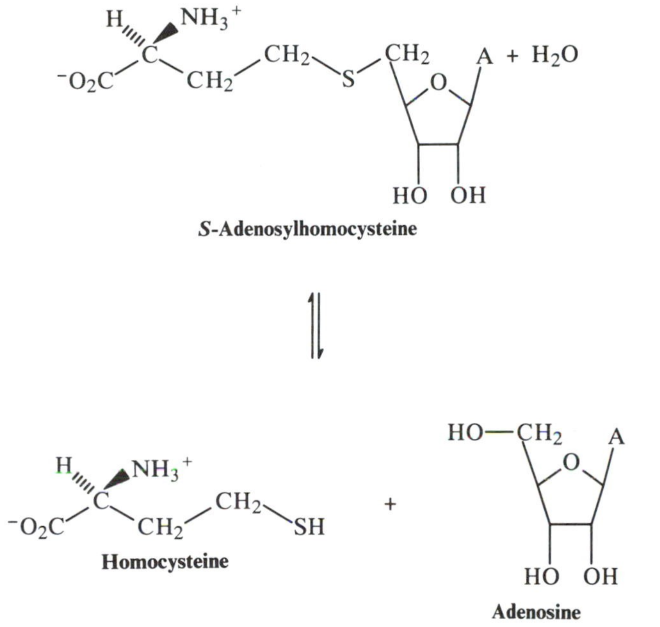

S-Adenosylhomocysteine + H_2O ⇌ Homocysteine + Adenosine

Figure 25-8

The Hydrolysis of *S*-Adenosylhomocysteine by *S*-Adenosylhomocysteinase.

Biosynthesis of Methionine

The next problem to be considered is how methionine is synthesized. For many organisms, including mammals, the methionine requirement can be met in part by the diet. However, most organisms can also synthesize methionine. In mammals and some microorganisms, but not in plants, the synthesis of methionine requires a cofactor derived from vitamin B_{12}. The structure of vitamin B_{12} is shown in Chapter 20 (Figure 20-4).

Two coenzymes that are derived from vitamin B_{12}, are known: adenosylcobalamin, which is sometimes called B_{12} coenzyme and was described in Chapter 20, and methylcobalamin. Methylcobalamin is a compound in which a methyl group is covalently bonded to the "top" position of the cobalt atom. This coenzyme contains a covalent carbon–cobalt bond, as does vitamin B_{12} coenzyme. The discovery and characterization of vitamin B_{12} is described in Chapter 20 (Box 20-2). Methylcobalamin is the cofactor that is required for methionine biosynthesis.

Figure 25-9

The Biosynthesis of *S*-Adenosylmethionine. This reaction is catalyzed by *S*-adenosylmethionine synthase. ATP reacts as an alkylating agent: the sulfur of methionine is the nucleophile, and triphosphate is the initial leaving group. The enzyme also catalyzes the hydrolysis of triphosphate to pyrophosphate and phosphate, which makes the reaction thermodynamically more favorable. Thus, the enzyme utilizes the energy of two phosphoanhydride bonds to drive this reaction to completion.

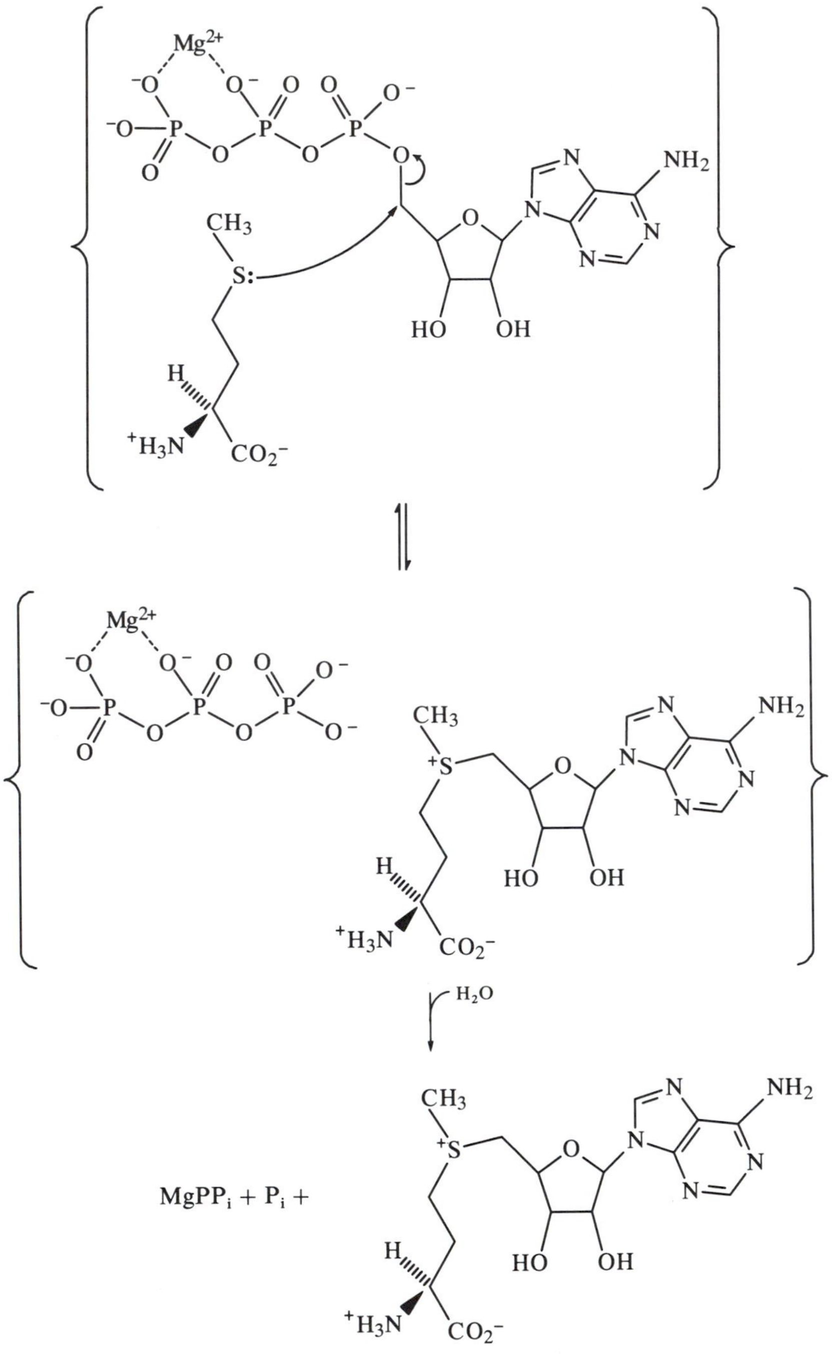

$MgPP_i + P_i +$

***S*-Adenosylmethionine**

The biosynthesis of methionine begins with the transfer of a methyl group from methylcobalamin to homocysteine according to Figure 25-10. This gives rise to methionine and a species known as vitamin $B_{12(s)}$, in which cobalt is in the $+1$ oxidation state (Co^{+1}). Vitamin $B_{12(s)}$ then reacts with 5-methyl tetrahydrofolate to regenerate methylcobalamin and produce tetrahydrofolate. The tetrahydrofolate is then converted back into 5-methyl tetrahydrofolate by the series of reactions in Figure 25-2.

The preceding discussion suggests a possible biochemical consequence of B_{12} deficiency (pernicious anemia). The inability to synthesize methionine cannot be the major problem, because administration of methionine does not cure the disease. However, the inability to synthesize methionine does have a

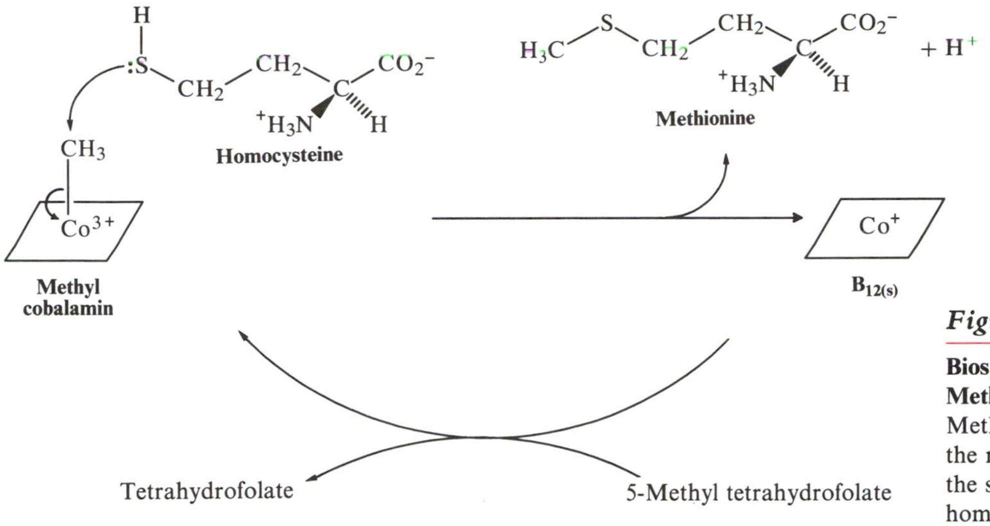

Homocysteine

Methionine

$+ H^+$

Methyl cobalamin

Co^{3+}

Co^{+}

$B_{12(s)}$

Tetrahydrofolate

5-Methyl tetrahydrofolate

Figure 25-10

Biosynthesis of Methionine by Methionine Synthase. Methionine synthase catalyzes the nucleophilic attack by the sulfhydryl group of homocysteine on the methyl group of methyl cobalamin, which displaces vitamin $B_{12(s)}$ and produces methionine. The vitamin $B_{12(s)}$ remains bound to the enzyme and itself attacks the N^5-methyl group of 5-methyl tetrahydrofolate to regenerate methylcobalamin at the active site. The tetrahydrofolate produced is converted back into 5-methyl tetrahydrofolate by other enzymes according to Figure 25-2.

consequence: 5-methyl tetrahydrofolate cannot be converted into tetrahydrofolate. There appears to be no other pathway by which this can be accomplished. Consequently, all of the body's folic acid accumulates as 5-methyl tetrahydrofolate and, in effect, a folic acid deficiency develops. This folic acid deficiency may well account for the symptoms of pernicious anemia. It should be pointed out that, although a secondary folic acid deficiency may be a consequence of B_{12} deficiency, which unquestionably has serious consequences, it is by no means certain that this is the only problem. It is possible that B_{12} has other functions that are as yet unknown.

One might suppose, in view of the foregoing discussion, that the administration of folic acid should cure pernicious anemia. In the early stages of the disease, administration of folic acid ameliorates it. However, folic acid cannot permanently prevent pernicious anemia. Perhaps the high concentration of 5-methyl tetrahydrofolic acid has undesirable consequences that cannot be overcome by folic acid.

Effect of N_2O on Methionine Synthase

Nitrous oxide (N_2O), or laughing gas, is a widely used anesthetic. It has been reported that people subjected to prolonged exposure to N_2O show some of the symptoms that are usually associated with pernicious anemia. Recall that pernicious anemia is a disease resulting from the inability to absorb vitamin B_{12} from the intestinal tract (Chapter 20, Box 20-2). The discovery that N_2O inactivates methionine synthase provided a possible explanation for the toxicity of N_2O. The inactivation of methionine synthase results from the reaction of vitamin $B_{12(s)}$, formed in the catalytic cycle, with N_2O. N_2O oxidizes vitamin $B_{12(s)}$ to its +2 oxidation state, vitamin $B_{12(r)}$(Co^{2+}); inactivation of the enzyme is believed to result from chemical modification of the apoprotein,or of vitamin B_{12} bound to methionine synthase, or from radicals that are generated in the reaction of vitamin $B_{12(s)}$ with N_2O.

N_2O turns out to be a valuable research tool in the investigation of the biochemical basis of pernicious anemia. In the study of metabolic disorders, it is useful to have animal models of the disease available. It is, however, very difficult to induce vitamin B_{12} deficiency in animals because the nutritional requirements are met by small amounts of the vitamin, and vitamin B_{12} is produced by intestinal bacteria. One can, however, obtain an animal model of pernicious anemia by maintaining pigs on a vitamin B_{12} deficient diet and exposing them daily for several hours to N_2O. It has become clear from these studies that a major cause of many of the symptoms of pernicious anemia is a failure of methionine synthesis.

Summary

Tetrahydrofolate is the central cofactor for supplying one-carbon units in metabolism. Tetrahydrofolate arises through reduction of the vitamin folic acid. One-carbon units at three oxidation levels are carried by tetrahydrofolate, and each is a source of one-carbon units for biosynthesis. 10-Formyl tetrahydrofolate is the source of formyl groups in the biosynthesis of purines. Methylene tetrahydrofolate carries its one-carbon unit at the oxidation state of formaldehyde, and it is the source of the methyl group in thymine of DNA. Thymidylate synthase catalyzes the reaction of dUMP with methylene tetrahydrofolate to form TMP and dihydrofolate, which is reduced back to tetrahydrofolate by NADPH and dihydrofolate reductase. 5-Methyl tetrahydrofolate carries its one-carbon unit at the methanol level of oxidation, and it is the source of the methyl group for methionine biosynthesis. Methionine synthase catalyzes the reaction of 5-methyl tetrahydrofolate with homocysteine to form methionine and tetrahydrofolate. The cofactor is methylcobalamin. The three one-carbon derivatives of tetrahydrofolate are produced from serine and formate. Serine transhydroxymethylase catalyzes the transfer of a formaldehyde-equivalent from serine to tetrahydrofolate to form methylene tetrahydrofolate. The latter can be reduced to 5-methyl tetrahydrofolate or oxidized to methenyl tetrahydrofolate. Methenyl tetrahydrofolate is hydrolyzed to 10-formyl tetrahydrofolate by cyclohydrolase. Formyl tetrahydrofolate synthase catalyzes the reaction of formate with ATP and tetrahydrofolate to produce 10-formyl tetrahydrofolate, ADP, and P_i.

The most-general methyl group donor in metabolism is *S*-adenosylmethionine, which is produced enzymatically from methionine and ATP. *S*-adenosylmethionine is the source of methyl groups for the biosynthesis of methyl ethers, methyl esters, and *N*-methyl compounds such as epinephrine.

ADDITIONAL READING

CHALLENGER, F. 1951. *Advances in Enzymology* 12: 429.
SMITH, E. L. 1960. *Vitamin B_{12}*. London: Methuen; New York: Wiley.
MATTHEWS, R. 1990. *Chemical Reviews* 90: 1275.

PROBLEMS

1. Write a detailed mechanism for:

ATP + 5-Formyl tetrahydrofolate ⟶ Methenyl tetrahydrofolate + ADP + P_i

Trace ^{18}O in the formyl group. Write all possible or expected exchange reactions.

2. Write mechanisms for:

(a) Nonenzymatic exchange of ^3H from ^3H_2O with H at position 5 of UMP in the presence of a thiol (RSH).

(b) The enzymatic conversion of 5-hydroxymethyl-dCMP and tetrahydrofolate into dCMP and methylene tetrahydrofolate.

(c) Irreversible inactivation of the enzyme in part b by 5-trifluoromethyl-dUMP, which forms a stable, covalent bond to the enzyme.

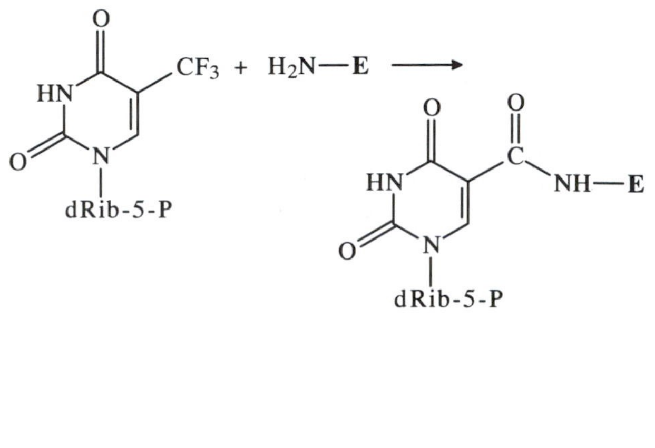

CF_3 + H_2N—E ⟶ C(=O)—NH—E

dRib-5-P dRib-5-P

Nitrogen Metabolism: Amino Acids, Urea, and Heme

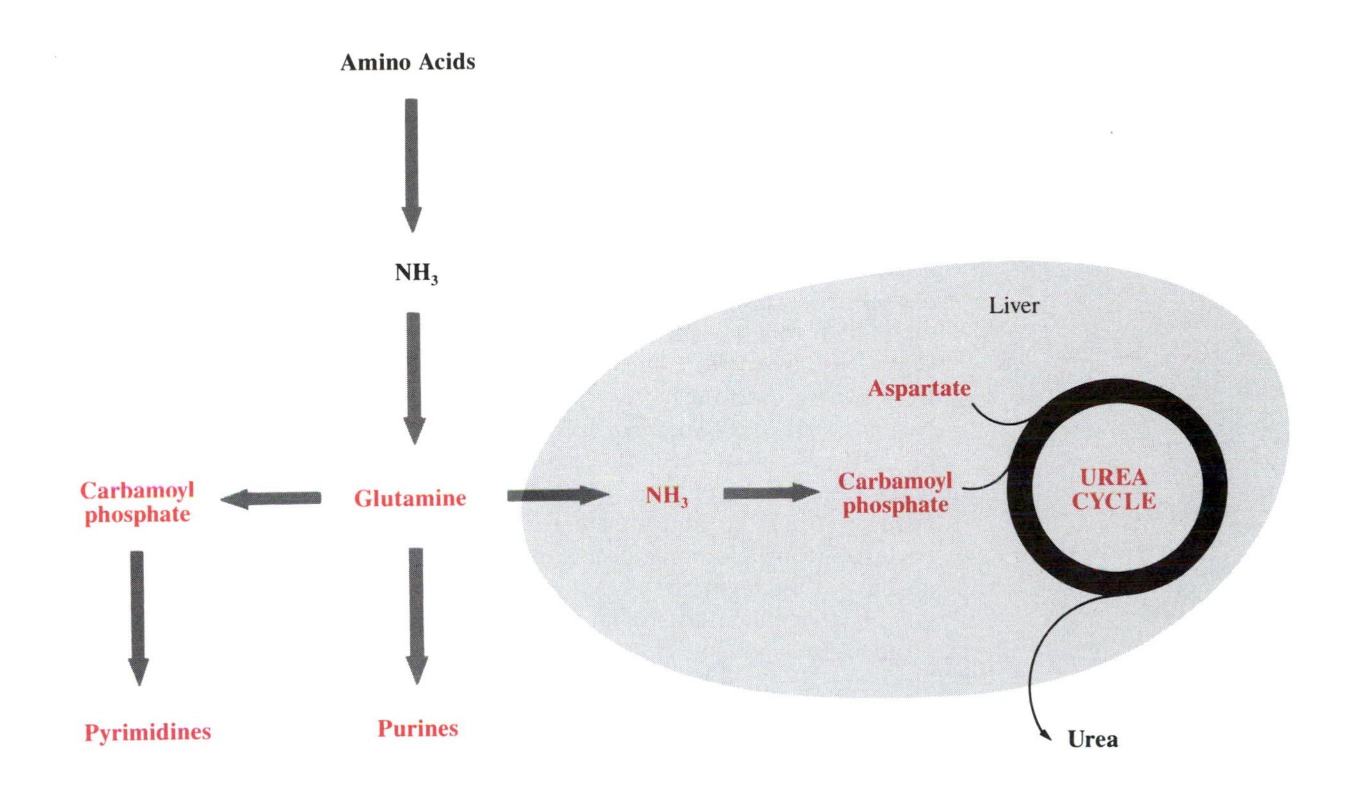

This diagram illustrates the central role of glutamine and carbamoyl phosphate in mediating the flow of nitrogen from amino acids to the biosynthesis of nucleotides. In ureatelic animals, these compounds also mediate the flow of excess nitrogen into urea.

Most biological molecules contain nitrogen. Important nitrogen-containing compounds in addition to amino acids, nucleotides, proteins, nucleic acids, phospholipids, and complex carbohydrates are neurotransmitters and heme, which are biosynthesized from amino acids. Nitrogen metabolism is an enormous subject that includes the biosynthesis of the twenty common amino acids, the degradation and interconversion of amino acids, the biosynthesis of the five heterocyclic bases in nucleic acids, the interconversion of nucleosides and nucleotides, the biosynthesis of neurotransmitters, the biosynthesis of heme, and the conversion of nitrogen from biodegradation into compounds that are excreted. Representative examples of the biosynthesis and biodegradation of amino acids are presented in this chapter, as well as the biosynthesis of heme. Most mammals excrete excess nitrogen in the form of urea, and the urea cycle also is described in this chapter. The metabolism of nucleotides is presented in Chapter 27.

Sources of Nitrogen for Cells

Cells obtain nitrogen for the biosyntheses of macromolecular building blocks such as amino acids, the heterocyclic bases of nucleic acids, and *N*-acetylhexosamines in ways that vary among organisms. A guide to these sources for various classes of organisms is given in Table 26-1. Bacteria and yeast can biosynthesize all of the amino acids and nucleotides from ammonia and simple carbon skeletons derived from carbohydrate metabolism. They can absorb ammonia from their growth media for the biosynthesis of amino acids and nucleic acid bases. Alternatively, they can absorb oxidized forms of nitrogen such as nitrate or nitrite, which are reduced to ammonia by the actions of nitrate or nitrite reductases. A few bacteria contain the nitrogen-fixing enzymes that reduce nitrogen from the atmosphere to ammonia. These are the nitrogen-fixing bacteria. Many microorganisms can also absorb amino acids from their growth media and use them directly for protein biosynthesis.

Higher animals derive most of their nitrogen from the digestion of nutrient proteins to amino acids, which are absorbed by cells and used in protein biosynthesis. Higher animals cannot biosynthesize all of the amino acids and must derive some of them from their nutrients. Those amino acids that animals can biosynthesize are known as *nonessential amino acids* and those that animals cannot biosynthesize are known as *essential amino acids*. The essential and nonessential amino acids are listed in Table 26-2.

Amino acids also serve as biosynthetic precursors of the nucleotides. They are the sources of the nitrogen and much of the carbon in the nucleic acid bases, which are biosynthesized in animal cells, as well as in microorganisms and plants. Heme also is biosynthesized from amino acids.

Table 26-1

Sources of nutrient nitrogen for several classes of organisms

Organism	Nitrogen form	Source
Microorganisms	NH_4^+, NO_2^-, NO_3^-, N_2, amino acids	Growth medium
Animals	Amino acids, proteins, nucleic acids	Nutrients
Plants	NH_4^+ (NO_2^-, N_2 bacterial symbiosis)	Soil

Table 26-2

Essential and nonessential amino acids

Essential	Nonessential
Lysine	Arginine
Phenylalanine	Tyrosine
Methionine	Cysteine
Threonine	Serine
Valine	Glycine
Leucine	Alanine
Isoleucine	Glutamine
Histidine	Asparagine
Tryptophan	Aspartic acid
	Glutamic acid
	Proline

Degradation of Amino Acids

Other important aspects of nitrogen metabolism are the utilization of carbon skeletons of amino acids as energy sources and the excretion of excess nitrogen. In higher animals, nitrogen excretion is an important part of metabolism, because it is required to maintain nitrogen balance. A serious imbalance can otherwise occur in carnivorous animals that derive most of their energy from proteins, because amino acids not used in biosynthesis are broken down into energy-producing molecules such as acetyl CoA, pyruvate, succinyl CoA, fumarate, and α-ketoglutarate. The nitrogen from amino acid degradation is released as ammonia, and excess ammonia, which is very toxic, is converted into urea by many animals and excreted.

The degradation of all amino acids proceeds in two stages beginning with removal of the amino group, generally to form an α-ketoacid or an unsaturated carboxylic acid and ammonia. In the second stage, the carbon skeleton is metabolized to compounds that can either be converted into fatty acids or enter the tricarboxylic acid cycle.

Deamination of Amino Acids

The first step in the metabolism of most amino acids is transamination with α-ketoglutarate to produce glutamate and the α-ketoacid with the same carbon skeleton as the amino acid (equation 1).

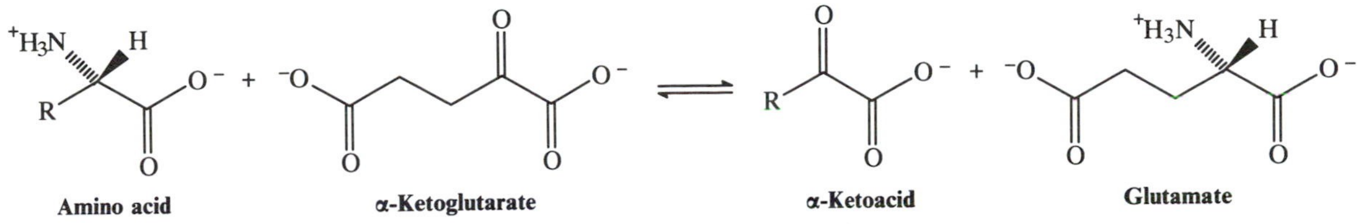

$$\text{Amino acid} + \alpha\text{-Ketoglutarate} \rightleftharpoons \alpha\text{-Ketoacid} + \text{Glutamate} \qquad (1)$$

These reactions are catalyzed by transaminases (aminotransferases), which utilize the coenzyme pyridoxal phosphate, through mechanisms described in Chapter 18.

Glutamate is then oxidized by NAD^+ (or $NADP^+$) in a reaction catalyzed by glutamate dehydrogenase according to equation 2.

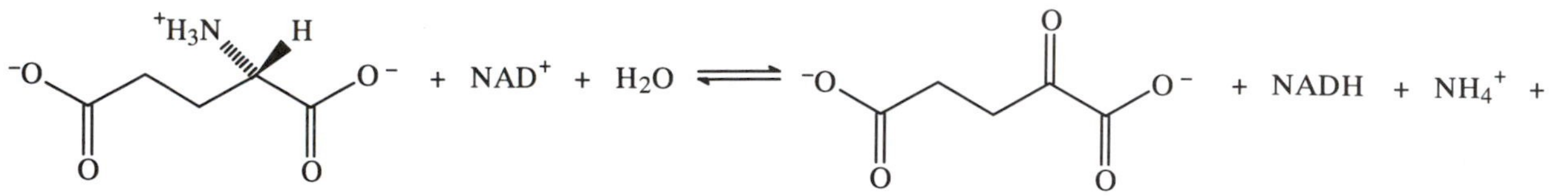

$$+ NAD^+ + H_2O \rightleftharpoons \quad + NADH + NH_4^+ + H^+ \qquad (2)$$

The dehydrogenation of glutamate proceeds in two discrete chemical steps at the active site of glutamate dehydrogenase (Figure 26-1). In the first step, glutamate is dehydrogenated by NAD^+ to α-ketoglutarate imine and NADH. In the second step, α-ketoglutarate imine undergoes hydrolysis to α-ketoglutarate and ammonia. The reaction is reversible and can produce glutamate under appropriate conditions.

Two other enzymatic reactions produce ammonia from amino acids. One is the hydrolysis of glutamine catalyzed by glutaminase (equation 3).

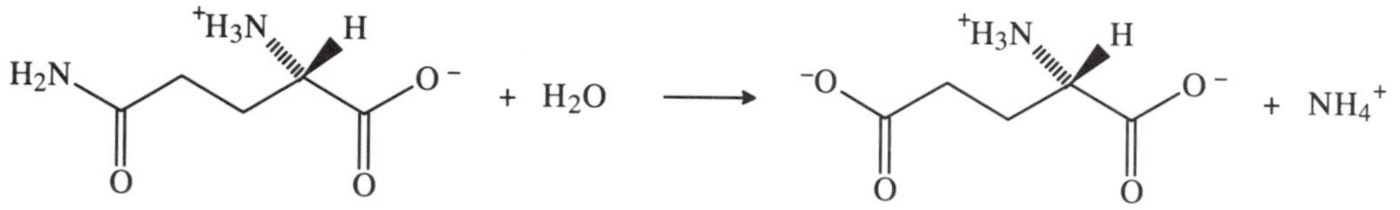

$$+ H_2O \longrightarrow \quad + NH_4^+ \qquad (3)$$

The glutamate generated is then dehydrogenated to α-ketoglutarate and ammonia by the action of glutamate dehydrogenase. The dehydration of serine and threonine catalyzed by dehydratases produces ammonia from each of these amino acids according to equation 4.

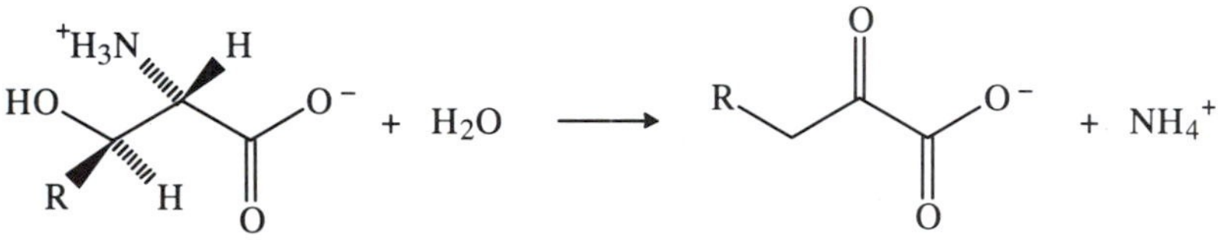

$$+ H_2O \longrightarrow \quad + NH_4^+ \qquad (4)$$

$R = H$ (Ser), CH_3 (Thr)

The mechanisms of these reactions and the function of pyridoxal phosphate are described in Chapter 18.

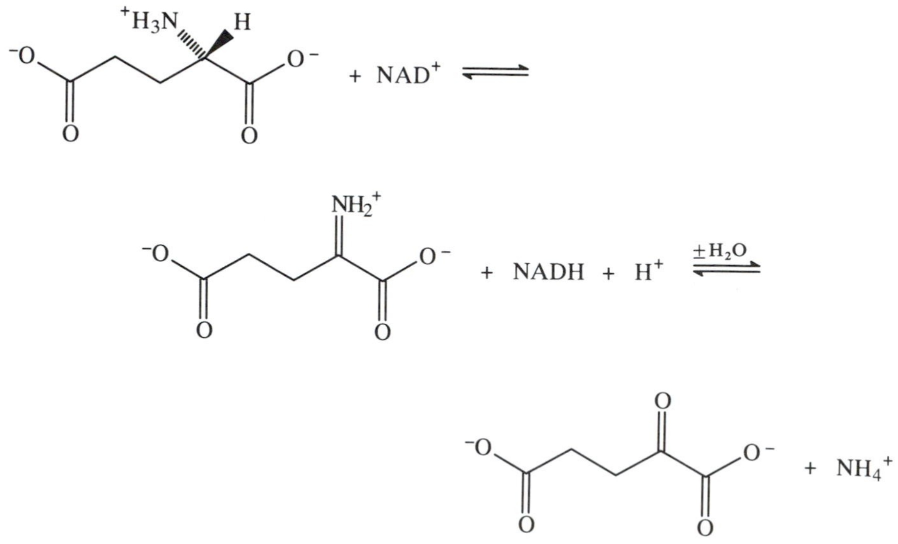

$$+ NAD^+ \rightleftharpoons$$

$$+ NADH + H^+ \overset{\pm H_2O}{\rightleftharpoons}$$

$$+ NH_4^+$$

Figure 26-1

Chemical Steps in the Dehydrogenation of Glutamate by Glutamate Dehydrogenase.

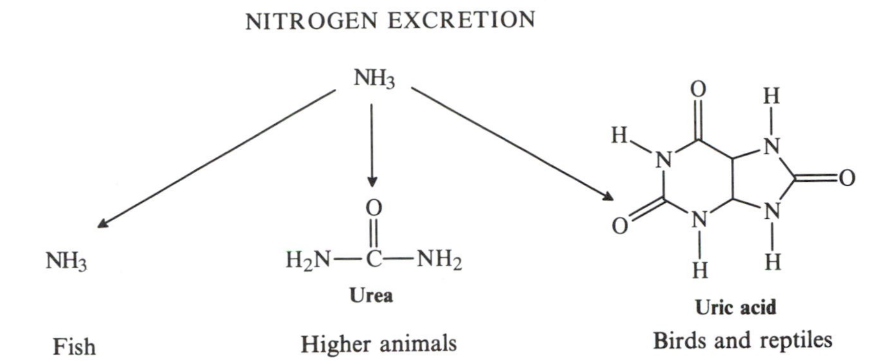

NITROGEN EXCRETION

NH_3

NH_3 — Fish

$H_2N—C(=O)—NH_2$ — Urea — Higher animals

Uric acid — Birds and reptiles

Figure 26-2

The Chemical Forms in Which Excess Nitrogen Is Excreted by Animals.

Ammonia produced in the degradation of amino acids has several fates depending on the metabolic requirements of cells. Ammonia is toxic at high concentrations, and that produced in amino acid degradation is quickly converted into other molecules such as glutamine and carbamoyl phosphate and used either for biosynthesis or for the production of molecules that are excreted. For example, the amido nitrogen of glutamine is used in the biosynthesis of nucleotides.

Excess ammonia must be eliminated from the organism because it is toxic; it is excreted in different chemical forms by different animals, as illustrated in Figure 26-2. Most higher animals convert excess ammonia into urea in the urea cycle, as described in a later section, and excrete it. Birds and reptiles convert ammonia into uric acid and excrete that compound, whereas fish excrete ammonia.

The Fate of Some Carbon Skeletons

We shall not describe the metabolism of all of the amino acids but will limit ourselves to a few typical examples. The three-carbon amino acids, alanine, cysteine, and serine, can be converted into pyruvate, either by transamination (alanine) or through an elimination reaction (cysteine and serine). Pyruvate is then metabolized in the TCA cycle.

Several amino acids, including isoleucine and threonine, are converted into propionyl CoA, which is then converted into methylmalonyl CoA and ultimately into succinyl CoA, a component of the TCA cycle. The pathway for the conversion of isoleucine into acetyl CoA and propionyl CoA is illustrated in Figure 26-3. The process begins with the transamination of isoleucine to 2-keto-3-methylvalerate in step 1, followed by dehydrogenation and decarboxylation to 2-methylbutyryl CoA in step 2. The latter reaction is catalyzed by the branched-chain α-ketoacid dehydrogenase complex, which is similar in composition and function to the pyruvate dehydrogenase complex described in Chapter 17. 2-Methylbutyryl CoA undergoes a sequence of dehydrogenation, hydration, dehydrogenation, and β-ketothiolase reactions in steps 3 through 6, similar to those in the degradation of fatty acids by β-oxidation (Chapter 19). The final products are acetyl CoA and propionyl CoA.

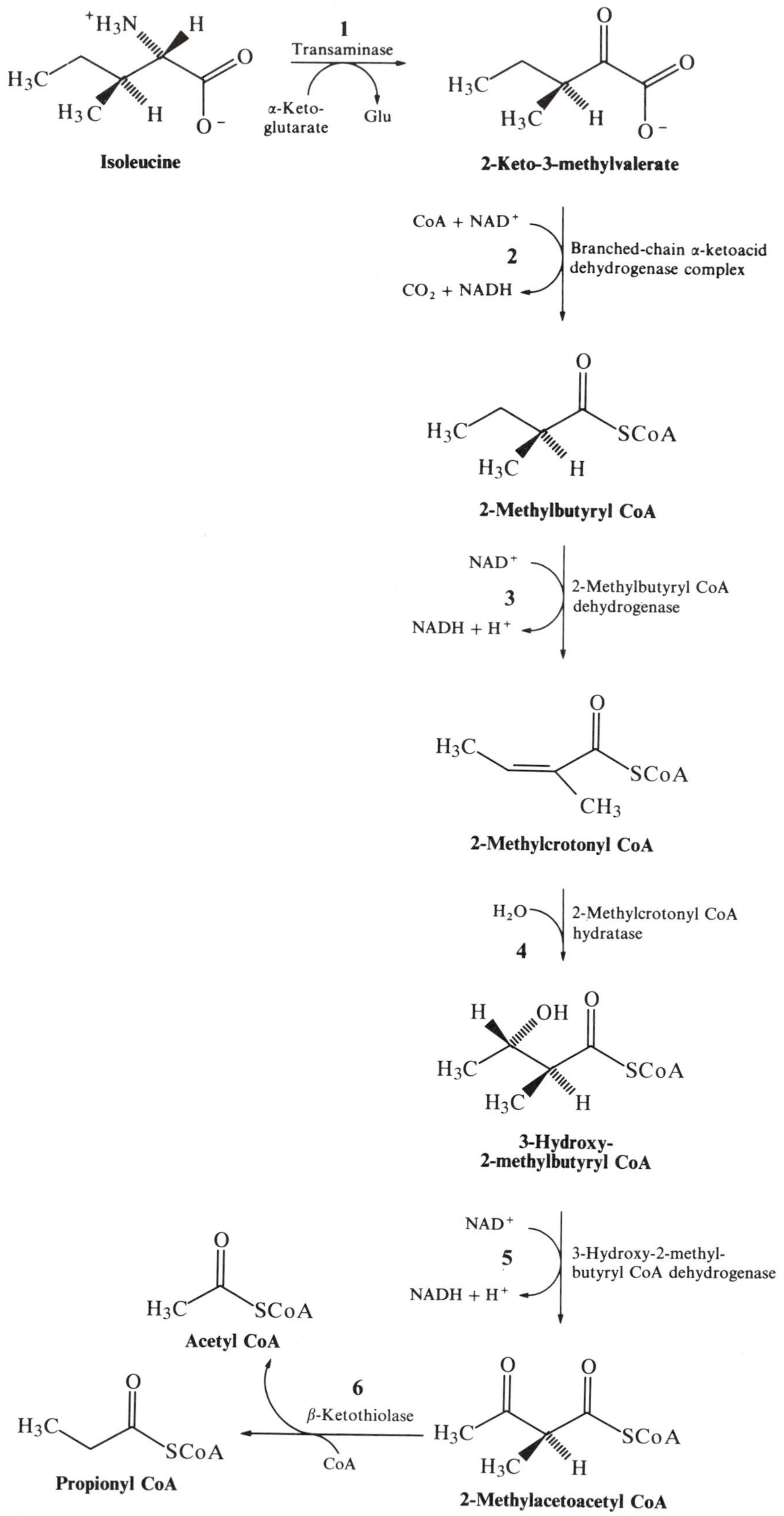

Isoleucine

1 Transaminase

α-Keto-glutarate → Glu

2-Keto-3-methylvalerate

$CoA + NAD^+$

2 Branched-chain α-ketoacid dehydrogenase complex

$CO_2 + NADH$

2-Methylbutyryl CoA

NAD^+

3 2-Methylbutyryl CoA dehydrogenase

$NADH + H^+$

2-Methylcrotonyl CoA

H_2O

4 2-Methylcrotonyl CoA hydratase

3-Hydroxy-2-methylbutyryl CoA

NAD^+

5 3-Hydroxy-2-methylbutyryl CoA dehydrogenase

$NADH + H^+$

2-Methylacetoacetyl CoA

6 β-Ketothiolase

CoA

Acetyl CoA

Propionyl CoA

Figure 26-3

Degradation of Isoleucine to Acetyl CoA and Propionyl CoA.

The conversion of propionyl CoA into succinyl CoA begins with carboxylation to methylmalonyl CoA, which is catalyzed by propionyl CoA carboxylase according to equation 5.

$$\underset{\textbf{Propionyl CoA}}{CH_3{-}CH_2{-}\overset{O}{\overset{\|}{C}}{-}SCoA} + HCO_3^- + ATP \xrightarrow{Mg^{2+}} \underset{\textbf{(R)-Methylmalonyl CoA}}{(H_3C)(H)({}^-O_2C)C{-}\overset{O}{\overset{\|}{C}}{-}SCoA} + ADP + P_i \quad (5)$$

Propionyl CoA carboxylase is one of the biotin-dependent carboxylases (Chapter 17). Carboxylation of propionyl CoA produces methylmalonyl CoA with the *R*-configuration at carbon-2. The enzyme methylmalonyl CoA epimerase catalyzes the interconversion of this isomer with that having the *S* configuration at carbon-2, as shown in equation 6.

$$\underset{\textbf{(R)-Methylmalonyl CoA}}{(H_3C)(H)({}^-O_2C)C{-}\overset{O}{\overset{\|}{C}}{-}SCoA} \rightleftharpoons \underset{\textbf{(S)-Methylmalonyl CoA}}{(H_3C)({}^-O_2C)(H)C{-}\overset{O}{\overset{\|}{C}}{-}SCoA} \quad (6)$$

The *S* stereoisomer is the substrate for a very unusual rearrangement to succinyl CoA catalyzed by methylmalonyl CoA mutase (equation 7).

$$(H_3C)({}^-O_2C)(H)C{-}\overset{O}{\overset{\|}{C}}{-}SCoA \rightleftharpoons \underset{\textbf{Succinyl CoA}}{{}^-O_2C{-}CH_2{-}CH_2{-}\overset{O}{\overset{\|}{C}}{-}SCoA} \quad (7)$$

Succinyl CoA is oxidized in the TCA cycle.

In the rearrangement of equation 7, the group—COSCoA migrates to the adjacent methyl group, and one of the hydrogen atoms from the methyl group takes the place of —COSCoA. (The hydrogen migration in equation 7 is not intended to show the stereochemical course of the transfer, which actually occurs with retention of configuration.) This reaction requires adenosylcobalamin, a coenzyme form of vitamin B_{12}, which is generally required for rearrangements of this type. (See Chapter 20 for the discovery of vitamin B_{12} and the mechanism of action of adenosylcobalamin.)

Breakdown of Aromatic Amino Acids

The metabolic breakdown of phenylalanine is outlined in Figure 26-4. Phenylalanine is first hydroxylated to tyrosine by phenylalanine hydroxylase, a monooxygenase. Tyrosine is then transaminated to *p*-hydroxyphenylpyruvate. The conversion of *p*-hydroxyphenylpyruvate into homogentisate is a complex reaction catalyzed by a single enzyme. The enzyme catalyzes the introduction of a hydroxyl group into the phenyl ring, as well as a decarboxylation and the migration of the methylene group to form homogentisate. Through the action of a dioxygenase, the aromatic ring of homogentisate is opened to 4-maleylacetoacetate, and an isomerase converts the Z-isomer into the E-isomer. Finally, through a retro-aldol condensation, 4-fumarylacetoacetate is cleaved to fumarate and acetoacetate. Acetoacetate is activated to acetoacetyl CoA and cleaved by β-ketothiolase to two moles of acetyl CoA. Both 4-fumarylacetoacetate and acetoacetyl CoA have the grouping —CO—CH_2—CO— that facilitates carbon–carbon bond cleavage (see Chapter 2).

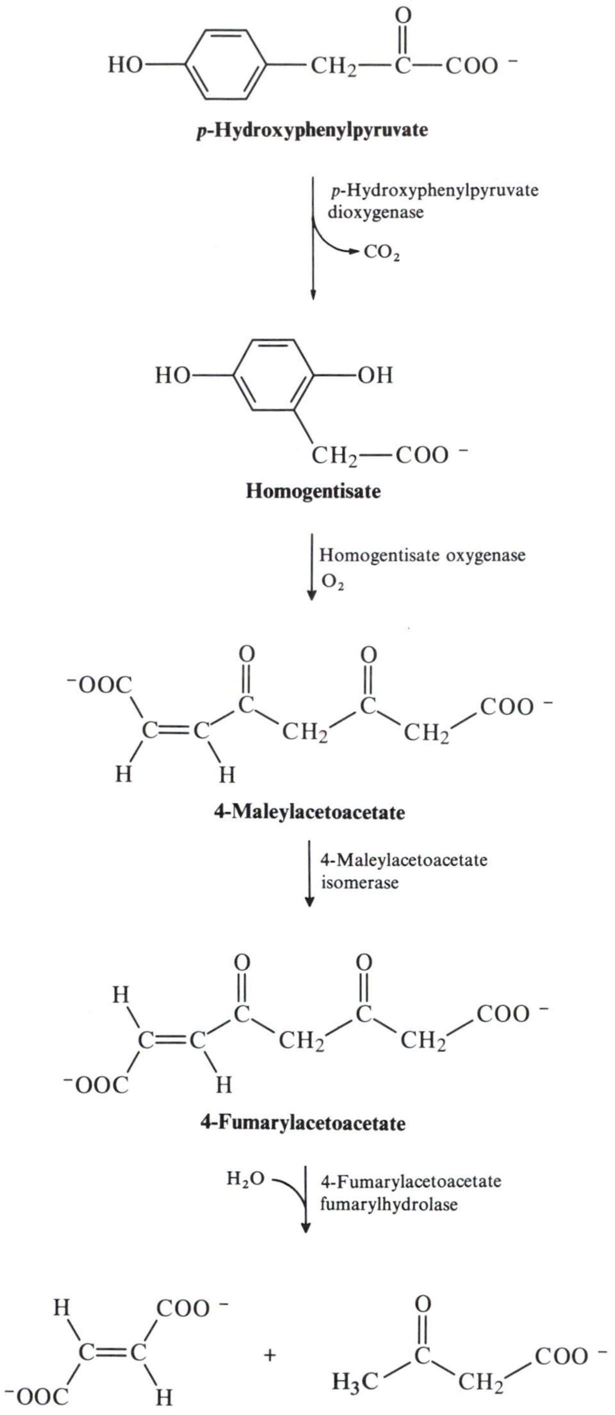

Figure 26-4

The Metabolism of Phenylalanine and Tyrosine. Phenylalanine is converted into tyrosine by hydroxylation catalyzed by phenylalanine hydroxylase. Tyrosine is then transaminated to *p*-hydroxyphenylpyruvate, which is subsequently degraded by the reactions shown.

HO—C_6H_4—CH_2—C(=O)—COO^-

***p*-Hydroxyphenylpyruvate**

↓ *p*-Hydroxyphenylpyruvate dioxygenase; → CO_2

Homogentisate

↓ Homogentisate oxygenase; O_2

4-Maleylacetoacetate

↓ 4-Maleylacetoacetate isomerase

4-Fumarylacetoacetate

↓ H_2O; 4-Fumarylacetoacetate fumarylhydrolase

^-OOC—CH=CH—COO^- + H_3C—C(=O)—CH_2—COO^-

Metabolic Disorders

Genes control the synthesis of enzymes. In nature, genes are subject to mutations that can alter the structures and expression of enzymes. Structural mutants of enzymes may exhibit decreased activities or no activities at all, and altered expression of enzymes may lead to the absence of particular enzymes in cells. These genetic defects affect the metabolism of the organism and will be passed from generation to generation. Such defects are known as genetic diseases or inborn errors of metabolism, an example being galactosemia, which was described in Chapter 11 and is caused by a defect in the enzyme galactose-1-phosphate uridylyltransferase. The consequences of genetic mutations in enzymes range from benign to fatal. Mental retardation is frequently a most-obvious consequence in human beings.

One of the recognized metabolic diseases is *alcaptonuria*, first described by the physician Lusitamus in 1649. In this disease, homogentisate oxidase is missing. Homogentisate accumulates, undergoes oxidation and polymerization, and gives urine a black color. The disease, though easily detected, appears to be benign. In 1902, Garrod recognized that this disease is transmitted as a single recessive Mendelian trait. He concluded that "the splitting of the benzene ring in normal metabolism is the work of a special enzyme, which in congenital alcaptonuria is wanting."

Phenylketonuria, described in Chapter 24, is a more-severe disease caused by the absence, or deficiency, of phenylalanine hydroxylase. Consequently, phenylalanine and its transamination product phenylpyruvate, as well as other toxic compounds derived from phenylpyruvate, accumulate. If untreated, patients suffering from phenylketonuria are severely mentally retarded by the age of one year. The biochemical basis for mental retardation in phenylketonuria is unknown. Phenylketonuria appears in about 1 in 20,000 newborns.

Lactic aciduria is caused by a defect in the pyruvate dehydrogenase complex. The inactivity of this complex prevents the conversion of pyruvate and NAD^+ into acetyl CoA, CO_2, and NADH. Pyruvate is instead reduced by NADH to lactate, which is excreted. The defect is usually in the pyruvate dehydrogenase or E_1 component of the complex, but in some cases it is in E_2 or E_3. The defect severely limits the energy value of carbohydrates in the diet because they cannot be oxidized to CO_2 with the production of NADH for use in the terminal electron-transport pathway and oxidative phosphorylation. However, affected persons still derive sufficient energy from the TCA cycle if they have protein in their diets, because the degradation of amino acids leads to the production of acetyl CoA, succinate, fumarate, oxaloacetate, and α-ketoglutarate, all of which are intermediates in the TCA cycle.

Excretion of Nitrogen and the Urea Cycle

The ammonia from amino acid degradation is used in most tissues to produce glutamine from glutamate and in liver to produce carbamoyl phosphate. Glutamine and carbamoyl phosphate are nitrogen sources in the biosynthesis of nucleotides, and excess glutamine and carbamoyl phosphate are used in the production of urea and uric acid for excretion. Most mammals excrete excess nitrogen in the form of urea (Figure 26-2) produced in the urea cycle. The production of uric acid for excretion by birds and reptiles is described in Chapter 27.

Carbamoyl Phosphate as a Source of Nitrogen in Urea

The starting point for the conversion of ammonia into urea in the liver is its reaction with bicarbonate and two moles of ATP to form carbamoyl phosphate, which is catalyzed by carbamoyl phosphate synthase I according to equation 8.

$$NH_4^+ + 2\,ATP + HCO_3^- \xrightarrow[N\text{-Acetylglutamate}]{Mg^{2+}} \underset{\textbf{Carbamoyl phosphate}}{H_2N{-}C({=}O){-}OPO_3^{2-}} + 2\,ADP + P_i \quad (8)$$

The enzyme is activated by *N*-acetylglutamate, which does not participate directly in the chemical reaction but is required to maintain the enzyme activity. Carbamoyl phosphate is a high-energy molecule formed by the elimination of one molecule of water between ammonia and bicarbonate ion and the elimination of another molecule of water between carbamate and ATP. For this reason, the reaction requires two moles of ATP. The mechanism by which two moles of ATP are used corresponds to the two stages of water elimination. The process almost certainly begins with the phosphorylation of bicarbonate by ATP at the enzymatic active site to produce phosphocarbonate, an ionized form of carbonic phosphoric anhydride, at the active site according to equation 8a.

$$HOCO_2^- + ATP \longrightarrow \left[{}^-O{-}C({=}O){-}O{-}P({=}O)(O^-){-}O^- \right] + ADP \quad (8a)$$

Phosphocarbonate is probably formed as an intermediate but, owing to its intrinsic reactivity, has never been isolated and characterized. It immediately reacts with ammonia at the active site in equation 8b, generating inorganic phosphate and carbamate.

$$NH_3 + \left[{}^-O{-}C({=}O){-}O{-}P({=}O)(O^-){-}O^- \right] \longrightarrow \underset{\textbf{Carbamate}}{H_2N{-}C({=}O){-}O^-} + P_i \quad (8b)$$

Carbamate is then phosphorylated at the active site by a second molecule of ATP to produce carbamoyl phosphate according to equation 8c.

$$H_2N{-}C({=}O){-}O^- + ATP \longrightarrow H_2N{-}C({=}O){-}O{-}P({=}O)(O^-){-}O^- + ADP \quad (8c)$$

All of these reactions take place at the active site, and neither phosphocarbonate nor carbamate is released from the active site. Only the final products are released as they are produced.

The formation of carbamoyl phosphate by this process is thermodynamically downhill because of the utilization of two moles of ATP per mole of ammonia and bicarbonate. Activation by phosphorylation in this reaction is similar to other acyl activations described in Chapter 10, and the product carbamoyl phosphate is itself activated for acyl group transfer.

Only about 1% of the ammonium ion is present as ammonia at the physiological pH of 7.3, but carbamoyl phosphate synthase is sufficiently reactive with ammonia to extract it and use it as a substrate at pH 7.3. The physiological concentration of ammonium ion is somewhat below K_m for carbamoyl phosphate synthase. This means that the enzyme normally operates under k_{cat}/K_m conditions; that is, under conditions in which the rate is proportional to the concentration of ammonium ion. Therefore, the enzyme responds immediately to an increase in ammonium ion with an increase in the rate of carbamoyl phosphate formation. This is metabolically significant because, if the enzyme were saturated or nearly saturated, an increase in rate could be brought about only by increasing the amount of enzyme in the cell, a process that would require a few minutes for protein biosynthesis.

In tissues other than liver, ammonia formed in metabolism is used to produce glutamine through the action of glutamine synthetase according to equation 9.

$$\text{Glutamate} + \text{ATP} + NH_4^+ \longrightarrow \text{Glutamine} + \text{ADP} + P_i \qquad (9)$$

Glutamine is then transported through the blood stream and absorbed by the small intestine, where it is hydrolyzed by glutaminase to glutamate and ammonia. Most of the ammonia is transported through the portal vein to the liver and used for urea production. Some of the ammonia is converted into citrulline in the small intestine and transported to the liver, where it enters the urea cycle (Figure 26-5).

The amido nitrogen of glutamine is also an important source of NH_2 groups in biosynthetic reactions, particularly in the biosynthesis of nucleotides (Chapter 27). One such use is in carbamoyl phosphate formation by carbamoyl phosphate synthase II, which uses glutamine rather than ammonia as the source of nitrogen. This reaction and others requiring glutamine as a source of NH_2 groups are described in Chapter 27. Carbamoyl phosphate synthase II produces carbamoyl phosphate primarily for pyrimidine biosynthesis (Chapter 27).

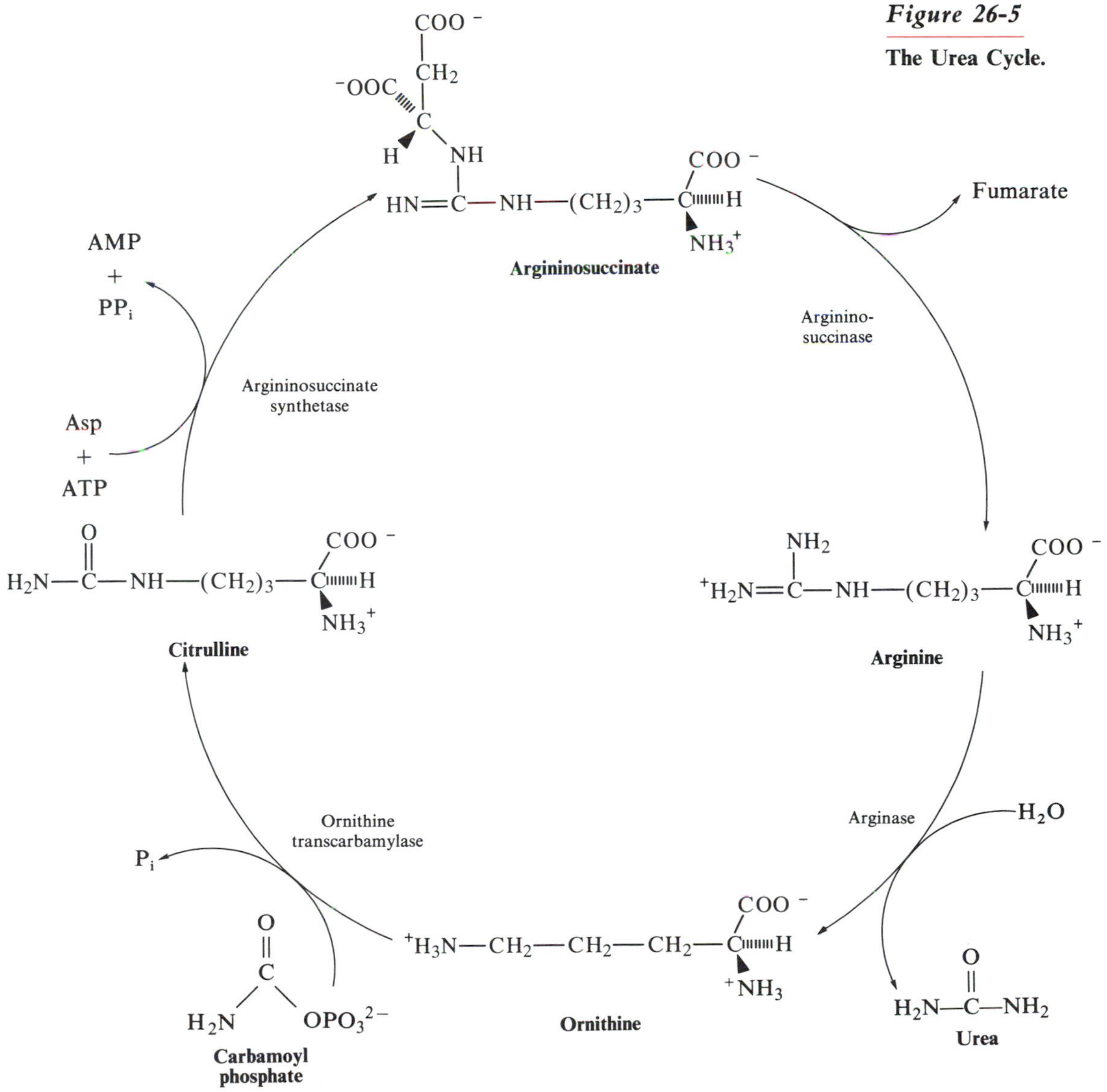

Figure 26-5

The Urea Cycle.

Urea Production in the Urea Cycle

The amido group of carbamoyl phosphate in the liver is used in the urea cycle to produce urea through the enzymatic reactions shown in Figure 26-5. Each revolution produces one molecule of urea, and carbamoyl phosphate is one of two sources of nitrogen for urea. The other nitrogen source is aspartate. The chemistry by which this is accomplished also converts aspartate into fumarate, which enters the TCA cycle.

In the first step of the urea cycle, ornithine transcarbamoylase catalyzes the transfer of the carbamoyl group in carbamoyl phosphate to ornithine to form citrulline. This is an acyl group transfer reaction analogous to those described in Chapter 10. Citrulline then reacts with aspartate and ATP in a reaction catalyzed by argininosuccinate synthase to produce argininosuccinate, AMP, and PP_i. This reaction involves two group transfers, the first being an AMP transfer to activate the ureido oxygen according to equation 10a.

$$H_2N(O{=})C{-}NH{-}(CH_2)_3{-}CH(CO_2^-)(NH_3^+) + ATP \xrightarrow{Mg^{2+}} HN{=}C(O{-}AMP){-}NH{-}(CH_2)_3{-}CH(CO_2^-)(NH_3^+) + PP_i \qquad (10a)$$

Citrulline

The AMP group in the product activates the ureido oxygen of citrulline for displacement in the next step, in which the α-amino group of aspartate displaces AMP from the AMP-derivative of citrulline to form argininosuccinate according to equation 10b.

$$^-O_2C{-}CH_2{-}CH(CO_2^-)(NH_2) + HN{=}C(O{-}AMP){-}NH{-}(CH_2)_3{-}CH(CO_2^-)(NH_3^+) \xrightarrow{Mg^{2+}}$$

Aspartate

$$^-O_2C{-}CH_2{-}CH(CO_2^-){-}NH{-}C({=}NH){-}NH{-}(CH_2)_3{-}CH(CO_2^-)(NH_3^+) + AMP \qquad (10b)$$

Argininosuccinate

Both steps take place at the active site, and the intermediate adenylylcitrulline (AMP-citrulline) produced in equation 10a remains bound to the enzyme throughout the reaction. In the next step, argininosuccinase catalyzes the elimination of fumarate from argininosuccinate to form arginine. This is an α, β-elimination reaction analogous to those described in Chapter 18. Hydrolysis of arginine by the action of arginase completes the formation of urea and regenerates ornithine, which continues with another round through the cycle. Each cycle consumes one carbamoyl phosphate, one aspartate, and one ATP and produces one urea and one fumarate.

Compartmentation of Enzymes

Urea is formed in the cytosol of liver cells in higher animals, but the cycle itself operates in both the cytosol and the mitochondrial matrix. Carbamoyl phosphate synthase and ornithine transcarbamoylase are located in the mitochondrial matrix, whereas argininosuccinate synthase, argininosuccinase, and arginase are in the cytosol. Therefore, transport between the matrix and cytosol is required. These transport systems lie within the inner mitochondrial membrane and relocate ammonia, bicarbonate, ornithine, and citrulline. The operations of a transport system are illustrated in Figure 26-6.

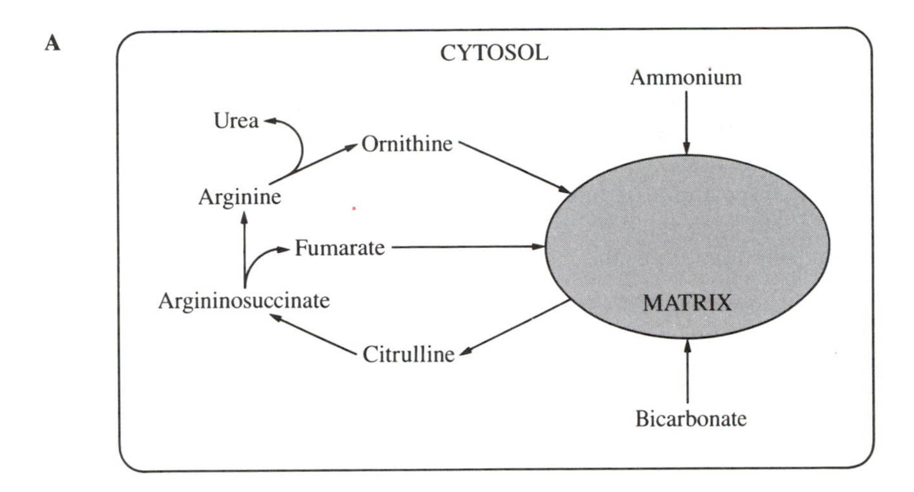

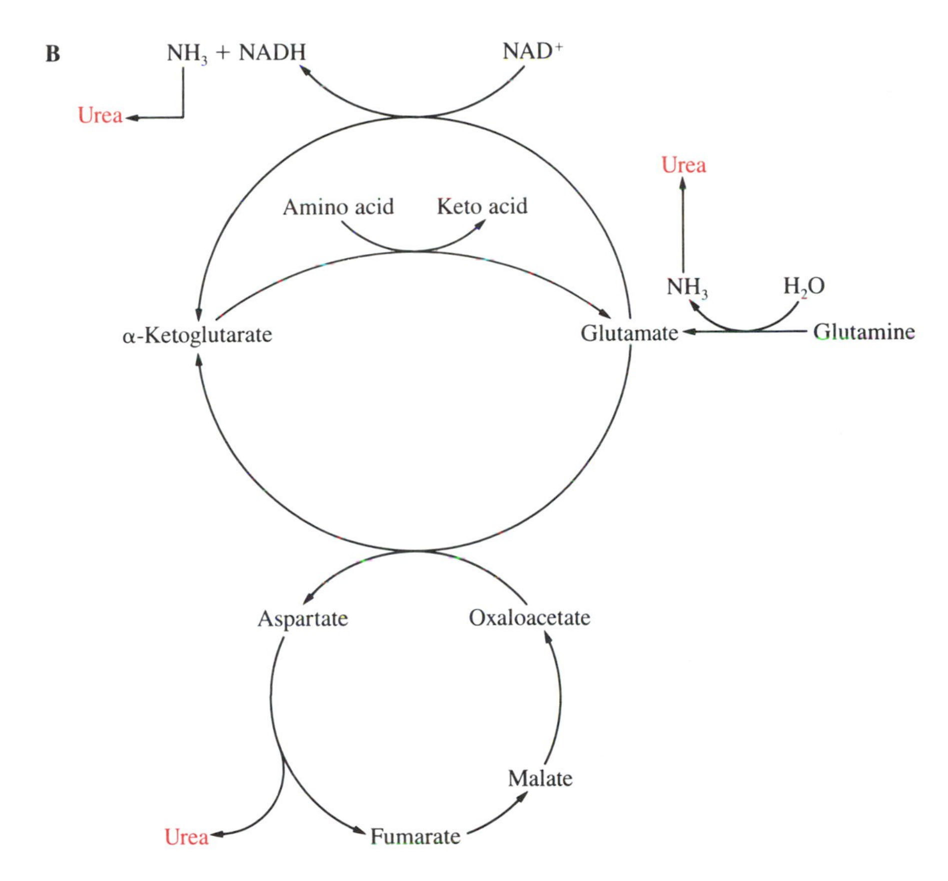

Figure 26-6

Intracellular Transport in the Operation of the Urea Cycle. A. Ammonia and bicarbonate must be taken into the mitochondrion, where carbamoyl phosphate synthetase and ornithine transcarbamoylase produce citrulline. Citrulline must be exported from the mitochondrion into the cytosol, where the production of urea and fumarate takes place. Ornithine is transported into the mitochondrion where it is again carbamoylated by carbamoyl phosphate.
B. Fumarate derived from aspartate is also recycled in urea production. It is hydrated to malate by fumarase and oxidized to oxaloacetate by malate dehydrogenase in the mitochondrion. Transamination with glutamate produces aspartate for another round of urea production. Glutamate also serves as a source of ammonia for urea through the action of glutamate dehydrogenase, and glutamate is regenerated by transamination of α-ketoglutarate with other amino acids.

Biosynthesis of Amino Acids

All of the common amino acids are biosynthesized in plants and microorganisms, but higher animals, including human beings, can synthesize only some of them. The essential amino acids must be included in the diet, and the nonessential amino acids are derived directly or ultimately from intermediates of the major metabolic pathways; that is, glycolysis, the TCA cycle, and the pentose phosphate pathway. In the following sections, we will describe the biosynthesis of a few amino acids as representative of amino acid biosynthetic pathways.

Amino acids derived directly from TCA intermediates are glutamate from α-ketoglutarate, aspartate from oxaloacetate, and alanine from pyruvate. These amino acids are synthesized by transamination and the glutamate dehydrogenase reaction, both of which you have seen before in the degradation of amino acids. These reactions are reversible and can lead to biosynthesis of the amino acids. Glutamine and asparagine are derived from aspartate and glutamate by amidation reactions catalyzed by glutamine synthase and asparagine synthase according to equations 9 and 11.

$$\text{Aspartate} + \text{ATP} + NH_4^+ \longrightarrow \text{Asparagine} + \text{AMP} + PP_i \qquad (11)$$

Both reactions are examples of acyl activation by ATP of the type described in Chapter 10.

Biosynthesis of Phenylalanine and Tyrosine

A complex and interesting reaction sequence leads to the biosynthesis of phenylalanine and tyrosine in microorganisms and plants. The building blocks are erythrose-4-phosphate and phosphoenolpyruvate. The aromatic amino acid biosynthetic pathway begins with the condensation of erythrose-4-phosphate with phosphoenolpyruvate to form the seven-carbon sugar 3-deoxy-D-*arabino*-heptulosonate-7-phosphate according to equation 12.

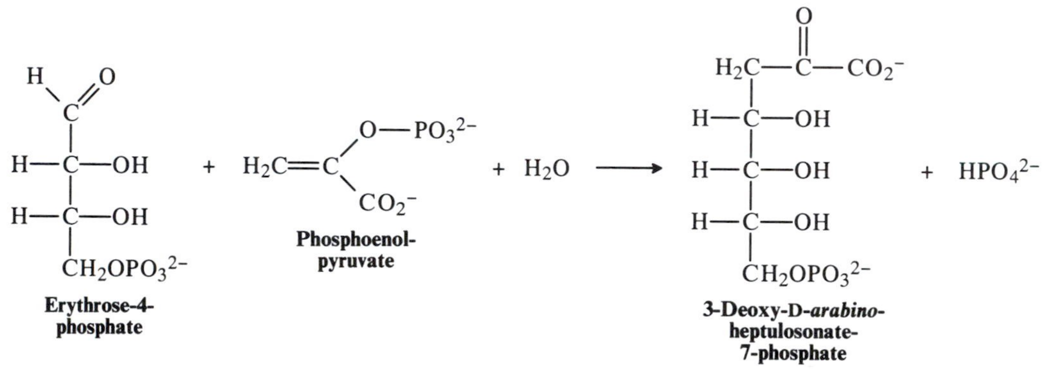

$$\underset{\textbf{Erythrose-4-phosphate}}{HC(=O)-CH(OH)-CH(OH)-CH_2OPO_3^{2-}} + \underset{\textbf{Phosphoenolpyruvate}}{H_2C{=}C(O-PO_3^{2-})CO_2^-} + H_2O \longrightarrow \underset{\textbf{3-Deoxy-D-arabino-heptulosonate-7-phosphate}}{H_2C(-C(=O)-CO_2^-)-CH(OH)-CH(OH)-CH(OH)-CH_2OPO_3^{2-}} + HPO_4^{2-} \qquad (12)$$

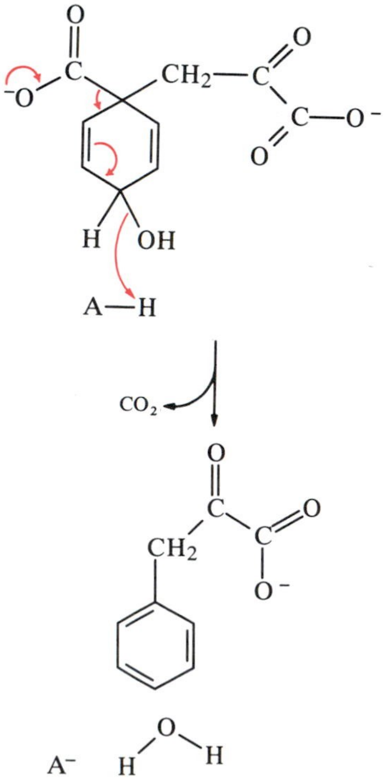

This is an aldol reaction catalyzed by 3-deoxy-D-*arabino*-heptulosonate-7-phosphate aldolase, in which phosphoenolpyruvate provides the enolate of pyruvate by undergoing hydrolytic dephosphorylation at the active site. However, as explained in Box 26-1, the reaction mechanism is complicated.

The conversion of 3-deoxy-D-*arabino*-heptulosonate-7-phosphate into chorismate is outlined in Figure 26-7. Dehydroquinate synthase catalyzes the cyclization of 3-deoxy-D-*arabino*-heptulosonate-7-phosphate to 3-dehydroquinate by an interesting mechanism in which carbon-7 is added to the 2-keto group and phosphate is eliminated. Dehydration of 3-dehydroquinate followed by reduction produces shikimate, which is phosphorylated to 3-phosphoshikimate. Condensation of 3-phosphoshikimate with phosphoenolpyruvate produces 5-enolpyruvylshikimate-3-phosphate. Elimination of phosphate leads to chorismate, the central compound from which phenylalanine, tyrosine, and tryptophan are produced.

The conversion of chorismate into phenylalanine and tyrosine is shown in Figure 26-8. Chorismate first undergoes isomerization to prephenate through a Claisen rearrangement catalyzed by chorismate mutase. The transformation of prephenate to phenylpyruvate entails a decarboxylation coupled to the elimination of H_2O, as illustrated in the margin. Decarboxylations coupled to elimination—in this case, the elimination of H_2O—are rare in biochemistry. Another example of this unusual process is the decarboxylation of mevalonate-5-pyrophosphate in cholesterol biosynthesis (Chapter 28), in which phosphate is eliminated instead of water. For tyrosine biosynthesis, prephenate is converted into *p*-hydroxyphenylpyruvate in a decarboxylation coupled to the reduction of NAD^+, a reaction type that was described in Chapter 16. The mechanism in Figure 26-9 is analogous to the decarboxylation of a β-ketoacid, but with an intervening double bond.

BOX 26-1

REACTIONS OF PHOSPHOENOLPYRUVATE IN AROMATIC AMINO ACID BIOSYNTHESIS

Phosphoenolpyruvate donates the enolpyruvyl group in equation 12 and in one reaction of Figure 26-7, but the mechanisms by which these reactions occur are poorly understood. The condensation of erythrose-4-phosphate and phosphoenolpyruvate appears to be a simple aldol condensation, as shown here.

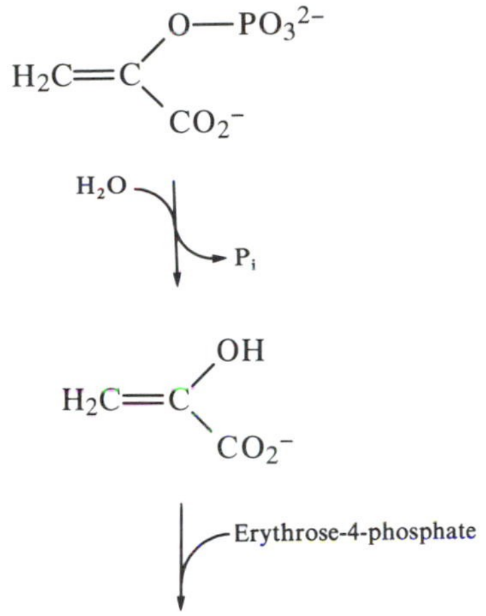

3-Deoxy-D-*arabino*-heptulosonate-7-phosphate

These reactions would take place at the active site; that is, free enolpyruvate would not take part. If the reaction occurred by this mechanism, then the phosphate produced should not contain the oxygen that originally bridged carbon-2 and the phosphoryl group in phosphoenolpyruvate. However, when the reaction was carried out with ^{18}O-labeled phosphoenolpyruvate, in which ^{18}O was attached to carbon-2 and phosphorus, the ^{18}O appeared in phosphate and not in 3-deoxy-D-*arabino*-heptulosonate-7-phosphate; that is, with C–O bond cleavage. The phosphate did not contain oxygen from water; therefore, the foregoing mechanism is *incorrect*.

The mechanism of this reaction is not known. According to one possible mechanism, a nucleophile at the active site adds to the double bond of phosphoenolpyruvate, and carbon-3 of phosphoenolpyruvate adds to the carbonyl group of erythrose-4-phosphate (RCHO) to give the enzyme-bound adduct (**2**). Water then displaces E-X to form compound **3**, which readily eliminates $HOPO_3^{2-}$ to yield 3-deoxy-D-*arabino*-heptulosonate-7-phosphate.

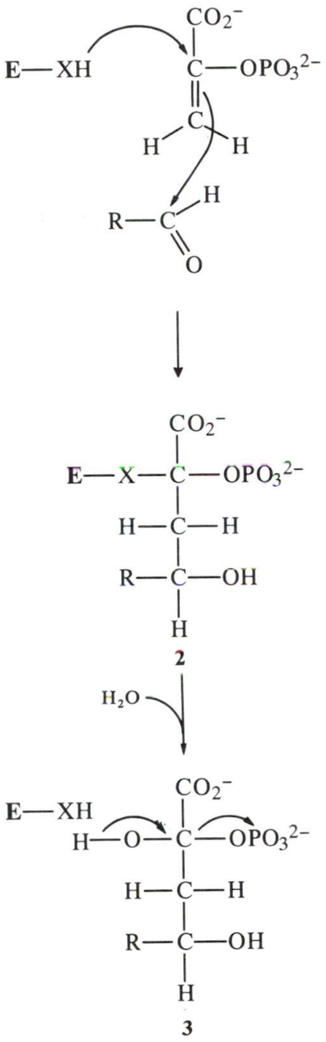

The reaction of shikimate-3-phosphate with phosphoenolpyruvate catalyzed by 5-enolpyruvylshikimate-3-phosphate synthase also is complex. When this reaction is carried out in ^3H_2O, the product contains tritium in the methylene group, as shown in Figure A (on page 706). This implies that the methylene group ($—CH_2—$) is reversibly converted into a methyl group ($—CH_3$) at some point in the reaction. In addition, phosphate formed in the reaction contains the oxygen originally present in the position bridging carbon and

BOX 26-1 (continued)

phosphorus. In other words, the reaction proceeds with C–O bond cleavage, as does the reaction catalyzed by 3-deoxy-D-*arabino*-heptulosonate-7-phosphate synthase. Figure A shows a mechanism that accounts for these facts. The intermediate resulting from the addition of shikimate-3-phosphate to phosphoenolpyruvate has been isolated from the enzyme and characterized. It has also been shown to be converted into the product at the active site of the enzyme at a rate that corresponds to the overall rate of the enzymatic reaction; that is, it is a true intermediate.

Figure A

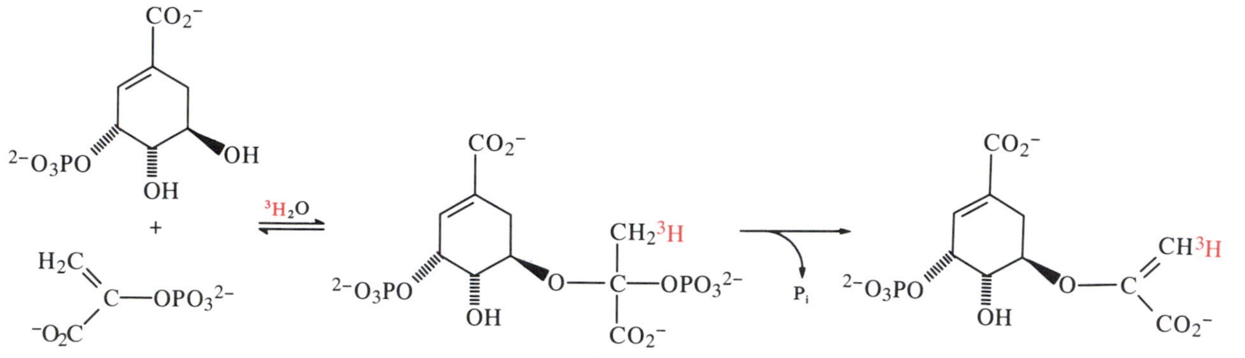

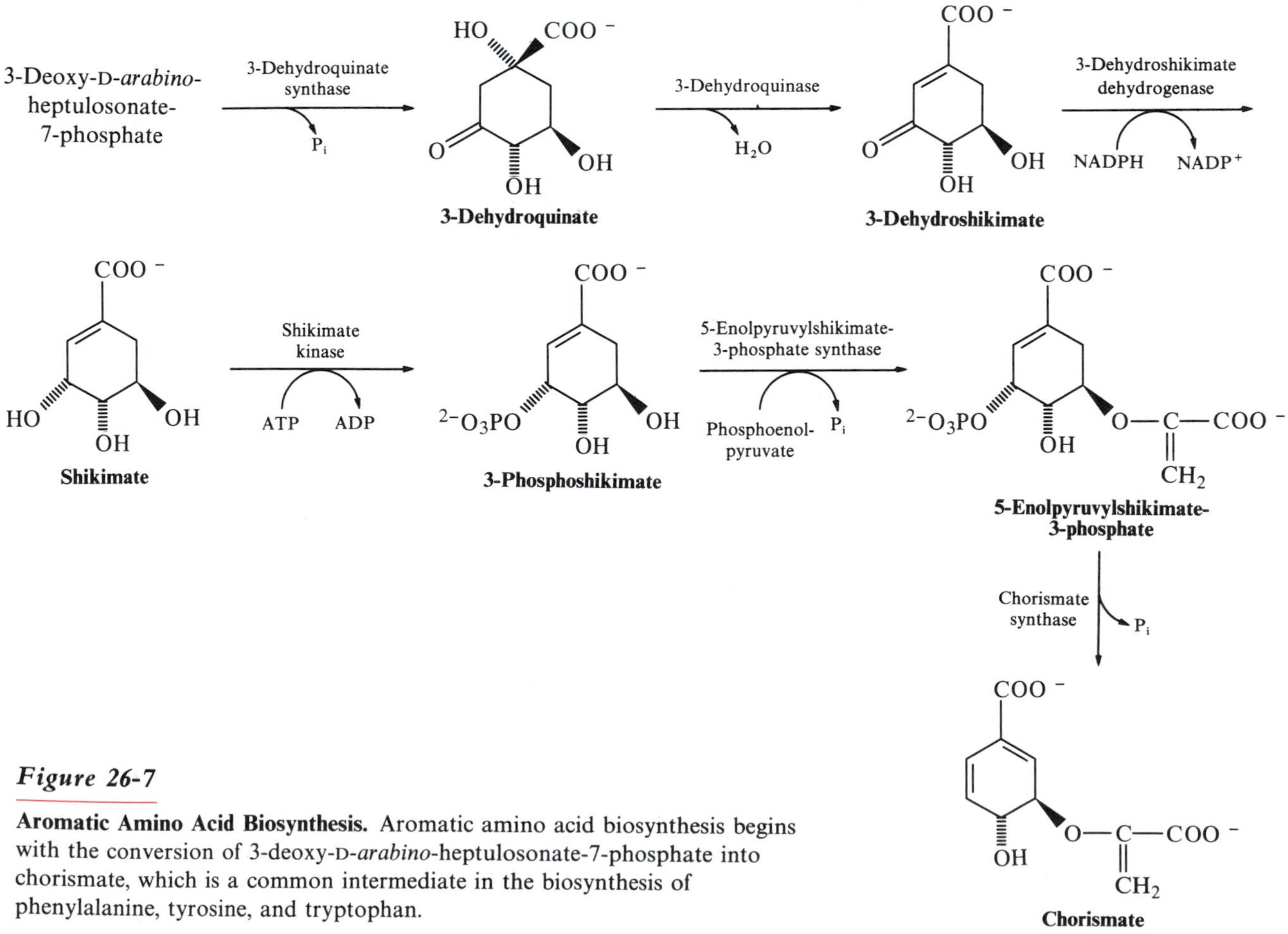

Figure 26-7

Aromatic Amino Acid Biosynthesis. Aromatic amino acid biosynthesis begins with the conversion of 3-deoxy-D-*arabino*-heptulosonate-7-phosphate into chorismate, which is a common intermediate in the biosynthesis of phenylalanine, tyrosine, and tryptophan.

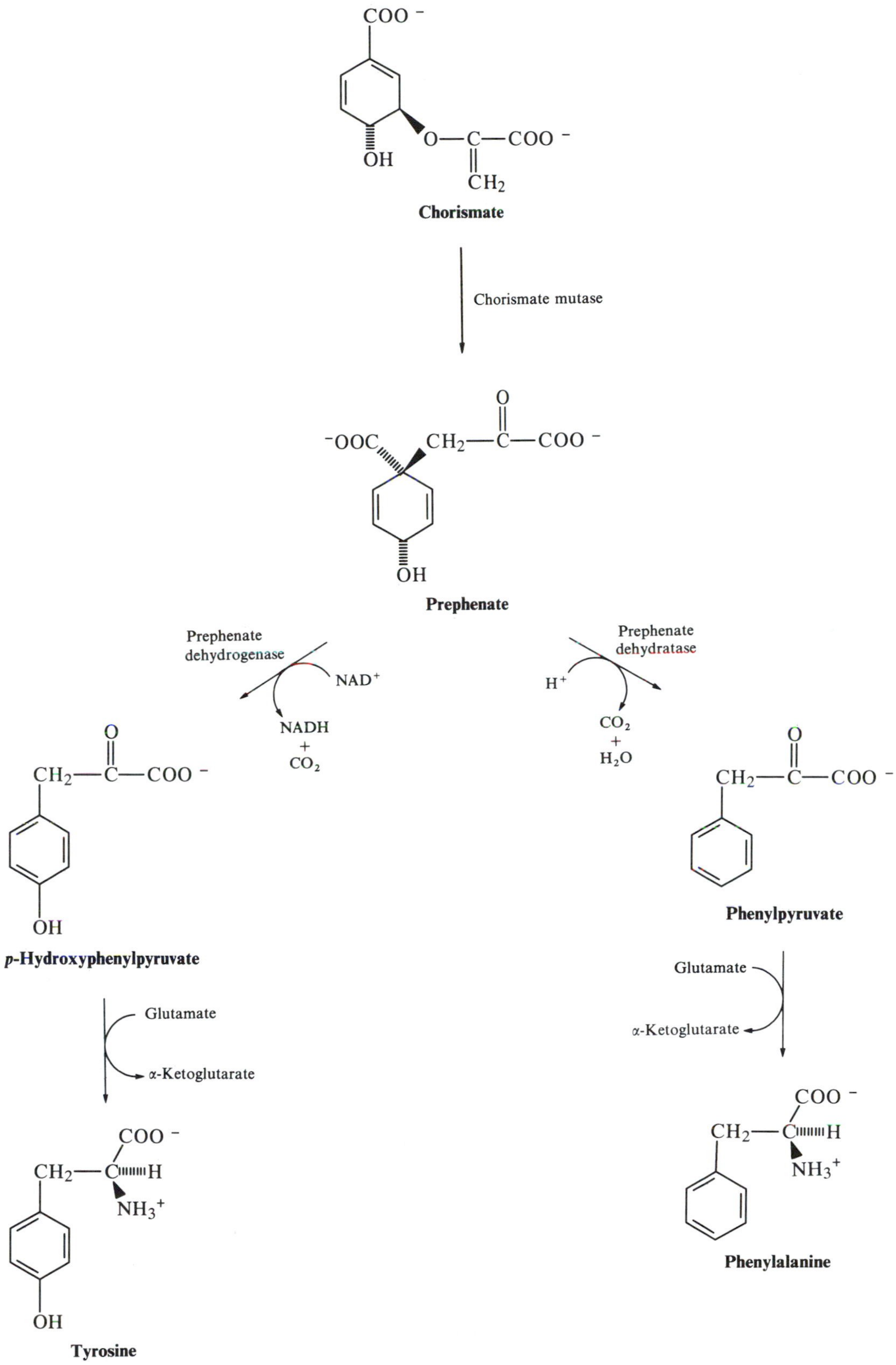

COO^-

OH

O—C—COO^-

CH_2

Chorismate

Chorismate mutase

^-OOC CH_2—C(=O)—COO^-

OH

Prephenate

Prephenate dehydrogenase

NAD^+

NADH + CO_2

Prephenate dehydratase

H^+

CO_2 + H_2O

CH_2—C(=O)—COO^-

OH

***p*-Hydroxyphenylpyruvate**

CH_2—C(=O)—COO^-

Phenylpyruvate

Glutamate

α-Ketoglutarate

COO^-

CH_2—C—H

NH_3^+

OH

Tyrosine

Glutamate

α-Ketoglutarate

COO^-

CH_2—C—H

NH_3^+

Phenylalanine

Figure 26-8

The Biosynthesis of Phenylalanine and Tyrosine. Shown here are the pathways for the conversion of chorismate into phenylalanine and tyrosine.

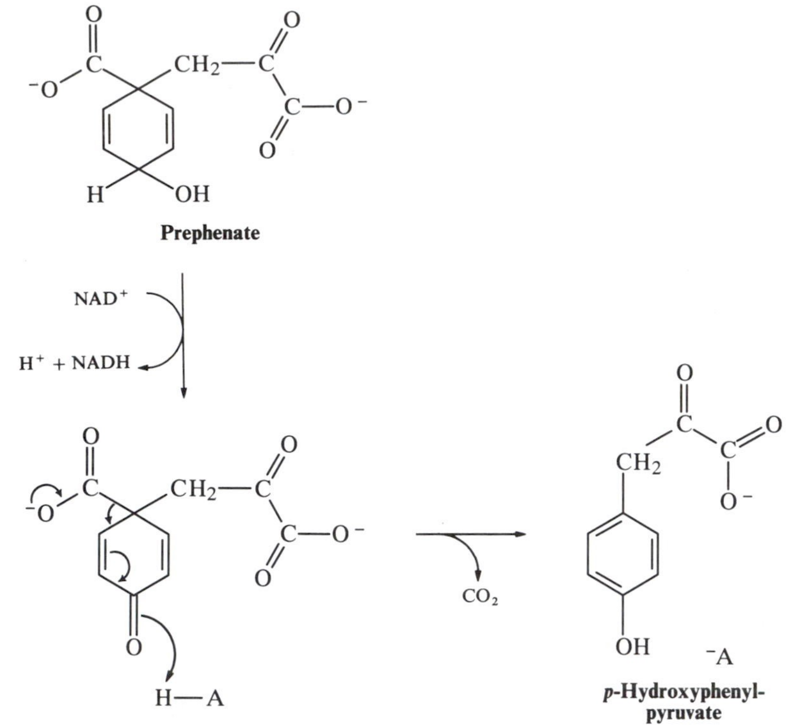

Figure 26-9

The Mechanism by Which Prephenate Is Converted into *p*-Hydroxyphenylpyruvate.
Prephenate is dehydrogenated by NAD^+ to introduce a carbonyl group at carbon-4 of the ring. This group is conjugated to the carboxylate group through the double bond in the ring; in the decarboxylation, this conjugation allows the carbonyl group to act as an electron sink, with protonation of the oxygen by a general acid at the active site. Decarboxylation leads to the formation of an aromatic phenol ring of *p*-hydroxyphenylpyruvate.

Biosynthesis of Tryptophan

The conversion of chorismate into tryptophan proceeds through another interesting reaction sequence outlined in Figure 26-10. Chorismate is initially converted into anthranilate in a complex reaction in which the amido group of glutamine serves as the "NH_3" donor. Anthranilate then reacts with phosphoribosyl pyrophosphate (PRPP) to form the *N*-ribosyl bond in phosphoribosyl anthranilate. PRPP is an effective ribosyl donor in this and other reactions in which ribosyl bonds are formed because PP_i is an excellent leaving group. Phosphoribosyl anthranilate then undergoes an interesting rearrangement that amounts to an intramolecular oxidation-reduction, with reduction of carbon-1′ and oxidation of carbon-2′ of the ribose ring. (Figure 26-11 shows how this rearrangement is related to an aldose-ketone isomerization, in which the aldehyde is replaced by an aldimine.) There follows a poorly understood reaction catalyzed by indole-3-glycerol phosphate synthase, in which decarboxylation accompanies ring closure and elimination of water between carbons 1 and 2 of ribulose to form the indole ring.

In the last step, tryptophan synthase catalyzes the reaction of serine with indole-3-glycerol phosphate to form tryptophan and glyceraldehyde-3-phosphate. The carbon atoms of serine replace those of glycerol phosphate in indole-3-glycerol phosphate. Tryptophan synthase from *E. coli* is a four-subunit

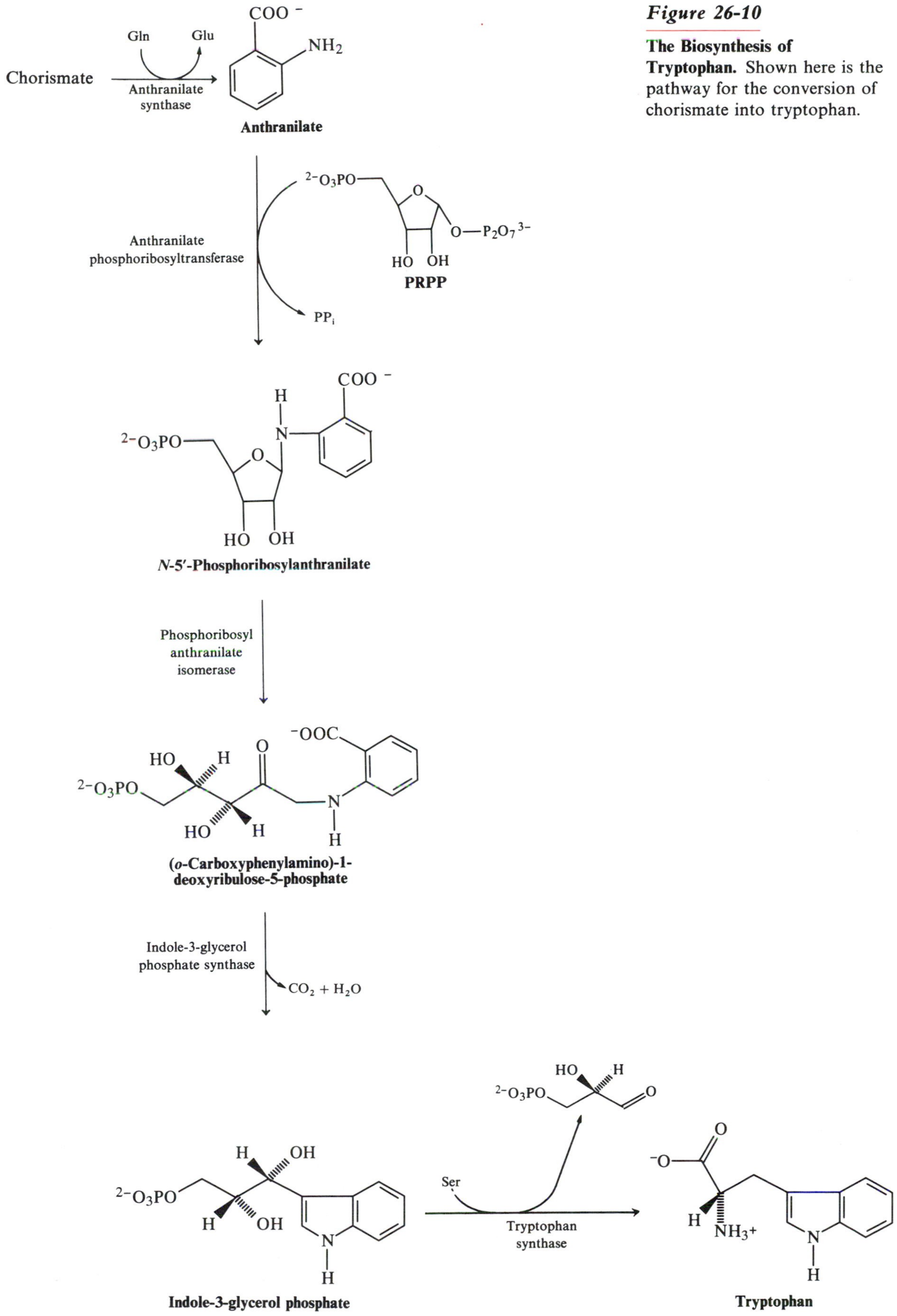

Chorismate

Gln → Glu

Anthranilate synthase

COO^-

NH_2

Anthranilate

Anthranilate phosphoribosyltransferase

$^{2-}O_3PO$

O

$O—P_2O_7^{3-}$

HO OH

PRPP

PP_i

COO^-

H

N

$^{2-}O_3PO$

O

HO OH

***N*-5′-Phosphoribosylanthranilate**

Phosphoribosyl anthranilate isomerase

^-OOC

HO H O

$^{2-}O_3PO$

N

HO H H

(*o*-Carboxyphenylamino)-1-deoxyribulose-5-phosphate

Indole-3-glycerol phosphate synthase

$CO_2 + H_2O$

H OH

$^{2-}O_3PO$

H OH

N

H

Indole-3-glycerol phosphate

Ser

HO H

$^{2-}O_3PO$

O

Tryptophan synthase

O

^-O

H NH_3^+

N

H

Tryptophan

Figure 26-10

The Biosynthesis of Tryptophan. Shown here is the pathway for the conversion of chorismate into tryptophan.

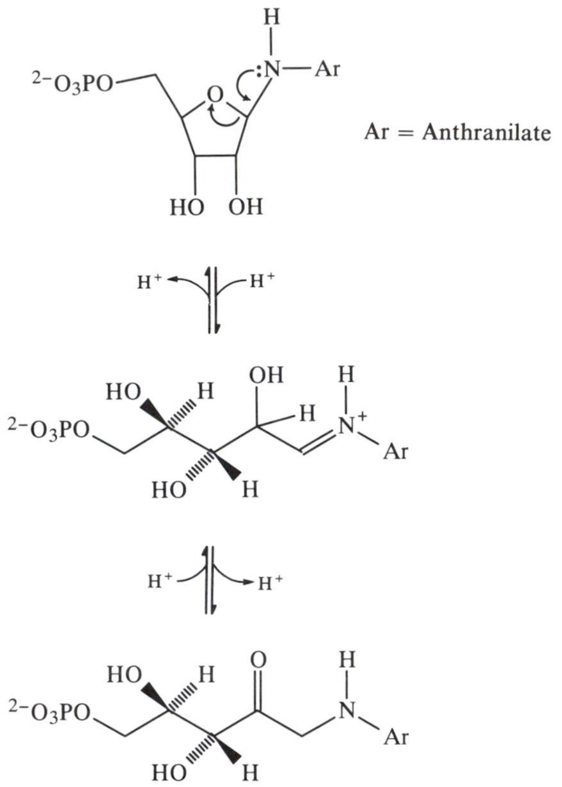

Ar = Anthranilate

Figure 26-11

Chemical Steps in the Conversion of *N*-5′-Phosphoribosyl Anthranilate into (*o*-Carboxyphenylamino)-1-deoxyribulose-5-phosphate.

enzyme, consisting of two α- and two β-subunits ($\alpha_2\beta_2$). The subunits separately catalyze the reactions of equations 13 and 14,

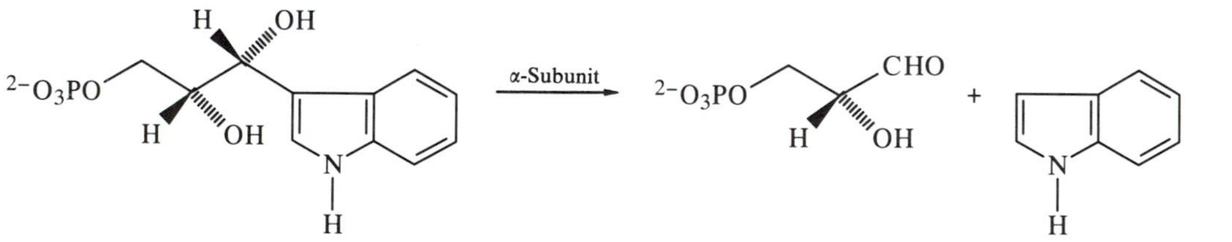

α-Subunit (13)

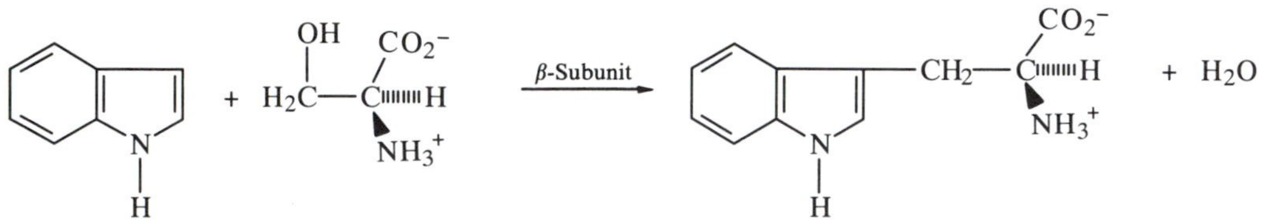

β-Subunit $+ H_2O$ (14)

the two reactions required to convert indole-3-glycerol phosphate into tryptophan. The reaction catalyzed by the α-subunit, equation 13, is a reverse aldol reaction analogous to others described in Chapter 2. That carried out by the β-subunit, equation 14, is analogous to other β-elimination reactions carried out by pyridoxal 5′-phosphate dependent enzymes described in Chapter 18. The β-subunit is a pyridoxal phosphate dependent enzyme. In the enzyme complex ($\alpha_2\beta_2$), the two subunits carry out the conversion of indole-3-glycerol phosphate and serine into tryptophan and glyceraldehyde-3-phosphate, and no free indole (nonenzyme bound) appears in the course of the reaction. The rates of the reactions within the complex are appreciably faster than the rates of reactions 13 and 14 catalyzed by the separate subunits, in which indole dissociates from the α-subunit in equation 13 and binds to the β-subunit in equation 14. Indole normally reacts too fast within the complex to dissociate from it. In the three-dimensional structure of tryptophan synthetase, there is a "tunnel" between the active sites of the α-subunit and β-subunit. The tunnel is large enough for the passage of indole, which is thought to diffuse through the tunnel from the α-subunit to the β-subunit.

The complete aromatic biosynthetic pathways are present in microorganisms and plants. Phenylalanine and tryptophan are essential amino acids for animals because animals cannot biosynthesize them. Tyrosine is not an essential amino acid because animals have phenylalanine monooxygenase, which converts nutrient phenylalanine into tyrosine.

Discovery of the Aromatic Amino Acid Biosynthetic Pathway

The first clues to the nature of the aromatic amino acid biosynthetic pathway were obtained from mutations. Certain mutants of *E. coli* were generated that could grow only if aromatic amino acids were present in the growth medium. It was found that these mutants could also grow in the presence of shikimic acid, a natural product isolated from plants, in place of aromatic amino acids. This suggested that shikimic acid or a molecule derived from it is a precursor of aromatic amino acids. All of these mutants apparently had their aromatic amino acid biosynthetic pathway blocked at points preceding shikimic acid in the biosynthetic pathway. The addition of shikimic acid cured the mutants and enabled them to produce the aromatic amino acids. This evidence indicated that shikimic acid is an intermediate in the biosynthesis of all the aromatic amino acids; that is, it is a *common* intermediate in the biosyntheses of phenylalanine, tyrosine, and tryptophan.

In other experiments, mutants of *Neurospora* that were generated by mutagenesis with ultraviolet light also required aromatic amino acids when grown alone. However, certain combinations of these mutants could grow together in the same culture without the addition of aromatic amino acids. These mutants were blocked at various points in the biosynthetic pathway, so that an intermediate produced and accumulated by one mutant could satisfy the growth requirement for another. It was possible to identify many of the intermediates of aromatic amino acid biosynthesis by determining which compounds were accumulated in the medium during the growth of the individual mutants.

Biosynthesis of Porphyrins

Porphyrins are important components of essential biological compounds, including hemoglobin and cytochromes, as well as chlorophyll. The classic experiments of D. Shemin and his coworkers in the 1940s made major contributions to the elucidation of the biochemical reaction sequence by which porphyrins are synthesized. An extract prepared from duck erythrocytes was found to contain the enzymes required for the synthesis of *protoporphyrin IX*, the iron-free precursor of heme. Its structure, with numbered carbon atoms, is given in Figure 26-12. Various heavy-isotope-labeled (^{14}C, ^{15}N) small molecules were added to the erythrocyte extract to determine whether they were incorporated into protoporphyrin IX and, if so, which atoms became labeled. The location of any isotopic label could be determined by a chemical degradation procedure that allowed each atom of protoporphyrin IX to be isolated in a

Figure 26-12

The Structure of Protoporphyrin IX.

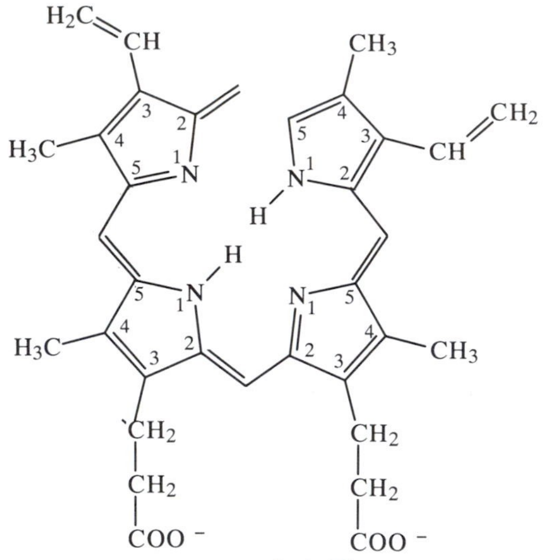

Protoporphyrin IX

known small molecule that could be analyzed for its isotopic composition. The results led to a picture of the biogenesis of protoporphyrin IX that greatly facilitated the elucidation of the biosynthetic pathway and the identification of enzymes required for the assembly of this molecule.

Porphobilinogen

By the use of [^{15}N]glycine to label protoporphyrin IX, it became clear that all of the nitrogen atoms of protoporphyrin IX are derived from glycine. Even though the pyrrole rings of protoporphyrin IX are not equivalent, ^{15}N from [^{15}N]glycine was found to be equally distributed among all four rings. This observation suggested that all the pyrrole rings are derived from a common precursor. It was also found that [2-^{14}C]glycine labels carbon-2 of the pyrrole rings, as well as the four methine carbon atoms that link the four pyrrole rings in protoporphyrin IX. However, no label derived from carboxyl-labeled glycine ([1-^{14}C]glycine) appeared in protoporphyrin IX.

In other experiments, [2-^{14}C]acetate and [1-^{14}C]acetate were used as precursors for protoporphyrin IX. The label was found to be distributed as shown in Figure 26-13. The most-striking relations in the labeling were that the specific radioactivities were equal at carbons 6 and 9, at carbons 4 and 8, and at carbons 5 and 3. It was apparent from this labeling pattern that the "methyl" side and "vinyl" sides of the pyrrole rings are derived from the same precursor, which must be either a three-carbon or a four-carbon compound produced from acetate. The specific radioactivities differed from one pair of carbons to the next because the crude extracts contained biosynthetic intermediates that differentially "diluted" the radioactivities at these paired positions.

It was further noted that the specific radioactivity of carbon-9, which was derived from [2-^{14}C]acetate, was equal to that of carbon-10, which had been found to be derived from [1-^{14}C]acetate. This constituted strong evidence that carbon-1 and carbon-2 of acetate were incorporated together into these two positions in the biosynthetic pathway—that is, carbons 9 and 10 must have been *equally* diluted during biosynthesis, which suggested that they were incorporated together into a precursor of the pyrrole ring that became the propionyl side chain. Finally, it was concluded that the four-carbon precursor must have been asymmetric, because carbons 3 and 10 did not have the same specific radioactivity in protoporphyrin IX. At the time of the experiments, it

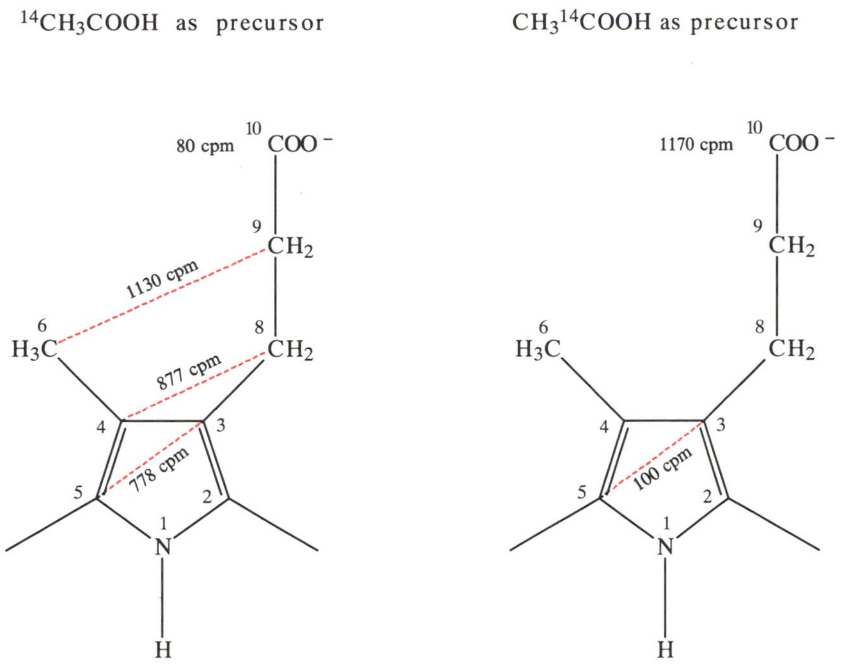

$^{14}CH_3COOH$ as precursor

80 cpm (10) COO^-; (9) CH_2; 1130 cpm; (6) H_3C; (8) CH_2; 877 cpm; 778 cpm; 1 N H

$CH_3{}^{14}COOH$ as precursor

1170 cpm (10) COO^-; (9) CH_2; (6) H_3C; (8) CH_2; 100 cpm; 1 N H

Figure 26-13

Labeling Pattern in the Pyrrole Rings of Protoporphyrin IX with the Use of [^{14}C]Acetates as the Precursors. The atoms connected by dashed red lines contain equal radioactivities, as shown by the numbers of cpm after incubation of the labeled acetates with the duck erythrocyte extract.

was proposed that the asymmetric precursor was an intermediate in the TCA cycle and was derived from α-ketoglutarate.

We now know that the four-carbon precursor is succinyl CoA and that the isotope distribution in the porphyrin ring is explained by the TCA cycle. Table 26-3 shows the distribution of ^{14}C in succinyl CoA that is obtained upon entry of [^{14}C]acetyl CoA from [^{14}C]acetate into the TCA cycle. The distribution depends on the number of cycles that have occurred by the time succinyl CoA is analyzed after the addition of a pulse of [^{14}C]acetate. The amount of label in carbons 2, 3, and 4 tends to become equal after a large number of cycles. Initially, the radioactivity is highest in carbon-2. The distribution of radioactivity in protoporphyrin IX conforms to this pattern, in which carbon-9 of the porphyrin corresponds to carbon-2 of succinyl CoA. The specific activities of carbons 9, 8, and 3 are nearly equal. Carbon-9 tends to have a somewhat higher specific activity. This distribution of radioactivity corresponds to that expected

Table 26-3

Labeling pattern of succinyl CoA by [^{14}C]acetate as a function of the number of TCA cycles

$HO_2\overset{1}{C}-\overset{2}{C}H_2-\overset{3}{C}H_2-\overset{4}{C}OSCoA$ Succinyl CoA	From [2-^{14}C]acetate (radioactivity per 10 cpm)				From [1-^{14}C]acetate (radioactivity per 10 cpm)			
NUMBER OF TURNS THROUGH THE TCA CYCLE	1st	2d	3d	∞	1st	2d	3d	∞
C-1 COOH	0	0	0	0	10	10	10	10
C-2 CH_2	10	10	10	10	0	0	0	0
C-3 CH_2	0	5	7.5	10	0	0	0	0
C-4 COSCoA	0	5	7.5	10	0	0	0	0

in succinyl CoA. When carboxy-labeled acetate enters the TCA cycle, the label remains in carbon-1 of succinyl CoA. Consistent with this is the observation that carboxy-labeled acetate leads to labeling exclusively at carbon-10 of protoporphyrin IX.

From these data, the precursors of the pyrrole ring could be defined as two molecules of glycine and two molecules of succinyl CoA. It was concluded that the reaction of glycine and succinyl CoA gives δ-aminolevulinic acid, two molecules of which combine to give porphobilinogen, a precursor of heme, as shown in Figure 26-14. By 1955, the conversion of labeled δ-aminolevulinic acid into porphobilinogen had been demonstrated with enzyme systems by several laboratories.

The enzymes have been further characterized. δ-Aminolevulinic acid synthase catalyzes the condensation of succinyl CoA and glycine. The enzyme utilizes pyridoxal phosphate as the cofactor to facilitate the formation of the glycine α-carbanion. The carbanion adds to the thioester group of succinyl CoA, leading to decarboxylation and, ultimately, the formation of δ-aminolevulinic acid, as shown in Figure 26-15. δ-Aminolevulinic acid dehydratase catalyzes the formation of porphobilinogen by condensation and cyclization of two molecules of δ-aminolevulinic acid, with elimination of water, according to the sequence of steps in Figure 26-16.

Porphyrin Biosynthesis

Figure 26-17 shows how four molecules of porphobilinogen are assembled to form a porphyrin ring. Uroporphyrinogen is the precursor of other porphyrins, including chlorophyll and heme. It is also a precursor of cobalamin, vitamin B_{12}. The condensation reaction requires the two enzymes porphobilinogen deaminase and uroporphyrinogen III cosynthase. The first enzyme catalyzes the

Figure 26-14

Biosynthesis of Porphobilinogen.

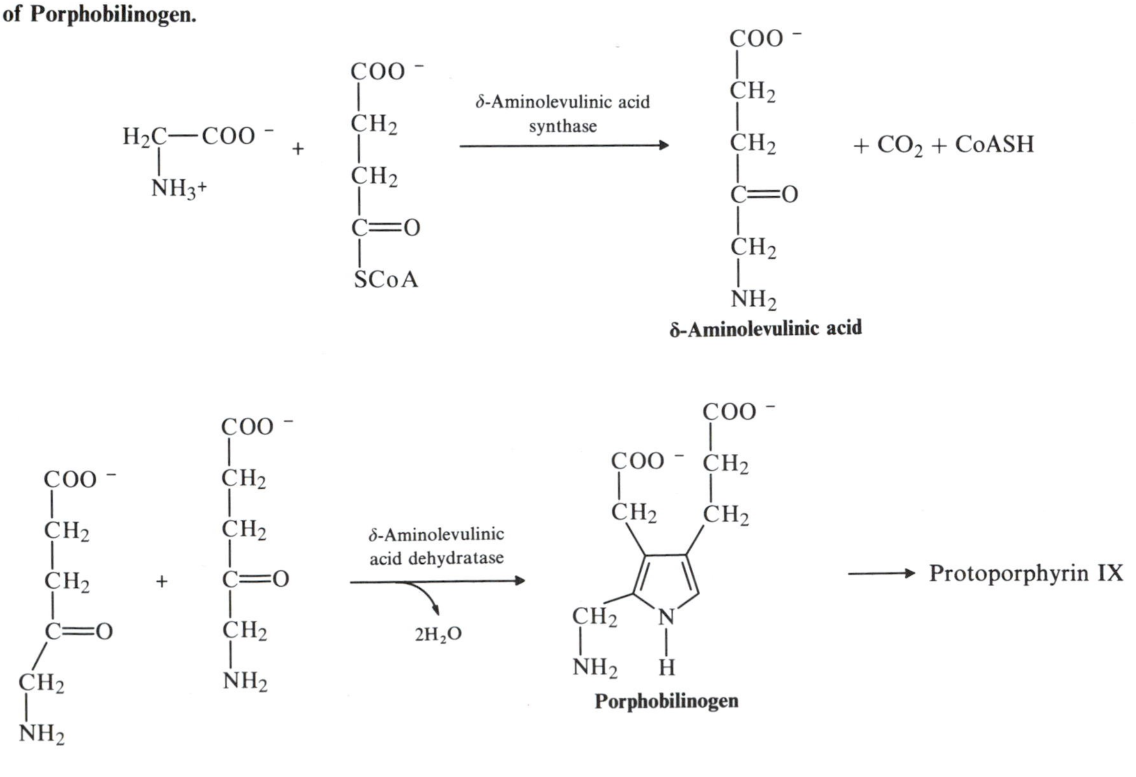

$$H_2C(NH_3^+)\text{—}COO^- + {}^-OOC\text{—}CH_2\text{—}CH_2\text{—}C(=O)\text{—}SCoA \xrightarrow{\delta\text{-Aminolevulinic acid synthase}} {}^-OOC\text{—}CH_2\text{—}CH_2\text{—}C(=O)\text{—}CH_2\text{—}NH_2 + CO_2 + CoASH$$

δ-Aminolevulinic acid

$$2\ {}^-OOC\text{—}CH_2\text{—}CH_2\text{—}C(=O)\text{—}CH_2\text{—}NH_2 \xrightarrow[2H_2O]{\delta\text{-Aminolevulinic acid dehydratase}} \text{Porphobilinogen} \longrightarrow \text{Protoporphyrin IX}$$

Porphobilinogen

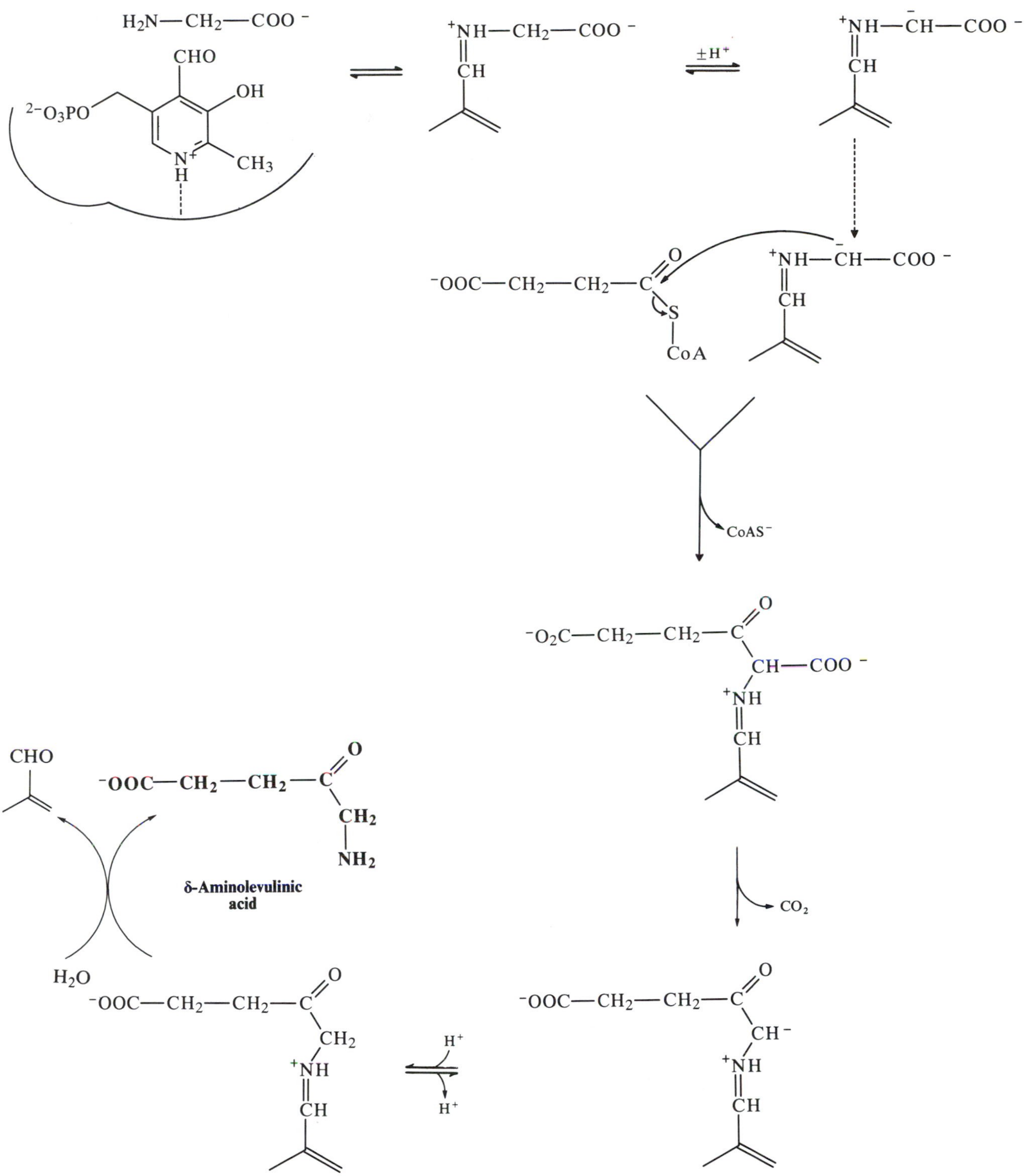

$H_2N—CH_2—COO^-$

CHO

OH

$^{2-}O_3PO$

N^+

H

CH_3

$^+NH—CH_2—COO^-$

CH

±H⁺

$^+NH—\bar{C}H—COO^-$

CH

$^-OOC—CH_2—CH_2—C$

O

S

CoA

$^+NH—\bar{C}H—COO^-$

CH

$CoAS^-$

$^-O_2C—CH_2—CH_2—C$

O

$CH—COO^-$

^+NH

CH

CO_2

$^-OOC—CH_2—CH_2—C$

O

CH^-

^+NH

CH

H⁺

H⁺

$^-OOC—CH_2—CH_2—C$

O

CH_2

^+NH

CH

H_2O

CHO

$^-OOC—CH_2—CH_2—C$

O

CH_2

NH_2

δ-Aminolevulinic acid

Figure 26-15

A Mechanism for the Formation of δ-Aminolevulinic Acid.

formation of the linear tetrapyrrole. The mechanism is not known, but a possible reaction sequence is shown in Figure 26-18. The cosynthase catalyzes ring closure, with concomitant rearrangement of one of the pyrrole rings. (Note the side-chain pattern.) Oxidations and decarboxylations lead to protoporphyrin IX, which is converted into heme by insertion of Fe.

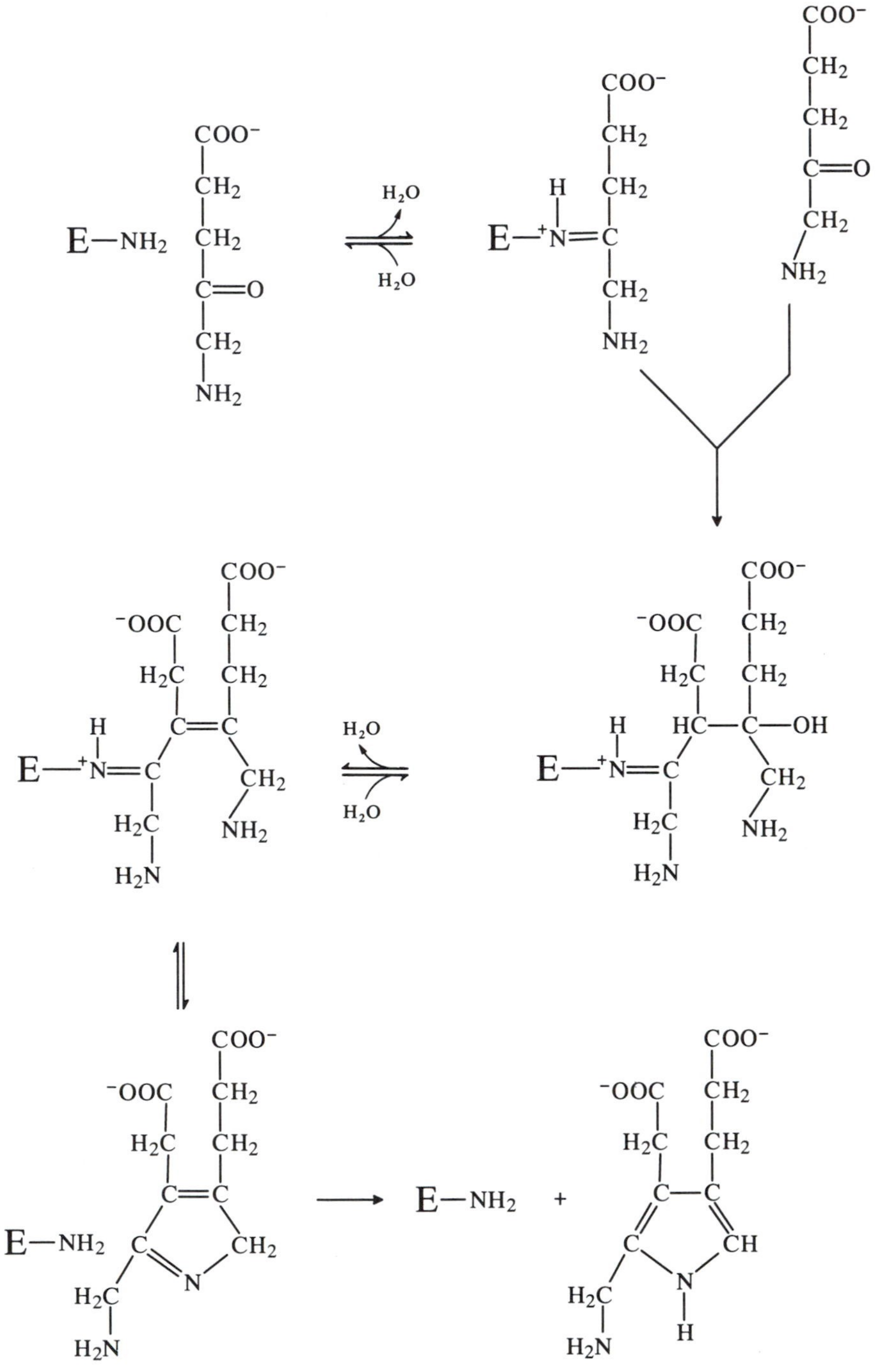

Figure 26-16

Mechanism of the Condensation Catalyzed by δ-Aminolevulinic Acid Dehydratase.

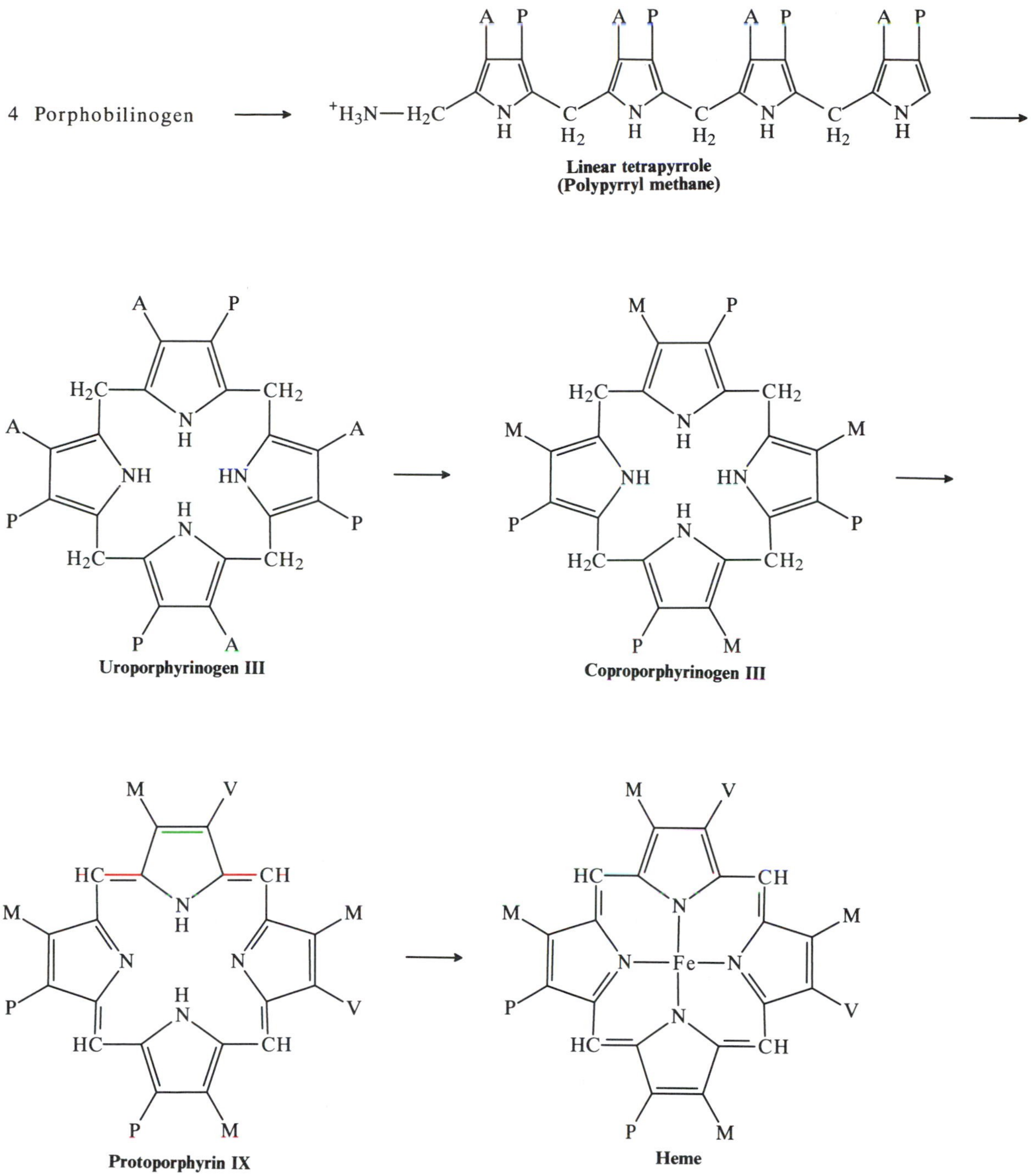

4 Porphobilinogen →

A P A P A P A P

$^{+}H_3N—H_2C$ N H C H_2 N H C H_2 N H C H_2 N H →

Linear tetrapyrrole
(Polypyrryl methane)

A P
H_2C N H CH_2
A NH HN A
P H N P
H_2C CH_2
P A

Uroporphyrinogen III

→

M P
H_2C N H CH_2
M NH HN M
P H N P
H_2C CH_2
P M

Coproporphyrinogen III

→

M V
HC N H CH
M N N M
P H N V
HC CH
P M

Protoporphyrin IX

→

M V
HC N CH
M N—Fe—N M
P N V
HC CH
P M

Heme

Figure 26-17

The Assembly of Heme from Porphobilinogen. Heme is iron protoporphyrin IX, which is produced from four molecules of porphobilinogen. The principal intermediates are shown, beginning with the tetrapyrrole and continuing with cyclized forms. Cyclization to uroporphyrinogen III is by an intramolecular reaction analogous to the intermolecular coupling processes illustrated in Figure 26-16. The other steps to protoporphyrin IX are side-chain processing reactions that oxidize and decarboxylate two propionic acid substituents. The abbreviations are: A, acetate; P, propionate; M, methyl; V, vinyl.

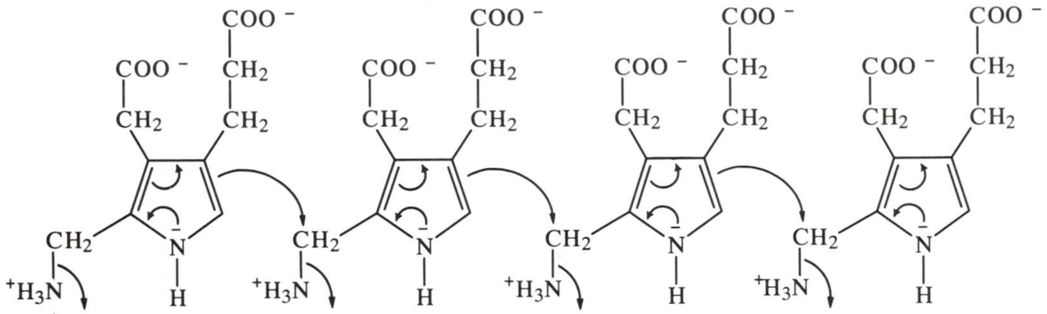

Figure 26-18

The Formation of a Tetrameric Pyrrole from Porphobilinogen. The scheme shows how the bonds linking four porphobilinogen molecules might be formed by elimination of ammonia between linked molecules.

Summary

Nitrogen is an important constituent of most biological molecules. Microorganisms and plants derive nitrogen from their growth medium and use it for the biosynthesis of amino acids and nucleotides. Higher animals derive most of their nitrogen from proteins in their diets. Animals can biosynthesize the nonessential amino acids from other amino acids and the carbon skeletons of intermediates of the TCA cycle and glycolysis. The essential amino acids must come from the diets of higher animals. Amino acids are metabolized by being broken down into ammonia and intermediates of the TCA cycle. Glutamine synthetase catalyzes the reaction of ammonia with glutamate to form glutamine, which serves as a source of NH_2 groups for biosynthesis. In most higher animals, excess ammonia is converted into urea by the urea cycle, and urea is excreted. The second nitrogen in urea is derived from aspartate in the urea cycle.

All amino acids are produced by plants and microorganisms in chemically complex biosynthetic pathways. The aromatic amino acids are produced from phosphoenolpyruvate and erythrose-4-phosphate by a pathway in which shikimic acid is a central intermediate and chorismate is the common intermediate at which the pathways for phenylalanine and tyrosine diverge from that for tryptophan. Protoporphyrin IX for heme is biosynthesized from glycine in a chemically complex pathway in which succinyl CoA and δ-aminolevulinic acid are key intermediates.

ADDITIONAL READING

Battersby, A. R., Fookes, G. W., Matcham, G. W. J., and McDonald, E. 1980. Biosynthesis of the pigments of life: formation of the macrocycle. *Nature* 285: 17.

Umbarger, H. E. 1978. Amino acid biosynthesis and its regulation. *Annual Review of Biochemistry* 47: 533.

Meister, A. 1965. *Biochemistry of the Amino Acids,* vols. 1 and 2. New York: Academic Press.

Meister, A. 1989. Mechanism and regulation of the glutamine-dependent carbamyl phosphate synthetase of *Escherichia coli.* In *Advances in Enzymology and Related Areas of Molecular Biology,* vol. 62. New York: Wiley-Interscience, p. 315.

PROBLEMS

1. The nonessential amino acid serine is derived from 3-phosphoglycerate in a multistep process. Suggest a sequence of enzymatic reactions that would lead to serine from 3-phosphoglycerate.

2. Proline is a nonessential amino acid. Considering the structures of other nonessential amino acids, which of them is the most-obvious precursor for proline. Suggest how proline could arise from this amino acid.

3. Arginine is biosynthesized in the urea cycle. If it is used as an amino acid for protein biosynthesis, which other amino acid should be depleted?

4. *N*-Acetylglutamate activates carbamoyl phosphate synthase I in the liver. Suggest an enzymatic reaction by which *N*-acetylglutamate could be synthesized.

5. Consider the reaction catalyzed by δ-aminolevulinic acid synthase. A mechanism is shown in Figure 26-15. Suggest an alternative reaction pathway by which pyridoxal phosphate might facilitate this reaction. What experiment might enable one to show that one or the other mechanism must be incorrect?

6. Suggest a metabolic pathway for the metabolism of threonine to propionyl CoA. How is propionyl CoA further metabolized as a source of energy?

7. The cyclization of 3-deoxy-D-*arabino*-heptulosonate-7-phosphate to 3-dehydroquinate catalyzed by dehydroquinate synthase is an interesting reaction, in part because the mechanism is not obvious. The enzyme contains a molecule of tightly bound NAD^+ that participates in the mechanism by undergoing transient reduction in the presence of the substrate. Suggest a mechanism for this cyclization.

8. The mechanism of action of anthranilate synthase is not well understood. Suggest a reaction sequence that might be catalyzed at the active site of the enzyme and that would account for the formation of anthranilate from chorismate and glutamine.

Nitrogen Metabolism: Nucleotides

27

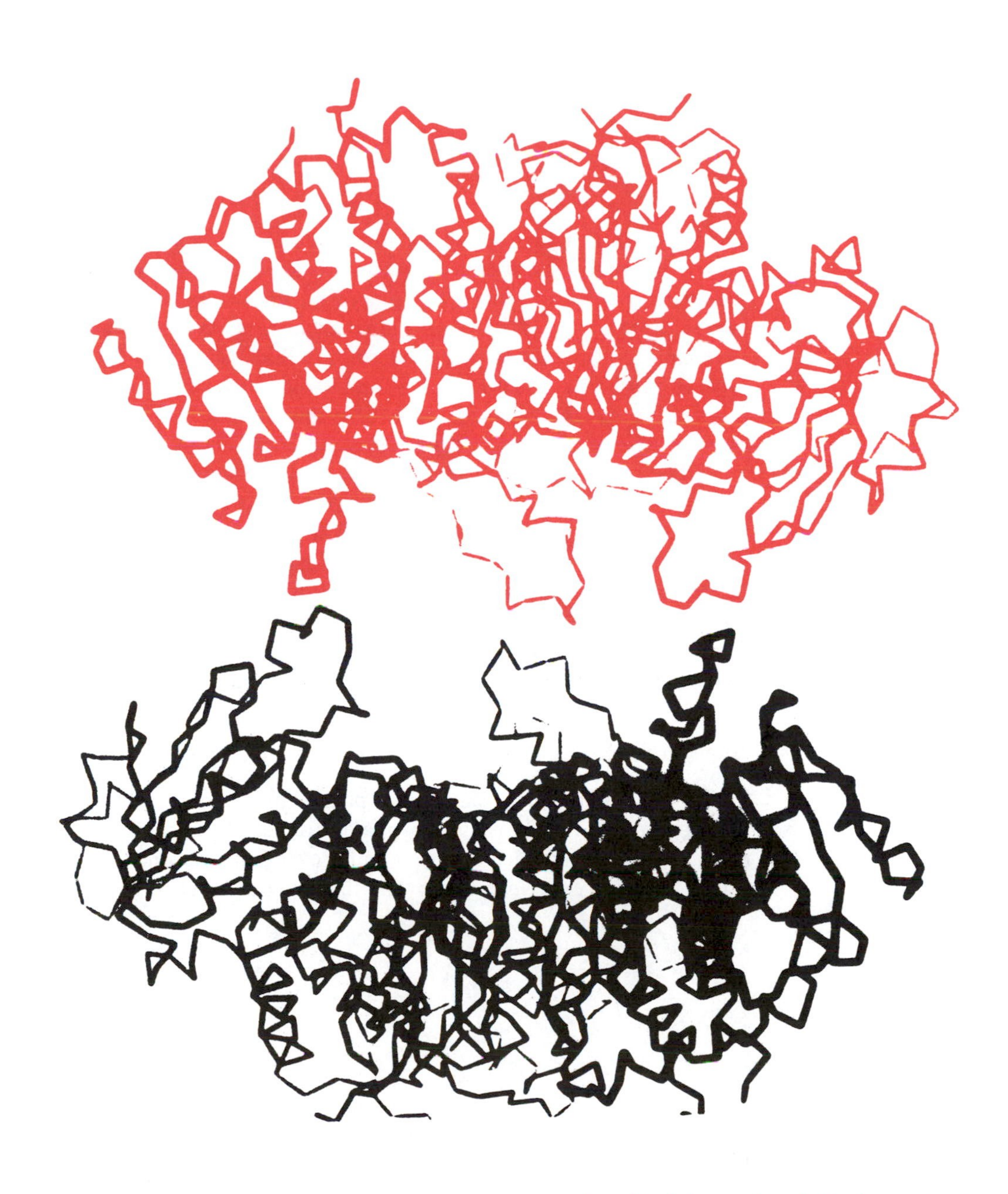

A model of the α-carbon chains in aspartate transcarbamoylase from Escherichia coli. The enzyme consists of two trimers of catalytic subunits and three dimers of regulatory subunits. (Copyright © Irving Geis.)

Nucleotides are among the most-important molecules in a cell. They are precursors of DNA and RNA; ATP is the "currency" for biochemical energy; nucleotide sugars are the glycosyl donors for the synthesis of most complex carbohydrates; nucleotides are component parts of coenzymes such as NAD^+, FAD, and CoA; and nucleotides such as 3′,5′-cyclic AMP are metabolic regulators. The structures of nucleosides and nucleotides were introduced in Chapter 10 and appear in twenty-eight of the thirty chapters in this textbook. In keeping with their fundamental importance to life, nucleotides are biosynthesized by all normal eukaryotic and prokaryotic cells. The only exceptions are a few mutant microorganisms that lack one or another enzyme of nucleotide biosynthesis. These mutants have been artificially generated for the express purpose of studying nucleotide biosynthesis. Thus, there are no "essential" nucleotides in the sense that there are nutritionally essential amino acids and fats. In the following section, we describe the biosyntheses of the most-important nucleotides.

Biosynthesis of Purine Nucleotides

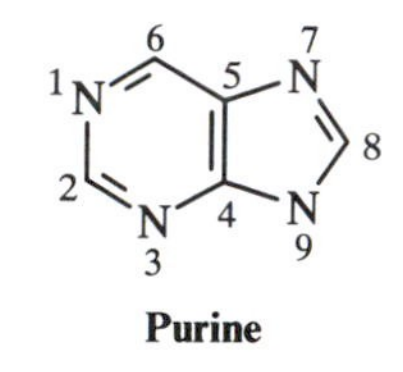

Purine

Uric acid

The biosynthetic origins of all the atoms in a purine ring are indicated in Figure 27-1. The numbering system for purines is shown in the margin, together with the structure of uric acid, which played an important role in determining the biosynthetic origins of the atoms in the purine ring. As mentioned in Chapter 26, birds and reptiles excrete excess nitrogen in the form of uric acid rather than as urea. Because uric acid is derived from the oxidation of adenine and guanine, its purine ring has the same biosynthetic origin as that of its precursors, and the origin of the ring atoms in uric acid is the same as in adenine and guanine.

The origins of the purine atoms shown in Figure 27-1 were determined in experiments using heavy isotopes of carbon (^{14}C) and nitrogen (^{15}N), which were incorporated into compounds fed to pigeons. The availability of heavy isotopes as by-products of the nuclear industry, together with the inexhaustible supply of pigeons, facilitated this research. About 80% of the solids excreted by pigeons consists of uric acid, so that crystalline uric acid was readily obtained. The uric acid was chemically degraded and analyzed for heavy-isotope labeling to determine the derivation of each atom in the purine ring.

Biosynthesis of Inosine-5′-phosphate (IMP)

Purine biosynthesis proceeds in two phases: first, the attachment of a nitrogen atom to carbon-1 of ribose-5-phosphate, and, second, construction of the purine ring around this nitrogen atom, so that it eventually becomes N-9 of the purine ring. The initial experiments showing that the purine ring is built up through a

"C-1" = 10-Formyl tetrahydrofolate

Figure 27-1

Biosynthetic Origins of Atoms in the Adenine Ring. The amide nitrogen of glutamine is the source of nitrogen for all but N-1, which is from the amino group of aspartate, and N-7, which is from the amino group of glycine.

series of ribotide intermediates again utilized pigeons and isotopic tracers, but this time with preparations of pigeon livers. The ribose-5-phosphate part of the intermediates is derived from PRPP (5-phosphoribosyl pyrophosphate), which was described earlier in a discussion of the biosynthesis of tryptophan and in Chapter 15. PRPP arises from the pyrophosphorylation of ribose 5-P by ATP, which is catalyzed by PRPP synthase according to equation 1.

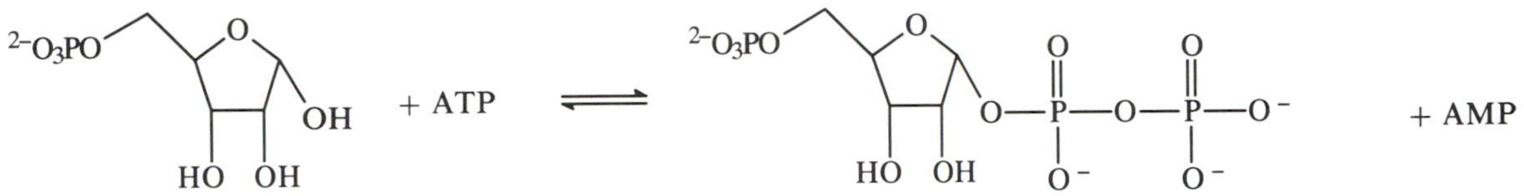

$^{2-}O_3PO$ … O … OH, HO OH + ATP ⇌ $^{2-}O_3PO$ … O … O—P(=O)(O⁻)—O—P(=O)(O⁻)—O⁻, HO OH + AMP (1)

PRPP

The next step is the displacement of PP_i from PRPP by NH_3, which is derived from glutamine, to give 5-phosphoribosyl-1-amine (equation 2).

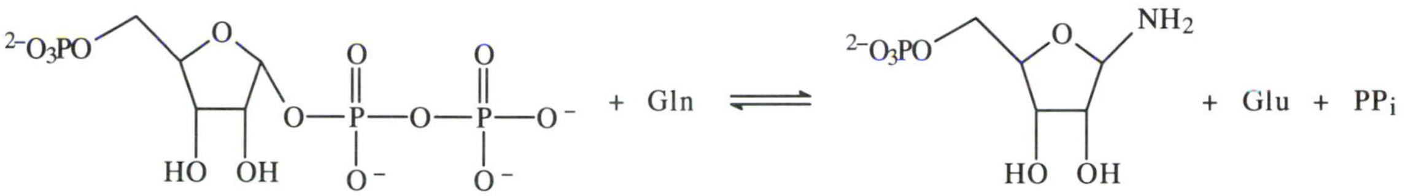

$^{2-}O_3PO$ … O … O—P(=O)(O⁻)—O—P(=O)(O⁻)—O⁻, HO OH + Gln ⇌ $^{2-}O_3PO$ … O … NH_2, HO OH + Glu + PP_i (2)

PRPP **5-Phosphoribosyl-1-amine**

We have encountered glutamine as an NH_3 donor in other reactions. This is the first "committed" step in the purine biosynthetic pathway; that is, it is the first step that is unique to purine biosynthesis. The amino group of 5-phosphoribosyl-1-amine eventually becomes N-9 of the purine ring.

5-Phosphoribosyl-1-amine is a highly reactive molecule that rapidly undergoes isomerization to a mixture of α- and β-anomers in solution and is labile to hydrolysis. In a cell, it is captured for purine biosynthesis much faster than it decomposes.

The construction of the purine ring requires nine additional reactions. The five carbon atoms are derived from glycine, CO_2, and the "C-1" units of 10-formyl tetrahydrofolate (see Figure 27-1). The metabolism of one-carbon compounds, including 10-formyl tetrahydrofolate, was presented in Chapter 25. Two of the nitrogen atoms in the purine ring itself are derived from glutamine, as is the exocyclic amino group of the adenine ring in Figure 27-1. The other two nitrogens are derived from aspartate and glycine.

The assembly of the five-member ring is outlined in Figure 27-2. In step 1, 5-phosphoribosyl-1-amine reacts with the carboxyl group of glycine to form an amide. This is an ATP-dependent reaction that is analogous to peptide-bond formation and is catalyzed by N^1-(5-phosphoribosyl)glycinamide synthase (also known as glycinamide ribotide synthase). The function of ATP is to activate the carboxyl group of glycine and facilitate amide formation with the amino group of 5-phosphoribosyl-1-amine. In the next reaction (2), the amino group of glycinamide is formylated by 10-formyl tetrahydrofolate by the transformylation mechanism described in Chapter 25. The N^1-(5-phosphoribosyl)-*N*-formylglycinamide is then converted into an amidine in step 3, an ATP-dependent reaction that is analogous to amide formation; it will be discussed in a later section in this chapter. In the last reaction (4) of Figure 27-2, the amidino group and *N*-formyl group of N^1-(5-phosphoribosyl)-*N*-formylglycinamidine react in an ATP-dependent dehydration to form the imidazole ring of N^1-(5′-phosphoribosyl)-5-aminoimidazole.

Figure 27-3 shows the sequence of reactions that complete the purine ring, all of which are familiar reaction types. The first is a carboxylation (5), in which an enamine (the nitrogen analog of an enol) is carboxylated. The reaction is therefore completely analogous to other carboxylation reactions in Chapter 17.

Figure 27-2

Biosynthesis of Purines. Shown here is the pathway for the conversion of 5′-phosphoribosyl-1-amine into 5-phosphoribosyl-5-aminoimidazole, an intermediate in purine biosynthesis.

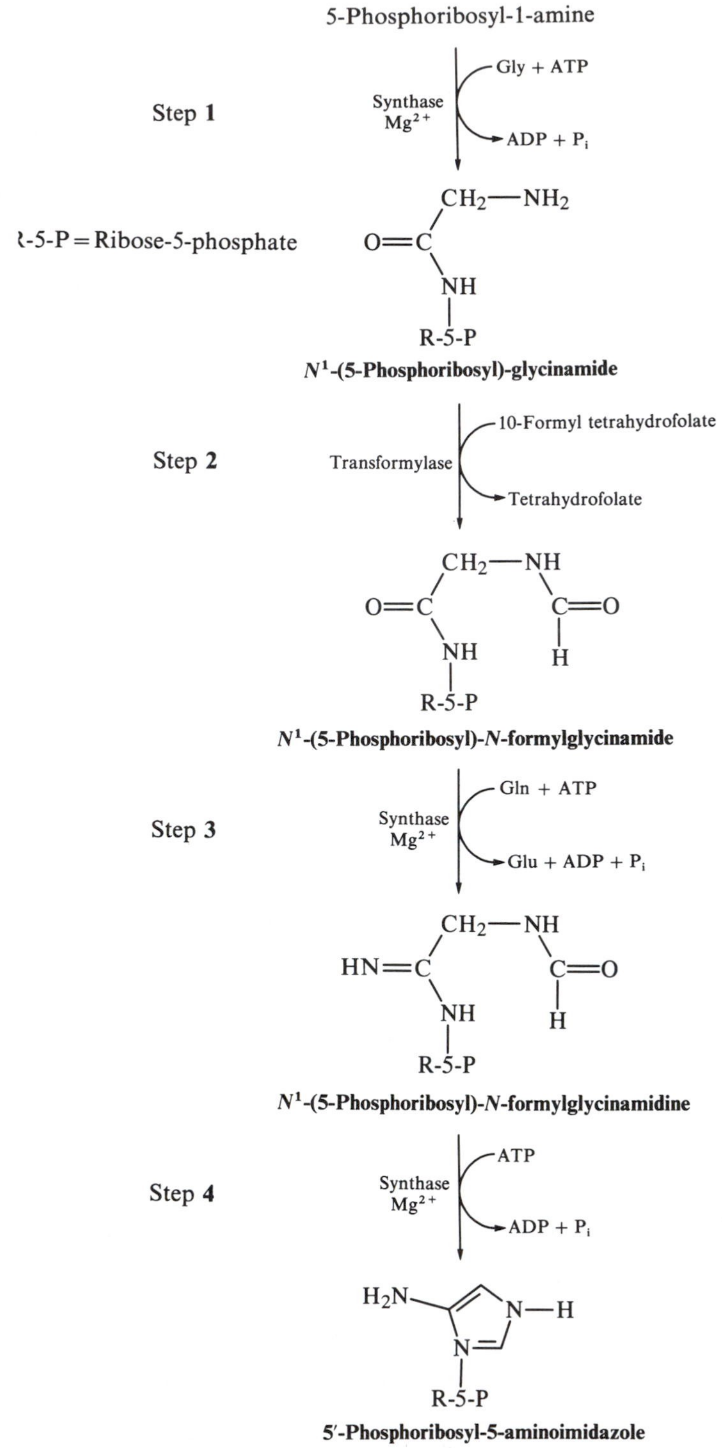

5-Phosphoribosyl-1-amine

Step **1** — Synthase, Mg^{2+}; Gly + ATP → ADP + P_i

R-5-P = Ribose-5-phosphate

CH_2—NH_2 / O=C / NH / R-5-P

N^1-(5-Phosphoribosyl)-glycinamide

Step **2** — Transformylase; 10-Formyl tetrahydrofolate → Tetrahydrofolate

CH_2—NH—C(=O)H / O=C / NH / R-5-P

N^1-(5-Phosphoribosyl)-*N*-formylglycinamide

Step **3** — Synthase, Mg^{2+}; Gln + ATP → Glu + ADP + P_i

CH_2—NH—C(=O)H / HN=C / NH / R-5-P

N^1-(5-Phosphoribosyl)-*N*-formylglycinamidine

Step **4** — Synthase, Mg^{2+}; ATP → ADP + P_i

H_2N—imidazole ring (N—H, N=) / R-5-P

5′-Phosphoribosyl-5-aminoimidazole

In the next step (6), an amide bond is formed between the carboxylate group introduced in the preceding step and the amino group of aspartate, with activation by ATP. This is followed by elimination of fumarate from the aspartyl moiety (7). The elimination of fumarate from an aspartyl moiety is also important in the urea cycle, as described in Chapter 26, and it will appear again in adenine biosynthesis; elimination reactions in general are described in Chapter 18. A second formyl transfer from 10-formyl tetrahydrofolate (8) introduces a formyl carbon onto the 5-amino group. This will become carbon-2 of the purine ring. Finally, inosine-5′-phosphate is formed by dehydration (9),

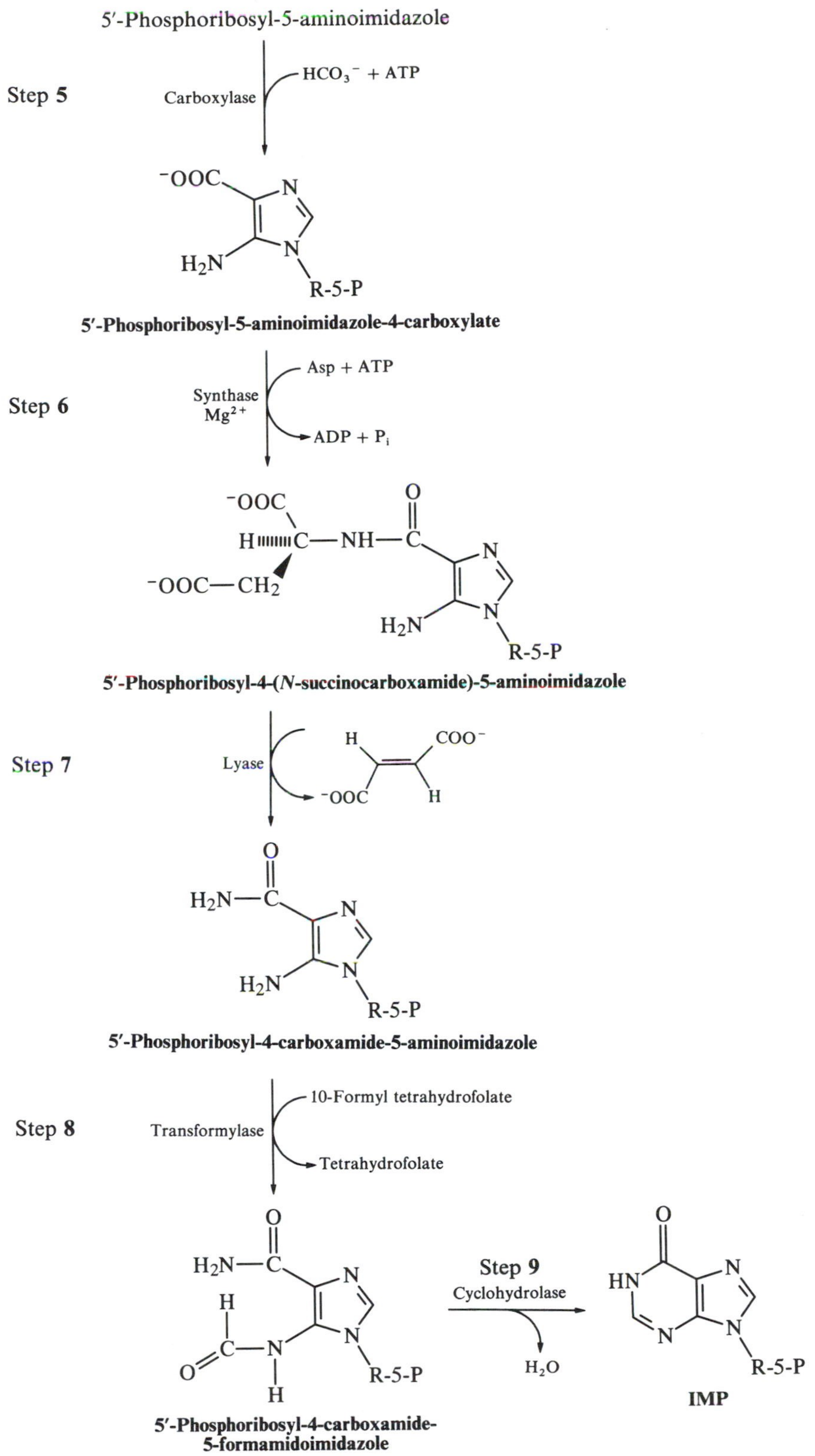

5′-Phosphoribosyl-5-aminoimidazole

Step **5** Carboxylase $HCO_3^- + ATP$

5′-Phosphoribosyl-5-aminoimidazole-4-carboxylate

Step **6** Synthase Mg^{2+} Asp + ATP → ADP + P_i

5′-Phosphoribosyl-4-(*N*-succinocarboxamide)-5-aminoimidazole

Step **7** Lyase → fumarate

5′-Phosphoribosyl-4-carboxamide-5-aminoimidazole

Step **8** Transformylase 10-Formyl tetrahydrofolate → Tetrahydrofolate

5′-Phosphoribosyl-4-carboxamide-5-formamidoimidazole

Step **9** Cyclohydrolase → H_2O

IMP

Figure 27-3

Biosynthesis of Purines. Shown here is the pathway for the conversion of 5′-phosphoribosyl-5-aminomidazole into IMP.

with extrusion of water between the formyl and carboxamide groups, to complete the six-member ring. This reaction does not require ATP to make the dehydration favorable, because it is intramolecular. Cyclization reactions are frequently sufficiently favorable energetically to allow dehydration without activation by ATP. However, the final cyclization of the amidine to 5′-phosphoribosyl-5-aminoimidazole in Figure 27-2 does require ATP.

AMP and GMP

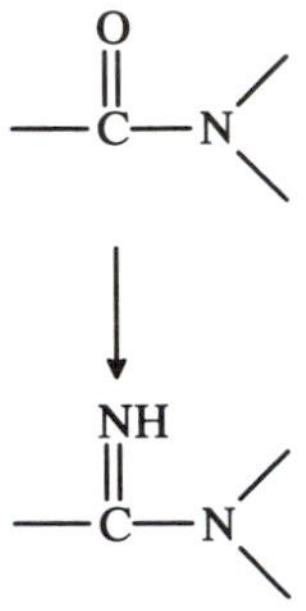

The purine biosynthetic pathway initially produces IMP, which is not a component of DNA or RNA. IMP is converted into AMP and GMP by the pathways shown in Figure 27-4. Again, the enzymes that catalyze these reactions are typical of those that we have seen before, with some special adaptations. One reaction in each pathway leads to the replacement of the oxygen in an amide group with nitrogen to form an amidine, as illustrated in the margin. A similar reaction takes place in the conversion of glycinamide into glycinamidine (Figure 27-2) and in the urea cycle (Chapter 26, Figure 26-5).

In AMP biosynthesis, aspartate is the source of nitrogen; whereas, in GMP biosynthesis, it is the carboxamide group of glutamine. These are difficult reactions, because they require the formation of intermediates in which the carbon is bonded to both the oxygen and the incoming nitrogen in a tetrahedral intermediate, followed by the elimination of oxygen. Oxygen elimination is difficult because oxide (O^{2-}) and hydroxide (HO^-) are poor leaving groups; therefore, ATP and GTP are used to convert oxygen into a good leaving group.

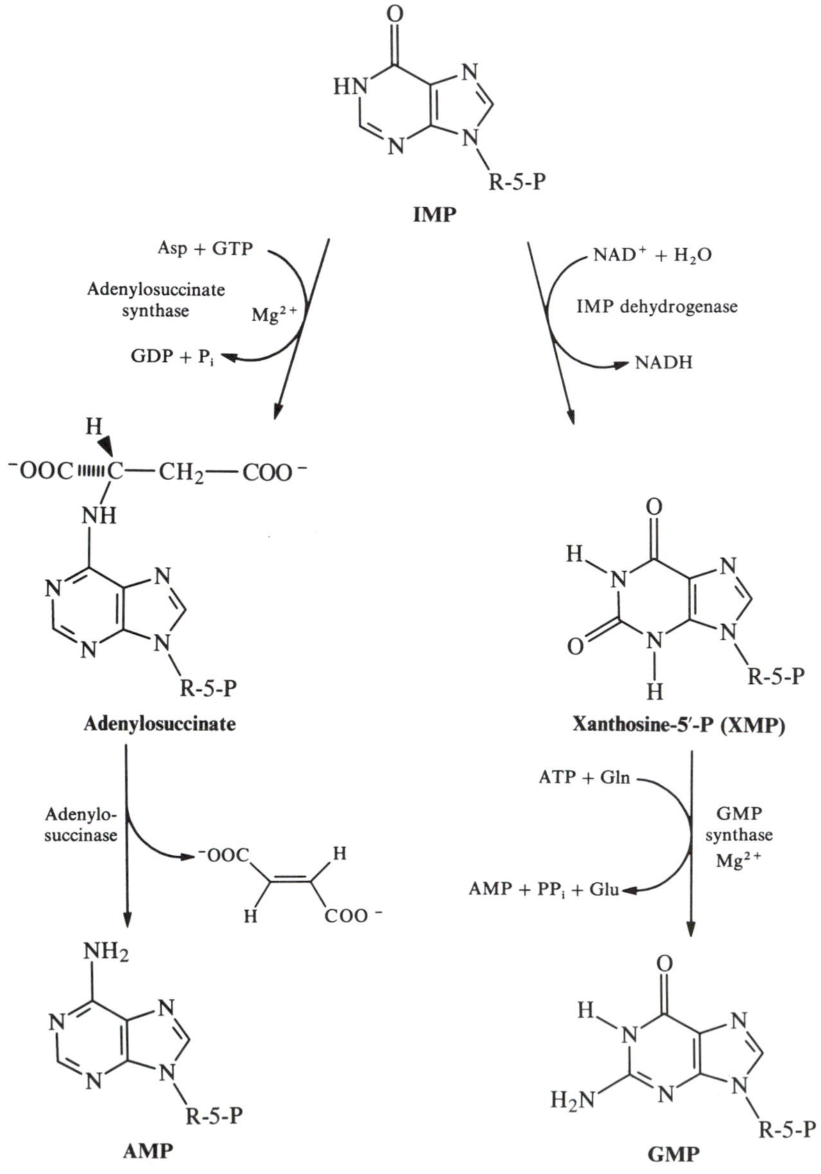

Figure 27-4

Biosynthesis of Purine Nucleotides. IMP is the precursor for AMP and GMP, which are produced by amination and oxidation of the inosine ring in IMP.

In ATP biosynthesis, GTP is used to activate the oxygen as a phosphate group ($—OPO_3^{2-}$); whereas, in GMP biosynthesis, ATP is used to activate oxygen as an adenylate group ($—OPO_3^{-}$-Ado).

We shall consider the conversion of IMP into adenylosuccinate in Figure 27-5 as a specific example of this type of reaction. Initially, GTP phosphorylates the oxygen atom of IMP. GTP is used in place of ATP because adenine nucleotides are presumably in short supply under conditions in which AMP biosynthesis is required by the cell. Nucleoside diphosphate kinase can produce GTP from GDP and any pyrimidine nucleoside triphosphate (Chapter 11). Phosphorylation creates a phosphoimidate ester, a reactive intermediate. The amino group of aspartate reacts with the phosphoimidate, presumably through the tetrahedral adduct shown in Figure 27-5, to give the product. Evidence supporting this mechanism includes the fact that the phosphate produced contains an oxygen atom derived from IMP.

The conversion of xanthylate (XMP) into guanylate (GMP) is catalyzed by GMP synthase and proceeds in an analogous manner (Figure 27-4), except that the phosphotransfer is replaced by an adenylyltransfer and the source of nitrogen is glutamine rather than aspartate. Glutamine is an NH_3 donor in many biochemical reactions. The mechanism of NH_3 formation in these reactions is interesting. Enzymes that utilize glutamine to produce ammonia consist of two types of subunits. In essence, they are multienzyme complexes. One subunit binds glutamine and can catalyze its hydrolysis to glutamate and

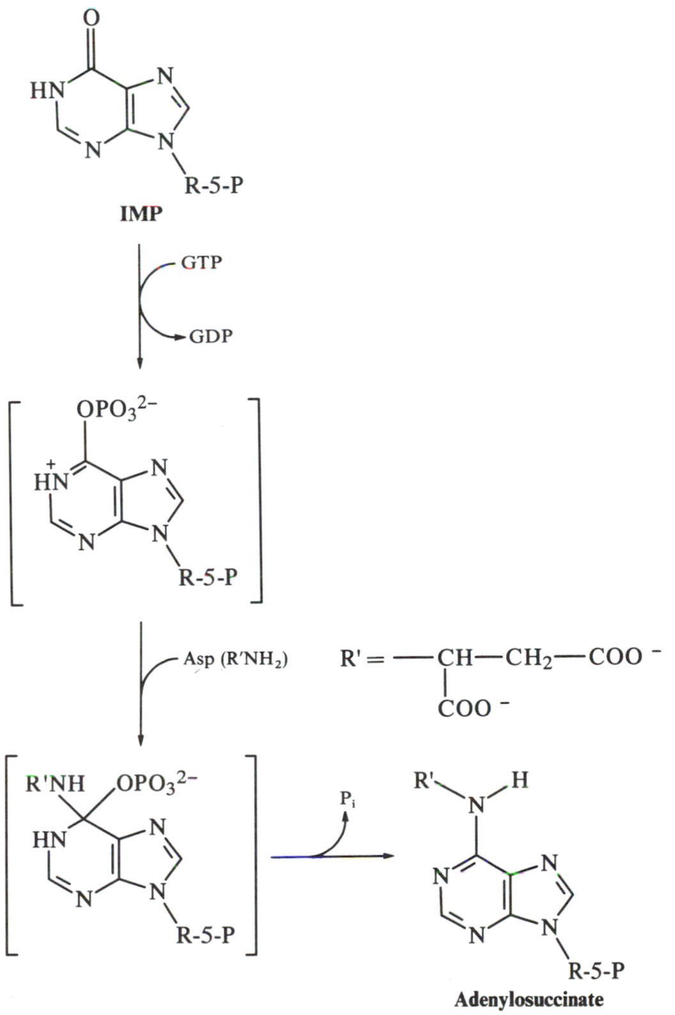

Figure 27-5

The Mechanism of the Adenylosuccinate Synthase Reaction. The bracketed species designate intermediates that are exclusively enzyme bound.

ammonia. The second subunit utilizes ammonia generated from glutamine by the first subunit. Ammonia is utilized without ever dissociating from the enzyme complex. These enzymes can generally utilize ammonium ion (NH_4^+) as a substrate in place of glutamine, but only at very high concentrations or at very high pH, where ammonium ions are substantially dissociated to ammonia.

Biosynthesis of Pyrimidine Nucleotides

The pyrimidine ring is contained within the purine ring, which is a fusion of pyrimidine and imidazole rings. From a purely chemical standpoint, one might expect the biosyntheses of these rings to be related; however, the biosynthesis of the pyrimidine ring is not similar to the biosynthesis of the purine ring. The purine ring is built up in a step-by-step assembly starting with PRPP, and all the intermediates are ribotides, or derivatives of ribose-5-phosphate. In contrast, the pyrimidine ring is built up from aspartate and carbamoyl phosphate and is attached to ribose-5-phosphate in a later step.

UMP and CTP

Figure 27-6 shows the biosynthetic origins of the atoms in the uracil ring, the biosynthetic precursor of all pyrimidines. Nitrogen-1 and carbons 4, 5, and 6 are derived from aspartate. Carbon-2 is derived from HCO_3^- and nitrogen-3 is from the amide nitrogen of glutamine.

Bicarbonate and glutamine are initially converted into carbamoyl phosphate by carbamoyl phosphate synthase II according to equation 3.

$$\text{Glutamine} + HCO_3^- + 2\,\text{ATP} \longrightarrow \text{Carbamoyl phosphate} + \text{Glutamate} + 2\,\text{ADP} + P_i \quad (3)$$

A very similar reaction takes place in the liver of animals, in which carbamoyl phosphate synthase I catalyzes the formation of carbamoyl phosphate from ammonia and bicarbonate when it is activated by *N*-acetylglutamate. This enzyme is limited to the liver and functions primarily in the urea cycle. Carbamoyl phosphate synthase II catalyzes the reaction of equation 3 in most other cells and is not subject to regulation by *N*-acetylglutamate. Carbamoyl phosphate carries the amide nitrogen of glutamine and the carbon of bicarbonate into the pyrimidine ring.

H_2N–C(=O)–O–P(=O)(O^-)–O^-

Carbamoyl phosphate

The carbamoyl group of carbamoyl phosphate is activated for acyl group transfer. Aspartate transcarbamoylase catalyzes its reaction with the α-amino group of aspartate to form *N*-carbamoyl aspartate in step 1 of the pathway for the biosynthesis of UMP in Figure 27-7. In step 2, dihydroorotase catalyzes the dehydration of *N*-carbamoyl aspartate by elimination of water between the amido group of the carbamoyl moiety and the β-carboxyl group of the aspartate moiety in an intramolecular reaction to form the pyrimidine ring of 5,6-dihydroorotate. The dehydrogenation of this molecule by NAD^+ in step 3 is

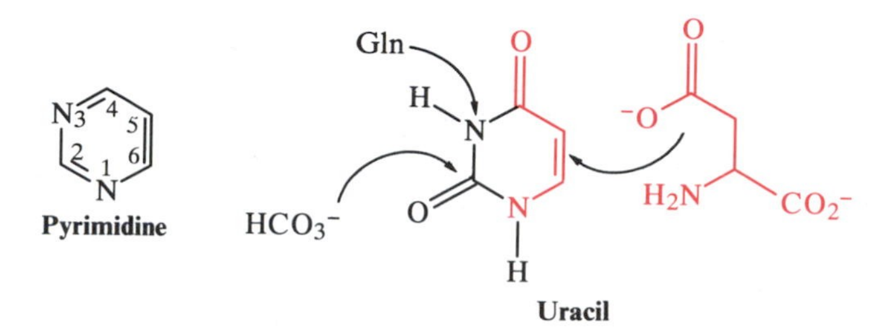

Figure 27-6

Biosynthetic Origins of the Atoms in the Pyrimidine Ring of Uracil. The uracil ring is the precursor of the other pyrimidine rings, cytosine and thymine, in nucleotides.

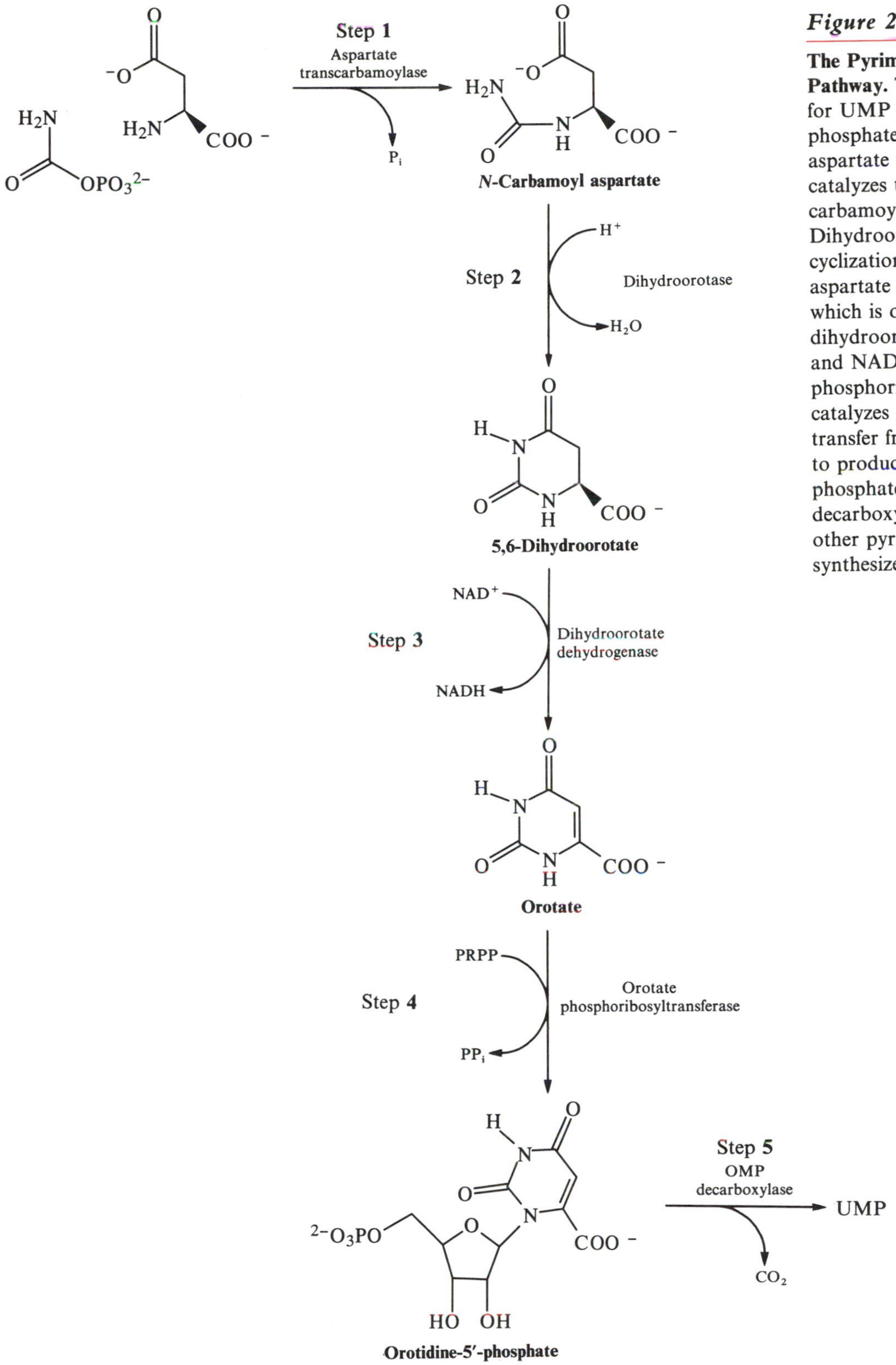

Step 1
Aspartate transcarbamoylase
P_i
N-Carbamoyl aspartate

Step 2
Dihydroorotase
H^+
H_2O
5,6-Dihydroorotate

Step 3
Dihydroorotate dehydrogenase
NAD^+
NADH
Orotate

Step 4
Orotate phosphoribosyltransferase
PRPP
PP_i
Orotidine-5′-phosphate

Step 5
OMP decarboxylase
CO_2
UMP

Figure 27-7

The Pyrimidine Biosynthetic Pathway. The building blocks for UMP are carbamoyl phosphate and aspartate; aspartate transcarbamoylase catalyzes their condensation to carbamoyl aspartate. Dihydroorotase catalyzes the cyclization of carbamoyl aspartate to 5,6-dihydroorotate, which is oxidized by dihydroorotate dehydrogenase and NAD^+ to orotate. Orotate phosphoribosyltransferase catalyzes phosphoribosyl transfer from PRPP to orotate to produce orotidine-5′-phosphate (OMP), which is decarboxylated to UMP. The other pyrimidine nucleotides are synthesized from UMP.

catalyzed by dihydroorotate dehydrogenase, a typical NAD-linked flavoprotein that catalyzes dehydrogenation α,β to a carboxyl group (Chapter 16). Analogous reactions appear in fatty acid biosynthesis and degradation (Chapter 19) and in amino acid degradation (Chapter 26). Orotate phosphoribosyltransferase catalyzes the reaction of orotate with PRPP in step 4 to form orotidine-5′-phosphate

(OMP) in a typical PRPP-dependent phosphoribosyl transfer of the type described for purine biosynthesis and in Chapter 26 for tryptophan biosynthesis. Finally, OMP is decarboxylated in step 5 by the action of OMP decarboxylase to form UMP. This reaction is *atypical* of the enzymatic decarboxylations that were described in Chapter 17, because there is no obvious means to stabilize the carbanion that would be generated by the expulsion of the carboxylate group as CO_2. The mechanism of this reaction is not known.

Nucleoside monophosphate and nucleoside diphosphate kinases catalyze the phosphorylation of UMP to UDP and UTP, which are converted into other pyrimidine nucleotides. The cytidine nucleotides arise from UTP in a reaction that is chemically similar to others that we have encountered in the production of AMP and GMP from IMP. CTP synthase catalyzes the amination of the 4-oxo group of the uracil ring in UTP to form CTP according to equation 4.

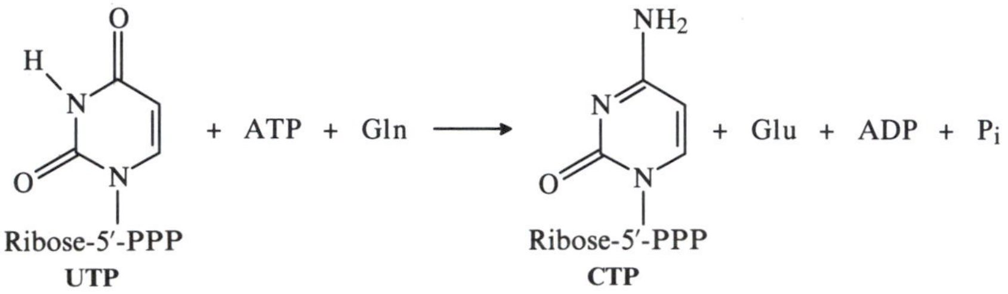

$$\text{UTP (Ribose-5'-PPP)} + \text{ATP} + \text{Gln} \longrightarrow \text{CTP (Ribose-5'-PPP)} + \text{Glu} + \text{ADP} + \text{P}_i \qquad (4)$$

Thymidylate (dTMP)

Thymine is present in DNA but not in RNA, and the thymine ring is derived from dUMP rather than from UMP. dUMP is derived from dCMP produced by ribonucleotide reductase, as described in a later section of this chapter. Thymidylate synthase is an important and interesting enzyme that catalyzes the conversion of dUMP into dTMP. The process is illustrated in Figure 27-8. This reaction proceeds by a mechanism presented in Chapter 25, in which 5, 10-methylene tetrahydrofolate functions as the source of a methyl group. The

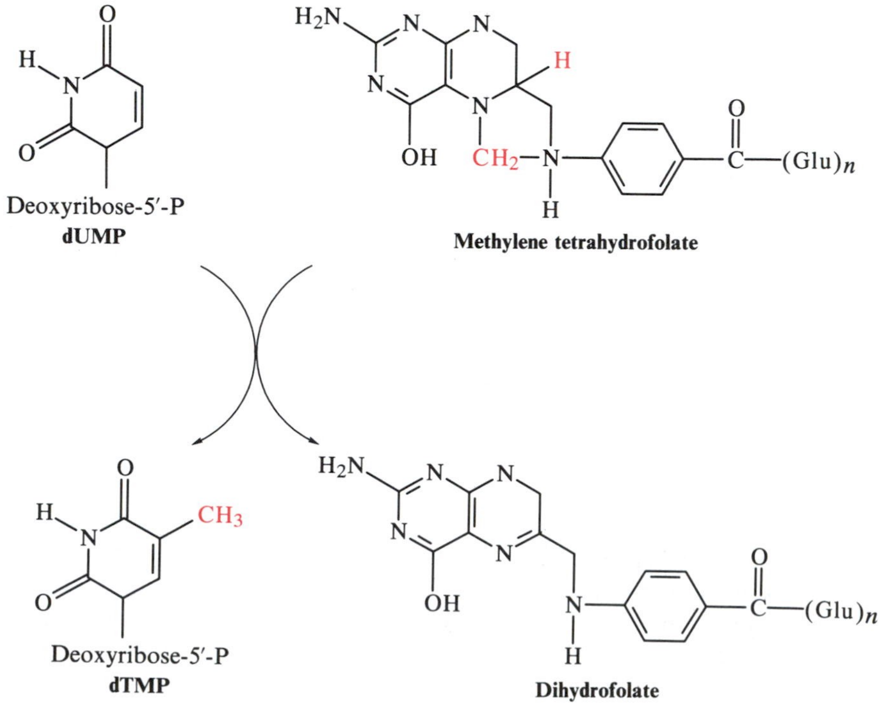

Figure 27-8

Methylation of the Uracil Ring in Thymine Nucleotide Biosynthesis. The methylene group of methylene tetrahydrofolate is transferred by thymidylate synthase to carbon-5 of dUMP, and the methylene group is reduced to the methyl group of thymine by tetrahydrofolate. Therefore, dihydrofolate is the product of thymidylate synthase. The mechanism of this important reaction is described in Chapter 25.

methylene group in the coenzyme is transferred and in the process is reduced to a methyl group at the expense of tetrahydrofolate, which is converted into dihydrofolate. The best folate substrates carry one or more glutamyl groups attached to the carboxyl group of the *p*-aminobenzoyl moiety. Dihydrofolate produced in this reaction must be reduced back to tetrahydrofolate, and this is catalyzed by dihydrofolate reductase. Inhibitors of thymidylate synthase and dihydrofolate reductase are important chemotherapeutic anticancer drugs. The mechanisms by which methotrexate and fluorouracil inhibit these enzymes are described in Chapter 25.

Regulation of Nucleotide Biosynthesis

The nucleotide biosynthetic pathways are subject to control by end-product inhibition; that is, by high levels of nucleoside triphosphates. The end products inhibit biosynthesis by acting as allosteric inhibitors of enzymes early in the biosynthetic pathways. In this way, a pathway can be shut down by an end product that acts on only one or two enzymes.

In purine biosynthesis, high levels of purine nucleotides inhibit PRPP synthase (equation 1); this regulates the level of PRPP, which is a substrate for the formation of 5-phosphoribosyl-1-amine in the first step of the purine biosynthetic pathway. Phosphoribosyl-1-amine synthase (equation 2) also is inhibited by high levels of purine nucleotides. Thus, the first two enzymes required for purine biosynthesis are inhibited by high levels of purine nucleotides. This ensures that, whenever purine nucleotides are present at high levels, 5-phosphoribosyl-1-amine will not be produced, and the biosynthesis of purine nucleotides (Figure 27-2) cannot proceed.

Pyrimidine biosynthesis is regulated differently in animals and in bacteria. In animal cells, CTP inhibits carbamoyl phosphate synthase (equation 3). Carbamoyl phosphate is a substrate for aspartate transcarbamoylase, which catalyzes the first step in UMP biosynthesis (Figure 27-7), so that at high levels of CTP the production of carbamoyl phosphate is attenuated and further biosynthesis of pyrimidines is inhibited. In bacteria, the control point is aspartate transcarbamoylase. CTP inhibits aspartate transcarbamoylase from *E. coli,* which blocks pyrimidine biosynthesis.

Aspartate transcarbamoylase from *E. coli* is an interesting example of a regulated enzyme. It is the first enzyme that was found to consist of catalytic and regulatory subunits. The catalytic subunits catalyze the reaction, and the regulatory subunits control the activity of the catalytic subunits. The enzyme contains twelve subunits of two types. Six are catalytic subunits (molecular weight = 34,000) and six are regulatory subunits (molecular weight = 17,000). The isolated catalytic subunits exist as trimers (α_3), whereas the regulatory subunits are dimers (β_2), so that the overall composition is $(\alpha_3)_2(\beta_2)_3$. The enzyme is inhibited by CTP and activated by aspartate in classical allosteric and cooperative fashion; that is, with a sigmoid dependence of the rate on the concentration of aspartate. However, the reaction of the enzyme with *p*-hydroxymercuribenzoate, which reacts selectively with the —SH group of cysteine, causes the catalytic and regulatory subunits to dissociate from each other and facilitates their separation by chromatography. The separated catalytic subunits are fully active, are not inhibited by CTP, and do not exhibit cooperative kinetic behavior with aspartate. Furthermore, the CTP-binding sites are on the regulatory subunits. Thus, the allosteric regulatory effects are transmitted from the regulatory to the catalytic subunits by conformational changes in the proteins and binding interactions between the catalytic and regulatory subunits.

Reduction of Ribonucleotides to Deoxyribonucleotides

Deoxyribonucleotides are produced by reduction of ribonucleotides in reactions catalyzed by ribonucleotide reductases. The most widely distributed ribonucleotide reductases catalyze the reduction according to Figure 27-9. These enzymes do not exhibit absolute specificities for the heterocyclic bases, and the reductase in a particular species accepts ADP, GDP, and CDP as substrates for reduction to dADP, dGDP, and dCDP. dUMP for dTMP biosynthesis arises from the hydrolysis of dCMP catalyzed by dCMP deaminase. Ribonucleotide reductases are specific for ribonucleoside diphosphates and not for monophosphates or triphosphates in most species, with some exceptions (see Box 27-1). The cellular source of reducing equivalents for ribonucleotide reductases is not known. Dihydrolipoate will serve as the reducing agent *in vitro*. The small redox protein *thioredoxin* (molecular weight = 12,000) will also serve as a reductant. It contains two redox-active thiol groups (Td_{red}) that are highly reactive and exhibit a reduction potential comparable to that of dihydrolipoate. Ribonucleotide reduction oxidizes thioredoxin to a disulfide form (Td_{ox}). The reduced form is regenerated by the action of thioredoxin reductase, a flavoprotein that catalyzes the reduction of thioredoxin by NADPH. Bacterial mutants lacking thioredoxin are viable and contain active ribonucleotide reductase. Therefore, thioredoxin is not required for ribonucleotide reduction, and it may not be the reductant *in vivo;* it is certainly not the *compulsory* reductant in the cell.

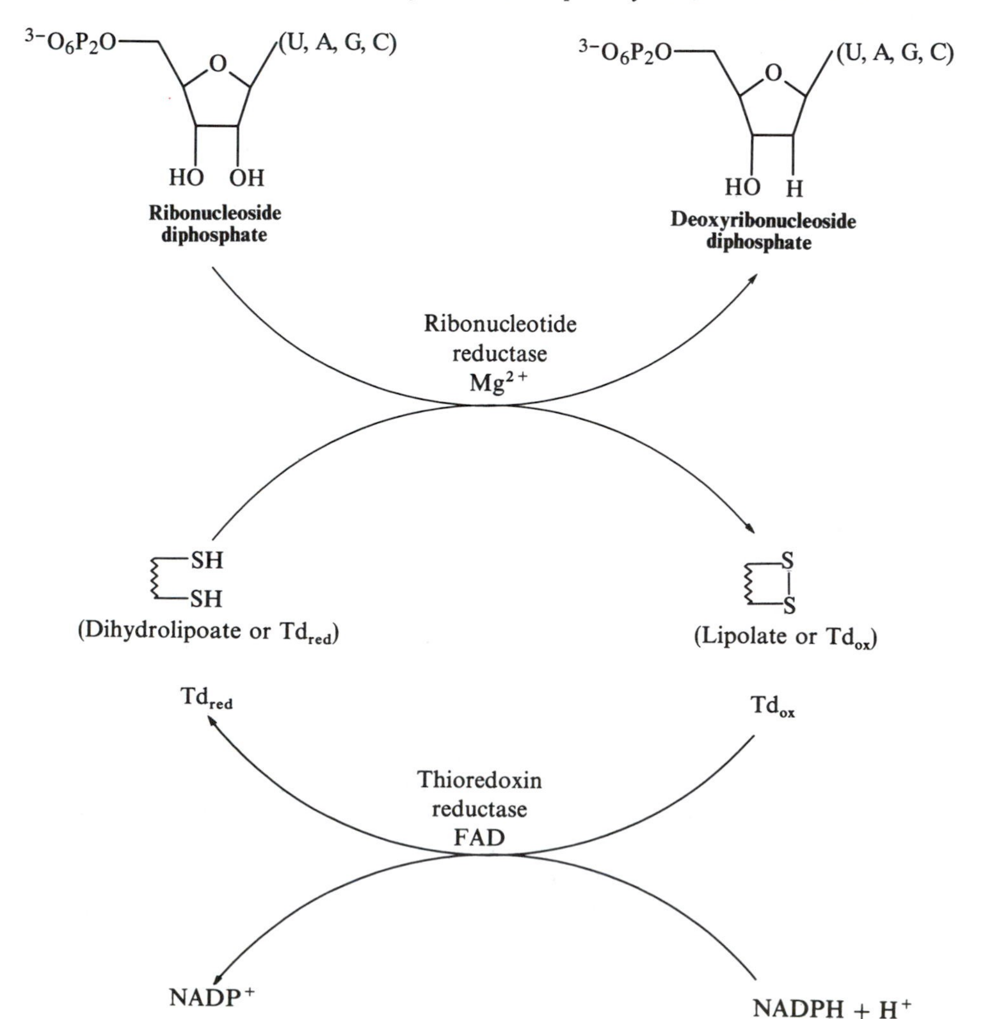

Figure 27-9

Reduction of Ribonucleotides to Deoxyribonucleotides. The reduction of ribonucleoside diphosphates to deoxyribonucleoside diphosphates is catalyzed by ribonucleotide reductase. The reducing agent is a dithiol such as dihydrolipoate or reduced thioredoxin, a small protein that contains a dithiol group that is active as a reducing agent. Thioredoxin reductase catalyzes the reduction of thioredoxin by NADPH.

BOX 27-1

THE MECHANISM OF ACTION OF RIBONUCLEOTIDE REDUCTASES

The mechanism by which ribonucleotide reductases catalyze the reduction of the 2′-carbon atom in nucleotides is of particular interest because of the difficulty of this type of reaction; no closely analogous chemical reactions are known. The reductase from *E. coli* consists of two types of subunits, one of which contains an iron cofactor consisting of two Fe ions bridged by an oxygen, a μ-oxo-Fe_2 complex, and a tyrosyl radical.

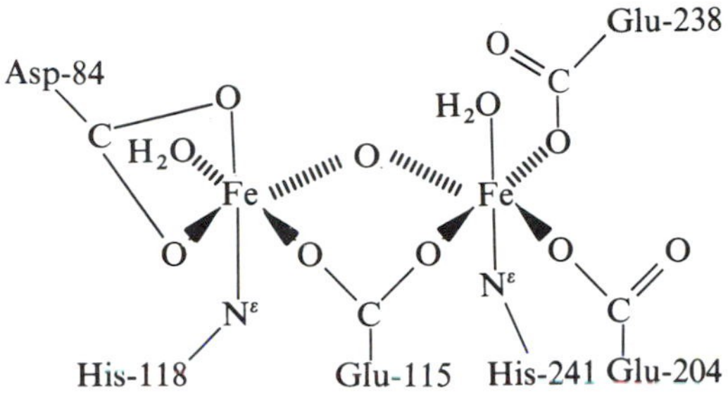

μ-Oxo-Fe_2

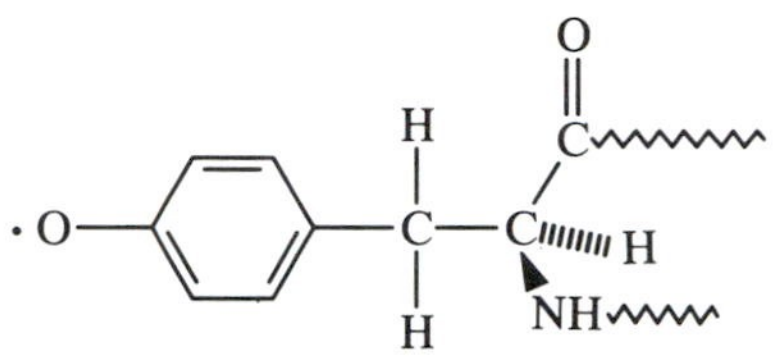

Tyrosyl radical

The majority of the ligands to Fe in the complex are provided by the protein. The μ-oxo-Fe_2 complex generates the tyrosyl radical in a reaction that requires O_2. The tyrosyl radical is required for activity, because the enzyme is inactivated by radical quenching agents; moreover, a phenylalanine mutant of the enzyme, generated by site-directed mutagenesis of the specific tyrosine residue, is inactive and cannot be activated by O_2. Two sulfhydryl groups also function in the reduction.

The available evidence indicates the reaction mechanism in Figure A. The tyrosyl radical abstracts the 3′-hydrogen atom from a substrate molecule to form tyrosine and the 3′-radical of the substrate. The substrate radical then undergoes dehydration, with catalysis by an acidic group on the enzyme, to form a radical cation. The radical cation is reduced by the dithiol (a hydride equivalent) to form the 3′-radical of the product. The product radical abstracts a hydrogen atom from the tyrosyl residue to generate the deoxynucleotide and regenerate the tyrosyl radical. Evidence for this mechanism includes the observation of a kinetic isotope effect for the reaction of 3′-tritiated substrates, which shows that the 3′-hydrogen atom is transiently removed in the mechanism. The 3′-hydrogen atom is conserved in the reaction, and the 2′-hydrogen atom in the product arises from water.

Figure A

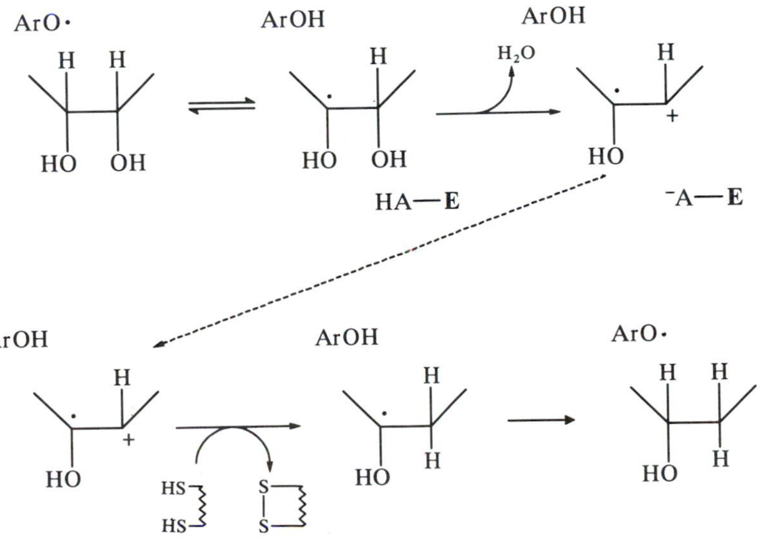

Recent evidence indicates that the mechanism may be more complex, in that the tyrosyl radical seems to act through other protein functional groups. It appears that the tyrosyl radical reacts reversibly with another protein group to form a secondary radical such as —S·, and this reacts directly with the substrate according to the mechanism in Figure A.

The cofactors of the mammalian ribonucleotide

BOX 27-1 (continued)

THE MECHANISM OF ACTION OF RIBONUCLEOTIDE REDUCTASES

reductases are similar to those of the enzyme in *E. coli.* Certain bacteria require different cofactors. The ribonucleotide reductase from *Lactobacillus leichmanii* utilizes only the ribonucleoside triphosphates as substrates and is activated by adenosylcobalamin. In this case, the organic radical initiator is the 5′-deoxyadenosyl radical resulting from homolytic cleavage of the cobalt–carbon bond (Chapter 20).

Ribonucleotide reductases from anaerobic bacteria cannot use the same cofactors as the *E. coli* system because oxygen is required to generate the tyrosyl radical. Many anaerobic bacteria use the adenosylcobalamin system. Even *E. coli* grown under anaerobic conditions must use other cofactors, which have not yet been identified.

Degradation and Salvage of Nucleotides

Nucleic acids are subject to metabolic degradation in several circumstances. DNA and RNA that are foreign to the organism or cell are broken down by digestion to nucleotides. Foreign nucleic acids most often are introduced into organisms in their food sources or through infection by viruses or bacteria. Nonforeign nucleic acids in a cell or organism also can be degraded to nucleotides. The chromosomal DNA within a cell is metabolically fairly stable for reasons presented in Chapter 10. Significant exceptions are that, upon the death of an animal cell, the DNA is degraded to produce nucleotides for other cells. Also, in the event of damage to DNA, the intracellular repair process requires a limited degradation of DNA in the vicinity of the damage. Cellular RNA, in contrast to DNA, is metabolically active and turns over fairly rapidly, especially mRNA. In fact, regulation of the turnover of mRNA is an important means by which the biosynthesis of specific proteins is regulated.

Nucleotides arising from degradation of foreign or nonforeign nucleic acids are either salvaged to produce required cellular constituents or degraded further for excretion. Salvage of the nucleotides is an energy-efficient means for maintaining the required cellular concentrations of nucleotides and nucleotide derivatives, such as ATP, GTP, UDP-glucose, and CDP-diacylglycerol. Nucleotides and their derivatives are required for most cellular functions, including mRNA biosynthesis; complex carbohydrate biosynthesis; the biosynthesis of coenzymes such as NAD^+, FAD, CoA, and adenosylcobalamin; the biosynthesis of complex lipids; and the operation of all metabolic pathways.

Degradation of Nucleic Acids and Nucleotides

The degradation of nucleic acids to nucleotides is catalyzed by a variety of nucleases, such as ribonuclease A (Chapter 12). Deoxyribonucleases degrade DNA to either deoxynucleoside-5′-phosphates or deoxynucleoside-3′-phosphates. Many of these nucleotides are simply used to produce coenzymes or newly synthesized nucleic acids. Significant amounts are broken down further to nucleosides and the heterocyclic bases. A general scheme for the breakdown is given in Figure 27-10. The breakdown of nucleotides to nucleosides is not strictly vertical. Rather it entails certain interconversions of nucleotides into nucleosides that precede their hydrolysis catalyzed by nucleotidases. AMP is deaminated by hydrolysis to IMP in a reaction catalyzed by AMP deaminase (equation 5).

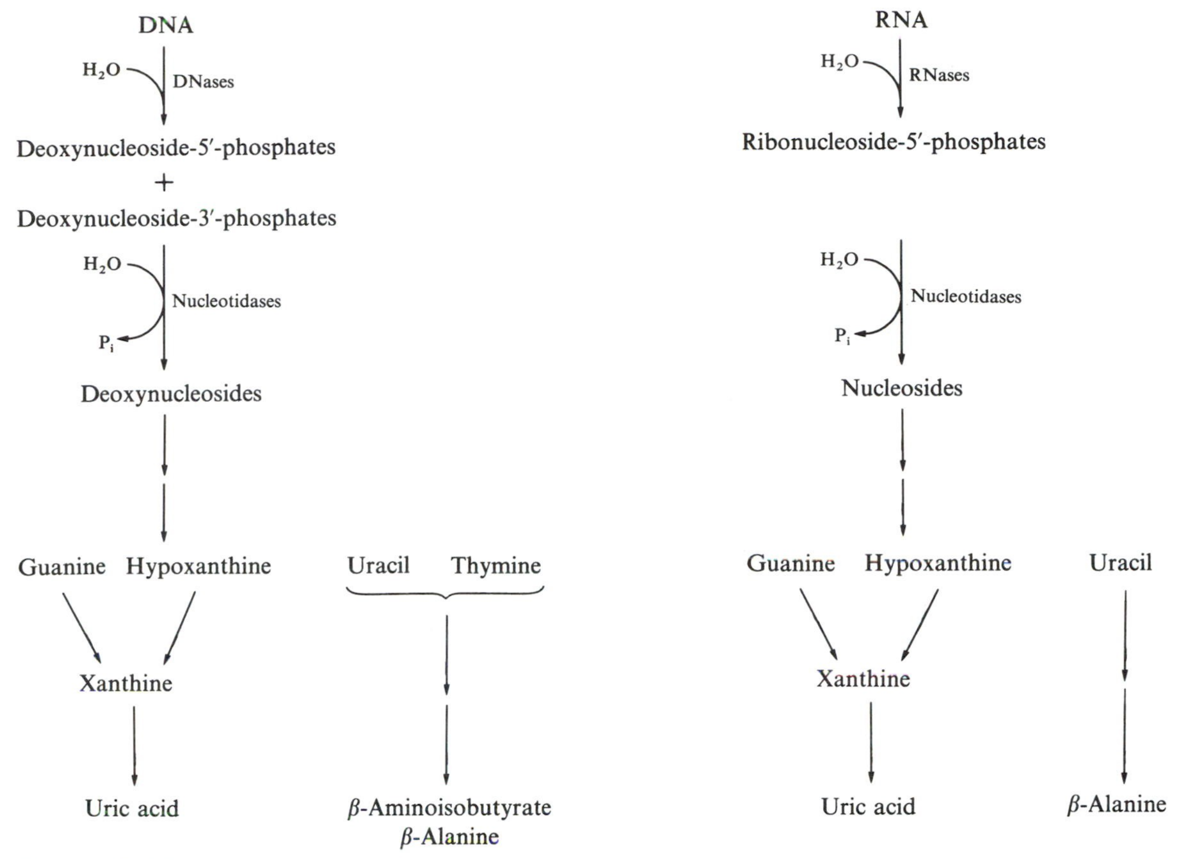

Figure 27-10

Pathways for Degradation of Nucleotides. Many nucleotides are salvaged and reused before being completely degraded by these pathways. See Figure 27-11 for the specific steps in uric acid formation.

NH_2 (adenine ring: N, N, N, N) + H_2O ⟶ O (hypoxanthine ring: N, NH, N, N) + NH_3 (5)

Ribose-5′-phosphate Ribose-5′-phosphate

The breakdown of nucleosides to heterocyclic bases also is not vertical. Cytidine is hydrolyzed to uridine by cytidine deaminase before being cleaved to uracil. Adenosine deaminase catalyzes the deamination of deoxyadenosine to deoxyinosine, which is further oxidized to uric acid.

IMP and GMP are hydrolyzed to inosine and guanosine by a nucleotidase. Inosine and guanosine are cleaved to hypoxanthine and guanine, respectively, by phosphorolytic cleavage of the *N*-ribosyl linkage catalyzed by purine nucleoside phosphorylase. This is illustrated in Figure 27-11, which also shows the further oxidation of hypoxanthine to xanthine and then to uric acid by xanthine oxidase. Guanine is hydrolyzed to xanthine and oxidized to uric acid.

Uric acid is the excreted form of purines and is the primary form in which birds and reptiles excrete excess dietary nitrogen from amino acids and proteins. The sources of nitrogen atoms in uric acid are aspartate, glycine, and the amide

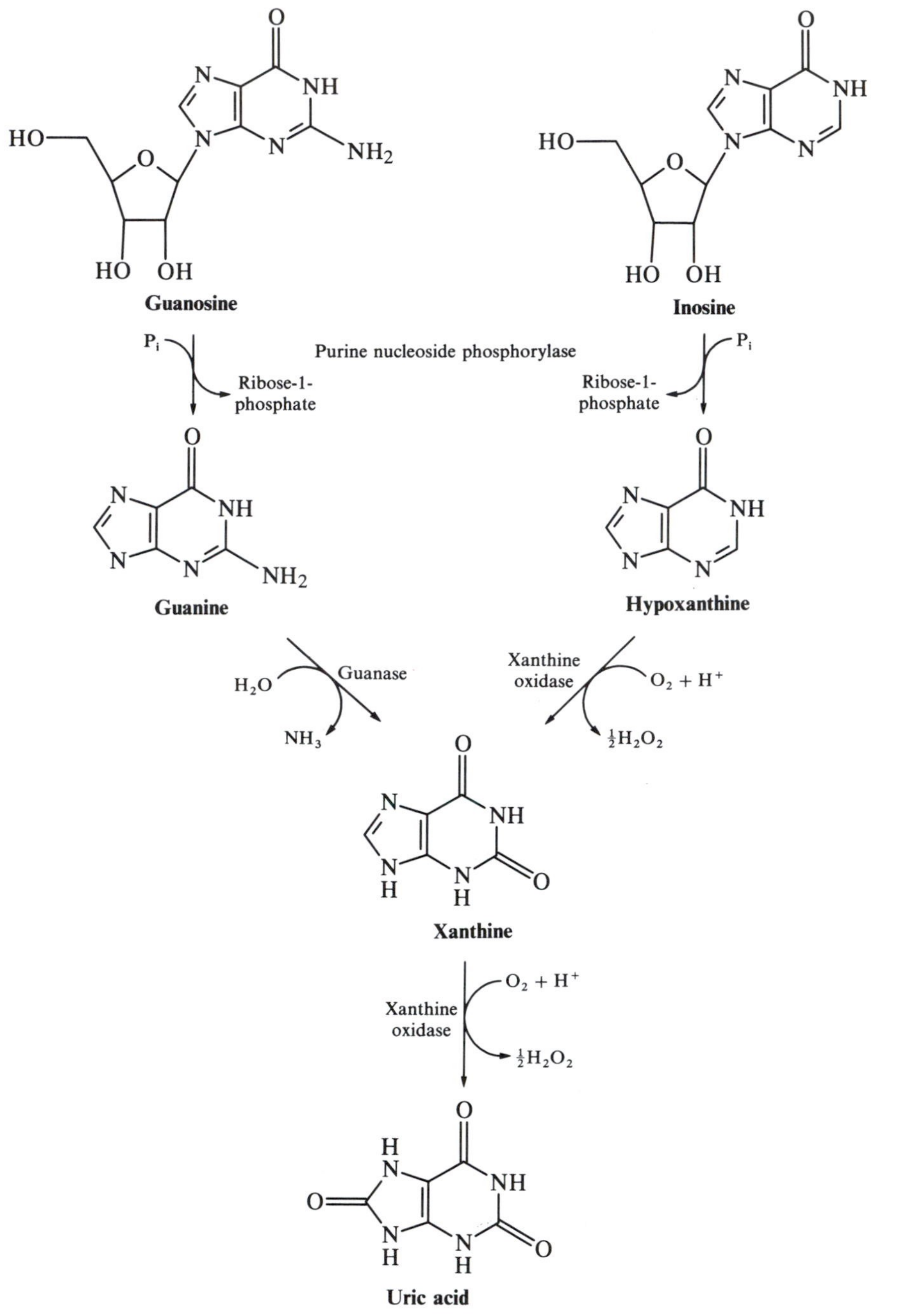

Figure 27-11

Degradation of Guanosine and Inosine to Uric Acid. Guanosine and inosine arise by nucleotidase-catalyzed hydrolysis of GMP and IMP. IMP is the central nucleotide intermediate in the purine biosynthetic pathway (Fig. 27-3). It is also derived from AMP by the action of AMP deaminase (equation 5).

nitrogen of glutamine (Figure 27-1). The amide group of glutamine is generated from ammonium ions by the action of glutamine synthetase (Chapter 26). Thus, nitrogen from amino acids is available for purine biosynthesis and uric acid formation. The glutamate generated from glutamine in purine biosynthesis is recycled by glutamine synthetase.

Salvage of Purine Nucleotides

The degradation of DNA and RNA leads to the formation of purine bases that can be used to synthesize nucleotides. Salvage of bases obviates the need to resynthesize them. The bases are salvaged by phosphoribosyl transferases that catalyze equations 6 through 8. Adenine phosphoribosyl transferase catalyzes reaction 6,

$$\text{Adenine} + \text{PRPP} \longrightarrow \text{AMP} + \text{PP}_i \quad (6)$$

and hypoxanthine-guanine phosphoribosyl transferase catalyzes reactions 7 and 8.

$$\text{Hypoxanthine} + \text{PRPP} \longrightarrow \text{IMP} + \text{PP}_i \quad (7)$$

$$\text{Guanine} + \text{PRPP} \longrightarrow \text{GMP} + \text{PP}_i \quad (8)$$

IMP can be converted into either AMP or GMP by the reactions in Figure 27-4, so that the reactions of equations 6 through 8 allow for the salvage of purines as AMP and GMP.

Summary

The biogenetic molecules for the synthesis of the purine rings of adenine and guanine nucleotides are CO_2, glycine, 10-formyl tetrahydrofolate, glutamine, and aspartate. Glycine provides carbons 4 and 5 and nitrogen-7, CO_2 is the source of carbon-6, the formyl group of 10-formyl tetrahydrofolate is the source of carbons 2 and 8, nitrogens 3 and 9 are derived from glutamine, and nitrogen-1 originates with the α-amino group of aspartate. The purine nucleotide IMP is the precursor of all purine nucleotides and is biosynthesized in ten steps beginning with the reaction of glutamine with PRPP to form 5-phosphoribosyl-1-amine. The purine ring is then built up from the amino group, which becomes N-9 of IMP, in nine steps. IMP is converted into AMP and GMP, which are phosphorylated to the triphosphates.

Purine

Pyrimidine nucleotides are biosynthesized from bicarbonate, glutamine, and aspartate. Aspartate is the source of nitrogen-1 and carbons 4, 5, and 6; bicarbonate is the source of carbon-2, and nitrogen-3 originates with the amide group of glutamine. Carbamoyl phosphate synthase II catalyzes the reaction of bicarbonate, glutamine, and two moles of ATP to form carbamoyl phosphate, glutamate, and two moles of ADP. Aspartate transcarbamoylase catalyzes the transfer of the carbamoyl group from carbamoyl phosphate to the α-amino group of aspartate to form carbamoyl aspartate. In the next step, dehydroquinase catalyzes the cyclization of carbamoyl aspartate to 5,6-dihydroorotate in a reaction that entails the elimination of water between the aspartyl β-carboxyl group and the amido nitrogen of the carbamoyl group. Oxidation of dihydroorotate to orotate and 5-phosphoribosyl transfer from PRPP produces orotidine-5′-phosphate, which undergoes decarboxylation to UMP. Phosphorylation of UMP to the diphosphate and triphosphate supplies the substrates for enzymes that catalyze the conversion of the uracil ring into cytidine or thymine.

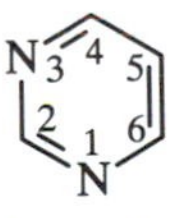

Pyrimidine

The deoxyribonucleotides are produced by reduction of ADP, GDP, CDP, and UDP by the action of ribonucleotide reductase. Thymidylate synthetase catalyzes the reaction of dUMP with 5,10-methylene tetrahydrofolate to form dTMP and dihydrofolate.

Excess purine nucleosides are degraded to uric acid. This process begins with phosphorolytic cleavage to the purine bases and ribose-1-phosphate catalyzed by purine nucleoside phosphorylase. The purines are then oxidized to uric acid by xanthine oxidase.

ADDITIONAL READING

HOFFEE, P. A., and JONES, M. E., eds. 1978. *Purine and Pyrimidine Nucleotide Metabolism.* Methods in Enzymology, vol. 51. New York: Academic Press.

REICHARD, P. 1988. Interactions between deoxyribonucleotides and DNA synthesis. *Annual Review of Biochemistry* 57: 349.

REICHARD, P., and EHRENBERG, A. 1983. Ribonucleotide reductase: a radical enzyme. *Science (Washington, D.C.)* 221: 514.

STUBBE, J. 1989. Protein radical involvement in biological catalysis. *Annual Review of Biochemistry* 58: 257.

KANTROWITZ, E. R., and LIPSCOMB, W. N. 1988. *Escherichia coli* aspartate transcarbamylase: the relation between structure and function. *Science* 241: 669.

PROBLEMS

1. A primary source of nitrogen for uric acid formation (and for urea formation) is aspartate. As a common amino acid, aspartate must not be permanently lost in nitrogen excretion. Explain how aspartate that is used for excretion of excess nitrogen can be regenerated to maintain its required balance for protein biosynthesis.

2. The mechanism of the OMP decarboxylase reaction is unknown. Explain the difficulty with this reaction. How might this reaction be catalyzed by the enzyme in such a way as to overcome the barrier to the simple expulsion of CO_2 from OMP?

3. Uracil and thymine are degraded to β-alanine and β-aminoisobutyrate, respectively. Suggest enzymatic reaction sequences leading to the formation of these degradation products.

4. Ribonucleotide reductase is inactivated by 2′-chloro-ADP, but 2′-chloro-ADP is not reduced to dADP. The inactivation is accompanied by the production of PP_i, adenine, and the compound shown below. This compound also inactivates ribonucleotide reductase. Suggest an explanation for the production of PP_i and adenine from 2′-chloro-ADP and for the inactivation of ribonucleotide reductase.

H_2C O O

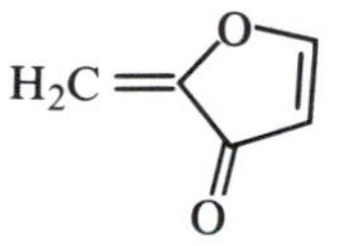

5. Consider the carboxylation of 5′-phosphoribosyl-5′-aminoimidazole, step 5 in Figure 27-3. The reaction is ATP-dependent and bicarbonate is probably the carboxylation substrate. Propose a mechanism for this reaction. Show the structures of the intermediates. How is this carboxylation related to the more-common examples of carboxylation described in Chapter 17?

28

Complex Lipids

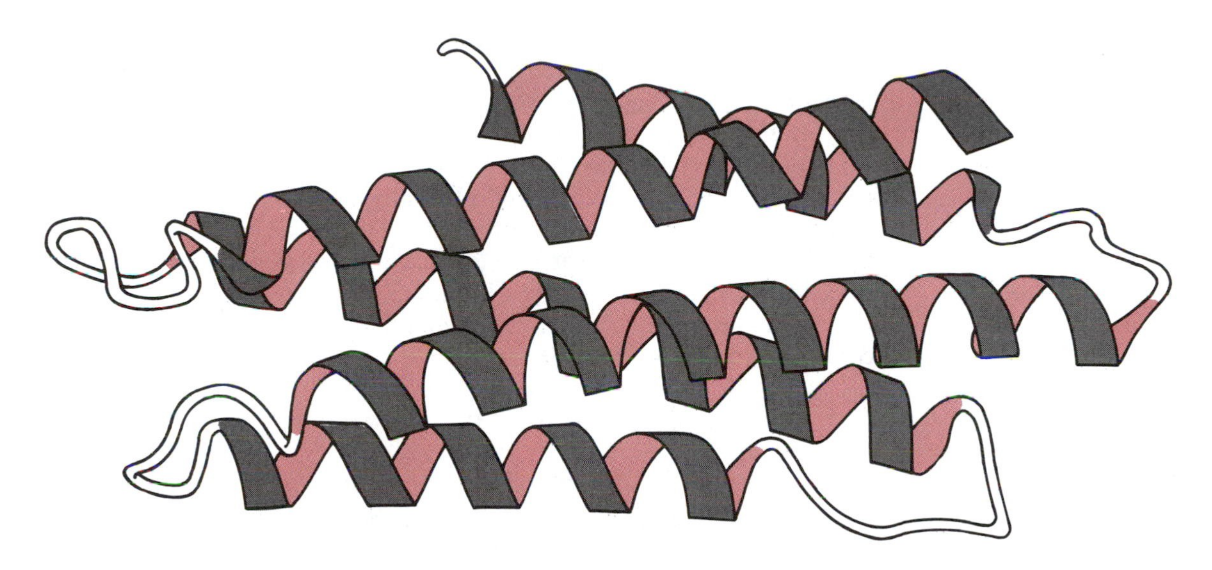

A ribbon diagram of the polypeptide chain of an insect lipoprotein. (Adapted from a computer-generated drawing from Hazel Holden.)

Lipids are crucially important to the structure and function of cells. Complex lipids, including phospholipids and, in eukaryotic cells, cholesterol, are the principal structural constituents of all cellular membranes. The simplest lipids are the fatty acids which are synthesized from acetyl CoA (Chapter 19). The phospholipids are biosynthesized from fatty acyl CoA, glycerol, and ATP, as described in Chapter 11; their association into bilayers is the principal physicochemical phenomenon in the assembly of biological membranes, which are described in Chapter 29.

In multicellular organisms, including higher animals, there are added dimensions to the functions of complex lipids. The transport of triglycerides and cholesterol from organs to tissues in animals is a complex process, owing to the insolubility of these lipids in water. The process requires the assembly and metabolism of lipoproteins, which are complexes of proteins and complex lipids. The steroid hormones are derived from cholesterol, and other hormones such as prostaglandin also are complex lipids. The prostaglandins, thromboxanes, and leukotrienes are biosynthesized from unsaturated fatty acids, as described in Chapter 24. Finally, the structures of nerve tissues include complex lipids in addition to those that compose the cellular membranes.

In this chapter, we describe the biosynthesis of cholesterol, the metabolism of lipoproteins, the conversion of cholesterol into bile salts and steroid hormones, and the biosynthesis of sphingolipids. Like fatty acids, most complex lipids are synthesized from acetyl CoA. In fact, the first steps in cholesterol biosynthesis are closely related to fatty acid biosynthesis. However, in the final steps, the chemistry of allyl carbocations becomes dominant. This type of reaction is rare in biochemistry; it is found in the synthesis of cholesterol and complex hydrocarbons such as terpenes.

The Biosynthesis of Cholesterol

Cholesterol is a component of eukaryotic membranes and a precursor of steroid hormones and bile salts. The structure of cholesterol is shown in Figure 28-1, together with the biosynthetic origin of each carbon atom. Most of the enzymatic reactions in the biosynthesis of cholesterol are now reasonably well

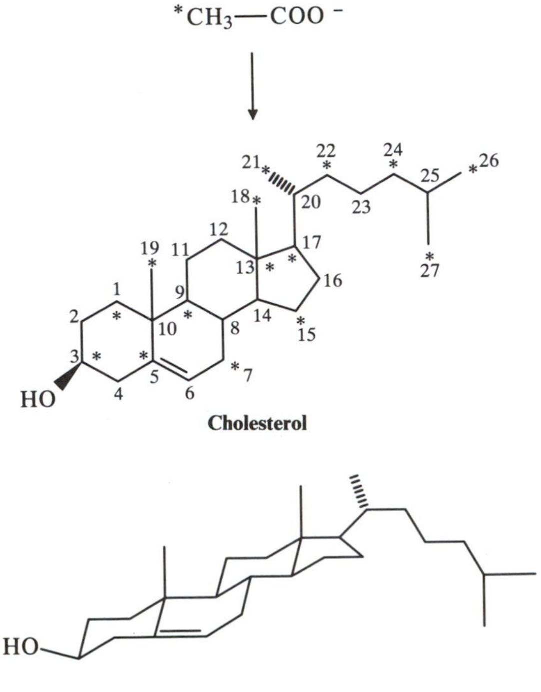

$$^{*}CH_3—COO^-$$

Cholesterol

Figure 28-1

The Biogenesis of Cholesterol. All of the carbon atoms in cholesterol are derived from acetate. The carbon atoms marked with asterisks in the upper structure are derived from the methyl carbon of acetate, and the other carbons are derived from the carboxyl carbon. The conformation and stereochemistry of cholesterol are illustrated at the bottom of the figure.

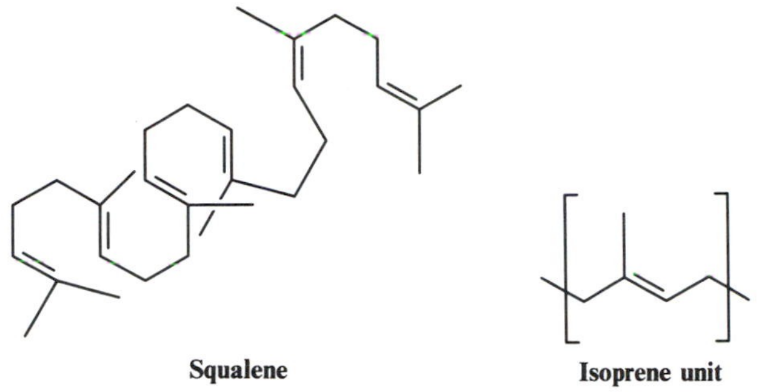

Figure 28-2

The Structure of Squalene and Its Relation to That of the Isoprene Unit.

understood through the work of K. Bloch, F. Lynen, G. Popjak, and J. Cornforth. Bloch proved that all carbon atoms of cholesterol are derived from acetate. By feeding animals acetate labeled with ^{14}C either in the methyl group or carboxyl group, he was able to establish which carbon atoms of cholesterol are derived from the methyl group of acetate and which are derived from the carboxyl group.

The discovery of squalene provided the next clue to the biosynthesis of cholesterol. Squalene is a hydrocarbon of thirty carbon atoms consisting of six isoprene units, as shown in Figure 28-2. Squalene was first found in shark liver, but small amounts are also present in the livers of other animals. Rats fed [^{14}C]acetate produce [^{14}C]squalene in the liver, and incubation of this labeled squalene with liver slices leads to the formation of [^{14}C]cholesterol.

An understanding of how the isoprene units are synthesized came from studies of bacterial metabolism, which led to the discovery of mevalonate, and it became apparent that mevalonate could give rise to the isoprene units making up squalene (Figure 28-3). Experiments with animals confirmed that [^{14}C]acetate gives rise to [^{14}C]mevalonate which in turn is a precursor of [^{14}C]squalene, according to the sequence in Figure 28-3. Subsequent research elucidated further details of the reaction sequence.

Mevalonic acid is derived from acetyl CoA through two condensation reactions, followed by a reduction, as shown in Figure 28-4. The first step is catalyzed by β-ketothiolase, the mechanism of which was described in Chapter 2. The second step is an addition of acetyl CoA to the ketonic group of acetoacetyl CoA by a mechanism analogous to that of citrate synthase, which was also described in Chapter 2. The reduction of 3-hydroxy-3-methylglutaryl CoA (HMG CoA) to mevalonate is unusual in that it requires a four-electron reduction and, therefore, requires two molecules of NADH. The reduction is carried out by a single enzyme, HMG CoA reductase. There is evidence suggesting that the aldehyde derived from HMG CoA is an intermediate that remains bound to the enzyme throughout the reaction (Figure 28-5).

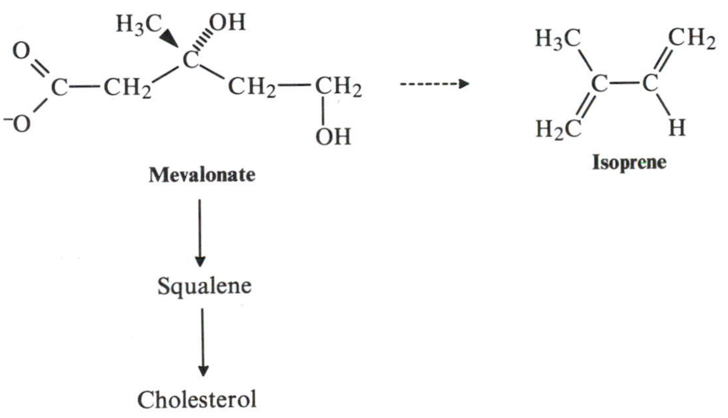

Figure 28-3

Mevalonate as a Biosynthetic Precursor of Squalene and Cholesterol. The isoprene unit from which squalene is assembled can be obtained through the decarboxylation and dehydration of mevalonate, as illustrated at the top of the figure. These are two steps in the assembly of squalene, and both require activation by ATP as shown in Figure 28-6.

Figure 28-4

Conversion of Acetyl CoA into Mevalonate in Cholesterol Biosynthesis. The mechanism of the condensation of two molecules of acetyl CoA to acetoacetyl CoA by β-ketothiolase is described in Chapter 2. The mechanism of the addition of a third molecule of acetyl CoA to form HMG CoA by HMG CoA synthase is analogous to that of citrate synthase, which also is described in Chapter 2. The reduction of HMG CoA to mevalonate by two molecules of NADH is an unusual reaction that requires the intermediate formation of the aldehyde as an intermediate. The aldehyde is reduced by NADH much faster than it can dissociate; therefore, it is not a free intermediate but rather an enzyme-bound intermediate.

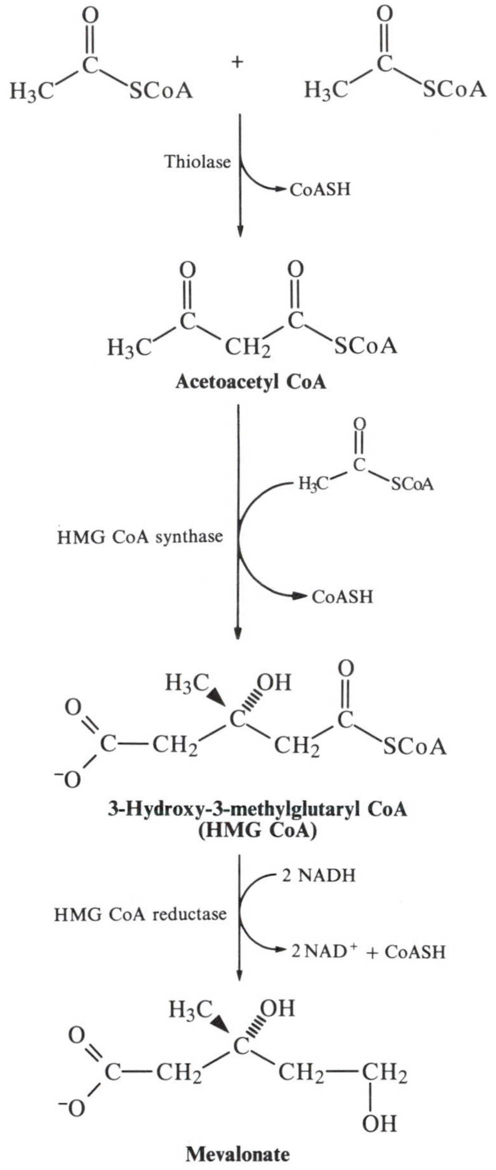

Mevalonate is converted into isopentenyl pyrophosphate (isopentenyl-PP) in three steps according to Figure 28-6. Isopentenyl-PP is the basic unit from which squalene and cholesterol are derived. The first two steps require ATP and are phosphoryl group transfer reactions that produce mevalonate-5-pyrophosphate (mevalonate-5-PP), which then undergoes decarboxylation in a reaction that also requires ATP. The bracketed compound in Figure 28-6, 3-phosphomevalonate-5-PP, is postulated to be an intermediate, although it has not been isolated. Phosphorylation of the 3-hydroxyl group of mevalonate makes it a better leaving group, thereby facilitating the decarboxylation.

The condensation of three molecules of isopentenyl-PP leads to the formation of farnesyl-PP (Figure 28-7). In the first step, isopentenyl-PP isomerase catalyzes the migration of the double bond to form dimethylallyl-PP. A molecule of dimethylallyl-PP donates a dimethylallyl group to isopentenyl-PP to form geranyl-PP and PP_i. This is a carbon–carbon bond-formation reaction that appears to proceed through a carbocationic intermediate. Elimination of PP_i from dimethylallyl-PP may generate a dimethylallyl carbocation that forms a bond with the π electron pair of isopentenyl-PP. Loss of a proton

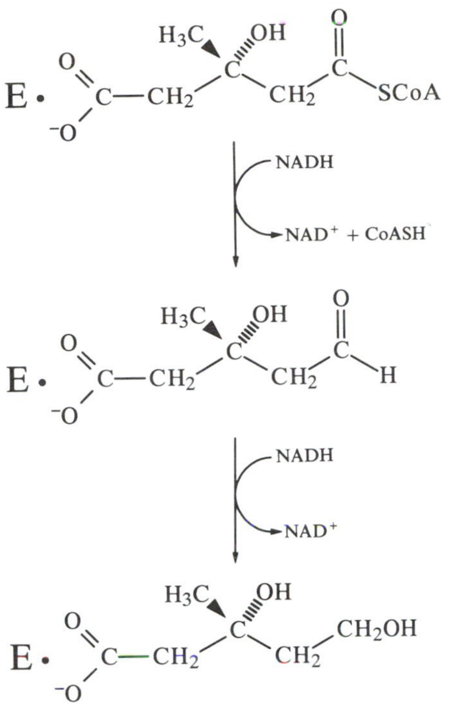

Figure 28-5

Steps in the Enzymatic Reduction of HMG CoA to Mevalonate. The reduction of HMG CoA to mevalonate requires two molecules of NADH and proceeds in two steps. In the first step, HMG CoA is reduced to the aldehyde, followed by the release of NADH and CoASH from the enzyme. The mevalonaldehyde is reduced to mevalonate by another molecule of NADH before it dissociates from the enzyme.

from carbon-3 of isopentenyl-PP leads to geranyl-PP. Another allyl transfer, a geranyl transfer, to isopentenyl-PP leads to farnesyl-PP.

Squalene synthase catalyzes the formation of squalene from two molecules of farnesyl-PP and NADPH. The reaction probably proceeds according to Figure 28-8 through the intermediate presqualene-PP, which is reduced by NADPH to squalene. Little is known about the enzymes that catalyze these reactions, primarily because they are associated with the membrane of the endoplasmic reticulum of eukaryotic cells. Membrane-bound enzymes are much more difficult to purify and study than soluble enzymes of the cytosol.

The mechanisms of the reactions in Figure 28-8 are not fully known; however, they almost certainly proceed through carbocationic intermediates. Elimination of pyrophosphate from farnesyl-PP would produce an allylic carbocation; the formation of allylic carbocations is favored by delocalization of the positive charge over several atoms through resonance, as shown in the margin. An allylic carbocation can undergo addition to a double bond, and the carbocationic adduct resulting from addition of the farnesyl allylic cation to a molecule of farnesyl-PP can undergo subsequent rearrangements and loss of a proton by a pathway leading to presqualene pyrophosphate (presqualene-PP) according to Figure 28-8. The exact mechanism of this process is not known; however, carbocations are known to undergo carbon-skeletal rearrangements, 1,2-hydride shifts, and deprotonations that can lead to presqualene-PP. The active site of the enzyme presumably interacts with the substrate in such a way as to facilitate the required isomerizations of the intermediates, while preventing other possible rearrangements from leading to the formation of other products. This could be accomplished by an active site in which a base is oriented in a position to accept only the correct proton in the last step of the reaction.

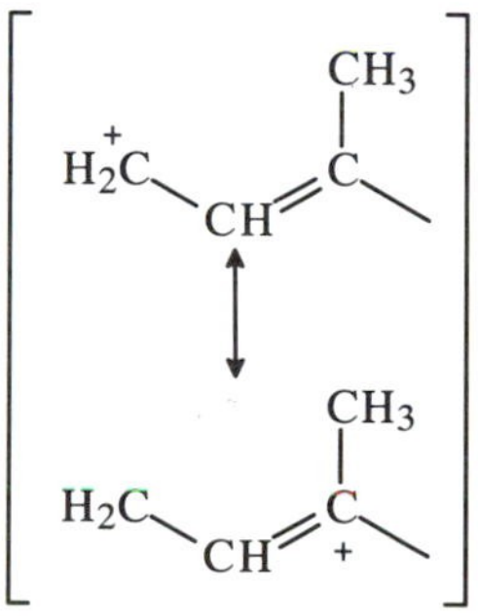

Allylic carbocation

The mechanism by which presqualene-PP is reduced to squalene by NADPH and a reductase is not known. Reduction of presqualene-PP itself by NADPH is unlikely to lead to squalene in a single step because this would require the cleavage of a carbon–oxygen bond, the cleavage of a carbon–carbon bond, the formation of a new carbon–carbon bond, and the formation of a new carbon–hydrogen bond, all through a single transition state. It is more likely

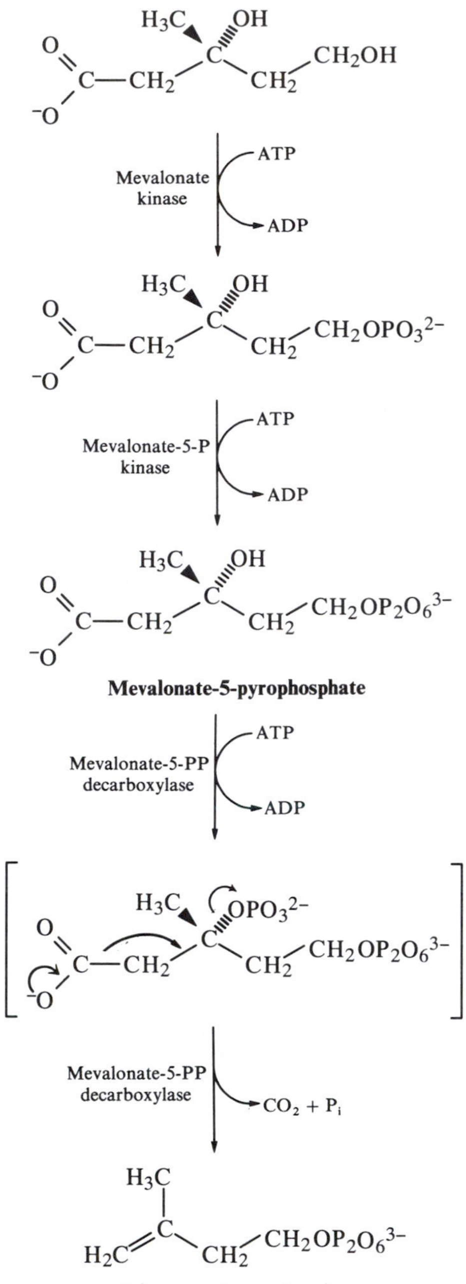

Figure 28-6

Steps in the Conversion of Mevalonate into Isopentenyl-PP. Mevalonate is phosphorylated in two steps by two different kinases to mevalonate-5-PP. The decarboxylation of mevalonate-5-PP is catalyzed by a single enzyme in two steps, phosphorylation to 3-phosphomevalonate-5-PP followed by decarboxylation with elimination of phosphate. 3-Phosphomevalonate-5-PP is presumed to be an intermediate that is not released from the enzyme, but it has never been positively identified.

Mevalonate kinase: ATP → ADP

Mevalonate-5-P kinase: ATP → ADP

Mevalonate-5-pyrophosphate

Mevalonate-5-PP decarboxylase: ATP → ADP

Mevalonate-5-PP decarboxylase: → $CO_2 + P_i$

Isopentenyl pyrophosphate

that the reduction of presqualene-PP (Figure 28-8) proceeds through a series of carbocationic intermediates such as those shown below.

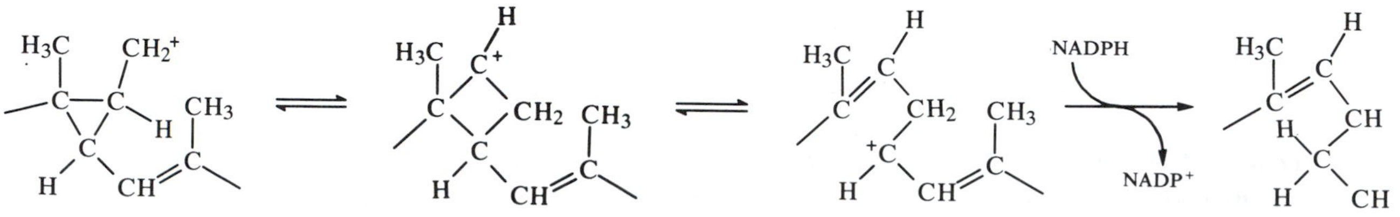

NADPH → NADP⁺

The cyclopropylcarbinyl cation resulting from the elimination of pyrophosphate from presqualene-PP could undergo a series of carbon-skeletal rearrangements to an allylic carbocation, the reduction of which by NADPH would produce squalene. The ring expansion from the cyclopropylcarbinyl carbocation to the

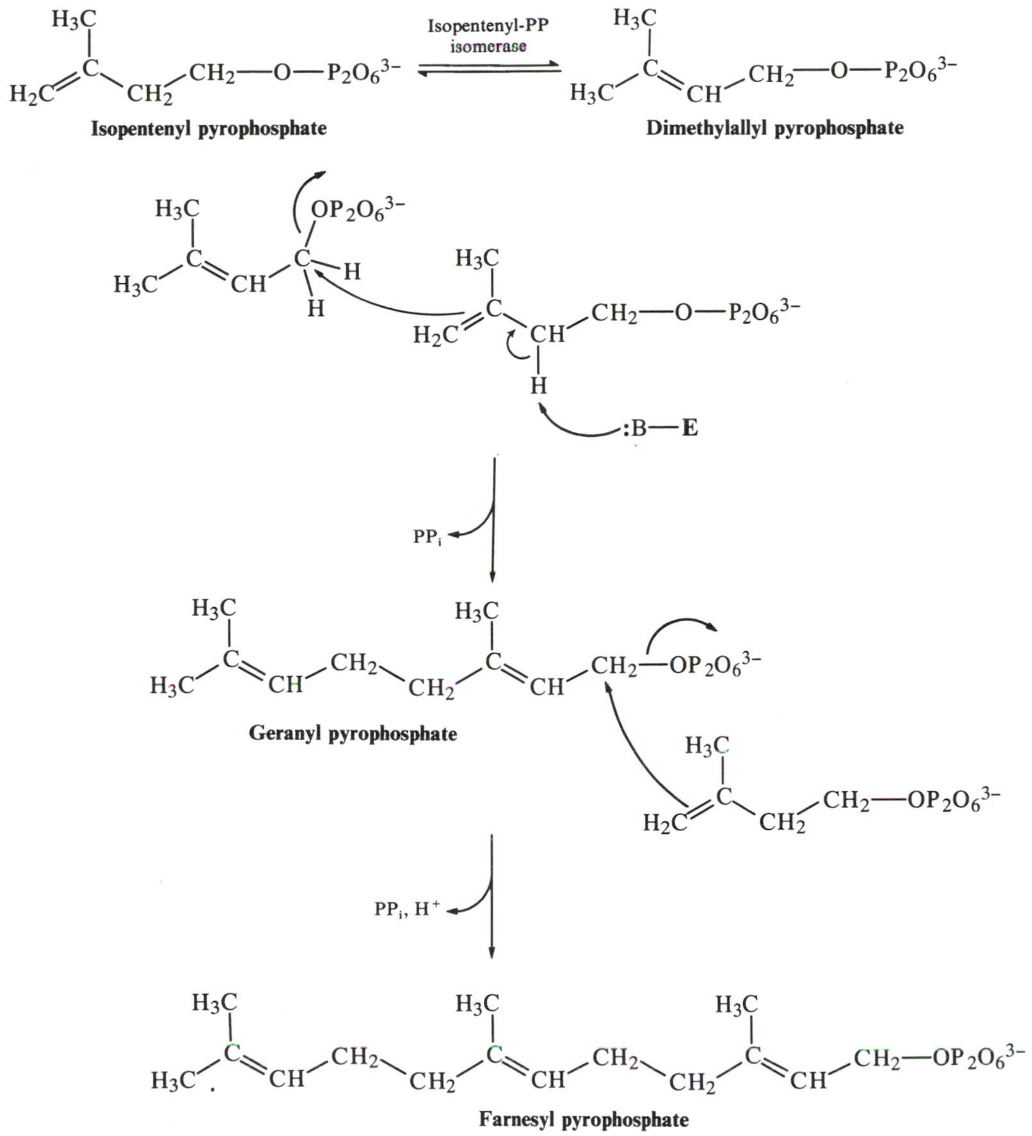

Isopentenyl pyrophosphate ⇌ (Isopentenyl-PP isomerase) Dimethylallyl pyrophosphate

Geranyl pyrophosphate

Farnesyl pyrophosphate

Figure 28-7

Steps in the Conversion of Isopentenyl-PP into Farnesyl-PP. Isopentenyl-PP isomerase catalyzes the conversion of isopentenyl-PP into its equilibrium mixture with dimethylallyl-PP, as described in Chapter 20. A carbon–carbon bond is then formed between dimethylallyl-PP and isopentenyl-PP by the transfer of the dimethylallyl group to the double bond of isopentenyl-PP. This is catalyzed by an allyl transferase in a reaction in which pyrophosphate is eliminated from dimethylallyl-PP. A second allyl transfer, a geranyl transfer, in the reaction of geranyl-PP with a second molecule of isopentenyl-PP produces farnesyl-PP.

cyclobutyl carbocation shown above is well precedented in the chemistry of carbocations. However, the mechanism of this interesting reaction is still not known.

Squalene undergoes cyclization to lanosterol according to Figure 28-9. Epoxidation of squalene is the first step, and the epoxide gives rise to a carbocation that undergoes cyclization through the formation of four carbon–

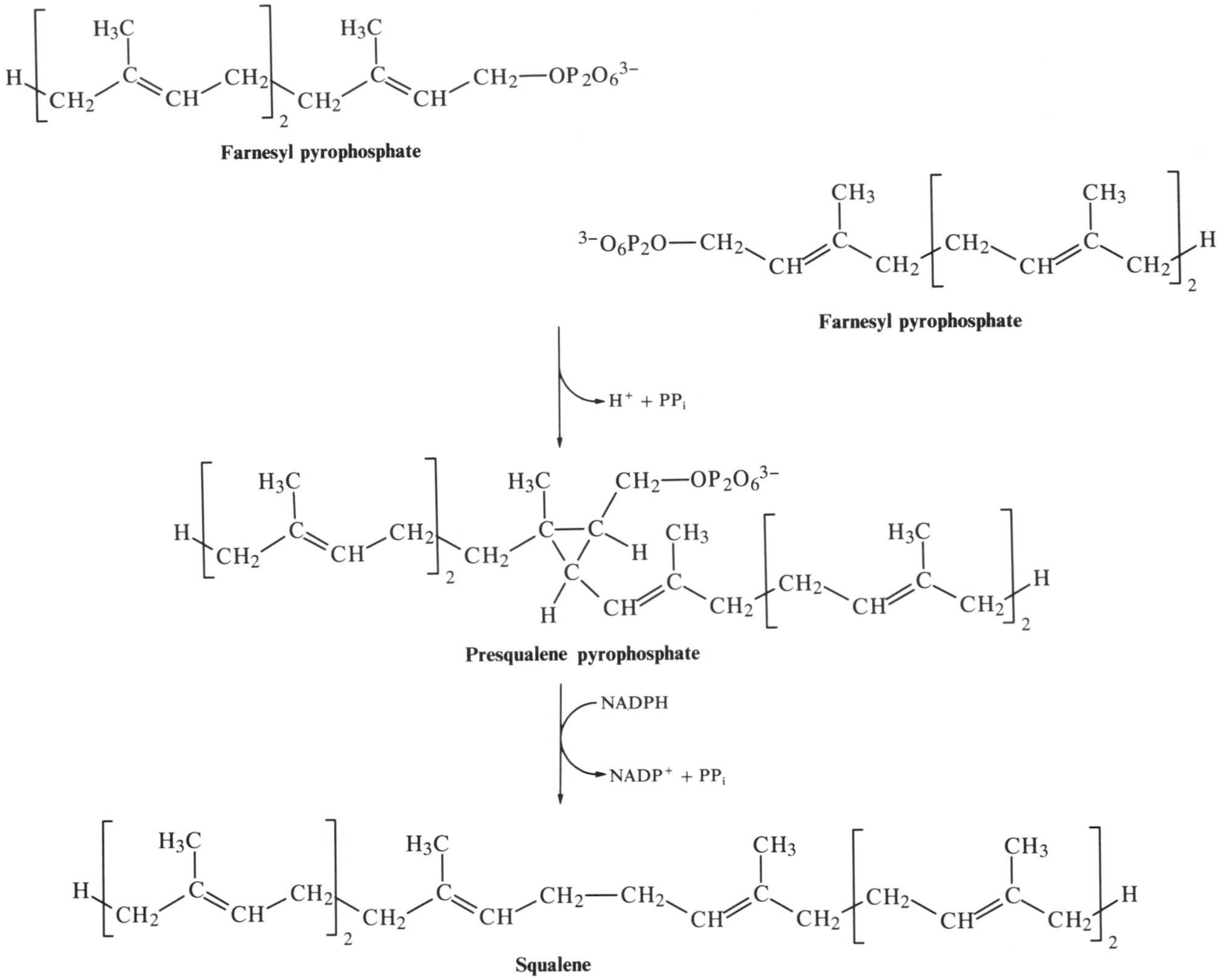

Farnesyl pyrophosphate

Farnesyl pyrophosphate

$H^+ + PP_i$

Presqualene pyrophosphate

NADPH

$NADP^+ + PP_i$

Squalene

Figure 28-8

Synthesis of Squalene from Two Molecules of Farnesyl-PP. Squalene is produced in the head-to-head coupling of two molecules of farnesyl-PP. The process requires two enzymes and the formation of presqualene-PP as an intermediate. The first enzyme catalyzes the elimination of PP_i from a molecule of farnesyl-PP and the coupling of the farnesyl allylic carbocation with the 2,3-double bond of the second molecule of farnesyl-PP to form presqualene-PP. The second enzyme catalyzes the elimination of PP_i, the skeletal rearrangement of the carbocationic intermediate, and reduction by NADPH to squalene.

carbon bonds to give a new cation. This rearrangement produces the basic carbon skeleton of cholesterol, and the new carbocation formed in the cyclization provides the driving force for the series of hydride and methyl migrations through which lanosterol is formed in the last step of Figure 28-9. It is not known whether initial cyclization of the protonated form of squalene-2,-3-oxide shown in Figure 28-9 proceeds in a single concerted step or in two or more steps. The carbocation shown as the initial cyclized species undergoes a series of 1,2-hydride shifts and 1,2-methyl group shifts culminating in the loss of a proton. These latter processes almost certainly proceed in a series of steps, with a new carbocationic intermediate formed in each step. The proton loss ends the process with the formation of lanosterol, a stable compound.

Lanosterol is converted in the endoplasmic reticulum into cholesterol through a series of about twenty enzymatic reactions. Three methyl groups are removed, double bonds are rearranged, and the 24, 25-double bond is reduced in this process. In demethylation reactions, the methyl groups are oxygenated to carboxylic acids and removed through decarboxylation. Recall that carbon

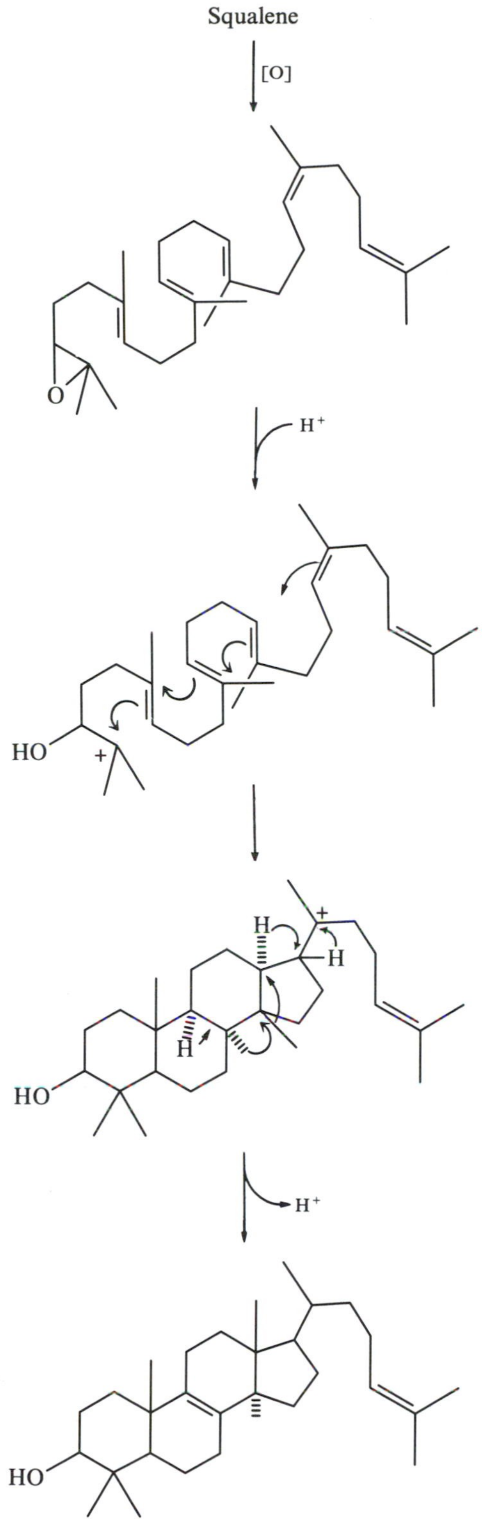

Squalene

[O]

H^+

H^+

Lanosterol

Figure 28-9

Steps in the Conversion of Squalene into Lanosterol. The cyclization of squalene begins with epoxidation of a double bond by squalene monooxygenase to form squalene 2,3-epoxide. The epoxide then undergoes cyclization to lanosterol. Protonation of the epoxide leads to a tertiary carbocation that cyclizes to a second tertiary carbocation in a process that leads to the formation of all four sterol rings. The second carbocation undergoes a series of 1,2-hydride and 1,2-methyl transfer reactions and finally eliminates a proton to form lanosterol.

dioxide can be eliminated from a carboxylic acid under physiological conditions only when there is a means for stabilizing the electron pair within the substrate (Chapter 17). Therefore, in the demethylation of lanosterol, carbonyl groups are introduced sequentially into the molecule in coordination with the oxidation of methyl groups, and these carbonyl groups facilitate decarboxylation.

Table 28-1

Composition of human lipoproteins

	PERCENTAGE OF DRY WEIGHT			
Type	**Protein**	**Cholesterol***	**Triglyceride**	**(%)Phospholipid**
Chylomicron	1–2	2–4	85–95	4–6
VLDL	6–10	19–29	45–65	15–22
LDL	18–22	50–58	3–9	16–25
HDL	45–55	18–25		26–32

*Total cholesterol, including esterified cholesterol.

Regulation of Cholesterol Biosynthesis

The first committed step of cholesterol biosynthesis is the reduction of HMG CoA to mevalonate. This is the major point at which the rate of cholesterol biosynthesis is controlled in higher animals and human beings. A high concentration of cholesterol in the cell inhibits the activity of HMG CoA reductase by two means: it decreases the rate at which the enzyme is synthesized by the cell, and it increases the rate at which the enzyme is degraded by the cell. Both phenomena reduce the activity of HMG CoA reductase, thereby reducing the rate at which cholesterol is produced.

The degradation of HMG CoA reductase is also regulated by phosphorylation and dephosphorylation of the enzyme. A specific protein kinase, HMG CoA reductase kinase, is activated through cAMP-dependent phosphorylation by ATP. The activated phosphorylated kinase catalyzes the phosphorylation of HMG CoA reductase by ATP. Phosphorylated HMG CoA reductase is very labile to degradation, so that a sudden increase in the concentration of cAMP and ATP in the cell will reduce the activity of HMG CoA reductase and the rate of cholesterol biosynthesis within a few minutes.

Plasma Lipoproteins

The major lipids found in blood include cholesterol, triglycerides, and phospholipids. The bloodstream transports cholesterol and triglycerides to organs and tissues; these molecules are almost insoluble in water and exist in blood as soluble complexes with proteins and phospholipids known as *lipoproteins.* The classes of lipoproteins differ with respect to protein constituents and lipid content. In this section, we describe the compositions and transformations of the major plasma lipoproteins.

Composition of Lipoproteins

Table 28-1 gives the abbreviated names of the major types of plasma lipoproteins, together with their compositions. Each lipoprotein is a complex of proteins, cholesterol, esterified cholesterol, triglycerides, and phospholipids. The types designated in Table 28-1 are chylomicron; VLDL, for very low density lipoprotein; LDL, for low-density lipoprotein; and HDL, for high-density lipoprotein. The densities of these particles are listed in Table 28-2 and range from above to below the density of water ($1.00\,g\,ml^{-1}$).

Table 28-2

Buoyant densities of lipoproteins

Lipoprotein	**Density ($g\,ml^{-1}$)**
Chylomicron	<0.96
VLDL	0.96–1.006
LDL	1.019–1.063
HDL	1.063–1.210

The lipoproteins are spherical or ellipsoidal particles containing proteins, esterified cholesterol, and triglycerides encased within a monolayer of phos-

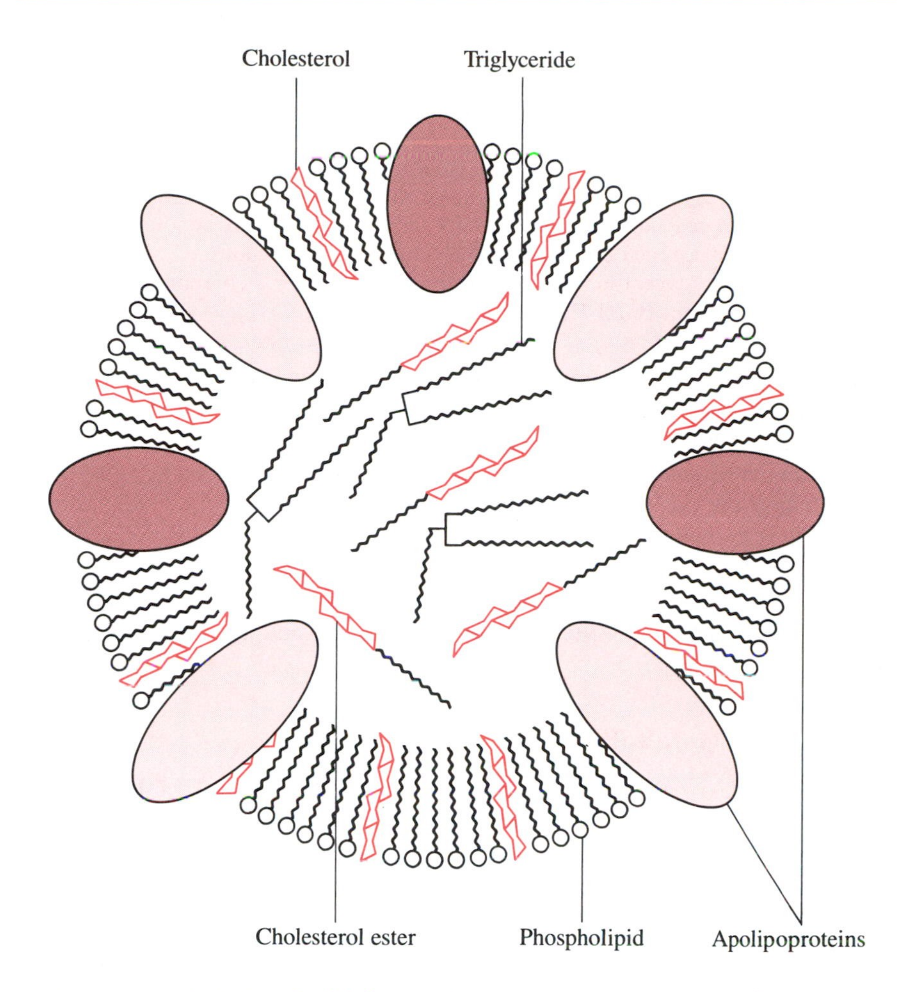

Figure 28-10

Composition and Organization of Lipoproteins. The lipoproteins vary in composition, as shown in Table 28-1. The organization of a lipoprotein, as shown here, consists of a spherical or oblate ellipsoidal outer envelope of phospholipids and cholesterol encasing the apolipoproteins, cholesterol esters, and triglycerides in the interior. Chylomicrons contain only traces of protein and cholesterol, and VLDL contains slightly more protein and some cholesterol. In human beings, LDL contains considerable protein and a great deal of cholesterol but little triglyceride, and HDL contains mainly protein and a small amount of cholesterol.

pholipids and cholesterol, which constitute the outer boundary (Figure 28-10). They differ from one another in composition; the lower density types contain high percentages of triglyceride and lower percentages of protein. Lipoproteins also differ with respect to the proteins that they contain. The purified proteins are known as *apolipoproteins* and range in molecular weight from 6500 to 550,000 (Table 28-3). Each lipoprotein contains a characteristic complement of

Table 28-3
Protein components of lipoproteins

Apolipoprotein	Lipoprotein	Molecular Weight	Plasma Concentration ($mg\ dl^{-1}$)
A-I	HDL	28,300	90–130
A-II	HDL	17,000	30–50
B-100	LDL	550,000	80–100
B-48	Chylomicrons	264,000	–
C-I	Chylomicrons/VLDL	6,500	4–7
C-II	Chylomicrons/VLDL	8,800	3–8
C-III	Chylomicrons/VLDL	8,750	8–15
D	HDL	33,000	10
E	Chylomicrons/VLDL	35,000	3–6
Cholesterol ester transfer protein	HDL	74,000	–

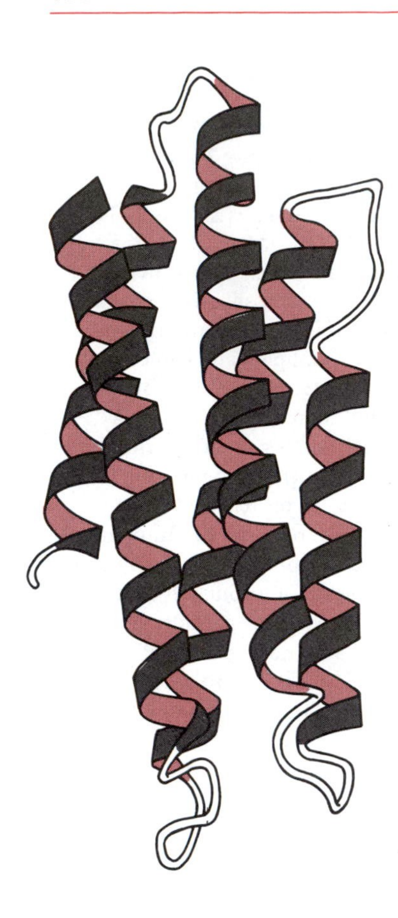

Figure 28-11

The Structure of an Apolipoprotein. The three-dimensional structure of apolipoprotein III from the insect *Locusta migratoria* consists of α-helices in a bundle. The bundle is illustrated here by a ribbon diagram. The helices are amphipathic; that is, each has polar amino acid side chains on one helix face and apolar amino acid side chains on another helix face. The helix bundle is formed through the association of apolar helical faces. The structure of apolipoprotein E has recently been found to be similar. (After a computer-generated drawing from Hazel Holden.)

apolipoproteins. The very small amount of protein in chylomicrons includes lipoproteins of types A, B-48, and C. VLDL contains types B-100, A, and C. LDL contains mainly apolipoprotein B-100, with a trace of type E, a receptor recognition factor. HDL contains mainly apolipoproteins of type A, with traces of type C and the receptor recognition factor type E.

Little is known about the structures and properties of lipoproteins, which are difficult to study in purified form in aqueous media. The three-dimensional structures are beginning to become known through X-ray crystallography. A ribbon drawing of the α-carbon chain of an apolipoprotein is shown in Figure 28-11.

Lipoprotein Metabolism

Triglycerides are carried into the bloodstream from the liver in VLDL and from the intestine in chylomicrons. The triglycerides are transported to other organs and tissues by these lipoproteins. Triglycerides in chylomicrons are largely dietary, so that the number of chylomicrons increases after eating fats and decreases afterward. Triglycerides in VLDL originate in the liver and so are derived from fatty acids carried into the liver by the bloodstream or from

Figure 28-12 (on facing page)

The Role of Chylomicrons and VLDL in Transporting Triglycerides. The most-important function of chylomicrons and VLDL is the transport of triglycerides (TG) from the intestine and liver to peripheral cells through the blood stream. Part A shows that free fatty acids (FA) and monoglycerides (MG) are absorbed from the intestine by cells of the intestinal mucosa and converted into triglycerides (TG). These cells produce apolipoproteins A and B-48 and assemble them in the Golgi complex with triglycerides into chylomicrons. The chylomicrons enter the bloodstream and are carried to capillaries in the tissues, where lipoprotein lipase catalyzes the hydrolysis of triglycerides to free fatty acids, which are transported into peripheral cells, and to complexes of fatty acids and albumin in the bloodstream. The remnant, depleted of triglycerides and now containing a trace of apolipoprotein E, is absorbed by liver cells through the B-100:E receptors and degraded by liver lysosomes. The function of VLDL is similar to that of chylomicrons. As shown in part B, VLDL is produced in liver cells from triglycerides; cholesterol and cholesterol ester (Ch); and apolipoproteins B-100, C, and E. VLDL enters the bloodstream, migrates to tissues, and undergoes the same triglyceride degradation process at capillary surfaces shown in part A for chylomicrons. Some of the free fatty acids are absorbed by cells and some are bound by albumin in the bloodstream. The remnants are depleted of triglycerides and apolipoprotein C, which seems to dissociate when triglycerides are degraded. The VLDL remnants are absorbed from the bloodstream by liver B-100:E receptors and converted into LDLs, which are released into the bloodstream. LDL is the primary carrier of cholesterol and cholesterol ester in the bloodstream of human beings; the LDL is absorbed by B-100:E receptors of peripheral cells and the liver, and it is degraded to cholesterol and amino acids by lysosomes.

biosynthesis in the liver. The primary function of VLDL and chylomicrons is to carry triglycerides to tissues, where they are hydrolyzed to free fatty acids and can be used as energy sources.

Figure 28-12 outlines the pathways through which chylomicrons and VLDL are produced and degraded. Intestinal mucosal cells absorb monoglycerides (MG) and free fatty acids (FA), convert them into triglycerides (TG), and assemble them with apolipoprotein B-48 and types A, C, and E to form chylomicrons, which are released into the bloodstream. Lipoprotein lipase in the capillaries of tissues hydrolyzes triglycerides of chylomicrons to free fatty acids,

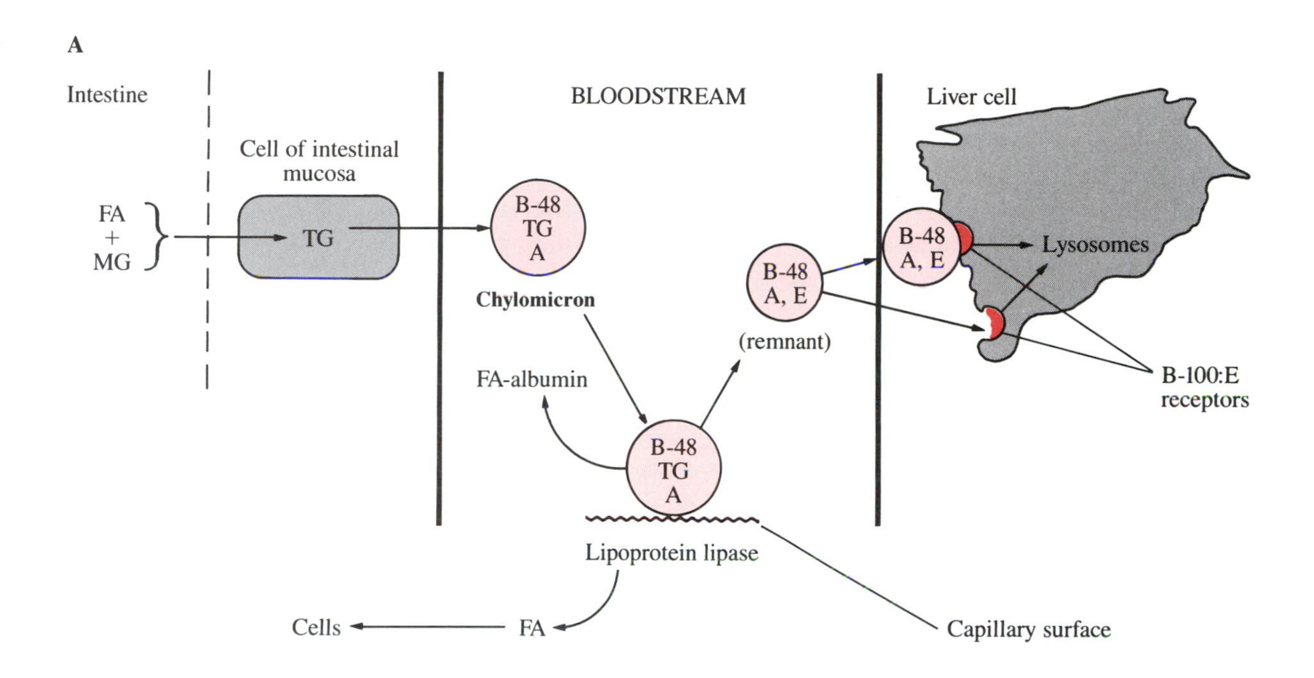

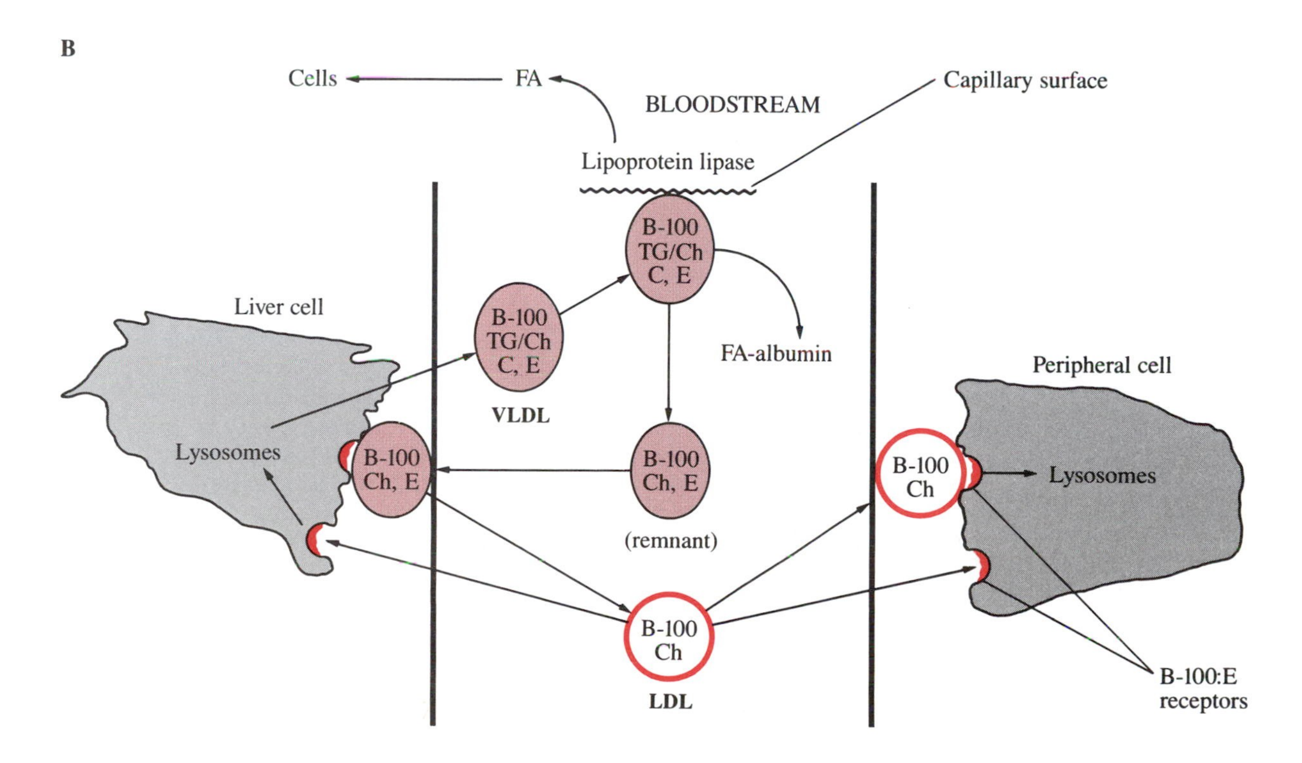

which are complexed with albumin and absorbed by cells. The remnants of chylomicrons in the bloodstream are absorbed by liver cells and degraded by lysosomes. VLDL is produced in the liver and follows a pathway of triglyceride distribution and degradation similar to that of chylomicrons. However, the remnant resulting from depletion of triglycerides has two fates: it can either be absorbed by the liver and degraded or it can be further degraded to lipoprotein B-100 and converted into LDL. Thus, VLDL can be a precursor of LDL by virtue of having the same B-type of lipoprotein (B-100).

The main function of human LDL is to carry cholesterol in the bloodstream. (In animals such as the rat and cow, HDL is the major carrier of cholesterol.) Note in Table 28-1 that, among human lipoproteins, LDL is the richest in cholesterol. Peripheral cells absorb LDL by means of receptors for lipoproteins B-100 and E, and the absorbed LDL is degraded by lysosomes. Peripheral cells do not degrade cholesterol; however, in adrenal cells it is used in the biosynthesis of steroid hormones.

The dynamics of LDL metabolism include the absorption and degradation of excess LDL by the liver (Figure 28-13). About two-thirds of normal LDL degradation occurs in the liver through the LDL receptor-mediated pathway, which M. Brown and J. Goldstein discovered. They received the Nobel Prize in 1985 for their work on LDL receptors (see Box 28-1). The liver LDL receptors are located in regions of the plasma membrane known as "coated pits," which are enriched in the protein clathrin. The receptor binds lipoprotein B-100, and this is followed by endocytosis of the LDL. The receptor dissociates from the endocytic vesicle and returns to the membrane, where it participates in

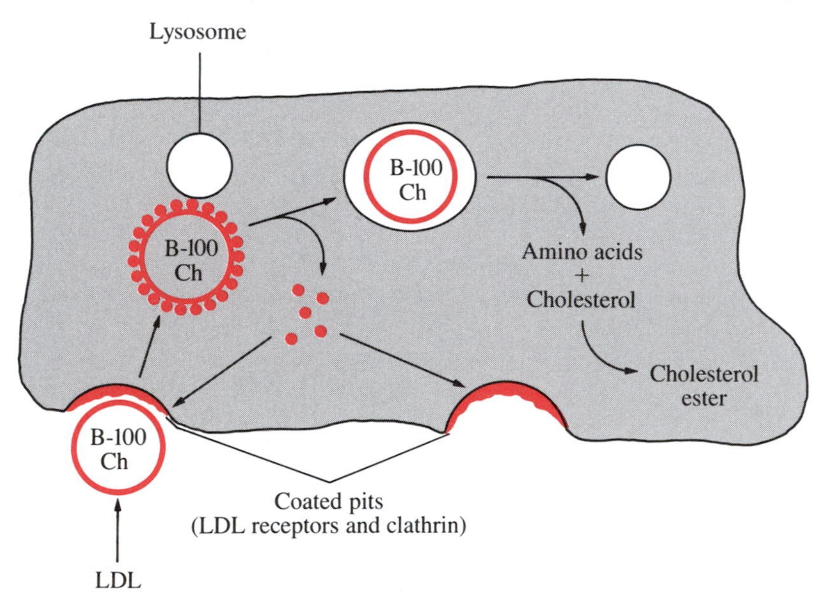

Figure 28-13

The Absorption and Degradation of LDL by Liver Cells. The LDL receptors of liver cells are in localized areas known as coated pits, which are formed by association of the protein clathrin. The receptors bind the LDL protein B-100. The bound LDL is encapsulated by the membrane and other components of the coated pit to form a vesicle within the cell. The LDL receptors are released from the vesicle, which undergoes fusion with a lysosome, and the receptors return to the coated pits. The enzymes within the lysosomes degrade LDL to amino acids and cholesterol, which are released within the cell. Cholesterol released within liver cells has several fates. Part of it is metabolized to bile salts, which are transported along with unmetabolized cholesterol into the gall bladder for excretion. Part of the cholesterol is used in the initial assembly of HDL (Figure 28-14).

BOX 28-1

FAMILIAL HYPERCHOLESTEROLEMIA

Atherosclerosis is the development of fibrous plaques in the arteries. These plaques become enlarged and calcified and eventually cause a narrowing and occlusion of the artery. Cholesterol is a major constituent of plaques, the formation of which can be caused by abnormally large amounts of cholesterol and other lipids in the blood, by hypertension, by smoking, and by combinations of these factors. Atherosclerosis is the most-common cause of coronary heart disease.

Grossly excessive cholesterol in the diet can lead to modest increases in blood cholesterol. However, very high blood levels of cholesterol often result from genetic factors. Excess saturated fats in the diet may exacerbate a person's genetic predisposition to overproduce or accumulate cholesterol. The molecular basis of all the genetic abnormalities are probably not yet known; however, some important ones were uncovered by Brown and Goldstein in their work on LDL receptors and hypercholesterolemia. These included the absence of coated pits, the absence of receptors, and the inability of coated pits to undergo endocytosis.

Familial hypercholesterolemia is caused by deficiencies or lack of LDL receptors in the liver. People who are homozygous for this deficiency have no functional hepatic LDL receptors, and their blood LDL levels of cholesterol are from four to six times greater than normal. Such people are born with atherosclerosis and have coronary heart disease in childhood. People who are heterozygous have from two to four times the normal blood LDL and are at very high risk of heart attack; 85% of them suffer a heart attack by the age of 60. One in five-hundred people is heterozygous for familial hypercholesterolemia.

The molecular defect in familial hypercholesterolemia is in the liver LDL receptor. A homozygous person produces defective receptors; in a heterozygous person, half of the receptors are normal and half are defective.

A number of drugs have recently been developed that are highly effective inhibitors of HMG CoA reductase and inhibit cholesterol biosynthesis. Most of these drugs are structurally similar to compactin, which is produced by certain fungi.

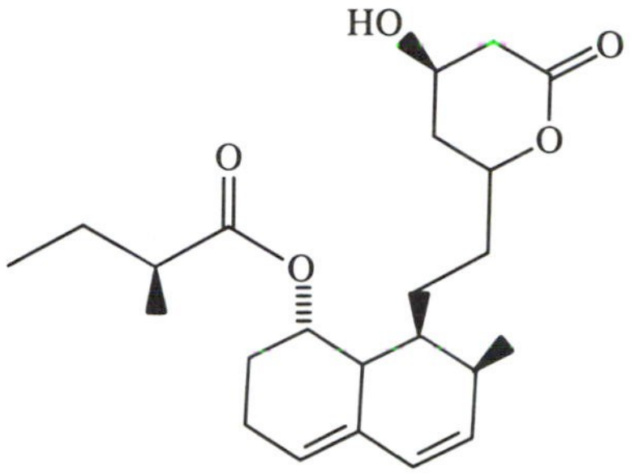

HO O O O O

Compactin

Lovastatin acid differs from compactin by the presence of an additional methyl group, and an opened lactone ring. It is in use for the treatment of patients with hypercholesterolemia.

HO C O OH OH O O

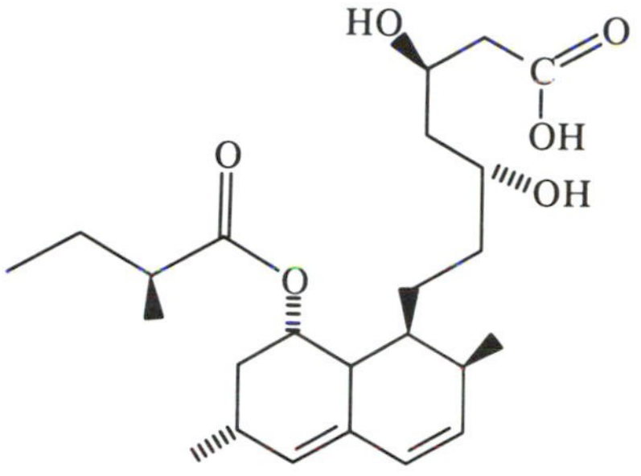

Lovastatin acid

Inhibition of β-hydroxymethylglutaryl CoA reductase by these drugs reduces the rate of cholesterol biosynthesis, and this tends to decrease blood cholesterol levels. The drugs also increase the number of liver LDL receptors and thereby increase the rate at which cholesterol is removed from the blood and excreted. The dual effects of these drugs lead to significant reductions in blood cholesterol, even for people who are heterozygous for familial hypercholesterolemia. Homozygous people have no functioning liver LDL receptors and cannot remove excess cholesterol from the blood by endocytosis into liver cells. The compactinlike drugs cause a modest reduction in cholesterol levels, but these people can clear cholesterol from the blood through the liver only after receiving liver transplants.

the formation of new coated pits. The vesicle fuses with a lysosome, which degrades LDL to amino acids and cholesterol. The released cholesterol is used as an essential component in the biosynthesis of membranes.

The concentration of cholesterol in the blood and tissues is very closely regulated. Excess cholesterol in cells exerts three regulatory effects on cholesterol metabolism:

1. It inhibits the activity of 3-hydroxy-3-methylglutaryl CoA reductase in cells by inhibiting the biosynthesis of this enzyme and enhancing the rate at which it is degraded.
2. It also inhibits the production of LDL receptors.

3. It stimulates the reesterification of cholesterol. Reesterification is catalyzed by the enzyme acyl CoA cholesterol acyl transferase, which is activated by free cholesterol.

Thus, excess cholesterol absorbed by cells reduces the intracellular production of cholesterol and the absorption of cholesterol from the bloodstream, while stimulating its esterification. A low level of cholesterol stimulates cholesterol biosynthesis. In this way, the cholesterol biosynthetic and absorptive capacities of cells are held in balance.

The degradation of LDL by liver microsomes includes the hydrolysis of cholesterol esters to cholesterol catalyzed by esterases. In addition to inhibiting cholesterol biosynthesis and uptake by liver cells, the excess cholesterol is converted in the liver into bile salts such as glycocholate and taurocholate (Chapter 6, Figure 6-2), which are stored in the gall bladder. Bile salts are detergents that have the capacity to solubilize cholesterol, and excess cholesterol is excreted with bile salts through the gall bladder. Thus, the liver excretes excess cholesterol acquired through the absorption and digestion of LDL. The liver is the only internal organ that can both degrade and excrete cholesterol; however, significant amounts of cholesterol are excreted through the skin.

High-density lipoproteins are assembled in liver and intestinal cells from cholesterol released in the degradation of LDL. Type-A apolipoproteins are the principal components of HDL, which also contains apolipoprotein D and the cholesterol transfer protein. Lecithin: cholesterol acyltransferase in the bloodstream catalyzes the esterification of cholesterol in HDL by acyl group transfer from the phospholipids in HDL according to equation 1.

$$
\begin{array}{l}
R_2{-}\overset{\overset{\displaystyle O}{\|}}{C}{-}O{-}\overset{\overset{\displaystyle H_2C-O-\overset{O}{\overset{\|}{C}}-R_1}{|}}{\underset{\underset{\displaystyle H_2C-O-\underset{\underset{\displaystyle OCH_2CH_2N^{+}(CH_3)_3}{|}}{\overset{\overset{O}{\|}}{P}}-O^{-}}{|}}{C}}{-}H \quad + \ HO{-}\text{cholesterol} \ \rightleftharpoons \\[2em]
HO{-}\overset{\overset{\displaystyle H_2C-O-\overset{O}{\overset{\|}{C}}-R_1}{|}}{\underset{\underset{\displaystyle H_2C-O-\underset{\underset{\displaystyle OCH_2CH_2N^{+}(CH_3)_3}{|}}{\overset{\overset{O}{\|}}{P}}-O^{-}}{|}}{C}}{-}H \quad + \ R_2{-}\overset{\overset{\displaystyle O}{\|}}{C}{-}O{-}\text{cholesterol} \qquad (1)
\end{array}
$$

Figure 28-14 illustrates the transformations of HDL. The initial aggregate of phospholipids, apolipoproteins A_1, A_2, and D, and cholesterol is disk-shaped HDL. Apolipoprotein A_1 stimulates the activity of lecithin: cholesterol acyltransferase, which converts the cholesterol into cholesterol ester. Upon esterification, the cholesterol ester is transferred from the outer lipid envelope into the lipid core of the particle, and the lipoprotein becomes spherical HDL.

The fate of HDL is not understood, but it is a subject of intensive investigation. A high level of HDL in the blood is correlated with the absence of atherosclerosis, and one hypothesis is that HDL somehow prevents plaque formation, perhaps by taking up free cholesterol. HDL is known to extract cholesterol from cells and other lipoproteins in the presence of lecithin: cholesterol acyltransferase. After sequestering cholesterol ester, HDL may facilitate its degradation and the excretion of excess cholesterol.

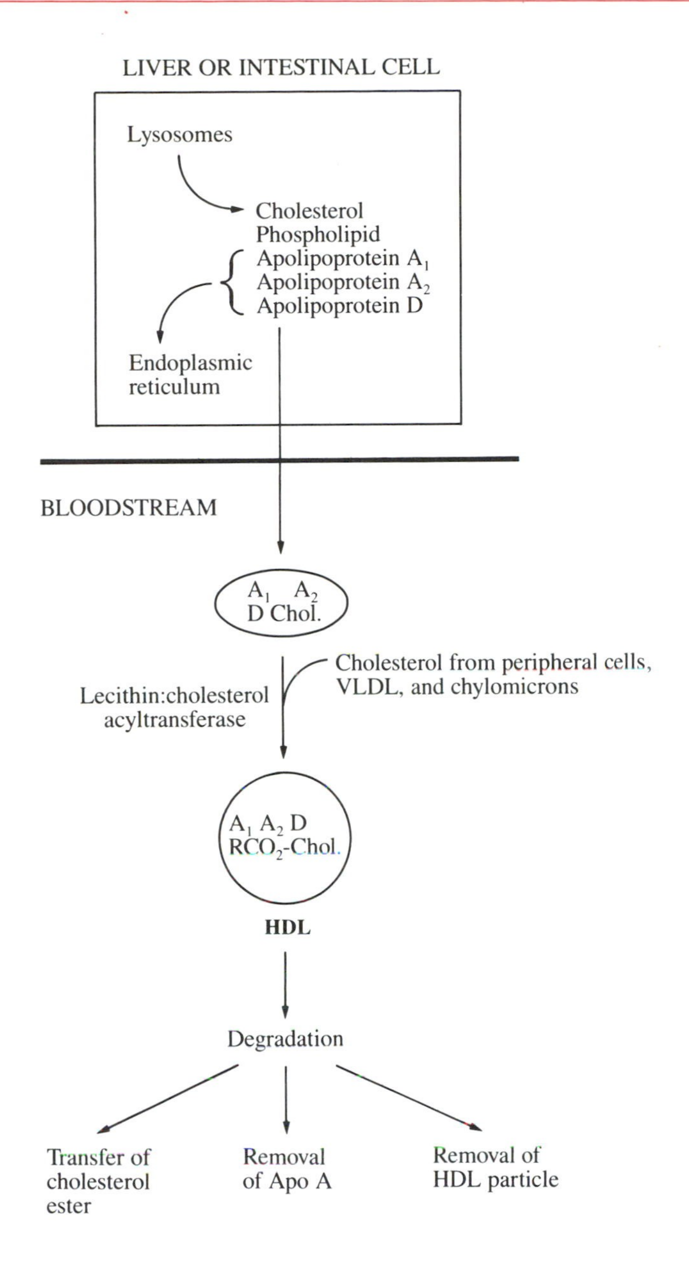

Figure 28-14

Steps in the Processing of HDL. HDL is initially assembled from cholesterol and apolipoproteins A and D within liver and intestinal cells and released into the bloodstream as discoid structures. Apolipoprotein A_1 in HDL activates lecithin: cholesterol acyltransferase in the blood stream, the enzyme that catalyzes fatty acyl group transfer to cholesterol from phospholipids in the outer envelope of the particle. Cholesterol ester (RCO_2-chol.) enters the central lipid core of the particle. This process is accompanied by a morphological change to spherical HDL. HDL can absorb and esterify cholesterol from cells and other lipoproteins. The details of the fate of HDL are unknown. HDL may supply cholesterol to cells and may return excess cholesterol to the liver for excretion.

Metabolism of Cholesterol

Production of Bile Acids

Cholesterol is converted into bile salts in the liver, and the bile salts are excreted through the gall bladder. The enzymology of this process is not fully known. It is largely an oxidative process that probably requires several mixed-function oxidases of the type introduced in Chapter 24. The chemistry of the transformations also requires several other types of chemical reactions, including carbon–carbon bond cleavages.

Figure 28-15 shows some of the steps in the conversion of cholesterol into glycocholate, a bile salt. The process requires hydroxylation of carbon atoms 7 and 12, inversion of configuration at carbon-3, and reduction of the double bond to form the first intermediate shown. Further steps remove carbon atoms 25, 26, and 27 and convert carbon-24 into the carboxyl group of cholic acid, the second

Figure 28-15

Intermediates in the Conversion of Cholesterol into the Bile Salt Glycocholate.

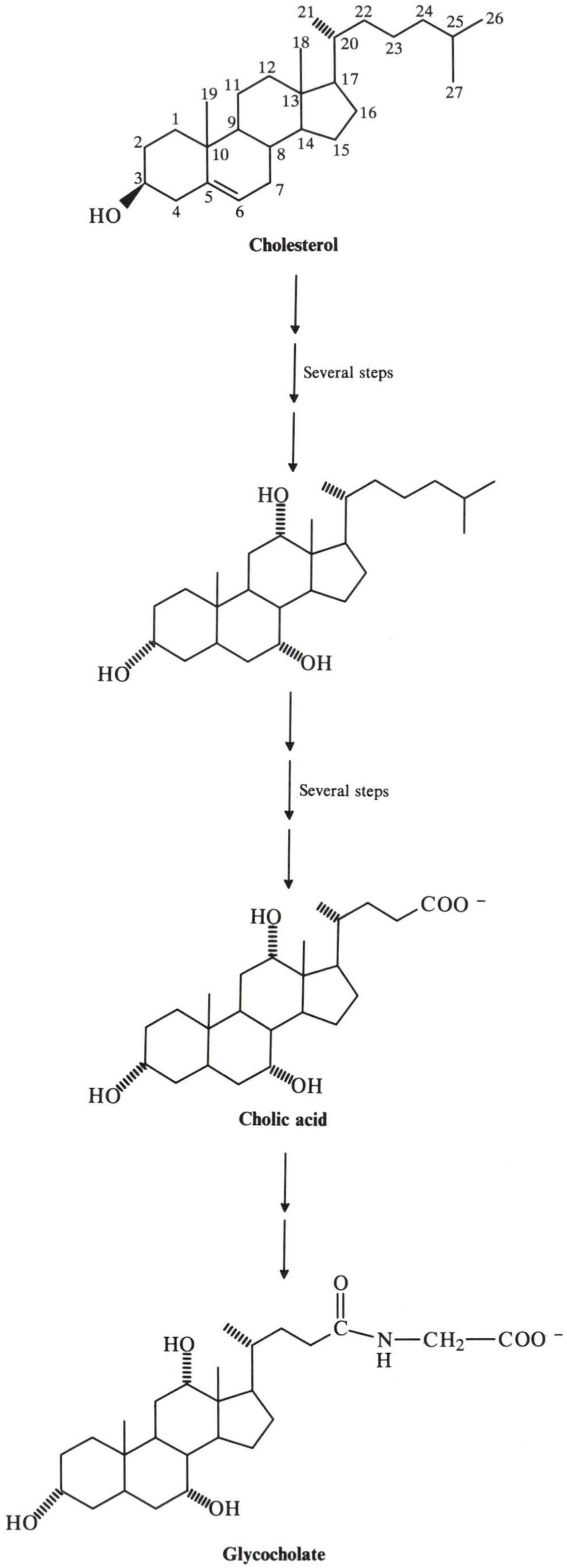

Cholesterol

Several steps

Several steps

Cholic acid

Glycocholate

intermediate in Figure 28-15. Cholic acid is converted into glycocholate in two steps that require CoASH, ATP, and glycine. ATP is cleaved to AMP and PP_i in this process.

Biosynthesis of Steroid Hormones

Steroid hormones are produced from cholesterol in the adrenal glands of higher animals. Figure 28-16 shows some of the early reactions and intermediates in the

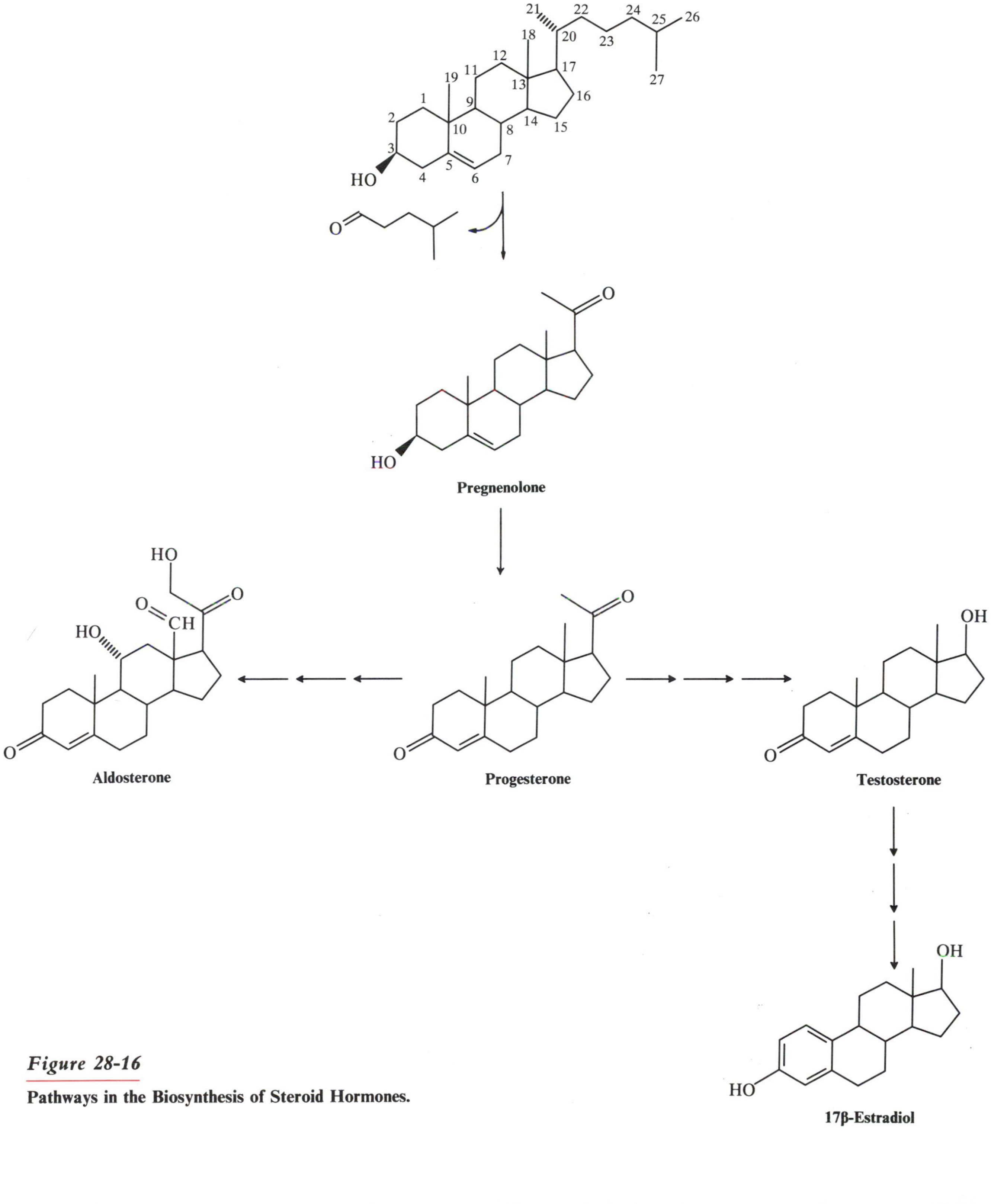

Pregnenolone

Aldosterone

Progesterone

Testosterone

17β-Estradiol

Figure 28-16

Pathways in the Biosynthesis of Steroid Hormones.

production of steroid hormones from cholesterol. The process requires several mixed-function oxidations (hydroxylations); the hydroxyl groups introduced by these reactions facilitate other transformations leading to the hormones.

The mitochondria of adrenal cells contain the enzyme desmolase, which catalyzes the cleavage of carbons 22 through 27 from cholesterol, producing pregnenolone and isocaproic aldehyde. Pregnenolone is dehydrogenated to progesterone in the endoplasmic reticulum of adrenal cells. Progesterone is the precursor of aldosterone, testosterone, and estradiol. The conversion of progesterone into aldosterone in the adrenal cortex requires a series of three hydroxylation reactions and a dehydrogenation. The conversion of progesterone into testosterone requires hydroxylation at carbon-17 in the adrenal glands, and the resulting 17α-hydroxyprogesterone is further processed to testosterone, the male sex hormone, in the gonads. The conversion of testosterone into estradiol, the female sex hormone, is catalyzed by a complex of enzymes known as the aromatase system, which is present only in the endoplasmic reticulum of ovaries and placenta.

Sphingolipids

Sphingolipids are found in most eukaryotic cells but not in prokaryotes. They are major lipids in the myelin sheath of nerve fibers; they are also minor constituents of other lipid structures in eukaryotes, including the plasma lipoproteins of animals. The sphingophospholipids are structural analogs of phospholipids, in which a sphingosine replaces glycerol as the esterifying component for the fatty acyl and phosphoryl groups. Two important sphingosines are 4-sphingenine and sphinganine. 4-Sphingenine is commonly known as sphingosine, and sphinganine is known as dihydrosphingosine. In sphingolipids, the terminal hydroxymethyl group is esterified to phosphorylcholine, and fatty acyl groups are bonded to the amino group and secondary hydroxyl group.

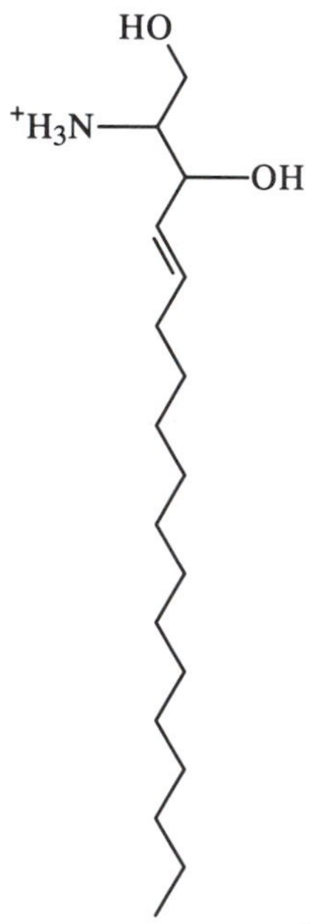

HO | $^{+}H_3N$ | OH

4-Sphingenine

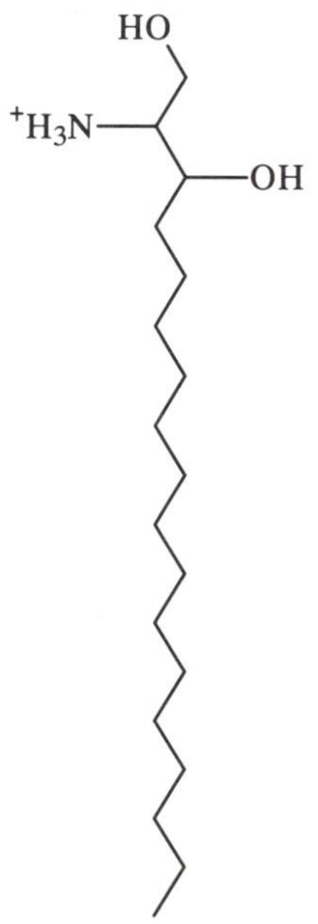

HO | $^{+}H_3N$ | OH

Sphinganine

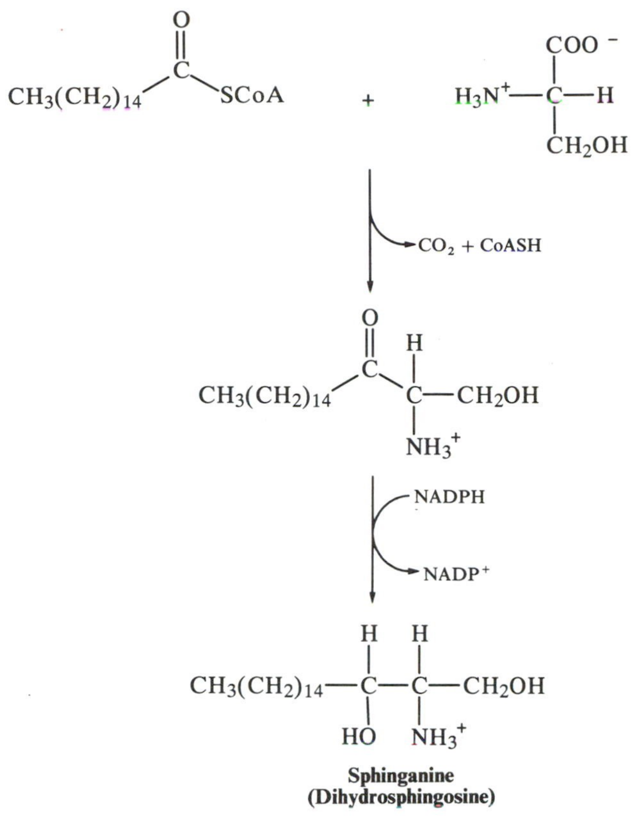

$CH_3(CH_2)_{14}-C(=O)-SCoA$ + $H_3N^+-CH(COO^-)-CH_2OH$

$\downarrow$ → CO_2 + CoASH

$CH_3(CH_2)_{14}-C(=O)-CH(NH_3^+)-CH_2OH$

$\downarrow$ NADPH → $NADP^+$

$CH_3(CH_2)_{14}-CH(OH)-CH(NH_3^+)-CH_2OH$

Sphinganine
(Dihydrosphingosine)

Figure 28-17

Biosynthesis of Sphinganine. Palmitoyl CoA and serine are the building blocks from which sphinganine (dihydrosphingosine) is assembled in two steps. 3-Ketosphinganine synthase catalyzes the condensation of these molecules, and 3-ketosphinganine reductase catalyzes the reduction to sphinganine.

The biosynthesis of sphinganine begins with the condensation of palmitoyl CoA and serine to 3-ketosphinganine, as shown in Figure 28-17. 3-Ketosphinganine is reduced to sphinganine. In some organisms, the sequence in Figure 28-17 starting with *trans*-2-palmitoleyl CoA leads directly to 4-sphingenine (sphingosine), but in other organisms an *N*-acylsphinganine is dehydrogenated to the corresponding *N*-acylsphingosine.

Sphingomyelin is produced from 4-sphingenine in two steps as shown in Figure 28-18. *N*-Acylation of 4-sphingenine by a fatty acyl CoA produces a ceramide, which accepts the choline phosphoryl group from phosphatidyl choline to form sphingomyelin and a diacylglycerol.

Ceramides are also intermediates in the synthesis of glycosphingolipids, as well as sphingomyelins. The glycosphingolipids are ceramides glycosylated with oligosaccharides of galactose, glucose, *N*-acetylgalactosamine, *N*-acetylneuraminic acid, and related compounds, some of which may be sulfated. The substrates for glycosyl transfer to the hydroxymethyl group of ceramides are the UDP-sugars, and the glycosyl units are added one at a time to build up the oligosaccharides (Chapter 15). These reactions take place in the Golgi complex of eukaryotic cells.

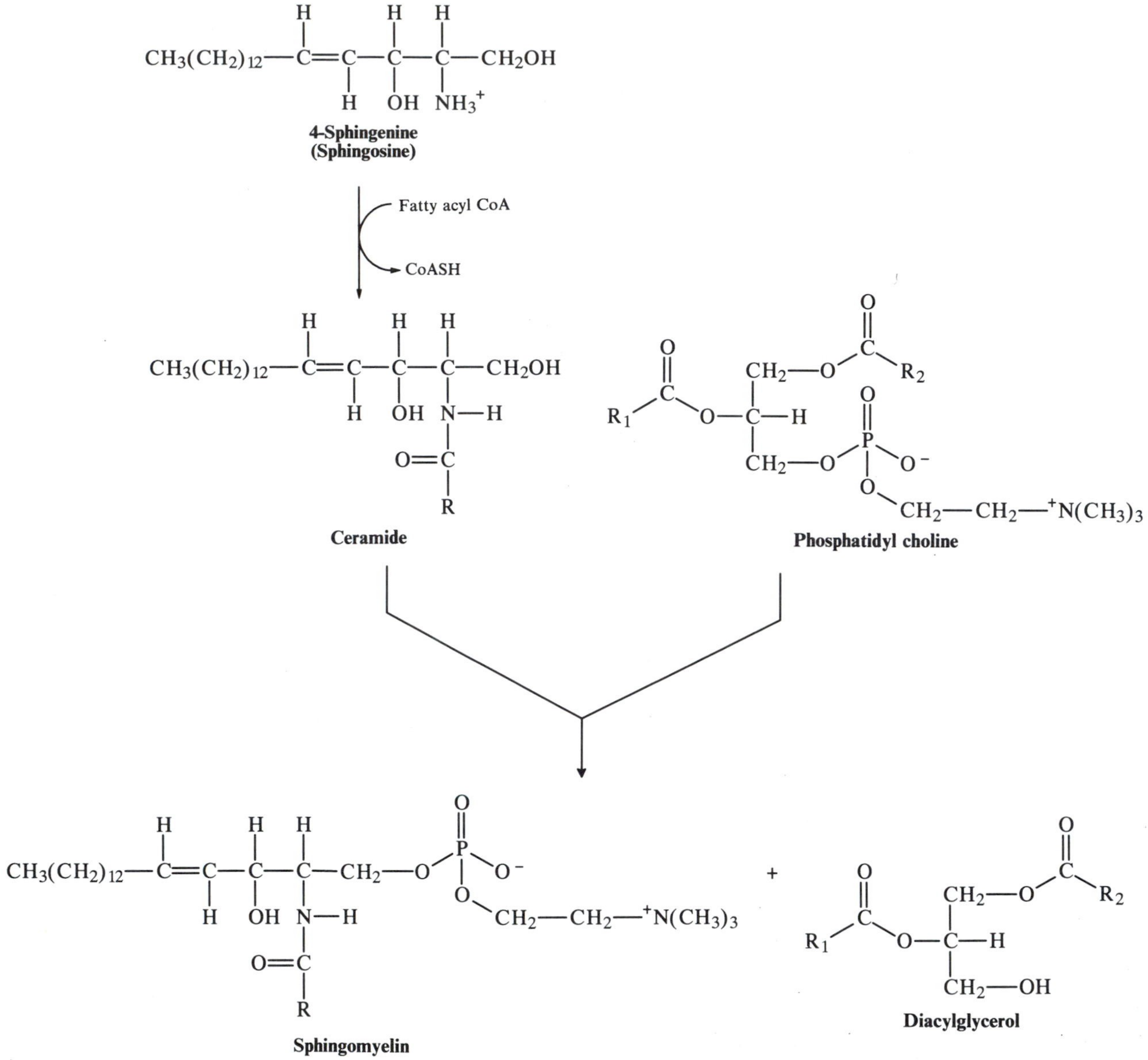

$CH_3(CH_2)_{12}{-}CH{=}CH{-}CH(OH){-}CH(NH_3^+){-}CH_2OH$

4-Sphingenine (Sphingosine)

Fatty acyl CoA → CoASH

$CH_3(CH_2)_{12}{-}CH{=}CH{-}CH(OH){-}CH(NH{-}C({=}O){-}R){-}CH_2OH$

Ceramide

$R_1{-}C({=}O){-}O{-}CH(CH_2{-}O{-}C({=}O){-}R_2)(CH_2{-}O{-}P({=}O)(O^-){-}O{-}CH_2{-}CH_2{-}{}^+N(CH_3)_3)$

Phosphatidyl choline

↓

$CH_3(CH_2)_{12}{-}CH{=}CH{-}CH(OH){-}CH(NH{-}C({=}O){-}R){-}CH_2{-}O{-}P({=}O)(O^-){-}O{-}CH_2{-}CH_2{-}{}^+N(CH_3)_3$

Sphingomyelin

+

$R_1{-}C({=}O){-}O{-}CH(CH_2{-}O{-}C({=}O){-}R_2)(CH_2{-}OH)$

Diacylglycerol

Figure 28-18

Conversion of Sphingosine into Sphingomyelin.

Summary

Cholesterol biosynthesis begins with the self-condensation of acetyl CoA into acetoacetyl CoA, which is catalyzed by β-ketothiolase. Acetoacetyl CoA undergoes a second condensation with acetyl CoA to give β-hydroxymethylglutaryl CoA (HMG CoA), which is reduced to mevalonate by HMG CoA reductase. Mevalonate is phosphorylated to mevalonate-5-PP, which is dehydrated and decarboxylated to isopentenyl-PP in an ATP-dependent reaction catalyzed by mevalonate-5-PP decarboxylase. Isopentenyl-PP undergoes reversible isomerization to dimethylallyl-PP, and these two molecules are then condensed by farnesyl-PP synthase to farnesyl-PP by a mechanism that requires the elimination of PP_i from dimethylallyl-PP to form an allylic carbocation intermediate. Additional allylic carbocationic condensations eventually lead to squalene. Squalene monooxygenase catalyzes the oxygenation of squalene to squalene epoxide, which is cyclized to lanosterol by squalene cyclase. Lanosterol is converted into cholesterol, which is an important component of membranes and a precursor of steroid hormones.

Triglycerides and cholesterol are transported through the bloodstream by the chylomicrons, very low density lipoproteins (VLDL), low-density lipoproteins (LDL), and high-density lipoproteins (HDL). Chylomicrons and VLDL are primarily responsible for transporting triglycerides from the intestine and liver to peripheral cells, and, in human beings, LDL is primarily responsible for transporting cholesterol to the liver for degradation to bile salts and excretion through the gall bladder. In some animals, this is the primary function of HDL.

Sphingolipids are assembled from sphingosine (4-sphingenine), which is biosynthesized from palmitoyl CoA and serine. Acylation of the amino group by fatty acyl CoA, and acylation of the hydroxymethyl group by phosphatidylcholine leads to sphingomyelin, an important lipid in the myelin sheath of nerves.

ADDITIONAL READING

POULTER, C. D. 1990. Biosynthesis of non-head-to-tail terpenes: formation of 1′-1 and 1′-3 linkages. *Accounts of Chemical Research* 23: 70.

BROWN, M. S., and GOLDSTEIN, J. L. 1986. A receptor-mediated pathway for cholesterol homeostasis. *Science (Washington D.C.)* 232: 34–47.

HAVEL, R. J., GOLDSTEIN, J. L., and BROWN, M. S. 1980. Lipoproteins and lipid transport. *Metabolic Control and Disease,* ed. Bondy and Rosenberg. pp. 393–494.

GODBOURT, U., and NEUFELD, H. N. 1986. Genetic aspects of arteriosclerosis. *Arteriosclerosis* 6: 357–377.

PROBLEMS

1. Suggest a mechanism for the conversion of two molecules of farnesyl pyrophosphate into presqualene pyrophosphate through the intermediate formation and rearrangements of carbocationic intermediates. Carbocations can undergo 1,2-hydride shifts, 1,2-alkyl group shifts, 1,3-alkyl group shifts, and loss of a β-proton.

2. Suggest a reasonable series of enzymatic reactions to convert cholesterol into the intermediate of cholic acid biosynthesis, shown in Figure 28-15. Describe the substrate and cofactor requirements for each step in your hypothetical biosynthetic sequence. (Hint: see Chapters 16 and 24.)

3. The intermediate of cholic acid biosynthesis shown in Figure 28-15 is converted into cholic acid in a series of enzymatic steps. Suggest a sequence of enzymatic reactions that would lead to this transformation, and describe the substrate and cofactor requirements for each step. (Hint: see Chapters 16, 17, and 24.)

4. In the biosynthesis of steroid hormones, outlined in Figure 28-16, cholesterol is first converted into pregnenolone by an enzyme known as desmolase in mitochondria of adrenal cells. In this reaction, carbons 22 through 27 are cleaved away in the form of isocaproic aldehyde. The mechanism of this reaction is not known. Suggest how this transformation might proceed by proposing the structures of intermediates that could lead to pregnenolone. (Hint: see Chapters 17 and 24.)

5. Suggest a sequence of enzymatic reactions that would lead to the conversion of pregnenolone into progesterone and then into aldosterone. Describe the substrate and cofactor requirements for each reaction. (Hint: see Chapters 16, 20, and 24)

6. Suggest a sequence of enzymatic reactions that would lead to the conversion of progesterone into testosterone and then into estradiol. Describe the substrate and cofactor requirements for each reaction. (Hint: see Chapters 16, 17, and 24.)

7. Figure 28-17 shows the formation of 3-ketosphinganine from palmitoyl CoA and serine. What cofactor do you expect to be required for this reaction? What is the mechanism? (Hint: see Chapter 18.)

Biological Membranes

29

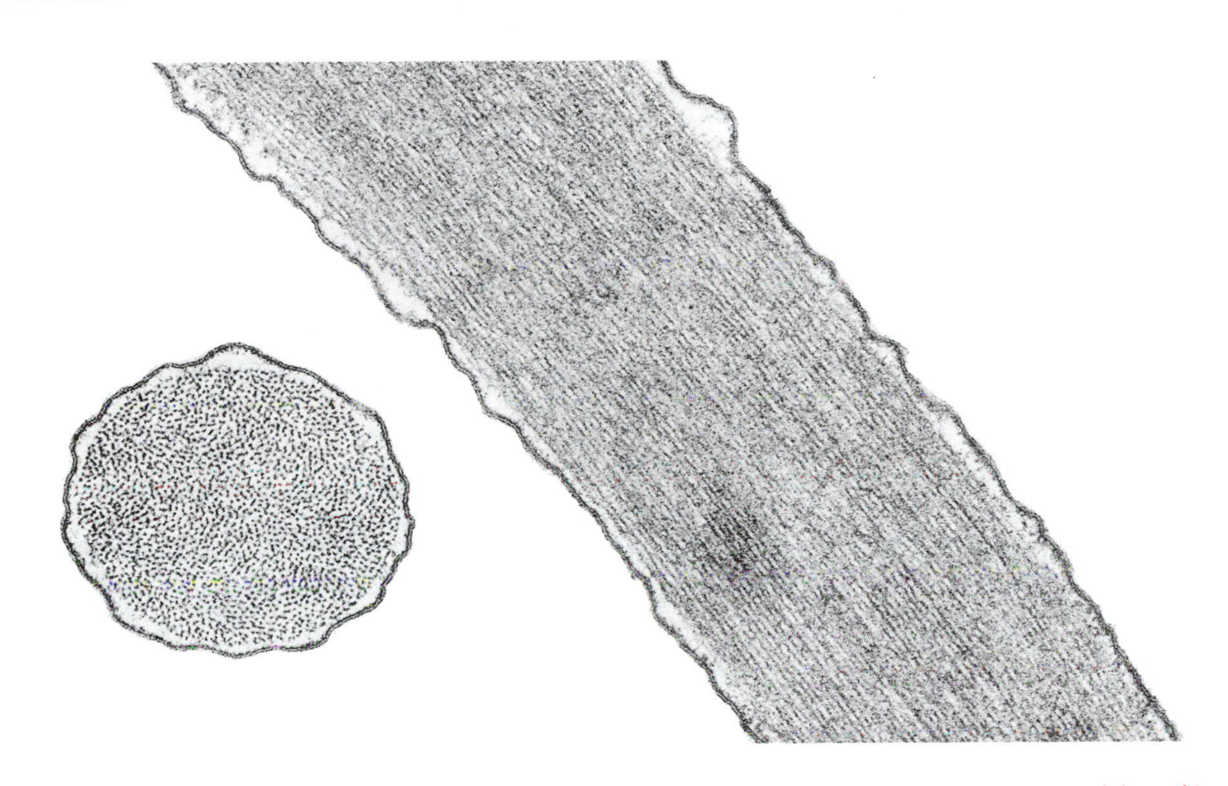

The stereocilium, seen here in transverse and long thin sections, is a membrane-encased bundle of actin filaments that projects from the apical surface of the hair cells of the inner ear. Stereocilia play a key role in transducing sound-induced fluid movement into changes in membrane potential, which is how we hear. (Electron micrographs courtesy of L. G. Tilney.)

Every cell, and every intracellular organelle, is covered by a membrane. Until quite recently, biological membranes were thought to be rather passive structures: impermeable bags for keeping macromolecules and metabolites locked inside the correct cellular compartments. But we now know that this view is completely mistaken. Cell membranes are active elements of biochemical metabolism; they tightly regulate the entry and exit of specific solutes, they serve as energy stores for powering many kinds of biochemically useful work, and they transfer information about the extracellular environment to the internal workings of the cell. In this chapter, we will examine first how biological membranes are built. Then we will consider some of the major functions that are carried out by cell membranes.

The Structure of Biological Membranes

Lipid Bilayers

For nearly a century it has been known that the membranes covering cells are composed of fatty material. With the realization more than fifty years ago that the membrane is exceedingly thin, on the order of a few nanometers, the idea of the bilayer lipid membrane gained general acceptance. There is now overwhelming evidence that all biological membranes are built on the lipid bilayer structure. To understand this structure, we must first consider the molecules from which it is built: the lipids.

Lipids form a broad class of compounds whose common feature is the spatial separation of chemically distinct parts: a hydrophilic region containing charged or hydrogen-bonding groups, and a hydrophobic region rich in hydrocarbons. All molecules with this characteristic are termed *amphiphilic,* because they have the capacity to interact with both polar and nonpolar solvents. Thus, phosphatidylcholine and other phospholipids, which are the major components of many biomembranes, carry two long fatty acyl chains esterified to the 1- and 2-positions of glycerol, whereas the 3-position contains a hydrophilic phosphate diester anion (Figure 29-1). Phospholipids are derivatives of phosphatidic acid (Chapter 11). Similarly, the hydrophobic parts of certain glycolipids and sphingomyelin (Figure 29-1) are composed of fatty acyl chains attached to sphingosine, whereas sugars or phosphorylcholine form the hydrophilic part (see Chapter 28).

When phospholipids are suspended in an aqueous medium, lipid bilayers form spontaneously (Figure 29-2). This structure solves the thermodynamic problem facing these molecules: how to keep the hydrophilic region hydrated, while at the same time removing the nonpolar hydrocarbon region from contact with water. The bilayer configuration is extremely stable in water, and no special biochemical mechanisms are required to maintain the structure; it is the great unfavorability of the lipid hydrocarbon chain's being in contact with water that keeps the bilayer membrane together. This unfavorability of placing a nonpolar residue into water is termed the *hydrophobic interaction* (Chapter 8).

To see just how strong this interaction is, let us calculate the probability of a lipid molecule leaving the bilayer and entering water. The Gibbs energy for the transfer of a hydrocarbon from a hydrocarbon phase into water is approximately +1 kcal per mole of methylene groups. Thus, to move a typical phospholipid, containing two 20-carbon fatty acid chains, out of the bilayer into water would cost about 40 kcal, a value corresponding to a probability for a single molecule leaving the bilayer of approximately 10^{-30}. In other words, if we had a membrane consisting of a mole of lipid molecules, we would almost never

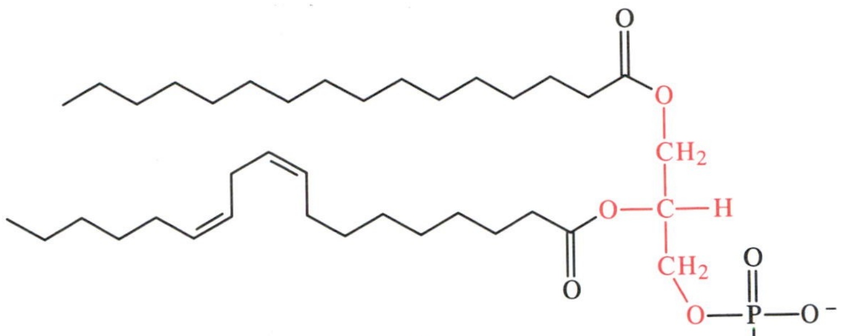

Phosphatidic acid

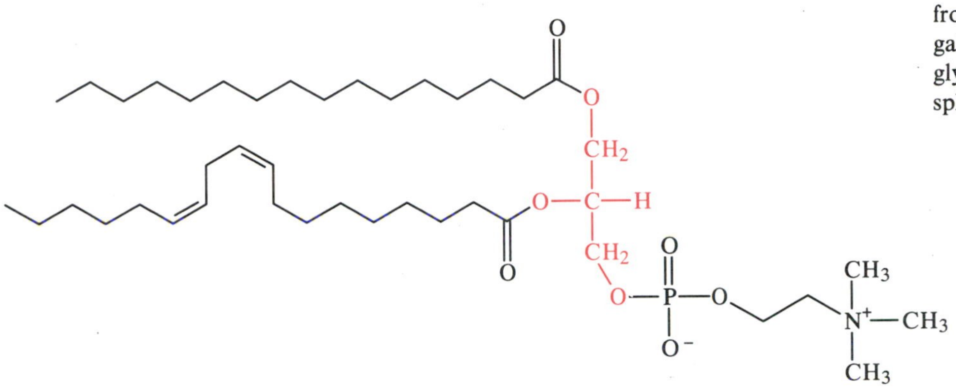

Phosphatidylcholine

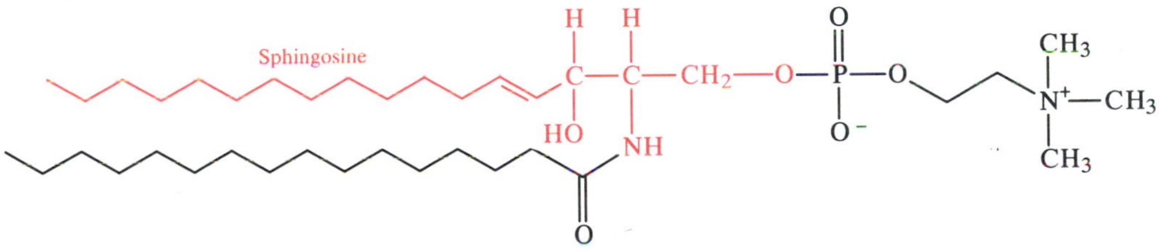

Sphingosine

Sphingomyelin

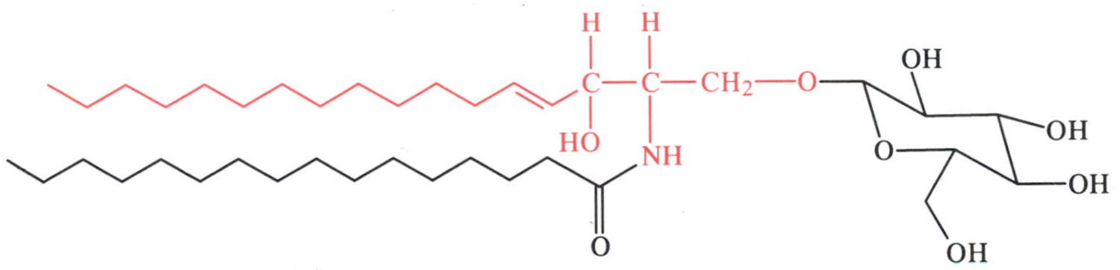

Cerebroside

Figure 29-1

Structures of Phospholipids and a Cerebroside in Membranes. Bilayer membranes are formed from phospholipids, such as phosphatidylcholine (lecithin). The hydrocarbon chains form an impermeable lipid bilayer and the charged groups interact with the aqueous medium on the two sides of the membrane. Membranes may also be formed from gluco- and galactocerebrosides, which are glycolipids that contain sphingosine instead of glycerol.

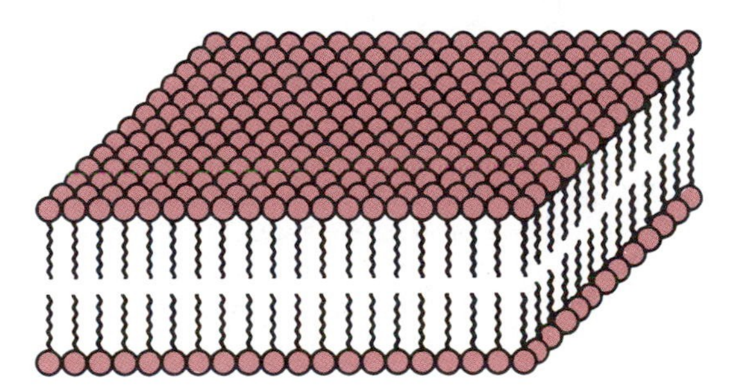

Figure 29-2

A Graphic Representation of a Lipid Bilayer. The hydrophobic tails of the lipid constituents in each layer are shown facing one another, with the hydrophilic heads constituting the two surfaces.

observe even a single phospholipid molecule in the aqueous phase. It is important to realize that this tendency of lipid hydrocarbon tails to stay together is not due to an especially strong affinity of these groups for one another; rather, it is the strong tendency of hydrocarbons to stay away from water that keeps the tails together in the bilayer interior (Chapter 8).

Furthermore, it is thermodynamically unfavorable to transfer a hydrophilic solute into the hydrocarbon interior of the bilayer. Because of the high energies that are required to move electrical charges or dipoles into a medium of low dielectric constant, ions, sugars, and other polar substances are virtually insoluble in hydrocarbons. Therefore, the lipid bilayer forms a permeability barrier to such solutes. Conversely, cell membranes are freely permeable to nonpolar molecules that easily dissolve in hydrocarbons, such as steroids, prostaglandins, free fatty acids, oxygen, and carbon dioxide.

Thus, the physical-chemical properties of hydrocarbons account for two fundamental characteristics of the lipid bilayer membrane: the stability of the bilayer structure in an aqueous medium, and the passive permeability of the membrane to many biologically important solutes. But we should not think that the strength of the hydrophobic interaction implies that the membrane is static. On the contrary, lipid molecules are in constant motion in the *plane* of the membrane. Lipids diffuse laterally along the membrane plane, and in this sense the lipid membrane may be viewed as a two-dimensional fluid. Indeed, it has been found that the lateral diffusion of lipids in the bilayer is rapid, although it is about 100-fold slower than the extremely fast diffusion of small solutes in water. The viscosity of the bilayer interior is therefore about 100-fold higher than that of water; it "feels" something like heavy machine oil. We will see later that the fluidity of the membrane and the free mobility of its constituents have important consequences in certain biochemical reactions.

Membrane Proteins

To a first approximation, we can view lipids as passive constituents of biological membranes; they form the basic bilayer structure of the membrane and provide a permeability barrier to polar solutes. Biological membranes are only about 50% lipid by weight, however. The other half is composed of protein. Figure 29-3 illustrates the prevailing view of the *fluid mosaic* cell membrane: it is an ocean of lipids in which proteins are embedded. Some of these proteins are peripheral; that is, they are exposed to the aqueous phase on only one side of the membrane, whereas others, the *integral membrane proteins,* span the bilayer and are thus

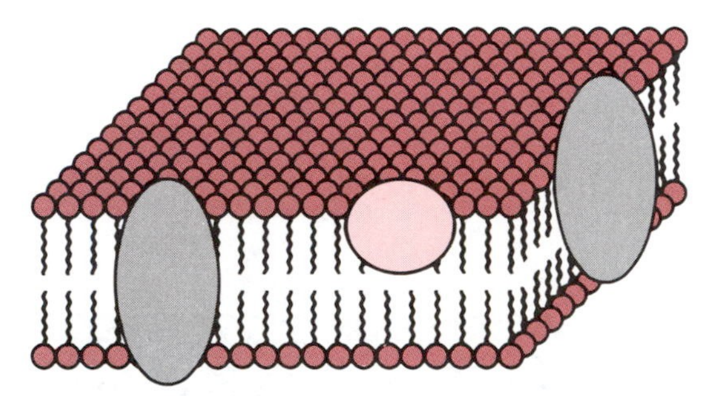

Figure 29-3

Transmembrane and Peripheral Proteins Embedded in a Lipid Bilayer. Shown are graphic representations of two proteins that span the lipid bilayer of Figure 29-2 and one protein that penetrates only part of the bilayer. All three proteins move freely within the plane of the bilayer; that is, within the lipid phase of the bilayer.

tightly locked into it. We now know that the many specific functions of biological membranes are due to the operation of membrane proteins and that the lipids act essentially as a two-dimensional solvent for these proteins.

What must the structure of proteins that reside in a lipid bilayer be? What special properties must such proteins possess? It is clear from the outset that they must be built in a fundamentally different way from cytosolic, water-soluble proteins. In the first place, membrane proteins do not exist in a single medium; rather, they have two "solvents" to contend with: some regions of the protein surface are exposed to water, whereas others reside in a hydrocarbon environment. Thus, these proteins must have two types of surfaces: nonpolar regions in the bilayer, and polar regions that are exposed to the aqueous phases. Furthermore, the regions of differing polarity must be well matched to the dimensions of the lipid bilayer.

We will consider four membrane proteins of increasing complexity, about which we have some biochemical understanding, in order to illustrate the different strategies that are used to cope with the basic physical constraint of being confined to this dual-solvent environment.

Gramicidin A. The simplest membrane protein for which both structure and function are reasonably well understood is gramicidin A, a peptide antibiotic that is synthesized by the bacterium *Bacillus brevis*. This peptide is small and extremely hydrophobic; it is composed of only fifteen amino acids, all of which (except for glycine) carry nonpolar side chains, as shown in Figure 29-4. A remarkable and eccentric feature of this protein, and one that is absolutely essential for its function, is the alternation of L- and D-amino acids along the peptide sequence. If the peptide backbone is stretched out straight, as shown in Figure 29-4B, it is clear that the alternation of L- and D-amino acids causes all of the individual side chains, R, to project toward *one side* of the chain (out of the plane of the page), leaving the hydrogen-bonded backbone groups facing the other side. In fact, when it is exposed to lipid bilayers, the peptide curls up into a helical structure (Figure 29-4C) such that all of the hydrophobic residues point outward, in contact with the hydrocarbon interior of the bilayer, and the polar NH–CO groups form the inner wall of the helix. This structure is *not* the familiar α-helix; this *π-helix* is much wider, with approximately six residues per turn. Consequently, a cylindrical pore of 0.4 nm diameter is formed through the center of the helix. This pore is filled with water and is wide enough to accommodate small ions, such as Na^+ and K^+.

The gramicidin helix is only 1.3 nm in length, which is not long enough to span the hydrocarbon region of the lipid bilayer. The peptide functions as a pore-former, however, by dimerizing lengthwise with hydrogen bonds between the *N*-formyl residues on the ends of the two chains. When two molecules that are embedded in opposite sides of the membrane come together, they form the dimer, as shown in Figure 29-4C. This dimeric pore is 2.6 nm in length, so that it can span the lipid bilayer to form a hydrophilic pathway that allows small ions to leak across the membrane.

This small peptide, though in detail atypical of integral membrane proteins (which are made of L-amino acids exclusively), illustrates a solution to the fundamental thermodynamic problems that must be solved by any protein that resides in a lipid bilayer: the necessity of exposing nonpolar groups on the protein surface that is in contact with the bilayer, while keeping polar groups associated with each other or with water. Gramicidin solves the first problem by the alternation of L- and D-amino acids with hydrophobic residues in its helical structure. The peptide solves the second problem by having a helix that is wide enough to accommodate water (something that an α-helix cannot do), while at the same time gaining an ion-transporting function.

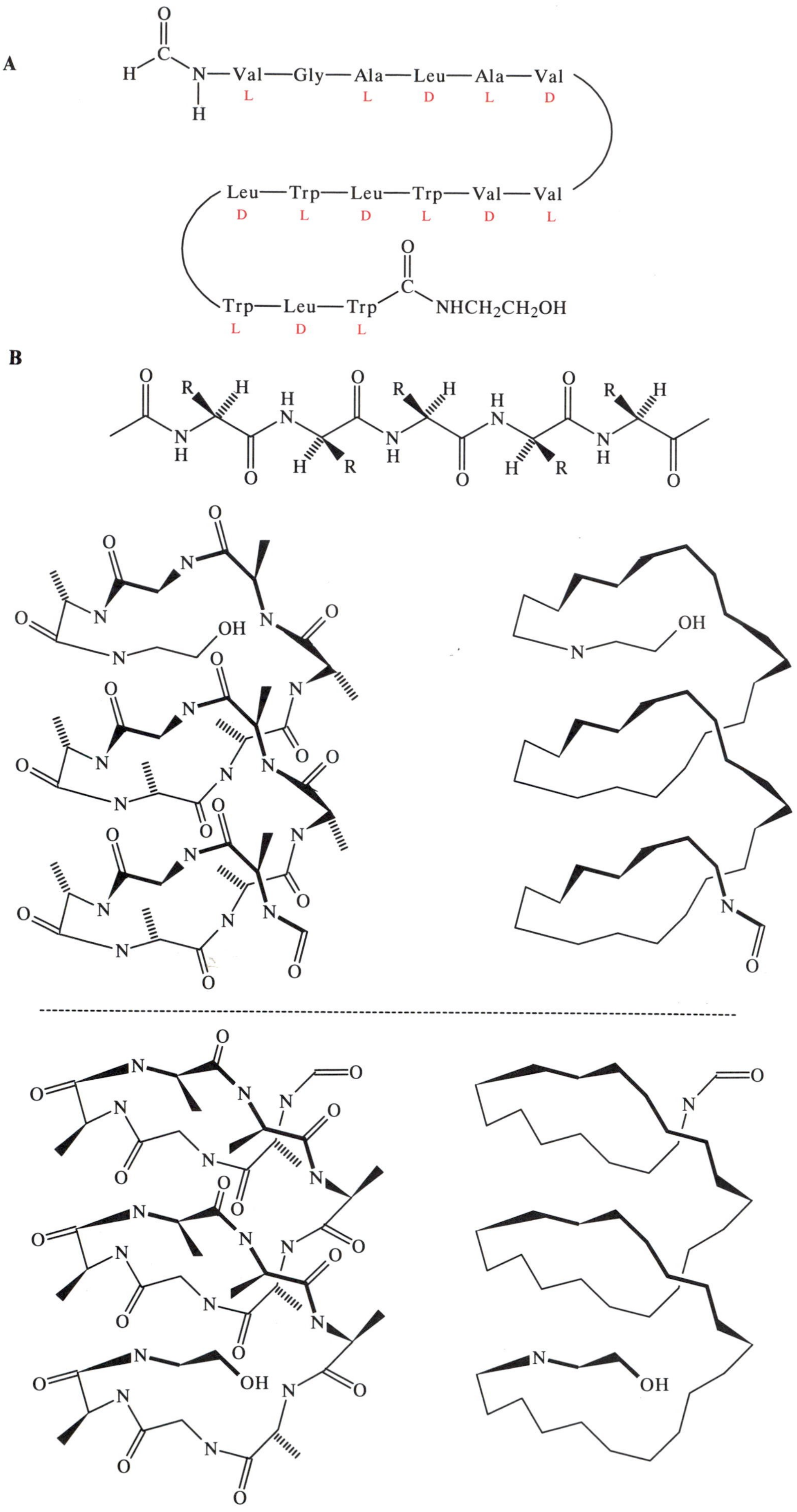

A

HC(=O)–NH–Val—Gly—Ala—Leu—Ala—Val
L L D L D

Leu—Trp—Leu—Trp—Val—Val
D L D L D L

Trp—Leu—Trp–C(=O)–$NHCH_2CH_2OH$
L D L

B

C

Figure 29-4 (on the facing page)

The Structure of Gramicidin A. Part A: Gramicidin A is an oligopeptide composed of fifteen amino acids. The N-terminus is *N*-formyl valine and the C-terminal tryptophan is β-hydroxyethyl tryptophanamide. The amino acid sequence consists of alternating L- and D-amino acids, except for one glycine residue. Part B: The effect of alternating L- and D-amino acids in a peptide is that all the amino acid side chains project in the same direction from the extended chain. Part C shows two π-helical molecules of gramicidin A associated head to head along a plane defined by the dashed line. The peptide nitrogen atoms, the peptide carbonyl groups, and the orientations of the amino acid side chains are shown at the left. At the right, these features have been omitted to show the main chain only. The length of the dimer is about 3.0 nm, about the same as the thickness of a typical lipid bilayer. The dimers are thought to span membranes, making them permeable to cations, which pass through the helices. [Adapted from Y. Ovchinnikov, *Eur. J. Biochem.* 94 (1979): 321.]

Cytochrome b_5. The endoplasmic reticulum membrane in many cells contains a redox protein named cytochrome b_5. This membrane-associated protein functions as an electron shuttle, accepting electrons from a variety of NADH-linked redox enzymes and donating them to several of the enzymes that participate in the metabolism of lipids, such as stearoyl CoA desaturase. The amino acid analysis of this protein does not appear extraordinary; it is not particularly rich in hydrophobic residues. However, a striking feature becomes apparent if we examine the amino acid sequence. If we plot an empirical index of the hydrophobicity of each residue against the residue number, as shown in Figure 29-5, we find that the C-terminal end of the protein contains a highly hydrophobic stretch of about thirty-five residues. No charged residues are found in this hydrophobic region.

It is now known that cytochrome b_5 is built as a normal globular enzyme, with a hydrophobic "tail" that serves to anchor the protein in the membrane (Figure 29-6). The protein's globular head may be cleaved from the membrane-associated tail with proteolytic enzymes; the globular part of the protein, which contains most of the molecular mass, is water soluble and retains its electron-shuttling activity. The function of the hydrophobic tail is to confine the

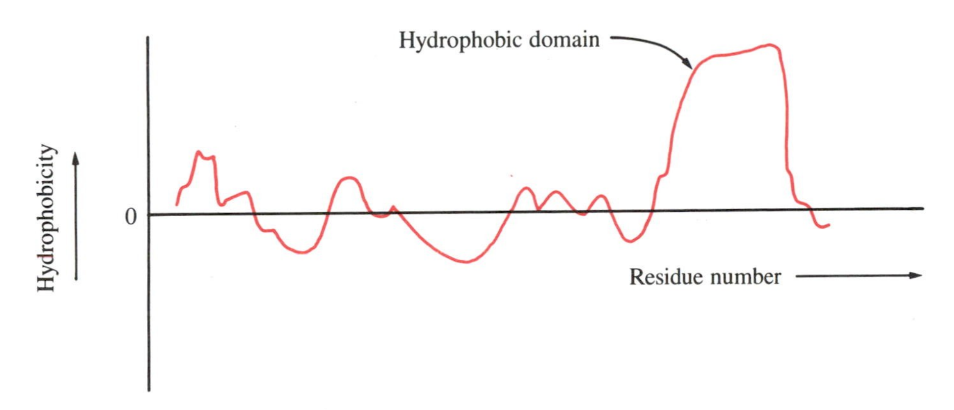

Figure 29-5

A Hydrophobicity Plot for the Amino Acid Sequence of Cytochrome b_5. This is a plot of a hydrophobicity parameter against residue number for the amino acid sequence of cytochrome b_5. Positive values of the parameter indicate hydrophobicity. The C-terminal residues of cytochrome b_5 are largely hydrophobic and constitute the hydrophobic domain of the molecule.

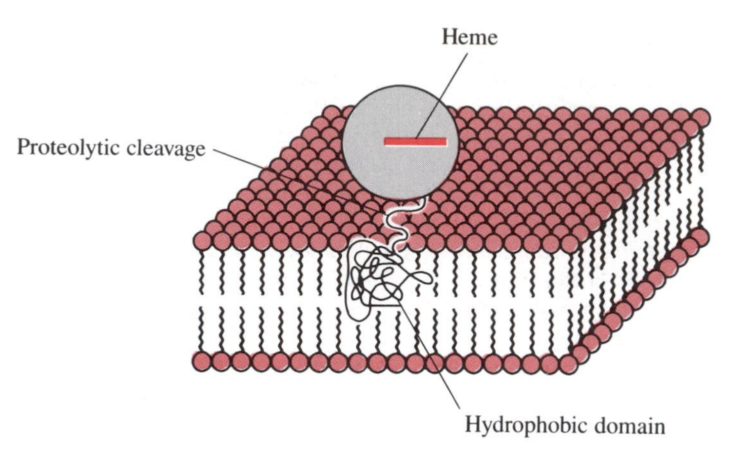

Figure 29-6

A Graphic Representation of Cytochrome b_5 Anchored in the Membrane. Cytochrome b_5 is represented as two domains, one typically globular and the other hydrophobic, linked through an unordered segment. The hydrophobic domain is buried in the membrane, and the catalytic, heme-bearing domain projects from one surface of the bilayer. The unordered segment linking the hydrophobic domain with the catalytic domain is typically more labile to proteolysis than are the two domains, which can be separated from each other by treatment with a proteolytic enzyme.

cytochrome to the two-dimensional membrane surface. This raises its effective concentration relative to the electron donor and acceptor enzymes with which it must interact because the enzyme is confined to the relatively small surface of the membrane, rather than the entire three-dimensional volume of the cell.

It is also known that mobility of the protein in the membrane is necessary for its efficient operation. If the temperature is decreased, so that the membrane fluidity is frozen out, then the cytochrome can no longer reduce other membrane-bound target proteins. Thus, cytochrome b_5 literally diffuses in the two dimensions of the membrane, accepting and donating electrons when it collides with the correct target enzymes, such as cytochrome b_5 reductase. Diffusion in only two dimensions restricts the number of directions in which it can move, so that the probability of its colliding with another molecule in the membrane in a given period is greatly increased, relative to that in a three-dimensional system. The simple strategy that makes this possible is to graft a very hydrophobic sequence of amino acids onto a conventional water-soluble cytochrome, which anchors the protein in the membrane. In this particular case, the anchor is a tail of hydrophobic amino acids that loops deeply into the lipid bilayer. Other systems are known in which fatty acids or other hydrophobic molecules are attached to water-soluble proteins to serve as hydrocarbon anchors.

Cytochrome b_5 is an example of a metabolically active protein that is confined to the membrane but interacts readily with small molecules in solution. Thus, the membrane can be regarded as a two-dimensional compartment of the cell, and this compartment is responsible for many of the most-important metabolic and regulatory functions of the cell.

Integral Membrane-Transport Proteins: Bacteriorhodopsin. Many integral membrane proteins act as specific transporters for solutes, so that they must span the lipid bilayer. The task of these proteins is to move hydrophilic solutes across the membrane. In many cases, the solute movement that takes place is thermodynamically "uphill," as with ATP-driven pumps (Chapter 30), and so chemical energy must somehow be coupled to drive the solute movement. Because of these complexities in mechanism, integral transport proteins are large and more complex in structure than the simple examples discussed so far. For the same reason, we do not know the three-dimensional structure of most of these proteins, which are very difficult to crystallize. But a few cases currently exist in which enough information is available to give us a picture of the folding of the peptide chains within the membrane.

One of the best studied of such transport proteins is bacteriorhodopsin, a protein with a molecular weight of 26,000 that acts as a light-driven proton pump for certain bacteria (Chapter 23). The protein has been crystallized and its structure has been determined at ~0.4 nm (4 Å) resolution. This resolution is high enough to show the overall shape of the protein and the arrangement of transmembrane helices. The peptide chain of the protein snakes back and forth across the lipid bilayer seven times, with an α-helix spanning the bilayer in each crossing (Figure 29-7A). These seven α-helices are arranged in a bundle, as shown in the ribbon drawing of Figure 29-7B (see also Figure 23-14). The hydrocarbon of the lipid bilayer is on the outside of the bundle, and the retinal chromophore is covalently attached to a lysine residue on one of the helices on the inside of the bundle, approximately in the center of the bilayer. Most of the

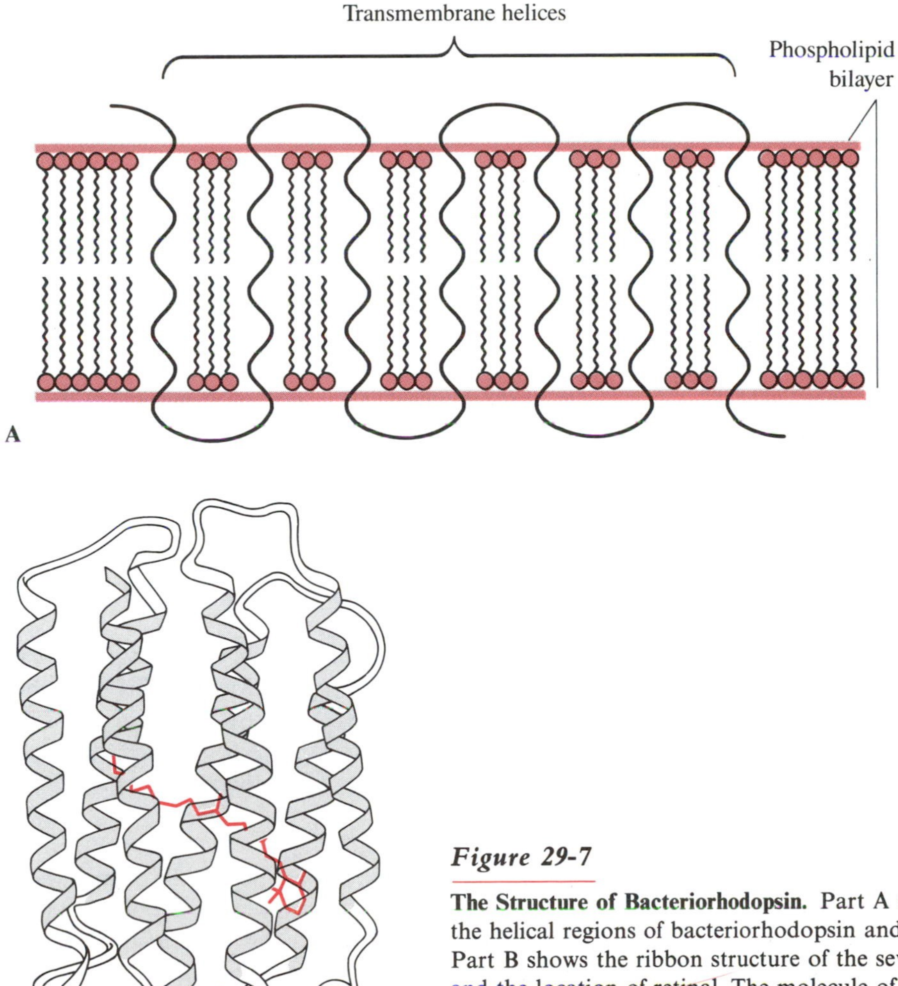

Figure 29-7

The Structure of Bacteriorhodopsin. Part A shows a schematic representation of the helical regions of bacteriorhodopsin and their orientation in the membrane. Part B shows the ribbon structure of the seven α-helices in bacteriorhodopsin and the location of retinal. The molecule of retinal, shown in red, is attached to the protein as an imine formed between retinal and the ε-amino group of a lysine. (Part B adapted from a ribbon drawing courtesy of R. Henderson.)

amino acid residues in the α-helices are hydrophobic, but polar amino acids are found at regular intervals in the sequence, so that the bundle of helices forms a channel through which ions can be transported. The segments of the peptide chain that connect the α-helices are rich in charged and polar residues and are almost certainly exposed to the aqueous medium.

The theme of multiple membrane-spanning domains in membrane-transport proteins has now been verified with every integral membrane transporter that has been examined to date. In every case, the protein is maintained in the membrane by bundles of hydrophobic membrane-spanning stretches that are composed either of α-helical or of β-sheet structures.

Although the membrane-spanning domains are highly hydrophobic, they are not completely so. We must remember that only the outer surfaces of these bundles come into direct contact with lipid hydrocarbon; the inner regions of the membrane-spanning sequences may contain hydrophilic groups that allow many of these complex membrane proteins to serve as hydrophilic channels through which the transported substrates can move.

In only a single case do we know the structure of a membrane protein at near-atomic resolution, the case being the photosynthetic reaction center complex of the photosynthetic bacterium *Rhodopseudomonas viridis.* The schematic diagram of this membrane-spanning protein on page 633 shows that the general rules outlined earlier for simpler proteins are verified here. The protein contains eleven membrane-spanning domains, all α helical. The outer surfaces of these helices are completely composed of nonpolar residues, whereas numerous polar groups on the inward-pointing surfaces make specific contacts with each other and with other molecules in the complex.

Membranes as Energy-conserving Machines

Before considering particular examples of Gibbs energy storage and utilization by solute gradients, we will explain more precisely what we mean by "the free energy that is stored in a solute gradient." Suppose that we have a membrane separating two solutions, "in" and "out," that contain a solute, s. Let us further suppose that the concentration of the solute in the inside solution, $[s]^{in}$, is higher than it is in the outside solution, $[s]^{out}$. If the membrane is permeable to the solute, we know from experience that it will spontaneously leak outward. This tendency is simply an expression of the Gibbs energy of dilution of a solute, which is always negative because solute dilution leads to an increase in the entropy of the system (see Chapters 5 and 9). Therefore, it requires work to separate a solution of ions into two parts that contain high and low concentrations of ions.

The driving force for movement, or for reaction, of an ion or molecule can be described by its *chemical potential,* μ, which is proportional to the logarithm of its concentration or, more exactly, its activity. If the chemical potentials of the solute in the two solutions are denoted by μ_s (in) and μ_s (out), then, when Δn moles of solute flow from in to out, the Gibbs energy of the system changes by:

$$\Delta G = \Delta n\ [\mu_s(\text{out}) - \mu_s(\text{in})] \tag{1}$$

(We are assuming here that the amount of solute flowing across the membrane is small enough so that its concentrations on the two sides do not change appreciably.) Thus, the Gibbs energy change is directly proportional to the difference in chemical potential across the membrane, the *chemical potential gradient.* The chemical potential for a solute in an ideal solution is:

$$\mu_s = \mu_s^\circ + RT \ln [s] \tag{2}$$

in which μ_s° is the value of the chemical potential of the solute, s, at an arbitrarily chosen standard concentration (usually 1 M). We can now calculate the change in Gibbs energy (equation 1; the standard chemical potential term cancels out).

The change in Gibbs energy for each mole of solute that is transferred across the membrane, also called the chemical potential gradient, $\Delta\mu_s$, is:

$$\Delta\mu_s = RT \ln ([s]^{out}/[s]^{in}) \tag{3}$$

Because the solute is at a higher concentration inside, this change in Gibbs energy is negative, which means that in principle we can perform useful work by "capturing" it with a suitable device.

There is an important complication in the foregoing treatment that almost always applies to biological membranes. Nearly all cell membranes are electrically polarized; that is, voltage differences exist across such membranes. (We should not worry here about *how* they are polarized; this is accomplished by a variety of mechanisms, some of which we will describe later.) Just as a chemical potential difference represents a *chemical* driving force for a solute, so does a voltage difference represent an *electrical* driving force if the solute carries a net electric charge. Therefore, for charged solutes, we must consider both the chemical and the electrical gradients in calculating the free energy of moving the solute across the membrane.

We can now define the *electrochemical potential gradient* by equation 4, when there is a voltage difference, V, across the membrane:

$$\begin{aligned}\Delta\mu_s &= \mu_s(\text{out}) - \mu_s(\text{in}) + zFV \\ &= RT \ln ([s]_{out}/[s]_{in}) + zFV\end{aligned} \tag{4}$$

in which z is the electrical charge on the solute and F, the Faraday constant, converts electrical units (voltage) into the Gibbs energy units (kcal mol^{-1}; Chapter 9). The electrical sign convention is always to report the voltage relative to the outside solution, which is taken as zero.

Henceforth, when we speak of a "solute gradient," we will be referring to the *electrochemical potential gradient* of the solute across the membrane. This is a direct measure of the Gibbs energy of dilution of the solute, as it falls down its thermodynamic hill from one side of the membrane to the other.

To illustrate how the electrical and chemical parts of the electrochemical potential gradient add together, Figures 29-8 and 29-9 depict the directions and magnitudes of the electrochemical gradients of several solutes across some

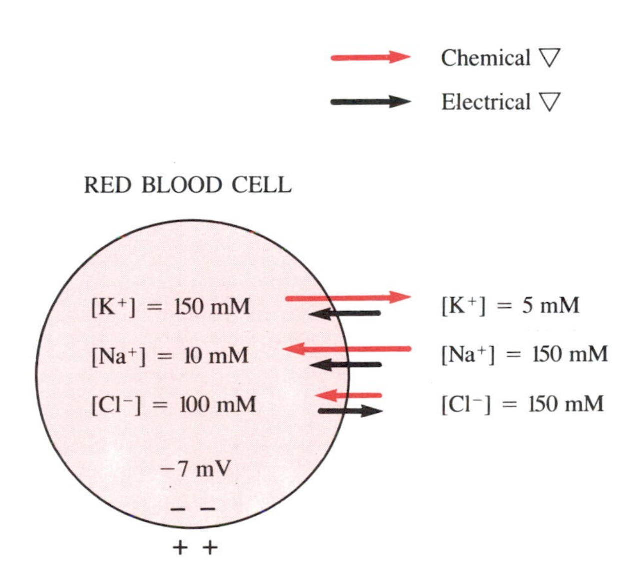

Figure 29-8

Electrochemical Gradients Across the Red Blood Cell Membrane. Electrochemical gradients across the red blood cell membrane arise from the different concentrations of potassium, sodium, and chloride ions on the two sides. The concentrations are millimolar. The red arrows indicate the direction of the chemical driving force, from the ion gradients, and the black arrows indicate the electrical driving force. There are additional anions and cations on both sides of the membrane that are not shown.

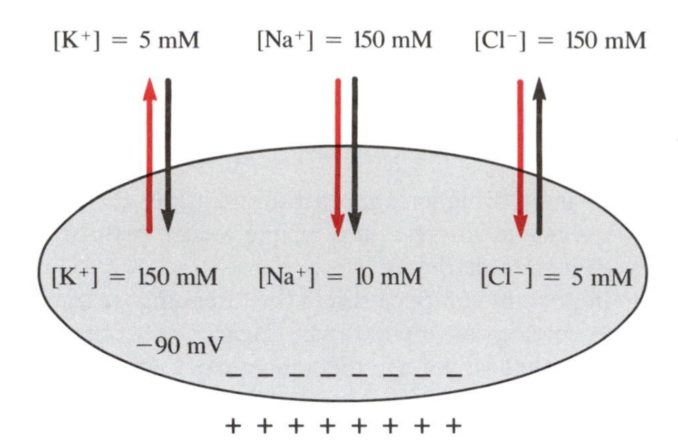

Figure 29-9

Electrochemical Gradients Across the Muscle Cell Membrane. The concentrations outside the muscle cell are the same as for the red blood cell (Figure 29-8), but the chloride concentration inside a muscle cell is much lower.

"typical" biological membranes. The human red blood cell contains a high K^+ and a low Na^+ concentration inside, as do most cells, as well as a relatively high concentration of internal Cl^-. The voltage across the red-cell membrane is low, only about -7 mV. Thus, the purely electrical driving forces on the three monovalent ions are low; for K^+ and Na^+, this electrical driving force is directed inward; and, for the negatively charged Cl^- ion, outward. The electrical driving force thus subtracts only slightly from the large, outwardly directed chemical gradient for K^+, and adds slightly to the large inwardly directed gradient for Na^+. But the same magnitude of electrical gradient for Cl^- almost exactly cancels the rather small inwardly directed concentration gradient for Cl^-. The Cl^- ion is at equilibrium across the red-cell membrane.

The situation is quite different in muscle. Here, the chemical gradients for Na^+ and K^+ are the same as in the red cell, but there is a very large inwardly directed chemical gradient for Cl^- as well (Figure 29-9). In this cell, the membrane potential is high, about -90 mV with negative charge on the inside of the membrane. Therefore, now both the K^+ gradient and the Cl^- gradient are exactly cancelled by the membrane voltage; both of these ions are very close to equilibrium across the membrane. However, the Na^+ electrochemical gradient is huge, because now the large chemical and electrical gradients are both directed inward. The Na^+ ion is kept very far away from equilibrium across this membrane.

The Creation of Solute Gradients: Primary Active Transport

In order for a biological membrane to harvest the Gibbs energy of dilution of a solute, a thermodynamic gradient of the solute must be established across the membrane. All biological membranes establish solute gradients by the action of *primary active transporters.* These are defined as integral membrane proteins that directly utilize the favorable free energy of a chemical reaction to "pump" a given solute against its concentration gradient across the membrane. We currently have identified many types of active transporters that are powered by that most common of biochemical energy sources, ATP hydrolysis. For example, the ATP-driven pumps that bring about the active transport of Na^+, K^+, Ca^{2+}, H^+ and amino acids have been extensively studied in a great variety of biological membranes. But numerous other sources of energy for solute transport are also known, such as the oxidation of substrates in the TCA cycle by O_2 to pump protons (Chapter 23), the decarboxylation-driven pumping of

Na^+, and the photochemically powered transport of solutes in light-harvesting membranes. We will briefly describe two such active transporters to illustrate how membranes capture the free energy of metabolic reactions and create energy reservoirs in the form of transmembrane solute gradients. The mechanism by which the uphill movement of a solute is coupled to the favorable Gibbs energy of a chemical reaction will be described in more detail in Chapter 30.

The Na,K ATPase

Virtually all cells maintain large concentration gradients of the two most-abundant biological solutes, Na^+ and K^+. As we have seen, K^+ is kept at a much higher concentration inside the cell than in the external medium, whereas Na^+ is kept lower inside than in the external medium. In a human erythrocyte, for example, these ion gradients are large: cytoplasmic K^+ is about 150 mM, whereas external K^+ is only 5 mM; internal Na^+ is typically 10 mM, whereas the plasma Na^+ concentration is about 150 mM.

In animal cells, these gradients are brought about by ATP and a single integral membrane enzyme: the Na,K ATPase, which is also called the "Na^+ pump." This enzyme catalyzes the reaction:

$$\mathrm{ATP} + 3\,\mathrm{Na^+(in)} + 2\,\mathrm{K^+(ex)} \rightleftharpoons \mathrm{ADP} + \mathrm{P_i} + 3\,\mathrm{Na^+(ex)} + 2\,\mathrm{K^+(in)} \quad (5)$$

Again, considering the erythrocyte to be our "typical" cell and neglecting effects of voltage, which are small, we can write down the total Gibbs energy change for each mole of ATP that is hydrolyzed by this enzyme when it is pumping Na^+ and K^+ against concentration gradients of these ions:

$$\Delta G = \Delta G^{\circ\prime}(\mathrm{ATP}) + RT\ln([\mathrm{ADP}][\mathrm{P_i}]/[\mathrm{ATP}]) + 3RT\ln([\mathrm{Na^+}]_{ex}/[\mathrm{Na^+}]_{in}) + 2RT\ln([\mathrm{K^+}]_{in}/[K^+]_{ex}) \quad (6)$$

Here, $\Delta G^{\circ\prime}$(ATP) is the Gibbs energy change for ATP hydrolysis in the standard state at physiological pH, and the first two terms of equation 6 represent the actual molar Gibbs energy change for ATP hydrolysis under ambient cellular conditions; that is, at the concentrations of ATP, ADP, and P_i that are found in a cell. Typically, this represents a large Gibbs energy change of about $-13\ \mathrm{kcal\ mol^{-1}}$, which is the amount of Gibbs energy that is available from the hydrolysis of ATP to do work in the cell (Chapter 9).

Let us now consider two special cases, to see where this available energy goes when the Na,K ATPase hydrolyzes ATP:

Case 1: No ion gradients. Suppose that we experimentally bring about nonphysiological conditions in the red cell by loading the cell with Na^+ and depleting it of K^+, such that now the internal concentrations of ions are equal to the external concentrations (Figure 29-10A). We now calculate the Gibbs energy of the Na,K ATPase reaction under these conditions:

$$\Delta G = -13\ \mathrm{kcal} + 0 + 0 = -13\ \mathrm{kcal} \quad (7)$$

In this case, the Gibbs energy change for the reaction is the same as the Gibbs energy of simply hydrolyzing ATP in solution, without "coupling" the hydrolysis to anything. To be sure, Na^+ is being pumped out of the cell, and K^+ in, but there are no concentration gradients against which the pumping occurs. Ions are merely transferred from one side of the membrane to the other, between solutions with identical concentrations of each ion. No work is done in such a transfer, and so all of the Gibbs energy of ATP hydrolysis is "lost" to the environment.

Case 2: Physiological ionic conditions. In this case, shown in Figure 29-10B, we consider the pumping of Na^+ and K^+ against the gradients that normally exist

A NO ION GRADIENTS

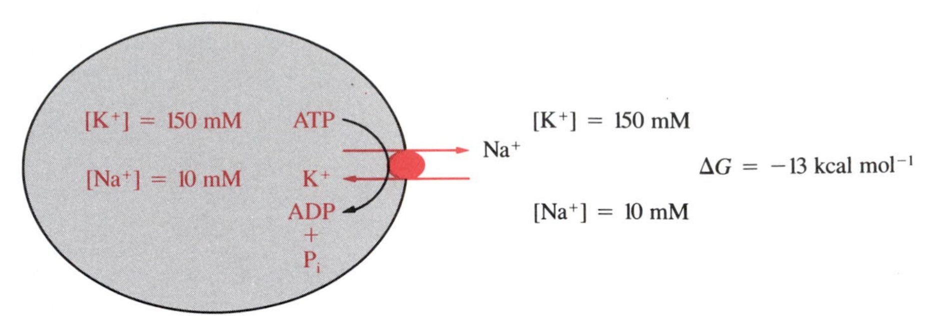

B PHYSIOLOGICAL GRADIENTS

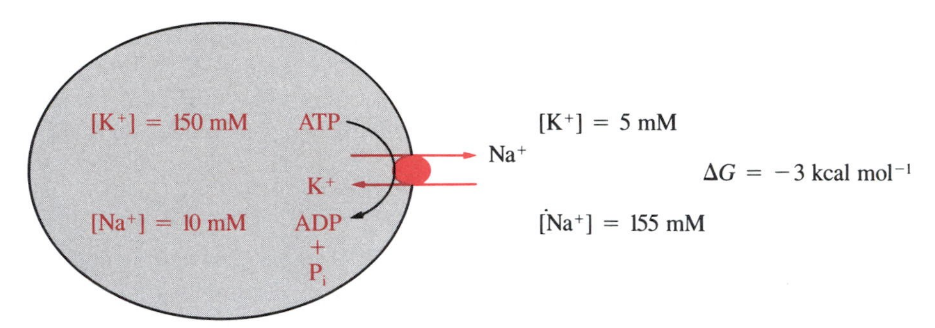

Figure 29-10

Transport of Na^+ and K^+ by Na,K ATPase. Part A shows that the transport of Na^+ and K^+ ions across the cell membrane by the Na,K ATPase results in the hydrolysis of ATP but that no work is done when there is no ion gradient across the membrane. Part B shows that, under physiological conditions, Na^+ and K^+ ions are transported across the cell membrane against a concentration gradient, so that most of the energy from the hydrolysis of ATP is conserved in the ion gradient.

in the cell. Now, equation 1 shows that the molar Gibbs energy change of hydrolyzing ATP is:

$$\Delta G = -13 \text{ kcal} + 4.8 \text{ kcal} + 5.2 \text{ kcal} = -3 \text{ kcal} \tag{8}$$

This is a very different answer. Transporting Na^+ and K^+ against ion gradients leads to a much smaller Gibbs energy change than the total Gibbs energy of ATP hydrolysis. Where has this extra energy gone? The answer is that a full 10 kcal of Gibbs energy has been *stored* in the ion gradients, because now the ions are being transported "uphill" against their concentration gradients. Thus, most of the Gibbs energy of ATP hydrolysis, which is simply lost to heat when no ion gradients are present, is now captured in the ion gradients. In exactly the same way that the chemical energy of combustion of gasoline running a pump can be captured in the gravitational potential energy of water pumped up into a storage tank, so is the chemical energy of ATP hydrolysis stored in the transmembrane gradients of Na^+ and K^+. This stored free energy can later be used to do other kinds of work in the cell, as we will see.

The Mitochondrial Respiratory Chain

We have seen in Chapter 22 that the mechanism of the mitochondrial respiratory chain is immensely complicated. The chain is composed of at least twenty different proteins working in a multienzyme complex to transfer electrons from various substrates to O_2. But, thermodynamically, the respiratory chain is beautifully simple; it is a pump for protons that is driven by the highly exergonic oxidation of NADH or other reduced substrates by molecular oxygen:

$$NADH + \frac{1}{2}O_2 + H^+ \longrightarrow NAD^+ + H_2O \tag{9}$$

In the mitochondrion, each NADH oxidized is used to pump six protons outward—that is, from the intramitochondrial space toward the cytoplasm—creating a proton gradient across the membrane that is equivalent to about 3 pH units. This proton gradient is then used to perform work of supreme biological importance: the synthesis of ATP.

The Utilization of Solute Gradients

One of the most-important reasons for the evolutionary development of nature's panoply of active transporters is to create solute gradients that can be used to do some kind of useful work. What do these solute gradients actually do for the cell? The answer is that there is no single answer: transmembrane solute gradients are used in a multitude of different ways as sources of free energy to power energy-requiring reactions. In the same way that ATP is a common "energy currency" for cellular metabolism, solute gradients serve as energy sources for many membrane-localized processes that require input of energy. In this section, we will consider a few of the types of useful work carried out by membranes that are energized by gradients of various solutes.

The Role of Solute Gradients

One of the most-important functions of a membrane in biological systems is to act as a machine that *utilizes* the Gibbs energy that is stored in solute gradients to drive many of the activities of a cell. We have seen some of the ways that solute gradients are built up, and we now consider how these gradients can be converted into useful work.

Aaron Katchalsky constructed a simple machine that illustrates how a difference in the concentrations of ions in two different solutions can be utilized to perform work, in this case mechanical work. This model is described in Box 29-1. It has little or no resemblance to any physiological process, but it is an interesting illustration of the principle that ion gradients can provide a driving force to bring about movement.

Oxidative Phosphorylation

The immediate source of energy for the synthesis of nearly all the ATP in aerobic organisms is the free energy of dilution of protons across the mitochondrial membrane. As we have seen in Chapter 22, the function of the respiratory chain is to set up a proton gradient across the mitochondrial membrane, by primary active transport, and this proton gradient, in turn, drives ATP synthesis. Although the detailed mechanism of the ATP synthase complex is not yet known, in principle the process is very simple. The proton-pumping ATPase in the mitochondrial membrane is an active transporter that operates in reverse (Figure 29-11). It allows protons to "fall" down their transmembrane gradient and uses this energy to synthesize ATP.

Let us consider the energy balance of the proton-pumping ATPase as we have the other primary active transporters. The reaction that this enzyme catalyzes is:

$$ATP + 3\,H^+(\text{in}) \rightleftharpoons ADP + P_i + 3\,H^+(\text{out}) \tag{10}$$

BOX 29-1

THE KATCHALSKI MACHINE

Many years ago, Aaron Katchalsky designed and built a remarkable "perpetual motion" machine, which is shown in part A of the adjoining illustration. This machine is constructed from a continuous belt of a charged polymeric material that is stretched tightly around the three-pulley arrangement shown. The belt dips into two aqueous solutions under the lower pulleys; one of these solutions contains a high concentration of salt, KCl, and the other pure water. The top pulley is actually two pulleys of different radii rigidly mounted together. The machine works because the belt is made from a material containing many charged groups, so that it is an ion-exchange resin. The belt contracts when it binds salt, because salt decreases the repulsive electrostatic forces between the fixed charged groups on the resin, and it expands when the salt is removed.

The machine works as follows. Consider a small section of the belt just dipped into the high-salt solution (part B of the illustration). This section will contract, which produces equal forces on the two sides of the upper, compound pulley. Because of the different radii on the two sides of this pulley, however, these equal forces give unequal torques, and a net torque tends to turn the belt so that the next section dips into the high-salt solution. Note that, as the belt turns, it relaxes as it dips into the low-salt side of the machine and the ions dissociate, so that no counterforces are placed on the compound pulley. A net torque is thus continuously maintained, and the belt keeps turning and draws the next section of belt into the high-KCl pot. The top pulley can be hooked up to the external world to perform work—to lift a weight, for example.

Is this a perpetual motion machine? Of course not, but we must understand exactly why not. What makes it run? What is the fuel powering this "Katchalsky machine"? As the machine turns, salt enters the polymeric belt in the high-salt solution and is carried along until it washes out as the belt dips into the low-salt solution. In other words, as the machine turns, salt is transferred from the high-salt solution to the low-salt solution. We know that dilution of a solute is a spontaneous process; that is, it has a negative free energy change. But the Gibbs energy change of any process is a measure of the maximum work that could be extracted out of that process, if only we were clever enough to invent a machine to capture the Gibbs energy. The Katchalsky machine is such a device; it extracts work out of the negative Gibbs energy of dilution of a solute—KCl in this case. Eventually, the machine stops turning as the concentration of salt in the high-salt pot diminishes and that in the low-salt pot rises.

What does this have to do with biological membranes? A great deal. In fact, a major function of biological membranes is precisely to act as a Katchalsky machine; that is, to perform useful work by utilizing the free energy of dilution of various physiological solutes. All cell membranes use metabolic energy to establish concentration gradients for different solutes; once established, these gradients supply energy to do other kinds of work according to the cell's needs. The solute gradients thus act as energy reservoirs for the cell, and membranes contain the machines that interconvert these different forms of free energy.

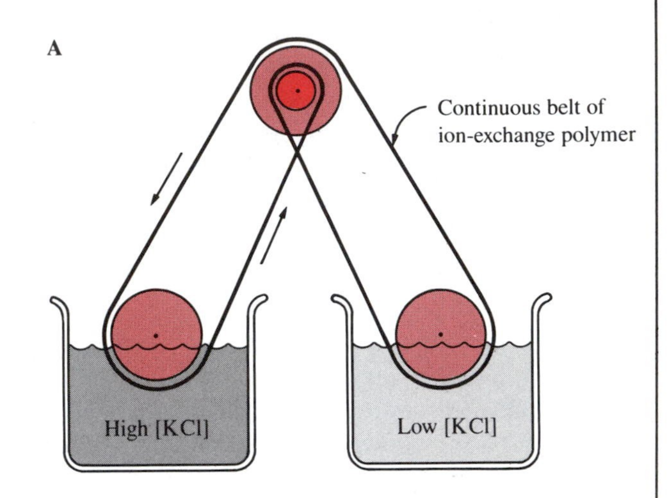

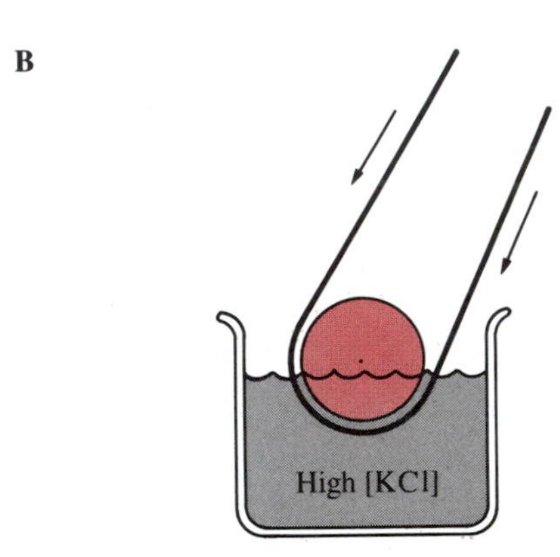

The Katchalsky Machine. The belt of ion-exchange polymer contracts when it binds salt, because of decreased electrostatic repulsion of the charges of the polymer (B). This causes the belt to move as shown in part A because a larger force is exerted by the large pulley than by the small pulley at the top. The salt dissociates in the right-hand solution. The cycle continues until the difference in the salt concentrations in the two solutions becomes too small to drive the pulley.

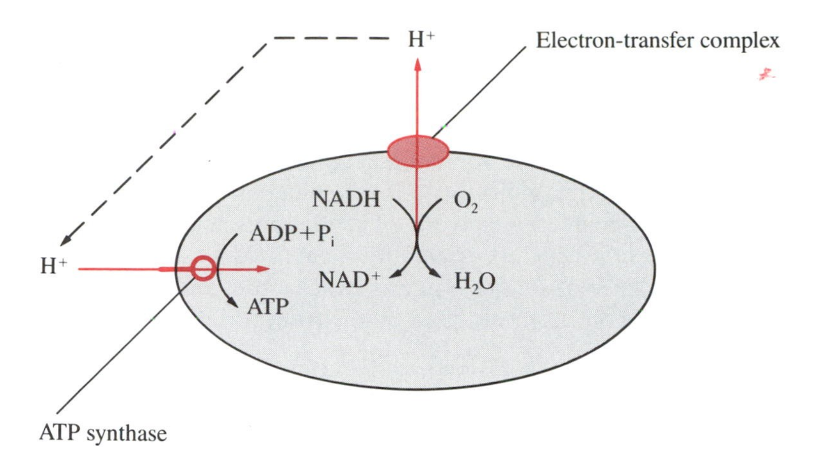

Figure 29-11

ATP Synthesis Driven by a Proton Gradient in Oxidative Phosphorylation. Oxidative phosphorylation utilizes the proton gradient that is generated from the oxidation of metabolites and the transfer of protons out of the mitochondrion to generate ATP. The energy released by the re-entry of protons into the mitochondrion through the F_1F_0ATPase generates ATP (Chapter 22).

As before, the Gibbs energy change for each mole of ATP that undergoes hydrolysis is:

$$\Delta G = \Delta G(\text{hyd}) + 3RT \ln([\text{H}^+]_{\text{out}}/[\text{H}^+]_{\text{in}}) \qquad (11)$$

Under normal conditions, ΔG(hyd) is -13 kcal mol^{-1}, as before, and the proton gradient is equivalent to approximately 3.3 pH units. Therefore, the Gibbs energy change of the reaction is:

$$\Delta G = -13\,\text{kcal} + 14.1\,\text{kcal} = +1.1\,\text{kcal} \qquad (12)$$

The fact that the Gibbs energy change in the reaction of equation 10 is positive means that the reaction goes in the reverse direction: protons enter the mitochondrion, flowing down the proton gradient *through* the ATPase. The Gibbs energy that is required in order to form the high-energy phosphate bond in ATP comes from the thermodynamically favorable dilution of protons as they fall down their steep electrochemical potential gradient across the mitochondrial membrane.

It is important to remember that the proton gradient is set up and maintained by the continual operation of the respiratory chain. The reason that this system works is that the chemical reactions of the respiratory chain that pump protons out are more favorable than the ATP hydrolysis reaction. The outwardly directed ATP-driven proton pump cannot overcome the very large proton gradient that is produced by the respiratory chain, so that it is forced to run backward. Thus, the mitochondrial membrane can be regarded as a Katchalsky machine, in which the movement of protons down a concentration gradient is utilized to bring about the useful work of ATP formation.

Concentrative Uptake of Nutrients: Symporters

The mitochondrial synthesis of ATP serves as an example of chemical work that is driven by proton transport across the membrane. Another type of useful work that the cell needs to do is to concentrate nutrients from its environment. Free-living bacteria, for example, are able to concentrate sugars and amino acids by many orders of magnitude greater than their concentrations in the environment.

The best-studied example of such a nutrient-accumulation system is the lactose transporter of *E. coli*. For many years, it was known that, in the presence of energy sources, this bacterium is able to accumulate lactose in the cytoplasm at concentrations that are at least 1000-fold higher than the concentration in the external medium. It was thought that a primary active transporter of some sort

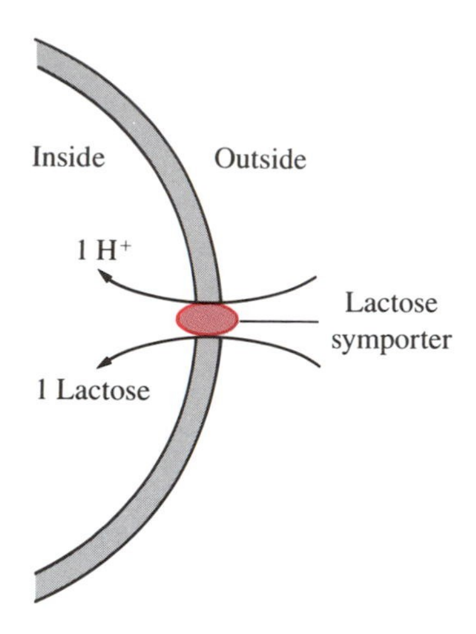

Figure 29-12

The Proton-coupled Lactose Symporter.

was responsible for this transport—perhaps an ATPase or an oxidation-reduction pump—but the search for the direct energy source of this hypothetical transporter remained fruitless for years.

Then, in the early 1970s, Mitchell proposed the radical idea that there is no direct chemical energy source for lactose accumulation—that this is not a primary active transporter, but rather a totally different type of mechanism that he called a "proton-coupled *symporter*." The idea was that the cell membrane contains a transporter, which is essentially a passive shuttle that catalyzes the movement of *two* different solutes—in this case, lactose and protons, as shown in Figure 29-12. The reaction catalyzed by the transporter was proposed to be:

$$\text{Lactose (out)} + \text{H}^+\text{(out)} \rightleftharpoons \text{Lactose (in)} + \text{H}^+\text{(in)} \qquad (13)$$

It is easy to show that the equilibrium condition for this reaction requires a relation between the proton gradient and the lactose gradient ($F = 96{,}500$ coulombs is the Faraday and V is the voltage difference across the membrane):

$$[\text{Lac}]_{\text{in}}/[\text{Lac}]_{\text{out}} = [\text{H}^+]_{\text{out}}/[\text{H}^+]_{\text{in}} \exp(-FV/RT) \qquad (14)$$

Mitchell further proposed that a proton gradient is set up across the bacterial membrane by the bacterial respiratory chain acting as an outwardly directed proton pump, in analogy with the situation for mitochondria. As long as the respiratory reactions proceed, this proton gradient is maintained, and protons will tend to fall into the cell. Those protons falling through the lactose transporter, because of the transporter's "rule" that requires a proton and a lactose molecule to move simultaneously, will force the "uphill" movement of lactose. If the respiratory chain maintains a proton gradient equivalent to 3 pH units, then lactose will be accumulated to a concentration ratio of 1000-fold. Thus, the thermodynamically favorable reaction driving lactose accumulation is the dilution of protons across the membrane. Once again, the membrane acts as a simple machine that utilizes a proton gradient to do useful work for the cell.

Mitchell's proposal was thought to be ridiculous by many at the time, but it has turned out to be correct. We now know that the lactose transport protein catalyzes exactly the reaction that was originally proposed. One of the early experiments, carried out in Mitchell's laboratory, is shown in Figure 29-13. Here, bacteria are placed in an anaerobic medium in the total absence of any chemical energy sources, and the pH in the medium is measured. If lactose is suddenly added to the medium, protons are observed to move into the bacteria and the pH of the solution increases rapidly. This is expected if the lactose transporter catalyzes the movement of lactose and protons together in a stoichiometric fashion. Eventually, the protons dragged in by the lactose slowly leak out, and the pH returns to the original value. Furthermore, the amount of protons transported is proportional to the amount of lactose added, and a stoichiometry of one proton per molecule of lactose was calculated. Experiments carried out later showed the converse effect, that a proton gradient artificially established across the bacterial membrane will drive the concentrative uptake of lactose in the absence of any other energy source.

It is now well established that this is the mechanism by which many organisms accumulate sugars, amino acids, and other nutrients. In bacteria, the nutrient accumulation is most often driven by a primary proton gradient. The same principle holds in animal cells as well, except that the primary energy source is often a gradient of Na^+ ions. The Na^+ gradient in animal cells is established by the Na,K ATPase, as already discussed; indeed, one of the important reasons for expending ATP to set up a Na^+ gradient is precisely to establish a transmembrane "Gibbs energy reservoir" that is used for nutrient accumulation.

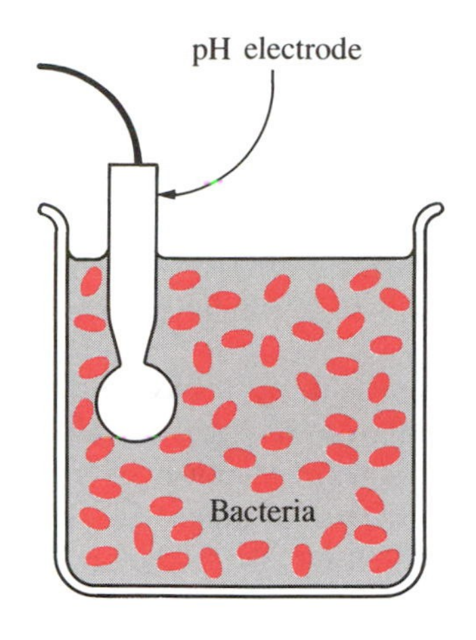

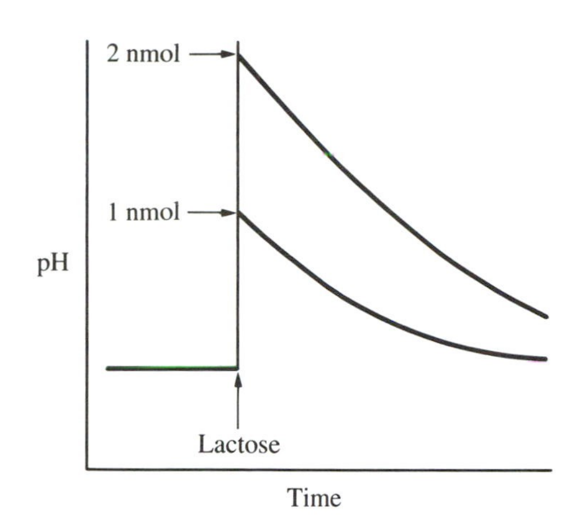

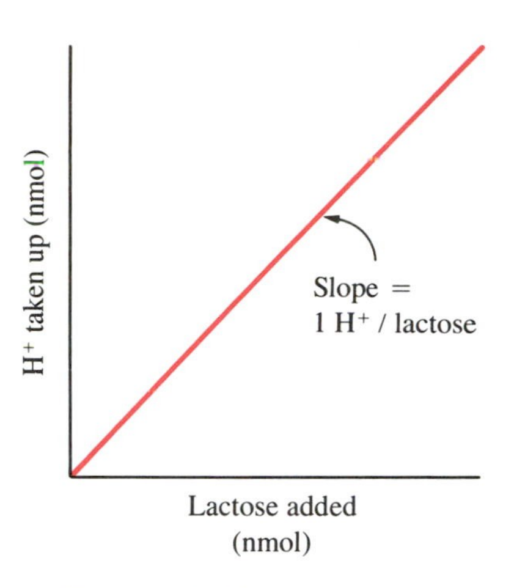

Figure 29-13

Mitchell's Demonstration of the Transport of Lactose into Bacteria by a Proton-driven Symport Mechanism. The addition of small amounts of lactose to a suspension of bacteria causes an increase in pH because protons are taken up with the lactose as it is transported into the bacteria. The right-hand graph shows that one proton is taken up for each lactose molecule that is added to the solution and taken up by the bacteria.

Na/Ca Exchanger: Antiporters

Not all coupled transporters move solutes in the same direction as the molecule or ion that provides the driving force for transport; not all transporters are symporters. We now know many examples of *antiporters,* which are transporters that catalyze the obligatorily coupled movement of two solutes, but in opposite directions across the membrane. One of the best studied of these, and one of prime importance in mammalian physiology, is the Na^+/Ca^{2+} exchanger. In many types of animal cells, this protein catalyzes the reaction:

$$3\,Na^+(out) + 1\,Ca^{2+}(in) \longrightarrow 3\,Na^+(in) + 1\,Ca^{2+}(out) \quad (15)$$

As with the lactose transporter, it is straightforward to deduce the equilibrium condition for this reaction:

$$[Ca]_{in}/[Ca]_{out} = \{[Na]_{in}/[Na]_{out}\}^3 \exp(FV/RT) \quad (16)$$

Equation 16 shows that, with an animal cell with a Na^+ gradient of 10-fold and a membrane potential of, say, -60 mV, the antiporter can pump Ca^{2+} out and establish a concentration ratio of 10^4 on the two sides of the membrane. The reason that the Ca gradient can be so much larger than the Na gradient is the high stoichiometric ratio of Na^+ to Ca^{2+}; for each Ca^{2+} that is transported out, three Na^+ ions fall inward, each of them down a 10-fold concentration gradient. Furthermore, the stoichiometry of the reaction is such that for every Ca^{2+} ion that is transported out, a net positive charge moves inward. This provides an additional favorable driving force because most cells maintain a membrane voltage that is internally negative. Again, the energy that drives this outward movement of Ca^{2+} against its gradient is the tendency of Na^+ to fall into the cell down its preestablished electrochemical gradient. This is indeed an example of biologically useful work that is driven by the Na^+ gradient: it is the ability of the cell to maintain a very low level of cytoplasmic calcium that allows this ion to be used as a regulatory signal for a large number of biochemical processes (see, for example, the regulation of muscle contraction in Chapter 30).

Electrical Work: Ion-Channel Proteins

Many types of cell membranes have the capability of generating electrical signals, and this also is a kind of work that is driven by transmembrane ion gradients. The β cells of the mammalian pancreas, for example, release insulin in response to an increased concentration of glucose in the blood stream, and the signal for the cell to do this is electrical in character. When the level of blood glucose is low, the membrane of the β cell maintains a resting transmembrane voltage of about $-50\,\text{mV}$. As the concentration of blood glucose rises, the membrane voltage goes into electrical oscillation, as shown in Figure 29-14; the voltage exhibits "bursts" of activity interspersed by quiet periods. Insulin is released only during the electrical burst, as a direct consequence of this electrical activity. The insulin then signals cells to take up more glucose, which lowers the blood glucose.

Likewise, sensory cells such as photoreceptors in the eye convert the sensory input, the absorption of a photon in this case, into an electrical signal: a change in voltage across the cell's plasma membrane. The mechanism of this process will be described in the next section. The electrical signal in the photoreceptor is then transmitted to the optic nerve and, ultimately, to the brain as electrical signals carried by neurons. Electrical signalling is necessary in situations in which speed is required, as in the transmission of nerve impulses, and when synchronous activity of many cells is necessary, as in the beating of the heart.

We now know that biological electrical signals are generated by a single class of membrane transporters—the ion channels—which use the Gibbs energy that is stored in transmembrane ion gradients. Ion channels are membrane-spanning proteins with a common feature: a hydrophilic pore that runs through the protein and thus across the membrane. This pore forms a permeability pathway through which ions can diffuse. Therefore, ion channels are nothing more than "leaks" across cell membranes. How can a membrane, which is constructed from an impermeable lipid bilayer and is filled with a variety of solute pumps to establish solute gradients, tolerate the existence of these "leak proteins"? It would seem that an ion-channel leak is exactly what a membrane would want to avoid at all costs.

The ion channels exhibit two key features that make them tolerable, and even useful, to cell membranes. First, they are able to select, with high specificity, which ions can permeate through the pore and which are forbidden to permeate. Second, ion-channel pores can open and close very rapidly, on the millisecond time scale, in response to external signals.

Let us consider the property of ionic selectivity in order to see how channel-mediated ionic leaks can set the voltage across the membrane (Figure 29-15). Consider a typical cell, with the usual high K^+ and low Na^+ inside, but let us suppose that there is absolutely no leakiness to ions. Imagine that we can measure the voltage across the membrane of this cell. Initially, there is no voltage difference across the membrane. Now, suppose that we suddenly create a leak pathway that is absolutely specific for K^+ ions. Because of the large concentration gradient for K^+, this ion will begin to leak out of the cell. But, in

Figure 29-14

Regulation of Insulin Release by the β Cells of the Pancreas. An increase in the concentration of glucose in the blood causes an increase in the membrane potential of the cell and, eventually, oscillation of the voltage. This stimulates the release of insulin by the cell.

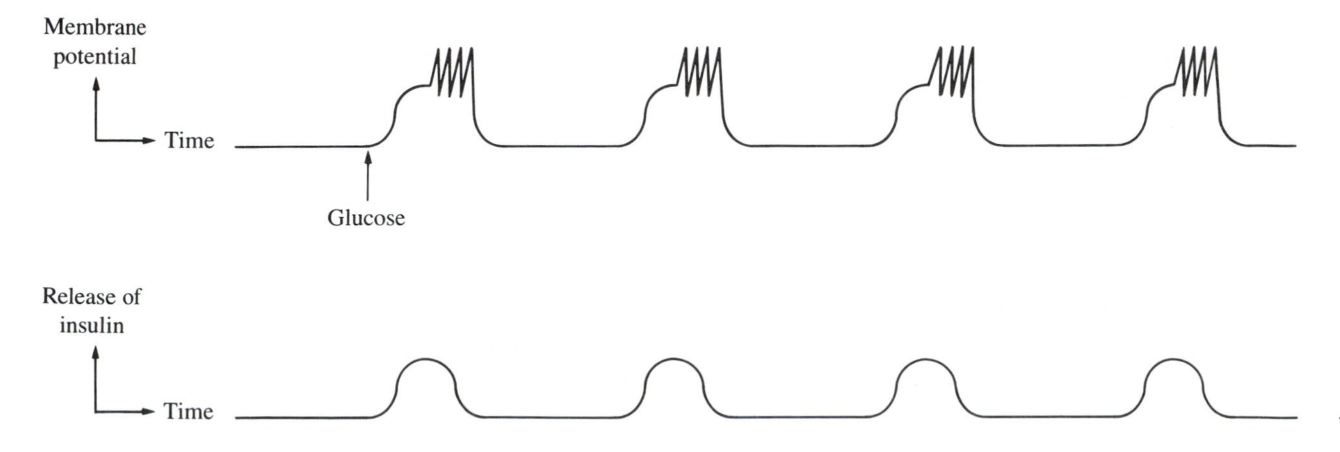

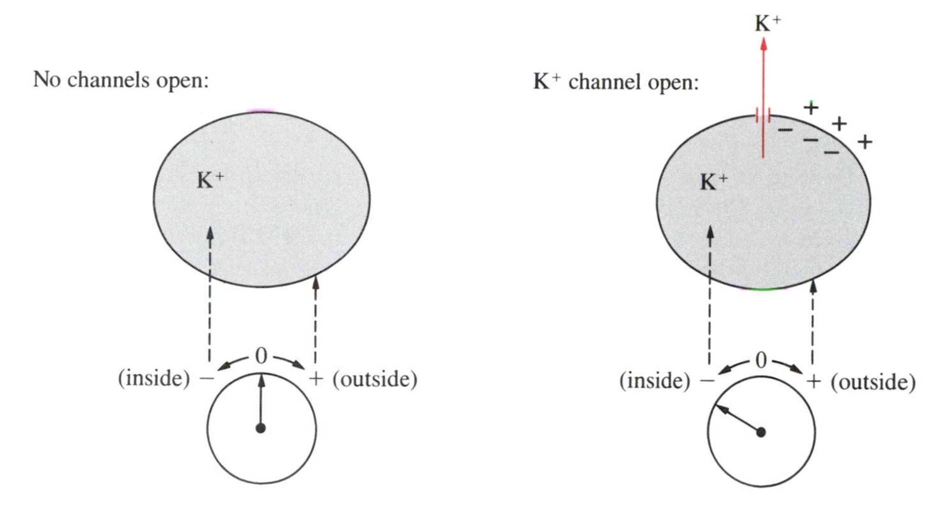

Figure 29-15

Operation of a Potassium Channel. The opening of a potassium channel results in the flow of a small number of potassium ions out of the cell, which causes the development of a gradient of charge, a voltage, across the membrane. This voltage prevents further efflux of potassium.

doing so, a positive charge will begin to build up on the outside of the membrane, because K^+ is electrically charged, leaving a net negative charge inside. This gradient-driven separation of electrical charge now causes us to record a voltage—a difference in electrical potential—on our transmembrane voltmeter.

Will K^+ continue to leak out of the cell indefinitely in this fashion? No: the voltage that is built up from the leakage of K^+ provides a thermodynamic force that counters further leakage. In fact, K^+ will stop leaking out of the cell as soon as the electrical gradient—the transmembrane voltage—exactly balances the concentration gradient (equation 4).

$$\begin{aligned}\Delta\mu_s &= \mu_s(\text{out}) - \mu_s(\text{in}) + zFV \\ &= RT\ln([\text{s}]_{\text{out}}/[\text{s}]_{\text{in}}) + zFV \qquad (4)\end{aligned}$$

This balance occurs when the voltage builds up to the value of the *Nernst equilibrium potential, V,* which is determined by the concentration ratio of K^+ on the two sides of the membrane:

$$V = \frac{RT}{F}\ln\frac{[\text{K}^+]_{\text{out}}}{[\text{K}^+]_{\text{in}}} = 59\ \text{mV} \times \log_{10}\frac{[\text{K}^+]_{\text{out}}}{[\text{K}^+]_{\text{in}}} \qquad (17)$$

Thus, a 10-fold ratio of K^+ concentrations across the membrane leads to a voltage difference of about -59 m*V*, in which the inside of the cell is electrically negative compared with the outside. This is the potential we would actually measure with our intracellular voltmeter.

The situation just described—that of a cell that is impermeable to everything but K^+—is not a bad approximation for many types of electrically active cells at "rest." For such cells, the K^+ ion is far more permeable than any other ion, mainly because of the existence in the cell membrane of "resting K^+ channels." For this reason, many cells exhibit internally negative membrane voltages that are sensitive to the transmembrane concentration gradient of K^+.

Thus, ion channels that are specific for K^+ serve to "polarize" cell membranes. In reality, cells are not totally impermeable to other ions. For example, Na^+ will slowly leak inward and allow the permeant K^+ ion to leak outward. But the operation of the energy-consuming Na,K ATPase can easily compensate for these slow ionic leaks and maintain the K^+ and Na^+ gradients.

A membrane that is polarized by a K^+ channel, as described in Figure 29-15, is in an equilibrium state. It maintains a voltage, but it does not dissipate the K^+ gradient; it does not use up any free energy that is stored in the gradient.

The transmission of electrical signals by biological membranes relies on a *change* in membrane voltage in response to a stimulus. This process, which does dissipate energy and has the capability of carrying information, relies on the second property of ion channels: the ability to switch rapidly between "open" and "closed" conformations.

To see this, consider a simplified version of an invertebrate photoreceptor cell in *Limulus,* the horseshoe crab (Figure 29-16). This cell is polarized in the dark, because of efflux of K^+ through resting K^+ channels. But, when photons are absorbed by the visual pigment, *Limulus* rhodopsin, a series of biochemical events occur that lead to the massive opening of other channels that are selective for Na^+. When this happens, the Na^+ permeability suddenly dominates the membrane, Na^+ ions flow down their gradient into the cell, and the membrane voltage swings from negative to positive. The Na^+ channels then close again, as "dark" conditions are re-established, and the K^+-dominated membrane once again returns to its negative resting potential as K^+ ions flow outward until the equilibrium K^+ potential is reached. It turns out that the number of ions that must leak across the cell membrane to change the membrane voltage is minuscule compared with the total content of ions in the cell, so that the internal ionic concentration changes only very slightly as a result of this single electrical signal.

What has happened during this sequence of electrical events following the absorption of light? When the Na^+ channels open, a small amount of Na^+ flows into the cell down the pre-established Na^+ gradient; when the Na^+ channels close again, a little K^+ flows out of the cell down its own gradient and re-establishes the resting potential. The ionic gradients have been dissipated slightly by the single photon event, and work must be done, in the form of ion pumping by the Na,K ATPase, to re-establish the original conditions. But the ionic gradients have been dissipated for a reason: the change in voltage that results from the action of these K^+ and Na^+ channels carries information. This

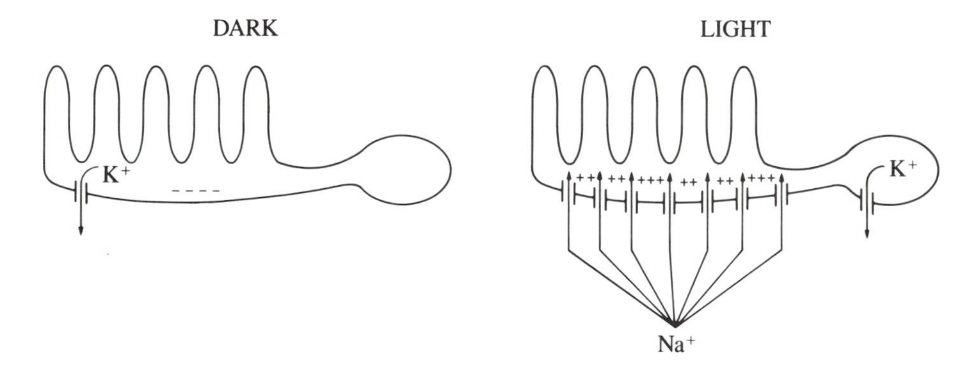

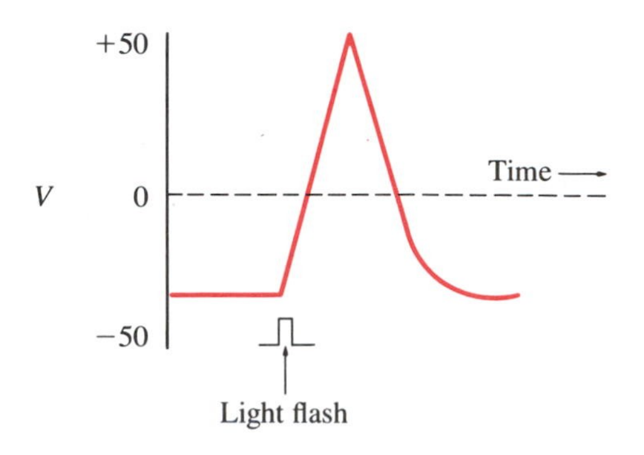

Figure 29-16

The Response to Light of a *Limulus* Photoreceptor. In the dark, the efflux of K^+ ions produces a negative voltage across the cell membrane. Light triggers a sudden influx of Na^+ ions, which reverses the membrane potential and gives rise to an electrical signal that results in vision.

information is transmitted to the brain by additional electrical signals that are mediated by the ion channels in neurons.

Ionic gradients are an absolute precondition for this signalling capacity. With no gradients present, the Na^+ permeability would change in response to light, but the membrane voltage would always stay at zero and the light would go undetected. The ability of the Na^+ channel to switch on and off in response to the light stimulus also is a necessary condition for the generation of an electrical signal.

This description is deliberately oversimplified and qualitative. The important point is that electrical signalling is a useful kind of work done by certain specialized cells. The energy source for this work is the Gibbs energy of dilution of charged solutes, Na^+ and K^+ in this case. Finally, the special characteristics of ion channels—ion-selective pores that can be switched open and closed—underlie the ability of these membranes to capture the free energy that is available from the ionic gradients and convert it into an electrical signal. In no case do we have a precise molecular picture that explains either the ion selectivity or the conformational changes that lead to the opening and closing of a channel. To gain such a picture is a challenge for future membrane biochemists.

The Biochemistry of Vision in Vertebrates

The *principles* of the visual cycle in vertebrates are similar to those in invertebrates. However, the detailed mechanism is different.

The vertebrate retina contains two types of photoreceptor cells: rods and cones. The rods are responsible for night vision or vision under conditions of low light intensity. The cones are responsible for color vision and are used under conditions of high light intensity—for example, during daylight hours. Rod cells are much larger and more numerous than are cone cells. Consequently, much more is known about the biochemistry of rod cells.

The rod photoreceptor cell is so well designed that it is capable of responding to a single photon of light. It is an elongated sensory neuron that is composed of three different segments: the *synaptic terminus;* the *inner segment* containing the nucleus, mitochondria, and the biosynthetic machinery of the cell; and, finally, the *outer segment* (Figure 29-17A). The outer segment is a highly specialized cylindrical structure composed of a plasma membrane that surrounds about a thousand free-floating membranous discs. Each disc is packed with the visual pigment rhodopsin. Rhodopsin within these discs first initiates photodetection by absorbing light that falls on the retina.

Rhodopsin is an integral membrane protein that is composed of a polypeptide chain, *opsin,* and a chromophoric molecule that enables the protein to absorb light of visible wavelengths. It is somewhat similar to bacteriorhodopsin (Chapter 23) but functions very differently. The polypeptide is 348 amino acids in length with α-helical transmembrane segments that span the lipid bilayer of the plasma membrane seven times. The chromophore is 11-*cis*-retinal, which is a derivative of all-*trans*-retinol, or vitamin A. A dietary deficiency of vitamin A causes night blindness and, eventually, degeneration of the retina. The 11-*cis*-retinal is covalently bound to the polypeptide chain by means of a protonated imine, or Schiff base, linkage to a lysine residue present in the seventh transmembrane helix of the protein opsin to form rhodopsin. Binding of the chromophore to the protein causes a shift in the wavelength maximum of the absorption spectrum from 440 nm to 500 nm.

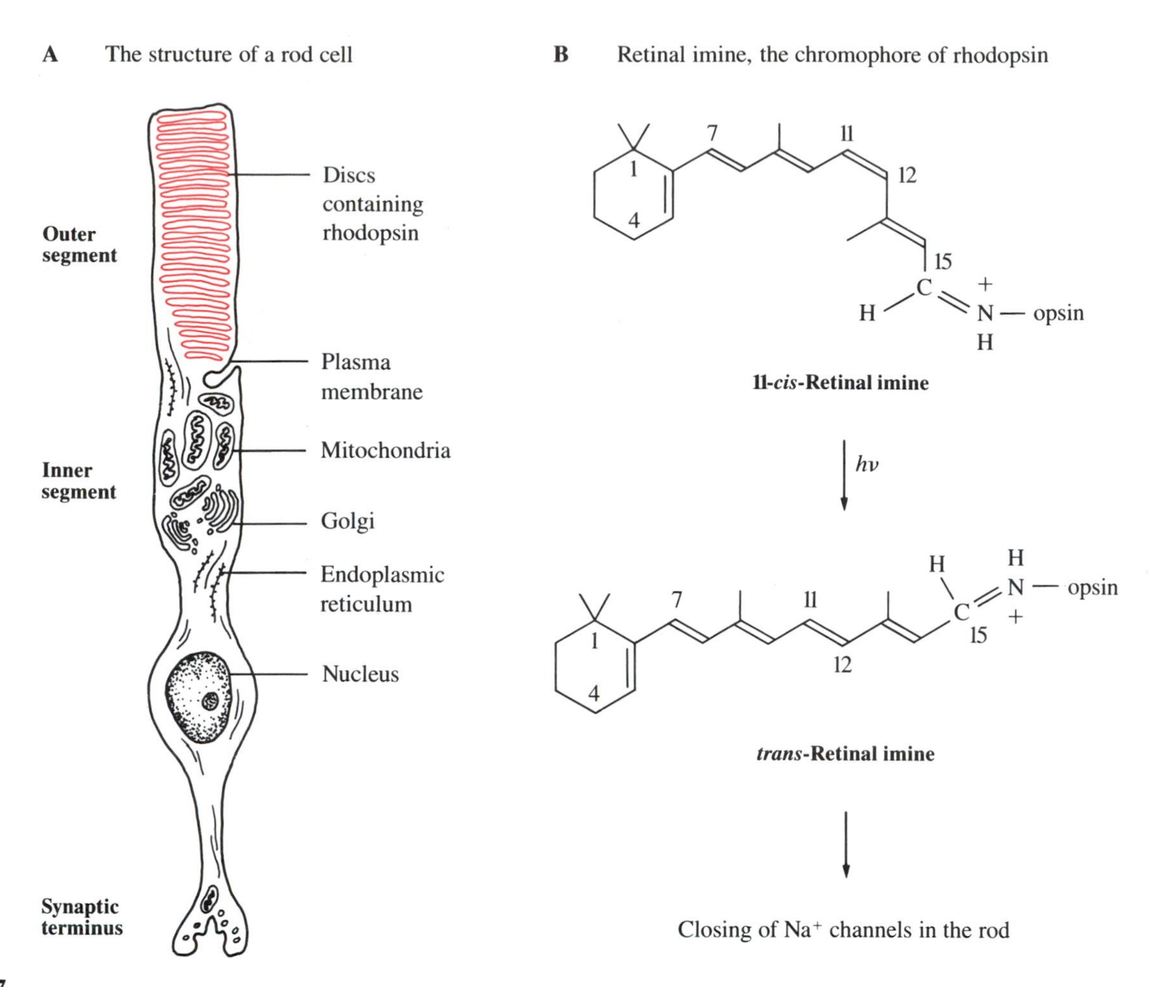

Figure 29-17

Components of Light-sensing Cells of the Eye. A: The structural organization of a rod cell in vertebrates. B: Retinal is bound to opsin as a protonated imine in rhodopsin. Light causes isomerization of 11-*cis*- to all *trans*-retinal, which initiates vision.

Upon absorption of light, the chromophore undergoes isomerization to the all-*trans* form, as shown in Figure 29-17B. This is the primary photochemical event in which light energy is transduced into chemical energy. The isomerization takes place within a matter of a few picoseconds, even at liquid nitrogen temperatures, and initiates a series of conformational changes in the protein that ultimately result in hyperpolarization of the plasma membrane. In the dark, sodium ions flow through channels in the plasma membrane of the rod outer segment and give rise to a dark current. The electrochemical gradient is maintained by a sodium pump, an active Na^+-K^+ ATPase that is present in the plasma membrane of the rod inner segment. Upon exposure of the rod to light, the sodium channels in the outer segment close. Sodium ions then accumulate in the extracellular space, which causes the cell to hyperpolarize. This inhibits the release of neurotransmitter from the synaptic terminus and initiates an electrical signal that is transmitted to the brain.

Let us now consider the mechanism by which the sodium channels in the plasma membrane are controlled by light-activated rhodopsin in the disc membranes (Figure 29-18). This system provides another example of a mechanism for response to an external stimulus (Chapter 21) in which the stimulus, light in this case, initiates a series of events that produce a time-dependent activation controlled by a G-protein and the hydrolysis of GTP. As already noted, the rod cell is an exquisitely sensitive neuron that is capable of responding to the absorption of a single photon of light. This response is brought about by a highly amplified enzymatic cascade that begins with absorption of light by

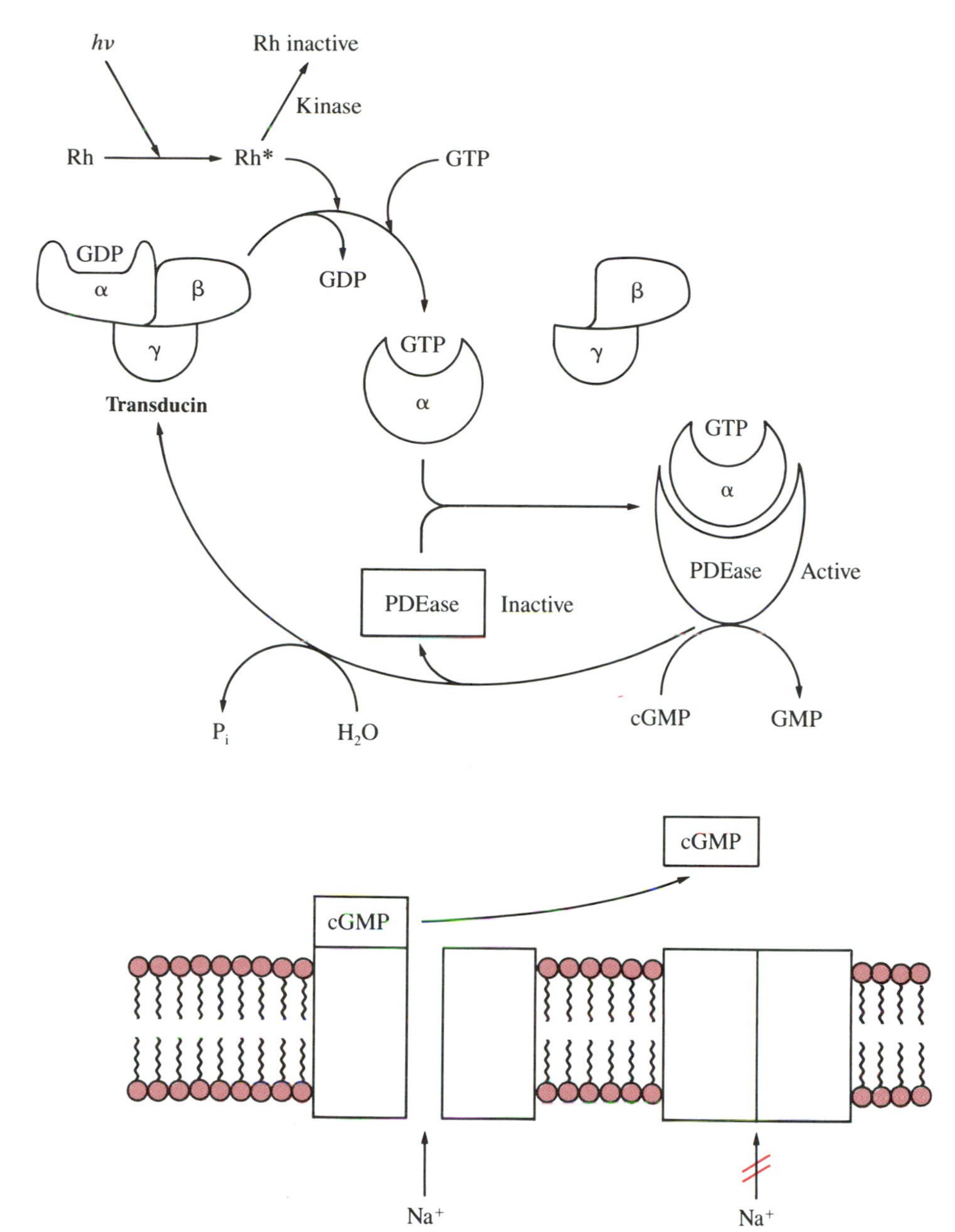

Figure 29-18

The Visual Cycle of Vertebrates. The absorption of light by rhodopsin initiates a series of reactions that ultimately decreases the permeability and increases the polarization of the plasma membrane; this initiates a signal to the brain that is responsible for vision. The individual steps of the reaction cycle are described in the text.

rhodopsin and includes activation of several other proteins from the rod outer segment. Once activated by light, rhodopsin binds to transducin, a guanine nucleotide-binding regulatory protein (G-protein) that is present on the cytoplasmic surface of the outer-segment discs.

Transducin is a heterotrimer composed of α,β, and γ subunits. The guanyl nucleotide, which is GDP in the resting state, is bound to the α subunit of transducin. Photoactivated rhodopsin catalyzes the exchange of the bound GDP for a molecule of GTP. This results in a change in the conformation of transducin and its dissociation into α-subunits and β/γ-subunits. The dissociated, GTP-bound α-subunit is the active form of transducin. The signal is amplified because one rhodopsin molecule catalyzes the activation of about five hundred transducin molecules.

The GTP-bound α-subunit of transducin then binds stoichiometrically to inactive cGMP phosphodiesterase and activates it to catalyze the rapid hydrolysis of cGMP to GMP, with a turnover number of about 4000 s^{-1}. In this manner, a single photon of light can cause the hydrolysis of about 100,000 molecules of cGMP. As the cGMP concentration in the outer segment decreases, cGMP dissociates from regulatory binding sites on the Na^+ channel in the plasma membrane, and this causes the channel pore to close. The cessation of Na^+ influx causes hyperpolarization of the plasma membrane, which inhibits the release of neurotransmitter. This provides the signal that is transmitted to the brain in vision.

The system is returned to the resting dark state by an intrinsic GTP hydrolysis activity in the α-subunit of transducin that regenerates transducin-GDP. The activity of photoactivated rhodopsin is also regulated by the phosphorylation of several serine residues near the carboxyl-terminal end of the peptide chain by a specific kinase, which decreases the activity of the pigment.

The rhodopsin is returned to the active, dark state by a complex series of reactions (Figure 29-19), in which the *trans*-retinal is removed from the protein, reduced to the corresponding alcohol, retinol, and esterified by transesterification with a fatty acid from a phospholipid in the membrane. The ester is then isomerized back to 11-*cis*-retinol, with concomitant hydrolysis of the ester bond. This prevents reversal of the reaction. The 11-*cis*-retinol is then oxidized back to the aldehyde and incorporated into a molecule of rhodopsin.

Membranes as Information-Transfer Devices

In addition to interconverting Gibbs energy among different forms, membranes have the general task of transmitting information about the external world to the metabolic machinery on the inside of the cell. Cells live in a constantly changing environment. The concentration of nutrients in the blood varies during the day, making it desirable to store sugar as glycogen in times of plenty and to release it in times of want. Cells in the kidney need to "know" when there is plenty of water available or when desiccation threatens, and so regulate the amount of water excreted into the urine. Even free-swimming cells must be aware of their environment; a sea-urchin egg, for instance, must "know" when the first sperm cell has fertilized it so that penetration by additional sperm cells

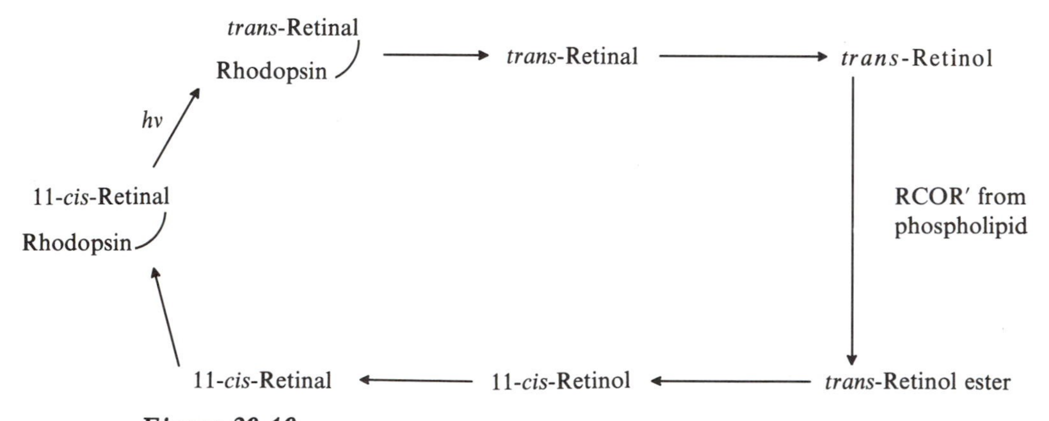

Figure 29-19

The Return of Excited Rhodopsin to Its Resting State. After excitation by light and isomerization of the 11-*cis*-retinal of rhodopsin, the *trans*-retinal is dissociated from the rhodopsin and reduced to retinol, esterified with a fatty acid from the membrane, isomerized to 11-*cis*-retinol with cleavage of the ester bond, oxidized to 11-*cis*-retinal, and combined with the protein opsin to give active rhodopsin.

is prevented. There are innumerable examples like these, in which metabolic events that take place inside the cell must respond to signals from outside the cell. The transmission of extracellular signals to cellular metabolism is a subject that might be called "transmembrane biochemistry."

At the basis of transmembrane biochemistry lies a single class of integral membrane proteins: the membrane receptors. Proteins of this class have the ability to bind a specific extracellular molecule that serves as a signal from the environment; this binding leads to a conformational change that is expressed on the *inside* of the membrane. The altered protein conformation on the internal side can then serve as the initiation point for additional intracellular changes. Because receptors are membrane-spanning proteins, there is no magic about all this; binding of a ligand at one location (extracellular) causes a conformational change that leads to a conformational change at another location (intracellular). Membrane receptors mediate the cellular responses to peptide hormones, neurotransmitters, immune-system antigens, and many other kinds of molecular signals that the cell encounters.

Summary

Biological membranes consist of bilayers that are composed of phospholipids and related molecules that have long hydrocarbon chains with polar groups on one end. The hydrocarbon chains resist exposure to water by self-association in the bilayer, and the polar groups interact strongly with water at the surface of the bilayer. Membrane proteins have nonpolar, hydrophobic regions that keep them embedded in the lipid bilayer. Many of these proteins, such as cytochrome b_5, undergo two-dimensional diffusion to interact with other proteins and substrates in the bilayer.

Some membrane proteins contain channels with varying specificities that bring about transfer of ions and metabolites across the membrane. Active transport across the membrane can be mediated by the transport of other molecules, through symport and antiport mechanisms, as in the symport of lactose and protons, or by coupling with the hydrolysis of ATP, as carried out by the Na,K ATPase and other ATP-dependent pumps. Bacteriorhodopsin is a transmembrane protein that uses the energy of light to transport protons or sodium ions across the cell membrane against a concentration gradient. These processes build up an electrochemical potential gradient across the membrane that can be utilized to do work by driving chemical reactions or the transport of other molecules and ions. Specific ion channels can generate electric potentials across membranes that are used as the driving force for other processes, or signals. Examples include the development of a membrane potential for nerve conduction, which mediates the transmission of visual signals to the brain after the stimulation of photoreceptors. The visual systems of vertebrates utilize a G-protein, transducin, to transmit the visual signal that is detected by rhodopsin, which closes sodium channels in the plasma membrane. This results in increased polarization of the cell membrane, inhibition of neurotransmitter release, and a signal that is transmitted to the brain.

ADDITIONAL READING

GENNIS, R. B. 1989. *Biomembranes: Molecular Structure and Function*. Springer-Verlag. An excellent, comprehensive summary of modern problems in membranes.

NICHOLLS, D. G. 1982. *Bioenergetics*. New York: Academic Press. A good summary of the Mitchell chemiosmotic theory, with examples taken from many types of energy coupling systems.

HOUSLAY, M. D., and STANLEY, K. K. 1984. *Dynamics of Biological Membranes.* New York: John Wiley. An introduction to membranes, with a strong focus on the physical properties of lipids in bilayers.

PROBLEMS

1. One detail of the Mitchell-Moyle experiment, not mentioned in the chapter, was the necessity of having a K^+ ionophore, valinomycin, present. This small molecule makes K^+ ions freely permeable to the bacterial membrane. To observe lactose-driven H^+ uptake by the bacterium, why is it necessary to have valinomycin present? Why would the experiment fail if a nonselective cation ionophore were used instead of valinomycin, which is highly specific for K^+?

2. The Na/Ca exchanger described in this chapter is mainly considered to be an outwardly directed Ca^{2+} "pump" used to maintain the very low concentration of Ca^{2+} inside the cell. But it can also act as a Ca^{2+} *influx* pathway triggered by changes in the cell's membrane potential. Convince yourself of this by calculating the overall Gibbs energy change for moving Ca^{2+} across the cell membrane by means of this transporter if:

(a) $\Delta V = -90\,\text{m}V$
(b) $\Delta V = -20\,\text{m}V$

3. γ-Aminobutyric acid, GABA, is an important neurotransmitter released at certain nerve terminals in the mammalian central nervous system. The cell has a major problem: to remove GABA released into the synaptic cleft in less than a millisecond after a nerve impulse and thus terminate neurotransmission. It does this by placing a high density of Na-GABA cotransporters in the nerve terminal. This "reuptake transporter" obligatorily cotransports three Na^+ ions with one GABA zwitterion. Given that the resting membrane potential is $-60\,\text{m}V$, calculate the maximum concentration gradient of GABA that can be generated by this transporter. How well would this work in a red blood cell?

4. One topic of current controversy concerns the forces determining the close packing of transmembrane helices in integral membrane proteins. A guiding principle of this subject is the idea that "once you're inside a membrane, hydrophobic forces no longer exist." What is the meaning of this aphorism?

5. Using the values of F and R in standard international units, do the arithmetic to show that a 10-fold gradient of a monovalent ion is energetically equivalent to $59\,\text{m}V$ or $1.4\,\text{kcal mol}^{-1}$ or $5.6\,\text{kJ mol}^{-1}$. (Assume that temperature = 25°C; $F = 96{,}500$ coulombs.)

6. Recalculate the Gibbs energy of the Na,K ATPase reaction described in the chapter under extreme ionic conditions, in which $[Na^+] = 1\,\text{mM}$, $[K^+] = 150\,\text{mM}$ inside the cell, and $[Na^+] = 150\,\text{mM}$ and $[K^+] = 1\,\text{mM}$ outside. Your answer should be a positive number; what does this mean? What is going on here?

Coupled Vectorial Processes: Muscle Contraction and Active Transport

30

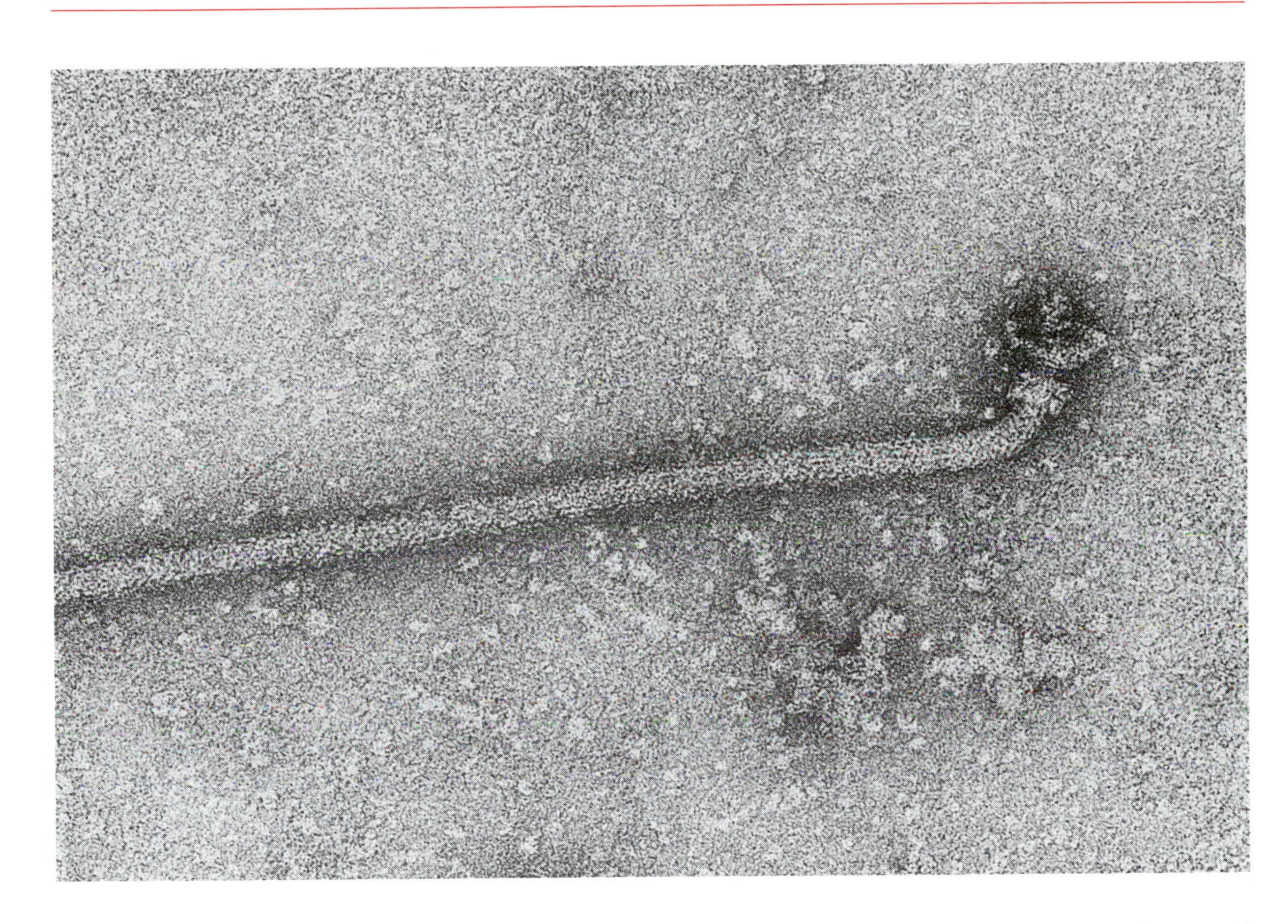

An electron micrograph of the proximal part of a bacterial flagellum (see Box 30-3). (Courtesy of Noreen Francis and David DeRosier.)

One of the most-important products of the metabolic dynamo (Figure 30-1) is *work*. This includes the mechanical work of muscle contraction, the osmotic work of active transport, and the building up of a membrane potential for the electrical work that is involved in many cellular activities, such as nerve conduction and the generation of electricity in the electric eel. Chemical energy is converted into work through *coupled vectorial processes*. The reactions are *vectorial* because they bring about the movement of a molecule or an organelle in a particular direction, and they are *coupled* because they connect chemical energy to the performance of work.

Coupled vectorial processes can be described by equation 1

$$\text{ATP} \rightleftharpoons \text{ADP} + \text{P}_i + \text{Work} \tag{1}$$

or by a similar equation with a different source of chemical energy. The problem is to understand this equation, in which a *physical* process, movement, is the product of a *chemical* reaction. Many of these processes are reversible; we have already seen in Chapters 22 and 23 that ATP synthesis is driven by the movement of protons across a membrane in oxidative phosphorylation and photophosphorylation. The *coupling* means that ATP is not hydrolyzed unless work is performed, in the forward direction. In the reverse direction, the energy that is available from an ion gradient or some physical process is not lost unless ATP is synthesized, if the reaction remains coupled in both directions. To understand this, it does not help to compare equation 1 to the chemistry and physics that are responsible for the irreversible conversion of chemical energy into work by a steam engine.

In this chapter, we will describe a few examples from the large number of coupled vectorial reactions that play a central role in the operation of every living system. We are just now beginning to understand how to think about reactions of this kind, and this area of study is one of the most interesting and challenging frontiers of biochemistry today.

We have already described the coupled transport of metabolites and ions by simple symport and antiport systems in Chapter 29. In this chapter, we will describe coupled vectorial processes that use the negative change in Gibbs free energy from the hydrolysis of ATP, a downhill process, to provide the energy for an uphill process—for example, the mechanical work of muscle contraction or the active transport across a membrane of ions and molecules, such as H^+, Na^+, Ca^{2+}, and sugars. These transport processes can occur against a concentration gradient and are somehow coupled to each other, as indicated by the circle connecting them in Figure 30-2, so that neither one can occur unless the other one also occurs. We want to understand the nature of this coupling. Figure 30-2

Figure 30-1

Lipmann's "Metabolic Dynamo."
The energy that is available from the oxidation of the carbon–hydrogen bonds of foodstuffs to carbon dioxide and water is used to form an energy-rich bond of ATP from ADP and inorganic phosphate, P_i. This energy is utilized to drive the different reactions and processes that take place in a living organism. Creatine phosphate, Cr~P, provides a reservoir of energy-rich phosphate bonds that can rapidly regenerate ATP from ADP. (Lipmann's "metabolic dynamo" is from his classic 1941 review on energy-rich compounds—see Chapter 9.)

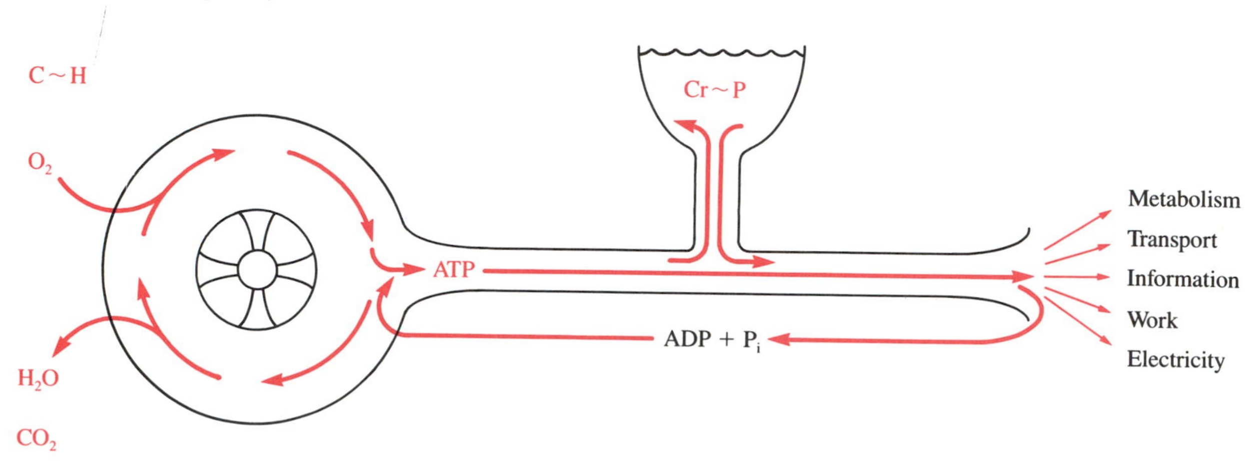

is vague, like many descriptions of coupled vectorial processes; we need a more-specific way of approaching the problem. If we can write a series of equations that describe the partial reactions of a cycle and understand what these equations represent, we can say that we have some understanding of the coupled vectorial process.

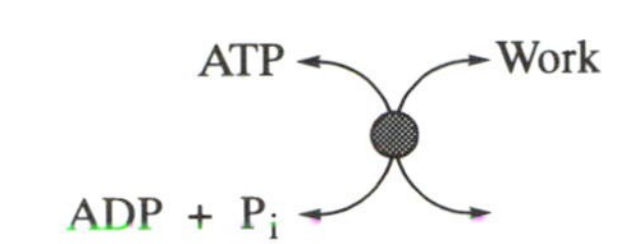

Figure 30-2

Coupling of the Hydrolysis of ATP and Work.

Muscle Contraction

The conversion of the chemical energy that is released upon the hydrolysis of ATP into the mechanical work of muscle contraction is certainly the best-known and most thoroughly studied coupled vectorial process. The history of the study of muscle contraction, like that of many biological processes, is long and complex, with many experiments that led only into blind alleys. A large fraction of the research that was carried out on the problem did not help to develop our understanding because it was directed at the wrong questions or was based on the wrong models.

Albert Szent Gyorgyi set the stage for our present understanding of muscle contraction in 1953 when he wrote that "Contraction in muscle is essentially a reaction of actomyosin, ATP, and ions."* This seemed a revolutionary statement at the time because it asserted that a physical process in a living system could be understood in terms of a chemical reaction. Most research on muscle at that time was concerned with its macroscopic mechanical properties, oxygen uptake during contraction, efficiency, and stimulation. Muscles were known to work with a high efficiency that was said by different workers to range between 40% and 90%; this is far higher than the 10% efficiency of automobiles. More-detailed research was based largely on the idea that contraction is due to the shortening of some molecule or fiber, as in a stretched rubber band.

Energy

It is now known that ATP is the immediate source of energy for muscle contraction, but this was demonstrated only quite recently. Intact muscle fibers contain the complete chain of enzymes that lead to energy production and storage, including creatine kinase and enzymes for oxidative phosphorylation and glycolysis. They also contain a high concentration of creatine phosphate, a high-energy phosphate compound that can produce ATP by reaction with ADP in the presence of creatine kinase. Because ATP and creatine are in equilibrium with ADP and creatine phosphate in the presence of creatine kinase, it was difficult to identify the most-direct energy source for contraction. ATP was finally shown to be the immediate source of energy by experiments in which creatine kinase was inhibited, so that the ADP that is formed in contraction is not immediately converted back into ATP by reaction with creatine phosphate. Furthermore, contraction was studied in a partially purified system of muscle fibrils that had been soaked in glycerol-water for a long enough time for the small molecules and soluble enzymes to diffuse away. When MgATP and calcium were added to these fibrils, they were found to contract (Figure 30-3). This experiment shows that ATP can provide the energy that is needed for muscle contraction in a simple system. The calcium stimulates contraction by a mechanism that we will examine in a later section in this chapter.

CH_3–N($CH_2CO_2^-$)–$C(=\overset{+}{N}H_2)NH_2$

Creatine

CH_3–N($CH_2CO_2^-$)–$C(=\overset{+}{N}H_2)NH$–PO_3^{2-}

Creatine phosphate

**Chemical Physiology of Contraction in Body and Heart Muscle.* New York: Academic Press, 1953, p. 4.

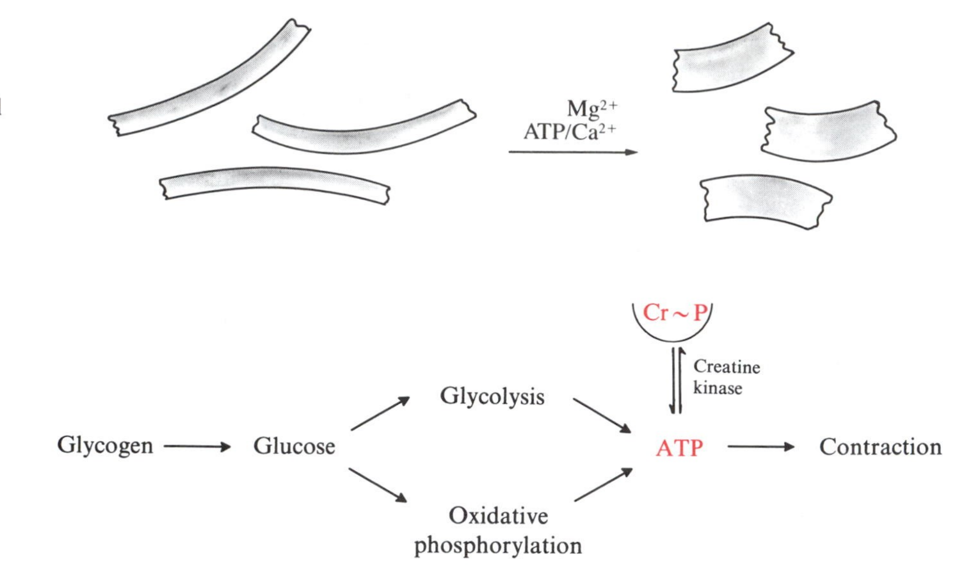

Figure 30-3

Muscle fibrils that have been extracted with aqueous glycerol shorten upon the addition of MgATP and calcium ions.

Figure 30-4

The Flow of Metabolic Energy for Muscle Contraction.

The flow of metabolic energy for muscle contraction is summarized in Figure 30-4. The role of creatine phosphate is to create a large reservoir of ~P (high-energy phosphate) in intact muscle. Glycolysis can provide ~P rapidly to replenish the pool of ~P, and oxidative phosphorylation provides ~P for sustained activity. These processes are indicated by the wheel and reservoir of Figure 30-1. The reserve stores of glycogen in the body can supply glucose for these processes until it is used up. For example, glycogen reserves in male runners are depleted after some twenty-six miles of running. It has been reported that women have a particular ability to utilize subcutaneous fat for energy, so that female runners can continue to mobilize energy after their glycogen is gone and can run for even longer distances.

Anatomy

Our present understanding of muscle contraction developed from a remarkable combination of anatomy and biochemistry. The science of anatomy went through a period of low popularity when biochemistry began to provide a clear description in chemical terms of how living systems work. However, in recent years, the importance of anatomy has again been recognized because coupled vectorial processes and other systems at the interface between biochemistry and the anatomy of the cell are becoming understood.

It has been known for a long time that there are two main classes of muscles:

1. *Striated muscle,* which is generally responsible for voluntary muscular activities.
2. *Smooth muscle,* as in the digestive system and the walls of arteries, which is generally controlled by the autonomic nervous system. It is ordinarily not subject to direct voluntary control, but some degree of control may be developed with appropriate training.

Cardiac muscle is a variant of striated muscle that is not under direct voluntary control.

Most of the research on the mechanism of muscle contraction has been on striated muscle. It is often assumed that smooth muscle works by the same or a similar mechanism but has a less-regular structure. Striated muscle has been known for many years from the examination of muscle fibers, or myofibrils, with the light microscope and later with electron microscopy. It is possible to distinguish A and I bands and a Z line through the center of the I band of a myofibril (Figure 30-5). The I, or isotropic, band brings about little rotation of polarized light and stains relatively weakly with dyes that bind to proteins. The A, or anisotropic, band rotates polarized light and has a higher protein concentration, so that thin sections for histologic examination stain more strongly. The location of enzymes in tissues can often be determined on a microscopic scale by *histochemical staining*. A substrate is added to a thin section of tissue, the enzyme is allowed to catalyze the reaction for some time, and then a reagent is added that causes the product of the reaction to precipitate or to give a color that can be observed under the microscope. Histochemical staining has shown that there is a high concentration of creatine kinase in the A band, which suggests that ATP is utilized and phosphoryl groups are transferred from creatine phosphate to ADP to regenerate ATP in this region of the muscle fiber.

Our understanding of the mechanism of muscle contraction was made possible by examination of striated muscle at higher magnifications with electron microscopy, as shown in Figure 30-6A and B. Figure 30-6C shows the arrangement of the thick and thin filaments in a cross section through the fibril. Figure 30-7 is a schematic picture of the A and I bands of relaxed muscle and contracted muscle. It is apparent that the A band that was observed at lower magnification (Figure 30-5) actually consists of a set of parallel thick filaments. These filaments interdigitate between thin filaments that extend into the A band from the I band. The I band constitutes the space between the A bands; its thin filaments are attached to the Z lines. Additional, less distinct, bands can be observed by electron microscopy. These depend on the amount of overlap of the thick and thin filaments, because there is more protein in the region in which the thick and thin filaments overlap. The cross section in Figure 30-6C shows a typical arrangement of thin filaments around the thick filaments.

H. E. Huxley and A. F. Huxley suggested independently in 1954 that muscle contraction is not caused by a shortening of contractile molecules or filaments, as had generally been assumed. Instead, they made the revolutionary suggestion that it is caused by the *sliding* of of these thick and thin filaments past each other. This proposal of the *sliding-filament model* would hardly have been

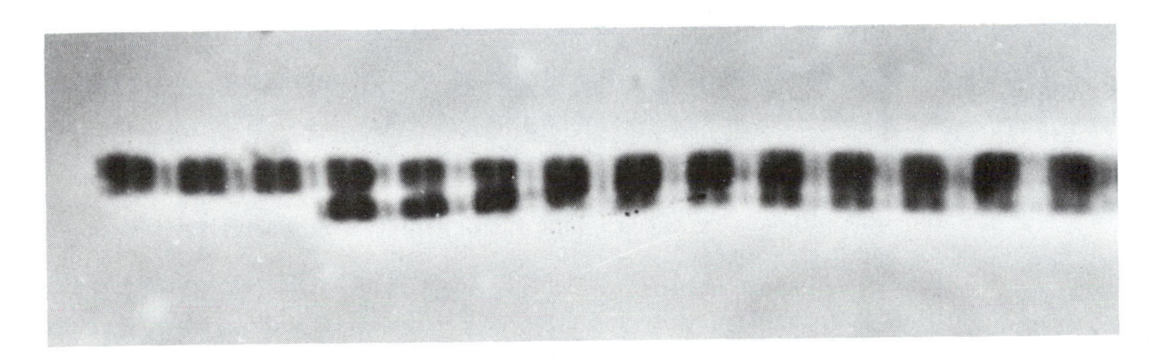

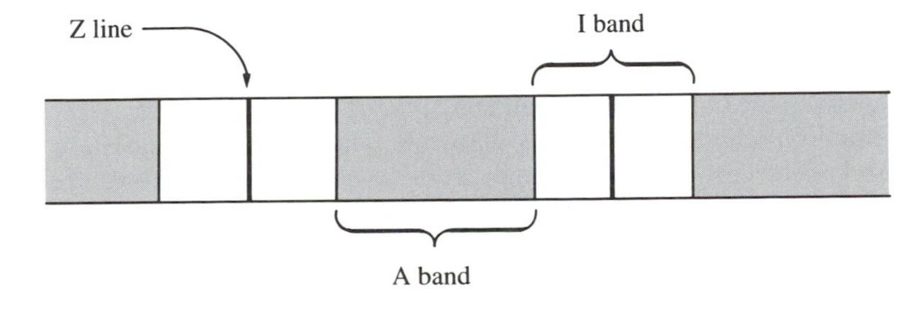

Figure 30-5

Striated Muscle at High Magnification. The diagram shows the anisotropic A band, the isotropic I band, and the Z line of the myofibril. (Micrograph courtesy of Dr. Hugh Huxley.)

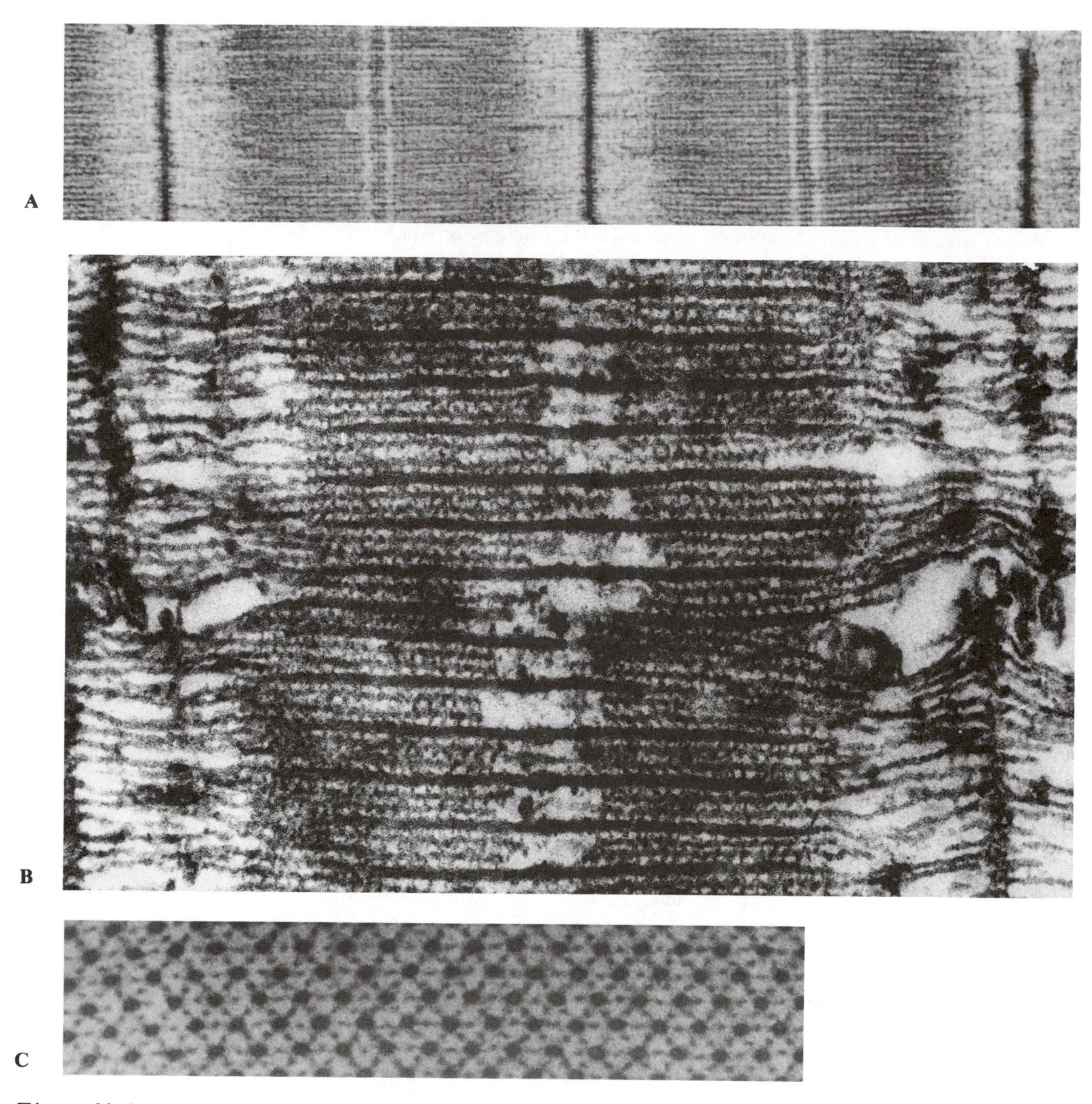

Figure 30-6

Striated Muscle, as Seen with the Electron Microscope. Two magnifications, A and B, show the thick and thin filaments, part B showing two thin filaments between the thick filaments. Part C is a cross section through the fibril. (Micrographs courtesy of Dr. Hugh Huxley.)

possible without a knowledge of the microanatomy of muscle. It is now virtually certain to be correct.

The important experimental observation is that electron microscopy of thin sections taken before and after the contraction of muscle shows that the contraction results from shortening of the I band; the length of the A band remains constant, at least in the initial stages of contraction, as shown in Figure 30-7B. There is no shortening of either the thick or the thin filaments. Instead, the thick and thin filaments slide past each other; this sliding pulls the Z lines together. *Cross bridges* from the thick filaments of the A bands can attach to the thin filaments and pull them past the thick filaments. These cross bridges can be seen by electron microscopy at high magnification, as shown in Figure 30-8. The movement has some resemblance to that of a caterpillar that is climbing up a string bean. The problem, then, is to understand how the feet, or cross bridges, sequentially attach, move, and then detach in order to cause the observed movement of the filaments.

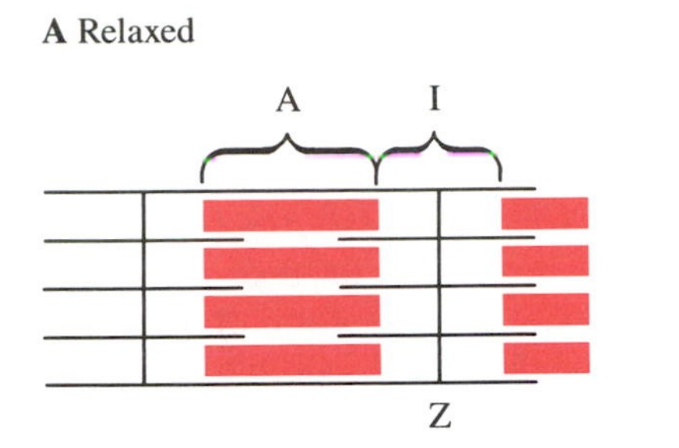

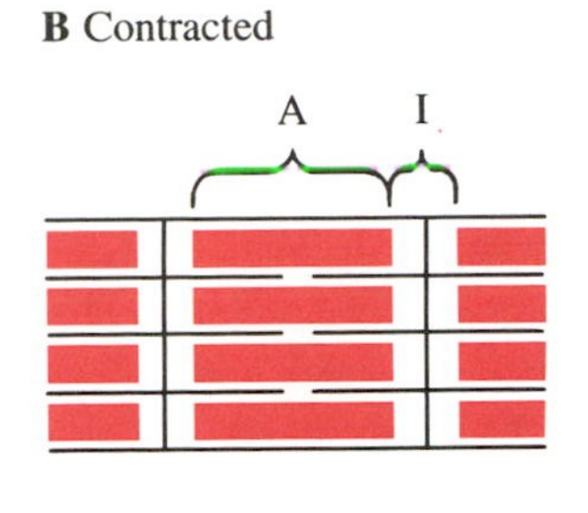

Figure 30-7

The Sliding-Filament Model for Muscle Contraction. Part A represents relaxed muscle; part B, contracted muscle, with shortening of the I bands but not the A bands.

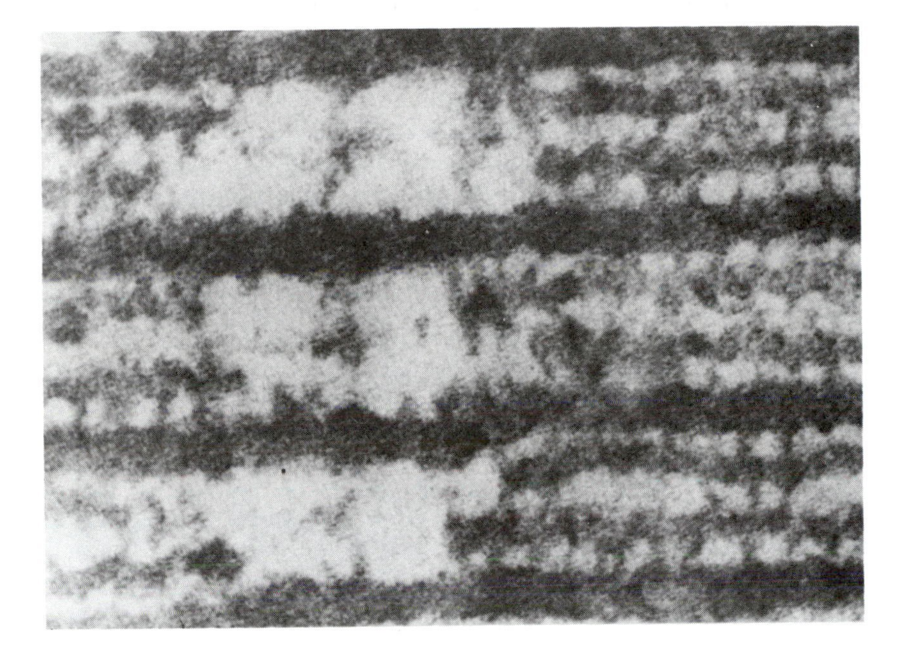

Figure 30-8

Electron Micrograph of Striated Muscle at High Magnification, Showing the Thick and Thin Filaments and the Cross Bridges that Connect Them. (Micrograph courtesy of Dr. Hugh Huxley.)

Biochemistry

Our present understanding of this problem came from a biochemical dissection of the muscle fiber. Brief extraction of muscle with 0.6 M potassium chloride gives a viscous solution of protein and removes the A band; it leaves behind the membranes and the thin filaments, which are still attached to the Z line (Figure 30-9). The extracted protein is myosin, which is sometimes called myosin A. This process is reversible, at least to some extent. The ability of muscle fibrils to contract is lost when the myosin is extracted by glycerol; it is regained when

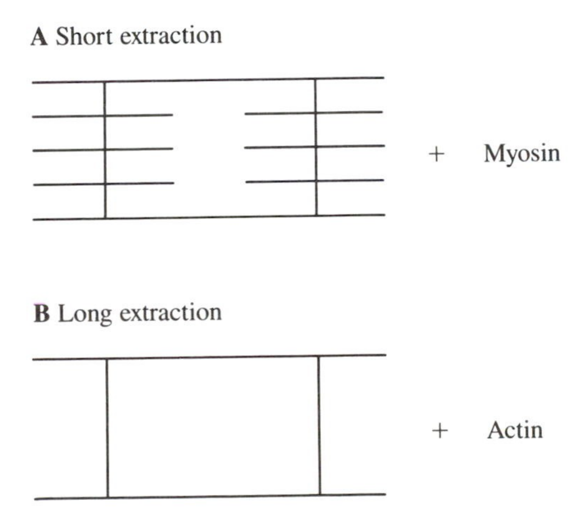

Figure 30-9

Extraction of Myosin and Actin from Striated Muscle. A short extraction of muscle sections with 0.6 M potassium chloride removes myosin, which constitutes the thick filaments of muscle. A longer extraction removes actin, which constitutes the thin filaments.

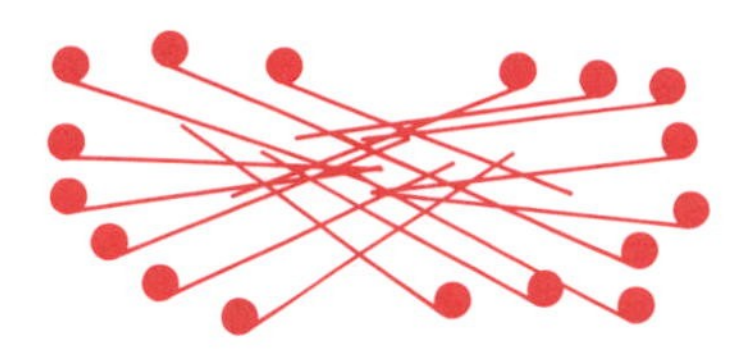

Figure 30-10

Schematic Drawing of Aggregated Myosin Molecules.

myosin is added back, although the structure is not so nicely arranged as in the original fiber. It has also been shown that antibodies to myosin bind specifically to the A band.

Myosin is a long, fibrous protein molecule; its dimensions have been estimated by electron microscopy to be approximately 20 × 1600 Å. It shows birefringence of flow: if a solution of myosin is forced through a capillary tube, the long molecules align themselves parallel to the direction of flow and rotate polarized light, just as in the intact A band. Myosin is soluble in concentrated salt solutions, such as 0.6 M potassium chloride. It comes out of solution at low ionic strength to give a structure that resembles a double-ended sheaf of wheat, with projections on both ends and sides (Figure 30-10). These projections are the cross bridges that connect the thick filaments to the thin filaments in intact muscle (Figure 30-8).

The molecular weight of myosin is now agreed to be close to 450,000. Earlier measurements gave molecular weights that ranged from 350,000 to 500,000. This range is large because it is very difficult to determine the molecular weight of a long, asymmetric molecule that tends to aggregate, undergo denaturation, and behave in a nonideal manner upon ultracentrifugation.

Myosin that is isolated from muscle can be dissected into its constituent parts in three different ways, which should not be confused with each other:

1. Myosin contains four *light chains,* each having a molecular weight of approximately 20,000. These relatively small molecules bind tightly to myosin but are not connected by covalent bonds. The function of these light chains in vertebrate skeletal muscle is still not well understood.
2. The bulk of myosin consists of two *heavy chains* of about 200,000 daltons each. These chains can be separated in solutions containing concentrated guanidine hydrochloride, a powerful protein solvent and denaturing agent (Chapter 8). Each of these chains ends in a head, which constitutes the cross bridge that can connect the myosin to the thin filament.
3. Brief treatment of myosin with trypsin cleaves it into two parts: *light meromyosin,* LMM, with a molecular weight of 140,000; and *heavy meromyosin,* HMM, with a molecular weight of about 340,000, as shown in Figure 30-11.

   Light meromyosin is almost entirely an α-helical rod; it presumably plays a structural role in the thick filament. Examination of light meromyosin by optical rotatory dispersion (ORD) gives a value of $b_0 = -660$, which is characteristic of an α helix.

   Heavy meromyosin consists of two heads and a stem. Further treatment with proteolytic enzymes, such as papain, cleaves HMM into subfragments consisting of two molecules of S_1, a 120,000 dalton fragment, which constitute the heads of myosin, and S_2, a 100,000 dalton chain that connects S_1 with the LMM chain (Figure 30-11).

Myosin has ATPase activity, which, as will be seen, is of great importance. It is remarkable that all of this activity is conserved in HMM and the S_1 subfragments, in spite of the proteolytic digestion.

Prolonged extraction of muscle fibers with 0.6 M KCl removes the thin filaments and gives a residual structure that contains membranes and the Z lines (Figure 30-9B). The solution contains a long *fibrous* protein, *F-actin,* that makes up the thin filaments. However, extraction of muscle directly with water at low ionic strength gives a solution of *G-actin,* which is a different form of the same protein. G-Actin is a *globular* protein having a molecular weight of 45,000 and containing one molecule of bound ATP and a calcium ion.

In the presence of 0.1 M KCl and a low concentration of magnesium ion, G-actin undergoes a dramatic change in its properties: it polymerizes into filaments of F-actin, the fibrous form. This interesting reaction is followed by the

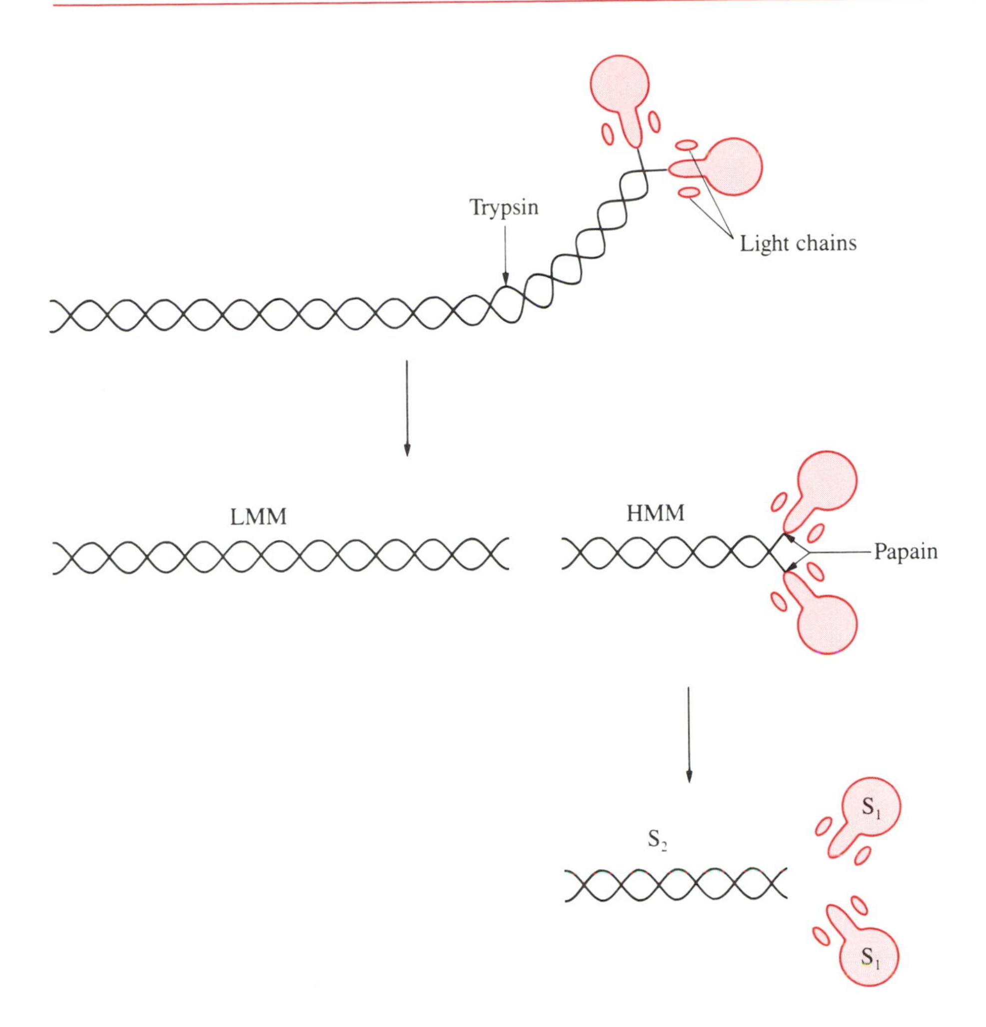

Figure 30-11

Diagram of a Myosin Molecule Before and After Limited Digestion with the Proteolytic Enzymes Trypsin and Papain. Digestion with trypsin cleaves myosin into light meromyosin (LMM) and heavy meromyosin (HMM), and further digestion with papain cleaves HMM into two subfragments, S_1 and S_2.

hydrolysis of ATP and the release of P_i. The reaction can be reversed to give monomeric G-actin by dialysis of F-actin against water that contains ATP (equation 2).

$$\underset{Ca^{2+}}{\overset{ATP}{\text{G-Actin}}} \xrightarrow[Mg^{2+}]{0.1\ M\,KCl} \underset{Ca^{2+}}{\overset{ADP}{\text{F-Actin}}} + P_i \quad (2)$$

(reverse: H_2O, ATP)

This sequence of reactions is of interest because it represents a change in the physical structure of a protein that is associated with the hydrolysis of ATP. At one time, it was thought that this might be the primary process in muscle contraction.

This is not the case, but the polymerization of actin has recently been shown to be one of the most-important processes in the maintenance of cell structure and function. The *microfilaments* of cells, which are part of the cytoskeleton (Chapter 1), are composed of F-actin. Actin is usually present at a higher concentration than any other cytoplasmic protein and microfilaments are found in all eukaryotic cells. These filaments form a dense network under the cell membrane that helps to control cell shape; in red blood cells, this network is connected to an anchoring protein, *spectrin,* and gives the cells the flexibility that they need in order to squeeze through small capillaries. Microfilaments are also responsible for the movement of cells and of cell surfaces, the ingestion of particles into the cell by phagocytosis, the separation of the two daughter cells in the course of cell division, the clustering of receptors on the cell surface, and the retraction of clots that is mediated by platelets.

We are just beginning to understand the mechanisms of these processes. It is very likely that myosin or myosinlike proteins, which are widely distributed in cells, interact with the actin filaments to cause movement. In fact, muscle contraction is almost certainly a comparatively newfangled process that has evolved from the widespread systems in which actin and myosin interact.

The behavior of actin microfilaments in the cell is controlled by a series of regulatory and structural proteins. These include spectrin, which was mentioned earlier, and *profilin,* which binds to the end of the actin filament and prevents the addition of actin monomers and elongation of the filament. The action of several of these regulatory or "capping" proteins is controlled by the calcium concentration in the cell.

Thus, actin filaments not only play an important role in maintaining the structure of of the cell, but are also responsible for many of the activities of the cell related to movement. This movement is made possible by the polarity, or directionality, of the assembly and disassembly of the microfilaments, which is controlled by the hydrolysis of ATP to ADP and P_i upon polymerization. This polarity leads to the preferential addition of actin at one end of the filament and loss from the other end, which results in what has been called "treadmilling."

Actin-ADP typically dissociates from one end of a microfilament with a first-order rate constant of $6\,s^{-1}$, which is about tenfold larger than the rate constant for the dissociation of actin-ATP from the other end of the filament. Therefore, there is a tendency for G-actin-ATP to bind at one end of the filament and for actin-ADP to dissociate from the other end. This results in treadmilling (Figure 30-12). The relative rates for the binding of actin-ATP and the dissociation of actin-ADP control the length of the filament. The ATP on G-actin-ATP undergoes hydrolysis after it has added to the filament and becomes F-actin-ADP.

The thin filaments of actin consist of two chains of F-actin that are arranged in an extended double helix. The myosin heads that project from the thick filaments of muscle cells can attach to these actin subunits. In order for a myosin head to detach from actin and find another actin monomer that faces in the same direction, the myosin has to move relative to the actin filament by 37 nm, which corresponds to seven actin monomers in the filament.

The product of prolonged extraction of muscle is actomyosin, a mixture of actin and myosin. Actomyosin is not a stable muscle protein. It is formed briefly when the myosin head attaches to the actin filament, but it dissociates rapidly when ATP binds to myosin. During contraction, actomyosin is formed and dissociates repeatedly, as the thick filament pulls the thin filament past it.

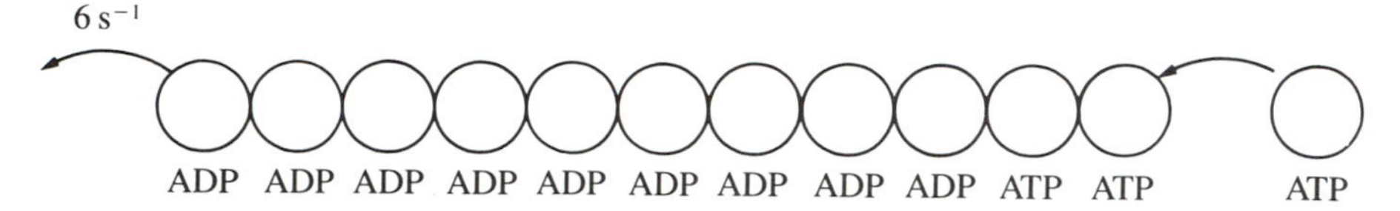

Figure 30-12

The loss of actin · ADP from one end and the addition of actin · ATP to the other end of an actin filament results in "treadmilling."

Enzymology

The key to our present understanding of the mechanism of muscle contraction came from combining this microanatomy and protein chemistry with enzymology. It had been known for a long time that the ATPase activity of myosin has peculiar properties. There is an initial *burst* of activity that is followed by slow steady-state turnover, just as in the hydrolysis of esters catalyzed by chymotrypsin. However, several years passed before it was established that this burst of activity corresponds to the hydrolysis of one mole of ATP for each myosin head; it took even longer until its implications for the mechanism of muscle contraction were realized.

The burst represents the hydrolysis of one mole of ATP to give ADP and P_i, which remain tightly bound to the enzyme; this is shown in the first two steps of equation 3.

$$\text{ATP} + \text{M} \underset{}{\overset{k_1}{\rightleftharpoons}} \text{M}\cdot\text{ATP} \underset{k_{-2}}{\overset{k_2}{\rightleftharpoons}} \text{M}{:}^{\text{ADP}}_{\text{P}_i} \xrightarrow[\text{}]{\overset{\text{Slow}}{k_3}} \text{M} + \text{ADP} + \text{P}_i \qquad (3)$$

(A Mg^{2+} ion is bound to the ATP and participates in this and subsequent reactions but is omitted from equation 3 for simplicity.) Several other possible explanations for such a burst were excluded. It does not represent the formation of a covalent enzyme-phosphate intermediate, $E\text{-}PO_3^{2-}$, that is hydrolyzed in acid during analysis. If there were a covalent $E\text{-}PO_3^{2-}$ intermediate, acid precipitation of the myosin in $H_2{}^{18}O$ immediately after the burst would give $H^{18}OPO_3^{2-}$ as the product from the hydrolysis of $E\text{-}PO_3^{2-}$, however, no incorporation of ^{18}O from water into the product was observed when this experiment was carried out.

The equilibrium constants for the steps of equation 3 are very curious. The binding of MgATP to myosin is extremely strong, with an association constant of $\sim 10^{11}\ M^{-1}$. Even more remarkable is the fact that the hydrolysis of ATP at the active site of myosin is readily reversible, with an equilibrium constant of only about 10 for hydrolysis. The reversibility of ATP hydrolysis was shown by using ATP that was labeled with ^{18}O in the terminal phosphate group and measuring the incorporation of oxygen atoms from water into the inorganic phosphate product. Figure 30-13 shows that each reversal of ATP hydrolysis incorporates an oxygen atom from water. The fact that all three ^{18}O-labeled oxygen atoms in the terminal phosphate group of ATP can be lost during hydrolysis means that the hydrolysis reaction must go back and forth many times; it is much faster than the dissociation of ADP and phosphate from the enzyme. This oxygen exchange also requires that the bound phosphate can rotate, so that different oxygen atoms are replaced by unlabeled oxygen from water as the reversible hydrolysis proceeds. This situation is similar to the behavior of the F_1 ATPase of oxidative phosphorylation (Chapter 22). We will return to these equilibrium constants shortly.

Because the $M\cdot ADP\cdot P_i$ complex accumulates after the burst, the slow, rate-limiting step for turnover of the ATPase must be the release of the products, ADP and P_i, in the k_3 step of equation 3. In the presence of magnesium, product release is very slow, with a rate constant of $\sim 0.02\ s^{-1}$. This is very much slower than the rate constant of $k_2 \sim 100\ s^{-1}$ for the hydrolysis step. However, in the presence of actin, the rate constant for ATP hydrolysis is greatly increased, to $\sim 20\ s^{-1}$. *This means that the actin must combine with* $M\cdot ADP\cdot P_i$ *and increase the rate of product release by at least 1000-fold.* Both the combination of myosin with actin and the hydrolysis of ATP are part of the contraction cycle, according to the sliding-filament model, so that the combination of $M\cdot ADP\cdot P_i$ with actin greatly accelerates a slow step in the process and makes free myosin available for the next turnover. This accounts for the high ATPase activity that is observed in

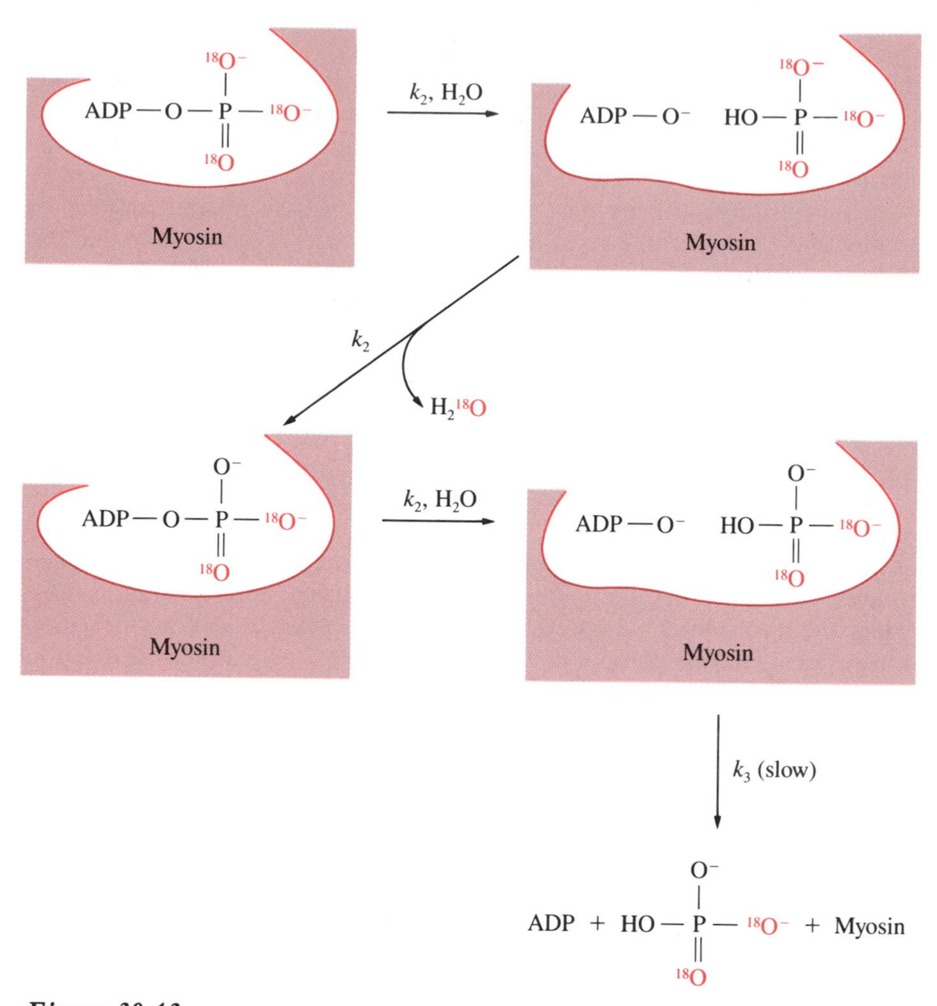

Figure 30-13

The reversible hydrolysis and resynthesis of ATP at the active site of myosin results in the loss of ^{18}O from the terminal phosphate of [γ-^{18}O]ATP.

the presence of actin, the so-called actomyosin ATPase. It also accounts for a large decrease in the amount of phosphate-water exchange (Figure 30-13) during ATP hydrolysis in the presence of actin. The increase in the rate of product dissociation (k_3) in the presence of actin gives less opportunity for the resynthesis of ATP (k_{-2}) and, therefore, less oxygen exchange.

The immediate product of this reaction after the release of P_i is AM · ADP, which rapidly loses ADP to give actomyosin, AM. Only one more step is needed to construct a cyclic process in which the formation and cleavage of actomyosin is accompanied by ATP hydrolysis: this is the dissociation of actomyosin into actin and myosin. This dissociation is brought about by ATP at a rate that is too fast to measure accurately; the rate constant is $\sim 1000\,s^{-1}$. The immediate product of this reaction is M · ATP, after which ATP hydrolysis occurs rapidly. All of the steps of the cycle can now be put together as shown in Figure 30-14.

Lymn and Taylor have, in fact, put these steps together and suggested that the reaction cycle of Figure 30-14 provides a reasonable mechanism for the contraction cycle of muscle. It agrees with the known enzymatic and structural properties of myosin and actin. There is only one more requirement for the cycle in order that it should give movement and do mechanical work. This is that the M · ADP · P_i does not combine with the same actin monomer from which the

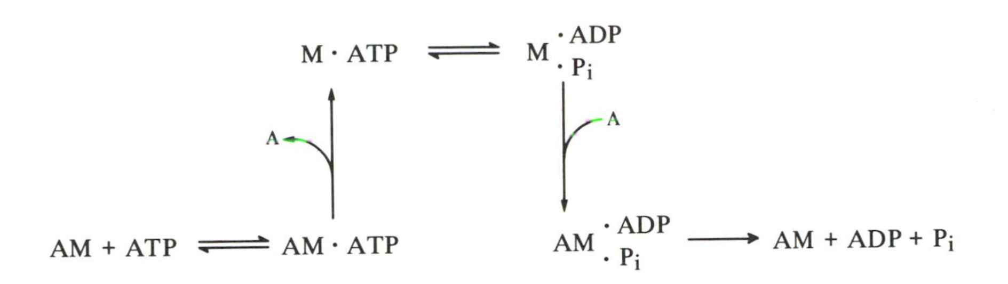

Figure 30-14

The Reaction Cycle of Myosin ATPase (M) in the Presence of Actin (A).

M·ATP dissociated; instead it must combine with an actin monomer that is further along on the thin filament. If this requirement is met, each reaction cycle results in movement of the two filaments relative to each other and shortening of the muscle fiber.

This process is illustrated schematically in Figure 30-15. We start with a myosin head on a thick filament that is attached to actin monomer 1 on a thin filament. The binding of ATP to myosin causes AM·ATP to dissociate rapidly, giving A and M·ATP. The M·ATP then undergoes rapid hydrolysis and the complex M·ADP·P_i combines with a *different* actin monomer, A_3. This causes the hydrolysis products, ADP and P_i, to dissociate. The attached myosin then undergoes a conformational change that pulls myosin and the thick filament to the right, relative to the actin on the thin filament. This step provides the driving force for movement of the thick filament relative to the thin filament and is called the "power stroke."

This process corresponds to *isotonic contraction,* in which there is a constant force but changing length of the muscle. Isotonic contraction can do work, such as lifting a suitcase. *Isometric contraction,* in which there is no net change in the length of a muscle under tension, presumably consists of many such small movements, but they are accompanied by an equivalent amount of backsliding of the detached segments of filament. Therefore, there is no net movement and isometric contraction does no work (it may, however, use up a lot of energy, as is well known to anyone who has held up a heavy suitcase for some time).

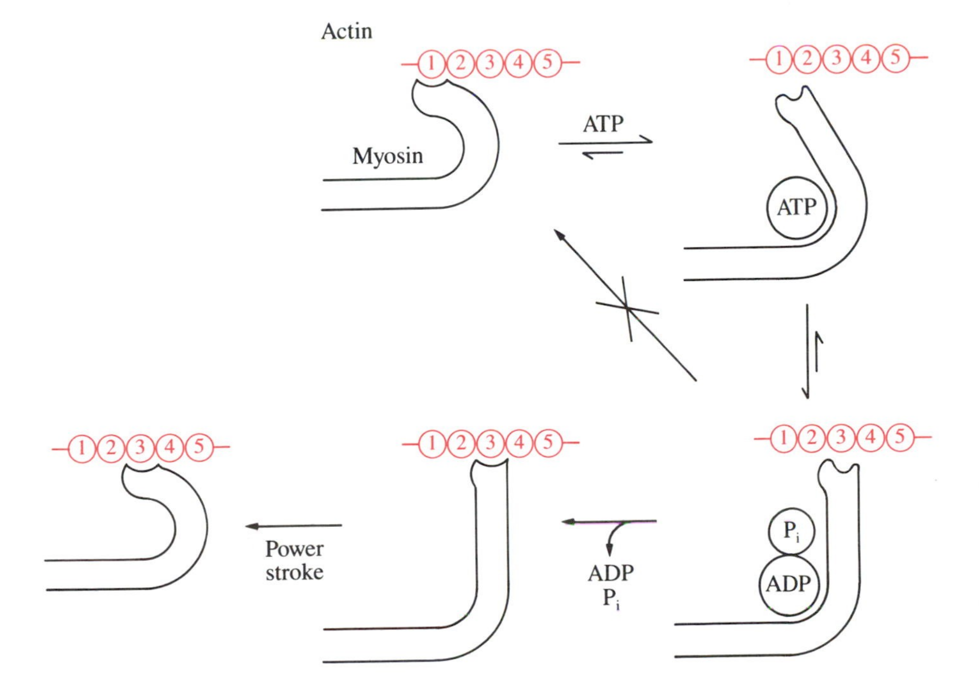

Figure 30-15

Reaction Cycle for Muscle Contraction. To perform work, the myosin molecule must reattach to the actin filament at a position different from its original position.

Coupling of the Hydrolysis of ATP and the Contraction of Muscle

We can now ask, What is it about the reaction cycle of Figure 30-15 that makes it move in the forward direction and carry out mechanical work? The reaction cycle results in the hydrolysis of ATP and performs work, so that it is a coupled vectorial process. What is responsible for the coupling between the hydrolysis of ATP and the performance of work?

The most-obvious requirement for doing work in the cycle of Figure 30-15 is that the $M \cdot ADP \cdot P_i$ must combine with A_3, not A_1. If the myosin-products complex combined with the original actin monomer, A_1, it would release products but there would be no net movement. The reaction cycle would then hydrolyze ATP to ADP and P_i but would not perform work; it would be uncoupled.

The mechanism by which this uncoupling is prevented has not yet been established. It may result from a conformational change in myosin that is brought about by the binding of ATP and moves the myosin head away from A_1 and toward A_3. This could occur either before or after the dissociation of actomyosin and hydrolysis; there is evidence for conformational changes in these steps. Another possibility is that the power strokes of other myosin molecules that are attached to actin pull the thick filament along while the myosin-product complex is detached, so that the myosin head recombines with A_3 rather than with A_1.

Thus, the coupling between the hydrolysis of ATP to free ADP and P_i and the production of work in the mechanism of Figure 30-15 depends on binding and dissociation steps. It does not depend directly on the hydrolysis step itself at the active site, which is rapid and reversible.

The basic requirements that account for coupling in this system can be described by four rules:

1. The dissociation and binding of actin and myosin occur only in the presence of nucleotides.
2. Myosin binds and dissociates nucleotides and P_i only in the presence of actin.
3. The binding (or dissociation) of ATP occurs only with A_1M.
4. The dissociation (or binding) of ADP and P_i occurs only with a different actin monomer—for example, A_3M—that is further away from the Z line.

If these four rules are followed, the muscle will contract and perform mechanical work in each reaction cycle.

Energies and Interaction Energies of the Contraction Cycle

The energetics for one reaction cycle of this system are shown in Figure 30-16. The ΔG values are based on physiological concentrations of reactants (see Box 30-1). The dissociation of the products, ADP and P_i, is strongly downhill in the last step and accounts for more than half of the total ΔG for the hydrolysis of ATP. Therefore, this step provides a large amount of energy that can be used to bring about mechanical work. It is energetically favorable because the binding of myosin to actin is very strong, whereas the binding energies for ADP and P_i are weak.

The dissociation of AM that is brought about by ATP is made possible by the very strong binding of ATP to M and by a conformational change that destabilizes the AM complex when ATP binds. This destabilization causes actin

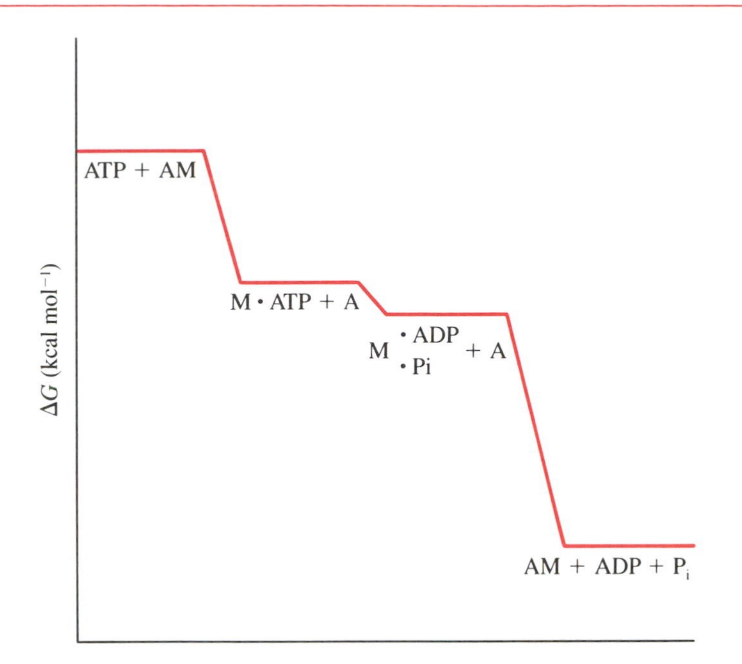

Figure 30-16

Energy Diagram Showing How the Free Energy of Hydrolysis of ATP Is Released in the Different Steps of Muscle Contraction. The diagram is based on estimated physiological concentrations of the reactants, which are not known with certainty.

to dissociate from myosin in less than 1 millisecond. The binding and destabilization are described by the box of reactions shown in Figure 30-17. The Gibbs energy changes for these reactions are related as shown in equations 4 and 5,

$$\Delta G_1 + \Delta G_2 = \Delta G_3 + \Delta G_4 \tag{4}$$

$$\Delta G_2 - \Delta G_3 = \Delta G_4 - \Delta G_1 = \Delta G_I \tag{5}$$

because the difference in energy between M + A + ATP and AM•ATP is the same when either the upper-right or the lower-left path in Figure 30-17 is followed.

The difference between the Gibbs energy changes on any two opposite sides of this box of reactions is the *interaction energy*, ΔG_I (equation 5, see also Chapters 5 and 22). The interaction energy is the amount by which ATP

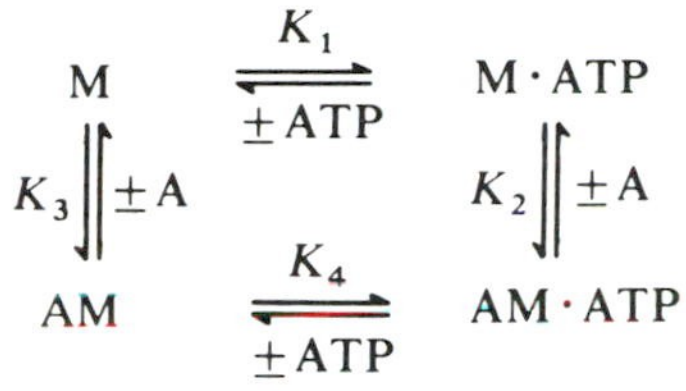

M $\underset{\pm ATP}{\overset{K_1}{\rightleftharpoons}}$ M•ATP

K_3 ⇅ ±A K_2 ⇅ ±A

AM $\underset{\pm ATP}{\overset{K_4}{\rightleftharpoons}}$ AM•ATP

Figure 30-17

Thermodynamic Box Describing the Equilibrium Constants for the Binding of ATP and Actin to Myosin.

BOX 30-1

GIBBS ENERGY CHANGES IN MUSCLE

Changes in Gibbs free energies that are based on a standard state of 1.0 M give a misleading picture of the energy balance in a physiological system because the concentrations of reactants and products in the cell are nowhere near 1 M. It is more useful to describe the Gibbs energy changes by using standard states that correspond to physiological concentrations of reactants, or "basic" Gibbs energy changes, as described in Chapter 9. The numbers in Figure 30-16 are estimated Gibbs energy changes for the concentrations of reactants that are found in muscle. They are uncertain because some of the equilibrium constants are not known accurately. In particular, actin and myosin are not ordinary reagents in solution; they are held near each other in the thick and thin filaments and the concentrations that determine the amount of binding to each other are not known.

Equations 10 and 11 of Chapter 9 show how the Gibbs free energy change of a reaction depends on the concentrations of reactants and products. Under physiological conditions, the free energy change of a reaction in which one molecule of reactant gives two molecules of product, such as the hydrolysis of ATP to ADP and inorganic phosphate, is much more negative than the standard free energy change for a standard state of 1 M. This is so because the physiological concentrations of reactants are usually much less than 1 M. Consider, for example, a reaction in which one molecule is at equilibrium with two molecules of product and the equilibrium constant is 1 M^{-1}. The fraction of the reacting molecules present as product is much larger when the reactant is 0.01 M and both of the products are 0.1 M than when both the reactants and the products are 1 M.

decreases the strength of the binding of myosin to actin ($\Delta G_2 - \Delta G_3$), which is equal to the amount by which actin decreases the strength of the binding of myosin to ATP ($\Delta G_4 - \Delta G_1$).

These relations are shown in the energy diagram of Figure 30-18. The binding of free M to A and to ATP is very strong; in both cases, it is steeply downhill and essentially irreversible under physiological conditions. However, the binding of A to M · ATP and of AM to ATP is much weaker and is readily reversible. The difference represents the interaction energy, or mutual destabilization, of A by ATP and of ATP by A when both are bound to myosin in AM · ATP. This interaction energy makes the system work, because each compound favors the dissociation of the other. If there were no interaction energy, the binding of A to M would be just as tight in the AM · ATP complex as in AM and dissociation would not occur. This is shown by the dashed lines in Figure 30-18. The interaction energy, ΔG_I, is the difference between this situation (dashed lines) and the actual situation (solid lines), as shown in the figure.

A similar interaction energy is responsible for the readily reversible hydrolysis of ATP at the active site of myosin (Figure 30-13), as it is also in the ATPases of oxidative phosphorylation and photophosphorylation (Chapters 22 and 23). This is shown for myosin in Figure 30-19. The binding of ATP to M is very strong, almost irreversible, whereas the binding of ADP and P_i is weak. The result is that the synthesis of ATP, which does not occur spontaneously in solution (left side of Figure 30-19), occurs easily at the active site of myosin (right side). The difference in these Gibbs energy changes corresponds to an interaction energy of $\Delta G_I = 13\,\text{kcal mol}^{-1}$. This raises the Gibbs energy of the complex M · ADP · P_i to a level that is high enough to allow it to form the complex M · ATP easily. The very strong binding of ATP stabilizes M · ATP and pulls the reaction toward ATP synthesis.

This interaction energy means that there is a large destabilization of the complex M · ADP · P_i relative to M · ATP, compared with the corresponding

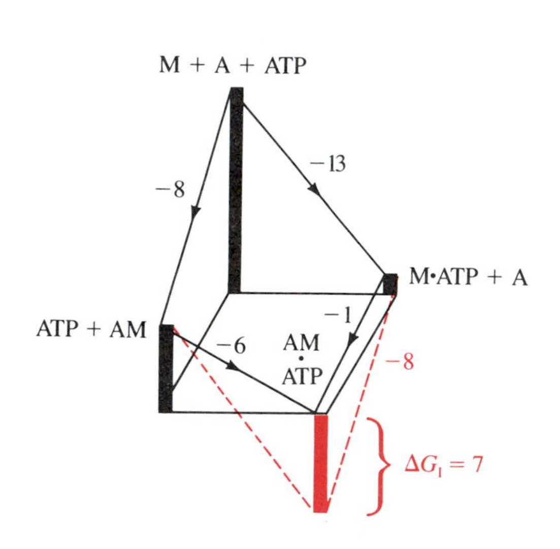

Figure 30-18

Energy Diagram for the Dissociation of Actomyosin by ATP. The binding of ATP weakens the binding of myosin to actin, and the binding of actin weakens the binding of myosin to ATP, by approximately $7\,\text{kcal mol}^{-1}$, the interaction energy. The plane is the Gibbs energy of AM · ATP.

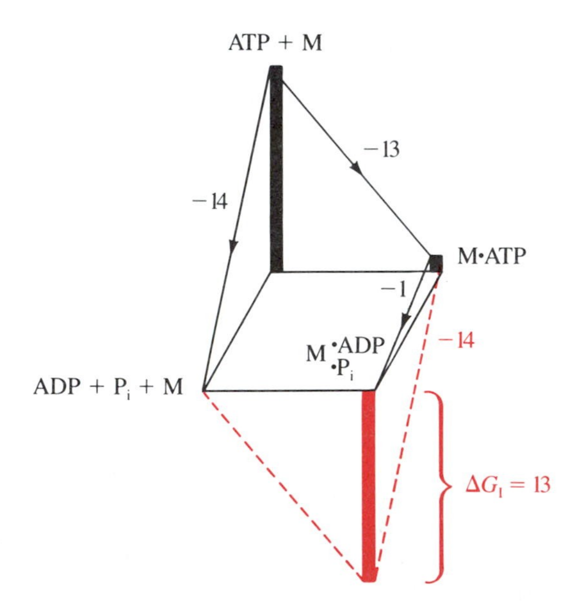

Figure 30-19

Energy Diagram for the Synthesis and Hydrolysis of ATP at the Active Site of Myosin and in Solution. The tighter binding of ATP to myosin, compared with that of ADP and P_i, favors the reaction on the enzyme by the interaction energy of approximately $13\,\text{kcal mol}^{-1}$.

reaction in the absence of myosin. There is presumably some compression, distortion, desolvation, or other destabilization of bound ADP·P_i that is relieved when ATP is formed. Binding of ADP and P_i certainly brings about a loss of entropy in the complex M·ADP·P_i, which corresponds to a positive ΔG_I (Chapter 5). The synthesis of ATP from ADP and P_i is much more favorable within the complex than it is in solution because the reaction of the bound reactants at the active site of myosin is intramolecular rather than bimolecular; it is much more probable and requires a smaller loss of entropy once ADP and P_i are bound in the correct position to react. We have already seen how the same advantage of intramolecular reactions provides an important contribution to the synthesis of ATP from ADP and P_i by the F_1F_0ATPase of oxidative phosphorylation (Chapter 22).

Control of Muscle Contraction: Stimulation and Relaxation

It is just as important that a muscle be able to relax as to contract. The absence of relaxation, after death, is a state of rigor mortis.

Many years of research went into the search for a soluble factor that causes relaxation and, in fact, several publications claimed the identification of such a factor. However, the results were not always reproducible and did not account for the relaxation that is observed under physiological conditions. Finally, it became clear that the reason the relaxing factor was hard to isolate is that it is not a chemical compound; it represents the *absence* of calcium ions rather than the presence of a soluble factor. The relaxing factor is analogous to phlogiston—the relaxing factor is the absence of calcium, just as phlogiston referred to the absence of oxygen in the period before oxygen was identified.

Conversely, muscle contraction is brought about by the *presence* of calcium. The stimulation of a voluntary muscle is initiated by the arrival of a nerve impulse at the motor end plate on the surface of a myofibril, which stimulates the release of acetylcholine at the neuromuscular junction. A stimulus is then transmitted along the membrane and a system of *transverse tubules,* which causes the opening of a gated calcium channel, the *ryanodine* receptor. This channel is located in the membrane of the *terminal cisternae* of the *sarcoplasmic reticulum* (Figure 30-20), which contains a high concentration of

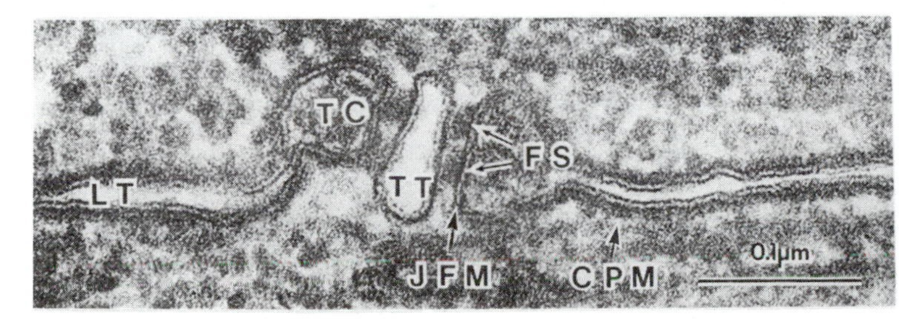

Figure 30-20

Electron Micrograph of a Muscle Fiber Showing the Transverse Tubules and Sections of the Longitudinal Tubules of the Sarcoplasmic Reticulum. The longitudinal tubules (LT) end in terminal cisternae (TC) that are connected to the transverse tubules (TT) by a junctional face membrane (JFM) and foot structures (FS). The sarcoplasmic reticulum is surrounded by the calcium pump membrane (CPM). The terminal cisternae contain calcium channels that open when a signal is received from the transverse tubule and initiate contraction by releasing Ca^{2+} into the cytoplasm. (Micrograph courtesy of Sidney Fleischer.)

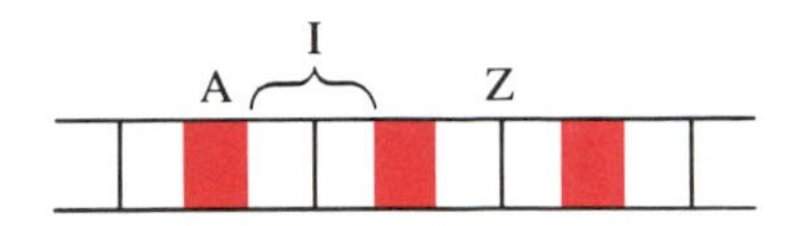

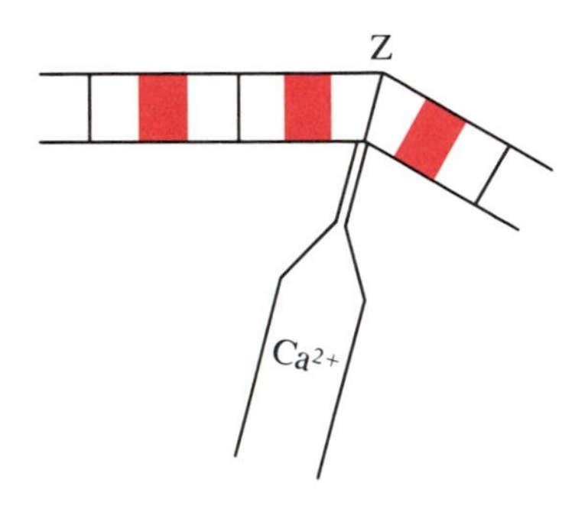

Figure 30-21

Local Contraction of a Myofibril When Calcium Is Released Near the Z Line and Transverse Tubule System.

calcium. Opening of the channel results in a rapid release of calcium to give a concentration of $\sim 10^{-5}$ M in the muscle cell. This release of calcium into the cytoplasm initiates contraction. The sarcoplasmic reticulum is an interconnected series of compartments that extends throughout much of the muscle cell.

The role of calcium in initiating contraction can be shown directly by adding calcium to isolated myofibrils. A dramatic demonstration of this is the localized contraction that occurs when calcium is released from a micropipette at the Z line of a microfibril (Figure 30-21).

In resting muscle, the calcium concentration is only $\sim 10^{-7}$ M and myosin exists largely as the $M \cdot ADP \cdot P_i$ complex, which is not attached or is only weakly attached to actin. The combination of $M \cdot ADP \cdot P_i$ and A, which would cause the release of products and contraction, is prevented by a long protein molecule, *tropomyosin,* that is bound to the actin filament and prevents the productive interaction of myosin with the actin in the thin filament. Tropomyosin, in turn, is controlled by the troponin system, a complex of three proteins, as shown schematically in Figure 30-22.

Calcium binds to one of these proteins, troponin C, and causes a change in its conformation. This, in turn, brings about a change in the conformation of at least one other troponin molecule and causes movement of the tropomyosin to a different position. This allows $M \cdot ADP \cdot P_i$ to combine with A in such a way as to cause the release of P_i, which is immediately followed by the power stroke (Figure 30-15). The contraction cycle then continues for as long as there is enough Ca^{2+} in the cytoplasm to keep the troponin-tropomyosin system in the activated state.

The activation of muscle contraction by calcium is brought about by different mechanisms in other kinds of muscle. Two of these mechanisms are described briefly in Box 30-2.

The Ca^{2+} Pump of the Sarcoplasmic Reticulum

The decrease of Ca^{2+} concentration in the cytoplasm that causes muscle relaxation is brought about by the active transport of Ca^{2+} into the sarcoplasmic reticulum. This process can lower the Ca^{2+} concentration in the cytoplasm to between 10^{-7} and 10^{-8} M. The Ca^{2+} transport is brought about by calcium ATPase, a Ca^{2+} pump that uses the energy from the hydrolysis of ATP to transport Ca^{2+} across the sarcoplasmic reticulum membrane against a gradient of as much as 10^4 in Ca^{2+} concentration. The calcium pump is a single protein molecule, with a molecular weight of 110,331, that is composed of 1001 amino acids. It is present in high concentration in the sarcoplasmic reticulum membrane (Figure 30-23).

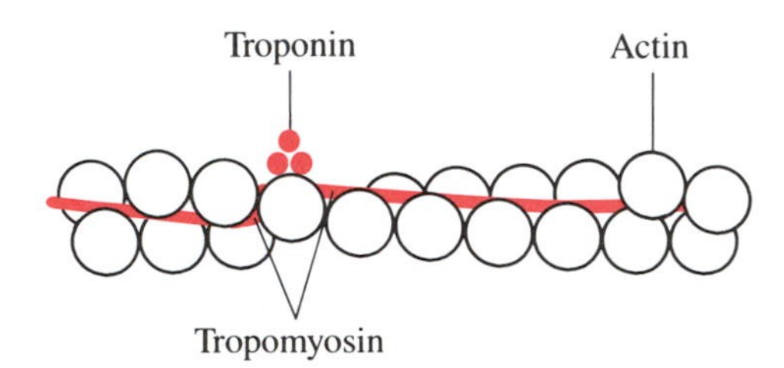

Figure 30-22

Tropomyosin and the Three Subunits of Troponin Bound to a Segment of an Actin Filament.

BOX 30-2

ACTIVATION OF SMOOTH AND MOLLUSCAN MUSCLE

Curiously, although the release of Ca^{2+} inside the cell is the universal signal for the initiation of muscle contraction (as it is also for many other activities of a cell), Ca^{2+} initiates contraction by different mechanisms in different kinds of muscle.

The contraction of most smooth muscles is initiated by a specific light chain protein kinase, which catalyzes the phosphorylation by ATP of a regulatory light chain that is bound to the myosin molecule. This allows the myosin cross bridges to attach to actin and the normal contraction cycle to proceed. The kinase, in turn, is activated by the binding of Ca^{2+} to *calmodulin,* which is a regulatory protein for a number of cellular activities. It undergoes a conformational change when Ca^{2+} binds; the complex then binds to protein kinase and activates phosphorylation, as shown in Figure A.

$$\text{Calmodulin} + Ca^{2+} \rightleftharpoons \text{Calmodulin}^* \cdot Ca^{2+}$$

$$\text{Calmodulin}^* \cdot Ca^{2+} + \text{Protein kinase} \longrightarrow \boxed{\text{Calmodulin} \cdot Ca^{2+} \; \text{Protein kinase}^*}$$

$$\xrightarrow{\text{ATP}}$$

$$\text{Light-chain myosin} \longrightarrow \text{Light-chain—P myosin}$$

$$\downarrow$$

$$\text{Actin} \longrightarrow \text{Actomyosin} \longrightarrow \text{Contraction}$$

Figure A

The muscles of scallops and other molluscs are activated by still another mechanism, in which Ca^{2+} binds directly to the myosin-products complex and induces the release of phosphate; this allows myosin to combine with actin, which initiates the power stroke and the contraction cycle, as shown in Figure B.

$$M \cdot ADP \cdot P_i + Ca^{2+}$$

$$\downarrow$$

$$Ca^{2+} \cdot M \cdot ADP \cdot P_i$$

$$\downarrow \; (\rightarrow P_i)$$

$$Ca^{2+} \cdot M \cdot ADP$$

$$\downarrow \; A$$

$$Ca^{2+} \cdot M \cdot ADP \cdot A$$

$$\downarrow$$

$$\text{Contraction}$$

Figure B

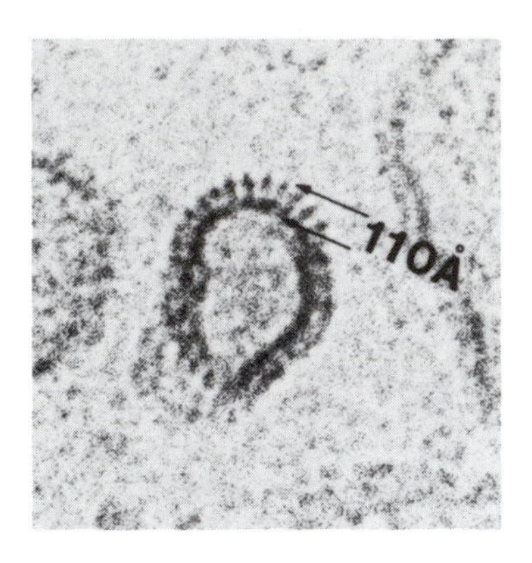

Figure 30-23

Electron Micrograph Showing Molecules of the Calcium ATPase of Sarcoplasmic Reticulum. (Courtesy of Sidney Fleischer.)

The calcium pump is a member of a family of cation pumps that use the energy from ATP hydrolysis to transport ions from the cytoplasm of a cell to the extracellular space or into vesicles within the cell. Crystals of the enzyme that are suitable for structure determination by high-resolution X-ray diffraction have not yet been obtained, but models of the overall structure have been constructed by MacLennan, Green, and others, as shown in Figure 30-24. It is likely that several other ion-pumping ATPases that have similar properties also have similar structures. The DNA of the calcium enzyme has been cloned and expressed, and site-directed mutagenesis has shown that different functions of the enzyme exist on different domains of the molecule. The amino acid sequence suggests that there are ten α helices, which probably form a bundle that crosses the membrane and provides a channel for the transport of Ca^{2+}. Removal of aspartate (D) and glutamate (E) residues in these helices by site-directed mutagenesis prevents the binding of calcium to the enzyme, which suggests that

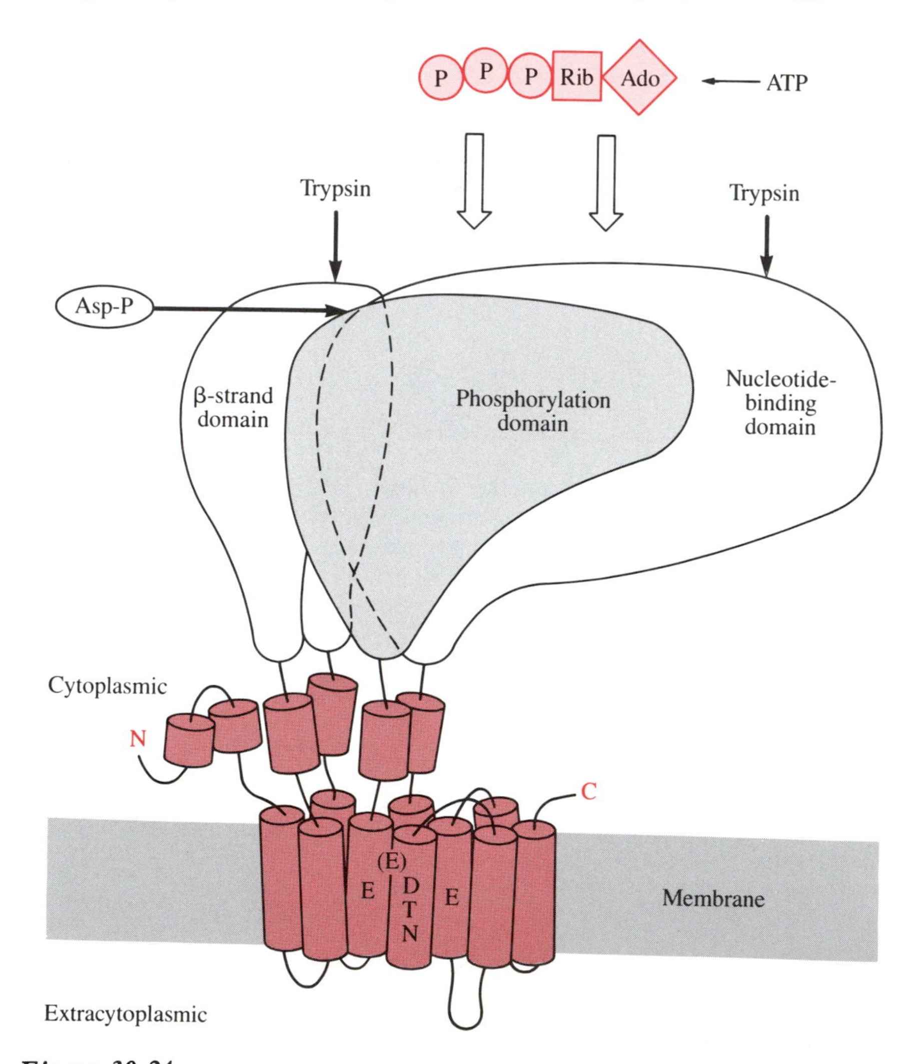

Figure 30-24

A Schematic Model for the Calcium ATPase of Sarcoplasmic Reticulum and Related Ion Pumps. Calcium ions are transported from one side of the sarcoplasmic reticulum membrane through a channel that is formed by the group of α helices at the bottom of the figure. However, the phosphorylation by ATP occurs in the cytoplasmic domain at some distance from the channel. Two sites at which the enzyme is rapidly cleaved by trypsin are indicated. (After a drawing by Jonathon Lytton and David H. MacLennan.)

the calcium-binding sites are in this channel. The binding of ATP to the enzyme is prevented by fluorescein isothiocyanate, FITC, which reacts with a different part of the protein that has been assigned to the nucleotide-binding domain. ATP phosphorylates the carboxylate group of an aspartate residue in a different part of the peptide chain that is assigned to the phosphorylation domain. Another domain has a high content of β strands, and short stretches of amino acids that are usually found in α helices connect these domains to the membrane helices. Two regions of the enzyme that are especially susceptible to cleavage by trypsin are also noted in the figure.

The calcium pump is of special interest because it catalyzes a relatively simple coupled vectorial process. For many years, it was not clearly understood how the chemical reaction of ATP hydrolysis could be coupled to a vectorial reaction, such as the movement of two calcium ions from one side of a membrane to the other. Furthermore, the reaction is reversible: the efflux of calcium ions from a sarcoplasmic reticulum vesicle that contains a high concentration of calcium will drive the conversion of ADP and P_i into a molecule of ATP.

Our understanding of the mechanism of pumping by the calcium ATPase was made possible, first, by the demonstration that Ca^{2+} transport is reversible and stoichiometric, with two Ca^{2+} ions transported for each ATP hydrolyzed (equation 6) and, second, by the use of high concentrations of enzyme that made it possible to examine the individual steps of the cycle.

$$2\,Ca^{2+}{}_{out} + ATP \rightleftharpoons 2\,Ca^{2+}{}_{in} + ADP + P_i \qquad (6)$$

In the first step of the transport cycle shown in Figure 30-25, the enzyme binds two Ca^{2+} ions from the cytoplasmic side of the sarcoplasmic reticulum membrane with a very high affinity to form $E\cdot Ca_2$. This activates the catalytic activity of the enzyme, so that it is rapidly phosphorylated by ATP to give the intermediate $E{\sim}P\cdot Ca_2$. ATP is present and can bind to the enzyme before Ca^{2+} binds, but it cannot phosphorylate the enzyme until after the Ca^{2+} is bound. Two Ca^{2+} ions are then released from $E{\sim}P\cdot Ca_2$ into the inside of the vesicle, to give the phosphorylated enzyme, E–P. The dissociation of Ca^{2+} activates the phosphoenzyme to react with water, which releases inorganic phosphate (P_i) and regenerates the free enzyme. This completes the reaction cycle.

This reaction cycle brings about the overall reaction of equation 6. In one direction, it uses the energy from the hydrolysis of ATP to pump two calcium ions into the vesicle against a large concentration gradient; and, in the other direction, it will synthesize ATP from ADP and phosphate if Ca^{2+} ions flow out

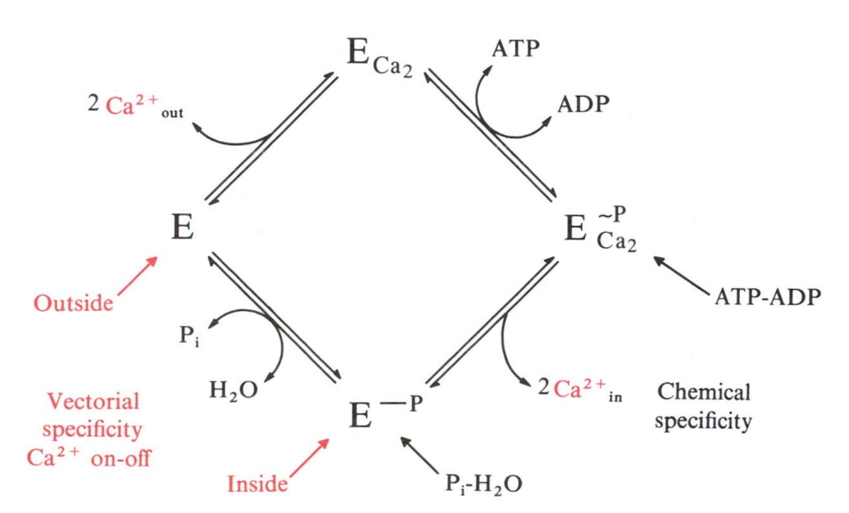

Figure 30-25

Reaction Cycle of the Calcium Pump of Sarcoplasmic Reticulum.

of vesicles that have been loaded with calcium. This can occur when a chelating agent is present to trap the calcium ions in the medium. Thus, the enzyme catalyzes the coupled vectorial process of equation 6 in both directions.

Coupling Between ATP Hydrolysis and Calcium Transport

What is it about this cycle that makes it work? The coupling between the hydrolysis of ATP and the transport of Ca^{2+} can be described by a set of "rules." These rules represent changes in the *specificity* of the enzyme for catalysis in different states of the enzyme, as shown in Figure 30-25. If the rules are followed, the enzyme will transfer Ca^{2+} across the membrane and hydrolyze (or synthesize) ATP, according to equation 6.

Most enzymes have only a single specificity for catalysis, but the calcium ATPase and related enzymes have different specificities that depend on what is bound to the enzyme. The rules that define the chemical specificity for catalysis by the Ca ATPase are:

1. The calcium-bound species, $E \cdot Ca_2$, reacts only with ATP to give $E \sim P \cdot Ca_2$, which reacts only with ADP in the reverse direction.
2. The free enzyme, E, reacts only with P_i to give E–P, which reacts with water to hydrolyze E–P in the reverse direction.

If these two rules were not followed, ATP would be hydrolyzed without the transport of Ca^{2+} and the system would be uncoupled.

The rules for vectorial specificity are:

1. The free enzyme, E, binds and dissociates Ca^{2+} only on the cytoplasmic side of the membrane (the outside of the sarcoplasmic reticulum vesicle).
2. The phosphorylated enzyme, E–P, binds and dissociates Ca^{2+} only on the inside of the vesicle.

If either of these rules were to be violated, Ca^{2+} would leak out of the vesicle and the system would be uncoupled.

Thus, the *chemical specificity* for catalysis is determined simply by whether or not Ca^{2+} is bound to the enzyme, and the *vectorial specificity* is determined by whether or not the enzyme is phosphorylated.

The coupling can be explained in an even simpler way. We want to couple the hydrolysis of ATP, which is a downhill process, to the pumping of Ca^{2+} into the vesicle, which is an energetically uphill process under physiological conditions (Figure 30-26A). This is accomplished by the Ca^{2+} ATPase simply by dividing both the vectorial process (the transport of Ca^{2+}) and the chemical process (the hydrolysis of ATP) into two steps. The steps for the chemical and vectorial processes *alternate* in the reaction cycle, as shown in Figure 30-26B. Therefore, neither the hydrolysis of ATP nor the transport of Ca^{2+} can occur unless the other process also occurs.

This sequence of reactions is shown schematically in Figure 30-27. The first half of the vectorial reaction of calcium transport (step 1) is the binding of two calcium ions from the cytoplasm to the cytoplasmic side of the vesicle membrane. This signals the enzyme to catalyze the first half of the chemical reaction, the phosphorylation of $E \cdot Ca_2$ by ATP (step 2). This is the critical vectorial step of calcium pumping because it changes the vectorial specificity of the enzyme: calcium can dissociate only to the cytoplasmic side of the membrane from the free enzyme, but it can dissociate only into the inside of the vesicle from the phosphorylated enzyme. This dissociation of calcium provides the second half of the vectorial reaction and signals the enzyme to change its specificity for

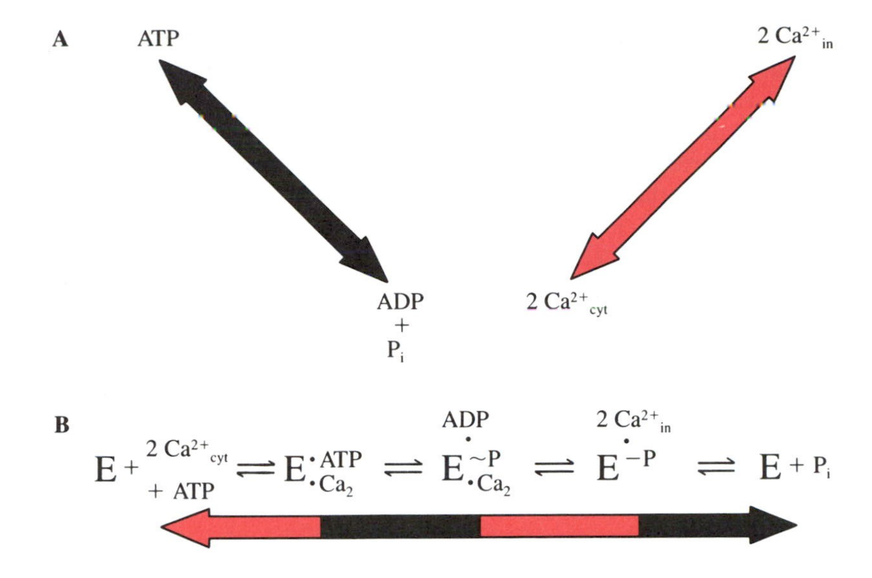

Figure 30-26

A. Calcium ATPase catalyzes the hydrolysis of ATP, a downhill chemical reaction, and calcium transport between the cytoplasm and the inside of a sarcoplasmic reticulum vesicle, an uphill vectorial reaction. B. The enzyme divides each reaction into two halves and catalyzes the alternating half-reactions. Thus, neither the chemical nor the vectorial reaction can go to completion unless the other reaction also takes place.

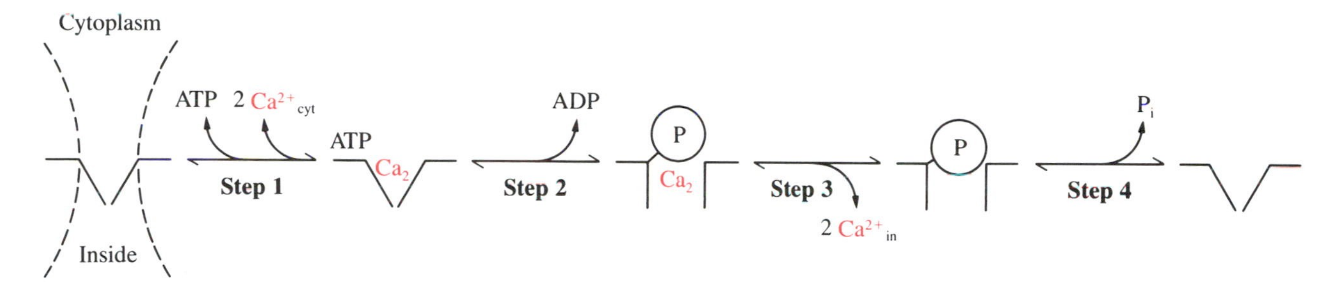

Figure 30-27

Schematic Representation of the Steps in Pumping of Ca^{2+} by the Calcium ATPase.

catalysis. This allows the enzyme to catalyze the rapid hydrolysis of the phosphoenzyme, which provides the second half of the chemical reaction and changes the vectorial specificity so that calcium can bind to the enzyme from the cytoplasm. The reaction cycle is now completed. Most, or all, of these steps are accompanied by conformational changes of the enzyme that change its chemical and vectorial specificities in the reaction cycle.

The coupling between ATP hydrolysis and movement in the contraction cycle of muscle can be explained similarly, by the alternation of chemical and vectorial steps. The vectorial steps result in the dissociation and association of actin and myosin, whereas the chemical steps include the binding of ATP and the dissociation of ADP and P_i, the products of hydrolysis. The binding of ATP, the first step of ATP hydrolysis, activates the dissociation of actomyosin, a vectorial step; and the binding of actin to myosin·ADP·P_i, a vectorial step, activates dissociation of the hydrolysis products, the final chemical step. Both the hydrolysis of ATP and the power stroke are fast steps that are not directly responsible for the coupling of ATP hydrolysis with movement.

Energetics of the Calcium ATPase

In order for calcium ATPase, or any other enzyme, to turn over substrate rapidly, there should not be intermediate species with a very high energy or low energy along the reaction path. An intermediate with a high Gibbs energy under physiological conditions introduces a barrier along the reaction path (Figure 30-28A), and an intermediate with a low energy produces an energy well from which it is difficult for the enzyme to escape (Figure 30-28B). The energetics of enzyme-catalyzed reactions are generally balanced in such a way that the intermediates along the reaction path have similar Gibbs energies under physiological conditions. Figure 30-28C shows that the intermediate species on the reaction path of the calcium pump of sarcoplasmic reticulum have similar energies in the presence of physiological concentrations of reactants and products.

On the other hand, an enzyme that catalyzes a coupled vectorial process, such as the transport of Ca^{2+}, must bring about large changes in the properties of the molecules or ions that are bound to it under different conditions. For example, the phosphoenzyme intermediate of the Ca^{2+} ATPase, E–P, is at equilibrium with inorganic phosphate, so that the cycle can proceed easily in the

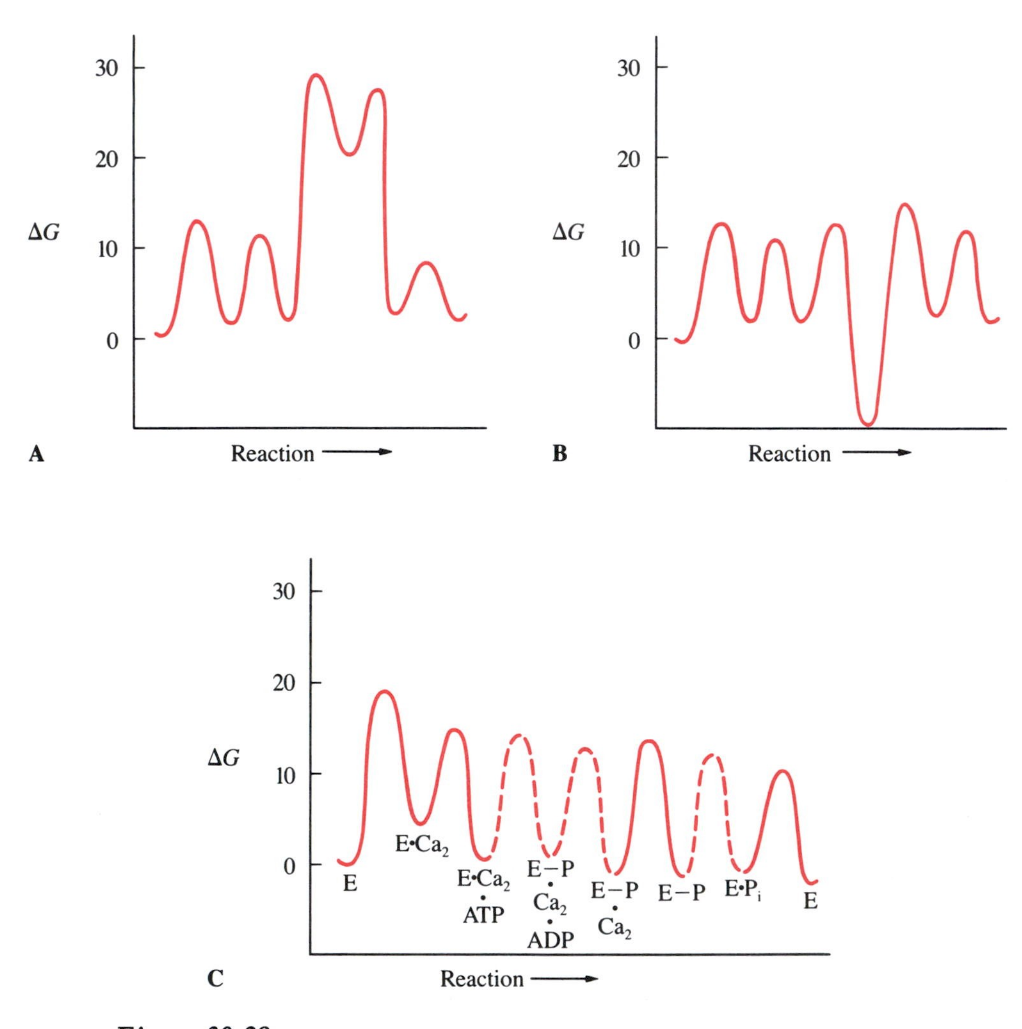

Figure 30-28

A. An unstable intermediate along the reaction path creates a barrier that makes rapid turnover difficult. B. A very stable intermediate also makes turnover difficult, because it is difficult for this intermediate to react further. C. The energy profile for the steps of the sarcoplasmic reticulum ATPase reaction cycle, calculated for *physiological concentrations of reactants* and a calcium gradient of 10^4 between the inside and the outside of the vesicle.

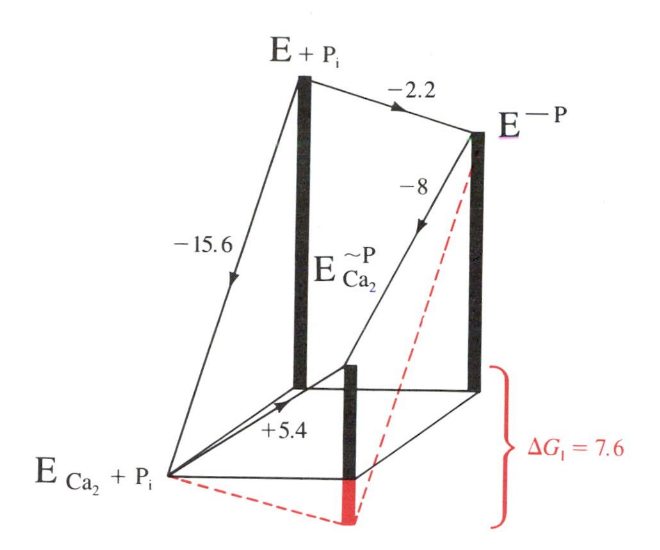

Figure 30-29

Energy Diagram for the Binding of Calcium and Phosphate to the Calcium ATPase. The binding of calcium and phosphate separately is favorable, but when they are bound together there is a mutual destabilization, or interaction energy, ΔG_I, of 7.6 kcal mol^{-1}.

reverse direction. Therefore, E−P is a *low-energy* phosphate compound, by definition. However, the phosphoenzyme intermediate with bound calcium, $E \sim P \cdot Ca_2$, must be a *high-energy* phosphate compound that is at equilibrium with ATP.

Furthermore, the enzyme must bind Ca^{2+} ions very tightly in order to pick up Ca^{2+} from the low concentration that exists in the cytoplasm of the muscle cell and form $E \cdot Ca_2$, but it must also be able to dissociate bound calcium easily into the inside of the vesicle, where the Ca^{2+} concentration is very much higher. These requirements mean that the escaping tendencies of the bound phosphate and Ca^{2+} must undergo large changes as the enzyme passes through different states of the reaction cycle, in order that turnover can occur easily under physiological conditions.

This behavior is described by an interaction energy, as shown in Figure 30-29. Figure 30-29 describes the box of reactions shown in Figure 30-30. The formation of phosphoenzyme from inorganic phosphate is slightly favorable and the binding of calcium to the free enzyme to form $E \cdot Ca_2$ is very strong, as it must be in order that the enzyme can reduce the calcium concentration in the cytoplasm to less than micromolar. However, the binding of calcium to the phosphoenzyme is weaker by 7.6 kcal mol^{-1}. This represents a *destabilization,* or interaction energy, of $\Delta G_I = 7.6$ kcal mol^{-1}. The interaction energy also raises the Gibbs energy of the $E \sim P \cdot Ca_2$ intermediate, compared with E−P, by the same amount, $\Delta G_I = 7.6$ kcal mol^{-1}. The interaction energy represents a *mutual destabilization* of the bound phosphate and Ca^{2+} in the $E \sim P \cdot Ca_2$ intermediate. Phosphorylation of $E \cdot Ca_2$ increases the escaping tendency of the bound Ca^{2+} and the binding of Ca^{2+} to E−P increases the escaping tendency of the phosphate by the same amount. This makes it possible for the bound Ca^{2+} to dissociate into the concentrated pool of calcium on the inside of the vesicle and for the phosphate to be transferred to ADP to make ATP. The consequence of this is that the Gibbs energies of all of the intermediates in the reaction cycle are similar under physiological conditions, so that there are no large barriers or energy wells along the reaction path (Figure 30-28C).

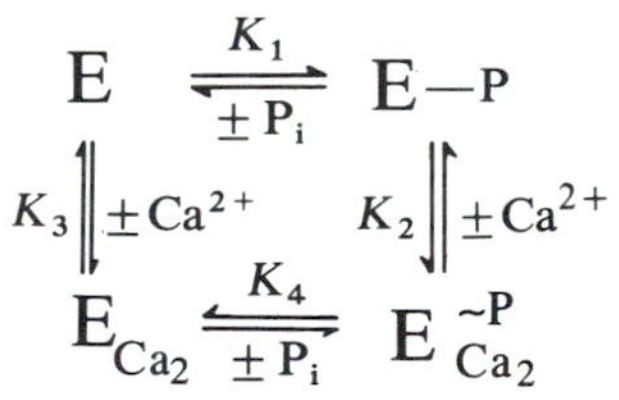

$$
\begin{array}{ccc}
E & \xrightleftharpoons[\pm P_i]{K_1} & E{-}P \\
K_3 \Big\Updownarrow \pm Ca^{2+} & & K_2 \Big\Updownarrow \pm Ca^{2+} \\
E_{Ca_2} & \xrightleftharpoons[\pm P_i]{K_4} & E^{\sim P}_{Ca_2}
\end{array}
$$

Figure 30-30

Thermodynamic Box Describing the Binding of Phosphate and Calcium to the Calcium ATPase.

Related Ion Pumps

The *sodium-potassium ATPase* of the cell membrane uses the energy of ATP to pump three Na^+ ions out of the cell and two K^+ ions into the cell, according to equation 7.

$$3\,Na^+_{in} + 2\,K^+_{out} + ATP \rightleftharpoons 3\,Na^+_{out} + 2\,K^+_{in} + ADP + P_i \quad (7)$$

As pointed out in Chapter 29, this enzyme has the important function of maintaining the very different internal and external ionic environments of a cell: it builds up a high concentration of potassium and low concentration of sodium inside the cell although the extracellular fluid contains a high concentration of sodium and a low concentration of potassium. These differences in concentration are used to generate an electric potential across the cell membrane and are responsible for the action potential of nerve conduction. They are also used as a source of energy for several other important processes. These processes include the transport of other molecules and ions that are driven by the movement of sodium ions into the cell down a concentration gradient (Chapter 29). In some cells, these ion gradients provide a major source of energy for physiological processes.

The properties of the Na-K enzyme (Chapter 29) are very similar to those of the Ca^{2+}-transporting ATPase, except that three Na^+ instead of two Ca^{2+} ions are transported from the cytoplasm to the outside of the cell for each molecule of ATP that is hydrolyzed (Figure 30-31). Another difference is that two K^+ ions are transported into the cell from the extracellular fluid in the lower part of the cycle.

This enzyme follows essentially the same coupling rules as does the calcium ATPase, but with Na^+ instead of Ca^{2+}. The binding of both sodium and potassium proceeds through an ordered mechanism, with binding to E–P on one side of the membrane and to free E on the other side. Three Na^+ ions bind to the enzyme from the cytoplasm and activate it for phosphorylation by ATP. After phosphorylation, they dissociate from the phosphorylated enzyme to the outside of the cell membrane. Two K^+ ions then bind to the phosphorylated enzyme from the extracellular fluid and activate the reaction of the phosphorylated enzyme with water. The K^+ ions dissociate to the cytoplasm after the phosphoenzyme is hydrolyzed. The energy that drives Na^+ and K^+ transport is provided by the hydrolysis of ATP, which drives the reaction cycle of Figure 30-31 in a clockwise direction.

Another closely related ATPase transports H^+ across the gastric mucosa to produce the high acidity of the stomach, which has a pH ranging from 1 to 2. It can also transport K^+ across the membrane in the reverse direction into the cell. Similar enzymes exist in other tissues. The proton-transporting ATPases of oxidative phosphorylation and photophosphorylation operate through an entirely different mechanism, as described in Chapters 22 and 23.

The gastric H^+ pump and related enzymes also form a phosphoenzyme intermediate and follow a reaction cycle similar to that of the Ca and Na-K pumps; all three enzymes have considerable homology in their amino acid

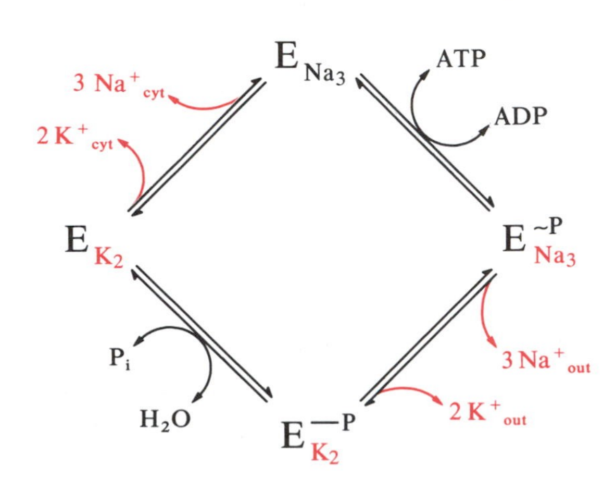

Figure 30-31

Reaction Cycle of the Sodium-Potassium Transporting ATPase.

sequence. Closely related proton and calcium pumps are found in the membranes of many cells and play an important role in the regulation of intracellular pH and calcium concentration.

Lysosomes and other intracellular vesicles contain a different proton pump, the *vacuolar ATPase*. This enzyme maintains an increased acidity, with a pH of about 5, inside these organelles. This pump does not form a phosphoenzyme intermediate; it has several similarities to the proton pump of oxidative phosphorylation and is probably a related enzyme. The low pH inside the vesicle is optimal for many of the reactions that occur within the vesicle and can act as a signal to cause the dissociation of protein complexes that are brought into the vesicle.

Still another mechanism for active transport has recently been discovered that utilizes the favorable free energy of decarboxylation to drive the transport of sodium ions. Enzymes have been found that couple the transport of sodium and the decarboxylation of oxaloacetate to form pyruvate (equation 8) or of methylmalonyl CoA to form propionyl CoA.

$$\begin{array}{c} COO^- \\ | \\ CH_2 \\ | \\ C{=}O \\ | \\ COO^- \end{array} + Na^+ + H^+ \longrightarrow \begin{array}{c} CH_3 \\ | \\ C{=}O \\ | \\ COO^- \end{array} + Na^+ + CO_2 \qquad (8)$$

The mechanism of action of these interesting ion pumps has not yet been determined.

Tubulin

Tubulin is a dimeric protein that is composed of α and β subunits and assembles into *microtubules*, which provide another major component of the cytoskeleton (Chapter 1). Microtubules are larger than actin microfilaments and, as indicated by their name, consist of tubes with a central hole, the function of which is unknown. They radiate from the centrosome of a cell to attachment sites just under the cell membrane and from mitotic spindles to chromosomes that are undergoing separation during cell division.

Although the structures of tubulin and microtubules are very different from those of actin and microfilaments, the behavior of the two systems is remarkably similar in many respects. The polymerization of tubulin monomers into microtubules is similar to the polymerization of G-actin into F-actin filaments, except that is is associated with the hydrolysis of bound GTP, rather than ATP. The microtubules have a polarity that favors assembly and dissociation of tubulin molecules at different ends of the microtubule. Their behavior is regulated in large part by microtubule-associated proteins, MAP. The mechanism of this regulation is just beginning to be understood at this time.

Microtubules tend to grow at one end, by the addition of GTP-tubulin, and shorten at the other end, by the dissociation of GDP-tubulin. However, one end of the tubulin is often connected to a binding site, which may be on a centrosome or a mitotic spindle. If both ends of the microtubule are connected to cell structures, it is stable for a considerable period.

Colchicine, which is used in the treatment of gout, and certain other drugs inhibit the polymerization of tubulin into microtubules. High doses of these drugs will stop cell division by preventing development of the mitotic spindle.

The movements of some types of cilia and flagellae are brought about by bundles of microtubules that interact with *dynein* and ATP in a coupled vectorial process. Dynein is a multisubunit ATPase; one of the polypeptide chains is 450,000 daltons and is the largest single polypeptide chain that is known. This reaction cycle bears an extraordinary resemblance to the myosin-actin contraction cycle of muscle, with dynein playing the role of myosin. Dynein contains two or three heads, which correspond functionally to the cross bridges of muscle and bind tightly to tubulin. Binding of ATP to the dynein-tubulin complex initiates a cycle of dissociation, hydrolysis, recombination of the dynein head with another tubulin unit, product dissociation, and movement. The remarkable similarity of this system to the myosin-actin system extends to the rate and equilibrium constants for most of the individual steps of the reaction cycle, which are almost the same in the two systems. There appears to be no structural or evolutionary relationship between these two systems, so that the close similarity suggests that the reaction cycle works so well that it has developed independently through evolution on two different occasions.

Microtubules also bring about movement of intracellular vesicles and organelles from one part of the cell to another, by a different mechanism. The most-dramatic example of this movement is the transport of vesicles along the axon of a nerve, from the cell body to the synapse that connects with another nerve. This corresponds to movements of as much as several feet along the human spinal cord, and even further in a giraffe or a blue whale.

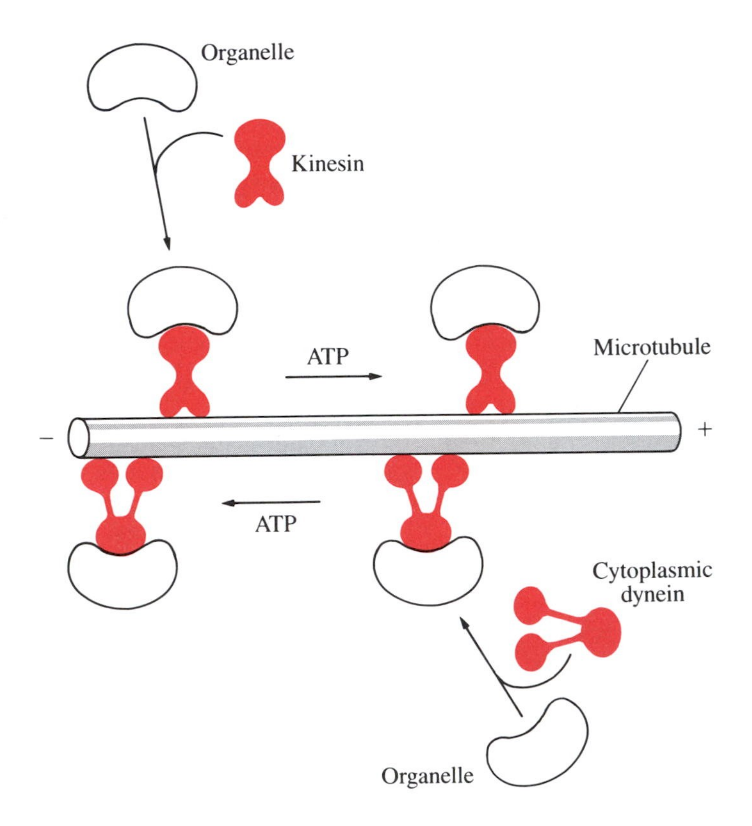

Figure 30-32

Kinesin and dynein move organelles along microtubules in opposite directions by ATP-driven reactions.

This transport mechanism has recently been studied in cell-free preparations by measuring the movement of small particles of gold or plastic along microtubules. These particles refract light and can be observed with a powerful microscope. The movement requires ATP and a large, multisubunit protein, *kinesin*. One end of kinesin binds to a particle, whereas the other end interacts with tubulin to produce movement. Several molecules of kinesin react with a single particle and move it along a "track" of tubulin. In an intact cell, this movement is toward the periphery of the cell.

ATP is required for the movement of kinesin along tubulin and is hydrolyzed during movement. The reaction cycle leads to successive dissociation and movement of the kinesin to different binding sites on tubulin monomers that are further along the microtubule. There are two binding sites, probably one on each of the two subunits of kinesin, that allow the kinesin to "walk" along the microtubule in such a way that it always stays attached to the microtubule and moves in one direction, toward the end of the microtubule indicated by the plus sign in Figure 30-32.

Transport along the microtubule in the opposite direction (retrograde transport) is driven by cytoplasmic dynein (Figure 30-32), but the mechanism of this process is not yet understood.

An entirely different mechanism for bringing about movement is found in gram-negative bacteria in which a motor propels the bacterium by bringing about a rotary motion of flagellae. This is described in Box 30-3.

BOX 30-3

THE FLAGELLAR MOTOR

The movement of some bacteria is brought about by a remarkable rotary engine in the cell membrane that is attached to long, helical flagellae. When this motor rotates in one direction, the movement of the flagellae propels the bacterium through the solution. However, when it rotates in the other direction, no force is exerted in the forward direction and the bacterium tumbles around in the solution, with little overall movement in any direction.

This movement serves to bring the bacterium closer to sources of food. Receptors in the cell membrane detect the presence of certain organic molecules and alter the duration of the forward motion and the tumbling phases of flagellar rotation in such a way that there is net movement of the bacterium toward the source of potential nourishment. It appears that, when the bacterium moves for a short distance, it is able to sense the small gradient in concentration of the organic molecule and correct its flagellar rotation so as to move in the correct direction. The motion of the flagellae can be observed directly with a microscope by attaching the flagellae to a glass surface and watching the rotation of the cell body.

The rotation of the motor is brought about by the transport of protons between the medium and the cytoplasm of the cell. In some species, a sodium gradient rather than a proton gradient is a major energy source for cellular activities and drives the flagellar motor.

The motor consists of two rings that move relative to each other, as illustrated in the drawing on page 820. The outer ring is fixed in the cell wall and the inner ring in the cytoplasmic membrane is attached to a rod that runs through the cell wall and is connected to the flagellar filament by a hook. Movement of protons between the medium and the cytoplasm causes these rings to rotate relative to each other, just as steam rotates the blades of a turbine as it passes through.

However, the exact mechanism of this extraordinary machine is not understood. It almost certainly is not identical with that of a turbine. There must be channels for the flow of protons across the membrane and cell wall, and a mechanism that couples this flow to rotation of the flagellae. A barrier that prevents the free flow of protons is presumably constructed in such a way that protons can pass through the membrane in a step-by-step process only when the rings move. Thus, a proton can pass through a channel in the cell wall but not through a channel on the inner ring, because the two channels are not aligned with each other. Movement of the inner ring then allows the proton to enter the cell. Repetition of this process many times will cause rotation of the inner ring and flagellum.

The structure of the flagellum is such that rotation in one direction causes smooth swimming in one direction, but rotation of the flagellum in the other direction causes

BOX 30-3 (continued)

THE FLAGELLAR MOTOR

a tumbling motion of the bacterium. Tumbling bacteria do not undergo translational motion but remain in one location. Thus, when bacteria are attracted to a molecule, their flagellae will rotate in the locomotive direction and they will swim into a concentration gradient of that molecule. The bacteria will tumble when there is no concentration gradient.

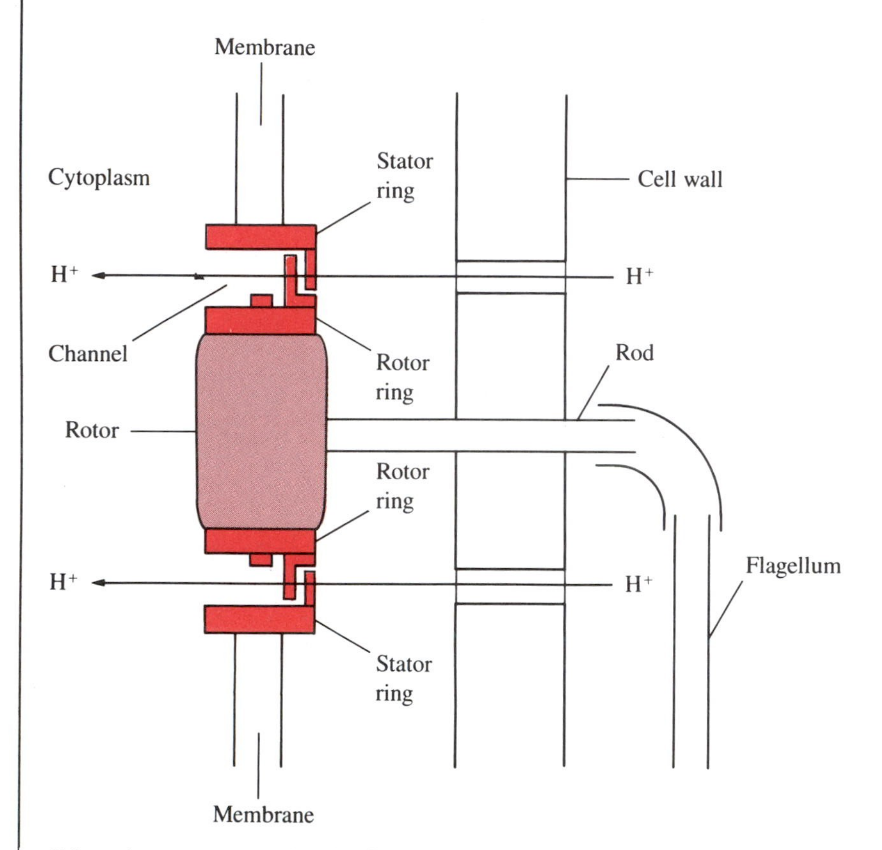

Schematic Cross Section of a Flagellar Motor That Is Driven by the Transfer of Protons Across the Bacterial Membrane. The motor is driven by coupling the movement of protons from the medium into the cell with rotation of a disk that is attached to the flagellum. A *stat*ionary ring or *stator*, is embedded in the membrane and a *rot*ating ring, or *rotor*, is attached to the flagellum and causes its rotation. The detailed mechanism by which proton transfer causes rotation is not known, but it has been suggested that protons can diffuse through the membrane only when the rotating ring moves a short distance in one direction. The rings are surrounded by a circlet of small particles that are believed to contain the proton channels. However, each channel allows proton transfer only part way across the membrane. Movement of the rotor in one direction allows the proton to escape from the other side of the membrane through a channel in the rotor.

Electron Micrographs of the Basal Body, or Rotor, of the Bacterial Flagellar Motor. These computer-averaged micrographs show the rings and bushings of the rotor. The flagellum is attached by a curved sleeve to the upper part of the rotor and the bushings are visible in the lower part of the structure. The lower image includes the "switch proteins" at the bottom of the structure which are required for reversing the direction of rotation of the motor. (Courtesy of D. Thomas and D. DeRosier.)

Summary

Coupled vectorial processes utilize the energy of a chemical reaction, usually the hydrolysis of ATP, to bring about movement. The chemical reaction and the movement are coupled, so that neither the chemical nor the physical reaction can occur unless the other also occurs. The contraction of striated muscle is initiated by the release of Ca^{2+} from sarcoplasmic reticulum vesicles. This causes a conformational change in troponin that allows myosin · ADP · P_i in the thick filaments to bind to actin in the thin filaments. This releases the bound ADP and P_i, and the myosin undergoes a conformational change that causes movement in a "power stroke." Binding of ATP then causes dissociation of myosin from the actin filament, the bound ATP is hydrolyzed, and the cycle is repeated. The calcium ATPase pumps Ca^{2+} into sarcoplasmic reticulum vesicles and decreases the calcium concentration in the muscle to $<1\ \mu M$; this causes the troponin complex to inhibit the interaction of the myosin-products complex with actin and results in relaxation because the contraction cycle is blocked. The calcium ATPase undergoes a cycle of phosphorylation by ATP, Ca^{2+} internalization, and hydrolysis of the phosphoenzyme that internalizes two Ca^{2+} ions. The Na,K ATPase maintains the Na^+-K^+ gradient across cell membranes by a similar mechanism by pumping 3 Na^+ out and 2 K^+ into the cytoplasm for each ATP that is hydrolyzed; the gastric ATPase uses essentially the same mechanism to pump protons into the stomach. A vacuolar proton ATPase maintains the pH at ~5 in lysosomes and other intracellular particles. Tubulin grows by the addition of GTP-tubulin to one end of a microtubule, and it shortens by dissociation of GDP-tubulin. The interaction of tubulin with dynein and ATP causes movement of cilia and flagellae, and intracellular organelles are transported along tubulin by kinesin and dynein. Bacterial flagellae are made to rotate by a motor that is driven by proton transport between the medium and the cell.

ADDITIONAL READING

Szent-György, A. 1953. *Chemical Physiology of Contraction in Body and Heart Muscle.* New York: Academic Press.

Sugi, H., and Pollack, G. H., eds. 1988. *Molecular Mechanism of Muscle Contraction.* New York: Plenum.

De Meis, L. 1981. *The Sarcoplasmic Reticulum.* New York: Wiley-Interscience.

Jencks, W. P. 1989. How does a calcium pump pump calcium? *Journal of Biological Chemistry* 264: 18855–18858.

Skou, J. C. 1990. The energy coupled exchange of Na^+ for K^+ across the cell membrane: the Na^+,K^+-pump. *FEBS Letters* 268: 314–324.

Macnab, R. M. 1987. Flagella. In *Escherichia coli and Salmonella typhimurium: Cellular and Molecular Biology*, vol. 1, ed. F. C. Neidhardt. Washington, D.C.: American Society of Microbiology, pp. 70–83.

Mitchison, T. J., and Kirschner, M. 1984. Dynamic instability of microtubule growth. *Nature* 212: 237–242.

PROBLEMS

1. Why was it important to study the contraction of muscle fibrils that had been soaked in aqueous glycerol solutions?

2. Explain how the length of actin filaments is controlled.

3. Explain isometric contraction of muscle. Why is it hard work to hold a 50 pound weight 1 foot above the floor?

4. Sodium ion binds to the sodium-transporting ATPase with a high affinity, but the affinity is decreased after phosphorylation of the enzyme. How can this be brought about? Why is it important?

5. Describe what would happen that would cause uncoupling if each of the rules for coupling ATP hydrolysis and Ca^{2+} transport were violated.

6. Show how the coupling between muscle contraction and ATP hydrolysis can be accounted for by alternating chemical and vectorial steps, so that neither ATP hydrolysis nor muscle contraction can occur unless the other also occurs.

7. Binding of Ca^{2+} to troponin C causes changes in the conformation of the troponin system that allow the myosin · ADP · P_i complex to combine with actin and bring about muscle contraction. Explain with a thermodynamic box and an energy diagram how the binding of calcium causes this conformational change.

8. How large must the interaction energy be in the troponin system to cause a 100-fold change in the equilibrium constant for the conformational change when Ca^{2+} binds?

9. Describe a set of coupling rules that will account for the coupling between the hydrolysis of ATP and the transport of Na^+ and K^+ by the Na,K ATPase.

Appendix: pH, pK, and Buffers in Biochemistry

The use of pH, pK, and buffers is described in elementary chemistry courses. Nevertheless, it can be useful to review these subjects and provide a brief reference to them.

The concentration of hydrogen ions in different fluids of living systems varies over an enormous range (Table A). The concentration of hydrogen ions in blood is 4×10^{-8} M, whereas gastric juice secreted into the stomach can have a hydrogen ion concentration of 0.1 M or more. Lysosomes, which are organelles that digest materials that are taken into the cell, contain approximately 10^{-5} M H^+, as well as proteolytic enzymes that have optimal activity at this acidity.

It is important to consider the concentrations of hydrogen ions when examining the processes that take place within cells, but it is awkward and inconvenient to describe the acidity in the different compartments or tissues of living systems by the concentrations of hydrogen ions because of the very large range of hydrogen ion concentrations that are found in living systems. Instead, the concentration of hydrogen ions is usually described by the pH, which is the negative logarithm of the hydrogen ion concentration (equation 1).

$$\mathrm{pH} = -\log[\mathrm{H}^+] \tag{1}$$

(This statement is not quite correct because the pH depends upon hydrogen ion activity, rather than concentration. Activities will be described in the last section of this appendix.) Thus, gastric fluid that contains 0.1 M hydrochloric acid has a pH of $-\log 0.1 = 1.0$. Hydrochloric acid is a strong acid, so that it is almost completely dissociated into H^+ and Cl^- ions. The pH of lysosomes is approximately 5.0 and the pH of blood is 7.4.

The dissociation of liquid water into hydrogen and hydroxide ions is usually described by the ion product, K_w (equation 2).

$$K_w = [\mathrm{H}^+][\mathrm{OH}^-] = 10^{-14}\,\mathrm{M}^2 \tag{2}$$

Table A

The pH and concentration of hydrogen ions in some body fluids and compartments

Fluid or Organ	pH	$[H^+]$
Blood	7.4	4×10^{-8} M
Sweat	5.0–7.5	3×10^{-8}–10^{-5} M
Tears	7.0–7.4	4×10^{-8}–10^{-7} M
Lysosomes	~5.0	10^{-5} M
Stomach	1–2	0.01–0.1 M
Cytoplasm	~7.0	10^{-7} M

Liquid water is generally assigned a dimensionless apparent concentration (or activity) of 1.0, rather than its true concentration of 55.5 M. It is more convenient for most purposes to use this value of 1.0 rather than 55.5 M.

The dissociation constant of an acid, K_a, is described by equations 3 and 4.

$$AH \xrightleftharpoons{K_a} A^- + H^+ \tag{3}$$

$$K_a = \frac{[H^+][A^-]}{[HA]} \tag{4}$$

The range of dissociation constants of weak acids is very large, so that it is also convenient to describe the dissociation constant of an acid by its pK_a, the negative logarithm of its dissociation constant, K_a, (equation 5).

$$pK_a = -\log K_a = -\log \frac{[H^+][A^-]}{[HA]} \tag{5}$$

Thus, the pK_a of acetic acid is 4.7, which is the negative logarithm of its dissociation constant of $K_a = 2 \times 10^{-5}$ M (equation 6).

$$CH_3COOH \xrightleftharpoons{K_a} CH_3COO^- + H^+ \tag{6}$$

The pK_a of protonated imidazole is 7.0, which corresponds to a dissociation constant of $K_a = 10^{-7}$ M (equation 7).

$$\text{HN}\overset{+}{\frown}\text{NH (imidazolium)} \xrightleftharpoons{K_a} \text{HN}\frown\text{N (imidazole)} + H^+ \tag{7}$$

The dissociation constants and pK_a values of some other acids are shown in Table B.

If we convert equation 4 into a logarithmic form, we obtain the simple relation shown in equation 8.

$$pH = pK_a + \log \frac{[A^-]}{[HA]} \tag{8}$$

This is known as the Henderson-Hasselbalch equation. It is one of the most widely used equations in the fields of biochemistry and chemistry. It allows the prediction of the pH of a solution if the buffer ratio, $[A^-]/[HA]$, and the pK_a of HA are known and the acid is weak. If the acid is moderately strong, the

Table B

Dissociation constants and pK_a values of some acids

Acid	K_a	pK_a
Oxalic	5.0×10^{-2}	1.3
Phosphoric	7.6×10^{-3}, 6.3×10^{-8}, 5.0×10^{-13}	2.1, 7.2, 12.3
Formic	1.6×10^{-4}	3.8
Fumaric	1.0×10^{-3}, 3.2×10^{-5}	3.0, 4.5
Succinic	2.9×10^{-5}, 3.2×10^{-6}	4.2, 5.5
Carbonic (apparent pK_a)	4.5×10^{-7}, 4.7×10^{-11}	6.4, 10.3
NH_4^+	5.8×10^{-10}	9.2
Glycine	4.6×10^{-3}, 2.5×10^{-10}	2.3, 9.6

dissociation of HA to form H^+ and A^- may decrease the concentration of HA significantly, so that the ratio of $[A^-]/[HA]$ in this solution will not be equal to the ratio of the concentrations of A^- and HA that were added to the solution.

Buffers

It is extremely difficult to maintain a pH of 7.4 in water. Even the uptake of carbon dioxide from the atmosphere will quickly decrease the pH of pure water significantly. Some of the carbon dioxide is hydrated to give carbonic acid, which has an apparent pK_a of 6.4. It takes very little acid to lower the pH of water from 7.4 to 7.0. (You might calculate how much the concentration of hydrogen ion increases with this change.) However, the fluids in cells and blood contain significant concentrations of weak acids and bases that react with added acids and bases and prevent large changes in the pH of these fluids under normal conditions. These compounds are called buffers. For example, the proteins of the blood contain ionizing groups that will neutralize small amounts of acid and base that enter the blood stream or tissue fluids, and there are other acids and bases in blood and the cytoplasm of cells that can neutralize added bases and acids. One such compound is carnosine, or β-alanylhistidine, which is found in brain and muscle and has a pK close to 7.0 for its protonated imidazole group.

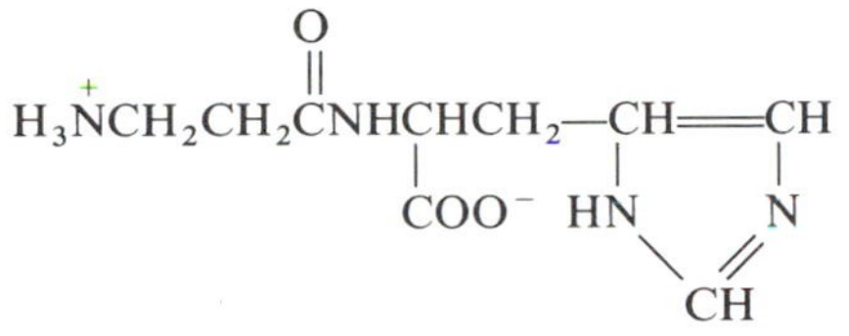

```
                      O
   +                  ||
H3NCH2CH2CNHCHCH2—CH══CH
                  |          |         |
               COO⁻  HN        N
                           \    //
                             CH
```

Carnosine
(β-Alanylhistidine)

The only known function of this interesting compound is to act as a buffer.

The capacity of a buffer to minimize pH changes is greatest at pH values near the pK_a of the buffer. The reason for this can be seen by reexamining the Henderson-Hasselbalch equation. When the pH of a solution is equal to the pK_a of the buffer, $\log([A^-]/[HA])$ is zero; that is, $[A^-] = [HA]$. Under this condition, the buffering capacity is maximal, because the concentrations of both A^- and HA are high, and the addition of a small amount of either an acid or a base will not change the ratio $[A^-]/[HA]$ very much. A small amount of an acid or base will not change the pH very much either because, as shown by the Henderson-Hasselbach equation, the pH changes with $\log([A^-]/[HA])$. At a pH that is far from the pK_a of the buffer, so that $[A^-]/[HA] = 0.01$ or 100, most of the buffer is in the form of HA or of A^- and the concentration of A^- or HA is very small. Addition of a small amount of a strong acid or base will have a large effect on the ratio $[A^-]/[HA]$ and on the pH. As a general rule, a buffer is regarded as effective at pH values within one unit on either side of the pK_a value for the buffer.

Activities and Activity Coefficients

It is frequently found that the pH of a solution containing a buffer is different from the value calculated from the buffer ratio and the pK_a. These differences are usually small, but they can be significant.

Differences of this kind can be due to changes in the properties of the solvent that are caused by the addition of organic compounds or salts. Salts will generally increase the dissociation of an uncharged acid, such as acetic acid (equation 9), because they stabilize the charged products.

$$CH_3COOH \rightleftharpoons H^+ + CH_3COO^- \tag{9}$$

Salts may also destabilize uncharged reactants because nonpolar organic molecules are likely to interact unfavorably with the charges of salts. In fact, organic molecules, including proteins, often can be forced out of solution by adding high concentration of salts. This is called "salting out" and is widely used for the precipitation and fractionation of proteins. Increasing concentrations of ammonium sulfate are added to an aqueous solution of proteins, and the proteins precipitate from the solution over a characteristic range of ammonium sulfate concentration.

Thus, salts can destabilize nonpolar molecules and stabilize charged molecules in aqueous solution. These changes may be described quantitatively by changes in the *activity* or *activity coefficient* of the molecules when their environment changes. An increase in activity or activity coefficient means that the molecule is less stable; it is more likely to react with another molecule or to fall out of the solution. A decrease in activity or activity coefficient means that it is more comfortable in the solution: its escaping tendency decreases and its solubility increases. The addition of alcohol to water will decrease the activity and activity coefficient of a hydrocarbon but will increase the activity and activity coefficient of most salts. Alcohols are less polar than water and are therefore poorer solvents for salts.

Thus, the behavior of a molecule in solution is determined by its activity, not its concentration. The relation between activity, a, and concentration, c, is defined by the activity coefficient, f, according to equation 10.

$$a = fc \tag{10}$$

The activity coefficient corrects for the change in activity of a compound in the presence of salt or upon the addition of an organic solvent to water. It provides a correction factor to account for changes in the stability and reactivity of a molecule when the nature of the solvent is changed. (Some disrespectful biochemists have claimed that f stands for a "fudge factor," but we would never make such a statement.)

The equilibrium constant of a reaction is normally defined in terms of the activities, not the concentrations, of the reactants and products (equation a). This equilibrium constant is indeed *constant*; it does not change when the concentrations of the reactants change enough to change the activity coefficients of the reacting molecules. However, it is also possible to define an equilibrium constant in terms of concentrations (equation b). This concentration equilibrium constant is not exact; it will change if the concentrations of the reactants change enough to cause changes in their activity coefficients.

$$K = \frac{a_B}{a_A} = \frac{f_B c_B}{f_A f_B} \tag{a}$$

$$K_{app} = \frac{c_B}{c_A} \tag{b}$$

Answers and Solutions to Problems

Many of the problems at the end of chapters in this book have specific answers that are given here. Other problems are more open ended in that there may be more than one solution. In these cases, a sample solution is provided, but the student is encouraged to find others. For some of the open-ended problems, there is a *set* of solutions, each of which refers to an individual biochemical system, and several related systems are known. For some problems, the answers may not be known with certainty and will have to be determined by further investigation. In these cases, several reasonable solutions can be formulated, and the one given here is only a sample. Some problems of this last type can be found in the later chapters. If a problem has more than one solution or answer, the answer is marked by an asterisk to indicate that other solutions or answers also are possible.

Chapter 1: Biochemistry and Life

1. Explanations:

(a) An organism is a life form that is capable of self-maintenance and replication—in general, an animal or plant.

(b) ATP is the most-important source of biochemical free energy in organisms and cells.

(c) An enzyme is a catalyst for a biochemical reaction in a cell.

(d) A prokaryote is a cell that contains no nucleus—in general, a bacterium.

(e) A mitochondrion is an intracellular eukaryotic organelle that couples the production of ATP to the oxidation of two-, three-, and four-carbon metabolites.

(f) A peroxisome is an intracellular eukaryotic organelle that contains oxidizing enzymes that degrade foreign materials and cell debris to simple molecules.

(g) A cell wall is the tough, outer coating of a plant or microrganismic cell that protects the plasma membrane from its environment.

(h) A cytoskeleton is the system of microfilaments and microtubules within a cell that maintains the structure of the cell and determines its morphology.

(i) An organized cellular structure in a higher animal, an organ consists of cells that carry out specialized functions in support of the whole animal.

(j) A cell is the basic unit of life and is able to maintain itself and reproduce.

(k) DNA is the deoxyribonucleic acid in a cell and has the function of storing and transmitting genetic information.

(l) A protein is a biological macromolecule composed of amino acids.

(m) A plasma membrane is the lipid, outer covering of a cell.

(n) A ribosome is a complex of ribonucleic acids and proteins that catalyzes the polymerization of amino acids into proteins.

(o) The endoplasmic reticulum is a membranous organelle within a eukaryotic cell; protein biosynthesis and a few other specialized biochemical reactions take place in the endoplasmic reticulum.

(p) A metabolite is a cellular compound that is an intermediate in the biosynthesis of an end product or in the biodegradation of a nutrient.

(q) A virus is a particle consisting of a nucleic acid and a few proteins that can enter a cell and undergo replication by controlling the cellular replication apparatus. It is a parasite of a cell.

(r) The endocrine system consists of the glands in an animal that produce hormones on command from the brain.

(s) A eukaryote is a cell that contains a nuclear envelope and organelles such as mitochondria, endoplasmic reticulum, lysosomes, peroxisomes, Golgi complex, and nucleolus.

(t) RNA is ribonucleic acid and has the function of transcribing the information from DNA and facilitating its translation into the functional molecules of a cell.

(u) A hormone, which is produced by a gland, is a molecule that migrates to another organ and stimulates it to carry out some function.

(v) A lysosome is an intracellular eukaryotic organelle that contains hydrolytic enzymes that degrade foreign matter and cellular debris.

(w) A lipid is a major component of cell membranes.

(x) A nucleus is the organelle in a eukaryotic cell that contains the chromosomes and enzymes that are required to replicate chromosomes.

(y) mRNA is messenger RNA, which carries the information for the biosynthesis (amino acid sequence) of a protein from DNA to a ribosome.

(z) A bacteriophage is a virus that infects bacteria.

2. Infection of a cell by a lytic virus begins with the attachment of the virus particle to the cell membrane or wall. The process continues with the chemical breaching of the membrane (and wall) and the entry of the particle or its nucleic acid into the cell. The cellular nucleic acid replication machinery and protein biosynthetic machinery are then recruited to produce hundreds or thousands of copies of the virus. The cell eventually dies and releases the replicated viruses. A lysogenous virus infects a cell but does not undergo self-replication immediately. Instead, it incorporates the information in its nucleic acid into the genome of the cell, where it may lie dormant for a time until it directs the production of virus particles.

3. Coenzymes have special chemical properties that are not found in the amino acids of enzymes; they combine with enzymes to form complexes that catalyze biological reactions.

4. Prokaryotes do not have the nuclear envelopes found in eukaryotic cells.

5. Signal molecules such as hormones carry messages from one organ to another. They are necessary to coordinate the functions of organs in a complex organism.

6. Lysosome loading is the accumulation of indigestible foreign matter in lysosomes. It is important because it can lead to cellular malfunctions and death.

7. Viruses are not generally regarded as forms of life because they cannot replicate or repair themselves independently of cells.

8. All cells must absorb nutrients and oxidize them to produce ATP; they must biosynthesize their molecules; they must repair themselves; and they must undergo replication. Specialized cells carry out additional functions characteristic of the cell type, such as the biosynthesis of hormones, muscular contraction, photosynthesis, the production of blood cells, and so forth.

9. An intracellular organelle of a eukaryotic cell, the Golgi complex directs the processing of proteins produced in the endoplasmic reticulum.

10. Biochemistry is the chemistry of life. Biology is the study of the properties and functions of living cells and organisms. Chemistry is the study of the transformations of molecules. Biochemistry is the study of the chemical transformations of biological molecules.

11. Cellular shape is maintained by the cytoskeleton of a cell.

12. Vitamins are molecules other than proteins and carbohydrates that are required by animals in their diets to maintain good health. Most vitamins are converted into coenzymes in animals.

13. DNA is important because it contains the genetic information that is required to specify the structures of the molecules in a cell. It stores the information and transmits it to the biosynthetic machinery, and it transmits the information from one generation of cells to the next. RNA mediates the expression of the genetic information in DNA. (The genetic information of viruses is contained within either DNA or RNA.)

14. Lysosomes and peroxisomes degrade foreign matter that adventitiously enters cells. They also degrade cellular debris that results from injury to the cell or from normal turnover of molecules in the cell. Lysosomes contain hydrolytic enzymes, whereas peroxisomes contain oxidizing enzymes.

15. Signal transduction is the transfer of information from a signal molecule to an apparatus that responds by carrying out some chemical process in a cell. It is important in regulating and coordinating the functions of cells.

Chapter 2: Carbon-Carbon Bond Formation and Cleavage

1. (a)

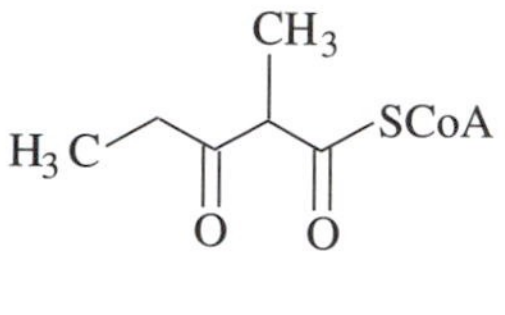

CH_3 H_3C SCoA   O   O

(b)

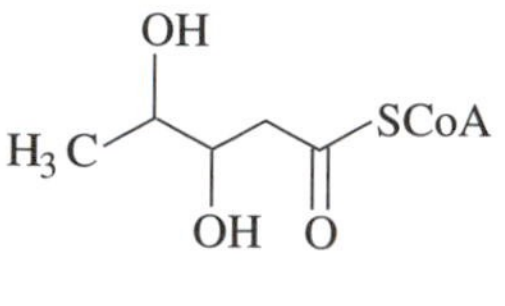

OH H_3C SCoA   OH   O

(c)

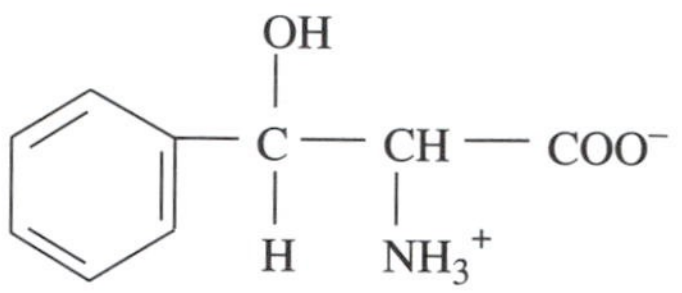

OH   C — CH — COO^- H NH_3^+

(d)

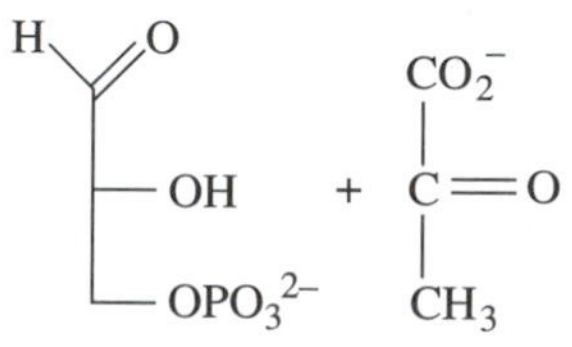

H   O   OH OPO_3^{2-} + CO_2^- C=O CH_3

(e)

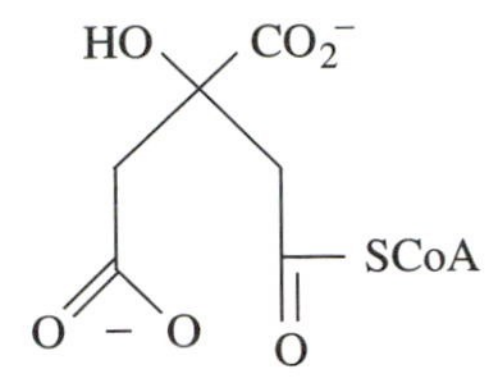

HO CO_2^-

SCoA

O – O O

2. The most likely cofactor is pyridoxal phosphate.

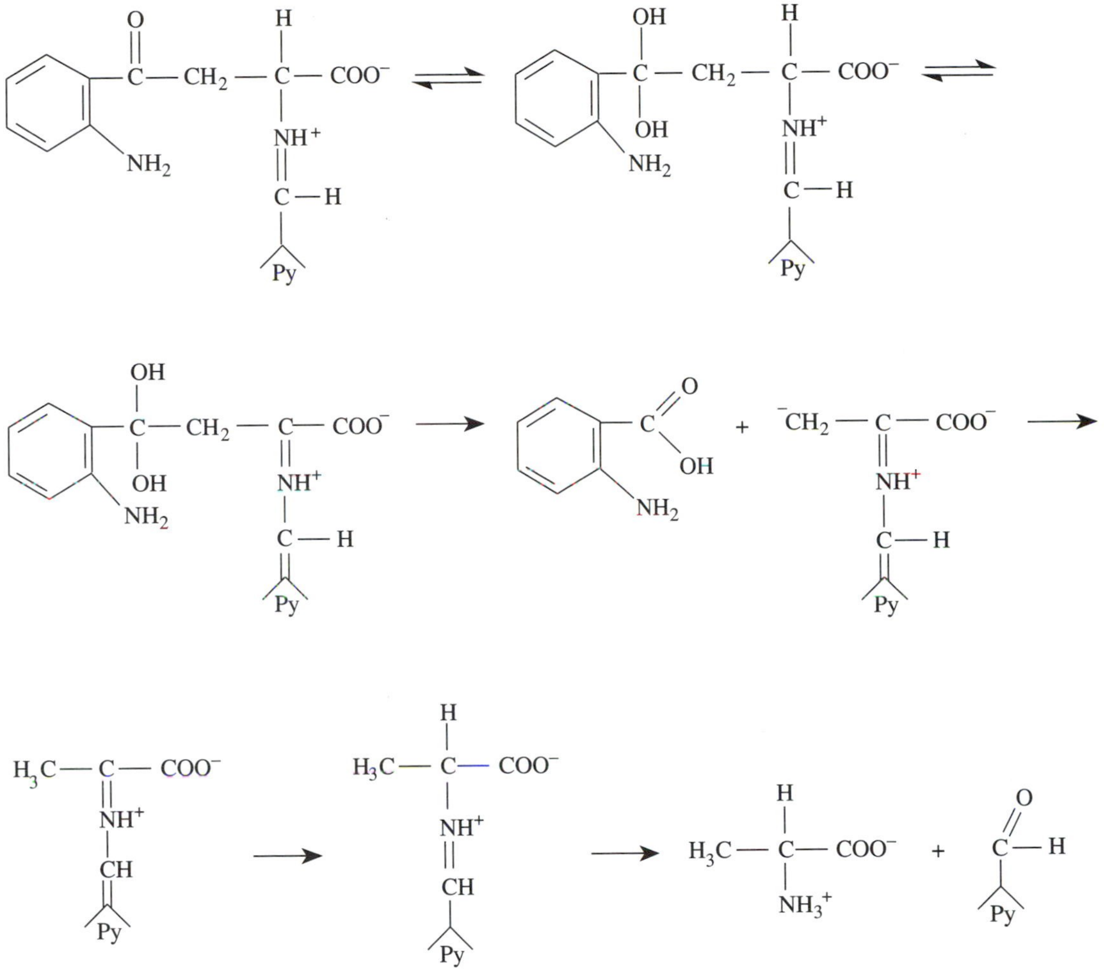

C—CH₂—C—COO⁻ ⇌ C—CH₂—C—COO⁻ ⇌

NH_2 NH^+ C—H Py OH OH NH_2 NH^+ C—H Py

→ + $^-CH_2$—C—COO⁻ →

H_3C—C—COO⁻ → H_3C—C—COO⁻ → H_3C—C—COO⁻ + C—H

NH^+ CH Py NH^+ CH Py NH_3^+ Py

3.

Enz — Nu: H_2C CO_2H → Enz— Nu: CO_2H + Br^-

Br

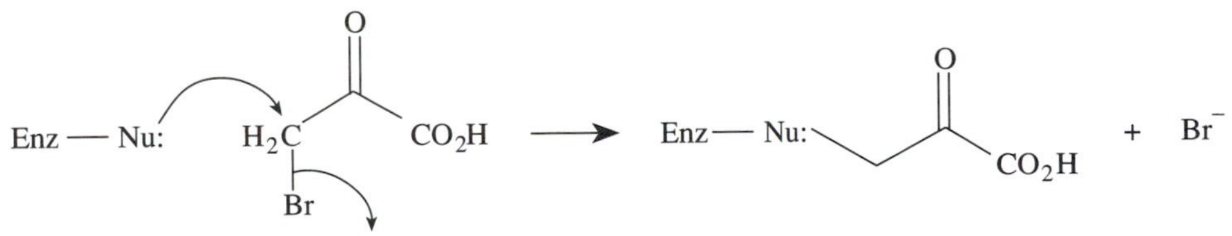

4. (a) There is a greater concentration of the (reactive) neutral species. Therefore, the imine-forming reaction is more likely to occur (faster).

(b) The amine could be in a hydrophobic environment. There could be an adjacent cation (another "amine," ionized).

5.

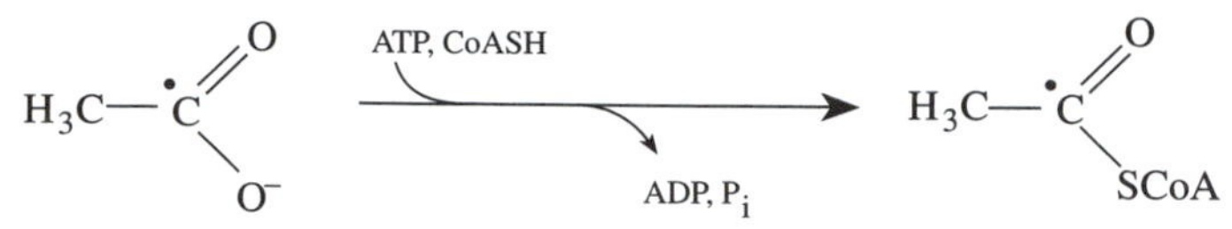

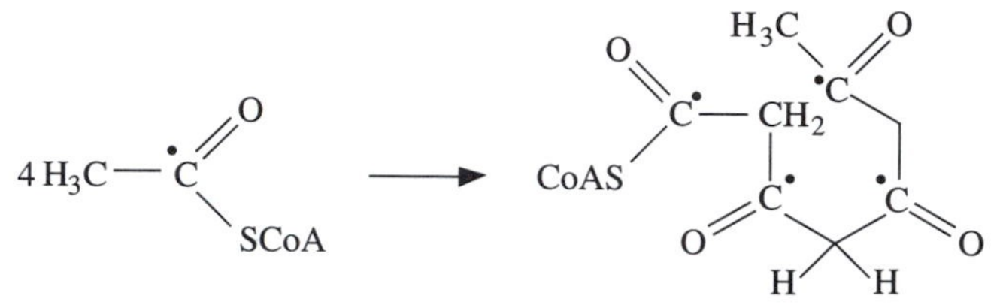

$H_3C—\dot{C}(=O)O^-$ $\xrightarrow[\text{ADP, P}_i]{\text{ATP, CoASH}}$ $H_3C—\dot{C}(=O)SCoA$

$4\ H_3C—\dot{C}(=O)SCoA \longrightarrow$ CoAS—$\dot{C}$(=O)—CH (ring, with CH_2, $H_3C—\dot{C}$=O, $\dot{C}$=O, $\dot{C}$=O, H, H)

A

This involves aldol condensation:

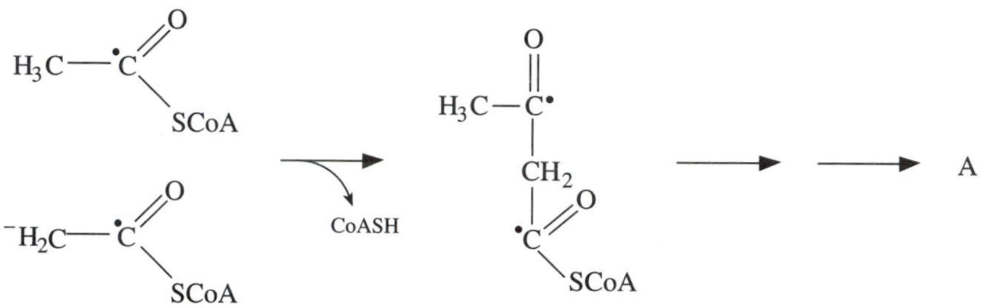

$H_3C—\dot{C}(=O)SCoA$ + $^-H_2C—\dot{C}(=O)SCoA$ $\xrightarrow[\text{CoASH}]{}$ $H_3C—\dot{C}(=O)—CH_2—\dot{C}(=O)SCoA \longrightarrow \longrightarrow$ A

Another aldol condensation:

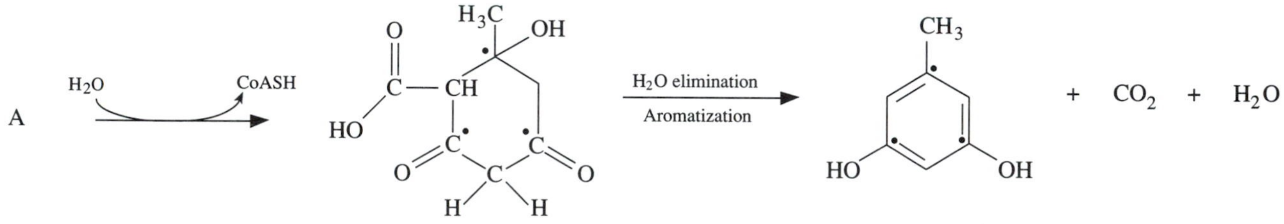

A $\xrightarrow[]{\text{H}_2\text{O} \quad \text{CoASH}}$ HO—C(=O)—CH (ring, with H_3C, OH, $\dot{C}$=O, $\dot{C}$=O, H, H) $\xrightarrow[\text{Aromatization}]{\text{H}_2\text{O elimination}}$ 3,5-(HO)$_2$-C_6H_3-CH_3 (CH$_3$, HO, OH) + CO_2 + H_2O

6. Product:

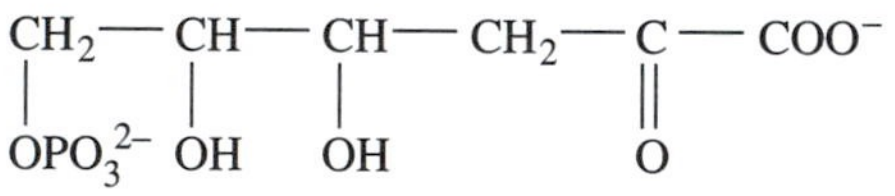

$CH_2(OPO_3^{2-})—CH(OH)—CH(OH)—CH_2—C(=O)—COO^-$

Schiff-base mechanism:

$E—NH_2 \xrightarrow{CH_3—C(=O)—COO^-} E—\overset{+}{N}H=C(COO^-)CH_3 \xrightarrow{CH_2(OH)—CH(OH)—C(=O)H} E—\overset{+}{N}H=C(COO^-)—CH_2—C(HO)(H)—C(H)(OH)—CH_2OH \xrightarrow{H_2O}$ Product

$E—\overset{+}{N}H=C(COO^-)CH_3 \xrightarrow{NaBH_4} E—NH—C(COO^-)(H)—CH_3$

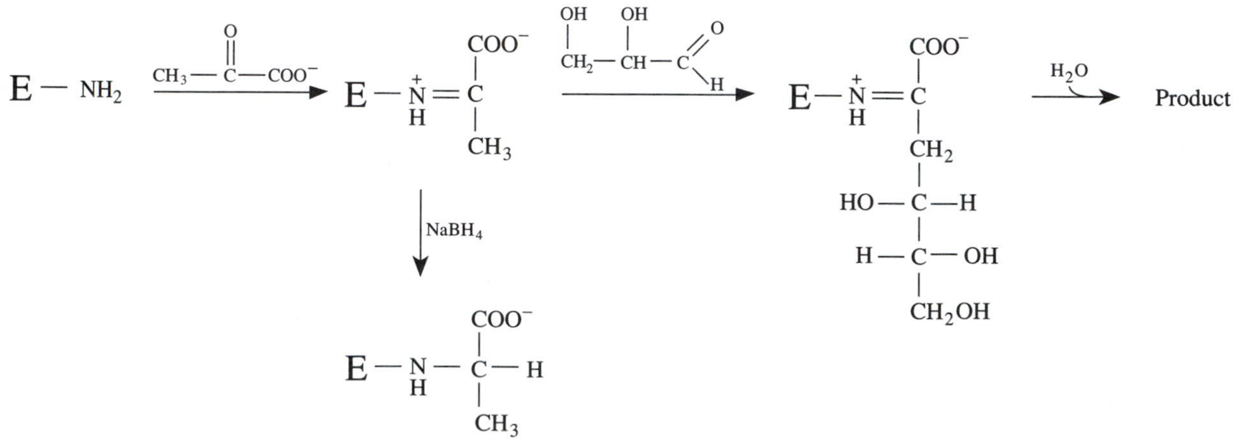

Inactive

Chapter 3: Introduction to Chymotrypsin: A Case Study for Catalysis By Enzymes

1. For the model

$$E + S \underset{k_{-1}}{\overset{k_1}{\rightleftharpoons}} ES \xrightarrow{k_2} E + \text{Product}$$

(a) The velocity is given by

$$v = k_2[ES]$$

and we will show that

$$v = \frac{k_2[E]_{tot}}{\frac{K_m}{[S]} + 1} \quad (a)$$

The conservation equation is

$$[E]_{tot} = [E] + [E \cdot S] \quad (b)$$

The observed initial rate in the steady state is

$$v = k_2[E \cdot S] \quad (c)$$

The steady-state requirement is

$$k_1[E][S] = (k_{-1} + k_2)[E \cdot S] \quad (d)$$

Define K_m as

$$K_m = \frac{k_{-1} + k_2}{k_1} \quad (e)$$

From equations d and b

$$k_1([E_{tot}] - [ES])[S] = (k_{-1} + k_2)[ES] \quad (f)$$

$$k_1[E_{tot}][S] - k_1[ES][S] = k_{-1}[ES] + k_2[ES] \quad (g)$$

Solve for [ES]:

$$k_{-1}[ES] + k_2[ES] + k_1[ES][S] = k_1[E]_{tot}[S] \quad (h)$$

$$[ES] = \frac{k_1[E]_{tot}[S]}{k_{-1} + k_2 + k_1[S]} = \frac{[E]_{tot}[S]}{\frac{k_{-1} + k_2}{k_1} + [S]} = \frac{[E]_{tot}[S]}{K_m + S} \quad (i)$$

From equation c:

$$v = k_2[ES] = \frac{k_2[E]_{tot}[S]}{K_m + [S]} = \frac{k_2[E]_{tot}}{\frac{K_m}{[S]} + 1}$$

(b) At high [S] ([S] $\gg K_m$):

$$k_{-1} \gg k_2 \qquad v = k_2[E_{tot}]$$

$$k_2 \gg k_{-1} \qquad v = k_2[E_{tot}]$$

At low [S] ([S] $\ll K_m$):

$$k_{-1} \gg k_2 \qquad v = \frac{k_1 k_2}{k_{-1}}[S][E_{tot}]$$

$$k_2 \gg k_{-1} \qquad v = k_1[E]_{tot}[S]$$

2. $k_{cat} = 0.16\ s^{-1}$, $V_{max} = 1.6\ \mu M\ s^{-1}$, $K_m = 0.13$ M

3. A pentapeptide with a reactive group that will be bound in position to react with the serine hydroxyl group (Figure 3-5) should be a strong inhibitor. The reactive group could be

$$-\overset{H}{C}{=}O,\ -\overset{H}{C}{=}S,\ -C{\equiv}N,\ -\overset{O}{\overset{\|}{C}}CH_2Cl$$

or other structures that can react with ROH. The pentapeptide provides the noncovalent binding energy to facilitate strong binding of the reactive group with little additional loss of entropy.

The two most-serious problems with such inhibitors or inactivators are: (1) they may react with other serine proteases, or other enzymes, that are important in the normal functioning of the organism, so that they are themselves toxic; (2) they may be metabolized rapidly, by hydrolysis or oxidation, so that they are not available for inhibition for a long enough time to be useful.

4.

$$\begin{array}{ccc} E + S & \xrightarrow{k_p} & \text{Product} \\ \pm S \Big\updownarrow K_s & & \\ E \cdot S & & \end{array}$$

The rate of the reaction will show saturation behavior with increasing [S] because of the formation of the unreactive complex, E·S. When [S] is $\gg K_s$, most of the enzyme exists as E·S. An increase in substrate concentration will then have no effect on the rate because the increase in k_p[S] will be balanced by a decrease in [E] when E forms more E·S.

Note that, according to the scheme, the reaction does not proceed through the E·S complex. The predominant E·S complex is unreactive; it is called a *nonproductive complex.* Of course, the reaction *does* proceed through an E·S complex, the *productive complex,* but this is present in very low concentration and is not responsible for the leveling off of the rate at high substrate concentration.

5. The velocity is given by

$$v = k_2[E^*] = k_1[E \cdot S] \quad (a)$$

because each turnover must proceed through these two steps and the steps are irreversible.

The conservation equation is

$$[E]_{tot} = [E] + [E \cdot S] + [E^*] \quad (b)$$

The concentration of ES is

$$[E \cdot S] = \frac{[E][S]}{K_m} \quad (c)$$

The steady-state requirement is

$$v = k_1[ES] = k_2[E^*]$$
$$= k_2[E_{tot}] - k_2[ES](1 + K_s / [S]) \qquad (d)$$

$$ES = \frac{k_2[E_{tot}]}{k_2\left(1 + \frac{K_s}{[S]}\right) + k_1}$$
$$= \frac{[E_{tot}]}{1 + \frac{K_s}{[S]} + \frac{k_1}{k_2}} \qquad (e)$$

$$v = \frac{k_1[E_{tot}]}{1 + \frac{K_s}{[S]} + \frac{k_1}{k_2}} \qquad (f)$$

When $k_1 \gg k_2$ and the enzyme is saturated with substrate, the rate-limiting step is k_2 (because $k_1 \gg k_2$) and the enzyme accumulates as E*. Thus, there is no substrate bound to the enzyme at saturating substrate concentrations.

6. See the inside front cover. Both threonine (T) and glutamate (E) have polar side chains, but the anionic propionate group of glutamate is much more hydrophilic than the ethanol side chain of threonine. Both phenylalanine (F) and leucine (L) have nonpolar side chains, but the aromatic toluene side chain of phenylalanine is more reactive chemically than the isobutyl side chain of leucine. The propylguanidinium side chain of arginine (R) is much more polar and hydrophilic than the *p*-cresol side chain of tyrosine (Y), but the aromatic phenolic group is much more reactive and susceptible to chemical modification. The methylethyl sulfide side chain of methionine (M) is nonpolar but can be methylated to give a reactive methylating agent in *S*-adenosylmethionine (see page 686 of the text), whereas the ethanol side chain of serine provides a hydroxyl group that is acylated or phosphorylated in the active site of many enzymes and other proteins; these reactions frequently regulate physiological functions.

7. The specificity pocket of chymotrypsin binds nonpolar side chains of amino acids and cleaves peptides at these amino acids, whereas the aspartate carboxylate group in the specificity pocket of trypsin binds the cationic side chains of lysine and arginine and cleaves peptides at these amino acids (see Figures 3-3 and 3-4).

Chapter 4: Catalysis by Chymotrypsin: The Catalytic Pathway

1. Chymotrypsin will attack the acyl group of phenylalanine to give a covalent phenylalanyl-enzyme (k_f), which normally reacts with water to give hydrolysis products. In the presence of alanine amide, however, the acyl-enzyme will react with alanine amide to regenerate starting materials (k_r), so that the rate of formation of hydrolysis products is decreased.

Gly-PheC(=O)-NH-Ala-NH_2 + HO-Enz

$\uparrow k_r \;\; \downarrow k_f$

Gly-PheC(=O)-O-Enz + H_2N-Ala-NH_2

$\downarrow H_2O$

Gly-PheC(=O)-O^- + HO-Enz + H^+

2. k_{cat} is the first-order rate constant (units of s^{-1} or min^{-1}) for hydrolysis of the acetyl-enzyme. It involves only one reacting species. This step is rate limiting at high substrate concentrations, because it is much slower than the rate of acylation of serine in the enzyme-substrate complex (k_1).

CH_3C(=O)OPhNO_2 + HO-Enz
PNPA

$\uparrow\downarrow K_s$

CH_3C(=O)OPhNO_2
HO-Enz

Fast $\downarrow k_1$

CH_3C(=O)-Enz + $^-$OPhNO_2 + H^+

Slow $\downarrow k_2 = k_{cat}$

CH_3C(=O)O^- + HO-Enz + H^+

The k_{cat} / K_m is the second-order rate constant (units of M^{-1} s^{-1} or M^{-1} min^{-1}) for reaction of PNPA with the enzyme to give products. It is equal to k_1 / K_s, and acylation (k_1) is rate limiting. The k_2 step is not significant at very low substrate concentration because the k_1 step is irreversible. At the very low substrate concentrations at which k_{cat} / K_m is measured, almost all of the enzyme molecules are not acylated during turnover and the rate increases linearly with increasing concentration of substrate. The K_m value is the concentration of substrate at which half of the enzyme molecules are acylated (CH_3COO-Enz) and half are not (HO-Enz). The observed rate of hydrolysis is half of that at substrate saturation (k_{cat}), at which all of the enzyme is acylated.

3. First, determine its substrate specificity by measuring the rate of hydrolysis of synthetic peptide substrates of known structure or determine the positions of bond cleavage by analyzing the C-terminal amino acid that is formed upon hydrolysis of a peptide or protein substrate.

Second, determine values of k_{cat} and k_{cat} / K_m for a peptide and an ester substrate.

Third, determine the dependence on pH of k_{cat}/K_m and k_{cat}. This could provide an indication for histidine at the active site.

Fourth, determine if the enzyme is inactivated by DFP (diisopropylphosphorofluoridate). (**Caution:** DFP is highly toxic; use this reagent only under the direction of an experienced investigator.) Inactivation by DFP is characteristic of enzymes that have serine at the active site.

Fifth, determine whether the activity is inhibited by reagents that react with the sulfhydryl group of cysteine residues. Is the enzyme protected against inactivation in the presence of a good substrate, because the substrate acylates a cysteine residue at the active site?

Finally, (1) add chelating agents for metals to determine whether there is a dissociable metal ion at the active site, and (2) analyze the purified protein for metal ions.

4. Graph A: The acylation step is rate limiting. When hydroxylamine is added, the acyl-enzyme reacts with hydroxylamine, instead of water, so that the hydroxamic acid is the product. However, the total rate does not increase because acylation is rate limiting.

Graph B: Deacylation is predominantly rate limiting. The rate of deacylation is increased by hydroxylamine, which reacts with the acyl-enzyme to form the hydroxamic acid. There is some decrease in the rate of hydrolysis because acylation is partially rate limiting.

The ratio of acid and hydroxamic acid formation is 1.0 at approximately 1.6×10^3 M [NH_2OH] in both A and B. This is expected if the same acyl-enzyme is formed from the amide and the ester; it would be unlikely if hydroxylamine and water reacted directly with the two different substrates at the active site.

The same maximum rate of hydrolysis of three different esters will be observed if the deacylation step is rate limiting. Different substrates usually have different acylation rates because they have different reactivities and the enzyme is likely to catalyze their hydrolysis at different rates.

5. The almost identical values of k_{cat} suggest that deacylation is predominantly rate limiting at substrate saturation for all of the esters. Acylation may be partly rate limiting for the ethyl ester, which has a much smaller chemical reactivity than the phenyl esters. The decrease in K_m, or increase in $1 / K_m$, with increasing chemical reactivity of the esters is caused by a faster rate of acylation by the more-reactive esters. All of the enzyme is converted into the acyl-enzyme at a relatively low concentration of the more-reactive esters because acylation is faster with these esters. This accounts for the increase in $1/K_m$ (or decrease in K_m) with the more-reactive esters.

Chapter 5: Catalysis by Chymotrypsin: The Importance of Binding Energy

1.

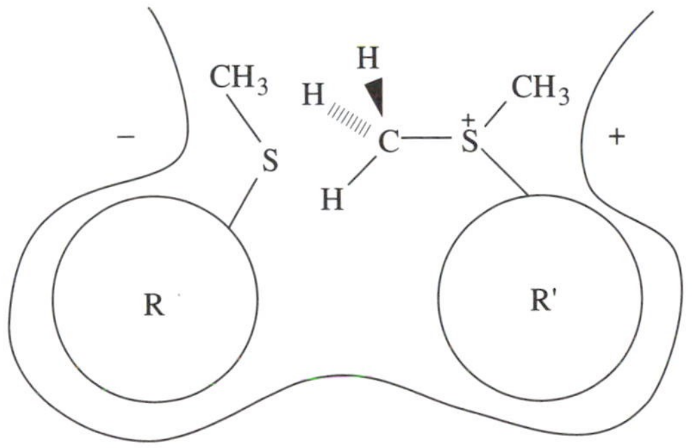

(a) The active site will hold the reactants firmly to form a structure that is as close to the structure of the transition state for bimolecular substitution as possible. This will bring about loss of entropy, so that little additional entropy will be lost in reaching the transition state.

(b) The bound reactants will be destabilized relative to the transition state, which will be stabilized. This can be brought about by a low polarity of the active site that destabilizes the positive charge on the sulfonium reactant and a negative charge that stabilizes the developing positive charge on the sulfonium ion product.

(c) The reactants will be compressed when they are held firmly against each other. However, this effect will be small because an enzyme cannot exert a strong enough force to bring about significant compression of a covalent bond between carbon atoms. The compression acts mainly to increase loss of entropy.

Nucleophilic catalysis is not likely to be important because the intermediate formed by reaction with a group on the enzyme is likely to be less reactive than the sulfonium ion itself.

2.

$$AcO^- + Ca^{2+} \underset{}{\overset{K_a}{\rightleftharpoons}} AcO^- \cdot Ca^{2+} \qquad K_a = 10\ M^{-1}$$

$$AcO^- + AcO^- \cdot Ca^{2+} \underset{}{\overset{K_a^*}{\rightleftharpoons}} AcO^- \cdot Ca^{2+} \cdot {}^-OAc \qquad K_a^* = 10\ M^{-1}$$

$$K_1 = \frac{[AcO^- \cdot Ca^{2+}]}{[AcO^-][Ca^{2+}]} = 10\ M^{-1}$$

Assume that a second acetate ion binds with the same association constant as the first acetate ion.

$$K_2 = \frac{[AcO^- \cdot Ca^{2+} \cdot {}^-OAc]}{[AcO^-][AcO^- \cdot Ca^{2+}]} = 10\ M^{-1}$$

$$K_{12} = K_1 K_2 = \frac{[AcO^- \cdot Ca^{2+} \cdot {}^-OAc]}{[AcO^-][AcO^-][Ca^{2+}]} = 100\ M^{-2}$$

$$K_{malonate} = \frac{[Mal^{2-} \cdot Ca^{2+}]}{[Mal^{2-}][Ca^{2+}]} = 500\ M^{-1}$$

$$\text{Effective molarity} = \frac{K_{malonate}}{K_1 K_2} = \frac{500\ M^{-1}}{100\ M^{-2}} = 5\ M$$

The effective molarity for chelation is usually not very large because the formation of a metal chelate requires less loss of entropy than does the formation of covalent bonds. An ion pair has much more freedom for movement, and therefore a more-positive entropy, compared with a covalent bond to carbon.

3. E = Effector, such as a hormone

R = Inactive form of receptor

R* = Active form of receptor

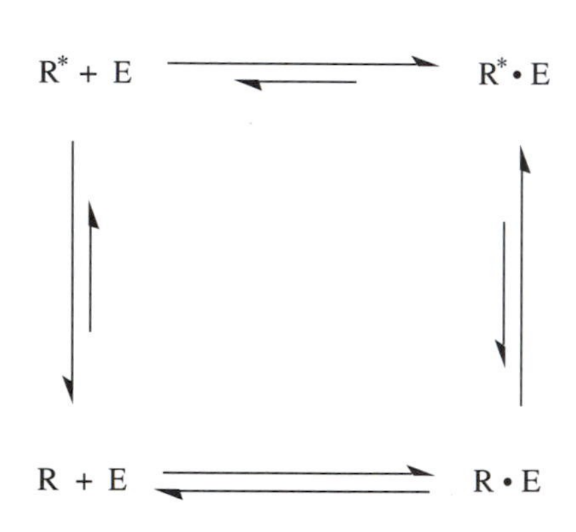

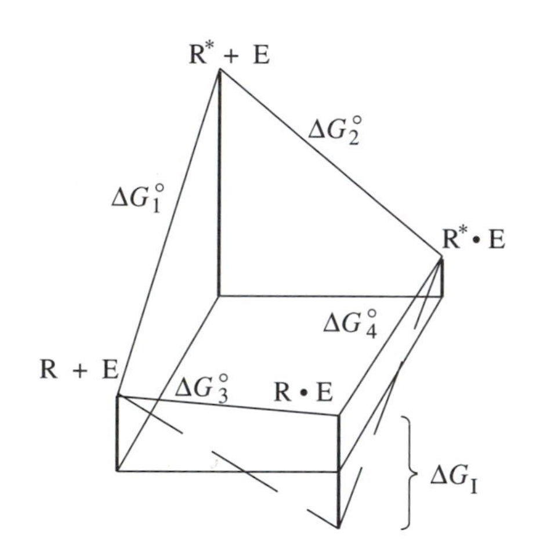

4. The acetic anhydride reacts with the amino group of Ile-16, which is present largely as the reactive free base at pH 9. This prevents protonation of the lysine amino group and formation of the Lys—$NH_3^+ \cdot {}^-OOC$—Asp ion pair in the active form of chymotrypsin (Figure 5-4).

5. Ethyl acetate lacks the alkylammonium group, $(CH_3)_3N^+CH_2$—, of acetylcholine. This group provides binding energy that is utilized not only to bind acetylcholine at the active site, but also to stabilize the transition state relative to the ground state of the reaction by noncovalent binding interactions. This interaction involves loss of entropy of the bound substrate and contributions from destabilization of the bound substrate relative to the transition state by noncovalent interactions that may include desolvation and nonbinding interactions (strain) that are relieved in the transition state.

Chapter 6: More Hydrolytic Enzymes, Blood Clotting, and the Antigen-Antibody Response

1. Most of the enzyme is in the active form at pH 3, with one carboxylate group protonated and one unprotonated. This allows both proton donation and proton removal to occur in the transition state for hydrolysis. One possible mechanism for this is shown in Figure 6-11. The addition or loss of a proton will prevent catalysis of proton removal or donation, respectively.

2. Lipase attacks the ester bonds of insoluble fats, but has relatively little activity toward soluble substrates. Increasing the concentration of phenyl propionate beyond its solubility and shaking the suspension would form an emulsion that could be attacked by the enzyme. The addition of a nonionic detergent would form micelles that could be an effective form of the substrate. (However, anionic detergents may denature the enzyme.)

3. The most-likely reason is oxidation of a sulfhydryl group in the active site. Bromelin is, indeed, a thiol protease. This oxidation can be prevented by storing the enzyme in the cold or in a dilute solution of a thiol, preferably in the absence of oxygen. Activity of the oxidized enzyme may be regained by the addition of a thiol or borohydride to reduce the oxidized thiol. The enzyme may also lose activity by denaturation, which may be accelerated by adsorption on the glass surface or the air-water interface.

4. Incubate solutions containing several very low concentrations of proenzyme and measure the rate of activation. If the half-time for activation is constant and independent of the enzyme concentration, the enzyme is activating itself. If it decreases with increasing enzyme concentration and with time, the active enzyme molecules are activating inactive molecules.

5. Calcium is required for clotting, and chelating agents for calcium will greatly decrease the rate of clotting. Blood is usually stored in plastic containers that provide a smooth surface, which is less likely to initiate clotting than is glass. Heparin may also be added at low concentration to inhibit clotting.

6. Multiple binding sites can greatly increase the binding constant of the antibody for its specific antigens because the formation of binding interactions after the first one is formed is an intramolecular reaction, which is much more favorable than the initial bimolecular reaction. Formation of the polymeric antigen-antibody complex serves as a signal for further reactions of the immune system.

7. There are many possibilities for inducing the formation of catalytic antibodies in this way. The molecule should resemble the transition state of the reaction that is being catalyzed. A condensation, such as a Diels-Alder reaction, should be especially favorable for such catalysis because correct approximation of the reactants (entropy loss) provides a major part of the barrier for the reaction. No proton transfers or additional reaction steps are required.

Chapter 7: The Structure of Proteins

1. A pH-gradient is set up in a polyacrylamide gel and the isoelectric point of a protein is estimated from the position to which it moves relative to the positions of proteins with known isoelectric points when an electric field is applied to the gel, as described on page 198.

2. Negative charges on the protein interact with positive charges on DEAE cellulose and slow the movement of the protein through the column. Increasing the pH will increase the number of negative charges and decrease the number of positive charges on the protein, so that it will bind more strongly to the DEAE and move more slowly through the column.

3. SDS electrophoresis depends on the size of the unfolded, denatured protein, which determines the resistance to movement through the polyacrylamide gel; larger unfolded proteins are more likely to be held back by the gel. In molecular exclusion chromatography, on the other hand, the native protein follows a pathway through the gel that depends on the size of the protein. Small proteins can travel through many interstices of the gel and follow a long path; large proteins can pass only through the larger channels, so that they follow a shorter path and come off the gel sooner.

4. The isoelectric point is the pH value at which there is no net charge on the protein. The isoionic point, which is close to the isoelectric point, is the pH value obtained after prolonged dialysis of the protein against water, so that all small ions are removed and the positive and negative charges on the protein balance each other.

5. Salting out of proteins is not well understood quantitatively. A salt with a high charge density interacts strongly with water. Insertion of a relatively nonpolar molecule into water displaces some of this salt from a fraction of the solution volume. High concentrations of salt occupy much of the available volume of the solution, and the addition of a large protein molecule decreases the volume that is available to the salt. This is energetically unfavorable, so that there is a tendency for the protein to fall out of solution so as to provide more volume in the solution that the salt can occupy.

6. The β-turn usually contains a hydrophilic amino acid, such as aspartate, serine or lysine, which interacts with the surrounding solvent; a proline or hydroxyproline residue, which facilitates the sharp, 180° angle of the turn; and a glycine, which has no side chain to interfere with the side chains of other amino acids.

7. Collagen has covalent cross-links that provide great strength and stability to the collagen triple helix. The triple helix is made possible by the high content of glycine residues, which permits the three chains to fit into the stable, closely packed helix without interference from too many side chains. The hydroxyl group of hydroxyproline probably stabilizes the helix by hydrogen bonding to other residues.

Chapter 8: Forces and Interactions in Aqueous Solution

1.
$$K_{denat} = 30\% / 70\% = 0.43$$
$$\Delta G_{denat} = -RT \ln K$$
$$= -1.98 \times 298 \times 2.303 \log K$$
$$= -1.98 \times 298 \times 2.303 \log(30/70)$$
$$= -1359 \log 0.42 = -1359(-0.37)$$
$$= 500 \text{ cal mol}^{-1}$$

2. Molecules of active chymotrypsin slowly digest other molecules of chymotrypsin because native chymotrypsin molecules occasionally undergo partial unfolding and are attacked by active molecules of chymotrypsin. This autodigestion is faster at higher temperature because more molecules are partially unfolded and are subject to digestion. However, if the temperature is increased very rapidly, it may be possible to unfold all of the molecules before digestion can occur.

3.
$$\Delta S = k \ln w \text{ (equation 1)}$$
$$50 = 1.98 \times 2.303 \log w$$
$$\log w = 50 / (1.98 \times 2.303)$$
$$= 10.97$$
$$w = 10^{11} \text{ states}$$

4. $\Delta G = 0$ when the protein is 50% denatured ($K_{denat} = 1.0$)

$$\Delta G = \Delta H - T\Delta S = 0$$
$$= 37{,}000 - 111\ T$$
$$T = 37{,}000 \div 111 = 333 \text{ K}$$
$$= 333 - 273 = 60°\text{C}$$

5. The temperature will drop slightly. Folding of a protein usually has a favorable enthalpy change ($-\Delta H$) and gives off heat; denaturation then takes up heat. However, this is often not true at low temperatures, and some proteins denature in the cold.

6.
$$\Delta G = \Delta H - T\Delta S$$
$$= 54{,}000 - (273 + 40) \times 200$$
$$= 54{,}000 - 62{,}000$$
$$= -8600 \text{ cal mol}^{-1}$$

The negative value of ΔG means that the protein is denatured.

$$\Delta G = -RT \ln K$$
$$= -1.98 \times 313 \times 2.303 \log K$$
$$\log K = -8600 / (-1427) = 6.0$$
$$K_{denat} = 1{,}000{,}000$$

7. The effects of changing solvent and temperature on the stability of the native structures of micelles and proteins could be compared. If proteins are similar to micelles, both should lose their native structure under similar conditions.

They do behave similarly under some conditions. Both have a tendency to lose their native structure at high temperature; they often also become less stable at low temperatures. The native structures of both are less stable and may be lost in the presence of denaturing agents, such as alcohols and urea.

The entropy changes for unfolding a protein and dispersing a micelle could be compared. The protein will undergo a large increase in rotational entropy when the bonds along the peptide chain are allowed to rotate in the unfolded protein. The micelles have much less structure than a protein but can gain a large amount of translational entropy when the micelles break up and the molecules are free to move independently in solution.

Chapter 9: Energy

1.
$$\Delta G^{\circ\prime} = -RT \ln K$$
$$= -1.98 \times 298 \times 2.303 \log 400$$
$$= -1359 \times 2.60$$
$$= -3536 \text{ cal mol}^{-1}$$
$$= -3.54 \text{ kcal mol}^{-1}$$

Note that K for $\Delta G^{\circ\prime}$ is based on the total concentration of each species. The temperature and pH should be specified; the calculation is for 25°C = 298 K.

2.
$$\Delta G = \Delta G^{\circ\prime} + RT \ln \frac{[\text{ADP}][\text{P}_i]}{[\text{ATP}]}$$
$$\Delta G^{\circ\prime}(\text{pH } 7.0) = -8500 \text{ cal mol}^{-1} \text{ (p. 240)}$$
$$= -8500 + 1.98 \times 310 \times 2.303 \log \frac{0.001 \times 0.04}{0.0001}$$
$$= -8500 + 1414 \times \log 0.4$$
$$= -9062 \text{ cal mol}^{-1} \text{ or } 9.0 \text{ kcal mol}^{-1}$$
$$= -38{,}060 \text{ joules mol}^{-1} \text{ or } 38.1 \text{ kJ mol}^{-1}$$

Calculated for T = 37°C= 310 K.

3. Calculate the pH-independent value of $\Delta G°$ for the hydrolysis of ATP for ionic species that are specified in the problem. Then calculate the pH-dependent value of $\Delta G°'$ for the hydrolyses of ATP at pH 6.5.

At pH 7.0:

$$\Delta G = \Delta G°'$$
$$= -8500$$
$$= \Delta G°' + RT \ln \frac{\text{fr } P_i^- \times \text{fr ADP}^{2-}}{\text{fr ATP}^{3-}}$$
$$\Delta G°' = -7658 \text{ cal mol}^{-1}$$

(fr = fraction)

Then calculate ΔG for 1 M total reactants (except H^+) at pH 6.5.

$$\Delta G°' = \Delta G° + 1359 \log \frac{\text{fr } P_i^- \times \text{fr ADP}^{2-}}{\text{fr ATP}^{3-}}$$
$$= -7249 \text{ cal mol}^{-1}$$

4.

$$\Delta G°' = -RT \ln K_{app} = -7249 \text{ cal mol}^{-1} \text{ at pH 6.5}$$
$$1.98 \times 310 \times 2.303 \log K_{app} = 7249$$
$$\log K_{app} = 5.13$$
$$K_{app} = 1.35 \times 10^5 \text{ M}$$

5.

$$K_2 = \frac{[\text{RCOOH}][\text{HOEt}]}{[\text{RCOOEt}]} = 22 \text{ M}$$
$$\Delta G = \Delta G°'(\text{pH } 7)$$
$$= \Delta G° + RT \ln = \frac{\text{fr RCOOH} \times 1.0}{\text{fr RCOOEt}}$$
$$= -1900 + 1420 \log \frac{10^{-4.7} \times 1}{0.91} \quad \text{(p. 247)}$$
$$= -1900 + 1420 \times (-4.7 + 0.04)$$
$$= -1900 + 1420 \times (-4.66)$$
$$= -1900 - 6617 = -8520 \text{ cal mol}^{-1}$$

$$K_3 = \frac{[\text{RCOO}^-][\text{H}^+][\text{HOEt}]}{[\text{RCOOEt}]}$$
$$= 0.11 \text{ M}^2 \qquad \Delta G° = +5720 \text{ cal mol}^{-1}$$
$$\Delta G = \Delta G°'(\text{pH } 7)$$
$$= \Delta G° + RT \ln \left(\frac{1 \times 10^{-7}}{.91} \right)$$
$$= \Delta G° + RT \ln 1.10 \times 10^{-7}$$
$$= 1360 + 1420 \times (-6.98)$$
$$= -9883 + 1360 = 8520 \text{ cal mol}^{-1}$$

6. The reaction is pH-independent below pH 1 and from pH 8 to pH 9. The value of $-\Delta G°'$ increases by -1420 cal mol^{-1} for each unit increase in pH in the range of pH 3–7 and above pH 11 because a proton is released upon hydrolysis in these regions of pH.

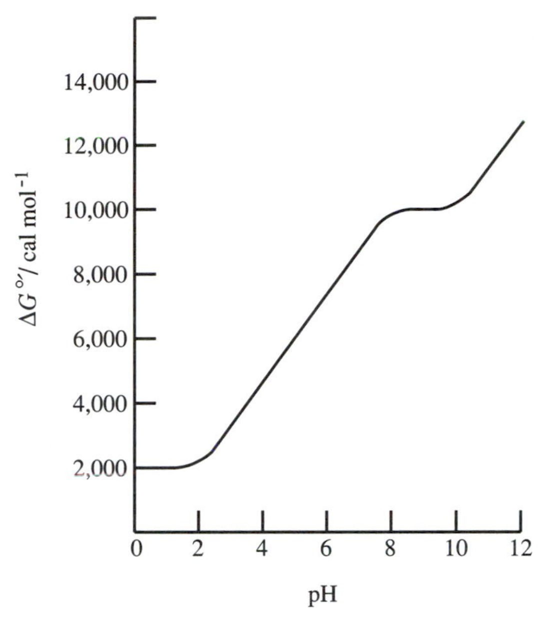

7. The values of $\Delta G°'$ (pH 7.0) are -8.5 kcal mol^{-1} for ATP and -4.7 kcal mol^{-1} for glucose-1-phosphate; the difference is -3.8 kcal mol^{-1}.

8. The values of $\Delta G°'$ (pH 7.0) are $-14{,}000$ kcal mol^{-1} for phosphoenolpyruvate and $-8{,}500$ kcal mol^{-1} for ATP. The difference of -5.5 kcal mol^{-1} for transfer of phosphate from phosphoenolpyruvate to ADP corresponds to an equilibrium constant of 7413 from $\Delta G°' = -RT \ln K$.

$$\Delta G°' = -RT \ln K = -5500 = -1423 \log K$$
$$\log K = 5500 / 1423 = 3.87$$
$$K = 7413$$

Chapter 10: Protein Biosynthesis and Acyl Activation

1. The mechanism of the acetyl CoA synthetase reaction is:

$$\text{ATP} + \text{Acetate} \rightleftharpoons \text{Acetyl-AMP} + \text{PP}_i$$

$$\text{Acetyl-AMP} + \text{CoASH} \rightleftharpoons \text{Acetyl CoA} + \text{AMP}$$

Acetyl CoA synthetase will catalyze the exchange of radioactive CoASH into acetyl CoA in the presence of AMP and the absence of PP_i. The exchange takes place through the reversal of the second step of the mechanism.

2.

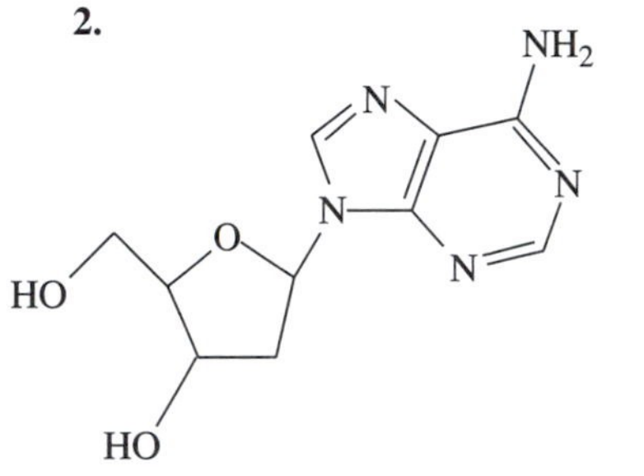

dAdo

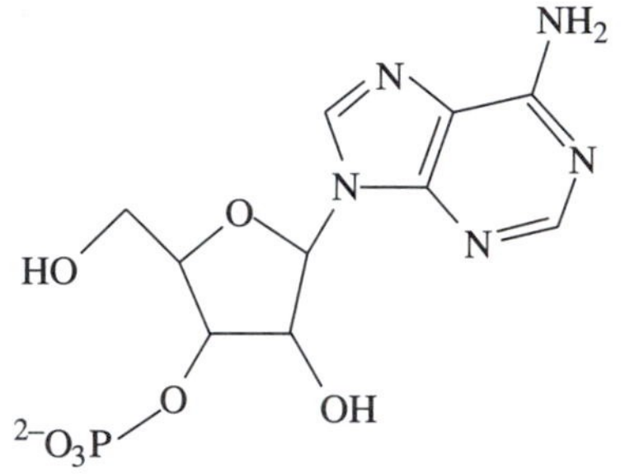

3'-AMP

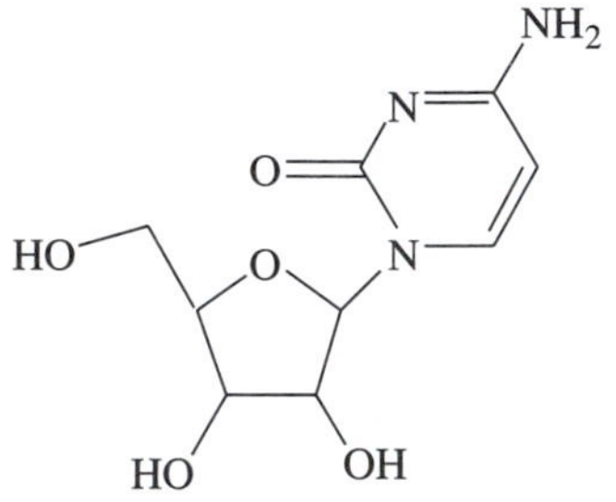

Cyd

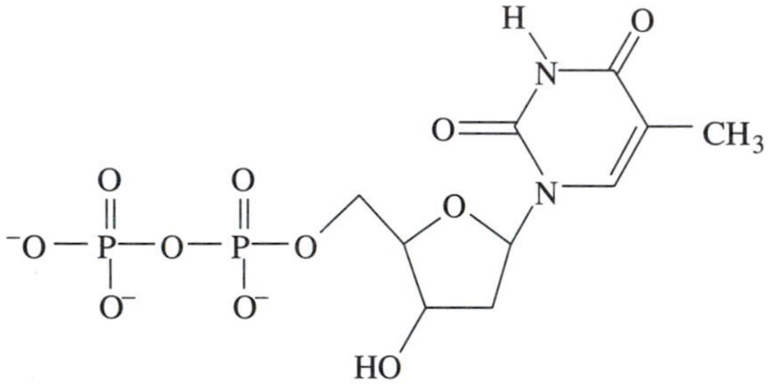

dTDP

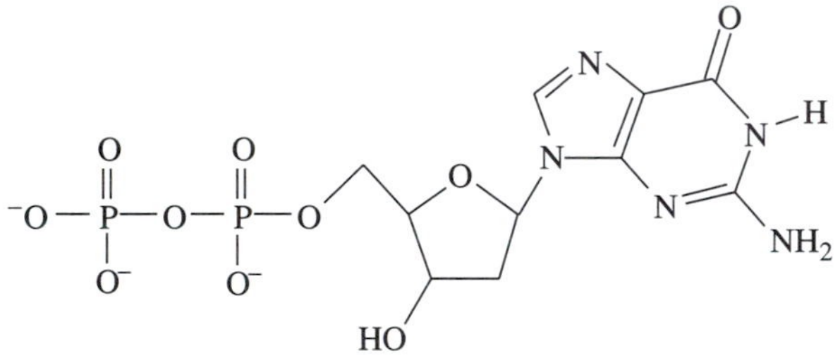

dGDP

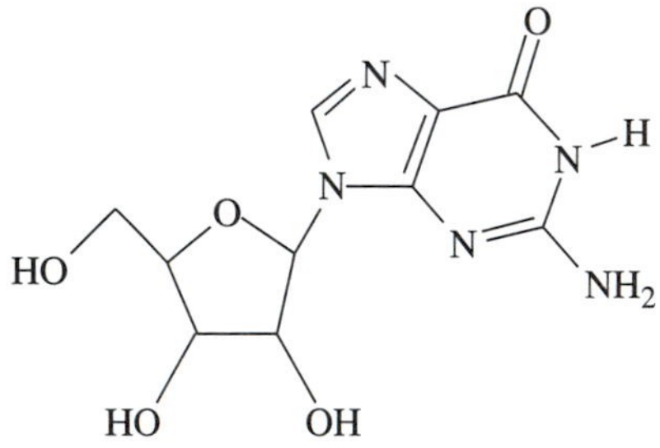

Guo

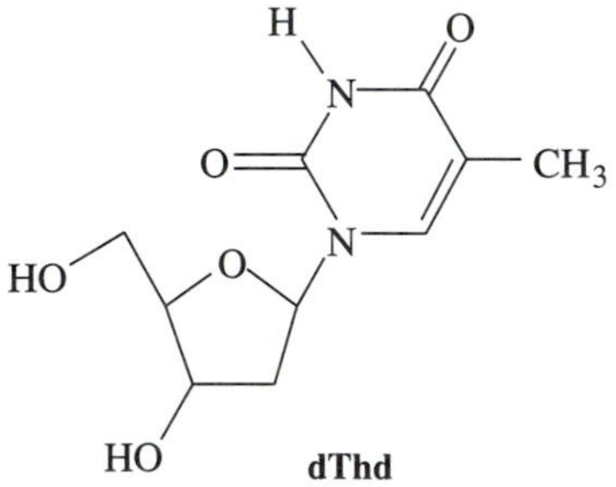

dThd

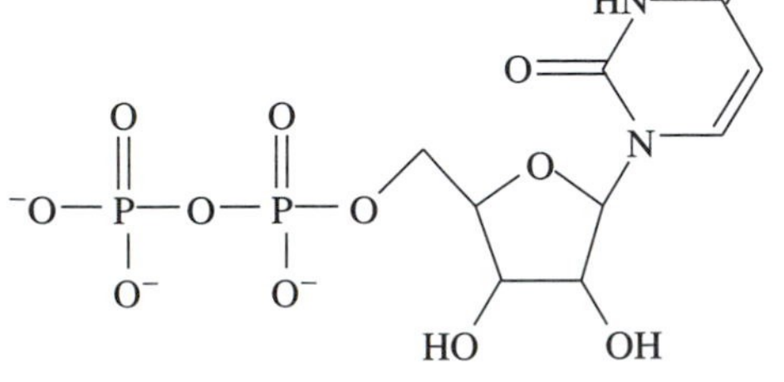

UDP

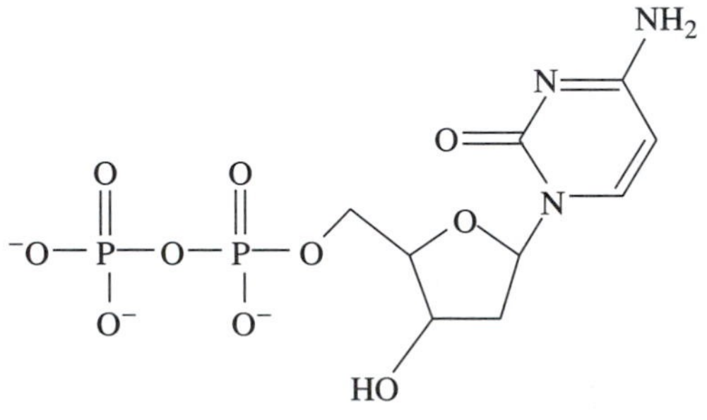

dCDP

3. An aminoacyl tRNA is less chemically reactive than an aminoacyl adenylate. It will undergo fewer side reactions and decompose more slowly in the cell than an aminoacyl adenylate.

4. Definitions:

(a) A nucleoside is a ribosyl derivative of a heterocyclic base. The base is ordinarily one of the common bases in nucleic acids, and the sugar is ordinarily ribose or 2-deoxyribose. However, closely related compounds containing an altered sugar or base also are known as nucleosides. Examples are nucleoside antibiotics.

(b) CDP is cytidine-5′-diphosphate.

(c) TTP is 2′-deoxythymidine-5′-triphosphate.

(d) The genetic code determines which amino acid will be incorporated into a particular position of a polypeptide in protein biosynthesis. It consists of codons of three bases that appear in mRNA and are read in the 5′-to-3′ direction. Each amino acid is encoded by one or more codons.

(e) Codons are the code words in the genetic code. Each codon is a three-base sequence that corresponds to a particular amino acid for incorporation into a protein.

(f) Initiation is the process by which initiation factors facilitate the assembly of an initiation complex for protein biosynthesis. The initiation complex consists of the assembled ribosome, formylmethionyl $tRNA^{fMet}$, mRNA, and elongation factors.

(g) Transcription is the process by which a species of RNA is produced that contains a nucleotide sequence that is complementary to that of a segment of one strand in DNA. The sequence in DNA is said to be *transcribed* into RNA.

(h) Translation is the process by which the transcribed nucleotide sequence in mRNA is translated into the amino acid sequence of a protein.

(i) An anticodon is a three-base sequence in the anticodon loop of tRNA that is complementary to a codon in mRNA. The anticodon binds to mRNA in the initiation and elongation complexes of protein biosynthesis.

(j) Elongation is the process by which a polypeptide is elongated though peptide-bond formation by an initiation complex formed from mRNA, a ribosome, aminoacyl tRNAs, and elongation factors. It is the peptide-elongation stage of translation.

(k) A ribosome is a ribonucleoprotein particle that carries out peptide-bond formation in protein biosynthesis.

(l) A nucleotide is a phosphorylated nucleoside.

(m) A 5′-end refers to the end of a nucleic acid chain that terminates with the substituent on the 5′-carbon of ribose or deoxyribose.

(n) D loop refers to a loop in tRNA that contains dihydrouracil.

(o) EF-T refers to a protein that is an elongation factor in protein biosynthesis.

(p) IF-3 refers to a protein that is an initiation factor in protein biosynthesis.

(q) P-site refers to the site in a ribosome at which peptide-bond formation takes place.

(r) 16s rRNA is the smaller of the two large species of rRNA found in the *E. coli* ribosome.

(s) Activation refers to the chemical process by which the kinetic and thermodynamic reactivity of a group is increased. In acyl activation, a carboxylate group is derivatized so that the negative charge on the carboxylate oxygen is abolished through the attachment of another group.

(t) $tRNA^{Leu}$ is the species of tRNA that will activate leucine for protein biosynthesis.

(u) mRNA is messenger RNA. It contains a nucleotide sequence that is complementary to that of a segment of DNA, and it encodes the sequence with which amino acids are incorporated into a protein.

(v) Termination is the process by which the elongation of a peptide chain is stopped at the end of translation.

(w) Puromycin is an antibiotic that blocks protein biosynthesis by binding to the amino acid site of a ribosome and forming a peptide bond with the growing end of a growing polypeptide, preventing further elongation.

(x) Mutation is any process by which the nucleotide sequence in a gene is subverted though alteration.

5. Aminoacyl tRNA synthetases play important roles in translation and activation. They activate amino acids for protein biosynthesis by converting them into aminoacyl tRNAs, and they participate directly in translation by matching up the amino acids with their respective anticodons.

6. Owing to the degeneracy of the genetic code, many sequences will code for the oligopeptide shown. One mRNA sequence is: 5′-AUGCGUGUUGUUGCUUGUUCUUUUGAUCCUCAUAAU-3′.

The number of possible sequences is 1 multiplied by the codon degeneracies of each amino acid in the peptide. The nucleotide sequence in the DNA from which the preceding mRNA was transcribed is: 3′d(TACGCACAACAACGAACAAGAAAACTAGGAGTATTA)-5′.

7. Cooperative oxygen binding by hemoglobin is the process through which the binding of a molecule of oxygen to a molecule of hemoglobin increases the affinity of the hemoglobin for second, third, and fourth molecules of oxygen. It leads to a sigmoid shape in the oxygen-binding curve. The increased affinity is brought about through a conformational change in one subunit of hemoglobin upon binding an oxygen molecule. The conformational change to a high-affinity state destabilizes the conformations of the other subunits through intersubunit interactions. These contacts reduce the amount of energy required to change the conformations of the other subunits and tend to transform them into their high-affinity conformations, in which they exhibit a high affinity for oxygen. Myoglobin does not bind oxygen cooperatively because it has only one binding site per molecule.

8. Heme is iron protoporphyrin IX, which is shown on page 288. One of its biological functions is to bind to hemoglobin and carry oxygen.

9. The binding of oxygen to Fe^{2+} in hemoglobin pulls the iron into the plane of the heme ring, and in the process it also pulls the transaxial ligand, the imidazole ring of a histidine, along with it. This touches off a protein conformational change that is transmitted through the network of van der Waals contacts and hydrogen bonds defined by the structure of the protein. This conformational change alters the intersubunit interactions and destabilizes them. The destabilization is relieved by complementary changes in the conformations of other subunits into the high-affinity conformation. The energy for this complementary conformational change is supplied by binding the first oxygen, and the affinity for the first oxygen is reduced by the amount of energy required to drive the conformational change. The second, third, and fourth oxygens can then bind to the high-affinity form.

10. Cooperative oxygen binding makes hemoglobin highly sensitive to the partial pressure of oxygen in tissues and the lungs. The steepness of the binding curve allows hemoglobin to undergo large changes in its oxygen content over a narrow range of oxygen partial pressures.

Chapter 11: Phosphoryl Group Transfer

1. UDP-galactose inhibiting versus UDP-glucose, competitive; UDP-galactose inhibiting versus galactose-1-P, noncompetitive; glucose-1-P inhibiting versus UDP-glucose, noncompetitive; glucose-1-P inhibiting versus galactose-1-P, competitive.

2. Kinetic mechanisms:

(a) Sequential binding of substrates to form ternary complexes.

(b) Sequential binding of substrates to form ternary complexes.

(c) Sequential binding of substrates to form ternary complexes.

(d) Ping-Pong kinetics with a free, covalent phosphoryl-enzyme intermediate.

(e) Sequential binding of substrates to form ternary complexes.

In reactions a, b, c, and e, the group-acceptor substrates are sterically and electrostatically different, so that they are unlikely to bind at the same subsite. By the principle of economy in the evolution of binding sites, two acceptor sites are required and ternary complexes are formed. In reaction d, the group acceptors are sterically and electrostatically similar, so they are likely to bind productively at the same subsite. By the principle of economy in the evolution of binding sites, both acceptors bind at the same site, and the enzyme works by a double-displacement mechanism and through a covalent phosphoryl-enzyme intermediate. The best way to distinguish these mechanisms is through a kinetic and stereochemical analysis.

3. Importance of enzymes in cells:

(a) Pyruvate kinase and 3-phosphoglycerate kinase produce ATP from ADP, phosphoenolpyruvate, and 1,3-diphosphoglycerate in cells. The acyl phosphates arise in metabolism.

(b) Creatine kinase and arginine kinase catalyze the production of ATP from ADP and the phosphagens creatine phosphate and phosphoarginine. The phosphagens are reservoirs of high-energy phosphorylated compounds.

(c) Hexokinase and glycerokinase catalyze the phosphorylation of glucose and glycerol to phosphoesters, which are trapped within cells by the charges on the phosphoryl groups. The phosphoesters are metabolized.

(d) UDP-glucose pyrophosphorylase produces UDP-glucose from glucose-1-P and UTP. UDP-glucose is a source of activated glucosyl groups for use in biosynthesis. Choline kinase produces choline phosphate from choline and ATP.

(e) Adenylate kinase maintains the balance of ATP, ADP, and AMP in the cell. Nucleoside diphosphate kinase maintains pools of nucleoside triphosphates by catalyzing the phosphorylation of nucleoside diphosphates by ATP, which is produced continuously by oxidation phosphorylation. The other nucleoside triphosphates are not produced by oxidative phosphorylation.

4. A dissociative transition state is one in which there is little bonding between the phosphoryl group in flight (at the transition state) and either the phosphoryl donor or the phosphoryl acceptor; that is, there is less bonding in the transition state than in the ground state. An associative transition state is one in which there is more bonding between the phosphoryl group donor in flight and the group acceptor and group donor than there is in the ground state.

5.

Single displacements	Double displacements
Sequential kinetics	Ping-Pong kinetics
Stereochemical inversion	Stereochemical retention
No covalent enzyme-substrate intermediate	Covalent enzyme-substrate intermediate
No enzymatic exchange reactions	Enzymatic substrate exchange reactions

6. The requirement for phosphate in the catalysis of the exchange of protons in pyruvate with protons in the medium is an example of substrate synergism. Phosphate at the active site stabilizes an active form of the enzyme by mimicking the phosphoryl group of phosphoenolpyruvate.

Chapter 12: Nucleic Acids

1. The structures of proteins and DNA are similar in a few respects. The hydrophobic parts of the molecules are associated in the interiors of the structures, whereas the polar and ionic parts are exposed on the surfaces to the aqueous medium. The structures are determined by the hydrophobic interactions, hydrogen bonding, ionic interactions, and van der Waals interactions. Secondary structure is important in both molecules. Both proteins and DNA undergo conformational changes; the functional importance of conformational changes in DNA are less well known than are those in proteins.

Proteins and DNA differ in several respects. The structural units in DNA are deoxyribonucleotides, whereas in proteins they are amino acids. Every DNA molecule contains two subunits with complementary primary structures, whereas a protein may consist of fewer or more subunits, which may be identical or different but which do not have a regular complementarity in their structures. All secondary structures in DNA are helical, whereas proteins have both helical and beta-sheet structures. Proteins have a definite tertiary structure, whereas tertiary structure in DNA is less definite.

2. RNA differs from DNA in that it is composed of ribonucleotides rather than deoxyribonucleotides, and it never consists of complementary subunits in the sense of the two chains in DNA. Species such as rRNA and tRNA, like proteins, have a specific tertiary structure in addition to secondary structure. mRNA is structurally dynamic in that it undergoes changes in secondary structure fairly rapidly. Some of the secondary structures probably have functional importance in recognition phenomena that are important in translation. RNA is composed of nucleotides rather than the amino acids of proteins; however, in other structural and dynamic respects, RNA and proteins have more in common structurally than do DNA and proteins.

3. There are essentially no degrees of freedom within the structures of the heterocyclic bases. Torsion is possible between the heterocyclic bases and ribosyl or deoxyribosyl rings by rotation about the *N*-glycosyl bond. Within the ribosyl or deoxyribosyl ring, the important freedom is that between the 2′-endo and 3′-endo conformations of these rings. There is torsional freedom about the bonds connecting carbons 4′ and 5′ and between carbon-5′ and a phosphoryl group.

4. Both methods depend on fragmentation of a segment of DNA by some controlled means and on end-group analysis of the fragments. In the Maxam-Gilbert method, the fragmentation methods are chemical—that is, nonenzymatic—and result in selective cleavages at A, G, A + G, C, T, and C + T sites in the chain. These cleavages allow the end groups of fragments separated on gels to be identified. In the Sanger method, the fragmentation is accomplished by systematically interrupted synthesis of complementary fragments by the use of DNA polymerase. The interruption in the synthesis of complementary fragments is controlled in such a way as to allow the end groups of separated fragments to be identified. The same gel electrophoretic techniques are used in both methods to separate the fragments.

5. DNA primase is a DNA-dependent RNA polymerase that produces short segments of complementary RNA that serve as primers for DNA polymerase III. DNA polymerase III catalyzes the polymerization of a complementary strand of DNA, the leading strand. DNA polymerase III also catalyzes the polymerization of short segments of DNA on RNA primers in the lagging strand. DNA polymerase I catalyzes the hydrolytic removal of the RNA primers in the lagging strand; it also catalyzes polymerization to fill the gaps left by the removal of primers. DNA ligase catalyzes phosphodiester-bond formation between segments of DNA in the lagging strand. Helicases separate the strands of DNA, and single-strand-binding proteins bind to single strands of DNA at the replication fork and stabilize them as single strands; this allows DNA primase and DNA polymerase III to act on single strands. DNA topoisomerases relax supercoils in DNA that are introduced by the unwinding of the double helix at the replication fork.

6. DNA is a better molecule for storing genetic information than RNA because it is chemically more stable. RNA undergoes hydrolysis, especially base-catalyzed hydrolysis, much more rapidly than does DNA. Thus, RNA would not preserve the genetic information as effectively as DNA. In this regard, it is interesting that the genetic information in RNA viruses in much more subject to change than that in DNA viruses.

7. RNA polymerase must bind to DNA and find a promoter. It must then separate the DNA strands locally and initiate polymerization by catalyzing phosphodiester-bond formation between the 3′-hydroxyl group of an ATP or GTP at the start site and the α-phosphoryl group of a second nucleoside triphosphate. It must then move along the DNA, separating the strands locally and catalyzing polymerization of RNA that is complementary to one strand of the DNA. Eventually, it must terminate the transcription and dissociate from the DNA.

8. 5′-Capping and 3′-polyadenylylation insulate the ends of mRNA from the actions of exonucleases that would digest the RNA to nucleotides.

9. A- and B-DNA differ most fundamentally in the conformation of the deoxyribosyl rings. In the 3′-endo conformation, the separation between the heterocyclic bases in the helix is smaller than it is in the 2′-endo conformation. The diameter of the helices is determined mainly by the lateral dimensions of the base pairs, not by the spacing between them.

10. Nucleases degrade nucleic acids to nucleotides in cells. Nuclease action has functional importance in DNA replication, where the RNA primers must be removed by DNA polymerase I acting as an exonuclease. Proofreading to correct adventitious mistakes in polymerization also requires the action of DNA polymerase as a nuclease. Turnover of mRNA is an important means of regulating the rate of protein biosynthesis, and turnover requires the action of nucleases to degrade the mRNA. Repair of DNA that has been damaged in some way requires the action of nucleases to remove the damaged segments. Any damaged and nonfunctional nucleic acid in cells can be recycled by the action of nucleases, which converts them into nucleotides for use in the resynthesis of nucleic acids. Certain nucleases carry out essential functions in processing various species of RNA, which are initially transcribed as pre-RNA and must be reduced in length by the action of nucleases. Some nucleases can protect a cell against foreign DNA.

Chapter 13: Gene Regulation

1. A very low level of constitutive synthesis of the enzymes in the *lac* operon is an essential part of the regulation of this operon. The natural inducer is *allo*-lactose, which binds to the repressor and causes it to dissociate from the operator, thus exposing the genes to transcription. *allo*-Lactose arises from lactose and is produced by an isomerization reaction that is catalyzed by β-galactosidase, an enzyme in the *lac* operon. Therefore, the induction of the enzymes in the *lac* operon requires the actions of those very enzymes. A low level of lactose permease is required to allow some lactose to enter the cell, and a low level of β-galactosidase is required to convert some of the transported lactose into *allo*-lactose, which derepresses the operon and leads to the increased production of the enzymes in the *lac* operon.

2. An enzyme that could excise a U from mRNA could regulate gene expression by introducing a start codon in mRNA where one did not exist (. . . AUUG . . . → . . .AUG. . .). It could also abolish a start codon (. . . AUG . . . → . . . AG . . .). Such an enzyme could also change the reading frame for translation. An alteration in reading frame could determine whether a protein is produced or not, because an alteration not only changes the amino acid sequence but can lead to the introduction of an early termination codon (. . . AUG_{init} . . . CGUUUAAAU . . . → . . . AUG_{init} . . . $CGUUAA_{term}$AU . . .).

3. A Shine-Delgarno sequence might be introduced through evolution at a position that is proximal to an initiation site. The high affinity of these sequences for ribosomes can increase the probability for the formation of the initiation complex.

4. A deletion in the *Z* gene could lead to a species of mRNA that contains a false initiation site, which would lead to a change in reading frame for translation. This could lead to a failure to recognize the termination codon for β-galactosidase and the initiation codons for permease and transacetylase.

Chapter 14: Recombinant DNA and Protein Engineering

1. The staggered cut introduced by a single restriction endonuclease is self-complementary in a vector. In attempts to clone fragments into such a site, there is always a competition between structures in which a single fragment has been hybridized to the vector and other species. A prominent and favored competing species is the circularized nicked vector, in which no fragment is present. This form is favored by being intramolecular—that is, by entropy—and the nick is easily sealed by DNA ligase. Therefore, in such a cloning experiment, circularization of the vector is always an important side reaction. It can be overcome by using a large ratio of fragments to linearized vector. But this ratio must be adjusted empirically to minimize the tandem incorporation of more than one fragment into the site.

2. Blunt-ended fragments can be modified by terminal deoxynucleotidyltransferase. For example, the enzyme extends primers by polymerization of dATP to 3′ ends. Thus, a set of fragments could have their 3′ ends extended with poly dA tails by the use of this enzyme. A cloning vector could be cut with a restriction endonuclease that introduces blunt ends, and these ends could be extended with poly dT tails. The vector should then capture the extended fragments through the complementarity of the tails, and the fragments could be cloned into the vector by the use of DNA polymerase, to fill in any gaps in the complementary tails, and DNA ligase to seal the nicks. The cloned fragments could be isolated and subjected to sequencing or further processing.

3. A source of carbon other than glucose or a sugar that can be transformed into glucose should be used in the growth medium. This is because the regulatory elements of the *lac* operon are sensitive to catabolite repression by glucose. The growth medium should also contain an inducer such as IPTG to maximize the transcription of the gene. If a plasmid with a high copy number is used, high levels of expressed enzymes can sometimes be produced even without the use of inducers, which are often expensive. The proteins are expressed because the concentration of the chromosomal repressor is often not high enough to saturate all of the operators on from 50 to 100 plasmids.

4. Synthetic Shine-Delgarno sequences that are engineered into the mRNA for a cloned gene can increase the probability of initiation-complex formation, which will enhance the translation process and the level of expression.

5. This experiment implicates H164 and H166 as essential residues in the action of uridylyltransferase. However, it gives no information about the functions of these residues. It is very likely that one of them is the nucleophilic catalyst in the mechanism of the reaction. Asparagine and glutamine are regarded as conservative replacements for histidine for two reasons: (1) the side chains are approximately the same size as that of histidine, and (2) the carboxamide group has hydrogen-bonding properties that are similar to those of the imidazole ring.

6. The principal conclusion is that H166 of the wild-type enzyme is the nucleophilic catalyst in the mechanism of action of this enzyme. In the mutant protein G166, the imidazolate group of the artificial substrate fits into the cavity left by the deletion of the imidazole ring of H166, and the mutant enzyme reacts with the artificial substrate as if it were the intermediate of the normal reaction—that is, the complex of the mutant enzyme with UMP-Im (E_{G166}–Im-UMP) is a mimic for the UMP-enzyme intermediate (E_{H166}–Im-UMP) of the wild-type enzyme. The mutant G164 cannot interact with UMP-Im in this way. No function can be assigned to

H164 on the basis of the negative evidence obtained in mutagenesis experiments. The function of the mutant G166 in catalyzing the model reaction is a positive result that supports the assignment of nucleophilic function to H166.

Chapter 15: Glycosyl Group Transfer

1.

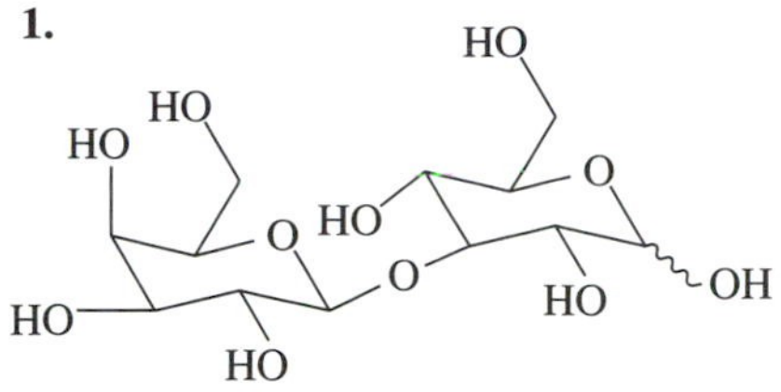

(ß-1, 3)-Galactosyl glucose

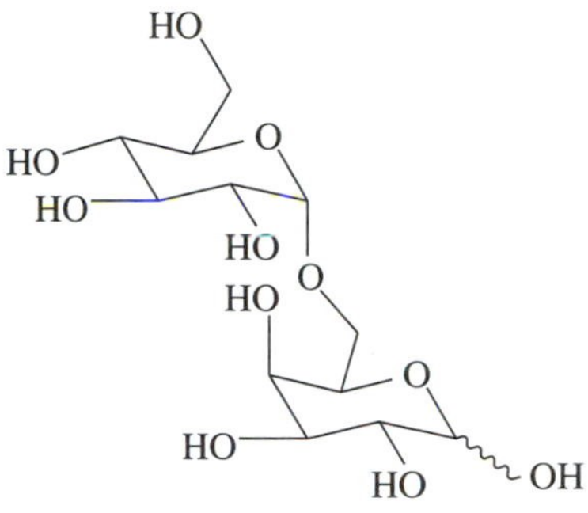

(α-1, 6)-Glucosyl galactose

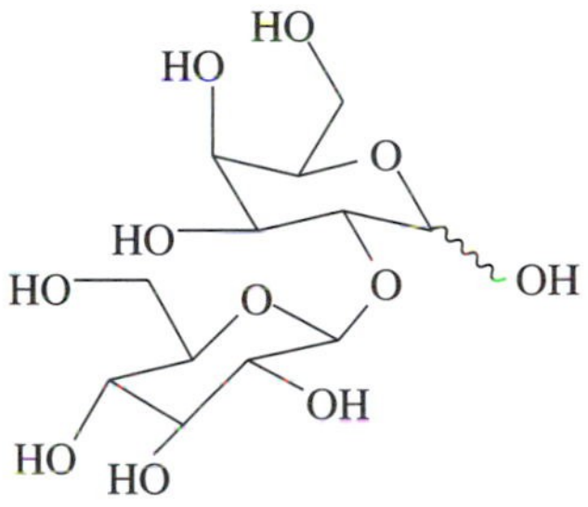

(ß-1,2)-Glucosyl galactose

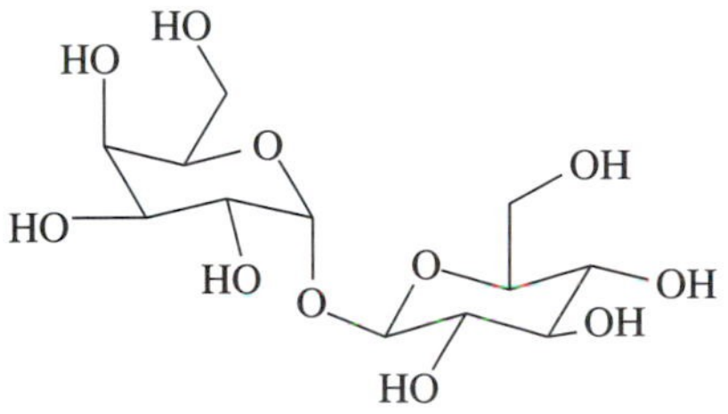

(α-1, 1)-Galactosyl ß-glucose

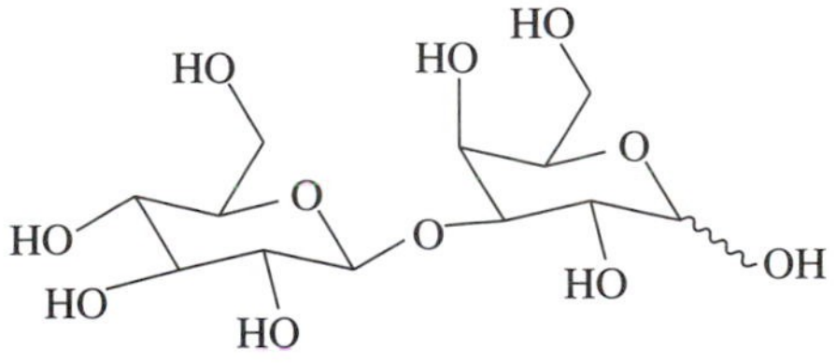

(ß-1, 3)-Glucosyl galactose

Among the possible glycosidic disaccharides composed of glucose and galactose in structures consisting of pyranose rings, glucose has four hydroxyl groups on carbons other than carbon-1 that could bond to a galactosyl group, and either the α- or β-anomeric configuration of galactose is possible. So there are eight galactosyl glucoses of this type, and the same number of glucosyl galactoses of this type are possible. There are four possible disaccharides in which carbon-1 of one sugar is bonded to carbon-1 of the other that differ in anomeric configuration: α-, α-; β-, β-; α-, β-; and β-, α-. A total of twenty pyranosyl disaccharides of glucose and galactoses are possible. Many more are possible if one considers the furanosyl forms of these sugars, and even more disaccharides with ether linkages are possible. These are not important in biochemistry.

2. A glycosyl bond cannot undergo dissociation of a proton; so base catalysis is of no use in providing an anion to drive off a leaving group. Direct attack of a hydroxide ion on carbon-1 with displacement of a sugar anion is very slow, probably for several reasons. First, at very high concentrations of hydroxide ions, there is a tendency for some of the hydroxyl groups in a disaccharide to be ionized, making the disaccharide an anion that tends to repel an attacking hydroxide ion. Second, the hydroxide ion in aqueous solution is very highly solvated and not as reactive as a nucleophile as might be expected from its basicity. Third, an alcoholate anion, $R—O^-$, would be the leaving group in a displacement by hydroxide, and such anions are poor leaving groups. Acid catalysis is effective because acids can donate a proton to the glycosyl group. This facilitates the leaving of the alcoholic group as R—OH, which is a good leaving group. The positive charge that develops on the glycosyl carbon in the transition state for the departure of the leaving group is stabilized by the ether oxygen in the glycosyl ring.

3. The hydrate is electronically and sterically similar in stability to either cyclic form of glucose. Its formation is much less favored, however, despite the fact that the concentration of water in a dilute solution of glucose is near 55 M. This is because the formation of the hydrate requires the binding of water to glucose; that is, it is a bimolecular process that takes place with a significant loss of entropy. The same electronic stabilization of the aldehyde group is accomplished by the intramolecular addition of the 5-hydroxyl group to the aldehyde, and this takes place with no induction of steric strain and, because of being intramolecular, little loss of entropy.

4.

Transition state for carbocationic mechanism

Transition state for displacement mechanism

The two mechanisms differ in the orientation of the carboxylate group relative to the galactosyl carbon and in that there is covalent bond formation between the galactosyl carbon and the carboxylate group in the transition state for displacement. In the transition state for carbocation formation, there is electrostatic stabilization by the carboxylate but no covalent bond formation. The carbocationic mechanism could take place if the carboxylate group were to be held in-plane with the carbocation so that it could not form a covalent bond to it.

5. The ether linkages of the methylated hydroxyl groups are stable to acids, whereas the glycosidic groups are labile to acid-catalyzed hydrolysis. One could hydrolyze the methylated glycogen in acid and characterize the products. The hydrolytic release of a methylated glucosyl group from an unbranched segment of methylated glycogen will produce 2,3,6-trimethylglucose, whereas 2,3-dimethylglucose will be released from a branch point. The ratio of the trimethylglucoses will show what fraction of glucose residues are at branch points in the polymer.

6.

Chapter 16: Biological Oxidation: NAD^+ and FAD

1. Hydride mechanism leads to the transfer of ^2H (G = glutathione):

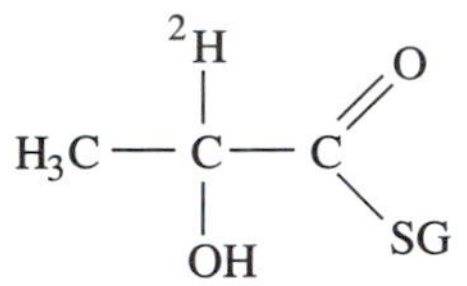

$$H_3C-C(^2H)(OH)-C(=O)SG$$

Enol mechanism:

$$H_3C-C(=O)-C(OH)(^2H)-SG \quad B_1$$

$$\rightleftharpoons$$

$$H_3C-C(OH)=C(OH)-SG \quad B_1{}^2H^+$$

$$\downarrow$$

$$H_3C-C(OH)(H)-C(=O)SG \quad B_1$$

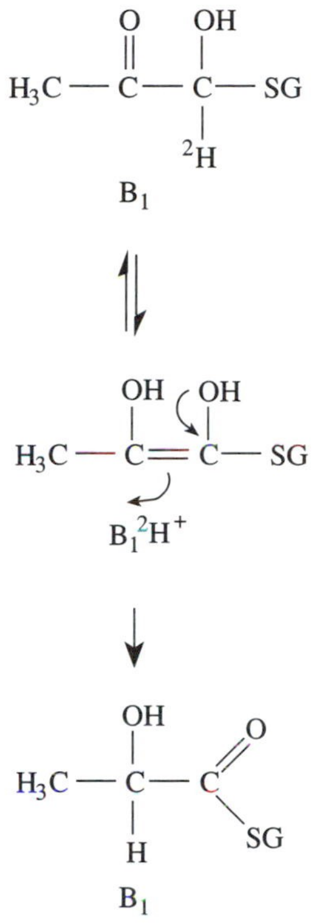

If $B_1{}^2H^+$ exchanges much faster than it transfers H^+ to the enol, no ^2H would be in the product. If the exchange is slow, ^2H will be found in the product.

2. Because glyceraldehyde phosphate dehydrogenase catalyzes

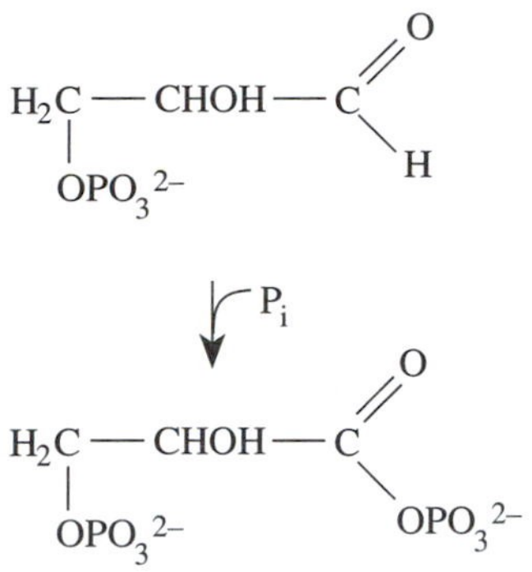

$$H_2C(OPO_3{}^{2-})-CHOH-C(=O)H \xrightarrow{P_i} H_2C(OPO_3{}^{2-})-CHOH-C(=O)OPO_3{}^{2-}$$

There is a P_i binding site. Phosphate of the inhibitor can bind to the P_i site. The aldehyde probably forms a thiohemiketal

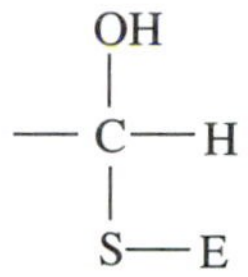

$$-C(OH)(H)-S-E$$

with the active-site thiol. The inhibitor thus binds at two points and, owing to the "chelation" effect, binds tightly.

3.

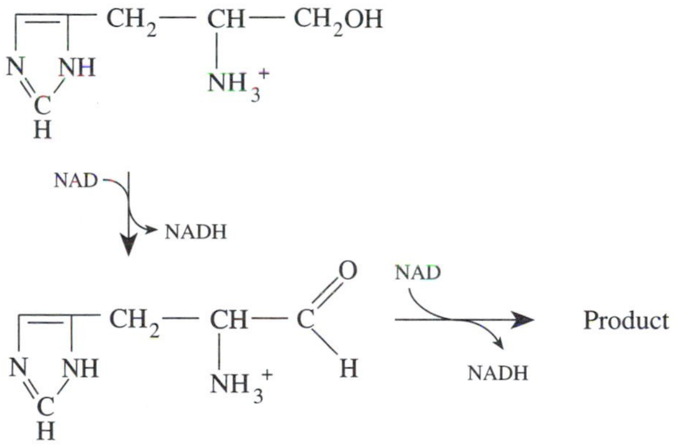

$$\text{Imidazolyl}-CH_2-CH(NH_3{}^+)-CH_2OH$$

$$\downarrow \; NAD \rightarrow NADH$$

$$\text{Imidazolyl}-CH_2-CH(NH_3{}^+)-C(=O)H \xrightarrow[NADH]{NAD} \text{Product}$$

The intermediate aldehyde is not released from the enzyme.

4.

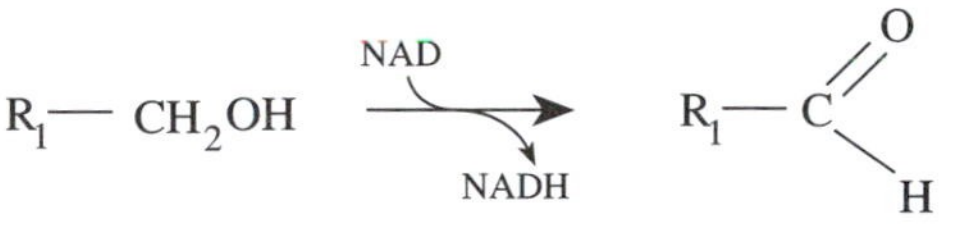

$$R_1-CH_2OH \xrightarrow[NADH]{NAD} R_1-C(=O)H$$

Assume that the enzyme that acts on R_1CH_2OH is "A" specific. Isolate NADH and oxidize to NAD with a "B" specific enzyme.

Chapter 17: Decarboxylation and Carboxylation

1.

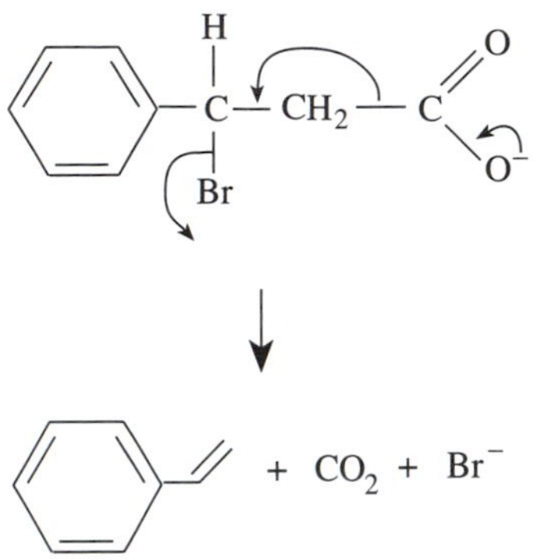

H
C — CH_2 — C
O
O^-
Br

$$\longrightarrow \quad + \ CO_2 + Br^-$$

2. The proton from the acid can move to the α-carbonyl group and activate decarboxylation of the remaining $-COO^-$ group.

3.

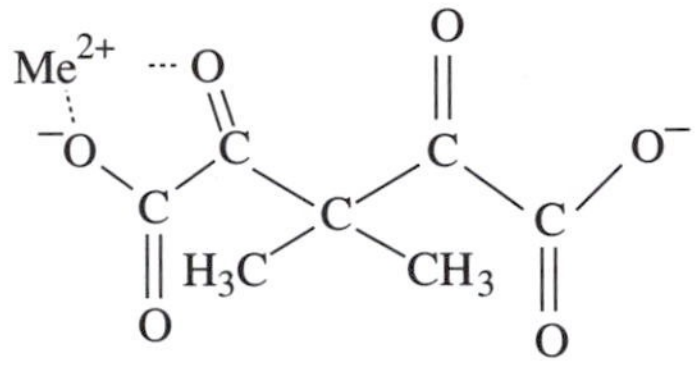

Me^{2+} ··· O
O
^-O C C C C O^-
H_3C CH_3
O O

Me^{2+} can form a complex with the negative charge of **3** but not with **1.** Me^{2+} stabilizes the charge on the carbonyl group that develops in the transition state.

4. The steady-state concentration of the acetone-derived imine is much higher than that derived from acetoacetate. The negative charge on the acetoacetate-derived imine repels BH_4^-.

5.

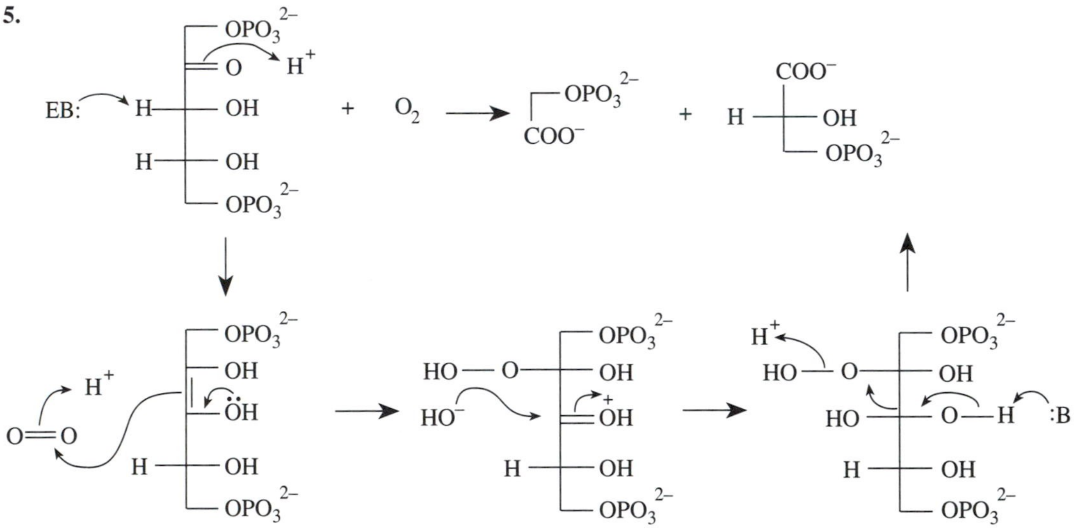

OPO_3^{2-}
O H^+
EB: H — OH + O_2 ⟶ OPO_3^{2-} / COO^- + H — OH / OPO_3^{2-} (COO^-)
H — OH
OPO_3^{2-}

↓ ↑

OPO_3^{2-} / OH / $\ddot{O}H$ / H — OH / OPO_3^{2-} ; O=O, H^+
⟶ HO — O — OH / HO^- / $\overset{+}{O}H$ / H — OH / OPO_3^{2-} (OPO_3^{2-})
⟶ H^+ / HO — O — OH / HO — O — H :B / H — OH / OPO_3^{2-} (OPO_3^{2-})

6. First partial reaction:

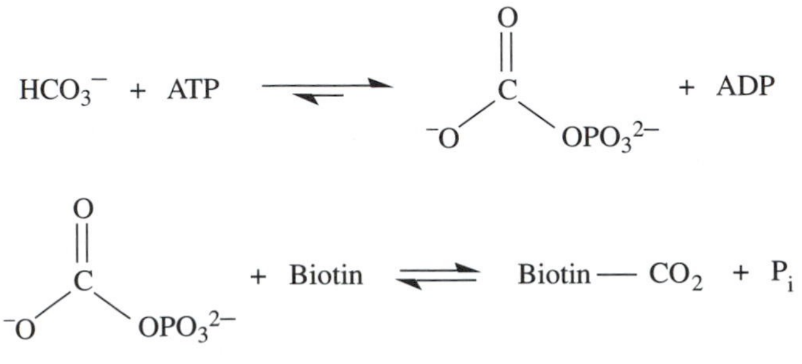

$$HCO_3^- + ATP \rightleftharpoons {}^-O-C(=O)-OPO_3^{2-} + ADP$$

$${}^-O-C(=O)-OPO_3^{2-} + \text{Biotin} \rightleftharpoons \text{Biotin}-CO_2 + P_i$$

Exchange ADP into ATP in the presence of HCO_3^-, but not biotin.

7. (a)

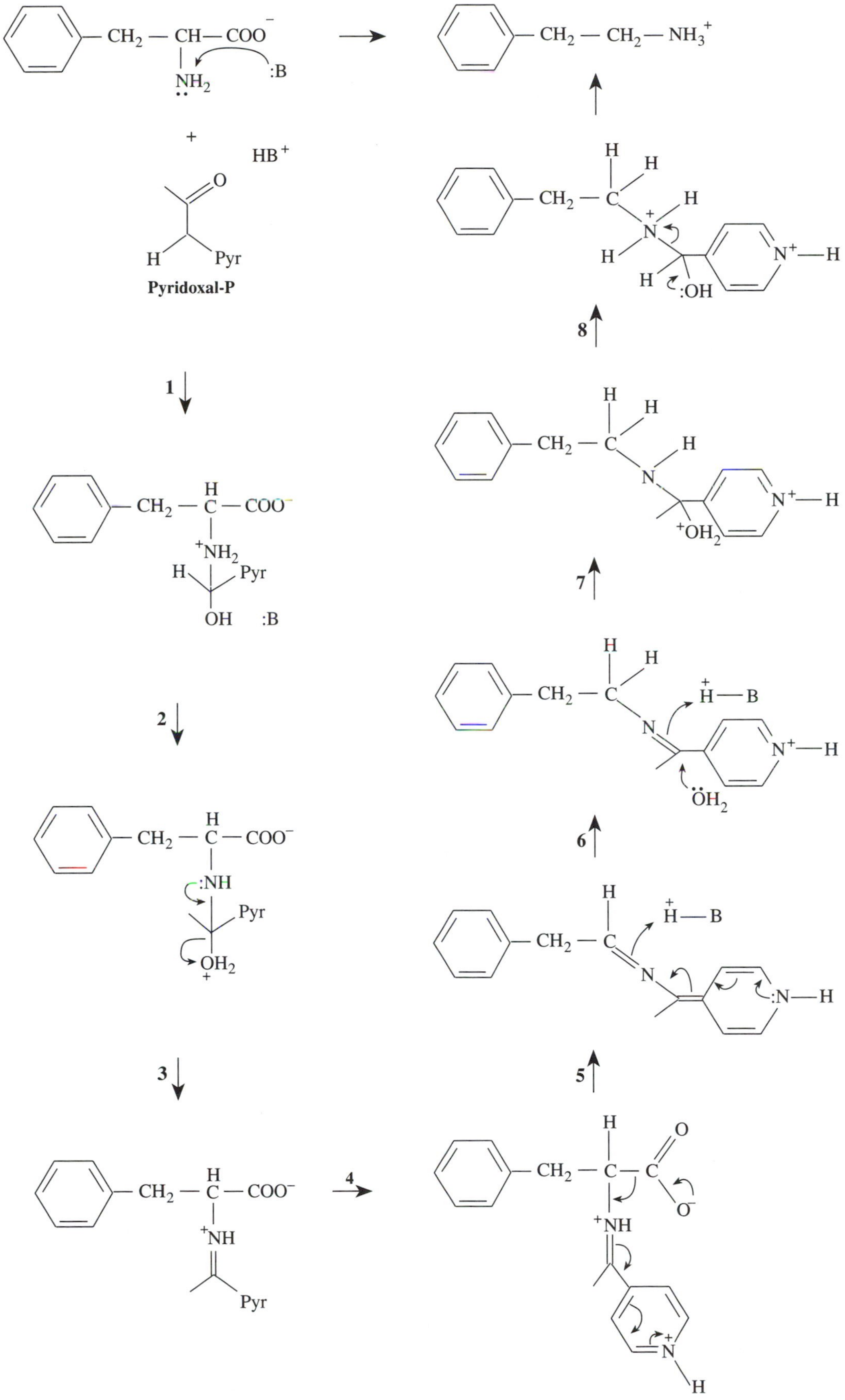

C_6H_5—CH_2 — CH— COO^- → C_6H_5—CH_2 — CH_2— NH_3^+
$\ddot{N}H_2$:B

\+ HB^+

O
H Pyr
Pyridoxal-P

1 ↓

CH_2 — C — COO^-
H
$^+NH_2$
H Pyr
OH :B

2 ↓

CH_2 — C — COO^-
H
:NH
Pyr
OH_2
\+

3 ↓

CH_2 — C — COO^-
H
^+NH
Pyr

4 →

H
CH_2 — C — C
O
O^-
^+NH
N^+
H

5 ↑

H
CH_2 — C
N
$\overset{+}{H}$—B
:N—H

6 ↑

H H
CH_2 — C
N
$\overset{+}{H}$—B
N^+—H
$\ddot{O}H_2$

7 ↑

H H
CH_2 — C
N H
N^+—H
$^+OH_2$

8 ↑

H H
CH_2 — C
H
$\overset{+}{N}$
H
H
N^+—H
:OH

↑

(b) See the scheme in part a. When phenylethylamine is added to the enzyme, reversal of steps 1 through 5 occurs and a deuterium from the solvent can wash in to C-1. The protonation is stereospecific; hence, the same proton is removed and added, giving only one deuteron at C-1 when D_2O is the solvent.

8.

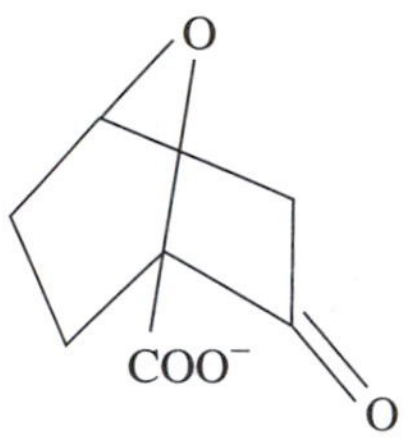

O

COO⁻

O

Decarboxylation violates Bredt's rule of no double bond at a bridge position.

9. It activates $HO-C(=O)-O^-$ to $^-O-C(=O)-OPO_3^{2-}$ so that carboxylation of biotin can occur.

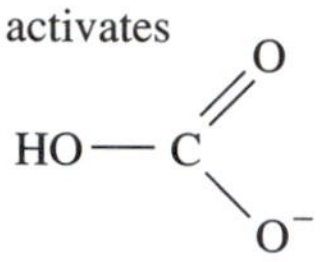

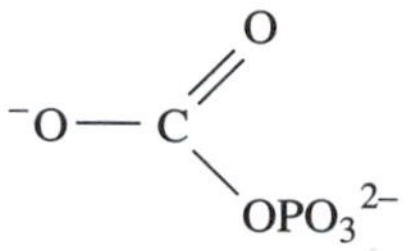

10.

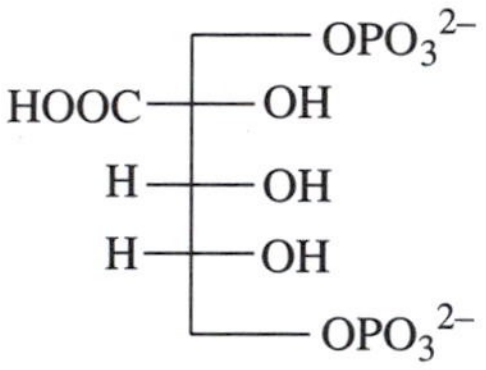

OPO_3^{2-}
HOOC—OH
H—OH
H—OH
OPO_3^{2-}

---

11. (a) Reaction I:

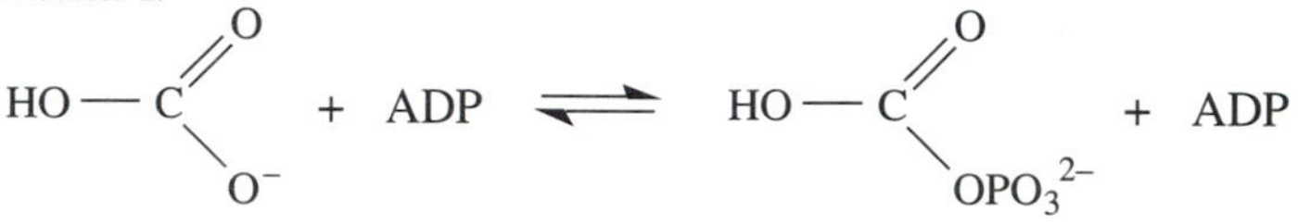

$$HO-C(=O)-O^- + ADP \rightleftharpoons HO-C(=O)-OPO_3^{2-} + ADP$$

Reaction II:

$$HO-C(=O)-OPO_3^{2-} + \text{Biotin} \rightleftharpoons \text{Biotin-}CO_2 + P_i$$

Reaction III:

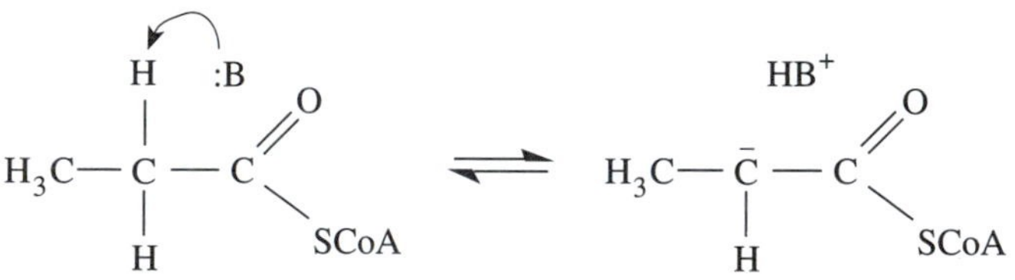

$$H_3C-CH_2-C(=O)-SCoA \; (+\,:B) \rightleftharpoons H_3C-\bar{C}H-C(=O)-SCoA \; (+\,HB^+)$$

Reaction IV:

$$H_3C-\bar{C}H-C(=O)-SCoA + \text{Biotin-}CO_2 \rightleftharpoons H_3C-CH(COO^-)-C(=O)-SCoA + \text{Biotin}$$

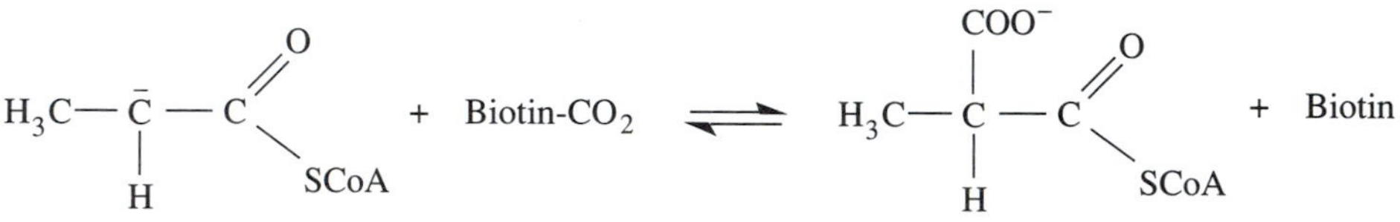

(b) We could expect incorporation of ^2H in reactions I through III. However, if reaction III does not occur until biotin-CO_2 is formed and if reaction IV is fast, no exchange would occur.

12. (a)

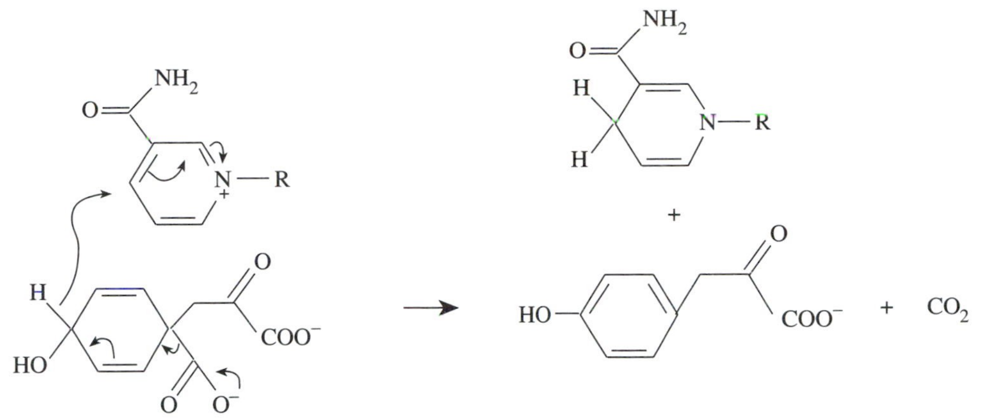

NH_2, O, N—R (+), H, HO, O, COO^-, O, O^- → NH_2, O, H, H, N—R + HO, O, COO^- + CO_2

(b)

H, HO, H^+, O, COO^-, O, O^- → O, COO^- + H_2O + CO_2

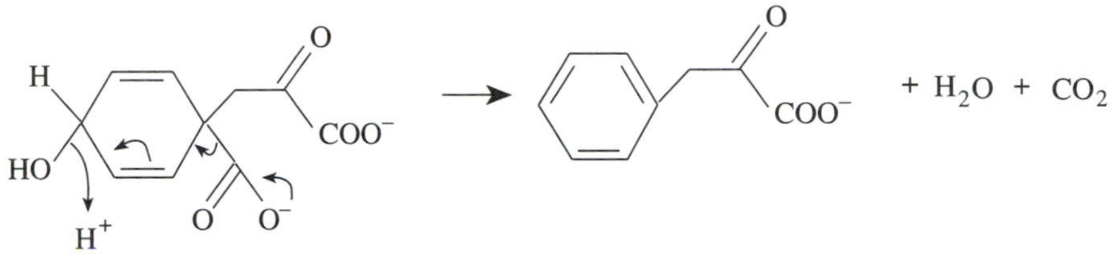

Chapter 18: Addition, Elimination, and Transamination

1. E1cB mechanism:

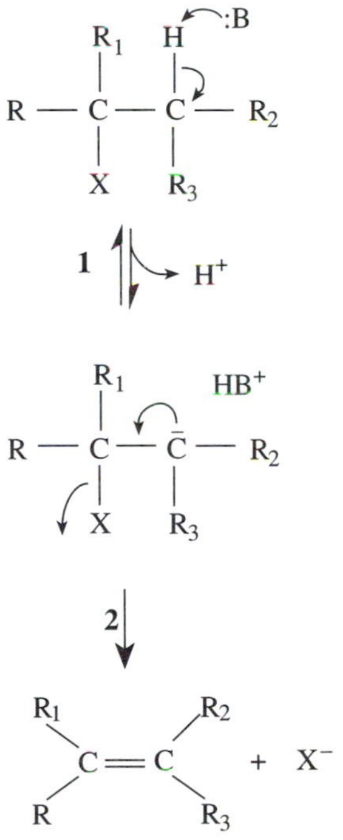

$$R-C(R_1)(X)-C(H)(R_2)(R_3) \quad (:B)$$

1 ⇅ H^+

$$R-C(R_1)(X)-\bar{C}(R_2)(R_3) \quad HB^+$$

2 ↓

$$R_1(R)C{=}C(R_2)(R_3) + X^-$$

Run the reaction in isotopic H_2O, either ^2H or ^3H.

(a) If step 2 is rate determining ($k_2 < k_{-1}$), the intermediate carbanion will return to starting material more often than to product. Isotope (^2H or ^3H) incorporation into starting material will occur.

(b) If step 1 is rate determining ($k_2 > k_{-1}$), no return of carbanion to starting material will occur and no isotope exchange will occur.

2. In the final stage of the reaction, this conversion occurs:

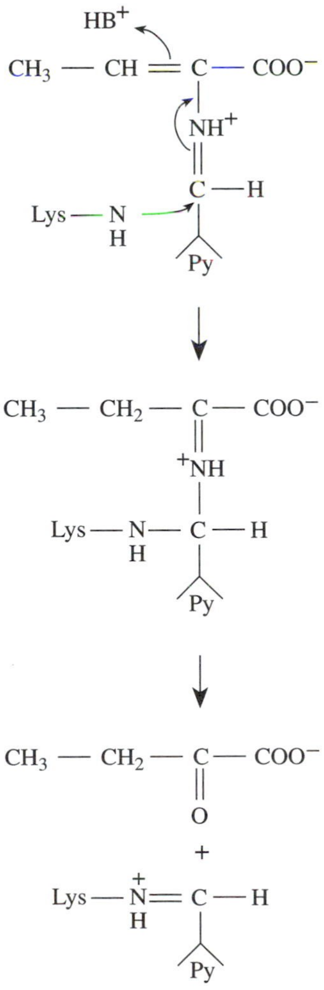

HB^+

$CH_3 — CH = C — COO^-$, NH^+, C—H, Lys—NH, Py

↓

$CH_3 — CH_2 — C — COO^-$, ^+NH, Lys—N(H)—C—H, Py

↓

$CH_3 — CH_2 — C(=O) — COO^-$

+

Lys—$\overset{+}{N}$(H)= C—H, Py

In $^{\bullet}H_2O$, the product acquires $^{\bullet}H$:

$$H_3C-C(^{\bullet}H)(H)-C(=O)-COO^-$$

If protonation occurs on the enzyme, the product will probably be chiral (*R* or *S*). If it occurs in solution, it will be racemic.

3. $RSH = HS-CH_2-CHNH_3^+COO^-$

$$CH_3-C(=O)-O-CH_2-CH_2-C(H)(COO^-)-NH^+=CH-Py \quad (B_1) \underset{1}{\rightleftharpoons} CH_3-C(=O)-O-CH_2-CH_2-\bar{C}(COO^-)-NH^+=CH-Py \quad (HB_1)$$

$$\underset{2}{\rightleftharpoons} CH_3-C(=O)-O-CH_2-C(H)(H)-C(COO^-)=NH^+-CH=Py \quad (B_2, HB_1)$$

$$B_2H \quad CH_3-C(=O)-O-CH_2-\bar{C}(H)-C(COO^-)=NH^+-CH=Py \quad (HB_1) \underset{3}{\rightleftharpoons} (CH_3-COOH) \quad CH_2=C(H)-C(COO^-)=NH-CH=Py \quad (RS^-, HB_1) \underset{4}{\rightleftharpoons}$$

$$RS-CH_2-CH_2-C(COO^-)=NH-CH=Py \xrightarrow{5} RS-CH_2-CH_2-C(H)(COO^-)-NH=CH-Py \quad (Lys-NH_2) \xrightarrow{6}$$

$$RS-CH_2-CH_2-CH(NH_2)-COO^- + Lys-N=CH-Py$$

Isotopic H would be at C-3 and C-4 of the product:

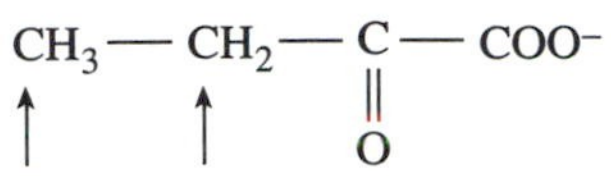

$$\underset{\uparrow}{CH_3}-\underset{\uparrow}{CH_2}-\underset{\underset{O}{\|}}{C}-COO^-$$

Isotopic H could be in the starting material if the intermediates can revert to starting material.

4.

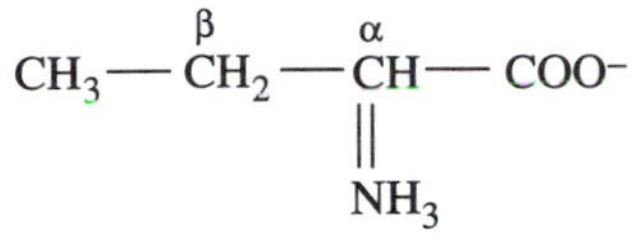

$$CH_3-\overset{\beta}{CH_2}-\overset{\alpha}{\underset{\underset{NH_3}{\|}}{CH}}-COO^-$$

Beta-hydrogens exchange because

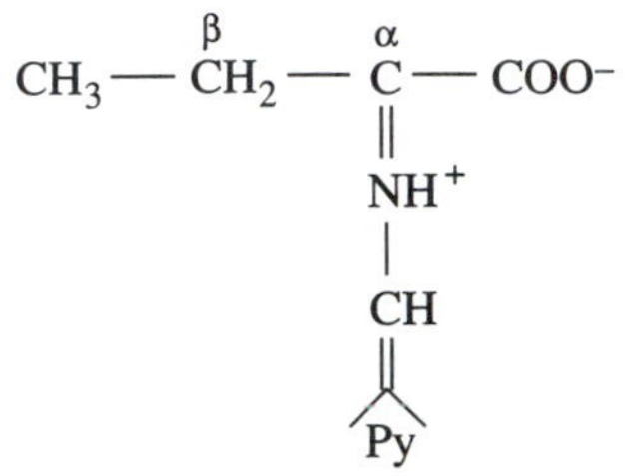

$$CH_3-\overset{\beta}{CH_2}-\overset{\alpha}{\underset{\underset{\underset{\underset{\underset{Py}{\|}}{CH}}{|}}{NH^+}}{\|}}{C}}-COO^-$$

is formed. The two β-hydrogens are not equivalent but, if rotation can occur about $C_\alpha-C_\beta$, both can exchange. Apparently the interaction of CH_3 with the enzyme is not sufficiently strong to prevent rotation.

When

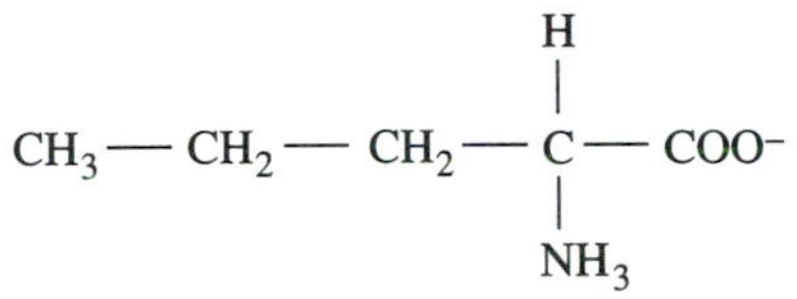

$$CH_3-CH_2-CH_2-\overset{\overset{H}{|}}{\underset{\underset{NH_3}{|}}{C}}-COO^-$$

is the substrate, an ethyl group is attached to the β-carbon. More binding energy is available to prevent rotation.

5. Alanine transaminase uses a Ping-Pong mechanism. The first stage is:

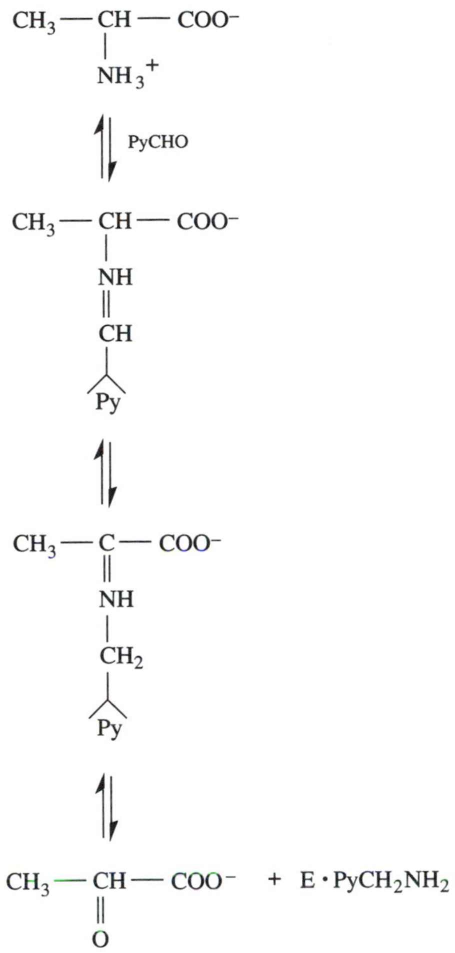

$$CH_3-\underset{\underset{NH_3^+}{|}}{CH}-COO^-$$

$$\Big\updownarrow \text{PyCHO}$$

$$CH_3-\underset{\underset{\underset{\underset{\underset{\underset{Py}{|}}{CH}}{\|}}{NH}}{|}}{CH}-COO^-$$

$$\Big\updownarrow$$

$$CH_3-\underset{\underset{\underset{\underset{\underset{\underset{Py}{|}}{CH_2}}{|}}{NH}}{\|}}{C}-COO^-$$

$$\Big\updownarrow$$

$$CH_3-\underset{\underset{O}{\|}}{CH}-COO^- \quad + \quad E\cdot PyCH_2NH_2$$

The two-step mechanism states that (1) $E\cdot PyNH_2$ (pyridoxamine) and pyruvate are formed; (2) $Py\text{-}NH_3^+$ reacts with ketoglutarate to regenerate PyCHO and glutamic acid. The exchange reaction occurs through the first half-reaction.

6.

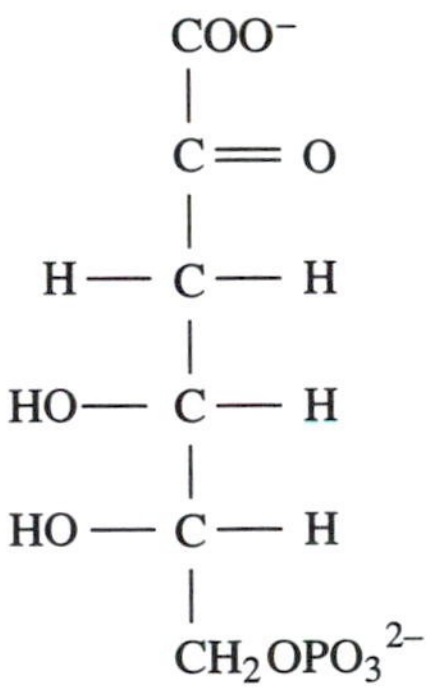

$$\begin{array}{c} COO^- \\ | \\ C=O \\ | \\ H-C-H \\ | \\ HO-C-H \\ | \\ HO-C-H \\ | \\ CH_2OPO_3^{2-} \end{array}$$

Cleavage between C-3 and C-4 (Schiff-base mechanism) yields

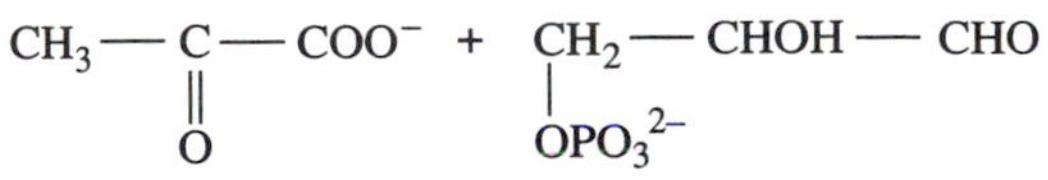

$$CH_3-\underset{\underset{O}{\|}}{C}-COO^- \quad + \quad \underset{\underset{OPO_3^{2-}}{|}}{CH_2}-CHOH-CHO$$

Glyceraldehyde-3-P can enter glycolysis and is converted into acetate + CO_2.

Chapter 19: Fatty Acids

1. Alternative mechanism:

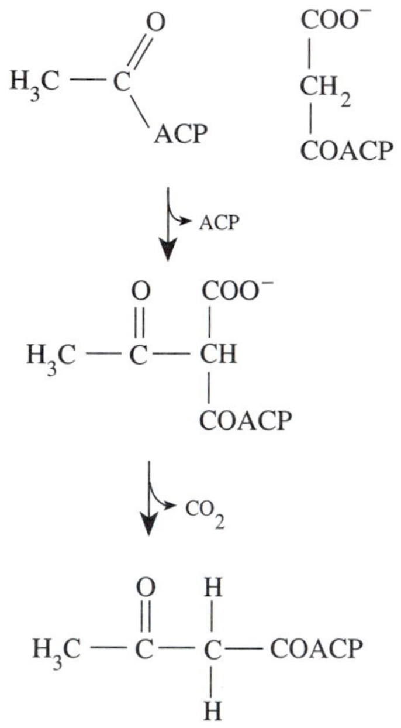

This mechanism results in ˙H incorporation from $˙H_2O$. The other mechanism does not, because the carbanion is derived from the decarboxylation of malonyl CoA.

2. Carnitine helps to maintain the compartmentalization of compounds. CoA esters are used in many reactions. The impermeability of the mitochondrial membranes to CoA esters prevents random diffusion of CoA esters.

3. Same answer as that for problem 2.

4. Methylmalonyl CoA could be incorporated into fatty acids in place of malonyl CoA. This in turn would lead to the production of branched fatty acids. The incorporation of branched fatty acids into membrane lipids could adversely affect membrane function.

5. Fatty acid biosynthesis begins with acetyl CoA. Propionyl CoA, derived from propionate, could interfere with the initiation of fatty acid biosynthesis.

6. Phosphopantetheine is attached to ACP, which provides a flexible arm that can reach all of the components participating in synthesis.

Chapter 20: Isomerization

1.

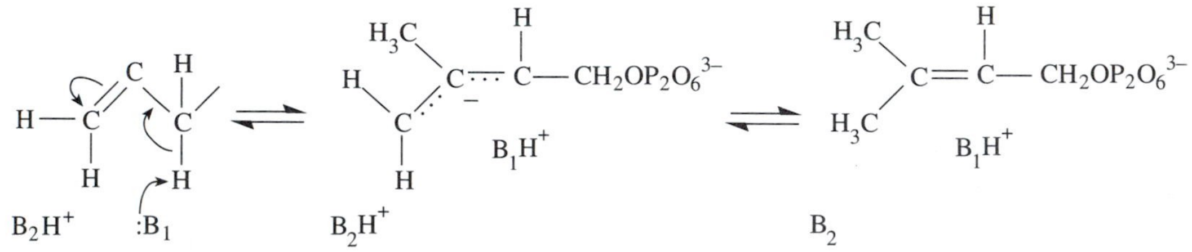

2.

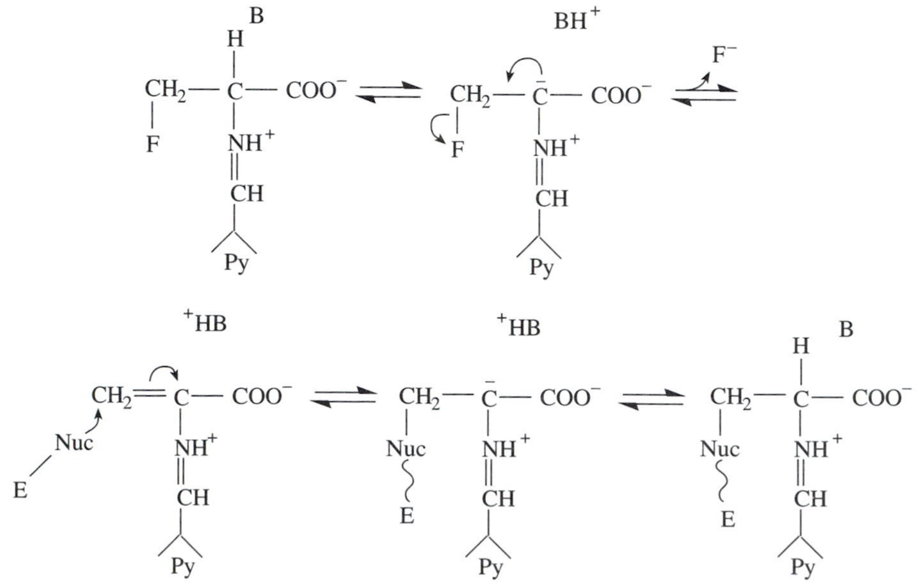

F^- is a reasonably good leaving group; therefore elimination occurs. This results in the formation of a Michael acceptor. A nucleophile—probably lysine—on the enzyme adds and forms a covalent bond.

3.

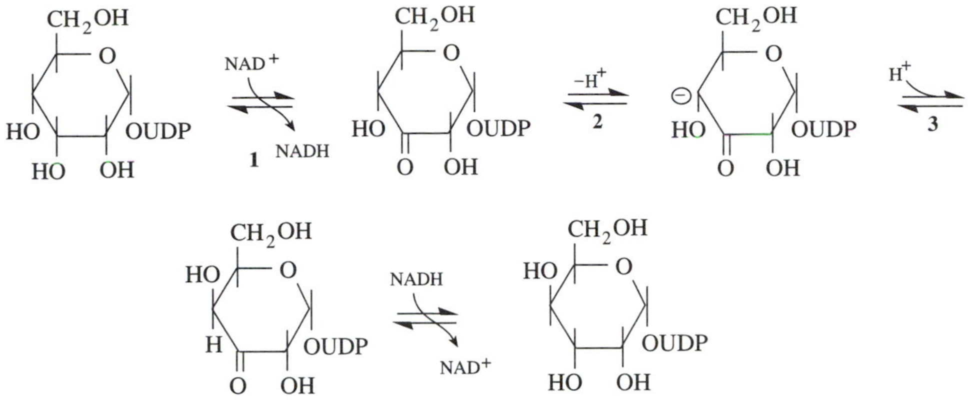

CH_2OH O HO OUDP HO OH $\underset{1\ \text{NADH}}{\overset{\text{NAD}^+}{\rightleftharpoons}}$ CH_2OH O HO OUDP O OH $\underset{2}{\overset{-H^+}{\rightleftharpoons}}$ CH_2OH O ⊖ HO OUDP O OH $\underset{3}{\overset{H^+}{\rightleftharpoons}}$

CH_2OH HO O H OUDP O OH $\underset{\text{NAD}^+}{\overset{\text{NADH}}{\rightleftharpoons}}$ CH_2OH HO O OUDP HO OH

Step 1: Oxidize at C-3 instead of C-4. This will stabilize the carbanion at C-4.
Step 2: Abstract proton at C-4, invert.
Step 3: Reduce carbonyl group at C-3.

4. (a)

$$A \rightleftharpoons B \rightleftharpoons C$$

(b)

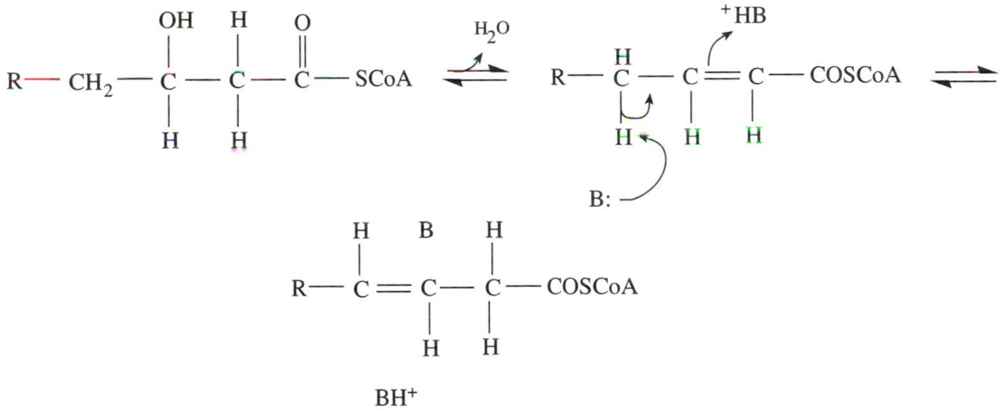

$R—CH_2—C(OH)(H)—C(H)(H)—C(=O)—SCoA \overset{H_2O}{\rightleftharpoons} R—C(H)(H)—C(H)=C(H)—COSCoA$ (${}^+HB$, B:) $\rightleftharpoons$

$R—C(H)=C(B)(H)—C(H)(H)—COSCoA$

BH^+

5. Probably similar to proline racemase.

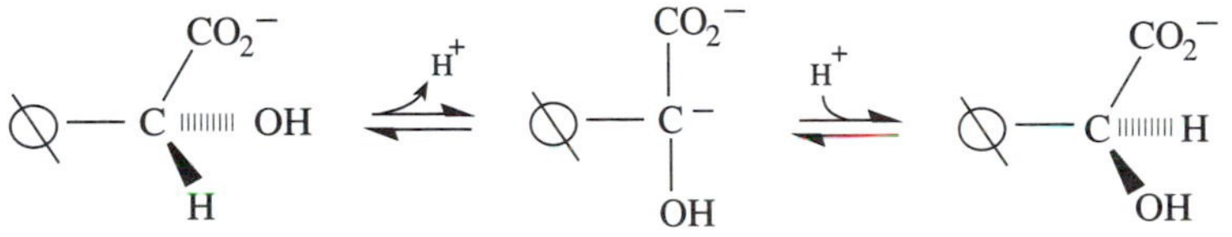

$\phi—C(CO_2^-)(OH)(H) \overset{H^+}{\rightleftharpoons} \phi—C^-(CO_2^-)(OH) \overset{H^+}{\rightleftharpoons} \phi—C(CO_2^-)(H)(OH)$

Chapter 21: Glycolysis and the Tricarboxylic Acid Cycle

1. No ATP is obtained from the oxidation of glyceraldehyde-3-P. This reduces the ATP yield to zero.

2.

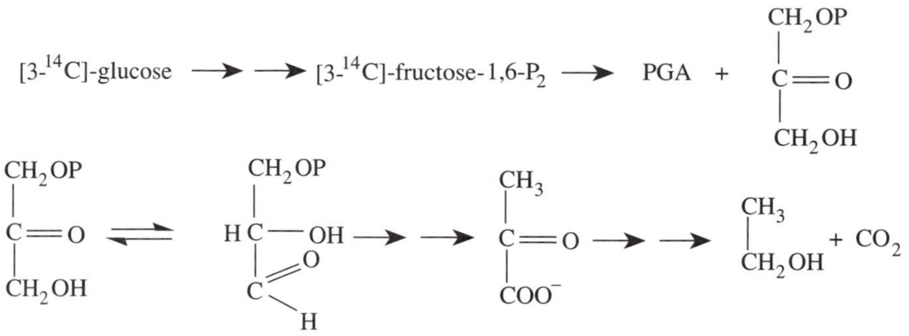

[3-^{14}C]-glucose ⟶ ⟶ [3-^{14}C]-fructose-1,6-P_2 ⟶ PGA + CH_2OP—C=O—CH_2OH

CH_2OP—C=O—CH_2OH ⇌ CH_2OP—HC—OH—C(=O)H ⟶ ⟶ CH_3—C=O—COO^- ⟶ ⟶ CH_3—CH_2OH + CO_2

3.

	Once	Three times	∞
COOH			
CH_2	x	x	x
CH_2		x	x
CO		x	x
COOH		x	x

x = ^{14}C label

4. The dicarboxylic acids are intermediates in the TCA cycle; therefore, the rate of operation of the cycle increases, as does the rate of CO_2 production.

***5.** There are many possible approaches to solving this problem, as well as many different opinions about the correct approach to use.

6. Probably not. A drug must be specifically directed toward a definite target. An analog of c-AMP would affect many enzyme systems.

***7.** Phosphopentose isomerase:

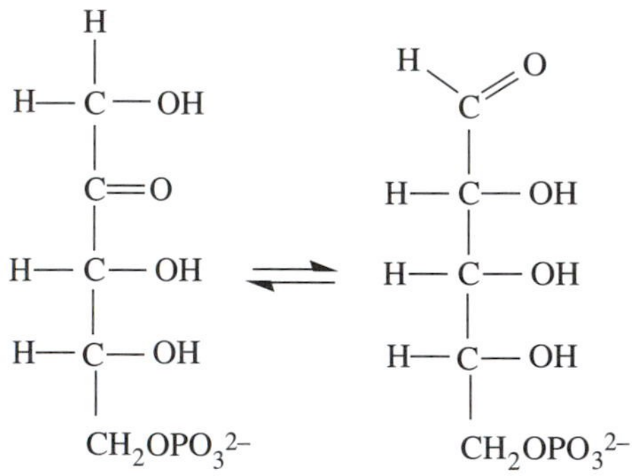

H—C(H)—OH / C=O / H—C—OH / H—C—OH / $CH_2OPO_3^{2-}$ ⇌ H—C=O / H—C—OH / H—C—OH / H—C—OH / $CH_2OPO_3^{2-}$

Possible mechanism:

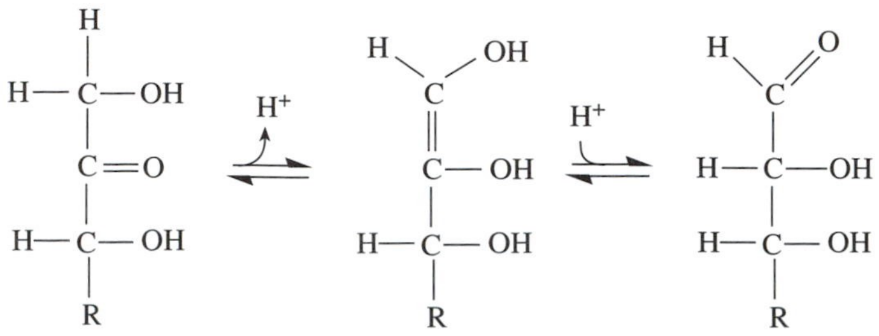

H—C(H)—OH / C=O / H—C—OH / R ⇌ (H^+) H—C—OH / ‖ / C—OH / H—C—OH / R ⇌ (H^+) H—C=O / H—C—OH / H—C—OH / R

Carry out reaction in isotopic H_2O and determine if aldose contains labeled H at C-2.

Chapter 22: The Electron-Transport Pathway and Oxidative Phophorylation

1. Participation in the production of ATP:

(a) Electron-transport complex I is NADH dehydrogenase in the inner mitochondrial membrane; it catalyzes the reduction of ubiquinone to dihydroubiquinone in a process that leads to the translocation of protons from the matrix into the intermembrane space. This proton gradient drives the synthesis of ATP from ADP and P_i by ATP synthase. Electron-transport complex II is succinate dehydrogenase, also in the inner mitochondrial membrane, which catalyzes the reduction of ubiquinone by succinate; however, it does not catalyze the translocation of protons.

(b) Ubiquinone is the electron acceptor for NADH dehydrogenase and succinate dehydrogenase in the inner mitochondrial membrane. It is a hydrophobic quinone that, in its oxidized and reduced forms, remains integrated into the membrane and can diffuse within the membrane from one electron-transfer complex to another and mediate electron transfer between the electron-transfer complexes. The ubiquinone/dihydoubiquinone system mediates electron transfer from complexes I and II to complex III.

(c) ATP synthase is a multiprotein complex consisting of F_1ATPase and F_0, which, when assembled in the inner mitochondrial membrane, is known as F_0F_1ATPase. The complex is anchored in the membrane by the component F_0, with the component F_1ATPase protruding into the matrix. The passage of protons through this complex from the intermembrane space into the matrix is driven by the pH gradient established by electron-transfer complexes I, III, and IV, and this translocation of protons drives the production of ATP from ADP and P_i.

(d) Cytochrome oxidase is electron-transfer complex IV in the inner mitochondrial membrane; it catalyzes the oxidation of four molecules of reduced cytochrome *c* by a molecule of oxygen to produce two molecules of water in a process that is accompanied by the translocation of protons from the mitochondrial matrix into the intermembrane space. The pH and electrochemical gradients so established drive the production of ATP by ATP synthase.

(e) Conformational coupling is a hypothesis that explains how ATP synthase couples proton translocation to the production of ATP from ADP and P_i. According to this hypothesis, ATP synthase catalyzes the formation of ATP from ADP+P_i bound at the active site, with an equilibrium constant for the enzyme-bound species that is near 1. One conformation of the complex has a high affinity for ADP and a low affinity for ATP, whereas another conformation has a low affinity for ADP and a high affinity for ATP. Proton translocation drives the transformation of one conformation into the other unidirectionally, so that ATP synthase continually releases ATP and binds ADP.

(f) Cytochrome *c* is an electron carrier in the inner mitochondrial membrane. Cytochrome *c* is reduced by complex III, diffuses within the membrane to complex IV, and reduces the electron-transfer cofactors in complex IV.

2. An electron pair introduced into the electron-transport pathway by NADH passes through complexes I, III, and IV, each of which couples electron transfer with the transport of protons from the matrix to the intermembrane space. This allows the production of as many as three molecules of ATP coupled to the reentry of protons through ATP synthase. An electron pair introduced by $FADH_2$ passes through complexes III and IV, so that fewer proton translocations take place in the oxidation of $FADH_2$ and only two molecules of ATP are produced.

3. The energy of pH gradients and electrochemical potentials can be used in place of ATP to drive active-transport processes, such as the transport of ions and molecules against concentration gradients.

4. Dinitrophenol prevents ATP formation in mitochondria by facilitating the transport of protons across the membrane, so that ATP synthase is by-passed. Both the dinitrophenolate anion and dinitrophenol can pass through the membrane rapidly, and their passage equalizes the pH on the two sides of the membrane and prevents a pH gradient from being created by the electron-transport pathway. Because of this, there is no resistance to electron transport, and the rate increases.

5. Both molecules blocked both ATP synthesis and electron transport. ATP synthase was inhibited by oligomycin, whereas dicyclohexylcarbodiimide irreversibly inactivated ATP synthase. These molecules did not inhibit any of the electron-transfer complexes, but they blocked electron transport in the mitochondrial membrane. Electron transport in the membrane could be restored when dinitrophenol was added, although ATP synthesis was not restored and ATP synthase remained inactive. This proved that the connection between ATP synthase and electron transport is broken by the action of dinitrophenol. The connection proved to be proton translocation.

6. Communication between the mitochondrial matrix and the cytosol must include the ADP/ATP transporter, the glycerophosphate shuttle, the malate shuttle, transport systems for pyruvate, fatty acyl carnitine, and citrate. In addition, there must be other transport systems to allow molecules required for protein biosynthesis to enter the mitochondrion, because the mitochondrial DNA encodes only a few mitochondrial proteins, including some of the subunits of cytochrome oxidase.

7. The protons translocated by the electron-transfer complexes create a pH differential and electrochemical potential on the two sides of the inner mitochondrial membrane. The energy associated with this differential is used by ATP synthase to produce ATP. ATP synthase can dissipate the pH differential by allowing protons to reenter the mitochondrial matrix, and the enzyme uses the energy made available by this process to convert ADP and P_i on the matrix side of the membrane into ATP.

8. For each molecule of acetyl CoA oxidized within the mitochondrion, one molecule of NADH is produced by isocitrate dehydrogenase, one by ketoglutarate dehydrogenase complex, and one by malate dehydrogenase, for a total of three molecules of NADH. In addition, one molecule of $FADH_2$ is produced by succinate dehydrogenase. The oxidation of three molecules of NADH and one molecule of $FADH_2$ can lead to the production of eleven molecules of ATP. To this must be added the one molecule of GTP(ATP-equivalent) produced by succinyl CoA synthetase, for a total of twelve. Of these twelve, eleven are produced by ATP synthase.

9. The main components are F_0 and F_1ATPase. F_0 anchors the ATP synthase in the mitochondrial membrane and constitutes the channel though which protons are translocated unidirectionally from the intermembrane space to the matrix. F_1ATPase is normally bound to F_0 and catalyzes the interconversion of ADP and P_i into ATP at its active sites. It releases ATP when protons are translocated from the intermembrane space into the matrix. F_1ATPase can be released from F_0 and becomes a soluble enzyme that catalyzes the hydrolysis of ATP.

10. Some of the energy from the oxidation of glucose that is not used to produce ATP is dissipated as heat, and some of it is used in place of ATP to drive membrane transport processes.

Chapter 23: Photosynthesis

1. Photosynthesis in green plants uses twelve molecules of water to form six molecules of oxygen for each molecule of glucose that is formed, as is described in equation 1 of the text. Two water molecules are used to form an oxygen molecule for each of the six carbon dioxide molecules that are used to form the molecule of glucose. This results in the utilization of twelve water molecules for each glucose. Six molecules of water are released because one water molecule is released when ribulose-1,5-bisphosphate is cleaved into two molecules of phosphoglycerate, as shown in equation 13 of the text. Additional water molecules are released when ATP is synthesized from ADP and phosphate during photophosphorylation, but this is an indirect result of photosynthesis.

2. The reduction of ferricyanide to ferrocyanide in the Hill reaction shows that the photosynthetic system can utilize light to form reducing equivalents. These reducing equivalents are used to reduce carbon dioxide to carbohydrate. Reoxidation of the carbohydrate to carbon dioxide drives ATP synthesis and provides the predominant source of useful energy for life on earth.

3. Accessory pigments ensure that most of the light that reaches a plant can be used to excite a pigment that can give rise to photosynthesis. The energy from this excitation can be transferred to the chlorophyll dimer and used to reduce carbon dioxide to carbohydrate. Light of relatively low wavelength is most effective because excitation energy increases with decreasing wavelength and the excitation has to have sufficient energy that it can excite chlorophyll, which has an absorption maximum at 453 nm.

4. An active photosynthetic system is illuminated with very brief flashes of light, so that only a single excitation can occur in each flash. Flashes are increased in intensity until there is no further increase in the yield of carbohydrate or of oxygen. The number of oxygen molecules released under these conditions is 1/8 of the number of chlorophyll molecules that can absorb light and transfer it to the active chlorophyll dimer to form oxygen, because eight quanta of light are required to form one oxygen molecule.

5. The formation of oxygen and of NADPH requires two excitations of chlorophyll, which is made possible by the Z scheme. Human beings require oxygen, as well as carbohydrate, so that human beings could not live in a world in which the photosynthetic reactions utilize only a single quantum of light and do not form oxygen.

6. Ascorbic acid is an effective reducing agent. It can reduce the P700 photosystem I after the pigment has been excited and used to form NADPH. This allows photosynthesis to occur without the participation of photosystem II, so that it can increase the quantum yield of photosynthesis at low light intensity.

Chapter 24: Oxidations and Oxygenations

1. There are several reasons that aromatic side chains on amino acids would make these amino acids likely candidates for compounds that regulate physiological functions by binding to receptors. Aromatic rings are rigid and provide a relatively large amount of binding energy that facilitates strong binding of the regulatory compound to its receptor. This binding energy is also needed to bring about a change in the conformation of the receptor from its stable ground state to the structure that releases a signal that initiates a physiological response. The aromatic ring is also stable under most conditions, but it can readily be hydroxylated or subjected to other reactions of the aromatic ring in order to synthesize a wide range of active regulatory molecules. See, for example, Figure 24-9.

2. Hydrolysis of phospholipids results in the release of arachidonic acid, which is converted into leukotrienes and prostaglandins. Both of these compounds are highly reactive, both in their physiological effects and in their reactivity. They activate a wide variety of physiological responses, including vasodilation and inflammation.

3. Aspirin blocks cyclooxygenase, which catalyzes steps in the synthesis of prostaglandins and thromboxanes, which tend to cause vasodilation and inflammation. These responses can be useful for defense against microorganisms and other environmental challenges, but can also lead to pain or discomfort, such as headaches. Prostaglandins and thromboxanes are responsible for much of the inflammation in rheumatoid arthritis and rheumatic fever. Aspirin inhibits cyclooxygenase by transferring an acetyl group to a serine hydroxyl group on the enzyme (equation 14 of the text).

4. An oxygenase catalyzes the transfer of an oxygen atom from molecular oxygen to the substrate, An example is cytochrome P450, which can insert oxygen into a carbon–hydrogen bond (Figure 26-6). This can be demonstrated directly by use of $^{18}O_2$. Monooxygenases frequently utilize 5,6,7,8-tetrahydrobiopterin as a cofactor to provide two reducing equivalents in a simple hydroxylation (see page 662 of the text). An oxidase acts to remove electrons from the substrate. It may incorporate oxygen into the product from water. This oxidation is often brought about by the coenzyme FAD. The FADH product is regenerated by oxidation in which molecular oxygen is frequently the reducing agent. FADH-acyl CoA dehydrogenase is a typical oxidase (page 657 of the text).

5. Some oxygenases utilize α-ketoglutarate to provide the reducing equivalents that are needed by an oxygenase. The oxidative decarboxylation of α-ketoglutarate to succinate and carbon dioxide is an irreversible two-electron oxidation that provides a strong driving force for reactions such as those catalyzed by prolyl and lysyl hydroxylases. Prolyl and lysyl hydroxylases introduce hydroxyl groups into collagen and elastin that provide hydrogen bonding and sites for glycosylation of collagen and elastin. These groups may participate in the cross-linking of collagen and elastin in connective tissue.

6. Phenylpyruvate is formed from phenylalanine by transamination. This reaction is not of great significance unless there is a deficiency in phenylalanine hydroxylase, which converts phenylalanine into tyrosine. A deficiency of this enzyme causes phenylketonuria, but the remaining phenylpyruvate in the tissues accumulates and causes mental retardation.

7.

NH_2 C—O^- O; $O_2 + 2\,H^+$; NADPH; $NADP^+$ + H_2O; NH_3^+ O C—O^- O; NH_3 CO_2; H OH O; OH OH

8. Molecular oxygen has four oxidizing equivalents, but only two oxidizing equivalents are utilized to hydroxylate proline by the substitution of OH for an H atom on the proline ring. The other two oxidizing equivalents are utilized by oxidative decarboxylation of α-ketoglutarate; one of the oxygen atoms is incorporated into the carboxylate group of the succinate product (see page 671 of the text).

9. The decarboxylation of 5-hydroxytryptophan to serotonin is catalyzed by a pyridoxal-containing enzyme, aromatic amino acid decarboxylase. This is a typical pyridoxal-catalyzed decarboxylation of an amino acid.

10. The children should be treated with γ-aminobutyric acid (GABA), which is an inhibitory neurotransmitter. Pyridoxal phosphate is required for the decarboxylation of glutamate to GABA, and a deficiency of pyridoxal phosphate can lead to hyperexcitability of the central nervous system and convulsions because of a decrease in the rate of GABA formation from glutamate.

Chapter 25: One-Carbon Metabolism

1.

H N N—R H C=^{18}O H; ADP; $^-$O—P(O)(O$^-$)—O—P(O)(O$^-$)—O—P(O)(O$^-$)—rib-Ad; H N$^+$ NHR H—C—^{18}O—P(O)(O$^-$)—O$^-$; H^+; N H C$^+$ NR H + ^{18}O—P(O)(O$^-$)—OH

Exchange reaction:

$$ATP/ADP^* \xrightarrow[\text{5-Formyl THF}]{E} ATP^*/ADP^*$$

2. (a)

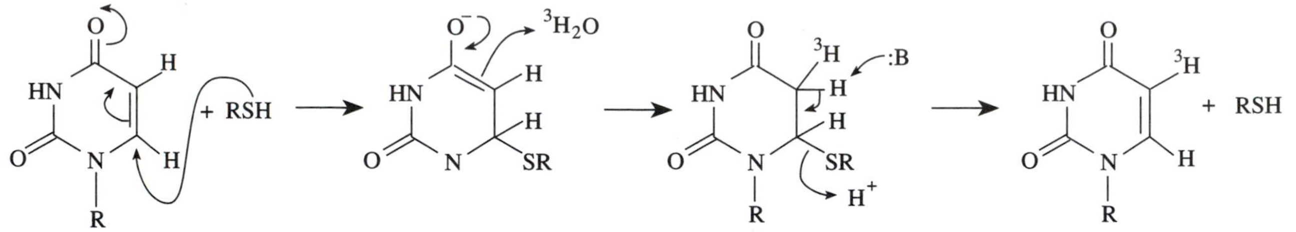

(b)

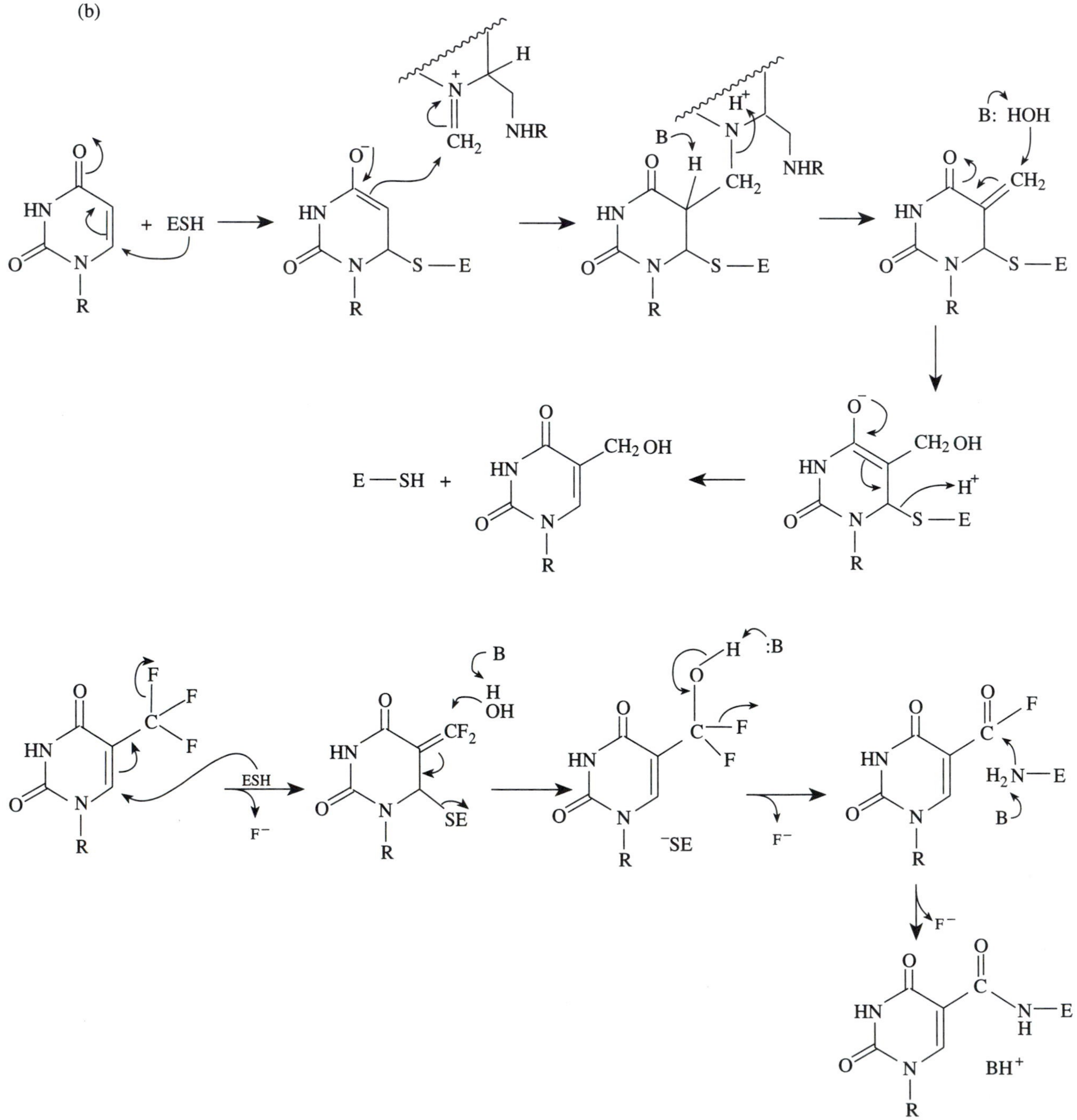

Chapter 26: Nitrogen Metabolism: Amino Acids, Urea, and Heme

***1.**

3-Phosphoglycerate + NAD^+ ⟶ 3-Phosphohydroxypyruvate + NADH + H^+

3-Phosphohydroxypyruvate + Glutamate ⟶ *O*-Phosphoserine + 2-Ketoglutarate

O-Phosphoserine + H_2O ⟶ Serine + P_i

Other sequences also are possible. The phosphatase step can be the first or second step, and the transamination can be substituted by a reductive amination similar to glutamate dehydrogenase.

***2.** Glutamate is a possible precursor because of the fact that it has five carbon atoms and an amino group, as does proline. If the 5-carboxyl group of glutamate were to be reduced to form glutamate 5-semialdehyde, the molecule could form an internal cyclic imine, which could be further reduced to proline. The reducing agent could be NADH, and the carboxyl group would probably be activated before it is reduced.

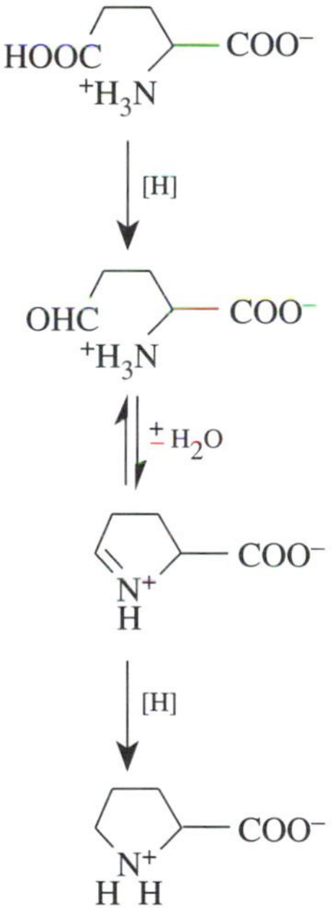

HOOC–(chain)–COO^- (^+H_3N)
↓ [H]
OHC–(chain)–COO^- (^+H_3N)
⇅ ± H_2O
(cyclic imine, N^+H)–COO^-
↓ [H]
(pyrrolidine, N^+ H H)–COO^-

***3.** Aspartate is used in the urea cycle as the source of the second nitrogen in urea and the second nitrogen in the guanidino group of arginine. To the extent that net arginine synthesis takes place, aspartate would be depleted. However, aspartate is easily produced from oxaloacetate by transamination.

***4.**

Acetyl CoA + Glutamate ⟶ *N*-Acetylglutamate + CoA

***5.** If the mechanism shown below is correct, the enzyme should catalyze the decarboxylation of glycine in the absence of succinyl CoA. If it does not, then either the mechanism in Figure 26-15 is correct or there is substrate synergism. If there is substrate synergism, the enzyme might catalyze the decarboxylation of glycine in the presence of an unreactive analog of succinyl CoA, such as malonyl CoA or propionyl CoA or another ester of succinate. The reduction products produced by inactivating the enzyme with sodium borohydride would also indicate whether the decarboxylation precedes or follows the attack of the carbanion on succinyl CoA.

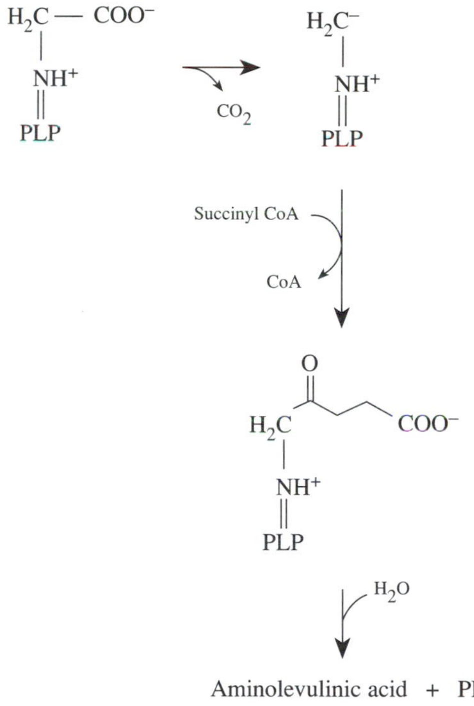

H_2C—COO^- | NH^+ ‖ PLP ⟶ ($-CO_2$) H_2C^- | NH^+ ‖ PLP
↓ Succinyl CoA → CoA
H_2C–C(=O)–CH_2CH_2–COO^- | NH^+ ‖ PLP
↓ H_2O
Aminolevulinic acid + PLP

***6.** The following is one possible sequence of enzymatic reactions that would lead to propionyl CoA. The first enzyme is PLP-dependent and catalyzes a β-elimination reaction. The second enzyme is an α-ketoacid dehydrogenase complex. The second reaction is catalyzed by the branched-chain α-ketoacid dehydrogenase complex.

H_3C—CH(OH)—CH(NH_3^+)—COO^- —(E · PLP; $-NH_4^+$)⟶ H_3C—CH_2—C(=O)—COO^- —(NAD^+ + CoA ⟶ NADH + CO_2)⟶ Propionyl CoA

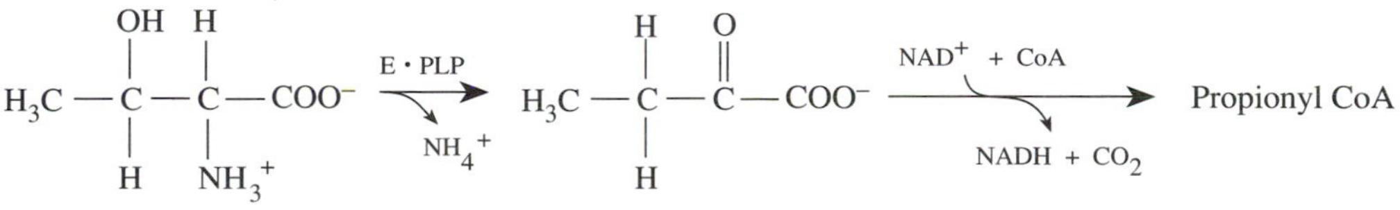

Propionyl CoA can be carboxylated to methylmalonyl CoA, which is isomerized to succinyl CoA, and succinyl CoA is metabolized through the TCA cycle.

***7.** The following mechanism showing the function of NAD^+ is supported by considerable experimental evidence.

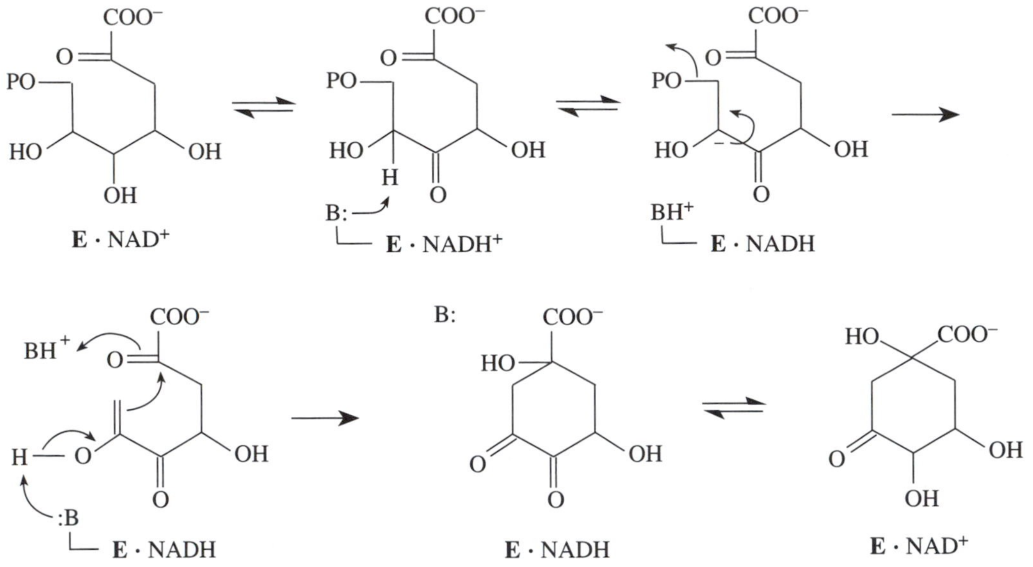

E · NAD^+ ⇌ E · $NADH^+$ ⇌ E · NADH → E · NADH → E · NADH ⇌ E · NAD^+

***8.** The enzyme should produce ammonia or the chemical equivalent of ammonia at the active site through the hydrolysis of glutamine, and the ammonia should undergo addition to the ring with elimination of water, as shown below.

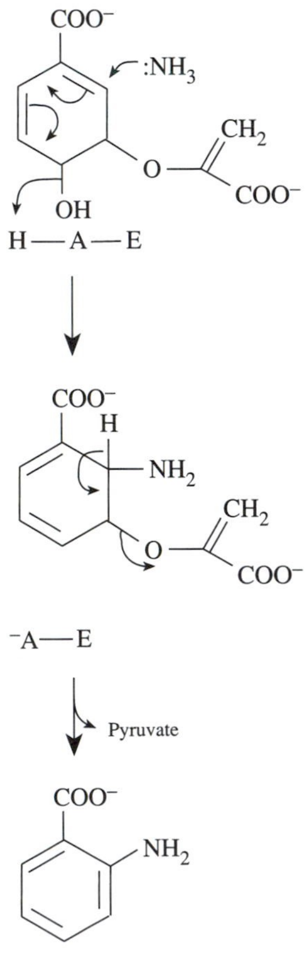

Pyruvate

The addition of ammonia and the elimination of water may proceed in steps, possibly with the departure of water to form a delocalized carbocation as an intermediate followed by the addition of ammonia. Loss of the proton from carbon-2 in the resulting intermediate should be an easy process when accompanied by the elimination of enolpyruvate; this is because of the concomitant aromatization of the ring, which is a highly favored reaction leading to a stable product.

Chapter 27: Nitrogen Metabolism: Nucleotides

1. The product carbon skeleton of aspartate resulting from its donation of an amino group in the formation of uric acid (and urea) is fumarate. Fumarate may be converted back into aspartate through action of the TCA cycle and glutamate-oxaloacetate transaminase. The hydration of fumarate by fumarase produces malate, which is dehydrogenated by malate dehydrogenase to oxaloacetate. Glutamate-oxaloacetate transaminase catalyzes the reaction of glutamate with oxaloacetate to form aspartate and 2-ketoglutarate. This regenerates aspartate. Glutamate is regenerated from 2-ketoglutarate by the action of glutamate dehydrogenase, which catalyzes the reaction of 2-ketoglutarate with ammonia and NADH to form glutamate and NAD^+.

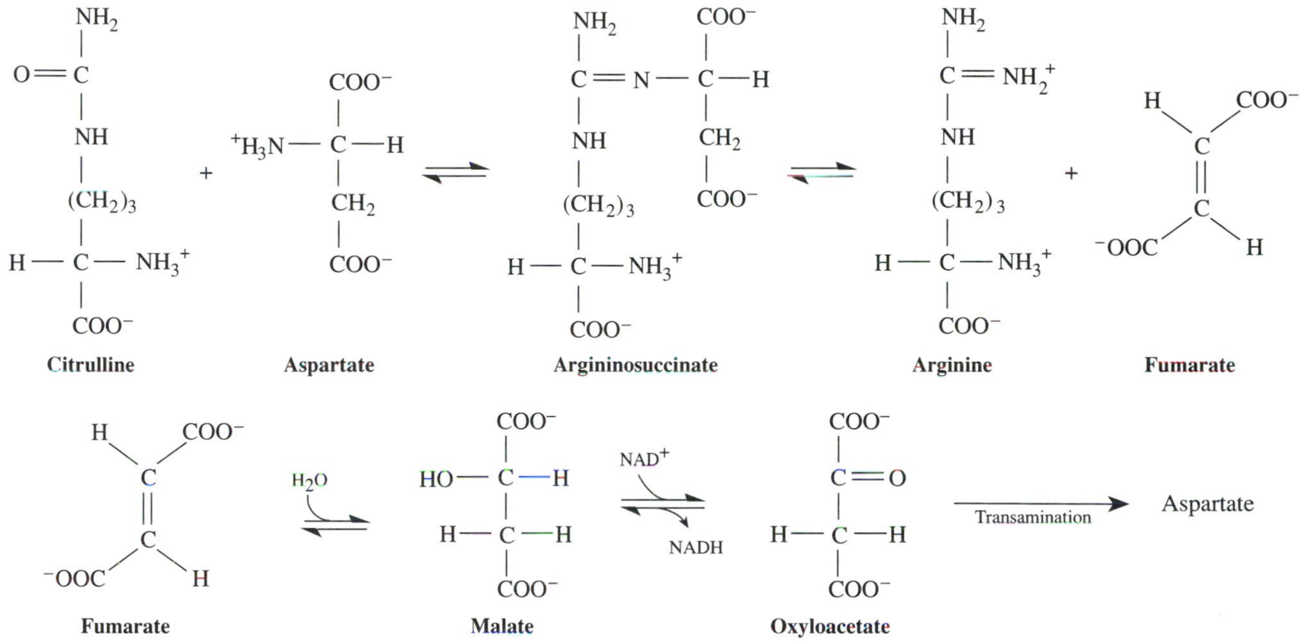

Citrulline + **Aspartate** ⇌ **Argininosuccinate** ⇌ **Arginine** + **Fumarate**

Fumarate $\xrightarrow{H_2O}$ ⇌ **Malate** ⇌ (NAD^+ → NADH) **Oxyloacetate** → (Transamination) Aspartate

***2.** The difficulty is the provision of a good means for stabilizing the incipient carbanion resulting from the decarboxylation of OMP. One hypothesis for obviating this problem was for OMP to undergo an addition of an enzymatic nucleophile such as a thiol group to carbon-6. This would allow an elimination mechanism, as illustrated below.

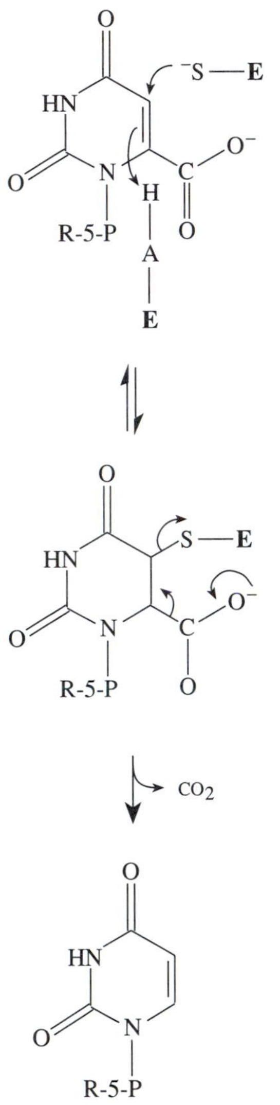

O HN O N R-5-P $^{-}$S—E C O^{-} O H A E

O HN O N R-5-P S—E C O^{-} O

CO_2

O HN O N R-5-P

Recent evidence indicates that an addition-elimination mechanism does *not* take place. Chemical model experiments indicate that the carbanion resulting from decarboxylation is sufficiently stabilized by the electronegative nitrogen atoms and oxygen substituents of the uracil ring to allow decarboxylation to take place.

***3.** Degradation of uracil and thymine:

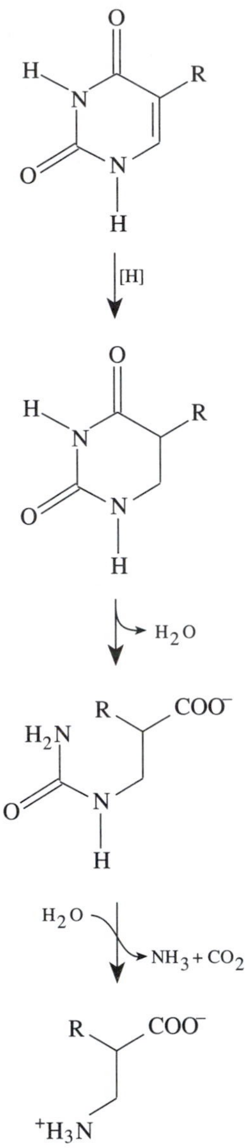

O H N R O N H

[H]

O H N R O N H

H_2O

R COO^- H_2N O N H

H_2O $NH_3 + CO_2$

R COO^- ^+H_3N

R = H; Uracil degradation to β-alanine.

R = CH_3; Thymine degradation to β-aminoisobutyrate.

***4.** It is likely that 2′-chloro-ADP reacts as if it were a substrate in the initial steps of the ribonucleotide reductase reaction. From the ribonucleotide reductase mechanism, the following is a reasonable hypothesis for the formation of the ribose product shown in the text.

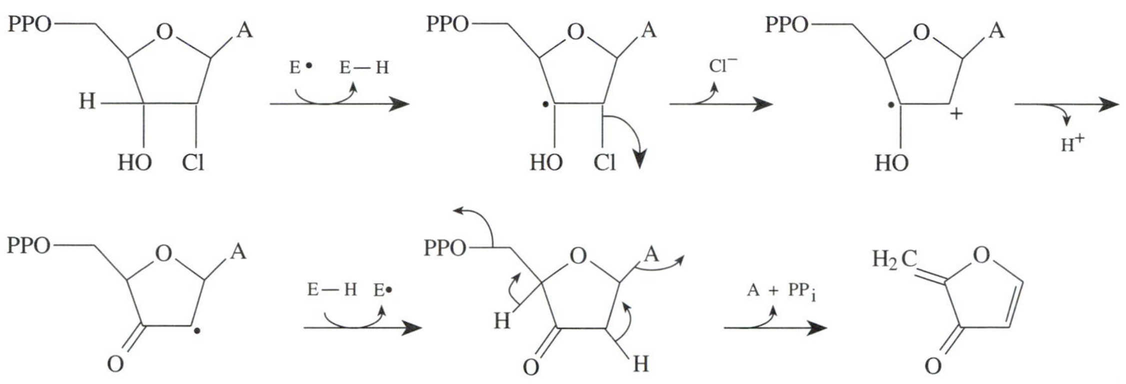

The product is a highly electorphilic species that readily alkylates sulfhydryl groups in proteins, including the essential cysteine residues of ribonucleotide reductase, through Michael addition reactions.

***5.**

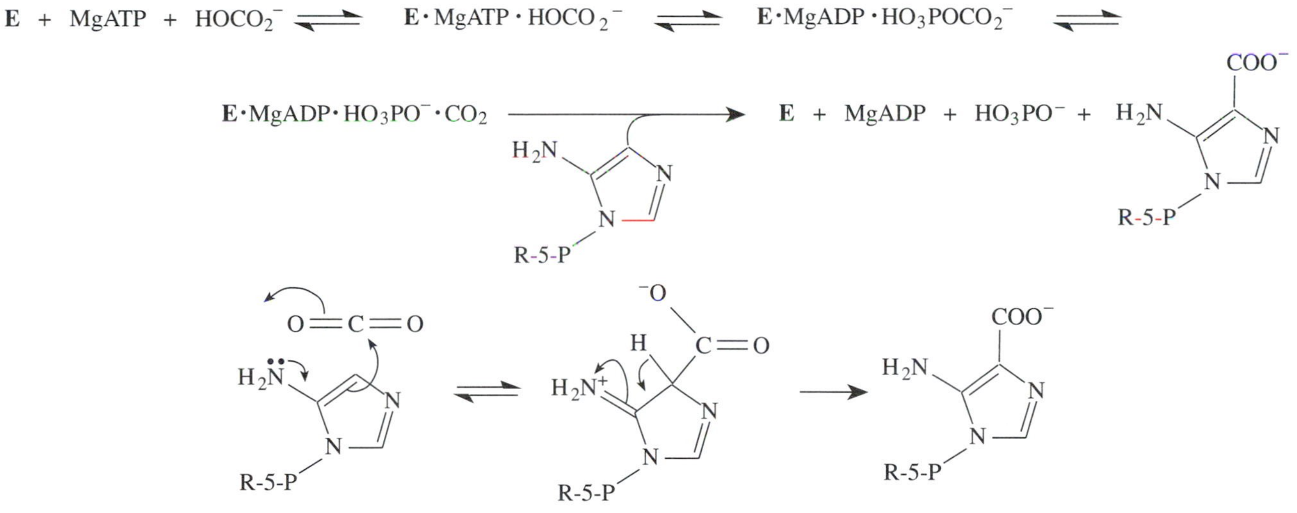

$$E + MgATP + HOCO_2^- \rightleftharpoons E \cdot MgATP \cdot HOCO_2^- \rightleftharpoons E \cdot MgADP \cdot HO_3POCO_2^- \rightleftharpoons$$

$$E \cdot MgADP \cdot HO_3PO^- \cdot CO_2 \longrightarrow E + MgADP + HO_3PO^- + \text{(4-carboxy-5-aminoimidazole ribonucleotide)}$$

The activation of bicarbonate by ATP is probably similar to that of other carboxylases. However, 5′-phosphoribosyl-5-aminoimidazole is not an enolic or carbanionic species. The carboxylation is facilitated by the electron pair on the 5-amino group to form a 4-carboxylated intermediate, from which the 4-proton must be lost in a second step. In other carboxylation reactions, there is no intermediate in which a proton must be lost by cleavage of a carbon–hydrogen bond. The carbon–hydrogen bond is normally broken in enol formation before the carboxylation step. (The sequence of substrate-binding steps shown here has not been established and is chosen to illustrate the chemical transformation. The same chemical transformations will be required if 5′-phosphoribosyl-5-aminoimidazole binds to the enzyme in an earlier step.)

Chapter 28: Complex Lipids

***1.**

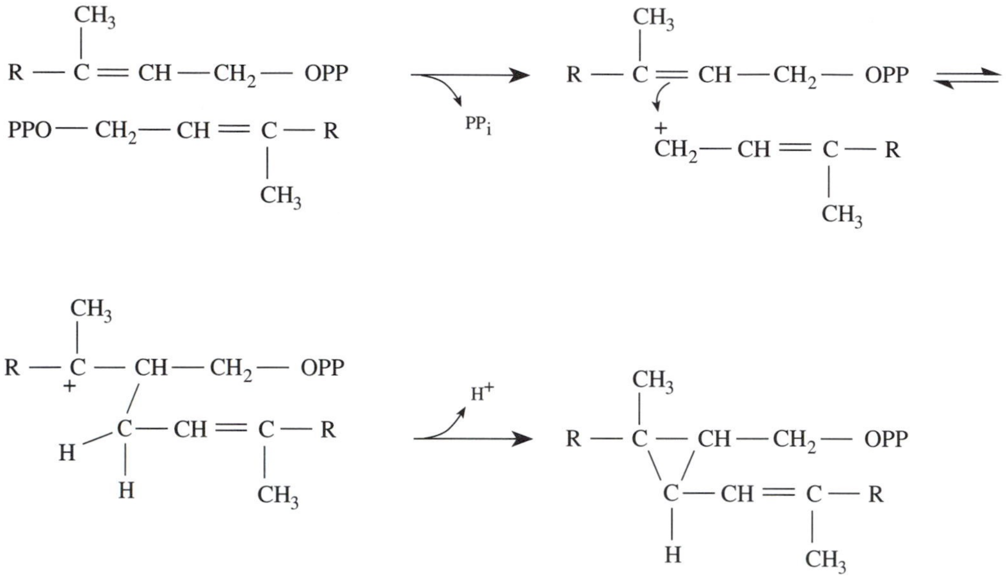

***2.** The first four steps must take place in the sequence shown. The cytochrome P450 steps can take place in any sequence.

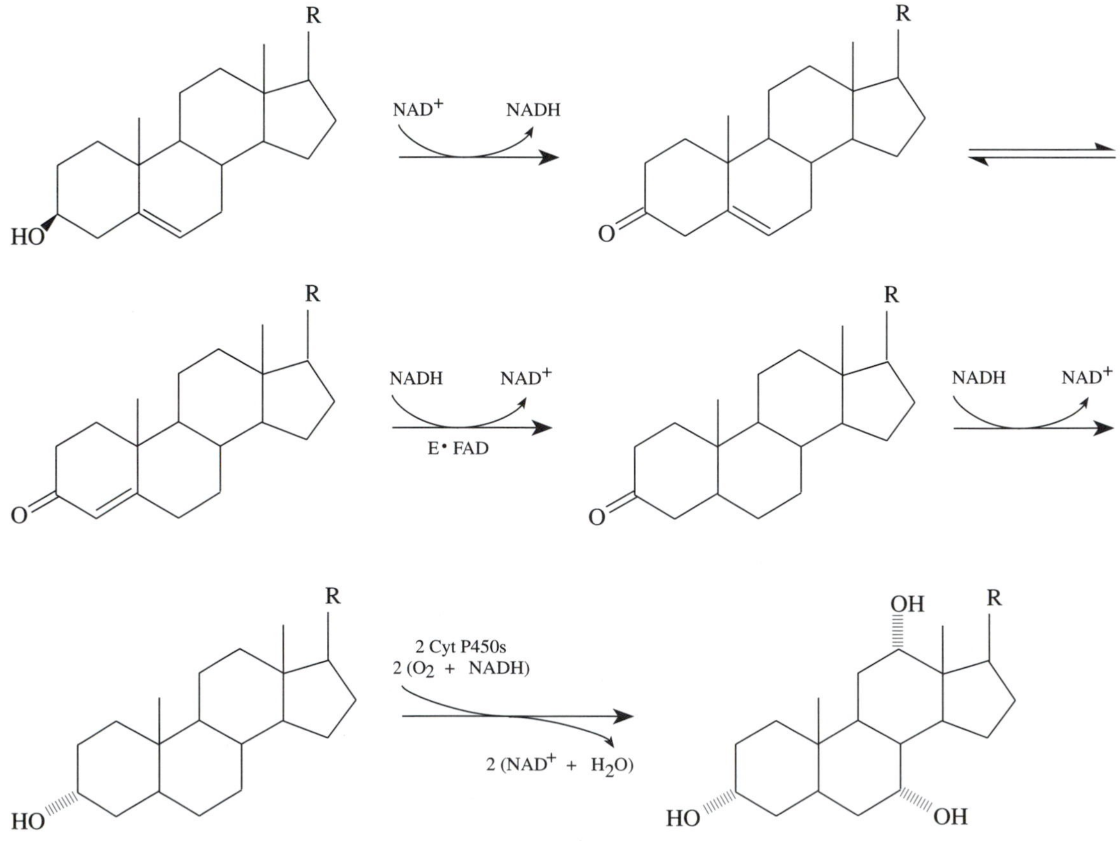

***3.** A number of sequences are possible. The following one is a short sequence that utilizes the chemical principles presented in Chapters 2, 16, and 24. Other possible sequences would require more dehydrogenation steps and would require decarboxylation steps.

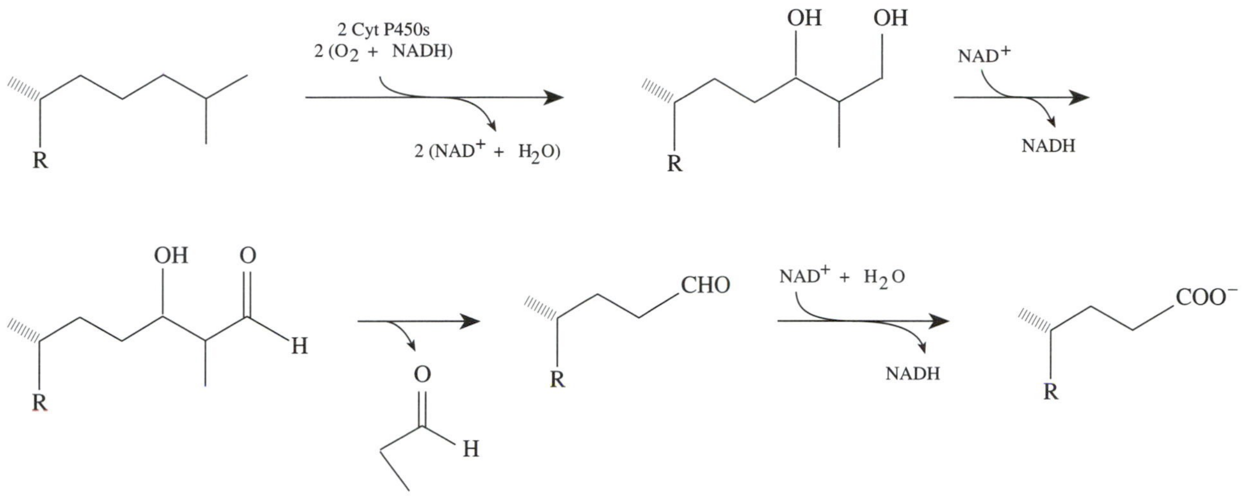

***4.** The last step of the following sequence is one that is similar to the carbon–carbon bond cleavages catalyzed by transketolases (Chapters 21 and 23) and phosphoketolase. These reactions require thiamine pyrophosphate, which is discussed in Chapters 5 and 17. Can you think of another way?

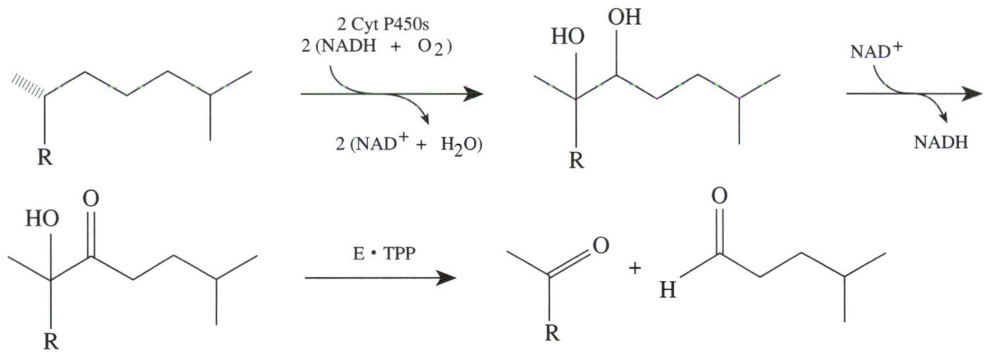

*5.

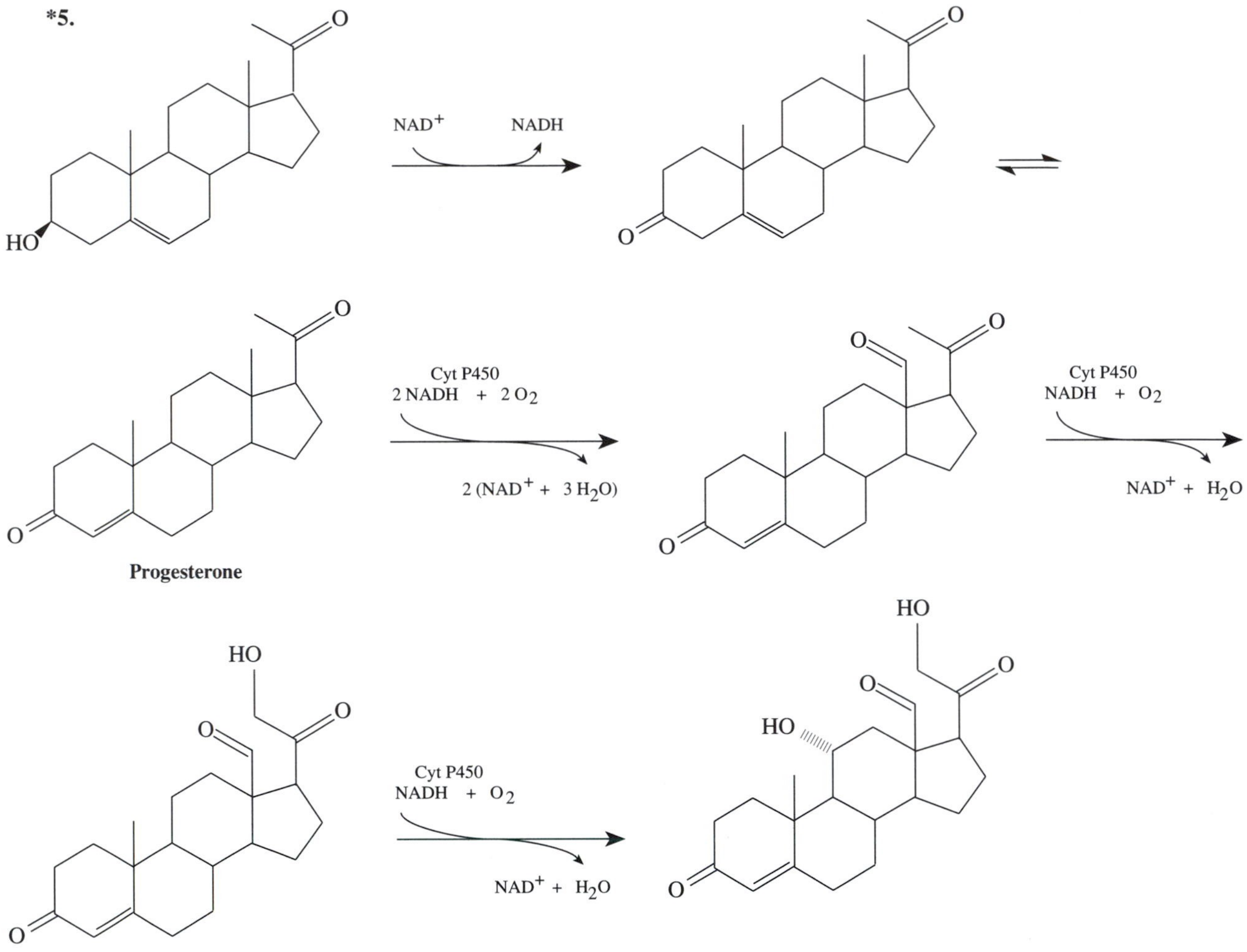

NAD^+ → NADH

Progesterone

Cyt P450, 2 NADH + 2 O_2 → 2 (NAD^+ + 3 H_2O)

Cyt P450, NADH + O_2 → NAD^+ + H_2O

Cyt P450, NADH + O_2 → NAD^+ + H_2O

The first oxygenation step may be catalyzed by one cytochrome P450 that produces a gem-diol, an aldehyde hydrate, which undergoes dehydration to the carboxaldehyde. The next two oxygenation steps are probably catalyzed by two different cytochrome P450s.

***6.** The following is a reasonable sequence, although the steps may take place in a different order. The side-chain cleavage is TPP-dependent; this is reasonable, but alternatives may be possible.

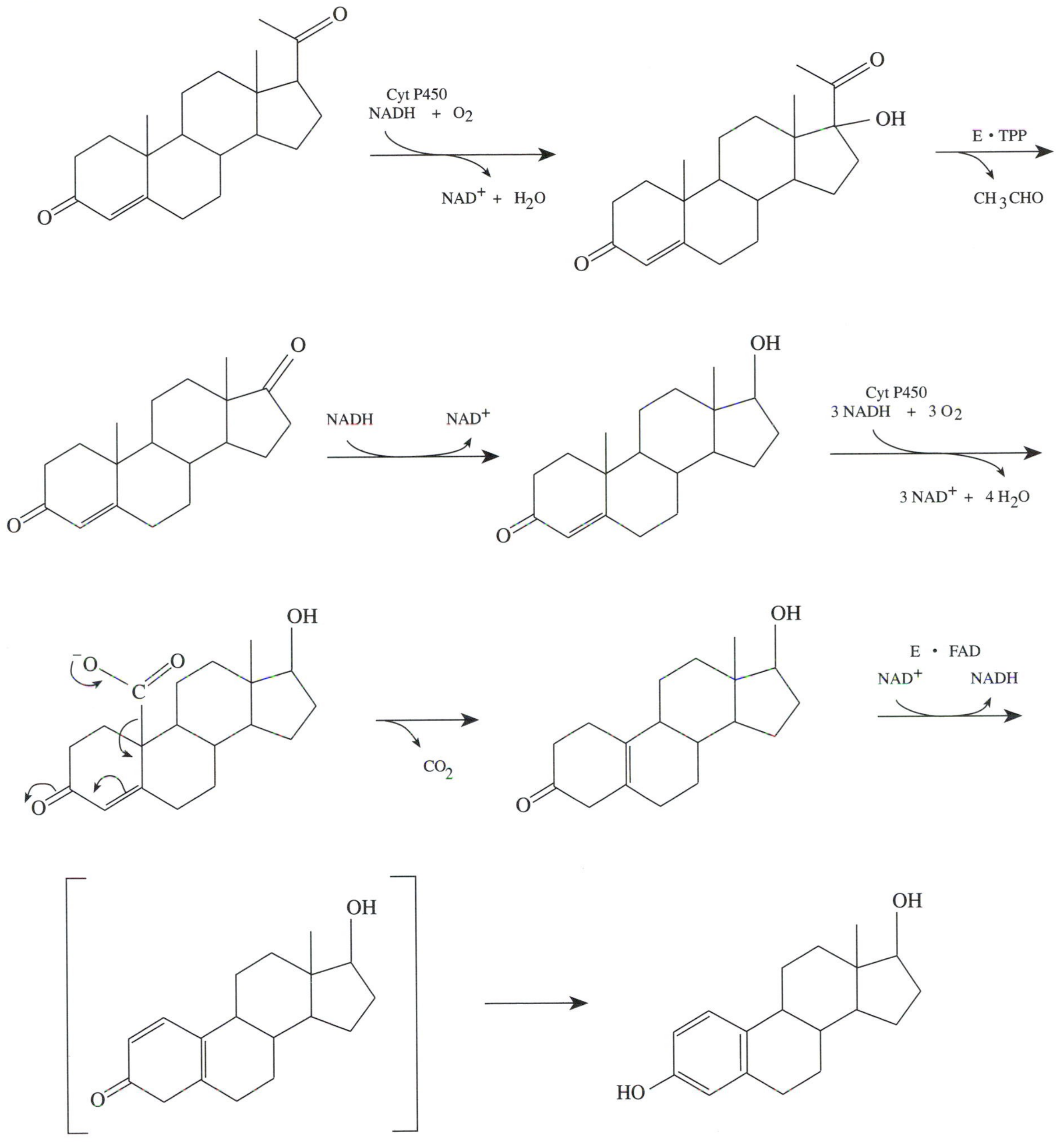

Cyt P450, NADH + O_2 → NAD^+ + H_2O

E • TPP → CH_3CHO

NADH → NAD^+

Cyt P450, 3 NADH + 3 O_2 → 3 NAD^+ + 4 H_2O

→ CO_2

E • FAD, NAD^+ → NADH

***7.** The cofactor is pyridoxal phosphate (PLP), and the mechanism of action of PLP is analogous to that postulated in Chapter 26 for the reaction of glycine with succinyl CoA to produce aminolevulinic acid.

Chapter 29: Biological Membranes

1. The experiment follows lactose-driven H^+ uptake into lipid vesicles by measuring the disappearance of H^+ from the external solution. For net H^+ movement to occur, in such an experiment, the electric charge carried by H^+ must be compensated by movement of another charged species. Otherwise, the coupled movement of lactose and H^+ would be self-inhibited by the buildup of a large transmembrane voltage difference (inside positive). Valinomycin, by making the vesicle membrane specifically leaky to K^+ ions, prevents this buildup of voltage, and thus permits net movement of lactose and H^+ to the vesicle interior.

2. The reaction catalyzed is:

$$3\,\mathrm{Na}_{out} + 1\,\mathrm{Ca}_{in} \rightleftharpoons 3\,\mathrm{Na}_{in} + 1\,\mathrm{Ca}_{out}$$

For this reaction going to the right,

$$\Delta G = RT\ln\frac{[\mathrm{Ca_{out}}]}{[\mathrm{Ca_{in}}]} + 3\,RT\ln\frac{[\mathrm{Na_{in}}]}{[\mathrm{Na_{out}}]} + \Delta z(\Delta V)F$$

$$\frac{\Delta G}{RT} = \ln\left(\frac{[\mathrm{Ca_{out}}]}{[\mathrm{Ca_{in}}]}\right)\cdot\left(\frac{[\mathrm{Na_{in}}]}{[\mathrm{Na_{out}}]}\right)^3 + \frac{\Delta zF\Delta V}{RT}$$

$$\mathrm{Ca_{out}}/\mathrm{Ca_{in}} = 10^4$$
$$\mathrm{Na_{in}}/\mathrm{Na_{out}} = 10^{-1}$$
$$\Delta z = 1$$

(a) For $\Delta V = -90$,

$$\Delta G/RT = \ln 10 - 3.8 \quad < 0$$

(b) For $\Delta V = -20$,

$$\Delta G/RT = \ln 10 - 0.8 \quad > 0$$

3. The reaction is:

$$3\,\mathrm{Na}_{out} + 1\,\mathrm{G}_{out} \rightleftharpoons 3\,\mathrm{Na}_{in} + 1\,\mathrm{G}_{in}$$

At maximum concentration gradient, reaction is at equilibrium, and

$$3\,\Delta\tilde{\mu}_{\mathrm{Na}} + \Delta\tilde{\mu}_{\mathrm{Gaba}} = 0$$

$$3\,RT\ln\frac{\mathrm{Na_{in}}}{\mathrm{Na_{out}}} + 3\,F\Delta V + RT\ln\frac{\mathrm{G_{in}}}{\mathrm{G_{out}}} = 0$$

$$\therefore\ \frac{\mathrm{G_{in}}}{\mathrm{G_{out}}} = \left(\frac{\mathrm{Na_{out}}}{\mathrm{Na_{in}}}\right)^3 e^{-3F\Delta V/RT}$$

(a) For $\Delta V = -60$ mV and $\mathrm{Na_{out}}/\mathrm{Na_{in}}$,

$$\mathrm{G_{in}}/\mathrm{G_{out}} = 10^6$$

(b) For red blood cell, $\Delta V \approx -5$ mV

$$\mathrm{G_{in}}/\mathrm{G_{out}} \approx 2\times 10^3$$

4. Hydrophobic forces arise from the tendency of water to be thermodynamically inhospitable to nonpolar solutes; thus, nonpolar groups tend to partition into the membrane interior. But, by the same logic, once nonpolar solutes are inside the nonpolar environment of the membrane, there is no water present to create hydrophobic interactions between them.

5. The thermodynamic equivalence between chemical and electrical potentials is:

$$2.3\,RT\log\frac{\mathrm{X_{in}}}{\mathrm{X_{out}}} = zF\Delta V$$

for $F, R, T = 300$ K and $\mathrm{X_{in}}/\mathrm{X_{out}} = 10$,

$$\Delta V = \frac{2.3\,RT}{zF}\ln\frac{\mathrm{X_{in}}}{\mathrm{X_{out}}}$$

for $\mathrm{X_{in}}/\mathrm{X_{out}} = 10$,

$$\Delta V = 58\ \mathrm{mV}$$

$$2.3\,RT\ln\frac{\mathrm{X_{in}}}{\mathrm{X_{out}}} = 1.4\ \mathrm{kcal\ mol^{-1}}$$

6. For extreme conditions:

$$\Delta G = \Delta G_{\mathrm{ATP}} + RT\ln\left(\frac{\mathrm{Na_{out}}}{\mathrm{Na_{in}}}\right)^3\cdot\left(\frac{\mathrm{K_{in}}}{\mathrm{K_{out}}}\right)^2$$

$$= -13\ \mathrm{kcal\ mol^{-1}} + RT\ln(150)^3\cdot(150)^2$$

$$\approx -13 + 15 = +2\ \mathrm{kcal\ mol^{-1}}$$

The fact that ΔG is > 0 means that the reaction will tend to run "backward"; that is, with Na^+ flowing into the cell, K^+ out of the cell, and ATP being synthesized. This "pump reversal" occurs because the Gibbs energy contained in the ion gradients is larger than the Gibbs energy of ATP hydrolysis.

Chapter 30: Coupled Vectorial Processes: Muscle Contraction and Active Transport

1. When the muscle fibrils are soaked in glycerol-water, the small molecules and ions in the fibrils, such as ATP, ADP, P_i, creatine phosphate, Mg^{2+}, and Ca^{2+}, are washed out. By adding these compounds back in various combinations and at different concentrations, it was possible to determine which compounds are needed to stimulate contraction and to provide the energy for contraction of the fibril, as well as the concentrations of each compound that are optimal for supporting contraction.

2. The length of the actin filament depends on the relative rates of elongation and shortening of the filament at the two ends. There is a tendency for molecules of actin-ATP to add to one end and for actin-ADP to dissociate from the other end. Structural and regulatory proteins that bind to one end of the actin filament, such as spectrin and profilin, can prevent elongation or shortening at that end.

3. Because there is no change in the length of the muscle in isometric contraction, there must be a continual shortening of fibrils in the muscle to balance the lengthening of other fibrils that are stretched by the weight. A heavier weight produces more lengthening, and more contraction events are necessary in order to counteract the lengthening and maintain a constant length.

4. The weaker binding means that phosphorylation brings about a change in the structure of the sodium-binding sites that provides less-favorable binding. A change in the structure of the protein that is brought about by phosphorylation at the ATP-binding site is transmitted to the aspartate residues in the transmembrane channel so as to cause movement of the $-COO^-$ groups of aspartate residues that results in weaker interaction with the Na^+ ions. This is necessary because Na^+ ions must bind with a high affinity on cytoplasmic-facing sites, owing to the low concentration of sodium in the cytoplasm, and they must bind with a much lower affinity at the extracellular sites, so that they can be discharged into the extracellular fluid in spite of the high concentration of Na^+, more than 0.1 M, in the extracellular fluid.

5. Rule 1: If $E{\cdot}Ca_2$ could react with inorganic phosphate to give phosphoenzyme, then the phosphoenzyme must react with water in the reverse direction (equation 1). ATP will phosphorylate the enzyme to give phosphoenzyme (equation 2), and hydrolysis of this phosphoenzyme (equation 1 in reverse) would result in ATP hydrolysis.

$$E{\cdot}Ca_2 + P_i \rightleftharpoons E\text{-}P{\cdot}Ca_2 + H_2O \qquad (1)$$

$$E{\cdot}Ca_2 + ATP \rightleftharpoons E\text{-}P{\cdot}Ca_2 + ADP \qquad (2)$$

Rule 2: If E could react with ATP to give the phosphoenzyme, E-P, the ATP would hydrolyze rapidly because E-P reacts with water to give E and P_i.

Rule 3: If E could bind Ca^{2+} from the inside of the vesicle, Ca^{2+} would leak out of the vesicle because $E{\cdot}Ca_2$ dissociates Ca^{2+} to the outside of the vesicle.

Rule 4: If E-P could bind and dissociate Ca^{2+} on the outside of the vesicle, Ca^{2+} would bind to E-P from the inside of the vesicle and then dissociate to the outside, so that Ca^{2+} would leak out of the vesicle.

6.

$$A_1M + ATP \xrightarrow[\text{(dissociation)}]{\text{Vectorial}} M \cdot ATP + A_1$$

$$M \cdot ATP \overset{\text{Chemical}}{\rightleftharpoons} M \cdot {}^{ADP}_{P_i}$$

$$A_2M + ATP + P_i \xleftarrow[A_2]{\text{Vectorial}} M \cdot {}^{ADP}_{P_i}$$

$$A_2M \xrightarrow[\text{Chemical}]{ATP} A_2M \cdot ATP$$

$$A_2M \cdot ATP \xrightarrow{\text{Vectorial}} M \cdot ATP + A_2$$

7.

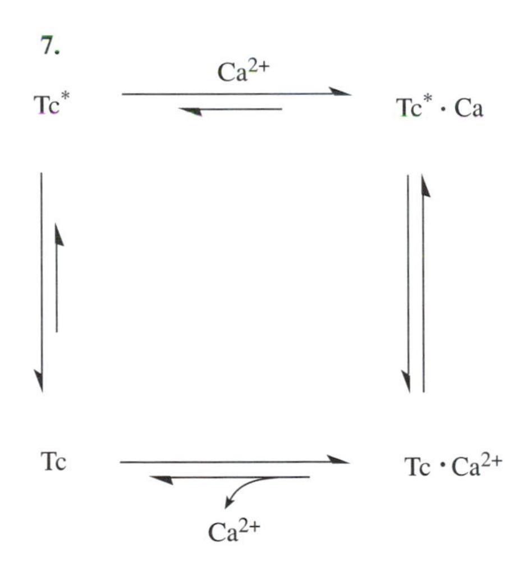

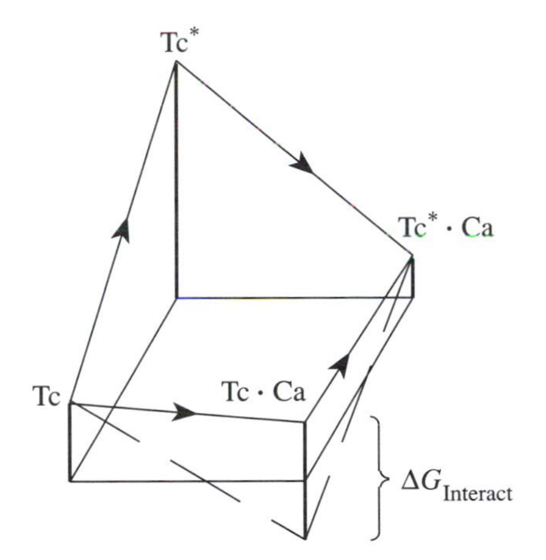

8. $\Delta G = -RT \ln K$

$$= -1423 \log K$$

For an increase in K of 100-fold (10^2), ΔG decreases by $2 \times 1423 = 2.85$ kcal mol^{-1}

9. Rule 1: Reversible phosphorylation of the enzyme by ATP occurs only with E_{Na_3}:

$$ATP + E_{Na_3} \rightleftharpoons ADP + E_{Na_3}^{\sim P}$$

Rule 2: Reversible dephosphorylation of the enzyme occurs only with E_{K_2}:

$$H_2O + E_{K_2}^{-P} \rightleftharpoons P_i + E_{K_2}$$

Rule 3: Three Na^+ bind from the cytoplasm (and dissociate) only with the free enzyme:

$$E + 3\,Na^+_{cyt} \rightleftharpoons E_{Na_3}$$

Rule 4: Three Na^+ dissociate (and bind) at the extracellular surface only with the phosphoenzyme:

$$E_{Na_3}^{\sim P} \rightleftharpoons E{-}P + 3\,Na^+_{ext}$$

Rule 5: Two K^+ bind (and dissociate) at the extracellular surface only with the free phosphoenzyme:

$$E{-}P + 2\,K^+ \rightleftharpoons E_{K_2}^{-P}$$

Rule 6: Two K^+ dissociate (and bind) at the cytoplasmic surface only with the free enzyme:

$$E_{K_2} \rightleftharpoons E + 2\,K^+_{cyt}$$

Index

Index

References to structural formulas are given in **boldface** type.